COMPARISON OF METRIC AND ENGLISH UNITS OF MEASUREMENT

LENGTH

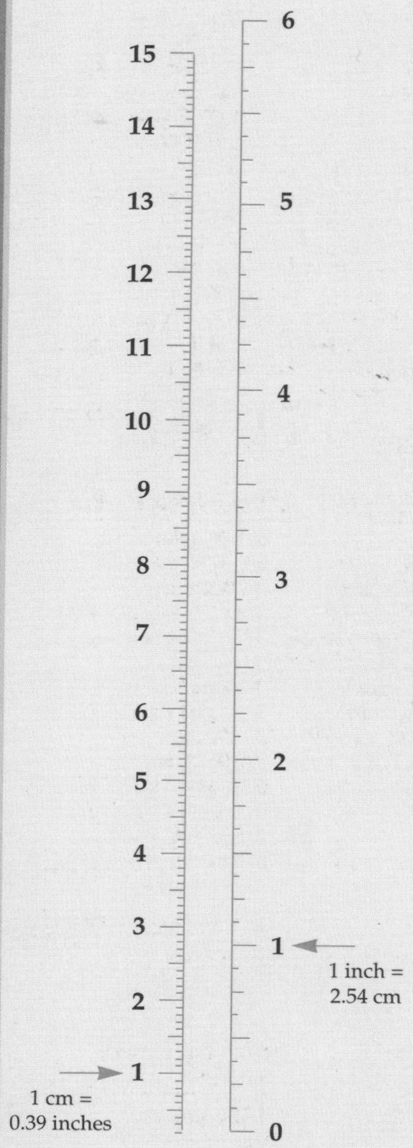

A speed limit of 88 kph is about 55 mph

A kilometer is about 0.6 mile

A meter is a little more than a yard (39 inches)

A centimeter is about the width of your index fingernail

1 inch = 2.54 cm

1 cm = 0.39 inches

Length Conversion

1 in	=	2.54 cm
1 ft	=	30.5 cm
1 yd	=	.91 m
1 mi	=	1.61 km
1 mm	=	.04 in
1 cm	=	.39 in
1 m	=	39.4 in
1 m	=	1.1 yd
1 km	=	.62 mi

WEIGHT

A large man weighs 90 kilograms

A small woman weighs 50 kilograms

This book weighs about 2 kilograms

A pound of cheese weighs about half a kilogram

A teaspoon of sugar weighs about 4 grams

A grain of salt weighs about 1 milligram

Weight Conversion

1 oz	=	28 g
1 lb	=	.45 kg
1 g	=	.035 oz
1 kg	=	2.2 lb

VOLUME

A large auto gas tank holds about 100 liters

One liter is a standard bottle of soda, milk, or wine

A cup of coffee is about 150 milliliters

5 milliliters is a standard teaspoon of medicine

A drop of water is about 50 microliters

Volume Conversion

1 tsp	=	5 ml
1 tbsp	=	15 ml
1 fl oz	=	30 ml
1 cup	=	.24 l
1 pt	=	.47 l
1 qt	=	.95 l
1 gal	=	3.8 l
1 ml	=	.03 fl oz
1 l	=	2.1 pt
1 l	=	1.06 qt
1 l	=	.26 gal

SECOND EDITION

Human Physiology

SECOND EDITION

Human Physiology

Rodney Rhoades, Ph.D.
Professor and Chairman
Department of Physiology and Biophysics
Indiana University School of Medicine

Richard Pflanzer, Ph.D.
Associate Professor of Biology and
Physiology/Biophysics
Indiana University School of Medicine
Purdue University School of Science

Saunders College Publishing
Harcourt Brace Jovanovich College Publishers
Fort Worth   Philadelphia   San Diego   New York   Orlando   Austin
San Antonio   Toronto   Montreal   London   Sydney   Tokyo

Text Typeface: Palatino
Compositor: York Graphic Services
Acquisitions Editor: Julie Levin Alexander
Developmental Editor: Cathleen Petree
Managing Editor: Carol Field
Project Editor: Margaret Mary Anderson
Copy Editor: Judy Patton
Manager of Art and Design: Carol Bleistine
Art Director: Christine Schueler
Art Assistant: Caroline McGowan
Text Designer: Tracy Baldwin
Cover Designer: Lawrence R. Didona
Text Artwork: Rolin Graphics
Layout Artist: Tracy Baldwin
Director of EDP: Tim Frelick
Production Manager: Charlene Squibb
Marketing Manager: Marjorie Waldron

Cover Artwork: Greg Purdon

Printed in the United States of America

HUMAN PHYSIOLOGY, SECOND EDITION

0-03-072616-6

Library of Congress Catalog Card Number: 91-058038

34 032 9876543

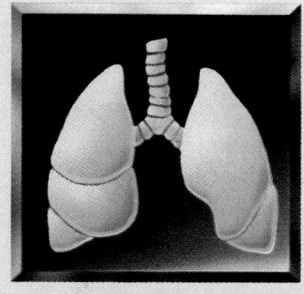

Preface

*T*ake a look at the cover of this textbook: like the illustration of the integrated circuit and the icons representing component systems, the body systems function in an integrated manner to maintain a relatively stable internal environment. This stability is requisite for good health. The maintenance of a stably functioning internal environment despite external influences—such as temperature, pathogens, or trauma—defines the concept of homeostatic balance. And, homeostasis is the pervasive theme of *Human Physiology, Second Edition.*

Physiology (from the Greek: physio = nature; logy = the study of) originally meant an inquiry into nature and included the systematized study of both living and nonliving matter. Today, the focus of the study of physiology is the integration and coordination of the systems of living organisms for the purpose of explaining and predicting homeostatic (homeo = same; stasis = standing) mechanisms that operate in the survival of any organism.

The human body is a complex organism comprising several major systems—nervous, skeletal/motor, endocrine, circulatory, respiratory, gastrointestinal, and reproductive, to mention a few. The study of physiology, therefore, is most logically presented by the "systems" approach. However, we believe it is important, especially for the beginning student of physiology, to realize that the body does not function as a set of independent systems, and that no single system is more important than another in maintaining life.

Human physiology is a fascinating discipline. Most of us, professor and student alike, have a natural curiosity about how our bodies work. The recent explosion of scientific information has afforded the life and health sciences tremendous opportunities and tools for better understanding the world within us. As educators, scientists, and authors, we face important challenges in presenting to students a vast amount of scientific information as a cohesive body of unifying principles and concepts, instead of as a catalogue of disconnected facts and mysteries. *Human Physiology, Second Edition,* develops and explains these underlying principles and concepts while conveying the excitement of unsolved research problems.

Student Audience

We have written this text for students who are planning careers in the life or health sciences. This text should be very useful to students of diverse backgrounds and levels of preparation. We have not presumed, for instance, that all students have had extensive experience with math, chemistry, physics, and biology. The background information on basic chemistry, physics, molecular and cell biology in the first six chapters will help those students who have had little or no college-level science course work. Students who have stronger backgrounds in the sciences will discover the basic science coverage to be a useful review.

In addition, underlying principles of human anatomy have been introduced where a knowledge of basic structure is necessary for understanding function. The emphasis of this text, however, is on function rather than on structure.

Our objectives in writing this text are the following:

- to develop concepts and principles of physiology in a logical, clear, and concise manner;
- to improve student ability to reason scientifically;
- to promote understanding and appreciation of normal homeostatic body functions;
- to engender excitement about physiological research and other related career interests.

The organization and approach of *Human Physiology, Second Edition* have evolved through many years of teaching the subject to students of various backgrounds and abilities. Our research, along with that of many colleagues and other scientists, helped us a great deal in shaping the content of this text. We sincerely hope that this presentation of human physiology will capture the interest of your students and stimulate them to pursue the discipline beyond the limits of this book.

Organization

The book is organized so that topics progress from the molecular–cellular level to integrated organ function. Chapter 1, designed to develop an appreciation of science, explains the scientific method, introduces basic concepts such as homeostasis, and provides a historical perspective. This chapter lays the foundation for the theme of Part I, which is basic cellular function and includes the essential chemistry, biochemistry, physics, and mathematical principles required to enhance understanding of fundamental physiological concepts.

Part II focuses on the body's internal environment and analyzes the nature of biological control systems including the properties of specialized cells from the nervous and endocrine systems that regulate body function.

Part III discusses the coordinated functions of the systems of circulation, digestion, respiration, and reproduction that all stem from the integration of specialized tissue function.

Features

- Chapter 1 provides a historical perspective of the science of physiology. This helps to define the discipline and offers the student an exciting sense of direction for mastering the text.
- Chapter 1 clearly explains the scientific method and offers a review of basic computations, measurements, graphic analysis, and anatomical terminology that serve as tools for learning physiology.
- Early coverage of molecular biology and cell physiology (Chapters 2–6) provides a strong basis for understanding systemic functions. The concepts that are derived from cellular and molecular biology are applied throughout the text.
- Physiological control systems (Part II) are presented before organ functions (Part III) to establish the concept of homeostasis early.
- Metabolism and energy transformation are covered early (in Chapter 6) in order to present intermediary metabolism as a fundamental concept rather than as a process that is only specific to nutrition.
- The artwork has been developed with the use of flowcharts to clarify complex physiological processes (for examples, see Figure 6–9 on page 211 and Figure 20–25 on page 725).
- There are separate chapters on the Autonomic Nervous System, Respiration in Unusual Environments, Exercise Physiology, Temperature Regulation, Calcium and Phosphate Metabolism, and Sexual Physiology. These chapters provide unique, in-depth coverage of these topics, which are typically not found in competitive textbooks.
- Most chapters contain focus boxes that highlight clinical physiological problems and issues. (For example, see the box on Alzheimer's Disease in Chapter 7.) These are designed to provide relevancy and to encourage career interest.

- End-of-chapter review questions assist in developing conceptual understanding.

New to the Second Edition

- Photographs and full-color line art have been effectively paired to better illustrate structure and function (for examples, see Figure 3–14 on page 106 and Figure 5–6 on page 161).
- Quantitative problems have been added to the end-of-chapter problem sets to help students more effectively apply concepts. Answers to all problems are provided in a new Appendix A.
- In addition to the clinical box featured in most chapters, a research box has been provided for most chapters. The research topics demonstrate the exciting ongoing activity of physiological research. For example, see the box on Space Shuttle Columbia, which is the first mission completely dedicated to life science research (Chapter 22).
- An extensive glossary has been added for student reference that includes phonetic pronunciations and the page numbers on which new terms are first defined in the text.
- The Nervous System (Chapters 7–11) has been completely updated and restructured to offer a clearer presentation of the synaptic transmission process and of nerve fiber development and regeneration.
- Chapter 17, "The Blood," has been rewritten and significantly expanded to update the information about plasma proteins and to modulate and clarify the immunity information.

Learning Aids

Many pedagogical aids have been incorporated to help the student learn concepts in physiology. They are the following:

- Icons have been developed to represent each system. The individual icons open each chapter. In addition, each part-opening design will utilize the icons to highlight the systems that are to be discussed in the upcoming chapters. This unity of design visually emphasizes the integrated nature of physiology.
- Each chapter begins with an outline of the chapter's content and ends with a summary.
- Each chapter concludes with a set of review questions (including, in most chapters, some

quantitative problems), a list of suggested future readings, and key terms that are referenced to specific pages within the chapter.
- Throughout the text, boldface type is used to emphasize important new terms.
- The artwork and photographs emphasize and reinforce important physiological concepts, and numerous flowcharts are used to reinforce the conceptual approach of both text and artwork.
- Three appendices offer the following: detailed answers to support the chapter review questions; tables of normal physiological values (e.g., blood pressure, blood gases, body water content); and medical terminology, which includes Latin and Greek prefixes, suffixes, and terms for body parts for the purpose of building a medical vocabulary.
- The index has been greatly expanded and provides extensive cross-referencing.
- A large glossary with phonetic pronunciations for many of the most difficult terms appears at the back of the text.
- Front endpapers contain tables of Decimal Factors and Prefixes, Conversion Factors and Equivalents for the Metric and English systems of measurement.
- The rear endpapers contain an alphabetized table of commonly used clinical abbreviations.

Ancillaries

A number of excellent supporting aids accompany the text. These include: an Instructor's Manual with Text Bank written by David Cotter of Georgia College; 150 carefully selected overhead transparencies of illustrations from the text; a Laboratory Manual written by Thomas H. Dietz and Joseph Woodring, both of Louisiana State University; and a Study Guide written by Florence C. Ricciuti of Albertus Magnus College. In addition, the Test Bank is computerized for the IBM and MacIntosh computers. We also offer a hypercard stack as well as a wide selection of NOVA videotapes. A selection of human cadaver dissection videotapes are available illustrating the respiratory, digestive, and circulatory systems.

Acknowledgments

We want to express our deepest thanks and appreciation to all the contributing authors without whose expertise, talent, and effort this book would not have been possible. We would also like to give thanks for a job well-done to the editorial staff and management of Saunders College Publishing for their help and guidance in significantly improving the first edition of this book. Special thanks go to our Developmental Editor, Cathleen Petree, whose patience and editorial talents were absolutely essential to the successful completion of this book. We are indebted, as well, to Margaret Mary Anderson, our Project Editor, Christine Schueler, Art Director, and to Ray Tschoepe, our Art Developmental Editor, for their advice and guidance during the production phase. We also thank Martha Colgan, the Developmental Editor of our first edition, for her hard work in expanding and greatly improving the index for this edition. Finally, we would like to thank Julie Levin Alexander, Senior Acquisitions Editor, for her support and commitment to this book, and for sharing with us her administrative talent in managing the human and material resources of this very complex project.

A very special thanks goes to Nancy Christian for typing portions of the manuscript and for her administrative assistance without which we surely would have suffered. Her assistance made our work as authors and editors much easier.

Lastly, we thank our families and friends for their tolerance of our preoccupation with the textbook for extended periods of time. Our very warm gratitude goes to our wives, Diane Pflanzer and Judy Rhoades, without whose love, understanding, and support, this project could not have been completed.

Reviewers

We express our sincere appreciation to the following faculty members and scientists for their conscientious reviews and assessments of the manuscript. Their suggestions, comments, and insights have been invaluable.

Reviewers of the First Edition:

Jerome Yochim, *University of Kansas*
Arnold J. Sillman, *University of California, Davis*
Rick Turnquist, *Augustana College*
Andy Anderson, *Utah State University*
Byron A. Schottelius, *University of Iowa*
John T. Fales, *University of Missouri, Kansas City*
John P. Harley, *Eastern Kentucky University*
L. Stephen Whitley, *Eastern Illinois University*
S.J. Coward, *University of Georgia*
David Noyes, *The Ohio State University*
Pegge Alciatore, *University of Southwest Louisiana*
Stephen Williams, *Glendale Community College*
Kenneth H. Bynum, *University of North Carolina, Chapel Hill*

Reviewers of the Second Edition

Steve Simasko, *SUNY at Buffalo*
Jack Wood, *Western Michigan University*
Ralph Ferges, *Palomar College*
John Scheide, *Central Michigan University*
Mark Biedebach, *California State University/Long Beach*
Daniel Gibson, *Worcester Polytechnic Institute*
Lin Aanonsen, *MacAlester College*
Mary Lynne Stephanou, *Santa Monica College*
Sunny Boyd, *University of Notre Dame*
Pegge Alciatore, *University of Southwest Louisiana*
Audrey Mackey, *Austin Community College*
Carl Hammen, *University of Rhode Island*
John G. Moner, *University of Massachusetts/Amherst*
Katherine Mechlin, *Wright State University*
Fred D. Hinson, *Western Carolina University*
Sheldon Lustick, *Ohio State University*
Harry Bernheim, *Tufts University*
Lawrence Wit, *Auburn University*
Leon Goldstein, *Brown University*
Robert Slechta, *Boston University*
Barbara Howell, *SUNY at Buffalo*
Stuart Coward, *University of Georgia*
Pinapka Murthy, *Fayetteville State University*
Edmund Tong, *Wheaton College*
Leonard Lundquist, *Lander College*

Rodney A. Rhoades
Richard G. Pflanzer
November, 1991

Contributing Authors

Reynaldo S. Elizondo, Ph.D.
Dean, College of Science
University of Texas at El Paso
Chapters 28, 29

Janice C. Froehlich, Ph.D.
Assistant Professor of Physiology/Biophysics and
   Medicine
Indiana University School of Medicine
Chapters 30, 31, 32

Joe R. Haeberle, Ph.D.
Associate Professor of Physiology/Biophysics
University of Vermont
Chapters 18, 19

Stephen A. Kempson, Ph.D.
Associate Professor of Physiology and Biophysics
Indiana University School of Medicine
Chapters 1, 2, 3, 4, 5, 6

Leon K. Knoebel, Ph.D.
Professor of Physiology and Biophysics
Indiana University School of Medicine
Chapter 23

Walter C. Low, Ph.D.
Associate Professor of Neurosurgery Physiology and
   Neuroscience
University of Minnesota Medical School
Chapters 1, 7, 8, 9, 10, 11

Richard A. Meiss, Ph.D.
Professor of Physiology/Biophysics and Obstetrics/
   Gynecology
Indiana University School of Medicine
Chapters 1, 16, 18

Daniel E. Peavy, Ph.D.
Associate Professor of Physiology and Biophysics
Indiana University School of Medicine
Research Chemist
R. L. Roudebush VA Medical Center
Chapters 1, 12, 13, 14, 15, 27

Richard G. Pflanzer, Ph.D.
Associate Professor of Biology and Physiology/
   Biophysics
Indiana University School of Medicine
Purdue University
Chapters 1, 17

Rodney A. Rhoades, Ph.D.
Professor and Chairman
Department of Physiology and Biophysics
Indiana University School of Medicine
Chapters 1, 20, 22

George A. Tanner, Ph.D.
Professor of Physiology and Biophysics
Indiana University School of Medicine
Chapters 1, 24, 25, 26

Wiltz W. Wagner, Jr., Ph.D.
V. K. Stoelting Professor of Anesthesiology
Professor of Physiology/Biophysics, Anesthesiology
   and Pediatrics
Indiana University School of Medicine
Chapters 21, 22

Contents Overview

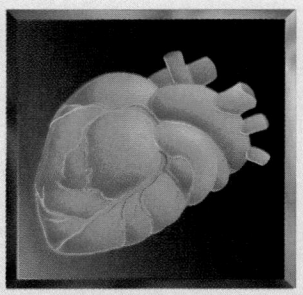

Contents

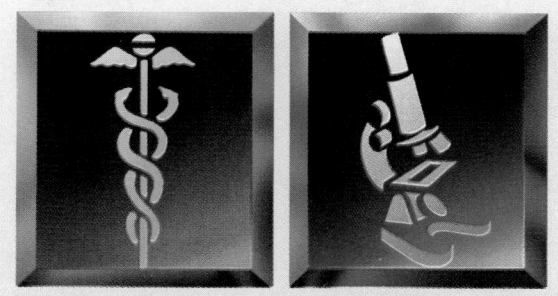

These focus boxes are intended to apply basic concepts and illustrate how diseases are characterized by altered functional changes. For most chapters, there are two types of focus boxes, clinical and research.

Focus Boxes

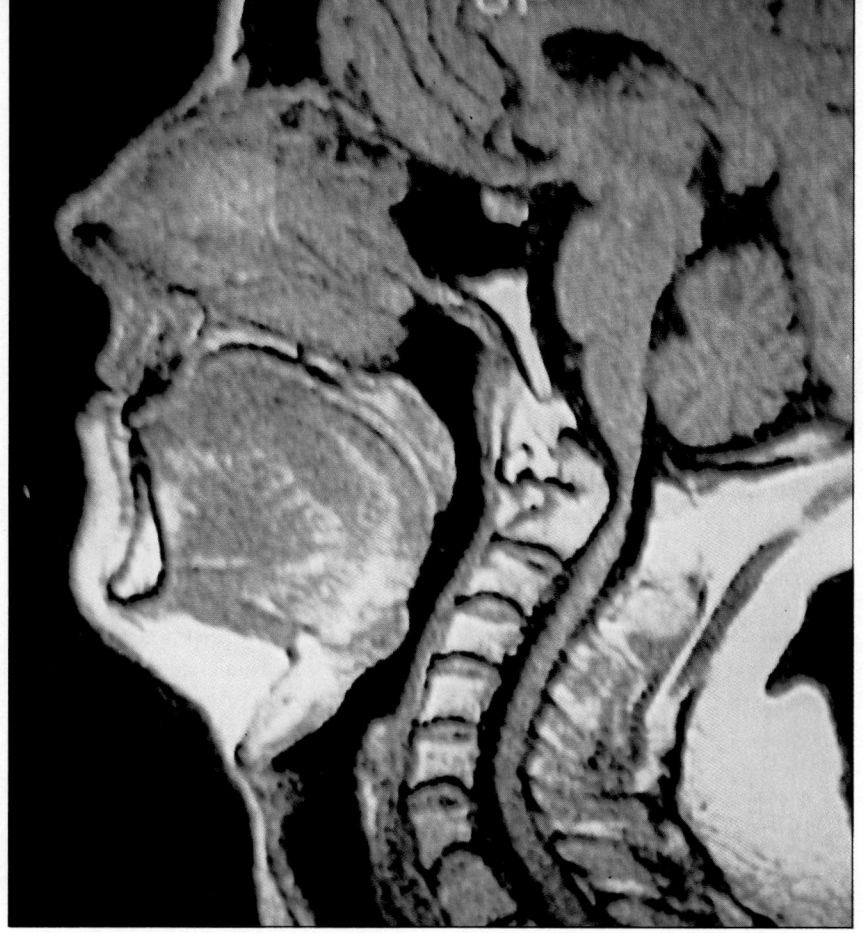

The Science of Physiology

*T*he ancient Greeks considered the universe to be composed of four elemental substances: air, water, earth, and fire. The Greeks further believed each human being was a miniature universe in whom these substances appeared in the form of four "humors," or fluids: *blood* (corresponding to air); *phlegm* (representing water), *black bile* (corresponding to earth), and *choler* or *yellow bile* (representing fire) (Fig. 1–1).

According to the Greek view, the fluids were not present in each person in equal amounts; one fluid usually predominated and characterized the individual's temperament. For example, one temperament was the "sanguine" type. We still use this word today to describe someone with a cheerful, confident personality. In ancient times, such a personality would have been attributed to a predominance of blood over the other three humors within that person's body, as evidenced perhaps by a ruddy complexion (*sanguine* comes from the Latin *sanguineus*, meaning "bloody"). A "melancholy" person was thought to have more black bile than any of the other humors, whereas a "choleric" per-

son was considered to have a predominance of yellow bile (the humor associated with fire), which produced an excitable, easily angered temperament. A "phlegmatic" person generally was calm and unemotional, even sluggish, supposedly because of a predominance of phlegm, the cold, waterlike humor. The Greeks thought that the particular quantities of each humor and the resulting proportion, or balance, determined a person's physical and psychological makeup.

According to the classical scheme, while a normal predominance of one fluid within each person's unique humoral balance produced a characteristic personality, an abnormal lack or excess of one or more humors resulted in disease. Doctors from ancient times until well into the Renaissance shaped their therapeutic methods around attempts to restore the proper balance of humors within the sick person's body. Purging, bloodletting, and similar treatments were all intended to adjust humors.

In contrast to the ancient Greek philosophers and physiologists, today we organize the universe of matter into 105, rather than 4, elemental substances, and we attribute each person's uniqueness to genetic makeup rather than to relative quantities of bodily fluids. Yet despite the revolutions in our conceptions of human physiology, some classical principles influence modern science. One enduring principle is the idea that the human body, if not a world in miniature, is composed of elements from the universe that obey the laws of nature. Another "modern" idea cherished by the ancient philosophers is that of a harmonious balance within the body. While today we speak of homeostasis and recognize it as a self-regulatory process involving not only fluids but dynamic processes, we share the classical view that health is the evidence of balance among body functions and substances. Finally, we have inherited the Renaissance attitude—which echoed the ancient Greek belief—that the human body is worthy of the most rigorous study and eloquent description.

The History of Physiology

The story of the development of physiological understanding in modern culture is fascinating. The following brief survey of the last 2000 years of physiological theory and research is meant to provide a

Figure 1–1
The four bodily "humors" of classical and medieval philosophy and physiology. The Greeks thought that the elements of the universe were represented in each human body by blood, phlegm, choler, and black bile. According to this scheme, a predominance of one humor produced each person's temperament.

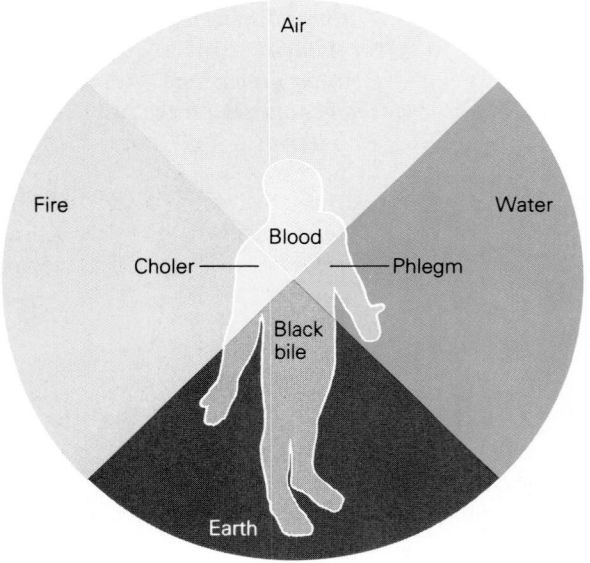

context for the study of twentieth-century physiology in all its complexity and sophistication. Further, the discussion communicates a sense of wonder at the vast distances physiologists have come since human beings first began to record observations and theories about the workings of the human body.

Physiological Knowledge in the Classical and Medieval Periods

Since so many Western attitudes about science, philosophy, and the place of human beings in the universe have been influenced by the traditions of the classical Greeks, that period seems a logical starting point to examine the development of physiology as we understand it today. The studies of the Greek philosophers and the investigations of the Greek physiologists at the great museum in Alexandria influenced European medical and physiological practice until well into the Renaissance because of the mediating influence of Galen. Galen's writings preserved the ancient studies and, in turn, were preserved through the medieval period for examination by Renaissance scholars.

The Observations of Aristotle

Although he is most often thought of as a philosopher, Aristotle (384–322 B.C.) did important work in biology and was one of the first to describe the blood vessels as a system with the heart at the center. Unlike other investigators of his time, he used a bloodless method of killing animals for dissection, finding that the vessels were easier to see and trace when the blood remained.

Some of Aristotle's concepts were corrected through the discoveries of later investigators. Aristotle thought of the heart as both the seat of the intellect and as a furnace that heated the blood to provide needed warmth. (Since warmth disappeared from the body so soon after death, it was thought to be the cause or source of life). Just as the blood needed a source of heat, so too the furnace needed ventilation. According to Aristotle, this was the purpose of breathing; the lungs were a ventilation system and the air a cooling agent.

Aristotle correctly observed that blood flowed between arteries and veins, but he could not see the microscopic capillaries that we now know provide the connection between them. He explained the flow by saying, in essence, that the vessels were themselves made of blood and that they disappeared when blood flow stopped.

. . . the channels of the blood vessels may be compared to the mud which a running stream deposits, they are as it were deposits left by the current of blood in the blood vessels. . . . Thus just as in the irrigation system the biggest channels persist whereas the smallest ones quickly get obliterated by the mud, though when the mud abates they reappear; so in the body the largest blood vessels persist while the smaller ones become flesh in actuality, though potentially they are blood vessels as ever before.

The Theories of Herophilus and Erasistratus

At the Museum at Alexandria, a famous center of Greek culture and learning in what is now Egypt, two men pursued a number of biological investigations in a continuation of Aristotle's search for an understanding of blood flow. Physiological research flourished at Alexandria partly because it was the only place that permitted dissection of human cadavers for the study of anatomy and physiology.

Herophilus (c. 335–280 B.C.), who is considered by some to be the first physiologist in the Western tradition, identified the brain as the seat of the intellect and demonstrated that the walls of the arteries are thicker than those of the veins. Using the newly invented water clock, he measured the pulse and showed that it varied when disease was present.

Erasistratus (310–240 B.C.) began his physiological studies as an assistant to Herophilus. Erasistratus believed that blood was made in the liver from food and was delivered to the organs of the body by an "ebb and flow" in the veins. He thought of the arteries as air vessels, not blood vessels. The air, called *pneuma*, was thought of as a living force taken in through the lungs to the left side of the heart, where it was transformed into a "vital spirit" and then transported, in "airy" form, through the arteries to the rest of the body. (To explain why blood, not air, flowed from a cut artery, in apparent contradiction with his theory, Erasistratus postulated the existence of tiny connections between veins and arteries that sometimes opened to let blood flow into the arteries.)

The "Pneumatology" of Galen

In a simple experiment with a goose feather, Galen (c. A.D. 130–201) demonstrated that blood, not air, flowed through the arteries. Galen created a hollow tube by cutting off the ends of the feather. Then he inserted the tube into a tied-off artery, from which blood immediately flowed into the tube. Since, by being tied off, the artery was supposedly separated from the veins, the blood that flowed from it had to have been inside the artery from the beginning.

Galen was a Greek who studied physiology both in Greece and at Alexandria. His practice as a surgeon to gladiators in his native city of Pergamum allowed him to observe the internal structure of the body at a time when dissections for the purpose of study were forbidden. Later he was appointed physician to a Roman emperor. Galen is known for his voluminous writings on philosophy, medicine, and physiology, in which he commented and expanded on the work of the investigators who proceeded him. Even though many of his ideas have been disproven, we still consider Galen an important figure because of the intellectual sophistication of his physiological schemes, the breadth and scope of his explanations, and the length of time and extent to which his teachings were accepted. His was the prevailing view of human physiology from his own time until the Renaissance.

Like Erasistratus, Galen believed that the blood was produced from food, in the liver (Fig. 1–2). He thought that the blood took on "natural spirits" and carried them through the veins to the bodily organs,

Figure 1–2
In the "pneumatology" of Galen, blood was made in the liver from food and carried "natural spirits" to the bodily organs, traveling both ways in blood vessels. *Pneuma* made from air in the lungs met blood in the heart and was transformed into "vital spirits" and then "animal spirits."

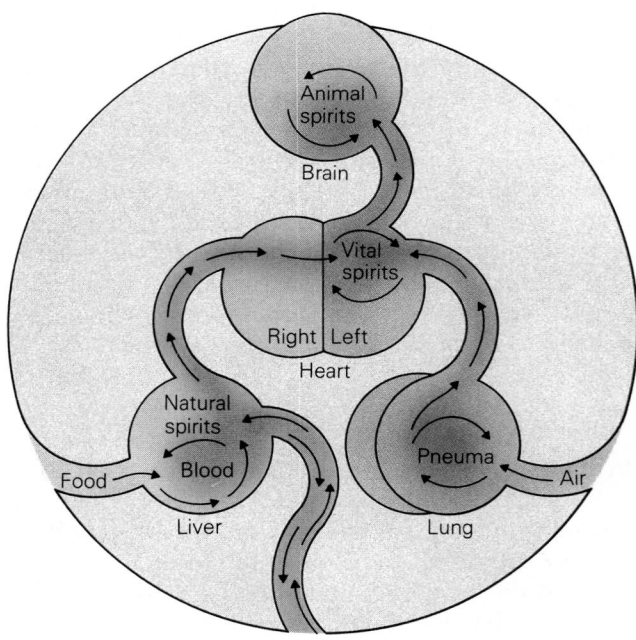

which needed the spirits to perform different functions. After its supply of spirits was depleted, the blood returned along the same venous pathways to the liver to be resupplied. (This idea that the blood flowed both ways in the vessels became one of the most revered of Galen's teachings. When Renaissance anatomists correctly challenged it because of their knowledge of valves, they found it very difficult to overthrow.)

According to Galen, some of the blood containing the "natural spirits" went first to the right side of the heart and then to the left side, where it contacted *pneuma*. *Pneuma* was a substance produced from air in the lungs and carried into the left side of the heart. When the "natural spirits" contacted *pneuma*, they were transformed into "vital spirits," a higher form of *pneuma*. These "vital spirits" were carried on up to the brain, where they were further transformed into yet a higher form of *pneuma* called "animal spirits." These various forms of *pneuma* drove the various processes of the body, according to Galen.

Despite his insistence on the importance of observation in the study of physiology, Galen often made incorrect assumptions about human anatomy based on his dissections of animals. As mentioned previously, however, the logic and elegance of his theories about bodily functions, which unified both the traditions he inherited and the knowledge available to his contemporaries, were convincing. His copious writings further established his authority, and throughout the late classical and medieval periods Galen's physiology was accepted. It prevailed largely unquestioned until the technological advances and humanistic spirit of the Renaissance permitted a fresh look at the body.

The Discoveries of the European Renaissance Physiologists

In the revival of classical learning that characterized the European Renaissance, Galen's writings about anatomy, physiology, and medicine first began to be widely translated from the Greek and Arabic into Latin. Although the writings were known to exist, and the general outlines of Galen's physiology had dominated medieval and early Renaissance medicine, only in the sixteenth century did the most important writings become available for close scrutiny by European physicians. By this time, the Galenic physiology had already taken hold of the imagination and come to be regarded almost as religious doctrine. Thus, Renaissance professors of medicine taught the anatomy of Galen, and anyone who departed from Galen's concepts was considered a secular heretic. To explain the discrepancies between the anatomy described by Galen and that visible in a dissected corpse, a professor might claim that the body simply had changed since Galen studied it.

The task of the Renaissance physiologists was thus to study Galen's physiology, discover its errors, and replace it with a modern scheme. Many independent-minded investigators contributed to the revolution in physiological knowledge that occurred in the late Renaissance, although they too defaulted to many of Galen's tenets whenever they could not observe or explain for themselves certain structures and functions. In his teachings, writings, and drawings, Andreas Vesalius corrected many of the inaccuracies of Galen's anatomy while perpetuating others and retaining much of Galen's erroneous physiology. William Harvey, in his discovery of the circulation, finally revised Galen's ancient notion that the air and the blood met in the heart, though he still viewed the lungs, as had Galen, as cooling organs.

The Drawings of Vesalius

Andreas Vesalius (1514–1564) was educated as a doctor and, at the age of 23, became a professor of surgery and anatomy at the great medical school in Padua, in what is now Italy. He had been taught the physiology of Galen and in turn began teaching it to his students. Unlike other professors of his time, however, who taught almost exclusively from Galen's writings, Vesalius centered his lectures not on the ancient texts but on human cadavers that he himself dissected as he lectured. (Previous instructors might not have used a human cadaver or might have lectured from afar while another person conducted the dissection.)

Another pedagogical device that Vesalius introduced into the study of physiology was the anatomical drawing. Galen's texts had been published with few or no illustrations; Vesalius created six anatomical "tables" that showed labelled parts of the body and allowed students to follow his lectures.

Despite his new approach to the teaching of anatomy, Vesalius considered himself part of the tradition of Galen and initially intended to simply explicate Galen's teachings. Hence, in the beginning he unwittingly reproduced some of Galen's erroneous assumptions about human anatomy. For example, Galen had thought that an organ called a "rete mirabile," similar to one he saw in hoofed animals, existed at the base of the human brain. This organ was supposedly where "vital spirits" were changed into "animal spirits." Vesalius showed this organ in his anatomical drawings, although, like Galen, he had never seen one in a human body.

Gradually, however, Vesalius began to realize that Galen's physiological dogma had been based not on an earlier version of the human body but on the bodies of apes, pigs, and other animals. He decided to write a new anatomy text that would reflect his own personal observations of a human body. The result was *De Humanis Corporis Fabrica* ("The Structure of the Human Body"), often simply called the *Fabrica* and now considered the first modern anatomy textbook. Published in 1543, the text contained plentiful illustrations and corrected many of Galen's errors.

Despite Vesalius's insistence on verification of anatomical details and his intention to correct Galen's anatomy, he perpetuated some of Galen's mistaken ideas. One ancient notion that reappeared in Vesalius's text was that the blood passed directly from the right to the left ventricle of the heart. Vesalius did not recognize what we now know as the pulmonary circulation (the movement of blood between the lungs and the heart) despite evidence of this movement that was becoming available in his time. He claimed, as had Galen, that air traveled from the lungs to the left ventricle of the heart, where it cooled the heart furnace and endowed the blood in the heart with "vital spirits."

The Theories of William Harvey

The picture of the blood flow handed down to the early Renaissance physiologists showed the heart as both a furnace and a meeting place for blood and air. We know now that the heart is a pump, not a furnace, and that the lung, not the heart, is where blood and air meet. Renaissance anatomists thought that blood moved directly from the right ventricle of the heart to the left to be imbued with "vital spirits" brought from the lungs by the *pneuma*. Modern theory explains that before blood can get from the right ventricle to the left it must first leave the heart and pass through the lungs. Before Harvey, Renaissance physiologists accepted Galen's view that blood returned along the same pathways it had come in its travels through the body. We now know that valves prevent this from happening and that blood flows only one way through each vessel. When it has gone as far as it can in one direction through the arteries, it crosses over into the veins to return to the heart.

Various pieces of this new picture were available to the Renaissance physiologists even as they taught the traditional scheme of Galen's *pneuma* and spirits. They knew, for example, that valves existed in the veins, and that the large vessels that connect the lungs to the heart contained flowing blood, not air. Only William Harvey, however, was able to make sense of these details and deduce from them the notion of the heart as a pump that maintains a *circular* movement of blood.

Born in England, William Harvey (1578–1657) was educated at Cambridge University and then at Padua, where he attended the lectures of the famous anatomy professor Fabricius (Fig. 1–3). Fabricius was renowned for his construction of a special amphitheater that allowed large groups of students to observe dissections. He is noted for his observation of valves as well, which later played a critical role in his famous student's conception of the circulation (Fig. 1–4). For his part, Fabricius dismissed the valves as merely slowing, not preventing, the backflow of blood through the veins. He held firmly to the physiology of Galen while patronizing the student whose physiological theories would soon turn Galen's views upside down.

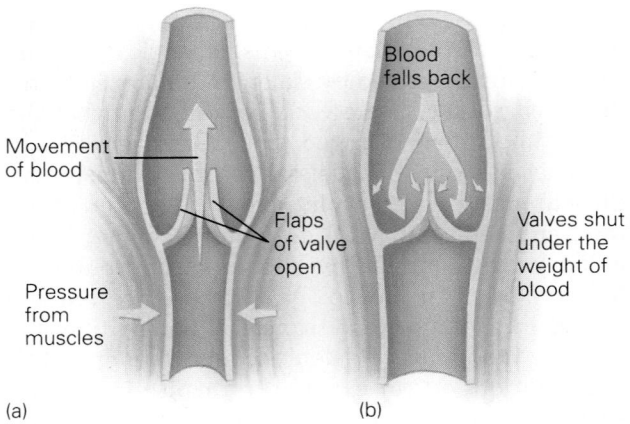

Figure 1–4
Valves in veins. The diagram shows how valves open to allow blood to flow in one direction but then close to prevent blood flow in the opposite direction.

Figure 1–3
William Harvey, the discoverer of the circulation (*The Bettman Archive*).

After completing his course at Padua, Harvey returned to England and began a career as a medical doctor (serving as royal physician) and lecturer. He also wrote a brief study of the circulation that was to become a classic in the history of science, revolutionizing the Western view of the movement of blood. *De Motu Cordis et Sanguinis in Animalibus* ("On the Motion of the Heart and of Blood in Animals"), published in 1628, described his search for the true role of the heart and the blood pathways.

The key features of Harvey's theories of blood flow that marked clear breaks from the physiology of Galen, Vesalius, and Fabricius were (1) the heart was a pump that worked by contracting to expel blood, rather than by expanding to fill with blood; (2) the blood went from the right ventricle of the heart first to the lungs, not directly to the left ventricle; (3) the air stayed in the lungs, where it met the blood, rather than coursing to the heart to meet the blood; and (4) the blood moved in a circular path through the body, taking different routes coming and going rather than traveling both directions in the same vessels.

Fifteen centuries earlier, Galen had used a goose feather to refute Erasistratus's claim that arteries were air vessels. Now Harvey used a similar experiment to refute Galen's claim that blood flowed both ways in a vessel (Fig. 1–5). He tied off the upper part of his arm to allow blood to flow into the arm through the arteries but to prevent blood already in the arm from flowing back out through the veins. As a result, blood coming into the arm through the arteries poured into the veins below the block but could not pass through them, causing them to swell. If, as Galen had claimed, blood could

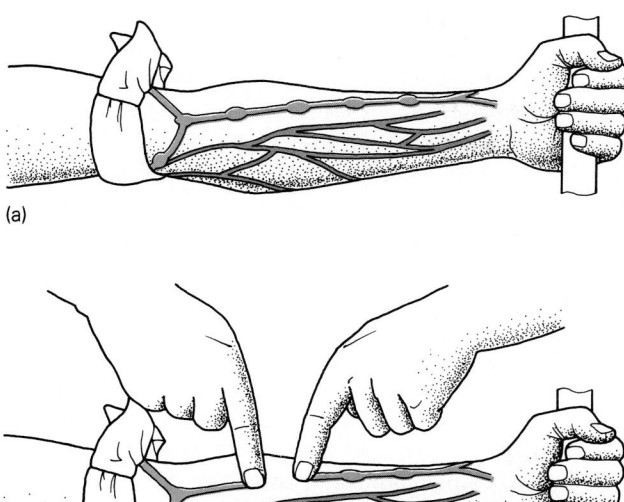

(a)

(b)

Figure 1–5
William Harvey's experiment showing direction of blood flow in veins and the connection between arteries and veins. A tourniquet allowed blood to flow through the arteries into the lower arm but prevented blood from flowing out of the veins. The veins below the tourniquet swelled up, indicating that blood was flowing into them from the arteries. When the swellings were removed, they reappeared in the end of the vein farthest from the heart, showing that blood was flowing into the veins from the arteries instead of from the heart.

have flowed back through the arteries through which it had entered the arm, the veins would not have swelled.

Thus, Harvey's discovery of the circulation came about through a combination of simple experiment and creative speculation. Not only did he acknowledge the evidence he saw; he went on to conceive an overall scheme that made sense of that evidence (Fig. 1–6). While others of his time knew of valves and the pulmonary circulation, usually they tried to fit these details into Galen's traditional physiology. Here Harvey describes how he combined direct observation with creative reasoning to break with the traditional view:

> If I started from the root of these vessels and tried with all the skill that I could muster to pass a probe in the direction of the small vessels, I was unable to do so over any great distance because of the obstacles provided by the valves; on the other hand, it was very easy to pass a probe from without inwards, that is, from the small branches towards the root of the veins. . . .

> Since calculations and visual demonstrations have confirmed all my suppositions . . . to wit, that the blood is passed through the lungs and the heart by the pulsation of the ventricles, is forcibly ejected to all parts of the body, therein steals into the veins and the porosities of the flesh, flows back everywhere through those very veins from the circumference to the centre . . .

> I am obliged to conclude that in animals the blood is driven round a circuit with an unceasing, circular sort of movement, that this is an activity of function of the heart which it carries out by virtue of its pulsation, and that in sum it constitutes the sole reason for that heart's pulsatile movement.

Another unique aspect of Harvey's studies was his use of *quantitative* methods, which included the use of the metric system for his measurements. He calculated that the total amount of blood pumped over the course of an hour, for example, was much greater than what the body could hold or produce from food. He reasoned that, instead of each beat pumping a new quantity of blood, the same blood must be passing through the heart repeatedly. This idea reconciled the ceaseless beating of the heart with the small amount of blood that Harvey determined to be present in the body.

Harvey's discovery of the circulation was critical for our modern conception of the workings of the body, but alone he could not supply all of the details

Figure 1–6
A simplified diagram of the circular movement of the blood as envisioned by Harvey.

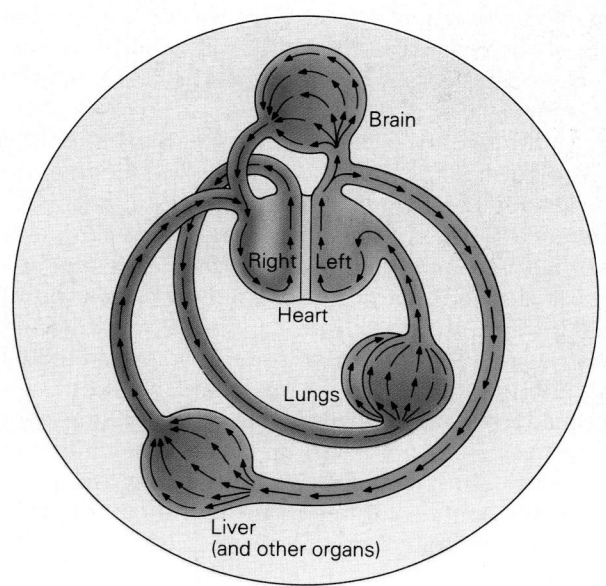

of the picture. While recognizing the pulmonary circulation, Harvey did not know its true purpose, gas exchange. He still thought of the lungs as a cooling system and continued to view blood as an elemental, irreducible substance and an innately vital force that provided the impulse for the heartbeat. We now know that the autonomic nervous system governs the beating of the heart and that blood comprises various types of cells and other matter. The existence of capillaries, which Harvey suspected but could not confirm, was discovered by the Italian physiologist Malpighi.

The Microscopic Studies of Malpighi

Harvey's explanation of the flow of blood required the existence of some physical connection between the arteries and veins so that, when blood reached the end of its course at the "circumference" of the body, it could immediately enter the veins for its return to the "centre." Harvey was unable to see the capillaries that we now know provide this connection. The invention of the microscope and the brilliant experiments of Marcello Malpighi (1628–1694) provided our first glimpse of these structures, in the lungs of a frog.

Modern Advances in Physiological Understanding

By the end of the seventeenth century, the picture of blood flow and heart action that we accept today had been set forth by Harvey and completed by Malpighi. Several other major discoveries awaited the work of eighteenth- and nineteenth-century chemists and physiologists, who would explore the role of the lungs and the substance of air, the existence and processes of cells, and the phenomenon of homeostasis.

In the twentieth century, explorations of bodily control systems, genetic processes, and bioenergetics remain the frontiers of physiological research. The pace of science has quickened. One example is the discovery of atrial natriuretic factor. In 1981 scientists showed that if part of the heart (the atrium) was ground up, made into an extract, and injected into rats, a large amount of sodium was excreted by the kidneys. This was an important breakthrough because some mechanism of this type had long been hypothesized to explain certain aspects of water and salt balance in the body. Within three years, atrial natriuretic factor had been purified, its amino acid sequence defined, assay methods developed, and synthesis by recombinant DNA technology accom-

plished. Scientists learned where it was made, how it was made, how it was metabolized, and what it did. The astonishing rapidity of this work was possible because of the rapid communication facilitated by international meetings and publications, the availability of sophisticated research methods, and the large number of scientists working in this promising area.

The Study of What We Breathe

The Greeks thought that the air was an elemental, irreducible, and living substance drawn into the lungs to cool the heart and to become *pneuma*. Galen taught that this *pneuma* was transformed into a succession of "spirits" that drove the life processes.

We now recognize that air is made up of several gases. It is not transformed in the heart but contributes oxygen to the blood in the lungs. We can trace this modern understanding to the work of seventeenth- and eighteenth-century physicists and chemists. A series of experiments by European physiologists and physicists (including Torricelli, Borelli, Boyle, Hooke, and Lower) had shown, by the mid-seventeenth century, that (1) despite its intangibility, air had weight and therefore substance; (2) part, not all, of air was necessary for both breathing and combustion; (3) the purpose of breathing was not to cool the body's "furnace," wherever it might have been, but simply to fill the lungs with air; (4) air did not move to the heart but stayed in the lungs, where it met blood that had traveled through the pulmonary circulation from the heart; and (5) in the lungs, air transformed blood in some way, as evidenced by the color change of blood from blue to red when air contacted it.

An English chemist, Joseph Priestley (1733–1804), reaffirmed the earlier findings of John Mayow that the actions of both combustion and breathing used up or removed some part of air. Using an experiment similar to Mayow's, Priestley placed a burning candle in a sealed jar (Fig. 1–7). The candle soon went out and could not be re-lit while still in the jar. Priestley then placed a green plant in the jar with the candle. After several days, he was able to re-light the candle. He repeated the test with a mouse. After a period in the sealed jar, the mouse died. After a green plant was left in the jar for a few days, however, another mouse was able to live in the jar for a longer period. Priestley deduced that some substance present in air was depleted by the processes of burning and breathing and that the substance could be resupplied by a plant. (This observation led to the later discovery of photosynthesis.)

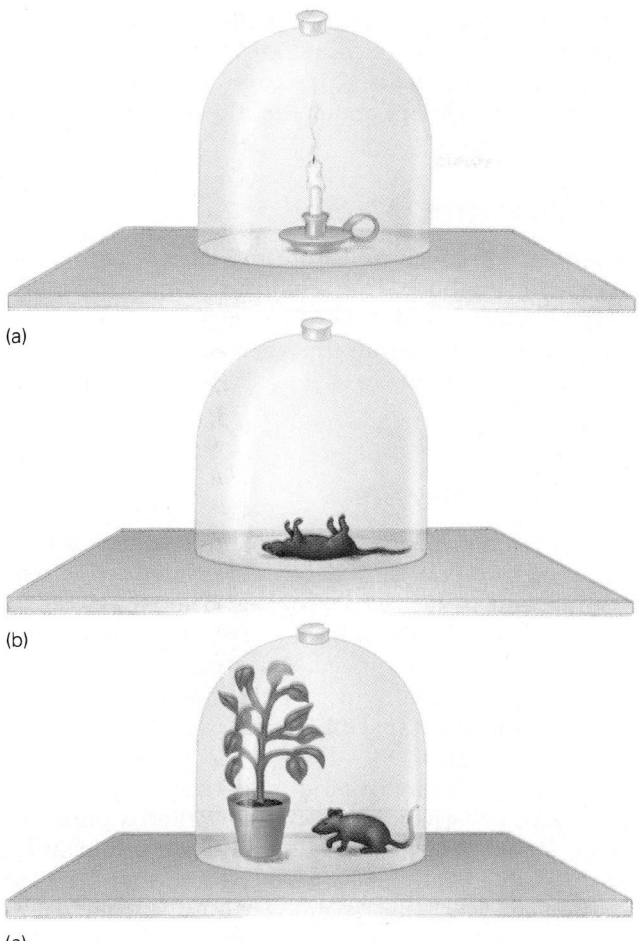

Figure 1–7
Joseph Priestley's experiment with a candle and a mouse, showing that something was removed from air during combustion and breathing. These observations later led to the discoveries of oxygen and photosynthesis.

At the time, Priestley did not recognize the meaning of his findings. Although he succeeded in isolating the pure form of oxygen from a mercury compound, he called it "dephlogisticated air," according to a prevailing theory than an element called "phlogiston" was present in air and other combustible matter and was destroyed by combustion. Priestley shared his findings, however, with the French scientist Antoine Lavoisier (1743–1794), who then identified the essential substance as oxygen.

Lavoisier thought of breathing as a physiological version of combustion because both processes required the same element. He claimed that the purpose of oxygen was to sustain the body's innate heat by allowing blood, the "combustible element," to burn. Lavoisier considered body heat the source of life and the lungs both a ventilation system and the place where combustion occurred. Later chemical advances would show that oxygen was involved in a complex trail of chemical reactions distributed throughout all the cells of the body.

The Advancement of Cell Theory

The Renaissance English scientist Robert Hooke (1635–1703), working with an early microscope, first detected the presence of small compartments he called "cells" in a piece of cork. Although he coined the term we now use to describe the basic unit of all life, it was not until two centuries after Hooke that the theory of the cell as we know it today arose in the writings of the German biologists Matthias Schleiden and Theodor Schwann. The cell theory is the broad basis of all modern biology and physiology and includes the following key concepts: (1) all organisms are made up of cells and their products; (2) new cells arise only from pre-existing cells; (3) all cells have the same fundamental chemical makeup and metabolic processes; and (4) the activities and processes of the organism as a whole result from the interdependent and cooperative workings of groups of cells. The concept that cells originate only from other cells, and not from nonliving matter as had been thought previously, was stated first by Rudolf Virchow (1821–1902) and then confirmed by Louis Pasteur's (1822–1895) experiments with bacteria.

By the late nineteenth century, chemists knew that the breaking of chemical bonds that occurred when substances such as air and food were taken into the body produced chemical energy. Louis Pasteur's research into the process of fermentation helped increase chemists' understanding of the energy-releasing processes that go on in each cell. Researchers since have identified high energy phosphate (ATP) as the repository for this energy in the cell. Twentieth-century scientists now recognize that oxygen is the substance that is linked to the utilization of ATP. The release of energy from ATP allows cell function to continue. Thus, ATP, not blood (as Lavoisier thought) is the substance whose "ignition" fuels the life of cells.

The Idea of Homeostasis

Before the invention of the thermometer, physicians and physiologists thought that body temperature varied from one individual to the next, and that, within each individual, the temperature could vary with such factors as age, location, climate, and time of day. Early scientists thought that the temperature of the body was subject to external manipulation and thus was unstable. The advent of the thermometer allowed scientists to measure each person's temperature against the same scale and observe that, apart from minute differences and variations due to abnormal (disease) states, the internal temperature of all human beings was the same regardless of what climate they lived in.

This realization led the French physiologist Claude Bernard (1813–1878) to develop the idea of the *"milieu intérieur,"* a constant internal state in which stable conditions of temperature and chemical conditions prevail despite changes in the environment. The American physiologist Walter Cannon (1871–1945) termed this process *homeostasis.* Other twentieth-century scientists since have done much research into the mechanisms by which the body maintains the optimal conditions for its own functioning.

The Study of Physiology

In a broad sense, science is a systematic method for discovering truth. Regardless of the scientific discipline (e.g., chemistry, physics, biology, or physiology), all scientists use a similar approach in their research. This approach employs a rigorous thought process in a rational, systematic, and error-minimizing way to discover new knowledge. Science is systematic because scientists attempt to organize their knowledge so that anyone who wishes to build

upon their work can obtain the necessary information. By using a set of commonly accepted rules and procedures, scientists strive to provide precise and accurate knowledge about the universe.

The Scientific Method

Scientists discover new information by using the **scientific method.** The scientific method involves posing questions and then searching for answers to those questions. Using the scientific method to obtain answers requires scientists to devise experiments.

Reasoning Processes

The thought processes that scientists use can be grouped into two general categories: **deductive reasoning** and **inductive reasoning.** The two types of reasoning can be thought of as opposites. In deductive reasoning, one proceeds from a general rule to a specific conclusion, whereas in inductive reasoning one goes from a specific observation to a general conclusion (Fig. 1–8).

In deductive reasoning, a scientist begins with supplied, or "given" information, called a **premise,** often stated in the form of an absolute rule using a word such as "all" or "always." From that premise or general rule, the scientist draws a conclusion about a specific example that matches the premise.

Figure 1–8
A schematic view of the deductive and inductive reasoning processes. In deductive reasoning, one proceeds from a premise or general law to a specific conclusion. In inductive reasoning, one goes from specific observations to a general conclusion, which can be disproved by a single exception.

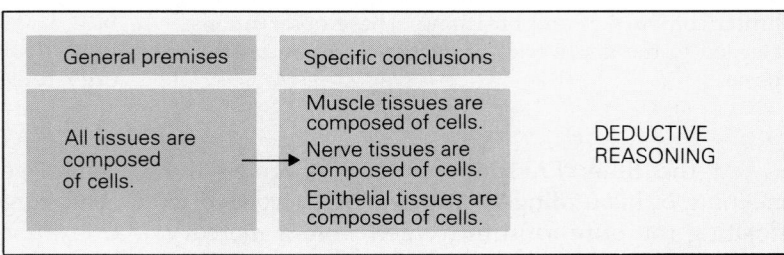

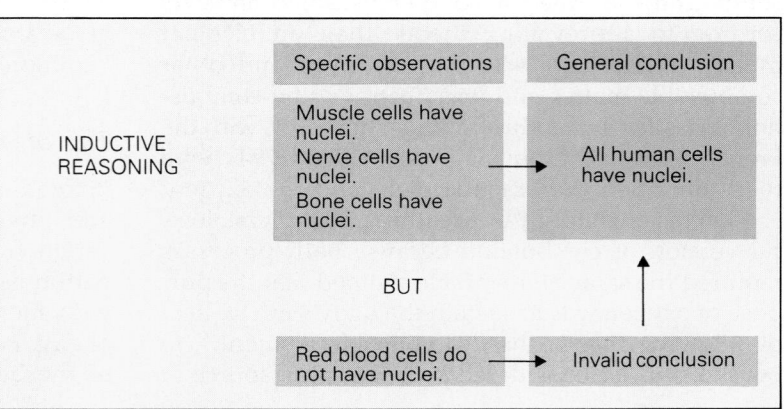

Frequently, the premise is actually a set of related premises, one general and one more specific. Following is an example of deductive reasoning:

Premise(s): (1) All tissues are composed of cells.
(2) Muscle is a type of tissue.
Conclusion: Muscle is composed of cells.

In inductive reasoning, a scientist proceeds from one or more specific observations to a general conclusion. A conclusion drawn by inductive reasoning is true only until a specific observation is found that disproves it; one specific exception is enough to invalidate the conclusion. Following is an example of inductive reasoning:

Observation(s): (1) Muscle cells have nuclei.
(2) Epithelial cells have nuclei.
(3) Nerve cells have nuclei.
(4) Muscle, epithelial, and nerve cells are all human cells.
Conclusion: All human cells have nuclei.

However, there is one exception that invalidates this conclusion. Human red blood cells lack nuclei; therefore, not all human cells have nuclei. This one example is enough to require that the conclusion be discarded.

Steps in the Scientific Method

The scientific method can be thought of as a sequence of steps, broadly classified as (1) observation, (2) hypothesis, (3) experiment, (4) conclusion, and, sometimes, (5) retesting and (6) publication.

Observation. Usually the scientific method begins with a set of observations that are not understood. For example, a neurophysiologist who is doing vision research observes a high incidence of cataracts (opacity in the lens of the eye that causes cloudiness and light scattering) in people who have been exposed to sunlight for long periods. Next, the scientist most likely begins to mull over the observation and wonder *why* cataracts form in such individuals. Usually at this stage, the scientist also talks to colleagues about the finding and searches the scientific literature for reports of the same observation. Next, the neurophysiologist may speculate about various possible causes of the cataracts, wondering, for example, whether ultraviolet rays (part of sunlight) cause the higher incidence of cataracts.

Hypothesis. Following this speculation, the neurophysiologist will form a **hypothesis,** which is a statement that tentatively explains the observation. In formulating the hypothesis, the scientist attempts to make a statement that is *testable,* realizing that it may not be true or that it can lead to new hypotheses. Following is an example of a hypothesis:

Hypothesis: Exposure to ultraviolet light causes cataracts in the lens of the eye.

Such a statement will lead to the next logical step: testing the hypothesis.

Experiment. A hypothesis is tested in an **experiment,** in which carefully specified conditions are created deliberately to produce the observations in question. The purpose of an experiment is to determine whether a cause-and-effect relationship exists between an observation and a variable (the hypothesized cause of the observation).

Ideally the experiment is designed in such a way that only one explanation exists for the observation. To accomplish this goal, most experimenters set up two sets of conditions. One set is the experimental set, in which the variable is present. The other set of conditions, called the **control,** is identical to the experimental set of conditions except that the variable is omitted. In essence, the control provides a standard to which the experimental group can be compared so that the effect of the variable can be observed. The control must contain conditions identical to the experiment; any variations or differences that creep into the design are termed **uncontrolled variables.**

Figure 1–9 illustrates an example of a controlled experiment used to test the cataract hypothesis.

Figure 1–9
A controlled experiment to test the hypothesis that ultraviolet light causes cataracts. The experimental group of animals is exposed to ultraviolet light for a certain period of time. The control group is exposed to light from which ultraviolet rays have been filtered out. All other conditions are the same for both the control and the experimental groups.

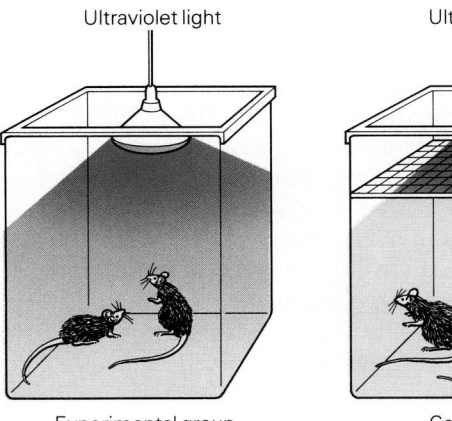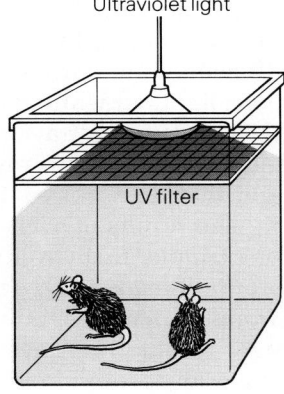

Experimental group   Control group

Since the hypothesis states that ultraviolet rays cause cataracts, ultraviolet rays are considered the variable. One group of animals, the experimental group, is exposed to light containing the variable, ultraviolet rays. Another group, the control group, is exposed to light from which the ultraviolet rays have been filtered, or removed. Apart from this difference, all other conditions are the same for the two groups: time of day, temperature of the environment, length of exposure, and so on. If one group were to be exposed inadvertently to the light for a longer period, the length of exposure would become an uncontrolled variable.

After the experiment is completed, results are collected. Results in this experiment might show, for example, that cataracts appeared in 20% of the animals in the experimental group, but in none of the control group.

Conclusion.   From the results, a **conclusion** can be drawn in which the original hypothesis is either accepted or rejected. The conclusion can be firm or tentative, depending on the results. For example, the results of the experiment with ultraviolet light point to a conclusion that mice exposed to ultraviolet light have a greater incidence of cataracts than mice that are not exposed to ultraviolet light. This conclusion in turn allows the neurophysiologist to accept the original hypothesis that ultraviolet rays cause cataracts.

Retesting.   To be able to accept the hypothesis more confidently, the physiologist can repeat the experiment, introducing another variable to determine whether any other conditions in addition to ultraviolet rays cause cataracts. For example, the researcher might dilate the pupils of the experimental mice, thereby increasing the area of the lens that is exposed to the ultraviolet rays. The rationale for this procedure would be that such an increase in exposure would cause a greater incidence of cataracts if ultraviolet light does in fact cause cataracts.

In such an experiment, four groups might be tested: Group I, the control group, not exposed to ultraviolet rays; Group II, a group exposed to ultraviolet light but without dilation of the pupils; Group III, exposed to ultraviolet light with pupils dilated; and Group IV, with pupils dilated but exposed to light that does not contain ultraviolet rays.

The results of such an experiment might show, for example, that Groups I and IV had no cataracts, Group II had a 20% incidence of cataracts, and Group III had a 30% incidence of cataracts. From these results, the physiologist could conclude that greater exposure to ultraviolet light causes a higher incidence of cataracts. This conclusion strengthens the original hypothesis that ultraviolet rays cause cataracts.

Publication.   Finally, the neurophysiologist is likely to compile the data from these experiments and submit a paper describing them to a scientific journal. The publication process is an important part of the scientific method. The sharing of new information is consistent with the goals of modern science, which strive to maintain an ongoing dialogue among scientists in order to foster greater advancement of knowledge. Many experiments have important ramifications for the health of human beings. For example, the finding that ultraviolet light causes cataracts has prompted government agencies to place warnings on sunglasses that do not filter out ultraviolet rays because sunglasses cause the pupils to dilate, thereby increasing the wearer's exposure to the harmful rays.

Not all investigations produce such clearcut and dramatic results. Many experiments fail because of uncontrolled variables, equipment malfunction, or poor design. In other cases, experimental results do not support the hypothesis, and a new hypothesis and experiment must be formulated. When experiments fail, scientists often talk to colleagues and return to the literature, reviewing the work of other investigators for helpful clues. Often, a finding in a related field provides insights not available from a study of the scientist's own specialty. The history of science contains many examples of experiments that seemed at first to have failed, only to lead to a major breakthrough as the researcher pondered the reasons for the failure.

Measurement, Computation, and Graphic Analysis

In the study of physiology, **quantitative methods** are invaluable for understanding bodily functions and processes. A quantitative approach aided William Harvey in his discovery of the circulation of the blood, revealing that the total amount of blood pumped by the heart was too great for the body to hold unless the same blood was pumped over and over again. Often in physiology it is the comparison of amounts, rates, times, and other measurable evidence of the body processes that reveals relationships between function and structure.

Units of Measurement

The metric system is almost exclusively the system used for physiological measurements. Length is expressed in **meters,** mass in **grams,** and volume in

liters. A length of 1 meter is equivalent to 39.37 inches. A mass of 1 gram is equivalent to 0.035 ounces, and a volume of 1 liter is the same as 0.26 gallons. These metric terms usually are abbreviated for convenience to **m** for meters, **g** for grams, and **L** for liters. A length of 1000 meters would be written as 1000 m.

Prefixes are used to indicate multiples or fractions of these basic units, and the prefixes are abbreviated also. The prefix **kilo,** abbreviated as **k,** indicates a multiple of 1000 of the basic unit. A length of 2000 meters (m), for example, is equivalent to 2 × 1000 m, or 2 kilometers (km). The combinations of prefixes and units that are most commonly used in physiological measurements are summarized and defined in Table 1–1.

These prefixes are necessary in order to accommodate the enormous range of sizes encountered in living organisms. A human being with a height of 1.83 m (equivalent to 6.0 feet) is composed of cells that for the most part have a diameter of about 10 μm. The cells can be discerned only with the aid of a microscope because the human eye cannot see objects smaller than about 100 μm. The cells contain amino acid molecules that are only 1 nm in diameter, and the individual atoms of the amino acids may have diameters that are less than 0.1 nm.

Table 1–1

Commonly Used Units of Measurement

The meter (m) is the unit of length.
1 kilometer (km) = 1×10^3 = 1000 m
1 decimeter (dm) = 1×10^{-1} = 0.1 m
1 centimeter (cm) = 1×10^{-2} = 0.01 m
1 millimeter (mm) = 1×10^{-3} = 0.001 m
1 micrometer (μm) = 1×10^{-6} = 0.000001 m
1 nanometer (nm) = 1×10^{-9} = 0.000000001 m

The gram (g) is the unit of mass.
1 kilogram (kg) = 1×10^3 = 1000 g
1 milligram (mg) = 1×10^{-3} = 0.001 g
1 microgram (μg) = 1×10^{-6} = 0.000001 g
1 nanogram (ng) = 1×10^{-9}
  = 0.000000001 g
1 picogram (pg) = 1×10^{-12}
  = 0.000000000001 g
1 femtogram (fg) = 1×10^{-15}
  = 0.000000000000001 g

The liter (L) is the unit of volume.
1 deciliter (dL) = $1 \times 10^{-1} = 0.1$ L
1 milliliter (mL) = 1×10^{-3} = 0.001 L
1 microliter (μL) = 1×10^{-6} = 0.000001 L
1 nanoliter (nL) = 1×10^{-9} = 0.000000001 L

Scientific Notation

In physiology, as in many of the sciences, it is often awkward or inconvenient to express measurements using standard digits (e.g., the diameter of a blood vessel as 10,000 μm, a blood glucose concentration of 0.005 moles per liter). Instead of writing out all the "zeroes," scientists use a system of **scientific notation,** in which quantities are expressed as products of a number and a power of ten. Powers of ten are expressed with **exponents,** which are small numbers written above and to the right of the base number, ten. The exponent indicates the number of times the base should be used as a factor in the expression. For example, $10^2 = 10 \times 10 = 100$. Thus,

10,000 μm = 1×10^4 μm
0.005 moles/liter = 5×10^{-3} moles/liter
0.001 grams = 1×10^{-3} grams

(Note that negative exponents indicate the number of times 10 is used as a *divisor* for quantities less than one. For example, $10^{-3} = 1/10 \times 1/10 \times 1/10$.) Numbers without zeroes can also be expressed in scientific notation (e.g., $46,782 = 4.6782 \times 10^4$).

Following are some simple rules for converting a number expressed in scientific notation back to the original number:

1. For positive exponents: Shift the decimal point to the right by the number of places indicated by the exponent.
2. For negative exponents: Move the decimal point to the left by the number of places indicated by the exponent.
3. When multiplying exponential quantities that have the same base, add the exponents ($5^3 \times 5^2 = 5^5$).

Ratios and Proportions

A **ratio** is an expression that compares two numbers or quantities by division. For example, to compare the numbers 100 and 10, we divide 100 by 10 to get 10/1. If we are describing the number of experimental subjects that were exposed to high concentrations of ozone, we might say that of 100 rats exposed to ozone, 10 died. The ratio of animals that died to the total number tested was 10/100 or 1/10 ("one to ten"). In other words, 1 out of every 10 animals died from the ozone exposure. Ratios can be expressed in several different ways with the same meaning:

1:250 means 1 part to 250 parts
1/5 means 1 part out of 5 parts
2 cats to 6 dogs means a ratio of 1 cat to every 3 dogs

A **proportion** is a mathematical statement of the equality of two ratios. By arbitrarily using the letters *A, B, C,* and *D* to express quantities, we can state a proportion in the following way:

A is to *B* as *C* is to *D*

This is equivalent to saying:

A times *D* equals *B* times *C*

For example, if $A = 15$, $B = 25$, $C = 3$, and $D = 5$, then

$$\frac{A}{B} = \frac{C}{D} = \frac{15}{25} = \frac{3}{5}$$

$$A \times D = B \times C; \; 15 \times 5 = 25 \times 3$$

Therefore, it follows that if three of the quantities are known, the value of the fourth can be determined.

For example, assume that an electrocardiogram is being recorded on a moving strip of paper (Fig. 1–10). The speed of the moving paper is 25 mm/sec. If each repeating cycle of the electrocardiogram represents one heartbeat, how many heartbeats are occurring each minute? The problem can be solved as follows:

1. The distance between cycles is 20 mm (as measured from the record).
2. The time interval between cycles is unknown; it equals *x* sec.
3. The ratios relating distance to time can be expressed as a proportion:

$$\frac{25 \text{ mm}}{1 \text{ sec}} = \frac{20 \text{ mm}}{x \text{ sec}}$$

4. $25x = 20$; $x = 0.8$ The interval between cycles is 0.8 seconds.
5. The number of cycles (beats) occurring each minute = *y* cycles (beats):

$$\frac{1 \text{ beat}}{0.8 \text{ sec}} = \frac{y \text{ beats}}{60 \text{ sec}} \quad 0.8y = 60$$

$$y = 75 \text{ beats/minute}$$

Interpretation of Graphic Data

Much of the study of physiology involves learning about relationships between variables. For instance, there is a definite relationship between the heart rate and the rate of exercise. There is a similar relationship between the body's oxygen consumption and the rate of exercise. Often such relationships can be expressed and understood through graphs.

A **graph** is a diagram that expresses a relationship between two or more quantities. In some cases there is a definite cause-and-effect relationship

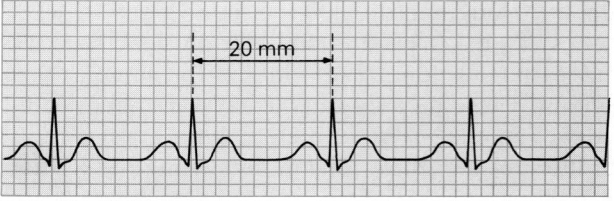

Paper speed: 25 mm/sec

Figure 1–10
A proportion based on the speed of the paper and the distance between points on this electrocardiogram can be used to determine the number of heartbeats per minute.

whereas in others the association is not as direct, but may be due to a third factor. Graphic presentation of data may not explain the reason for the relationship, but it can provide clues by illustrating the shape of it. A graph puts into visual form abstract ideas or experimental data, so that their relationships become apparent.

Variables. The related quantities displayed on a graph are called **variables.** The simplest sort of graph uses a system of coordinates or axes to represent the values of the variables. Usually the relative size of the variable is represented by its position along the axis, and numbers along the axis allow the reader to estimate the values. If the relationship being plotted is one of cause and effect, the variable that expresses the cause is called the **independent variable.** Usually this is represented by the horizontal axis (also sometimes called the *x*-axis or the **abscissa**). The variable that changes as a result of changes in the independent variable is the **dependent variable.** It is usually represented on the vertical axis (also called the *y*-axis or **ordinate**). The two axes are arranged at right angles to each other and cross at a point called the **origin** (Fig. 1–11).

To show the relationship between two variables that are directly related (at some specific value, such as point *A* in the figure), the value on the *x*-axis (X_1) is extended vertically, and the corresponding value on the *y*-axis (Y_1) is extended horizontally. The point *A* at which these lines cross is determined by their relationship. If another pair of points (X_2 and Y_2) is chosen, their point of intersection on the graph also can be plotted; this is point *B*. A line drawn between points *A* and *B* can then give information about how all other *x* and *y* values on this graph should relate to each other.

Types of Relationships. As you may have guessed, this explanation represents a very simple case in which some important assumptions were made. We

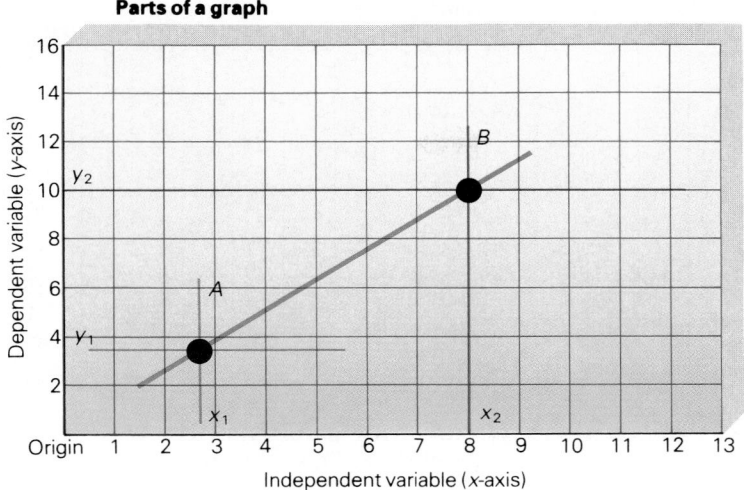

Parts of a graph

Figure 1–11
In a typical line graph, the vertical axis (the *y*-axis or ordinate) shows the values for the dependent variable, whereas the horizontal axis (the *x*-axis or abscissa) shows values for the independent variable. The two axes meet at a point called the origin.

first assumed that for every *x*-value there was only one *y*-value, and we further assumed that all of the *y*-variables were directly related to all of the *x*-variables. This is the simplest kind of relationship that a graph can represent. It is called a **direct relationship:** the *y*-values get larger as the *x*-values get larger. An example of this type of graph is shown in Figure 1–12. The data plotted here could have come from an experiment in which various concentrations of an enzyme were used to study how fast a particular chemical reaction happened at each concentration.

Some of the imperfections of the "real world" are evident here. The data points show that there is not a perfect relationship between each *x*-value and its corresponding *y*-value. We would say that the data show "scatter." (Sometimes this sort of graph is called a **scatter diagram.**) In many cases a mathematical procedure may be carried out to determine the "best-fitting" line to describe the relationship. This is called the **ideal line** in the figure. A relationship that can be described by a straight line is called a **linear relationship.**

Inverse relationships are also common. In such relationships, the *y*-values get *smaller* as the *x*-values get larger. Such relationships may still be linear if they are described by straight lines. Other inverse relationships may be curvilinear, as in Figure 1–13. The graph shown here summarizes the experimental finding that a muscle exerting a large force must contract more slowly than one exerting only a small force. A graph of this sort is adequate for the process of interpolation within the range of data, but it is poor for extrapolation outside that range.

Interpolation and Extrapolation. If an experimenter had enough confidence in the reliability of the data, two kinds of predictions could be made

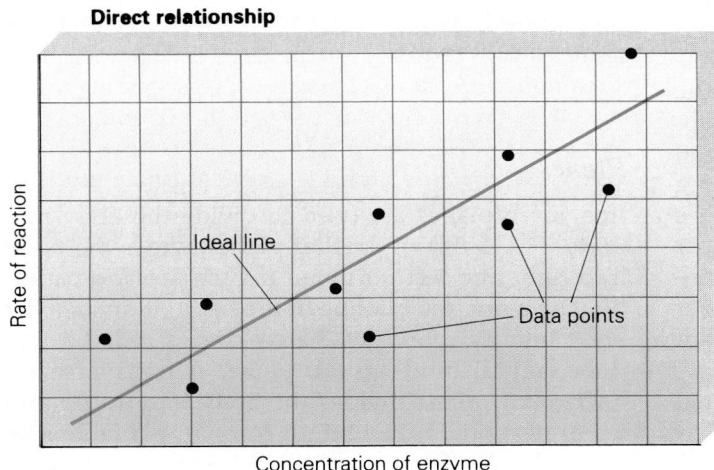

Direct relationship

Figure 1–12
In a direct relationship, *y*-values increase as *x*-values increase. Data points frequently do not fall exactly on a straight line but are "scattered" about an ideal line, which is drawn to show the general shape of the relationship. A linear relationship can be shown by a straight line.

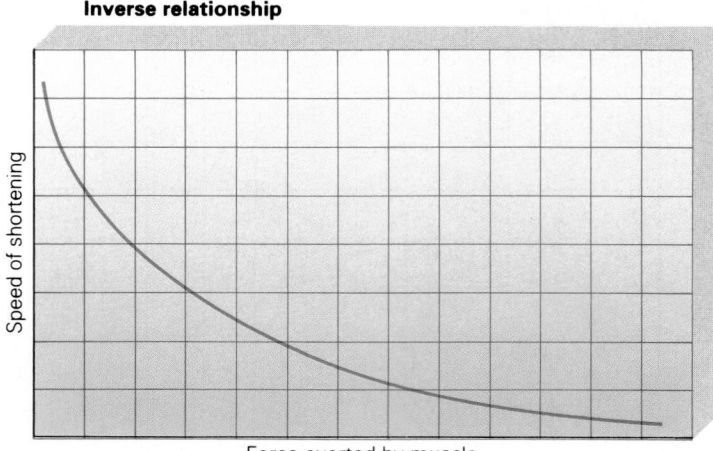

Inverse relationship

Speed of shortening

Force exerted by muscle

Figure 1–13
In an inverse relationship, *y*-values decrease as *x*-values increase, thus producing a downward-sloping line or curve.

from such a graph. Predicting data values that fall "between the points" is called **interpolation.** This process is useful if the curve is to be used as a guide for interpreting or testing the reliability of newly obtained data. A more risky procedure is **extrapolation,** which involves extending the ideal line into ranges where experimental data are not present. If there is good reason to believe that the same relationship should hold outside this range, then this prediction could be valid and might permit useful information to be gained.

However, most relationships found in nature are not simple. Over some ranges, a relationship may be direct (and linear, as in the example in Fig. 1–12), and then it may change to another form as a wider range of variables is considered. In this case, extrapolation from a limited amount of data could lead to a wrong conclusion.

Take a few moments to flip through the pages of this book. You will see many graphs. Some express simple relationships over their entire range of data, whereas others are more complicated, expressing several relationships at once. Some, especially those that show the way in which an important variable such as blood pressure varies with time during the course of a heartbeat, present a very complex appearance. In some cases, there are several lines (and possible extra axes) on the graph, each describing some aspect of the idea being presented. Some graphs illustrating other vital processes appear much simpler. Some use bars or parts of a circle instead of lines to illustrate relationships. Whatever their form, all these graphs are designed to present important relationships in the clearest possible way. When you learn to interpret information presented graphically, you are well on your way to understanding the language of physiology.

Anatomical Planes and Positions

In order to study functions of the human body, one must have a basic knowledge of anatomical terminology since function is often described in terms of bodily structures. Conventional lay terms such as *top, bottom, above, below, behind, under,* and *on the side of* can be ambiguous when used in complex descriptions of anatomy. To ensure precise understanding, anatomists have devised a special vocabulary that is part of the language of anatomy and physiology. The following survey of anatomical terms is by no means exhaustive. It is meant to introduce the correct meaning and use of common anatomical terms that appear in subsequent chapters. When encountering new terms, you will find a medical dictionary an invaluable aid for mastering meaning and usage.

Structures of the body are always described with reference to a standard static position called the **anatomical position** (Fig. 1–14). In this position, the body is erect, the face and feet are directed forward, and the arms are straight and directed toward the ground, with the hands rotated so that the palms face forward.

Planes

Imaginary planes are used to divide the body into parts (Fig. 1–14). A **sagittal** plane divides the body into right and left portions. If such a plane passes directly along the midline of the body, it divides the body into right and left halves and is called a **median sagittal (midsagittal) plane.** A **transverse,** or **horizontal, plane** divides the body into upper and lower portions. A **frontal,** or **coronal, plane** divides the body into front and back portions.

Positions

Terms of position are used to locate a structure relative to other structures. **Anterior** means "nearer to the front of the body," whereas **posterior** means "nearer to the back of the body." (Sometimes **ventral** and **dorsal** are used in place of **anterior** and **posterior,** respectively.)

Medial means "nearer to the midsagittal plane," whereas **lateral** means "further from the midsagittal plane." **Superior** means "nearer to the head," whereas **inferior** means "nearer to the lower end of the body." (Sometimes the terms **cranial** and **caudal** are used instead of **superior** and **inferior,** respectively.)

Internal means "nearer the center of an organ, cavity, or part of the body"; **external** means "farther from the center." **Superficial** means "nearer to the surface" of the body; **deep** means "farther from the surface."

Figure 1–14
A diagram showing basic anatomical planes.

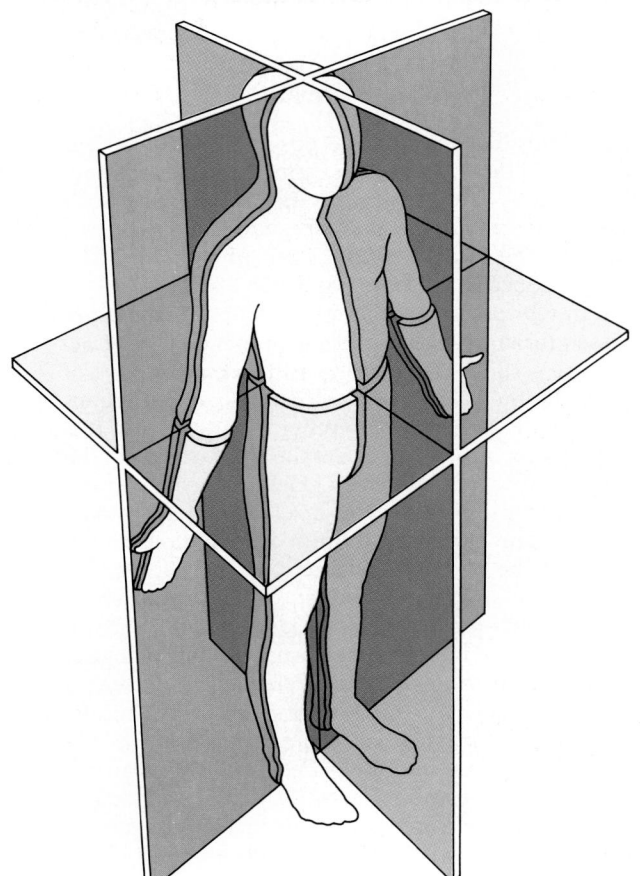

Two special terms are used in describing extremities or structures related to their long axes. **Proximal** means "nearer to the origin or point of attachment," whereas **distal** means "farther from the origin or point of attachment." For example, the humerus (the bone in the upper arm) is attached at its proximal end to the shoulder and at its distal end to the elbow.

General Organization of the Body

Living matter is made up of **protoplasm,** a complex mixture of chemicals that displays the attributes of life. These attributes include (1) organization into specific kinds of structural units; (2) the ability to enter into chemical activities that include the transformation of energy and the maintenance or synthesis of protoplasm; (3) the ability to respond to changes in the environment; and (4) the ability to grow and reproduce.

Protoplasm is organized into cells, which are the basic structural and functional units of the human body and of all life. A **cell** can be defined as a microscopic bit of organized protoplasm surrounded by a membrane called the **plasma membrane.** The adult human body comprises about 100 trillion cells, all of which function collectively to maintain an individual's life.

Each human life begins as a single cell, a fertilized ovum, which then divides to form two cells, four cells, eight cells, and so on. In addition to undergoing numerous cell divisions during development, cells also begin to exhibit specialized functions. There are about 200 different cell types in the body, as determined by differences in both structure and function.

Cell Differentiation and Specialization

The process by which a single cell type (the fertilized ovum) develops into many different cell types is known as **cell differentiation.** As you will see, these different cell types are arranged first into **tissues,** or groups of related cells. Tissues in turn are arranged in groups to form organs, and organs in turn form systems. An analogy can be drawn between the cells of the body and a society of individuals. Just as mail carriers, police, doctors, and teachers have specific roles for the common good of the entire society, so too do specialized cells such as muscle, nerve, connective tissue, and epithelial cells serve specific functions to promote the survival of the organism. The body can be thought of as a society or social order of cells.

In a multicellular organism such as a human being, labor is divided among groups of specialized cells, each group performing one principal function such as movement, digestion of food, or reproduction. Specialization of cells has been an essential factor in the development of the large size of multicellular animals. The different groups of cells work in concert to maintain the life of the organism, and considerable interdependence exists among the different cell types. For example, most of the cells of large animals depend on the red blood cells to provide them with oxygen, while at the same time the red blood cells depend on the pumping action produced by the muscle cells of the heart to propel them through the body. Such a division and sharing of labor have allowed multicellular organisms to grow to great size, whereas a unicellular organism must carry out for itself all the life processes, and can exist independently only by remaining microscopically small.

Cells and Tissues

Because multicellular organisms require division of labor and specialization among cells, the cells in an individual's body differ in size, shape, internal architecture, and function. Some cells are spherical; others are shaped like cubes, discs, stars, or cylinders. Each cell manifests the form and structure best suited to its function (Fig. 1–15). The neuron (a nerve cell), for example, bears numerous cellular extensions that act to receive and transmit the chemical and electrical signals that form the basis for nervous function. A skeletal muscle cell has a long cylindrical shape, which helps the cell move other parts of the body when it contracts. The "bi-concave disc" shape of a red blood cell helps it in its work of transporting oxygen through the body.

When one or more types of specialized cells become closely associated, they form a tissue. There are four classes of tissues in animals: epithelial tissue, connective tissue, muscle tissue, and nervous tissue (Table 1–2).

R E S E A R C H   F O C U S

Imaging of Internal Structures

The spectrum of electromagnetic radiation contains wavelengths that are suitable for use in imaging techniques. Visible light, wavelength 10^{-7} to 10^{-6} m, is the most familiar form of electromagnetic radiation. It bounces off the surface of solid structures, allowing us to see them. In contrast, X-rays (wavelength 10^{-11} to 10^{-8} m), gamma rays (10^{-18} to 10^{-10} m), and radio waves (10^{-1} to 10^{4} m) pass through most solid objects and can be used to study the internal structure of the human body. Some of the specific techniques are discussed briefly here.

Radiography is a technique for studying internal structures using X-rays. The rays are passed through the body onto a photographic plate to produce a two-dimensional photograph. Some parts of the body, such as bone, absorb more of the X-rays than other parts and can be distinguished in the photograph. The photograph, usually called an X-ray, used to be known as a **roentgenogram** because X-rays were discovered by Wilhelm Roentgen, a German physicist, at the end of the 19th century. Roentgen used X, the symbol for an unknown quantity, because he did not understand completely the phenomenon he had found. In modern times radiography has become a very useful diagnostic tool. **Contrast radiography** is useful for viewing certain soft tissues which are not seen easily in normal X-rays. For example, the upper

gastrointestinal tract can be observed if the patient swallows a drink containing barium prior to an X-ray. Since X-rays cannot pass through barium, the position of the gastrointestinal tract, now lined with barium, can be located in the X-ray photograph. The **computerized axial tomograph (CAT scan)** is a system in which X-rays from many different angles are passed through a selected plane of the body in a fraction of a second. Detectors measure the absorption of X-rays by the various body structures and all the information is fed into a computer. The computer displays a reconstructed image of the cross-section of the body on a television screen. Hundreds of different cross-sections, like the slices in a loaf of bread, can be obtained and combined to form a three-dimensional image of the contents of large cavities in the body, such as the abdomen or thorax. The **positron emission tomogram (PET scan)** produces very clear images and can identify chemical and physiological changes. It uses gamma rays released by chemicals made especially for this purpose. These chemicals, which are similar to those normally present in the body, are injected into the patient. The PET scan detects the gamma rays released inside the patient and uses them to form an image. This image provides information about where the chemicals are used in the body. One use of the technique has been in studies of

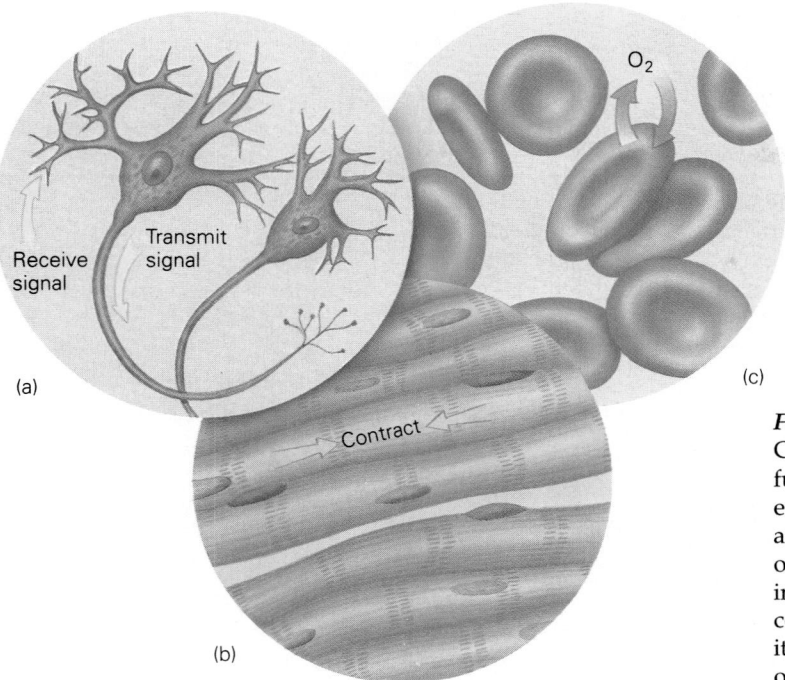

(a)

(c)

(b)

Figure 1–15
Cell shapes are designed to allow optimal functioning of specialized cells: (a) The long extensions of the neuron enable it to receive and transmit nervous signals. (b) The fibers of a muscle cell allow it to change shape during contraction and relaxation. (c) The biconcave disc shape of a red blood cell assists it in carrying oxygen to the tissues and cells of the body.

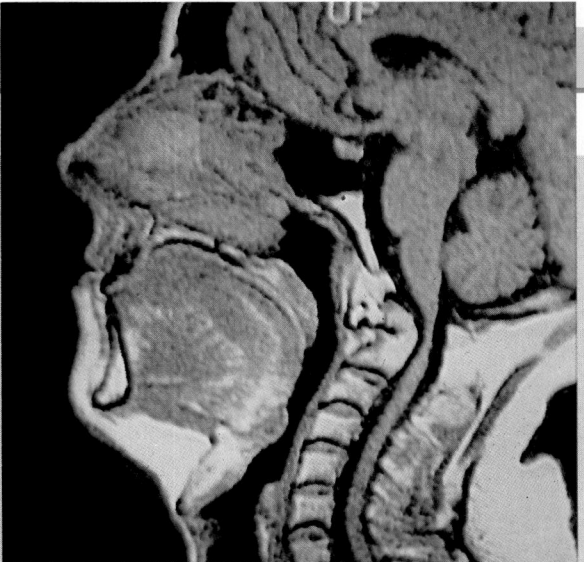

MRI of a herniated brain. (© Howard Sochurek/ Medichrome/The Stock Shop.)

the brain to gain information about blood flow and utilization of glucose, an important source of energy.

Magnetic resonance imaging (MRI) is also known as nuclear magnetic resonance (NMR). It uses low energy radio waves and a strong magnetic field, and can produce excellent cross-sectional views without exposing the patient to high energy gamma rays or X-rays, or to a contrast medium such as barium. MRI also can provide information about structure/ function changes as well as the location of specific chemicals in vivo.

Ultrasonography is based on a different principle. It uses high frequency sound waves (**ultrasound**) to view internal anatomy and is better than X-rays for imaging soft tissues. The equipment consists of a transducer that produces and receives the silent high frequency sound waves. The transducer is placed against the body and slowly passed over the area to be examined. Sound waves pass through the skin into the body. As they strike various organs, they send echoes back to the transducer. Different kinds of tissue, bone, blood, and other fluids produce different echoes, which are separated and identified by the transducer. The transducer forms a visual representation of the varying echoes as it changes sound waves into electrical energy. This electrical signal is converted into an image on a television screen. Ultrasound is particularly useful for visualizing the pregnant uterus and for detecting abnormalities of fetal organs. It can be repeated as often as required without harm. **Echocardiography** is a form of ultrasound used to examine the pumping action of the heart, especially the position and movement of valves and other structures within the heart.

All of these techniques are painless and are described as **noninvasive** because they allow the internal structures of the body to be studied without physical intervention such as surgery.

Table 1–2

Examples of Cell Types	Examples of Cell Functions	Examples of Tissue Functions	Tissue	Subtypes	Major Tissue Type
squamous cells cuboidal cells columnar cells	formation of protective layers	lining of body surfaces	simple epithelia stratified epithelia	EPITHELIAL SHEETS	
lacrimal gland cells	secretion of tears				
sweat gland cells	secretion of sweat				
intestinal gland cells	secretion of digestive juices	secretion of substances through ducts	exocrine glands		
liver cells	secretion of bile				
sebaceous gland cells	secretion of sebum				EPITHELIAL TISSUE
parathyroid cells	secretion of parathyroid hormone			GLANDS	
thyroid cells	secretion of thyroxine				
hypothalamic cells	secretion of oxytocin				
adrenal cells pituitary cells	secretion of epinephrine secretion of human growth hormone	secretions of hormones directly into blood	endocrine glands		
ovarian cells	secretion of estrogen				
testicular cells	secretion of testosterone				
fibroblasts	production of polymers that form connective tissue fibers	firm support and transmission of mechanical forces	dense regular connective tissue dense irregular connective tissue	DENSE CONNECTIVE TISSUE	
macrophages (histocytes) lymphocytes mast cells	protection against infection immunity secretion of heparin	soft support and protection	areolar connective tissue adipose connective tissue reticular connective tissue	LOOSE CONNECTIVE TISSUE	
osteoblasts	secretion of bone matrix				

Table 1–2 continued

Examples of Cell Types	Examples of Cell Functions	Examples of Tissue Functions	Tissue	Subtypes	Major Tissue Type
osteocytes	bone metabolism	rigid support of body structures		BONE	
osteoclasts	removal of old bone				CONNECTIVE TISSUE
chondrocytes	secretion of intercellular substance	flexible support of body structures		CARTILAGE	
chondroblasts	production of chondrocytes				
bone marrow cells	production of all red and most white cells	production of blood cells	myeloid tissue	BLOOD-FORMING TISSUE	
lymph node cells	production of some white cells		lymphoid tissue		
red blood cells	transport of oxygen	circulation of oxygen and immune factors		BLOOD	
white blood cells	protection against foreign substances				
skeletal muscle fiber	voluntary movement of bones			SKELETAL MUSCLE	
smooth muscle fiber	involuntary movement of blood vessels, ducts, and other structures			SMOOTH MUSCLE	MUSCLE TISSUE
cardiac muscle fiber	rhythmic movement of heart			CARDIAC MUSCLE	
unipolar neurons	transmission of sensory impulses				
bipolar neurons	transmission of sensory impulses	transmission of nervous impulses		NEURONS	
multipolar neurons	transmission of motor impulses				

(Table continues on p. 22)

Table 1–2 continued

Examples of Cell Types	Examples of Cell Functions	Examples of Tissue Functions	Tissue	Subtypes	Major Tissue Type
astrocytes	support of neurons		neuroglia of the central nervous system		NERVOUS TISSUE
oligodendrocytes	formation of myelin sheaths	support and protection of neurons		NEUROGLIA	
microglia	response to injury		neuroglia of the peripheral nervous system		
Schwann cells	formation of myelin sheaths				
satellite cells					

Epithelial Tissue

Epithelial tissue consists of closely packed cells arranged in flat sheets ranging from one to several layers in thickness. This type of tissue covers exposed body surfaces and lines body cavities, tubes, and organs. Epithelial tissue protects the body parts and absorbs, filters, and secretes substances. The skin is made of epithelial tissue.

Connective Tissue

The cells that make up **connective tissue** are somewhat loosely arranged and separated by an intercellular **matrix** (ground substance). The matrix material, made by the connective tissue cells themselves, frequently contains cellular products such as fibers, soluble proteins, and crystalline complexes. Connective tissue is the most widely distributed tissue in the body. It supports, protects, binds, and partitions nearly all bodily components. Specific examples are cartilage, bone, ligaments, blood, and adipose (fat) tissue.

Muscle Tissue

The cells of **muscle tissue** are elongated and contain many parallel contractile fibers called **myofibrils.** In general, muscular tissue is specialized for contraction and functions to accomplish movement of the organism. There are three types of muscle tissue:

skeletal muscle, cardiac muscle, and smooth muscle. **Skeletal muscle** is attached to the bones and is used for conscious, voluntary movements of the body, such as walking. **Cardiac muscle** is present only in the walls of the heart and produces the autonomic pumping contractions of the organ. **Smooth muscle** is found in hollow internal organs, such as the blood vessels, and allows them to undergo autonomic, "involuntary" contraction and expansion.

Nervous Tissue

The chief components of **nervous tissue** are **neurons,** cells that are specialized for receiving and conducting electrical impulses. The characteristic shape of a neuron is an enlarged main cell body with hairlike extensions that receive and transmit the nerve impulses. **Glia,** sometimes called **glial cells** or **neuroglia,** provide neurons with structural and metabolic support and protection and also form part of nervous tissue. This type of tissue is the primary component of the brain and spinal cord. In addition, bundles of nervous tissue called simply **nerves** are found in all parts of the body.

Organs and Systems

Combinations of the primary tissues make up the organs of the body. An **organ** is defined as a group of tissues that have been combined for the performance of a specific function or series of related functions. The stomach, for example, contains all four types of primary tissues, organized for the purpose of receiving, storing, and digesting food and for moving partially digested food into the small intestine for further processing. No single tissue can perform as well as several tissues working together.

In a similar manner, organs and other structures that share in the performance of related tasks are grouped into **systems.** The urinary system, for example, consists of the kidneys, ureters, urinary bladder, and urethra. These organs work toward the common purpose of removing waste products from the blood and eliminating them from the body.

The grouping of organs and related structures into systems provides a logical approach for a textbook such as this. Keep in mind as you proceed through the various chapters, however, that although many levels of organization and function exist in the human body, all of the body's components function collectively to maintain an individual's integrity and life.

Bodily Fluid Compartments

The body can be thought of as mostly a watery solution. Approximately 60% of its weight is water. The body fluids contain various mineral ions (e.g., sodium, potassium, chloride) and organic substances (e.g., proteins, glucose) dissolved in the body water. Bodily fluids usually are grouped into two major divisions or compartments: **intracellular fluid** and **extracellular fluid.**

The two compartments are separated by the plasma membranes of each of the body's cells, and the chemical compositions of the two compartments differ strikingly. The differences in chemical composition are maintained by the cells themselves and have important effects on the cell membrane potential, cell excitability, cellular metabolism, and the ability of the cell membrane to transport substances in and out of the cell. Much of cell physiology is concerned with the mechanisms for maintaining differences in the ionic composition of intra- and extracellular fluids and the consequences of these differences for cell function.

Intracellular Fluid

The intracellular fluid is the fluid within cells. It is not, therefore, one single continuous compartment, but rather a conglomeration of trillions of cellular subcompartments. Intracellular fluid has a relatively low concentration of sodium ions and a high concentration of potassium ions.

Extracellular Fluid

The extracellular fluid is the fluid outside of the cells. Extracellular fluid, in contrast to intracellular fluid, has a high concentration of sodium ions and a low concentration of potassium ions. The extracellular fluid that bathes our cells is a medium of exchange between one cell and another and between cells and the external environment.

Extracellular fluid consists of three types: interstitial fluid, plasma, and lymph. **Interstitial fluid** is the fluid found between body cells. **Plasma** is the fluid portion of the blood, and **lymph** is the fluid in lymphatic vessels. Together these extracellular fluids make up the **internal environment** of the body, whose constant regulation is the purpose of the complex physiological processes, as we are about to discover.

Homeostasis: Control of the Internal Environment

In a complex, multicellular organism such as a person, most of the living cells of the body are not directly exposed to the gaseous external environment (the atmosphere) but exist in the liquid internal environment of the extracellular fluid that consists of lymph, plasma, and interstitial fluid. Conditions in this extracellular fluid must be maintained within certain narrow limits in order to permit our cells to live and function. Conditions such as oxygen tension, pH, osmotic pressure, temperature, and the concentrations of various metabolic substrates, hormones, and waste products are all closely regulated by the cooperative workings of tissues, organs, and systems. At a certain temperature, the human body will function well; at another temperature, its processes will fail. When certain concentrations of ions exist in its internal environment, the body will thrive; at lower or higher concentrations, it will be unable to live or function well.

In many diseases, the composition and/or volume of the internal environment becomes abnormal. Consider the person with pulmonary disease, for example, in whom arterial blood is not oxygenated adequately. The oxygen tension in such a person's internal environment will be reduced. Consider the person with renal failure, who cannot get rid of metabolic waste products at an adequate rate. In such a person, toxic waste materials and hydrogen ions will accumulate in the extracellular fluid; essentially, the body will be poisoned by its own internal environment. The functioning of body cells will be disturbed under such conditions as these.

The organism will be forced to lead a more restricted existence.

The famous nineteenth-century French physiologist Claude Bernard was the first to point out that **constancy,** or stability, in the internal environment—the extracellular fluid—of an organism is the requisite condition for a free and independent existence. Coining the phrase *milieu intérieur,* "internal environment," Bernard paved the way for our current understanding of **homeostasis,** the stable state of the internal environment that is maintained by physiologic processes. Despite the many changes that may occur in the external environment, the temperature and composition of the extracellular fluid remain constant. Factors such as ion concentrations, oxygen content, nutrient and waste product concentrations, and temperature normally do not change appreciably, either minute by minute or day by day.

Indeed, much of the study of physiology is taken up with the analysis of the regulatory processes by which the body maintains the constancy of the internal environment. Some of these regulatory processes are simple, whereas others are highly complex. As you read about each of the various mechanisms in later chapters of this book, bear in mind that each process serves to accomplish, at least in part, one common goal, which is to maintain homeostasis of the internal environment of the body. Thus, for example, although the regulation of breathing in the lungs and the secretion of digestive enzymes into the intestine seem to have little in common, each of these processes serves in the end to maintain homeostasis.

The Advantages of an Internal Environment

In a single-celled, primitive organism (e.g., an amoeba living in a pond), there is no internal environment, simply because there is no *extra*cellular fluid that is still part of the organism. The entire environment of the unicellular creature is outside it, beyond its borders and therefore beyond its control. Such an organism is subject to the vagaries of the external world. Changes in temperature, light, and concentrations of various chemicals in the external environment dictate its existence. It has little freedom since it has no internal environment that it can manipulate and use as a buffer or mediator between itself and the external world. If the water (its external environment) freezes, the amoeba freezes also. If it does not die, it is at least immobilized.

A more highly developed organism, such as a frog, has a greater degree of independence from external conditions. By virtue of its multicellularity, it has both an external *and* an internal environment. Its internal environment is the extracellular fluid that surrounds its cells; this fluid is *outside* the cells but still *inside* (and therefore part of) the frog. The frog, however, is unable to regulate its own body temperature. When the weather turns cold, it may not freeze, but it will be forced into inactivity. Reptiles such as snakes and alligators must spend a great deal of time simply sitting in the sun after periods of exposure to cold air or water, in order to raise the temperature of their internal environment to a level that permits life processes to continue at full speed.

In contrast to reptiles, mammals have developed a great deal of freedom from external conditions by evolving sophisticated mechanisms for maintaining a stable internal environment. Most noticeably, their internal body temperatures remain virtually constant under a wide range of external temperature conditions, by virtue of physiological and behavioral mechanisms that counteract the effects of cold or heat in the external environment. Mammals are able to keep their body temperature close to 37°C whether the air around them is at 0°C or 45°C. Furthermore, unlike frogs, many mammals are fully active under all climatic conditions; they do not require periods of physiological inactivity while waiting for their bodies to warm.

Homeostasis and Protein Conformation

Intuitively, one can appreciate that cells, as living things, can flourish and prosper in certain sets of conditions and not in others. By providing a constant environment for the cells of the body, the various homeostatic mechanisms cooperate to create optimal conditions for cell function.

Beyond this, however, one might ask, "What is the molecular basis of this need for a constant internal environment?" The answer to this question lies in the fact that the primary components or building blocks of cells are macromolecules called **proteins.** Proteins serve a variety of functions in cells. They form enzymes, which are responsible for the various metabolic processes by which cells obtain energy and synthesize other macromolecules in the cell. They also serve as structural components in the cell interior and the cell membrane.

Every protein in the cell has its own distinctive **conformation,** or shape, that allows it to best carry out its own particular function. The shape, and therefore the function, of proteins are affected by such factors as ion concentrations, temperature, and pH. Changes in these factors can change protein conformation. With the right conditions of ion composition, temperature, and pH, proteins assume their proper conformations and carry out their tasks perfectly. Under abnormal chemical, temperature, and pH conditions, however, they lose their shape and their ability to function. Protein malfunction in turn prevents the cells as a whole from carrying out their own specialized tasks, and the malfunctions accumulate until they become manifest in a disease condition in the organism.

But proteins function *inside* cells. The internal environment is *extra*cellular fluid. What is the connection? As we shall see in Chapter 4, the connection is everywhere, anywhere cells and their membranes are found. The plasma membranes of cells serve as the boundary between the two body fluid compartments—the extracellular fluid and the trillions of subcompartments of the intracellular fluid. At the same time, they serve as a transport system between the two compartments, constantly permitting or encouraging movement of substances between cells and the extracellular fluid. The composition of that fluid determines whether or not the cells in turn can obtain from it what they need to make proteins and other molecules essential for their existence.

Thus, the molecular basis for the homeostatic process is, in essence, the maintenance of protein conformation. By constantly ensuring that proper conditions exist in the extracellular fluid, homeostatic mechanisms allow proper conditions inside cells, thereby permitting cell proteins to assume their proper shapes and carry out their special functions. The optimal functioning of specialized cells, in turn, ensures the coherent functioning of tissues, organs, and organ systems and the health of the body as a whole.

Control Systems

The mechanisms by which the body maintains homeostasis are known collectively as **control systems.** Control systems share several features. These systems use mechanisms called sensors to *detect* conditions in the body and effectors to *change* conditions in the body.

Sensors and Effectors

In order to monitor and control a variable of the extracellular fluid such as temperature, the body must first detect the variable. This is accomplished most often by **sensors,** which determine the level of the variable with respect to a **reference point.** For example, in monitoring a physiological parameter such as concentration of calcium in the extracellular fluid, the sensor obtains information about the concentration of calcium present at a particular time and compares it to the level of calcium that *should* be present. If too much or too little calcium is present, this information is sent from the sensors to the effectors, which then attempt to change the parameter to the desired level. Thus, control systems detect discrepancies between *desired* levels or quantities and *actual* levels or quantities.

Two general classes of control systems are active in the body: negative feedback systems and positive feedback systems. **Negative feedback systems** work to restore normal values of a variable and thus exert a stabilizing influence. **Positive feedback systems** destabilize and thus have limited value in maintaining homeostasis.

The Nervous System in Homeostatic Control

The nervous system is one of the major homeostatic control systems in the body. For example, it monitors the temperature of the body and attempts to maintain it at 37°C. It monitors the mean arterial blood pressure and tries to keep it at approximately 90 mm Hg. It monitors the concentration of hydrogen ions in the extracellular fluid and attempts to maintain a pH of 7.4.

The sensors that are part of the homeostatic function of the nervous system are associated with the autonomic nervous system (see Chapter 10). Sensors called **baroreceptors** are nerve endings that wrap themselves around blood vessels to monitor blood pressure. The information they transmit to the brain causes the blood pressure either to rise or to fall. This adjustment is accomplished by the activation or inactivation of effector systems such as the heart, causing it to change its rate and force of contraction, and the smooth muscles, causing them to dilate or constrict blood vessels.

Chemical sensors, also called **chemoreceptors,** of the autonomic nervous system monitor the concentrations of hydrogen ions and carbon dioxide in the extracellular fluid. These chemicals affect the pH of the body, which in turn affects the ability of cells to carry out essential enzymatic reactions. Alterations in the levels of hydrogen ions and carbon dioxide are detected by chemoreceptors in the brain and peripheral autonomic nervous system. The information these chemoreceptors transmit to the brain eventually affects the ways in which the lungs and kidneys work to maintain a constant pH in the body.

Sensors within the brain are also capable of monitoring body temperature. These **thermoreceptors,** located in the hypothalamus, detect the temperature of the blood flowing in the brain. When the temperature falls below 37°C, effector mechanisms are activated. These mechanisms reduce the amount of blood that flows to the skin, thus minimizing heat loss. Heat is retained in the body, thereby raising the temperature. In addition, hormones are released to increase the rate of cellular metabolism, which releases energy in the form of heat. Moreover, neural mechanisms go into action to produce shivering, which generates even more heat. Together these mechanisms generate and conserve heat in the body until the temperature is raised to 37°C. If too much heat is retained, different effector mechanisms take action to dissipate heat. Blood flow to the skin increases, and sweat glands are activated. Heat is lost from the body, and the temperature falls to 37°C.

C L I N I C A L   F O C U S

Positron Emission Tomography

Positron emission tomography (PET) is an imaging method based upon the measurement of positron emission from radionuclides such as oxygen-15, fluorine-18, nitrogen-13, and carbon-11. The imaging method is similar to that used for Computerized Axial Tomography scanning (CAT scanning). With this method it is possible to generate cross-sectional images of the body which show the distribution of a particular radiolabeled substance. Depending on the radiolabeled compound used, it is possible to create a map of a variety of physical and biochemical processes which occur in the body. For example, the distribution of $^{13}NH_3$ is related to the distribution of blood flow in organs such as the brain or the heart. Changes in regional metabolism can be monitored by measuring the distribution of ^{18}F-2-deoxyglucose. Since these images can be generated in real time, it is possible to image changes in regional metabolism and blood flow in the brain, for example, in response to changes in mental activity. Studies of this nature dramatically demonstrate the process of blood flow autoregulation in response to altered cellular activity and metabolism.

PET has been used to monitor the destruction of tumor cells following radiation treatment. Changes in blood flow and metabolic rate can be measured to distinguish actively growing cells from those which have been destroyed by radiation. Patients with Alzheimer's disease (dementia) have a fairly typical pattern on a PET scan which can be used for diagnostic purposes. Recent experimental studies have used PET as a tool to investigate the physiology of psychiatric disorders such as schizophrenia and epilepsy. PET has also been used to detect coronary artery disease and to distinguish between normal and nonfunctional myocardium following a heart attack.

Because of the very short half-life of most positron emitters (20–100 min), it is necessary to generate these isotopes on site. This requires an on-site cyclotron as well as a laboratory for rapid preparation of isotopically-labeled compounds under sterile conditions. In spite of the substantial cost of building and operating such facilities, PET may provide a relatively safe and economical alternative to many currently available diagnostic procedures. Experimental studies using PET are providing a wealth of new information concerning the biochemical basis of numerous physiological processes. PET is proving to be a powerful and unique tool for the investigation of normal and pathological brain physiology by providing investigators a means of directly imaging changes in regional brain metabolism associated with specific types of mental activity.

SUMMARY

The History of Physiology

Aristotle and the investigators of Alexandria searched for explanations of blood flow, temperament, and other aspects of physiology. Galen's "pneumatology," based on these early studies, dominated the Western view of body processes until well into the Renaissance.

Renaissance scientists corrected many errors in Galen's teachings. William Harvey's discovery of the closed circulation marked the end of Galen's dominance and the beginning of modern physiology.

Scientists of the eighteenth and nineteenth centuries studied the breathing process and the nature of the cell. Twentieth-century frontiers in physiology include genetics, bioenergetics, and control systems.

The Study of Physiology

The scientific method, which emphasizes experiment, quantitative results, retesting, and information exchange, forms the basis of all modern scientific research. In conducting investigations, scientists use both inductive and deductive reasoning.

Through measurement, computation, and the use of graphs, scientists apply quantitative methods to the study of physiology.

Physiologists use standard terms of reference to describe planes and positions of the body.

General Organization of the Body

The body consists of protoplasm organized into cells of different types and functions.

Cells that are similar in shape and function form groups called tissues. Muscle, nerve, epithelial, and connective tissues are the four major tissue types in the body.

Groups of tissues that work together form organs; organs in turn combine their functions to maintain systems.

The body comprises two fluid compartments: the fluid within each plasma membrane, the intracellular fluid; and the extracellular fluid (plasma, lymph, and interstitial fluid) that surrounds the cells.

Homeostasis: Control of the Internal Environment

The extracellular fluid forms the "internal environment" of the body, allowing it to maintain constant conditions such as temperature.

Through homeostasis, or self-regulation, the body protects the shapes of proteins, allowing them to function optimally in cells.

Many control systems work to maintain homeostasis. They employ sensors to detect changes in bodily conditions and effectors to adjust those conditions.

REVIEW QUESTIONS

Identify Each Term

1. The units that Harvey used to measure blood pressure.
2. The notation used by scientists to express numbers by a power of ten.
3. The system of measurements used by scientists.
4. The term used to describe fluid outside of the cell.
5. The term used for receptors that sense chemical changes in the body.
6. The term for a group of specialized cells that become closely associated to perform a specific function.
7. The term used in a line graph to represent the horizontal axis.

Define Each Term

8. homeostasis	11. proximal
9. hypothesis	12. distal
10. scientific method	13. y-axis

Choose the Correct Answer

14. Which of the following *is not* part of the scientific method?
   a. observation
   b. hypothesis
   c. experimentation
   d. speculation

15. The imaginary plane that divides the body into right and left portions is the:
   a. transverse plane
   b. sagittal plane
   c. frontal plane
   d. horizontal plane

16. The individual who first described blood vessels as a system with the heart at the center was:
   a. Aristotle
   b. Vesalius
   c. Harvey
   d. Galen

17. The individual who discovered oxygen was:
    a. Bernard
    b. Hooke
    c. Lavoisier
    d. Priestley
18. The individual who discovered the circulation of the blood was:
    a. Galen
    b. Harvey
    c. Malpighi
    d. Vesalius
19. The individual who discovered capillaries was:
    a. Erasistratus
    b. Galen
    c. Herophilus
    d. Malpighi
20. The individual who coined the term homeostasis was:
    a. Aristotle
    b. Bernard
    c. Cannon
    d. Lavoisier

Calculate

21. Express the following in scientific notation.
    a. 1,000,000 cells
    b. 0.0005 mm
22. Convert the following to ordinary numbers.
    a. 5×10^{-3} moles/liter
    b. 1×10^{-4} μm
23. Convert 12 mg to kilograms.
24. Of 100 rabbits exposed to ultraviolet light, 10 developed cataracts. What is the ratio of rabbits that developed cataracts to those that did not develop cataracts?

Answer Each Question in One or Two Sentences

25. What are the major concepts that the cell theory introduced?
26. What advantages does the application of the scientific method have over earlier approaches to the study of physiology?
27. Why is homeostasis so important in cell function?
28. How do graphs aid the study of physiology?
29. What is the purpose of scientific notation?
30. What are the major types of control systems in the body?

SUGGESTED READINGS

Boas, M. *The Scientific Renaissance 1450–1630*. New York: Harper & Row, 1962.

Broad, W., and Wade, N. *Betrayers of the Truth*. New York: Simon & Schuster, 1982.

Comroe, J.H., Jr. *Retrospectroscope—Insights into Medical Discovery*. Menlo Park, Calif.: Von Gehr Press, 1977.

Davis, W. *The Serpent and the Rainbow*. New York: Simon & Schuster, 1985.

Dyson, F. *Disturbing the Universe*. New York: Harper & Row, 1981.

Feynman, R.P. *Surely You're Joking, Mr. Feynman!* New York: Bantam Books, 1986.

Fishman, A.P., and Richards, D.W., eds. *Circulation of the Blood—Men and Ideas*. New York: Oxford University Press, 1964.

Knight, B., M.D. *Discovering the Human Body*. New York: Lippincott and Crowell, 1980.

Miller, J. *The Body in Question*. New York: Random House, 1978.

Poncins, G. *Kabloona*. New York: Reynal and Hitchcock, Inc., 1941.

Rothschuh, K.E., M.D. *History of Physiology* (Guenter, B.R., M.D., ed.). New York: Robert E. Krieger Publishing, 1973.

Sagan, C. *Cosmos*. New York: Random House, 1980.

Saunders, J.B. deC.M., and O'Malley, C.D. *The Anatomical Drawings of Andreas Vesalius*. New York: Bonanza Books, 1982.

Scott, L.M., and Waterhouse, J.M. *Physiology and the Scientific Method*. New Hampshire: Manchester Press, 1987.

Simon, T. *The Heart Explorers*. New York: Basic Books, Inc., 1966.

Thomas, L. *The Youngest Science*. New York: Bantam Books, 1984.

KEY TERMS

anatomical planes (p. 16)

blood vessels (p. 6)

cell differentiation (p. 17)

cell theory (p. 9)

control systems (p. 25)

data interpretation (p. 14)

experimental design (p. 11)

homeostasis (p. 10)

hypothesis (p. 11)

negative feedback (p. 25)

organ (p. 22)

physiology (p. 2)

positive feedback (p. 25)

protoplasm (p. 17)

scientific measurements (p. 12)

scientific method (p. 10)

Cellular Functions

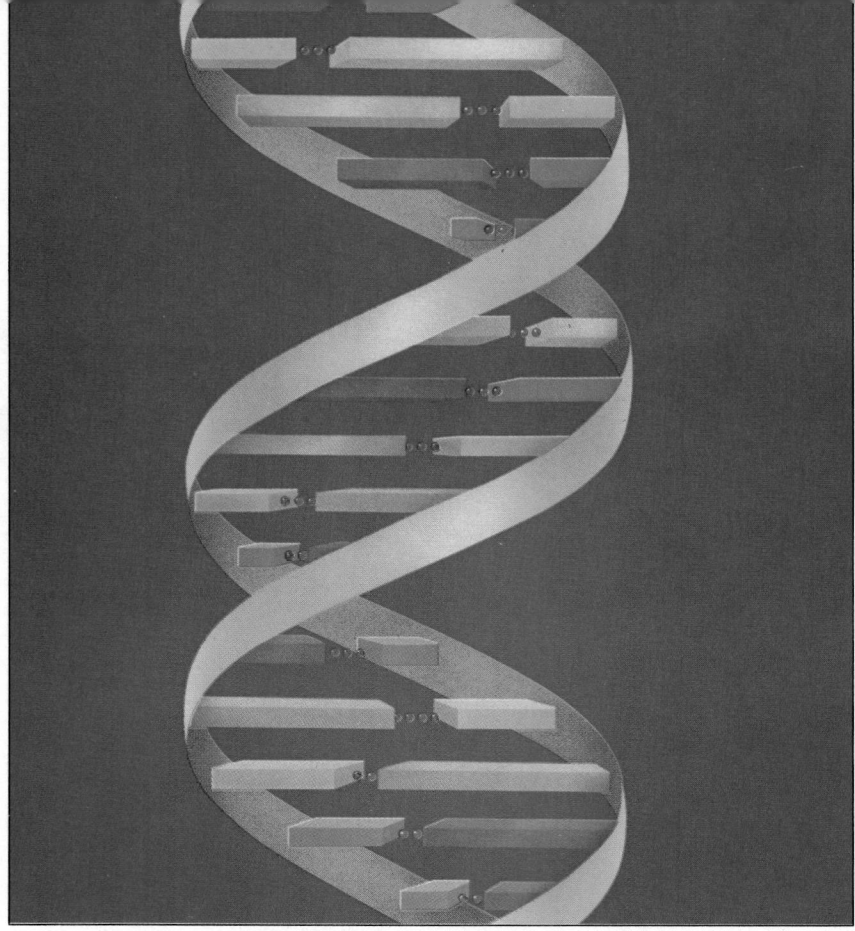

CHAPTER 2

Chemical and Physical Principles

*L*iving organisms are composed of **matter.** Matter is anything that occupies space, has mass, and can assume the form of either a solid (e.g., stone), a liquid (e.g., water), or a gas (e.g., oxygen).

We currently organize the matter of the universe into elements. An **element** is a substance that cannot be broken down into simpler substances by chemical reactions. There are 92 elements that occur naturally, although scientists have made more than a dozen new elements. In nature as a whole, the four most abundant elements are oxygen, silicon, aluminum, and iron. Together, they account for 87% of the atoms in the earth's crust. Elements are commonly referred to by their abbreviations, called **chemical symbols** (e.g., "O" for oxygen and "Fe" for iron).

Elements are composed of atoms. An **atom** is the smallest unit of an element that retains the properties of the element. Atoms of the same element combine in specified ways to form different varieties of the same element. Atoms of different elements combine in specified ways to form compounds. A **compound** is a substance formed by the union of the atoms of two or more elements. Many substances, whether made of one element or more than one element, have molecules (not atoms) as their smallest stable units. A **molecule** is formed by the union of two or more atoms, whether they are the same elements or not. Oxygen atoms, for example, in samples of pure oxygen, form pairs called oxygen (O_2) molecules rather than existing singly. Water, in contrast, is a compound substance; one water molecule consists of two hydrogen atoms combined with one oxygen atom. A molecule or other regular group of atoms is designated by a **molecular formula**, such as H_2O, which shows the number of each type of atom in the group. The subscript "2" after the H, for example, means that two hydrogen atoms are present in a molecule of water. If there is no subscript following a chemical symbol, the symbol represents a single atom.

In living cells, molecules combine further to form large molecules known as **macromolecules.** These macromolecules all contain carbon, whose unique structure allows macromolecules to assume huge proportions and a great variety of shapes. Proteins, sugars, lipids, and nucleic acids are the chief macromolecules found in most living cells. The specific properties of these macromolecules and the organization of their many interactions within the cell form the subcellular structures and facilitate the chemical reactions that are the basis for all physiological processes and for life itself.

Overview: Matter and Energy in the Living Cell

The predominant elements in the matter of the universe are oxygen, silicon, aluminum, and iron. In contrast, the matter of the human body is largely hydrogen (H), oxygen (O), carbon (C), and nitrogen (N). Together they account for 99% of the atoms in the body. The marked difference between the chemical compositions of the human body and its external environment indicates that hydrogen, oxygen, carbon, and nitrogen must be uniquely suited for the processes that maintain the living state. In addition to these 4 most abundant elements, 23 others have been found to be essential components of living organisms. The 27 elements essential for life are listed in Table 2–1.

Table 2–1
Essential Elements of Living Organisms*

Element	Chemical symbol
Hydrogen	H
Oxygen	O
Carbon	C
Nitrogen	N
Phosphorus	P
Sulfur	S
Sodium	Na
Potassium	K
Magnesium	Mg
Calcium	Ca
Chlorine	Cl

Trace elements (present in very small amounts)
Manganese (Mn), Iron (Fe), Cobalt (Co), Copper (Cu), Zinc (Zn), Boron (B), Aluminum (Al), Vanadium (V), Molybdenum (Mo), Iodine (I), Silicon (Si), Tin (Sn), Nickel (Ni), Chromium (Cr), Fluorine (F), and Selenium (Se).

*Not all of these elements are essential for every species.

The matter of a living organism is highly organized. It also undergoes frequent transformation. The maintenance of this organized state and the processes of transformation require a constant input of energy. **Energy** is the capacity to produce change. It is measured by the amount of work performed to achieve a given change. In the universe at large, the energy used for work comes from various sources, such as heat, light, electricity, and chemical reactions. In the cell, the energy used for work is mostly **chemical energy** released by reactions that take place within cells. In the cell, the work to be done consists of moving matter from one place to another and changing it from one form to another—that is, breaking down or building up molecules from other molecules.

The different forms of energy also exist in two different states: **Potential energy** is stored energy; **kinetic energy** is energy that produces motion. Both states of energy enter into the metabolic processes of cells.

The chemical basis of the matter of widely diverse forms of life is strikingly similar. All living cells have fundamentally similar constituents and activities. Thus, both a complex multicellular organism such as a human being and a simple unicellular bacterium such as *Escherichia coli (E. coli)* use the same simple molecules to synthesize, or make, larger molecules. Both human cells and bacterial cells store energy in the form of adenosine triphosphate (ATP). Both cells store genetic information in deoxyribonucleic acid (DNA).

R E S E A R C H   F O C U S

Scanning Tunneling Microscope: A new research tool for mapping the molecular landscape of the cell

The scanning tunneling microscope (STM) was developed in 1982 by Drs. Gerd Binnig and Heinrich Rohrer of IBM Zurich. The development of the STM attracted much attention because it allowed scientists to image individual molecules and atoms. The potential of the STM was recognized early on in its development, and the Nobel Prize was awarded four years later to Drs. Bining and Heinrich.

In biology, the STM offered another advantage: it can image surface details down to the atomic level without using a destructive probe and eventually will redefine the concept of microscopy. In some cases the STM will replace techniques such as scanning electron microscopy (SEM) and transmission electron microscopy (TEM).

The STM works by tracing an ultrafine conductive tip over a sample, and detecting topography through tiny variations in a quantum-mechanical "tunneling current" that seeps between the atoms at the surface and the tip (Figure 1). The business end of an STM consists of a piezoelectric transducer that scans the tip over the surface while a computer-controlled feedback loop monitors the tunneling current and applied scan voltage (Figure 1). The feedback loop adjusts the vertical position of the tip to keep the tunneling current constant. The adjustment signal versus the lateral position of the tip provides a way of positioning the tip in three dimensions that allows topographical images to be generated in which individual atoms can be resolved (Figure 2). An STM image is made up of a series of line scans and displays the path of the tip followed over the surface. A major disadvantage of the STM is that most biological materials are not sufficiently conductive enough to allow STM images to be made. Recently, this problem has been resolved by the invention of the Atomic Force

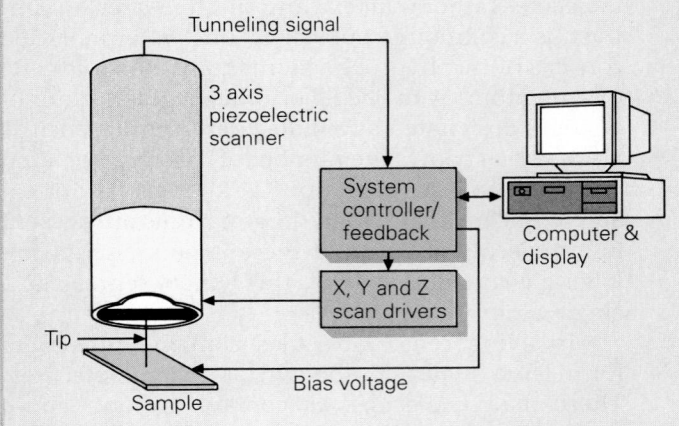

Figure 1.
Block diagram of the STM. An extremely sharp conductive tip can be moved in three dimensions with an x, y, z piezoelectric translator to trace the contours of a surface.

The chemical content of an *E. coli* cell has been analyzed in detail. Its composition is 70% water, 15% proteins, 7% nucleic acids, 2% carbohydrates, 2% lipids, 1% inorganic ions, and 3% other small molecules. The cell is mostly water, and the major macromolecules present are proteins and nucleic acids. The chemical composition of the typical mammalian cell is markedly similar to that of this bacterial cell.

To determine why certain chemical structures dominate all forms of life so regularly, the current chapter provides a survey of general chemical and physical principles. Then it focuses more closely on the specific principles that influence the life processes. The chapter begins with the simplest components of matter—atoms and small molecules—and the chemical and physical laws that govern their behavior. It then proceeds to the major classes of giant molecules such as sugars, proteins, lipids, and nucleic acids. A look at their components and properties provides a sound basis for the description, in later chapters, of their unique roles in the life of cells.

Atoms and Elements

An **atom** is the smallest unit of an element that retains all the chemical properties of the element. In solid matter, atoms are closely packed. In liquid and gaseous mater, atoms are more loosely arranged. The same element can appear in different forms

---

Microscope (AFM), which can image biological material and other substances that do not readily conduct electrons. The STM and AFM are very similar with their scanning, feedback, and display systems. However, instead of a conductive tip, the AFM utilizes a tip that detects interatomic forces between the tip and the sample.

*Hansma, P.K., Elings, V.B., Marti, O., and Bracker C.E. Scanning tunneling microscopy and atomic force microscopy: Application to biology and technology. Science 24:209–216, 1988.

†Thompson, M. and Elings, V. Scanning tunneling and atomic force microscopy: Applications in life sciences. Am. Laboratory 23:(No. 6):36–42, 1991.

(a)

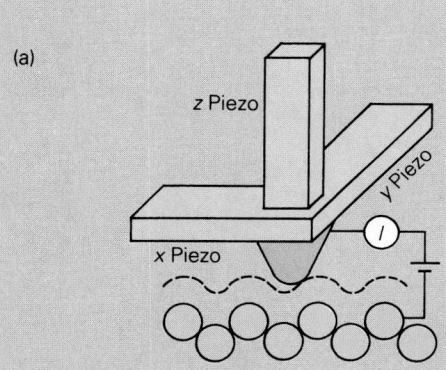

(b)

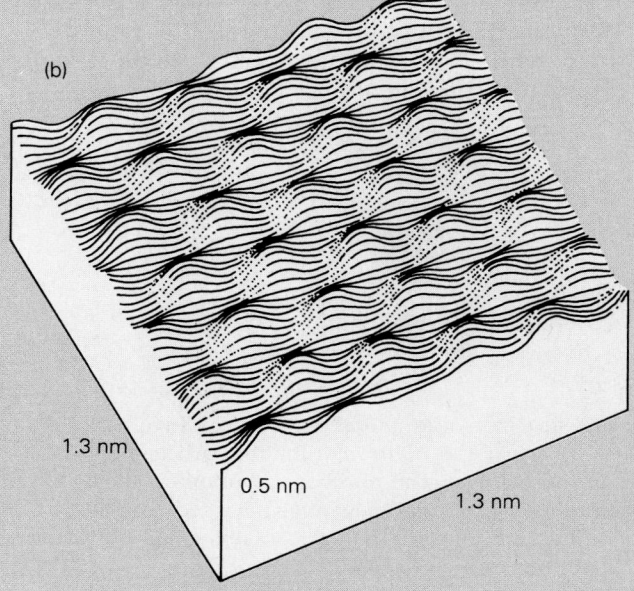

1.3 nm

0.5 nm

1.3 nm

Figure 2.
(*a*) Schematic view of the STM tip. The tip is mounted on a piezoelectric *x*, *y*, *z* scanner. As the tip scans the surface, atomic contours are revealed. (*b*) An image generated by STM slows carbon atoms at the surface.

depending on the temperature and pressure of its surroundings. For example, at room temperature and pressure, the element carbon is a solid because its atoms are packed closely together. At higher temperatures, carbon becomes liquid; at still higher temperatures, it takes the form of a gas as its atoms become separated from each other.

Subatomic Particles

Each atom, regardless of its element, is composed of certain subatomic particles. Three types of particles are important to an understanding of the behavior of atoms in life processes: neutrons, protons, and electrons. **Neutrons** are electrically neutral, whereas **protons** are positively charged and **electrons** negatively charged. Protons and neutrons stay in the **nucleus**, or innermost area, of the atom, whereas electrons constantly orbit the nucleus (Fig 2–1). The electrical attraction between the positively charged protons in the nucleus and the negatively charged electrons in orbit around the nucleus stabilizes the atom's structure.

Because atoms are so small, they most often occur together in large groups. In order to count large numbers of atoms conveniently, scientists have come up with a counting method called the **mole.** Just as in everyday speech we talk of a dozen eggs, a dozen pencils, or a dozen pages—all the time meaning 12 eggs, 12 pencils, or 12 pages—scientists speak of a mole of atoms to mean 6.022045×10^{23} atoms. The number 6.022045×10^{23} is called **Avogadro's number.** A mole is a convenient and accurate way to express the same number of atoms of any substance. Thus a mole of sodium atoms would be 6.022045×10^{23} sodium atoms; a mole of chlorine atoms would be 6.022045×10^{23} chlorine atoms.

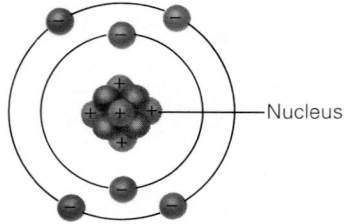

Figure 2–1
The basic structure of an atom. Protons (positively charged particles) and neutrons (neutral particles) are found in the nucleus. The number of protons is the atomic number. Electrons (negatively charged particles) orbit the nucleus. A neutral atom contains equal numbers of protons and electrons.

Uniqueness of Elements: Protons and Atomic Number

Atoms of different elements are distinct from one another because of the different numbers of each subatomic particle they contain. In fact, the essential identity of an atom is found in the number of protons in its nucleus, called the **atomic number.** In the uncharged or electrically neutral atom, the number of electrons is the same as the number of protons. An atom of the element hydrogen has one proton in its nucleus and one electron in orbit around the nucleus. An oxygen atom typically has 8 protons and 8 electrons. A carbon atom has 6 protons and 6 electrons, a nitrogen atom has 7 of each particle, and a uranium atom has 92 of each particle.

Variation Within Elements: Neutrons, Mass Number, and Isotopes

The actual mass of an atom is too small to be measured or expressed conveniently. Instead, scientists refer to the **mass number** of an element. The mass number is the total number of protons and neutrons in the nucleus. Since the number of protons is different for every element, the atomic mass is also different for every element. For example, the atomic mass of the most common form of hydrogen is 1 because the hydrogen atom has one proton (p) and no neutrons (n) in its nucleus (Fig. 2–2). Oxygen has

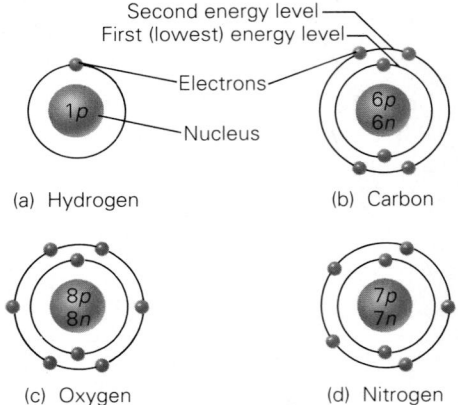

Figure 2–2
The electron configurations of four important elements. Hydrogen needs one more electron to complete its only shell. Carbon needs four electrons, oxygen two, and nitrogen three to complete their outer shells. The electron orbits are depicted as concentric circles for the sake of simplicity, but their true movements are more complex.

8 protons and 8 neutrons, so its mass number is 16. Nitrogen has 7 protons and 7 neutrons, so its mass number is 14.

Moles and grams (the metric unit of mass) can be related in the following way: one mole of an element, or 6.022045×10^{23} atoms, contains x grams of the element, x being equal to the mass number of the element. For example, one mole of sodium atoms (6.022045×10^{23}) weighs 23 grams, 23 being the mass number of sodium. One mole of chlorine atoms weighs 35 grams, since the mass number of chlorine is 35.

While all atoms of the same element have the same atomic number (number of protons), atoms of the same element may have differing mass numbers because they may have differing numbers of neutrons. For example, hydrogen exists in three forms: one form contains no neutrons in its nucleus, another contains one neutron, and another two neutrons. These three forms of hydrogen are called **isotopes** of hydrogen. Since the mass number of an atom takes into account the number of neutrons, different isotopes of an element have different mass numbers.

An isotope is designated with both the atomic number (number of protons) and the mass number (number of protons plus neutrons) written to the left of the chemical symbol. For example, the isotope of hydrogen that has no neutrons has an atomic number of 1 (because all hydrogen atoms have 1 proton) and a mass number of 1, written as $^{1}_{1}\text{H}$. The isotope of hydrogen that has one neutron (deuterium) would be written as $^{2}_{1}\text{H}$; the isotope with two neutrons (tritium) as $^{3}_{1}\text{H}$. Isotopes are important in biomedical research, as will be discussed in Chapter 3.

Ions: Atoms with Electrical Charges

A neutral, or uncharged, atom has the same number of protons (positive charges) and electrons (negative charges). When a neutral atom gains or loses an electron, it acquires an electric charge and becomes an *ion*. An atom that loses one or more electrons acquires a positive charge because it now has more protons (positively charged particles) than electrons (negatively charged particles). Such an atom is called a **cation** (CAT-eye-on). An atom that gains one or more electrons acquires a negative charge and is called an **anion.**

For example, a potassium cation is formed when neutral potassium loses one electron. The single charge is denoted by a plus sign written to the

upper right of the chemical symbol for potassium: K^{+}. The chloride anion, formed when chlorine gains an electron, is written as Cl^{-}. A calcium ion, formed by the loss of two electrons, would be denoted by a number 2 and a plus sign: Ca^{2+}. An oxide ion, formed by the gain of two electrons, is written as O^{2-}.

Ions have a strong impact on the ability of certain substances to dissolve in water. The potassium ion is a very important ion in the study of physiology because it is the major cation found inside most cells, while the sodium ion (Na^{+}) is the major cation present outside cells in fluids such as plasma. Chloride ions are the most abundant anions in plasma.

Electron Configuration and the Outer Shell

The electrons, in their orbit of the nucleus, move in a specified **electron configuration,** consisting of a series of **shells** that can be thought of as concentric circles (although this is a very simplified view of the way electrons actually move). The shells represent different energy levels and can hold different numbers of electrons. The shell with the lowest energy level, located closest to the nucleus, can accommodate two electrons. The shell with the next highest energy level can hold eight electrons. The electron configurations of hydrogen, oxygen, carbon, and nitrogen atoms are illustrated in Figure 2–2. Larger atoms have more electrons occupying several different energy levels. For example, the 92 electrons in the uranium atom are arranged in seven different shells, each with a different energy level.

The number of electrons in the outermost shell of an atom determines many of the chemical properties of the element. An atom may be stable and neutral even if its outermost shell is not filled to capacity, but it can fill this outermost shell by interacting in certain ways with other atoms both of the same element and of different elements. An atom of carbon, for example, has four electrons in its outermost shell. Since this is the second shell from the nucleus, it can hold up to eight electrons. The carbon atom will try to fill that shell with four more electrons.

An atom can fill its outermost shell in one of three ways. It can share electrons with one or more atoms that do not have complete outer shells. It can obtain electrons outright from other atoms. Or it can donate all of the electrons in its currently incomplete outer shell to other atoms so that its next-lower shell, which is filled to capacity, becomes its outermost shell.

Molecules, Groups, and Compounds

The process of sharing, gaining, or losing electrons leads to the creation of a **chemical bond** between atoms. A chemical bond represents potential energy, energy being stored in the molecule. The release of this potential energy in chemical reactions for use by the cell is an important aspect of cellular physiology.

Atoms combine in definite proportions to form variations of elemental substances or new substances called compounds. Two atoms from the same element may join to form a "diatomic" molecule (see Fig. 2–3). Oxygen and hydrogen, for example, rarely exist as single atoms; instead, they are found as O_2 and H_2, pairs of atoms bonded together.

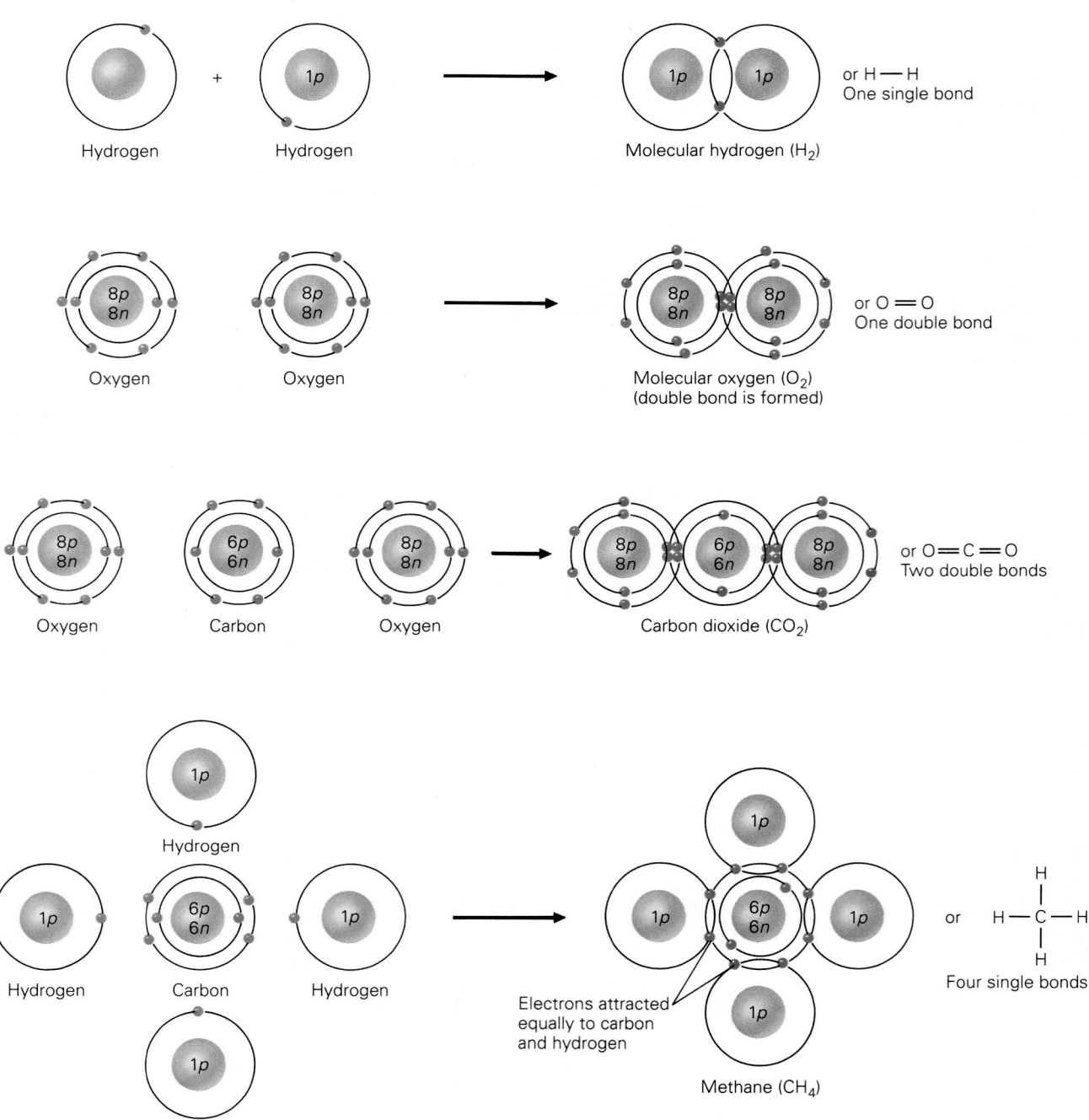

(a)

Figure 2–3

Covalent bonds are formed when two or more atoms share electrons. (*a*) Nonpolar covalent bonds are formed when the shared electrons are attracted equally to the nuclei of the original atoms. Hydrogen (H_2) and oxygen both exist as diatomic molecules formed by single nonpolar bonds. Carbon dioxide (CO_2) and methane (CH_4) are compounds by double and single nonpolar covalent bonds, respectively. (*b*) Polar bonds are formed when the shared electrons are attracted more strongly to one nucleus than to another. In ammonia (NH_3), the three shared pairs are more strongly attracted to the nitrogen nucleus, creating a partial negative charge there (symbolized by δ^-) and a slight positive charge (symbolized by δ^+) in the region of the molecule near the three hydrogen nuclei. A similar polarity, formed by two polar bonds, exists in the water molecule.

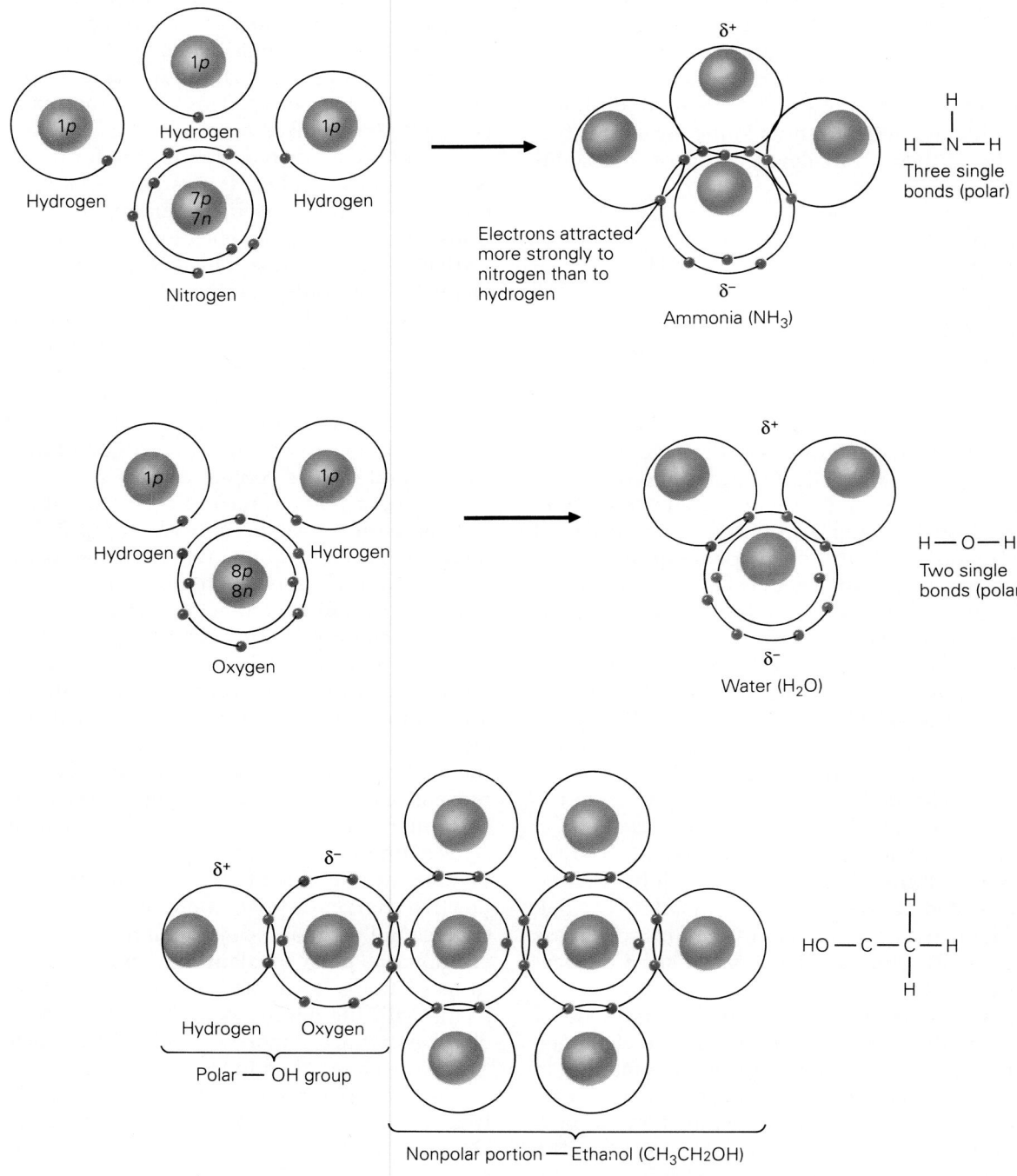

Two or more atoms from different elements may come together to form a new compound. Depending on the type of bond they form, the result may be a molecule (a discrete group of atoms) or a larger structure known as a **crystal lattice,** which is made up of atoms arranged in precise patterns. Whether or not separate molecules are formed, atoms bond in definite proportions, which are indicated by a molecular formula for the compound. For example, even though sodium chloride exists as a crystal lattice made up of many alternating sodium and chloride ions, rather than discrete molecules, the compound has the formula NaCl, indicating that sodium and chloride are always present in a ratio of 1:1.

The molecular formula of a compound shows the relative quantities of atoms; the **structural formula** shows their arrangement in space. for example:

	water	methane
Molecular formula	H_2O	CH_4
Structural formula	H—O—H	H—C—H (with H above and H below)

A **chemical group** is a small cluster of bonded atoms that functions almost as a single atom. Chemical groups behave in distinctive ways and have special names. Examples of chemical groups are —OH (the "hydroxide group"), —COOH (the "carboxyl group"), and —PO_4 (the "phosphate group"). The dash written before the chemical symbols indicates that the group is not a free molecule but is attached to a larger molecule.

Just as atoms are counted in groups called moles, so are molecules. A mole of molecules, regardless of the substance, is 6.022045×10^{23} molecules. A mole of oxygen is 6.022045×10^{23} molecules. A mole of water is 6.022045×10^{23} H_2O molecules. The **molecular mass** of a compound is the sum of the atomic masses, or mass numbers, of its elements. Molecular mass is expressed in units known as **daltons** (symbol Da), after John Dalton, who proposed the atomic theory of matter. The dalton is defined as 1/12 of the mass of carbon (mass number = 12). Thus, methane (CH_4) has a molecular mass of 12 + 4 = 16 Da.

Regardless of whether a molecule is made of atoms from the same element or different elements, the type of bond formed depends on the manner in which the atoms interact. If they complete their outermost shells by sharing electrons, the bond formed is a covalent bond. If the atoms gain or lose electrons outright, the bond formed is an ionic bond.

Covalent Bonds: Polar and Nonpolar Molecules

Chemical bonds formed when two or more atoms share electrons are known as **covalent bonds.** These bonds are very stable and strong and are the type of bond most frequently found in biological compounds. Hydrogen, oxygen, carbon, and nitrogen, the four elements most prevalent in living cells, readily form covalent bonds with each other (Fig. 2–3). Atoms of these four elements all have an incomplete outer shell: hydrogen lacks one electron, oxygen two, carbon four, and nitrogen three. Not only can they form single bonds by sharing one pair of electrons; they can form double and (in the case of carbon) triple bonds. This ability increases their versatility in chemical reactions.

In addition, carbon atoms can form covalent bonds with each other. If a carbon obtains each of the four electrons it needs by sharing a pair of electrons with four other carbon atoms, it will form four distinct covalent bonds. The ability to bond with as many as four other carbon atoms allows carbon atoms to form the enormous variety of linear, branched, and cyclic structures that serve as the backbones of large molecules (Fig. 2–4). More than a million compounds are known to contain carbon, and many thousands of them are vital for the maintenance of life. A very large proportion of the molecules found in cells contain carbon.

Covalent bonds and the molecules they form may be either polar or nonpolar (Fig. 2–3*b*). In a **nonpolar covalent bond,** the electrons in the pair are shared equally by each atom in a pair, so the molecule overall is completely neutral. In general, compounds formed with carbon and hydrogen are nonpolar covalent compounds. Their overall shapes are square or rectangular. In a **polar covalent bond,** a pair of electrons is shared unequally; that is, the electrons in the pair are attracted more strongly to the nucleus of one bonded atom than they are to the nucleus of the other bonded atom. The electrons can be thought of as spending more time near the nucleus of one atom, creating a partial negative charge, or pole, at one end of the molecule and a partial positive charge, or pole, at another end. The small charges are symbolized as δ^- and δ^+. Water is an example of a **polar covalent compound.** The two hydrogen atoms form an asymmetrical, triangular shape with the one oxygen atom, with the shared electrons attracted more strongly to the nucleus of the oxygen atom than to the nuclei of the hydrogen atoms. This polarity of certain molecules, and the lack of it in others, has important consequences for chemical reactions, as the section on solutions will show.

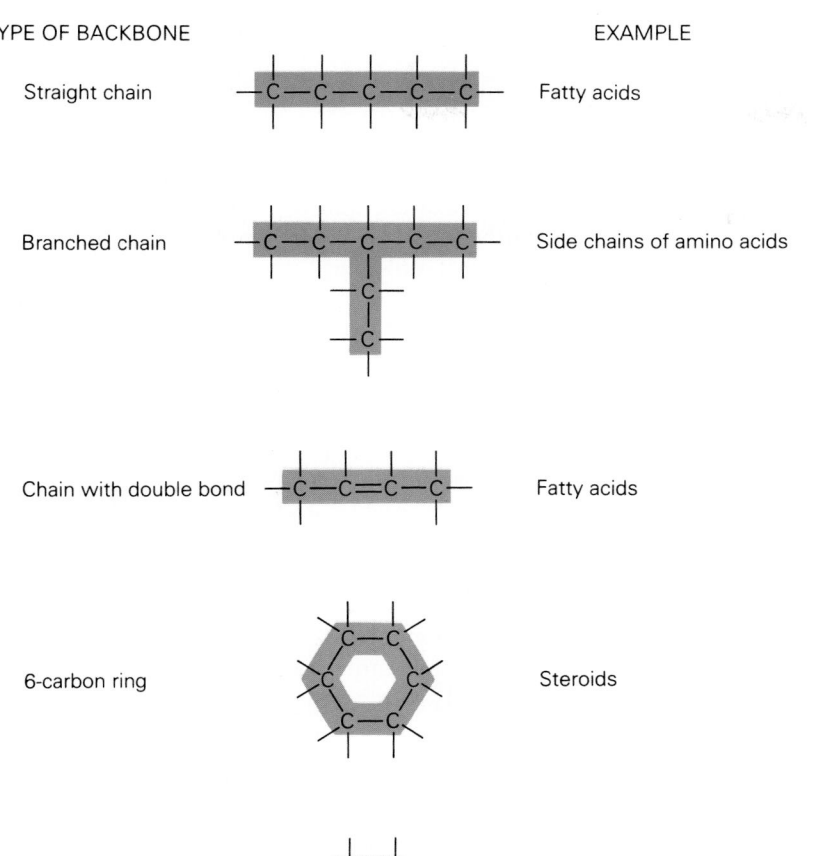

TYPE OF BACKBONE

Straight chain — Fatty acids

Branched chain — Side chains of amino acids

Chain with double bond — Fatty acids

6-carbon ring — Steroids

6-carbon ring with double bonds — Side chains of amino acids

EXAMPLE

Figure 2–4
The ability of carbon to form bonds with four other carbon atoms at once allows it to form a large variety of different linear, branched, and ring structures that serve as the backbones for organic macromolecules.

Ionic Bonds and Ionic Compounds

When atoms complete their outer shells by either gaining or losing electrons instead of sharing them, they form **ionic bonds** to create **ionic compounds.** When two atoms form an ionic bond, one atom acquires one or more electrons and therefore one or more negative charges. The other atom, by losing the electrons to the first atom, acquires an equal number of positive charges. The equal and opposite electrical charges of the two atoms attract them and hold them together; this attraction is the basis of the ionic bond. A specific example (Fig. 2–5) is the formation of sodium chloride. The neutral sodium (Na)

Figure 2–5
Sodium (Na) forms an ionic bond with chlorine (Cl) by donating the single electron in its outer shell to the chlorine atom. The sodium atom then has a complete inner shell, but it has become a cation because it has one more proton than electron. The chlorine atom also has a complete outer shell after gaining the electron from sodium, but it has become an anion because it has one more electron than proton.

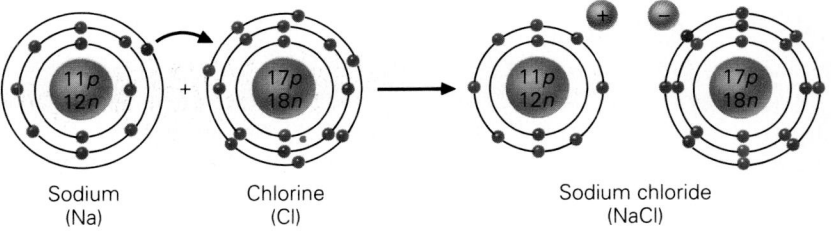

Sodium (Na)   Chlorine (Cl)   Sodium chloride (NaCl)

atom has 11 protons and 11 electrons. After donating the lone electron in its outermost shell, it has only 10 electrons left—one less than the number of protons—and a net electrical charge of +1. The neutral chlorine (Cl) atom, with 17 protons and 17 electrons, is able to fill its outermost electron shell by accepting the single electron the sodium atom has given up. The chlorine atom then has 18 electrons—one more electron than protons—and a net electrical charge of −1. An ionic bond has been formed between the oppositely charged sodium and chlorine atoms, and the chemical compound sodium chloride (NaCl) results.

Ionic bonds between atoms do not produce molecules per se; rather, they make crystal lattices, or orderly patterns of cations and anions. As long as the cations and anions are bound together in the lattice, the compound overall is neutral. If the ions become dissociated, as in a solution (see Fig. 2–9a), they exist as charged particles.

Van der Waals Bonds

Another type of attractive force develops when any two atoms are close together, about 0.3 to 0.4 nm apart. The **van der Waals bond** forms when the negatively charged electron "cloud" formed by an asymmetrical orbit of electrons in one atom is attracted to the positively charged nucleus of a neighboring atom (Fig. 2–6). The electrons do not leave one atom for another, however. Van der Waals bonds are a much weaker attractive force than covalent and ionic bonds. The amount of energy needed to break a van der Waals bond is less than 1% of the energy necessary to disrupt a covalent bond between hydrogen and oxygen.

Although individual van der Waals bonds are relatively weak, they become important when large numbers of such bonds form at the same time. This situation often occurs in large molecules when the specific structure of the molecule permits the close proximity of many of the atoms. The van der Waals bonds help to stabilize the three-dimensional shape, which is often important for the function of the molecule.

Hydrogen Bonds

A **hydrogen bond** is similar to a van der Waals bond but slightly stronger. Hydrogen bonds occur among polar covalent molecules, such as water, made up of hydrogen and certain elements such as nitrogen, oxygen, and fluorine. In water, for example, the partial positive charge of the hydrogen atom of one water molecule is attracted to the partial negative charge of the oxygen atom of another water molecule (Fig. 2–7). Hydrogen bonds can occur between molecules or within the same molecule.

Hydrogen bonds are still relatively weak compared to covalent bonds. The input of energy needed to disrupt the hydrogen bonds between water molecules is only 4% of that needed to break the covalent bond between the hydrogen and oxygen in the same molecule. Hydrogen bonds form from the hydrogen atoms in many biological compounds. They are important for maintaining the shapes of large molecules such as proteins and nucleic acids.

Chemical Reactions: Basic Principles

All atoms and molecules are moving continuously and therefore frequently collide with one another. These frequent collisions sometimes disrupt the electron configurations of the atoms involved, thereby breaking the covalent or ionic bonds the atoms have formed with other atoms. When bonds are broken, compounds break down into their elements; when these elements recombine in new ways, new compounds are formed. This process is called a **chemical reaction.** According to the Law of Conservation of Mass, the total number of atoms is the same before and after a chemical reaction; the

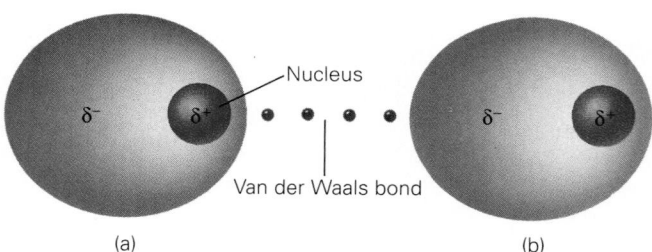

Figure 2–6
Van der Waals bonds form between two atoms because, if the two atoms are close enough, asymmetry in the distribution of electrons in atom (a) allows the positively charged nucleus to attract negatively charged electrons in atom (b).

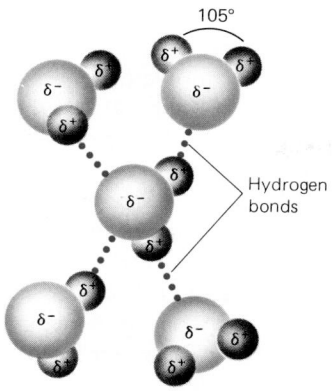

Figure 2–7
The polarity of the water molecule allows it to form hydrogen bonds with other water molecules.

atoms are simply rearranged into different groups, which represent new compounds with different chemical properties than the original compounds.

A chemical reaction is represented by a **chemical equation,** which shows the relative quantities of each compound in terms of its molecular formula. A simple example is the reaction of perchloric acid ($HClO_3$) and potassium carbonate (K_2CO_3) to form potassium perchlorate ($KClO_3$), carbon dioxide (CO_2), and water (H_2O). The two original compounds are called **reactants;** the resulting compounds are called **products.** Note that the same number of atoms of each element are present both before and after the reaction. This is ensured by providing two molecules (indicated by the number 2 to the left of the molecular formula) of the reactant perchloric acid for each molecule of potassium carbonate.

Reactants			Products		
$2HClO_3$	+	K_2CO_3 $\longrightarrow$	$2KClO_3$	+	CO_2	+	H_2O
perchloric acid		potassium carbonate	perchlorate		carbon dioxide	water

total atoms	total atoms
2 H	2 H
2 Cl	2 Cl
9 O	9 O
2 K	2 K
1 C	1 C

Sometimes, the reaction proceeds in one direction, giving stable products. Some reactions, however, form unstable products. Either they break down and recombine into the original reactants or break down into new products altogether. An example of this is the reaction of carbon dioxide (CO_2) and water to form carbonic acid (H_2CO_3). The prod-

uct, H_2CO_3, is unstable, breaking down into the original reactants. Hence, the reaction is called *reversible* and is indicated by a set of opposite arrows. Many metabolic reactions that occur in the cell are reversible.

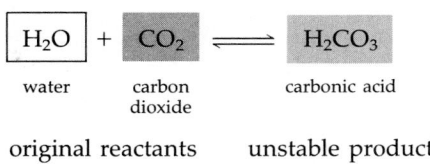

H_2O	+	CO_2	$\Longrightarrow$	H_2CO_3
water		carbon dioxide		carbonic acid

original reactants   unstable product

When a reaction reaches a point at which the formation of products is exactly balanced by the reformation of the original reactants, the reaction is said to be in **equilibrium.**

The rate or speed at which a reaction occurs depends on a number of factors. Some factors that affect rates of reaction are temperature, concentration, and physical orientation of molecules.

Temperature. An increase in temperature generally increases the rate of a chemical reaction by increasing the velocity at which the reactant molecules and atoms are moving. The particles collide more frequently, increasing the possibility that electron configurations will be disrupted. The rate of many reactions doubles when the temperature increases by only 10°C. The metabolic reactions that occur in cells occur at 37°C, a temperature higher than most environments. A further increase in temperature would be harmful to cells because reaction products such as proteins would become unstable, breaking down and losing the ability to perform their functions.

Concentration. At least initially, an increase in the concentration of reactants usually increases the rate of reaction. The presence of more reactant molecules in the same space will increase the frequency with which moving particles collide, thereby disrupting more bonds. Eventually, however, the rate will decrease because the reactants are quickly used up in the formation of products. In the cell, the concentrations of molecules are carefully controlled so that metabolic reactions proceed at optimal rates.

Orientation. The way in which molecules are oriented, or arranged in space, often affects the rate at which their atoms will break off and form new bonds. The atoms also must be near enough for the reaction to occur. Certain important bonds occur only between specific chemical groups (e.g., peptide bonds that form between amino acids in protein molecules).

Some chemical reactions fall into a special category known as **oxidation-reduction reactions,** or

redox reactions. An example is the reaction in which iron changes to ferric oxide (rust):

$$4\ Fe\ +\ 3\ O_2\ \longrightarrow\ 4\ Fe^{3+}\ +\ 6\ O_2^-\ \longrightarrow\ 2\ Fe_2O_3$$

iron   oxygen   ferric oxide (rust)

In this reaction, the iron loses electrons, a process known as **oxidation.** Each iron (Fe) atom loses three electrons, leaving an excess of three positively charged protons and creating iron cations (Fe^{3+}). Each oxygen atom accepts two electrons from the iron atoms; the extra electrons confer two negative charges on each oxygen atom, creating oxide anions (O^{2-}). The process of gaining electrons is known as **reduction.** The charged iron and oxygen atoms then form ionic bonds and produce two molecules of the compound ferric oxide. Three iron atoms lose a total of 12 electrons, which are gained by a total of six oxygen atoms. The iron is being **oxidized** at the same time that the oxygen is being **reduced.** Oxidation and reduction always occur simultaneously, one reactant accepting the electrons removed from another.

Water and Aqueous Solutions

Water is essential for living organisms and is the major chemical compound in most cells, accounting for up to 70% of cell weight. Many biologically important compounds are soluble in water; in fact, the chemical reactions they undergo require an **aqueous** (watery) medium or environment. Here we will consider some special properties of water that make it an essential substance for living cells.

Polarity of the Water Molecule

The water molecule is formed by the covalent bonding of two hydrogen atoms to one oxygen atom. To complete its outer shell, oxygen needs two electrons. It obtains these electrons by sharing an electron with each of two hydrogen atoms. These bond to the oxygen atom to form a triangular structure. The covalent bonds formed between the atoms are polar bonds; that is, the shared electrons are attracted more strongly to the oxygen nucleus than to the hydrogen nuclei. This creates a small negative charge in the area of the water molecule near the oxygen nucleus and a small positive charge near the two hydrogen nuclei. Because it has these oppositely charged areas at opposite ends, the water molecule is said to be **polar** (see Fig. 2–3).

Arrangements of Water Molecules

The polarity of water molecules leads them to form hydrogen bonds with each other when the partial positive charge at one end of one molecule attracts the partial negative charge at the opposite end of another molecule (Fig. 2–7). Hydrogen bonds are responsible for the higher melting and boiling points of water compared to other substances with similar structures that do not form hydrogen bonds (e.g., ammonia [NH_3]). In solid water (ice), for example, the molecules are arranged in a regular three-dimensional pattern so that each molecule is hydrogen-bonded to four other water molecules. When ice melts by heating, and liquid water forms, only about 15% of the hydrogen bonds are broken. Thus, liquid water contains clusters of molecules that have retained a highly organized structure. Energy in the form of heat is required to change ice to liquid water and liquid water to a gas because these changes require the disruption of the hydrogen bonds between water molecules. Similar changes in the physical state of ammonia require less energy because there are no hydrogen bonds to be broken.

Water as a Solvent

In chemistry, a **solution** is a type of mixture in which one substance, called the **solute,** is uniformly distributed throughout another substance, usually a liquid, called the **solvent.** A solvent dissolves, or breaks up, the solute without reacting chemically with it: that is, electrons are not shared or exchanged between the two substances. The polarity and hydrogen-bonding tendencies of water make it an excellent solvent of both ionic compounds and polar covalent compounds.

Any substance that dissolves in water is said to be **hydrophilic** ("water-loving"). Ionic compounds and polar covalent compounds are generally hydrophilic. Nonpolar covalent compounds and groups are not easily soluble in water; they are said to be **hydrophobic** ("water-hating").

Solutions of Ionic Compounds

Water weakens the ionic bonds that occur in ionic compounds. The ionic bonds in potassium chloride, for example, are readily broken when solid potassium chloride is added to water because the polar water molecules are able to form shells around the potassium and chlorine ions (Fig. 2–8a). Some water molecules orient their positively charged hydrogen atoms toward the negatively charged chlorine

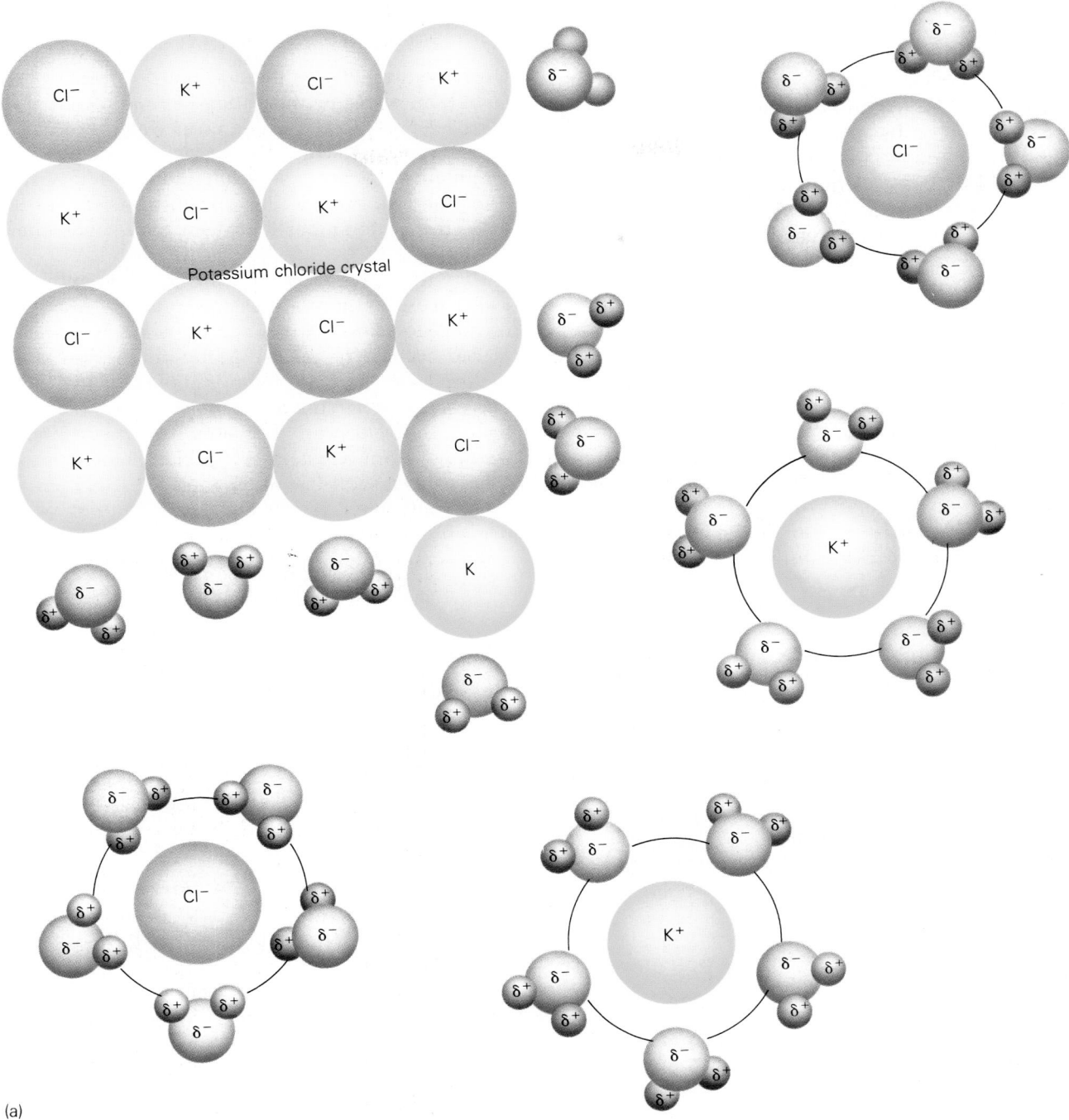

(a)

(b)

Figure 2–8

(a) Compounds such as potassium chloride (KCl) that form crystal lattices stabilized by ionic bonds dissolve readily in water because the small polar water molecules can slip in between and surround the K^+ and Cl^- ions. Some water molecules point their positively charged poles toward the Cl^- anions, forming shells around them. Other water molecules point their negatively charged poles toward the K^+ cations, forming shells around them.
(b) The ethanol molecule consists mostly of nonpolar covalent bonds, but the —OH (hydroxide) group is polar, so it forms hydrogen bonds with water. These bonds allow ethanol and other polar covalent compounds to dissolve in water.

atoms; others point their negatively charged oxygen atom toward the positively charged potassium atoms. The net force of all the water molecules attracting each ion overcomes the bonds between the ions. As a result, the potassium and chloride ions break away, or **dissociate,** from one another. However, in the process they do not take or give back the electrons they lost or donated in the first place, so they still have unequal numbers of protons and electrons. Therefore, they remain ions in solution as they were in the crystal lattice.

Solutions of Polar Covalent Compounds

Uncharged polar molecules also dissolve readily in water because they can form hydrogen bonds with water molecules. Ethanol is an example (Fig. 2–8*b*). The hydrogen atom in the —OH group of ethanol has a small positive charge because the shared electrons in the covalent bond are more strongly attracted to the oxygen atom. The polar —OH group provides the hydrogen atom for the hydrogen bond, while the water provides the oxygen atom.

Solutions of Nonpolar Compounds: Hydrophobic Interactions

Uncharged molecules that are not polar are poorly soluble in water because they have no positive or negative poles to attract the polar water molecules. Thus, water molecules have no effect on the covalent bonds that hold the solute molecules together. The polar water molecules can form neither shells around the atoms nor hydrogen bonds with them (Fig. 2–9*a*).

Hydrophobic substances are common in biological processes. Inportant examples are the chains of carbon atoms with only hydrogen atoms attached (called *hydrocarbon chains*). Many important macromolecules are partially composed of these hydrocarbon chains. When the macromolecules are placed in water, the hydrocarbon chains—the hydrophobic portions—form clusters or "pockets" to minimize their contact with the water molecules (Fig. 2–9*b*). This arrangement causes the least disruption of hydrogen bonds in the water. By creating such collections of molecules, hydrophobic interactions play an

Figure 2–9
(*a*) Nonpolar covalent molecules and groups, such as those formed from only carbon and hydrogen (hydrocarbons), do not dissolve in water because they lack positive and negative poles. The polar water molecules are unable to attract their atoms away from each other since they cannot form hydrogen bonds with them or shells around them. (*b*) Instead, nonpolar molecules cluster together in hydrophobic "pockets"; these reactions are called hydrophobic interactions. Such pockets allow water to retain its hydrogen-bonded structure.

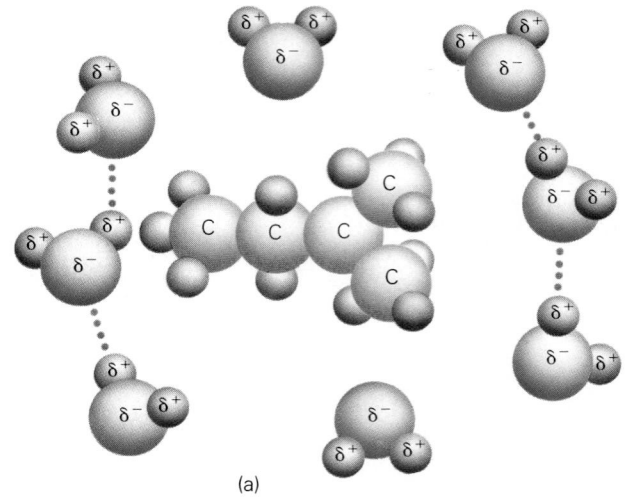

(a)

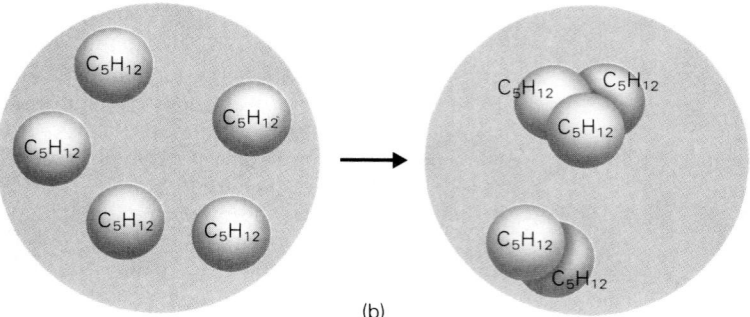

(b)

important role in maintaining the structures of certain cell organelles and of the plasma membrane that surrounds the cell. The strong tendency to cluster means that these hydrophobic groups can create a sturdy mass.

Concentrations of Solutions

We have said that some chemical reactions require an aqueous medium in order to occur and that the concentration of reactants can be an important factor in determining the rate of reaction. A solution may contain several different solutes, which may then react with one another; indeed, the purpose of their being in solution is to take part in important metabolic reactions. The concentration of reactants in these metabolic processes is determined by their concentration in solutions. Scientists have devised convenient ways of expressing these concentrations in order to understand the rates at which certain reactions occur both in the laboratory and in the living cell.

The concentration of a solution can be expressed in several ways. One way is to compare the number of moles of solute to the kilograms of solvent: this is known as **molality.** In aqueous solutions, this is the same as comparing moles of solute to liters of solution, since one kilogram of water is one liter. For example, if one mole (58.5 g) of sodium chloride (the solute) is dissolved in one liter of water, concentration is expressed as 1 mole per kilogram (abbreviated as 1 mol/kg). This is a "1 molal" solution, abbreviated as "1 m" solution.

An alternative way to express concentration is **molarity.** Molarity is a unit of concentration, based on the number of moles of solute per liter of solution. A solution with a concentration of 1 mol/L is called a "1 molar" solution, abbreviated as "1 M" solution. The solution is said to have a molarity of 1. A solution with a concentration of 2 mol/L is a 2 M solution.

Very low concentrations can be expressed with the same prefixes used in metric measures. A 1 mM solution of sodium chloride contains 1 millimole (58.5 mg) of sodium chloride per liter of solution. The solutions that occur in living organisms have a variety of concentrations, but most are in the millimolar range. Human blood plasma, for example, contains sodium and potassium at concentrations of approximately 140 mM and 4 mM, respectively.

Osmolarity is another expression of concentration that considers the total number of solute **particles** rather than the relative weights of specific solutes. Osmolarity is discussed in the section on osmosis.

The pH Scale

Water (H_2O or HOH) has a slight tendency to dissociate into hydrogen ions (H^+) and **hydroxide** (OH^-) **ions.** This dissociation process can be represented as:

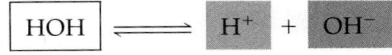

At the same time, the ions tend to reassociate, forming new water molecules. When the rate of dissociation equals the rate of reassociation, the reaction is said to be in **equilibrium.** At this point, the net concentrations of HOH molecules and of H^+ and OH^- ions remain constant, even while individual molecules and ions may be breaking down and reforming. The dissociation of water reaches equilibrium very rapidly, so that only very few of the water molecules are dissociated at any instant.

Ideally, the concentrations of H^+ and OH^- ions in water are exactly equal and very small, making the water a neutral compound. In reality, however, the concentrations of these ions are not equal because of the presence of solutes in the water that, by dissociating in it, increase or decrease the amount of H^+ ions relative to the number of OH^- ions. A solution in which the H^+ concentration is greater than the OH^- concentration is said to be **acidic.** A solution in which the number of OH^- ions is greater than the number of H^+ ions is a **basic** solution.

We can express the acidity or basicity of a solution by specifying the concentration of H^+ ions. Because this concentration is very small, scientists have devised a logarithmic method of expressing it, called **pH.** This expression is the negative of the base-10 logarithm (log) of the H^+ concentration:

$$pH = -\log [H^+]$$

where [ ] indicates the H^+ concentration in moles per liter.

In one liter of neutral water, there are 1×10^{-7} moles of H^+ ions (and 10^{-7} moles of OH^- ions). The pH of neutral water then is:

$$
\begin{aligned}
pH &= -\log (1 \times 10^{-7}) \\
&= -[\log (1.0) + \log (10^{-7})] \\
&= -[(0.00) + (-7)] \\
&= 7
\end{aligned}
$$

Suppose the concentration of H^+ ions were to increase 10-fold, to 1×10^{-6} mol/L. The pH of the solution would then be:

$$
\begin{aligned}
pH &= -\log (1 \times 10^{-6})] \\
&= -[\log (1.0) + \log (10^{-6})] \\
&= -[(0.00) + (-6)] \\
&= 6
\end{aligned}
$$

The solution is more acidic, and the pH has decreased. Conversely, if the H^+ concentration were to decrease, the solution would become more basic, and the pH would increase.

The pH scale goes from 0 to 14. A solution with a pH of 0 is highly acidic; one with a pH of 14 is highly basic. In terms of pure numbers, the scale is small, so that a 1000-fold increase in H^+ concentration produces a pH decrease of only 3 units. It is important to remember, therefore, that a *small decrease* in pH represents a *large increase* in the H^+ concentration, and vice versa.

Acids and Bases in Solution

An **acid** is any substance whose dissociation in water releases H^+ ions. The hydrogen ion does not contain any neutrons and can be thought of as essentially equivalent to a proton. Therefore, an acid is often referred to as a **proton donor** because it gives off protons in the form of H^+ ions. A **base** is a **proton acceptor,** a substance that accepts H^+ ions and thereby decreases the H^+ concentration in a solution.

Acids: Proton Donors

The addition of an acid to a solution greatly increases the concentration of free H^+ in the solution, producing a more acidic solution. The dissociation of an acid, HA, can be represented symbolically by the following equation:

$$HA \longrightarrow H^+ + A^-$$

where A^- represents the anion that remains when the hydrogen cation is formed. The strength of an acid is a function of its tendency to give up H^+ ions. In the case of a strong acid, such as HCl, all the molecules will dissociate in water and the dissociation will not be reversible:

$$HCl \longrightarrow H^+ + Cl^-$$

hydrochloric acid
(strong acid)

Weak acids, in contrast, give up H^+ ions less readily and do not dissociate completely in water. Thus, a solution of a weak acid contains both dissociated ions and complete molecules. Most of the acids produced in living organisms are weak acids.

Bases: Proton Acceptors

Usually a base gives off hydroxide ions (OH^-) in the process of dissolving in water and these decrease the H^+ concentration by accepting free H^+ ions. Sodium hydroxide (NaOH) and ammonium hydroxide (NH_4OH) are examples of strong bases that dissociate completely in water to yield free OH^- ions. The hydroxide anions then accept H^+ ions, and the result is water:

$$NaOH \longrightarrow Na^+ + OH^-$$

sodium hydroxide

$$OH^- + H^+ \rightleftharpoons H_2O$$

This reaction occurs when NaOH is added to water. The OH^- ions combine with the free H^+ ions already present in the water, decreasing the concentration of H^+ in the solution. In the case of a weak base, such as the bicarbonate ion (HCO_3^-), H^+ also is used up as follows:

$$H^+ + HCO_3^- \rightleftharpoons H_2CO_3 \rightleftharpoons H_2O + CO_2$$

Bases use up H^+ ions and decrease the H^+ concentration, thereby producing basic solutions.

If a weak base such as $NaHCO_3$ is present in solution with a strong acid such as HCl, the base will absorb the H+ ions released by the acid. The following four processes will be occurring:

Dissociation of the strong acid

$$HCl \longrightarrow H^+ + Cl^-$$

Dissociation of the weak base

$$NaHCO_3 \rightleftharpoons Na^+ + HCO_3^-$$

Formation of a weak acid

$$H^+ + HCO_3^- \rightleftharpoons H_2CO_3$$

Formation of a neutral salt

$$Na^+ + Cl^- \longrightarrow NaCl$$

These processes can be summarized in the following equation:

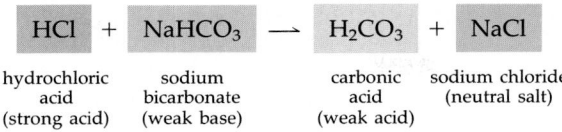

hydrochloric acid (strong acid) + sodium bicarbonate (weak base) → carbonic acid (weak acid) + sodium chloride (neutral salt)

Buffers

The pH of blood is controlled very closely and does not vary outside a small range, usually 7.38 to 7.42. This is a remarkable achievement because the chemical reactions in the cells of the body produce more than 10,000 nmol of H^+ ions each day. An average blood pH of 7.4 represents an H^+ concentration of only 40 nmol/liter, indicating that highly efficient mechanisms must exist to absorb the H^+ ions added to the blood. These mechanisms will be addressed in detail in later chapters. A general discussion of buffers will serve as an introduction to the "buffer" systems of the blood.

The purpose of a **buffer system** is to minimize the change in pH that occurs when an acid or base is added to any solution. For example, the dissociation of carbonic acid, a weak acid, occurs as follows:

$H_2CO_3 \rightleftharpoons H^+ + HCO_3$

carbonic acid (weak) → bicarbonate ion (base)

If the H^+ concentration of this solution were to be decreased by the addition of a strong base such as sodium hydroxide (NaOH), the equilibrium of the solution will be disturbed because the base would use up H^+ ions. The equilibrium would be re-established by dissociation of some more carbonic acid molecules (H_2CO_3), yielding additional H^+ ions to replace those consumed by the base. The overall reaction can be summarized as follows, using NaOH as an example of a strong base:

NaOH + H_2CO_3 → NaHCO$_3$ + H_2O

sodium hydroxide (strong base) + carbonic acid (weak acid) → sodium bicarbonate (weak base) + water

This is actually a combination of two reactions:

NaOH → Na^+ + OH^-

$H_2CO_3 \rightleftharpoons H^+ + HCO_3$

The sodium cations, Na^+, combine with the bicarbonate anions, HCO_3^-, and the H^+ ions combine with the OH^- ions, producing sodium bicarbonate and water. Thus, the strong base NaOH is converted to a weak base, NaHCO$_3$, and the reaction prevents a marked change in pH. The carbonic acid in this reaction is the buffer because its presence produces new H^+ to replace used H^+, thus maintaining the H^+ concentration.

Addition of an acid also disturbs the equilibrium of a solution because the H^+ concentration is increased. Bicarbonate ions will combine with the added H^+ to form carbonic acid, and in this way the total H^+ concentration is reduced to the level present before the acid was added. The overall reaction with HCl, a strong acid, can be summarized as:

HCl + NaHCO$_3$ → H_2CO_3 + NaCl

hydrochloric acid (strong acid) + sodium bicarbonate (weak base) → carbonic acid (weak acid) + sodium chloride (neutral salt)

The bicarbonate ion produced by the dissociation of the weak base NaHCO$_3$ uses up excess H^+ ions released by the dissociation of the strong acid HCl. The strong acid HCl is converted to the weak acid H_2CO_3 and excess H^+ ions are used up to minimize the pH change. As NaHCO$_3$ is consumed, carbonic acid is generated.

It is through these reactions that a mixture of NaHCO$_3$ and H_2CO_3 functions as the bicarbonate buffer system that prevents harmful changes in blood pH when acids or bases are added to blood. Destruction of one component of the buffer system always regenerates the other component, so that the buffer capability is maintained. This aspect of a buffer system distinguishes it from a simple neutralization reaction. A neutralization reaction also prevents a pH change, but the product of the reaction is a neutral **salt,** a compound in which the hydrogen atom of an acid (e.g., HCl) has been replaced by a different cation (e.g., NaCl). The salt NaCl is produced by the following neutralization reaction:

HCl + NaOH → NaCl + H_2O

hydrochloric acid (strong acid) + sodium bicarbonate (strong base) → sodium chloride (neutral salt) + water

NaCl cannot serve as a buffer. The resulting solution, although at neutral pH, has no capacity for dealing with further threats to the pH.

The effect of a buffer system is illustrated in Figure 2–10. Addition of 1 mmol (10^{-3} mol) of a strong acid, which dissociates completely in water, to

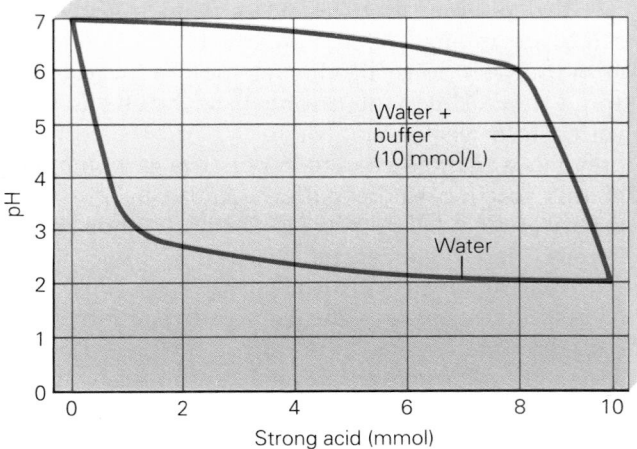

Figure 2–10
The presence of a buffer in a solution maintains a stable pH in spite of the sudden increase in H^+ concentration caused by the dissociation of a strong acid.

1 liter of pure water increases the concentration of H^+ from 10^{-7} mol/liter to about 10^{-3} mol/liter. The pH will fall from 7.00 to 3.00. If an appropriate buffer is present at 10 mmol/liter, the addition of 1 mmol strong acid will decrease the pH of the solution only to 6.90, a change of only 0.10 unit compared to the change of 4.00 units when no buffer was present. The buffer minimizes the decrease in pH until about 8 mmol of acid have been added. At this point, all the free buffer anions have been consumed by combining with the added H^+ ions. The buffer system no longer has an effect and the pH declines quickly. The buffer in this example was most effective in the range from 6.00 to 7.00. Most chemical buffers work best over a limited pH range, and each buffer has its own range.

The Behavior of Gases

When a liquid is heated sufficiently, it begins to evaporate; its molecules break apart and begin moving about rapidly in space. The substance has passed into the gas phase. Scientists describe four physical properties of any gas: (1) the **pressure** (P) or the force the gas molecules exert on their surroundings; (2) the **volume** (V) or the amount of space the gas molecules occupy; (3) the **amount,** or number (n) of moles of gas; and (4) the **temperature** (T) expressed in Kelvins (273 plus the Celsius temperature).

The Ideal Gas Laws

Three factors can affect the volume of a gas: pressure, temperature, and amount. At a constant temperature, the same amount of gas will decrease in volume with an increase in pressure. At constant pressure, the same amount of gas will increase in volume with an increase in temperature. At a constant temperature and pressure, the same amount of gas will remain at constant volume; the only factor that would change the volume would be an increase in the number of moles of gas present.

These relationships can be summarized as follows: (1) volume is inversely proportional to pressure; (2) volume is directly proprotional to temperature; and (3) volume is directly proportional to amount. They are further summarized in an equation known as the **Ideal Gas Law:**

$$PV = nRT$$

in which R is a constant, known as the **ideal gas constant,** equal to 0.082 liter · atm/deg · mol for any gas.

Partial Pressures in Gaseous Solutions

Unlike solids and liquids, which do not all readily dissolve in one another, all gases are completely soluble in all other gases. Air is a solution of several gases, including oxygen, nitrogen, carbon dioxide, and water vapor. According to the **law of partial pressures,** the total pressure in a container holding a solution of gases is the sum of the partial pressures of the component gases. **Atmospheric (or barometric) pressure,** which is the pressure exerted by air on the surface of the earth, is the sum of the partial pressures of all the gases that are present in air.

The partial pressure of a gas is proportional to the amount, in moles, of the gas solution it makes up. Oxygen, for example, makes up 21% of air. Atmospheric pressure is 760 mm Hg (so determined because, at sea level, air causes liquid mercury, Hg, to rise 760 mm in a closed tube). The partial pressure of oxygen, or P_{O_2}, is 21% of 760 mm Hg, or 160 mm Hg. If all other gases were removed from a container of air, with only oxygen remaining, it would still exert a pressure of 160 mm Hg. (Atmospheric pressure is used as a standard unit of pressure called the "atmosphere." One atmosphere— 1 atm—is equal to 760 mm Hg.)

Partial pressures are designated as P_{O_2}, P_{CO_2}, P_{N_2}, P_{H_2O}, and so on. The idea of partial pressure is very useful for dealing with concentration gradients and other aspects of physiology. Because partial pressures are directly proportional to the amount of gas present, they can be used to determine the number of moles of any constituent gas in a solution of gases.

The Kinetic Theory of Gases

According to the *kinetic theory of gases*, gas molecules are in continuous, random motion, colliding with one another and with their surroundings. The pressure of the gas arises from the forces of the collisions with the surroundings. The speed of the molecules is directly related to the temperature; if the temperature increases, the molecules move faster.

Gradients and the Movement of Substances

A **gradient** is a difference between two quantities divided by the physical distance (measured in units of length) between them. Gradients are responsible for the movement of various substances throughout the body, including the movement of blood through the circulatory system, air into lungs, and ions through membranes of cells. These gradients take a variety of forms and manifest themselves as pressure gradients, electrical gradients, or concentration gradients. Concentration gradients are part of a larger topic, diffusion and osmosis.

Pressure Gradients

The movement of blood through arteries and veins is a result of a **pressure gradient.** Just as water will flow through a pipe only if the pressure at one end is greater than the pressure at another end, pressure in one region of the circulatory system must be greater than that at another region in order for blood to flow between them.

The larger the difference in pressure between two points, the greater the flow. In other words, blood flow is directly proportional to the difference in pressure between the two points. The difference in pressure between any two points can be symbolized as $(P_1 - P_2)$, where P_1 is the pressure at the starting point and P_2 the pressure at the final point. (This difference, or pressure change, can also be expressed as ΔP, where the Greek letter Δ or "delta" symbolizes a change.) The relationship between flow, F, and pressure can be represented as:

$$F \propto (P_1 - P_2)$$

where $\propto$ means "is proportional to."

Flow is met with **resistance** in the form of veins and arteries, or, more precisely, their diameters. Vessels with small diameters impede or resist the flow of blood to a greater degree than vessels with large diameters. Other sources of resistance are the length of the vessel and the viscosity of the fluid. Flow is inversely related to resistance, that is, the greater the resistance, the smaller the flow.

$$F \propto 1/R$$

We can express the combined effects of pressure and resistance on flow as:

$$F = (P_1 - P_2)/R$$

The movement of air into the lungs is subject to the same influences. Gradients established by pressure differences between the external environment and the interior of the lung cause air to rush into or out of the lungs. The rate of air flow is greatly affected by the lung airway resistance. In people with asthma, for example, the constriction of the airways, or decrease in their diameters, creates resistance to the flow of air, and breathing becomes difficult.

Voltage Differences and Ion Flow

Gradients of an electrical nature are responsible for the movement of charged ions through cell membranes. **Electrical gradients** are also known as **potentials** or **potential differences.** These potential differences result from differences in the number of positive and negative charges on either side of a membrane. Typically, the interior of a cell is more negatively charged than the outside. The resulting electrical gradient causes positively charged ions in the extracellular fluid to be attracted to the negatively charged intracellular space. Likewise, negatively charged ions in the intracellular fluid are attracted to the positively charged extracellular space.

Ions can move across cell membranes by way of specific channels. The diameter of a channel in relation to the size of the ion passing through it helps determine the amount of resistance the ion encounters. The flow of ions is referred to as ionic current, I. The relationship of ionic current flow to potential differences and resistance can be expressed as:

$$I = (V_1 - V_2)/R$$

in which $(V_1 - V_2)$ is the potential difference or electrical gradient and R is the resistance of the channel. This expression (like that for pressure) is a variation of **Ohm's Law,** which relates flow to gradients and resistances in a variety of physical and biological systems.

Concentration Gradients: Diffusion and Osmosis

Diffusion is a process that involves the movement of particles such as ions and molecules in solution. An understanding of these processes in general is essential for an understanding of the movement of particles into and out of cells. In diffusion, a substance such as solute moves through and gradually becomes evenly mixed with another substance such as a solvent. Osmosis is a special kind of diffusion in which water moves through a membrane.

Diffusion of Solutes in Liquid Solvents

In solutions as well as elsewhere in nature, molecules, ions, and other particles move from one location to another by random thermal motion sometimes called **Brownian motion.** The magnitude and direction of the net movement of particles are determined by **concentration gradients.** A concentration gradient exists when two adjacent regions have different concentrations of particles. The larger the difference in concentrations, the greater the net movement between them. Particles will tend to move from regions of higher concentration to regions of lower concentration, that is, "down" the concentration gradient. If there is no concentration gradient (a situation that will occur when particles have dispersed throughout the solution), then the net move-

ment of particles will be zero, even though individual particles may move back and forth between the two regions. Such a solution is said to be in **dynamic equilibrium.**

The movement of sugar molecules in water illustrates the effects of a concentration gradient (Fig. 2–11). In Figure 2–11a, the sugar concentration in the lower region of the beaker is higher than elsewhere in the beaker. Although the sugar molecules are moving constantly, and some are moving toward the bottom of the beaker, most will move up and away from the region of concentration until they are evenly dispersed throughout the water (Fig. 2–11d). Even at this point, individual particles will still be moving in all directions, but there will be no net movement and no net change in concentration. The concentration gradient that existed when the particles were more concentrated in one region has been abolished by diffusion.

The driving force for solute diffusion is simply the difference in solute concentration, that is, the concentration gradient, between two regions or compartments. Diffusion of solutes abolishes the concentration differences. At this point, there is no net movement because there is no driving force.

Flux is a general term used to describe the rate of movement of solute molecules. It is expressed in terms of amount of solute moving per unit time, for example, mmol/minute. **Fick's Law** of diffusion states that the flux (J) of an uncharged solute is di-

Figure 2–11
Diffusion occurs when one substance moves randomly throughout another and becomes evenly mixed with it. Diffusion occurs only when a concentration gradient exists, such that the substance to be mixed is more concentrated in one region than another. Here sugar molecules move from a region of high concentration to regions of low concentration in a beaker of water. When they are fully dispersed, a gradient no longer exists and diffusion ceases. Although the individual sugar particles are still moving in all directions, there is no net movement in any one direction because there is no gradient.

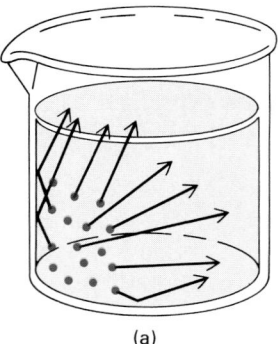
(a)

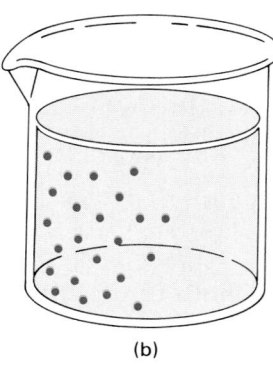
(b)

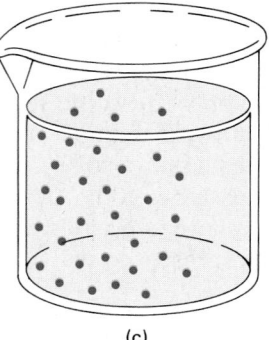
(c)

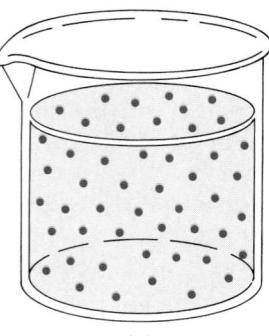
(d)

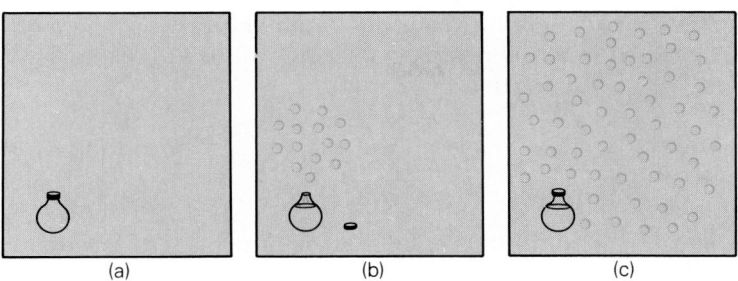

Figure 2–12
A model for the diffusion of gas molecules down a concentration gradient. When a perfume bottle is opened in a closed space, some of the liquid evaporates, and the gaseous perfume molecules pass into the surrounding space. A concentration gradient then exists because the molecules are more concentrated near the bottle than in the rest of the room. When the bottle is recapped, the perfume molecules slowly diffuse throughout the room until they are evenly distributed, and the concentration gradient no longer exists.

rectly proportional to the driving force (dc/dx) and the area (A) through which diffusion can occur. Thus:

$$J = -DA \cdot dc/dx$$

D is the **diffusion coefficient,** which considers the mobility of the solute molecules in solution. This mobility will vary with both the specific solute and the specific solvent. The driving force (dc/dx) is the concentration gradient—literally, the change in concentration relative to the change in distance. The minus sign is a standard way of showing that the direction of net solute movement is always from the region of high concentration to the region of low concentration, that is, "down" the concentration gradient.

Diffusion of Gases in Gaseous Solutions

Gas molecules diffuse in the same way as do solid or liquid solute particles. Imagine a room in which there is no movement of air molecules (Fig. 2–12). If a bottle of perfume is opened, some of the liquid

perfume evaporates. The gas molecules come out of the bottle into the room. Slowly, the perfume molecules spread and eventually reach an equilibrium in which the distribution of the scent is even throughout the room. The perfume spreads by diffusion, owing to the concentration gradient that exists because initially the molecules are more concentrated near the bottle than in the rest of the room. The molecules move from this region of high concentration to regions of low concentration.

The movement of gas along a concentration gradient can be pictured as the movement of molecules in two boxes with a common wall (Fig. 2–13a). Suppose one box is filled with oxygen molecules. The molecules move randomly about in the box, bouncing off the walls and ricocheting off each other. Now, suppose a hole is opened in the common wall between the boxes (Fig. 2–13b). A few molecules will travel immediately through the hole into the second box and begin bouncing around. More and more of the molecules will accumulate in the second box. As the concentration builds in the second box, some of the molecules will ricochet back through the

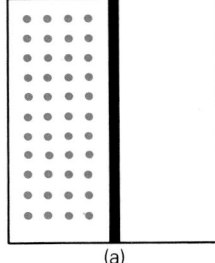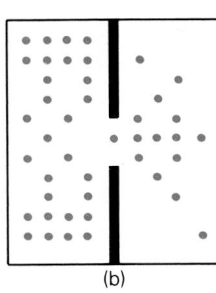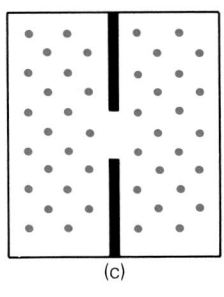

Figure 2–13
In response to a concentration gradient, gas molecules move through a hole in a wall between two boxes until they are evenly distributed.

hole into the first box. Eventually, the same number of molecules will be traveling in both directions through the hole, and the number of molecules will be the same in both boxes. At that time, the gas in the two boxes will be in dynamic equilibrium (Fig. 2–13c).

This diffusion process can take place even if the overall pressure is the same in each box. It can also occur if there are other kinds of molecules in each box. For example, nitrogen, carbon dioxide, and water vapor might be present in each box in equilibrium. If oxygen is introduced, it will move along its own concentration gradient as if it were the only type of molecule present (Fig. 2–14a, b, and c).

Likewise, different gases can be present in different concentrations in each box. For example, there can be a lot of carbon dioxide in one box and a lot of oxygen in the other. Even though the total pressure is the same in each box, the gases will diffuse simultaneously along their respective concentration gradients and move in opposite directions between boxes as soon as the hole is opened (Fig. 2–14d, e, and f).

One kind of molecule can move through the spaces between other molecules. There is mostly empty space between the molecules present in air. The movement of perfume molecules between air molecules is analogous to the movement of oxygen molecules through the hole in the wall. In air, there is a very large number of spaces through which molecules can move, and this speeds up any diffusion processes that occur. The wall in the box is analo-gous also to various membranes in the body through which gas molecules diffuse in response to concentration gradients.

Osmosis: Diffusion of Water Through a Membrane

When the molecules that are moving in diffusion are water molecules, the movement is known as **osmosis**. Osmosis is the flow of water across a barrier in response to a difference in solute concentration on either side of the barrier. Water flows from the side of the barrier with the lower concentration of solutes (the more dilute region) to the side of the barrier with the higher concentration (the less dilute region). The water molecules simply move down their own concentration gradient, like any other type of molecule. More water molecules are present in the dilute solution than in the concentrated solution, which is why there is net movement of water toward the region of high solute concentration.

Osmosis is involved in the preservation of meat by salting. When bacteria contact salt on the surface of meat, a concentration gradient exists because the salt is more highly concentrated outside the bacterial cell than inside. The plasma membrane of the cell is impermeable to the salt; salt cannot diffuse into the cell to abolish the gradient. The membrane is permeable to water, however, and water flows out of the cell toward the region in which the salt is more concentrated. The bacterial cell dies before it can infect the meat.

Figure 2–14
(a, b, and c) In a solution of gases, one constituent gas will diffuse independently along its own concentration gradient to become evenly distributed throughout the solution. (d, e, and f) Two concentration gradients can exist in two adjoining spaces. Molecules of each gas will move along their own concentration gradients at the same time until both gases are evenly spread throughout the entire space.

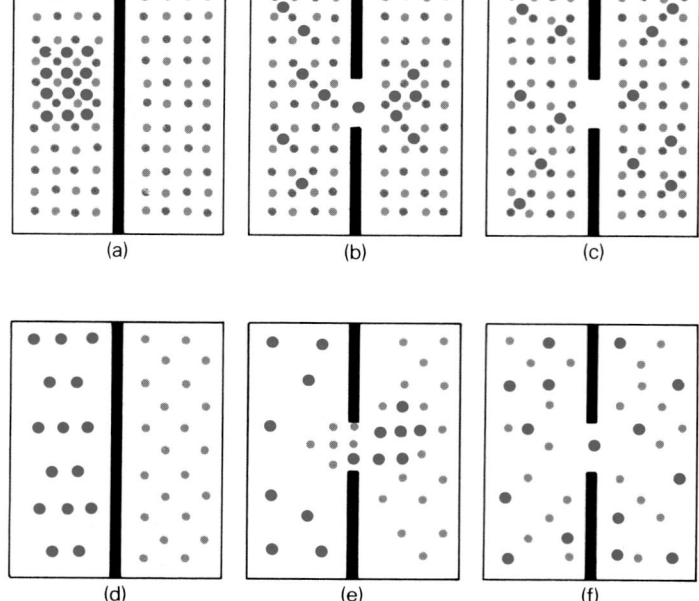

Osmotic Concentration.   When discussing osmosis, we use a special measure of concentration in order to better understand the relationship of water and solute particles. The number of solute particles a solute releases in solution is not related to the mass of the solute. Chemical concentration, e.g., molality or molarity, is based essentially on weight and does not always reveal the extent to which solute particles are present in the solution. For example, in a solution of glucose in water, the glucose molecules ($C_6H_{12}O_6$) stay intact rather than dissociating into atoms. In a solution of sodium chloride in water, by contrast, the Na^+ and Cl^- ions dissociate completely and this doubles the total number of particles that are present in the solution.

Osmotic concentration is expressed in units called **osmoles (Osm)**. For physiological solutions, we use **milliosmoles (mOsm)** since the concentrations are generally low.

Just as molarity expresses moles of solute per liter of solution, osmolarity expresses osmoles of solute per liter of solution. The relationship between molarity and osmolarity can be summarized as:

$$\text{osmoles} = n \times \text{moles}$$
$$(\text{milliosmoles} = n \times \text{millimoles})$$

where n is the number of particles derived from each molecule of solute when in solution. Note that this relationship applies only to dilute solutions that are assumed to behave ideally. Deviation from ideal behavior occurs as the concentration of the solution increases.

For example, the osmotic concentration of a glucose solution with a chemical concentration of 60 mmol/liter is 60 mOsm/liter because n equals 1 for glucose. Glucose exists as single, intact molecules in water, producing only one solute particle per molecule.

In contrast, the osmotic concentration of a solution of NaCl whose chemical concentration is 60 mmol/liter is 120 mOsm/liter because $n = 2$. Each molecule of NaCl produces two solute particles (Na^+ and Cl^- ions) and both become osmotically active solute particles. A $CaCl_2$ solution with a chemical concentration of 20 mmol/liter will have an osmotic concentration of 60 mOsm/liter because each $CaCl_2$ group dissociates completely in solution into three ions: one Ca^{2+} ion and two Cl^- ions. A 20 mM solution of $CaCl_2$ is said to have an osmolarity of 60 mOsm.

Osmotic Pressure.   **Osmotic pressure** is a property of a solution that is proportional to the solute concentration. A dilute solution will have a lower osmotic pressure than a concentrated solution. Thus, water will move by diffusion from a region of low osmotic pressure to a region of high osmotic pressure.

The osmotic pressure difference between two solutions of different solute concentrations is illustrated in Figure 2–15. A solution of water containing 1.0 M sucrose is placed inside an inverted thistle funnel. The thistle funnel is placed in a bath of pure water. The water and the sucrose solution are separated by a membrane placed across the mouth of the funnel. The membrane allows only pure water, not sugar molecules, to pass through. The apparatus is arranged so that, initially, the level of the sucrose solutions is the same as the level of the water in the bath.

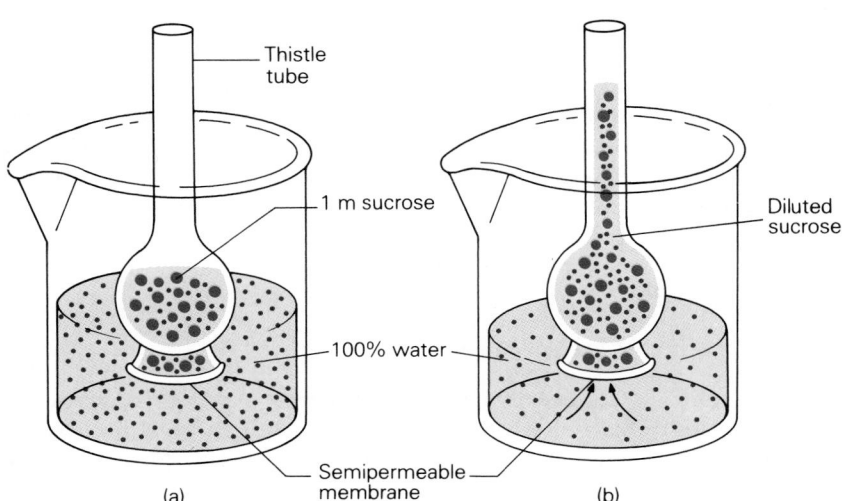

(a)   (b)

Figure 2–15
When a thistle tube containing a sucrose solution is placed in a bath of pure water, the water moves across a membrane up into the tube, diluting the sucrose solution and raising the level of the solution in the tube. No sucrose flows from the tube into the bath because the membrane is impermeable to sucrose.

The sucrose solution has a higher osmotic pressure than the water because it has a higher solute concentration. Consequently, water will begin to move from the pure water bath across the membrane into the thistle funnel. The movement of water into the sucrose solution gradually dilutes the solution and raises the level of the solution inside the thistle tube.

Eventually, the increased fluid level in the funnel generates a new pressure, hydrostatic pressure, which now opposes the movement of more water into the tube. Soon there is enough hydrostatic pressure to completely stop the osmosis of water into the thistle tube. At this point, the hydrostatic pressure reflects the difference in osmotic pressure between water and the original sucrose solution.

The increase in the fluid level in the funnel can be prevented by applying pressure down the funnel at the start of the experiment. This pressure blocks the osmotic movement of water into the sucrose solution. The pressure applied is the osmotic pressure of 1.0 M sucrose.

If we know the solute concentration of a solution, it is relatively simple to calculate the osmotic pressure. Osmotic pressure is related to the solute concentration by the **van't Hoff law,** which is described by the equation:

$$\pi = RTC$$

in which π is the osmotic pressure, R is the ideal gas constant (0.082 liter · atm/degree · mole), T is the absolute temperature in Kelvins (°C + 273), and C is the total solute concentration expressed as osmoles of solute per liter of solution. For example, the osmotic pressure of a solution of 1.0 Osm/liter at 0°C is calculated as follows:

$$\pi = RTC$$
$$= 0.082 \frac{\text{liter} \cdot \text{atm}}{\text{degree} \cdot \text{mole}} \times 273° \times 1.0 \frac{\text{Osm}}{\text{liter}}$$
$$= 22.4 \text{ atm}$$

Thus, a 1.0 Osm solution exerts an osmotic pressure of 22.4 atm at 0°C. A similar calculation reveals that human blood plasma, which is normally a 0.31 Osm solution, will have an osmotic pressure of 7.9 atm at 37°C. The osmotic pressure of a solution depends on the total osmolarity of the solution and not on the nature of the specific solute or solutes. The value of C for pure water containing no dissolved solutes is zero; hence, the osmotic pressure of pure water is zero.

Since R and T are constants if there is no temperature change, the osmotic pressure will be directly related to osmolarity. In this case, there is no need to calculate osmotic pressure in order to determine which of several solutions has the highest osmotic pressure. A comparison of the osmolarity of the solutions will suffice; the solution with the highest osmolarity will have the highest osmotic pressure.

The osmolarity of a solution can be directly determined by use of a freezing point **osmometer,** even if the identity of the specific solute is unknown. The freezing point of pure water is depressed when solutes are present, and the extent of the depression is proportional to the total amount of all the solutes present. Most osmometers are calibrated to give a readout directly in osmoles or milliosmoles, and they provide a rapid and accurate method for determining the osmolarity of solutions.

Two solutions that have the same osmolarity are said to be **iso-osmotic.** Thus, 300 mOsm NaCl is iso-osmotic with 300 mOsm sucrose. A solution with a higher osmolarity than another is said to be **hyperosmotic;** a solution with a lower osmolarity is said to be **hypo-osmotic** relative to one with a higher osmolarity.

Osmotic pressure is only one property of a solution that is affected by the presence of solute particles. Dissolved solutes lower the vapor pressure and increase the boiling point of solutions. They also lower the freezing point. When salt is spread on a wet road in winter, it dissolves in the water, preventing the water from freezing at 0°C, the point at which pure water would normally freeze. $CaCl_2$ is used instead of NaCl because it releases more solute particles (three rather than two) per molecule and has a greater effect on the freezing point.

Macromolecules

The unique properties of the carbon atom allow it to form four bonds at once and lead to a huge variety of molecules of different shapes and enormous sizes. These molecules are called **macromolecules.** These substances play important roles in the life processes of all cells and organisms. The most important macromolecules in cell physiology are carbohydrates, lipids, proteins, and nucleic acids. The following brief survey provides a glimpse of these compounds and a review of their chief functions in cells.

Carbohydrates

Carbohydrates are often referred to as **sugars.** All **carbohydrates** are composed of carbon, hydrogen, and oxygen atoms. Carbohydrates are classed in three broad groups: monosaccharides, disaccha-

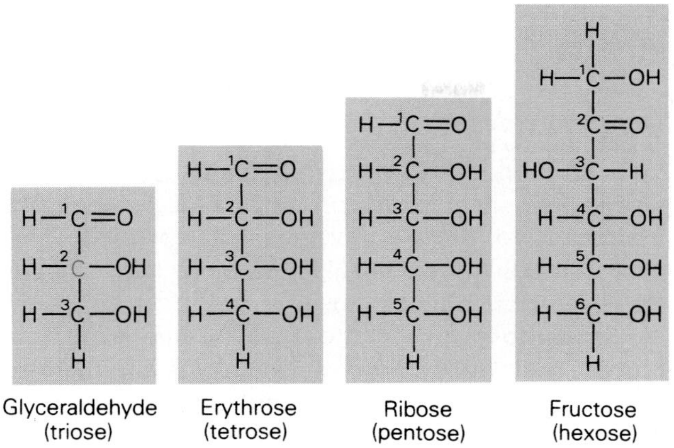

Figure 2–16
The structures of the monosaccharides glyceraldehyde, erythrose, ribose, and fructose.

rides, and polysaccharides. In addition, they form compounds with proteins and lipids to form glycoproteins and glycolipids.

Monosaccharides

The simplest type of carbohydrates are called **monosaccharides.** The general chemical formula for a monosaccharide is $(CH_2O)_n$, where n is a whole number indicating the number of carbons. A monosaccharide with three carbons, such as glyceraldehyde, is a **triose** (Fig. 2–16). A monosaccharide with four carbons is a **tetrose;** one with five carbons a **pentose,** and one with six carbons a **hexose.**

In structural formulas of sugars and other "organic" molecules, the carbon atoms often are numbered for identification. Specific carbons often have specific roles based on their position in the molecule and on the other atoms attached to them. A carbon atom, such as the second carbon atom ("carbon 2") in the glyceraldehyde molecule, is described as **asymmetrical** when it is linked covalently to four different chemical groups. Carbon 2 in glyceraldehyde (Fig. 2–16) is linked to a hydrogen (H), an aldehyde (—CHO) group, a hydroxide (—OH) group, and —CH_2OH.

Ring Structures. Monosaccharides can have different forms, either linear chains or rings. In solutions, monosaccharides most often form rings. One common hexose is **glucose,** $C_6H_{12}O_6$, or $(CH_2O)_6$. In glucose, ring formation occurs when the aldehyde (—CHO) group on carbon 1 reacts with the hydroxyl (—OH) group on carbon 5 to form a stable covalent bond, giving rise to a six-member ring structure (Fig. 2–17).

Although drawn as a flat structure, the plane of the ring is perpendicular to the plane of the page, and the hydrogen atoms and hydroxyl groups lie above and below the plane of the ring. The edge of the ring nearest the reader is indicated by a thicker line between the carbon atoms. In simplified drawings of ring structures, the letter C indicating a carbon atom is omitted, but the carbons are understood to be present wherever two "sides" of the ring meet. Furthermore, the letter H indicating hydrogen atoms is also omitted but understood to be present wherever needed to supply each carbon with a total of four bonds. In the most simplified drawings, only the —OH groups are identified by letters (Fig. 2–17).

Figure 2–17
The linear and ring structures of the sugar glucose, a common hexose. When the —OH group on carbon 5 comes near the =O group on carbon 1, the two carbons are joined by the —O— atom, and a ring is formed.

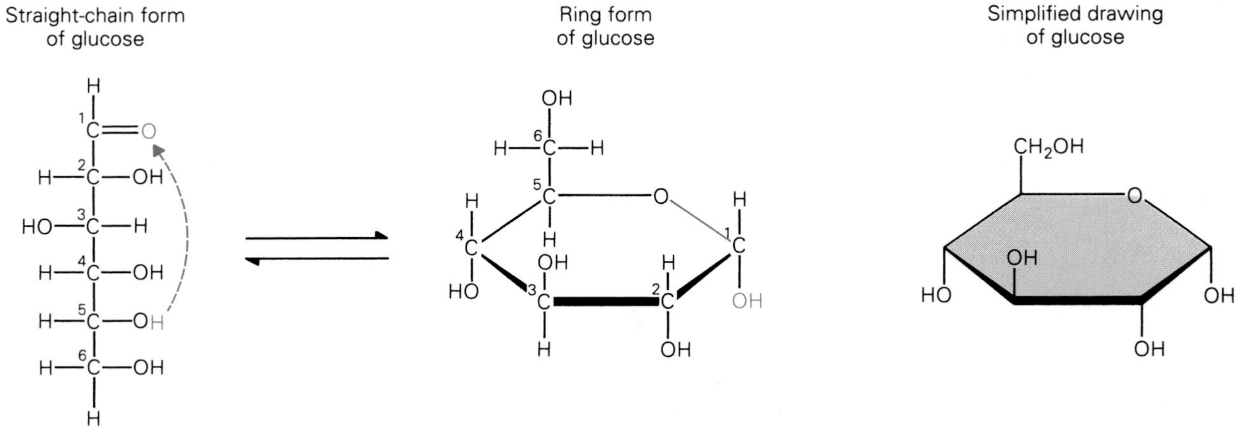

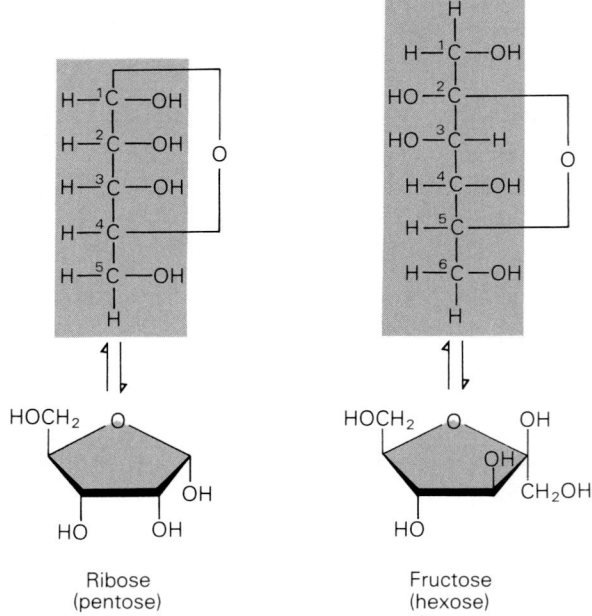

and nucleic acids can assemble only in linear, unbranched chains.

Disaccharides

Two monosaccharides can combine to form a **disaccharide.** The most common disaccharides are **maltose** (two glucose molecules), **lactose** (one glucose molecule plus one galactose molecule), and **sucrose** (glucose plus fructose).

Isomaltose, like maltose, is formed from two glucose molecules, and both isomaltose and maltose have the same chemical formula. However, they have different structures because their monosaccharide units join at different carbon atoms. In maltose, the glucose molecules join at carbons 1 and 4; in isomaltose, they join at carbons 1 and 6 (Fig. 2–19). Maltose and isomaltose are called **isomers,** molecules with the same chemical formula but different structural arrangements of the atoms.

Polysaccharides

Most of the carbohydrates found in nature occur as **polysaccharides,** molecules composed of many monosaccharide units arranged in branched chains. Polysaccharides are distinguished from one another mainly by their monosaccharide units and the length and amount of branching of their chains. Glucose is the most common monosaccharide unit of polysaccharides. Glucose is the fuel that most animal cells require, and it can be stored in certain tissues in the form of the polysaccharide **glycogen.** This important polysaccharide is especially abundant in the liver, where it can account for up to 10% of the wet weight of the tissue. It consists of a chain of glucose units that are linked by chemical bonds between carbons 1 and 4 of adjacent glucose units. Branches occur about every 10 glucose units, arising from a bond between carbons 1 and 6 (Fig. 2–20).

Figure 2–18
The cyclic forms of ribose (a pentose) and fructose (a hexose) both contain five-member rings.

Pentoses form five-member ring structures, as do some hexoses such as fructose (Fig. 2–18). Two important pentoses are ribose and deoxyribose, which are present in nucleic acids.

When a hexose forms a five-member ring, carbon 1 is left outside the ring because the ring forms when the carbonyl (C=O) group on carbon 2 reacts with the hydroxyl (—OH) group on carbon 5 (Fig. 2–18).

Monosaccharides have a unique ability to form chemical bonds at more than one carbon atom. One monosaccharide can combine with several others to create branched chains of monosaccharide units. In contrast, other macromolecules such as proteins

Figure 2–19
Maltose and isomaltose are isomers. Both are formed from two molecules of glucose; both have the same molecular formula. They have different structural formulas, however, and therefore constitute two different disaccharides.

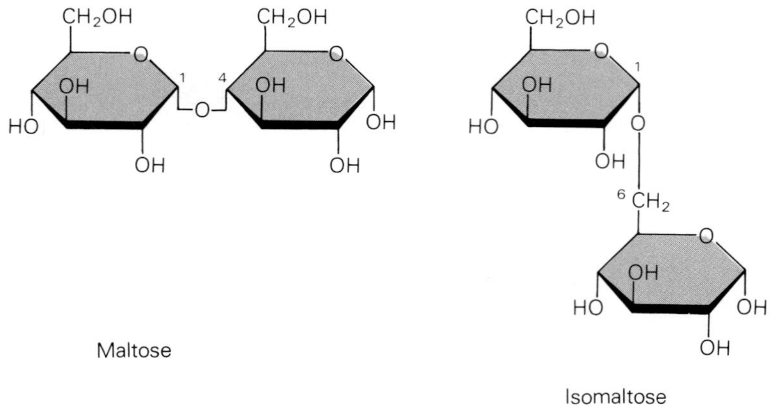

Maltose

Isomaltose

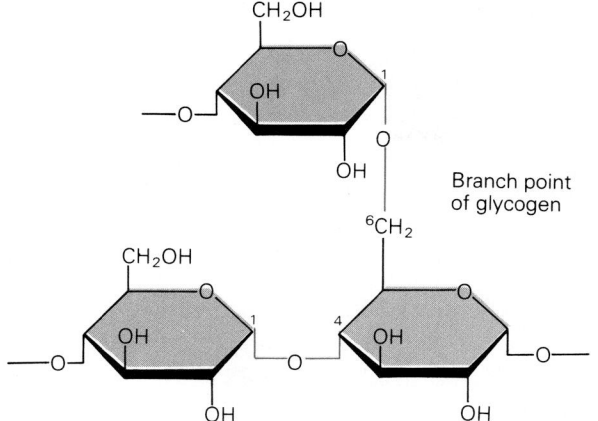

Figure 2–20
A branch point in the polysaccharide glycogen.

The molecular mass of glycogen has been estimated to be in the range of 3×10^5 to 3×10^6. This means that the number of glucose units present in a glycogen molecule could be as high as 16,667.

Sugar Combinations

Many of the polysaccharides present in living organisms occur in covalent combination with proteins or lipids. **Glycoproteins** and **glycolipids** are present in cell membranes. The polysaccharide chains in these compounds tend to be short but highly branched, containing several different monosaccharide units arranged in complex sequences. They are thought to be important in recognition processes such as those involved in the action of hormones. A hormone released into the blood will be distributed to all the cells in the body, but will affect only those cells that have an external site that the hormone "recognizes" and binds. The carbohydrate portions of membrane glycoproteins and glycolipids are thought to be important parts of these binding sites for hormones.

Lipids

Lipids are sometimes called **fats** and are distinguished from the other cellular macromolecules because they are poorly soluble in water and highly soluble in organic lipids such as ether and chloroform. They are composed of carbon, hydrogen, and oxygen atoms, the same elements that make up carbohydrates; however, the proportions of these elements differ in lipids. The lipids contain only small amounts of oxygen relative to carbon and hydrogen.

Three major classes of lipids will be considered here: fatty acids, lipids containing glycerol, and lipids that do not contain glycerol.

Fatty Acids

A **fatty acid** is a long chain of covalently linked carbon atoms to which only hydrogen atoms are attached, except for a carboxylic acid (—COOH) group at one end. The long hydrocarbon chain is hydrophobic, or poorly soluble in water, whereas the —COOH group is highly water soluble and reacts readily with other chemical groups, forming covalent bonds. Most of the fatty acids in animal cells are covalently linked to other molecules via the carboxylic acid group. Two of the most abundant fatty acid molecules in animal cells are **palmitic acid** and **stearic acid** (Fig. 2–21).

Figure 2–21
The structures of stearic acid and palmitic acid, two common saturated fatty acids. The molecules consist mostly of hydrophobic hydrocarbon chains. The carboxylic acid (—COOH) group at one end, however, is water-soluble or hydrophilic. Fatty acids are sometimes given the general symbol RCOOH, R representing the hydrocarbon chain.

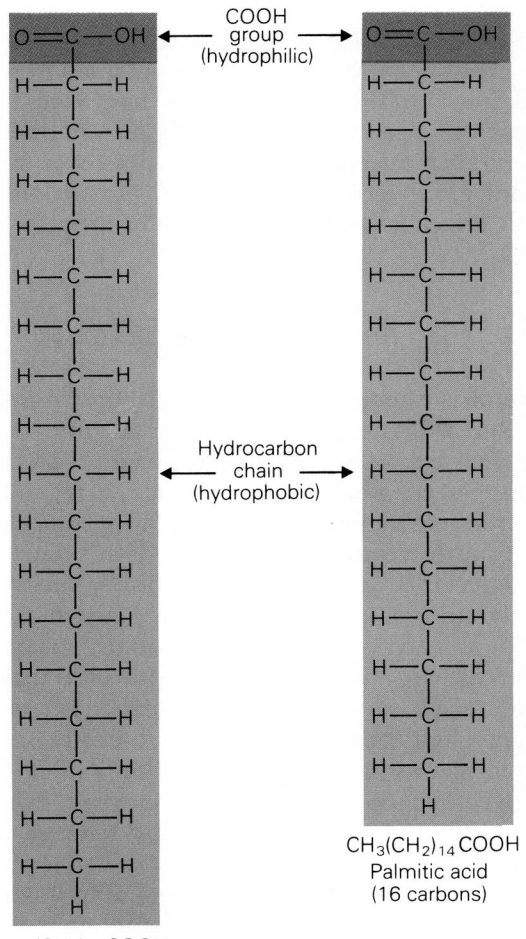

$CH_3(CH_2)_{16}COOH$
Stearic acid
(18 carbons)

$CH_3(CH_2)_{14}COOH$
Palmitic acid
(16 carbons)

Specific fatty acids are distinguished from one another by the number of carbon atoms in their hydrocarbon chains and the number of carbon-carbon double bonds. When only single bonds exist between all the carbons in a fatty acid chain, the fatty acid is said to be **saturated,** i.e., saturated with hydrogen, because as many hydrogen atoms as possible are bonded to each carbon. When one or more double bonds are present between carbon atoms, the fatty acid is **unsaturated.** The most abundant unsaturated fatty acids in animal cells are oleic acid, linoleic acid, and arachidonic acid, which contain one, two, and four double bonds, respectively:

oleic acid $CH_3(CH_2)_7CH{=}CH(CH_2)_7COOH$

linoleic acid $CH_3(CH_2)_4CH{=}CHCH_2CH{=}CH(CH_2)_7COOH$

arachidonic acid $CH_3(CH_2)_4(CH{=}CHCH_2)_4(CH_2)_2COOH$

Fatty acids are a valuable source of fuel for cells. Considerably more energy can be obtained from the breakdown of 1 g of a fatty acid than from 1 g of carbohydrate. Arachidonic acid is an intermediate structure produced during the synthesis of an important group of molecules called **prostaglandins,** which affect a wide variety of physiological processes.

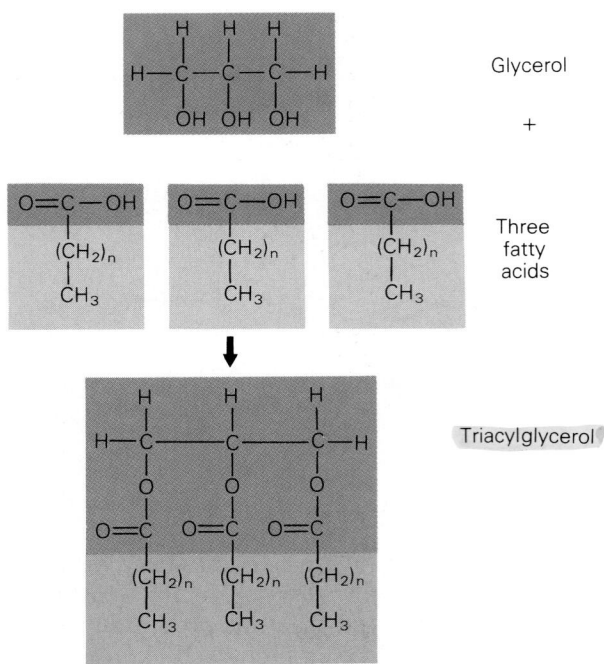

Figure 2–22
Steps in the formation of a triacylglycerol, also termed a triglyceride, from glycerol and three fatty acids. The fatty acids may be identical or different. *n* stands for a whole number, the number of carbon atoms in the parentheses representing all but the last carbon in the fatty acid hydrocarbon chain. For stearic acid, *n* is 16 (see Fig. 2–21).

Glycerol Compounds and Phospholipids

Glycerol is a molecule with three hydroxyl groups (Fig. 2–22). These hydroxyl groups can form bonds with the carboxylic acid groups of fatty acids. The attachment of one, two, or three fatty acids to glycerol produces a **monoacylglycerol, a diacylglycerol,** or a **triacylglycerol,** respectively. Both mono- and diacylglycerols are important natural compounds, but the triacylglycerols (Fig. 2–22) are the most abundant and represent an important means by which cells are able to store fatty acids until they are needed. Cells can convert excess carbohydrate and protein to fatty acids and then store the fatty acids in triacylglycerols. The continued accumulation of triacylglycerol deposits in the body, however, can lead to obesity. Triacylglycerols often are referred to as **triglycerides.**

Another group of glycerol-containing lipids are the **phospholipids.** The parent compound of this group is **glycerol-3-phosphate,** a glycerol molecule to which a phosphate (—PO₄) group has been at-

tached (Fig. 2–23*a*). A phospholipid is formed from this basic structure by (1) the addition of fatty acids to the free hydroxyl groups on carbons 1 and 2 to give **phosphatidic acid** and (2) the addition of another chemical group to the phosphate on carbon 3. In Step 1, the fatty acid added to carbon 1 of the glycerol unit is usually saturated and contains 16 to 18 carbon atoms, whereas the fatty acid added to carbon 2 is usually unsaturated and contains 16 to 20 carbon atoms. The most common groups added in Step 2 are choline, ethanolamine, serine, and inositol. The phospholipids that result are called phosphatidylcholine,   phosphatidylethanolamine, phosphatidylserine,   and   phosphatidylinositol, respectively (Fig. 2–23*b*).

At physiological pH (7.40), the phospholipids are in the charged forms shown, and one end of the molecule is a polar (hydrophilic) region. In contrast, the long hydrocarbon chains of the two fatty acid groups constitute a highly hydrophobic region of the molecule (Fig. 2–23*c*). The dual nature of these phospholipids—hydrophilic at one end and hydro-

An Ether Phospholipid

Platelet-activating factor (PAF) is an ether phospholipid which has striking and diverse physiological effects. It is an analog of phosphatidylcholine (Fig. 2-23c). Unlike phosphatidylcholine, PAF has a long hydrocarbon chain only on carbon 1. At carbon 2, instead of a second hydrocarbon chain, PAF has an acetyl group (see diagram below). This increases the water solubility of the molecule and allows it to function in an aqueous environment. It is termed an ether phospholipid because the hydrocarbon chain at carbon 1 is attached by an ether linkage (—C—O—C—).

$$
\begin{array}{c}
\overset{O}{\underset{\parallel}{\text{H}_3\text{C}-\text{C}}}-\text{O}-\text{C}^2-\text{H} \quad\quad \overset{\text{H}_2\text{C}^1-\text{O}-\text{CH}_2-(\text{CH}_2)_{14}-\text{CH}_3}{} \\
\text{C}^3\text{H}_2-\text{O}-\overset{O}{\underset{O^-}{\overset{\parallel}{P}}}-\text{O}-\text{CH}_2-\text{CH}_2-\text{N}^+(\text{CH}_3)_3
\end{array}
$$

platelet-activating factor

PAF is of interest because it is the first known example of a biologically active phospholipid. Its identity was discovered in 1979. Prior to this period it was assumed that the principal role of phospholipids was to provide a physical barrier as a component of plasma membranes and intracellular membranes.

PAF was found originally in the blood of animals undergoing a severe allergic reaction and was shown to be a factor which activated blood **platelets.** Platelets are blood cells which participate in the formation of blood clots (Chapter 17). However, the term *platelet-activating factor* is now considered a misnomer because PAF affects the cells of many different tissues. In some situations the action of PAF is achieved at very low concentrations such as 1.0 pM (10^{-12} M). PAF has received most attention for its role in **asthma** and **endotoxin shock.** PAF is one of a group of substances produced during an asthmatic reaction and its presence leads to inflammation and narrowing of air passages. Release of PAF is stimulated also during a bacterial invasion. **Toxic-shock syndrome,** for example, occurs because fragments of destroyed bacteria act as endotoxins and induce production of PAF. The latter lowers blood pressure and decreases the amount of blood pumped by the heart, leading to shock. Both asthma and endotoxin shock, when severe, can lead to death.

In addition to these adverse effects in abnormal situations, PAF also may have important beneficial roles in normal physiological processes such as reproduction. For example, PAF secreted by the fertilized egg is helpful for implanting the egg in the wall of the uterus. In the late stages of pregnancy, PAF is produced in increasing amounts by the fetal lung. This may help to stimulate production of fetal lung **surfactant,** a substance which prevents lung collapse in newborns (see Chapter 20). Release of PAF into the fluid surrounding the fetus may contribute to the initiation of labor because PAF is a potent stimulator of uterine contractions.

A major development in the field occurred in 1991 when the structure of the PAF **receptor** was identifed. The receptor is a small protein of 342 amino acids which is present in the plasma membrane of many types of cells. Attachment of PAF to the receptor is a necessary first step in the sequence of events by which PAF changes the behavior of a cell. Researchers are hopeful that the design and development of new drugs to block binding of PAF to its receptor could lead to new treatments for asthma and endotoxin shock, and to new ways for preventing conception and premature birth.

phobic at the opposite end—plays an important structural role in the plasma membrane surrounding all cells. The membrane is made up of a "lipid bilayer": two layers of phospholipid molecules oriented so that their hydrophobic hydrocarbon chains point toward each other, away from the aqueous surroundings both inside and outside the cell (Fig. 2–23d). The hydrophilic parts of the phospholipids in each layer are oriented outward. This arrangement represents a very stable physical structure with unique properties, as will be discussed in Chapter 4.

Sphingomyelin is the only phospholipid not derived from glycerol. The basic unit of sphingomyelin is **sphingosine:**

$$\text{CH}_3(\text{CH}_2)_{12}-\text{CH}=\text{CH}-\text{CHOH}-\text{CHNH}_3^+-\text{CH}_2\text{OH}$$

Sphingosine contains an unsaturated hydrocarbon chain and is converted to sphingomyelin by the addition of a long-chain fatty acid, a phosphate group, and a choline group. Overall, the structure of sphingomyelin is very similar to that of phospha-

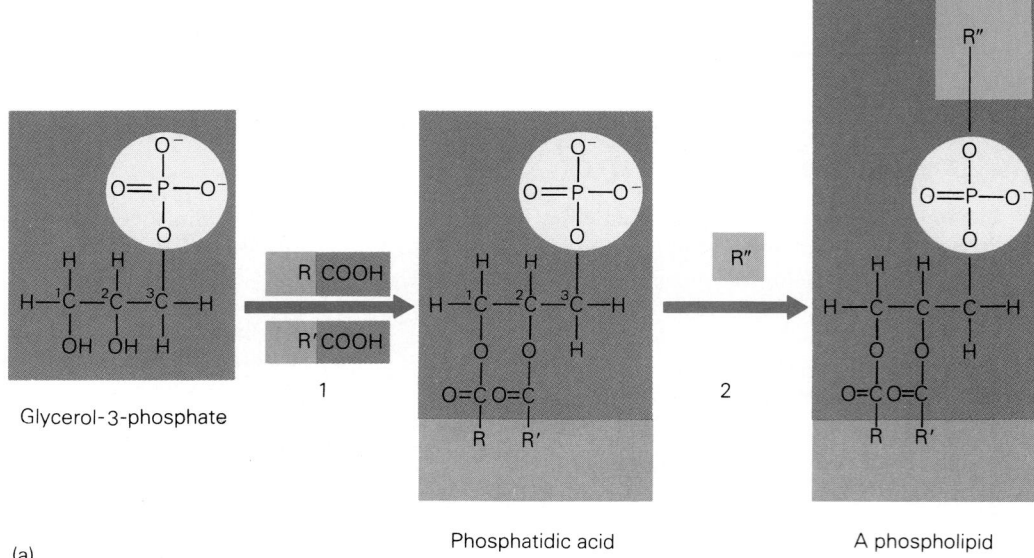

Glycerol-3-phosphate

Phosphatidic acid

A phospholipid

(a)

Figure 2–23

Steps in the formation of a phospholipid. (*a*) Two different fatty acids, symbolized by RCOOH and R'COOH, join with glycerol-3-phosphate to form phosphatidic acid. *R* and *R'* represent the long, hydrophobic hydrocarbon chains of the fatty acids. (*b*) Another specific group, either choline, ethanolamine, serine, or inositol, is added to phosphatidic acid to form a specific phospholipid. (*c*) An example of a specific phospholipid: phosphatidylcholine, formed by the addition of choline to phosphatidic acid. (*d*) In water, phospholipid molecules become oriented with their hydrophobic tails as far from the water as possible, forming either a sphere or a double layer.

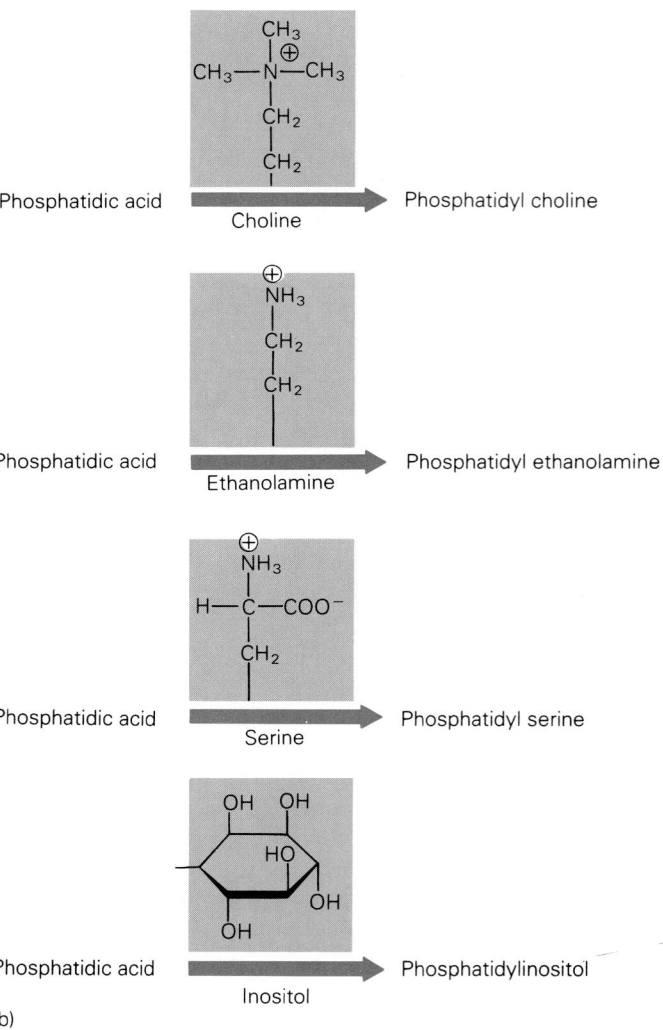

Phosphatidic acid →(Choline)→ Phosphatidyl choline

Phosphatidic acid →(Ethanolamine)→ Phosphatidyl ethanolamine

Phosphatidic acid →(Serine)→ Phosphatidyl serine

Phosphatidic acid →(Inositol)→ Phosphatidylinositol

(b)

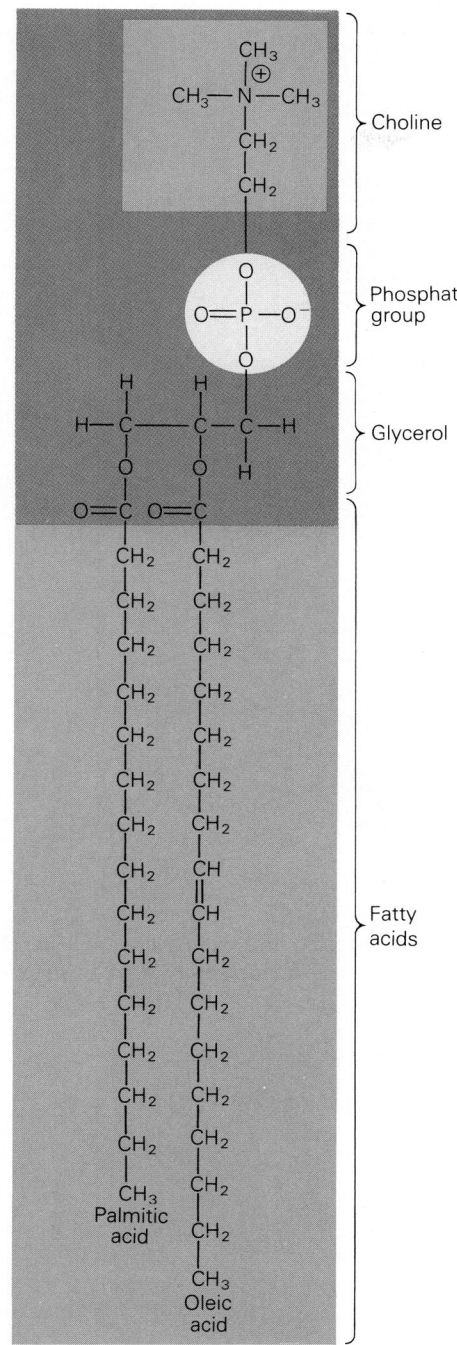

(c)

Choline

Phosphate group

Glycerol

Fatty acids

Palmitic acid

Oleic acid

(d)

Hydrophilic

Hydrophobic

Hydrophilic

Water

Water

tidylcholine (Fig. 2–23c), since it has two hydrophobic hydrocarbon chains and a hydrophilic phosphorylcholine group.

Steroids

Some of the most important lipids do not contain glycerol; these are the **steroids.** The basic structure of a steroid is four interlocking rings of carbon atoms (Fig. 2–24) and is quite different from that of other lipids. The male and female sex hormones and some hormones secreted by the adrenal glands, such as aldosterone, are steroids. The term **sterol** is often used when a hydroxyl group is present. **Cholesterol** is an example of a sterol that plays an important role in the structure of plasma membranes. Like the phospholipids, cholesterol has a polar hydrophilic region (the hydroxyl group) and a nonpolar hydrophobic region (the rest of the molecule).

Glycolipids

Glycolipids are lipids, also derived from sphingosine, that contain monosaccharide units. They have two hydrocarbon chains, as sphingomyelin does, but between one and seven monosaccharide units usually are present instead of the phosphorylcholine group. The monosaccharides make up a hydrophilic region of the molecule. Glycolipids occur in cell membranes.

Proteins

Proteins are some of the largest macromolecules in the cell. The molecular mass of a protein is usually between 5000 and several million. Proteins carry out remarkably diverse functions in cell. For example, enzymes, the molecules that catalyze specific biochemical reactions, are proteins. Carrier molecules that transport small molecules and ions are proteins. Many hormones are proteins, as are antibodies that combine with foreign bacteria. In addition, proteins are a major component of the muscles, allowing coordinated movement.

Two key aspects of protein molecules are important for an understanding of cell function: (1) the nature of the amino acids themselves, based on their side chain groups, and (2) the shape of the protein macromolecule, called its *conformation*.

Amino Acids

The basic units of protein molecules are **amino acids.** The general structure of an amino acid is illustrated in Figure 2–25. The "alpha" (α) carbon atom

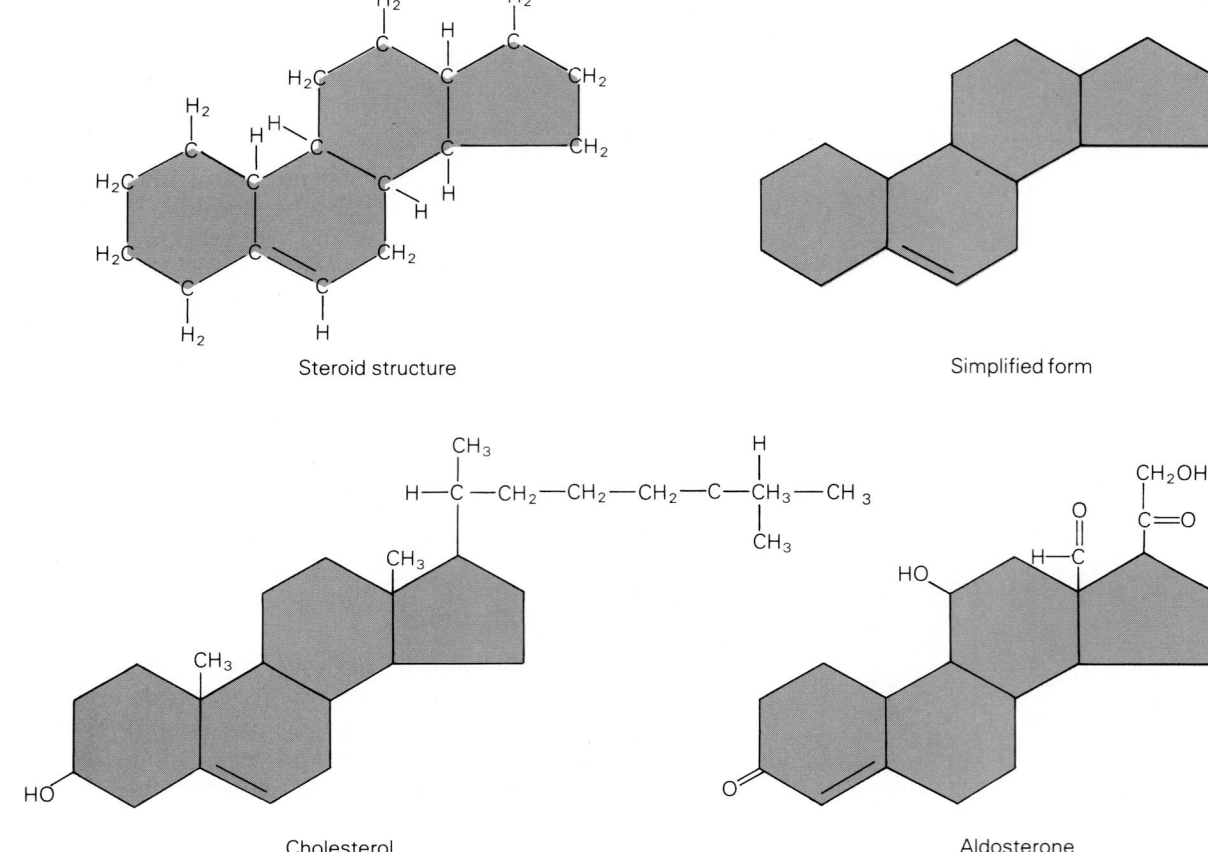

Steroid structure

Simplified form

Cholesterol

Aldosterone

Figure 2–24
Basic structure of steroids, and examples: cholesterol and aldosterone are both formed from a basic structure of four interlocking rings.

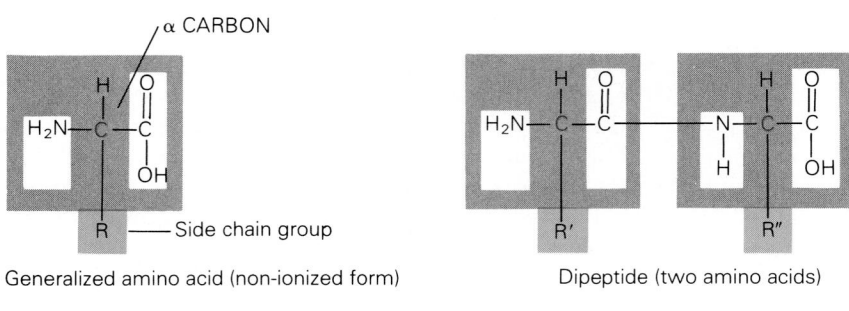

Generalized amino acid (non-ionized form)

Dipeptide (two amino acids)

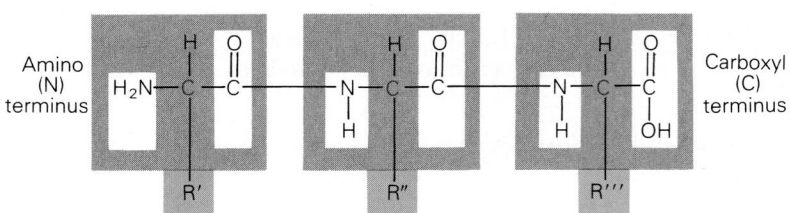

Tripeptide (three amino acids)

Figure 2–25
The general structure of an amino acid. A dipeptide is formed from the attachment of two amino acids, a tripeptide from three. Proteins are chains called polypeptides, made up of many amino acids, *R*, *R'*, *R''*, and *R'''* symbolize the specific side chain groups that distinguish the amino acids from each other.

is asymmetrical, bonded to four different chemical groups. It is called the **alpha-carbon** to distinguish it from carbons that may be present in the **side chain (R) group.** Twenty different amino acids are available to make proteins, and the same 20 amino acids occur in all proteins whether they are made by bacterial, plant, or animal cells.

Amino acids are distinguished from one another by their side chain groups. These groups have important properties that determine the properties of the protein molecules they make up.

Proteins are formed from linear unbranched chains of amino acids. The amino acids are covalently linked to one another via **peptide bonds,** formed between the alpha-carboxyl (—COOH) group of one amino acid and the alpha-amino (—NH₂) group of an adjacent amino acid (Fig. 2–25). A **dipeptide** is a chain of two amino acids, a **tripeptide** a chain of three, and a **polypeptide** a chain of eight or more amino acids. Protein molecules typically are polypeptide chains of between 50 and 2500 amino acids. The exact number of amino acids and the sequence in which they combine are characteristic features of a specific protein that determine its properties and functions.

The amino acids are divided into three broad groups, based on the properties of their side chains at physiological pH (Fig. 2–26): amino acids with ionic polar side chains, those with nonionic polar side chains, and those with nonpolar side chains.

Ionic Polar Side Chains. Ionic side chains found in amino acids contain terminal —COOH or —NH₂ groups, which donate or accept, respectively, H^+ ions in solution at pH 7.4 to become COO^- or —NH_3^+. Side chains that donate H^+ ions are, of course, acids (e.g., aspartic acid). Side chains that accept H^+ ions are bases (e.g., lysine and histidine). Note that the —NH₂ and —COOH groups on the alpha carbon are involved in formation of peptide bonds and cannot donate or accept H^+ ions. The ionic side chains are found on the external surface of protein molecules, where they can interact with the polar water molecules.

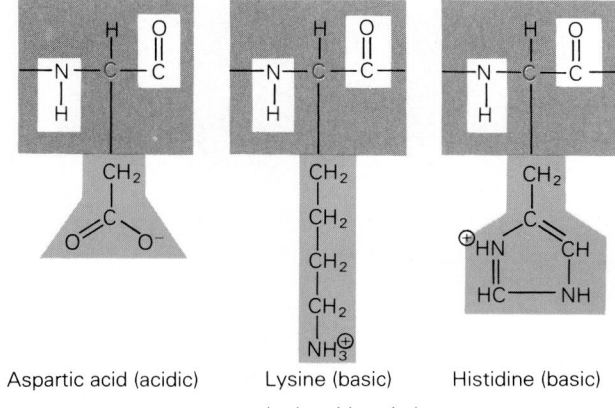

Aspartic acid (acidic)   Lysine (basic)   Histidine (basic)

Ionic side chains

Figure 2–26

Structures of some amino acid side chain groups. Ionic side chains such as those that produce the amino acids aspartic acid, lysine, and histidine are hydrophilic (water-soluble), as are nonionic polar side chains such as those that form cysteine, serine, and glutamine. Nonpolar groups such as those that form leucine, phenylalanine, and methionine are not water-soluble. (The —NH₂ and —COOH groups of the alpha-carbon are shown as if they were involved in the formation of peptide bonds.)

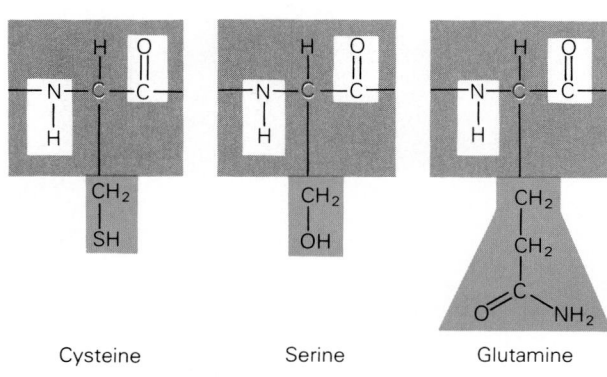

Cysteine   Serine   Glutamine

Nonionic polar side chains

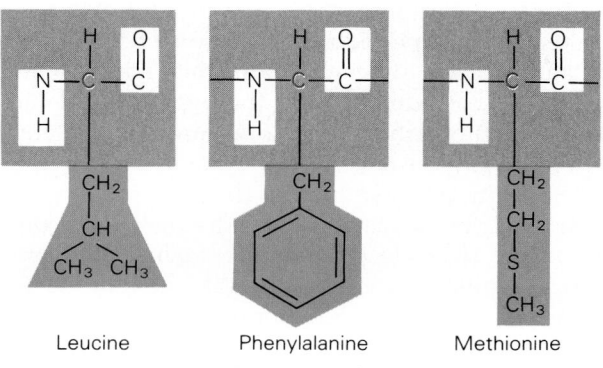

Leucine   Phenylalanine   Methionine

Nonpolar side chains

Nonionic Polar Side Chains. Some groups found in amino acids are polar but have no net charge at physiological pH. Examples are the —NH_2 group of glutamine, the —OH group of serine, and the —SH group of cysteine. These groups may be exposed on the protein surface, as are ionic side chains, or they may be buried within the protein molecule, where they are able to form hydrogen bonds. The side chain —SH groups of adjacent cysteines also interact to form a covalent bond known as a **disulfide bridge:**

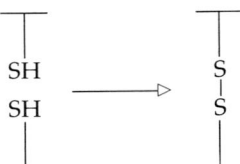

Nonpolar Side Chains. Some side chain groups are nonpolar and therefore hydrophobic. They are usually "hidden" from water by sequestration inside the protein structure where they cluster together via hydrophobic interactions. The side chains of leucine, methionine, and phenylalanine are good examples.

Protein Structure

There are different levels of organization in the structure of a protein molecule. These levels are important to understand because protein shape affects protein functioning in cells. The sequence in which the amino acids are linked together determines the primary structure of a protein; the way in which the resulting chain bends or folds is the secondary structure. Further bending and folding create the tertiary structure, and in some complex proteins, the physical relation of separate chains makes up the quaternary structure.

Primary Structure. The chemical bonds that are important for maintaining the **primary structure,** or the sequence of amino acids, in a protein molecule are the covalent peptide bonds between the adjacent amino acids. By convention, the amino- or *N*-terminus is considered to be the beginning of the chain, and the carboxyl- or *C*-terminus is the end (see Fig. 2–25).

The hormone insulin was the first protein for which the primary structure was discovered. It consists of two separate polypeptide chains, one containing 21 amino acids and the other containing 30 amino acids. Two disulfide bridges hold the chains together, and a third disulfide bond joins two cysteines in the shorter chain (Fig. 2–27). Eighteen different amino acids are present in the molecule.

Secondary Structure. A polypeptide chain is usually flexible and, under biological conditions, becomes folded into a specific shape or **conformation.** This conformation is called the **secondary structure** of the protein.

The folding occurs spontaneously because the side chain groups of the different amino acids react in a variety of ways with each other and with water. For example, hydrophobic interactions develop because nonpolar side chains tend to be pushed together in the interior of the molecule, where they can avoid contact with the aqueous environment. Hydrogen bonds develop when two peptide bonds come close together as the polypeptide folds (Fig. 2–28). Any polar side chains buried in the interior

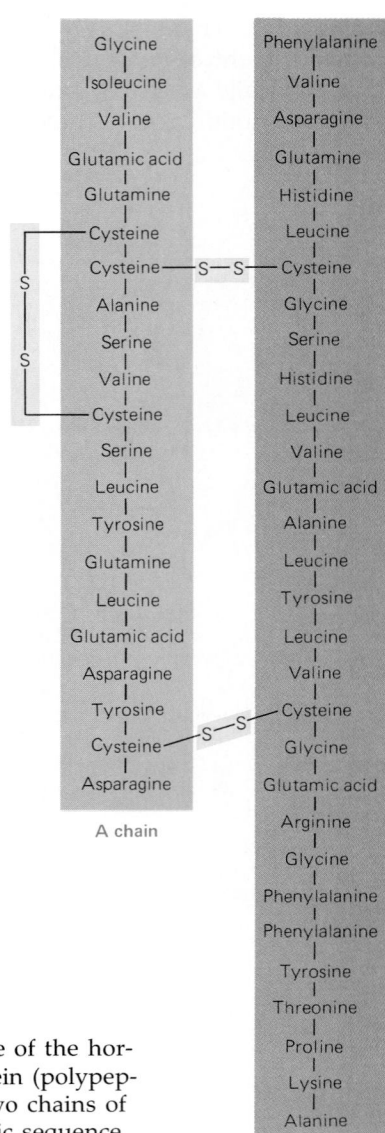

Figure 2–27
The primary structure of the hormone insulin, a protein (polypeptide), consisting of two chains of amino acids in specific sequence joined by two disulfide bridges.

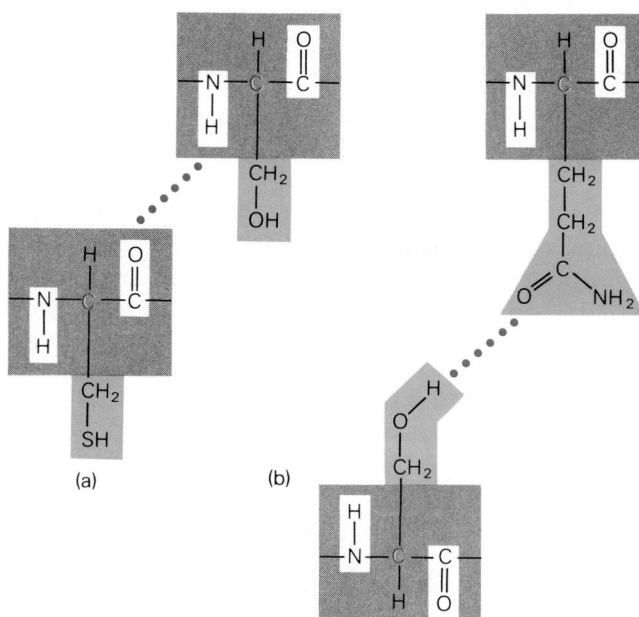

(a)

(b)

Figure 2–28
(*a* and *b*) The hydrogen bonds formed within a folded polypeptide chain are important for stabilizing the secondary structure. Hydrogen bonds can form between —NH and —CO groups attached to the alpha carbon or between specific side chain groups. (*c* and *d*) The secondary structure of a protein can be either an alpha-helix or a beta-sheet. (*e*) Tertiary structure of a hypothetical protein, maintained by several different types of bonds. (*f*) The quaternary structure of hemoglobin.

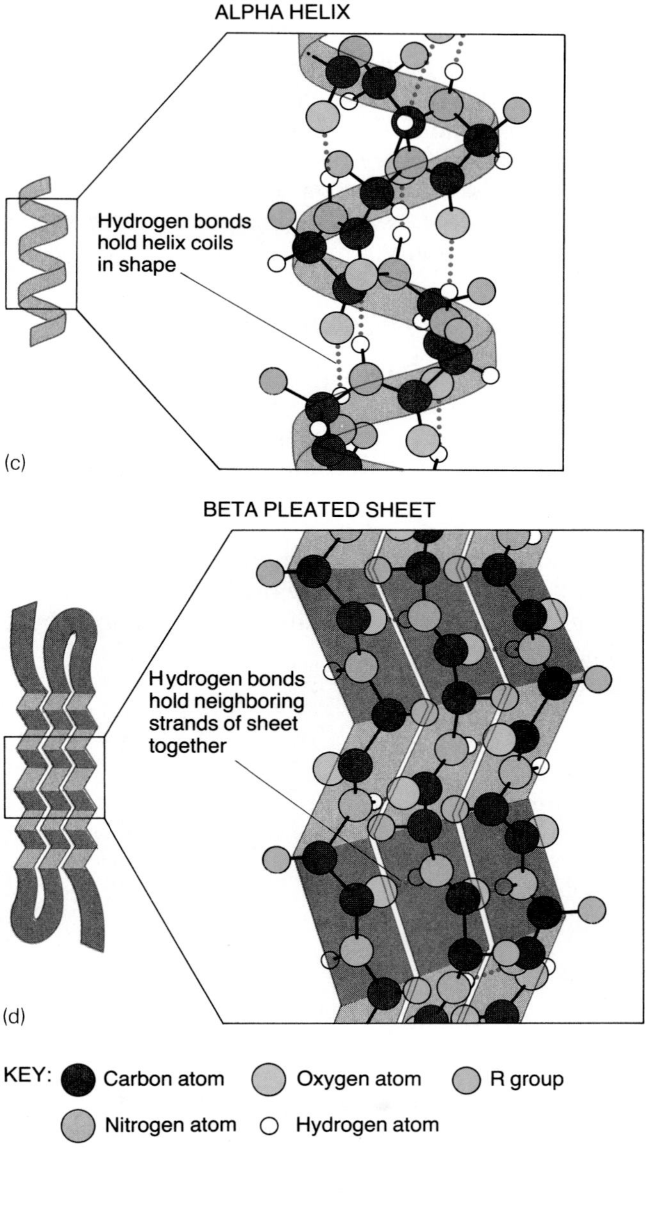

ALPHA HELIX

Hydrogen bonds hold helix coils in shape

(c)

BETA PLEATED SHEET

Hydrogen bonds hold neighboring strands of sheet together

(d)

KEY:
- ● Carbon atom
- ◯ Oxygen atom
- ◯ R group
- ◯ Nitrogen atom
- ○ Hydrogen atom

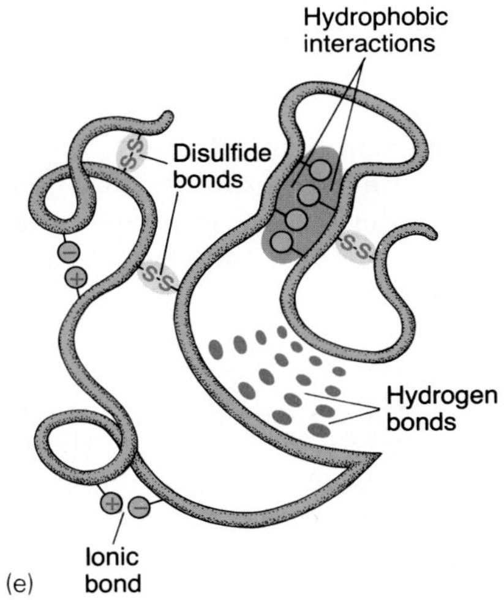

Hydrophobic interactions

Disulfide bonds

Hydrogen bonds

Ionic bond

(e)

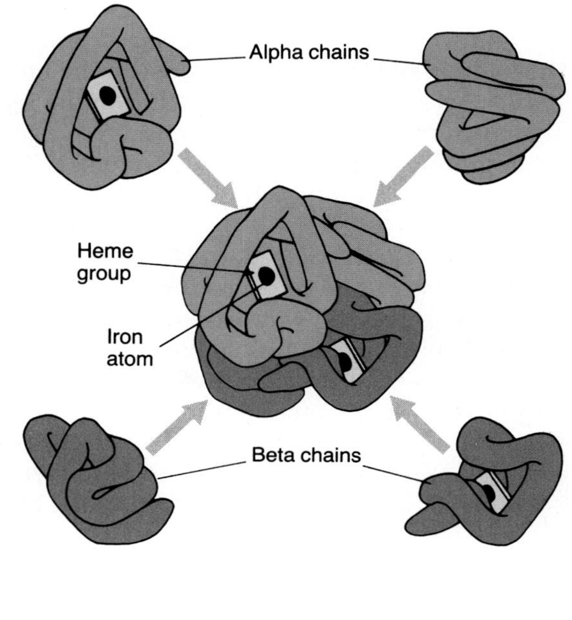

Alpha chains

Heme group

Iron atom

Beta chains

(f)

also become involved in hydrogen bond formation (Fig. 2–26). The ionic side chains and many of the polar side chains tend to be arranged near the surface of the molecule, where they can interact with the polar water molecules. Finally, covalent disulfide bridges between cysteine-SH groups frequently stabilize the folded conformation. These various interactions depend on the specific amino acids that are present and the sequence in which they are arranged in different proteins. They produce a unique secondary structure for each protein.

All of the specific conformations, however, can be broadly classed as either alpha-helices or beta-sheets (Fig. 2–28c and d). An **alpha-helix** is a type of conformation that results when a single polypeptide chain coils about itself to form a uniform cylindrical structure. Each complete turn of the helix is occupied by 3.6 amino acids. This structure is stabilized mainly by hydrogen bonds between the atoms of the peptide bonds. A **beta-sheet** results when the polypeptide chain runs back and forth upon itself. It is stabilized, like the alpha-helix, by hydrogen bonds between atoms of peptide bonds in neighboring segments of the chain.

Alpha-helixes and beta-sheets are stable structures that are more rigid than extended polypeptide chains. Both of these types of conformation may occur within the same protein molecule, and they may be joined by unfolded segments that adopt neither conformation.

Tertiary Structure. The next level of protein structure, the **tertiary structure** (Fig. 2–28c), is formed by the folding of the alpha-helix upon itself or by stable associations of beta sheets and alpha-helices so that a beta-sheet is covered on both sides by alpha-helices. Ionic and polar side chains of amino acids tend to be on the outside of the structure, whereas nonpolar side chains tend to be buried on the inside. The tertiary structure is stabilized primarily by the hydrophobic interactions between the nonpolar groups; hydrogen bonds (formed as in the secondary structure) and ionic bonds between charged side chain groups are also important. Proteins with tertiary structures are typically globular.

Quaternary Structure. Many large globular proteins consist of two or more separate polypeptide chains. The specific way in which the individual folded chains or subunits fit together to form the biologically active protein is known as the **quaternary structure. Hemoglobin,** the protein in red blood cells that binds and carries oxygen, is comprised of 574 amino acids arranged in four separate subunits (Fig. 2–28d). The subunits are held to-

gether by noncovalent interactions. One molecule of oxygen can bind reversibly to each of the four subunits.

The final conformation, or shape, of the intact protein depends on the interactions among the amino acid side chain groups. These in turn depend on the primary structure, or amino acid sequences, of the polypeptide chains. A change in just one amino acid in the polypeptide chain can alter the physiological function of the protein. The disease **sickle-cell anemia** results when a single amino acid is changed in two of the subunits of hemoglobin. The function of the altered hemoglobin is abnormal and seriously threatens the survival of the individual.

Nucleotides and Nucleic Acids

A **nucleotide** is a chemical structure consisting of a pentose sugar, one of a set of special groups called bases, and one to three phosphate groups. Nucleotides exist both as free molecules and as subunits of **nucleic acid** molecules. As free molecules, they are important in cell metabolism. As subunits of **ribonucleic acid (RNA)** and **deoxyribonucleic acid (DNA),** they are essential elements of genetic processes and heredity.

Components of Nucleotides

A nucleotide is composed of three smaller subunits: a sugar (either ribose or deoxyribose), a phosphate group or groups, and a specific chemical group called a **base.** The sub-subunits are linked by covalent bonds.

Sugars. The nucleotides of ribonucleic acids contain the pentose **ribose;** the nucleotides of deoxyribonucleic acids contain the pentose **deoxyribose** (Fig. 2–29).

Figure 2–29
The structures of ribose (the pentose sugar that forms part of ribonucleic acid or RNA) and deoxyribose (the pentose sugar that forms part of deoxyribonucleic acid or DNA).

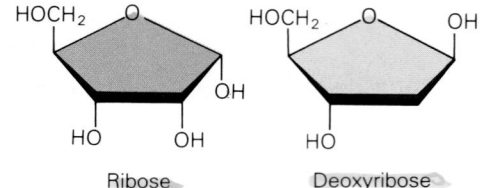

PYRIMIDINE BASES

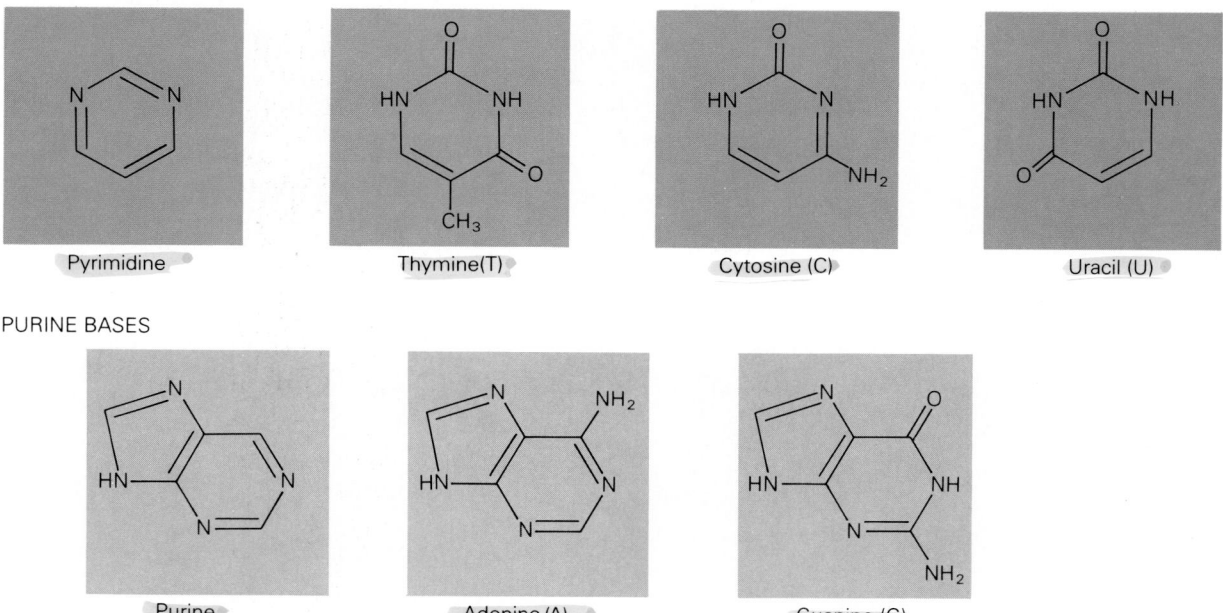

Figure 2–30
Structures of the pyrimidine- and purine-derived bases that are present in nucleotides. The single-letter abbreviations for the bases are shown in parentheses.

Bases.   The bases that form a part of nucleotides are nitrogen-containing ring compounds derived from either **purine** or **pyrimidine** (Fig. 2–30). They are called bases because they are proton acceptors. In DNA, the principal pyrimidine-derived bases are **thymine** and **cytosine;** in RNA, **uracil** and **cytosine.** The principal purine-derived bases in both DNA and RNA are **adenine** and **guanine.**

Phosphate Groups.   A free nucleotide molecule can have either one, two, or three phosphate groups.

Formation of Nucleotides and Their Roles in Cell Metabolism

The basic structure of a nucleotide can be symbolized as follows:

> Nucleotide = Base + Sugar + Phosphate

The base-plus-sugar group that makes up part of a nucleotide is called a **nucleoside.** Thus the basic formula for a nucleotide can also be represented as:

> Nucleotide = Nucleoside + Phosphate

The base adenine becomes a nucleoside—**adenosine**—through the formation of a covalent bond between one of its ring nitrogen atoms and carbon 1 of ribose (Fig. 2–31). The nucleoside adenosine is converted to the nucleotide **adenosine monophosphate (AMP)** when a phosphate group is attached covalently to adenosine at carbon 5 of the ribose unit.

The addition of phosphate groups to adenosine monophosphate produces two new nucleotides: **adenosine diphosphate** or **ADP** (when one phosphate group is added to AMP) and **adenosine triphosphate** or **ATP** (when two phosphate groups are added to AMP) (Fig. 2–31).

Other bases besides adenine can form nucleosides and nucleotides. Some examples are shown on page 64.

Nucleotides are important subunits for nucleic acids, the genetic storage molecules, but free nucleotides have important roles in the cell energy processes as well. Adenosine triphosphate (ATP) is critical as the storage form of the chemical energy required for many of the biochemical reactions in the cell. GTP is required for protein synthesis, UTP for glycogen synthesis, and CTP for phospholipid synthesis.

In addition, some compounds derived from nucleotides fulfill important functions in the cell. One example is **cyclic AMP** (Fig. 2–32), which medi-

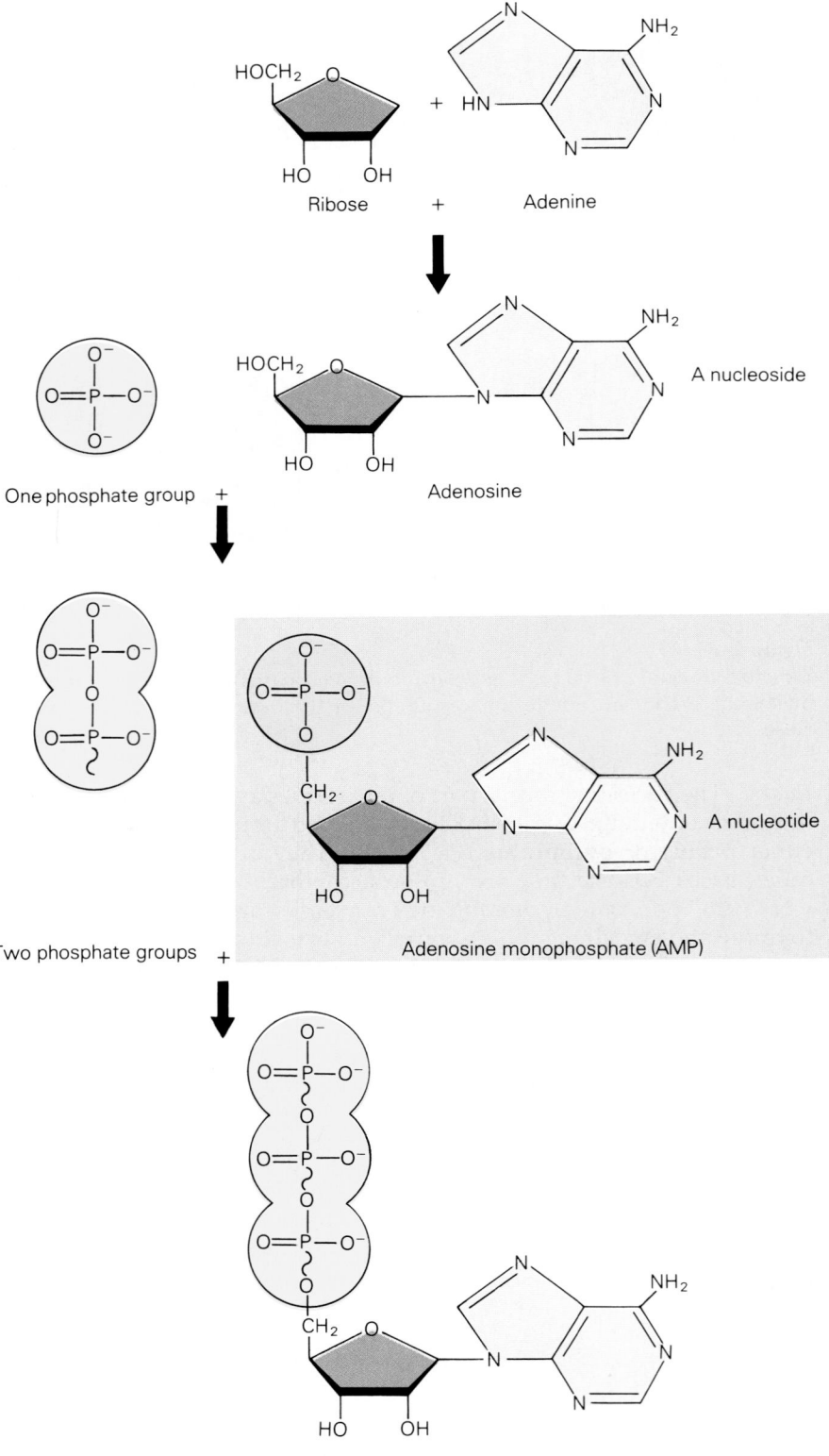

Figure 2–31
Assembly of the units of a nucleotide. First, the pentose sugar ribose (or deoxyribose) attaches to a base such as adenine to form a nucleoside (in this case, adenosine). Next, a phosphate group (—PO₄) attaches to the nucleoside to form a nucleotide, adenosine monophosphate (AMP). The attachment of two more phosphate groups produces a different nucleotide, adenosine triphosphate (ATP).

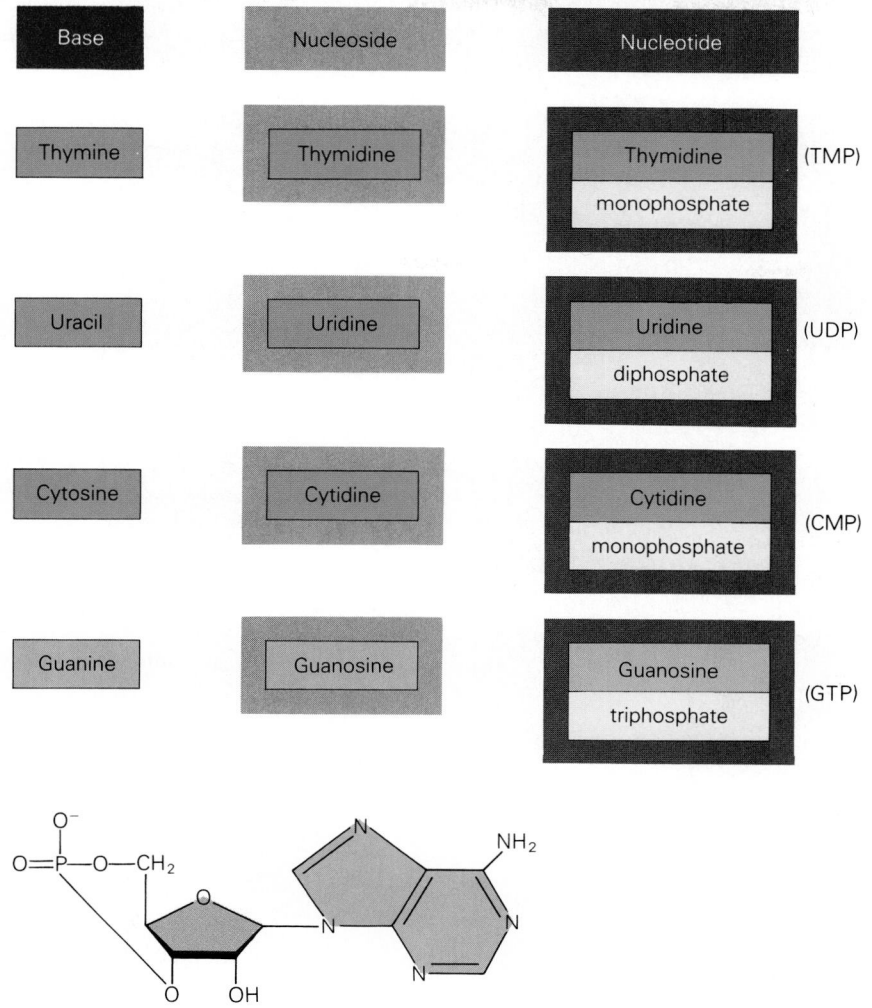

Base	Nucleoside	Nucleotide	
Thymine	Thymidine	Thymidine monophosphate	(TMP)
Uracil	Uridine	Uridine diphosphate	(UDP)
Cytosine	Cytidine	Cytidine monophosphate	(CMP)
Guanine	Guanosine	Guanosine triphosphate	(GTP)

Cyclic adenosine monophosphate (cyclic AMP)

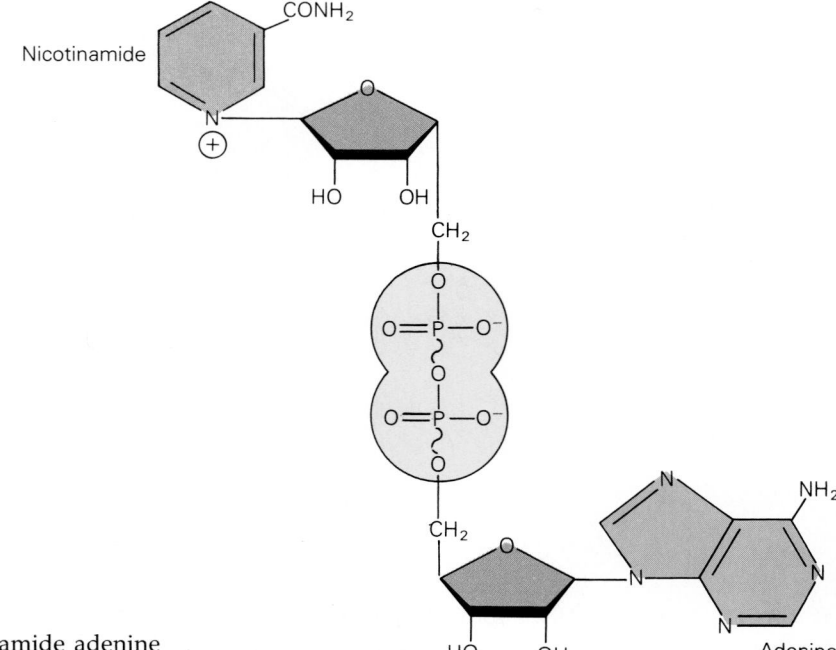

Nicotinamide adenine dinucleotide (NAD)

Figure 2–32
The structures of cyclic AMP and nicotinamide adenine nucleotide (NAD), two important derivatives of adenine nucleotides.

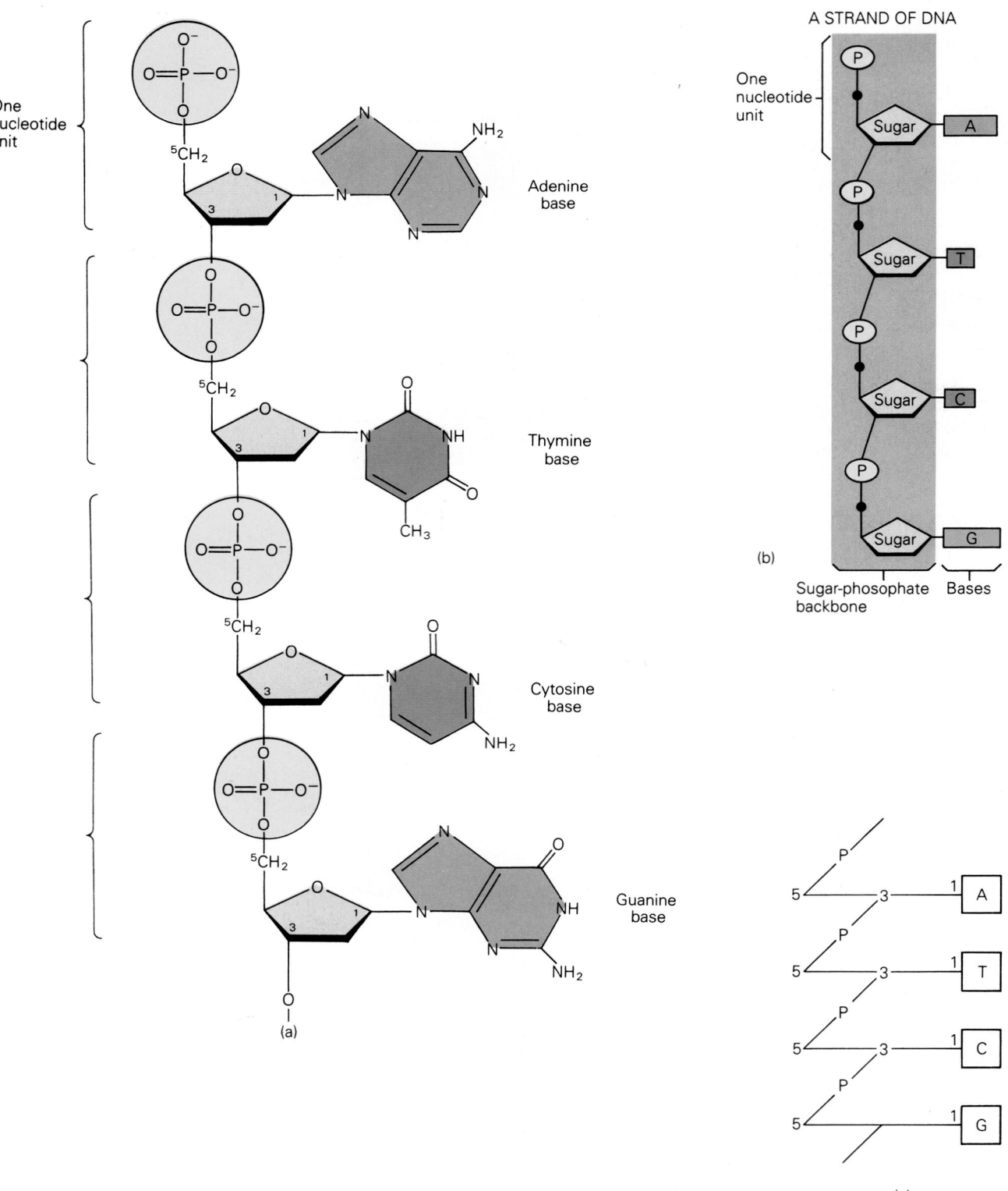

Figure 2–33

(*a*) A single strand of DNA is formed by monophosphate nucleotides linked together via the sugar (deoxyribose) and phosphate groups. (*b*) The "sugar-phosphate backbone" is distinct from the bases, which are not directly linked to each other. (*c*) The shorthand notation for a DNA single strand.

ates the action of many hormones in cells. Adenine nucleotides also are components of many **coenzymes,** compounds required for many biochemical reactions. An important coenzyme is **nicotinamide adenine dinucleotide,** or **NAD** (Fig. 2–32).

Nucleic Acids

Nucleic acids are composed of monophosphate nucleotide units covalently linked in long unbranched chains. The different nucleotides are linked by covalent bonds between the sugar of one nucleotide and the phosphate of the next (Fig. 2–33a). The phosphate group of the first nucleotide is attached to the hydroxyl group on carbon 3 of the pentose sugar of the next nucleotide. Thus, each sugar in a nucleic acid has a base attached to carbon 1 and phosphate groups attached to carbons 3 and 5. The sugar-phosphate chain, analogous to the peptide bond in proteins, is a constant feature throughout any nucleic acid molecule and is known as the **sugar-phosphate backbone** of the nucleic acid (Fig. 2–33b). The variable part of the nucleic acid, analogous to the variable amino acid side chain groups in a protein, is the sequence of bases.

The precise sequences of the different bases in both DNA and RNA molecules represent the genetic information of the cell and of the organism as a whole. DNA and RNA interact in complex processes that transmit this genetic information to cellular systems responsible for the synthesis of all proteins involved in cell activities. The specific base sequences of DNA and RNA determine the specific amino acid sequences in the protein molecules, thereby dictating protein shape and function. These factors in turn affect the functioning of cells and the shape and characteristics of the organism as a whole.

The Structure of DNA

Certain aspects of the DNA molecule distinguish it from RNA and will become important in later chapters that discuss the replication of genetic material and cell reproduction. The DNA molecule differs from the RNA molecule in that it contains the sugar deoxyribose rather than ribose; it also contains certain bases that RNA does not have. In addition, DNA differs from RNA in that it consists of two separate nucleotide chains, or "strands," instead of one.

The Double Helix. The two nucleotide strands in DNA are coiled around each other in a configuration called a **double helix** (Fig. 2–34). Each turn of the double helix contains 10 bases.

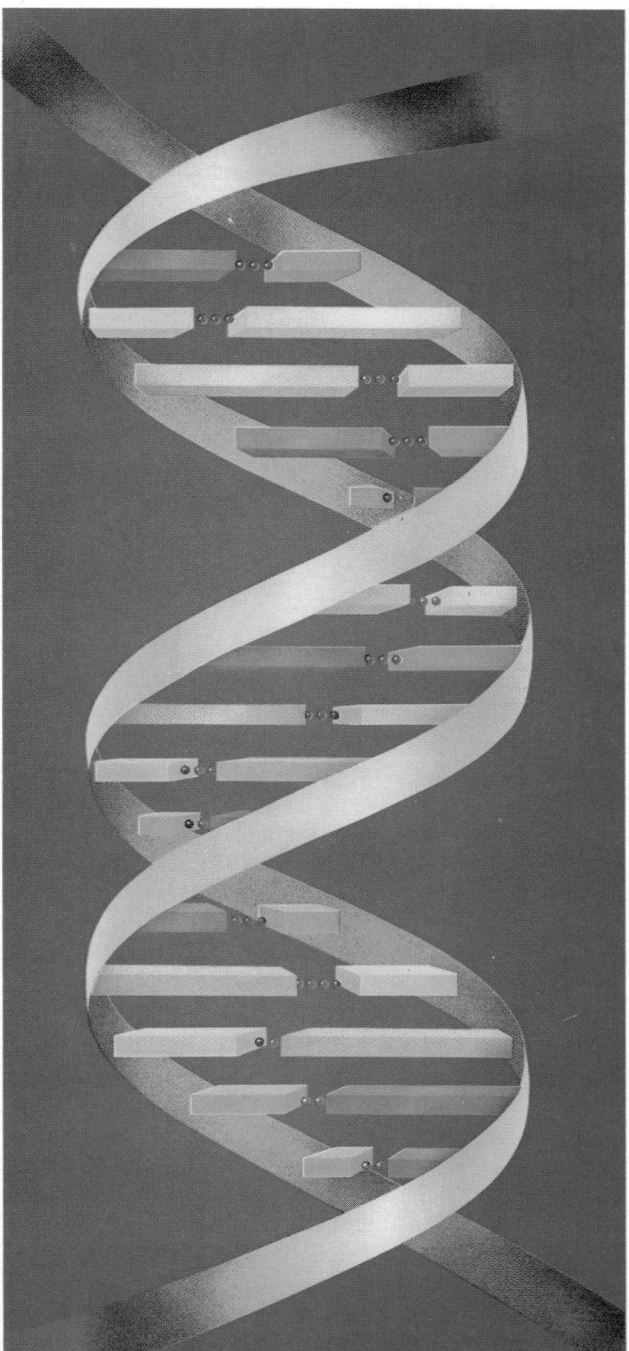

Figure 2–34
The double helix formed when two complementary strands of DNA coil around each other.

The two nucleotide strands that form the double helix are oriented so that their bases are inside the column of the helix, with the bases from one strand situated close to the bases from the other strand. As a result, hydrogen bonds form between a base in one strand and a base in the other. These hydrogen bonds help to hold the strands together in the helix.

Base-pairing.   A hydrogen bond will form only between a purine-derived base and a pyrimidine-derived base. Hence, these interactions, called **base-pairings,** are highly specific and are the most important aspect of the DNA double helix. Adenine (a purine) always forms two hydrogen bonds only with thymine (a pyrimidine) (Fig. 2–35*a*). Guanine (a purine) always forms three hydrogen bonds only with cytosine (a pyrimidine) (Fig. 2–35*b*).

Complementary Strands.   As a result of the base-pairing, the two DNA strands in the double helix are not identical; instead, they are **complementary** (Fig. 2–35*c* and *d*). Adenine in one strand always lies opposite thymine in the other strand. Cytosine in one strand always lies opposite guanine in the other strand.

Antiparallel Strands.   The two nucleotide strands are **antiparallel** as well as complementary. This term refers to the arrangement of the sugar-phosphate backbone within each strand of DNA. An antiparallel arrangement is depicted in Fig. 2–35*d*. Nucleoside *T* at the top of the left strand is linked to nucleoside *C* below it by a phosphate group attached to carbon 3 on the sugar of *T* and carbon 5 on the sugar of *C*. In contrast, nucleoside *A* at the top of the right strand is linked to nucleoside *G* below it by a phosphate group attached to carbon 5 on the sugar of *A* and carbon 3 on the sugar of *G*. Both the complementary and the antiparallel properties of the DNA strands play an essential role in the mechanism that passes on identical copies of DNA to offspring during cell division.

Length of DNA Molecules.   A striking characteristic of DNA molecules is the length of the nucleotide chains. The bacterium *Escherichia coli*, for example, contains a single molecule of DNA that is 1.2 mm long. It contains 3.4 million nucleotides and has a molecular mass of 23×10^8. In a typical animal cell, the DNA double helixes are themselves further coiled into tight bundles. One DNA molecule in an animal cell might contain 1 meter of DNA, with 3×10^9 nucleotides.

Thermodynamic Principles

Living organisms are highly organized, or **ordered,** systems. Maintaining this ordered state requires constant work. The capacity to do work is provided by **energy,** which can exist in several different forms, such as radiant energy (e.g., light and heat), mechanical energy, chemical energy, and electrical energy. Various physiological processes use different energy forms. For example, some of the energy released during metabolism is heat that helps to maintain normal body temperature. Mechanical energy is involved in moving a part of the body, such as an arm, by the contraction of skeletal muscles. Chemical energy is released or absorbed during breakage or formation of chemical bonds in the molecules involved in metabolic reactions. Electrical energy results from the flow of charged particles such as ions and is essential for the transmission of nerve impulses. As elsewhere in nature, energy in the body can be converted from one form to another.

The unit used to measure energy is the **kilocalorie (kcal).** One kilocalorie is defined as the amount of heat that will raise the temperature of 1 kg of water by 1°C. Most forms of energy are readily converted to heat and measured in kilocalories.

The study of energy is known as **thermodynamics** (literally, "heat changes"). **The First Law of Thermodynamics** states that *energy can neither be created nor destroyed;* there is a constant supply of energy in the universe. **The Second Law of Thermodynamics** states that the *disorder* (also called *randomness* or *entropy*) *in the universe is always increasing.*

It is not difficult to appreciate that even simple tasks such as assembling a new bicycle require work to create the necessary order. The parts of a bicycle, when shaken from the package, do not fall to the ground in just the right position and sequence needed for spontaneous assembly of the bicycle. Assembly always requires some work.

Similarly, a living organism, which must impose order on its internal functions as well as on its surroundings, must do work to resist the universal tendency toward disorder. Much of this work takes place in the individual cells of the organism, in the form of reactions that build up macromolecules to support the structures and functions of the organism. This "biological" work requires energy; however, energy does not simply exist in the cell. The cell must transform the potential energy held in the bonds of other molecules into a form it can use to do work.

Free Energy and Chemical Reactions

The bonds that hold atoms together in a molecule or other chemical group represent potential energy, or stored energy that can be released to do work. According to the laws of thermodynamics, however,

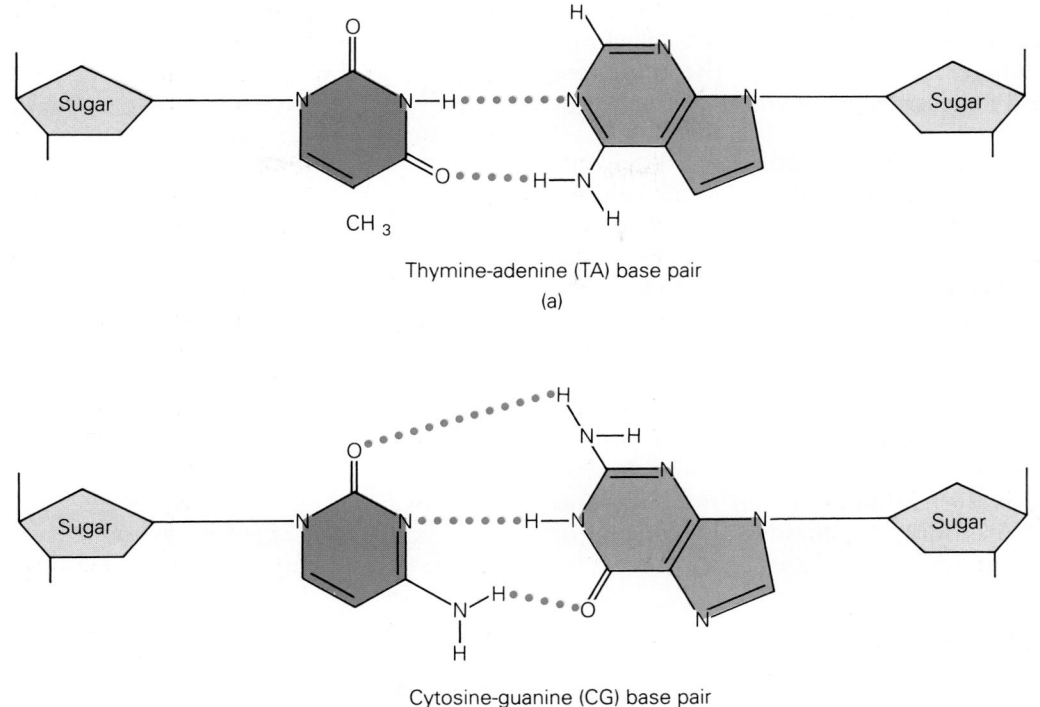

Thymine-adenine (TA) base pair

(a)

Cytosine-guanine (CG) base pair

(b)

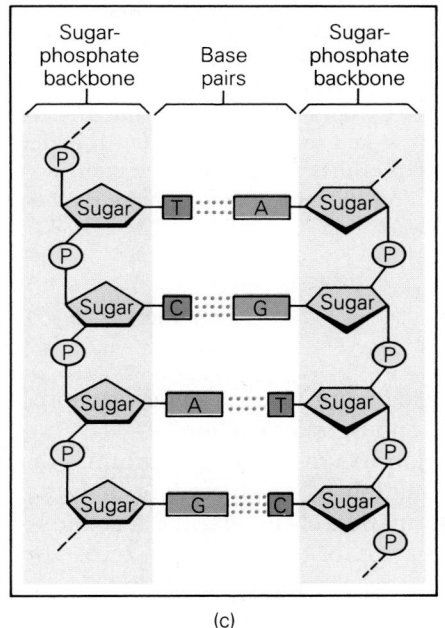

(c)

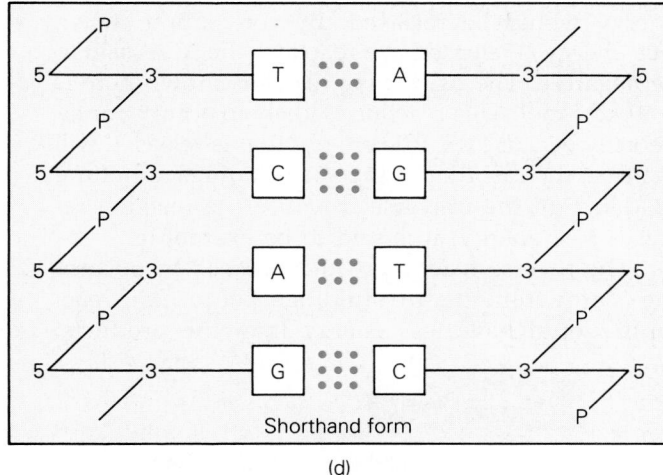

Shorthand form

(d)

Figure 2–35

Base-pairing between bases of separated DNA strands. (*a*) Thymine and adenine form one base pair. These two bases always form two hydrogen bonds between them. (*b*) Cytosine and guanine form the other base pair. These two bases always form three hydrogen bonds between them. (*c*) The two strands are complementary because specific bases on one strand form hydrogen bonds only with specific bases on the other strand. (*d*) The shorthand notation indicating complementary and antiparallel strands.

not all of the stored energy can be used; some will always be lost because of the tendency toward disorder everywhere in the universe. The term **free energy** is used to describe the energy that can be put to work.

Free energy is given the symbol G and, as with all energy, is measured in units called kilocalories (kcal). While it is difficult to determine the free energy content of a particular molecule, it is relatively easy to calculate the *change* in free energy that occurs in the course of a chemical reaction. The change in free energy is symbolized as **ΔG** (the Δ symbol is the Greek letter "delta").

The concept of free energy is useful for predicting the direction in which a reaction will tend to proceed. Consider the following hypothetical reaction, in which the G values of the reactants and products are used only to illustrate the concept of **free energy change:**

$$A + B \rightarrow C + D$$

Total G	Total G	$\Delta G = -40$ kcal/mol
100 kcal/mol	60 kcal/mol	

The total free energy of molecules C and D (expressed as kcal per mole) is less than the total free energy of molecules A and B. If the reaction were to proceed from **left** to **right**, then 40 kcal/mol of free energy would be released. By convention, when free energy is released in a reaction, the ΔG is said to be **negative.** The ΔG for the reaction shown here is -40 kcal/mol. This reaction is likely to occur spontaneously because the free energy that is released will increase the disorder of the surroundings, a natural tendency in the universe. Because the reaction releases free energy, it is said to be **exergonic.**

The reverse reaction would proceed by absorbing energy from the surroundings, because the reactants would have less energy than the products:

$$C + D \rightarrow A + B$$

Total G	Total G	$\Delta G = +40$ kcal/mol
60 kcal/mol	100 kcal/mol	

The change in free energy would be considered **positive,** $+40$ kcal/mol. Reactions that must absorb energy from the surroundings before proceeding will not occur spontaneously; they are considered **energetically unfavorable** and are termed **endergonic.** Figure 2–36 compares these two types of reactions diagrammatically. Many reactions that take place in the cell are endergonic; they proceed by being coupled with other reactions that are exergonic. The

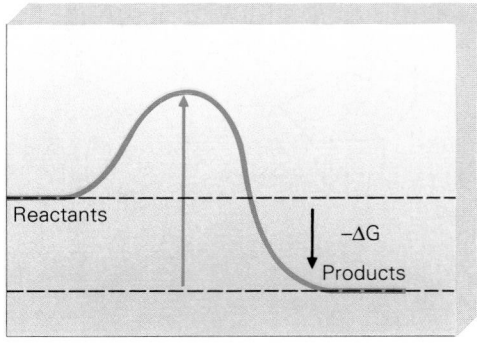

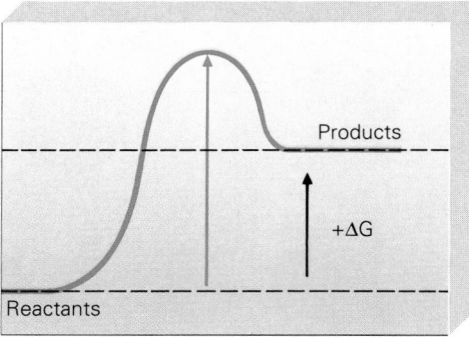

Figure 2–36
Diagrams comparing the free energy changes in endergonic and exergonic reactions. (*a*) In exergonic reactions the products have less free energy than the reactants. Energy has been released in the reaction, and the free energy change (ΔG) is said to be negative. (*b*) In endergonic reactions, the products have more free energy than the reactants, and the free energy change is said to be positive.

free energy released from an exergonic reaction is used to drive an endergonic reaction.

When a reaction has reached equilibrium—that is, when the formation of products is exactly balanced by the reformation of the original reactants—the free energy change for the reaction is zero. Consider the following reaction at equilibrium:

$$AB \rightleftharpoons A + B$$

$$\Delta G = 0$$

The rate at which AB breaks down is balanced by the rate at which A and B recombine to form AB. The reaction can be "driven," or caused to proceed, to the "right" or toward the formation of products A and B, by the removal of product B as soon as it becomes available. This removal prevents the new product *B* from combining with product *A*. The removal of *B* upsets the equilibrium of the reaction, which will be re-established by the further breakdown of *AB*. This continual attempt to maintain equilibrium by providing new product *B* causes the reaction to proceed until all of the *AB* is broken down. In effect, the constant removal of *B* keeps the free energy of the products (mostly *A*) low. Since the reaction tends to proceed spontaneously in the direction from high to low free energy, it proceeds in the direction of the products *A* and *B*. Many metabolic reactions take place with a very small free energy change. They can be driven in either direction easily when one of the reactants or one of the products is removed by involvement in a separate reaction. This provides a cell with the capacity to reverse a metabolic chain of reactions and to respond quickly to specific demands for either reactants or products of the metabolic pathway.

Activation Energy and Enzyme-Catalyzed Reactions

Regardless of whether it is endergonic or exergonic, a reaction requires a certain amount of energy to begin. This energy, which is the initial energy required to break the existing bonds and form new ones, is called the **activation energy.** This can be thought of as an energy barrier that reactant molecules must overcome before they can participate in the reaction. The disruption of existing bonds is illustrated in Figure 2–36 by the initial increase in free energy that occurs in both exergonic and endergonic reactions. In a chemical reaction, only those reactant molecules with sufficient internal energy will actually react. The lower the activation energy barrier of a reaction, the more likely it is to proceed because more reactant molecules will have enough energy.

One way to increase the rate of a reaction, as we have said, is to increase the temperature. In the human body, increasing the temperature sufficiently to enable every reaction to occur at a high enough rate would not be feasible. The body must maintain a constant (and relatively low) temperature, too low for many of the reactions necessary for life.

Another way to increase the rate of reaction is to ensure that the reactant molecules approach each other in the proper orientation for specific bonds to occur. But in the body, as elsewhere, molecules are in random, disorderly motion. How then does the body ensure that these reactions occur?

The solution lies in molecules called **enzymes,** which act as natural catalysts to increase the rate of chemical reactions in the body by lowering the activation energy levels for those reactions. They physically bind reactant molecules and hold them in place so that specific reactions can occur.

The Catalytic Role of Enzymes

A **catalyst** is a substance that increases the rate of a chemical reaction without actually participating in it and that is recovered unchanged when the reaction is completed. Catalysts do not influence the direction of a reaction, the final concentrations of the reactants and products, or the free energy change. They merely speed up a reaction that is already thermodynamically feasible.

The degree by which a catalyst can speed up a reaction is very impressive. It is not unusual for a reaction to occur one million times faster than normal when a catalyst is present. The increase is made possible in part because, in many reactions, a single enzyme molecule can catalyze the reaction of 10,000 reactant molecules every second. Such a number is called the **turnover number** for the catalyst. Typically, only a few catalyst molecules are needed to convert a relatively large number of reactant molecules into product.

Catalysts are widely used in industry to increase the rates of reactions and thereby the quantities of products that can be manufactured in a given amount of time. In the cell, natural catalysts called enzymes control the rates of many important reactions, speeding them up when necessary and slowing them down when necessary.

The Mechanics of Enzyme Action

Enzymes are proteins, and each enzyme has a characteristic three-dimensional shape and surface structure resulting from the folding of the polypeptide chain (Fig. 2–37). Enzymes facilitate the reactions of metabolism in cells by providing binding sites on their surfaces for the reactant molecules. These sites, which may be indentations on the surface of the enzyme, are known as **active sites.** The reactant molecules that bind to the active sites are known as the **substrates** of the enzyme.

We know of more than 1000 different enzymes. About 90% of the proteins present in a cell are enzymes. Given these large numbers, it is remarkable that a key aspect of enzyme action is **specificity.** A particular enzyme will catalyze only a very few reactions, and sometimes only one reaction, because its active sites will accept and bind only certain reactants.

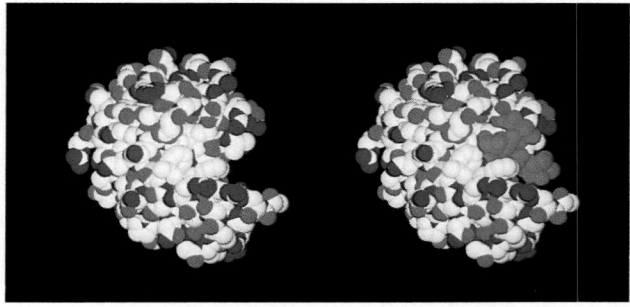

Figure 2–37
A computerized model of an enzyme (protein) that is unattached to its substrate (left) and one that is attached (right). (© Visuals Unlimited.)

Although individual enzymes are "tailored" for specific substrate molecules, most enzymes can be classified according to the general types of reactions they catalyze. A **dehydrogenase**, for example, catalyzes the removal of hydrogen from a molecule (oxidation), whereas an **oxidase** catalyzes the addition of oxygen. A **transferase** mediates the transfer of groups of atoms, and an **isomerase** rearranges the atoms in a molecule. A **hydrolase** splits chemical bonds by adding the constituents of water (H and OH) to the separated atoms, and a **ligase** joins ("ligates" or "ties") molecules together by forming new bonds in reactions coupled to breakdown of ATP. Note that the name for each class of enzymes always ends in "-ase."

An enzyme (E) works by forming a temporary **enzyme-substrate complex** (E-S) with its substrates via hydrogen bonds at its active sites. This complex holds the substrate molecules in the proper orientation for the reaction to occur. When the reaction is complete, the E-S complex breaks apart to yield the reaction products (P) and the free, *unchanged* enzyme:

$$E + S \longrightarrow E - S \longrightarrow E + P$$

This equation is represented in diagram form in Figure 2–38, which shows the "lock-and-key model" of enzyme action. Evidence suggests, however, that not all active sites have the rigid conformation shown in this figure. In some cases, the active site may assume a shape complementary to the desired reaction only after the substrates are bound, a process known as **induced fit** (Fig. 2–39).

Formation of an E-S complex increases the frequency with which the reactant molecules collide in the proper orientation for the reaction to occur. This is one of the reasons that an enzyme accelerates the rate of a reaction. In the absence of enzymes, reactant molecules do not bind specifically to each other; they collide randomly. Each reactant molecule can potentially react with any number of other molecules that are present. The particular reactant molecules that must come together for a specific reaction are not likely to meet spontaneously at a high rate. By binding only the reactant molecules specific to a desired reaction, an enzyme "favors" that reaction over the other possible reactions in which the molecules could participate.

To further encourage reactions, some enzymes bind a substrate molecule in a manner that distorts the structure of the substrate. This may place a strain on the existing chemical bond—helping to break it—or it may facilitate a chemical reaction between groups that are not normally close enough together to react. These effects contribute to a higher reaction rate.

From a thermodynamic standpoint, enzymes in effect decrease the activation energy necessary for the reaction to occur—that is, the energy needed to break old bonds and form new ones. More reactant molecules have the required energy to overcome this activation energy barrier and can therefore react to form products. Again, enzymes act specifically, decreasing activation energy only for the desired reaction. In the cell, products of reactions are often

Figure 2–38
The "lock-and-key" model of enzyme action. The substrates fit tightly into the active sites on the enzyme molecule. When they have formed products, the enzyme releases them and is unchanged, free to catalyze another reaction.

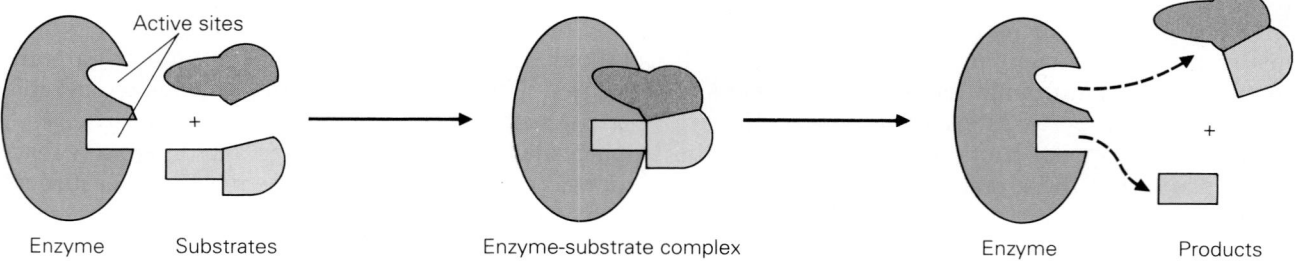

Enzyme   Substrates   Enzyme-substrate complex   Enzyme   Products

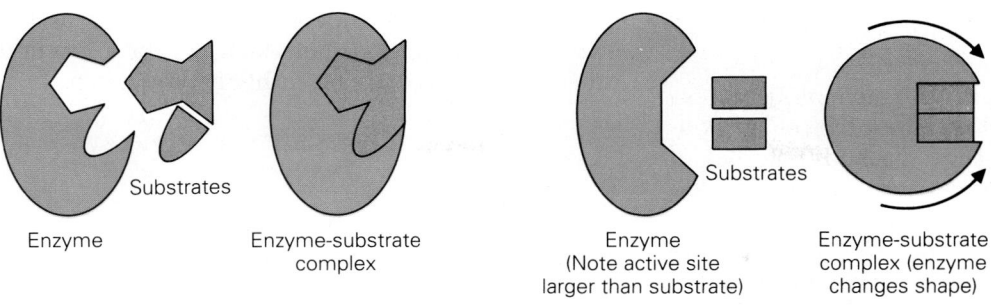

Figure 2–39
The "lock-and-key" (*a*) model has been revised for some enzymes, in which "induced fit" (*b*) is a more likely scheme. According to this model, the active sites are somewhat larger than the substrates. When the substrates contact them, the active sites change shape to bind the substrates.

intermediate products in long chains of reaction. At each step, a product must be directed down one of several alternative pathways. Because an enzyme decreases the activation energy only for the specific reaction that it catalyzes (Fig. 2–40), the reactant molecules are more likely to attain this level of activation energy and undergo the reaction catalyzed by the enzyme. They are less likely to undergo other possible reactions that require a higher activation energy.

Figure 2–40
Most products of chemical reactions in the body are intermediate products in complex metabolic pathways. Enzymes direct such intermediates down a specific reaction pathway by lowering the activation energy for only the desired reaction. The intermediate product is more likely to participate in a reaction with a lower activation energy than a higher one.

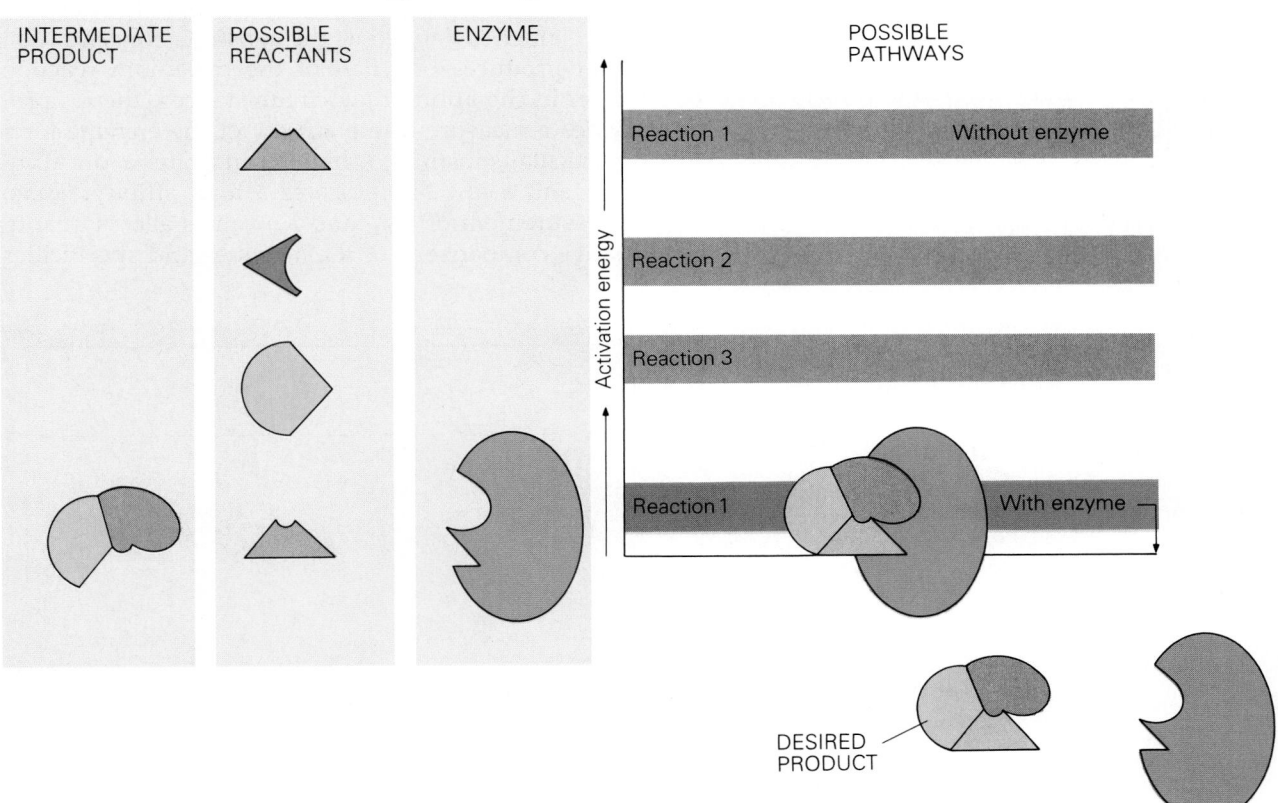

Enzyme Kinetics

Researchers can study reaction rates and other characteristics of enzyme-catalyzed reactions using enzymes that have been isolated from the cell and purified. These in vitro (outside the body) studies of enzyme activity allow comparisons of different enzymes under the same conditions.

The relation between reaction rate and substrate concentration, at a constant enzyme concentration, is a particularly useful measure and is expressed graphically in Figure 2–41. Initially, the reaction rate increases rapidly as the concentration of substrate increases because at low substrate levels the enzyme is not working at full capacity. Any substrate molecules present are rapidly converted to reaction products. Eventually a maximum reaction rate ($V_{\max}$) is reached when the substrate concentration is high enough. Further increases in substrate concentration produce no change in the reaction rate. A $V_{\max}$ is achieved because the enzyme molecule has a limited number of active sites. When substrate molecules occupy all of the sites on the enzyme, it is said to be "saturated" and is working at maximum capacity. The addition of more substrate cannot increase the rate of reaction. The reaction rate in the absence of the enzyme is shown for comparison. It is much slower and increases linearly without exhibiting a $V_{\max}$ over the same range of substrate concentration (Fig. 2–41).

Knowledge of the $V_{\max}$, the maximum amount of substrate that can be utilized and turned into reaction product in a given time, allows calculation of the turnover number of the enzyme from the relationship:

$$V_{\max} = k[E]$$

in which k is the turnover number and $[E]$ is the enzyme concentration. For example, for a reaction catalyzed by the enzyme carbonic anhydrase:

$$CO_2 + H_2O \xrightarrow{\text{Carbonic anhydrase}} H_2CO_3$$

k can be calculated as follows: When carbonic anhydrase is fully saturated with substrate, just 10^{-6} mol $[E]$ of this enzyme will catalyze the formation of H_2CO_3 at a rate of 0.6 mol/sec ($V_{\max}$). Hence, the turnover number k is 0.6 divided by 10^{-6}, which is 600,000/sec. This is one of the largest known turnover numbers for any enzyme. In contrast, the turnover number for DNA polymerase is only 15/sec, which is very small. Note that the $V_{\max}$ and the turnover number are determined by laboratory experiments and provide a measure of the maximum activity of an enzyme. However, most enzymes in the cell are not saturated by substrate and do not operate at their $V_{\max}$. The advantage this provides to the cell is the capacity to deal with fluctuations in the concentrations of its molecules. Thus, when the demand for glucose catabolism is increased rapidly by physical exercise, the enzymes involved have the capacity to respond by processing more substrate molecules in the same amount of time.

Another useful term that can be obtained from the plot in Figure 2–41 is the **Michaelis constant, K_m,** which is defined as the substrate concentration that produces one half of the maximum reaction rate. In the appropriate circumstances, the K_m provides a measure of the affinity of the enzyme for a particular substrate. A high K_m indicates a low affinity, and a low K_m indicates a high affinity. Thus, measurement of V_{max} and K_m values allows quantitative comparisons of the activities and specificities

Figure 2–41
When an enzyme is present, a reaction proceeds initially at very high rate, reaching a maximum rate ($V_{\max}$) when the enzyme is saturated with substrate. The Michaelis constant (K_m) is the concentration of substrate that will produce one half of the maximum reaction rate.

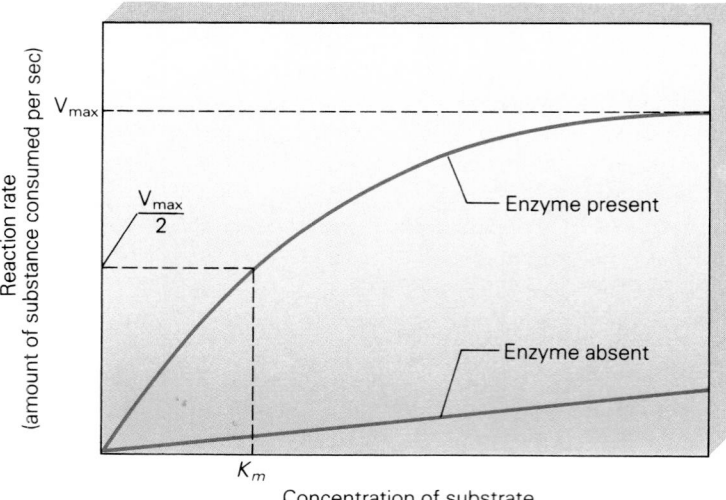

of different enzymes for various substrates. This kind of analysis has greatly aided scientists' understanding of the reactions of cellular metabolism.

The Effects of pH and Temperature

In addition to concentration of substrate, enzyme activity depends on both the temperature and pH of the medium (surroundings). An increase in temperature will increase the rate of most enzyme-catalyzed reactions until temperatures in the range of 50°C to 60°C are reached. At this point, the temperature is high enough to **denature,** or break down the tertiary structure of, the enzyme protein. When this occurs, the enzyme rapidly loses its catalytic activity.

The pH range over which most enzymes can function is relatively narrow but varies from one enzyme to another. **Pepsin,** for example, which breaks down proteins in the stomach, works best at a pH of 2, the **pH optimum** for this enzyme. In contrast, **trypsin,** secreted by the pancreas, is most effective in breaking down proteins at a pH of 8, and **alkaline phosphatase** removes phosphate groups from substrates most effectively when the pH is close to 10. Changes in pH result in addition or removal of hydrogen ions (H^+) from the enzyme, thereby changing the number of positive and negative charges present at the active site and altering the catalytic efficiency of the enzyme.

Enzyme Cofactors: Coenzymes

A **cofactor** is a nonprotein component that some enzyme molecules need to function as catalysts. For example, several enzymes involved in glucose catabolism require a magnesium ion for their activity. Other enzymes use iron, copper, or zinc ions. In some instances, the metal ion actually carries out the catalytic reaction, although the reaction is facilitated by the presence of the enzyme protein. In other cases, the metal ion serves to maintain the enzyme molecule in the structure needed for biological activity. A change in pH can alter the binding of a cofactor and result in a change in the catalytic activity of the enzyme. It is difficult to remove metal cofactors that are tightly bound to an enzyme without disturbing the protein structure and denaturing the enzyme.

An enzyme cofactor that is an organic molecule is referred to as a **coenzyme** (Fig. 2–42). These types of cofactors can serve as donors or acceptors of the atoms or functional groups added to or removed from the substrate.

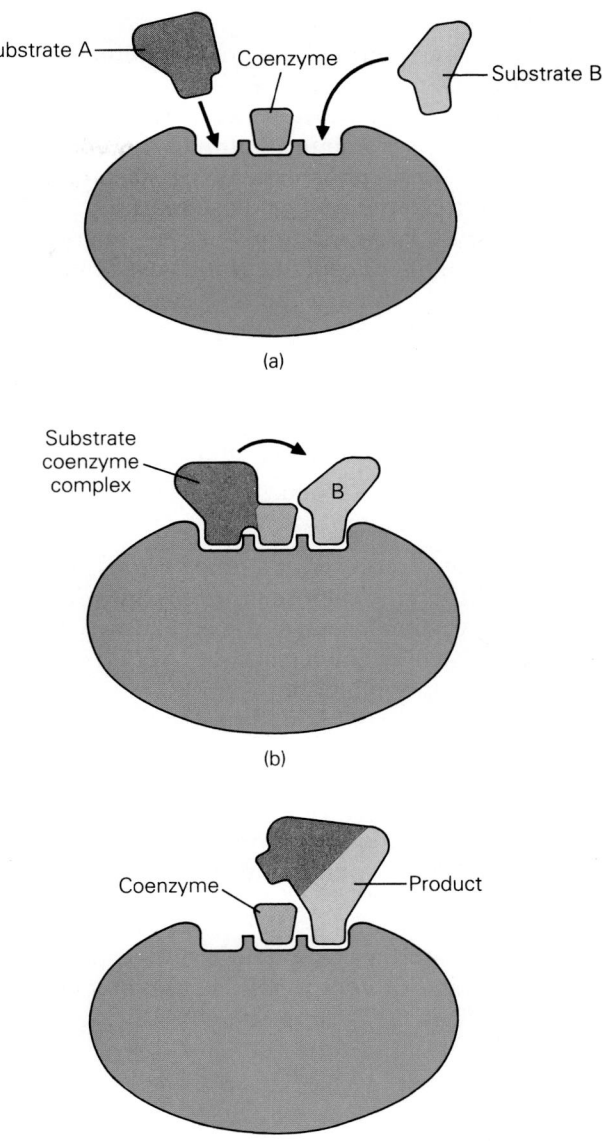

Figure 2–42
Coenzymes are nonprotein molecules that aid enzymes in the attachment of substrates.

Energy Exchange in the Cell: ATP and ADP

The intricate processes by which cells transform chemical energy to do work will be examined in detail in Chapter 6. A general overview at this point, however, will reveal the basic relationship—and an important molecule—linking the endergonic and exergonic reactions in the cell.

We have learned that the macromolecule ATP (adenosine triphosphate) is a nucleotide consisting of three phosophate (P_i) groups, a ribose ring, and adenine. When a phosphate group is removed, ATP becomes ADP; when another is removed, ADP becomes AMP. The removal of one phosphate group from ATP releases a large amount of free energy; the attachment of one phosphate group absorbs a large amount of free energy. For these reasons, ATP acts as a store of large amounts of free energy. The bonds among its phosphate groups are often represented by wavy lines in the ATP structural formula (see Fig. 2–31) and are known as "high-energy bonds."

Exergonic reactions that occur in the cell produce free energy that is used to **phosphorylate** (add a phosphate group to) AMP and ADP:

$$AMP + P_i + Energy \longrightarrow ADP$$
$$ADP + P_i + Energy \longrightarrow ATP$$

These reactions represent a storage process in which the cell temporarily deposits free energy released in exergonic reactions, such as the breakdown of nutrient molecules, in molecules of ATP (Fig. 2–43) through phosphorylation.

This energy is not stored for long, however, for other reactions that need large amounts of free energy to begin are also occurring: these endergonic reactions are mainly those of biosynthesis. They obtain the free energy they need from the ATP produced by the exergonic reactions:

$$ATP \longrightarrow Energy + ADP + P_i$$

leaving ADP and a phosphate group to be used in the next round of energy release and storage.

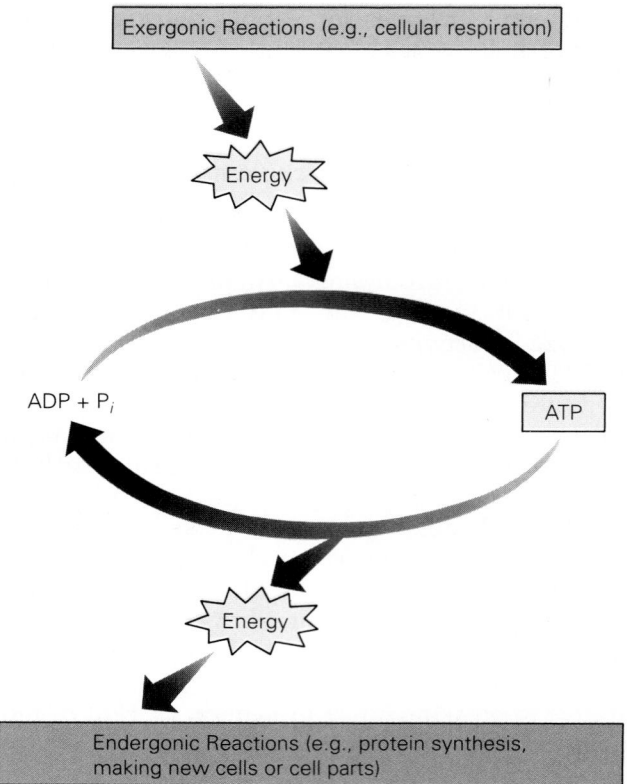

Figure 2–43
Both endergonic and exergonic reactions occur in the cell. Endergonic reactions obtain the free energy they need to proceed from the free energy spontaneously released from exergonic reactions. This free energy is exchanged in the form of ATP.

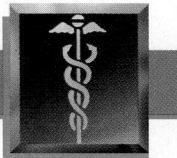

Black Death

One quarter of the population of England was killed by **bubonic plague** in 1348. A characteristic feature of this disease is the **bubo,** a painful swelling due to inflammation of lymph glands. The affected glands are located principally in the groin, armpit, head, and neck. The disease came to be known as the **Black Death,** probably because bleeding from small blood vessels in the skin produced discoloration. Inadequate oxygenation of the blood due to associated pneumonia also contributed to the bluish-black appearance of the skin. Bubonic plague is a disease of rodents, and rats are the main source of human infection. The disease, transmitted from rats to humans by fleas, is caused by the bacterium *Yersinia pestis.* The recent discovery that this bacterium contains an enzyme very similar to one normally found in mammalian cells may help to explain why the infection progressed so rapidly and had such lethal consequences for humans in the Middle Ages.

Addition of phosphate groups (**phosphorylation**) to tyrosines on intracellular proteins may be a critical step in regulating the growth and development of cells. For example, the action of several hormones, including insulin, is accompanied by tyrosine phosphorylation on certain proteins. Specific enzymes which add phosphate groups to the tyrosine amino acids of proteins are known as **protein tyro-**

sine kinases. The phosphate groups can be removed by another class of enzymes known as **protein tyrosine phosphatases.** Thus, the balance between these two groups of enzymes with their opposing actions may play a critical role in the control of important events in cell physiology.

The bacterium *Yersinia* was found to contain mammalian protein tyrosine phosphatases. This finding was unexpected because the bacterium contains no tyrosine phosphate groups, the normal substrates for these enzymes. It has been speculated that abnormally high levels of protein tyrosine phosphatases in mammalian cells could disrupt normal cell functions. When *Yersinia* invades a human cell and multiplies, it leads to a rapid increase in protein tyrosine phosphatase activity inside the cell. An uncontrolled increase in this enzyme activity results in the sudden removal of phosphate groups from a wide variety of intracellular proteins. This disruption of normal intracellular controls may contribute to the rapid deterioration and death of the infected person.

Fortunately, the development of modern antibiotics (see CLINICAL FOCUS in Chapter 5) has provided an effective means of treating *Yersinia* infections. Although outbreaks still occur in parts of Asia and South America, bubonic plague has been virtually eliminated from the western world.

SUMMARY

Overview: Matter and Energy in the Living Cell

Matter is composed of elements, the smallest units of which are atoms. Atoms combine to form molecules, and elements combine to form groups and compounds. Carbon, hydrogen, nitrogen, and oxygen are the four predominant elements in the human body. Living things require a constant supply of energy to maintain order in the matter that comprises them.

Atoms and Elements

All atoms are composed of the same subatomic particles. Atoms of the same element have the same number of protons in their nuclei. Uncharged atoms have equal numbers of protons and electrons. Isotopes are atoms of the same element with different numbers of neutrons.

Ions are atoms with electrical charges due to unequal numbers of electrons and protons.

The electron configuration of an atom, especially the arrangement of electrons in its outer shell, determines its reactivity. An atom will try to fill its outer shell with electrons.

Molecules, Groups, and Compounds

Atoms share, gain, or lose electrons to form bonds with other atoms, producing molecules, groups, and compounds. Atoms that share electrons form polar or nonpolar covalent molecules.

Atoms that form bonds by gaining or losing electrons produce ionic compounds, often arranged as crystal lattices.

Van der Waals bonds develop between atoms that are close together, owing to the attraction of the nucleus of one atom to the electrons of another.

Hydrogen bonds form between hydrogen atoms of polar molecules and groups of small atoms such as oxygen and nitrogen of other polar molecules.

Chemical Reactions: Basic Principles

Chemical reactions occur when compounds react with other compounds to form new products. The rate of a chemical reaction depends on such factors as temperature, concentration, and orientation.

Water and Aqueous Solutions

The polarity of the water molecule makes water a strong solvent of ionic and polar covalent compounds. These compounds are therefore called "hydrophilic," in contrast to nonpolar compounds, which are "hydrophobic," or insoluble in water.

Because of hydrogen bonds, water molecules form three-dimensional patterns in the solid and liquid phases.

Water molecules dissolve ionic compounds by forming shells around the ions, causing them to dissociate

from the crystal lattice. Water dissolves polar covalent compounds by forming hydrogen bonds with them. Nonpolar covalent compounds do not dissolve in water; instaed, the nonpolar molecules undergo hydrophobic reactions, clustering together in pockets in the water.

Concentration of a solution is measured in terms of molality (moles of solute per kilograms of solvent) or molarity (moles of solute per liter of solution).

The pH scale is a means of measuring the relative acidity or basicity of a solution by expressing logarithmically the H^+ concentration.

An acid is a substance that dissociates in solution to release H^+ ions, or protons. A base is a substance that accepts H^+ ions in solution, often by releasing OH^- ions. An acidic solution is one with a high H^+ concentration; a basic solution is one with a low H^+ concentration. Buffers are weak acids or bases added to solutions to minimize changes in pH due to the presence of a strong acid or base.

The Behavior of Gases

Gases can be described in terms of pressure, temperature, volume, and amount (in moles).

These four properties are related in a constant way for all gases, as described in the Ideal Gas Equation, $PV = nRT$.

In a solution of gases, the total pressure of the solution is equal to the sum of the partial pressures of the constituent gases.

According to the kinetic theory of gases, gas molecules are moving in continuous random motion, colliding with each other and with their surroundings, and increasing in speed as the temperature increases.

Gradients and the Movement of Substances

A gradient is a difference in two quantities divided by the distance between them. Some substances move through the body in response to pressure gradients.

Other substances move in response to voltage gradients, or differences in electrical charges between regions.

Concentration gradients are responsible for the movement of substances by diffusion and osmosis. Solute particles in both liquid and gaseous solutions spread evenly throughout solvents by means of diffusion. Water molecules pass through selectively permeable membranes by means of osmosis in order to dilute regions that have high concentrations of solute particles.

Macromolecules

The unique covalent bonding capabilities of the carbon atom allow it to form molecules of huge proportions and intricately varied structures. These macromolecules are important in all cell processes. One class of macromolecules, called carbohydrates, is made up of smaller units called monosaccharides. Ring structures are common among carbohydrates. Monosaccharides combine in linear and branched chains to form di- and polysaccharides. Sugars are important sources of chemical energy in cells.

Lipids are macromolecules that often consist of long hydrocarbon tails that are not soluble in water. The hydrophobic behavior of phospholipids in water makes them ideal structural components of cell membranes. Lipids are also important sources of chemical energy in cells. The major classes of lipids are fatty acids, glycerol compounds, and nonglycerol compounds.

Proteins are complex macromolecules that carry out diverse and important functions in cells. Proteins consist of amino acids arranged in four levels of structure. The sequence of amino acids and the shape of protein macromolecules are highly specific features of proteins that determine their functions in the cell.

Nucleotides are composed of a sugar, a phosphate group, and a special group called a base. Nucleic acids (DNA and RNA) are composed of nucleotides linked together in long strands. They store genetic information.

Thermodynamic Principles

Thermodynamic principles play an important role in cell physiology. All chemical reactions involve a free energy change between the reactants and the products.

Enzymes increase the rates of specific reactions by bringing reactant molecules together and straining existing bonds, thus lowering the initial activation energy barrier for the reaction.

Endergonic reactions in the cell absorb the free energy they need from exergonic reactions.

REVIEW QUESTIONS

Identify Each with Term

1. A reaction which releases free energy.

2. A macromolecule which contains adenine, guanine, thymine, and cytosine.

3. The maximum amount of substrate that can be turned into product in a given time during an enzyme-catalyzed reaction.

4. A macromolecule which contains both protein and carbohydrate.

5. The rate of movement of solute molecules.

Define Each Term

6. Molecular formula

7. Atomic number

8. Covalent bond

9. Solution

10. Hexose

Choose the Correct Answer

11. Adenine forms a base-pair with:
   a. thymine
   b. guanine
   c. cytosine
   d. adenine

12. Attachment of serine to phosphatidic acid forms a:
   a. dipeptide
   b. glycolipid
   c. phospholipid
   d. nucleoside

13. A dehydrogenase enzyme catalyzes:
   a. joining of molecules by forming new bonds
   b. addition of oxygen to a molecule
   c. rearrangement of atoms in a molecule
   d. removal of hydrogen from a molecule

14. Nucleic acids are long chains of:
   a. amino acids
   b. nucleosides
   c. fatty acids
   d. nucleotide monophosphates

15. Adenosine triphosphate (ATP):
   a. contains guanine
   b. is a nucleoside
   c. acts as an intracellular store of free energy
   d. is an enzyme that adds phosphate groups to substrates

Calculate

16. What is the molecular mass of water?

17. What is the concentration of hydrogen ions in a solution at pH 5.0?

18. What is the osmolarity of 25 mM K_2SO_4?

Answer Each Question in One or Two Sentences

19. How is a peptide bond formed between two amino acids?

20. What are the bonds which help to stabilize the secondary structure of a protein?

21. What does it mean when the nucleotide strands in DNA are described as complementary?

22. Why do solid crystals of potassium chloride dissolve easily in water?

23. How does an enzyme increase the rate of a reaction?

SUGGESTED READINGS

Cloud, P. "The biosphere." *Scientific Ameircan*, 249:176–189, 1983.

Felsenfeld, G. "DNA." *Scientific American*, 253:58–66, 1985.

Gross, J. "Collagen." *Scientific American*, 204:120–138, 1961.

Hill, T.L. "Biochemical cycles and free energy transduction." *Trends in Biochemical Sciences*, 2:204–207, 1977.

Karplus, M., and McCammon, J.A. "The dynamics of proteins." *Scientific American*, 254:42–51, 1986.

Koshland, D.E. "Control of enzyme activity and metabolic pathways." *Trends in Biochemical Sciences*, 9:155–159, 1984.

Scott, J.E. "Molecules for strength and shape." *Trends in Biochemical Sciences*, 12:318–321, 1987.

Sharon, N. "Glycoproteins." *Scientific American*, 230:78–92. 1974.

Smith, E.L., Hill, R.L., Lehman, I.R., Lefkowitz, R.J., Handler, P., and White, A. *Principles of Biochemistry: General Aspects*, 7th ed. New York: McGraw-Hill, 1983.

Stryer, L. *Biochemistry*, 3rd ed. San Francisco: Freeman, 1988.

KEY TERMS

acid (p. 48)
activation energy (p. 77)
amino acid (p. 63)
atom (p. 33)
base (p. 48)
buffer (p. 49)
carbohydrate (p. 56)

chemical reaction (p. 42)
compound (p. 33)
covalent bond (p. 40)
diffusion (p. 52)
electron (p. 36)
element (p. 33)
enzyme (p. 77)

equilibrium (p. 47)
fatty acid (p. 59)
kinetic energy (p. 80)
lipid (p. 59)
macromolecule (p. 56)
nucleic acid (p. 68)
osmolarity (p. 47)

osmosis (p. 54)
oxidation-reduction (redox) reaction (p. 43)
protein (p. 63)
solution (p. 44)

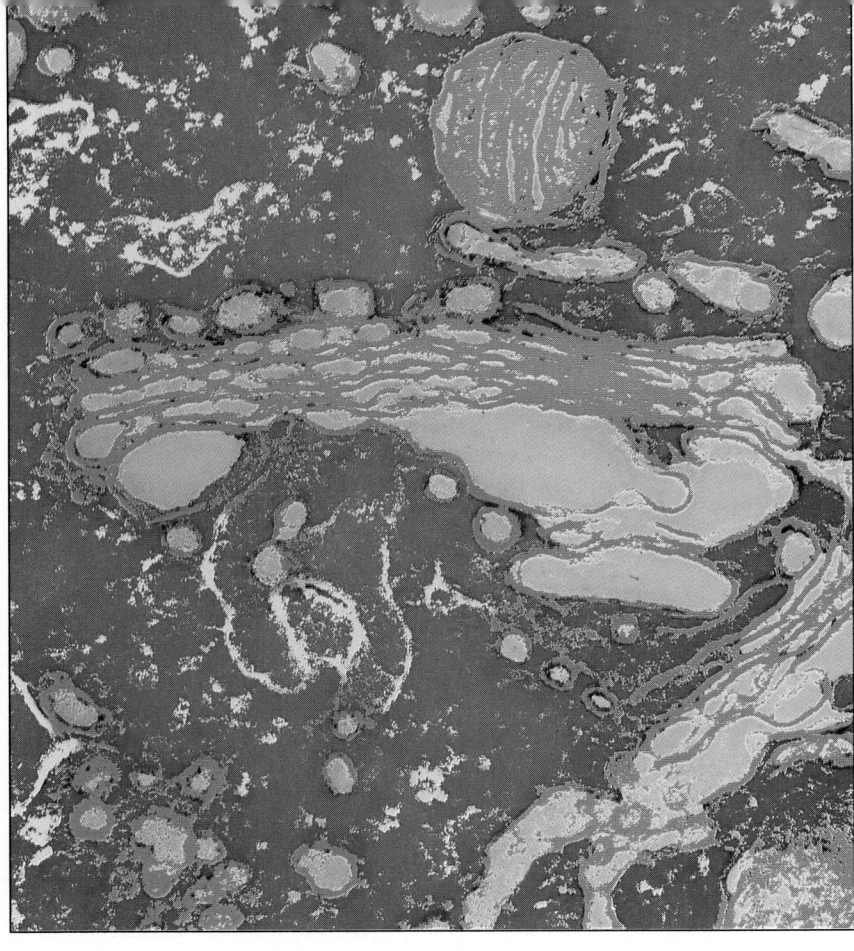

The Structure and Functions of Cells

CHAPTER OUTLINE

According to the **cell theory,** which forms the basis of all modern biology and physiology, (1) all organisms are made up of cells and their products; (2) new cells arise only from pre-existing cells; (3) all cells have the same fundamental chemical makeup and metabolic processes; and (4) the activities and processes of the organism as a whole result from the interdependent and cooperative workings of groups of cells.

In all of nature there are two broad classes of cells: eukaryotic cells and prokaryotic cells. The primary distinction between these cell types is the presence or absence of a membrane-bounded compartment surrounding the genetic material, the DNA. In **prokaryotic cells,** the DNA is not in a separate compartment from the rest of the cell, whereas all **eukaryotic cells** have a well-defined **nucleus,** or central area, contained within its own nuclear membrane.

Organisms in turn are broadly classified according to whether they have prokaryotic or eukaryotic cells. Many bacteria are prokaryotic organisms, for example, while all mammals are eukaryotic organisms. Eukaryotic cells are generally larger (10 to 100 μm in diameter) than prokaryotic cells (1 to 10 μm).

In addition to a membrane-enclosed nucleus, all eukaryotic cells contain many small subcellular structures, most of which are also enclosed in a membrane. These structures are grouped under the general term **organelles.** They are distributed throughout the **cytosol,** the fluid surrounding the nucleus. Together, the organelles and the cytosol make up the **cytoplasm.** The cytoplasm and the nucleus make up the two major compartments of the cell (Fig. 3–1). They are enclosed by a cell membrane called the **plasma membrane.**

Overview: Cell Processes and Products

A **cell** is the smallest distinct unit of life capable of an independent existence. Three features of cells permit independent life: (1) a set of genes that contain the information the cell needs both to direct its own activities (primarily the synthesis of macromolecules) and to pass on to new cells; (2) the machinery for obtaining energy from foodstuffs to sustain growth, reproduction, and movement; and (3) a cell membrane that provides a physical boundary between the cell and its environment.

Cell processes can be thought of in two broad categories: (1) the **synthesis,** or creation from precursors, of macromolecules both for use by the cell itself and for export to other cells; and (2) **degrada-**

Figure 3–1
The cell has two major compartments: the nucleus and the cytoplasm. The cytoplasm contains the major cell organelles and a fluid called cytosol.

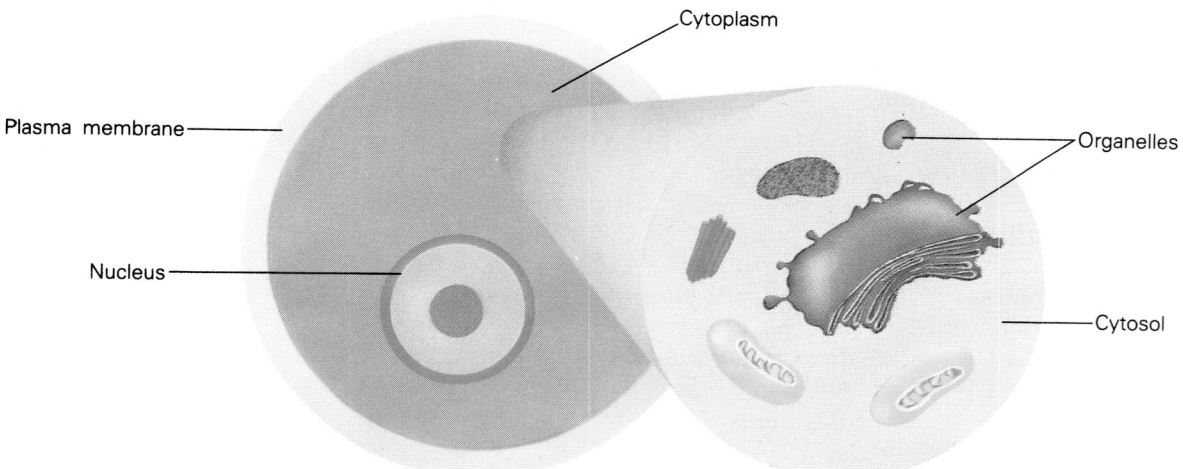

tion, or breakdown, of other macromolecules for the purpose of obtaining energy. These processes together make up the cell's **metabolism.**

Just as each organ of the body has a unique structure that permits it to perform a special function in a system of the body, each organelle of the cell has a unique structure and function that enables it to participate in the life processes of the cell and allows the cell, in turn, to function in the body. The most important aspect of each cell's role in the body is the type of proteins it synthesizes.

Every cell in the body can potentially make any protein in the body. However, certain cell types make only certain proteins. For example, some cells in the pancreas have as their chief function the production of insulin for use by other cells. Insulin is a type of protein called a **hormone.** In order to produce insulin, the cell needs instructions. To ensure that future generations of pancreatic cells know how to produce insulin, the cell must preserve those instructions by copying them.

Even if they do not secrete proteins for use by other cells, all cells need certain proteins for their own internal functions. Many of these proteins are **enzymes.** As described in Chapter 2, they assist many important chemical reactions in the cell, controlling both synthetic and degradative processes.

The instructions that tell each cell how to make the specific proteins it needs are contained in the DNA and RNA in the nucleus. The machinery for assembling the proteins is found outside the nucleus, however. When the cell needs to make a protein, the instructions are brought via RNA from the nucleus to the organelles that make the proteins. When the cell needs to divide, the instructions are duplicated in the nucleus so that each new nucleus will have a copy.

Following is a general summary of the major cell processes and products, and the organelles that carry them out. A typical, generalized animal cell with all its organelles is depicted in Figure 3–2(a). Refer also to the flow chart in Figure 3–3, which is a schematic depiction of the interlocking cell processes.

The Process of Controlling Cell Functions

Genetic material, as mentioned previously, is necessary for the independent existence of most cells. Contained in the macromolecules of DNA, genetic material is essentially a set of instructions that the cell uses to (1) make the specific proteins and other macromolecules it needs to carry out its own desig-

nated function in the body; and (2) to reproduce itself. A set of related organelles in the cell work together to "express" the information contained in the DNA by synthesizing macromolecules according to its instructions.

The **nucleus** is the site where **DNA** is made, stored, and duplicated. This DNA is used not only in cell division but throughout the life of the cell. Recall from Chapter 2 that the DNA molecule is a double strand of base-paired nucleotides. A **gene** is a portion of a DNA molecule containing a specific sequence of nucleotides that serve as a code dictating the sequence of amino acids to be followed when a particular polypeptide is assembled in the cytoplasm.

Directing Synthesis

The nucleus is also the site of **RNA** manufacture. RNA carries genetic information (in the form of instructions about how to make proteins) from the DNA in the nucleus to the synthesizing organelles in the cytoplasm. RNA travels through **nuclear pores,** or openings, in the nuclear membrane to get from the nucleus to the cytoplasm.

On the endoplasmic reticulum, RNA forms a **ribosome.** Ribosomes are organelles that make proteins, according to the instructions carried by RNA, by joining amino acids found in the cytosol. Once made, many of the proteins are transported to the **Golgi complex.** There they are joined with other molecules and transported through the cytosol. Some of the proteins travel to the cell membrane for transport to the outside of the cell. Some proteins stay in the cell to participate in the internal cell processes. For example, some may be enzymes used to digest food for the cell, while others become working parts of membranes in the cell.

Preserving Information

Many cells in the body are continually dividing. Unlike a human being, which can produce offspring and still continue to exist, a "parent" cell is completely replaced by its daughter cells. Therefore, any information to be passed on to the new cells must be prepared ahead of time, before cell division takes place.

The information about synthesis of macromolecules that must be passed on is contained within DNA. This molecule replicates, or copies, itself. Two copies of the same DNA then exist, each to become the nuclear DNA of one of the new cells. DNA replication is discussed in detail in a later chapter.

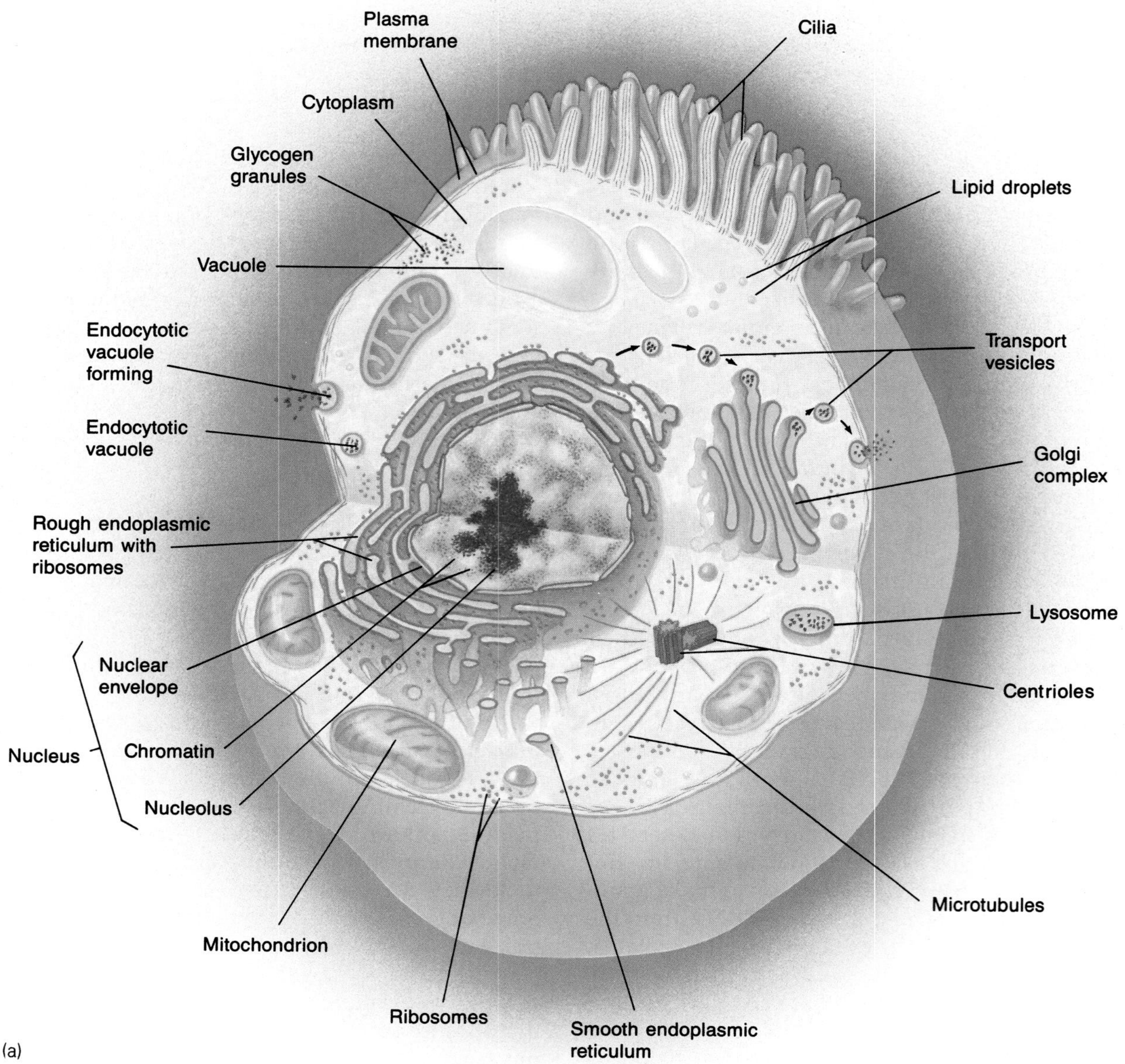

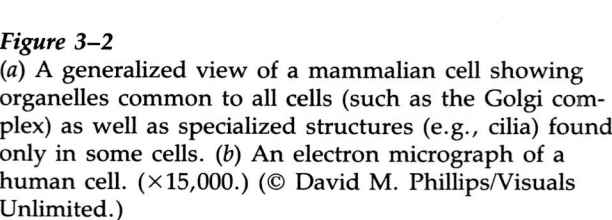

(a)

(b)

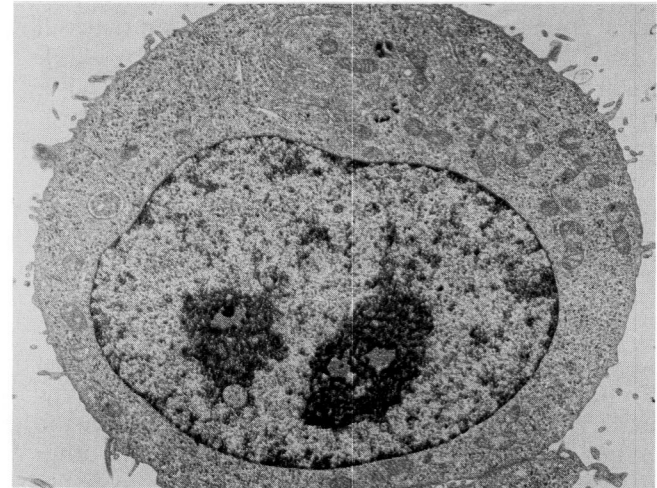

Figure 3–2
(a) A generalized view of a mammalian cell showing
organelles common to all cells (such as the Golgi complex) as well as specialized structures (e.g., cilia) found
only in some cells. (b) An electron micrograph of a
human cell. (×15,000.) (© David M. Phillips/Visuals
Unlimited.)

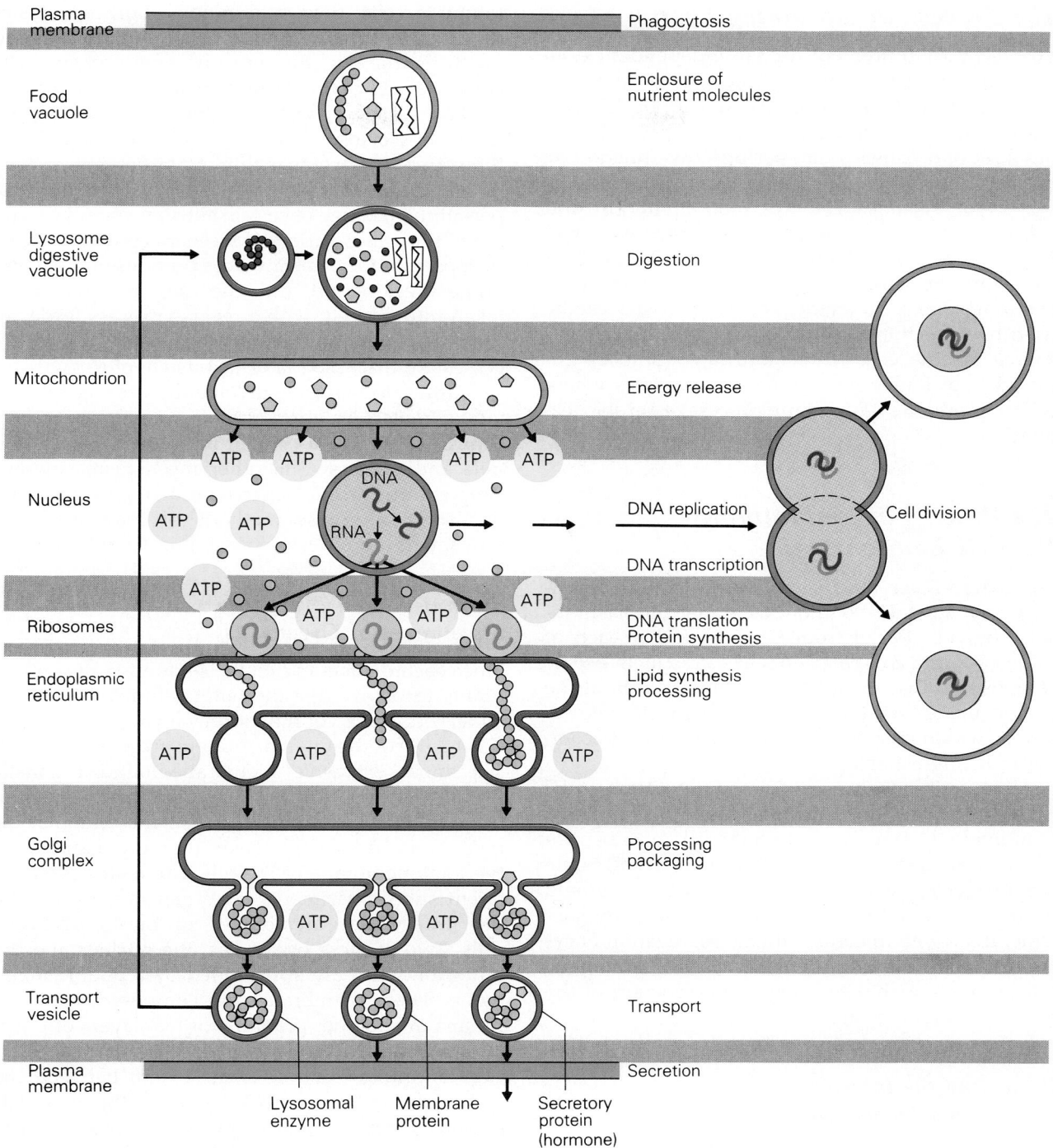

Figure 3–3
A flow chart summarizing the flow of matter and energy in the metabolism of a cell. Reactions in the cytosol and the mitochondria release energy (in the form of ATP) from nutrient molecules. The energy is used to synthesize other molecules, such as proteins and lipids, which perform various functions in cells. Some proteins are secreted; others form parts of membranes; and still others are enzymes the cell needs. Some lipids are used to form parts of membranes.

The Process of Obtaining Energy

The second requirement for the independent existence of a cell is the ability to take energy from food and convert it into a useful form to support life processes, including growth and reproduction. The cell has a set of organelles that perform this function by supporting a complex chain of chemical reactions. A system of **vacuoles** conveys food matter in and waste matter out of the cell. **Lysosomes** are small bags of enzymes that surround food molecules that come into the cells in vacuoles. The enzymes in the lysosomes break down the molecules so they can be used by mitochondria to produce ATP, the energy source for the cell. **Mitochondria,** organelles enclosed in a double membrane, are the sites where ATP is produced through the process of aerobic respiration.

The Process of Maintaining the Cell Environment

The extracellular fluid surrounding the cell is primarily an aqueous solution, as is the cytoplasm inside the cell. Recall from Chapter 1 that each of these two fluid compartments has a distinctive ionic composition that must be maintained. Without the plasma membrane that separates them, the two solutions would merge and their contents mix, becoming indistinguishable.

Just as it is necessary for the cytoplasm to be segregated from the extracellular fluid, the contents of individual organelles within the cell must be protected from each other and from the cytosol. Some cell products, such as digestive enzymes, would destroy other cell products, such as secretory proteins, if allowed to contact them. Segregation of cytoplasm from the extracellular fluid, and of various cell products within the cell, is the function of the cell's membrane system.

Protecting the Interior: The Plasma Membrane

The outer **cell membrane,** called the **plasma membrane,** is composed of two layers of phospholipid molecules. Recall from Chapter 2 that phospholipids consist of a long hydrophobic, or water-insoluble, tail, and a hydrophilic, or water-soluble, portion. Because of hydrophobic interactions, the phospholipid molecules making up the cell membrane act the way any such molecules would act when placed in an aqueous solution such as that found in cells: their hydrophobic portions cluster together, leaving their hydrophilic portions exposed on the surface to the aqueous environment. The clustering of lipid molecules creates a sturdy barrier to the passage of water-soluble particles, keeping the watery cytosol from flowing out of the cell and the watery extracellular fluid from flowing in.

At the same time, the plasma membrane must permit the passage of certain substances both in and out of the cell. A variety of mechanisms exist for this purpose. Endocytosis involves portions of the membrane breaking off and enclosing material to be taken into the cell, with a reverse process (exocytosis) used to let material out of the cell. These processes are discussed later in this chapter.

Other mechanisms involve the passage of individual molecules through the membrane in very controlled ways via protein molecules called **channels** and **carriers.** Still other mechanisms involve simple diffusion such as that occurring in any aqueous solution. These methods of transmembrane movement are covered in detail in Chapter 4.

Separating Cell Processes: Membranes of Organelles

Since each distinct organelle is bounded by a membrane, it follows that the outer cell membrane of a eukaryotic cell accounts for only a minor proportion (2% to 5%) of the total membrane present in the cell. The most extensive membrane-bound organelle present in most cells is the endoplasmic reticulum (see Fig. 3–2). The membrane of the endoplasmic reticulum of a liver cell makes up more than 50% of the total membrane of the cell. There are seven different subcompartments, or organelles, formed by the intracellular membranes of most eukaryotic cells. Two of these organelles, the nucleus and the mitochondrion, are surrounded by a double membrane. These membranes are composed, as is the plasma membrane, primarily of two layers of phospholipid molecules.

Continuity exists to some extent between the subcompartments in that some of the materials present in one compartment may be passed on to another, but direct physical links between the compartments are probably very limited. Each of the compartments within the cell has a unique role, carried out at the molecular level by the enzymes that are either packed inside the organelle or components of the membrane of the organelle. Since each organelle has a different function, the set of enzymes within a specific organelle will not be found in any of the other organelles. **Catalase,** for example, is found only in organelles called peroxisomes. The enzyme that breaks down glycogen into glucose molecules occurs only in the cytosol. The enzymes that catalyze the sequence of reactions in the citric acid cycle are present only within mitochondria.

Enzymes found in lysosomes, called **lysosomal enzymes,** are essential to normal cell function. The absence of just one of these enzymes, as occurs in some genetic diseases, has severe consequences for the individual. However, this group of **degradative enzymes** can break down all of the macromolecules responsible for the structure and function of the cell. Clearly, these enzymes cannot be allowed direct contact with other organelles or with the components of the cytosol because they would cause irreversible, lethal damage. The cell solves the problem very neatly by keeping the enzymes within lysosomes so that they are always available within the cell. They are allowed access to structures and molecules targeted for degradation (breakdown) only when the lysosomes fuse with endocytotic vesicles containing material to be broken down. This process ensures that the degradative process is confined locally within a specific compartment, the secondary lysosome (see Fig. 3–11).

Similarly, the presence of catalase within peroxisomes ensures that the toxic hydrogen peroxide produced within these organelles is destroyed on the spot before it has a chance to leak into the cytosol, where it would cause serious damage.

The presence of intracellular compartments also permits separation of the processes of synthesis and degradation of molecules. The advantage derived from physically separating these processes is that they can occur simultaneously within the cell and be controlled independently. The metabolism of fatty acids provides a good example. These molecules are continually synthesized by the cell because they are components of the phospholipid molecules needed for the structure of cell membranes. At the same time, the cell is also degrading other fatty acids because they are an excellent source of chemical energy; in fact, they yield more energy than equal weights of carbohydrates such as glucose. The goal of synthesis is the opposite of the goal of degradation, and the two processes would not be compatible if both occurred openly in the cytosol. Eukaryotic cells solve the problem by locating the reactions that degrade fatty acids in the mitochondria and those that synthesize fatty acids in the cytosol.

Finally, compartmentalization of cellular functions allows for intracellular processing and "sorting" of some of the synthesized molecules. We will discuss shortly some of the sorting and processing that occurs in the endoplasmic reticulum (ER) and the Golgi complex. For example, proteins destined to become components of membranes are trapped within the ER membrane during synthesis, whereas proteins destined for export from the cell are discharged into the lumen of the ER, where they are attached to carbohydrates to become glycoproteins. From there, they are transported to the Golgi complex, where the carbohydrate portion of the molecules is chemically modified. The Golgi complex also carries out another sorting step based on the specific structures present in the glycoprotein molecules. It distinguishes between the molecules that must be packaged inside secretory vesicles for export from the cell, and the molecules that are destined to become lysosomal enzymes and must be directed into primary lysosomes for use inside the cell.

Cell Compartments and Their Functions

The Nucleus: The Library of the Cell

The nucleus (Fig. 3–4) is the site where DNA and RNA are made. The information stored in DNA and RNA molecules directs the synthesis of proteins that determine the shape and function of a cell. (The red blood cell is the only common mammalian cell that lacks a nucleus, and it survives for only a few months.) The nucleus also contains proteins. The approximate composition of the nucleus in a liver cell, expressed as a percentage of the dry weight of the nucleus, is 80% protein, 15% DNA, and 5% RNA.

DNA: The Permanent Files

Recall from Chapter 2 that DNA (deoxyribonucleic acid) is an enormously long, unbranched, double strand of nucleotides arranged in an alpha-helix. A typical animal cell contains more than 1 meter of DNA, all of which must fit inside a nucleus with a diameter of only 0.000006 meter (6 μm). Each section of the chromosomal DNA that carries the information necessary for synthesis of a single polypeptide is known as a **gene.** The sum of all the chromosomal genes is termed the **genome** of the cell. All cells in the body, except the reproductive cells, normally contain the same DNA. The DNA molecules remain unchanged throughout the life of the organism unless they are damaged in some way.

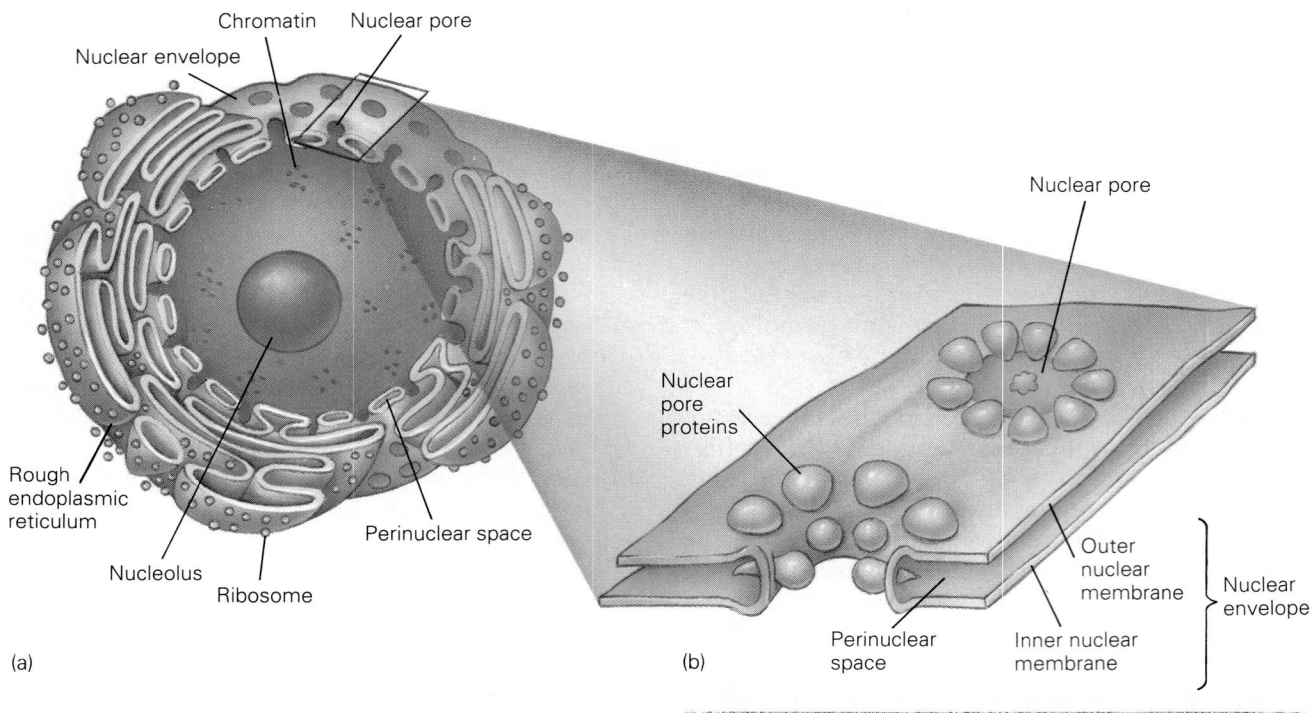

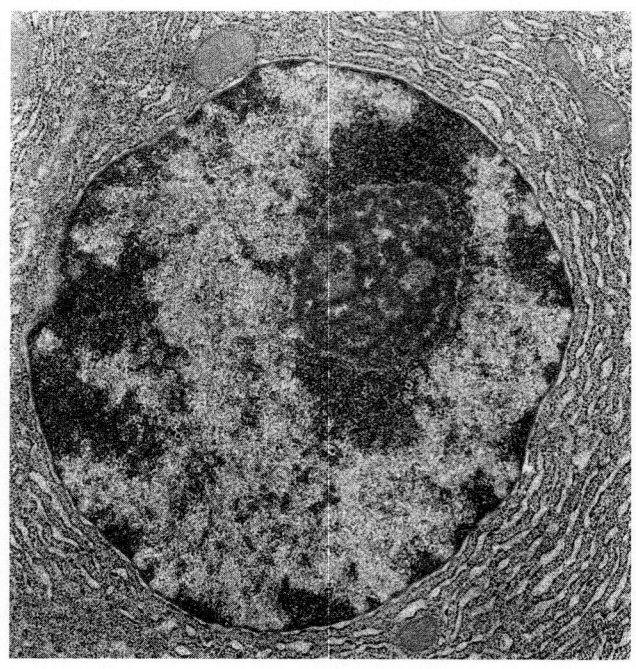

Figure 3–4
(*a*) The cell nucleus is enclosed in a double membrane called the nuclear envelope. Pores in the envelope permit the passage of molecules in and out of the nucleus. The outer layer of the nuclear envelope is continuous with the endoplasmic reticulum, so that the lumen of the ER is continuous with the perinuclear space. In the nondividing nucleus, DNA is visible as chromatin. The nucleolus plays a role in the synthesis of ribosomes from RNA. (*b*) A nuclear pore is formed from the fusion of the two layers of the nuclear envelope. Proteins are thought to be located in the pores. (*c*) An electron micrograph of the nucleus and nucleolus of a glandular cell. (© Don W. Fawcett/Photo Researchers.)

The tight packing of DNA (Fig. 3–5) is achieved by specialized proteins called **histones,** which bind to the DNA and organize its structure in the nucleus. Histones are by far the most common protein found in the nucleus. They are rich in amino acids with basic (positively charged) side chains, such as lysine and arginine (see Fig. 2–26). These positively charged side chain groups may be able to interact with and bind to the negatively charged phosphate groups in DNA (see Fig. 2–33), regardless of the specific nucleotide sequence.

Chromatin. Most of the DNA is associated with histones and other nuclear proteins, giving rise to a

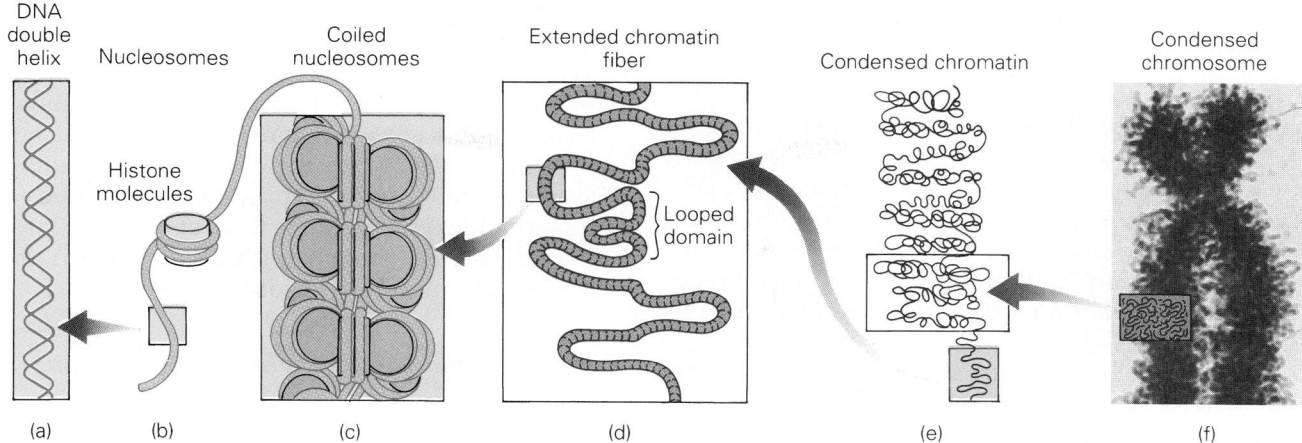

Figure 3–5
The DNA is packed in a series of stages beginning with nucleosomes. In a nondividing cell, DNA appears as chromatin; in a dividing cell, it appears in the more tightly coiled form of chromosomes.

complex known as **chromatin.** The basic packing unit of chromatin is the **nucleosome** (Fig. 3–5b), in which parts of the DNA double helix are wrapped around histone molecules and linked by stretches of "free," or **linker,** DNA. The nucleosome is still a relatively extended form of chromatin, and it is unlikely that much of the chromatin in the nucleus is present in this form.

A higher level of organization is produced by the packing together of nucleosomes in a regular array known as a **chromatin fiber.** Loops in these fibers give rise to **looped domains,** which reduce the initial length of the DNA double helix about 500-fold (Fig. 3–5d).

Chromosomes. Further levels of folding must occur, perhaps involving close packing of the loops, in order to condense the DNA into a single **chromosome** (Fig. 3–5e, f). Thus, each chromosome of a eukaryotic cell consists of a single very large molecule of DNA arranged so that the double helix is folded in a compact and highly organized way. Each species has a characteristic number of chromosomes; human cells, for example, have 23 pairs. When cells are not dividing, the chromosomes are in a relatively uncoiled state and are not readily identifiable as individual units. Only when cell division is about to occur do the chromosomes become apparent.

RNA: Selective Retrieval of Information

The RNA (ribonucleic acid) present in the nucleus is destined for export into the cytoplasm, where the machinery for protein synthesis is located. In cells such as rat liver cells, less than 10% of the total RNA in the cell is located at any one time in the nucleus.

An RNA molecule consists of a single strand of nucleotides that represent a copy of a limited region of the nucleotide sequence of one of the strands of DNA. Three major functional types of RNA interact in the cytoplasm to synthesize proteins according to the information stored in DNA: **messenger RNA (mRNA); ribosomal RNA (rRNA);** and **transfer RNA (tRNA).** For now, the related roles of these three types of RNA can be summarized as follows (more detailed discussions will follow in Chapter 5): Ribosomal RNA joins with certain proteins in the cytosol to form ribosomes. These are the protein-making structures that occur in the cytosol or are attached to the endoplasmic reticulum. Messenger RNA contains the information concerning which amino acids should be joined together, and in which specific sequence, on the ribosomes. Transfer RNA seeks out the required amino acids in the cytosol and brings them to the ribosome to be assembled into polypeptide chains.

One of the most obvious structures within the nucleus is the **nucleolus,** which is particularly prominent in cells that synthesize a lot of protein. The nucleolus is not limited by a membrane; it contains proteins and large loops of DNA from which rRNA is rapidly synthesized. Cells begin assembling rRNA and proteins into ribosomes in the nucleolus. However, ribosomes are not mature, or completely formed, until they reach the cytoplasm. This aspect of ribosome formation will be discussed in Chapter 5.

The Nuclear Envelope: A Porous Double Membrane

The interior of the nucleus is separated from the cytoplasm by the **nuclear envelope,** a set of two separate membranes. These membranes are separated by a space about 10 to 50 nm wide termed the **perinuclear space.** The outer membrane of the nuclear envelope often has attached ribosomes that are engaged in protein synthesis. This outer membrane appears to be continuous with the endoplasmic reticulum (Fig. 3–4*a*).

Much exchange occurs between the interior of the nucleus and the cytoplasm. Since DNA and RNA are synthesized only in the nucleus, the individual nucleotides and other parts required for their synthesis must be obtained from the cytoplasm. Conversely, RNA molecules made in the nucleus must move out to the cytoplasm in order to carry out their roles in protein synthesis. The rapid exchange of materials between the nuclear and cytoplasmic compartments probably occurs through the **nuclear pores.** These are open channels in the nuclear envelope that form in areas where the inner and outer membranes are pinched together (Fig. 3–4*b*). A mammalian cell has about 3000 to 4000 pores per nucleus. The diameter of each pore is 9 nm. This size permits the passage of small water-soluble molecules but excludes large macromolecular structures.

The Cytoplasm: Synthesis and Energy Release

The normal structure, growth, and function of a cell require a constant supply of macromolecules, which are made from simpler molecules, or **precursors.** Synthesis of precursors such as amino acids or nucleotides, for example, is achieved by a set of chemical reactions that are part of the cell metabolism. Also important to intermediary metabolism are the reactions that break down small molecules and trap their chemical energy as ATP, as discussed in the previous section. All the reactions of metabolism take place outside the nucleus (Fig. 3–1).

The Cytosol: A Wealth of Enzymes and Precursors

The cytosol of a mammalian cell, such as a liver cell, is a large compartment. It represents about 50% of the total volume of the cell and contains the small molecules that are precursors for macromolecules. It also contains thousands of enzymes, which account for its high protein content (about 20% by weight). Many of the enzymes are catalysts for the reactions of cell metabolism. Thus, the cytosol resembles a highly viscous fluid rather than a watery solution.

The cytosol contains some structures that are not organelles but are worthy of mention, including glycogen granules and lipid droplets. **Glycogen,** a polysaccharide, is the storage form of carbohydrate. The lipid droplets contain **triglycerides,** which are the storage form of fatty acids (see Chapter 2).

Ribosomes and the Endoplasmic Reticulum: Synthesis of Proteins and Lipids

The endoplasmic reticulum (ER) is a membrane sheet, folded repeatedly to form a large interconnected system (see Fig. 3–2). Two regions of the ER are structurally and functionally different, however: the **rough,** or **granular, ER** and the **smooth,** or **agranular, ER** (Table 3–1). The rough ER is so called because ribosomes are bound to its outer surface; smooth ER has no ribosomes attached.

The ER makes new proteins and lipids, including molecules destined to be released outside the cell by a process known as **secretion.** Rough ER is particularly extensive in cells that synthesize and secrete proteins, such as certain hormones. Smooth ER is well developed in cells specializing in lipid synthesis or the synthesis and secretion of steroid hormones. Muscle cells have a highly specialized smooth ER known as the **sarcoplasmic reticulum.** An important function of this structure is to remove calcium ions from the cytosol.

Protein Synthesis on the Rough ER. Individual amino acids are assembled into polypeptides on ribosomes. Some of the ribosomes occur freely in the cytosol, but many are attached to the ER. The sequence of the amino acids in each polypeptide is dictated by mRNA. Polypeptide chains formed in

Table 3–1

Structural and Functional Differences Between Rough and Smooth Endoplasmic Reticulum (ER)

	Rough ER	Smooth ER
Ribosomes present	Yes	No
Protein synthesis	Yes	No
Lipid synthesis	Yes	Yes
Steroid synthesis	No	Yes

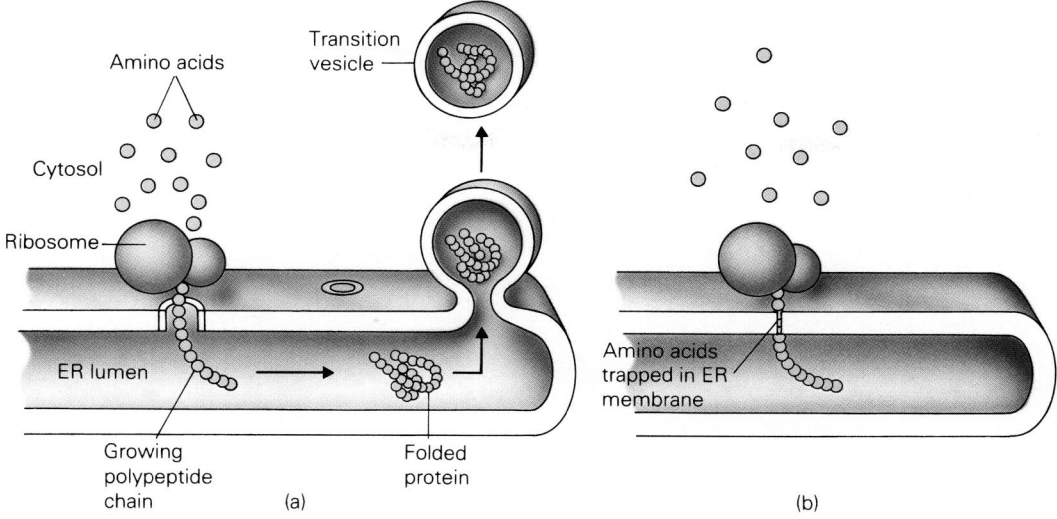

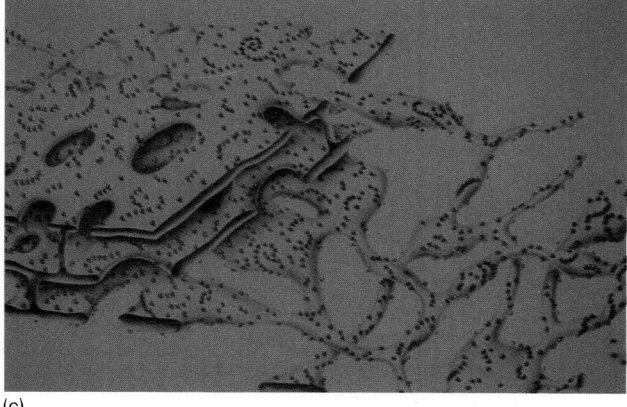

Figure 3–6
In the rough endoplasmic reticulum, a ribosome synthesizes a polypeptide from amino acids found in the
cytosol. (*a*) A protein destined to be a secretory product
or an enzyme enters the ER lumen as it is assembled,
folding into its various levels of structure. It is enclosed
in a transition vesicle for transport to the Golgi complex. (*b*) A protein destined to be membrane protein is
partially trapped in the ER membrane as it is synthesized. (*c*) An electron micrograph of the rough endoplasmic reticulum. (© Don W. Fawcett/Visuals Unlimited.)

the ribosomes are directed into the lumen of the ER
and separated from the cytosol by the ER membrane
(Fig. 3–6). Most of the proteins synthesized in the
rough ER become modified by the addition of monosaccharide units to form glycoproteins. In contrast,
the proteins synthesized on free ribosomes in the
cytosol are excluded from the ER lumen and cannot
be modified by addition of carbohydrate.

The proteins that are destined to become structural components of membranes (see Chapter 4) are
trapped within the rough ER membrane during synthesis. The mechanism for trapping the polypeptide
chain may be a specific sequence of amino acids that
interacts strongly with the interior of the ER membrane and prevents further movement of the polypeptide into the ER lumen (see Fig. 3–6). The part of
the polypeptide that is synthesized after the trapping sequence is left on the cytosolic side of the ER
membrane. Movement of ER membrane to other
membrane-bounded organelles and to the plasma
membrane will be discussed shortly.

Lipid Synthesis in the Smooth ER. The endoplasmic reticulum in many cells is the major site of the
synthesis of lipids, including triacylglycerols, phospholipids, and steroids. Lipids from each of these
three general classes form complexes with proteins
in the ER lumen. The resulting structures are the
compounds known as **lipoproteins.** The lipoproteins that are secreted from the cell and that enter
the blood have a critical role in the transport and
distribution of lipids, such as cholesterol, that are
absorbed by the intestine.

Lipids such as phospholipids and cholesterol
are also important structural components of membranes. The successive steps leading to the formation of a phospholipid occur on the ER membrane
because the enzyme proteins that catalyze the synthesis are part of the membrane. Thus, synthesis of
phosphatidylcholine (see Fig. 2–23*c*) occurs on the
cytosolic surface of the ER membrane; the molecules
of phosphatidylcholine that are produced will remain within the ER membrane (Fig. 3–7). Clearly,

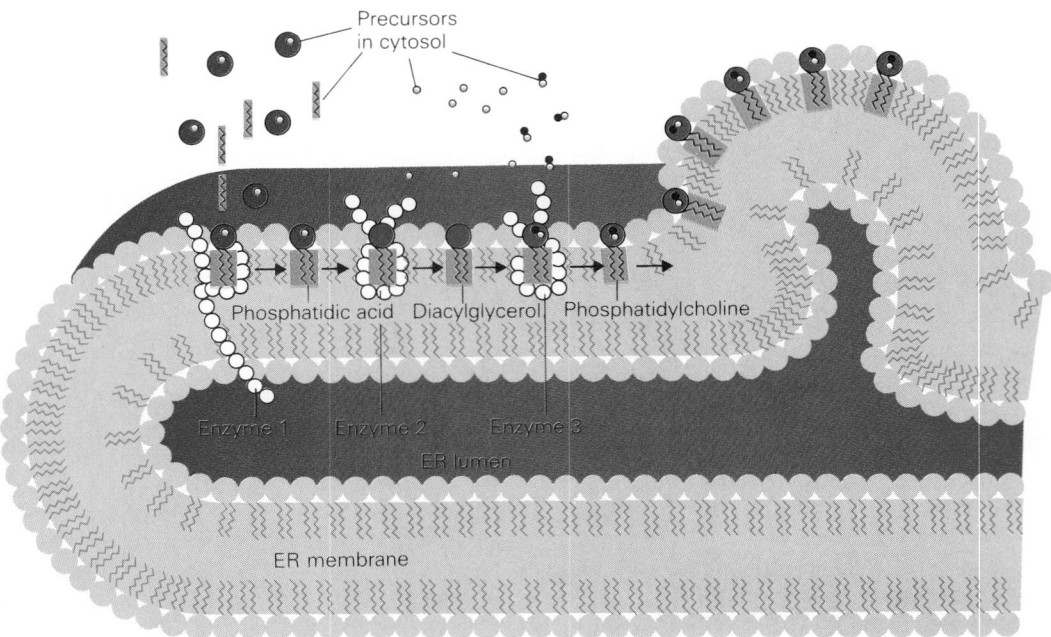

Precursors
in cytosol

Phosphatidic acid   Diacylglycerol   Phosphatidylcholine

Enzyme 1   Enzyme 2   Enzyme 3

ER lumen

ER membrane

Figure 3–7
In the smooth ER, a phospholipid is assembled from precursors in the cytosol through a series of reactions catalyzed by enzymes located in the ER membrane. The developing phospholipid molecule moves through the layer of existing phospholipid molecules that make up one half of the ER membrane.

synthesis of the proteins and lipids that are required for production of new cellular membrane structures is another important function of the ER.

A macromolecule that is to be secreted is first formed in the ER and then moves to the Golgi complex. This occurs via a transition vesicle formed from a portion of the ER membrane. In the Golgi complex, the macromolecule is processed and packaged for export out of the cell or for storage or use within the cell.

The Golgi Complex: Processing and Packaging

The function of the Golgi complex is to modify the macromolecules synthesized in the ER. This is especially important in secretory cells. The Golgi complex processes and packages the macromolecules inside small membrane bags, or **vesicles,** ready for transport to the cell surface or the cytosol.

The Golgi complex is a collection of smooth membranes that look like flattened sacs. Each sac is called a **cisterna.** Commonly, about six cisternae will be stacked together to form a structure known as a **Golgi stack** (Fig. 3–8). Small vesicles surround and radiate away from the Golgi stack.

The **cis,** or **forming, face** of the stack is oriented toward and closely associated with rough ER. The two structures are thought to be connected by the small membrane-bounded **transition vesicles** that shuttle newly synthesized macromolecules from the rough ER to the Golgi complex (see Fig. 3–8).

The **trans,** or **mature, face** of the Golgi complex, on the other hand, is oriented toward the plasma membrane and away from the nucleus, especially in secretory cells. Membrane-bounded **secretory vesicles** appear to arise from the trans face (see Fig. 3–8) to transport newly processed and packaged macromolecules from the Golgi complex to the plasma membrane for export.

There is good evidence that the Golgi complex modifies the macromolecules destined for secretion. One of its important activities involves modification of the polysaccharide chains that were added covalently to the proteins in the rough ER. The processing of these glycoproteins within the Golgi complex involves a considerable amount of "trimming" as well as the addition of new monosaccharide units.

The form of the Golgi complex can vary considerably. In some cell types it may be compact and limited, whereas in others it may be more spread

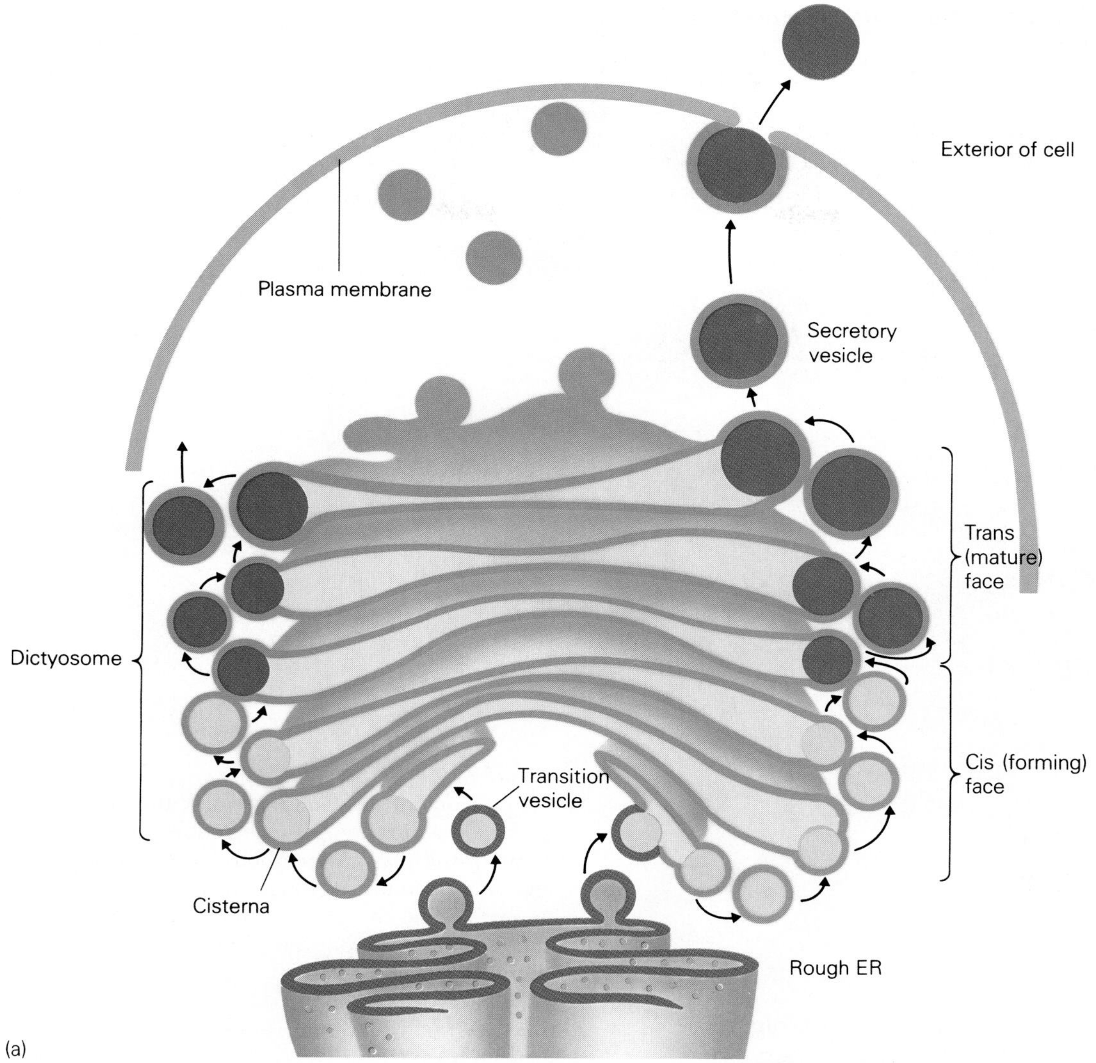

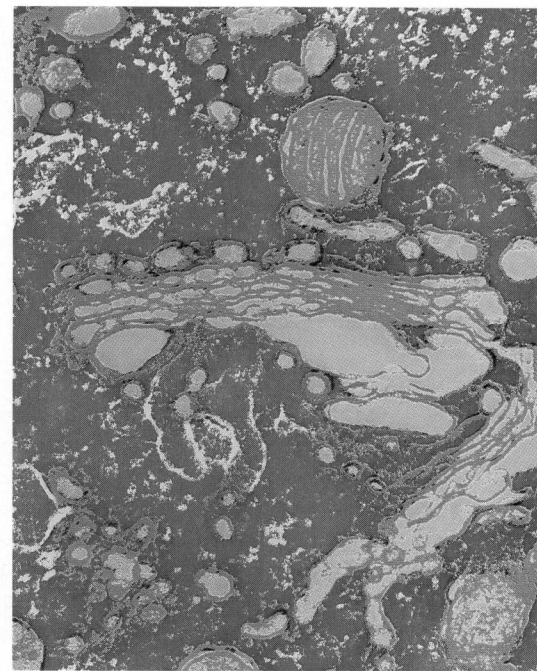

Exterior of cell

Plasma membrane

Secretory
vesicle

Trans
(mature)
face

Dictyosome

Cis (forming)
face

Transition
vesicle

Cisterna

Rough ER

(a)

(b)

Figure 3–8

(*a*) One unit of the Golgi complex, called a Golgi stack, is composed typically of about six folded sacs called cisternae. Macromolecules formed in the ER enter the cis face of the Golgi complex and pass through to the trans face via a series of vesicles. In the process, the lipids or proteins are modified. They leave the Golgi complex in secretory vesicles. (*b*) A false-color transmission electron micrograph of a section through one dictyosome of the Golgi apparatus. The dictyosome is the stack of paired membranes (green outlined in red) in the middle of the image. The large red sphere near the top is a mitochondrion. (×31,800.) (© Dr. Gopal Murti/ SPL/Photo Researchers, Inc.)

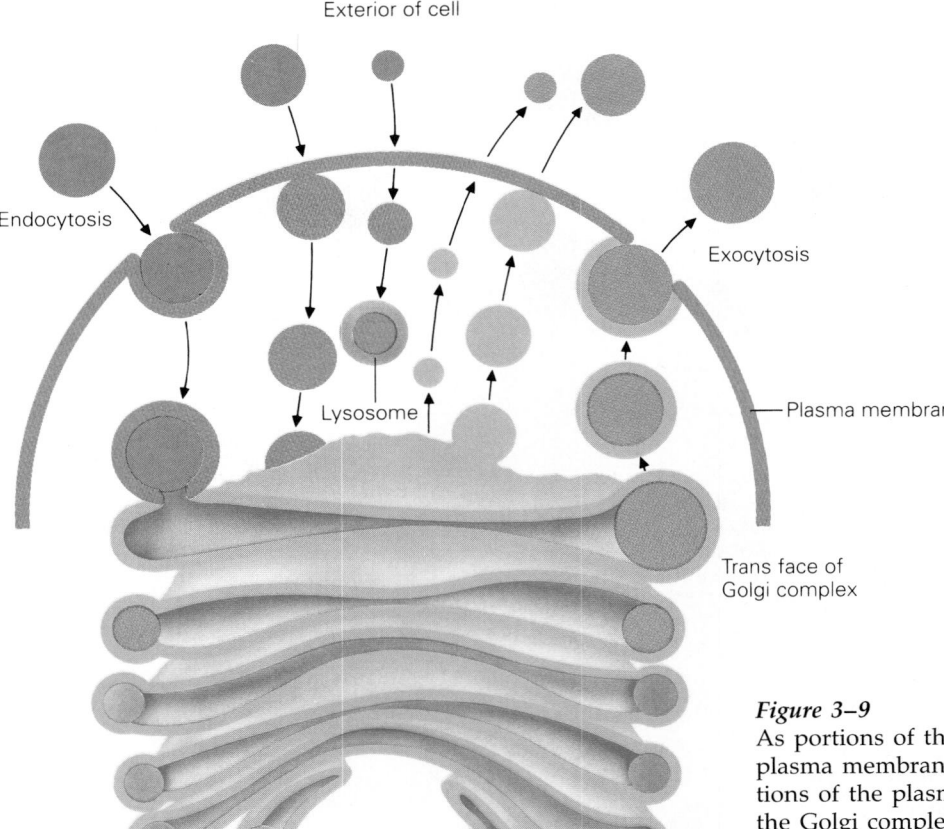

Exterior of cell

Endocytosis

Exocytosis

Lysosome

Plasma membrane

Trans face of
Golgi complex

Figure 3–9
As portions of the Golgi complex fuse with the plasma membrane in the secretion process, portions of the plasma membrane are returned to the Golgi complex through various endocytotic events. Thus, the plasma membrane maintains the same size despite the continual addition of membrane from secretory vesicles.

out, almost like a net. The position of the Golgi complex within the cytosol may also vary. In nerve cells it often surrounds the nucleus, whereas in secretory cells it is usually found between the nucleus and the side of the cell that is engaged in the secretory process. The number of Golgi stacks in a cell varies considerably with the type of cell. The Golgi complex in a liver cell accounts for about 7% of the total cell membrane, and in this cell type is a much less extensive structure than the ER.

Secretory Vesicles: Exocytosis and Membrane Recycling

A secretory vesicle containing a packaged macromolecule moves from the inner face of the Golgi complex to the plasma membrane. The vesicle membrane fuses with the plasma membrane in a process known as **exocytosis,** in which the secretory vesicle is allowed to open and expel its contents into the extracellular fluid. Some secretory vesicles are designed so that the exocytotic step occurs very rapidly whenever the plasma membrane is stimulated from the outside by a signal such as a specific hormone.

At the same time that the products of ER synthesis and Golgi complex packaging are conveyed to the exterior of the cell by exocytosis, the membrane of the secretory vesicle becomes part of the plasma membrane. Recall that some proteins have been synthesized in the ER for use as membrane proteins. They have also been packaged in the Golgi complex and form the membranes of the same secretory vesicles that carry the proteins destined for export. In this way, the new membrane proteins and lipids synthesized and trapped within the ER membrane can be transported to the cell surface and inserted into the plasma membrane, by the fusion step, to replace damaged or "worn out" proteins and lipids.

Not all of the membrane lost from the trans face of the Golgi as secretory vesicles is replaced by new membrane arriving at the cis face in the form of transition vesicles from the ER. Some of the replacement membrane is derived from pieces of the plasma membrane that are internalized and delivered to the trans face of the Golgi by processes such as **endocytosis,** which is the reverse of exocytosis (Fig. 3–9). The internalized membrane may be returned to the plasma membrane by a subsequent exocytotic step.

Exchange of membrane in this manner between the Golgi and the plasma membrane conserves the membranes of the transport vesicles, which may be highly specialized. It is economical for the cell to use them for several series of transport events rather than to synthesize new vesicle membranes each time. This **membrane recycling** mechanism can provide the cell with reusable raw materials because some of the internalized membrane is directed to **lysosomes.** There the component proteins and lipids are taken apart and deposited in the cytosolic pool of precursor molecules to await a new round of synthesis. Membrane recycling also helps to avoid an increase in the surface area of secretory cells when large numbers of secretory vesicles fuse with the plasma membrane.

Lysosomes and Peroxisomes: Digestion of Food and Waste

Lysosomes and peroxisomes are small membrane-bounded bags, about 0.5 μm in diameter. Together they account for less than 1% of the total membrane in a cell, but they are found in almost all cells in large numbers. One mammalian liver cell contains about 300 to 400 of each type of organelle.

Lysosomes. The **lysosome** is an organelle that contains about 50 different digestive enzymes. As a group, these enzymes allow cells to completely break down any of the macromolecules in the body (Fig. 3–10a). Since these enzymes would destroy the cells themselves, they are kept within a specific

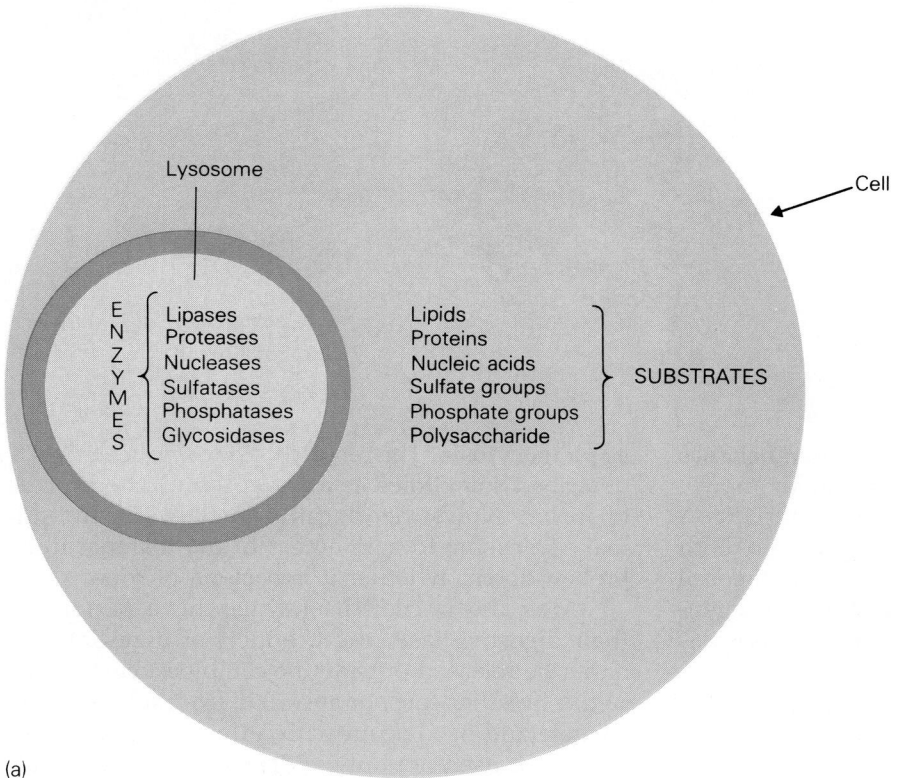

(a)

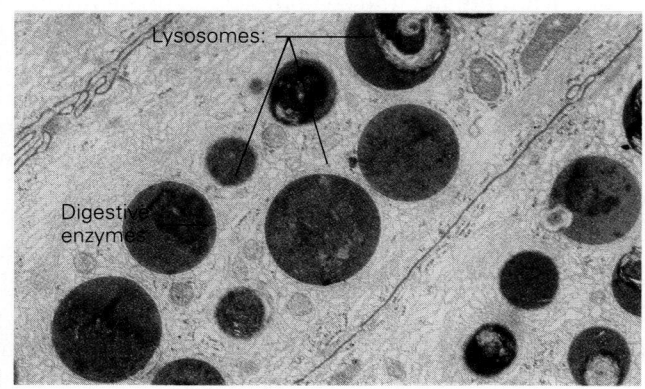

(b)

Figure 3–10
(*a*) Lysosomes protect the cell contents from degradation by keeping within their membranes enzymes that can break down all of the macromolecules in the cell. The only substrates allowed to contact the lysosomal enzymes are those within food vacuoles or other vacuoles containing material meant to be digested. (*b*) A transmission electron micrograph of lysosomes. (© Don W. Fawcett/Visuals Unlimited.)

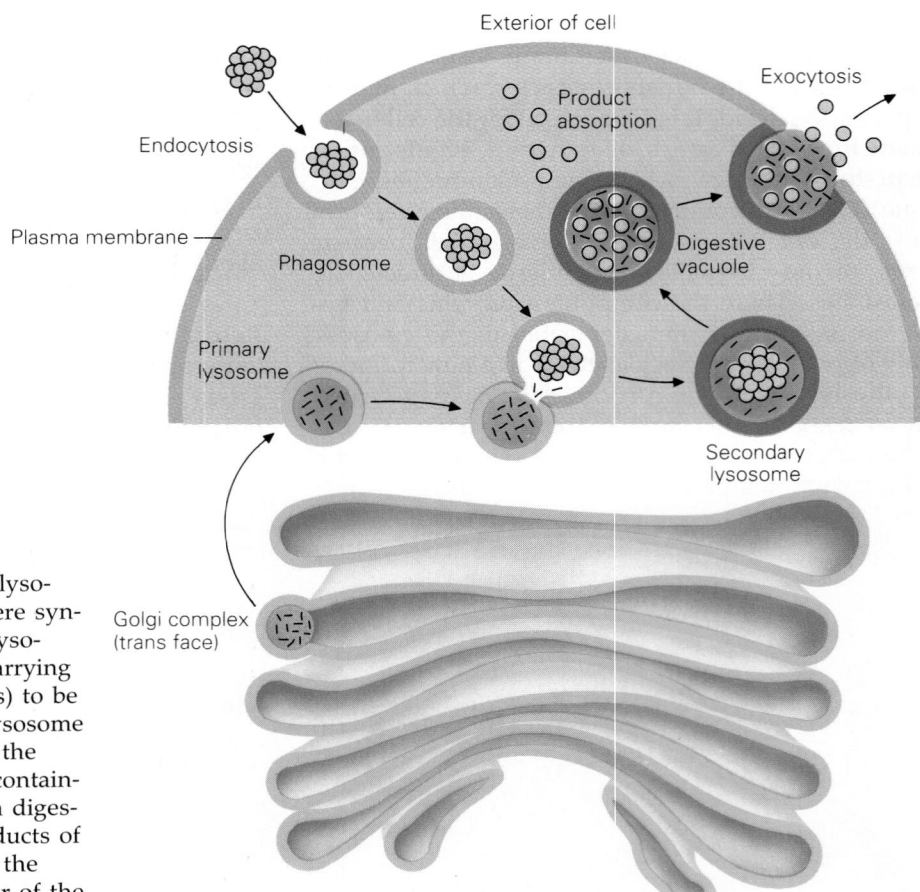

Figure 3–11
Primary lysosomes contain only lysosomal enzymes (proteins that were synthesized in the ER). A primary lysosome fuses with a phagosome carrying
material (e.g., nutrient molecules) to be
digested, creating a secondary lysosome
that contains both enzymes and the
nutrient substrates. The vesicle containing digested material is termed a digestive vacuole; from there the products of
digestion are either absorbed by the
cytosol or expelled to the exterior of the
cell by exocytosis.

membrane-bounded organelle. Two general classes
of lysosomes have been distinguished: newly
formed **primary lysosomes** contain only the digestive enzymes; **secondary lysosomes** contain both
enzymes and molecules (substrates) that the cell
must break down. (Recall that "substrate" is a general term for any material on which an enzyme is
designed to work.)

Primary lysosomes are thought to arise by bud-
ding from the trans face of the Golgi complex
(Fig. 3–11). Most of the lysosomal enzymes they
contain are glycoproteins that originate in the rough
ER. When they enter the Golgi complex, the polysaccharide parts of the enzyme molecules present a
unique signal to the Golgi complex, which responds
by packaging these enzymes into the vesicles that
become primary lysosomes.

There are various types of secondary lysosomes. Among the largest are the **digestive vacuoles** that arise after a primary lysosome fuses with
an endocytotic vesicle containing particulate material such as a bacterium. Internalization of particulate material is a special type of endocytosis known

as **phagocytosis.** The resultant endocytotic vesicle is
described sometimes as a **phagosome.** The process
of fusion allows the digestive enzymes of the primary lysosome to gain access to the material to be
broken down, whether it is bacteria or food.

After the lysosomal enzymes have performed
their digestive task, the products of digestion can
either be released outside the cell by exocytosis, for
re-use by other cells, or absorbed from the digestive
vacuole and used again within the same cell. Lysosomes are used not only to digest material originating from outside the cell but also to break down the
cell's own organelles and other matter in a process
known as **autophagy.** This process can be a response to cell injury, or it can occur during "remodeling" of the cell or during adverse nutritional conditions. The secondary lysosomes formed during
autophagy are called **autophagic vacuoles.**

Peroxisomes. The **peroxisome** also contains specialized enzymes. These enzymes are synthesized in
the cytosol and transported into peroxisomes as
they form, probably as buds from the smooth ER.

The enzyme **catalase** is contained almost exclusively within the peroxisomes. Other enzymes within peroxisomes use oxygen (O_2) to carry out oxidative reactions. One of the products of these reactions is hydrogen peroxide (H_2O_2), which is potentially harmful to the cell. Catalase helps to degrade the hydrogen peroxide to harmless water by the following reaction:

$$2\ H_2O_2 \xrightarrow{\text{Catalase}} 2\ H_2O + O_2$$

The peroxisomes in many body cells are important in detoxifying a number of molecules.

Mitochondria: Energy Transformation

The mitochondrion is thought of as a "powerhouse" for the cell because its metabolic activities provide the energy needed to sustain the life of mammals and other **aerobic** (oxygen-breathing) organisms.

The principal fuels for most life forms are carbohydrates. After digestion, carbohydrates are stored in cells in the form of glycogen. In the cytoplasm, glycogen is broken down into glucose. The primary mechanism for processing these molecules and extracting the energy stored in them is **aerobic respiration**. This metabolic pathway is completed within the mitochondrion by a process that uses most of the oxygen that mammals take in with each breath. During aerobic respiration, about 36 molecules of adenosine triphosphate (ATP) are produced for each molecule of glucose that is broken down. Without the participation of the mitochondrion, only two molecules of ATP are produced. Clearly, the mitochondrion is the major site of ATP production within the cell. ATP is the form in which the cell is able to trap chemical energy and transfer it to sites where it is needed to drive many synthetic reactions and other energy-requiring processes. These processes are described in detail in Chapter 6.

Mitochondria are much smaller than the nucleus. They are usually cylindrical and about 0.5 to 1.0 μm in diameter. The number of mitochondria per cell varies with the cell type but animal cells usually contain large numbers. A liver cell may have as many as 1700 mitochondria, which would account for one fifth of the cell volume and 40% of the total cell membrane.

The mitochondrion, like the nucleus, is bounded by two separate membranes. The **outer mitochondrial membrane** is smooth and unfolded, but the **inner mitochondrial membrane** is distinguished by numerous well-developed infoldings called **cristae,** which project deep into the interior of the organelle and increase its surface area considerably (Fig. 3–12). The two membranes divide the mitochondrion into two compartments, the narrow **intermembrane space** and the much larger **matrix.**

The inner mitochondrial membrane is a highly impermeable structure, unlike the outer membrane. Most of the "work" of the mitochondrion is carried out by enzymes and other molecules located in the inner membrane and the matrix. The matrix compartment also contains ribosomes and 5 to 10 small double-helical DNA molecules that the mitochon-

Figure 3–12
(a) The mitochondrion, the cell's "powerhouse," contains a double membrane. The numerous infoldings of the inner membrane, called cristae, create a large surface area on which the reactions of cellular respiration can take place. (b) An electron micrograph of a mitochondrion from an adrenal cortex cell. (© Don W. Fawcett/Visuals Unlimited.)

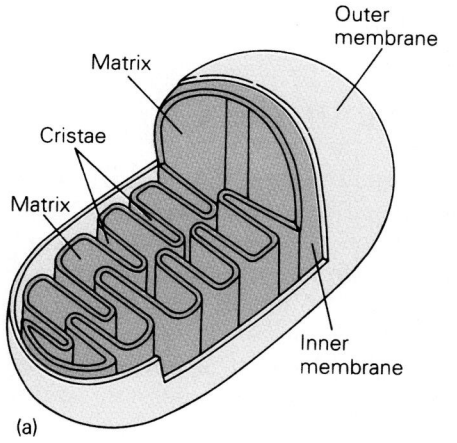

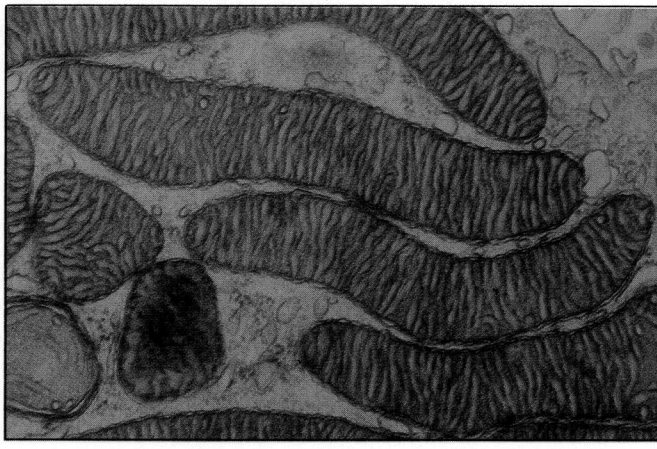

dria use to synthesize some of the mitochondrial proteins. Mitochondrial DNA has a circular structure. In mammalian cells, it makes up less than 1% of the total cellular DNA.

The major component of mitochondria, besides water, is protein. Most (67%) of the protein of liver cell mitochondria is located in the matrix, and about 21% is in the inner membrane. Many of these protein molecules are specific enzymes; about 120 different enzymes are present. Some of the enzymes in the matrix participate in a series of reactions known as the **citric acid cycle,** also referred to as the **Krebs cycle** or the **tricarboxylic acid cycle.** This cycle is the final common pathway for the breakdown of all the fuel molecules in the cell, not only carbohydrates but also amino acids and fatty acids. These processes will be discussed in detail in Chapter 6.

The Cytoskeleton: Shape and Structure

Specialized proteins in the cytosol are organized into the **cytoskeleton** of the cell, which is responsible for maintaining the cell shape and the positions of the internal organelles, and for mediating movement of the cell itself or of the organelles within the cell. Specific examples of organelle movement that have been discussed already in this chapter are (1) the movement of secretory vesicles from the Golgi complex to the plasma membrane, and (2) the movement of endocytotic vesicles or phagosomes from the plasma membrane to the cell interior.

The cytoskeleton is made up of a dense network of different types of protein fibers, the **microtubules, intermediate filaments,** and **microfilaments.** Their structures and sizes are compared in Table 3–2. Microtubules and intermediate filaments are found in deeper regions of the cell, whereas microfilaments generally are located just beneath the plasma membrane.

Microfilaments also extend into cell processes. Good examples are the fingerlike projections called **microvilli,** which cover the inner surfaces of the intestine and the kidney proximal tubule. Figure 3–13 (drawn to scale) gives an indication of the huge size of the cytoskeletal protein fibers relative to the protein molecules present in the membrane of the microvillus.

Microtubules and microfilaments can be described as dynamic structures because their formation is readily (and often) reversed, and because the distribution of these structures within a cell is subject to change as the physiological conditions change. In contrast, the intermediate filaments are much more stable structures, and their formation may be irreversible. Most of the cytoskeletal fibers are made up of large numbers of a single type of protein molecule or subunit (see Table 3–2).

Myosin filaments (also known as **thick filaments**) are a fourth type of fiber found in almost all cell types. They are most abundant in muscle cells where, together with the thin actin filaments, they form part of the contractile machinery of these cells. The presence of **actin** and **myosin** molecules in nonmuscle cells suggests that nonmuscle cells may possess a contractile mechanism such as that of muscle cells and use it for some internal movements.

The organization of the cytoskeleton is complex. It has been suggested that the major fibers are linked by another set of thin filaments, the **microtrabecular lattice.** These filaments appear to hold the fibers and the internal organelles in their places (Fig. 3–14). They even hold ribosomes in suspension. However the microtubules may have the major role in the overall design of the cytoskeleton because they appear to influence the distribution of microfilaments and intermediate filaments. They may provide the framework on which the rest of the cytoskeleton is erected. The molecular mechanisms by which the cytoskeleton can mediate the movement of cells and the movement of organelles within cells have not been determined.

The Life Cycle of the Cell

The many different cell types present in the human body have different patterns and rates of division. Cells such as red blood cells and nerve cells stop dividing when they reach maturity, whereas certain epithelial cells in the intestine and skin divide rapidly and continuously.

The **cell cycle** is the period of time from the beginning of one cell division to the beginning of the next division (Fig. 3–15). Cell cycle times vary from

Table 3–2
Cytoskeletal Fibers

	Diameter (nm)	Structure	Protein Subunit
Microtubules	24	Hollow	Tubulin
Intermediate filaments	10	Hollow	Several types
Microfilaments	6	Solid	Actin
Myosin filaments	15	Solid	Myosin

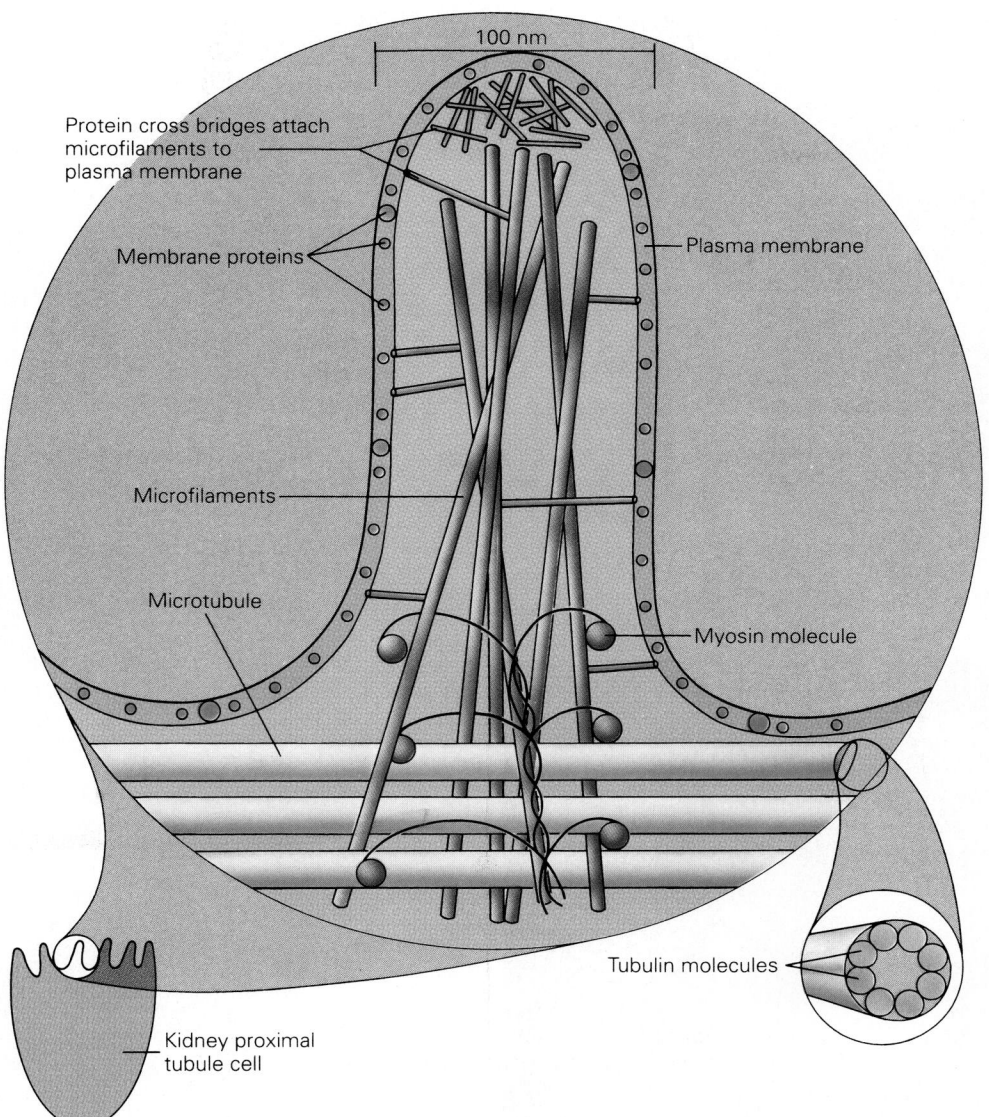

100 nm

Protein cross bridges attach
microfilaments to
plasma membrane

Membrane proteins

Plasma membrane

Microfilaments

Microtubule

Myosin molecule

Tubulin molecules

Kidney proximal
tubule cell

Figure 3–13
Microvilli are found in cells of the digestive system and kidney. The internal structure of a
microvillus, drawn to scale, shows that the microtubules and microfilaments that create its
fingerlike shape are relatively large structures in the cell. They are composed of and connected by protein molecules (e.g., actin, myosin, tubulin) specialized for contraction and
support. For example, a microtubule is formed by a circular collection of tubulin molecules.
Tubulin is a protein that can be detected through the use of fluorescent antibodies (see
Fig. 3–17).

just eight hours for rapidly dividing cells to more
than 100 days for rarely dividing cells.

The cell cycle consists of two main phases: interphase and cell division. **Interphase** is the period between the end of one cell division and the beginning
of the next, during which the cell carries out all the
normal cell processes of growth and metabolism,
including the replication of DNA. **Cell division** is
the period during which the nucleus and then the

cytoplasm divide to form two new cells, each containing one of the copies of the original DNA produced during the preceding interphase.

The process of cell division, sometimes called
the **"M" phase,** involves two steps: mitosis and cytokinesis. In **mitosis,** the nucleus divides into two
new nuclei. In **cytokinesis,** the cytoplasm divides,
each half taking with it one of the two new nuclei to
form a new cell.

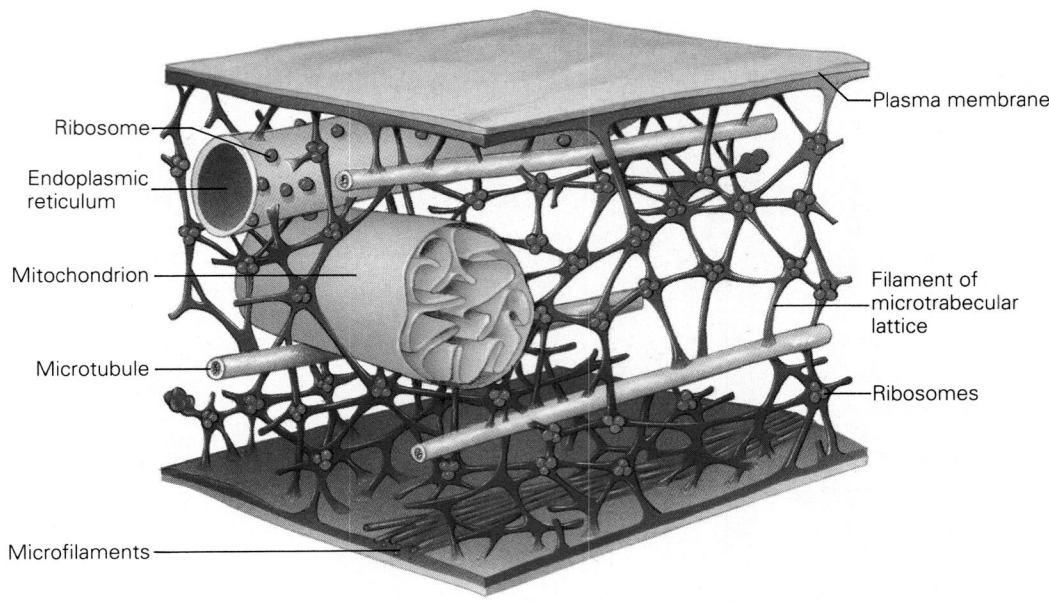

Ribosome

Endoplasmic reticulum

Mitochondrion

Microtubule

Plasma membrane

Filament of microtrabecular lattice

Ribosomes

Microfilaments

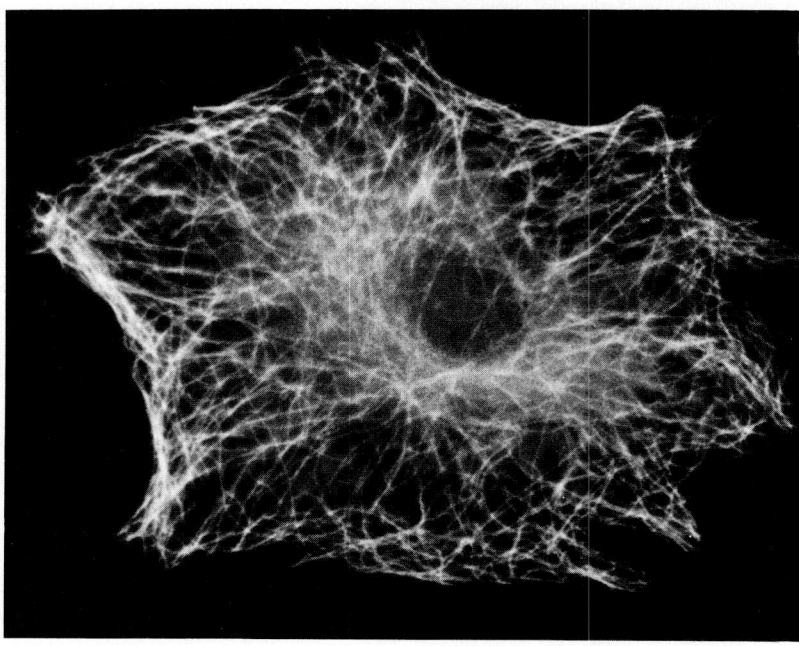

(b)

Figure 3–14
(*a*) Microtubules and filaments are found throughout the cell, not only in microvilli. They may be linked by the microtrabecular lattice. (*b*) Fluorescent antibodies show the position of microtubules in a human cell (see also Fig. 3–17).

Interphase: Growth and DNA Replication

Much of the preparation for cell division occurs during **interphase** (see Fig. 3–15). It occupies 90% or more of the cell cycle. Interphase itself is divided into three phases: the synthesis, or S, phase and two gap, or G, phases.

The period of interphase that directly follows cell division is the **first gap,** or **G_1, phase.** During this period the new cell begins its growth as the organelles it inherited from the parent cell resume their biosynthetic activities (which slowed during the M phase).

The **S,** or **synthesis, phase** follows the first gap phase. This is the part of interphase during which

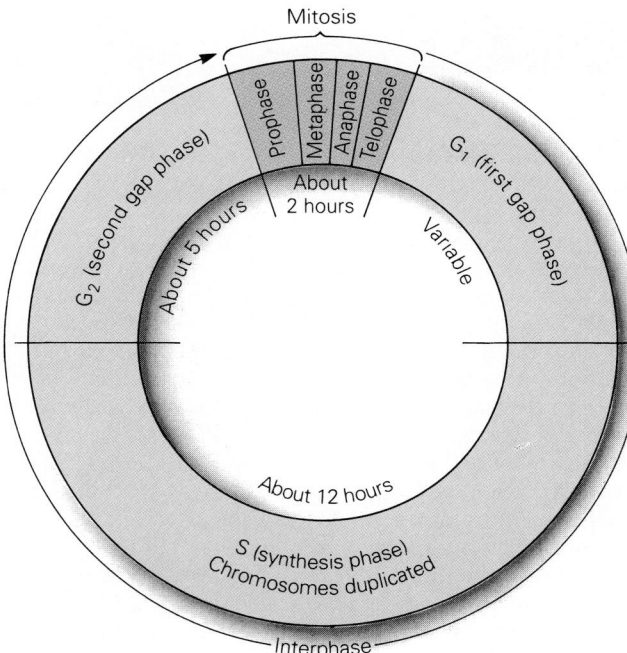

Figure 3–15
The life cycle of the cell contains two phases: interphase and mitosis. During interphase, the cell carries out normal metabolic activities and duplicates DNA. During mitosis, the cell divides into four steps called prophase, metaphase, anaphase, and telophase.

the cell synthesizes DNA in the nucleus in order to duplicate the DNA in preparation for cell division.

Directly following the S phase is the **second gap phase (G_2),** during which the cell increases protein synthesis in final preparations for division.

The time taken to progress from the beginning of the S phase to the end of the M phase is markedly similar in all cell types. The principal difference between cells with short cell cycles and those with long cell cycles is the amount of time spent in the G_1 phase; slowly dividing cells stay in G_1 for days or years.

Cell Division: Mitosis and Cytokinesis

The process of mitosis occurs in four stages: prophase, anaphase, metaphase, and telophase. Each of the stages merges smoothly with those coming before and after. The stages are distinguished mainly by the different movement of the DNA.

Prophase. In the first state of mitosis, called **prophase,** the DNA duplicated during the preceding S phase condenses into chromosomes. The nuclear membrane breaks down, and the contents of the nucleus become mixed with the cytoplasm.

Metaphase. The duplicated chromosomes, not yet separated, line up at the center of the cell and attach to the **mitotic spindle,** which is formed from microtubules.

Anaphase. The third stage begins when the duplicated chromosomes separate. The two sets, each representing a complete copy of the original DNA, are pulled to opposite ends of the cell.

Telophase. During the final stage of mitosis, a new nuclear membrane forms around each set of chromosomes, and they begin to uncoil. **Cytokinesis,** or division of the cytoplasm into two daughter cells, also occurs during this stage. Following cytokinesis, the new cells begin interphase.

Abnormal Cell Cycles

The rate of cell division in normal tissues is controlled so that the cells will divide only when new cells are needed. Removal of part of the liver, for example, will stimulate rapid division and growth of the remaining cells. This new growth ceases when the normal liver mass has been restored. The normal functions of tissues and organ systems would be destroyed rapidly if cell division were uncontrolled.

Cancer cells are the cells of any malignant **neoplasm.** "Neoplasm" means "new growth." The term *malignant neoplasm* is used because the cells divide and multiply rapidly and behave abnormally. In contrast to the highly organized growth of normal cells, cancer cells grow upon one another, invading normal tissues and threatening their functions. Cancer cells do not respond to the normal controls of cell division and growth, and the body is simply a very favorable environment to support their growth.

A cancer cell probably develops when the DNA of a cell experiences **mutation,** or alteration, after exposure to radiation, chemicals, or viruses. When the cell divides, all the cells derived from it contain copies of the altered DNA and inherit the defect. An accumulation of several DNA mutations arising independently from various insults may complete the transformation of a normal cell into a cancer cell.

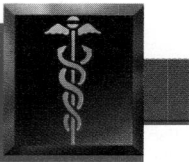

C L I N I C A L F O C U S

Cancer

In the discussion of cell cycles in this chapter, it is pointed out that different cell types have different rates of cell growth and division. Mature nerve cells, for example, grow no further and cell division ceases. In contrast, epithelial cells such as those in the intestine and skin have a very rapid cell cycle. This produces new cells which are needed to replace the continuous loss of older cells.

Cancer cells also grow and divide rapidly but their growth is out of control. A mass of cells growing out of control is known as a **neoplasm,** more commonly referred to as a **tumor.** The study of tumors **(oncology)** has revealed that division and growth of the cells are not controlled effectively by the normal mechanisms. Also, tumor cells range from well to less **differentiated** compared to normal cells. A tumor cell which is less differentiated will be one which has lost some or all of the specialized properties of the normal tissue cells from which it is derived. **Benign** tumors are relatively slow growing and remain in a localized area of the body. They are usually removed easily by surgery.

The term *cancer* is used most often to refer to **malignant,** life-threatening tumors. Malignant neoplasms grow more rapidly than benign growths. They also are capable of spreading rapidly to other parts of the body, a process known as **metastasis.** One way this can occur is by some cells breaking away from the original tumor and traveling via the bloodstream or lymphatic system (Chapter 19) to a distant site where they grow to produce a secondary tumor. Cancer of the breast, for example, frequently spreads to bone, lung, or brain tissues. Normal cells and organs are destroyed by the pressure exerted by an expanding malignant tumor. Direct pressure on adjacent nerves produces pain. Life for the normal cells is made even more difficult because the tumor cells compete for the supply of nutrients and other important molecules.

At the present time, cancer is the second leading cause of death in the United States. It is second only to heart disease. The current "war" on cancer began with the National Cancer Act in 1971. At that time, hopes were high for a quick cure. This view was over-optimistic but was very helpful in obtaining the necessary research funds. Twenty years later, the overall death rate from cancer has not changed, although this statistic obscures the fall in death rates in specific cases such as stomach and uterine cancers in patients under age 65. Major advances have been made, however, by researchers who are trying to understand the origins of cancer, or what turns a normal cell into a malignant one. We now know about **oncogenes,** which transform normal cells into malignant ones and trigger tumor formation, and **tumor-suppressor genes,** which retard growth and division of transformed cells. It is thought now that some cancers arise because mutations activate oncogenes, thus stimulating cell growth and division. These processes continue unchecked because additional mutations inactivate tumor-suppressor genes in the same cell. One of the current goals of cancer research is to learn how to control oncogenes and tumor-suppressor genes. The challenge for the future will be to convert these advances in basic medical research into treatment and prevention of cancer in the clinical setting.

Research Techniques in Cell Physiology

The information about cell structure and function described in the preceding paragraphs has been obtained through a variety of research methods, some as simple as observation through a microscope, others involving the sophisticated analysis of complex chemical reactions. Following is a brief survey of the research methods still being used to elucidate the details of cell physiology.

Cell Cultures

Cells taken out of mammalian tissues will grow and multiply in the laboratory if provided with appropriate nutrients and other factors. Human cells have been used almost routinely for **cell culture** since 1952. Many types of cells can be grown as a single layer attached to glass or plastic surfaces and covered with a liquid **growth medium,** or mixture of nutrients. Under these conditions, the cells are ideally suited for direct study by various techniques of microscopy.

Alternatively, cells can be grown in suspension in liquid media for the purpose of producing large numbers of cells. Special conditions are used to avoid bacterial growth in the liquid medium, and antibiotics are frequently added as an extra precaution. The growth medium contains glucose, various salts, amino acids, and vitamins and is often supplemented with serum derived from the blood of a horse or calf. (**Serum** is the fluid that remains in the blood after the blood cells have clotted together.) The precise composition of serum is always uncertain, but it is an adequate supplement to the growth medium for routine cell culture.

Cells that are transferred directly from the tissues of an animal to a plastic dish containing artificial growth medium are known as *primary* cultures. As they grow and multiply, by dividing to produce daughter cells, they can be **subcultured**—that is, the cells can be removed and dispersed among many new culture dishes to produce a large number of cells.

One purpose of growing cells in culture is to obtain cells that still exhibit the specific functions of the tissue from which they were removed. Ideally, a culture contains only one specific cell type. Thus, cell cultures have an advantage over studies of intact tissues, in which several different cell types, products, and processes are usually at work, complicating the picture. When they contain only the pertinent cells, cultures are extremely useful for studies of unique cellular functions because the cells are growing under carefully controlled conditions and are far more accessible than when part of an intact tissue inside a living animal.

Most normal cells will divide only a limited number of times in culture before dying. Some researchers believe that studies of this built-in "senescence" will provide important clues about the aging process in humans and other animals. Occasionally, however, a few apparently normal cells will grow indefinitely in culture: they are known as a **cell line.** A cell line, and also a primary culture, can be stored frozen in liquid nitrogen ($-200°C$) for long periods of time, even years. When warmed to $37°C$, some of the cells will resume growth and division. This ability is very useful once a suitable group of cultured cells has been obtained. Some of them can be stored frozen for studies in the future, creating an almost inexhaustible supply of cells.

Among their many research purposes, cell cultures are used in investigations of the interactions that allow similar cell types to "recognize" one another and to form the close associations that lead to the development of a specific tissue. When cells are cultured on a plastic surface, they grow and multiply only to the point at which most of the surface is covered by a monolayer of cells. Further growth is inhibited by close contact with other cells (so-called **contact inhibition**), and the monolayer is said to be **confluent.** Interestingly, cancer cells that grow in an uncontrolled way in the body exhibit the same uncontrolled growth pattern in culture. They are much less sensitive to contact inhibition.

Finally, the study in culture of **mutant** cells, abnormal cells that cannot synthesize a specific protein, often reveals a great deal of information about the role of that protein in normal cells. The specific functions that protein normally direct are absent from the cultured mutant cell. The altered behavior of the mutant cell is a strong clue to the function of the protein in normal cells of the same type.

A word of caution: Cells in culture—especially cell lines that have been cultured for a number of years—are not growing in their normal physiological environment. Therefore, despite the usefulness of the technique, the behavior of cells in culture may be different from those in an intact tissue. The observations made about cultured cells must be checked eventually against cells in their normal situation.

Microscopy and Cytochemistry

First used to study cells and tissues in the seventeenth century, the microscope in its modern form is essential in physiological research. A significant advantage of microscopic studies is that the researcher can see the complete cell and its organelles in a relatively undisrupted state.

The Light Microscope

The **light microscope** (Fig. 3–16) is familiar to all students of science. It is extremely useful for examining basic sizes and shapes of cells, and it can be used to detect the presence of nuclei and mitochondria when the cells are stained with dyes that change the color of only these organelles. The tissue of interest is usually **fixed** with a solution that preserves the cell structure and **embedded** in a hard wax to provide support so that thin slices can be prepared for viewing.

Students and researchers are often concerned that these treatments may distort the appearance of the cells. This problem can be avoided by using a different optical system, such as that used in a **phase contrast microscope.** This instrument enhances the contrast between the internal organelles

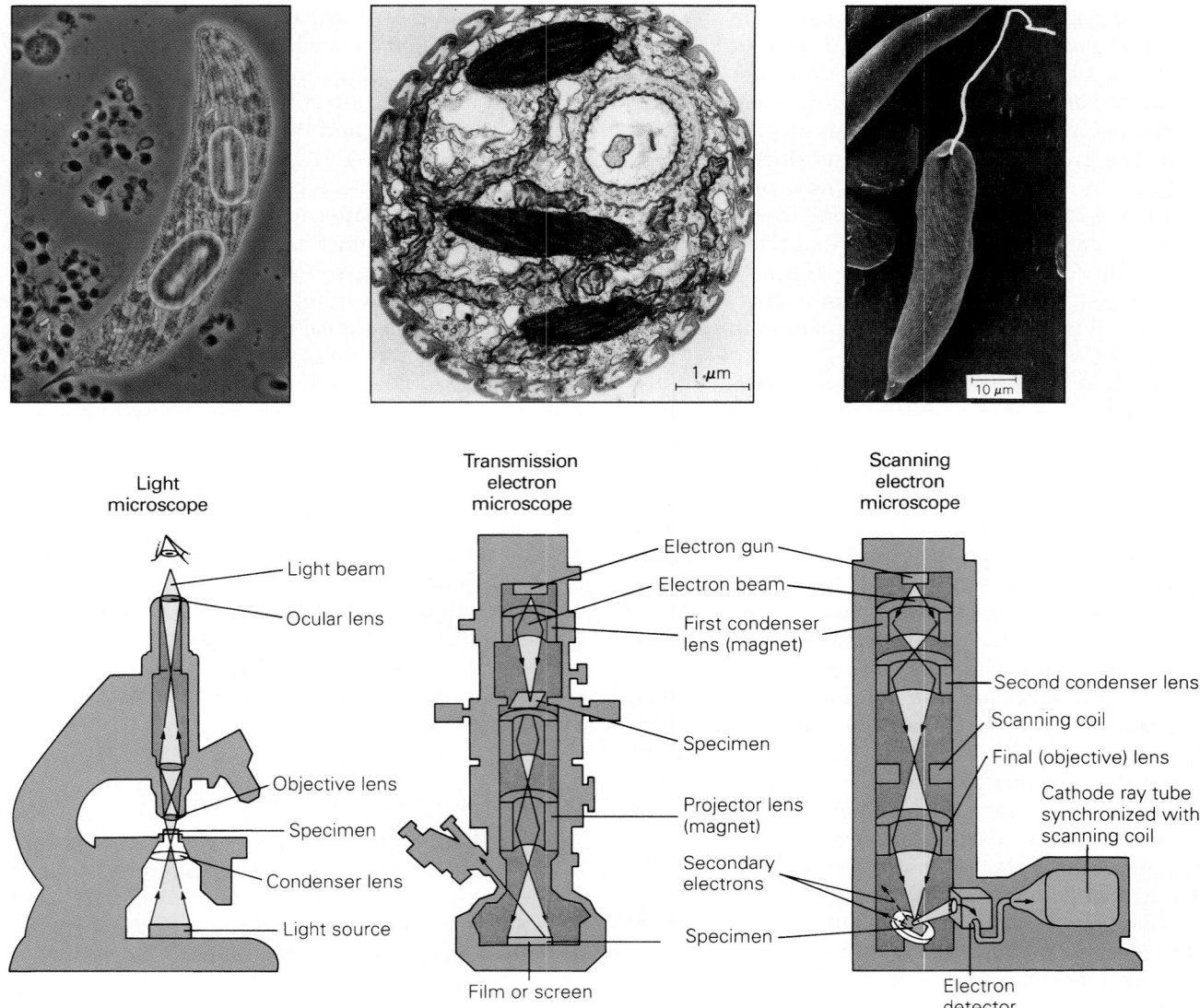

Figure 3–16

Whereas a light microsope focuses a beam of light through curved glass to produce an image, a transmission electron microscope focuses a beam of electrons using an electromagnet. The electrons pass through the specimen to produce an image that can be made visible to the human eye on a screen or photographic film. A scanning electron microscope images the deflection of electrons by the specimen. The photographs show views of the unicellular organism *Euglena.* The light microscope shows the whole cell; the transmission electron microscope shows a thin cross section; the scanning electron microscope shows the cell's outer surface with its pattern of fine ridges.

so that mitochondria, lysosomes, chromosomes, nucleoli, and other structures can be seen quite clearly in living, unstained, and non-fixed cells. Light microscopes are rarely used for magnifications greater than 1000 to 1500 times normal size because beyond this point the **resolution,** or clarity of the image, is poor.

The Electron Microscope

The development of the **transmission electron microscope** (see Fig. 3–16) in the 1940s and 1950s made greater magnifications (more than 100,000 times normal size) possible. The use of this type of microscope has provided a wealth of information about the organization of cells and the structure of their organelles.

In an electron microscope, magnetic lenses focus a beam of electrons in much the same way that glass lenses in ordinary microscopes focus a beam of light (see Fig. 3–16). To be seen by an electron microscope, specimens must be fixed and then stained with solutions that introduce atoms such as lead or uranium into the various structures of the cell. The presence of these large atoms permits fewer electrons to pass through the specimen and gives rise to a dark image of the specimen on a bright fluorescent screen like a television screen. This image can then be recorded on a photographic plate to provide a permanent record.

The **scanning electron microscope** (see Fig. 3–16) is especially useful for examining the surface of a specimen. These microscopes image the deflection of electrons by the specimen. The **high-voltage electron microscope** uses an energy source of one million volts and stands about 32 feet high. It allows much thicker tissue sections to be viewed with high resolution and in considerable depth, producing an almost three-dimensional view of the cell interior. The use of this technique led to the concept of the microtrabecular lattice in the cytosol.

Cytochemical Studies

Cytochemistry is used in combination with microscopy to produce color or enhanced contrast at the sites of specific molecules. Specific stains for DNA and RNA have been used to demonstrate the location of these molecules within the cell. If the molecule being studied is an enzyme, chemicals can be added to sections of a tissue to cause the enzyme to generate an insoluble reaction product. This material will be deposited wherever the enzyme is located. Researchers can identify these sites precisely by looking at the section through a light microscope or electron microscope. They can see the reaction product readily because it is too dense to allow the passage of light or electrons.

Radioactivity in Cell Research

Some isotopes are unstable and tend to break down to a more stable form. Tritium, an isotope of hydrogen, is a specific example. This process is termed **radioactive decay** because it releases high-energy electrons or other particles known as **radiation.** Unstable isotopes that undergo radioactive decay are **radioactive isotopes,** or **radioisotopes.** Although radioisotopes are rare in nature, radioisotopes of phosphorus, iodine, sulfur, carbon, cal-

cium, and hydrogen are readily manufactured for use in cell research.

Since radiation can be detected with extreme sensitivity, the amounts required in the laboratory are very small and well within the accepted safety limits. The use of radioactive molecules (molecules to which a radioisotope has been attached) is a powerful technique that allows researchers to follow the course of almost any process in a cell. A radioactive molecule performs identically to the naturally occurring molecule. Its biochemical properties are unaltered, and the cell reacts as if it were the natural nonradioactive form. However, because it is "labeled" with a radioisotope, any changes in its chemical form or its location can be followed as the molecule is processed within the cell. Researchers worked out the complex metabolic pathways of cells largely with the aid of this technique.

The location of radioactive molecules within cells or tissues is best determined by the technique of **autoradiography.** After incubating the cells with an appropriate radioactive compound to allow its incorporation into the macromolecules of the cells, the researcher washes the cells or slices of tissue to remove unincorporated radioactivity, fixes and embeds them, and cuts them into thin sections for either light or electron microscopy. Each preparation is then overlaid with a film of photographic emulsion and left in the dark. The radioactivity in the preparations reacts with the emulsion in much the same way that photographic film reacts to light. After developing the film, the researcher can determine the position of the radioactivity in the cells from the position of the dark spots on the emulsion when viewed in the microscope. Autoradiography has been very useful in discovering the secretory pathways in cells and in determining the role of the Golgi complex in this process.

The Immune System and Cell Studies

The mammalian **immune system** is an important defense mechanism that recognizes and destroys invading microorganisms and cells not normally present in blood. An important phase of this reaction is the synthesis and release of special proteins, known as **antibodies,** into the bloodstream. A molecule on the surface of the foreign cell triggers their production. A foreign molecule, such as a protein, polysaccharide, or nucleic acid, can elicit a similar response. Any molecule that stimulates the production of antibodies is termed an **antigen.** The antibodies that are released are highly specific molecules that recognize and bind only to the antigen.

This reaction may inactivate the cell bearing the antigen, or, alternatively, the antibody-antigen complex on the surface of the cell may stimulate removal of the cell from the bloodstream by phagocytic cells specifically designed for this purpose.

Immunocytochemistry

Mammals can produce an enormous variety of antibodies. The specificity of these molecules makes them powerful research tools because they can be used to pick out specific molecules within a cell. For example, antibodies to tubulin can be used to study the organization of microtubules in a cell. A researcher labels the antitubulin antibodies by attaching a fluorescent molecule. The antibodies then are allowed to react with thin sections of cells or with intact cells that have been treated to allow them to enter. The antibodies will bind only to microtubules, since only the microtubules contain tubulin. Because the antibodies are fluorescent, they can be traced in a microscope fitted with an ultraviolet light source. Only the fluorescent label will be seen, but it suggests the presence of the antitubulin antibodies, which in turn indicate the microtubules to which they are bound (Fig. 3–17*a* and *b*). Thus, the fluores-

cent sites within the cell indicate the location of microtubules. If the antibodies to an antigenic molecule are labeled with **ferritin,** an iron-protein complex that does not permit the passage of electrons, the location of the antigen within cells can be seen in great detail in the electron microscope.

Radiolabeled Antibodies

In a method similar to using fluorescent markers, a radioactive isotope can be attached to an antibody to give a **radiolabeled** antibody. The radiolabeled antibody can be used in combination with autoradiography to pinpoint the location of specific molecules in the cell. Researchers have used this approach to follow the movement and fate of the plasma membrane that is internalized by cells during endocytosis. In this case, the antibody used is one that binds to molecules that are components of the plasma membrane.

Radioimmunoassays

Another use of radioisotopes and immunochemistry provides the means to measure very small amounts $(1 \times 10^{-12}$ mole) of certain molecules with great ac-

Figure 3–17

In the fluorescent-labeled antibody technique, an antibody to tubulin, which is the protein that makes up microtubules, is injected into a cell. The antibody has been labeled with a fluorescent marker. (*a*) The antibody attaches to the tubulin molecules as they form microtubules throughout the cell. A special microscope allows a view of the fluorescent markers, which indicate the presence of the antibodies, in turn showing the presence of microtubules (*b*).

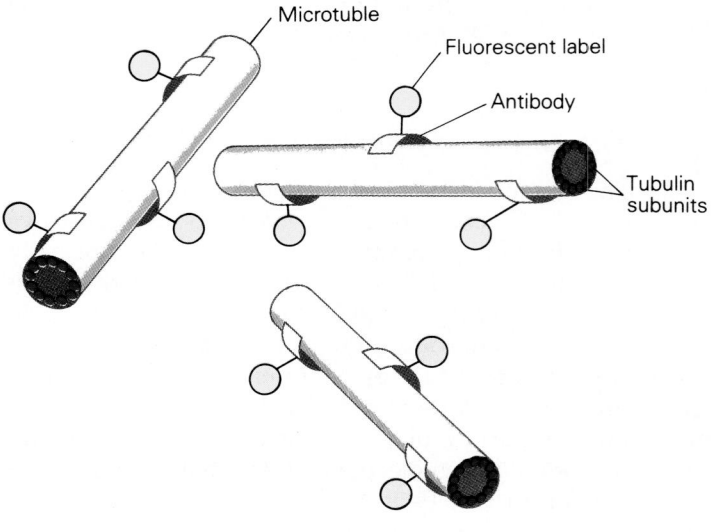

(a)

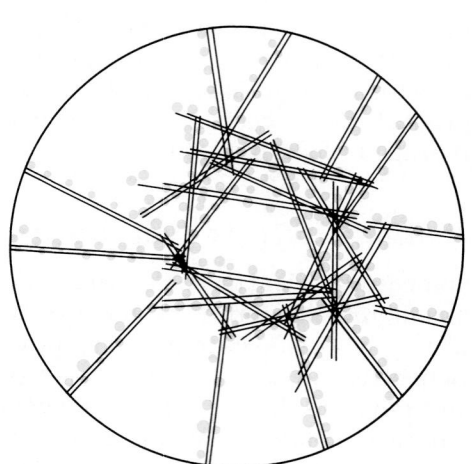

(b)

curacy. The measuring procedure is known as a **radioimmunoassay.** A specific example of its use is the measurement of cyclic AMP in various tissues, in blood, and in urine.

The basic principle of the method is to extract the cyclic AMP and allow it to bind to an anticyclic AMP antibody. A trace amount of radiolabeled cyclic AMP is added to the mixture, which is then set aside for several hours. The unlabeled and labeled cyclic AMP are identical molecules that compete for binding to the antibody. The amount of radiolabeled cyclic AMP present is always the same. The only variable is the amount of unlabeled cyclic AMP in the sample; this determines how much labeled cyclic AMP can be bound to the antibody. When there is very little cyclic AMP in the sample, a lot of **labeled** cyclic AMP can bind to the antibody because only a few unlabeled cyclic AMP molecules will compete for binding. If there is a lot of cyclic AMP in the sample, however, much more will bind to the antibody and only a small amount of labeled cyclic AMP will be able to bind (Fig. 3–18). A researcher can determine the amount of radiolabeled cyclic AMP bound by the antibody by isolating the antibody-antigen complex and measuring the radioactivity present. The amount of radiolabeled cyclic

Figure 3–18
A radioimmunoassay to determine the relative quality of cyclic AMP present in a solution. (*a*) Radioactive (orange) cyclic AMP molecules and antibodies (blue) specific to cyclic AMP are added to a solution containing an unknown amount of (white) cyclic AMP. When the antibody-antigen complex is isolated, the presence of radioactive cyclic AMP is relatively low, revealing that a large amount of cyclic AMP was present in the original solution. (*b*) The same process is followed in another solution, in which the amount of cyclic AMP is also unknown. The same numbers of radioactive molecules and antibodies are used, but the isolated antigen-antibody complex has a high proportion of radioactive cyclic AMP. This indicates that the original solution had a low level of cyclic AMP.

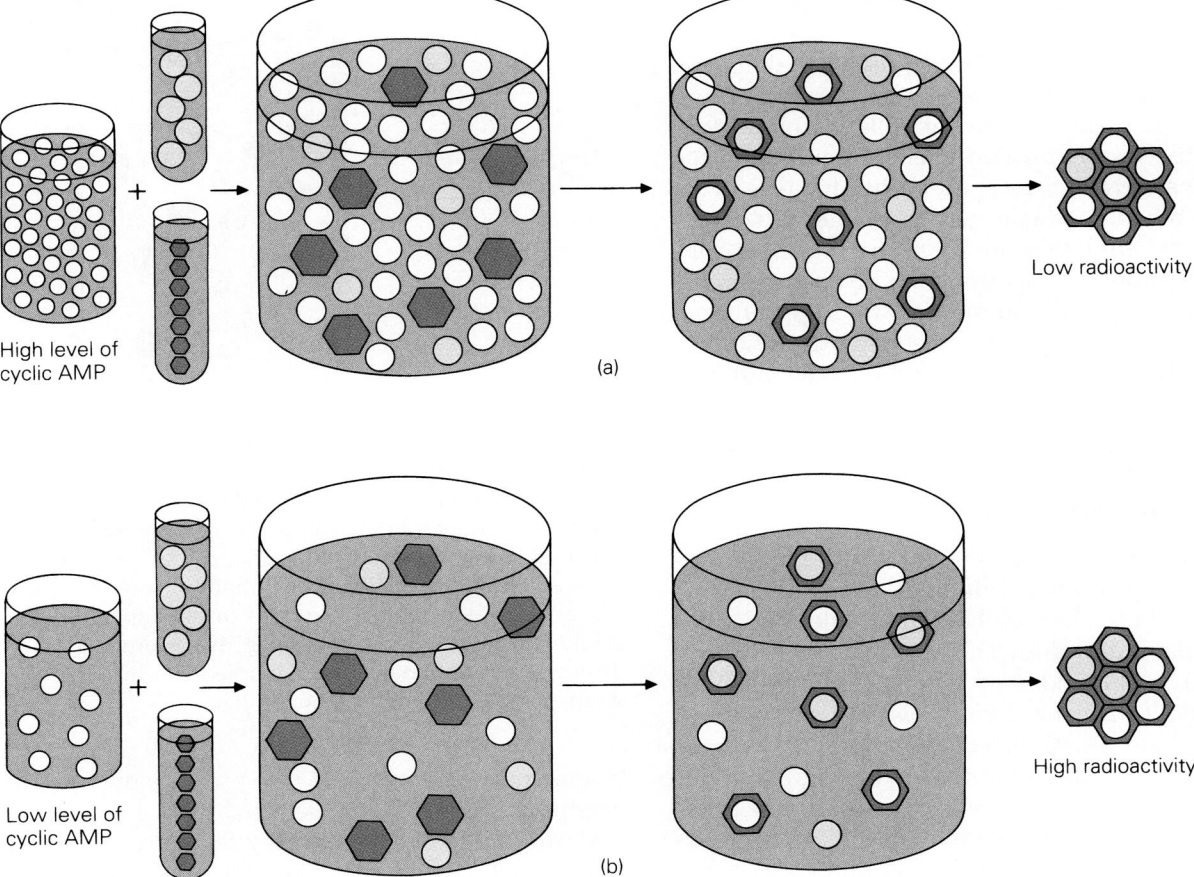

High level of cyclic AMP

(a)

Low radioactivity

Low level of cyclic AMP

(b)

High radioactivity

AMP bound is inversely related to the amount of cyclic AMP present in the sample. The latter is easily calculated.

Affinity Chromatography

Another technique that uses the specificity of antigen-antibody reactions is **affinity chromatography.** Following is an example of how this technique might be used to study a particular antibody. A researcher produces the antibody to be studied by injecting a purified sample of an antigen several times into an animal (such as a goat or a rabbit) in which the antigen is a foreign molecule. When the researcher draws blood from the animal, it now contains the antibodies the researcher wants to use as well as proteins and other antibodies normally present in the blood. The antibody in question must be separated out. The researcher allows the blood sample to clot in order to remove red blood cells. The remaining serum is passed through a column of an insoluble carbohydrate matrix to which an antigen to the antibody is attached. The antibody will bind to the antigen and will be retained on the column while the other serum proteins pass straight through (Fig. 3–19). The antibody can then be washed off the column and recovered separately in a relatively pure form.

A disadvantage of this technique is that antibodies to a single injected antigen are produced by a variety of antibody-producing cells in the animal. Each of these cells produces its own antibody that recognizes and binds to a part of the structure of the antigen, but it is usually not the same part recognized by the antibodies produced by the other cells. Thus, the antibodies specific for a given antigen are not a singular molecular species, and the different antibody molecules vary widely in the efficiency with which they bind to the antigen. The antibody preparation is described as **heterogeneous,** and it has limited use.

Monoclonal Antibodies

The problem of heterogeneity was solved in 1976 by the use of **hybridoma cell lines** to produce large quantities of identical antibody molecules known as **monoclonal antibodies.** The cells that produce antibodies have a limited lifetime outside the body. But researchers can make them grow and multiply indefinitely in a culture dish by joining them to cancer cells, which have an unlimited ability to grow under these conditions. The two joined cells become a single cell that still produces antibodies but now has an unlimited lifetime; the cell line is known as a **hy-**

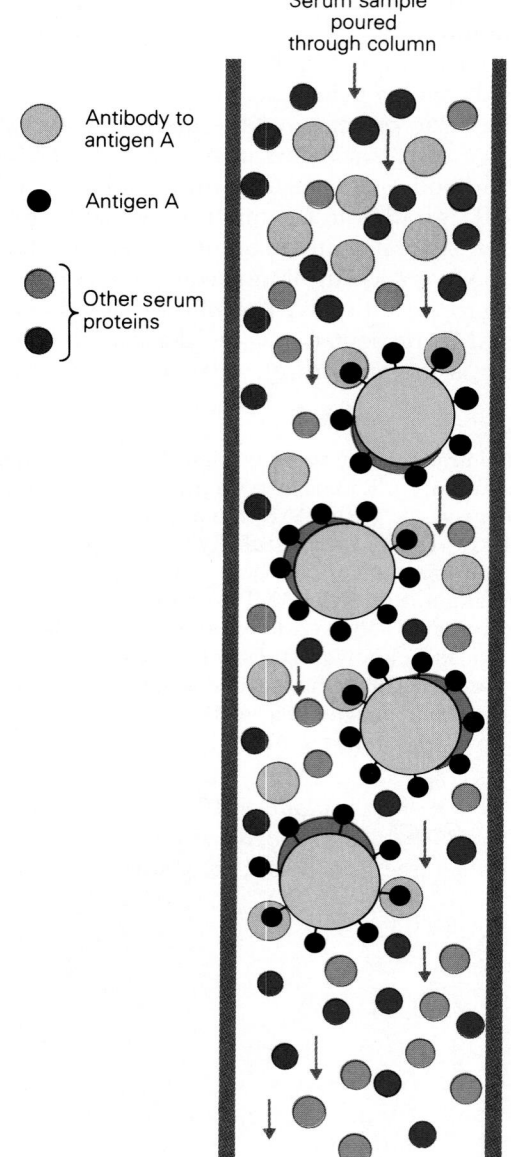

Figure 3–19
Affinity chromatography. Antibodies are produced when an antigen is injected into the blood of an animal such as a goat or a rabbit. To isolate these antibodies, researchers draw a blood sample, collect the serum, and then pour the serum through a column that contains an antigen specific to the antibody being studied. The antibodies being studied bind to the antigen in the column, whereas other proteins in the serum pass through. The antibodies being studied can then be washed from the column and studied.

bridoma. The hybrid cells that produce the specific needed antibody can be selected, and a single hybrid cell is used to start the growth of large numbers of new cells, a procedure known as **cloning.** Since all the new cells were originally derived from the same hybrid cell, the antibody molecules that the new cells produce will be identical.

An additional advantage is that the technique allows a selection step to identify the cell producing the required antibody. This means that researchers do not need to purify the antigen used at the start of the process because cells producing antibodies against the impurities can be eliminated at the selection step. Initial purification of the antigen is a critical step in the conventional procedure for antibody production, as described previously. The high specificity of monoclonal antibodies makes them tools that are as sensitive as radioisotopes for detecting specific molecules in a complex mixture or in a section of tissue.

Amino Acid Sequencing and HPLC

When a protein has been isolated, scientists often wish to determine the exact sequence of amino acids that make it up. Various methods exist for **amino acid sequencing.** Researchers usually begin by breaking the protein into shorter peptide chains, using enzymes that act to split peptide bonds only between specific amino acids. The peptide fragments can then be analyzed by various chromatographic means. One recent advance in this area is the development of **high-performance liquid chromatography (HPLC).** Solutions of amino acids are poured into columns containing densely packed separation substances, which separate amino acids based on size or other factors. In contrast with other column separation techniques, however, HPLC involves the use of great pressures to force the solution through the column in a very short time. HPLC also provides a greater degree of resolution than do other chromatographic techniques.

Fractionation of Cells

The purpose of **cell fractionation** is to separate the different organelles of the cell in order to study each organelle on its own in a strictly controlled, "cell-free" system. This isolation permits direct, detailed analysis of the structure and the function of the organelle.

Fractionation is begun by a physical disruption of the tissue and its cells, a process that can be controlled to minimize damage to the internal organelles. Before being broken up, the tissue usually is placed in a solution of sucrose, which helps to protect organelle structure. A common method of tissue disruption is **homogenization,** in which the tissue and its cells are broken by being forced between

a rotating pestle and the walls of a glass vessel (Fig. 3–20). The clearance between the two surfaces is small enough to break cells but too large to harm most of the internal organelles. Mitochondria, lysosomes, and nuclei remain intact, while large structures such as the endoplasmic reticulum are broken into small pieces, as is the plasma membrane. The resultant thick suspension of broken cells in sucrose is called an **homogenate.**

Once the tissue and its cells have been broken up, centrifugal force is used to separate the organelles. The most widely used technique for separating the cell organelles present in a mixture such as an homogenate is **differential centrifugation.** The homogenate is placed in a special tube that can be spun at high speed in a machine called a **centrifuge.** The centrifugal force generated by these machines can exceed 100,000 times the force of gravity (g). Under the influence of a centrifugal force, the structures present in the homogenate begin to move toward the bottom of the centrifuge tube (see Fig. 3–20). The large and dense structures (e.g., the nuclei) move at the fastest rate and collect as a "pellet" at the bottom of the tube. The centrifugal force can be adjusted so that only the nuclei collect during one centrifugation, whereas the other organelles remain suspended in the solution or "supernatant."

The supernatant is then poured into a new tube, which is spun at a higher centrifugal force. A different group of smaller organelles is collected as a separate pellet. In this way, the homogenate can be divided by centrifugal force into several "fractions," each containing a specific organelle. The total centrifugal force applied to the homogenate depends not only on the force used but also on the time for which it is applied. The sequence of different forces and times that is commonly used to fractionate tissue homogenates is as follows:

Step	Centrifugation	Pellet Contents
1. Homogenate	1000 × g for 10 min	Nuclei, large pieces of plasma membrane
2. Supernatant from 1	10,000 × g for 20 min	Mitochondria, lysosomes, peroxisomes
3. Supernatant from 2	100,000 × g for 1 hr	Endoplasmic reticulum, small pieces of plasma membrane

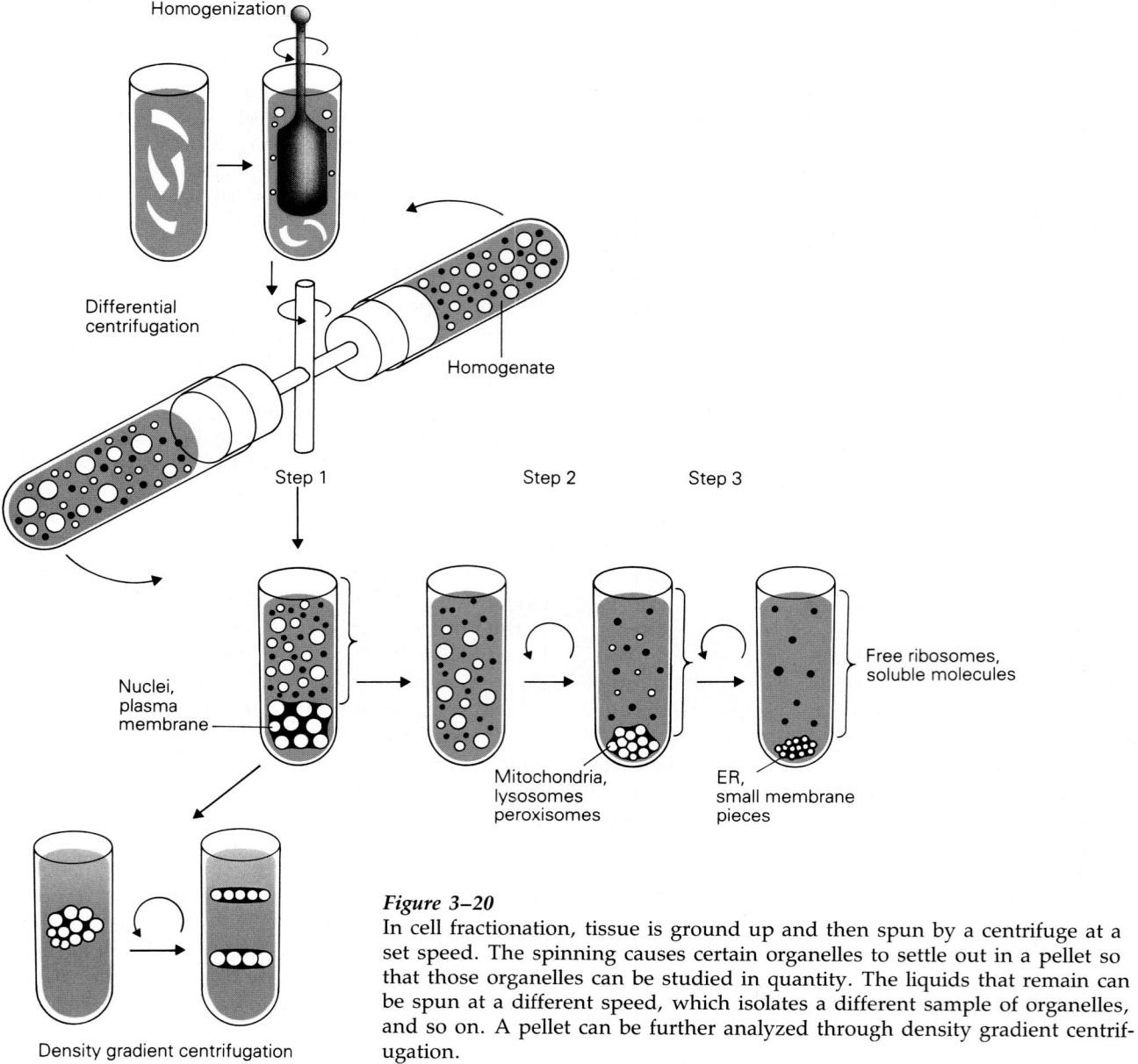

Figure 3–20
In cell fractionation, tissue is ground up and then spun by a centrifuge at a set speed. The spinning causes certain organelles to settle out in a pellet so that those organelles can be studied in quantity. The liquids that remain can be spun at a different speed, which isolates a different sample of organelles, and so on. A pellet can be further analyzed through density gradient centrifugation.

4. Supernatant from 3 contains free ribosomes and soluble molecules.

Although the pellet fractions are enriched in certain organelles, they are not pure because more than one type of organelle is usually present. This is one of the limitations of differential centrifugation. Different organelles in each pellet can be separated by resuspending the pellet material and placing it as a thin layer on top of a sucrose solution that increases in density from the top to the bottom of the tube (see Fig. 3–20). This technique, **density gradient**

centrifugation, allows the separation of organelles that have the same size but different densities. Upon application of a centrifugal force, the organelles with different densities will separate from one another because they will move at different rates through the gradient. At the end of the centrifugation the gradient usually contains two to three separate "zones" or bands of material, each one representing a relatively pure fraction of a specific organelle (see Fig. 3–20). The purity and integrity of a mitochondrial fraction, for example, can be determined by electron microscopy and by checking to see whether enzymes characteristic of lysosomes, peroxisomes, endoplasmic reticulum, and other structures are present.

Isolated fractions of cell organelles are used widely to study the processes that occur within cells, because the process of interest can be isolated and studied without interference from the complex side reactions that occur in the intact cell. The use of a cell fraction that could synthesize polypeptides when provided with mRNA was a crucial step in understanding the mechanism of protein synthesis.

Genetic Engineering

Recall that a **gene** is a section of a DNA molecule that contains all the information needed for the synthesis of a single polypeptide chain. The techniques of genetic engineering allow scientists to form new DNA molecules by combining pieces of DNA from markedly different organisms. The new DNA is known as **recombinant DNA.** The cell that contains this new DNA will use it to synthesize a new protein, one that the cell does not normally produce.

One of the early commercial successes of genetic engineering was the insertion of the gene for human insulin, a polypeptide, into the DNA of the bacterium *Escherichia coli (E. coli)* so that the bacterium synthesized human insulin in addition to its own proteins. Large amounts of insulin now can be produced at relatively low cost when the *E. coli* are grown on a large scale. The method is one of the most powerful tools for understanding how the information stored in chromosomal genes is used and how it is regulated. Here we briefly review some of the basic techniques, terminology, and potential applications of genetic engineering.

The first step in the preparation of recombinant DNA is to break both strands of the native DNA at specific sites, using enzymes known as **restriction endonucleases.** Next, the broken strands of two different DNA molecules are joined together by a different enzyme called **DNA ligase** (Fig. 3–21). The combined use of these enzymes allows any DNA fragment to be grafted into the DNA of a bacterium.

If the inserted DNA fragment represents a single gene, these recombination experiments can pro-

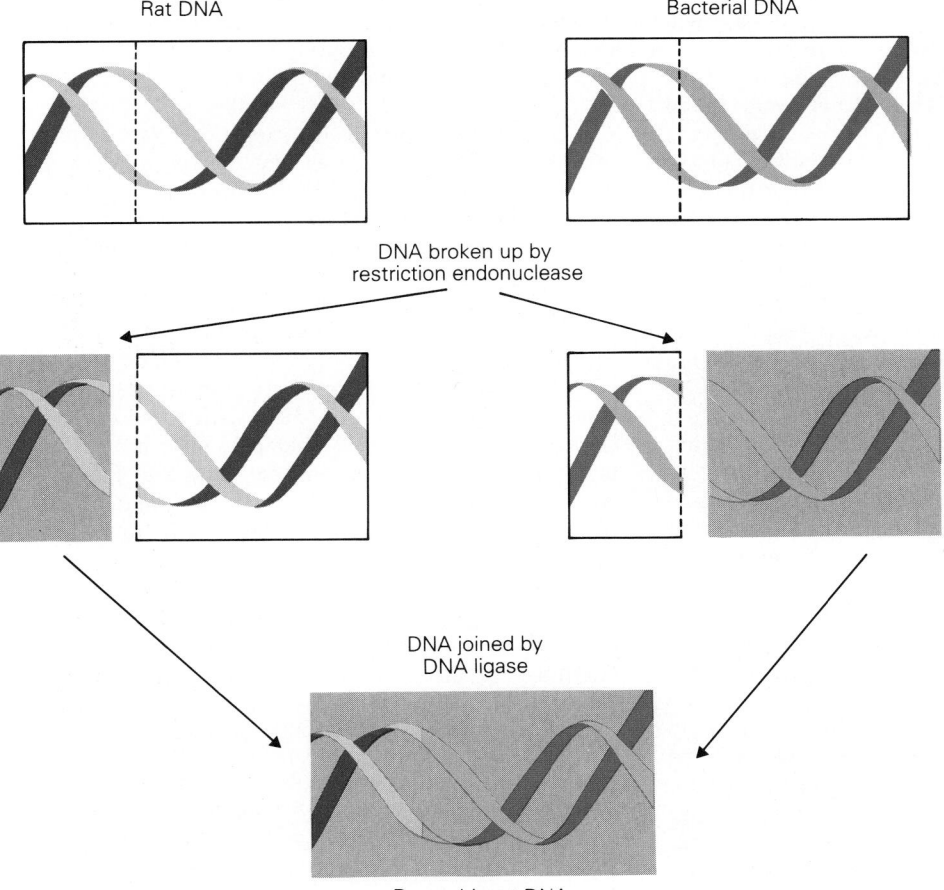

Rat DNA

Bacterial DNA

DNA broken up by restriction endonuclease

DNA joined by DNA ligase

Recombinant DNA

Figure 3–21
Researchers prepare recombinant DNA by splicing together DNA strands from two different organisms.

R E S E A R C H   F O C U S

Gene Therapy

About 1% of all babies are born with an inherited genetic defect. We know of more than 4000 different types of inherited disorders, many of which lead to early death because there is no fully effective treatment. Due to the recent advances in recombinant DNA technology, discussed briefly in this chapter, scientists are on the verge of repairing certain genetic defects by introducing the normal undamaged genes into patients. The first two human gene therapy trials in the USA were approved in September 1990 by the National Institutes of Health (NIH) Recombinant DNA Advisory Committee. This marks the start of a new era for biomedical science, in which genetic engineering will be used to alleviate the suffering caused by genetic diseases and perhaps even serious illnesses such as cancer.

The best candidates for gene therapy are disorders caused by a single damaged gene rather than by multiple genes (recall that every chromosome contains thousands of genes). Ideally the new gene will exactly replace the damaged one and the disease will be cured for life with a single treatment and no side effects. (Random insertion of a new gene anywhere in one of the chromosomes is less ideal because of the risk of activating a cancer-inducing gene, or *oncogene*.) In reality, however, it is extremely difficult to control the fate of DNA introduced into a cell. For this reason, one of the approaches currently being studied is the introduction of a healthy gene into a cell without removal of the defective gene. In this way the healthy gene supplies the missing product of the defective gene and the risk of side effects is reduced. The disadvantage of this approach is that several treatments would be required throughout life, and so it is not a cure. The procedure, known as *gene augmentation*, involves removing cells from the patient, introducing the healthy gene into the cells, and returning the altered cells to the patient. Genetic flaws do not have to be corrected in all of the body's cells, provided that the altered cells can supply enough of the missing product. Bone marrow and skin and liver cells are the best candidates for gene therapy. They withstand removal from the body, can be returned without great difficulty, and survive for a long time after replacement within the body.

An important strategy for introducing a healthy or "therapeutic" gene into the isolated cells uses the native ability of a *virus* to enter cells. The most promising system uses a *retrovirus*, a type of virus that uses RNA as genetic material. Once inside a human cell, the retrovirus converts its RNA to DNA and inserts

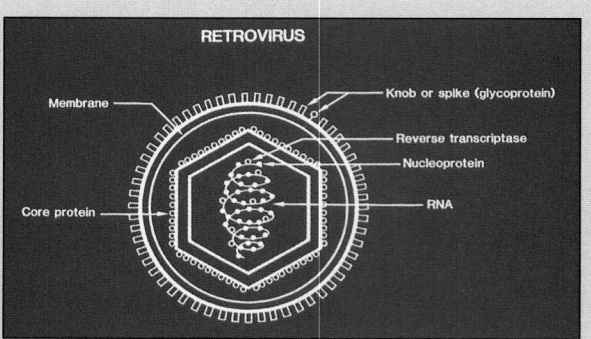

A schematic of the AIDS retrovirus. (© Peter Arnold/Centers for Disease Control.)

the DNA into a chromosome, causing the cell to begin making viral proteins. Many retroviruses have been engineered to serve primarily as gene delivery vehicles. Investigators substitute a therapeutic gene for viral genes and delete the instructions for making new viral proteins. The modified retrovirus still enters cells and inserts the therapeutic gene into cellular DNA, but now it has lost the ability to reproduce.

One of the NIH trials is focused on a rare condition known as *severe combined immuno-deficiency*, which often arises because the function of white blood cells, part of the body's defense against disease, is impaired by a damaged gene. The patients in this study are infused with their own white blood cells, which have been previously removed and modified (using the retrovirus technique) to contain the therapeutic gene. The other NIH trial recognizes that gene therapy can also be used to enhance the ability of cells to fight serious diseases such as cancer. A class of white blood cells that penetrate tumors is being modified to contain a new gene that will increase their anti-cancer activity. Their therapeutic potential will be evaluated in patients with advanced cancer.

Gene therapy is an exciting development but is not without problems, both technical and ethical. There is dramatic success in some patients while others show no improvement. The technique will require extensive research in a high technology environment for several more years. Intentional modification of human genes also is disconcerting for many people. However, the consensus among scientists is that human gene therapy can be beneficial if carefully scrutinized and is worth pursuing if it proves successful in combating human disease.

duce large amounts of the gene because it is copied and reproduced as the bacteria rapidly divide and multiply. This process is known as **DNA cloning.** Researchers recover the gene from the bacterial DNA using the same restriction endonuclease used to break open the original DNA. The amount of recombinant DNA that is now available is sufficient for detailed studies of the precise sequence of nucleotides.

The bacterial DNA used for DNA cloning is not the chromosomal DNA; rather, it is a small accessory DNA molecule known as a **plasmid.** Researchers use these molecules because they are easily separated from the chromosomal DNA and, once the gene of interest has been attached to them, they can be put back inside living bacterial cells (Fig. 3–22). Plasmids that are used in DNA cloning studies are often referred to as **cloning vectors.**

Once the nucleotide sequence of a specific gene has been identified, it is possible to check the accuracy of this analysis by chemically synthesizing the gene from its constituent nucleotides. The gene can be incorporated into a cloning vector and inserted into bacteria for cloning and to see if the bacteria produce a functional protein from the synthetic gene.

The ability to use synthetic genes broadens the scope of these techniques enormously. For instance, after sequencing a gene that produces a functional protein, researchers can synthesize one that is slightly different, a **mutant** gene, or one that makes a nonfunctional protein. If this mutant gene is inserted into a mammalian cell to replace a normal gene, the cell then synthesizes a nonfunctional protein instead of a functional one. Changes in the behavior of cells that contain the mutant gene provide information about the biological role of the normal protein within the cell.

At present it is extremely difficult to introduce genes into mammalian DNA, but ultimately it may be possible to genetically repair a mammalian cell that is nonfunctional because of the presence of a mutant gene. In this case, a chemically synthesized fragment of DNA that contains the normal gene would be inserted into the DNA of the cell.

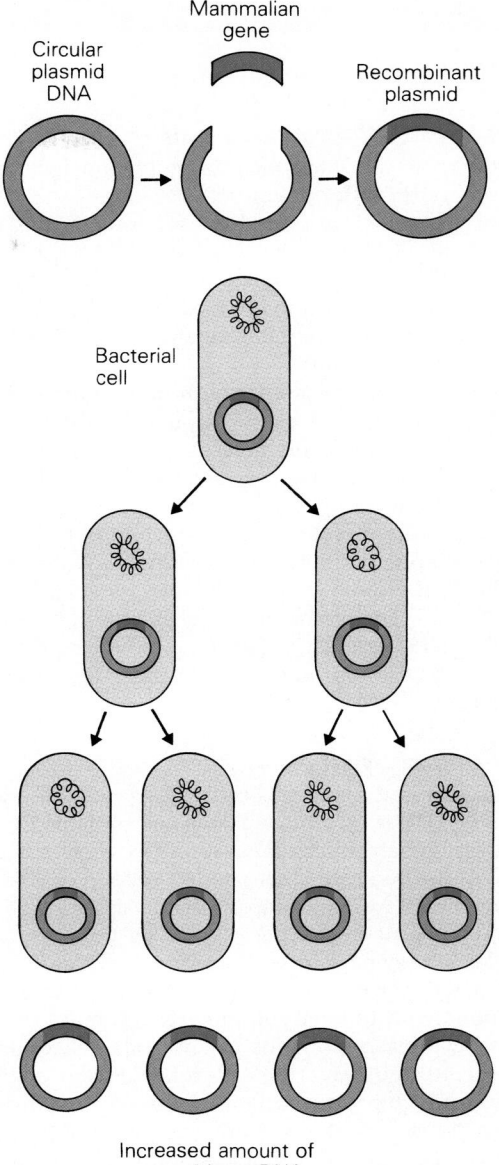

Figure 3–22
Researchers can produce large quantities of recombinant DNA by placing them in bacterial cells, which duplicate the recombinant molecule each time they divide.

SUMMARY

Overview: Cell Processes and Products

Cell functions are controlled from the nucleus, where DNA stores the information the cell uses to synthesize proteins. Proteins are important macromolecules in the cell because most are enzymes that catalyze specific cell reactions. Others are secretory proteins that enable the cell to perform its specific function in the body. The nucleus is also the site where information is duplicated to be passed on to new cells.

The cell must obtain energy from food to perform its various functions. It does so through a system of food vacuoles, lysosomes, and digestive vacuoles. The products of digestion are used by the mitochondrion to trap energy in the form of ATP.

The plasma membrane that surrounds the cell protects it from fluctuations in the external environment but also allows the movement of specific substances in and out of the cell. The membranes that surround the individual cell organelles ensure that their activities proceed independently, without endangering other organelles and activities.

Cell Compartments and Their Functions

The nucleus is the site where nucleic acids are synthesized. The DNA in the nucleus of a cell contains the same permanent genetic information found in every other non-sex cell in the body. RNA is a coded version of DNA that contains the information stored in DNA in a form that can be used by protein-making organelles outside the nucleus. Molecules pass in and out of the nucleus through pores in the nuclear envelope. The outer membrane is continuous with the endoplasmic reticulum.

The cytoplasm contains the various cell organelles distributed throughout the cytosol. Cytosol, a viscous fluid with a high protein content, is the site of many metabolic reactions. It contains many enzymes and precursors for macromolecules. Proteins are synthesized both on ribosomes found free in the cytosol and on those attached to the ER. Proteins synthesized on the rough ER become glycoproteins that are destined to be membrane components or secretory products. Lipids are synthesized within the ER membrane. Products of ER synthesis are shuttled to the Golgi complex and then to their final destination, whether outside the cell, in cell organelles, or cell membranes. Lysosomes and peroxisomes contain specialized enzymes that digest food and waste material in the cell. Mitochondria, bounded by double membranes, use much of the oxygen inhaled with each breath to produce ATP from the products of digestion. The cytoskeleton, consisting of three types of protein fibers organized into an extensive microtrabecular lattice, gives shape and support to the cell and its organelles.

The Life Cycle of the Cell

The life cycle of the cell is divided into two main phases, interphase (the growth phase) and cell division (mitosis). In interphase, which occupies most of the cell cycle, the cell conducts normal activities such as protein synthesis and replicates its DNA.

During mitosis, the nucleus and cytoplasm divide in four phases, with each of the two new daughter cells taking an exact copy of the original DNA.

Cancerous cells divide abnormally—that is, very rapidly—creating neoplasms. They do not respond to the normal controls of cell division and growth.

Research Techniques in Cell Physiology

Research techniques in cell physiology include cell cultures, in which cells are grown artificially in the laboratory in dishes or in suspension in a liquid growth medium. Cells grown in culture are very accessible and allow unique studies to be carried out.

Light and electron microscopes can be used in conjunction with staining to reveal important structures and functions of the cell.

Radioactive isotopes can be used to great advantage in cell research because they can be added to normal cell macromolecules without affecting their behavior, yet are easily detected. Radioactive molecules can be used to follow the course of almost any metabolic activity in the cell.

Researchers make great use of the immune system to study various proteins in the cell. Immunocytochemical techniques involve the use of fluorescent antibodies to illuminate cell structures, such as microtubules, made of proteins. Antibodies can be labeled with radioactive markers for easier detection, and various chromatographic techniques can also be used to isolate samples of antibodies for study. The precise specificity of monoclonal antibodies provides a powerful technique for locating specific molecules within the cell.

Once an antibody is isolated, its amino acid sequence can be determined through a variety of methods, including high-performance liquid chromatography.

To isolate cell organelles for concentrated study, researchers break up tissue samples and spin them in a centrifuge so that cell organelles of different weights collect in a series of pellets. Each organelle can then be studied on its own in a controlled, cell-free system.

In genetic engineering, scientists create recombinant DNA by splicing together portions of DNA molecules from different organisms and then injecting the new DNA into bacteria, which replicate the DNA as they divide. These techniques allow scientists to study the functions of chromosomes—that is, how they store genetic information and how the cell uses this information.

REVIEW QUESTIONS

Identify Each with Term

1. The fluid surrounding the nucleus of a cell.
2. The complex formed between DNA and nuclear proteins such as histones.
3. The process by which secretory vesicles fuse with the plasma membrane and expel their contents outside the cell.
4. The structure which maintains cell shape and mediates cell movements.
5. Division of the cell cytoplasm to form two daughter cells after completion of mitosis.

Define Each Term

6. Eukaryotic cell
7. Genome
8. Autoradiography
9. Density gradient centrifugation
10. Recombinant DNA

Choose the Correct Answer

11. The site of protein synthesis is the:
    a. Nucleus
    b. Lysosome
    c. Nucleolus
    d. Rough endoplasmic reticulum
12. Which of the following organelles is surrounded by a double membrane?
    a. Golgi complex
    b. Nucleus
    c. Peroxisome
    d. Smooth endoplasmic reticulum
13. Intermediate filaments are structural components of the:
    a. Cytoskeleton
    b. Mitochondrion
    c. Ribosome
    d. Chromosome
14. Monoclonal antibodies are produced by:
    a. Cytochemistry
    b. Hybridoma cell lines
    c. Radioimmunoassay
    d. Cloning vectors
15. The growth phase of the cell cycle is known as:
    a. Mitosis
    b. Metaphase
    c. Interphase
    d. M phase
16. The endoplasmic reticulum of a liver cell accounts for what % of the total membrane of the cell?
    a. 2–5
    b. 10–20
    c. more than 50
    d. 100

Answer Each Question in One or Two Sentences

17. List 3 advantages conferred by the presence of intracellular compartments.
18. Explain how 1 meter of DNA can fit inside the nucleus of a typical animal cell.
19. What are the functions of the three different types of RNA?
20. Why are mitochondria important for aerobic respiration?
21. How do cells from a primary culture differ from cells from a cell line?

SUGGESTED READINGS

Alberts, B., Bray, D., Lewis, J., Raff, M., Roberts, K., and Watson, J. D. *Molecular Biology of the Cell* (2nd edition). New York and London: Garland Publishing Inc., 1989.

Avers, C. J. *Molecular Cell Biology*. Reading, MA: Addison-Wesley, 1986.

Bainton, D. "The discovery of lysosomes." *J. Cell Biology*, 91:66s–76s, 1981.

De Duve, C. "Exploring cells with a centrifuge." *Science*, 189:186–194, 1975.

De Duve, C. "Microbodies in the living cell." *Scientific American*, 248 (May 1983):74–84.

Everhart, T. E., and Hayes, T. L. "The scanning electron microscope." *Scientific American*, 226:54–69, 1972.

Farquhar, M., and Palade, G. "The Golgi apparatus (complex) from artifact to center stage." *J. Cell Biology*, 91:77s–103s, 1981.

Fulton, A. B. "How crowded is the cytoplasm?" *Cell*, 30:345–347, 1980.

Johnson, R. S. "Fingers-crossed celebration of gene therapy." *FASEB Journal*, 4:3063–3064, 1990.

Palade, G. "Intracellular aspects of the process of protein synthesis." *Science*, 189:347–358, 1975.

Porter, K. R., and Tucker, J. B. "The ground substance of the living cell." *Scientific American,* 244:56–67 (March 1981).

Rosenberg, S. A. "Adoptive immunotherapy for cancer." *Scientific American,* 262:62–69 (May 1990).

Verma, I. M. "Gene therapy." *Scientific American,* 263:68–84 (November 1990).

Watson, J. D., Tooze, J., and Kurtz, D. T. *Recombinant DNA: A Short Course.* New York: W. H. Freeman & Co., 1983.

KEY TERMS

antibody (p. 111)
antigen (p. 111)
chromatin (p. 95)
chromosome (p. 95)
cytokinesis (p. 105)
cytoplasm (p. 96)
cytoskeleton (p. 96)
cytosol (p. 96)

deoxyribonucleic acid (DNA) (p. 89)
endoplasmic reticulum (p. 93)
exocytosis (p. 100)
gene (p. 93)
genome (p. 93)
Golgi complex (p. 89)

hormone (p. 89)
lysosome (p. 92)
metabolism (p. 89)
mitochondrion (p. 92)
mitosis (p. 105)
nucleolus (p. 95)
nucleus (p. 88)
organelle (p. 88)

phagocytosis (p. 102)
plasma membrane (p. 88)
prokaryotic cell (p. 88)
recombinant DNA (p. 117)
ribonucleic acid (RNA) (p. 89)

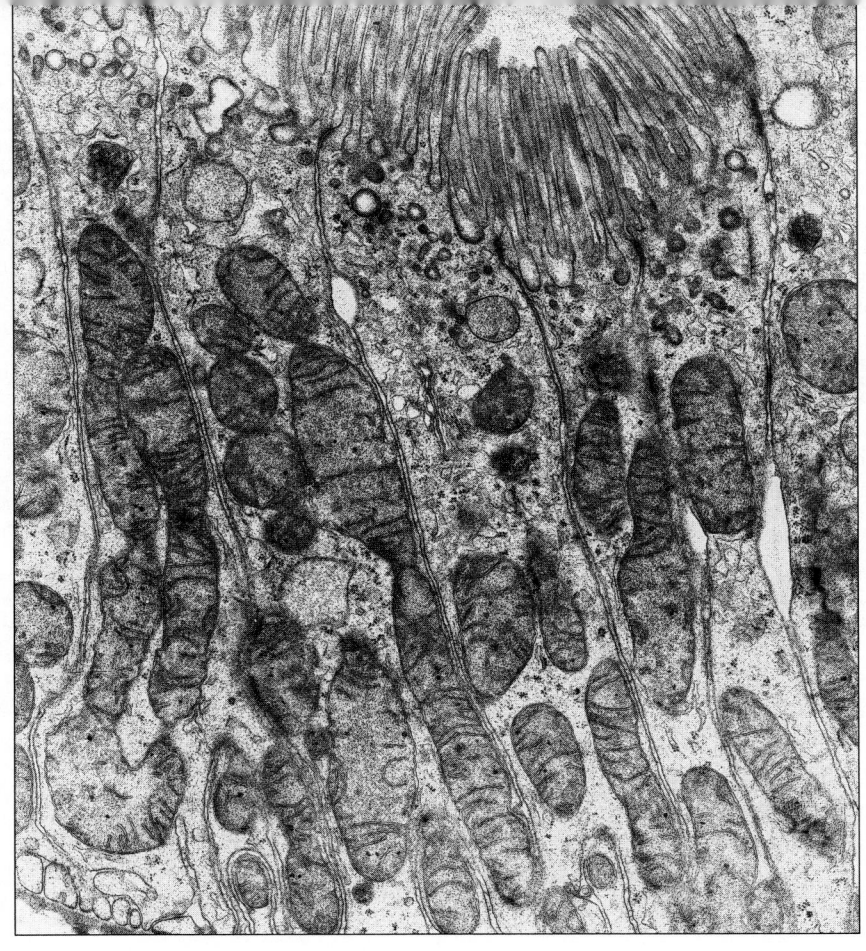

Transport Through the Cell Membrane

CHAPTER OUTLINE

In this chapter we will consider first the structure and organization of the main components of plasma membranes. Later sections will discuss the mechanisms by which water, small ions, and solutes can move across membranes and epithelial cells. Although this chapter focuses primarily on the membrane that surrounds each cell, keep in mind that the membranes surrounding the various cell organelles are similar to the plasma membrane in composition and behavior.

Overview of Plasma Membrane Functions

Every cell is surrounded by a plasma membrane that separates the contents of the cell from the environment. The plasma membrane is a highly selective filter that permits nutrients to enter and waste products to leave the cell. This selective nature allows the composition of the cell cytosol to be very different from its surrounding environment. For example, the concentration of potassium ions inside most cells is about 25 to 30 times higher than the concentration of potassium outside, in the extracellular fluid. Other important differences must be maintained between the two fluid compartments (Fig. 4–1). In addition to this function as a selective barrier, the plasma membrane also has many sites on its external surface which bind specific chemical signals produced by other cells. Thus, the plasma membrane has an important role in cell to cell communications.

The Components and Structure of the Membrane

The following observations on the behavior of the major membrane components have been integrated into a generalized model of the structure of cell plasma membranes (Fig. 4–2). It is often referred to

Figure 4–1
A simplified animal cell and its plasma membrane, which separates the intracellular fluid (the cytosol) from the extracellular fluid.

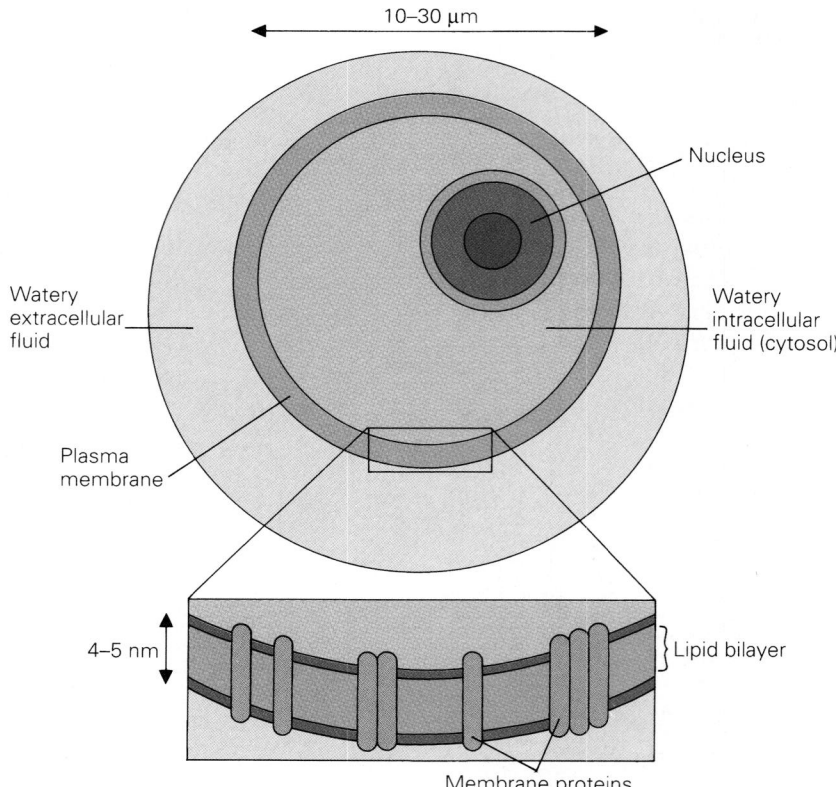

10–30 μm

Nucleus

Watery extracellular fluid

Watery intracellular fluid (cytosol)

Plasma membrane

4–5 nm

Lipid bilayer

Membrane proteins

The current fluid-mosaic model of the structure of plasma membranes.

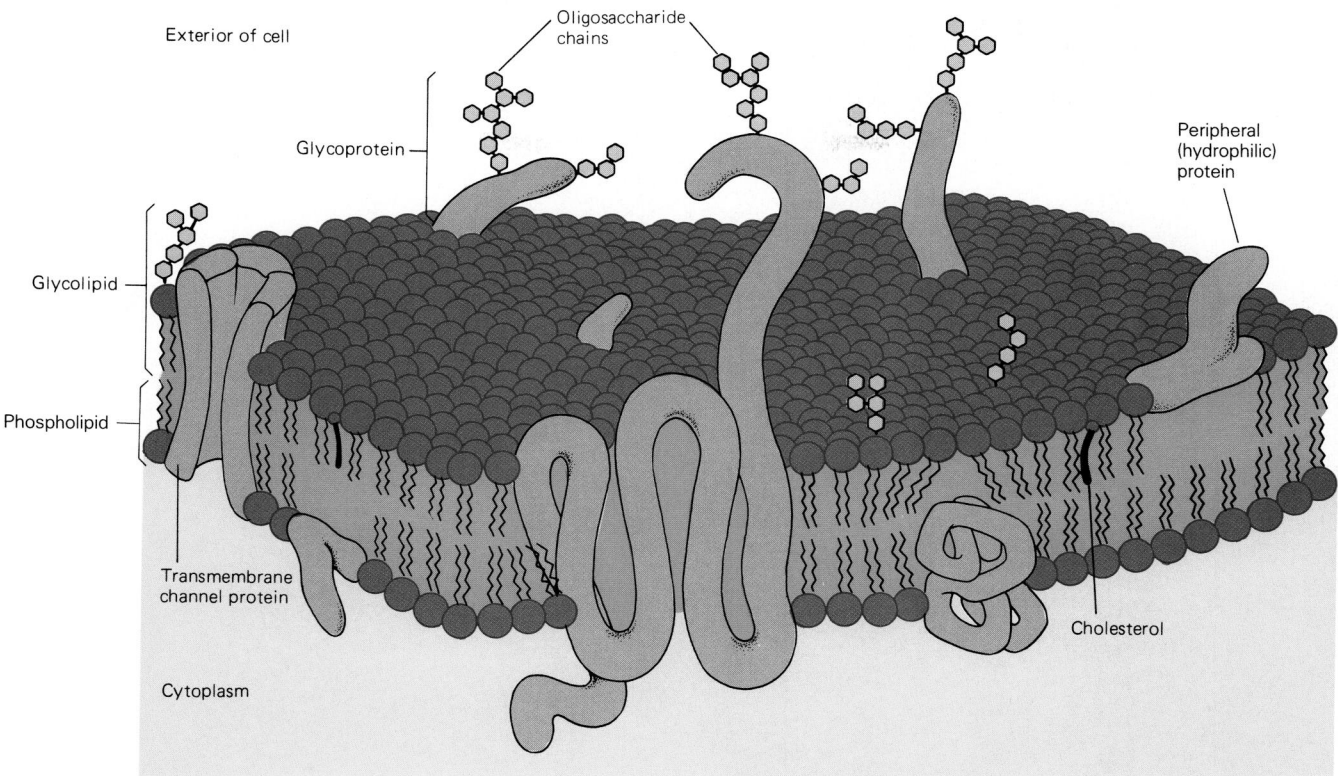

Figure 4–2
The current fluid-mosaic model of the structure of plasma membranes.

as the **fluid mosaic model** or the **Singer-Nicolson model,** the latter being the names of the scientists who first put forward the concept. Although Figure 4–2 is designed to represent the structure of plasma membranes, the structure of intracellular membranes is very similar.

The plasma membrane consists primarily of lipids and proteins. The core structure of the plasma membrane is a double layer of phospholipid molecules known as the **lipid bilayer** (see Fig. 4–1). Lipids can account for as much as 50% of the mass of a plasma membrane.

Many of the functional properties of the membrane, however, are due to the presence of specific proteins that are closely associated with, and often embedded in, the lipid bilayer. Typically, proteins also account for about 50% of the mass of cell membranes, although this figure can vary from 18% in the membrane of the Schwann cell (a supporting cell in the nervous system) to 76% in the inner membrane of a mitochondrion (Fig. 4–3).

The relative quantities of lipid and protein in a plasma membrane usually reflect some aspect of the function of the cell or organelle that the membrane protects. Thus, the high lipid content of the Schwann cell membrane suits the role of this membrane as an insulator around certain nerve fibers. Conversely, the high protein content of the inner

Figure 4–3
The proportions of membrane protein and lipid vary as the function of the membrane changes from primarily an insulator (myelin) to a complex variety of enzymatic and transport functions (mitochondrial inner membrane).

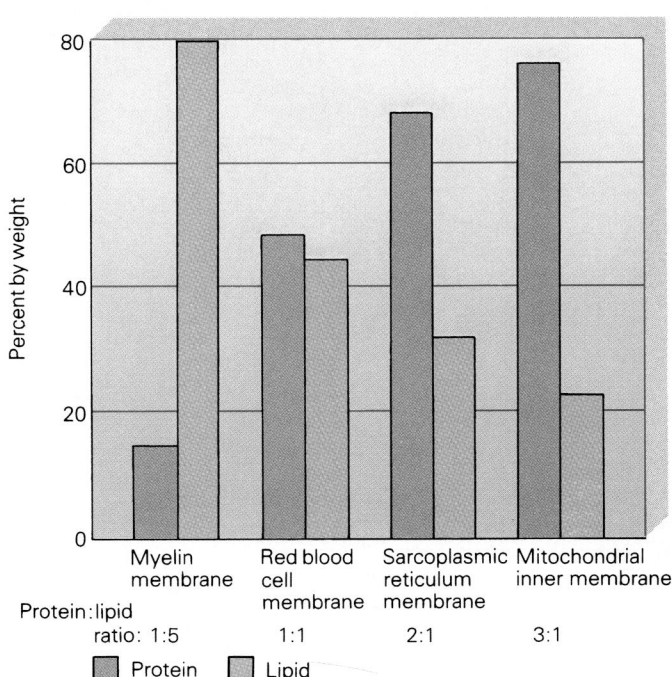

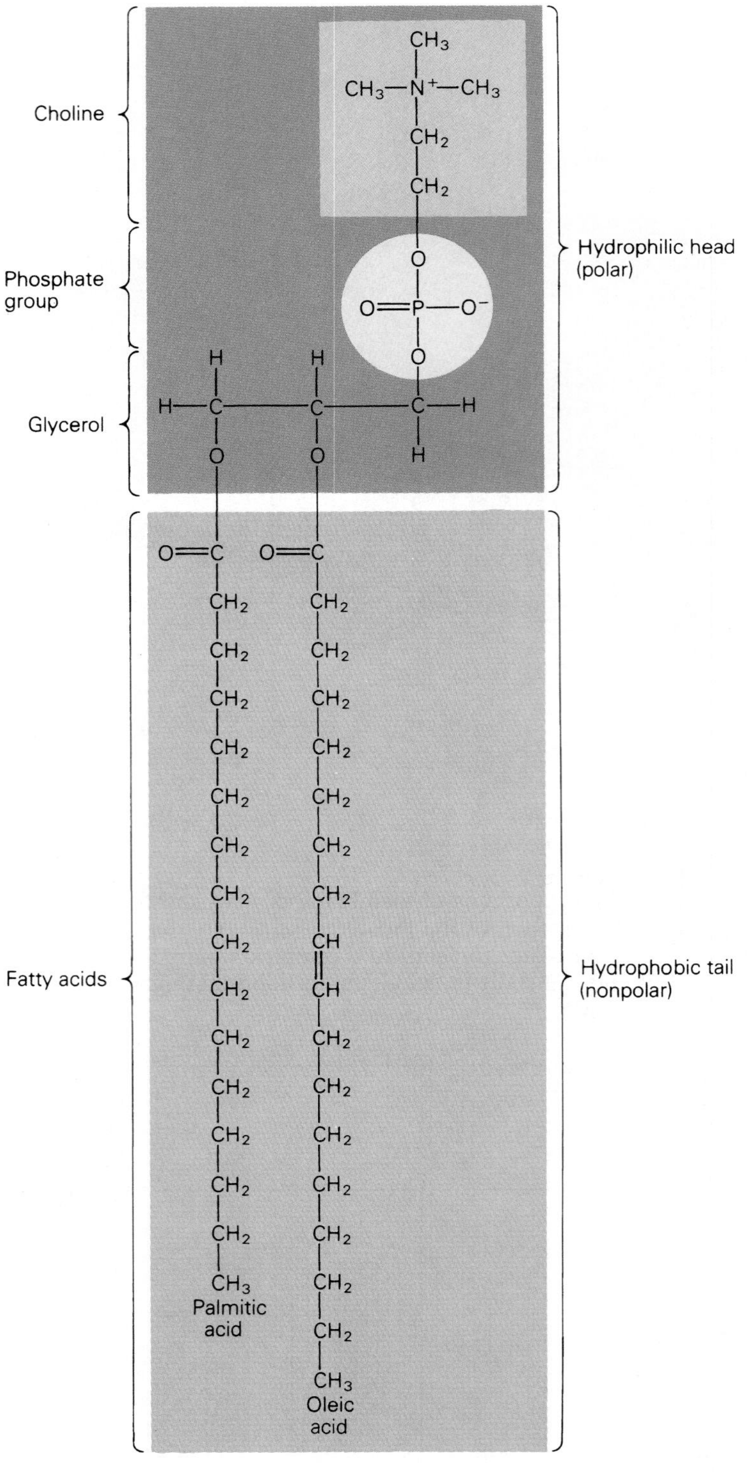

Choline {

Phosphate group {

Glycerol {

$$CH_3$$
$$CH_3—N^+—CH_3$$
$$CH_2$$
$$CH_2$$

Hydrophilic head (polar)

$$O$$
$$O=P—O^-$$
$$O$$

$$H \quad\quad H$$
$$H—C—C—C—H$$
$$O \quad\quad O \quad\quad H$$

Fatty acids {

$$O=C \quad\quad O=C$$
$$CH_2 \quad\quad CH_2$$
$$CH_2 \quad\quad CH_2$$
$$CH_2 \quad\quad CH_2$$
$$CH_2 \quad\quad CH_2$$
$$CH_2 \quad\quad CH_2$$
$$CH_2 \quad\quad CH_2$$
$$CH_2 \quad\quad CH$$
$$\quad\quad\quad\quad ||$$
$$CH_2 \quad\quad CH$$
$$CH_2 \quad\quad CH_2$$
$$CH_2 \quad\quad CH_2$$
$$CH_2 \quad\quad CH_2$$
$$CH_2 \quad\quad CH_2$$
$$CH_3 \quad\quad CH_2$$
Palmitic $\quad\quad CH_2$
acid $\quad\quad\quad CH_2$
$$\quad\quad\quad\quad CH_3$$
$$\quad\quad\quad\quad Oleic$$
$$\quad\quad\quad\quad acid$$

Hydrophobic tail (nonpolar)

Figure 4–4
The structure of a typical phosphatidylcholine molecule.

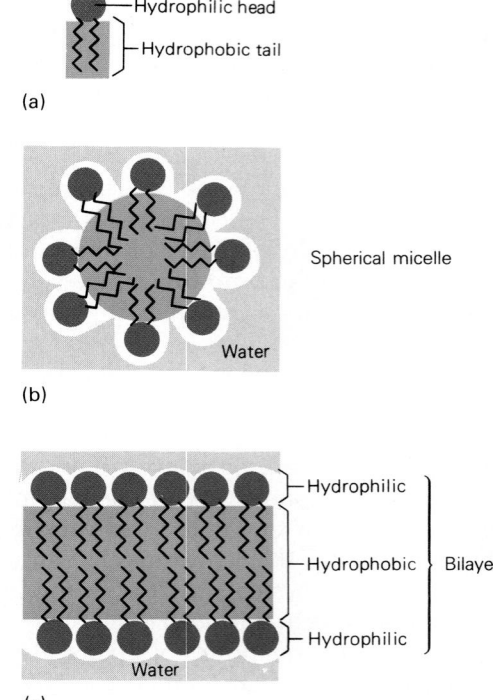

(a)

Spherical micelle

(b)

Hydrophilic — Hydrophobic — Hydrophilic — } Bilayer

Water

(c)

Figure 4–5
Phospholipids form either spherical micelles or bilayers in aqueous solution.

membrane of a mitochondrion reflects the rich complement of enzymes and other proteins vital for the processes of energy transformation that occur in these organelles.

The Lipid Bilayer of Amphipathic Molecules

The specific types of lipids found in membranes are **phospholipids, cholesterol,** and **glycolipids.** All of these molecules are insoluble in water. However, they all have hydrophilic, or water-soluble, regions and hence are called **amphipathic** molecules (*amphi* = "both").

Look at the structure of **phosphatidylcholine** (Fig. 4–4) for an example of a typical amphipathic phospholipid. Phosphatidylcholine has a hydrophilic head group and two long hydrophobic fatty acid tails, each of which may contain 14 to 24 carbon atoms in an unbranched chain. (The 16- to 18-carbon fatty acids are the most common fatty acid chains found in phospholipids; typical examples are **palmitic acid** and **oleic acid.**)

Recall from Chapter 2 that nonpolar molecules and groups, when placed in aqueous solution, cluster together in hydrophobic pockets. Since they consist mainly of hydrophobic fatty acid chains, phospholipids form clusters in order to "hide" their hydrophobic tails from water molecules. They form either spheres, called **spherical micelles,** or **bimolecular sheets (bilayers)** (Fig. 4–5). In both types of clusters, the hydrophobic tails of the phospholipids are buried in the interior of the structure, and only the hydrophilic head groups are exposed to the water.

The micelles are rather small structures, less than 20 nm in diameter, whereas a bimolecular sheet or bilayer is a much more extensive structure. When enough phospholipid molecules occur together, they are more likely to form a bilayer than a micelle. This is because the interior of a micelle has room for only a limited number of bulky hydrophobic fatty acid tails. A bilayer is a more efficient way for the phospholipids to protect their hydrophobic portions because it tends to close in on itself (forming a circle) to eliminate the free edges where the hydrophobic tails would be exposed to water. For the same reason, it reseals itself if punctured. Thus, a bilayer is the favored structure in an aqueous solution.

The Movement of Phospholipids Within the Bilayer

Studies on the movement of individual phospholipid molecules within a lipid bilayer have revealed that the bilayer is a dynamic structure. Phospholipid molecules move laterally along one layer of the membrane, but crossing from one layer to the other is much more difficult.

Movement Within Molecules: Flexing and Rotation. Within each phospholipid molecule there is a certain amount of movement. The phospholipid molecules are able to rotate about their axes (Fig. 4–6a). In addition, at physiological temperatures (37°C), the hydrocarbon tails are mobile and undergo rapid flexing movements. The rate of this movement is higher if the hydrocarbon tail is short and contains double bonds.

Stabilization by Cholesterol. In eukaryotic cell membranes, as much as 20% of the total lipid may be cholesterol. Cholesterol stabilizes the lipid bilayer. Like the phospholipids, cholesterol is an amphipathic molecule; however, its steroid ring structure is more rigid than the long fatty acid tails of the phospholipids. When present in the lipid bilayer,

Figure 4–6
Phospholipid molecules (*a*) rotate on their axes and (*b*) undergo flexing motions. The rigid steroid ring structures of cholesterol stabilize certain portions of the bilayer.

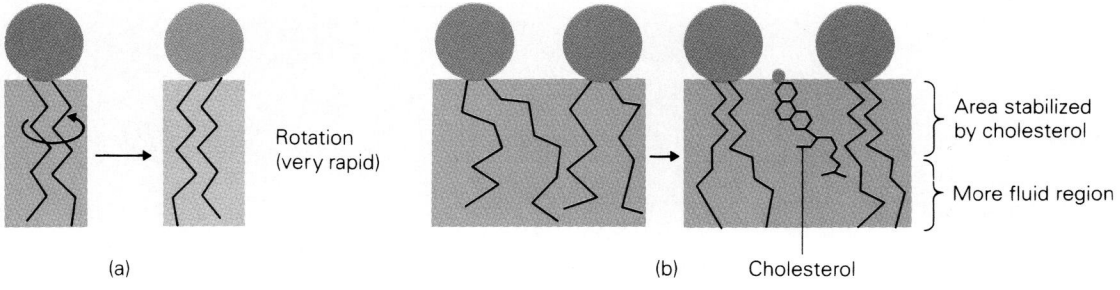

Rotation (very rapid)

(a)

(b)

Cholesterol

Area stabilized by cholesterol

More fluid region

the ring structure of cholesterol can interact with and reduce the motion of the neighboring phospholipid hydrocarbon chains (Fig. 4–6*b*).

Movement Within Layers. Phospholipid molecules exchange positions with their neighbors in the same half of the bilayer by very rapid lateral diffusion (Fig. 4–7*a*). By this process a lipid molecule takes only 1 sec to move about 2 μm, the entire length of a large bacterial cell.

Movement Between Layers. Lipids also move from one half of the lipid bilayer to the other in what is called "**flip-flop.**" This motion requires that the polar head groups of the phospholipids in one half of the bilayer pass through the hydrophobic interior of the bilayer to reach the other half of the bilayer (Fig. 4–7*b*). This is a difficult task; hence, flip-flop movement is extremely slow.

Lipid Asymmetry in the Bilayer

Phosphatidylcholine is not the only phospholipid in biological membranes. Other important membrane phospholipids are **phosphatidylethanolamine, phosphatidylserine,** and **phosphatidylinositol** (all de-

rived from glycerol), and **sphingomyelin** (derived from sphingosine). These different phospholipids are not distributed equally between the two halves of the lipid bilayer. A good example is the plasma membrane of the red blood cell, in which most of the phosphatidylcholine and sphingomyelin is present in the outer half of the bilayer, whereas most of the phosphatidylethanolamine and all of the phosphatidylserine are present in the inner half (Fig. 4–8).

The persistence of this **lipid asymmetry** in membranes is good evidence that the rate of phospholipid flip-flop is very slow. If flip-flop were rapid, the various phospholipids would probably mix randomly and thus be distributed equally between the bilayers. The phospholipid asymmetry must originate during the initial assembly of the bilayer, but scientists do not understand the mechanism and possible advantages of this arrangement.

The Texture of the Plasma Membrane

The net result of the movements of the phospholipids and their interactions with cholesterol is that, at 37°C, the interior of the lipid bilayer of biological membranes is more like a viscous fluid such as olive

Figure 4–7
(*a*) Phospholipid molecules diffuse laterally very rapidly within one layer of the plasma membrane.
(*b*) Phospholipids move very slowly from one layer to the other.

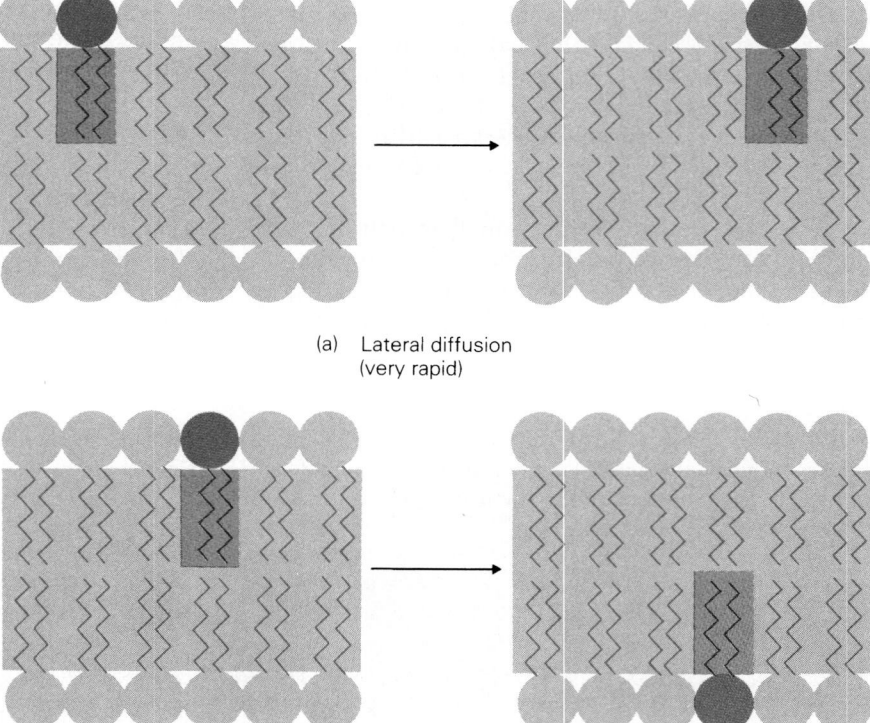

(a)   Lateral diffusion
(very rapid)

(b)   "Flip-flop"
(very slow)

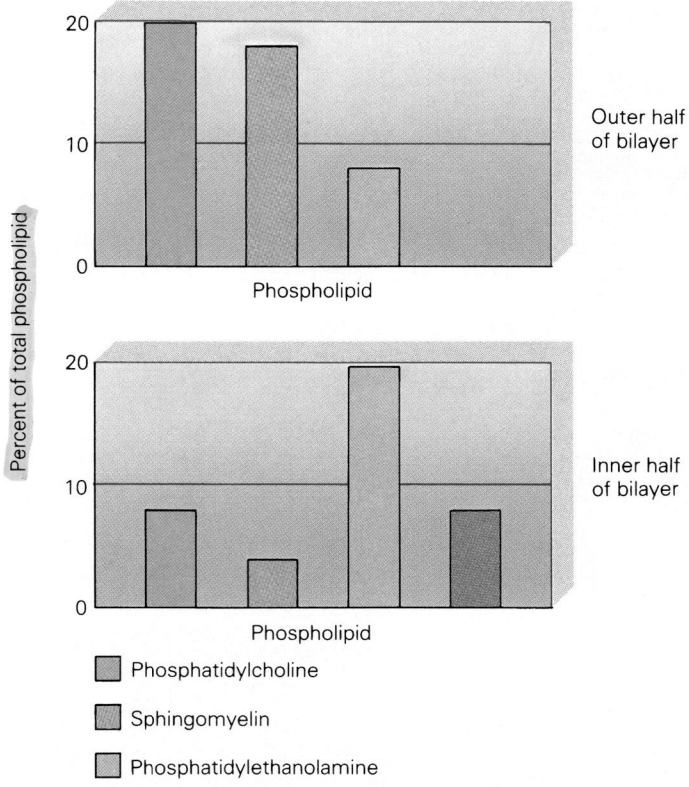

Outer half
of bilayer

Inner half
of bilayer

■ Phosphatidylcholine

▨ Sphingomyelin

□ Phosphatidylethanolamine

▨ Phosphatidylserine

Figure 4–8
Asymmetrical distribution of lipids in the erythrocyte membrane. Note that each half of the bilayer contains 50% of the total lipids present in the membrane.

oil than a rigid crystalline matrix. This fluidity of the membrane structure is probably very important for the normal function of many membrane proteins.

Membrane Proteins

Proteins are an important component of most biological membranes. Some proteins, called extrinsic proteins, are found only on the surface of the membranes, whereas others, called intrinsic proteins, partially or completely penetrate the lipid bilayer.

Intrinsic Proteins: Amphipathic Molecules

The proteins that penetrate the bilayer, called **intrinsic** (or **integral**) **proteins,** account for about 70% of total membrane protein. Most of them probably span the bilayer completely, so that parts of the protein are exposed on both sides of the bilayer.

Intrinsic proteins are usually amphipathic molecules. Their hydrophobic regions cluster in the interior of the bilayer with the hydrophobic portions of the lipids, whereas their hydrophilic regions remain exposed to the aqueous medium outside the bilayer (Fig. 4–9).

The three-dimensional structure or conformation (see Chapter 2) of much of the hydrophobic part of the protein is usually an alpha-helix. Any polar (hydrophilic) amino acid side chains in this portion of the molecule are sequestered within the alpha-helix. Only the nonpolar (hydrophobic) side chains are exposed to the hydrophobic interior of the bilayer. In contrast, in regions of the protein that project outside the lipid bilayer, the nonpolar side chains are buried in the interior of the protein conformation, whereas the polar and ionic side chains are exposed to the hydrophilic environment.

Many of the intrinsic membrane proteins display the same dynamic properties as the membrane lipids, such as rapid lateral diffusion and rotation about an axis perpendicular to the plane of the membrane. The fluidity of the lipid bilayer makes these movements possible. However, flip-flop of intrinsic membrane proteins is even more difficult than it is for phospholipids; in fact, it probably does not occur.

The lateral mobility of these membrane proteins can sometimes be prevented by interactions with peripheral components of the membrane (such as other proteins) or with components of the cell ctyoskeleton (microtubules and microfilaments).

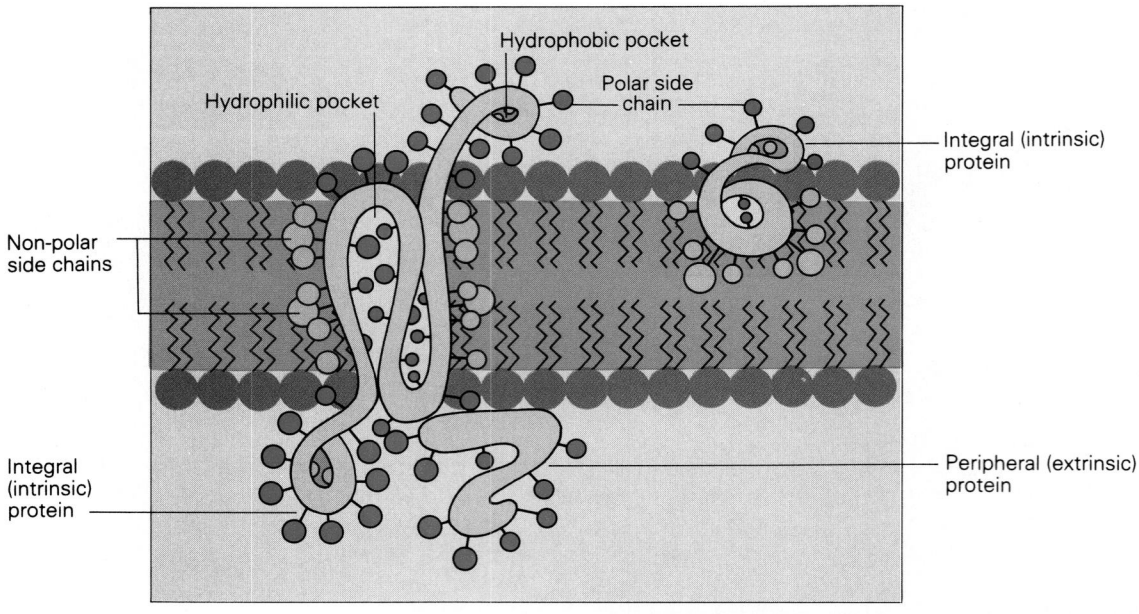

(a)

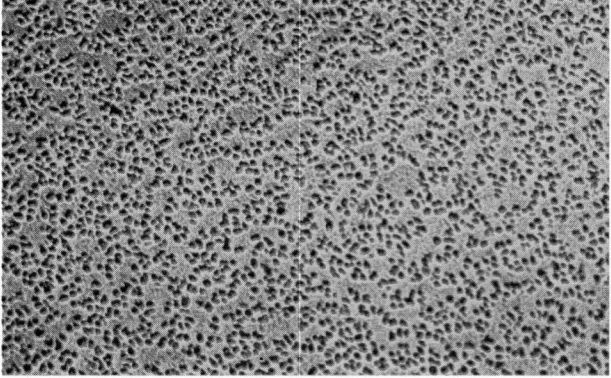

(b)

Figure 4–9
(*a*) The arrangement of integral and peripheral proteins in a biological membrane. (*b*) A freeze-fracture electron micrograph of the plasma membrane of a red blood cell. (© Harold H. Edwards, Visuals Unlimited.)

Extrinsic Proteins: Hydrophilic Molecules

Membrane proteins that do not penetrate the lipid bilayer are termed **extrinsic** (or **peripheral**) **proteins** (Fig. 4–9). They are usually not amphipathic molecules. They are surrounded by the aqueous environment of the cell and are associated with the membrane primarily through interactions (electrostatic and hydrogen-bonding) with the polar portions of intrinsic membrane proteins and with the polar head groups of the phospholipids. These are relatively weak interactions. Consequently, extrinsic proteins can be easily extracted from membranes by relatively simple procedures.

Protein Asymmetry in the Membrane

Both extrinsic and intrinsic membrane proteins, like the phospholipids, are often distributed unequally between the two halves of the bilayer. That is, a specific protein may be associated with only one side of the bilayer. Specific advantages of protein asymmetry are better understood than those of phospholipid asymmetry.

This asymmetry is related to cell function. In the membrane of the red blood cell, for example, the extrinsic protein called **spectrin** is found only on the cytoplasmic (inner) side of the membrane. It is part of a filamentous network inside the cell that may be important in maintaining the biconcave shape while being flexible enough to allow the cell to change shape as necessary during its passage through narrow capillaries.

In contrast, many receptors for specific chemical signals such as hormones are located only on the external surface of the cell membrane and often represent part of an intrinsic membrane protein. If such a protein spans the lipid bilayer, a potential mechanism exists for converting an extracellular signal into an intracellular response. For example, binding of a hormone to the receptor could trigger a conformational change in the protein, especially if the protein is composed of several separate subunits. This conformational change could open a hydrophilic channel between the protein subunits and allow a specific ion to cross the plasma membrane.

Carbohydrate on the Surface of the Membrane

Carbohydrate is present in the plasma membrane of all eukaryotic cells and accounts for 2% to 10% of the mass of the membrane. In terms of quantity, carbohydrate is a minor component of the plasma membrane, yet it has important functions in membrane physiology.

Carbohydrates, like the membrane phospholipids and proteins, are distributed unequally between the two layers of the membrane. The carbohydrate molecule is a hydrophilic structure that does not penetrate the lipid bilayer but is found only on the surface of the membrane (see Fig. 4–2). Most of the membrane carbohydrate is bound either to intrinsic membrane proteins, forming glycoproteins, or to lipids in the bilayer, forming glycolipids.

Glycolipids are present only in the external half of the bilayer. The glycoproteins are more abundant than the glycolipids. Most of the proteins that are exposed on the external surface of the membrane have carbohydrate attached, but fewer than 10% of the lipid molecules in the external half of the bilayer have carbohydrate attached.

The amount of carbohydrate present in an individual glycoprotein may be large relative to the amount of protein. An extreme example is **glycophorin,** an intrinsic glycoprotein in the red blood cell membrane. This protein is composed of 131 amino acids and about 100 monosaccharide units. The monosaccharides together make up approximately 60% of the mass of the glycophorin molecule.

The location of membrane carbohydrates on the surface of the plasma membrane is probably an important factor in cell-to-cell recognition processes. These carbohydrates also probably serve as specific membrane receptors for extracellular messengers such as hormones. Unlike amino acids, which form only linear chains, monosaccharide units can be assembled in branching chain structures. This property provides almost limitless possibilities for construction of membrane receptor sites to which only one specific extracellular molecule can bind.

The Types of Movement Through the Membrane

Two types of movement occur as substances pass through the plasma membrane of a cell: (1) **passive,** or **spontaneous,** processes that occur in response to gradients and produce a *less* concentrated state overall; and (2) **active,** or **energy-requiring,** processes that occur in opposition to gradients and produce a *more* concentrated state where a high degree of concentration already existed. **Passive transport** processes include both **simple diffusion,** such as **osmosis,** and **facilitated diffusion. Active transport** processes include **primary** and **secondary active transport** (Fig. 4–10). Note that the same substance—sodium ions, for example—can move one way through the membrane by a passive process and the other way by an active process.

Passive Transport: Spontaneous Movement down a Gradient

Spontaneous, random movements of substances into or out of the cell include simple diffusion and facilitated diffusion. **Simple diffusion** takes place when a substance, usually one that can dissolve in lipids, passes directly through the lipid interior of the plasma membrane. **Facilitated diffusion** occurs in the presence of a protein that enables the passage of the substance. However, both processes are passive and spontaneous movements that occur in response to a gradient.

Simple Diffusion of Lipid-Soluble Substances

Diffusion in free solution or any other type of mixture occurs because of the random movement that all particles exhibit. This motion is an expression of the thermal energy of the molecules. As a result of diffusion, a substance such as a solute will tend to become distributed in an equal concentration throughout the entire space available.

A Selectively Permeable Membrane. Chapter 2 discussed the diffusion of solute particles in free solution, from a region of high solute concentration to an adjacent region of low concentration. There was no physical division between these two regions; they were more or less continuous.

Diffusion can also occur in the presence of a **selectively permeable membrane.** If a solute is highly concentrated on one side of a membrane but able to

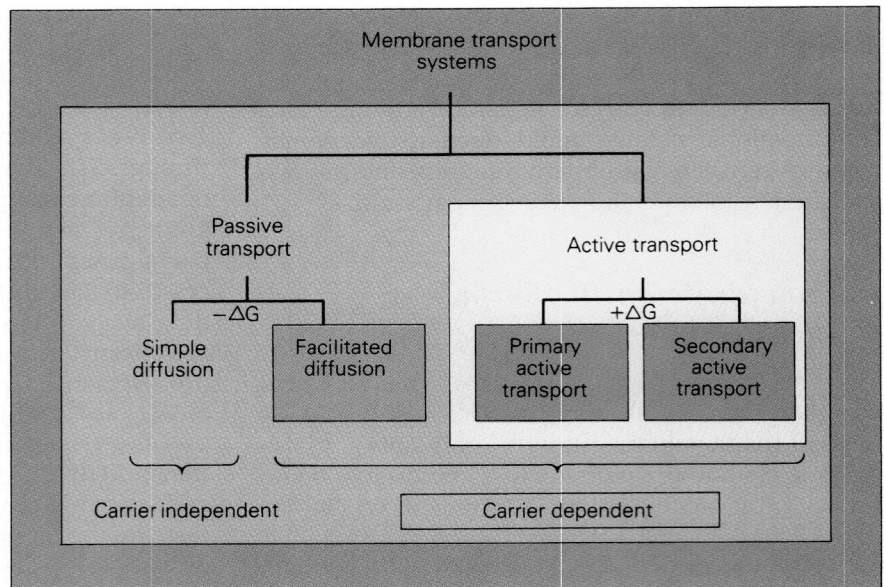

Figure 4–10
An overview of membrane transport systems.

pass through the membrane, it will do so, diffusing into the adjoining region of low concentration by the same process that would occur if the membrane were not there (Fig. 4–11). If another solute were present to which the membrane were not permeable, that solute would remain highly concentrated on one side of the membrane.

The plasma membrane, as a mixture of lipids and proteins, represents a barrier to the diffusion of some solutes into or out of the cell. An inherent, important property of any lipid bilayer is that it is nearly impermeable to ions and most polar molecules (water is a notable exception to this general rule because it readily crosses lipid bilayers even though it is a polar molecule). In contrast, a lipid bilayer is permeable to nonpolar molecules—that is,

molecules that can dissolve in its nonpolar interior. The lipid bilayer of the plasma membrane is selectively permeable; it readily permits some substances to cross it but prevents or retards the free passage of others.

Consider the diffusion of glycerol, a polar molecule, from the plasma to the interior of a red blood cell. The diffusion coefficient D is a measure of the mobility of a specific solute molecule within a specific solute. D for glycerol in water is 1.0×10^{-5} cm^2/sec. This is several orders of magnitude larger than the value of D for glycerol inside the lipid phase of the plasma membrane: 1.7×10^{-10} cm^2/sec. This large difference reflects the fact that the mobility of glycerol is greatly diminished once it enters the membrane.

Figure 4–11
Diffusion can occur in the presence of a selectively permeable membrane. As long as solute particles can pass through the membrane, they will diffuse from a region of high concentration to one of low concentration.

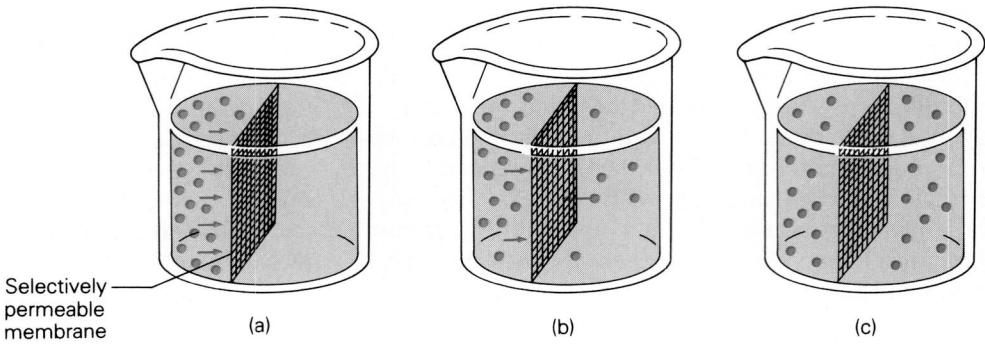

Selectively permeable membrane

(a)   (b)   (c)

The Permeability Coefficient. The rate of diffusion of solute through a cell membrane is governed by the thickness of the membrane and by the lipid solubility of the solute. The combined effect of these two factors and the diffusion coefficient can be expressed as the **permeability coefficient P.** The value of P varies with the specific solute and the specific cell membrane.

In spite of the barrier represented by the cell membrane, an uncharged solute will still tend to move across it if the solute concentration inside the cell is different from the solute concentration outside—that is, if there is a concentration gradient. In simple diffusion, the solute will always move down the concentration gradient, from the region of high solute concentration to the region of low concentration, just as in free solution. Fick's Law (see Chapter 2) can be adapted to represent the rate of solute diffusion across a membrane as follows:

$$J = -PA[C_1 - C_2]$$

C_1 and C_2 are the solute concentrations on either side of the membrane. Thus, when C_1 is greater than C_2, there is net movement of solute from Side 1 to Side 2 (Fig. 4–12). As the solute concentration builds up on Side 2, the rate of movement back to Side 1 increases. Eventually the solute concentration gradient will be abolished, and these opposing fluxes will become equal. At this point, the system will be in a dynamic equilibrium—that is, individual solute particles will continue to move from one side of the membrane to the other, but the net movement of solute will cease. This can also be seen from the equation: When $C_1 = C_2$, the value of J is zero.

The Electrochemical Gradient. When the solute is electrically charged, as are Na^+ and Cl^- ions, for example, the flux of solute is influenced not only by a concentration gradient but also by an electrical gradient. A typical animal cell has a small negative potential (about -70 millivolts) on the inside relative to the outside. This difference tends to cause positively charged solutes to move across the plasma membrane and into the cell while it opposes the entry of negatively charged solutes. This electrical gradient represents an additional factor in diffusion that must be considered when the solute bears a charge. The total gradient is now the sum of the chemical (concentration) and electrical components and is termed the **electrochemical gradient.**

Considered separately, the chemical and electrical gradients often favor solute movement in the same direction. For example, the passive entry of Na^+ into cells is favored by both the concentration gradient (low Na^+ concentration inside the cell) and the electrical gradient (negative potential inside the cell).

On the other hand, electrical and chemical gradients can oppose each other. For example, the passive leak of K^+ ions out of cells goes against, or "up," the electrical gradient (there is a positive potential outside the cell). However, the concentration gradient is so large—because of the high K^+ concentration inside the cell—that the electrochemical gradient force favors the exit of K^+ from the cell.

Overton's Rules. The principal barrier to solutes that cross cell membranes by simple diffusion is the lipid bilayer. The rate at which a solute diffuses

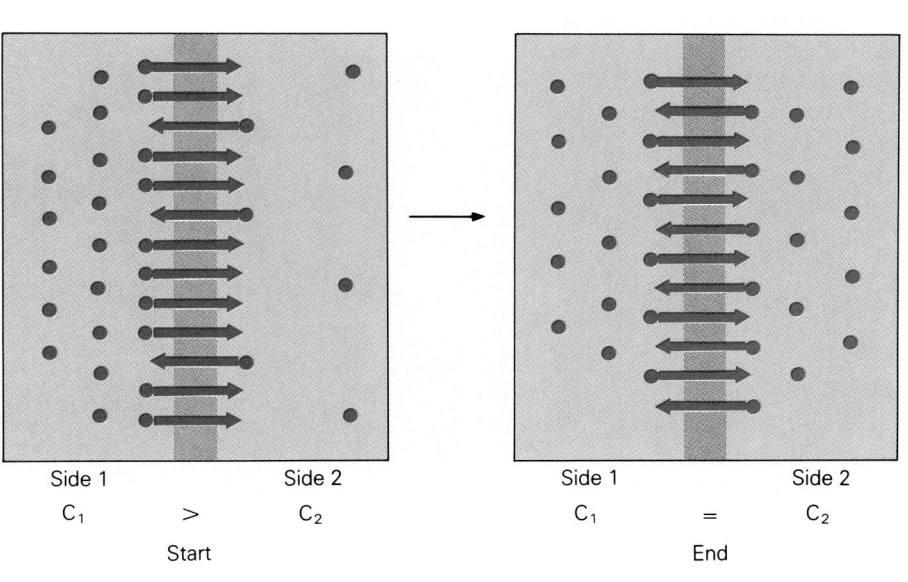

Figure 4–12
Diffusion of solute molecules across a biological membrane. At the start there are 25 molecules on Side 1 and only 7 on Side 2, so there will be net movement of solute from Side 1 to Side 2. The end point is reached when each side has 16 molecules. At this point, there is no net movement of solute across the membrane because equal numbers of molecules are moving in each direction.

Side 1		Side 2
C_1	>	C_2
	Start	

Side 1		Side 2
C_1	=	C_2
	End	

through the membrane depends in large part on two factors: its ability to dissolve in lipids and its molecular size. In general, a nonpolar, and therefore lipid-soluble, solute will diffuse across a cell membrane much more rapidly than a polar (water-soluble) solute of the same size. The effect of the size of the solute molecule is particularly important for polar solutes. Membrane permeability for these molecules, already limited by their polarity, decreases further as their molecular size increases.

These relationships are summarized in **Overton's Rules,** as follows: (1) the permeability of cell membranes to small nonpolar solutes is directly proportional to the lipid solubility of the solute; and (2) the permeability of cell membranes to polar solutes is inversely proportional to the molecular size of the solute. Thus, a small, highly lipid-soluble molecule has the best chance of rapidly penetrating cell membranes by simple diffusion.

An exception to these general statements is water itself. Despite its small size, a water molecule is highly polar (lipid-insoluble) and yet it crosses cell membranes very rapidly. The precise mechanism is not fully understood, but it has been suggested that membranes contain small nonspecific pores that allow water and small water-soluble solutes to cross the bilayer without entering the lipid phase. The diffusion of water will be considered in the section on osmosis.

An electrical charge on the solute is another factor that decreases its lipid solubility. Thus, membranes are almost impermeable to ions, even though an ion represents a solute of small size. Any ionic diffusion that occurs probably uses specific channels in proteins that span the lipid bilayer (see the section on facilitated diffusion later in this chapter).

Diffusion of Acids and Bases. The very low permeability of cell membranes to solutes bearing an electrical charge has important consequences for the diffusion of **weak acids** and **weak bases.** These compounds, unlike strong acids and bases, do not dissociate completely when dissolved in water (see Chapter 2). Recall that an acid is a proton donor while a base is a proton acceptor. We can write the following general equations to represent the behavior in solution of a weak acid (HA) and a weak base (B):

$$HA \rightleftharpoons H^+ + A^- \text{ (weak acid)}$$
$$B + H^+ \rightleftharpoons BH^+ \text{ (weak base)}$$

A solution of a weak acid or weak base will contain a mixture of the undissociated form and the dissociated form, the proportions of these two forms varying with the pH of the solution. In the case of the weak base, for example, a low pH (high H^+ concen-

tration) will shift the equilibrium to the right, resulting in more BH^+. Conversely, a high pH (low H^+ concentration) will shift the equilibrium to the left, resulting in more B and more H^+. The pH at which the concentration of the associated form is exactly equal to that of the dissociated form is defined as the *pK* and is a characteristic property of the weak base or the weak acid.

Diffusion Trapping. The neutral form of a weak base diffuses across a cell membrane whereas the charged form does not, which gives rise to the phenomenon of **diffusion trapping.** This is illustrated in Figure 4–13 using ammonia as an example of a weak base. Ammonia is produced in the cells of the kidney tubules as an end product of the metabolism of amino acids, especially glutamine. Ammonia is neutral and lipid-soluble and easily diffuses out of the cell, moving down its concentration gradient and into the fluid present in the lumen of the tubule. Typically, this fluid has a low pH; it is slightly acidic relative to the cell cytosol. When the ammonia contacts the excess H^+ in this acidic solution, it gains a proton to form the ammonium ion. The ammonium ion is positively charged and lipid-insoluble and cannot diffuse back into the tubule cell. It is effectively "trapped" in the lumen and eventually excreted in the urine.

Protonation of available ammonia in the tubular fluid maintains a low concentration of ammonia out-

Figure 4–13
Diffusion trapping of a weak base. NH_3 can cross the plasma membrane because it is a neutral molecule. When it gains a proton in acidic solution, it becomes the charged ammonium ion NH_4^+, which cannot cross the membrane.

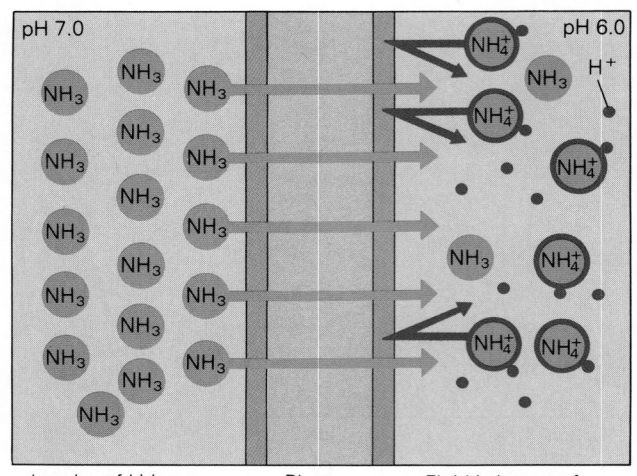

| Interior of kidney tubule cell | Plasma membrane | Fluid in lumen of kidney tubule |

side the tubule cells, and continued production of ammonia by cellular metabolism maintains a relatively high concentration of ammonia inside the tubule cells. The combination of these processes preserves the concentration gradient that favors the passive diffusion of ammonia out of the cells.

Simple Diffusion of Water: Osmosis

Recall that **osmosis** is the term used to describe simple diffusion when the molecules that are moving are water molecules. Osmosis is defined as the flow of water across a membrane in response to a difference in solute concentration on each side of the membrane. The water always moves from the dilute solution (the region of high water concentration) to the concentrated solution (the region of low water concentration).

Changes in Cell Volume: The Erythrocyte.

Nearly all cell membranes are permeable to water and must regulate the amount of water that enters the cell. Plant cells and many bacteria have a rigid cell wall that surrounds the plasma membrane and prevents an increase in the volume of the cell due to an influx of water. Animal cells lack an exterior cell wall and have developed other mechanisms to deal with osmotic stress.

We will use the red blood cell, or **erythrocyte,** to illustrate the changes that occur in animal cells when the osmolarity of the external medium changes. The erythrocyte is a favored experimental tool of cell physiologists because of its simple structure. It has no intracellular membranes and is available in large numbers.

Previously in this chapter, we stated that many biological membranes are permeable to solutes as well as to water. This is certainly true of the erythrocyte plasma membrane, but the net movement of the major solutes, such as Na^+, K^+, and Cl^- ions, will be relatively small during osmotic experiments of short duration.

The osmolarity of the erythrocyte cytosol is close to 300 mOsm. Suppose an erythrocyte is placed in an iso-osmotic solution of 300 mOsm NaCl. The total osmolarity (and hence the osmotic pressure) of the solution inside the cell will be the same as that of the solution outside. There will no concentration gradient favoring osmotic water flow and no net movement of water into or out of the cell; the intracellular volume will not change.

Next, suppose the erythrocyte is placed in hyperosmotic 400 mOsm NaCl. The erythrocyte will no longer be in osmotic equilibrium because the external solution has a higher osmolarity than the cell cytosol. In response to the higher solute concentration outside the cell, water flows out of the cell (Fig. 4–14). However, solutes remain in the cell, and therefore the osmolarity of the cell cytosol increases. Eventually the osmolarity of the cytosol will increase to the point where it is equivalent to that of the external solution, and osmotic flow of water will cease. The cell will return to an osmotic equilibrium with the external medium, but this will occur at the expense of part of the intracellular volume. The decrease in volume is not accompanied by a decrease in surface area, and the cell assumes a spiky, or **crenated,** shape (Fig. 4–14).

When a normal erythrocyte is placed in a hypo-osmotic solution of 200 mOsm NaCl, the osmolarity of the cytosol is higher than that of the external solution. Water will enter the cell, leading to an increase in intracellular volume (Fig. 4–14). Osmotic flow of water into the cell will continue until the osmolarity of the cytosol has been diluted to 200 mOsm. At this point net movement of water into the cell will cease, and osmotic equilibrium will be re-established, but with an increase in cell volume (Fig. 4–14).

The erythrocyte has a unique characteristic that enables it to accommodate a considerable increase in cell volume. The normal cell is bi-concave, not spherical. When water enters from a hypo-osmotic solution, the cell simply fills out to become more like a sphere (Fig. 4–14). When the cell has become a perfect sphere, the intracellular volume will have increased by 67% over the volume of the original bi-concave cell. In this way, the erythrocyte can undergo a large increase in volume without experiencing a change in the surface area of the cell. Most other types of animal cells are not bi-concave and cannot respond in this way to osmotic entry of water. The plasma membrane is not elastic—it cannot stretch—and so the cells can accommodate only a very small increase in intracellular volume. If osmotic entry of water continues, the cells undergo osmotic **lysis**—that is, holes develop in the plasma membrane and the intracellular contents are lost. The cells literally burst open.

Osmotic Fragility.

An erythrocyte will burst open, or **lyse,** if it is placed in a medium that is very dilute (e.g., 100 mOsm NaCl). The difference in osmolarity between the inside and the outside of the erythrocyte is now so large that the cell cannot achieve osmotic equilibrium. Water will continue to enter even when the cell has become a sphere, and so the cell will lyse.

Osmotic fragility is a term used to characterize how a sample of erythrocytes responds to osmotic stress. It represents the osmolarity of NaCl that

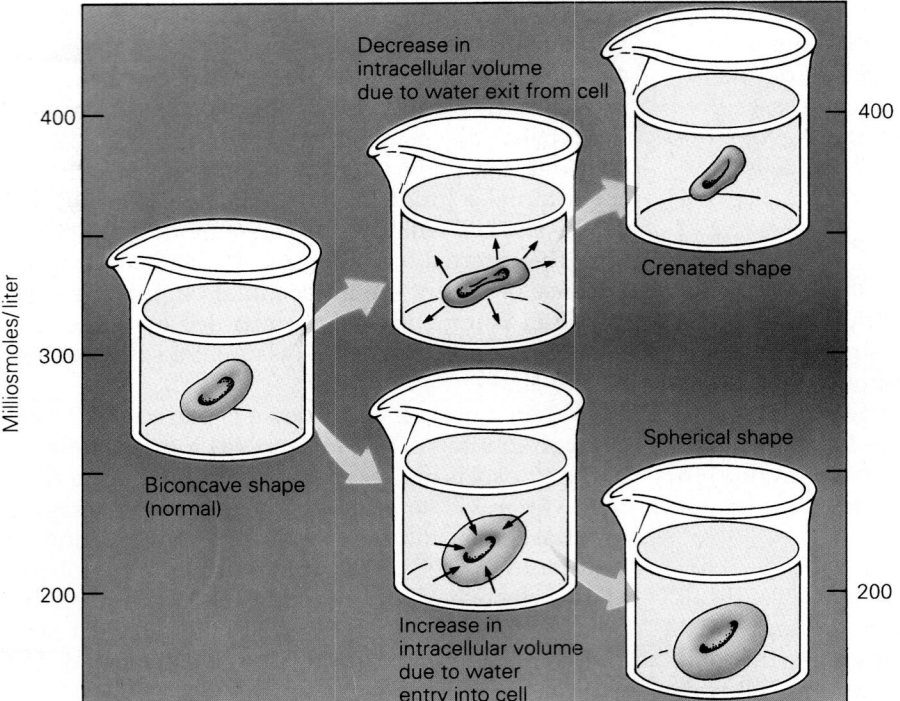

Figure 4–14

Response of a human erythrocyte to changes in the osmolarity of the extra-cellular medium. The normally bi-concave disc becomes crenated when water flows out of the cell because of a greater osmolarity in the surrounding fluid. The cell becomes a sphere when water flows into it from solutions with a lower osmolarity.

causes 50% of the erythrocytes to lyse. In the case of normal erythrocytes, 50% of the cells lyse at an os-molarity of 150 mOsm (Fig. 4–15). Some cells, pre-sumably the older ones, are more fragile and lyse at a slightly higher osmolarity than 150 mOsm. Other cells, presumably the younger ones, are less fragile and lyse at a slightly lower osmolarity. However, almost all of the population of cells lyse over a fairly narrow range of osmolarities, as shown by the steep slope of the curve in Figure 4–15.

Osmotic fragility can be an important test for diseases that involve a change in erythrocyte behav-ior. When the erythrocytes become more resistant to osmotic stress, they are said to be less fragile, and the curve (Fig. 4–15) shifts to the right. When the population is less resistant to osmotic lysis, the cells are said to be more fragile, and the curve shifts to the left.

A specific example of a fragile population are the erythrocytes from mice with **hereditary sphero-cytosis.** These cells tend to be spherical rather than bi-concave. As a result, the cells can no longer ac-commodate the volume increase caused by osmotic

Figure 4–15

Abnormal erythrocytes are more fragile and lyse in so-lutions of higher osmolarity compared to normal eryth-rocytes.

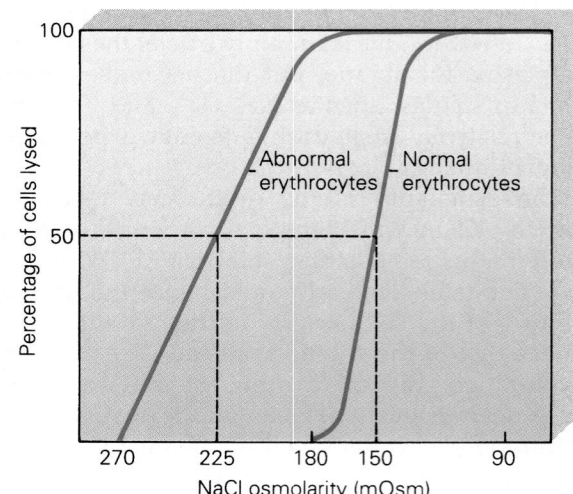

flow of water into the cells. As seen in Figure 4–15, some of the cells lyse when the osmolarity of the NaCl medium decreases only slightly. Fifty percent of the cells lyse at an osmolarity of 225 mOsm, which represents a considerable change from the behavior of the normal cell population. The erythrocytes in outdated blood from a blood bank provide another example of increased osmotic fragility. These cells are less resistant to osmotic stress than erythrocytes in fresh drawn human blood.

After a discussion of some specialized transport mechanisms, we will return to the concept of osmosis and the way in which the cell transports materials to maintain appropriate cell volume.

Facilitated Diffusion of Lipid-Insoluble Substances

A lipid bilayer is essentially impermeable to ions and large water-soluble molecules such as D-glucose, amino acids, and other metabolites. Still, these substances diffuse across the membrane in response to electrochemical gradients. They do so by a process of **facilitated diffusion,** so called because their passage is facilitated by intrinsic membrane proteins that act as **carriers** or **channels.** Ions and other lipid-insoluble substances normally would not be able to pass through the nonpolar interior of the membrane. The proteins enable them to get from one side of the lipid bilayer to the other, in part by shielding them from the lipid molecules.

The molecular mechanisms of action of membrane carriers and channels are the subjects of intense investigation.

Simple diffusion and facilitated diffusion both occur spontaneously and passively in response to an electrochemical gradient. Similarly, both processes cease once the gradient has been abolished. Since facilitated diffusion is mediated by a carrier or channel protein, however, a number of characteristics distinguish it from simple diffusion: (1) facilitated diffusion allows a very high rate of solute transport, (2) it is a saturable process; there is a limit to the transport rate that cannot be exceeded, (3) it is a highly specific process, and (4) solute transport can be blocked by competitive inhibitors.

Carrier Proteins.   A hypothesis to explain the operation of a carrier is depicted in Figure 4–16. This model proposes (1) that the carrier protein is composed of at least two subunits, and (2) that binding of solute to a site exposed to the cell exterior triggers an intramolecular rearrangement in the structure of the protein such that the solute binding site is now exposed to the cell interior. The solute is always surrounded by hydrophilic regions of the protein subunits and is never exposed to the hydrophobic lipid phase of the membrane. Once the solute is exposed to the interior of the cell, where the solute concentration is low, it dissociates from the binding site. This step allows the carrier protein to revert back to its original conformation, with the empty binding

Figure 4–16

A possible mechanism for carrier-dependent transport. (*a*) Solute concentration is high outside and low inside the cell. A solute molecule binds to the carrier protein at a specific site that is exposed on the exterior surface of the cell membrane. (*b*) Binding of the solute to the carrier triggers a change in conformation, and the solute binding site becomes exposed to the cell interior. (*c*) The solute easily dissociates from the carrier because of the low concentration of solute inside the cell. This dissociation allows the carrier protein to revert to its original conformation. The solute binding site is again exposed to the exterior of the cell and the cycle can be repeated. Probably the changes in protein conformation are much more subtle than depicted here.

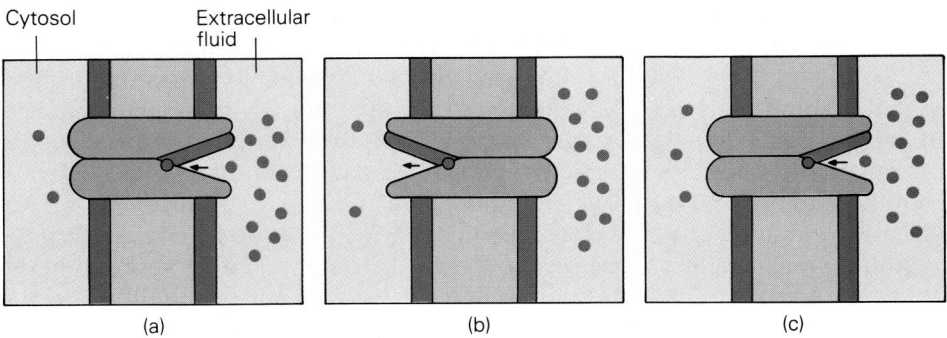

Cytosol   Extracellular fluid

(a)   (b)   (c)

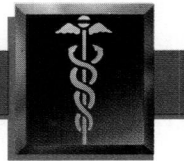

Cystic Fibrosis

A disorder that is specifically linked to cell transport and secretion is **cystic fibrosis (CF).** CF is an inherited genetic disorder occurring in children in which the secretions of the body glands are abnormally thick and sticky. They clog up the glands and prevent certain organs from working properly. The pancreas cannot function, and without pancreatic enzymes it is difficult to absorb certain types of food. The liver and lungs also become clogged with sticky mucous, which in turn leads to cystic and fibrotic changes in these organs. In addition, children with cystic fibrosis have abnormally high concentrations of electrolytes (salt) in their sweat and saliva.

The most severe effect of CF is upon the lung. Sticky mucous not only blocks the air passages but also serves as a culture (growth) medium for harmful bacteria. Consequently, the small airways become obstructed, and the bacteria or their toxins cause respiratory infections that often lead to pneumonia. The repeated infection causes scarring and promotes fibrosis of the lung.

Another complication in CF patients, as a result of the clogged airways, is the low concentration of oxygen in the lungs. This decreases the amount of oxygen that gets into the blood and also triggers a response that causes the pulmonary arteries to constrict. The constriction causes pulmonary hypertension (an abnormal high blood pressure in the lungs) and puts an increased strain on the heart. In fact, 98% of all CF patients die from cardiopulmonary complications.

Prior to the 1940s, babies with CF died early from respiratory failure. In the 1950s, with advances in antibiotic therapy, half of all CF infants lived to be 3 to 5 years old. Today, with stong antibiotics and proper therapy, half of all children with CF live into their twenties, which for everyone else is the prime of life. The disease is an autosomal recessive disorder in

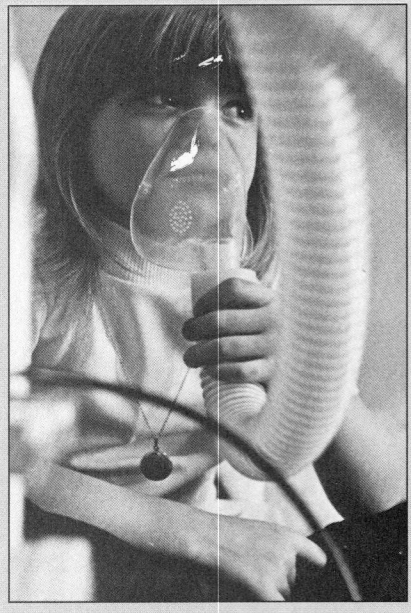

A child with cystic fibrosis using a nebulizer which disperses medications into a fine mist that can be inhaled. (Courtesy of the Cystic Fibrosis Foundation.)

Caucasian children and occurs in about one in 1,500 live births. CF is rare in black children and essentially absent among Asiatics.

Scientists are rapidly closing in on the disease process of cystic fibrosis. They have discovered that an abnormal gene leads to a defective transport of chloride ion in the mucous gland. If the defect in chloride transport is the cause of the abnormal mucous in CF patients, then someday it may be possible to treat CF patients with gene therapy (see FOCUS in Chapter 3) designed to restore normal chloride transport.

site again exposed to the cell exterior. The cycle can then be repeated to move another solute molecule into the cell.

Channel Proteins and Gating. A channel mechanism is particularly important for the rapid transmembrane movement of ions such as Na^+ and K^+. One hypothesis is that a hydrophilic channel across the membrane is formed between the subunits of a protein. The open channel permits a much higher rate of transport than that provided by a carrier. For example, some channels enable ions to pass through them at the rate of 10^8 ions/sec, whereas the fastest carriers move solutes across membranes at a rate no greater than 10^5 molecules/sec.

It is unlikely than an ion moves through a channel by simple diffusion down the electrochemical gradient. Most likely the ion undergoes specific interactions with charged groups along the sides of the channel. These reactions help to move the ion through the channel. The types of charged groups present in the channel may also be important in determining the specificity of the channel for one ion type.

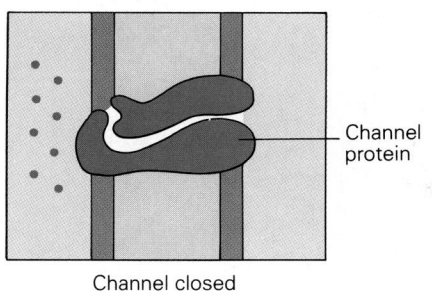
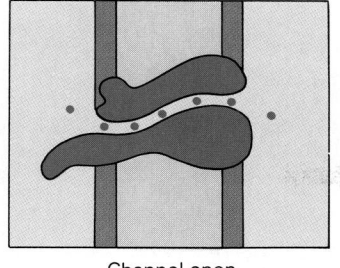

Channel closed   Channel open

Figure 4–17
A highly simplified view of the gating mechanism for regulating ion transport through a membrane channel protein.

Another difference between a channel and a carrier, in addition to the overall rate of transport, is that a channel can be kept closed by a "**gate**" (Fig. 4–17). A closed channel will not permit movement of ions even though the electrochemical gradient may favor it. The mechanisms that regulate the opening and closing ("**gating**") of ion channels will be discussed later in Chapter 7.

The High Rate of Facilitated Diffusion. A solute that crosses a cell membrane by facilitated diffusion does so at a rate that is much more rapid than that of a solute crossing by simple diffusion, even though both solutes are of the same molecular size and have the same lipid solubility. Consider a hypothetical example in which there are 17 solute molecules outside a cell and only 1 inside (Fig. 4–18); assume that

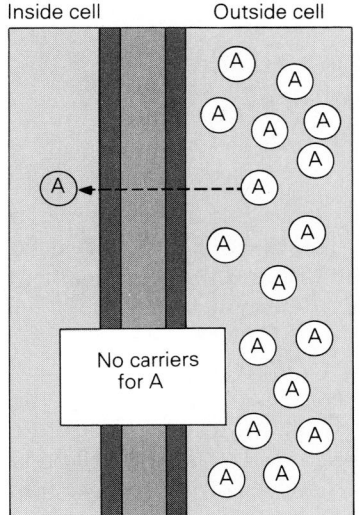
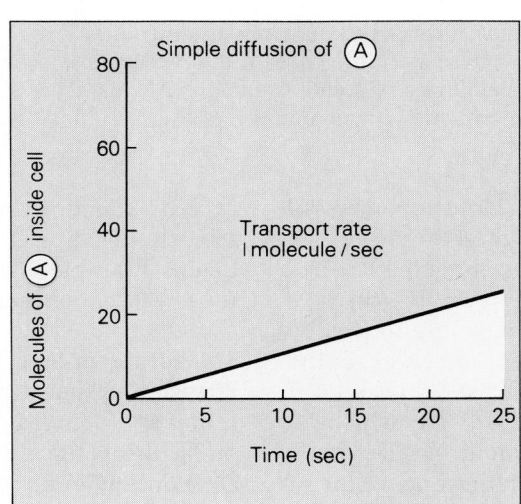

Figure 4–18
The rate of net transmembrane movement of Solute B, which occurs by facilitated diffusion, is much higher than the rate of simple diffusion of Solute A, even though Solutes A and B have similar molecular sizes and lipid solubilities.

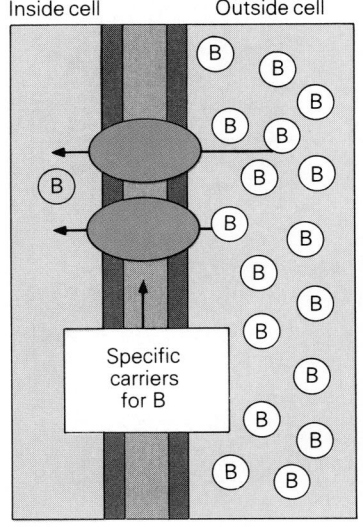
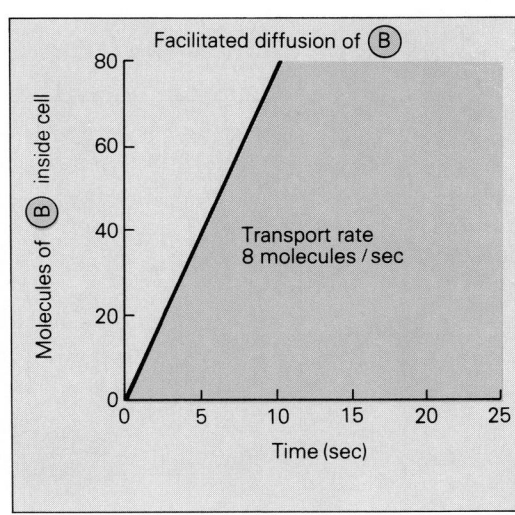

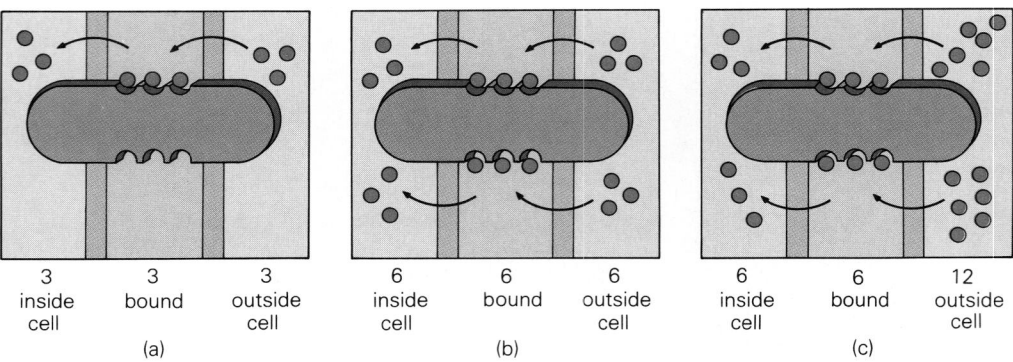

3 inside cell	3 bound	3 outside cell	6 inside cell	6 bound	6 outside cell	6 inside cell	6 bound	12 outside cell
(a)			(b)			(c)		

Figure 4–19

The saturation of a membrane carrier with solute. (*a*) Assume that a carrier protein in a membrane has specific binding sites for 6 solute molecules. If 3 solute molecules outside a cell arrive at the carrier, all 3 molecules will be transported inside the cell in one cycle of the carrier. (*b*) If 6 molecules outside the cell arrive at the carrier, all 6 molecules will be moved inside the cell in one cycle. That is, the rate of net transmembrane transport of solute doubles when the amount of solute doubles. (*c*) When 12 solute molecules arrive at the carrier, the rate of solute movement will not increase because the carrier can bind and transport only 6 molecules per cycle. Once the carrier is saturated with solute, any further increase in solute concentration outside the cell will have no effect on the overall rate of solute transport. At this point the rate of solute transport has reached a maximum.

the solute is electrically neutral. Net movement of solute inside the cell will cease when 8 more molecules have entered—that is, when there are 9 molecules on each side of the membrane. It will take 8 sec to achieve this balance if the solute moves by simple diffusion at a rate of 1 molecule entering the cell per second. In contrast, it will take only 1 sec if the solute moves by facilitated diffusion, which allows solute entry at a rate of 8 molecules/sec.

This example illustrates that the end point of both processes is the same but is achieved faster by facilitated diffusion. In this way the carrier protein serves a function analogous to that of an enzyme in a biochemical reaction. The rates of transport used in this example are arbitrary and are not meant to represent experimentally determined values.

Saturation in Facilitated Diffusion. A facilitated diffusion system can become saturated because the number of binding sites on the carrier for solute molecules is limited. The rate of transport will be at a maximum when all the sites are occupied by solute (Fig. 4–19). A graph drawn to relate transport rate to solute concentration (see Fig. 4–20) shows that ini-

Figure 4–20

Graphical representation of the saturability of facilitated diffusion. The maximum rate of transport possible is termed the V_{max}, and the K_m is the concentration of solute necessary to reach one half of the V_{max}. Although simple diffusion is not saturable, the rate of transport is much slower at the physiological range of solute concentration.

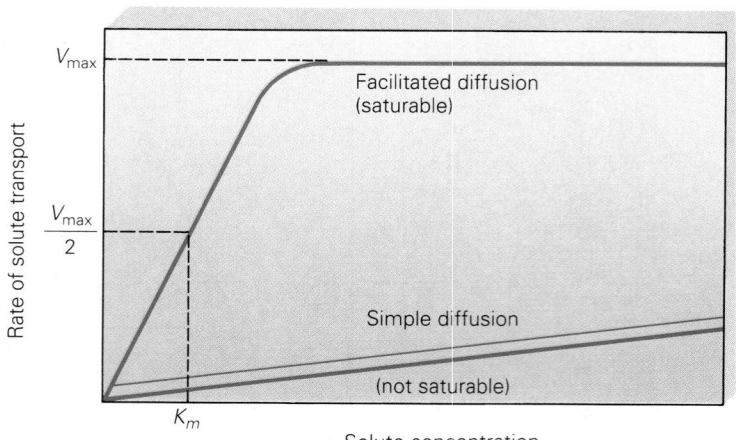

tially the rate of solute transport increases in proportion to the solute concentration. Once the solute concentration has reached a level at which the carrier binding sites are always full (saturation), however, any further increases in solute concentration will produce no change in the solute transport rate. The rate of transport is now at its maximum; this is termed the V_{max}. The solute concentration that gives rise to one half of the V_{max} is known as the K_m. These kinetic parameters are a very useful way of comparing the characteristics of different transport systems. It is worth adding that transport by simple diffusion is not saturable. The rate of transport increases in proportion to the solute concentration (see Fig. 4–20). The limitation posed by carrier saturation is rarely a problem because the overall rate of transport by this mechanism is so much faster than that by simple diffusion. The V_{max} for transport by facilitated diffusion is analogous to an enzyme-catalyzed reaction, which proceeds at a maximum rate once the enzyme is fully saturated with substrate (see Chapter 2). Any further increase in substrate concentration will have no effect on the reaction rate. The analogy with enzymes is limited, however, because the solute is not permanently altered as a consequence of binding to the carrier protein.

The Specificity of Facilitated Diffusion.

Unlike simple diffusion, facilitated diffusion is a highly specific process. Since transport of the solute involves binding to a carrier protein, it is likely that there are specific carriers in the cell membrane for each of the solutes that enter or leave the cell by facilitated diffusion. The specificity of the carrier is not absolute. For example, structurally related amino acids are often able to share the same carrier. The transport of sugars across the erythrocyte plasma membrane provides another example. The transport system works best when glucose is the solute being transported, but related sugars such as galactose can utilize the same system to cross the membrane.

Competitive Inhibition in Facilitated Diffusion.

Structurally related compounds will compete for the same binding site on a membrane carrier, giving rise to the phenomenon of **competitive inhibition** of solute transport, as characterized by an increase in the K_m without a change in the V_{max} of the transport system. Thus, amino acid A, which shares the same transport system as amino acid B, will act as a competitive inhibitor of the transport of B if both are present simultaneously (Fig. 4–21). The inhibitory effect of A can be overcome by an increase in the concentration of B so that more molecules of B are

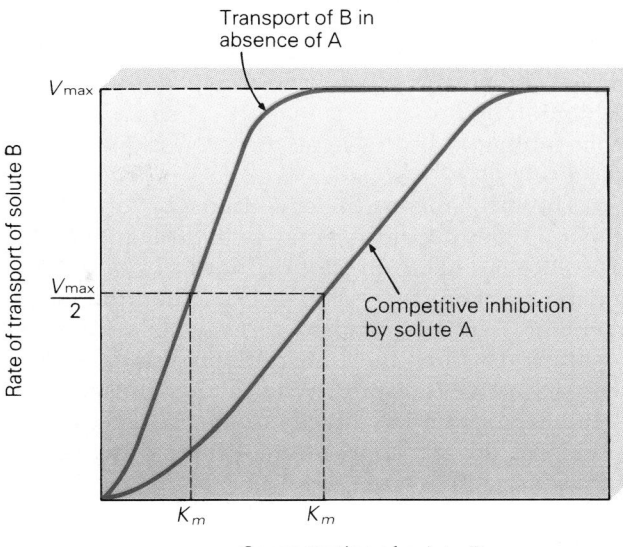

Figure 4–21
Competitive inhibition of facilitated diffusion of Solute B by the structurally similiar Solute A. In the presence of A, the concentration of B must increase to reach the same V_{max} as that attained in the absence of A.

available to compete with A for the carrier binding sites. In this way the rate of transport of B can be increased to the same V_{max} that was achieved in the complete absence of A, but a higher concentration of B (i.e., an increased K_m) will be required (see Fig. 4–21).

Active Transport: Energized Movement Against a Gradient

Thus far we have discussed passive transport processes, in which molecules and ions move in or out of the cell from regions of high to regions of low concentration. These movements, while sometimes requiring the presence of a special protein molecule, do not require energy. They occur in response to electrochemical gradients. The result is always a *decrease* in concentration of a once highly concentrated region.

Frequently, however, the cell requires an increase in concentration of an already highly concentrated solute. The solute must move from a less concentrated region outside the cell to a more concentrated region inside—that is, *against* the prevailing gradient. This process requires input of energy. The transport system will continue to operate for as long as the energy supply and the solute are available. Active transport, like facilitated diffusion, requires carrier proteins. The energy supply for active transport is derived from cellular metabolism; hence, active transport systems are susceptible to inhibition by compounds that act primarily as metabolic inhibitors and have no direct effect on the carrier proteins.

The mechanisms of active transport are of two general types depending on whether the movement of solute is linked directly or indirectly to energy-yielding reactions (see Fig. 4–10). When the movement of solute is coupled directly to an energy-yielding reaction, the transport process is termed **primary active transport.** When active transport of a solute is not coupled directly to the energy-yielding reactions, the transport mechanism is described as **secondary active transport.**

Primary Active Transport

The best known example of a primary active transport system is probably the **sodium pump.** This pump is present in the plasma membrane of most animal cells and is also known as the **Na$^+$/K$^+$-ATPase pump** because Na$^+$ is pumped out of the cell at the same time as K$^+$ is pumped in, and because the ion movements are coupled directly to breakdown of ATP. Removal of one of the phosphate groups from ATP releases free energy that is used to drive the movement of both Na$^+$ and K$^+$ against their respective electrochemical gradients. It is chiefly through this active transport system that the cell is able to maintain a high K$^+$ concentration and a low Na$^+$ concentration in the cytosol.

The breakdown of ATP is tightly coupled to the transport of Na$^+$ and K$^+$; the one cannot occur without the other. For every molecule of ATP that is split into ADP and phosphate, 3 Na$^+$ are pumped out of the cell and 2 K$^+$ are pumped in (Fig. 4–22). The molecular events that link ATP breakdown to ion transport are not fully understood, but it seems clear that the enzyme protein is temporarily phosphorylated by attachment of the phosphate group released from the ATP. The phosphorylation step is Na$^+$-dependent—that is, it occurs only when Na$^+$ is present, and presumably leads to a change in protein conformation that moves bound Na$^+$ out of the cell. The subsequent dephosphorylation (removal of phosphate) is K$^+$-dependent and probably allows the protein to return to its original conformation, moving bound K$^+$ into the cell. These steps are depicted in the scheme in Figure 4–23. The Na$^+$/K$^+$-ATPase is represented as a protein with four subunits. This active transport system is so important to the cell that about one third of the cell's energy supply is used to maintain its activity.

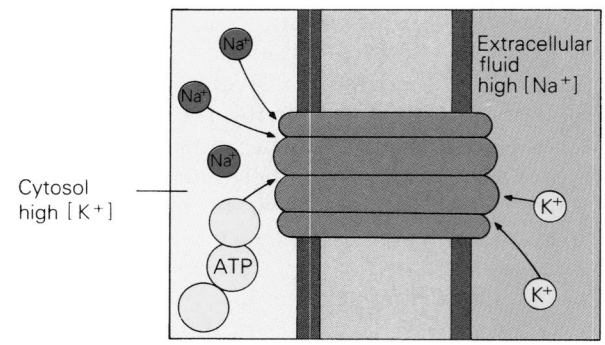

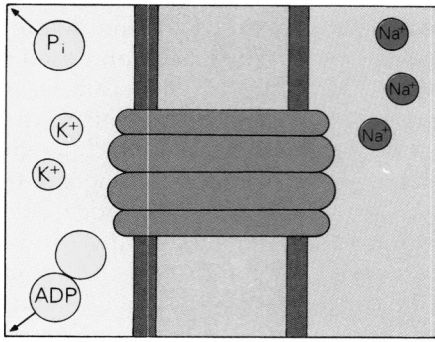

Figure 4–22
An overview of primary active transport by Na$^+$/K$^+$-ATPase. The free energy derived from hydrolysis of ATP is used to drive transport of Na$^+$ out of the cell and K$^+$ into the cell. Both ions are moved against their electrochemical gradients. Three Na$^+$ exit for every two K$^+$ that enter the cell.

Secondary Active Transport

When transport of a solute is not coupled directly to energy-yielding reactions, it is described as **secondary active transport.** A common example of this kind of mechanism is a transport system that is driven by the energy stored in the electrochemical gradient for another solute. In animal cells, this solute is usually Na$^+$, which enters cells passively by moving down a very favorable electrochemical gradient.

Symport and Antiport. Recall that there is a low sodium concentration and a negative potential inside the cell relative to the outside. Many different solutes are transported across cell membranes against their electrochemical gradients by coupling to the movement of Na$^+$. Movement of the solutes in the same direction as Na$^+$ is known as **symport** or **cotransport;** movement in the opposite direction is known as **antiport** or **exchange** (Fig. 4–24). Solute transport systems that are coupled to Na$^+$ movement have a very specific requirement for Na$^+$;

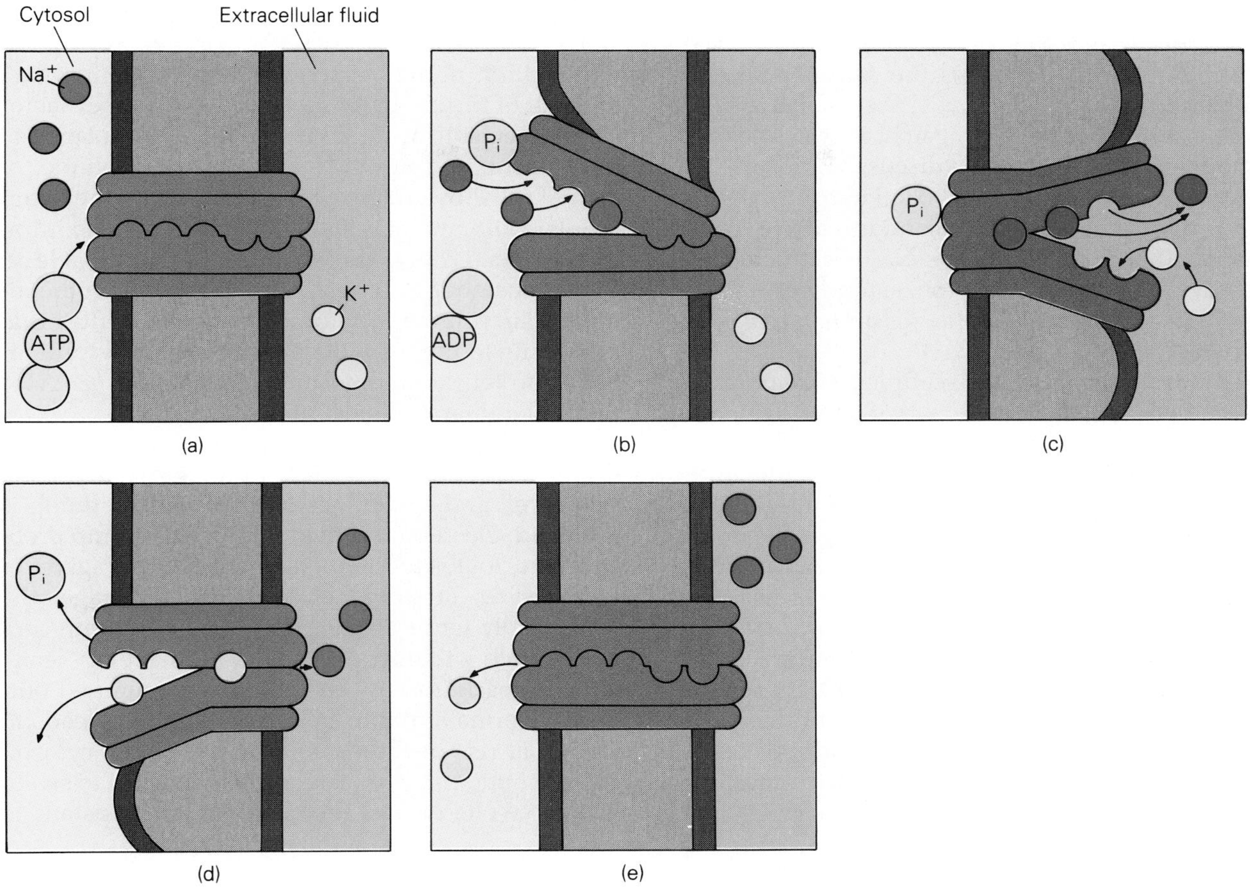

Cytosol   Extracellular fluid

Na$^+$

ATP

K$^+$

(a)

P$_i$

ADP

(b)

P$_i$

(c)

P$_i$

(d)

(e)

Figure 4–23
Hypothetical scheme of action of Na$^+$/K$^+$-ATPase. (*a, b*) In the presence of Na$^+$ inside the cell, ATP binds to and phosphorylates one of the protein subunits. Na$^+$ then binds to specific sites on the phosphorylated protein. (*c, d*) These events trigger a conformational change that exposes the bound Na$^+$ to the cell exterior, and Na$^+$ dissociates from the protein. The binding site for K$^+$ is now exposed, and external K$^+$ binds. (*e*) Binding of K$^+$ promotes dephosphorylation (removal of the phosphate) of the protein, which allows it to revert to the original conformation. The conformational change moves the bound K$^+$ inside the cell, where it dissociates from the protein. The cycle now can be repeated. The conformational changes are greatly simplified in the drawing.

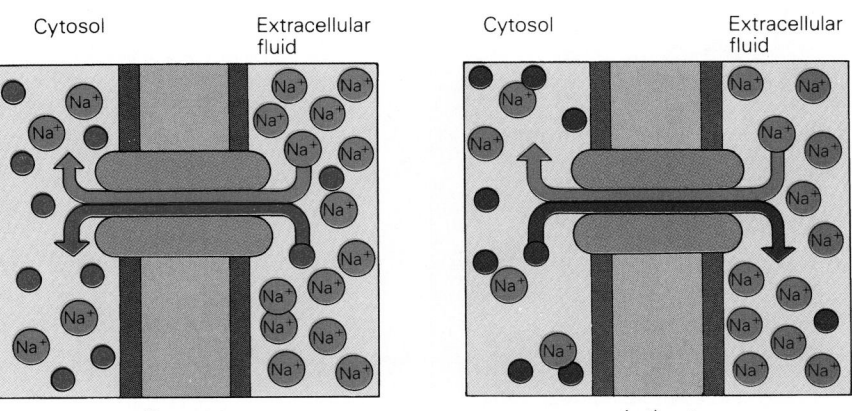

Cytosol   Extracellular fluid

Na$^+$

Symport

Cytosol   Extracellular fluid

Na$^+$

Antiport

Figure 4–24
Secondary active transport occurs when solute transport is driven by the electrochemical gradient for Na$^+$. Both solute and Na$^+$ bind to the membrane carrier. The Na$^+$ gradient is maintained by primary active transport of Na$^+$.

other ions of similar size such as K^+ and Li^+ cannot be used in place of Na^+.

The Na^+ gradient across the cell membrane is maintained by the Na^+/K^+-ATPase, which actively pumps Na^+ out of the cell against its electrochemical gradient. As we have just discussed, this process is driven directly by a metabolic energy supply in the form of ATP. Thus, secondary active solute transport that is driven by the Na^+ electrochemical gradient depends ultimately on a supply of ATP but is not linked directly to the reactions that break down ATP.

The mechanism of Na^+-coupled transport systems is believed to involve a two-site carrier that binds both Na^+ and the solute. A possible mechanism is illustrated in Figure 4–25. Na^+ outside the cell binds readily to the carrier and encourages binding of solute to the carrier even though the solute concentration is low outside the cell. A conformational change in the carrier protein now exposes the bound Na^+ and solute to the cell interior. Na^+ readily dissociates from its site because the intracellular Na^+ concentration is very low. Loss of Na^+ from the carrier may alter the binding of solute to the carrier and cause the solute to dissociate from the carrier even though there is already a high intracellular concentration of the same solute.

Transport Processes and Cell Volume

The sodium pump, which uses ATP to pump Na^+ ions out of the cell in exchange for K^+, is a key factor in the mechanism for controlling the osmolarity of the cytosol, and hence the intracellular volume, of animal cells. By effectively adding Na^+ to the extracellular fluid and simultaneously keeping the intracellular Na^+ concentration low, the cell is able to prevent osmotic entry of water and maintain the intracellular volume. The other principal solutes that contribute to the osmotic pressure of cell cytosol include proteins, organic phosphates such as ATP, metabolic intermediates, and K^+ ions.

The importance of the sodium pump in the control of cell volume is evident in the fact that animal cells swell and sometimes burst when the pump is inhibited. Decreased activity of the Na^+ pump probably contributes to the increased osmotic fragility of erythrocytes in stored blood. During storage, the ATP supply tends to fall. Since the Na^+ pump cannot operate without ATP, any Na^+ ions that "leak" (diffuse) back into the cell will not be pumped out. This abnormal buildup in intracellular Na^+ leads to a gradual rise in the osmolarity of the cell cytosol, with osmotic entry of water causing the cell to swell. In this swollen state, the cells are less resistant to

Figure 4–25
Hypothetical scheme of action of a two-site carrier for Na^+-coupled solute symport. (*a*) Binding of Na^+ outside the cell to the carrier increases the affinity for the solute, which also binds even though it is present at a low concentration. (*b*) A conformational change in the carrier protein exposes the binding sites to the cell interior. Na^+ dissociates from the carrier because the Na^+ concentration is low inside the cell. This dissociation step leads to a decrease in carrier affinity for solute, so the latter also dissociates from the carrier even though the solute concentration is high inside the cell. The conformational changes are probably much more subtle than those shown in these diagrams.

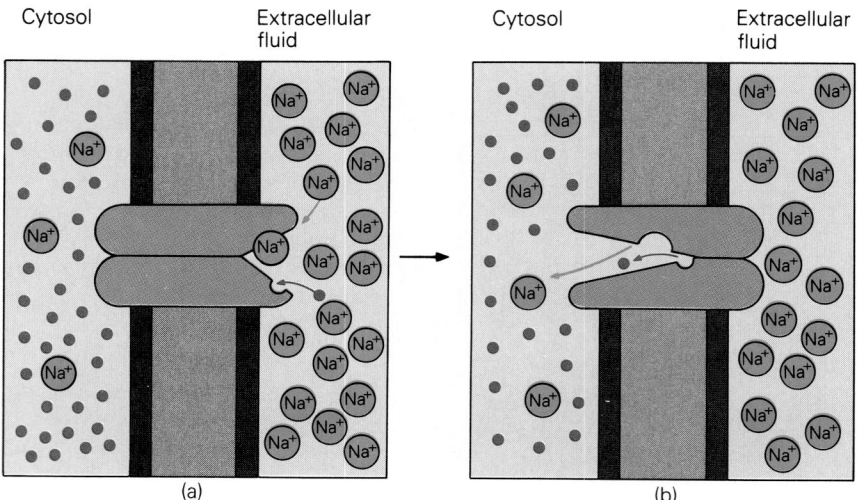

(a)   (b)

osmotic stress than normal biconcave erythrocytes because they cannot accommodate the same increase in intracellular volume.

Typically, animal cells are exposed to a high concentration of external Na^+. There is no net increase in intracellular Na^+ because the rate at which Na^+ leaks into the cell is closely matched by the rate at which Na^+ leaves the cell via the Na^+ pump. This mechanism prevents the external and internal Na^+ concentrations from reaching equilibrium even though Na^+ constantly enters and leaves the cells.

Thus, Na^+ outside the cell can be considered to behave as if it were a nonpermeant solute. Similarly, the rate at which K^+ is pumped into the cell matches the rate at which K^+ leaks out, so the high internal K^+ concentration and the low external K^+ concen-

tration are maintained. Thus, K^+ inside the cell behaves effectively as a nonpermeant solute because, as with Na^+, there is no net movement of K^+ into or out of the cell even though the internal and external concentrations are markedly different.

Some solutes, such as urea and glycerol, penetrate the plasma membranes of an animal cell very rapidly. The cell does not have a mechanism for pumping these solutes back out again. Such **permeant solutes** exert only a transient effect on cell volume, however. Unlike Na^+ and K^+, their external and internal concentrations rapidly equilibrate.

Consider an erythrocyte placed in a large volume of a solution containing 300 mOsm NaCl and 60 mOsm urea (Fig. 4–26). Initially the cell will experience a hyperosmotic environment. Water will

Figure 4–26
Permeant solutes such as urea exert only a transient effect on the volume of an erythrocyte because their intracellular and extracellular concentrations rapidly equilibrate. In reality, Steps I and II probably overlap considerably.

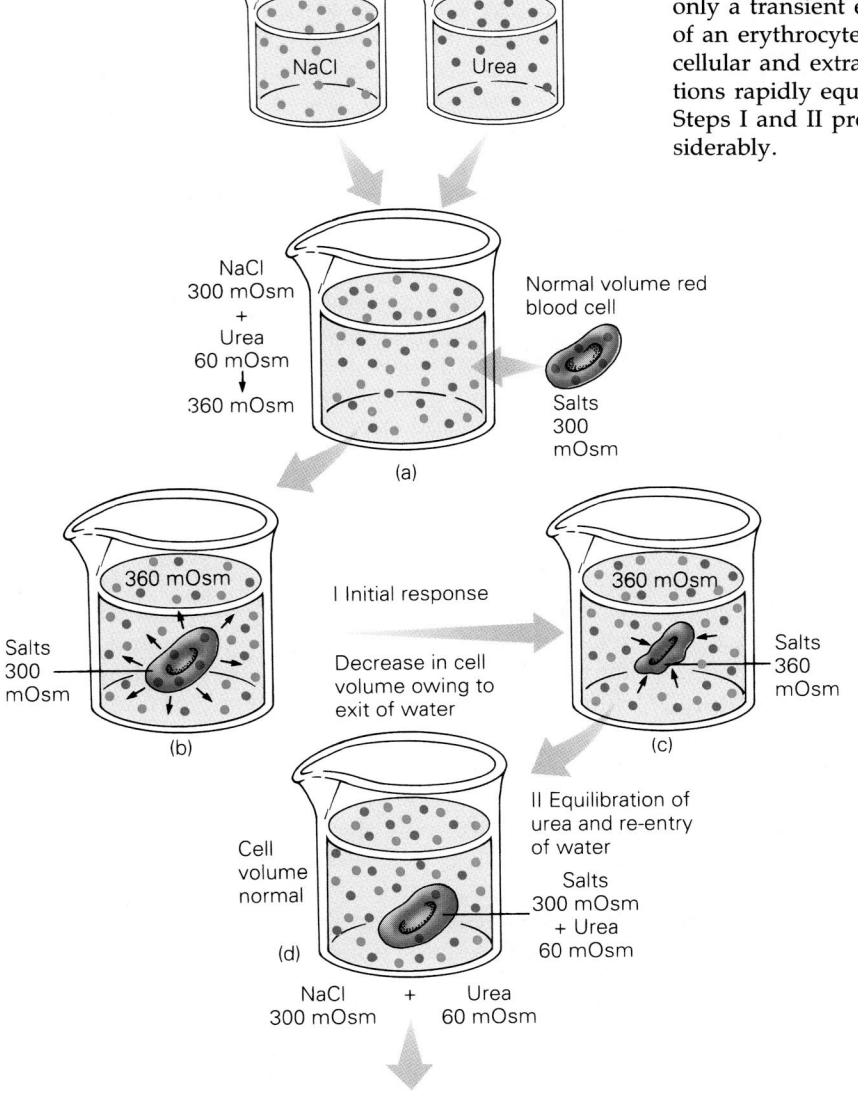

flow out, and the cell volume will decrease. However, as the urea rapidly enters the cell and the intracellular urea concentration increases toward 60 mOsm, water also will enter and the cell will begin to swell. The final volume of the cell at equilibrium is determined only by the osmolarity of the impermeant solutes in the extracellular medium. In this case, the impermeant solutes (NaCl) have a total osmolarity of 300 mOsm, the same as the cell cytosol, and the cell will assume its original volume.

Na^+ is the most abundant solute present in blood plasma. As an effectively impermeant solute, it is the major determinant of intracellular volume in the animal, even though plasma contains many membrane permeable solutes. A drop in plasma Na^+ concentration will reduce plasma osmolarity, and osmotic flow of water into cells will cause them to swell. Conversely, a rise in plasma Na^+ concentration will lead to cell shrinkage, owing to loss of intracellular water. This is one of the reasons that plasma Na^+ concentration is closely regulated. A change in Na^+ concentration induces a series of rapid responses designed to return it to the normal range.

Close control of the plasma Na^+ level means that most normal human cells are not subjected to large shifts in plasma osmolarity. However, challenges to the internal osmolarity can occur in cells designed to move water and specific solutes from one side of the cell to the other. This is true particularly for epithelial cells (see next section). Changes in internal osmolarity cause water to flow into or out of the cell and lead to changes in intracellular volume. Many cells have mechanisms for rapidly correcting the volume changes, and these mechanisms do not utilize the Na^+/K^+-ATPase pump. Most cells respond to an increase in cell volume by activating pathways that increase the efflux of K^+ and Cl^- from the cell. The loss of solute decreases intracellular osmolarity, which in turn stimulates efflux of water from the cell and leads to a return to normal cell volume. This response is termed a **regulatory volume decrease.** Some, but not all, cells can respond to a decrease in normal cell volume by a complementary process known as a **regulatory volume increase.** In this case, shrinkage of the cell activates pathways which increase influx of solutes such as Na^+ and Cl^-. Gain of solutes increases intracellular osmolarity, causing an influx of water and a return to normal cell volume. These rapid volume regulatory responses may help epithelial cells maintain a near-constant volume even with large changes in the amount of water and solutes passing through the cells.

Epithelial Transport

The transport systems in the plasma membrane of the cell are important for supplying nutrients and removing waste products. In addition to these roles, the specific arrangement of these transport systems in epithelial cells permits the net movement of solutes and water from one side of the epithelium to the other. This organization is crucial for the function of epithelial cells such as those lining the lumen of the small intestine and the renal proximal tubule, tissues that are specialized for reabsorption. These cells allow **transcellular transport** of water and solutes because the entry and exit pathways are on opposite sides of the cell (Fig. 4–27).

The entry step is at the apical membrane. This is the membrane where the Na^+-coupled symport systems for amino acids, glucose, and phosphate are located. The exit step is across the basolateral membrane where the Na^+/K^+-ATPase is located. The Na^+ gradient across the apical membrane is maintained because the ATPase in the basolateral membrane constantly pumps Na^+ out of the other side of the cell to keep intracellular Na^+ low.

Junctional Complexes. The asymmetrical distribution of transport systems between the apical and basolateral plasma membranes persists because of the **junctional complexes** between the cells. The junctional complexes occur at the point where the plasma membranes of neighboring cells come into close contact, and they prevent lateral diffusion of the proteins in the cell membrane. Thus, the carrier proteins in the apical membrane are physically prevented from leaving the apical membrane and mixing with the carrier proteins in the basolateral membrane.

In some types of epithelia, these junctions are relatively permeable and allow water and ions to move between the epithelial cells, providing an additional or **paracellular** pathway across the epithelium (see Fig. 4–27). This type of epithelium is termed a "**leaky**" epithelium because water and some solutes can leak between the cells in addition to moving through the transcellular pathway. A good example of this type of epithelium is the proximal tubule of the kidney. In other epithelia, such as the small intestine and the collecting tubule of the kidney, the cell junctions are impermeable; the only route by which water and solutes can cross the epithelia is the transcellular pathway. These types of epithelia are termed "**tight**" epithelia because they do not permit leakage between the cells.

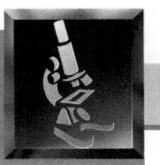
Single Channel Recording

The flow of ions such as Na^+ and K^+ across nerve cell membranes plays a central role in transmission of nervous impulses (see Chapter 7). The ion flow occurs through integral membrane proteins known as **channels.** Each nerve cell has several different types of ion channels, and as many as one million channel proteins of each type may be present in the plasma membrane. Ion flow through large populations of channels, and the resulting electrical activity, has been studied in nerve cells for many years. In 1976, a new method known as the **patch-clamp** technique was introduced. It was developed in Germany by E. Neher and B. Sakmann and is now used in a large number of laboratories. The great advantage of this technique is that it allows the study of a **single** ion channel.

The technique is used in the following way: A clean glass pipette is pressed against an intact cell to form a tight seal. The tip of the pipette has a diameter of about 1 μm and is polished to avoid damage to the cell membrane. Slight suction allows formation of a very stable glass-membrane seal. The ion channels underneath the tip of the pipette can now be studied to determine their regulation by intracellular molecules. The patch of membrane underneath the pipette tip can also be pulled away from the cell by withdrawing the pipette from the cell surface. The membrane patch remains undamaged and functional in the aperture of the pipette. The membrane patch can be detached from the cell because the glass-membrane interactions are stronger than the phospholipid interactions that maintain the integrity of the plasma membrane. The great advantage of using a detached patch is that the plasma membrane surface that faced the inside of the intact cell now becomes accessible to experimental manipulations.

In essence, whenever a channel protein opens within the membrane spanning the pipette tip, there will be a flow of ions from one side of the membrane to the other along their concentration and electrical gradients. This flow of ions generates a tiny electric current that can be measured, in picoamperes (pA), by very sensitive electronic measuring equipment. These currents can be detected because the patch-clamp instrumentation eliminates electrical noise that, if present, would obscure the small single channel currents. Patch clamping isolates a small area of membrane so that only a few channels are present. This allows the contribution from each channel to be separated and analyzed.

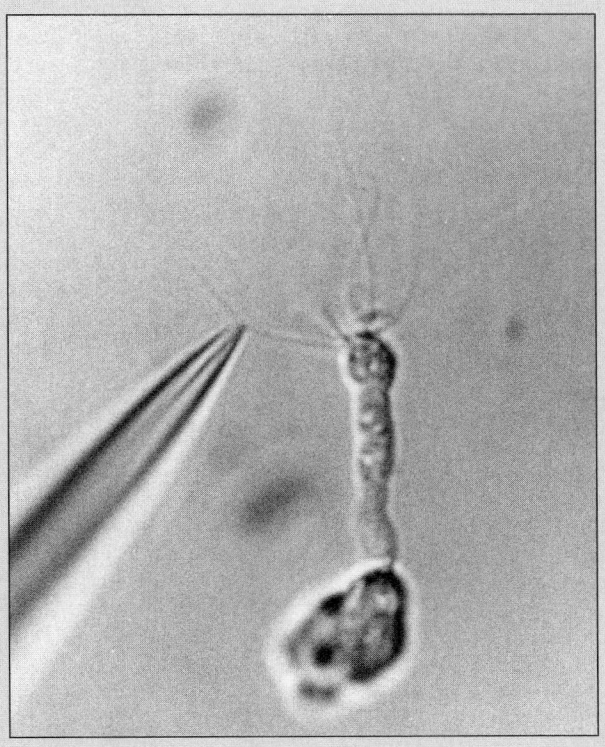

A photograph showing the view in a light microscope of a patch clamp pipette pressed against the plasma membrane of a cilium which extends from a cell derived from the olfactory epithelium of a toad. Most of the pipette is not in focus and can be seen to the left of the cell. ($\times$ 2,300)

Recording single channel activity has allowed new understanding of the action of channel proteins in nerve and other cells. Several unique properties of ion channels have been discovered: (1) Ion channels behave in a simple open-shut manner, i.e. they are either fully open or fully closed. They open for a period of time that is often less than 1 millisecond. (2) Channels open in a random way and so current flows for varying lengths of time. The average duration of current flow is a quantity used to characterize the specific type of channel. (3) The ease with which current flows (known as **conductance**) when a channel opens is also characteristic of the type of channel. (4) Most channels exhibit selectivity with regard to the ions that can pass through them. Some channels allow passage of Na^+ but not K^+, for example. In contrast, others allow flow of K^+ but not Na^+.

SUMMARY

The Components and Structure of the Membrane

The core structure of a plasma membrane is a double layer of lipid molecules known as the lipid bilayer. This structure allows the passage of small lipid-soluble solutes but is highly impermeable to charged or polar solutes.

Most membrane proteins are embedded in the lipid bilayer, but a few are only loosely associated with one side of the bilayer. The proteins carry out most membrane functions by acting as specific transporters, receptors, or enzymes.

Carbohydrates are a relatively minor component of the membrane and occur as both glycoproteins and glycolipids. The carbohydrate portion of the molecule is located exclusively on the exterior surface of the membrane.

The Types of Movement Through the Membrane

A solute crosses a membrane by simple diffusion in response to an electrochemical gradient for the solute. Lipid-soluble, nonpolar molecules cross the plasma membrane this way. Specific membrane proteins, most likely spanning the lipid bilayer, bind charged or polar solutes and "facilitate" their diffusion across the membrane. Cells respond to changes in the osmolarity of their surroundings by gaining or losing water, sometimes lysing in the process.

Active transport is the movement of solute against its electrochemical gradient, achieved by input of energy (usually in the form of ATP) derived from cellular metabolism. The solute is moved across the membrane by a specific membrane protein carrier. This step can be linked either directly or indirectly to the energy-yielding reactions.

Many animal cells resist osmotic entry of water and, hence, an increase in intracellular volume by pumping Na^+ ions out so that the intracellular Na^+ concentration remains low relative to the extracellular concentration.

Epithelial Transport

The different types of transport mechanisms act in concert in epithelial cells to achieve directional movement of solutes and water from one side of the cell to the other. This process is aided by the structural and functional organization of the cells.

REVIEW QUESTIONS

Identify Each Term

1. General term for a membrane protein which penetrates the phospholipid bilayer.

2. The process of simple diffusion when the molecules in motion are water molecules.

3. Carrier dependent transport of solute in response to the electrochemical gradient.

4. Coupled movement of two solutes in opposite directions across a membrane.

5. The process by which a swollen cell returns rapidly to normal size.

Define Each Term

6. Amphipathic molecule.

7. Overton's rules.

8. Diffusion trapping.

9. Secondary active transport.

10. Tight epithelium.

Choose the Correct Answer

11. Which of the following processes requires a supply of metabolic energy?

    a. Simple diffusion

    b. Facilitated diffusion

    c. Primary active transport

    d. Osmosis

12. Simple diffusion differs from facilitated diffusion because it:

    a. is much slower

    b. can be blocked by competitive inhibitors

    c. reaches a maximum rate

    d. requires expenditure of energy

13. The plasma membrane of a cell separates:

    a. organelles from cytosol

    b. cytosol from chromosomes

    c. cytosol from mitochondrial matrix

    d. extracellular fluid from cytosol

14. In the phospholipid bilayer of a plasma membrane:

    a. phospholipids are distributed asymmetrically

    b. proteins are distributed asymmetrically

    c. both phospholipids and proteins move laterally

    d. all the above are true

15. The plasma membrane Na$^+$ pump
   a. moves K$^+$ out of the cell
   b. exchanges 3 Na$^+$ for 2 K$^+$
   c. uses energy released by ADP breakdown
   d. moves Na$^+$ into the cell

16. A membrane which is permeable only to water separates two solutions of glucose dissolved in water. On one side of the membrane (side A) the glucose concentration is 0.1 g/ml. On the other side (side B) the glucose concentration is 0.5 g/ml. Initially, the rate of water flow will be:
   a. most rapid from side A to side B
   b. most rapid from side B to side A
   c. the same in both directions
   d. zero (no flow in either direction)

Answer Each Question in One or Two Sentences

17. List four of the principal features of the fluid mosaic (Singer-Nicolson) model of the plasma membrane.

18. Carbohydrate on the external surface of the plasma membrane is probably important for membrane receptors and cell-cell recognition. Why are carbohydrates particularly suited for these functions?

19. Explain why the erythrocyte can increase its volume by 67% without rupturing.

20. Explain the term regulatory volume decrease.

21. What distinguishes a leaky epithelium from a tight epithelium?

SUGGESTED READINGS

Armstrong, W. McD. "The cell membrane and biological transport." *Physiology*, 5th ed. Selkurt, E.E., ed. Boston: Little, Brown, 1984. pp. 1–26.

Finean, J.B., Coleman, R., and Michell, R.H. *Membranes and Their Cellular Functions*, 2nd ed. New York: John Wiley & Sons, Inc., 1978.

Graves, J.S., ed. *Regulation and Development of Membrane Transport Processes*. New York: John Wiley & Sons, Inc., 1985.

Handler, J.S. "Overview of epithelial polarity." *Annual Review of Physiology*, 51:729–740, 1989.

Hille, B. *Ionic Channels of Excitable Membranes*. Sunderland, MA: Sinauer Associates Inc., 1984.

Keynes, R.D. "Ion channels in the nerve-cell membrane." *Scientific American*, 240:126–135, March, 1979.

Lewis, S.A., and Donaldson, P. "Ion channels and cell volume regulation: chaos in an organized system." *News in Physiological Sciences*, 5:112–119, 1990.

Pederson, P.L., and Carafoli, E. "Ion motive ATPases: I. Ubiquity, properties, and significance to cell function." *Trends in Biochemical Sciences*, 12:146–150, 1987.

Quinton, P.M. "Cystic fibrosis: a disease in electrolyte transport." *FASEB Journal*, 4:2709–2717, 1990.

Rothman, J.E., and Lenard, J. "Membrane asymmetry." *Science*, 195:743–753, 1977.

Singer, S.J., and Nicolson, G.L. "The fluid mosaic model of the structure of cell membranes." *Science*, 175:720–731, 1972.

Stevens, C.F. "Studying just one molecule: single channel recording." *Trends in Pharmacological Sciences*, 5:131–134, 1984.

Walmsley, A.R. "The dynamics of the glucose transporter." *Trends in Biochemical Sciences*, 13:226–231, 1988.

Welsh, M.J. "Abnormal regulation of ion channels in cystic fibrosis epithelia." *FASEB Journal*, 4:2718–2725, 1990.

KEY TERMS

active transport (p. 131)
carrier protein (p. 131)
channel gating (p. 138)
cholesterol (p. 127)

diffusion coefficient (p. 132)
facilitated diffusion (p. 131)
fatty acid (p. 127)
fluid mosaic model (p. 125)

junctional complex (p. 146)
lipid (p. 125)
lysis (p. 135)
permeability coefficient (p. 133)

phospholipid (p. 127)
simple diffusion (p. 131)
sodium pump (p. 142)
tight epithelium (p. 146)

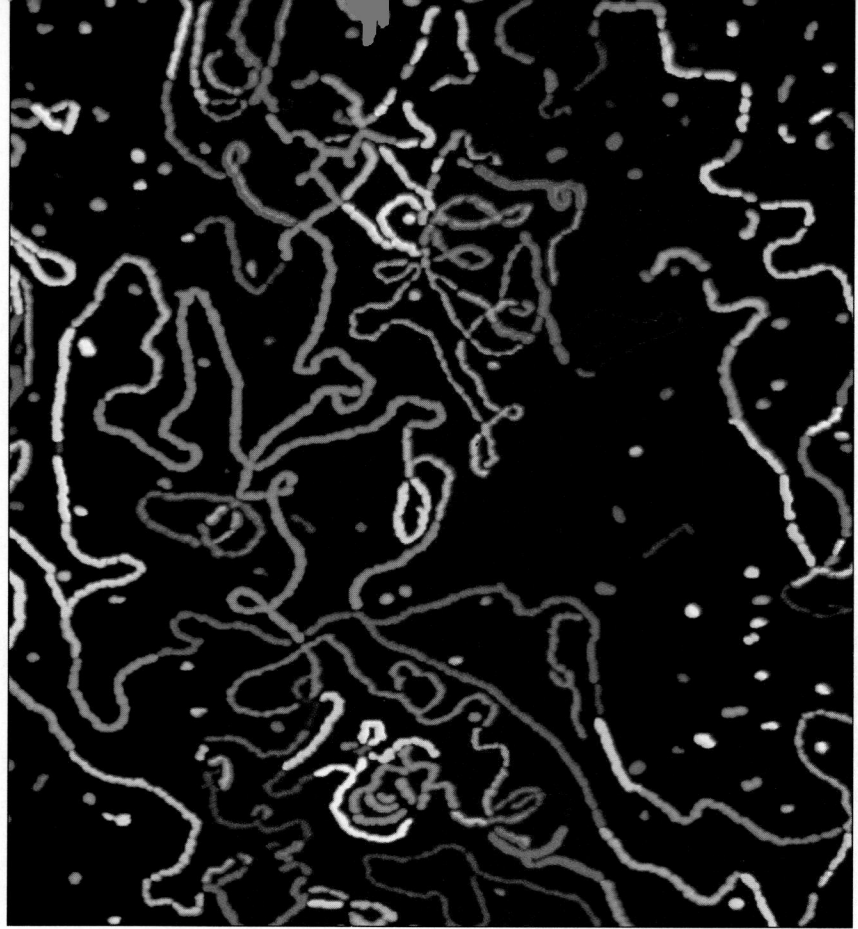

Cellular Control Mechanisms

CHAPTER OUTLINE

*M*uch of the success of multicellular organisms is enabled by the organization of groups of differentiated cells into specialized tissues and organs. The tissues coordinate to carry out all the functions necessary for the organism's survival.

The specialization and coordination of cell functions require very effective communication, both within each cell and among the different tissues. A good example of the importance of efficient communication is the process involved in avoiding a predator. When the tissues responsible for sight, sound, or smell detect the presence of a predator, the survival of the organism often depends on the almost-instantaneous response of other tissues that can carry out rapid evasive movement. Mammals have systems that provide communication and coordination among the different organs. These are the **nervous system** and the **endocrine system.** A large section of the book is devoted to descriptions of these very sophisticated systems.

This chapter emphasizes the ways in which some of the intracellular structures and macromolecules described in the preceding chapters participate in specific control systems that reside and operate within a cell, as well as ways in which these controls are influenced by signals from other cells. We begin by discussing the use of the genetic information stored in DNA, perhaps the fundamental **intracellular control system.** We will learn how this system is subject to regulation primarily by intracellular influences. The latter part of the chapter introduces the concept of **extracellular control systems,** or **cell-to-cell signaling,** the ways in which one cell communicates with and influences the function of another.

An Intracellular Control System: DNA, RNA, and Protein Synthesis

The cells of a complex eukaryotic organism such as a human being have developed a variety of structural and functional specializations, giving rise to different types of tissues, such as epithelial, connective, muscle, and nervous tissue. As previously mentioned in the brief discussion of the functions of the cell nucleus in Chapter 3, the molecular basis for the differentiation of functions and structures of cells is the structure of their proteins. In Chapter 4 we learned, for example, that different cells have different types and amounts of proteins in their membranes. These proteins are distributed in specific ways in the membrane in order to effectively perform their roles as carriers, channels, receptors, or shaping agents. The red blood cell produces hemoglobin, whereas a nerve cell makes none of this protein. All cells in the body make and use Na^+/K^+-ATPase, but different cells make different amounts of it. Most cells in the body produce myosin (a contractile protein), but only muscle cells make it in the abundance necessary to produce the type of motion we associate with visible body movements.

Overview: Protein Synthesis as Genetic Expression

With few exceptions, the markedly different types of cells contain the same **genome** in their nuclei—that is, each cell, regardless of tissue type, has the same DNA molecules, the same "permanent files" or master set of instructions for protein synthesis. These DNA molecules are in groups called **chromosomes.** Every cell in the human body, except the reproductive cells, has the same number (46) of chromosomes containing the same DNA molecules (Fig. 5–1a). These DNA molecules are made up of **genes** (Fig. 5–1b). Each gene consists of a certain sequence of nucleotides that dictates the sequence of amino acids in a specific protein produced by the cell. Every cell in the body, except the sex cells, can potentially make every protein found in the body.

It would seem that if all cells in the body contained the same genetic information, they would all make the same proteins in the same amounts. How then do cells become **differentiated,** producing some proteins and not others? How does one cell type, such as a pancreatic cell, know to produce an abundance of insulin, whereas another cell, such as a red blood cell, knows *not* to produce insulin (even though it possesses the gene for insulin), but to produce hemoglobin instead? How does a cell in the kidney know to produce myosin in the controlled amount required for its own shape and internal movements, but not in the abundant amounts that would produce movement of the entire organ? How does a skeletal muscle cell know to produce myosin in abundance, enough to produce movement of other tissues?

Genetic Expression: Selective Use of the Permanent Files

The answer to the previous questions lies in the phenomenon of **genetic expression** or, more precisely, the **control of genetic expression.** An individual cell has much more genetic information than it ever needs. Most of the genetic information in the cell is unused, or **suppressed.** At any given time in a particular cell, only a small part of the DNA is being used as a source for information about protein synthesis. In other words, only certain genes are being tapped for information about how to make the specific protein they describe; only certain genes in each cell are **expressed** in the form of **protein products.**

Thus, gene expression is the principal factor that determines the structural, functional, and behavioral characteristics of a cell. In complex eukary-

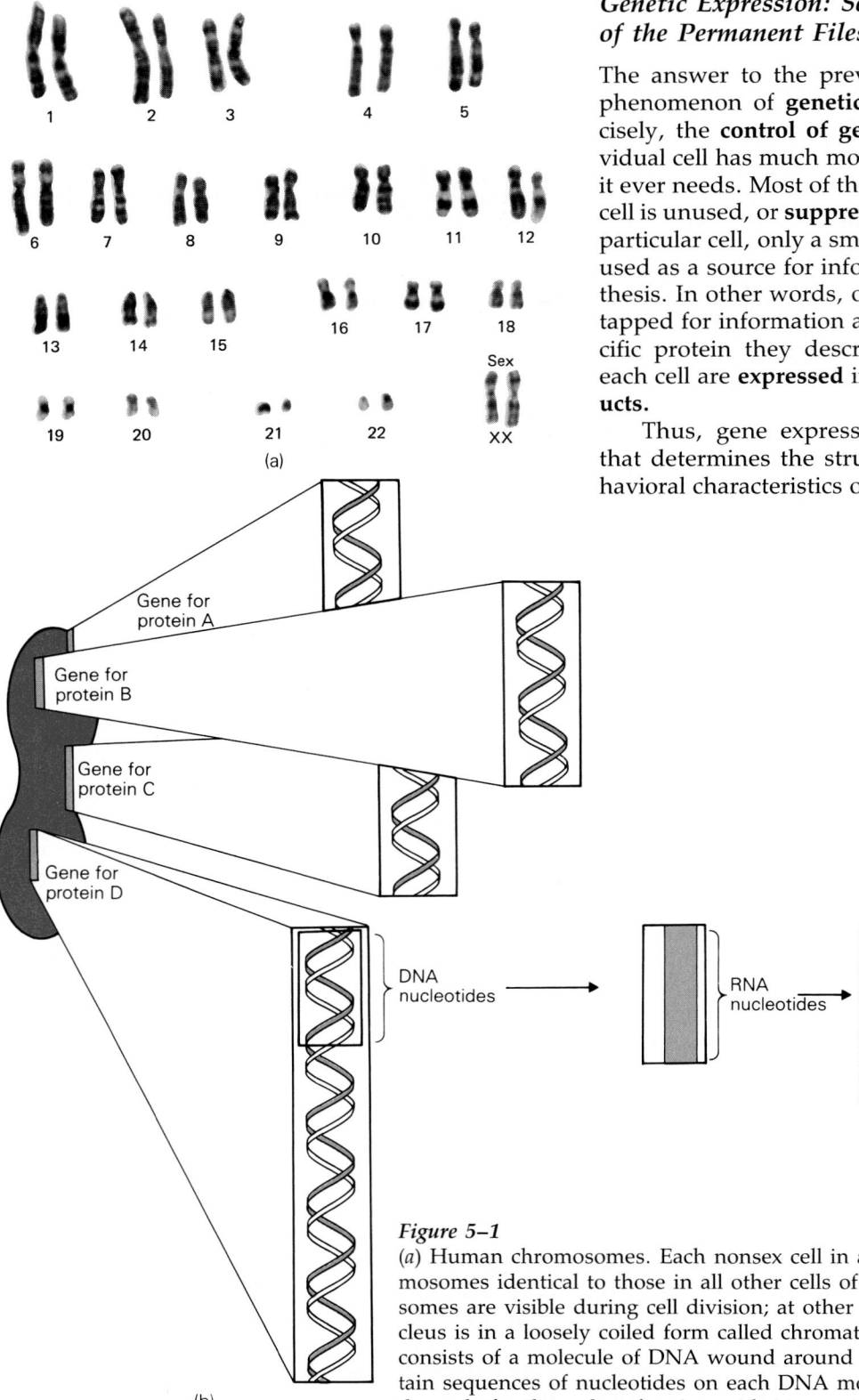

(a)

(b)

Figure 5–1

(*a*) Human chromosomes. Each nonsex cell in a human body has 46 chromosomes identical to those in all other cells of the same body. Chromosomes are visible during cell division; at other times, the DNA in the nucleus is in a loosely coiled form called chromatin. (*b*) Each chromosome consists of a molecule of DNA wound around proteins called histones. Certain sequences of nucleotides on each DNA molecule, called genes, contain the code for the order of amino acids in a protein molecule to be made by the cell. The DNA nucleotide sequences are transcribed into RNA nucleotide sequences, which are then translated into amino acid sequences to form a polypeptide.

otes, different genes are being expressed as the characteristic protein products of each different cell type. Furthermore, in a given cell, certain genes may be expressed only at certain times (e.g., when the cell receives the appropriate stimulus, possibly a hormone). Some of the mechanisms that control the expression of genes in a cell are discussed in detail in a later section.

The Genetic Code: From Nucleic Acids to Amino Acids

The term **genetic code** refers to the way in which information about protein synthesis is stored in DNA, "translated" into RNA, and then used to make proteins. The information in a DNA molecule specifies the sequence in which amino acids must be assembled to form a functional polypeptide molecule. A relationship exists between the nucleotides of RNA and DNA and the amino acids of a polypeptide chain, so that the order of nucleotides dictates the order of amino acids.

DNA and RNA are each composed of only four different nucleotides, whereas polypeptides can contain up to 20 different amino acids. Since there are only four nucleotides in each type of nucleic acid to specify the 20 different amino acids, we assume that **combinations** of nucleotides, rather than single nucleotides, are used to describe the amino acids. The genetic code must specify, in the sequence in which the four nucleotides are arranged, the sequence in which the 20 common amino acids should be arranged. Think of DNA, RNA, and proteins as three different languages, each with a different alphabet and vocabulary. Protein synthesis then is a process of translation, from the language of nucleic acids to the language of proteins.

The Four Letters of the DNA and RNA Alphabets. DNA molecules are formed from **deoxyribonucleotides,** four different molecules distinguished by four different **bases:** adenine, cytosine, guanine, and thymine (Fig. 5–2a). The DNA "alphabet" consists of four letters—A, C, G, and T—representing the four DNA bases (Table 5–1). These four letters can be combined in a total of 64 different "words," or **triplets,** such as CGA, CGT, CGG, and so on.

RNA molecules are formed from **ribonucleotides**—again, four different molecules distinguished by four different bases: adenine, cytosine, guanine, and uracil (see Fig. 5–2a). Thus, the RNA alphabet also contains four letters—A, C, G, and U (Table 5–1). These can also be combined in groups of three to make a total of 64 different "words," called **codons,** found on a special type of RNA called mRNA. As we will learn, the codons on an RNA molecule are **complementary** to the triplets on the DNA molecule from which they were transcribed. For example, the DNA triplet CGA has the complementary mRNA codon GCU (uracil is the RNA base complementary to adenine). Each codon represents a "word" in the language of mRNA, translated from the triplet "word" in the language of DNA.

The Twenty Letters of the Protein Alphabet. Each mRNA codon in turn refers to one of the 20 "letters" in the protein alphabet, one of the 20 common amino acids (Table 5–1). Scientists have deduced the individual codons that represent the links between the nucleic acid strands and the polypeptide chains. The DNA triplets and complementary mRNA codons for each of the 20 amino acids are listed in Table 5–2. Figure 5–2b gives examples of the way in which the relationship among the three languages results in an RNA molecule and a polypeptide.

Of the 64 different codons, 61 specify a particular amino acid (Table 5–2). The codon for **methionine (AUG)** has a dual role because it also serves as a **start codon,** which is involved in initiating **translation** of the nucleotide triplets in mRNA. The remaining 3 codons do not specify amino acids; instead, they are used as **stop codons** to terminate translation.

Table 5–1
DNA, RNA, and Protein "Alphabets"

DNA Bases	RNA Bases	Amino Acids				
A	A	Alanine	Cysteine	Histidine	Methionine	Threonine
C	C	Arginine	Glutamic acid	Isoleucine	Phenylalanine	Tryptophan
G	G	Asparagine	Glutamine	Leucine	Proline	Tyrosine
T	U	Aspartic acid	Glycine	Lysine	Serine	Valine

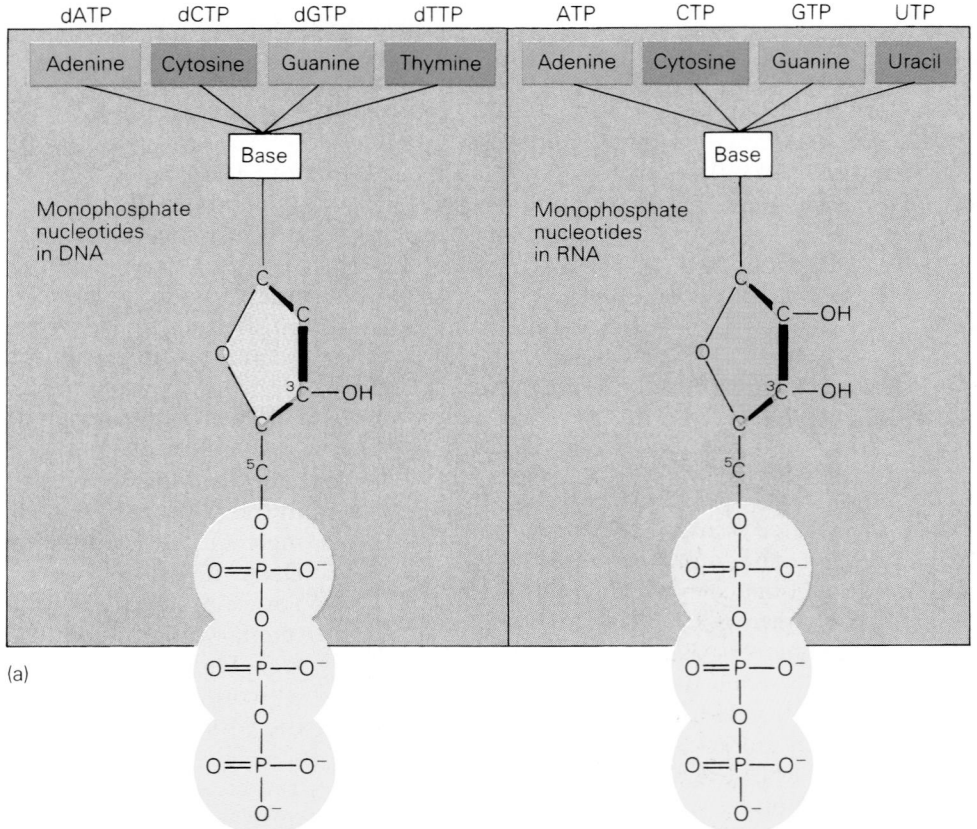

Figure 5–2

(*a*) The DNA "alphabet" consists of four nucleotides, derived from triphosphate nucleotides and distinguished by their bases: adenine, cytosine, guanine, and thymine. The RNA "alphabet" contains nucleotides formed from a different sugar molecule than that in DNA and distinguished by a slightly different set of bases: adenine, cytosine, guanine, and uracil. (*b*) The genetic code on DNA is copied in a complementary strand of RNA, and then a matching chain of amino acids is formed. Each set of three bases on a DNA strand, called a triplet, is transcribed into a complementary set of RNA bases, called a codon. Each codon refers to a specific amino acid.

The Vast Language of Proteins.   Consider the enormous number of words that can be generated from our alphabet of 26 letters. Then consider the huge number of different polypeptide chains that could be constructed from nearly that many amino acids, without the constraints on "word" length found in verbal language. This will give you some sense of the variety of proteins that can be synthesized in the cell and the great possibilities for specialization of function that this variety allows. Approximately 2000 different proteins are thought to exist in the body.

Think of the difference in meaning one letter change makes in a word—for example, the change from *deride* to *decide*. This is similar to the difference

that one amino acid can make in the nature of a protein. No other type of macromolecule allows this great variety and specificity of structure and function. Compared with lipids, carbohydrates, and nucleic acids (which have only a few subunits to arrange), proteins (with 20 different subunits) have an enormous evolutionary advantage because of the variety of different functional macromolecules that can be assembled. Indeed, it is likely that the molecular basis of the evolution of species lies in the ability of cells to make new proteins in response to changes in their environment.

A Consistent and Universal Code.   The genetic code is said to be **consistent** because each codon specifies

Table 5–2
The Genetic Code

DNA Triplets	mRNA Codons		Amino Acids	DNA Triplets	mRNA Codons		Amino Acids
AAA	UUU	1	Phenylalanine	GTA	CAU	34	Histidine
AAG	UUC	2		GTG	CAC	35	
AAT	UUA	3	Leucine	GTT	CAA	36	Glutamine
AAC	UUG	4		GTC	CAG	37	
GAA	CUU	5		TTA	AAU	38	Asparagine
GAG	CUC	6		TTG	AAC	39	
GAT	CUA	7		TTT	AAA	40	Lysine
GAC	CUG	8		TTC	AAG	41	
TAA	AUU	9	Isoleucine	CTA	GAU	42	Aspartic acid
TAG	AUC	10		CTG	GAC	43	
TAT	AUA	11		CTT	GAA	44	Glutamic acid
CAA	GUU	12	Valine	CTC	GAG	45	
CAG	GUC	13		ACA	UGU	46	Cysteine
CAT	GUA	14		ACG	UGC	47	
CAC	GUG	15		ACC	UGG	48	Tryptophan
AGA	UCU	16	Serine	GCA	CGU	49	Arginine
AGG	UCC	17		GCG	CGC	50	
AGT	UCA	18		GCT	CGA	51	
AGC	UCG	19		GCC	CGG	52	
GGA	CCU	20	Proline	TCT	AGA	53	
GGG	CCC	21		TCC	AGG	54	
GGT	CCA	22		TCA	AGU	55	Serine
GGC	CCG	23		TCG	AGC	56	
TGA	ACU	24	Threonine	CCA	GGU	57	Glycine
TGG	ACC	25		CCG	GGC	58	
TGT	ACA	26		CCT	GGA	59	
TGC	ACG	27		CCC	GGG	60	
CGA	GCU	28	Alanine	TAC	AUG	61	Methionine START
CGG	GCC	29					
CGT	GCA	30		ATT	UAA	62	STOP
CGC	GCG	31		ATC	UAG	63	
ATA	UAU	32	Tyrosine	ACT	UGA	64	
ATG	UAC	33					

only a single amino acid. It is essentially **universal** because it is the same in diverse types of organisms including plants and humans. Because there are 64 different codons for only 20 amino acids, some amino acids are specified by more than one codon. For example, there are 2 mRNA codons for phenylalanine, 6 for leucine, 3 for isoleucine, etc. (Table 5–2). For this reason, the code is said to be **degenerate.**

A gene and its polypeptide product are usually regarded as **colinear molecules.** This term means that the codon representing the beginning of the gene determines the amino acid present at the beginning of the polypeptide, while the codon representing the second triplet in the gene determines the second amino acid in the polypeptide, and so on. Many eukaryotic genes are interrupted with nucleotide sequences known as **introns,** the functions of which have yet to be determined. Introns will be discussed again later in this chapter.

Let us now look at how the nucleic acids themselves are assembled and translated into proteins. Figure 5–3 gives a simplified overview of the flow of information during these processes. We will discuss **DNA synthesis** (also known as **DNA replication**) and **RNA synthesis** (also known as **DNA transcrip-**

tion). We will also look at methods of altering newly synthesized RNA, known as **RNA processing,** before it leaves the nucleus, methods that result in the production of functional RNA molecules. Finally, we will examine the process of **protein synthesis** (also known as **RNA translation**), in which the information in RNA is translated into a polypeptide chain.

Synthesis of Nucleic Acids

As discussed in Chapter 2, the macromolecules deoxyribonucleic acid (DNA) and ribonucleic acid (RNA) are each composed of molecules called nucleotides. Nucleotides contain a ribose sugar and one to three phosphate groups, in addition to one of five bases. DNA and RNA are distinguished by the type of ribose sugar and bases their nucleotides contain. All nucleotides forming part of a nucleic acid strand have one phosphate group.

Synthesis of DNA: Replication of the Permanent Files

When a cell divides into two, the contents of both the cytoplasm and the nucleus divide (Fig. 5–4).

Figure 5–3
A simplified overview of the flow of information in the expression of the genetic code. The information on a DNA molecule is transcribed, or reproduced in coded form, on an RNA molecule. After the RNA molecule is processed, it leaves the nucleus and travels to a ribosome, where the coded information is translated into a sequence of amino acids that forms a polypeptide chain.

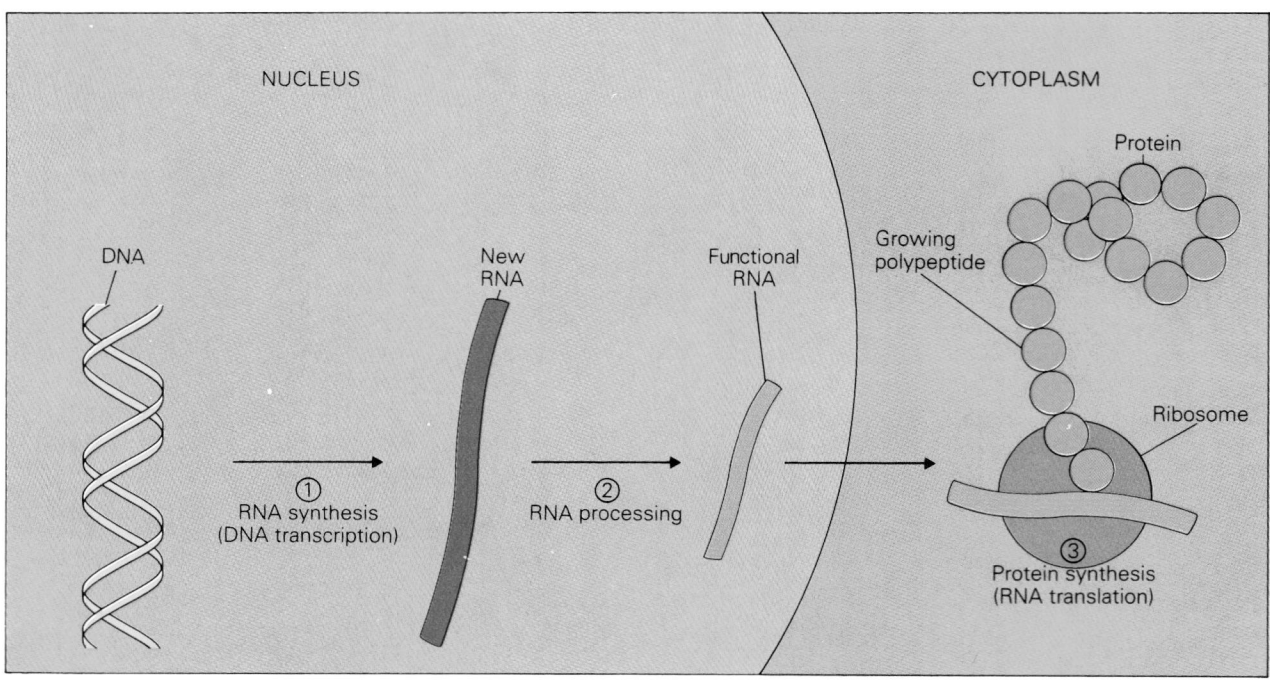

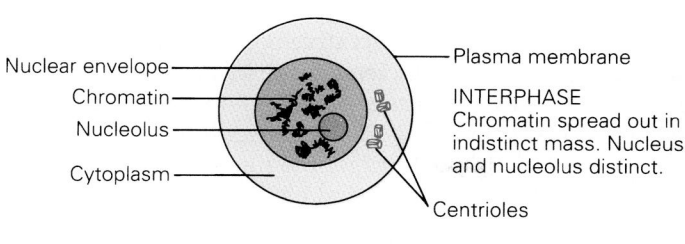

Nuclear envelope — Plasma membrane

Chromatin

Nucleolus

Cytoplasm

Centrioles

INTERPHASE
Chromatin spread out in indistinct mass. Nucleus and nucleolus distinct.

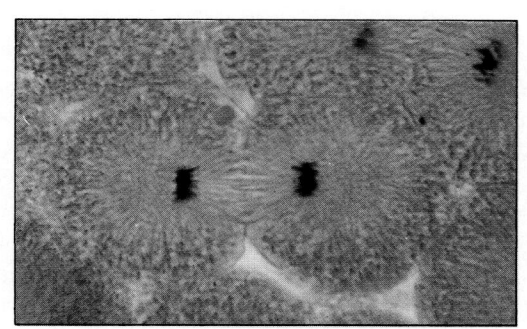

MITOSIS

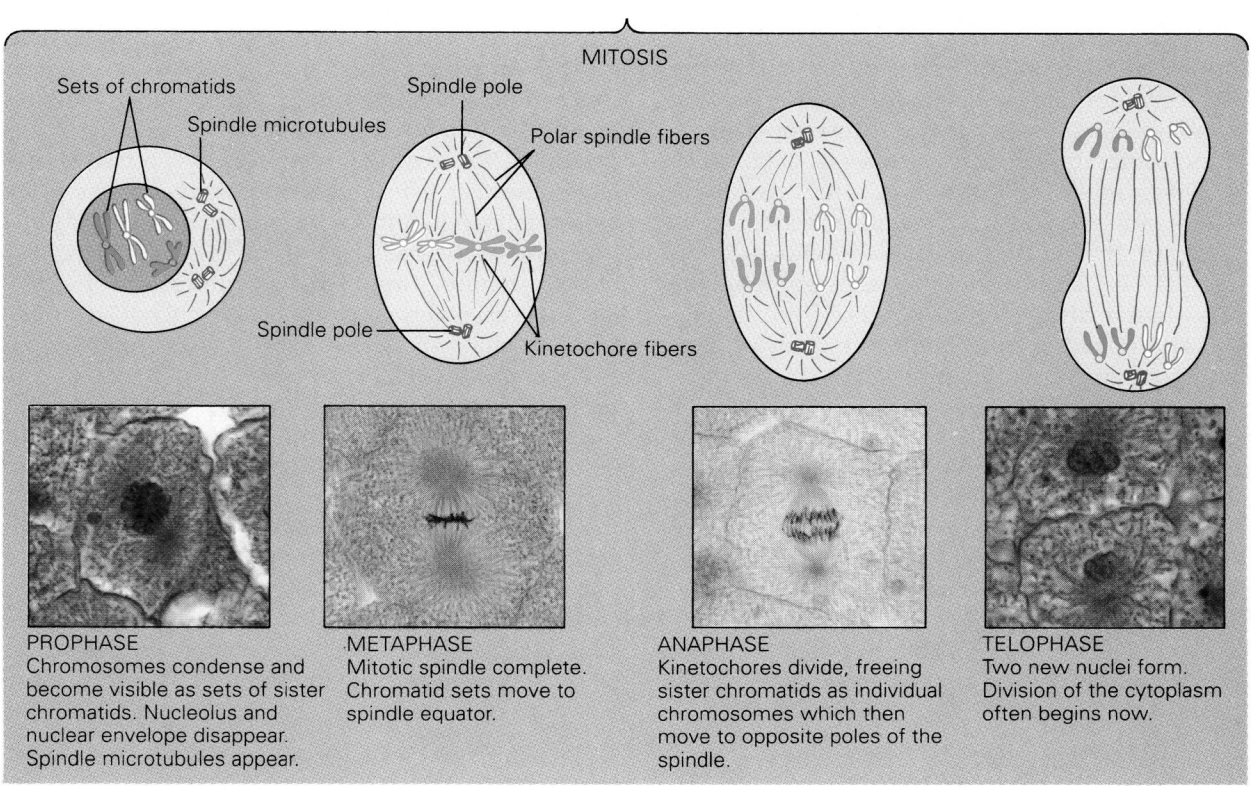

Sets of chromatids

Spindle microtubules

Spindle pole

Polar spindle fibers

Spindle pole

Kinetochore fibers

PROPHASE
Chromosomes condense and become visible as sets of sister chromatids. Nucleolus and nuclear envelope disappear. Spindle microtubules appear.

METAPHASE
Mitotic spindle complete. Chromatid sets move to spindle equator.

ANAPHASE
Kinetochores divide, freeing sister chromatids as individual chromosomes which then move to opposite poles of the spindle.

TELOPHASE
Two new nuclei form. Division of the cytoplasm often begins now.

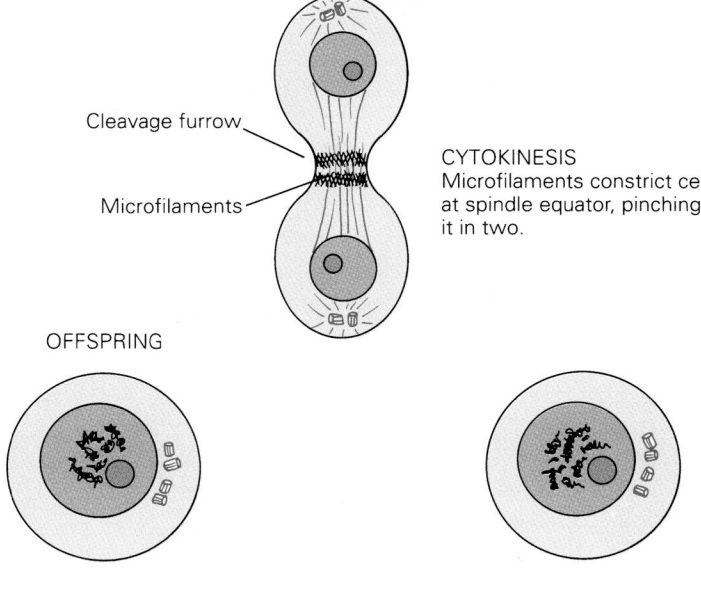

Cleavage furrow

Microfilaments

CYTOKINESIS
Microfilaments constrict cell at spindle equator, pinching it in two.

OFFSPRING

Figure 5–4
(*a*) During the interphase portion of the life cycle of a cell, the chromatin form of the DNA in the nucleus replicates, or copies itself. (×450.) (© M. Abbey/Photo Researchers, Inc.) (*b*) In mitosis, the chromatin forms the tightly coiled bundles known as chromosomes, with the pairs of identical chromosomes linked together to form chromatids (×450.) (© M. Abbey/Photo Researchers, Inc.) (*c-e*) The chromatids separate, after lining up at the spindle equator, such that one copy of each chromosome becomes part of one of the nuclei of the two new cells. Thus, each new cell has the same genetic information contained in the parent cell. (×315.) (© John D. Cunningham/Visuals Unlimited.)

Division of the nucleus, a process known as **mitosis** (see Chapter 3), ensures that both of the new cells have an identical copy of the DNA of the parent cell. Thus, the new cells will be identical in form and function to the parent.

Before the nucleus can divide, the DNA molecules within the nucleus (in the form of **chromatin**) must be copied, or **replicated.** The replication process must be accurate so that the genetic information is not altered, and it must be fast so that it can occur in time for a cell to divide. DNA replication, as mentioned in Chapter 3, occurs during the interphase preceding mitosis. Each chromosome, a long DNA molecule, is copied to produce two identical DNA molecules called **chromatids** (see Fig. 5–4).

A DNA molecule consists of the two long strands of nucleotides arranged in a double helix (see Chapter 2). The two strands are held together by hydrogen bonds between the bases of the nucleotides. Adenine always forms bonds with thymine, and guanine always forms bonds with cytosine. The specific base pairing produces two strands that complement each other.

Semiconservative Replication. During the process of DNA replication, the two strands of the DNA molecule are first separated. Each one then acts as a mold, or **template,** upon which a new complementary strand of nucleotides is assembled. In this way, two new molecules of DNA are produced. Each molecule contains one strand that was the template derived from the parent molecule and one newly synthesized strand (Fig. 5–5). The process is de-

scribed as **semiconservative** because the structure of the parent DNA molecule is disrupted but the structure of the subunits—that is, the nucleotide sequence of each strand—is conserved for many generations. Maintenance of the nucleotide sequence of DNA is a critical step that ensures that the new cells receive an exact copy of the genetic information in the parent cell.

Separating the Strands of the Double Helix. Before replication of DNA can take place, the double helix must be unwound and "unzipped" to separate the two strands. Unwinding is aided by an enzyme known as a **DNA topoisomerase,** and unzipping (breaking the hydrogen bonds between bases) by a **DNA helicase.** The separated strands are stabilized by proteins that bind to the single DNA strands and therefore assist the opening of the helix. They are known as **helix-destabilizing proteins.** Each strand then acts as a template for the assembly of a complementary chain. Separation of the two original DNA strands produces a structure called a **replication fork** at the site of active synthesis of the new strands (Fig. 5–6a). Many replication forks can exist on a DNA molecule at once (Fig. 5–6b).

Replicating the Strands: DNA Polymerase. DNA replication is catalyzed by an enzyme called **DNA polymerase,** which takes instructions from the DNA strand that is acting as the template for the new DNA strand. The specific reaction catalyzed by DNA polymerase is the joining together of the various nucleotides that are components of DNA. The

Figure 5–5
DNA replication is known as semiconservative replication because each new DNA molecule contains one new strand and one original strand. Thus, single strands of DNA are preserved and passed on for many generations.

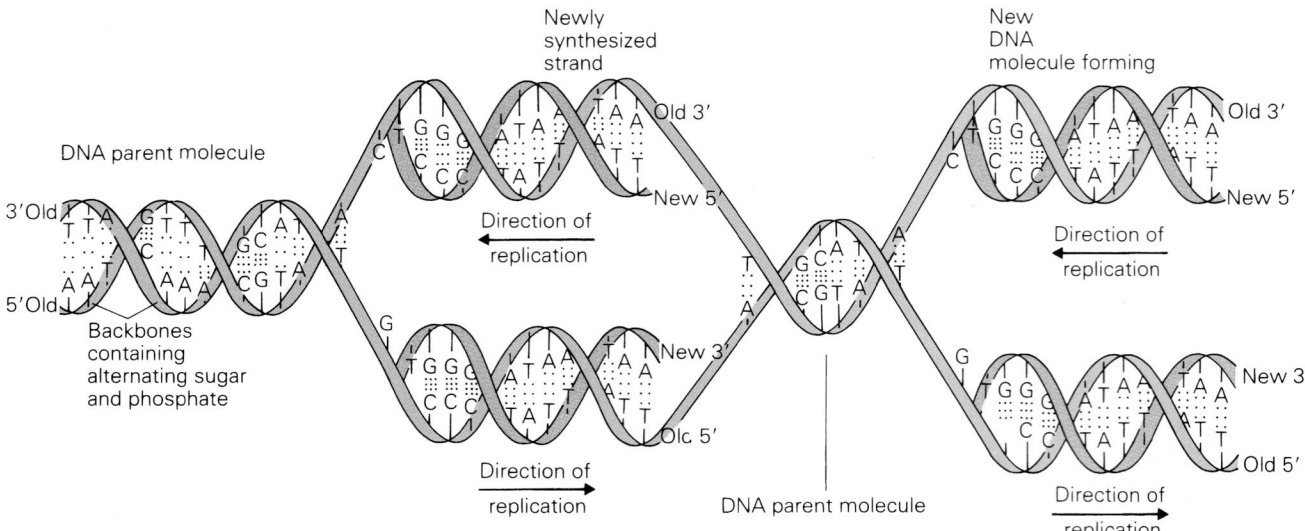

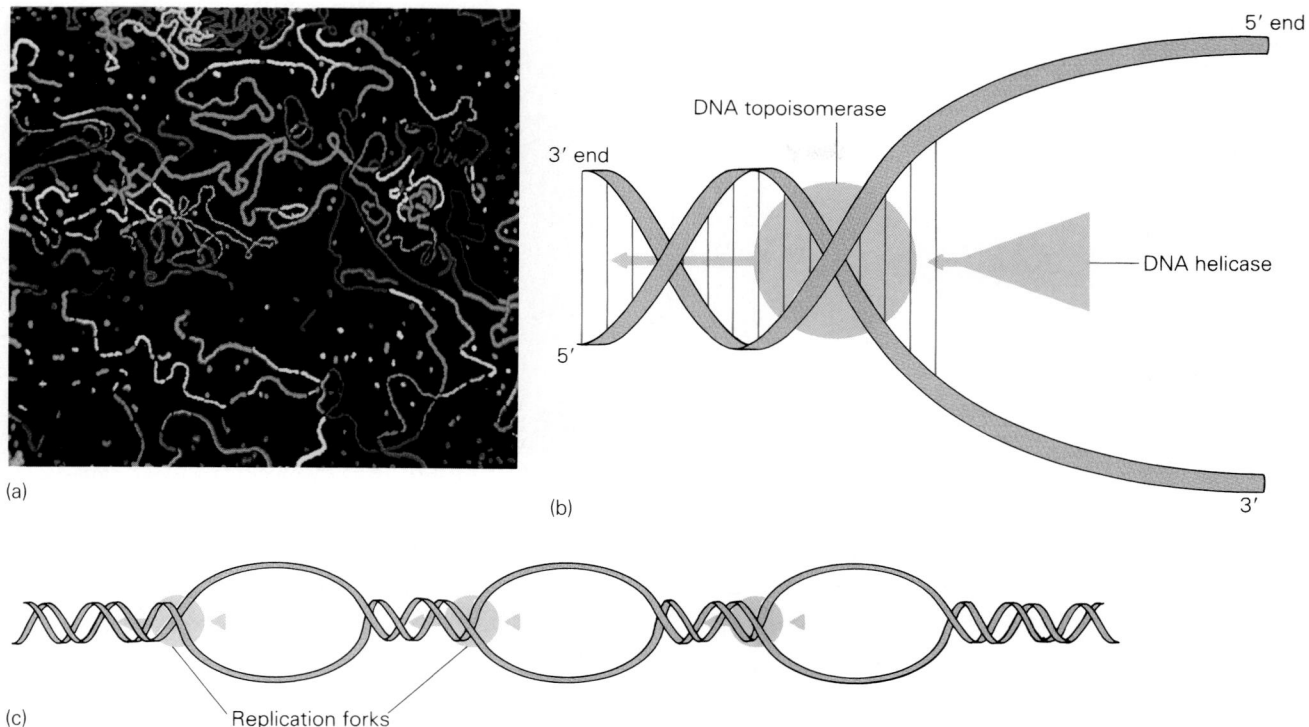

Figure 5–6
(*a*) An electron micrograph of an unwound human DNA molecule. (© John A. Olschowka, Ph.D/Richard Insel, M.D./Dr. Louise Chow/Phototake. (*b*) Before DNA replication can begin, the double helix must be unwound and the two strands separated. Unwinding is accomplished by an enzyme called DNA topoisomerase, and another enzyme, DNA helicase, breaks the hydrogen bonds holding the base pairs—and, hence, the two strands of the molecule—together. (*c*) DNA topoisomerase and DNA helicase may be active at several sites on the same DNA molecule, thereby forming many replication forks.

individual nucleotides must contain deoxyribose and three phosphate groups; otherwise the enzyme cannot use them. Thus, DNA polymerase takes the nucleotides

> deoxyadenosine triphosphate (dATP)
> deoxycytidine triphosphate (dCTP)
> deoxyguanosine triphosphate (dGTP)
> deoxythymidine triphosphate (dTTP),

removes two of their phosphate groups to form the monophosphate nucleotides

> dAMP
> dCMP
> dGMP
> dTMP,

and joins them in the order specified by the complementary bases on the DNA template to form a new strand of DNA. Recall that the base C always forms a pair with the base G, and A with T. Hence, a DNA strand of nucleotides with bases in the order ACAGTG will serve as a template for a set of nucleotides with bases in the order TGTCAC. Base-pairing

ensures that the order of bases in the new DNA strand is identical to the order in the original strand, from which the template has recently separated.

Figure 5–7*a* depicts the end of a DNA strand as it is synthesized on a DNA template. The incoming nucleotide dCTP has been selected for attachment to the end of the strand because its base cytosine (C) is the only base that can form hydrogen bonds with guanine (G), the base of the next nucleotide on the DNA template. The preliminary hydrogen bonding step serves to line up the individual nucleotide molecules on the DNA template. DNA polymerase then catalyzes a reaction that covalently attaches dCTP to the end of the new strand. During the reaction, the two phosphate groups on the end of the dCTP molecule are removed, and the rest of the molecule is attached via the remaining phosphate to the hydroxyl group on carbon 3 of the deoxyribose ring of the nucleotide at the end of the new strand (see Fig. 5–7*a*). DNA polymerase can add nucleotides to the DNA chain at a rate of more than 20 per second, and each one is always attached to carbon 3 of the sugar of the last nucleotide. Thus, the new DNA chain

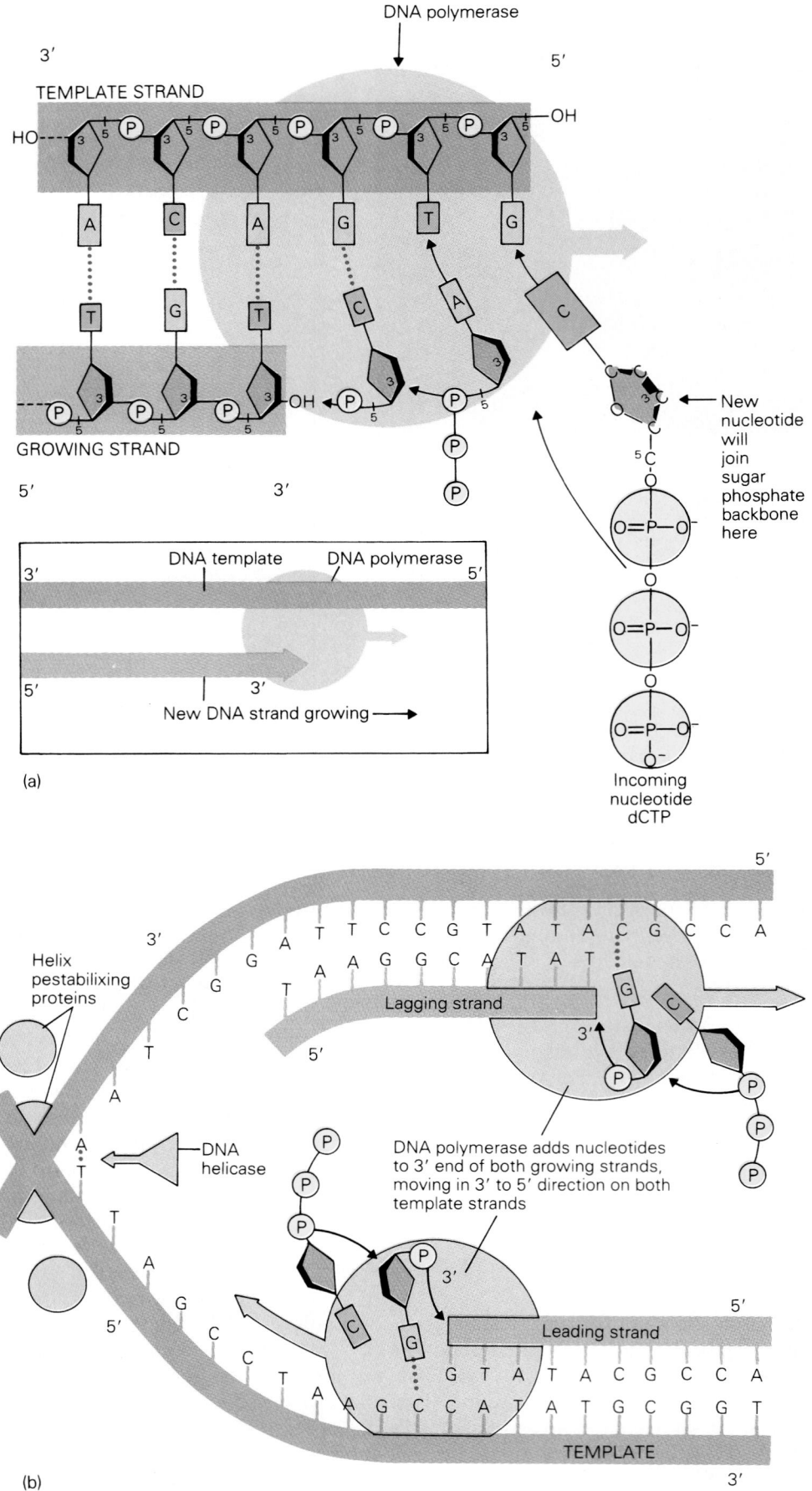

(a)

(b)

grows from the end where there is a free hydroxyl group on carbon 3.

Synthesis in the 5' to 3' Direction.

A special numbering system is commonly used for nucleotides in order to distinguish between the two ring structures that are present in the molecule—that is, the sugar and the base. By convention, the numbers of the carbon atoms of the sugar are given a superscript (e.g., 3' or 5'), whereas the carbon atoms of the base are numbered in the usual way. A DNA strand is formed by linking nucleotides together via phosphate groups between carbon 3' of one nucleotide and 5' of another. The nucleotide at the beginning of the DNA strand has a free phosphate group on the 5' carbon. This end of the strand is usually called the **5' ("five prime") end.** The nucleotide at the other end of the strand has a free hydroxyl on the 3' carbon. This is termed the **3' ("three prime") end** of the strand (Fig. 5–7).

In theory, a new nucleotide could be attached to either the 5' end or the 3' end of the chain. In practice, however, DNA polymerase attaches nucleotides only to the 3' end of a strand, so that the new strand always grows from the **5' to 3' direction.**

Antiparallel Strands: Leading and Lagging Strands.

In the DNA molecule, the 3' ends of the two complementary nucleic acid strands are at opposite ends of the double helix (see Fig. 5–5); they are said to be **antiparallel.** A DNA template and its complementary growing strand are also antiparallel. Hence, a template with the nucleotide sequence 3'—

TGGCGTATAC—5' will direct the synthesis of the complementary and antiparallel strand 5'—ACCGCATATG—3' (see Fig. 5–7b). This complicates the process of DNA replication because DNA polymerase moves along the *template* strands in the 3' to 5' direction only, and synthesizes new DNA by adding nucleotides only to the 3' end of the growing strand.

In the simultaneous replication of both strands of the DNA double helix, one of the new DNA strands, the so-called **leading strand,** is synthesized in a straightforward manner because it can grow in the 5' to 3' direction on the template strand (see Fig. 5–7b). It is synthesized continuously and "follows" the DNA polymerase as the polymerase moves from the 3' end of the template strand.

The other new strand of DNA, the **lagging strand,** is formed by a process of **discontinuous synthesis,** in a series of small sections (**Okazaki fragments**) (see Fig 5–7b). Each section of the lagging strand is assembled in exactly the same way as a section of the leading strand, in the 5' to 3' direction by a DNA polymerase moving from 3' to 5' on the template. By the time one section is finished, the replication fork has opened up to expose more of the template, allowing another DNA polymerase to synthesize a second fragment of new DNA behind the first (Fig. 5–8a and b). However, DNA polymerase does not join the segments of the lagging strand. This task is carried out by another enzyme, called **DNA ligase** (Fig. 5–8c).

In summary, the replication forks depicted in Figure 5–8a are always moving to the left, and the overall growth of both of the new DNA strands is from right to left following the movement of the fork. However, the synthesis of both of the new strands is exclusively in the 5' to 3' direction, which means that the lagging strand is actually synthesized in the direction opposite to the movement of the replication fork. As already pointed out (see Fig. 5–5), this mechanism of DNA replication produces two double-stranded DNA molecules, each consisting of one nucleic acid chain from the "parent" and one newly synthesized chain.

Proofreading in DNA Replication.

Replication of DNA is a remarkably accurate process. There is probably only one error per 10^9 base pairs. This efficiency is well within the limits required to maintain the genome of a mammalian cell (about 3×10^9 base pairs). One of the reasons for the high degree of accuracy is that the polymerase is able to check or "proofread" the new chain as it is synthesized, meaning that the polymerase responds to the incorporation of an incorrect nucleotide by immediately removing it and inserting the correct one before

◀ *Figure 5–7*

(a) DNA polymerase moves from the 3' end to the 5' end of a DNA template, joining each new nucleotide first to the complementary base on the template via hydrogen bonds and then to the previous nucleotide on the growing strand at the 3' carbon. DNA polymerase adds nucleotides only to the 3' end of the growing strand (so called because it has a free hydroxyl group attached to the 3' carbon of the sugar). (b) DNA works on both strands of the parent DNA molecule simultaneously. However, because the parent strands are antiparallel, and because DNA polymerase travels in the 3' to 5' direction on a DNA template, in effect it moves in opposite directions. When DNA polymerase is moving in the same direction as DNA helicase, the growing strand is termed the *leading strand.* When DNA polymerase is moving away from the replication fork, the growing strand is called the *lagging strand.*

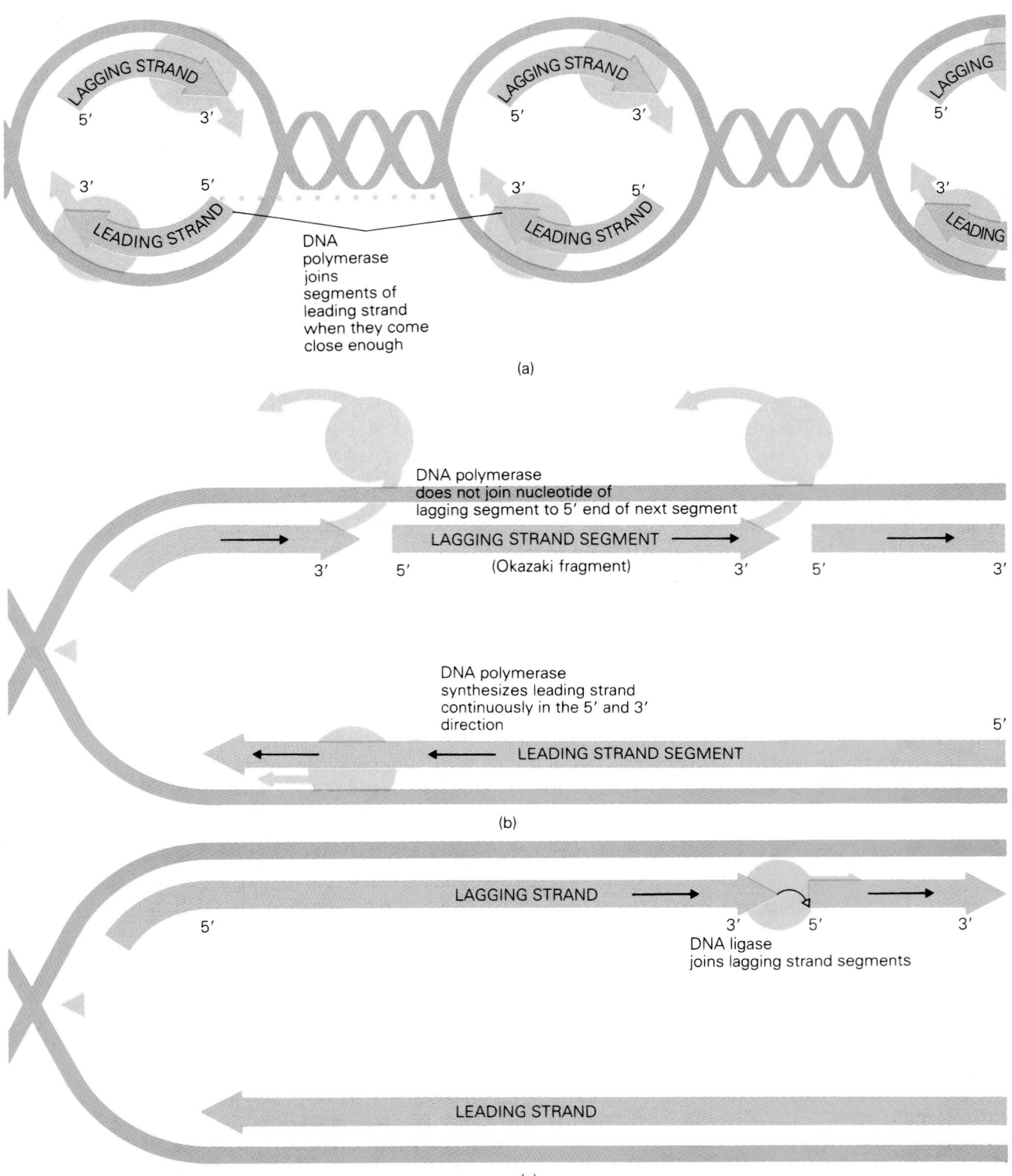

(a)

DNA polymerase joins segments of leading strand when they come close enough

(b)

DNA polymerase does not join nucleotide of lagging segment to 5' end of next segment

LAGGING STRAND SEGMENT
(Okazaki fragment)

DNA polymerase synthesizes leading strand continuously in the 5' and 3' direction

LEADING STRAND SEGMENT

(c)

LAGGING STRAND

DNA ligase joins lagging strand segments

LEADING STRAND

Figure 5–8
(*a*) Throughout a DNA molecule, several DNA polymerases may be working at many sites simultaneously. (*b*) The leading strand is synthesized continuously; the lagging strand is synthesized in small segments. DNA polymerase cannot join these segments. (*c*) Another enzyme, DNA ligase, joins lagging strand segments, known as Okazaki fragments.

moving on to the next. DNA replication is also remarkably rapid. The average human DNA chain contains more than 10^8 nucleotides linked together. Such a chain can be replicated completely within a few hours. This rapidity is possible because many replication forks are present on each DNA molecule that is undergoing replication, so that many different sections of the huge molecule can be replicated simultaneously.

Synthesis of RNA: Selective Transcription of the Permanent Files

DNA does not provide protein-synthesizing information directly to the ribosomes; the information in DNA is first transcribed into RNA. In fact, RNA is synthesized in a process known as **DNA transcription.**

Three kinds of RNA are synthesized: **messenger RNA (mRNA), ribosomal RNA (rRNA),** and **transfer RNA (tRNA).** All three kinds of RNA molecules interact closely in the mechanism whereby a nucleotide sequence in DNA is translated into a specific protein. All three types of RNA are transcribed from a DNA template; all three types of RNA consist of ribonucleotides complementary to DNA nucleotides. The nucleotides on mRNA molecules are translated directly into amino acids, while the nucleotides on tRNA and rRNA molecules provide other information crucial to protein synthesis. The interactions of the three RNA molecules will be discussed in more detail soon.

A portion of one strand of DNA serves as a template for each new RNA strand. The base sequence in the RNA is complementary to the sequence in the DNA strand, and the base pairs formed ensure that the correct base sequence is copied. Synthesis of RNA differs from synthesis of DNA in four principal ways that are summarized in Table 5–3.

Transcribing the DNA Template: RNA Polymerase. During transcription of DNA, the new chain of RNA is synthesized by the enzyme **RNA polymerase.** There are three distinct types of RNA polymerase enzymes in eukaryotic cells, and all three are located in the nucleus. **RNA polymerase I** synthesizes the major rRNA molecules in the nucleolus. The other polymerases are active in the portion of the nucleus that is outside the nucleolus. **RNA polymerase II** synthesizes mRNA, and **RNA polymerase III** synthesizes tRNA and small rRNA molecules.

Promoter and Terminator Sites. Transcription of a given gene begins when the RNA polymerase binds to a **promoter** site on a DNA template, a discrete sequence of nucleotides immediately preceding the 3′ end of the gene (Fig. 5–9). Polymerase action stops when the enzyme reaches the **terminator** site, which is a sequence of nucleotides immediately following the 5′ end of the gene (Fig. 5–9). At this point, the polymerase releases both the finished RNA strand and the DNA template. The promoter and terminator act as recognition sites only; they are not transcribed. The promoter sequence also determines which strand of the double-stranded DNA will be transcribed by the RNA polymerase.

Synthesis in the 5′ to 3′ Direction: Antiparallel Strands. During synthesis of RNA, the RNA polymerase moves along the DNA template in the 3′ to 5′ direction, adding ribonucleotides to the 3′ end of the growing RNA strand. Thus, the new RNA strand grows in the antiparallel 5′ to 3′ direction (Fig. 5–10). A portion of the DNA template with the base sequence 3′—CCAGTAAC—5′ will direct the incorporation of the sequence 5′—GGUCAUUG—3′ into RNA (see Fig. 5–10). On a single DNA template, several RNA molecules may be synthesized at the same time (Fig. 5–11).

Table 5-3
Principal Differences Between DNA Synthesis and RNA Synthesis

	DNA Synthesis	RNA Synthesis
Sugar required	Deoxyribose	Ribose
Bases required	Adenine	Adenine
	Cytosine	Cytosine
	Guanine	Guanine
	Thymine	Uracil
Fate of DNA template	Incorporated into reaction product	Not incorporated into reaction product
Type of product	Double strand of nucleotides	Single strand of nucleotides

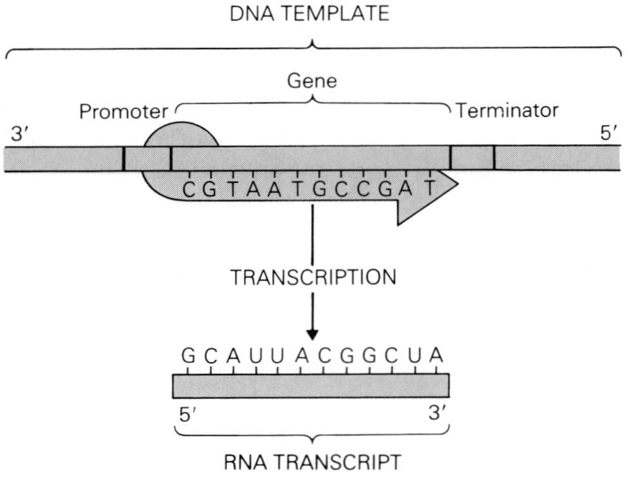

Figure 5–9

The synthesis of a molecule of RNA begins when the enzyme RNA polymerase encounters a sequence of DNA nucleotides called a promoter. Ribonucleotides are assembled in an order complementary to the order of nucleotides on the DNA template in a process known as transcription, producing an "RNA transcript." Transcription stops when the RNA polymerase encounters a terminator sequence on the DNA template.

Figure 5–10

RNA polymerase works in the 3' to 5' direction on a DNA template, adding ribonucleotides to the 3' end of a growing RNA strand. Ribonucleotides initially form hydrogen bonds with complementary bases on the DNA template, but when transcription is completed, the RNA transcript is released from the template.

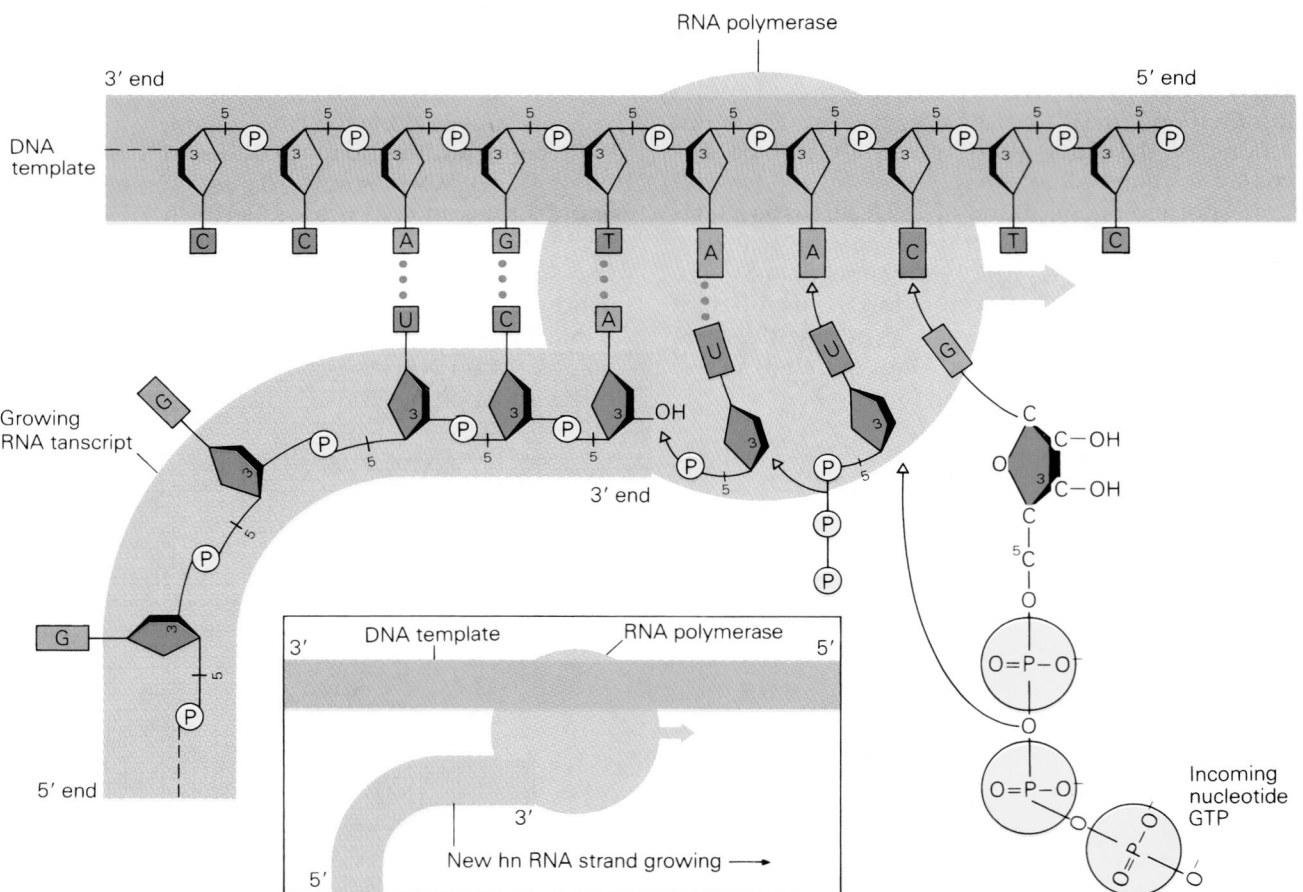

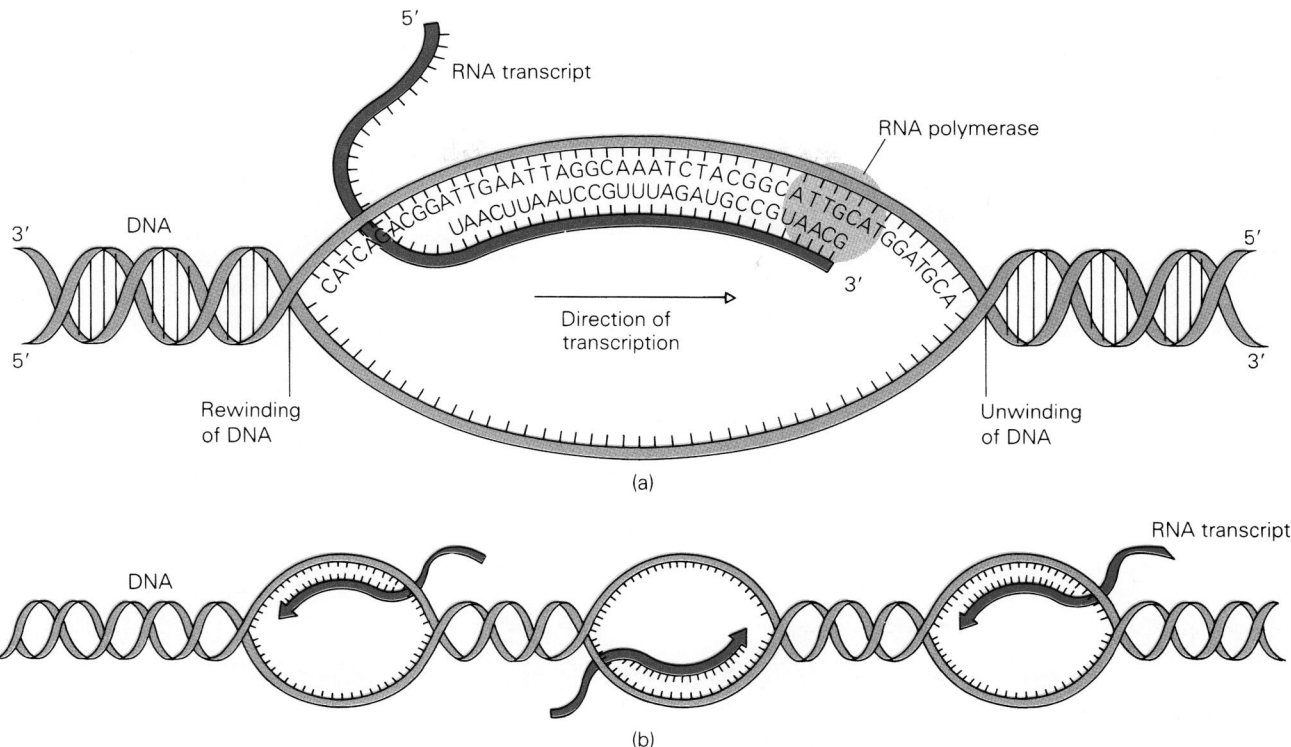

Figure 5–11
A single DNA molecule may provide templates for several different RNA transcripts growing simultaneously.

RNA Processing: Post-transcriptional Modification

Before they leave the nucleus to direct protein synthesis, newly synthesized RNA molecules are covalently modified in a series of reactions known as **RNA processing** or **post-transcriptional modification.** Several examples are discussed below.

The RNA that is synthesized by the action of RNA polymerase II is known as **heterogeneous nuclear RNA (hnRNA).** The synthesis of hnRNA accounts for more than 50% of the total RNA synthesis in a typical mammalian cell. Many of these hnRNA molecules are destined to leave the nucleus in the form of much smaller mRNA molecules.

The 5' Cap and the Poly(A) Tail. Processing of hnRNA helps to provide the resultant mRNA with characteristics that distinguish it from rRNA and tRNA, the products of the action of the other RNA polymerases. These characteristics are a **5' cap** and a **poly(A) tail.** The 5' cap is formed by addition of **7-methylguanosine** and three phosphate groups, represented in shorthand notation as **Gppp** (Fig. 5–12), to the nucleotide at the 5' end of the RNA tran-

script. This is the end of RNA that is synthesized first, and it is capped immediately—before the rest of the molecule is finished. The cap is required later to initiate binding of mRNA to ribosomes in the cytoplasm. The poly(A) tail consists of 100 to 200 adenine residues that are added to the 3' end of the RNA after transcription is completed (see Fig. 5–12). The function of the poly(A) tail is unknown, but it may serve to stabilize the mRNA structure and/or assist its export from the nucleus.

RNA Splicing: Introns Out. Newly synthesized RNA molecules in eukaryotic cells contain intervening nucleotide sequences known as **introns,** nucleotides transcribed from DNA nucleotide sequences that contain no useful information (noncoding segments). These sequences are reproduced in the hnRNA during transcription and must be removed before the RNA leaves the nucleus. After addition of the 5' cap and the poly(A) tail, the introns are cut out in order to convert the RNA to mature mRNA that can be used to direct the synthesis of a complete protein. This is another example of RNA processing, and it is known specifically as **RNA splicing** (see Fig. 5–12) because the remaining segments,

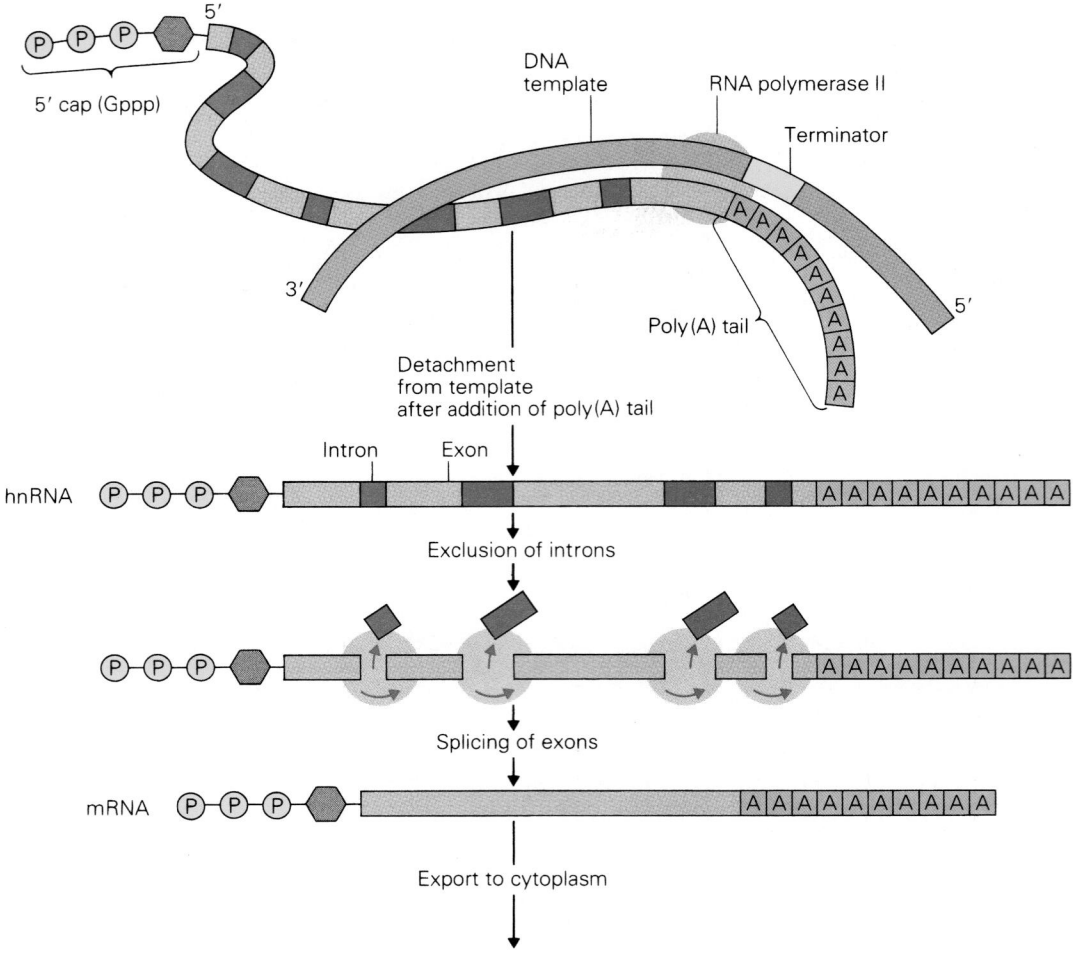

Figure 5–12
As part of the processing of newly synthesized RNA molecules, a "5' cap"
and a "poly(A) tail" are added at either end. In addition, introns (transcripts
of meaningless segments of the DNA template) are excised, with the remaining RNA segments, known as exons, spliced together.

called **exons,** are joined together to form functional
mRNA.

RNA processing removes most of the mass of
the hnRNA. The mature mRNA that is produced by
these reactions accounts for only 3% of the total
RNA present in a cell.

Formation of Functional tRNA.   A molecule of
tRNA is relatively small, usually 70 to 80 nucleotides
long, compared to the other types of RNA. The secondary structure adopted by a tRNA molecule is
represented in Figure 5–13. Transfer RNA molecules have two important features. One, called the
amino acid accepting end or the **aminoacyl attachment site,** is where an amino acid destined to join a
growing polypeptide chain attaches for transport to

a ribosome. The other, called the **anticodon,** is a set
of three nucleotides complementary to a codon on
an mRNA molecule. Each amino acid has a specific
tRNA molecule designed to convey it to a ribosome
and link it in the correct order to another amino
acid, in a process to be described.

In a recent major advance in the understanding
of the genetic code, molecular biologists have determined that the identity of a tRNA molecule, by
which it is assured of linking up with the correct
amino acid, is found in a few specific bases in each
tRNA molecule. In some molecules, these "identity
sites" include the bases that form the anticodon as
well as other bases in the molecule. In others, the
identity sites do not include the anticodon. Experiments have shown that if the identity elements are

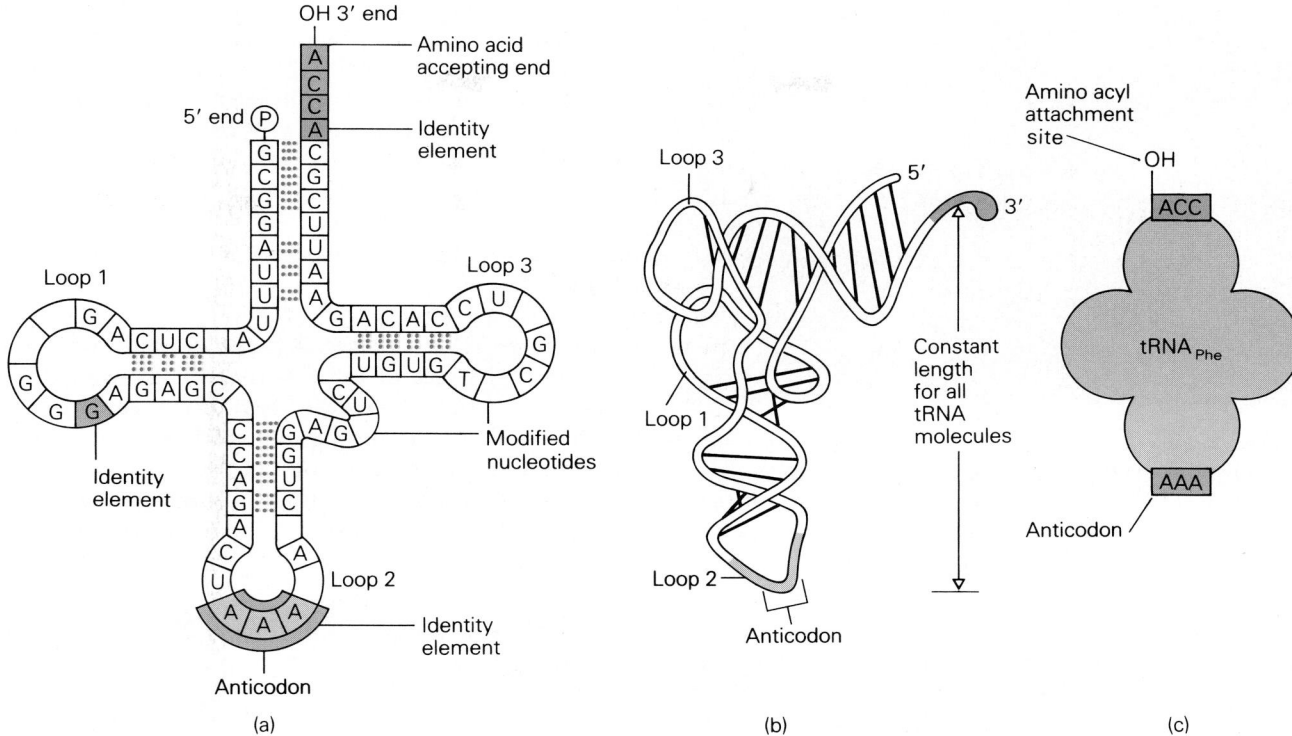

Figure 5–13
A molecule of tRNA contains two important sites: the aminoacyl attachment site, where an amino acid attaches to the molecule, and the anticodon, a triplet of bases complementary to a specific codon on a molecule of mRNA. In addition, equally important "identity sites," sometimes including the anticodon, ensure that the molecule attaches to the correct amino acid.

removed from a tRNA molecule, the molecule no longer recognizes its amino acid. It may link up with another amino acid or simply fail to function altogether.

Formation of Ribosomal Subunits from rRNA. Ribosomal RNA is formed in the nucleolus, an area of the nucleus. There, rRNA combines with special proteins to form the large and small subunits that will later combine to produce ribosomes. The subunits do not combine in the nucleolus; they are exported into the cytoplasm and remain separated except when they are participating in protein synthesis.

Synthesis of Proteins: Translation of mRNA

After export from the nucleus, all three kinds of RNA cooperate in protein synthesis in the cytoplasm. Ribosomal RNA makes up part of the ribo-

some, the organelle responsible for protein synthesis. Messenger RNA carries to the ribosome the codons specifying the amino acids to be used in the polypeptide chain. Transfer RNA obtains the amino acids in the cytosol and brings them to the ribosome, where the polypeptide is assembled.

RNA in the Cytoplasm: A Three-Way System

Although all three types of RNA are complementary transcripts of a DNA template, only mRNA nucleotides are directly translated into amino acids. The nucleotides on rRNA and tRNA serve other purposes in the protein synthesizing process, interacting with mRNA to produce a polypeptide. Figure 5–15 gives an overview of the role each type of RNA plays in protein synthesis.

Ribosomal RNA: The Protein-Synthesizing Organelle. The several steps involved in translation of mRNA and synthesis of a polypeptide chain take place on and are coordinated by **ribosomes.** Eukary-

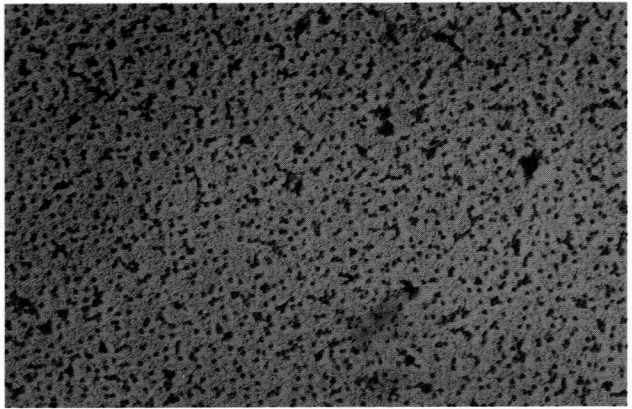

Figure 5–14
An electron micrograph of eukaryotic ribosomes.
(× 121,000.) (© David M. Phillips/Visuals Unlimited.)

otic ribosomes are composed of approximately equal weights of rRNA and protein (Fig. 5–14). The molecular weight of these ribosomes is about 4.5×10^6. Each ribosome is composed of two subunits. Some ribosomes are found free in the cytosol or in clusters called **polysomes** (see Fig. 5–19); other ribosomes are attached to the endoplasmic reticulum.

Transfer RNA: Procuring Amino Acids.   A preliminary step in protein synthesis involves coupling amino acids to the tRNA molecules, a process driven by the energy derived from ATP and requiring the catalytic action of an enzyme called **amino-acyl-tRNA synthetase (AAS):**

amino acid + tRNA + ATP ⟶
aminoacyl-tRNA + AMP

The amino acid is attached to the ribose ring of the nucleotide at the 3′ end (the amino acid-accepting end) of the tRNA.

Each of the 20 amino acids has one specific synthetase enzyme and one specific tRNA. The synthetase must be able to "recognize" both the amino acid and the tRNA that is specific for that amino acid (Fig. 5–16). Free amino acids cannot participate in protein synthesis because they are unable to form peptide bonds (—CONH—) spontaneously. Participation is mediated by the tRNA molecules to which the amino acids are covalently linked.

The tRNA molecules help to maintain the accuracy of translation by making sure that the correct amino acid is inserted at the correct location in the growing polypeptide chain. The basis of this mecha-

nism is complementary base-pairing between the anticodons in the tRNA molecules (see Fig. 5–16) and the codons in mRNA. Thus, it is the tRNA molecule, not the attached amino acid, that recognizes the codon in mRNA (see Fig. 5–16). Only one of the many kinds of tRNA molecules in the cell can form base-pairs with a specific codon in mRNA; this is how the amino acid specified by the codon can be selected accurately and added to the polypeptide chain.

Messenger RNA: Conveying the Amino Acid Sequence.   The role of mRNA in the three-way RNA system is to bear the codons that dictate the order of amino acids in the new polypeptide product. Messenger RNA is the focal point of protein synthesis, a meeting place for rRNA and tRNA and the mediator between the information stored in DNA and the final polypeptide product.

Initiation of Protein Synthesis: Formation of the Ribosome

In eukaryotic cells, the synthesis of a polypeptide is initiated at the AUG start codon that is nearest to the 5′ cap of the mRNA. The other AUG codons in the mRNA do not act as start codons. They specify incorporation of methionine into the polypeptide. The AUG start codon allows the ribosome to bind to the mRNA (Fig. 5–17). The ribosome itself is not able to "recognize" the AUG start codon. This process is mediated by a molecule of **initiator tRNA** bound to the ribosome. Specialized proteins called **initiation factors** are also involved in the complex initiation process.

A ribosome has two binding site for tRNA molecules, denoted as the **P-site** and the **A-site** (see Fig. 5–17). When a tRNA molecule binds to either of these sites, it is held close to the mRNA strand that is also bound. The bound tRNA is oriented so that the nucleotides in its anticodon form complementary base pairs with the nucleotides of the appropriate codon in the mRNA. Thus, the aminoacyl-tRNA molecule cannot fit into the ribosomal binding site unless the appropriate mRNA codon also is present.

During elongation of the polypeptide chain (see the following section), when both tRNA binding sites are occupied by tRNA molecules, the two tRNA molecules form base pairs with adjacent codons in the mRNA (Fig. 5–18). The tRNA in the P-site is attached to the amino acid at the growing end of the polypeptide chain, while the tRNA in the A-site carries the next amino acid that will be added to the polypeptide.

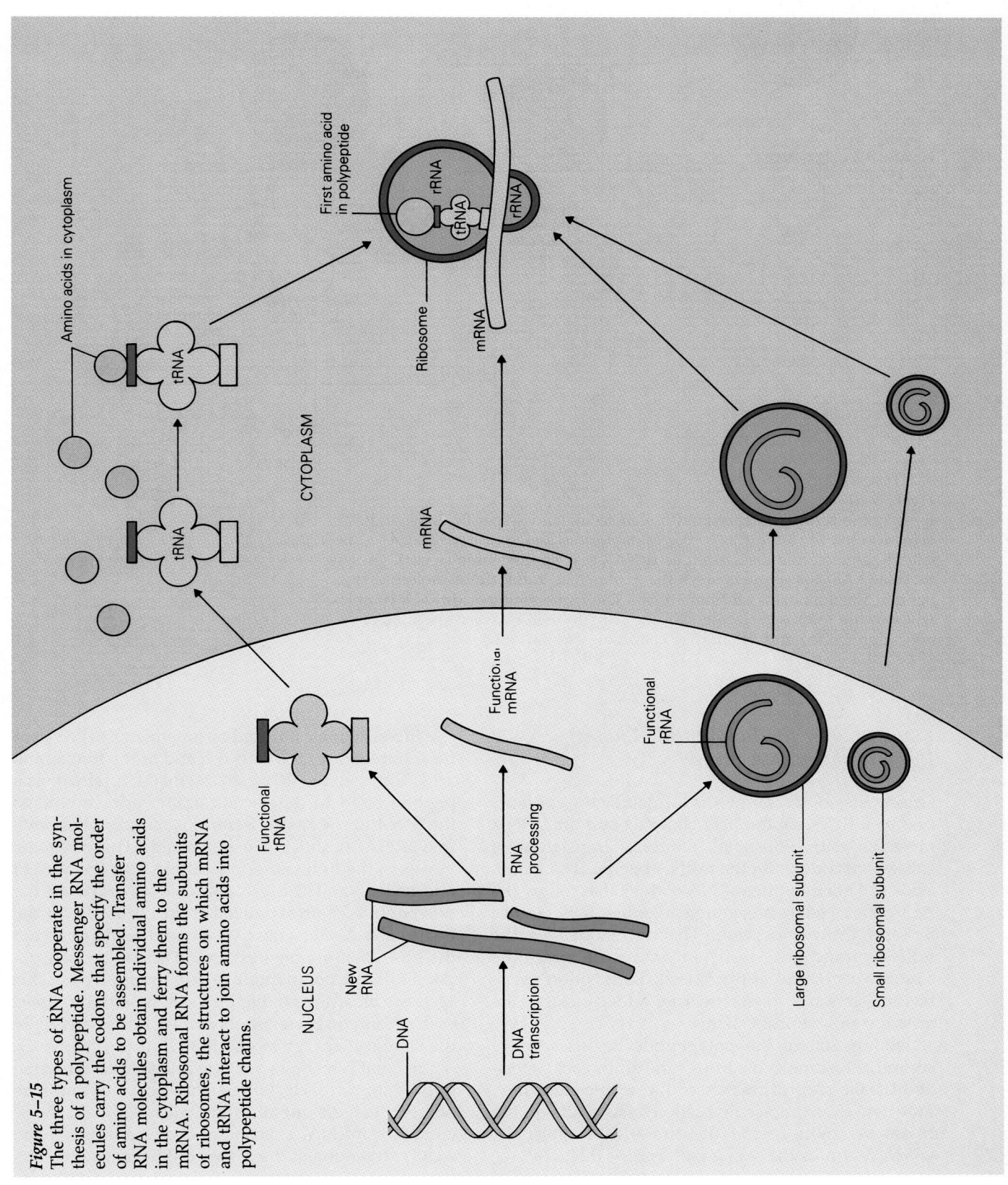

Figure 5–15
The three types of RNA cooperate in the synthesis of a polypeptide. Messenger RNA molecules carry the codons that specify the order of amino acids to be assembled. Transfer RNA molecules obtain individual amino acids in the cytoplasm and ferry them to the mRNA. Ribosomal RNA forms the subunits of ribosomes, the structures on which mRNA and tRNA interact to join amino acids into polypeptide chains.

Amino acids in cytoplasm

tRNA

tRNA

CYTOPLASM

First amino acid in polypeptide

rRNA

tRNA

tRNA

rRNA

Ribosome

mRNA

mRNA

Functional mRNA

Functional rRNA

Large ribosomal subunit

Small ribosomal subunit

Functional tRNA

RNA processing

DNA transcription

New RNA

NUCLEUS

DNA

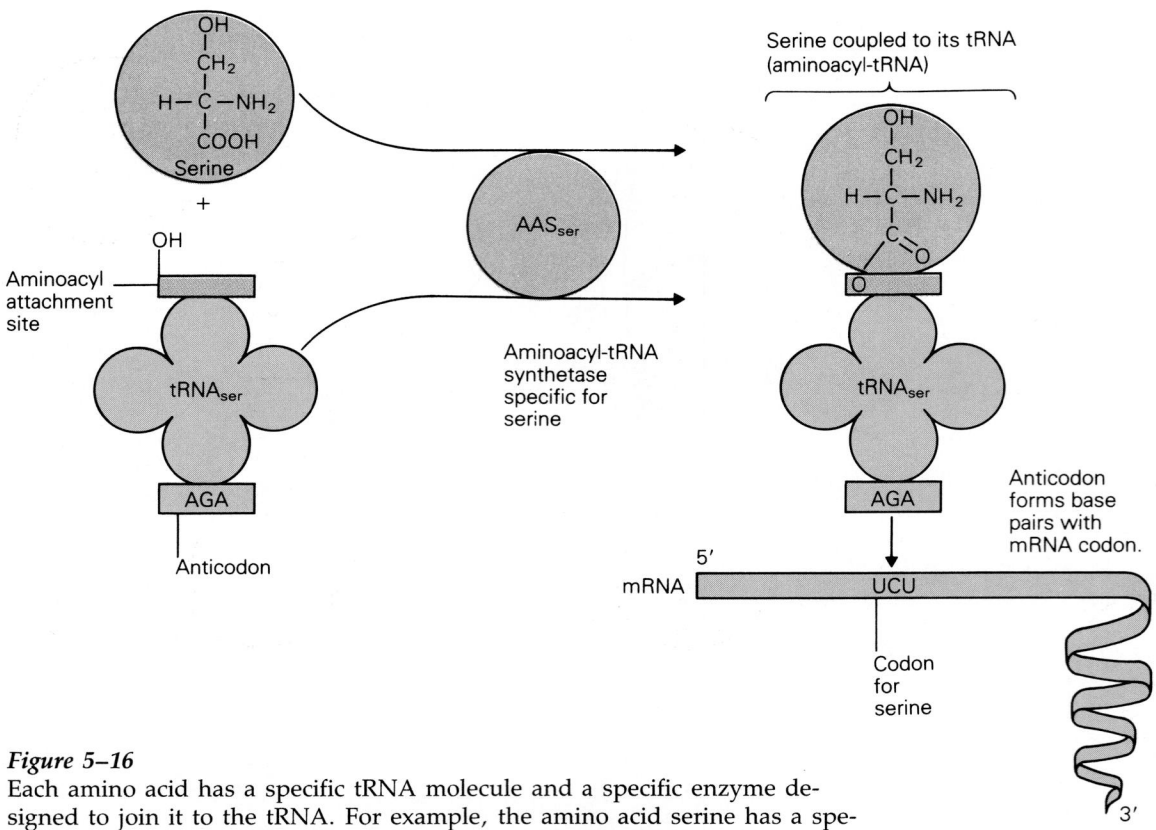

Figure 5–16
Each amino acid has a specific tRNA molecule and a specific enzyme de-signed to join it to the tRNA. For example, the amino acid serine has a spe-cific tRNA molecule designated as tRNA$_{SER}$. A specific aminoacyl tRNA syn-thetase (AAS) for serine couples the amino acid to the tRNA molecule via a bond at the aminoacyl attachment site. Later, the amino-acid-tRNA complex will join the mRNA molecule via base pairs formed between the tRNA$_{SER}$ anticodon (AGA) and the mRNA codon for serine (UCU).

Elongation of the Polypeptide Chain: Reading the mRNA Strand

Once a ribosomal complex is formed by the conjunc-tion of a tRNA-amino acid complex and an mRNA molecule with ribosomal subunits, translation of successive codons on the mRNA begins. The amino acids specified by the mRNA codons that follow the AUG start codon are joined together in sequence to form the polypeptide. The nucleotides on the mRNA strand are "read" in groups of three, the codons, beginning at the 5' end. Each codon speci-fies the amino acid that will be added next to the growing polypeptide chain.

Elongation of the polypeptide begins when a second tRNA molecule brings the amino acid speci-fied by the second mRNA codon to the ribosome. This new tRNA molecule occupies the A-site on the ribosome, right beside the first tRNA molecule, which occupies the P-site (see Fig. 5–18b).

In a reaction catalyzed by an enzyme called **pep-tidyl transferase,** the carboxyl end of the first amino acid, bound to the tRNA on the P-site, is joined via a peptide bond to the second amino acid, bound to the tRNA at the A-site. After this reaction, the now-free tRNA at the P-site is ejected. The ribosome moves to the right along the mRNA by a distance of one codon, and the tRNA at the new end of the polypeptide chain is moved from the A-site to the P-site. The A-site is now vacant and ready to accept the tRNA bearing the next amino acid to be incorpo-rated into the polypeptide. The various steps of this cycle are illustrated in Figure 5–18. It occurs 20 times each second on bacterial ribosomes, and is driven by the chemical energy stored in GTP (see Chapter 2).

Several ribosomes may be attached to the same mRNA strand, each synthesizing a copy of the same polypeptide. As one ribosome moves toward the 3' end of the mRNA, a second ribosome attaches to the AUG start codon at the 5' cap and begins synthesiz-

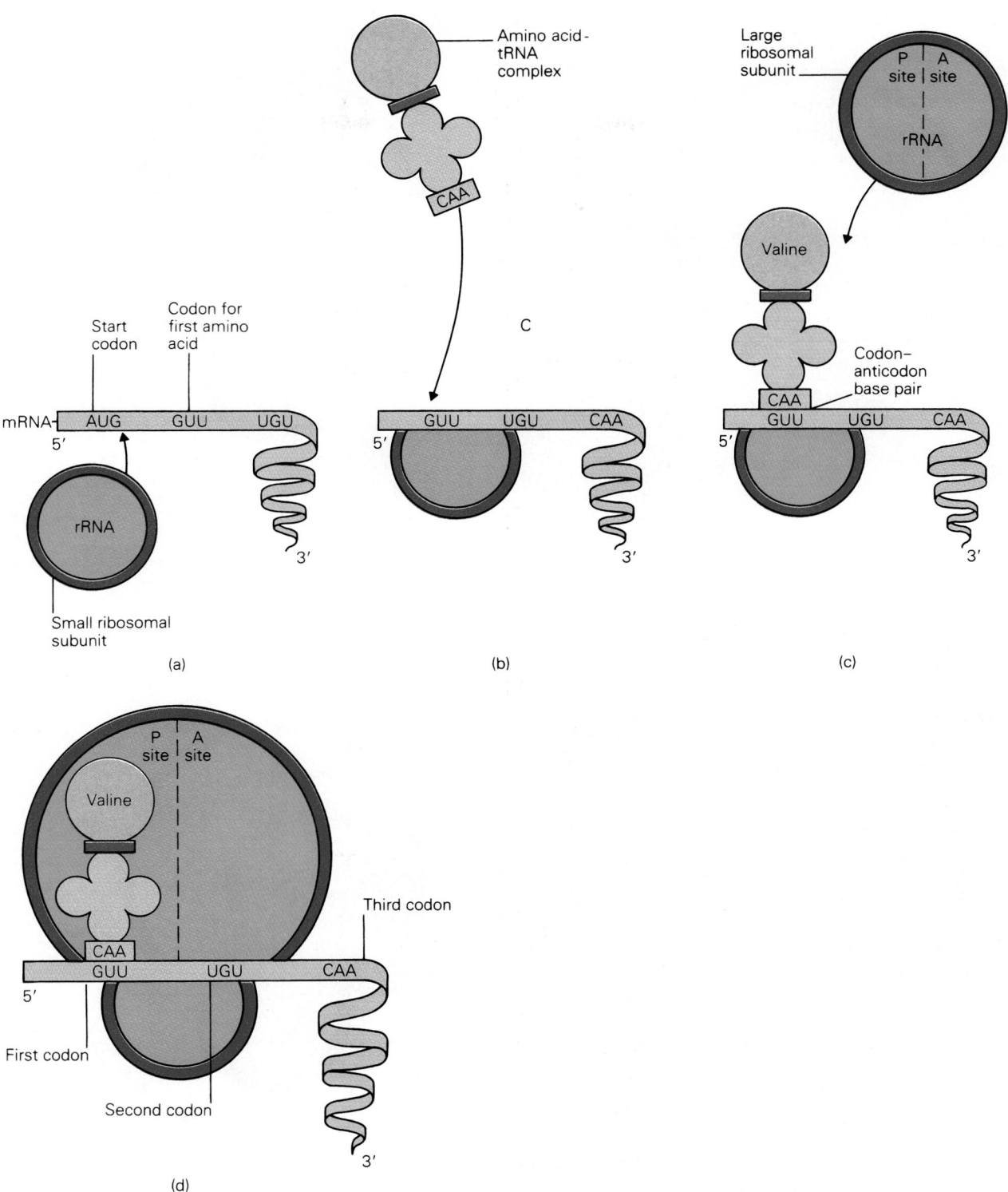

Figure 5–17
Protein synthesis is initiated when a small ribosomal subunit attaches to a
start codon (AUG) on an mRNA molecule (*a*). An aminoacyl complex bearing
an amino acid (e.g., valine) attaches to the mRNA strand via hydrogen
bonds between the tRNA anticodon and the mRNA codon (*b*). A large ribosomal subunit then attaches, forming a complete ribosome (*c*). The ribosome
contains two sites for tRNA molecules to occupy, the A site and the P site (*d*).

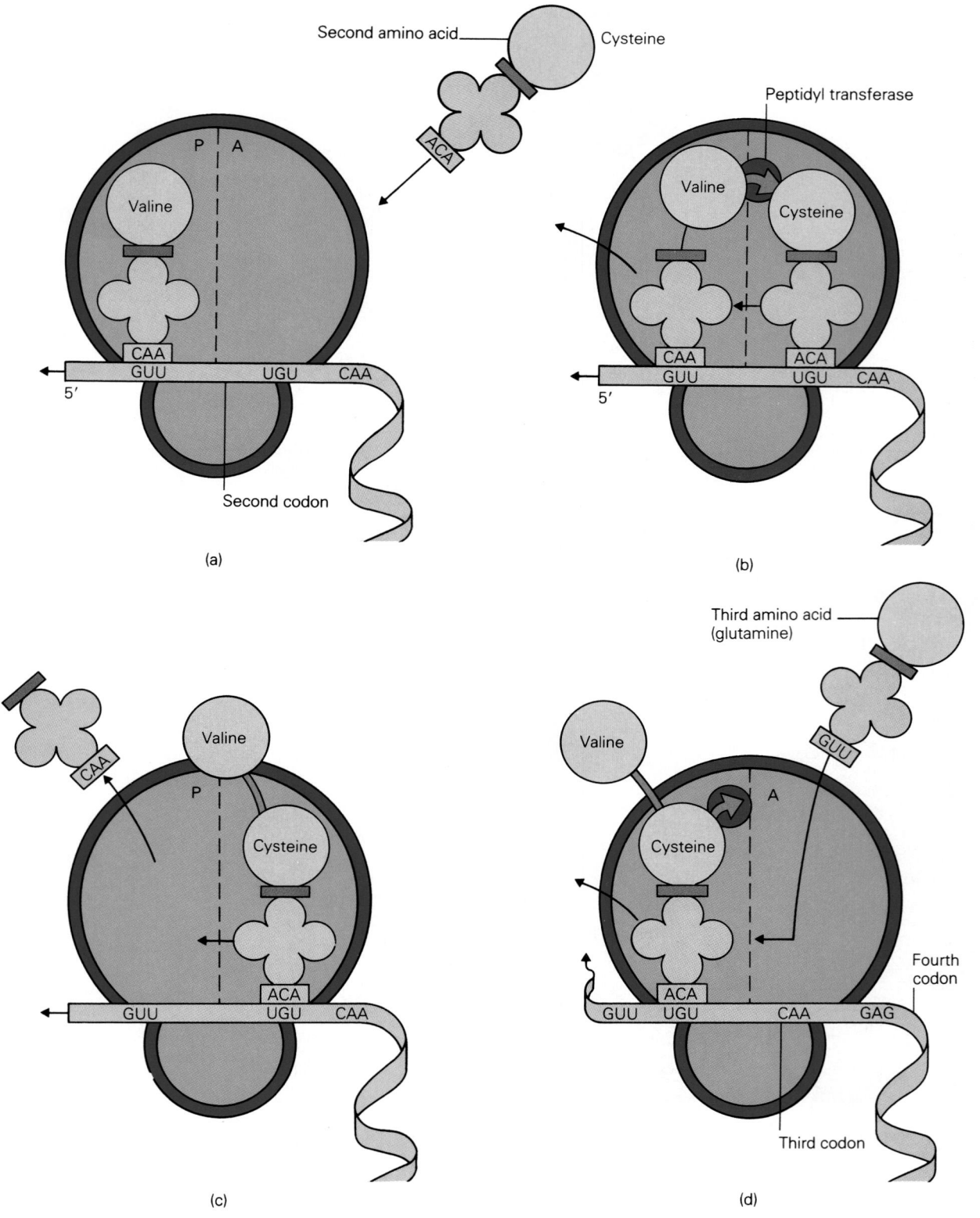

Figure 5–18
Elongation of a polypeptide begins when a tRNA molecule bearing another amino acid enters the A site of the ribosome. The amino acid is specific to the second codon on the mRNA strand (*a*). Aided by the enzyme peptidyl transferase and energy from GTP, a peptide bond forms between the two amino acids that now sit side by side on the ribosome. At the same time, the first amino acid dissociates from its tRNA molecule (*b*). The free tRNA molecule leaves the P site, and the ribosome moves to the right along the mRNA strand, moving the second codon and attached tRNA-amino acid complex from the A site to the P site (*c*). A third amino acid-tRNA complex, corresponding to the third codon, now enters the vacant A site (*d*). A peptide bond forms

174

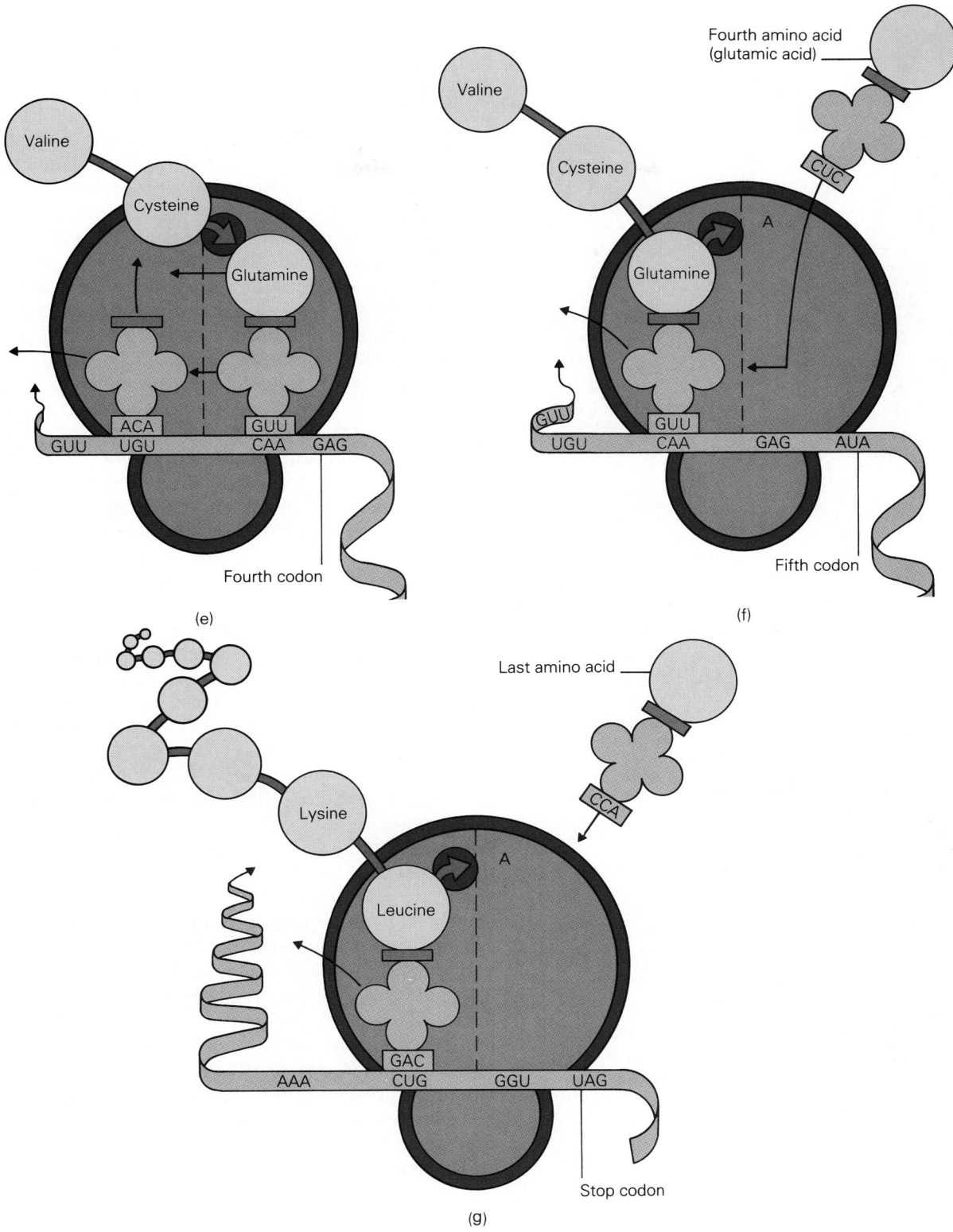

(e)

(f)

(g)

between the second and third amino acids, and the second tRNA molecule vacates the A site (*e*). The third amino-acid-tRNA complex is moved into the P site, and another amino acid complex approaches (*f*). This cycle repeats until the entire mRNA strand has been translated into amino acids (*g*). When the ribosome reaches a stop codon, the ribosomal complex dissociates.

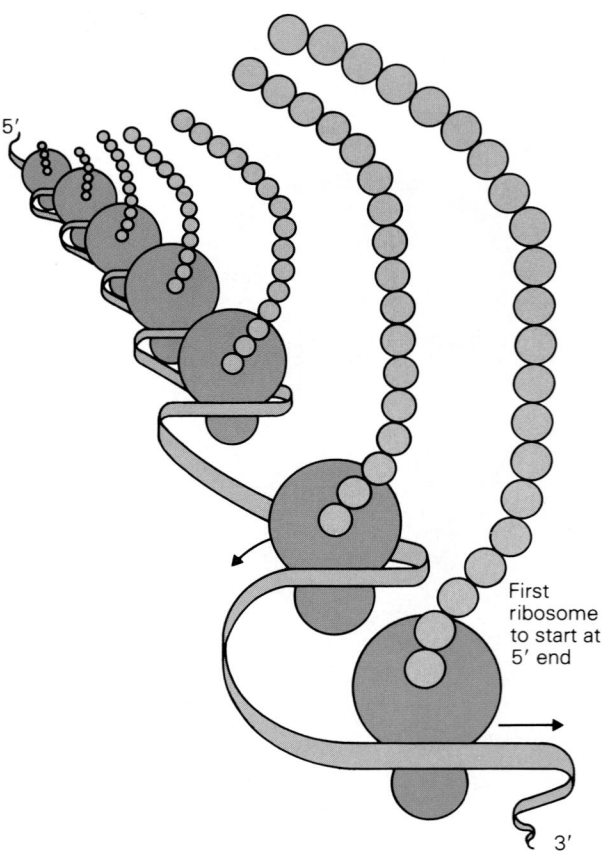

First
ribosome
to start at
5' end

Figure 5–19
Several ribosomes may be at work at the same time on
the same mRNA strand, creating a structure known as
a polysome.

ing a second copy of the polypeptide. As more ribosomes become attached in this way to a single strand of mRNA, a structure known as a **polysome** (or **polyribosome**) is formed (Fig. 5–19).

Translation of the mRNA ceases when a **stop codon** is encountered. At this point, the completed polypeptide is released, and the ribosome dissociates from the mRNA.

Control of Gene Expression

The expression of prokaryotic genes is regulated primarily at the level of transcription, rather than translation (see Fig. 5–15). A specific **repressor** protein can bind to the DNA at a site next to a gene and can block transcription of the gene by RNA polymerase. Repression of the gene can be reversed by an **inducer** molecule that forms a complex with the repressor and prevents binding of the repressor to the

DNA, thus allowing normal transcription of the gene by the polymerase. This kind of a mechanism readily explains how the expression of different genes can be turned on or off at different times.

Control at the level of transcription also may be the most important way of regulating gene expression in eukaryotes, but the system is complex and many of the mechanisms still unclear. The action of steroid hormones on the expression of specific genes is one control system in which many of the steps have been worked out through extensive investigations. Steroid hormones include estrogens (the female sex hormones produced by the ovary and the placenta), testosterone (the male sex hormone from the testis), and mineralocorticoids such as aldosterone from the adrenal glands.

The general action of these hormones is as follows. On arriving at the cell, the hormone crosses the plasma membrane and binds to a specific receptor molecule inside the cell. The steroid-receptor complex is able to enter the nucleus and bind to the DNA at a specific site. The binding of this complex either increases or decreases the transcription of specific genes. This changes the amount of mRNA and leads, in turn, to an increase or decrease in the specific protein products of the genes. It is these few specific proteins that produce the changes in cellular function that represent the characteristic response of the cell to the hormone.

While not a general mechanism for regulating gene expression, **amplification** of genes has been well documented in certain cells such as the developing eggs of a species of African clawed toad. In the nuclei of the egg cells, the genes that code for rRNA are repeated about 450 times on the chromosomes. This repetition, or amplification, allows the cell to make a huge amount of one gene product very quickly. The accumulation of rRNA in the toad eggs allows the formation of large numbers of ribosomes. These structures facilitate the very rapid protein synthesis that will be needed when the egg is fertilized and starts to divide and grow rapidly. About 10^{12} ribosomes are needed per cell. Without gene amplification, the cell would need several hundred years to make this number of ribosomes.

In some eukaryotic cells, gene expression can be regulated at steps following transcription—for example, by modification of the RNA molecules themselves (see Fig. 5–12), or by a control system at the level of translation. The latter provides a means for synthesizing large amounts of a single protein without amplifying the gene that codes for that protein. This occurs when transcription of the gene produces a very stable mRNA molecule that remains in the cytoplasm for days and can be used repeatedly for

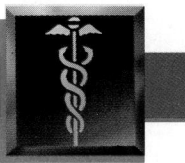

Antibiotics

Antibiotics are molecules produced by various species of microorganisms such as bacteria and fungi. An antibiotic released by one species blocks the growth of other species and may eventually destroy them. Hundreds of antibiotics differing in physical and chemical properties and mechanism of action have been found. More than 50 of these are used clinically to control and cure human diseases caused by invading microorganisms. The specific antibiotic selected for use is usually determined by the identity of the microorganism.

Penicillin and closely related molecules are widely used to treat many infectious diseases. Alexander Fleming discovered penicillin in 1928 when he noticed that growth of bacteria was inhibited by a contaminating fungus of the Penicillium family. Penicillin blocks the synthesis of the tough bacterial cell wall that normally keeps bacterial cell membranes intact despite their high internal osmolarity. Other antibiotics, such as *chloramphenicol* and the *tetracyclines*, act on bacterial ribosomes to inhibit protein synthesis. *Aminoglycosides* also act at the ribosomal level, where they interfere with the initiation of protein synthesis. An additional effect of aminoglycosides is to induce misreading of the mRNA code so that abnormal polypeptides are produced. *Rifampin*, a drug used to treat tuberculosis, inhibits RNA polymerase, which is the enzyme responsible for synthesis of RNA from DNA. Other key steps in bacterial metabolism are inhibited by the *sulfonamides*.

A serious medical problem arises when invading bacteria acquire the genes that allow them to resist the antibiotics normally used to destroy them. The resistant cells in the bacterial population multiply unimpaired and eventually predominate because the growth of the drug-sensitive cells is suppressed. The time taken for this to occur varies greatly among various microorganisms and drugs. Antibiotic resistance requires a stable genetic change that can be inherited by each new generation. *Mutation* is often the cause, and the alterations that occur are remarkable. Bacterial mutants have developed resistance to penicillin by producing an enzyme that destroys the penicillin. Mutants that are highly resistant to rifampin were found to contain a slightly altered RNA polymerase, which fulfills its normal function but is no longer sensitive to inhibition by the drug. Antibiotic resistance can also be acquired by the transfer of genetic material from one bacterial cell to another.

Bacteriophages are viruses that enter bacterial cells; some of these "phages" can carry fragments of bacterial DNA. If the DNA contains a gene for drug resistance that the bacterial cell can transcribe, then the bacterium may become drug resistant and capable of passing on this property to its offspring. The same results can occur when genes for drug resistance are passed from one bacterium to another via a direct bridge between the cells, a process known as *conjugation*. The emergence of antibiotic-resistant bacteria has created a constant need for new drugs.

---

translation and synthesis of the specific protein product. In other words, the mRNA molecules that are involved in the production of the major proteins of a cell tend to be long-lived. A specific example is the synthesis of the protein hemoglobin in the immature red blood cell. As red blood cells mature, their nuclei are lost and new RNA cannot be synthesized, yet hemoglobin synthesis continues for weeks. This is possible because the mRNA molecules that code for this protein have a much longer lifetime than the other mRNA molecules in the cell. These observations suggest that controlling the lifetime of mRNA molecules in the cytoplasm may be a mechanism for regulating the expression of certain genes.

The final step at which gene expression may be regulated is that following translation. Many newly synthesized polypeptides undergo one or more types of covalent alterations known collectively as **post-translational modification** (not to be confused with the post-transcriptional modification of RNA discussed earlier [see Fig. 5–12]). Some of the covalent modifications of polypeptides were discussed earlier in Chapters 2 and 3—for example, the attachment of carbohydrate to produce a glycoprotein and the regulation of the biological activity of a protein by alternate cycles of attachment and removal of a phosphate group. Other examples include the permanent covalent attachment of a coenzyme molecule essential for the activity of an enzyme protein,

RNA Scissors

In 1982 Thomas Cech showed that RNA can catalyze specific intracellular reactions. Pre-ribosomal RNA was found to have the ability to cut and splice itself, thus removing sequences not needed for its cellular function. These findings were complemented by the work of Sidney Altman, who found that transfer RNA could be cut by another RNA molecule. The discovery that RNA could act as an enzyme, as a type of "molecular scissors," led to the 1989 Nobel Prize in Chemistry for Cech and Altman. This discovery overturned the long-held principle that the nucleic acids (RNA and DNA) act solely as the informational molecules of a cell, while the proteins (such as enzymes) act solely as the functional molecules. It is now clear, at least in some cases, that RNA has the dual role of information carrier and enzyme catalyst.

The implications of this relatively new information are only beginning to be understood. One obvious implication of the presence of RNA enzymes in a cell is that, in contrast to previous ideas, it cannot be assumed that all catalytic activities in a cell are due to proteins. For example, some of the steps which produce a functional RNA molecule are catalyzed by RNA. There is another implication for the function of *ribosomes*, the structure on which proteins are synthesized. Ribosomes are composed of RNA and several proteins. It may be one of the ribosomal RNA molecules, rather than one of the ribosomal proteins, which catalyzes formation of peptide bonds. Finally, there are evolutionary implications because it has been suggested that nucleic acids and proteins evolved together. This statement is based on the previous assumption that only proteins could serve as functional molecules in a cell. However, it is now possible to believe that when life originated neither DNA nor proteins were required because RNA may have been able to fulfill all the important functions.

Using RNA enzymes, researchers are learning to manipulate RNA with the hope of altering it as easily as DNA is now manipulated during cloning procedures (see Chapter 3). The potential practical applications of RNA enzymes are considerable. A specific example is their potential use to fight infections by *viruses*. A virus is a packet of nucleic acid surrounded by a protective coat. It is unable to generate metabolic energy or synthesize proteins. Viruses can be thought of as cellular parasites because they reproduce by infecting a cell and directing the host cell machinery to synthesize more viral material. Viruses cause many diseases in humans and also in crop plants. Some of the human diseases include *AIDS*, *hepatitis*, and *herpes*. Many viruses lack DNA and store their genetic information in RNA molecules. Thus it may be possible to destroy viruses by using specific RNA enzymes as scissors to cut the viral RNA. This will prevent the virus from reproducing after it has invaded a host cell. RNA enzymes are remarkably specific, and the key in this approach will be to use one which will recognize and cut out a specific base sequence in the foreign RNA. The RNA fragments can then be disposed of by other enzymes in the host cell. The AIDS virus is a specific example of a virus which uses RNA as its genetic material. Some initial studies indicate that the AIDS virus grown in culture under experimental conditions can be inactivated by a specific type of RNA enzyme. Other RNA scissors may prove to be useful for treating the skin lesions caused by the herpes virus.

RNA scissors work well in the research laboratory but some technical problems must be overcome before clinical applications can be tested. Methods have to be developed for introducing the pieces of catalytic RNA into cells, and for ensuring that the RNA remains chemically stable inside the cell in order to carry out its enzymatic action. Careful validation of the precision of RNA scissors and their lack of effect on the host cell RNA also will be required. This will be the next challenge.

and the production of a biologically active polypeptide from a much larger inactive polypeptide molecule (known as a **precursor**) through the activity of enzymes that selectively remove small sections. The latter mechanism is involved in the production of the hormone **insulin,** a polypeptide of 51 amino acids, from an inactive precursor called **proinsulin,** a polypeptide of 84 amino acids.

Finally, although the function is not understood, the side chain groups of certain amino acids in the polypeptide may be altered. A hydroxyl (—OH) group is commonly attached to the side chain of the amino acid proline to produce hydroxyproline, and a methyl (—CH_3) group may be attached to the side chain of amino acids such as lysine and glutamic acid.

Repair of Damaged DNA

The genetic information stored in DNA must remain essentially unchanged as it is passed on from cell to cell. The replication process is designed to faithfully reproduce molecules that are identical to the parent molecule, but it cannot correct alterations to DNA that occur in the intervals between the replication steps. DNA is very sensitive to damage by factors commonly present in the environment. In fact, it has been estimated that several thousand bases in the genome of a mammalian cell are damaged every day. Changes on this scale in the structure of DNA could have drastic consequences, leading to mistakes in genetic information, malignant cells, and a threat to the long-term continuance of the species. Fortunately, cells have developed repair systems to correct almost any kind of damage, and very few stable changes are allowed to accumulate in the genome of each cell.

Environmental factors that cause spontaneous damage to DNA are examples of extracellular signals that are potential modifiers of an internal cellular control system. Ultraviolet light is an environmental factor that damages DNA. Other factors include a number of common chemicals and the radiation produced by the decay of radioactive isotopes.

The repair mechanisms that correct the damage from ultraviolet light rays (wavelength 20 to 200 nm) are the best understood. When exposed to ultraviolet light, pyrimidine bases, such as thymine, that are adjacent to one another on the same DNA strand tend to become linked together by covalent bonds. The resultant structure, a **thymine dimer,** causes distortion of the DNA chain and blocks replication (Fig. 5–20). A number of enzymes recognize this structure and can repair it.

Photoreactivation

The simplest method of repair is by **photoreactivation,** a process whereby a specific enzyme binds to the DNA section containing the thymine dimer and uses the energy of visible light (wavelength 300 to 600 nm) to break the covalent bonds that hold the

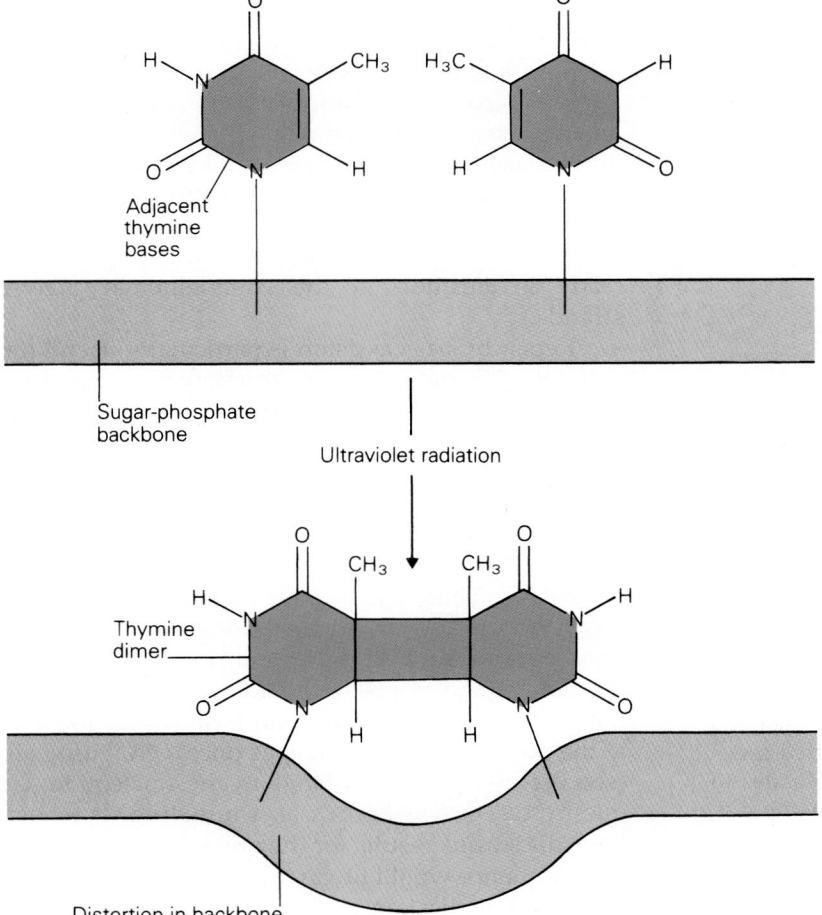

Adjacent thymine bases

Sugar-phosphate backbone

Ultraviolet radiation

Thymine dimer

Distortion in backbone

Figure 5–20
Ultraviolet radiation may cause adjacent thymine bases on a DNA strand to join together, forming a thymine dimer and distorting the shape of the sugar-phosphate backbone.

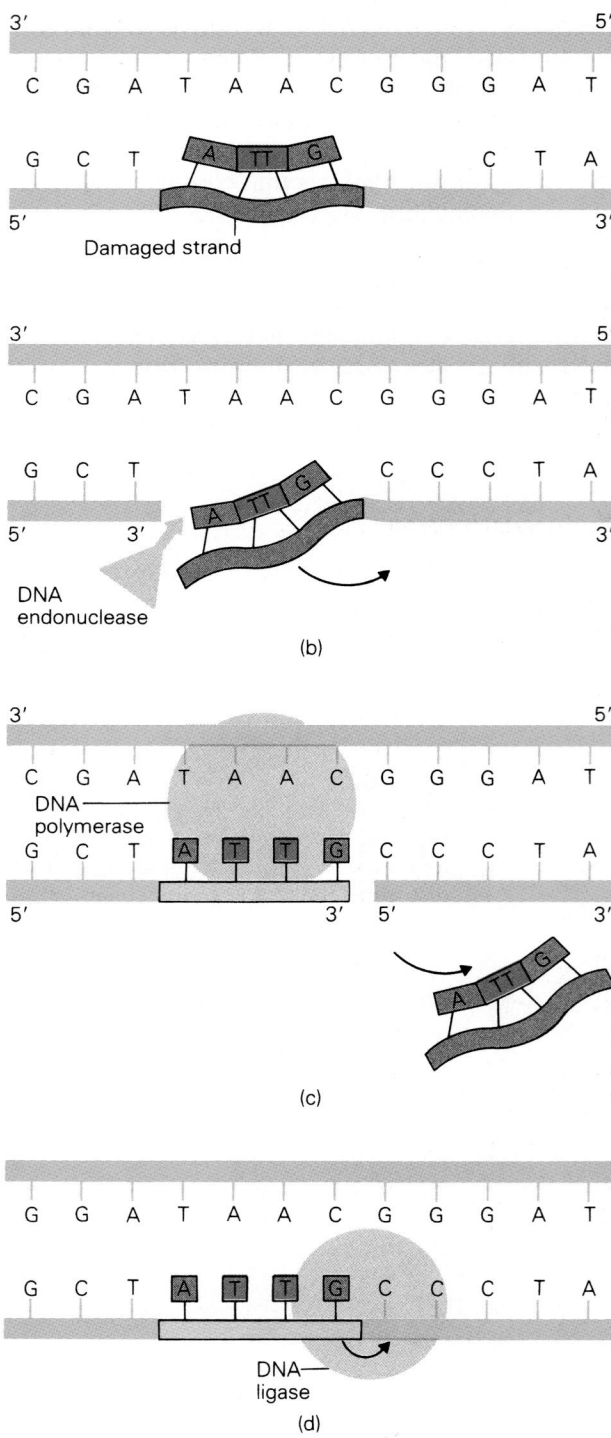

(b)

(c)

(d)

Figure 5–21
The damage caused by a thymine dimer is repaired when DNA endonuclease excises the distorted segment and DNA polymerase joins new nucleotides to replace those removed. DNA ligase joins the new segment to the undamaged portion.

dimer together. The DNA reverts to its original structure containing adjacent thymine bases, and the bound enzyme is released.

Excision Repair

Other enzymes act by selectively cutting out or **excising** the section of DNA that contains the thymine dimer. The remaining gap can be filled relatively easily because the complementary strand has all the information for the correct sequence of nucleotides needed to fill the gap. The process is known as **excision repair** and consists of essentially three steps (Fig. 5–21). Repair is initiated by a specific **endonuclease** enzyme that breaks the DNA chain containing the dimer. Next, DNA polymerase simultaneously cuts out the thymine dimer and replaces it with a new segment that is complementary to the undamaged strand. Finally, the 3' end of the new segment is joined to the rest of the chain by **DNA ligase.**

Base Excision

In a different type of repair mechanism, any nucleotide containing an altered base can be removed. In **base excision, DNA glycosylase** enzymes "recognize" the altered bases and remove them. The remaining sugar-phosphate portions of the nucleotides are also removed and replaced with the correct intact nucleotides by the stepwise actions of an endonuclease, DNA polymerase, and DNA ligase, in a sequence similar to that depicted in Figure 5–21. The correct nucleotide can be inserted because the nucleotide with which it will form a base-pair is already in position in the complementary DNA strand.

Repair by base excision is particularly useful for correcting changes due to spontaneous **deamination,** a term used to describe the loss of NH_3. Deamination of cytosine produces uracil (Fig. 5–22b), which is a base not normally present in DNA. The introduction of uracil causes a problem because it resembles the structure of thymine so closely that it can form hydrogen bonds with adenine, just as thymine does. Thus, an AU base pair will form during replication of the strand that contains uracil. After another replication step, a thymine will be placed in the new strand opposite the adenine so that in the new DNA molecule an AT pair base will be present at the position where there was once a GC base pair (see Fig. 5–22c). This change in the nucleotide sequence of DNA is known as a **mutation.** It would have disastrous results for the organism because the new sequence would be carefully copied during further DNA replication and passed on to each of the new cells. The mutation must be corrected by prompt removal of the uracil by base excision.

These are a few examples of the variety of DNA-repairing enzymes that recognize and replace ab-normal nucleotides, ensuring that the genetic information stored in the nucleotide sequence of DNA remains unaltered during the life of a cell. A built-in proofreading mechanism ensures that the nucleo-

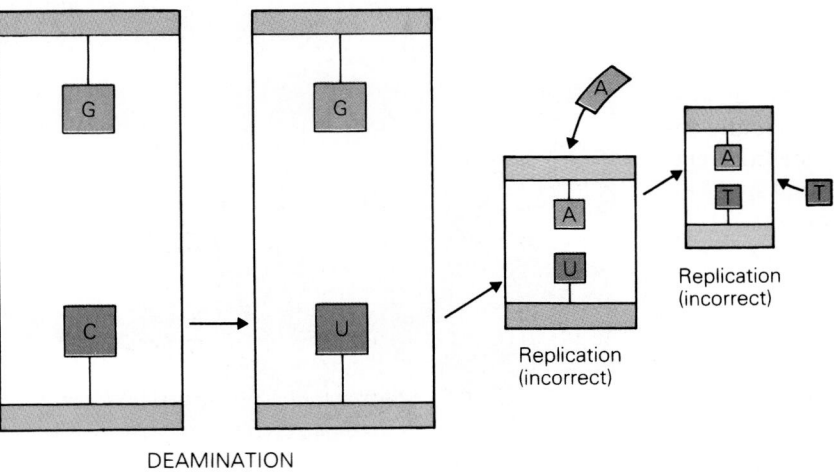

Figure 5–22
In normal DNA replication, the same base pair is copied indefinitely (*a*). A mutation can occur, however, when the base cytosine undergoes deamination to become uracil (*b*). When the strand containing the erroneous uracil is replicated, adenine forms a base pair with uracil. When the strand containing the erroneous adenine is copied, thymine forms a base pair with the adenine (*c*). Thus, a GC base pair has been changed to an AT base pair, changing the basic genetic code for that cell.

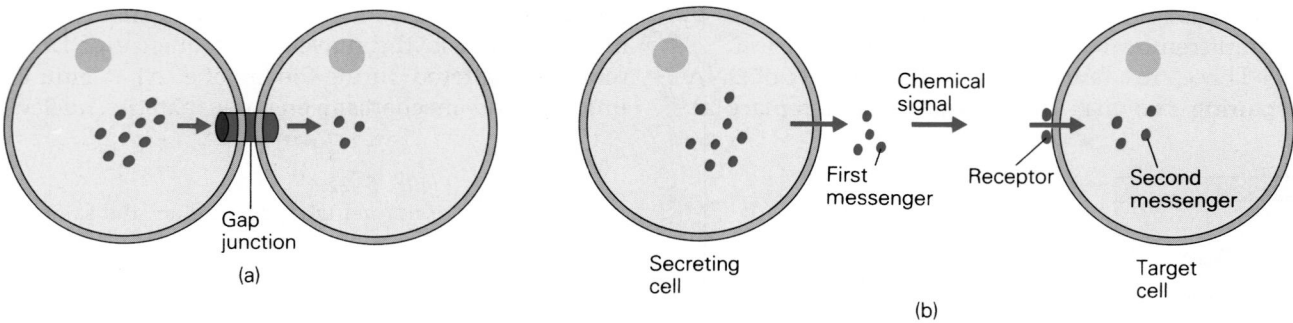

Figure 5–23
The two broad types of communication between cells: (*a*) gap junctions (between adjacent cells) and (*b*) chemical signals (between distant cells). Chemical signals usually involve an extracellular signal produced by one cell, a receptor in or on the target cell, and an intracellular signal in the target cell.

tide sequence is copied very accurately during replication of DNA, so that an identical store of information is passed on to the new cells produced by cell division.

Cell-to-Cell Signaling: Hormones, Neurotransmitters, and Other Molecules

The coordination of the different functions of the many tissues in a multicellular animal is achieved because cells have developed communication systems. Cells that are in direct physical contact with one another have structures known as **gap junctions** in their plasma membranes (Fig. 5–23*a*). These junctions allow small molecules to pass directly from the cytosol of one cell to the cytosol of the adjacent cell. Cells some distance apart are able to communicate by sending **chemical signals,** molecules secreted by one cell that bind to the plasma membrane or enter the cytoplasm of the distant cell (see Fig. 5–23*b*). Chemical signaling systems usually involve three separate components: extracellular signals, called first messengers; receptors; and intracellular signals, called second messengers.

Interactions Between Adjacent Cells: Gap Junctions

Gap junctions are common structures in tissues of all animals. They consist of a group of small pores or channels, each about 1.5 nm in diameter, that allow small water-soluble molecules with molecular

weights under 1000 to move directly from the cytosol of one cell to the cytosol of an adjacent cell (Fig. 5–24). This includes ions, sugars, amino acids, and nucleotides, but excludes the macromolecules.

The major advantages of this direct communication system are its speed and reliability. The "message" must travel only a short distance, with little chance of intervention by outside forces because it

Figure 5–24
A gap junction forms direct connections between the cytoplasm of adjacent cells, via proteins that create a tunnel-like structure in the plasma membrane of each cell. Two tunnels of adjacent cells connect to allow materials to pass between them.

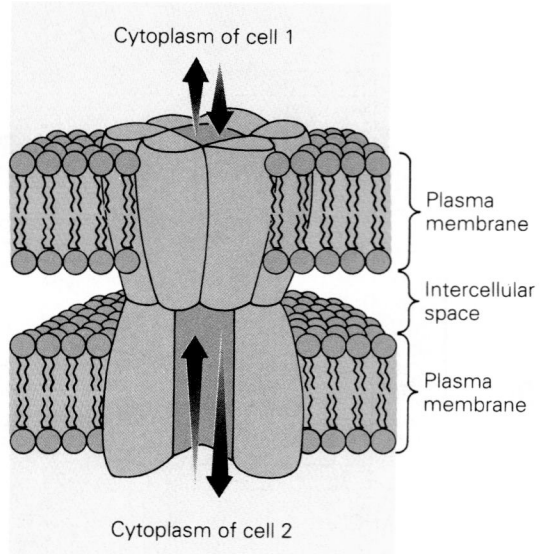

does not leave the cytosol. In principle, this system could also allow **metabolic coupling** between different cells present in a tissue. This means that molecules made by only one group of cells within the tissue could be shared with the other cell types present by being passed through the gap junctions.

Gap junctions also may allow electrical communication between cells because they permit the rapid flow of ions. Electrical communication can occur in this way between certain nerve cells. Normally, nerve cells communicate chemically with each other through substances called **neurotransmitters,** which cross from one nerve cell to the other through junctions called **synapses.** The nerve impulse from one nerve cell is passed to the next cell by the release of the chemical neurotransmitter, which crosses the synapse and initiates the nerve impulse in the next nerve cell (this is discussed in more detail later in the chapter).

Occasionally, however, nerve cells must communicate with more speed than chemical neurotransmitters can provide. In these situations, a gap junction can act as a special kind of synapse that transmits the nerve impulse electrically rather than chemically. Electrical transmission is much faster than chemical transmission, and sometimes this difference is very important. An electric eel can generate a powerful electric shock only because all the cells in its electrical organ are coupled via electrical synapses and can be synchronized to "fire" at the same time. Chemical transmission would be too slow to achieve the necessary degree of precision. **Electrical coupling** is also important in mammalian tissues such as the heart, for example, in which all the muscle cells must contract simultaneously in order for the heart to pump blood efficiently.

Interactions Between Distant Cells: Chemical Signals

Compared with the study of communication between adjacent cells, which is difficult to observe or measure, it is relatively easy to study communication systems that use chemical signals. These signals take the form of specific molecules released by one cell and sent to produce a change in a **target cell.** Consequently, much more is known about such systems, and the remainder of this chapter is concerned with how they operate.

Chemical signaling systems generally involve three components: (1) the signal molecules, also called **extracellular signals** or **first messengers;** (2) **receptors,** molecules in either the membrane or cytoplasm of the target cell; and (3) **intracellular signals** or **second messengers.** Extracellular signals travel from the cell that produced them to the target cell, where they react with receptor molecules. The receptor molecules in turn produce intracellular signals. The cumulative effect of these interactions is to elicit the desired physiological response in the target cell.

Extracellular Signals: First Messengers

Cells use various types of chemical signals to communicate with other cells. **Local chemical mediators** affect only nearby cells and are rapidly destroyed if they are not used. **Neurotransmitters** are produced by nerve cells and also affect cells that are nearby. Finally, **hormones** are sent into the blood by secretory cells to affect target cells, which may be nearby or far away from the hormone-producing cells. Figure 5–25 summarizes these three types of signaling systems.

With the exception of the prostaglandins and related compounds, most of these chemical signals are synthesized by specialized cells where they are stored until the appropriate stimulus triggers their release. Specific examples of chemical signals are listed in Table 5–4, which summarizes the wide variety of molecules used.

These groups of chemical signals can be classed into two large groups: **water-soluble signals** and **lipid-soluble signals.** Many of the hormones, such as parathyroid hormone and vasopressin, and all of the neurotransmitters, are water-soluble. These molecules are removed or destroyed very rapidly (within seconds, or a few milliseconds for neurotransmitters) after being released. Steroid hormones and thyroid hormones (such as thyroxine), in contrast, are lipid-soluble molecules that can persist in the bloodstream for several hours, or even days in the case of thyroid hormones. Thus, water-soluble signal molecules are used to produce responses that are needed rapidly but only for a short time, whereas lipid-soluble signal molecules are used to elicit responses that may develop more slowly but will last for a long time.

A common feature of the chemical communication systems is the presence of specific **receptors** that allow the signaling molecule to recognize and bind to the target cell. Receptor function is discussed in more detail in a later section.

Local Chemical Mediators. Some signal molecules released by a cell affect only the cells in the immediate vicinity, either because the molecules are destroyed rapidly once released or because they readily bind to specific **receptors** that are present on the

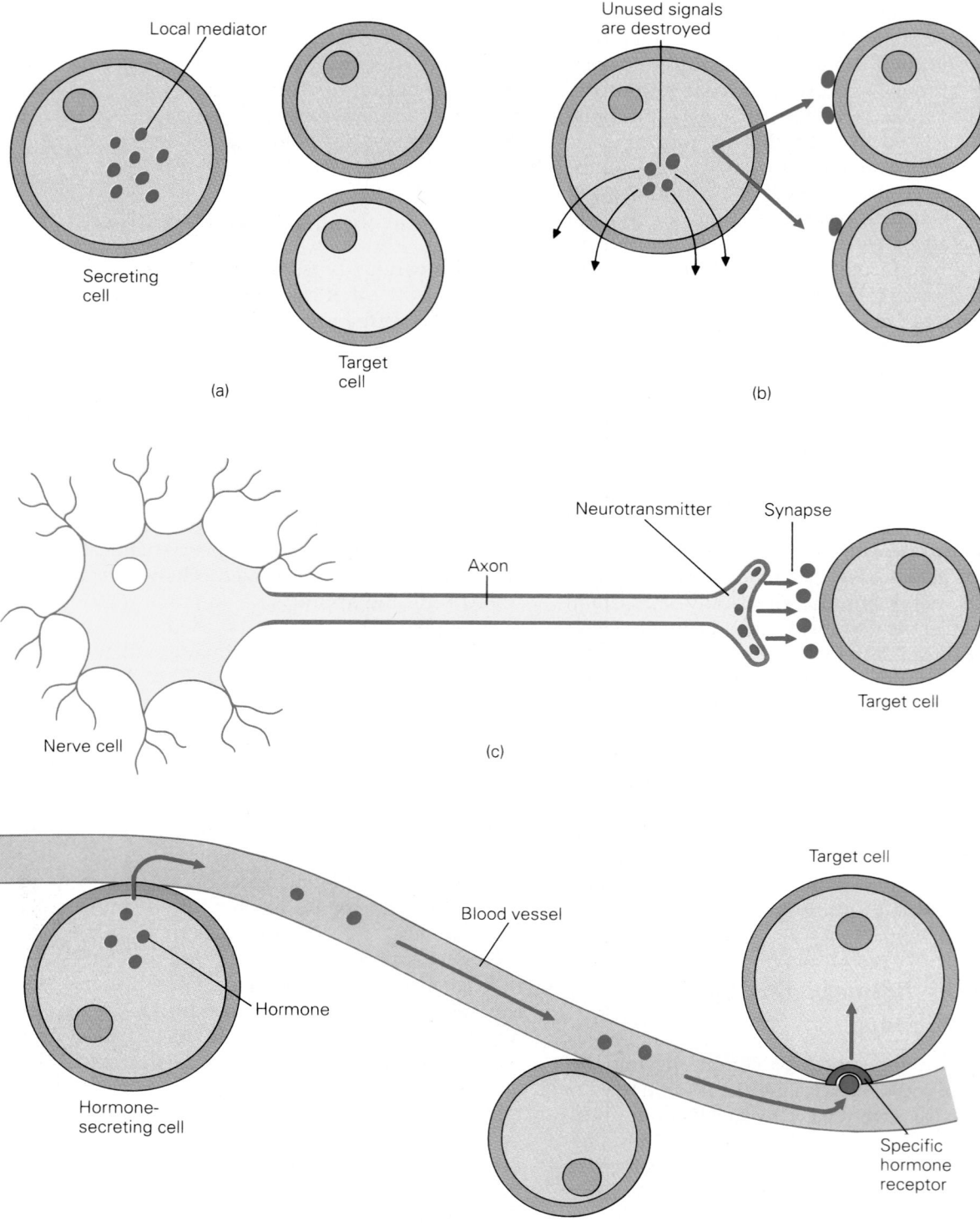

Figure 5–25
The three basic types of chemical signaling systems. (*a* and *b*) Local chemical mediators act only on cells near them and are quickly destroyed if they are not used. (*c*) Neurotransmitters are produced by nerve cells and are delivered via a synapse at a target cell, which may be a long distance away. (*d*) Hormones travel via the blood from hormone-secreting cells to target cells that have specific receptors for them.

Table 5–4

Examples of Molecules That Serve as Chemical Signals

Type of Molecule	Local Mediator	Neurotransmitter	Hormone
Peptides	—	Neuropeptides	Vasopressin
Polypeptides	—	—	Insulin
Amino acids and derivatives	Histamine	Glycine	Epinephrine
Steroids	—	—	Testosterone
Fatty acid derivatives	Prostaglandins	—	—
Other small molecules	—	Acetylcholine	—

first cells that they encounter nearby (see Fig. 5–25*a* and *b*). Destruction and cell binding of these molecules occur so rapidly that only a minor fraction of the amount released will reach the bloodstream. The molecules are called **local chemical mediators;** specific examples include **histamine** and the **prostaglandins.** Histamine is derived from the amino acid histidine, and is stored in the specialized **mast cells** present in connective tissues.

Histamine is released rapidly in response to injury, infection, and allergic reactions. One of its principal actions is to cause expansion and leakiness of local blood vessels. This effect allows antibodies and white blood cells access to the site of injury, where they perform their tasks of destroying foreign substances.

Prostaglandins were named after the prostate gland, which was once thought to be their only source. It is now recognized that prostaglandins are formed by most tissues, rather than by specialized cells, and that they are released continuously instead of being stored. They are synthesized from essential fatty acids containing 20 carbon atoms, such as arachidonic acid (see Chapter 2).

Prostaglandins have diverse effects on different tissues. Smooth muscles are major target tissues that either contract or relax, depending on the specific type of prostaglandin. Prostaglandins also mediate inflammatory responses. The effectiveness of aspirin in reducing fever and inflammation is due largely to its ability to inhibit prostaglandin synthesis. Many of the actions attributed originally to the prostaglandins may be due to their highly unstable but potent relatives known as **prostacyclins** and **thromboxanes.** Another metabolic pathway that begins with arachidonic acid leads to formation of the **leukotrienes,** another group of locally acting chemicals that appear to mediate allergic and inflammatory reactions.

Neurotransmitters. A second group of chemical signals, known as **neurotransmitters,** is used by

nerve cells to influence their target cells. Although the target cell may be many centimeters from the nerve cell body, the neurotransmitter is delivered directly to the target cell by a long slender extension of the nerve cell called an **axon.** At the **synapse** (the gap between the end of the axon and the target cell), the neurotransmitter is released by the axon and diffuses across to the target cell (see Fig. 5–25*c*). The gap it must cross is very small (only a few nanometers), and it reaches the target cell in less than 1 msec. Receptor molecules that are present on the surface of the target cell in the region of the synapse provide binding sites for the neurotransmitter.

Delivery of the neurotransmitter directly to the target cell, and the presence of specific receptors there, ensure that the effects of the chemical signal are restricted to a very small area, even though the target cell may be many centimeters from the main body of the nerve cell.

A variety of molecules serve as neurotransmitters at chemical synapses. The amino acid **glycine** and small peptides known as **neuropeptides** are important in the central nervous system. Synapses where the target is a muscle cell use **acetylcholine** as the neurotransmitter. The nervous system is a unique communication system because, in addition to transmitting information, it can store a vast amount of information in the cells of the brain.

Hormones. The cells of a few specialized tissues synthesize and secrete chemicals known as **hormones** into the blood, which then carries them to target cells that are far away from the hormone-secreting cells (see Fig. 5–25*d*). This type of communication system relies on the flow of blood and requires the chemical signal (the hormone) to travel distances that are much greater than the distance a neurotransmitter must travel in crossing a synapse. Consequently, hormone-secreting cells transmit information much more slowly than nerve cells. Unlike the systems discussed previously, most hormones are not delivered directly to the target by the secreting cell. Furthermore, the effects of the hormone are much more diffuse.

Once released, many hormones affect more than one type of cell or tissue in the body. **Parathyroid hormone,** for example, raises the level of calcium ions in the blood by acting on bone to release Ca^{2+} and by acting on the kidney to decrease excretion of Ca^{2+} from the body. The secreted hormone must be able to "recognize" the target cells and must not affect the other cells that it encounters. Accurate recognition is achieved because the target cells have receptor molecules that the hormone can bind to. Nontarget cells lack the specific receptor and are not recognized by the hormone.

The different types of hormone-secreting cells make up the **endocrine system** of the body. This system is influenced in part by the nervous system. One result of this influence is that the secretion of several specific hormones is subject to overall regulation by the nervous system.

Hormones are as varied a group of molecules as the other types of chemical signals. Insulin and parathyroid hormone are polypeptides. Vasopressin (also called antidiuretic hormone) is a small peptide of nine amino acids. Epinephrine is derived from the amino acid tyrosine. The female and male sex hormones (estrogens and testosterone, respectively) are steroids.

Receptors: Mediating Between First and Second Messengers

When a chemical signal arrives at a target cell, it ultimately changes processes such as enzyme-catalyzed reactions, transport of ions, and transcription of genes. The initial event, however, is the interaction of the signal molecules with specific receptors located either on the surface of the target cell or on its cytoplasm.

The Nature of Receptors.   A **receptor** is a macromolecule, usually a protein, to which the signal molecule can bind with high specificity. This means that other molecules do not bind easily to the receptor. This specificity may arise because the specific shape of the signal molecule may facilitate its binding to the receptor. Other molecules with slightly different shapes may be unable to fit into the binding sites on the receptor. This point is illustrated in Figure 5–26. In addition to specificity, the signal molecule binds to the receptor with **high affinity,** which means that the binding is strong.

The combined purpose of these receptor properties is to ensure that only the signal molecule will bind and remain attached to the receptor. These properties are particularly important for hormone receptors because a hormone that is delivered to a

Figure 5–26
Receptor molecules have specific shapes designed to discriminate among various chemical signals. Although chemical signals A, B, and C are very similar in shape, only B fits the shape of the receptor exactly, so only B binds with the receptor and affects the target cell.

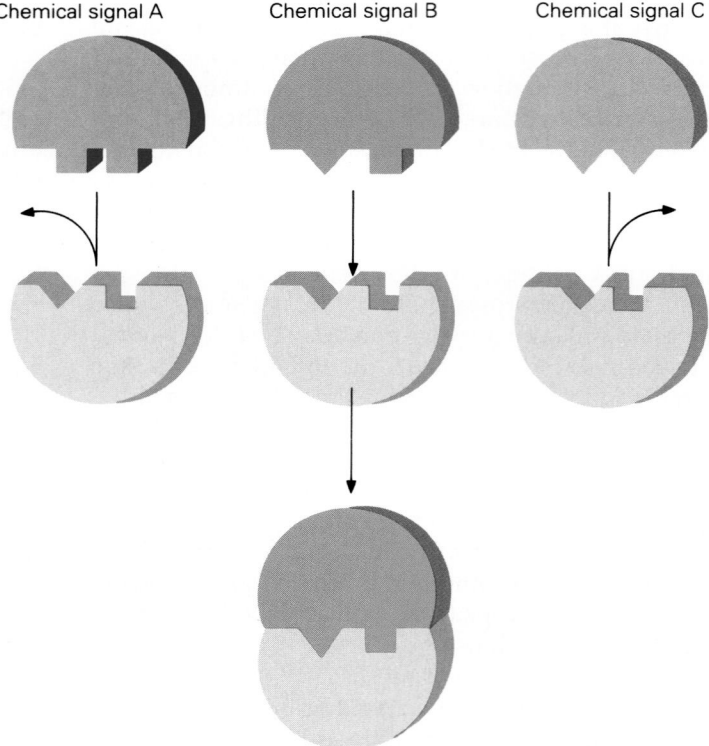

cell via the bloodstream is usually present in a low concentration (10 nM or less) and is only one of the many different types of molecules in the blood that will be "seen" by the receptor.

Binding of a hormone to its receptor in a target cell triggers a sequence of events that leads ultimately to the changes that represent the cellular response to the hormone. The cellular location and action of hormone receptors vary with the type of hormone.

Receptors in the Cytoplasm. The receptors for all steroid hormones and for **vitamin D** (a sterol that promotes calcium absorption in the intestine) are soluble proteins located in the cytosol of the cell (Fig. 5–27). The hormones must penetrate the cell membrane in order to gain access to the receptors. This process is facilitated by the lipid-soluble nature of the hormones. Binding of the hormone to the receptor is followed by movement of the hormone-receptor complex into the nucleus, where it binds to specific acceptor sites on the DNA and regulates the transcription of specific genes.

Thyroid hormones also exert their effects via gene transcription and through a similar mecha-

nism, except that the receptors for these hormones are concentrated in the nucleus even in the absence of hormone. Regulation of gene expression at the level of transcription was discussed earlier.

Receptors in the Plasma Membrane. Polypeptide hormones and **catecholamines** such as epinephrine initiate a completely different series of events upon binding to their receptors. The receptors for these hormones are located on the plasma membrane of the target cell. The hormones bind to the receptors, producing a response within the cell without gaining entry to it (see Fig. 5–27). The effects of the hormones are mediated by second messengers that are generated within the cell when the hormones bind to their receptors. (The actions of second messengers are discussed in detail later.)

In summary, a plasma membrane receptor is more than just a binding site. It acts more like a **transducer.** It can convert an extracellular event—the binding of a signal molecule—into an intracellular response that modifies the behavior of the target cell.

Plasma membrane receptors are intrinsic membrane proteins that usually account for less than 1%

Figure 5–27
Receptors can be located in either the plasma membrane or the cytoplasm of a target cell. Receptors for water-soluble signals are often intrinsic membrane proteins, whereas receptors for lipid-soluble signals, which can diffuse through the plasma membrane, are often found inside the cell.

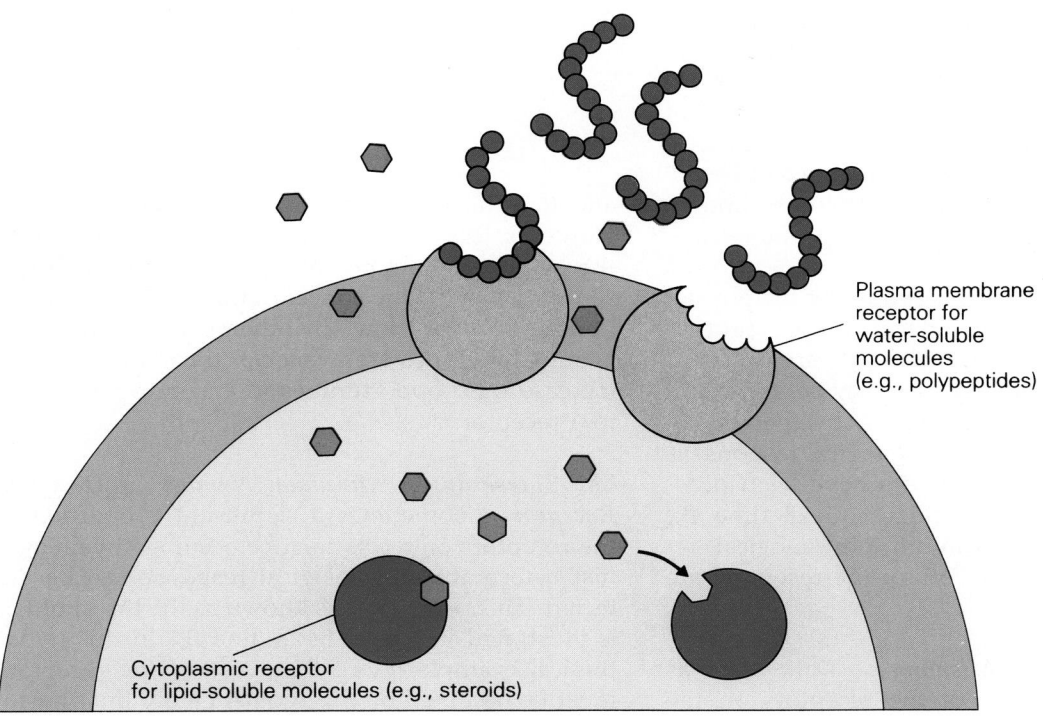

Plasma membrane receptor for water-soluble molecules (e.g., polypeptides)

Cytoplasmic receptor for lipid-soluble molecules (e.g., steroids)

of the total mass of protein present in the membrane. For these reasons, they are very difficult to identify, isolate, and study. The study of receptors has been aided considerably by the development of techniques for attaching radioisotopes (see Chapter 3) to signal molecules without interfering with the physiological activity of the molecule. Receptors bind the labeled molecules, making it possible to follow some aspects of receptor function, such as numbers, distribution, and the fate of the bound molecules.

Ligands: Agonists and Antagonists. Any compound, such as a signal molecule, that binds with high specificity to a receptor is termed a **ligand.** Any ligand capable of both binding to a plasma membrane receptor and evoking a physiological response is known as an **agonist.** A ligand that binds with high affinity to a receptor but evokes no response is called an **antagonist** because, by occupying receptors, it interferes with (or "antagonizes") the action of an agonist.

Antagonists of hormone receptors can be of great therapeutic importance in medicine. Certain hormones called **catecholamines** act on the heart to increase both the heart rate and the amount of blood it pumps, with the net effect of increasing blood pressure. Too much catecholamine action can cause high blood pressure and angina (chest pain). A catecholamine antagonist called **propranolol** is used to block these effects of catecholamines.

Another example is **spironolactone,** which is an antagonist of the steroid hormone **aldosterone.** Aldosterone increases the retention of Na^+ by the kidneys; hence, water is also retained (see Chapter 25). Treatment with spironolactone blocks the action of aldosterone and produces an increase in the excretion of water, which is important in cases of severe fluid retention and edema. Spironolactone is known as a **diuretic,** a substance that increases water excretion.

In some instances, the magnitude of the physiological response to an agonist is directly proportional to the number of receptors that are occupied (Fig. 5–28, line A). The response may be, for example, the uptake of a specific solute by the cell. In other cases, the maximum physiological response is achieved before all the receptors have been occupied. This is illustrated in Figure 5–28 (line B), where 100% of the maximum physiological response occurs when only 75% of the receptors are occupied.

"Spare Receptors": A Misnomer. Data such as these suggest that some of the receptors on the plasma membrane of a given cell are not needed,

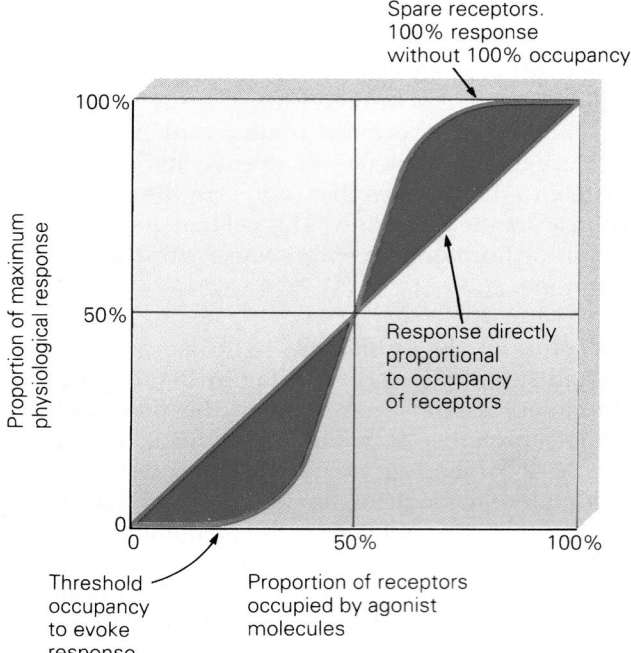

Figure 5–28
The response of a target cell to a chemical signal shown in graphic form. Sometimes the magnitude of the response by the target cell is directly proportional to the number of receptors occupied rather than to the quantity of signal molecules (line A). In other cases a minimum number of receptors, known as the threshold, must be occupied before the target cell shows any response to the signal (line B). Not all receptors need be occupied for the cell to show the maximum response to the signal. Some "spare" receptors may remain unoccupied.

and they have been described as **spare receptors.** However, this is not an accurate description because the extra receptors provide the cell with increased sensitivity when the concentration of the agonist or hormone is very low. In other words, when few agonist molecules are available, the extra receptors increase their opportunities to encounter and bind to a receptor.

The Threshold: A Minimum Number of Occupied Receptors. Sometimes a significant proportion of the receptors on a cell must be occupied by an agonist before any physiological response can be detected. This proportion is known as the **threshold;** it is illustrated in Figure 5–28 (line B). In this hypothetical example, more than 25% of the receptors must be occupied by the agonist before the physiological response will occur.

Variations in Receptor Quantities. One of the original hypotheses regarding receptors was that they were present in cells at fixed concentrations. Quantitative studies of receptor numbers using radiolabeled ligands have revealed, however, that the concentration of many receptors in cells can be regulated. For example, a sustained elevation of the concentration of an agonist circulating in the blood often leads to a decrease in the number of receptors present in target cells. This phenomenon is known as **down regulation** of receptor number and is accompanied by a decrease in the physiological response **(desensitization)** of the cells to the circulating agonist. When the agonist concentration falls, the receptor number rises again. This phenomenon makes many cells particularly sensitive to a **change** in the normal concentration of a chemical signal, as opposed to its absolute concentration.

The mechanism by which agonists can regulate the number of their receptors in target cells remains unclear, but it may involve any one of a number of procedures, including inactivation of receptors, and changes in the rate of synthesis or rate of degradation of the receptor proteins. In the specific case of plasma membrane receptors, binding of agonist molecules to receptors can stimulate internalization of the receptor-agonist complex by an endocytotic process.

Since few receptors have been isolated and studied in great detail, little is known about the interactions that stabilize the binding of signal molecules. The binding is noncovalent and reversible and probably involves contributions by hydrophobic interactions, hydrogen bonds, and ionic bonds (see Chapter 2). The remainder of this chapter examines some of the ways by which the extracellular signal—binding of an agonist molecule to a plasma membrane receptor—is converted (transduced) to an intracellular response.

Intracellular Signals: Second Messengers

It is generally accepted that the conformation (three-dimensional shape) of a receptor molecule is altered when an extracellular signal (first messenger) molecule binds to it. The change in receptor conformation generates an intracellular signal molecule known as a **second messenger.** This process also increases or **amplifies** the signal because several second messenger molecules are generated for each extracellular signal molecule that is bound to a receptor (Fig. 5–29). The second messengers initiate

Figure 5–29
When an extracellular signal binds to a receptor molecule, the receptor changes its conformation. This change triggers the activation of intracellular signal molecules, which amplify the extracellular signal and produce the physiological response of the target cell.

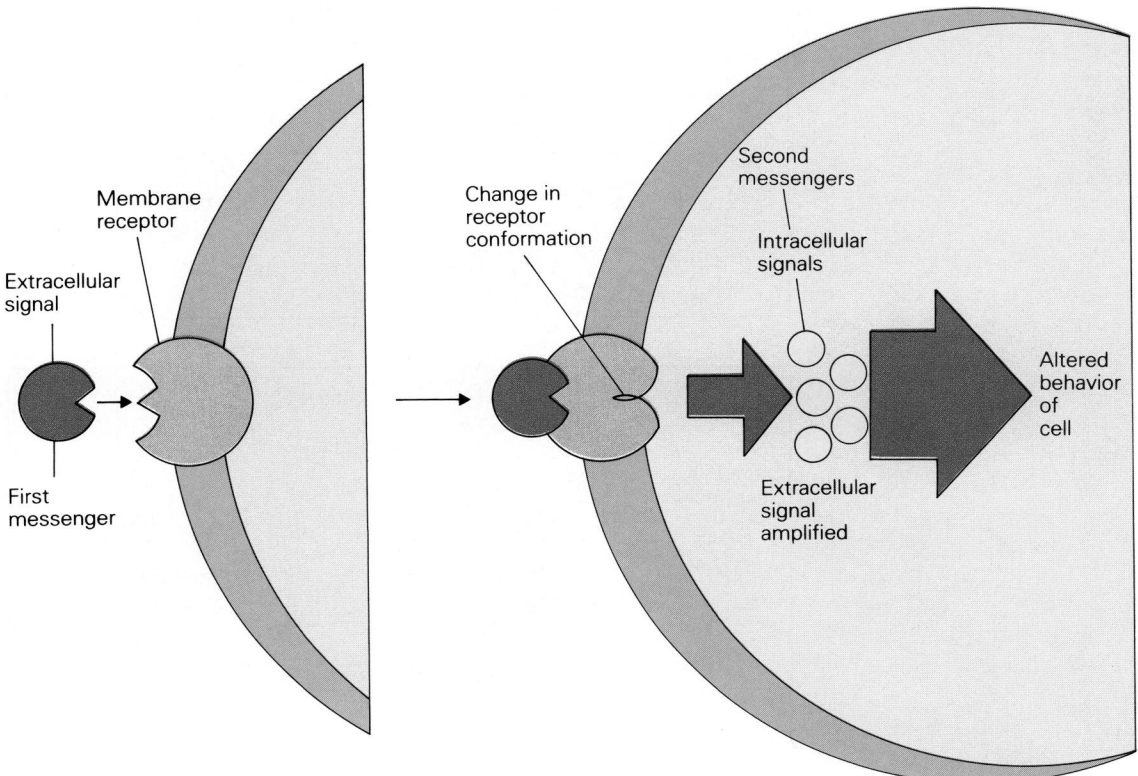

the intracellular reactions that modify the behavior of the target cell.

Three principal mechanisms are thought to produce second messenger molecules inside cells when an external signal molecule binds to a plasma membrane receptor. These are summarized in Figure 5–30.

Cyclic AMP as a Second Messenger. One of the best understood mechanisms is the production of a second messenger, known as **cyclic AMP,** from ATP by a reaction catalyzed by **adenylate cyclase** (see Fig. 5–30a). The latter is a membrane-bound enzyme that is stimulated by binding of the signal molecule to the receptor. This mechanism will be discussed again in more detail in Chapter 12.

Calcium Ions as Second Messengers. The conformational change in the receptor caused by an extracellular signal could open a **membrane channel,** which in turn could lead to two possible events. First, a small transient influx of ions into an electrically excitable nerve or muscle cell will briefly change the electrical potential difference between the inside and outside of the cell. This change can trigger an **action potential** (see Chapter 7) that spreads rapidly throughout the membrane of the target cell. Many neurotransmitters work in this fashion. Second, a major influx of a specific ion could raise the intracellular concentration of that ion to the point at which it acts as a second messenger to stimulate an intracellular response. **Calcium ions** are likely candidates for this role. The extracellular concentration of

Figure 5–30
Three examples of intracellular signals, or second messengers. (*a*) An extracellular signal, when it binds to a receptor, may activate adenylate cyclase, generating cyclic AMP from ATP. (*b*) An extracellular signal may cause a channel to open in the plasma membrane, causing an influx of Ca^{2+} ions. (*c*) An extracellular signal may activate phospholipase C, which converts phosphatidylinositol bisphosphate (a membrane phospholipid) into diacylglycerol and inositol trisphosphate.

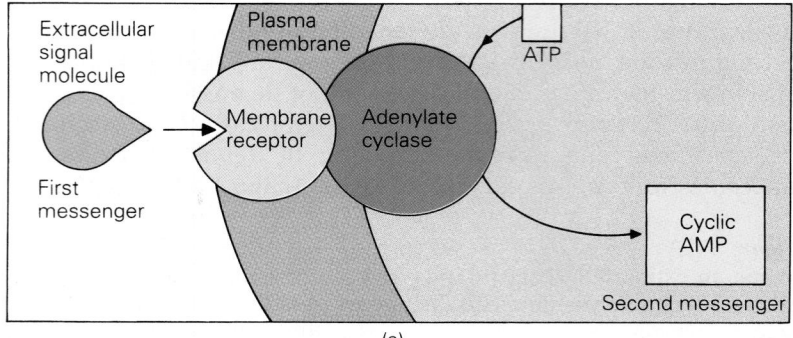
(a)

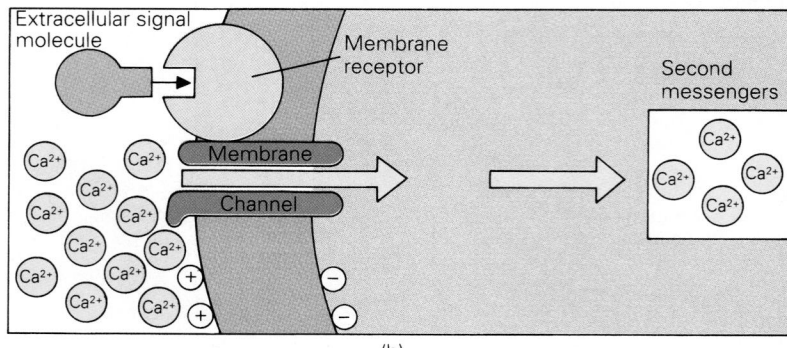
(b)

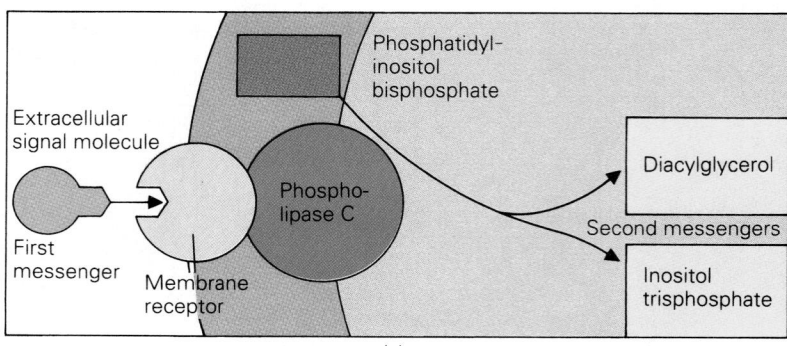
(c)

Ca^{2+} is not large (1000 μM), but since the intracellular concentration of free Ca^{2+} is kept so low (0.1 μM)—in part because it is potentially toxic to the cell—a highly favorable concentration gradient exists for passive entry of Ca^{2+}. The electrical gradient across the membrane also favors entry of the positively charged Ca^{2+} (see Fig. 5–30b).

Inositol Trisphosphate and Diacylglycerol as Second Messengers. More recently, a third system for producing second messengers has been identified. It involves the breakdown of the phosphorylated form of a membrane phospholipid called **phosphatidylinositol bisphosphate.** This compound is produced by the covalent attachment of two phosphate groups to phosphatidylinositol (see Chapter 2). Binding of the signal molecule to the membrane receptor leads to activation of a specific **phospholipase C** enzyme

that catalyzes the breakdown of the phosphorylated lipid into **inositol trisphosphate** and **diacylglycerol,** a pair of intracellular second messenger molecules (see Fig. 5–30c).

Action of Second Messengers. The mode of action of second messengers is summarized in Figure 5–31. Cyclic AMP modifies the behavior of a cell by activating enzymes called **cyclic AMP-dependent protein kinases.** When activated by cyclic AMP, these kinases catalyze reactions that transfer a phosphate group from ATP to the side chain of an amino acid such as serine in specific protein molecules. Covalent attachment of a phosphate group (phosphorylation) to a protein usually modifies its biological activity (see Fig. 5–31). If the protein is an enzyme, for example, the result of phosphorylation could be either activation or inactivation of its enzymatic ca-

Figure 5–31
Second messengers may activate certain enzymes that catalyze the phosphorylation of certain proteins, which in turn produce the physiological response of the cell to the extracellular signal.

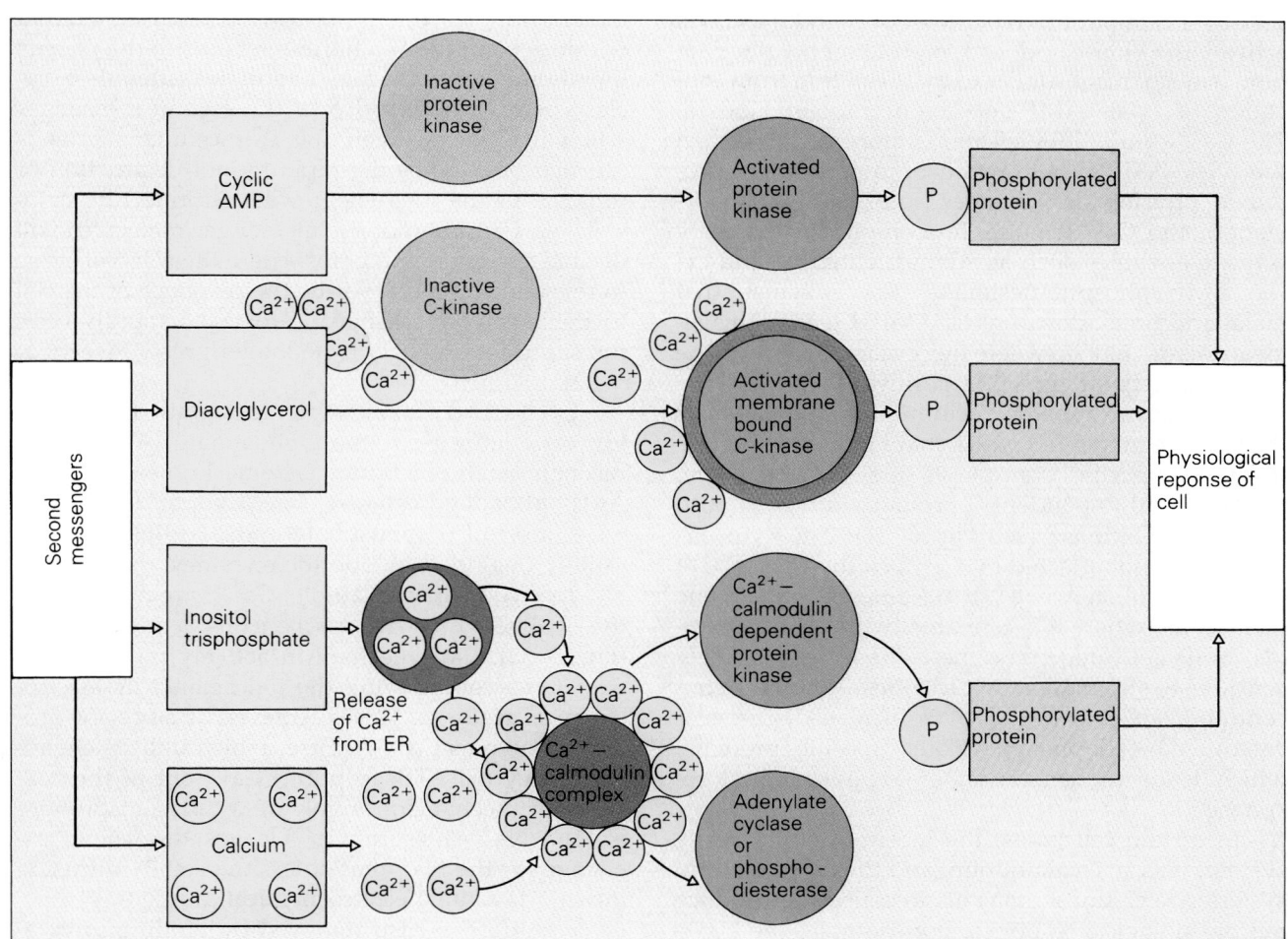

pacity. Such a change could lead to the physiological response of the target cell.

Inositol trisphosphate and diacylglycerol are important parts of receptor systems that operate primarily by increasing the level of Ca^{2+} in the cell cytosol. Inositol trisphosphate is involved in the release of Ca^{2+} from intracellular stores such as the endoplasmic reticulum. This effect, coupled with the influx of extracellular Ca^{2+}, can raise the cytosolic concentration of free Ca^{2+} from the resting level of 0.1 μM to somewhere in the range of 1 to 10 μM (see Fig. 5–31). The increased Ca^{2+} becomes bound with high affinity to **calcium-binding proteins** and produces a change in the conformation of these proteins and, hence, a change in their function. There are two types of these proteins.

The first type is the true receptor proteins that have no enzymatic activity but, upon binding Ca^{2+}, alter their association with other proteins. This alteration leads to a change in the function of the other proteins. Specific examples of this type of receptor protein are **troponin C,** found in skeletal and cardiac muscle cells, and **calmodulin,** which is found in almost all cells. Binding of a Ca^{2+}-calmodulin complex by a calmodulin-dependent protein kinase will activate the kinase and lead to protein phosphorylation, a mechanism similar to that resulting from activation of cyclic AMP-dependent protein kinases. The Ca^{2+}-calmodulin system is more versatile than the cyclic AMP system because, in addition to producing physiological changes via protein phosphorylation, the Ca^{2+}-calmodulin complex can directly activate enzymes such as adenylate cyclase and cyclic AMP phosphodiesterase, the enzymes that make and break down cyclic AMP. Clearly, this allows interaction between the cyclic AMP and the Ca^{2+} intracellular signaling pathways.

The second type of calcium-binding proteins is the Ca^{2+}-regulated enzymes that bind Ca^{2+} ions directly. A specific example is the Ca^{2+}-sensitive, phospholipid-dependent protein kinase, also known as **C-kinase** (see Fig. 5–31). The cytosolic C-kinase is not affected by Ca^{2+}. In the presence of diacylglycerol, however, it becomes bound to the membrane, where it is activated by the phospholipids. In its activated state, the C-kinase is extremely sensitive to stimulation by Ca^{2+} ions. When the concentration of cytosolic Ca^{2+} is increased, the activated C-kinase phosphorylates specific proteins, which leads to generation of a physiological response.

In certain cell types, the inositol trisphosphate system, via Ca^{2+}-calmodulin, and the diacylglycerol system, via C-kinase, may act in concert to produce the physiological response. For example, the Ca^{2+}-regulated contraction of smooth muscle occurs in two phases, an initial rapid component followed by a slower but more sustained component. It has been proposed—but not yet proven—that the rapid phase is due to release of Ca^{2+} from intracellular stores, leading to muscle contraction via protein phosphorylation by a Ca^{2+}-calmodulin–dependent protein kinase. In contrast, the slower sustained contraction could be mediated by influx of extracellular Ca^{2+} and activation of a C-kinase, which also leads to muscle contraction via protein phosphorylation.

Removal of Second Messengers. The concentrations of second messengers in the cytosol of an unstimulated cell are usually very low because these molecules are continuously and rapidly destroyed or removed from the cytosol (Fig. 5–32). The cytosolic concentration of a second messenger is altered principally by changes in its rate of synthesis or influx into a cell. This ensures that an increased rate of synthesis or influx will rapidly increase the cytosolic concentration of the messenger, allowing the target cell to respond quickly to the extracellular signal (see Fig. 5–32). The concentration of cyclic AMP is usually only 1 μM, but this can be increased fivefold within seconds after a hormone binds to the plasma membrane receptor and stimulates adenylate cyclase. Synthesis or influx of the second messenger will fall to zero when the extracellular signal is "turned off." Since the rapid rate of destruction or removal of the messenger will continue, the cytosolic concentration of the messenger molecules will decline rapidly to the point where the cell will cease to respond. In other words, the response of the cell to the second messenger is reversed rapidly once the second messenger is no longer present (see Fig. 5–32).

Cyclic AMP, for example, is degraded very rapidly to adenosine 5'-monophosphate (AMP) by a reaction catalyzed by an enzyme known as **cyclic AMP phosphodiesterase** (see Chapter 12). Likewise, inositol trisphosphate and diacylglycerol are rapidly degraded by specific enzymes.

To give another example, Ca^{2+} is removed from the cytosol by several mechanisms acting together (Fig. 5–33). One mechanism actively pumps Ca^{2+} out of the cell, moving the ion against its electrochemical gradient. This is achieved by a specific protein, known as Ca^{2+}-ATPase, which utilizes chemical energy (as ATP) to pump Ca^{2+} out of the cell. Another mechanism, involving a different protein, is Ca^{2+}/Na^+ antiport. Ca^{2+} leaves the cell in exchange for the Na^+ that enters the cell by diffusing down a favorable electrochemical gradient.

Free Ca^{2+} is also removed by binding to cytosolic calcium-binding proteins and other molecules,

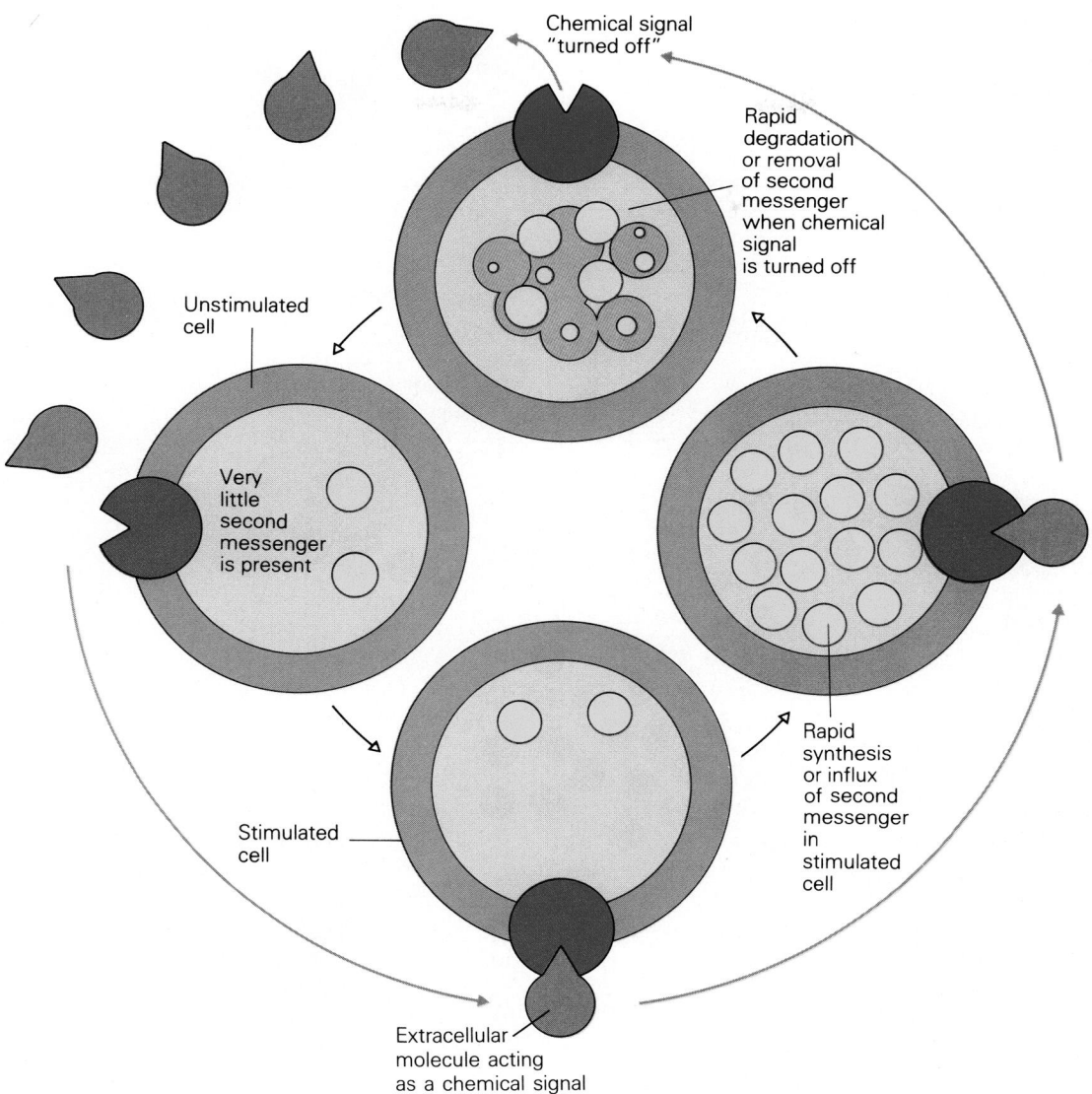

Figure 5–32
A second messenger is rapidly synthesized when a first messenger molecule (an extracellular signal) binds to a receptor. When the extracellular signal is removed, however, the second messenger is rapidly destroyed or removed from the cell. Thus, the next time a signal binds to the receptor, the second messenger must be synthesized anew.

and by sequestration inside intracellular organelles such as mitochondria and the endoplasmic reticulum. The large number of different mechanisms that can regulate cytosolic Ca^{2+} levels may be, at least in part, a defense system to guard against possible toxic effects of uncontrolled concentrations of this ion.

Case Study: Glycogen Metabolism and Protein Phosphorylation. Protein phosphorylation is one example of a way in which the production of second messengers can mediate a change in cell function that represents the cellular response to a chemical

signal. The metabolism of **glycogen** in skeletal muscle cells provides clues to a possible sequence of events. Recall from Chapter 2 that glycogen is a polysaccharide, the form in which glucose is stored until its chemical energy is needed by the muscle cell. The hormone **epinephrine** stimulates glycogen breakdown and shuts off glycogen synthesis—the net effect being to provide the cell with an adequate supply of glucose for strenuous activity.

Epinephrine works by stimulating adenylate cyclase in the plasma membrane and elevating intracellular cyclic AMP. The cyclic AMP activates a protein kinase, initiating a sequence of phosphorylation

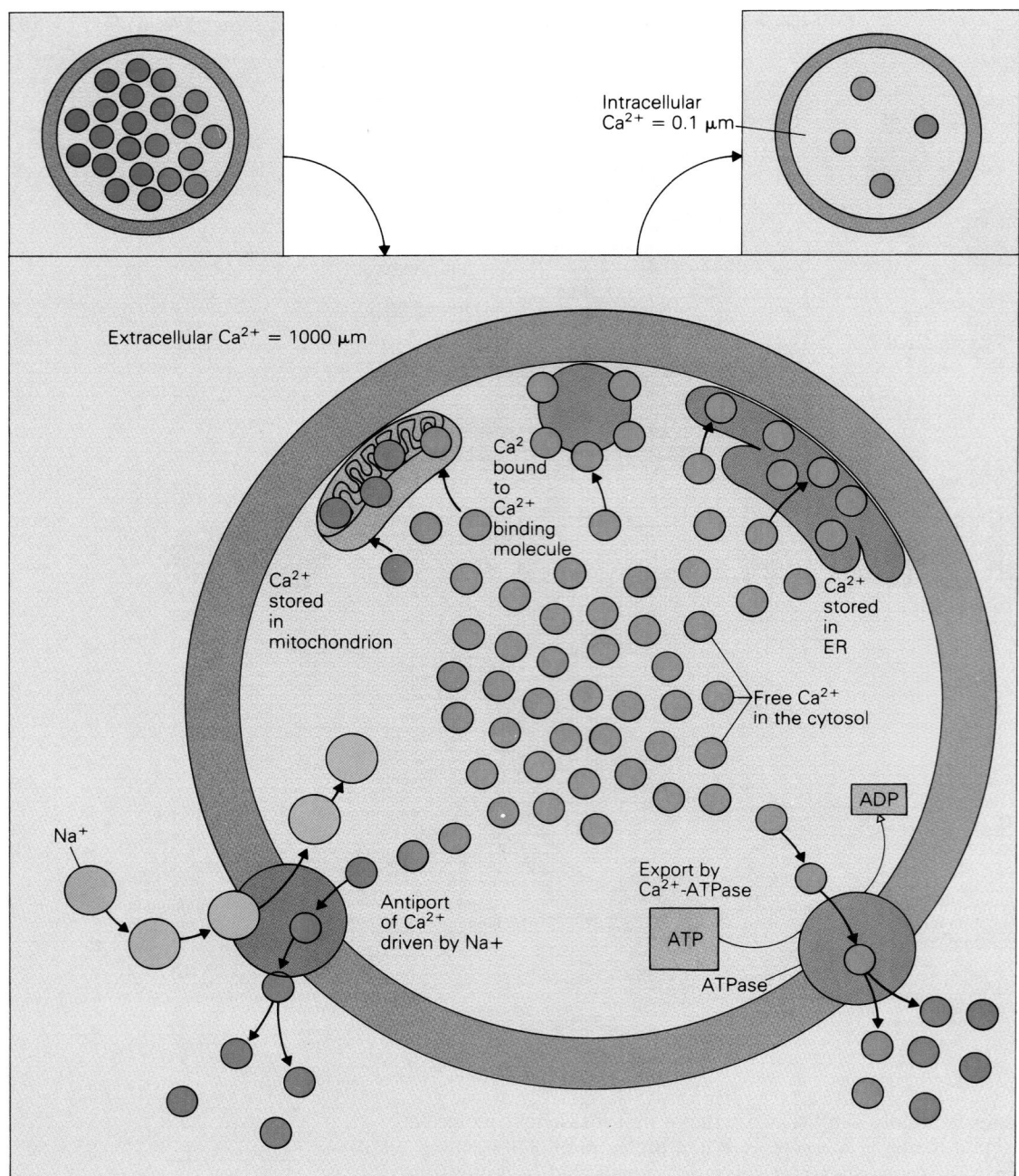

Intracellular
Ca²⁺ = 0.1 μm

Extracellular Ca²⁺ = 1000 μm

Ca²
bound
to
Ca²⁺
binding
molecule

Ca²⁺
stored
in
mitochondrion

Ca²⁺
stored
in
ER

Free Ca²⁺
in the cytosol

Na⁺

Antiport
of Ca²⁺
driven by Na+

Export by
Ca²⁺-ATPase

ADP

ATP

ATPase

Figure 5–33
Examples of the ways in which a second messenger is removed from a cell.
The second messenger calcium ions in the cytosol may be removed by active
transport, by antiport, by sequestration inside cytoplasmic organelles, or by
binding to special molecules.

reactions (Fig. 5–34*a*). Phosphorylation by the protein kinase of Enzyme 1 activates this enzyme; once activated, Enzyme 1 phosphorylates Enzyme 2, which begins to remove individual glucose molecules from glycogen. The same protein kinase phosphorylates a third enzyme (Enzyme 3) that is involved in synthesis of glycogen from glucose. In

contrast to Enzymes 1 and 2, phosphorylation of Enzyme 3 causes inactivation.

The overall effect of these protein phosphorylations is to make available a maximum supply of glucose by simultaneously stimulating breakdown of glycogen into glucose molecules while inhibiting the use of glucose for synthesis of glycogen.

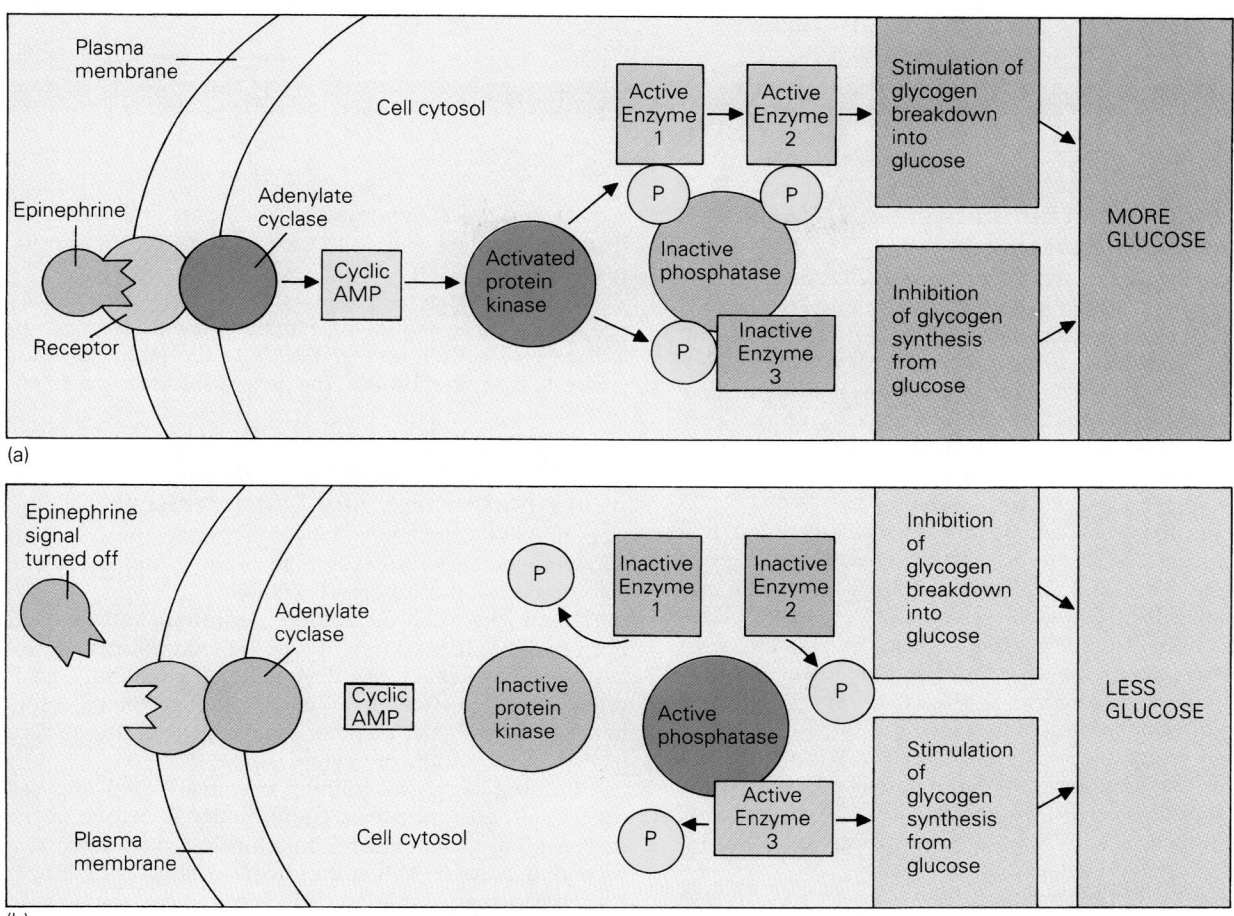

(a)

(b)

When the epinephrine signal is turned off, the phosphorylation steps just described are rapidly reversed by reactions catalyzed by the enzyme **phosphoprotein phosphatase** (see Fig. 5–34b). In other words, the phosphate groups on each of the three key enzymes are removed by the phosphatase, with the net effect of inhibiting glycogen breakdown (Enzymes 1 and 2 are inactivated) and stimulating glycogen synthesis (Enzyme 3 is activated). The opposing action of the phosphoprotein phosphatase does not interfere with the action of cyclic AMP because the phosphatase is inhibited while the level of cyclic AMP is elevated.

Regulation of both breakdown and synthesis of glycogen is an important feature of this cyclic AMP control system. It increases the response to an elevated level of cyclic AMP in the cell, the specific response being a readily available supply of glucose molecules. We now know of more than 25 different enzymes regulated by a reversible phosphorylation process. These and other types of protein modifications are so abundant that most of the proteins that mediate physiological processes are probably regulated in this way.

Figure 5–34
An example of the response of a cell to an extracellular signal and the reversal of that response once the signal is turned off. In this example, the desired effect of the chemical signal epinephrine on the target cell is an increase in available glucose as opposed to its storage in the form of glycogen. (*a*) When epinephrine binds to the receptor, adenylate cyclase produces the second messenger cyclic AMP. This, in turn, activates protein kinase, which causes the phosphorylation of three enzymes. Phosphorylated Enzymes 1 and 2 stimulate glycogen breakdown, while Enzyme 3, which normally stimulates glycogen synthesis from glucose, is inactivated by phosphorylation. The net result is the desired response of the target cell, an increase in the availability of glucose. (*b*) When the epinephrine signal is removed, all three enzymes are dephosphorylated by phosphoprotein phosphatase. As a result, glycogen synthesis is resumed and less glucose is available.

SUMMARY

An Intracellular Control System: DNA, RNA, and Protein Synthesis

Genetic information, the instructions a cell uses to synthesize proteins, is stored in both DNA and mRNA by a code formed from the specific linear sequence in which the four different nucleotides are arranged. Each triplet of DNA nucleotides is transcribed into a codon of RNA nucleotides; each codon specifies one amino acid in a polypeptide. Most of the total genetic information in a cell is suppressed. Different cell types arise because the genes expressed in these cells are different.

When a parent cell divides to yield two daughter cells, each of the daughter cells has an identical copy of the parent cell's DNA. DNA is copied by a replication process catalyzed by DNA polymerase. RNA is synthesized, or transcribed, from DNA in a process catalyzed by RNA polymerase. Messenger RNA carries the genetic information stored in the nucleus to the cytoplasm, where it is used to direct synthesis of proteins. Ribosomal RNA is an important constituent of ribosomes, and transfer RNA mediates the incorporation of individual amino acids into a polypeptide. Newly synthesized RNA is converted to mature functional RNA by several processing reactions that occur in the nucleus.

Amino acids cannot participate directly in synthesis of a polypeptide; first they must be linked to a molecule of tRNA. There is a different tRNA molecule for each amino acid. Translation of the information in mRNA occurs on and is coordinated by the ribosomes. Both mRNA and tRNA are bound to these structures during assembly of a polypeptide. The amino acid linked to the tRNA is added to the end of the growing polypeptide chain by the formation of a peptide bond, in a reaction catalyzed by peptidyl transferase, a ribosomal enzyme.

At any given time, only a small part of the DNA is being transcribed and expressed in the form of specific polypeptides.

DNA is very sensitive to damage by factors normally present in the environment. Highly efficient repair mechanisms replace the damaged parts of the molecule, thus ensuring that alterations in the genetic information do not persist.

Cell-to-Cell Signaling: Hormones, Neurotransmitters, and Other Molecules

Gap junctions are formed by membrane proteins that allow movement of substances from the cytoplasm of one cell directly to that of an adjacent cell.

Local chemical mediators, neurotransmitters, and hormones are three different types of extracellular control systems that use chemical signals. A chemical signal, called a first messenger, interacts with specific receptors in the target cell. The receptor can be either on the surface of the cell or within the cytoplasm.

Binding of an extracellular chemical signal to a cell surface receptor induces a conformational change in the receptor that generates an intracellular signal known as a second messenger. When the chemical signal is no longer present, the concentration of the second messenger falls rapidly, owing to enzymatic destruction or other mechanisms that remove the messenger from the cytosol. Some second messengers produce changes in the function of a cell by stimulating a kinase enzyme to catalyze phosphorylation of specific proteins. In the absence of the second messenger, the phosphorylation is readily reversed by a reaction mediated by a phosphatase enzyme.

REVIEW QUESTIONS

Identify Each with a Term

1. The term applied to the genetic code to indicate that many amino acids are specified by more than one codon.

2. The term used to describe DNA replication when each new DNA molecule contains one nucleotide strand derived from the parent DNA.

3. A nucleotide sequence in newly synthesized RNA that contains no useful information.

4. A structure that forms a direct connection between the cytoplasm of two adjacent cells.

5. General term for any compound that binds with high specificity to a receptor.

Define Each Term

6. Start codon.

7. RNA polymerases.

8. Post-translational modifications.

9. Antagonist.
10. Cyclic AMP phosphodiesterase.

Choose the Correct Answer

11. Which of the following describes the function of a nucleolus?
    a. Produces ATP
    b. Synthesizes DNA polymerase
    c. Produces ribosomal subunits
    d. Stores genetic information for protein synthesis.

12. Amino acids are carried to the ribosomes in the cytoplasm by:
    a. messenger RNA
    b. mitochondrial RNA
    c. ribosomal RNA
    d. transfer RNA

13. Unwinding the DNA double helix prior to replication is aided by:
    a. DNA polymerase
    b. DNA topoisomerase
    c. DNA endonuclease
    d. DNA ligase

14. Chemical signaling systems generally involve:
    a. an extracellular signal molecule
    b. a receptor molecule
    c. an intracellular signal molecule
    d. all of the above

15. The ion which acts as a second messenger is:
    a. calcium
    b. magnesium
    c. potassium
    d. sodium

16. Inositol trisphosphate is an example of a:
    a. second messenger
    b. membrane receptor
    c. first messenger
    d. calcium channel

17. The enzyme responsible for making cyclic AMP is:
    a. cyclic AMP phosphodiesterase
    b. cyclic AMP-dependent protein kinase
    c. adenylate cyclase
    d. DNA glycosylase

Answer Each Question in One or Two Sentences

18. How can only four different nucleotides be used to construct a genetic code that will specify 20 different amino acids?

19. List five principal differences between DNA synthesis and RNA synthesis.

20. If all the cells in an animal contain the same genetic information, how do cells with such marked structural and functional differences arise?

21. What is the difference between an agonist and an antagonist?

22. List the three principal types of extracellular chemical signals.

23. How do second messengers produce a change in the function of a cell?

SUGGESTED READINGS

Abdel-Latif, A.A. "Calcium-mobilizing receptors, polyphosphoinositides, and the generation of second messengers." *Pharmacological Reviews*, 38:227–272, 1986.

Avers, C.J. *Molecular Cell Biology*. Reading, MA: Addison-Wesley, 1986.

Berridge, M.J. "Inositol triphosphate and diacylglycerol: two interacting second messengers." *Annual Review of Biochemistry*, 56:159–193, 1987.

Carafoli, E., and Penniston, J.T. "The calcium signal." *Scientific American*, 253:70–78, November 1985.

Caskey, T.H. "Peptide chain termination." *Trends in Biochemical Sciences*, 5:234–237, 1980.

Cech, T.C. "RNA as an enzyme." *Scientific American*, 255:64–75, November 1986.

Chein, S., and Gargus, J.J. "Molecular biology in physiology." *FASEB Journal*, 1:97–102, 1987.

Darnell, J.E. "RNA." *Scientific American*, 253:68–86, October 1985.

Darnell, J., Lodish, H., and Baltimore, D. *Molecular Cell Biology* (2nd edition). New York: Scientific American Books (Freeman), 1990.

Felsenfeld G. "DNA." *Scientific American* 253:58–67, October 1985.

Gilman, A.G. "G proteins: transducers of receptor-generated signals." *Annual Review of Biochemistry*, 56:615–649, 1987.

Lake, J.A. "The ribosome." *Scientific American* 245:84–97, August 1981.

Lindahl, T. "DNA repair enzymes." *Annual Review of Biochemistry*, 51:61–87, 1982.

Moldave, K. "Eukaryotic protein synthesis." *Annual Review of Biochemistry*, 54:1109–1150, 1985.

Ptashne, M. "How gene activators work." *Scientific American*, 260:41–47, January 1989.

Radman, M., and Wagner, R. "The high fidelity of DNA duplication." *Scientific American*, 259:40–46, August 1988.

Rasmussen H. "The cycling of calcium as an intracellular mesenger." *Scientific American,* 261:66–73, October 1989.

Rodbell, M., Birnbaumer, L., Pohl, S.L., and Krans, H.M.J. "The glucagon-sensitive adenyl cyclase system in plasma membranes of rat liver. An obligatory role of guanyl nucleotides in glucagon action." *Journal of NIH Research,* 2:63–70, 1990.

Schramm, M., and Selinger, Z. "Message transmission: receptor controlled adenylate cyclase system." *Science,* 225:1350–1356, 1984.

Schulman, L.H., and Abelson, J. "Recent excitement in understanding tRNA identity." *Science,* 240:1591–1593, 1988.

Snyder, S.H. "The molecular basis of communication between cells." *Scientific American,* 253:132–140, October 1985.

KEY TERMS

antibiotic (p. 177)

anticodon (p. 168)

codon (p. 155)

cyclic AMP (p. 190)

DNA polymerase (p. 160)

DNA replication (p. 158)

DNA transcription (p. 158)

gap junction (p. 182)

genetic code (p. 155)

genetic expression (p. 154)

ligand (p. 188)

messenger RNA (p. 165)

receptor (p. 186)

ribosomal RNA (p. 165)

transfer RNA (p. 165)

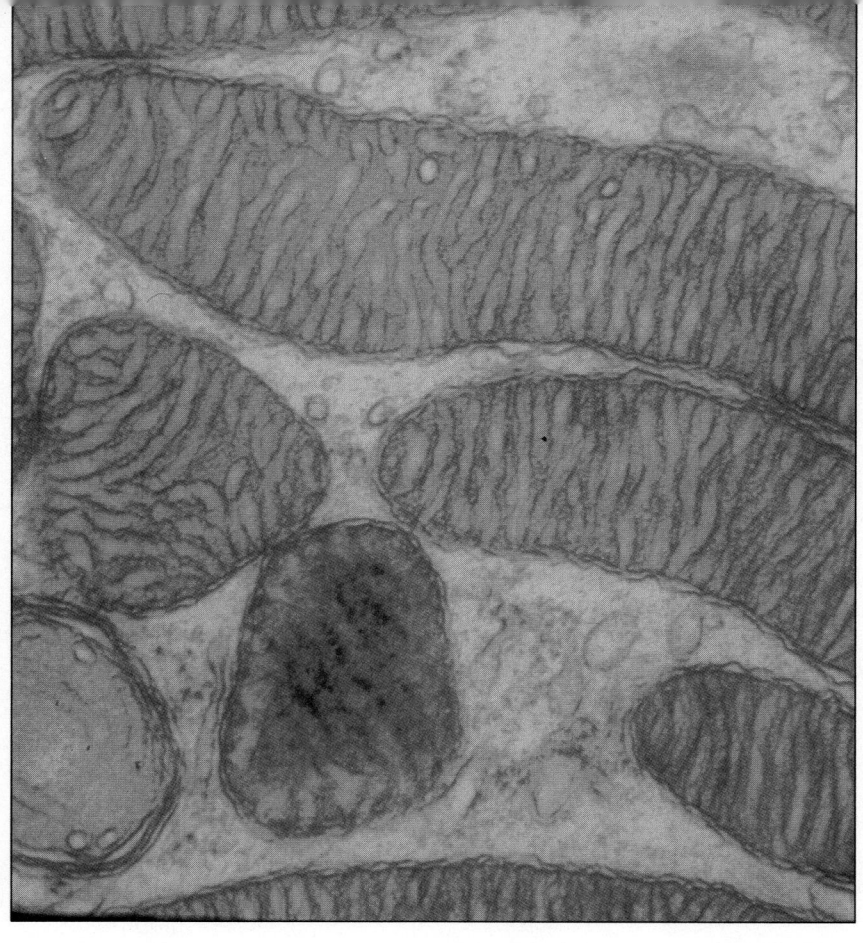

Energy and Cellular Metabolism

M etabolism is the term used to describe the highly organized and integrated set of chemical reactions and energy transformation that sustain a living organism. These reactions are myriad; at least 1000 different reactions can take place within the unicellular bacterium *Escherichia coli*.

Two distinct aspects of the metabolism of a cell are **energy extraction** and **energy use.** In a series of steps known as **catabolism,** the cell extracts energy from its environment, the extracellular fluid. Catabolism is usually achieved in animals by the breakdown of the molecules present in food. This process has the additional purpose of converting complex macromolecules into simple molecules that can be used as "building blocks" to make new macromolecules. In a series known as **anabolism** or, more descriptively, **biosynthesis,** the energy extracted in catabolism is used to assemble the building block molecules into needed proteins, nucleic acids, lipids, and other new macromolecules for the cell.

Every cell is always engaged in both catabolism and biosynthesis, and normally these processes are carefully balanced. The supply of a specific macromolecule is matched closely to the demand for it; ideally, there is no shortage or excess. The fine control that this balance requires is remarkable, both at the cellular level and, on a larger scale, at the level of the whole animal. Consider, for example, that the average adult human can eat about six tons of food over a span of 40 years without losing or gaining weight and without changing the composition of the body.

Although cellular metabolism is a highly complex process, the central pathways are not difficult to understand. Moreover, they are essentially similar in most animal cells. This chapter examines some of the basic principles that drive all metabolic processes and their specific involvement in the metabolism of carbohydrates, fats, and proteins.

Basic Principles of Energy Transformation in the Cell

Cells obtain the energy they need to carry out work from their environment and, ultimately, release some of that energy back to the environment in the form of heat. Some of the ways by which living cells obtain and transform energy are depicted in Figure 6–1.

Photosynthesis: From Radiant Energy to Organic Products in Plants

Plants use light from the sun, a form of radiant energy, for the **photosynthesis** of ATP, the most versatile form of chemical energy (see Chapter 2). The ATP drives biosynthesis within plant cells, including the series of reactions known as **CO_2 fixation,** in which CO_2 from the atmosphere is used in the creation of carbohydrates and other high-energy organic compounds. They are considered high-energy compounds because their many covalent bonds represent solar energy stored as chemical energy. Photosynthesis requires CO_2 and produces O_2 as a byproduct. The process can be summarized as:

$$CO_2 + H_2O + \text{radiant (solar) energy} \longrightarrow$$
low-energy
compounds

$$(CH_2O)_n + O_2$$
organic
molecule
(high-energy
compound)

Plants convert some newly synthesized carbohydrates, by a series of metabolic reactions, to the other types of high-energy organic molecules that they need.

Cellular Respiration: From Plant Products to Chemical Energy in Animals

The organic molecules made by the plants are a source of chemical energy and building block molecules for the animals that eat the plants. Unlike plants, animals cannot use the energy from sunlight directly. They can obtain this energy only after its transformation by plants.

The organic "fuel" molecules that animals ingest when they consume plants are catabolized by reactions that require atmospheric O_2, and are termed **cellular respiration.** They produce CO_2 and water as byproducts and can be thought of as the reverse of photosynthesis:

$$(CH_2O)_n + O_2 \longrightarrow CO_2 + H_2O + \text{chemical energy}$$
high-energy   low-energy   (ATP)
organic   compounds
compounds

These complex reactions are directed by **enzymes** and designed to release the chemical energy of the

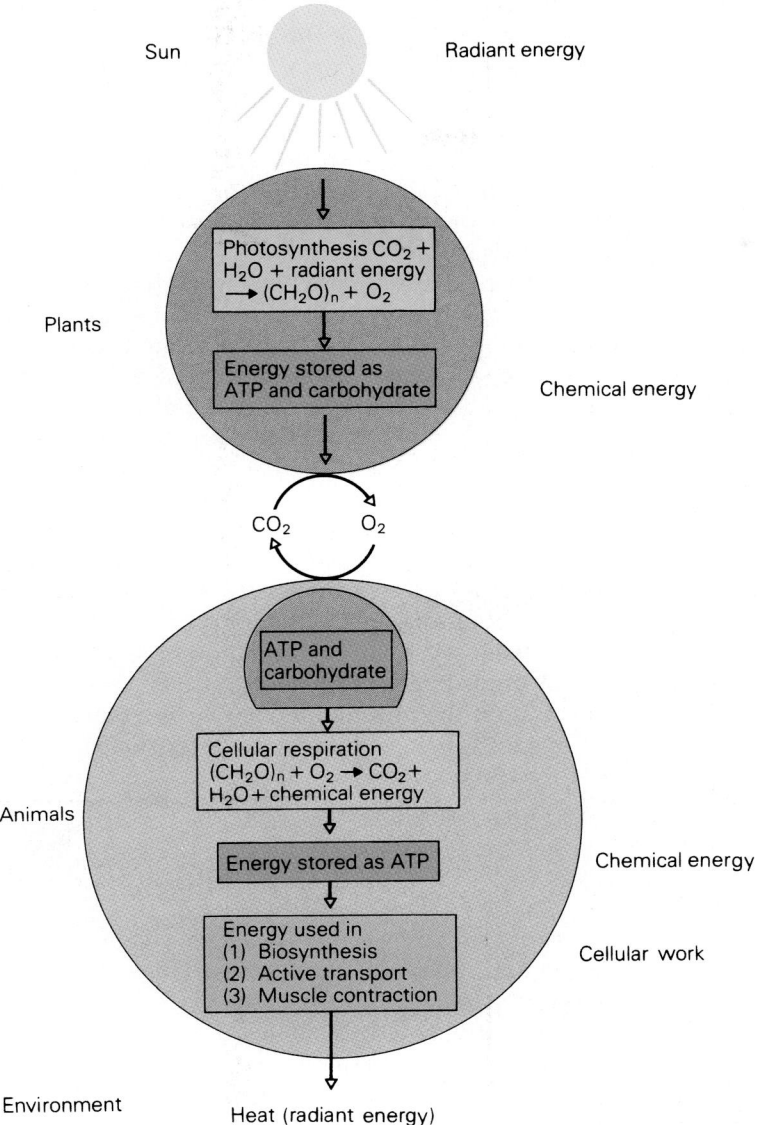

Figure 6–1
The flow of energy. Plants use radiant energy directly from the sun to photosynthesize high-energy organic molecules. Animals obtain these molecules by eating plants, and then transform their energy into ATP, which they use to do biological work.

organic molecules in a controlled fashion so that it can be "trapped" and stored as ATP. This is a form of chemical energy that cells can use for biosynthesis, active transport, and the generation of force and movement required for muscle contraction.

Energy Loss and the Constant Demand for ATP

The transformation of energy by animals cells is not a perfect process. Not all of the energy released from a fuel molecule can be used to form ATP. The untrapped energy is lost to the environment in the form of heat. Because they constantly need ATP and are continually losing heat energy, living cells require constant input of energy in order to survive. ATP is not a long-term storage form of chemical energy. It is used almost as soon as it is formed; the "turnover" of an ATP molecule in a cell is very rapid. At any moment, less than 1 g of ATP is present in the human body. However, in a person who is resting with minimal activity, about 45,000 g of ATP will be formed and used each day. Much more ATP will be required if the person undergoes strenuous physical activity during the day.

Chemical Reactions in Cells: Some Special Considerations

Free Energy Change in Cell Reactions

Recall from Chapter 2 that some reactions, called **exergonic reactions,** are highly likely to proceed spontaneously. The driving force for such reactions is the difference in total free energy between reactants and products. If a large difference exists—one in which the reactants contain much more free energy than the products—a large amount of useful energy will be released when the reaction occurs.

Other reactions, called **endergonic reactions,** cannot proceed without input of free energy. Both types of reactions occur in the cell. Endergonic reactions (in which ΔG is said to be positive) obtain the free energy they need by being **coupled** to exergonic reactions (in which ΔG is negative).

One of the first steps in glucose catabolism, for example, is to add a phosphate group (Pi) to carbon atom 6 in the glucose molecule, an endergonic reaction called a **phosphorylation:**

Glucose + Pi $\longrightarrow$ Glucose-6-Pi

$$\Delta G = +3 \text{ kcal/mol}$$

This reaction takes place because the Pi that is added comes from a molecule of ATP. The Pi group has been removed from ATP by a strongly exergonic reaction:

ATP $\longrightarrow$ ADP + Pi $\Delta G = -7$ kcal/mol

In the coupling of these two reactions, the 7 kcal/mol of free energy that ATP gives up by losing a Pi is used to drive the endergonic phosphorylation of glucose. A summary equation shows the overall reaction and the net change in free energy:

Glucose + ATP $\longrightarrow$ Glucose-6-Pi + ADP

$$\Delta G = -4 \text{ kcal/mol}$$

The ΔG for the overall reaction is negative, indicating that the phosphorylation of glucose by this mechanism is thermodynamically feasible.

Reversible Reactions in the Cell

Reactions that are highly exergonic will proceed in one direction, toward the formation of products. Recall from Chapter 2, however, that many reactions are **reversible;** that is, the products may immediately recombine by the reverse reaction. The equation below describes a reversible reaction:

$AB \rightleftharpoons A + B$ $\Delta G = 0$

Such a reaction is said to be in **equilibrium** when the rate of formation of the products ($A + B$) is exactly balanced by the rate at which the products recombine to yield the reactant (AB). In this situation there will be no net change in free energy. The smaller the difference in free energy between reactants and products, the less likely the reaction is to proceed to completion; that is, the more likely it is to be in equilibrium.

In an equilibrium reaction, it might be difficult to obtain enough of the Product B, since it tends to recombine with A as soon as it is formed. One way to ensure that such a reaction proceeds "to the right," toward "completion" or the formation of A and B, is to remove any B as soon as it is formed. Then AB will continue to break down, since there will be no B to recombine with A.

Similarly, a metabolic reaction can be driven to completion even though there is no difference in the free energy of the reactants and the products; that is, even though the formation of products is almost exactly balanced by the reformation of reactants. In the cell, metabolic reactions are driven to completion because the product of one reaction (such as B) is constantly being "removed" by participating in another reaction:

(1) $AB \longrightarrow A + B$
                    $\downarrow$
(2) $B + C \longrightarrow D$

Many reactions of metabolism, while not at equilibrium, take place with a very small change in free energy and can be driven in either direction relatively easily. This fact allows a cell to reverse a pathway of metabolism according to the specific demands of any given moment. For example, the synthesis of glucose from precursors takes place by reversal of many of the reactions that are used to catabolize it.

These last points emphasize one of the characteristic features of cellular metabolism: that no single reaction is taking place independently or in isolation. Hundreds of reactions are occurring in the cell at the same time, and most of them are interrelated; the products of one reaction are being used for the next reaction in a metabolic pathway. Even the so-called final products of one pathway are likely to be used in another pathway.

Oxidation–Reduction Reactions in the Cell

Recall from Chapter 2 that **redox reactions** are combinations of **oxidation reactions** (involving loss of electrons) with **reduction** reactions (involving gain of electrons). A compound becomes "reduced" by gaining electrons.

$$Ae^- + B \longrightarrow A + Be^-$$

| electron donor | electron acceptor | has been oxidized | has been reduced |

Hydrogen: The "Electron" in Biological Redox Reactions.

Organic compounds do not give up electrons easily. Removal of an electron during oxidation is achieved only by loss of an entire atom. The specific atom given up is almost always hydrogen, and so the oxidation process is actually a **dehydrogenation** reaction and the reduction process a **hydrogenation** process. In general, compounds that are highly reduced—contain many hydrogen atoms—are high-energy compounds. Oxidized compounds contain few hydrogen atoms and are low-energy compounds.

$$AH + B \longrightarrow A + BH$$

| electron donor | electron acceptor | has been oxidized | has been reduced |
| high energy | low energy | low energy | high energy |

Photosynthesis: A Reduction Process.

The formation of an organic molecule through photosynthesis can be thought of as a reduction process, in which CO_2 becomes reduced by gaining electrons in the form of hydrogen atoms:

$$H_2O + CO_2 + \text{radiant energy} \longrightarrow$$

| electron donor | electron acceptor |

$$O_2 + (CH_2O)n$$

| has been oxidized | has been reduced |
| | high-energy organic molecule |

At the same time, the H_2O becomes oxidized, losing electrons. The resulting O_2, an oxidized molecule, is a low-energy molecule.

Cellular Respiration: An Oxidation Process.

Conversely, the breakdown of an organic molecule in cellular respiration can be thought of as an oxidation process (the removal of electrons in the form of hydrogen atoms) that produces the highly oxidized, low-energy compound CO_2:

$$(CH_2O)_n + O_2 \longrightarrow CO_2 + H_2O + \text{energy}$$

| electron donor (high-energy organic compound) | electron acceptor | has been oxidized | has been reduced |

Note that the electrons (H atoms) removed from the organic molecule are accepted by the O_2 molecule, forming water. Oxygen is therefore called the electron acceptor. (Since oxygen is the most common

electron acceptor, the loss of electrons is called oxidation.) As we have said, organic fuel molecules such as glucose ($C_6H_{12}O_6$) are highly *reduced* compounds because they contain many hydrogen atoms. Thus, the catabolism, or breakdown, of these molecules is essentially an *oxidative* process, involving the removal of electrons in the form of those many hydrogen atoms. The hydrogen atoms often combine with oxygen to form water. For example, the complete catabolism of one glucose molecule uses up six molecules of oxygen and can be summarized as:

$$C_6H_{12}O_6 + 6\ O_2 \longrightarrow 6\ CO_2 + H_2O$$

Before combining with water, however, the hydrogen atoms that glucose gives up during oxidation are transferred temporarily to molecules called coenzymes, chiefly NAD, which thereby become reduced. The reduced forms of NAD and other coenzymes in turn donate electrons to the **electron transport chain,** known also as the **respiratory chain**, which is embedded in the inner mitochondrial membrane (see Chapter 3). The flow of electrons along the respiratory chain occurs with the release of free energy that is used to drive endergonic synthesis of ATP (Fig. 6–2). This process will be discussed later in the chapter.

Electron Acceptors: Key Molecules in Cellular Respiration.

Electrons cannot simply be released but must be accepted by another molecule. Oxygen and coenzymes play critical roles in cellular respiration by accepting the electrons given off as organic molecules are oxidized. Coenzymes accept electrons in

Figure 6–2

Energy transformation occurs in the mitochondria of animal cells through a series of oxidation-reduction reactions, in which electrons in the form of hydrogen atoms are transferred from highly reduced compounds such as NADH to highly oxidized compounds such as O_2. The energy released in these reactions is used to phosphorylate ADP to give ATP, the energy storage molecule.

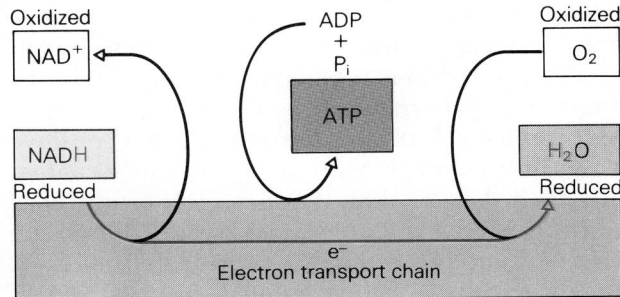

intermediate steps in the process. In the final steps, oxygen accepts electrons from the coenzymes.

Coenzymes: Intermediate Electron Acceptors. An example of a biological redox reaction that involves a coenzyme is the conversion of lactic acid into pyruvic acid:

$$
\begin{array}{ccc}
\text{COOH} & & \text{COOH} \\
| & \xrightarrow{\text{oxidation}} & | \\
\text{CHOH} & & \text{C=O} \quad + \ 2\,\text{H} \\
| & & | \\
\text{CH}_3 & & \text{CH}_3 \\
\text{lactic acid} & & \text{pyruvic acid} \\
\text{electron} & & \text{electrons} \\
\text{donor} & &
\end{array}
$$

The freed hydrogen atoms are usually transferred to a coenzyme (so called because it works closely with an enzyme). The coenzyme most frequently employed as a hydrogen acceptor in metabolic reactions is **nicotinamide adenine dinucleotide,** or NAD (see Chapter 2; Fig. 6–8).

The formation of pyruvic acid by this reaction is a reversible process in the cell. Pyruvic acid can be converted back to lactic acid by the addition of two hydrogen atoms; by accepting the hydrogen atoms, the pyruvic acid becomes reduced:

$$
\text{Pyruvic acid} + 2\text{H} \xrightarrow{\text{reduction}} \text{Lactic acid}
$$
$$
\begin{array}{cc}
\text{electron} & \text{electrons} \\
\text{acceptor} &
\end{array}
$$

The hydrogen atoms for this reaction are provided by the same coenzyme (NAD) involved in the oxidation of lactic acid. Note that during the interconversion of lactate and pyruvate, the NAD is donating or accepting hydrogen atoms from the reactants, thereby becoming either oxidized or reduced. The process can be summarized as:

$$
\begin{array}{cccc}
\text{Lactic} + \text{Oxidized} & \rightleftharpoons & \text{Pyruvic} + \text{Reduced} \\
\text{acid} \quad\ \ \text{NAD} & & \text{acid} \quad\quad \text{NAD}
\end{array}
$$

This illustrates the point that oxidation and reduction of the organic molecules involved in cellular metabolism are true redox reactions. The oxidation of one molecule, such as lactic acid, is coupled to the reduction of another, such as NAD.

In some oxidative reactions involving removal of hydrogen atoms from an organic compound, the protons and electrons of the hydrogen atoms depart separately or become separated after removal of the atoms. Recall from Chapter 2 that the hydrogen atom contains 1 proton bearing a single positive charge and 1 electron with a single negative charge. Thus, a hydrogen atom that has given up its electron will have a single positive charge, due to the remaining proton, and can be represented as H^+. A single electron can be represented as e^-. By way of

example, the oxidation of hydroquinone can be represented in two stages as follows:

reduced   two protons removed   two electrons removed   oxidized

Oxygen: The Ultimate Electron Acceptor. Organic molecules vary in their tendencies to give up or accept electrons. Their affinity for electrons can be measured in terms of their oxidation-reduction potential, better known as the **redox potential.** The redox potential of molecules is expressed on a scale in which the redox potential of hydrogen gas (H_2) is 0 millivolts (mV). A molecule that has a negative redox potential on this scale has a low affinity for electrons. A specific example is NADH, the reduced form of NAD. This molecule, with a redox potential of -320 mV, has a low affinity for electrons and will donate them to another electron carrier, thereby reverting to its oxidized form, NAD^+.

At the other end of the scale of redox potentials is molecular oxygen, which, with a redox potential of $+820$ mV, has a high affinity for electrons and will readily accept them, becoming reduced to water in the process. Molecular oxygen is the ultimate aceptor of the electrons donated by NADH to the mitochondrial respiratory chain (see Fig. 6–2).

High-Energy Phosphate Compounds

Earlier in this chapter, we discussed the use of the chemical energy stored in a molecule of ATP to drive an endergonic reaction such as the phosphorylation of a glucose molecule. ATP is an example of a special group of compounds known as **high-energy phosphate compounds** that have a strong tendency to transfer their phosphate group to an acceptor molecule. This transfer is accompanied by the release of a large amount of free energy.

The ΔG for a given reaction is influenced by the concentrations of the reactants and products; it is not a very useful term for comparing the energetics of different reactions. For comparisons of reactions, the free energy difference between the reactants and the products is considered under a set of "standard conditions": temperature of 25°C, pressure of 1 atmosphere, and a concentration of 1.0 M for all reactants and products present. Under these conditions, the free energy change for a reaction is called the **standard free energy change** and is given the symbol $\Delta G°$. This term is a constant for a specific reaction. For example:

$$ATP \rightarrow ADP + Pi \qquad \Delta G° = -7 \text{ kcal/mol}$$

In the cell, however, the "standard conditions" are not present. The temperature is 37°C, and the concentration of ATP is very high relative to ADP and Pi. Under these conditions, the free energy change (ΔG) for ATP breakdown by the above reaction is about -12 kcal/mol. The large difference between $\Delta G°$ and ΔG in this example indicates that it is the ΔG that must be used to predict the direction of reactions within the cell.

The standard free energy change ($\Delta G°$) can be used to compare the tendency of different high-energy phosphate compounds to transfer their phosphate group to an acceptor molecule. The $\Delta G°$ values for the phosphorylated compounds found in cells range from about -2 to -13 kcal/mol. ATP is unique in that it occupies a central position in this scale (Fig. 6–3). All of the high-energy phosphate compounds for which $\Delta G°$ is more negative than ATP have a lower affinity for the phosphate group and can form ATP by donating their phosphate group to ADP:

$$ADP + Pi \rightarrow ATP$$

Compounds for which $\Delta G°$ is less negative than ATP have a greater affinity for the phosphate group and serve as acceptors for phosphate groups donated to them by ATP. Specific examples are given in Figure 6–3. Phosphoenolpyruvate and 1,3-diphosphoglycerate are intermediates formed during glucose catabolism. Phosphate groups removed from these compounds are transferred to ADP to form ATP. By a similar mechanism, phosphocreatine is used to replenish the ATP supply in skeletal muscle during periods of intense activity when the metabolic production of ATP cannot meet the demand. In fact, phosphocreatine is the major phosphorylated compound present in the resting muscle cell. The amount of phosphocreatine is three to eight times greater than the amount of ATP.

The direct formation of a high-energy phosphate compound such as ATP, by a chemical reaction that incorporates a phosphate group, is a process known as **substrate-level phosphorylation.** We shall see that this reaction does not require oxygen, in contrast to the mechanism for ATP formation by the electron transport chain in the mitochondrion (see Fig. 6–2). Phosphorylation of ADP in the mitochondrion is coupled to loss of electrons (oxidation) by the components of the electron transport chain and is termed **oxidative phosphorylation.** Most of the ATP supply for the cell is generated by oxidative phosphorylation. While ADP is the only molecule that can be phosphorylated by mitochondrial oxidative phosphorylation, substrate-level phosphorylation can generate other high-energy phosphate compounds such as GTP (by phosphorylation of GDP) and 1,3-diphosphoglycerate (by phosphorylation of glyceraldehyde-3-phosphate).

Figure 6–3
The standard free energy of hydrolysis of some phosphate compounds important in metabolism. Arrows indicate that phosphate groups can be transferred from high-energy donors to low-energy acceptors.

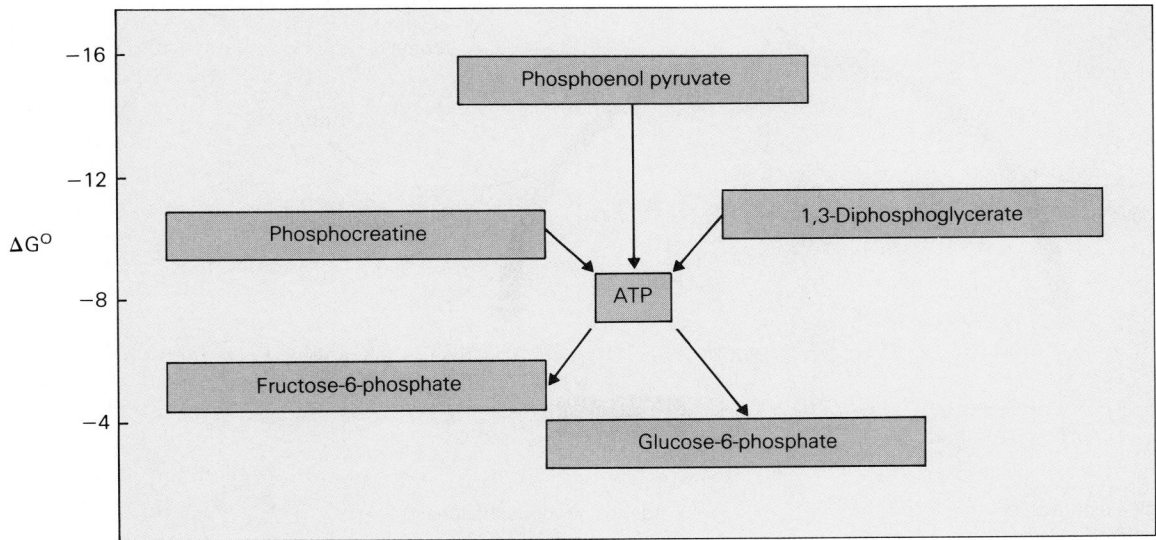

The large release of free energy that occurs upon removal of a phosphate group from a high-energy phosphate compound is a property of the whole molecule; the energy does not reside solely in the covalent bond by which the phosphate group is attached. There is nothing special about this bond, even though it may be referred to as a "high-energy bond." Several factors account for the free energy that is released. In the specific case of ATP, one important factor is the reduction of **electrostatic repulsion** that is achieved by loss of a phosphate group (Fig. 6–4). Removal of the terminal phosphate group yields ADP, while the removal of two phosphate groups (which occurs in some reactions) yields AMP. In either case the free energy released is the same.

Enzymes in Cellular Metabolism

In Chapter 2 we learned that an enzyme facilitates reactions by recognizing specific reactants and bringing them close together by binding them to its active sites. For many reactions, the physical orientation of the reactant molecules determines the rate of reaction; certain bonds such as peptide bonds cannot form unless the reacting molecules (e.g., amino acids) are oriented in a specific way (Fig. 6–5). By binding reactants in specific ways, enzymes ensure that they meet in the most favorable orientation for reactions to occur. For example, ATP and glucose are both bound by the enzyme **hexokinase,** which facilitates the phosphorylation reaction in which a phosphate group is added to glucose. The enzyme in effect couples an exergonic reaction (removal of a phosphate group from the ATP) to an endergonic reaction (addition of a phosphate group to the glucose). Glucose-6-phosphate and ADP are the reaction products.

Enzymes are also important in directing various products of cell reactions down the correct pathways. In Chapter 2 we discussed the ways in which enzymes lower the activation energy for specific reactions. Enzymes lower the activation energy only for specific desired reactions, so that at each step reactants are more likely to participate in those reactions than in others that might be possible.

Figure 6–4
Release of phosphate (Pi) or pyrophosphate (PPi) groups reduces the electrostatic repulsion between negatively charged phosphate groups in adenosine triphosphate (ATP).

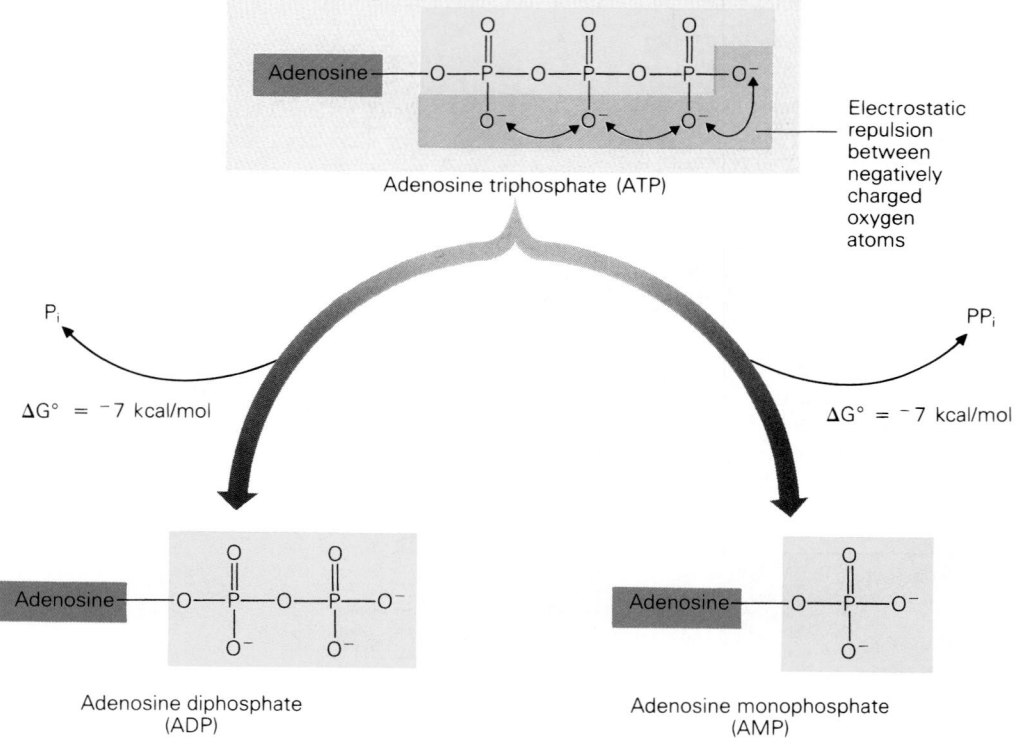

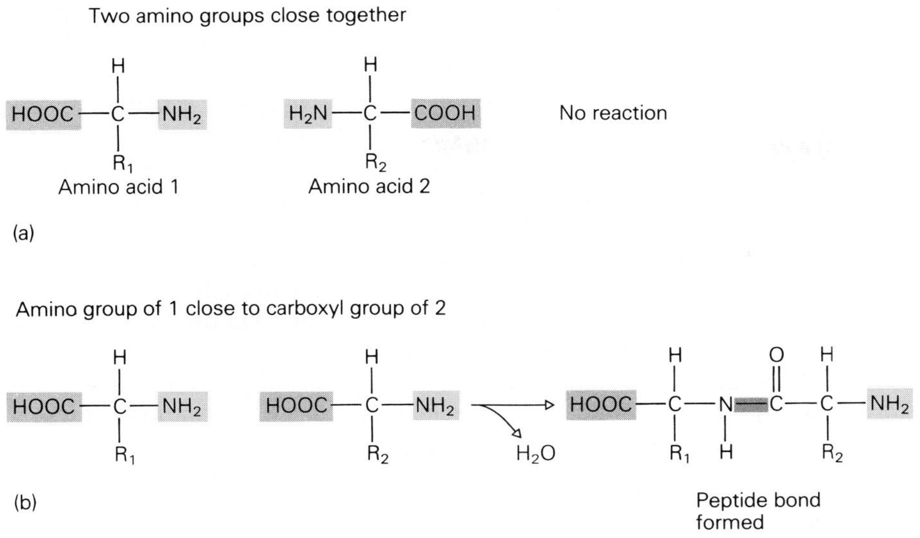

Two amino groups close together

$$HOOC—\overset{\overset{\displaystyle H}{|}}{\underset{\underset{\displaystyle R_1}{|}}{C}}—NH_2 \qquad H_2N—\overset{\overset{\displaystyle H}{|}}{\underset{\underset{\displaystyle R_2}{|}}{C}}—COOH \qquad \text{No reaction}$$

Amino acid 1   Amino acid 2

(a)

Amino group of 1 close to carboxyl group of 2

$$HOOC—\overset{\overset{\displaystyle H}{|}}{\underset{\underset{\displaystyle R_1}{|}}{C}}—NH_2 \qquad HOOC—\overset{\overset{\displaystyle H}{|}}{\underset{\underset{\displaystyle R_2}{|}}{C}}—NH_2 \quad\longrightarrow\quad HOOC—\overset{\overset{\displaystyle H}{|}}{\underset{\underset{\displaystyle R_1}{|}}{C}}—\overset{\overset{\displaystyle H}{|}}{\underset{\underset{\displaystyle H}{|}}{N}}=\overset{\overset{\displaystyle O}{\|}}{C}—\overset{\overset{\displaystyle H}{|}}{\underset{\underset{\displaystyle R_2}{|}}{C}}—NH_2$$

H_2O

(b)   Peptide bond
                                                                 formed

Figure 6–5
Certain reactions require reactants to have a specific physical orientation. Peptide bonds, for example, will form only when the amino group of one amino acid is near the carboxyl group of another amino acid.

Coenzymes. **Coenzymes,** which have already been mentioned as intermediate electron acceptors in cell metabolism, are important features of enzyme action in cell processes. Coenzymes are nonprotein organic molecules that some enzymes require before they can catalyze reactions. **Coenzyme A,** for example, feeds 2-carbon acetyl groups ($CH_3CO—$) from glucose into the **citric acid cycle,** and NAD serves as a carrier of hydrogen atoms. Many coenzymes, such as NAD, are separated easily from the enzyme. NAD and its phosphorylated version NADP are derivatives of the vitamin **niacin,** while the coenzyme flavin adenine dinucleotide (FAD) is derived from another vitamin called **riboflavin.**

The cell must maintain all of the various metabolic substrates in the required concentrations, ultimately by controlling the rates of the chemical reactions that produce and consume them. All of these reactions are believed to require catalysis by enzymes, and one of the great advantages of an enzyme catalyst is that its activity in the cell can be controlled. One control mechanism, discussed already in Chapter 5, is covalent modification of the enzyme protein by attachment of a phosphate group, a reaction that is itself catalyzed by a specific **kinase** enzyme. Phosphorylation of an enzyme will either activate or inactivate it, and this change can be reversed by removal of the phosphate group with a **phosphoprotein phosphatase** enzyme (Fig. 6–6).

Allosteric Inhibition. A specific enzyme usually catalyzes each of the distinct reactions of a pathway of metabolism. A control mechanism that regulates the output of an entire pathway is **end-product inhibition,** also known as **allosteric inhibition.** Typically an allosteric inhibitor is the final product of a

metabolic pathway that "feeds back" to inhibit the enzyme catalyzing the *first step* of that pathway (Fig. 6–7a). When the first enzyme (e.g., Enzyme *A*) in the pathway is inhibited, the first reaction will not occur. Hence, no substrate will be available for the next reaction, catalyzed by Enzyme *B*, and so on. In

Figure 6–6
The phosphorylation of certain enzymes can activate them, whereas the removal of a phosphate group inactivates them.

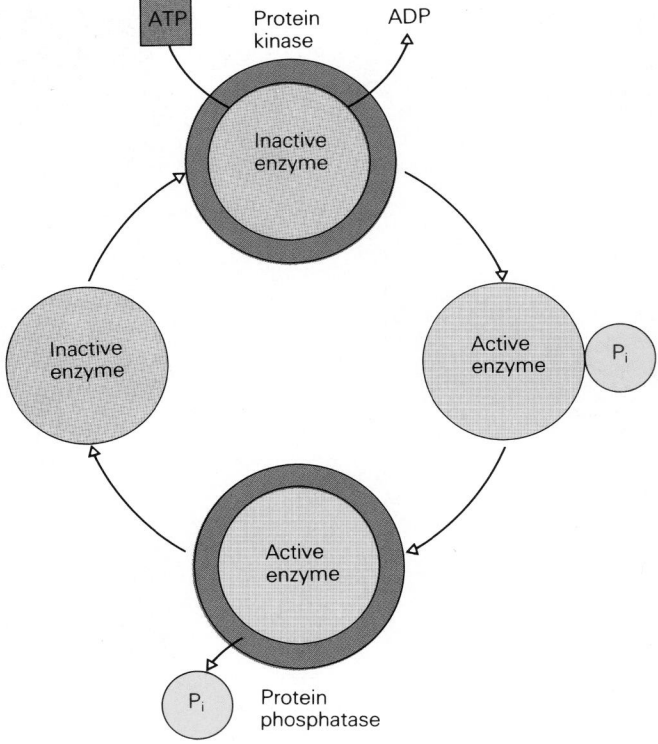

this way, the rate of formation of the final product X will rapidly decrease when the first step is inhibited.

An allosteric inhibitor inhibits an enzyme by binding to it at a site that is separate from the active site, causing a conformational change in the protein that impairs the function of the active site (see Fig. 6–7b).

Allosteric inhibition by the end product of a metabolic pathway is a reversible process. As formation of Product X declines, owing to inhibition of Enzyme A (see Fig. 6–7), the concentration of X in the cell will also decline because now it is used and consumed faster than it is produced. The concentration of X will fall until there is no longer enough of X to inhibit Enzyme A. When the inhibitor of Enzyme A is removed, the pathway leading to production of X will be "turned on" again.

Cells can also control enzyme-catalyzed reactions by increasing or decreasing synthesis of the enzyme proteins, a process that changes the amount of enzyme present in the cell. If a cell requires more of Product X produced by the pathway

depicted in Figure 6–7, for example, it can synthesize more of Enzyme A, Enzyme B, and the other enzymes involved in this pathway. This type of control is relatively slow to take effect; it may take several hours or days for the enzyme levels to change significantly. It does not provide the rapid and fine control possible with covalent modification or allosteric inhibition.

Most enzymes in the cell are not saturated by substrate and do not operate at their V_{max}. This provides the cell with the capacity to deal with fluctuations in the concentrations of its metabolites. Thus, when physical exercise rapidly increases the demand for glucose catabolism, for example, the enzymes involved have the capacity to respond by processing more substrate molecules in the same time.

Redox Reactions and Coenzyme Action. The oxidized forms of the coenzymes NAD and NADP have a single positive charge (Fig. 6–8) and, when reduced, they accept electrons and protons from

Figure 6–7
End-product (allosteric) inhibition of the initial steps in two metabolic pathways (*a*). Allosteric inhibitors typically bind to a site on the enzyme molecule that is different from the active site. The binding causes the active site to change shape, preventing the enzyme from catalyzing the first reaction in the pathway (*b*).

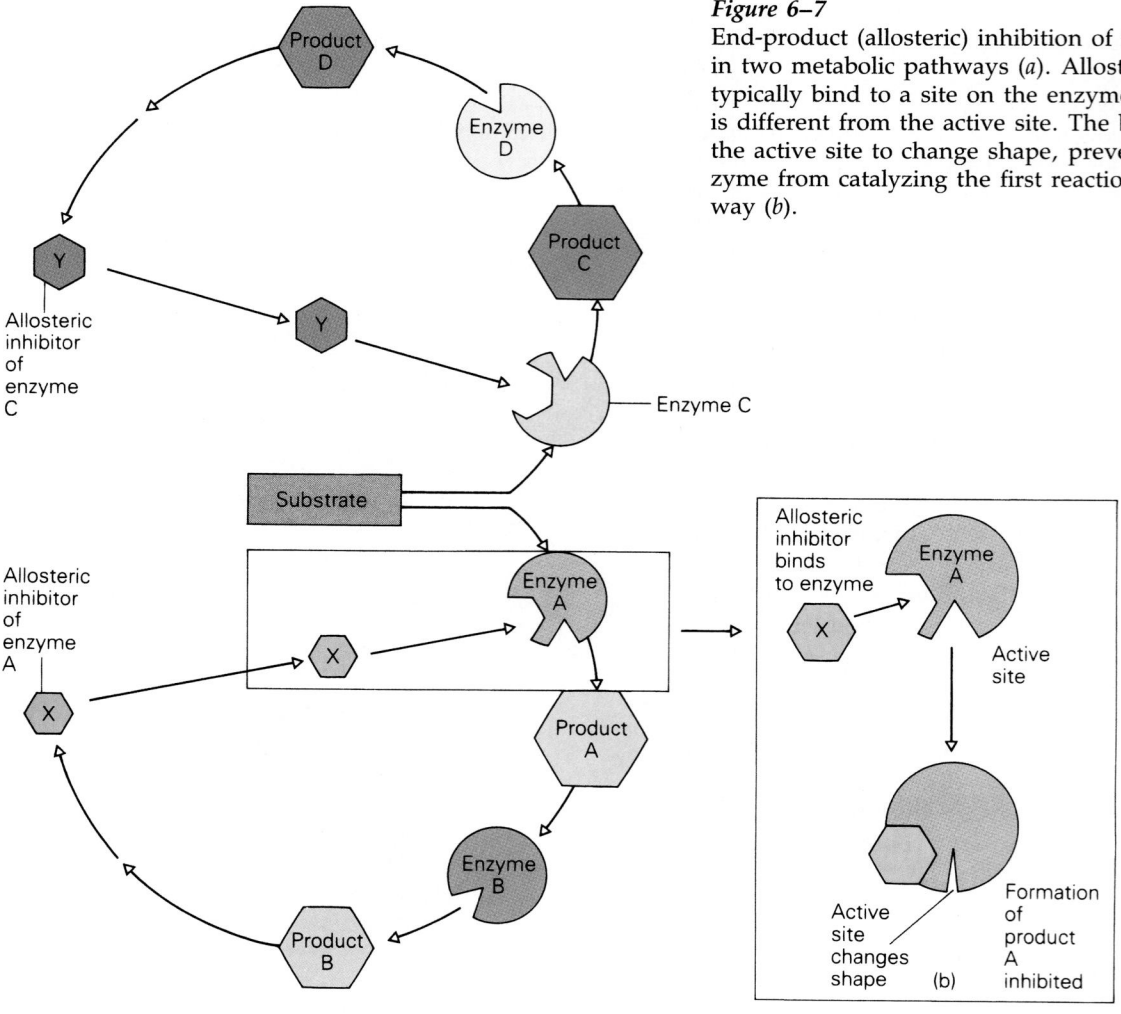

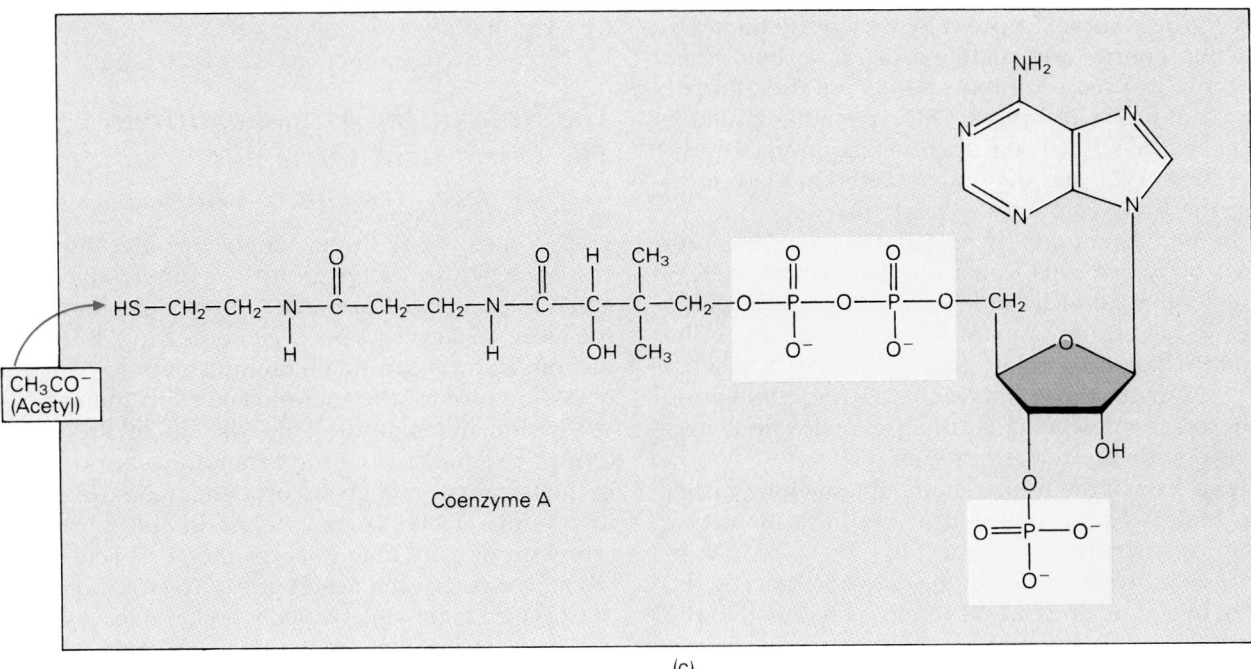

Figure 6–8
The structures of common coenzymes. Arrows indicate the points that accept
hydrogen atoms (NAD⁺, NADP⁺, and FAD) or acetyl groups (coenzyme A).

two hydrogen atoms. Since a hydrogen atom consists of one electron and one proton, two hydrogen atoms are equivalent to two electrons and two protons. NAD^+ and $NADP^+$ accept both electrons but only one of the protons:

$$NAD^+ + 2H^+ + 2e^- \longrightarrow NADH + H^+$$

oxidized   reduced

FAD, like NAD^+, can accept two electrons, but it also accepts two protons. In other words, FAD can accept two hydrogen atoms when it is being reduced (Fig. 6–8):

$$FAD + 2H^+ + 2e^- \longrightarrow FADH_2$$

oxidized   reduced

Recall that the reduction of these coenzymes is coupled to oxidation of a molecule such as glucose. NAD^+ and FAD are important coenzymes used in catabolic pathways such as the oxidation of glucose and other fuel molecules. $NADP^+$ is important in biosynthetic pathways of metabolism such as the synthesis of fatty acids.

An Overview of Catabolism: Dietary Proteins, Carbohydrates, and Fats

The simple act of consuming food is only a preliminary to the complex process by which the body obtains the energy and matter it needs to live. Even the complex process of digestion is not the ultimate destiny of food substances. Only the individual cells can enact the specialized reactions required to transform the energy and matter contained in food into a form the body can use.

Before examining in detail the complex processes by which cells transform matter and energy, let us look at all of them in general and discuss the ways in which they relate. We will first review the way in which cells obtain food molecules and what they do with those molecules before delivering them to the mitochondria, the organelles finally responsible for the release of energy.

The cells of the human body obtain energy from food that has been ingested in the form of dietary proteins, carbohydrates, and fats. These food substances are broken down in the digestive tract by the action of certain enzymes. Proteins consumed in the form of meat, for example, are broken down into amino acids. Carbohydrates eaten in the form of fruits, vegetables, and grains are broken down into simple sugars (monosaccharides) such as glucose.

Dietary fats such as margarine and oils are degraded into fatty acids and glycerol. After digestion, these macromolecules, the products of digestion, circulate in the blood to all cells in the body (Fig. 6–9).

Digestion in the Cytoplasm: Deamination, Glycolysis, and Fatty Acid Oxidation

The chemical bonds that hold together these various macromolecules represent a store of energy the cells need to tap. They do so by obtaining the macromolecules from the blood and breaking them down. Each type of food molecule is broken down by a specific process. Amino acids obtained by the cell are broken down by **deamination,** which separates nitrogen atoms and carbon atoms. Monosaccharides (most commonly glucose) are broken down by **glycolysis.** These reactions occur in the cytosol of the cell. Fatty acids are broken down by a process known as **fatty acid oxidation** in the mitochondrial matrix (see Fig. 6–9). All of these processes are regulated and facilitated by enzymes.

Most of the molecules that result from the breakdown of amino acids, glucose, and fatty acids are converted to **acetyl (CH_3CO—) groups** in the matrix of the **mitochondrion,** the "powerhouse" of the cell. This step is preliminary to the citric acid cycle, the chief function of the mitochondrion and the central processing point for all food molecules (see Fig. 6–9).

The Role of the Mitochondrion: The Citric Acid Cycle and the Electron Transport Chain

The products of all three cytoplasmic digestive processes, whether acetyl groups or other molecules, go on to participate in a series of reactions known as the **citric acid cycle.** This cycle, occurring in the mitochondrion, is the final common pathway for the breakdown of all the fuel molecules in the cell.

In the mitochondrial matrix, all of the acetyl groups produced from food molecules are coupled to the terminal—SH group of coenzyme A (Fig. 6–8) to produce **acetyl coenzyme A.** In this form, the acetyl groups are able to enter the citric acid cycle, which converts each acetyl group to two molecules of CO_2 (see Fig. 6–9). This conversion releases high-energy electrons that are temporarily trapped by the coenzymes NAD and FAD. The coenzymes later feed the electrons into the electron transport chain, which is a series of electron carriers, many of them

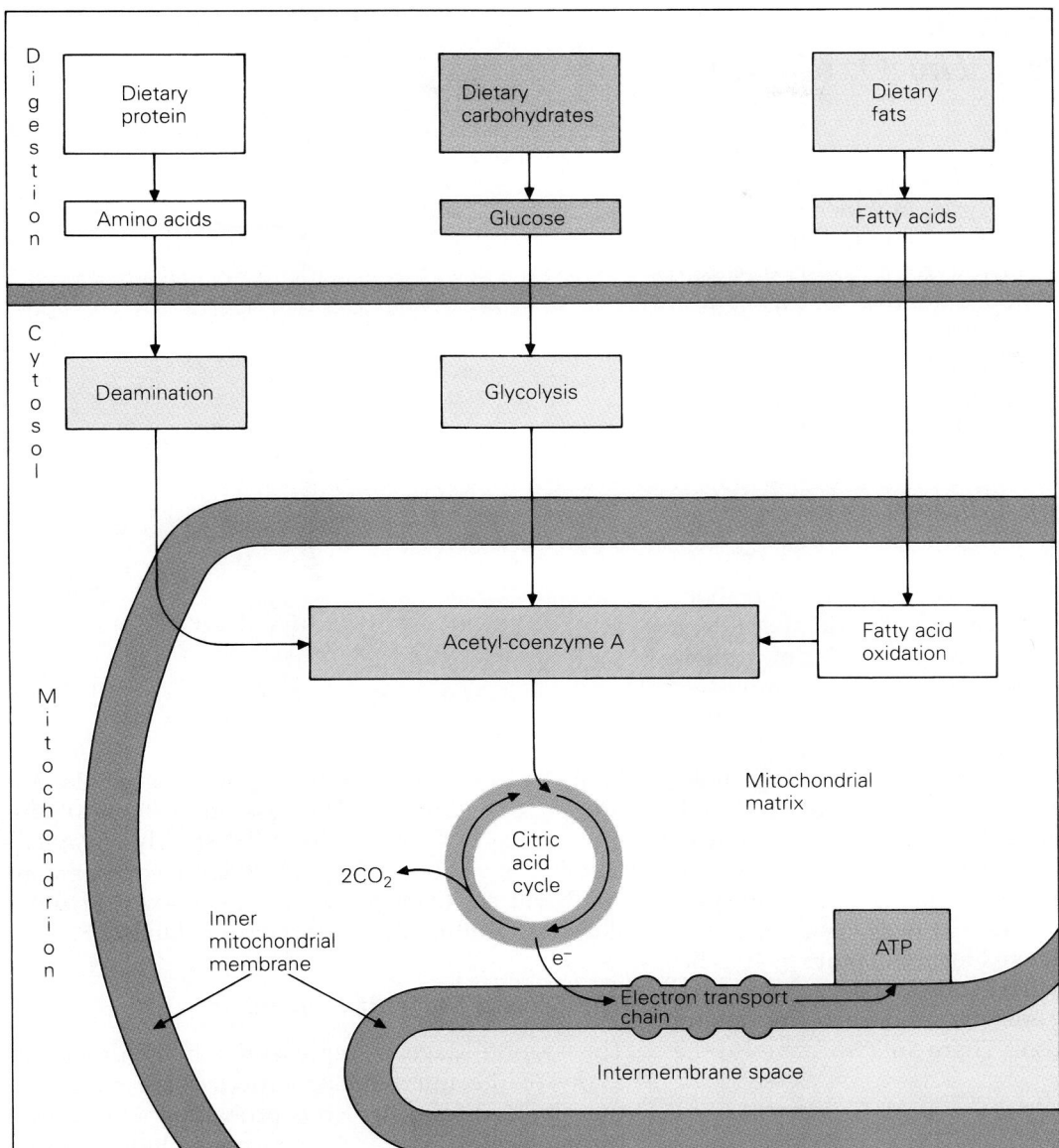

Figure 6–9
An overview of catabolism, which shows the locations of important processes. (The outer membrane has been omitted.)

proteins, embedded in the inner mitochondrial membrane. The redox reactions of the electron carriers release the energy used to synthesize ATP.

The final step in this process requires oxygen, making cellular respiration an **aerobic (oxygen-requiring) process.** The electron transport chain uses oxygen as the final acceptor of the electrons, a step that converts the oxygen to water. As the electrons pass from one electron carrier to the next, their energy is released in several small bursts that are used to drive the energy-dependent synthesis of ATP from ADP and inorganic phosphate (Pi). Thus, the ADP becomes phosphorylated; this step is tightly coupled to the flow of electrons. The electron flow will not occur unless the phosphorylation step can proceed. The loss of electrons, as we have said, is defined as oxidation. Since electron transport consists of successive oxidations and is coupled to ADP phosphorylation, the entire process is known as **oxidative phosphorylation.**

The Metabolism of Carbohydrates: The Oxidation of Glucose

The metabolism of carbohydrate involves three main processes: (1) **glycolysis**, in which glucose obtained from food is converted to pyruvic acid; (2) **the citric acid cycle,** in which pyruvic acid is oxidized, releasing electrons that are trapped by coenzymes; and (3) **oxidative phosphorylation,** also called **chemiosmotic synthesis** of ATP, in which electrons held by coenzymes are transferred to oxygen, releasing free energy that is stored in ATP.

Glycolysis: From Glucose to Pyruvic Acid

The overall reaction that occurs in this metabolic pathway is the breakdown of a **glucose** molecule, a 6-carbon compound, into two molecules of **pyruvic acid,** a 3-carbon compound. This is accomplished with the net release of enough free energy to phosphorylate two molecules of ADP to ATP by substrate-level phosphorylation. In addition, two molecules of the coenzyme NAD^+ are reduced, providing the potential for more ATP synthesis when the reduced NAD molecules later give up electrons to the mitochondrial electron transport chain (see Fig. 6–2). The pathway of glycolysis is represented in condensed form in Figure 6–10. This pathway consists of 10 reaction steps, each one catalyzed by a specific enzyme, and each one taking place in the cell cytosol. There are two stages to the pathway.

The First Stage of Glycolysis

The first stage (Fig. 6–11) consumes two ATP molecules for phosphorylating and converting the glucose molecule to a form that can be used to drive net synthesis of ATP. There are two phosphorylation steps (Steps 1 and 3), and neither can be reversed under the conditions that are present in the cell. The second phosphorylation (Step 3) is catalyzed by **phosphofructokinase** and is the primary point for controlling the rate of glycolysis.

Phosphofructokinase is regulated allosterically by both ADP and ATP, which have opposite effects: ADP stimulates the enzyme and ATP inhibits it. The significance of these opposing effects is as follows. When ATP is converted to ADP during biosynthesis, the concentration of ADP will be high relative to

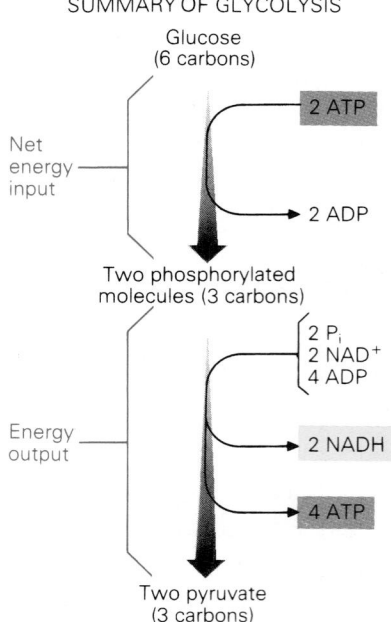

SUMMARY OF GLYCOLYSIS

Figure 6–10
A summary of the pathway of glycolysis.

ATP. The ADP will stimulate phosphofructokinase and increase the rate of glucose breakdown to generate more ATP. When the ATP supply is replenished, the relatively high ATP level compared to ADP will act to decrease the breakdown of more glucose by inhibiting phosphofructokinase.

The Second Stage of Glycolysis

The second stage is represented in Figure 6–12. Many of the intermediate products are shown in their ionized form, which is probably the form that the molecules assume in aqueous solution. Thus, pyruvic acid is shown as the **pyruvate** ion. The 6-carbon molecule from the first stage (fructose-1, 6-bisphosphate) is split into two 3-carbon molecules called **dihydroxyacetone phosphate** and **glyceraldehyde-3-phosphate** (Step 4). The molecule of dihydroxyacetone phosphate undergoes a rearrangement **(isomerization)** of its atoms and becomes glyceraldehyde-3-phosphate (Step 5). Thus, two molecules of glyceraldehyde-3-phosphate are ultimately derived from fructose-1,6-bisphosphate. Each molecule of glyceraldehyde-3-phosphate is converted first to *1,3-diphosphoglycerate* (Step 6) and them to *phosphoenolpyruvate* (Steps 7 to 9). These are high-energy phosphate compounds that can donate a phosphate group to ADP to generate ATP (see Fig. 6–3). Formation of 1,3-diphosphoglycerate in Step 6

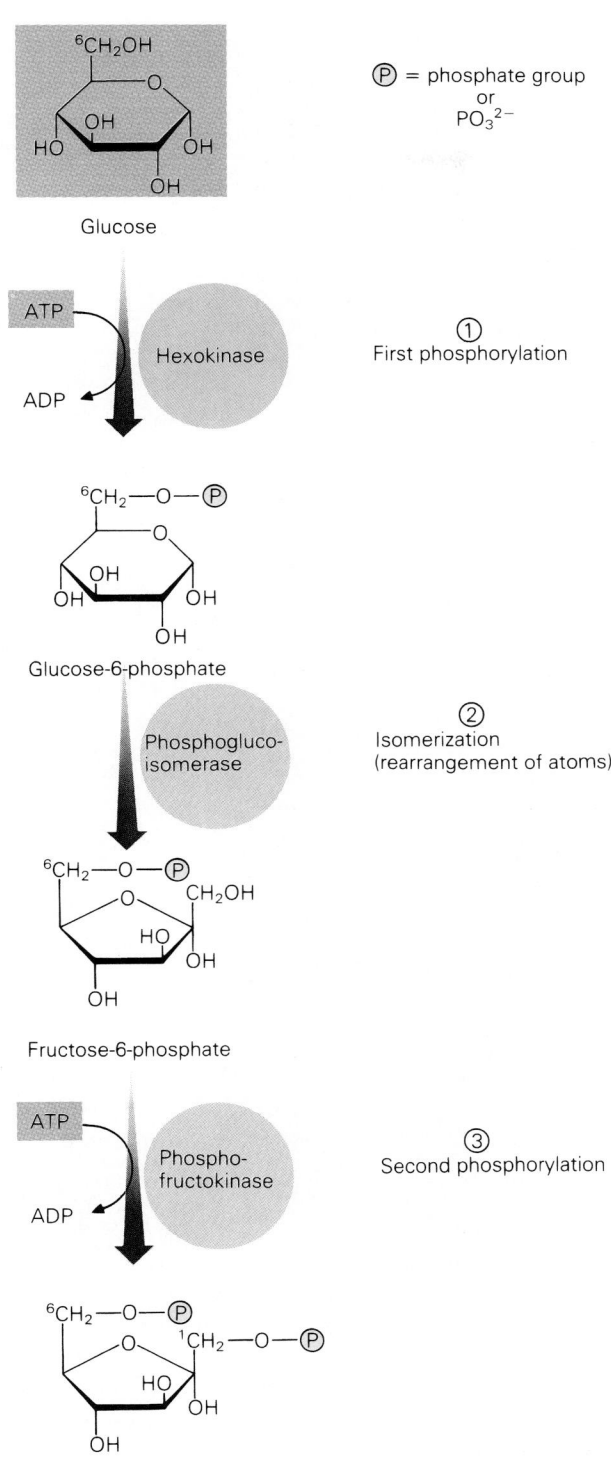

Glucose

(P) = phosphate group
or
PO_3^{2-}

① First phosphorylation

Glucose-6-phosphate

② Isomerization
(rearrangement of atoms)

Fructose-6-phosphate

③ Second phosphorylation

Fructose1,6-bisphosphate

Figure 6–11
The first stage of glycolysis.

is not only a substrate-level phosphorylation; it is also an oxidation reaction. The protons (H⁺) and electrons that are removed from glyceraldehyde-3-phosphate during oxidation are accepted by NAD⁺, which is reduced to NADH. The phosphorylation of ADP by phosphoenolpyruvate in Step 10 is essentially irreversible under the conditions present in the cell.

The second stage of glycolysis produces a total of two NADH molecules and four ATP molecules from each glucose molecule that is metabolized. Since two ATP molecules are consumed in the first stage, there is a net gain by the cell of two ATP molecules as a result of glycolysis.

The yield of free energy from glucose as it is catabolized by the glycolytic pathway to pyruvic acid is low. The standard free energy change ($\Delta G°$) is about −20 kcal/mol, and only two molecules of ATP are gained from the process. Compare this with the $\Delta G°$ of −686 kcal/mol when glucose is fully oxidized to CO_2 and water:

$$C_6H_{12}O_6 + 6 O_2 \longrightarrow 6 CO_2 + 6 H_2O$$

Glycolysis yields less than 3% of the maximum amount of free energy that can be obtained from glucose. However, it is important to realize that this pathway does not require molecular oxygen—that is, glycolysis can proceed under **anaerobic conditions**. This capability is important in tissues such as skeletal muscle when the demand for ATP rapidly increases at a time when oxygen is in low supply, such as during strenuous exercise. The skeletal muscle cells are able to meet their increased requirement for ATP by increasing the rate of glycolysis, producing ATP despite the fact that little or no oxygen is available.

However, under these anaerobic conditions, the pyruvic acid produced by glycolysis is converted to lactic acid (Fig. 6–13), which accumulates bacause it cannot be metabolized further. When exercise stops, the lactic acid in the muscle cells is transferred to the liver, where it is converted back to pyruvic acid and broken down by aerobic metabolism to CO_2 and water. This requires additional oxygen, which is obtained by continued excessive breathing after muscular activity has ceased, a phenomenon known as **oxygen debt**.

Remember that two molecules of pyruvic acid are generated from every molecule of glucose that enters the glycolytic pathway (Fig. 6–10). If the oxygen supply is plentiful—as is usually the case in most cells—the pyruvic acid produced by glycolysis is fed rapidly into the **citric acid cycle**, a pathway that takes place in the **mitochondrial matrix**.

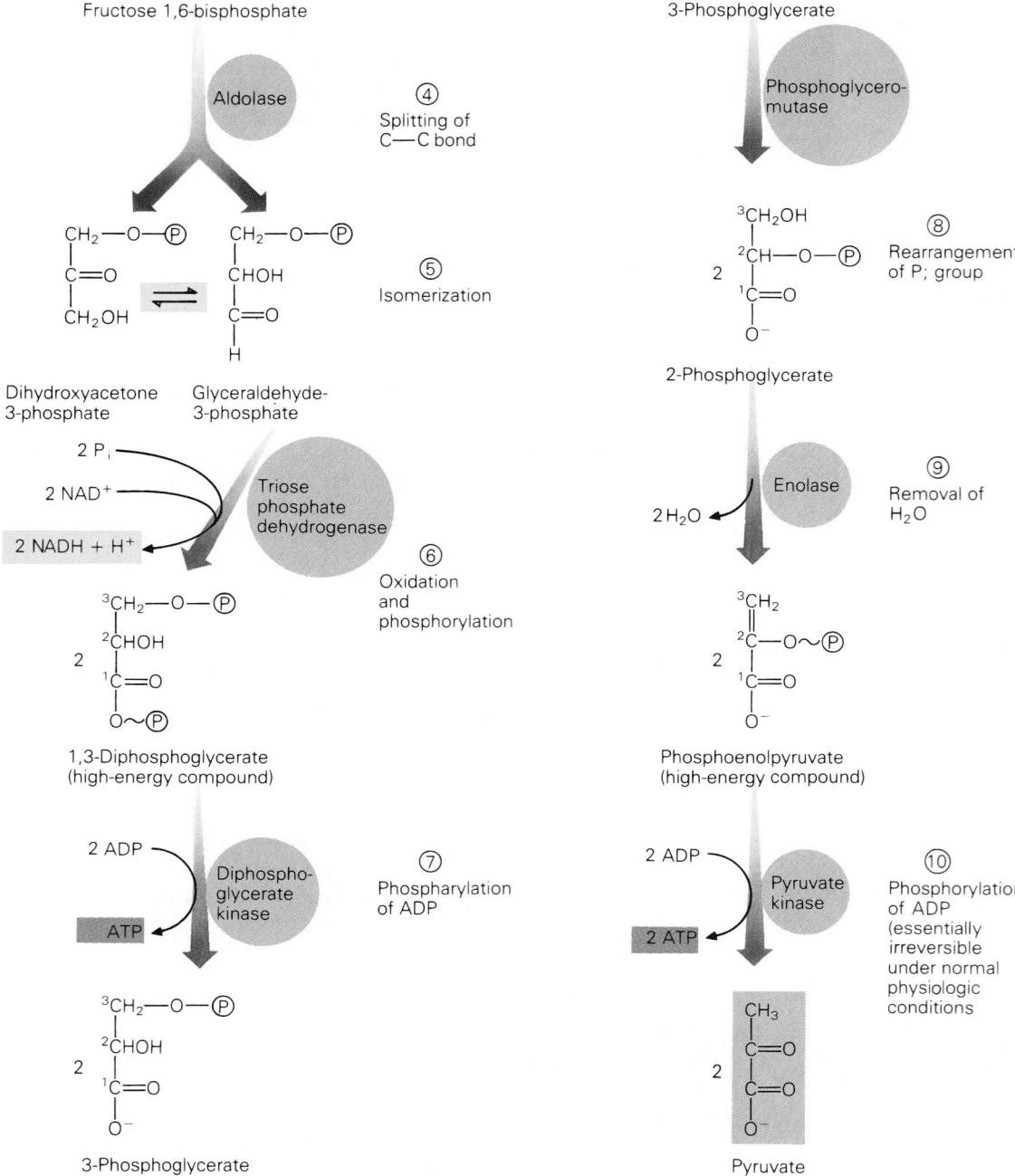

Figure 6–12
The second stage of glycolysis.

The Citric Acid Cycle: From Pyruvic Acid to Reduced Coenzymes

Pyruvic acid produced in glycolysis next enters the mitochondrial matrix, where it is coupled to **coenzyme A** in preparation for processing via the **citric acid cycle** (also called the **Krebs cycle** or the **tricarboxylic acid cycle**). The first step in the cycle is the formation of **acetyl coenzyme A.**

The Preparation of Acetyl Coenzyme A

In a reaction catalyzed by **pyruvate dehydrogenase** (see Fig. 6–13), an acetyl group ($CH_3CO—$) is transferred from pyruvic acid to the —SH group of coenzyme A to form **acetyl coenzyme A.** The acetyl group contains two of the three carbons present in pyruvic acid. (The third carbon of pyruvic acid is lost as CO_2, which is ultimately exhaled via the

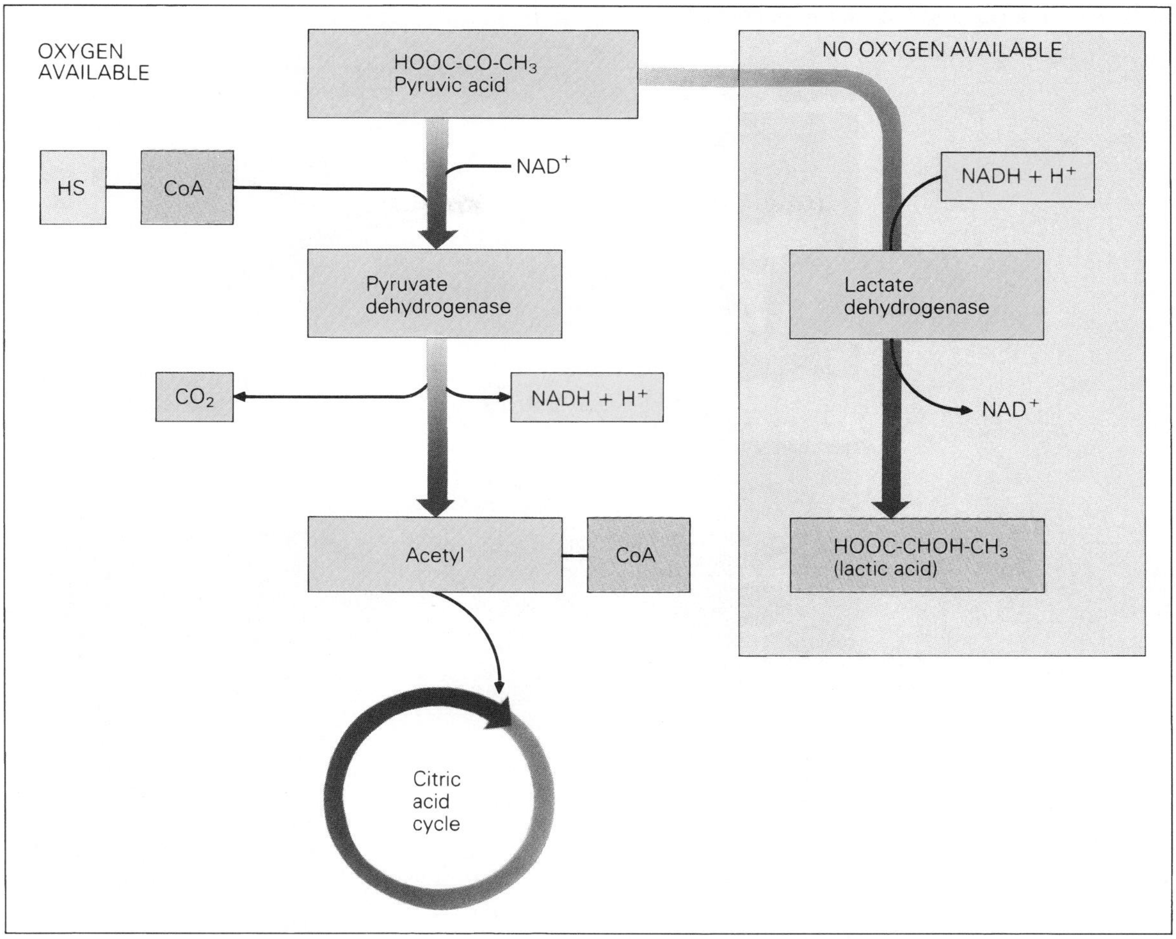

Figure 6–13

The metabolic fate of pyruvic acid. When oxygen is present, pyruvic acid reaches the citric acid as acetyl-coenzyme A. When oxygen is absent, pyruvic acid is converted to lactic acid. (CoA = coenzyme A.)

lungs.) The reaction also removes protons and electrons from pyruvic acid and coenzyme A; these are accepted by NAD^+, which is reduced to NADH. The net result of the pathway up to this point is that glucose, a compound containing 6 carbon atoms, is fed into the citric acid cycle as two 2-carbon units in the form of acetyl groups (Fig. 6–14).

The 2-carbon acetyl groups enter the citric acid cycle in a reaction catalyzed by **citrate synthetase,** which transfers the acetyl group from acetyl coenzyme A to a 4-carbon compound called **oxaloacetic acid.** One product of this reaction is the 6-carbon compound **citric acid** (hence the name for the cycle). The other reaction product is free coenzyme A, which becomes available to accept another acetyl group from another pyruvic acid:

oxaloacetic acid + acetyl coenzyme A ⟶

  4 C atoms   2 C atoms

        citric acid + coenzyme A

          6 C atoms

The pathway of the citric acid cycle is a true cyclic process because the last reaction regenerates the 4-carbon oxaloacetic acid, which is free to accept an incoming acetyl group from acetyl coenzyme A.

Figure 6–14

The formation of acetyl coenzyme A from pyruvic acid and coenzyme A (CoA).

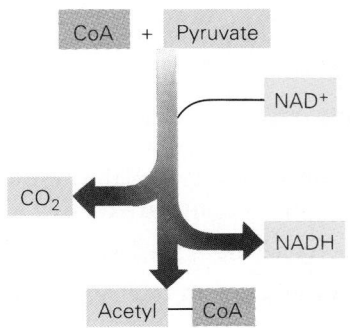

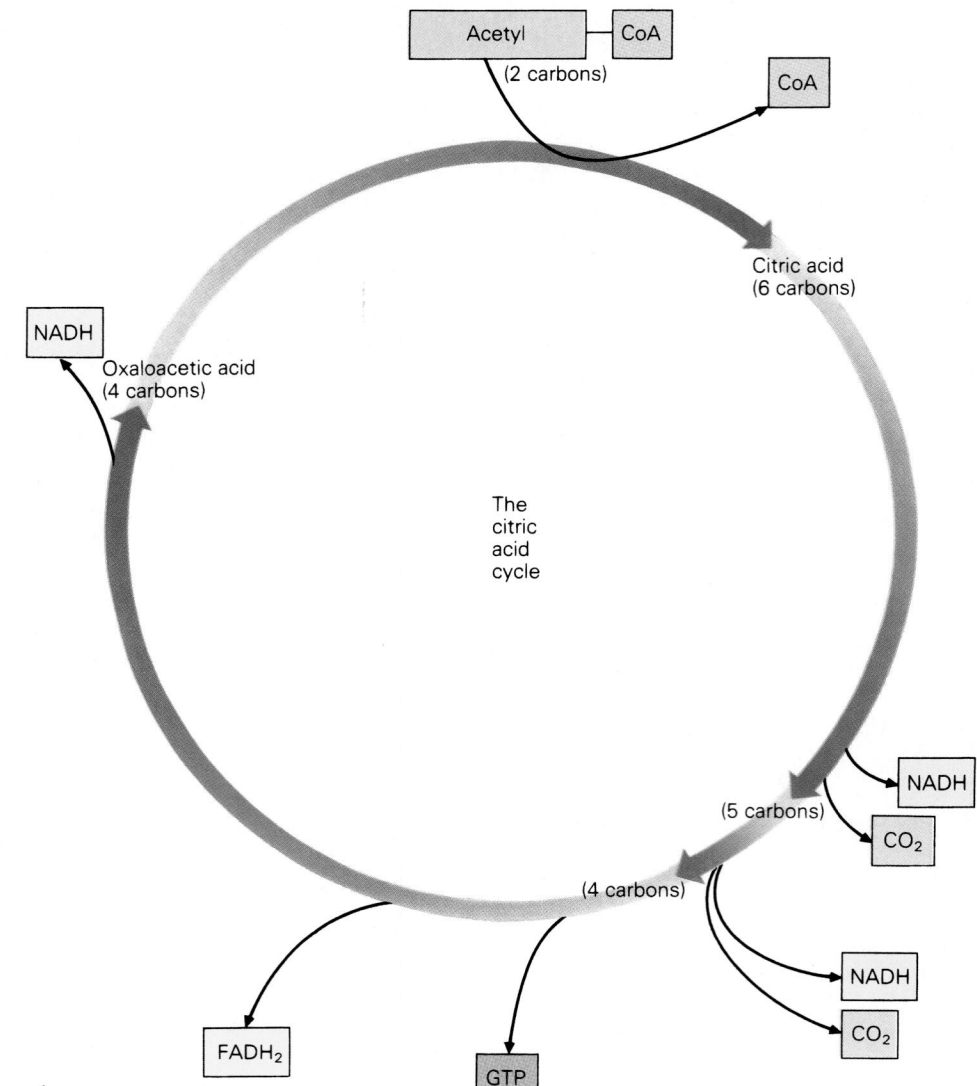

Figure 6–15
A summary of the citric acid cycle.

These steps are summarized in a simplified view of the citric acid cycle in Figure 6–15.

The Reduction of NAD⁺ and FAD

The reactions of the citric acid cycle regenerate free coenzyme A and produce two molecules of CO_2 for every 2-carbon acetyl group that enters the cycle. The process can be summarized as:

$$CH_3CO—\text{coenzyme A} \rightarrow 2\ CO_2 + \text{coenzyme A}$$

The carbon atoms lost in one round of the cycle, in the form of CO_2 exhaled by the lungs, are those that entered as an acetyl group in a previous round. Since CO_2 is much more oxidized than the $CH_3CO—$ (acetyl) group, the latter must become oxidized during the cycle. These oxidation steps are coupled to the reduction of the electron-carrying coenzymes NAD⁺ and FAD. There are four such reactions that generate, from each acetyl group entering the cycle, a total of three molecules of NADH and one molecule of $FADH_2$. In addition, one high-energy phosphate molecule (GTP) is formed by substrate-level phosphorylation (see Fig. 6–15).

Each reaction of the citric acid cycle is shown in detail in Figure 6–16. There are nine consecutive steps beginning with the entry of an acetyl group (Step 1). In Steps 2 and 3, H and OH are interchanged to form **isocitrate**. Steps 4 and 5 are oxidation-reduction reactions. The net result is that the 6-carbon isocitrate is oxidized to the 4-carbon **succinyl coenzyme A** with the release of two CO_2 molecules. At the same time, two molecules of NAD^+ are reduced to NADH. Step 4 is catalyzed by **isocitrate dehydrogenase**. Step 6 is a substrate-level phosphorylation reaction that utilizes the free energy released by breakdown of succinyl coenzyme A to drive phosphorylation of GDP to GTP:

$$\text{succinyl coenzyme A} + \text{GDP} + \text{Pi} \longrightarrow$$
$$\text{succinate} + \text{GTP} + \text{coenzyme A}$$

This is the only reaction in the cycle that produces a high-energy phosphate compound directly. The

GTP formed in this reaction is used to phosphorylate ADP to produce ATP:

$$\text{ADP} + \text{GTP} \longrightarrow \text{ATP} + \text{GDP}$$

The next three reactions involve the 4-carbon compounds of the cycle and culminate in the regeneration of oxaloacetic acid. There are two oxidation reactions (Steps 7 and 9) coupled to the reduction of one molecule of FAD to $FADH_2$ and one molecule of NAD^+ to NADH.

Molecular oxygen is not required for the reactions of the citric acid cycle, but it is essential for oxidation of the reduced coenzymes by the mitochondrial electron transport chain (see Fig. 6–2). A supply of oxidized coenzymes is needed for continued operation of the cycle. Hence the cycle cannot operate under anaerobic conditions because under

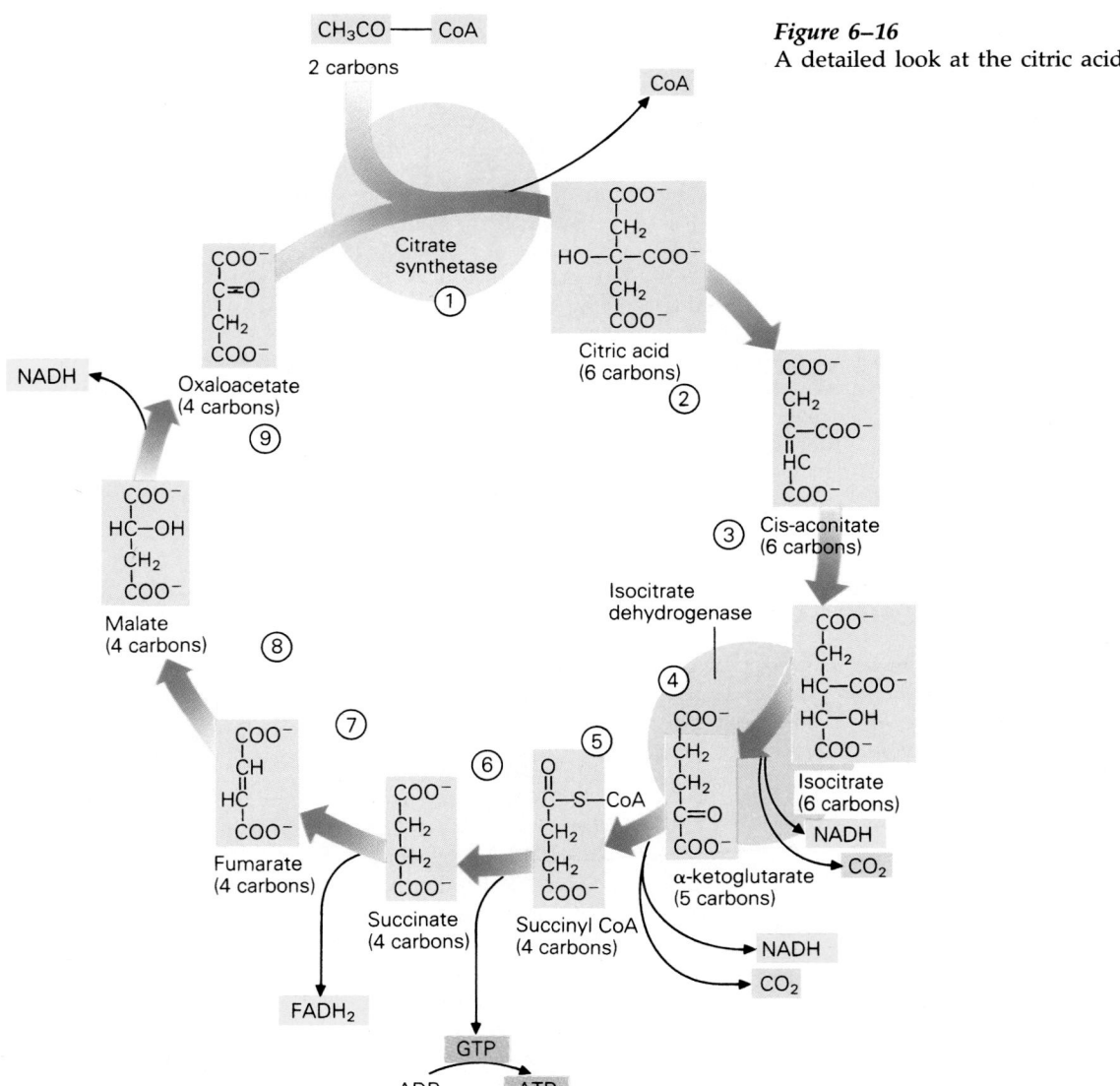

Figure 6–16
A detailed look at the citric acid cycle.

such conditions all the coenzyme molecules remain in a reduced form. In contrast, a supply of NAD^+ is always available for anaerobic glycolysis in the cytosol because the reduction of pyruvate to lactate is coupled to oxidation of NADH to NAD^+ (see Fig. 6–13).

Allosteric Inhibition in the Citric Acid Cycle

The citric acid cycle contains several control points. The cycle is slowed when the ATP supply is adequate and accelerated when the ATP supply is low. When the supply of ATP is high, ATP itself acts as an allosteric inhibitor of citrate synthetase to reduce the rate of formation of citrate (see Step 1, Fig. 6–16) and slow down the cycle. When the supply of ATP is low and the concentration of ADP relatively high, ADP acts as an allosteric activator of isocitrate dehydrogenase to increase the rate of production of alpha-ketoglutarate (Step 4) and speed up the cycle. These and other mechanisms cooperate to match precisely the rate of the reactions of the cycle to the requirement of the cell for ATP.

The reactions of the citric acid cycle generate two molecules of ATP (from GTP) when the two acetyl groups from one glucose molecule are processed. There is also a net gain of two molecules of ATP during glycolysis (see Fig. 6–10), which gives a

R E S E A R C H   F O C U S

The Role of Oxygen in Tissue Injury

In most mammalian cells, oxygen is used largely by the mitochondria, the site where most of the ATP is synthesized. Oxygen is essential for efficient production of ATP, and during this production process **(cellular respiration)** most of the oxygen is converted to water. A small but significant amount (1% to 2%) of the oxygen consumed during cellular respiration is converted to highly reactive **oxygen intermediates**. These intermediates can injure and even kill cells, and most cells have developed protective mechanisms. The margin of safety is a narrow one, however, and the cellular defenses can be overwhelmed if the generation of oxygen intermediates is increased. It has been proposed that reactive oxygen intermediates may contribute to many illnesses, including inflammatory diseases, toxic reactions, cancer, and aging disorders.

Molecular oxygen is the terminal electron acceptor in the electron transport chain of the inner mitochondrial membrane (Fig. 6–17). The transfer of 4 electrons (and 4 protons) to molecular oxygen produces complete reduction to water, a safe product:

$$O_2 + 4e^- + 4 H^+ \longrightarrow 2 H_2O$$

In contrast, partial reduction generates hazardous compounds. For example, transfer of a single electron to oxygen produces the **superoxide anion:**

$$O_2 + e^- \longrightarrow O_2^-$$

Superoxide is highly reactive because addition of the electron to one of the oxygen atoms gives the oxygen an unpaired electron in its outer electron shell (see Chapter 2).

One of the mechanisms for protecting the cell from the action of superoxide anions involves the enzyme **superoxide dismutase**, which is present in all aerobic cells. This enzyme is referred to as a **scavenger** of oxygen intermediates because it converts the superoxide anion to hydrogen peroxide and oxygen, as follows:

$$2 O_2^- + 2 H^+ \longrightarrow H_2O_2 + O_2$$

The hydrogen peroxide produced by this reaction is scavenged by the enzyme **catalase,** present in peroxisomes (Chapter 3). The products of catalase action are water and oxygen:

$$2 H_2O_2 \longrightarrow 2 H_2O + O_2$$

Peroxidase, an enzyme less common than catalase, scavenges hydrogen peroxide by using a reducing agent (AH_2):

$$H_2O_2 + AH_2 \longrightarrow 2 H_2O + A$$

Hydrogen peroxide also may create the extraordinarily reactive **hydroxyl radical,** usually written as ·OH. This nomenclature denotes that there is an unpaired electron in the outer shell of the oxygen atom but that, unlike superoxide, there is no excess of electrons compared to protons, and hence the structure has no overall negative charge. The hydroxyl radical differs from the hydroxide ion because although hydroxide has an excess of electrons (hence an overall negative charge) there are no unpaired electrons.

total of four ATP molecules synthesized directly during the complete oxidation of glucose ($C_6H_{12}O_6$) to six molecules of CO_2. Most of the chemical energy of the glucose molecule is conserved, by transfer of electrons, in the reduced coenzymes. Two NADH are produced by aerobic glycolysis of glucose to two pyruvates (see Fig. 6–10). Two more NADH are produced during formation of two acetyl coenzyme A molecules (see Fig. 6–14), and six NADH are produced when two acetyl groups are fed into the citric acid cycle (see Fig. 6–15). Thus, the complete oxidation of a single molecule of glucose generates 10 molecules of NADH; in addition, two molecules of $FADH_2$ are formed. The subsequent transfer of electrons from these reduced coenzyme molecules to O_2 drives the synthesis of 32 more molecules of ATP by **oxidative phosphorylation.**

The Electron Transport Chain: From Reduced Coenzymes to ATP

The transfer of electrons from reduced coenzymes to O_2 takes place on the electron transport chain, which is an integral part of the inner membrane of the mitochondrion. These oxidation steps (loss of electrons) are coupled to phosphorylation of ADP to

---

Production of hydroxyl radicals from hydrogen peroxide can arise from interaction with compounds containing iron, as follows:

$$H_2O_2 + Fe^{2+} \longrightarrow \cdot OH + OH^- + Fe^{3+}$$

In the normal liver cell, the average concentrations of these toxic molecules have been estimated to be 10^{-12} to 10^{-11} M for O_2^-, 10^{-9} to 10^{-7} M for H_2O_2, and 10^{-15} to 10^{-14} M for $\cdot OH$. Both hydrogen peroxide and superoxide can diffuse from their intracellular production sites and affect the entire cell. Hydrogen peroxide can also cross biological membranes. In contrast, the highly reactive $\cdot OH$ probably only interacts near its production site.

Unless destroyed by scavengers, these oxygen-derived molecules can injure and kill cells. They damage cell membranes by reacting with phospholipids, destroy enzymes and other proteins by oxidizing sulfhydryl groups, and induce breaks in the strands of DNA molecules.

Reactive oxygen intermediates have been shown to play a role in the development of injury to several different tissues, including the kidney. Studies on the cellular mechanisms of **acute renal failure**, defined as an abrupt decrease in kidney function, have revealed the possibility of increased production of reactive oxygen intermediates. A small proportion of patients who receive **gentamicin,** an antibiotic used to treat infection, develop acute renal failure. Studies in animals have revealed that gentamicin increases the kidney's production of superoxide and hydrogen peroxide. Furthermore, the presence of scavengers blocks

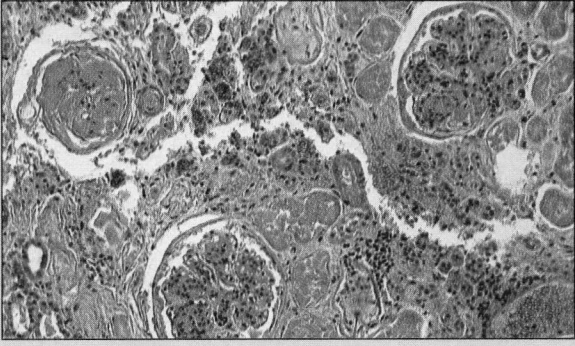

A micrograph of kidney cells from a diabetic with renal damage. ($\times$ 100.) (© Ida Wyman/Phototake.)

the development of renal failure during gentamicin treatment. Thus, reactive oxygen intermediates appear to play an important role in acute renal failure induced by gentamicin.

Oxygen-derived radicals are not always bad news. Some cells in the body make use of these intermediates to carry out their normal functions. For example, **phagocytic cells,** an important part of the body's defense against invading organisms, make use of the toxic action of these intermediates to kill the invaders. The generation of reactive oxygen intermediates is markedly enhanced in phagocytes that have surrounded and engulfed a foreign organism. Furthermore, defective production of reactive oxygen intermediates by phagocytic cells results in the body's increased susceptibility to infections.

form ATP. The process produces most of the ATP supply for the cell and, since molecular oxygen is an essential reactant, it is an aerobic pathway called **oxidative phosphorylation.** The overall process by which oxidative phosphorylation generates ATP is called the **chemiosmotic synthesis of ATP.**

Cytochromes: Electron Carriers

The driving forces for oxidative phosphorylation are the negative redox potentials of NADH and $FADH_2$, which allow these molecules to readily donate electrons to proteins serving as **electron carriers** in the electron transport chain. Many of these electron carriers are proteins called **cytochromes,** and their redox potentials increase sequentially from the beginning to the end of the chain.

The importance of this increase in redox potentials is that each cytochrome has a greater affinity for electrons than the previous one in the chain. This increase facilitates the passage of electrons from one cytochrome to the next and, ultimately, to oxygen. Each cytochrome becomes reduced when it accepts electrons and oxidized when it donates the electrons to the next cytochrome. The components of the electron transport chain are organized into three major complexes (Fig. 6–17).

The flow of electrons from NADH to molecular oxygen represents an oxidation-reduction reaction that can be summarized as:

$$1/2 \ O_2 + NADH + H^+ \longrightarrow H_2O + NAD^+$$

The standard free energy change $\Delta G°$ for this reaction is −53 kcal/mol. It is an exergonic reaction; the free energy released is used to drive the endergonic phosphorylation of ADP to produce ATP (see Fig. 6–4). Under standard conditions, about 7 kcal/mol of free energy can be stored in one ATP molecule, so the free energy released by the above reaction could be stored in seven molecules of ATP.

Figure 6–17
The oxidation of NADH and $FADH_2$ by the electron transport chain. Each of three electron transport complexes has one phosphorylation site where enough free energy is released to synthesize one molecule of ATP. The ATP is not synthesized directly. The free energy released at each site is used to pump protons (H^+) across the inner mitochondrial membrane.

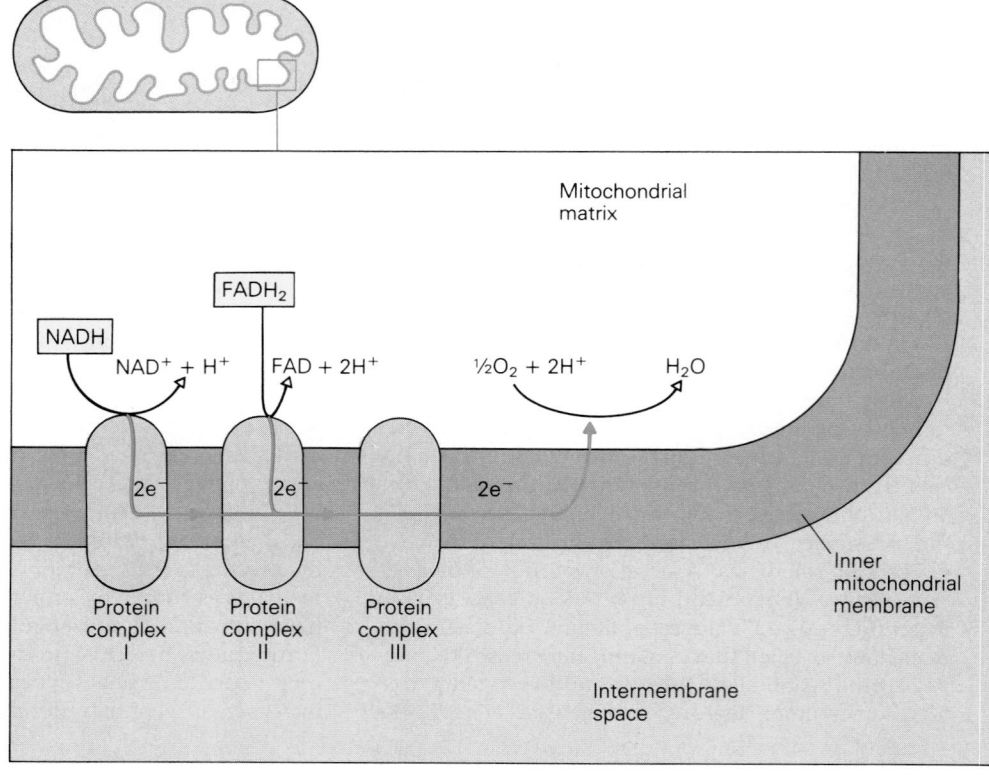

Phosphorylation Sites

The electron transport chain ensures that the free energy is not released all at once, but even so, not all of this free energy can be used to synthesize ATP. This is because under physiological conditions there are only three sites in the chain where the transfer of electrons releases enough free energy to synthesize a molecule of ATP. These sites are termed **phosphorylation sites,** and one is present in each of the three electron transport complexes (see Fig. 6–17). Thus, in the cell, three molecules of ATP are synthesized for each molecule of NADH that is oxidized in the electron transport chain.

The redox potential of $FADH_2$ is less negative than that of NADH, and it cannot donate electrons to the beginning of the electron transport chain in Complex I. $FADH_2$ donates its electrons to the chain at a point between Complexes I and II; they do not pass through the first phosphorylation site (see Fig. 6–17). Just two ATP molecules are synthesized when each molecule of $FADH_2$ is oxidized.

The Flow of Protons: The Chemiosmotic Mechanism

The free energy released at the phosphorylation sites is harnessed by the electron transport complexes in order to pump protons (H^+) from the mitochondrial matrix to the intermembrane space (Fig. 6–18). The movement of positively charged protons, unaccompanied by negatively charged anions, sets up an electrical gradient across the inner mitochon-drial membrane. In addition, the accumulation of protons in the intermembrane space builds a chemical gradient. The result of the net electrochemical gradient is a passive flow of protons from the intermembrane space back to the matrix. This flow can occur only through a specific protein in the inner membrane that bears the enzyme **ATP synthetase.** Just as the flow of sodium ions down an electrochemical gradient can be used to do "work" such as moving another solute against its electrochemical gradient (secondary active transport, Chapter 4), the flow of protons is coupled by ATP synthetase to drive the endergonic synthesis of ATP from ADP and a phosphate group (see Fig. 6–18). In other words, the ATP synthetase transforms the energy of the proton electrochemical gradient into chemical energy in the form of ATP.

The proton pumps use the free energy released by electron transport to maintain the proton gradient, which in turn maintains the flow of protons through ATP synthetase. This is known as the **chemiosmotic mechanism** of ATP synthesis because it links chemical reactions to transport processes. The newly formed ATP is transferred out of the mitochondrion and used to drive metabolic reactions in the rest of the cell.

A "balance sheet" summarizing the pathways involved and the yield of 36 ATP molecules from complete oxidation of one glucose molecule is presented in Figure 6–19. Only one additional comment is needed: the inner mitochondrial membrane is impermeable to NADH (and NAD^+), so the cytosolic "pool" of this coenzyme never mixes with the

Figure 6–18
The chemiosmotic mechanism of ATP synthesis. Protons (H^+) are pumped from the matrix to the intermembrane space, then returned to the matrix by passive flow down the electrochemical gradient. ATP synthetase uses the passive proton flow to drive the synthesis of ATP.

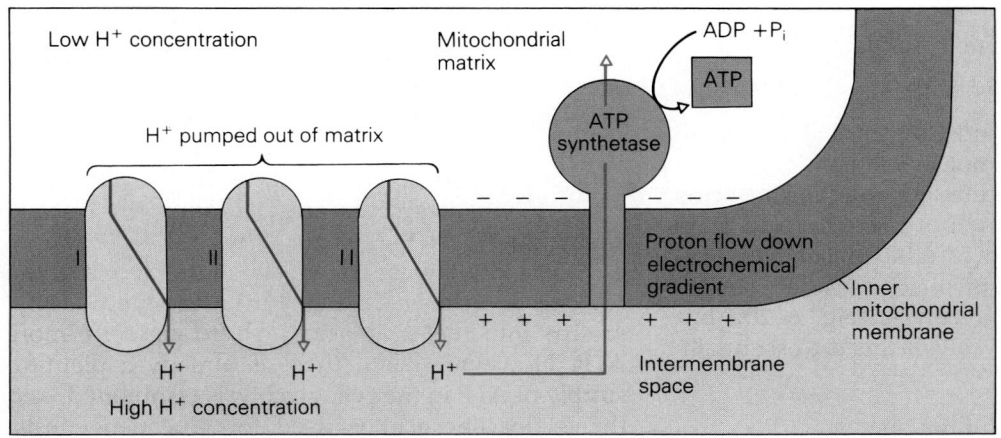

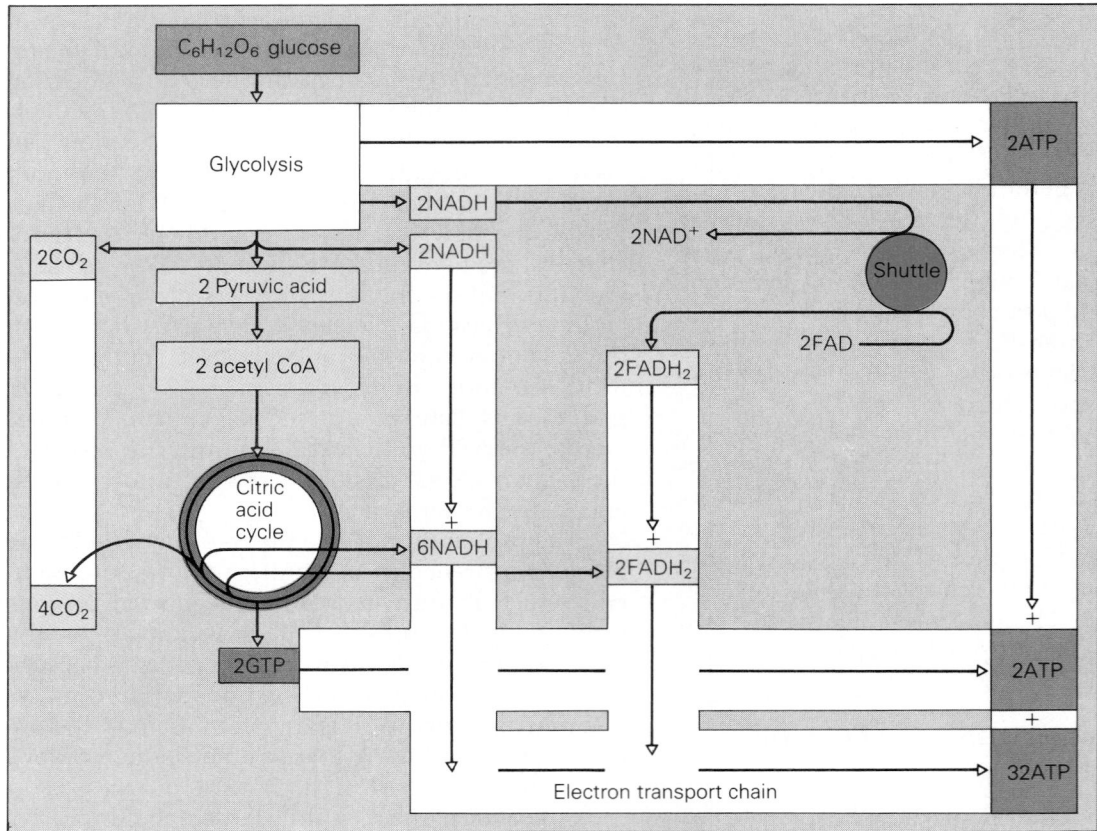

Figure 6–19
The production of 36 ATP molecules by complete oxidation of 1 glucose molecule.

mitochondrial pool. Cytosolic NADH produced by glycolysis is oxidized by a shuttle mechanism that, in many cell types, transfers electrons and protons to FAD inside the mitochondrion. As a result of this mechanism, each cytosolic NADH molecule ultimately produces only two ATP molecules.

Gluconeogenesis: The Synthesis of Glucose

Gluconeogenesis is the pathway by which glucose can be synthesized from molecules that are not carbohydrates. These molecules include **amino acids** produced by the breakdown of proteins, and **glycerol** from the breakdown of fats (triacylglycerols). Many of the reactions of gluconeogenesis are simply the reverse of those of glycolysis (Fig. 6–20), but there are three steps of glycolysis that are essentially irreversible:

   1. Glucose $\rightarrow$ Glucose-6-Phosphate

   2. Fructose-6-Phosphate $\rightarrow$ Fructose-1, 6-Bisphosphate
   3. Phosphoenolpyruvate $\rightarrow$ Pyruvate

These glycolytic reactions are bypassed with a different reaction mechanism during gluconeogenesis.

Lactate and some amino acids (e.g., alanine, glycine, serine) enter the gluconeogenic pathway by conversion to pyruvate. Other amino acids (e.g., aspartic acid, asparagine) enter in the form of oxaloacetic acid (see Fig. 6–20). Note that oxaloacetic acid, an intermediate in the synthesis of phosphoenolpyruvate from pyruvic acid during gluconeogenesis, is also an intermediate in the citric acid cycle, where it accepts acetyl groups from acetyl coenzyme A (see Fig. 6–15). When the cell's supply of ATP is low, gluconeogenesis is inhibited, and the oxaloacetic acid is used primarily to funnel acetyl groups into the citric acid cycle to generate more ATP. However, when there is already a plentiful supply of ATP in the cell, glycolysis is inhibited, and the oxaloacetic acid is used for gluconeogenesis.

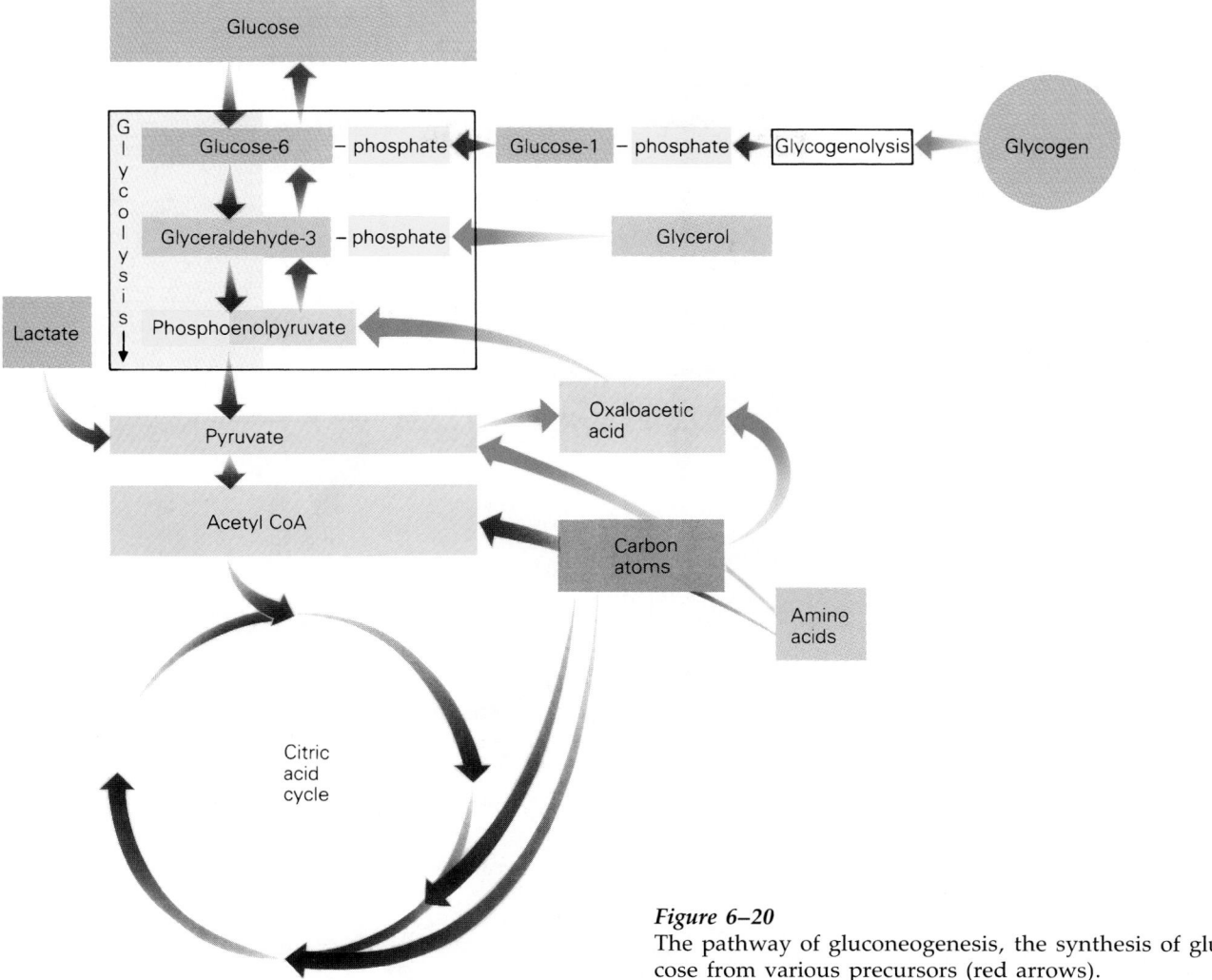

Figure 6–20
The pathway of gluconeogenesis, the synthesis of glucose from various precursors (red arrows).

This is economical for the cell because ATP, and also GTP, are needed to drive the synthesis of glucose from pyruvate:

$$2 \text{ pyruvate} + 4 \text{ ATP} + 2 \text{ GTP} + 2 \text{ NADH} \longrightarrow$$

$$\text{glucose} + 4 \text{ ADP} + 2 \text{ GDP} + 2 \text{ NAD}^+$$

Glycerol from fat breakdown feeds into the gluconeogenic pathway at the level of glyceraldehyde-3-phosphate. Fat breakdown also produces fatty acids, which are degraded to acetyl coenzyme A (see later in this chapter). However, acetyl coenzyme A cannot be converted to pyruvate, so no glucose is produced from fatty acids. Glycogen, the storage form of glucose in tissues such as the liver, is broken down in the process of **glycogenolysis** to units of glucose-1-phosphate, which enter gluconeogenesis after conversion to glucose-6-phosphate (see Fig. 6–20). In this way, liver glycogen is converted to glucose that can be supplied via the blood-

stream to other tissues needing a source of energy. As with lactate and the amino acids, both glycerol and glycogen are used for glycolysis when the ATP supply is low.

The Pentose Phosphate Pathway: An Alternative Glucose Oxidation Path

The purpose of glucose oxidation, via glycolysis and the citric acid cycle, is to produce ATP as an energy source for biosynthesis. Another pathway for the oxidation of glucose exists in the cytosol, called the **pentose phosphate pathway** or the **hexose monophosphate shunt.** The main purpose of the pentose phosphate pathway is to generate NADPH, the reduced form of the coenzyme $NADP^+$ (see Fig. 6–8), to drive the oxidation-reduction reactions in various biosynthetic pathways such as the synthesis of fatty acids and steroids. It also provides the pentose sug-

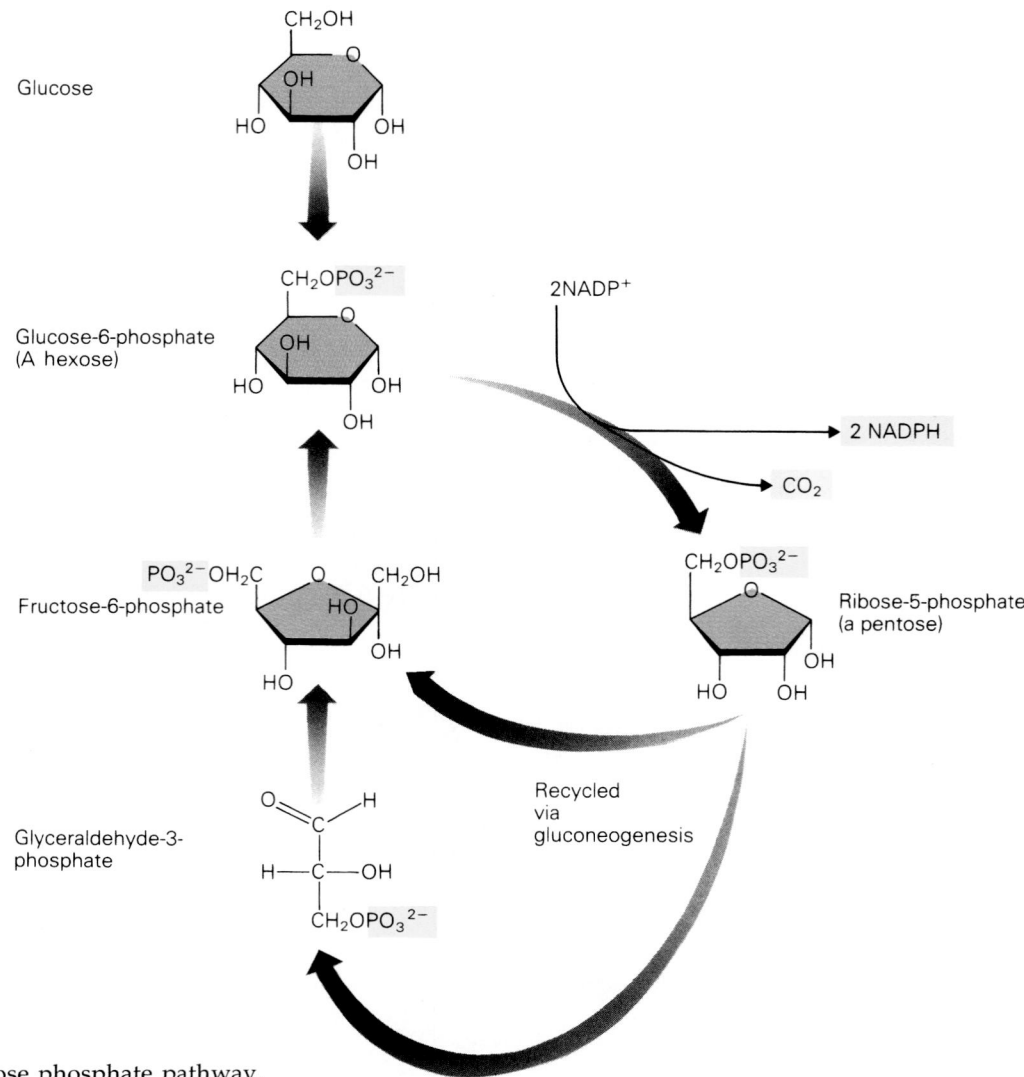

Figure 6–21
A summary of the pentose phosphate pathway.

ars that are essential components of nucleotides and nucleic acids (see Chapter 2). This process uses different enzymes from those used in glycolysis.

The pathway starts with glucose-6-phosphate, a hexose. After several reactions, the final result is the reduction of two molecules of $NADP^+$ to NADPH and the formation of one molecule of ribose-5-phosphate, a pentose (Fig. 6–21). The carbon atom removed from glucose-6-phosphate is lost as CO_2. Excess ribose-5-phosphate not needed for biosynthetic reactions is recycled by conversion to fructose-6-phosphate and glyceraldehyde-3-phosphate, two of the intermediates of the glycolytic pathway. These intermediates are used to synthesize glucose-6-phosphate by the enzymes of the gluconeogenic pathway (see Fig. 6–21), and the cycle begins again. Six rounds of the cycle result in the complete oxida-

tion of one molecule of glucose-6-phosphate to six molecules of CO_2 with the generation of NADPH:

$$\text{glucose-6-phosphate} + 12\ NADP^+ \longrightarrow$$
$$6\ CO_2 + 12\ NADPH$$

In contrast to aerobic oxidation of glucose, the pentose-phosphate pathway does not use ATP, oxygen, or the citric acid cycle. It does not produce ATP, and the only electron acceptor involved is $NADP^+$. Which of these pathways is used for glucose catabolism seems to depend on whether or not the cell is heavily engaged in biosynthesis. Skeletal muscle, for example, catabolizes glucose solely via glycolysis and the citric acid cycle. Adipose and liver tissue, on the other hand, catabolize a large fraction of glucose via the pentose-phosphate pathway. These tissues are important sites of fatty acid synthesis, a reaction pathway that consumes a large amount of NADPH and is located in the cytosol where the pentose phosphate pathway also occurs.

Note the markedly different roles of NADH and NADPH. NADH is oxidized by the electron transport chain to generate ATP, whereas NADPH is oxidized directly in biosynthetic reactions.

Metabolism of Fats

Fats stored in adipose tissue provide a rich energy source for the body because they can be oxidized to produce ATP. The body can store more fat than carbohydrate, and, as we shall see, much more energy can be obtained from fats than from carbohydrates.

The amount of ATP produced by a given weight of fat exceeds that produced by the same weight of carbohydrates. Since a mobile animal, unlike a stationary plant, must carry its energy store with it, the amount of ATP available per unit weight of the energy store is an important factor. This is a major reason that animals use fat as an energy store.

Catabolism of a neutral fat, or **triacylglycerol,** molecule begins with the release of the glycerol backbone and the three long-chain fatty acids. Glycerol is readily converted to glyceraldehyde-3-phosphate (see Fig. 6–20), which can be used for either glycolysis or gluconeogenesis. The fate of the fatty acids is illustrated in Figure 6–22, using palmitic acid as an example. Coenzyme A is attached to the carboxyl (—COOH) group and, in a series of com-

Figure 6–22
The catabolism of fatty acids to acetyl coenzyme A (acetyl CoA). The long hydrocarbon chains of fatty acids are represented by the symbols R_1, R_2, and R_3.

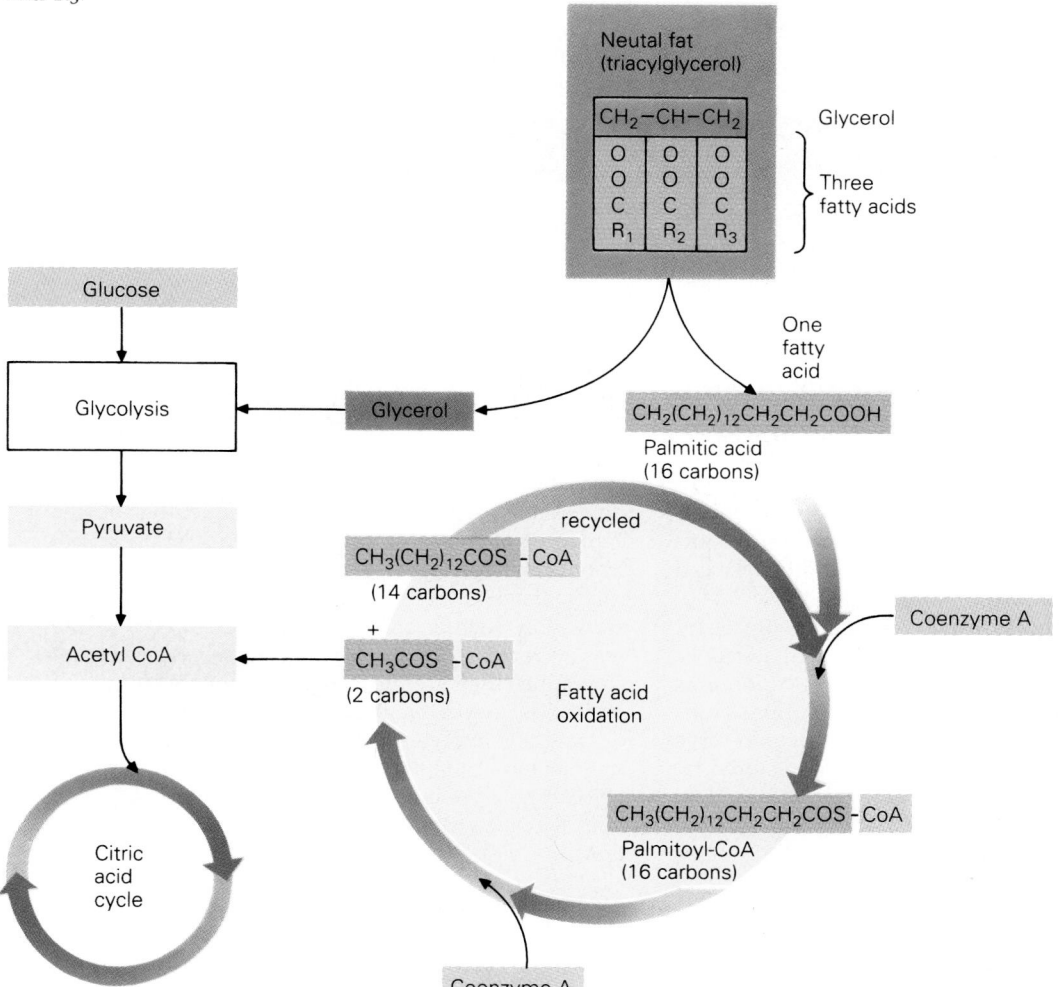

plex reactions within the matrix of a mitochondrion, a fragment of two carbon atoms is removed. This fragment, which also contains coenzyme A, is actually a molecule of acetyl coenzyme A (see Fig. 6–22). The 16-carbon chain of palmitic acid is reduced to 14 carbons by the removal of acetyl coenzyme A. The 14-carbon fatty acid is recycled through the same series of reactions and another two carbon atoms are removed as acetyl coenzyme A. In this way, the original long-chain fatty acid is converted into several molecules of acetyl coenzyme A. While NADH and $FADH_2$ are produced by this series of reactions, no ATP is generated. All the ATP derived from fatty acid catabolism is produced when the acetyl coenzyme A groups are fed into the citric acid cycle (see Fig. 6–22). The catabolism of palmitoyl-coenzyme A, the activated form of palmitic acid, can be summarized as follows:

$$palmitoyl\text{-}coenzyme\ A + 7\ FAD + 7\ NAD^+$$
$$+ 7\ coenzyme\ A \longrightarrow$$
$$8\ acetyl\ coenzyme\ A + 7\ FADH_2 + 7\ NADH$$

When this yield of acetyl coenzyme A and reduced coenzymes is processed via the citric acid cycle and electron transport, it can be calculated that 131 ATP molecules will be generated. Since two high-energy phosphate bonds are used to activate the fatty acid (attachment of coenzyme A), the net gain from oxidation of 1 molecule of palmitic acid is 129 molecules of ATP. This exceeds by far the 36 ATP molecules generated by oxidation of one molecule of glucose, reinforcing the point that fats are a much richer energy source than carbohydrates.

Fatty acids are synthesized in a different subcellular compartment (the cytosol) and by a different pathway from that used for their catabolism. The pathway begins with acetyl coenzyme A, and the carbon chain is elongated by a succession of reactions that add two carbon atoms at a time. The energy source for these reactions is ATP, with NADPH from the pentose-phosphate pathway providing reducing "power" for oxidation-reduction reactions. Synthesis of palmitic acid can be summarized by these equations:

$$8\ acetyl\ coenzyme\ A + 7\ ATP + 14\ NADPH \longrightarrow$$
$$palmitic\ acid + 7\ ADP + 14\ NADP^+ +$$
$$8\ coenzyme\ A$$

C L I N I C A L   F O C U S

Starvation

A well-nourished human male weighing 70 kg has considerable fuel reserves amounting to a total of about 160,600 kcal. Most (84%) of these reserves are fat, and the remainder are protein (15%) and carbohydrate (1%). If he were to remain completely inactive, the energy requirement to sustain life for 24 hours would be 1,600 kcal. Thus, the stored fuels could meet his energy needs for about 3 months during complete starvation. This time would be reduced if any activity occurred.

The carbohydrate reserve is mainly glycogen, which is converted to glucose, the principal metabolic fuel for the brain. The brain requires a continuous supply of glucose, and metabolic adaptations must occur during starvation because the glycogen reserve is emptied within 1 day. The fatty acids of stored fat molecules (triacylglycerols) cannot be converted to glucose. Initially, the brain receives sufficient glucose through gluconeogenesis, which uses amino acids produced by breakdown of protein, principally in muscle tissue. Serious depletion of muscle tissue would impair movement, so an additional metabolic adaptation occurs in order to diminish the requirement for glucose. Triacylglycerols are mobilized to yield acetyl coenzyme A, which cannot enter the citric acid cycle because oxaloacetate, the entry point, is depleted by gluconeogenesis. Consequently, there is a marked increase in the synthesis of *ketone bodies* from acetyl coenzyme A by the liver. Ketone bodies can be used as fuel by the brain. After three days of starvation, ketone bodies provide about 30% of the energy required by the brain; after several weeks of starvation, they are the major metabolic fuels for the brain. The need for glucose is reduced and muscle degradation is decreased. It is clear from the nature of the metabolic adaptations that the size of the triacylglycerol reserves is a major factor in determining how long a person can survive during complete starvation.

Metabolism of Proteins

The building blocks of proteins are amino acids. In contrast to glucose and fatty acids, which can be stored until needed, excess amino acids cannot be stored and instead are used as a metabolic fuel. Many of the amino acids are catabolized by removal of the amino group ($—NH_2$) in the form of ammonia, a process known as **deamination**. Ammonia is potentially toxic and so is converted to urea and excreted from the body in the urine (Fig. 6–23).

The carbon atoms of the amino acids can enter the pathway of glucose oxidation, where they are converted to CO_2 and water with the generation of ATP. Some amino acids enter the pathway as pyruvate, some as acetyl coenzyme A, and others as intermediates in the citric acid cycle (see Fig. 6–23).

About half of the basic set of 20 amino acids required by humans are described as **essential amino acids** because they cannot be synthesized; they must be ingested as food. The other needed amino acids are termed **nonessential amino acids** because they can be synthesized by relatively simple pathways.

Some of the steps by which amino acids are incorporated into a polypeptide chain during protein synthesis require a source of energy. ATP provides the energy needed for attachment of each amino acid to its tRNA molecule, and GTP is the energy source required to move a ribosome from one mRNA codon to the next (see Chapter 5).

In addition to their use in protein synthesis, certain amino acids are the precursors for a number of important molecules. These molecules include the **purines** and **pyrimidines** used for synthesis of DNA and RNA. **Histamine** is a powerful locally acting vasodilator synthesized from the amino acid histidine. The hormones **epinephrine** and **thyroxine** are formed from the amino acid tyrosine, and the nicotinamide ring of NAD^+ is derived from the amino acid tryptophan.

Figure 6–23
During catabolism, carbon atoms from amino acids are used to form pyruvate, acetyl groups, and various products of the citric acid cycle. Nitrogen atoms from amino acids are eventually excreted from the body in the urine.

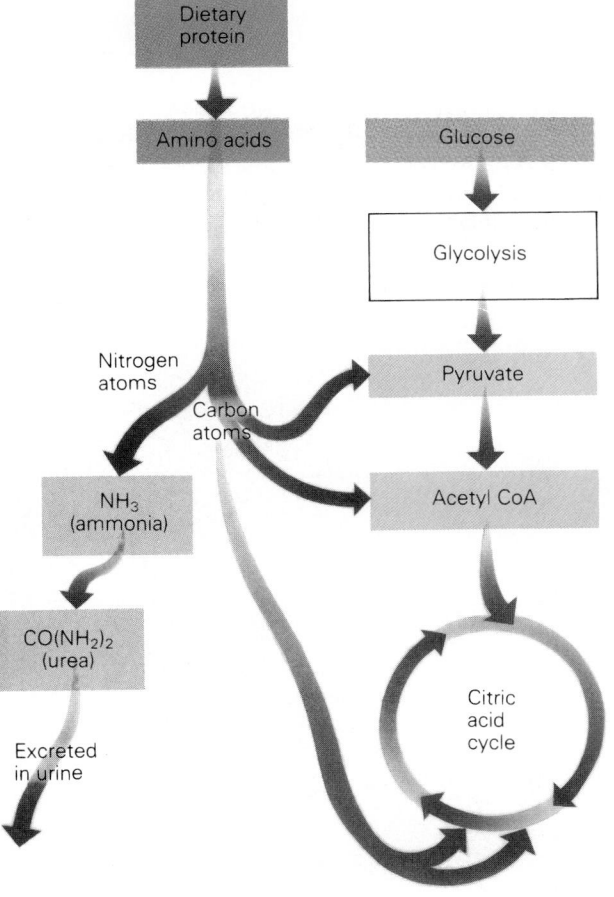

Acetyl Coenzyme A: Mediating Between Catabolism and Biosynthesis

It should be possible to appreciate from this discussion of cellular metabolism that the citric acid cycle, together with mitochondrial electron transport, is the principal pathway for supplying the cell with ATP. Very little of the chemical energy present in the macromolecules of the cell is made available until small units of these macromolecules are fed into the citric acid cycle (Fig. 6–24).

Although the carbon fragments of several amino acids can enter the citric acid cycle directly, almost all the other intermediate compounds produced by catabolism are fed into the cycle as acetyl groups coupled to coenzyme A. The acetyl groups are also the 2-carbon precursors used in the synthesis of fatty acids and cholesterol. Steroid hormones are derived from cholesterol. Thus, coenzyme A is required not only to derive energy from fuel molecules, but also to provide the many carbon atoms required for biosynthesis of a large variety of or-

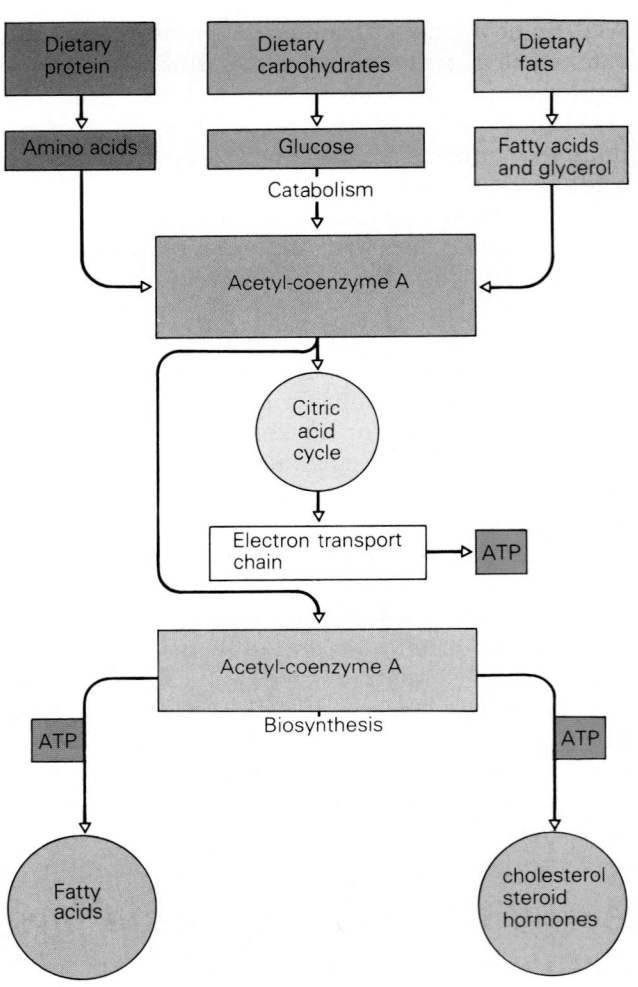

ganic molecules. Carbon atoms derived from catabolism of fats and proteins can be shunted through acetyl coenzyme A and utilized in biosynthesis (see Fig. 6–24), depending on the metabolic demands at any given time. In other words, acetyl coenzyme A provides a place for the different metabolic pathways of the major macromolecules of the cell to intersect. Note that in mammals, acetyl coenzyme A cannot be converted to pyruvate; hence fats cannot be converted to carbohydrates.

Figure 6–24
An illustration of the central role of acetyl coenzyme A in cellular metabolism of mammals.

SUMMARY

Basic Principles of Energy Transformation in the Cell

Plants transform radiant energy from the sun into carbohydrates, a form of chemical energy that animals can use.

Using catabolism, animals transform the chemical energy in carbohydrates into ATP. They use ATP to drive biosynthetic reactions.

Energy is stored in an ATP molecule only for a short time, and cells are constantly losing energy to the environment in the form of heat.

Endergonic reactions in the cell are driven by free energy produced by exergonic reactions. Many cell reactions are reversible, driven one way or the other depending on which products the cell needs at a given time. When oxidation–reductions occur in the cell, electrons are donated and accepted in the form of hydrogen atoms. Most energy-releasing processes in animal cells are oxidative processes, removing hydrogen atoms from highly reduced, high-energy organic compounds (compounds that contain many hydrogen atoms). Molecules that accept the electrons released by oxidation are important to cell reactions. The coenzymes NAD^+ and FAD are intermediate electron acceptors, whereas oxygen is the final electron acceptor in the cell. High-energy phosphate compounds have a strong tendency to transfer their phosphate group to an acceptor molecule. All of the chemical reactions in the cell are catalyzed by enzymes. The catalytic activity of some enzymes depends on coenzymes. Some enzymes are subject to allosteric inhibition by the end products of the metabolic pathways they initiate.

An Overview of Catabolism: Dietary Proteins, Carbohydrates, and Fats

After digestion, dietary fats, proteins, and carbohydrates are carried to the cells by blood in the form of amino acids, monosaccharides such as glucose, and fatty acids.

The citric acid cycle and the electron transport chain take place in the mitochondria and make up the central metabolic pathway for all fuel molecules, whether proteins, carbohydrates, or fats.

The Metabolism of Carbohydrates: The Oxidation of Glucose

The reactions of glycolysis split one glucose molecule into two molecules of pyruvic acid and generate two molecules of ATP.

Each molecule of pyruvic acid is fed into the citric acid cycle as a 2-carbon acetyl group attached to coenzyme A. The reactions of the cycle generate three NADH, one $FADH_2$, and one GTP molecule.

The process of oxidative phosphorylation couples the oxidation of reduced coenzymes, by the electron transport chain, to the phosphorylation of ADP to form ATP. About 36 ATP molecules are generated from 1 glucose molecule by this process.

Glucose can by synthesized from molecules that are not carbohydrates by a series of reactions known as gluconeogenesis.

The pentose phosphate pathway is important in cells that are carrying out biosynthesis of molecules such as fatty acids. The pathway starts with glucose-6-phosphate and generates NADPH, which is used to drive redox reactions in biosynthetic pathways.

Metabolism of Fats

Fats are a much richer energy source than carbohydrates. About 129 molecules of ATP are generated from complete oxidation of 1 molecule of palmitic acid.

Metabolism of Proteins

The carbon atoms of amino acids are fed into the pathway for glucose oxidation; the nitrogen atoms are excreted in the urine in the form of urea. Amino acids are the precursors of proteins as well as other physiologically important molecules.

Acetyl Coenzyme A: Mediating Between Catabolism and Biosynthesis

The citric acid cyle and electron transport chain supply almost all of the ATP needed by the cell. Acetyl coenzyme A is the central molecule that connects the different metabolic pathways in the cell.

REVIEW QUESTIONS

Identify Each with a Term

1. The general term for the energy-requiring process by which simple molecules are assembled into new macromolecules needed by the cell.

2. A chemical reaction that cannot proceed without input of free energy.

3. The coenzyme used most frequently as a hydrogen acceptor in metabolic reactions.

4. The metabolic pathway by which glucose can be synthesized from molecules that are not carbohydrates.

5. The compound excreted in urine that represents the means for disposing of nitrogen produced by deamination of amino acids.

Define Each Term

6. Chemiosmotic mechanism
7. Pentose phosphate pathway
8. Oxidative phosphorylation
9. Allosteric inhibitor
10. High-energy phosphate compound

Choose the Correct Answer

11. When normal metabolic production cannot meet the demand for ATP in a skeletal muscle cell (e.g., during intense activity) new ATP is provided by another mechanism using:
    a. phosphocreatine
    b. GTP
    c. glucose-6-phosphate
    d. lactate dehydrogenase

12. Complete oxidation of carbohydrates such as glucose requires:
    a. oxidative phosphorylation
    b. glycolysis
    c. citric acid cycle
    d. all of the above

13. An organic molecule that functions in metabolism as an acceptor of acetyl groups is:
    a. NADP
    b. FAD
    c. coenzyme A
    d. pyruvate

14. How many molecules of ATP are synthesized for each molecule of NADH oxidized in the electron transport chain?
    a. 1
    b. 2
    c. 3
    d. 7

15. Which of the following are produced by catabolism of stored fat molecules (triacylglycerols) when the supply of oxaloacetate is depleted?
    a. Pyruvate
    b. Ketone bodies
    c. Glucose
    d. NADP

Calculate

16. How many molecules of ATP are gained from the complete metabolism of one molecule of pyruvic acid (A) under anaerobic conditions and (B) under aerobic conditions?

17. How many molecules of (A) NADH and (B) $FADH_2$ are gained from the complete oxidation of one molecule of palmitic acid, a 16-carbon fatty acid?

Answer Each Question in One or Two Sentences

18. What is the fundamental difference between the roles of the reduced coenzymes NADH and NADPH?

19. In what metabolic processes does acetyl coenzyme A participate?

20. How does ATP slow down the citric acid cycle?

21. What are the major steps in the process by which animals are able to use energy from the sun to synthesize the macromolecules they need?

22. Why do animals use fat as an energy store?

SUGGESTED READINGS

Brown, M.S., and Goldstein, J.L. "How LDL receptors influence cholesterol and atherosclerosis," *Scientific American*, 251:58–66, November 1984.

Capaldi, R.A. "Mitochondrial myopathies and respiratory chain proteins." *Trends in Biochemical Sciences*, 13:144–148, 1988.

Cross, C.E., Halliwell, B., Borish, E.T., Pryor, W.A., Ames, B., Saul, R., McCord, J., and Harman, D. "Oxygen radicals in human disease." *Annals of Internal Medicine*, 107:526–545, 1987.

Fersht, A. *Enzyme Structure and Mechanism*, 2nd ed. New York: Freeman, 1985.

Fothergill-Gilmore, L.A. "The evolution of the glycolytic pathway." *Trends in Biochemical Sciences*, 11:47–51, 1986.

Halliwell, B. "Oxidants in human disease: some new concepts." *FASEB Journal*, 1:358–364, 1987.

Hinkle, P.C., and McCarty, R.E. "How cells make ATP." *Scientific American*, 238:104–123, March 1978.

Koshland, D.E. "Switches, thresholds and ultrasensitivity." *Trends in Biochemical Sciences*, 12:225–229, 1987.

Pardee, A.B. "Molecular basis of biological regulation: origins of feedback inhibition and allostery." *Bioessays*, 2:37–40, 1985.

Racker, E. "From Pasteur to Mitchell: a hundred years of bioenergetics." *Federation Proceedings*, 39:210–215, 1980.

Schlenk, F. "The ancestry, birth and adolescence of adenosine triphosphate." *Trends in Biochemical Sciences*, 12:367–368, 1987.

Schulman, R.G. "NMR spectroscopy of living cells." *Scientific American*, 248:86–93, January 1983.

Shah, S.V. "Role of reactive oxygen metabolites in experimental glomerular disease." *Kidney International*, 35:1093–1106, 1989.

Smith, E.L., Hill, R.L., Lehman, I.R., Lefkowitz, R.J., Handler, P., and White, A. *Principles of Biochemistry: General Aspects*, 7th ed. New York: McGraw-Hill, 1983.

Stryer, L. *Biochemistry*, 3rd ed. San Francisco: Freeman, 1988.

KEY TERMS

adenosine triphosphate (ATP) (p. 221)
aerobic (p. 211)
anabolism (biosynthesis) (p. 200)
anaerobic (p. 213)
catabolism (p. 200)

citric acid cycle (Krebs cycle, tricarboxylic acid cycle) (p. 207)
coenzyme, coenzyme A (p. 207)
cytochrome (p. 220)
electron carrier (p. 220)

electron transport chain (p. 219)
essential amino acids (p. 227)
fat (p. 225)
gluconeogenesis (p. 222)
glycolysis (p. 210)
metabolic pathway (p. 223)

nonessential amino acids (p. 227)
oxidative phosphorylation (p. 205)
photosynthesis (p. 200)
standard free energy change (p. 204)

Physiological Control Systems

The following systems are discussed in Part II:

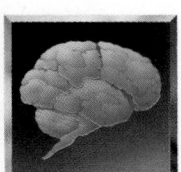

Nervous

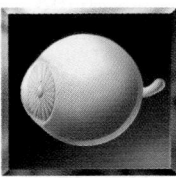

Sensory

Muscular

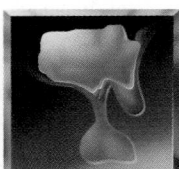

Thermoregulatory

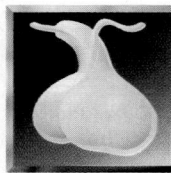

Endocrine

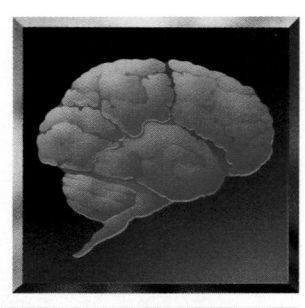

CHAPTER
7

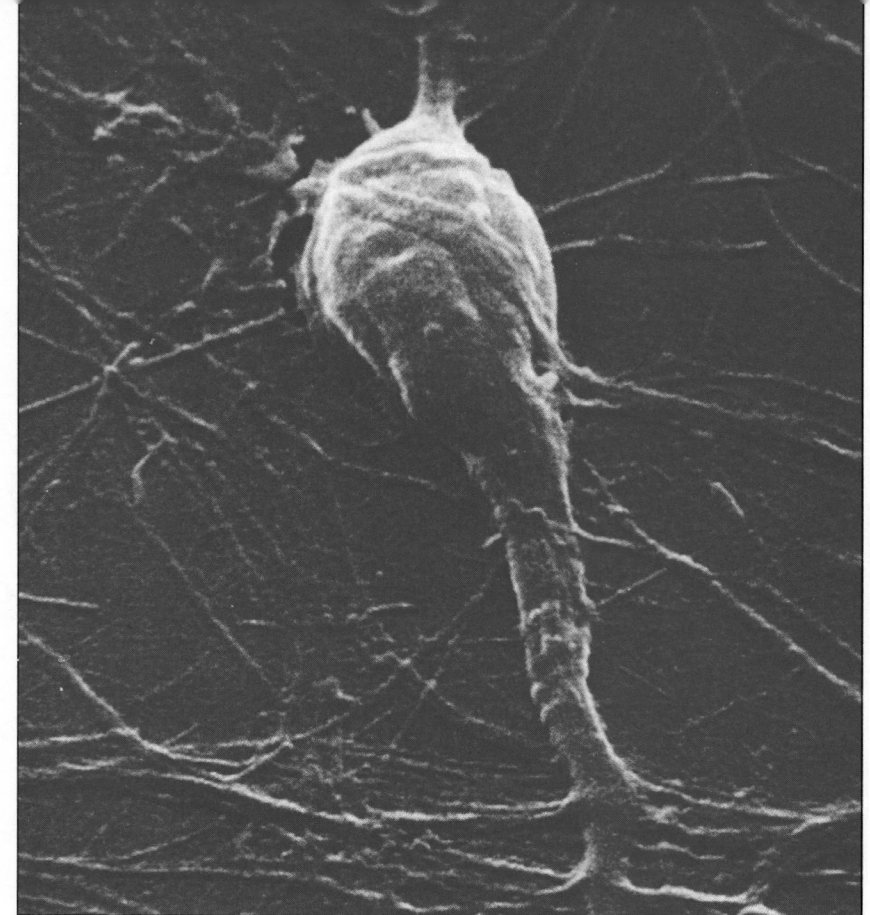

Functional Organization of the Nervous System

CHAPTER OUTLINE

The nervous system is one of the major control systems in the body. It coordinates the activity of other organ systems so that they operate in concert with one another. During exercise, for example, the nervous system activates selective muscle groups for movement; it causes the heart to pump faster to supply more blood to various parts of the body via the circulatory system, and it causes the respiratory system to increase the rate and depth of breathing to provide the body with more oxygen. These aspects of nervous system function represent the motor output to skeletal, cardiac, and smooth muscles of target organs.

The nervous system also detects and transmits sensory information from the external environment to the brain. We respond to our environment based on what we see, hear, touch, smell, and taste. In general, these sensations are processed in separate pathways within the nervous system. At higher levels, however, there is a convergence of various sensory inputs. Such is the case with information from the visual and auditory systems, which converges in areas of the brain involved in the interpretation of language.

The picture that emerges from the study of the nervous system is that specific areas of the brain are responsible for discrete functions. In this chapter we will focus on the functional organization of the brain and how the cellular components of the nervous system transduce and transform neural information as it is transmitted from one area of the brain to another.

Overview: The Central and Peripheral Nervous Systems

The nervous system is composed of the **brain, brain stem, spinal cord,** and **peripheral nerves** (Fig. 7–1). The brain is encased in the skull and, when exposed, it appears as a convoluted mass of tissue that makes up the **neocortex**. The neocortex is proportionally larger in humans than in most other species. It is functionally divided (Fig. 7–2) into areas that control movement **(motor cortex),** areas that respond to sensations felt on the surface of the skin **(somatic sensory cortex),** visual images seen by the eye **(visual cortex),** and sounds detected by the ear **(auditory cortex).** Other areas of the neocortex are responsible for language **(Wernicke's area** and **Broca's area),** and personality **(frontal cortex).**

Beneath the neocortex are subcortical areas involved in a variety of different functions. The **thalamus** (Fig. 7–3a and 7–3b), for example, relays information from the sense organs to specific sensory cortical areas. Collectively, the sense organs, thalamus, and sensory cortices form the **sensory systems.** These systems will be discussed in Chapter 8. Other subcortical areas such as the **hypothalamus** (Fig. 7–3a) control the secretion of hormones for growth, reproduction, and metabolism; and others such as the **striatum** (Fig. 7–3b) are involved in the initiation and coordination of movement.

The neocortical and subcortical components of the brain rest upon the **brain stem** (Fig. 7–1)—an area that plays important roles in several functions. By receiving information from the motor cortex and transmitting it to the spinal cord, the brain stem aids

Figure 7–1
The human nervous system, consisting of the brain, spinal cord, and nerves that extend from each.

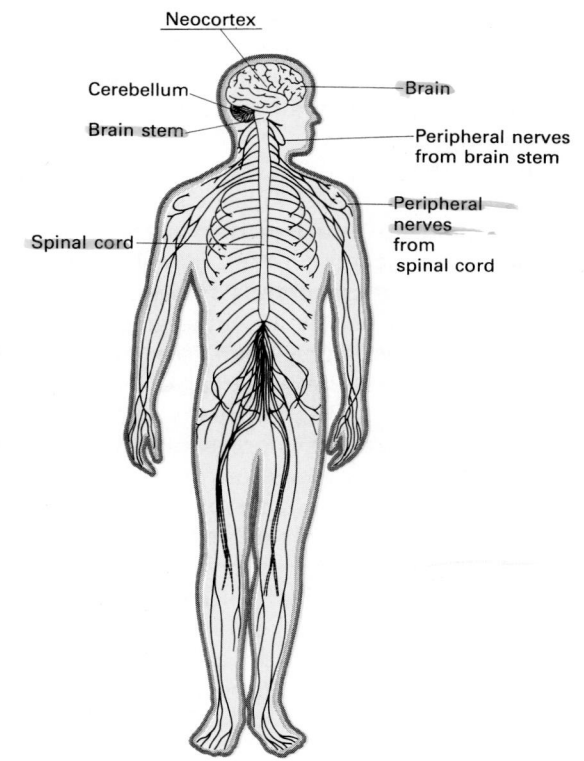

Neocortex

Cerebellum

Brain stem

Brain

Peripheral nerves from brain stem

Peripheral nerves from spinal cord

Spinal cord

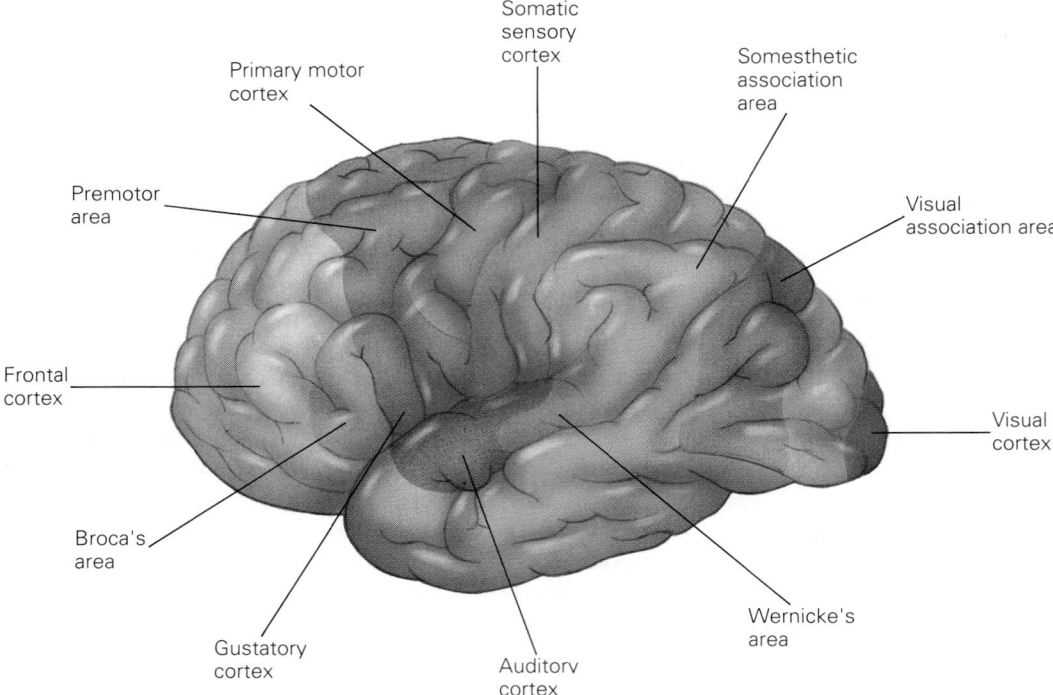

Figure 7–2
The functional areas of the neocortex.

in maintaining body posture. It also transmits information to and from the **cerebellum,** another area involved in locomotor coordination. Collectively, these areas form the **motor system,** which will be discussed in Chapter 9.

Areas within the brain stem also control the breathing and heart rates. These changes occur automatically and are not typically under conscious control. These functions of the brain stem are thus part of the **autonomic nervous system,** which will be discussed in Chapter 10. Emerging from the brain stem are cranial nerve fibers (see Fig. 7–1) connected to the body's internal organs. These cranial nerves are sensory and motor fibers that innervate the face and head. Sensory nerves transmit information *from* the internal organs conveying information about their condition or state of function. Motor nerves transmit information *to* the internal organs to control how they operate. They also receive and transmit information to the muscles and skin of the head and neck. The **spinal cord** is located caudal to the brain stem (Fig. 7–1), and it also transmits information to and receives information from the body's internal organs, as well as from skeletal muscles and skin.

The brain, brain stem, and spinal cord make up what is known as the **central nervous system (CNS),** while the nerve fibers that enter and exit the brain stem and spinal cord comprise the **peripheral nervous system (PNS)** (Fig. 7–4). The peripheral nervous system is comprised of a **somatic** portion and an **autonomic** portion. Within the **somatic nervous system,** sensory nerves transmit information from sense receptors to the central nervous system, and motor nerves transmit instructions about appropriate responses to skeletal muscles. Within the **autonomic nervous system,** sensory nerves transmit information about the condition of internal organs to the central nervous system, and motor nerves instruct glands and involuntary muscles about ap-

Figure 7–3 ▶
(*a*) A medial view of the left half of the brain, showing some of the functional areas of the neocortex and the subcortical structures. (*b*) A horizontal section through the cerebrum at the level of the basal ganglia and thalamus, showing the striatum. (*c*) The striatum comprises the caudate nucleus, putamen, and globus pallidus.

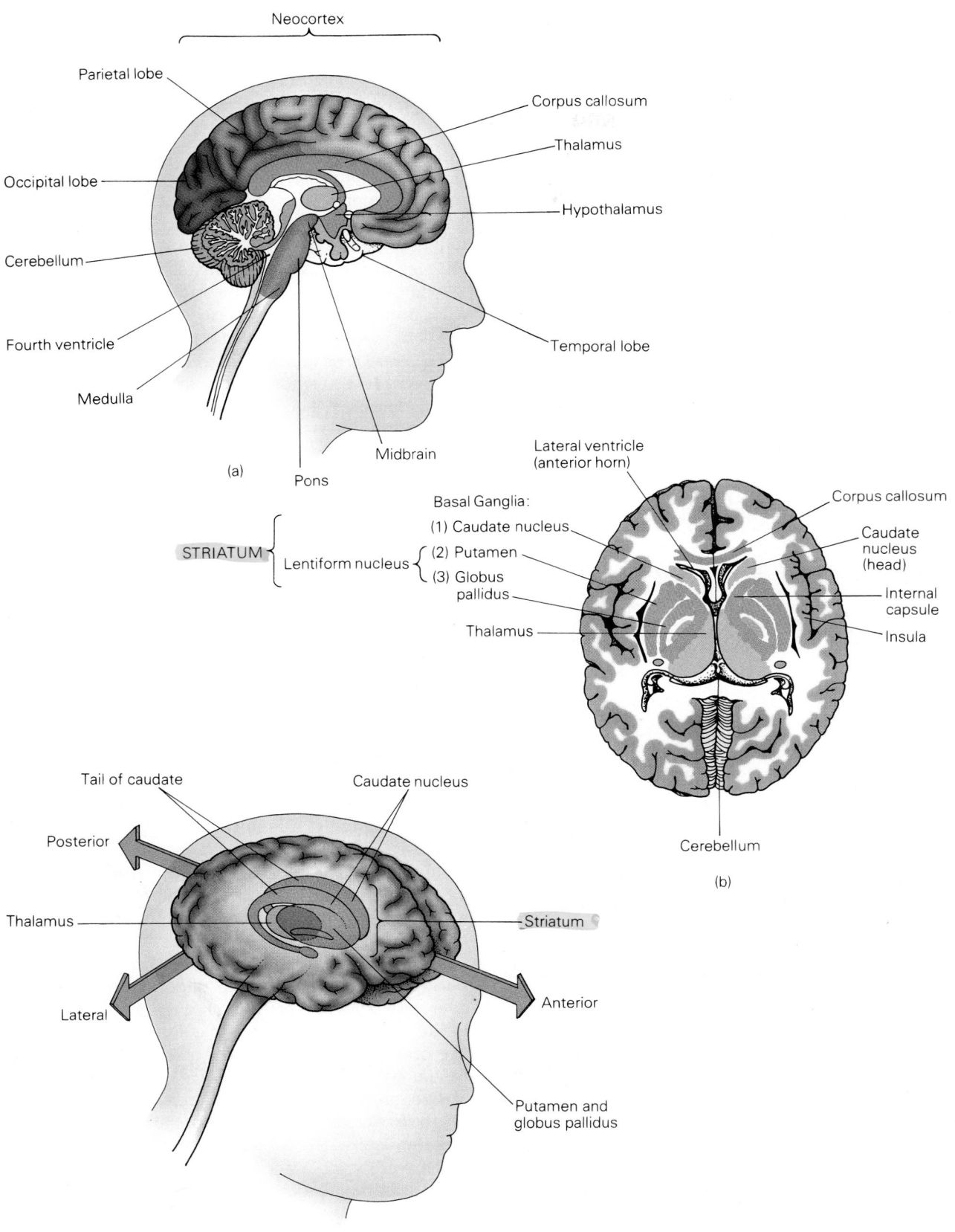

Neocortex

Parietal lobe

Corpus callosum

Thalamus

Occipital lobe

Hypothalamus

Cerebellum

Temporal lobe

Fourth ventricle

Medulla

Pons

Midbrain

(a)

Lateral ventricle (anterior horn)

Corpus callosum

Basal Ganglia:

(1) Caudate nucleus

Caudate nucleus (head)

STRIATUM

Lentiform nucleus

(2) Putamen

(3) Globus pallidus

Internal capsule

Insula

Thalamus

Cerebellum

(b)

Tail of caudate

Caudate nucleus

Posterior

Thalamus

Striatum

Lateral

Anterior

Putamen and globus pallidus

(c)

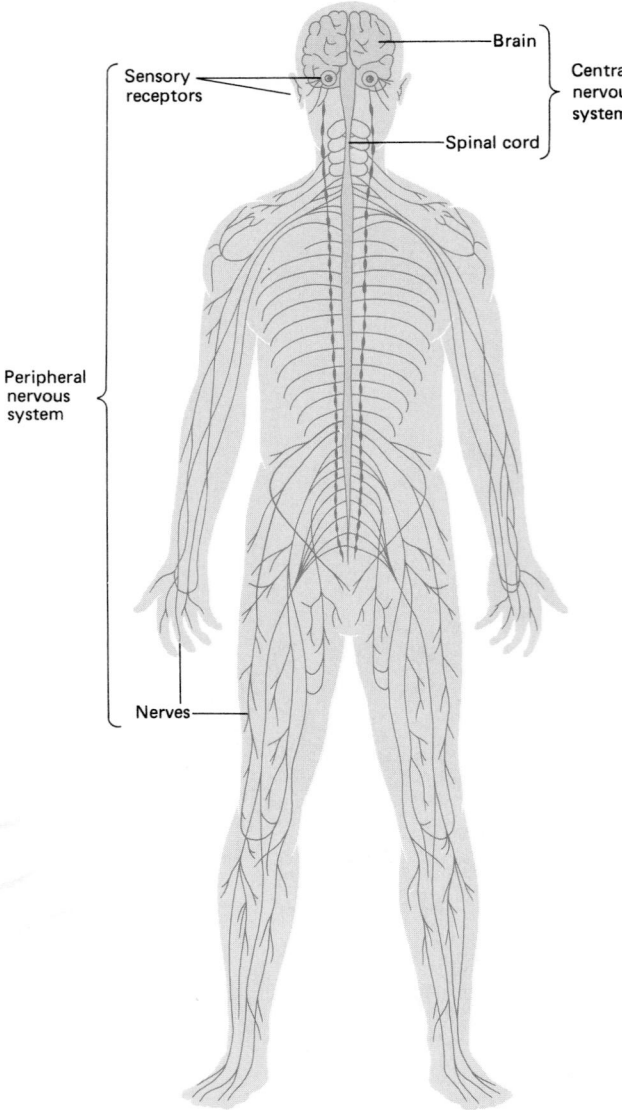

Figure 7–4
The nervous system has two main divisions: the central nervous system and the peripheral nervous system.

organs systems (such as our skeletal muscles) to initiate the appropriate responses (Fig. 7–5).

Cells of the Nervous System

The nervous system contains a complex organization of over a trillion cells. Many cells rapidly communicate with each other by producing electrical impulses that are sent from one cell to another. We will now direct our attention in the remainder of this chapter to the basic mechanisms by which cells in the nervous system operate and communicate with each other.

The nervous system is composed of two types of cells: neurons and glia (or glial cells). **Neurons** are designed to transmit information rapidly from one cell to another. They are responsible for the functions normally associated with the nervous system, such as sensory perception and movement. **Glia,** on the other hand, help maintain the environment surrounding neurons and aid in their ability to transmit information rapidly.

Glia: Multifunctional Cells

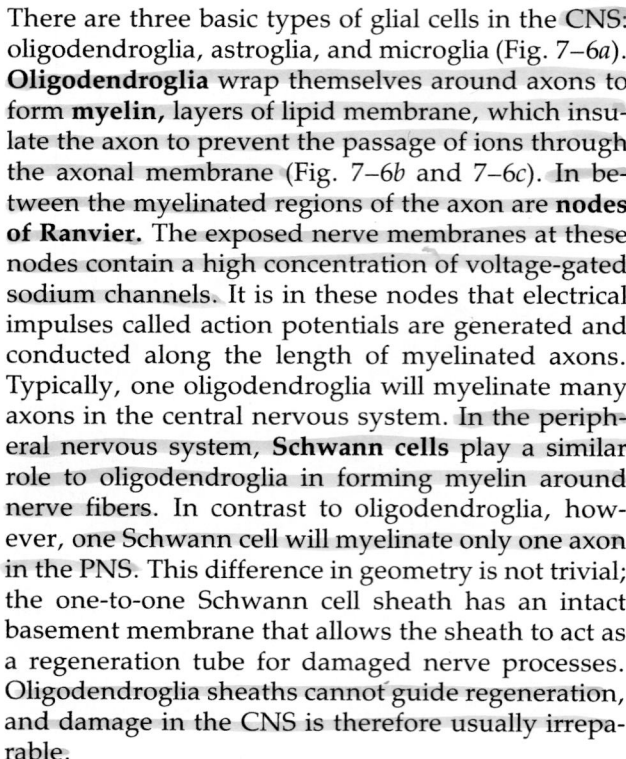

There are three basic types of glial cells in the CNS: oligodendroglia, astroglia, and microglia (Fig. 7–6*a*). **Oligodendroglia** wrap themselves around axons to form **myelin,** layers of lipid membrane, which insulate the axon to prevent the passage of ions through the axonal membrane (Fig. 7–6*b* and 7–6*c*). In between the myelinated regions of the axon are **nodes of Ranvier.** The exposed nerve membranes at these nodes contain a high concentration of voltage-gated sodium channels. It is in these nodes that electrical impulses called action potentials are generated and conducted along the length of myelinated axons. Typically, one oligodendroglia will myelinate many axons in the central nervous system. In the peripheral nervous system, **Schwann cells** play a similar role to oligodendroglia in forming myelin around nerve fibers. In contrast to oligodendroglia, however, one Schwann cell will myelinate only one axon in the PNS. This difference in geometry is not trivial; the one-to-one Schwann cell sheath has an intact basement membrane that allows the sheath to act as a regeneration tube for damaged nerve processes. Oligodendroglia sheaths cannot guide regeneration, and damage in the CNS is therefore usually irreparable.

Astroglia are star-shaped cells which exhibit the greatest diversity of function among glial cells in the CNS. One type of astroglia, the **fibrous astrocyte,** is

propriate responses. Motor nerves in the autonomic nervous system are either sympathetic or parasympathetic in function. **Sympathetic nerves** generally prepare the body for stressful situations, while **parasympathetic nerves** prepare the body for such processes as food consumption and digestion. In general, the nervous system is organized to receive information from sensory organs (such as the eyes and ears), interpret this information, and respond to these sensory inputs by transmitting information to

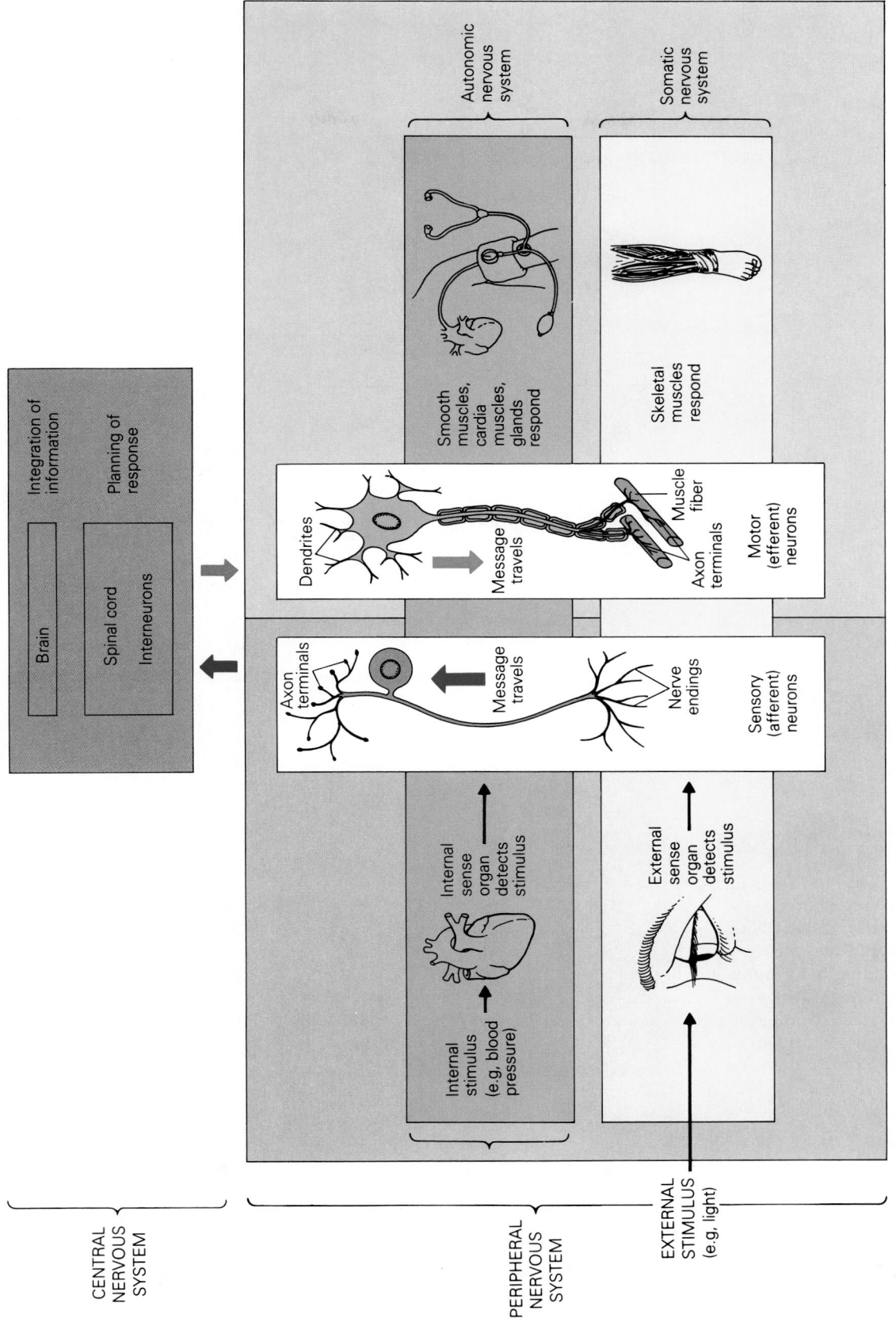

Figure 7-5
The functions of, and flow of information through, the various components of the nervous system.

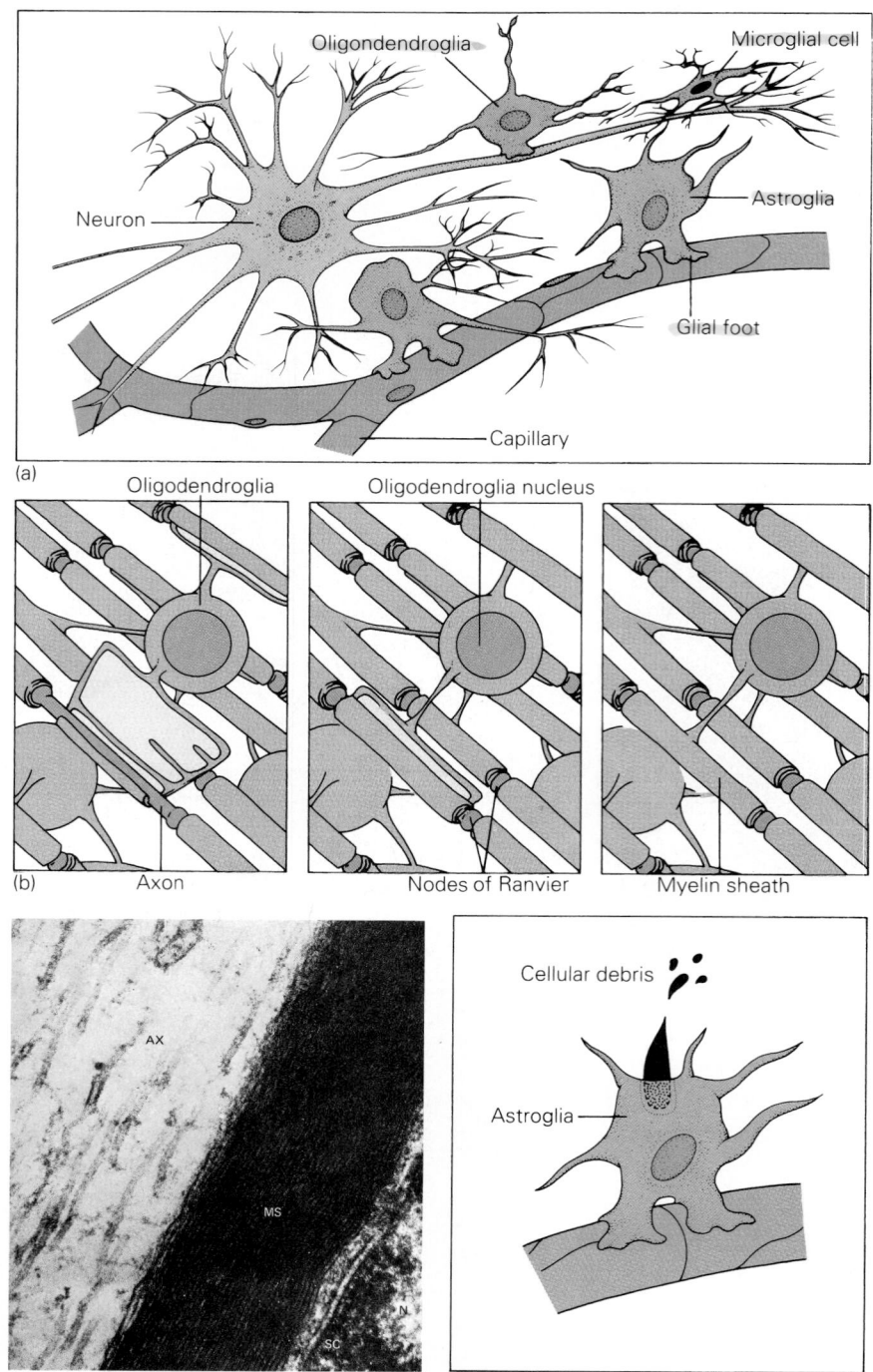

Figure 7–6
Types of glial cells and their functions in protecting neurons.
(*a*) Astroglia make contact with blood vessels and neurons.
(*b, c*) Oligodendroglia form myelin sheaths around axons. AX = axon; MS = myelin sheath.
(*d*) Astroglia remove cellular debris from the central nervous system.

found in areas containing predominantly nerve fibers. It is called fibrous because of the large number of intermediate filaments found in glial cells. **Protoplasmic astrocytes** are similar to fibrous astrocytes but have fewer filaments. They are found in areas containing predominantly nerve cell bodies, dendrites, and synapses. Both types form "glial end-feet" on blood vessels (Fig. 7–6*a*). The anatomical relationship between glial end-feet and blood vessels was once thought to represent the blood-brain-barrier that prevented unwanted substances in the circulatory system from penetrating the brain. However, more recent studies have found that the tight junctions between the endothelial cells that make up the blood vessels are the key cellular components of the blood-brain-barrier.

During injury to the brain, another type of astrocyte called a **reactive astrocyte** appears. It removes degenerating debris by the process of phagocytosis. Reactive astrocytes proliferate after the degenerated neurons have been removed and, in conjunction with fibroblasts, form **glial scars** that are typically seen after brain injury.

Astrocytes regulate the concentration of K^+ ions in the extracellular space by absorbing and redistributing them to other astrocytes. This is accomplished by the transport of K^+ ions through the gap junctions that link networks of glial cells. This type of spatial buffering is needed to maintain the constant extracellular environment of K^+ that is critical to the function of neuron-generated electrical impulses.

Astrocytes also regulate the concentration of chemicals called **neurotransmitters** that are released by nerve cells during the process of **synaptic transmission.** This uptake by glial cells is needed to terminate synaptic transmission and prevent the flooding of neurotransmitters between nerve cells.

Astrocytes are also thought to be involved in the maintenance of interstitial pH, which is required for the optimal function of many cellular processes. As in other organ systems, the enzyme **carbonic anhydrase** plays a critical part in acid-base balance. In the central nervous system, this enzyme is found mainly in astrocytes and oligodendroglia.

Microglia are found in the nervous system near blood vessels. Typically their occurrence is somewhat rare in healthy brain tissue. After injury, however, they are found to migrate to the site of damage, where their main function is the removal of cellular debris by phagocytosis. Unlike other glial cells, recent studies have demonstrated that microglia originate from outside of the brain, most likely from bone marrow.

Neurons: Conductors of Nervous Impulses

Neurons are also called **nerve cells.** They have four functionally distinct regions: the soma, dendrites, axon, and axon terminals (Fig. 7–7). The **soma** is the cell body of a neuron and its metabolic center. It contains the components necessary to fabricate and package proteins used in other parts of the cell to maintain a variety of cellular functions. **Dendrites** are fiber-like structures that branch out from the cell body. Both dendrites and soma form the surface of a neuron, which receives information from other nerve cells.

The **axon** is a long fiber-like structure that extends from the soma to make contact with other nerve cells. The **initial segment** of the axon is a sensitive region that triggers the discharge of electrical impulses. In many nerve cells, sections of the axon are surrounded by myelin. The areas in between the segments of insulation are called **nodes of Ranvier.** The axon ends in many **axon terminals,** which make contact with other nerve cells at junctions called **synapses** in order to transmit information to other cells.

Types of Neurons

Neurons can be classified in two ways: by their structure, as multipolar neurons, bipolar neurons, or unipolar neurons; and by function, as sensory neurons, motor neurons, or interneurons.

Classification by Structure. A **multipolar neuron** has several short dendrites and one long axon (Fig. 7–7). A **bipolar neuron** has one axon and one dendrite, on opposite sides of the soma. A **unipolar neuron** has one short branch that gives off an axon and a dendrite. (A **nerve** is a bundle of axons from different neurons.)

Classification by Function. A **sensory neuron,** also called an **afferent** ("going toward") **neuron,** carries information about external or internal stimuli from sensory receptors to the central nervous system. A **motor neuron,** also called an **efferent** ("leading away") **neuron,** carries movement instructions to muscle from the central nervous system, in response to sensory stimuli. An **interneuron,** found in the intermediate zone of the spinal cord, for example, connects sensory and motor neurons and integrates their functions. Motor neurons are typically multipolar neurons, whereas sensory neurons are often unipolar.

Specialized Neuronal Structures

As with other types of cells, the nucleus of nerve cells contains the genetic information needed for the synthesis of specific proteins. Some proteins, for example, are involved in the formation of new functional connections with other cells as a result of signals from afferent inputs or during the process of regeneration.

In neurons the nucleus is relatively quite large. A substantial portion of the genetic information contained within the genome is continually transcribed by neurons. Based on hybridization studies, it is estimated that $\frac{1}{3}$ of the genome is actively transcribed,

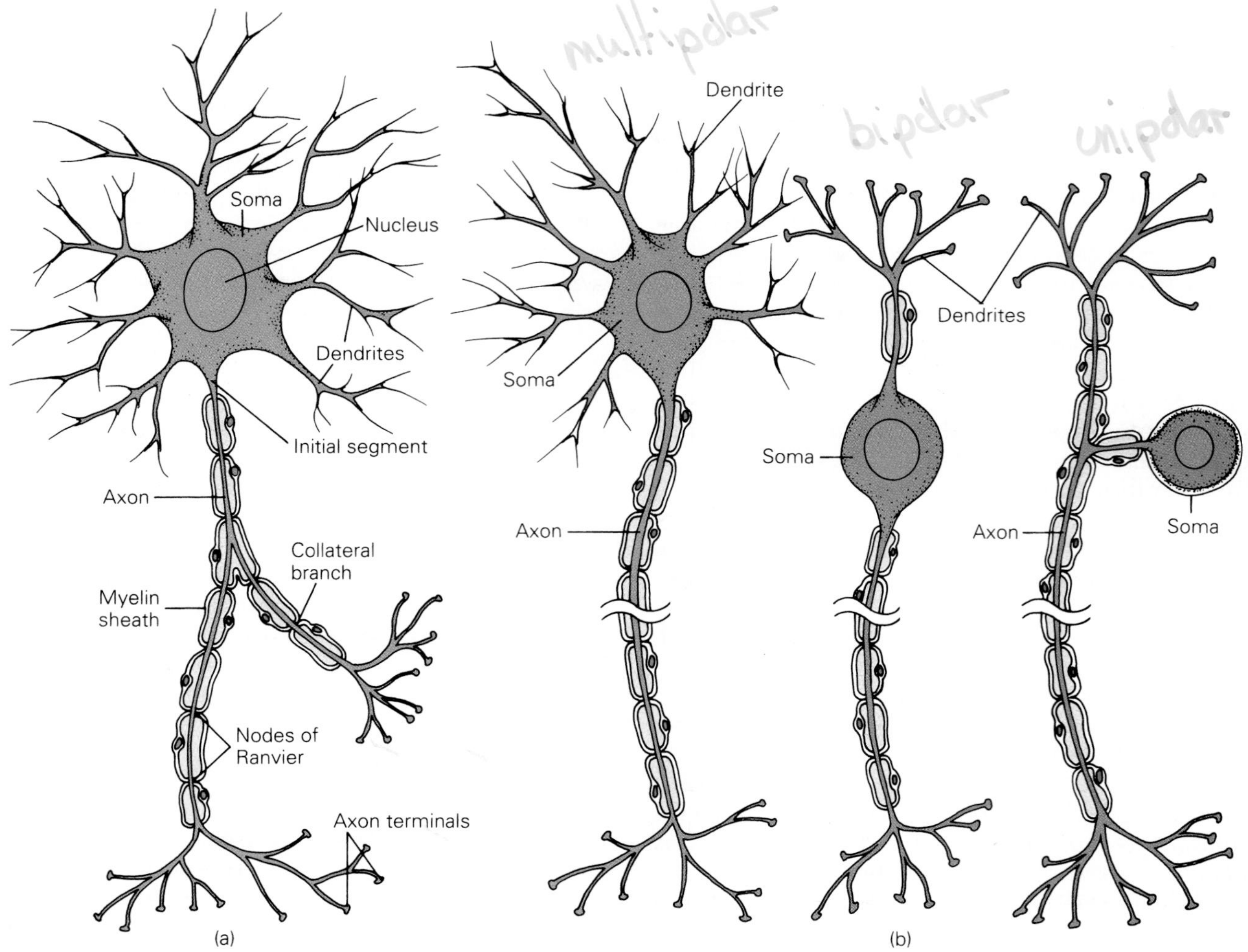

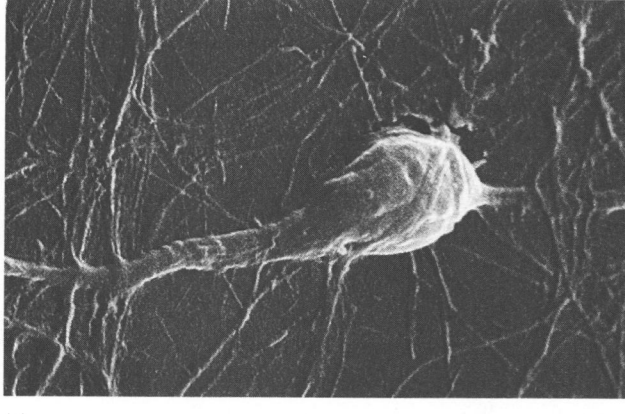

(a)   (b)   (c)

multipolar   *bipolar*   *unipolar*

Figure 7–7
(a) The general structure of a neuron. (b) Types of neurons. Multipolar neurons have several short dendrites and one long axon, while bipolar neurons have one axon and one dendrite. Unipolar neurons have one short branch that gives off an axon and a dendrite. (c) SEM of a human neuron in tissue culture. (© Petit Format/Science Source.)

producing more mRNA than any other organ in the body. Because of the high level of transcriptional activity, the nuclear chromatin is in the form of **dispersed chromatin** (Fig. 7–8). In contrast, the chromatin in non-neuronal cells in the brain such as glia is found in clusters on the internal face of the nuclear membrane. The neuronal nucleus also contains one or two nucleoli, which often have attached electron-dense bodies. One of the dense bodies associated with the nucleolus in females is the Barr body, which is the condensed chromatin of the inactive X-chromosome.

Most of the proteins formed by free ribosomes and polyribosomes remain within the cell, while proteins formed by rough endoplasmic reticulum are exported. Polyribosomes and rough endoplasmic reticulum (RER) are found predominantly in the soma of neurons. Axons contain no RER and are unable to synthesize proteins. Dendrites, on the other hand, contain ribosomes and RNA, and are

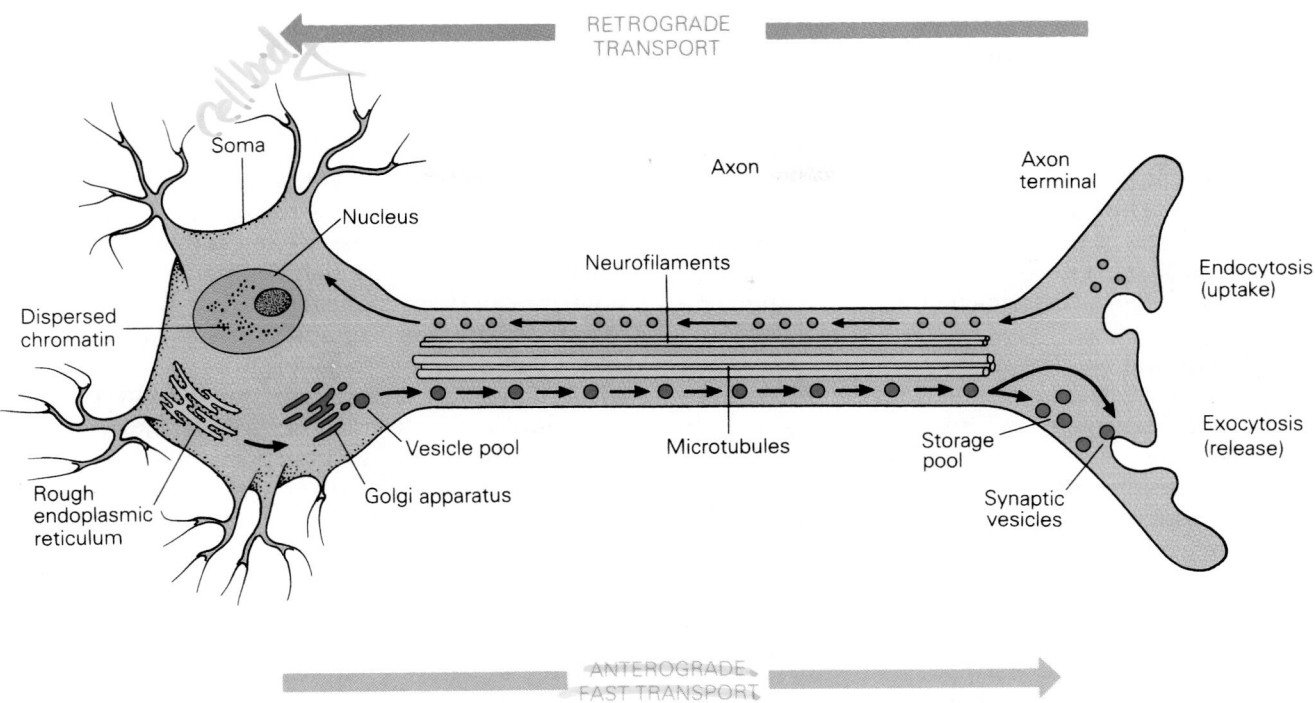

Figure 7–8
Types of axonal transport. Materials move by anterograde fast transport from the soma to the axon terminal and by retrograde transport from the axon terminal to the soma.

thought to play an important part in the modulation of dendritic function.

The smooth endoplasmic reticulum is involved in the intracellular storage of calcium. Smooth endoplasmic reticulum within neurons binds calcium and maintains the intracellular cytoplasmic concentration at a low level of 0.1 μmoles. Prolonged elevation of intracellular calcium has been shown to lead to neuronal death and degeneration.

In neurons the Golgi apparatus is found in the soma. As in other types of cells, this structure is engaged in the terminal glycosylation of proteins synthesized in the RER. The Golgi apparatus takes proteins produced for exportation in the RER and forms protein-containing vesicles. These vesicles are released into the cytoplasm, where some are carried by axoplasmic transport to the axon terminals.

The Neuronal Cytoskeleton. The transport of proteins from the Golgi apparatus and the highly specialized form of the neuron are dependent on the internal framework of the cytoskeleton. The neuronal cytoskeleton is made of microfilaments, neurofilaments, and microtubules (Fig. 7–9). **Microfilaments** are composed of actin, a contractile protein commonly found in muscle. They are 4–5 nm in diameter, and in neurons they are found in dendritic

spines and growth cones. **Neurofilaments,** on the other hand, are found in axons and dendrites and are thought to provide structural rigidity. They are not found in the growing tips of axons and dendritic

Figure 7–9
The cytoskeleton structure of an axon, consisting of microtubules and neurofilaments that aid in the transport of organelles and secretory products to and from the axon terminal.

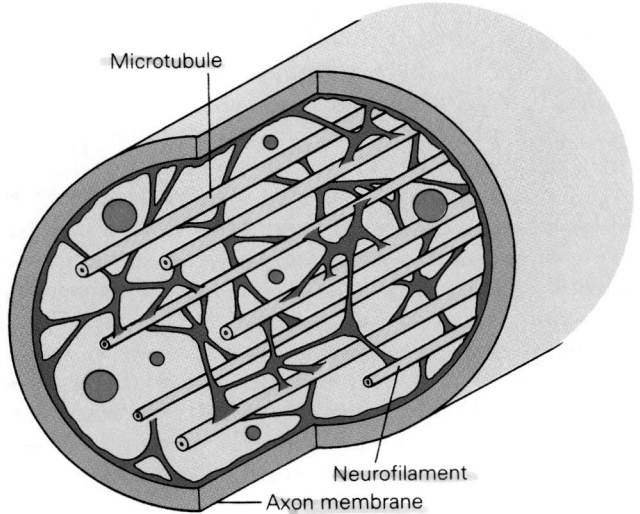

spines, which are more dynamic structures. Neurofilaments are about the size (10 nm diameter) of the intermediate filaments found in other types of cells. However, intermediate filaments in other cell types consist of one protein, while neurofilaments are composed of three proteins (70 kd, 140 kd, and 220 kd in size). The core of the filament consists of the 70 kd protein, similar to intermediate filaments in other cells. The two other neurofilament proteins are thought to be side arms that interact with microtubules.

Microtubules are responsible for the rapid movement of material in axons and dendrites. They are 23 nm in diameter and are composed of tubulin with molecular weights of 52 and 56 kd. In neurons, microtubules have accessory proteins called **microtubule-associated proteins (MAPS)**. Dendrites have high molecular weight MAPS, while axons have low molecular weight MAPS; these two types of MAPS are thought to determine whether material is distributed to dendrites or to axons.

In neurons, mitochondria are highly concentrated in the region of axon terminals. They produce adenosine triphosphate (ATP), which is required as a source of energy for many cellular processes. In the axon terminal the mitochondria not only provide a source of energy for the process of synaptic transmission but also provide substrates for the synthesis of certain neurotransmitter chemicals such as the amino acid glutamate. In addition, enzymes involved in the degradation of other types of neurotransmitter chemicals are embedded in the outer membrane of the mitochondria. Thus the role of mitochondria in the neuron is multifunctional and varied.

Axonal and Dendritic Transport Mechanisms. The shapes of most cells in the body are relatively simple in comparison to the complexity of neurons, with their elaborate axonal and dendritic processes. Because of the length of nerve cell processes, however, neurons have developed specialized mechanisms to transport proteins, organelles, and other cellular material along the length of axons and dendrites needed for the maintenance of the cell. These transport mechanisms are capable of moving cellular components along fiber processes in an **anterograde** direction away from the soma, or in a **retrograde** direction toward the soma (see Figure 7–8). **Kinesin,** a microtubule associated protein, is involved in anterograde transport of organelles and vesicles via the hydrolysis of ATP. Retrograde transport, on the other hand, is mediated by another microtubule associated protein called **dynein.**

In the axon, anterograde transport occurs at both a slow and a fast rate. The rate of **slow axonal transport** is 1 to 2 mm/day. Structural proteins such as actin, as well as neurofilaments and microtubules, are transported at this speed. The rate of **fast axonal transport** is 400 mm/day. Fast transport mechanisms are utilized by organelles, vesicles, and membrane glycoproteins needed at the synaptic terminal. Another feature that distinguishes fast from slow transport mechanisms is that fast transport requires Ca^{2+}, glucose, and ATP (i.e., it is dependent upon oxidative metabolism).

In dendrites, anterograde transport occurs at a rate of 0.4 mm/day, and like fast transport, it also requires ATP. Dendritic transport appears to be involved in the movement of ribosomes and RNA, suggesting that protein synthesis occurs within dendrites.

In retrograde axonal transport, material is moved from terminal endings to the cell body. This provides a mechanism for the cell body to sample the environment around its synaptic terminals. In some neurons, maintenance of synaptic connections depends on the **transneuronal** transport of trophic substances, such as nerve growth factor (NGF), across the synapse. After transport to the soma, NGF activates mechanisms for protein synthesis.

Nerve Fiber Development and Regeneration. One of the major features distinguishing nerve cell differentiation and growth from that of other cell types is the outgrowth of the axon from the nerve cell body, in a specific direction and along a specific pathway, to form synaptic connections with specific targets. The growth of axons is determined largely by interactions between the growing axon and the tissue environment. At the leading edge of a growing axon is the growth cone. **Growth cones** are structures that give rise to protrusions called filopodia. Growth cones contain actin, a contractile protein, and are quite motile, with filopodia extending and retracting at a rate of 6 to 10 μm/min. Newly synthesized membranes in the form of vesicles are also found in the growth cone and fuse with it as it extends. As the growth cone elongates, microtubules and neurofilaments are added to the distal end of the fiber and partially extend into the growth cone. They are transported to the growth cone via the process of slow axonal transport.

The direction of axonal growth is directed in part by **cell adhesion molecules (CAMs)**. CAMs are glycoproteins that are expressed on cell surfaces to promote cell adhesion. Neuron-glia-CAM (Ng-CAM) is expressed in postmitotic neurons and is particularly prominent in growing neurites (axons and dendrites); these neurites migrate along certain types of glial cells that provide a guiding path to target sites. The secretion of tropic factors by target cells also influences the direction of axon growth. Once the proper target site is reached, and synaptic connections are formed, the processes of growth cone elongation and migration are terminated.

During development, axon terminals are found on target cells in excess of the numbers seen after the nervous system has matured. Cells in the lateral geniculate nucleus of the visual system, for example, receive both ipsi- and contralateral inputs during development. In the adult, however, these cells no longer receive inputs from both eyes but only from either the ipsilateral or contralateral input. This loss of synaptic contacts is a result of a selection process whereby the most active inputs predominate and survive at the expense of less active synaptic contacts.

Growth cones are found not only during development but also during the regeneration of nerve fibers. When a nerve fiber is cut, the distal end degenerates while the fiber segment proximal to the soma develops growth cones for elongation and extension. Regeneration occurs at a rate of 1 mm/day, the rate of slow axonal transport.

The Ion Channels of Nerve Cell Membranes

The electrical impulse that initiates synaptic transmission is called an **action potential** (to be discussed in detail later) and is dependent on the properties of the axon membrane. As are all cell membranes, this membrane is made up of a double layer of lipids, with specialized proteins penetrating the double layer. These proteins regulate the movement of ions across the membrane, which in turn creates the action potential. Certain ions (Na^+, K^+, Cl^-, and Ca^{2+}) can cross the membrane only through protein pores in the membrane that form **ion channels.** These channels typically allow only specific ions to pass, while blocking others because of their size, charge, or state of hydration. There are three basic types of ion channels: passive, chemically activated, and voltage-activated channels (Fig. 7–10).

Passive ion channels are found in membranes throughout all areas of the nerve cell. Each passive channel is identified according to the specific ion it allows through (e.g., Na^+ channel, K^+ channel, Cl^- channel, and Ca^{2+} channels).

Chemically activated ion channels are located predominantly on dendrites and the soma. These channels are generally closed by "gates" to prevent the flow of ions through the membrane. Chemical transmitters bind to sites on these protein channels and open the gate across the channel to permit the flow of ions through the channel. These chemically activated channels are also known as **receptors.**

Voltage-activated ion channels, found in the membranes of axons and soma, are opened when they detect a certain voltage. They are responsible for generating and propagating the action potential.

The Membrane Potential and the Action Potential

The movement of ions across their channels is caused by a difference in the electrical charge or differences in the concentration of ions between one side of the membrane and the other. The imbalance in ionic concentration is called the **concentration gradient.** The resulting imbalance in electrical charge is called the **membrane potential.**

Resting cells, cells not in the process of conducting nervous impulses, have a stable membrane potential. The onset of an **action potential,** in which the membrane potential temporarily changes, indicates a cell in the process of nervous conduction.

The Electrical Potential Across the Membrane of a Resting Cell

The membrane potential exists because of a difference in the number of positive and negative charges across the membrane. These charges are attributed to the **cations** (positively charged ions) and **anions** (negatively charged ions) that occur on each side of the membrane. When there is a difference in the net charge between the inside and outside of a cell, an **electrical potential (E)** is established. This electrical potential is measured in units called **volts (V)** and is determined by the charge difference between two points. The charge at two points in the extracellular fluid is the same, for example, so the charge difference and electrical potential are therefore zero. In contrast, since there is a charge difference between a point in the extracellular space and a point within the cell, an electrical potential is established. In many nerve cells, this electrical potential difference across the membrane, E_m, is approximately -0.060 V or -60 mV (Fig. 7–11). The convention that has been adopted for measuring the voltage differences across the membrane is to use the extracel-

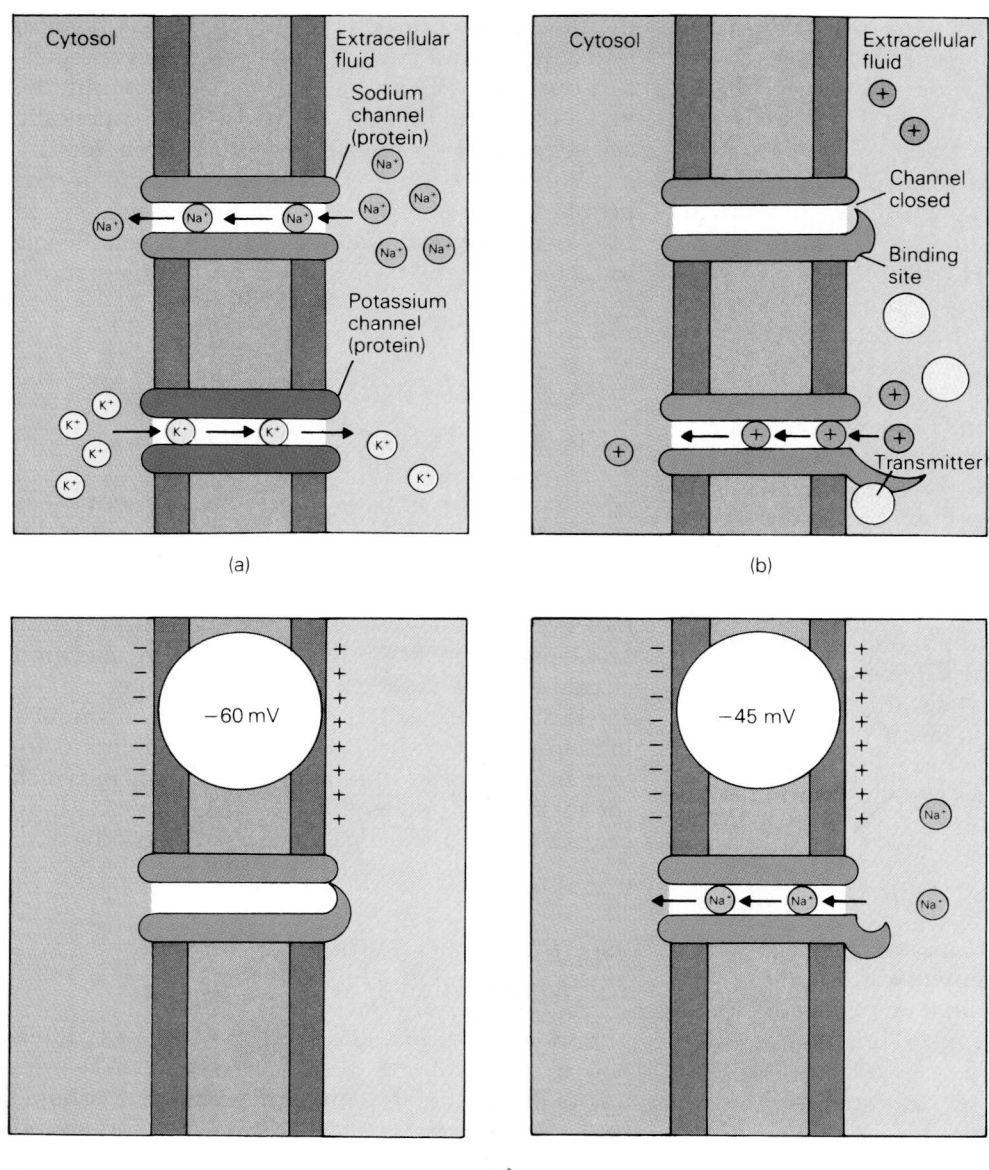

Figure 7–10
Types of ion channels. (*a*) Passive channels are specific to individual ions. (*b*) Chemically activated channels, also known as receptors, depend on activation by a chemical transmitter. (*c*) Voltage-activated channels open or close when they detect a change in the membrane potential.

lular voltage as a reference. A minus sign thus indicates that the inside of the cell is more negative than the extracellular fluid.

The electrical potential can be viewed as the electrical equivalent to pressure. Differences in pressure between two points cause currents of water and wind to flow. In a similar fashion, an electrical potential results in the movement of ions or elec-

trons. In physical systems, the movement of electrons in a copper wire is the basis of **current flow.** In biological systems, current flow is due to the movement of ions. This movement occurs when anions are attracted to positively charged regions and cations to negatively charged regions. The flow of ions is called **ionic current (I)** and is measured in units called **amperes (A).** The ionic sodium current, I_{Na^+},

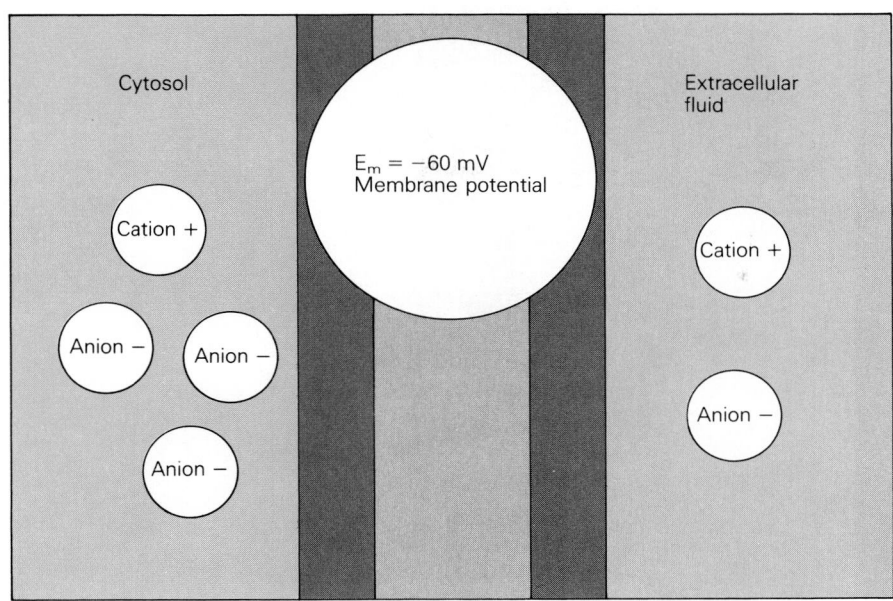

Figure 7–11
The membrane potential (E_m) of −60 mV indicates that the inside of the cell is more negative than the extracellular fluid.

that moves through a single channel, for example, is approximately 12×10^{-12} A.

The ease with which ions can flow through an ion channel is called **conductance (G)** and is measured in dimensional units called **siemens (S).** Several factors affect the conductance of an ion through a channel. The first is the size of the ion relative to that of the channel. If the channel is large in comparison to the size of the ion, the ion will pass with relative ease and the channel therefore has a high conductance. A smaller channel increases the probability of the ion colliding with the walls of the channel, making it more difficult for the ion to pass. Smaller channels therefore have a lower conductance. Other factors, such as the charge of an ion and its state of **hydration** (the addition of water molecules), can also affect its conductance through the channel. Each type of channel has a specific conductance for its associated ion. Sodium ion channels, for example, have a conductance of 15×10^{-12} S, while calcium ion channels have a conductance of 1×10^{-12} S. Consequently, a sodium ion is able to move through its channel more easily than calcium ions can move through a calcium channel.

The relationship between electrical potential, ionic current, and conductance is given by **Ohm's Law.** This law states that the rate of ion flow through a channel is directly proportional to the conductance of the channel and the magnitude of the electrical potential, $I = G \times E$. Larger electrical potentials and greater channel conductances produce a faster flow of ions across the membrane. This is similar to the flow of water through a tube. Larger pressures and tubes with bigger diameters will allow more water to flow.

The overall effect of all of the channels for a particular ion is called the **membrane conductance** for that ion. Membrane conductance of an ion, therefore, depends on the number or density of a particular type of channel within a region of the membrane (Fig. 7–12). Membrane conductance can vary from one region of a nerve cell to another. The membrane conductance for sodium ions, for instance, is greater in the region of the initial segment of the axon and nodes of Ranvier than in other areas of the cell. As we will learn, this regional difference in the number of Na^+ channels aids in the initiation and propagation of the action potential, the basis of nervous impulses.

Concentration Gradients and Diffusion

The distribution of ions in the extra- and intracellular spaces of nerve cells is similar to that of most cells in the body. The inside of nerve cells has a relatively high concentration of K^+ ions and a low concentration of Na^+. In contrast, the extracellular fluid contains a relatively low concentration of K^+ ions and a high concentration of Na^+. The high intracellular concentration of K^+ ions in comparison to the extracellular compartment produces a concentration gradient that causes K^+ to diffuse out of the nerve cell through the passive K^+ channels (Fig. 7–13). Similarly, the high extracellular concentration of Na^+ causes Na^+ to diffuse into the cell through passive Na^+ channels. If the diffusion of these ions

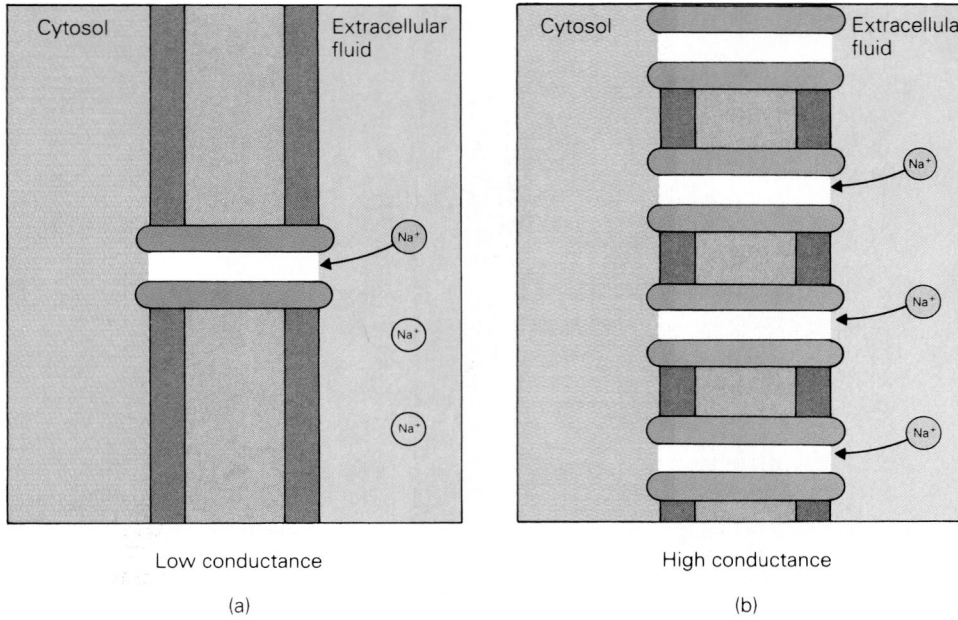

Low conductance   High conductance
(a)   (b)

Figure 7–12
Membrane conductance for a particular ion depends on the number of ion channels that occur in the membrane.

continued, the concentration gradients would eventually equilibrate, and the intracellular concentrations of K$^+$ and Na$^+$ would be equal to their respective extracellular concentrations. The action of the

Figure 7–13
Concentration gradients for sodium and potassium. Sodium (Na$^+$) ions are more highly concentrated outside than inside the cell, so they tend to passively diffuse into the cytosol, whereas K$^+$ ions (more concentrated inside the cell) tend to diffuse into the extracellular fluid.

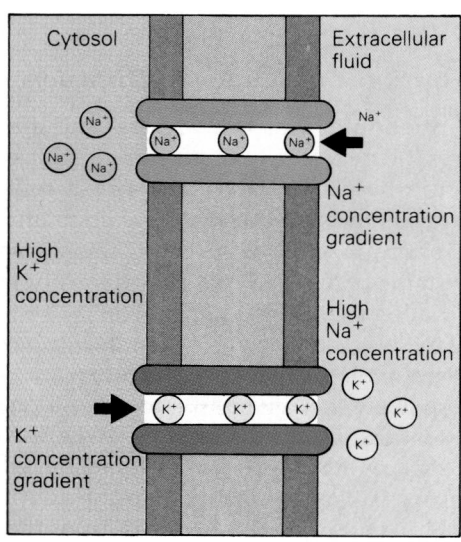

Na/K-ATPase pumping mechanisms, however, works to transport Na$^+$ ions back out to the extracellular compartment and K$^+$ back into the intracellular compartment (see Chapter 4). Consequently, the Na/K pumping mechanisms maintains the Na$^+$ and K$^+$ gradients despite the continued diffusion of the ions across the membrane.

The Membrane Potential of a Hypothetical Nerve Cell: The Equilibrium Potential

To understand how the membrane potential is established, let us first consider a hypothetical cell that contains only K$^+$ ions and large negatively charged organic anions, represented as A$^-$ (Fig. 7–14). These anions are charged amino acids and proteins that are too large to pass through the ion channels of the membrane. Assume that inside a nerve cell are relatively high concentrations of K$^+$ and A$^-$ ions, and that initially these ions do not exist in the extracellular fluid. As we discussed earlier, the establishment of concentration gradients would cause K$^+$ and A$^-$ to diffuse out of the cell through passive channels. In our hypothetical cell, however, only K$^+$ ions can diffuse across the membrane. If K$^+$ and A$^-$ are initially contained within the cell in equal concentrations, then before diffusion starts there will be no net charge (Fig. 7–14a). As K$^+$ begins to diffuse outward, fewer K$^+$ ions are available inside

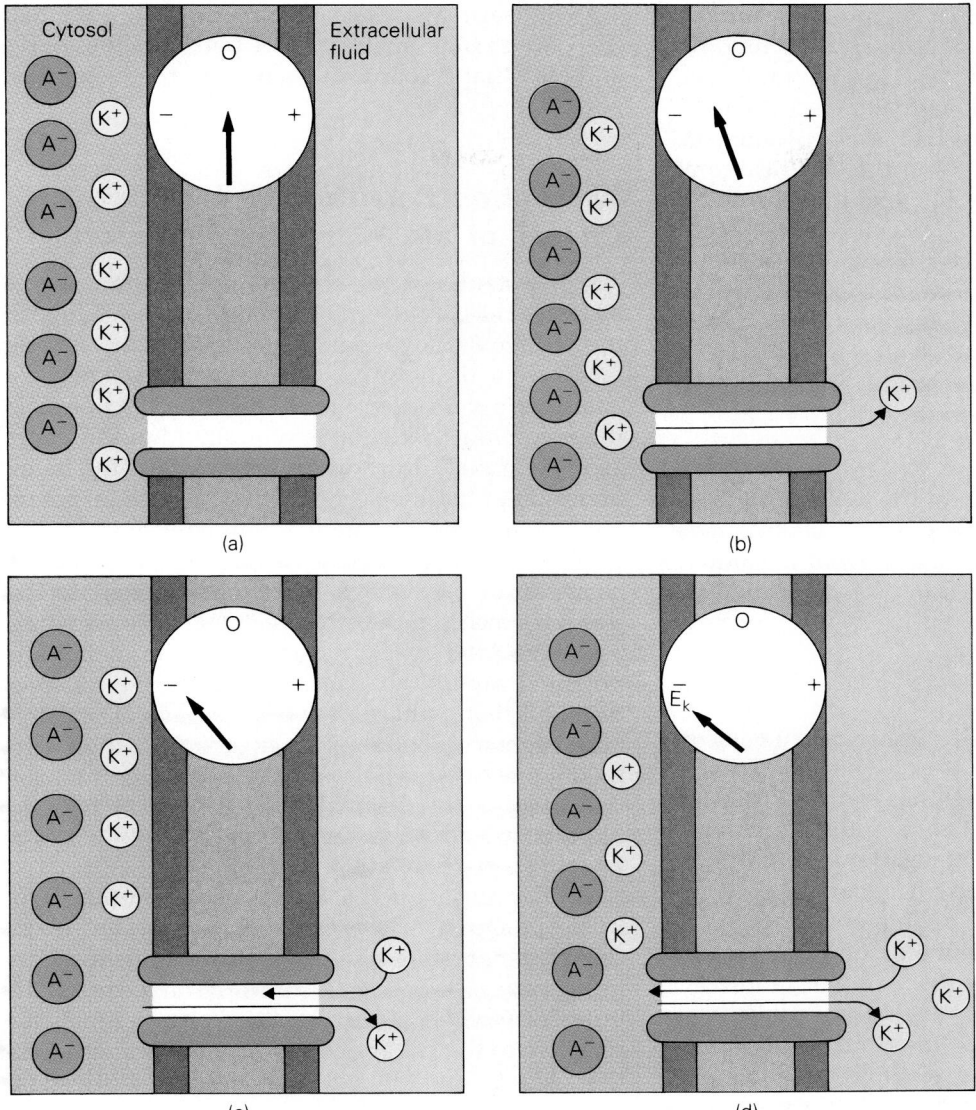

Cytosol   Extracellular
                           fluid

(a)

(b)

(c)

(d)

Figure 7–14
A membrane potential develops when a difference in electrical charge between the
cytosol and the extracellular fluid arises, owing to the movement of ions down a concentration gradient. (a) Initially, there is no net charge within the cell because the
number of anions and cations is equal. (b) K$^+$ ions flow out of the cell down their
concentration gradient, while anions must remain inside because the membrane is
impermeable to them. The cytosol now has a net negative charge. (c) K$^+$ ions next
flow into the cell because they are attracted to the negative charge of the cytosol, although other K$^+$ ions are still flowing out, owing to the concentration gradient.
(d) Equal numbers of K$^+$ ions flow in and out of the cell when the intracellular
charge is very negative. The membrane potential when this occurs is the equilibrium
potential (E_K) for K$^+$.

the cell to balance the negatively charged A$^-$ ions.
Consequently, the interior of the cell becomes more
negative than the outside of the cell (Fig. 7–14b). As
we learned earlier, when the cytosol and the extracellular fluid have different numbers of charged
ions, an electrical potential develops and the

charged ions are attracted to the region that is oppositely charged. The electrical force established by the
negatively charged interior of the cell attracts positive ions, causing them to flow from the extracellular fluid across the membrane to the cytosol. Therefore, even though the concentration gradient for K$^+$

alone would continue to force K$^+$ ions outward, an electrical force gradually develops, attracting K$^+$ ions back into the cell (Fig. 7–14c). As more and more K$^+$ ions leave the cell, a greater electrical force develops. Eventually, because of the opposing concentration gradient (Fig. 7–14d), the electrical force becomes sufficient to draw K$^+$ ions into the cell at the same rate that they leave the cell. The voltage at this particular electrical force is known as the **equilibrium potential.** The equilibrium potential can be expressed as the product of several constants and the logarithm of the ratio of the extra- and intracellular concentration of a particular ion, and can be described by the **Nernst equation.**

$$E_{ion} = \frac{(RT)}{(zF)} \times \log_e \frac{[ion]_o}{[ion]_i}$$

where R is the gas constant, T is the absolute temperature, z is the ion valance, and F is the Faraday constant. At 20°C this expression for K$^+$ is reduced to

$$E_K = 58 \times \log_{10} \frac{[K^+]_o}{[K^+]_i}$$

Typically the ratio of $[K^+]_o/[K^+]_i$ results in a value of E_K of about −75 mV.

Thus, for this simple cell that contains only K$^+$ and A$^-$ and allows only K$^+$ to cross the membrane, the membrane potential is equal to the equilibrium potential of K$^+$. Nerve cells in general, however, contain other ions that affect the membrane potential of the cell. The two other major ions that contribute to the membrane potential are Na$^+$ and Cl$^-$. Both Na$^+$ and Cl$^-$ have very high extracellular concentrations in comparison to their intracellular concentrations and therefore passively diffuse into the cell. Na$^+$ and Cl$^-$ do not diffuse across the cell membrane as easily as K$^+$ ions. When the nerve cell is at "rest"—that is, when it does not generate an action potential—its membrane is most permeable to K$^+$. This is the reason that the resting membrane potential is near the equilibrium potential for K$^+$. (Typically, the K$^+$ equilibrium potential, E_{K^+}, is approximately −75 mV; the resting membrane potential, E_m, is −60 mV; and the Na$^+$ equilibrium potential, E_{Na^+}, is +55 mV.) The factors that affect the resting membrane potential, E_m, can be expressed in terms of the intra- and extracellular concentrations of Na$^+$, K$^+$, and Cl$^-$ ions, as well as their membrane **permeabilities,** P_{Na}, P_K, and P_{Cl}. This relation is given by the **Goldman equation.**

$$E_m = \frac{(RT)}{(zF)} \times$$
$$\ln\left[\frac{P_K[K^+]_o + P_{Na}[Na^+]_o + P_{Cl}[Cl^-]_i}{P_K[K^+]_i + P_{Na}[Na^+]_o + P_{Cl}[Cl^-]_o}\right]$$

We will learn later that during the action potential, there are changes in the relative permeabilities of K$^+$ and Na$^+$ that account for changes in the membrane potential.

The Action Potential: A Temporary Change in the Membrane Potential

Action potentials are electrical impulses transmitted by nerve cells. They can be thought of as a change in the voltage of the membrane potential that causes it to go from its negative resting state to a positive value for a very brief time. This type of change in the membrane potential is typically found in nerve axons. Figure 7–15 shows the changes in the membrane potential that characterize the action potential. The membrane potential is initially at its resting level of −60 mV. As the membrane potential becomes more positive (a process called **depolarization**), it reaches a **threshold** value at approximately −45 mV. After reaching threshold, the membrane potential rapidly changes to more positive values during a **rising phase.** It reaches a **peak** at +25 mV and begins to **repolarize.** In repolarization, the membrane becomes **hyperpolarized,** or more negative than in the original resting state. It remains hyperpolarized for a time before returning to the resting level of −60 mV.

What causes such a drastic change in the membrane potential? The voltage changes seen in the action potential are a result of the opening and closing of voltage-sensitive ion channels that control the influx of Na$^+$ and efflux of K$^+$. These channels are sensitive to the voltage across the membrane. When they detect that the voltage has reached a threshold

Figure 7–15
A graph showing the various components of the action potential.

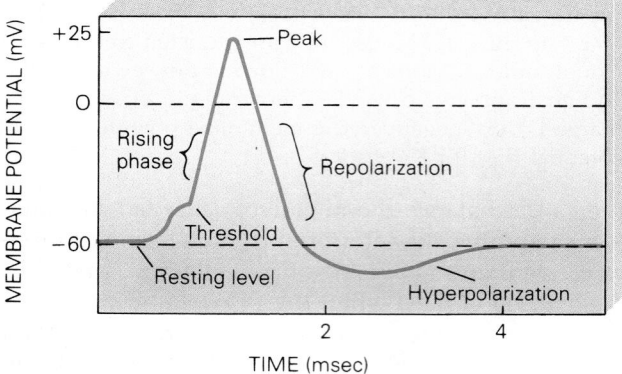

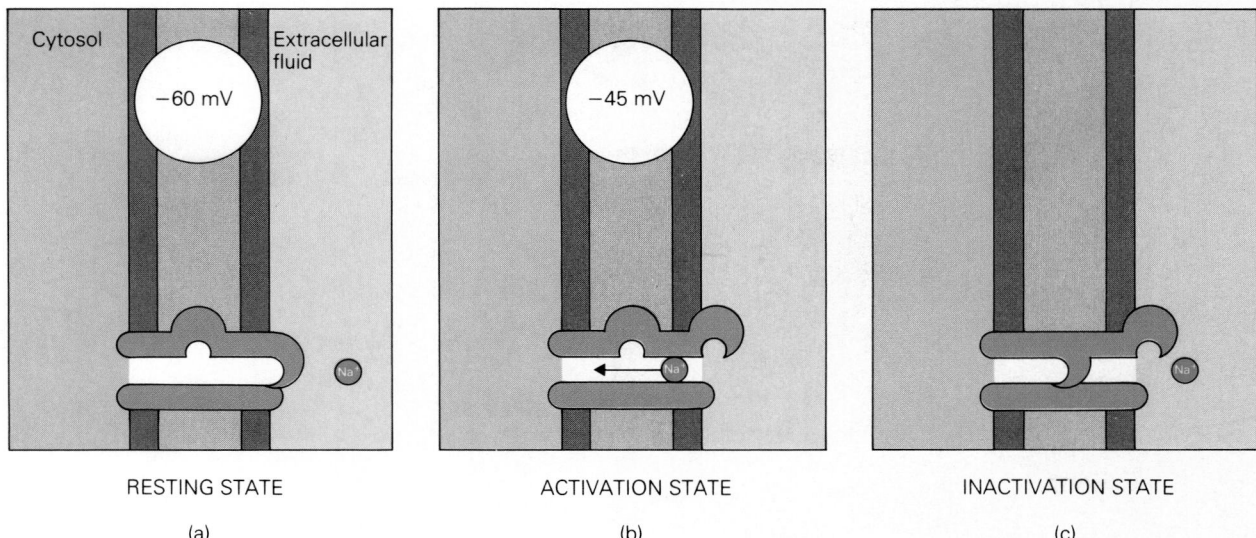

Cytosol   Extracellular
                              fluid

−60 mV   −45 mV

RESTING STATE   ACTIVATION STATE   INACTIVATION STATE

(a)   (b)   (c)

Figure 7–16
Mechanism of the voltage-sensitive Na⁺ channel. (*a*) In the resting state, the
voltage-activated gate is closed. (*b*) In the activation stage, triggered by a
change in the membrane potential, the voltage-activated gate is open, allowing Na⁺ ions to flow into the cell. (*c*) In the inactivation state, the time-
dependent gate is closed.

level, they open their gates to allow the passage of
these ions. The sodium channel appears to have
three states of operation, a **resting state, an activating state,** and an **inactivating state** (Fig. 7–16).
When the nerve cell is at rest, the sodium channel is
in its resting state, and its gates are closed to prevent the influx of Na⁺ ions. If the nerve cell is stimulated and the membrane potential is depolarized to
the threshold value of approximately −45 mV, the
sodium channel switches from its resting state to the
activating or open state. Sodium ions flow into the
cell until the inactivating gate closes the channel.
The closing of the inactivating gate is a time-dependent phenomenon and is independent of the opening of the activation gate. Thus, Na⁺ channel activation and inactivation can be thought of as two
separate gates that open and close independently.
The inactivating gate remains closed for a period of
time, after which both gates return to their resting
state. The properties of the sodium channel, therefore, are dependent on voltage and time. Once the
threshold value of voltage is reached, it triggers this
series of events that is carried out sequentially.

A threshold value of voltage also opens voltage-
sensitive K⁺ channels, but the process is entirely
different from the opening of Na⁺ channels (Fig.
7–17). Unlike the Na⁺ channel, there is no inactivation gate in the K⁺ channel. Also, the K⁺ channels
open more slowly than the Na⁺ channels and re-

main open until they sense a particular voltage. The
closing of these channels is not a time-dependent
process; it depends only on the voltage, the membrane potential. When the membrane is depolarized, the K⁺ channels remain in the open state;
when the membrane potential repolarizes, the K⁺
channels close.

Ion Channels and the Action Potential

The influx of Na⁺ ions and efflux of K⁺ ions produce the action potential. In the resting state,
voltage-sensitive Na⁺ and K⁺ ion channels are
closed (Fig. 7–18). The simultaneous activation of
many sodium channels in the membrane of an axon
causes an influx of Na⁺ ions. This influx of positive
charges causes the membrane potential to become
more positive, producing a gradual depolarization
of the membrane. Once the threshold value of the
membrane potential has been reached (typically
−45 mV at the initial segment), a series of events is
triggered, leading to the **initiation** or **generation** of
the action potential. At the threshold level of the
membrane potential, more voltage-sensitive sodium
channels are activated, resulting in a greater influx
of Na⁺ ions. This influx of positive charges depolarizes the membrane still further, leading to a further
influx of Na⁺ and consequently a further depolarization.

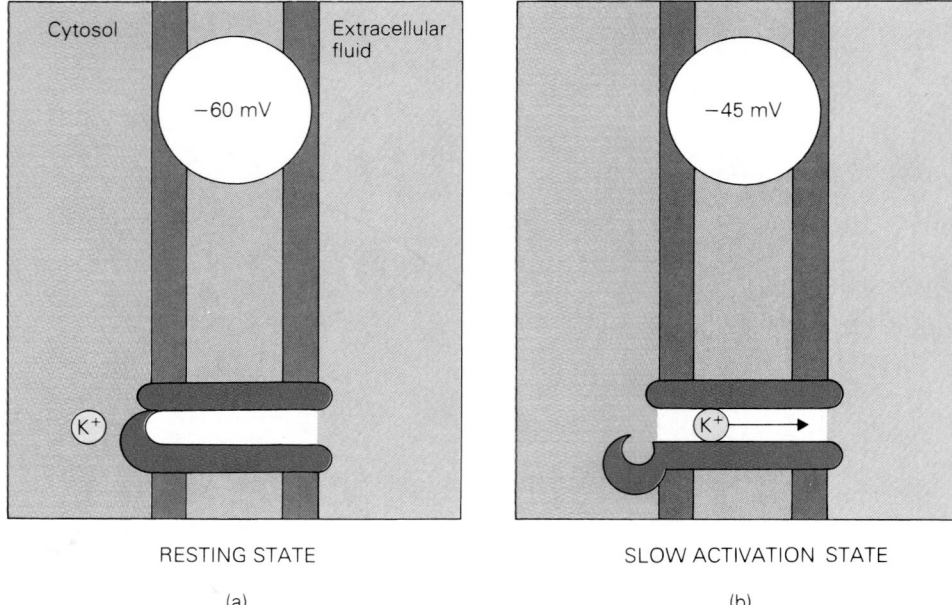

Figure 7–17
Mechanism of the voltage-sensitive K⁺ channel. The opening of the K⁺
channel is entirely voltage-dependent and is slower than the Na⁺ channel.

This **regenerative process** (Fig. 7–19) is the basis for the rising phase of the action potential once threshold has been reached. The regenerative process can occur at different speeds depending on how much the membrane is depolarized and how quickly the membrane potential achieves the threshold value. Greater depolarizations of the membrane potential above the threshold can lead to the opening of more sodium channels, in turn depolarizing the membrane even faster.

Note that not all depolarizations will result in an action potential. If the depolarization is too small, too few voltage-sensitive sodium channels will be activated to perpetuate the regenerative process and no action potential will be produced. Thus, the action potential is an *"all-or-none"* phenomenon.

At the peak of the action potential (Fig. 7–18) the membrane is much more permeable to Na⁺ than to K⁺; consequently, the value of the membrane potential is closer to the Na⁺ equilibrium potential than to the K⁺ equilibrium potential. After the peak has been reached the inactivation channels close, Na⁺ influx is blocked, and the membrane potential begins to repolarize. The membrane potential repolarizes because at this point the membrane conductance for K⁺ is greater than that for Na⁺. As the sodium channels begin to become inactivated, the potassium channels begin to become activated. This increase in potassium conductance causes the membrane potential to become more negative and contributes to the repolarization phase of the action potential. Finally, the prolonged opening of K⁺ channels causes a continued efflux of K⁺ ions. This removal of positive charges from the cell in turn causes the membrane potential to remain hyperpolarized briefly before returning to the resting level.

Refractory Periods

For a period of time just after the initiation of an action potential, the axon is unable to generate a second action potential regardless of how much the membrane is depolarized. This phase is known as the **absolute refractory period** and typically lasts several milliseconds after the onset of the first action potential (Fig. 7–20). After this period, an axon is capable of initiating a second action potential, but only if the membrane is depolarized to a large degree. During this **relative refractory period,** the threshold voltage needed to initiate an action potential becomes much greater.

What causes this increase in threshold? In the absolute refractory period, the Na⁺ inactivation gates are still in their closed state and have not been reset. Consequently, no matter what the voltage difference across a sodium channel, the channel will not open. In the relative refractory period, many

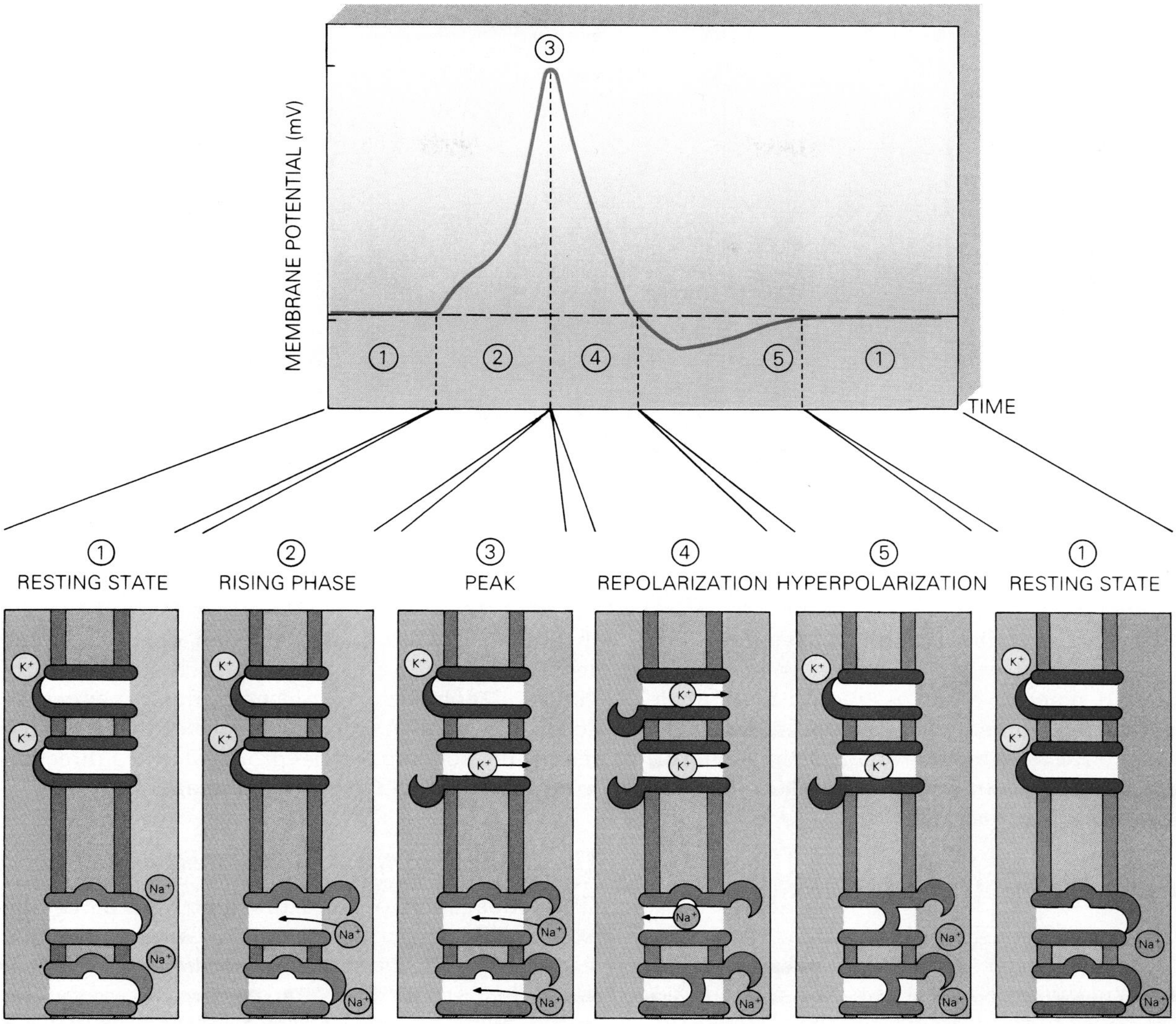

MEMBRANE POTENTIAL (mV)

③

① ② ④ ⑤ ①

TIME

① RESTING STATE ② RISING PHASE ③ PEAK ④ REPOLARIZATION ⑤ HYPERPOLARIZATION ① RESTING STATE

K+ K+ K+ K+ K+ K+
K+ K+ K+ K+ K+ K+

Na+ Na+ Na+ Na+ Na+ Na+
Na+ Na+ Na+ Na+ Na+ Na+

Figure 7–18
The phases of the action potential and corresponding
movements of ions and ion channels.

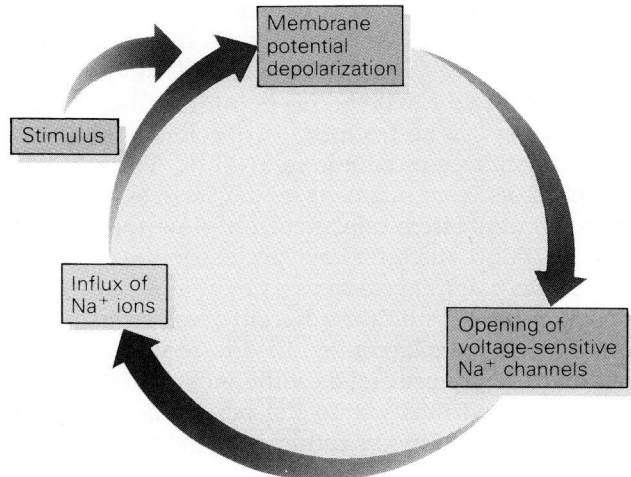

Figure 7–19
The regenerative nature of the process that initiates the
action potential.

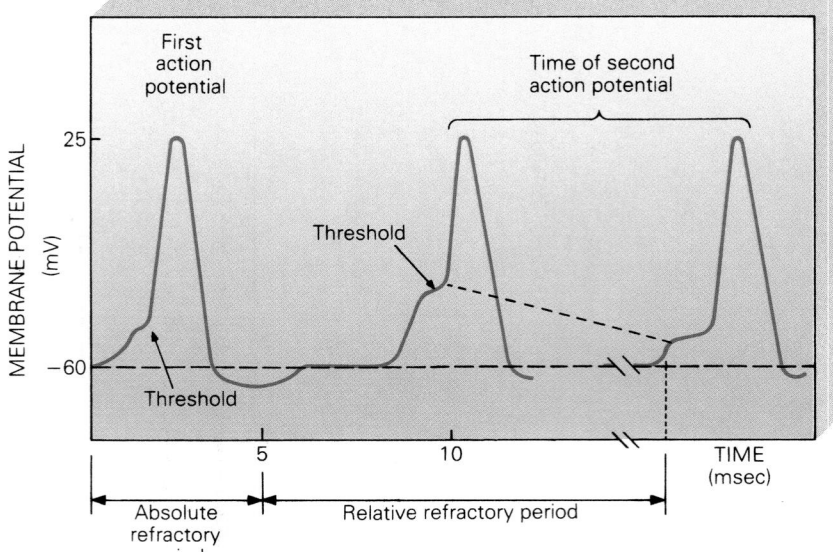

Figure 7–20
A graph showing the absolute and relative refractory periods that follow the action potential.

(but not all) of the sodium channels have now been reset. Thus a greater depolarization is needed to trigger the regenerative process required to initiate the second action potential. The significance of the absolute refractory period is that its time interval determines the fastest frequency at which an axon can generate action potentials.

Propagation of the Action Potential

Thus far we have considered how an action potential is initiated. After initiation it **propagates,** or moves, along the axon from the region of the initial segment down to the terminal endings. How does this propagation occur?

Let us first consider the axon as a tube with many membrane regions (Fig. 7–21). In its resting (nonconducting) state, concentrations of various ions are maintained by various active transport systems (Fig. 7–21a). Assume that the first membrane region generates an action potential so that Na+ ions flow into the axon at this point (Fig. 7–21b). The sodium ions come from the extracellular fluid surrounding adjacent membrane regions. The removal of Na+ ions from adjacent regions causes these regions to depolarize. Since current flow follows the path of a loop, the influx of Na+ ions in the region of the action potential is followed by the movement of positive charges into the adjacent areas of the membrane. The depolarization of adjacent membrane regions is therefore aided by the movement of positively charged ions within the nerve fiber along its length. If this depolarization is sufficient to reach

threshold, it will initiate an action potential in the second region. In this manner, each successive region is depolarized to threshold and generates an action potential. As a consequence of these mechanisms, the action potential propagates in a direction from the soma to the axon terminal.

The Velocity of the Action Potential

The propagation of an action potential along the axon occurs at a constant speed. In unmyelinated nerve fibers, the conduction velocity is proportional to the diameter of the axon. The larger the diameter of the axon, the greater the speed of propagation. This is because axons with large diameters do not offer as much resistance to the flow of ions along the length of the axon. In myelinated axons, the velocity of propagation is determined not only by the diameter of the axon but also by the distance between the nodes of Ranvier.

The formation of myelin around an axon prevents the penetration of ions needed for the conduction of the action potential. Between the segments of myelin, however, are nodes of Ranvier, where the membrane of the nerve axon is exposed and contains large numbers of Na+ channels. It is in the nodes of Ranvier that the membrane is depolarized and action potentials are generated (Fig. 7–22). The generation of an action potential in one node of Ranvier causes the membrane potential in the adjacent node of Ranvier to depolarize and also to generate an action potential. In this way, the propagation of an action potential along a myelinated nerve axon

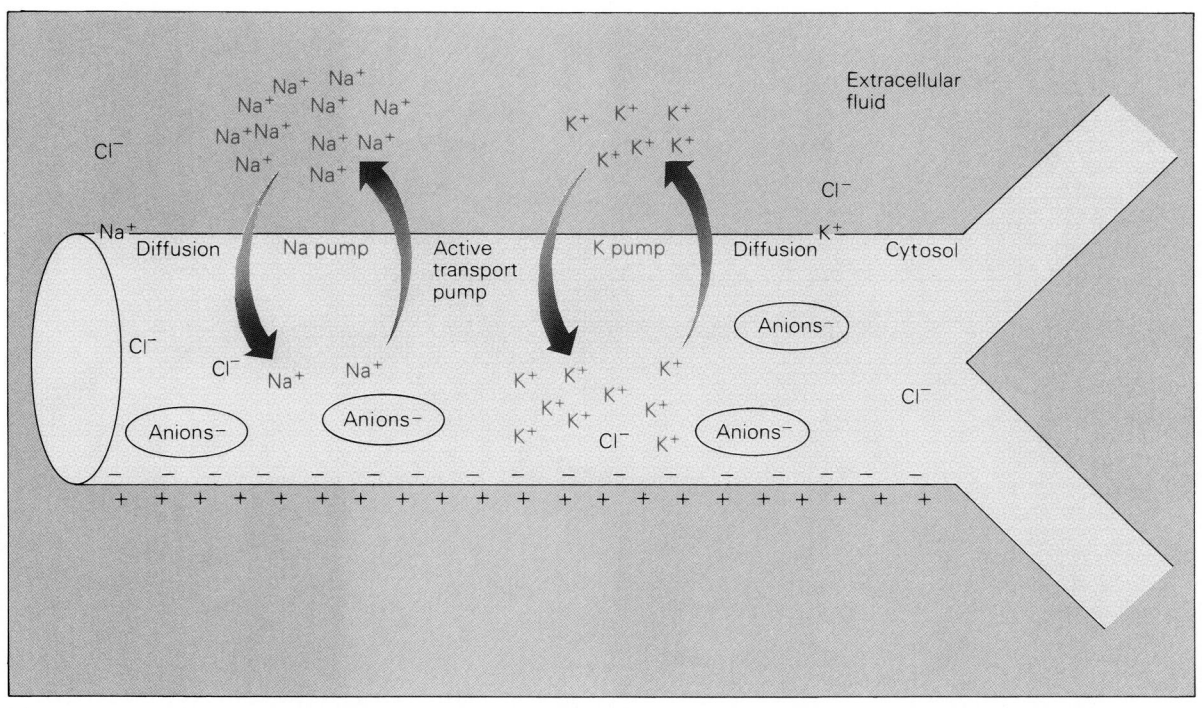

(a)

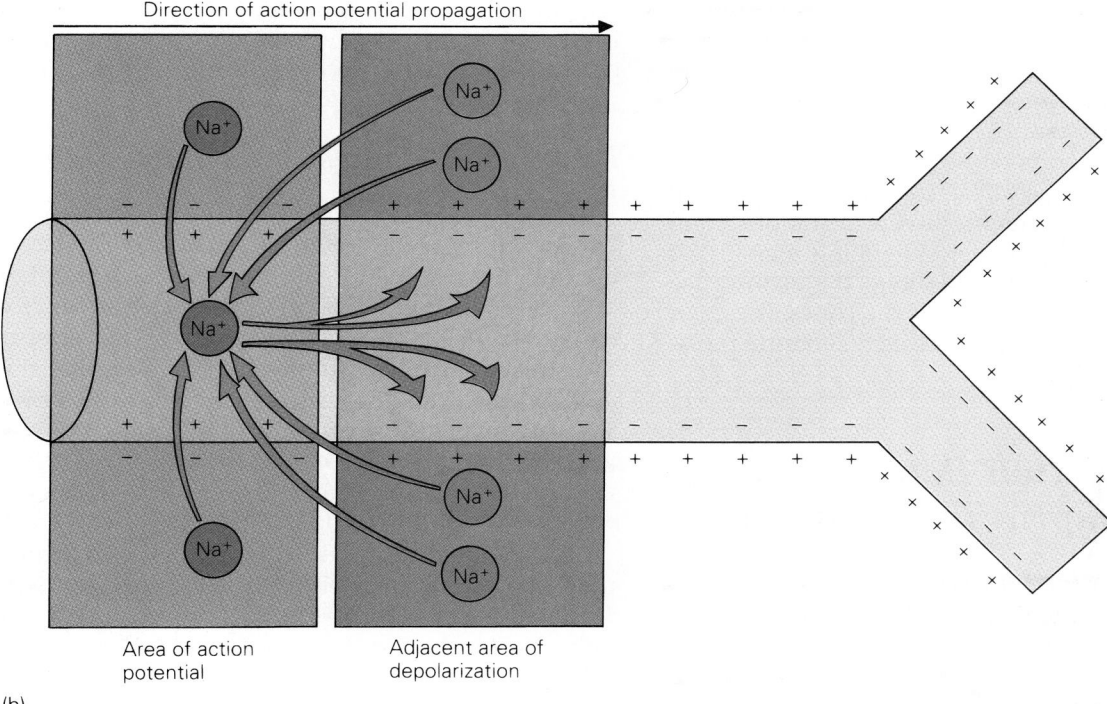

(b)

Figure 7–21

The propagation of the action potential. (*a*) In a resting axon, ion concentrations are maintained by active transport. (*b*) The action potential travels as adjacent regions of the axon are successively depolarized by the movement of Na$^+$ ions.

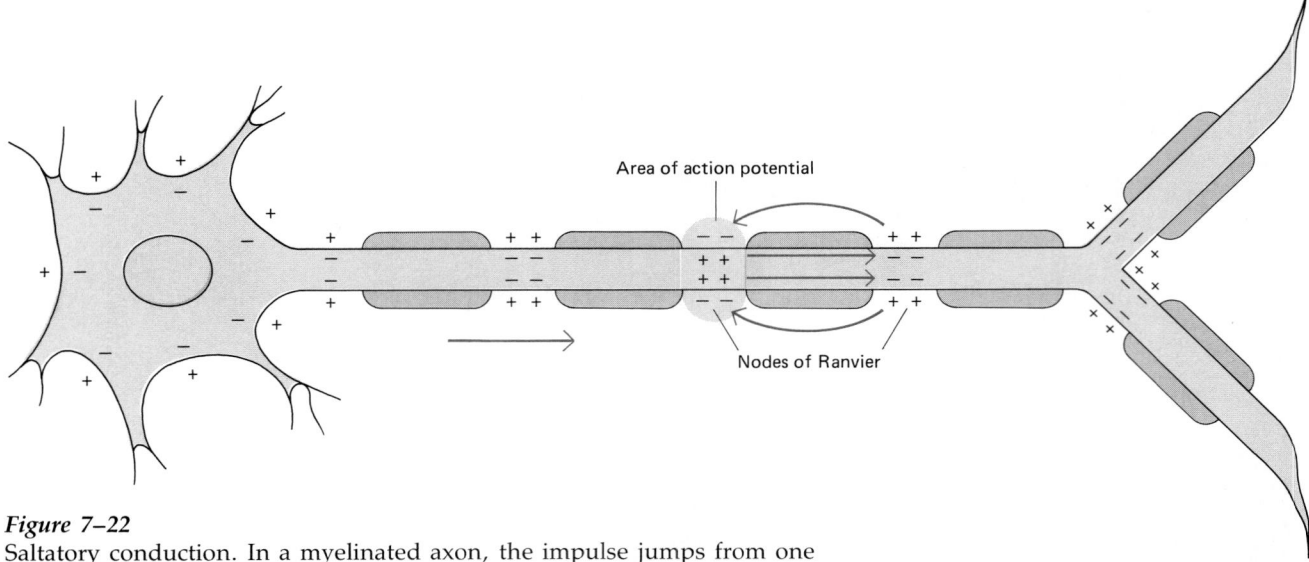

Figure 7–22
Saltatory conduction. In a myelinated axon, the impulse jumps from one
node of Ranvier to the next.

appears to jump (*saltus* in Latin) from one node of
Ranvier to the next in the process of **saltatory conduction.** Thus the greater the distance between
nodes of Ranvier, the greater the velocity of action
potential propagation.

When the action potential reaches the end of the
axon, it invades the *synaptic terminal.* This structure
is one of the key elements in the communication
between nerve cells. At the synaptic terminal, the
action potentials transmitted along the axon cause
the release of chemicals used in the communication
between neurons in a process called **synaptic transmission.**

Communication Between Nerve Cells: Synaptic Transmission

The communication between nerve cells occurs at
junctions called **synapses.** In the nervous system,
there are two types of synapses, **electrical synapses**
and **chemical synapses.** Electrical synapses consist
of **gap junctions** that permit the direct passage of
ions and other small molecules from one cell to another. Electrical synapses transmit information in
both directions. In chemical synapses, on the other
hand, there is a space, called a **synaptic cleft,** between one cell and the other that prevents the direct
passage of ions. In order for ions to flow into the

second cell, chemical transmitters must be released
from the first cell into the synaptic cleft, where they
bind with a receptor located on the second cell.

Electrical Synapses: Communication Between Adjacent Cells

The structure of the electrical synapse is similar to
the bridge junctions or gap junctions seen in other
cells (see Chapter 3). Under the electron microscope, the membranes of two adjoining cells appear
to be fused together. Studies have shown that the
pre- and postsynaptic cells are connected by a protein channel called a **connexon** that spans the gap
between them (Fig. 7–23). The connexon is made of
six protein subunits called **connexin** that are arranged into a hexagonal assembly.

Electrical synapses are found between axons
and soma, axons and dendrites, dendrites and dendrites, and soma and soma. These synapses are
pathways for the direct conduction of ions from one
cell to another. In addition, the channels are large
enough to allow the passage of molecules such as
cAMP, sucrose, and small peptides. Thus, these
"electrical" synapses serve as channels for both
electrical and metabolic communication.

In terms of electrical communication, these synapses synchronize the activity of many adjoining
cells as well as provide a pathway for a rapid com-

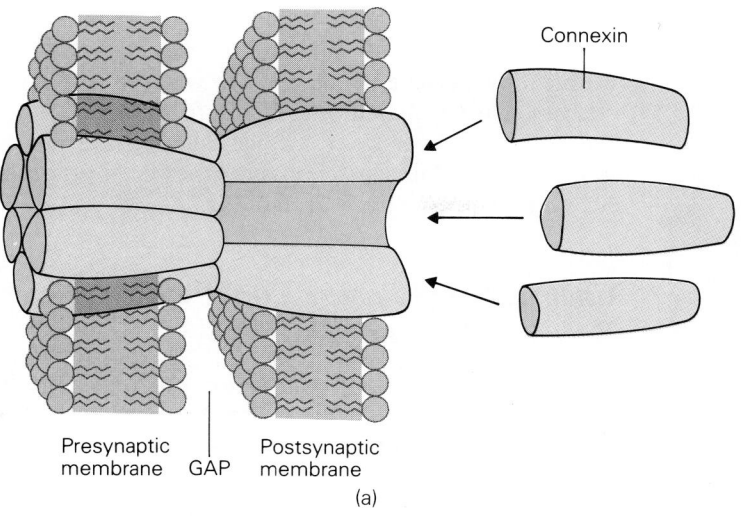

Connexin

Presynaptic
membrane   GAP   Postsynaptic
membrane

(a)

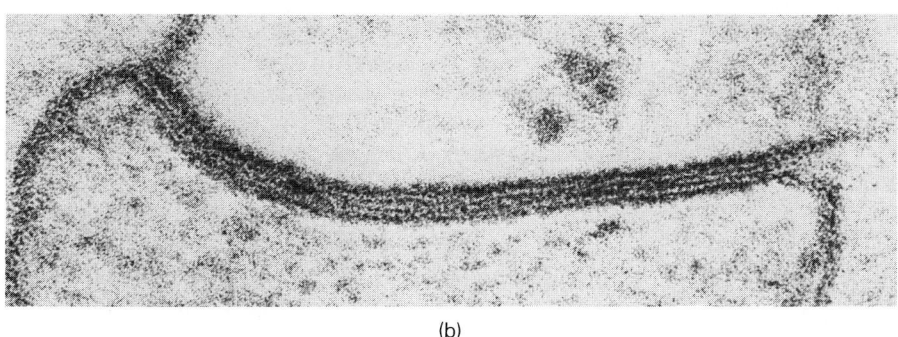

(b)

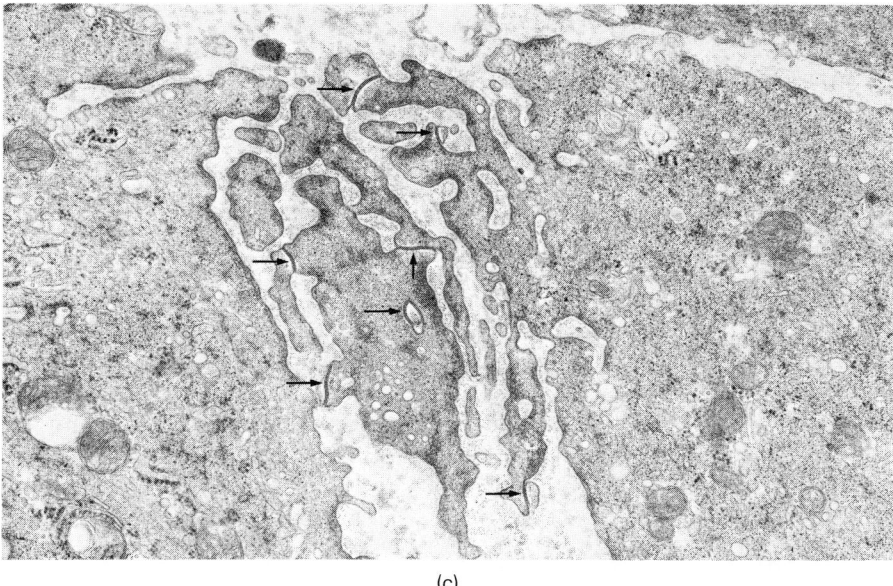

(c)

Figure 7–23
Types of synapses. An electrical synapse (*a*) consists of a gap junction formed by proteins in adjacent cell membranes. The individual subunits that make up each half of the junction are known as connexins. (*b*) Electron micrograph of a gap junction between smooth muscle cells in the wall of the uterus. (*c*) Several gap junctions shown at a lower magnification (approximately × 36,000). (Electron micrographs courtesy of Puri and Garfield, from *Biology of Reproduction*, 27:967–975, 1982).

(continued on next page)

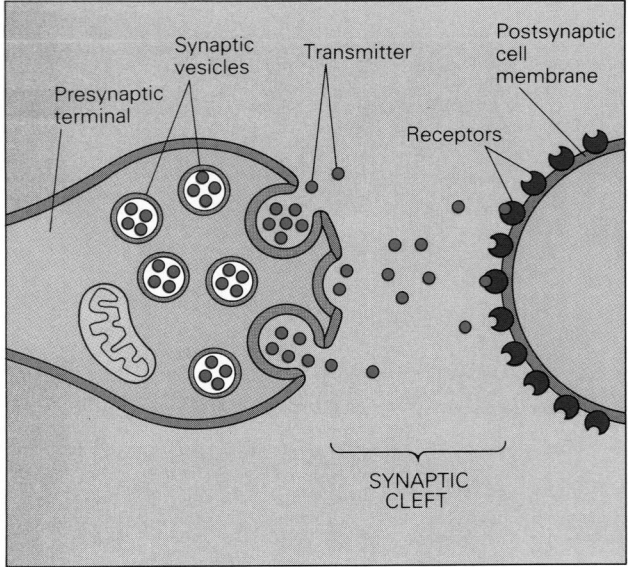

(d)

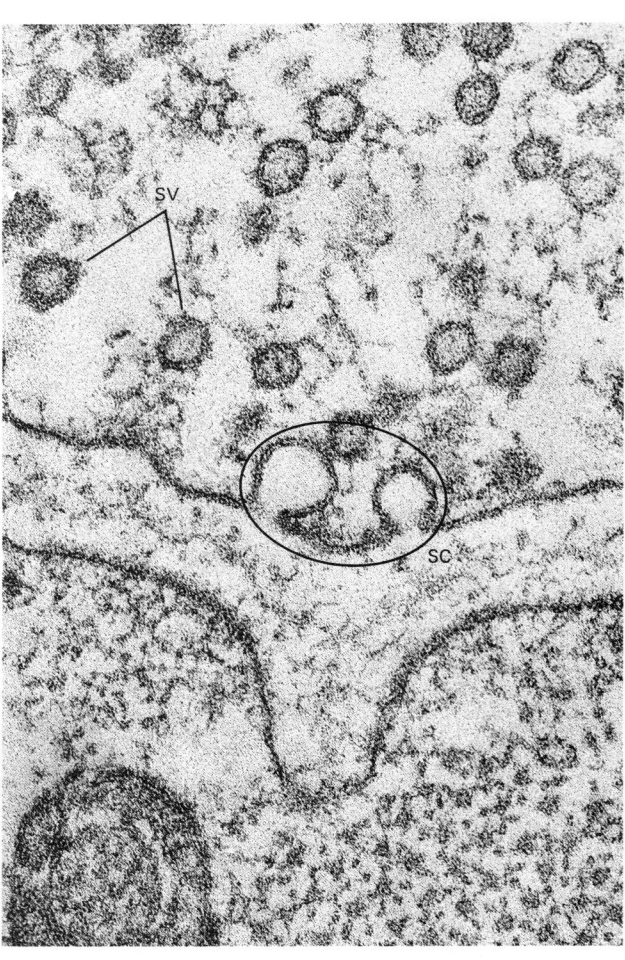

(e)

munication between cells. As we will see in a later chapter, electrical synapses in the form of gap junctions are also found in cardiac cells of the heart, many smooth muscle cells, and other cells that display a synchronization of activity.

Chemical Synapses: Communication Between Distant Cells

Although electrical synapses are found in many areas of the nervous system, the predominant type of synapse is the chemical synapse. In chemical synapses, at least two cells participate—the cell producing the chemical signal, called the **presynaptic** cell, and the target cell receiving the signal, called the **postsynaptic cell.** The presynaptic component of the synapse consists of the terminal ending, which contains vesicles called **synaptic vesicles** (Fig. 7–23d). These vesicles, filled with chemicals referred to as **neurotransmitters** (or **transmitters**), fuse with the presynaptic membrane. The fusion of the vesicle with the presynaptic cell membrane causes the release of the chemical neurotransmitters into the **synaptic cleft.** The transmitters in turn act upon receptors located in postsynaptic cell membranes. Synaptic contacts can be formed at different sites on the postsynaptic cell. Those between axons and dendrites are called **axodendritic synapses.** Contacts between axons and somata are called **axosomatic synapses,** and contacts between two axons are called **axoaxonic synapses.** In the peripheral nervous system, the postsynaptic cell can also be a muscle or a glandular cell.

Presynaptic Mechanisms of Chemical Transmission

The process of synaptic transmission involves four phases: the synthesis of the transmitter, the storage and release of the transmitter, the interaction of the transmitter with the postsynaptic receptor, and the termination of synaptic transmission (Fig. 7–24).

Figure 7–23 (continued)
(*d*) Chemical synapses consist of a presynaptic axon terminal, a synaptic cleft, and a postsynaptic cell membrane. Synaptic vesicles release transmitter molecules that diffuse across the synaptic cleft and bind to receptors on the target (postsynaptic) cell membrane. (*e*) Electron micrograph of a chemical synapse, showing fusion of a vesicle with a presynaptic membrane, approximately × 125,000. SV = synaptic vesicles; SC = synaptic cleft. (Courtesy of Dr. John Heuser.)

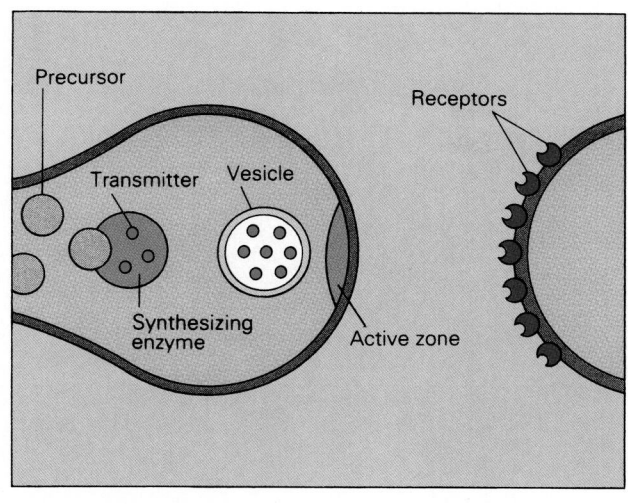

(a)

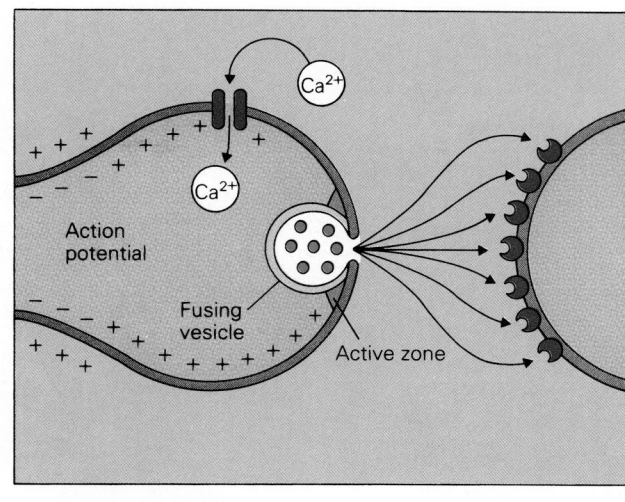

(b)

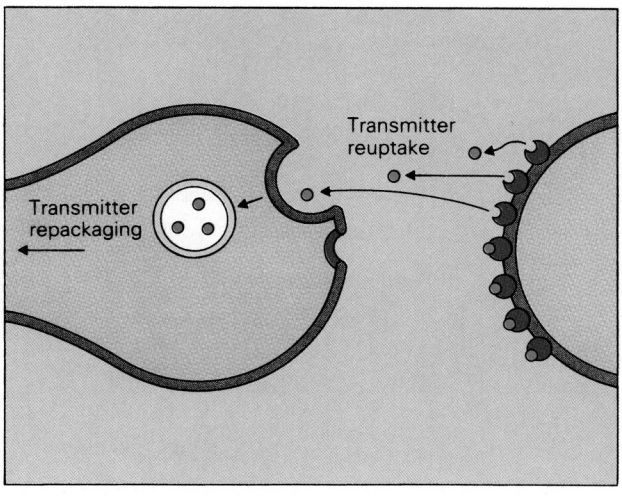

(c)

Figure 7–24
In the presynaptic terminal, chemical transmitter molecules are synthesized by various enzymes and packaged into secretory vesicles (*a*). When calcium ions flow into the cell, the secretory vesicles fuse with the cell membrane, and the transmitters are released into the synaptic cleft (*b*). Some transmitter molecules bind to receptor proteins on the postsynaptic membrane, whereas others are pumped back into the presynaptic terminal by a re-uptake mechanism (*c*).

The biochemical mechanisms for synthesis differ for each neurotransmitter and will be considered in detail later in the chapter. In general, however, each type of transmitter is packaged and stored in vesicles, which are then positioned for the release of their chemical transmitters into the cleft.

The release of the neurotransmitter is produced by the action potential when it invades the terminal ending. The resulting change in the membrane potential activates voltage-sensitive Ca^{2+} channels, causing an influx of Ca^{2+} ions into the terminal. By mechanisms not yet clearly understood, the Ca^{2+} ions cause the synaptic vesicles to fuse with the presynaptic membrane and release transmitter molecules into the synaptic cleft by the process of **exocytosis**.

Each synaptic vesicle releases a fixed, or **quantum**, number of neurotransmitter molecules. Acetylcholine-containing vesicles, for example, have approximately 10,000 molecules per vesicle. The number of synaptic vesicles that fuse with the synaptic membrane to release their chemical transmitters depends on the concentration of Ca^{2+} ions within the synaptic terminal. Greater concentrations of Ca^{2+} within the terminal cause more vesicles to release their contents into the synaptic cleft.

The amount of Ca^{2+} within the terminal is regulated by the activity of the cell (that is, the number of action potentials it has generated). As action potentials invade the synaptic terminal with greater frequency, there is a residual increase in Ca^{2+} ions within the terminal. This increase in Ca^{2+} leads to a

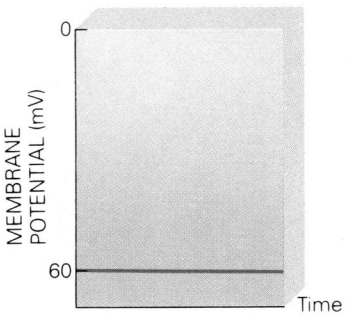

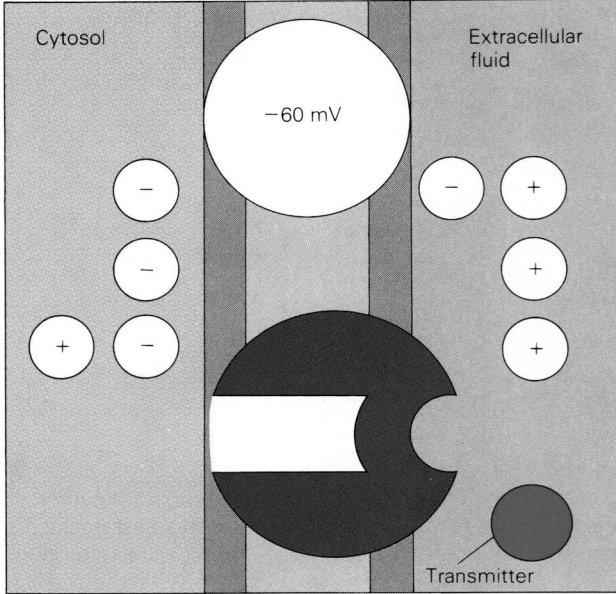

RESTING MEMBRANE
POTENTIAL

(a)

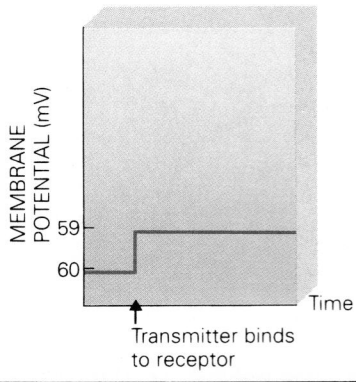

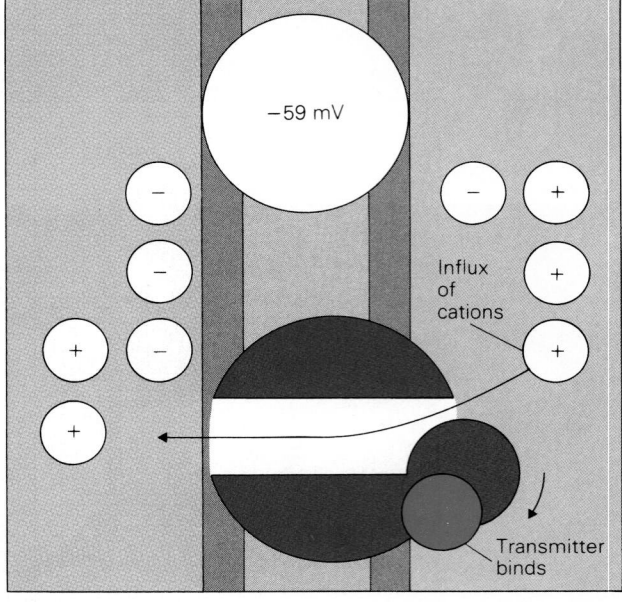

RECEPTOR-ACTIVATED MEMBRANE
POTENTIAL DEPOLARIZATION

(b)

Figure 7–25
Ion flow and postsynaptic potentials. (*a*) Resting membrane potential. (*b*) When a transmitter binds to a receptor, the receptor depolarizes the membrane by allowing cations to flow into the cell. Chemical transmitters can also produce hyperpolarization by causing cations to flow out (*c*) or anions to flow in (*d*) to the cell.

greater release of the neurotransmitter from the synaptic terminal. In this manner, the frequency of action potentials produced by the nerve cell can govern the amount of chemical transmitter it releases.

The amount of transmitter released can also be affected by extrinsic processes such as synaptic inputs from other cells that alter the calcium levels within the terminals. The mechanisms of these axoaxonic inputs will be discussed later in this chapter.

Postsynaptic Mechanisms of Chemical Transmission

After a chemical transmitter is released into the synaptic cleft, it can bind to a receptor located on the postsynaptic cell. The coupling of the transmitter with the receptor causes the receptor to open, permitting the passage of specific ions through the membrane. The receptor can directly activate the ion channel or indirectly activate it through a second messenger such as cAMP, cGMP, or IP_3.

The movement of charged ions into or out of a postsynaptic cell can affect its membrane potential. For instance, the activation of a neurotransmitter-sensitive ion channel that permits the influx of Na^+ into the cell results in more positive charges inside

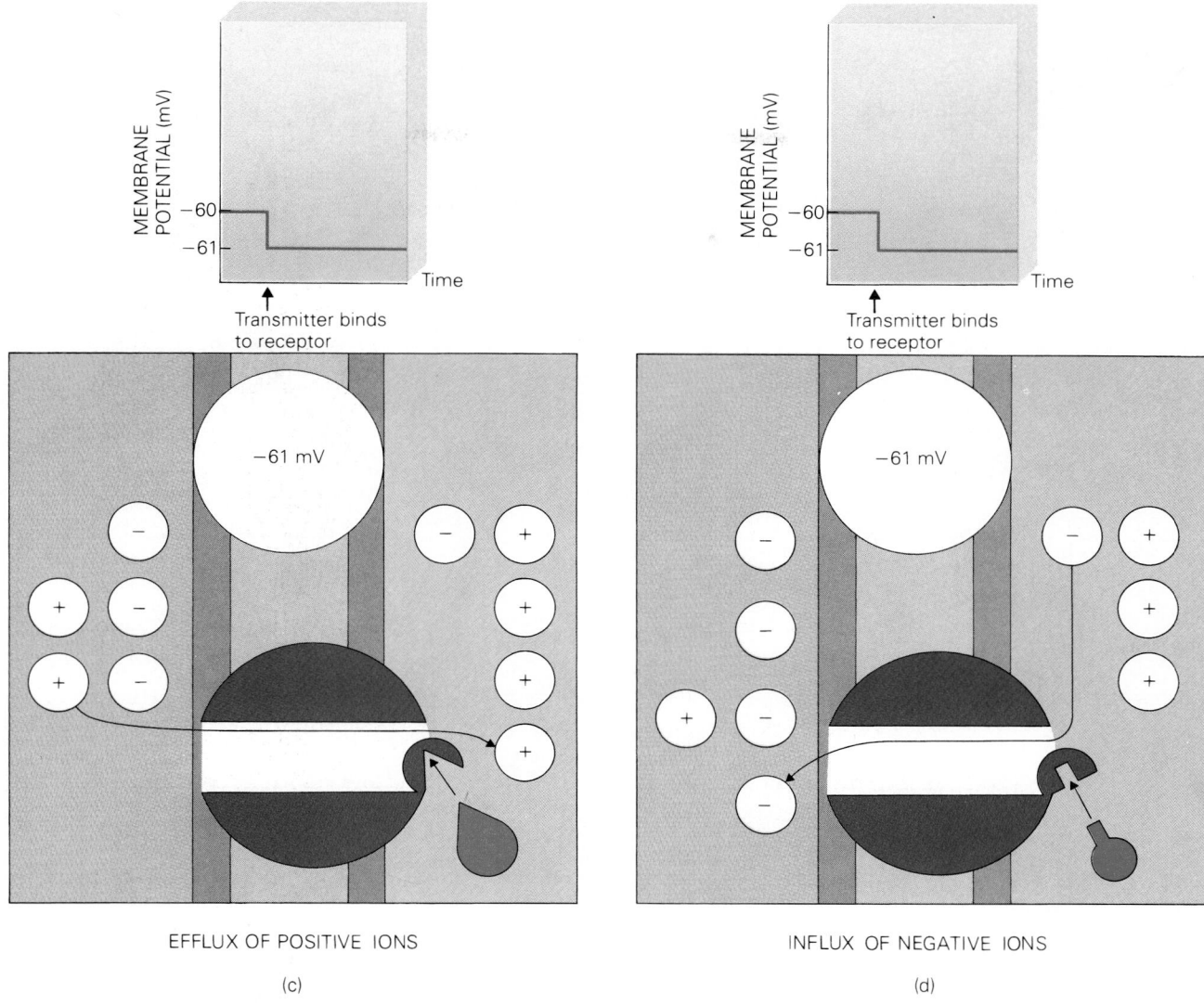

EFFLUX OF POSITIVE IONS

(c)

INFLUX OF NEGATIVE IONS

(d)

RECEPTOR-ACTIVATED MEMBRANE-POTENTIAL HYPERPOLARIZATION

of the cell. This in turn depolarizes the cell, making the membrane potential more positive (Fig. 7–25). Likewise, the efflux of positive charges from the interior of the cell or the influx of negative charges into the cell hyperpolarizes it, causing the membrane potential to become more negative.

The properties of these chemically activated ion channels are quite different from those of the voltage-activated ion channels responsible for the action potential. First of all, the chemically activated channels remain open as long as the transmitter is bound to the receptor. Secondly, the chemically activated channels are generally not sensitive to changes in the membrane potential.

The activation of a single receptor results in the influx or efflux of one or more types of ion. The number of ions that flow through the channel per unit time is referred to as the **single channel current.** The **synaptic current** is made up of all the single channel currents through the membrane at the synapse.

The flow of synaptic current resulting from the release of transmitters from one vesicle produces a change in the membrane potential referred to as a **unitary postsynaptic potential** (Fig. 7–26). These postsynaptic potentials can summate, or act together, to depolarize or hyperpolarize the membrane. Postsynaptic potentials that depolarize the membrane tend to excite the nerve cells to discharge action potentials and are therefore called **excitatory postsynaptic potentials (EPSPs).** In contrast, postsynaptic potentials that hyperpolarize the mem-

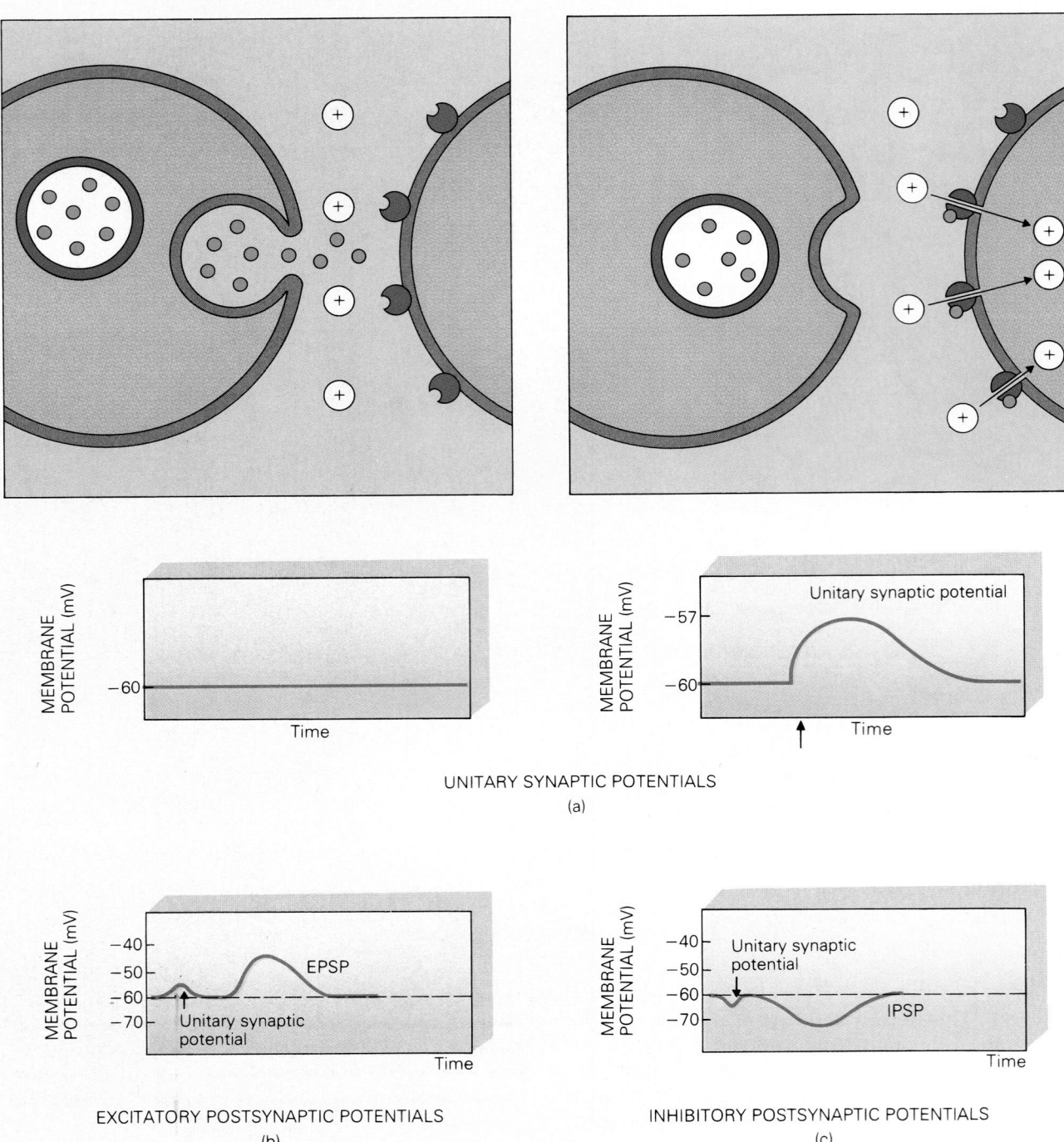

Figure 7–26
Postsynaptic potentials. (*a*) Unitary synaptic potentials. (*b*) Excitatory post-synaptic potentials (EPSPs). (*c*) Inhibitory postsynaptic potentials (IPSPs).

brane potential tend to prevent or inhibit the nerve cell from generating an action potential and are therefore called **inhibitory postsynaptic potentials (IPSPs).**

Excitatory Postsynaptic Potentials. A number of ions are capable of producing EPSPs. We will first focus on the mechanisms involved in producing EPSPs by Na⁺ ions alone. The synaptic activation of

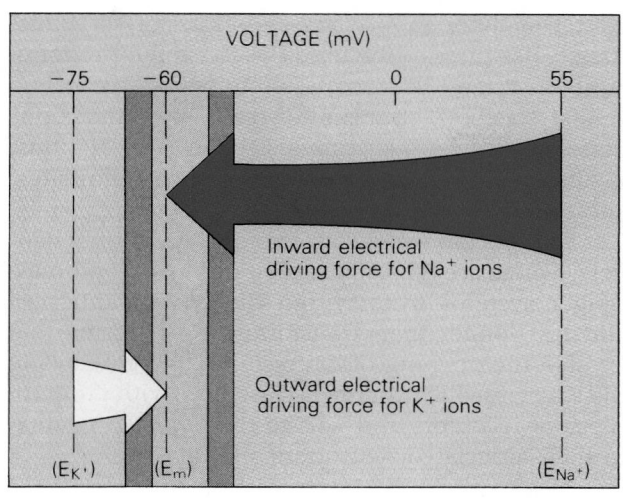

(a)

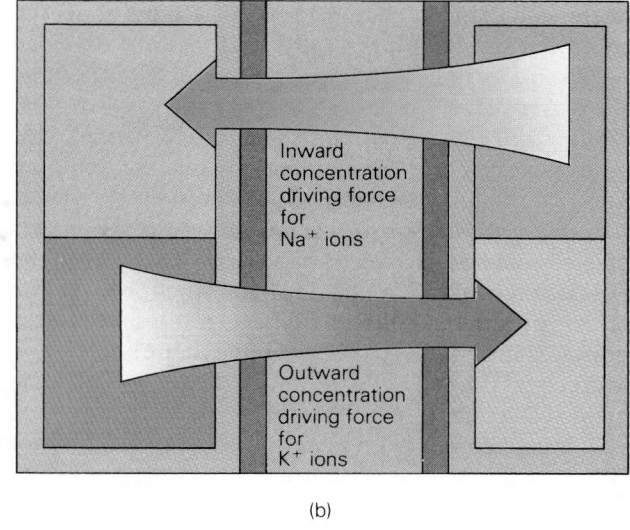

(b)

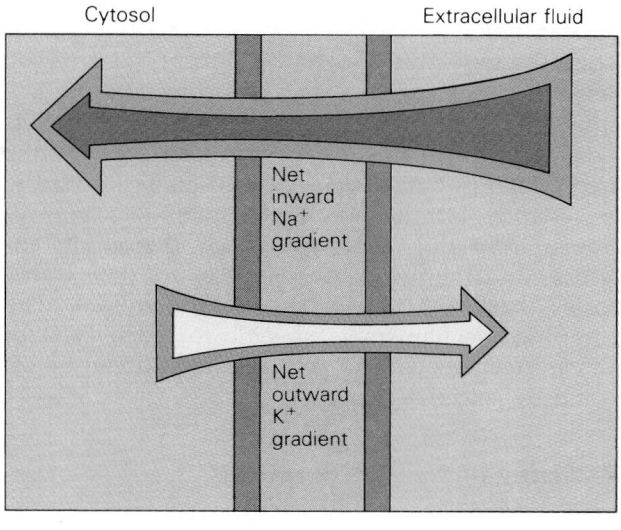

(c)

Figure 7–27
The net gradient for ion flow comprises the electrical gradient (*a*) and the chemical (concentration) gradient (*b*). The net gradient for Na^+ flow into the cell is greater than the net gradient K^+ flow out of the cell (*c*).

a receptor that selectively allows Na^+ to flow results in the influx of Na^+ into the interior of the cell. The addition of more positive charges into the cell therefore depolarizes the membrane. The influx of Na^+ ions is due to two factors: a concentration gradient and an electrical gradient. As we learned earlier, the concentration of Na^+ ions is greater outside of the cell than inside; this concentration gradient moves Na^+ ions into the cell. In addition, a membrane potential of -60 mV establishes a strong electrical gradient for the influx of Na^+ (Fig. 7–27). The equilibrium potential for Na^+ is typically $+55$ mV. Recall that the equilibrium potential is the membrane potential at which the movement of ions resulting from electrical forces is equal to the movement of ions—in the opposite direction—resulting from the concentration gradient. Since the value of the rest-

ing membrane potential is much less than the equilibrium potential for Na^+, the electrical force will not be sufficient to counteract the inward flux of Na^+ ions. In fact, the electrical driving force for Na^+ aids in the inward influx of Na^+ ions.

The electrical driving force acting on a particular ion is the difference between the membrane potential and the equilibrium potential for that ion ($E_m - E_{ion}$). In the case of Na^+, this driving force is $[-60$ mV $- (+55$ mV$)]$ or -115 mV. The negative value of the electrical driving force for Na^+ aids in attracting Na^+ ions into the cell. Thus, the concentration gradient and the electrical gradient both act to produce an influx of Na^+.

Excitatory postsynaptic potentials can also be produced by the opening of channels that are somewhat nonselective and result in the simultaneous flow of both Na^+ ions into the cell and K^+ ions out of the cell. This is characteristic of a type of acetylcholine receptor found at the junction between nerve and muscle cells. How can an ion channel that allows positively charged ions such as K^+ to flow out of a cell cause the membrane to depolarize? This occurs because more Na^+ ions flow into the cell than do K^+ ions out. At a resting membrane potential of -60 mV, the driving forces that influence the influx of Na^+ ions are greater than those that affect the efflux of K^+ ions. Like the Na^+ ion, the forces acting on K^+ produce K^+ fluxes that act in the same direction, but in the case of K^+ we have an outward flux. Moreover, the outward K^+ flux is not as great as the inward Na^+ flux.

The electrical force resulting from the difference between the resting potential and the equilibrium potential for K^+ is [-60 mV$-(-75$ mV)] or $+15$ mV. The positive value of the driving force results in an efflux of K^+ ions. Together, the concentration and electrical forces act to move K^+ ions from the interior of the cell outward. From the calculation of electrical driving forces, however, we can see that the resulting magnitude of the K^+ efflux is much smaller than the magnitude of the Na^+ influx (Fig. 7–27). Thus, when this type of acetylcholine receptor is activated, the total influx of Na^+ ions will be substantially greater than the efflux of K^+ and will therefore result in the depolarization of the membrane.

Inhibitory Postsynaptic Potentials. As with the EPSPs, there are a number of ways that an IPSP can be produced. One mechanism is the synaptic activation of a receptor that opens its channel to K^+ ions. Because the concentration and electrical gradients for K^+ cause it to move outward, the interior of the cell loses positive charges and becomes more negative. Consequently, the membrane potential hyperpolarizes, and an inhibitory postsynaptic potential develops. Other receptors that produce IPSPs are those that permit Cl^- to flow across the membrane. The increase in the flow of negative charges into the cell also acts to hyperpolarize the membrane potential.

Closing of Ion Channels by Receptors. Our discussion of transmitter-receptor interactions has thus far focused only on the opening of ion channels. There is also a class of receptors, however, that close their ion channels when coupled to a transmitter. The closing of these chemically activated ion channels can also produce depolarization or hyperpolarizing changes of the membrane potential.

For example, the closing of Na^+ channels that are normally open prevents the influx of Na^+ ions. As a consequence, the efflux of K^+ is the predominant effect, and the membrane hyperpolarizes. In a similar fashion, transmitter-receptor interactions that close K^+ channels prevent the efflux of K^+ ions. In this situation, the influx of Na^+ predominates, and the membrane depolarizes.

The mechanisms of transmitter-activated channel closings are typically carried out by second messenger systems from within the postsynaptic cell and last longer than transmitter mechanisms that involve the opening of ion channels. The details of the ways in which channels close vary with different types of receptors and will be discussed in relation to their associated neurotransmitters.

Termination of Synaptic Transmission

The termination of synaptic transmission occurs when the transmitter is removed from the synaptic cleft. This process is accomplished in most neurotransmitter systems by the transport of the transmitter back into the presynaptic terminal. A **re-uptake** mechanism pumps most transmitters back into the presynaptic terminal. Other transmitters are removed from the synaptic cleft by degrading enzymes, and the metabolic products are then transported back into the presynaptic terminal. The pumping mechanism is specific for each type of transmitter substance or metabolite and can be affected by selective drugs.

Recycling of Neurotransmitters

After a transmitter has been transported back into the presynaptic terminal, it is again packaged into vesicles for storage in preparation for release. In cases in which the metabolite is transported into the presynaptic terminal, the transmitter is resynthesized from the metabolic precursor and then packaged and stored into vesicles. The repacking of the neurotransmitter is an energy-dependent process requiring ATP. The details of this process will be discussed later in relation to the neurotransmitter, norepinephrine.

Diffuse and Discrete Chemical Synapses

Chemical synapses have been shown to have either discrete or diffuse actions. In **discrete synapses,** the chemical neurotransmitter is released from restricted areas of the presynaptic terminal, called **active zones,** into a small synaptic cleft; only 30 nm

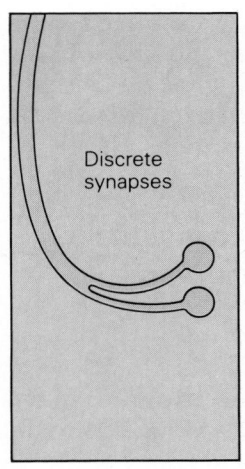

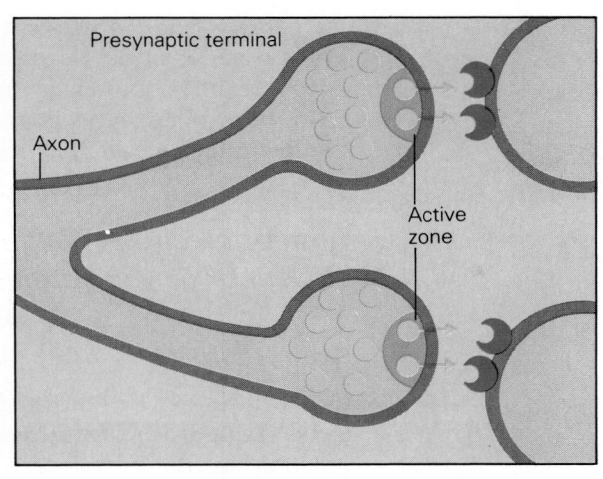

(a)

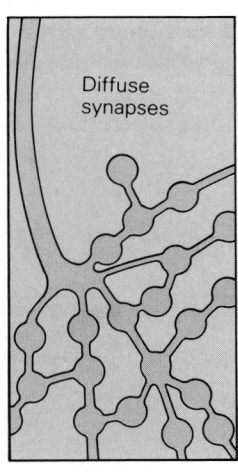

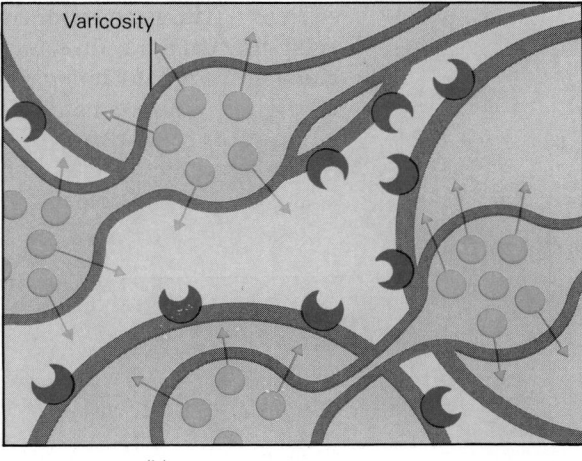

(b)

Figure 7–28
(*a*) In discrete synapses, transmitters are released only through active zones of the axon terminal and travel across very small clefts to target cells. (*b*) In diffuse synapses, axon terminals have varicosities that release transmitters over a wider range.

separate the pre- and postsynaptic membranes (Fig. 7–28). An example of a discrete synapse is the junction between nerve and striated muscle cells, which tend to activate a small region on the muscle fiber.

In **diffuse synapses,** on the other hand, transmitter release is not limited to specific active zones, and the distance between the presynaptic and postsynaptic membrane can be as much as 150 nm. These synapses have the form of beads on a chain, called **varicosities.** The varicosities are an extension of the axon and form an overlapping network of synapses **en passant,** or synapses in passage. As the action potential invades each varicosity, vesicles fuse with the presynaptic membrane, and the action potential continues on to the next varicosity. The effect of this type of synapse is to activate a large surface area of one cell or a large number of cells in a diffuse manner. The diffuse synapse is typical of nerve terminals of the sympathetic autonomic ner-

vous system and of nerve cells containing catecholamines within the central nervous system.

Neurotransmitters and Associated Receptors

Chemical neurotransmitters can be broadly classified into two groups: **low-molecular-weight transmitters** and **neuropeptide transmitters** (see Tables 7–1 and 7–2). The low-molecular-weight transmitters are synthesized within the presynaptic terminal. The necessary synthesizing enzymes are produced within the soma and transported to the terminal region. Neuropeptides, on the other hand, are fabricated in the soma and carried via axonal transport mechanisms to the synaptic terminal.

Table 7–1

Low-Molecular-Weight Transmitters

Acetylcholine

$$CH_3-\overset{\overset{\displaystyle O}{\|}}{C}OH_2CH_2\overset{\oplus}{N}(CH_3)_3$$

Dopamine

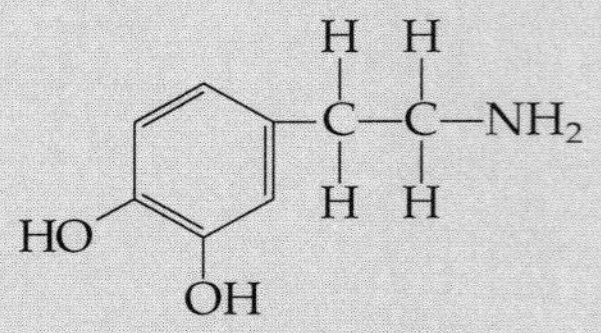

Norepinephrine

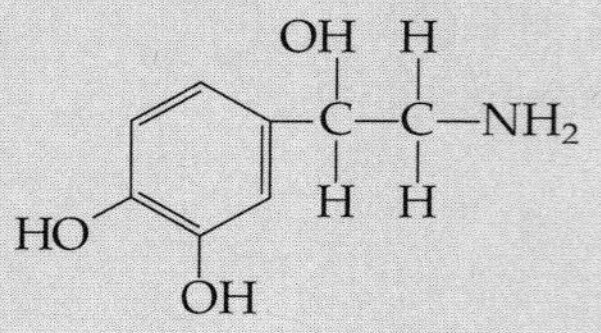

Serotonin

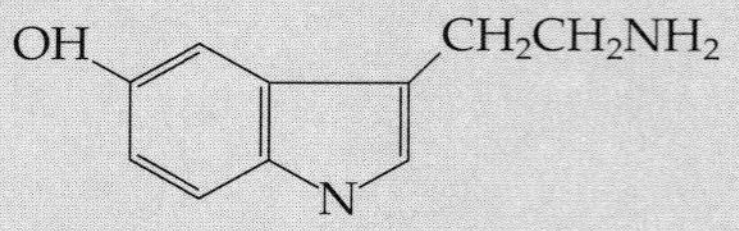

Glutamate $\quad HOOCCH_2CH_2\overset{\overset{\displaystyle NH_2}{|}}{C}HCOOH$

GABA $\quad HOOCCH_2CH_2CH_2NH_2$

Although the processes of transmitter synthesis, release, degradation, uptake, and physiological effect are not completely understood for the peptides, they are fairly well characterized for many of the low-molecular-weight transmitters.

Low-Molecular-Weight Transmitters: Products of the Axon Terminal

Acetylcholine

One of the first low-molecular-weight transmitters to be studied was **acetylcholine (ACh).** It is found not only in the central nervous system but also in the peripheral nervous system, where it acts as the chemical transmitter between nerve and muscle. ACh is synthesized from acetyl CoA and choline with the catalytic enzyme, choline acetyltransferase (Fig. 7–29). ACh is packaged and stored in vesicles within the presynaptic terminal, where is it positioned for release near the active zones. The fusion of the synaptic vesicle with the presynaptic membrane causes the release of ACh into the synaptic

Table 7–2

Neuropeptides

Neuroactive Peptides: Mammalian Brain Peptides Categorized According to Principal Tissue Localization

Hypothalamic-releasing hormones	Gastrointestinal peptides
Thyrotropin-releasing hormone	Vasoactive intestinal polypeptides
Gonadotropin-releasing hormone	Cholecystokinin
Somatostatin	Gastrin
Corticotropin-releasing hormone	Substance P
Growth hormone–releasing hormone	Neurotensin
	Methionine-enkephalin
Neurohypophyseal hormones	Leucine-enkephalin
Vasopressin	Insulin
Oxytocin	Glucagon
Neurophysin(s)	Bombesin
	Secretin
Pituitary peptides	Somatostatin
Adrenocorticotropin	Motilin
β-Endorphin	
α-Melanocyte-stimulating hormone	Others
Prolactin	Angiotensin II
Luteinizing hormone	Bradykinin
Growth hormone	Sleep peptide(s)
Thyrotropin	Calcitonin
	CGRP (calcitonin gene-related peptide)
	Neuropeptide Y

Table 7–3

Types of Neurotransmitters and Receptors

Neurotransmitter	Receptor	Membrane Conductance	Membrane Potential	2nd Msgr
glutamate	kainate	increase g_{Na}, g_K	EPSP	
	quisqualate	increase g_{Na}, g_K	EPSP	
	NMDA	increase g_{Ca}	EPSP	
acetylcholine	nicotinic	increase g_{Na}, g_K	EPSP	
	muscarinic M1	decrease g_K	EPSP	IP$_3$ and DAG
	muscarinic M2	increase g_K	IPSP	cAMP
serotonin	5HT-1A	increase g_K	IPSP	cAMP
	5HT-1B			
	5HT-1C	increase g_{Cl}	IPSP	IP$_3$
	5HT-1D			
	5HT-2	decrease g_K	EPSP	IP$_3$
	5HT-3	increase g_{Na}, g_K	EPSP	
GABA	GABA-A	increase g_{Cl}	IPSP	
	GABA-B	increase g_K	IPSP	cAMP?
dopamine	D-1		EPSP?	cAMP
	D-2		IPSP	cAMP
norepinephrine	alpha-1	increase g_K	IPSP (CNS)	
			[contraction (PNS)]	IP$_3$, DAG
	alpha-2	decrease g_{Ca}		cAMP
	beta-1		[heart acceleration]	cAMP
	beta-2		[dilation (PNS)]	cAMP

cleft. It then diffuses across the cleft and binds with receptors on the postsynaptic membrane.

In the central and peripheral nervous systems, there are two types of receptors for acetylcholine: nicotinic receptors and muscarinic receptors. Nicotinic acetylcholine receptors are sensitive to nicotine, while muscarinic receptors respond to the drug muscarine.

Nicotinic Acetylcholine Receptors. Nicotinic acetylcholine receptors have been well characterized and found to consist of five subunits: beta, gamma, delta, and two alphas. The five components are thought to be arranged so that an ion channel is formed at the central core of the receptor (Fig. 7–29b). The active binding sites for the nicotinic receptor are on the two alpha subunits. Both binding sites must be occupied by an acetylcholine molecule in order for the receptor to become activated. Once activated, the receptor opens its gate to permit the simultaneous influx of Na$^+$ ions and the efflux of K$^+$ ions (see Table 7–3).

As we have seen, the driving forces for Na$^+$ influx are much greater than for K$^+$ efflux. Consequently, the excess positive charge movement into the cell causes the membrane to depolarize. The ion channel of the nicotinic receptor remains in the open state until ACh uncouples from its receptor. After ACh dissociates from its receptor, the receptor channel closes and Na$^+$ and K$^+$ are no longer able to pass through the channel. ACh is then free to diffuse within the synaptic cleft, where it binds with the membrane-bound enzyme acetylcholinesterase, **AChE.** The AChE enzyme degrades ACh by hydrolysis to produce choline and acetate. Choline is then taken up into the presynaptic terminal by a high-affinity uptake mechanism and recycled to be used again to synthesize ACh. This process of cholinergic nicotinic transmission is found at the neuromuscular junction in addition to various locations in the central nervous system.

Muscarinic Acetylcholine Receptors. The structure and function of the second type of acetylcholine receptor, the muscarinic receptor, are quite different from those of the nicotinic acetylcholine receptor. Two types of muscarinic receptors, M1 and M2, have been identified. Both are composed of 7 membrane-spanning domains and both exert their actions through a G protein. The activation of M1 receptors, however, results in a decrease of K$^+$ conductance via phospholipase C, while M2 receptor activation causes an increase in K$^+$ conductance

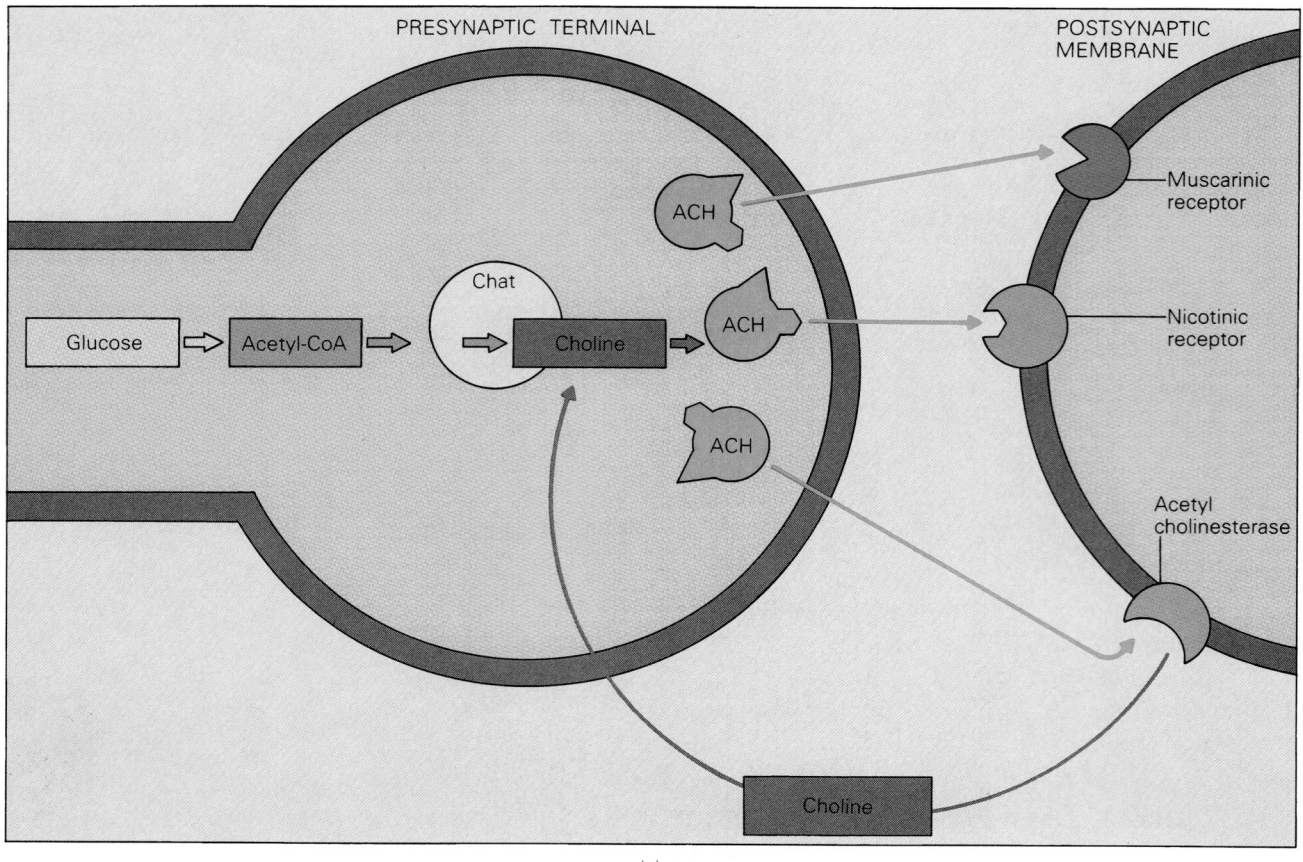

(a)

Figure 7–29
(*a*) Acetylcholine is synthesized from acetyl CoA in the axon terminal and binds to two types of receptors: muscarinic and nicotinic receptors. After binding to acetylcholinesterase, acetylcholine is degraded to choline and taken up into the presynaptic terminal for resynthesis of ACh. (*b*) The nicotinic acetylcholine receptor consists of five subunits. When acetylcholine binds to two of the subunits, the protein opens to form an ion channel into the postsynaptic cell.

via the inhibition of adenylate cyclase. As a consequence, when ACh binds to an M1 receptor it depolarizes the membrane, and when it binds to an M2 receptor it causes a hyperpolarization.

Biogenic Amines

Biogenic amines are a class of low-molecular-weight transmitters characterized by the presence of an amine group. A subgroup of the biogenic amines, the catecholamines, have a catechol ring. This subgroup includes the transmitters **dopamine, norepinephrine,** and **epinephrine.** The synthesis of each catecholamine follows a similar biochemical pathway that starts with the synthesis of dopamine.

Dopamine. Dopamine is synthesized from the amino acid tyrosine, which is converted to DOPA by the enzyme tyrosine hydroxylase (TH). Tyrosine hydroxylase is the controlling enzyme that regulates the overall synthesis of dopamine. DOPA, in turn, is converted to dopamine by the enzyme DOPA decarboxylase (Fig. 7–30).

Based on their pharmacological properties, two subtypes of dopamine receptors have been identified: D1 and D2 (Fig. 7–30). **D1 receptors** are cou-

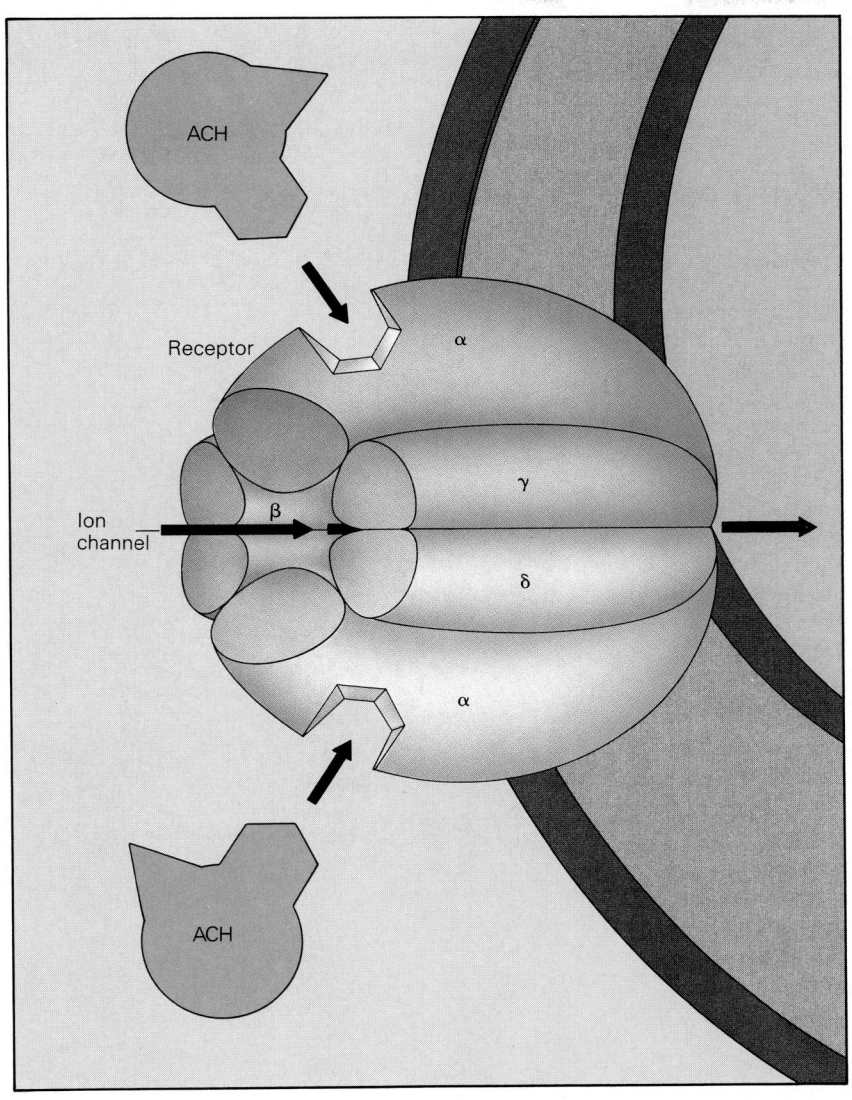

(b)

pled to G_s proteins, leading to the activation of adenylate cyclase (AC), while **D2 receptors** are coupled to G_i proteins, which cause a decrease in AC activity. The activation of D2 receptors has been shown to hyperpolarize the postsynaptic membrane by increasing potassium conductance. After uncoupling with the receptor, dopamine is transported by a re-uptake pumping mechanism back into the presynaptic terminal, where it is repackaged into synaptic vesicles. Eighty percent of the dopamine within the synaptic cleft is transported back to the presynaptic terminal. Thus, the re-uptake process is the main

mechanism by which dopamine transmission is terminated. The remaining 20% of dopamine is degraded within the cleft by the enzyme catechol-O-methyl transferase (COMT).

The synaptic transmission of dopamine is greatly affected by commonly used illegal drugs. Cocaine, for example, inhibits the re-uptake of dopamine into the presynaptic terminal, and amphetamine increases the release of dopamine into the synaptic cleft. Both drugs effectively increase the levels of dopamine within the cleft to activate dopamine receptors.

CLINICAL FOCUS

Alzheimer's Disease: The Loss of Acetylcholine Neurons and Memory

Alzheimer's disease is characterized by the loss of recent memories. Initially, a person's more distant memories are preserved, but as the disease progresses even the earliest of memories are lost. Eventually, the victim is unable to recognize even close family members. As the disease progresses further, other deficits begin to appear. Along with loss of memory, speech is often impaired; gradually, difficulties in reading, writing, and performing complex movements occur. In later stages, Alzheimer victims are unable to perform even simple tasks such as eating, dressing, and caring for themselves. Most patients die within 5 to 15 years after the onset of the disease.

Postmortem studies of brain tissue from Alzheimer's patients have revealed a selective loss of neurons that synthesize acetylcholine. These acetylcholine neurons are located at the base of the brain (the nucleus basalis of Meynert, the diagonal band of Broca, and the medial septal nucleus) and send axons to all areas of the neocortex and hippocampal formation. These areas of brain are responsible for sensory perception, movement, speech, and language, as well as learning and memory (see Chapters 8, 9, and 11). The loss of these neurons is thought to cause the gradual deterioration of memory, language, and motor functions.

Of all individuals with Alzheimer's, 15% to 20% have inherited the disorder. Research indicates that the genetic form of this disease follows an autosomal dominant pattern of inheritance and that offspring have a 50% chance of inheriting the disease from a parent.

Environmental factors also may play a role in Alzheimer's disease. Studies have shown abnormally high concentrations of aluminum in degenerating neurons of Alzheimer's patients; however, the role of aluminum still needs to be clarified. In addition, infectious agents such as viruses may be involved. Slow-acting viruses have been found to cause other forms of dementia in humans. At this time, there is not treatment to prevent or slow the progression of Alzheimer's disease. Drugs that increase the amount of acetylcholine have been found to temporarily improve memory in Alzheimer's patients. This approach, however, is useful only when a sufficient number of acetylcholine-producing neurons remain.

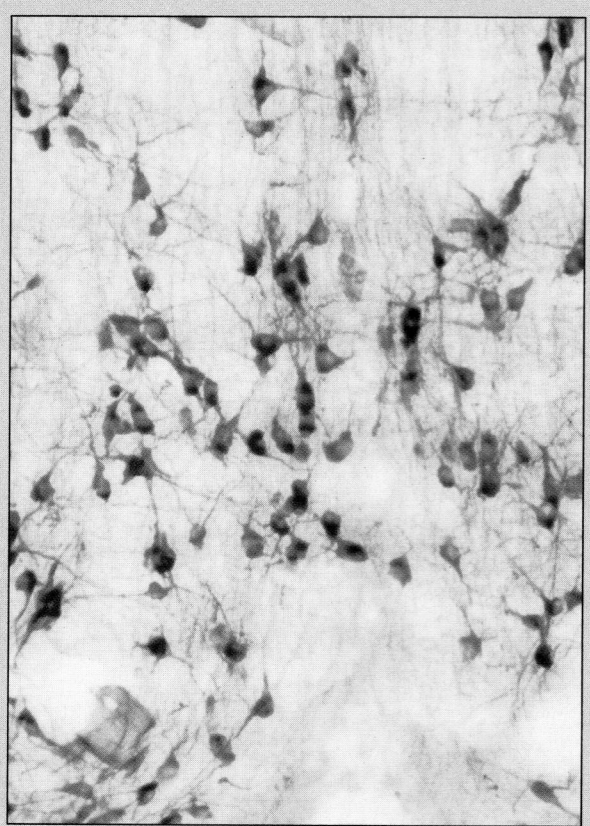

ChAT immunocytochemistry of cholinergic neurons. (Courtesy of Dr. Walter Low.)

As the median age of our population increases, more people will succumb to this age-related disorder. Within the next 50 years, nearly 20% of our population will be older than 65 years. Up to 6% of this group will be afflicted with Alzheimer's disease, while approximately 15% will suffer from a milder form of the disorder. The cost of caring for these individuals currently exceeds $10 billion per year and is expected to exceed $40 billion within the next few decades.

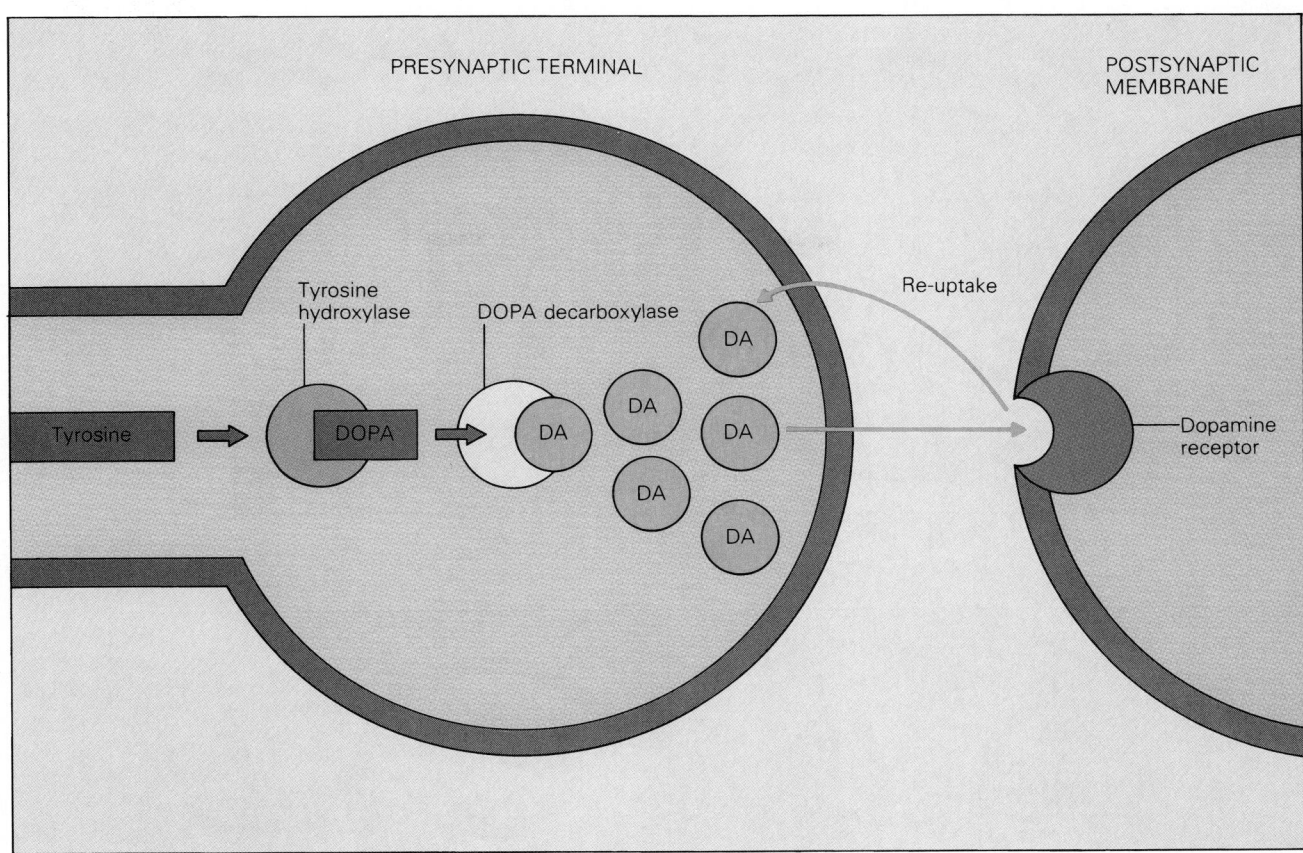

Figure 7–30
Dopamine is synthesized in the axon terminal and released to the synaptic cleft, to be pumped back into the presynaptic terminal.

Norepinephrine. Norepinephrine (NE) is another member of the catecholamine family that is found throughout the central nervous system and also at the junction between nerves and smooth muscles in the autonomic nervous system. Norepinephrine is synthesized from dopamine by the enzyme dopamine-beta-hydroxylase (DBH) (Fig. 7–31a).

The formation of NE is quite labile and is regulated by mechanisms that can change the enzymatic activity associated with NE synthesis over either a short or long period of time. The short-term changes involve the modulation of tyrosine hydroxylase. Typically the activity of TH is regulated by both dopamine and norepinephrine by the process of end-product inhibition (see Chapter 5). When the enzyme is phosphorylated, however, the process of end-product inhibition is less effective. In addition, the phosphorylation of tyrosine hydroxylase increases the affinity of the enzyme to the tyrosine substrate. Thus the phosphorylation of the tyrosine hydroxylase can increase its enzymatic activity. The phosphorylation of tyrosine hydroxylase has been shown to occur by cAMP-dependent protein kinase and Ca^{2+}/calmodulin mechanisms (see Chapter 5). The short-term increases in the synthesis of NE occur within minutes and are easily reversible.

Long-term increases in NE can be caused by factors such as stress, since stress can lead to increases in tyrosine hydroxylase and dopamine beta-hydroxylase enzymes. Environmental factors can therefore play a significant role in modifying nerve cell function on a long-term basis.

The binding of NE to alpha receptor in the central nervous system opens K^+ channels and causes a hyperpolarization of the postsynaptic cell. In the peripheral nervous system, two types of norepinephrine receptors have been identified: alpha receptors and beta receptors. Alpha-1 receptors are found on the smooth muscles of blood vessels. When they are activated, an increase in Ca^{2+} ion

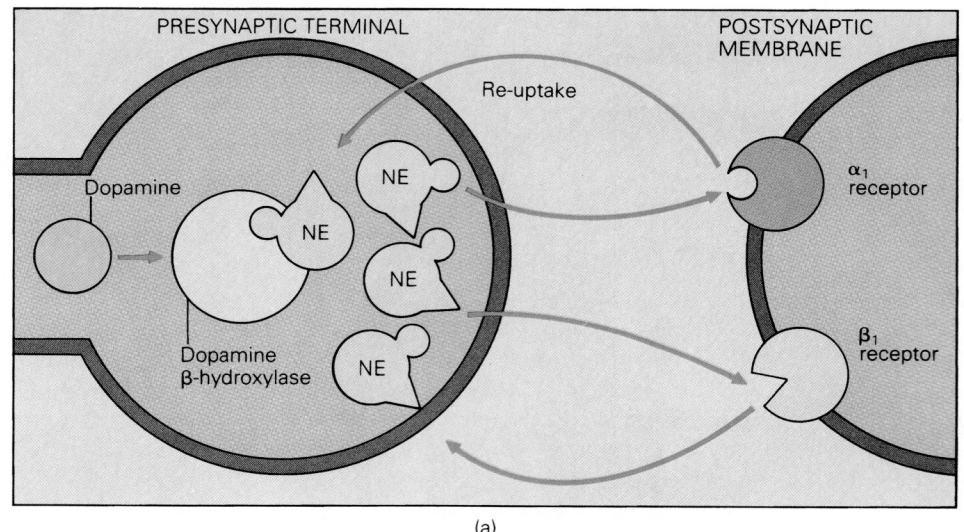

(a)

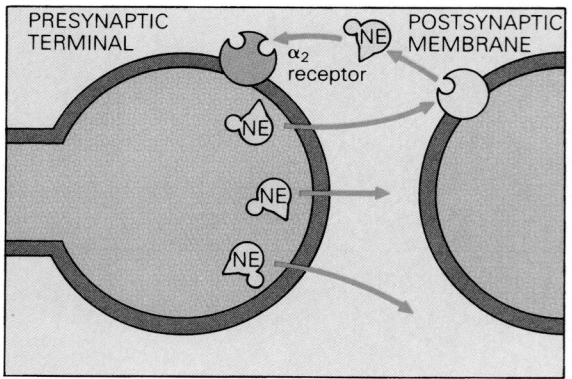

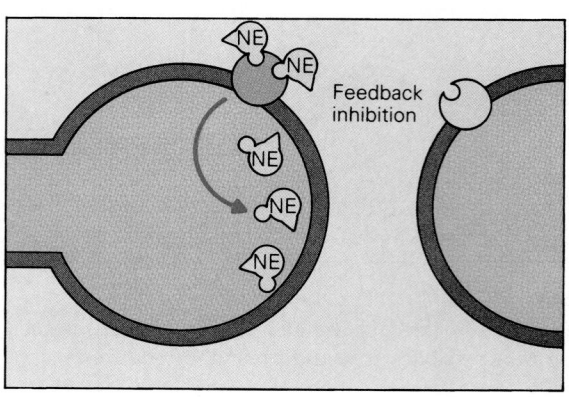

(b)

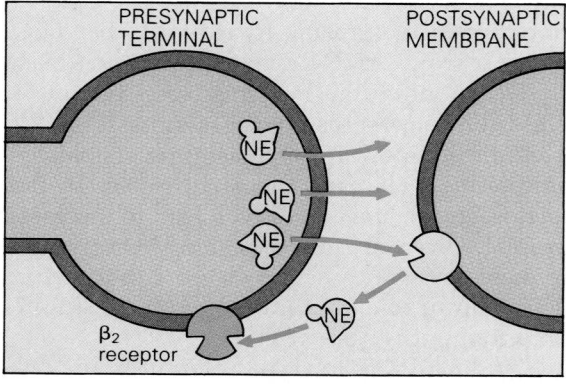

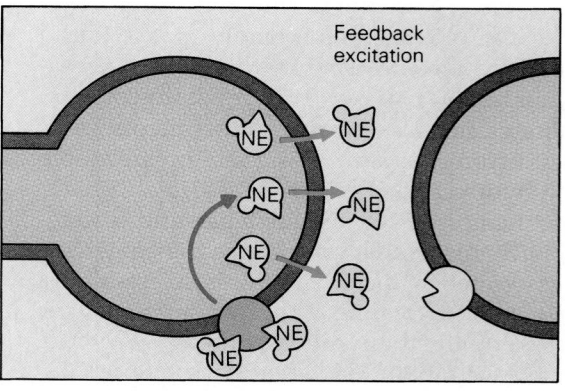

(c)

Figure 7–31

(*a*) Norepinephrine (NE) binds to two types of receptors on the post-synaptic membrane: α_1 and β_1. On the presynaptic membrane, binding of NE to α_2 receptors causes norepinephrine release to be inhibited (*b*), while binding of NE to β_2 receptors causes more NE to be released (*c*).

influx occurs, which in turn causes the contraction of smooth muscles. Alpha receptors also stimulate the hydrolysis of phosphatidylinositol, a lipid in the postsynaptic membrane that activates the intracellular messenger diacylglycerol. Diacylglycerol in turn activates protein kinase C, which initiates a variety of cellular functions (see Chapter 5).

Beta-2 receptors are also found on smooth muscles. Their activation, however, results in a relaxation of smooth muscles by mechanisms that remain to be determined. In addition to activating ionic channels, the binding of NE to beta-2 receptors activates metabolic pathways by converting ATP to cAMP. As previously discussed, this second messenger in turn activates a cAMP-dependent protein kinase, which regulates a number of important cellular functions (see Chapter 5). Beta-1 receptors are found in the heart, kidney, and adipose tissue, where they are responsible for the acceleration of heart rate, renin secretion, and lypolysis, respectively.

Alpha-2 and beta-2 receptors are located on the membrane of the presynaptic terminal and appear to regulate the amount of NE that is released (Fig. 7–31b). These receptors are often referred to as **autoreceptors.** As the amount of NE within the synaptic cleft is increased, more alpha-2 autoreceptors become activated. These autoreceptors then inhibit the release of NE from the presynaptic terminal in a process of **feedback inhibition.** The activation of beta-2 receptors, on the other hand, increases the release of NE in a process called **feedback excitation.**

After uncoupling from its receptor, NE is taken back up into the presynaptic terminal, where it is repackaged into vesicles in preparation for release again into the synaptic cleft. As with dopamine, this re-uptake process removes approximately 80% of the NE from the synaptic cleft and is the main mechanism that terminates NE synaptic transmission. The remaining NE is degraded within the cleft by the enzyme COMT.

Serotonin. Another common biogenic amine is **5-hydroxytryptamine (5-HT),** or **serotonin.** It is found throughout the brain, but is primarily synthesized in the region of the brain stem. The synthesis of 5-HT begins with the amino acid tryptophan, which is converted to 5-hydroxytryptophan (5-HTP) by the enzyme tryptophan hydroxylase (Fig. 7–32a). In turn, 5-HTP is converted to serotonin by 5-HTP decarboxylase.

After its release into the synaptic cleft, serotonin can interact with several types of serotonin receptors. The best characterized serotonergic receptors are the 5HT-1A, 5HT-1C, 5HT-2, and 5HT-3 receptors. Activation of either the 5HT-1A or 5HT-1C receptors results in IPSPs. The inhibitory potential of 5HT-1A receptor activation, however, is mediated by an increase in K^+ conductance via cAMP as the second messenger, while that produced by 5HT-1C activation is carried out by an increase in Cl^- conductance via IP_3. Activation of either the 5HT-2 or 5HT-3 receptors results in the generation of EPSPs. The excitatory potential produced by the activation of the 5HT-2 receptor is due to a decrease in K^+

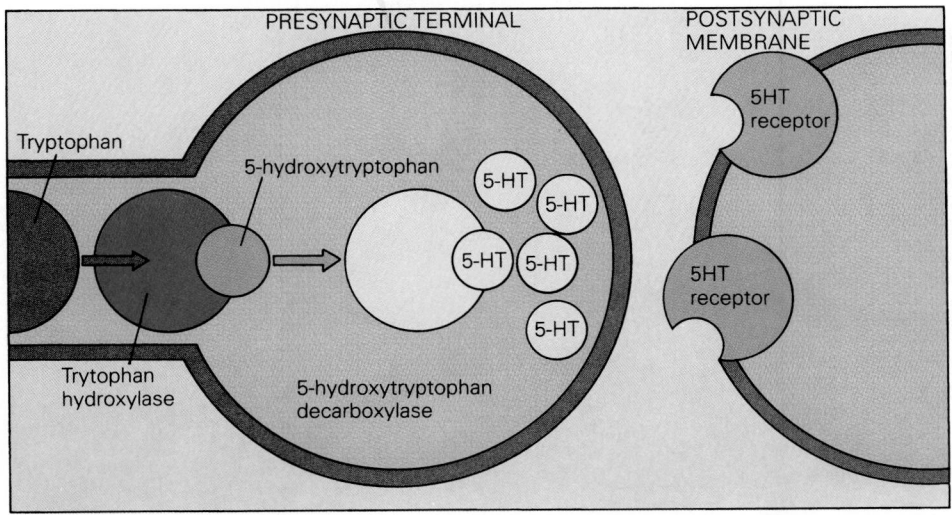

Figure 7–32
Serotonin neurotransmission.

conductance via IP$_3$, while the 5HT-3 receptor is directly coupled to an ion channel that allows the influx of Na$^+$ ions and the efflux of K$^+$ ions.

After uncoupling from its receptor, serotonin is transported back into the presynaptic terminal. As with NE and DA, approximately 80% of the serotonin within the synaptic cleft is removed by this re-uptake process. The remainder of the serotonin is degraded by an enzyme called monoamine oxidase (MAO).

Amino Acids

A variety of amino acids also satisfy the conditions necessary to be classified as neurotransmitters. We will consider only glutamate and gamma amino butyric acid, since they are found in great abundance throughout the nervous system.

Glutamate. **Glutamate** is synthesized from alpha-ketoglutarate by way of the citric acid cycle (Fig. 7–33). It is one of the most potent excitatory neuro-

transmitters in the nervous system. There are three subtypes of glutamate receptors: kainate, quisqualate, and N-methyl-D-aspartate (NMDA). Activation of the *kainate* and *quisqualate receptors* produces excitatory postsynaptic potentials by opening ion channels that increase Na$^+$ and K$^+$ conductance. *NMDA receptor* activation results in an increase in Ca^{2+} conductance. This receptor, however, is blocked by Mg^{2+} when the membrane is in the resting state and becomes unblocked when the membrane is depolarized. Thus the NMDA receptor can be thought of as both a ligand and a voltage-gated channel.

Synaptic transmission by glutamate is terminated in part by its re-uptake into the presynaptic terminal. Much of the glutamate within the synaptic cleft, however, is transported into glial cells. There, it is converted to glutamine by the enzyme glutamine synthase. The glutamine in turn is transported back into the presynaptic terminal, where it is then reconverted to free glutamate and repackaged into vesicles. Glial cells, therefore, play a critical role in regulating glutamate neurotransmission.

Figure 7–33
Glutamate is synthesized from a product of the citric acid cycle and is converted to glutamine in a glial cell before re-uptake.

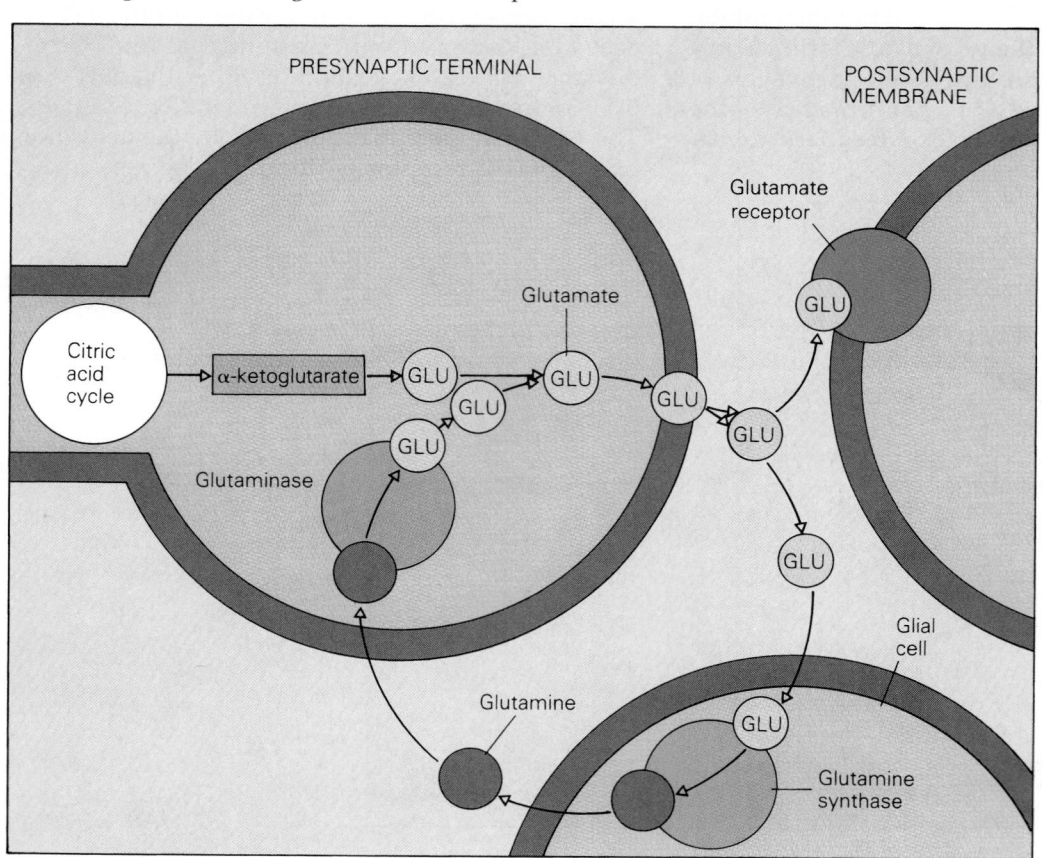

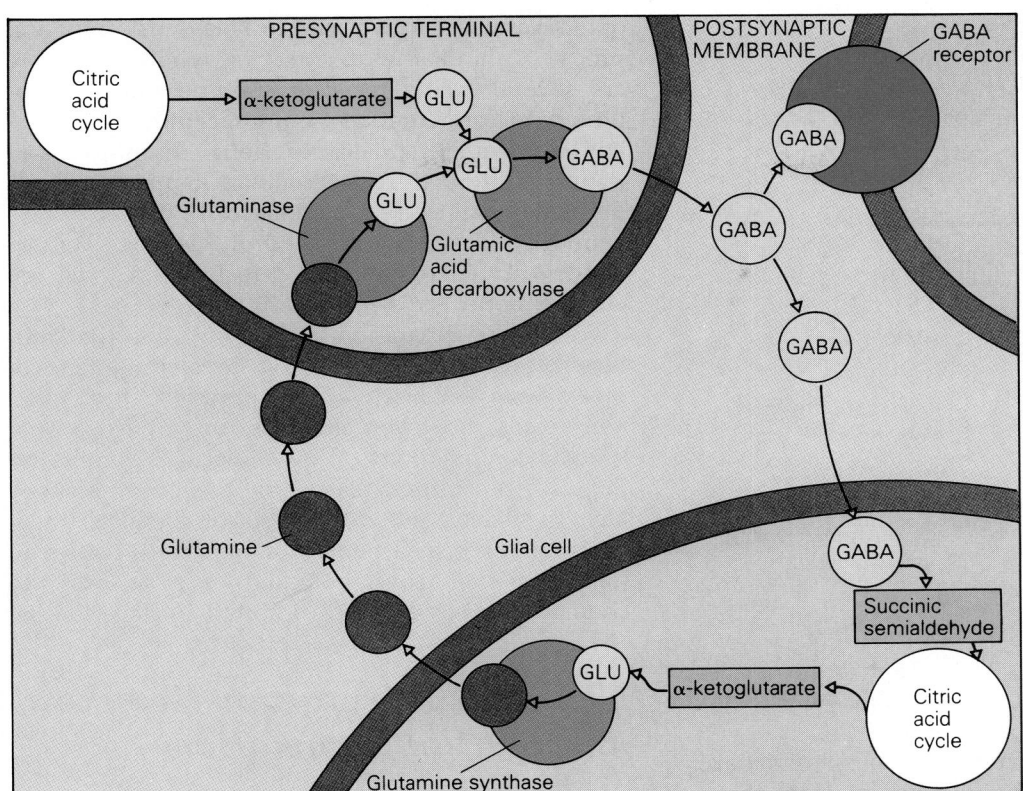

Figure 7–34
The GABA synapse.

Gamma Amino Butyric Acid. Another amino acid neurotransmitter found throughout the central nervous system is **gamma amino butyric acid (GABA).** GABA is a potent inhibitory neurotransmitter synthesized from glutamate by the enzyme glutamic acid decarboxylase (GAD) (Fig. 7–34).

There are two types of GABA receptors: $GABA_A$ and $GABA_B$. The **$GABA_A$ receptor** is a ligand-gated Cl^- channel, and its activation produces inhibitory postsynaptic potentials by increasing the influx of Cl^- ions. This receptor is composed of 2 alpha and 2 beta subunits, each of which spans the membrane 4 times. Its activation requires the binding of GABA to the beta subunits. The increase in Cl^- conductance is facilitated by drugs called **benzodiazepines,** which bind to the alpha subunits. Benzodiazepines such as librium are used as anticonvulsants and sedatives.

The activation of the **$GABA_B$ receptor** also produces inhibitory postsynaptic potentials. With this receptor, however, the IPSP results from an increase in K^+ conductance via the activation of a G protein. The synaptic transmission of GABA is terminated by its re-uptake into the presynaptic terminal and by its transport into glial cells. The mitochondria within glial cells convert GABA into succinic semial-dehyde by the enzyme GABA-T. At the same time, this enzyme is coupled to the conversion of alpha-ketoglutarate to glutamate. Glutamate in turn is converted to glutamine by glutamine synthase and is then transported to the presynaptic terminal. Within the presynaptic terminal, glutamine is converted into glutamate and subsequently into GABA to be packaged into synaptic vesicles.

Neuropeptides: Products of the Soma

Neuropeptides are chemical transmitters that consist of chains of amino acids. The processing of neuropeptides differs considerably from that of low-molecular-weight transmitters. Neuropeptides are synthesized in the soma of neurons rather than in the synaptic terminal. In addition, neuropeptides are created when large proteins, or **polyproteins,** are broken down. The various peptides are packaged within secretory vesicles and carried to the terminal area by mechanisms of fast axonal transport. Within the synaptic terminal, vesicles containing peptides are found to co-exist with vesicles containing low-molecular-weight transmitters. Table 7–4 lists the low-molecular-weight and neuropeptide transmitters that have been found within the same

Table 7–4

Colocalization of Low-Molecular-Weight Transmitters and Neuropeptides

Low-Molecular-Weight Transmitter	Neuropeptide
Acetylcholine	Vasoactive intestinal peptide
Norepinephrine	Somatostatin Enkephalin Neurotensin
Dopamine	Cholecystokinin Enkephalin
Adrenalin	Enkephalin
Serotonin	Substance P Thyrotropin-releasing hormone

synaptic terminals. It is thought that the released neuropeptide modulates the actions of the low-molecular-weight neurotransmitter.

At the terminal ending, the process of synaptic transmission of peptides is different from that of low-molecular-weight transmitters. Once they are released into the synaptic cleft, there are no re-uptake mechanisms to recycle the neuropeptides. Therefore the process of peptide transmission cannot be sustained as it is for the low-molecular-weight transmitters.

Neuropeptides can be classified in one of several families of peptides based on their amino acid sequence and function (Table 7–5). The family of **neurohypophyseal peptides** is structurally similar to those found in the posterior pituitary and function to regulate plasma osmolarity and lactation. **Secretins** and **gastrins** are peptides that are structurally similar to the peptides and hormones found in the gastrointestinal system (see Chapter 26). The family of **insulins** is similar in structure to the insulin hormone and is responsible for the growth and maintenance of nerve cells, while the **somatostatins** are structurally similar to growth hormone.

Opiates are peptides that bind to opioid receptors. They appear to be involved in the regulation of pain information. Opioid peptides include met-enkephalin, leu-enkephalin, dynorphin, and beta-endorphin (see Table 7–5). Structurally, they share homologous regions consisting of the amino acid sequence Tyr-Gly-Gly-Phe. The opiates are derived from three propeptides: **proenkephalin, pro-opiomelanocortin,** and **prodynorphin.** Proenkephalin gives rise to met- and leu-enkephalin; pro-

opiomelanocortin gives rise to beta-endorphin; and prodynorphin is the precursor of dynorphin. There are several opioid receptor subtypes. Beta-endorphin binds preferentially to **mu receptors;** enkephalins bind preferentially to **delta receptors,** and dynorphin binds preferentially to **kappa receptors.** The enkephalins are metabolized by two enzymes: aminopeptidase, which hydrolyzes the Tyr-Gly bond, and enkephalinase, which hydrolyzes the Gly-Gly bond.

From our discussion of chemical neurotransmitters, it can be seen that nerve cells are capable of producing a variety of transmitters. Individual nerve cells, however, are able to synthesize only specific combinations of low-molecular-weight and peptide transmitters based on the enzymes they have available. This defined set of chemical transmitters is used by a neuron at all of its synapses to transduce the action potentials it generates into chemical signals that are detected by target cells.

Neuronal Integration: Temporal and Spatial Summation

The dendrites and somata of nerve cells have many different types of receptors embedded within their membranes. The synaptic activation of these receptors can simultaneously produce excitatory and inhibitory responses. If a nerve cell receives both excitatory and inhibitory information from many

Table 7–5

Families of Peptides

Opioid: opiocortins, enkephalins, dynorphin, FMRF amide

Neurohypophyseal: vasopressin, oxytocin, neurophysins

Tachykinins: substance P, physalaemin, kassinin, uperolein, eledoisin

Secretins: secretin, glucagon, vasoactive intestinal peptide, gastric inhibitory peptide, growth hormone–releasing factor, peptide histidine isoleucineamide

Insulins: insulin, somatomedins, relaxin, nerve growth factor

Somatostatins: somatostatins, pancreatic polypeptide

Gastrins: gastrin, cholecystokinins

Brain Slices and Long-Term Potentiation

Physiological studies of the central nervous system have been greatly advanced by the development of new methods in which tissue from the brain is removed, cut into thin slices, and placed into incubation chambers for experimentation. In this **brain slice** technique, nerve cells are maintained in an extracellular environment similar to that found in a normal brain. The slices can also be perfused with various chemical compounds to determine their effects on neuronal activity and cell-to-cell communication. Another advantage of the slice technique is that it maintains the synaptic organization between nerve cells within the slices of tissue (see figure).

The development of the brain slice technique has enabled scientists to study various cellular mechanisms of neuronal communication. One cellular phenomenon called **long-term potentiation (LTP)** has been extensively studied using brain slices. LTP is the long-lasting enhancement of synaptic strength and was first observed in the **hippocampal formation,** an area of the brain long known to be involved in processes of learning and memory.

LTP in the hippocampus can be induced by a brief high-frequency stimulation of nerve fibers that synapse upon target pyramidal neurons. The resulting enhancement is specific only to those synapses that have been activated.

In the hippocampal formation, LTP occurs at synapses that use glutamate as the neurotransmitter. The underlying basis of LTP appears to be a persistent alteration in the **quisqualate** (Q) receptor function, most likely either an increase in the number of Q receptors within the synapse, or a change in single channel properties. Whatever the changes in Q receptor function, they occur as a result of a sequence of events. The first step in the induction of LTP is the depolarization of the postsynaptic cell. This is a result of the initial activation of Q receptors by glutamate. In contrast, the **N-methyl-D-aspartate** (NMDA) receptor is not initially activated by glutamate since the channel is still blocked by Mg^{2+} ions when the membrane potential is at the resting level (see text). With greater depolarizations of the postsynaptic membrane, however, Mg^{2+} ions are removed from the

(a) Hippocampal formation and brain slice. (© Dennis Kunkel/CNRI/Phototake.) (b) Glutamate receptors quisqualate (Q) and N-methyl-D-aspartate (NMDA) involved in long-term potentiation.

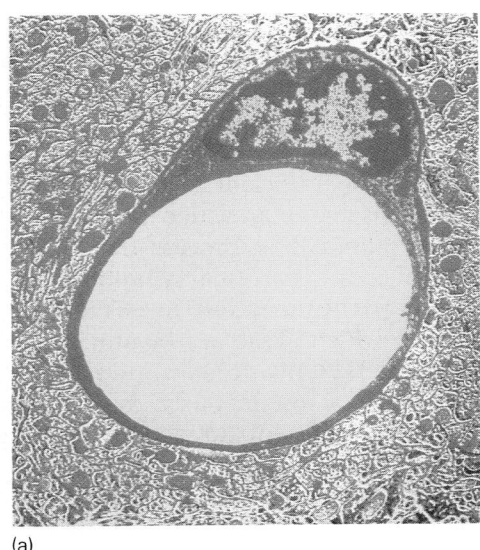

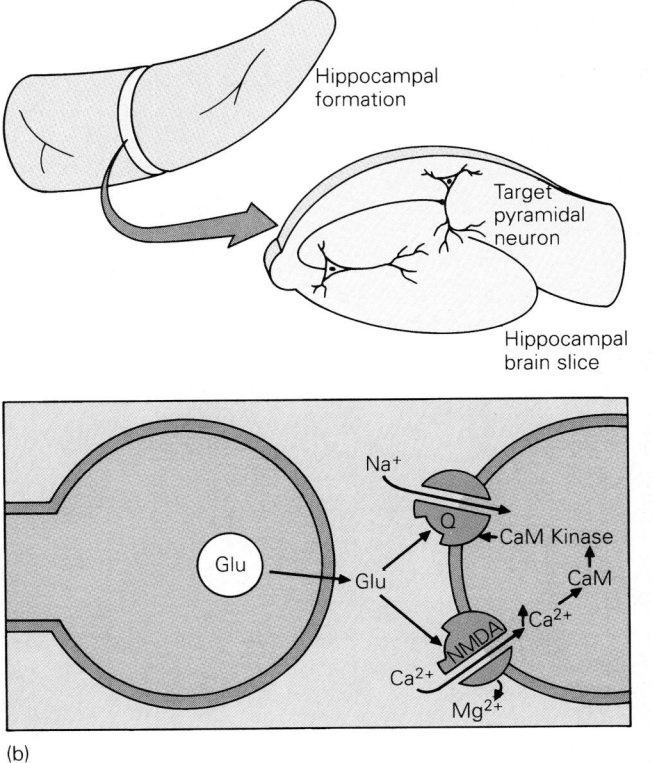

(a)

(b)

channel, permitting the influx of Ca^{2+} ions through the NMDA receptor channel as a result of glutamate binding to this receptor.

The importance of the NDMA receptor is seen in studies where it is blocked by the antagonist **D-2-amino-5-phosphonovalerate** (APV). Perfusions of APV into the incubation chamber can prevent the induction of LTP in hippocampal slices. In addition, the removal of Ca^{2+} from the incubation medium, and hence the removal of extracellular Ca^{2+}, also prevents LTP.

Influx of Ca^{2+} activates a number of possible intermediate steps that lead to LTP. The most important biochemical pathway is the activation of **Ca^{2+}/calmodulin** (CaM)-dependent protein kinase II (CaM

kinase II). Perfusion of drugs such as trifluoperpazine, which block the action of CaM, also block LTP. Moreover, the intracellular injection of synthetic peptides that inhibit Ca^{2+}/CaM-dependent substrate phosphorylation by CaM kinase II prevents the induction of LTP within the injected cell but not in adjacent hippocampal neurons.

The hippocampal slice technique has also revealed that LTP has associative properties. This has been shown by stimulating a weak pathway that by itself fails to produce LTP. It does produce LTP when stimulated at the same time that an independent pathway ending on the same population of neurons is activated.

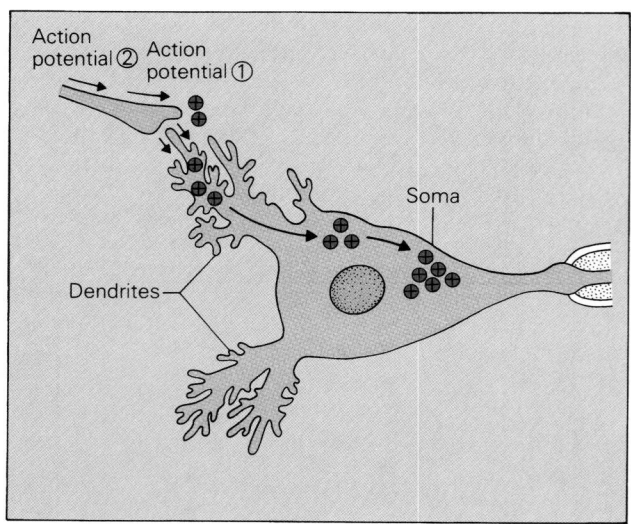

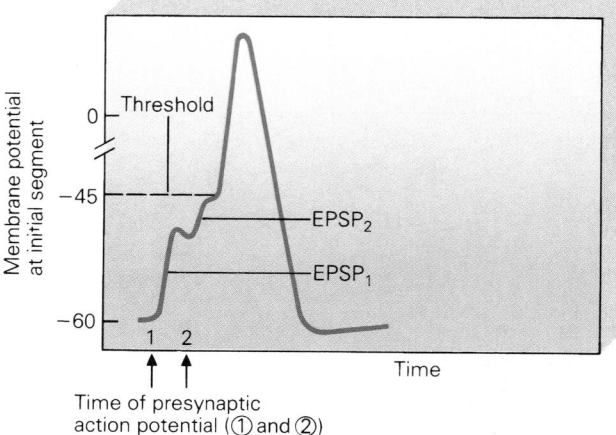

Figure 7–35
Temporal summation of EPSPs.

synaptic inputs, how does it decide, so to speak, whether or not to initiate an action potential? The decision is made by adding all of the postsynaptic potentials, using two different processes: temporal summation and spatial summation. If the addition of all of the synaptic potentials results in a depolarization of the membrane potential to approximately −45 mV near the initial segment or trigger zone of the axon, then an action potential will be generated.

Temporal summation involves the addition of postsynaptic potentials that arise from the activation at one synaptic input (Fig. 7–35). The first time that the synapse is activated, it may produce a depolarization that is not sufficient to cause the membrane potential to reach threshold. This depolarization by itself would decay back to the resting level. If, however, a second action potential invades the terminal, the release of neurotransmitters causes the postsynaptic membrane to again depolarize. Moreover, if the time between the first and second action potentials is short enough, the two postsynaptic membrane potentials together will produce a depolarization that may be sufficient to reach the threshold voltage and discharge an action potential.

Another summation process that involves the activation of more than one synaptic input is the process of **spatial summation.** In this case, two synaptic inputs, which alone are insufficient to cause the membrane potential to reach threshold (Fig. 7–36), are activated simultaneously, thus resulting in a depolarization sufficient to achieve a threshold value and discharge an action potential. The action potential discharge by a nerve cell, therefore, is based on the summation of excitatory and inhibitory postsynaptic potentials that occur in time over the space of the neuron's receptive surface.

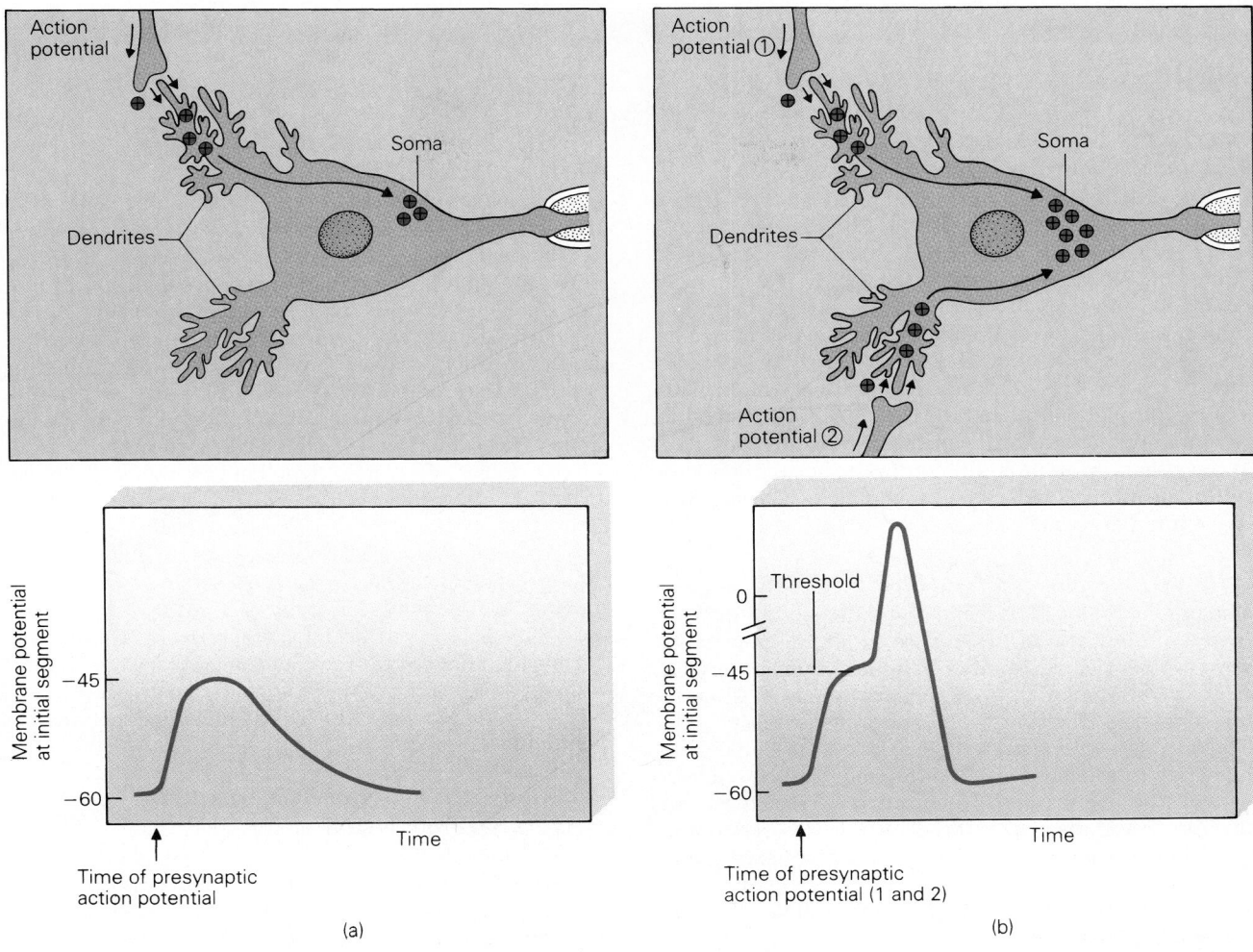

Figure 7–36
Spatial summation of EPSPs.

SUMMARY

Overview: The Central and Peripheral Nervous Systems

The nervous system is composed of the brain, brain stem, spinal cord, and peripheral nerves. The first three make up the *central nervous system (CNS)*, whereas the latter constitute the *peripheral nervous system (PNS)*. The brain is subdivided into functional regions involved in sensory input from the skin, eye, ears, or other sensory organs. Other areas are involved with locomotor activity as well as other functions. This functional partitioning of brain areas is found throughout the nervous system. A general organizational theme of the nervous system is that the brain receives sensory input and provides the appropriate output for such functions as speech, movement, or the control of internal organs.

Cells of the Nervous System

Glia, or glial cells, aid in maintaining the environment of the brain. They also act as scavengers in the brain to remove cellular debris. Another important function of glial cells is the formation of myelin around the axon of nerve cells to increase the conduction and transmission of electrical impulses. *Neurons* are the basic cellular elements in the nervous system involved in the transmission of information. The membranes of neurons have structural and functional properties specialized for the integration and rapid transmission of information from one nerve cell to another.

Neurons have extensive processes called dendrites and axons that are involved in the reception and transmission of information. The cytoskeletal structure of these processes is made of microtubules and neurofilaments. In addition to providing the structural framework for the neuron, these components are involved in the transport of cellular material from the somal to the outer reaches of axonal and dendritic processes. The fast axonal transport process will carry organelles and vesicles at a rate of 400 mm/day, while slow axonal transport will carry material at a rate of 1.0 to 2.0 mm per day. These processes play important roles in the maintenance and regeneration of nerve fibers. The membranes of nerve cells are composed of passive, voltage-activated, and chemically activated ion channels. These channels are made of proteins embedded within the membrane and permit specific ions to move through the membrane. Chemically activated ion channels, also known as receptors, are opened (or closed) when specific chemicals bind to receptor sites on the channel. Voltage-activated ion channels open (or close) when the membrane is depolarized. Passive ion channels remain in the open state and play a role in establishing the membrane potential.

The Membrane Potential and the Action Potential

A difference in the number of positive and negative ions between the inside and outside of a cell produces the membrane potential. In general, the inside of a cell is more negatively charged than the extracellular space, and thus establishes an electrical driving force for the movement of ions across the membrane. Differences in the concentration of ions between the inside and outside of cells establish concentration gradients, which act as chemical driving forces for the movement of ions across the membrane. The summated effects of electrical and chemical driving forces determine whether an ion will move through the membrane in an inward or outward direction.

The axon of nerve cells transmits information by generating action potentials. Action potentials are initiated and propagated along the axon as a result of the influx and efflux of Na^+ and K^+ ions, respectively. These ionic currents that flow across the axonal membrane are a result of the opening and closing of voltage-sensitive ion channels for Na^+ and K^+, which are embedded within the membrane.

Communication Between Nerve Cells: Synaptic Transmission

Action potentials that reach the axon terminal initiate the process of synaptic transmission. At electrical synapses, the cytoplasm between two cells is in direct contact. Thus ions will move directly from one cell to another.

At chemical synapses, synaptic transmission occurs when an action potential depolarizes the membrane within the synaptic terminal to activate voltage-sensitive Ca^{2+} channels. The opening of these channels causes the influx of Ca^{2+} into the terminal, which in turn leads to the fusion of synaptic vesicles with the preterminal membrane. As a consequence, chemical transmitters contained within the vesicles are released into the synaptic cleft, where they eventually bind to receptors located on the membrane of postsynaptic cells.

Neurotransmitters and Associated Receptors

The binding of chemicals called neurotransmitters with their receptors affects the ionic conductances of the target cells. These receptors are chemically activated ion channels that permit the passage of specific ions. Well-characterized neurotransmitters include acetylcholine, dopamine, norepinephrine, serotonin, glutamate, and gamma amino butyric acid.

Less well understood transmitters known as neuropeptides are synthesized in the soma.

Neuronal Integration: Temporal and Spatial Summation

The influx and efflux of ions through the chemically activated channels produce excitatory or inhibitory postsynaptic potentials. These responses are integrated by the postsynaptic cell to produce a summated response. If the summation of these alterations in the postsynaptic membrane causes a depolarization that reaches threshold levels, an action potential will be generated in the postsynaptic nerve cell. The transmission of information between nerve cells, therefore, is an electrochemical process that features the conduction of an electrical impulse coupled to the secretion of chemical transmitters.

REVIEW QUESTIONS

Identify Each with a Term

1. The term used for a positively charged ion.
2. The term used for a negatively charged ion.
3. The term used when the interior of the nerve membrane becomes more positively charged than when in its resting state.
4. The term used when the interior of the nerve membrane becomes more negatively charged than when in its resting state.

Define Each Term

5. membrane potential
6. equilibrium potential
7. action potential
8. saltatory conduction

Choose the Correct Answer

9. What is the direction of the driving forces for the movement of Na^+ ions when a nerve cell is at rest?
   a. inward electrical gradient
   b. outward electrical gradient
   c. inward chemical gradient
   d. both (a) and (c)
10. The repolarization of the action potential is a result of:
   a. an increase in sodium ion conductance
   b. a decrease in sodium ion conductance
   c. an increase in potassium ion conductance
   d. both (b) and (c)
11. A peripheral nerve fiber will regenerate at a rate of:
   a. 0.1 mm/day
   b. 0.4 mm/day
   c. 1.0 mm/day
   d. 400 mm/day

12. Which of the following statements is *not* true about glial cells?
   a. Astroglia remove degenerating debris.
   b. Each oligodendroglia cell forms myelin around one axon.
   c. Microglia remove degenerating debris.
   d. Glial cells transport glutamate from the synaptic cleft into their cytoplasmic compartment.

Calculate

13. If the extra- and intracellular concentrations for K^+ are 4 mM and 120 mM, respectively, what is the equilibrium potential for K^+?
14. The forearm of a patient is electrically stimulated by two electrodes to activate the motor nerves that cause the thumb to contract. The first electrode is placed 15 cm from the tip of the thumb while the second electrode is placed 30 cm away. The first electrode causes the thumb to contract 2.5 msec after the electrical stimulus. The second electrode causes the thumb to contract 4.5 msec after the stimulus. What is the conduction velocity of the motor nerves that innervate the thumb?

Answer Each Question in One or Two Sentences

15. Describe the membrane conductance changes that occur during the absolute refractory period of the action potential.
16. What membrane conductance changes can produce EPSPs?
17. What membrane conductance changes can produce IPSPs?
18. Describe the process of temporal summation.
19. What effects do the concentration and electrical gradients have on the movement of ions across the nerve membrane?

SUGGESTED READINGS

Cajal, S.R. "A new concept of the histology of the central nervous system," 1892. D.A. Rottenberg (trans.). In D.A. Rottenberg and F.H. Hochberg (eds.) *Neurological Classics in Modern Translation*, New York: Hafner, 1977.

Dale, H.H., Feldberg, W., and Vougt, M. "Release of acetylcholine at voluntary motor nerve endings." *Journal of Physiology*, 86:353–380, 1936.

Eccles, J.C., Fatt, P., and Koketsu, K. "Cholinergic and inhibitory synapses in a pathway from motor-axon collaterals to motoneurons." *Journal of Physiology*, 126:524–562, 1954.

Hodgkin, A.L., and Huxley, A.F. "A quantitative description of membrane current and its application to conduction and excitation in the nerve." *Journal of Physiology*, 117:500–544, 1952.

Kandel, E.R., and Schwartz, J. *Principles of Neural Science.* Amsterdam: Elsevier Press, 1985.

Katz, B., and Miledi, R., "The timing of calcium action during neuromuscular transmission." *Journal of Physiology*, 189:535–544, 1967.

Kuffler, S.W. "Neuroglial cells: Physiological properties and a potassium mediated effect of neuronal activity on the glial membrane potential." *Proceedings of the Royal Society of London [Biology]*, 168:1–210, 1967.

Neher, E., and Sakmann, B. "Single-channel currents recorded from membrane of denervated frog muscle fibres." *Nature*, 260:799–802, 1976.

Sherrington, C.S. *The Integrative Action of the Nervous System*, 1906. Reprint, New Haven: Yale University Press, 1947.

Zalutsky, R.A. and Nicoll, R.A. "Comparison of two forms of long-term potentiation in single hippocampal neurons." *Science*, 248:1619–1624, 1990.

KEY TERMS

acetylcholine (p. 266)
action potential (p. 245)
autonomic nervous system (p. 236)
axon (p. 241)
axonal transport (p. 244)
brain stem (p. 235)

central nervous system (p. 236)
cerebellum (p. 236)
dendrite (p. 241)
depolarization (p. 250)
electrical potential (p. 245)
glia (p. 238)
ion channel (p. 245)

membrane potential (p. 245)
microfilament (p. 243)
microtubule (p. 244)
myelin (p. 238)
neocortex (p. 235)
nerve (p. 241)
neuron (p. 241)
neurotransmitter (p. 241)

norepinephrine (p. 271)
peripheral nervous system (p. 236)
Schwann cell (p. 238)
spinal cord (p. 236)
synapse (p. 256)
thalamus (p. 235)
visual cortex (p. 235)

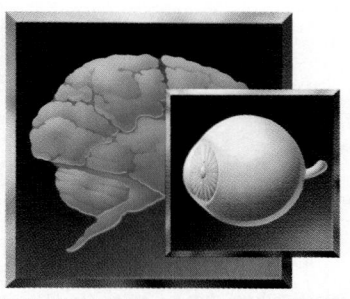

Sensory Systems

CHAPTER OUTLINE

From the moment we awaken in the morning until we fall asleep at night, our bodies are bombarded with information from the outside world. Most of us awake to the sound of an alarm clock that is detected by our **auditory system.** Our eyes focus on the clock to send information to the **visual system** and confirm that it is time to get up. We then climb out of bed into the warmth of a hot shower that is sensed by nerve endings of the **somatic sensory system.** At the breakfast table the smell of toast and coffee comes wafting to our noses and **olfactory system,** and the flavor of a rich cup of coffee stimulate the taste buds of our **gustatory system.**

Throughout the day we are continually exposed to information from the external environment that must reach the brain so that we can respond accordingly. The detection and transmission of this information are the functions of our **sensory systems.**

Information about our physical environment is detected by **sensory receptor cells** located within various **sense organs** (see Table 8–1). Sensory information is transmitted to the neocortex by different pathways depending on the type of sensation (Fig. 8–1). Somatic sensation from the body and limbs, for example, is transmitted from receptor cells to the spinal cord, where it then ascends to the higher lev-

Figure 8–1
Sensory information travels to the brain via specialized pathways for each type of sensation.

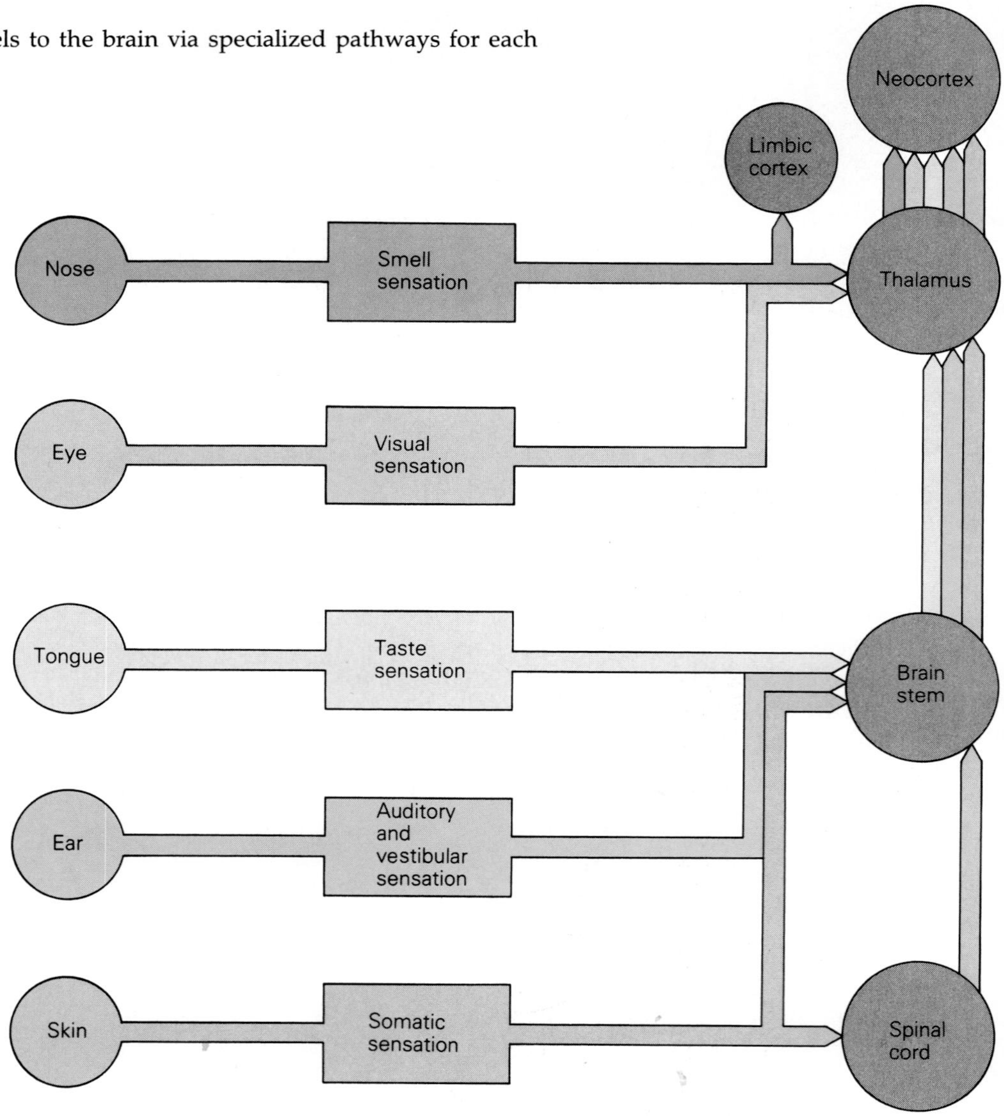

Table 8–1
Sensory Modalities and Receptor Cells

Sensory Mode	Receptor	Sense Organ
Vision	Rods and cones	Eye
Hearing	Hair cells	Ear (organ of Corti)
Rotational acceleration	Hair cells	Ear (semicircular canals)
Linear acceleration	Hair cells	Ear (utricle and saccule)
Smell	Olfactory neurons	Olfactory mucous membrane
Taste	Taste receptor cells	Taste buds
Touch—pressure	Nerve endings	Skin
Warmth	Nerve endings	Skin
Cold	Nerve endings	Skin
Pain	Naked nerve endings	Skin
Joint movement and position	Nerve endings	Various
Muscle length	Nerve endings	Muscle spindle
Muscle tension	Nerve endings	Golgi tendon organ

els of the nervous system. Auditory information detected by the ears, on the other hand, is transmitted from sensory receptor cells to the brain stem, where it ascends to eventually reach the neocortex. In general, a sensory receptor will respond only to a specific type of stimulus. After detecting the stimulus, it transduces this information into electrical impulses that encode the stimulus intensity. In this chapter we will consider the mechanisms by which sensory receptors detect and transduce information for touch, vision, hearing, taste, and smell.

General Principles of Sensory Transduction

Sensory receptors are activated when they detect a specific stimulus. This specific stimulus is called an **adequate stimulus** (Fig. 8–2) and is unique to each sensory receptor. In the visual system, for example, the photoreceptors detect colors of light but are insensitive to frequencies of sound. Likewise, the auditory receptors in the ear respond only to sound and are insensitive to light.

An adequate stimulus will produce a change in the membrane potential of the sensory receptor cell. This change in the membrane potential is called a **generator potential.** In some sensory receptors, such as somatic sensory receptors, the generator potential is a **depolarization** of the membrane. In others, such as the photoreceptors of the eye, it is a **hyperpolarization** of the membrane. The generator potential, in turn, produces an action potential or a series of action potentials. These action potentials can be generated by the receptor cell itself or by a neuron connected to the receptor cell. They transmit information about the nature of the stimulus to the central nervous system.

Figure 8–2
All sensory transductions involve an adequate stimulus, a generator potential resulting from a membrane conductance change, and one or more action potentials. The red "spikes" represent action potentials that have been compressed in time to reveal a pattern of frequency of occurrence.

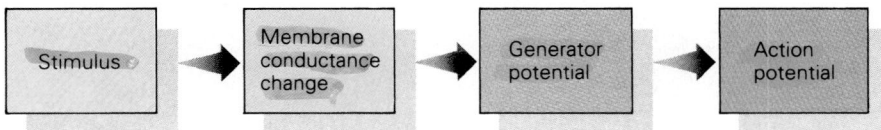

The transduction mechanisms of all sensory systems, therefore, involve (1) an adequate stimulus, (2) a generator potential, and (3) the initiation of action potentials. In order to examine these mechanisms in greater detail, let us consider a somatic sensory receptor called a **Pacinian corpuscle** that responds to touch, or **tactile,** stimuli. This type of receptor is located beneath the skin and consists of a free nerve ending encapsulated by layers of connective tissue (Fig. 8–3). The nerve ending is wrapped with myelin along the length of the fiber. The fiber itself extends into the spinal cord, where it forms synapses with other nerve cells.

The adequate stimulus for the Pacinian corpuscle is **pressure** applied to the skin. This pressure causes the layers of connective tissue and free nerve endings to compress. The compression of the free nerve ending causes an opening of Na$^+$ channels embedded within the nerve membrane (see Fig. 8–

3*b*). The resulting influx of Na$^+$ ions, in turn, depolarizes the membrane and produces a generator potential. If the generator potential is of sufficient amplitude, it will then depolarize the membrane in the region of the first node of Ranvier to threshold and initiate an action potential (Fig. 8–3*c*).

The Intensity of a Stimulus: How Nerves Inform the Brain

The number of action potentials generated per unit time is a function of the intensity of the pressure. Greater pressures produce a greater number of action potentials in a single fiber. This is known as the **frequency code of stimulus intensity** (Fig. 8–4*a*). The increase in the number of action potentials generated with greater pressure is due to the continued opening of the Na$^+$ channels and continued depolarization of the generator potential above the

Figure 8–3
(*a*) A Pacinian corpuscle, an example of a sensory receptor for touch, consists of a neuron wrapped in layers of connective tissue, with a myelinated nerve ending (axon). (*b*) A tactile (pressure) stimulus causes the neuron within the connective tissue to change shape, allowing Na$^+$ to enter the cell. (*c*) Pressure stimulus produces a depolarizing generator potential and an action potential.

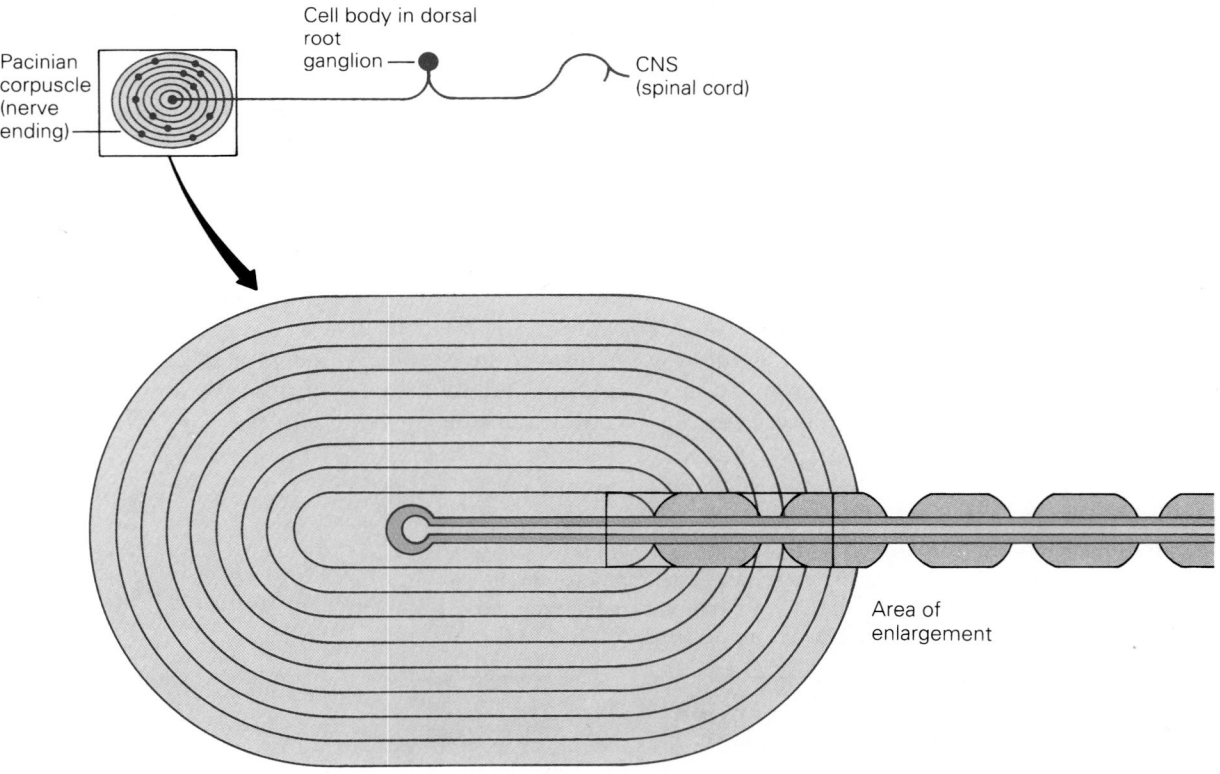

(a)

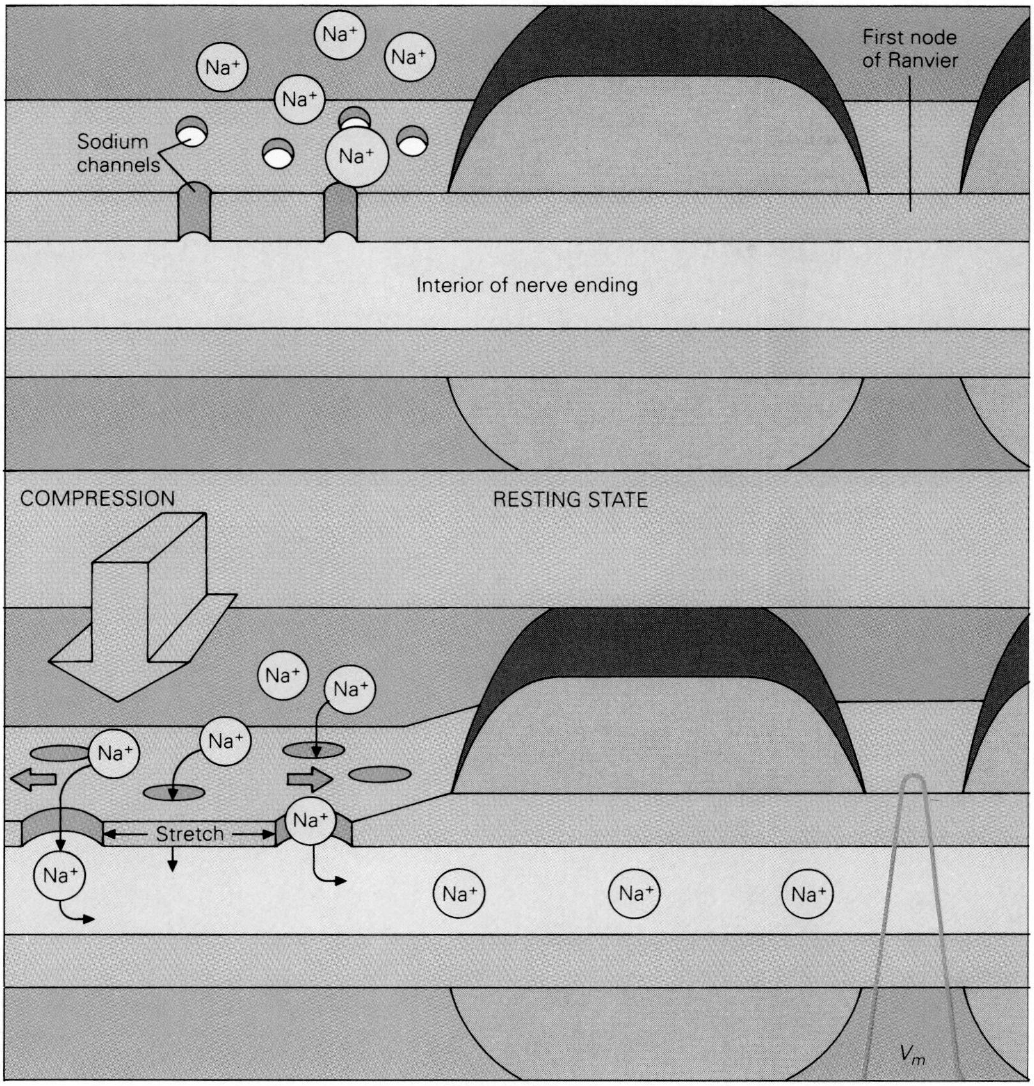

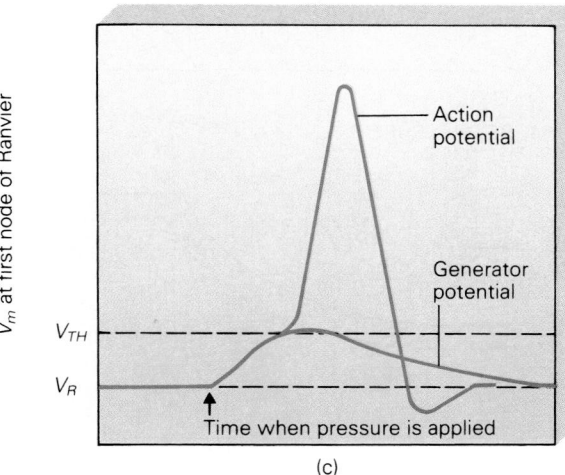

Sodium channels

Na⁺

Interior of nerve ending

First node of Ranvier

COMPRESSION

RESTING STATE

Stretch

V_m

(b)

V_m at first node of Ranvier

Action potential

Generator potential

V_{TH}

V_R

Time when pressure is applied

(c)

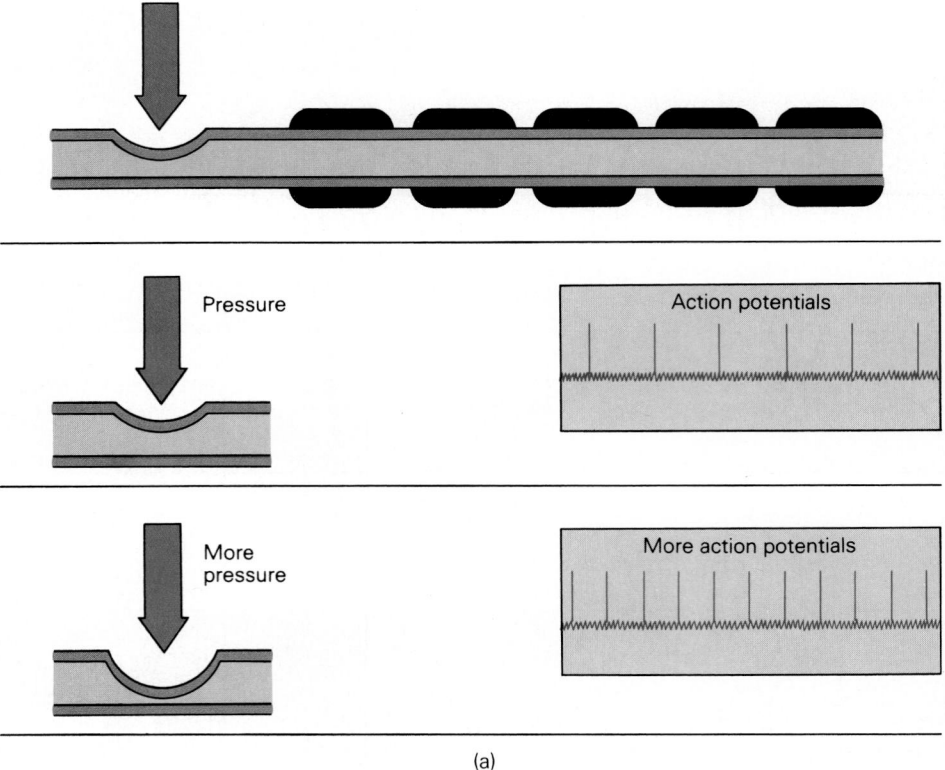

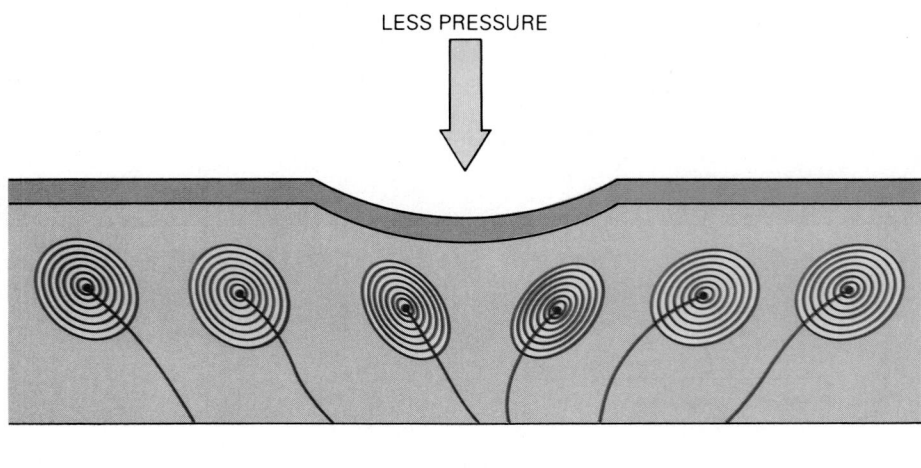

Figure 8–4
(*a*) According to the frequency code of stimulus intensity, more pressure produces more action potentials from the same receptor. (*b*) According to the population code of stimulus intensity, more pressure produces more action potentials because more receptors are affected.

threshold value. As a consequence, after the repolarization of one action potential, another is generated.

A second method of informing the brain about the intensity of the stimulus is a **population code.** With this method of coding, more sensory receptors of the Pacinian corpuscle type are activated as the pressure becomes greater. This occurs because the increased pressure affects a greater area beneath the skin (Fig. 8–4b).

The Quality of a Stimulus: How Nerves Send Clear Messages

The type of information sent to the brain is also coded in the way that nerve fiber pathways leading to the brain are arranged physically. The skin, for example, contains sensory receptors for temperature in addition to pressure. The information about the temperature of the skin reaches the brain by a different nerve fiber pathway than does information about pressure. In this way, information about the quality, or type, of stimulus is maintained within each pathway without becoming mixed with other types of stimuli. This aspect of the nervous system, the mechanisms of coding for the **type** of stimulus

detected by a sensory receptor, is called the **labeled-line code of stimulus quality** (Fig. 8–5).

Sensory Adaptation: How Nerves Adjust to Repetitive Stimuli

Often, when a stimulus is continuously applied, the brain at some point no longer consciously perceives it. This adjustment happens, for example, with background noise such as the ticking of a clock. After a period of time it is unnoticed. This phenomenon is called **sensory adaptation.** The adaptation to a sensory stimulus can be caused by mechanisms either within the brain or at the receptor site. Adaptation mechanisms that work at receptor sites are the most clearly understood and best illustrated by the Pacinian corpuscle.

When pressure applied to the Pacinian corpuscle is continuously maintained, the free nerve ending eventually reverts back to its original shape (Fig. 8–6) even though the layers of connective tissue remained deformed. Since the Na^+ channels embedded within the nerve membrane are opened only when the nerve fiber is deformed, they close when the form of the nerve is again circular. Consequently, even though pressure on the Pacinian cor-

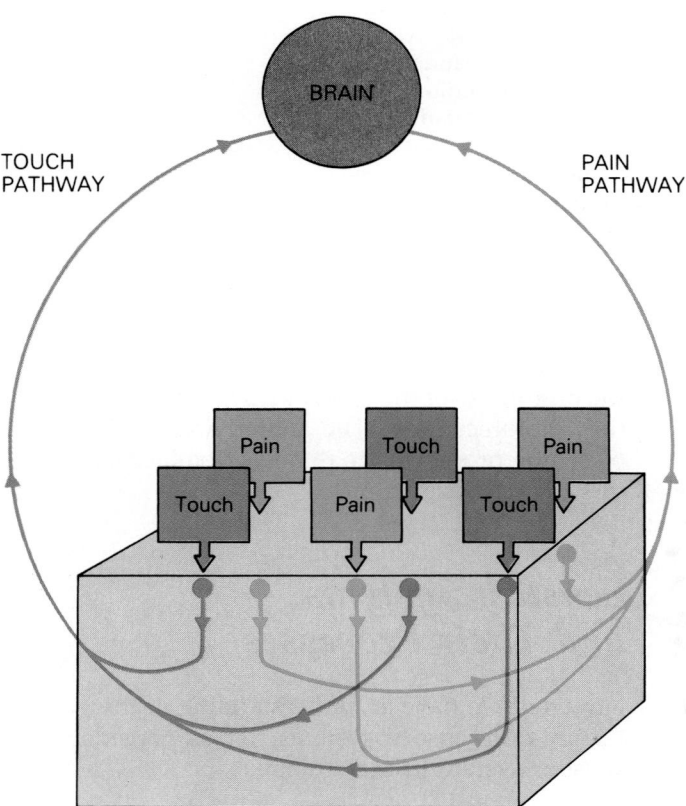

TOUCH PATHWAY

PAIN PATHWAY

Figure 8–5
The labeled line code of stimulus quality. Although different stimuli affect the skin, each type of stimulus has its own pathway to the brain, ensuring clarity of sensory messages.

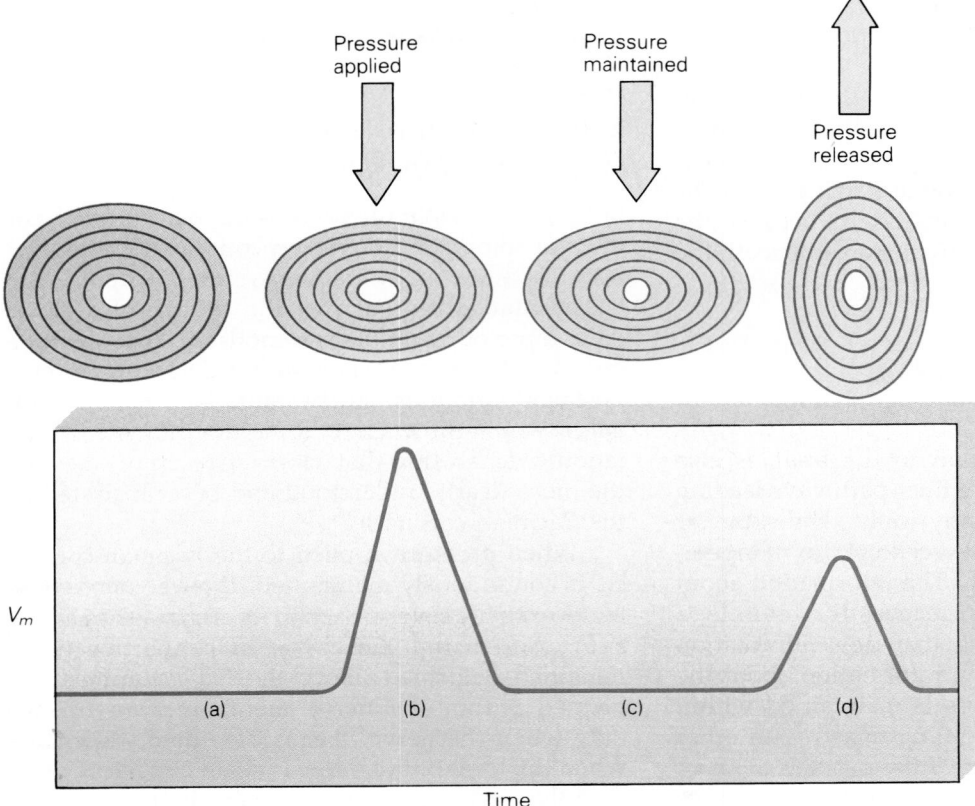

Figure 8–6

The process of sensory adaptation in a Pacinian corpuscle. (*a*) When not stimulated, the nerve ending inside the connective tissue layers has a rounded shape. (*b*) When pressure is applied, the entire corpuscle is pressed down and flattened, including the nerve ending inside. The elongation opens Na^+ channels into the nerve ending as shown in Figure 8–3. (*c*) As pressure is maintained, the corpuscle remains flattened, but the nerve ending inside resumes its resting shape, and Na^+ channels close; hence, no action potential is generated. (*d*) When pressure is released, the entire corpuscle, including the nerve ending, "springs back," becoming compressed in a different direction.

puscle is still maintained, it ceases to produce a generator potential when it resumes its normal shape and thus no longer transmits information about the pressure of the stimulus.

Mechanisms of adaptation are also found at the molecular level. Chemical receptors on membranes, for example, are internalized (see Chapter 12) and removed from the surface of the membrane after continued exposure to drugs. With the removal of these chemical receptors, higher levels of the drug are necessary to achieve the same effect. This may be why, after repeated use, the body develops a tolerance to drugs such as heroin. A similar molecular adaptation mechanism is found in receptors for vi-

sion. During sunny days, our eyes quickly adapt to the brightness of the outdoors by the "bleaching" of the photoreceptors. The underlying basis of this bleaching process is the removal of molecular receptors that capture light.

Sensory Systems and Their Processes

Now that we have an understanding of the general features of sensory systems, let us consider each sensory system in more detail.

The Somatic Sensory System

The **somatic sensory system** is sensitive to external stimuli that affect the skin or surface of the body. This sensory system also conveys information about the position of the body's limbs in relation to the rest of the body.

Types of Somatic Sensory Receptors

The major types of somatic sensory receptors are (1) **tactile receptors** activated by the mechanical stimulation of the body's surface, (2) **thermal receptors** activated by changes in temperature on the surface of the body, (3) **pain receptors** activated by noxious (harmful) stimuli, and (4) **proprioceptive** receptors activated by the movement of the limbs. Each type of sensation is detected by a variety of sensory receptors.

Tactile Receptors: Touch, Pressure, and Vibration.
Tactile receptors are called **mechanoreceptors.** These receptors are responsible for detecting touch, pressure, and vibrations applied to the skin (Fig. 8–7). We have already examined the functional prop-

erties of one type of pressure-sensing mechanoreceptor, the **Pacinian corpuscle.** In addition to pressure, however, the Pacinian corpuscle is capable of detecting vibrations. The mechanisms that allow it to detect vibrations are extensions of its ability to rapidly adapt to applied pressures.

Recall that when pressure is applied to the Pacinian corpuscle for a length of time, it quickly stops generating action potentials. As we discussed earlier, this occurs when the Pacinian nerve ending reverts back to its original shape (see Fig. 8–6). As the applied pressure is released, the layers of connective tissue surrounding the nerve ending spring back to their original circular form and then become compressed again in the opposite direction because of the elastic nature of the connective tissue (see Fig. 8–6d). This recompression squeezes the nerve ending and again causes Na$^+$ channels to open and produce a generator potential and action potentials. In this way, both the application and release of pressure result in the discharge of action potentials. This oscillation of pressure is the basis of a vibratory stimulus. It should be clear from this example that the Pacinian corpuscles must adapt very quickly. This ability to quickly adapt to a stimulus classifies these corpuscles as **rapidly adapting receptors.**

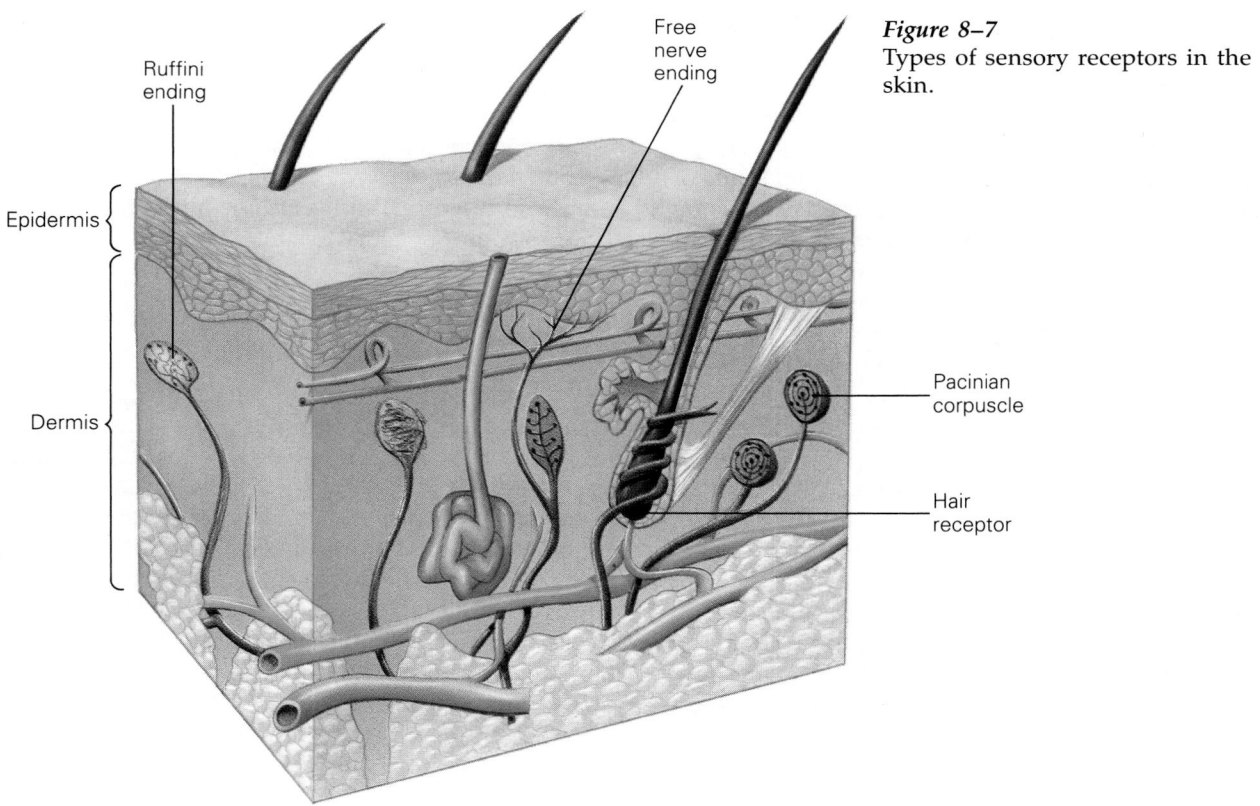

Figure 8–7
Types of sensory receptors in the skin.

Another rapidly adapting tactile mechanoreceptor is the **hair receptor.** The hair receptor is a hair follicle on the surface of the skin that has a nerve fiber wrapped around its base (see Fig. 8–7). The bending of the hair cell such as that caused by a slight breeze produces a mechanical displacement of the nerve fiber at the base of the follicle. This displacement, in turn, opens Na$^+$ channels, resulting in a generator potential and the initiation of action potentials. This type of receptor is very sensitive to a gentle touch across the surface of the skin.

Another group of tactile receptors, called **slowly adapting tactile mechanoreceptors,** continue to generate action potentials as long as the stimulus is applied. An example of a slowly adapting mechanoreceptor is the **Ruffini capsule** (see Fig. 8–7). This receptor is sensitive to stretching of the skin such as that produced during a massage.

Thermal Receptors: Heat and Cold. Sensory receptors that respond to temperature are called **thermal receptors.** There are two types of thermal receptors: one type responds to temperatures below 30°C, while the other type responds to temperatures above 30°C. These thermoreceptors are capable of adapting to the external environment. When exposed to cold weather, for example, the cold thermoreceptors do not become activated until temperatures lower than 30°C are reached, while warm temperatures become activated well before 30°C.

Nociceptors: Painful Touch and Temperatures. **Nociceptors** are free nerve endings that detect painful stimuli (Fig. 8–7). In general, there are two types of nociceptors: **mechanical nociceptors** activated by intense mechanical stimulation such as a knife cut on the arm or a slap to the head, and **heat nociceptors** that respond to temperatures above 45°C. The transduction of pain is activated by a chemical process that also involves the immune system.

Proprioceptors: Position and Movement. The position of the body's limbs is detected by **proprioceptors.** One type of proprioceptor detects the stationary position of the limbs in space with respect to the other parts of the body. Other proprioceptors transmit dynamic information about limb movement to convey the sense of movement, or **kinesthesia.** The brain needs this information to determine where the arms and legs are located in order to calculate how much further they need to go to complete a certain movement.

The operation of these receptors can be tested by trying to bring the tips of your left and right index fingers together with your eyes closed. If your proprioceptors are working properly, your fingers should touch without your having to watch and visually guide them together.

The sense of stationary or static position is transmitted to the brain by mechanoreceptors located in joint capsules, cutaneous mechanoreceptors, and mechanoreceptors in muscles that are specialized to transduce the stretching of the muscle. Extremes of joint angles are sensed by **joint capsules,** while intermediate angles are transduced by the **muscle spindle receptor.** The functional properties of these receptors and muscle tension receptors will be discussed in detail in Chapter 9, in conjunction with the motor system.

The static proprioceptors produce a continuous frequency of action potentials in response to different joint positions. If the joint is left in one particular position, the receptor will generate action potentials at one specific frequency (see Fig. 8–8a). This type of action potential response is called a **tonic discharge.** The dynamic proprioceptor generates action potentials only with a change in direction of movement. The burst of action potentials produced is very brief (Fig. 8–8b). This type of action potential response is called a **phasic discharge.**

The information from somatic sensory receptors is transmitted to the central nervous system by several different pathways organized according to three general principles: (1) Each type of somatic sensation has its own pathway; (2) most pathways cross over from one side of the brain to the other; (3) the nerve cells within each nucleus are topographically organized according to the location of their sensory receptors on the surface of the body. As we shall see, these three principles also apply to the organization of other sensory systems.

The Dorsal Root Ganglion: A Cluster of Neurons

Each of the somatic sensory receptors previously described is associated with a peripheral nerve axon of a particular diameter. The cell bodies of these sensory receptors are located within a cluster of cells immediately outside the spinal cord called the **dorsal root ganglion** (Fig. 8–9). The fibers with the largest diameters are those associated with touch, pressure, and vibration sense. These are heavily myelinated fibers that range from 13 to 22 μm in diameter and have action potential conduction velocities of 70 to 120 meters per second (see Table 8–2). These fibers release the neurotransmitter glutamate at their terminal endings. The fibers with the smallest diameters are the lightly myelinated and unmyelinated fibers that convey pain and tempera-

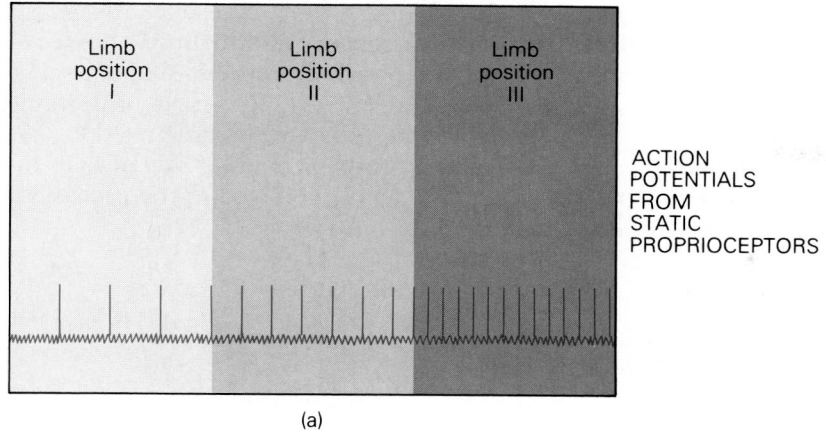

(a)

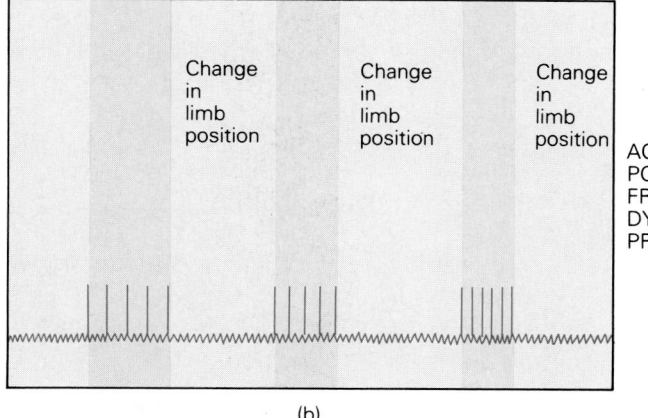

(b)

Figure 8–8
Action potential patterns produced by static (*a*) and dynamic (*b*) proprioceptors.

Figure 8–9
The dorsal root ganglion is a cluster of neurons located just outside the spinal cord on the dorsal side.

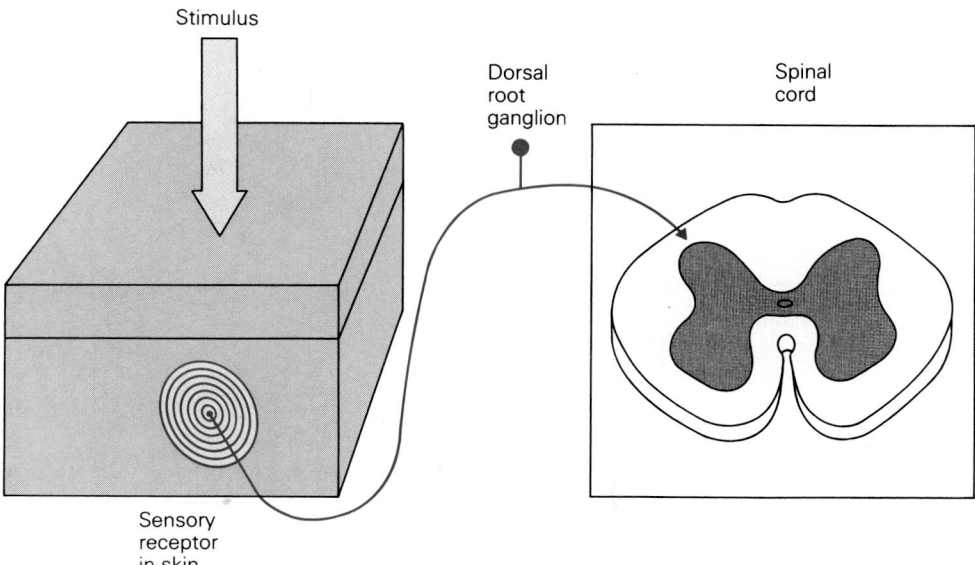

ture information. These fibers range from 1 to 5 μm in diameter and have action potential conduction velocities of 2 to 15 meters per second. These fibers release the neuropeptide substance P at their terminal endings.

Two major pathways transmit somatic sensory information from the surface of the body to the neocortex: the **dorsal column pathway** and the **antero-** **lateral pathway** (Fig 8–10*c*). These pathways sequentially transmit sensory information to nerve cell nuclei in the **spinal cord, brain stem, thalamus,** and **neocortex** (see Fig. 8–1). In order to understand how these pathways operate, we first need to consider the functional organization of the areas of the nervous system that are involved in the processing of somatic sensory information.

Figure 8–10
(*a*) The spinal cord is described frequently in terms of its transverse orientation (i.e., dorsal and ventral). (*b*) A cross-section of the spinal cord reveals gray and white areas as well as dorsal and ventral horns. (*c*) Two major pathways through the spinal cord that transmit sensory information to the brain are the dorsal column pathway and the anterolateral pathway.

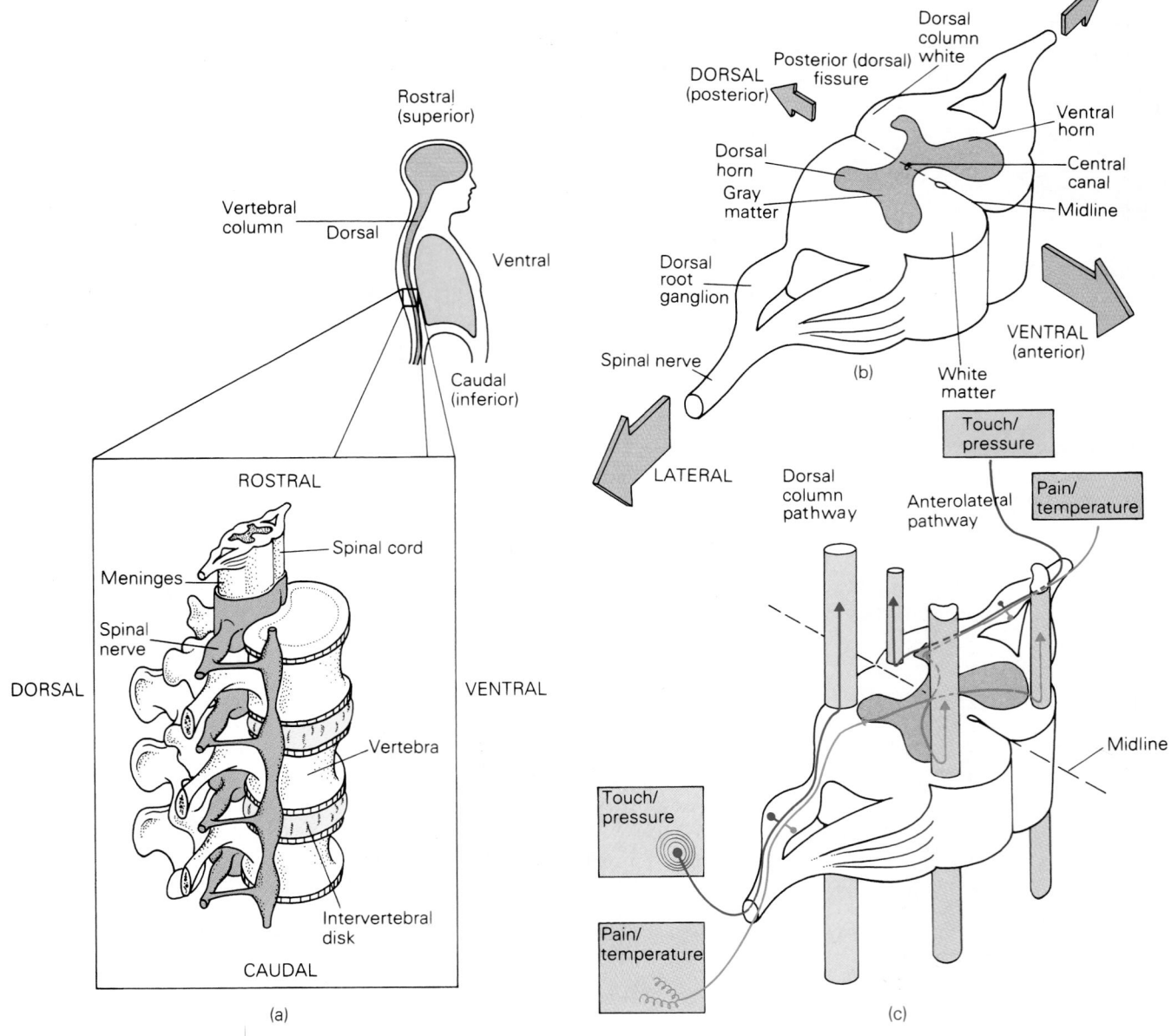

Table 8–2
Sensory Fibers

Type	Diameter (µm)	Conduction Velocity (m/sec)	Sensory Information
I_a (A$_\alpha$)	22	120	Primary muscle spindle
I_b (A$_\alpha$)	22	120	Primary touch and pressure
II (A$_\beta$)	13	70	Secondary muscle spindle, touch, and pressure
III (A$_\delta$)	5	15	Lightly myelinated fibers for touch, pressure, and pain
IV (C)	1	2	Unmyelinated fibers for pain and temperature

The Spinal Cord in Somatic Sensory Function

The spinal cord is shaped like a tube that runs from the base of the head to the small of the back (Fig. 8–10a). This longitudinal orientation is called the **rostro-caudal axis** (*rostral* meaning "toward the nose" and *caudal* meaning "toward the tail" in Latin). The spinal cord is also oriented so that one part of the tube is facing the back and the other is facing the stomach. This transverse orientation is called the **dorso-ventral axis** (*dorsal* meaning "toward the back" and *ventral* meaning "toward the belly" in Latin).

Four other terms also used quite often to describe the location of nuclei and fiber tracts in the brain are: *medial* meaning "close to the midline" of the body; *lateral* meaning "to the side" of the body; *anterior* meaning "toward the front" of the body; and *posterior* meaning "toward the back" of the body.

A transverse section of the spinal cord shows that it has a central gray area and a peripheral region that is white (see Fig. 8–10b). The gray matter is composed of nerve cells of the spinal cord, whereas the white matter is composed of axons. The myelination of axons is what makes the peripheral area appear white. The gray region has four protrusions shaped like butterfly wings that are symmetrical about the midline. The dorsal protrusions, called the **dorsal horns**, contain nerve cells involved in the processing of sensory information. The ventral protrusions are called the **ventral horns** and contain nerve cells that innervate skeletal muscles and are thus involved in motor function. This latter group will be discussed in Chapter 9.

Nerve fibers that carry sensory information in the spinal cord follow one of the two major pathways already mentioned: the dorsal column pathway or the anterolateral pathway (see Fig. 8–10c). Nerve fibers carrying information about touch and pressure traverse the dorsal column pathway, whereas those carrying information about pain and temperature traverse the anterolateral pathway. When pain and temperature fibers enter the spinal cord, they cross over to the midline; consequently, the anterolateral pathway transmits information from the opposite (contralateral) side of the body. In contrast, fibers of the dorsal column pathway do not cross over until they reach the medulla.

Segments of the Spinal Cord. The spinal cord is subdivided into 31 segments (Fig. 8–11). Each spinal cord segment receives sensory information from particular areas of the skin called **dermatomes.** These dermatomes contain receptors for all of the somatic sensations. The loss of sensory information by certain dermatomes can thus be used as a clue to the location of damage to the spinal cord.

In the uppermost region of the cord are 8 cervical segments (C1 to C8). These segments receive sensory information from the back of the head, neck, shoulders, part of the arms, and hands. Next are 12 thoracic segments (T1 to T12), which receive sensory inputs from parts of the arms and hands and trunk of the body. Following the thoracic segments are 5 lumbar segments (L1 to L5), which receive sensory inputs from the waist, thighs, upper and lower legs, and parts of the feet. The remaining segments of the spinal cord are five sacral segments (S1 to S5) and one sacro-coccygeal segment. These receive sensory input from the back of the legs, buttocks, and anus.

Sensory fibers of the dorsal column system enter the spinal cord and ascend toward the brain in

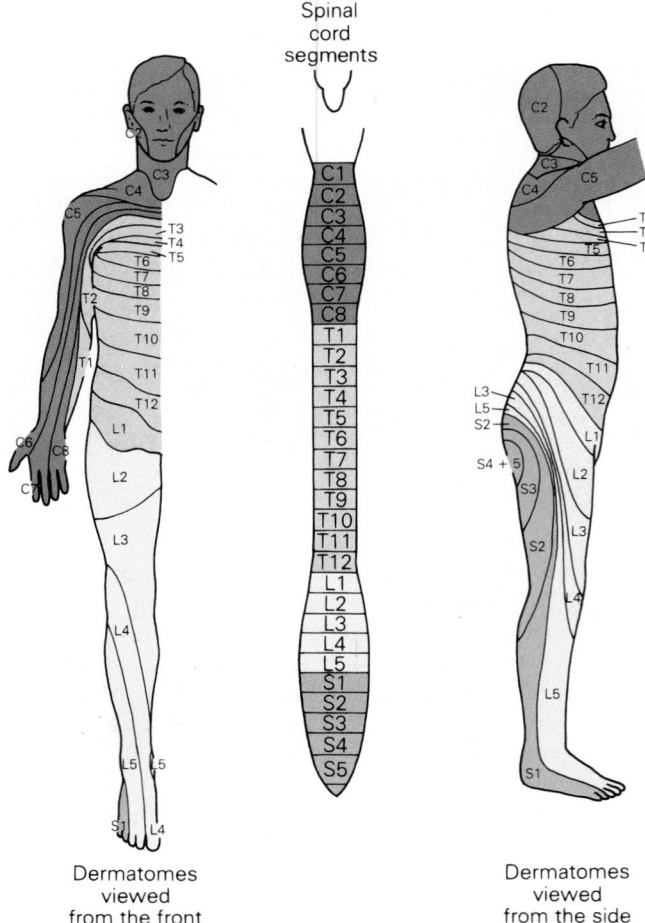

Spinal cord segments

Dermatomes viewed from the front

Dermatomes viewed from the side

Figure 8–11
The 31 segments of the spinal cord are linked to areas of the skin known as dermatomes.

the dorsal portion of the spinal cord (Fig. 8–12). These fibers continue to the medulla oblongata, where they form synapses with neurons in the **dorsal column nuclei.** Fibers from the dorsal column nuclei, in turn, cross the midline and become part of the **medial lemniscal** pathway. These fibers reach the **thalamus** on the opposite side of the brain. Neurons in the thalamus, in turn, extend their fibers to the cortex.

Lesions of the Spinal Cord. In accidents that damage the spinal cord, the loss of cutaneous sensation appears in patterns that are characteristic of the level of the lesion. For example, an accident that severs half of the spinal cord on the right side of the body at the waist causes a loss of pain and temperature sensation on the left leg and a loss of touch and vibration sense on the right leg (Fig. 8–13). This pattern occurs because tactile information is transmit-

ted in the spinal cord along the same side of the body that receives information about touch and pressure. Pain and temperature information, on the other hand, crosses the midline once it enters the spinal cord (Fig. 8–13). Obviously, accidents that transect all or half of the spinal cord will also result in some form of paralysis. This loss of motor control after spinal cord lesions will be considered in Chapter 9.

The Thalamus in Somatic Sensory Function

The **thalamus** is one of the brain areas that receives most types of sensory information. It receives not only somatic information, but also visual, auditory, and taste information. The region of the thalamus that receives somatic sensory information is the **ventral posterior lateral (VPL) nucleus.** Neurons in the VPL of the thalamus then transmit sensory information to the neocortex.

Figure 8–12
The dorsal column, medial lemniscal pathway.

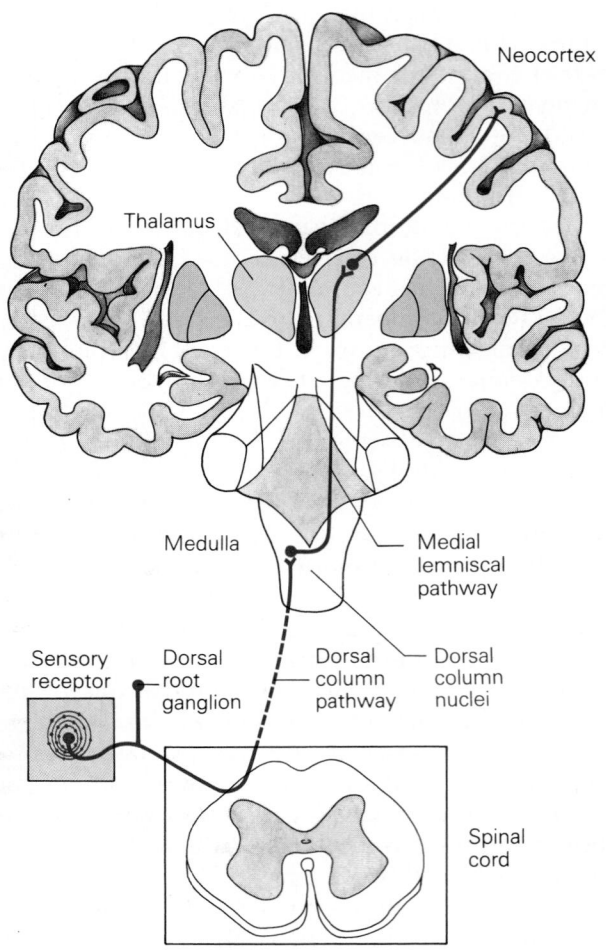

Neocortex

Thalamus

Medulla

Medial lemniscal pathway

Sensory receptor

Dorsal root ganglion

Dorsal column pathway

Dorsal column nuclei

Spinal cord

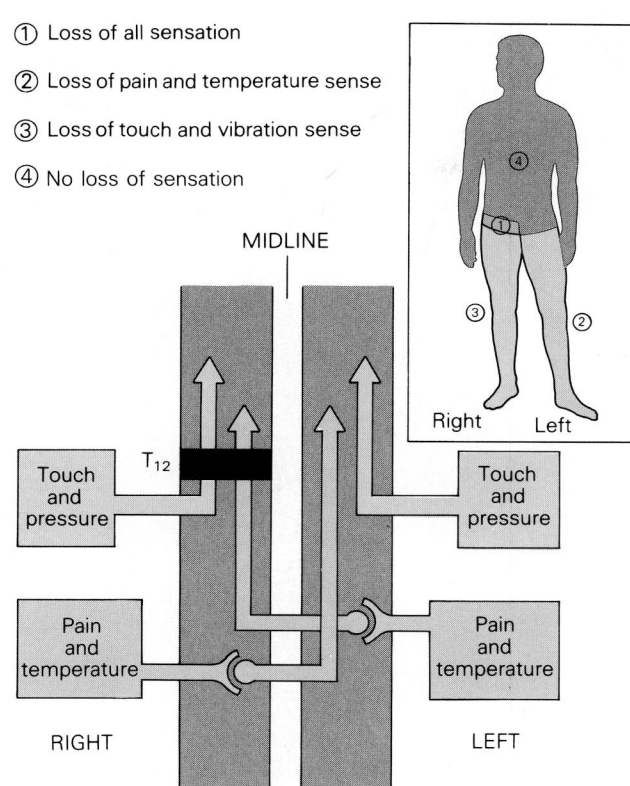

① Loss of all sensation

② Loss of pain and temperature sense

③ Loss of touch and vibration sense

④ No loss of sensation

The Somatic Sensory Cortex

The **neocortex** is the part of the brain that evolved most recently (*neo* meaning new). It is divided into discrete areas that receive somatic, visual, auditory, and gustatory sensations as well as areas for the control of movement and other functions (Fig. 8–14). The area of the neocortex that receives somatic sensations is called the **somatic sensory cortex.**

A cross-section of the somatic sensory cortex shows that it is organized like a map of the body (Fig. 8–15). Sensory information from the legs is sent to the medial portion of the cortex. Adjacent but more lateral regions of the cortex receive sensory information from the torso. The most lateral regions of the cortex receive sensory information from the arms, hands, and face.

Figure 8–13
Some effects of spinal cord lesions. Damage to the spinal cord on the right side of the body near the waist causes the left leg to lose pain and temperature sensation, while the right leg loses the sense of touch and vibration.

Figure 8–14
The primary somatic sensory cortex and other areas of the neocortex.

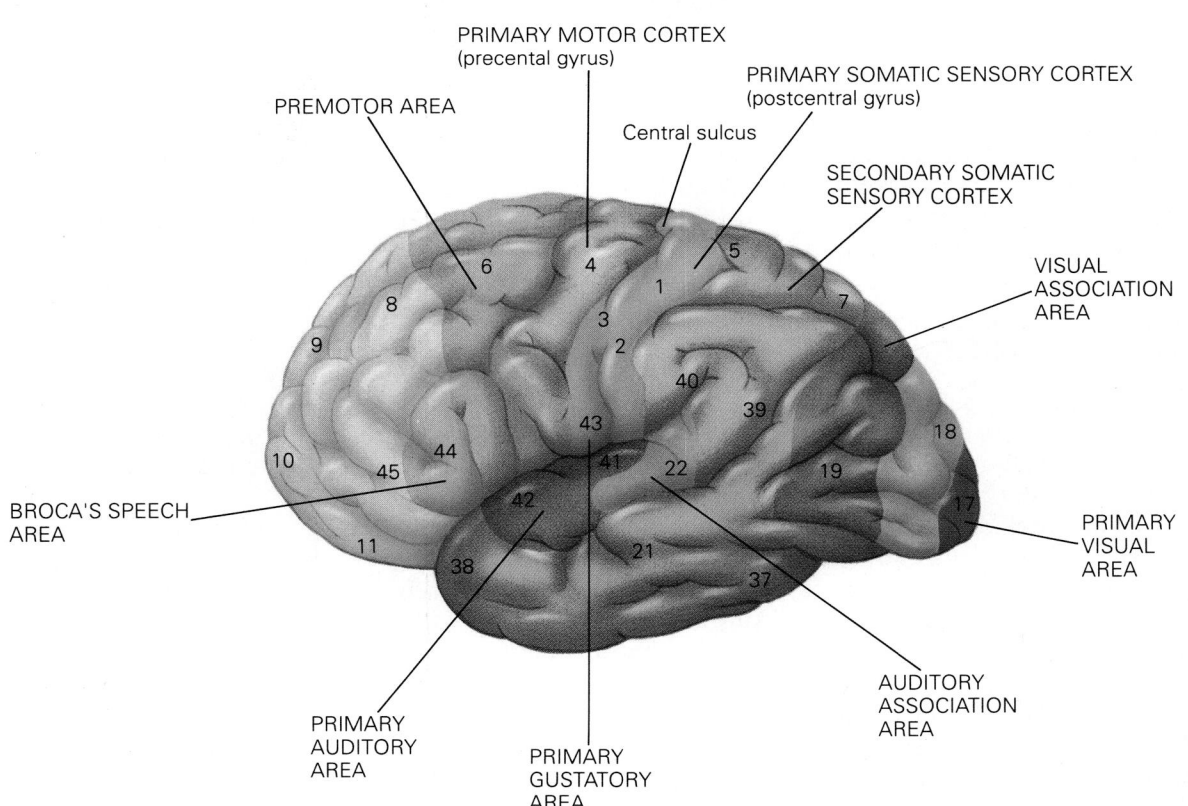

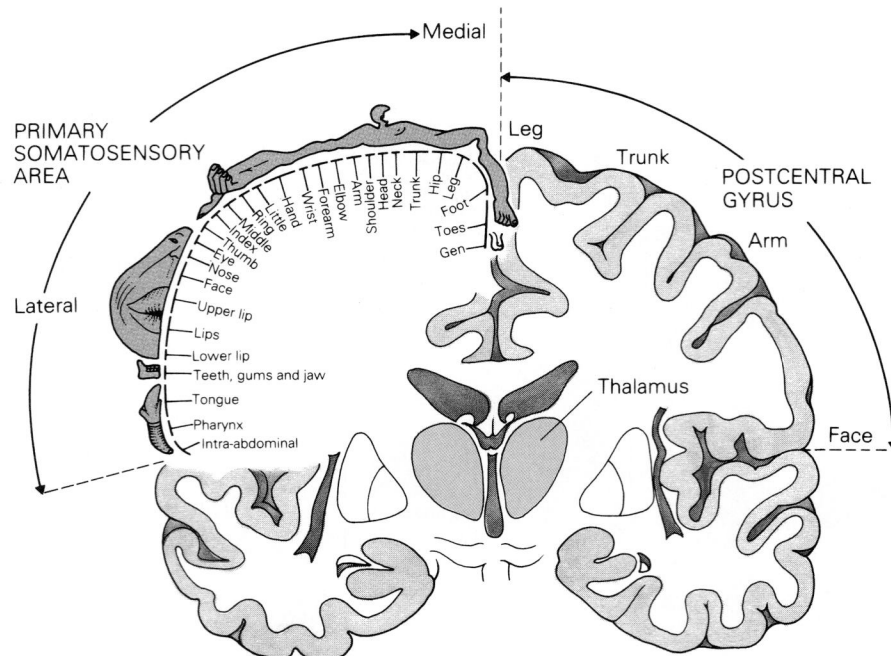

Figure 8–15
The somatotopic organization of
the primary somatic sensory
cortex.

This type of topographical organization of sensory information from various regions of the body on to the sensory cortex is called a **somatotopic organization.** The map that is created in the somatosensory cortex is a distortion of the true proportions of the body. The regions of the body that have the greatest density of sensory receptors are represented by a greater proportion of the cortex. The areas in the cortex representing the thumb and fingers of the hand, for instance, occupy substantially more space than that represented by the torso.

Receptive Fields of Neurons in the Somatic Sensory System

Stimuli detected by a sensory receptor are transmitted to various groups of nerve cells. The most effective location of a stimulus, however, differs for each cell within a cluster. For instance, pressure applied to an area of the skin directly above a Pacinian corpuscle will activate the associated dorsal root ganglion cell (Fig. 8–16). If pressure is applied to another location on the surface of the skin, another dorsal root ganglion cell is activated; however, the first dorsal root ganglion cell is not affected by pressure applied to this second location on the skin. Thus, only a small area on the receptive surface of the skin can activate a nerve cell in the dorsal root

ganglion. This small area of skin represents the **receptive field** for that cell. The receptive field can be described as a circular area on the skin where pressure is applied to excite the cell. We will denote the ability to excite the cell with a "+," and the receptive field area of the skin by a circle.

Nerve cells in the dorsal column nuclei, the next cluster of cells along the dorsal column pathway, have different receptive fields than those in the dorsal root ganglia. Dorsal column nerve cells generate a low rate of spontaneous action potentials. This is their tonic baseline activity. When pressure is applied to an area of the skin, a dorsal column nerve cell (cell #1 in Fig. 8–17a) is excited to produce a higher frequency of action potential discharge than its baseline rate of activity. However, when the pressure is applied to a different location, the frequency of the action potentials is decreased to less than the baseline rate.

This inhibition, "–," of the dorsal column nerve cell, is a result of activating a second Pacinian corpuscle, which in turn activates a different nerve cell (cell #2) within the dorsal column. The activation of this second dorsal column nerve cell leads to the activation of an **inhibitory interneuron.** This inhibitory neuron, in turn, prevents the excitation of the first dorsal column neuron. There are also other areas of the skin that, when pressed, will inhibit the

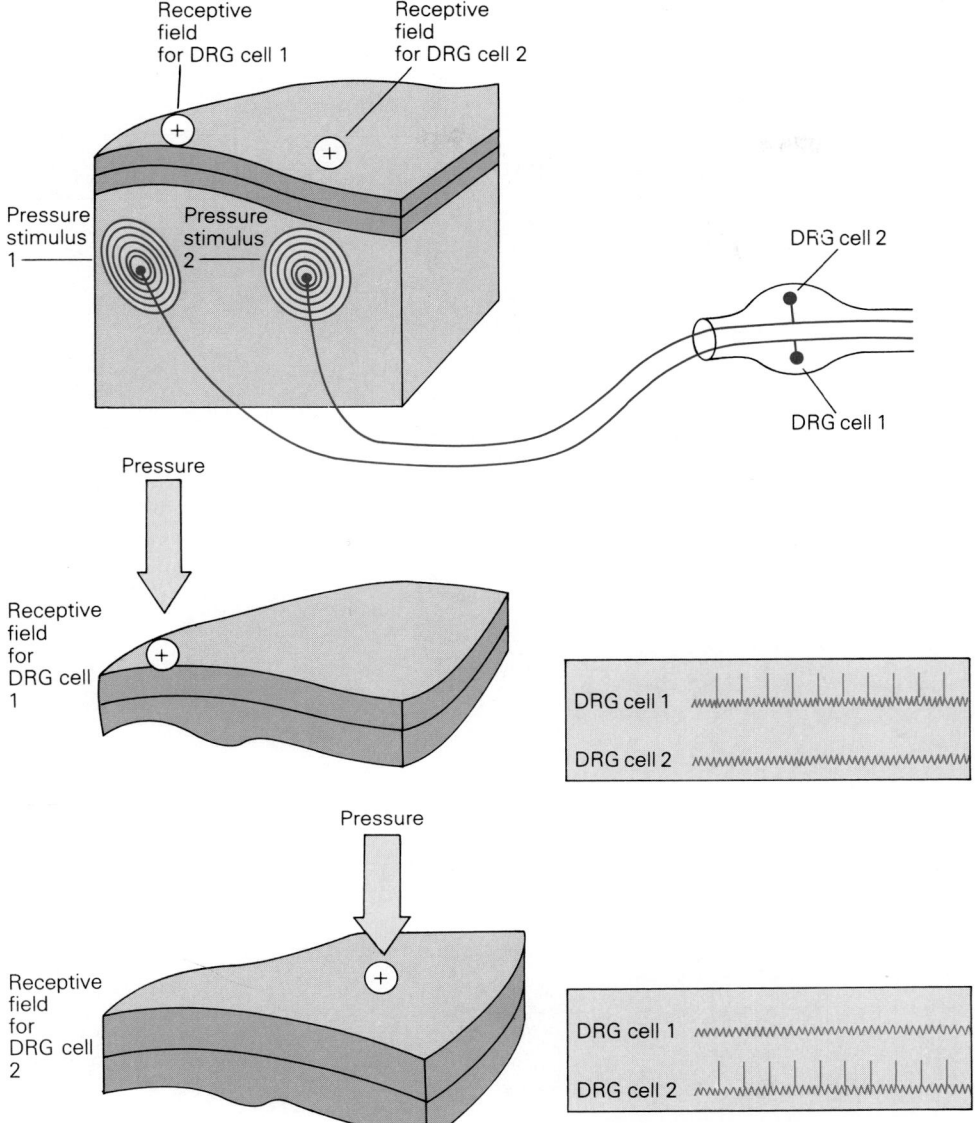

Figure 8–16
A cell in the dorsal root ganglion has a specific receptive field on the skin.

response of dorsal column nerve cell #1 (see Fig. 8–17b). When the excitatory and inhibitory areas on the skin are mapped for nerve cell #1, an annulus or doughnut-shaped receptive field is revealed with an excitatory center and an inhibitory ring (Fig. 8–17c).

This annular receptive field is also found for nerve cells in the thalamus and certain areas of the somatosensory cortex. The cells in the somatosensory cortex that exhibit the annular receptive field are all located in area 3 of the **primary somatosensory cortex** (see Fig. 8–15). This is a narrow strip of cortical tissue that is also called **Brodmann area 3** (named for the neuroanatomical studies by K. Brodmann in 1909). In Brodmann areas 1 and 2, the receptive fields are no longer annular. The stimulus that most effectively activates nerve cells in areas 1 and 2 is pressure applied to the skin in the form of a rectangle (Fig. 8–18) rather than a ring. The effec-

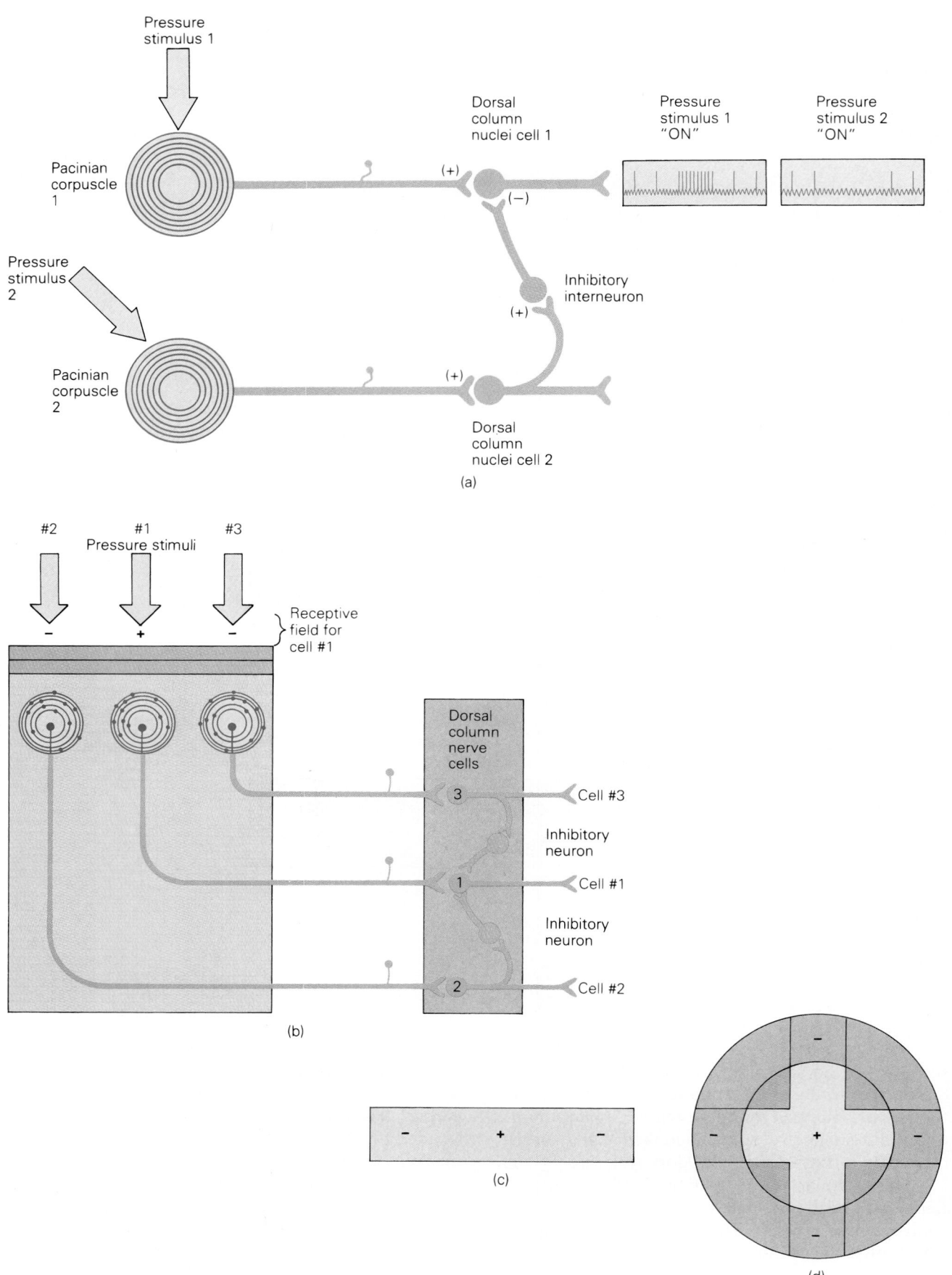

(a)

(b)

(c)

(d)

◀ *Figure 8–17*

Annular receptive fields of neurons in the dorsal column nuclei are formed by inhibitory interneurons. (*a*) Activation of Pacinian corpuscle #1 excites dorsal column cell #1, while activation of Pacinian corpuscle #2 inhibits dorsal column cell #1 via an inhibitory interneuron (red). (*b*) Mapping of receptive field on the surface of the skin for dorsal column cell #1. Pressure stimulus #1 activates this cell, while pressure stimuli #2 and #3 inhibit dorsal column cell #1. (*c*) One dimensional view of receptive field for dorsal column cell #1. (*d*) Two dimensional view of receptive field for dorsal column cell #1.

Figure 8–18

The different types of receptive fields for cells of different nuclei.

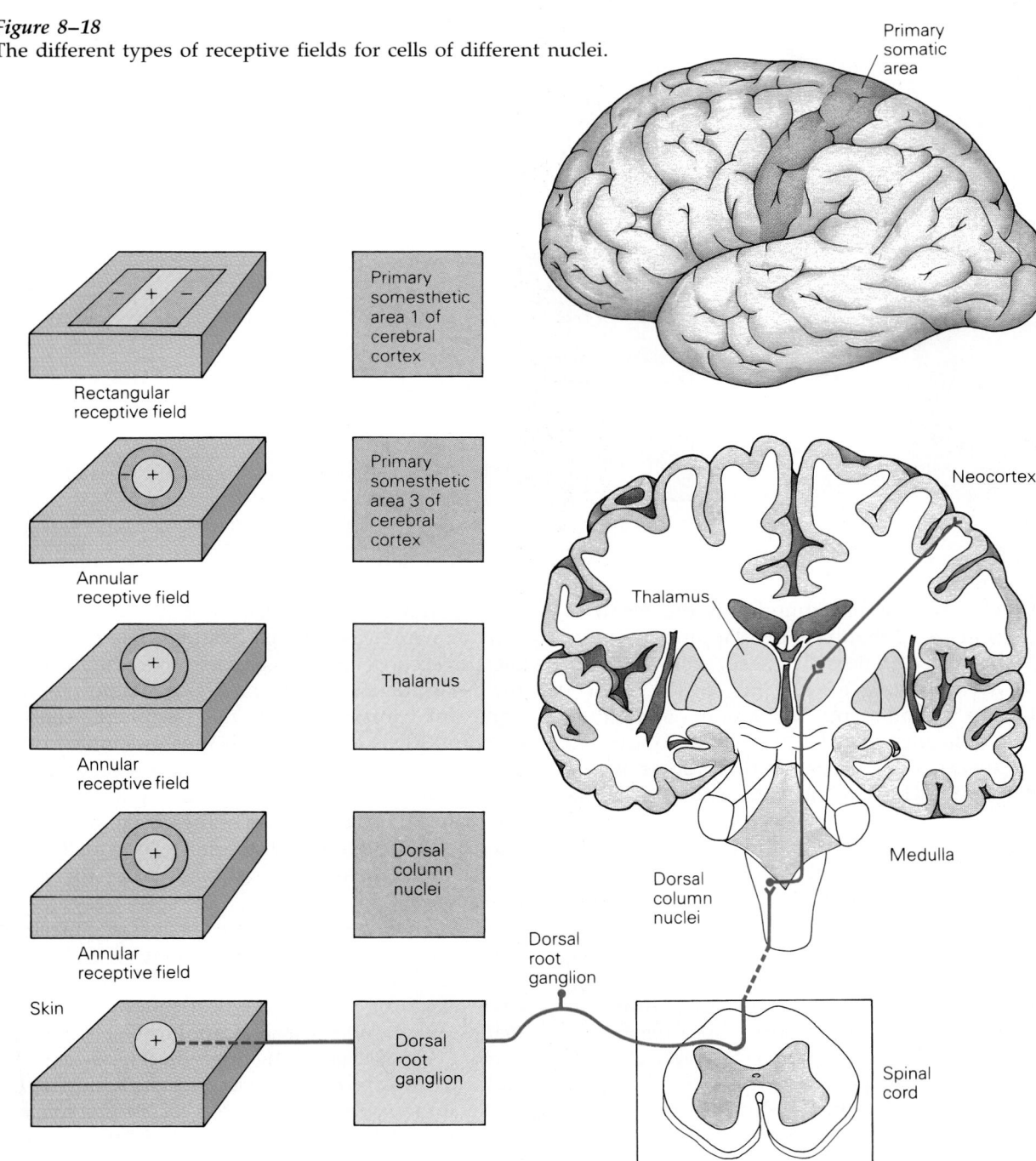

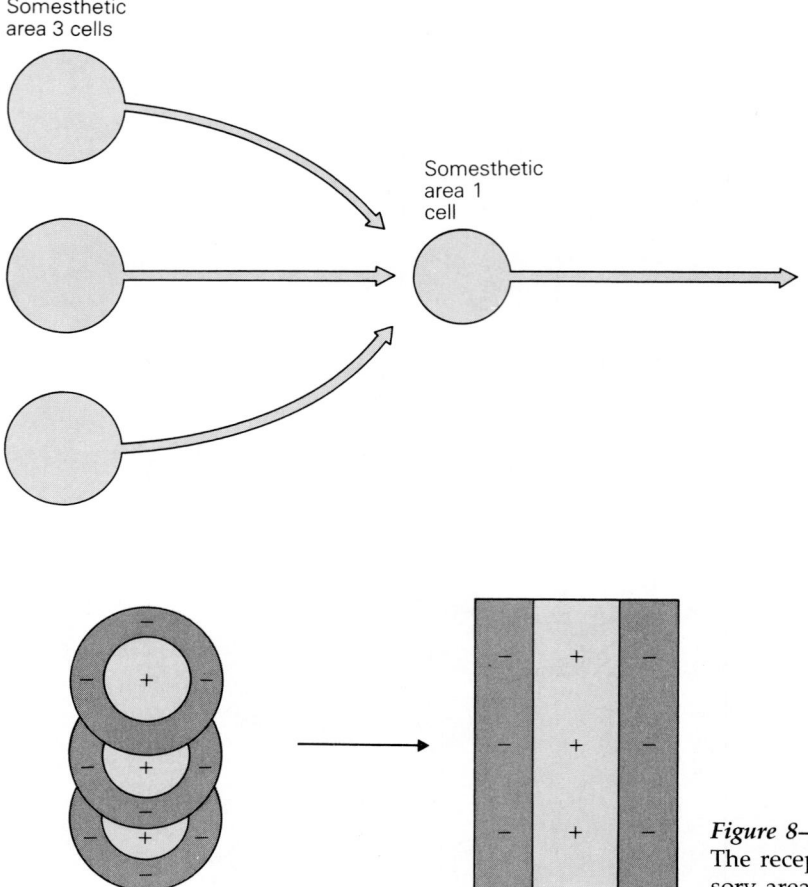

Somesthetic area 3 cells

Somesthetic area 1 cell

Figure 8–19
The receptive fields in the secondary somatosensory areas are rectangular.

tiveness of a rectangular stimulus is the result of nerve cells in area 3 that have overlapping receptive fields. The output fibers of these cells converge upon a nerve cell on area 1 (Fig. 8–19). Consequently, the simultaneous activation of all three input cells is needed in order to excite the one cell in area 1. This excitation occurs by the process of spatial summation. This process is described in Chapter 7.

Although sensory information of a particular type of stimulus is transmitted from area 3 to 1, and area 1 to 2, in any one Brodmann area of the somatosensory cortex, information from one type of somatic sensation tends to predominate. The nerve cells in Brodmann area 1, for example, respond to rapidly adapting cutaneous receptors; those in area 2 respond to deep pressure; those in area 3a respond to stretch receptors in skeletal muscle; and those in area 3b respond to rapidly and slowly adapting cutaneous receptors. In this way the somatic sensory cortex has several somatotopic maps of the body.

The most effective shapes of stimuli needed to activate nerve cells in area 2 can become quite complex. In essence, the physical features of an object in the environment are first broken down into its fundamental components by the peripheral sensory receptors. This information is then assembled to reconstruct the holistic features of the object that are capable of activating nerve cells in higher centers of the brain. In our discussion of the visual system, we will see that a similar synthesis and reconstruction of the external environment takes place for visual information.

The somatic sensory cortex is also organized so that all of the cells that respond to one type of sensation are located together within vertical columns. Each nerve cell within a vertical column of the somatic sensory cortex, for example, is activated by Pacinian corpuscles in one fingertip (Fig. 8–20). The location of the receptive fields for cells within a vertical column is also similar. *Thus, the vertical columns in neocortex represent basic units of sensation-specific and location-specific function.*

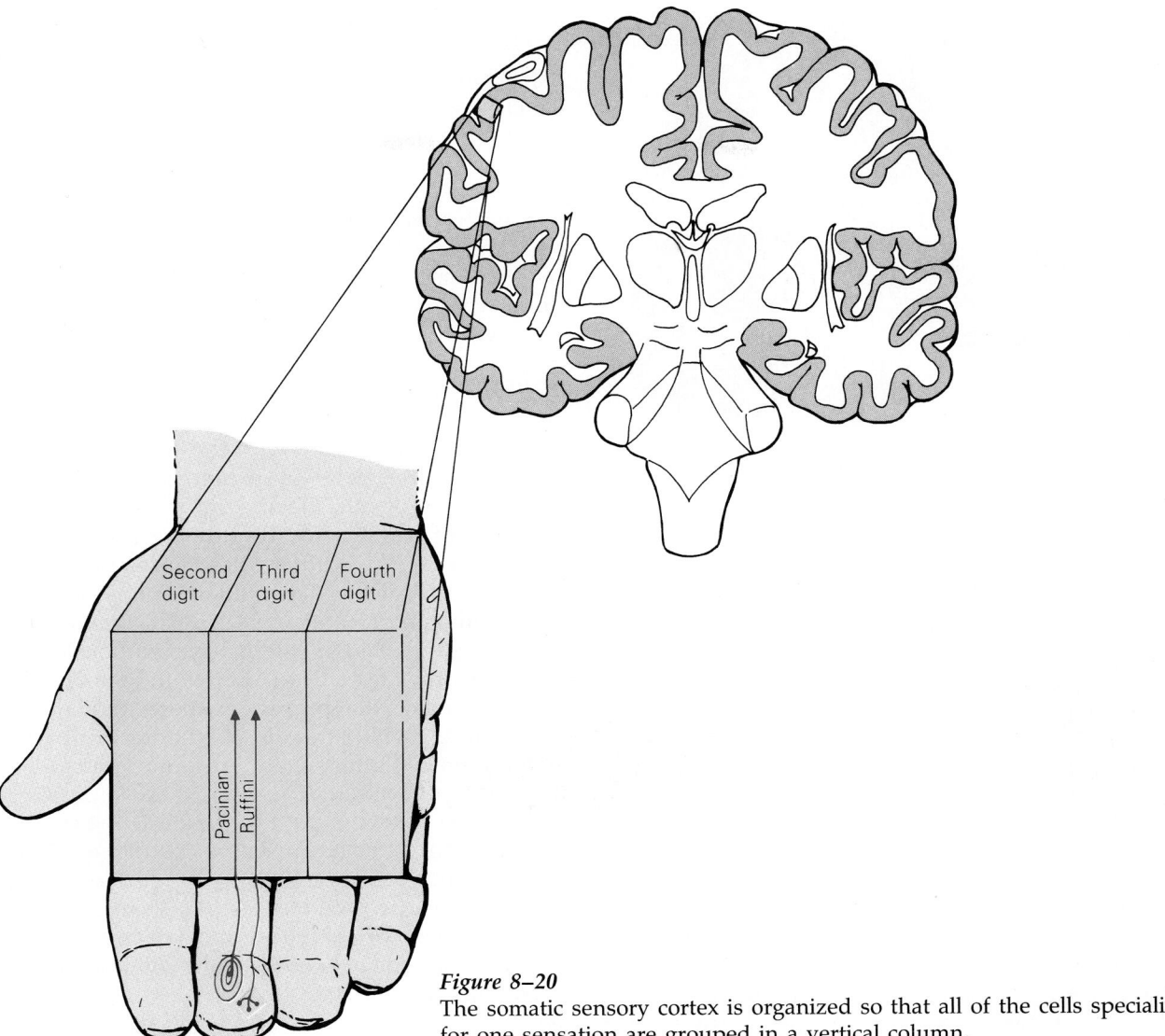

Figure 8–20
The somatic sensory cortex is organized so that all of the cells specialized for one sensation are grouped in a vertical column.

The Anterolateral Pain and Temperature Pathway

As do the nerve fibers that transmit tactile sensation, the fibers that send pain and temperature information enter the dorsal portion of the spinal cord (Fig. 8–21). After entering the spinal cord, however, the majority of them form synaptic connections with nerve cells located in the dorsal horn. This group of nerve cells is collectively called the **substantia gelatinosa.** These cells, in turn, cross the midline and ascend toward the brain as a bundle of fibers along the anterior and lateral portion of the spinal cord, forming the **anterolateral pathway.** These fibers have inputs to the reticular formation of the brain stem and the thalamus. The reticular formation serves as an alerting system and activates other areas of the brain.

Inputs to the thalamus terminate in the **ventrobasal (VB)** nuclei. This information in turn is sent to the primary somatic sensory cortex in a somatotopic fashion, and to the frontal cortex. The pain information that reaches the primary sensory cortex is associated with acute pain, whereas the information that reaches the frontal cortex is associated with chronic pain.

The perception of pain can be altered by other sensory inputs from the periphery or controlled by descending inputs from the brain stem and other higher brain centers. This gating of pain information

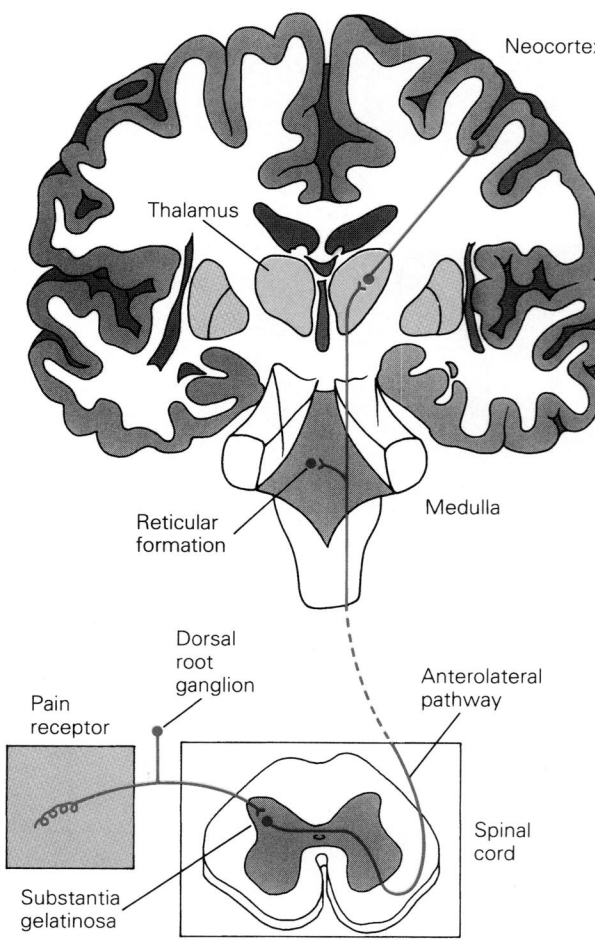

Figure 8–21
Pain and temperature information travels to the brain via the anterolateral pathway.

is carried out by an inhibitory interneuron located in the dorsal horn of the spinal cord (Fig. 8–22). As previously discussed, the pain fibers from the periphery release the peptide called substance P as the neurotransmitter. In the absence of substance P release, pain cannot be detected by the brain. The interneuron acts to inhibit the release of substance P by a mechanism called **pre-synaptic inhibition.** The axon terminal of the interneuron forms a synapse upon the terminal of pain fibers. The terminals of the interneurons release the peptide called **enkephalin,** which in turn inhibits the release of substance P from the terminals of the pain fibers.

Thus, the interneuron is the key element in altering the sensation of pain. It can be activated by sensory fibers associated with touch, pressure, and vibration, or by fibers from higher brain centers. The influence of other tactile inputs is seen in the reduction of pain caused by rubbing the skin surrounding a region of injury; this causes the pain to subside more quickly. The influence of higher brain centers on pain perception is seen in cases in which individuals involved in emergency situations do not realize that they have been injured until much later. The mechanisms that prevent the perception of pain under these highly stressful conditions are part of an autonomic "fight or flight" response that will be discussed in Chapter 10.

Another aspect of pain sensation is the perception that pain actually caused by the distress of internal organs is coming from the surface of the body. This is often seen in cases of heart attacks in which the area of the chest and left arm become painful. This type of **referred pain** will also be considered in Chapter 10.

Figure 8–22
The perception of pain can be inhibited by the action of an interneuron.

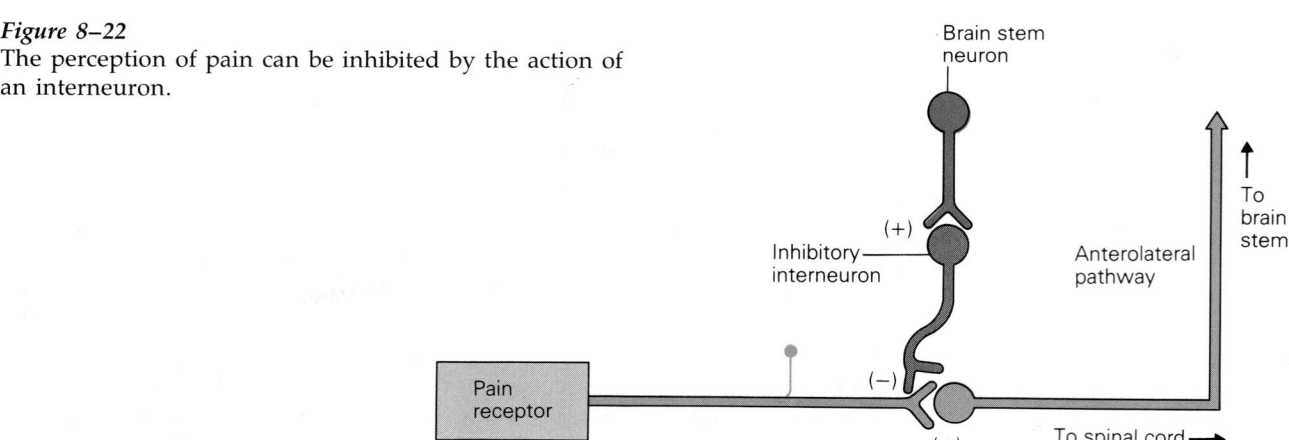

The Visual System

The visual system allows us to determine the shape and color of objects such as books, and the form of the letters printed within its pages. Parts of the visual system also allow us to recognize the grouping of letters in the form of words, and the grouping of words to represent ideas. These latter functions will be discussed in Chapter 11 when we consider the localization of language centers in the brain. In this section, we will consider how the visual system detects the shape and color of objects and the movement of objects in the external environment. The features of shape, color, and movement are processed by different cell groups within the visual system, beginning with those found in the eye.

The eye is the sensory organ of the visual system that transduces light into electrical impulses, which the nervous system then uses to transmit information to the brain. The light perceived is part of the continuum of **electromagnetic radiation** that ranges from radio waves to X-rays (Fig. 8–23). Each type of electromagnetic radiation is characterized by a **wavelength.** Radio waves have long wavelengths whereas X-rays have very short wavelengths. The colors of light that we see have wavelengths in between radio waves and X-rays, and range from 400 to 700 nm. These wavelengths correspond to the range of colors seen in a rainbow.

In addition to its wave-like nature, light can also be thought of as a composite of particles called **photons.** These photons have wavelengths and energies associated with different colors. All objects have the capability of selectively absorbing certain photons of light, and those photons not absorbed are reflected back into the environment. The reflected photons are those that give an object its color and that interact with the eye.

Anatomical Organization of the Eye

Figure 8–23 is an illustration of a cross-section of the eye. It shows the cornea, iris, lens, vitreous humor, and retina. Together the **cornea** and **lens** act like the lens of a camera. They operate to bend the light rays and focus them on the region of the fovea. Focusing a camera changes the distance between the lens and the film. Our eyes accomplish this feat by changing the shape of the lens; **ciliary muscles** tighten and flatten the lens to focus on objects that are far away, or relax it to focus on objects that are near by. The **retina** is the part of the eye containing receptor cells called **photoreceptors** that absorb photons. As the illustration shows, photons must first pass through the cornea and lens before reaching the retina. The most sensitive part of the retina is a region called the **fovea;** consequently, the lens attempts to focus the image of an object onto the fovea. The image projected onto the retina, however, is upside down. The **inversion** of the image is similar to that produced by a camera lens onto a roll of film. Even though the retina detects the form of objects in an upside-down fashion, the brain reorganizes this information so that we interpret the outside world as right-side-up.

The Nerve Cells of the Retina. We can best understand how the eye transduces the form of an object by first examining the functional organization of the nerve cells within the retina. In addition to photoreceptors, the retina contains other types of nerve cells called **ganglion cells, bipolar cells,** and **horizontal cells** (see Fig. 8–23). Photoreceptors, either **rods** or **cones,** are responsible for absorbing the photons and are situated at the back portion of the retina, behind these other types of cells. Photons must therefore pass through several layers of cells before reaching the layer of photoreceptors.

Rods are cylindrically shaped photoreceptors (Fig. 8–23c) that contain one type of pigment capable of absorbing photons representing a broad range of wavelengths. Rods display a high degree of convergence onto other cells and thus as a group are very sensitive to low levels of illumination. They therefore operate well during the night.

Cones are conically shaped and absorb photons associated with a more narrow range of wavelengths for color. Cones display a low degree of convergence onto other cells and therefore as a group are not as sensitive as rods but have better spatial resolution. There are three types of cones, each responsive to photons of a different range of wavelengths. These cones form the basis of color perception and are used for day vision.

Photoreceptors and Phototransduction. The photoreceptor pigment that absorbs photons is called **visual pigment.** This pigment is embedded within the membranes of **discs,** lamellar structures in the **outer segment** of the photoreceptor. Also present in the disc membrane is a protein called **transducin** and an enzyme called **phosphodiesterase** (Fig. 8–24a). The pigment in a rod cell disc is **rhodopsin.** When the rhodopsin molecule absorbs a photon, its three-dimensional form changes. In its new form, rhodopsin is now capable of activating transducin. Transducin, in turn, is now able to activate phosphodiesterase, which hydrolyzes cyclic GMP

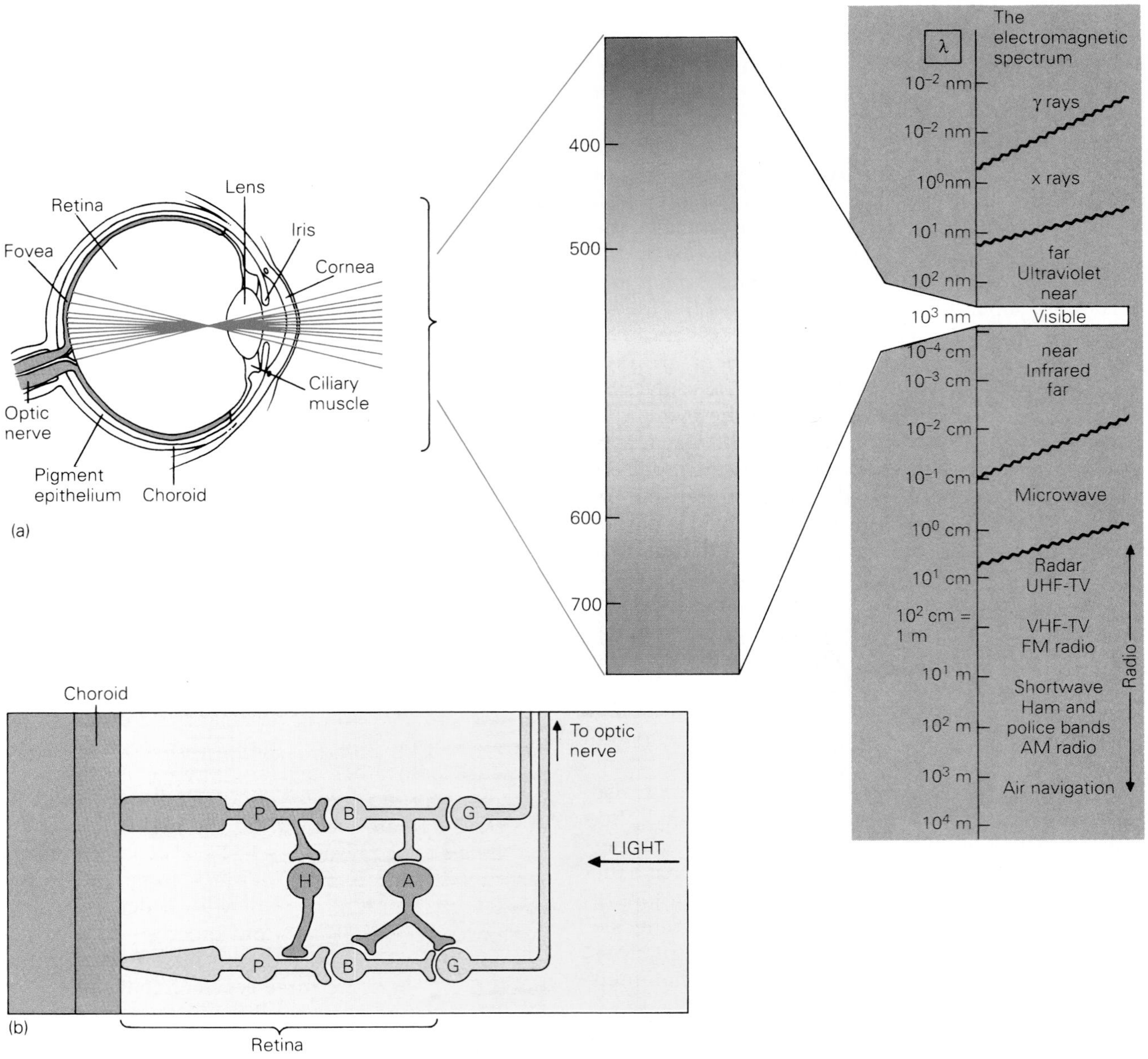

The electromagnetic spectrum

| λ |

10^{-2} nm	γ rays
10^{-2} nm	x rays
10^0 nm	
10^1 nm	far
10^2 nm	Ultraviolet near
10^3 nm	Visible
10^{-4} cm	near Infrared
10^{-3} cm	far
10^{-2} cm	
10^{-1} cm	Microwave
10^0 cm	
10^1 cm	Radar UHF-TV
10^2 cm = 1 m	VHF-TV FM radio
10^1 m	Shortwave Ham and police bands AM radio
10^2 m	
10^3 m	Air navigation
10^4 m	

400
500
600
700

(a)

Lens
Retina
Iris
Fovea
Cornea
Optic nerve
Ciliary muscle
Pigment epithelium
Choroid

Choroid

To optic nerve

LIGHT

(b)

Retina

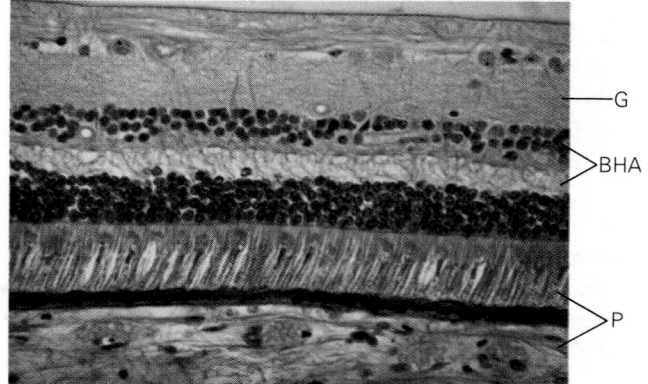

(c)

G
BHA
P

Figure 8–23

(*a*) General structure of the eye and the visible portion of the electromagnetic spectrum. (*b*) Organization of photoreceptors (P, rods and cones), bipolar (b), horizontal (H), amicrine (A), and ganglion (G) cells in the retina. (*c*) The general structure of a rod cell and a cone cell. Note the membranous discs within the outer segments of both cells. The membranes of these discs contain the pigment that, when activated by photons, produces visual sensation. (*d*) Photomicrograph of the different cell types of the retina (× 100). (© Ed Reshke/ Peter Arnold, Inc.)

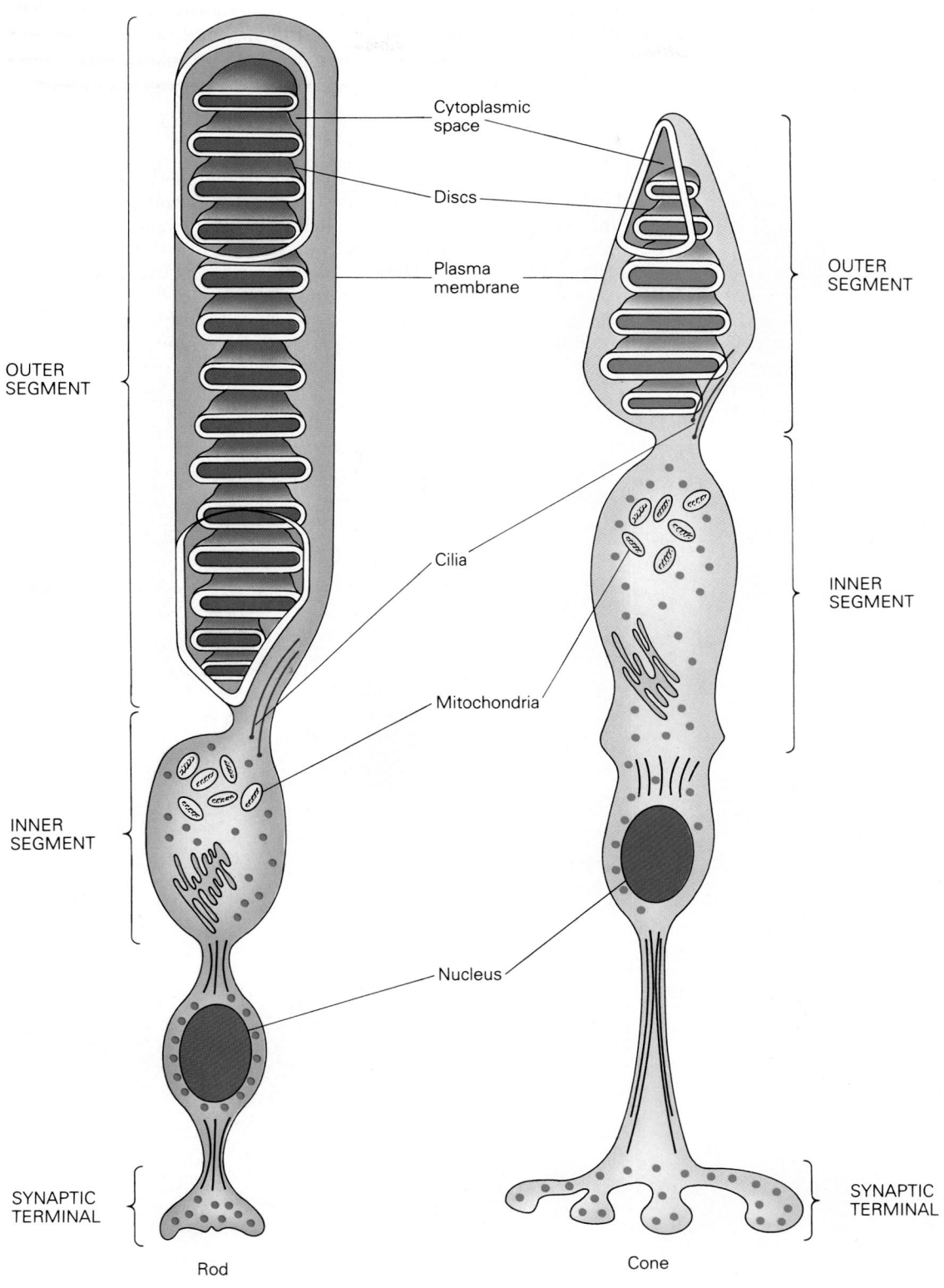

OUTER
SEGMENT

INNER
SEGMENT

SYNAPTIC
TERMINAL

Cytoplasmic
space

Discs

Plasma
membrane

Cilia

Mitochondria

Nucleus

OUTER
SEGMENT

INNER
SEGMENT

SYNAPTIC
TERMINAL

Rod

Cone

(d)

(cGMP) to 5'-GMP. In this process, the net effect is the reduction in the amount of intracellular cGMP.

What is the significance of decreasing the cGMP concentration? Normally, cGMP opens Na^+ channels in the photoreceptor plasma membrane. When cGMP is decreased, Na^+ channels begin to close, thus reducing the influx of positive ions into the cell. This results in a hyperpolarization of the membrane. Thus, the photoreceptor transduces the photon of light into a more negative membrane potential (Fig. 8–25). In order to understand how this hyperpolarization is transformed into a series of action potentials, we must first consider the organization and response of nerve cells in the retina.

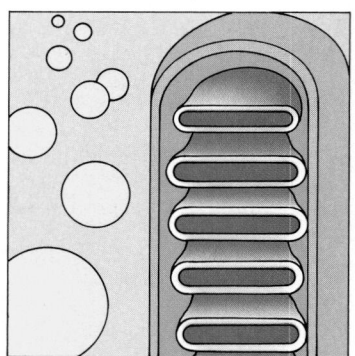

Figure 8–24

(*a*) In the absence of light, sodium channels in the plasma membrane of red cells are kept open by the presence of cGMP, and sodium ions enter the cell. (*b*) When a photon of light enters a rod cell, it causes a rhodopsin (pigment) molecule to activate transducin, which in turn activates phosphodiesterase. The enzyme hydrolyzes cytosolic cGMP of the rod to 5'-GMP. The reduction in the amount of cGMP causes Na^+ channels in the rod plasma membrane to close, reducing the amount of Na^+ entering the cell and producing a hyperpolarization (*b*).

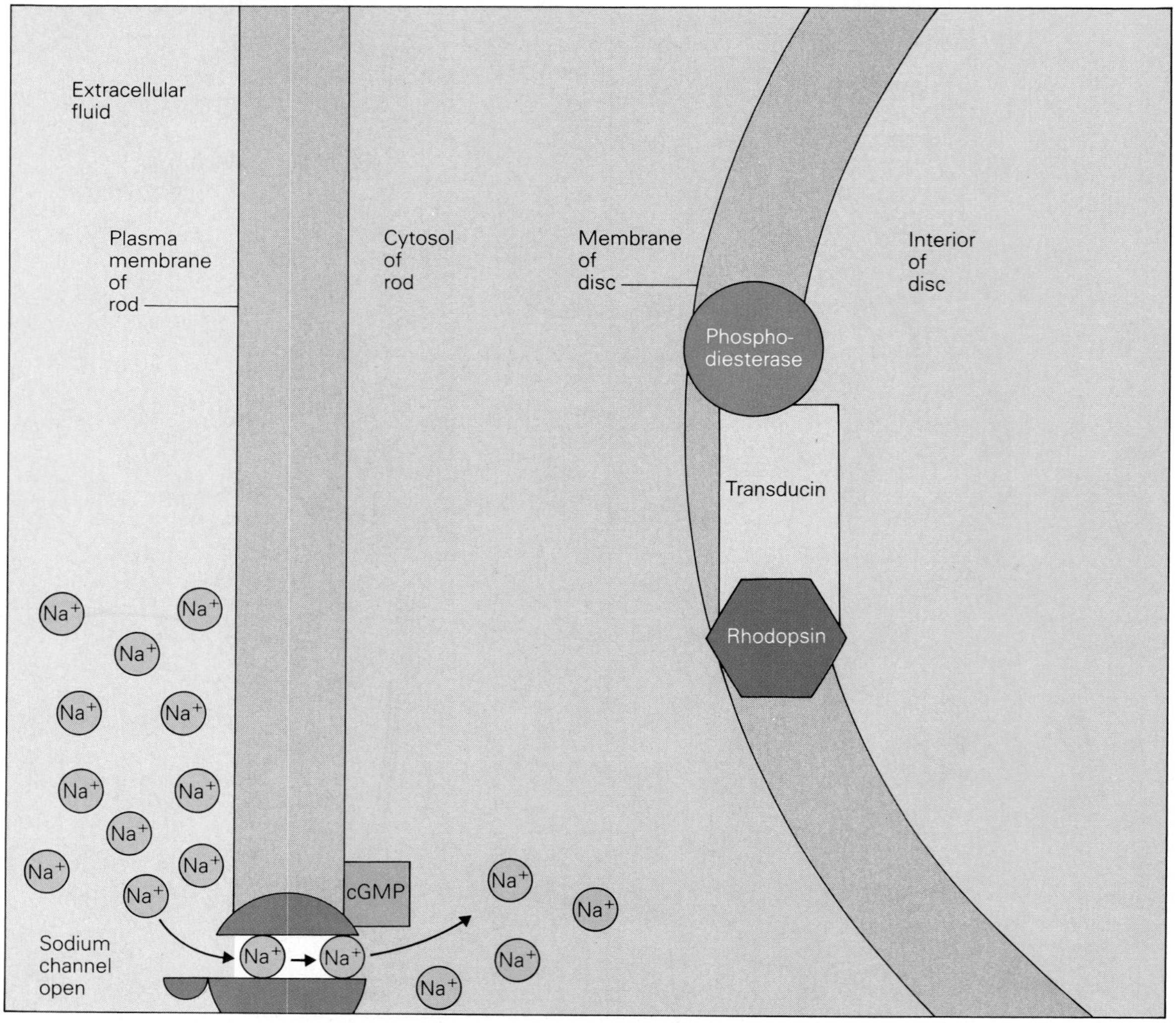

Extracellular fluid

Plasma membrane of rod

Cytosol of rod

Membrane of disc

Interior of disc

Phospho-diesterase

Transducin

Rhodopsin

Na^+

Na^+

Na^+

Na^+

Na^+

Na^+

Na^+

Na^+

Na^+

Na^+

Na^+

cGMP

Na^+

Na^+

Sodium channel open

$Na^+ \rightarrow Na^+$

Na^+

Na^+

(a)

Interaction of Cells in the Retina. We can best understand the way in which visual information is processed in the retina by considering the interactions between photoreceptors, bipolar cells, horizontal cells, and ganglion cells. Photoreceptors form synaptic connections with **bipolar cells** (Fig. 8–26). Some bipolar cells, in turn, form synaptic connections with **ganglion cells.** These cells generate action potentials even in the absence of input from the bipolar cells. The bipolar cells, however, can depolarize the ganglion cells, resulting in a greater frequency of action potentials. Ganglion cell activity can be inhibited by a fourth type of nerve cell found in the retina, the **horizontal cell.** This cell receives its inputs from photoreceptors located in the retina lateral to the region of excitation (Fig. 8–26*b*). When these photoreceptors detect light, they inhibit the horizontal cell.

The inhibition of the horizontal cell, in turn, causes a depolarization of adjacent photoreceptors (most likely by preventing the release of an inhibitory transmitter) and the subsequent inhibition of ganglion activity. Depending on where light impinges on the retina, ganglion cells can thus be activated or inhibited.

The areas of the retina that excite or inhibit a ganglion cell make up the receptive field for that ganglion cell. Therefore—as in the case of the dorsal column nuclei cells of the somatosensory pathway—the retinal ganglion cells have receptive fields that are shaped like an annulus. For some ganglion cells, the center of the annulus is excitatory, while the periphery is inhibitory (Fig. 8–27*d*). These neurons are called **"on" center** and **"off" surround cells.** Other ganglion cells have **"off" centers** and **"on" surrounds.**

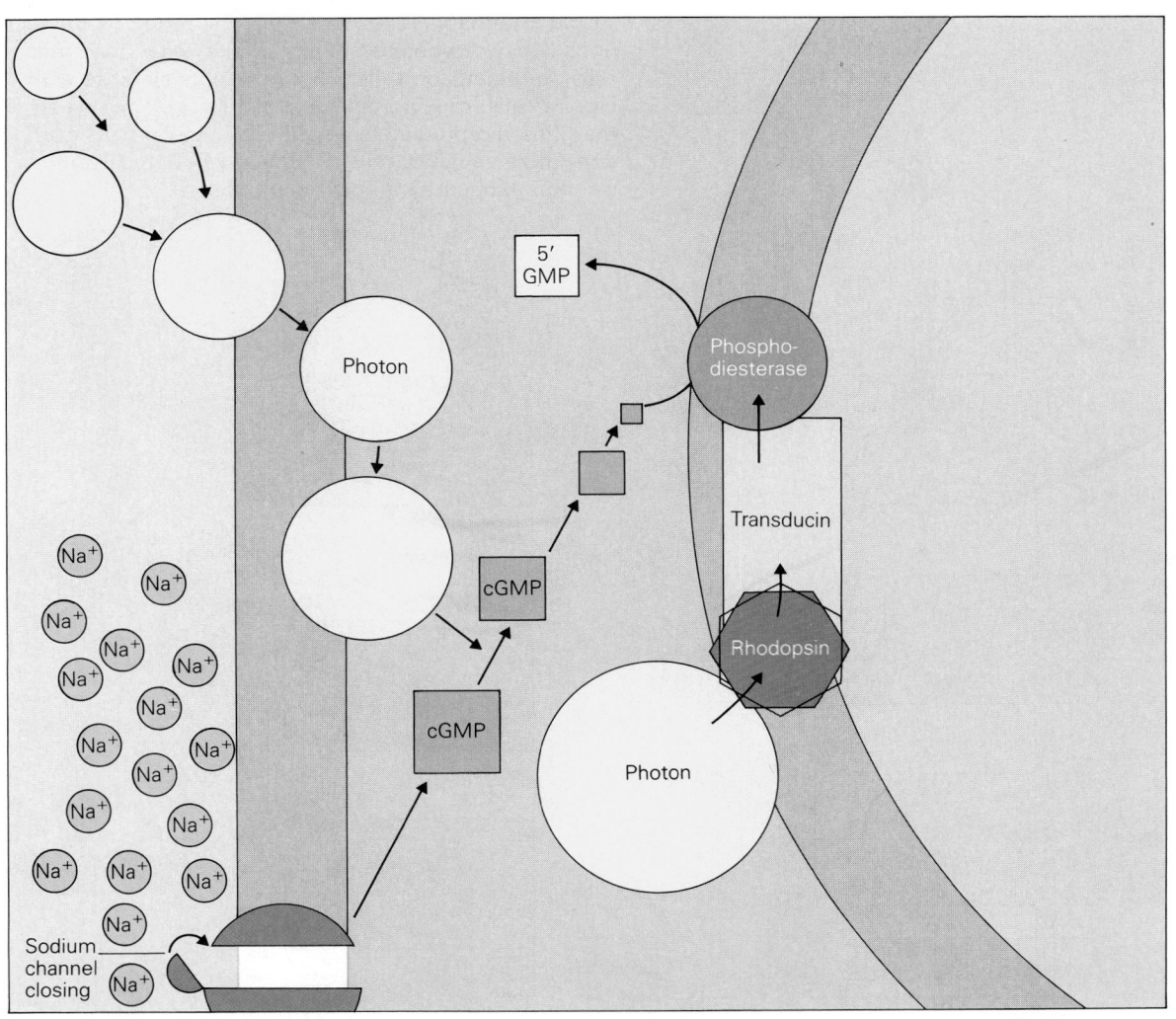

(b)

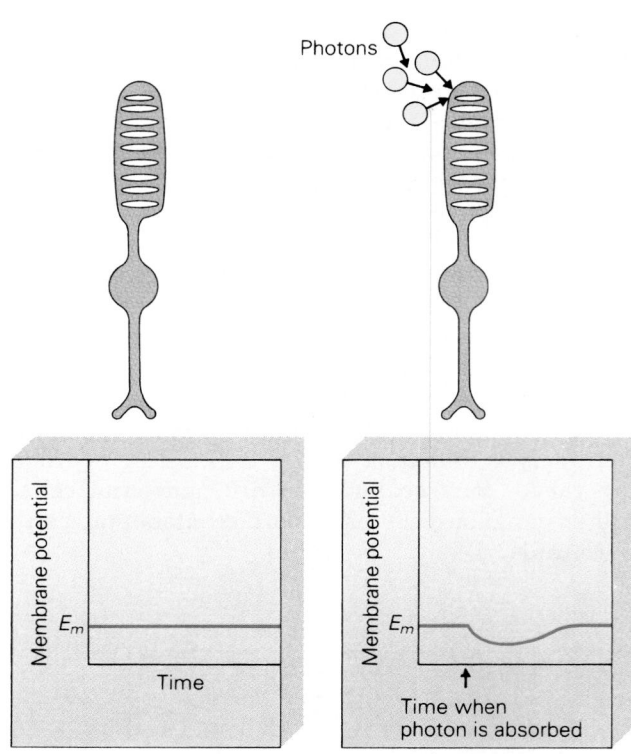

Figure 8–25
The hyperpolarization of a rod cell membrane caused by a photon is eventually transduced into action potentials that travel to the optic nerve.

Figure 8–26
Visual information is processed in the retina by interactions between photoreceptors, bipolar cells, horizontal cells, and ganglion cells. Photoreceptors synapse with bipolar cells, causing bipolar cells to depolarize when the photoreceptors detect light. The bipolar cells can depolarize ganglion cells. Ganglion cell depolarization can be inhibited by a horizontal cell.

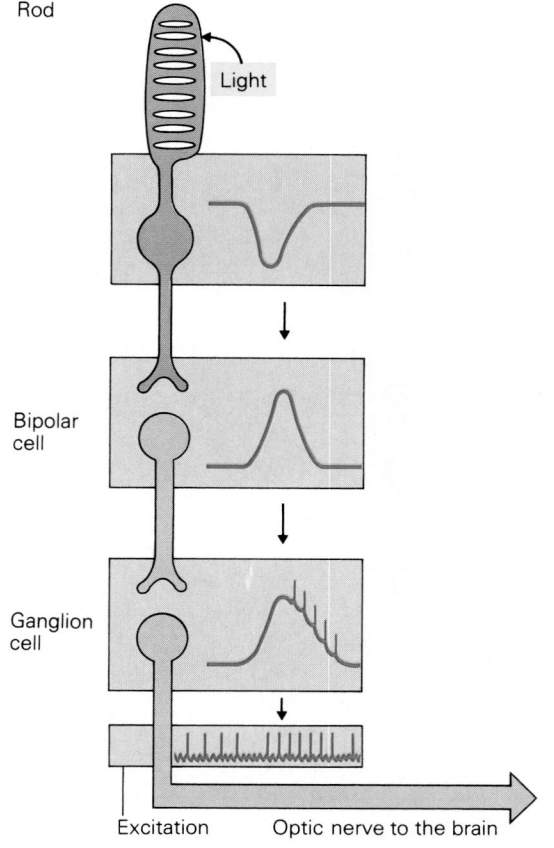

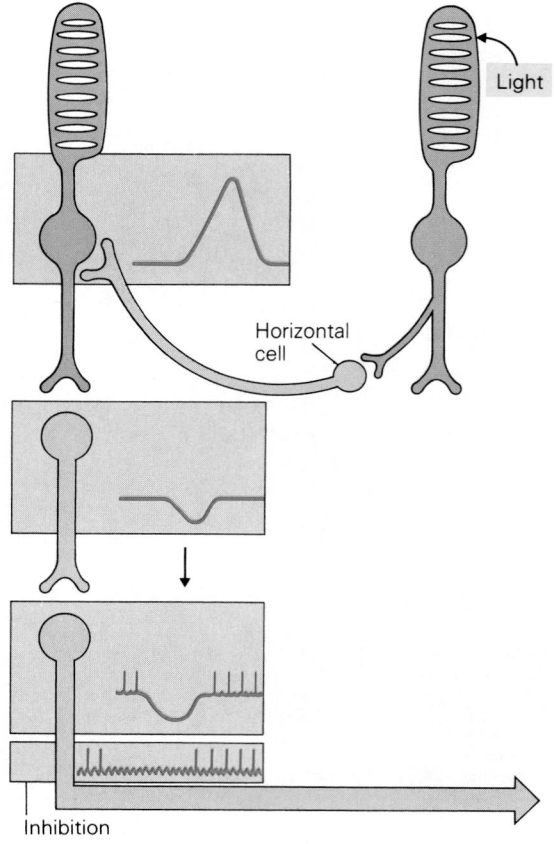

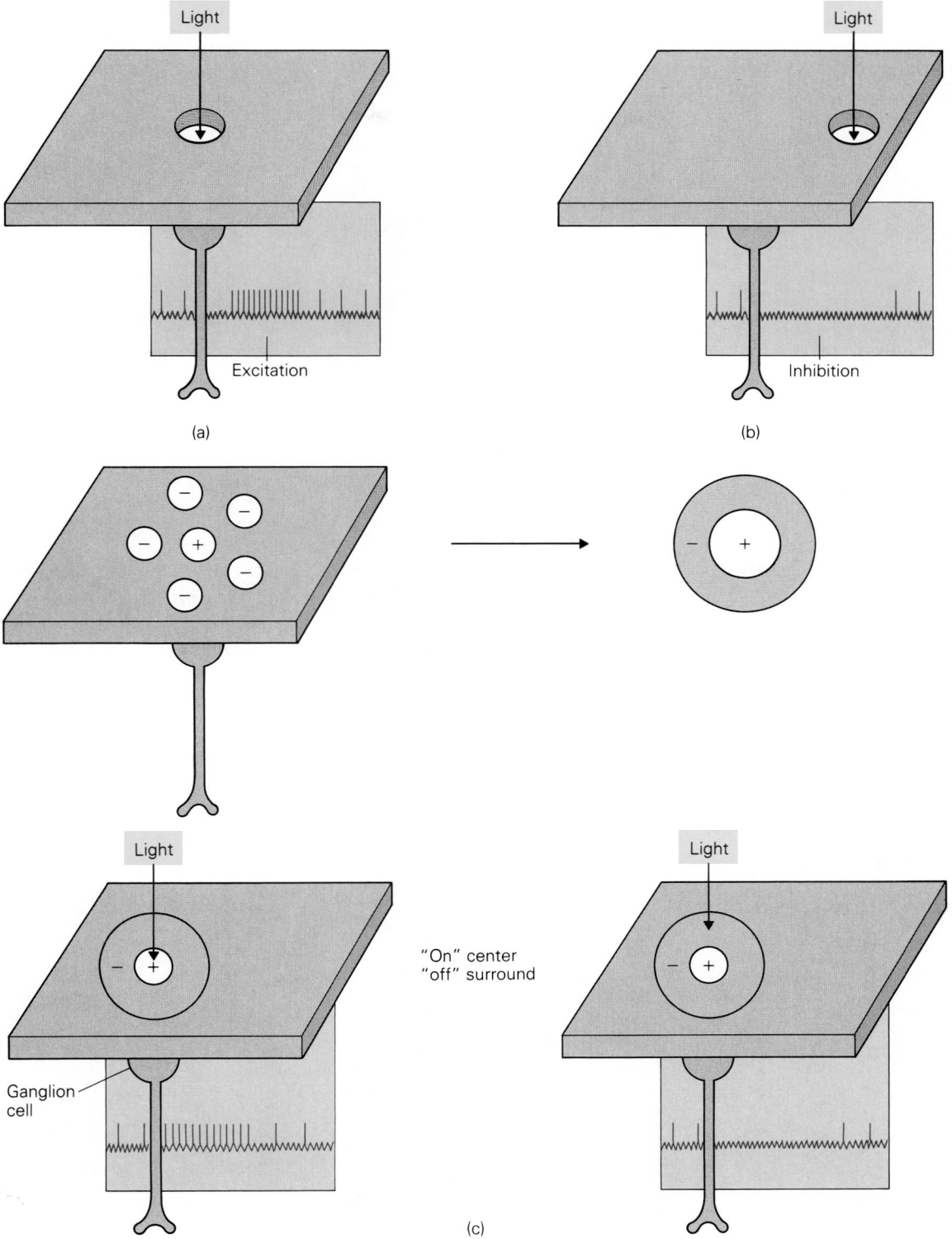

Figure 8–27
The effects of light on ganglion cell activity. (*a*) Light near the region of a
ganglion cell produces depolarization and action potentials, whereas light in
regions farther from the cell may inhibit the formation of action potentials
(*b*). (*c*) A ganglion cell with a receptive field whose center is excitatory and
whose surrounding ring is inhibitory is an "on-center and off-surround" cell.

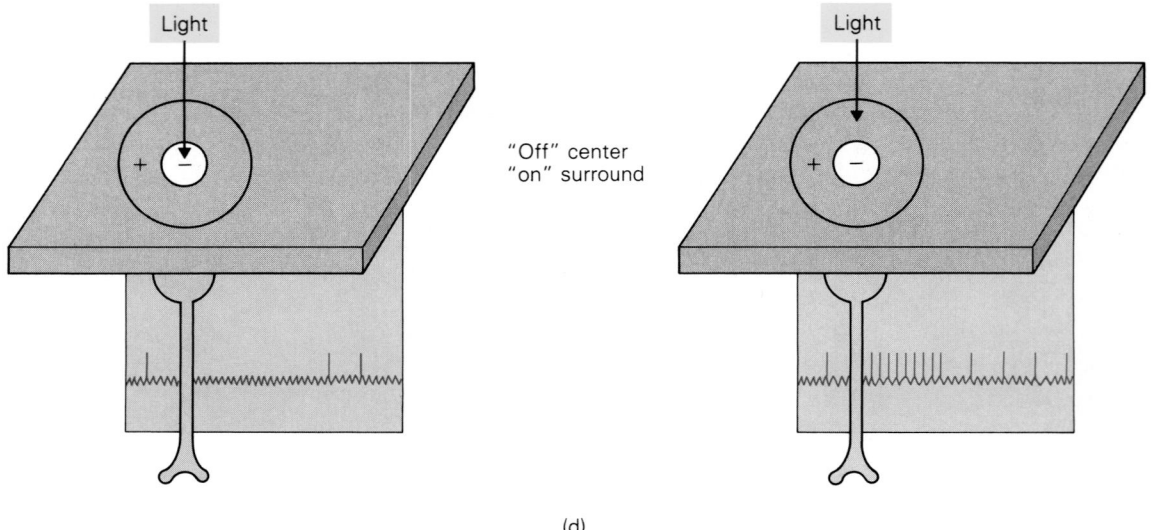

"Off" center
"on" surround

(d)

Figure 8–27 (cont.)
(*d*) Other ganglion cells have "off" centers and "on" surrounds.

The spatial pattern of activated ganglion cells in the two-dimensional plane of the retina represents the form of an object whose image is projected onto the surface of the retina. The image of a pencil, for example, would activate a selective array of ganglion cells (Fig. 8–28). The image of the pencil is thus encoded by the spatial location of activated ganglion cells within the retina.

Pathways for Visual Information

Other groups of neurons in the visual pathway have receptive fields that differ from those of retinal ganglion cells. In order to fully appreciate the visual receptive fields that are exhibited by neurons in other areas of the brain, we first need to consider the anatomical pathways that connect these regions together.

Figure 8–28
The location of activated ganglion cells encodes the image of an object.

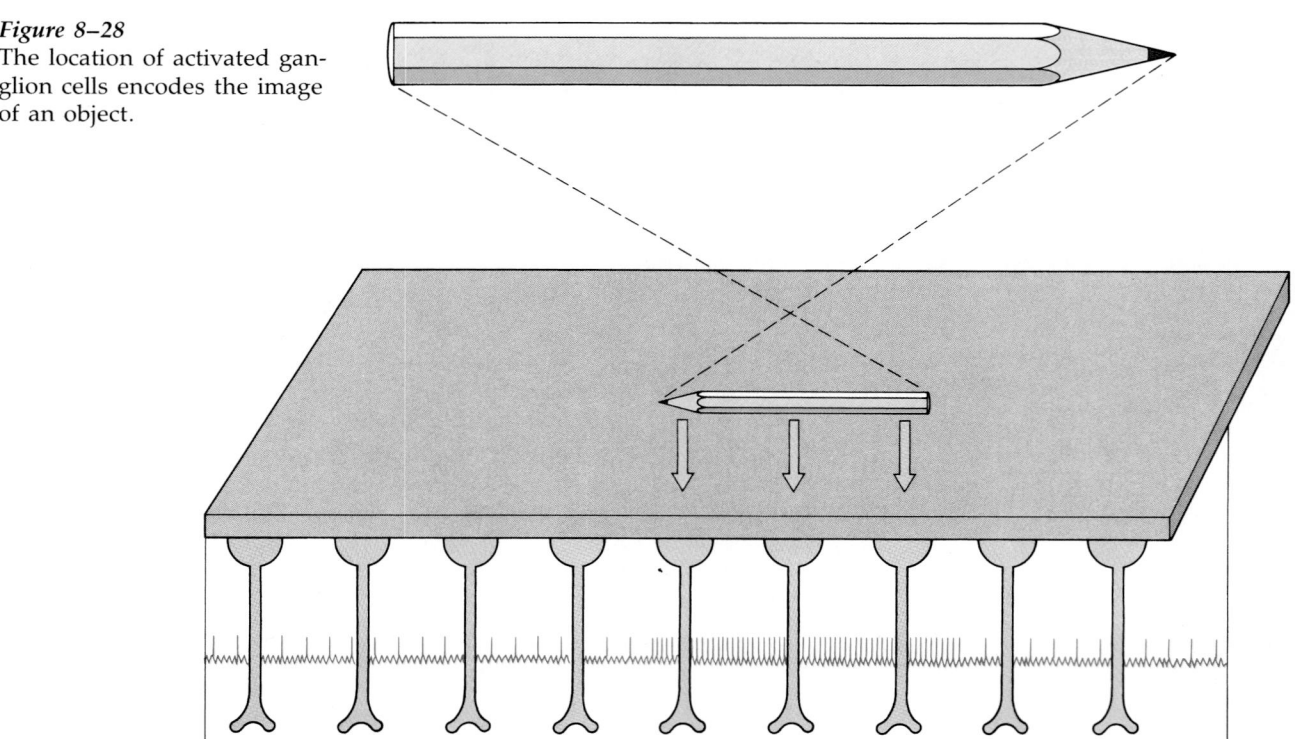

The ganglion cells of the retina are the only cells that send axons outside of the eye (Fig. 8–29*a*). These fibers make synaptic connections with neurons in the lateral geniculate nucleus of the thalamus. Fibers from ganglion cells situated in the retina near the nose (nasal region) cross the midline and form connections with the lateral geniculate nucleus on the opposite side of the brain. In contrast, fibers from ganglion cells in the retina located farthest away from the nose (temporal region) do not cross the midline and thus form connections with the lateral geniculate nucleus on the same side of the brain.

The lateral geniculate nucleus contains six distinct layers of neurons (Fig. 8–29*b*). The first, fourth, and sixth layers contain cells that respond to light impinging upon the eye on the opposite side of the brain. The second, third, and fifth layers contain cells that respond to light impinging upon the eye on the same side of the brain. Nerve cells in these layers, in turn, have axons that synapse upon neurons in the primary visual cortex (Brodmann area 17).

The cells of the visual cortex are organized in the form of a map of the external visual field (see Fig. 8–29*c*). An area in the center of the visual field

Figure 8–29
(*a*) Ganglion cells extend their axons outside of the eye through the optic nerve. Ganglion cell fibers in the temporal retina cross the midline at the optic chiasma and form connections with cells in the lateral geniculate nucleus on the opposite side of the brain. (*b*) In the lateral geniculate nucleus, visual information is sorted according to which eye it originated in and then sent to the visual cortex, where cells are arranged in ocular dominance columns, as well as in the form of a map of the external visual field (*c*).

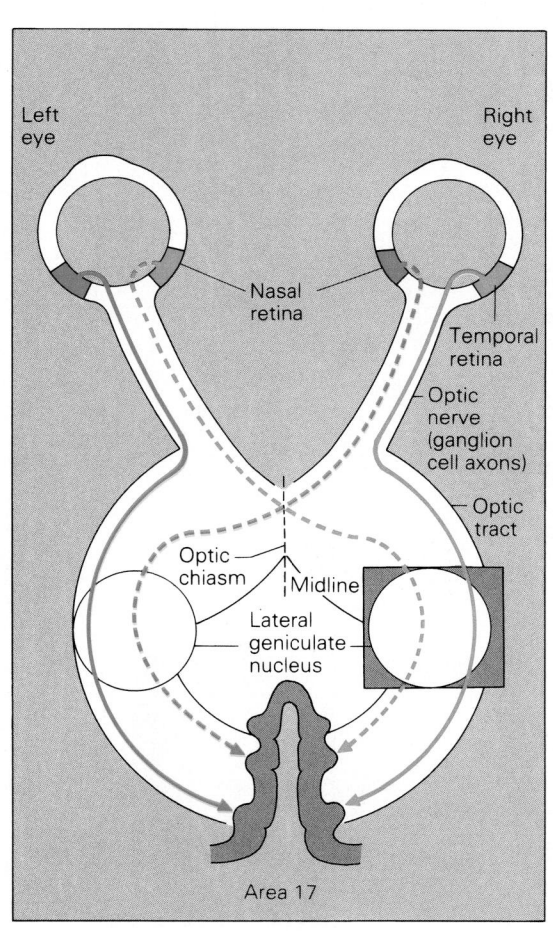

(a)

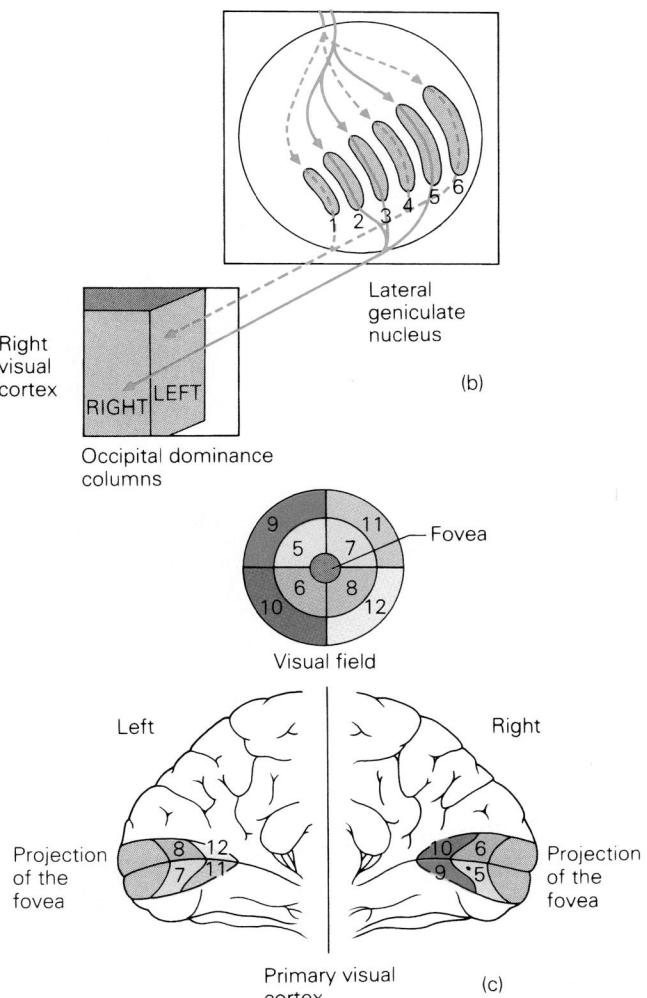

(b)

(c)

projects to the fovea of the retina. Since the fovea is the region in the retina of greatest acuity and contains the greatest density of cones, the associated areas in the visual cortex also occupy the largest amount of cortical space. Regions of the visual field that are adjacent to the center are represented in the visual cortex in areas next to the representation of the fovea. Likewise, the cortical representation of the peripheral visual field is found adjacent to the para-foveal cortical representation. In this manner, the visual field is precisely mapped onto the visual cortex and is referred to as a **visuotopic organization.** Since the visual field is also precisely mapped onto the retina, the visual cortex can also be thought of as a **retinotopic map.**

Within each region of the visual cortex, the cells are organized into functional columns. Each column of cells responds to light impinging upon either the right or left eye. These columns are therefore referred to as **ocular dominance columns.**

Receptive Fields of Neurons in the Lateral Geniculate Nucleus and Visual Cortex

The receptive fields exhibited by cells in the lateral geniculate nucleus are similar to those of the ganglion cells—that is, annular with either excitatory centers and inhibitory rings (or with inhibitory centers and excitatory rings). In the primary visual cortex, however, the visual stimulus that most effectively activates the neurons in this region of the brain is light that impinges upon the retina in the form of a rectangular bar (see Fig. 8–30).

Why have the receptive fields changed from annular to rectangular in going from cells in the LGN to those in the primary visual cortex? The cells in the primary visual cortex receive inputs from cells in the LGN that have overlapping annular receptive fields. Figure 8–30*a* shows three neurons in the LGN that synapse onto a nerve cell in the visual cortex. The activation of one cell in the LGN alone is

Figure 8–30
(*a*) The cells in the primary visual cortex receive information from LGN cells with overlapping receptive fields; thus, the visual cortex cells have rectangular receptive fields. (*b*) Some visual cortex cells respond only to stimuli with the correct orientation. (*c*) Orientation-sensitive cells are arranged in columns in the visual cortex. (*d*) Some examples of complex stimuli needed to activate neurons in the secondary visual cortices.

Neurons of
the lateral geniculate nucleus

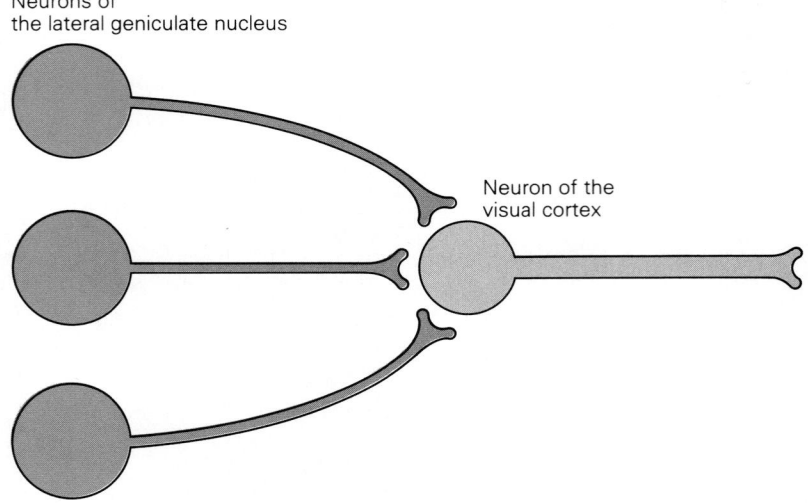

Neuron of the
visual cortex

Receptive fields

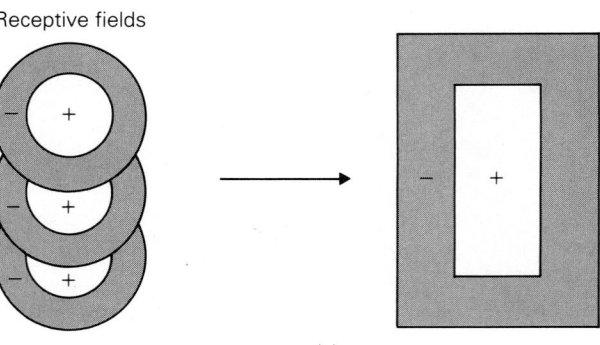

(a)

not sufficient to activate the neurons in the primary visual cortex; the simultaneous activation of all three is required to turn on the visual cortex cell. Since the three cells in the LGN have annular receptive fields that overlap in a linear array, the combination of these receptive fields has the form of a rectangular bar.

The nerve cells in the primary visual cortex are also sensitive to the orientation of the rectangular bar of light in addition to its form. Some columns of cells are most sensitive to a rectangular bar of light oriented vertically (see Fig. 8–30b). With this orientation, they generate the greatest frequency of action potentials. Other columns of cells are most responsive to a rectangular bar of light oriented horizontally, and still other columns of cells are sensitive to bars of light oriented at different angles. These columns of cells are referred to as **visual orientation columns** (see Fig. 8–30c). Cells in vertical orientation columns, for example, would be acti-

vated by the letter "I," whereas cells in horizontal orientation columns would be activated by the horizontal component of the letter "H." Studies with laboratory animals have shown that young animals raised in visual environments with vertical bars but without horizontal bars do not perceive horizontal bars when tested later in life. These results indicate that the horizontal orientation columns do not exist in the visual cortex of these animals and demonstrate the importance of the visual environment in the development of visual perception.

Stimuli more complex than rectangular bars are required to activate neurons in the secondary visual cortices (areas 18 and 19). The most effective forms of visual stimuli capable of activating neurons in visual areas 18 and 19 have complex shapes that are combinations of simple rectangular bars (see Fig. 8–30d). Cells in these areas of the visual cortex combine the features of the receptive fields from the cells in the primary visual cortex. In this way, infor-

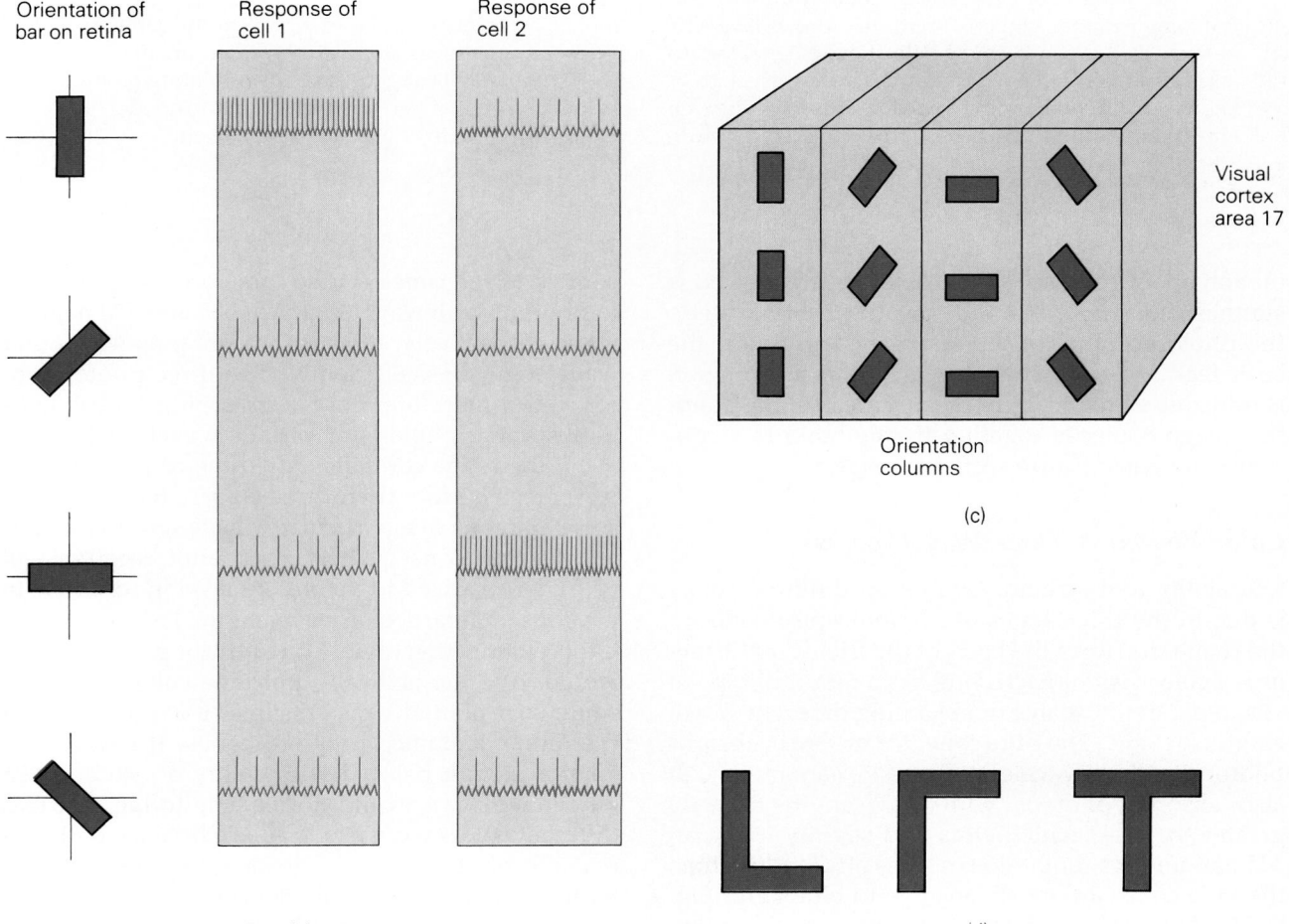

Orientation of bar on retina

Response of cell 1

Response of cell 2

Visual cortex area 17

Orientation columns

(c)

(b)

(d)

R E S E A R C H F O C U S

Environments That Enhance Brain Development

Research with laboratory animals reveals that exposure to sensory-enriched environments can change the structure and chemistry of the brain. Rats raised in environments containing toys, stimulating objects, and other rats exhibit an increased thickness of their neocortex. In addition, the cell bodies and nuclei of neurons are larger, dendrites are longer, and the area of synaptic contacts is greater. These rats also have more protein, more glial cells, larger capillaries, and an increased ratio of RNA to DNA.

The most prominent change in these stimulated animals is found in the visual association cortex (area 18); however, most areas of the cortex, as well as the cerebellum and hippocampus (Chapters 9 and 11), are also affected. Except for the frontal cortex, where the right hemisphere is affected more than the left, both hemispheres appear to be equally influenced by the enriched environment.

The more varied the enriched environment and the longer the rat remains there, the longer it retains its increased cortical dimensions after being moved to a less stimulating environment. These effects have been found in both young and very old rats. Research also indicates that rats exposed to enriched environ-

Rat in a sensory-enriched environment, Columbia University. (© Hank Morgan/Science Source/Photo Researchers, Inc.)

ments perform maze tasks significantly better than those raised in less stimulating environments.

These results suggest that a stimulating environment is essential for optimal growth and development of the brain, especially in children.

---

mation about the form of the object being viewed is synthesized. Thus, like the somatosensory system, the photoreceptors of the visual system detect the basic features of a visual object. As this information is transmitted to the higher visual areas of the brain, the image is pieced together by the brain to reconstruct the basic features of the image.

Color Vision: A Three-Way Process

Our ability to discriminate between different colors is due to the three types of photoreceptor cones in the retina that form the basis of the **trichromatic theory** of color vision. Each cone has a different type of visual pigment capable of absorbing different wavelengths of light. The **blue cone,** for instance, absorbs photons with a wavelength of 443 nanometers. It also absorbs photons with wavelengths slightly greater than 443 nanometers and slightly less than 443 nanometers, but it does so less efficiently. Thus, the blue cones are most sensitive to blue light (Fig. 8–31). The other two types of cones are the **green**

cones and **red cones.** These cones are most sensitive to photons with wavelengths of 535 and 570 nanometers, respectively; they absorb photons with other wavelengths less efficiently. The three photoreceptor cones, therefore, have overlapping probabilities of absorbing photons of various wavelengths.

If the retina contains only three cones, how can we perceive more than three colors? Each cone absorbs photons of a particular wavelength to a different degree. Consequently, each photoreceptor will be hyperpolarized to a different level in response to photons of a particular wavelength. The differential hyperpolarization of only one type of photoreceptor would give the ability to perceive colors, but in a somewhat limited way. Assume, for example, that the retina contained only red cones. If we were to view a picture painted with colors of 540 and 605 nanometers, we would not be able to tell that two different colors were used. This is because both colors emit photons that are absorbed by the red cone with equal probability (Fig. 8–32*a*). The addition of a green cone, however, results in a greater absorption

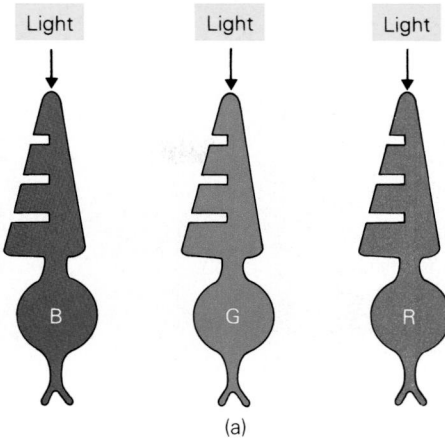

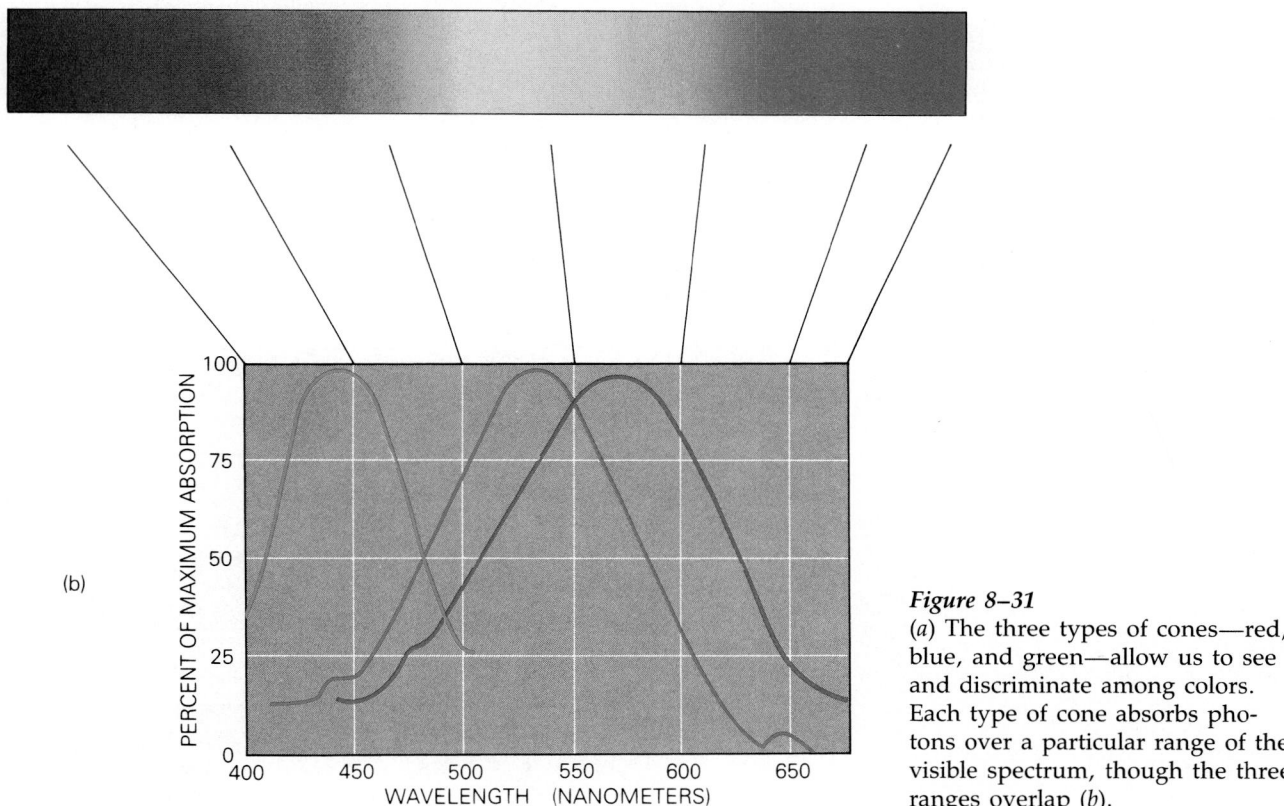

Figure 8–31
(*a*) The three types of cones—red, blue, and green—allow us to see and discriminate among colors. Each type of cone absorbs photons over a particular range of the visible spectrum, though the three ranges overlap (*b*).

of the 540 nanometer photons, and thus allows us to distinguish between the two colors. With three photoreceptor cones, our ability to distinguish between different colors becomes even better.

Color vision that relies on one type of cone is called **monochromatic color vision;** color vision that relies on two types of cones is called **dichromatic color vision.** Most people exhibit **trichromatic color vision.** Some people, however, are **color blind** and are not able to tell the difference between certain colors. Color blindness is a sex-linked trait that affects males predominantly. These males are usually **dichromats,** possessing only two types of cones, and represent approximately 2% of the population. Half of these dichromats do not have green cones, and the other half do not have red cones. These in-

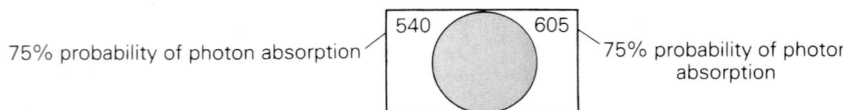

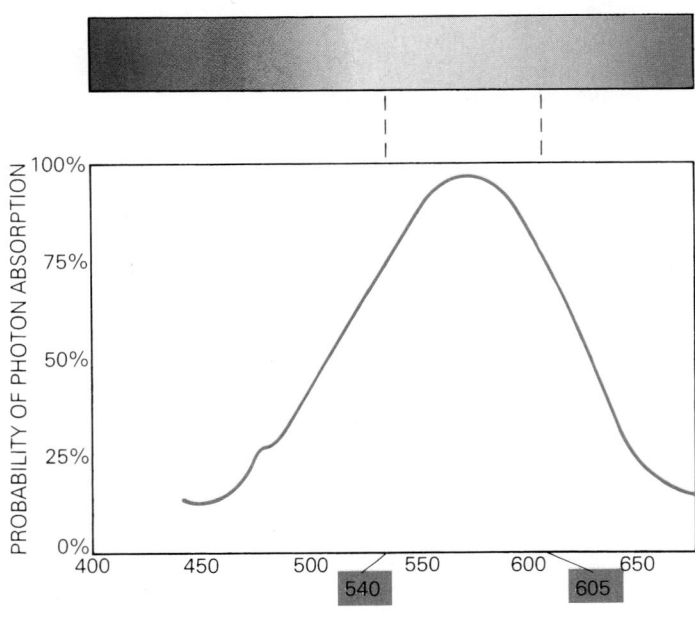

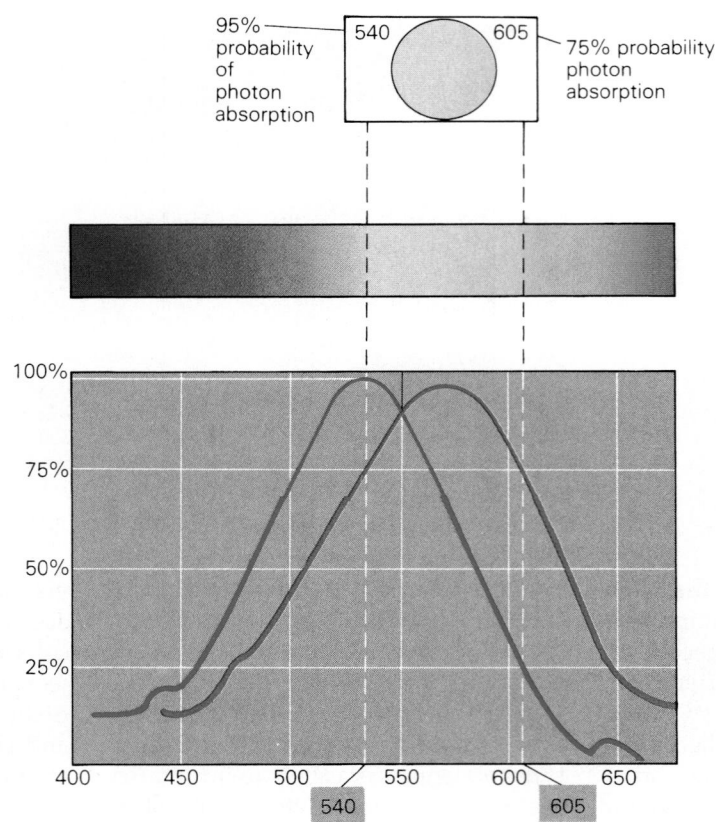

Figure 8–32
(*a*) A person with only red cones may not be able to distinguish two colors for which the rod has the same probability of absorbing photons.
(*b*) The presence of a second type of cone allows the two colors to be distinguished by increasing the probability that the rod will absorb photons at one wavelength.

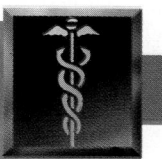

Ocular Dominance and Critical Periods of Development

In the lateral geniculate nucleus (LGN) and visual cortex are clusters of nerve cells that respond to the stimuli of light impinging upon either the left or right eye. The cells in the LGN are arranged in several layers, with each layer specific for inputs from one eye. The number of layers in the LGN is species dependent. Rodents, for example, have two layers, while monkeys have six. The segregation of left and right eye inputs emerges gradually during development from an initial pattern where inputs from both eyes are intermingled. In cats in which this phenomenon has been studied in detail, ganglion cell fibers from the contralateral eye enter the LGN at embryonic day 32 (about half of a cat's gestational period). The ipsilateral inputs arrive a few days later. During the next four and a half weeks the inputs from the two eyes gradually segregate; by birth the eye-specific layers are present as a result of a retraction of fibers from the non-dominant input.

Laboratory studies have shown that competition between the ipsilateral and contralateral fibers is involved in the segregated pattern of eye inputs. The loss of one eye, for example, results in the expansion of fibers from the preserved eye into those layers typically occupied by inputs from the lost eye. Other studies have demonstrated that the segregated patterns are a result of activity-dependent processes. This has been shown by blocking, with the drug tetrodotoxin, the initiation and propagation of action potentials by ganglion axons. The administration of this drug in prenatal cats results in a loss of eye-specific layers within the LGN.

In the visual cortex the pattern of left-eye and right-eye inputs takes the forms of columns called

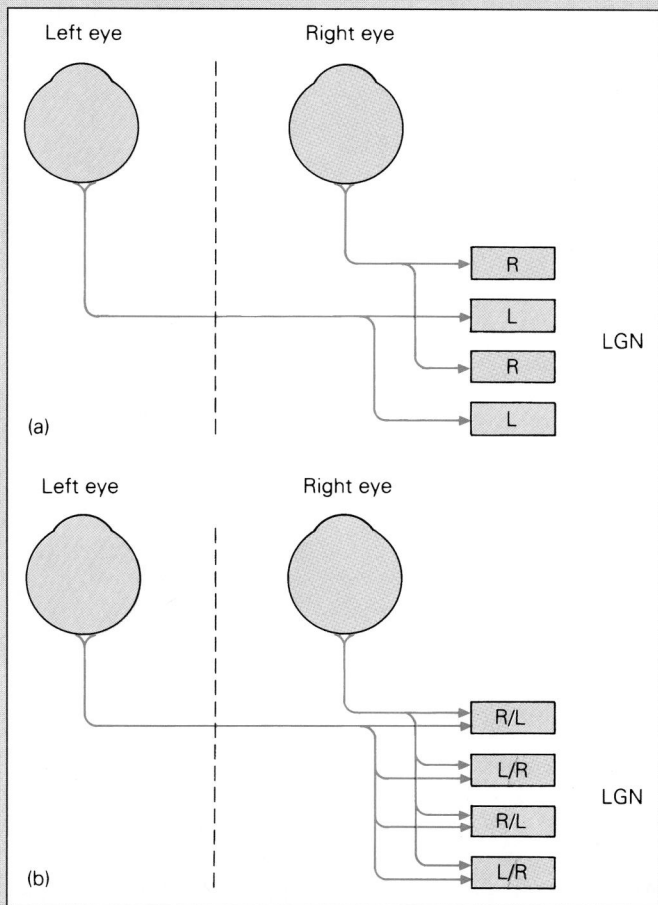

Eye-specific layers of the lateral geniculate nucleus. (*a*) Normal pattern of eye-specific layers in the adult cat. (*b*) Overlap of eye-specific layers in the prenatal cat.

ocular dominance columns. These columns exhibit alternative left-eye and right-eye inputs. The width of each column is about 0.5 mm in the monkey

The closure of one eye at various times after birth alters the pattern of ocular dominance columns. Closure of one eye at birth in monkeys greatly reduces the width of the associated ocular column in visual cortex; closure at 4 weeks after birth or later causes no alterations in the ocular dominance columns. These results reveal the modifiable nature of ocular dominance columns and the fact that their formation depends on a postnatal critical period of development.

Visual stimuli of various types have also been shown to alter the functional properties of cells in the visual cortex. Kittens raised in an environment of vertical black and white stripes develop cells in the visual cortex that respond preferentially to vertical bars of light. Animals raised in visual environments that minimize movement perception have fewer movement selective cells. These studies suggest that exposure to a variety of visual stimuli is essential for the proper development of connections in the visual pathways.

Figure (cont.)
(c) Adult pattern of eye-specific layer in the cat after prenatal removal of one eye.
(d) Adult pattern of eye-specific layers after prenatal administration of TTX to both eyes.

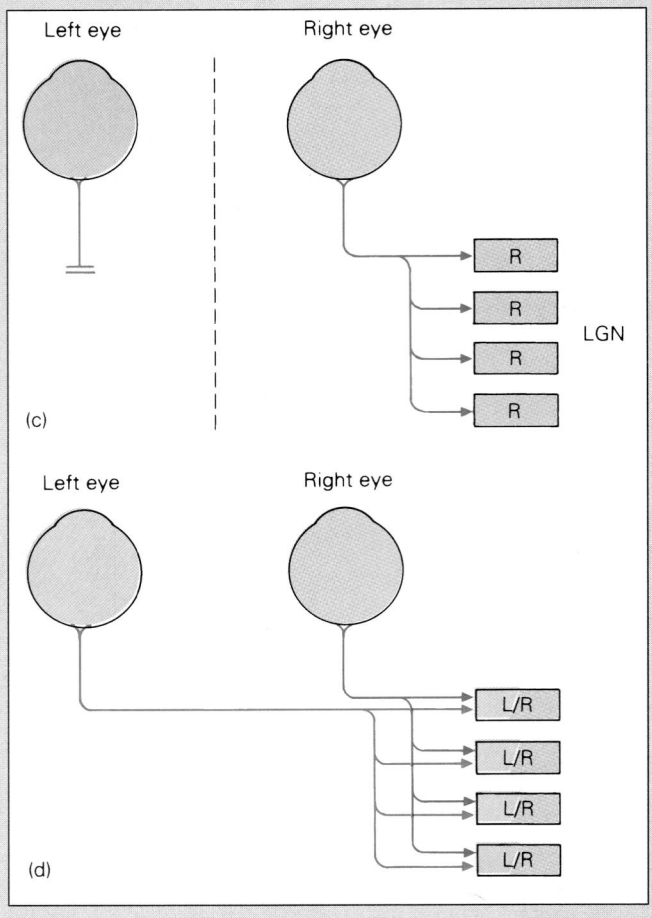

dividuals cannot see the colored numbers embedded within the design of Figure 8–33.

The trichromatic theory of color vision alone is not sufficient to explain completely how we perceive color. After the selective absorption of photons by cones, cellular responses that are also responsible for the perception of color come into play.

Ganglion cells, for example, differ in the way they process information from cones. Small ganglion cells are able to distinguish information about different colors. They in turn transmit this color information to small cells in the lateral geniculate nucleus (Fig. 8–34). These so-called **parvocellular neurons** in the geniculate transmit information to neurons in

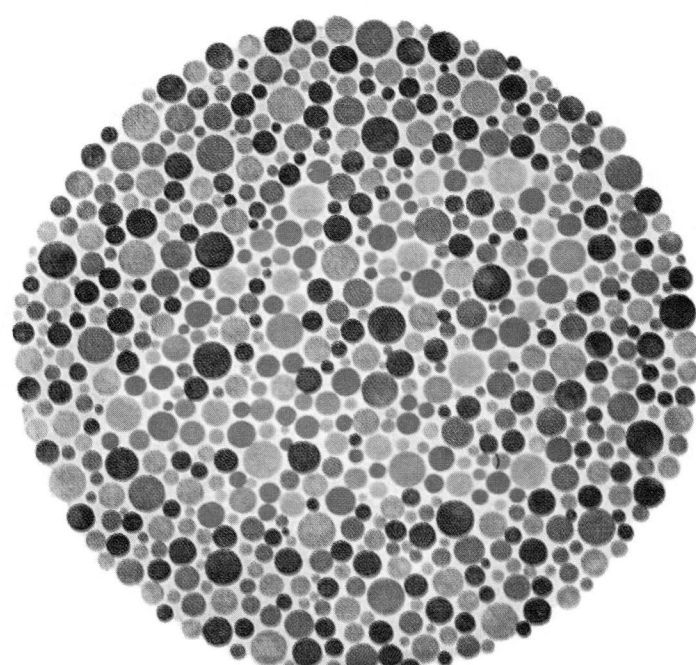

Figure 8–33

A rough guide to red-green color-vision defects (Ishihara, 1968). More severe than green weakness, the phenotype green blindness is due to the absence of the green color-vision pigment in cone cells. The two green defects are controlled by alleles. A parallel situation relates the two defects of the red visual pigment. What you see in the patterns of dots is the diagnostic criterion:

Perception	Phenotype
42	Normal
4 (and less clearly a 2)	green-weak
4 only	green-blind
2 (and less clearly a 4)	red-weak
2 only	red-blind

This color plate is not intended for use in the diagnosis of color blindness. A diagnosis of color blindness must be based on the original plates, which appear in Ishihara, S.: *Tests of Colour-Blindness*, 38 plate edition, Kanehara & Co., LTD, Tokyo, Japan (1968).

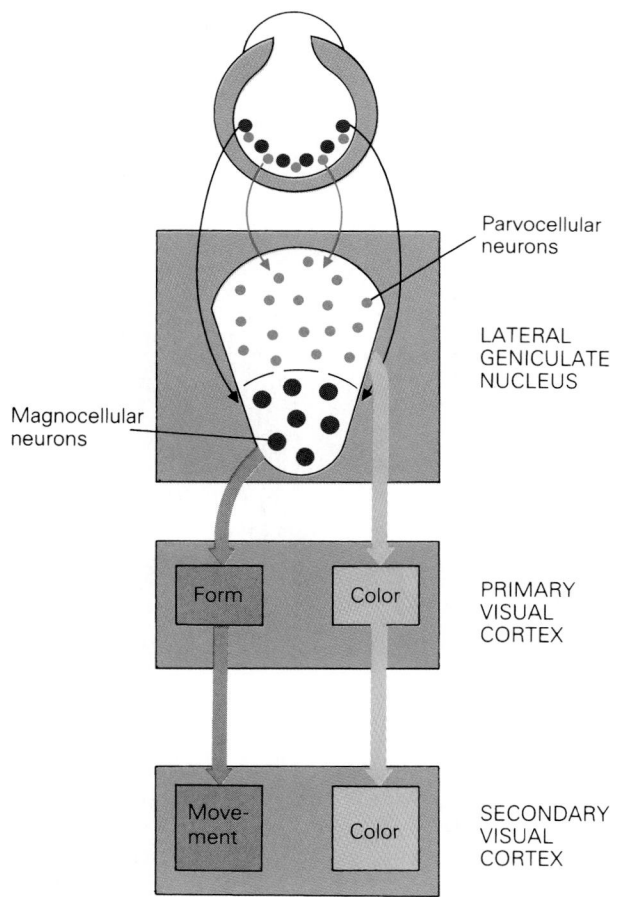

the visual cortex that respond selectively to color information. Large ganglion cells, on the other hand, process information from cones in such a way that they lose their ability to respond selectively to specific colors. They in turn transmit information to large cells in the lateral geniculate nucleus. These so-called **magnocellular neurons** transmit information to the primary and secondary visual cortices, where it is analyzed to extract features about an object's movement.

Neurons that selectively respond to color are divided into three different groups: (1) **broad-band cells,** (2) **single-opponent cells,** and (3) **double-opponent cells** (Fig. 8–35). Broad-band cells are excited by a single color presented to the center of the receptive field and inhibited when that color is in the periphery. Single-opponent cells are also excited when one color of light is in the center of the receptive field; however, they are inhibited when another color is in the periphery. For example, one type of

Figure 8–34

Small ganglion cells distinguish among different colors and transmit the information to small cells, or parvocellular neurons, in the lateral geniculate nucleus. Large ganglion cells transmit information about an object's movement to large, or magnocellular, neurons in the lateral geniculate.

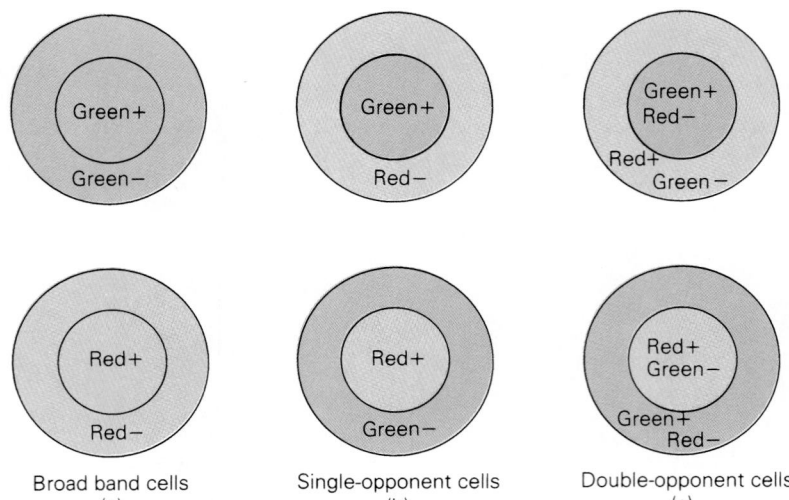

Figure 8–35
Neurons that respond selectively to color are classified as broad-band, single-opponent, or double-opponent cells, depending on the effects of different colors.

Broad band cells
(a)

Single-opponent cells
(b)

Double-opponent cells
(c)

cell excited by red in the center of the receptive field is inhibited by green in the periphery. The opposing colors are usually paired in combinations of red and green, blue and yellow, or black and white, and oppose one another across the compartmental boundaries of the receptive field.

Double-opponent cells differ from single-opponent cells in that the color pairs oppose each other within the boundaries of the receptive field. In this type of cell, for example, red in the center of the field might excite the cell, while green within the center would inhibit the cell. In addition, green in the peripheral boundary of the field might excite the cell, while red in this region would be inhibitory.

The double-opponent cell can also be thought of as transmitting information about the contrast be-

Figure 8–36
The annular receptive fields of some cells in the visual cortex overlap to give bar-shaped receptive fields for color.

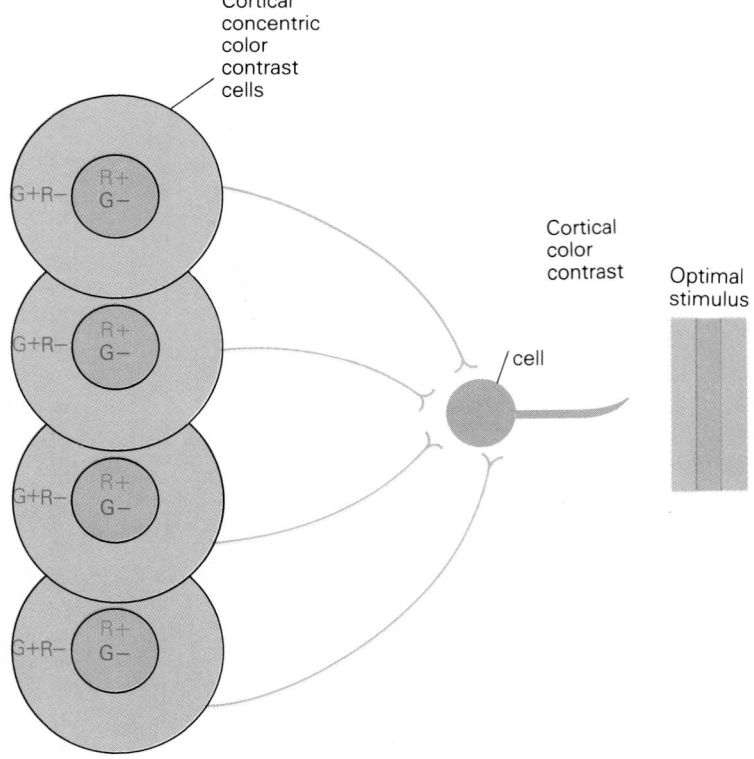

Cortical concentric color contrast cells

Cortical color contrast

cell

Optimal stimulus

tween two colors. In the previous example, the cell would become maximally excited when red is present in the center of the field and green is in the periphery. The activation of this type of cell would therefore communicate to the brain that two contrasting colors are adjacent to one another in the receptive field.

Some cells in the visual cortex display the characteristics of double-opponent cells with annular receptive fields. Other cells have rectangular receptive fields. The response properties of these cells are a summation of the annular properties displayed by converging neurons (Fig. 8–36). The annular patterns of these converging neurons overlap and thus give rise to a bar-shaped receptive field. These cells, however, do not appear to be associated with orientation columns and respond to rectangular forms of color in any orientation. The color-sensitive cells are clustered in functional columns that are parallel to orientation columns (Fig. 8–34). This arrangment suggests that the color and form of an object are analyzed independently. In support of this theory are observations of stroke patients who are unable to determine the color of an object even though they recognize its form.

Disorders of Visual Sensation

As mentioned earlier, the most sensitive part of the retina is the **fovea** (Fig. 8–37). In this part of the retina, the ganglion cell fibers are sparse so that photons can easily reach the photoreceptors. In addition, all of the photoreceptors in the fovea are cones. The fovea, therefore, is the region of the retina with the greatest spatial resolution and the region where the lens of the eye attempts to focus the incoming images of light. Some individuals, however, have lenses that cannot properly focus the light onto the fovea. Lenses without enough focusing power cause the image to be focused at some point behind the retina. Individuals with this type of lens disorder are farsighted and require biconvex eye glasses or contacts to adjust the focal point. In some cases, the lens focuses the image in front of the fovea. Individuals with this type of lens disorder are nearsighted and require biconcave lenses to extend the focal length.

Visual disturbances can also occur when accidents or disease damage parts of the visual pathway. These disturbances are characterized by a loss of visual perception in some part of the **visual field** (Fig. 8–38). The left and right eyes have overlapping visual fields; this overlapping region is called the nasal portion of the visual field. Images in the nasal

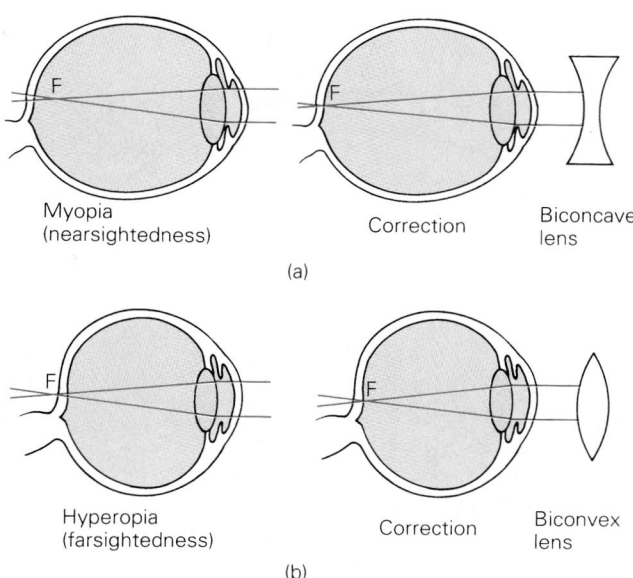

Figure 8–37
In a nearsighted (myopic) person, the lens focuses an image in front of the retina, requiring correction with biconcave lenses. In a farsighted (hyperopic) person, the lens focuses the image behind the retina, requiring a biconvex lens for correction.

portion of the visual field are projected to the temporal portion of the retina in the left and right eye. Images in the temporal portion of the right visual field project to the nasal portion of the retina of the right eye only. Likewise, images in the temporal portion of the left visual field project to the nasal portion of the retina of the left eye.

Trauma such as a blow to the eye can sometimes damage the optic nerve. Damage to the optic nerve of the right eye, for example, results in a loss of the ability to see objects in the right visual field. Disorders such as pituitary tumors can damage nerve fibers in the optic chiasm, which results in an inability to perceive objects in the left and right visual fields. The optic tract on one side of the brain can also be damaged. Damage of the optic tract on the right side of the brain results in defects in the left temporal and right nasal visual field.

Lesions can also occur in the secondary visual cortices (areas 18 and 19), resulting in **visual agnosia.** Patients with visual agnosia have difficulty recognizing the faces of friends, familiar objects, or colors (see Table 8–3).

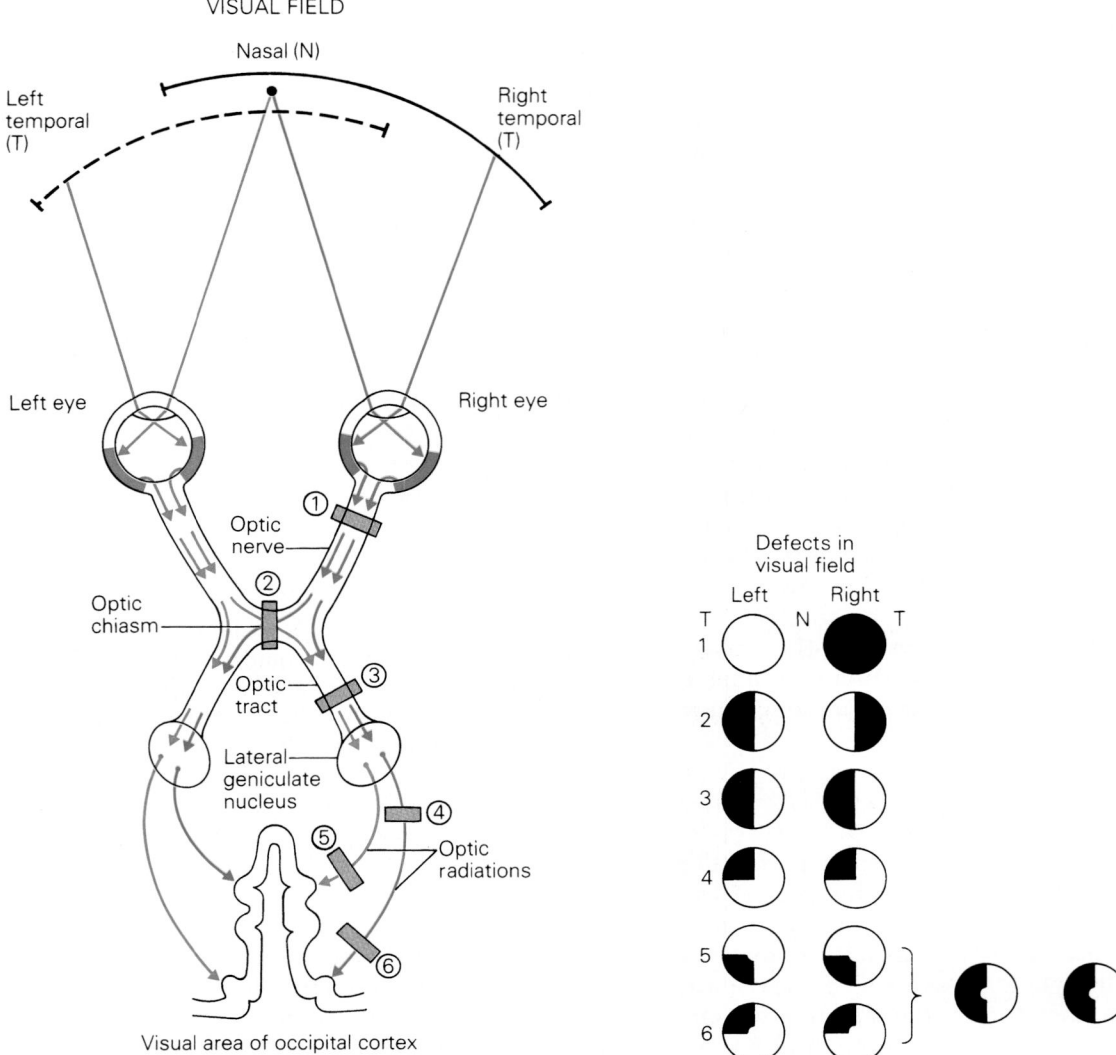

Figure 8–38
Damage to some parts of the visual pathway may result in disturbances in
the visual fields of the eyes.

Table 8–3	
Visual Agnosia	
Type of Agnosia	*Visual Defects*
Agnosia for color	
Color agnosia	Associating colors with objects
Color anomia	Naming colors
Agnosia for form and pattern	
Object agnosia	Naming objects
Agnosia for drawings	Recognizing drawn objects
Prosopagnosia	Recognizing faces
Agnosia for depth and movement	
Visual spatial agnosia	Stereoscopic vision
Movement agnosia	Discerning movement of objects

The Auditory System

The auditory system detects complex sounds such as spoken words and breaks them into their basic sound frequencies. The fundamental frequency components are transduced into action potentials by receptor cells in the ear and transmitted to the brain. Auditory areas of the brain subsequently recombine the basic sound features of the words, which are then interpreted in areas of the brain responsible for language (see Chapter 11).

Sound Transduction and the Functional Organization of the Ear

Sound is the adequate stimulus of the auditory system and is transmitted through the air by the compression and expansion of air molecules in the form of air pressure waves. The sense organ for the transduction of sound is the ear. The ear consists of the **outer ear, middle ear,** and **inner ear** (Fig. 8–39). The outer ear channels the sound waves to the **ear drum,** located at the interface between the outer and middle ear. The sound waves cause the ear drum to vibrate much like the vibration of the speakers of a stereo system. The vibration of the ear drum, in turn, causes the small bones or **ossicles** within the middle ear to move. These bones, called the **mal-**leus, incus, and **stapes** (also known as the **hammer, anvil,** and **stirrup),** are connected in series and transfer the vibrations of the ear drum to the inner ear.

The stapes, the last in the series of ossicles, is connected to the inner ear at the **oval window,** a membranous structure that vibrates with the movement of the ossicles. The inner ear consists of a fluid-filled chamber called the **cochlea.** The cochlea is a cone-shaped structure wrapped in a coil much like the shape of a snail's shell. Along the length of the cochlea is a membrane called the **basilar membrane** (Fig. 8–40c). This membrane is a flexible structure that vibrates when sound impinges upon the ear drum. The vibrations of the basilar membrane are in the form of traveling waves (see Fig. 8–40c) that deform the shape of the membrane. As the traveling wave progresses down the length of the membrane, the amplitude of the deformation becomes greater and greater until it reaches a peak. As the traveling wave progresses further, its amplitude diminishes. The location of the peak amplitude of the traveling wave along the basilar membrane depends on the frequency of the sound. High-frequency sounds cause the traveling wave to reach its peak amplitude near the base of the cochlea. Low-frequency sounds cause the traveling wave to reach its peak near the apex of the cochlea. Interme-

Figure 8–39
(*a*) The general anatomy of the ear.

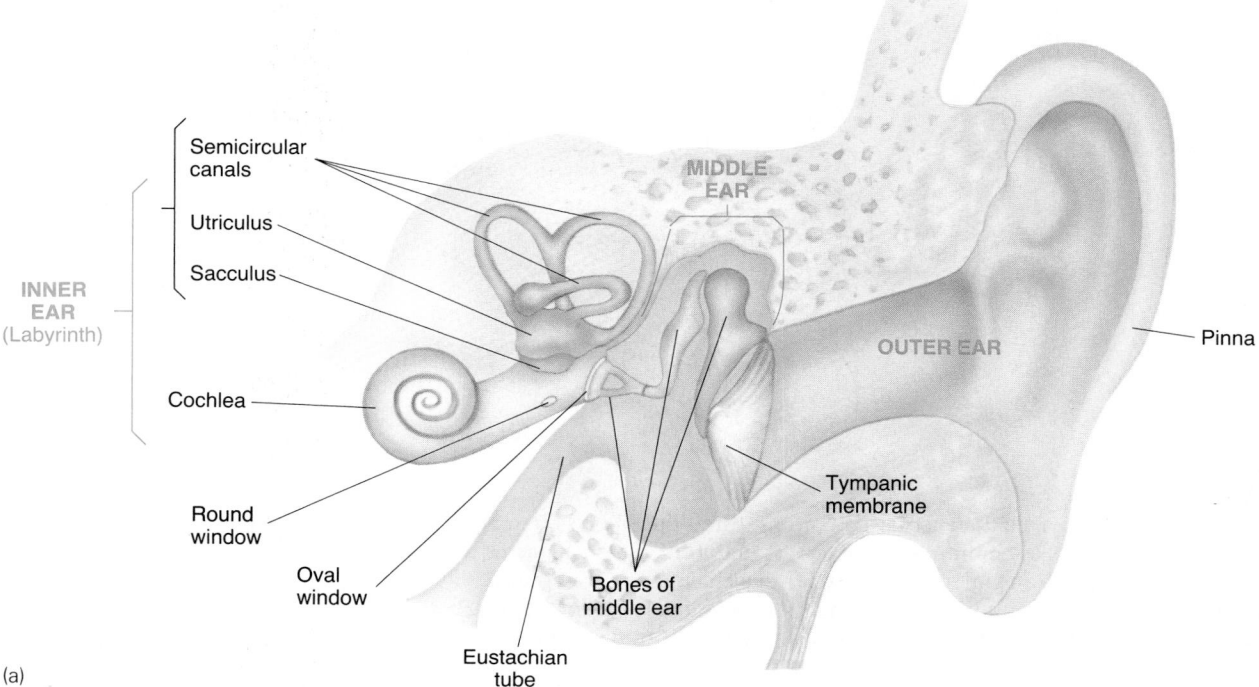

Semicircular canals

Utriculus

Sacculus

INNER EAR (Labyrinth)

Cochlea

Round window

Oval window

Eustachian tube

MIDDLE EAR

OUTER EAR

Pinna

Tympanic membrane

Bones of middle ear

(a)

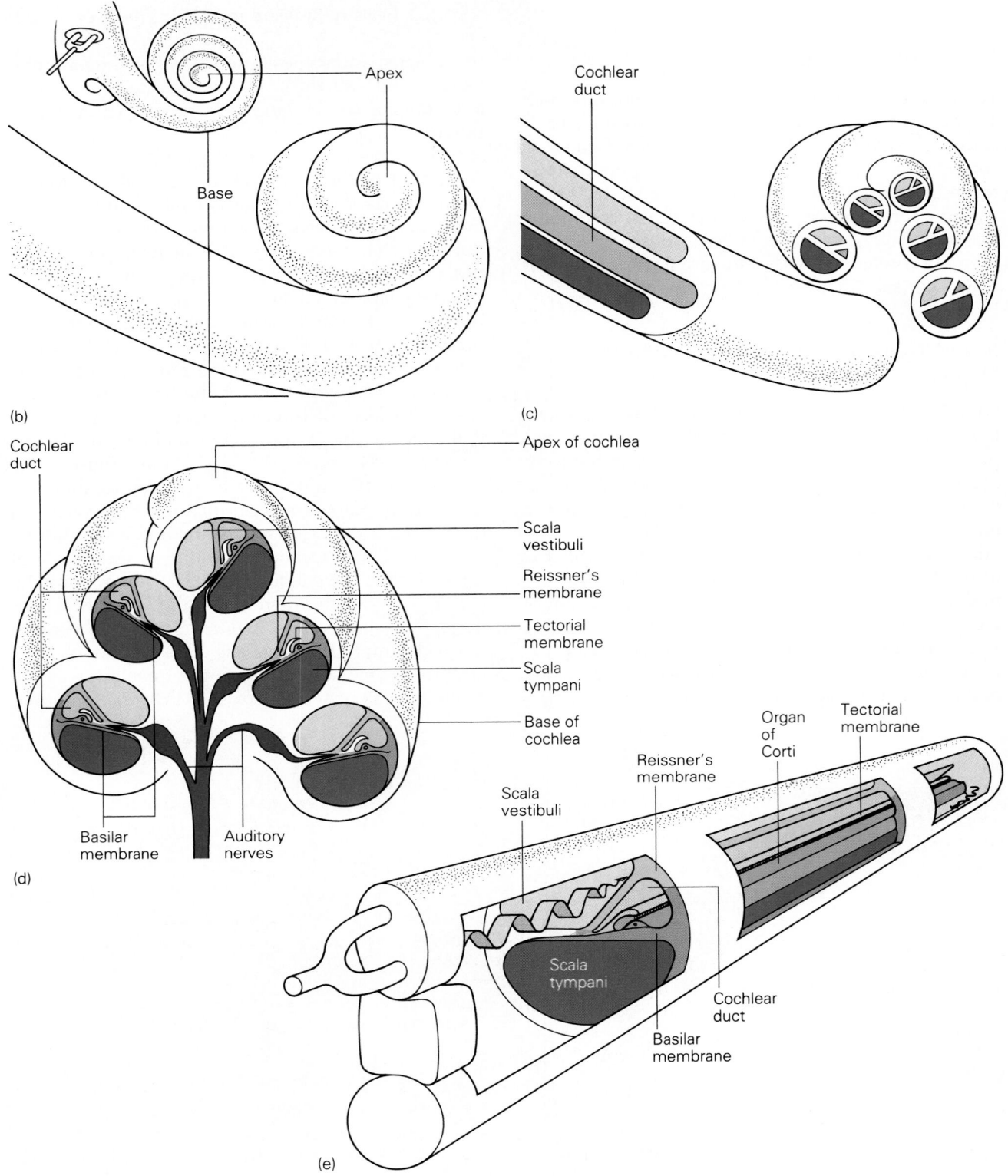

(b)

(c)

(d)

(e)

Figure 8–39 (cont.)
(*b, c*) The cochlea is a coiled structure containing three fluid-filled chambers formed by a system of membranes. (*d*) A cross-section of the cochlea showing the fluid chambers and membranes. (*e*) The cochlea as if "unwound," showing the path of a sound pressure wave from the oval window through the scala vestibuli, around the apex of the cochlea, into the scala tympani, and back to the round window at the base of the cochlea. (Some pressure waves do not go all the way to the apex.)

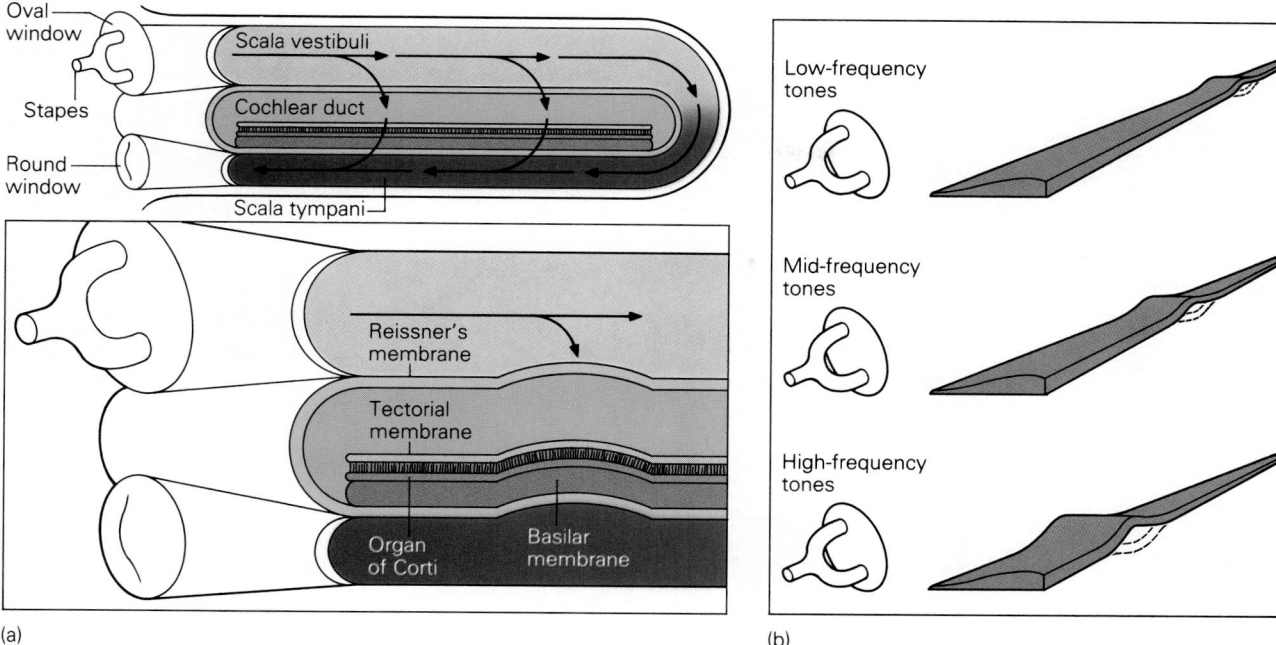

Figure 8–40
(*a*) The cochlear membranes shown in a different cross-sectional view. Sound causes the fluid in the scala vestibuli to exert pressure on Reissner's membrane, which in turn causes the basilar membrane to bulge. This movement sets up vibrations in the form of waves in the basilar membrane (*b*). The waves peak at different locations on the membrane depending on the frequency of the sound.

diate frequencies of sound cause peak amplitudes to occur at locations along the membrane between the base and apex of the cochlea. Thus, the basilar membrane is designed mechanically to respond to different frequencies of sound by changing its form.

The deformations of the basilar membrane are transduced into action potentials by **hair cells.** The hair cells sit upon the basilar membrane and have fiberlike elements called **stereocilia** that protrude from the cell (Fig. 8–41*a*). These cilia are embedded in another membranous structure called the **tectorial membrane.** When the basilar membrane vibrates and reaches a peak amplitude, the stereocilia are bent in relation to the hair cell (Fig. 8–41*b*). This bending of the stereocilia causes ion channels of the hair cells to open, leading to a depolarization of the membrane potential. This depolarization is analogous to the generator potential described for the Pacinian corpuscle of the somatosensory system. Unlike the Pacinian corpuscle, however, the hair cell is not capable of producing its own action potential. When the hair cell is depolarized, Ca^{2+} ions flow into the cell and release transmitters that in turn depolarize auditory nerve fibers (see Fig. 8–

41*c*). The auditory nerve fiber then generates action potentials that are transmitted to the brain to convey information relating to frequencies of sound. Hair cells are also responsible for transducing information regarding rotational and linear acceleration needed for balance and equilibrium (see Table 8–1). These hair cells are part of the vestibular system and will be discussed in Chapter 9.

Auditory Pathways and Receptive Properties of Auditory Nerve Cells

The peripheral fibers of the auditory nerve make contact with the cochlear nucleus in the brain stem (Fig. 8–42). Information from the cochlear nucleus is transmitted to both sides of the brain and eventually reaches the medial geniculate nucleus of the thalamus. From the medial geniculate nucleus, information is transmitted to the primary auditory cortex in the temporal cortex.

The receptive field of neurons along the auditory pathways is characterized by the use of tones with different frequencies and intensities. A tone is a sound wave of one frequency. The sound pro-

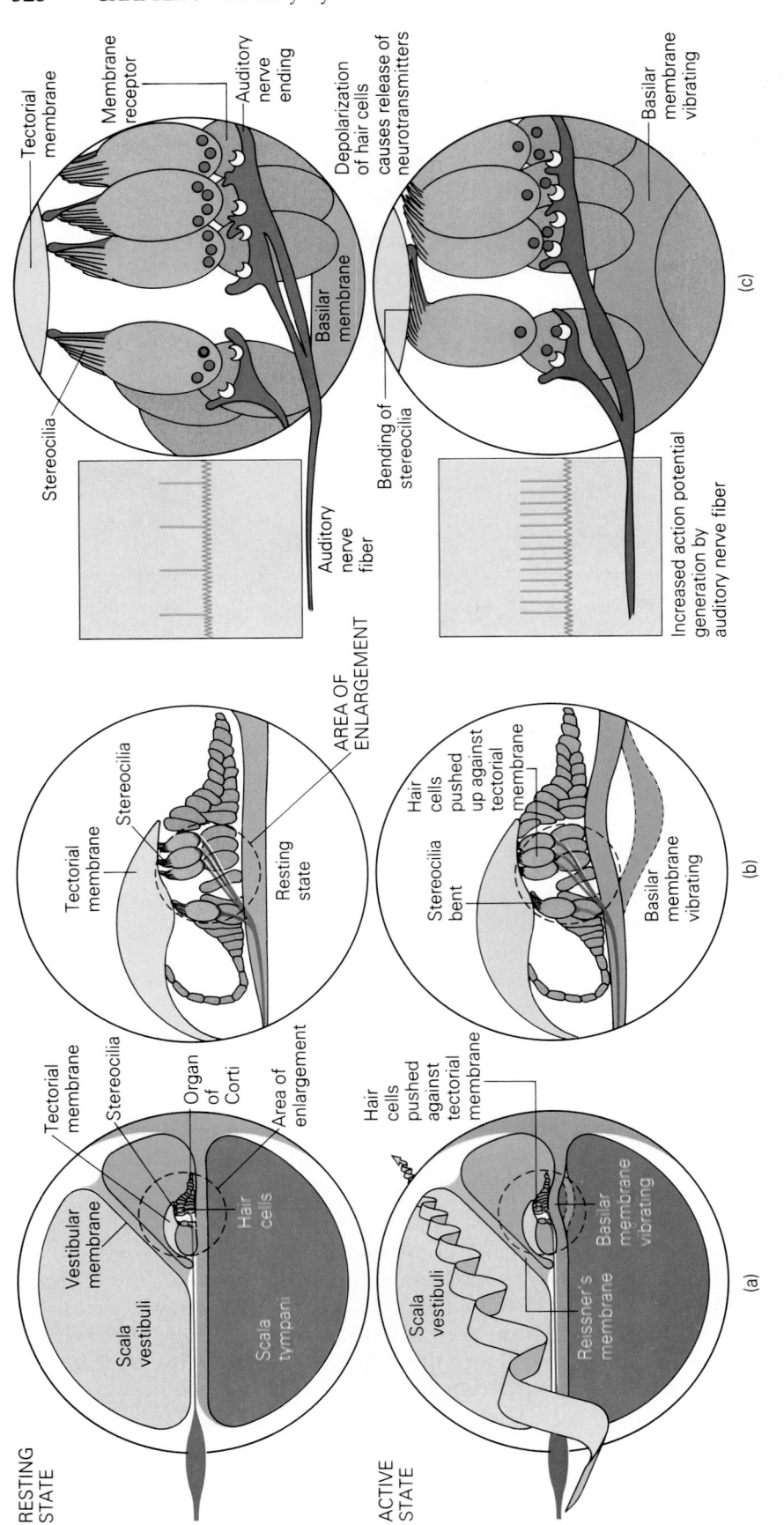

Figure 8–41

(*a*) When the basilar membrane changes shape as waves travel along it, hair cells above the membrane, which form the organ of Corti, are pushed closer to the tectorial membrane so that (*b*) stereocilia that protrude from the hair cells bend (*c*). The bending causes hair cell ion channels to open, leading to membrane depolarization, inflow of Ca^{2+}, and release of a neurotransmitter that causes action potentials to arise in auditory nerve fibers. (*d*) SEM of inner hair cells of the human ear. ($\times$ 1705). (© Dr. G. Bredberg/SPL/Photo Researchers, Inc.)

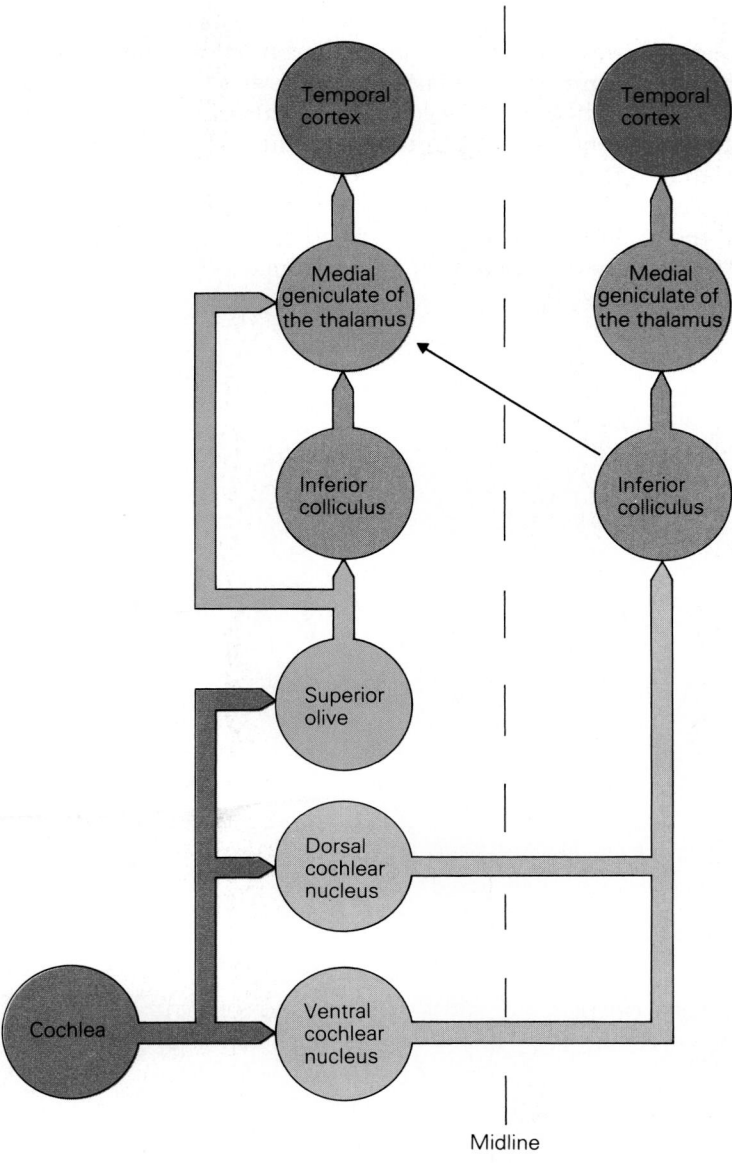

Figure 8–42
Auditory pathways, originating in the cochlea on one side of the brain and terminating bilaterally in the temporal cortices.

duced by playing a single note on the piano, for example, produces a pure tone. This tone, in turn, produces a sound pressure wave that causes the compression and expansion of air molecules at one frequency (Fig. 8–43). If many notes are played simultaneously, the sound produced is a combination of all of the individual frequencies. Tone can be represented graphically by an oscillating waveform; the peak of the waveform represents the compression of the air molecules, and the valley of the waveform represents the expansion of the molecules. The difference between the peak and the valley is the **amplitude** (A_o), or intensity of the sound wave. The time between the peaks is the **time period** of the tone (**T**) and is measured in units of seconds. Its inverse is the **frequency** (**f**) of the tone and is measured in units of **Hertz** (**Hz**). The amplitude and frequency of a tone can be used to characterize the receptive properties of nerve cells in the auditory pathway. If the frequency of the tones is varied, one frequency will be found that most effectively activates a particular auditory nerve fiber. This frequency is known as the **characteristic frequency** (f_o) for the nerve fiber. Different auditory nerves have different characteristic frequencies. These auditory nerve fibers have a **tonotopic organization** such that those with a high characteristic frequency innervate hair cells near the base of the cochlea, while those with a low characteristic frequency innervate hair cells near the apex of the cochlea.

Neurons of the auditory system also respond in a variety of different ways to a particular tone. Some neurons generate action potentials only when the tone is on; other neurons are prevented from generating action potentials with the onset of the tone. Some neurons generate action potentials when the frequency of the tone is suddenly changed, whereas others generate action potentials only when the amplitude changes. The variety of responses suggests that nerve cells of the auditory system extract the basic features of a sound from its complex waveform. The fundamental features of a sound are reassembled by the convergence of neurons. These neurons converge onto other neurons that respond only to the reassembled complex auditory stimuli. For example, certain neurons in the auditory cortex of monkeys respond only to specific sounds used by that species of animal. Animal studies of auditory

Figure 8–43
(*a*) A single tone produces a sound pressure wave that alternately compresses and diffuses air molecules at a particular frequency and amplitude. The characteristic frequency of a nerve fiber is the frequency that most effectively activates the fiber (*b*). In this example, f_2 would be the characteristic frequency. (*c*) Some nerve fibers generate action potentials only in response to a change in the frequency or amplitude of a tone.

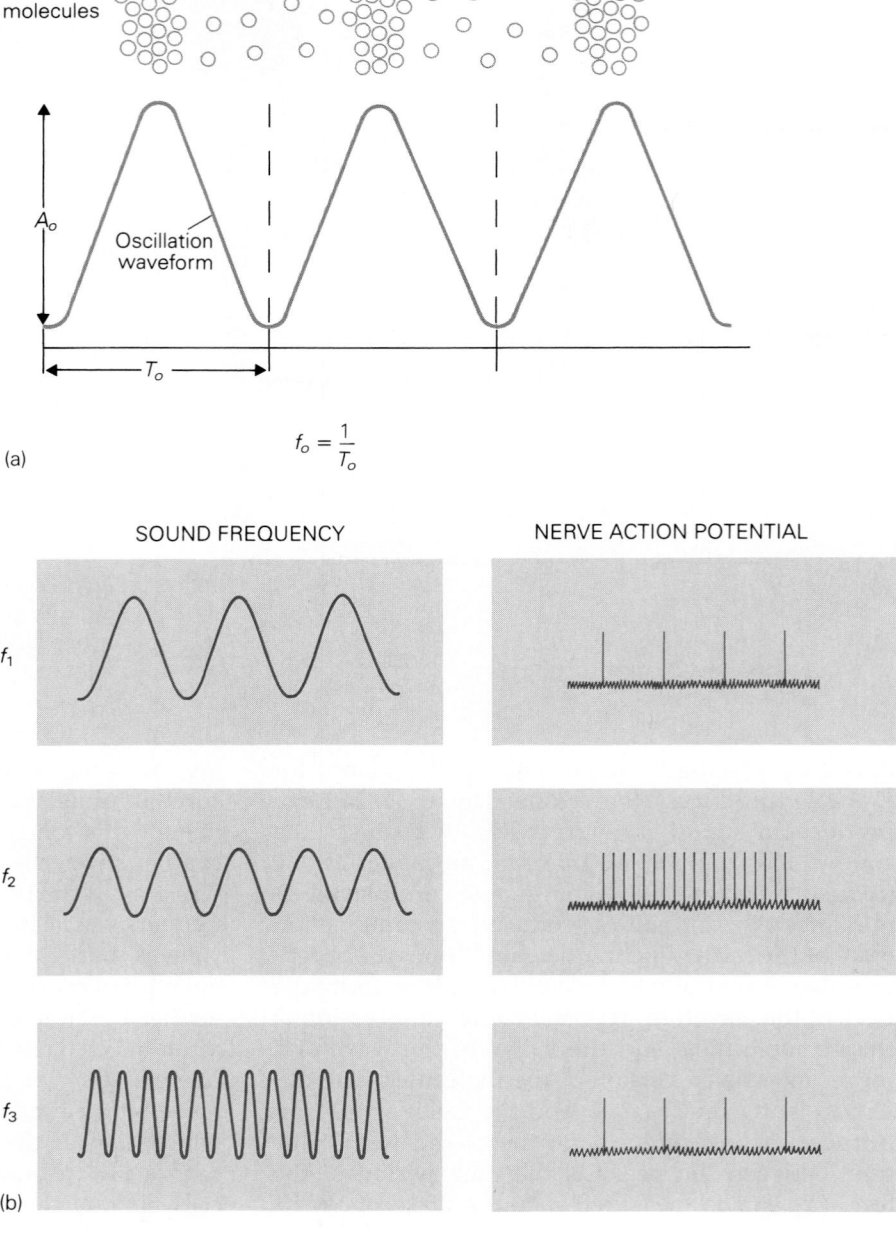

responses suggest that there are neurons in the human auditory cortex that respond only to the word *go,* and others that respond to the word *stop.*

Disorders of Auditory Sensation

A loss of hearing can occur in a variety of ways. Hair cells can be destroyed when the eardrum is suddenly exposed to a loud noise such as an explosion, which is thought to crush hair cells between the basilar and tectorial membranes. The loss of hair cells is associated with frequencies of sound that make up the noise. The resulting hearing disorder is called **sensory neural hearing loss.** Another disturbance of hearing of the sensory neural form is called **tinnitus,** a condition in which a constant ringing in the ear is perceived even in the absence of sound waves. This is due to the inappropriate discharge of action potentials by particular nerve fibers associated with specific frequencies. The exact cause of this disorder is unknown.

Another type of hearing disorder is **conductive hearing loss.** Conductive hearing loss results from damage to the middle ear. This can result from an inflammation of the small bones in the middle ear, or from abnormal bone growth caused by a process called **otosclerosis.** Conductive hearing loss produces a general reduction in hearing ability. Hearing by conduction through the skull, however, is still functional and can be tested by placing a tuning fork over the bony structure behind the ear to elicit the sensation of sound.

Central hearing losses are typically due to brain tumors and lesions of the auditory pathway within the central nervous system. Lesions in the primary auditory cortex (area 41) on one side of the brain have little effect because of the bilateral projection of sound information. However, lesions in the auditory association cortex (area 22) or Wernicke's area produce **auditory sensory aphasia,** a condition in which voices can be heard but the words appear to have no meaning.

Hearing disorders in infants can lead to lifelong language disorders. Ear infections, for example, that chronically impair the ability of infants to hear spoken words hinder the development of auditory nerve cell connections needed for hearing and speech. If these auditory connections are not formed within the first few years after birth, they will never be made. This developmental disorder illustrates the importance of the auditory environment and the formation of critical nerve cell connections underlying language skills.

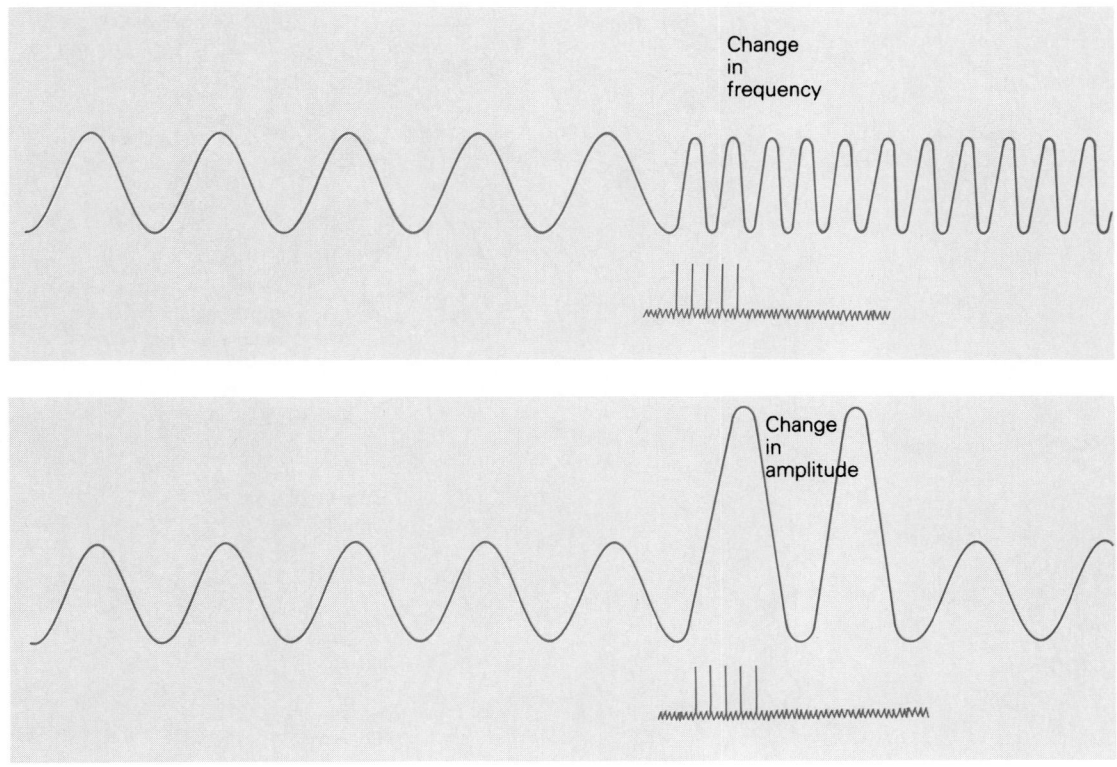

(c)

The Gustatory System

Our ability to taste food is the result of the activation of taste receptor cells in the tongue. These receptors detect chemicals associated with four basic taste qualities: sweet, salty, sour, and bitter. The tip of the tongue contains receptors that are sensitive to sweet substances (Fig. 8–44*a*). The lateral region near the tip of the tongue is sensitive to salty substances; the lateral region near the back of the tongue is sensitive to sour substances. The middle region at the back of the tongue is sensitive to bitter substances. Taste receptors are also located in the palate and pharynx.

Taste receptors are called **gustatory receptors** and are a type of epithelial cell clustered with other cells called **supporting** and **basal** cells (see Fig. 8–44*b*). Together these cells compose the **taste bud.** Taste receptors degenerate every 10 days and are replaced by new receptor cells derived from the differentiation of basal and supporting cells.

Transduction of Taste Sensations

A taste receptor can respond to several types of substances; however, each receptor cell responds best to only one type of taste quality. The most effective chemical stimuli for a salt receptor, for example, are Na^+ and Cl^-. Taste receptors for sourness are best activated by protons (H^+) from acids such as vinegar, receptors for sweetness are activated most effectively by sugar, and receptors for bitterness are

Figure 8–44

(*a*) Taste buds sensitive to the various tastes are grouped in specific regions of the tongue. (*b*) Taste buds consist of taste cells surrounded by supporting and epithelial cells. (*c*) Pathways for the transmission of gustatory sensation.

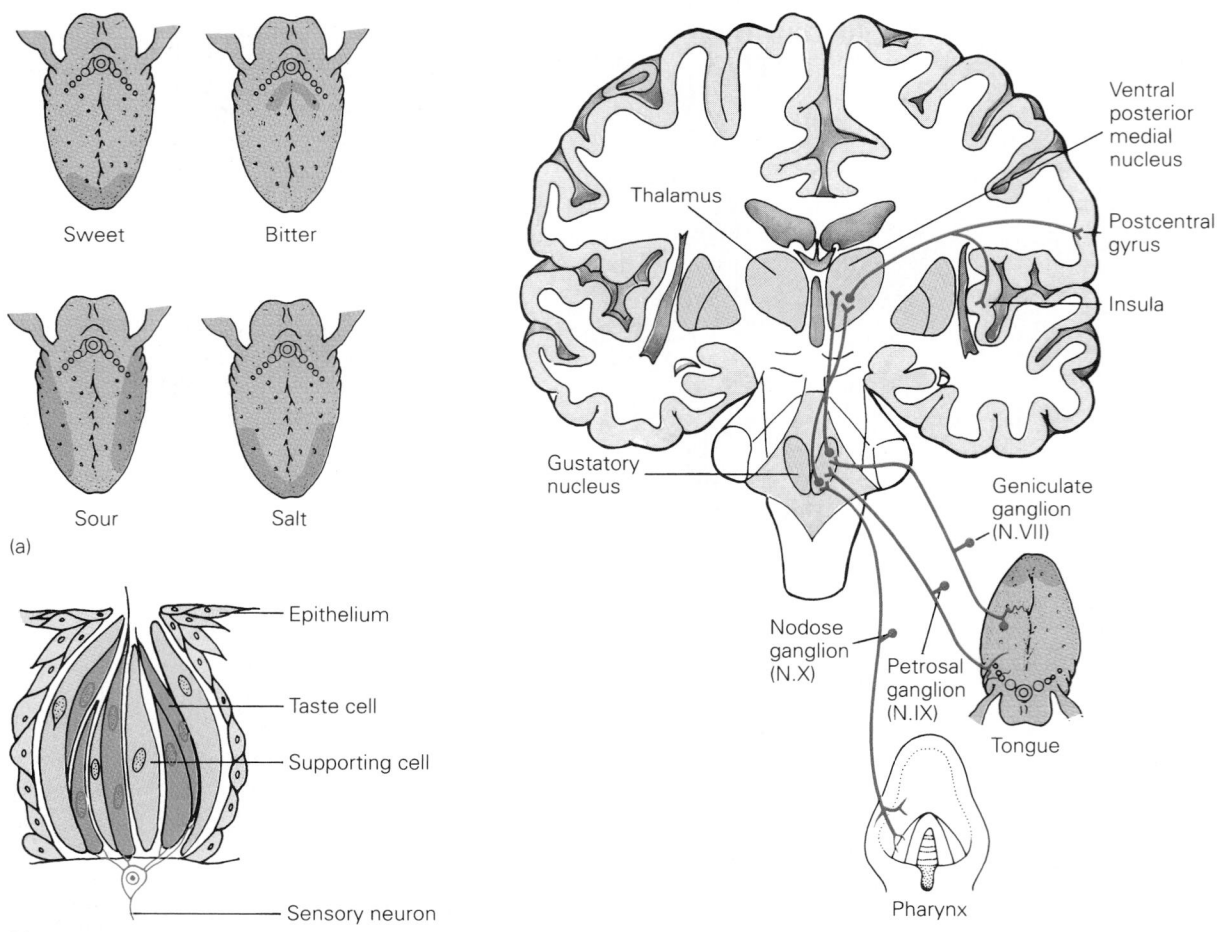

best activated by alkaloids such as quinine found in tonic water.

The binding of a chemical compound to a molecular receptor on a taste cell produces a depolarizing generator potential. The generator potential is thought to cause the release of transmitters, resulting in the initiation of action potentials in the afferent fiber. Afferent fibers branch many times before they innervate a taste bud; consequently, one fiber often innervates several taste buds, and the action potentials from a single fiber may therefore represent the activation of many taste buds.

A single fiber can respond to stimuli from different categories of chemicals, but as noted earlier, it responds best to one type of chemical group. This property of taste fibers is similar to the responses of fibers in the auditory pathway for sound frequencies and visual pathways for colors of light. As in the other sensory pathways, the higher brain centers of the taste pathway use this type of information to make comparisons and contrast the taste qualities of consumed substances.

Gustatory Pathways

Taste receptor cells are innervated at their base by nerve fibers that form part of the VII, IX, or X cranial nerve (see Fig. 8–44c). Taste buds in the front of the tongue are innervated by the VII cranial nerve; those in the back of the tongue are innervated by the IX cranial nerve; and those in the pharynx are innervated by the X cranial nerve. Thus, information about bitter tastes is transmitted by way of a different pathway than information about sweet, sour, and salty tastes. Fibers innervating the taste buds in the palate are part of the X cranial nerve.

All afferent fibers synapse on cells in the brain stem in the **gustatory nucleus.** Neurons in the gustatory nucleus, in turn, send their axons to the thalamus on the same side of the brain and terminate in the **ventral posterior medial (VPM) nucleus** of the thalamus. From the VPM, taste information is transmitted to the gustatory area of the neocortex just in front of the somatic sensory region of the tongue.

Alternate pathways for taste sensation transmit information to the limbic system and hypothalamus (see Chapter 11). These pathways are responsible for the affective and emotional aspects of taste. For instance, some individuals become nauseated by the taste of certain foods. This can happen even if the food was at one time a favorite dish. The consumption of spoiled food, for example, can lead to food poisoning. In some cases, a single exposure to the spoiled food can condition an individual to avoid it in the future.

The Olfactory System

Our ability to smell odors is a result of the activation of **olfactory receptor cells** in the nose. Olfactory receptor cells are contained within the nasal cavity of the nose and surrounded by supporting and basal cells (Fig. 8–45a). As do the receptor cells in taste buds, the olfactory receptor cells degenerate and are replaced by basal cells that become new receptor cells. The regenerative cycle for olfactory cells, however, is 60 days.

The receptor cells have fiber processes that extend from the cell body into the mucosa of the nose (see Fig. 8–45b); it is thought that these cilia interact with odor-producing molecules. Human beings can detect seven general types of odors: camphoraceous, musk, floral, peppermint, ethereal, pungent, and putrid. The binding of an odor molecule to the membrane of the receptor cell produces a generator potential. When threshold is reached, an action potential is initiated. Individual olfactory receptors respond best to one type of odor, but can also respond to others to a lesser degree. Thus, the olfactory system also exhibits the contrast mechanisms of the other sensory systems.

Olfactory receptor cells innervate **mitral cells** located in the **olfactory bulb.** The bundle of fibers from mitral cells forms the **olfactory tract.** These fibers directly innervate areas of cortex known as the **limbic system,** and the orbitofrontal cortex indirectly via the thalamus (see Fig. 8–45b). The olfactory innervation of the limbic system is exceptional in that it represents the only sensory system that directly innervates areas of cortex.

The olfactory system plays an obvious role in stimulating appetite and eating behavior. The activation of eating behavior is carried out by a motivational area of the brain called the **hypothalamus** and will be considered in Chapter 11, on central integrative systems.

The olfactory system also plays a role in mating behavior. Perfumes and colognes are examples of the use of artificial odors to attract members of the opposite sex. In animals such as mice, the odor of a female stimulates the release of testosterone in males and enhances the growth rate of the male sexual organ in adolescents. The odor of a female also causes the release of hormones in the adult male that in turn activate the sebaceous glands to give off a male odor to signal its readiness for mating. Male odors also increase the rate of sexual maturation in adolescent female mice and accelerate the estrous cycle of adult females.

In humans, the olfactory system is poorly developed; thus, visual, tactile, and auditory stimuli play more important roles in human mating behavior.

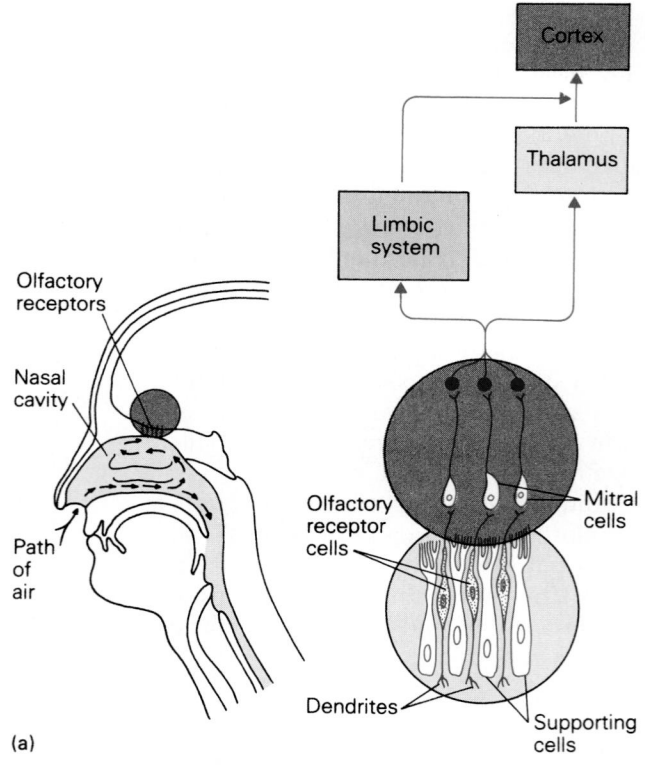

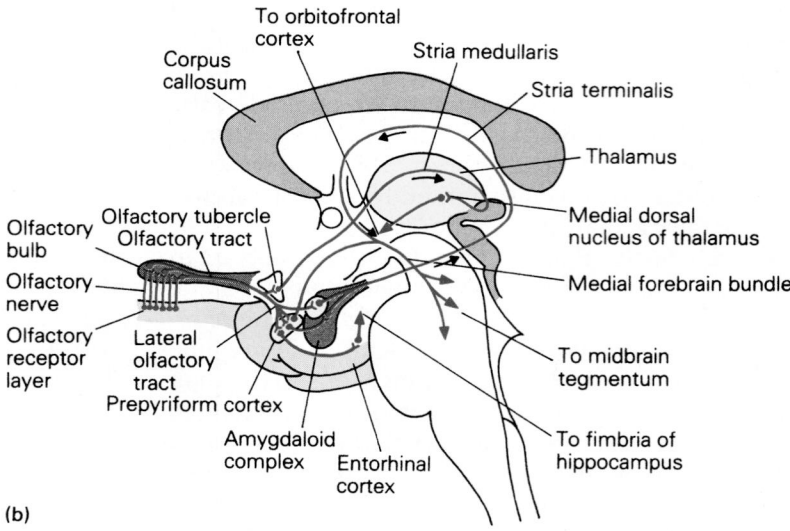

Figure 8–45
(*a*) Olfactory receptor cells in the nasal cavity contact mitral cells in the olfactory bulb. Mitral cell fibers form the olfactory tract, which innervates the limbic system and the orbitofrontal cortex. (*b*) Olfactory pathways.

SUMMARY

General Principles of Sensory Transduction

We detect information from the external environment with receptors that transduce sensory stimuli into action potentials, which are then conducted to the central nervous system. Sensory systems detect stimuli from the outside world in the form of light, sound, heat, pressure, chemical substances, and odors. Each type of stimulus is capable of activating a specific type of sensory receptor. The intensity of a stimulus is coded by the number of action potentials generated in a frequency code and by the number of receptors activated in a population code.

The quality of a sensory stimulus is coded in the labeled-line code of stimulus quality.

In sensory adaptation, receptors stop responding to a stimulus after a time.

Sensory Systems and Their Processes

Somatic receptors on the surface of the skin detect pressures applied to the skin as well as temperature changes and painful stimuli. Pressure sensation is transmitted ipsilaterally in the spinal cord along the dorsal column pathway. Pain and temperature sensation are transmitted contralaterally along the anterolateral pathway of the spinal cord. Relay stations that process pressure sensation are located in the dorsal column nuclei of the medulla and the thalamus. In contrast, pain and temperature information are transmitted directly to the thalamus. From the thalamus, somatic information is sent to the somatosensory cortex. This cortex is organized in the form of a somatotopic map of the body with the lower limbs located in the medial region of the cortex and the upper extremities and face in the lateral region. Nerve cells that process somatic information from the periphery to the somatosensory cortex have receptive fields. These receptive fields are areas of the skin that, when stimulated, can either excite or inhibit a particular nerve cell. The pattern of excitation or inhibition forms unique shapes that are characteristic of neurons at different levels along the somatic sensory pathway. Sensory receptor cells respond to the basic features of a complex stimulus. In this way, they decompose the stimulus into its fundamental components. Contrasts and comparisons are made of the basic features by groups of nerve cells in each sensory pathway. These features are eventually reassembled by nerve cells in the sensory areas of the neocortex.

Sensory receptors for vision are called photoreceptors and are located in the retina of the eye. Rod photoreceptors are used for night vision while cones are involved in color vision. There are three types of cones used in the process of color. These are red, green, and blue cones because of their sensitivity to specific wavelengths or colors of light. Light in the form of photons is absorbed by photoreceptors, causing a hyperpolarization of the photoreceptors. This information is then transmitted to ganglion cells via bipolar cells. Visual information is then transmitted from ganglion cells in the retina to the lateral geniculate nucleus of the thalamus. Information from the lateral geniculate is then transmitted to the visual cortex. The receptive fields of ganglion cells are annular with "on" centers and "off" surrounds or with "off" centers and "on" surrounds. This type of receptive field is also exhibited by neurons in the lateral geniculate. In the primary visual cortex, neurons are more sensitive to bar-shaped stimuli. Ganglion cells detect the basic features of a visual image. These features are assembled to reconstruct the total image at the higher levels of the visual system.

The auditory system transforms sounds into action potentials. The external ear channels sound to the ear drum, causing the ear drum to vibrate. The ear drum is connected to a series of small bones called the malleus, incus, and stapes, which also oscillate in response to sound pressures. The mechanical displacement of the stapes is transferred to a fluid-filled structure called the cochlea. The cochlea contains sensory receptor cells called hair cells. The movement of fluid within the cochlea also displaces the basilar membrane. The displacement of the basilar membrane activates the hair cell, which in turn excites fibers of the auditory nerve. Each auditory nerve responds to a narrow range of sound frequencies. In addition, the nerves are tonotopically organized so that those innervating hair cells near the base of the cochlea respond best to high frequencies while those innervating the apex respond best to low frequencies. At the higher levels of the auditory system, neurons respond best to more complex sound frequencies.

Taste information is detected by the gustatory system using taste buds located in the tongue. Taste buds near the back of the tongue respond best to bitter tastes, those in the middle respond best to sour and salty tastes, and those on the tip respond best to sweet tastes. The chemical activation of a taste receptor in turn excites afferent fibers that innervate the receptor. This information is eventually transmitted to the gustatory area of the neocortex near the somatic sensory region associated with the tongue.

The sense of smell is conveyed by olfactory receptors in the nose. These receptors are innervated by mitral cells whose axons constitute the olfactory tract. This pathway provides inputs to the limbic system and thalamus. Information reaching the thalamus is then transmitted to the frontal cortex. The sense of smell is evoked when odor-producing molecules interact with the cilia of the olfactory receptors. Humans can detect camphoraceous, musk, floral, peppermint, ethereal, pungent, and putrid odors. In humans, the sense of smell is poorly developed in comparison to other animals.

REVIEW QUESTIONS

Identify Each with a Term

1. The term used to describe the functional column of cells in the visual cortex that respond to visual input to one eye.

2. The term used to describe the sensory receptor that detects limb position.

3. The term used to describe a sensory receptor that discharges tonically in response to a sensory stimulus.

4. The term used to describe the functional organization of sensory inputs from the surface of the body to the neocortex.

Define Each Term

5. generator potential
6. conductive hearing loss
7. fovea
8. labeled-line code of stimulus quality

Choose the Correct Answer

9. The constant exposure to loud, high-frequency noise near 20,000 Hz would result in the loss of hair cells on the basilar membrane:
   a. near the base of the cochlea
   b. near the apex of the cochlea
   c. at a location midway between the base and the apex

10. The sensory information carried by fibers in the optic tract on the right side of the brain is derived from:
    a. the entire right visual field
    b. the nasal component of the left and right visual fields
    c. the temporal component of the left and right visual fields
    d. the nasal component of the right visual field and the temporal component of the left visual field

11. The absorption of a photon of light by a photoreceptor results in:
    a. the depolarization of the photoreceptor
    b. the inhibition of the phosphodiesterase enzyme
    c. the conversion of cGMP to 5'-GMP
    d. all of the above

12. A patient involved in a sailing accident has lost the ability to sense the flow of air across his left arm and leg and to perceive pain on the right side of his body. These symptoms are suggestive of:
    a. a hemisection of the right side of the brain stem above the level of the medulla
    b. a hemisection of the left side of the spinal cord at the cervical level
    c. a hemisection of the right side of the spinal cord at the thoracic level
    d. a hemisection of the left side of the spinal cord at the thoracic level

Answer Each Question in One or Two Sentences

13. What are the general mechanisms of sensory transduction?
14. How is stimulus intensity encoded by the nervous system?
15. What is the concept of a receptive field?
16. What is the trichromatic theory of color vision?

SUGGESTED READINGS

Hubel, D.H., and T.N. Wiesel. "Receptive fields, binocular interaction and functional architecture in the cat's visual cortex." *Journal of Physiology*, 160:106–154, 1962.

Hubel, D.H., T.N. Wiesel, and S. LeVay. "Plasticity of ocular dominance columns in monkey striate cortex." *Philosophical Transactions of the Royal Society of London*, 278:377–409, 1977.

Hudspeth, A.J. "Transduction and tuning by vertebrate hair cells." *Trends in Neuroscience*, 6:366–369, 1983.

Moulton, D.G. "Spatial patterning of response to odors in the peripheral olfactory system." *Physiological Reviews*, 56:578–593, 1976.

Mountcastle, V.B. "Modality and topographic properties of single neurons of cat's somatic sensory cortex." *Journal of Neurophysiology*, 20:408–434, 1957.

Penfield, W., and T. Rasmussen. *The Cerebral Cortex of Man. A Clinical Study of Localization of Function*. New York: Macmillan, 1950.

Shatz, C.J., "Impulse activity and the patterning of connections during CNS development." *Neuron*, 5:745–756, 1990.

Stewart, W.B., J.S. Kauer, and G.M. Shepherd. "Functional organization of rat olfactory bulb analysed by the 2-deoxyglucose method." *Journal of Comparative Neurology*, 185:715–734, 1979.

von Bekesy, G. *Experiments in Hearing*. New York: McGraw-Hill, 1960.

KEY TERMS

auditory system (p. 284)
bipolar cell (p. 305)
cochlea (p. 325)
dermatome (p. 295)
dorsal root ganglion (p. 292)
ear drum (p. 325)
gustatory system (p. 284)

hair receptor (p. 292)
inner ear (p. 325)
mechanoreceptor (p. 292)
middle ear (p. 325)
nociceptor (p. 292)
olfactory system (p. 333)
outer ear (p. 325)

pain receptor (p. 291)
photoreceptor (p. 305)
proprioception (p. 292)
retina (p. 305)
rhodopsin (p. 305)
sensory system (p. 284)

somatic sensory system
 (p. 284)
thermal receptor (p. 291)
visual field (p. 323)
visual system (p. 284)

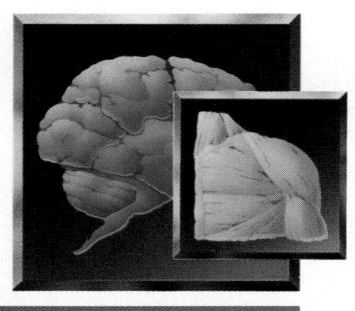

C H A P T E R
9

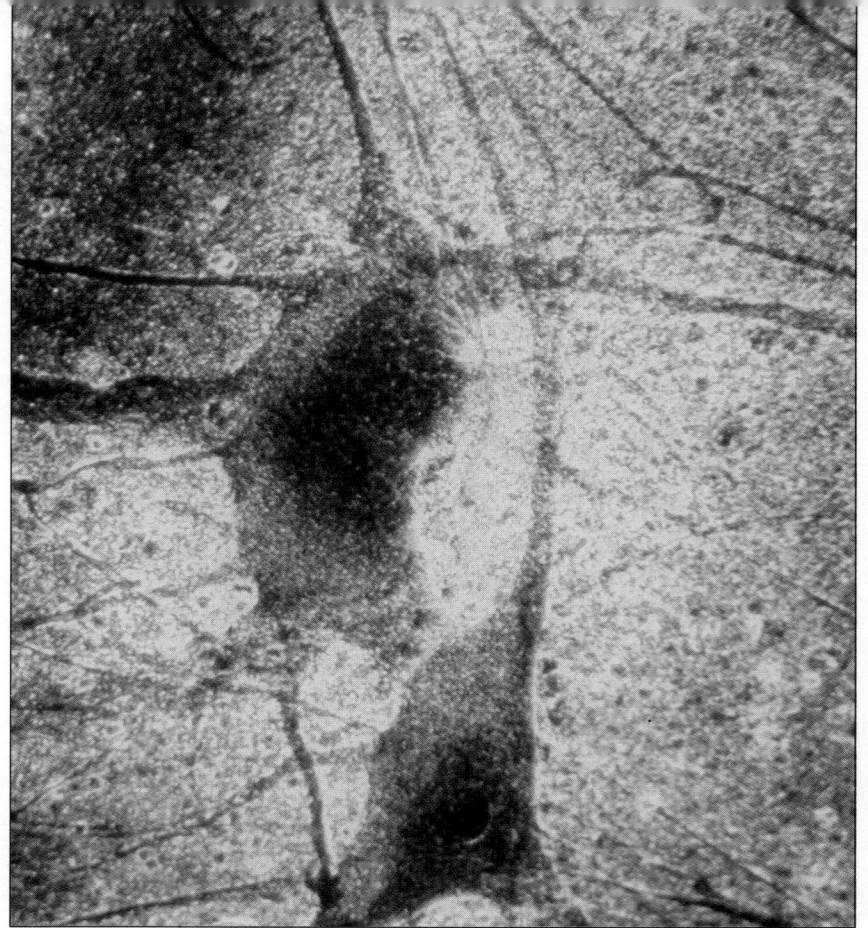

Motor Systems

*T*he act of moving requires the coordinated activation and inhibition of various muscle groups. This is seen in activities such as reaching, walking, running, jumping, and typing. Each activity uses different sets of muscle groups to produce a movement in a particular direction with specific velocity and force. The planning, initiation, execution, and termination of these movements are carried out by the **motor system.**

The repertoire of motor responses ranges from **reflex** (automatic, involuntary) **actions** such as the knee jerk to complex, **voluntary** activities such as playing the piano. Reflex responses are controlled in the spinal cord, while more complex motor behaviors are organized in other areas of the CNS. The range of complexity of motor behaviors reflects the **hierarchical** (or **serial**) **organization** of the motor system (Fig. 9–1). The spinal cord is the lowest level of the hierarchy; it is followed in order by the **brain stem, motor cortex (Brodmann area 4),** and the **premotor and supplemental areas** of the cortex (Brodmann area 6). Other non-cortical areas that play important roles in motor function are the cerebellum and basal ganglia (Fig. 9–2).

In the hierarchical organization, the highest motor levels influence the levels below them. For instance, the premotor area has inputs to the motor cortex; the motor cortex, in turn, provides inputs to the brain stem; and the brain stem provides inputs to the spinal cord. This hierarchical arrangement connects different areas of brain together serially in a descending fashion.

In addition to this serial organization, the motor system also has a **parallel organization.** In this arrangement, information from the motor cortex and brain stem, for example, can simultaneously and independently affect the spinal cord (see Fig. 9–1). With both the parallel and serial organization of the motor system, the targets of the efferent (outgoing) information are the **motor neurons** of the spinal cord. These motor neurons are the nerve cells that directly activate or inhibit the skeletal muscles responsible for movement.

The body contains three major types of muscle: skeletal muscle, cardiac muscle, and smooth muscles. **Cardiac muscles** are the muscles of the heart; they pump blood through the circulatory system (see Chapter 18). **Smooth muscles** surround certain

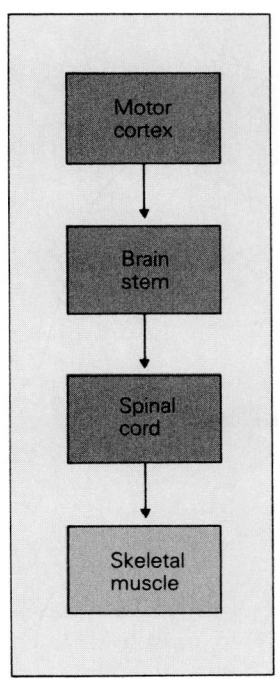

(a)

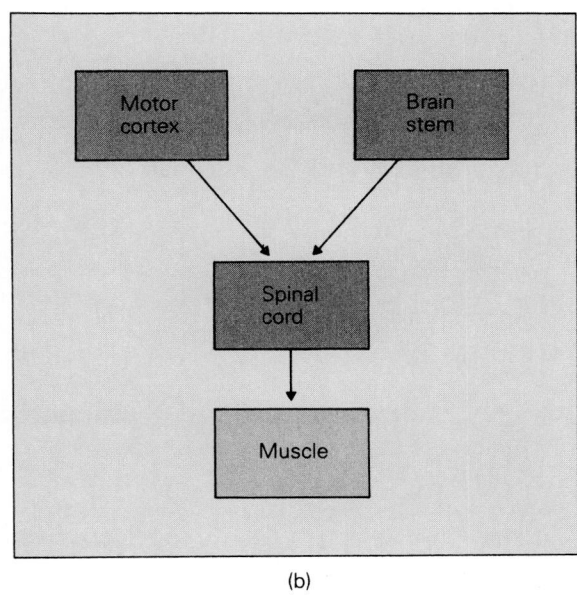

(b)

Figure 9–1
The motor system is organized in both hierarchical and parallel fashions.

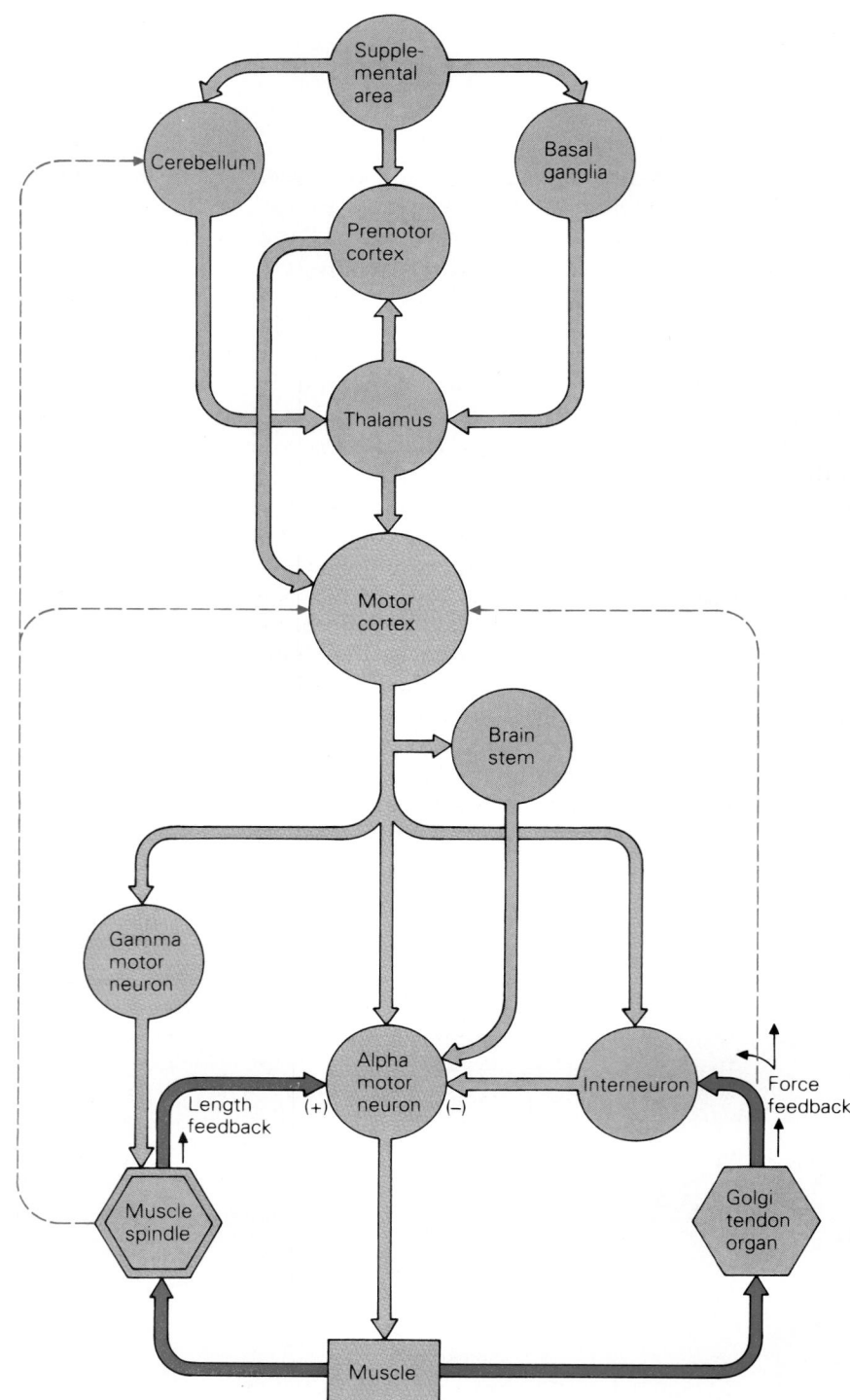

Figure 9–2
An overview of the motor system
and the flow of information
among the various components.

blood vessels to control blood pressure and are also associated with a variety of other organ systems. In general, they maintain the homeostatic state of the body (see Chapter 19). In this chapter, we will focus on the areas of brain that control the **skeletal muscles,** also called **striate** or **striated muscles,** which are responsible for locomotion, fine movement, and the maintenance of balance and posture.

Muscles Controlled by Nerves

Overview of the Skeletal Muscle System

Although the skeletal muscle system is discussed in detail in Chapter 16, we will review some general aspects of muscle structure and function here in order to better understand the interactions of muscles and neurons to be discussed shortly.

General Structure of Skeletal Muscle

A skeletal muscle is made up of **muscle cells,** which, because of their elongated shape, are called **muscle fibers.** Muscle cells are **multinucleate;** one muscle cell contains many nuclei. A **muscle** comprises many muscle fibers arranged in a parallel way. Two types of muscle fibers make up a skeletal muscle: **extrafusal fibers** and **intrafusal fibers** (Fig. 9–3). Extrafusal fibers are the strongest and most com-

mon fibers and form the exterior of the muscle. Inside the muscle are the intrafusal fibers, arranged parallel to the extrafusal fibers. Both extra- and intrafusal fibers are attached to the tendon at either end of the muscle. By a process to be discussed in more detail in Chapter 16, muscle fibers shorten when activated, resulting in contraction of the entire muscle and producing movement of a bone.

Types and Actions of Skeletal Muscle

Most skeletal muscles are attached to bones at two points: an **origin** and an **insertion** (Fig. 9–3), by strong connective tissue structures called **tendons.** As skeletal muscles **contract**—typically meaning that they shorten, although sometimes they remain the same length—they move bones, frequently causing joints to open or close. Skeletal muscles are usually arranged in **antagonistic** pairs or groups such that contraction of one muscle or muscle group will cause lengthening of another. In fact, all movements are controlled by antagonistic muscles working in opposite ways.

Antagonistic muscles are arranged in parallel fashion across a joint where two bones meet. The muscle that, by contracting, causes the joint to stretch out is called an **extensor,** while the muscle that, by contracting, causes the joint to close up is called the **flexor.** A flexor for a joint is said to be **synergistic** to another flexor for the same joint, while a flexor and an extensor for the same joint are said to be **antagonistic.**

Skeletal muscles produce movements of the bones of the **axial skeleton,** consisting of the skull, backbone, and ribs. These muscles are important in maintaining posture. Other skeletal muscles move the bones of the **distal skeleton,** primarily the limbs.

Skeletal muscles can be categorized as either fast muscles or slow muscles. **Fast skeletal muscles** contract rapidly. They are poorly vascularized (poorly supplied with blood vessels). Operating under anaerobic conditions, they fatigue rapidly. These muscles are best suited for physical activities such as a sprinting. In contrast, **slow skeletal muscles** contract slowly and are more resistant to fatigue. They operate under aerobic metabolism. These muscles are best suited for activities such as maintaining posture.

Innervation of Skeletal Muscle

The contraction of skeletal muscles necessary for movement is controlled by nerve cells in the spinal cord called **alpha motor neurons** (see Fig. 9–2). These neurons form synaptic connections with skel-

Figure 9–3
A skeletal muscle consists of extrafusal and intrafusal muscle fibers attached to tendons, which connect the muscle to a bone. Skeletal muscles work in antagonistic pairs of flexors and extensors.

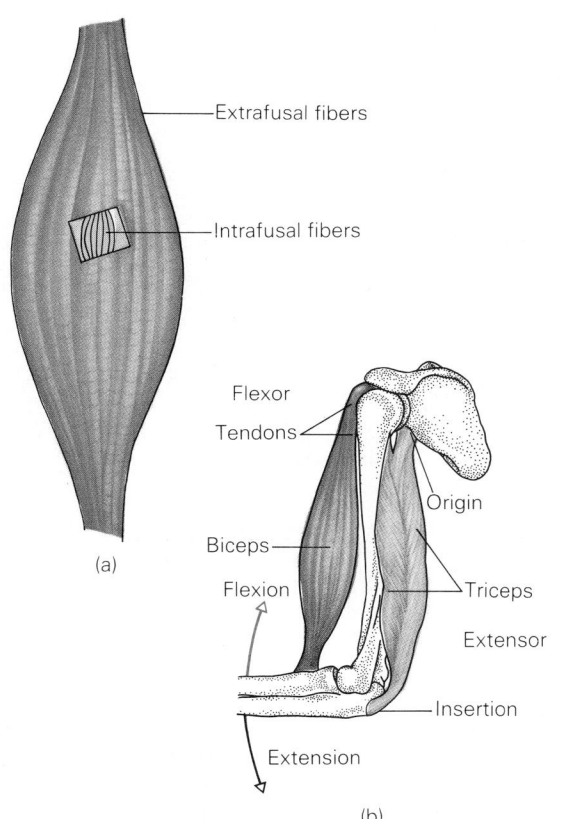

etal muscles at **neuromuscular junctions,** where the neurotransmitter **acetylcholine** is released from the nerve fiber terminal. Within the neuromuscular junction, acetylcholine binds to cholinergic **nicotinic receptors** that are located on the surface of the skeletal muscle membrane. As in the CNS, activation of the nicotinic receptor causes a depolarization of the membrane. In the case of skeletal muscles this depolarization leads to a sequence of events resulting in the contraction of the muscle (see Chapter 16).

The motor neurons that innervate fast and slow types of muscles have different functional properties. Motor neurons that innervate fast skeletal muscles discharge action potentials with high frequencies and have rapid action potential conduction velocities. In contrast, motor neurons that innervate slow skeletal muscles have low rates of action po-

tential discharge and exhibit slow conduction velocities.

Motor Neurons in Skeletal Muscle

The Motor Unit: The Myoneural Junctions of a Single Motor Neuron. The motor neurons that innervate skeletal muscles and cause them to contract have cell bodies located in the ventral horn of the spinal cord. Their axons leave the spinal cord through the ventral root and find their way to the appropriate skeletal muscle.

One motor neuron can form synaptic connections with many muscle fibers. An individual motor neuron and all the muscle fibers that it innervates form a **motor unit** (see Fig. 9–4). *The motor unit is the functional unit of the **neuromuscular system.***

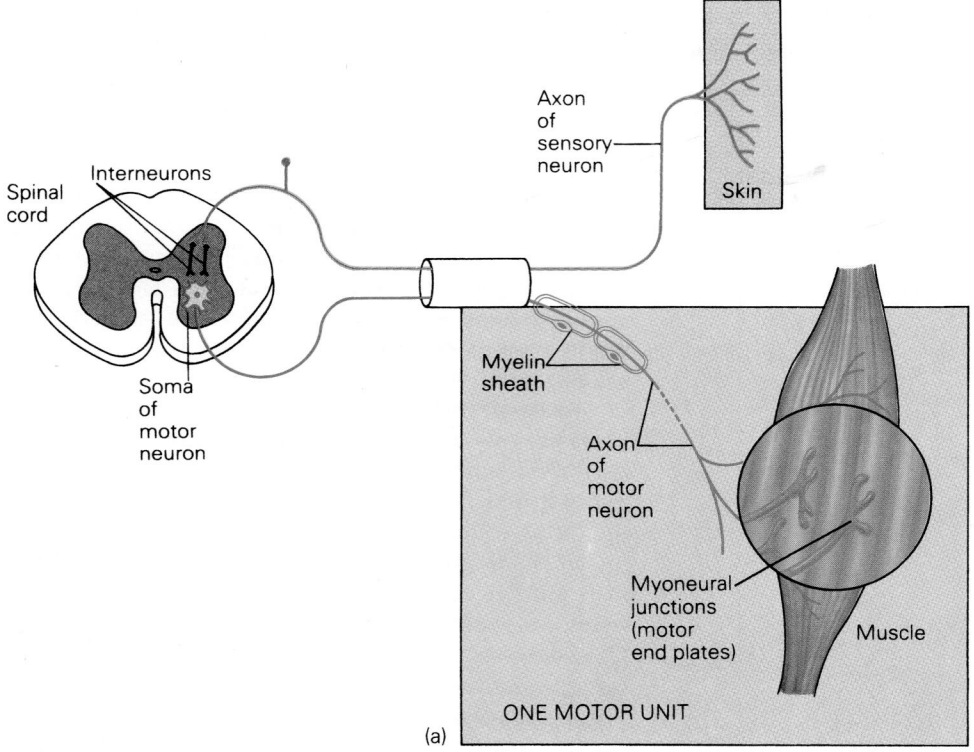

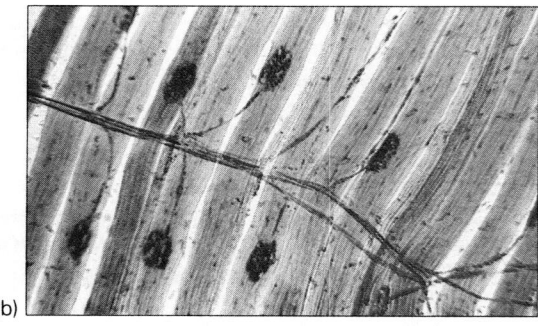

Figure 9–4
(*a*) The motor unit consists of a motor neuron and all the muscle fibers it innervates through myoneural junctions. (*b*) Photomicrograph of a nerve fiber innervating a skeletal muscle. (© J. & L. Weber/Peter Arnold, Inc.)

Motor units come in various sizes. A small motor unit consists of a motor neuron that innervates few muscle fibers. This type of motor unit is responsible for the precise control of very fine movements such as contraction of a finger or darting movements of the eyes. A large motor unit consists of a motor neuron that innervates many muscle fibers. Such a unit is responsible for gross movements such as contraction of the legs or the maintenance of posture. The concept of motor unit size parallels that of the size of sensory receptive fields. *Just as small receptive fields are associated with greater resolution of stimuli, small motor units are associated with greater precision of fine movement.*

The Coding of Contractile Force. The amount of force that the muscle generates during contraction is controlled by the nervous system in two ways. First, a motor neuron can control the tension developed by a single muscle fiber with the frequency of action potentials it generates (Fig. 9–5). This code is the motor system's **frequency code** of contractile force. Secondly, more motor neurons and therefore more motor units can be activated to increase a muscle's force of contraction. This code is the motor system's **population code** of contractile force.

The Spinal Cord in the Motor Response System

The spinal cord plays a crucial role in the neuromuscular system. Some automatic body movements, called reflex movements, are controlled solely by the spinal cord and have no control from higher structures. Moreover, locomotion, or walking, can be produced by spinal cord neurons alone.

Organization of Motor Neurons in the Spinal Cord

Motor neurons are organized in the spinal cord in two ways: (1) the motor neurons in the dorsal area of the ventral horn are responsible for flexor movements, while cells in the ventral region are responsible for extensor movements (Fig. 9–6); and (2) those in the dorsolateral region of the ventral horn innervate muscles in the extremities while those in the ventromedial region of the ventral horn innervate the axial muscles of the body to maintain posture.

Motor neurons that innervate a single muscle are functionally grouped in the spinal cord in **motor neuron pools.** Motor neurons within a single pool are located in several adjacent segments of the spi-

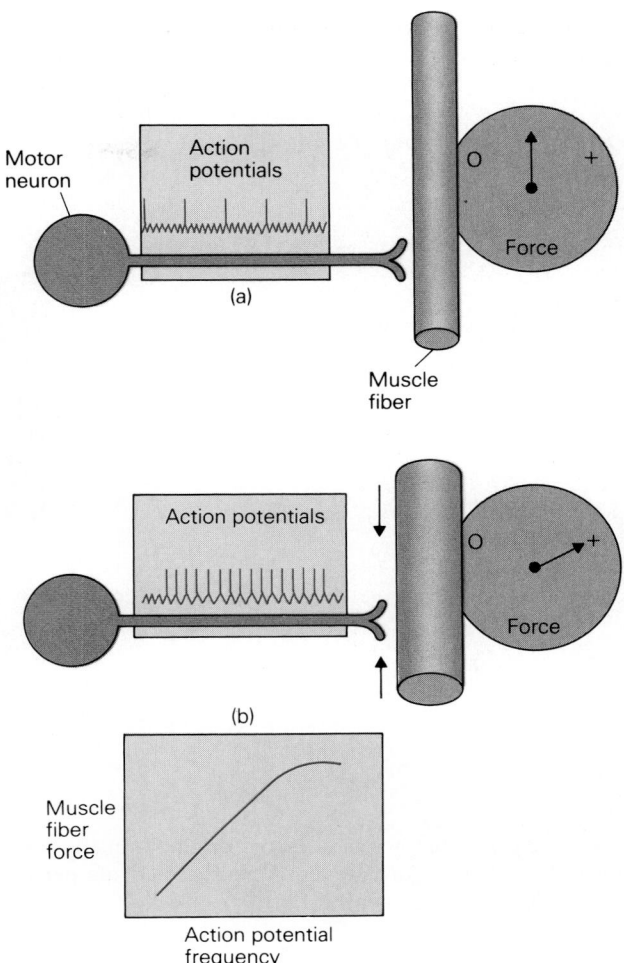

Figure 9–5
A motor neuron generates action potentials that cause a muscle fiber to contract and exert force.

nal cord (Fig. 9–7). The activation of a motor neuron pool thus coordinates the contractile action of the muscle. Motor neurons within a pool are those recruited to increase contractile force by the population code.

Interneurons of the Spinal Cord

The interneurons of the spinal cord also play an important role in movement. These cells are located in the intermediate zone of the spinal cord. Those located in the lateral part of this zone have axons that synapse **ipsilaterally** (on the same side of the body) with motor neurons that innervate distal limb muscles. Interneurons lying closer to the midline have axons that synapse with motor neurons on both sides of the spinal cord that control muscles for posture.

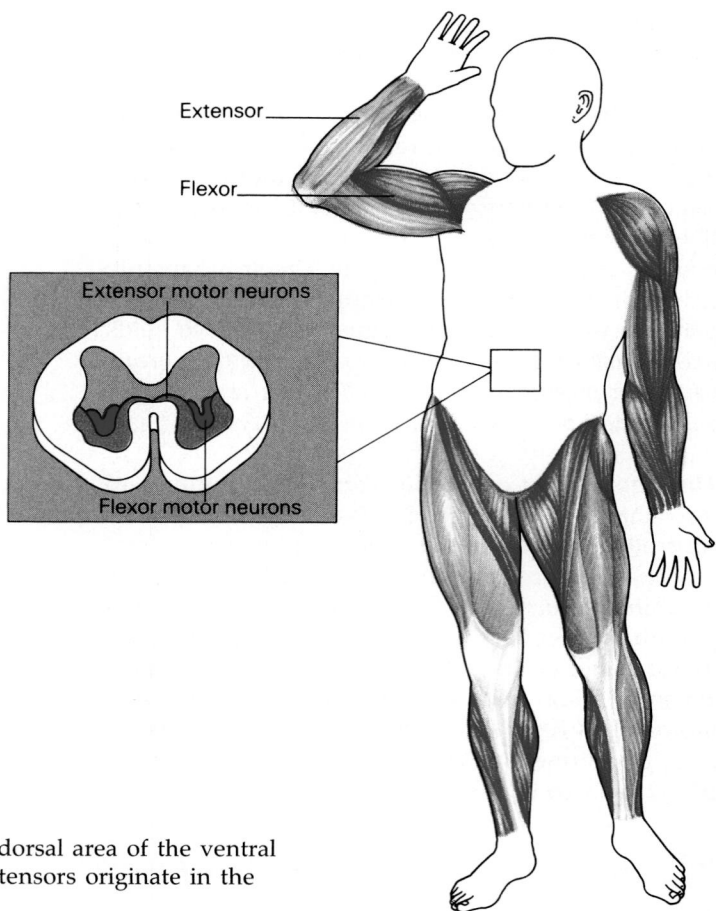

Figure 9–6
Motor neurons that activate flexors originate in the dorsal area of the ventral horn of the spinal cord, while those that activate extensors originate in the ventral region of the ventral horn.

Interneurons also send their axons up and down the spinal cord to form synaptic connections in the regions several cord segments away from their cell bodies. They connect with motor neurons that control the contraction of several muscle groups. In this way, the activities of muscle groups are coordinated in maintaining posture and for reaching movements.

Control of Locomotion

The act of walking requires the coordinated contraction and relaxation of flexor and extensor muscles of the legs. This rhythmic alteration between flexion and extension is produced entirely by the interconnections of neurons in the spinal cord. These neurons and their interconnections compose the **locomotion generators** for walking. Animals whose spinal cords have been severed can walk without inputs from higher centers. This phenomenon is seen in chickens, for example, which remain capable of running for a short period after decapitation. Experimental animals with spinal cord transections

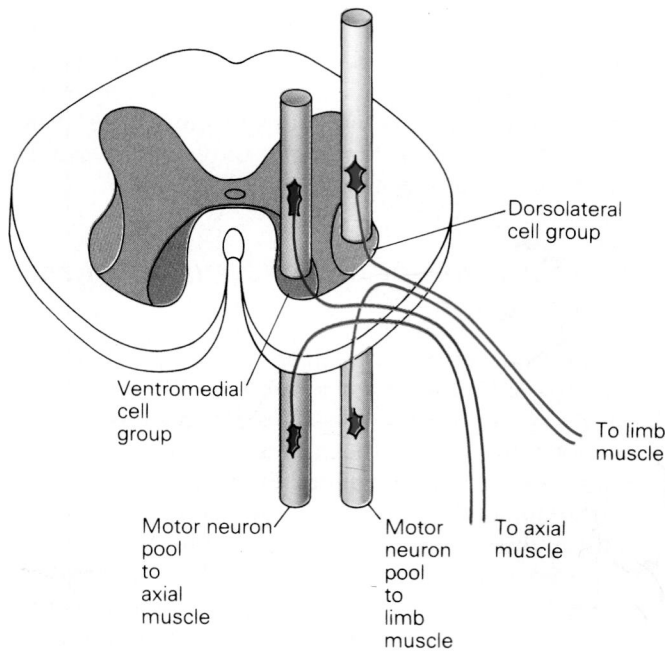

Figure 9–7
A motor neuron pool is a group of motor neurons that innervate a single muscle.

are also capable of walking on treadmills. The speeds at which they walk are dictated by the speed of the treadmill. In these experimental animals, if one limb is prevented from walking, the other limb will continue to walk. The locomotor generator for each limb, therefore, does not require activity from the other limb. However, when all limbs are active, the generators for limb movement are coupled to produce a coordinated response (Fig. 9–8).

The locomotor generators in the spinal cord are under the control of a "locomotor command center" in the brain stem. Electrical stimulation of this command center causes animals with spinal cord lesions to walk on a treadmill. Weak stimulation causes walking while stronger stimulation can cause the animal to run.

During the walking process in normal animals, the locomotor generators coordinate the flexion and extension of leg muscles to carry out the **stance phase** and the **swing phase** of walking. In the stance phase, the foot is on the ground and supporting the weight of the body. In the swing phase, the foot is off the ground and swinging forward. If sensory feedback from muscle spindles and Golgi tendon organs is prevented from reaching the spinal cord during the stance phase, the step cycle can be

stopped during the extension of the leg. In other phases of the walk cycle, however, the absence of sensory information does not inhibit the process of locomotion.

Sensory information can also change the activation of one motor response to another. The activation of tactile receptors on the top of the foot during the swing phase of walking, for example, results in a flexion response that is appropriate for stepping over an object. Sensory information to the spinal cord is therefore capable of changing ongoing motor programs and reflexes.

Sensory Fibers in Skeletal Muscle

Sensory neurons also have fiber processes within skeletal muscles. They transmit information about the state of contraction and tension in muscle fibers to the CNS so that it can revise its instructions as necessary to the motor neurons producing the contraction or tension. Sensory neurons that innervate intrafusal muscle fibers form what is known as a muscle spindle, while sensory neurons that innervate tendons form Golgi tendon organs.

The Muscle Spindle: A Length Detector. The sensory receptors that detect the length of the muscle and its velocity of contraction are called **muscle spindles** (Fig. 9–9). These structures are formed by sensory neurons that entwine intrafusal muscle fibers.

The muscle spindle is composed of two types of intrafusal fibers: **nuclear bag fibers** and **nuclear chain fibers** (see Fig. 9–9a). The nuclear bag fiber is innervated by **Type Ia nerve fibers,** which transmit information about muscle length and velocity of contraction to the CNS. Nuclear chain fibers are innervated by **Type II nerve fibers,** which transmit information about muscle length.

The muscle spindle acts as a **stretch receptor**— that is, it increases its discharge of action potentials when the intrafusal fibers stretch. Conversely, they decrease their discharge when the fibers shorten.

Gamma motor neurons innervate contractile elements at the poles of the muscle spindles (see Fig. 9–9a). The activation of these contractile elements causes the central region of the muscle spindle to stretch as a rubber band would. In this manner, the gamma motor neurons can regulate the sensitivity of the Type Ia sensory nerves of the muscle spindles. Figure 9–9c shows the action potential activity of Type Ia muscle spindle sensory nerves as a function of the length of the extrafusal muscle fiber. In the absence of gamma motor neuron activa-

Figure 9–8
Locomotor generators and the locomotor command center.

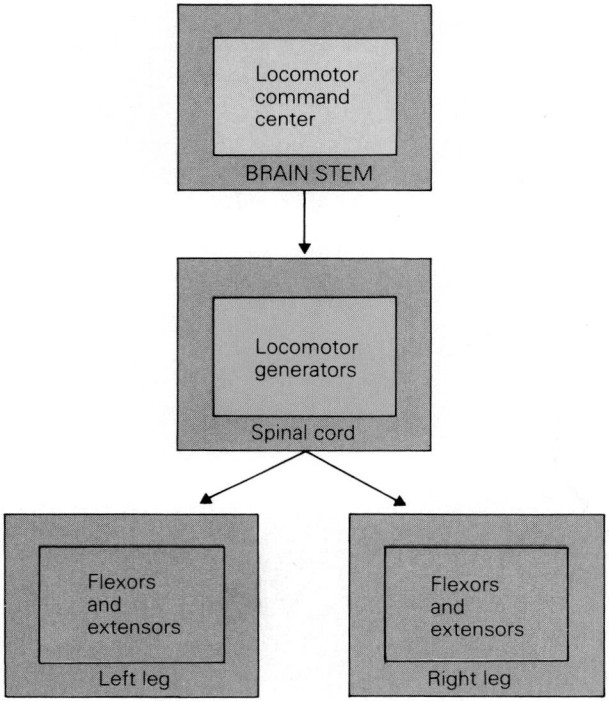

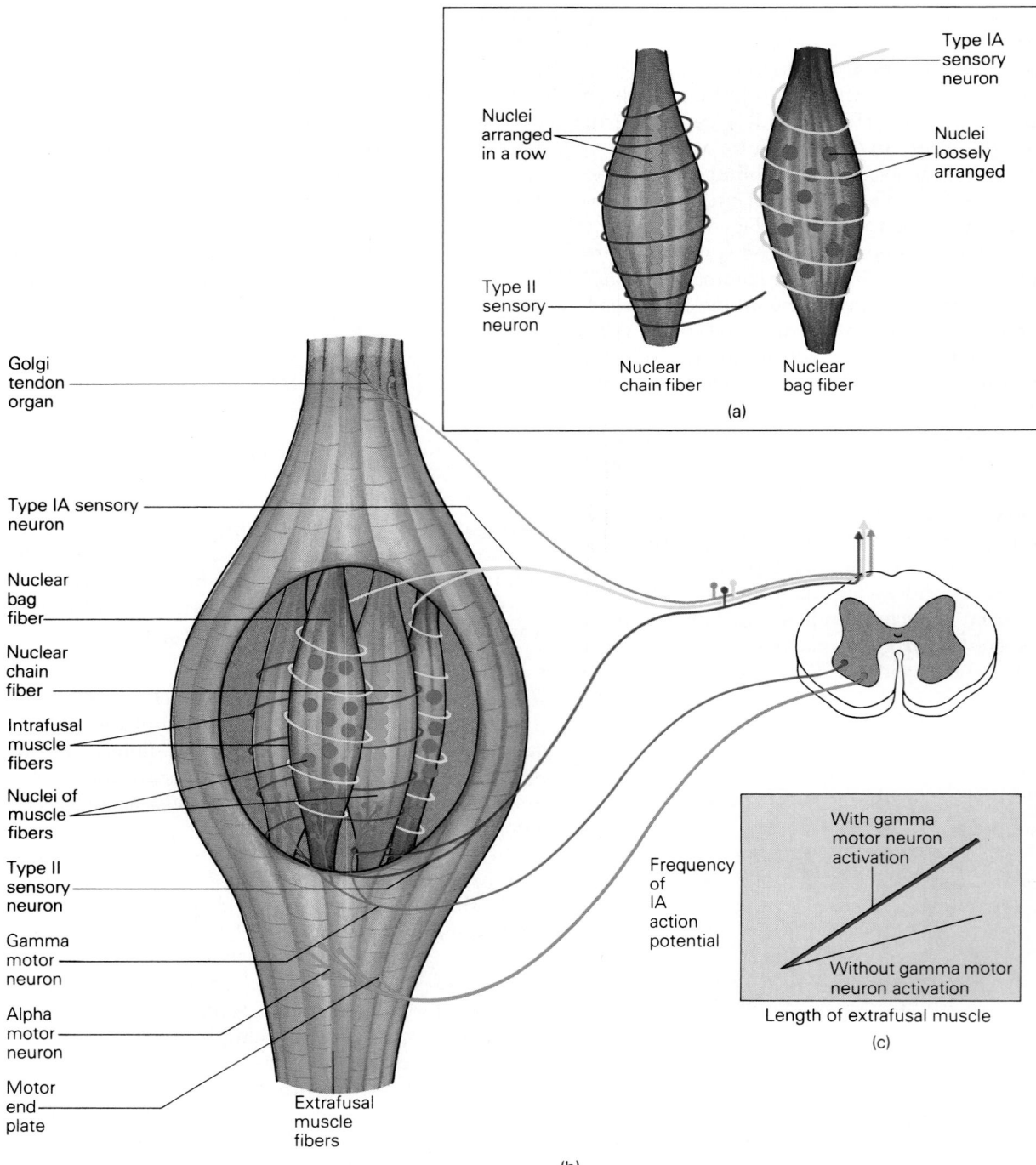

Figure 9–9
(*a*) Intrafusal muscle fibers consist of nuclear bag fibers and nuclear chain fibers. (*b*) These fibers and the sensory neurons that innervate them make up a muscle spindle, a receptor for muscle length and state of contraction. Golgi tendon receptors innervate tendons and transmit information about the force exerted by the muscle. Gamma motor neurons activate intrafusal fibers, while alpha motor neurons activate extrafusal muscle fibers. (*c*) When activated by gamma motor neurons, Type Ia sensory neurons generate action potentials in response to changes in extrafusal muscle length more frequently than in the absence of gamma motor neurons.

tion, a change in the length of the extrafusal muscle fibers results in a slight change in the frequency of action potentials generated by the sensory neurons. With gamma motor neuron activation, however, a slight change in the length of the extrafusal muscle fibers causes a great change in the frequency of generated action potentials. In this way, the activation of gamma motor neurons maintains the sensitivity of the muscle spindles even as the extrafusal muscle fibers shorten during contraction.

Golgi Tendon Organs: Tension Receptors. A second type of muscle receptor is the **Golgi tendon organ.** This type of receptor is located in series with the extrafusal muscle fiber (see Fig. 9–9b) and transmits information about the force or tension produced by the contraction of the muscle. As greater tension develops when the muscle contracts, the Golgi tendon organ generates more action potentials. Since the Golgi tendon organ and muscle spindle are somatic sensory receptors, they transmit their information about muscle force, velocity, and length to the spinal cord, where it is then forwarded to the somatic sensory cortex or used within the spinal cord for reflex action.

Spinal Cord Reflexes

Spinal cord reflexes represent the most basic of motor responses. These reflexes are carried out entirely within the spinal cord and are modified by inputs from high centers to generate complex movements. They are also used to help diagnose disorders of the motor system.

The Stretch Reflex

The **stretch reflex** is commonly called the **knee jerk reflex.** A stretch reflex, for example, is activated by tapping the **patellar tendon** below the knee to stretch the muscle spindles. Action potentials conducted along the muscle spindle sensory neurons enter the spinal cord via the dorsal roots (Fig. 9–10). The fiber from the muscle spindle branches after entering the spinal cord, with the ascending branch joining the dorsal column pathway. The other branch forms synaptic connections with motor neurons in the ventral horn.

These motor neurons activate muscle fibers, causing the leg to extend quickly or kick out. In order for the leg to extend, however, the antagonistic muscles that cause the leg to flex must be inhibited. This inhibition comes about by the activation of inhibitory interneurons within the ventral horn. These interneurons receive information from muscle spindle sensory neurons and in turn form inhibitory synaptic connections with motor neurons of flexor muscles. Consequently, the activation of these interneurons prevents the contraction of the flexors. By inhibiting flexors and activating extensors, the interconnections of nerve cells within the spinal cord produce well-coordinated motor responses.

The Inverse Myotatic Reflex

A reflex that involves the Golgi tendon organ is the **inverse myotatic reflex.** This reflex is seen when a person attempts to lift more weight than he or she can actually carry. Weightlifters, for example, often add increasing amounts of weight to bar bells when lifting the weights in a maneuver called the "curl." In this maneuver, a person lifts the bar bells by contracting the bicep muscles. If the weight is too great for the biceps to lift, however, they suddenly relax, and the bar bells are dropped.

The inhibition of the biceps occurs because the Golgi tendon organ detects excessive force, which might damage the muscle. With increasing force, the Golgi tendon organ begins to discharge action

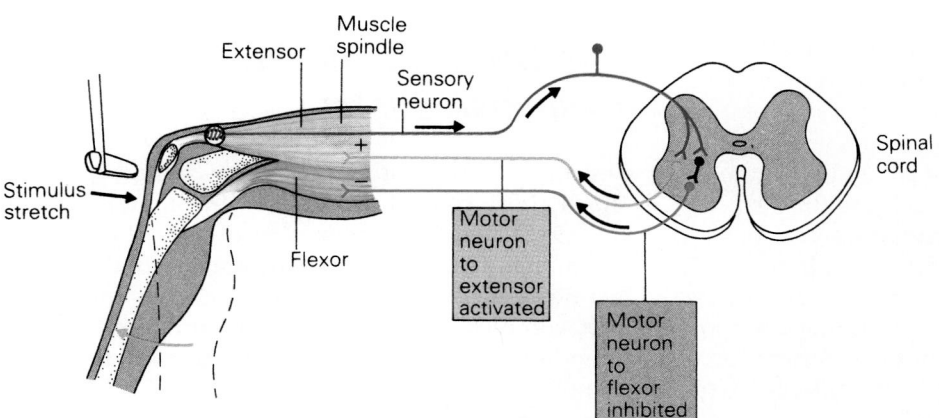

Figure 9–10
The stretch reflex occurs when a muscle spindle detects a tap on the knee and informs the spinal cord, which activates the leg extensor and inhibits the flexor, causing the knee joint to open up.

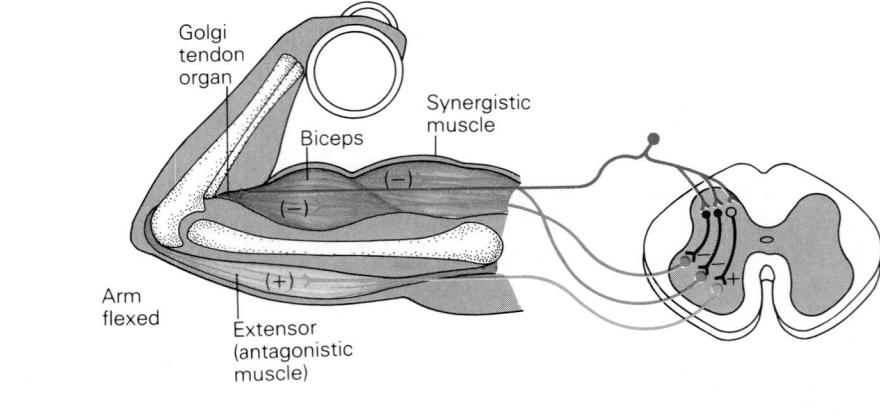

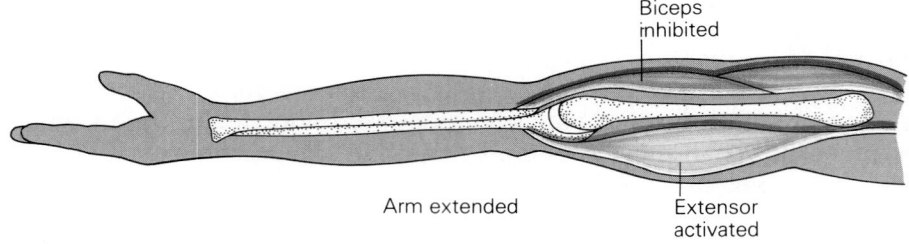

Figure 9–11

The inverse myotatic reflex causes a flexed bicep muscle to relax and an extensor in the arm to contract so that the arm extends and drops a weight that might be harmful.

potentials (Fig. 9–11). As with the muscle spindles, these action potentials are transmitted to the dorsal column nuclei and subsequently to the somatic sensory cortex. In the spinal cord, however, information about excessive muscle tension is transmitted to inhibitory interneurons that turn off the motor neuron innervating the biceps. At the same time, interneurons turn off any muscles (flexors) synergistic to the biceps and turn on any muscles (extensors) that are antagonistic to the biceps, so that the arm extends and drops the barbells. This reflex thus protects the muscle from damage that might occur when a person attempts to lift too heavy a weight.

In addition to regulating muscle tension, this reflex is also thought to involve the smooth onset and termination of muscular contraction in walking. In the motion of stepping, the flexors of one leg begin to contract while its extensors relax. The flexion of one leg needs to be coordinated with the extension of the opposite leg. To achieve this, sensory information from the Golgi tendon organ from the leg being flexed is transmitted to the other side of the spinal cord to activate the extensors and inhibit the flexors of the opposite leg. Although sensory inputs play an important role in the coordination of walking, the motor "program" for walking is contained within the spinal cord.

The Flexor Withdrawal Reflex

A reflex that involves cutaneous receptors is the **flexor withdrawal reflex.** An example of this reflex is activated by pain receptors on the bottom of the foot (Fig. 9–12). When you step on a tack, for instance, pain information is transmitted to the spinal cord and causes a contraction of the flexors to remove the foot from the tack. At the same time, extensors of the leg that would normally keep the foot on the tack are inhibited. In order to support the rest of the body, however, the extensor muscles in the other leg are activated while its flexors are inhibited.

The Brain in the Motor Response System

Several important structures in the brain represent the highest control centers for the motor system (Fig. 9–2). The brain stem controls spinal cord output in maintaining posture and balance of the body. Various areas of the neocortex organize and interpret complex sensory input and program the responding movements. The cerebellum and the basal ganglia also plan and program specific types of motor responses.

Comparisons are made between the intended movement, as designed by the motor program, and the actual movement. Deviations from the intended movement are then corrected by transmitting another set of instructions to the motor neurons in the spinal cord.

The Brain Stem: Control of Posture and the Spinal Cord

One of the primary roles of the brain stem is maintaining the body's posture and balance. Nerve cells within different clusters in the brain stem send axons that terminate in the spinal cord. Three pathways from the brain stem provide input to the motor neurons of the spinal cord: (1) the ventromedial pathway, (2) the lateral reticulospinal pathway, and (3) the rubrospinal pathway.

The Ventromedial Pathway

The **ventromedial pathway** impinges on motor neurons that control axial muscles. This pathway has three major components, all of which descend along the ventral and medial region of the spinal cord (see Fig. 9–13a). The first component is the **vestibulospinal tract,** which originates in the vestibular nucleus and carries information for the reflex control of equilibrium (discussed in a later section). The second component of the ventromedial pathway is the **tectospinal tract,** which originates in the **tectum,** a structure involved in the coordinated control of head and eye movements. The third component is the **medial reticulospinal tract,** which originates in the **reticular formation,** a structure involved in maintaining posture by the activation of extensor muscles.

The Lateral Reticulospinal Tract

The second major descending pathway from the brain stem to the spinal cord is the **lateral reticulospinal tract.** Nerve fibers in this pathway are derived from the lateral reticular nucleus and descend the spinal cord in the lateral region of the cord (see Fig. 9–13b). These fibers innervate flexors in the control of posture.

The Rubrospinal Tract

The rubrospinal tract has fibers that originate in the red nucleus of the brain stem. These fibers descend along the dorsal and lateral border of the cord to innervate motor neurons that control distal flexor muscles (see Fig. 9–13c). The effects of the major

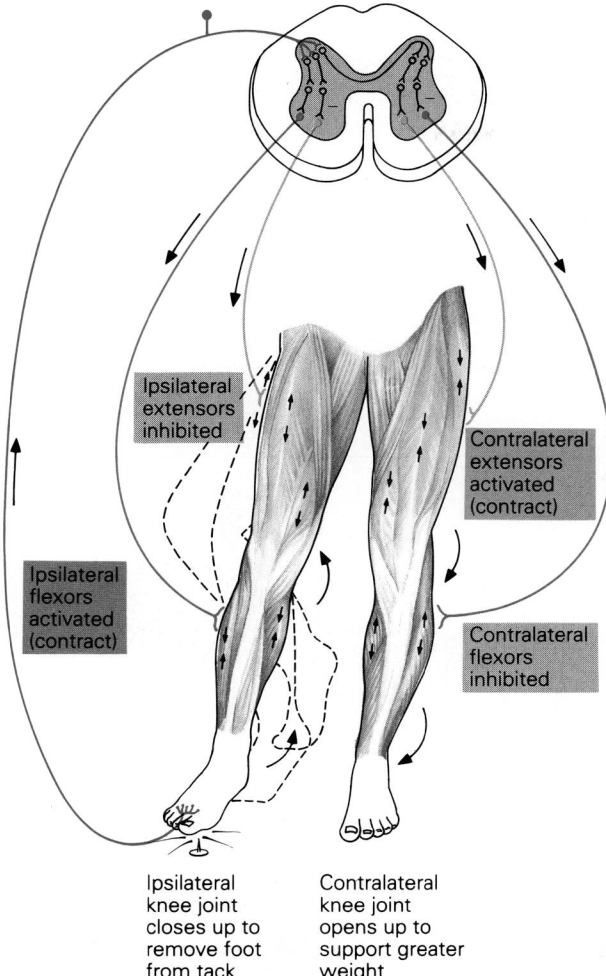

Ipsilateral extensors inhibited

Contralateral extensors activated (contract)

Ipsilateral flexors activated (contract)

Contralateral flexors inhibited

Ipsilateral knee joint closes up to remove foot from tack

Contralateral knee joint opens up to support greater weight

Figure 9–12
The withdrawal reflex in one leg requires the cooperation of the opposite leg, arranged by interneurons in the spinal cord.

These components of the motor system interact with each other to carry out specific functions. The desire to move activates the supplemental motor cortex, basal ganglia, and cerebellar components of the motor system. Information from these areas is transmitted to the premotor cortex, where movements are planned and programmed. From the premotor cortex, information about the planned movement is transmitted to the motor cortex. The motor program is then carried out when neurons in the motor cortex instruct those in the spinal cord to activate the skeletal muscles. The force and velocity of contraction, and the length of muscle shortening, are monitored by Golgi tendon organs and muscle spindles. This sensory information is fed back to the motor neurons at the spinal level and also transmitted back to the motor cortex and cerebellum.

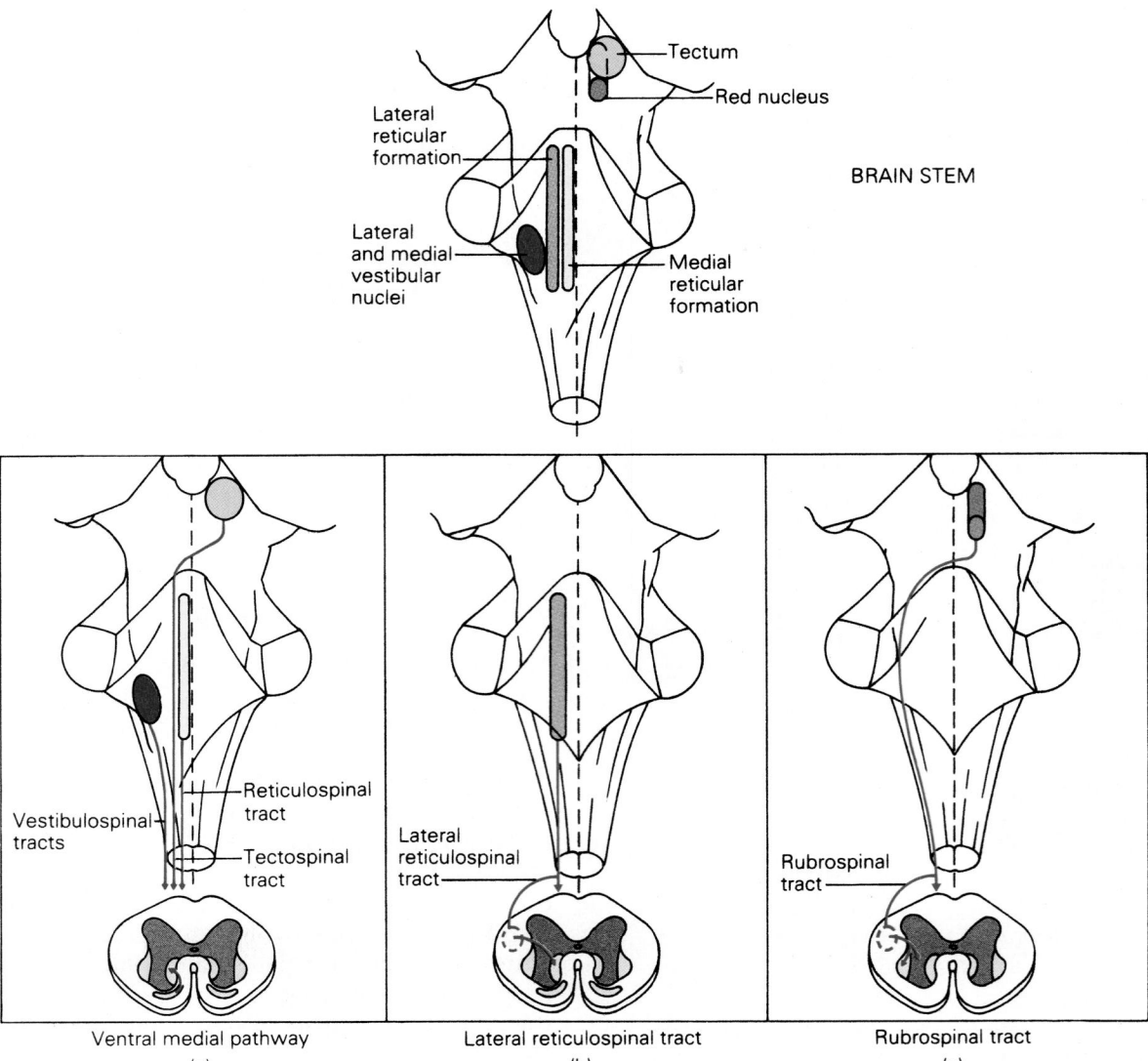

Figure 9–13
Pathways from the brain stem to the spinal cord: (*a*) the ventromedial pathway, (*b*) the lateral reticulospinal tract, and (*c*) the rubrospinal tract.

descending brain stem pathways on flexor and extensor muscles in the maintenance of posture and equilibrium are summarized in Figure 9–14.

Motor Cortex Control of Reaching and Fine Voluntary Movement

The control of reaching and fine voluntary movement is due to instructions transmitted via descending pathways to the spinal cord from the **motor cortex.** The motor cortex occupies a cortical region rostral to the somatic sensory cortex (Fig. 9–15). The cells in the motor cortex are organized in a somatotopic manner similar to that of the somatic sensory cortex (Fig. 9–16). Cells in the medial region of the motor cortex cause the contraction of muscles in the leg. Nerve cells located in more lateral regions of the motor cortex activate the muscles of the torso, arm, hand, and face.

The parts of the body involved with fine movement, such as the fingers, occupy more space in the motor cortex than parts of the body involved in gross movement, such as the torso. This relationship is in keeping with the general principle that the amount of cortical space devoted to different parts of the body is in proportion to its sensory sensitivity or degree of fine motor control.

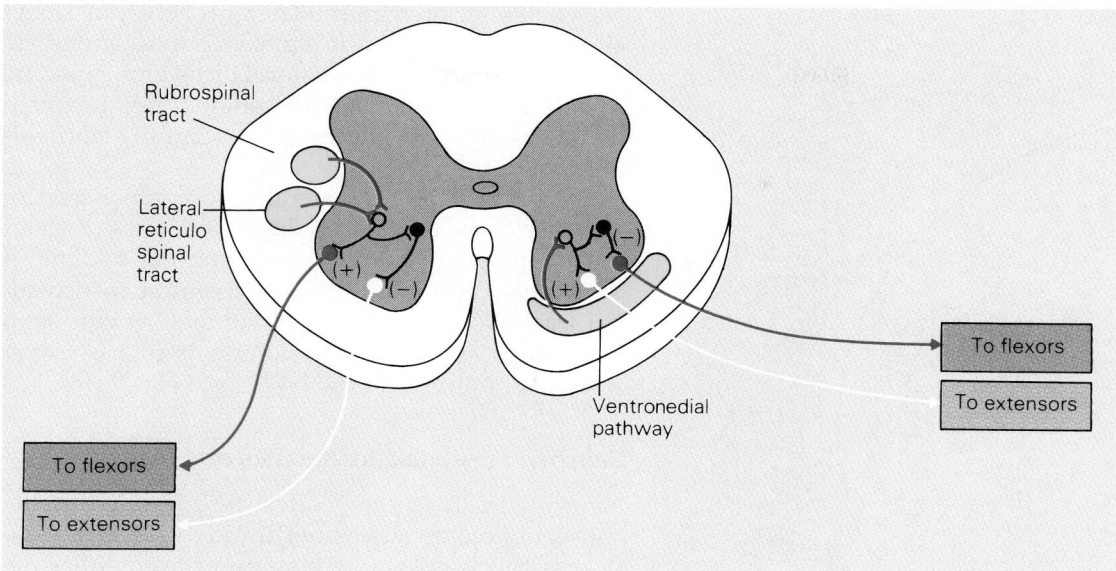

Figure 9–14
A summary of the effects of the brain stem on spinal motor neurons.

Organization of the Motor Cortex

The cells of the motor cortex are organized in functional columns called **cortical efferent zones (CEZ).** All of the cells within one efferent zone are involved in the contraction of a given muscle. Neurons in the motor cortex are also organized horizontally into six different layers (Fig. 9–17). Those located in layer V provide the output from the motor cortex. Pyramid-shaped nerve cells in this layer send axons to the spinal cord to synapse with spinal motor neurons and interneurons. These cortical neurons are involved primarily in controlling distal muscles. They excite both alpha and gamma motor neurons and thus allow the muscle spindles to remain sensitive to changes in muscle length.

Descending Pathways from the Motor Cortex

Information from the motor cortex is transmitted to the spinal cord and brain stem by the **corticospinal pathway** and **corticobulbar pathway,** respectively. Inputs to the brain stem ultimately influence axial muscles near the midline for the maintenance of posture, while inputs to the spinal cord control the distal limb muscles. Approximately 30% of the corti-

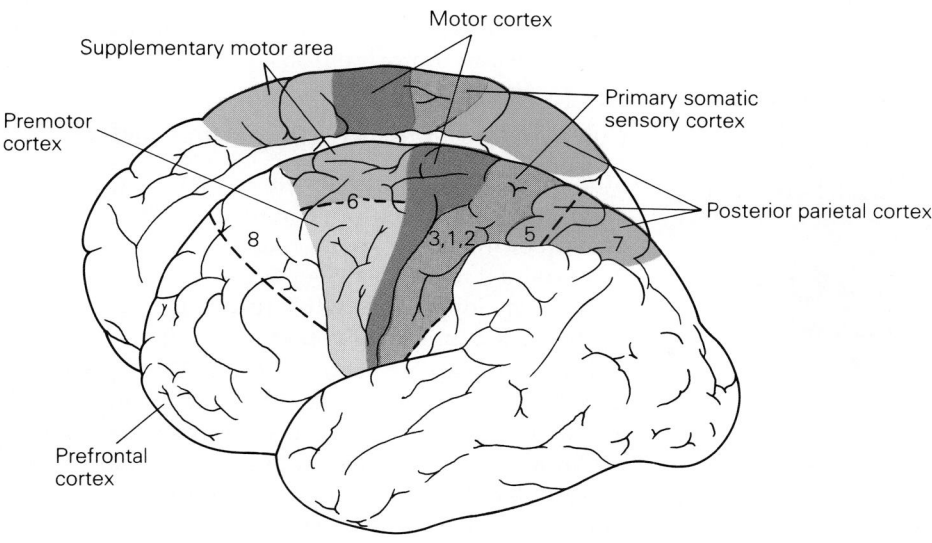

Figure 9–15
The functional areas of the cerebral cortex involved in motor control.

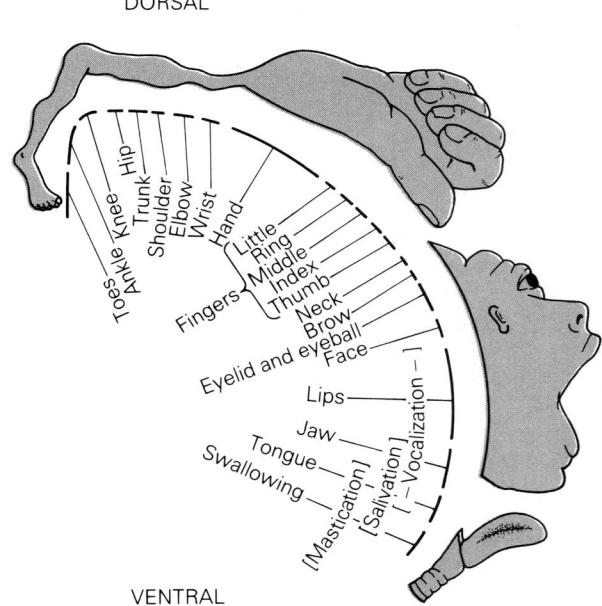

DORSAL

VENTRAL

Figure 9–16
The somatotopic organization of the motor cortex.

cospinal and corticobulbar fibers are from neurons in area 4; another 30% originate from the premotor cortex (area 6). The remaining fibers are derived from the somatic sensory cortex.

As these axons descend from the motor cortex, they form the **pyramidal tract.** When the pyramidal tract reaches the level of the brain stem, the majority of the fibers that continue to the spinal cord cross the midline of the body and continue their descent along the **lateral corticospinal tract** (Fig. 9–18).

Figure 9–17
The cell layers of the motor cortex.

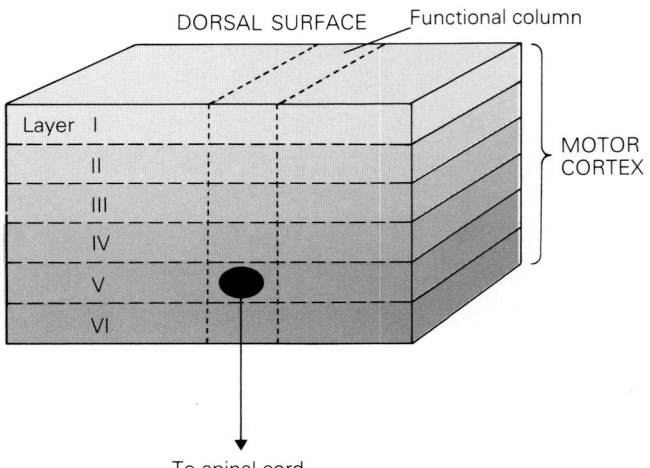

DORSAL SURFACE   Functional column

Layer I

II

III

IV

V

VI

MOTOR CORTEX

To spinal cord

Nerve fibers exit from the lateral corticospinal tract at various levels of the spinal cord to innervate motor neuron pools that control distal limb muscles. The direct connections from motor cortex to spinal cord thus permit the independent control of individual muscles.

In addition to the lateral corticospinal tract, a minority of the fibers from the motor cortex descend to the spinal cord without crossing the midline of the body. These fibers form the **ventral corticospinal tract** and primarily innervate motor neurons in the medial region of the ventral horn associated with axial muscles of the body (see Fig. 9–18).

Sensory Feedback to the Motor Cortex

Neurons in the motor cortex are informed of the consequences of movement through sensory feedback pathways. They receive input from either the muscles they project to or from areas of skin surrounding the muscle. This long loop of sensory feedback results in the alteration of information from the motor cortex to the spinal cord to correct any deviations from the intended movement. Sensory feedback to the motor cortex occurs by way of the somatic sensory cortex (Fig. 9–19).

The nerve cells in the sensory cortex are connected to those in the motor cortex in a topographic manner. Cells in the sensory cortex receiving proprioceptive input from muscles in the thumb, for example, are connected with cells in the cortical efferent zone responsible for the contraction of these same muscles. In this way, feedback is provided by the sensory system to inform the cells in the motor cortex whether the instructions they have transmitted for the muscular contraction have been faithfully executed. If not, then the sensory information that is sent back to the motor cortex causes the cells in the CEZ to modify their activity and thus modify the activity of the motor neurons.

Cortical Coding of Reaching Movements

Studies correlating nerve cell activity in the motor cortex with reaching movements have been carried out predominantly in monkeys. These monkeys were trained to move a lever to different locations in two-dimensional space while recordings were made of the action potential activity of individual nerve cells in motor cortex (Fig. 9–20a and 9–20b). During a reaching motion, the activity of a cell in motor cortex is best correlated with a specific direction of reaching in space. This preferred direction can be represented by a **directional cell vector,** which indicates not only direction but also the frequency of

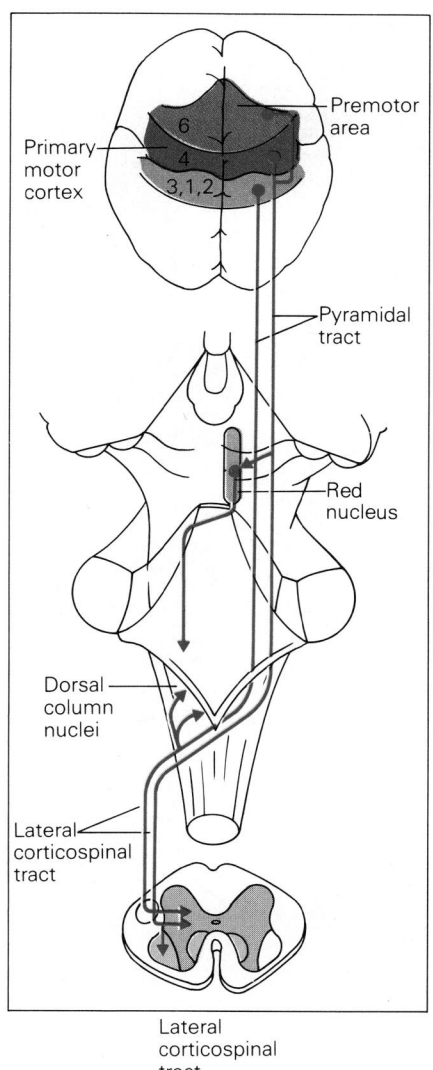

Lateral
corticospinal
tract

(a)

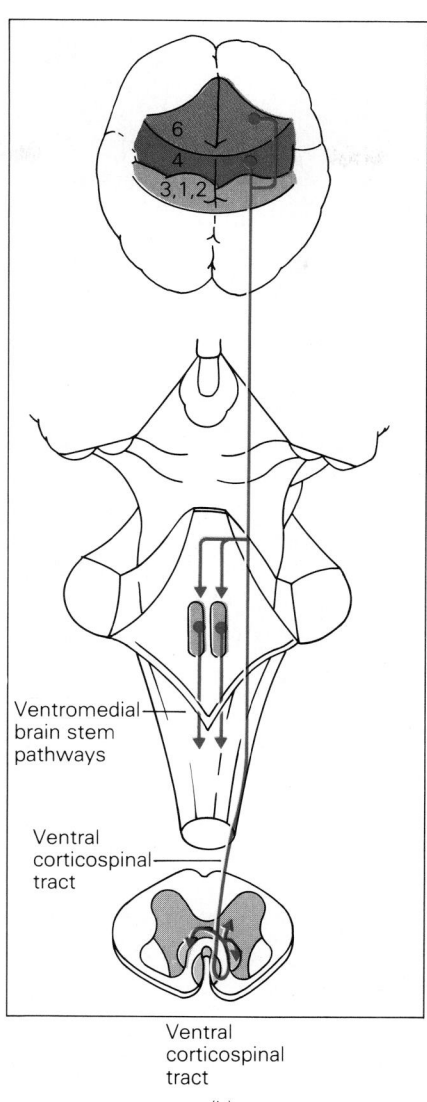

Ventral
corticospinal
tract

(b)

Figure 9–18
Descending pathways from the
motor cortex: the lateral and ventral corticospinal tracts.

action potential discharge (Fig. 9–20c). From the direction vector of individual cells, a **population vector** can be generated, representing the preferred direction of a cluster of cells in motor cortex. This
population vector is highly correlated with direction
of movement (Fig. 9–20d) and represents a *distributed code* such that each neuron carries part of the
information for movement and that the activity of
the cells as a group uniquely determines the direction of movement in space.

The activity of nerve cells in motor cortex encode not only directional information but also information about the force required to make a particular
reaching movement. With an increase in force there
is a corresponding increase in the frequency of
action potentials produced by individual nerve cells
in motor cortex. This results in an increase in the

length of the cell vector and also the population vector. Thus these vectors are most likely representa-
tions of the direction movement and the force of
muscular contraction needed to carry out a particular movement.

Coding of contractile force is carried out by two
types of neurons. First, dynamic neurons code for
the *rate of force development*. These neurons change
their activity when there is an increase in the level
of applied force. A second type of neuron codes for
steady-state force. These neurons discharge action
potentials throughout the entire period during
which force is maintained.

The pyramidal cells in the motor cortex begin
generating action potentials approximately 50 milliseconds before a movement. Nerve cells in other
areas of the brain involved in motor control become

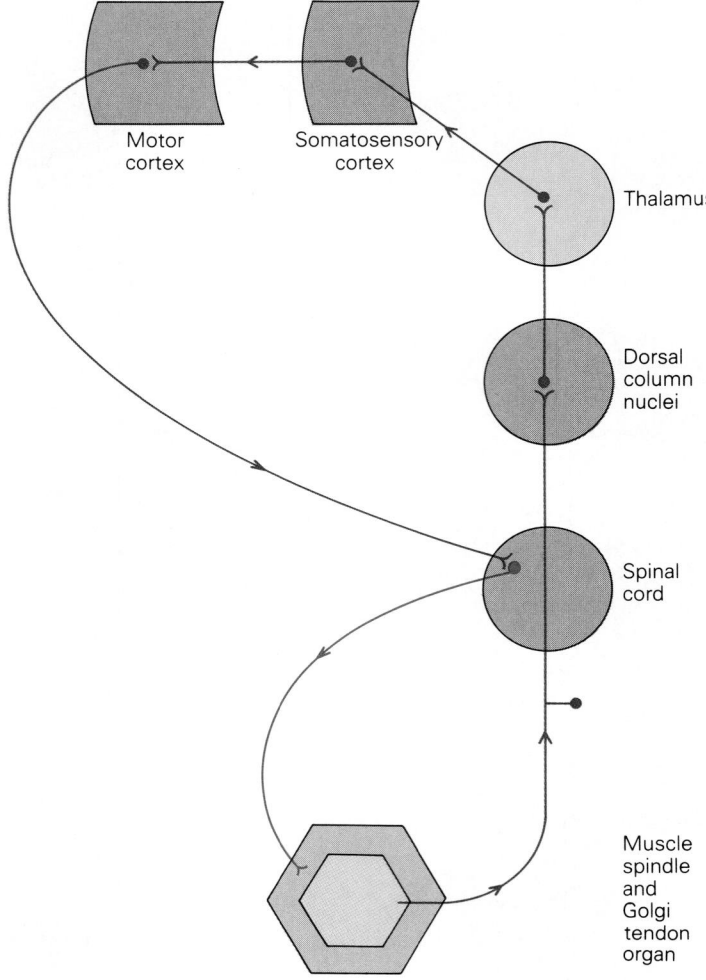

Figure 9–19
Sensory feedback pathways to the motor cortex.

active much earlier than the motor cortex. These areas of the brain, called the **supplemental and premotor cortex, posterior parietal cortex, cerebellum,** and **basal ganglia,** are involved in the planning and programming of movement.

The Supplemental and Premotor Areas: Programming Movement

The motivational factors that cause specific movements most likely originate in subcortical areas of the brain such as the **hypothalamus.** This area of the brain is involved with processes such as hunger, thirst, and sex, and will be discussed in Chapter 11. The motivational areas of the brain that respond to the need to accomplish a particular task transmit this information to the **supplemental and premotor cortex** (Brodmann area 6), which designs the necessary movements to accomplish the desired task.

In the act of reaching for a glass of water to bring to the lips, the nervous system must determine which motor program will activate certain muscles, and when and how much they need to be contracted. The components of the motor program are thought to be developed by supplemental and premotor cortices (area 6).

Stimulation of the supplemental and premotor areas produces movements more complex than those produced by stimulation of the motor cortex. Activation of the premotor area causes movements of the torso or the opening and closing of the hand. In contrast, stimulating the motor cortex produces small, twitching movements. The supplemental area has direct inputs to the motor cortex for the control of distal limb muscles. However, it also has inputs to motor neurons in the spinal cord that control axial muscles involved with the maintenance of posture.

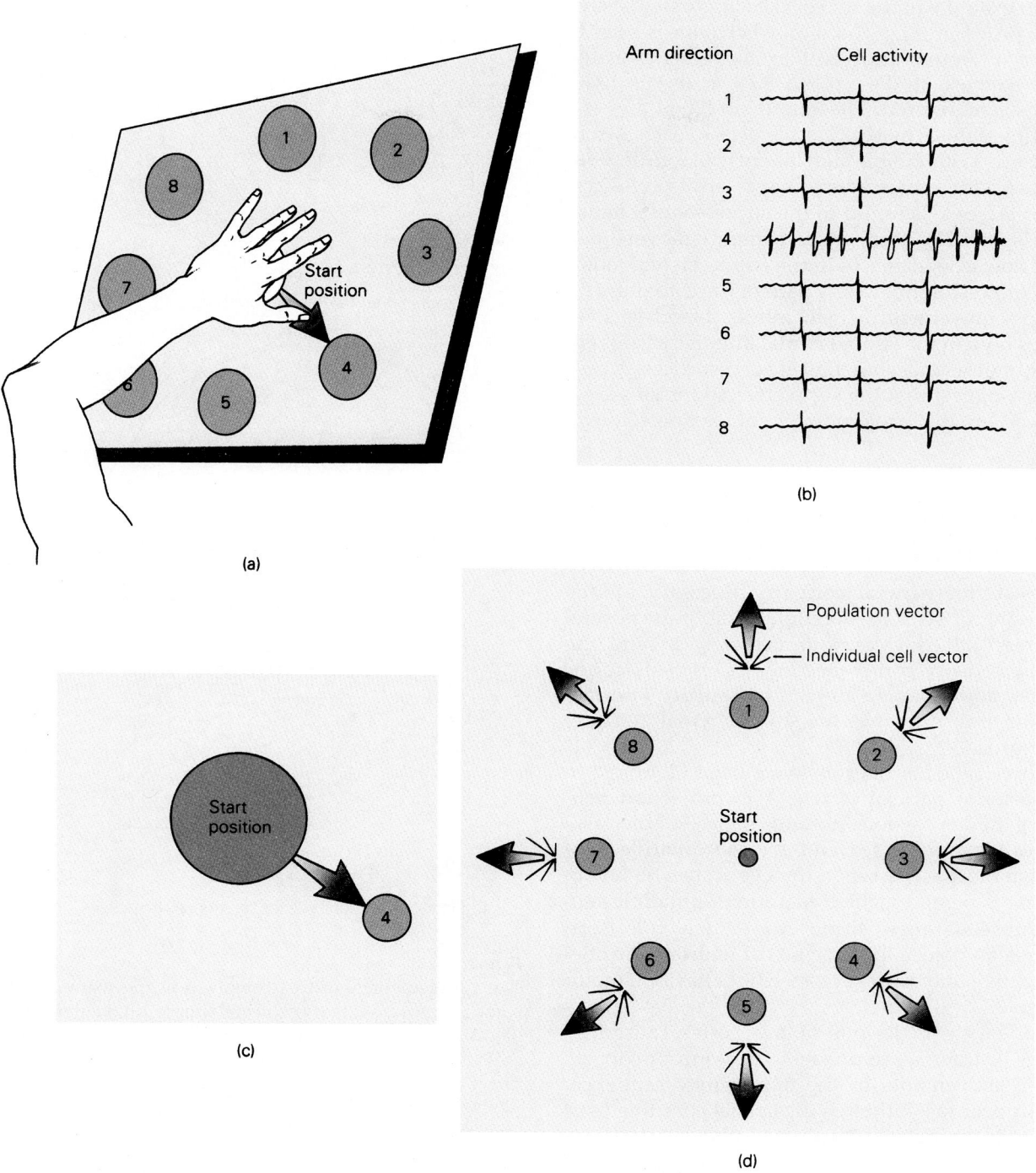

Figure 9–20
(*a*) Eight directions of possible arm movements in a 2-dimensional plane, starting from a center position. (*b*) Activity of single neuron in motor cortex correlated with arm reaching direction. (*c*) Vector for single cell indicating preferential direction and intensity of activity. (*d*) Population vector representing the summation of individual cell vectors that encodes the direction of arm movement.

An example of the role of the supplemental area in the programming of a motor behavior is seen in studies of simple and complex movements of the hand. In these studies, blood flow to certain brain regions is monitored. The amount of blood flow is a measure of the activity of nerves cells in each area of the brain. With simple movements, blood flow increases in the motor and somatic sensory cortices, but no increase is found in the supplemental motor area (Fig. 9–21). With more complex movements, blood flow also increases in the supplemental motor area. Interestingly, when patients are told to rehearse a movement in their minds, blood flow increases only in the supplemental motor area and not in the motor and somatic sensory cortices. These studies suggest that the supplemental motor area is involved in the programming of complex movements.

The Posterior Parietal Cortex: Sensory Stimuli and Purposeful Movement

The **posterior parietal cortex** is necessary for the processing of sensory stimuli leading to purposeful movement. It is located immediately behind the somatic sensory cortex (see Fig. 9–15). This cortex receives both somatic and visual sensory information and transmits it to the supplemental and premotor areas.

Three types of neurons have been identified in the posterior parietal cortex. **Arm projection neurons** generate action potentials when the arm reaches for an object. **Hand-eye coordination neurons** are most active when the eye fixates on an object that is being touched; **hand manipulation neurons** increase their firing rate when the hand explores an object. These types of neurons thus become active only with very specific behavioral motor responses.

Patients with lesions of this cortex caused by strokes or trauma are unable to perform previously learned movements in the appropriate sequence. They appear to synthesize the spatial coordinates of objects in abnormal ways and behave as if their movements are not in accord with the coordinates of the objects in space. For example, when drawing a clock, a patient with a posterior parietal lesion places all of the numbers on one half of its face.

The Cerebellum and the Basal Ganglia

The basal ganglia and cerebellum (Fig. 9–22) are the major subcortical components of the motor system. They both receive inputs from the neocortex and

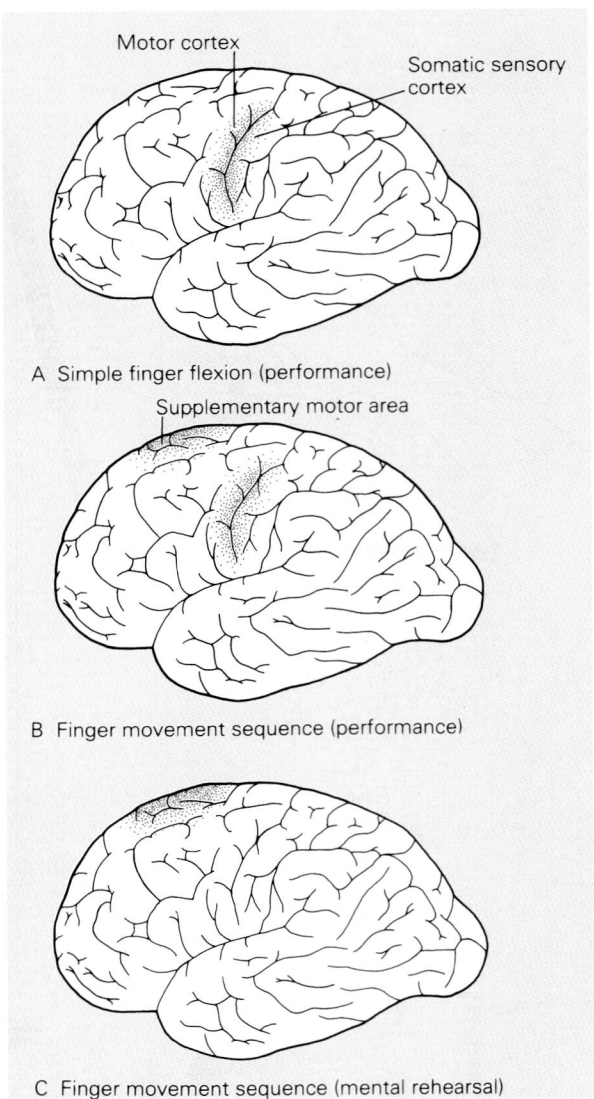

A Simple finger flexion (performance)

B Finger movement sequence (performance)

C Finger movement sequence (mental rehearsal)

Figure 9–21
The supplemental cortex is involved in the planning of complex movements. (*a*) During simple finger flexion, blood flow (indicated by red stippling) increases in the motor cortex, while (*b*) during a complex sequence of finger movements, blood flow is also increased in the supplemental cortex. (*c*) When the same sequence is mentally rehearsed, blood flow increases in the supplementary cortex but not in the primary motor cortex.

transmit information back to the cortex by way of the thalamus. The inputs to the basal ganglia, however, are from the entire cortex, while those of the cerebellum are primarily from sensory and motor areas. In addition, the basal ganglia do not receive direct sensory information from somatic receptors, nor do they transmit descending information di-

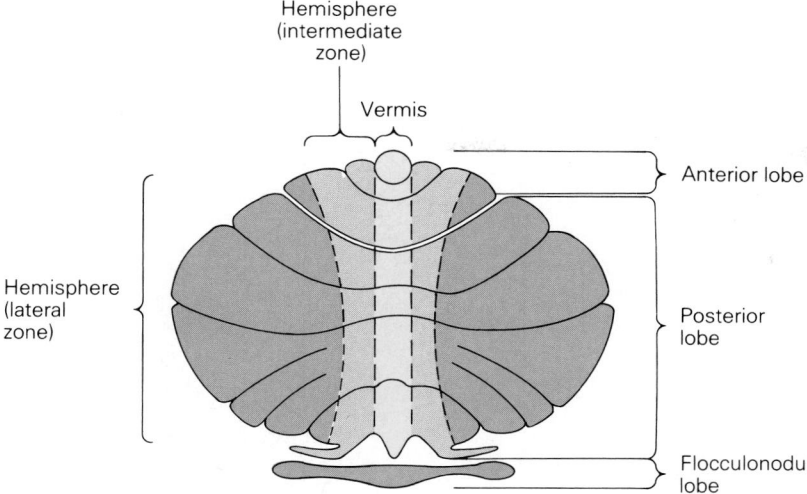

Hemisphere (intermediate zone)

Vermis

Hemisphere (lateral zone)

Anterior lobe

Posterior lobe

Flocculonodular lobe

Figure 9–22
A dorsal view of the cerebellum.

rectly back to the spinal cord as does the cerebellum. These differences suggest that the basal ganglia are involved in more complex motor functions, while the cerebellum is more involved with the control of movement that requires constant monitoring by sensory feedback.

The Cerebellum: Planning, Coordination, and Posture

The cerebellum ("little brain" in Latin) has three basic functions: (1) the planning of a movement, (2) the control of posture and equilibrium, and (3) the control of smooth limb movement. The cerebellum accomplishes the latter two functions by comparing information concerning an intended movement with sensory feedback about the actual movement, and adjusting its output to compensate for differences between the two. It participates in the planning of a movement by receiving information from motor (supplemental and premotor) and parietal cortices and then uses these inputs to initiate a planned movement.

These cerebellar functions are associated with specific anatomical locations. The cerebellum has three major lobes: the **anterior lobe,** the **posterior lobe,** and the **flocculonodular lobe** (see Fig. 9–22). The flocculonodular lobe is involved in the maintenance of equilibrium and posture, while the midline and intermediate regions of the anterior and posterior lobes are involved in limb movement. The planning and initiation of motor programs are the responsibility of the lateral region of the anterior and posterior lobes.

The most prominent nerve cell in the cerebellum is the **Purkinje cell,** found throughout the cerebellum and packed tightly into a single cellular layer

(Fig. 9–23). The axons of Purkinje cells form synaptic connections with neurons in one of three deep cerebellar nuclei, the **dentate nucleus,** the **interpositus nucleus,** or the **fastigial nucleus.** At the synapses between Purkinje cells and the neurons of the deep cerebellar nuclei, the transmitter GABA is released, causing a hyperpolarization of the postsynaptic neurons. The effect of Purkinje cells in the deep cerebellar nuclei is therefore inhibitory.

The deep nuclei receive information from Purkinje cells located in different regions of the cerebellum. The fastigial nucleus receives information from Purkinje cells in the flocculonodular lobe and is therefore involved in the maintenance of balance and posture. The fastigial nucleus transmits its information to the vestibular nucleus, which in turn controls motor neurons that innervate axial muscles (see Fig. 9–23).

The interpositus nucleus receives information from Purkinje cells in the medial and intermediate regions of the anterior and posterior lobes of the cerebellum and is therefore involved in the control of limb muscles. The interpositus sends its information to the thalamus and red nucleus of the brain stem. The red nucleus in turn controls motor neurons innervating distal limb muscles.

The dentate receives information from Purkinje cells in the lateral regions of the anterior and posterior lobes of the cerebellum and is therefore involved in the planning and initiation of movement. The dentate transmits its information to the thalamus, which in turn forwards information to the motor and premotor cortex. The dentate also provides inputs to the red nucleus.

Since the outflow of information from the cerebellum must eventually pass through the deep cerebellar nuclei, these nuclei constitute the output of

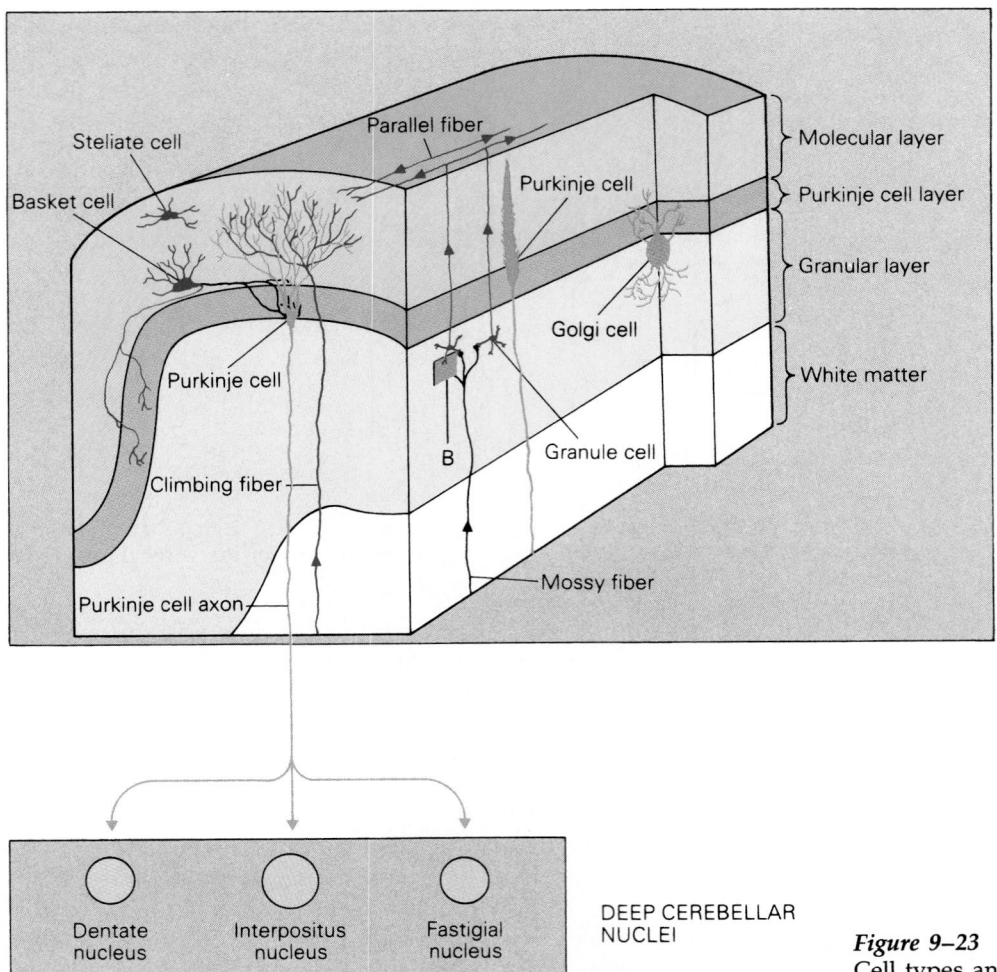

Figure 9–23
Cell types and circuits in the cerebellum.

the cerebellum. Research indicates that the input of information to the cerebellum is simultaneously transmitted to the deep cerebellar nuclei (Fig. 9–24). After the incoming information is processed by the cerebellum, it is transmitted by way of the Purkinje cells to the deep nuclei. The outputs from the deep nuclei are thus modulated by the inhibitory actions of the Purkinje cells in the comparison of intended and actual movements. Neurons of the deep nuclei then transmit information to correct for any deviation from the intended motion.

The inputs to the Purkinje cells of the cerebellum traverse two routes: the climbing fiber and parallel fiber pathways. Climbing fibers originate in the inferior olivary nucleus (Fig. 9–25*a*). Axons from neurons in this nucleus form synaptic connections around the soma and dendrites of Purkinje cells. Each Purkinje cell receives synaptic connections from only one climbing fiber. The activation of this input causes the Purkinje cell to generate complex patterns of action potentials (Fig. 9–25*b*).

The parallel fiber input to the Purkinje cell originates from **granule cells.** These fibers synapse with

Figure 9–24
Information input to the cerebellum is simultaneously transmitted to the deep cerebellar nuclei.

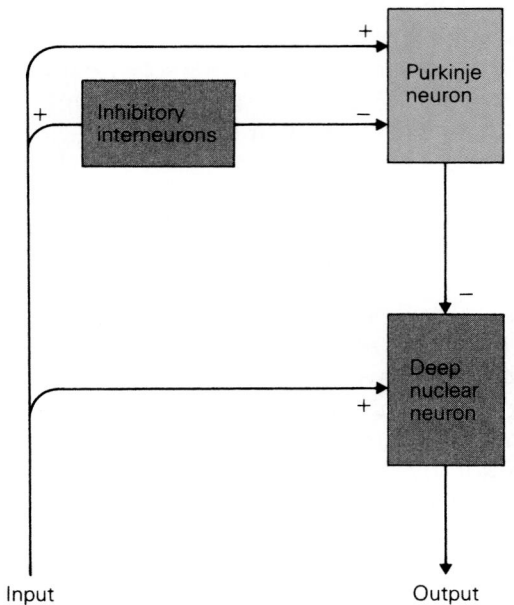

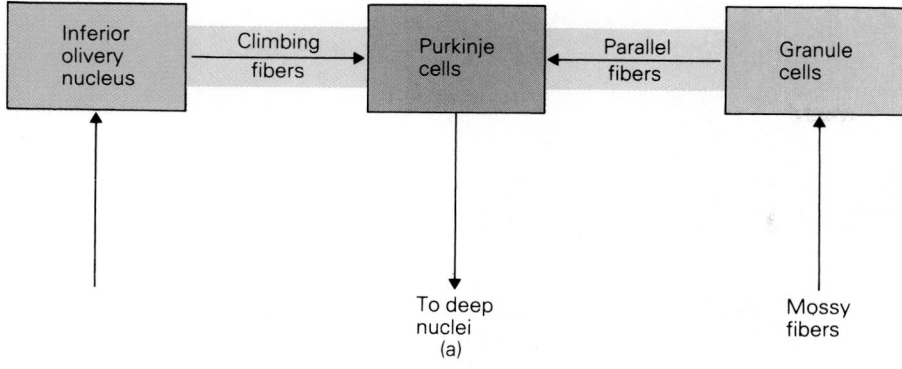

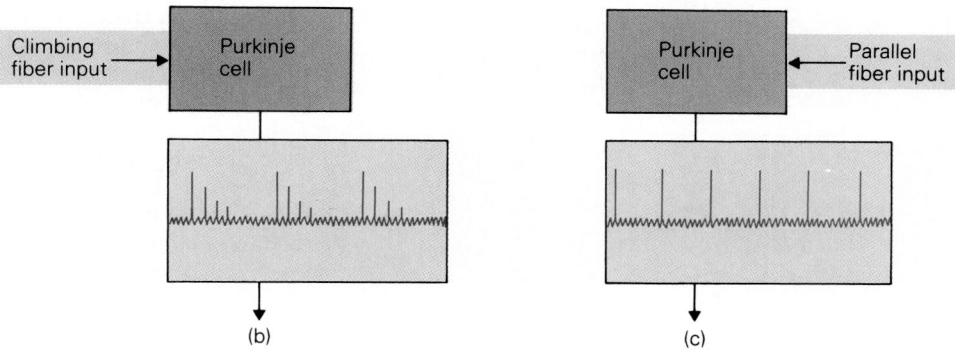

Figure 9–25
Climbing fibers to the cerebellum originate in the inferior olivary nucleus, while parallel fibers originate in the granule cells (*a*). Activation of climbing fibers generates complex pattern of action potentials (*b*), while activation of parallel fibers generates simple action potential patterns (*c*).

the dendrites of Purkinje cells. Each Purkinje cell receives inputs from many granule cells, and the activation of the parallel fibers causes the Purkinje cells to generate a simple pattern of action potentials.

Inputs to the inferior olivary nucleus come from pathways originating in supplemental and premotor areas of the cortex as well as from the spinal cord, while inputs to granule cells are derived from so called **mossy fibers.** These mossy fibers also originate in the neocortex as well as in the spinal cord. The mossy fibers and olivary climbing fibers, however, respond quite differently during movement. Sensory stimuli and voluntary movements enhance the activity of mossy fibers but have little effect on climbing fiber activity. Climbing fibers are therefore thought to modulate the responsiveness of Purkinje cells to mossy fiber and hence to parallel fiber inputs. This is seen most clearly during the learning of motor behaviors.

When monkeys are trained to maintain a lever in one position with their hands, Purkinje cells generate a simple pattern of action potentials (Fig. 9–

26). The pattern is characteristic of that produced by Purkinje cells activated by parallel fibers. When different weights are placed on the lever to change its position, the animal must compensate for this added force in order to maintain its original position. When more weight is added, Purkinje cells begin to produce complex (rather than simple) patterns of action potentials. These complex patterns of Purkinje cell action potentials are due to their activation by climbing fibers. As the animal learns to adapt to the new weight, the complex pattern of action potentials diminishes and the simple pattern re-emerges. Thus, the Purkinje cells are capable of changing their pattern of activity during the adaptation of new motor responses.

In addition to information concerning the force of muscle contraction, the activity of nerve cells in the cerebellum can encode the direction of reaching. Similar to that described for cells in the motor cortex, individual cells in the cerebellum become activated when a particular direction of movement is made. This preferential direction of movement gives rise to directional vectors for cells and population

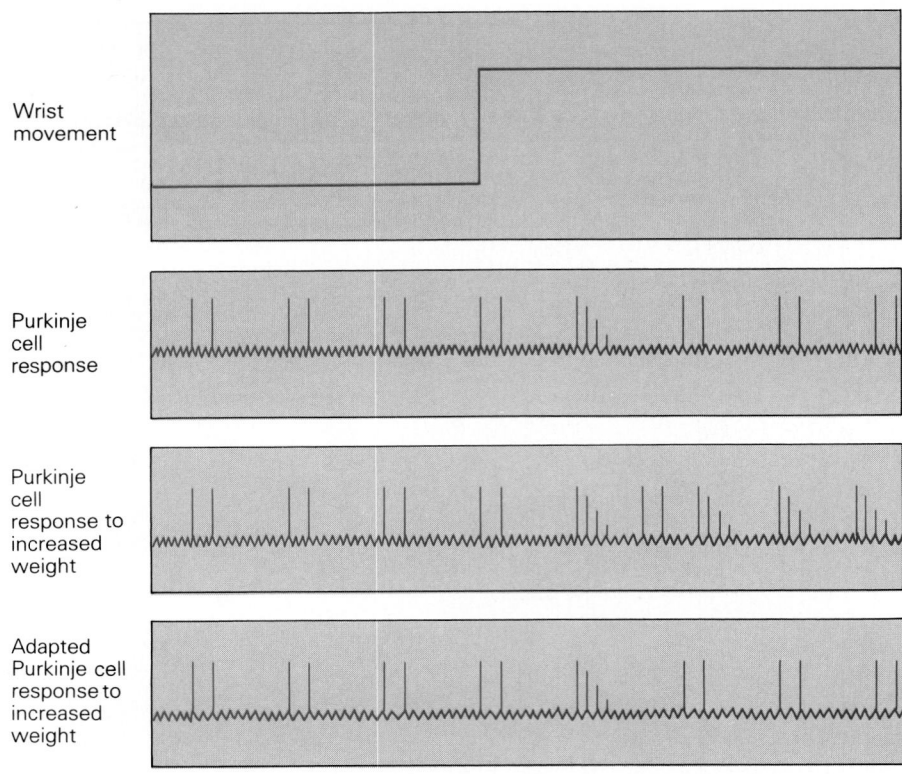

Wrist movement

Purkinje cell response

Purkinje cell response to increased weight

Adapted Purkinje cell response to increased weight

Figure 9–26
Purkinje cells can change their pattern of activity during adaptation to new motor responses. Complex action potential patterns show the involvement of climbing fibers in the learning of a new motor skill. As an animal adapts to the skill, simpler action potential patterns emerge.

vectors for cell clusters within the cerebellum. These cerebellar vectors therefore represent the neural code for the direction of limb movement.

The Basal Ganglia: Planning of Movements

Other major areas of the brain involved in motor control are the basal ganglia. The basal ganglia include the **caudate, putamen,** and **globus pallidus** (Fig. 9–27). The caudate and putamen receive inputs to the basal ganglia, while the globus pallidus provides the output. Inputs to the basal ganglia are from the entire neocortex, thalamus, and substantia nigra of the brain stem. The primary input, however, is from the neocortex. The extensiveness of this input suggests that the basal ganglia are involved in other functions besides motor activities. In fact, diseases of the basal ganglia often produce cognitive abnormalities.

The major output of the basal ganglia is to the prefrontal and premotor cortices by way of the thalamus (Fig. 9–27*c*). Through this pathway, the basal ganglia can modulate the descending components of the motor system. The basal ganglia also have outputs to the substantia nigra. The interconnections between the basal ganglia and substantia nigra play a prominent role in diseases of the motor system. Nerve fibers from the substantia nigra that terminate in the basal ganglia release dopamine as the neurotransmitter. The degeneration of these dopamine fibers is responsible for the motor disorder called Parkinson's disease. Patients with parkinsonism exhibit rigidity, a rhythmic tremor at rest, and difficulty in initiating movement. These symptoms can be alleviated, in part, with drugs such as L-dopa, which act as precursors to increase the synthesis of dopamine.

Vestibular Control of Equilibrium and Balance

The vestibular system aids in maintaining the body's balance by detecting the position and motion of the head in space. The sensory organs of the vestibular system are the **semicircular canals** (Fig. 9–28). Three semicircular canals make up the vestibular apparatus: the **superior, inferior,** and **horizontal semicircular canals.** Each canal is situated perpendicular to the other two. With this arrangement, each semicircular canal detects the angular acceleration in one of three planes in space. The vestibular apparatus also consists of the **utricle** and **saccule**

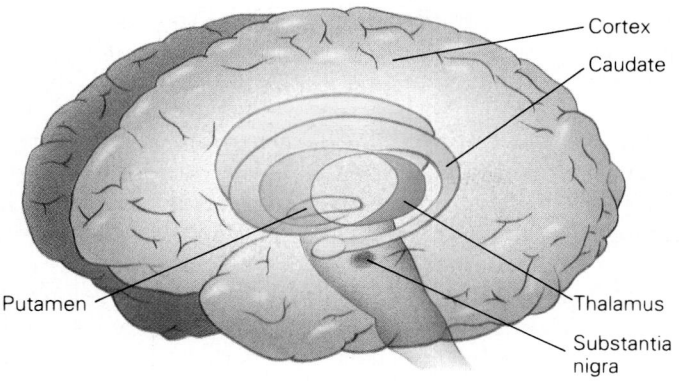

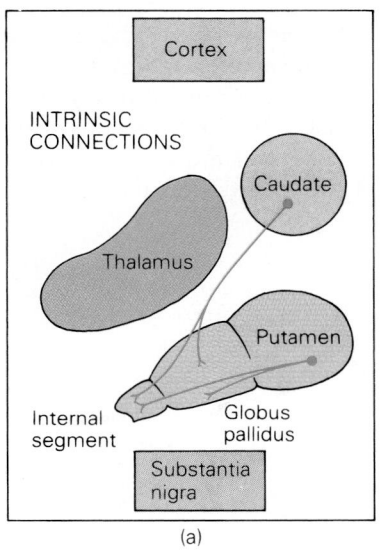
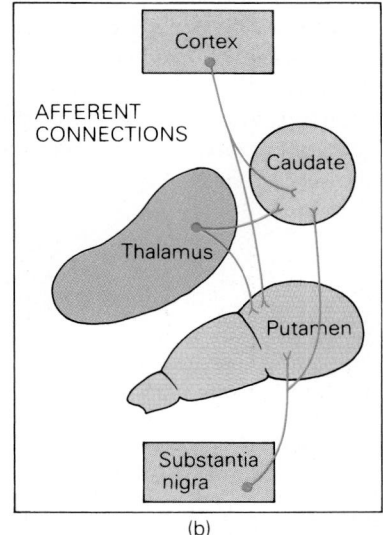
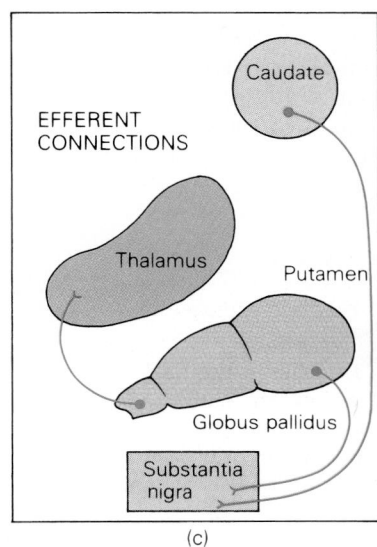

Figure 9–27
The caudate and putamen of the basal ganglia receive inputs, while the globus pallidus provides the output. The basal ganglia receive inputs from the neocortex, thalamus, and substantia nigra.

Figure 9–28
The vestibular apparatus consists of the semicircular canals, including the inferior, superior, and horizontal canals.

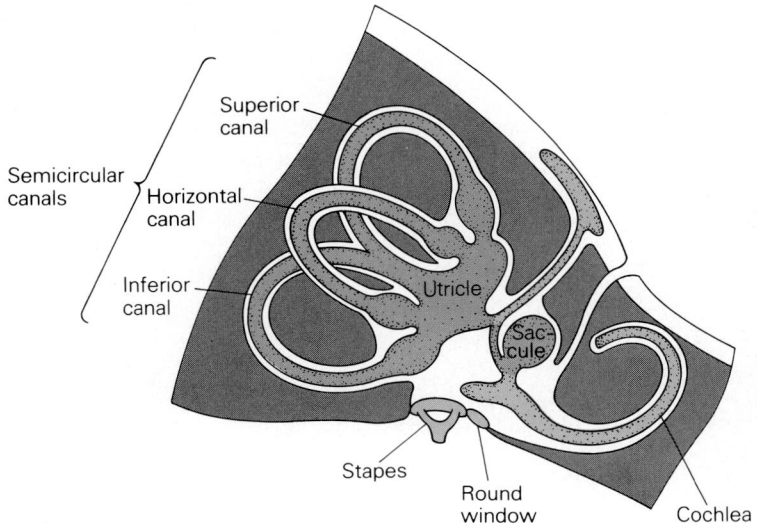

(see Fig. 9–28). These are so-called **otolithic organs;** they detect linear acceleration.

The semicircular canals and otolithic organs are filled with the same endolymph fluid that surrounds the hair cells of the cochlea. As does the cochlea, the semicircular canals and otolithic organs have hair cells responsible for the transduction of sensory stimuli. The appropriate stimulus for these vestibular hair cells is the acceleration of the head. As the head begins to move, the inertia of the endolymph fluid causes the stereocilia to bend (Fig. 9–29). This bending of the stereocilia causes the hair cell either to depolarize or to hyperpolarize.

The type of membrane potential change in the hair cell depends on the direction in which the stereocilia bend. The stereocilia of hair cells in the vestibular apparatus are aligned according to size, with the smallest cilium placed at one end of the group and the largest cilium, called the *kinocilium,* at the opposite end (see Fig. 9–29). If the stereocilia are bent toward the kinocilium, the hair cell will depolarize. If the stereocilia are bent toward the smallest cilium, the hair cell will hyperpolarize.

These hair cells, like those in the cochlear nucleus, are not capable of generating action potentials. They do, however, excite or inhibit the **vestibular nerve fibers** that innervate them. As the hair cell depolarizes, it releases a neurotransmitter that excites the innervating vestibular nerve (see Fig. 9–29). The hyperpolarization of a hair cell causes the vestibular nerve to produce fewer action potentials. This may be due to a decrease in the amount of an excitatory transmitter that is tonically being released from the hair cell.

How do these characteristics of vestibular hair cells allow them to detect the accelerated movement of the head? The hair cells are arranged in a polarized manner on both sides of the brain. In the horizontal semicircular canals, for example, the stereocilia on each hair cell are positioned so that the

Figure 9–29
The cilia in the hair cells of the vestibular apparatus are arranged in order of increasing length, with the largest cilium, the kinocilium, at one end. When the cilia are bent in the direction of the kinocilium, the hair cell membrane becomes depolarized; when the kinocilium bends the other cilia toward the smallest cilium, the membrane becomes hyperpolarized.

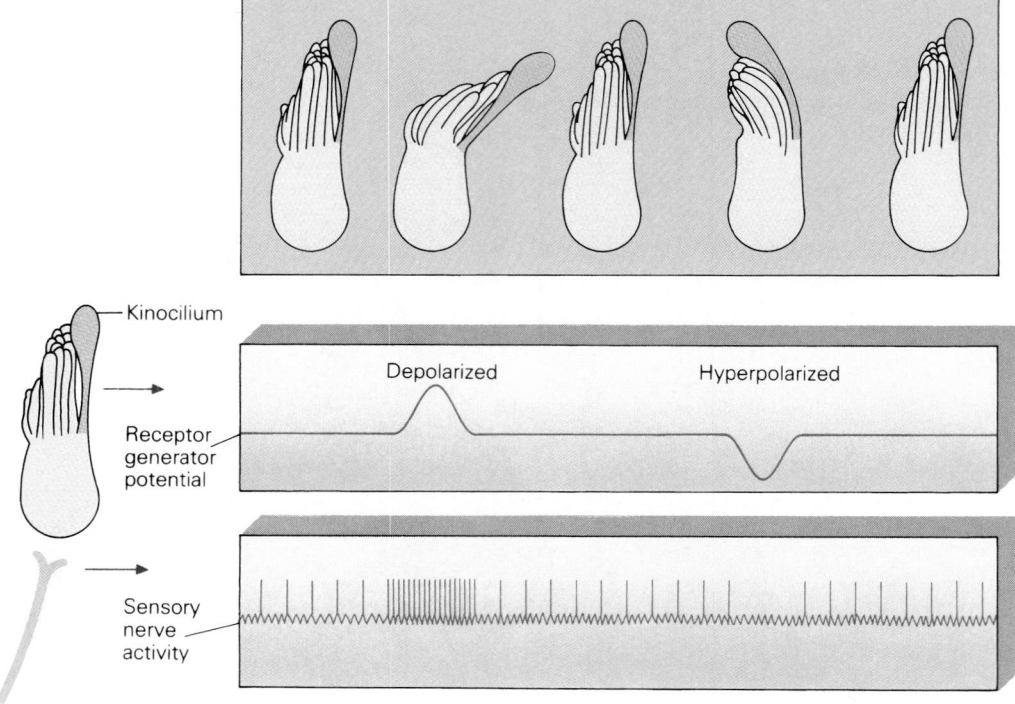

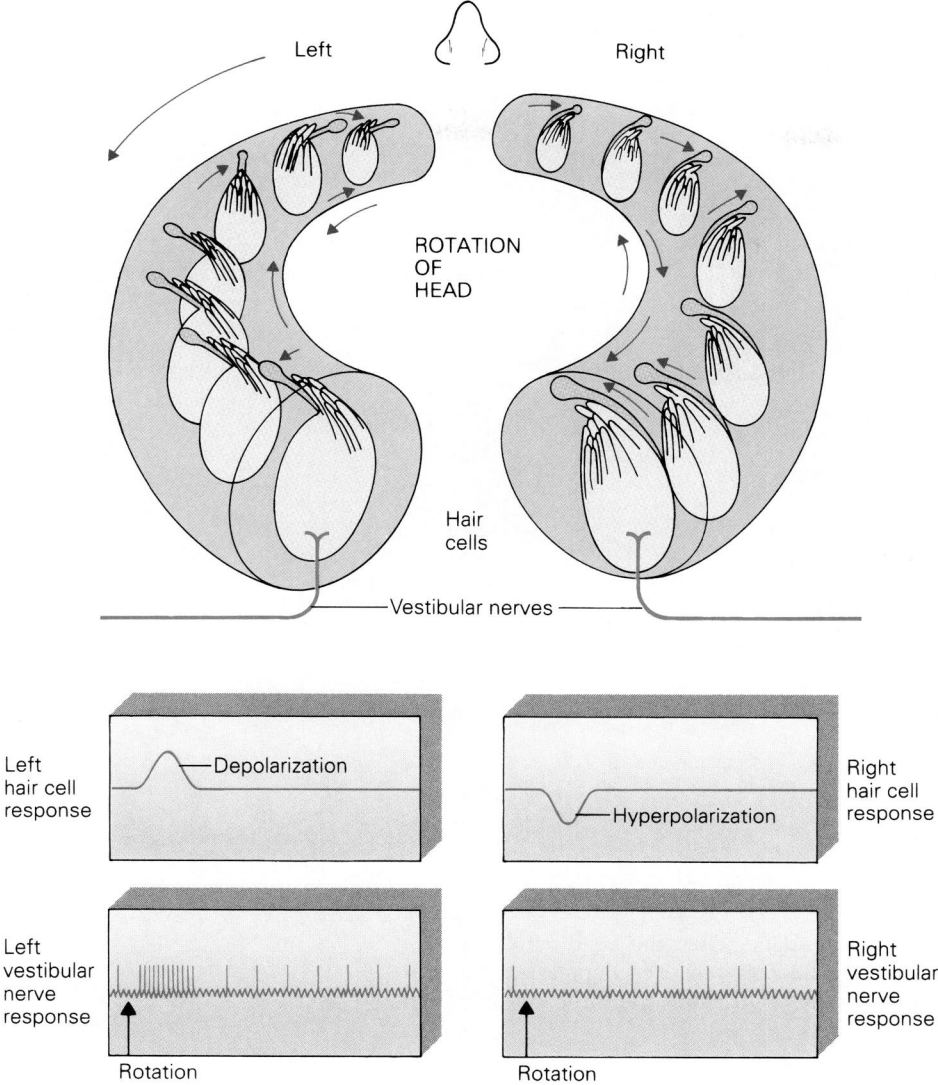

Left

Right

ROTATION
OF
HEAD

Hair
cells

Vestibular nerves

Left
hair cell
response

Depolarization

Right
hair cell
response

Hyperpolarization

Left
vestibular
nerve
response

Rotation

Right
vestibular
nerve
response

Rotation

Figure 9–30
In the horizontal semicircular canal, the kinocilia of the hair cells are oriented toward the nose. When the head rotates in a counterclockwise direction, the endolymph in effect flows in the opposite direction relative to the cilia. On the left side of the brain, the hair cells depolarize because the fluid motion causes the cilia to bend toward the kinocilium, while the opposite effect takes place on the right side. The vestibular nerves detect the depolarization and hyperpolarization and thereby inform the brain about the direction of rotation.

kinocilium is nearest to the nose or front of the head, while the smallest cilium is nearest the back of the head (Fig. 9–30). As the head rotates from right to left (counterclockwise), the inertia of the endolymph makes it appear to be moving from left to right (clockwise) with respect to the stereocilia. The fluid's relative motion causes the stereocilia on the left side of the brain to depolarize the hair cells, while the stereocilia on the right side of the brain hyperpolarize the hair cells. Correspondingly, the vestibular nerves on the left side of the brain increase their rate of action potential generation, while the nerves on the right side of the brain decrease their rates of action potential generation. This

information is thus transmitted to inform the brain that the head is rotating clockwise.

The information from the vestibular nerves is transmitted to the vestibular nucleus located within the brain stem. From there, it is sent down to the spinal cord to act on nerve cells that regulate movement, and up to the cerebellum and somatosensory cortex. The cerebellum and somatosensory cortex use this information to coordinate muscular activity in order to maintain balance.

In addition to the transduction of information about the movement of the head, the vestibular apparatus also transduces information about the stationary position of the head in space with respect

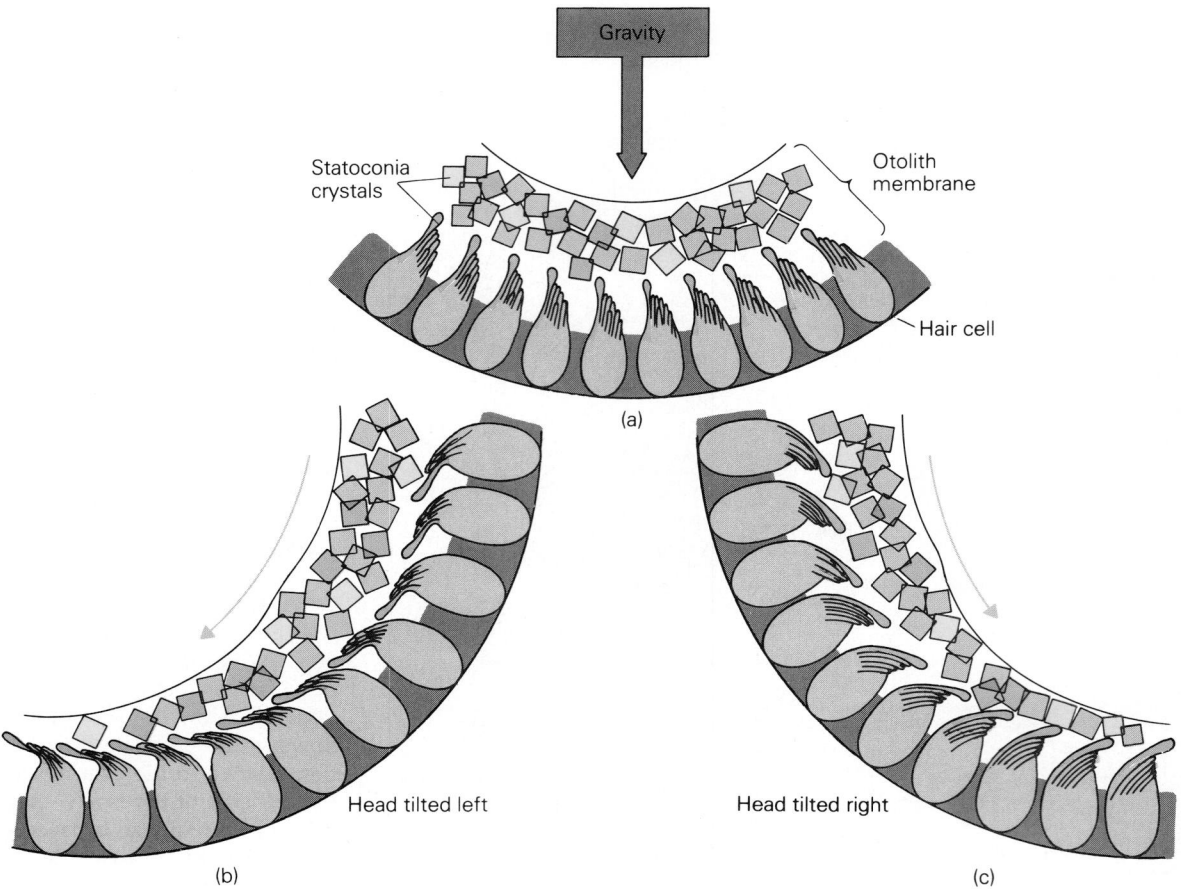

Figure 9–31
The stereocilia of the hair cells of the utricle bend when the head bends to the right or left under the weight of statoconia crystals. The depolarization or hyperpolarization of hair cells informs the brain about the direction of bending.

to gravity. The stereocilia of the hair cells in the utricle are embedded in an **otolithic membrane** that contains calcium carbonate crystals called **statoconia** (Fig. 9–31). The statoconia give the otolithic membrane mass, so that when the head is tilted to the left or right, the gravitational force causes the weighted otolithic membrane to bend the stereocilia. The stereocilia are arranged on the hair cells so that when the head is tilted to the left, the hair cells depolarize, and when the head is tilted to the right, the hair cells hyperpolarize.

Disorders of the Motor System

Because the different parts of the motor system have such distinctive roles, lesions in these areas result in characteristic movement disorders. These lesions often occur because of trauma, stroke, or tumors.

Lesions of the Corticospinal Tract

Upper-motor-neuron lesions result from damage to descending pyramidal cell fibers (Fig. 9–32). These lesions are characterized by paralysis on the side of the body opposite the site of the lesion, increase in muscle tone, hyperactive reflexes, and the extension of the big toe and fanning of the other toes in response to stroking the bottom of the foot. This latter sign is called the **Babinski sign.** Muscle atrophy associated with this type of lesion is not prominent.

Corticospinal tract lesions occur within the spinal cord and disrupt the inputs to motor neurons. These lesions are characterized by loss of strength and movement in muscle groups, loss of strength in voluntary muscle contraction, and the Babinski sign. These effects are ipsilateral to (occurring on the same side of the body as) the side of the lesion.

Lower-motor-neuron lesions can occur with spinal cord injuries that directly affect the motor

Huntington's Disease

Huntington's disease (HD) is a neurological disorder characterized by depression, dementia, psychiatric problems and a progressive loss of motor control. In 1872 George Huntington described the motor components of this disorder as an "irregular and spasmodic action of certain muscles, as of the face, arms, etc. These movements gradually increase, when muscles hitherto unaffected take on the spasmotic action, until every muscle of the body becomes affected." Ultimately, the "patient presents a spectacle which is anything but pleasing to witness."

In addition to the motor abnormalities, Huntington made key observations regarding its hereditary nature. HD is passed from generation to generation by an autosomal dominant mode of genetic transmission. Both males and females are equally susceptible and each offspring of an affected parent has a 50% chance of inheriting the gene. Symptoms of HD can begin at any time; the average age of onset, however, is about 40 years of age. At present there is no treatment that can cure the disease or slow its course of action. Symptoms gradually worsen and death occurs an average of 16 years after onset.

The cause of the abnormalities seen in HD is the progressive loss of neurons in the caudate and putamen of the neostriatum. These neurons receive nerve fiber inputs from the neocortex. They, in turn, innervate the globus pallidus and use GABA as their neurotransmitter. In HD, therefore, a major inhibitory input to the globus pallidus is missing.

The reason for the degeneration of neurons in the caudate and putamen is still unknown. One theory speculates that fiber inputs from neocortex, which release the transmitter glutamate onto neurons in the caudate and putamen, may be the source of the problem. Glutamate is an excitatory transmitter that when released in excess can damage postsynaptic neurons. Exactly how the HD gene may be involved in this process remains to be determined.

Although the precise role of the HD gene in the destruction of caudate and putamen neurons is not known, the recent application of molecular tech-

Pedigree of an HD family.

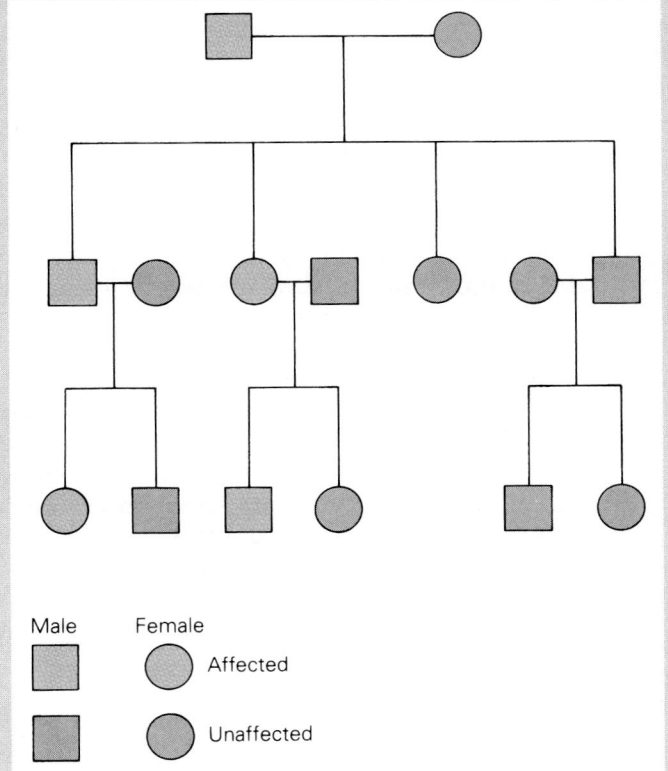

Male　Female

■ ○ Affected

■ ○ Unaffected

niques has determined the chromosomal location of the gene. Using genetic linkage analysis and DNA markers, James Gusella, Michael Conneally, and their collaborators have localized the HD gene to the short arm of chromosome 4. In genetic linkage analysis, two traits are said to be linked when they are inherited together by several members of a family. This linkage typically occurs when the two genes lie close together on the same chromosome. The linkage of two genes is considered proven when the odds in favor of linkage are greater than 1000:1. These odds are generally expressed as the log of the odds (LOD), so that odds of 1000:1 give an LOD score of 3.0.

Human DNA has many areas in which nucleotide sequences may vary from person to person. These polymorphic areas can be used as a source of genetic markers and are detected by restriction enzymes that cut specific nucleotide sequences. The resulting restriction fragment length polymorphisms **(RFLPs)** can be used as markers not only for the HD gene but for other genetic disorders as well.

Using an RFLP marker for chromosome 4, it was found that the marker was tightly linked to the HD gene with an LOD score of 6, indicating a million to one odds in favor of linkage. Ultimately, by using other markers for chromosome 4, the exact site of the HD gene will be located and hence its nucleotide sequence. Once the nucleotide sequence has been determined, the HD gene product can also be deduced. Knowledge of the gene product will then open the door to experiments addressing the issue of how the product can lead to the degeneration of caudate and putamen neurons.

Since the genetic markers used to localize the HD gene and the gene itself are closely localized, the markers can also be used for diagnostic purposes. Present studies indicate that markers for the HD gene can be used to determine with 99% accuracy whether or not an individual or a fetus at risk will eventually develop the disease.

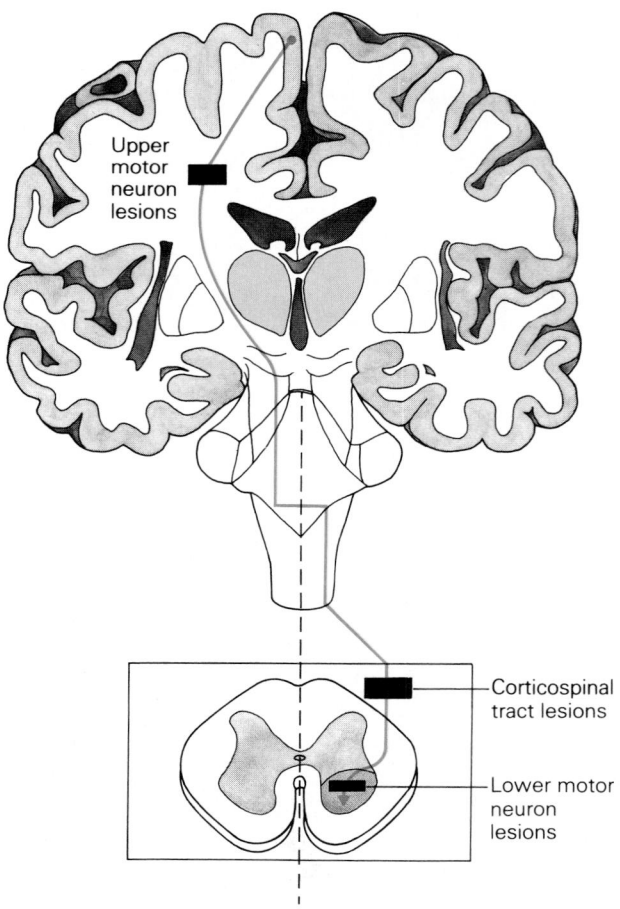

neurons rather than their inputs. These lesions are characterized by ipsilateral hypoactive reflexes, paralysis limited to specific groups of muscles, and flaccid (limp) muscles with prominent atrophy (decrease in size).

Lesions of the Cerebellum

Lesions of the cerebellum cause ipsilateral disturbances. Lesions occurring in the lateral cerebellum result in a lack of limb coordination. Lesions occurring in the vermis produce ataxia (loss of coordination); those localized in the flocculonodular lobe produce a disturbance in equilibrium as well as ataxia. Diseases of the cerebellum are also characterized by inaccurate range and direction of movement, diminished resistance to passive movement, inability to rapidly move a limb and to stop sharply, and tremor with movements. Interestingly, chronic alcoholism also affects the cerebellum. Excessive, long-term consumption of alcohol leads to the degeneration of Purkinje cells. As a consequence, chronic alcoholics often exhibit a wide base ataxic gait.

Figure 9–32
Lesions of the motor system result in characteristic motor disorders.

Parkinson's Disease and Brain Cell Transplants

Parkinson's disease is a disorder of the motor system characterized by slowness of movement, tremor of the fingers and hands (most prominent at rest), absence of facial expression, slowness of speech, stooped body posture, and a shuffling gait. In extreme cases, patients are bedridden and incapable of feeding and caring for themselves.

This disorder frequently appears in persons between the ages of 50 and 65 years. Currently in the United States, approximately 1 million patients are afflicted with parkinsonism, with 50,000 new cases reported each year.

Examination of brain tissue from Parkinson's patients has revealed a loss of dopamine-producing neurons in an area called the substantia nigra, located in the brain stem nucleus. These neurons send axons to form connections with other neurons in the neostriatum, an area of brain involved in locomotor activity. The resulting depletion of dopamine in the neostriatum appears to be the major cause of this disorder.

The reason for the degeneration of dopamine cells in the substantia nigra is unknown. A form of juvenile parkinsonism is thought to be genetically inherited. Evidence for a genetic component for adult parkinsonism is more controversial. Epidemiological studies reveal a strong correlation between parkinsonism and agricultural communities. It is thought that pesticides may be a contributing factor in the development of Parkinson's disease.

One form of treatment for Parkinson's disease—the administration of drugs such as L-dopa that become precursors for the synthesis of dopamine—has been found to be an effective form of therapy for many patients. Some patients, however, do not respond to drug therapies, or experience debilitating side effects such as hallucinations.

In animal models of Parkinson's disease, scientists have taken a new approach in which nerve cells that produce dopamine are transplanted directly into the neostriatum. These transplanted neurons form synaptic connections with those of the host. In addition, they are capable of correcting the locomotor abnormalities associated with Parkinson's disease.

The sources of these dopamine neurons in laboratory animal experiments are fetuses of the same species. Because of moral and ethical issues concerning the use of human fetal tissue, however, medical researchers currently are exploring the possible use of alternative sources of dopamine cells for transplantation. One possibility is the use of a patient's own cells from the adrenal medulla. Preliminary studies in parkinsonian patients with adrenal medulla transplants have resulted in impressive results at several medical centers. These studies await replication by other centers, but offer the promise that such an approach may work for parkinsonian patients who do not respond to drug therapies.

Dopamine neurons in the substantia nigra. (*a*) Substantia nigra neurons in a normal mouse. (*b*) Substantia nigra in a mouse model of Parkinson's disease. Note degeneration of dopamine neurons. (Courtesy of L. C. Triarhou, B. Ghetti, and W. C. Low.)

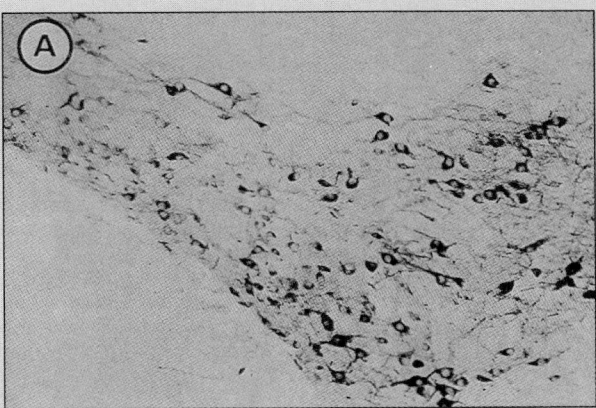

SUMMARY

Muscle: Controlled by Nerves

The motor system controls locomotion and fine movement, and maintains body posture and equilibrium, by acting on motor neurons in the spinal cord that innervate skeletal muscles. The components of the motor system (the spinal cord, brain stem, and motor cortex) interact in both hierarchical (serial) and parallel fashion.

A motor neuron and the muscle fibers that it innervates (by forming myoneural junctions) constitute a motor unit. A motor neuron causes a muscle to contract or relax.

The Spinal Cord in the Motor Response System

Motor neurons have cell bodies located in the gray matter of the spinal cord, in the ventral horn.

The spinal cord contains interneurons, which play a role in coordinating the responses of antagonistic muscles.

Walking movements are created by locomotor generators in the spinal cord. These locomotor programs are driven by the locomotor command center in the brain stem and can be modified by sensory feedback.

Sensory fibers within skeletal muscles detect muscle length, force, and velocity of muscle shortening.

Reflex actions are initiated when sensory information is sent from muscle, tendon, or skin receptors to the spinal cord, which causes a reflexive movement in the appropriate muscles. Reflexes include the stretch, myotatic, and withdrawal reflexes.

The Brain in the Motor Response System

Complex motor behaviors require inputs from the motor cortex, supplemental and premotor cortices, cerebellum, and basal ganglia. The maintenance of posture and balance is carried out by brain stem inputs to the spinal cord. Three pathways provide motor input to the spinal cord: the ventromedial, lateral reticulospinal, and rubrospinal pathways.

The motor cortex is responsible for fine movements. It is a central motor command center that instructs the motor neurons to activate their respective skeletal muscles. The motor cortex is organized in a somatotopic fashion similar to the somatic sensory cortex. Regions of the body containing small motor units represent the largest portion of the motor cortex. The cortex contains neuronal populations with summated activities (population vectors) responsible for the direction of limb movement.

The supplemental and premotor cortices are involved in the planning of movements; they transmit this information to the motor cortex.

The posterior cortex is involved in the processing of sensory information leading to purposeful movement. This cortex receives somatic and visual sensory information and transmits this information to the supplemental and premotor areas.

The cerebellum has three motor functions: the initiation and planning of movement, the coordination of limb movement, and the maintenance of posture and equilibrium. The cerebellum carries out these functions in concert with the basal ganglia. The basal ganglia are also involved in the planning of movements. They have inputs from all neocortical areas in addition to the motor cortex and are involved in functions such as cognition and movement.

Vestibular Control of Equilibrium and Balance

The vestibular system provides sensory information to maintain body posture and balance. The semicircular canals detect changes in angular movement, while the otolithic organs detect changes in linear movement. Sensory information from the semicircular canals and otoliths is transmitted by the vestibular nerve to the vestibular nucleus of the brain stem and then to motor neurons in the spinal cord, which activate the appropriate muscles to maintain balance.

Disorders of the Motor System

Lesions in different areas of the motor system produce unique movement disorders. Upper-motor-neuron lesions of the corticospinal tract result in bodily paralysis and increased muscle tone, with hyperactive reflexes and the extension of the big toe on the side opposite to the lesion. Corticospinal lesions in the spinal cord produce loss of strength and movement in specific muscle groups, loss of strength in voluntary contraction, and extension of the big toe on the same side as the lesion. Lower-motor-neuron lesions of the spinal cord result in paralysis restricted to specific groups of muscles, hypoactive reflexes, and flaccid muscles.

Cerebellar lesions are characterized by a lack of coordination and equilibrium.

REVIEW QUESTIONS

Identify Each with a Term

1. The term used to describe the type of muscle that decreases the angle of a joint.

2. The term used to describe a functional group of nerve cells in the motor cortex that activate the same skeletal muscles.

3. The type of sensory receptor that detects changes in muscle length and velocity of muscle shortening.

4. The sensory organ that helps maintain balance and equilibrium by sensing when the body undergoes angular acceleration.

Define Each Term

5. extensor

6. alpha motor neuron

7. Babinski sign

8. utricle

Choose the Correct Answer

9. If fibers of gamma motor neurons are cut, the contraction of extrafusal muscle fibers will cause:

   a. a decrease in type Ia nerve fiber activity

   b. a decrease in type Ib nerve fiber activity

   c. an increase in type II nerve fiber activity

   d. both (a) and (b)

10. Which of the following statements about the cerebellum is *not* true?

   a. Purkinje cells of the cerebellum release GABA as their neurotransmitter.

   b. Lesions on one side of the cerebellum cause a contralateral disturbance.

   c. Nerve fibers from the cerebellar dentate nucleus project to thalamus on the contralateral side of the body.

   d. Lesions of Purkinje cells near the midline of the cerebellum result in a disturbance in balance and equilibrium.

11. The alpha motor neurons in the spinal cord responsible for maintaining posture are located in the:

   a. medial aspect of the dorsal horn

   b. lateral aspect of the dorsal horn

   c. medial aspect of the ventral horn

   d. lateral aspect of the ventral horn

12. An example of a large motor unit is the innervation of:

   a. muscles of the thumb

   b. muscles responsible for the movement of the eyes

   c. axial muscles controlling posture

   d. muscles responsible for speech

Answer Each Question in One or Two Sentences

13. How do alpha motor neurons increase the force of muscle contraction?

14. What are the functional properties of motor neurons that innervate fast and slow muscles?

15. What is the role of gamma motor neurons in regulating the sensitivity of muscle spindle activity?

16. Compare and contrast motor units versus motor neuron pools.

SUGGESTED READING

Asanuma, H. "Cerebral cortical control of movement." *Physiologist*, 16:143–166, 1973.

Evarts, E.V. "Pyramidal tract activity associated with a conditioned hand movement in the monkey." *Journal of Neurophysiology*, 29:1011–1027, 1966.

Georgopoulos, A.P., Kettner, R.E., and Schwartz, A.B. "Primate motor cortex and free arm movements to visual targets in three-dimensional space. II. Coding of the direction of movement by a neuronal population." *Journal of Neuroscience*, 8:2928–2937, 1988.

Grillner, S., and Shik, M.L. "On the descending control of the lumbosacral spinal cord from the 'mesencephalic locomotor region.'" *Acta Physiological Scandinavia*, 87:320–333, 1973.

Grillner, S., and Zangger, P. "How detailed is the central pattern generation for locomotion?" *Brain Research*, 88:367–371, 1975.

Fortier, P.A., Kalaska, J.F., and Smith, A.M. "Cerebellar neuronal activity related to whole-arm reaching movements in the monkey." *Journal of Neurophysiology*, 62:198–211, 1989.

Liddell, E.G.T., and Sherrington, C. "Recruitment and some other features of reflex inhibition." *Proceedings of the Royal Society of London*, 97:488–518, 1925.

Lundberg, A. "Integration in a propriospinal motor centre controlling the forelimb in the cat." In *Integration in the Nervous System*, edited by H. Asanuma and V.J. Wellson, 47–64. Tokyo: Igaku-Shoin, 1979.

Matthews, B.H.C. "Nerve endings in mammalian muscle." *Journal of Physiology*, 78:1–53, 1933.

Mitchell, S.J., Richardson, R.T., Baker, F.H., and DeLong, M.R. "The primate globus pallidus: Neuronal activity related to direction of movement." *Experimental Brain Research*, 68:491–505, 1987.

Mountcastle, V.B., Lynch, J.C., Georgopoulos, A., Sakata, H., and Acuna, C. "Posterior parietal association cortex of the monkey: Command functions for operations within extrapersonal space." *Journal of Neurophysiology*, 38:871–908, 1975.

Sherrington, C. *The Integrative Action of the Nervous System*, 2nd ed. New Haven: Yale University Press, 1947.

KEY TERMS

Babinski sign (p. 364)
basal ganglia (p. 360)
caudate (p. 360)
cerebellum (p. 357)
extensor (p. 341)
flexor (p. 341)

hierarchical motor
  organization (p. 339)
motor cortex (p. 350)
motor system (p. 339)
muscle (p. 339)
muscle fiber (p. 341)

muscle spindle (p. 345)
myoneural junction (p. 342)
reflex (p. 347)
semicircular canals (p. 360)
skeletal muscle (p. 341)
smooth muscle (p. 339)

striate muscle (p. 340)
vestibular system (p. 360)
vestibulospinal tract (p. 349)

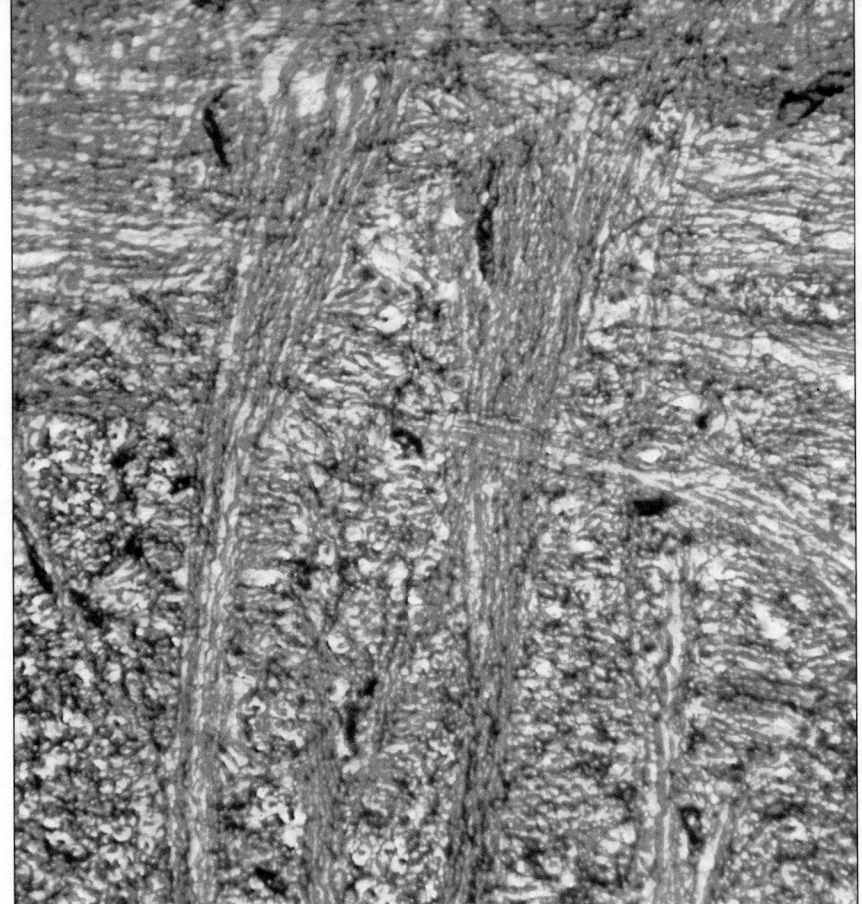

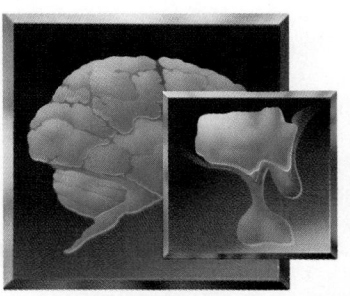

The Autonomic Nervous System

CHAPTER OUTLINE

The **autonomic nervous system (ANS)** coordinates bodily functions needed for survival. It senses the need for food and activates the digestive processes; it senses the need for water and activates processes for the retention and consumption of fluids. The ANS aids in removing waste products from the body and it prepares the body for the stress of life-threatening conditions.

The ANS is also a regulatory system that maintains the environment necessary for cells to function properly. It maintains the body's temperature at 37°C, mean blood pressure at 90 mm Hg, pH at 7.4, and blood plasma osmolarity at 300 mOsm. These are the optimal conditions of the **internal environment** for the functioning of the organs and their cells.

The actions of the ANS are not under direct voluntary control. Unlike the **somatic nervous system,** with which we can consciously command skeletal muscles to move (e.g., the muscles in the fingers of the hand to pick up a pencil), the actions of the ANS are all subconscious. It is difficult to consciously will the heart to beat faster or command the blood pressure to increase. Instead, autonomic processes transmit sensory information from **"visceral" organs** to the central ANS, which in turn sends instructions to the smooth muscles and other cells of those organs, producing an appropriate automatic response.

This chapter examines the functional organization of the ANS, and the ways in which the ANS controls other organ systems to supply the necessary nutrients to the body, to remove waste products of metabolism, to coordinate the response to stress, and to regulate the internal environment.

Organization of the Autonomic Nervous System

The ANS has both central and peripheral components. The central components of the ANS are the **hypothalamus,** the **brain stem,** and the **spinal cord** (Fig. 10–1). The peripheral components consist of the nerves that innervate the organs of the body, classified as either parasympathetic or sympathetic nerves. Autonomic nerve fibers include both afferent (sensory) and efferent (motor) neurons.

The Peripheral Autonomic Nervous System

The peripheral ANS is functionally composed of the **sympathetic nervous system** and the **parasympathetic nervous system.** The sympathetic division of the ANS coordinates the body's response to stress, while the parasympathetic division coordinates the body's more "vegetative" activities such as digestion. The sympathetic and parasympathetic nervous systems not only have distinct anatomical differences, but also release different neurotransmitters at their target sites. The sympathetic nervous system simultaneously activates more organ systems than its parasympathetic counterpart. It plays an important role during stress responses when there is a need to coordinate increases in heart rate, blood pressure, and respiration rate. By contrast, in parasympathetic responses, different organ systems are activated more independently.

Nerve fibers of the sympathetic nervous system emerge from the spinal cord at the thoracic and lumbar levels (Fig. 10–1b). The cell bodies of these fibers are located in the **interomedial lateral nuclei (IML)** of the spinal cord (Fig. 10–2). These cells are called **preganglionic nerve cells** and have short axons that innervate cells in **ganglia** located near the spinal cord. The cells within the ganglia are **postganglionic nerve cells.** The postganglionic cells have long axons and synapse with cells in other organ systems.

Fibers from the parasympathetic nervous system exit from the brain stem and also leave the spinal cord at the level of the sacrum (Fig. 10–1b). These preganglionic fibers have long myelinated and unmyelinated axons and innervate postganglionic nerve cells clustered near or in the target organ. As a consequence, the postganglionic parasympathetic nerve cells have very short axons.

As Figure 10–1b illustrates, most organs are innervated by both sympathetic and parasympathetic nerve fibers. The effects of each system on the target organ, however, are quite different. One system increases the activity of the target organ, while the opposing system decreases its activity. For example, the sympathetic innervation of the heart increases the heart rate, while parasympathetic innervation causes the heart rate to decrease. In contrast, sympathetic innervation of the small intestines decreases intestinal motility, while parasympathetic

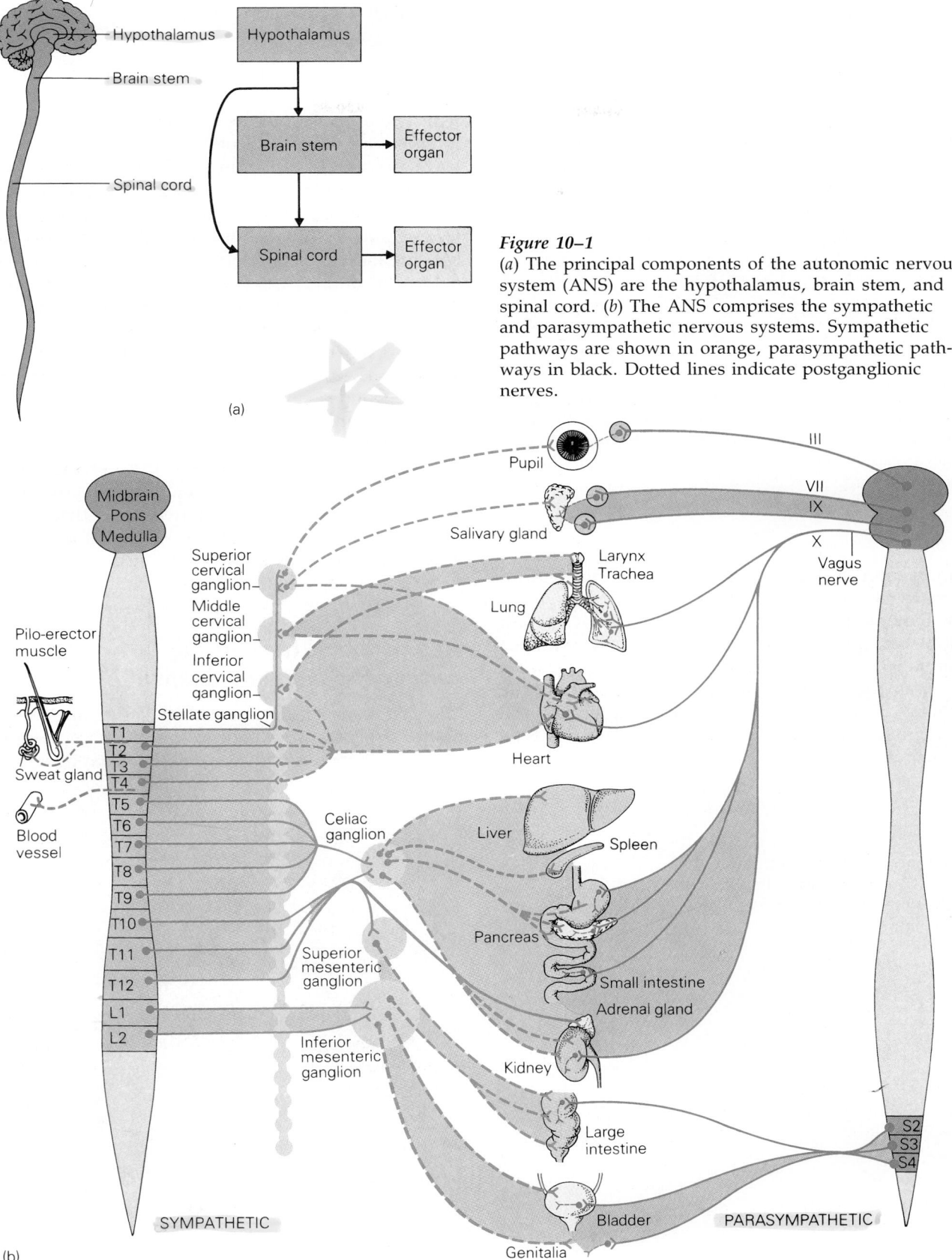

Figure 10–1

(*a*) The principal components of the autonomic nervous system (ANS) are the hypothalamus, brain stem, and spinal cord. (*b*) The ANS comprises the sympathetic and parasympathetic nervous systems. Sympathetic pathways are shown in orange, parasympathetic pathways in black. Dotted lines indicate postganglionic nerves.

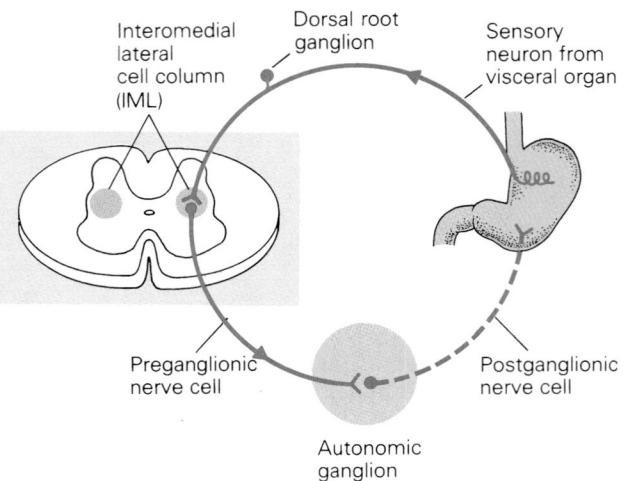

Figure 10–2
Nerve fibers of the sympathetic nervous system originate in the interomedial lateral nuclei (IML) of the spinal cord. These preganglionic nerve cells innervate postganglionic nerve cells located within ganglia near the spinal cord.

innervation causes it to increase. From this description, the actions of the sympathetic and parasympathetic nervous systems seem to oppose each other. However, the sympathetic and parasympathetic inputs to a target organ are not active at the same time; as one system is activated, the activity of the other is diminished. Later in this chapter, we will examine the mechanisms that coordinate this type of activity.

The Central Autonomic Nervous System

The organization of the ANS in the spinal cord is somewhat similar to that of the somatic nervous system. The preganglionic nerve cells are located in the interomedial lateral cell column and have fibers that leave the spinal cord from the ventral roots, while sensory fibers that carry information from the target organ enter the spinal cord through the dorsal roots (see Fig. 10–2). The cell bodies of these sensory fibers are located in the dorsal root ganglia. This spinal organization is identical for the sympathetic and the parasympathetic components of the autonomic nervous system. Sympathetic fibers, however, emerge from the thoracic and lumbar segments of the spinal cord, while parasympathetic fibers emerge from sacral segments of the cord.

The organization of the autonomic nervous system in the brain stem is different from that in the spinal cord. The autonomic sensory fibers coming into the brain stem are segregated from the somatic sensory fibers, which enter more laterally than autonomic sensory fibers (Fig. 10–3). On the other hand, autonomic efferent nerves leave the brain stem in more lateral regions than somatic efferents.

Neurotransmitters in the Autonomic Nervous System

The two major neurotransmitters in the autonomic nervous system are **acetylcholine** and **norepinephrine**. Acetylcholine is released at the synaptic termi-

Figure 10–3
Autonomic sensory and motor nerves enter and leave the brain stem from different regions than somatic nerves do.

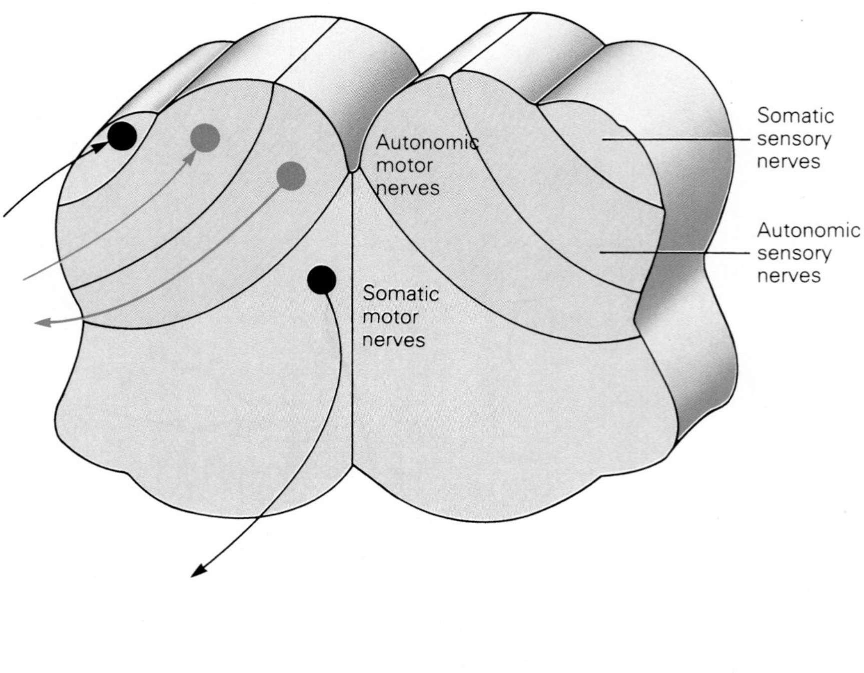

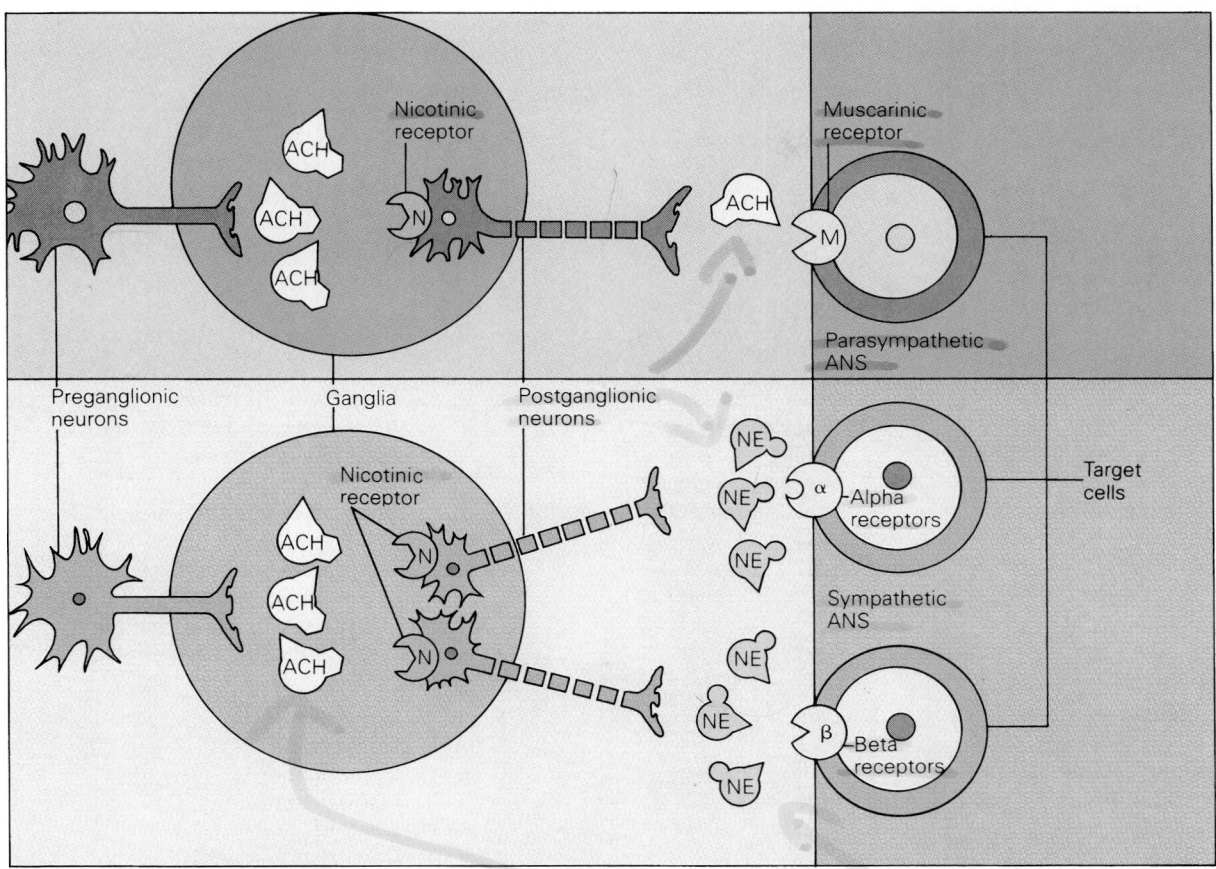

Figure 10–4
All preganglionic neurons release acetylcholine (ACH). Postganglionic parasympathetic nerve cells also release ACH, while postganglionic sympathetic cells release norepinephrine (NE). Exceptions to this, however, are fibers from sympathetic nerve cells that innervate sweat glands, and blood vessels of skeletal muscles that release ACH. Receptors for ACH are either muscarinic (M) or nicotinic (N); receptors for NE are either alpha or beta receptors.

nals of all preganglionic neurons (Fig. 10–4). In postganglionic nerve cells, norepinephrine is found in the terminals of sympathetic fibers, while acetylcholine is found in the terminals of parasympathetic fibers. Exceptions to this rule are postganglionic sympathetic fibers that innervate sweat glands, and blood vessels of skeletal muscles that release acetylcholine. Nerve fibers that transmit acetylcholine are sometimes called **cholinergic** fibers; those that transmit norepinephrine are sometimes referred to as **adrenergic** fibers.

The acetylcholine receptor on the cell body of postganglionic neurons is different from the acetylcholine receptor on target cells of the parasympathetic nervous system. Cholinergic receptors are classified as either **nicotinic** or **muscarinic** acetyl-choline receptors, based on whether they are activated by the drug nicotine or the drug muscarine. The acetylcholine receptor on the postganglionic neuron can be activated by nicotine and is thus a nicotinic receptor, while the acetylcholine receptor on target cells such as smooth muscle is muscarinic. Norepinephrine activates two types of receptors, **alpha receptors** and **beta receptors.** Beta receptors are distinguished from alpha receptors by their greater sensitivity to isoproterenol, a drug similar to norepinephrine.

As discussed in Chapter 7, the alpha adrenergic receptors are subdivided into alpha-1 and alpha-2 subtypes based on their anatomical location. The alpha-1 receptor subtype is located on postsynaptic target cells of sympathetically innervated organs,

Table 10–1

Summary of Cholinergic and Adrenergic Influences on Effector Organs*

Effector Organs	Cholinergic Impulses	Adrenergic Impulses	
	Response	*Receptor type†*	*Response*
Eye			
Radial muscle of iris	—	α	Contraction (mydriasis)
Sphincter muscle of iris	Contraction (miosis)		—
Ciliary muscle	Contraction for near vision	β	Relaxation for far vision
Heart			
S–A node	Decrease in heart rate	β_1	Increase in heart rate
Atria	Decrease in contractility, and (usually) increase in conduction velocity	β_1	Increase in contractility and conduction velocity
A–V node and conduction system	Decrease in conduction velocity	β_1	Increase in conduction velocity
Ventricles	—	β_1	Increase in contractility, conduction velocity, automaticity, and rate of idiopathic pacemakers
Blood vessels			
Coronary	Dilation	α	Constriction
		β_2	Dilation
Skin and mucosa	—	α	Constriction
Skeletal muscle	Dilation	α	Constriction
		β_2	Dilation
Cerebral	—	α	Constriction (slight)
Pulmonary	—	α	Constriction
		β_2	Dilation
Abdominal organs	—	α	Constriction
	—	β_2	Dilation
Kidney	—	α	Constriction
Salivary glands	Dilation	α	Constriction
Lung			
Bronchial muscle	Contraction	β_2	Relaxation
Bronchial glands	Stimulation		Inhibition (?)
Stomach			
Motility and tone	Increase	α_2, β_2	Decrease (usually)
Sphincters	Relaxation (usually)	α	Contraction (usually)
Secretion	Stimulation		Inhibition (?)
Intestine			
Motility and tone	Increase	α_2, β_2	Decrease

and alpha-2 receptors are located on the presynaptic terminal of noradrenergic nerve fibers.

The beta receptor can also be subdivided into subtypes beta-1 and beta-2. At the beta-1 receptor, epinephrine and NE are more or less equipotent. In contrast, beta-2 receptors respond better to epinephrine than to NE, suggesting that they are normally activated by circulating hormones rather than by sympathetic nerve stimulation. Table 10–1 lists different organs and the effects of acetylcholine and norepinephrine on them.

Control of Bodily Organs by the Sympathetic and Parasympathetic Systems

The actions of the sympathetic and parasympathetic inputs to an organ have opposite effects. These actions operate to control various parameters in the body. For instance, when the blood pressure becomes too high, the parasympathetic inputs are activated to decrease the heart rate. They accomplish

Table 10–1 continued

Summary of Cholinergic and Adrenergic Influences on Effector Organs*

Effector Organs	Cholinergic Impulses Response	Adrenergic Impulses Receptor type†	Response
Sphincters	Relaxation (usually)	α	Contraction (usually)
Secretion	Stimulation		Inhibition (?)
Gallbladder and ducts	Contraction		Relaxation
Urinary bladder			
Detrusor	Contraction	β	Relaxation (usually)
Trigone and sphincter	Relaxation	α	Contraction
Ureter			
Motility and tone	Increase (?)	α	Increase (usually)
Uterus	Variable‡	α, β_2	Variable‡
Male sex organs	Erection		Ejaculation
Skin			
Pilomotor muscles	—	α	Contraction
Sweat glands	Generalized secretion	α	Slight, localized secretion§
Spleen capsule	—	α	Contraction
Adrenal medulla	Secretion of epinephrine and norepinephrine		—
Liver	Glycogen synthesis	α, β_2	Glycogenolysis
Pancreas			
Acini	Secretion		
Islets	Insulin secretion	α	Inhibition of insulin secretion
		β_2	Insulin secretion
Salivary glands	Profuse, watery secretion	α	Potassium and water secretion
Lacrimal glands	Secretion	β	Amylase secretion
Nasopharyngeal glands	Secretion		—
Adipose tissue	—	α, β_1	Lipolysis
Kidney	—	β_2	Renin secretion

*Modified from Goodman & Gilman, *The Pharmacological Basis of Therapeutics*. 4th ed. New York: Macmillan, 1970.

†Where adrenergic receptor type has been established, α receptors are all α_1 unless indicated otherwise.

‡Depends on stage of menstrual cycle, amount of circulating estrogen and progesterone, and other factors. Responses of pregnant uterus different from those of nonpregnant.

§On palms of hands and in some other locations ("adrenergic sweating").

this by releasing acetylcholine. When acetylcholine binds to muscarinic receptors located on **pacemaker cells,** which are cardiac cells that regulate heart rate, the cells decrease their rhythmic rate of action potential discharge. This occurs because the muscarinic receptors, once acetylcholine binds to them, allow the K^+ ions to escape from the cells, causing them to become hyperpolarized. This slows their pacemaker activity, decreases the heart rate, and ultimately decreases the blood pressure (see Ch. 18).

When the blood pressure is too low, the sympathetic nervous system steps in. The sympathetic inputs to the heart release norepinephrine. The binding of norepinephrine to beta receptors on cardiac cells causes the cells to depolarize faster, resulting in an increase in heart rate. In addition, norepinephrine also causes the heart to contract with greater force. These mechanisms result in a compensatory increase in blood pressure (see Chapter 19).

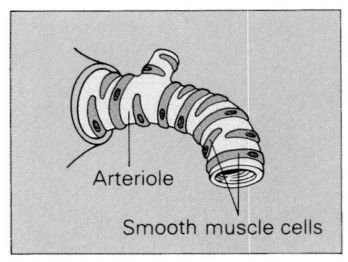

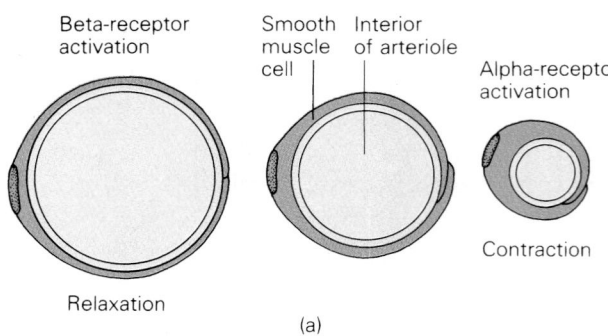

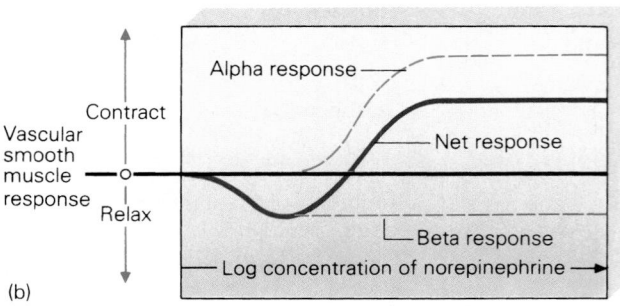

Figure 10–5
Alpha-1 norepinephrine receptors cause smooth muscle cells lining blood vessels to contract, reducing the inner space in the vessel and thereby increasing blood pressure. Beta-1 norepinephrine receptors cause smooth muscle cells to relax, thus widening vessels and decreasing blood pressure (*a*). The net response, however, is a contraction of smooth muscle cells (*b*).

Blood pressure is also controlled by the contraction and relaxation of smooth muscles that surround blood vessels (Fig. 10–5). These smooth muscles respond to both norepinephrine and acetylcholine that are released by nerve fibers. In addition, circulating norepinephrine and epinephrine released by the adrenal medulla can also affect the contraction of smooth muscle. Alpha-1 norepinephrine receptors cause contraction of smooth muscles, which increases blood pressure (see Chapter 19). Norepinephrine beta-1 receptors, on the other hand, cause relaxation of the smooth muscle and decreased blood pressure. The relaxation of some blood vessels in smooth muscle can also occur by the activation of muscarinic acetylcholine receptors of the

parasympathetic nervous system. Thus, smooth muscles are relaxed by the activation of either muscarinic or beta-1 receptors. Reduction of alpha-1 receptor activation can also produce relaxation. This is the mechanism typically found in the increase in blood flow caused by a decrease in the vasoconstriction of blood vessels.

Receptors for norepinephrine and acetylcholine also exist in the membranes of the presynaptic terminals. The alpha-2 receptors are found on synaptic terminals that release acetylcholine. These receptors inhibit the release of acetylcholine. Muscarinic receptors are found on synaptic terminals that release norepinephrine. Activation of these receptors inhibits the release of norepinephrine (Fig. 10–6). These receptors therefore allow the ANS to enhance the effect of either the sympathetic or parasympathetic actions while inhibiting the opposite action by local mechanisms at target organs. For example, when sympathetic inputs to the heart release norepinephrine, causing the heart to beat faster, the norepinephrine also binds to presynaptic alpha-2 receptors located on acetylcholine nerve terminals. The activation of the alpha-2 receptors inhibits the release of acetylcholine, which would normally decrease the heart rate. Likewise, when parasympathetic inputs to the heart release acetylcholine and reduce the heart rate, acetylcholine also binds to muscarinic receptors located on norepinephrine terminals. The activation of these muscarinic receptors decreases the amount of norepinephrine released, which would typically increase the heart rate. Thus, the antagonistic actions of the sympathetic and parasympathetic nervous systems are controlled at the site of the target organ.

A more complex coordination of sympathetic and parasympathetic responses is found at the central levels of the ANS. At the level of the spinal cord and brain stem, these responses are viewed as **autonomic reflexes,** while at the hypothalamic level the responses are viewed as **autonomic control systems.**

Reflexes Governed by the Autonomic Nervous System

The Spinal Autonomic Reflex Arc

Earlier in this chapter, we discussed the sensory connections from the target organ to the spinal cord. This pathway transmits information about the status of the target organ. Within the spinal cord, this information is then transmitted up to the brain and to neurons within the IML of the spinal cord. Since

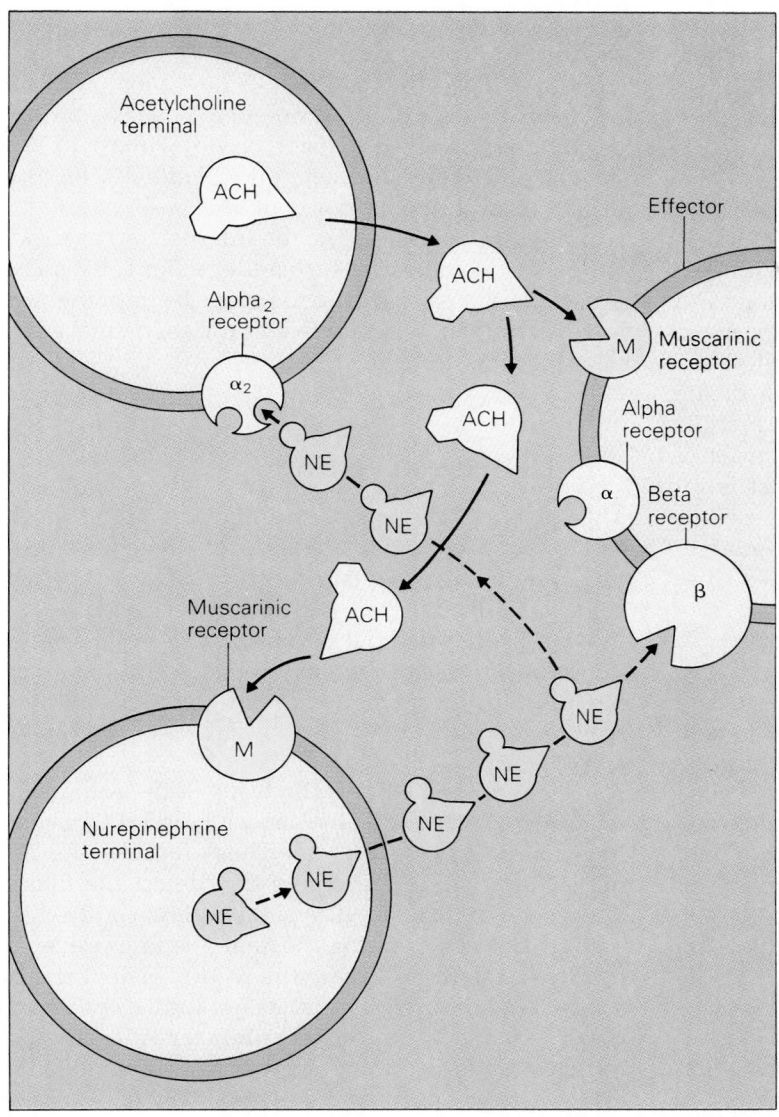

Figure 10–6
Receptors for norepinephrine and acetyl-choline located in the presynaptic membranes can inhibit the release of these neurotransmitters, allowing the ANS to enhance the effect of a sympathetic or parasympathetic action.

the cells in the IML provide the output from the ANS to the target organs, a loop of nerve cell activity is therefore established; this forms the **spinal autonomic reflex arc** (Fig. 10–7). The formation of this neuronal circuit starts with the detection of visceral information by autonomic sensory receptors in the organ. Sensory fibers in turn activate interneurons within the spinal cord. These neurons synapse with the cells of the IML. The IML cells integrate incoming information from many sensory neurons that enter the spinal cord within one spinal cord segment and discharge to activate postganglionic neurons. Finally, the postganglionic cells convey information to the target organ, causing it to make the appropriate responses. This reflex arc is quite similar to other reflex circuits such as the knee-jerk reflex of the somatic nervous system (see Chapter 9).

Figure 10–7
A spinal autonomic reflex arc.

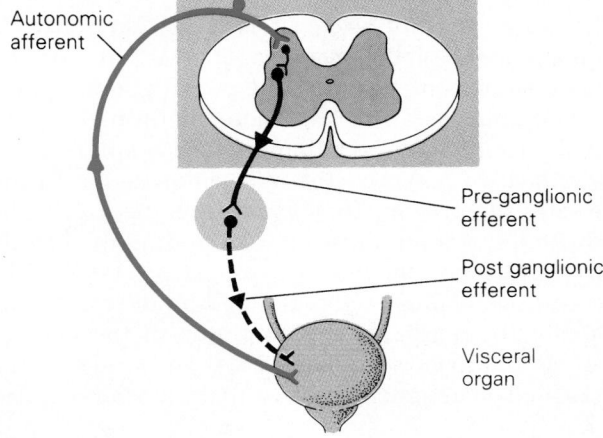

The particular response in the autonomic reflex system varies with the target organ. In the bladder, this reflex is responsible for urination; in the colon and rectum, it is responsible for defecation.

The Urinary Bladder Reflex

The bladder stores urine produced by the kidneys. It fills at a rate of approximately 50 ml per hour and and can store about 150 to 200 ml before the urge to urinate occurs. The ability to keep urine in the bladder is called **continence,** and the emptying of the bladder is called **micturition.**

The bladder (Fig. 10–8*a*) contains a wall of smooth (involuntary) muscle (detrusor) that is innervated by sensory nerve endings. These nerves function as stretch receptors, sensing the enlargement of the bladder. Two groups of muscles close off the exit from the bladder and prevent urine from leaving. The first group is made of smooth muscle and is called the **internal sphincter.** These muscles are innervated by parasympathetic nerve fibers in the pelvic nerves and by sympathetic nerve fibers in the hypogastric nerves. The second group of muscles are striated (voluntary) muscles and are called the **external sphincter.** They are innervated by motor neurons that form the pudendal nerve. These motor neurons are under voluntary control of higher brain centers.

As urine begins to fill and distend the bladder, the stretch receptors become activated (Fig. 10–8*b*) and discharge action potentials. The action potentials from sensory fibers activate preganglionic parasympathetic nerve cells that result in the contraction of the smooth muscles surrounding the wall of the bladder. These actions constitute the urinary bladder reflex in the spinal cord.

In order for the bladder to empty, however, the internal and external sphincter muscles must be placed in a relaxed state. This involves the activation of a "micturition center" in the brain stem by inputs from the sensory stretch receptors that sense the enlargement of the bladder. The output from the micturition center inhibits the preganglionic sympathetic neurons and alpha motor neurons that normally cause the internal and external sphincters to contract. In addition, the micturition center further excites the preganglionic parasympathetic cells that cause the smooth muscles surrounding the bladder to contract and the internal sphincter to relax. Thus, the spinal bladder reflex is aided by the micturition center to coordinate the emptying of the bladder.

Individuals who have been in accidents that crushed their spinal cords can still control the bladder reflex. If the damage is above the level of the sacrum, the spinal reflex is intact. The reflex is initially lost, however, as a result of "spinal shock," but recovers over the next one to five weeks. Urination after spinal cord damage occurs with great frequency and is initiated when the bladder is slightly filled. This occurs because of the absence of descending inputs that cause the internal and external sphincters to contract. The bladder reflex can also be initiated after spinal cord damage by tapping the lower abdomen, which activates the segmental cutaneous-visceral reflex.

The Defecation Reflex

The colon and rectum store unabsorbed food products or fecal (waste) material from the small intestine. The ability to retain fecal material within the colon and rectum is called **continence,** and the removal of fecal material is called **defecation.** Continence is controlled by spinal reflexes and descending inputs from higher brain centers, and is maintained until approximately 2 liters of feces have accumulated.

As with the bladder, the walls of the colon and rectum are surrounded by smooth muscles that contract to expel fecal material. These muscles are innervated by stretch receptors that detect the filling of the rectum (Fig. 10–9). A set of internal and external muscles surrounding the anal opening prevents fecal material from leaving the rectum. The internal sphincters are smooth muscles not under voluntary control, while the external sphincters are striated muscles, under voluntary control and innervated by motor neurons from spinal segments S2 to S4. Typically, both internal and external sphincters are closed. The tonic contraction of the external sphincter is maintained by the activation of the motor neurons from sensory nerves from the skin and from the external sphincter muscle itself.

As the rectum fills with fecal material, stretch receptors become activated and transmit this information to the spinal cord via the pelvic nerve. Within the spinal cord, these sensory inputs excite preganglionic parasympathetic neurons that in turn excite postganglionic nerve cells innervating the smooth muscles of the colon and rectum. This causes the smooth muscles to contract, forcing fecal material to move toward the anal opening. The sensory stretch receptors also transmit information to higher brain centers to initiate the urge to defecate. These higher centers activate descending inputs to the parasympathetic preganglionic nerve cells to facilitate the defecation reflex. In order to further

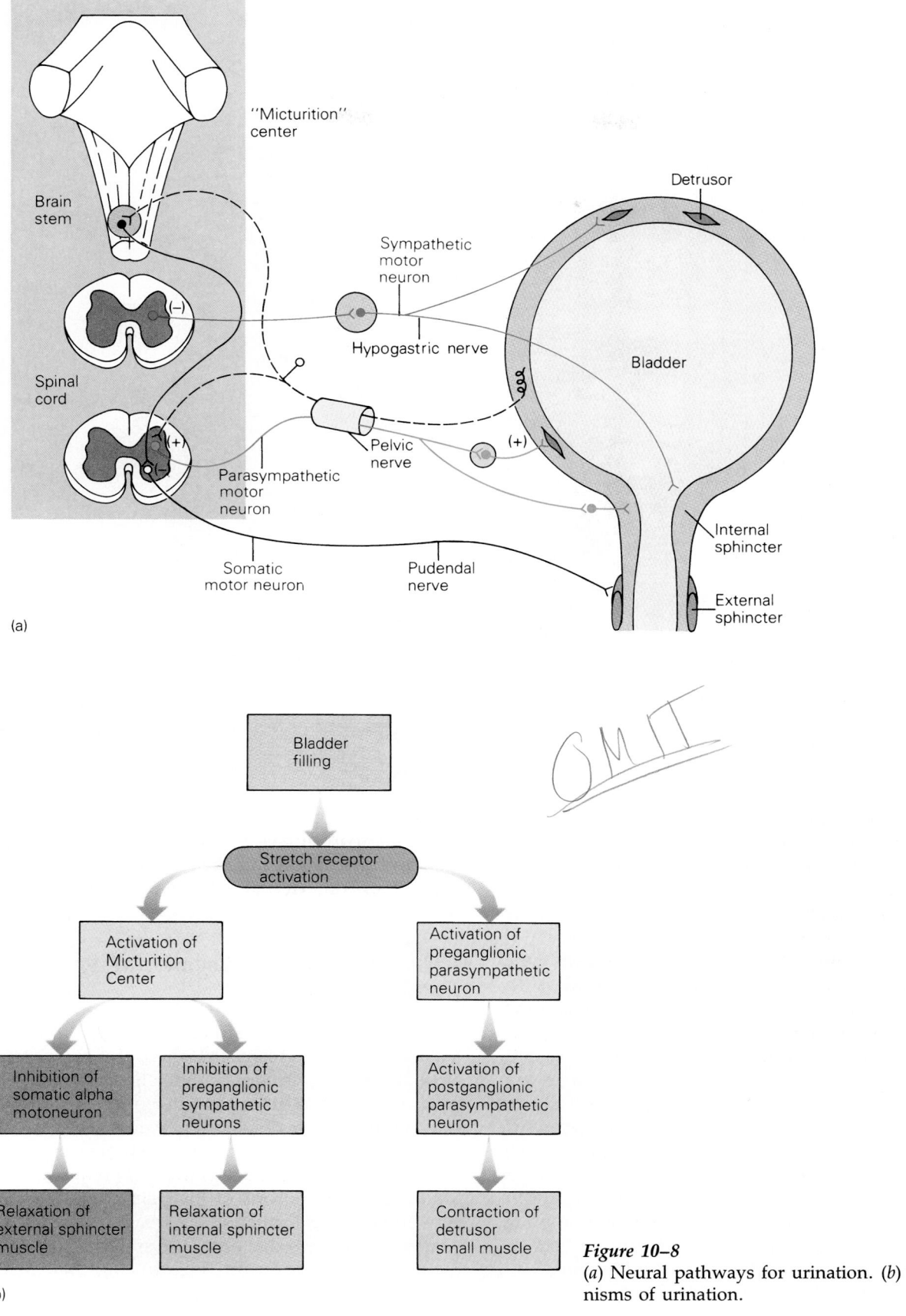

(a)

(b)

Figure 10–8

(*a*) Neural pathways for urination. (*b*) Mechanisms of urination.

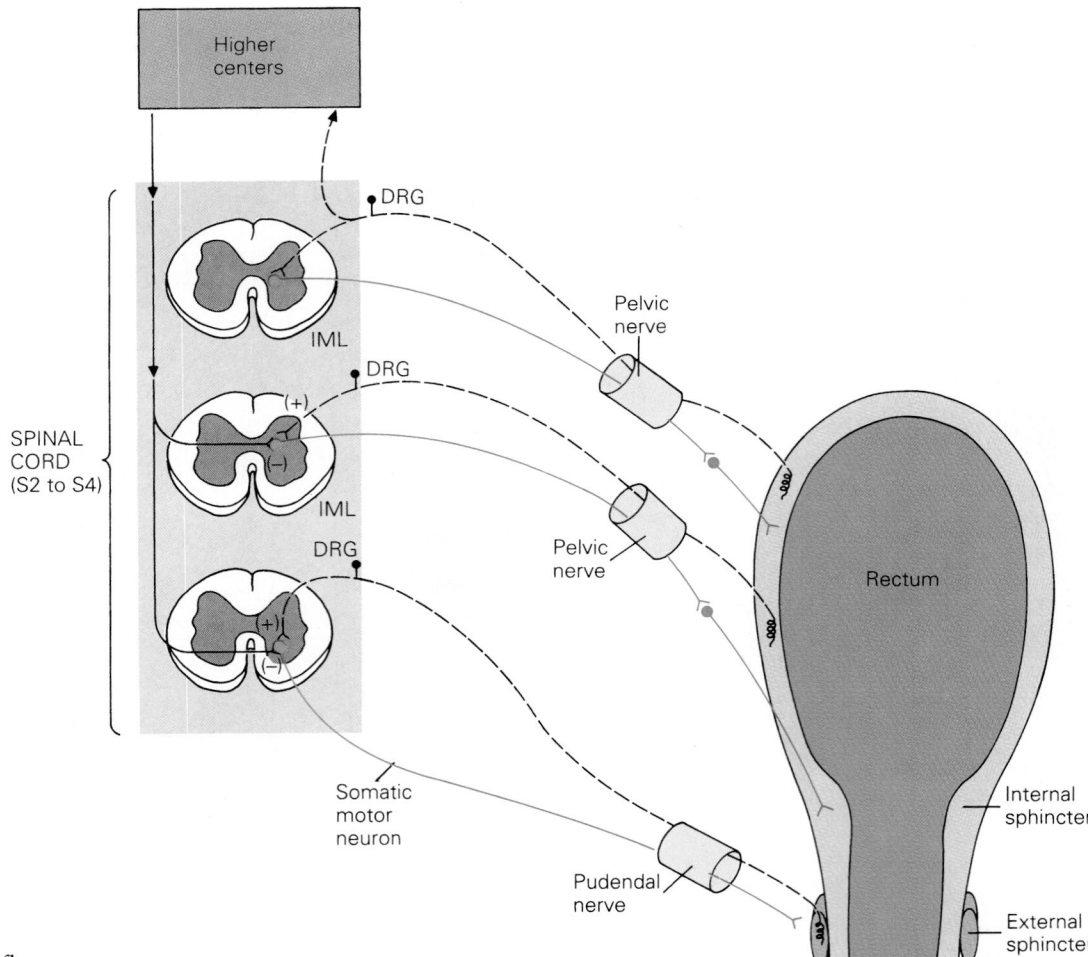

Figure 10–9
The defecation reflex.

coordinate the emptying of the rectum, the descending inputs activate preganglionic parasympathetic neurons that lead to the relaxation of the internal sphincters and inhibit the motor neurons of the pudendal nerve that innervates the striated muscles of the external sphincters.

These actions alone, however, are not sufficient for defecation to proceed. The force necessary for the expulsion of feces also comes from the contraction of the abdominal muscles. This contraction causes an increase in the intra-abdominal pressure, which squeezes the contents inside of the rectum and colon. This increase in pressure works in concert with the contraction of the smooth muscles surrounding the colon and rectum, and together they are sufficient to generate the forces necessary for fecal expulsion. Thus, descending inputs from sensory-motor areas of cortex orchestrate defecation by coordinating both autonomic and somatic actions.

The defecation reflex is lost in individuals who have been in accidents that destroy the sacral seg-ments of their spinal cord. However, transections of the spinal cord above the sacrum preserve the reflex, and defecation can occur with the manual expansion of the external and internal sphincters.

Referred Pain and Segmental Spinal Reflexes

People who experience stomach pains and heart attacks commonly perceive that their pain comes from the surface of their bodies. Since the pain is attributed to a location other than its source, it is called **referred pain.** This displaced perception of pain is a result of nerve fibers from pain receptors from internal organs and the surface of the body that converge onto the same neurons of the pain pathway. This convergence of information occurs with visceral and somatic pain fibers that enter the spinal cord within the same spinal segment (Fig. 10–10). Pain fibers from the heart, for example, enter the spinal cord at

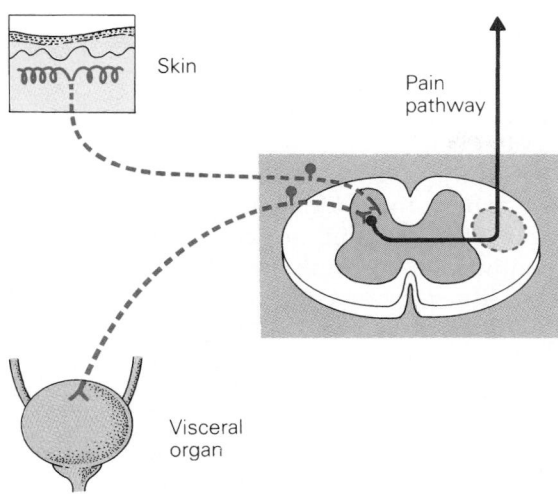

Figure 10–10
An example of a referred pain pathway.

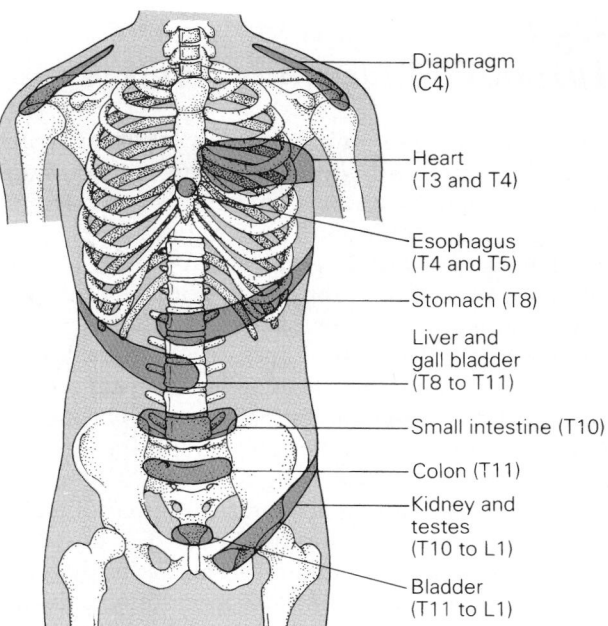

Figure 10–11
Hyperalgesic zones for pain originating in the major internal organs.

Figure 10–12
An example of a segmental spinal autonomic reflex, in which reddening of the skin signals internal inflammation.

thoracic segments T1 through T3 and synapse with neurons in the **substantia gelatinosa** of the pain pathway (see Chapter 8). Pain information from the left chest and the upper portion of the left arm also synapse with these same neurons. Consequently, since the brain has come to associate pain information from spinal segments T1 through T3 with the left chest and arm, it continues to do so, and thus heart attacks are often perceived as pain in the left arm and chest.

The painful areas on the surface of the body that are associated with inflammations of internal organs are called **hyperalgesic zones.** Figure 10–11 shows the hyperalgesic zones for pain originating in the heart, stomach, kidney, liver, gallbladder, and other internal organs. The hyperalgesic zones are useful in diagnosing disorders of visceral organs.

In addition to the increased pain associated with the hyperalgesic zone, the inflammation of internal organs often causes the skeletal muscles in this zone to tighten and the skin to redden (Fig. 10–12). These signs are often used to help determine which organ is in distress. The reddening of the skin and the tightening of the muscles are results of reflex connections. These connections occur between sensory fibers from the internal organ that synapse with neurons activating skeletal muscles to cause contractions. The inputs from sensory fibers also inhibit preganglionic sympathetic neurons and thus decrease vasoconstriction and increase blood flow to the skin. The segmental spinal reflex connections, therefore, occur within the appropriate spinal cord segments where visceral sensation is received and transmitted out through somatic and autonomic pathways.

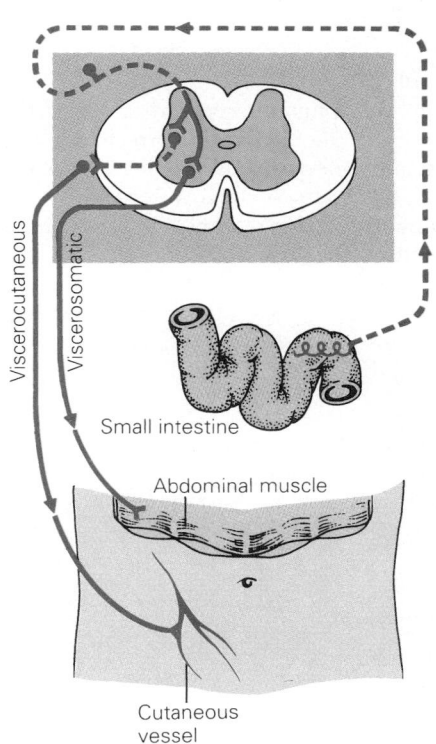

Regulatory Systems of the Autonomic Nervous System

Actions of the autonomic nervous system involving the brain stem and hypothalamus are part of **autonomic regulatory systems** that are more complex than spinal cord reflexes. These regulatory systems maintain a constant internal environment for the optimal functioning of the various organs and their cells. Regulatory systems of the body have characteristics similar to those found in a modern home. For example, home thermostats have sensors to determine if the ambient temperature is warmer or colder than the selected temperature. If the ambient temperature is warmer than the desired temperature, the **sensor system** activates the air conditioner to introduce cooler air into the house. On the other hand, if the ambient temperature is colder than the desired temperature, the sensor system activates the heater to provide warmer air. The air conditioner and heater in this example are **effector systems** that work to change the air temperature. The change in air temperature provided by the effector systems always attempts to minimize the difference between the actual and desired temperatures. This type of regulatory system is referred to as a **negative feedback system** and is the basis of control systems in the body (see Chapter 1).

Regulation of Thirst, Osmolarity, and Water Volume

A large portion of the human body consists of water. Water is lost in the form of sweat and urine. It also leaves the body as water vapor in air exhaled from the lungs. In order to maintain a constant amount of water within the body, thirst mechanisms motivate us to drink water; in addition, the ANS and kidneys act in concert to retain water that would normally be excreted as urine. The amount of water in the blood affects its **osmolarity** (see Chapter 2). Consequently the regulation of thirst, water volume, and osmolarity is interrelated.

The osmolarity of the blood is a critical factor in maintaining the internal environment of the cells in the body. As discussed in Chapter 4, hyper- and hypo-osmotic fluids cause cells to shrink or swell and alter their function. The desired osmolarity for blood plasma is 300 mOsm. Increases in plasma osmolarity above this value can occur from excessive dehydration and sweating. This increase in osmolarity is detected by osmoreceptors in the **supraoptic nucleus** and the **paraventricular nucleus** of the hypothalamus (Fig. 10–13; see also Fig. 11–1). In response to increases in osmolarity, these nuclei secrete the hormone **vasopressin,** also called **antidiuretic hormone (ADH),** into the blood. Eventually this hormone acts on the kidneys and causes them

Figure 10–13
The hypothalamus detects increases in the osmolarity of the blood and corrects them by activating mechanisms that increase the water in the body, such as water consumption and water retention by the kidney, controlled by the secretion of ADH.

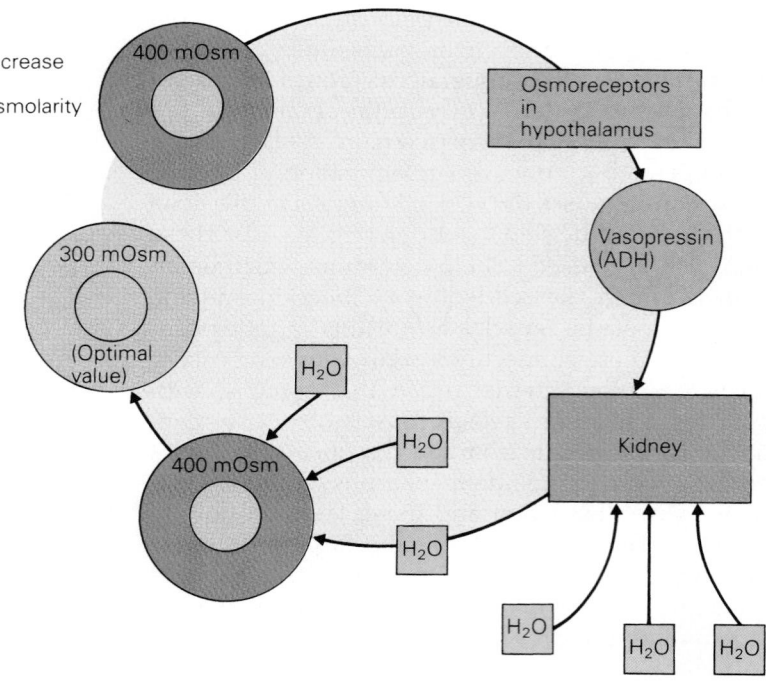

to reabsorb more water. The retention of water reduces the plasma osmolarity to maintain it at 300 mOsm.

The activation of hypothalamic osmoreceptors also initiates thirst mechanisms that involve somatic responses leading to the consumption of water. The intake of water in turn reduces plasma osmolarity.

In addition to central osmoreceptors, peripheral stretch receptors act as **water-volume sensors.** These sensory receptors are located in large veins near the heart. Water volume can decrease dramatically if blood is lost through hemorrhages or other accidents. As the water volume decreases, the volume receptors activate centers in the brain stem (Fig. 10–14). These centers activate sympathetic neurons associated with the renal nerve and cause the kidneys to release the enzyme **renin** into the blood stream. In the circulatory system, renin converts the peptide angiotensinogen to angiotensin I (AI). AI in turn is converted to **angiotensin II (AII),** which circulates to the brain and causes the hypo-

thalamus to release ADH into the circulatory system. Eventually, ADH reaches the kidneys to increase water reabsorption. Release of hypothalamic ADH by either osmoreceptors or AII is therefore the key pathway leading to the retention of water by the kidney.

As the feeling of thirst is satisfied, water consumption subsides. This happens, however, long before the water is absorbed into the circulatory system and long before plasma osmolarity is restored to normal. The feeling of being quenched is thought to be triggered by stretch receptors in the stomach. How this information is transmitted to the brain has not yet been determined.

Regulation of Breathing

Cellular production of energy from glucose requires oxygen. As cells become more active, their need for oxygen and energy increases. Thus, the amount of oxygen available to each cell must be closely regu-

Figure 10–14
Water volume deficiencies are detected by sensors near the heart, setting off a series of reactions and mechanisms that increase the water content of the body.

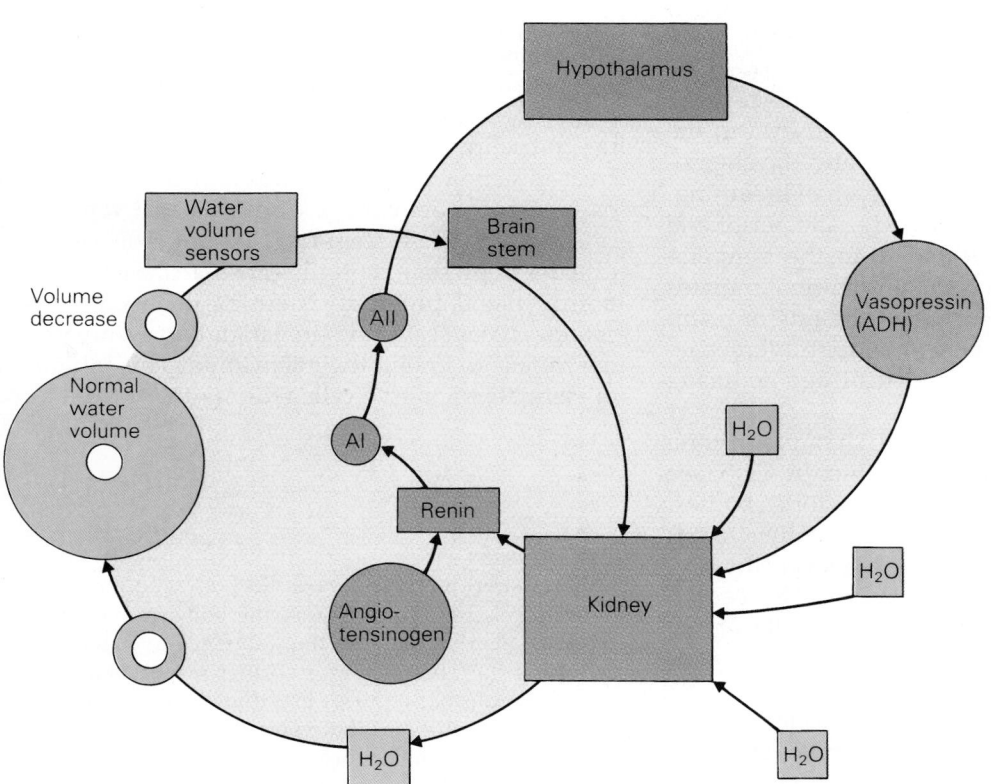

lated to match metabolic demands. In addition, for cells to function properly, the waste products of metabolism, CO_2 and H^+, must be removed from the internal environment. The supply and regulation of oxygen to the body and the removal of CO_2 (and H^+ indirectly) are primary functions of the lungs. These functions are carried out in concert with the ANS.

Receptors called **chemoreceptors** in the ANS monitor the amount of O_2 available to the cells and the amount of CO_2 they produce. In the peripheral nervous system there are two types of chemoreceptors: **carotid bodies** and **aortic bodies** (Fig. 10–15*a*). These chemoreceptors are sensitive to levels of O_2 and CO_2. The respiratory muscles respond to the activation of the carotid and aortic bodies by increasing the frequency and depth of breathing. In this way, more oxygen is supplied to the body and more CO_2 is removed (see Chapter 20).

The peripheral chemoreceptors indirectly activate motor neurons that control the muscles for breathing. Peripheral chemoreceptors have inputs to neurons of the respiratory center in the brain stem (see Fig. 10–15*b*) that generate the rhythm of **inspiration** (breathing in) to bring oxygen to the body and **expiration** (breathing out) to remove carbon dioxide. The rhythm-generating neurons in turn have inputs to the motor neurons that control the muscles for respiration (see Fig. 10–15*c*). The muscles used in inspiration are the diaphragm and external intercostal muscles (see Chapter 20).

The carotid and aortic bodies are also sensitive to blood flow. Decreases in blood flow result in a decrease in the amount of oxygen delivered to the chemoreceptors. This decrease activates the chemoreceptors, increases the rate and depth of breathing, and brings more oxygen to the body. The peripheral chemoreceptors are also sensitive to the temperature of the blood. A rise in temperature stimulates the chemoreceptors and increases the rate of respiration. In some animals, this is an important mechanism for dissipating (releasing) heat and reducing the body's temperature.

In addition to the peripheral chemoreceptors, chemoreceptors are also located within the CNS (see Fig. 10–15*d*). As are the carotid and aortic bodies, these central chemoreceptors are sensitive to increases in carbon dioxide and H^+; however, they are not sensitive to decreases in oxygen. The central chemoreceptors are located in the **medulla** of the brain stem and are in contact with the **cerebrospinal fluid (CSF)**. As discussed in Chapter 7, the CSF bathes the brain and provides the extracellular environment for the brain cells. Changes in the chemical composition of the CSF alter the function of nerve cells within the brain.

The CSF is isolated from the blood and circulatory system by the blood-brain barrier (see Chapter 7). However, gases such as oxygen, carbon dioxide, and other lipid-soluble substances are able to penetrate the barrier, going from blood to brain or from brain to blood. The body can therefore regulate carbon dioxide levels in the CSF by removing it in respiration.

The maintenance of sufficient oxygen in the body and the removal of carbon dioxide are thus regulated by chemoreceptors of the autonomic nervous system. Peripheral chemoreceptors detect decreases in oxygen; both peripheral and central chemoreceptors detect increases in carbon dioxide and H^+. Peripheral and central chemoreceptors act together to coordinate the intake of oxygen and removal of carbon dioxide based on the metabolic demands of the body.

Regulation of Blood Pressure

The maintenance of blood pressure is necessary for blood to flow and provide oxygen and nutrients to the organs of the body. The desired average arterial pressure is 90 mm Hg. Blood pressure is sensed by stretch receptors called **baroreceptors** located on structures in the heart called the **aortic arch** and **carotid sinus**. Pressures above 90 mm Hg cause the stretch receptors to expand and discharge more action potentials. Pressures below 90 mm Hg cause the receptors to contract and discharge fewer action potentials.

The response to changes in blood pressure is complex, involving several areas of the brain. When the pressure is greater than 90 mm Hg, the increased action potential activity in the fibers from baroreceptors activates a cardiovascular regulatory center in the brain stem (Fig. 10–16). This center is called a **depressor center** because it causes a reduction in blood pressure. Neurons in the depressor center synapse with preganglionic sympathetic nerve cells to inhibit their activation. This inhibition of sympathetic nerve cells reduces the heart's force

Figure 10–15
(*a*) Chemoreceptors in the carotid body and aortic arch detect levels of O_2 and CO_2 in the body and initiate mechanisms that increase the frequency and depth of breathing. (*b*) The respiratory center responds to a variety of stimuli that regulate breathing. (*c*) Rhythm-generating neurons in the respiratory center influence the activity of motor neurons that control muscles for respiration. (*d*) Chemoreceptors located within the CNS also influence the rate and depth of respiration.

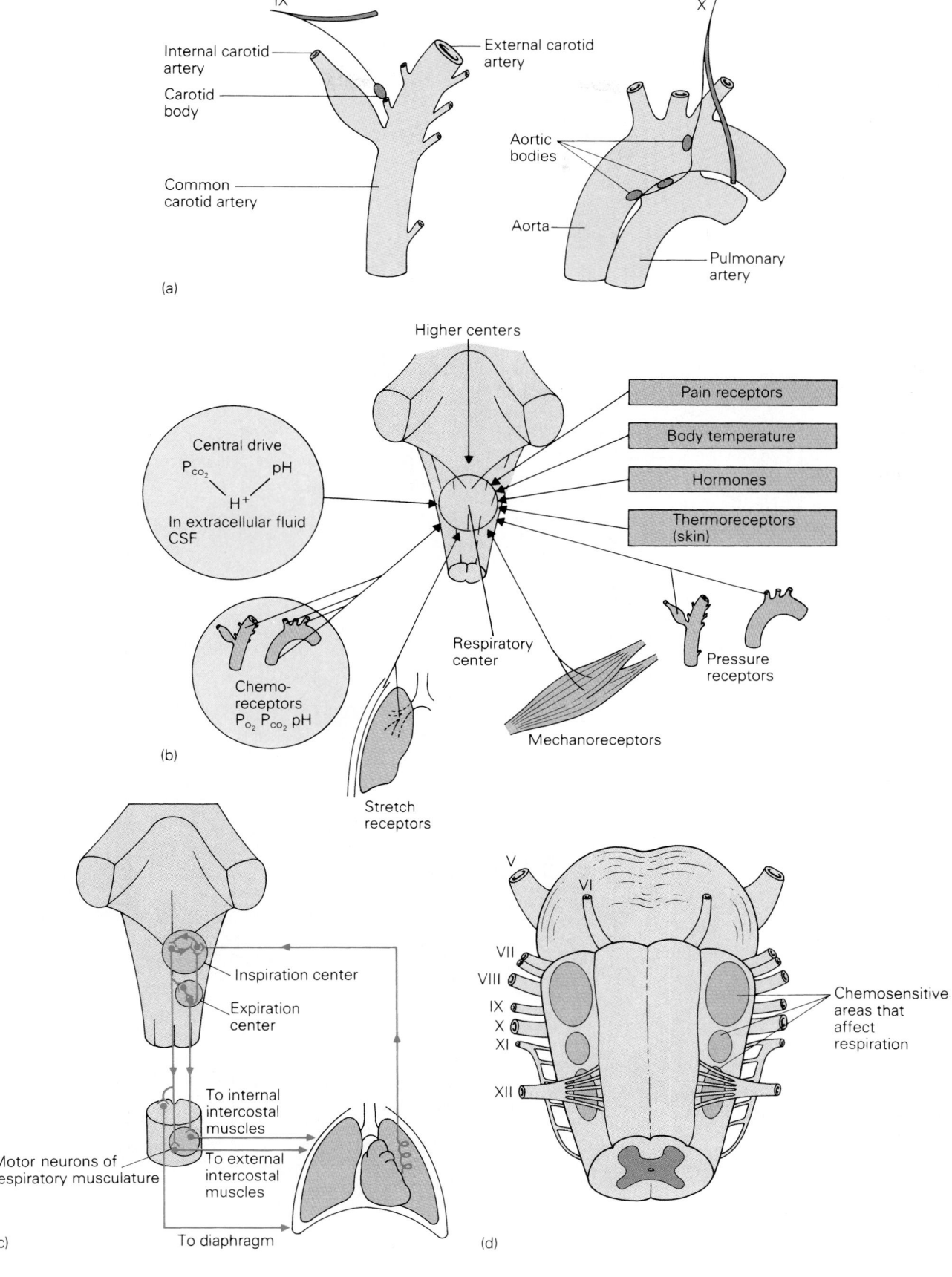

(a)

IX

Internal carotid artery

Carotid body

Common carotid artery

External carotid artery

X

Aortic bodies

Aorta

Pulmonary artery

(b)

Higher centers

Central drive

P_{CO_2} pH

H^+

In extracellular fluid CSF

Chemoreceptors
P_{O_2} P_{CO_2} pH

Stretch receptors

Respiratory center

Mechanoreceptors

Pain receptors

Body temperature

Hormones

Thermoreceptors (skin)

Pressure receptors

(c)

Inspiration center

Expiration center

Motor neurons of respiratory musculature

To internal intercostal muscles

To external intercostal muscles

To diaphragm

(d)

V

VI

VII

VIII

IX

X

XI

XII

Chemosensitive areas that affect respiration

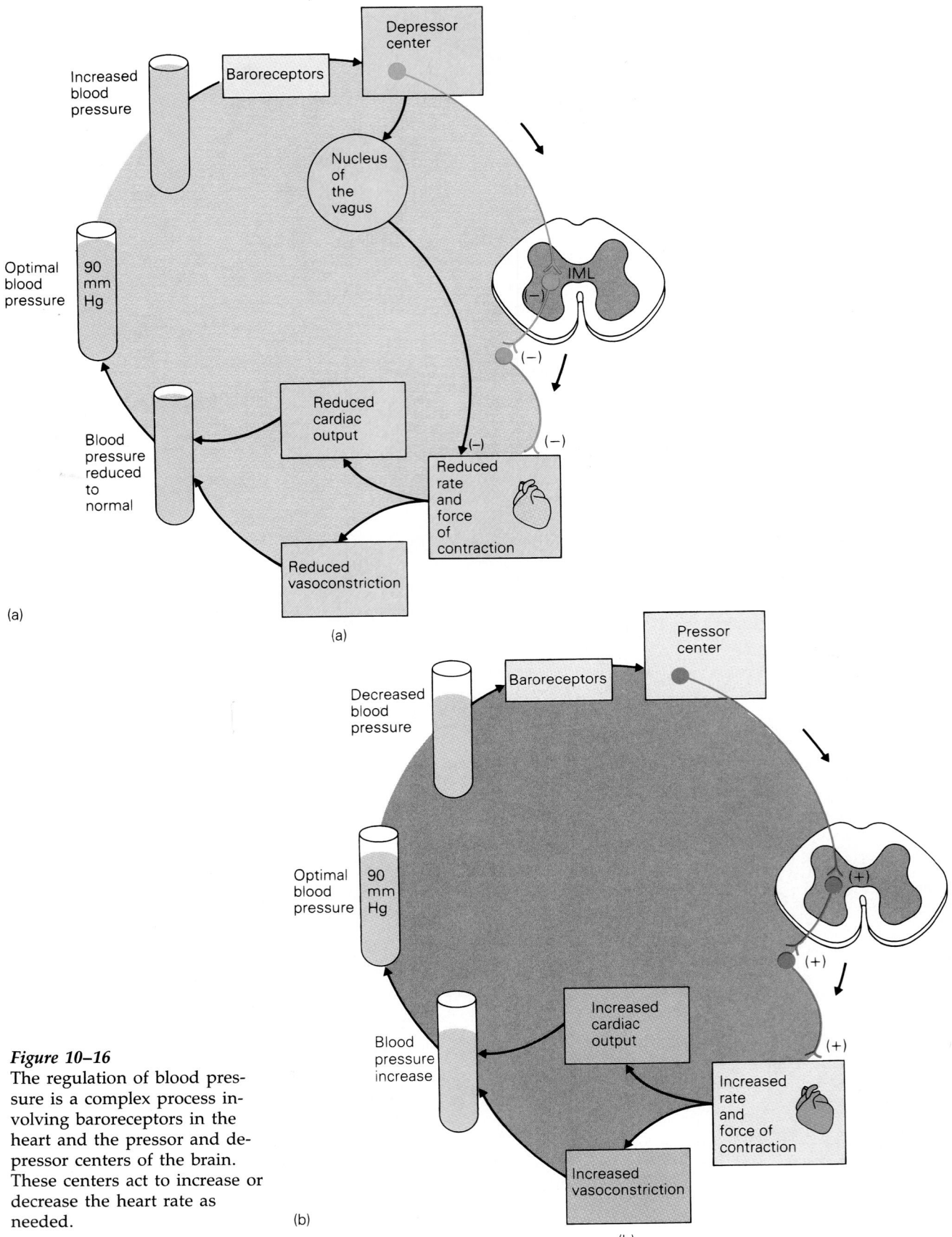

(a)

(a)

(b)

Figure 10–16
The regulation of blood pressure is a complex process involving baroreceptors in the heart and the pressor and depressor centers of the brain. These centers act to increase or decrease the heart rate as needed.

(b)

of contraction and rate of contraction, and reduces the vasoconstriction of blood vessels. The depressor center also activates another region of the brain stem called the **nucleus of the vagus,** which directly reduces the heart rate. The net result of these actions is a reduction in blood pressure.

Decreases in blood pressure below 90 mm Hg will cause the baroreceptors to activate the cardiovascular **pressor center,** located in the brain stem. The pressor center attempts to increase blood pressure in response to sudden decreases such as those due to hemorrhages. The pressor center in turn stimulates the preganglionic sympathetic nerve cells to increase heart rate and the force of the heart's contraction. In addition, it causes blood vessels to constrict. The net effect of these actions is to increase blood pressure in response to sudden decreases detected by the baroreceptors.

The hypothalamus affects the regulation of blood pressure by its excitatory inputs to the pressor center and preganglionic sympathetic neurons to increase blood pressure. This action of the hypothalamus, however, is only one of many functions in its regulation of autonomic activity.

C L I N I C A L   F O C U S

The Relaxation Response

The relaxation response is characterized by a decrease in the activity of the sympathetic nervous system, a decrease that results from conditioning and training. During this response, there is a decrease in oxygen consumption, heart rate, blood pressure, and respiration rate, and an increase in the alpha waves of the electroencephalogram (Chapter 11). The relaxation response is not just simple relaxation. In simple relaxation, changes in the rate of respiration, oxygen consumption, and alpha wave activity do not occur.

Learning to induce the relaxation response is a common practice associated with various Eastern religions such as Zen Buddhism. One can induce the relaxation response by sitting in a comfortable position, focusing on an object, and blocking out all other distractions while repeating a word or phrase.

The relaxation response is thought to modify the way in which stressful stimuli affect the sympathetic nervous system. A temporary activation of the sympathetic system is important in preparing the body to respond appropriately to stressful situations by the release of substances such as norepinephrine. The prolonged release of these substances, however, can have deleterious effects that include abnormally high blood pressure and a decrease in immune system function. Research indicates that the relaxation response reduces the body's response to norepinephrine secreted into the circulatory system.

The relaxation response produces changes in the sympathetic nervous system that extend outside of the training period. Clinical studies have shown that the relaxation response can alleviate heart beat disorders, reduce pain, and lower blood pressure in hypertensive patients. If practiced daily, it may alleviate physiological conditions arising from undesirable increases in sympathetic nervous system activity, and thus improve and maintain overall health.

Buddhist monks during morning meditation, Eiheiji Monastery, Japan. (© Dieter Blum/Peter Arnold, Inc.)

Autonomic Functions of the Hypothalamus

The activation of different regions of the hypothalamus produces a variety of coordinated autonomic responses. The activation of the **dorsal hypothalamus,** for example, increases blood pressure, intestinal motility, and intestinal blood supply but decreases the blood supply to the skeletal muscles (Fig. 10–17). These responses are associated with feeding behaviors. Activation of the **ventral hypothalamus** increases blood pressure and the blood supply to the skeletal muscles but decreases intestinal motility and blood flow to the intestines. These responses are associated with defense behaviors or the so-called **"fight or flight responses"** (see Table 10–2). This response prepares the body for the stress of threatening situations. For example, at the onset of intense pain or sudden fright, blood vessels to the skeletal muscles dilate, and heart rate increases along with an increase in the force of heart contraction. These actions increase blood flow to the skeletal muscles in preparation for running or fighting. In addition, energy sources are mobilized to provide more fuel for the skeletal muscles and heart. The liver produces more glucose, for example; adipose tissue is converted to free fatty acids; and the rate and depth of respiration are increased to provide more oxygen to the cells in the body.

Hypothalamic Control of the Adrenal Gland

The coordinated actions of the hypothalamus during the stress response are mediated in part by the activation of the adrenal gland. As shown in Figure 10–18, the adrenal gland secretes catecholamines (epinephrine and norepinephrine) and glucocorticoids in response to stress. The mechanisms by which the hypothalamus causes the release of these adrenal hormones, however, are quite different. The adrenal gland is composed of an outer cortical area called the **adrenal cortex** and a central area

Figure 10–17
The hypothalamus coordinates the functions of different organs. Activation of the dorsal regions of the hypothalamus results in "feeding" behaviors, while activation of ventral regions produces the "fight or flight" response to a threat.

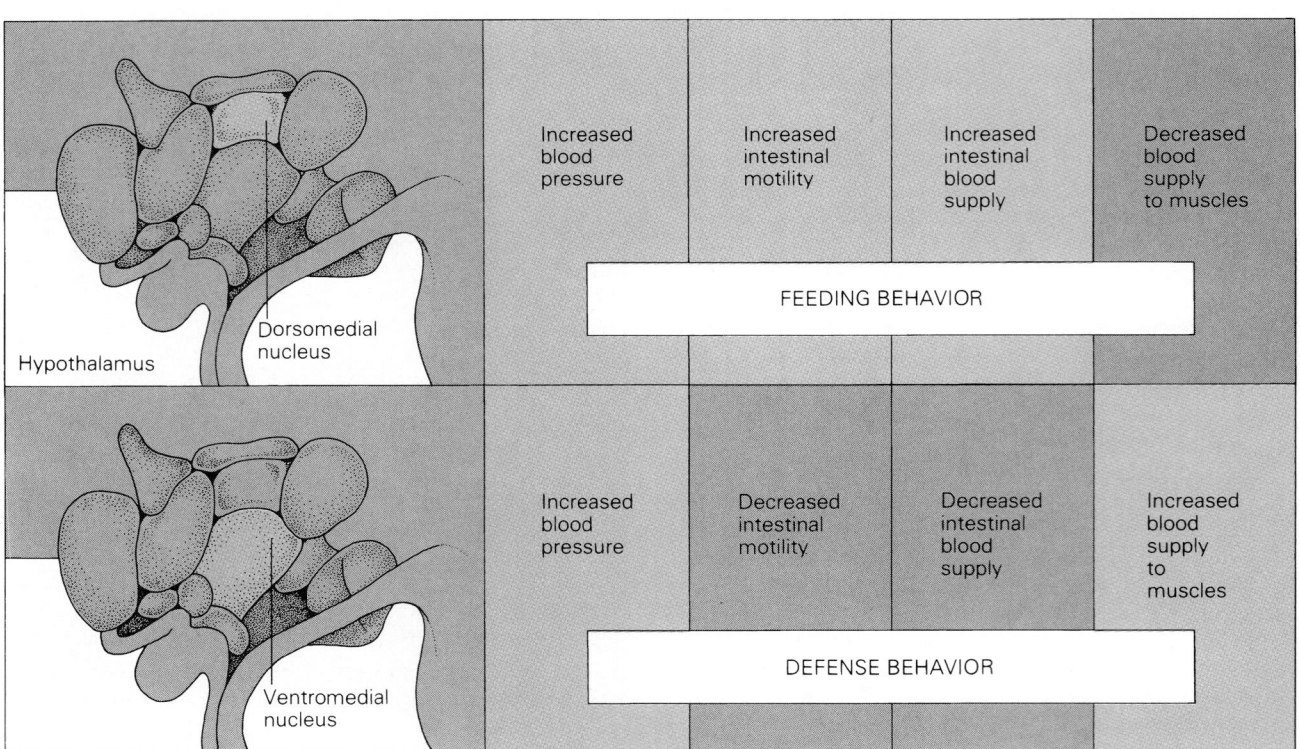

Table 10–2

Physiological Changes in the Fight or Flight Response

Fight or Flight Response

1. Increased blood pressure
2. Increased heart rate
3. Increased force of heart contraction
4. Increased heart conduction velocity
5. Increased depth and rate of respiration
6. Shift of blood flow distribution away from the skin and splanchnic regions and more to skeletal muscles and heart (and, of course, brain)
7. Mobilization of liver glycogen to glucose (glycogenolysis)
8. Mobilization of free fatty acids from adipose tissue (lipolysis)
9. Contraction of spleen capsule (increased hematocrit)
10. Mydriasis (widening of pupil)
11. Accommodation for far vision (relaxation of ciliary muscle)
12. Widening of palpebral fissure (eyelids open wide)
13. Piloerection
14. Inhibition of gastrointestinal motility and secretion, contraction of sphincters
15. Sweating ("cold" sweats as skin blood vessels are constricted)

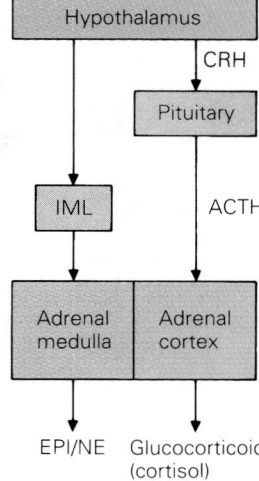

Figure 10–18
Hypothalamic control of the adrenal gland during the stress response.

called the **adrenal medulla** (see Chapter 14). The cells in the adrenal cortex secrete **glucocorticoids** (as well as other steroids), while the medullary cells secrete catecholamines. The secretion of glucocorticoids is mediated by a cascade of hormones, beginning with the release of **corticotropin-releasing hormone (CRH)** from the hypothalamus. CRH, in turn, causes the release of **adrenocorticotropic hormone (ACTH)**, which then causes the secretion of glucocorticoids from the adrenal cortex. In contrast, the release of catecholamines from the adrenal medulla is carried out by the direct connection of nerve fibers from the hypothalamus to the IML, and by nerve fiber from the IML to cells in the medulla. The catecholamine-secreting cells of the adrenal medulla therefore function as postganglionic sympathetic cells. The cells of the adrenal medulla are homologous to postganglionic neurons since they are di-

rectly innervated by preganglionic sympathetic nerve fibers. As previously mentioned, the cells of the adrenal gland, when activated, release the catecholamines **epinephrine** and **norepinephrine** into the circulatory system. (Epinephrine-synthesizing cells represent 80% of the cells in the adrenal while the remaining 20% synthesize norepinephrine.)

The catecholamines that circulate in the blood affect the same receptors and target organs as postganglionic sympathetic nerve cells (Fig. 10–19). Because they enter the blood, however, their actions are not as discrete as those generated by sympathetic nerve cells. Circulating catecholamines cause the air passages in the lungs to dilate and increase the supply of oxygen to the skeletal muscles, heart, and brain. In the cardiovascular system, circulating catecholamines cause the heart to beat faster and with greater force; they also cause the constriction of arteries in the skin, mesentery, and kidney. In contrast, the arteries of the skeletal and coronary (heart) muscles dilate. The net effect of these actions on the cardiovascular system is to provide more blood to skeletal muscles, heart, and brain and to reduce blood flow to the vegetative organs in the body. Since one of the primary functions of blood is to carry oxygen from the lungs to the other organs of the body, the increase in blood flow also provides more oxygen.

Circulating catecholamines also increase the supply of fuel for cells in the form of glucose, free fatty acids, and lactate. The action of catecholamines on the liver causes an increase in **glycogenolysis,** a

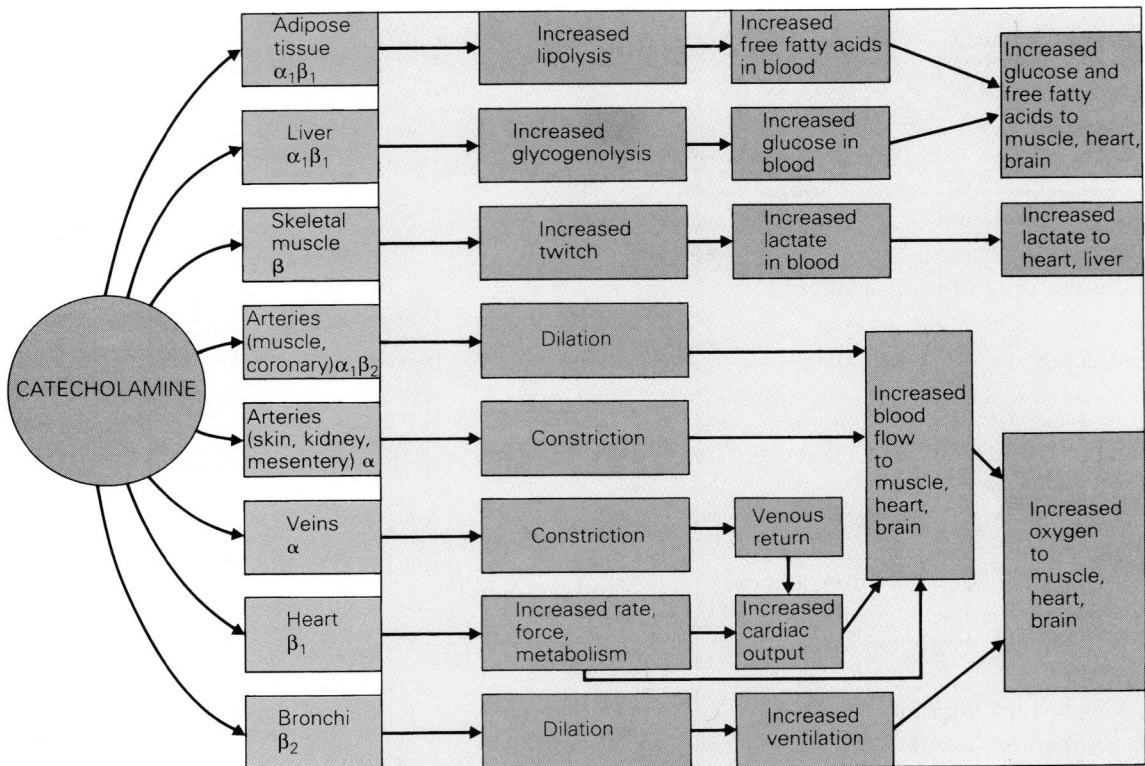

Figure 10–19
Circulating catecholamines produced by the adrenal medulla affect the same
receptors and target organs that postganglionic sympathetic neurons do.

process that increases the amount of glucose in the blood (see Chapter 14). Circulating catecholamines cause adipose or fat tissue to release free fatty acids. They also cause skeletal muscles to twitch and produce lactate. The net effect of these metabolic actions is to increase the substrates for cells in the skeletal muscles, heart, and brain to maintain their increased activity.

Several factors influence the function of the adrenal medulla. One is an individual's emotional state. As you will see in Chapter 11, the brain center for emotion is the hypothalamus. Stress-induced increases in nerve cell activity in the hypothalamus activate preganglionic sympathetic neurons. These preganglionic neurons in turn innervate and activate the adrenal medulla, causing the secretion of epinephrine and norepinephrine into the blood stream. Other factors such as the extreme loss of blood, hypothermia, hypoglycemia, and hypoxia will also activate the adrenal medulla in attempts to correct functional disorders and maintain a constant internal environment (see Chapter 14).

Chronic Stress and the Loss of Brain Cells

Glucocorticoids are steroids released by the adrenal cortex during stress. The work of Robert Sapolsky and his colleagues at Stanford University has shown that prolonged elevation of glucocorticoids can lead to the degeneration of nerve cells in rodents and monkeys. The area of brain most sensitive to glucocorticoid damage is the hippocampal formation, a region long known to be involved in learning and memory. Earlier studies by Bruce McEwen and co-workers at the Rockefeller University revealed that the hippocampus is a major target site in the brain for circulating glucocorticoids.

In studies of vervet monkeys housed in primate colonies in Kenya, Sapolsky and his co-workers found that socially subordinate monkeys exhibited gastric ulcers, high incidences of bite wounds, and hyperplastic adrenal cortices suggestive of sustained glucocorticoid release. Moreover, these characteristic signs of social stress were associated with a severe degeneration of nerve cells within the hippocampal formation.

In experiments to determine whether glucocorticoid secretion alone could lead to neuronal loss, glucocorticoid-containing pellets were implanted directly into the hippocampi of non-socially stressed monkeys. These animals ultimately displayed the same type of nerve cell loss as socially stressed monkeys.

Other studies have shown that chronic stress causes premature aging of hippocampal brain activity. Conversely, reducing the exposure of glucocorticoids can protect against the age-related loss of hippocampal neurons. Furthermore, glucocorticoids in concentrations that are not in themselves toxic can impair the ability of hippocampal nerve cells to survive neurological insults such as reductions in blood flow and oxygen to the brain, and seizures. Together these studies suggest that prolonged exposure to glucocorticoids or stress can accelerate the loss of hippocampal neurons as well as increase their susceptibility to neurological insults.

Effects of chronic stress on brain cells in monkeys. (*a*) Normal nerve cells of the hippocampal formation. (*b*) Degenerating nerve cells from chronically stressed monkey. (© Dr. Dennis Kunkel/Phototake.)

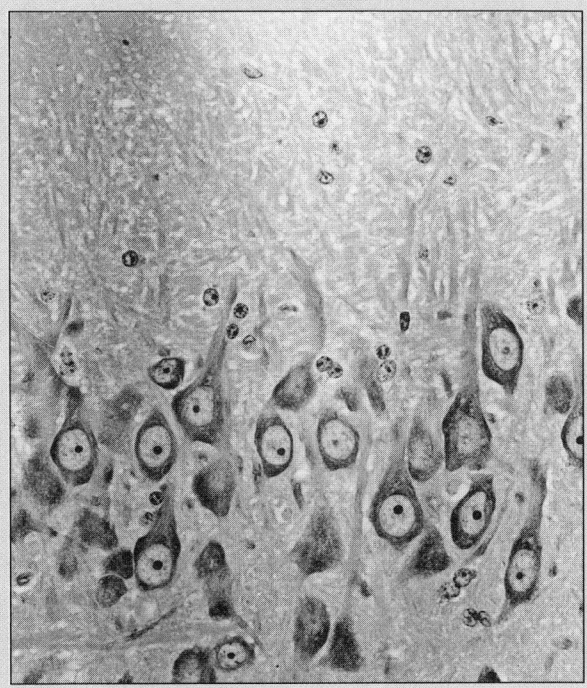

(a)

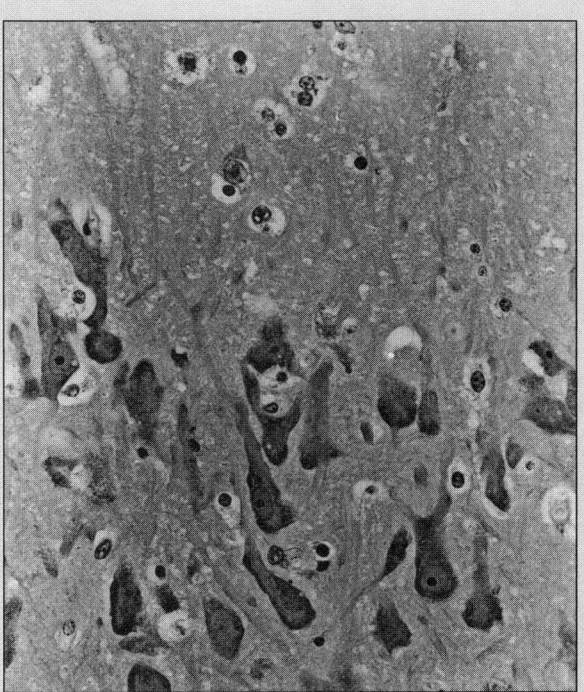

(b)

SUMMARY

Organization of the Autonomic Nervous System

The autonomic nervous system (ANS), a regulatory system that attempts to maintain a constant internal environment, has both central and peripheral components.

The central component of the ANS consists of the hypothalamus, brain stem, and spinal cord. The peripheral ANS can be further subdivided into the *sympathetic* and *parasympathetic* nervous systems. The sympathetic nervous system prepares the body to respond to stressful conditions, while the parasympathetic nervous system prepares the body for vegetative functions such as digestion. The sympathetic and parasympathetic nervous systems are anatomically quite different.

Both preganglionic sympathetic nerve fibers and preganglionic parasympathetic fibers release acetylcholine from their nerve terminals. Postganglionic sympathetic nerve fibers typically release norepinephrine from their nerve terminals, while postganglionic parasympathetic nerve fibers release acetylcholine.

Control of Bodily Organs by the Sympathetic and Parasympathetic Systems

Organs with vegetative functions are activated by parasympathetic inputs and inactivated by sympathetic inputs. Organs involved in stress responses are activated by sympathetic inputs and inactivated by parasympathetic inputs.

Reflexes Governed by the Autonomic Nervous System

Autonomic reflexes are carried out primarily in the spinal cord.

Reflexes such as the urinary bladder reflex and the defecation reflex are activated by stretch receptors and play important roles in coordinating the actions of smooth muscles and sphincters.

The convergence of pain information from internal organs and regions on the surface of the body is the basis of referred pain.

Regulatory Systems of the Autonomic Nervous System

Osmoreceptors that detect plasma osmolarity are located in different parts of the hypothalamus and regulate thirst, osmolarity, and water volume.

Brain stem responses involved in respiration are activated by chemoreceptors sensitive to oxygen, carbon dioxide, and H^+.

Brain stem responses that control blood pressure are activated by stretch receptors that sense pressure changes.

Autonomic Functions of the Hypothalamus

The hypothalamus is the center for the "fight or flight" response. It coordinates the variety of bodily changes needed for survival during life-threatening situations. It increases blood flow and oxygen to the brain, heart, and skeletal muscles while reducing blood flow to the organs involved with the digestion of food.

Hypothalamic Control of the Adrenal Gland

The adrenal gland coordinates the body's response to stress by releasing epinephrine, norepinephrine, and glucocorticoids into the circulatory system. Epinephrine and norepinephrine in the blood stream have effects similar to the activation of sympathetic nerve fibers; the onset of their actions, however, is delayed, since they must course through the circulatory system.

REVIEW QUESTIONS

Identify Each with a Term

1. The phrase used to describe the coordinated physiological changes seen in response to stress.

2. The hormones released from the adrenal medulla during stress.

3. The neurotransmitter released by postganglionic parasympathetic nerve terminals.

Define Each Term

5. sympathetic nervous system
6. parasympathetic nervous system

7. referred pain
8. viscera

Choose the Correct Answer

9. Which type of receptor is found on the target cells of postganglionic parasympathetic nerve fibers?
    a. nicotinic receptor
    b. muscarinic receptor
    c. alpha receptor
    d. beta receptor

10. A drug that enhances the release of norepinephrine from nerve terminals in the heart would:

a. increase heart rate.

b. decrease heart rate.

c. increase the force of contraction.

d. both (a) and (c)

11. An increase in plasma osmolarity above 300 mOsm

   a. is detected by osmoreceptors in the aortic arch.

   b. causes the release of norepinephrine by the hypothalamus.

   c. increases the reabsorption of water by the kidney.

   d. both (a) and (c)

12. A drug that activates alpha receptors would:

   a. constrict blood vessels in skeletal muscles.

   b. dilate coronary blood vessels.

   c. cause renin secretion from the kidney.

   d. contract bronchial muscles in the lung.

Answer Each Question in One or Two Sentences

13. Describe the conditions of the extracellular environment needed for optimal cellular function, and how these conditions are maintained by the autonomic nervous system.

14. Describe the role of the autonomic nervous system in the process of urination.

15. How does the autonomic nervous system regulate thirst?

16. What are the physiological changes that occur during the "fight or flight" response?

SUGGESTED READINGS

Appenzeller, O. *The Autonomic Nervous System*, 3rd ed. New York: Elsevier, 1982.

Ciriello, J., Calaresu, F.R., Renaud, L.P., and Polosa, C. *Organization of the Autonomic Nervous System—Central and Peripheral Mechanisms*. New York: Alan R. Liss, Inc., 1987.

Gabella, G. *Structure of the Autonomic Nervous System*, London: Chapman and Hall, 1976.

Gebber, G.L., and Barman, S.M. "Lateral tegmental field neurons of cat medulla: A potential source of basal sympathetic nerve discharge." *Journal of Neurophysiology*, 54:1498–1512, 1985.

Hilton, S.M., and Smith, P.R. "Ventral medullary neurones excited from the hypothalamic and midbrain defence areas." *Journal of the Autonomic Nervous System*, 11:35–42, 1984.

Janig, W. "The Autonomic Nervous System." In *Human Physiology* 2nd ed., edited by R.F. Schmidt and G. Thews, 333–370. Berlin: Springer-Verlag, 1989.

Reis, D.J., Ruggiero, D.A., and Morrison, S.F. "The C1 area of rostral ventrolateral medulla: A critical brainstem region for control of resting and reflex integration of arterial pressure." *American Journal of Hypertension*, 2:363S–374S, 1989.

Sapolsky, R.M., Uno, H., Rebert, C.S., and Finch, C.E. "Hippocampal damage associated with prolonged glucocorticoid exposure in primates." *Journal of Neuroscience*, 10:2897–2902, 1990.

Schramm, L.P. "Spinal factors in sympathetic regulation." In *The Molecular Basis for the Central and Peripheral Regulation of Vascular Resistance*, edited by A. Magro, 303–352. New York: Plenum Press, 1986.

Uno, H., Tarara, R., Else, J.G., Suleman, M.A., and Sapolsky, R.M. "Hippocampal damage associated with prolonged and fatal stress in primates." *Journal of Neuroscience*, 9:1705–1711, 1989.

KEY TERMS

adrenocorticotropic hormone (ACTH) (p. 391)

alpha receptor (p. 375)

baroreceptor (p. 386)

beta receptor (p. 375)

carotid sinus (p. 386)

chemoreceptor (p. 386)

corticotropic-releasing hormone (CRH) (p. 391)

epinephrine (p. 391)

"fight or flight" response (p. 391)

glucocorticoids (p. 391)

hypothalamus (p. 390)

interomedial lateral nuclei (IML) (p. 372)

medulla (p. 386)

parasympathetic nervous system (p. 372)

referred pain (p. 382)

sympathetic nervous system (p. 372)

vagus (p. 373)

vasopressin (p. 384)

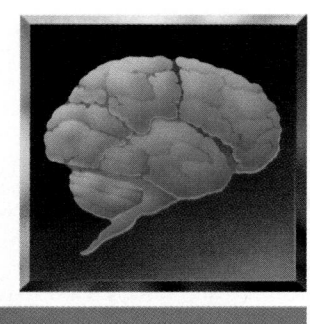

C H A P T E R
11

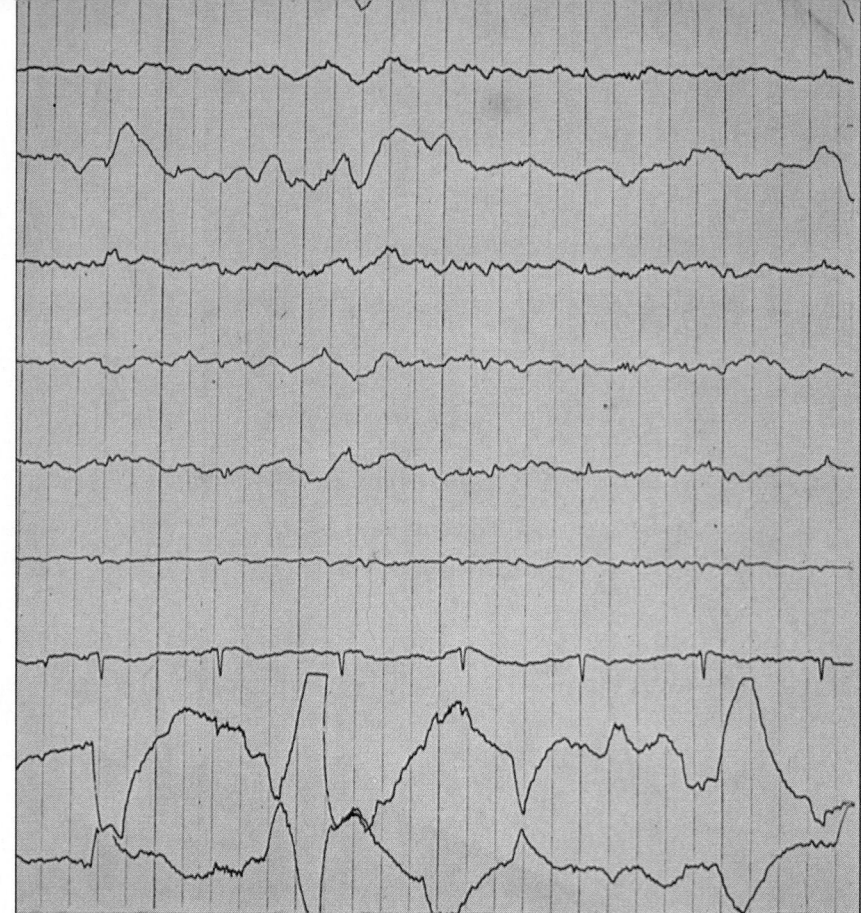

Central Integrative Systems

In the previous chapters, we saw how the nervous system detects and processes sensory information from the external environment and how the motor system puts out information to skeletal muscles for locomotion and fine movement. These sensory and motor functions were described one by one, suggesting that they occur in isolation. In reality, however, human behavior is highly coordinated in order to carry out purposeful actions. The brain areas involved in these coordinated behaviors are part of the **central integrative systems.** Central integrative systems can be subdivided into several broad categories. Motivational systems, for example, are those responsible for our drive to satisfy such basic needs as hunger and thirst. Other systems carry out functions such as learning and memory, which endow us with the ability to acquire and store new information. Still others endow us with the ability to communicate in the form of written and spoken language. In this chapter, we will consider the brain areas and the mechanisms responsible for these integrative functions. We will also examine how the brain brings together and integrates information from its two hemispheres, and how

functions are localized on one side of the brain. We begin with how the brain controls bodily rhythms by acting as a master clock.

Bodily Rhythms, Cyclic Systems, and the Hypothalamus

The previous chapter described the role of the hypothalamus in maintaining the body's internal environment, and its role in neuroendocrine functions will be discussed in Chapter 13. In this chapter, we will consider the role of the hypothalamus in controlling biological rhythms and motivational systems. The hypothalamus is composed of many different cell clusters (Fig. 11–1), several of which are involved in regulation of body rhythms.

Rhythms of the body vary on daily or monthly cycles. **Monthly cycles** are those seen in the process of ovulation, in which the female body prepares itself for reproduction by the release of gonadotropic hormones from the hypothalamus (see Chapter 30).

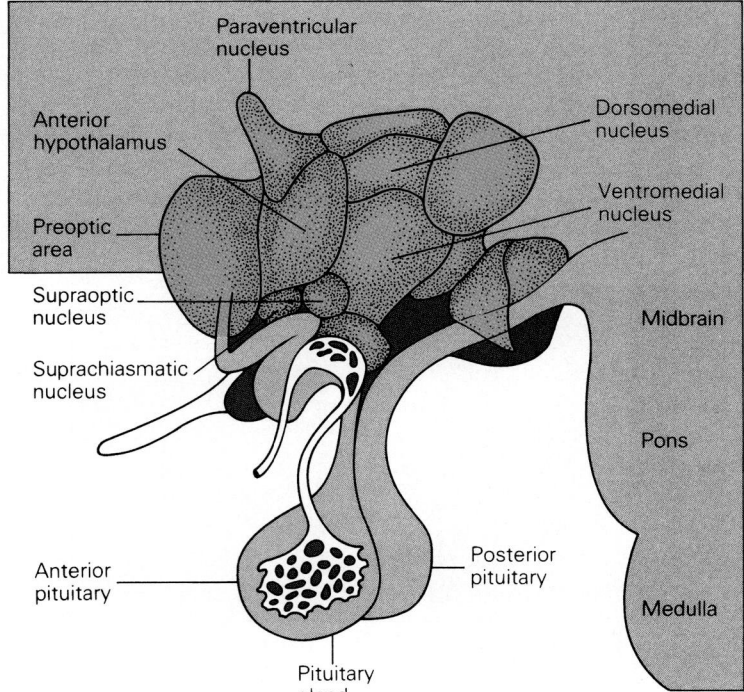

Figure 11–1
The hypothalamus comprises several different cell clusters or nuclei, several of which are involved in central integrative systems of the body.

Diurnal (daily) or **circadian** (*circa* = "approximately" and *dies* = "day") **rhythms** are those that vary on a daily basis.

Circadian Rhythms and the "Master Clock"

Well over 100 bodily functions and biochemical processes have been found to vary in accordance with a 24-hour schedule. Body temperature, for example, decreases during the night and increases during the day (Fig. 11–2). This rhythmic variation is controlled by the hypothalamus, which influences the secretion of the hormone **thyroxin** by the thyroid gland. This hormone in turn speeds up the rate at which cells consume oxygen, metabolize glucose, and produce heat. The control by the hypothalamus of other hormones such as **ACTH** and **cortisone** has a similar circadian rhythm (see Chapter 14).

Many of the body's rhythms have periods that are longer or shorter than 24 hours. The sleep-wake cycle, for instance, has a 25-hour period. The body's rhythms, however, are entrained and synchronized to external factors such as the 24-hour light-dark cycle, or to our work schedules.

The "master clock" thought to be responsible for coordinating various biological rhythms is the **suprachiasmatic nucleus** of the hypothalamus (see Fig. 11–1). this nucleus receives information about light and darkness directly from the retina of the eyes. By this pathway, the light-dark cycle entrains the body's biological clock. The destruction of this nucleus in experimental animals disrupts their sleep-wake cycle, along with other bodily functions.

The uncoupling of the body's rhythms can result in fatigue and poor mental and physical performance. On long airplane flights, most of us have experienced the fatigue of "jet lag." When we cross different time zones, our bodily functions are no longer synchronized with the day-night cycle; hormonal fluctuations, body temperature, and the sleep-wake cycle are desynchronized. Eventually, the synchrony of these cycles is reestablished and the body's rhythms again function in harmony.

Sleep and Wakefulness

Despite research efforts of many scientists, the need for a sleep-wake cycle remains a mystery. Sleep has various stages in which the nerve cells are very active. According to some theories, the increased activity of certain nerve cells during sleep creates the internal stimuli necessary for the maturation of the brain during development. Other theories suggest that sleep and dreams strengthen the nerve cell connections associated with the memories that were formed during the day.

The Stages of Sleep

During one night of sleep, the brain displays several different stages of electrical activity (Fig. 11–3). This activity can be monitored by electrodes placed on the scalp that produce a record, called an **electroencephalogram,** or **EEG,** of brain activity. The EEG of a person who is awake exhibits electrical activity that is low in amplitude and high in frequency. During the onset and progression of sleep, the fre-

Figure 11–2
Body temperature shows a rhythmic variation throughout the day and night.

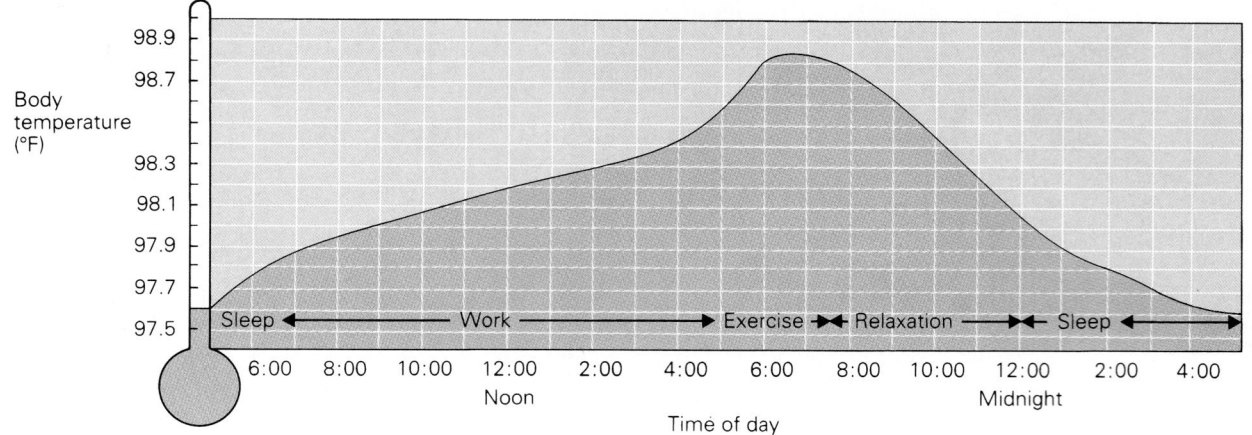

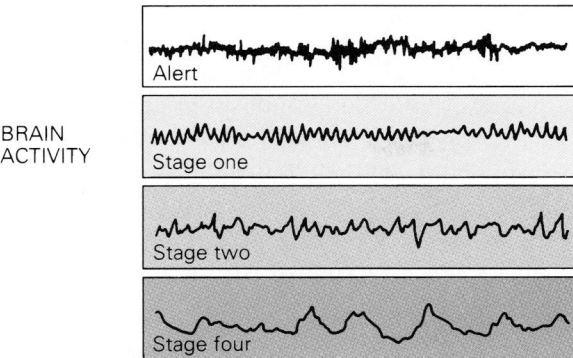

BRAIN ACTIVITY

Alert

Stage one

Stage two

Stage four

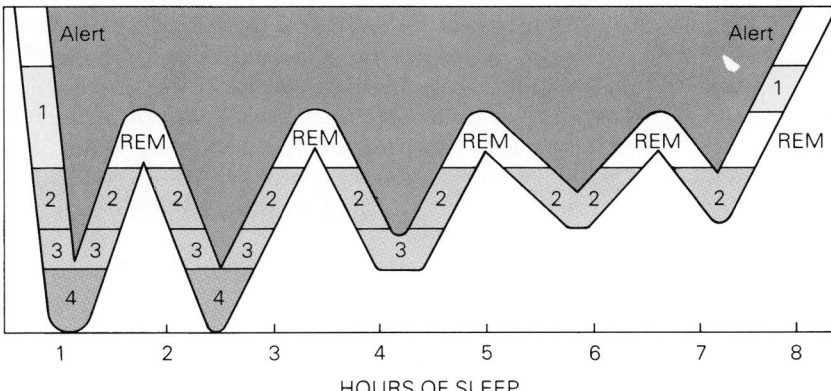

HOURS OF SLEEP

Figure 11–3
(*a*) The electrical activity of the brain during the various stages of sleep can be recorded by means of electroencephalograms. (*b*) During one night's sleep, a person goes through three to five cycles, each cycle including several sequential sleep stages.

quency of the electrical activity decreases while the amplitude increases.

Stages 1 to 4. Based on the changes of frequency and amplitude of the brain's electrical activity, four distinct stages of sleep have been identified. Stage 1 of sleep has the lowest amplitude and highest frequency of electrical activity, while Stage 4 has the highest amplitude and lowest frequency. An individual passes through each stage in sequence an average of 3 to 5 times during the night. In the first 45 minutes of sleep, the brain progresses from Stage 1 to Stage 4. During this progression, the skeletal muscles are relaxed, but a sleeper frequently tosses and turns to readjust the sleeping position. In addition, heart rate and blood pressure decrease during Stage 4, while gastrointestinal motility increases. (These latter functions are associated with the parasympathetic activation of the autonomic nervous system, discussed in Chapter 10).

REM Sleep. During the next 45 minutes of sleep, the electrical activity of the brain reverts back from Stage 4 to Stage 1. In Stage 1 of this sleep cycle, the

skeletal muscles, except those that control the movement of the eyes, are completely inhibited from moving. In fact, eye movement becomes very rapid during this phase of sleep and it is thus referred to as **rapid eye movement (REM)** sleep. During REM sleep, heart rate and blood pressure also increase, while gastrointestinal motility decreases. These changes are associated with the activation of the sympathetic division of the autonomic nervous system (ANS).

In general, REM sleep is associated with visual dreaming. One night's sleep contains about 3 to 5 REM episodes. Each successive episode is longer and is characterized by increases in the electrical activity of the brain and the number of visual images. Dreams are primarily visual; however, people who are born blind have auditory dreams. Individuals who lose their sight eventually lose their ability to dream visually.

The deprivation of REM sleep in humans has no apparent long-term effects on behavior. When permitted to sleep undisturbed after REM deprivation, subjects initially exhibit more REM sleep. Normal sleeping patterns, however, are soon reestablished.

Variations with Age. The amount of sleep and the proportion of REM and non-REM sleep vary with age. In general, the amount of total sleep required decreases with age, with the greatest changes occurring for REM and Stage 4 sleep (Fig. 11–4). During the last few weeks of gestation, for example, REM sleep occupies about 80% of the total sleep time. This amount decreases to 50% for full-term newborns and levels off at 25% by 10 years of age. The amount of total sleep time spent in Stage 4 also decreases rapidly with age. Unlike REM sleep, however, Stage 4 sleep often disappears after 60 years of age.

Figure 11–4
Total sleep time decreases with age, as does the percentage of total sleep time spent in REM sleep and in Stage 4.

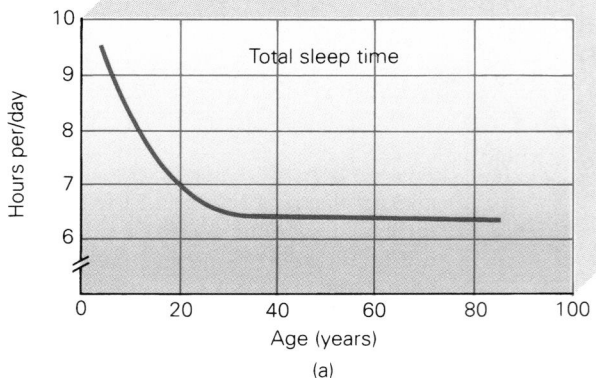

(a)

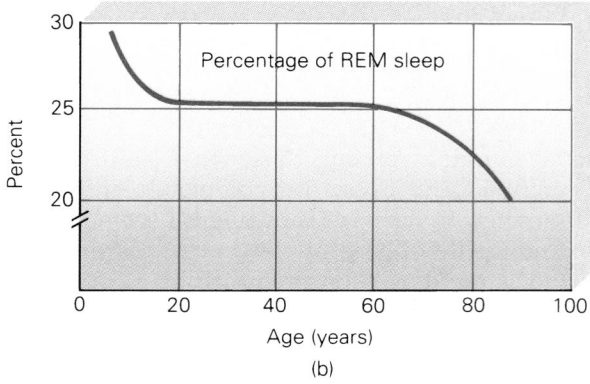

(b)

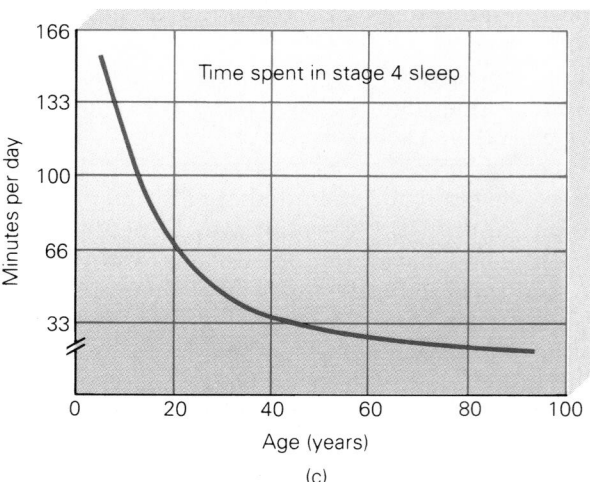

(c)

Neural Mechanisms of Sleep

The areas of brain responsible for the sleep-wake cycle are the hypothalamus and brain stem. Electrical stimulation of the **preoptic area** (Fig. 11–5) of the hypothalamus induces non-REM sleep, and the destruction of this area of brain causes insomnia in laboratory rats. The preoptic area is located near the body's biological clock, the **suprachiasmatic nucleus** (Fig. 11–5), which receives direct input from the retina to provide information about the light-dark cycle. Thus, the suprachiasmatic nucleus is thought to provide inputs to the preoptic area to entrain the sleep-wake cycle.

Sleep also can be induced by injections of serotonin into the preoptic nucleus. A possible internal source of this serotoninergic innervation is the **raphe nucleus** (Fig. 11–5) of the brain stem. The destruction of the raphe causes insomnia in laboratory animals. With small lesions of the raphe, non-REM sleep gradually returns, but REM sleep does not. Thus, it is thought that the raphe is responsible for the generation of REM sleep. The descending fibers from the raphe nucleus innervate the spinal cord and may be responsible for inhibiting the activity of motor neurons of skeletal muscles. In this way, the motor programs that are planned and programmed during dreaming and REM sleep are normally prevented from being carried out.

Disorders of Sleep

Insomnia is a general term for the inability to sleep. Normal causes of insomnia include disruptions of the body's circadian rhythms such as those due to the travel across time zones. **Narcolepsy** is a severe disorder in which an individual suddenly lapses into sleep during the middle of the day. Narcolepsy is thought to result from the activation of the REM-sleep generator. Narcoleptic attacks that occur during the waking state have all of the characteristics of REM sleep. These attacks often last from 90 to 100 minutes, the period of time normal for REM sleep. In addition, they are accompanied by vivid hallucinations similar to REM dreams. Finally, narcoleptic attacks are associated with a lack of muscle tone, except for eye movement.

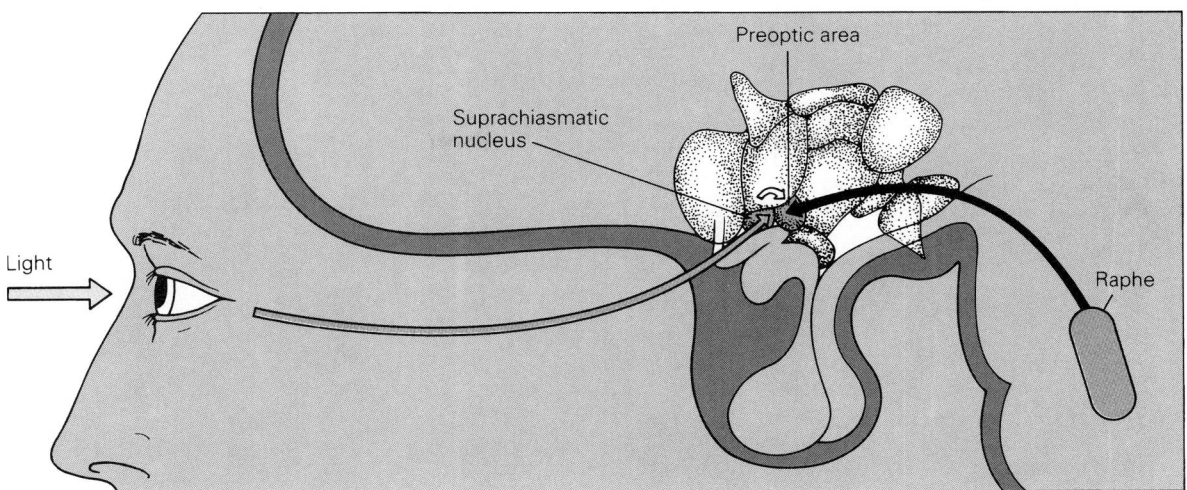

Figure 11–5
The suprachiasmatic nucleus, the body's "biological clock," receives information about light directly from the retina and is thought to regulate the sleep-wake cycle with the cooperation of the preoptic nucleus. The raphe nucleus of the brain stem is thought to be involved in the generation of REM sleep.

Narcolepsy is an inherited disorder; it is thought that the defect is the inability to inhibit the REM generator. The primary inhibitory inputs to the REM generator are noradrenergic nerve fibers. Consequently, drugs such as amphetamines that enhance norepinephrine release are used in treating this disorder.

Motivational Systems: Maintaining an Internal Balance and Ensuring Survival

Motivational systems initiate behaviors that satisfy bodily needs. These systems typically operate to ensure the survival of an individual or species. In some cases, motivational systems are driven by signals that represent the imbalance of certain chemicals within the body. Hunger and thirst, for example, are activated by internal signals that communicate the need for food and water in order to provide energy for the body and to maintain a constant level of osmolarity within the body's internal environment. Other motivational systems, such as those responsible for sexual drive, are activated by external signals from members of the opposite sex as well as internal signals from our own bodies. In this section, we will consider the motivational systems of hunger, thirst, and sexual drive, and the brain's so-called "reward system."

Hunger: Maintaining Levels of Glucose and Triglycerides

The consumption of food is required for the growth, maintenance, and repair of all organ systems, and for the production of energy required by all cells in the body. The hypothalamus plays a central role in regulating food intake by acting as a sensor of glucose availability (Fig. 11–6). This process is part of the motivational system involved in the short-term regulation of food intake.

The long-term regulation of food intake appears to involve the maintenance of body weight. Research has shown that laboratory animals forced to gain weight by eating an excessive amount of food will reduce their food consumption and revert back to their original body weight when left to eat at their own pace. Animals that are forced to lose weight by starvation will also eventually gain weight back to their original level when allowed free access to food. These studies suggest that the body may have a "set-point" for weight, which it attempts to maintain.

The maintenance of body weight is now thought to operate in an attempt to maintain a constant amount of **adipose tissue,** or **fat cells.** Fat cells store **triglycerides,** a major source of energy for the body (see Chapter 2). In adults, the amount of adipose tissue remains relatively constant, and there is a good correlation between body weight and the amount of adipose tissue. The variable that is

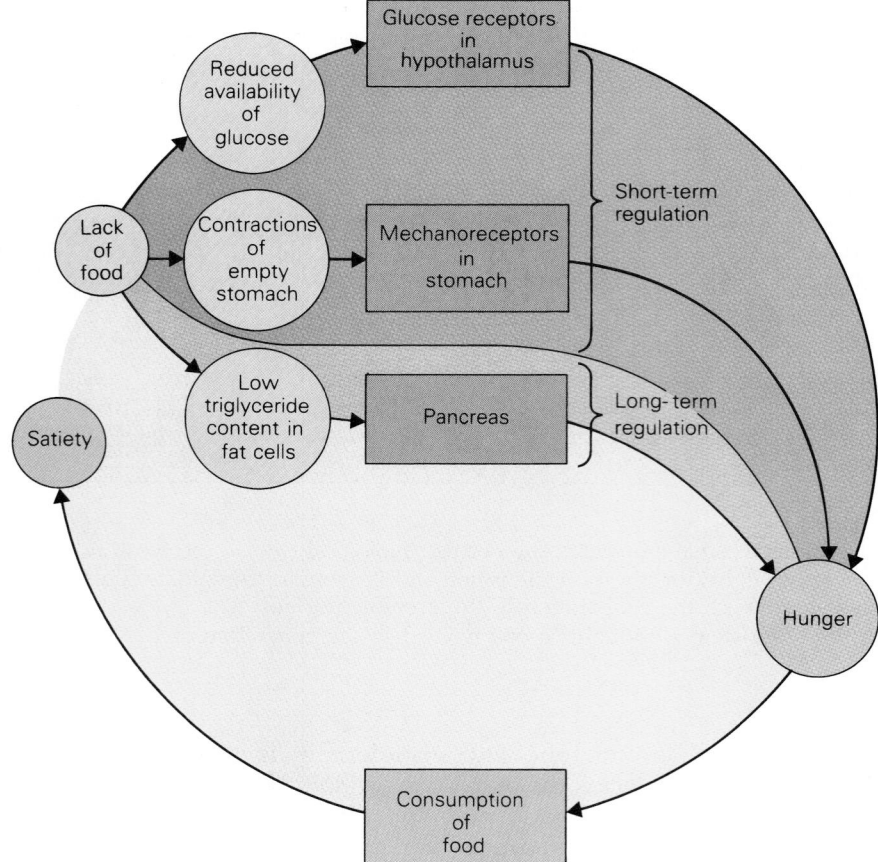

Figure 11–6
Three monitoring devices cooperate to produce the feeling of hunger and resulting consumption of food: (1) glucose receptors in the hypothalamus detect the lack of available glucose; (2) mechanoreceptors detect the contractions of an empty stomach; and (3) the pancreas detects low levels of triglycerides in fat cells.

thought to be regulated in the maintenance of adipose tissue is the volume of individual fat cells. These cells are thought to release a chemical signal in amounts proportional to their triglyceride content. This adipose signal affects the level of **insulin,** a hormone produced by the pancreas, in the circulatory system. Research has shown that basal insulin levels are proportional to body adipose tissue. Circulating insulin, in turn, penetrates the blood-brain barrier and binds to receptors in the ventral hypothalamus. The activation of these hypothalamic insulin receptors lowers food intake and, subsequently, body weight. Thus, insulin is involved not only in the control of circulating glucose after a meal (see Chapter 26), but also in the long-term regulation of body fat and hence of body weight (see Fig. 11–6). What remains to be determined about the long-term control of adipose tissue is how adipose cells regulate the levels of insulin in plasma.

The other aspect of the control of hunger is **satiety,** or the feeling of fullness after a meal. The mechanisms of satiety involve the ventromedial hypothalamus and were discussed in Chapter 10.

Thirst: Maintaining Osmolarity of the Blood

The human body consists of approximately 60% water. Water escapes from the body in the form of sweat and urine. It also leaves the body as water vapor in air exhaled from the lungs. In order to maintain a constant amount of water within the body, thirst mechanisms motivate us to drink water; in addition, the ANS and kidneys act in concert to retain water that would normally be excreted as urine. The amount of water in the body affects the amount of water in the blood, its **osmolarity** (see Chapter 2), which must be maintained at a constant level in order for blood cells to function. Consequently, the regulation of thirst, water volume, and osmolarity is interrelated.

The osmolarity of the plasma is a critical factor in the internal environment of the cells in the body. As explained in Chapter 4, hyper- and hypo-osmotic fluids cause cells to shrink or swell, altering their function. The desired osmolarity for blood plasma is 300 mOsm. Increases above this value in plasma osmolarity can occur from excessive dehydration and sweating. This increase is detected by osmoreceptors in the **supraoptic** nucleus and **paraventricular nucleus** of the hypothalamus (Fig. 11–7). In response to increases in osmolarity, these nuclei secrete the hormone **vasopressin,** also called **antidiuretic hormone (ADH),** into the blood. Eventually, this hormone acts on the kidneys, causing them to reabsorb more water. This reabsorption of additional water reduces the plasma osmolarity to 300 mOsm.

The activation of hypothalamic **osmoreceptors** also initiates thirst mechanisms that involve somatic responses leading to the consumption of water. The intake of water in turn reduces plasma osmolarity.

As the feeling of thirst is satisfied, water consumption subsides. This happens, however, long before the water is absorbed into the circulatory system and long before plasma osmolarity is restored to normal. It is thought that the cessation of water consumption is triggered by stretch receptors in the stomach. How this information is transmitted to the brain has not yet been determined.

Figure 11–7
Osmoreceptors in the hypothalamus detect water deficiency and spark the release of ADH (which leads the kidneys to retain water) as well as the feeling of thirst (leading to the consumption of water). These two mechanisms decrease the osmolarity of the blood to normal levels.

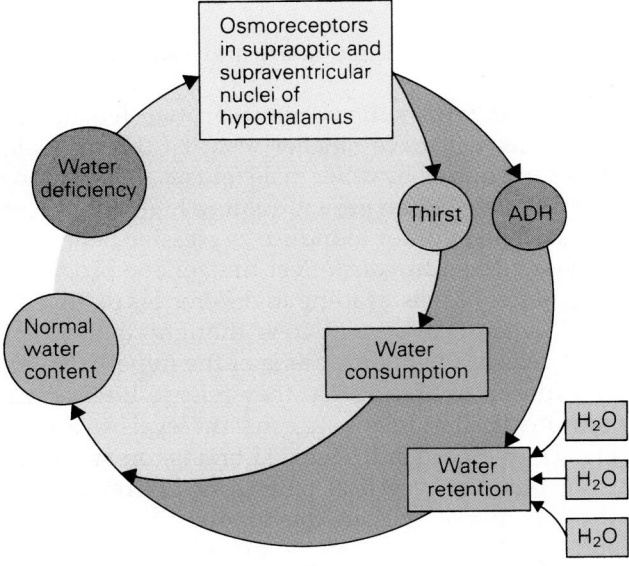

The Sexual Drive: Perpetuating the Species

Mating is a basic drive common to all mammals. It is essential for reproduction and the continuation of the species. The sexual behaviors of the males and females of a species are quite different. Yet among different species there is a great deal of similarity in the mating posture of the animals. During mating, males display a **mounting** posture, and the female adopts a receptive posture called **lordosis.** In studies of laboratory animals, the frequency of these two postures is used as a measure of sexual drive. In both males and females, the **anterior hypothalamus** (see Fig. 11–1) appears to be one of the brain centers for sexual drive. Lesions of this area of brain in experimental animals reduce their sexual activity.

The stimulus for activating sexual behavior in many species is the release of chemicals called **pheromones.** Ovulating female monkeys (those who are prepared to conceive), for example, emit pheromones along with other vaginal secretions. These chemical odors are detected by the olfactory system of male monkeys and enhance their sexual drive. Sensory information from the male olfactory system is transmitted to the pyriform cortex and amygdala and indirectly to the hypothalamus (Chapter 8). In lower animals, pheromones play a prominent role in sexual behavior. In humans, other social cues come into play. The use of perfume or cologne as a sexual attractant, however, is still a common practice.

Mounting and lordosis are also used as indicators of differences in sexual behavior or sexual phenotype. Under certain circumstances, both the male and female animals can adopt the sexual behavior of the opposite sex.

The development of a particular pattern of sexual behavior is a result of the presence or absence of particular steroid hormones called **androgens** during **critical periods** of development. The presence of androgens during the fetal and early postnatal periods causes the brain to develop male sexual-behavioral characteristics, while the absence of these steroids results in the development of female characteristics (Fig. 11–8). In laboratory mice, for example, the removal of the testis (the primary source of androgens) from males soon after birth causes these animals to develop lordotic behavioral responses as adults. If given ovaries, these males will ovulate as would a normal female. Likewise, the sexual behavior of a female can also be altered if the hormonal environment is altered during critical periods of development. Injections of androgens into newborn female mice cause them to develop

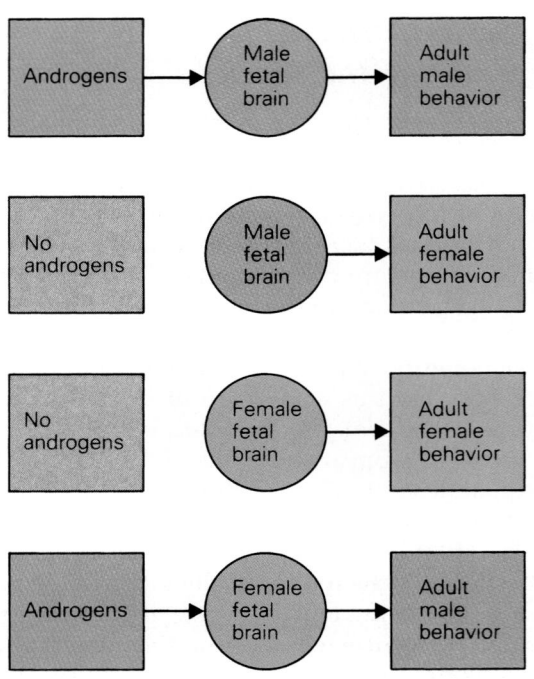

Figure 11–8
In laboratory experiments involving mice, exposure (or lack of it) to androgens during fetal development has been shown to affect adult sexual behavior. If a male fetal brain is exposed to androgens during critical periods of development, the adult will exhibit male behavior. If androgen exposure does not occur, the adult male will exhibit female behavior. If a female fetal brain is not exposed to androgens during critical periods, the adult will exhibit female behavior, but if the female fetal brain is exposed to androgens, the adult will exhibit male behavior.

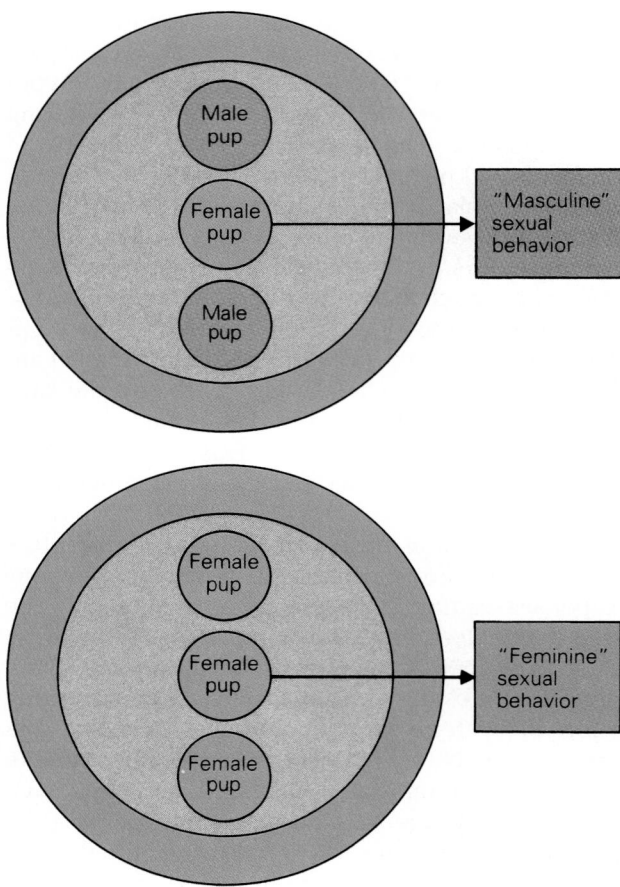

Figure 11–9
Fetal environment can affect the adult sexual behavior of laboratory animals. A female fetus that develops between two males will show male behavior as an adult, while a female surrounded by other females in the fetal stage will develop as a female.

male mounting behaviors as adults. Research has shown that the critical time period for the exposure (or lack of exposure) to androgens in order to affect brain-sex alterations in laboratory animals is the first few days after birth. In humans, androgen activity is elevated during the 12th and 22nd weeks of gestation and during the first six weeks after birth.

The fetal environment of these laboratory animals also influences the sexual behavior that these mice will display as adults. The positioning of a fetal mouse in the uterus is a random process. A female fetus, for instance, could be placed in between two male fetuses, two females, or in between a male and a female. Research has shown that female fetuses placed between two male fetuses are exposed to higher androgen levels than if placed between two female fetuses (Fig. 11–9). As adults, these "androgenized" females are more aggressive; they exhibit male mounting behaviors, display irregular estrus

(ovulation) cycles, mate later, and cease ovulating earlier.

A male fetus situated between two females in the uterus will have smaller testicles than a male fetus surrounded by other male fetuses. In addition, these animals, when grown, require higher doses of testosterone in order to induce aggressive behavior.

How do androgens affect the genetic programming of nerve cells, leading to differences in "male" and "female" behavior? It is thought that when nerve cells in the preoptic area of the hypothalamus are exposed to androgens, they release **luteinizing hormone (LH)** at relatively constant levels as adults (Fig. 11–10). Nonandrogenized brains, on the other hand, develop a cyclical pattern of LH release. In addition, nerve cells in the preoptic area of nonandrogenized female brains are sensitive to estro-

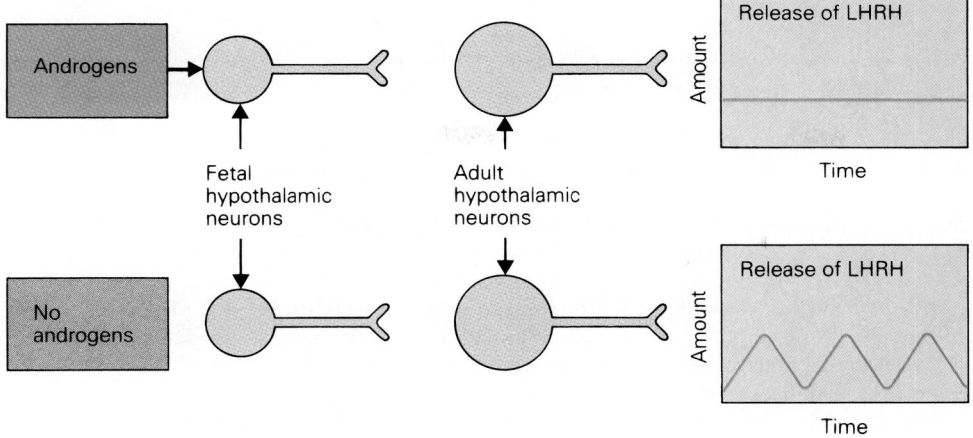

Figure 11–10
When nerve cells in the preoptic area of the hypothalamus are exposed to androgens, they release luteinizing hormone releasing hormone (LHRH) at relatively constant levels as adults, while preoptic nerve cells of animals not exposed to androgen show a cyclic pattern of LHRH release.

gen secreted by ovarian follicles. Androgenized brains are not affected by estrogen. Thus, the function of nonandrogenized preoptic neurons has been altered not only in their release of LH, but also in their sensitivity to estrogen.

Alterations in sexual behavior are also observed in humans. In a condition known as **androgen insensitivity,** males are incapable of responding to androgens. This disorder is a genetic defect transmitted by the mother. Males with this defect develop as females in their external appearance as well as in their psychosexual preference.

Drugs can also affect the androgenic environment of the developing brain. Barbiturates, pesticides such as DDT, and hormones such as **diethylstilbestrol (DES)** (at one time given as a "morning after" birth control pill and also to prevent premature births) have been shown to masculinize the brain.

Learning and Memory Systems

Prior to our birth, and from the moment we are born, we acquire information and knowledge about our environment. In the womb, babies learn to distinguish the sounds of their mothers' voices. Soon after birth, they are conditioned to orient their heads to their mothers' breasts in order to receive milk. Eventually, babies are trained to use a toilet in the excretion of body waste. The acquisition of knowledge is called **learning,** and the retention of knowledge is called **memory.**

Associative and Nonassociative Learning

Research has shown that there are two types of learning: **associative learning** and **nonassociative learning.** In nonassociative learning, we learn about a single type of stimulus and adapt to it according to its relevance to our survival. The sound of a ticking clock, for example, may initially draw our attention, but after a period of time we ignore it. This type of behavior, in which we learn to decrease our response to a repeated stimulus, is called **habituation.** In contrast, we can become more sensitized to certain other sounds. In school, for example, we learn to recognize the sound of a fire alarm and respond accordingly. **Sensitization** is thus an increase in response to a stimulus.

In one type of associative learning, an animal acquires an understanding about the relationship or association between one stimulus and another. This learning process is called **classical conditioning.** An example of classical conditioning is the salivation of a dog in response to a stimulus of light. Normally, the dog salivates only in response to a piece of food. After a light stimulus (the **conditioned stimulus** or **CS**) has been paired with the presence of a piece of meat (the **unconditioned stimulus** or **UCS**) in several conditioning trials, however, the light alone is

R E S E A R C H   F O C U S

Drugs of Abuse and the Reward System of the Brain

The reward system of the brain was first discovered when stimulating electrodes were placed into the brains of laboratory animals, which were then allowed to control the amount of electrical current passing through the electrodes. By pressing a bar, animals were able to deliver current to the electrodes. When the electrodes were placed in specific areas of the brain, these animals would press the bar well over 100 times per minute (Figure 1). When given a choice of receiving food instead of the electrical current, animals would choose the electrical stimulus, starving as a result. The reward areas of the brain thus constitute a powerful motivational system that overrides drives such as hunger, sex, and, in some cases, the avoidance of pain and punishment.

The areas of the brain that are associated with the reward system are shown here (Figure 2). They include the **substantia nigra, hypothalamus, nucleus accumbens, caudate nucleus,** and the **frontal cortex.** The common thread linking these areas together is the **dopamine nerve fiber pathway** that originates in the substantia nigra. This pathway contains axons that form synaptic connections with each of the reward areas of the brain.

The release of dopamine by these synaptic connections is critical to the reward process. Drugs such as amphetamine and cocaine that enhance the release of dopamine also augment the rate of self-stimulation. Drugs that block dopamine release or its ability to bind to its receptor decrease the rate of self-stimulation. The reward system may therefore play a role in the process of addiction to certain abused drugs.

Figure 1.
Rat with a brain-stimulating electrode.
(© Yoav/Phototake.)

Figure 2.
The reward system of the brain involves the substantia nigra, the hypothalamus, the nucleus accumbens, the caudate nucleus, and the frontal cortex.

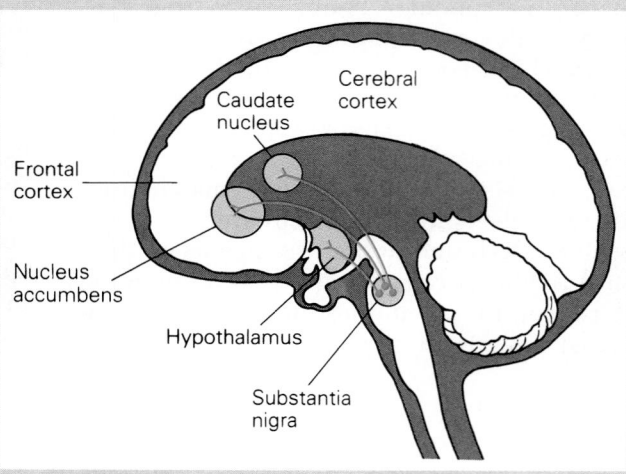

sufficient to cause the dog to salivate (salivation is the **conditioned response** or **CR**). The dog has been **conditioned** to salivate in response to light.

The other type of associative learning is called **operant conditioning.** In this type of learning, animals acquire an understanding of the relationship between their own behavior and subsequent rewards (or punishment). With this type of learning, animals tend to repeat behaviors that are rewarded and avoid behaviors that are punished.

In both classical and operant conditioning, a critical factor is the time interval between the appearances of the two stimuli, or of a behavior and its reward. With classical conditioning, if the time between the appearances of a conditioned and unconditioned stimulus is long, the conditioned response is poor. Likewise, with operant conditioning, if the reward is delayed, the conditioned behavior is weak.

Only certain types of stimuli are capable of producing a conditioned response. The most effective stimuli are those relevant to the survival of the animal. Food poisoning, for example, is a case of conditioning in which nausea acts as a form of punishment. Subsequent exposure to the taste of the food causes an animal to avoid it. The conditioning of food aversion is most effective when taste is used as the stimulus. If other stimuli such as sounds or lights instead of taste are paired with nausea, the animal does not develop an aversion to these stimuli. Consequently, the suitability of a stimulus depends largely on the nature of the response to be learned.

Short- and Long-term Memory

Memory has different components and progresses through different stages. Information is first stored briefly in **short-term memory.** Short-term memory is used, for example, in the action of dialing a telephone number after looking it up in the telephone book. After the phone call, we typically forget the number if we no longer need it. In contrast, if the telephone number happens to be one we use often, we will most likely retain it in **long-term memory.**

Short-term and long-term memory have different characteristics. Short-term memory is easily disrupted and short-lived, while long-term memory is stable, less volatile, and long-lasting. To be stored in long-term memory, information must first be transferred from short-term memory. In addition, protein synthesis is required for the formation of long-term memory. If the information is disrupted while it is in short-term memory, it will not be permanently registered in long-term memory. The process of forming long-term memory from short-term memory is called **consolidation.**

Memory Circuits: Neuronal Pathways for Learning

Learning and memory systems are found in several areas of the brain. These systems are associated with the interconnection of nerve cells in the form of neuronal **memory circuits** that are necessary for learning and memory to occur. Memory circuits have been identified for activities such as the learning of motor skills, conditioning of cardiovascular responses, and visual discrimination, as well as for other types of learning.

The learning of skilled movements such as playing the piano and typing are examples of motor learning. These types of behaviors, however, are very complex and therefore difficult to study in terms of memory circuits. A simpler paradigm that can be used to study motor learning is the conditioning of the eye-blink reflex. In this paradigm, which has been used in the rabbit, a puff of air blown on the cornea of the eye causes the eyelid to blink (an unconditioned stimulus producing an unconditioned response). If the puff of air is paired with a tone, the sound of the tone alone will ultimately cause the eye to blink (a conditioned stimulus producing a conditioned response). The neural pathways used in this learning paradigm are (1) somatic sensory pathways that transmit information about the puff of air, and (2) motor pathways that cause the eyelid to blink (Fig. 11–11). In addition, during the conditioning process, the sound of the tone becomes a predictor of the puff of air and eventually conditions the same eye-blink response. Consequently, information from the auditory pathway must ultimately affect the motor pathways to result in the eye blink. The critical memory circuits involved in this conditioning process are the connections between the inputs to and the nerve cells within the cerebellum (see Fig. 11–11).

Information about the puff of air is transmitted through unconditioned stimulus pathways. In the cerebellum, the pathway used by an unconditioned stimulus is the climbing fiber inputs to the Purkinje cells. The conditioned stimulus, the tone, is transmitted through conditioned stimulus pathways. In the cerebellum, this pathway is the mossy fiber input.

The output of the conditioned responses from the cerebellum to motor nuclei that control eye blinking is by way of the interpositus and red nucleus (see Fig. 11–11). The interpositus is one of the three deep cerebellar nuclei that provide the major output from the cerebellum (see Chapter 9).

What changes occur in the memory circuit to account for learning? During the conditioning process, activation of the climbing fiber pathway (the US pathway) causes the Purkinje cells to generate a complex spiking pattern of action potentials (Fig. 11–12). As conditioning progresses, fewer complex spikes are produced in response to activation of the US inputs. Well trained animals display a decrease in the number of simple spikes generated in response to activation of the mossy fiber inputs. The net effect is a reduction in Purkinje cell activity. As we saw in Chapter 9, the Purkinje cell terminals release GABA, an inhibitory neurotransmitter. Consequently, a reduction in Purkinje cell activity in es-

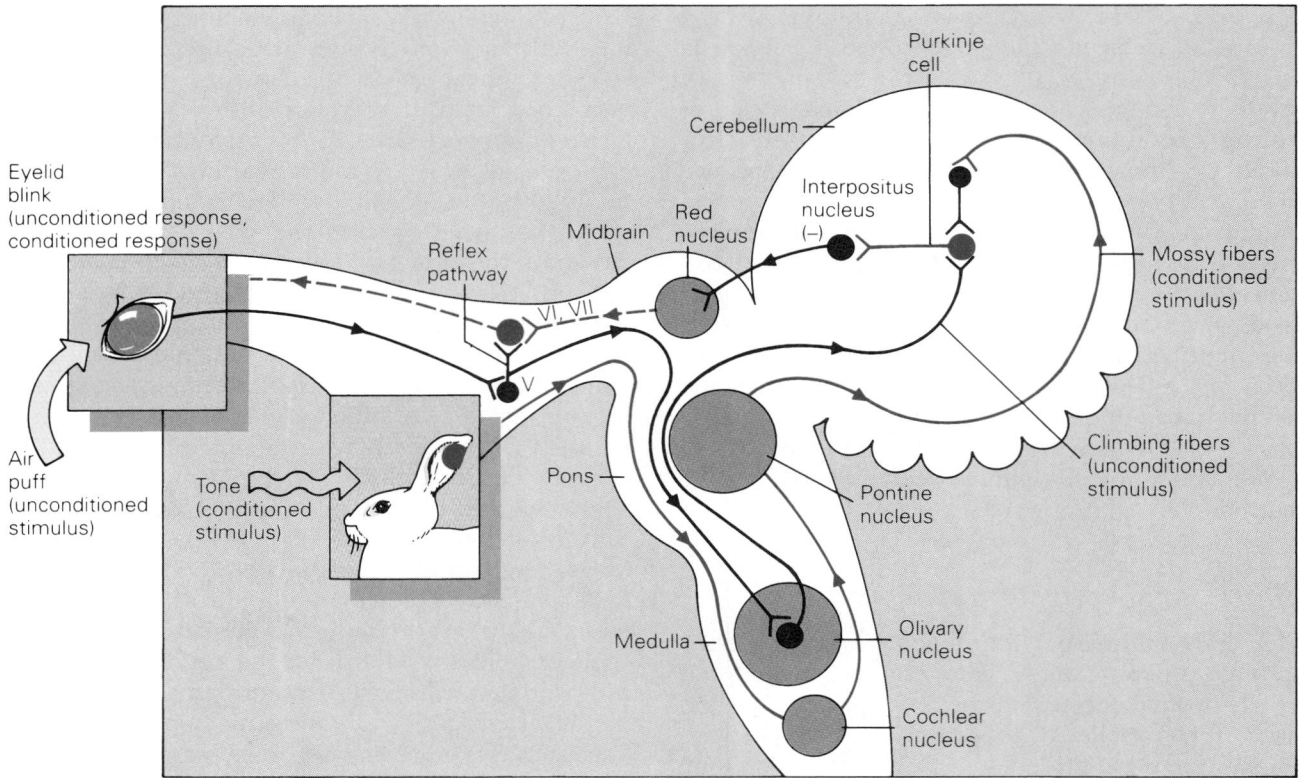

Figure 11–11
A learning paradigm for the eye-blink reflex in a rabbit. Somatic sensory pathways transmit information about a puff of air (an unconditioned stimulus), which causes the eyelid to blink through the reflex pathway. This response eventually becomes linked to the sound of a tone (a conditioned stimulus). The auditory-motor pathways then cause the eyelid to blink.

sence causes the nerve cells in the interpositus nucleus to become more active.

The critical dynamic changes that occur during learning take place in the Purkinje cells or within inputs that control the activity of Purkinje cells. The decrease in Purkinje cell activity is a form of conditioned habituation. The possible cellular mechanisms of these alterations are discussed later in this chapter.

Other memory circuits that have been identified in the brain include the connections of the hypothalamus and amygdala in learning a conditioned cardiovascular response, and of the primary and secondary visual cortices in visual memory. Memory circuits in other areas of the brain are currently being studied in association with other memory functions, and a diversity of these circuits will most likely be identified throughout the brain.

Other areas of the brain, such as the temporal lobe and the hippocampus, also play a prominent role in learning and memory. Patients undergoing neurosurgical operations report that electrical stimulation of the temporal lobes causes them to "hear" melodies that they heard or learned in the past.

The hippocampus plays a role in the transformation of short- to long-term memory. Lesions of the hippocampus were at one time made in attempts to prevent seizures in patients with epilepsy. After the surgery, long- and short-term memory in these patients remained intact. However, their ability to convert short-term to long-term storage was lost. Patients who were taught a new task, for example, could perform the task as long as they were not interrupted. If they were temporarily distracted, however, and then asked to perform the task again, they were not able to do so. In addition, although they were able to recognize people they had known before the operation, these patients were unable to remember faces of new individuals to whom they were introduced after the operation. As a consequence, these patients continued to "live in the past" as they formed no new memories.

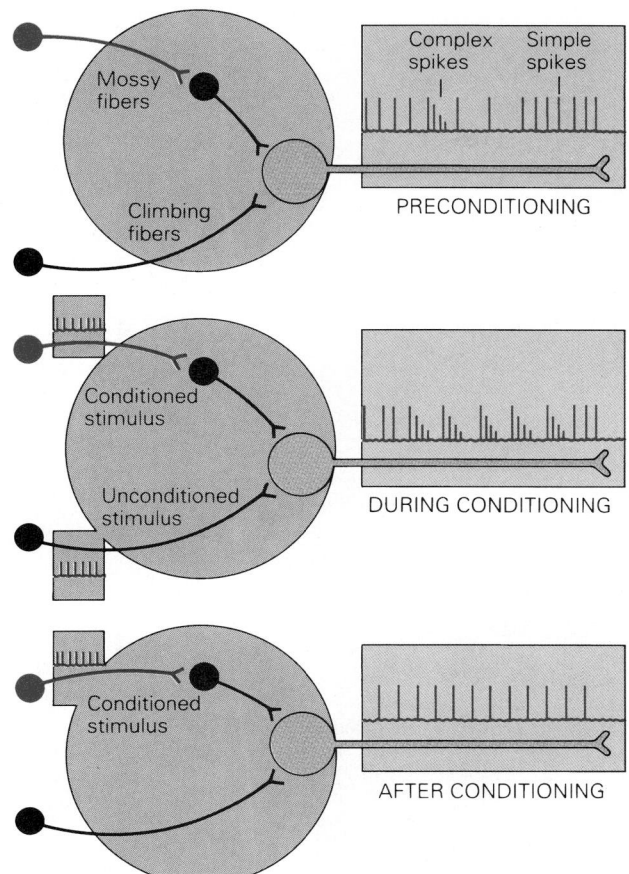

Figure 11–12
During the conditioning process, the Purkinje cells of the cerebellum show a complex spiking pattern of action potentials, which decrease as conditioning progresses.

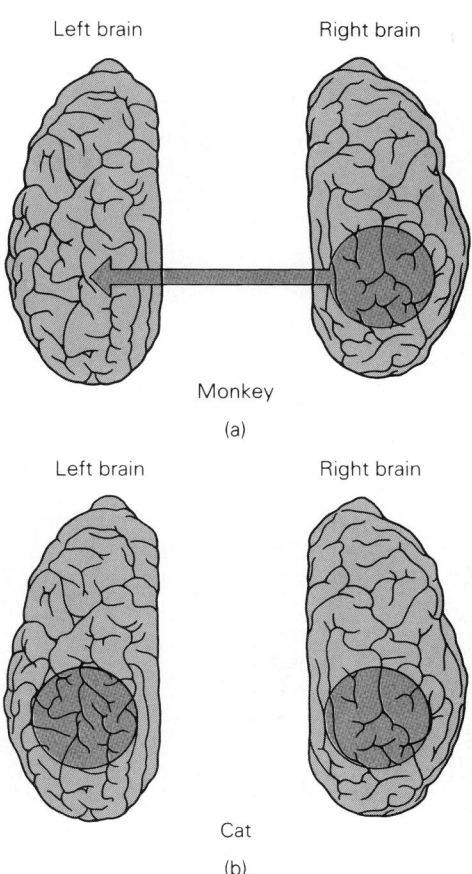

Figure 11–13
In monkeys, learning stored in one hemisphere of the brain is not accessible to the other if the two hemispheres have been disconnected, while in the brain of a cat, information is stored in both hemispheres and so is accessible to both even when disconnected.

Learning and the "Split Brain"

"Split-brain studies" of experimental animals reveal that information reaching the cerebral hemispheres is processed and stored differently for different species. Studies of monkeys demonstrate that learned responses stored in one hemisphere are not accessible by the other when the fibers connecting the two hemispheres are disconnected (Fig. 11–13). Thus, it is suggested that the hemispheres must normally work together in order to exchange stored information. In studies with cats, on the other hand, learned responses stored in one hemisphere are also transferred and stored in the other. Consequently, when the hemispheres are disconnected after storing learned information, each hemisphere is still capable of producing the correct response. Different species of mammals therefore approach the storage of information in different ways.

Cellular Mechanisms of Learning

Although it is not yet possible to study the cellular mechanisms of human learning and memory, studies of many different species are beginning to reveal common mechanisms that may also operate in humans. These cellular mechanisms involve alterations of synaptic function. These alterations can come about by changes in presynaptic or postsynaptic mechanisms of transmission, or by the formation of new synaptic connections.

Habituation: A Decrease in Synaptic Transmission

Habituation, or the decrease in a behavioral response to a repeated stimulus, can be explained at the cellular level in terms of a decrease in synaptic

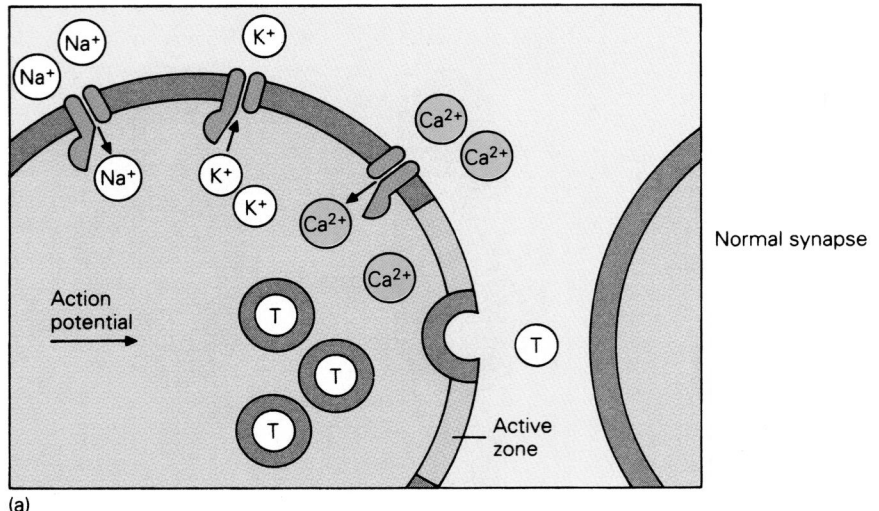

(a)

Normal synapse

Na⁺

Habituated synapse

(b)

Facilitated synapse

(c)

Figure 11–14
At the cellular level, habituation is thought to
occur because of a decrease in synaptic transmission due to a decrease in the influx of Ca^{2+}
(*b*). Sensitization at the cellular level is thought
to involve an increase in synaptic transmission,
initiated by an influx of Ca^{2+} (*c*). This influx
occurs as a result of action by a facilitating interneuron, which causes cAMP to phosphorylate a
potassium channel, thereby closing it and creating an electrochemical gradient favorable to the
influx of Ca^{2+}.

transmission as a result of the inactivation of Ca^{2+} channels in the presynaptic terminal (Fig. 11–14b). The reduction of Ca^{2+} influx in turn decreases the amount of neurotransmitter released into the synaptic cleft. In addition, the number of synapses that contain active zones for the release of vesicles decreases, as does the area of the active zones. Together, these mechanisms diminish the functional capacity of the synapse.

Sensitization: An Increase in Synaptic Transmission

In contrast to habituation, cellular mechanisms of **sensitization** enhance an animal's response to a particular stimulus. The underlying mechanism is called **presynaptic facilitation** and it involves interneurons that form synapses with axons of other neurons (see Fig. 11–14). These facilitating interneurons enhance the release of a transmitter by raising the levels of cAMP, which in turn phosphorylates and turns off K^+ channels. The reduction in K^+ efflux effectively prolongs the action potential and as a consequence allows the Ca^{2+} channels to stay open longer. With an increase in Ca^{2+} influx, more vesicles release their transmitter contents into the synaptic cleft. The sensitization process is also associated with an increase in the number of synapses with active zones and a doubling in the area of the active zones. These changes, in conjunction with the enhancement in Ca^{2+} influx, suggest that an overall increase in transmitter release is the basis of the sensitization process.

Classical Conditioning: A Sensitization Process

Although there may be a variety of different cellular mechanisms of classical conditioning for different memory circuits, one mechanism has been found to be similar to that of the sensitization process. This mechanism leads to an enhanced response to a conditioned stimulus. The activation of a conditioned stimulus by this mechanism not only excites the postsynaptic neuron, but temporarily raises the intracellular level of Ca^{2+} in the presynaptic terminal (Fig. 11–15b). As we saw in Chapter 5, an increase in intracellular Ca^{2+} levels can activate a calmodulin-mediated increase in adenylate cyclase activity. As a consequence, cAMP levels are amplified, leading to the phosphorylation and inactivation of K^+ channels and a prolongation of the action potential. As in the sensitization response, the prolongation of the action potential results in a greater Ca^{2+} influx at the terminals and an enhancement in transmitter release.

Thus, the pairing of the unconditioned stimulus and the conditioned stimulus activates cellular processes that alter the excitability of the postsynaptic cell. As consequence, the CS input alone is ultimately capable of activating the cell.

The cellular mechanisms of the classically conditioned eye-blink response previously described remain to be determined. Since the response is associated with a decrease in Purkinje cell activity, however, the cellular mechanism of this conditioning process would most likely be similar to that of habituation.

The Effects of Use and Nonuse of Neural Pathways

During muscle exercise, there is an eventual change in muscle tone and muscle mass. In a similar way, the use and nonuse of neural pathways can strengthen or weaken the connections between nerve cells. Animals raised in "enriched" environments have nerve cells that form more synaptic connections with other nerve cells than would normally appear. In contrast, animals raised in "impoverished" environments develop fewer synapses than normal.

The use and nonuse of pathways can also alter topographical maps in the cortex. The map of the somatic sensory cortex and the motor cortex, for example, can be modified when an animal is trained to use one finger more than the others. The amount of cortex representing the function of the exercised finger expands beyond that which is normally found. In contrast, if the nerve fibers innervating the finger are severed, the cortical space devoted to that finger is gradually eliminated and eventually incorporated into the maps of other fingers.

The results of these studies indicate that the brain is a dynamic organ system, capable of changing the strength of existing synaptic connections and in some cases forming new synapses as part of the process of learning and memory. Many of these alterations in synaptic function can take place only during certain critical periods of development, while other modifications can occur at all ages. These observations argue for continued learning and exposure to new challenges, regardless of age.

Language Systems

Language is a tool that allows the communication of thoughts, knowledge, feelings, and emotions in both written and spoken form. The deciphering of words from complex sounds or from complex visual

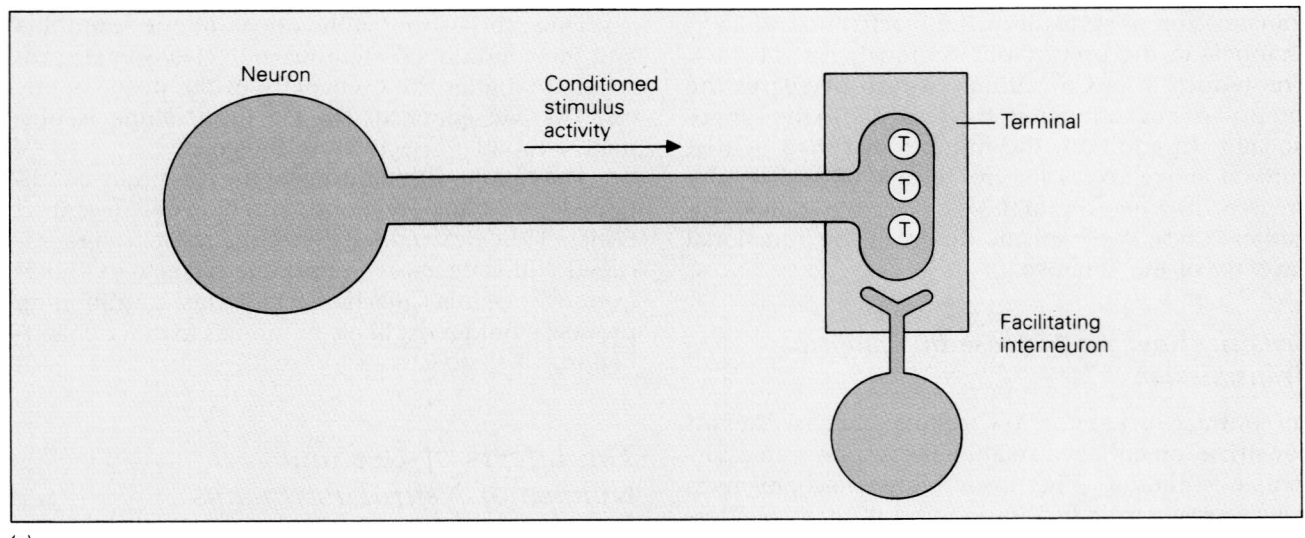

(a)

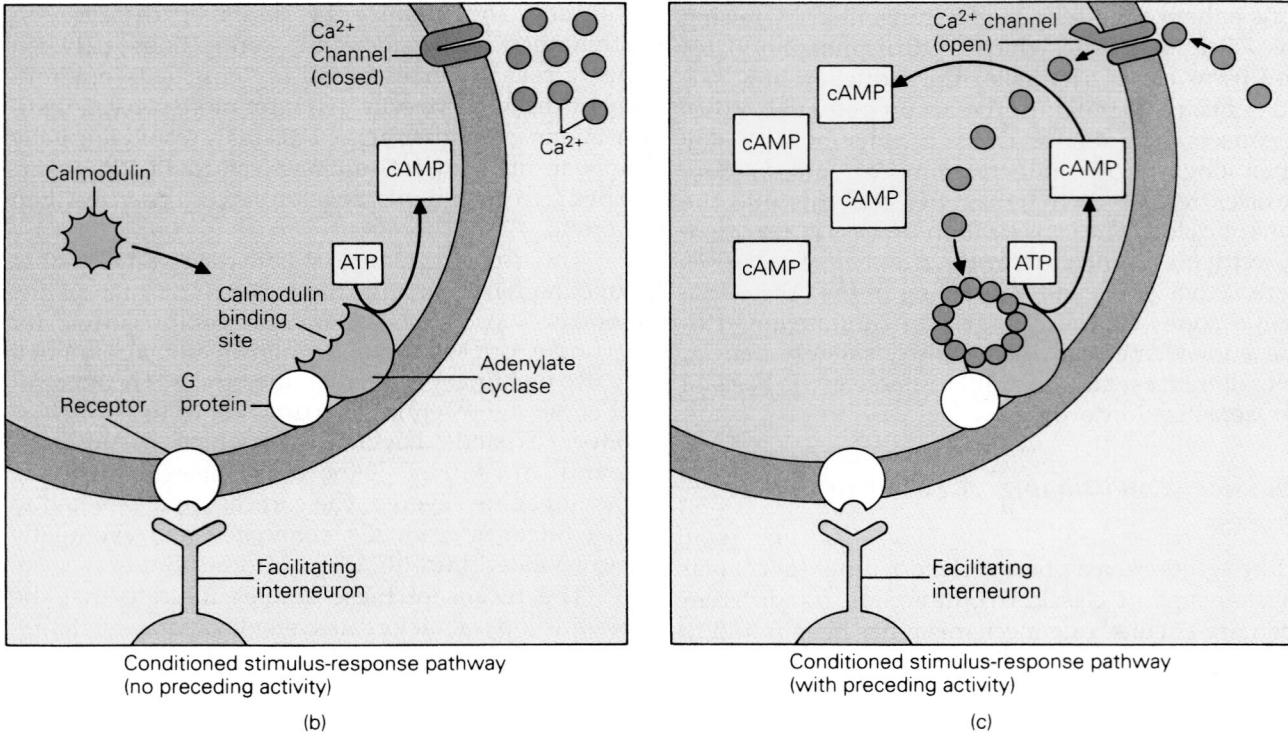

Conditioned stimulus-response pathway
(no preceding activity)

(b)

Conditioned stimulus-response pathway
(with preceding activity)

(c)

Figure 11–15

A mechanism for classical conditioning. (*a*) Interaction between a neuron along the conditioned stimulus-response pathway and a faciliting interneuron. (*b*) Enlargement of boxed area shown in figure above, illustrating the effect of no activity along the conditioned stimulus-response pathway. (*c*) Activation of conditioned stimulus-response pathway temporarily raises intracellular levels of Ca^{2+}, activating a calmodulin-activated increase in adenylate cyclase production of cAMP. This activity closes K^+ channels and prolongs the action potential.

symbols and the extraction of concepts from these words represent activities at the highest level of sensory integration. Written language allows us to transfer knowledge from generation to generation so that each generation can build upon the achievements of the past.

The development of language has thus played a critical role in the progress of humankind. It is difficult to determine when language was first used; however, fossil evidence suggests that the areas of the brain responsible for language existed over 500,000 years ago. Spoken language obviously developed before written language. Consequently, another leap in human evolution may have been the formation of connections between the visual and language areas of the brain, allowing people to develop symbolic writing as a method of communication.

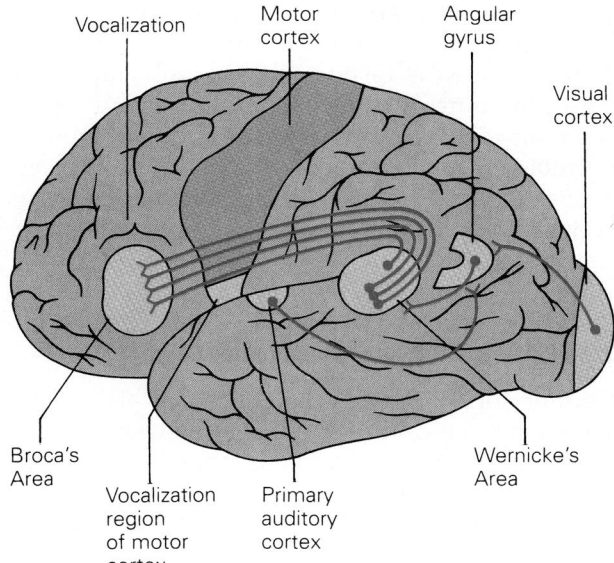

Figure 11–16
Language functions are localized in Wernicke's area and Broca's area. Other areas of the brain involved in language function are the primary auditory cortex, the visual cortex, the vocalization region of the motor cortex, and the angular gyrus.

The Dominant Hemisphere: Comprehension and the Motor Aspects of Language

The hemisphere in which the language centers are located is called the **dominant hemisphere.** Typically, the speech area in the dominant hemisphere is larger than the same region in the nondominant hemisphere. This difference in size is apparent in the human fetus by the 31st week of gestation and suggests that the development of language and speech may be an innate process. Indeed, infants are born with the ability to learn all human languages. This ability to learn other languages fluently after acquiring a first language, however, appears to diminish with age.

Language functions are located in regions of the brain called **Wernicke's area** and **Broca's area** (see Fig. 11–16). Wernicke's area is responsible for the comprehension of language, while Broca's area is responsible for motor aspects of speech and language. In right-handed individuals, these language centers are found in the left hemisphere. The majority of left-handed people also have language centers in the left hemisphere; however, many left-handed individuals have language centers in both hemispheres or only in the right hemisphere.

For our understanding of spoken and written language, both auditory and visual information must reach the language centers. The reading of a word on a page, for example, requires that sensory information about the shape and form of the letters be transmitted from the retina to the primary and secondary visual cortices (see Chapter 8). This information is then transmitted to an association cortex

called the **angular gyrus.** This area of brain is thought to be involved with the association of visual, auditory, and tactile sensations (Fig. 11–16). From the angular gyrus, information about the word is transmitted to Wernicke's area, where it is thought to be recognized or comprehended. From Wernicke's area, information about the word is transmitted to Broca's area, where the motor programs for speech reside. Broca's area in turn contains nerve fibers that transmit information to the regions of the motor cortex responsible for control of movements of the mouth and vocal cords for speaking and of the hand for writing.

The Nondominant Hemisphere: Intonation and the Emotional Aspects of Language

The regions of the nondominant hemisphere of the brain that reside in the location corresponding to the location of Wernicke's and Broca's areas in the dominant hemisphere are responsible for the **affective** aspects of language (aspects related to mood or emotion). The manner in which a person speaks depends largely on his or her mood, and a listener can interpret simple statements in different ways depending on the speaker's tone of voice. The region in the nondominant hemisphere comparable to Wernicke's area is responsible for the comprehension of the tone of voice in speech, while the nondominant region comparable to Broca's area is responsible for the expression of intonation. Damage to these areas of brain causes **aprosodias,** disorders producing the inability to understand or express intonation.

Disorders of Speech and Language

Disorders of speech and language can also be explained in terms of the connections between the language centers in the dominant hemisphere. Disorders of language resulting from damage to the brain are called **aphasias.** A congenital disorder affecting reading is **dyslexia.**

Aphasia: Damage to the Language Centers

Several types of aphasia occur. In **Broca's aphasia,** damage to Broca's area disrupts the motor programs that control speech and writing. Patients with this disorder lose their ability to speak fluently and grammatically and are unable to express their ideas in writing.

Damage to Wernicke's area produces an aphasia characterized by loss of language comprehension. Patients with this disorder can still speak and write but have difficulty understanding the spoken and written language.

Dyslexia: A Congenital Reading Disability

Dyslexia is more common among boys than among girls, and also more common among left-handers than among right-handers. Dyslexics have difficulty

R E S E A R C H   F O C U S

Dyslexia: A Reading Disorder with Neuroanatomical Abnormalities

Dyslexia is a reading disorder of developmental origin. Children with this disorder cannot learn to read with normal proficiency despite specialized instruction. Although these children appear to have normal intellectual ability and motivation, they have difficulty associating letters with their corresponding sounds, and distinguishing letters that are similar in form, such as "p" and "b."

Dyslexics represent about 2% to 16% of school-age children. Male dyslexics outnumber female dyslexics by 3:1, and the disorder is more common among left-handed children. Research indicates that inheritance plays an important role in the transmission of dyslexia.

Neurological signs often related to dyslexia include stuttering, abnormal electroencephalograms (measures of brain electrical activity), and problems involving manual dexterity of one or both hands.

Anatomical abnormalities are also apparent. The brain region required for the interpretation of language, Wernicke's area, is notably smaller in dyslexics, and the cell layers are disorganized. In addition, blood vessels are found in regions where they are not normally present. An abnormally high percentage of dyslexics have dominant right hemispheres.

Current theories argue that testosterone activity in dyslexics during the prenatal period is abnormally high. This alteration in testosterone preferentially affects the development and function of the left hemisphere. As a consequence, handedness and speech (stuttering) are affected, along with the manifestation of dyslexia.

A person with dyslexia learning to write. (© Will & Deni McIntyre/Science Source/Photo Researchers, Inc.)

associating letters with their corresponding sounds, and distinguishing letters that are similar in form, such as "p" and "b." Research has revealed that the **planum temporale,** a region that includes Wernicke's area, is reduced in size in dyslexics. In addition, the layers of nerve cells in people with dyslexia have not properly separated in this region, suggesting that the migration of neurons has been altered during development.

Laterality of Brain Function

The localization of language to one hemisphere is only one example of the laterality of brain function. Other functions, such as musical and spatial abilities, also exhibit a lateralization to one hemisphere. Tests show that in right-handed subjects, the left ear is better at recognizing music, while the right ear is better at recognizing verbal material.

Laterality of Spatial Abilities

Studies of so-called split-brain patients who have undergone surgery to disconnect the two hemispheres have shown that spatial abilities appear to reside in the right hemisphere. This surgical procedure cuts through the corpus callosum and is used as a last resort to control the spread of epileptic seizures. When lesions of the corpus callosum are made, the two hemispheres are no longer able to send information from one side to the other.

Patients with lesions of the corpus callosum who are asked to reassemble a jigsaw puzzle are unable to do so with their right hands, but are able to piece the puzzle together with their left hands. This is because the left hand is controlled by the right hemisphere, which houses the spatial abilities.

The spatial abilities of the right hemisphere are also seen in the processing of the written language. The Japanese, for example, use two writing systems: one consists of 71 letters that form the basis of phonetic (sound-related) symbols, and the second consists of over 40,000 **characters** or ideographic symbols, each one representing a separate idea or object. The first system uses the letters to aid in the pronunciation of a word, while the second uses "pictures" to represent the word. Although both the phonetic and ideographic systems of writing use the language centers of the left hemisphere, the ideographic system also uses the right side of the brain.

Lesions of the angular gyrus near the language centers in the left hemisphere impair reading in the phonetic system of writing but generally not in the ideographic system. Even in cases when ideographic reading is impaired, patients are able to explain the meaning of the word. For the same word written in the phonetic system, they lose this ability. Thus ideographic writing, which uses the language centers in the dominant hemisphere, also relies on the spatial abilities of the nondominant hemisphere.

Differences of Laterality in Male and Female Brains

Differences in the development of the lateralization of brain function exist between males and females. In tests that evaluate the development of spatial function, males exhibit a lateralization of spatial tasks to the right hemisphere by 6 years of age. With females, spatial function is equally developed in both hemispheres until the age of 13.

The absence of early hemispherical specialization imparts certain advantages to females. Damage to the left hemisphere in childhood, for example, impairs language development in males more than in females. It appears that the delay in hemispherical specialization allows the transfer of language function from the left to right hemisphere to occur more readily in females.

To some extent, differences in the laterality in brain function between males and females are also seen in adults. Damage to the left hemisphere is more strongly associated with language disorders in adult males than in adult females. Likewise, damage to the right hemisphere is more strongly associated with nonlanguage disorders in males than in females. These observations suggest that the female brain is more symmetrical in function than that of the male.

SUMMARY

Bodily Rhythms, Cyclic Systems, and the Hypothalamus

Many bodily functions fluctuate on a monthly and/or daily basis. Those that vary daily are called circadian rhythms. Many of these rhythms are entrained to the 24-hour light-dark cycle by the suprachiasmatic nucleus of the hypothalamus. The desynchronization of bodily rhythms can result in fatigue and poor mental and physical performance.

The cycle of sleep and wakefulness varies on a 25-hour schedule but is entrained to a 24-hour period. Sleep involves repeated cycles of brain activity, each having similar stages.

Motivational Systems: Maintaining an Internal Balance and Ensuring Survival

Hunger is the desire for food, a motivational system that operates to satisfy the energy needs of the body and the need for growth, development, and maintenance of the body's organs. Hunger is regulated by both short- and long-term mechanisms that lead to food consumption.

Thirst is a drive to maintain a constant level of osmolarity within the body's internal environment, through fluid retention and consumption.

Sexual drives are important for reproduction and the perpetuation of the species. They are activated by the interaction of social environment, hormones, and genetically programmed sexual behaviors. The development of a male or female sexual behavior is influenced by androgens during development.

The reward system of the brain is a motivational system that can override other motivational drives such as hunger and thirst.

Learning and Memory Systems

In nonassociative learning, an understanding of the properties concerning a single type of stimulus leads to either a habituation or sensitization to the stimulus. In associative learning, we acquire an understanding of the relationship either between two stimuli (called classical conditioning) or between a stimulus and a behavior (called operant learning).

The storage of information progresses from short-term memory to long-term memory. If information is disrupted while in short-term memory, it will not become consolidated into long-term storage.

Memory is stored in a variety of memory circuits found in different areas of the brain. Memory circuits have been identified for the learning of motor responses, cardiovascular responses, and visual perceptions.

In primates, information presented to one hemisphere is stored in that hemisphere. The opposite hemisphere has access to this information but is not capable of transferring this information for long-term storage.

Habituation is a decrease in response to a repeated stimulus. Sensitization is an increase in response to a repeated stimulus and is due to an increase in the release of neurotransmitters.

The connections between nerve cells can be altered by the amount of activity between the cells. The use of neural pathways can strengthen the connections between groups of nerve cells, while the nonuse of pathways weakens their connections.

Language Systems

Language centers are located in one or the other side of the brain depending on the individual. The hemisphere in which language is located in an individual is called the *dominant hemisphere*. Within the dominant hemisphere, Wernicke's area is responsible for the interpretation of language, and Broca's area for the motor aspects of speech.

The nondominant hemisphere plays a role in the affective aspects of language.

Disorders of speech and language include aphasias and dyslexia. Aphasias result from brain damage, while dyslexia is congenital.

Laterality of Brain Function

Spatial, artistic, and musical abilities are located predominantly in the right hemisphere.

Laterality of brain function develops earlier in males than in females. In the event of damage to one side of the brain, the opposite side is capable of taking over the functions of the damaged hemisphere.

REVIEW QUESTIONS

Identify Each with a Term

1. The term used to describe the hemisphere where language is localized.

2. The area of brain that exhibits an abnormal development in people with dyslexia.

3. The term for the sleep disorder in which individuals cannot prevent themselves from suddenly falling asleep during the day.

4. The area of brain responsible for REM sleep.

Define Each Term

5. synaptic facilitation
6. synaptic habituation
7. associative learning
8. REM sleep

Choose the Correct Answer

9. Which of the following statements about speech and language is *not* correct?

    a. Damage to Wernicke's area (22) produces a loss of comprehension for both the written and spoken language.

    b. Patients with damage to Wernicke's area (22) lose the ability to write.

    c. Damage to area 22 in the non-dominant hemisphere produces a loss in comprehending the intonation of speech.

    d. Patients with dyslexia have an abnormal development of the planum temporale.

10. In order to prevent seizure activity from spreading from one side of the brain to the other, a patient with a dominant left hemisphere undergoes surgical transection of the corpus callosum to disconnect the cerebral hemispheres. When tested after surgery, the patient can correctly identify the image of an object projected to the temporal aspect of his left visual field by:

    a. verbally naming the object

    b. selecting the object from among other objects with his right hand

    c. both a and b

    d. none of the above

11. Which of the following peptides is secreted by the supraoptic and paraventricular nucleus of the hypothalamus in response to an increase in blood osmolarity?

    a. prolactin

    b. oxytocin

    c. vasopressin

    d. beta-endorphin

12. Which of the following is *not* a cellular mechanism that leads to an enhanced response to a conditioned stimulus?

    a. activation of voltage-gated potassium channels

    b. increase in intracellular levels of calcium

    c. amplification of cAMP levels

    d. a calmodulin-mediated increase in adenylate cyclase activity

Answer Each Question in One or Two Sentences

13. What are the various stages of sleep? Which stage is typically associated with dreaming?

14. What are the brain areas involved in the control of sleep?

15. How is sexual behavior affected by androgens?

16. What are the differences between short- and long-term memory?

SUGGESTED READINGS

Dement, W., and Kleitmann, N. "Cyclic variations in EEG during sleep and their relation to eye movements, body motility, and dreaming." *Electroencephalography and Clinical Neurophysiology*, 9:673–690, 1957.

Domasio, A.R., and Geschwind, N. "The neural basis of language." *Annual Review of Neuroscience*, 7:127–147, 1984.

Gazzaniga, M.S., and LeDoux, J.E. *The Integrated Mind*. New York, NY: Plenum Press, 1978.

Goy, R.W., and McEwen, B.S. *Sexual Differentiation of the Brain*. Cambridge, Mass.: MIT Press, 1980.

Kandel, E.R., and Schwartz, J.H. "Molecular biology of learning: Modulation of transmitter release." *Science*, 218:433–443, 1982.

Olds, J. and Milner, P. "Positive reinforcement produced by electrical stimulation of septal area and other regions of rat brain." *Journal of Comparative Physiology and Psychology*, 47:419–427, 1954.

Sperry, R.W. "The great cerebral commissure." *Scientific American*, 210:42–52, 1964.

Steriade, M., and Hobson, J.A. "Neuronal activity during the sleep-wake cycle." *Progress in Neurobiology*, 6:155–376, 1976.

Thompson, R.F., Berger, T.W., and Madden IV, J. "Cellular processes of learning and memory in the mammalian CNS." *Annual Review of Neuroscience*, 6:447–491, 1983.

Zola-Morgan, S.M., and Squire, L.R. "The primate hippocampal formation: Evidence for a time-limited role in memory storage." *Science*, 250:288–290, 1990.

KEY TERMS

accumbens nucleus (p. 406)

adipose tissue (p. 401)

androgens (p. 403)

angular gyrus (p. 413)

aphasia (p. 414)

associative learning (p. 405)

Broca's area (p. 413)

circadian rhythm (p. 398)

classical conditioning (p. 405)

dominant hemisphere (p. 413)

dyslexia (p. 415)

electroencephalogram (EEG) (p. 398)

habituation (p. 409)

long-term memory (p. 407)

nonassociative learning (p. 405)

operant conditioning (p. 406)

REM sleep (p. 399)

sensitization (p. 411)

short-term memory (p. 407)

Wernicke's area (p. 413)

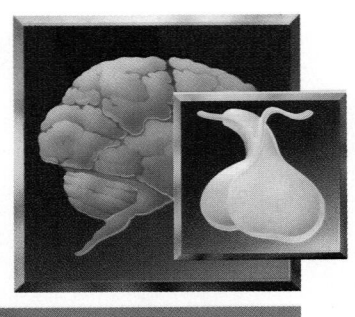

C H A P T E R
12

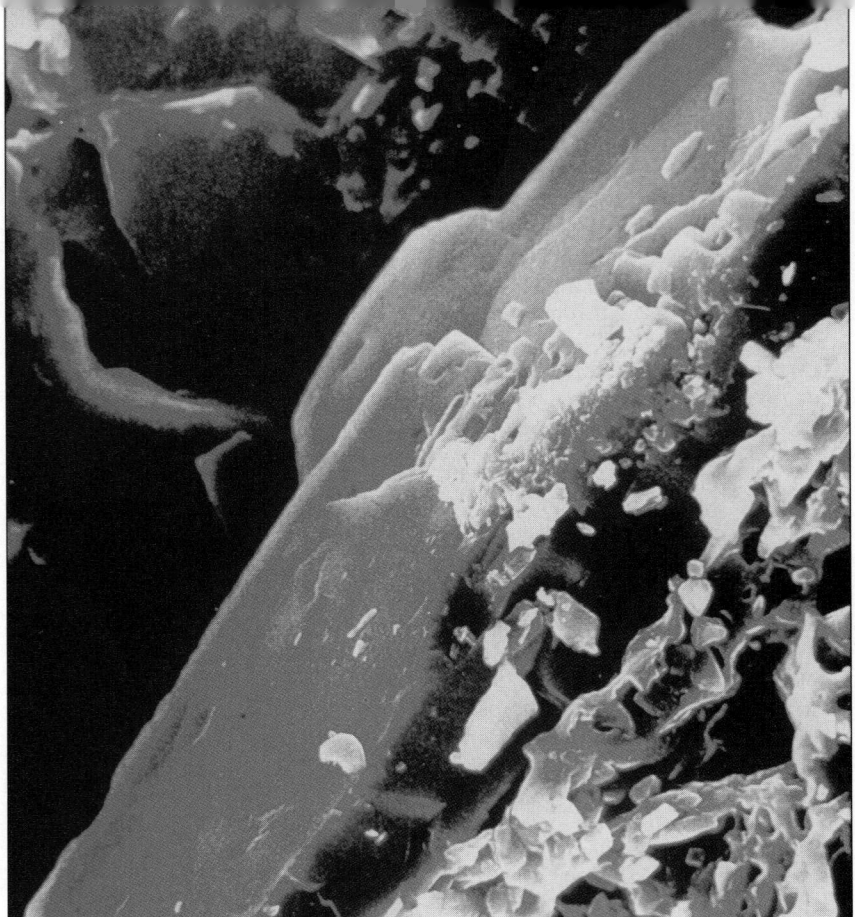

Endocrine Control Mechanisms

CHAPTER OUTLINE

*E*ndocrinology is the branch of biology concerned with the actions of hormones and the organs or glands in which the hormones are produced. The exact boundaries of the field are difficult to define, since endocrinology often overlaps with areas such as anatomy, biochemistry, and neurophysiology. Included within the field of endocrinology are such topics as (1) the anatomy and physiological function of the endocrine organs, (2) the chemical nature of the hormones these glands produce, (3) the cellular mechanisms by which the hormones produce their biological effect, and (4) clinical manifestations of abnormal endocrine function. This brief list is by no means complete, as there are numerous other exciting areas of study within the field of endocrinology. However, the majority of current research efforts in endocrinology can probably be placed into one of these four categories.

General Concepts of Endocrine Control

The word hormone is derived from the Greek *hormaein*, which means "to excite" or "to arouse." But what exactly is a hormone, and conversely, what is not a hormone? Is the glucose that circulates in the blood a hormone? What about the lactic acid produced by a vigorously exercising muscle, or the carbon dioxide produced by virtually all cells in the body? Why do we consider insulin, produced by the pancreas, and epinephrine (also known as adrenaline), secreted by the adrenal medulla, to be hormones?

Over the years, scientists have made many attempts to define a hormone, and much discussion has ensued over the propriety of any particular definition. As new hormones were discovered, definitions of hormones were rewritten to include the newest members of the "hormone family." Today, it is virtually impossible to write an accurate definition that includes the large numbers of known hormones or hormone-like substances, while still excluding other substances that do not function in a manner similar to that of a typical hormone.

The Nature of a Hormone

Despite the complexity involved in describing hormones, at this point in the discussion a broad, working definition is probably useful to help you gain a general appreciation of what hormones are and how they function in the body. Hormones are generally considered to be substances secreted into the blood in very small amounts by specialized cells or glands and carried by the bloodstream to other parts of the body, where they interact with specific receptors in target tissue cells to produce a particular biological response (Fig. 12–1*a*). This is a generalized definition of a hormone, not an absolute one. Several bona fide hormones would probably not fit this definition in its entirety, and likewise, several nonhormonal substances could be viewed as filling most of the criteria of this definition. (Some of the terms used in this definition, such as *receptor* and *target tissue*, will be described in greater detail later in this chapter.)

In recent years, our knowledge of the degree of the complexities involved in cell-to-cell communication mechanisms has increased dramatically. Virtually every cell in the body receives information in the form of chemical signals, from either nearby cells or distant cells or organs. We have only begun to characterize and understand the details of these very powerful control systems in many instances. A fuller understanding of these mechanisms may lead to important breakthroughs in the treatment of a wide variety of diseases.

Not all hormone-like substances are carried to their action sites by the blood. Some reach their target cells by diffusion and exert their effects by what are known as **autocrine** or **paracrine** mechanisms. In this case, the hormone product is not secreted directly into the blood to be carried throughout the entire body, but instead is released into the interstitial fluid surrounding the cells in which it is produced. The primary transfer mechanism from its production site to its action site is by **diffusion.** Autocrine regulation (see Fig. 12–1*b*) involves the secretion of a hormone product that exerts its effects on the cell in which it was produced, while paracrine regulation (see Fig. 12–1*c*) involves the

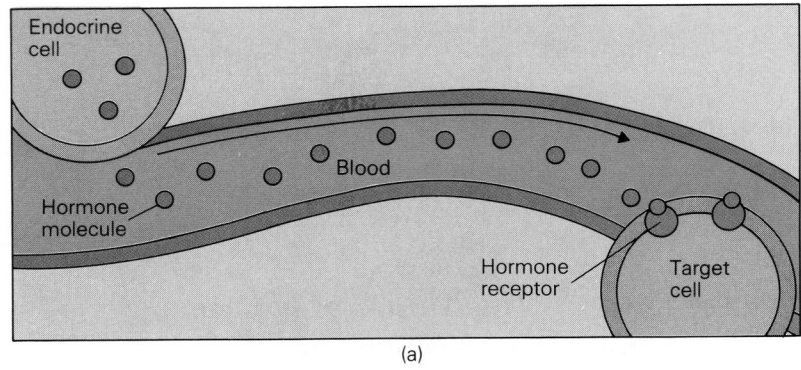

(a)

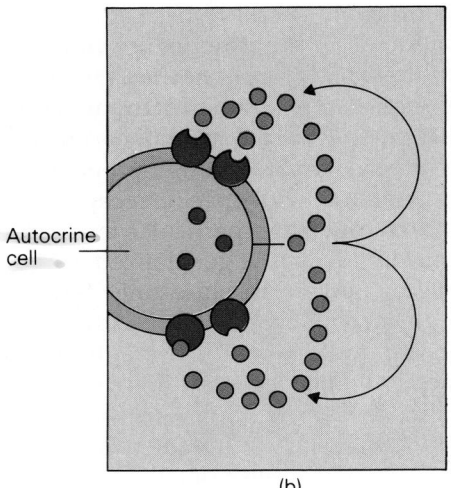

(b)

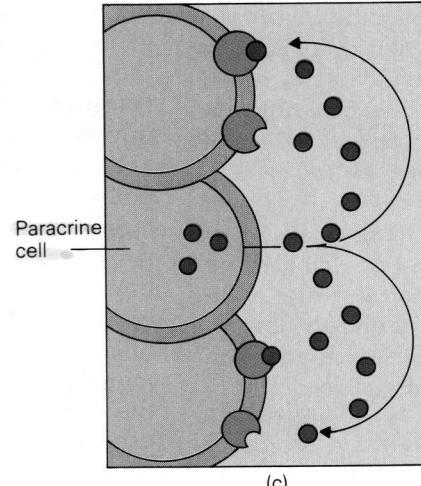

(c)

Figure 12–1
Comparison of endocrine (*a*), autocrine (*b*), and paracrine (*c*) mechanisms of cell-to-cell communication.

secretion of a hormone that diffuses and acts on adjacent cells. While our knowledge of classical endocrine control systems is extensive, information about the autocrine and paracrine systems is rather limited and remains to be more completely characterized. However, it is likely that these systems play a major role in the establishment and maintenance of a normal, healthy state.

The Relationship Between the Nervous System and the Endocrine System

As ancient single-cell organisms evolved into today's complex animals, it became necessary and advantageous for cells to specialize in their functioning. Specialization allowed the various cell types to perform unique functions in a cooperative manner and thus contribute to the overall survival of the organism. Along with this specialization came an increasing need for communication between cells or groups of cells as they became more interdependent. Such communication ensures that the activities and functions of various tissues and organs are carried out in a coordinated, purposeful manner.

In the human body, cells communicate through two major systems: the **nervous system,** discussed in recent chapters, and the **endocrine system.** There is a considerable degree of interaction between the two systems. This overlap between the nervous and endocrine systems constitutes an area of study known as **neuroendocrinology.**

Figure 12–2 illustrates the different mechanisms whereby neural, endocrine, and neuroendocrine cells receive and transfer information. Neurons, for example, receive information from other nerve cells in the form of chemical signals (neurotransmitters) that reach the cell by diffusion across synaptic clefts. The neuron in turn releases neurotransmitters at its synapses, which then activate subsequent cells in the neuronal chain. Thus, in the example of neurotransmission shown in Figure 12–2*a*, information in the form of an electrochemical signal is transferred from left to right.

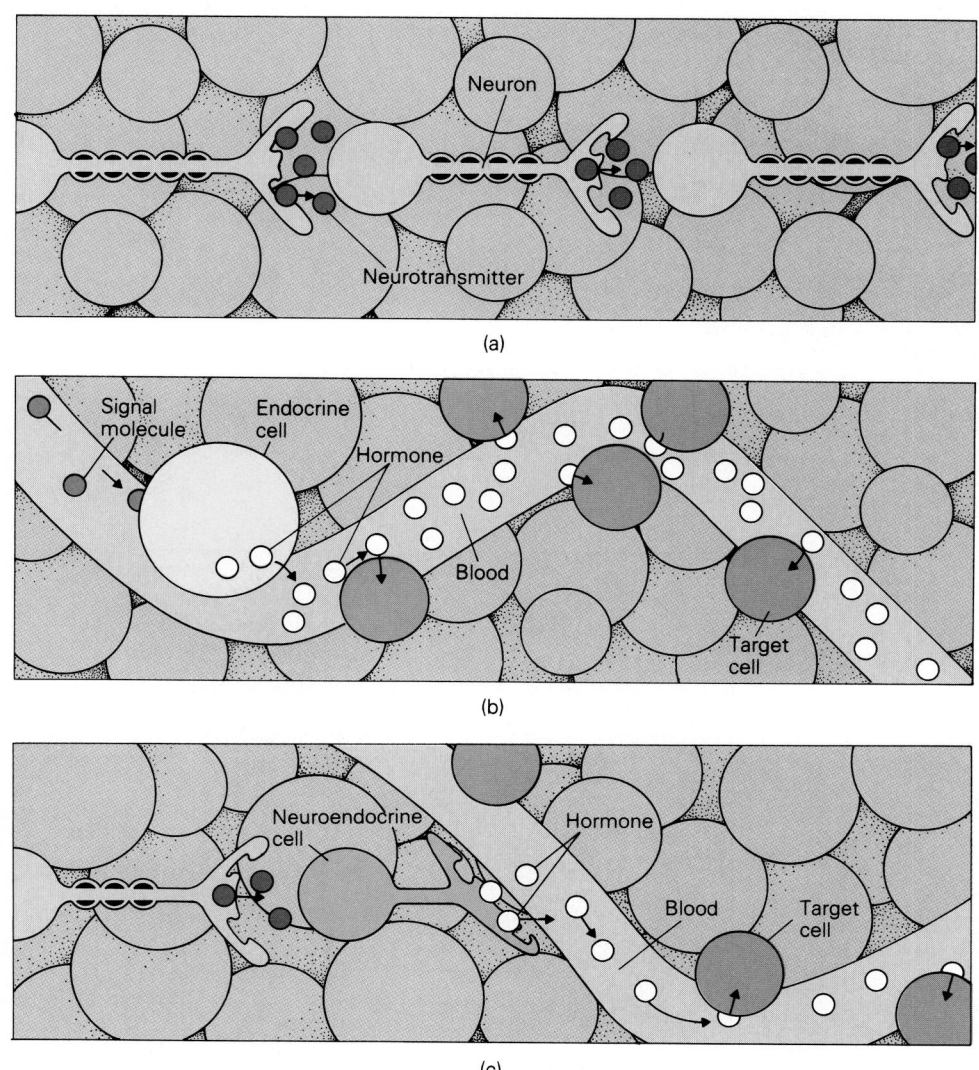

Figure 12–2
Basic mechanisms of neural (*a*), endocrine (*b*), and neuroendocrine (*c*) communication methods.

A typical endocrine cell, on the other hand, receives information in the form of chemical messages ("signal" molecules) from the blood, as illustrated in (Figure 12–2*b*. In response to the appropriate input signal, the endocrine cell then secretes its own specific hormonal signal into the blood. The endocrine cell may sense the concentration of another hormone present in the blood, or it may perceive the concentration of some particular ion or nutrient, such as sodium or glucose. When the endocrine cell detects a significant change in the concentration of that particular factor in the blood, it appropriately alters the rate of release of its own secreted hormone product.

Neuroendocrine cells function as an interface between the nervous and endocrine systems, providing a crucial link between these two communication systems. As shown in Figure 12–2*c*, a neuroendocrine cell is directly innervated by nerve cells but, unlike other neurons, in response to stimulation it secretes its product, a hormone, directly into the blood. Thus, neuroendocrine cells act as one-way translators. They convert **electrochemical signals** from the nervous system into **hormonal signals** in the blood.

Figure 12–3 illustrates a basic difference in the functional organization of the nervous and endocrine systems. The ability of one particular neuron

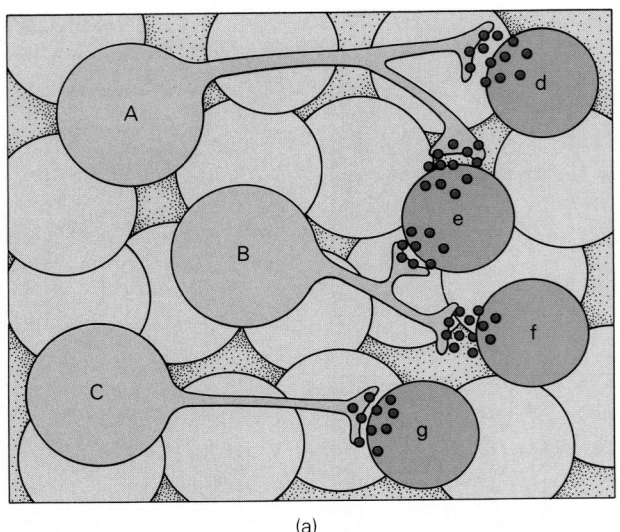

(a)

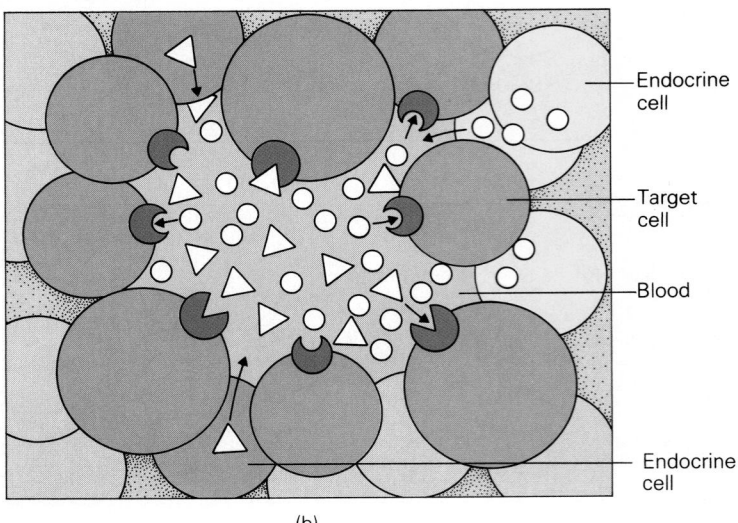

Endocrine cell

Target cell

Blood

Endocrine cell

(b)

Figure 12–3
Mechanisms of cellular communication in the nervous (*a*) and endocrine (*b*) systems.

to communicate with its target cells—whether they are other nerve cells, muscle cells, or other excitable cells—is determined by the anatomical connections within the system. A single neuron can transmit information—in this case, an electrochemical signal—only to those cells with which it makes synaptic contact. For example, in Figure 12–3*a*, nerve cell A can communicate only with cells d and e, nerve cell B only with cells e and f, and nerve cell C only with cell g. Thus, the flow of information within the nervous system is determined by the point-to-point anatomical connections made between different cells. We can make an analogy between the nervous system and an intercom system in an office building. Only those workers who are connected by the wiring of the intercom system can directly communicate with one another.

By contrast, once a hormone is secreted into the blood, it is carried to nearly every cell in the body.

This difference is illustrated in Figure 12–3*b*. Whether or not a certain cell will respond to a particular hormone molecule is determined by the presence or absence in the cell of a specific **receptor** for that hormone. A receptor is a molecule in the membrane of or within the cell that specifically recognizes and binds a particular hormone and, as a result, initiates the biological responses characteristic of that particular hormone. As depicted in Figure 12–3, all cells are exposed to the hormones present in the blood. Only those cells having the specific receptor for a hormone can respond to it. For example, only those cells having a "triangular" receptor can respond to the "triangular" hormone, and so on. In this case, we can draw an analogy between a hormone and a radio signal: a radio transmitter broadcasts its signal in every direction for all who care to listen; however, only those radios tuned to the appropriate station will detect that particular radio signal. Similarly, a hormone circulates throughout the body, but only those cells "tuned" to that particular hormone (i.e., that possess specific receptors for that hormone) will be able to sense its presence in the blood. The concept of receptor specificity will be discussed in greater detail later in this chapter.

Principles of Feedback Regulation in the Endocrine System

One of the hallmark features of the endocrine system is **feedback regulation.** Endocrine cells have the ability to manufacture and secrete a particular hor-

mone; in most instances, they are also equipped to detect or monitor the magnitude of the biological effect of that hormone. This permits the endocrine cell to adjust the rate of hormone secretion in an appropriate manner to achieve the desired effect, which is the maintenance of homeostasis.

In most instances, feedback regulation involves **negative feedback,** although some cases of **positive feedback** are also known. Figure 12–4a illustrates a very simple negative feedback system. In this case, the endocrine cell (A) secretes a hormone, which has an effect on one of its target cells (B). The hormone triggers a specific biological response in the target cell, and the size of the response is in turn sensed, or monitored, by the endocrine cell that first

Figure 12–4
(a) A simple negative feedback loop. (b) A more complex negative feedback system involving two different endocrine cells and two different hormones.

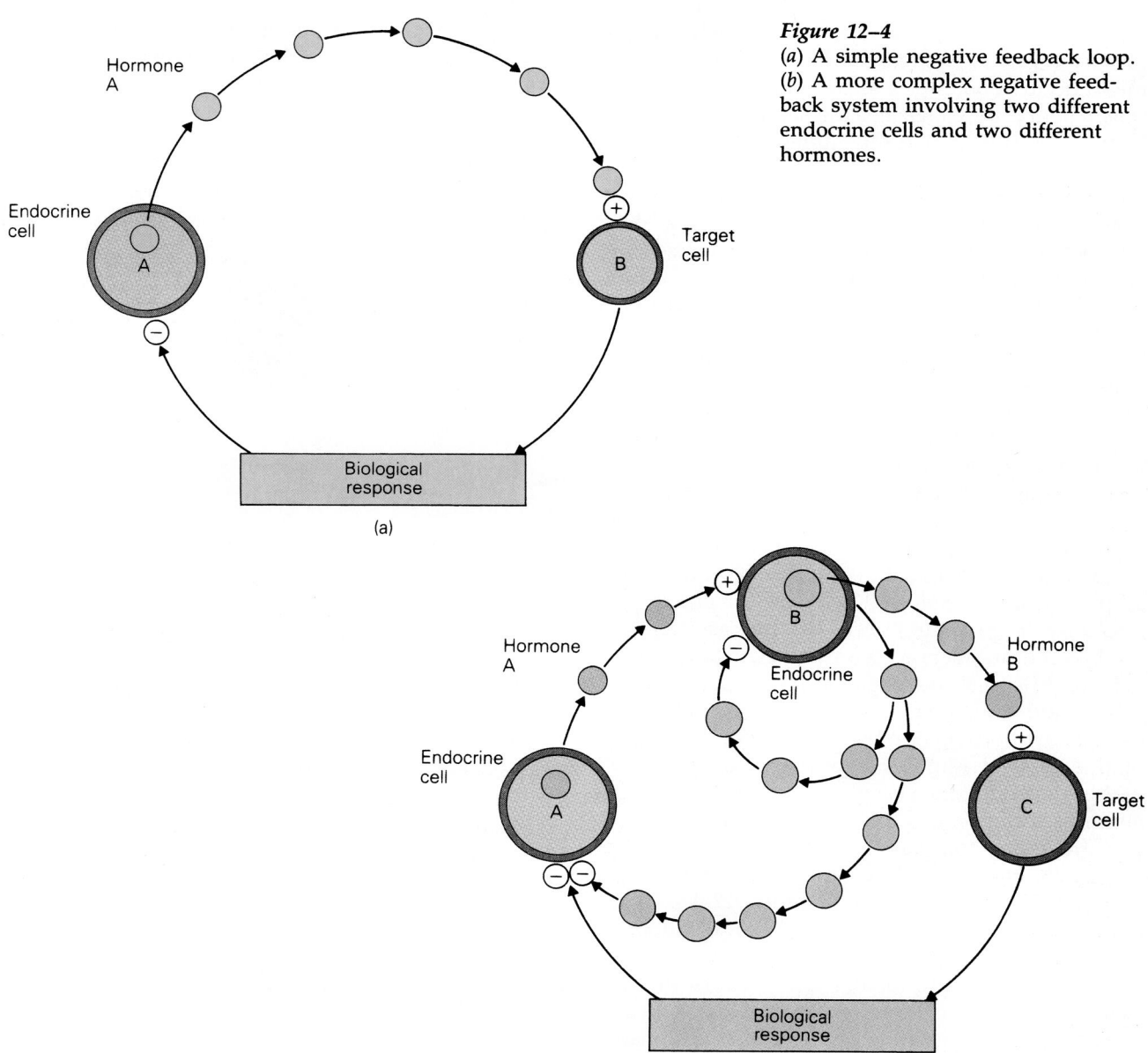

secreted the hormone. If the response is too small, the endocrine cell will produce and secrete more hormone in an attempt to elicit a larger, more normal response. Likewise, if the response of the target cell is too great, the endocrine cell will reduce the amount of hormone it secretes until the response decreases into the normal range. In this case, the particular response being produced can be said to have a negative feedback effect on the amount of hormone secreted by the endocrine cell.

An example of this simple type of negative feedback regulation is the regulation of the blood glucose concentration by the hormone **insulin,** which is produced by specialized endocrine cells within the pancreas. These cells are triggered to secrete insulin when the amount of glucose in the blood exceeds the normal concentration of about 100 mg of glucose per 100 ml of blood. Therefore, when blood glucose levels are increased above normal, such as might occur soon after a meal, the insulin-secreting cells respond by increasing the secretion of insulin. Insulin in turn stimulates the uptake of glucose from the blood by such tissues as muscle and adipose (fat). The overall response to insulin therefore is a reduction in the amount of glucose circulating in the blood. The stimulus for insulin secretion is therefore lessened, and the secretion rate decreases. (These effects of insulin will be discussed in greater detail in Chapter 15.)

More complex types of negative feedback systems are also prevalent within the endocrine system. An example of one such system is illustrated in Figure 12–4b. In this case, endocrine cell A secretes a hormone (A), which has as its target endocrine cell B. In response to hormone A, endocrine cell B increases its production of hormone B. This hormone in turn acts on target cell C to produce some particular biological response. In similar fashion to the example shown in Figure 12–4a, the response produced by the final target cell—in this case, cell C—inhibits the production of hormone A. In addition, hormone B itself has negative feedback effects on cell A, and it also has negative feedback effects to limit its own production by cell B. Several examples of this type of multilevel feedback can be found within the endocrine system, especially in the regulation of secretion of pituitary hormones.

The more complex type of feedback control shown in Figure 12–4b offers certain advantages over the simple type shown in Figure 12–4a. It permits a greater degree of fine-tuning of the endocrine cells with regard to their hormone-secreting activity. This in turn leads to biological responses that are closely matched to the original stimulus. A nearly constant internal environment results. Additionally,

these multilevel control systems provide some degree of back-up in the event that one segment of the overall system functions abnormally. In such a situation, a multilevel control system is better able to achieve the desired biological response than a simple system such as the one shown in Figure 12–4a.

The Primary Endocrine and Neuroendocrine Glands and the Hormones They Produce

Figure 12–5 shows the principal endocrine glands in the body and their approximate anatomical locations. It is clear from this figure that tissues or organs with endocrine functions can be found in many parts of the body. This figure is by no means complete, as there are a number of other tissues and organs within the body that are known to produce and secrete hormones. Table 12–1 gives a more com-

Figure 12–5
The anatomical location of the primary endocrine glands in the human body.

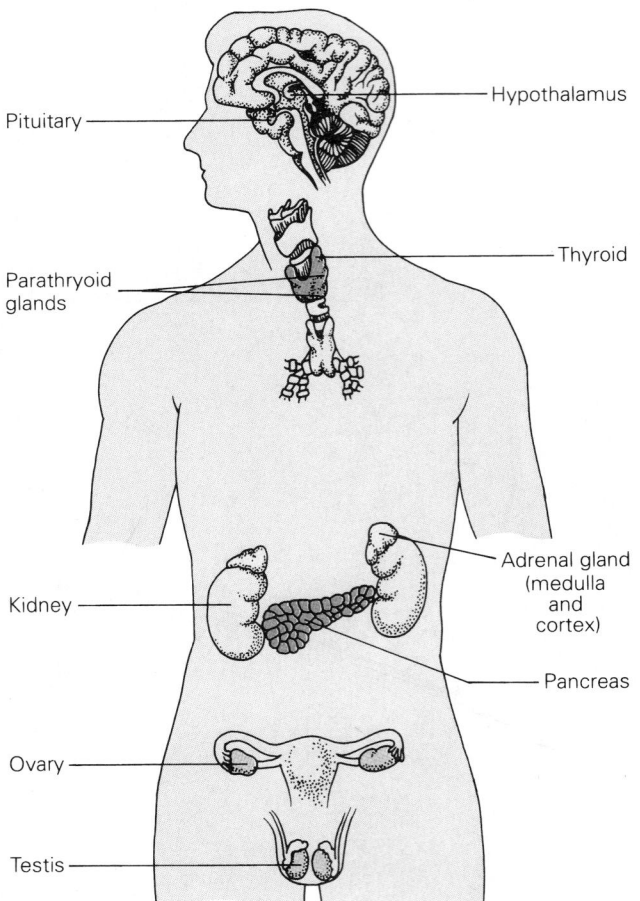

Table 12–1

Summary of the Primary Hormones in the Body

Hormone	Gland or Source	Target Cells	Primary Biological Effect
Release or release-inhibiting hormones	Hypothalamus	Anterior pituitary	Regulate hormone secretion
Oxytocin	Synthesized in hypothalamus, secreted from posterior pituitary	Breast Uterus	Triggers milk "let down" Stimulates uterine contraction
Vasopressin (antidiuretic hormone) (ADH)	Synthesized in hypothalamus, secreted from posterior pituitary	Kidney Blood vessels	Increases water reabsorption Causes constriction
Growth hormone (GH)	Anterior pituitary	Many; especially bone, fat, and liver	Stimulates growth of skeleton and muscle
Prolactin (PRL)	Anterior pituitary	Breast	Stimulates milk production
Adrenocorticotropic hormone (ACTH)	Anterior pituitary	Adrenal cortex	Promotes adrenal steroid production
Thyroid-stimulating hormone (thyrotropin) (TSH)	Anterior pituitary	Thyroid gland	Promotes thyroid hormone production
Follicle-stimulating hormone (FSH)	Anterior pituitary	Gonads (in females, ovarian follicle cells; in males, Sertoli cells of testes)	Stimulates growth and development
Luteinizing hormone (LH)	Anterior pituitary	Gonads; ovarian follicle cells Leydig cells of testes	Triggers ovulation Stimulates testosterone production
Insulin	Pancreas—islets of Langerhans	Primarily liver, muscle, and fat	Regulates metabolism and blood glucose
Glucagon	Pancreas—islets of Langerhans	Primarily liver	Regulates metabolism and blood glucose
Somatostatin	Hypothalamus Pancreas—islets of Langerhans	Anterior pituitary Other cells of islets	Inhibits growth hormone secretion Regulates insulin and glucagon secretion
Glucocorticoids	Adrenal cortex	Primarily liver, muscle, and fat	Regulate metabolism
Aldosterone	Adrenal cortex	Kidney	Regulates sodium excretion
Epinephrine	Adrenal medulla	Cardiovascular system Muscle, liver, and fat	Stimulates cardiovascular function Regulates energy metabolism
Angiotensin II	Formed by conversion steps involving kidney, blood, and lung	Adrenal cortex	Stimulates aldosterone production
Atrial natriuretic factor (ANF)	Atrial wall of heart	Primarily kidney	Regulates sodium excretion

(Table continues on p. 426)

Table 12–1 continued

Summary of the Primary Hormones in the Body

Hormone	Gland or Source	Target Cells	Primary Biological Effect
Thyroid hormones (T_3, T_4)	Thyroid gland	Many cell types	Regulate energy metabolism
Somatomedins			
Insulin-like growth factor I (IGF-I)	Primarily liver	Primarily bone	Promotes bone growth
Insulin-like growth factor II (IGF-II)	Primarily liver	Many tissues	Promotes tissue growth and repair
Parathyroid hormone (PTH)	Parathyroid glands	Bone, kidney	Regulates plasma calcium and phosphate
1,25 Dihydroxyvitamin D_3	Conversion in kidney from precursor formed in liver	Gastrointestinal tract; bone	Regulates plasma calcium and phosphate
Calcitonin (CT)	Parafollicular cells of thyroid gland	Bone	Regulates plasma calcium and phosphate
Gastrointestinal (GI) hormones Gastrin, secretin, cholecystokinin, gastric inhibitory peptide, and somatostatin	Various cells of the GI tract	GI tract; gallbladder and pancreas	Regulates digestive processes, including secretion and motility
Androgens	Primarily testes; also adrenal cortex	Reproductive tract	Stimulates growth and development
Estrogens	Ovaries and placenta	Reproductive tract; breasts	Stimulates growth and development
Progesterone	Ovaries and placenta	Uterus; breasts	Promotes proper development and function
		CNS	Inhibits ovulation
Placental hormones Chorionic gonadotropin, chorionic somato-mammotropin, estrogen, and progesterone	Placenta	Various reproductive tissues	Maintains pregnancy
Mullerian inhibitory factor (MIF)	Testes	Mullerian ducts	Causes regression of ducts in fetus
Inhibin	Testes	Hypothalamus and pituitary	Inhibits FSH secretion
Melatonin	Pineal gland	Reproductive system	Influences onset of sexual maturity
Erythropoietin	Kidney	Bone marrow	Stimulates red cell formation
Bradykinin	Kidney and other tissues	Blood vessels	Causes vasodilation

plete listing of the known hormones, their tissues of origin, and a brief description of the processes they regulate. The specific effects of individual hormones can generally be categorized as relating to the regulation of one of four physiological processes: (1) energy production, use, and storage; (2) salt and water metabolism; (3) growth and development; (4) reproductive function. A later section of this chapter describes details of the biochemical mechanisms involved in hormonal regulation of a variety of cellular processes.

The descriptions of target tissues and hormonal effects presented in Table 12–1 are considerably abbreviated and simplified. Most hormones have *several* different target tissues. In addition, many hormones produce multiple individual effects within their specific target cells. These multiple effects, when added together, produce the overall response characteristic of that particular hormone. The concept of multifunctional hormones is illustrated by hormones A and C in Figure 12–6. As an example,

the male sex hormone, **testosterone,** has several different effects in different tissues. It is required for normal sperm formation to occur in the testes and also promotes the growth and development of most parts of the male reproductive tract, such as the prostate gland and seminal vesicles. Testosterone is also responsible for the development of secondary male sexual characteristics at puberty, including beard growth and a deepening of the voice. Thus, this one single hormone has many different effects in a variety of different tissues.

Just as one hormone may influence many different processes, a single process may in turn be regulated by several different hormones, as depicted in Figure 12–6. For example, the process of converting blood sugar into glycogen in the liver is regulated by insulin, glucagon, epinephrine, adrenal glucocorticoids, and thyroid hormone. Several other hormones have minor influences as well. The overall effect is an integrated response reflecting the circulating levels of all these hormones.

Figure 12–6
Multiple hormone action. Hormones may have many effects in several different tissues. In addition, one process may be regulated by many different hormones.

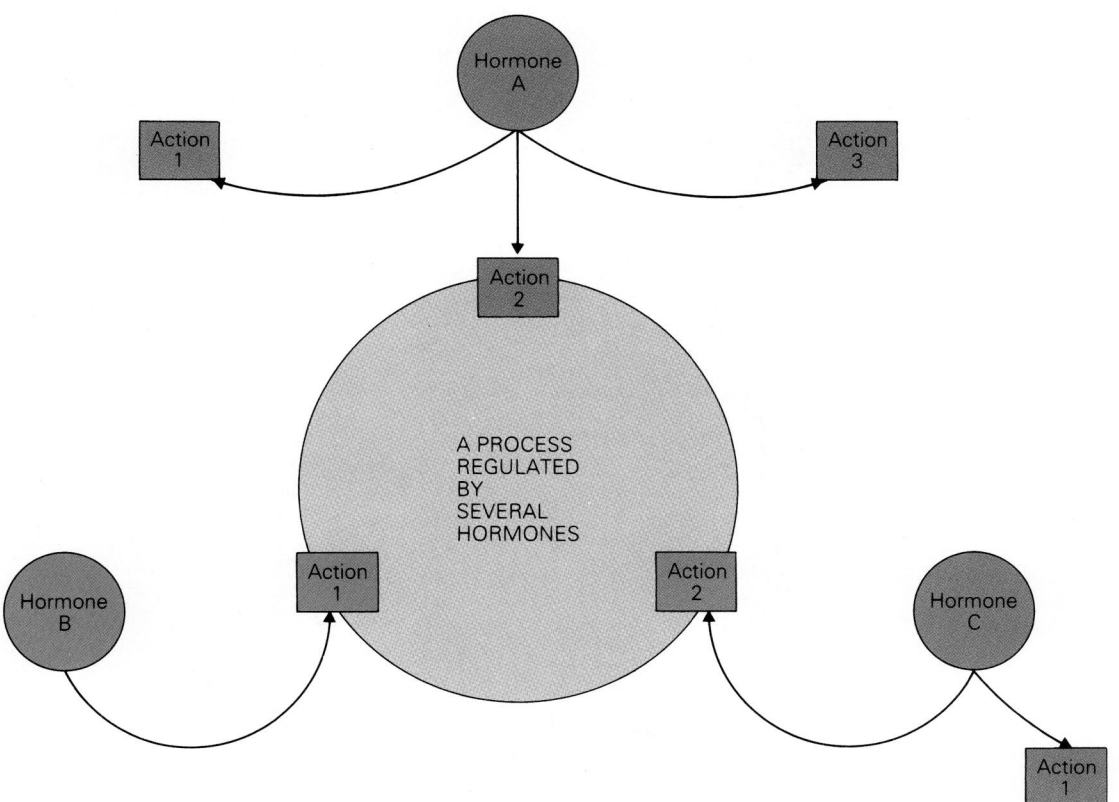

Thus, endocrine regulation is characterized by single hormones with multiple effects and by processes regulated by several hormones.

The Chemistry of Hormones

Up to this point, the discussion of hormones has been rather general in nature. Let us now begin to examine the hormones in more detail, beginning first with the chemistry of hormones.

Structures and Synthesis of Hormones

Hormones generally fall into one of three different chemical categories: (1) amino acid derivatives, (2) peptides or proteins, and (3) steroids. Figure 12–7 shows an example of each type of hormone.

Hormones Derived from Amino Acids

Some hormones are simple derivatives of amino acids either obtained from the diet or synthesized in the body. With just a few relatively simple chemical changes in the molecule, an ordinary amino acid can be converted into the powerful regulatory substance that is a hormone. Examples of hormones that are amino acid derivatives are **epinephrine, melatonin, and thyroxine;** their structures are shown in Figures 12–7 and 12–8. Epinephrine is derived from tyrosine, and melatonin is formed from the amino acid tryptophan. Thyroxine consists of four iodide molecules joined to a pair of tyrosines that have been coupled end to end.

Epinephrine and Melatonin: Amines. Epinephrine and melatonin also represent a subset of hormones

Figure 12–7
Examples of the three types of hormones.

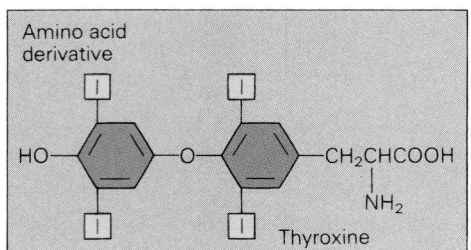

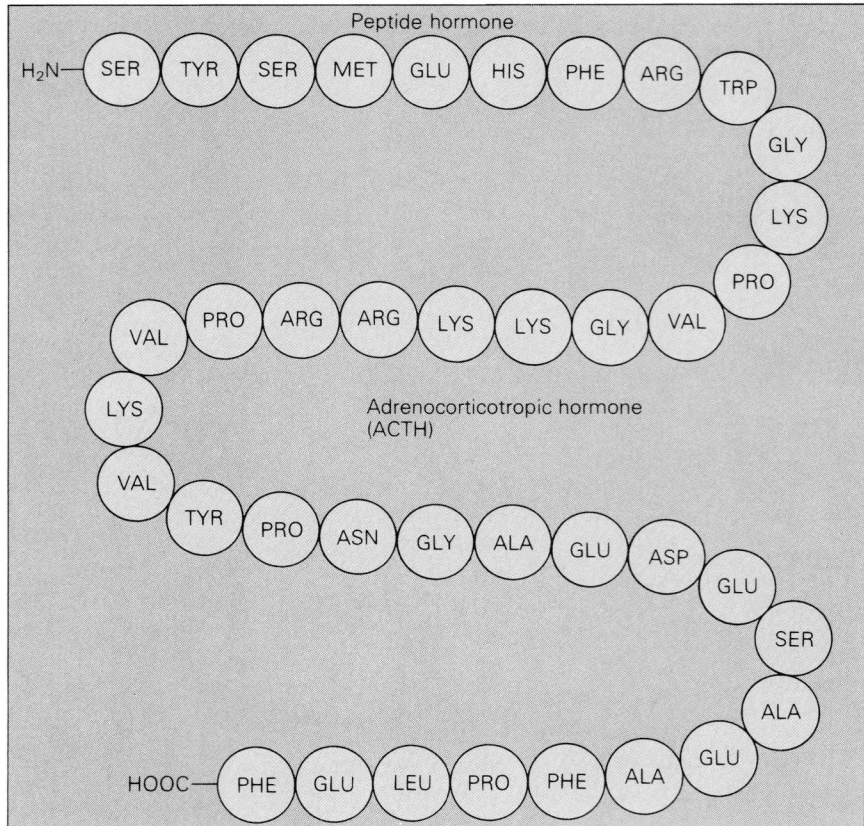

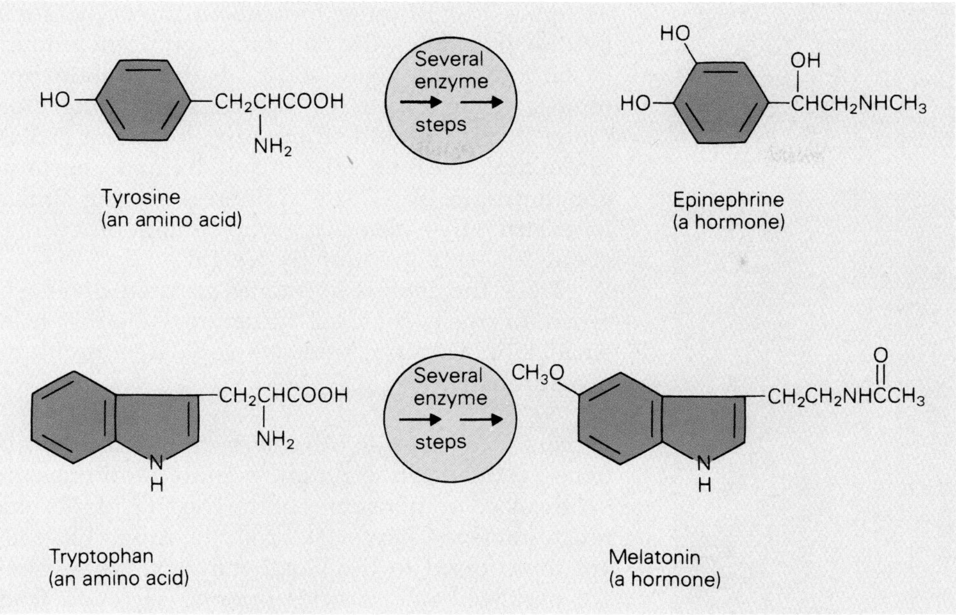

Figure 12–8
Examples of two hormones that are derivatives of amino acids.

known as **amines,** which have a free amino group. Most amines are synthesized, stored, and secreted in a similar way. Within the endocrine cell, an amino acid is converted to an amine by a sequence of specific enzyme-catalyzed steps. The amine is packaged into secretory granules in the cell, where it is held until it is discharged into the blood in response to the appropriate stimulus to the endocrine cell. Stimuli that trigger amine secretion not only cause discharge of preformed granules from the cell but also tend to increase the activity of key regulatory enzymes involved in formation of the hormone. Thus, most stimuli not only increase the rate of secretion of previously formed hormone but also accelerate the rate of formation of new hormone molecules.

Thyroxine: An Iodide Compound. Although it is an amino acid derivative, thyroxine is synthesized in a somewhat different manner from amines. Details of thyroxine formation will be presented in Chapter 13. Briefly, thyroxine is first formed as part of a much larger precursor protein, known as **thyroglobulin,** which is stored in relatively large amounts within the **thyroid gland.** Because iodide is not always available in the diet, thyroxine is the one hormone in the body that is not solely made from readily available dietary constituents. A system has therefore evolved that allows for the storage of large amounts of thyroxine in its precursor form, thus providing a reservoir of hormone in the event that

iodine is not available in the diet. Stimuli for thyroxine secretion result in the rapid breakdown of the thyroglobulin molecule to liberate thyroxine and stimulate the synthesis of new thyroglobulin protein.

Hormones That Are Peptides or Proteins

Many hormones are either peptides or proteins. Figure 12–9 shows the structures of several hormones in the peptide hormone family. As this figure shows, they can be very small peptides such as **thyrotropin releasing hormone (TRH),** which is a tripeptide produced by neuroendocrine cells of the hypothalamus. It plays a role in the regulation of thyroid hormone production. Peptide hormones can be intermediate in size, such as **adrenocorticotropic hormone (ACTH)** (see Fig. 12–7), which contains 39 amino acids, or **parathyroid hormone (PTH)** (see Fig. 12–9*b*), which contains 84 amino acids. Or they can be much larger peptides such as **growth hormone (GH),** which in humans is a chain of 191 amino acids. Some are simple, single-chain peptides such as the ACTH and PTH molecules, while others such as growth hormone contain disulfide (S—S) bonds that connect cysteine residues within different portions of the peptide chain. Still others, such as **follicle-stimulating hormone (FSH)** (see Fig. 12–9*d*), are composed of different peptide subunits joined together to form the active hormone molecule.

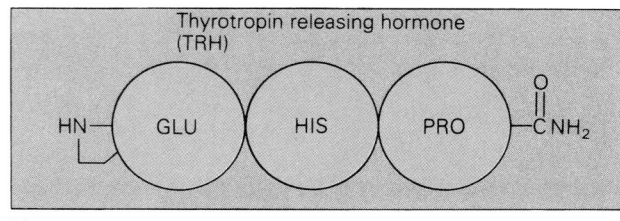

(a)

(b)

As is true of other proteins in the body, some peptide hormones also contain a significant amount of carbohydrate. One family of **glycoprotein hormones**, which includes **follicle-stimulating hormone (FSH), thyroid-stimulating hormone (TSH), luteinizing hormone (LH),** and **human chorionic gonadotropin (hCG),** in addition to having similar basic structures, also as a group contain on average about 20% carbohydrate by weight.

Thus, the peptide hormones are quite diverse in terms of their chemical structure, ranging from small, single-chain peptides to very large, multiunit glycoproteins.

Precursors of Peptide Hormones. As are other proteins destined for secretion, peptide hormones are synthesized on ribosomes of the rough endoplasmic reticulum (see Chapter 3). From the rough ER, they are transported to the Golgi complex, where they are packaged into vesicles prior to secretion from the cell. During these transport and packaging steps

Figure 12–9
Examples of four common peptide hormones.

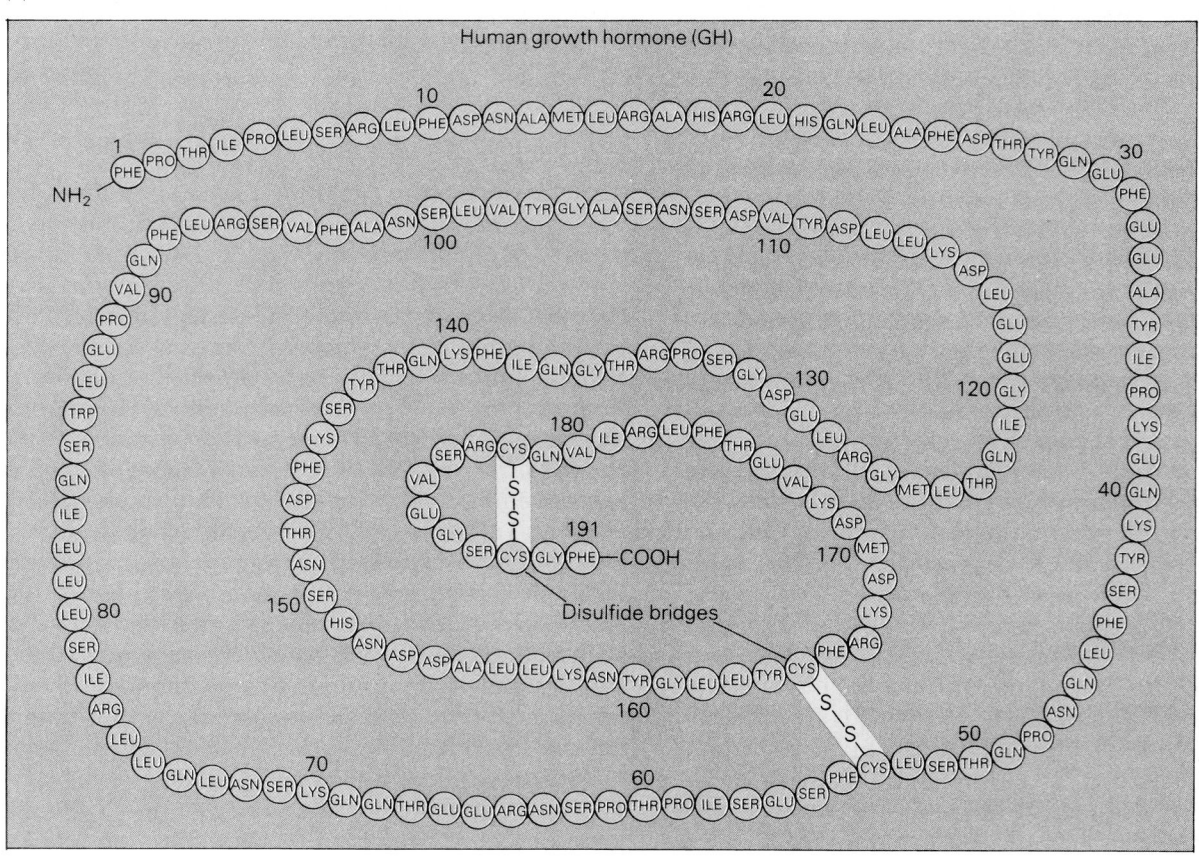

(c)

in the cell, some very important events take place. Many, if not most, peptide hormones are first synthesized in the form of much larger **precursor** molecules. In some cases, the precursor may even contain the amino acid sequence for several hormones or hormone-like substances. The precursor for ACTH, for example, contains the ACTH sequence as well as the sequences for at least four other bioactive peptides. The precursor for TRH contains several repetitions of the TRH sequence, so that a number of TRH molecules may be formed from a single precursor molecule. During the intracellular transport phase of hormone secretion, the precursor molecule is "clipped" by specific proteolytic ("protein-breaking") enzymes to produce the mature, active hormone molecules that will be secreted into the blood.

When an endocrine cell that produces a peptide hormone receives an appropriate stimulus, vesicles containing the mature hormone rapidly move toward the cell surface, and the contents are released by the process of exocytosis (see Chapter 3). Much as they do in other cell types, calcium ions play a key role in coupling a stimulus with secretion from the endocrine cell. In most cases, the amount of hormone available in preformed vesicles is somewhat limited and usually only enough to support high rates of secretion for a matter of minutes. To compensate for this, stimuli that promote peptide hormone secretion usually also promote increased rates of synthesis of the hormone or its precursor.

Hormones Derived from Cholesterol: Steroids

The third major category of hormones are the steroid hormones. Figure 12–10a shows the structure of cholesterol, which is the precursor for all of the steroid hormones. Differences in biological activity between the various steroid hormones are the result of different, sometimes very subtle alterations or substitutions in the cholesterol backbone. Figure 12–10a also shows the designation of each of the four rings (A to D) and the numbering convention used to designate each of the 27 carbon atoms in the cholesterol molecule. These are used when the various steroids are described according to strict chemical nomenclature. For simplicity, however, in most instances the steroids are referred to by their common names.

In the body, the major sites of steroid hormone production are the adrenal cortex, ovaries, testes, and the placenta. Hormonally active metabolites of vitamin D, which are also derived from cholesterol and have structures similar to the steroid hormones, are produced by sequential modifications occurring first in the liver and then in the kidney.

Types of Steroid Hormones. The steroid hormones can be grouped into five categories according to the type of biological process they regulate: (1) glucocorticoids, (2) mineralocorticoids, (3) androgens, (4) estrogens, and (5) progestins. Glucocorticoids, produced by cells in the cortex of the adrenal gland, are involved in regulating glucose and energy metabolism (see Table 12–1). Mineralocorticoids are the principal steroids produced by the outer portion of the adrenal cortex and are involved in regulating sodium balance by the kidneys. Androgens, the male sex hormones, are produced primarily in the testes of the male, although the adrenal cortex produces and secretes physiologically significant amounts of androgens as well. Estrogens, the female sex hormones, are secreted by the ovaries and the placenta. The progestins are involved in the

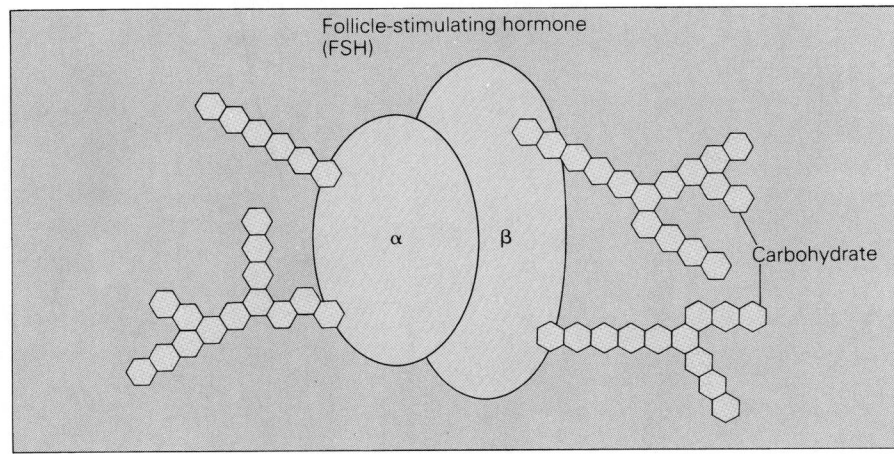

(d)

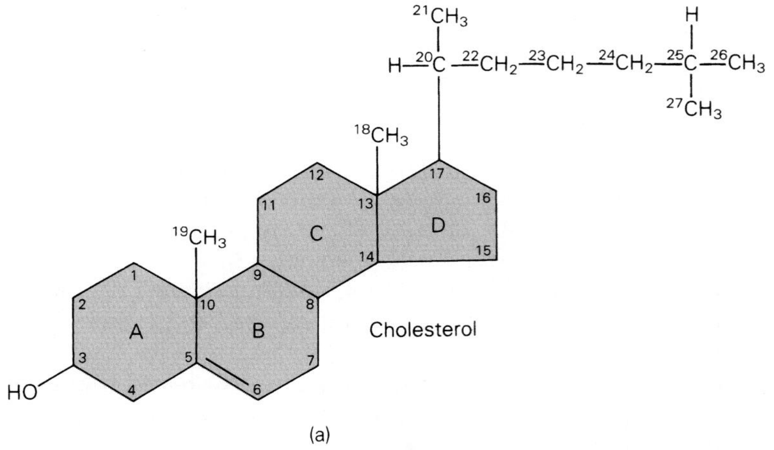
Cholesterol
HO
A
B
C
D
(a)

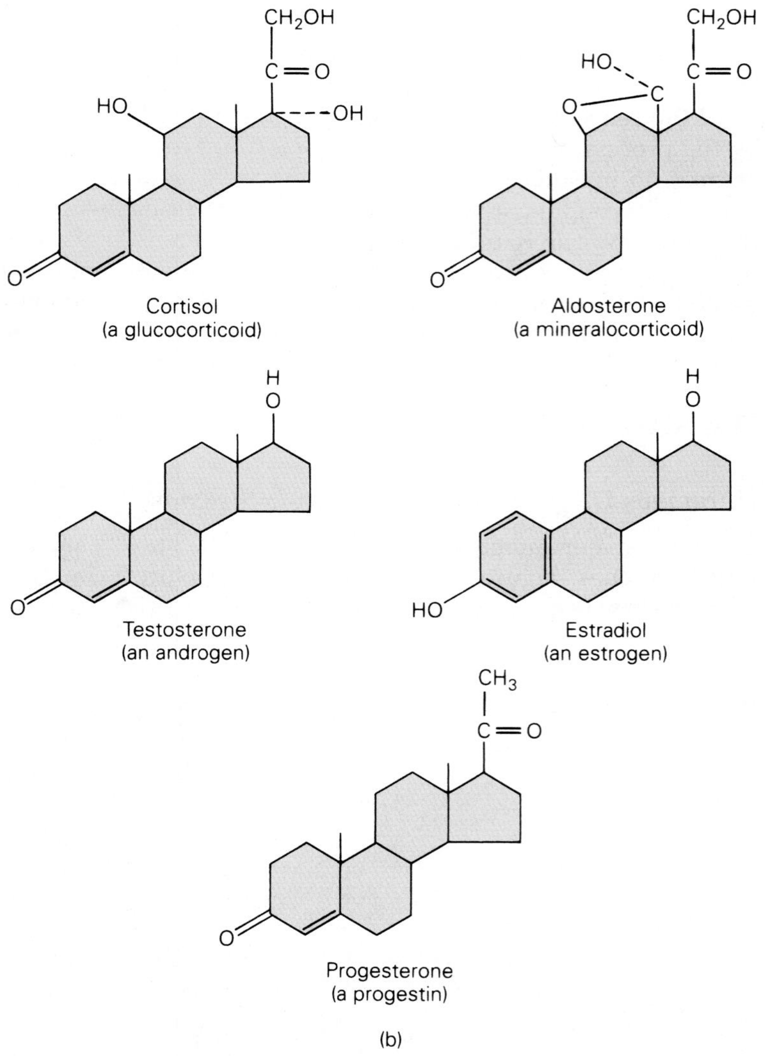
Cortisol
(a glucocorticoid)
Aldosterone
(a mineralocorticoid)
Testosterone
(an androgen)
Estradiol
(an estrogen)
Progesterone
(a progestin)
(b)

Figure 12–10
(*a*) Chemical structure of cholesterol. The letters A through D designate each of the four rings in the molecule. Numbers designate each of the 27 carbon atoms in cholesterol. (*b*) Examples of steroid hormone structure. Each steroid shown represents one of the five classes of steroid hormones.

Peptide Hormones in Medicine

In the past, there has been a limited supply of authentic human peptide hormones for therapeutic purposes. The only readily available materials were those obtained from animals. However, owing to differences in structure, the human immune system may view some peptide hormones from animals as foreign and produce an immune response in the human patient. In addition, if the three-dimensional structure of a particular animal hormone is markedly different from its human counterpart, the animal hormone may be totally ineffective for humans. In a commercial setting, many of the smaller peptides such as TRH can be manufactured readily in the authentic human form, but this is not feasible for intermediate-sized or larger peptide hormones. Physicians have therefore been limited in their capacity to treat patients with peptide hormones.

The animal and human forms of some peptide hormones are so similar in structure that in most people the animal hormone is quite effective and is not viewed by the immune system as foreign. Beef (bovine) insulin and pork (porcine) insulin, purified from cattle and pig pancreas tissue, are structurally very similar to human insulin and have been used successfully to treat human diabetics for many years. Other hormones, such as growth hormone (GH), have such different structures in animals and humans that only the human form is effective in treating human patients. In the past, the supply of human GH has been limited to the very small amount that could be obtained from human cadavers. The available quantities of this material were both extremely limited and very expensive. Many individuals who would have benefitted greatly from therapy with GH were not treated.

With the application of recombinant DNA and cloning techniques, it is now theoretically possible to produce the authentic human form of any peptide hormone in large amounts and at little expense. Both human insulin and human GH have been produced by these means and are available commercially. Most likely, many other human peptide hormones will be produced similarly and become available in the near future.

maintenance of pregnancy and are also produced by the ovaries and the placenta.

Figure 12–10b shows examples from each category of steroid hormone. This figure shows that differences in structure of these steroids are relatively small. In the body, however, these small changes in structure are translated into markedly different biological activities. This phenomenon is a result of the high degree of receptor specificity, causing them to recognize and respond to one particular hormone and not to another.

Pathways in Steroid Hormone Synthesis. Conversion of the cholesterol precursor into each of the steroid hormones occurs by a number of well-defined steps, each catalyzed by a specific enzyme within the steroid-producing cell. Figure 12–11 is a simplified diagram of the primary synthetic pathways involved in the formation of each of the major steroids. It should be evident from this figure that several common steps are involved in the formation of each of the five classes of steroids. For example, both cortisol and aldosterone are formed from pro-

gesterone, and thus the conversion of pregnenolone to progesterone is common to the synthesis of both steroids. Similarly, **testosterone,** an androgen, also serves as an intermediate in the formation of **estradiol,** an estrogen.

Given these common steps, what factors determine which hormone a particular steroid-producing endocrine gland will produce? As indicated previously, each step in steroid hormone biosynthesis is catalyzed by a specific enzyme. The relative amount of each of these enzymes in the endocrine cell determines the particular steroids produced. For example, cells in the outer portion of the adrenal cortex are the site of most aldosterone production in the body. These cells have relatively high amounts of the enzymes needed to convert pregnenolone to aldosterone, but have little or none of the enzymes needed to convert pregnenolone or progesterone to cortisol, testosterone, or estradiol. Similarly, the steroid-producing cells of the testes possess the enzymes needed to form testosterone, but not those needed to form significant amounts of cortisol or aldosterone. Likewise, they have relatively low

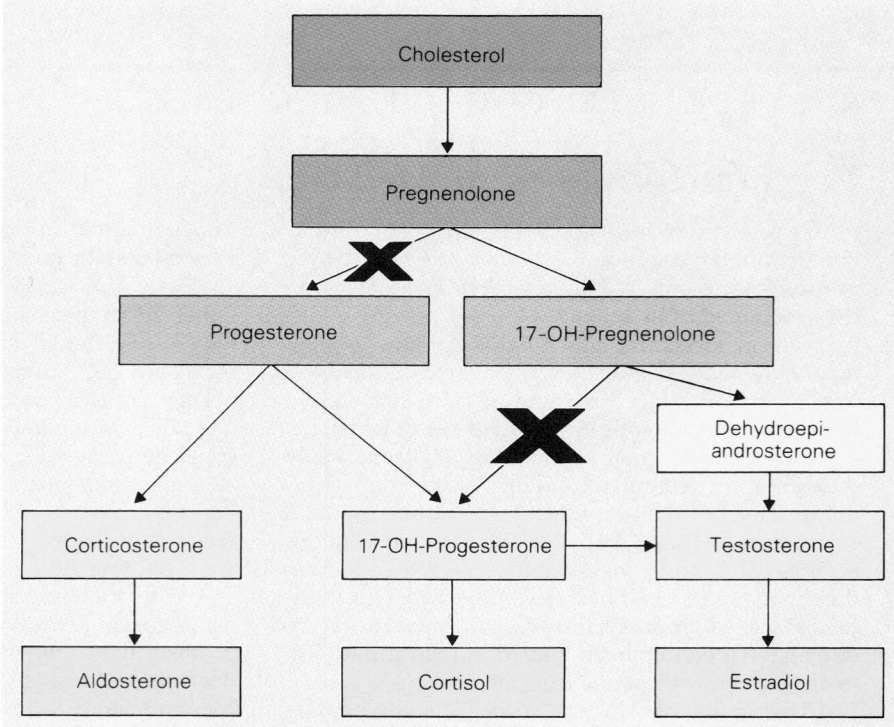

Figure 12–11
Simplified diagram of the pathways of steroid hormone synthesis. Note that the first step in synthesis of all steroids is the conversion of cholesterol to pregnenolone.

amounts of the enzyme needed to convert testosterone into estradiol. Estrogen-producing cells of the ovary, however, have a set of steroid-metabolizing enzymes similar to that of the testes, but they also have relatively high amounts of the enzyme needed to transform testosterone into estradiol. Thus, each of these tissues produces and secretes different steroid hormones because of the different enzymes they contain.

Abnormalities in Steroid Synthesis: Enzyme Deficiencies. In some disease states, genetic deficiencies of certain enzymes in the steroid biosynthetic pathway can result in major changes in the relative amounts of each of the steroids produced. This may result in inappropriately high secretion of some steroids. For example, one of the most common defects involves a deficiency of the enzyme responsible for both the conversion of pregnenolene to progesterone and the conversion of 17-OH-pregnenolone to 17-OH-progesterone within cells of the adrenal cortex. The Xs in Figure 12–11 indicate the steps in steroid synthesis that are absent in this particular enzyme deficiency state. Not only does this enzyme

deficiency prevent these cells from producing cortisol and aldosterone, leading to the clinical symptoms of cortisol and aldosterone deficiency, but it also causes a shunting of pregnenolene to dehydroepiandrosterone (see Fig. 12–11), which is then secreted in large amounts. Dehydroepiandrosterone itself has little biological activity, but it can be converted in peripheral tissues to form testosterone, resulting in masculinization and virilization. These effects can be particularly detrimental in a developing female fetus (see Chapter 30).

Several other endocrine abnormalities are known to result from genetic defects that lead to deficiencies in single enzymes of the steroid synthetic pathways.

Factors Affecting the Rate of Steroid Synthesis. Unlike peptide or protein hormones, which are stored in secretory vesicles, steroid hormones are synthesized and secreted only on demand. They are not stored in the endocrine cell prior to secretion, but exit the cell immediately after formation. The rate of steroid hormone synthesis therefore determines the rate of hormone secretion. As with other enzyme-catalyzed reactions, steroid hormone synthesis can be influenced by alterations in the activity of the "rate-limiting enzyme" (the enzyme that catalyzes the slowest step) in the pathway. The conversion of cholesterol to pregnenolone (see Fig. 12–11) is usually the rate-limiting step in the synthesis of all steroid hormones. Factors that increase the rate of steroid hormone secretion therefore do so primarily by increasing the activity of the enzyme that converts cholesterol to pregnenolone. In addition, to ensure that the amount of cholesterol inside endocrine cells is adequate to meet the needs for steroid hormone biosynthesis, factors that increase steroid synthesis also stimulate uptake of cholesterol from the blood and at the same time cause a release of free cholesterol from storage sites within the cell. Steroid hormone secretion is therefore increased by two mechanisms. One results in an increase in the uptake and availability of cholesterol inside the cell, and the other stimulates the conversion of cholesterol to pregnenolone, which is the slowest step in the pathway of steroid synthesis.

Transport of Hormones in the Blood

Once a hormone is released from the endocrine cell, it travels via the blood, in most cases, to reach its site of action in the body. This can present two problems. The first and most important is that several hormones are only slightly soluble in the blood. For example, because they are lipid in nature, steroid hormones have limited solubility in aqueous solutions. Thyroxine, the principal hormone secreted by the thyroid gland, is also relatively insoluble in the plasma. Therefore, some mechanism is needed to increase the solubility of these hormones in the aqueous environment of the blood. The second problem is that, because of the small size of several of the hormones, they would be readily lost from the kidneys into the urine if allowed to exist free in solution in the plasma. Large peptide or protein hormones, because they generally are freely soluble in water and large enough to not pass through the kidneys into the urine, do not present either of these problems.

Specific Transport Proteins

To solve the problems of solubility and excessive urinary loss, thyroxine and the steroids are bound to large "carrier" proteins for transport in the blood. Generally these hormones bind to two types of carrier proteins: (1) specific transport proteins that bind one particular hormone with a relatively high degree of specificity and a relatively high affinity; and (2) albumin and prealbumin, which bind hormones without high specificity or affinity.

An example of the first type of carrier protein is **cortisol-binding globulin,** or **CBG,** which primarily binds cortisol and relatively little, if any, of other steroids. Similarly, **thyroxine-binding globulin,** or **TBG,** binds and carries mainly thyroxine in the blood. These and other specific carrier proteins are listed in Table 12–2. These carrier proteins are synthesized and secreted by the liver, and both nutritional and endocrine factors are known to influence their rate of synthesis.

Albumin and Prealbumin

Steroid hormones and thyroxine also bind to other plasma proteins for transport, primarily **albumin** and **prealbumin,** which are also produced in the

---

Table 12–2
Circulating Carrier Proteins*

Name	Principal Hormone Carried
Specific carrier proteins	
Cortisol-binding globulin (CBG)	Cortisol, some aldosterone
Thyroxine-binding globulin (CBG)	Thyroxine (T_4)
Sex steroid-binding globulin (TeBG)	Testosterone and estradiol
General carrier proteins	
Albumin	Many steroids, thyroxine
Prealbumin	Thyroxine, some steroids

*The insulin-like growth factors are also carried in the blood bound to specific carrier proteins. These will be discussed in more detail in Chapter 13.

liver. Hormones generally bind to albumin and prealbumin with low specificity and often with low affinity. However, because these two proteins are abundant in the plasma, they have a relatively important role in the transport of some steroid hormones. For example, about 50% of the circulating aldosterone, and about 10% of the circulating cortisol, is carried by binding to albumin.

Free and Bound Hormone: A Dynamic Equilibrium

The binding of steroid or other hormones to these plasma proteins occurs according to the principles of a simple chemical equilibrium. Thus, the free and the bound forms of the hormone exist in the plasma in a dynamic equilibrium with one another. In most instances, the equilibrium favors the bound form, and the majority of the hormone therefore exists in the bound state. However, a finite amount will also be present in a free, or unbound, state. Only the hormone present in the free state is able to move from the blood across the walls of capillaries (the small blood vessels that bring blood into contact with most cells) to reach cells and interact with hormone receptors. The free hormone is therefore the physiologically active form of the hormone. As free hormone is lost from plasma, it is replaced by hormone newly released from carrier proteins. Similarly, if a sudden large burst of hormone secretion occurs, most of the newly secreted hormone molecules will become bound to the carrier proteins, and the free hormone concentration will increase only slightly. Thus, in addition to facilitating hormone solubility and transport, carrier proteins also serve as a reservoir and buffer system to help maintain a relatively constant concentration of the free hormone in the blood.

Peripheral Transformation, Degradation, and Excretion of Hormones

In most cases, the form of the hormone secreted by the endocrine cell is the form directly responsible for producing the biological effects of the hormone; hormones need not be transformed to be fully active. Two notable exceptions to this general rule are thyroxine and testosterone. Both hormones are converted "peripherally" (within some of their target tissues) to forms displaying biological potency many times greater than the parent hormone. Details of the conversion of thyroxine and testosterone to their more active forms will be presented in Chapters 13 and 30, respectively. Also possibly included on this list of peripherally transformed hormones are **angiotensin II** and **1,25 dihydroxyvitamin D₃,** both of which are formed by sequences of conversion steps involving several different tissues.

The concentration of any one hormone circulating in the blood is determined by both its rate of secretion and the rate at which it is removed or metabolized to an inactive form. For most hormones, the major pathways for removal or degradation occur in the liver or kidneys.

Steroid hormones are primarily degraded by the liver and the metabolites secreted by the liver into the bile, eventually being eliminated in the feces. The liver first converts the steroid into a relatively less active form and then, to make the molecule more water soluble, couples the steroid to a polar sulfate or glucuronide group in a process known as **conjugation.** The conjugated steroids are then primarily secreted by the liver into the bile, with smaller amounts being excreted in the urine. Other hormones, such as epinephrine, are inactivated by specific degrading enzymes circulating in the blood and are then rapidly excreted in the urine. Determination of the rates of urinary excretion of a variety of steroid hormones or epinephrine metabolites has proven to be a useful indirect index of the rate of secretion of the active hormone and has provided physicians with an important diagnostic tool.

Many of the larger peptide hormones, such as insulin and prolactin, are taken into their target tissue cells by a process known as **receptor-mediated endocytosis.** After the hormone binds to its receptor on the surface of target cells, the hormone-receptor complex is internalized or taken into the cell. Once internalized, the hormone is separated from its receptor. In most cases, the hormone molecule is then proteolytically degraded, and the majority of the internalized receptors are recycled back to the cell surface. Many peptide hormones are taken up and degraded by the kidneys, although the exact contribution made by the kidneys can vary considerably depending on the hormone.

Measurement of Hormone Concentrations in the Blood

As indicated earlier, hormones are present in extremely low concentrations in the blood, usually in the range of 10^{-9} to 10^{-12} M. To put these low concentrations into perspective, consider that a hormone concentration of 10^{-9} M corresponds roughly to one hormone molecule in the blood per 50 billion water molecules. Thus, exceedingly sensitive techniques are needed to allow the measurement of circulating hormone concentrations. The measurement of circulating hormone concentrations, either in the basal (resting) state or after an appropriate stimulus has been given, is an important step in the diagnosis of many different endocrine disorders. Because of the similarity in structure and activity of many of the hormones, hormone assays (measurements) must be highly specific if meaningful results are to be obtained.

In the late 1950s, two American scientists, Solomon Berson and Rosalyn Yalow, developed an assay technique that is both extremely sensitive and highly specific. The technique can be performed quite rapidly and inexpensively. The assay that Berson and Yalow developed is known as the **radioimmunoassay (RIA).** The basic technique, described in Chapter 3, has been modified and adapted in a number of different ways, but these newer assays are still based on the original principal of **competition binding,** in which the amount of radioactive hormone that binds to an antibody reveals how much nonradioactive hormone is present. This assay technique can be used to measure the concentrations of hormones in the plasma, as well as a variety of drugs and even vitamins. As a result, the radioimmunoassay has revolutionized the field of endocrinology, both clinically and in the area of basic research. For their efforts in the development of the technique, Berson and Yalow were awarded the Nobel Prize in Physiology and Medicine in 1977.

Mechanisms of Hormone Action

For convenience, hormones can be considered to use one of two different general mechanisms of action to produce their effects. One mechanism, used by amine and peptide hormones, involves membrane receptors and the generation of an intracellular signal or "second messenger" (see Chapter 5). These hormones take effect by altering (activating) proteins that already exist in the cell. The other mechanism, used by steroid hormones and thyroxine, uses intracellular receptors and involves alterations in the expression of particular genes within the nucleus of the cell. These hormones work by initiating production of new proteins in the cell. These two models of hormone action will be considered in more detail shortly. First, let us look at the nature of hormone receptors.

Properties of Hormone Receptors

As was described earlier, a receptor is a molecule inside the cell or the plasma membrane that specifically recognizes and binds one particular hormone. Binding of the hormone to the receptor initiates a sequence of events that ultimately results in the biological effects typical of that hormone. Therefore, specificity within the endocrine system is determined at the level of the hormone receptor.

Receptors are themselves proteins, or in some cases glycoproteins, and thus have definite three-dimensional structures. As illustrated in Figure 12–3, only those hormones with a structure complementary to the structure of a specific receptor can bind to and activate the receptor. A very simple key and lock analogy works here: the key functions like a hormone, and the lock like a receptor.

The ability of receptors to discriminate between hormones is not absolute, however. Similarities in structure between hormones can result in an overlap of hormonal activities. For example, cortisol primarily produces effects typical of a glucocorticoid, but it also interacts weakly with aldosterone receptors and therefore is considered a weak mineralocorticoid. This type of overlapping activity is fairly common among the steroids, owing to the numerous similarities in structure among these hormones. Similar, though fewer, examples of overlapping activities can be found among the peptide hormones. Growth hormone and prolactin have similar structures, and each displays a weak degree of the other's activity. Thus, while receptors are responsible for the specificity that exists within the endocrine system, the inability of the receptor to discriminate absolutely between hormones having similar structures leads to some overlap of activities. Normally, these weak secondary activities are of little or no physiological consequence. However, in an abnormal situation in which the circulating concentration of a particular hormone is markedly elevated, these secondary activities of the hormone may be strong enough to produce noticeable biological effects.

The binding of a hormone to its receptor involves weak chemical forces and is usually readily reversible. When a hormone binds to its receptor, the three-dimensional conformation of the receptor is altered. In this regard, a hormone functions in a manner similar to that of an allosteric modifier of an enzyme. The alteration in structure of the receptor is the first step in the sequence of events leading to the biological effects of the hormone. The magnitude of the biological response is therefore proportional to the number of receptors occupied by hormone molecules. However, receptors for any particular hormone are present only in a finite or limited number in their target tissue cells. It follows, therefore, that once a certain hormone concentration is reached at which all receptors are occupied, no greater response can be produced by further increases in the hormone concentration. Thus, for each biological effect of a hormone, there is a maximal response that can be produced by that hormone.

Current evidence indicates that all the biological effects of hormones are the result of interactions with their respective receptors. The hormones themselves do not directly affect other cellular constituents, such as enzymes, but instead use their receptors as intermediaries to produce the desired effects. The receptor therefore serves as a **signal transducer,** converting a specific extracellular hormonal signal into an intracellular signal that will redirect or alter cell function.

Given the role of the receptor in signal transduction, what is the nature of the intracellular signal generated by the receptor, and how does a single message or signal within the cell lead to a variety of different effects of the hormone?

The "Second Messenger" Model of Hormone Action

Peptide hormones are generally hydrophilic ("water-loving") and therefore not soluble in the lipid layer of the plasma membrane. As a result,

they are unable to readily penetrate the cell membrane and gain access to the interior of the cell. These hormones must therefore use mechanisms involving intracellular signals, or second messengers, to produce their effects. For this class of hormone, the receptor is an intrinsic membrane protein with its hormone-binding component exposed to the extracellular fluid. As illustrated in Figure 12–12, binding of the hormone to that portion of its receptor exposed to the outside of the cell results in generation of a second messenger. Many of the membrane receptors are glycoproteins oriented such that the carbohydrate portion of the molecule is exposed to the outside of the cell. As discussed in Chapter 5, several second messengers are known. Details of the mechanisms involved in the formation of each and the role they play in hormone action will be discussed shortly.

These surface receptors are not rigidly fixed in place in the membrane, as was once thought, but instead show considerable lateral mobility. In addition, as discussed in the earlier section on hormone degradation, a number of peptide hormones have been shown to undergo receptor-mediated endocytosis, in which both the receptor and the hormone are taken into the cell in an endocytotic vesicle. Binding of several peptide hormones to their receptors has also been shown to induce a clustering of the occupied receptors into patches in the plasma membrane. This clustering is believed to facilitate the endocytotic uptake of the hormone-receptor complexes.

As was indicated in Chapter 5, three intracellular second messenger systems have been identified and characterized: (1) the adenylate cyclase/cyclic

Figure 12–12
The general action mechanism of hormones with membrane receptors.

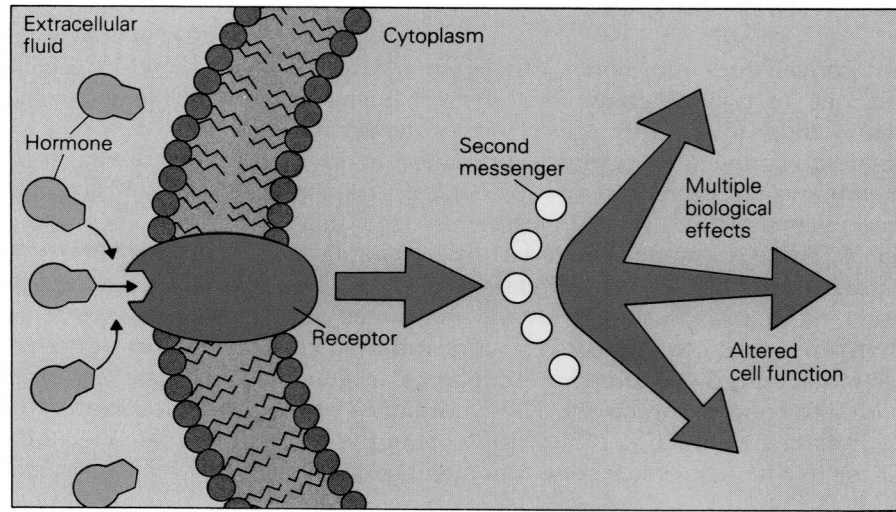

AMP system, (2) the phosphatidylinositol and diacylglycerol system, and (3) receptor-linked ion channels. For some peptide hormones, such as insulin, the mechanism of action has yet to be fully determined, so there may be additional, as yet unrecognized, second messengers.

Cyclic AMP as a Second Messenger

Cyclic 3′,5′-adenosine monophosphate, or more simply **cyclic AMP (cAMP),** was one of the first second messengers to be recognized and has been extensively studied and characterized. As shown in Figure 12–13, this cyclic nucleotide is generated from cellular ATP by the action of the enzyme **adenylate cyclase,** which is located on the inner surface of the cell membrane. Cyclic AMP is rapidly inactivated by conversion to AMP, a process catalyzed by one or more enzymes known as **phosphodiesterases.** As a result of its rapid inactivation, any increase in cytoplasmic cAMP will be short-lived, allowing for the rapid termination of any effects it mediates.

The activity of adenylate cyclase in coupled to membrane receptors by another group of integral membrane proteins known as **G-proteins,** so named because they bind and require **guanosine triphosphate (GTP)** to function. One type of G-protein stimulates adenylate cyclase, while a second type of G-protein inhibits it. These are referred to as the G_s and G_i **proteins,** respectively (Fig. 12–14). Binding of hormones to membrane receptors (designated as R_s in the figure) coupled to adenylate cyclase through G_s proteins results in stimulation of adenylate cyclase and an increase in intracellular cAMP concentrations. Similarly, binding of a hormone to receptors (R_i receptors) linked to the G_i proteins will result in inhibition of adenylate cyclase and a reduction in cAMP. Table 12–3 shows an abbreviated list of hormones that produce their effects at least in part as a result of changes in the intracellular concentration of cAMP. Most hormones act through a stimulatory receptor and therefore produce their effects by increasing intracellular cAMP. These changes in cellular cAMP occur within a matter of just a few seconds after the hormone binds to its receptor. Likewise, the intracellular effects of cAMP occur very rapidly; hence, the overall system is well suited to producing very rapid, acute effects within the cell.

Figure 12–13
Conversion of ATP to cAMP by the enzyme adenylate cyclase, and the subsequent conversion of cAMP to AMP by phosphodiesterase.

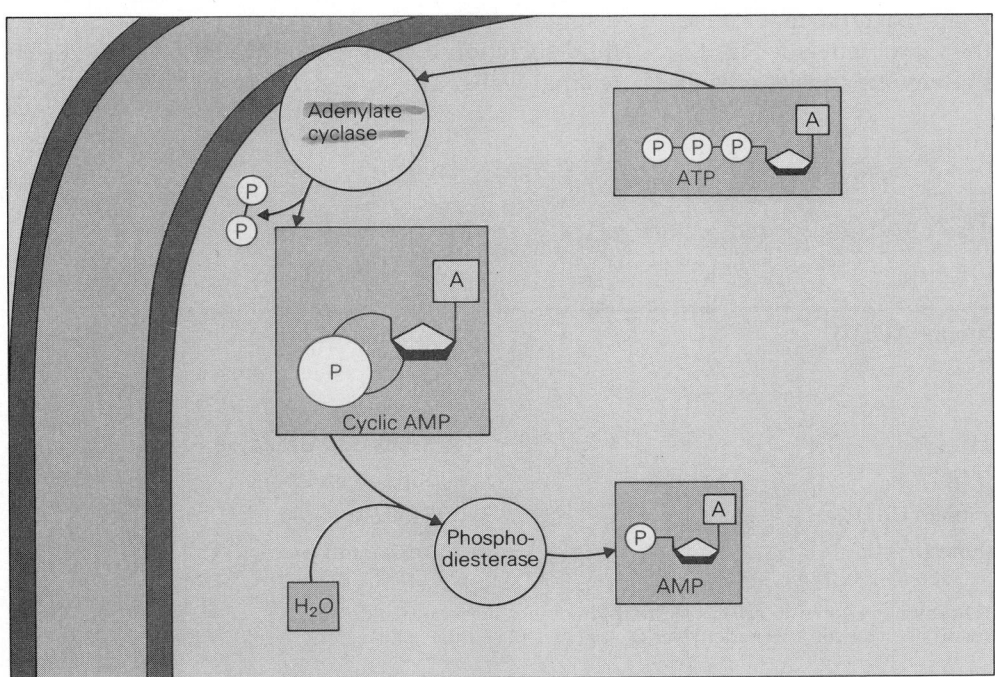

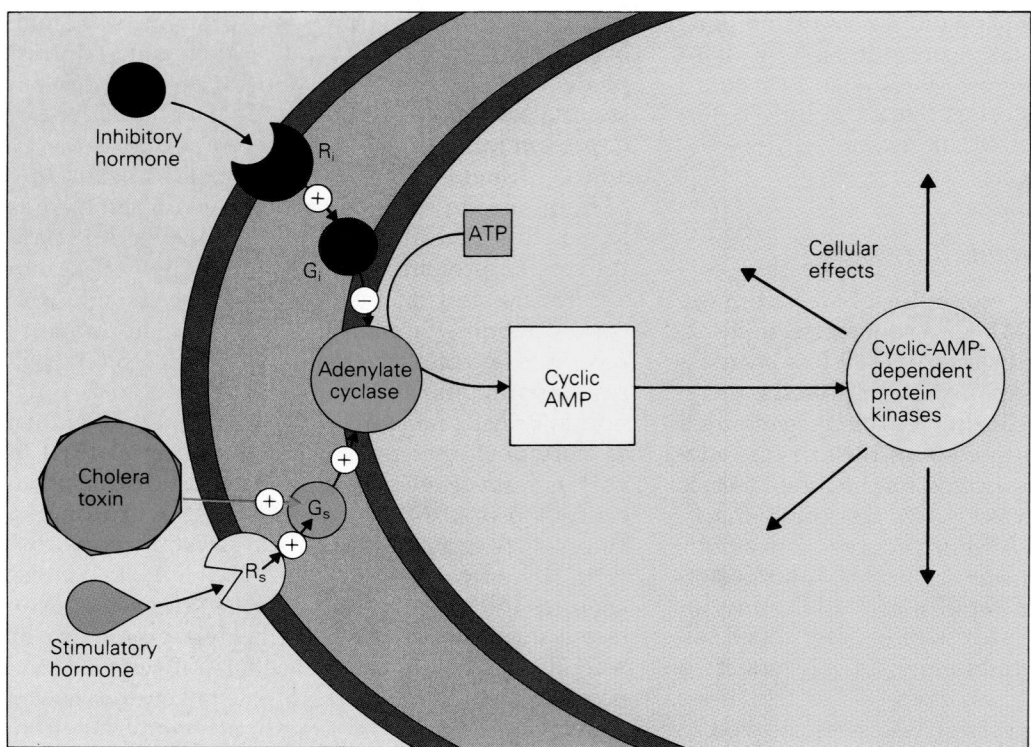

Figure 12–14
Illustration of the coupling of membrane receptors to adenylate cyclase by the G_s and G_i proteins.

A number of agents, including some toxins, are known to bypass the receptor and act directly on various components of the adenylate cyclase system. A toxin released by the bacteria that causes cholera directly activates the G_s protein (see Fig. 12–14). This in turn activates adenylate cyclase, result-

ing in the continual formation of cAMP, even in the absence of the external hormonal signal normally required for adenylate cyclase activation. In the intestine, cAMP is known to be a potent activator of fluid secretion by the mucosal cells that line the intestinal walls. The effect of cholera toxin on the G_s

Table 12–3

Some Hormones that Use cAMP as Second Messenger*

Hormone	Target Tissue
Adrenocorticotropic hormone (ACTH)	Adrenal cortex
Epinephrine	Heart, skeletal muscle, and adipose
Glucagon	Liver
Luteinizing hormone (LH)	Testis and ovary
Parathyroid hormone (PTH)	Kidney and bone
Thyroid-stimulating hormone (TSH)	Thyroid gland
Follicle stimulating hormone (FSH)	Testis and ovary

*This is only a partial list of hormones known to act via cAMP. Many other hormones and local mediators too numerous to list also stimulate adenylate cyclase and raise intracellular cAMP concentrations.

protein therefore accounts at least in part for the severe diarrhea characteristic of this disease. This is just one example of how the molecular basis of a number of disease states may lie in a disruption of normal events involved in cell-to-cell signaling.

Given that certain hormones act by altering cellular cAMP concentrations, how can a change in the concentration of this single compound then lead to the many different effects that some hormones are known to produce within one cell type? As was described in Chapter 5, cAMP is known to produce its effects by activating enzymes known as **cyclic AMP-dependent protein kinases.** Each particular protein kinase catalyzes the **phosphorylation** of (addition of a phosphate group to) a limited number of specific cellular proteins. Phosphorylation of these proteins results in alterations in their activities, which in turn leads to specific changes in cell function. As indicated in Figure 12–15, the activation of several different protein kinases within a cell by cAMP can result in a variety of different intracellular effects. In this way, binding of a hormone to its receptor on the surface of the cell and the generation of a single intracellular signal can regulate a variety of different processes within the cell. In addition, the particular set of kinases present in cells may vary from one tissue to the next, such that the responses elicited as

a result of the increase in cAMP vary from tissue to tissue.

Inositol Trisphosphate and Diacyglycerol as Second Messengers

Although the turnover of membrane lipids was first suggested as a possible mechanism for intracellular signaling in the 1950s, it is only within the past few years that the importance of this mechanism has become widely appreciated. **Phosphatidylinositol** is a minor phospholipid constituent of the plasma membrane and is primarily located in the inner half of the bilayer. Upon the donation of two phosphate groups from ATP, phosphatidylinositol is converted to **phosphatidylinositol 4,5-bisphosphate (PIP_2),** which serves as the immediate precursor of the generation of two intracellular second messengers. As shown in Figure 12–16, PIP_2 is split to form **diacylglycerol (DG)** and **inositol trisphosphate (IP_3).** The cleavage is catalyzed by a membrane-bound enzyme known as **phospholipase C.** Several other enzymes function to replenish the supply of membrane PIP_2, therefore preventing its depletion.

As with adenylate cyclase, the activity of phospholipase C is functionally coupled to membrane hormone receptors by a GTP-binding protein (G-

Figure 12–15
Example of how an increase in intracellular cAMP produced in response to a hormone can result in many different biological responses in the cell. The particular set of kinases shown in this figure is purely hypothetical. Some cells may have fewer kinases, some more, and some may have a different mixture altogether.

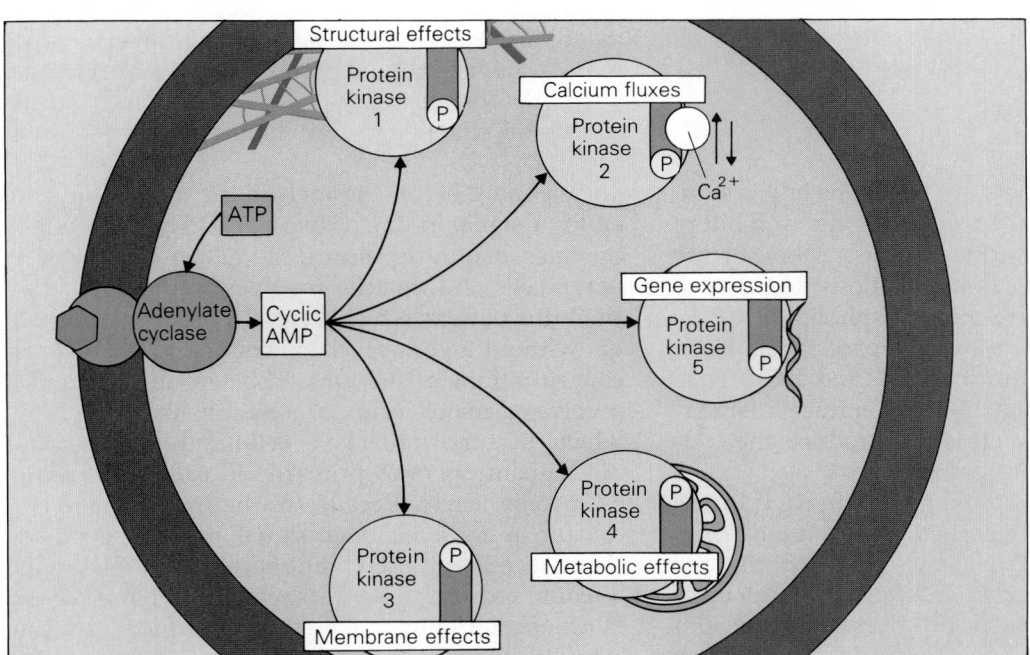

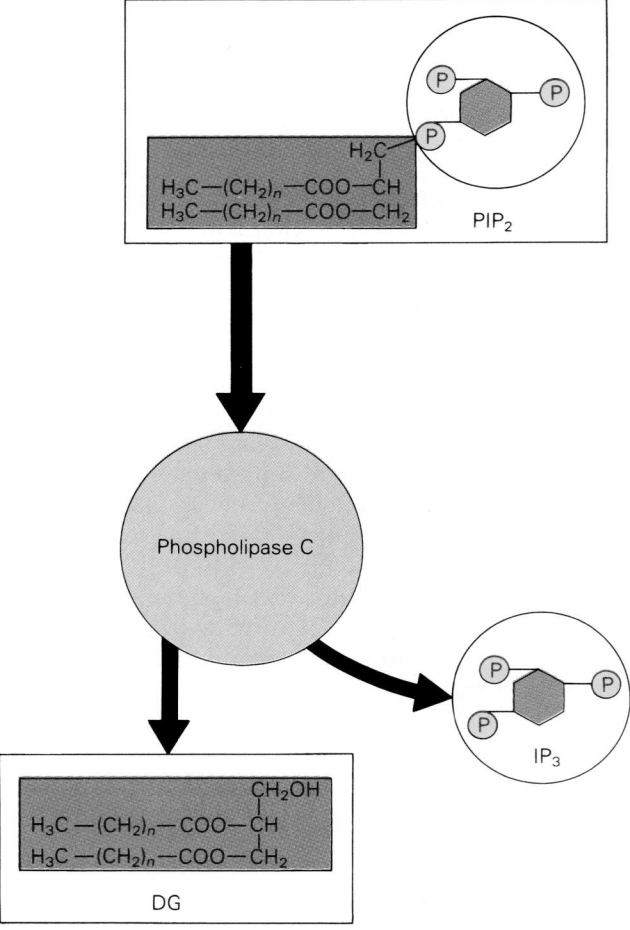

Figure 12–16
Pathway of conversion of phosphatidylinositol 4,5-bisphosphate (PIP$_2$) to inositol trisphosphate (IP$_3$) and diacylglycerol (DG) by phospholipase C.

mediator in a variety of different processes. Thus, the cellular effects of one branch of this mechanism are ultimately the result of increased cytoplasmic calcium concentrations.

The other product formed from PIP$_2$ hydrolysis, diacylglycerol (DG), remains associated with the membrane. Within the membrane, it activates a membrane-bound protein kinase known as **C-kinase.** In a manner analogous to that described for the cAMP-dependent kinases, C-kinase catalyzes the phosphorylation of a number of specific cellular proteins, thereby altering their activity and producing specific changes in cell function. Although many of the details of this two-branch IP$_3$/DG pathway have yet to be fully characterized and understood, it appears that the two mechanisms act together to regulate a number of different processes.

The biochemistry of inositol phosphate metabolism is in fact probably much more complicated than described in the preceding paragraphs. A large number of multiply-phosphorylated inositol phosphates, such as IP$_4$, IP$_5$, and IP$_6$, have now been identified in mammalian cells. However, details of the metabolic pathways leading to their formation and their postulated functions are not completely understood.

Ions as Second Messengers

A third signal transduction mechanism used by membrane receptors is that of receptor-linked ion channels. In certain hormone/receptor systems, the change in configuration of the receptor caused by hormone binding results in opening of specific, gated, ion channels within the membrane. Whether or not other membrane proteins are involved in coupling the changes in receptor configuration to the functional changes in ion channels is uncertain at present.

Opening of ion channels in the membrane generates a signal in one of two ways. The first, which operates mainly in electrically active cells such as nerve cells and muscle, involves a small and transient flux of ions, which alters the membrane potential without any appreciable change in cytoplasmic concentrations of the ions. The second mechanism involves a major influx of ions into the cytoplasm, which in turn initiates a cellular response. This mechanism operates primarily in cells that are not electrically active. Because of the important role of calcium as an intracellular signal, it is not surprising that a number of membrane receptors are linked to calcium channels, as illustrated in Figure 12–18. Hormone binding activates the receptors, causing calcium channels to open and allowing calcium to

protein). At present, there appears to be only a stimulatory G-protein, since as yet no inhibitory G-protein has been identified for this system. Figure 12–17 illustrates the functional relationship between the receptor, G-protein, and phospholipase C. As indicated, activation of phospholipase C results in the hydrolysis of PIP$_2$ to form IP$_3$ and DG. These two substances both play a role as intracellular second messengers, although they produce their effects by different mechanisms.

The IP$_3$ released from the membrane acts inside the cell by causing the release of calcium ions from storage sites in the endoplasmic reticulum, thereby raising the concentration of free calcium in the cytoplasm. As was described in Chapter 5, calcium serves an important role as an intracellular signal or

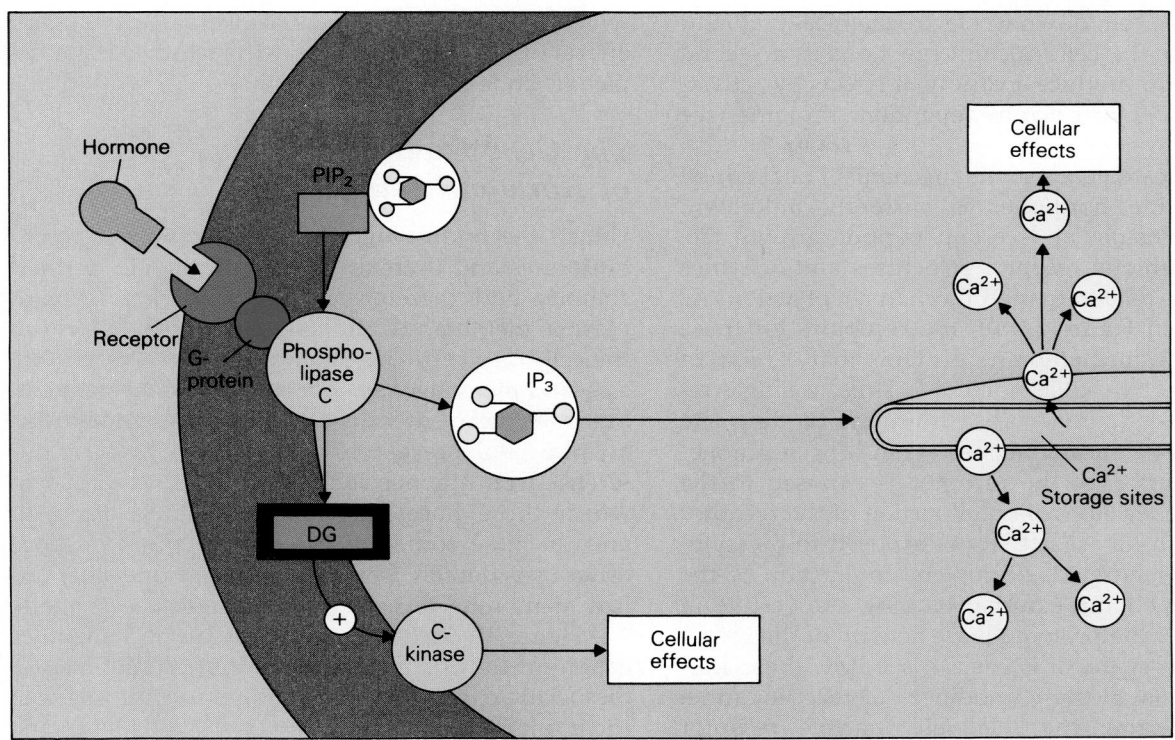

Figure 12-17
Diagram of the formation and the mechanisms of action of inositol trisphosphate (IP₃) and diacylglycerol (DG) as intracellular second messengers.

Figure 12-18
Coupling of membrane receptors to membrane
calcium channels.

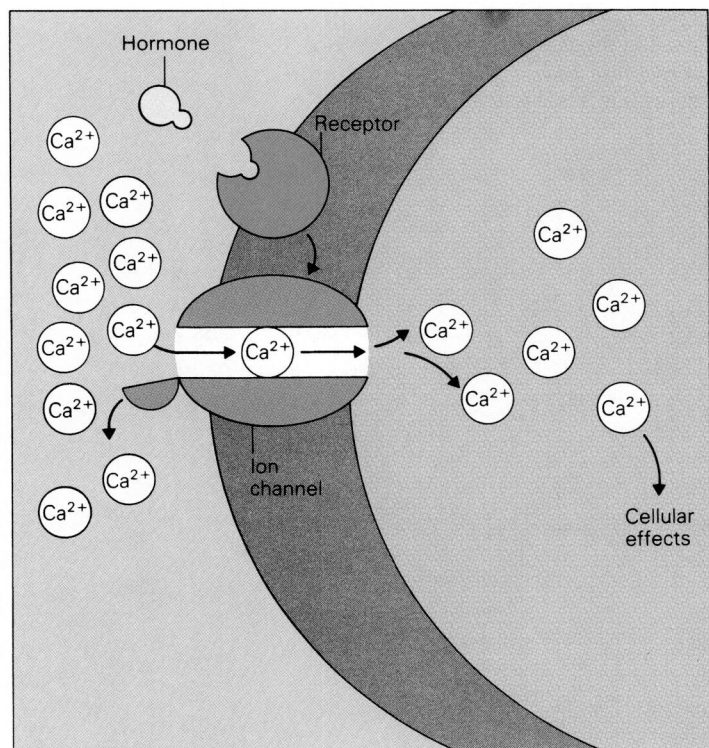

flow into the cell, down its electrochemical gradient. Once inside the cell, calcium functions as a second messenger to produce a variety of effects by activating a number of calcium-dependent proteins (see Chapter 5).

As indicated earlier, the mechanism of action of some peptide hormones is currently unknown. However, insulin and certain peptide growth factors have unique receptor structures and activities that may be related to their mechanism of action. As illustrated in Figure 12–19, the receptors for these agents are transmembrane proteins that exhibit an intrinsic protein kinase activity. Thus, a single receptor molecule has both a hormone binding site and an enzymatically active region. The hormone-binding portion of the receptor is exposed to the outside of the cell, while the portion of the receptor with protein kinase activity is exposed to the cytoplasm. As Figure 12–19 depicts, interaction of the hormone with the external binding site results in activation of the protein kinase activity of the receptor. In this regard, it is easy to see how these hormones act as allosteric modifiers of receptor function, converting the relatively inactive receptor kinase into a more active form. Presumably, the kinase catalyzes phosphorylation of cytoplasmic proteins, altering their activity and producing the desired biological effects. At present, the degree to

which these events are responsible for the biological effects of insulin and the growth factors is not completely understood.

The Gene Expression Model of Hormone Action

Unlike the hydrophilic peptide hormones, steroid hormones and thyroxine are hydrophobic, or lipid soluble, and therefore can readily pass through plasma membranes and other membranes within the cell. As a result, these hormones do not require a second messenger system to produce their effects, but instead can enter the cell themselves and bind to an intracellular receptor. In addition to this basic difference from the peptide hormones, the manner in which these hormones regulate cellular function and the time course of the changes they produce differ considerably from those of hormones that utilize membrane receptors and second messengers.

Figure 12–20 depicts the general mechanism of action for thyroid hormones and the steroids. After dissociating from its carrier protein in the plasma, the hormone freely moves across the cell membrane into the cytoplasm of the cell. For many years, scientists believed that receptors for steroid hormones existed free in the cytoplasm of target cells. However, recent evidence suggests that the receptors

Figure 12–19
Possible mechanism of action on hormones that have receptors with intrinsic protein kinase activity.

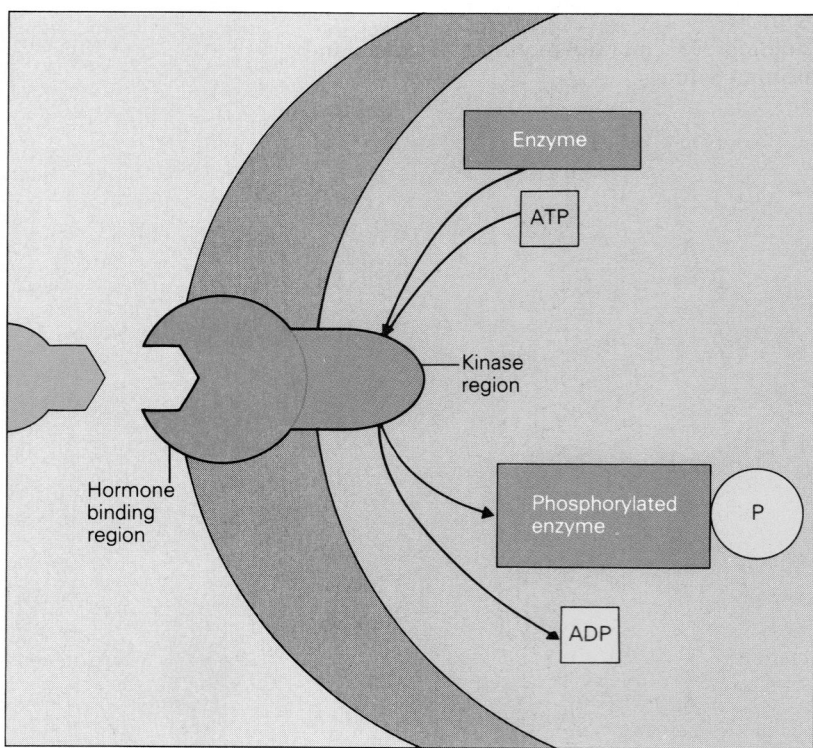

Growth Factors, Cellular Communication, and Oncogenes

In recent years, there has been a nearly exponential increase in the number of growth factors identified and at least partially characterized. As the name implies, exposure of certain cells or tissues to specific growth factors stimulates growth in those cells. In addition, growth factors also stimulate a variety of metabolic processes in their target tissues. A typical growth factor is peptide in nature and generally exerts its hormone-like actions via autocrine or paracrine mechanisms. In many instances, the signal transduction pathways utilized by growth factors are similar if not identical to the mechanisms used by peptide and protein hormones described in the accompanying sections of this chapter.

The area of cellular communication mechanisms and signal transduction has been the subject of active research for many years. Much attention has been focused on these processes with the hope that understanding the pathways responsible for regulating normal cell growth and function would lead to greater insight into conditions of uncontrolled growth, such as in cancer. In recent years, this assumption has been supported by the finding that altered versions of protein-encoding genes that function in signal transduction pathways related to cell growth are present in certain forms of animal cancer. These altered genes are known as *oncogenes* (from the Greek *onkos*, meaning mass or tumor). The oncogenes arise as a result of a virus picking up either a fragment or a corrupted copy of a normal gene from its host cell. As a result of normal viral activity, the gene fragment is inserted into subsequent host cells, where the oncogene product, or *oncoprotein,* is produced.

Normally, when a growth factor reaches its target cells, the signal transduction pathway leading to growth stimulation is only transiently activated. Presumably, growth-inhibitory signals are generated at some point to prevent cellular proliferation from proceeding indefinitely. In cells containing oncogenes, however, pathways are continuously activated, causing cells to endlessly grow and replicate. In many instances, oncogenes encode proteins (oncoproteins) that are very similar to the gene products found in normal cells, except that these proteins have lost important regulatory effects on their activity, and therefore do not require activation by an external signal. Thus, this situation is clearly one in which the molecular basis of disease results from a disruption of the normal events involved in cell-to-cell signaling and communication.

For the sake of simplicity, a hormone-signaling pathway can be subdivided into three components. The first component is the signal, or hormone itself. The second is the signal transducer, or receptor for that hormone. And third are the various intracellular post-receptor effectors, such as adenylate cyclase, G-proteins, and protein kinase C, involved in hormone action and growth-regulatory pathways. In theory, any component of a growth-regulating pathway could become oncogenic (tumor causing) if it were to be altered in some critical manner that prevented the regulation of that component.

In fact, oncogenes (and their oncoproteins) have been identified that are associated with each of the three components of hormone-signaling pathways described above. For example, epidermal growth factor, or EGF, binds to and activates a cell-surface receptor that is a transmembrane tyrosine kinase similar to that illustrated in Figure 12–19. The *erbB* oncogene (the name is an abbreviation of the virus in which it was discovered) has been shown to code for a protein analogous to the EGF receptor, but lacking the extracellular binding region of the normal EGF receptor. In the absence of the extracellular EGF binding domain, the intracellular tyrosine kinase activity of this altered receptor remains continuously activated, and therefore those processes that the receptor normally regulates are under continuous stimulation. Another example is the **sis** oncogene, which encodes a functionally active subunit of a growth factor known as PDGF (platelet-derived growth factor). Cells containing a sis oncogene manufacture the PDGF subunit constantly and at an inappropriately high rate, constantly stimulating adjacent target cells to proliferate. Numerous other examples are also readily evident: the **ras** oncogene encodes a protein analogous to a normal G-protein except that an amino acid substitution results in continuous activation of the counterfeit G-protein; the **erb-A** oncoprotein is an unregulated version of the thyroid hormone receptor; the **src** family of oncogenes code for a group of unregulated tyrosine kinases; and the **raf/mil** oncoproteins are cytoplasmic serine/threonine kinases that lack normal regulated function.

The study of oncogenes is likely to remain an area of intense research for many years. As a result of studying oncogene biology, we have learned much about structure/function relationships of various components of hormone action pathways. For example, studies of the erbB oncoprotein suggest that in a normal EGF receptor, the extracellular domain of the receptor continuously inhibits the cytoplasmic tyrosine kinase activity of the cytoplasmic portion of the receptor. Normally, when EGF binds to the receptor, this inhibitory effect is released, leading to activation of the cytoplasmic receptor kinase function. In addition to providing insight into normal signalling processes, scientists anticipate that as the mechanisms by which oncogene products disrupt normal growth are better understood, it may be possible to design pharmacological intervention to counteract their effects and restore normal growth.

exist either on the nuclear envelope or within the nucleus of the cell. Binding of the hormone to the receptor causes a conformational change in the receptor so that the hormone/receptor complex then interacts with specific "acceptor sites" on the nuclear DNA. This interaction initiates transcription of mRNA molecules that code for specific proteins. After transcription, the mRNA precursors are processed within the nucleus before moving into the cytoplasm of the cell (see Chapter 5). Because of the increased abundance of these particular mRNA molecules, the synthesis of a few specific proteins is

Figure 12–20
Mechanism of action of hormones with intracellular receptors, indicating the effects on gene expression and the possible ways in which cell function might be altered.

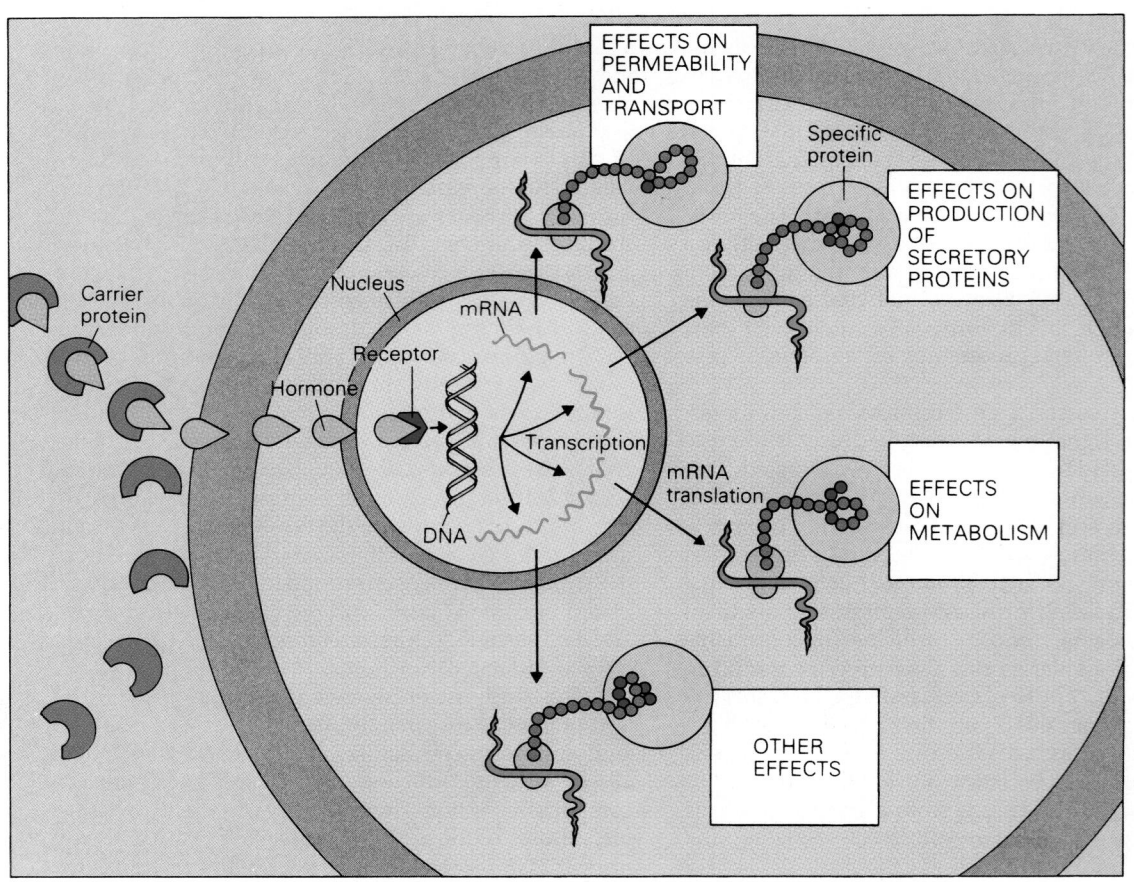

increased. Increased synthesis of these particular proteins leads to the altered cellular function characteristically produced by the particular hormone. As indicated in Figure 12–20, these newly synthesized proteins may be involved in a variety of different cellular functions.

The cellular response to a particular hormone acting via these mechanisms usually results from increased synthesis of several proteins. Although each of these proteins may serve a different role in the cell, each contributes in some way to producing the final coordinated biological effect. For example, aldosterone, a mineralocorticoid produced by the adrenal cortex, has the effect of increasing sodium transport across tubule cells of the kidney. As shown in Figure 12–21, sodium first enters the cell on the tubular side by diffusion through specific sodium channels and is then pumped out on the opposite side of the cell by an energy-requiring sodium pump. Aldosterone both increases the synthesis of several mitochondrial citric-acid-cycle enzymes involved in ATP production, making more energy available to power the sodium pump, and stimulates synthesis of sodium channel proteins,

Figure 12–21
Apparent mechanism of action of aldosterone to increase sodium transport across kidney tubule cells.

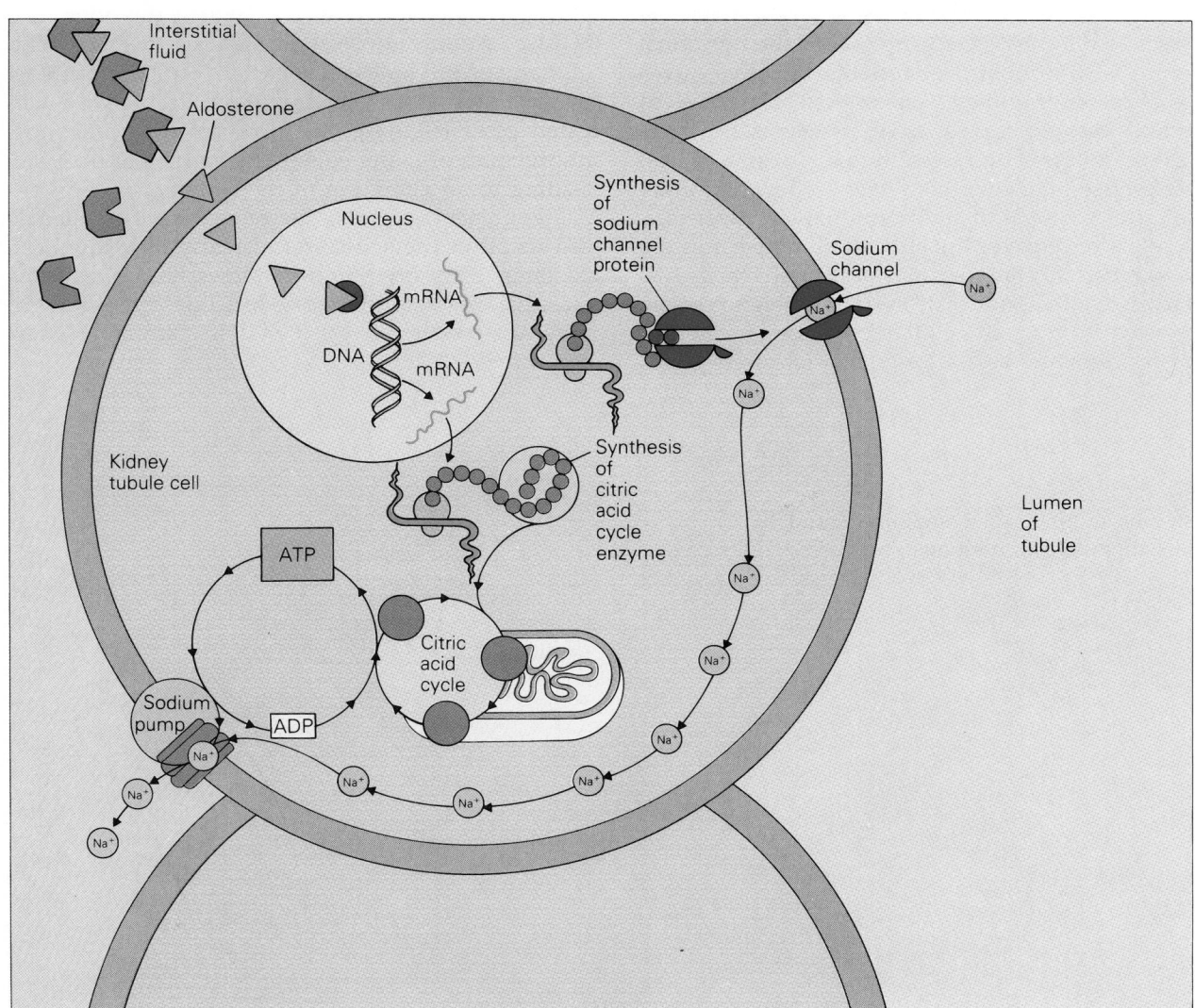

making more sodium available to the sodium pump. There is some evidence that aldosterone may also increase the synthesis of sodium pump proteins themselves. The overall effect is to increase the transfer of sodium across kidney tubule cells. Thus, as is typical of other steroid or thyroid hormones that take effect by regulating events within the nucleus, aldosterone alters the function of its target cells by changing the rate of synthesis and therefore the amount of several key proteins in the cell. Recall that, by contrast, hormones that act through membrane receptors using second messengers generally do so by modifying the activity of pre-existing proteins, using such mechanisms as phosphorylation or dephosphorylation.

Another basic difference between the cell-surface and intracellular receptor systems involves the time course over which responses occur. In general, responses to hormones that use the second-messenger system can be observed in a matter of seconds or minutes. In addition, once the hormonal stimulus is withdrawn, cell function rapidly returns to its basal (resting) state. In contrast, responses to steroid or thyroid hormones are usually not evident for many minutes or several hours. In some cases, the response may even take days to develop. Likewise, after removal of the hormonal signal, it may take hours or days for the effect to diminish and the cells to return to their original state. With these two mechanisms, the endocrine system is able to closely regulate a number of cellular processes on both a short-term (acute) and a long-term (chronic) basis.

Amplification of Hormone Signals

One characteristic feature within the endocrine system is the amplification or increase in signal strength. As a result of this amplification, binding of just a few hormone molecules to their receptors can result in a dramatic change in cell function. Signal amplification occurs at nearly every step in hormone action, from binding of the hormone to its receptor to the final biological effect produced by the hormone. An example of signal amplification is illustrated in Figure 12–22. It is easy to see that after just a few steps, the original signal consisting of one hormone molecule has now been multiplied many times over. The amplification occurs because each step in the sequence involves enzymes that, when active, are responsible for forming not just one but many single "signal" molecules. For example, when glucagon binds to its receptor, it activates adenylate cyclase. Adenylate cyclase catalyzes the formation of many cAMP molecules, each of which in turn activates a particular protein kinase. Each cAMP-activated protein kinase will now catalyze the phosphorylation of many copies of a particular enzyme, leading to an alteration in the activity of many enzyme molecules. Thus, just one glucagon molecule can result in the activation of many individual cell proteins. This amplification allows hormones to be effective regulators of cell function even though they are present in the blood in extremely small amounts.

Figure 12–22
Illustration of how signal amplification might occur with regard to hormone action.

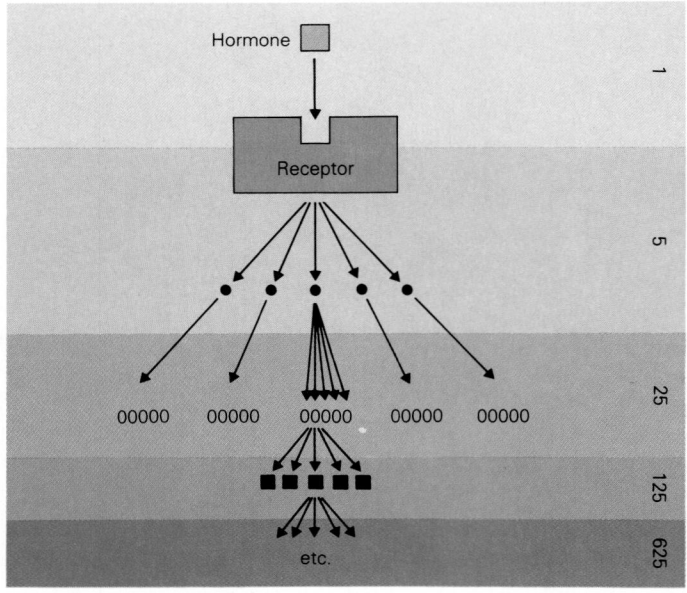

SUMMARY

General Concepts of Endocrine Control

Endocrinology is the study of hormones and the glands that produce them, including the synthesis, secretion, and action of hormones and the clinical abnormalities produced by hormone dysfunction. Most substances considered hormones are secreted in small amounts into the blood and then carried to other parts of the body, where they interact with receptors and produce a distinct biological response in the cells of their target tissues.

Neuroendocrine cells link the *nervous* and *endocrine* systems by converting electrochemical signals within the nervous system into hormonal signals in the blood.

Feedback regulation in the endocrine system usually involves *negative feedback*, although positive feedback can also occur.

The Chemistry of Hormones

In general, hormones regulate processes related to either energy metabolism, salt and water metabolism, growth and development, or reproduction. Most hormones act on several tissues and produce several different specific effects.

Chemically, hormones generally fit into one of three categories: amino acid derivatives, peptide or protein hormones, and steroid hormones, which are derived from cholesterol. Steroids can be further subdivided into glucocorticoids, mineralocorticoids, androgens, estrogens, and progestins.

Steroid hormones and thyroid hormones travel through the blood by binding to specific carrier proteins and to albumin and other plasma proteins. Peptide and protein hormones do not require carrier proteins for transport.

Most hormones are secreted in a biologically active form. Steroid hormones are primarily degraded by the liver. Many of the peptide hormones are proteolytically degraded by their target tissues. The kidneys also contribute to the proteolytic degradation of several peptide hormones.

The radioimmunoassay allows for the measurement of hormone concentrations in the blood. Development of the assay was an extremely important advancement in the field of endocrinology.

Mechanisms of Hormone Action

The receptor is the cell molecule that recognizes the hormone and initiates its biological effects. Receptors are responsible for the *specificity* that exists within the endocrine system, although the specificity is not absolute in every case.

Peptide and protein hormones generally act via plasma membrane receptors present on the cell surface. Binding of a hormone to its receptor on the cell surface results in the generation of an intracellular second messenger, which then alters, or regulates, cell function. Second messengers include cyclic AMP, inositol trisphosphate, diacylglycerol (DG), and ions, notably calcium.

Steroid hormones generally exert their effects via intracellular receptors located on or in the nucleus. The hormone/receptor complex interacts with the DNA to alter transcription of messenger RNA molecules coding for specific proteins. Alterations in the synthesis of these specific proteins lead to the particular cellular response characteristically produced by the hormone.

Signal amplification occurs at nearly every step in hormone action, allowing hormones to be effective regulators of cell function even though they are present in the blood in extremely small amounts.

REVIEW QUESTIONS

Identify Each with a Term

1. The terms for the mechanisms by which hormones reach their target cells solely by diffusion, rather than via the bloodstream.

2. The term for the cell type that links the nervous system and the endocrine system.

3. The precursor for synthesis of all steroid hormones.

4. The assay system commonly utilized to measure hormone concentrations in the blood.

5. The enzyme responsible for the formation of cAMP.

6. The names of the two second messengers formed as a result of the hydrolysis of phosphatidylinositol 4,5-bisphosphate (PIP_2) by phospholipase C.

Define Each Term

7. hormone receptor
8. neuroendocrinology
9. mineralocorticoids
10. conjugation of steroids
11. G_s protein
12. receptor-mediated endocytosis
13. phosphodiesterase

Choose the Correct Answer

14. Which of the following is *not* a steroid hormone?
   a. aldosterone
   b. epinephrine
   c. cortisol
   d. progesterone

15. Which of the following is *not* one of the five categories of steroid hormones?
   a. progestins
   b. estrogens
   c. prostaglandins
   d. glucocorticoids

16. Which of the following is a glycoprotein?
   a. parathyroid hormone (PTH)
   b. growth hormone (GH)
   c. thyroxine (T_4)
   d. follicle-stimulating hormone (FSH)

17. Which of the following does *not* appear to use cAMP as a second messenger?
   a. glucagon
   b. epinephrine
   c. thyroid-stimulating hormone (TSH)
   d. insulin

18. Which of the following exerts it effects primarily via the gene expression model of hormone action?
   a. growth hormone
   b. aldosterone
   c. parathyroid hormone
   d. luteinizing hormone (LH)

19. Hormones are involved in regulating processes related to:
   a. production, use, and storage of energy
   b. reproductive functions
   c. growth and development
   d. salt and water metabolism
   e. all of the above

20. The ion thought to be involved in stimulus-secretion coupling of hormone secretion by endocrine cells is:
   a. calcium
   b. sodium
   c. potassium
   d. hydrogen ion

21. Which of the following hormones is *not* synthesized from readily available dietary constituents?
   a. adrenocorticotropic hormone (ACTH)
   b. thyroxine
   c. progesterone
   d. growth hormone

22. Which of the following is thought to be transformed in its target tissues to a more active hormone form?
   a. insulin
   b. follicle-stimulating hormone (FSH)
   c. estradiol
   d. testosterone

23. Receptors for steroid hormones are located:
   a. within the nucleus
   b. in the cytoplasm of non-target cells
   c. always near the site of synthesis of the hormone
   d. in the nuclear membrane
   e. either A or D, not B or C

24. Which of the following is *not* a general characteristic of peptide hormones?
   a. are transported in the blood largely free in solution
   b. are synthesized and secreted on demand
   c. utilize cell membrane receptors and intracellular second messengers
   d. are usually first synthesized as a larger precursor protein

Answer in One or Two Sentences

25. What is the role of hormone receptors in hormone action, and how do receptors convey specificity within the endocrine system?

26. What is the general mechanism of action of steroid and thyroid hormones?

27. Why is signal amplification an important feature of hormone action and how does it occur?

28. Describe the adenylate cyclase/cAMP second messenger system and explain how an increase in intracellular cAMP can lead to different effects in different cell types.

29. Describe the phosphatidylinositol second messenger system.

SUGGESTED READINGS

Berridge, Michael J. ''The molecular basis of communication within the cell.'' *Scientific American*, 253:142–152, 1986.

Berridge, Michael J., and Irvine, Robin F. ''Inositol phosphates and cell signalling.'' *Nature*, 341:197–205, 1989.

Carafoli, Ernesto, and Pennington, John T. ''The calcium signal.'' *Scientific American*, 253:70–78, 1986.

Druker, Brian J., Mamon, Harvey J., and Roberts, Thomas M. ''Oncogenes, growth factors, and signal transduction.'' *New England Journal of Medicine*, 321:1383–1391, 1989.

Fregley, Melvin J., and Luttge, William G. *Human Endocrinology: An Interactive Text.* New York: Elsevier Biomedical, 1982.

Hedge, George A., Colby, Howard C., and Goodman, Robert L. *Clinical Endocrine Physiology.* Philadelphia: W.B. Saunders Company, 1987.

Lewis, Deborah L., Lechleiter, James D., Kim, Donghee, Nanavati, Chaya, and Clapham, David E. "Intracellular regulation of ion channels in cell membranes." *Mayo Clinic Proceedings* 65:1127–1143, 1990.

Smith, Emil L., Hill, Robert L., Lehman, I. Robert, Lefkowitz, Robert J., Handler, Philip, and White, Abraham. *Principles of Biochemistry: Mammalian Biochemistry,* 7th ed. New York: McGraw-Hill Book Company, 1983.

Snyder, Solomon H. The molecular basis of communication between cells. *Scientific American,* 253:132–141, 1986.

Tepperman, Jay, and Tepperman, Helen M. *Metabolic and Endocrine Physiology,* 5th ed. Chicago: Year Book Medical Publishers, Inc., 1987.

Walters, Marian R. "Steroid hormone receptors and the nucleus." *Endocrine Reviews,* 6:512–543, 1985.

KEY TERMS

adenylate cyclase (p. 439)
adrenal cortex (p. 431)
adrenal glands (p. 424)
aldosterone (p. 432)
autocrine (p. 419)
cortisol (p. 432)
cyclic AMP (p. 439)

diacylglycerol (DG) (p. 441)
endocrine system (p. 419)
estrogen (p. 431)
follicle-stimulating hormone (FSH) (p. 429)
glucocorticoid (p. 431)
hormone receptors (p. 437)

human chorionic gonadotropin (hCG) (p. 430)
inositol trisphosphate (IP_3) (p. 441)
neuroendocrine (p. 421)
paracrine (p. 419)

progesterone (p. 432)
steroid hormones (p. 431)
testosterone (p. 432)

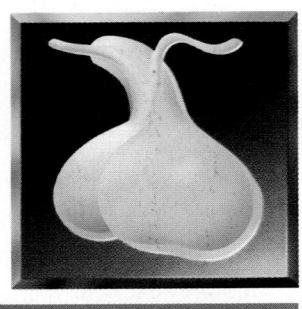

C H A P T E R
13

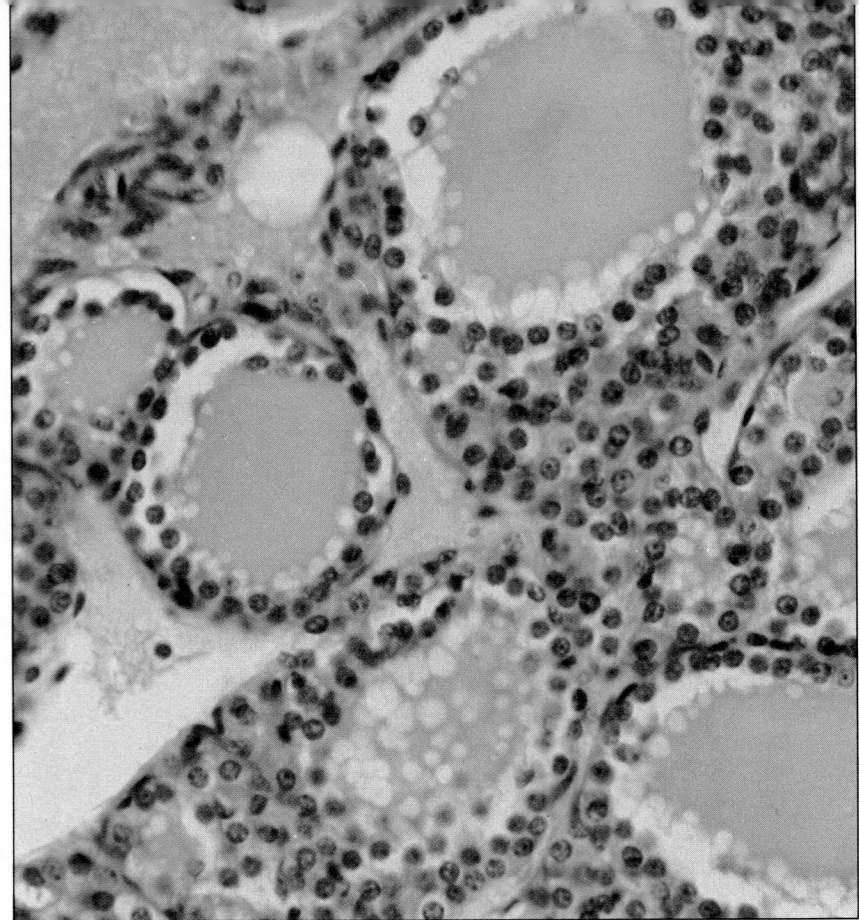

The Pituitary and Thyroid Glands

CHAPTER OUTLINE

The pituitary gland plays a central role in the hormonal regulation of a wide variety of processes, and thus has been one of the most extensively studied of all endocrine glands. As we will show, pituitary function is under the control of the hypothalamus. The interaction of the pituitary and hypothalamus is therefore an excellent example of the interaction between the nervous and endocrine systems.

We begin with a discussion of pituitary hormones and the regulation of pituitary function by the hypothalamus. Later sections of this chapter describe the actions of two pituitary hormones: growth hormone and thyroid-stimulating hormone.

Anatomy of the Pituitary Gland and Its Relation to the Hypothalamus

Gross Anatomy of the Pituitary Gland

In humans, the pituitary measures about 1 cm (1/2″) in diameter and is located just below the hypothalamus at the base of the brain (Fig. 13–1). As described in Chapter 10, the hypothalamus serves as an integrative center for a variety of different bodily processes. The hypothalamus also controls pituitary function and therefore serves as an important regulator for the endocrine system. The pituitary is not a single gland, but consists of three separate glands, or lobes: the **anterior lobe,** the **intermediate lobe,** and the **posterior lobe.** The pituitary is connected to the hypothalamus by a thin segment of tissue known as the **pituitary stalk.** The hormones secreted by each of the three different lobes of the pituitary are listed in Table 13–1. As this table indicates, the pituitary gland secretes at least nine different hormones. However, as will be described later, the pituitary does not itself synthesize all nine hormones.

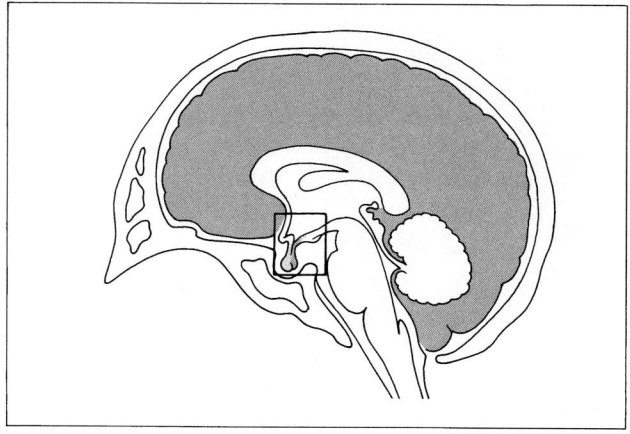

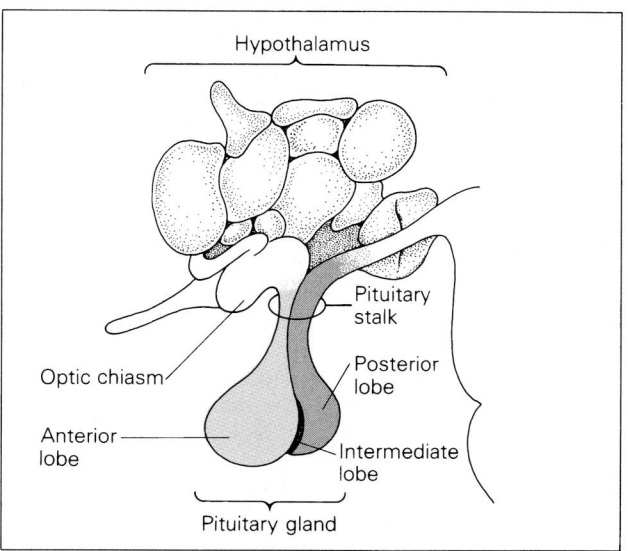

Figure 13–1
The location of the pituitary gland relative to the brain and hypothalamus.

Table 13–1
Hormones Secreted by the Pituitary Gland

Hormones secreted from the anterior lobe
    Luteinizing hormone (LH)
    Follicle-stimulating hormone (FSH)
    Prolactin (PRL)
    Adrenocorticotropic hormone (ACTH)
    Growth hormone (GH)
    Thyroid-stimulating hormone (TSH)

Hormones secreted from the intermediate lobe
    Melanocyte-stimulating hormone (MSH)

Hormones secreted from the posterior lobe
    Oxytocin
    Antidiuretic hormone (ADH, also known as vasopressin)

Regulation of hormone secretion by each of the lobes of the pituitary involves different mechanisms. Cells of the anterior pituitary function as true endocrine cells (see Fig. 12–1) in that they receive signals or information via the blood and in turn release their own hormones directly into the blood. The posterior pituitary, in contrast, is really an extension of the nervous system and functions like typical neuroendocrine cells. In humans, the intermediate lobe is rudimentary and does not exist as a

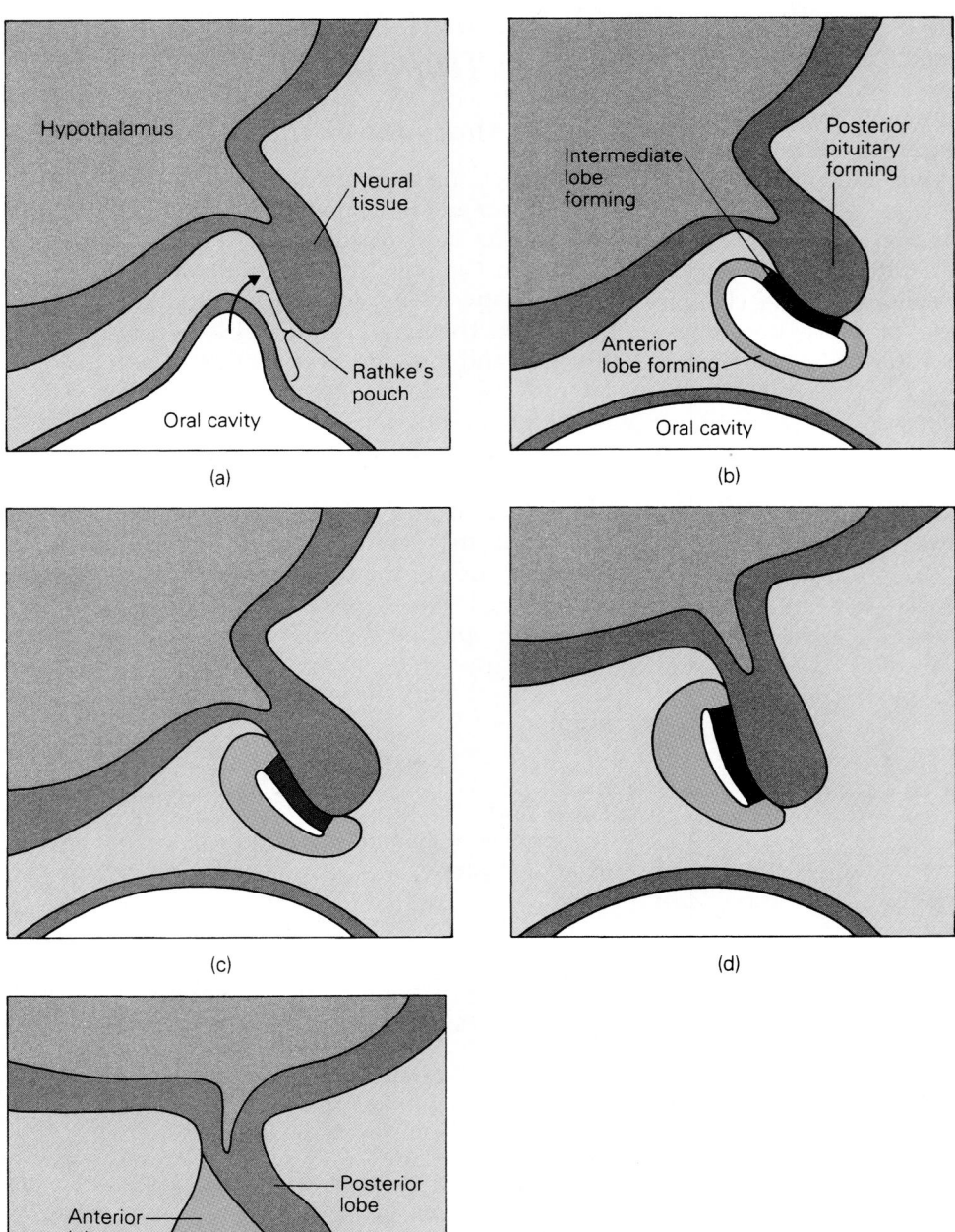

Figure 13–2
The formation of the anterior, posterior, and intermediate lobes of the pituitary during embryological development. The normal sequence of development is from (*a*) through (*e*). (*a*) and (*b*) Note that the posterior lobe arises from neural tissue, while the anterior lobe is formed from Rathke's pouch. The intermediate lobe forms from the portion of Rathke's pouch that makes contact with the neural tissue. The cleft (*e*) is the remnant of what was once the hollow interior of Rathke's pouch.

distinct lobe per se, but consists of only a few cells interspersed along the borders of the anterior and posterior lobes. Although the intermediate lobe serves an important role in a number of lower vertebrates, it has little or no endocrine function in humans.

Functional Anatomy of the Pituitary Gland

An examination of the embryologic development of the pituitary illustrates how both the endocrine and neuroendocrine portions of the gland are formed. As Figure 13–2 indicates, during early fetal development a pocket of cells known as **Rathke's pouch** pinches off and moves upward from the roof of the primitive oral cavity. At about the same time, a finger-like projection of neural tissue extends downward from the base of the hypothalamus. Rathke's pouch eventually meets this neural protrusion of the hypothalamus, and the two structures fuse to form the anterior and posterior lobes of the pituitary, respectively.

Neural connections between the hypothalamus and the posterior lobe of the pituitary are maintained during development. As Figure 13–3 shows, the posterior pituitary contains the axons and terminals of neurons that have their cell bodies within the hypothalamus. These neurons terminate close to numerous capillaries located throughout the posterior pituitary. As indicated in Figure 13–3, the cell bodies of these neurons are found within two distinct areas of the hypothalamus, the **supraoptic nucleus** and the **paraventricular nucleus.** The significance of these anatomical features will become evident in the discussion of the regulation of posterior pituitary hormone secretion that follows.

In contrast to the posterior pituitary, the function of the anterior pituitary is not regulated by direct innervation but instead by vascular (blood vessel) connections with the hypothalamus. As shown in Figure 13–3, arterial blood reaching the hypothalamus enters a specialized region known as the **median eminence,** where vessels branch into a network of primary capillaries. From these primary capillaries, the blood enters the **long portal veins** and is carried down through the pituitary stalk to the anterior lobe of the pituitary, where a second capillary network is located. From this second capillary network in the anterior pituitary, blood leaves the gland through the veins and then mixes with the systemic venous blood.

This complex of blood vessels, which consists of the primary capillary network, the secondary capillary network, and the vessels that connect them, is known as the **hypothalamic-pituitary portal system.** As will be described later, the hypothalamic-pituitary portal system carries factors from the hypothalamus to the anterior pituitary that serve to regulate the secretion of anterior pituitary hormones.

In addition to the long portal veins that connect the hypothalamus and anterior pituitary, a less prominent portal system connects the posterior and anterior lobes of the pituitary. This vascular system is referred to as the **short portal system.** Some evidence suggests that the short portal veins carry substances secreted by the posterior lobe to the anterior lobe, where they regulate hormone secretion from the anterior pituitary. The exact degree to which this system is involved in regulating anterior pituitary function is not well understood at this time, however.

Regulation of Pituitary Hormone Secretion by the Hypothalamus

The mechanisms by which the hypothalamus regulates the secretion of hormones from both the posterior and anterior lobes of the pituitary are shown in diagrammatic form in Figure 13–4. This figure illustrates the functional significance of the neural and vascular connections between the hypothalamus and the pituitary shown in Figure 13–3.

The Posterior Pituitary: A Neuroendocrine System

Neural connections between the hypothalamus and posterior pituitary are shown on the right side of Figure 13–4. As indicated previously, the cell bodies of the neurons that send projections to the posterior pituitary are located primarily in the supraoptic and paraventricular nuclei of the hypothalamus. As indicated in Table 13–1, the posterior pituitary secretes two hormones: **oxytocin** and **antidiuretic hormone (ADH,** or vasopressin). These two hormones are actually synthesized within the cell bodies of the neurons in the supraoptic and paraventricular nuclei of the hypothalamus. Oxytocin and ADH are then transported to the posterior pituitary through the axons of the same neurons that produce them. Generation of action potentials in these nerves results in the release of oxytocin or vasopressin from the nerve endings adjacent to capillary beds in the

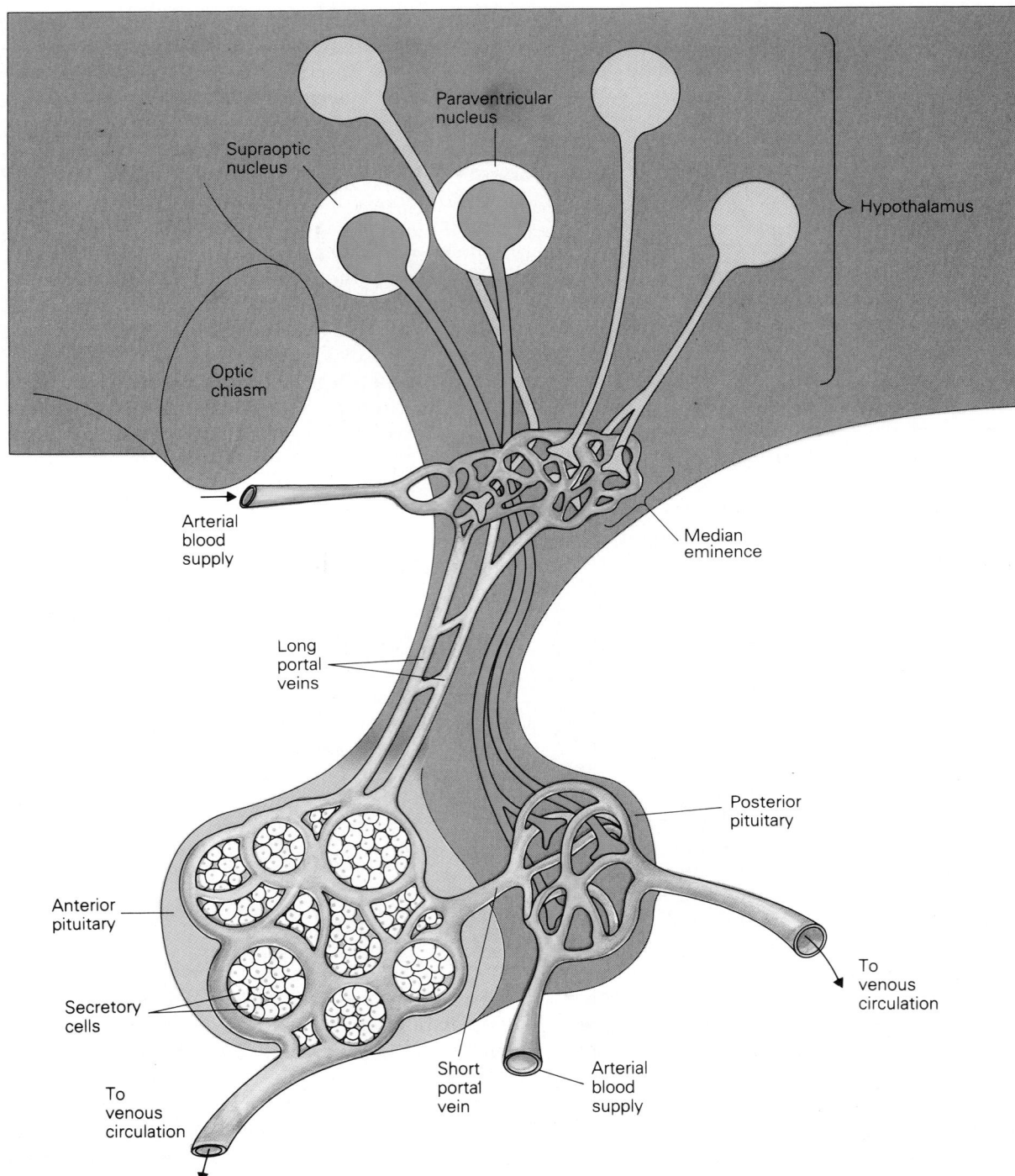

Figure 13–3
The neural and vascular connections between the hypothalamus and the anterior and posterior lobes of the pituitary.

posterior pituitary. The hormones enter the blood vessels and are then carried throughout the body. The cells that produce oxytocin and ADH are therefore good examples of neuroendocrine cells.

Neural inputs to the supraoptic and paraventricular nuclei from other areas of the hypothalamus or from other areas of the brain modulate the electrical activity of the neuroendocrine cells in these two

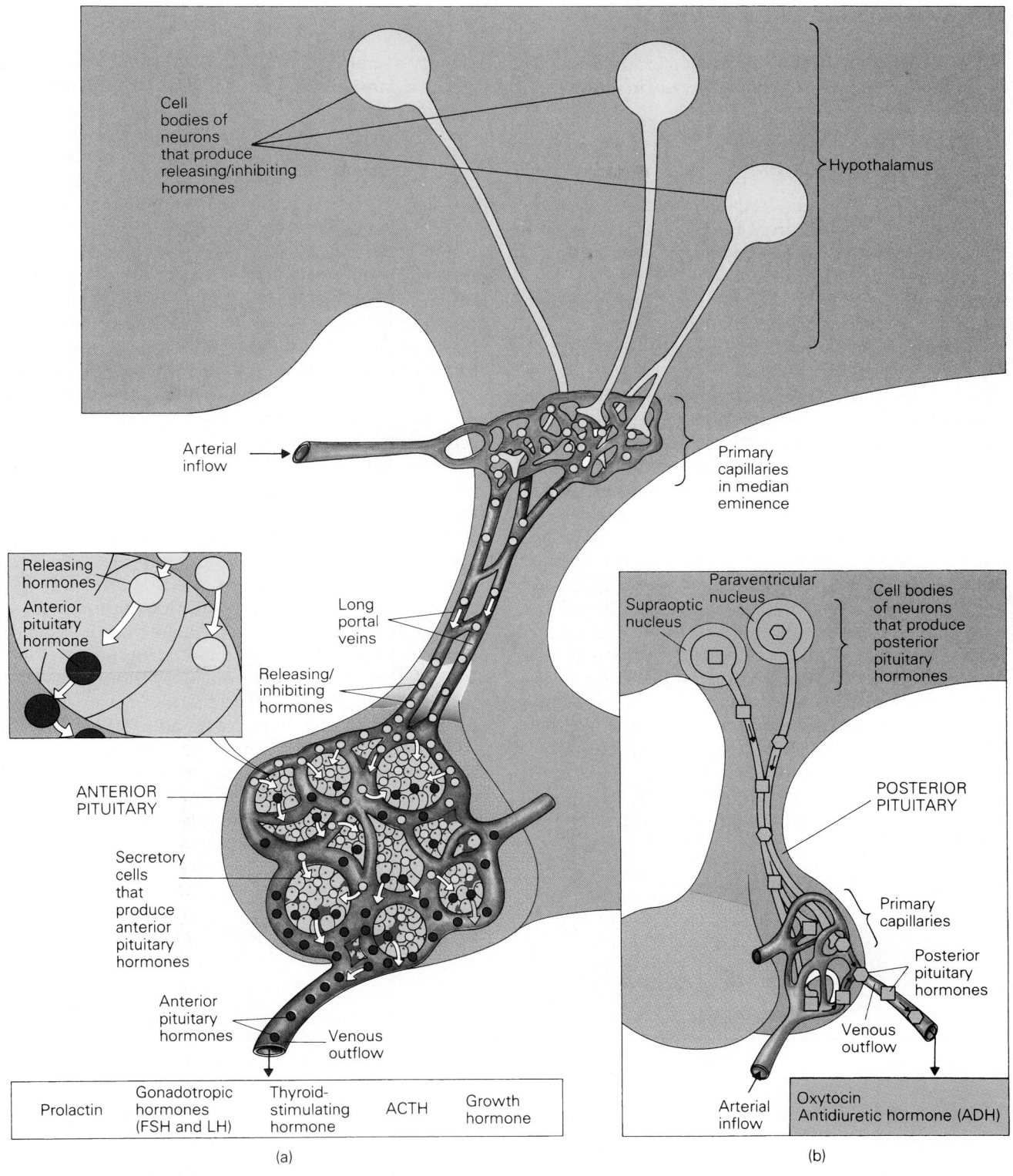

Figure 13–4
An illustration of the mechanisms involved in the regulation of anterior and posterior pituitary function. (*a*) Neuroendocrine cells of the hypothalamus transmit releasing/inhibiting hormones to the capillaries in the median eminence. These hormones travel via the long portal veins to the secretory cells of the anterior lobe, where they stimulate or inhibit the release of anterior pituitary hormones into the blood. (*b*) Neurons that originate in the supraoptic and paraventricular nuclei of the hypothalamus synthesize oxytocin and ADH and transmit them through axons that release the hormones into capillaries in the posterior pituitary. The hormones are then funneled into the circulation.

nuclei. In this manner, a variety of stimuli are capable of regulating the secretion of oxytocin and ADH.

The Anterior Pituitary: An Endocrine System

As Table 13–1 indicates, cells of the anterior pituitary synthesize six different hormones. Secretion of these hormones is under the control of several different **releasing/inhibiting hormones** produced in the hypothalamus. All of the releasing/inhibiting hormones that have been chemically characterized thus far are either small peptides or amino acid derivatives. As illustrated on the left side of Figure 13–4, these hormones are synthesized within the cell bodies of neuroendocrine cells located in a number of different sites in the hypothalamus. The axons of these neuroendocrine cells converge in the median eminence of the hypothalamus, where they terminate on or near the primary capillary network of the hypothalamic-pituitary portal system. Generation of action potentials in these neuroendocrine cells causes secretion of particular releasing/inhibiting hormones from their nerve terminals. Upon release, the releasing/inhibiting hormones enter the primary capillaries and travel via the long portal vessels to the anterior pituitary. The releasing/inhibiting hormones then exit from the secondary capillaries and act on the endocrine cells of the anterior pituitary to regulate (stimulate or inhibit) secretion of anterior lobe hormones.

Table 13–2 lists the hypothalamic releasing/inhibiting hormones involved in regulating anterior pituitary secretion. Note that some stimulate secretion of a particular hormone, while others inhibit hormone secretion. As a group, they are therefore referred to as "releasing/inhibiting hormones." The system would be relatively easy to understand if there were a simple one-to-one relationship between each anterior pituitary hormone and a single releasing hormone or a pair of releasing and inhibiting hormones. However, a single hypothalamic hormone may influence the secretion of several anterior pituitary hormones. For example, **somatostatin** regulates the secretion not only of growth hormone but also of **prolactin** and **TSH.** Similarly, GnRH regulates the secretion of both FSH and LH. The degree to which multiple activities of these hypothalamic releasing/inhibiting hormones regulate hormone secretion from the anterior pituitary has not yet been fully established. The general belief, however, is that secretion of any particular hormone from the anterior pituitary depends on the relative amounts of the various hypothalamic releasing/inhibiting hormones that reach the anterior pituitary.

Cell bodies of neurons that synthesize releasing/inhibiting hormones are not randomly distributed throughout the hypothalamus, but instead each releasing factor is synthesized by neurons located in different, discrete nuclei of the hypothalamus. For example, GnRH is produced primarily in one nucleus of the hypothalamus, while CRH is produced in another. Although they are synthesized in different regions of the hypothalamus, both hormones are ultimately released into the capillary network in the median eminence. Secretion of re-

---

Table 13–2
Releasing Hormones Involved in Regulating Anterior Pituitary Secretion

Hypothalamic-releasing Hormone	*Primary Effect on Anterior Pituitary**
Corticotropin-releasing hormone (CRH)	Stimulates ACTH secretion
Thyrotropin-releasing hormone (TRH)	Stimulates TSH secretion
Gonadotropin-releasing hormone (GnRH)	Stimulates LH and FSH secretion
Somatostatin	Inhibits GH secretion
Growth-hormone-releasing hormone (GHRH)	Stimulates GH secretion
Prolactin-releasing factor (PRF)**	Stimulates prolactin secretion
Prolactin-inhibitory hormone (PIH)	Inhibits prolactin secretion

*As described in the text, several of the releasing hormones influence the secretion of more than one anterior pituitary hormone. For simplicity, only the primary effects of the releasing hormones are presented here.

**The chemical structure of the substance which stimulates prolactin secretion has not been established. By convention, it is referred to as a "factor" until its structure is known.

leasing/inhibiting hormones from these hypothalamic neurons depends on the input they receive from neurons located in other areas of the hypothalamus or in higher brain centers.

Actions and Effects of the Pituitary Hormones

Now that we have examined the anatomical and physiological bases for regulating hormone secretion from the anterior and posterior lobes of the pituitary, a logical question might be, "What are the actions of the pituitary hormones?" The following sections describe the specific biological effects of the pituitary hormones. Note the wide variety of physiological processes that the pituitary hormones regulate.

Hormones of the Posterior Pituitary

Antidiuretic Hormone (ADH): An Increase in Water Retention

Antidiuretic hormone, or **ADH,** is one of two small peptide hormones secreted by the posterior pituitary. It is nine amino acids in length and is synthesized primarily in the supraoptic nucleus of the hypothalamus, although smaller amounts are produced in the paraventricular nucleus as well

The peripheral action mechanisms of ADH will be discussed in detail in Chapters 24 and 25, when the regulation of fluid balance by the kidney is described. Briefly, however, the primary effect of ADH is to increase water retention by the kidney. The net result of this effect is a decrease in urine volume and an increase in extracellular fluid volume. ADH can also act to constrict blood vessels (vasoconstriction). The other name for ADH, **"vasopressin,"** stems from this vasoconstrictor effect, but this response occurs only when the hormone is secreted in relatively large amounts.

The two primary physiological stimulators of ADH release are (1) an increase in osmolality of the blood or extracellular fluid, and (2) a large decrease in blood volume, as in a hemorrhage, or heavy bleeding. How do changes in osmolality or blood volume trigger ADH release? Specialized cells within the hypothalamus known as **osmoreceptors** are very sensitive to increases in osmolality of the extracellular fluid. Even a slight increase in osmolality will stimulate the osmoreceptor cells. The osmoreceptors in turn activate the neuroendocrine cells of the supraoptic nucleus and trigger ADH release from the posterior pituitary. As a result of promoting water retention by the kidney, ADH causes a dilution of the extracellular fluid, which returns plasma osmolality to normal.

A decrease in blood volume also stimulates ADH release. However, stimulation of ADH release requires a 10% or greater blood volume decrease, which is a severe loss of blood. As a result of increasing fluid retention by the kidneys, ADH acts to restore blood volume to normal. In addition, a large loss of blood can result in sufficient ADH secretion to cause constriction of blood vessels. By constricting blood vessels, this effect serves to lessen the decrease in blood pressure that would occur with severe hemorrhage.

In addition to changes in osmolality and blood volume, several other factors are known to influence ADH release. For example, alcohol is a strong inhibitor of ADH release. When ADH secretion is inhibited, urine volume increases, and excess fluid is lost from the body. This accounts, in part, for the increase in urine volume and the thirst experienced following the consumption of alcoholic beverages. Nicotine, barbiturates, and certain anesthetics are known to stimulate ADH release. The overproduction of ADH caused by these agents results in excess fluid retention.

In the absence of ADH secretion, which can occur as a result of either a genetic abnormality or damage to the posterior pituitary, the kidneys are unable to conserve water, and large quantities of fluid are lost in the urine (a process called *diuresis*). This condition is known as **diabetes insipidus.** Diabetes insipidus is distinctly different from what is commonly termed "diabetes," which is actually diabetes mellitus and will be discussed in Chapter 15.

Oxytocin: Regulation of Breast Milk Release

Oxytocin, the other hormone secreted by the posterior pituitary, is also a small peptide hormone (nine amino acids in length) and is very similar in structure to ADH. Like ADH, oxytocin is synthesized in the cell bodies of neurons in both the supraoptic and paraventricular nuclei of the hypothalamus, but its primary site of synthesis is in the paraventricular nucleus.

The principal action site of oxytocin is the female breast. In the breast, oxytocin stimulates contraction of specialized smooth muscle cells, which results in the transfer of milk from its site of synthesis in structures known as **alveoli** into the larger ducts of the breast. The net result of this effect is that the milk is made available for a nursing infant.

This action of oxytocin will be described in Chapter 31, when the hormonal control of lactation is discussed. In addition to its effects on the breast, oxytocin can also stimulate contraction of smooth muscle in the uterus.

The primary stimulus for oxytocin secretion in a female is suckling by a nursing infant. When an infant nurses, touch sensors in the mother's nipple are activated and neural signals are transmitted to her brain. This information is processed in the brain, and the oxytocin-producing cells in the paraventricular nucleus are stimulated, resulting in oxytocin release from the posterior pituitary.

A variety of psychological stimuli can also influence oxytocin release. The mere sound of a baby crying can trigger oxytocin release in nursing mothers, while fear or apprehension can strongly inhibit its secretion. Thus, in addition to the neural mechanism that is triggered by suckling, input from other areas of the brain is also important in regulating oxytocin secretion.

Hormones of the Anterior Pituitary

Five different hormone-secreting cell types are found in the anterior pituitary. These are listed in Table 13–3 along with the hormone(s) produced by each cell type. The **somatotrophs,** which produce growth hormone, are the most numerous cells of the anterior pituitary. Normally, about 50% of the cells present in the anterior pituitary are somatotrophs. **Corticotrophs,** the ACTH-secreting cells, compose about 20% of the cells, and the TSH-producing cells, the **thyrotrophs,** about 5%. **Lactotrophs** and **gonadotrophs** are the other two cell types in the anterior pituitary.

The size and activity of the latter two cell types vary under different physiological conditions. In lactating (nursing) females, the lactotrophs, which secrete prolactin, are large and numerous, reflecting

Table 13–3

Hormone-Producing Cells of the Anterior Pituitary

Name of Cell Type	Principal Hormone Produced
Somatotrophs	Growth hormone (GH)
Corticotrophs	ACTH
Thyrotrophs	TSH
Lactotrophs	Prolactin
Gonadotrophs	LH and FSH

the high rate of prolactin secretion in these individuals. In males or nonlactating females, fewer lactotrophs are present, and they are smaller as well. The gonadotrophs, which secrete both LH and FSH, account for only about 5% of the cell population of the anterior pituitary of males. In females, their size varies in accordance with the cyclical changes in LH and FSH secretion that occur during the monthly menstrual cycle.

As Table 13–3 indicates, these five cell types are responsible for the production of six hormones in the anterior pituitary. Gonadotrophs are responsible for the synthesis and secretion of *both* LH and FSH. As Chapter 30 will describe, the concentrations of LH and FSH in the blood do not always change in parallel during the monthly ovarian cycle. Thus, although they are produced within the same cell, their secretion is regulated independently. At present, the mechanisms by which a single releasing hormone (GnRH) is able to differentially regulate the secretion of two hormones (LH and FSH) are not well understood.

The following sections describe the secretion and actions of each of the anterior pituitary hormones. The physiology of LH, FSH, and prolactin will be presented in detail in Chapters 30 and 31. Similarly, the physiology of ACTH will be presented primarily in Chapter 14. Therefore, these hormones will only be discussed briefly here. Instead, we will emphasize the remaining two hormones, growth hormone and TSH.

Luteinizing Hormone (LH) and Follicle-Stimulating Hormone (FSH): Regulation of Reproductive Function

LH and FSH are collectively referred to as the **gonadotropins.** This name reflects the fact that the target tissue of these hormones is the **gonads,** or the ovaries and testes in females and males, respectively. In general, the gonadotropins have two primary effects: (1) to promote the development and maturation of sperm and egg, and (2) to stimulate production of sex steroid hormones by the gonads. The principal sex steroids in males and females are **testosterone** and **estradiol,** respectively. As Table 13–2 indicates, secretion of gonadotropins is under the control of GnRH from the hypothalamus.

Prolactin: Regulation of Milk Synthesis

Prolactin has no known function in males. In females, the primary effect of prolactin is to stimulate milk production by the breast. The target cells for prolactin are the milk-producing alveolar cells of the breast. (Specific mechanisms by which prolactin stimulates milk formation will be presented in Chapter 31.) Prolactin secretion is under the dual control of a **prolactin-releasing factor (PRF)** and a **prolactin release-inhibiting hormone (PIH).** Similar to the situation for oxytocin described previously, prolactin release increases in response to stimulation of the mother's nipple by a suckling infant.

Adrenocorticotropic Hormone (ACTH): Regulation of Cortisol

The primary effect of **adrenocorticotropic hormone (ACTH)** is to regulate the synthesis and secretion of the steroid hormone **cortisol** from the adrenal cortex (a portion of the adrenal gland, an endocrine gland located near the kidney). The cortisol produced by the adrenal gland in turn has numerous effects on intermediary metabolism in a variety of different tissues. As the name indicates, ACTH also has **trophic** or "growth-promoting" effects on cells of the adrenal cortex. As Table 13–2 indicates, the secretion of ACTH is regulated by **corticotropin-releasing hormone (CRH),** which is produced in the hypothalamus. The specific effects of ACTH on the adrenal cortex and the effects of adrenal steroids on metabolism will be covered in detail in the following chapter.

ACTH is actually synthesized as part of a much larger precursor protein that contains not only ACTH but also several other biologically active peptides. This precursor has been named **pro-opiomelanocortin peptide,** or **POMC peptide,** to reflect the fact that it contains the sequences for endogenous (internally synthesized) opioid compounds (see Chapter 7), the sequence for **melanocyte-stimulating hormone (MSH),** and the sequence for ACTH.

Growth Hormone (GH): Body Growth and Metabolism

As was indicated in Figure 12–9 in the previous chapter, **human growth hormone (hGH or GH)** is a peptide hormone consisting of 191 amino acids. The structure of GH is remarkably similar to that of a peptide hormone secreted by the placenta during pregnancy, **human chorionic somatomammotropin (hCS).** Growth hormone is also somewhat similar to prolactin, although to a much lesser extent. GH, hCS, and prolactin compose a family of related peptide hormones collectively referred to as the **somatomammotropic hormones,** named to indicate their effects on body growth and breast development. The sections that follow will focus on the ef-

fects of GH on body growth and metabolism. The physiology of hCS and prolactin is presented in later chapters.

Regulation of Growth Hormone Secretion. As Tables 13–2 and 13–3 indicate, GH secretion from the somatotrophs of the anterior pituitary is under the dual control of growth-hormone-releasing hormone (GHRH) and somatostatin, both of which are produced by the hypothalamus. As indicated previously in Table 12–1, somatostatin is also produced in the pancreas, where it is thought to play a local role in regulating pancreatic hormone secretion. This particular action of somatostatin will be described further in Chapter 15. At this point we should emphasize, however, that the somatostatin produced by the hypothalamus and that produced by the pancreas serve different physiological functions, and also that the secretion of somatostatin from each site is independently regulated.

As Figure 13–5 shows, growth hormone is not secreted in a steady continuous fashion throughout the day, but instead is secreted in a somewhat pulsatile manner. The most consistent period of GH secretion occurs about one hour after the onset of deep sleep (Fig. 13–5). This can occur during the normal sleep period at night or during a daytime nap. The phase of sleep associated with dreams, known as REM sleep (see Chapter 11), initiates the return of GH secretion to the basal pre-sleep levels. The exact significance of this sleep-related surge in GH secretion is not known, but it has been suggested that it may be important in stimulating processes of tissue growth and repair during sleep.

Figure 13–5
Typical 24-hr pattern of growth hormone secretion in a normal individual. As indicated, the person slept from 11:30 PM until 7:00 AM. The pink bars indicate periods of REM sleep. Note the sleep-induced increase in the plasma growth hormone concentration that occurs between midnight and 1:00 AM. Note also that with the onset of REM sleep, growth hormone secretion returns to the basal level.

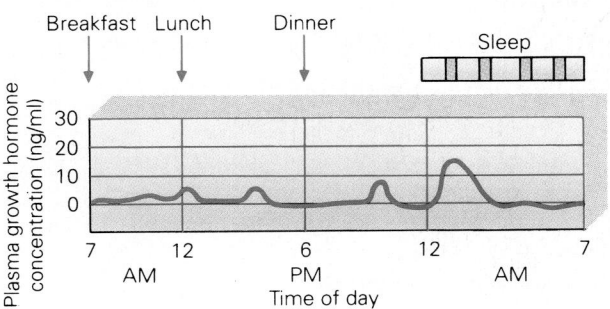

In addition to the sleep-induced increase in GH secretion, a series of pulses often occurs anywhere from two to four hours after a meal (see Fig. 13–5). The relative frequency and the size of these pulses tend to increase about the time of puberty. Some of the other agents or conditions known to influence GH secretion are listed in Table 13–4. Hypoglycemia (low blood glucose) stimulates GH secretion, while high blood glucose inhibits GH secretion. A useful test that physicians can perform to evaluate whether a patient's pituitary is capable of secreting GH involves the administration of a dose of insulin, followed by the monitoring of GH concentrations in the blood. As Chapter 15 will describe, insulin decreases the blood glucose concentration. In a normal individual in this test situation, the hypoglycemia caused by insulin triggers a pronounced increase in GH secretion. In contrast, the absence of increased GH secretion in response to insulin indicates to the physician that pituitary GH secretion is impaired. As Table 13–4 also shows, various types of stress, both physical and emotional, stimulate GH secretion. Several amino acids, especially arginine, also stimulate GH secretion.

How is such a wide variety of stimuli, such as those listed in Table 13–4, able to regulate GH secretion? As Figure 13–6 illustrates, GHRH and somatostatin serve as the final common pathway for regulating GH secretion. Inputs from various areas of the brain and hypothalamus regulate GHRH and somatostatin release in the median eminence, thus regulating GH secretion from the anterior pituitary.

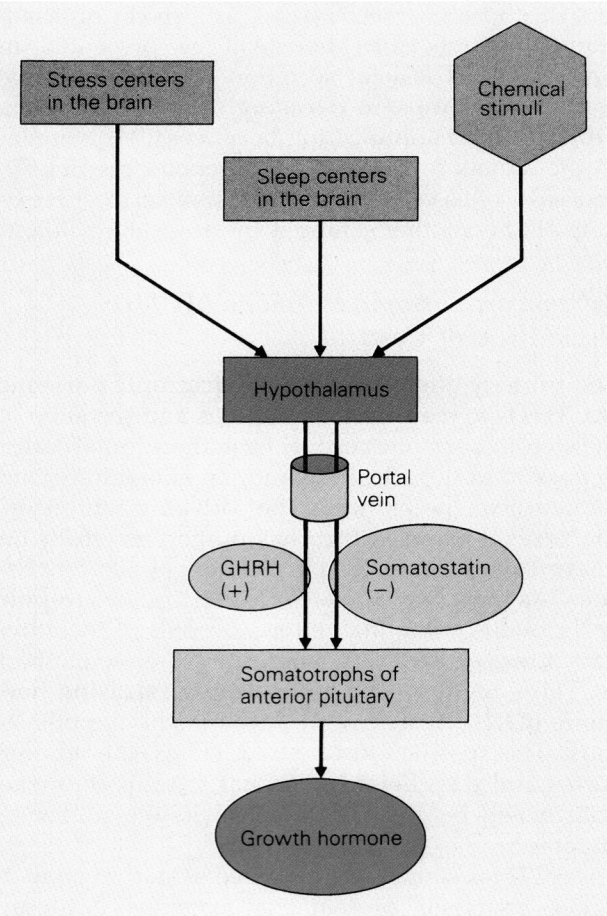

Figure 13–6

A variety of factors can influence growth hormone secretion by regulating the production of GHRH and somatostatin by the hypothalamus.

Table 13–4

Partial Listing of Factors or Conditions Known to Influence Growth Hormone (GH) Secretion

Stimulators
  Deep sleep*
  Low blood glucose concentration
  Stress
    Physical trauma
    Infection
    Psychological stress
  Amino acids, especially arginine

Inhibitors
  REM sleep*
  High blood glucose concentration

*See Chapter 11 of this text.

The Effects of Growth Hormone. The overall effect of GH is to promote tissue growth. In this regard, GH is considered to be an **anabolic hormone.** (Those hormones that tend to have the opposite effect, promoting breakdown of tissues, are known as **catabolic hormones.**)

Another important anabolic hormone in the body is insulin, a hormone produced by the pancreas. While many of the effects of GH are similar to those of insulin, some are exactly opposite those of insulin and actually antagonize or impair the actions of insulin. GH is therefore said to have both **insulin-like** effects and **anti-insulin** or **"diabetogenic"** effects.

The most striking and obvious effect of GH is that it stimulates linear growth of the skeleton. In addition, it also stimulates growth in a number of tissues in the body. Some of the effects are due to direct actions of the hormone on the tissue, while others result from the stimulation by GH of **somatomedin** production by the liver (Fig. 13–7a). A later section of this chapter discusses the physiology and actions of the somatomedins.

As Figure 13–7a shows, GH directly stimulates the uptake of amino acids from the blood into muscle cells and also stimulates protein synthesis in muscle. In addition to these effects, which stimulate muscle growth, GH stimulates liver protein synthesis as well. These effects on muscle and liver are similar to those of insulin and are known as the **insulin-like effects** of GH.

However, as indicated previously, some of the effects of GH antagonize those of insulin. In adipose tissue, GH decreases glucose uptake and stimulates the breakdown of fat stores (Fig. 13–7a), effects that are opposite those of insulin. Other effects of growth hormone, which decrease glucose uptake into muscle and increase gluconeogenesis in the liver, tend to raise the blood glucose concentration. These effects also antagonize those of insulin, which normally cause a decrease in the blood glucose concentration.

As indicated in Figure 13–7a, the net effect of GH on metabolism in adipose tissue and in muscle is to decrease fat storage and promote the accumulation of muscle protein. The overall effect of the hormone is therefore to promote the accumulation of lean body mass.

Growth Hormone Stimulation of Somatomedins. Early studies of GH recognized that not all of the effects of the hormone observed in the whole animal were the result of direct actions of GH on individual tissues. For example, although GH was known to promote bone growth when injected into young animals, it was without effect when directly added to pieces of bone incubated in a test tube. Later, scientists determined that GH stimulated the production of a group of peptide substances known as **somatomedins,** and that the somatomedins were directly responsible for some of the effects of the hormone.

The liver is the predominant, and probably most important, production site of somatomedins, although other tissues may produce them as well, especially during fetal development. The two most abundant somatomedins produced by the liver are known as **insulin-like growth factor I (IGF-I)** and **insulin-like growth factor II (IGF-II).** This particular nomenclature for the somatomedins grew from the observation that, in addition to insulin itself, the plasma of most animals contains peptide substances exhibiting many of the growth-promoting activities of insulin. IGF-I and IGF-II are very similar to each other in structure and also bear a remarkable resemblance to **proinsulin,** which is the precursor for insulin found in the pancreas (see Chapter 15). This similarity in structure between the two IGFs and proinsulin accounts for much of the insulin-like activity of the two peptides.

IGF-I production by the liver appears to be primarily determined by the concentration of GH in the blood. As a result, IGF-I is markedly decreased in the blood of individuals deficient in GH and markedly increased in many patients who have higher than normal concentrations of GH in their blood. In addition to GH, insulin and thyroid hormones also tend to promote IGF-I production. In contrast to IGF-I, the production of IGF-II by the liver is much less dependent on the concentration of circulating GH. A deficiency in circulating GH usually results in a decrease in IGF-II to only about half of the normal value.

The full spectrum of activities of the IGFs has not yet been fully characterized and investigated. In general, however, the IGFs can be thought of as having two types of effects, as indicated in Figure 13–7a. The first type of effect involves a stimulation of events leading to an increase in linear growth of the skeleton. Of the two growth factors, IGF-I is primarily responsible for producing this particular effect. IGF-I stimulates skeletal growth by increasing the formation of cartilage in specialized regions near the ends of growing bone known as **epiphyseal plates** (Fig. 13–7b). The cartilage deposited in the epiphyseal plates will eventually be replaced by bone mineral, and the bone will therefore increase in length. IGF-I promotes bone growth by stimulating those cells involved in cartilage formation, cells known as **chondrocytes.** Specifically, IGF-I stimulates the synthesis of collagen as well as total cell protein in chondrocytes, and promotes the proliferation of these cells as well (see Fig. 13–7a). The process of bone formation will be described in greater detail in Chapter 27.

The second type of IGF effect is related to processes involved in stimulating tissue growth and tissue repair. IGF-II appears to be primarily responsible for producing these effects. By stimulating a number of cellular processes such as protein synthesis and RNA synthesis, IGF-II is able to elicit a general anabolic response in a variety of different tissues and organs. This results in generalized tissue growth and an increase in organ size (see Fig. 13–7a).

Conditions of Growth Hormone Deficiency or Excess. Given the role of GH in stimulating skeletal and soft tissue growth, one might imagine that the

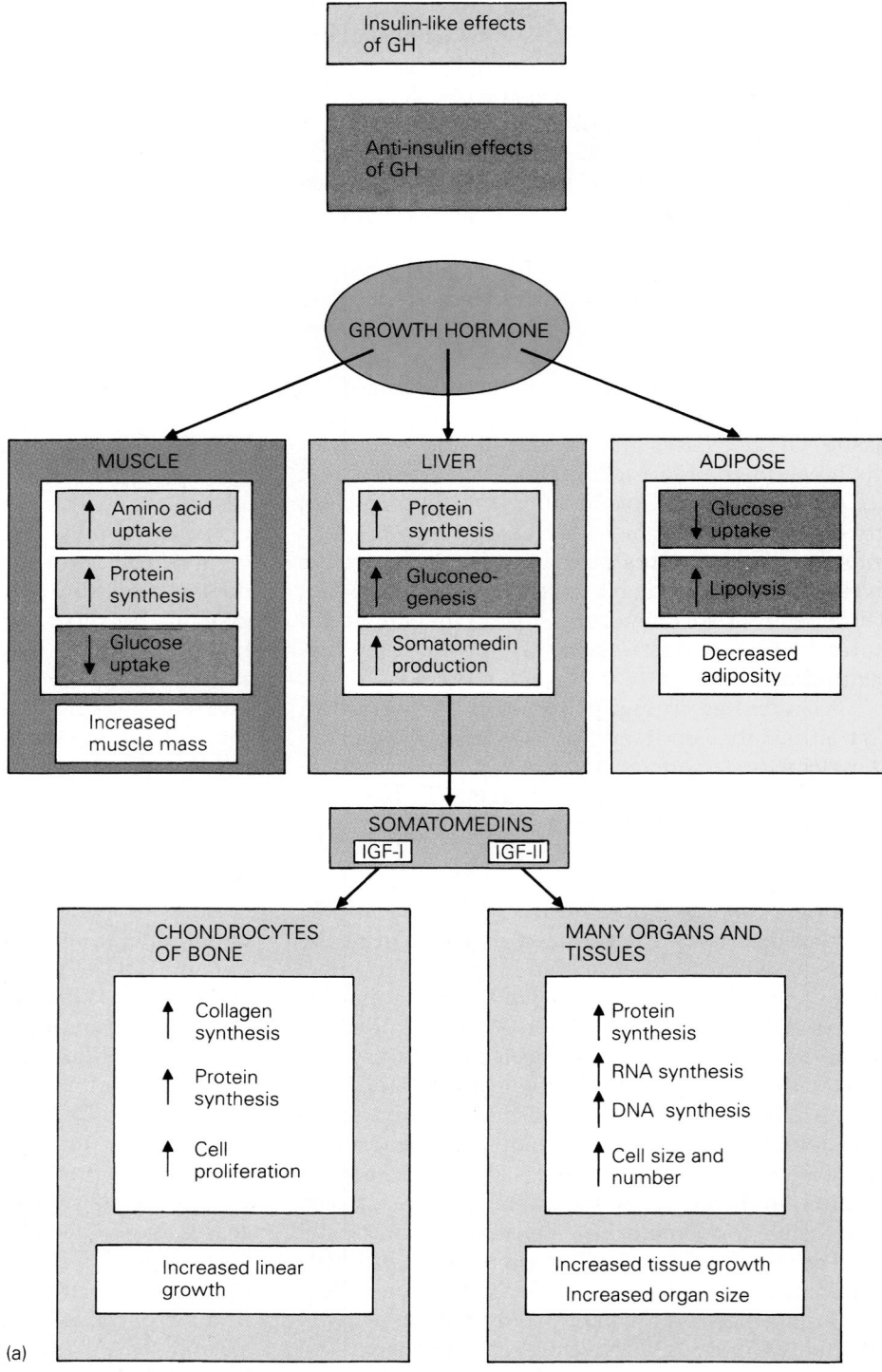

Figure 13–7
(*a*) Growth hormone has direct effects on muscle, liver, and adipose tissue, and indirect effects that occur as a result of somatomedin production. The overall effect of growth hormone is to promote skeletal growth and the accumulation of lean body mass.

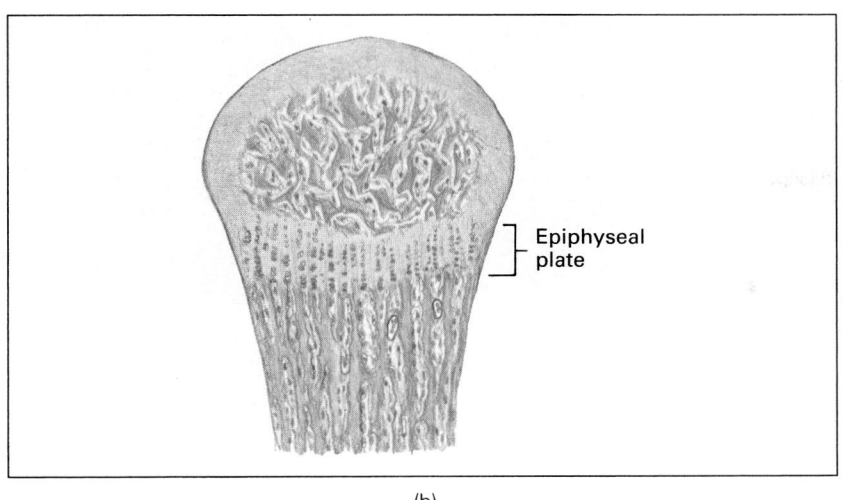

(b)

Figure 13–7 (continued)
(*b*) IGF-I stimulates bone growth by stimulating cells involved in cartilage formation at the epiphyseal plate.

Epiphyseal plate

consequences of abnormalities that involve either under- or oversecretion of the hormone would be significant. The particular clinical symptoms that occur, however, depend on the age of the individual when the deficiency or excess occurs.

If a deficiency in GH secretion exists during childhood, the condition of **dwarfism** results. Because the deficiency in GH secretion usually occurs as the result of a defect in pituitary function, this condition has come to be known as **pituitary dwarfism.** Scientists now realize, however, that the condition of pituitary dwarfism may result not only from a pituitary deficiency of GH production, but also from either an impairment in IGF-I production by the liver or by an inability of the target tissues to respond to the somatomedins that the liver produces. A tribe of Pygmies in Africa was recently studied and found to have normal GH concentrations in their blood but low circulating IGF-I concentrations due to a genetic impairment in growth factor production by their livers. Other well known forms of pituitary dwarfism appear to result from an impairment in the responsiveness of target tissues. Thus, dwarfism can result from abnormal function in either the anterior pituitary, the liver, or the target tissues. In contrast to the situation in children, however, if the deficiency in GH production occurs during adulthood after normal bone growth has occurred, generally no overt clinical symptoms will be evident.

Overproduction of GH can result in one of two clinically significant conditions: **gigantism** or **acromegaly.** Gigantism results when increased GH secretion occurs during childhood. In these individuals, the skeleton is stimulated to grow excessively, up to heights of 8 feet or more. A photograph of a

person displaying an extreme case of gigantism is shown in Figure 13–8. As evident in the figure, the long bones of the body are stimulated to grow disproportionately, and as a result these persons have relatively long arms and long legs.

Figure 13–8
The world's tallest woman, Sandy Allen (7' 7 1/4", 480 lbs.), pictured at home with her 12-year old brother.

If, on the other hand, overproduction of GH begins after the time of puberty, acromegaly results. Around the time of puberty, the growth centers, or epiphyseal plates, of the long bones of the body "close," meaning that they become unresponsive to hormonal stimulation. Thus, bones of the arms and legs cease to grow at this time. However, the bones of the hands, feet, skull, and lower jaw do not "close" and can still be stimulated to grow. As a result of the oversecretion of GH, these bones continue to grow and enlarge, leading to the distinctive physical appearance of acromegaly shown in Figure 13–9. Typically, these individuals have an enlarged lower jaw, large hands and feet, and also

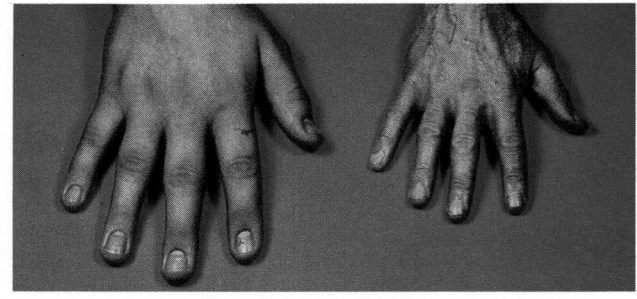

Figure 13–9
Hand of someone with acromegaly (left) placed next to a normal hand (right). (© Custom Medical Stock Photo.)

C L I N I C A L   F O C U S

Dwarfism in Animals

As indicated in the accompanying sections, the endocrine system can have a profound influence on growth and body size. Normal growth requires a sequence of events involving growth hormone (GH) secretion, GH action on the liver, IGF-I secretion by the liver, and IGF-I action on tissues to promote growth. In addition, the appropriate secretion of and response to various other hormones must also take place. An impairment at any one of these steps can lead to reduced growth and short stature.

Among mammals, the domestic dog exhibits perhaps the highest degree of variation in body size of any species. Selective breeding has resulted in both very large breeds, such as the St. Bernard, which may reach 200 lbs or more in adulthood, and very small breeds, such as the Chihuahua, which may weigh less than 2 lbs. Even within one particular breed, such as the poodle, there may be a large variation in size. The poodle breed is composed of three varieties—Standard, Miniature, and Toy—which are classified according to criteria set forth by the American Kennel Club.

Recently, a group of investigators addressed the question of whether the variation in size among the different varieties of poodle could be explained on the basis of their particular endocrine status. They measured both GH secretion and plasma IGF-I concentrations in groups of poodles belonging to the Standard, Miniature, and Toy varieties. All three groups of dogs showed similar GH secretion, suggesting that the differences in stature among groups were not due to differences in the capacity of their pituitary glands to produce and secrete GH. However, significant differences in plasma IGF-I concentrations were observed among groups. As shown in the figure, Standard poodles had significantly greater amounts of circulating IGF-I than did their Miniature or Toy counterparts. Relative sizes of the three poodle varieties are indicated by the size of their respective caricatures in this figure. If the body weight of each animal used in this study is graphed against the concentration of IGF-I in their blood, a highly significant correlation is obtained. That is, the larger the animal, the higher the circulating IGF-I concentration.

Although these results are consistent with the notion that the variation in stature among these dogs were the direct result of differences in circulating IGF-I, this has not yet been proven. There may well have been some other primary change that directly led to both smaller stature *and* lower IGF-I secretion. However, it is tempting to speculate that in breeding the different varieties of poodle, the breeders have in fact been selecting for differences in the ability of these animals to produce IGF-I. It will be interesting to see whether future investigations support this idea. (*Source:* Eigenmann, J.E., Patterson, D.F., and Froesch, E.R. *Acta Endocrinologica* 106:448–453, 1984.)

have generally large facial features due to the stimulation of soft tissue growth resulting from increased somatomedin production. In extreme situations, the diabetogenic effects of GH can also lead to overt cases of diabetes mellitus in individuals suffering from acromegaly.

The Thyroid Gland and Thyroid-Stimulating Hormone (TSH)

As its name implies, the primary target tissue of thyroid-stimulating hormone (TSH) is the thyroid gland. In the thyroid, TSH stimulates cell growth and the secretion of thyroid hormones. Thyroid hor-

mones in turn affect a variety of metabolic processes in the body. Therefore, this section provides a description not only of the physiology of TSH, but also of the thyroid gland and the actions of thyroid hormones.

Anatomy of the Thyroid Gland.   The gross anatomy of the thyroid gland is shown in Figure 13–10. The thyroid is one of the largest endocrine glands in the body, weighing approximately 20 grams in a normal adult. As Figure 13–10 indicates, the thyroid consists of two lobes that lie on either side of the trachea, just below the larynx. A thin band of tissue known as the **isthmus** connects the two lobes. The

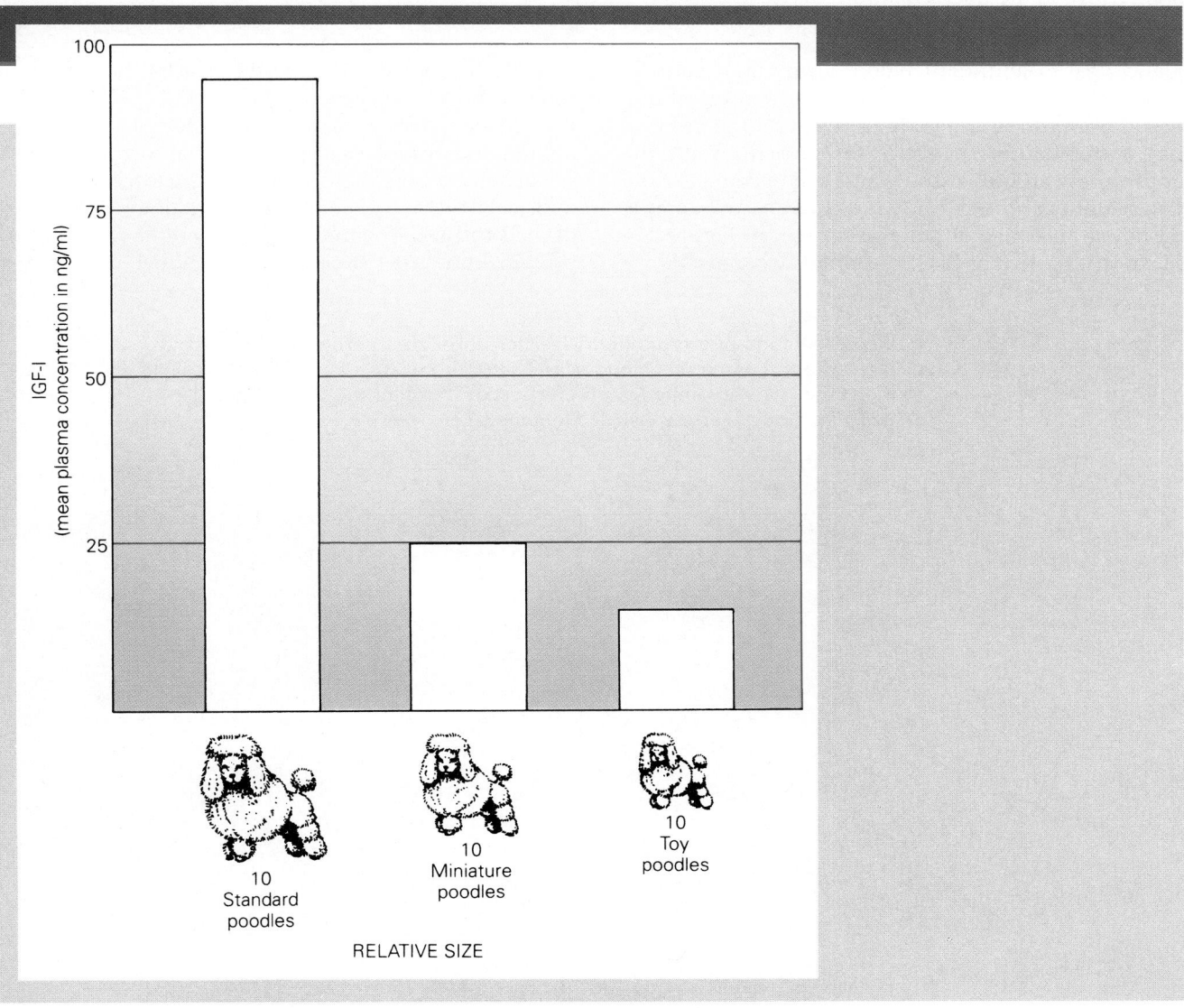

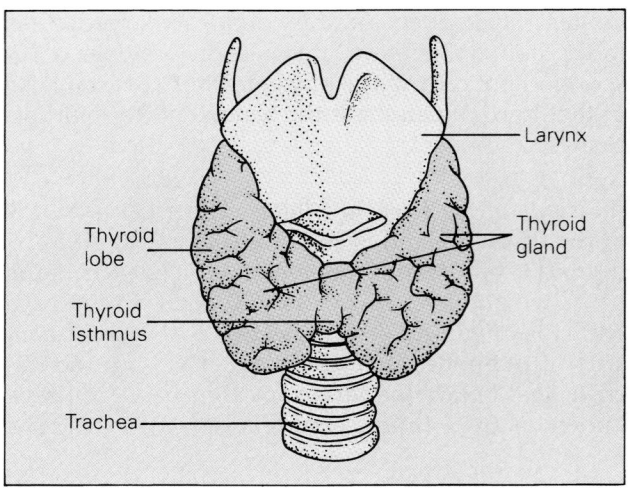

Figure 13–10
Gross anatomy of the thyroid gland showing its location relative to the trachea and the larynx.

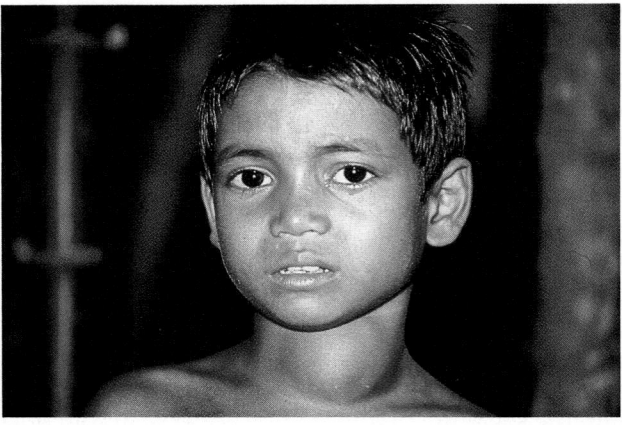

Figure 13–11
A boy with goiter. (© John Paul Kay/Peter Arnold, Inc.)

gland has an abundant blood supply and actually exhibits one of the highest blood flow rates of any tissue or organ in the body. The thyroid gland also has a tremendous capacity for growth. With the appropriate stimulus, the gland can become greatly enlarged, a condition known as **goiter.** An example of an extreme case of goiter is shown in Figure 13–11. In this particular case, the goiter was caused by a

dietary deficiency in iodide. A later section of this chapter describes the mechanism by which iodide deficiency causes goiter.

The thyroid gland consists of cells that are arranged in closely packed follicles (Fig. 13–12). The thyroid of a normal healthy adult contains about 3 million of these follicles. The cells that form a follicle are usually cuboidal in shape, and the interior of the follicle is filled with a protein-containing material termed **colloid.** The major component of colloid is a protein known as **thyroglobulin,** which serves as

Figure 13–12
(*a*) Representation of a typical cross-section through a portion of the thyroid gland. Note the numerous blood vessels and the relative abundance of colloid in the gland. (*b*) Photomicrograph of a section through the thyroid gland. (× 425.) (© Bruce Iverson/Visuals Unlimited.)

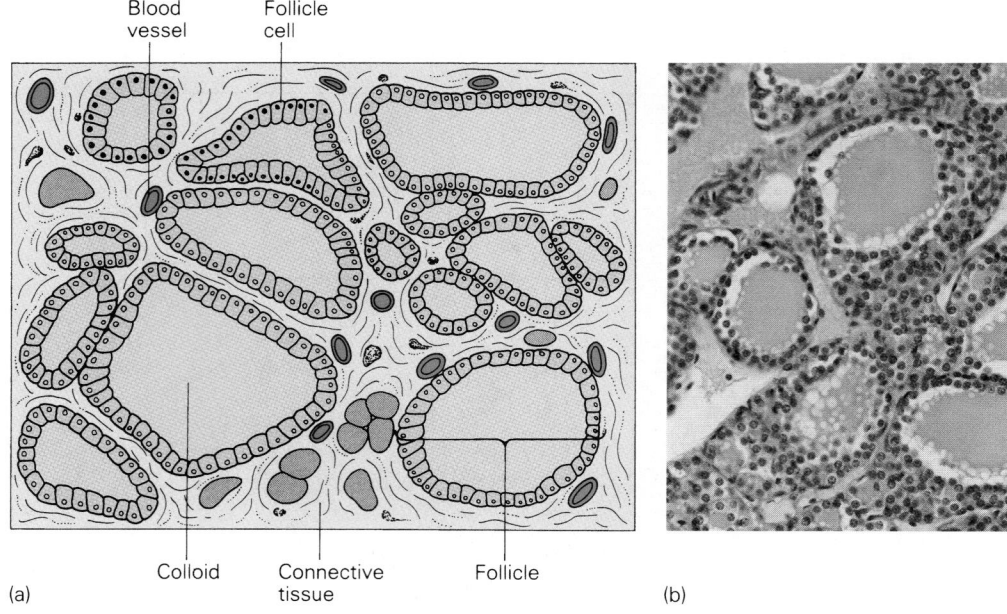

the precursor of thyroid hormones. A considerable amount of colloid is usually present within each follicle, so that colloid is normally the major constituent of the total mass of the thyroid gland. An extensive capillary network surrounds each follicle.

Synthesis and Secretion of Thyroid Hormones. The major compounds of interest relating to thyroid hormone production and metabolism are shown in Figure 13–13. **Thyroxine (tetraiodothyronine, or T₄)** is the primary hormone product of the thyroid gland. **Triiodothyronine (T₃)** is also produced by the thyroid gland but in lesser amounts. (In target tissues, T_4 is converted into T_3, as will be discussed later.) The other two iodinated compounds shown in Figure 13–13, **monoiodotyrosine (MIT)** and **diiodotyrosine (DIT),** are found primarily within the

thyroid follicle cell, although small amounts are also secreted into the blood.

Each of the compounds shown in Figure 13–13 contains iodine as an integral part of the molecule. In comparison to other hormones, thyroid hormones are unique in that their synthesis requires a constituent (iodine) not always present in the diet. As a result, mechanisms have evolved that permit the concentration of iodide in the gland and the storage of large amounts of thyroglobulin, the thyroid hormone precursor. Normally, enough thyroglobulin may be stored in the gland to sustain thyroid hormone secretion for 2 months in the absence of any further thyroglobulin synthesis.

Thyroglobulin itself is a large glycoprotein synthesized in the thyroid follicle cell (Fig. 13–14a). As with other glycoproteins that are secreted, the pro-

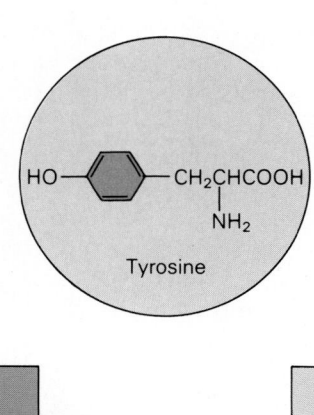

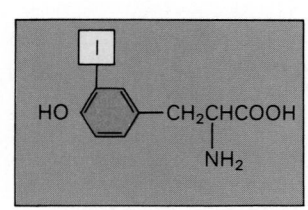

Monoiodotyrosine (MIT)

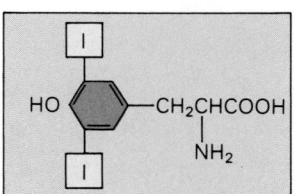

Diiodotyrosine (DIT)

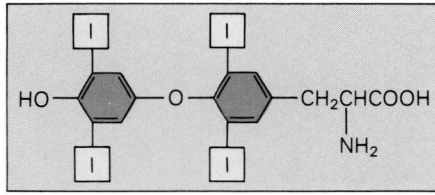

T₄ (thyroxine)

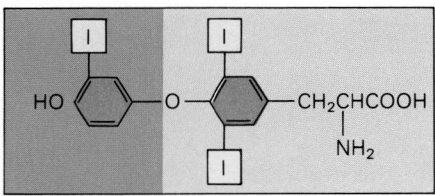

T₃

Figure 13–13
Structures of the primary iodinated compounds found in the thyroid gland and in the blood. Note the difference in number and location of the iodides between T_4 and T_3.

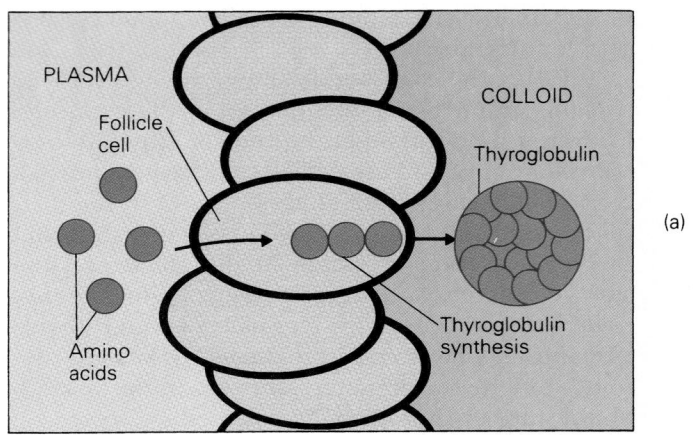

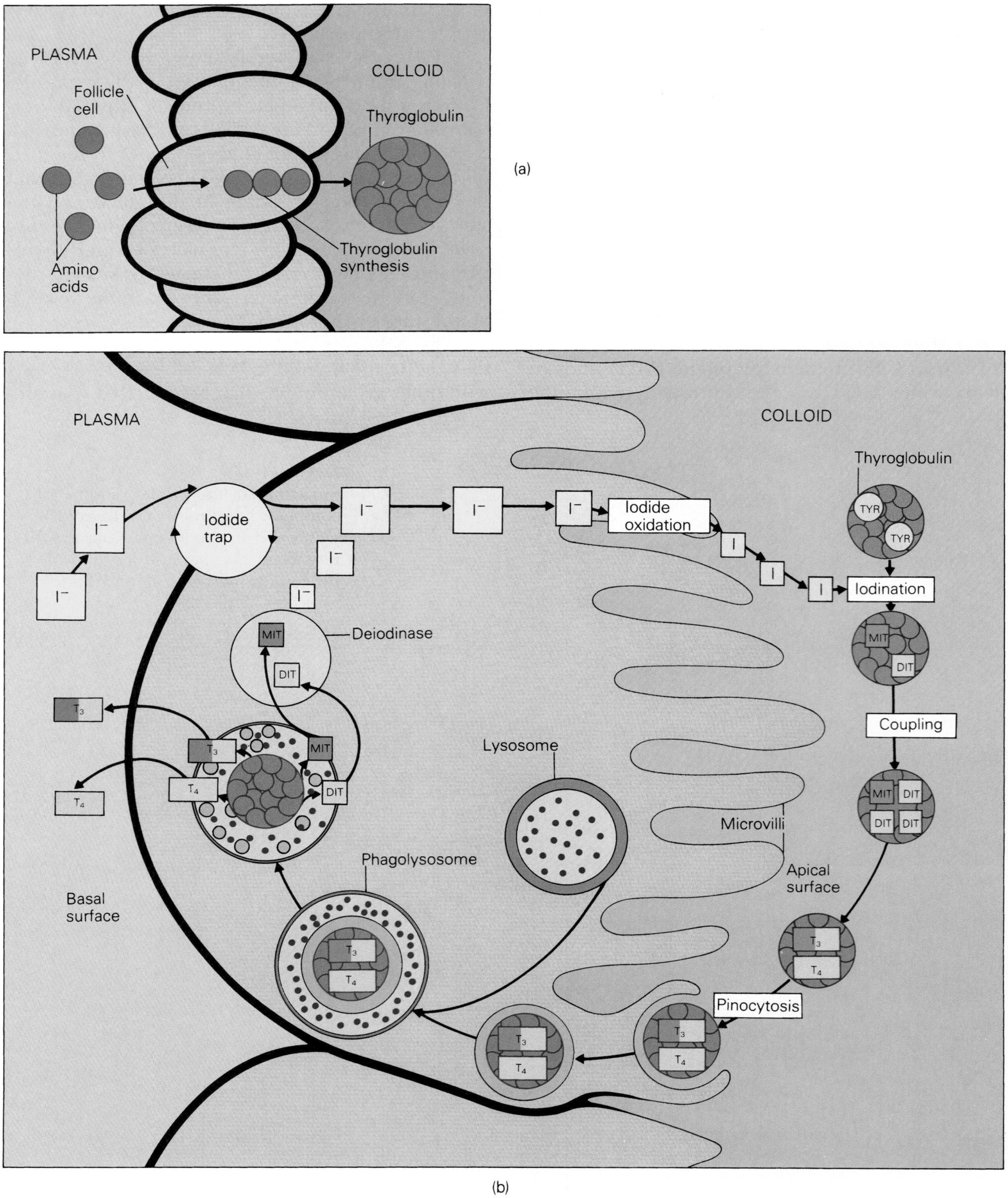

PLASMA

Follicle
cell

Thyroglobulin

COLLOID

Amino
acids

Thyroglobulin
synthesis

(a)

PLASMA

COLLOID

Thyroglobulin

Iodide
trap

I⁻

I⁻

I⁻

Iodide
oxidation

TYR

TYR

Iodination

Deiodinase

MIT

DIT

MIT

DIT

Coupling

T₃

MIT

MIT

DIT

DIT

DIT

Lysosome

Microvilli

T₄

T₃

DIT

T₄

Apical
surface

Basal
surface

Phagolysosome

T₃

T₄

T₃

Pinocytosis

T₄

T₃

T₄

(b)

Figure 13–14
(*a*) The synthesis of thyroglobulin occurs in the follicle cell. (*b*) The major steps involved in thyroid hormone formation and release. (*c*) A closer look at the iodination and coupling reactions that occur in the colloid.

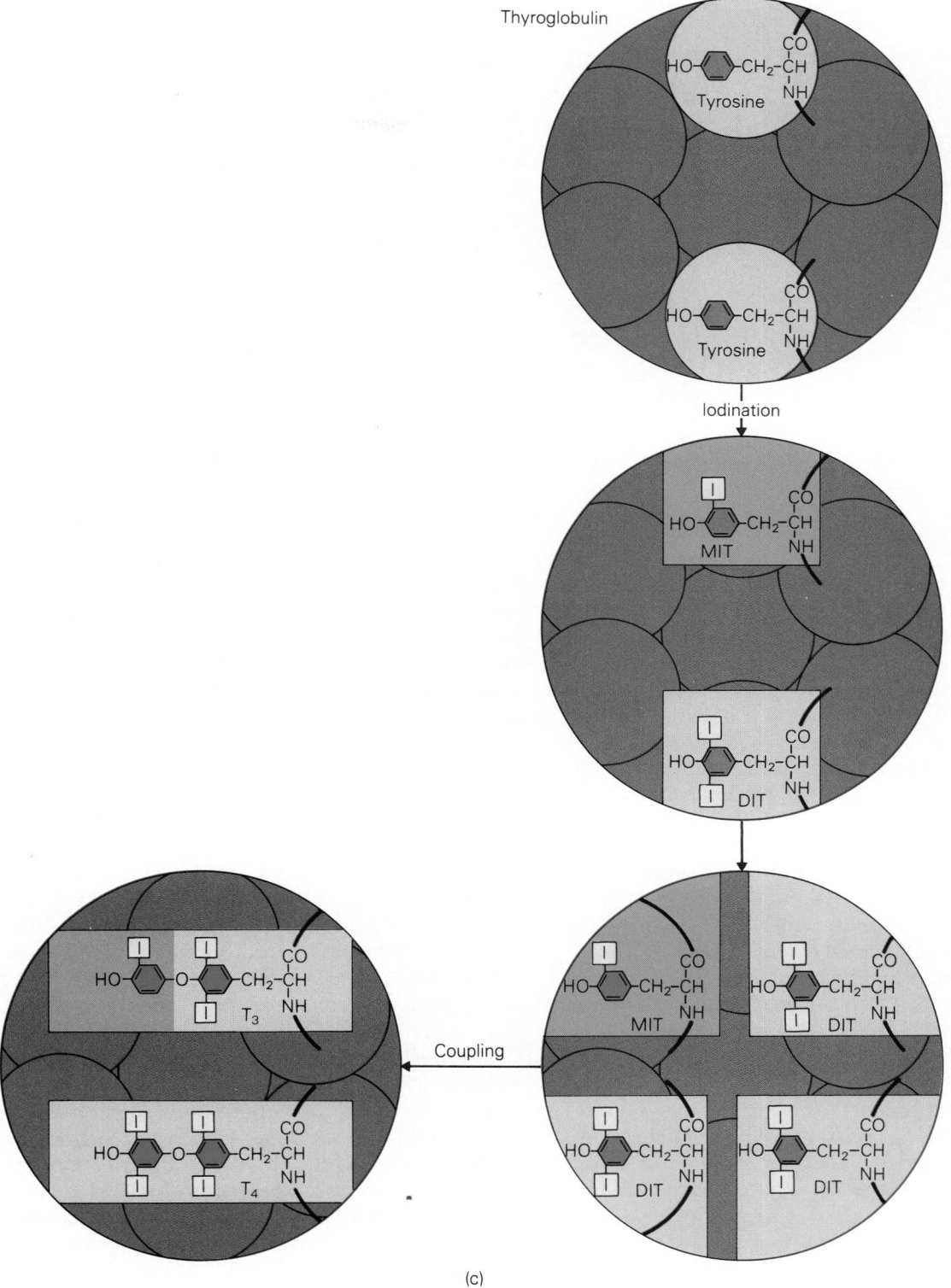

(c)

Figure 13–14 (continued)

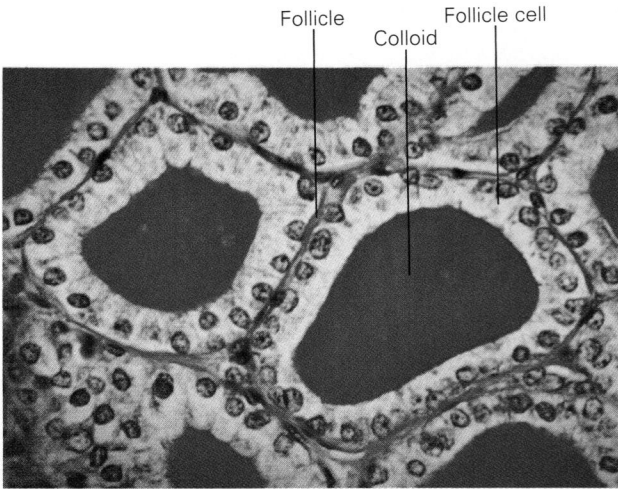

Follicle   Follicle cell
    Colloid

Figure 13–14 (continued)
(*d*) Micrograph of a follicular cell from the human thyroid gland. (© Biophoto Associates/Photo Researchers, Inc.)

tein component of thyroglobulin is synthesized in the rough endoplasmic reticulum of the cell, and the carbohydrate portion of the molecule is added in the Golgi complex. From the Golgi complex, thyroglobulin is then secreted on the apical side of the cell into the lumen of the follicle, where it is iodinated and stored as part of the colloid.

The first step in thyroid hormone formation involves the uptake of iodide (I^-, the ionic form of the element iodine) into the thyroid follicle cell across its basal membrane from the plasma (Fig. 13–14*b*). The uptake involves an active transport process that pumps iodide into the cells and can produce an intracellular iodide concentration 30 times higher than that in the extracellular fluid. This transport system is so effective at pumping iodide into the follicle cell that the process is often referred to as the "**iodide trap.**"

Once inside the cell, free iodide diffuses down a concentration gradient toward the apical surface of the cell, where it leaves the cell and enters the interior of the follicle. Soon after leaving the follicle cell, the iodide undergoes **oxidation** and **organification** reactions. Enzymes that rapidly oxidize the iodide and convert it to iodine (I) are present on the surface of the follicle cell membrane that faces the colloid. The iodine formed by these enzymes then serves as a substrate for other enzymes that catalyze the addition of iodine to tyrosine residues within the thyroglobulin protein. Tyrosine residues containing either one or two iodine molecules can be formed, and thus at this point thyroglobulin contains both MIT and DIT as part of its protein structure.

The next major step in thyroid hormone formation involves the coupling of two of these iodinated tyrosines within the thyroglobulin molecule to form either T_3 or T_4. If two DIT residues are coupled, then T_4 will be formed (refer to Fig. 13–14*c*). If, however, DIT and MIT residues of thyroglobulin are coupled, then T_3 is produced. It should be emphasized that at this point the T_3 and T_4 are *not* free, but are still present as part of the thyroglobulin peptide. In later steps to be described, the thyroglobulin is broken down to release the free hormones.

When follicle cells are stimulated to produce thyroid hormones, numerous microvilli are formed on the apical surface of the cells. These microvilli extend out from the cell into the colloid and engulf a portion of colloid by pinocytosis (see Fig. 13–14*b*). This small droplet of colloid is taken into the follicle cell and moved toward the basal surface of the cell. During this movement, the colloid droplet meets with lysosomes, and the structures fuse to form a **phagolysosome** (see Fig. 13–14*b*). The phagolysosome then continues to move toward the basal end of the cell, and as it does, the thyroglobulin molecule is broken down by proteolytic enzymes to produce free T_4 and T_3. The T_3 and T_4 are secreted into the interstitial fluid, where they are picked up by nearby capillaries and carried by the circulation throughout the body.

The proteolytic breakdown of thyroglobulin also results in the release of small amounts of free MIT and DIT, which are present due to incomplete coupling reactions. The freed MIT and DIT are subject to metabolism within the follicle cell by very active **deiodinases** present in the cytoplasm. The deiodinases remove iodide from MIT and DIT to produce iodide and tyrosine. The iodide can be recycled for further hormone synthesis, thereby conserving the iodide. Very little free MIT or DIT is actually released from the thyroid gland.

Regulation of Thyroid Hormone Synthesis and Secretion. The primary regulator of thyroid follicle cell activity and thyroid hormone secretion is **thyroid-stimulating hormone,** or **TSH,** which is produced by the thyrotrophs of the anterior pituitary (see Table 13–3). Specific receptors for TSH are present on the basal surface of the follicle cell. These receptors are coupled to adenylate cyclase, and hence the binding of TSH to its receptors leads to an increase in the intracellular concentration of cAMP. The specific effects of TSH on thyroid hormone production are listed in Table 13–5. Note that TSH stimulates virtually every aspect of thyroid hormone formulation. In addition to these specific effects on thyroid hormone formation, TSH also stimulates

Table 13–5

Effects of TSH on Thyroid Hormone Production

In the thyroid follicle cell, TSH stimulates:
1. Iodide uptake by active transport mechanisms
2. Thyroglobulin synthesis
3. Reactions resulting in the oxidation and organification of iodide
4. Microvilli formation and the engulfment of colloid at the apical cell surface
5. The rate of movement of lysosomes from the basal toward the apical surface of the follicle cell
6. The rate of movement of phagolysosomes from the apical toward the basal surface of the cell
7. The activity of the deiodinase enzymes

growth of the thyroid follicle cell. Thus, when TSH secretion is increased, the follicle cells are stimulated to grow and become elongated and more columnar in shape. Likewise, when TSH secretion is low, the cells regress and become somewhat flattened.

A diagram outlining the regulation of thyroid hormone secretion is presented in Figure 13–15. The relationship between the hypothalamus, anterior pituitary, and thyroid gland indicated in this figure is known as the **hypothalamic-pituitary-thyroid axis.** As described previously, thyroid hormone secretion is under the control of TSH from the pituitary. The secretion of TSH in turn is regulated primarily by TRH produced by the hypothalamus. The secretion of TRH can be influenced by a variety of factors, including inputs from higher centers within the CNS and from temperature regulatory centers in the hypothalamus. The temperature regulatory centers in turn receive information concerning changes in body temperature and changes in environmental temperature. Also indicated in Figure 13–15 is the important negative feedback effect that thyroid hormones exert upon the anterior pituitary to limit TSH secretion. As a result, thyroid hormones exert negative feedback effects that limit their own production. In addition, some of the biological effects of thyroid hormones result in increased heat production in the body (see the following), which also has the negative feedback effect of limiting further thyroid hormone formation.

Prior to the introduction of iodized salt into the diet, the occurrence of goiter, or enlargement of the

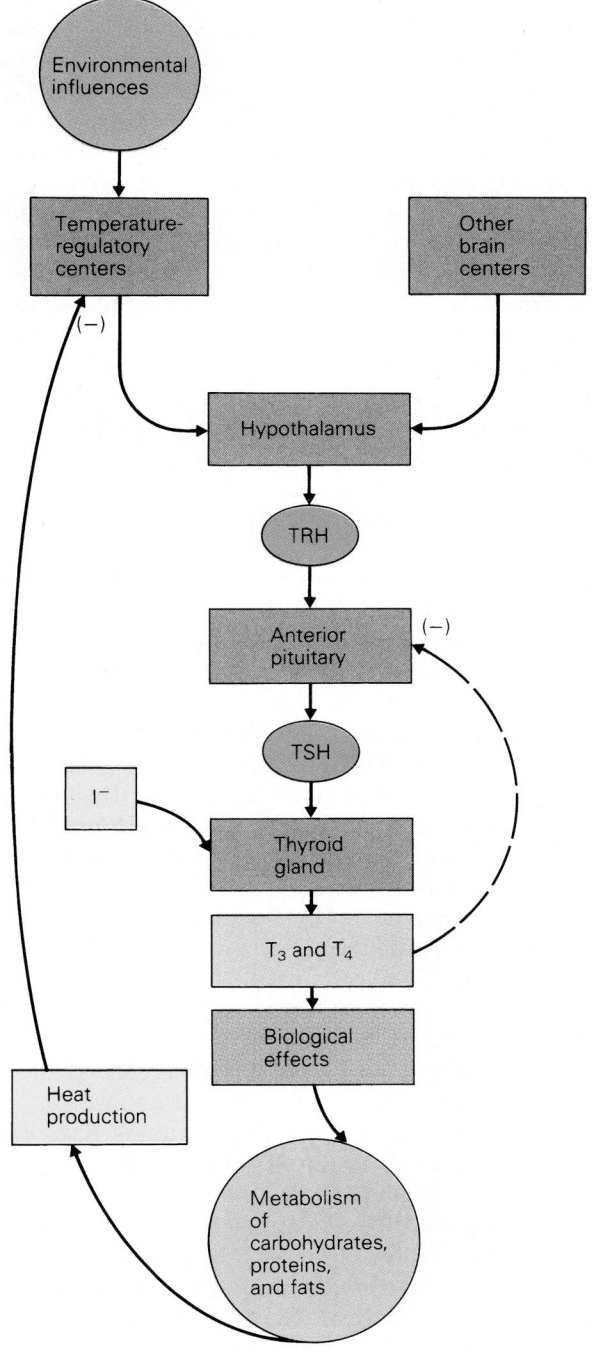

Figure 13–15
The primary steps involved in the regulation of thyroid hormone production.

thyroid gland, was fairly common. Figure 13–15 illustrates how a deficiency of dietary iodide leads to an enlargement of the thyroid gland. As described previously and indicated in Figure 13–15, iodide is a necessary component for thyroid hormone formation. In the absence of iodide, thyroid hormones cannot be produced. In the absence of thyroid hormone secretion, the negative feedback effects of T_4 and T_3 on TSH release are also absent (see Fig. 13–15). TRH and TSH secretion therefore increases, and the growth-promoting effects of TSH on thyroid follicle cells cause the gland to increase in size, despite the inability to actually synthesize active hormone. As a result, goiter is produced with a prolonged absence of dietary iodine. A number of agents known as **goitrogens** inhibit iodide uptake or organification reactions, resulting in a blockage of thyroid hormone synthesis and the formation of a thyroid goiter. Vegetables of the turnip family are known to contain goitrogens and can produce goiter if eaten raw in excessively large quantities; however, cooking destroys the activity of the goitrogens.

Thyroid Hormone Transport and Metabolism. After secretion from the thyroid gland, T_3 and T_4 are carried in the blood reversibly bound to a number of plasma proteins. The carrier proteins serve to increase solubility of the thyroid hormones and to buffer acute changes in secretion, as was discussed earlier in Chapter 12. Both T_4 and T_3 are associated primarily with **thyroxine-binding globulin (TBG),** but are also bound to a lesser extent to **albumin** (see Table 12–2). The relative affinity or strength with which the two hormones bind to these proteins differs greatly, however. T_4 has a much higher affinity and therefore binds more tightly to the plasma proteins than does T_3. As a result, T_3 is more rapidly degraded and removed from the plasma. Thus, not only is T_4 secreted from the thyroid gland in much greater amounts than T_3, but it is also removed from the blood much more slowly than T_3.

Target tissues for thyroid hormones have the capacity to convert T_4 into T_3. Although T_3 and T_4 are capable of producing exactly the same biological effects, their relative strengths or potencies differ considerably. Within a target cell, T_3 is much more active than T_4. It is currently thought that most, if not all, of the biological effects of T_4 are due to its conversion to T_3 within the target tissue cell.

Effects of Thyroid Hormones on Metabolic Processes. Thyroid hormones exert numerous effects on metabolic processes in a wide variety of different tissues and cells; these are summarized in Table 13–6. The effects of thyroid hormones are so widespread that virtually no tissue or organ system escapes the adverse effects that result from thyroid hormone deficiency or excess. In general, the effects of thyroid hormones on metabolic processes are

---

Table 13–6

Summary of the Effects of Thyroid Hormones

Stimulate calorigenesis in most cells

Increase cardiac output
    Increase rate of cardiac contractions
    Increase strength of cardiac contractions

Increase oxygenation of blood
    Increase rate of breathing
    Increase number of red blood cells in the circulation

Effects on carbohydrate metabolism
    Promote glycogen formation in liver
    Increase glucose uptake into adipose and muscle

Effects on lipid turnover
    Increase lipid synthesis
    Increase lipid mobilization
    Increase lipid oxidation

Effects on protein metabolism
    Stimulate protein synthesis

Promote normal growth
    Stimulate growth hormone (GH) secretion
    Promote bone growth
    Promote IGF-I production by liver

Promote development and maturation of nervous system
    Promote neural branching
    Promote myelinization of nerves

---

slow in onset and long-lasting, as compared to the effects of other hormones. Thus, thyroid hormones exert relatively long-term regulatory influences over metabolism.

One of the primary effects of thyroid hormones is the stimulation of **calorigenesis,** or heat production, in the body (see Table 13–6). This response occurs after a delay of several hours or days and is reflected by an increase in oxygen consumption. This particular effect is evident in most tissues, with brain, spleen, and testes being the most notable exceptions. To provide for an increase in oxygen consumption by the tissues, thyroid hormones also have effects on the cardiopulmonary system. In the heart, thyroid hormones increase cardiac output (the amount of blood being pumped) by increasing both the rate and the strength of cardiac contractions. The amount of oxygen carried by the blood is also enhanced by thyroid hormones as a result of increases in both the resting rate of breathing and the number of red blood cells present in the blood.

Thyroid hormones influence several aspects of carbohydrate metabolism, although many of the effects that they produce depend on or are modified by other hormones. For example, normal concentrations of thyroid hormones increase glycogen formation in the liver (see Table 13–6), but this response occurs only when insulin is also present. Thyroid hormones also increase the uptake of glucose from the blood into adipose tissue and muscle and thereby act to potentiate the effect of insulin in this regard. In addition, many of the effects of epinephrine are enhanced by thyroid hormones as a result of their enhancement of the responsiveness of the adenylate cyclase/cAMP system.

Thyroid hormones have a number of effects on lipid metabolism in both liver and adipose tissue (see Table 13–6). Thyroid hormones stimulate virtually all aspects of lipid metabolism, including the synthesis, mobilization, and oxidation of lipids. In general, however, the oxidation of lipids is affected more than is their synthesis, so that in conditions of hormone excess the net effect is a decrease in the size of most fat stores in the body.

Thyroid hormones exert important regulatory effects on protein metabolism in the body as well. When present in normal physiological concentrations, thyroid hormones increase protein synthesis and promote an overall accumulation of protein in the body (see Table 13–6). However, when thyroid hormones are present in excess, they tend to cause a decrease in protein synthesis and an increase in protein breakdown. The result of these two effects is an overall loss of protein from the body. The effects of thyroid hormones on protein metabolism are therefore described as **biphasic.** Thyroid hormones have an additional effect on the pituitary, one that influences protein metabolism indirectly. In the anterior pituitary, thyroid hormones are known to stimulate the somatotrophs and increase GH secretion. As a result, owing to their influence on pituitary GH secretion, thyroid hormones also affect protein metabolism.

In addition to specific effects on protein metabolism, thyroid hormones play a role in regulating the overall growth and development of several tissues. In humans, thyroid hormone is required for normal skeletal growth. This is due in part to the stimulatory effects of thyroid hormones on GH production and in part to thyroid hormone stimulation of maturation of the epiphyseal growth centers in bone. In addition, as was noted in the earlier section describing somatomedins, thyroid hormones influence the production of IGF-I by the liver, which further contributes to the promotion of normal growth. Thyroid hormones are required for the normal develop-

ment and maturation of the teeth, hair follicles, and skin.

Development and maturation of the CNS are markedly affected by thyroid hormones. This effect of thyroid hormones is particularly evident in their absence. The branching of axons and dendrites that normally occurs during fetal and early neonatal development occurs to only a limited degree in the absence of thyroid hormones. Normal myelinization of nerves is also impaired in the absence of thyroid hormones. As a result, severe mental retardation can occur if thyroid hormone is deficient during the period of fetal and early neonatal development. In adults, thyroid hormones influence mental alertness and responsiveness to external stimuli. The velocity of conduction of action potentials in peripheral nerves can also be shown to vary in response to an excess or deficiency of thyroid hormone secretion.

Conditions of Abnormal Thyroid Hormone Secretion. Many of the specific effects of thyroid hormones that were discussed previously are readily evident in conditions of either thyroid hormone deficiency or thyroid hormone excess.

Hypothyroidism is the condition that exists when there is a deficiency in thyroid hormone production. In most instances, this occurs as a result of a defect in the thyroid gland itself. However, in some cases the defect occurs in either the hypothalamus or pituitary, resulting in a deficiency in TSH production. Because of a slower metabolic rate and a reduced rate of heat production, hypothyroid individuals cannot tolerate cold temperatures. Water tends to accumulate in the skin of these individuals, leading to a condition termed **myxedema,** in which the skin has a thickened, puffy appearance. This change is most evident in the facial features. The heart rate and the strength of cardiac contractions are reduced, both effects contributing to an overall reduction in the cardiac output. There is also a general slowing of all intellectual functions, leading to a feeling of lethargy and also possibly some degree of speech impairment.

Excess secretion of thyroid hormones results in **hyperthyroidism.** When tissues are presented with excessive quantities of thyroid hormones, a complex of biochemical and physiological events occurs. Hyperthyroid individuals exhibit such symptoms as a markedly increased heart rate, an insensitivity to heat, and considerable weight loss owing to the lipid-mobilizing and protein-catabolic effects of excess thyroid hormones. In contrast to the lethargy that occurs in hypothyroidism, hyperthyroid individuals tend to be highly responsive to external stimuli. The most common form of hyperthyroidism is a condition known as **Grave's disease,** which occurs as a result of an interesting set of circumstances. In affected individuals, an abnormality in the immune system leads to the production of antibodies that recognize and bind to sites on the surface of thyroid follicle cells. When these antibodies bind to the follicle cell, the TSH receptor is activated, as though TSH itself were present. This results in a marked increase in thyroid hormone production, and a goiter forms as a result of the TSH-like effects of the antibodies on thyroid cell growth. The cause of the immune system abnormality and the mechanism by which the antibody activates the TSH receptor are not completely understood, however.

As indicated previously, a deficiency in thyroid hormone secretion during fetal development can have a dramatic impact, owing to the hormone's role in promoting normal development of the nervous system. A form of mental handicap known as **cretinism** can occur as a result of an untreated thyroid hormone deficiency. In addition to mental impairment, linear growth is also impaired in these individuals, so that dwarfism is also evident. If therapy with thyroid hormone is begun very soon after birth, severity of the impairment can be dramatically reduced. For this reason, most infants are checked very soon after delivery for the presence of adequate thyroid hormones in their blood.

SUMMARY

Anatomy of the Pituitary Gland and its Relation to the Hypothalamus

The pituitary, a small gland located just below the hypothalamus at the base of the brain, consists of two main lobes: the anterior lobe and the posterior lobe.

The posterior pituitary is derived from neural tissue and contains the terminals of neuroendocrine cells that originate in the hypothalamus. In contrast, cells of the anterior pituitary are derived from non-neural tissue. The hypothalamic-pituitary portal system carries blood from the hypothalamus to the anterior pituitary.

Regulation of Pituitary Hormone Secretion by the Hypothalamus

The posterior pituitary secretes two hormones, oxytocin and ADH. These two hormones are synthesized in the cell bodies of neuroendocrine cells in the hypothalamus and are then transported down their axons to the posterior pituitary where they are released into capillaries when the hypothalamus receives an appropriate stimulus.

Cells of the anterior pituitary synthesize and secrete at least six different hormones: LH, FSH, prolactin, ACTH, GH, and TSH. Secretion of each hormone is under the control of releasing/inhibiting hormones produced by neuroendocrine cells in the hypothalamus and released into the blood vessels of the median eminence. The releasing/inhibiting hormones travel to the anterior lobe via the hypothalamic-pituitary portal system, where they act to regulate secretion of the six anterior pituitary hormones.

Actions and Effects of the Pituitary Hormones

The primary actions of ADH are to increase water retention in the kidneys and to stimulate vasoconstriction. Changes in ADH secretion occur primarily in response to changes in plasma osmolality and blood volume. Oxytocin acts primarily on the breast and causes release of milk from the alveoli into the collecting ducts.

LH and FSH, collectively referred to as the *gonadotropins*, regulate development of sperm and egg in males and females, respectively, and also regulate production of sex steroids by the gonads. Gonadotropin secretion is regulated by GnRH produced in the hypothalamus. *Prolactin* stimulates milk production by the alveolar cells of the female breast. Prolactin secretion is regulated by a PRF and a PIH. Prolactin secretion increases in response to suckling stimuli. *ACTH* regulates the synthesis and secretion of cortisol from the adrenal cortex and also stimulates growth of the adrenal cortex. ACTH secretion is regulated by *CRH. Growth hormone* (GH) secretion from the anterior pituitary is regulated by *GHRH* and *somatostatin*. GH effects are generally anabolic, although specific effects can be either *insulin-like* or *diabetogenic*. GH affects metabolism primarily in muscle, liver, and adipose tissue, promoting accumulation of lean body mass. GH stimulates production of somatomedins by the liver. Somatomedins stimulate both skeletal growth and tissue growth. Abnormalities of GH secretion can result in dwarfism, gigantism, or acromegaly. *TSH* regulates synthesis and secretion of thyroid hormones and stimulates growth of thyroid follicle cells. Pituitary TSH secretion is regulated by TRH secreted by the hypothalamus. Thyroid hormones contain iodide, which thyroid follicle cells actively trap. Iodide is added to *thyroglobulin*, which is stored as *colloid* within the thyroid follicles. With TSH stimulation, follicle cells engulf and degrade colloid to produce T_4 and T_3. T_4 is secreted in the greatest amount, but its effects are due to conversion to T_3 in target tissues. One of the primary effects of thyroid hormones is the promotion of *calorigenesis* in most cells of the body. Other effects are produced within the cardiopulmonary systems. Thyroid hormones also influence carbohydrate, lipid, and protein metabolism. Thyroid hormones have an important influence on the maturation of the nervous system during fetal development. Conditions of hypothyroidism and hyperthyroidism reflect the absence or excess of thyroid hormone actions, respectively.

REVIEW QUESTIONS

Identify Each with a Term

1. The fetal structure that during development gives rise to the anterior pituitary.

2. The specialized region of the hypothalamus that contains the primary capillaries of the hypothalamic-pituitary portal system.

3. The hypothalamic factor that inhibits growth hormone secretion by the anterior pituitary.

4. The name of the condition resulting from a deficiency in ADH secretion or action.

5. The name of the condition resulting from excess GH secretion during adulthood.

6. The component of colloid that serves as the precursor of thyroid hormones.

Define Each Term

7. hypothalamic-pituitary portal system
8. corticotrophs
9. gonadotropins
10. somatomedins
11. "iodide trap"

Choose the Correct Term

12. Which of the following is *not* secreted by the anterior pituitary?
    a. oxytocin
    b. growth hormone
    c. ACTH
    d. FSH

13. Which of the following is *not* a stimulus for GH secretion?
    a. deep sleep
    b. high blood glucose
    c. infections
    d. amino acids

14. The insulin-like effects of GH include
    a. increased protein synthesis in muscle
    b. increased lipolysis in adipose tissue
    c. increased gluconeogenesis in liver
    d. all of the above

15. Which of the following are the most numerous hormone-producing cells of the anterior pituitary?
    a. gonadotrophs
    b. corticotrophs
    c. lactotrophs
    d. somatotrophs

16. Which "hormone:site of synthesis" pair listed below is incorrect?
    a. oxytocin:supraoptic nucleus
    b. corticotropin-releasing hormone:hypothalamus
    c. somatostatin:somatotrophs
    d. FSH:gonadotrophs

17. Overproduction of GH can result in:
    a. cretinism
    b. acromegaly
    c. gigantism
    d. either B or C

18. An abnormal enlargement of the thyroid gland is termed:
    a. myxedema
    b. thyroid isthmus
    c. goiter
    d. colloid

19. In the thyroid gland, the "iodide trap":
    a. is responsible for the conversion of thyroglobulin into T_3 and T_4
    b. removes iodide from MIT and DIT
    c. is located on the apical surface of the cell and is directly responsible for transporting iodide from the intracellular space into the colloid space
    d. is located on the basal surface of the cell and is responsible for pumping iodide from the extracellular space into the thyroid follicle cell

20. Intolerance to cold, myxedema, and reduced cardiac output are typically seen in:
    a. hyperthyroidism
    b. hypothyroidism
    c. Grave's disease
    d. acromegaly

Answer in One or Two Sentences

21. Describe the physiologically important neural and vascular connections between the hypothalamus and the pituitary.

22. Describe the processes involved in the synthesis and secretion of posterior pituitary hormones.

23. What are the primary stimuli for secretion and the primary actions of antidiuretic hormone?

24. Describe the hypothalamic-pituitary-thyroid axis.

25. What are the differences in physical appearance among dwarfism, gigantism, and acromegaly? What factors account for these differences?

26. What effects do thyroid hormones have on carbohydrate, lipid, and protein metabolism?

SUGGESTED READINGS

Fregley, Melvin J., and Luttge, William G. *Human Endocrinology: An Interactive Text.* New York: Elsevier Biomedical, 1982.

Ganong, William F. *Review of Medical Physiology*, 12th ed. Los Altos: Lange Medical Publications, 1985.

Goodman, H. Maurice. *Basic Medical Endocrinology.* New York: Raven Press, 1988.

Griffin, James E., and Ojeda, Sergio R. *Textbook of Endocrine Physiology.* New York: Oxford University Press, 1988.

Guyton, Arthur C. *Human Physiology and Mechanisms of Disease*, 4th ed. Philadelphia: W.B. Saunders Company, 1987.

Norman, Anthony W., and Litwack, Gerald. *Hormones.* Orlando: Academic Press, 1987.

Smith, Emil L., Hill, Robert L., Lehman, I. Robert, Lefkowitz, Robert J., Handler, Philip, and White, Abraham. *Principles of Biochemistry: Mammalian Biochemistry*, 7th ed. New York: McGraw-Hill Book Company, 1983.

Tepperman, Jay, and Tepperman, Helen M. *Metabolic and Endocrine Physiology*, 5th ed. Chicago: Year Book Medical Publishers, 1987.

KEY TERMS

acromegaly (p. 465)
anterior pituitary gland (p. 453)
chondrocytes (p. 463)
epiphyseal plates (p. 463)
gigantism (p. 465)

goiter (p. 468)
growth hormone (p. 461)
hypothalamic-pituitary portal system (p. 455)
median eminence (p. 455)
pituitary gland (p. 453)

posterior pituitary gland (p. 455)
somatomedin (p. 463)
thyroglobulin (p. 468)
thyroid hormone (p. 469)

thyroid-stimulating hormone (TSH) (p. 473)
triiodothyronine (T_3) (p. 469)

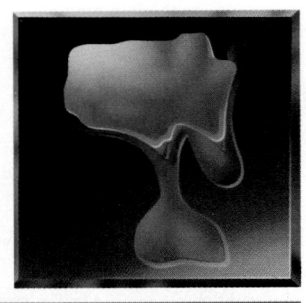

CHAPTER

14

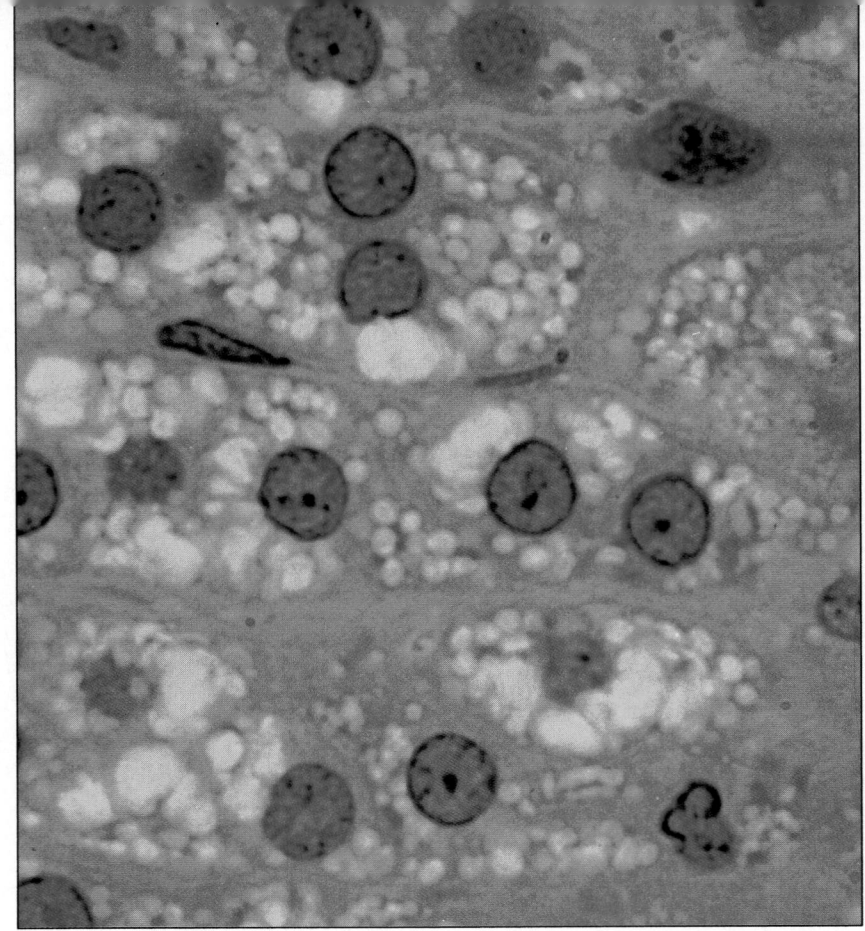

The Adrenal Glands

*T*his chapter discusses the physiology of the adrenal glands and the hormones they produce. Humans have two adrenal glands, each one composed of two distinct endocrine tissues that differ in cell type, hormone products, and regulation. The two endocrine components of the adrenal gland are the outer portion, or **cortex,** which constitutes about 80% of the gland, and the inner portion, or **medulla,** which forms the remaining 20%. The cortex produces and secretes several different steroid hormones, while the medulla produces and secretes catecholamines. The first section of this chapter focuses on the adrenal cortex, and a discussion of the adrenal medulla follows in the latter sections.

Gross Anatomy of the Adrenal Glands

The shape of each adrenal gland roughly resembles that of a pyramid, and each of the two adrenals in humans rests like a cap just above the upper end of each kidney (Fig. 14–1). There is no direct physical connection between each adrenal and kidney, however. The arterial blood supply entering each adrenal gland is also separate from that entering the kidneys. In a normal adult, each adrenal gland weighs about 3 to 4 grams and is about 5 cm (2 inches) across at its widest point.

The Adrenal Cortex

Cell Structure of the Adrenal Cortex

Based on the arrangement of cells as seen under the microscope, the adrenal cortex can be divided into roughly three different zones. Each zone secretes a different class of steroid hormone. The narrow, outermost region of the adrenal cortex is termed the **zona glomerulosa** (Fig. 14–2) and consists of cells arranged in "whorls" **(glomeruli).** Cells of the zona glomerulosa are responsible for the production of **mineralocorticoids.** The middle region of the cortex is the widest and is termed the **zona fasciculata** because of the arrangement of these cells in "fascicles" or cords separated by venous sinuses (small channels for venous blood) (Fig. 14–2). Cells of the zona fasciculata produce primarily **glucocorticoids.** The

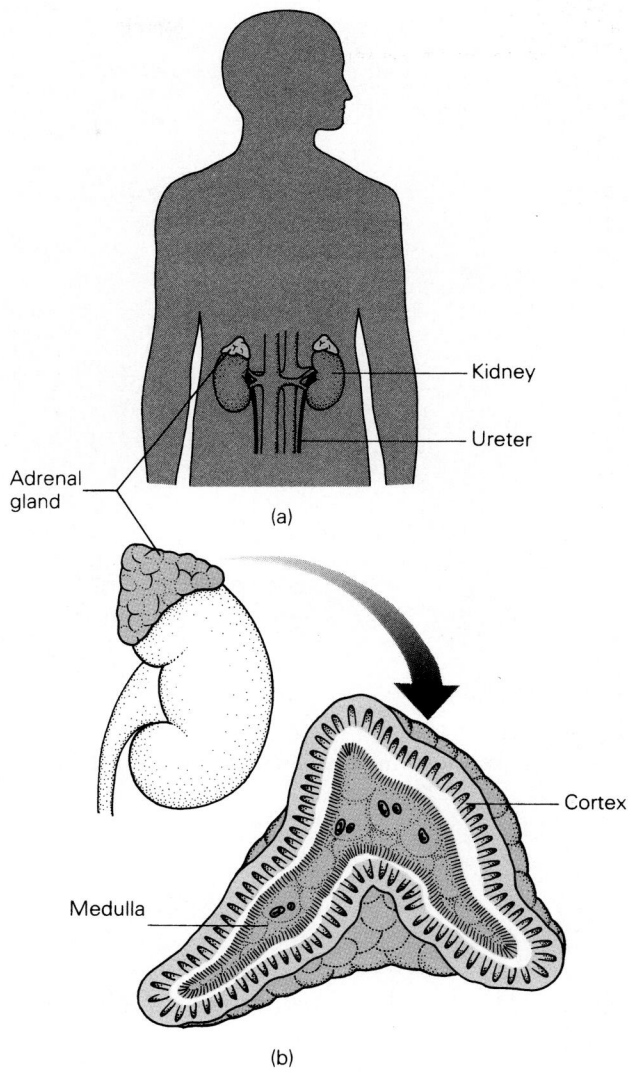

Figure 14–1
(*a*) Location of the adrenal glands relative to the kidneys. (*b*) The adrenal gland shown in cross-section, with both the cortex and medulla visible.

innermost region, which borders the medulla, is known as the **zona reticularis,** so named because the cells are arranged in the form of a network, or reticulum. This zone of the cortex produces and secretes primarily **androgens.**

Numerous lipid droplets are present within the cells of the adrenal cortex, especially in cells of the zona fasciculata. These lipid droplets represent stored cholesterol to be used for the synthesis of adrenal steroid hormones, as will be discussed later.

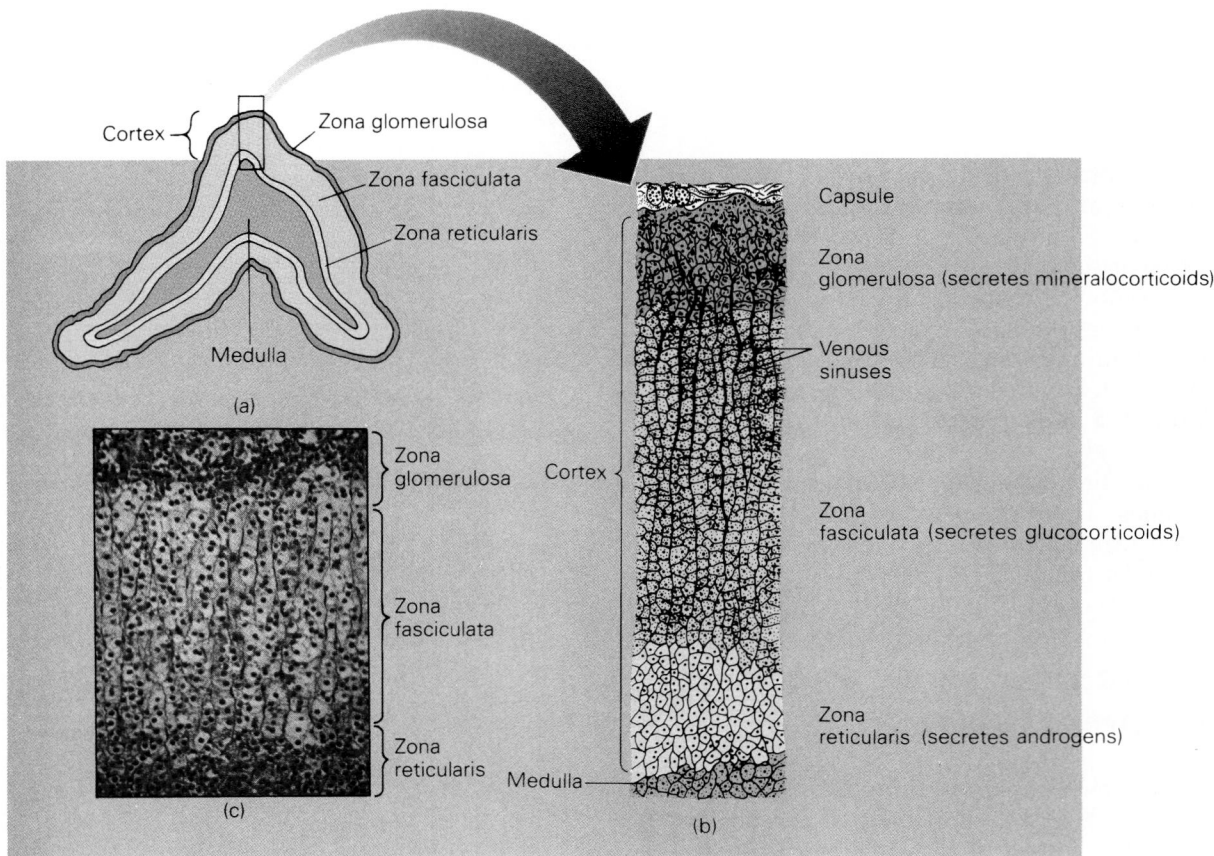

Figure 14–2
The adrenal cortex: three zones. (c) © Ed Reschke/Peter Arnold, Inc.

Regulation of the Functions of the Adrenal Cortex by ACTH

The activity of the adrenal cortex is under the control of adrenocorticotropic hormone (ACTH), which is produced by the corticotrophs of the anterior pituitary. As was described in Chapter 13, ACTH secretion is in turn regulated by corticotropin-releasing hormone (CRH), which is produced by neuroendocrine cells of the hypothalamus and carried to the anterior pituitary by the hypothalamic-pituitary portal vessels (refer to Figs. 13–3 and 13–4). This relationship between CRH, ACTH, and the adrenal cortex is illustrated in Figure 14–3 and is known as the **hypothalamic-pituitary adrenal axis.**

ACTH has both **tropic** effects and **steroidogenic** effects on cells of the adrenal cortex. The tropic effects of ACTH are the promotion of the general growth and function of all three zones of the adrenal cortex. Thus, in the absence of ACTH, cells of the adrenal cortex become smaller, and the production of all three classes of steroid hormone is dramatically reduced. The steroidogenic effect of ACTH involves the regulation of glucocorticoid secretion by cells of the zona fasciculata. As will be described later in this chapter, secretion of mineralocorticoids from the zona glomerulosa is primarily regulated by factors other than ACTH.

Biosynthesis of Steroid Hormones by the Adrenal Cortex

In all, some 30 to 50 different steroids have been isolated from the adrenal cortex. However, only three are produced and secreted in quantitatively

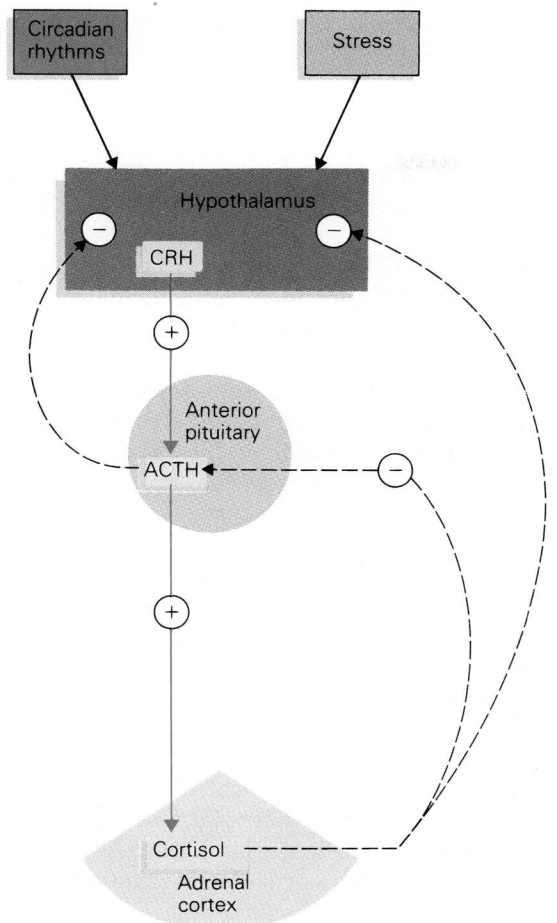

Figure 14–3
The hypothalamic-pituitary adrenal axis. Stimulatory effects are shown by the solid arrows; inhibitory effects by the dashed arrows.

and aldosterone, a mineralocorticoid, is produced in the least amount. This chapter focuses on the physiology of these two steroids. The third, dehydroepiandrosterone (DHEA), is itself a rather weak androgen, but it can be converted in peripheral tissues to testosterone, a much more powerful androgen (see Chapter 30). In males, androgen production by the adrenal is of little or no physiological significance, since far greater amounts are produced by the testes. In females, however, androgens produced by the adrenal gland account for most of that found circulating in the blood.

The general pathways involved in steroid hormone biosynthesis were previously described in Chapter 12 (see Fig. 12–11). For reference, a simplified diagram of the pathways is also presented here (Fig. 14–5). Recall that steroid hormones are synthesized and secreted on demand and are not stored within the cells in which they are produced. Therefore, the regulation of steroid hormone secretion occurs at the level of steroid hormone synthesis. As was described in Chapter 12, the first step in the synthesis of all steroid hormones begins with the enzymatic conversion of cholesterol to pregnenolone. In most instances, this step in steroid synthesis is the slowest, or rate-limiting, step. Once pregnenolone is formed, the subsequent steps proceed fairly rapidly. Therefore, the enzyme that catalyzes the conversion of cholesterol into pregnenolone represents the primary site of regulation of steroid synthesis (see Step A, Fig. 14–6).

Increasing the supply of cholesterol inside the cells also enhances steroid synthesis. Most of the cholesterol that cells of the adrenal cortex use to make steroids is taken up from the blood, where it is carried in the form of lipoproteins. Cells of the adrenal cortex can remove these lipoproteins from the blood and bring them into the cell, making the cholesterol available for steroid synthesis (see Step B, Fig. 14–6). These cells store any cholesterol not

significant amounts. These are **cortisol, aldosterone,** and **dehydroepiandrosterone (DHEA),** which are shown in Figure 14–4. Of the three, cortisol, a glucocorticoid, is produced in the greatest amount,

Figure 14–4
Chemical structures of the principal steroids of the adrenal cortex.

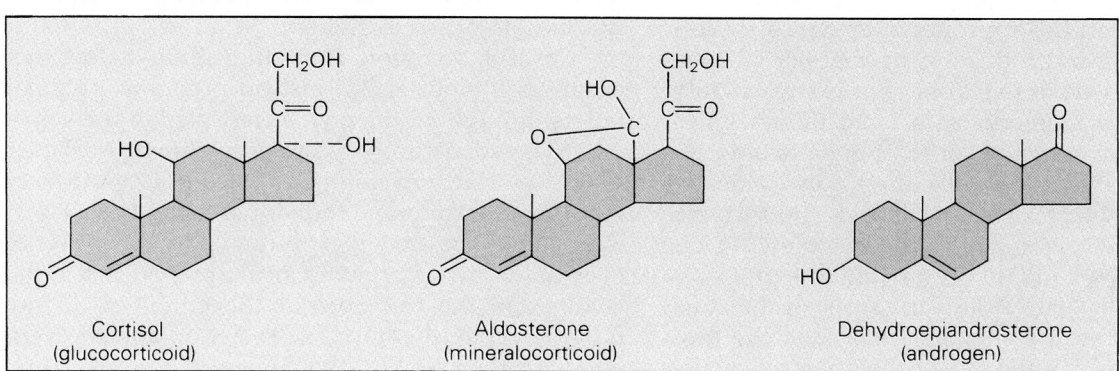

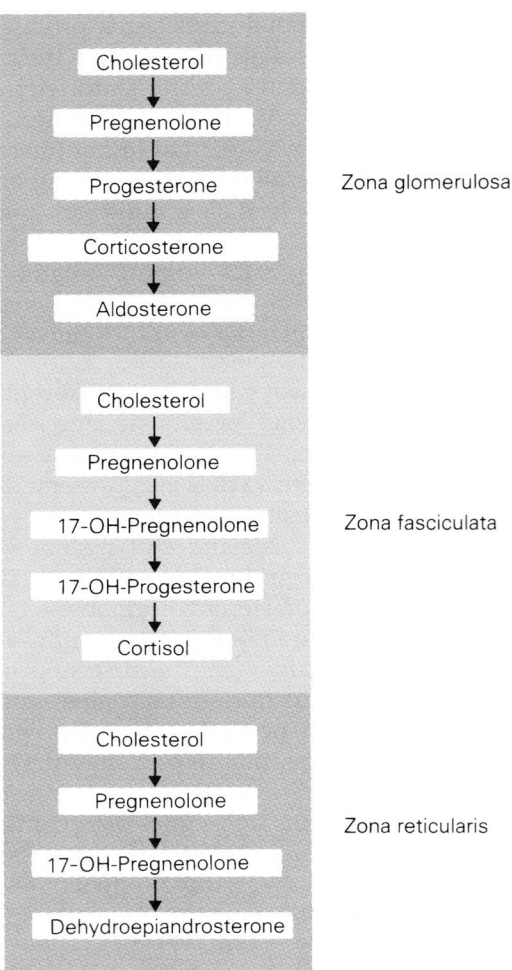

Figure 14–5
Pathways of steroid synthesis in the adrenal cortex.

immediately used for steroid synthesis in the form of lipid droplets. Later, the stored cholesterol can be rapidly mobilized and used for steroid synthesis when needed (see Step C, Fig. 14–6).

Cells within each of the three zones of the adrenal cortex contain the appropriate enzymes specifically suited to the production of the steroid hormone characteristic of that zone. For example, only cells of the zona glomerulosa contain the enzymes required for conversion of corticosterone to aldosterone, and thus only these cells can produce aldosterone (see Fig. 14–5). These cells lack the enzyme required for the conversion of pregnenolone to 17-OH-pregnenolone and so are unable to synthesize either cortisol or DHEA. Similarly, cells of the zona fasciculata contain the enzymes that favor the production of cortisol, and cells of the zona reticularis

exhibit high levels of the enzymes favoring the synthesis of DHEA.

Secretion and Actions of Glucocorticoids

Regulation of Glucocorticoid Secretion

As illustrated in Figure 14–3, secretion of glucocorticoids from cells of the adrenal cortex is regulated by a sequence of steps beginning with CRH secretion by the hypothalamus. CRH release by the hypothalamus is regulated by inputs from a variety of higher brain centers to the hypothalamus. Therefore, much of the regulation of glucocorticoid secretion is due to these inputs to the hypothalamus. In general, glucocorticoid secretion from the adrenal glands is responsive to two types of stimuli. The first type involves a pattern of secretion that varies over the 24-hour period of a day, and the second category involves increased secretion in response to a variety of specific stimuli.

The Circadian Rhythm of Glucocorticoid Secretion. Figure 14–7 shows the typical changes in plasma cortisol concentrations that might occur over a 24-hour period in a normal individual. As the figure indicates, the plasma cortisol concentration is highest around the time of awakening in the morning and lowest around midnight. Because, like the sleep-wake cycle, this general pattern of secretion is repeated approximately every 24 hours, it is termed a circadian rhythm. These changes in cortisol secretion are due to similar circadian variations in CRH and ACTH secretion. This rhythm in cortisol secretion has its origin in higher brain centers that in turn influence the hypothalamic-pituitary system. These circadian rhythms are related primarily to the sleep-wake patterns of the individual rather than by environmental light-dark cycles. Changes in a person's sleep-wake cycle, such as working nights and sleeping days, will result in a temporal shift in the daily rhythm of cortisol secretion.

Cortisol secretion from the adrenal does not occur continuously, but instead occurs in irregular bursts throughout the day, as reflected in the somewhat jagged nature of the curve shown in Figure 14–7. The size and timing of these pulses can vary considerably from one individual to another. However, the bursts are largely buffered by the presence of specific carrier proteins in plasma, so that rapid changes in the free cortisol concentration do not occur. Despite the bursts of secretion and the variability in their timing, the general pattern of secre-

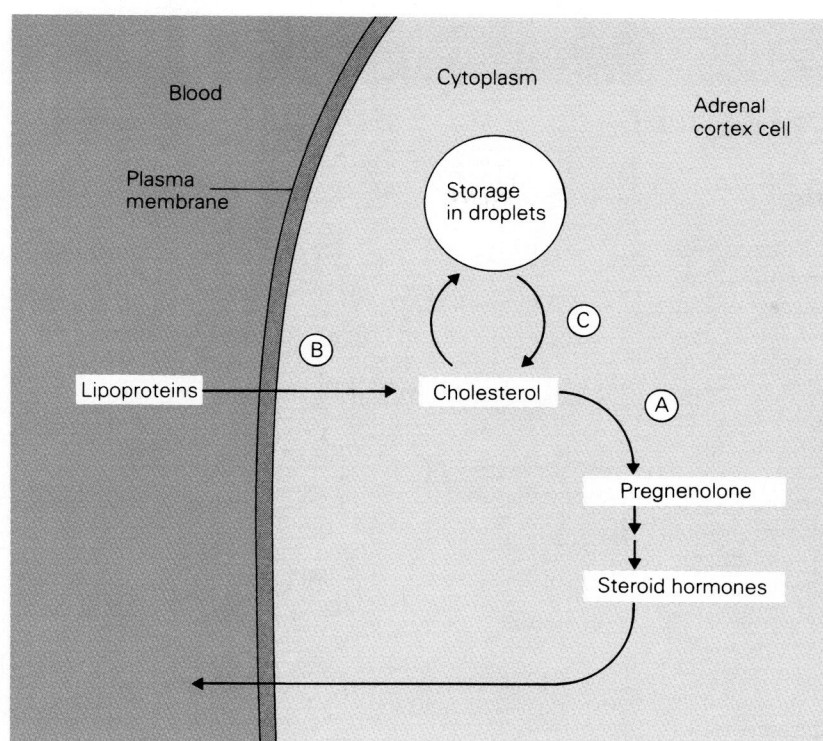

Figure 14–6
The regulatory steps in biosynthesis of adrenal cortex steroids. Step A, the conversion of cholesterol to pregnenolone, is the primary site of regulation of steroid synthesis. Steps B and C, the uptake of cholesterol from the bloodstream and its removal from intracellular storage sites, respectively, are secondary sites of regulation.

tion shown in Figure 14–7 is fairly consistent from one individual to the next.

Stress-Induced Glucocorticoid Secretion. A variety of different stress stimuli can also cause significant increases in cortisol synthesis and secretion (Fig. 14–3). Increased secretion in response to stress occurs in addition to the normal circadian rhythm of secretion described previously. Table 14–1 lists sev-

eral of the different types of stress known to cause increased glucocorticoid secretion. Generally these stresses can be either physical or psychological in nature. Hypoglycemia (low blood glucose) is an important stimulator of cortisol secretion and provides for important adaptive changes that occur

Figure 14–7
Graph of the daily rhythm of the plasma cortisol concentration.

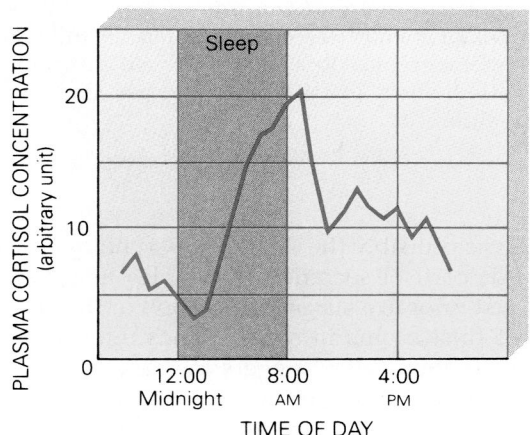

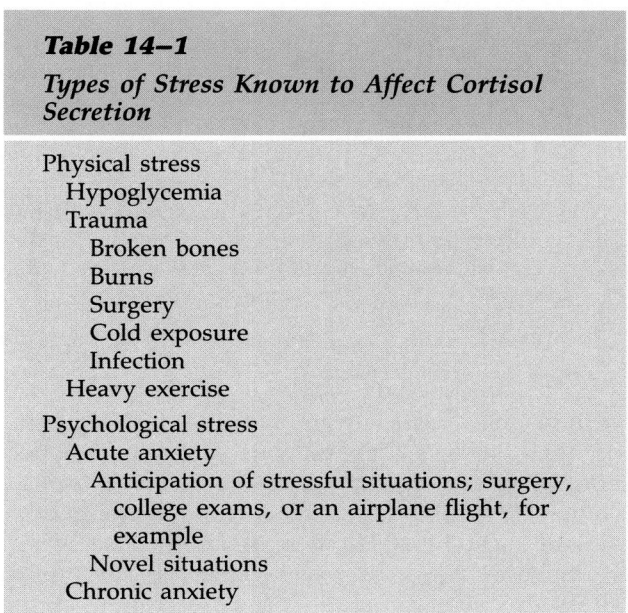

Table 14–1
Types of Stress Known to Affect Cortisol Secretion
Physical stress
Hypoglycemia
Trauma
Broken bones
Burns
Surgery
Cold exposure
Infection
Heavy exercise
Psychological stress
Acute anxiety
Anticipation of stressful situations; surgery, college exams, or an airplane flight, for example
Novel situations
Chronic anxiety

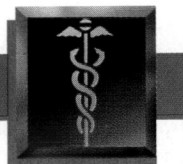

C L I N I C A L   F O C U S

Clinical Use of Steroids

One of the first adrenal steroids to be isolated and synthesized was cortisone, which is structurally related to cortisol. When cortisone was discovered, it was hailed as a possible wonder drug and was at first believed to have potential for widespread use in medicine. Although these initial expectations of cortisone did not materialize to the extent hoped, a variety of steroids have in fact been found extremely useful in medical practice.

Most commonly, corticosteroids are prescribed (1) to replace hormones in patients unable to secrete them in adequate quantities; (2) to suppress inflammatory reactions in situations in which inflammation would be undesirable; and (3) to minimize the immune response in cases of allergy or following organ transplantation. The requirements for steroid hormone action in each of these three cases differ, however, and dictate which type of corticosteroid should be used. Depending on the situation, a physician may want to treat a patient with a corticosteroid that has primarily glucocorticoid activity, while in another situation a corticosteroid with primarily mineralocorticoid activity would be more beneficial.

The pharmaceutical industry has produced numerous steroid derivatives, each of which has differing degrees of glucocorticoid or mineralocorticoid activity. Organic chemists have synthesized many analogs of adrenal steroids, hoping to produce compounds with certain desirable qualities exaggerated and other undesirable qualities reduced or eliminated.

The table below shows relative anti-inflammatory and mineralocorticoid potencies of four commonly prescribed steroid preparations. Dexamethasone and prednisone are synthetic steroids; cortisol and aldosterone are naturally occurring steroids that have been described in this chapter.

Note that the ratio of anti-inflammatory/mineralocorticoid potency of these four steroids varies considerably, from a nearly infinite ratio for dexamethasone to 0.00025 for aldosterone.

Steroid	Relative Anti-inflammatory Activity*	Relative Mineralocorticoid Activity*	Ratio of: Anti-inflammatory/ Mineralocorticoid Activity
Dexamethasone	30	0	infinite
Prednisone	4	0.7	6
Cortisol	1	1	1
Aldosterone	0.1	400	0.00025

* Activities are expressed relative to that of cortisol, which is arbitrarily set equal to 1.0.

In addition to the four steroids listed, literally hundreds of synthetic steroids have been produced. How does the physician choose which steroid to use? If the purpose of treatment is to reduce inflammation, then a steroid with a high ratio of anti-inflammatory activity to mineralocorticoid activity, such as dexamethasone, should be chosen. Large doses of the steroid could then be given to suppress inflammation while having little or no effect on mineral metabolism. If, however, the purpose of steroid therapy is to replace normal hormone levels in an individual with inadequate adrenal function, then a hormone such as cortisol, which has both glucocorticoid activity and a weak amount of mineralocorticoid activity, could be used.

In other cases, pharmaceutical derivatives of adrenal steroids that are very poorly absorbed by the skin have been formulated. These are useful in creams or ointments, which can be applied to the skin to produce a maximal local effect but have little or no generalized effect on the body because they are poorly absorbed.

with fasting. Practically any type of physical trauma or injury, such as a broken bone, severe burn, infection, or surgery, will result in increased cortisol secretion. Strenuous exercise, such as in competitive athletics, also results in increased cortisol secretion.

Several types of psychological or emotional stress can also stimulate cortisol secretion. Acute anxiety is probably the strongest psychological stimulus for cortisol secretion. Examples might be the time just prior to a surgical operation or the anticipation of final exams in school. Encountering a new social situation often triggers increased cortisol secretion. In contrast, chronic anxiety is considerably less important as a stimulus for cortisol secretion.

Recent experiments have suggested that ACTH may play a role in promoting learning. A common denominator in many of the psychological stresses that trigger increased ACTH and cortisol secretion is the involvement in a new or challenging situation. Increased ACTH secretion may therefore play a role in facilitating the learning of appropriate responses in new situations. As described in the previous chapter, ACTH is initially synthesized in the anterior pituitary as part of the larger POMC peptide. When ACTH secretion is stimulated, several other bioactive peptides, including the morphine-like beta-endorphin (see Chapter 7), are secreted along with ACTH. The secretion of peptides such as beta-endorphin during times of stress is beneficial since they function as an endogenous analgesic and increase the pain threshold.

Negative Feedback Relationships in Glucocorticoid Secretion. As is true of most other hormones, cortisol secretion is regulated by a series of negative feedback loops. The dotted lines in Figure 14–3 show that cortisol exerts dual negative feedback effects to inhibit the secretion of ACTH from the pituitary. First, cortisol decreases the activity of CRH-producing neurons in the hypothalamus, thereby decreasing the amount of CRH released. Second, cortisol decreases the sensitivity of the anterior pituitary corticotrophs to CRH. As a result, CRH is less effective in stimulating ACTH release when cortisol is present than when it is absent. Thus, the combination of a decrease in CRH secretion as well as a decrease in sensitivity to CRH contributes to reduced ACTH secretion, which ultimately results in decreased cortisol secretion. In addition, as indicated in Figure 14–3, ACTH itself exerts a negative feedback effect on the hypothalamus to reduce CRH secretion.

Transport of Cortisol in the Blood

Normally, about 80% of the cortisol in blood is bound to a specific carrier protein known as **corticosteroid-binding globulin.** Another 15% of the cortisol is bound to albumin, and the remaining 5% normally exists free in solution. As with other hormones, only free cortisol molecules are able to interact with receptors and produce the physiological effects characteristic of the hormone.

The Physiological Effects of Glucocorticoids

Glucocorticoids play an important role in our day-to-day survival. The actions of glucocorticoids enable us to survive the potentially damaging effects of wide fluctuations in environmental conditions, as well as a number of noxious or potentially noxious factors. As a result, glucocorticoids are considered essential for life. Even something as simple as a brief period of fasting would not be possible without the actions of cortisol. Before we look at the specific effects of glucocorticoids, let us briefly review the general mechanism of steroid action.

Glucocorticoids produce their effects via the general mechanism for steroid hormone action described in Chapter 12 (see Fig. 12–20). Recall that steroid hormones are lipid soluble; they readily pass through the plasma membrane and bind to receptors located in the nucleus of the cell. The hormone-receptor complex then binds to specific sites on the DNA, altering transcription of mRNA molecules that code for specific proteins. As a result, the synthesis of a few specific proteins is appropriately increased or decreased, and the desired changes in cell function occur. Given these mechanisms, what are the specific changes in cell function that are produced by cortisol?

Cortisol exerts its primary effects on three different tissues in the body: liver, skeletal muscle, and adipose tissue. The specific physiological effects of the hormone in these three tissues are summarized in Table 14–2. The overall effect of cortisol under most conditions can be summarized as being **catabolic,** owing to its promotion of protein breakdown in tissues. At very high concentrations, such as might be used therapeutically in medical treatment, glucocorticoids also produce effects that are **anti-inflammatory** and **immunosuppressive.**

Cortisol in the Liver: Increased Blood Glucose. In the liver, the net response to cortisol is increased

---

Table 14–2
Summary of Effects of Cortisol on Metabolism

Liver
    Increases gluconeogenesis
    Increases glycogen synthesis

Skeletal muscle
    Decreases protein synthesis
    Increases protein degradation
    Decreases glucose uptake

Adipose tissue
    Decreases glucose uptake
    Increases lipid mobilization

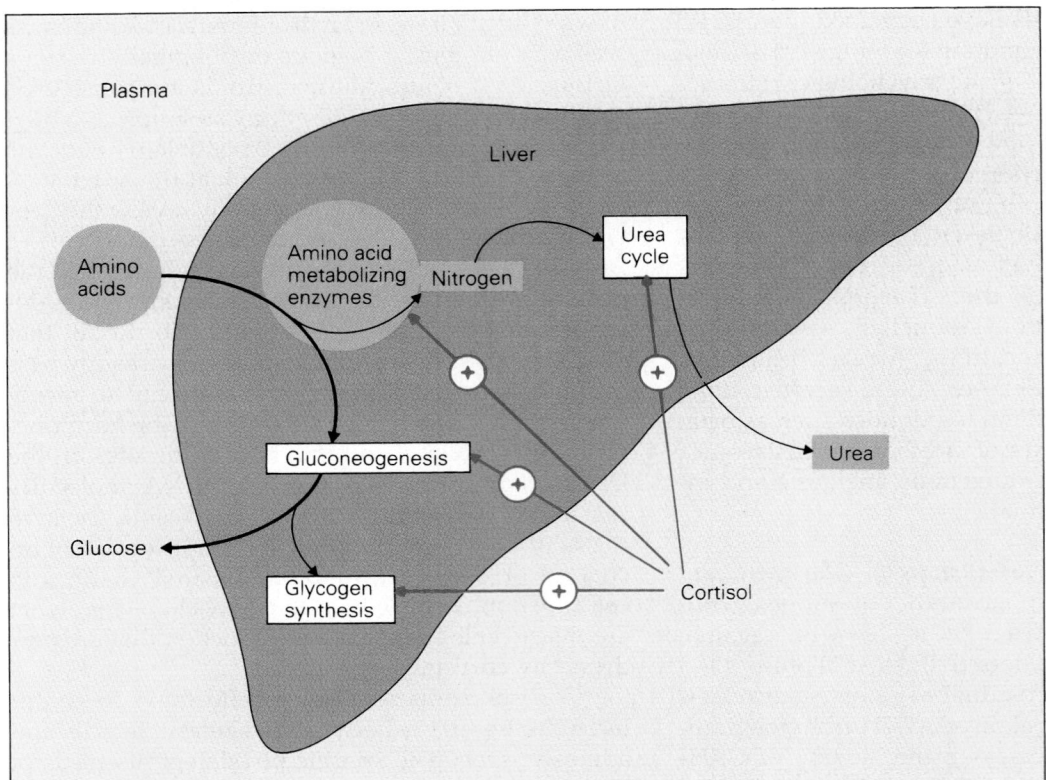

Figure 14–8
Summary of the effects of cortisol on liver metabolism.

gluconeogenesis (see Table 14–2), primarily as a result of increased conversion of amino acids into glucose (for a discussion of the process of gluconeogenesis, refer to Chapter 6). The increased rate of formation of glucose that occurs is the result of several specific effects of the hormone on the liver (Fig. 14–8). First, cortisol increases the activity of several enzymes that catalyze key steps in the gluconeogenic pathway. As a result, glucose formation from a variety of substrates is increased. Second, cortisol increases the activity of several enzymes involved in amino acid metabolism, facilitating the use of amino acids as substrates for gluconeogenesis. The third way in which glucocorticoids promote gluconeogenesis in liver is by stimulating the activity of enzymes of the urea cycle. This stimulation increases the ability of the liver to dispose of the nitrogen liberated during metabolism of amino acids. Together, these specific effects of cortisol permit increased conversion of amino acids into glucose (see Fig. 14–8). As a result, glucocorticoids tend to increase blood glucose. Glucocorticoids also tend to promote the formation of liver glycogen. This is due in part to increased glucose formation in the liver via gluconeogenesis and in part to a stimulation of enzymes involved in glycogen formation.

Cortisol in Skeletal Muscle: Decreased Protein. In skeletal muscle, cortisol has a catabolic action and tends to promote the net loss of protein. This occurs as a result of two different effects of the hormone (Table 14–2 and Fig. 14–9). First, cortisol inhibits muscle protein synthesis. As a result, the uptake of amino acids from the blood and their subsequent incorporation into muscle protein is reduced. Second, cortisol increases the degradation of existing muscle protein. The amino acids released during the degradation of muscle protein are then free to enter the blood. As a result of these two effects of cortisol, there is a net transfer of amino acids from muscle protein into the blood. The liver can then use these amino acids for the synthesis of glucose via gluconeogenesis. The combined effects of cortisol on liver and muscle result in the maintenance of blood glucose at the expense of muscle protein. These catabolic effects of cortisol on muscle occur in skeletal muscle *only*, not in cardiac muscle. Furthermore, the catabolic actions of cortisol and the anabolic actions of other steroids should not be confused. Anabolic steroids used to increase muscle mass consist primarily of androgens and not of glucocorticoids. Cortisol also decreases glucose uptake by skeletal muscle, an effect that opposes that of insulin.

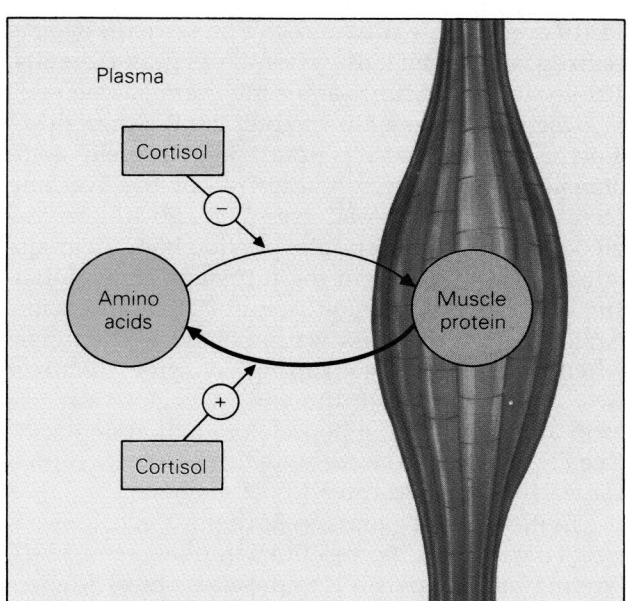

Figure 14-9
Summary of the effects of cortisol on muscle protein metabolism.

Cortisol in Adipose Tissue: Decreased Glucose Uptake. In adipose tissue, cortisol decreases glucose uptake (see Table 14-2). As in muscle, this effect is opposite to that of insulin and thus represents an anti-insulin effect of the steroid. Glucocorticoids also promote the mobilization of lipid from some of the adipose tissue stores in the body. Not all adipose tissue is affected equally, however. This differential sensitivity of adipose tissues to glucocorticoids can best be seen when the glucocorticoids are secreted in excess. When excess secretion occurs, fat stores in the legs and arms decrease, and the fat is redistributed to the trunk and shoulder-blade region. This leads to a very distinct physical appearance, which will be discussed in a later section that describes the effects of cortisol hypersecretion.

The Permissive Actions of Glucocorticoids. In a number of instances, although cortisol itself does not directly regulate a particular metabolic process or initiate events inside the cell, it must be present for other hormones to produce their full effect. Cortisol therefore tends to have an amplifying effect on the action of other hormones. These are known as the **permissive actions** of glucocorticoids. One example of a permissive action of glucocorticoid occurs in adipose tissue, where the stimulation of lipid breakdown by epinephrine is enhanced when cortisol is also present. Although numerous other permissive actions of glucocorticoids are known, the exact nature of the events resulting in these effects is not well understood.

Effects of Glucocorticoids on Blood Vessels and Blood Cells. In addition to its effects on carbohydrate, lipid, and protein metabolism, cortisol has other important effects in the body. One of the more important effects of cortisol is to enhance the responsiveness of blood vessels, or **vascular reactivity.** In the absence of cortisol, the diameter of arterioles (very small arteries) is not appropriately regulated during times of stress, and consequently blood pressure can fall dramatically, even resulting in death. Cortisol restores the responsiveness of the blood vessels and therefore is said to restore vascular reactivity.

Glucocorticoids also affect various blood cell types and lymphoid tissues. Cortisol decreases the number of eosinophils and basophils in the blood and increases the number of neutrophils, red blood cells, and platelets in the blood (see Chapter 17 for a further discussion of blood cell types). As will be discussed, high doses of glucocorticoids also influence inflammation and the immune response.

The Pharmacological Effects of Glucocorticoids

The effects of cortisol described in the foregoing paragraphs are those generally observed in response to physiological amounts of the hormone. At much higher concentrations, these effects of the hormone are magnified, and other effects are produced as well, some of which have proven very useful for medical purposes. These are known collectively as the **pharmacological effects** produced by glucocorticoids and other closely related steroids.

Perhaps the two most important pharmacological effects of glucocorticoids are their **anti-inflammatory** effects and their **immunosuppressive** effects. In large doses, glucocorticoids profoundly inhibit the inflammatory response to tissue injury or infection by inhibiting virtually every step in the inflammatory response, including the vasodilation, increased capillary permeability, and increased phagocytosis normally observed in inflammation. High doses of glucocorticoids also suppress the immune system, resulting in decreased antibody production. The number of circulating lymphocytes decreases, owing to increased lymphocyte destruction in lymphoid tissues, and the size of lymph nodes also decreases. As a result of these pharmacological effects, glucocorticoids are particularly useful in medicine for reducing the inflammatory response to injury or infection, or for suppressing the rejection response to transplanted organs. Although the immunosuppression produced by glucocorticoids is of great value following organ transplantation, it does leave the individual more susceptible to infections, necessitating treatment with antibiotics as well as with steroids.

Secretion and Actions of Mineralocorticoids

Regulation of Mineralocorticoid Secretion

Aldosterone is the principal mineralocorticoid produced in the zona glomerulosa of the adrenal cortex. The two primary regulators of aldosterone secretion are **potassium ion** and a peptide hormone known as **angiotensin II.** The kidneys play a key role in determining the plasma levels of each of these two factors, and are also the major action site of aldosterone in the body.

Regulation by Angiotensin II. **Angiotensin II** is formed by a sequence of steps that begin in the kidney and are collectively known as the **renin-angiotensin system.** These steps are outlined in Figure 14–10 and begin with the secretion of **renin** by specialized cells in the kidney known as **granular cells.** The location and function of the granular cells will be described in detail in Chapter 24. Renin is a proteolytic enzyme that acts upon a large protein, **angiotensinogen,** which is synthesized by the liver and secreted into the blood (see Fig. 14–10). In the blood, renin splits a specific peptide bond in angiotensinogen, resulting in the formation of **angiotensin I,** which is a peptide chain of 10 amino acids. Angiotensin I is carried via the blood to the lungs, where **angiotensin-converting enzyme** removes two amino acids from the end of angiotensin I to form **angiotensin II,** a peptide of eight amino acids (see Fig. 14–10). From the lungs, angiotensin II travels via the blood to the adrenal glands.

In the adrenal glands, cells of the zona glomerulosa contain specific receptors for angiotensin II. Binding of angiotensin II to these receptors triggers increased synthesis and secretion of aldosterone. The exact mechanisms whereby angiotensin II binding leads to increased aldosterone synthesis are not known, although it appears that the phosphatidylinositol pathway provides the second messengers that couple hormone binding with the cellular effects of increased steroid synthesis (refer to Fig. 12–17).

Figure 14–10
Summary of the renin-angiotensin system.

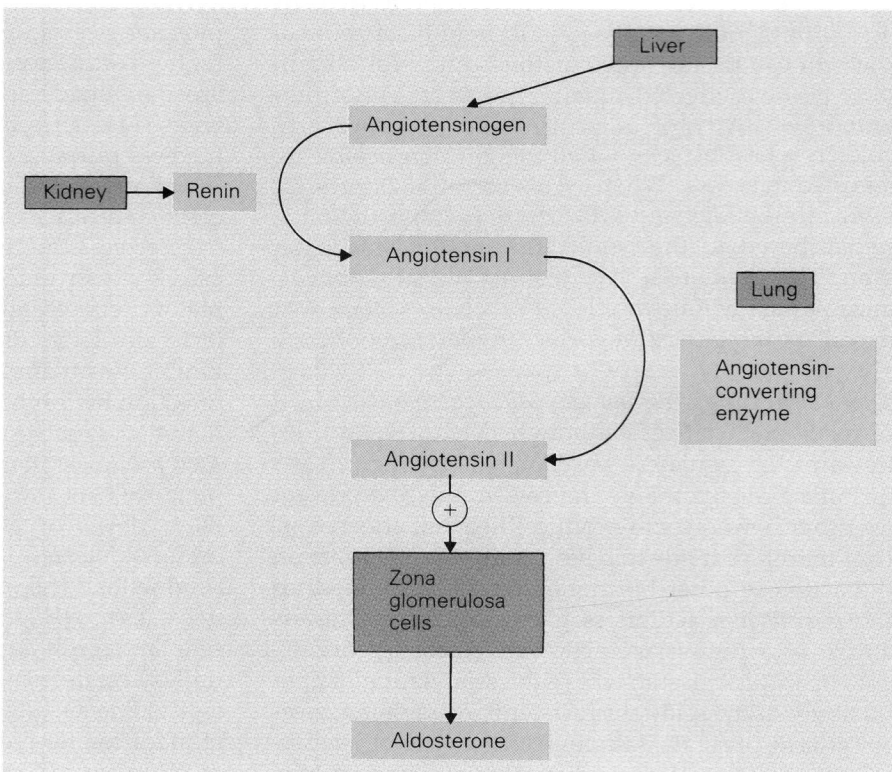

As indicated by the series of steps shown in Figure 14–10, the amount of aldosterone produced by the adrenal cortex is related to the amount of renin released from the kidney. Therefore, those factors that regulate **renin secretion** ultimately serve to regulate **aldosterone secretion** by the adrenal cortex. The primary stimulus for renin secretion is a decrease in blood pressure or blood flow to the kidneys. Conversely, an increase in blood pressure or blood flow to the kidneys inhibits renin release. In addition, sympathetic innervation of the granular cells modulates the response in renin secretion to these and other stimuli.

Regulation by Potassium Ion. The second major regulatory factor involved in the control of aldosterone secretion is the concentration of **potassium ion.** Aldosterone secretion is stimulated by an *increase* in the plasma potassium concentration. The potassium ion appears to directly affect the cells of the zona glomerulosa. These cells are extremely sensitive to very small changes in the plasma concentration of potassium; an increase in potassium of as little as 5% above normal is enough to increase aldosterone secretion. As the following section will describe, one of the actions of aldosterone is to promote potassium excretion by the kidney and thus to lower plasma potassium concentrations. Thus, aldosterone and the kidneys make up the two components of a feedback loop that regulates the plasma potassium concentration.

The Effects of Mineralocorticoids

The net effect of aldosterone is to increase extracellular fluid volume. This increase in extracellular fluid volume results in increased blood volume, which tends to raise blood pressure and blood flow. Thus, aldosterone indirectly affects blood pressure and blood flow in the cardiovascular system and is an important component of the processes involved in regulating fluid balance in the body (see Chapter 25).

The kidneys are the primary action site of aldosterone. In response to aldosterone, the kidneys retain more sodium in the body while excreting more potassium. Thus, in response to aldosterone, urinary sodium decreases, while urinary potassium increases. The specific mechanisms involved in these effects of aldosterone are presented in Chapter 24 and 25.

In addition to its effects on aldosterone secretion, angiotensin II has other important effects related to the maintenance of blood pressure and extracellular fluid volume. These include a stimulation of smooth muscle contraction in blood vessels, activation of thirst centers in the brain, and stimulation of ADH release from the posterior pituitary. Each of the respective changes that result (decreased blood vessel diameter, increased water intake, and increased fluid retention by the kidneys) contributes to the maintenance of a normal blood volume and blood pressure. These and other effects of angiotensin, as well as other characteristics of the renin-angiotensin system, will be described in Chapter 25, which discusses in detail the regulation of fluid and electrolyte balance.

Abnormalities of Adrenal Cortex Function

Several common diseases result from either deficiencies or oversecretion of adrenal cortex steroids. As with other endocrine disorders, the symptoms expressed are the direct result of either the absence or magnification of the effects of the hormones involved. A discussion of these conditions is therefore a useful way to review the actions of these hormones, as well as to become familiar with endocrine disease states that are fairly common in the general population.

Excess Secretion of Glucocorticoids

The most common disorder of adrenal cortex function involves increased secretion of glucocorticoids. This can occur either as a result of a primary abnormality in steroid hormone production by the adrenal cortex or as a result of overproduction of ACTH by the pituitary, causing excessive stimulation of the adrenal cortex. In either case, the physiological changes that occur are due to the exaggerated effects of cortisol. The condition that results from excess cortisol secretion is known as **Cushing's syndrome.** Figure 14–11 illustrates the general appearance of a patient with Cushing's syndrome. Note the thin arms and legs, due in part to the loss of muscle mass as a result of the protein-catabolic effects of excess cortisol. Fat is redistributed from the extremities to the trunk, also contributing to a thinning of the arms and legs. Loss of fat in the extremities is accompanied by increased fat in the face, the trunk, across the shoulder blades, and at the base of the neck. Accumulation of fat in the face gives it a characteristically rounded appearance. As a result of the protein-catabolic effects of cortisol, connective tissue is lost from the skin, causing it to become thinner. As a result, blood vessels are located closer to the surface, which is most often ap-

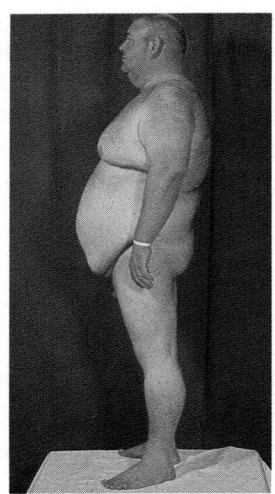

Figure 14–11
A man with Cushing's syndrome. (© Science VU/ Visuals Unlimited.)

parent in the red cheeks that characterize Cushing's patients. In addition, as the skin of the abdomen is stretched with increased fat deposition, it thins even further, and reddish-purple stretch marks, or **striae,** form, owing to the increased visibility of underlying blood vessels. Loss of connective tissue in small blood vessels makes them more fragile and easily broken than normal, resulting in an increased susceptibility to bruising.

Hypertension (high blood pressure) is also commonly present in persons with Cushing's syndrome and in many cases is the primary contributing factor leading to death of these individuals. These patients also suffer from increased susceptibility to infection and poor wound healing due to suppression of the inflammatory and immune systems caused by excess cortisol. Excess cortisol also impairs bone me-

R E S E A R C H   F O C U S

The Steroid and Thyroid Hormone Receptor Superfamily

According to the current model for steroid and thyroid hormone action, binding of the hormone to its receptor induces an allosteric change that allows the hormone-receptor complex to interact with DNA and alter transcription of a few specific genes. Considerable effort has been directed towards understanding the mechanism of action of these hormones, beginning in the early 1970's when hormone receptors were first functionally identified. Details of how the events involved in hormone action occur have yet to be fully resolved, but in the past several years, considerable insight in this area has been provided by the application of the techniques of molecular biology.

It is now known that the receptors for steroid and thyroid hormones comprise a family of proteins bearing considerable similarity with one another. The similarity among receptors for this group of hormones is illustrated in schematic form in the accompanying figure. Each receptor is thought to consist of three regions, or domains, each of which appears to serve a different function. The presumed function of each domain is based in part on our knowledge of structure/function relationships of other proteins, and also on the results of mutation and deletion studies with these receptors, using the methods of molecular biology. The region of greatest similarity among the receptors is the central domain, which is thought to be the DNA-binding portion of the protein. This region is rich in basic amino acids, whose positive

charges are likely to allow significant interaction with the negatively charged DNA. Amino acid sequences in this portion of the molecule also suggest that the peptide backbone may exist in the form of several loops, or "fingers," which are spaced to allow interdigitation with the DNA helices. The hormone-binding region of the molecule is thought to begin 50 to 70 amino acids downstream, towards the carboxyl-terminal end of the protein. The structural homology in this domain is generally more graded than in the DNA-binding domain and appears to parallel the relatedness of the hormones themselves. The amino terminal end of the molecule shows the greatest variability between receptors, in terms of both amino acid sequence and length. A definitive role for this region has not been demonstrated, although it may contribute to important functional differences between receptors.

Results of mutation studies have provided insight into the activation mechanism of these receptors that occurs as a result of hormone binding. Deletion of the hormone-binding region of the molecule results in a receptor that is constantly active. This suggests that in an intact receptor, the hormone-binding domain normally prevents the DNA-binding region from functioning; binding of hormone to an intact receptor relieves this inhibition. Another approach used by molecular biologists to characterize these receptors has been to exchange domains between re-

tabolism, and thus patients with Cushing's syndrome may also suffer from numerous bone fractures.

Many of the effects of cortisol antagonize the actions of insulin. Because a deficiency in insulin action results in diabetes, severe cases of Cushing's syndrome can actually produce diabetes if left untreated for a prolonged period of time.

Insufficient Secretion of Adrenal Cortex Steroids

An overall deficiency in steroid production by the adrenal cortex leads to the clinical condition termed **Addison's disease.** If not treated, this particular disease state leads to death. Addison's disease can be caused by primary disease of the adrenal cortex, leading to a loss of its steroid-secreting capacity, or

Table 14–3
Typical Findings in Addison's Disease

Low plasma sodium, high plasma potassium
Low blood pressure
Muscle weakness, fatigue
Vomiting, loss of appetite, dehydration
Low blood sugar
Excess pigmentation of skin in some patients

it can be due to loss of ACTH production by the pituitary gland. Loss of the trophic effects of ACTH prevents the adrenal cortex from producing adequate amounts of both cortisol and aldosterone.

Typical symptoms of Addison's disease (see Table 14–3) reflect the loss of both glucocorticoid

ceptor types. In one type of experiment, termed a "finger swap," the DNA-binding domain of one receptor type is substituted for that of another. The resulting hybrid receptor exhibits a switch in specificity that is determined by its DNA-binding domain. Thus, although specificity of hormone action is determined at the hormone-binding step, once this occurs the nature of hormone-receptor interaction with DNA appears to be determined by inherent structural features in the DNA-binding region of that particular receptor. Thus the hormone acts as an allosteric activator of the DNA-binding portion of the molecule.

It is clear that discovery of the steroid and thyroid hormone receptor superfamily has greatly expanded our knowledge of human physiology. It is also likely to have an impact on the strategies for treatment of certain diseases, since there is evidence to suggest that mutant receptors may contribute to some forms of malignancy. For example, the protein product of a particular oncogene proposed to be involved in the development of leukemia is a derivative of the thyroid hormone receptor in which hormonal control of the activity of the receptor has apparently been lost (see Chapter 12 for a further discussion of oncogenes and hormones). These mutant receptors remain constantly active, even in the absence of thyroid hormone. Understanding disease at this molecular level should permit the design of treatments that will specifically and directly alleviate the defect and restore normal function.

Schematic representation of the structure of several members of the steroid and thyroid hormone superfamily. The amino terminal end of the protein is to the left; the carboxyl terminus is to the right. The blue center region is the DNA-binding domain. The pink segment on the right is the hormone-binding domain, and the gray bar on the left is the highly variable amino terminal end of the protein.

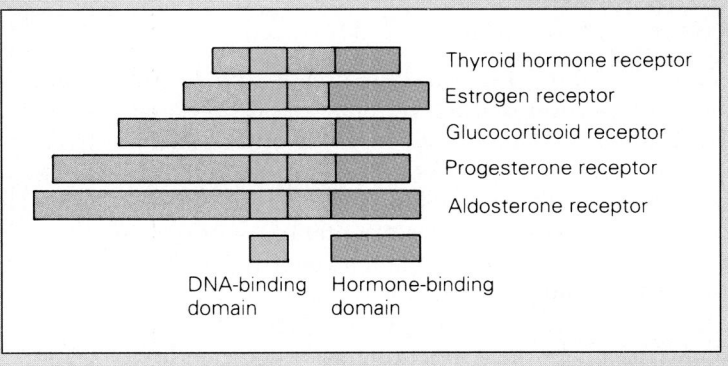

Thyroid hormone receptor
Estrogen receptor
Glucocorticoid receptor
Progesterone receptor
Aldosterone receptor

DNA-binding domain   Hormone-binding domain

and mineralocorticoid action. Due to loss of aldosterone, a person suffering from Addison's disease typically exhibits low plasma sodium and high plasma potassium levels. Low blood pressure is a common finding, in part due to electrolyte and water loss leading to reduced blood volume caused by the lack of aldosterone, and also due to the absence of the stimulatory effect of cortisol on vascular reactivity. Disturbances of electrolyte balance may lead to muscle weakness and easy fatigue as well as other symptoms, including vomiting and loss of appetite. Low blood glucose is also common in persons with Addison's disease, owing to an absence of the stimulatory effect of cortisol on gluconeogenesis in the liver.

In addition, hyperpigmentation of the skin and gums may occur in patients with Addison's disease when the abnormality results from a primary loss of adrenal cortical function (see Table 14–3). Recall from Figure 14–3 that in the absence of glucocorticoids, their negative feedback effect to inhibit ACTH secretion is lost, and plasma ACTH therefore increases dramatically. The structure of ACTH is similar to melanocyte-stimulating hormone (MSH), which increases pigmentation in the skin (refer to Table 12–1). Although the MSH-like activity of ACTH is weak, the greatly increased plasma concentrations of ACTH that occur in some forms of Addison's disease are sufficient to cause increased skin pigmentation.

The Adrenal Medulla

The adrenal medulla, which secretes the catecholamines epinephrine and norepinephrine, is an important component of the sympathetic nervous system. As a result, the sympathetic nervous system and the adrenal medulla are often referred to together as the **sympathoadrenal system.**

Anatomy of the Adrenal Medulla

As shown in Figure 14–1, the medulla is located in the interior portion of each adrenal gland. Within the adrenal gland, blood flows from the outer cortex inward to the medulla. As a result, blood that reaches the medulla contains high concentrations of glucocorticoids. This vascular orientation is important since it has been shown that glucocorticoids can influence hormone synthesis in the medulla.

The functional unit of the adrenal medulla is the **chromaffin cell,** which functions as a neuroendocrine cell. The name "chromaffin" originated from the fact that these cells are readily stained with solutions containing the element chromium. Functionally, the chromaffin cells of the adrenal medulla are analogous to **postganglionic fibers** of the sympathetic nervous system (see Fig. 14–12). Recall that in the sympathetic nervous system, preganglionic fibers form synapses with postganglionic fibers in the sympathetic chain ganglia or in one of the outlying sympathetic ganglia. The postganglionic fiber then projects to its destination in a particular target organ such as the heart, as illustrated in Figure 14–12. The arrangement is slightly different for the adrenal medulla, however. Preganglionic fibers to the medulla pass through the sympathetic chain without synapsing and directly innervate the chromaffin cells of the adrenal medulla. In response to stimulation, the chromaffin cells secrete epinephrine directly into the blood. The epinephrine is then carried to various effector organs in the body. Thus, although the chromaffin cells of the medulla are analogous to sympathetic postganglionic fibers, they differ in that the chromaffin cells influence the activity of a variety of effector organs throughout the body. In contrast, sympathetic postganglionic neurons generally influence the activity of single effector organs.

Synthesis and Secretion of Epinephrine

The primary product secreted by the adrenal medulla is epinephrine. The pathway of epinephrine synthesis is outlined in Figure 14–13. Note that norepinephrine, another important catecholamine, is an intermediate in the synthesis of epinephrine. However, the adrenal medulla secretes only small amounts of norepinephrine.

Epinephrine synthesis begins with the amino acid tyrosine (see Fig. 14–13). Tyrosine is converted to **DOPA** (dihydroxyphenylalanine) by the enzyme **tyrosine hydroxylase.** This is the rate-limiting (slowest) step in epinephrine synthesis, and thus alterations in the activity of this enzyme directly influence the rate of epinephrine synthesis. Input of sympathetic stimulation to the medulla increases epinephrine synthesis by increasing the activity of tyrosine hydroxylase. Subsequent steps resulting in the conversion of DOPA to norepinephrine proceed rapidly. The final step in epinephrine synthesis is the conversion of norepinephrine to epinephrine, which is catalyzed by the enzyme **phenylethanolamine-N-methyl transferase (PNMT).** This enzyme is stimulated by cortisol, and thus the high concentration of this steroid in blood flowing from the cortex to the medulla enhances epinephrine synthesis

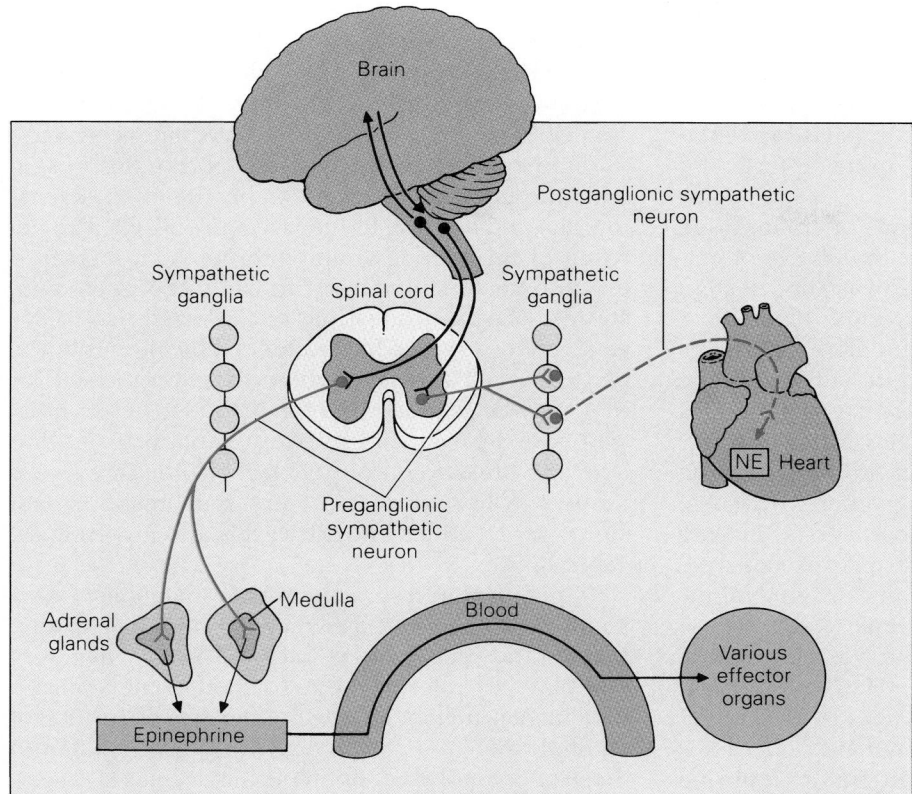

Figure 14–12
Figure illustrating the anatomical analogy between cells of the adrenal medulla and the sympathetic postganglionic neurons. Note that a particular postganglionic fiber (dotted orange line) has effects on one specific effector organ, such as the heart. In contrast, cells of the adrenal medulla, because they secrete epinephrine into the blood, are able to influence the activity of various effector organs throughout the body.

by increasing the activity of phenylethanolamine-N-methyl transferase.

Following synthesis, epinephrine is packaged and concentrated within the chromaffin cell into dense membrane-bounded vesicles known as **chromaffin granules.** Stimulation of the chromaffin cell as a result of sympathetic discharge causes the release of epinephrine from these granules into the blood.

The Actions of Epinephrine

Stimulation of the sympathetic nerves to the adrenal medulla results in the secretion of large amounts of epinephrine and smaller amounts of norepinephrine into the circulation. These hormones are then carried in the blood to all tissues of the body, where they produce the same effects as those resulting from direct sympathetic stimulation of the tissue, except for one important difference. Because epinephrine and norepinephrine are removed from the blood more slowly than from the area of the sympathetic nerve terminal, effects of the circulating hormones are more prolonged than effects resulting from direct sympathetic stimulation.

The adrenal medulla is part of a general response system in the body that triggers a sequence of events appropriate to **"fight-or-flight"** situations.

Figure 14–13
Pathways of epinephrine synthesis in adrenal chromaffin cells. Conversion of tyrosine to DOPA, catalyzed by tyrosine hydroxylase, is normally the rate-limiting step. The enzyme PNMT catalyzes the conversion of norepinephrine to epinephrine.

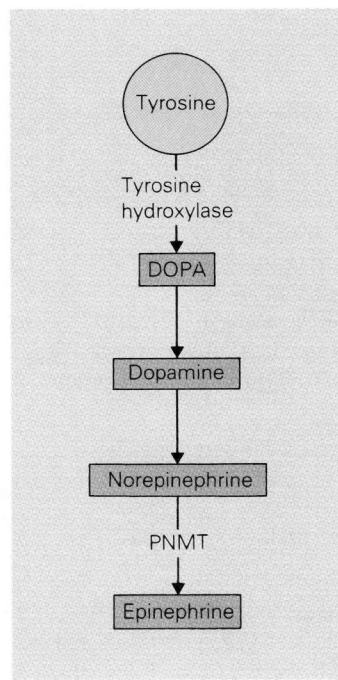

The net result of this response is to increase the capability of the body to perform vigorous muscle activity. Thus, if faced with a life-threatening situation, the body can respond more effectively, bettering the chances of survival.

Examples of some of the actions of epinephrine are listed in Table 14–4. Many of the effects of epinephrine occur within the cardiovascular system and serve to increase and redistribute blood flow between various tissues. Epinephrine increases both the heart rate and the strength with which the heart contracts; as a result, it increases the amount of blood pumped per minute (the cardiac output). Epinephrine also produces changes that affect the distribution of blood flow among different tissues. Epinephrine causes dilation of blood vessels in skeletal muscle while simultaneously causing constriction of blood vessels in the skin and in visceral organs such as the gastrointestinal tract. These changes in diameter of the blood vessels cause a shunting of blood away from the skin and internal organs to the muscles. These changes in blood flow are clearly advantageous to an animal faced with a "fight-or-flight" situation. In addition, epinephrine affects the CNS, increasing the state of mental alertness, which is also advantageous in a life-threatening situation.

Epinephrine also increases the ability of the blood to transport oxygen. By causing relaxation of smooth muscle in the airways of the lungs, epinephrine increases inflow of oxygen to the lungs. In some species, epinephrine also causes the spleen to contract, which increases the circulating blood volume and the number of red blood cells in the circulation. Both factors contribute to an increased oxygen-carrying capacity of the circulatory system.

Epinephrine affects metabolic processes in several different tissues, most notably the liver, skeletal muscle, and adipose tissue (see Table 14–4). The net result of the actions of epinephrine is an increased breakdown of stored fuels into metabolizable substrates. The substrates that are released may serve as an energy source for local metabolism within the particular cell or tissue where they are formed, or they may be released into the blood for general distribution. In liver and muscle, epinephrine stimulates the breakdown of glycogen, while in adipose tissue the hormone stimulates the breakdown of stored triglycerides into fatty acids and glycerol (see Table 14–4).

Epinephrine indirectly affects metabolism as a result of its effects on pancreatic hormone secretion. Insulin and glucagon are secreted by the pancreas and play key roles in the regulation of metabolism. Insulin has metabolic effects that oppose those of epinephrine, and glucagon has effects similar to those of epinephrine. Epinephrine decreases insulin secretion while increasing glucagon secretion. Thus, this effect of epinephrine on the pancreas serves to indirectly promote its own metabolic actions. Discharge of the sympathetic nervous system also leads to an increased basal metabolic rate as a result of increased cellular metabolism throughout the body. This effect is primarily due to release of norepinephrine from sympathetic nerve endings, rather than release of epinephrine from the adrenal medulla.

A summary of the metabolic effects of epinephrine is presented in Figure 14–14. In the liver, epinephrine stimulates glycogenolysis and thus increases glucose release from the liver into the blood. Epinephrine also stimulates glycogenolysis in muscle, but muscle cannot release free glucose into the blood. Instead, glucose is metabolized to lactate, which is then released. Lactate can be taken up by liver and converted to glucose via gluconeogenesis. As glucose is metabolized to lactate in muscle, ATP, which the muscle can use for contraction, is produced.

In adipose tissue, epinephrine stimulates **lipolysis,** which is the breakdown of triglycerides to fatty acids and glycerol. The fatty acids released can be utilized by other tissues to obtain energy. Heart and skeletal muscle in particular are able to fill a considerable amount of their energy needs by burning fatty acids; as a result, these tissues do not have to use glucose. This ability to use fatty acids as an alternative energy source is important, since glucose is the primary fuel used by nerves and the brain. Provision of fatty acids is therefore said to have a "spar-

---

Table 14–4
Summary of the Actions of Epinephrine

Effects on the cardiovascular system
  Increased cardiac output
    Increased heart rate
    Increased strength of cardiac contractions
  Vasodilation in skeletal muscle
  Vasoconstriction in internal organs and skin

Effects on other tissues
  Relaxation of smooth muscle in
  gastrointestinal tract, urinary bladder, airways
  of lung
  Increased mental alertness

Effects on metabolism
  Increased glycogenolysis in liver and muscle
  Increased lipolysis in adipose tissue
  Decreased insulin secretion
  Increased glucagon secretion

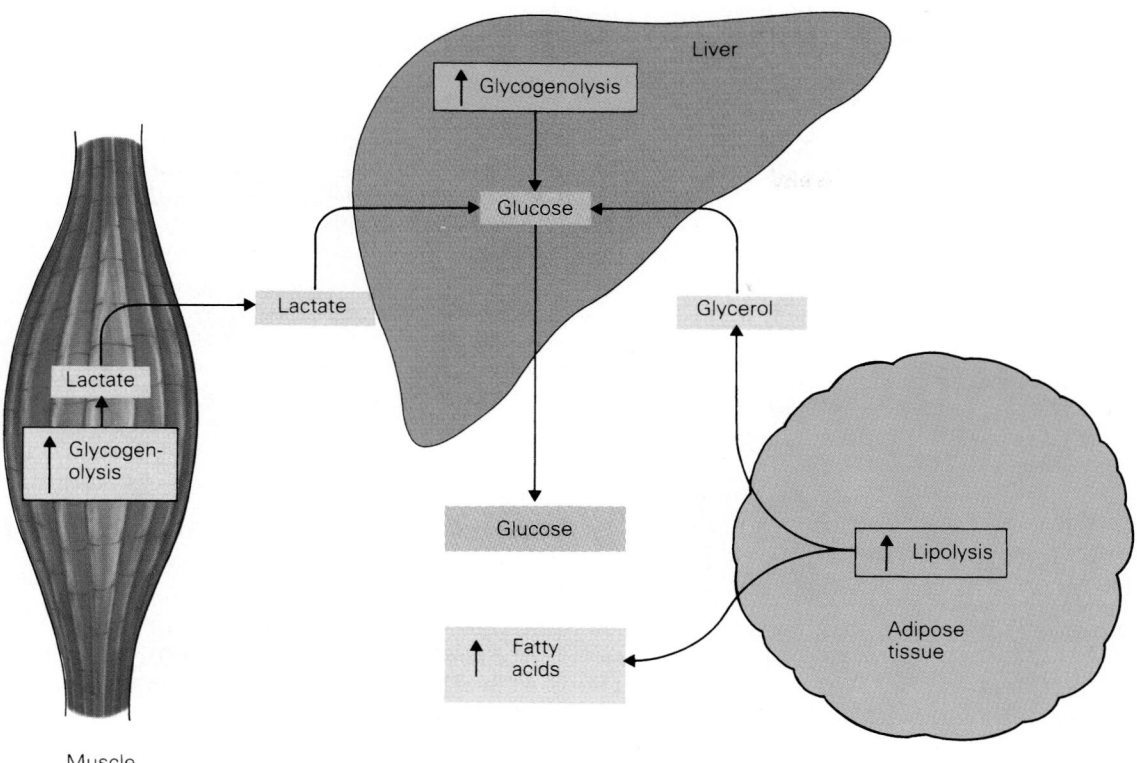

Figure 14–14
Summary of the principal effects of epinephrine on energy metabolism. The primary effects of epinephrine are to stimulate glycogenolysis in liver and muscle, and lipolysis in adipose tissue. The net result is an increase in blood glucose and blood fatty acid concentrations.

ing effect" on the blood glucose. Glycerol released from adipose tissue as a result of lipolysis can be taken up by the liver and converted into glucose.

The net effect of epinephrine on metabolism is increased blood glucose concentration. The increase occurs as a direct result of increased glycogenolysis in liver and increased gluconeogenesis from lactate and glycerol in the liver, and as an indirect result of the sparing effects of fatty acids on blood glucose. The effect of epinephrine to inhibit insulin secretion and stimulate glucagon secretion from the pancreas also plays an important part in the elevation of blood glucose.

Summary of Effects of Adrenal Hormones

When we consider together the effects of the hormones produced by the adrenal cortex and adrenal medulla, it is clear that the adrenal glands play an important part in the body's ability to effectively respond to stress. Epinephrine produces effects on metabolism and on the cardiovascular system and lungs that result in an increased supply of oxygen and metabolic fuels for the muscles and brain. Cortisol stimulates gluconeogenesis in the liver and thus helps to maintain blood glucose during prolonged periods of stress. As a result of either direct discharge of sympathetic nerves to the kidney or the release of epinephrine from the adrenal medulla, renin release can be increased during stress. An increase in renin results in increased aldosterone secretion. In response to increased aldosterone, the kidney retains more sodium and water, resulting in a greater blood volume. This is obviously beneficial when the body suffers blood loss. In addition, the constriction of blood vessels in the skin caused by epinephrine serves to limit blood loss in the case of superficial wounds. Thus, the hormones of the adrenal glands produce a variety of effects that enable the body to adapt to potentially harmful situations and therefore serve to promote survival.

SUMMARY

Gross Anatomy of the Adrenal Glands

Humans have two adrenal glands, one located just above each kidney. Each adrenal consists of two distinct endocrine glands, the adrenal cortex and the adrenal medulla. The cortex secretes steroids, and the medulla secretes catecholamines.

The Adrenal Cortex

The three zones of the adrenal cortex are the zona glomerulosa, the zona fasciculata, and the zona reticularis. These three zones produce, respectively, mineralocorticoids, glucocorticoids, and androgens.

The adrenal cortex is controlled by ACTH from the anterior pituitary, which is in turn regulated by *CRH* secretion by the hypothalamus. These form the *hypothalamic-pituitary adrenal axis.* ACTH has both tropic and steroidogenic effects on the adrenal cortex.

The adrenal cortex primarily produces three steroids: cortisol, aldosterone, and dehydroepiandrosterone. Cortisol secretion is regulated by ACTH.

Cortisol is secreted according to a circadian rhythm. Plasma cortisol levels are highest around 8 AM and are lowest around midnight. Both *physical* and *psychological* stress also stimulate cortisol secretion. Cortisol has negative feedback effects on both the hypothalamus and the anterior pituitary.

The overall effect of cortisol on metabolism is catabolic. In the liver, cortisol produces effects which tend to increase the blood glucose concentration. In skeletal muscle, cortisol promotes protein loss, and in adipose tissue, it decreases glucose uptake and promotes the mobilization and redistribution of fat stores. Cortisol enhances vascular reactivity and also has permissive effects with regard to the actions of several other hormones. In large doses, glucocorticoids have anti-inflammatory and immunosuppressive actions useful in medicine.

Aldosterone is the principal mineralocorticoid produced by the adrenal cortex. Aldosterone secretion is regulated by the renin-angiotensin system and by changes in plasma potassium concentrations. Aldosterone acts on the kidneys to increase sodium retention and potassium excretion.

When excess glucocorticoids are secreted, Cushing's syndrome results. Persons with Cushing's syndrome have a distinct physical appearance due to the magnified effects of cortisol on protein and fat metabolism. A deficiency in steroid production by the adrenal cortex leads to the condition known as Addison's disease. Persons with Addison's disease exhibit abnormalities in electrolyte balance and in blood pressure regulation.

The Adrenal Medulla

The adrenal medulla is an important component of the sympathoadrenal system. The chromaffin cells of the medulla are neuroendocrine cells analogous to postganglionic neurons in the sympathetic nervous system.

The adrenal medulla produces and secretes catecholamines. *Epinephrine* is the primary product of the adrenal medulla.

Epinephrine produces a variety of different effects in the body that together prepare the animal for a fight-or-flight situation. Epinephrine affects the cardiovascular and pulmonary systems to increase the delivery of oxygen to the muscles. Epinephrine affects metabolism in liver, muscle, and adipose tissue, increasing supplies of metabolic fuels such as glucose and fatty acids.

Summary of Effects of Adrenal Hormones

The overall effects of hormones produced by the adrenal cortex and adrenal medulla increase the likelihood of survival of an animal faced with harmful or potentially harmful situations.

REVIEW QUESTIONS

Identify Each with a Term

1. The zone of the adrenal cortex responsible for the synthesis of glucocorticoids.

2. The primary mineralocorticoid secreted by the adrenal cortex.

3. The name of the proteolytic enzyme secreted by the kidney that converts angiotensinogen into angiotensin I.

4. The condition that results from excess cortisol secretion.

5. The name of the enzyme that catalyzes the rate-limiting step in epinephrine synthesis.

6. The name of cells of the adrenal gland responsible for epinephrine synthesis.

Define Each Term

7. zona glomerulosa
8. corticotropin-releasing hormone
9. pharmacological effects of glucocorticoids
10. angiotension II
11. Addison's disease
12. lipolysis
13. renin-angiotensin system

Choose the Correct Answer

14. Dehydroepiandrosterone is primarily secreted from which region of the adrenal gland?

    a. zona fasciculata

    b. zona glomerulosa

    c. zona reticularis

    d. medulla

15. Of the following, when would plasma cortisol levels be highest?

    a. at the onset of deep sleep

    b. just prior to awakening

    c. when ACTH levels are lowest

    d. right after a meal, when blood glucose levels are high

16. Cortisol has a number of effects on carbohydrate, protein, and fat metabolism. Cortisol:

    a. decreases gluconeogenesis in liver

    b. increases glucose uptake into adipose tissue

    c. decreases urea production in liver

    d. decreases protein synthesis in muscle

17. Angiotensin converting enzyme is responsible for the conversion of:

    a. angiotensinogen into angiotensin I

    b. angiotensin I into angiotensin II

    c. renin into angiotensin I

    d. angiotensin II into aldosterone

18. Effects of epinephrine on the cardiovascular system include:

    a. reduced strength of cardiac contractions

    b. increased heart rate

    c. redistribution of blood flow to internal organs

    d. vasoconstriction in skeletal muscle

19. Which of the following is *not* a metabolic effect of epinephrine?

    a. increased glycogenolysis in liver

    b. decreased insulin secretion

    c. increased lipolysis in adipose tissue

    d. increased gluconeogenesis

20. Which of the following "hormone class: hormone" pairs is *incorrectly* matched?

    a. glucocorticoid:cortisol

    b. catecholamine:epinephrine

    c. mineralocorticoid:dehydroepiandrosterone

    d. all of the above pairs are correctly matched

Answer in One or Two Sentences

21. What is meant by the terms "tropic" and "steroidogenic" effects of ACTH?

22. How does the hypothalamic-pituitary-adrenal axis function?

23. What is the physiological basis for the excess pigmentation seen in some individuals with Addison's disease?

24. What is the relationship of the sympathetic nervous sysem and the adrenal medulla?

25. How do the characteristics of Cushing's syndrome relate to the physiological effects of cortisol?

SUGGESTED READINGS

Bolander, Franklyn F. *Molecular Endocrinology*. San Diego: Academic Press, 1989.

Evans, Ronald M. "The steroid and thyroid hormone receptor superfamily." *Science*, 240:889–895, 1988.

Fregley, Melvin J., and Luttge, William G. *Human Endocrinology: An Interactive Text*. New York: Elsevier Biomedical, 1982.

Ganong, William F. *Review of Medical Physiology*, 12th ed. Los Altos: Lange Medical Publications, 1985.

Hedge, George A., Colby, Howard D., and Goodman, Robert J. *Clinical Endocrine Physiology*. Philadelphia: W.B. Saunders Company, 1987.

Norman, Anthony W., and Litwack, Gerald. *Hormones*. Orlando: Academic Press, 1987.

Tepperman, Jay, and Tepperman, Helen M. *Metabolic and Endocrine Physiology*, 5th ed. Chicago: Year Book Medical Publishers, 1987.

Walters, Marian R. "Steroid hormone receptors and the nucleus." *Endocrine Reviews*, 6:512–543, 1985.

Wilson, Jean D., and Foster, Daniel W., eds. *Williams Textbook of Endocrinology*, 7th ed. Philadelphia: W.B. Saunders Company, 1985.

KEY TERMS

Addison's disease (p. 493)

angiotensinogen (p. 490)

corticosteroid-binding globulin (CBG) (p. 487)

Cushing's syndrome (p. 491)

hypothalamic-pituitary-adrenal axis (p. 482)

lipolysis (p. 496)

renin-angiotension system (p. 490)

sympathoadrenal system (p. 494)

tyrosine hydroxylase (p. 494)

urea cycle (p. 488)

zona fasciculata (p. 481)

zona glomerulosa (p. 481)

zona reticularis (p. 481)

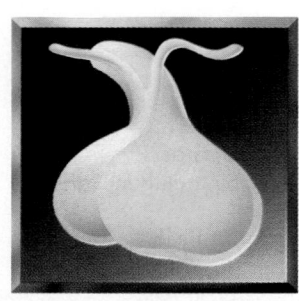

CHAPTER
15

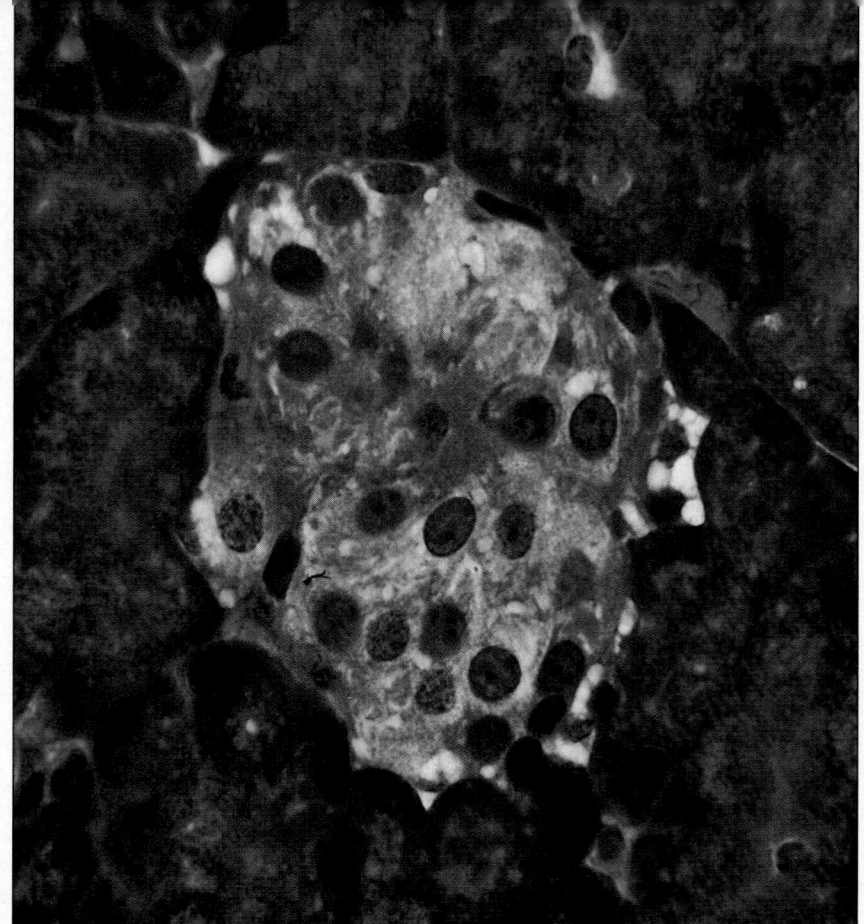

The Endocrine Pancreas

The pancreas is located in the abdominal cavity adjacent to the upper part of the small intestine, as shown in Figure 15–1. The pancreas serves two functions, which are carried out by different groups of cells within the organ. These two groups of cells are designated as the **exocrine** and **endocrine** portions of the pancreas.

The **exocrine pancreas** is responsible for the secretion of fluid and various enzymes involved in the process of digestion. These secretions of the exocrine pancreas empty into the upper part of the small intestine. The regulation and function of the exocrine pancreas will be described in Chapter 23.

The **endocrine pancreas,** the subject of this chapter, represents an anatomically small part of the pancreas. It secretes into the blood two hormones, insulin and glucagon, that are important regulators of energy metabolism and fuel homeostasis in the body. This chapter describes the physiology of those hormones. Although the exocrine and endocrine portions of the pancreas are usually discussed in separate chapters in physiology textbooks, as they will be here, bear in mind that the pancreas as a whole serves functions related to the overall process of digestion, uptake, and use of metabolic fuels.

The Islets of Langerhans: Functional Units of the Endocrine Pancreas

The endocrine pancreas consists of groups of cells known as the **islets of Langerhans.** Figure 15–2 shows a cross-section of the pancreas and the relationship of the islets to the exocrine portions of the gland. The islets of Langerhans are small groups of cells interspersed within the cells of the exocrine pancreas. The islets comprise only about 1% to 2% of the total mass of the pancreas, although the average human pancreas has about one million islets. Each islet is richly supplied with blood vessels, and the hormones the islet cells secrete enter these blood vessels directly. In contrast, the cells of the exocrine pancreas are arranged to form blind-ended sacs and secrete their products into a duct system that eventually empties into the small intestine.

The islets are composed of four major cell types, listed in Table 15–1. Each cell type synthesizes and secretes a different hormone. As later sections will describe, **glucagon** and **insulin,** which are produced

Figure 15–1
Location of the pancreas in the abdominal cavity.

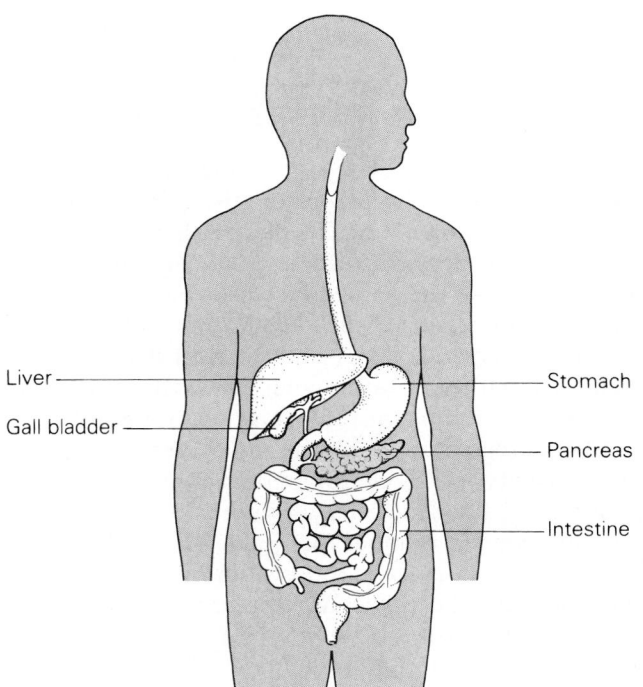

Liver
Gall bladder
Stomach
Pancreas
Intestine

Table 15–1		
Major Cell Types of the Islets of Langerhans and the Hormones They Produce		
Name	*Hormone Produced*	*Percentage of Total Islet**
Alpha cell	Glucagon	25
Beta cell	Insulin	60
Delta cell	Somatostatin	10
F cell	Pancreatic polypeptide	1

*The remaining 4% consists of connective tissue and blood vessels.

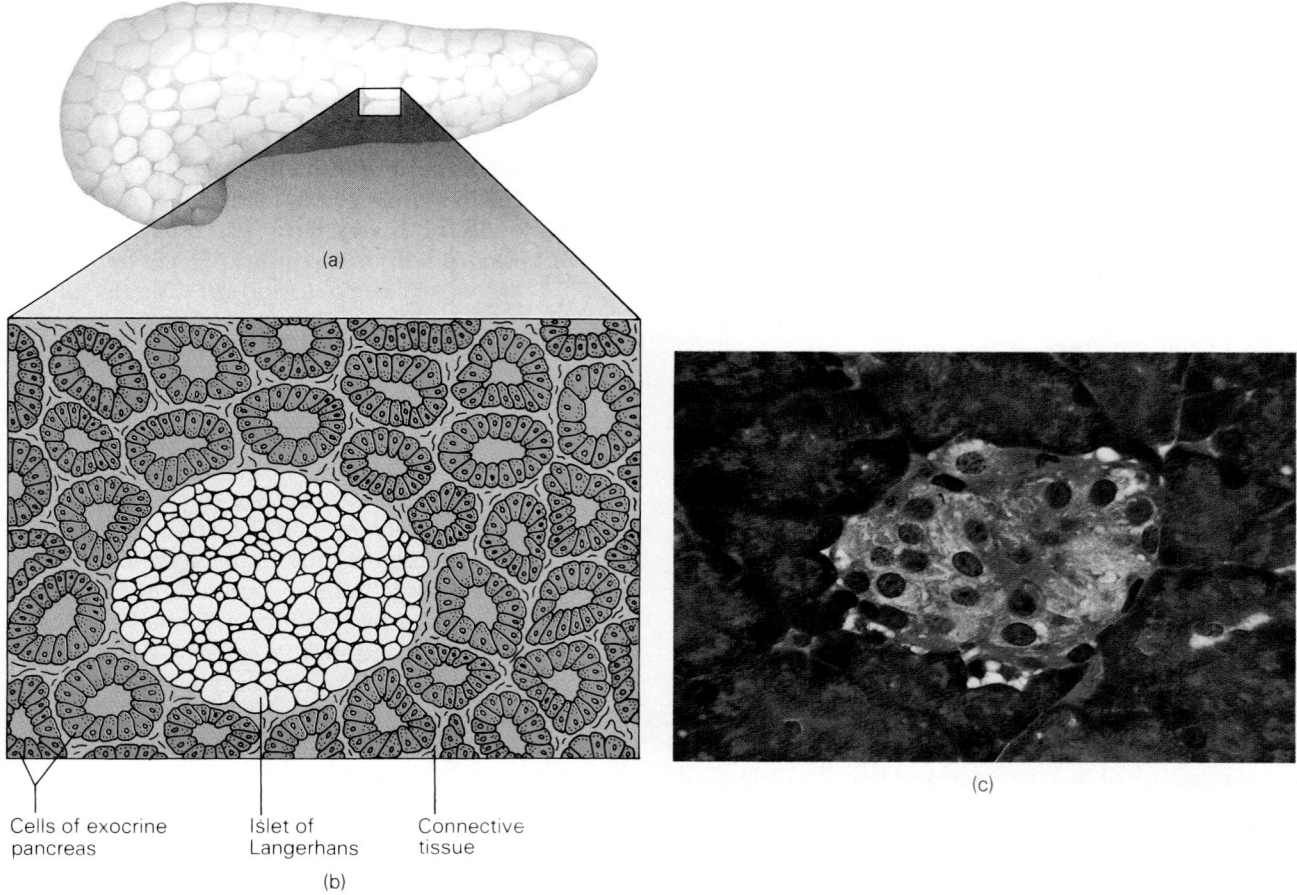

(a)

Cells of exocrine
pancreas

Islet of
Langerhans

Connective
tissue

(b)

(c)

Figure 15–2
(*a*) Relationship of the islets of Langerhans to the (*b*) exocrine cells of the
pancreas. (*c*) Photomicrograph of an islet of Langerhans. (© David M.
Phillips/Visuals Unlimited.)

in the **alpha cells** and **beta cells,** respectively, are
important hormones involved in the regulation of
blood glucose concentrations and fuel homeostasis.
The **somatostatin** produced by **delta cells** in the islets is identical to the somatostatin produced in the
hypothalamus. Recall that hypothalamic somatostatin functions as a release inhibiting hormone,
regulating secretion of growth hormone from the
anterior pituitary (Chapter 13). In the pancreas,
somatostatin may act to regulate hormone secretion
by the alpha and beta cells. The physiological function of **pancreatic polypeptide,** produced by the **F
cells,** has not been fully characterized at this time.

Each individual islet shows an orderly arrangement of the different cell types. As Figure 15–3 illustrates, the insulin-secreting beta cells tend to be located more toward the center of the islet and are
generally the most numerous cells in the islet (see

also Table 15–1). The less numerous glucagon-
secreting alpha cells are located toward the edges of
the islet, forming a rim. The delta cells that produce
somatostatin are scattered in between this rim of
alpha cells and the core of beta cells (see Fig. 15–3).
As will be described later, somatostatin can inhibit
hormone secretion from both alpha cells and beta
cells. Therefore, the arrangement of cells within the
islet as shown in Figure 15–3 may be important with
regard to regulating hormone secretion from the different cell types. F cells show roughly the same distribution as the delta cells, although they are considerably fewer in number (see Table 15–1).

Islets also receive innervation from both the
sympathetic and **parasympathetic** divisions of the
autonomic nervous system (ANS). The importance
of these neural inputs in regulating hormone secretion will be described in later sections.

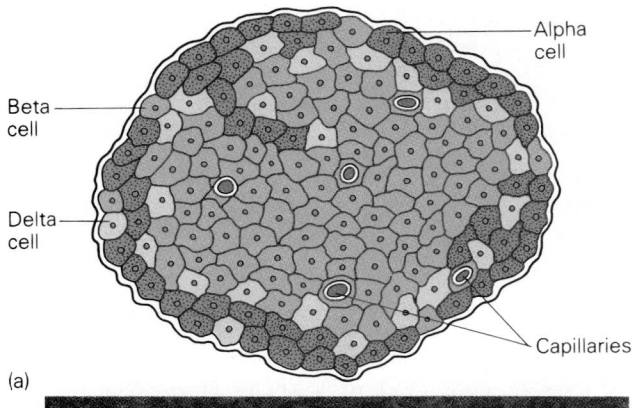

(a)

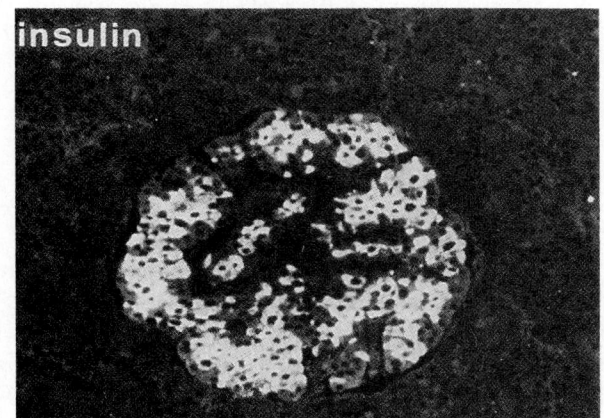

(b)

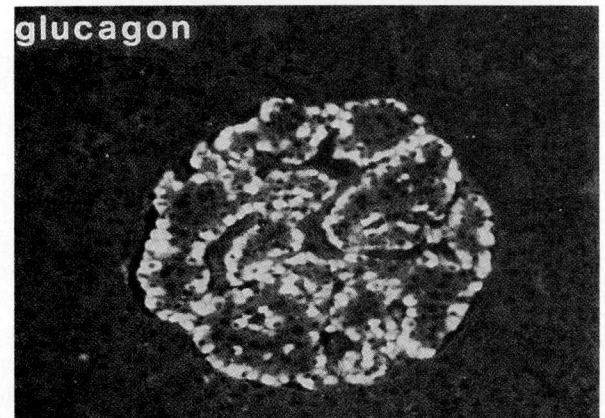

(c)

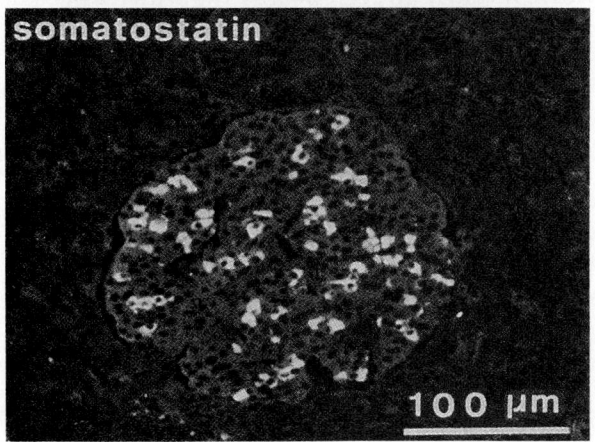

(d)

The Synthesis and Secretion of Pancreatic Hormones

As mentioned previously, the islets of Langerhans secrete three hormones whose actions have been well studied: **insulin,** produced by beta cells; **glucagon,** produced by alpha cells; and **somatostatin,** produced by delta cells. Insulin is secreted in response to a number of factors, most importantly increased blood glucose. It promotes the uptake of glucose from the blood into cells, thereby decreasing blood glucose. Conversely, glucagon is secreted when blood glucose levels are low; it acts to raise them. Somatostatin is secreted in response to several factors as well, including increased blood glucose and glucagon levels.

Insulin: A Response to Increased Glucose in the Blood

As indicated in Figure 15–4, insulin is a peptide hormone that consists of an A-chain and a B-chain held together by two disulfide (S—S) bonds. In addition, there is a third disulfide bond within the A chain. These disulfide linkages are essential for the biological activity of the hormone.

Synthesis of Insulin by Beta Cells of the Pancreas

Insulin is derived from a larger precursor molecule known as **proinsulin.** Proinsulin is first synthesized in the rough endoplasmic reticulum of the beta cells. As Figure 15–5 shows, proinsulin consists of a single peptide chain that contains both the A and B chains of insulin linked by a third connecting peptide segment. As the hormone is packaged into secretory vesicles within the beta cell, proinsulin is converted into insulin by proteolytic enzymes that clip the peptide chain in two places. The remaining peptide segment is known as **C-peptide** (Fig. 15–5).

Figure 15–3
(*a*) A typical islet of Langerhans, showing the anatomical relationship between the major hormone-producing cell types. (*b-d*) Consecutive serial sections from an islet of a normal human pancreas. Tissue sections were treated with antibodies to either insulin, glucagon, or somatostatin to determine the location of insulin-secreting beta cells (*b*), glucagon-secreting alpha cells (*c*), or somatostatin-secreting delta cells (*d*) by the technique of immunofluoresence microscopy. (Photos by L. Orci from *Williams Textbook of Endocrinology,* Seventh Edition, by Wilson and Foster, W.B. Saunders, 1985.)

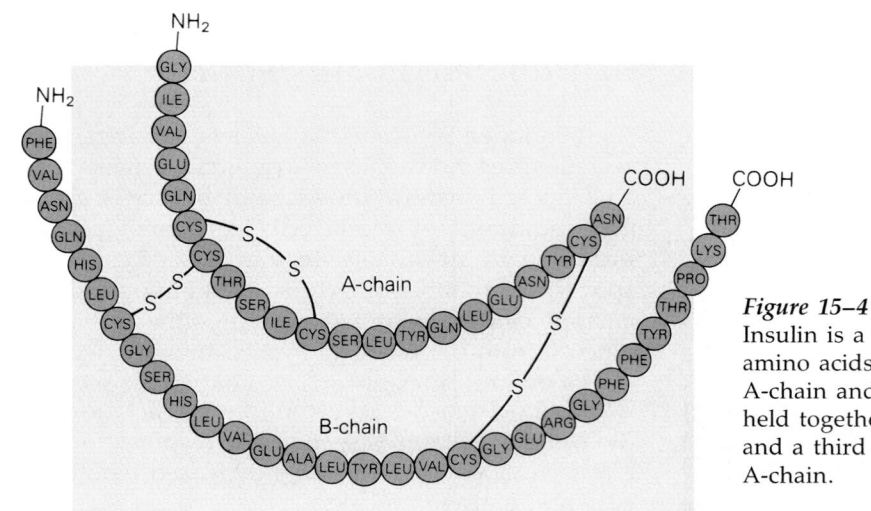

Figure 15–4
Insulin is a two-chain peptide consisting of 51 amino acids. There are 21 amino acids in the A-chain and 30 in the B-chain. The two chains are held together by a pair of disulfide (S—S) bonds, and a third disulfide bond is present in the A-chain.

When the beta cells secrete insulin into the blood, they also secrete an equal amount of C-peptide. However, C-peptide does not appear to have any biological function other than its role in insulin synthesis. In this regard, the initial synthesis of insulin as proinsulin appears to be necessary for the disulfide bonds in the hormone to form properly. Once these have formed, the C-peptide can be removed to yield the hormonally active insulin molecule. Under certain conditions, small amounts of proinsulin may be secreted from the beta cells. However, proinsulin is only about 1% as potent as insulin and thus appears to be of little or no physiological significance in the blood.

Secretion of Insulin in Response to Various Stimuli

Several factors cause the beta cells of the endocrine pancreas to secrete insulin (Table 15–2), including increased blood glucose, amino acids, fatty acids, gastrointestinal hormones, neural and pharmacological stimuli, and other hormones secreted by the pancreas. The exact mechanism by which each of the factors listed in Table 15–2 is able to influence insulin secretion is not known, although cAMP, Ca^{2+}, or both, appear to serve as intracellular signals for insulin secretion from the beta cells. When insulin secretion is stimulated, insulin-containing vesicles move toward the plasma membrane of the beta cell by a microtubule-microfilament system. The membrane of the vesicle fuses with the plasma membrane, and the contents of the vesicle (insulin and C-peptide) are released by exocytosis. In addition to stimulating the release of preformed insulin, factors such as glucose that stimulate insulin secre-

tion also increase proinsulin synthesis and the conversion of proinsulin to insulin.

After release from the beta cells into the surrounding capillary network, most of the insulin circulates free in the blood. A small amount is loosely associated with plasma proteins such as albumin, but the significance, if any, of this association, is not known.

Increased Blood Glucose. The most important physiological stimulus for insulin secretion from the pancreas is an increased concentration of glucose in the blood. Figure 15–6 shows the average changes in plasma concentrations of glucose and insulin in a group of normal individuals who consumed 75 grams of glucose (at the time indicated by the arrow) following an overnight fast. Note that with a normal overnight fasting blood glucose concentration of 80 to 90 mg/100 ml (prior to ingestion of the glucose), insulin secretion is low. As the plasma glucose concentration increases to above about 100 mg/100 ml, however, insulin secretion by the pancreas is stimulated considerably. Notice in Figure 15–6 that after glucose ingestion, the concentration of glucose in plasma rises rapidly and is closely matched in time by an increase in the concentration of circulating insulin. In this case, the plasma insulin concentration increases to a peak value approximately 10 times that observed before glucose ingestion. With administration of larger amounts of glucose, the blood glucose concentration can be driven even

Figure 15–5
Proinsulin is converted into insulin and C-peptide within the beta cell by proteolytic removal of two amino acids from each of the two sites indicated.

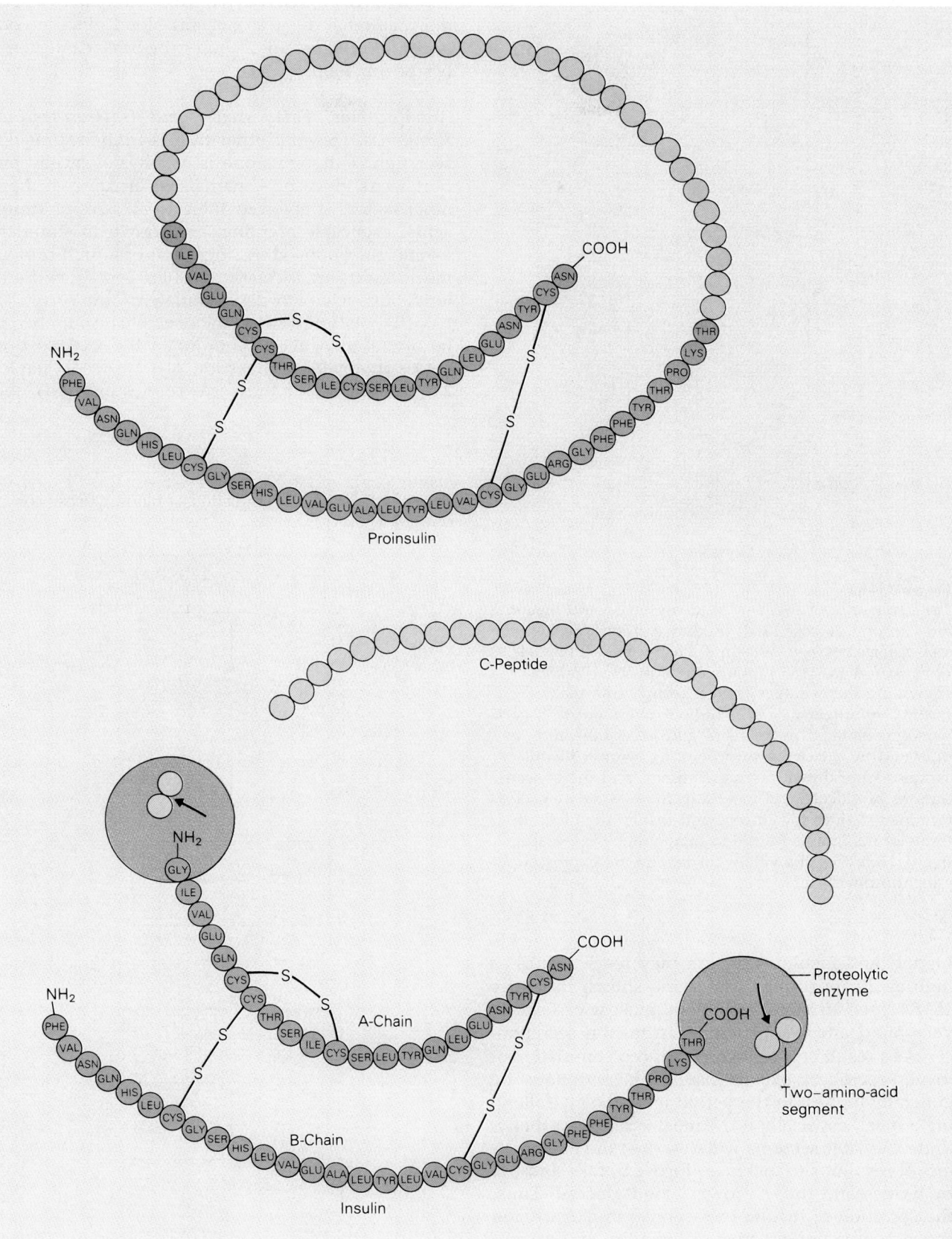

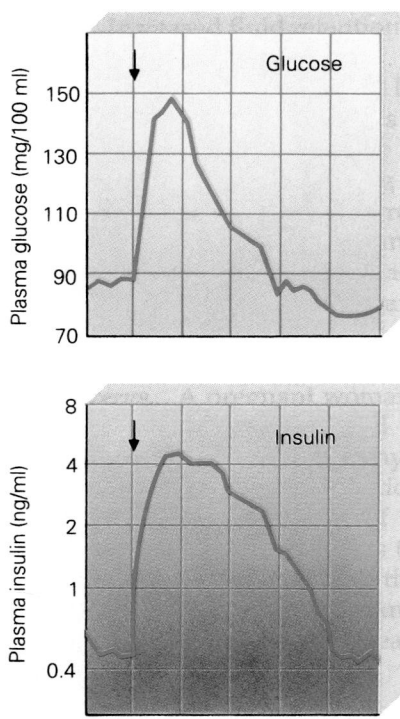

Figure 15–6
Time course of changes in plasma glucose and insulin after an oral glucose load. (Redrawn from *Williams Textbook of Endocrinology*, Wilson & Foster, eds. [Philadelphia: Saunders, 1985.] Figure 25–8, p. 996.) Values shown are the averages from a group of normal, healthy volunteers. After a 12-hour overnight fast, each person consumed 75 grams of glucose at time zero, as indicated by the downward-pointing arrows. Blood samples were drawn at various times, and the concentrations of glucose and insulin in the plasma were determined. Values for glucose and insulin are given in terms of milligrams (mg) and nanograms (ng), respectively. Note that the values for insulin are expressed as a logarithmic scale.

higher, and insulin secretion may reach values as high as 30 times the basal value shown in Figure 15–6. Thus, an increase in blood glucose can elicit a very rapid and large increase in insulin secretion.

As a result of the glucose-induced stimulation of insulin secretion, plasma insulin concentrations are generally highest in the period immediately following a meal, especially if the meal is rich in carbohydrate. As later sections will describe, the overall effect of insulin is to increase glucose uptake and use by tissues and thus to lower blood glucose. Thus, the secretion of insulin in response to glucose and the accompanying action of insulin to decrease

blood glucose are part of a feedback loop involved in the maintenance of a constant blood glucose concentration. Figure 15–7 shows the basic characteristics of this loop.

Amino Acids, Fatty Acids, and Gastrointestinal Hormones. Several other factors influence insulin secretion, although none is as physiologically important as glucose. A partial list of some of these other factors is given in Table 15–2. Several **amino acids,** especially arginine, are known to stimulate insulin secretion. Thus, ingestion of a high-protein meal results in increased insulin secretion. Similarly, fatty acids can also stimulate insulin secretion.

The gastrointestinal tract secretes a number of hormones that act to coordinate the various processes involved in digestion of food (see Chapter 23). A number of these gastrointestinal (GI) hor-

Figure 15–7
Role of insulin in the regulation of blood glucose concentration.

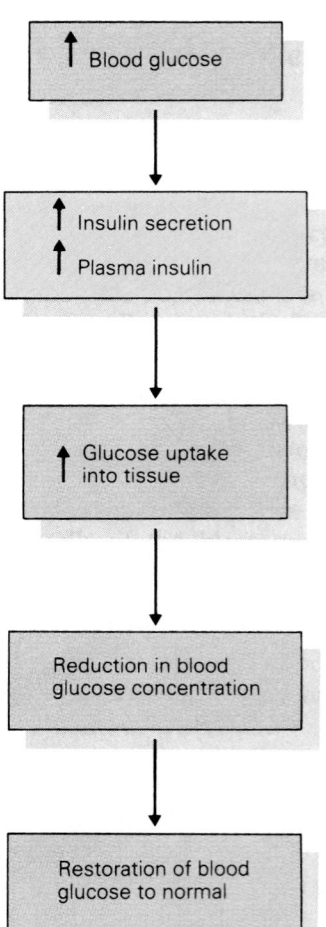

Table 15–2

Factors Affecting Insulin Secretion from the Pancreas

Stimulators	Inhibitors
Increased blood glucose	Somatostatin
Amino acids	Epinephrine
Fatty acids	Norepinephrine
Gastrointestinal hormones (gastrin, secretin)	
Acetylcholine	
Sulfonylureas	
Glucagon	

mones, including gastrin and secretin, also stimulate insulin secretion by the pancreas (Table 15–2). It is well known that the administration route of glucose influences the amount of insulin secreted by the pancreas. If glucose is given orally, a greater increase in insulin secretion occurs as compared to when the same amount of glucose is given intravenously. This difference is due to a stimulatory effect of GI hormones on insulin secretion. As Chapter 23 will describe, GI hormones are generally secreted in response to the presence of food in the GI tract. Soon after food enters the GI tract, the blood glucose concentration is expected to rise. The effect that the GI hormones have of stimulating insulin secretion therefore provides an anticipatory signal to the beta cells that the blood glucose concentration is likely to increase in the near future.

Parasympathetic Signals. Neural inputs to the islets appear to play a role in regulating insulin secretion. Note in Table 15–2 that acetylcholine stimulates insulin secretion. Recall from Chapter 10 that acetylcholine serves as the neurotransmitter for postganglionic fibers in the parasympathetic division of the autonomic nervous system. Activation of parasympathetic neurons to the islets, as might be expected to occur during digestion, also results in increased insulin secretion. This, too, provides an anticipatory signal to the pancreas that an increase in blood glucose is likely to occur. Table 15–2 also indicates that epinephrine and norepinephrine inhibit insulin secretion. Activation of sympathetic fibers to the islets or release of epinephrine from the adrenal medulla, as might occur in response to stress, results in an inhibition of insulin secretion.

The significance of the stress-induced inhibition of insulin secretion was discussed earlier in Chapter 14.

Drugs and Other Islet Hormones. A class of drugs known as **sulfonylureas** also promote insulin secretion from the beta cells (see Table 15–2). These drugs can be taken orally and are widely used in the treatment of some forms of diabetes mellitus, as will be discussed in a later section of this chapter.

Table 15–2 also indicates that two other islet hormones, glucagon and somatostatin, influence insulin secretion. Glucagon is known to augment glucose-stimulated insulin secretion. By contrast, somatostatin inhibits insulin secretion. The physiological significance of these effects is not clear, but they suggest that the alpha cells and the delta cells exert local regulatory influences on insulin secretion from the beta cells within each islet.

Glucagon: A Response to Decreased Glucose in the Blood

Synthesis of Glucagon by Alpha Cells of the Pancreas

Glucagon is a simple straight-chain polypeptide consisting of 29 amino acids. It is synthesized in the rough endoplasmic reticulum of the alpha cells of the islets, and, as is true of many other peptide hormones, it is first synthesized as part of a larger precursor molecule that is later converted into the active hormone. In this case, the immediate precursor, **proglucagon,** is about five times the size of glucagon itself. Interestingly, proglucagon is also synthesized in certain cells of the small intestine and in the hypothalamus. However, in these two tissues, proglucagon is not converted into glucagon. Instead, several other **glucagon-like peptides** are formed. The function of these peptides and the regulation of their production are not well understood at this time. It is clear, however, that the regulation of secretion of the glucagon-like peptides differs from that of glucagon secretion by the alpha cells of the pancreas.

Secretion of Glucagon in Response to Various Stimuli

Decreased Blood Glucose. As with insulin, the major physiological regulator of glucagon secretion by the alpha cells of the pancreas is the blood glucose concentration. However, in direct contrast to

Table 15–3

Factors Affecting Glucagon Secretion from the Pancreas

Stimulators	Inhibitors
Low blood glucose	Fatty acids
Amino acids	Somatostatin
Acetylcholine	Insulin
Norepinephrine	
Epinephrine	

insulin, a *decrease* in blood glucose concentration stimulates glucagon secretion (Table 15–3). Therefore, circulating glucagon concentrations are highest during periods of fasting, when the blood glucose levels tend to be lowest. Conversely, an increased blood glucose concentration, such as might occur after a meal, tends to inhibit glucagon secretion.

Thus, alterations in blood glucose either above or below normal result in opposite changes in insulin and glucagon secretion. Insulin and glucagon have opposite effects on metabolism, lowering and raising blood glucose, respectively. It is not surprising, therefore, that the regulation of secretion of these two hormones by glucose involves a reciprocal relationship.

Amino Acids, Fatty Acids, and the Autonomic Nervous System. In addition to glucose, a number of the substances that regulate insulin secretion also influence glucagon secretion (see Table 15–3). Amino acids also stimulate glucagon secretion; as with insulin, arginine is the most potent stimulus among these. As a result of the stimulation of insulin and glucagon secretion by amino acids, the plasma concentration of both hormones increases after a high-protein meal. The parallel increases in both insulin and glucagon secretion in response to amino acids serve as a protective mechanism to ensure that blood glucose levels are maintained following ingestion of meals rich in protein but low in carbohydrate, as will be discussed in later sections describing the actions of these two hormones.

Fatty acids also affect glucagon secretion. An increase in circulating fatty acids inhibits glucagon secretion, whereas a decrease in fatty acid concentrations increases glucagon secretion from the pancreas (see Table 15–3).

In addition, the ANS also has a role in regulating glucagon secretion from the alpha cells of the islets. In this case, both the sympathetic and parasympathetic divisions promote glucagon secretion.

Insulin. Insulin also affects glucagon secretion from the alpha cells (see Table 15–3). Normally, high blood glucose tends to inhibit glucagon secretion, but this effect depends on the presence of insulin. In the absence of insulin, the alpha cell cannot detect the elevated blood glucose, and glucagon secretion continues at a high rate. This fact is important in those forms of diabetes mellitus in which the insulin-secreting beta cells are destroyed. Not only is insulin secretion deficient; glucagon secretion is *inappropriately* high due to the absence of the inhibitory effect of insulin on glucagon secretion.

Somatostatin: A Response to Increased Glucose and Glucagon in the Blood

As described in Chapter 13, somatostatin is a small peptide hormone. As was the case for insulin and glucagon, somatostatin is also first synthesized as part of a larger precursor molecule. The immediate precursor, **prosomatostatin,** is then converted into the smaller somatostatin peptide. Factors regulating somatostatin secretion from the delta cells are similar to those regulating insulin and glucagon secretion. The primary stimuli of somatostatin secretion are (1) increased blood glucose, (2) increased plasma glucagon, and (3) amino acids.

As described previously and indicated in Tables 15–2 and 15–3, secretion of insulin, glucagon, and somatostatin by the islet cells is interrelated in that each is influenced to some degree by the other islet hormones. These relationships are summarized in Figure 15–8. As indicated earlier, the anatomical arrangement of cells within the islet and the possible paracrine effects of each hormone on the secretion of the other islet hormones (Fig. 15–8) suggest that these factors may be important in the fine tuning of islet cell secretions.

The Metabolic Effects of Pancreatic Hormones

Even while at rest or sleeping, the body is continually using energy to drive such vital processes as ion transport, the synthesis of various cellular proteins, and the mechanical activity involved in respiration

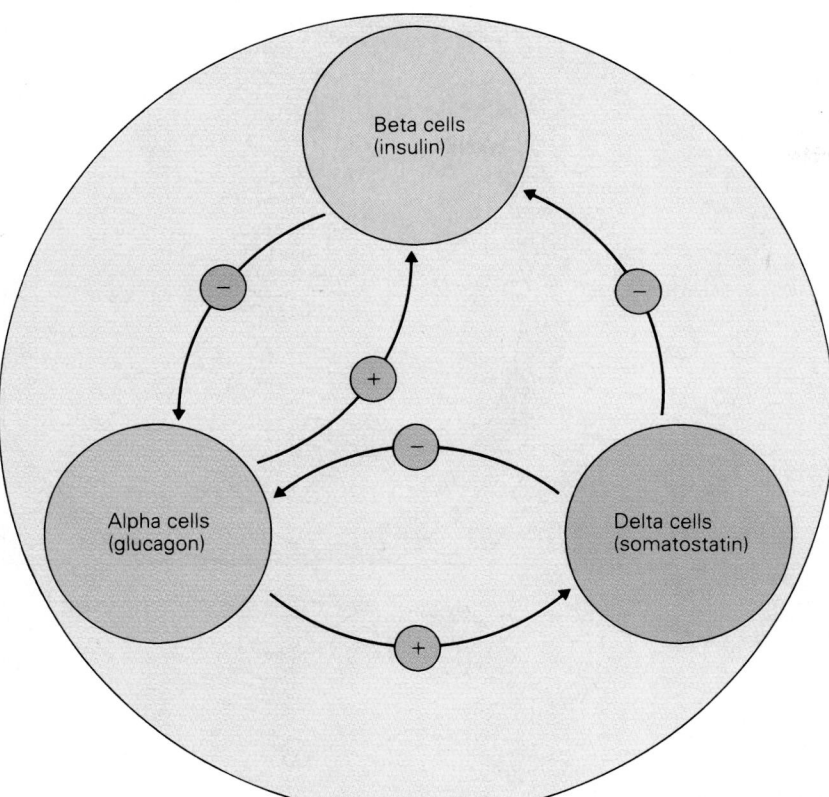

Figure 15–8
Possible paracrine interactions of hormones in the islets of Langerhans.

or cardiac contractions. Physical activity increases energy requirements above those in the basal (resting) state. However, although energy use is continuous, the intake of energy in the form of food is intermittent. Thus, excess fuels taken in with a meal must be stored for subsequent use between meals. As a result, mechanisms that coordinate and regulate fuel homeostasis have evolved. The following sections discuss the actions of insulin and glucagon, the two primary hormones involved in regulating fuel homeostasis in the body.

As described earlier, alterations in the blood glucose concentration provide the primary signals regulating insulin and glucagon secretion from the pancreas. Glucagon and insulin have reciprocal effects that tend to raise and lower the blood glucose concentration, respectively. Thus, these two hormones serve as key components of feedback loops whereby the blood glucose concentration is maintained within fairly narrow limits. As we have learned in previous chapters, other hormones such as cortisol, growth hormone (GH), and epinephrine also have effects that influence the blood glucose concentration. An appropriate question to ask at this point might be, "What is the physiological basis

for this tight regulation of blood glucose concentrations?"

The answer to that question lies in the fact that the central nervous system (CNS) relies almost solely on glucose for its metabolic needs. Although other tissues, such as muscle and liver, can use alternate substrates for fuel, the nervous system ordinarily must derive all of its energy from glucose. Thus, if the blood glucose concentration were to fall to low levels, the brain could not obtain energy, and death would soon follow. Conversely, although a marked elevation of blood glucose would not produce such immediate deleterious effects, it could lead to wasteful loss of glucose in the urine as well as concurrent losses of large volumes of water. In a normal individual, the blood glucose concentration following an overnight fast is usually in the range of 75 to 115 mg of glucose per 100 ml of blood. Following ingestion of a large amount of glucose, such as with a high-carbohydrate meal, the concentration may reach as high as 200 mg glucose per 100 ml of blood. However, as shown in Figure 15–6, in a normal individual the blood glucose concentration quickly returns toward the basal level, owing to the secretion of insulin and its effects on glucose use.

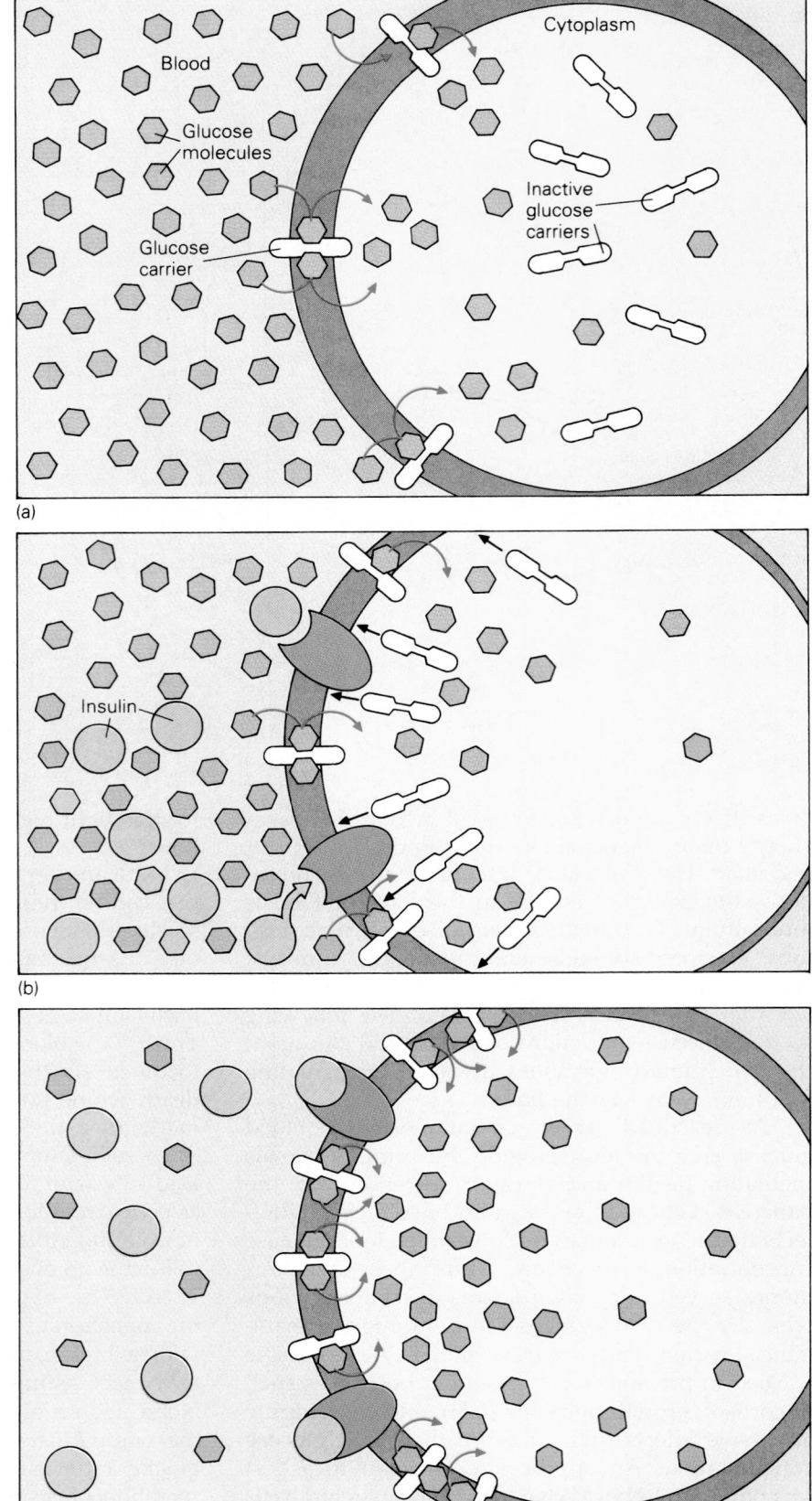

Figure 15–9
Illustration of the effect of insulin to stimulate glucose uptake into cells. This figure depicts the situations that might exist in the absence (*a*) or presence (*b*) of insulin. Note that insulin increases glucose uptake by increasing the number of glucose transporters in the cell membrane (*c*).

The Effects of Insulin on Metabolism

Insulin has been termed the **hormone of nutrient abundance.** When influx of nutrients is high, insulin provides the signal that directs the storage of excess fuels while at the same time it suppresses the mobilization of pre-existing fuel stores. Insulin therefore has **anabolic** effects.

The specific cellular mechanism by which insulin produces its effects has yet to be determined, despite considerable research efforts in this area in the past several years. The insulin receptor is located in the plasma membrane, as might be expected, since insulin is a peptide hormone. Although the structure of the receptor has been characterized biochemically, it is not currently known how the binding of insulin to its membrane receptor is transduced into the intracellular events that the hormone produces. Research has shown that the insulin receptor itself has protein kinase activity, and thus insulin may produce some of its effects by stimulating its receptor to phosphorylate specific regulatory proteins within its target cells. Thus, insulin may act via a mechanism similar to the one depicted earlier in Figure 12–19.

The major target tissues of insulin are skeletal muscle, adipose tissue (fat), and liver.

The Effects of Insulin on Carbohydrate Metabolism

The most obvious action of insulin is that it very rapidly and effectively lowers the blood glucose concentration. Most tissues in the body depend on insulin for the uptake of glucose from the blood. In the absence of insulin, glucose does not readily enter cells and therefore cannot be metabolized and used for energy. Two primary exceptions to this rule are the brain and the liver, whose cells are readily permeable to glucose even in the absence of insulin. In tissues such as muscle and adipose, glucose enters cells by both simple diffusion and facilitated diffusion. (Recall that facilitated diffusion differs from active transport in that it is not energy-dependent and cannot move glucose against a concentration gradient. Instead, facilitated diffusion involves specific carrier proteins that "shuttle" glucose across the membrane at a faster rate than would occur by diffusion alone.) As indicated in Figure 15–9a glucose transporters are located both in the plasma membrane and in intracellular sites. Insulin promotes the movement of transporters from their intracellular location to the plasma membrane (see Fig. 15–9b). By increasing the number of transporters present in the plasma membrane, insulin is able to increase the rate of glucose uptake into cells.

In addition to stimulating glucose uptake into cells, insulin also stimulates the metabolism and use of glucose. In most cells, particularly those in liver and muscle, insulin stimulates glycogen synthesis while inhibiting glycogen breakdown (Fig. 15–10). The net result is an increase in the amount of glycogen stored in the cells. Insulin also stimulates a number of the enzymes of the glycolytic pathway. Glucose metabolism via glycolysis therefore increases in most insulin-sensitive cells. In addition to stimulating glucose uptake and use by peripheral tissues such as muscle and adipose tissue, insulin

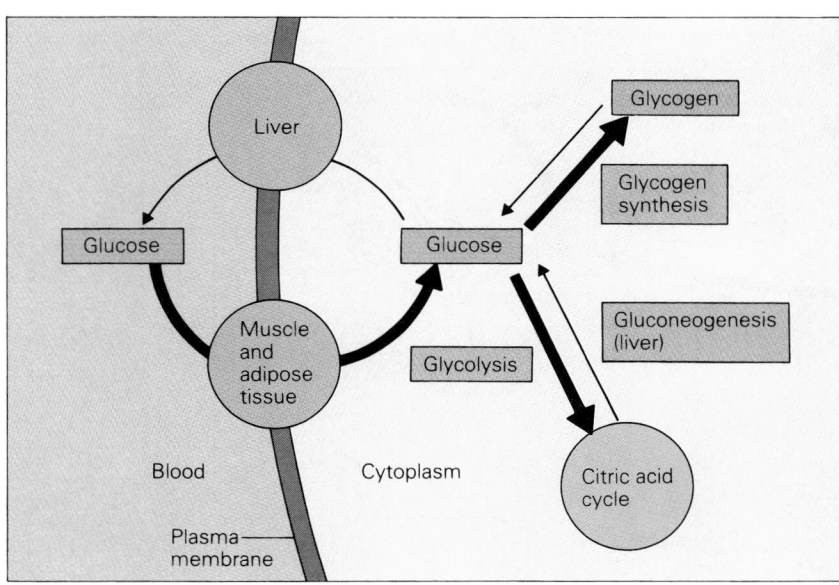

Figure 15–10
Summary of the effects of insulin on carbohydrate metabolism. Processes accelerated by insulin are shown with the heavy arrows; those slowed by insulin are shown with the light arrows.

decreases glucose output by the liver in two different ways. First, insulin decreases the activity of several key enzymes in the gluconeogenic pathway. (Recall that gluconeogenesis is the process by which nonglucose substrates such as amino acids are converted into glucose.) Second, insulin promotes the use of amino acids in peripheral tissues (see the following section on protein metabolism) and so decreases the supply of amino acids, which are normally the major substrate for gluconeogenesis in the liver.

The Effects of Insulin on Lipid Metabolism

As a result of several specific effects of insulin on lipid metabolism, the hormone promotes the storage of excess fuel as triglycerides. In both liver and adipose tissue, insulin stimulates synthesis of fatty acids. In adipose tissue, the fatty acids combine with glycerol-phosphate and are stored as triglycerides (Fig. 15–11). By contrast, the liver stores only a small amount of the fatty acids it produces. Instead, the majority of fatty acids are packaged as lipoproteins and released into the blood. The lipoproteins are taken up by adipose tissue, and the fatty acids are then converted and stored as triglycerides in the adipose tissue (see Fig. 15–11). In adipose tissue, insulin stimulates the activity of the enzyme **lipoprotein lipase,** which promotes the uptake of lipoproteins from the blood (see Fig. 15–11). The effect of insulin to stimulate glucose uptake by adipose tissue also tends to promote triglyceride storage. Inside the adipose cell, glucose can be either metabolized and used for fatty acid synthesis or used to form glycerol-phosphate. Glycerol-phosphate serves as the backbone to which fatty acids are attached to form triglycerides (see Fig. 15–11). Thus, by increasing glucose entry into the adipose cell, insulin promotes the synthesis of both the fatty acid

Figure 15–11
Summary of the effects of insulin on adipose cell metabolism. Processes accelerated by insulin are indicated with the heavy arrows; those slowed by insulin are indicated with the light arrows. The net effect of insulin is to promote triglyceride storage. (LPL = Lipoprotein lipase.)

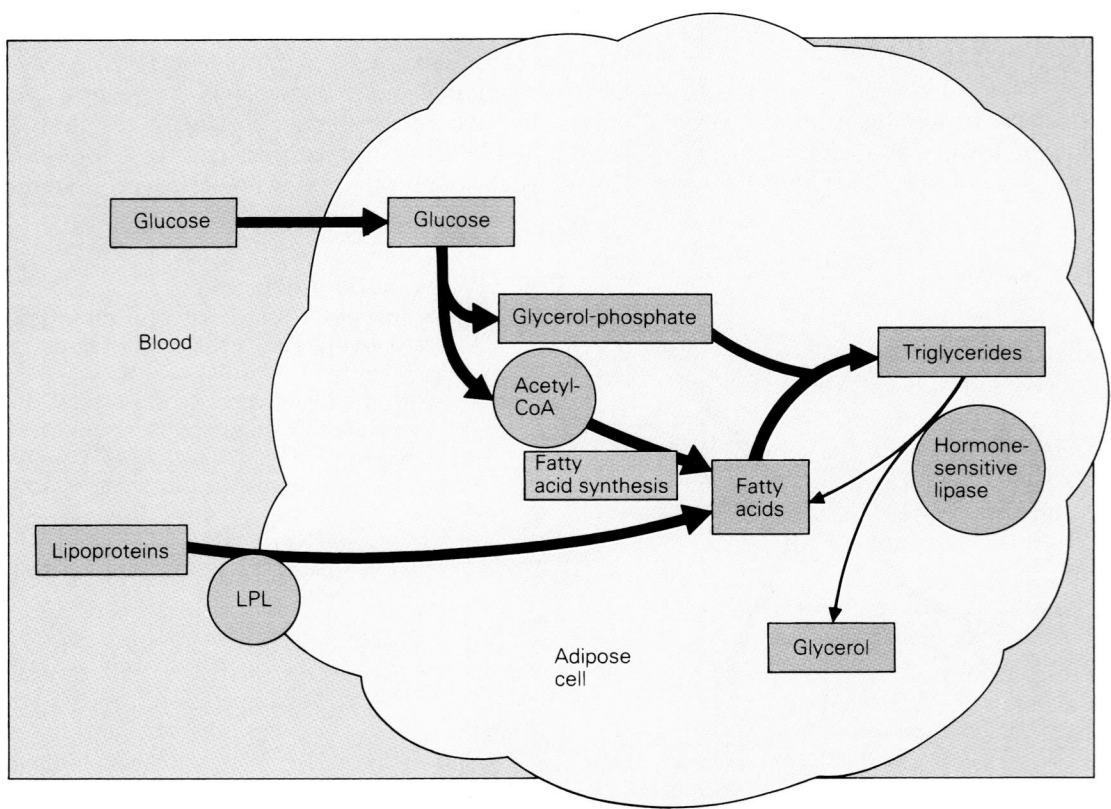

and glycerol-phosphate components of triglycerides.

In addition to its stimulatory effect on fat synthesis, insulin also inhibits lipolysis (the breakdown of triglycerides). The breakdown of triglycerides into their component fatty acids and glycerol is catalyzed by an enzyme known as **hormone-sensitive lipase** (see Fig. 15–11). Insulin inhibits the activity of this enzyme both in liver and in adipose tissue, thus decreasing lipolysis in these two tissues. Therefore, the overall effect of insulin on lipid metabolism is to promote triglyceride storage while inhibiting triglyceride breakdown.

The Effects of Insulin on Protein Metabolism

Insulin also has profound effects on protein metabolism in the body. These effects are most readily seen in skeletal muscle and liver, although the *anabolic* effects of insulin, promoting the accumulation of protein, occur in most tissues.

The effects that insulin has on muscle protein metabolism are of particular significance since approximately 40% of the total body protein is present in muscle. Therefore, much of the influence that insulin has on overall protein balance in the body results from its effects on muscle protein metabolism. In muscle, insulin stimulates the active transport of amino acids from the blood into individual muscle cells (see Fig. 15–12). As a result, more amino acids are available for muscle protein synthesis. Insulin also stimulates the process of protein synthesis in muscle (see Fig. 15–12), in part through an increased number of ribosomes in each muscle cell and in part due to increases in the activity of individual ribosomes. At the same time, insulin strongly inhibits protein degradation in muscle (see Fig. 15–12). Therefore, the overall effect of insulin in muscle is to promote the accumulation of muscle protein.

Insulin has similar effects on protein metabolism in the liver and adipose tissue; it increases protein synthesis and decreases protein degradation. In addition, as described earlier, insulin also inhibits gluconeogenesis in the liver. As a result, fewer amino acids are converted into glucose, and more amino acids are available for either muscle or liver protein synthesis.

As a result of its anabolic effects on protein metabolism, insulin produces a "positive nitrogen balance," meaning a net accumulation of protein in the body. By contrast, when insulin is deficient, as in diabetes mellitus, there is a net loss of protein, or a negative nitrogen balance. Insulin therefore serves as an important regulator of tissue growth.

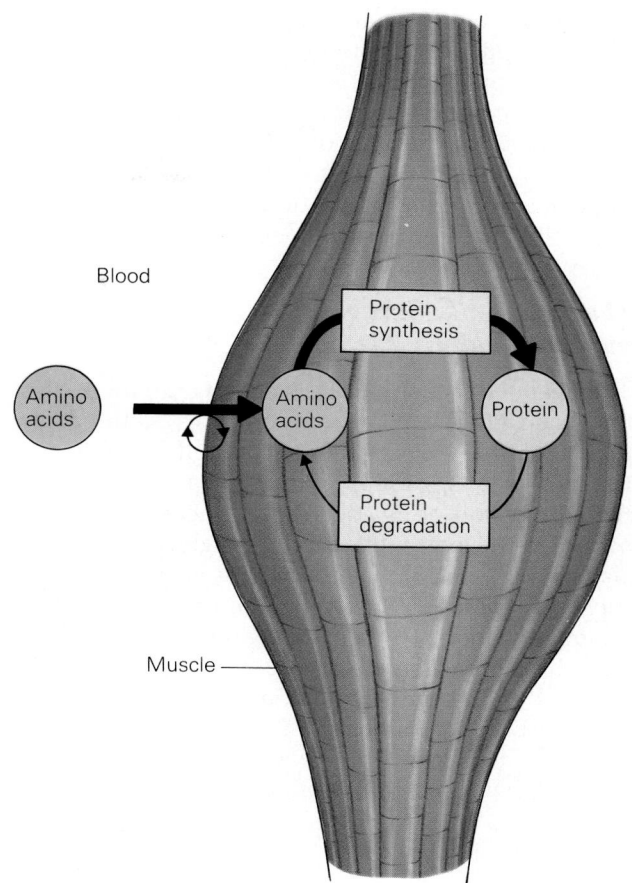

Figure 15–12
Summary of the effects of insulin on muscle protein metabolism. Processes accelerated by insulin are indicated with the heavy arrows; those slowed by insulin are indicated with the light arrows.

The Effects of Glucagon on Metabolism

As will be evident in the sections that follow, glucagon affects many of the same metabolic pathways that insulin affects. However, in each instance, the effects of glucagon are exactly opposite those of insulin.

The mechanism of glucagon action has been fairly well established. Since it is a peptide hormone, glucagon interacts with a receptor on the plasma membrane of its target cells. This receptor is coupled to adenylate cyclase and results in an increase in the concentration of cAMP inside the cell. As described earlier in Chapter 5, cAMP activates several protein kinases inside the cell, leading to changes in the phosphorylation state and activity of various regulatory proteins and enzymes.

The principal target site of glucagon is the liver, where its overall effect on metabolism is catabolic. Some effects of glucagon on adipose tissue have also been reported, but these usually require a relatively high concentration of the hormone and thus their physiological significance is doubtful.

The Effects of Glucagon on Carbohydrate Metabolism

The overall effect of glucagon action is an increase in the blood glucose concentration. This occurs as the result of three separate effects that glucagon has on the liver (see Fig. 15–13). First, glucagon stimulates the conversion of glycogen into glucose (**glycogenolysis**) in the liver while inhibiting glycogen synthesis (see Fig. 15–13). The net result is an increase in the conversion of liver glycogen into glucose, which enters the blood. The second means by which glucagon raises the blood glucose concentration is by stimulating the conversion of nonglucose substrates into glucose (**gluconeogenesis**) in the liver (see Fig. 15–13). Thirdly, as the following section will describe, glucagon affects lipid metabolism, resulting

in glucose "sparing," which tends to raise the blood glucose concentration. To illustrate the opposing effects of insulin and glucagon on carbohydrate metabolism, compare Figure 15–10 with Figure 15–13.

The Effects of Glucagon on Lipid Metabolism

The effects of glucagon on lipid metabolism also oppose those of insulin. In the liver, glucagon stimulates hormone-sensitive lipase and thereby stimulates lipolysis (Fig. 15–14). Therefore, in response to glucagon, more triglycerides break down into free fatty acids and glycerol. In addition to being oxidized and used directly for energy production in muscle, fatty acids can be partially metabolized and converted in the liver into **ketones** (also known as **ketone bodies** or **ketoacids**). The ketones, which are four-carbon compounds, are released by the liver and can be taken up and used by peripheral tissues for energy. Glucagon stimulates the formation of ketones (**ketogenesis**) in the liver. Skeletal muscle and the heart both derive a considerable amount of their energy needs from the oxidation of ketones. In fact, ketones are a major source of energy for the heart. By stimulating ketogenesis in the liver, gluca-

Figure 15–13
Summary of the effects of glucagon on liver carbohydrate metabolism. Processes stimulated by glucagon are shown with the heavy arrows; processes slowed by glucagon are shown with light arrows.

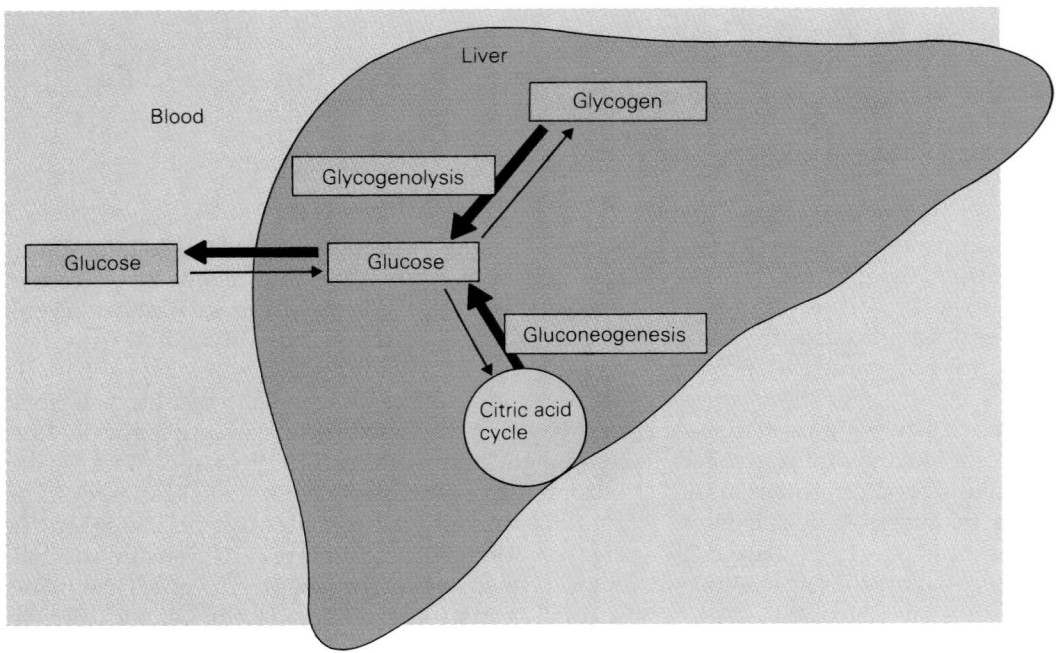

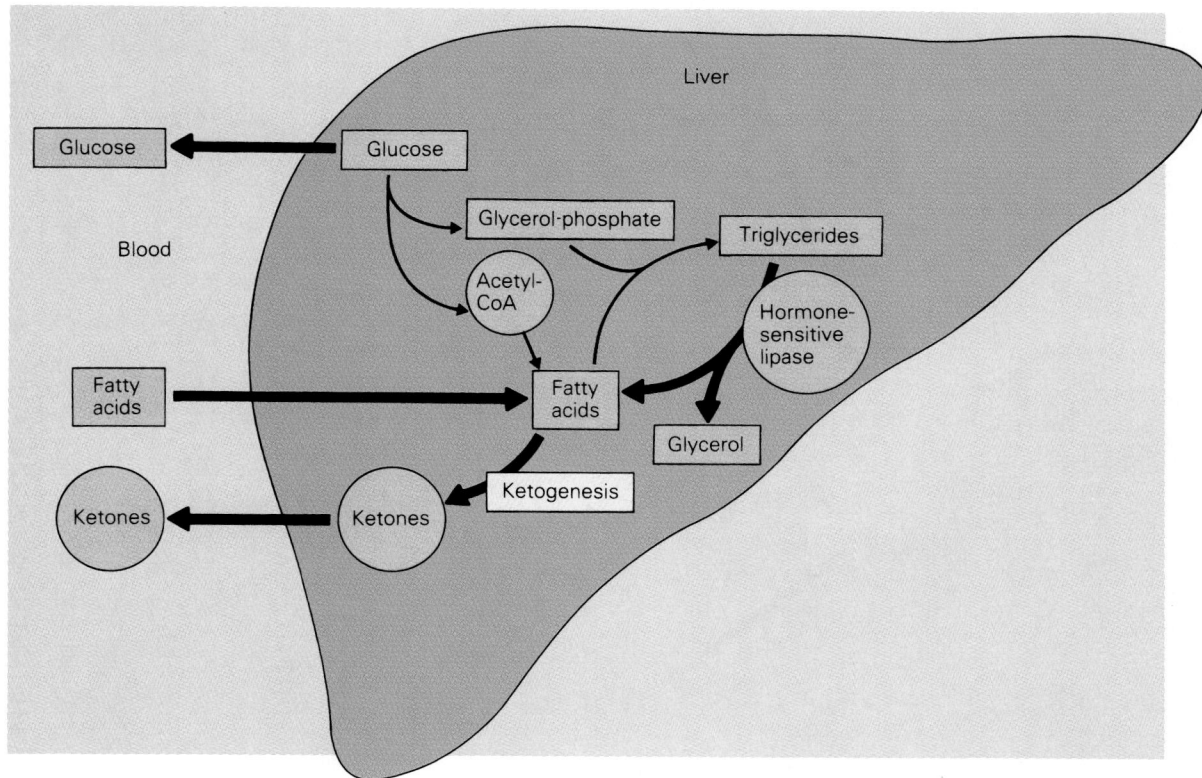

Figure 15–14
Summary of the effects of glucagon on liver fat metabolism. Processes stimulated by glucagon are shown with the heavy arrows; processes slowed by glucagon are shown with the light arrows.

gon provides increased amounts of ketones for use by muscle and heart. This has the effect of sparing glucose, which, when coupled with the other effects of glucagon, also serves to raise the blood glucose concentration.

The Effects of Glucagon on Protein Metabolism

In the liver, glucagon is a potent stimulator of protein degradation (Fig. 15–15). The free amino acids released as liver proteins are broken down and can be used for glucose synthesis via gluconeogenesis. Glucagon has an additional effect of increasing the supply of amino acids for gluconeogenesis by stimulating the transport of amino acids from the blood into liver. As more amino acids are converted into glucose, more nitrogen, a by-product of amino acid metabolism, must be disposed of. To accomplish this, glucagon also stimulates **urea synthesis** (see Fig. 15–15).

As indicated earlier, amino acids stimulate the secretion of *both* insulin and glucagon from the pancreas. In almost every case, the actions of insulin are opposite to those of glucagon. Depending on the particular metabolic state, it is desirable to have either insulin or glucagon present in high concentrations in the blood, but usually not both, since their opposing actions would cancel one another out. However, the apparent paradoxical effect of amino acids to stimulate *both* insulin and glucagon secretion serves an important purpose. When one consumes a meal high in protein, it is desirable to use the amino acids in the meal for protein synthesis in tissues, and thus insulin secretion is beneficial under these conditions. However, when the meal is also low in carbohydrate, very little glucose enters the blood from the GI tract, and the actions of insulin to decrease blood glucose would, if left alone, cause the blood glucose concentration to drop to very low levels. It is therefore beneficial for glucagon to also be secreted in this situation. The effects

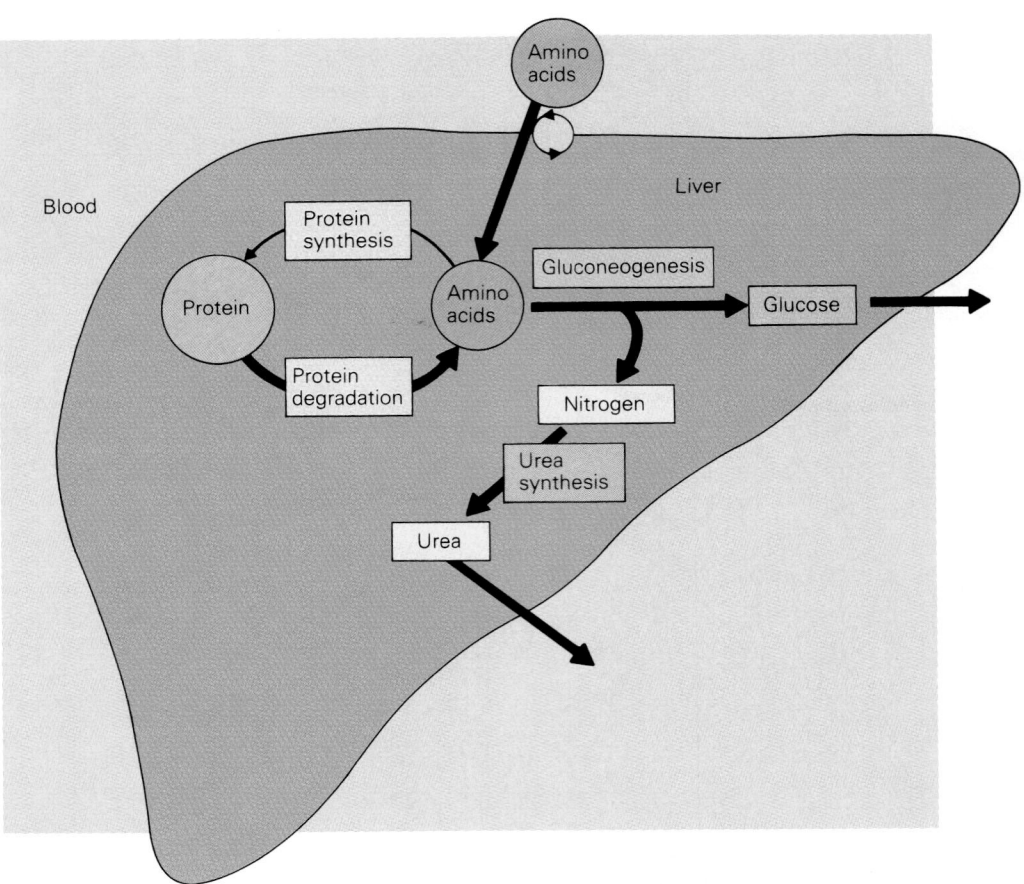

Figure 15–15
Summary of the effects of glucagon on liver protein metabolism. Processes
stimulated by glucagon are shown with the heavy arrows; processes slowed
by glucagon are shown with the light arrows.

of glucagon oppose the blood-glucose-lowering actions of insulin, and therefore the blood glucose concentration is maintained near normal, while at the same time insulin is able to promote the efficient use of amino acids for protein synthesis in various tissues.

The Importance of the Insulin/ Glucagon Ratio

Because of the opposing actions of insulin and glucagon, it has been suggested that the **insulin-to-glucagon (I/G) ratio,** rather than the absolute concentration of either hormone, is the primary factor determining the metabolic status in the body. Thus, when the I/G ratio is high (insulin high, glucagon low), the effects of insulin predominate, and an anabolic state exists. Excess fuels are stored as glycogen

and triglycerides, and tissue protein synthesis increases. Conversely, when the I/G ratio is low (insulin low, glucagon high), the effects of glucagon predominate, and the tissues are in a catabolic state.

Shortly after a high-carbohydrate meal, an individual's molar ratio of insulin to glucagon can be as high as 30. When one awakens in the morning after an overnight fast, the I/G ratio is about 2. If one fasts for a day or two, the I/G ratio may decrease to 0.5 or less. Thus, in going from a fully fed state to a fasted state, the I/G ratio may change by a factor of 50 to 60.

As an example of the importance of the I/G ratio in determining the relative metabolic state, consider the situation that can exist in those cases of diabetes mellitus in which insulin secretion is deficient. In the absence of insulin, the blood glucose concentration increases dramatically. Normally, a high blood glucose concentration inhibits glucagon secretion (refer to Table 15–3), but in the absence of insulin this inhibitory effect is lost, and therefore plasma glucagon levels increase. In such cases, the I/G ratio is extremely low. The catabolic effects of glucagon greatly predominate, leading to increased glycogenolysis, lipolysis, ketogenesis, and gluconeogenesis. Blood glucose concentrations increase to extremely high levels, and a negative nitrogen balance exists. Experimental treatment of diabetic patients with somatostatin, which increases the I/G ratio by inhibiting glucagon secretion, dramatically reduces the severity of the catabolic state (somatostatin also inhibits insulin secretion, but in this case insulin secretion is already extremely low). Thus, the ratio of the relative concentrations of insulin and glucagon is probably more important in determining the overall metabolic state than is the absolute concentration of either hormone.

Overview of Metabolic Regulation by Hormones: The Fed State Versus the Fasted State

As described earlier, the body's use of energy is continuous, whereas the intake of energy in the form of food is intermittent. This situation requires that fuels be stored in the body for use during between-meal periods or for periods when food is not readily available.

Fuel Storage Forms

Table 15–4 shows the chemical nature and tissue distribution of stored fuels in an average 150-lb person. Note that most calories in the body are stored in the form of triglycerides in adipose tissue. By comparison, relatively little of the total stored fuel

Table 15–4
Distribution of Stored Fuels in the Body

Storage Form	Site of Storage	Energy Stored (kcal)
Fat (triglyceride)	Adipose tissue	141,000
Glycogen	Muscle	480
Glycogen	Liver	280
Protein	Muscle	24,000

consists of glycogen in either liver or muscle. Since liver glycogen can be broken down directly into free glucose and released into the blood, one might expect that it would be a better form in which to store excess calories. However, compared to glycogen, triglycerides are considered a much more efficient means of storing excess fuel, for several reasons. First, one gram of pure fat contains about twice the calories of one gram of carbohydrate. In other words, triglycerides have about twice the **caloric density** of glycogen. Secondly, glycogen is very hydrophilic, whereas fat is hydrophobic. Thus, glycogen attracts a considerable amount of water, which adds a significant amount of weight to the tissues in which glycogen is stored. If we stored all of our excess calories as glycogen and not as fat, the average person would weigh about 700 lb. Obviously, this would be a very inefficient way in which to store fuel, since the transport of this excess fuel would represent a considerable burden.

Note also in Table 15–4 that muscle protein represents about 15% of the total stored fuels in the body. During a prolonged fast, the body can and does call upon its muscle protein reserves as a source of stored energy.

The Flow of Nutrients in the Fed State

By examining nutrient flow in both the fed and the fasted states, we can review the effects of insulin and glucagon on metabolism and at the same time begin to appreciate the flow of various nutrients among organs.

Figure 15–16 summarizes the metabolic state expected in the liver, muscle, and adipose tissue of someone who has just eaten a meal rich in carbohydrate. Remember that in this situation, the expected concentration of insulin in the blood is high, whereas glucagon is low. In the liver, insulin promotes the conversion of glucose into glycogen, so that liver glycogen stores are increased (see Fig. 15–16). Insulin also stimulates fatty acid synthesis from glucose in the liver. Some of the fatty acids are stored as triglycerides, but most are packaged as lipoproteins and exported to adipose tissue (see Fig. 15–16). In adipose tissue, lipoprotein lipase (LPL) is stimulated by insulin, resulting in increased uptake of lipoproteins from the blood. The entry of glucose into adipose tissue is also stimulated under these conditions, as is synthesis of fatty acids and glycerol-phosphate (see Fig. 15–16). Triglycercide formation from fatty acids and glycerol-phosphate is also stimulated, resulting in expansion of the fat stores in adipose tissue. In skeletal muscle, glucose

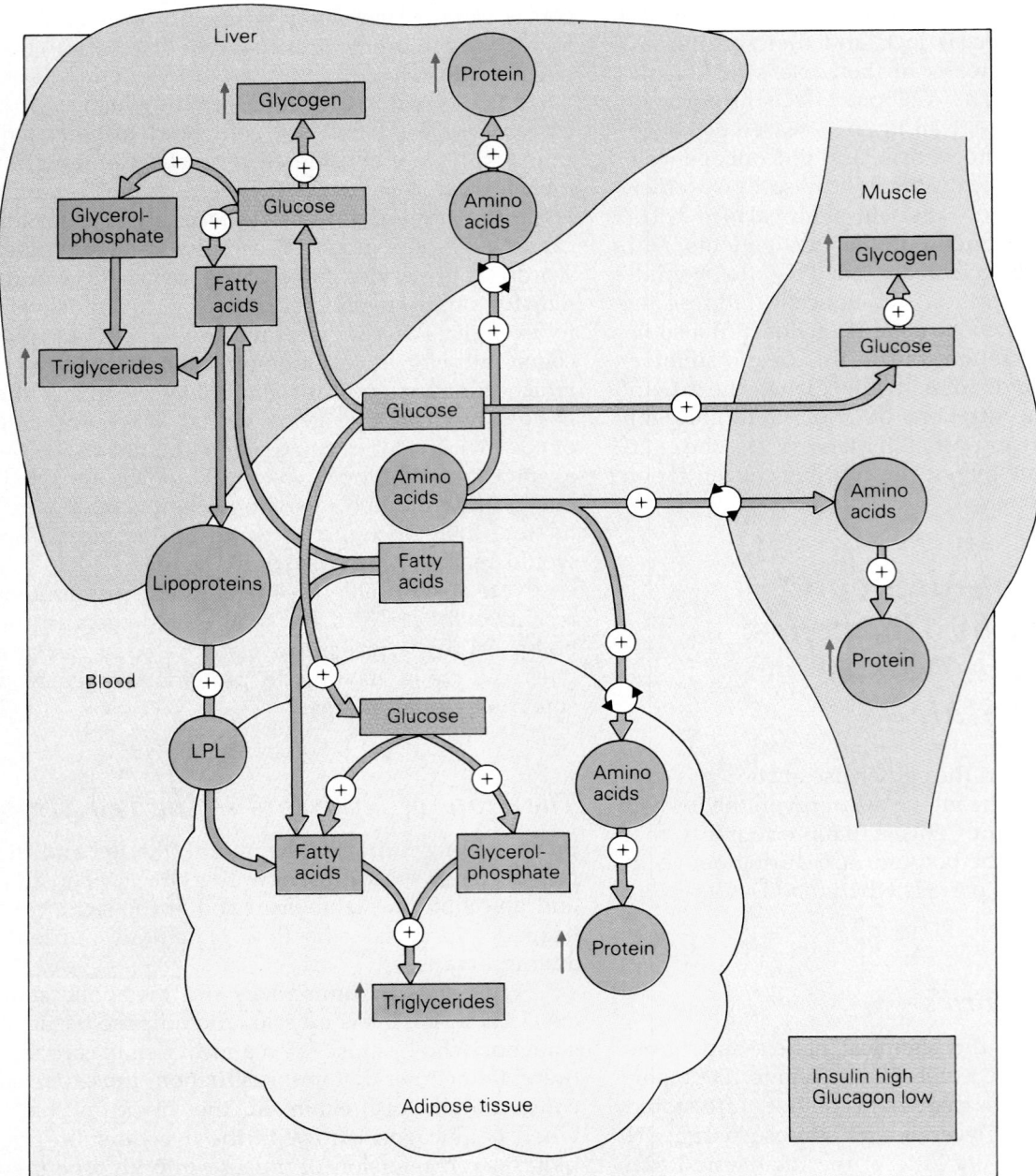

Figure 15–16
Summary of nutrient flow immediately after a meal. Those processes accelerated due to the prevailing high insulin/low glucagon concentrations are indicated by the +.

uptake is stimulated, as is glycogen synthesis (see Fig. 15–16). Thus, the amount of glycogen stored in muscle increases under these conditions. The uptake of amino acids into liver, muscle, and adipose tissue also increases in the fed state in response to insulin. In each tissue, insulin also stimulates the conversion of amino acids into protein while at the same time inhibiting protein degradation. Thus, the protein stores in each of the three tissues increase (see Fig. 15–16). Any fatty acids taken in with the meal are either taken up directly by the adipose tissue and stored as triglycerides or taken up by the liver and then packaged as lipoproteins before export to the adipose tissue.

Thus, as indicated in Fig. 15–16, in the period shortly after consumption of a high-carbohydrate meal (which most meals are), the combination of high insulin and low glucagon promotes fuel storage in liver, adipose, and muscle. Glycogen stores are increased in liver and muscle, triglyceride storage is increased in adipose tissue, and protein synthesis is promoted in all three tissues.

The Flow of Nutrients in the Fasted State

Figure 15–17 summarizes the state of nutrient flow expected during a brief period of fasting. In this situation, the expected blood insulin levels are low while glucagon levels are high. In liver, glycogen breakdown and gluconeogenesis are stimulated in

Figure 15–17
Summary of nutrient flow during fasting. Processes accelerated under the conditions of low insulin/high glucagon are indicated by the +. (LP = lipoproteins.)

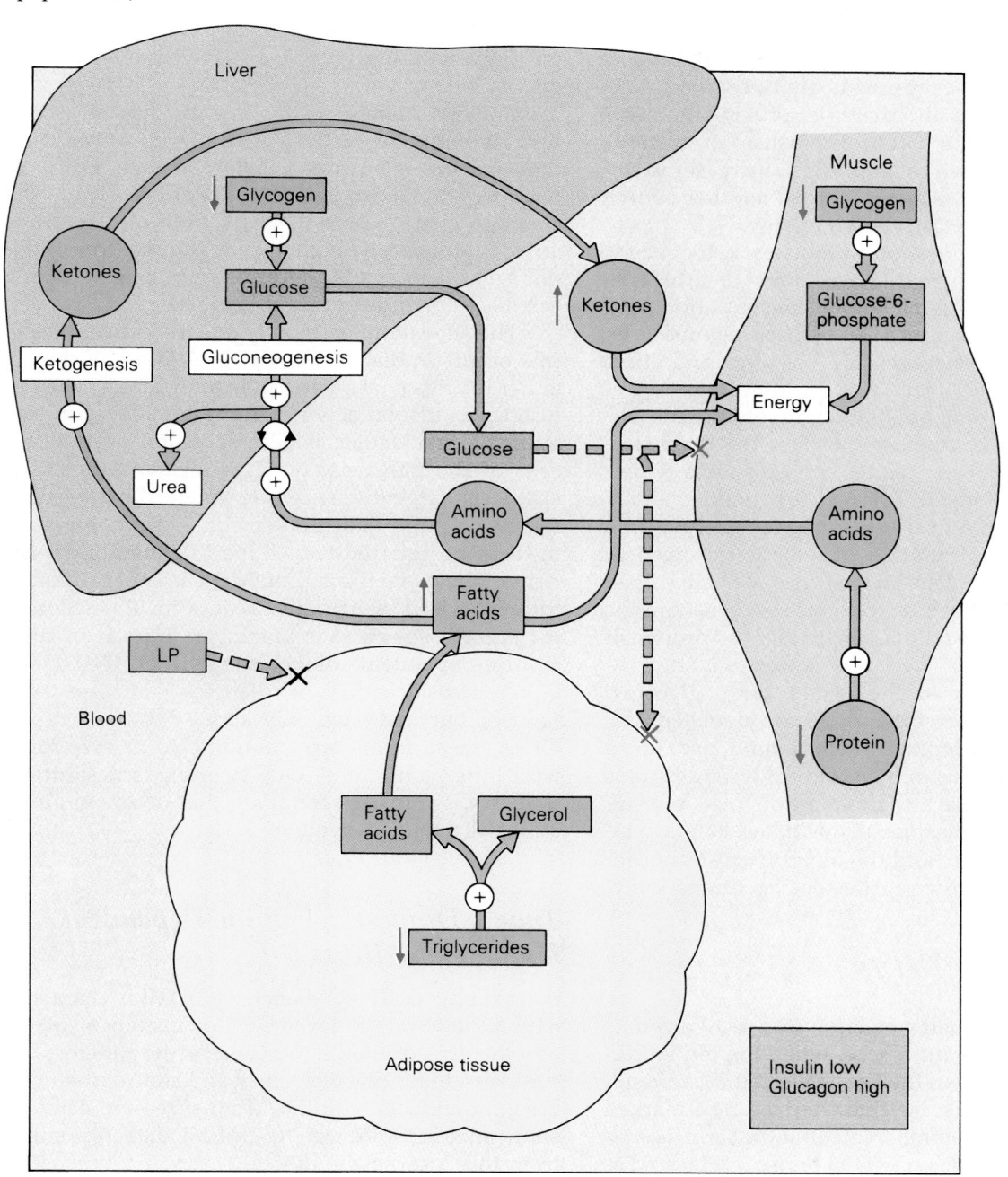

an attempt to maintain blood glucose levels near normal (see Fig. 15–17). In the absence of insulin, the breakdown of muscle protein is increased, and the amino acids released by muscle are used in the liver to synthesize glucose via gluconeogenesis. To facilitate the use of amino acids by liver, glucagon also stimulates amino acid uptake and urea synthesis in liver (see Fig. 15–17). In the absence of insulin, glucose entry into muscle and adipose tissue is greatly reduced, which also contributes to the maintenance of blood glucose levels. In adipose tissue, lipolysis predominates over lipid synthesis under these conditions. A result is increased release of fatty acids from adipose tissue, which leads to an increased concentration of fatty acids in the blood (see Fig. 15–17). The fatty acids can be used by muscle to obtain energy and therefore provide the muscle with an alternate fuel to use instead of glucose. Glycogen breakdown in muscle also increases in the absence of insulin, which provides another source from which muscle can derive energy.

In the liver, glucagon stimulates ketogenesis, and, as a result of increased conversion of fatty acids into ketones, the plasma ketone concentration rises dramatically. The ketones can be used by muscle as an energy source, which again has a sparing effect on plasma glucose. If the fast continues for several days, an important adaptation involving the ketones takes place. As stated earlier, the brain normally requires glucose as its energy source. However, with prolonged fasting, the brain actually adapts and begins to use ketones for energy. This adaptation is important, as it lessens the need to convert muscle protein into glucose. As Table 15–4 shows, considerably more calories are present in the body in the form of fat than are present as protein in muscle.

To summarize: in the fasted state, glycogen stores in both liver and muscle are depleted and muscle protein is broken down; amino acids used for glucose synthesis in liver and triglyceride stores in adipose tissue are also broken down. As a result of these changes, plasma fatty acid and ketone concentrations increase, and the blood glucose concentration is prevented from decreasing dramatically.

Diabetes Mellitus

The disease of diabetes mellitus was recognized as early as the first century A.D., when the term *diabetes*, which stems from the Greek word for "siphon," was applied to a disease characterized by a marked increase in urine volume. Note that the term *diabetes* is commonly used (as it will be here) to refer to dia-betes mellitus and should not be confused with the disease of diabetes insipidus, which is an entirely unrelated endocrine abnormality caused by completely different factors (refer to Chapter 13 for a discussion of diabetes insipidus).

Diabetes mellitus is by far the most common of all endocrine disorders and is a worldwide health problem. In the United States, about 6 million individuals have been diagnosed as having diabetes, and estimates suggest there may be another 4 to 5 million more who are borderline diabetics not yet diagnosed. Diabetes is the third leading medical cause of death and the second leading cause of blindness in the United States. Therefore, research on the causes and cures of diabetes is an area of intense interest.

Diabetes mellitus is not a single disease as was once thought. Instead, it is now well appreciated that diabetes comprises a heterogeneous group of disorders that differ in both cause and severity. The common characteristic of every form of diabetes is an inappropriately high blood glucose concentration, which in fact is the primary means by which the disease is most often diagnosed.

The elevation in blood glucose concentration that occurs in diabetes is the result of a deficiency, either relative *or* absolute, in insulin action. The deficiency in insulin action is most commonly due to either (1) inadequate insulin secretion by the beta cells of the pancreas, or (2) a relative lack of response by target tissue cells to the insulin present in the circulation. Inadequate insulin secretion results in the form referred to as **Type I** or **insulin-dependent diabetes mellitus (IDDM)**. The other common form is caused by a relative lack of insulin action in its target tissues and is known as **Type II** or **non-insulin-dependent diabetes mellitus (NIDDM)**. Other classifications of diabetes are also well known, but make up only about 2% of all cases. Often, these forms are secondary to or associated with other conditions. One example is **gestational diabetes,** a transient condition that occurs in about 3% of all pregnant women.

Type I Diabetes: Insulin-Dependent Diabetes Mellitus

As mentioned, Type I diabetes (IDDM) is characterized either by very low levels of insulin or by an absolute lack of insulin secretion by the pancreas. To compensate for this deficiency and control their disease, individuals with IDDM must receive daily insulin injections. About 10% of all diabetics suffer from this form of the disease.

Advances in Insulin Delivery

Although approximately 90% of all individuals with diabetes are Type II diabetics and generally do not require insulin therapy, some 20 million Type I diabetics worldwide do require insulin injections to remain alive. Insulin was first discovered and purified in Toronto, Canada, in 1921 and became available commercially a short time later, in 1923. Since its discovery, the insulin molecule has been the subject of several "firsts" in physiology and biochemistry. It was the first peptide or protein to have its amino acid sequence deduced. This achievement was a major event in the history of biology. Insulin was the first naturally occurring protein to be synthesized in the laboratory, and the discovery of proinsulin provided important information about the mechanisms of protein biosynthesis. More recently, insulin was one of the first mammalian peptides produced by the application of recombinant DNA technologies.

Four Canadians, F. C. Banting, C. H. Best, J. B. Collip, and J. J. Macleod, were instrumental in the discovery of insulin, although historically Banting and Best have received most of the credit for the work. As described in the accompanying chapter, the islets of Langerhans are surrounded by the exocrine tissue of the pancreas. The exocrine pancreas contains large amounts of proteolytic digestive enzymes, which normally are secreted into the small intestine. Before Banting and Best's work, other investigators had failed to isolate insulin because, when they ground up the pancreas to extract the protein, they also released digestive enzymes that destroyed the insulin. Banting and Best took advantage of situations in which the exocrine pancreas was relatively inactive and were thus able to extract and purify intact insulin.

Following the initial discovery of insulin, a second major advancement in the treatment of diabetes was the development in the late 1930s of insulin preparations with extended duration of action. Regular insulin is short-acting, and thus patients had to administer the hormone many times during the course of a day and night. The extended-action insulin formulations are designed to reduce the solubility of insulin in body fluids. These alterations allow the insulin to enter the circulation slowly. Thus a single injection can supply the body's needs over a longer period of time compared to regular insulin. Different types of extended-action insulins have been produced, each having a different duration of action. Some are effective for 6 to 12 hours, while others have their maximum effect 18 to 24 hours later. By tailoring the type and dose of different insulin preparations to an individual patient's needs, it is possible to limit the required injections to a few per day.

In the past few years, advancements in electronics and biotechnology have been applied to the field of insulin delivery as well. One example from this area is the development of an artificial pancreas. The essential elements of such a device include a glucose sensor to measure the blood glucose concentration, a pump and reservoir to deliver insulin to the bloodstream, and a computer to monitor information from the glucose sensor and make appropriate adjustments in the rate of insulin delivery by the pump. Initially, the equipment needed to perform these functions nearly filled a room. However, considerable progress has been made towards the miniaturization of this equipment. The hope is that someday an artificial pancreas small enough to be implanted and left in place for prolonged periods of time can be produced.

Another approach to the problem of insulin delivery that has shown considerable promise is that of islet transplantation. Although the method has been used only experimentally in humans, there is considerable hope that the technique will one day have widespread use, since in theory such a procedure could "cure" diabetes. The basic technique of islet transplantation involves separating the islets of Langerhans from the exocrine pancreas and then giving the recipient an injection containing a suspension of the isolated and purified islets. The primary obstacle to the adoption of this technique, as is the case with many transplant procedures, is rejection of the transplanted tissue by the immune system. As techniques for immunosuppression improve, the likelihood that islet transplantation will be used in the treatment of human diabetics also increases.

Finally, yet another major advance has been the production of insulin by the application of recombinant DNA technology. To accomplish this, researchers insert the human proinsulin gene into the DNA of bacteria. These altered bacteria produce human proinsulin and can be grown in large quantities in fermentation vats. The proinsulin is harvested, purified from the bacteria, and then converted into human insulin by the proteolytic removal of the C-peptide. Using this approach, virtually limitless amounts of authentic human insulin can be produced, ensuring the widespread availability of insulin despite the growing numbers of diabetics in our population.

The loss of beta cell function usually occurs during adolescence, although less frequently it occurs later in life. Recent investigations suggest that the loss of beta cell function is due to an autoimmune disorder that causes the immune system to attack and destroy the pancreatic beta cells. The causes of the autoimmune disorder are not well understood, but at least two major components appear to contribute to the disease state. The first is a genetic component by which certain individuals have an increased susceptibility to the disease. However, the genetic component is not by itself sufficient to cause the autoimmunity. The effects of an as yet unidentified environmental component are also required to produce the disease. What that agent might be is not currently known, although viruses have been suggested as likely candidates. The environmental component in some way triggers the autoimmune response, resulting in beta cell destruction.

Type II Diabetes: Non-Insulin-Dependent Diabetes Mellitus

About 90% of all diabetics suffer from the form of the disease known as **Type II diabetes** (NIDDM). Type II diabetes results when the target tissues of insulin, for reasons not currently understood, lose their responsiveness to the hormone. Normal or often higher-than-normal concentrations of insulin may be present in the blood, but the cells of target tissues such as adipose and muscle simply do not respond to the hormone as they normally would, a condition referred to as **insulin resistance.** In many cases, persons with Type II diabetes are also obese. The exact relationship between the obesity and insulin resistance is not clear, but the severity of the disease can often be reduced considerably if the patients are placed on a diet and lose weight. In addition to diet, persons with Type II diabetes are often treated with the oral hypoglycemic agents known as **sulfonylureas,** which, as indicated in Table 15–2, stimulate insulin secretion from the beta cells of the pancreas. In addition to stimulating insulin secretion, the sulfonylureas also appear to augment the actions of insulin in tissues and therefore help overcome the insulin resistance that exists in the target tissues.

The cause of the insulin resistance of Type II diabetes is poorly understood. As indicated previously, there does appear to be some link with obesity, but obesity per se is not sufficient to cause Type II diabetes. There does, however, appear to be a strong genetic component to Type II diabetes. Chances are nearly 100% that if one genetically identical twin develops Type II diabetes, the other will also, even if they are raised separately under entirely different environmental and socioeconomic conditions.

Complications of Diabetes Mellitus

Several immediate complications can arise if the glucose concentration in the blood is not checked by insulin; these problems include hyperglycemia, ketoacidosis, and electrolyte imbalance. In addition, long-term complications of the condition include vascular problems, vision problems, and impairment of nerve function.

Acute Complications of Diabetes

As indicated previously, in the absence of either insulin secretion or insulin action, the blood glucose concentration increases markedly, a condition named **hyperglycemia.** If the hyperglycemia becomes severe enough and the concentration of glucose in the blood exceeds the capacity of the kidneys to recapture the glucose by active transport, glucose will be excreted in the urine (a condition called **glucosuria**). Because of osmotic effects, the glucose present in the urine will also attract considerable amounts of water, which will be excreted along with the glucose. As a result, urine volume and frequency of urination will increase (a condition known as **polyuria**). As increased amounts of water are lost in the urine, **dehydration** of the individual can also occur.

In Type I (insulin-deficient) diabetics, the unopposed actions of glucagon result in increased ketone formation by the liver. The ketones are acids, and when they are produced in sufficiently large amounts, the normal acid/base balance in the body is considerably disturbed, resulting in the condition termed **ketoacidosis.** As with glucose, if ketones reach a high enough concentration in the blood, they will begin to "spill over" into the urine. As this spillover occurs, ketones also carry cations, such as sodium and potassium, with them into the urine. Thus, accompanying severe ketoacidosis is a loss of these ions, resulting in an **electrolyte imbalance** in the body. If left untreated, severe ketoacidosis accompanied by dehydration can rapidly result in coma and death.

Islet Amyloid Polypeptide: A Recently Discovered Beta Cell Secretory Product

Early in this century, investigators who studied the pancreases of human cadavers at the time of autopsy noted the presence of proteinaceous deposits in and about the islets of Langerhans in a large proportion of subjects who were known to be diabetics. What these deposits were remained a mystery until just recently, when the characterization of this material revealed a new beta cell secretory product, which has been given the name **islet amyloid polypeptide (IAPP),** also referred to as **amylin** by some. Although considerable experimental work must be done before the exact role of IAPP can be defined, there is reason to believe that it may play a role in normal physiology and in the pathophysiology of diabetes.

Amyloid is a general name given to the type of deposits seen in the islets, but which are also characteristic of several other disease states. IAPP is a 37-amino acid peptide structurally related to several peptides with neuroendocrine function and also to calcitonin, a hormone involved in the regulation of plasma calcium (see Chapter 27). IAPP is first synthesized as a larger precursor, in a manner much like that of other secretory peptides, suggesting that it too is a normal secretory product of the pancreas. Studies using antibodies that recognize IAPP have localized the protein to secretory granules of beta cells in normal islets. Thus, like insulin, IAPP appears to be synthesized primarily in beta cells and not in other tissues. It is presumed, but not yet proven experimentally, that IAPP is co-secreted along with insulin. If this presumption is correct, then the factors regulating insulin secretion must also regulate IAPP secretion. At present, due to the lack of a specific radioimmunoassay (RIA) for the peptide, there is little data regarding blood IAPP levels either in normal or diabetic individuals. Development of this assay, which is expected to be forthcoming, should yield answers to these and other important questions.

Assuming that IAPP is a normal secretory product of the beta cell, what might be its role in regulating homeostasis? Experiments conducted thus far suggest that IAPP may oppose or inhibit some of insulin's actions in skeletal muscle. By contrast, IAPP does not appear to have an effect on insulin action in adipose tissue. Thus, by causing tissue-specific insulin resistance, IAPP may serve to redirect nutrient storage under certain conditions. This is undoubtedly an overly simplified view, and much additional experimentation must be done before normal IAPP physiology can be described with any certainty.

Based on the finding that IAPP can induce insulin resistance, it is tempting to speculate that an abnormality of IAPP homeostasis, leading to excessive secretion of the peptide from the islets, may be an important mechanism contributing to the pathogenesis of NIDDM. While there is evidence consistent with this view, no cause and effect relationship has yet been shown. Similarly, since IAPP is synthesized in beta cells, destruction of these cells by the autoimmune disorder characteristic of IDDM would be expected to result in a deficiency in IAPP secretion as well as in insulin secretion. What effects this might have on homeostatic regulation or the development of secondary complications are also questions to which answers are actively being sought at this time. It will be interesting to follow the unraveling story of IAPP (amylin) physiology over the next few years.

Chronic Secondary Complications of Diabetes

In addition to the possible acute complications that may occur, several secondary complications usually accompany longstanding diabetes. These secondary complications usually involve gradual changes that develop over a period of years and often considerably shorten the life expectancy of these individuals. The most common secondary complications of diabetes are seen within the vascular system. Changes much like those seen with atherosclerosis lead to the narrowing of larger blood vessels in the brain, heart, and lower extremities. The resulting reduction in circulation to these areas may result in stroke, heart attack, or loss of limb, respectively. Lesions in the microvasculature (small blood vessels and capillaries) are also common in diabetics and are most detrimental in the kidney and in the retina of the eye. These changes can result in kidney disease and in the condition termed **diabetic retinopathy.** Diabetic retinopathy is due to the deterioration of blood vessels that nourish the retina. As the blood vessels in the retina break down, scar tissue is formed that eventually interferes with the light-sensing function of the retina. Each year, several thousand diabetics become legally blind as a result of diabetic retinopathy.

Another common secondary complication of diabetes involves impairment in nerve function (**diabetic neuropathy**). Fibers of the ANS are often involved, frequently resulting in abnormalities in bladder or gastrointestinal tract function, or in impotence. Diabetic neuropathy also frequently involves peripheral sensory nerves, resulting in loss of feeling in the lower limbs in particular.

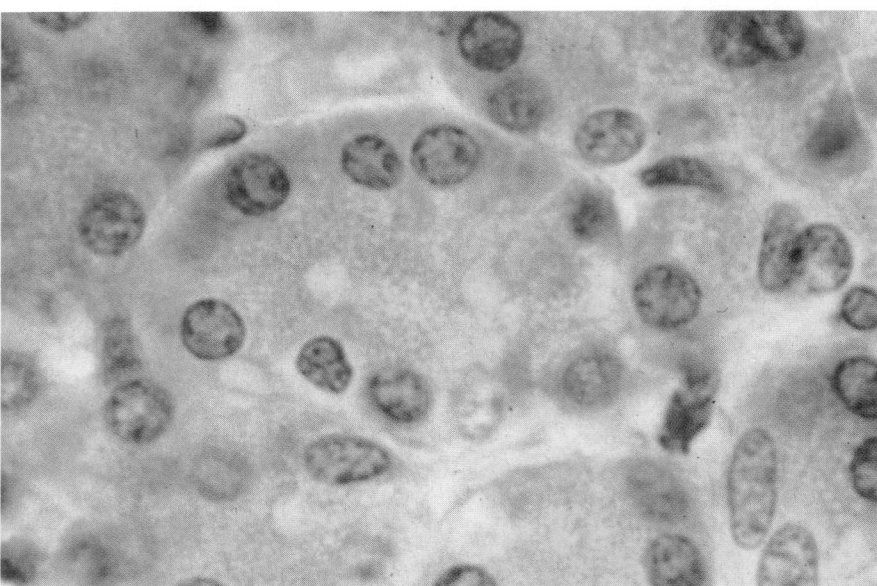

Figure 15–18
Histological stain of pancreas acinar cells. (© John D. Cunningham/Visuals Unlimited.)

SUMMARY

The Islets of Langerhans: Functional Units of the Endocrine Pancreas

The functional units of the endocrine pancreas are groups of cells known as the *islets of Langerhans*. The islets comprise four major cell types: the alpha cells, which secrete glucagon; beta cells, which secrete insulin; delta cells, which secrete somatostatin; and F cells, which secrete pancreatic polypeptide.

The Synthesis and Secretion of Pancreatic Hormones

Insulin consists of an A-chain and a B-chain held together by disulfide bonds. Insulin is first synthesized as proinsulin. The most important physiological stimulus for insulin secretion is an increased blood glucose concentration. Other factors that stimulate insulin secretion from the pancreas include amino acids, several of the GI hormones, acetylcholine, sulfonylureas, and glucagon. Inhibitors of insulin secretion include somatostatin, epinephrine, and norepinephrine.

Glucagon is a simple polypeptide first synthesized in the alpha cells as proglucagon. Glucagon secretion is stimulated by a decreased blood glucose concentration. Conversely, increased blood glucose inhibits glucagon secretion. Amino acids, fatty acids, and both branches of the ANS also stimulate glucagon secretion. Insulin and somatostatin both inhibit glucagon secretion.

Somatostatin is first synthesized in the delta cells as prosomatostatin. Somatostatin secretion is stimulated by increased blood glucose, glucagon, and amino acids.

The Metabolic Effects of Pancreatic Hormones

Insulin promotes fuel storage and protein synthesis and therefore has anabolic effects. Insulin rapidly lowers the blood glucose concentration by increasing glucose uptake into muscle and adipose tissue by facilitated diffusion. Insulin also stimulates glycogen formation and glycolysis in its target cells and decreases glucose output by the liver. In liver and adipose tissue, insulin *stimulates fatty acid synthesis* and promotes the *storage* of fatty acids as *triglycerides*. In adipose tissue, insulin promotes the uptake of lipoproteins into fat cells from the blood and inhibits lipolysis. Insulin stimulates the accumulation of protein in most tissues of the body, promoting a positive nitrogen balance.

Glucagon, which raises the intracellular concentration of cAMP, has catabolic effects on metabolism. The primary target tissue of glucagon is the liver. Glucagon increases the blood glucose concentration by stimulating glycogenolysis and gluconeogenesis in the liver. Glucagon also stimulates lipolysis and stimulates ketogenesis in liver. Fatty acids and ketones are alternative energy sources for many tissues and thus their increased availability has a glucose-sparing effect. Glucagon provides increased substrate for gluconeogenesis in liver by stimulating protein degradation and amino acid transport.

Because of the opposing actions of insulin and glucagon, the status of nutrient flow and metabolism in the body is determined by the relative amounts of each of these two hormones. A high I/G ratio exists in an anabolic state, and a low I/G ratio is found in a catabolic state.

Overview of Metabolic Regulation by Hormones: The Fed State Versus The Fasted State

The principal storage form of calories in the body is triglycerides (fat). Fat has a greater *caloric density* than glycogen and is hydrophobic. It is therefore a *more efficient* storage form than glycogen.

Shortly after a meal, the actions of insulin on metabolism predominate over those of glucagon. As a result, glycogen stores are increased in liver and muscle, triglyceride stores are increased in liver and adipose tissue, and protein synthesis is promoted in all three tissues.

In the fasting state, glucagon levels are high, insulin levels are low, and nutrient flow is exactly opposite to that in the fed state. Glycogen stores in liver and muscle are depleted, muscle protein is broken down, and the amino acids used for *gluconeogenesis* in liver and the *triglyceride* stores in adipose tissue are broken down. These changes occur in an attempt to maintain blood glucose concentrations near normal.

Diabetes Mellitus

Diabetes is a common endocrine disorder caused by either an *absolute* or *relative deficiency* in insulin action. Type I diabetes, or insulin-dependent diabetes mellitus (IDDM), is caused by destruction of insulin-secreting beta cells of the pancreas. Individuals with Type I diabetes generally require treatment with insulin.

Type II diabetes, or non-insulin-dependent diabetes mellitus (NIDDM), is caused by lack of responsiveness of insulin target tissues to the hormone, a condition termed insulin resistance. Usually, Type II diabetics can be treated by diet and by oral hypoglycemic agents.

Hyperglycemia, glucosuria, polyuria, and dehydration are common acute complications of diabetes. Type I diabetics are also prone to ketoacidosis. Chronic secondary complications of diabetes include diabetic retinopathy and impaired nerve function.

REVIEW QUESTIONS

Identify Each with a Term

1. The small clusters of cells that make up the endocrine portion of the pancreas.

2. The cells of the pancreas responsible for the secretion of insulin.

3. The name of the precursor molecule of insulin.

4. The name of the enzyme in adipose tissue that promotes the uptake of lipoproteins from the blood.

5. The name of the process (which occurs primarily in the liver) whereby non-glucose substrates are converted into glucose.

6. The type of diabetes characterized by a relative lack of insulin action in its target tissues.

7. The portion of the pancreas responsible for the secretion of fluid and digestive enzymes into the small intestine.

Define Each Term

8. C-peptide
9. hormone-sensitive lipase
10. glycogenolysis
11. ketogenesis
12. caloric density
13. Type I diabetes mellitus (IDDM)
14. sulfonylureas
15. hyperglycemia

Choose the Correct Answer

16. Which cell type is most numerous in the islets of Langerhans?
    a. alpha cell
    b. beta cell
    c. delta cell
    d. F cell

17. Which of the following is incorrectly paired?
    a. delta-cell:somatostatin
    b. alpha cell:pancreatic polypeptide
    c. beta cell:insulin
    d. all of the above pairs are correct

18. Insulin secretion from the pancreas increases in response to:
    a. ingestion of a high protein meal.
    b. decreased blood glucose concentration.
    c. activation of the sympathetic nervous system.
    d. somatostatin.

19. Several paracrine interactions are thought to be involved in regulating hormone secretion from the islets of Langerhans. These include:
    a. somatostatin stimulation of glucagon secretion.
    b. glucagon inhibition of insulin secretion.
    c. insulin stimulation of somatostatin secretion.
    d. insulin inhibition of glucagon secretion.

20. Among the effects of insulin on carbohydrate metabolism are:
    a. inhibition of glycogenolysis.
    b. stimulation of glycolysis.
    c. stimulation of glycogen synthesis.
    d. all of the above.

21. In the liver, glucagon:
    a. stimulates ketogenesis.
    b. inhibits glycogenolysis.
    c. stimulates glucose uptake.
    d. all of the above.

22. In adipose tissue, insulin:
    a. stimulates hormone-sensitive lipase and lipoprotein lipase.
    b. promotes lipolysis.
    c. stimulates urea synthesis.
    d. stimulates glucose uptake.

23. Which of the following is not associated with insulin-dependent diabetes mellitus?
    a. autoimmune disease
    b. high blood glucose concentrations
    c. ketoacidosis
    d. low blood glucagon

Answer in One or Two Sentences

24. What is the anatomical arrangement of the cell types of the islets of Langerhans, and how might the arrangement of cells be involved in the regulation of hormone secretion?

25. What are the effects of insulin on muscle protein metabolism?

26. Why are triglycerides more efficient than glycogen as a storage form of energy?

27. What are the basic differences in characteristics of Type I and Type II diabetes?

28. What are some of the more common secondary complications of diabetes?

29. Describe the mechanism whereby insulin stimulates glucose uptake into cells.

SUGGESTED READINGS

Diabetes in America. A report of the National Diabetes Data Group, U.S. Department of Health and Human Services. National Institutes of Health Publication No. 85–1468, Bethesda, 1985.

Fregley, Melvin J., and Luttge, William G. *Human Endocrinology: An Interactive Text*. New York: Elsevier Biomedical, 1982.

Ganong, William F. *Review of Medical Physiology*, 12th ed. Los Altos: Lange Medical Publications, 1985.

Hedge, George A., Colby, Howard D., and Goodman, Robert L. *Clinical Endocrine Physiology*. Philadelphia: W.B. Saunders Company, 1987.

Nishi, Masahiro, Sanke, Tokio, Nagamatsu, Shinya, Bell, Graeme I., and Steiner, Donald F. "Islet amyloid polypeptide. A new β cell secretory product related to islet amyloid deposits." *Journal of Biological Chemistry*, 265:4173–4176, 1990.

Olefsky, Jerrold M. "The insulin receptor. A multifunctional protein." *Diabetes*, 39:1009–1016, 1990.

Rifkin, Harold, and Porte, Daniel, Jr. *Ellenberg and Rifkin's Diabetes Mellitus: Theory and Practice*, 4th ed. New York: Elsevier, 1990.

Tepperman, Jay, and Tepperman, Helen M. *Metabolic and Endocrine Physiology*, 5th ed. Chicago: Year Book Medical Publishers, 1987.

Wilson, Jean D., and Foster, Daniel W., eds. *Williams Textbook of Endocrinology*, 7th ed. Philadelphia: W.B. Saunders Company, 1985.

KEY TERMS

caloric density (p. 517)

diabetes mellitus (p. 520)

endocrine pancreas (p. 501)

exocrine pancreas (p. 501)

fatty acid (p. 512)

glucagon (p. 501)

glucosuria (p. 522)

glycerol phosphate (p. 512)

hyperglycemia (p. 522)

insulin (p. 501)

islets of Langerhans (p. 501)

ketoacidosis (p. 522)

lipoprotein lipase (p. 512)

lipoproteins (p. 512)

pancreas (p. 501)

polyuria (p. 522)

proinsulin (p. 503)

Type I diabetes (p. 520)

Type II diabetes (p. 520)

PART *III*

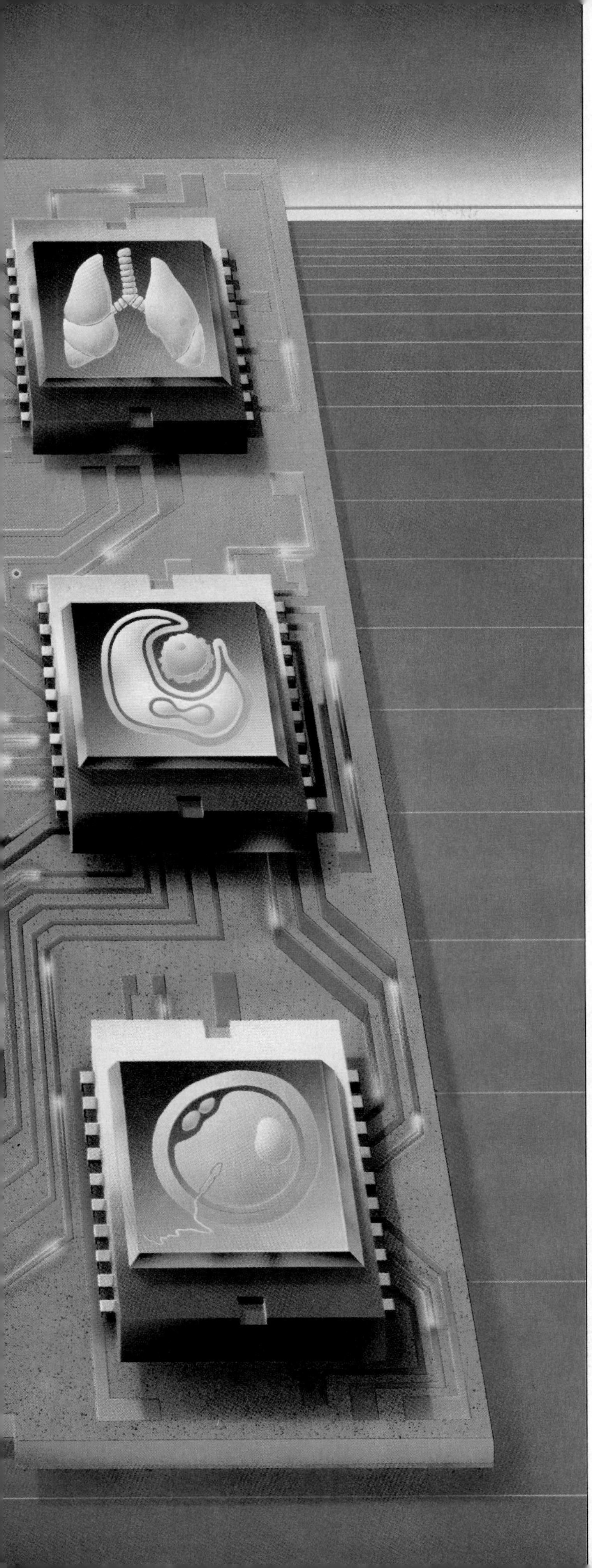

Integrative Organ Functions

The following systems are discussed in Part III:

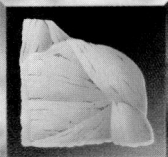

Muscular

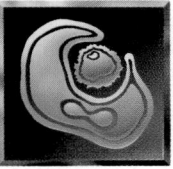

Defense

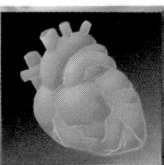

Cardiovascular

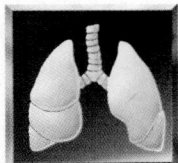

Respiratory

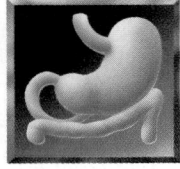

Digestive

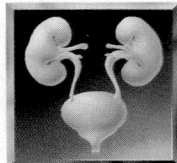

Renal System

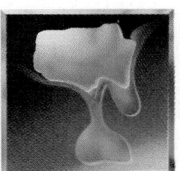

Thermoregulatory

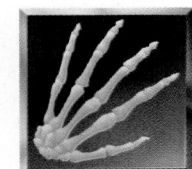

Skeletal

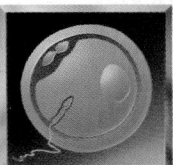

Reproductive

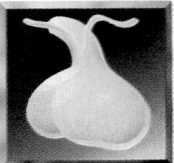

Endocrine

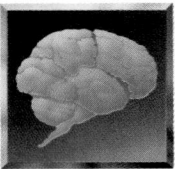

Nervous

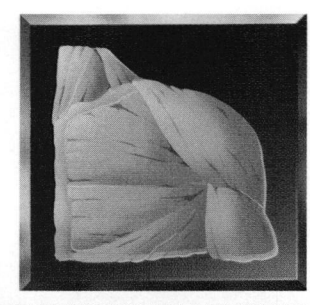

CHAPTER
16

Muscle

*T*here are many types of muscle in the body. While we are most familiar with those large muscles attached to bones and responsible for our ability to move, breathe, and manipulate our environment, many other muscles are used for bodily activities of which we are usually unaware. All of our muscles have important features in common. They may be similar in structure, function, and in various other ways. This chapter covers all of these aspects of muscle and elucidates the critical roles of muscle in the body economy. Because skeletal muscle is the type best understood, most of the chapter will be devoted to it.

Overview: The Roles and Types of Muscle

The range of activities that muscles carry out in the body is extremely broad, so it is not surprising that muscles show a wide range of functional adapta-

tions that specifically suit their many tasks in the overall function of the body.

Muscle as an Effector

As described in Chapter 10, our central and peripheral nervous systems function by gathering information and instructing muscles to take useful action in response. This coordination between nerve and muscle enables us to walk, talk, digest our food, defend our bodies, propagate our species, and do almost everything else that we do. In addition to effecting changes prescribed by the nervous system, muscles also help to regulate what goes into that system by adjusting the sensitivity of our sense organs. Some muscles are under neural control, while others are more independent. Very few muscles, however, are totally independent. The degree of neural control serves to adapt various muscles to their special roles. Some of the many functions of muscle in relation to the nervous system are shown in Figure 16–1.

Figure 16–1
Some of the more important functions of muscle. Most of the categories shown could be divided into even more functions.

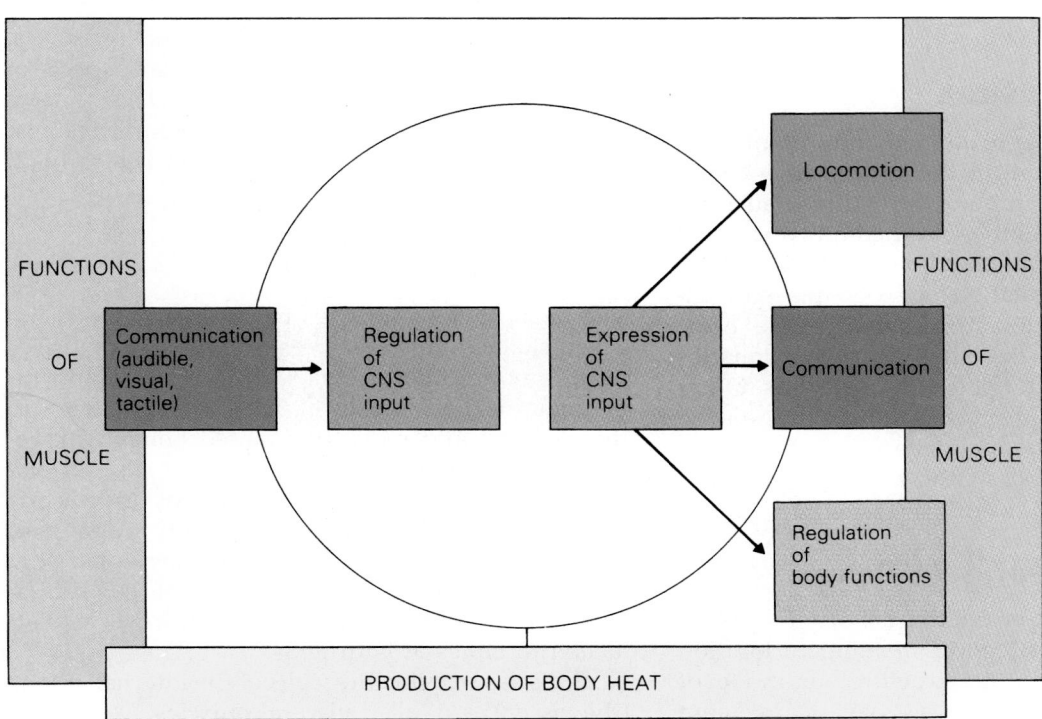

Our other important control system, the endocrine system, also has muscle as one of its important effectors, or means of expression, although this function may not be very obvious when our bodies are functioning well.

Muscle as a Motor

Muscles can do what they do because they are a sort of **biological motor.** Like mechanical or electrical motors, they consume fuel, do useful work, and waste some of the energy that they consume as heat. Muscles produce extra heat because they, like other motors, are not completely efficient. That is, they cannot turn all of the energy they consume into mechanical work. The heat they produce as a result of this inefficiency may at times be very important in overall bodily function. Under most conditions of activity, most of our body heat is produced as a result of the contraction of muscles. When the body temperature falls, brisk physical exercise (which may be relatively efficient) or shivering (which is very inefficient from a mechanical standpoint) provides the necessary warmth. During heavy exercise or under the stress of high environmental temperatures, our problem may involve getting rid of the excess heat produced by muscular contraction. Many of our conscious "cooling" mechanisms also involve the use of muscle, as may become apparent when the work of fanning to cool off produces more heating than cooling.

Muscle as a Regulator

Muscles also help to regulate many important body functions. Muscles control the movement of substances through the tubular structures (such as blood vessels and intestines) of the body and expel those substances from the body at the proper time. For example, the regulation of blood pressure involves a complex interaction between the heart muscle, which pumps the blood, and other muscles that control the diameter of the blood vessels. Other regulatory mechanisms in which muscles play a central role include the maintenance of an upright posture and the control of the body temperature.

Classification of Muscle Types

Several different sets of criteria are used to place muscles in separate categories relating to location and function, microscopic structure, or mode of control and action. These topics serve as convenient

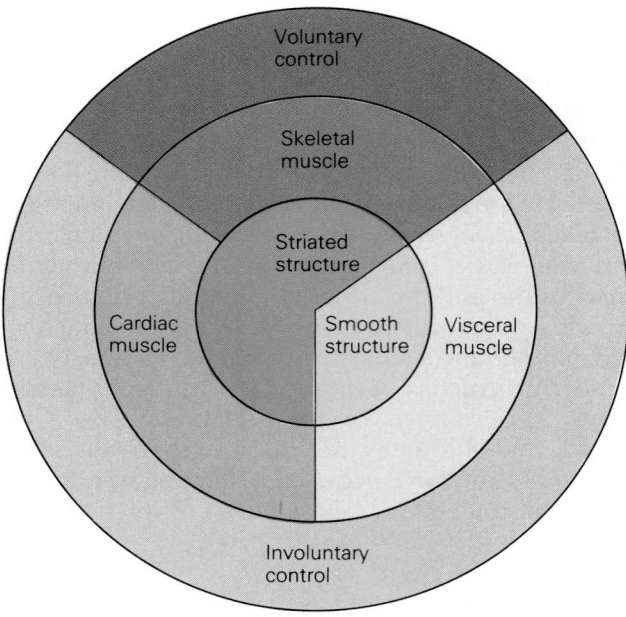

Figure 16–2
A simplified view of the classification of muscle. Although there may be some difficulties with each way of classification, muscle may be classified by three different means: by location (skeletal, cardiac, or visceral); by structure (smooth or striated); and by mode of control (voluntary or involuntary). In reality, some categories and functions overlap.

ways to describe and discuss particular aspects of muscle, but, as with any attempt at classification, some exceptions will have to be noted. The most usual sort of classification is shown in Figure 16–2.

Classification of Muscles by Location and Function

The most obvious muscles in the human body are called **skeletal muscles.** These are the large muscles used, among other things, for locomotion and for maintaining body posture. They usually attach to the skeleton and move the jointed bones with respect to one another. In some instances, the skeletal attachments may be indirect or may exist at only one end of the muscle (as is the case with the tongue, which is a skeletal muscle), or they may not exist at all (as is the case with the upper end of the esophagus). The bulk of skeletal muscles, however, are attached to bones at both ends.

Muscles that line the walls of the internal organs (the viscera) are called **visceral muscles.** Muscles of this type are also located in nonvisceral organs such as blood vessels and in sensory organs such as the eye, where they aid in focusing and adapting to different levels of lighting.

The location of **cardiac muscle** is obvious from its name; it makes up the pumping muscle mass of the heart. Cardiac muscle is found only in the heart (although it may extend a little way into the large vessels that enter and leave the organ). While confined to a relatively small anatomical region, its critical role in body function in health and disease gives great importance to its understanding and study.

Classification of Muscles by Structure

When viewed at the level of microscopic structure, muscle falls into two general categories: striated muscle and smooth muscle. The term *striated* means "striped" and refers to the regular cross-striped appearance of the muscle cells under a microscope. This appearance is an important clue to the way in which these muscles function on a molecular level, a topic that will be treated in some detail later in this chapter.

Skeletal muscle is **striated muscle.** Cardiac muscle is also striated muscle. While its cells are considerably smaller than those of skeletal muscle, its molecular function is similar in many respects. Many of the details of the function of skeletal muscle have been found to apply to cardiac muscle also. In addition, cardiac muscle also shares some functional similarities with visceral muscle.

Visceral muscle cells lack the striations found in the other muscle types and are therefore referred to as **smooth muscle.** Their structure provides fewer clues to their function than in the case of skeletal muscle, although many aspects of their internal processes are becoming better understood as research in this field progresses.

Classification of Muscles by Mode of Action and Control

Because most movements of skeletal (striated) muscle occur in response to a conscious (willed) effort, these muscles may be called **voluntary muscles.** However, many actions of skeletal muscle, such as the acts of breathing or walking, are almost automatic and are thus in a sense involuntary. The strictest basis for this type of classification is that skeletal muscles, in order to function, must receive a specific signal or signals from the central nervous system (CNS). In the event of an accident or illness that disconnects a skeletal muscle from its nerve supply, paralysis results.

On the other hand, the actions of most smooth (visceral) muscles are **involuntary.** In many cases these involuntary actions are initiated by the **autonomic nervous system (ANS)** in response to some internal reflex adjustment that may take place automatically. Smooth muscle is also capable of acting on its own, sometimes in response to hormonal or environmental factors. Voluntary control of some smooth muscle functions is also possible; for instance, specially trained persons using the techniques of "biofeedback" can cause blood vessels to dilate or constrict. In the majority of situations, however, the function of smooth muscle is completely involuntary.

Cardiac muscle is also an involuntary muscle, and no signals from the nervous system are required to start or maintain its contraction. The heart will continue to beat indefinitely after all nerve connections to it are removed, although the rate and strength of the heartbeat are no longer under external control. Because of the special nature of the cardiac muscle, it is not usually described as being either voluntary or involuntary.

Smooth muscle is itself subject to further classification in the area of its mode of control. Some smooth muscle is called **multiunit smooth muscle;** this muscle is closely controlled by the nervous system. Since the nervous elements involved in its control are part of the ANS, it is not subject to voluntary control. Other smooth muscle, that which makes up the bulk of our visceral organs, is called **unitary smooth muscle.** This sort of muscle is less strictly controlled by the nervous system, and large numbers of its cells function together as a single unit. The distinctions mentioned here should become clearer later in this chapter.

A summary of the three classifications is provided in Figure 16–2. When allowance is made for the exceptions noted previously, the classifications are useful for discussing and comparing the various muscle types. Within each type are important variations, and these will be taken up as the specific details of each muscle type are discussed.

The Structure of Muscle

The study of muscle structure can be done at two different levels. At one level, large-scale structural aspects of muscle can be observed and described without microscopic aid. The position, size, shape, and attachment points of a skeletal muscle give important information about its function. At another level, light microscopes and electron microscopes reveal details of the fine structure responsible for the actual mechanism of contraction. Muscle is an excellent example of the way in which the structure and function of a biological tissue are intimately related. For this reason we will discuss in detail the structure of striated muscle before going on to other muscle types. Striated muscle can serve as a basis of comparison for the description of other muscle types.

The Structure of Skeletal Muscle

Contraction, the major function of a skeletal muscle, involves its shortening and the resulting movement of parts of the skeleton. The large-scale structure of skeletal muscle reflects this function, and the organization of the cells and tissues reflects the structural and metabolic adaptations necessary for functions in which large forces are generated and large amounts of energy are expended.

Gross Structure of Skeletal Muscle

In mechanical terms, the skeleton works with the muscles as a **lever system.** The term *lever system* refers to the fact that groups of muscles working together with bones permit a larger range of movement than could muscles acting on their own. Muscles attach to bones by means of very strong connective tissue structures called **tendons** (Fig. 16–

3). The more stationary (and **proximal**) attachment is called the **origin** of a muscle, and the other (**distal**) end is called the **insertion.** The wide or thick central portion of a muscle is called the **belly.** Some muscles, the **biceps** for example, have a double origin (*biceps* means "two heads") and a single insertion, while others have a narrow origin and a very wide insertion. The variable anatomy of skeletal muscles represents adaptations to the specific functions of the individual muscles. Muscles with many parallel (side-by-side) fibers can exert a large force but cannot shorten rapidly or by a very great amount. On the other hand, long muscles with many fibers in series (end-to-end) are able to shorten rapidly but develop less force.

Because most skeletal muscles are attached to bones at either end, the range of motion of a muscle itself may be limited by the skeletal system. The needed range of motion must then be multiplied by the lever system provided by the skeleton. This arrangement has some important consequences. Large amounts of muscle shortening, as we shall see, can significantly affect the amount of force a muscle can exert, and limiting the range of muscle shortening helps to compensate for these effects. However, this arrangement also means that the muscle must exert a much greater force at its insertion than is actually produced at the end of the limb. For instance, in the human forearm, the lever is

Figure 16–3
Antagonistic muscle arrangements and the skeletal lever system. The major muscles that move the upper arm work in opposition to each other. The biceps flexes the arm, while the triceps extends it. The tendon of the biceps attaches at a point approximately 5 cm away from the fulcrum of the lever system (the elbow joint), and the hand is 35 cm away. Therefore, the upward force exerted by the muscle must be seven times the downward force supplied by a weight in the hand. However, any shortening of the muscle is multiplied by seven, meaning that the hand moves seven times as fast and as far as does the lower end of the biceps muscle.

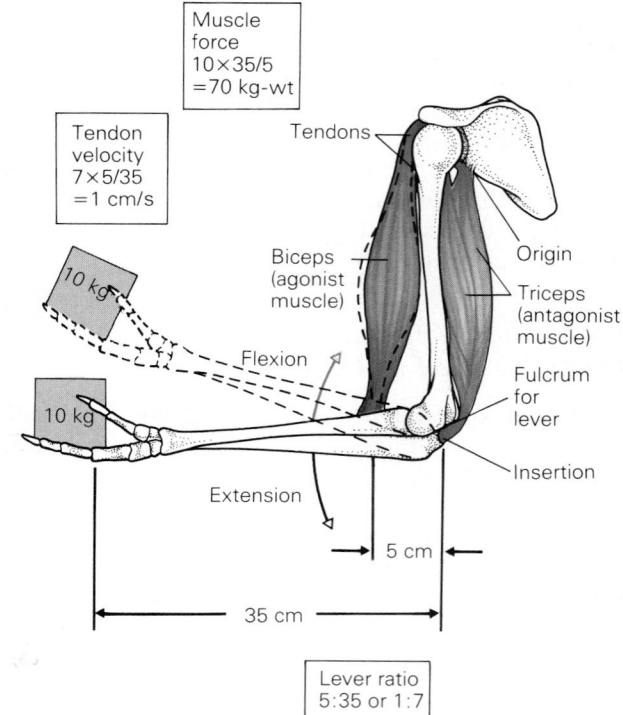

Muscle force
$10 \times 35/5$
$= 70$ kg-wt

Tendon velocity
$7 \times 5/35$
$= 1$ cm/s

Hand force
10 kg-wt

Tendons

Biceps (agonist muscle)

Origin

Triceps (antagonist muscle)

Flexion

Fulcrum for lever

Insertion

Extension

5 cm

35 cm

Lever ratio
5:35 or 1:7

hinged at the elbow. The hand is approximately seven times as far from the joint as is the insertion of the biceps. This means that if the hand is holding an object with a mass of 10 kilograms, the tendons of the biceps must exert seven times as much force (see Fig. 16–3). This situation can lead to serious muscle and tendon injury in athletes and in people who do heavy manual labor. On the other hand, it means that the hand can move seven times as fast as the biceps can shorten. Such an increase in speed is often valuable. For instance, it allows the leg muscles to propel the body at a high rate of speed, or to jump up in the air much higher than the distance over which the muscles themselves can shorten.

As mentioned in Chapter 10, in most situations, (especially in the limbs) skeletal muscles are arranged in so-called **antagonistic pairs.** When a muscle moves a joint in one direction (e.g., **flexion,** in which a joint closes and a limb assumes a more withdrawn position), another muscle or muscles move the joint in the opposite direction, toward a more extended position (i.e., **extension**). This arrangement is necessary because muscles can only *pull*; they cannot *push*.

Skeletal **muscle cells** (also called **muscle fibers**) are quite large compared to other cells. Typical diameters are on the order of 100 micrometers, while cells from some large muscles may be many centimeters in length. The way in which muscle cells are assembled into the complete muscle is diagrammed in Figure 16–4. In an intact muscle, shown in cross section in Figure 16–4a, the individual muscle fibers are surrounded by a delicate connective tissue sheet called the **endomysium.** Bundles of muscle fibers are grouped into **fasciculi** (shown in longitudinal section in Fig. 16–4b) by a connective tissue called the **perimysium,** and the entire muscle is covered by a connective tissue sheet called the **epimysium.** During physical exercise, muscle requires a plentiful supply of oxygen and nutrients, and must rapidly get rid of waste products. Accordingly, muscle is supplied with a plentiful network of blood vessels arranged among the fibers so that no muscle cell is very far away from a blood vessel.

Both the connective tissue sheets and the muscle fibers themselves are connected at either end into a very tough connective tissue extension of the muscle that makes up the tendon. Some tendons are thick and short, like those that provide the insertion of the biceps. Some tendons are very long; the muscles that move the fingers are actually located in the forearm, and their tendons run across the wrist joint and the back and palm of the hand to connect the muscles to the bones of the fingers.

Cellular and Molecular Structure of Skeletal Muscle

At the cellular level of organization, there is a remarkably clear relationship between the structure and function of muscle. The molecules that make up the contractile substance of muscle have a dual role. Not only are they the major structural elements of the muscle cell; they are also the enzymes that convert the biochemical energy (derived ultimately from the food we eat) into mechanical energy.

When viewed through a light microscope, skeletal muscle cells show a repeating pattern of dark and light crossbanding. This striped, or striated, pattern gives skeletal muscle one of its names; more importantly, it provides a clue to how striated muscle actually works. Analysis using electron microscopy and other methods has revealed that the dark and light crossbands, or striations, are composed of two types of protein filaments, with each type being arranged in sets side by side across the cell. These protein filaments are called **myofilaments,** and the two types differ from each other in their size and internal structure. The two sets overlap each other in a characteristic way that changes as the muscle is lengthened or shortened.

The Structure of a Sarcomere. The two sets of myofilaments are organized into a structure called a **sarcomere;** this is the basic unit of the contractile machinery of the skeletal muscle. Figure 16–5 shows both a cross-section and longitudinal section view of the microscopic structure of skeletal muscle. The center portion of the sarcomere (the **A-band**) is made up of **thick filaments,** and the lateral regions are made up of **thin filaments.** A structure called the **M-line** runs across the middle of the A-band and holds the thick filaments in side-by-side alignment. The thin filament array interdigitates (overlaps) with the thick filament array, and the depth of its penetration depends on the overall length of the muscle. The region where only thin filaments are present is called the **I-band.**

Both sets of filaments are packed in such a way that a cross-sectional cut through a muscle fiber shows a hexagonal pattern; this is the most compact way in which the filaments could be packed (Fig. 16–6a). If a cut is made in a region in which sets of thick and thin myofilaments overlap, then the spaces provided by the hexagonal array of the thick filaments will be occupied by an interlocking hexagonal array of the thin filaments. "Crossbridges" from the thick filaments will be seen to reach out directly toward the thin filaments.

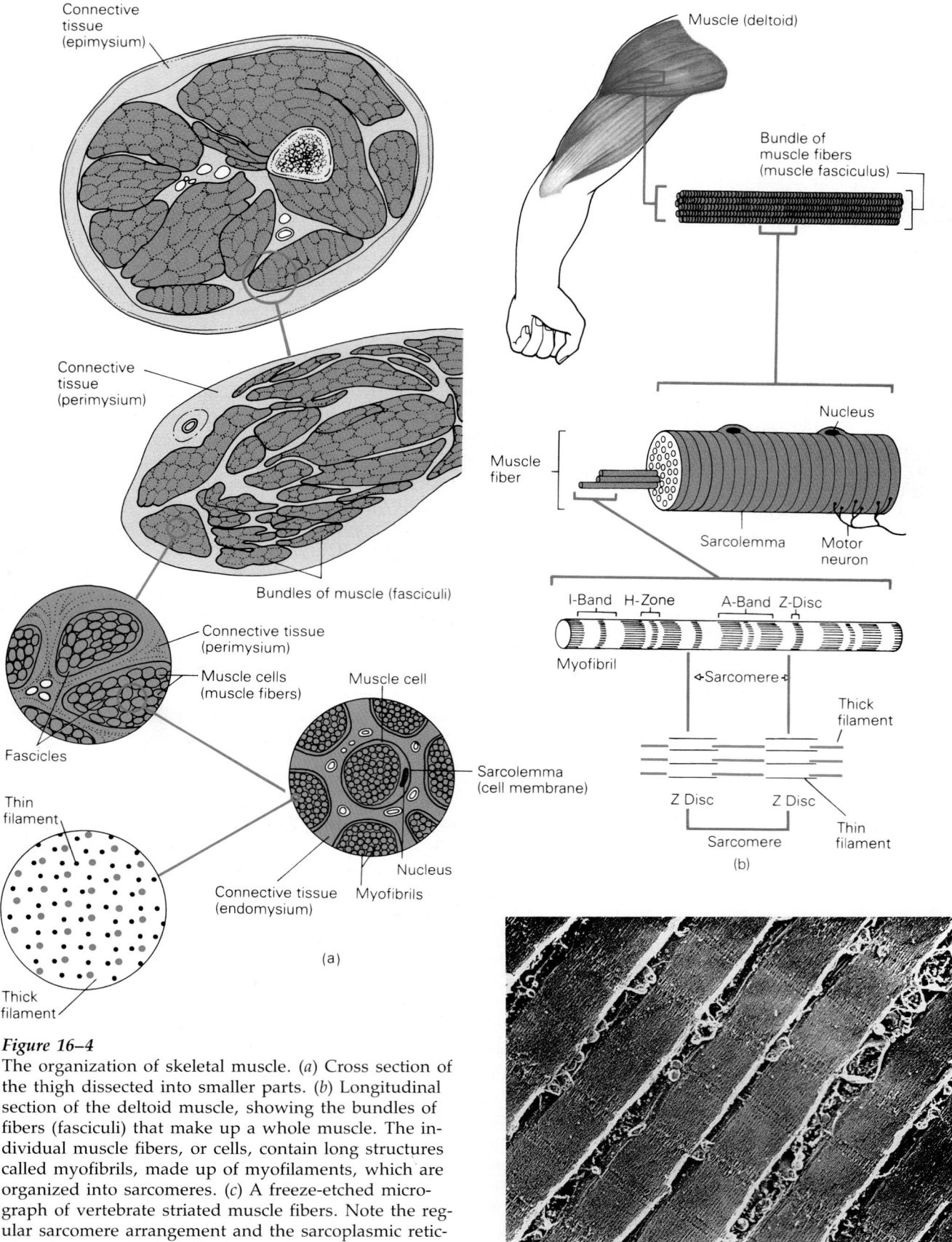

Figure 16–4
The organization of skeletal muscle. (*a*) Cross section of
the thigh dissected into smaller parts. (*b*) Longitudinal
section of the deltoid muscle, showing the bundles of
fibers (fasciculi) that make up a whole muscle. The individual muscle fibers, or cells, contain long structures
called myofibrils, made up of myofilaments, which are
organized into sarcomeres. (*c*) A freeze-etched micrograph of vertebrate striated muscle fibers. Note the regular sarcomere arrangement and the sarcoplasmic reticulum between the myofibrils. (© Clara Franzini-
Armstrong/Photo Researchers, Inc.)

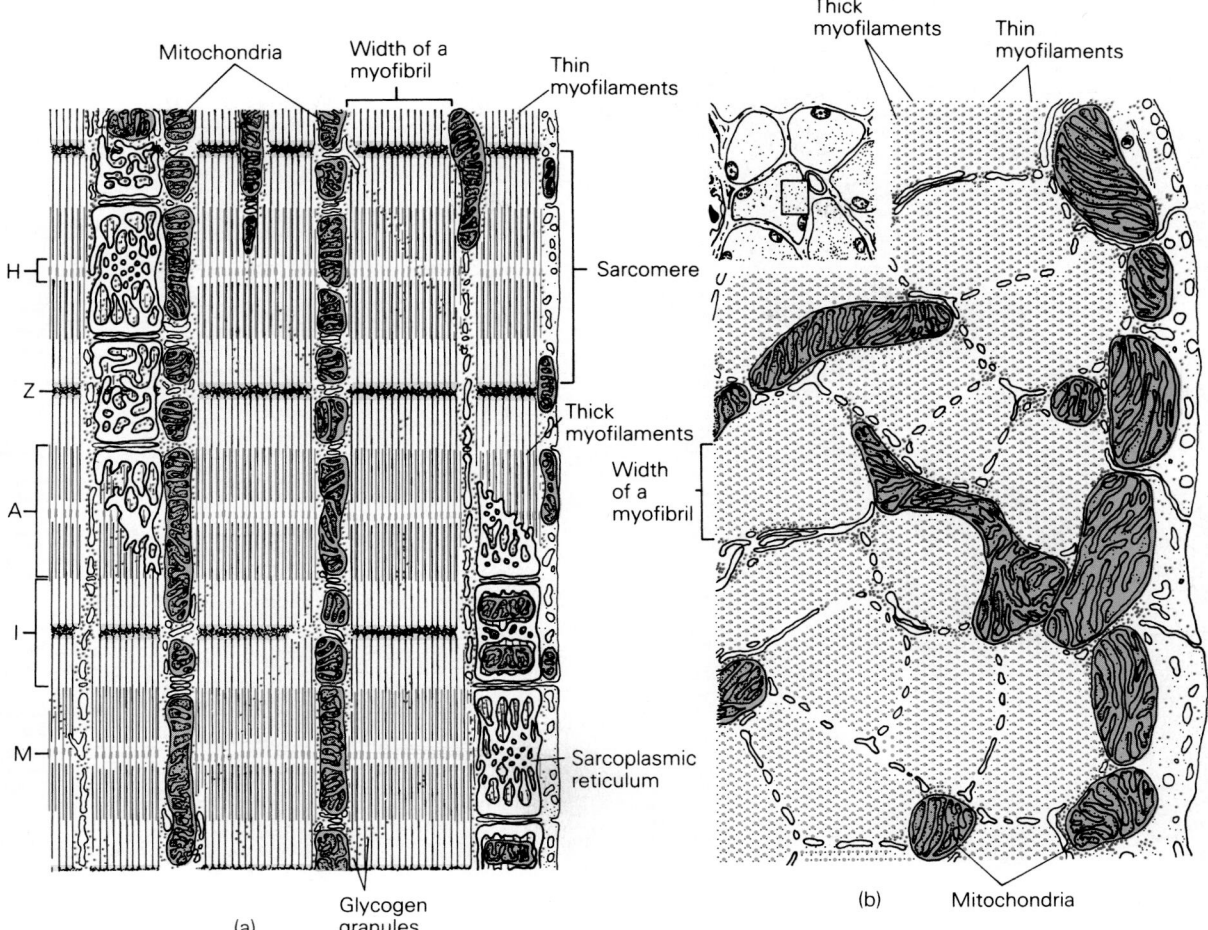

Figure 16–5
The microscopic structure of skeletal muscle. These drawings were made
from electron micrographs of skeletal muscle. (*a*) A longitudinal section
showing the arrangement of the sarcomeres. The letters H, Z, A, I, and M
refer to the parts of the sarcomere diagrammed in Figure 16–6. (*b*) A cross
section of the same tissue. Several cells are shown in the inset at the upper
left; the area in the rectangle is magnified in the rest of the figure. (Modified
from Krstic, R.V., *Ultrastruktur der Saugetierzelle*. Berlin: Springer-Verlag,
1976.)

The usual names given to the parts of the sarcomere are shown in Figure 16–6*b*. The center of each
I-band is marked by a structure called the **Z-line,**
which serves to mark the boundary of the sarcomere. (Sometimes this structure is called the **Z-disc**
in order to emphasize its two-dimensional nature.)
The function of the Z-line is to connect the ends of
the thin filaments from one sarcomere to those of
the next one in line. The Z-lines also have cross-
connections that keep the thin filaments in register
with each other in a lateral direction. In the center
region of the A-band, where the thin filaments of
the I-band have not penetrated, is a somewhat
lighter region called the **H-zone** (the M-line runs
through the center of the H-zone, since the filament
overlap is symmetrical).

During the process of contraction, in which the
muscle may shorten considerably, *neither the thick
nor the thin filaments change their length.* All of the
change in the length of the muscle is accounted for
by a greater or lesser degree of overlap between the
thick and thin filaments. This means that when the
muscle shortens, the length of the A-band stays the
same while that of the I-band decreases (see Fig.

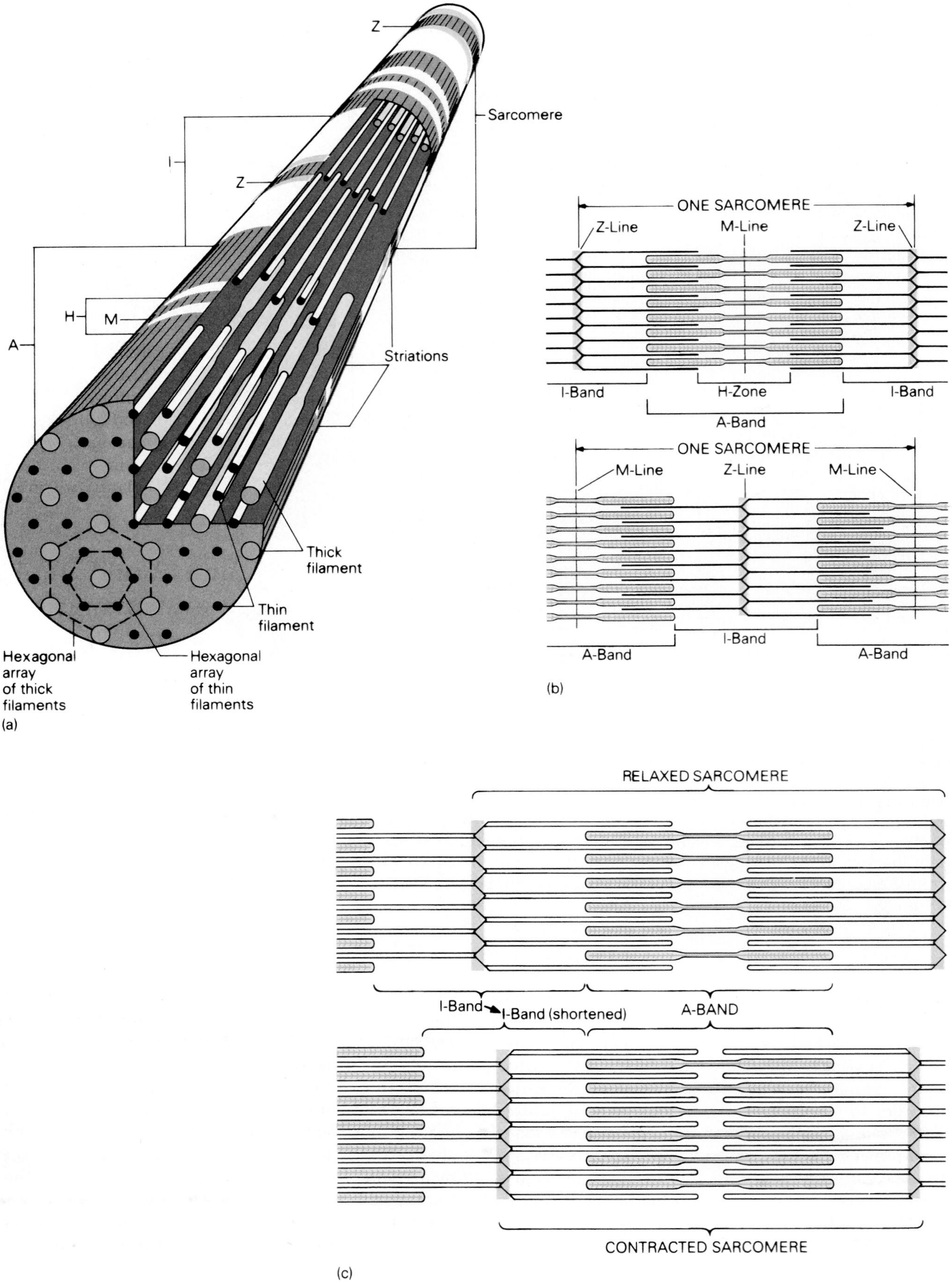

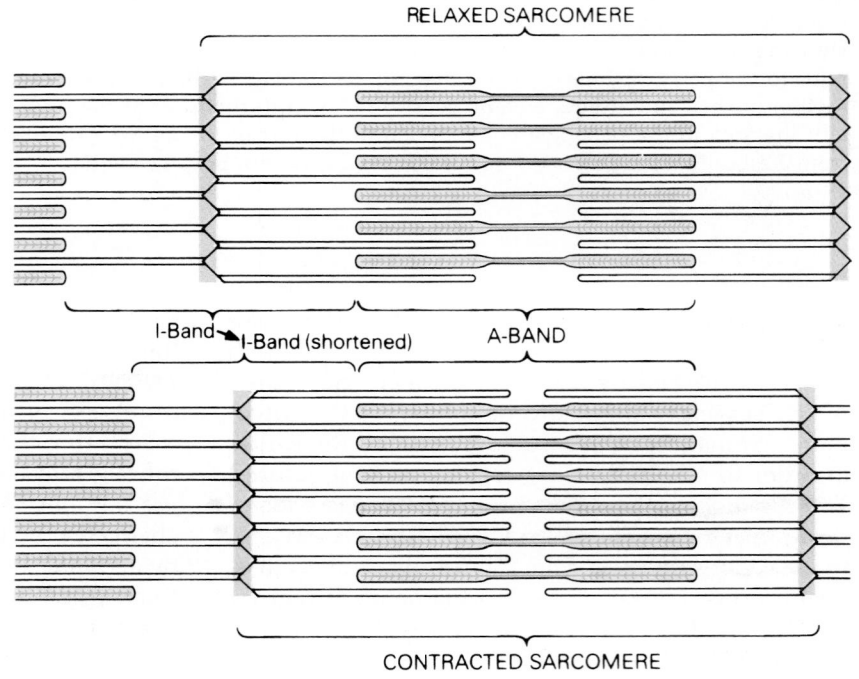

(a)

Z

Sarcomere

I

Z

H

M

A

Striations

Thick filament

Thin filament

Hexagonal array of thick filaments

Hexagonal array of thin filaments

(b)

ONE SARCOMERE

Z-Line M-Line Z-Line

I-Band H-Zone I-Band

A-Band

ONE SARCOMERE

M-Line Z-Line M-Line

A-Band I-Band A-Band

(c)

RELAXED SARCOMERE

I-Band I-Band (shortened) A-BAND

CONTRACTED SARCOMERE

16–6c). The thin filaments themselves do not change in length. Shortening of the muscle is brought about by a cyclic "rowing" motion of the crossbridges, in which they successively attach to the closest available thin filament site, change their shape, then let go of the actin and re-attach further along. This process will be discussed in more detail.

Hundreds of sarcomeres connected in series (end-to-end) form a functional unit called a **myofibril,** which runs for the length of the cell. Most muscle cells contain a large number of parallel myofibrils, and the intracellular spaces between them are occupied by **mitochondria,** and by molecules of glycogen, a source of chemical energy. Space between the myofibrils is also occupied by the **internal membrane system.**

Structure of the Myofilaments.   **Thick filaments** are made up of a very large protein called **myosin** (Fig. 16–7). A thick filament is composed of about 300 to 400 myosin molecules. Each myosin molecule has a long rod-shaped section (a structural portion sometimes called the **tail region**) and a double globular head at one end. The head portion of the molecule has the biochemical and enzymatic properties that participate in the actual contractile function. On the head portion are protein chains called **light chains** because of their molecular weights. These will be discussed in detail later.

When myosin molecules are packed together to form a thick filament, the tail regions of the molecules make up the structure of the filament itself, while the head portions project from the sides of the filament. The projections of the myosin heads are what form the structures called **crossbridges.** Because of the way the molecules are packed, the heads project from the filament in a radial fashion, and the projection makes a full turn for every six myosin molecules. The angle between the successive myosin heads is thus 60 degrees, and they project out of the filament every 14.3 nanometers. In this way, they form a helical pattern (like a spiral staircase) that winds down the length of the thick filament.

The tail regions of molecules at either end of the filament point toward its center, making the thick filament **symmetrical** about its center and **bi-directional.** It also means that there is a **bare zone** in the middle of each thick filament; this is an area containing only the tail portions of myosin molecules. These arrangements are critical to the way in which skeletal muscle functions.

A globular protein called **actin** makes up the **thin filaments.** The slightly oblong actin molecules are each about 5.5 nanometers in diameter and are joined end-to-end to form a long structure that resembles a bead-chain. Two such chains are wound about each other so that the pair makes a half-turn every seven actin monomers. This means that the thin filaments are also helical structures, with a half-pitch of approximately 35 to 37 nanometers (compared with the 42.9 nm pitch of the thick filaments). Each half-turn of seven actin molecules is also associated with a long fibrous protein molecule called **tropomyosin.** Each of these tropomyosin molecules bears a protein complex called **troponin.** The troponin is in turn composed of three subunits whose names are **troponin-I, troponin-T,** and **troponin-C.** Although it is the actin protein of the thin filaments that interacts with the myosin crossbridges of the thick filaments, the troponin-tropomyosin system has an important function in regulating the contraction of skeletal muscle.

The Internal Membranes of a Muscle Fiber.   These membrane structures participate in the control of contraction and relaxation. (See Fig. 16–8 for their anatomical arrangements.) The **plasma membrane** and a thin layer of connective tissue fibers form the outer covering of the cell, called the **sarcolemma.** Extending inward from the sarcolemma into the depths of the cell is a set of membrane-lined passageways, the **transverse tubules** (often called **t-tubules** for short). Depending on the type of muscle being considered, the t-tubules may enter the fiber either at the region where the A- and I-bands overlap, or at the level of the Z-discs. The t-tubules are an inward extension of the plasma membrane,

◀ *Figure 16–6*

The structure of a sarcomere. (*a*) Diagram showing the relationship of the longitudinal and cross-sectional arrangement of the myofilaments. (*b*) The A-band is made of thick filaments arranged side by side; the thin filaments extend from the A-line into the A-band. The I-bands are made up of the portions of thin filaments where there is no overlap between thick and thin filaments, and the H-zone is the region of the A-band into which the thin filaments do not extend. The M-line runs down the center of the A-band and appears to help in maintaining the uniform positions of the thick filament array. A sarcomere may also be defined as the distance between M-lines, with the Z-line at its center, since the sarcomeres repeat hundreds of times down the length of a myofibril. (*c*) During contraction, the width of the A-band stays constant, while the H-zones and the I-bands become narrower and the Z-lines move closer together.

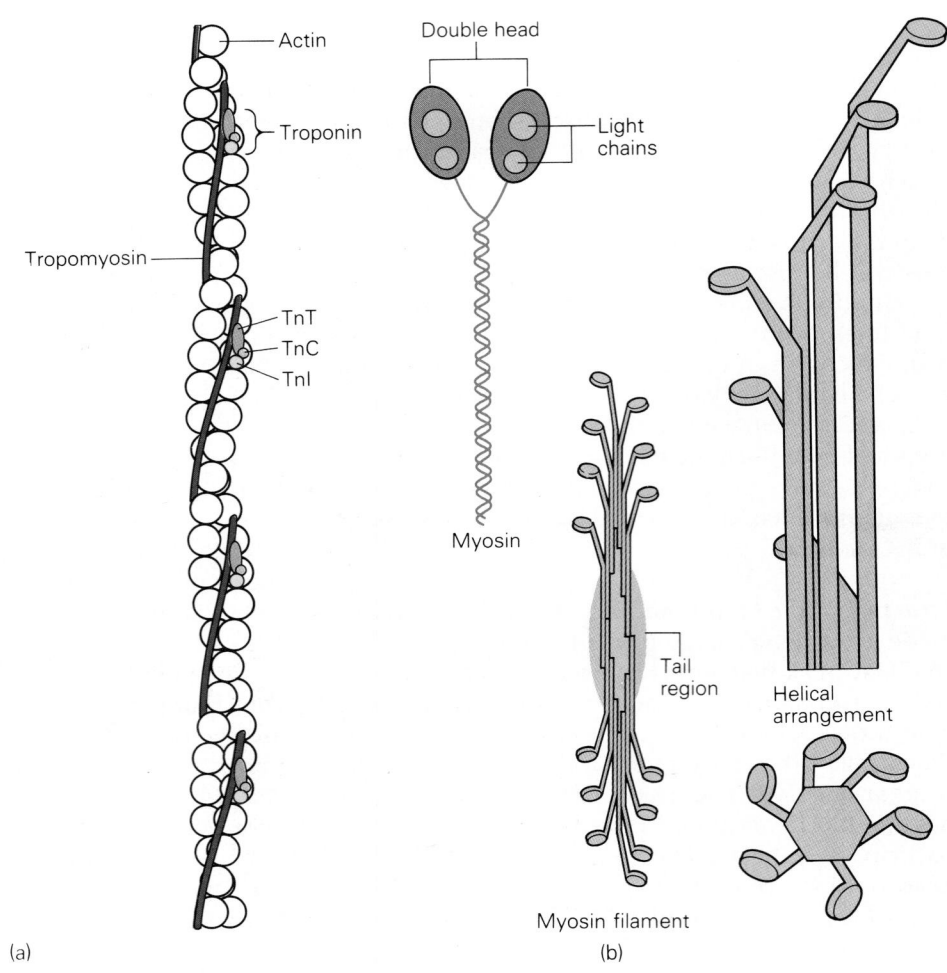

(a)

Double head

Light chains

Myosin

Tail region

Myosin filament

Helical arrangement

(b)

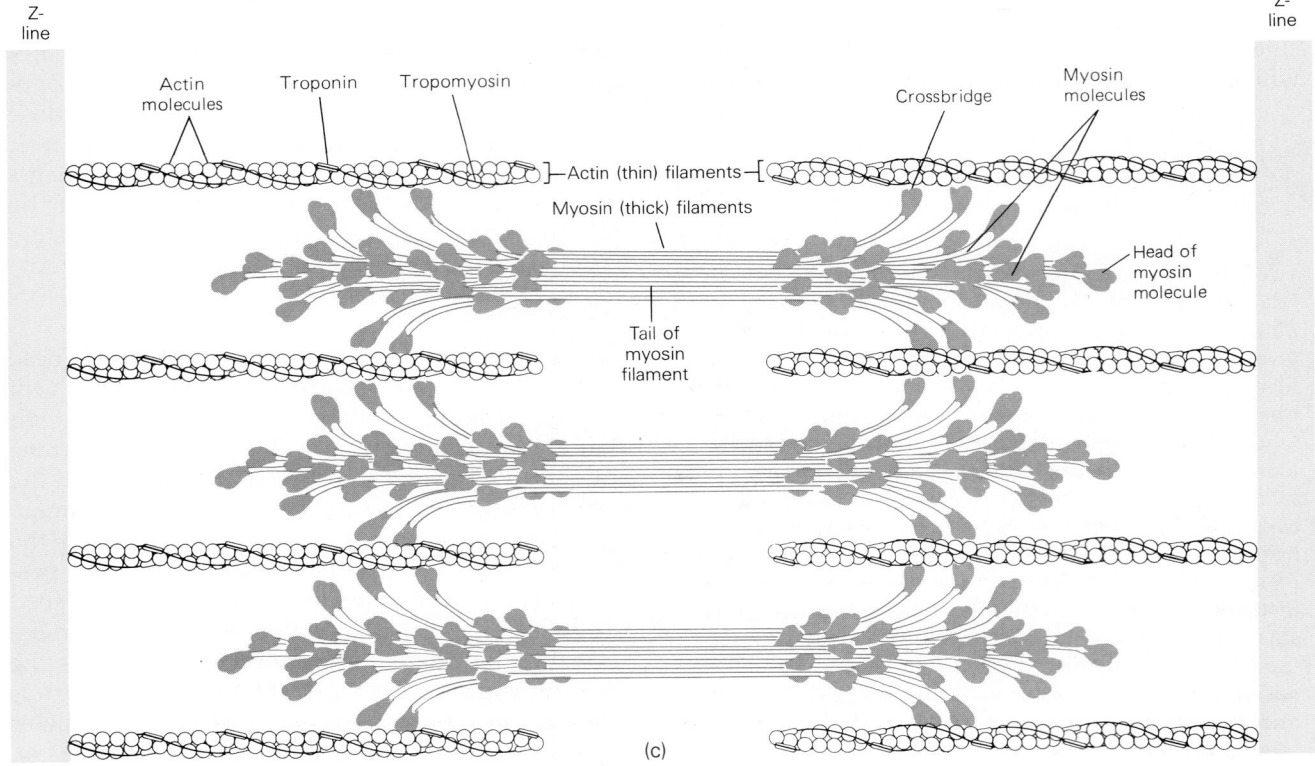

(c)

◀ *Figure 16–7*

The structure of the myofilaments. (*a*) The actin filaments are made up of two chains of action monomers twisted one-half turn for every seven monomers. The regulatory protein troponin (composed of three subunits) is associated with the fibrous protein tropomyosin, which extends along a group of seven monomers in the filament. (*b*) The structure of the individual myosin molecules. The myosin molecules are assembled into a two-ended filament with their heads protruding out to form crossbridges and their tail portions forming the main body of the filament. (*c*) Diagram showing how myosin and actin filaments are arranged in a sarcomere.

Figure 16–8

The internal membranes of skeletal muscle showing the structure of the t-tubule system and the sarcoplasmic reticulum. (Modified from Krstic, R.V., *Ultrastruktur der Saugetierzelle*. Berlin: Springer-Verlag, 1976.)

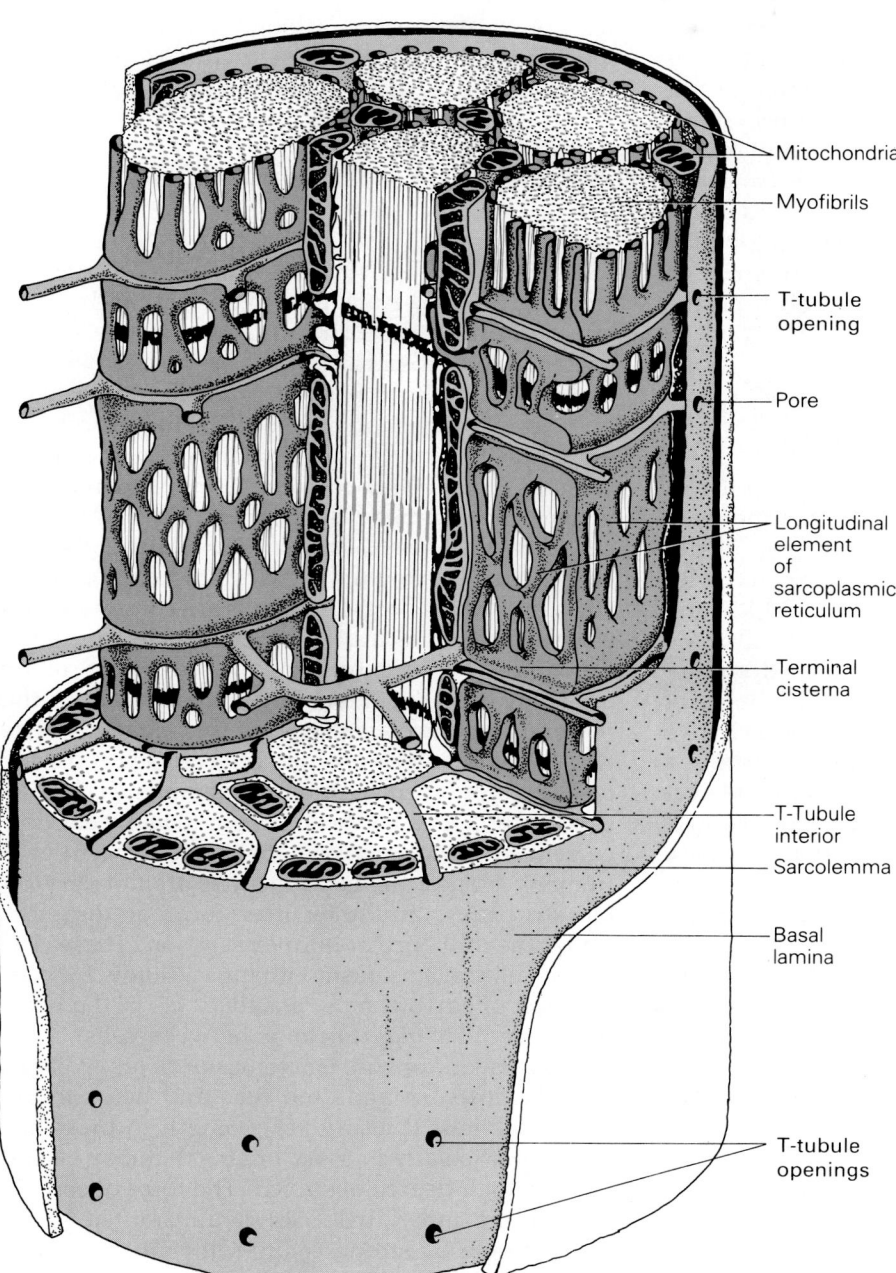

- Mitochondria
- Myofibrils
- T-tubule opening
- Pore
- Longitudinal element of sarcoplasmic reticulum
- Terminal cisterna
- T-Tubule interior
- Sarcolemma
- Basal lamina
- T-tubule openings

and the interior of the t-tubule system is continuous with the extracellular space. Closely associated with the t-tubular system is another set of membranes that run at right angles to the t-tubules and down the length of the sarcomere. This is the **sarcoplasmic reticulum.** Unlike the t-tubules, the sarcoplasmic reticulum is not connected to the extracellular space; however, it does come into close contact with the t-tubules as they course across the fiber. This region of contact, which consists of the t-tubules and the enlarged end portions of the sarcoplasmic reticulum (called **lateral sacs** or **terminal cisternae**), is called a **triad** in skeletal muscle because it is made up of one t-tubule and the terminal cisternae from two adjacent sarcomeres. Periodic structures (sometimes called "**feet**") appear to join the two membrane systems at the triad. These internal membrane systems function in the control of the contraction and relaxation of the muscle. (Details of this process and further aspects of the anatomy will be discussed later.) The function of the "feet" structures is not fully understood, but they are thought to play a role in communicating the signal for contraction from the t-system to the interior of the fiber.

The Structure of Cardiac Muscle

The cellular structure of heart muscle is similar in many ways to that of skeletal muscle, but there are also some very important differences (Fig. 16–9). The cells of heart muscle are small compared with those of skeletal muscle. They are from 5 to 15 micrometers in diameter and may be from 20 to 30 micrometers long. The cells are connected end-to-end and, to a lesser extent, side-by-side. Some of the cells may branch so that one end connects to two other cells. At the regions of connection between the cells is a specialized structure called the **intercalated disc.** This is an area of extremely close contact in which intercellular space is essentially absent; many projections from one cell extend into another, and the other cell has the mirror image of this arrangement. There is a strong mechanical connection between cells in this region, as well as a close electrical connection. The mechanical connection is provided by a cellular structure called a **desmosome,** and the electrical connection is made through a **gap junction** (a structure found also between smooth muscle cells). The specialized contact structure of the intercalated disc allows a tissue made up of small cells to function in many ways as though it were a single large cell, and this forms the cellular basis of the coordinated action of the muscle in the beating heart.

The intracellular organization of cardiac muscle is the aspect in which cardiac muscle is most like skeletal muscle. The structure and arrangement of skeletal and cardiac myofilaments are quite similar, as is their protein composition. The similar contractile structure gives rise to the same sort of striated appearance that skeletal muscle has, and many of the mechanical responses are similar. Because the muscle of the heart has a largely **aerobic** metabolism, it contains numerous mitochondria. A transverse tubular system is present in muscle from many areas of the heart, although the tubules are larger in diameter and usually make close contact with only one portion of longitudinal sarcoplasmic reticulum. This results in a structure called a **dyad,** rather than a **triad** as in skeletal muscle.

The Structure of Smooth Muscle

The Ultrastructure of Smooth Muscle

Although it is very similar to skeletal and cardiac muscle in the molecular structure of its contractile proteins, smooth muscle has a very different cellular and tissue structure. (Drawings made from electron micrographs are shown in Fig. 16–10.) Its cells are quite small, 5 to 15 micrometers in diameter and perhaps 200 to 500 micrometers long, depending on which type of muscle is examined. The cells have a single nucleus, centrally located, and the ends taper off to a small diameter. In many smooth muscle tissues, numerous portions of cells are connected very closely together by a structure called a **nexus,** or **gap junction.** This structure involves a fusion of the outer layer of the membrane of each cell so that there is no extracellular space at this region between the cells. The interiors of the cells are still separated by the inner layers of the cell membranes, but the gap junction area has been shown to allow electrical current to pass more easily between adjacent cells. In many tissues, the gap junctions are not very stable structures, and they may form or disappear under the influence of hormonal action. These close cell-to-cell contacts in smooth muscle allow for coordination of smooth muscle activity (as in the heart, which also has gap junctions between cells).

A network of connective tissue fibers (collagen and elastin) surrounds the cells and helps to link them together. It also gives strength to the whole tissue, particularly in those organs that may be subject to a high degree of stretch. The force of contraction of the individual cells is transmitted to the whole tissue by strong cell-to-cell connections and by the connective tissue network.

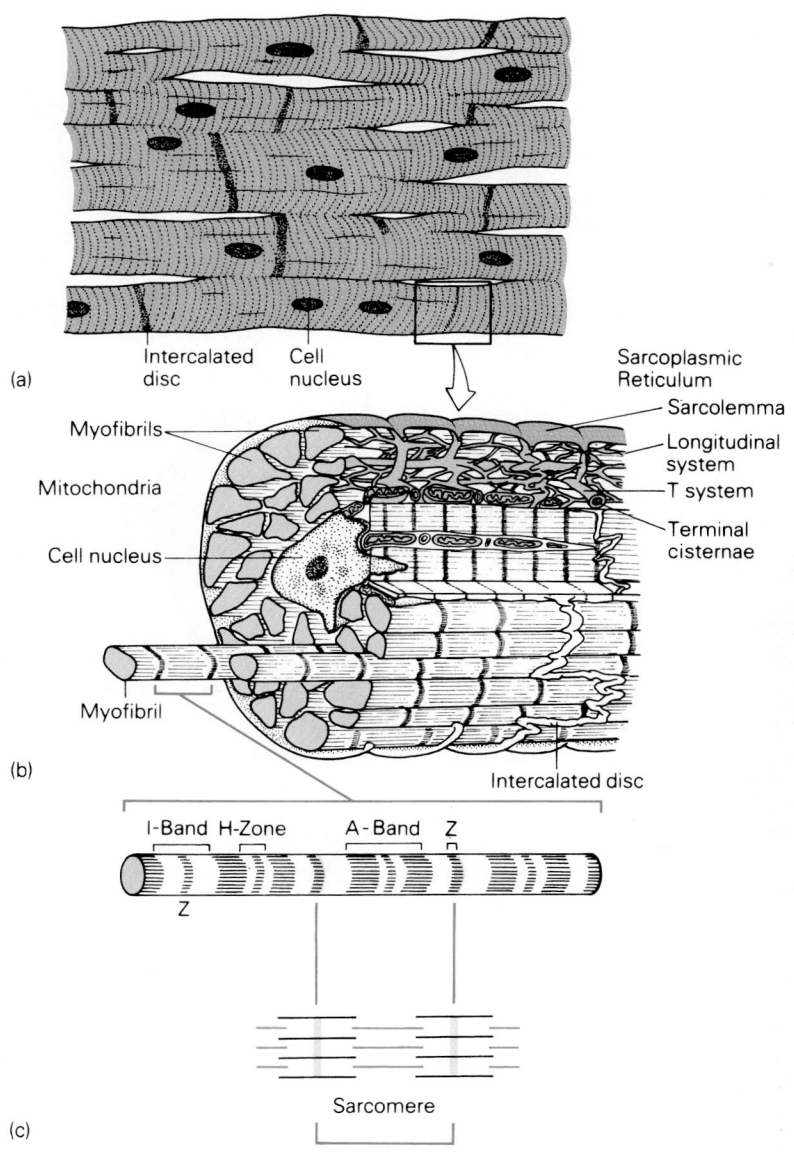

(a)

Intercalated disc   Cell nucleus   Sarcoplasmic Reticulum

Myofibrils   Sarcolemma

Mitochondria   Longitudinal system

Cell nucleus   T system

Terminal cisternae

Myofibril

(b)

Intercalated disc

I-Band   H-Zone   A-Band   Z

Z

Sarcomere

(c)

(d)

Figure 16–9

The structure of cardiac muscle. (*a*) The cells of the muscle are small, sometimes branched, and they attach end-to-end. (*b*) Seen at higher magnification, bundles of myofibrials run the length of the muscle, and a sarcoplasmic reticulum and t-system, which function much as they do in skeletal muscle, are present. (*c*) The sarcomere is organized in the same way that it is in skeletal muscle (Modified from Braunwald, E., Ross, J. and Sonnenblick, E. H. *Mechanisms of Contraction of the Normal and Failing Heart.* Boston: Little, Brown, 1968.) (*d*) Electron micrograph of the longitudinal section through an intercalcated disk of cardiac muscle. (© Biophoto Associates/Photo Researchers, Inc.) (*e*) Transmission electron micrograph of cardiac muscle showing the intercalated disks in perspective. (© Don W. Fawcett/ Visuals Unlimited.)

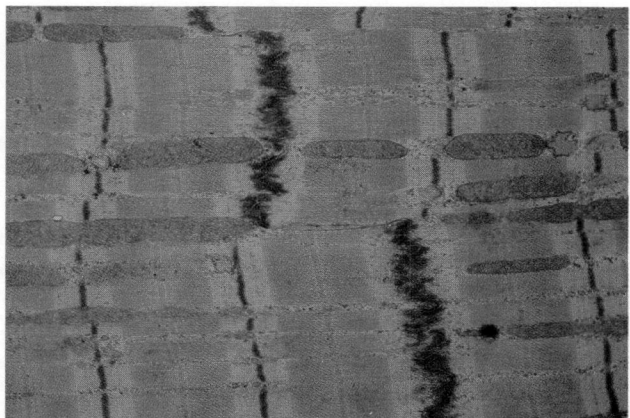

(e)

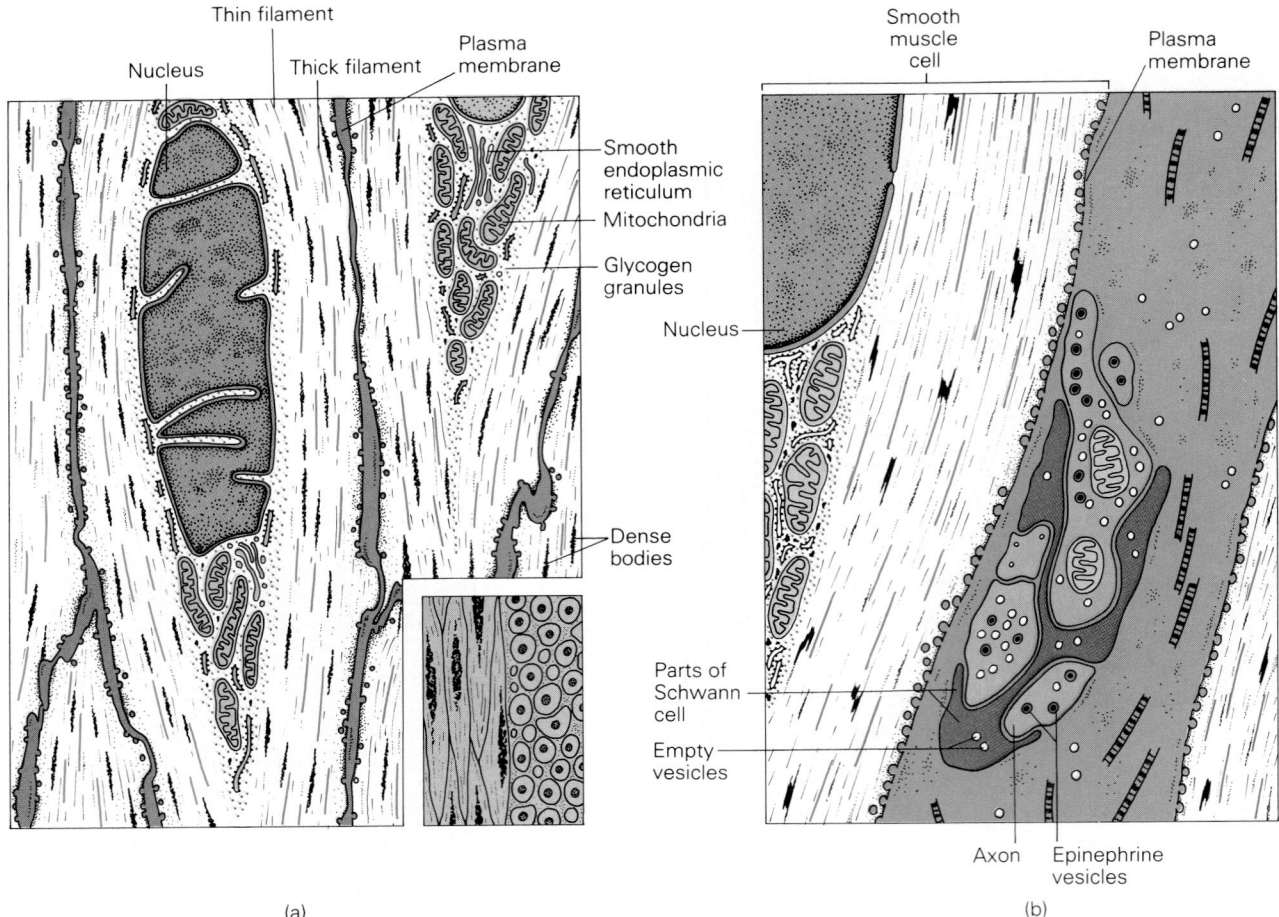

(a)

(b)

Figure 16–10
The structure of smooth muscle. (*a*) Drawing based on an electron micrograph of a longitudinal section of parts of several smooth muscle cells. It is the area enclosed in the rectangle in the inset portion at the lower right. Most of the cell organelles tend to cluster at the ends of the nucleus. (*b*) A nerve-muscle contact between a smooth muscle cell and a small bundle of autonomic nerve fibers. There is no specialized region of contact. (Modified from Krstic, R.V., *Ultrastruktur der Saugetierzelle.* Berlin: Springer-Verlag, 1976.)

Mechanical Arrangements of Smooth Muscle

Smooth muscle makes up much of the walls of organs such as the stomach, intestines, and uterus. It is generally arranged in layers. In the small intestine, for example, one layer (the inner one) is arranged in a circular fashion around the lumen, and the outer layer runs lengthwise (longitudinally). When the circular layer contracts, the intestine is reduced in diameter, and when the longitudinal layer contracts, the intestine becomes shorter. Unlike skeletal muscles, which usually have antagonistic muscles to stretch them out after contraction,

smooth muscles must return to their pre-contraction length in a different way. An increase in the volume of the contents of the organ re-stretches the muscles in the walls. This usually comes about by the entry of new fluid (e.g., by newly formed urine in the bladder) or by movement of contents from elsewhere along a tubular structure (as in the intestine). Coordinated movements of smooth muscle organs such as the intestine often involve the alternate contraction and relaxation of the two muscle layers. In some saclike organs such as the uterus, muscle fibers are oriented in several directions, reducing the overall volume of the organ when the muscle contracts. The musculature of most blood vessels is simpler, usually consisting only of a circular layer of muscle. In very tiny blood vessels (e.g., an arteriole, which is pictured in Fig. 16–11), a single cell may wrap completely around the circumference of the vessel. Contraction of this muscle cell leads to a narrowing of the blood vessel and a reduction in the amount of blood flow. Such smooth muscle plays a vital role in the regulation of blood flow and pressure.

The Function of Muscle

Most of what muscle does is obvious. We are aware that it shortens and develops force to lift an object, and there is some apparent relationship between how heavy a load is and how fast it can be lifted. These everyday impressions have been refined into a more precise set of concepts that are used to describe, quantify, and compare the way in which muscle functions. Before we approach this more exact analysis of muscle function, some basic definitions are necessary.

The Nature of Skeletal Muscle Contraction

First of all, the term **contraction** has a special meaning. In everyday usage, it means "to shorten"; in the terminology of muscle physiology, it means any form of muscle activity in response to stimulation, whether it involves shortening or not. For instance, in an **isometric contraction** (Fig. 16–12*a*) (*iso* = same; *metric* = measure), the muscle does not do any external shortening; instead, it develops **force,** or **tension,** while pulling against immovable attachments. In an **isotonic contraction** (*tonic* = tension), the force in the muscle remains constant while the muscle shortens. In an **auxotonic contraction,** the force continually increases as the muscle shortens (e.g., pulling back a bowstring or stretching a rubber band involves this sort of contraction). A **meiotonic** contraction, however, is one in which the force *lessens* as the muscle shortens. Many useful mechanical devices, such as gearshift levers and typewriter keyboards, allow this type of contraction in order to provide a clear indication that an action has been performed. These defined conditions represent special circumstances, and realistic contractions usually involve some combination of them. For purposes of analysis, however, contraction conditions can be set up to emphasize the specific properties desired.

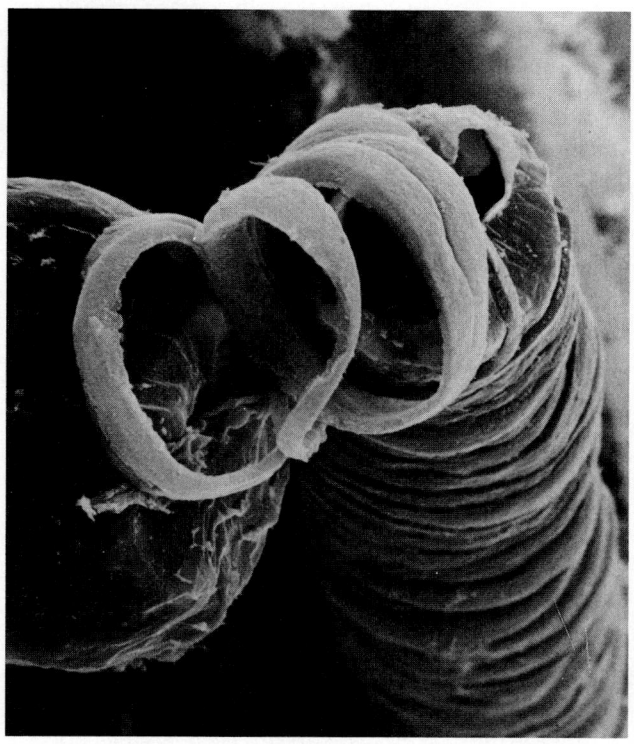

(a)

Blood cells   Smooth muscle

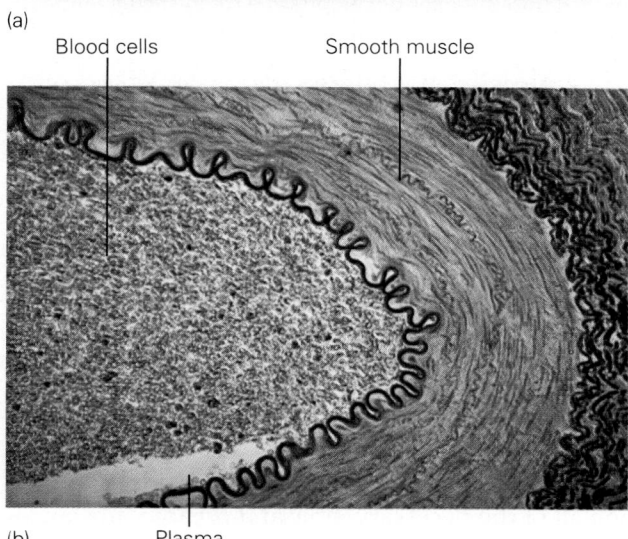

(b)   Plasma

Figure 16–11 (*a*) An example of the arrangement of smooth muscle cells in the wall of a blood vessel. This is a scanning electron micrograph of a small blood vessel called an arteriole. At the cut end of this vessel, some of the smooth muscle cells have been spread apart so that they may be seen better, while further down the vessel the cells are in their natural position. Contraction and shortening of the smooth muscle cells, which turn in a circular direction, would cause the arteriole to become smaller in diameter. It is by this means that smooth muscle can regulate the flow of blood to the tissues. (SEM courtesy of Dr. A. Evan.) (*b*) Photomicrograph of a cross section of an artery. Note the circular arrangement of the smooth muscle. (© Stan Elems/Visuals Unlimited.)

RESEARCH FOCUS

Modern Smooth Muscle Research

In the past, much of the research on smooth muscle has concentrated on comparing its function with that of skeletal muscle. This work has established the basic similarity of the processes of contraction and has also pointed out some of the highly specialized functions of smooth muscle. Because of the vital importance of these specializations in understanding human physiology in health and disease, much research is now focused on learning about the molecular adaptations of smooth muscle.

A wide variety of research tools are being applied in modern smooth muscle research. Improvements in electron and optical microscopy have provided a detailed look at the structure and arrangement of the contractile apparatus. Advances in biochemistry and molecular biology have allowed investigators to study the complex interactions between calcium as the activator of contraction and the processes regulating the contractile proteins. New techniques for observing the function of cell membranes have shown that there are a number of separate ways in which messages from the body (whether by neurotransmitters, circulating hormones, or locally-produced signaling substances) are translated into contraction and relaxation of smooth muscle. There have been many successful attempts to correlate mechanical events in a contraction with the biochemical events that control them; much of this work uses smooth muscle tissues whose cell membranes have been made permeable (with detergents or bacterial toxins), while leaving the contractile proteins and associated controlling molecules intact. Such experimental preparations, which contract and relax much like intact muscle, can be activated directly with calcium, and the internal concentrations of ATP and other substances can be closely controlled. The contractile function of single isolated smooth muscle cells, attached to highly sensitive force- and length-measuring equipment, has been studied in some detail, and recent advances in microscopy and immunological staining techniques have permitted the observation of the function of individual actin and myosin molecules completely removed from the cells. Important insight has been gained into the roles of smooth muscle in such health problems as hypertension (high blood pressure), vascular disease, and asthma, and new avenues for therapy and treatment of such diseases are being developed as a result of active research in the area of smooth muscle function.

Isometric Contractions in the Laboratory: Clues to the Sliding-Filament Hypothesis of Contraction

Under laboratory conditions, an isolated skeletal muscle (and other muscle types as well) can be made to produce contractions conforming to any of these definitions. The relationships between muscle and its external mechanical environment are usually studied by using some "standard" experimental conditions. The simplest type to setup (as diagrammed in Fig. 16–12) can record **isometric contractions.** Here the muscle is securely attached to a device called a **force transducer,** which produces an electrical signal that varies in direct proportion to the force the muscle exerts. The other end of the muscle is attached rigidly to a fixed support (whose position the experimenter may adjust). All connections, and the force transducer itself, are made very stiff so as to prevent any shortening of the muscle. The force signal from the transducer is fed to a device such as a **chart recorder,** which produces an ink-on-paper graph of the force as it changes with time. (An **oscilloscope,** which produces a similar, although temporary, record on a **cathode ray tube,** may also be used.) The muscle is kept alive by being immersed in a physiological saline solution containing all of the necessary ionic and organic substances (along with oxygen and an energy source) that it requires.

The Muscle Twitch. A single electrical stimulus to a skeletal muscle sets off a series of electrical and chemical events that result in a single, brief contraction called a **twitch.** On the graph drawn by the chart recorder, the force rises rapidly to a peak and then declines a bit more slowly back to the resting value. For a given stimulus sufficient to activate all of the muscle fibers, and under a constant set of temperature and metabolic conditions, all twitches made at the same muscle length will produce similar contraction records.

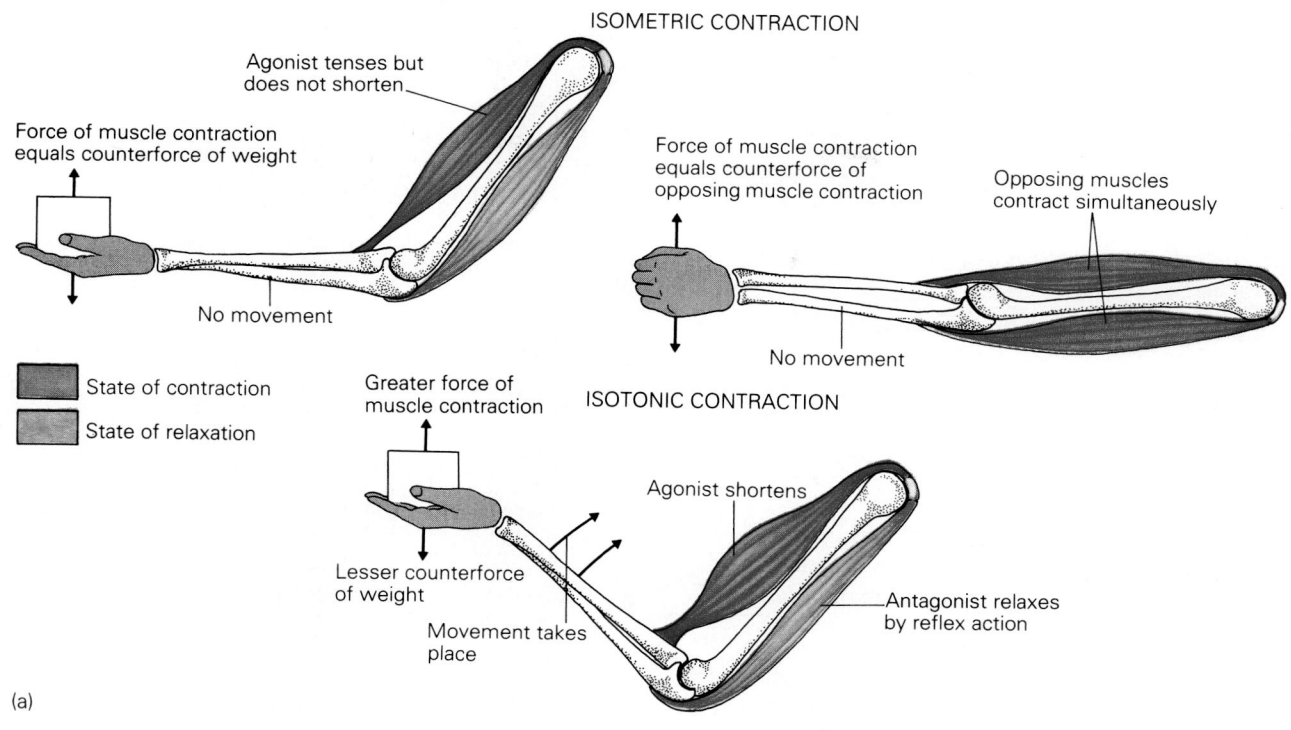

ISOMETRIC CONTRACTION

Agonist tenses but does not shorten

Force of muscle contraction equals counterforce of weight

No movement

Force of muscle contraction equals counterforce of opposing muscle contraction

Opposing muscles contract simultaneously

No movement

State of contraction

State of relaxation

Greater force of muscle contraction

ISOTONIC CONTRACTION

Agonist shortens

Lesser counterforce of weight

Movement takes place

Antagonist relaxes by reflex action

(a)

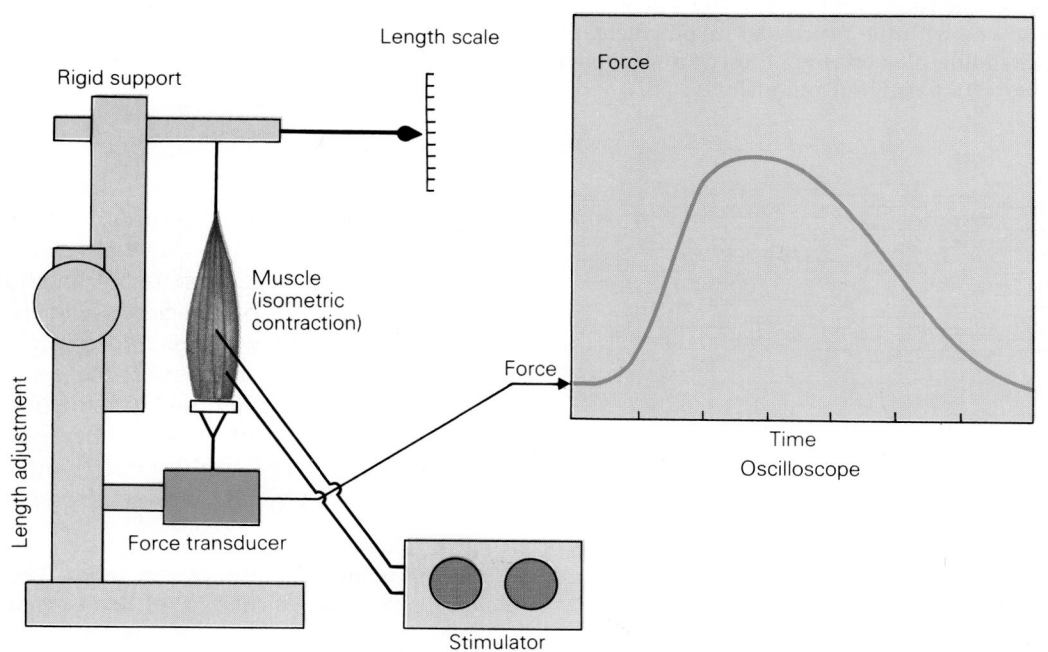

Length scale

Rigid support

Force

Muscle (isometric contraction)

Length adjustment

Force

Force transducer

Time
Oscilloscope

Stimulator

(b)

Figure 16–12
(*a*) Diagrams of isotonic and isometric contraction mechanisms. (*b*) Typical apparatus for measuring isometric contraction. The length of the muscle may be varied between contractions, and the record of the contraction is viewed on the face of an oscilloscope. The muscle is stimulated with an electric shock.

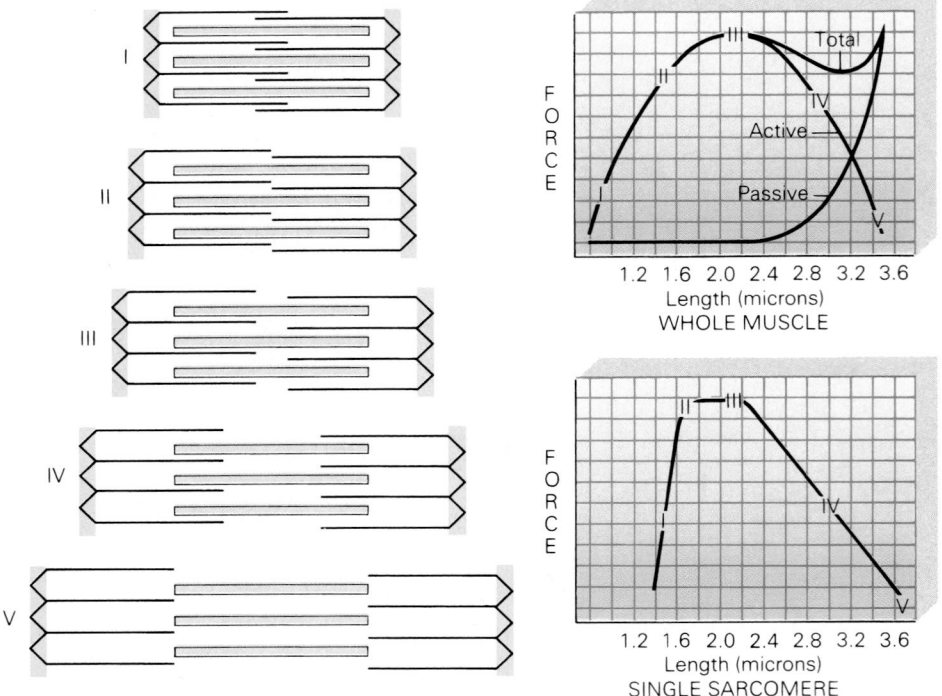

Figure 16–13
The length-tension curve. The upper portion shows how the resting length affects the tension (or force) produced by a muscle during an isometric contraction. The curve labeled "Total" represents the "Active" force plus the "Passive" force. The key to understanding the basis for this curve lies in the sarcomere diagrams to the left. They show the relative amounts of overlap in most of the sarcomeres of the muscle at the various resting lengths. This may be better seen in the lower graph, the length-tension curve of a single sarcomere. This curve is made up of straight lines that correlate very well with the overlap of myofilaments; this correlation is one of the main bits of evidence for the sliding-filament hypothesis. In the upper curve, the collective effect of millions of sarcomeres obscures the simple relationship.

(a)

(b)

The Isometric Length-Tension Curve. The mechanical factor with the most profound effect on an isometric contraction is the *length* at which the muscle is held. Figures 16–13 and 16–14 illustrate the relationships that will be described shortly. Making a series of isometric twitches, and changing the muscle length *between* (not during) the contractions, allows an important relationship to be observed. First the muscle is held at a very short length (shorter than it would ever be in the body) and then stimulated. If it is then progressively stretched out be-

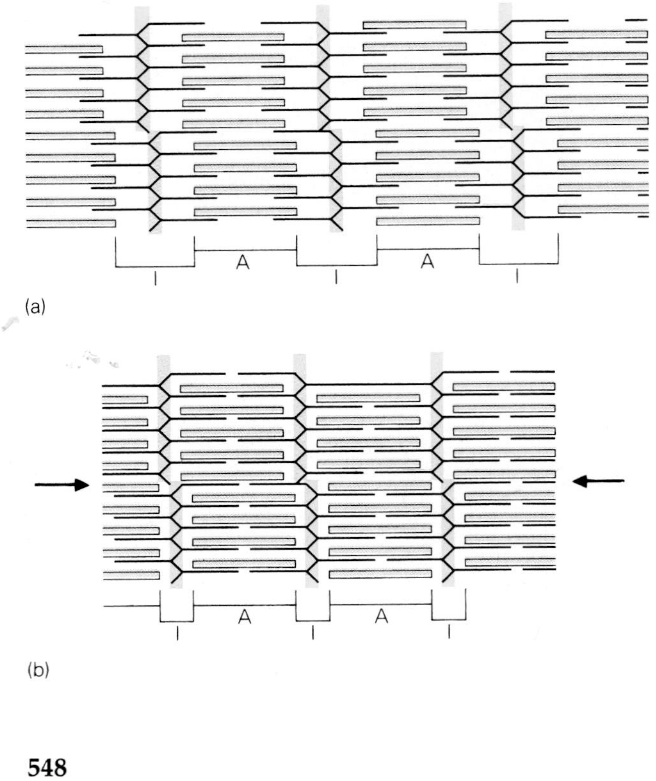

Figure 16–14
The shortening of sarcomeres in series. The upper portion represents a "snapshot" of two sarcomeres somewhere along a myofibril. There is about 50% overlap of thick and thin filaments at this length. The lower portion shows sarcomeres from a shortened muscle. Here the overlap is complete, and three sarcomeres now fit roughly in the space of two. The shortening of all of the sarcomeres in a muscle is added together, and the tiny amount of shortening of each sarcomere results in a very large shortening of the whole muscle.

tween contractions, the peak amount of twitch force will increase each time until some length is reached at which the twitch force is greatest. From this point on, further lengthening will result in two effects. First, force will begin to be measured in the muscle even before the stimulus is applied. This is called the **resting** (or **passive**) **force** and is due mostly to the connective tissue that holds the muscle together, acting much like a rubber band as it is stretched. Secondly, the amount of force (the so-called active force) that the muscle can develop over and above the resting force now diminishes as the muscle is lengthened further. At an extreme degree of stretch (provided that the muscle has not torn), there will be a very high resting force and very little active force in response to stimulation.

This relationship, in which the force capability of the muscle is greatest at some intermediate length, is called the **isometric length-tension curve**. Usually the maximum force capability occurs at a length that is close to the natural length of the muscle in the body. This length is often termed the **optimal length** or L_o, while the peak force at this length is designated P_o or F_o.

While this length-tension behavior is very important to an understanding of how muscle works, most skeletal muscles are limited in how far they can be stretched out by their attachments to the skeleton, and the length-tension curve is of relatively little importance in normal function. Heart muscle, however, does not have its length changes limited by skeletal attachments, and its resting length is set by the amount of blood that returns to fill the heart between beats. The length-tension curve thus allows the heart to make a more forceful contraction if it receives a larger filling of blood between contractions.

Optimal Length and the Overlap of Myofilaments. Why does the resting length have such an important effect on the contraction of the muscle? The answer to this question has provided some important clues to the basic physiology of skeletal muscle, clues that relate the structure of the muscle to its function. Careful measurements with the electron microscope have shown that the A-bands do not change in width as the muscle is lengthened or shortened, but the width of the I-bands increases as the muscle is stretched (and vice-versa). The spacing between the Z-lines also changes in proportion to the muscle length. Other measurements have shown that the length of the myofilaments themselves also does not change. What does change, however, is the amount of overlap between thick and thin filaments. Stretching the muscle decreases this amount of

overlap; in some muscles, the sarcomeres may be pulled completely apart. These observations (when compared with the length-tension curve of the living muscle) provided evidence that the overlap of the myofilaments controls the amount of force that the muscle can develop (see Fig. 16–13). During delicate experiments with single fibers of skeletal muscle in specially controlled isometric contractions, the sarcomere length of the living muscle was measured as the length of the muscle was varied. These studies showed that the amount of force developed was indeed directly related to the amount of myofilament overlap at lengths greater than L_o. Over a small region of the length-tension curve near L_o, the force was independent of the length, but as the muscle was shortened to lengths less than L_o, the force decreased.

Myofilament Overlap and the Amount of Force. These findings are interpreted in the light of the structural evidence for the existence of crossbridges and the biochemical evidence that it is the head (crossbridge) portion of the myosin molecule that interacts with actin and catalyzes the release of energy from ATP. When the muscle is stretched out beyond L_o, there is incomplete overlap between the thick and thin filaments, and not all of the crossbridges on the thick filaments have the chance to attach to an actin filament. Therefore, when the muscle produces force, the amount of force is proportional to the myofilament overlap. In the region of the curve near L_o, overlap is complete, and force is at its greatest. Because of the **bare zone** at the center of the thick filaments (where there are only myosin "tails" and no crossbridges), shortening the sarcomere does not allow any more crossbridges to attach to the thin filament as it slides further in. This is why the force does not change with length in this region. As the sarcomere shortens further, myofilaments from opposite sides of the sarcomere begin to interfere with each other, and less force can be produced. At the extreme shortening, where force has virtually disappeared, the thick filaments have been pushed against the Z-lines, and there is no further opportunity for shortening or developing force.

It has also been shown that the processes that activate the myofilaments do not function as well at short lengths, and this factor also reduces the force. In experiments with whole muscles, the regions of the length-tension curve are not as clear-cut, because the millions of sarcomeres making up the whole muscle are all at somewhat different places on this curve. For this reason, and because of the effects of the connective tissue, the length-tension curve for a whole muscle is a smooth curve without a flat region at its peak. Nevertheless, it is well established that the effects of length on muscle function do reflect a fundamental property of the molecular mechanisms responsible for muscle contraction.

The Sliding-Filament Model of Contraction.

The account of muscle function just given is usually called the **sliding-filament hypothesis.** This is a very well-established body of knowledge that integrates much of what is known about skeletal muscle into a consistent framework. Because cardiac muscle is similar to skeletal muscle in structure and function, the same basic hypothesis appears to be valid here as well. Less is known about the relationship between structure and function in smooth muscle because of the different cell arrangement and lack of obvious structural regularity. However, what is known about the details of smooth muscle function and structure is at least consistent with a sliding-filament hypothesis, and as more information becomes available such a mechanism will probably be experimentally supported.

Isotonic Contractions in the Laboratory: Clues to the Work of Muscle as a Motor

Analysis of isotonic contractions can reveal other aspects of muscle contraction. Because muscles shortening at a constant force are doing work, it is during isotonic contraction that the functions of muscle as a **motor** are best emphasized.

The equipment necessary to produce isotonic contraction from an isolated muscle is somewhat more complex than that used in isometric experiments. A force transducer is still used to measure the tension produced, but now the changes in muscle length must be measured, and a means of keeping the force constant during shortening must be provided. These requirements may be met by a **lever system** (Fig. 16–15). The muscle is attached to one end of a very light (but rigid) lever, and the desired load (such as a brass weight) is hung from the other end. The muscle is anchored at its other end to the force transducer. A support is placed under the weight so that when the muscle is at rest it is not stretched out. The position of this support sets the resting length of the muscle. If the muscle

Figure 16–15

A typical apparatus for measuring isotonic contraction. The equipment is a bit more complicated than that shown in Figure 16–12 because the muscle is now free to shorten if it develops enough force, and the length change must be recorded.

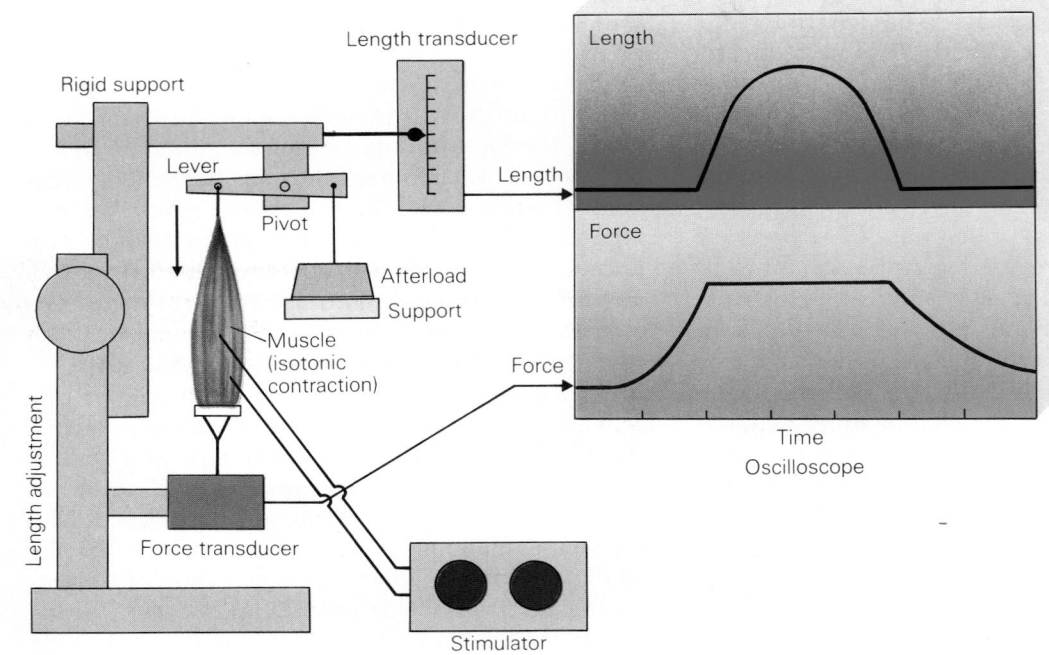

length is appropriately set (see Fig. 16–15), there will be no passive tension (recall this definition from the previous section). The force that the weight provides during contraction is called the **afterload,** because the muscle "feels" it only after beginning to contract. (If the muscle is stretched out by the attached weight so that some **resting tension** is present, this resting force is called the **preload.**)

The Mixed Contraction. When a stimulus is applied to an afterloaded muscle, the resulting contraction is at first **isometric.** It remains isometric until the muscle has developed an amount of force equal to the overload. Then the muscle begins to shorten and lift the load. Now the force will remain essentially constant and equal to the afterload as long as the load is lifted clear of its support. This is the **isotonic** portion of the contraction. As relaxation begins, the load is lowered (still isotonically) to the support. When the support prevents further lowering of the load, the muscle length no longer changes, and relaxation becomes isometric, just as the first portion of the contraction was. Only if no load at all were applied would the contraction be entirely isotonic. Under most circumstances, then, an isotonic contraction actually consists of an isometric portion followed by the isotonic shortening. Such a contraction is sometimes called a **mixed contraction.** With a large afterload, the isometric portion lasts relatively longer, because the muscle takes longer to build up enough force to lift the load.

Several important relationships can be expressed during isotonic (mixed) contractions (Fig. 16–16). First, the lighter the load, the sooner it can be lifted. (This information could actually also be obtained from the force record of an isometric contraction.) Secondly, the larger the load, the less the muscle is able to shorten. In fact, if the afterload is greater than what the muscle can lift, the contraction will be completely isometric.

The Force-Velocity Curve. Most importantly, the speed at which a muscle shortens is related to the amount of the afterload (see Fig. 16–16). This speed is measured at the very beginning of shortening, since the muscle continually slows down during most of the shortening. The lighter the afterload, the greater the speed (velocity) of shortening. As the afterload is made larger and larger (in subsequent contractions), the speed with which the muscle begins its shortening becomes less and less. The relationship is described by the **force-velocity curve.** The highest velocity of shortening, called V_{max}, oc-

curs with zero load. While useful contractions do not occur under conditions of no load, a measurement of V_{max} provides a good indication of the maximal rate at which the energy-converting processes of a muscle can operate. This information is often used for comparisons among different types of muscle. At the other extreme of loading, the lowest velocity of shortening is *zero,* and at this point the contraction is completely isometric.

The Power of a Muscle. The force-velocity curve emphasizes the function of muscle as a biological motor. It provides a basis for studying how our muscles interact with the loads and forces encountered in our environment. We use muscles to do **work,** which is defined as *moving a force through a distance.* A concept related to this is the notion of **power output,** defined as the **rate** of doing work. Analysis of the information expressed in the force-velocity curve shows that the value obtained by multiplying each force by the corresponding value of velocity is the power output of the muscle. A graph of the way in which the power output of muscle varies with the afterload (Fig. 16–17) force shows that at the maximum force that the muscle can attain, the power output is zero. Because the isometric muscle does not shorten, it does no work and therefore produces no power. A related situation occurs at V_{max}, the velocity at no load. Here shortening is the most rapid, but no force is exerted. Again, this means that the power output is zero. Between these two extremes, the power output passes through a maximum when the force is set to about 30% of its maximum value.

The Efficiency of Muscular Contraction. For many skeletal muscles, it is at about this point that the **efficiency** is the greatest. (Efficiency is defined as *the amount of work done for a certain input of energy.*) This relationship has some practical consequences for the way in which we do work (or play). An athlete (a long-distance runner, for instance) who needs to maximize endurance will run for most of the race at a pace that is less than "all out." Only in the final stages of the race will the runner strive for the greatest speed. This high-speed running is done much less efficiently than the running in the earlier part of the race and cannot be sustained as long. In human-powered machines, especially those designed to be operated for long periods of time, the force-velocity and power output properties of exercising muscle need to be considered. Machines such as bicycles (and now airplanes) that get their power from

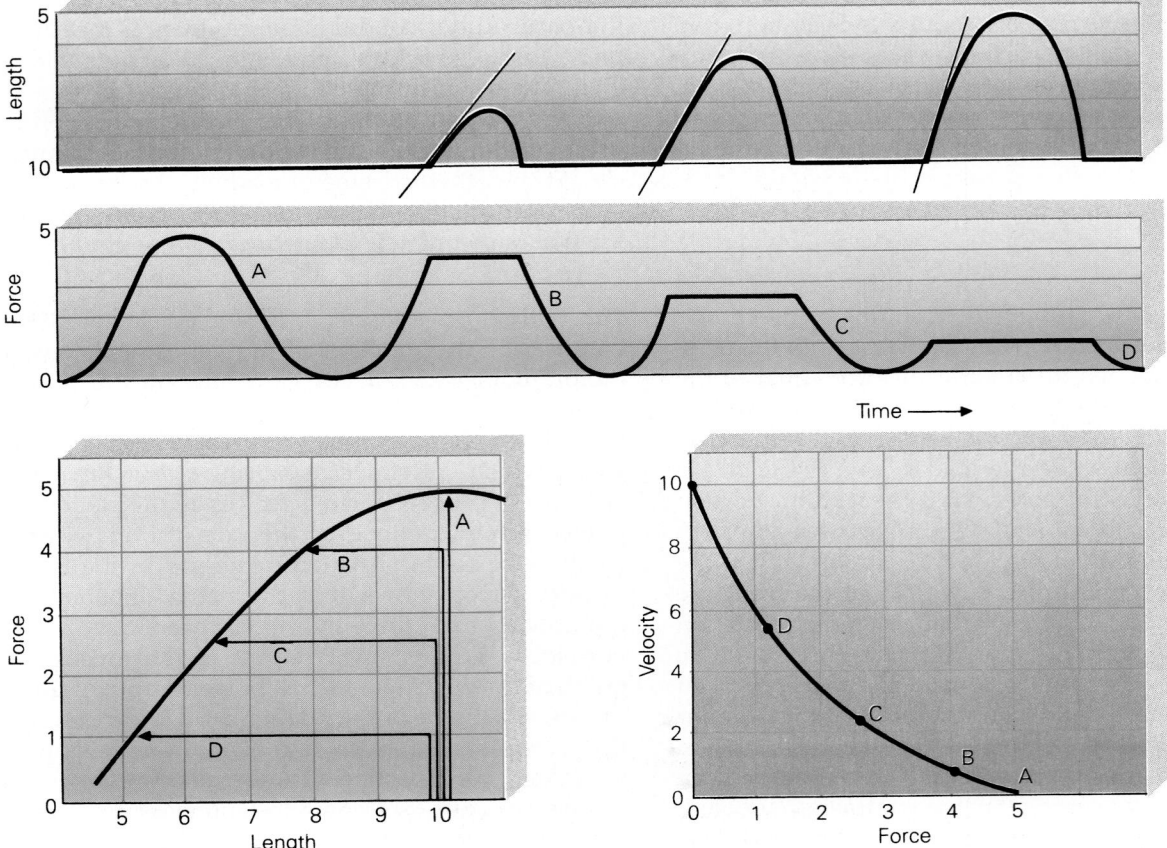

Figure 16–16
Force, velocity, length, and time in isotonic contraction. Isotonic contractions can be measured in several ways. (The letters A, B, C, and D refer to the same contraction in each of the viewpoints.) At the top of the figure is the usual display of force and length as they change with time. The same four contractions are on the length-tension curve axes at the lower left. Here the dimension of time is not shown, but it can be seen that the length-tension curve provides the limit to shortening at each of the different forces. The force-velocity curve at the lower right shows how the velocity of shortening (taken from the sloping lines in the uppermost data display) varies at different forces.

human muscle are designed so that the load that they present to the muscles is near to the optimum afterload on the force axis of the force-velocity curve. Many bicycles, in fact, are provided with a number of "gear ratios" so that the rider can match the muscle force to the leg-muscle force-velocity curves as the vehicle goes up and down hills, for example. While such considerations are not always important in everyday activities, all muscular activity is subject to the limitations described by the length-tension and force-velocity curves.

The Nature of Contraction in Cardiac Muscle and Smooth Muscle

The contraction of heart muscle also can be described by length-tension and force-velocity curves, but its mechanical situation in the body is much different from that of skeletal muscle. Because it is not attached to the rigid bones of the skeleton, it does not have such a strictly limited range of lengths over which it must operate. In addition, unlike skeletal muscle, which is set by its attachments to operate at

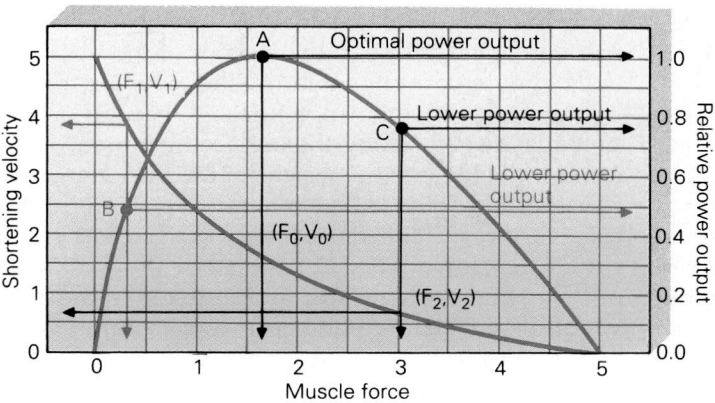

Figure 16–17
How the power output of muscle varies with the afterload. At the highest and lowest values of force, the power output is zero (see text). At an afterload of around one third the maximum (at A), the power output is maximal. Decreasing the force (at B) increases the shortening velocity but lowers the power output; increasing the force (at C) decreases the shortening velocity and also decreases the power output. The efficiency of the muscle is also reduced at the force-velocity combinations at which the power output is reduced.

lengths near L_o, cardiac muscle generally operates at lengths somewhat less than L_o. This gives it some length "reserve" with which to produce more forceful contractions when they are needed.

Cardiac muscle, in normal body function, also relaxes differently from skeletal muscle that has undergone isometric contraction. During isotonic relaxation, a skeletal muscle is stretched back to its original length by the load it has lifted. The relaxation is thus mostly isotonic. Ventricular muscle (a type of cardiac muscle), however, at its maximum shortening, has the afterload *removed* by the action of the aortic and pulmonary valves (see Chapter 18), and the relaxation is mostly *isometric*. The muscle is re-lengthened by the re-filling of the heart produced by returning blood. The mechanical consequences of such a cycle will become apparent when the function of the heart is discussed.

Because smooth muscle is usually found as a part of the makeup of the wall of a hollow organ or tube (rather than as a muscle stretched between two attachment points), it rarely can make a "typical" isotonic contraction. Most contractions in which smooth muscle shortens are done against a decreasing load as fluid or soft contents are propelled or contents of an organ are expelled. Smooth muscle of the blood vessels is maintained in a partially shortened condition in which force is also constant for long periods. At the other extreme, smooth muscle sphincters contract isometrically and stay contracted most of the time. Even though the physiological situation of smooth muscle is very different from that of skeletal and cardiac muscle, its actions too are described by force-velocity curves similar to those of skeletal muscle. The differences are mostly in speed (smooth muscle is much slower than skeletal muscle) and in the shape of the length-tension curve

(smooth muscle typically can function over a very wide range of lengths).

The Control of Muscle

Muscle is subject to a variety of bodily influences that control its function. Some of these controlling factors provide very precise and specific control of some types of muscle. Other control mechanisms may have a more subtle but still important control over the functions of various muscle types.

Control of Muscle by Cellular Membrane Systems

As with nerve cells, the cells of muscle are surrounded by an electrically excitable **plasma membrane.** The characteristics of this membrane vary widely among the muscle types. In the case of most skeletal muscles, the function of the membrane can be understood in much the same terms used to describe the way in which neuronal membranes behave. We encounter additional complexity in the membrane properties of heart muscle, and a wide variety of membrane functions are found in smooth muscle. However, most of the basic features of membrane control can be illustrated with skeletal muscle as an example.

Action Potentials in the Plasma Membrane

The surface membrane of most skeletal muscle cells can be excited to produce **action potentials.** As in neurons, the action potential is caused by time- and voltage-dependent changes in the membrane permeability to sodium and potassium ions. These changes may be started by a brief electrical shock to the membrane, but stimulation usually occurs by way of the **motor neurons** that normally innervate voluntary muscle. The details of neural activation will be discussed later on. At present we will simply consider that stimulation has taken place and that an action potential is present in a muscle cell plasma membrane. This action potential is called **propagated** because it arises at one location on the membrane and spreads rapidly along the fiber. As in a nerve fiber, the active region causes depolarization of resting membrane areas nearby. Since muscle cells lack the myelin sheath found in many nerves, the propagation of the activity is somewhat slower than that found in myelinated neurons. The relative slowness with which action potentials travel *along* the muscle cell plasma membranes is not especially important from a practical standpoint, however, since the muscle contraction itself is considerably slower than the speed of the action potential. A much more important problem involves the spread of activation *inward* from the surface plasma membrane to the depths of the fiber where the contractile material is located.

Action Potentials in the Transverse Tubular System

As cells go, those of skeletal muscle are quite large. The deepest parts of a muscle fiber may be as much as 50 micrometers away from the surface membrane. While this does not appear to be very far, it does pose an important problem. Suppose that muscle activation were due to the release of a chemical substance at the plasma membrane. This substance could then diffuse into the depths of the fiber. If this were actually the case, muscle contractions would have to be much slower than they are because diffusion over the distances involved would take a significant amount of time. If local electrical currents from the active plasma membrane flowed to the depths of the cell and caused its activation, they would not have sufficient strength to affect more than a few of the surface myofibrils.

In skeletal muscle these potential problems are overcome by the special conduction that takes place in the **transverse tubular system (t-system).** The membrane-lined tubules of this system, whose structure was outlined previously, penetrate the interior of the fiber at the level of each sarcomere. Recall that the insides of these t-tubules are open to the extracellular space, and their membrane linings are extensions of the surface membrane. When they enter each sarcomere of the cell, they effectively bring the extracellular space and the muscle plasma membrane very close to the interior of the fiber. As does the surface membrane, the t-tubule membranes conduct action potentials. When an action potential sweeps along the surface membrane, it is also conducted into the depths of the fiber along the t-system and arrives at the innermost portions of the cell with very little delay. This mechanism overcomes the slowness of diffusion and the weakness of the membrane currents, and this step in the activation of skeletal muscle becomes very rapid.

Calcium Ions in the Sarcoplasmic Reticulum

The t-tubule system does not actually open into the interior cytoplasm of the cell, since the tubules are completely lined with cellular membrane. Instead, they come into close contact with the membranes of the **sarcoplasmic reticulum (SR),** a closed membrane system. This means that it is a structure that maintains a separate region of the cell interior. This enclosed region is used to store and release the actual chemical substance that controls the contractile activity of the muscle proteins. This substance is **ionized calcium** (calcium ions), and in the resting muscle it is highly concentrated within the SR. It is contained within the saclike areas called **terminal** (or **lateral**) **cisternae.** The longitudinal elements of the SR communicate with the terminal cisternae at each end of the sarcomere and form a highly branched network surrounding the myofilament bundles. The interior of the longitudinal elements is continuous with the terminal cisternae.

As this description suggests, each sarcomere has its own separate portion of sarcoplasmic reticulum. The junction between a t-tubule and the terminal cisternae of two adjacent sarcomeres (the **triad**) is the place where the electrical signal from the outside of the fiber actually is communicated with the interior of the fiber. Just how the signal passes from the t-tubule to the SR at this point is not well understood, although it is assumed that the local movement of electrical charges plays a role. It is well known, however, that when this signal has been passed, the SR is caused to release a large portion of its stored calcium ions into the cytoplasm very near the myofilaments. These calcium ions free the myofilaments from their resting state (by a process that will be discussed shortly), and contraction takes place. Relaxation is associated with the movement of calcium ions back into the longitudinal elements of the SR. This ionic movement takes place up a steep concentration gradient via an ATP-dependent calcium ion pump present in this region of the SR membrane. While this mechanism (diagrammed in

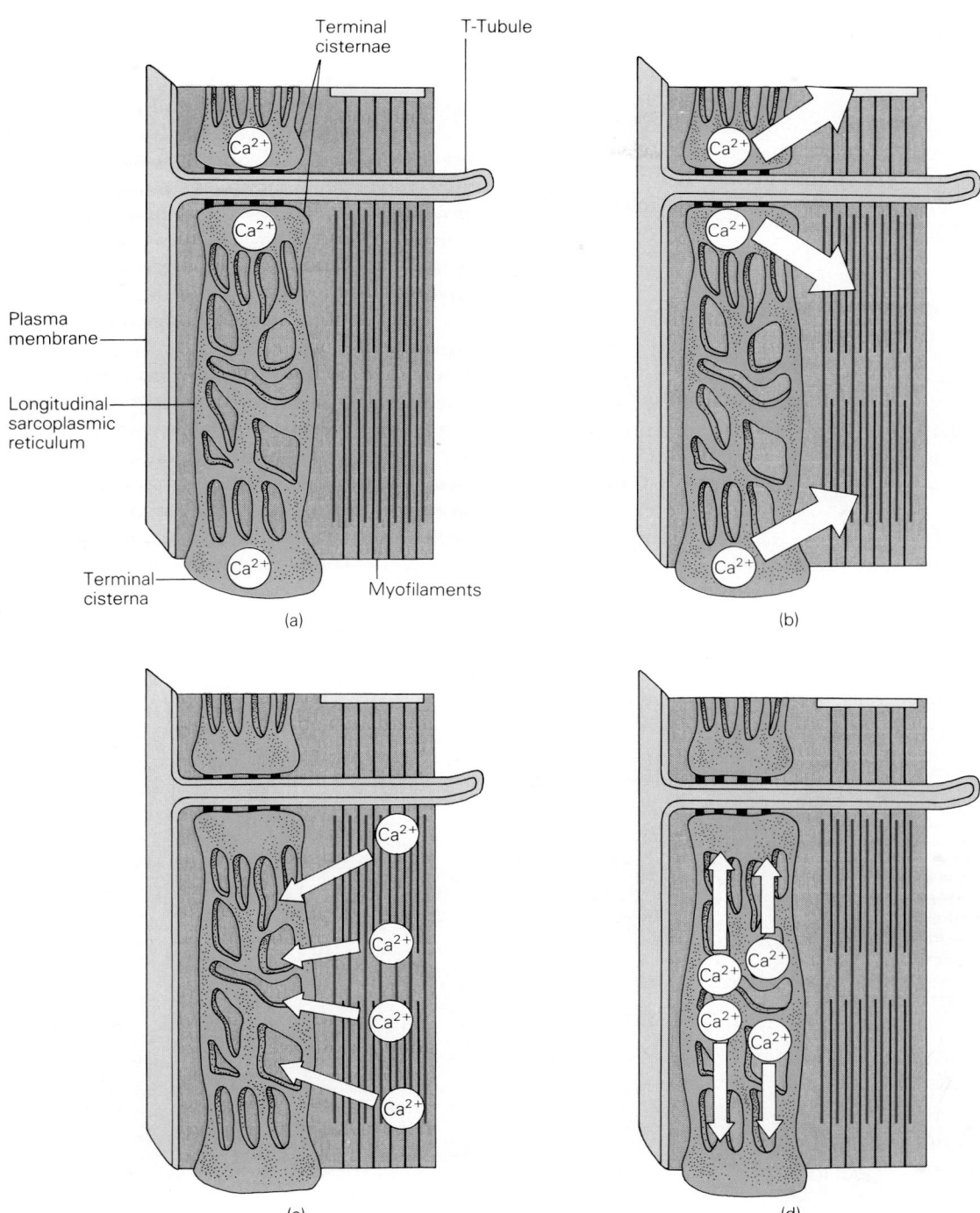

Terminal cisternae

T-Tubule

Ca²⁺

Ca²⁺

Plasma membrane

Longitudinal sarcoplasmic reticulum

Terminal cisterna

Ca²⁺

Myofilaments

(a)

Ca²⁺

Ca²⁺

Ca²⁺

(b)

Ca²⁺

Ca²⁺

Ca²⁺

Ca²⁺

(c)

Ca²⁺ Ca²⁺

Ca²⁺ Ca²⁺

(d)

Figure 16–18

How calcium moves during a contraction. (*a*) With the muscle at rest, the calcium ions are stored in the terminal cisternae. (*b*) As activation begins, calcium is released from the sarcoplasmic reticulum into the region of the myofilaments. (*c*) Activation ends as calcium is taken back up into the longitudinal elements of the sarcoplasmic reticulum. (*d*) Immediately after the contraction, the calcium moves back into the terminal cisternae.

Fig. 16–18) may seem complicated, it quite effectively overcomes the diffusion and conduction obstacles posed by the large size of skeletal muscle fibers. It is capable, in some muscles, of allowing activation to be repeated as many as 60 to 100 times per second. Muscles without such a mechanism would be limited to very slow rates of contraction.

Control of Muscle by Calcium Ions

The regulation of muscle contraction involves controlling the interaction of the actin and myosin myofilaments. In all types of muscle, this control is carried out through the action of calcium ions, though the actual process differs greatly between smooth muscle on one hand and cardiac and skeletal muscle on the other.

The Molecular Mechanisms of Muscle Contraction

When an action potential, through the sequence of events described previously, causes the SR of skeletal muscle to release a quantity of stored calcium ions, these ions diffuse to the region of the myofilaments. Here they bind to the **troponin-C (TnC)** subunit of the regulatory protein complex, troponin. The troponin complex is attached to the tropomyosin molecule on the thin (actin) myofilament. When calcium ions bind to troponin-C, the actin filament becomes able to interact with the myosin crossbridge. The contraction that results involves the repeated interaction of actin and myosin in a series of steps diagrammed in Figure 16–19. In resting skeletal muscle (with calcium ions not yet released), an ATP molecule is bound to each myosin molecule projection that forms a crossbridge. The ATP molecule has been "split," but its energy cannot be released until the actin is allowed to interact with the myosin. When the presence of calcium ions permits this interaction, the crossbridge attaches to the actin filament at an angle of approximately 90° and the phosphate ion that was split from the ATP molecule is released. As the energy of the ATP molecule is released, the angle of attachment becomes closer to 45°. This change in attachment angle (the so-called **power stroke**) causes the actin filaments to be pulled in toward the center of the sarcomere, and the ADP molecule is released from the myosin. Only when a new ATP molecule binds to the myosin head can the

Figure 16–19
The crossbridge cycle of skeletal muscle. The series of events goes on continuously as long as calcium and ATP are present.

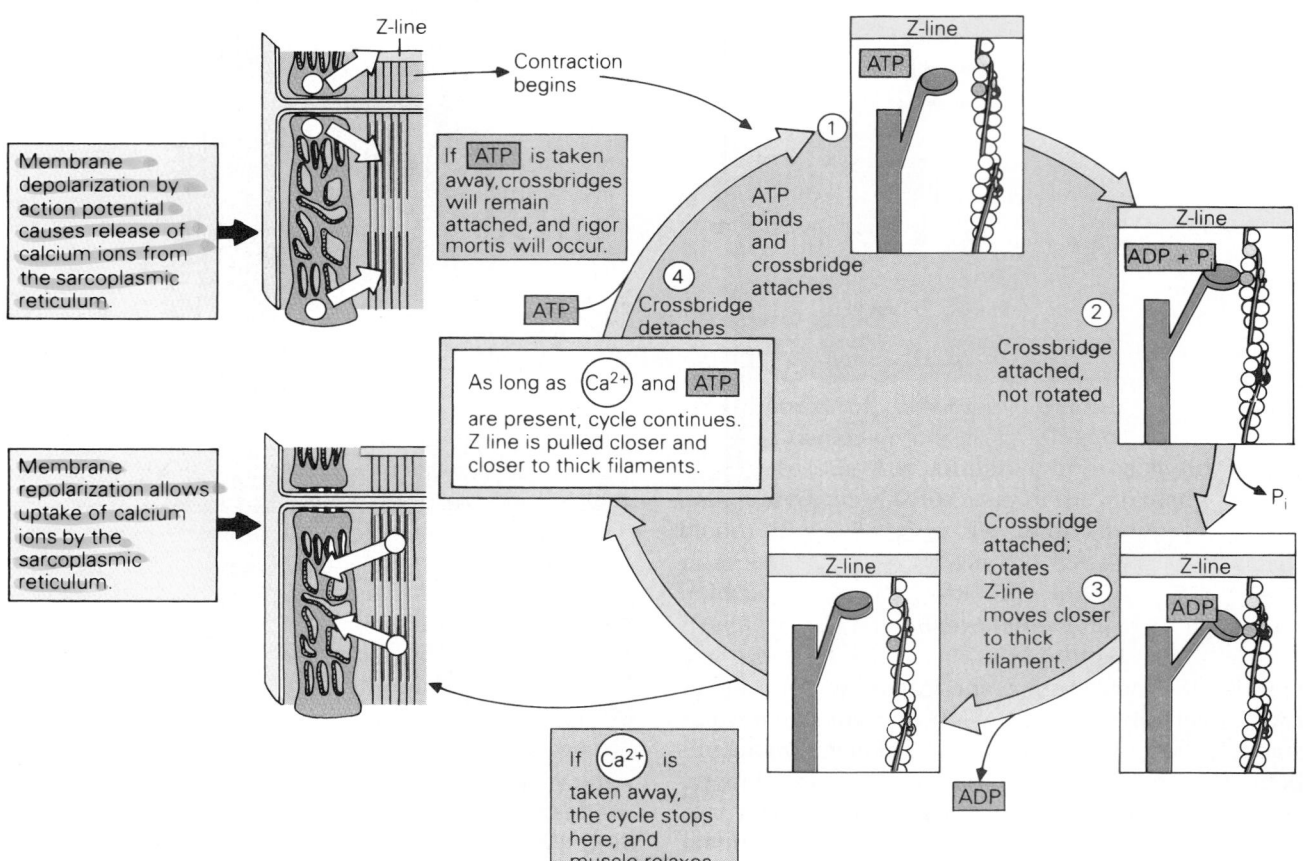

crossbridge detach itself from the actin and become ready for a new cycle.

As long as enough ATP and calcium ions are present, the cyclic process of attachment, angle change, ATP binding and hydrolysis, and detachment can take place. Repeated crossbridge cycles result in the actin filaments being pulled further into the myosin filament array, and the whole muscle becomes shorter. As was discussed previously, the force that the muscle exerts is important in determining how fast the crossbridge cycle will run. If the ends of the muscle are held stationary (as in isometric conditions), the crossbridge cycles will still continue to some extent, since some internal movement is allowed by the elastic nature of the muscle. When stimulation of the muscle is stopped, no additional calcium ions are released from the SR, and calcium ions that diffuse free from the troponin or that are free in the cytoplasm are taken back up by the SR. Crossbridges will not reattach, and relaxation takes place. If the supply of ATP becomes depleted, attached crossbridges will not be able to detach, and the muscle becomes stiff and unable to relax. An extreme example of this situation is the condition known as **rigor mortis,** which occurs after death when the supplies of ATP in the muscle become exhausted.

Actin-Linked Regulation of Muscle Contraction

Muscle regulation that is accomplished through some action on the **thin filaments** is called **actin-linked regulation.** This type of regulation is also found in heart muscle, where the steps involved are quite similar to those in skeletal muscle. In cardiac muscle, however, the amount of calcium ions released in response to an action potential is usually not sufficient to activate all of the possible actin molecules, and a maximal contraction is not obtained. This provides cardiac muscle with a "reserve capacity" for contraction, because there are several ways in which extra calcium ions can be released to meet additional requirements. In the specialized action potentials of cardiac muscle, one of the ions that flows into the cell is calcium. Because of the small size of cardiac muscle cells, it is possible for a significant amount of calcium ions to enter the cell through the plasma membrane during the course of an action potential. Factors that increase the duration of action potentials, or the number of action potentials per minute, or the amount of calcium ions entering with each action potential, can increase the force of contraction.

Myosin-Linked Regulation of Muscle Contraction

Regulation of contraction that takes place via **thick filaments** is called **myosin-linked regulation.** Smooth muscle is controlled in this way. Its thin filaments lack the troponin protein complex, but on the head of each myosin molecule are two protein chains that function in regulation. With molecular weights of 17,000 and 20,000 daltons, they are referred to as **light chains.** Only when a phosphate group (from ATP) is bound to the 20,000 light chain of myosin can interaction take place between actin and myosin. Since the thin filaments of smooth muscle lack the troponin regulatory complex, they are always ready for interaction with myosin, and it is the myosin, not the actin, that must be activated.

The activation and regulation processes of smooth muscle can be divided into three groups of reactions that take place more or less simultaneously. These are shown in Figure 16–20. The first group (area 1) is the step where control of contraction begins. Calcium ions are released into the cytoplasm (upper left), either from the sarcoplasmic reticulum or from outside the cell via the cell membrane. These ions bind to a protein called **calmodulin,** which is associated with an enzyme called **myosin light-chain kinase (MLCK).** This binding activates the MLCK and allows it to catalyze one of the reactions of the second group (area 2). These reactions are associated with the phosphorylation of the myosin light chains; when the terminal phosphate group of an ATP molecule is transferred to each myosin molecule, the myosin becomes activated and able to interact with actin. A phosphatase enzyme simultaneously acts to remove the phosphate group, but during activation the phosphorylation reaction occurs at such a high rate that the dephosphorylation reaction is relatively unimportant. The interaction between actin and activated myosin (third group) is shown in area 3. This is a crossbridge cycle similar to that of skeletal muscle, involving the cyclic attachment, rotation, and detachment of crossbridges to produce force and movement. This cycle continues as long as there is a supply of ATP from the cellular energy stores and as long as myosin remains in its activated state.

Relaxation takes place when calcium ions are taken up from the cytoplasm (upper right) by the action of calcium pumps in the sarcoplasmic reticulum or plasma membrane. The MLCK becomes inactive and can no longer catalyze the phosphorylation reaction. The phosphatase reaction is then able to dephosphorylate the myosin, and so it can no longer interact with actin. The crossbridge cycle

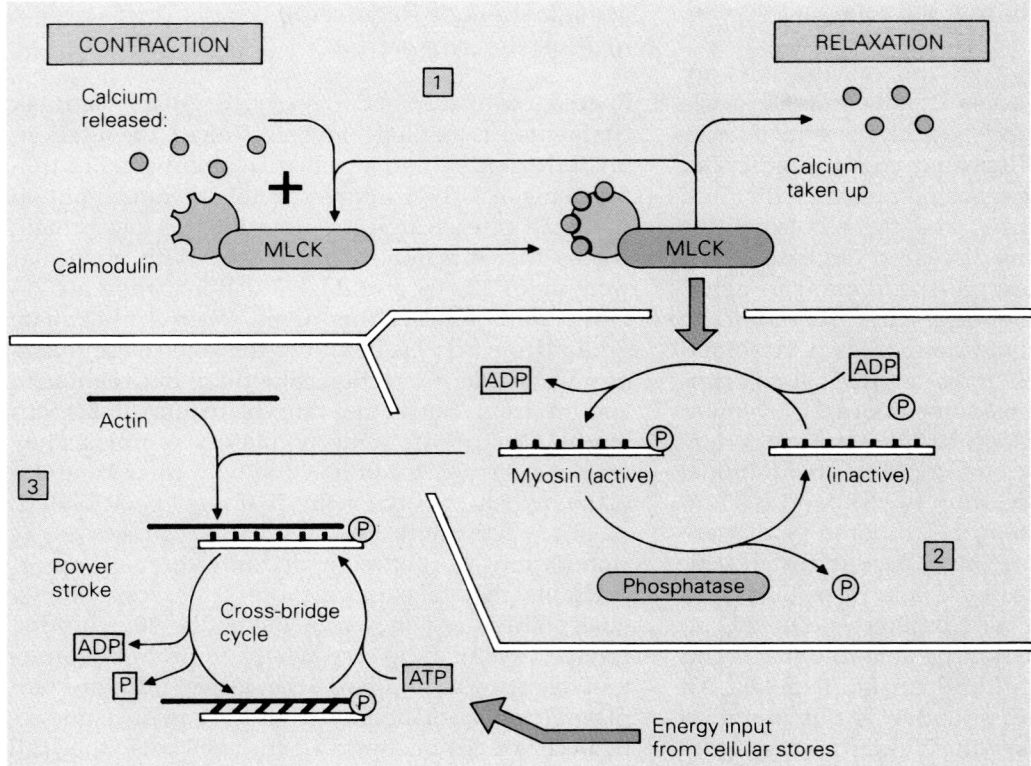

Figure 16–20
The major steps in the regulation of smooth muscle contraction. A contraction is initiated by the release of calcium into the cytoplasm, and the myosin molecules are activated by the steps in areas 1 and 2 so that the crossbridge cycle (shown in area 3) begins to operate. Relaxation begins when the calcium is taken back up and the phosphatase enzyme (area 2) is no longer working against the MLCK and can dephosphorylate (and thus deactivate) the myosin molecules.

comes to a halt and the muscle relaxes. In addition to these basic processes, smooth muscle also contains other regulatory processes responsible for its ability to maintain a contraction for a long period of time without consuming large amounts of energy. These processes appear to involve modifications in the properties of crossbridges; in some smooth muscles crossbridges become "**latch bridges**" that detach very slowly and therefore spend most of their time in the attached, force-producing state. Thus, while the details of regulation of smooth muscle are much different from the regulatory processes in skeletal muscle, the control over contraction and relaxation is still exercised by calcium ions.

Types of Stimuli that Activate Muscle Contraction

One of the important results of the way in which muscle is controlled is seen in the types of contractions produced in response to various types of stimuli. Many muscles (especially skeletal and cardiac) can respond to a single stimulus with a single action potential. This usually results in a single brief contraction. Such a response, already mentioned, is called a **twitch**. The rapid contraction and relaxation of skeletal muscle during a twitch are the result of two processes that take place at the same time. As we have seen, the result of the action potential is the sudden release of stored calcium ions from the SR. Even while these calcium ions are diffusing to the myofilaments and activating the contraction, the processes that take the calcium ions back up are also operating. Even when the amount of calcium ions released is potentially enough for a maximal contraction (as is usually the case in skeletal muscle), their rapid uptake limits the amount of force that can develop. In most cases, the muscle does not have a chance to develop its full amount of force before relaxation processes have set in. Crossbridge cycling, necessary for the buildup of force, takes time, and this time is not available.

Summation of Action Potentials: Producing a Sustained Contraction in Skeletal Muscle

This rapid relaxation and low force mean that twitch contractions are not particularly useful for everyday tasks. However, this limitation is readily overcome by a repetitive type of contraction. The action potential of twitch-type skeletal muscles is very brief compared with the duration of the twitch contraction. Even though the action potential has the usual absolute and relative refractory periods typical of excitable membranes in general, these also are brief compared to the twitch time. This means that the muscle can be restimulated before its relaxation is complete, and the force produced by the second stimulus can add to the force left from the first.

This process is called **summation,** and it results in a contraction, called a **tetanus,** that lasts longer and produces more force than a single stimulus could produce. What has happened inside the muscle is that the number of calcium ions available to activate the myofilaments has been increased by the second stimulus. Stimuli can be repeated at regular intervals in order to keep the internal calcium ions continuously available to the myofilaments. This continuous supply in turn keeps the muscle from relaxing completely, and the result is a sustained contraction that produces significantly more force than that developed in a twitch (Fig. 16–21). If the stimuli are spaced relatively far apart, the force will rise and fall somewhat between stimuli, but if the stimuli are close enough together, the developed force will be steady. This is called a **complete,** or **fused, tetanus.** Just how many times a second a muscle must be stimulated in order to produce a fused tetanus depends on the relative rates of contraction and relaxation of the muscle. The lowest frequency that will produce a fused tetanus is called the **tetanic fusion frequency.** It is around 20 to 60 times per second for most skeletal muscles.

Useful contractions—those that get us through daily life—are usually a mixture of twitches and partly fused tetanic contractions. These are distributed among the many individual fibers that make up a complete muscle, so that the contraction is "shared." The question of which fiber does what to produce a partial contraction of a muscle is a matter for the CNS to decide, as will be discussed shortly.

Distancing of Action Potentials: Producing Discrete Twitches in Cardiac Muscle

Cardiac muscle is a twitch-only type of muscle. A sustained and maximal tetanic contraction would be of little use in the task of pumping blood, and heart muscle has some special properties that guard against contractions of this sort. The action potentials of heart muscle are quite long; they typically last almost as long as the contraction itself. This means that the membrane is **refractory** during most of the twitch, so that restimulation is not possible while the muscle has any significant tension. At higher rates of stimulation (high heart rates), the duration of both the action potential and the contraction are reduced. This allows for closely spaced twitches, with time for complete relaxation between them.

Even though cardiac muscle cannot develop a partial tetanus, the properties of its contractions can be adjusted by other means. In contrast to the situation in skeletal muscle, the amount of calcium released in response to a single action potential is generally not sufficient to provide for full activation. A number of factors can change the amount of calcium made available for each contraction. Some drugs, those of the **digitalis** family, for example, increase the overall internal supply of calcium, and more is released from the SR with each beat. Because calcium is one of the ions that enters during the action potential, changes in this **calcium current** can cause changes in the contraction. Drugs called **calcium blockers** can reduce the strength of contractions, and the chemical **epinephrine** (both a drug and a hormone) can increase the strength of contraction by permitting the accumulation of internal calcium. Making the heart contract more times per minute (up to a point, of course) will also result in the accumulation of calcium and an increase in the strength of the contraction. This is known as the **force-frequency relationship.** All of the examples given here are agents that affect the **contractility** of the heart muscle. These influences are called **inotropic agents** (*ino* = muscle; *tropic* = having an effect). They permit changes in the properties of the con-

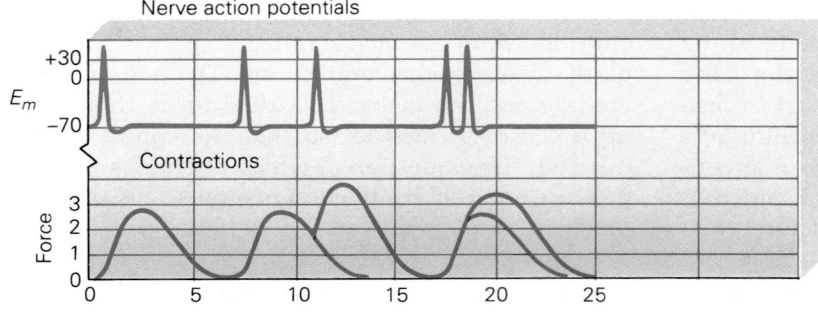

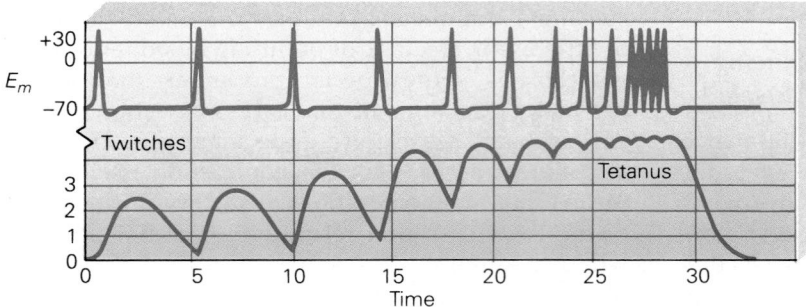

Figure 16–21

The relationship between twitches and a tetanus. Upper portion: A single motor nerve action potential produces a single twitch contraction (first event). Applying a stimulus during relaxation (second event) produces a contraction whose force adds to that of the first. Two stimuli delivered quite close together (third event) also produce an addition of force (possible because of the short refractory period of the muscle cell membrane). Lower portion: When action potentials are made to occur closer and closer together, less and less time is available for relaxation, and the forces of the individual twitches add up. When the rate of stimulation is high enough, a smooth tetanus can be produced. A gradual build-up of twitches is not necessary for a tetanic contraction; a tetanus will be produced immediately if the stimulus rate is high at the outset.

traction that do not depend on the possibility of a tetanic activation.

Along with the changes in contractility, changes in muscle length (as we have seen previously) can have an important effect on the strength of the contraction of cardiac muscle. (Recall the relationships expressed in the **length-tension curve.**) In the functioning heart, these two types of influences work together to produce a very effective degree of control of the pumping function of the heart.

Phasic and Tonic Contractions: A Variety of Responses in Smooth Muscle

Smooth muscle can contract in a variety of different ways. Some smooth muscles can give twitchlike contractions in response to single stimuli or brief bursts of stimuli. These contractions are called **phasic** and are often found in smooth muscles that propel or move materials. Other smooth muscles give long and sustained contractions in response to single or continued stimulation by nerves, drugs, or hormones. These contractions are called **tonic** and are likely to be found in smooth muscle organs, such as sphincters or bladders, that must remain contracted for long periods of time. Many smooth muscles are almost never in a state of complete relaxation, and the steady force that they generate is called **tonus.** The level of tonus is often set by the concentration of a specific hormone such as norepinephrine.

In many cases, the contraction of smooth muscle is associated with electrical activity of the cell membranes. Most phasic contractions are associated with **action potentials.** In some tissues, such as intestinal muscle, with rhythmic contraction patterns, there is a kind of electrical activity called the **slow wave.** The slow waves, which are many seconds in duration, periodically depolarize the membrane to the threshold for action potential generation; this mechanism acts to produce periodic contractions. Other smooth muscles show action potential activity that may be caused by a pacemaker cell and conducted to adjacent cells by means of the gap junctions previously mentioned.

Control of Muscle by the Central Nervous System

Almost all of the muscle in the body is under the direct or indirect control of the CNS, although smooth muscle and cardiac muscle do have a degree of autonomous function that is suited to their specialized roles. Because neural control is so highly developed in the case of skeletal muscle, most of the general process can be understood by a study of how the action of a motor nerve regulates the function of a skeletal muscle.

Nervous Control of Skeletal Muscle

Skeletal muscle depends on the somatic division of the CNS for the control of its contraction. As described in Chapter 10, under normal conditions, skeletal muscle will not contract unless it is stimulated by a **motor neuron,** which has its origin in the CNS. Connection between the motor nerve fibers, which are large myelinated axons, and the muscle fibers takes place at a structure called the **myoneural junction.** Because in some cases this structure spreads out somewhat on the muscle fiber, it is also called a **motor end-plate.**

The Myoneural Junction and the Motor Unit. The myoneural junction is a special case of a **synapse.** As in the vast majority of synapses in the CNS, the transmission of the impulse from the nerve to the muscle is by chemical means. The particular transmitter substance at the myoneural junction is **acetylcholine (ACh).** Unlike some CNS synapses, only **excitation** of the postsynaptic membrane is possible at the myoneural junction. Because many of the myoneural junctions in a muscle lie at its surface, they have been extensively studied, and the process of transmission is well understood. In fact, much of what we know about synapses in general has been learned by the study of this highly specialized synapse.

The structure of the myoneural junction provides keys to an understanding of its function. As a motor axon from the motor nerve bundle nears the muscle, it branches to innervate a number of individual muscle fibers. The assemblage of a motor axon, together with all of the muscle fibers that it innervates, is called a **motor unit.** Since all of the muscle fibers served by a single branched axon are activated together when an impulse passes down the axon, the fineness of control of a muscle depends on just how many muscle fibers a single axon controls. Muscles capable of fine and delicate movements, such as the **extraocular** muscles that move the eyeball, have only a few muscle fibers served by a single axon, while muscles used for rapid and relatively coarse movements, such as those of the posterior thigh, have many muscle fibers controlled by a single axon. Each muscle fiber normally receives innervation from only a single axon terminal, although some multiply-innervated muscle fibers do exist.

Synaptic Transmission at the Myoneural Junction. As the motor axon branches, the myelin sheath becomes reduced to the cytoplasm and cell membranes of a single **Schwann cell** (the type of cell that forms the myelin sheaths around motor axons), and this covers the axon as it makes its final approach to the muscle (Fig. 16–22). The terminal region of the axon lies in a "groove" on the surface of the muscle. This arrangement increases the amount of nerve and muscle cell portions that make up the junctional region. The motor axon terminal contains small vesicles that carry the transmitter substance. Also present in this presynaptic region is a large number of mitochondria. These provide the metabolic energy for the synthesis of the transmitter substance and the ionic pumping mechanisms necessary for the recovery processes that follow the passage of an impulse. The muscle plasma membrane making up the postsynaptic side of the junction is folded into a series of deep grooves called **junctional folds,** which are lined with the receptors for the ACh transmitter substance. The receptors are complex protein molecules with dual function. Each receptor has two binding sites for ACh and also functions as an ion channel. Normally, the receptor is not permeable to ions, but when the transmitter substance has attached to the binding sites, sodium and potassium ions can pass through the channel down their electrochemical gradients. When the ACh is no longer bound to the receptor, the channel again becomes impermeable to ions. The postsynaptic membrane also contains an enzyme called **acetylcholinesterase.** This enzyme is essential for breaking

PRESYNAPTIC PORTION
(motor axon)

POSTSYNAPTIC PORTION
(muscle fiber)

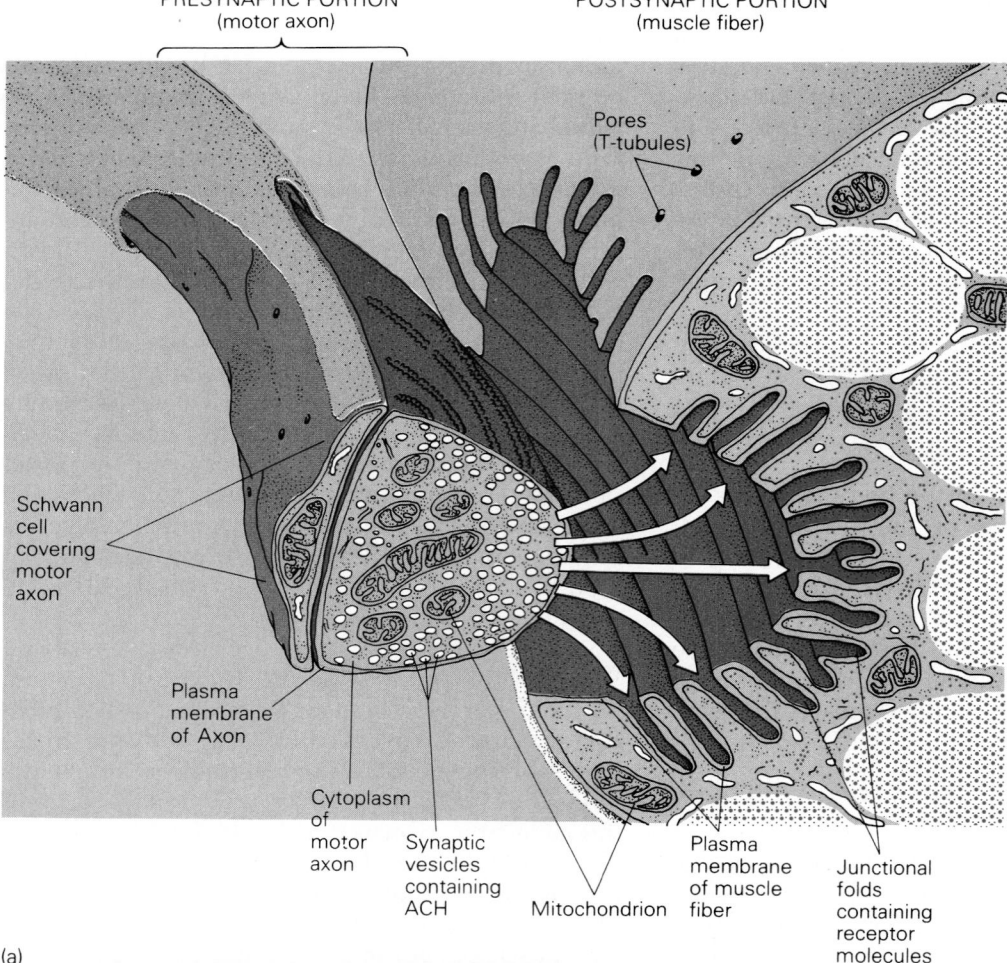

Pores
(T-tubules)

Schwann
cell
covering
motor
axon

Plasma
membrane
of Axon

Cytoplasm
of
motor
axon

Synaptic
vesicles
containing
ACH

Mitochondrion

Plasma
membrane
of muscle
fiber

Junctional
folds
containing
receptor
molecules

(a)

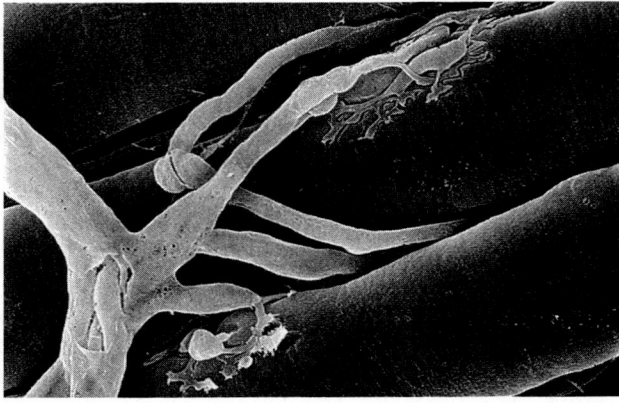

(b)

Figure 16–22
(*a*) An expanded view of the myoneural junction. This drawing shows the essential features of the myoneural junction of skeletal muscle. (Modified from Krstic, R.V., *Ultrastruktur der Saugetierzelle.* Berlin: Springer-Verlag, 1976.) (*b*) SEM of a region of (hamster) striated muscle fiber showing the branching motor axon and two myoneural junctions. (© Don Fawcett/Science Source).

down the transmitter to an inactive form once it has done its job of causing the channel to open.

Between the time when a motor nerve is stimulated and when the muscle contracts, a long series of events takes place (Fig. 16–23). The action potential travels down the motor axon by the mechanisms previously described. When the action potential spreads into the terminal regions of the axon branches, membrane calcium channels open in response to the depolarization, allowing the entry of calcium ions into the neuron. This increase in the intracellular calcium ions causes the synaptic vesicles containing the acetylcholine to move to the region of the presynaptic membrane. Here, the vesicle membranes fuse with the membrane. In the region of the fusion, the plasma membrane breaks down, and the contents of the vesicle are released into the synaptic cleft. They diffuse across the cleft to the postsynaptic region, where they attach to the specific binding sites on the receptor molecules. When

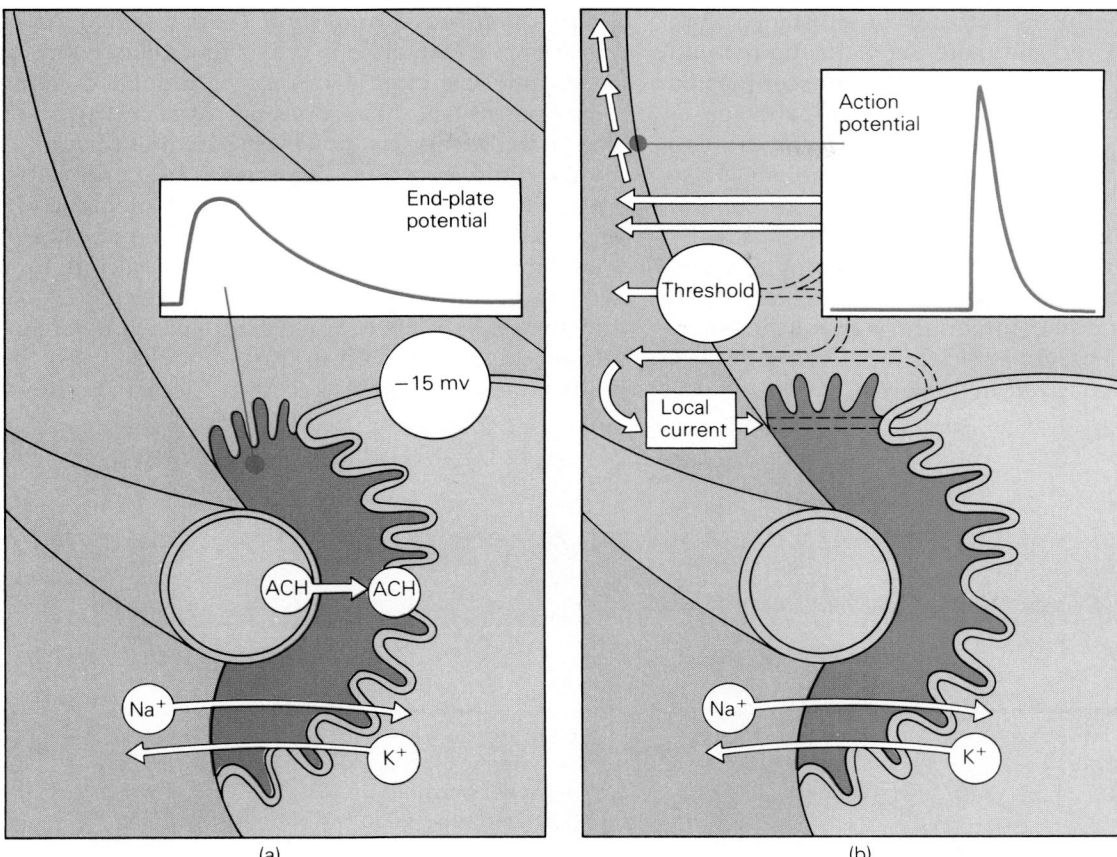

Figure 16–23
How the end-plate potential produces a muscle action potential. The region
of membrane (the postsynaptic membrane) immediately opposite the axon
terminal does not produce an action potential. Instead, its depolarization
causes local currents to flow through the interior of the muscle cell and
across the adjacent muscle cell membrane. This flow of current depolarizes
the membrane to a threshold value and produces an action potential. This
action potential then travels the length of the muscle cell membrane and enters via the t-tubule system to activate the contraction.

this binding takes place, the membrane channels
associated with the receptors become permeable
simultaneously to both sodium and potassium ions.
This causes the membrane potential to move away
from the resting potential (about -80 mV) and toward a new potential of about -15 mV, which is
called the **end-plate potential.** (This is the potential
expected from a membrane permeable to both sodium and potassium at the same time.) The postsynaptic membrane is unlike the rest of the plasma
membrane in one important aspect. Since its channels are chemically activated and both permeability
changes occur together, it cannot produce an action
potential. What the end-plate potential does, however, is to cause a small local current (the **end-plate
current**) to flow across the membrane at this region.
This inward current flows into the fiber and begins
to spread down its length. The corresponding outward current is forced to flow across adjacent resting areas of the general plasma membrane, and it
depolarizes these regions so that they reach their
threshold for action potential generation. The action
potentials produced by the end-plate potential then
propagate down the length of the fiber and activate
the contraction in ways that have already been described. After the passage of the action potential,
the muscle membrane returns to its resting condition.

If the action of the transmitter substance were not terminated, the end-plate depolarization would continue. This would prevent the muscle plasma membrane from being stimulated again, and subsequent synaptic transmission would fail. Hydrolysis of the ACh by the postsynaptic acetylcholinesterase terminates ACh action and produces its components, acetate and choline. The choline is actively taken back up into the presynaptic terminal, where it is made back into ACh. The acetate fragment, which is a chemical common throughout the body, is not saved by any special mechanism. These events are summarized in Table 16–1.

Factors that Block Myoneural Transmission.   All of these events take place in only a few milliseconds of time, and the muscle can be restimulated many times per second. The presence of so many steps critical to the process, however, indicates that many things could go wrong. **Myoneural blockade** is the term given to this transmission failure. Some blockade can take place presynaptically. The presence of too many magnesium ions or too few calcium ions near the axon terminal will prevent release of the transmitter substance. Some drugs, called **hemicholiniums,** interfere with uptake of choline, and the transmitter substance becomes depleted. **Botu-**

Table 16–1
Sequence of Events During Neuromuscular Transmission

Event	Possible Blocking Agent	Event	Possible Blocking Agent
Action potential arrives at axon terminal.	Nerve-blocking agents such as procaine or tetrodotoxin	End-plate region (postsynaptic membrane) becomes depolarized; end-plate potential is set up.	
Extracellular calcium ions enter axon terminal.	Low extracellular calcium or high magnesium concentration	Local end-plate currents cause adjacent muscle membrane to depolarize.	
Transmitter vesicles migrate to axon membrane and fuse with it.		Muscle action potential is triggered.	
Acetylcholine is released from the transmitter vesicles into synaptic cleft.	Botulinum toxin	Acetylcholine diffuses away from receptors and is hydrolyzed by cholinesterase enzyme.	Eserine or other cholinesterase inhibitors
Acetylcholine molecules diffuse across synaptic cleft.		Choline is taken up by presynaptic terminal (now repolarized.)	Hemicholinium drugs
Acetylcholine molecules bind to receptor proteins in postsynaptic membrane on muscle.	Curare, succinylcholine	Acetylcholine resynthesized and vesicles re-filled.	
Ion channels in postsynaptic membrane open and allow the flow of sodium and potassium ions.			

linum toxin, an extremely deadly poison produced by a bacterium (*Clostridum botulinum*) present in spoiled food, also causes a failure of transmitter release, and poisoning with this substance results in paralysis and death.

Postsynaptic blockade is also possible. The substance **curare** (used on blowgun darts by aboriginal hunters in South America) can bind to the acetylcholine receptors, in preference to acetylcholine. However, its binding does not produce a permeability change or depolarization. It is also resistant to breakdown by the acetylcholinesterase. When it is present in sufficient quantity, therefore, myoneural transmission cannot take place.

A close chemical relative of acetylcholine is the drug **succinylcholine,** which is used during surgical procedures to produce muscle relaxation. This drug is broken down only very slowly by acetylcholinesterase. Another possible site for blockade is the acetylcholinesterase enzyme. This enzyme is inhibited by drugs such as **physostigmine** and **eserine,** which are similar to the active (organophosphate) compounds in chemical warfare agents, the so-called nerve gases. When the action of cholinesterase is inhibited, the end-plate membranes remain depolarized, and the muscle action potential mechanism cannot be reset.

In all of the situations mentioned, the myoneural blockade would result in death from respiratory failure. The diaphragm, a skeletal muscle, would fail to respond to nerve impulses, and along with other skeletal muscles, it would be paralyzed. Carefully controlled doses of blockers such as succinylcholine can be used as surgical muscle relaxants as long as respiration is maintained artificially.

Other factors also affect the myoneural junction. As is true of other synapses, it is subject to **fatigue** due to depletion of the transmitter substance. However, the transmission normally has a large **safety factor.** That is, more transmitter is released than is necessary for transmission, and the end-plate potential is normally larger than necessary to stimulate the muscle membrane. In addition, the process of contraction in the muscle is subject to fatigue far sooner than is the process of myoneural transmission. An abnormal situation, though, is present in the disease known as **myasthenia gravis,** which means "severe weakness of the muscles." In this condition, it appears that an immune reaction against the acetylcholine receptors reduces their numbers, and the end-plate potentials may not be large enough to stimulate the muscle fibers. Weakness or paralysis results from this condition, but effective therapy, at least in the short term, is possible. Careful administration of cholinesterase inhibitors allows the acetylcholine to remain active in the synaptic cleft for a longer time, and what receptors there are may be bound to and activated several times in succession. This allows sufficient end-plate current to flow to permit near-normal muscle function.

Nervous Control of Smooth and Cardiac Muscle

The two general types of smooth muscle, multiunit and unitary, differ considerably in the ways they are controlled by the nervous system. Multiunit smooth muscles are generally very densely innervated by autonomic nerve fibers and have rudimentary motor end-plates on each cell in the tissue. Myoneural transmission in these muscles is much like that found in skeletal muscles. The structures involving this muscle are quite small. They include some very small blood vessels, the **arterioles,** and the muscles that make our hair stand on end, the **piloerector muscles.** Because of the small size and the plentiful innervation of the tissues, the fact that the cells do not communicate with one another is of little practical consequence.

Most tissues composed of unitary smooth muscle are also innervated by postganglionic fibers of the autonomic nervous system. There is nothing approaching a one-to-one connection between the nerve supply and the muscle cells. Instead of a nerve-muscle connection such as a motor end-plate, the innervating fibers course among the smooth muscle cells. Periodically the nerves have swellings (called **varicosities**) that hold vesicles containing various transmitter substances (e.g., acetylcholine or norepinephrine). Action potentials traveling down the nerves cause the release of the transmitter substance, which can then diffuse across the intercellular spaces to the vicinity of the muscle cells. Because there are so many cells in the tissue, only a few of them are directly affected by the transmitter substance. The cells that depolarize and produce action potentials, however, can affect neighboring cells through gap junctions. The gap junctions form low-resistance connections between adjacent cells; depolarizing one cell can cause an adjacent cell to depolarize by means of **local current flow.** This has the effect of spreading the activity throughout the tissue, and even though most individual cells are not directly innervated, the tissue tends to function as though they were. This is the source of the term *unitary,* and a tissue that behaves in this way is called a **functional syncytium** (*syn* = same; *cytium* = cells). Many unitary smooth muscle tissues are spontaneously active and can contract rhythmically

C L I N I C A L   F O C U S

Myasthenia Gravis

Myasthenia gravis (meaning "serious muscle weakness") is an illness associated with muscular weakness and extreme fatigue following moderate exercise. It stems from a failure of neuromuscular transmission rather than from any defects in the muscle contraction mechanism itself. While no cure has been found for the disease, several kinds of therapy can give patients some symptomatic relief.

Scientists now know that myasthenia gravis is one of the so-called **autoimmune diseases.** For unknown reasons, some people produce antibodies that react with the **acetylcholinesterase receptors** in the postsynaptic (motor end-plate) plasma membrane. This immune reaction destroys many of the receptors, and adequate **end-plate potentials** can no longer be produced. This means that a muscle action potential does not result from a nerve impulse, and so the muscle fails to contract. For a person afflicted with the condition, the result is that extreme effort may be required for even ordinary movements. In extreme cases, even breathing becomes very difficult.

Specific diagnosis of the condition involves study of the neuromuscular junction and immune system. Careful observation of the relationships between the nerve stimulation and muscle response can reveal delays in the transmission process even before it fails completely, and the presence of antibodies specific to the acetylcholine receptors may be detected in the blood of affected persons.

Several forms of treatment are possible. Drugs that inhibit the action of acetylcholinesterase allow the remaining postsynaptic receptors to function for a longer period of time, and adequate end-plate potentials can be produced. Unfortunately, the effect is only temporary, and the drug must be readministered repeatedly. This approach can lead to harmful side effects and, in any case, treats only a symptom of the underlying cause, with no effect on the general course of the disease.

Like other autoimmune diseases (such as rheumatoid arthritis), myasthenia gravis can undergo spontaneous remissions in its early stages, although the symptoms usually return. Removal of the thymus gland (which is known to play an important role in the immune system) results in substantial improvement of the condition, especially if it is done early in the course of the disease.

Other therapies, also directed at the immune system, may be applied. Administration of gamma globulin may produce temporary improvement. Steroid drugs can be used to reduce the severity of the problem for a time, but they cause harmful side effects in many cases. Immunosuppressant drugs reduce the function of the entire immune system and can produce an improvement in the symptoms, although the patient will then be at risk from other infections and diseases. The technique of **plasmapheresis,** in which the proteins (including antibodies) are removed from the blood by a special transfusion method, can produce temporary improvement. But the procedure, though relatively safe, is time-consuming and expensive.

While an understanding of the effects of the disease can be obtained through study of the neuromuscular junction, its actual cause lies elsewhere. With immunological methods, myasthenia gravis can be produced in experimental animals, especially in rodents and monkeys. Since the animal condition very much resembles the human disease, studies involving animal immune responses offer hope for development of a more satisfactory treatment.

---

on their own. In these cases, the effect of the autonomic innervation may be to regulate (increase or decrease) the ongoing function.

Heart muscle also receives an extensive autonomic innervation, but the function of the nerves in heart muscle is not to initiate contractions. The muscle of the heart does not contain motor end-plates, and its contractions begin directly in the muscle tissue, not in response to an external stimulus. As in some smooth muscles, the function of nerves in the heart is to regulate the ongoing timing and strength of contraction. Cell-to-cell communication is also of very great importance in the function of heart muscle, because, as is true of smooth muscle, it is composed of cells that are very small in relation to the size of the tissue. The process of the control of heart muscle contraction is sufficiently specialized to be treated separately in Chapter 18.

The Metabolism of Muscle Contraction

Because muscle does work, it must consume fuel in the form of energy-yielding biochemical compounds. The fuel most directly consumed by the actin-myosin contractile system is the universal high-energy compound, ATP. While muscle cells require energy for ion pumping, growth, and cell maintenance, as do other cells, their major function is that of contraction, and this function consumes more energy than does any other type of cell. Not surprisingly, the metabolic adaptations of muscle are specialized to provide an adequate and constant supply of ATP for the contractile process.

Energy Pathways in Muscle Contraction

As we have seen, ATP is broken down during the crossbridge cycle and provides the energy that is turned into mechanical work. If chemical inhibitors of the general cell energy metabolism are used to prevent replenishment of the ATP supply, a contracting muscle will shortly become exhausted and enter a state similar to **rigor mortis.** However, a muscle maintains activity for a longer period of time than simple chemical measurements of the ATP content would indicate. This is because of a reserve "energy pool" in the cell in the form of a compound called **creatine phosphate.** This high-energy compound is capable of transferring its energy to ATP very readily, so that as soon as the supply of ATP to the myofilaments begins to fall, it is rapidly replenished. The ADP from the contraction process is rephosphorylated to ATP, and the creatine phosphate becomes creatine. This creatine is itself rapidly rephosphorylated by ATP derived from the general metabolism of the cell. The creatine phosphate pool provides a "buffer" for the rapid supply of ATP for the work of contraction, as well as a link to the cellular sources of ATP. Depending on the type of muscle fiber, this cellular ATP is produced by one or both of two common biochemical pathways, as diagrammed in Figure 16–24.

Glycolysis: An Anaerobic Pathway

Glycolysis, the first of these pathways, yields a relatively small amount of ATP from the metabolic fuel (glucose) in a series of reactions that take place in the cytoplasm of the cell. Recall from Chapter 6 that since this reaction pathway takes place in the ab-

sence of oxygen, it is called **anaerobic.** For each molecule of glucose metabolized by this pathway, only two molecules of ADP are "recharged" to ATP.

The Citric Acid Cycle: Aerobic Metabolism

When sufficient oxygen is present, however, the end products of glycolysis (and those of fatty acid metabolism) can enter an **aerobic (oxidative)** reaction pathway, the citric acid cycle. These reactions, which take place in the mitochondria of the cell, yield an additional 36 molecules of ATP from a metabolized glucose molecule.

Given the much higher ATP production of aerobic metabolism, it is reasonable to ask why muscles would also have the less efficient anaerobic pathways. The answer lies in the rates at which energy is consumed and at which metabolic fuels can be supplied. Under conditions of normal activity, the blood supply to a muscle can bring the energy-supplying molecules (fatty acids glucose) to a muscle at a rate that allows aerobic metabolism to supply almost all of the ATP needed by the contractile system. During moderate exercise as well, the blood supply and the oxygen it delivers can keep up with the needs of the muscle.

The Need for an Anaerobic Pathway

When the level of activity reaches about 70% of the maximum possible, however, aerobic metabolism is no longer able to supply sufficient ATP. At this point, glycolytic (anaerobic) metabolism begins to take over to provide the ATP. The reactions of the glycolytic pathway run more quickly than the aerobic ones, and they can rapidly supply large amounts of ATP from glucose in the blood and from the glycogen (a polymer of glucose) stored in the muscle itself.

The Oxygen Debt

While rapid, however, these anaerobic reactions are not very efficient, and they produce large amounts of a temporary byproduct, **lactic acid.** When the bout of heavy exercise is over, the muscle is left with an overabundance of lactic acid and is deficient in its stores of creatine phosphate and glycogen. **Muscle fatigue** often results from prolonged exercise in which the energy supply fails to keep up with all of the energy demands. In a fatigued muscle, the maximum force that can be exerted decreases; after a period of rest, the capability of the muscle is restored. Paradoxically, chemical measurements

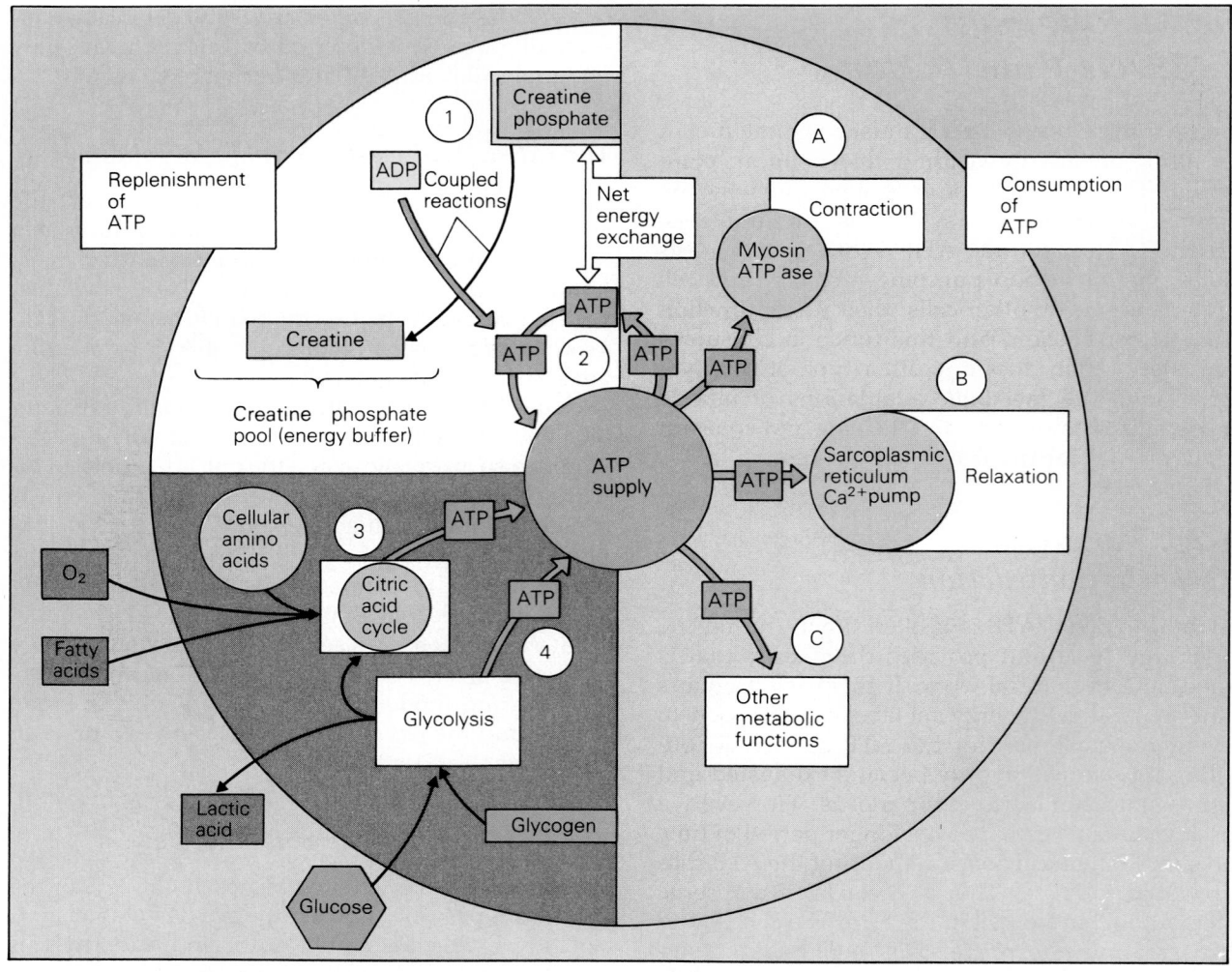

Figure 16–24
Summary of a cell's mechanisms for replenishing its supply of ATP, with the numbers indicating the order of replenishment. The cell uses ATP for a variety of functions. The letters A, B, and C indicate the order in which cellular functions use up ATP.

made on fatigued muscle show that the total ATP content is not severely reduced; instead, the failure to contract appears to lie somewhere in the processes that couple the membrane depolarization to the activation of the myofilaments. Whatever the defect, it is quickly restored.

The metabolic imbalances that follow contraction are corrected by the aerobic metabolism that occurs in the resting muscle after the activity. The lactic acid is oxidized to compounds that can provide ATP, and the glycogen stores are resynthesized from glucose. These processes are responsible for the continued high oxygen consumption of a resting muscle after exercise. This oxygen is required to run the aerobic reactions until the original state of the muscle is restored. The burden of lactic acid and the glycogen deficit that occur during exercise make up the so-called **oxygen debt,** which must be "paid back" quantitatively by postcontraction oxygen consumption.

Muscle Efficiency and Heat Production

Another important byproduct of muscle metabolism is **heat.** The biochemical reactions of contraction are only about 20% efficient; the other 80% of the energy is degraded into heat. This is the heat that the body must get rid of during heavy exercise; it is also the heat that warms the body by shivering or other exercise in the cold.

Types of Muscle Fiber

In light of the many functions that muscles must perform, it is not surprising that there are specialized types of fibers adapted to particular sorts of tasks. These adaptations involve structural and biochemical specializations that are interrelated and that fall into some fairly distinct categories. Table 16–2 gives the characteristics of these specialized muscle types.

Some muscle tasks, such as maintaining body posture, require that muscles maintain tension for long periods with a low expenditure of energy. Other tasks, such as sprinting in a race or running after a bus, require rapid contractions that may be made at a high energy cost. Between these extremes are muscles that require a mixture of metabolic and structural specializations. In general, muscle fibers fall into two broad categories, **white fibers** and **red fibers.** (For example, consider the light (white) and the dark (red) meat from poultry.)

Red Muscle Fibers and Aerobic Metabolism

The color differences among muscle fibers arise because of the differing content of a protein called **myoglobin,** which imparts a red color to the tissue. The presence of myoglobin also gives a clue to the most important distinction between the muscle classes. Myoglobin is a protein (similar to the protein **hemoglobin** found in red blood cells) that is capable of binding, storing, and releasing oxygen. It is found most abundantly in those muscle fibers that depend on oxidative (aerobic) metabolism, where it serves as an oxygen source in times of heavy demand.

Red fibers can be further classified as **fast twitch** and **slow twitch** types. They both have plentiful mitochondria and a rich blood supply. Their speeds of contraction are correlated with their rates of ATPase activity—that is, with the rate at which the actin-myosin contractile proteins can break down ATP and release its energy. The slow fibers

Table 16–2
Classification of Skeletal Muscle

| Feature | Fast | | Slow |
	White	Red	Red
Mechanical			
Contraction speed	Fast	Fast	Slow
Force capability	High	Medium	Low
S.R. Ca^{2+} pumping	High	High	Moderate
Motor axon velocity	100 M/sec	100 M/sec	85 M/sec
Biochemical			
ATPase activity	High	High	Low
Source of ATP	Anaerobic glycolysis	Oxidative phosphorylation	Oxidative phosphorylation
Glycolytic enzymes	High	Moderate	Low
Number of mitochondria	Low	High	High
Myoglobin content	Low	High	High
Glycogen content	High	Moderate	Low
Rate of fatigue	Fast	Moderate	Slow
Structural			
Diffusion distance	Large	Small	Moderate
Fiber diameter	Large	Moderate	Small
Number of capillaries	Few	Many	Many
Sarcomere structure	Very regular	Less regular	Irregular Z-lines
Functional			
Role in body	Rapid, powerful movements	Medium endurance	Postural endurance
Example	*Latissimus dorsi*	Found in mixed-fiber muscles such as *Vastus lateralis*	*Soleus*

are very resistant to fatigue when fulfilling their usual function of maintaining posture or working at lower rates. Attempts to use such muscles for rapid, sustained tasks will cause them to fatigue rather quickly as their ATP consumption outstrips the ability of the citric acid cycle to replenish the ATP supply. The fast fibers of the red group have a higher ATPase activity; they contract and relax more quickly and are suited to moderate endurance-type activities.

White Muscle Fibers and Anaerobic Metabolism

White muscle fibers, on the other hand, are adapted for fast and powerful contractions, but they fatigue quickly. They contain the enzymes necessary for glycolytic (anaerobic) metabolism. Glycolysis, a process that operates faster than oxidative metabolism, can use the stored glycogen at a high rate to produce ATP. Since their immediate contractile function does not depend heavily on their oxygen supply, these muscles are not as well supplied with a dense capillary network as are the red types. The white muscles show a high ATPase activity and very rapid contraction and relaxation. The rapid contraction appears to be due to specializations in the myosin molecules that permit them to convert the energy in ATP more rapidly. The rapid relaxation is due in part to a more aggressive calcium-pumping mechanism present in the sarcoplasmic reticulum. As might be inferred from the anaerobic character of their metabolism, the cells contain fewer mitochondria than do the red types.

Patterns of Innervation and Muscle Growth

These various characteristics are summarized in Table 16–2. A close look at the features compared there will reveal many correlations among function and biochemical makeup. Note that the characteristics listed are for fibers, not whole muscles. Most whole muscles contain a mixture of these fiber types, so they are not strictly confined to a single type of activity. Studies on athletes have shown that, in spite of particular types of training programs, the fiber composition of a given muscle remains relatively constant. For an individual, the specific mixture of fiber types begins to be established before birth and becomes set in childhood. In some newborn mammals, the fibers are predominantly the slow type. Patterns of innervation and usage during early development play an important role in deciding the ultimate distribution of fiber types. Experiments have been done with animals in

which the nerve supply to fast and slow muscles was surgically reversed. These studies have shown that the re-innervated muscles take on characteristics related to the type of nerve supplying them. That is, formerly slow muscles begin to take on the contraction characteristics of fast muscles.

The Development of Muscle in Response to Use

It is well known that the pattern of use is important in determining the size and strength of a muscle. An increase in the mass of a muscle (without an increase in the actual number of cells) is called **hypertrophy**. Such hypertrophy is most readily brought about by exercises that emphasize isometric or slow isotonic contractions repeated many times. Muscles subjected to such routines will increase in their cross-sectional area and in the amount of contractile protein they contain. The relative distribution of fiber types will remain relatively constant. A negative aspect of this type of muscle adaptation (to a specific set of exercise conditions) is that the vascular supply to the muscle does not increase in proportion to the increase in muscle mass, nor is there a large improvement in the function of the heart. This means that the endurance of the hypertrophied muscle is not greatly increased, and its improved function is largely limited to the sort of exercise that produced the hypertrophy.

Heart muscle is also subject to hypertrophy in response to increased demands. This occurs normally during athletic conditioning but may also be a result of some disease process. For example, if the improper functioning of a heart valve makes the heart do extra work to maintain the blood pressure, a so-called enlarged heart will result. Such enlargement can at least partially compensate for the condition that caused it, but it also increases the metabolic demands of the heart and may make it more prone to disturbances in the heartbeat rhythm (see Chapter 18).

Exercise that emphasizes isotonic contractions at moderate workloads produces a different type of modification in the muscle. The popular "aerobic" exercise regimens produce less marked increases in strength than the isometric exercises mentioned previously, but do result in an increased vascular supply to the muscle. This increases the ability to produce increased performance for extended periods of time, such as for the duration of a crosscountry race. Such exercises also provide conditioning for the heart muscle and can increase the efficiency of its pumping.

Adaptation in the opposite direction is also possible. Long-term immobilization of a muscle (such as might be produced by wearing a cast) causes **disuse atrophy**, in which the mass and strength of the muscle decrease. When a muscle is paralyzed by injury to its motor nerve and cannot contract on its own, it is also subject to atrophy. If there is a probability that nerve function may eventually be restored, periodic electrical stimulation of the muscle may prevent some of the atrophy.

Less severe atrophy occurs when a regular form of exercise is discontinued, and the muscle adapts to the lower level of use. This may occur during the "off-season" periods in the life of an athlete and is a factor in the "deconditioning" observed after long periods of weightlessness during space travel. Both of these situations can be at least partially reversed by a proper exercise program.

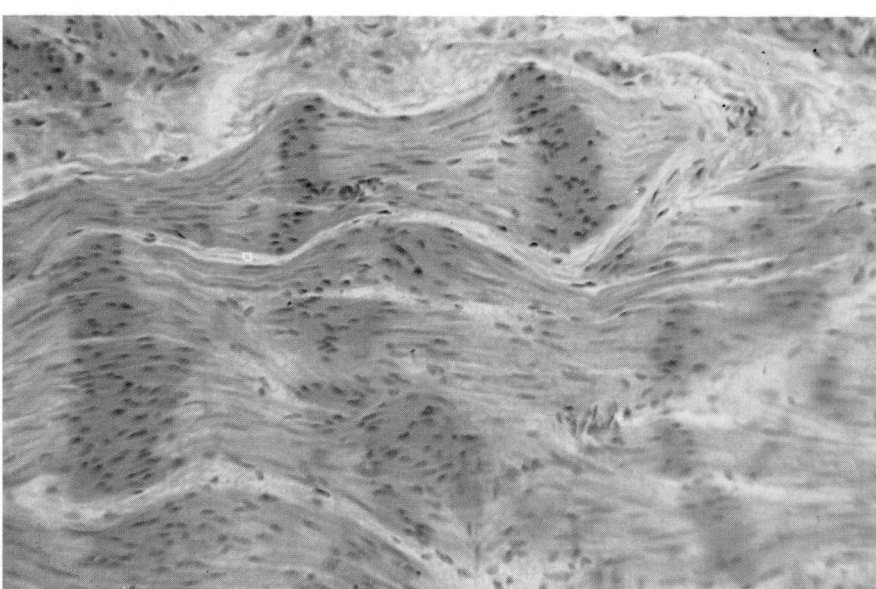

Figure 16–25
Contractile bands of smooth muscle. (© John D. Cunningham/ Visuals Unlimited.)

SUMMARY

Overview: The Roles and Types of Muscle

The many functions of muscle require that different types of muscle be specialized for different tasks. Muscle functions as an effector, a motor, and a regulator. It also regulates much of the input to the nervous system. The degree of neural control varies with the type of muscle.

Muscle may be classified in three ways: on the basis of its anatomical location (as skeletal, visceral, or cardiac); on the basis of its structure (as striated or smooth); and on the basis of its mode of control (as voluntary or involuntary). There is some overlap in these categories.

The Structure of Muscle

Most skeletal muscles are attached to bones at an origin and an insertion by tendons. As skeletal muscles shorten, they power the lever systems made up of the bones of the skeleton. Usually, skeletal muscles are arranged in antagonistic pairs or groups, and contraction of one muscle group will cause the lengthening of another. A whole skeletal muscle is made up of long and slender cells called fibers, which are bound up into the muscle by connective tissue. A skeletal muscle receives a plentiful blood supply.

Skeletal muscle fibers are composed of sarcomeres, which are the fundamental units of contraction. The sarcomeres are composed of thick (myosin) and thin (actin) myofilaments joined by crossbridges. Contraction of the individual sarcomeres results in shortening or force development of the whole muscle. The transverse tubule system and the sarcoplasmic reticulum function in controlling the contraction of the muscle.

Cardiac muscle is striated, and the sarcomeres are also similar to those of skeletal muscle. The small cells are closely coupled together both mechanically and electrically.

Smooth muscle is composed of small unstriated cells. They contain contractile proteins, although no obvious sarcomere arrangement is present, and the internal membrane systems are not very extensive. Cells are coupled together mechanically and electrically.

The Function of Muscle

Muscles can shorten and exert a constant force (under isotonic conditions), or they can exert a force that does not result in movement (isometric conditions). Most practical contractions involve a mixture of these two conditions. A single brief activation of a skeletal muscle results in a brief contraction called a twitch, and repeated activation results in a sustained contraction called a tetanus. The length of a muscle before contraction has an important effect on the amount of force the muscle can exert when stimulated, and there is a particular length for optimal force. This relationship is described by the length-tension curve. The interaction between the actin filaments and the myosin crossbridges produces mechanical motion. The speed of this process and the overall amount of force generated are influenced by muscle length and the force during shortening. The crossbridge cycle is controlled by presence or absence of calcium ions and requires ATP.

When muscle shortens, its speed is related to the amount of force exerted, a relationship described by the force-velocity curve. The power output of muscle also depends on the relationships presented in this curve. There is an optimal shortening speed for maximal power output, a speed also related to the efficiency with which the muscle does work.

Cardiac muscle contracts using a similar mechanism, although the external control is different. Smooth muscle also appears to function in much the same way.

The Control of Muscle

The transverse tubular systems link the action potential in the plasma membrane to the release of calcium ions from the lateral cisternae of the sarcoplasmic reticulum. Release of calcium in the region of the myofilaments allows the contractile proteins to interact in the crossbridge cycle. This action is exerted via control proteins on the actin (thin) filaments and is called actin-linked regulation.

Cardiac muscle also has a similar actin-linked regulatory process. Smooth muscle is under the control of a myosin-linked regulatory process, in which calcium ions activate the myosin filaments by promoting the phosphorylation of the myosin crossbridges.

Skeletal muscle contractions are controlled by the plasma membrane. Cardiac muscle is also controlled by its plasma membranes, but only twitch contractions are possible. Many smooth muscles are also controlled by the electrical activity of the membrane, but control is also exerted by substances (drugs, hormones) that can directly cause the release of calcium from internal storage sites. A long-lasting contraction of smooth muscle is often called a tonic contraction, or tonus.

The electrical activity of the skeletal muscle plasma membrane is controlled by the central nervous system (CNS). The link between a motor axon and a muscle fiber is called the myoneural junction, or motor end-plate. A nerve action potential causes the release of acetylcholine from the nerve terminals to the muscle end-plate membrane, producing the end-plate potential. This in turn causes the muscle membrane to produce an action potential, which travels along the muscle fiber and activates contraction. Substances that interfere with the release, removal, or action of acetylcholine will block myoneural transmission and paralyze the muscle. Some smooth muscles are controlled by the autonomic nervous system (ANS), although they do not have well-developed myoneural junctions. Cardiac muscle membranes function under the control of specialized cells in the heart, and the nervous system acts to regulate both the rate and strength of the heartbeat.

The Metabolism of Muscle Contraction

The energy for muscle contraction is supplied by glucose and fatty acids and is ultimately used by the muscles in the form of ATP. Glycolysis, an anaerobic process, produces relatively little ATP from each glucose molecule and produces lactic acid as a waste product. Such a pathway produces energy quickly but inefficiently and results in an oxygen debt in the form of lactic acid. This must be subsequently "paid off" by oxygen consumption during aerobic metabolism. Aerobic metabolism produces large amounts of ATP, but it is relatively slow.

Some skeletal muscles are specialized for powerful and rapid movements, but they become fatigued rapidly. These muscles have a primarily anaerobic metabolism, are composed of large fibers, and are light in color. They are termed fast, white muscles. Other muscles contract more slowly but can sustain activity for long periods of time. These muscles have a primarily aerobic metabolism, with many mitochondria and a rich blood supply. This group, which contains both slow and moderately fast muscle fiber types, is dark-colored; these are the red muscles. Many skeletal muscles are made up of combinations of these fiber types and are suited to a variety of functions.

Muscle responds to increased use by undergoing hypertrophy, which is an increase in the size, weight, and protein content of the muscle. Cardiac muscle may undergo hypertrophy as the result of increased blood pressure or improper function of a heart valve. Muscle that is immobilized or little used will undergo atrophy, a reduction in size and protein content.

REVIEW QUESTIONS

Identify Each with a Term

1. The basic contractile unit of skeletal muscle.
2. The protein strands that form the basic contractile mechanism of all types of muscle.
3. The type of muscle associated with blood vessels and internal organs.
4. The type of muscle activity in which force is developed without any change in muscle length.
5. The connection between a motor neuron and a skeletal muscle fiber.
6. The chemical fuel most directly consumed by the muscle contraction process.
7. The relationship between muscle force and speed of shortening.

Define Each Term

8. twitch
9. tetanus
10. sarcoplasmic reticulum
11. troponin
12. crossbridges
13. end-plate potential
14. acetylcholine esterase
15. rigor mortis
16. myoglobin
17. disuse atrophy

Choose the Correct Answer

18. In the *sliding-filament hypothesis* for muscle contraction:
    a. movement occurs due to shortening of the myofilaments.
    b. the cyclic interactions between myosin crossbridges and actin filaments causes the movement to occur.
    c. movement is caused by the action of calcium ions on the thick filaments.

19. In an isotonic contraction of skeletal muscle:
    a. no movement occurs.
    b. the muscle shortens at a constant force.
    c. force increases as the muscle shortens.

20. Increasing the length of a skeletal muscle prior to an isometric contraction:
    a. may either increase or decrease the force that can be developed, depending on how long the muscle was originally.
    b. has no effect on the force that the muscle can develop.
    c. can only decrease the force of contraction.

21. When a weight is held in the hand, the lever action of the bones of the arm causes the flexor muscles of the upper arm:
    a. to exert a force greater than that provided by the weight.
    b. to shorten more rapidly than the speed at which the weight itself is moved.
    c. to move the weight even though they exert less force than the weight provides.

22. The role of the transverse tubules (t-tubules) in the excitation of skeletal muscle contraction is:
    a. to provide an inward path for the spread of muscle action potentials.
    b. to serve as the storage site for calcium ions.
    c. to connect the sarcomeres end-to-end.

23. The primary role of calcium in the activation of *skeletal* and *cardiac* muscle is:
    a. to cause depolarization of the muscle cell plasma membrane.
    b. to remove the inhibition of the reaction between the actin filaments and the myosin filaments.
    c. to activate myosin molecules so that they can interact with actin filaments.

24. The primary role of calcium in the activation of *smooth muscle* is:

a. to cause depolarization of the muscle cell plasma membrane.

b. to remove the inhibition of the reaction between the actin filaments and the myosin filaments.

c. to activate myosin molecules so that they can interact with actin filaments.

25. During the process of transmission at the myoneural junction, the role of acetylcholine is:

a. to cause the depolarization of the presynaptic (nerve) cell membrane.

b. to prevent depolarization of the muscle cell membrane.

c. to cause the depolarization of the postsynaptic (end-plate) membrane of the muscle cell.

26. Which statement best describes the control of smooth muscle by the nervous system?

a. Smooth muscle is completely inactive unless it is stimulated by a motor nerve.

b. Nerve stimulation may either initiate mechanical (contractile) activity or it may modify ongoing activity.

c. Nerve stimulation is usually inhibitory and serves mostly to prevent contraction.

27. During the *aerobic* metabolism of exercising skeletal muscle:

a. the primary source of energy is glucose, and relatively little ATP is formed.

b. the production of ATP is associated with the consumption of oxygen.

c. large amounts of lactic acid are produced.

28. An oxygen debt in exercising skeletal muscle:

a. is due to the build-up of lactic acid produced by anaerobic metabolism.

b. can usually be "paid back" by anaerobic metabolism.

c. occurs only when the muscle is at rest.

29. Which set of characteristics is found in those skeletal muscle fibers best adapted to produce fast and powerful movements?

a. resistance to fatigue, red color, mostly aerobic metabolism

b. small diameter fibers, red color, irregular sarcomere structure

c. little resistance to fatigue, large fiber diameter, white color, mainly anaerobic metabolism

Answer in One or Two Sentences

30. What are the three types of muscle in the body?

31. What conditions can lead to hypertrophy and atrophy in skeletal muscle?

32. How does the speed of contraction of cardiac muscle compare with that of smooth and skeletal muscle?

33. How does the drug curare block transmission at the skeletal muscle myoneural junction?

34. Why does skeletal muscle produce more force during a tetanus than during a twitch?

35. What internal cellular event leads to relaxation in all types of muscle?

SUGGESTED READINGS

Aidley, D.J. *The Physiology of Excitable Cells*. Cambridge, England: Cambridge University Press, 1971.

Carlson, F.D., and Wilkie, D.R. *Muscle Physiology*. Englewood Cliffs, N.J.: Prentice-Hall, 1974.

Cohen, C. "The protein switch of muscle contraction." *Scientific American*, 233(5):36, 1975.

Jones, D.A. and Round, J.M. *Skeletal Muscle in Health and Disease. A Textbook of Muscle Physiology*. Manchester, England: Manchester University Press, 1990.

Katz, B. *Nerve, Muscle, and Synapse*. New York: McGraw-Hill, 1966.

Lester, H.A. "The response to acetylcholine." *Scientific American*, 236(2):106, 1977.

Murray, J.M., and Weber, A. "The cooperative action of muscle proteins." *Scientific American*, 230(2):58, 1974.

Ruegg, J.C. *Calcium in Muscle Activation*. New York: Springer-Verlag, 1986.

Shephard, R.J. *Physiology and Biochemistry of Exercise*. New York: Praeger Scientific, 1982.

Squire, J.M. *Molecular Mechanisms in Muscular Contraction*. Boca Raton: CRC Press, 1990.

Stein, R.B. *Nerve and Muscle: Membranes, Cells, and Systems*. New York: Plenum Press, 1980.

Woledge, R., Curtin, N.A., and Homsher, E. *Energetic Aspects of Muscle Contraction*. New York: Academic Press, 1985.

KEY TERMS

actin (p. 539)
afterload (p. 551)
antagonistic pair (p. 535)
auxotonic contraction
  (p. 545)
contractility (p. 559)
crossbridge (p. 539)
end-plate potential (p. 563)

force-velocity curve (p. 551)
hypertrophy (p. 570)
involuntary muscle (p. 533)
isometric contraction
  (p. 545)
isotonic contraction (p. 545)
length-tension curve
  (p. 560)

motor unit (p. 561)
muscle fiber (p. 569)
myofilament (p. 535)
myoglobin (p. 569)
myosin (p. 539)
myotonic contraction
  (p. 545)

sarcomere (p. 539)
sliding-filament
  hypothesis (p. 550)
tendon (p. 534)
tension: active, passive,
  resting (p. 548)
voluntary muscle (p. 533)

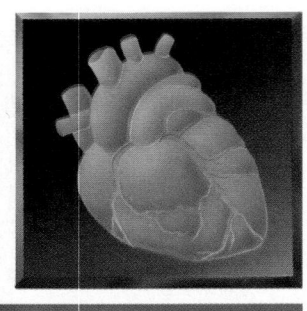

C H A P T E R
17

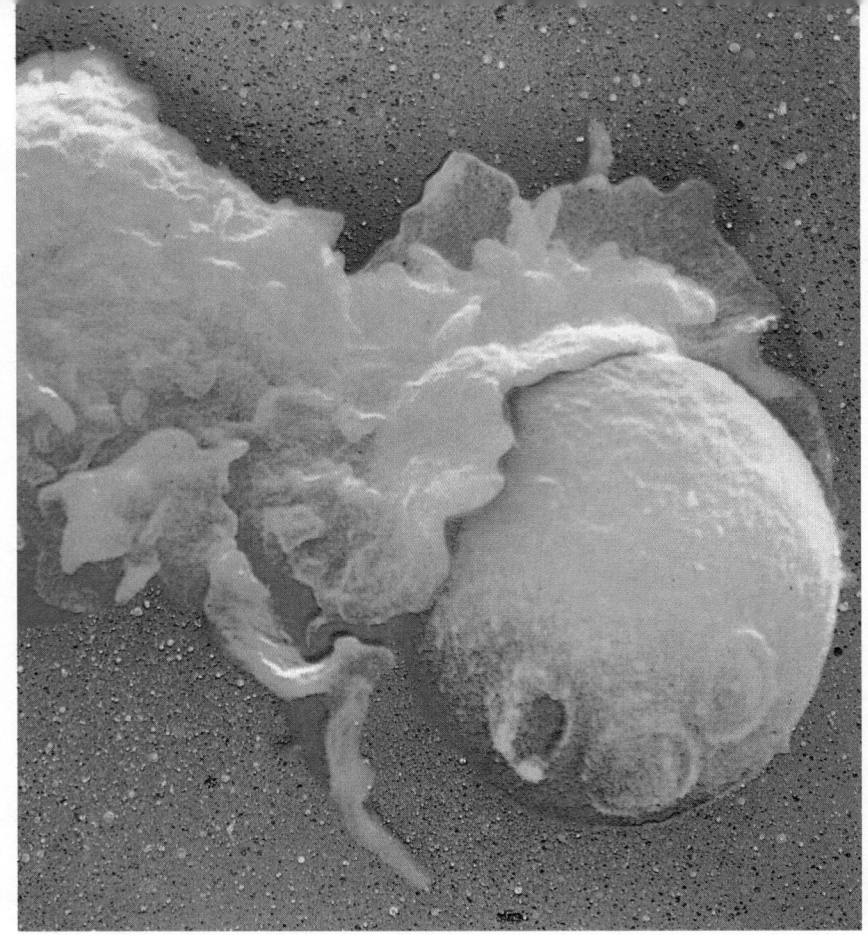

Functions of the Blood

For centuries, blood has been regarded as a vital fluid. Even primitive man associated the loss of blood with a decline in well-being and, if the loss was sufficient, death. The ancient Greeks considered blood to be one of four elemental substances (called fluids or humors) of the human body (see Chapter 1). Because of the life-sustaining qualities of blood, many magical properties have been accorded to it by civilizations of the past, a practice that continues in primitive societies and cults of the modern world.

Strictly speaking, blood is not a bodily fluid like tears, saliva, and urine; instead, it is a living connective tissue composed by volume of approximately 45% cellular elements and 55% intercellular fluid. Blood is normally confined to the blood vasculature and chambers of the heart, and is circulated throughout the body by the heart's pumping action. The lives of all cells depend on circulating blood because it helps maintain the relatively stable internal environment essential for normal cell function. When blood circulation ceases, cells of the brain, heart, and other organs begin to die within a few minutes.

In this chapter, we will examine some physical and chemical properties of blood, formation and functions of each type of blood cell, normal and abnormal chemistry of the blood, and why one person's blood differs from another's.

General Functions of Blood

Circulation of the blood through the vasculature provides a functional connection among the various cells, tissues, and organs of the body, and serves to maintain a relatively stable internal environment for optimal cellular activity. The role of the blood in homeostasis largely involves transport, defense against foreign agents, and the maintenance of internal temperature and pH.

Transport

Blood is the primary means by which substances are conveyed from one area of the body to another. Nutrients such as glucose, amino acids, and fatty acids, plus electrolytes and water, are absorbed from the gastrointestinal tract and carried by the blood to various body tissues. Oxygen is absorbed by blood as it passes through the lungs and is transported to respiring tissues. Carbon dioxide produced during cellular respiration is absorbed by the blood as it circulates through tissues and is transported to the lungs where carbon dioxide is excreted in expired air. In addition, other cellular waste such as urea, uric acid, and excess water is carried by the blood to organs of excretion. Hormones and other chemical messengers produced within the body are transported by blood from sites of production to target cells. In essence, blood serves as a vehicle for moving many different kinds of substances to various areas within the body for a variety of purposes.

Defense Against Foreign Agents

Some of the blood cells are phagocytic—that is, they are capable of ingesting and rendering harmless substances foreign or toxic to the body and microorganisms that are pathogenic or disease-producing. Other blood cells produce and release chemicals (antibodies) that react with and nullify the harmful effects of foreign agents, or chemicals that regulate blood flow and blood clotting during a defensive response to injury.

Maintenance of Internal Temperature and pH

Body metabolism produces a considerable quantity of heat as a by-product, and the excess heat must be dissipated into the environment (see Chapter 28) in order to maintain a stable internal temperature. Blood conducts heat from the body's core to the lungs, respiratory passageways, and skin, where dissipation of the heat occurs.

Blood plays a major role in the regulation of extracellular fluid pH. Chemical buffers present in blood convert strong acids and bases into weak acids and bases, thereby minimizing large shifts in pH during the course of daily metabolism. The blood also transports acidic and basic substances to organs of excretion (see Chapter 26). Normal blood is slightly alkaline, with a pH between 7.35 and 7.45.

Properties of Whole Blood

Composition, Volume, and Distribution

Whole blood is an opaque red liquid connective tissue consisting of microscopically visible elements—**red corpuscles** (erythrocytes), **white corpuscles** (leukocytes), and **platelets** (thrombocytes)—suspended in an amber intercellular fluid called **plasma.** The various components of whole blood can be separated using fractionation techniques such as centrifugation. These components, called **blood fractions,** include red corpuscles (packed RBCs), white corpuscles, platelets, and the protein fractions of the plasma (e.g., gamma globulin, the protein fraction that contains antibodies). A blood fraction, rather than whole blood, is often used clinically to alleviate symptoms of disease. For example, packed RBCs are sometimes used to reduce the effects of severe anemia, and gamma globulin is used to increase resistance to disease.

Blood accounts for approximately 6% to 8% of adult body weight. Normal adult **blood volume** ranges from 4.5 to 5.5 liters for a female and 5.0 to 6.0 liters for a male. Total blood volume varies, being altered by changes in body water balance (see Chapter 25), pregnancy, hemorrhage, and other factors. **Normovolemia** is a blood volume within normal range. An above-normal blood volume is called **hypervolemia** and a below-normal volume is **hypovolemia.**

The distribution of blood between the heart and the various types of blood vessels is not uniform. Generally, about 12% of the blood is in pulmonary vessels, 79% is in systemic vessels, and 9% is within the heart. Factors that alter the distribution of blood are discussed in Chapter 19.

Specific Gravity and Viscosity

The **specific gravity (SG)** of a liquid is the ratio between the weight of a given volume of the liquid and the weight of an equal volume of pure water. The specific gravity of water is 1.000; the specific gravity of normal adult blood ranges from 1.050 to 1.060. The specific gravity of blood depends on the number of blood cells and the concentration of chemicals normally dissolved in the plasma. Diseases that alter cell numbers or chemical concentrations of the blood may change the specific gravity.

Viscosity is a measure of a fluid's resistance to flow. The greater the viscosity the greater the fluid's resistance to flow. If we assume the viscosity of pure water to be 1.00, then whole blood viscosity by comparison is approximately 3.50 to 5.50 and that of blood plasma is 1.90 to 2.60; mean values for an adult are 4.5 and 2.2, respectively. Blood viscosity is determined by the number of cells present—cells resist being moved—in whole blood and the mutual attraction of molecules dissolved in the plasma. When the total number of cells increases to above normal, or when the plasma concentration of large molecules such as protein is elevated to above normal, the viscosity of the blood increases, forcing the heart to work more strenuously to maintain a normal rate of flow. A prolonged pathological increase in blood viscosity overloads the heart and may weaken its ability to pump. It also tends to elevate systemic blood pressure, which damages blood vessels and increases the possibility of hemorrhage, or reduces blood flow to the point where circulation through certain organs (e.g., the kidneys) becomes inadequate.

Sedimentation Rate of Blood

If a small quantity of blood is removed from the body, mixed with a chemical that prevents clotting (an anticoagulant), and placed in a vertical graduated cylinder, the blood cells, having a higher specific gravity than the plasma, will slowly sink to the bottom, leaving behind a transparent amber upper layer of plasma. The leukocytes are lighter than the erythrocytes; as a result, the leukocytes form a thin whitish layer (called the buffy coat) between the upper plasma and the lower sedimented erythrocytes (Fig. 17–1). The rate at which the erythrocytes

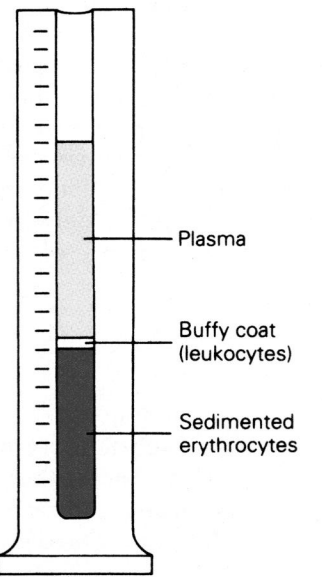

Figure 17–1
Sedimented blood.

Plasma

Buffy coat
(leukocytes)

Sedimented
erythrocytes

settle is called the **sedimentation rate (SR).** Normal sedimentation rates for adults are: males—2 to 8 mm per hour; females—2 to 10 mm per hour.

Sedimentation of erythrocytes proceeds in three phases: formation of rouleaux, rapid settling, and final packing of the red cell mass. The formation of rouleaux (clumping of red cells together like a stack of coins) does not occur with cells of abnormal shape such as those in sickle cell anemia, and the sedimentation rate is therefore slower. Increases in the concentration of plasma proteins tend to increase the sedimentation rate of blood by facilitating rouleaux formation. The sedimentation rate tends to be elevated in acute and chronic infections such as arthritis, tuberculosis, rheumatic fever, and toxemia. The sedimentation rate usually decreases in polycythemia (too many RBCs), allergies, and hyperglycemia (elevated blood sugar).

The Hematocrit

The percentage of the blood volume made up of erythrocytes is called the **hematocrit.** Normal adult values are: males—47.0 ± 5.0; females—42.0 ± 5.0.

Normal hematocrit values vary with age, sex, and the part of the body from which the blood sample is taken. The hematocrit value of arterial blood is slightly less than that of venous blood. The hematocrit varies from tissue to tissue, and the tissue hematocrit is generally lower than either the systemic arterial or systemic venous hematocrit.

To determine hematocrit, venous blood is drawn into a siliconized syringe (to minimize cell damage), rendered incoagulable with oxalate, deposited in a Wintrobe tube (Fig. 17–2), and centrifuged for 10 minutes at about 1500 g (g = relative acceleration due to gravity = 980 cm/sec^2). After centrifugation, the hematocrit is calculated by dividing the measured red cell volume by the measured whole blood volume and multiplying the result by 100 to obtain the percentage.

Clinically, it is often more convenient to use the **microhematocrit** method rather than the Wintrobe method to determine hematocrit. A small-bore (1 mm) heparinized (to prevent coagulation) capillary tube is filled by a drop of capillary blood taken from a fingertip puncture. The blood is then subjected to centrifugation and the hematocrit determined.

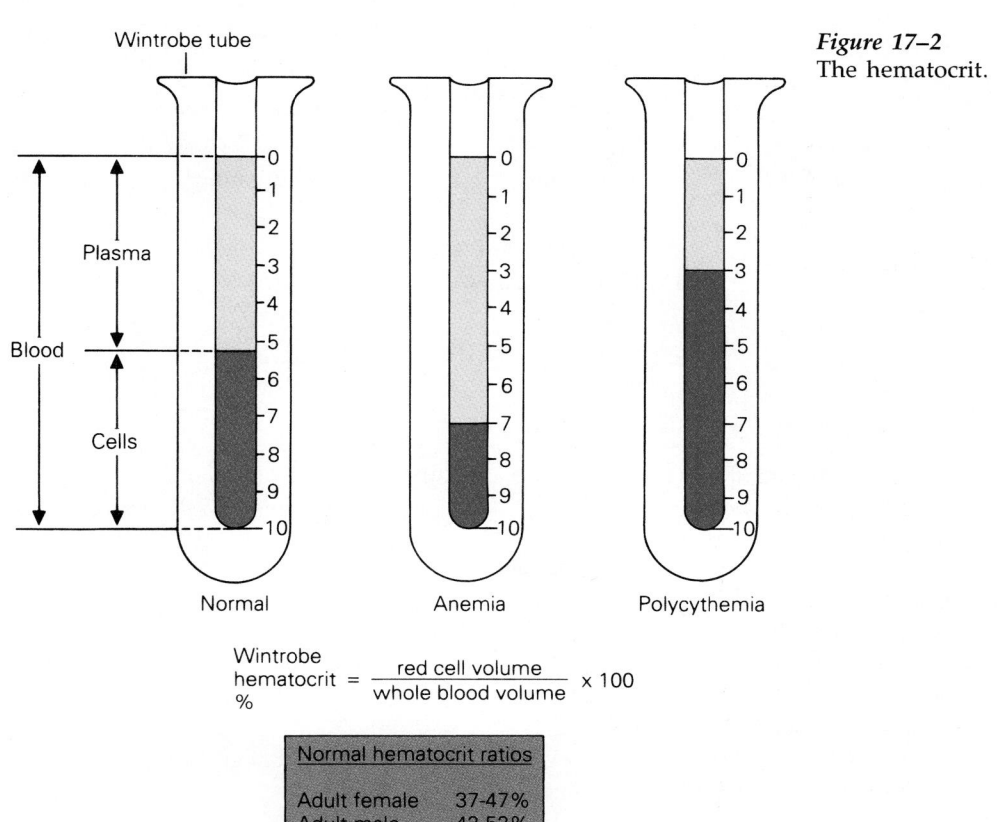

Figure 17–2
The hematocrit.

$$\text{Wintrobe hematocrit \%} = \frac{\text{red cell volume}}{\text{whole blood volume}} \times 100$$

Normal hematocrit ratios	
Adult female	37-47%
Adult male	42-52%

After centrifugation of the blood, about 4% of the plasma always remains trapped among the erythrocytes. To obtain a **corrected hematocrit,** the value of the **apparent hematocrit** is simply multiplied by 0.96.

Abnormal hematocrits occur in diseases that alter the number of circulating erythrocytes, the volume of plasma, or both. In anemias characterized by a reduction in the number of circulating erythrocytes, the hematocrit may fall to 25, whereas in polycythemia (too many erythrocytes), the hematocrit may approach 70. The hematocrit is useful in diagnosing and assessing progress in treatment of diseases that alter the number of circulating normal erythrocytes, or conditions that alter the normal ratio between erythrocyte and plasma volumes (e.g., dehydration).

Plasma

Plasma is the complex fluid in which blood cells and platelets circulate. Plasma is defined as whole blood minus the formed elements (cells and platelets). Plasma may be separated from whole blood by centrifugation, but if left to stand, it coagulates within minutes, forming a gel plus a clear fluid called **serum.** Serum is similar in composition to plasma, except that it lacks coagulation factors.

Table 17–1
Normal Constituents of Blood

Compound[a]	Concentration and Fraction[b]	Compound[a]	Concentration and Fraction[b]
Acetoacetic acid + acetone (ketone bodies)	0.2–2.0 mg/100 ml (S)	Glucose (true)	70–100 mg/dl (B)
		Iodine, protein-bound	3.5–8.0 μg/dl (S)
Adenosine diphosphate	100 μm (E)	Iron	50–150 μg/dl (S)
Adenosine monophosphate	13 μm (E)	Lactic acid	0.6–1.8 mEq/L (B); 1 mm (P)
		Lead	3 μm (E)
Adenosine triphosphate	1 μm (E)	Lipase	Below 2 units/ml (S)
Aldosterone	3–10 mg/dl (P)	Lipids, total	450–1000 mg/dl (S)
Alpha-amino acid nitrogen	3.0–5.5 mg/dl (P)	Magnesium	1.5–2.0 mEq/L; 1–2 mg/dl (S)
		Nitrogen, nonprotein	15–35 mg/dl (S)
Amino acid, total	300 mg/dl (P), 120 mg/dl (E)	Phosphatase, acid total[c]	Male: 0.13–0.63 sigma units/ml (S); female: 0.01–0.56 sigma units/ml (S)
Ascorbic acid	0.4–1.5 mg/dl (fasting B)		
Bicarbonate	25 mM		
Bilirubin	Direct: 0.1–0.3 mg/dl (S); indirect: 0.2–1.2 mg/dl (S)	Phosphatase, alkaline[c]	2.0–4.5 Bodansky units/ml (S)
		Phospholipids	145–225 mg/dl (S)
Calcium	8.5–10.5 mg/dl; 4.3–5.3 mEq/L (S)	Phosphorus, inorganic	3.0–4.5 mg/dl (adult, S); 0.9–1.5 mm/L (adult, S)
Carbon dioxide	26–28 mEq/L (S)	Potassium	4 mm (P); 111 mm (E)
Carotenoids	0.08–0.4 μg/ml (S)	Protein	
Ceruloplasmin	27–37 mg/dl (S)	Total	6.0–8.0 gm/dl (S)

Composition

Normal plasma is transparent and light yellow in color, and is about 93% water and 7% solutes. One liter of human plasma contains about 930 grams of water, 60 grams of protein, 8 grams of inorganic solutes such as Na^+, K^+, Cl^-, HCO_3^-, and Ca^{++}, and 2 grams of nonprotein organic substances such as glucose, glycerol, and fatty acids. Plasma contains dissolved gases (O_2, N_2, CO_2, hormones, enzymes, vitamins, pigments, minerals), and a variety of cell waste products (e.g., urea, uric acid) and cell nutrients such as amino acids. Major constituents of human plasma are shown in Table 17–1.

The composition of plasma varies slightly from minute to minute as substances are exchanged across capillary walls between plasma and interstitial fluid. Figure 17–3 is a diagram of the three major fluid compartments in the body: **the blood plasma, the interstitial fluid,** and **the intracellular fluid.** The interstitial fluid constitutes the immediate environment of all living cells of the body and is separated from the intracellular fluid by the plasma membrane. The capillary wall separates blood plasma from interstitial fluid. Normally, the three fluid compartments are in osmotic equilibrium. The exchange of water and small molecules, such as electrolytes, between plasma and interstitial fluid is

Table 17–1 continued
Normal Constituents of Blood

Compound[a]	Concentration and Fraction[b]	Compound[a]	Concentration and Fraction[b]
Chloride	100–106 mEq/L (S); 102 mm (P); 78 mm (E)	Albumin	3.5–4.5 gm/dl (S)
		Globulin	3.5–4.5 gm/dl (S)
Cholesterol	150–240 mg/dl (S)	Pyruvic acid	0–0.11 mEq/L (P)
Cholesterol esters	≅60–75% of total cholesterol (S)	Riboflavin	500 nanomol (E)
		Sodium	136–145 mEq/L: 310–340 mg/dl (S); 140 mm (P); 6 mm (E)
Cobalt	17 μm (E)		
Copper, total	100–200 μg/dl (S); 17 μm (E)		
Cortisol (17-hydroxycorticoids)	5–18 μg/dl (P)	Sulfate	0.5–1.5 mEq/L (S)
		Transaminase (serum glutamic oxaloacetic transaminase [SGOT][c])	10–40 units/ml (S)
Creatine	40 μm (P): 600 μm (E)		
Creatinine	0.7–1.5 mg/100 ml (S); 450 μm (P); 160 μm (E)	Urea nitrogen (blood urea nitrogen, BUN)	8–25 mg/dl (B)
2,3-Diphosphoglycerate (DPG)	4 mm (E)	Uric acid	3.0–7.0 mg/dl (S)
Glucose (folin)	80–120 mg/dl (fasting, B)	Vitamin A	0.15–0.6 μg/ml (S)

[a] The substances and the values listed for them in this table give an indication of the biochemical complexity of blood and are not to be construed as absolute. Blood levels of specific components can vary among normal individuals or in the same individual at different times. Furthermore, the normal values for specific blood constituents can differ greatly among different laboratories, depending on the technique used for their determination.

[b] Abbreviations used for the several blood fractions are: B = whole blood; P = plasma; S = serum; E = erythrocytes.

[c] Activities of over 50 specific enzymes have been detected in various blood fractions, only 3 of which are listed here.

Source: From D. Jensen, *The Principles of Physiology,* 2nd ed. New York: Appleton-Century-Crofts, 1980. Table 32–2.

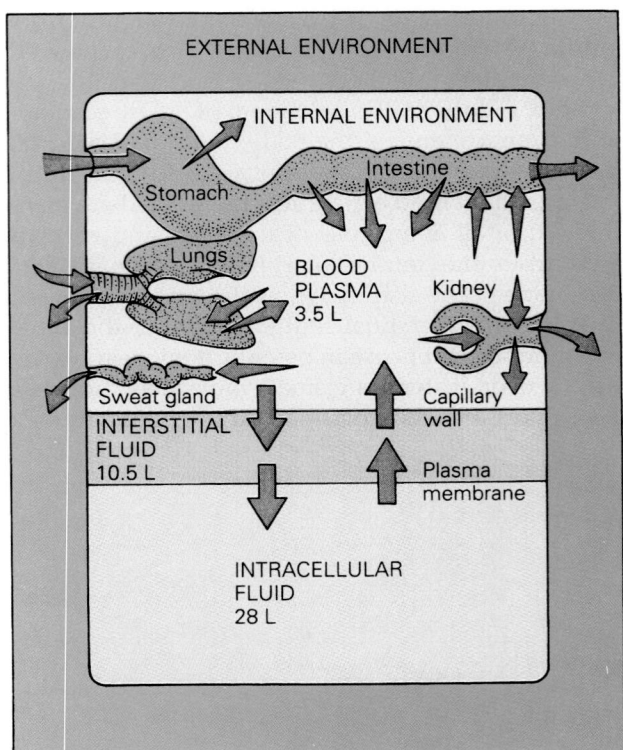

Figure 17–3
Fluid compartments in the body.

very rapid; about 70% of the plasma fluid is exchanged with the interstitial fluid in one minute. Because of this rapid exchange, the water and electrolyte concentrations of plasma and interstitial fluid are nearly identical and vary little despite considerable variation in the uptake and release of substances by cells (see Chapter 25). The major difference in concentration between plasma and interstitial fluid is that of protein, because protein molecules are generally too large to readily pass through the capillary wall. The greater concentration of proteins in the plasma establishes an osmotic gradient that, as we shall see, governs the distribution of water between the blood and the interstitial fluid.

Plasma Proteins

Types and General Functions

By means of electrophoresis, ultracentrifugation, immunoelectrophoresis, or other separation techniques, the plasma proteins can be divided into the following fractions: **albumin, globulins** (alpha-1, alpha-2, beta-1, beta-2, and gamma), and **fibrinogen** (Fig. 17–4). Plasma proteins are large molecules with high molecular weights (44,000 to 1,300,000) and configurations (arrangement of atoms) ranging from spherical (B-lipoprotein) to ellipsoid (albumin, globulins). Due to their large size, plasma proteins are classified as **colloids.** Plasma proteins collectively function in several different ways.

First, plasma proteins serve as an important reserve supply of amino acids for cell nutrition. If necessary, cells of the reticulo-endothelial system (macrophages in liver, gut, spleen, lung, and lymphatic tissue) can ingest plasma proteins, break them down, and release component amino acids into the blood so that other cells can use them for synthesis of new protein.

Second, plasma proteins serve as carriers for other molecules (see Fig. 17–4). Many types of small molecules bind to specific plasma proteins and are transported from absorptive organs (e.g., intestine) or storage organs (e.g., liver) to other tissues for utilization. Iron, for example, is carried in the blood combined with a transport protein called transferrin. Many of the ions in the blood (e.g., Ca^{++}) bind to plasma proteins, as do drugs, pigments, and hormones. Lipids (fats) absorbed into the lymph and blood by the intestine are water-insoluble; yet, when combined with proteins in lymph and blood, they can be made soluble for transport (see Chapter 23).

Third, plasma proteins act as buffers, helping to maintain a stable blood pH. The normal range for the pH of blood is 7.35 to 7.45; that is, normal blood is slightly alkaline. Proteins are amphoteric; they are able to bind H^+ or OH^- depending on whether they exist as weak acids or weak bases, which in turn depends on pH. Generally, plasma proteins function as weak bases, binding excess H^+, and thus help to keep the blood slightly alkaline (see Chapter 26).

Fourth, several plasma proteins, some of which are enzymes, interact in specific ways to cause blood to coagulate. Coagulation of blood is part of the body's response to vascular injury and helps protect against the loss of blood and invasion by microorganisms and viruses.

Fifth, plasma proteins produce a **colloid osmotic (oncotic) pressure** that governs the distribution of water between the blood and the interstitial fluid.

The total osmotic pressure of normal plasma is about 7.3 atm (1 atm = 760 mmHg), or 5550 mmHg. The total osmotic pressure is the osmotic pressure plasma would exert if separated

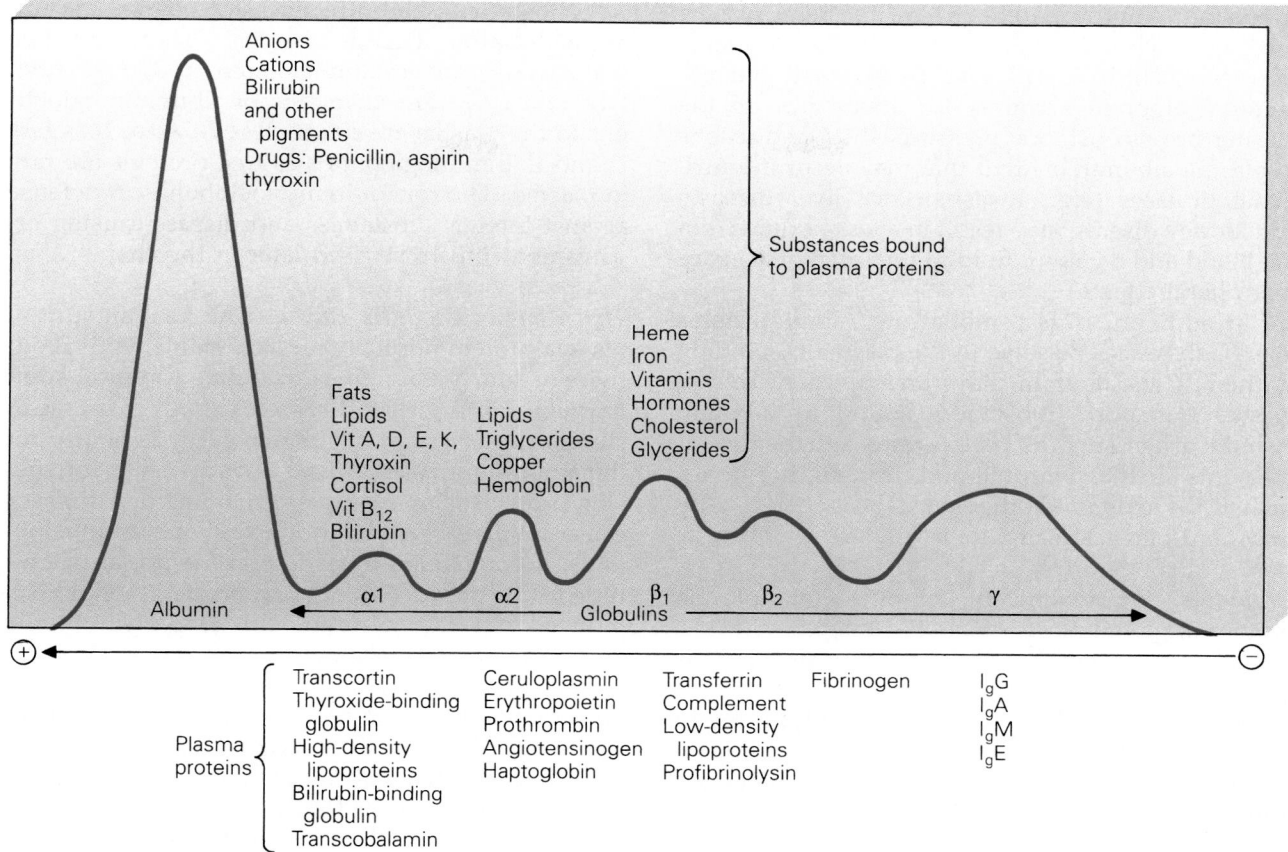

Figure 17–4
Electrophoretic survey of plasma proteins and substances bound to plasma proteins.

from pure water by a selectively permeable membrane. Solutions with the same osmotic pressure as plasma are called **isotonic.** An erythrocyte placed in an isotonic solution, such as 0.85% NaCl (physiological saline), will neither gain nor lose water. Solutions with a higher osmotic pressure than plasma are called **hypertonic,** and those with a lower osmotic pressure are called **hypotonic.** An erythrocyte will lose water and shrivel (crenate) if placed in a hypertonic solution, whereas it will gain water, expand, and hemolyze (lose hemoglobin) if placed in a hypotonic solution. Since plasma, interstitial fluid, and intracellular fluid are normally in osmotic equilibrium, homeostasis depends critically on regulation of the osmotic pressure of the plasma. Any deviation from normal extracellular fluid osmotic pressure causes cells to gain or lose volume, and hence possibly malfunction or die.

About 99.5% of the total osmotic pressure of plasma is due to small molecules such as electrolytes, urea, glucose, and others that readily pass through capillary membranes along with water.

Thus, the osmotic pressure due to these solutes is about the same in interstitial fluid as in plasma. The contribution of plasma proteins to the total osmotic pressure of plasma is about 0.5%; normal colloid osmotic pressure (oncotic pressure) of the plasma is 28 mmHg. Nevertheless, oncotic pressure plays an important role in regulating the distribution of water between plasma and interstitial fluid, because plasma proteins do not readily pass through capillary membranes. Small amounts of protein that do pass from blood to interstitial fluid do not collect in the interstitial fluid because of cell uptake and removal by the lymphatic system, and therefore a protein concentration gradient from blood to interstitial fluid exists, is maintained, and osmotically governs the movement of water and small molecules between plasma and interstitial fluid. Mechanisms of capillary exchange are presented in greater detail in Chapter 19.

In addition to functioning collectively as previously described, the individual types of plasma proteins have specific properties and functions.

Albumin.   Approximately 60% of the total plasma protein is albumin, one of the smallest proteins (mw = 69,000) in the plasma. Being small and numerous, albumin accounts for about 80% of the plasma protein osmotic pressure; thus, reductions in plasma albumin content that may occur in nutritional diseases (e.g., kwashiorkor), liver disease, and kidney disease may result in a loss of fluid from the blood and a gain of fluid in the interstitial space (peripheral edema).

In addition to its osmotic role, albumin helps many substances dissolve in the plasma by binding to them; thus albumin plays an important role in plasma transport. Substances bound by albumin include drugs such as barbiturates and penicillin, pigments such as bilirubin and urobilin, hormones such as thyroxin, and other substances such as fatty acids and bile acid salts.

Globulins.   Approximately 40% of the total plasma protein is globulin, of which 4% is alpha-1, 8% is alpha-2, 7% is beta-1, 4% is beta-2, and 17% is gamma.

Alpha-1 globulins are primarily **glycoprotein** (protein joined to carbohydrate) and small amounts of lipoprotein (protein joined to lipid). **High density lipoprotein** (HDL) functions in lipid transport and appears to prevent cholesterol from invading and settling in the walls of arteries (see Research Focus box). Other proteins in the alpha-1 group include **thyroxine-binding globulin, cortisol-binding globulin (transcortin),** and **vitamin B_{12}–binding globulin** (transcobalamin), all of which transport their respective substrates.

Alpha-2 globulins contain **haptoglobin,** a protein that combines with free hemoglobin to prevent its excretion by the kidney, and **ceruloplasmin,** a copper-containing oxidase enzyme. Other proteins in the alpha-2 group include: **prothrombin,** a protein involved in blood coagulation; **erythropoietin,** a hormone that stimulates erythrocyte production; and **angiotensinogen,** a hormone precursor involved in regulating blood pressure, body fluid, and electrolyte balance.

The **beta globulins** (beta-1 and beta-2) include most of the carrier proteins for lipids. Beta-1 lipoprotein, called **low-density lipoprotein (LDL),** favors the deposition of cholesterol in arterial walls and thus appears to play a role in disease of the blood vessels and heart (see Research Focus box). Other substances transported by beta globulins include phospholipids, glycerides, lipid-soluble vitamins (A, D, E, K), and metals such as copper and iron. **Transferrin,** a copper and iron transporting protein, is among the beta globulins.

The **gamma globulin** fraction contains the **immunoglobulins (I_g)** (antibodies). There are five major classes of immunoglobulin: I_gG, I_gA, I_gM, I_gD, and I_gE. More than 99% of all immunoglobulins in the plasma are either class G, A, or M. Class D and E immunoglobulins as free proteins are rare in plasma. The role of immunoglobulins in defense against foreign substances and disease-causing organisms will be discussed later in the chapter.

The Albumin/Globulin Ratio.   The total amount of plasma protein normally remains stable, with about twice as much albumin as globulin. A typical adult normally consumes and replaces about 15 grams of albumin and 5 grams of globulin every 24 hours, but the amounts vary according to body needs for specific proteins. For example, inflammatory diseases cause an increase in the production of immunoglobulins, accompanied by a decrease of equal magnitude in the production of albumin. The A/G ratio is reduced but the total amount of protein in the plasma remains stable (A + G).

Fibrinogen.   Fibrinogen is a soluble plasma protein manufactured in the liver. When blood clots, fibrinogen is converted into an insoluble protein called fibrin, which forms the foundation of the blood clot. Coagulation of blood will be discussed later in the chapter.

Hematopoiesis: The Formation of Blood Cells

The formed elements of the blood, sometimes referred to as corpuscles (little bodies), include the **erythrocytes** (red cells), **leukocytes** (white cells), and **thrombocytes** (platelets). Of the formed elements, only the leukocytes are true cells, possessing a nucleus and other organelles; however, we will follow convention and refer to the erythrocyte as though it were a true cell. Leukocytes include **agranular leukocytes** or **agranulocytes** (**monocytes** and **lymphocytes**), and **granular leukocytes** or **granulocytes** (**basophils, eosinophils,** and **neutrophils**).

All of these blood cells and structures originate from primitive cells known as **pluripotent uncommitted stem cells,** found mainly in the bone marrow (Fig. 17–5). (Specifically, they are found in the red bone marrow of certain bones, although other sites are possible, particularly in the fetus.) These uncommitted stem cells are called ''pluripotent'' and ''uncommitted'' because they are capable of developing into any of the various blood cells. In the pro-

Cholesterol, Triacylglycerols, and Lipoproteins

One of the primary risk factors for development of heart and vascular disease is an abnormally high level of cholesterol in blood serum. Excess cholesterol tends to be deposited, in the form of fatty plaques, on the lining of blood vessels (atherosclerosis), much the way mineral scale builds up on the inside of a water pipe. These deposits reduce blood flow and oxygen delivery to tissues downstream, and increase the chances of internal blood clotting (thrombosis) and heart attack. Excess cholesterol also accumulates inside and outside the cells that form the walls of arteries; this results in a loss of vascular elasticity and a hardening of the arterial wall (arteriosclerosis), thereby increasing the blood pressure and heart work.

Elevated blood triacylglycerols (triglycerides) also contribute indirectly to heart and vascular disease because triacylglycerols infuence the way in which the body processes cholesterol. For reasons not clear, very high levels of blood triacylglycerols also cause the pancreas to become severely inflammed (pancreatitis). Cholesterol and triacylglycerols are lipids which, by themselves, are insoluble in plasma and thus share a common system of transport in the blood: the lipoproteins.

Transport proteins, minus the triacylglycerols and cholesterol they normally carry, are called apolipoproteins. Apolipoproteins serve as a packaging for lipids so they can be transported in small particle form, and also as messengers, delivering the lipid to specific cells (e.g., adipose cells) for storage or use.

Four major categories of lipoprotein participate in the transport of triacylglycerols and cholesterol: (1) chylomicrons, (2) very low density lipoproteins (VLDL), (3) low density lipoproteins (LDL), and (4) high density lipoproteins (HDL).

Chylomicrons transport dietary cholesterol and triacylglycerols from the intestine to adipose tissues and liver. Chylomicrons contain mostly triacylglycerols and a little cholesterol.

VLDL transport triacylglycerols and cholesterol made by the liver to adipose and other tissues. Adipose cells remove the triacylglycerols from VLDL and leave behind the apolipoprotein and cholesterol that form LDL.

LDL transports cholesterol to peripheral tissues and regulates endogenous cholesterol levels in those tissues. High levels of LDL promote atherosclerosis and arteriosclerosis, and thus the cholesterol contained in LDL is sometimes referred to as "bad" cholesterol.

HDL transports cholesterol from peripheral tissues to the liver, where the cholesterol is removed from the blood. HDL thus help to protect against cholesterol-related heart and vascular disease. The cholesterol in HDL is sometimes referred to as "good" cholesterol.

Both the total amount of cholesterol in the blood and the ratio between LDL and HDL cholesterol are important in determining the risk of developing heart and vascular disease. Cholesterol management strategies are geared toward lowering total blood cholesterol to below 200 mg/dl, and LDL-cholesterol to below 130 mg/dl.

One way to lower LDL-cholesterol and raise HDL-cholesterol is to modify dietary intake of fatty acids. A diet rich in saturated fatty acids such as those found in animal fats (e.g., butter, cheese, cream, meats) is associated with high plasma cholesterol levels and high LDL. Dietary substitution of polyunsaturated fatty acids, such as those found in fish and vegetable oils, for saturated fatty acids tends to decrease total cholesterol and LDL-cholesterol levels, and to raise HDL cholesterol levels. The mechanisms involved are not yet clear and are the subject of much current research.

---

cess known as **hematopoiesis,** which is generally ongoing throughout life, these uncommitted stem cells become committed stem cells that can now develop into only one or more of the specific mature blood cells shown in Figure 17–5. Uncommitted stem cells initially develop into four types of committed stem cells: (1) those giving rise to erythrocytes, (2) those giving rise to megakaryocytes (platelet-forming cells), (3) those giving rise to monocytes and granulocytes, and (4) those giving rise to lymphocytes.

Hematopoiesis is a general term for the development of all types of blood cells. Figure 17–5 gives an overview of the process. **Erythropoiesis** is a term that specifically describes the development of red blood cells from uncommitted stem cells; **leukopoie-**

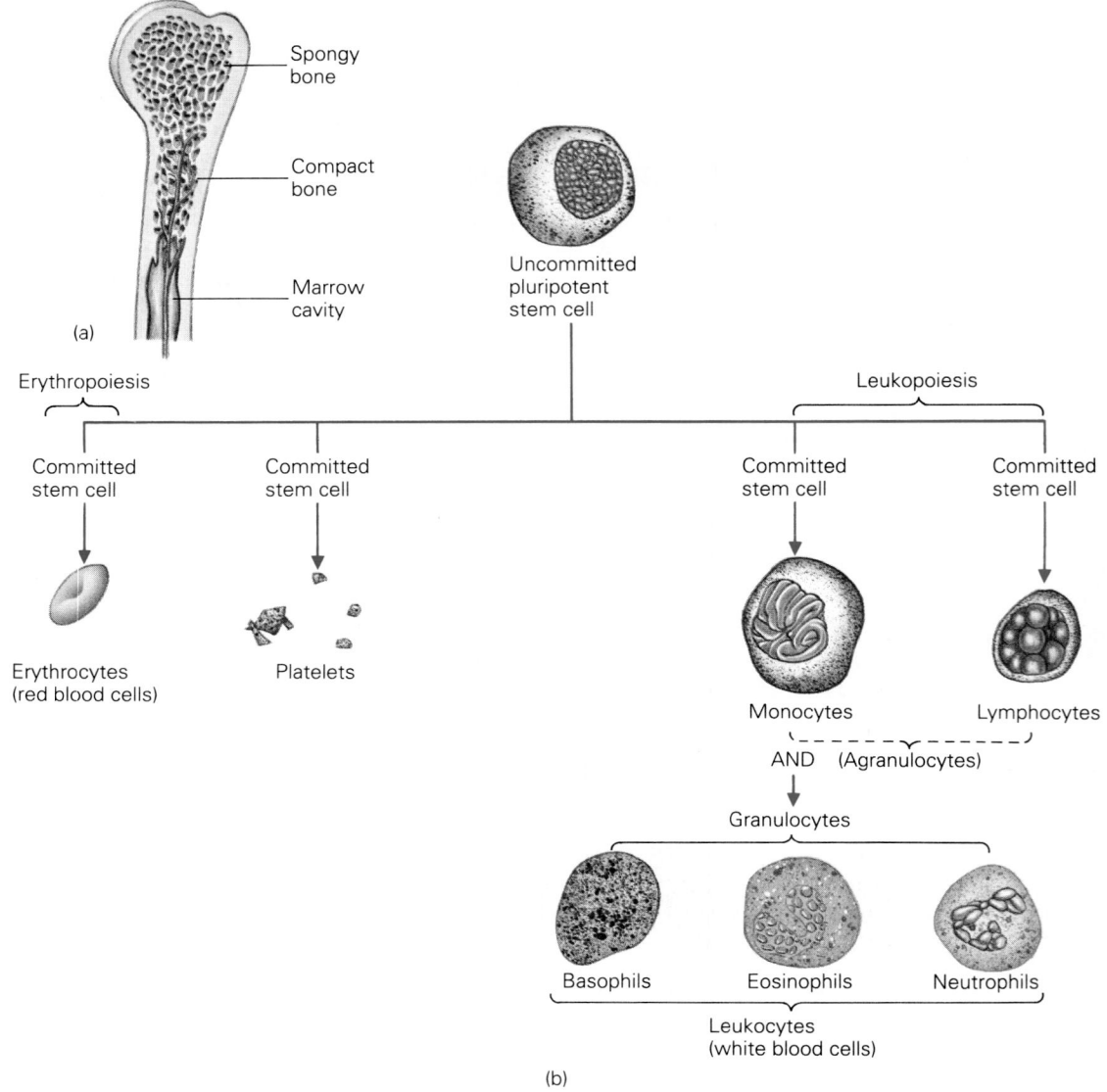

Figure 17–5
(*a*) Blood cells are generated from uncommitted stem cells primarily in the red marrow of bones. (*b*) An overview of hematopoiesis.

sis is the term for the development of the white blood cells. The specific steps in each of these processes are described and illustrated in the relevant upcoming discussions of each cell type.

Erythrocytes: Red Blood Cells

Characteristics of Erythrocytes

The erythrocyte, or **red blood cell (RBC),** is shaped like a biconcave disc with a diameter of about 8 μm and a maximum thickness of about 2 μm (Fig. 17–

6). Its shape results in a high ratio of surface-area to volume, favoring more efficient diffusion into and out of the cell. The principal function of the erythrocyte is to transport oxygen and carbon dioxide in the blood; as we will learn, its shape is ideally suited to its function.

Because it lacks a nucleus and other organelles, the erythrocyte is not a true cell. The interior of the erythrocyte consists of a homogeneous matrix containing a loosely arranged meshwork of fibrous and globular proteins that form a cytoskeleton. The cytoskeleton is anchored to the inner surface of the plasma membrane and so gives the erythrocyte its discoid shape. Erythrocytes are pliable, elastic cells

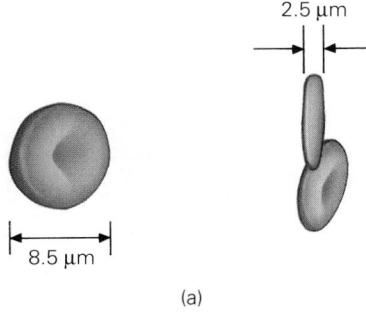

(a)

(b)

Figure 17–6
(*a*) Typical dimensions of erythrocytes, or red blood cells. (*b*) A scanning electron micrograph of a red blood cell. (From Bernstein, E., *Science* 173, 1971. Copyright 1971, the American Association for the Advancement of Science.)

that bend, twist, and flex as they pass through tiny blood vessels; after such changes, they normally regain their discoid shape very quickly. Genetic abnormalities (such as **hereditary spherocytosis** or **hereditary elliptocytosis**) that affect the synthesis of cytoskeleton proteins may result in red blood cells with abnormal spherical or elliptical shapes; such cells do not function normally and are destroyed at a faster-than-normal rate, resulting in anemia.

Normal human blood contains approximately 5 million red blood cells per cubic millimeter (mm^3) of whole blood. Normal adult ranges are: males— 5.4 ± 0.8 million/mm^3; females—4.8 ± 0.6 million/ mm^3.

Hemoglobin and the Oxygen-Carrying Capacity of Erythrocytes

Erythrocytes contain **hemoglobin,** a pigment molecule that gives the cell a pale color. Each erythrocyte contains 200 to 300 million molecules of hemoglobin, accounting for about one third of the mass of the red cell. Hemoglobin readily associates and dissociates with oxygen and carbon dioxide and is responsible for the erythrocyte's ability to transport these gases. Hemoglobin also plays an important role as a chemical buffer in the body's defense against changes in hydrogen ion concentration (pH).

The hemoglobin molecule (Fig. 17–7) consists of a **globin molecule** and four attached **heme groups.**

Each heme group contains an iron atom with which a molecule of oxygen may associate and dissociate; hence, each hemoglobin molecule can transport a maximum of four molecules of oxygen. When blood passes through the lungs, hemoglobin becomes saturated with oxygen, forming **oxyhemoglobin** and becoming bright red in color. Then, when this blood passes through body tissues, some of the oxygen dissociates from hemoglobin. **Reduced hemoglobin** is formed, and the blood turns a darker red.

Figure 17–7
A model of the hemoglobin A molecule.

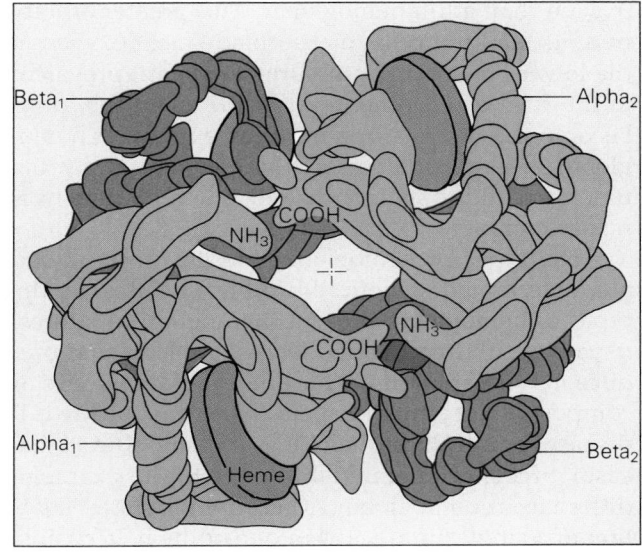

Normal blood contains approximately 15 grams of hemoglobin per deciliter (100 ml) of whole blood. Normal adult ranges are: males—16.0 ± 2.0 g/dl; females—14.0 ± 2.0 g/dl.

About 98% of the oxygen carried in the blood is in the form of oxyhemoglobin. The **oxygen-carrying capacity** of blood is defined as the amount of oxygen the hemoglobin in 100 ml of whole blood is capable of transporting. Each gram of hemoglobin can combine with 1.34 ml of oxygen. Since normal blood contains approximately 15 grams of hemoglobin per deciliter, the oxygen-carrying capacity of normal adult blood is approximately 20 ml of O_2 per deciliter.

Variations and Abnormalities in Hemoglobin Structure

Several forms of hemoglobin have been found in human blood. Two types (**type A** in adults and **type F** in the fetus) are normal hemoglobins, whereas others are considered abnormal. Differences in the globin molecules give rise to the various types (the heme groups of all types are alike).

The globin part of hemoglobin consists of two dissimilar pairs of peptide chains, each pair designated by the Greek letter alpha (α), beta (β), gamma (γ), or delta (δ). A heme group is bound to each of the four chains. 96% of adult hemoglobin is **type A$_1$,** consisting of two alpha chains and two beta chains. Two percent of adult hemoglobin is **type A$_2$,** consisting of two alpha and two delta chains, and two percent is type F, containing two alpha and two gamma chains.

Hemoglobin F is the predominant type in fetal blood. Fetal hemoglobin has a greater affinity for oxygen than adult hemoglobin. This greater affinity permits fetal red cells to absorb adequate oxygen at the lower P_{O_2} (partial pressure of oxygen) prevalent in the fetal environment. At birth, when the lungs become the infant's organ of gas exchange, hemoglobin A begins to replace hemoglobin F. By the time the child is six months old, the replacement is nearly complete.

The types of hemoglobin present in the blood are determined genetically. Abnormalities in the types of hemoglobin present **(hemoglobinopathies)** usually result from altered DNA templates that produce minor variations in the amino acid sequence or composition of either the beta chain (e.g., **sickle cell disease**) or the alpha chain (e.g., **hemoglobin H disease**). **Sickle cell hemoglobin (type S),** for example, differs from hemoglobin A because of the substitution of a single amino acid in one of the beta chains.

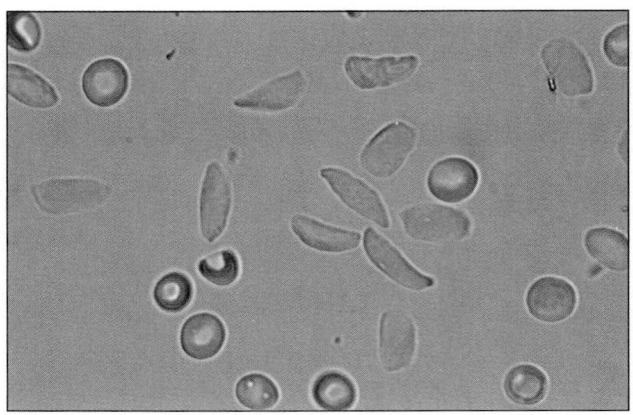

Figure 17–8
Blood from a patient with sickle cell anemia. (© Dr. Donald L. Rucknagel, Director of the Comprehensive Sickle Cell Center/Children's Hospital Medical Center/ Cincinnati, Ohio.)

Deoxygenation of hemoglobin S causes a reversible change in the red cell shape from a biconcave disc to a sickle shape (Fig. 17–8). Sickle cells hemolyze (lyze, or burst) readily, producing anemia. Sickle cells often become trapped in capillaries, blocking the flow of blood and producing tissue swelling, pain, and tissue hypoxia (inadequate oxygen supply). These conditions, if severe, can result in tissue death.

Erythropoiesis: The Formation of Red Blood Cells

Erythropoiesis is the formation of erythrocytes from uncommitted stem cells. In the fetus, red blood cells are formed from stem cells in the yolk sac, liver, spleen, and red bone marrow. In the adult, red cell production is confined to the red marrow of the humerus (upper arm bone), femur (thigh bone), ribs, sternum (breastbone), scapulae (shoulder blades), skull, vertebrae, and pelvis. In certain diseases during adult life, **extramedullary erythropoiesis** (red blood cell formation occurring outside the bone marrow) may be seen in the spleen and liver. Each day, the adult marrow produces about 230 billion red cells. If necessary, the marrow can increase its production of red cells five-fold.

An erythrocyte exists for approximately 4 months, after which it is removed from the blood and destroyed by macrophages in the spleen, liver, and marrow. Normally, the rate of destruction equals the rate of production, so that the number of circulating erythrocytes is fairly constant.

Erythropoiesis begins with the differentiation of **uncommitted stem cells** (Fig. 17–9) into **committed stem cells.** Those stem cells committed to erythropoiesis gradually progress through a series of developmental stages, during which they lose their nuclei and acquire hemoglobin.

Upon losing their nuclei, the stem cells become **reticulocytes.** Both mature erythrocytes and reticulocytes are released by the marrow into the blood. Released reticulocytes mature into erythrocytes about 1.5 days after entering the circulation. Normally, blood contains about 1 reticulocyte for every

Figure 17–9
The process of erythropoiesis, by which red blood cells develop from uncommitted stem cells.

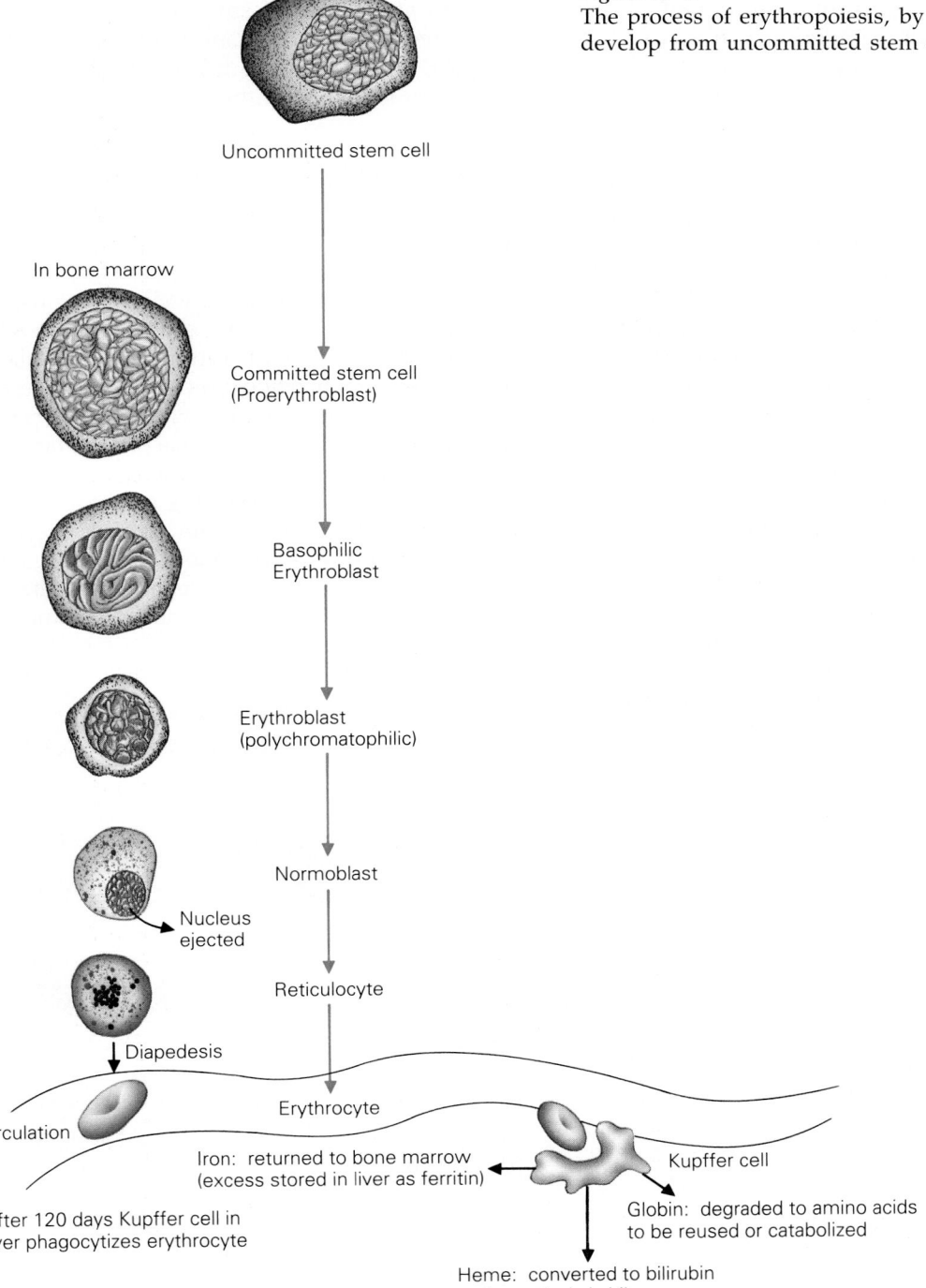

Uncommitted stem cell

In bone marrow

Committed stem cell
(Proerythroblast)

Basophilic
Erythroblast

Erythroblast
(polychromatophilic)

Normoblast

Nucleus
ejected

Reticulocyte

Diapedesis

Erythrocyte

In circulation

Iron: returned to bone marrow
(excess stored in liver as ferritin)

Kupffer cell

After 120 days Kupffer cell in
liver phagocytizes erythrocyte

Globin: degraded to amino acids
to be reused or catabolized

Heme: converted to bilirubin
and excreted via bile

100 erythrocytes, their production serving as an index of erythropoietic activity (i.e., the higher the reticulocyte proportion in the blood, the higher the rate of red cell production in the marrow).

Iron and Other Substances Required for Erythropoiesis

Several substances are essential for proper formation of red cells and hemoglobin: **amino acids, iron, copper,** and **vitamins B₂ (riboflavin), B₁₂ (cyanocobalamin), pyridoxine,** and **folic acid.** Deficiencies in any one of these substances may lead to anemia.

Iron is required for the production of heme; in fact, approximately two thirds of the body's iron content is found in hemoglobin. Most of the remaining iron is stored in the liver as **ferritin,** an iron-protein complex, and in the intestinal wall, spleen, and marrow. Stored iron is readily made available for hemoglobin synthesis via a protein called transferrin found in the plasma. Transferrin combines with iron that has been absorbed from the intestinal tract or released from storage. It then transfers the iron to erythrocytes developing in the bone marrow. Transferrin also transports iron released from worn-out red cells to the bone marrow to be reused for hemoglobin synthesis, or to the liver to be stored.

Nearly all body cells require iron, because iron-containing enzymes are required in oxidative metabolism. The shedding of cells by the lining of the intestinal and reproductive tracts as well as the skin, coupled with smaller amounts of iron in sweat and urine, results in an average daily loss of about 4 mg of iron. Iron loss is normally replaced by dietary intake. The recommended daily allowance is 15 mg, although typically only about 4 mg of iron per day is absorbed, primarily in the small intestine.

Iron is absorbed by active transport, a process that occurs until all the plasma transferrin is saturated with iron. When transferrin is saturated, iron absorption practically ceases, and the remaining dietary iron is excreted in the feces. Conversely, when iron stores are low, iron absorption increases until iron stores and plasma iron are replenished. Thus, a feedback mechanism involving absorption, transport, and storage maintains a stable supply of iron for hemoglobin synthesis.

Hormonal Control of Erythropoiesis

Erythropoiesis is controlled by a hormone called erythropoietin. A reduction in oxygen tension in systemic arterial blood stimulates the kidneys to release erythropoietin, which travels via the blood to the marrow. In the marrow, erythropoietin stimulates the differentiation of erythrocyte stem cells and shortens their maturation time, thereby increasing the rate of red cell formation and release (Fig. 17–10). Under conditions of maximal stimulation (such as when the tissues seriously lack oxygen, a condition termed severe **hypoxia**), the erythropoiesis can increase five-fold to correct the deprivation.

The synthesis of erythropoietin in turn is influenced by sex hormones. **Androgens** (male sex hormones) such as testosterone stimulate production of erythropoietin; in part, this stimulation may account for the higher RBC count normally seen in adult males.

Figure 17–10
The mechanism of erythropoietin action.

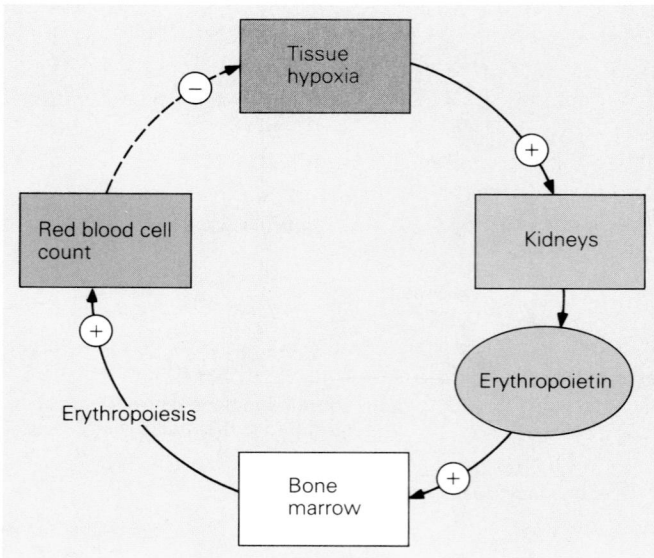

Hemolysis: Destruction of Erythrocytes

The erythrocyte has no nucleus, mitochondria, endoplasmic reticulum, or other organelles required for cell maintenance, and thus has a very limited lifespan. As erythrocytes age, changes in their plasma membranes make them susceptible to recognition and ingestion by **macrophages** in the spleen, liver, and marrow. About 230 billion red cells are destroyed each day.

When the red cell is destroyed by a macrophage, hemoglobin is broken down, and important components of the molecule are recycled within the body (Fig. 17–11). The peptide components are reduced to individual amino acids, which may be reused by other cells for protein synthesis. The heme portions are broken down into iron (Fe^{3+}) and **biliverdin,** a greenish pigment, which is reduced to **bilirubin** and released into the plasma. In the plasma, bilirubin binds to **albumin,** a carrier protein, and is transported to the liver. The liver cells take up the bilirubin, join it to **glucuronic acid,** and excrete the complex as a bile salt into the small intestine. Intestinal bacteria convert bilirubin into **urobilinogen,** most of which is eliminated in the feces in the form of **stercobilin,** a brown pigment. Some urobilinogen is absorbed from the intestine and excreted by the kidney into the urine, where it becomes oxidized to **urobilin,** the amber coloring matter in urine. The iron is released into the plasma, where it combines with transferrin and travels to other cells for storage or reuse.

Bilirubin is very toxic to the nervous system and, if allowed to accumulate in the body, can cause neural damage. An elevation of blood bilirubin above normally low levels is called **hyperbilirubinemia.** This condition may result from an excessively high rate of cell destruction, liver disease, or blockage of the liver biliary system. **Icterus,** or **yellow jaundice,** a symptom of hyperbilirubinemia, is caused by an elevation in plasma bilirubin to the point at which the eyes, skin, and mucous membranes appear yellow. Hyperbilirubinemia is common in newborn infants, who often become jaundiced after the first day of life. The chief cause is an immature liver unable to excrete bilirubin as fast as it is being formed. Fortunately, bilirubin is a photosensitive pigment that decomposes when exposed to near-ultraviolet light; thus, when the jaundiced infant's skin is exposed to the appropriate wave-

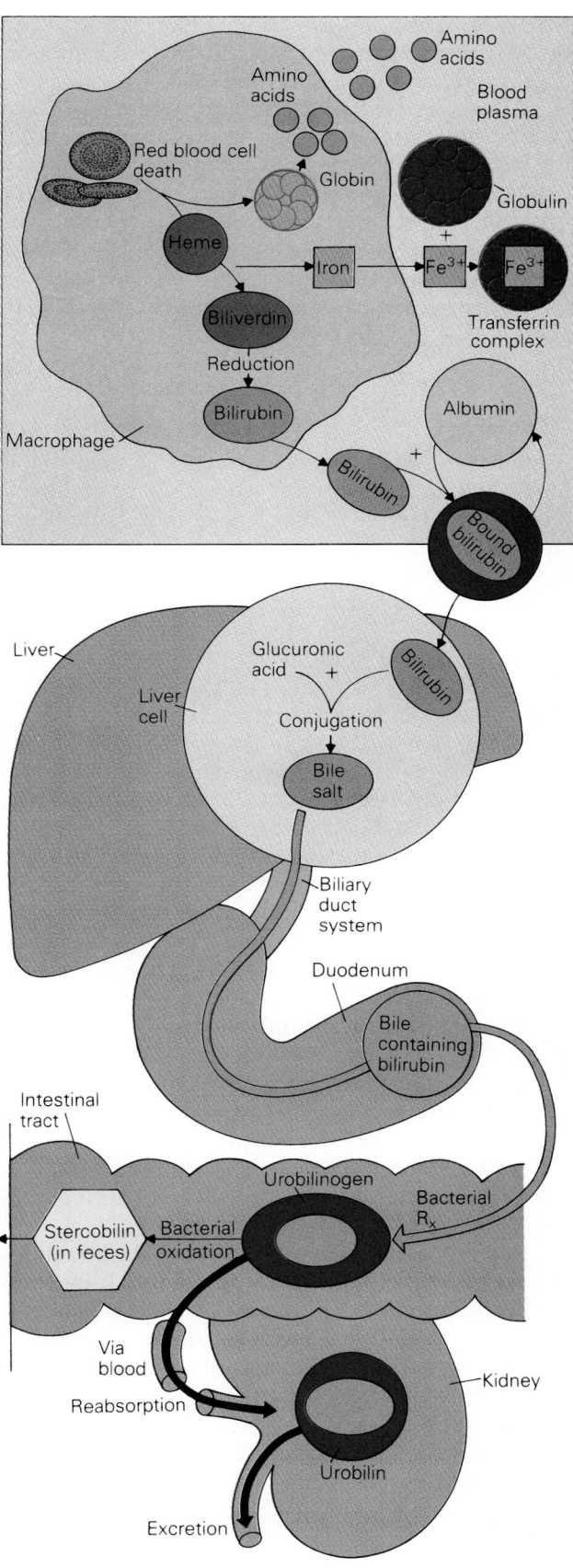

Figure 17–11
The metabolism of hemoglobin.

length of light in an incubator, blood bilirubin can be kept at low levels until the infant's liver matures enough to handle the bilirubin load.

Although most aged erythrocytes are destroyed by macrophages, a few actually disintegrate within the blood. When the red cell membrane is damaged, hemoglobin escapes into the blood, where it quickly becomes bound to a protein called **haptoglobin.** The resulting complex prevents the kidney from excreting the hemoglobin in the urine. Macrophages in the liver then remove and process the hemoglobin, as indicated in Figure 17–11.

Red Blood Cell Indices and Anemia

Anemia is a condition characterized by an abnormally low oxygen-carrying capacity of the blood, resulting from marked deficiencies in the number of circulating red cells, the amount of hemoglobin in the blood, or both. Anemia is considered present if the hemoglobin is less than 12 g/dl or the hematocrit is less than 37%. Because the oxygen-carrying capacity of their blood is reduced, anemic persons may experience physical weakness, shortness of breath, and difficulty in performing mental work. Chronic physical and mental fatigue are common symptoms.

Anemia may result from a failure of red cell production or an excessive rate of cell loss, or both. Anemias are classified on the basis of cell morphology (shape and size) and red cell indices.

With respect to morphology, red cells are described in specific terms. A **normocytic** erythrocyte is normal in size and shape; a **normochromic** erythrocyte is normal in color (and thus in hemoglobin content). A **microcytic** erythrocyte is smaller than normal; a **macrocytic** erythrocyte is larger than normal. A **hypochromic** erythrocyte is paler in color than normal (indicating insufficient hemoglobin); a **hyperchromic** erythrocyte is deeper in color than normal. **Anisocytosis** is a variation in size, **poikilocytosis** a variation in shape. The optimum situation is for erythrocytes to be normocytic and normochromic.

Red cell indices are used to detect abnormalities in erythrocyte size, shape, and color. Three commonly used indices are the **mean corpuscular volume** index **(MCV),** the **mean corpuscular hemoglobin (MCH),** and the **mean corpuscular hemoglobin concentration (MCHC).**

The MCV index is a quantitative assessment of the red cell volume reported in cubic micrometers. It is computed by the following formula:

$$MCV = \frac{\text{hematocrit} \times 10}{\text{RBC (millions per mm}^3)}$$

The normal adult value of MCV is $78 \pm 5 \ \mu m^3$. A low MCV indicates microcytosis, and a high MCV indicates macrocytosis.

The MCH index is used to determine the concentration of hemoglobin within the erythrocytes. It is reported in micrograms and is computed as follows:

$$MCH = \frac{\text{g hemoglobin} \times 10}{\text{RBC (millions per mm}^3)}$$

The normal adult value of MCH is $29 \pm 5 \ \mu g$. A low MCH indicates microcytosis, hypochromia, or both. A high MCH suggests hyperchromia.

The MCHC is an index of the proportion of hemoglobin measured (weight/volume) per average red cell in a sample of blood. It is calculated by the following formula:

$$MCHC = \frac{\text{g hemoglobin} \times 100}{\% \text{ hematocrit}}$$

The normal adult value for MCHC is $34 \pm 2\%$. A low MCHC means hypochromia, but a high MCHC indicates a loss of RBC volume without a proportionate loss of hemoglobin (microcytosis). Table 17–2 summarizes normal red cell values for adult males and females.

In addition to changes in red cell morphology and indices, most forms of anemia are characterized by an elevated **reticulocyte count.** Increased numbers of circulating reticulocytes indicate increased bone marrow activity with early release of reticulocytes. This is common in **hemolytic anemias,** for example, when red cells are destroyed faster than marrow can produce them.

Anemia is a sign of disease. There are as many types of anemia as there are underlying causes. Only a few of the more common anemias will be discussed here.

Table 17–2
Normal Red Cell Values (Adult)

RBC Measurement	Male	Female
Red cell count (millions/mm³)	5.4 ± 0.8	4.8 ± 0.6
Hemoglobin (gm/100 ml)	16.0 ± 2.0	14.0 ± 2.0
Hematocrit (%)	47.0 ± 5.0	42.0 ± 5.0
MCV (μm³)	87 ± 5	87 ± 5
MCH (μg)	29 ± 2	29 ± 2
MCHC (%)	34 ± 2	34 ± 2

Iron-deficiency anemia is caused by lack of dietary iron, inadequate absorption of iron, loss of iron stores, or abnormalities in iron utilization. Deficiencies in iron metabolism lead to a reduction in hemoglobin production and the release of microcytic, hypochromic red cells from the marrow. Good dietary sources of iron include liver, red meat, egg yolks, carrots, and spinach.

Pernicious anemia occurs because of a deficiency in the absorption of vitamin B_{12}. The lining of the stomach secretes a glycoprotein called the **intrinsic factor,** which is essential for the absorption of vitamin B_{12} in the small intestine. Vitamin B_{12} and the intrinsic factor combine to form the **erythrocyte maturation factor,** which is required for maturation of red cells in the marrow. Genetic lack of the intrinsic factor or inadequate dietary intake of vitamin B_{12} leads to the impairment of erythrocyte maturation. The marrow releases large nucleated red cell precursors called **megaloblasts** that contain more hemoglobin than normal red cells but do not function normally. Pernicious anemia is characterized by the presence in the blood of macrocytic, hyperchromic, immature red cell precursors and a reduction in the numbers of normal erythrocytes.

Aplastic anemia occurs as a result of a decrease in the production of all blood cells in the marrow (**aplasia** denotes faulty or incomplete development). Blood-forming cells do not mature at a normal rate in this disorder. The resultant anemia is characterized by reduced numbers of normal red cells. Aplastic anemia can be caused by radiation, cytotoxic drugs used for the treatment of malignant disease, exposure to chemicals (arsenic, DDT, benzene) that poison the marrow, or genetic failure of bone marrow development.

Hemolytic anemia occurs when the rate of red cell destruction exceeds the capacity of the marrow to replace lost cells. Hemolytic anemias are usually accompanied by jaundice and elevated serum bilirubin. Causes of hemolytic anemia include damage to the red cell or red cell defects (such as hereditary spherocytosis and sickle cell disease) that cause early destruction by macrophages, and various extracellular agents (toxins, antibodies) that cause hemolysis.

Thalassemias are a group of disorders caused by inheritance of one or more genes that limit the rate of synthesis of alpha or beta chains of hemoglobin. The impairment leads to a hypochromic, microcytic anemia, the severity of which is determined by the number and kind of genes affected. **Thalassemia major** is usually homozygous and most often affects beta chain synthesis (**B-thalassemia major, Cooley's anemia**). Beta-thalassemias occur most frequently in inhabitants of the Mediterranean area (*thalassa* = "sea"), and alpha-thalassemias are more frequent among Asians and Central Africans.

Hemorrhagic anemias occur after acute or chronic blood loss, such as that associated with hemorrhoids, bleeding ulcers, and excessive menopausal bleeding, and are usually characterized by normocytic normochromic erythrocytes.

Polycythemia: An Increase in Red Blood Cells

Polycythemia (*poly* = "many"; *cyte* = "cell") is either a relative or an absolute increase in the number of circulating red cells above 6.2 million per cubic millimeter. It may result from an increase in the concentration of erythrocytes without an increase in the total red cell mass (**relative polycythemia**) due to dehydration and hemoconcentration, or it may result from an increase in the concentration of erythrocytes accompanied by an increase in total cell mass (**polycythemia vera**) due to the hyperactivity of the bone marrow.

Secondary polycythemia (erythrocytosis) is a physiological polycythemia that occurs in response to arterial hypoxia and its stimulation of erythropoiesis. Exposure to chronic hypoxic (low oxygen) environments (e.g., at high altitudes) is often accompanied by secondary polycythemia. It may also occur in emphysema, pulmonary fibrosis, carbon monoxide poisoning, and other states in which oxygenation of the blood is reduced.

Polycythemia vera (*true polycythemia*) is a chronic, slowly progressive disease involving excessive production of all blood cells produced in the marrow (**hyperplasia**). In most cases the cause is unknown (**idiopathic**), while in others the polycythemia may be evidence of tumors in the bone marrow, kidney, or brain. Polycythemia vera is not associated with lack of oxygen in the blood.

Polycythemias can be life-threatening because they increase the viscosity of the blood, placing an increased load on the heart and blood vessels. They also make an individual more susceptible to thrombosis, the intravascular formation of blood clots.

Leukocytes: White Blood Cells

Leukocytes (*leuko* = "white"; *cyte* = "cells") are called **white blood cells (WBC)** because they lack pigmented (color) molecules such as hemoglobin. White blood cells must be stained in order for the different types to be distinguished and to be more visible when they are counted.

Types of Leukocytes

There are several types of leukocytes (Fig. 17–12). Each leukocyte possesses a nucleus and other organelles; some contain large cytoplasmic vesicles, or granules, while others do not. These granules are mostly lysosomes. Leukocytes are classified according to the shapes of their nuclei and the presence or absence of cytoplasmic granules.

Leukocytes containing large granules are called granular leukocytes, or granulocytes. Granulocytes also have a complex lobulated nucleus (a nucleus consisting of several lobes, or segments) and are therefore called **polymorphonuclear leukocytes.** The types of granulocytes, named according to the staining properties of their granules, are the neutrophil, the eosinophil, and the basophil.

Leukocytes without a granular cytoplasm are called agranulocytes. They have a simple, unlobulated nucleus and thus are designated as **monomorphonuclear leukocytes.** The two basic types of agranulocytes are the monocyte and the lymphocyte.

On average, whole blood contains 7000 to 10,000 leukocytes/mm^3; this figure represents the **combined white blood cell (WBC) count.** Because leukocytes have different functions, however, it is diagnostically useful to count the white blood cells according to type; this count is called the **differential white blood cell count.** When a differential count is performed, the number of each type of leukocyte is usually expressed as a percentage of the total or combined white blood cell count. A neutrophil count of 65%, for example, means that neutrophils make up 65% of all leukocytes present in a blood sample. Normal differential counts for the leukocytes are given in Figure 17–12.

Leukocytes function collectively to combat foreign substances that enter the body. They are part of a system of defensive cells that **phagocytize** (engulf and destroy) material, detoxify poisons, produce antibodies, and release chemical messenger, enzymes, and other substances. Each type of leukocyte performs different functions in body defense, but each function is necessary for an integrated and effective defense.

Leukocytes are not confined to the blood or lymph; they may also be found in other tissues, particularly in loose connective tissue. They are motile (moving) cells, moving (traveling) about via **ameboid movement** (movement similar to that of an amoeba, in which the cell projects protoplasmic extensions ahead and then follows them). They are capable of passing through the capillary wall in a process called **diapedesis** to enter the spaces around the blood vessels whenever necessary. Leukocytes are attracted to sites of injury, inflammation, or bacterial invasion when chemical substances are liberated by damaged cells or bacteria, and when immune complexes form during immune responses. The attraction and subsequent movement of leukocytes toward these chemical substances is known as **chemotaxis.**

Granulocytes: Granular Leukocytes

Neutrophils. The neutrophil functions as the body's first line of defense against invasion by **pyogenic** (pus-forming) **bacteria.** Neutrophils are the most numerous white cells. They are very motile and actively phagocytic. They can ingest and destroy several kinds of microbes, small particulate matter, and fibrin (from blood clots). Most granules

Figure *17–12*
Types of leukocytes.

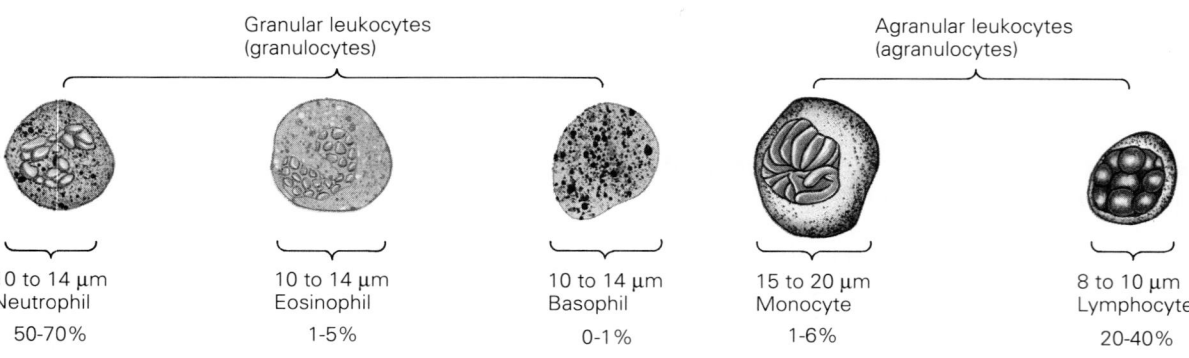

Granular leukocytes (granulocytes)			Agranular leukocytes (agranulocytes)	
10 to 14 μm Neutrophil 50-70%	10 to 14 μm Eosinophil 1-5%	10 to 14 μm Basophil 0-1%	15 to 20 μm Monocyte 1-6%	8 to 10 μm Lymphocyte 20-40%

Combined white blood cell count = 7000-10,000/mm^3

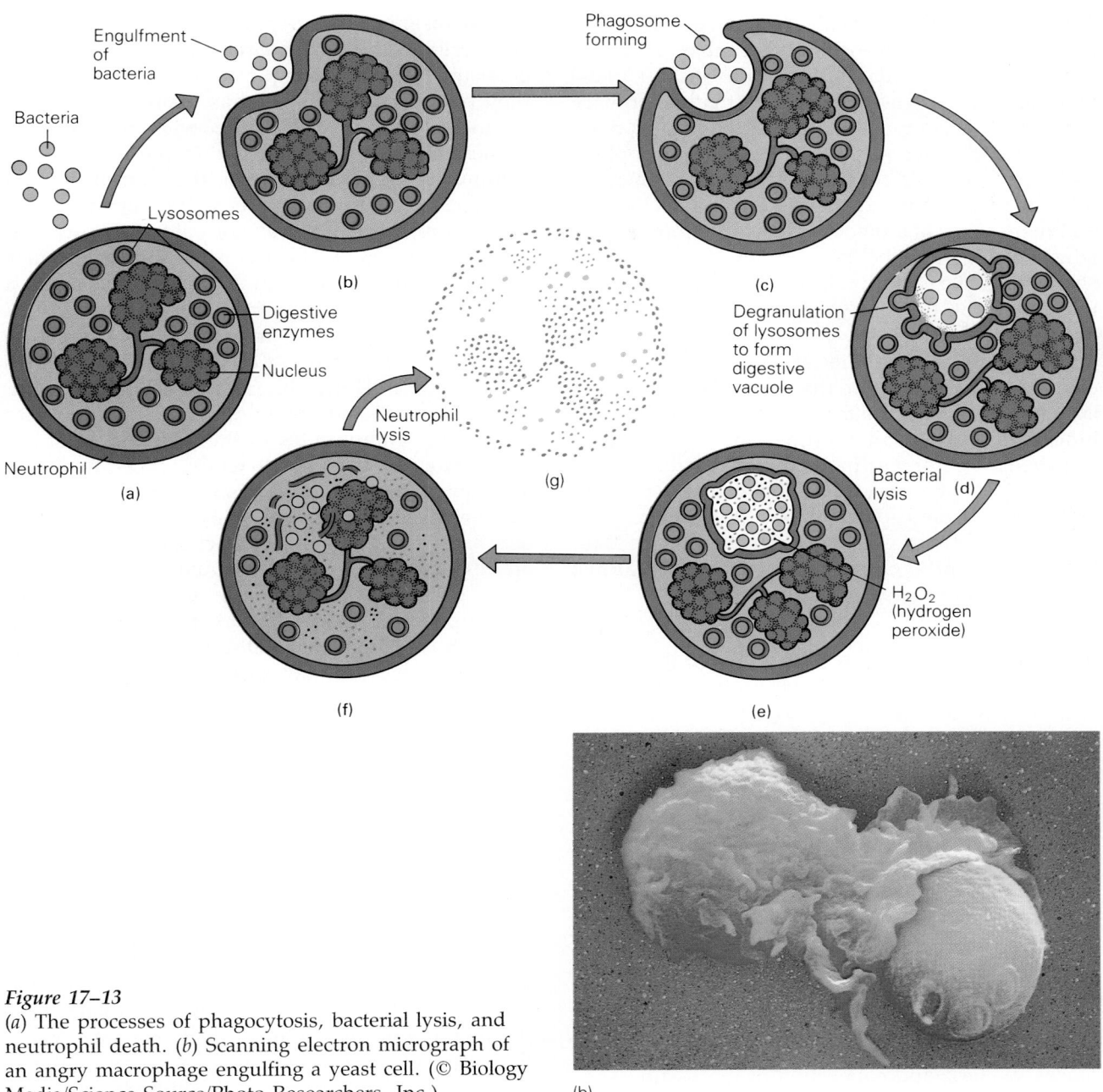

Bacteria

Engulfment
of
bacteria

Lysosomes

Digestive
enzymes

Nucleus

Neutrophil

(a)

(b)

Phagosome
forming

(c)

Degranulation
of lysosomes
to form
digestive
vacuole

(d)

Bacterial
lysis

H_2O_2
(hydrogen
peroxide)

(e)

Neutrophil
lysis

(g)

(f)

(b)

Figure 17–13
(*a*) The processes of phagocytosis, bacterial lysis, and
neutrophil death. (*b*) Scanning electron micrograph of
an angry macrophage engulfing a yeast cell. (© Biology
Media/Science Source/Photo Researchers, Inc.)

in the cytoplasm of neutrophils are lysosomes containing **hydrolytic enzymes** that digest ingested
matter. Other cytoplasmic granules contain
phagocytins, a group of antibacterial proteins. Neutrophils are usually the first white cells to arrive at
the site of a bacterial invasion, and they arrive in
large numbers. In the process of defending the body
against microbial invasion, many neutrophils die
(Fig. 17–13), forming what is called **pus** at the invasion site. Dying neutrophils also release digestive

enzymes into surrounding connective tissue. To a
limited extent, these enzymes damage tissue and
contribute to localized pain and swelling at a site of
inflammation. After the invading or traumatizing
factor has been controlled, many neutrophils remain to assist in healing.

Eosinophils. The eosinophil is less motile than the
neutrophil and, although capable of phagocytosis,
is not as active in this regard. The eosinophil does

not, for example, normally phagocytize bacteria. The lysosomes of the eosinophils contain many enzymes, such as **oxidases, peroxidases,** and **phosphatases,** whose presence indicates that the primary function of eosinophils is detoxification of foreign proteins and other substances. As do neutrophils, eosinophils exhibit chemotaxis; however, their attraction to traumatized tissues appears to depend on the presence of antibodies specific to foreign proteins. Antigen-antibody complexes have been shown to cause eosinophils to travel from the blood into connective tissues, where they phagocytize and destroy the complexes.

Normally, eosinophils are more abundant in connective tissues other than the blood—particularly the lung, mammary glands, omentum, and inner wall of the small intestine. They are observed to increase in number in allergic and autoimmune reactions, during decomposition of body protein, and in certain parasitic infections. It is believed that basophils and mast cells (connective tissue cells) release **eosinophilic chemotactic factors** that attract eosinophils to an area of tissue damage.

Basophils.   Basophils are the rarest of the leukocytes and are usually found in greater numbers outside the blood in loose connective tissue. They show little ameboid movement and do not contain lysosomes. Many of their cytoplasmic granules contain either **heparin,** an anticoagulant, or **histamine,** a vasodilator substance that also increases capillary permeability. Basophils release histamine in areas of tissue damage in order to increase blood flow, attract neutrophils, and facilitate the movement of leukocyte emigration from blood to interstitial fluids in the area of damage. The effects of histamine release have been experienced by everyone who has suffered from the common cold; invasion of the nasal tissue by viruses induces basophils to gather at the invasion site and release histamine, which causes swelling and congestion of the tissue. Antihistamines and vasoconstrictors contained in decongestant nasal sprays may provide relief but actually interfere with the body's normal defense, possibly prolonging infection.

Agranulocytes: Agranular Leukocytes

Monocytes constitute 1% to 6% and lymphocytes 20% to 40% of the circulating white blood cells. Lymphocytes are second in number only to the neutrophil. All lymphocytes appear similar, but they differ in their development, functions, and life span. There are two functionally distinct types of lymphocytes: **T-lymphocytes** and **B-lymphocytes.**

T-lymphocytes.   **T-lymphocytes** originate from stem cells in the bone marrow (Fig. 17–14). They enter the blood and complete their development as they pass through the **thymus gland,** a lymphoid organ in the thorax. T-lymphocytes are the most numerous type of lymphocyte in the blood. They continually migrate between the spleen, lymph nodes, and connective tissues. As T-lymphocytes pass through the thymus, receptors are inserted into them that allow the T-lymphocyte to recognize antigens (chemicals foreign to the body that evoke a protective response, the "immune response").

These lymphocytes are said to be "preconditioned" or "competent." Competent T-lymphocytes respond to antigens either by attacking them directly or by releasing chemicals called lymphokines that attract granulocytes to the area and stimulate B-lymphocytes and other T-lymphocytes. Activated T-lymphocytes may also release nonspecific toxins that neutralize the antigens. T-lymphocytes become activated when antigens contact them and bind to specific membrane receptors. Activated T-lymphocytes give rise, through cell division, to clones of lymphocytes responsive to the activating antigen, thus conferring **cell-mediated immunity.** T-lymphocytes also play important roles in helping B-lymphocytes function normally.

B-lymphocytes.   B-lymphocytes develop in the bone marrow and migrate by way of the blood to lymphoid tissues in the spleen, tonsils, lymph nodes, and walls of the small intestine, where they remain for variable periods of time before recirculating in the blood and lymph. When stimulated by an antigen, some B-lymphocytes become specialized to make and secrete **antibodies,** proteins that react with the antigens and help destroy them. Other B-lymphocytes become specialized to recognize the antigen should it re-enter the body at a later date; these B-lymphocytes are called **memory cells.** B-lymphocytes can live for many years and are the basis of the body's **humoral** or **antibody-mediated immunity.** Mechanisms of humoral and cell-mediated immunity will be discussed in greater detail later in the chapter.

Monocytes.   **Monocytes** are the largest of the leukocytes, some being nearly three times the size of an erythrocyte. Monocytes originate in the bone marrow, taking about three days to develop before being released into the blood. They remain in circulating blood for only a few days, after which they migrate to connective tissues in various organs. There they develop into **macrophages,** large phagocytic cells that form part of the body's defense against microorganisms and harmful chemicals.

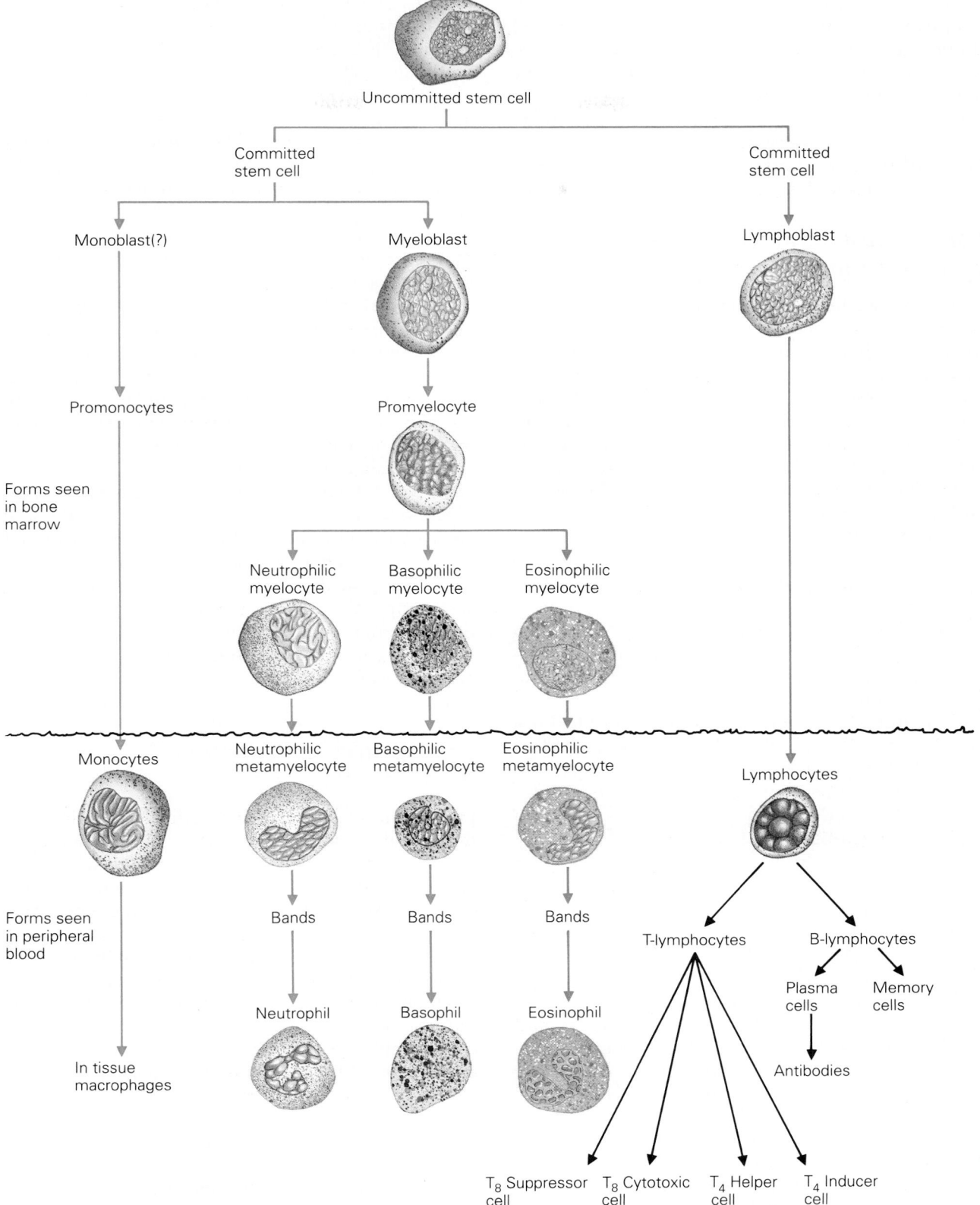

Figure 17–14
The processes of leukopoiesis.

Some tissue macrophages, such as those in the liver **(Kupffer cells)** or lung **(alveolar macrophages)** do not move about as much as do macrophages in loose connective tissue, but all macrophages display ameboid movement and are actively phagocytic. They contain a variety of digestive enzymes and are very effective in destroying bacteria, detoxifying harmful chemicals, and cleaning up cellular debris in areas of tissue damage.

Leukopoiesis: The Generation of White Blood Cells

Leukopoiesis, the formation of white blood cells, begins with a single type of uncommitted (primitive) stem cell in the marrow (see Fig. 17–14). Granulocytes and monocytes develop to maturity in the marrow. Lymphocytes usually begin their lives in the marrow, but most leave before maturing. Potential T-lymphocytes mature in the thymus and the spleen. Potential B-lymphocytes develop and mature in lymphoid tissue of the intestines, the spleen, and the bone marrow.

Uncommitted stem cells can leave the marrow and become **free stem cells** moving from organ to organ via the blood and lymph. Thus, hematopoietic centers in the marrow may serve to maintain lymphoid tissue via "seeding" with free stem cells. In turn, lymphoid organs may become sites of **extramedullary hematopoiesis** when the marrow fails to function normally.

Granulopoiesis (formation of granulocytes) has been shown to be stimulated by the presence of **leukocyte-inducing factor (LIF),** a specific factor in plasma that can be demonstrated when leukocytes are depleted. Various other chemicals released into the blood when tissue is damaged may also stimulate leukopoiesis, although the mechanisms are not clear.

The life span of a leukocyte depends on the role it plays in defense. Neutrophils and other actively motile and phagocytic leukocytes tend to live only a few days because they are continually defending the body against invasion by bacteria and dying in the process. On the other hand, B-lymphocytes and some T-lymphocytes may live and continue to function normally for years before being destroyed. Aged leukocytes are destroyed by the liver, spleen, marrow, and lymph nodes.

Leukocytosis: An Increase in White Blood Cells

Leukocytosis is an abnormal increase in the total number of circulating leukocytes—sometimes as many as 400,000 cells per cubic millimeter. Leukocy-tosis may have physiological or pathological causes. All combined white blood cell counts above 50,000 cells/mm^3 indicate the condition. **Physiological leukocytosis** (WBC count 10,000 to 20,000/mm^3) is nonpathological and occurs in newborn infants and in pregnancy, emotional disturbances, menstruation, and severe exercise. **Pathological leukocytosis** occurs in bacterial and viral infections, metabolic and hormonal disturbances, allergies, and malignancies.

Leukemia is a malignant disease characterized by the proliferation of hematopoietic cells that are immature and therefore functionally impaired. Environmental factors (e.g., radiation, benzene), genetic factors, and RNA viruses have been implicated. Leukemias are classified by the course and duration of the illness and the abnormal type of cells and tissue involved. **Acute leukemias** (e.g., **acute lymphocytic leukemia, acute myelogenous leukemia**) are characterized by rapid onset, massive numbers of immature leukocytes, severe anemia, and other signs. **Chronic leukemias** (e.g., **chronic granulocytic leukemia, chronic lymphocytic leukemia**) are characterized by gradual onset and increased numbers of mature leukocytes. Chronic leukemias may lead to acute leukemias. Although symptoms and complications vary with the type of leukemia, they generally include anemia, excessive bleeding due to impairment of thrombocytes, and local and systemic infection due to impaired granulocyte function.

Leukopenia: A Reduction in White Blood Cells

Leukopenia is an abnormal reduction in the total number of circulating leukocytes. It may result from bone marrow defects, arrest of cell development and maturation, or excessive destruction of leukocytes. **Agranulocytosis** is a type of leukopenia characterized by a marked reduction in neutrophils. Usually, it is caused by a depression in (slowed functioning of) the bone marrow due to irradiation or cytotoxic drugs. As a result of agranulocytosis, the body's defenses against bacterial invasion are considerably weakened and the chance of dying from bacterial infection greatly increased.

Thrombocytes (Platelets)

A **platelet** or **thrombocyte** (*thrombus* = "clot") is a cell fragment split from a large cell called a **megakaryocyte** (Fig. 17–15). One megakaryocyte gives rise to about 6000 thrombocytes. Since thrombocytes are not actually cells produced by a mitotic division of a

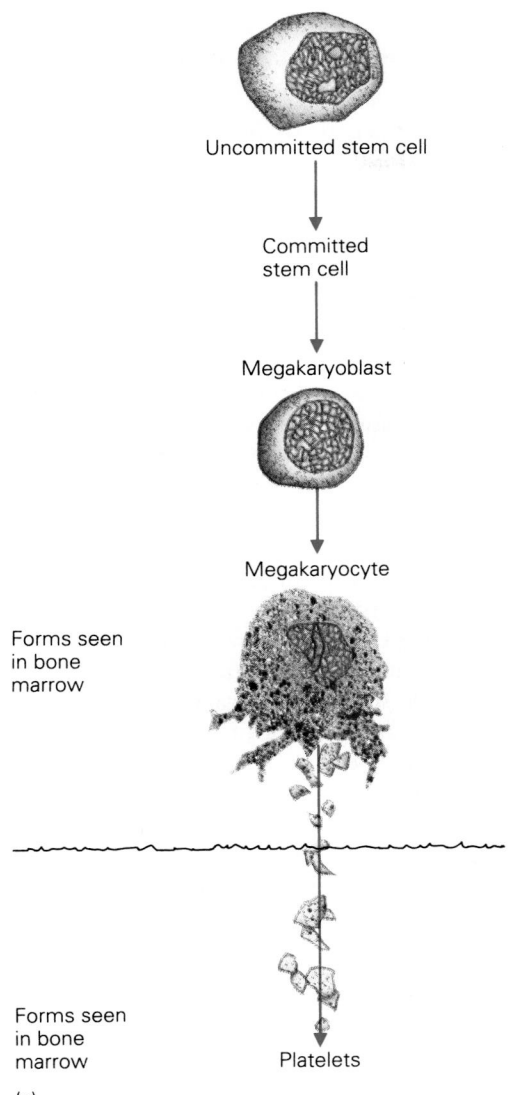

Figure 17–15
(*a*) The development of platelets. (*b*) Photomicrograph of thrombocytes (platelets). (© Manfred Kage/Peter Arnold, Inc.)

parent cell, the term **platelet** is more commonly used.

Most megakaryocytes remain in the marrow, releasing platelets into the circulating blood. Some megakaryocytes enter the blood and travel to other organs (particularly the lung) where they remain and produce platelets.

Platelets are small, membrane-bounded bodies without nuclei, about 2 to 4 μm in diameter. They are packed with granules, vesicles, microfilaments, microtubules, and occasionally mitochondria. They are normally present in numbers ranging from 240,000 to 340,000/mm^3 of blood and play several important roles in hemostasis. Following injury, chemicals contained within or absorbed by platelets are released to the lining of the blood vessels, where they stimulate contraction of the injured vessels to minimize blood loss. Because of their adhesive properties, they clump together at the site of vascular damage, effectively plugging the defect end. In addition, they participate in the formation of factors that initiate coagulation of the blood, and they release **platelet-derived growth factor (PDGF),** a protein that promotes repair after vascular injury. The role of platelets in hemostasis will be discussed later.

Circulating platelets have a lifespan of approximately one to two weeks. If not consumed in the process of blood coagulation, platelets eventually are destroyed by macrophages in the liver and spleen. The spleen is also an important storage organ for platelets. In certain kinds of stress, such as a hemorrhage or burns, sympathetic stimulation of the spleen results in the release of large numbers of stored platelets into circulating blood.

Hemostasis: Mechanisms to Prevent Blood Loss

The term **hemostasis** (*hemo* = "blood"; *sta* = "remain") refers to mechanisms that minimize or prevent the loss of blood when a blood vessel is opened. Hemostasis is vital because unchecked hemorrhage (blood loss) eventually leads to cardiovascular collapse and death. Four interrelated events constitute hemostasis: (1) local vasoconstriction, (2) formation of a platelet aggregate (clump),

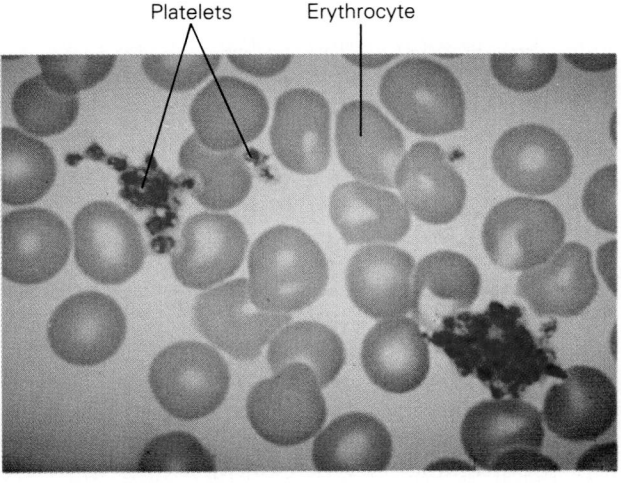

(b)

(3) formation of a blood clot, and (4) clot retraction and dissolution.

Vasoconstriction: An Autonomic Nervous System Response

When a blood vessel is injured, its immediate response is to constrict and thereby reduce blood flow. This initial vasoconstriction is due to local spasm of the smooth muscle in the wall of the blood vessel and to sympathetic reflexes. Vasoconstriction is maintained by the release of vasoconstrictor chemicals from platelets that begin to accumulate at the damaged site.

Platelet Aggregation: The Framework for Coagulation

Damaged cells of the injured blood vessel release **adenosine diphosphate (ADP),** which attracts platelets and causes them to clump together at the damage site. Platelets coming in contact with the torn, rough edges of the vascular wall **degranulate** (release stored chemicals), releasing **ADP, serotonin** (a vasoconstrictor), and **platelet factors** necessary for coagulation of blood. The release of ADP attracts more platelets and causes them to swell and become sticky; they adhere in increasing numbers to the damaged site and form a plug (a **platelet aggregate**). Activated platelets also produce **thromboxane A$_2$,** a powerful vasoconstrictor and platelet aggregator derived from platelet **prostaglandin H$_2$.** In addition, the spherical platelets extend **pseudopodia** (footlike extensions of their cytoplasm) to nearby exposed collagen, anchoring the platelet plug and establishing a framework upon which coagulation can proceed.

Coagulation: The Production of Fibrin

Coagulation is the process by which some of the blood loses its fluid consistency and becomes a **clot** (a semisolid mass similar in consistency to gelatin). In the formation of a clot, an enzyme called **thrombin** converts **fibrinogen,** a soluble plasma protein, into an insoluble plasma protein, **fibrin.** Fibrin is a threadlike protein that forms a meshlike network at the site of vascular damage, trapping red cells and plasma and thus forming a clot (Fig. 17–16). The complex sequence of chemical events that produce fibrin is divided into three phases or stages, designated as **Stage I, Stage II,** and **Stage III** (see Fig.

17–17). Common synonyms for blood clotting factors involved in each stage are given in Table 17–3.

Stage I: Formation of a Prothrombin Converting Factor

Stage I may begin when blood comes in contact with injured tissue (the **extrinsic pathway**) or it may be initiated in the absence of tissue damage (the **intrinsic pathway**).

When blood comes in contact with injured tissue, clotting is initiated via the extrinsic pathway, owing to the release of the **tissue lipoprotein thromboplastin** (Factor III). Tissue thromboplastin interacts with a plasma protein, **proconvertin** (Factor VII), and **calcium ions** to form an agent that activates the **Stuart factor** (Factor X). The activated Stuart factor, in the presence of **calcium ions,** forms complexes with **accelerin** (Factor V) on phospholipid micelles provided by tissue thromboplastin to form the **prothrombin-converting factor.**

Intrinsic coagulation may occur inside or outside of the body. In either case, the first step is the activation of the **Hageman factor** (Factor XII). In the body, activation of the Hageman factor may occur from collagen, fibrin, or platelet membrane during platelet aggregation. In addition, it can apparently be activated under conditions of stress, anxiety, fear, and other states. Outside the body (e.g., in a test tube), activation of the Hageman factor occurs when blood comes in contact with foreign substances whose common property appears to be a negative surface charge.

The Hageman factor activates the plasma enzyme **plasma thromboplastin antecedent (PTA,** Factor XI), which, in the presence of calcium ions, activates a plasma protein, the **Christmas factor** (Factor IX). Activated Christmas factor interacts with another protein, **antihemophilic factor** (Factor VIII), on the surface of phospholipids and in the presence of calcium to form a complex that activates the Stuart factor. The succeeding steps in the formation of prothrombin-converting factor are the same as for the extrinsic mechanism.

Stage II: Conversion of Prothrombin to Thrombin

Prothrombin is a plasma globulin manufactured by the liver and normally present in circulating plasma. It is the inactive precursor of an active enzyme called **thrombin.** Thrombin is not normally present in plasma unless blood is clotting. In the presence of calcium ions and the prothrombin-converting factor, prothrombin is enzymatically split into two

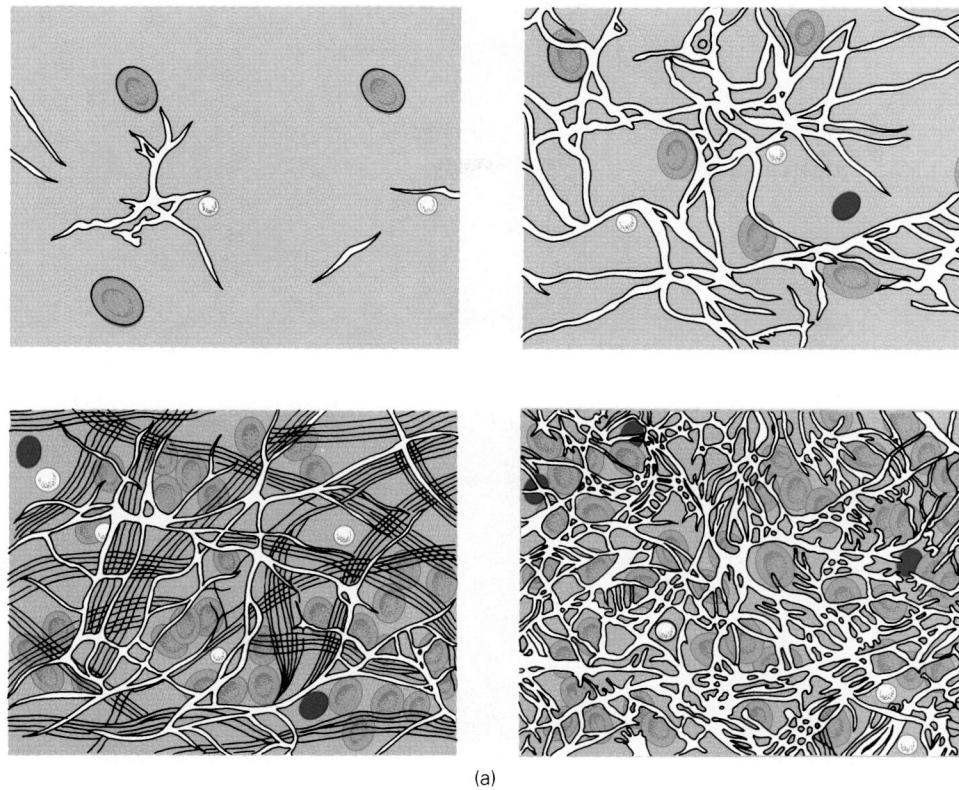

(a)

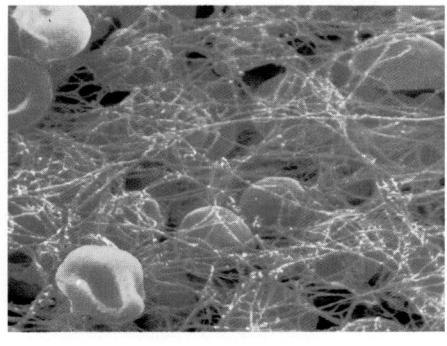

(b)

Figure 17–16
(*a*) Formation of a clot. Fibrin forms long threads in which blood cells, platelets, and plasma become trapped. (*b*) SEM of a blood clot. (© CNRI/SPL/Photo Researchers, Inc.)

fragments—one inert and the other possessing the properties of thrombin. Initially, the conversion of prothrombin proceeds too slowly to produce significant amounts of thrombin needed for coagulation. Thrombin itself, however, increases its own rate of formation by converting a labile plasma protein, **proaccelerin** (Factor V), into **accelerin,** which then accelerates the formation of thrombin. Thrombin also activates the antihemophilic factor and is needed to activate a fibrin-stabilizing factor in Stage III.

Stage III: Conversion of Fibrinogen to Fibrin
Fibrinogen is a soluble plasma protein produced by the liver and normally circulating in the plasma. Thrombin converts **fibrinogen** to **fibrin** by cleaving (breaking off) two pairs of small polypeptides (fibrinopeptides) from each fibrin molecule, leaving a **fibrin monomer** (a single link in a chain of molecules that make up fibrin). Fibrin monomers spontaneously link together (polymerize), forming an insoluble threadlike protein known as **fibrin.** In addition, thrombin activates another plasma en-

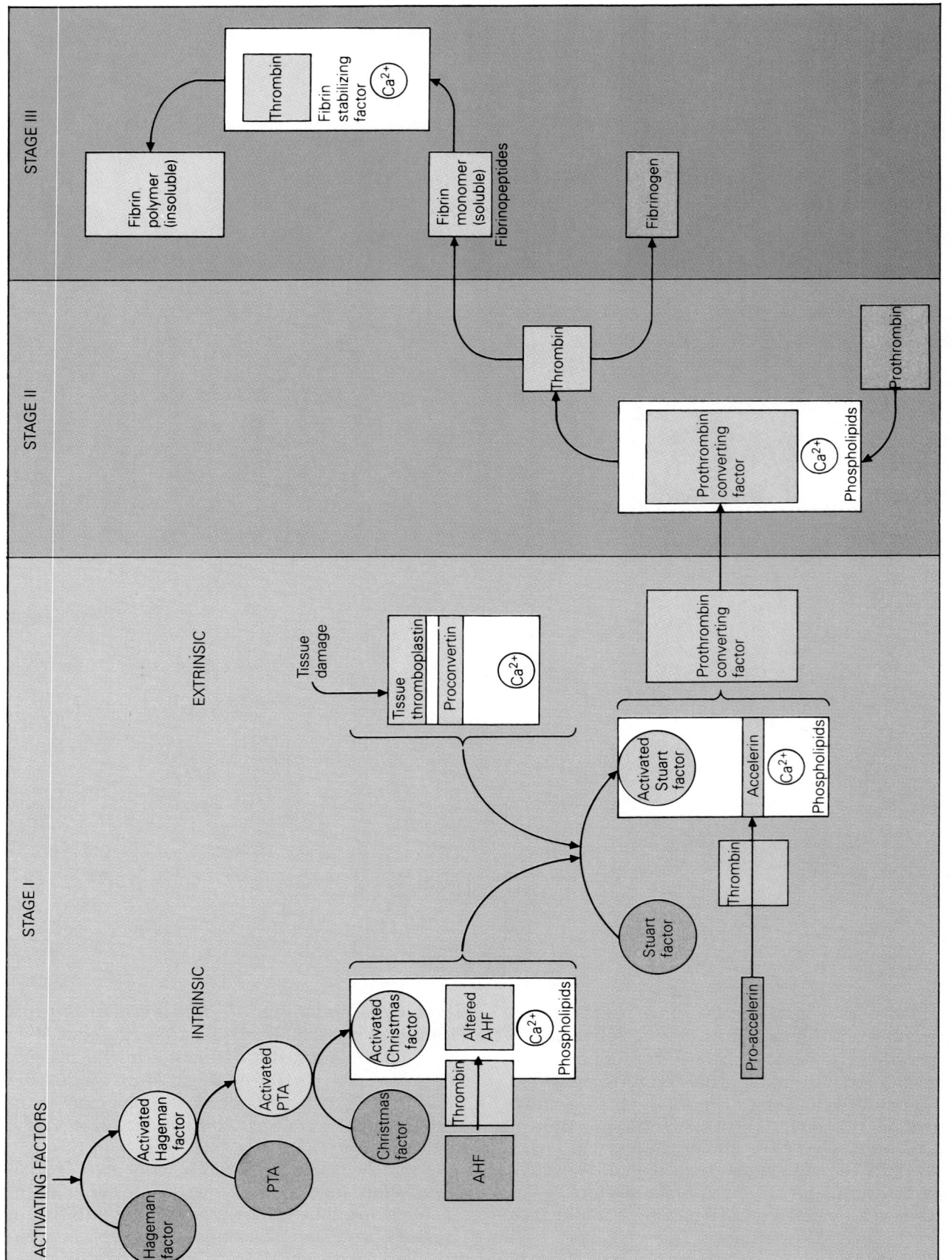

Figure 17-17
A schema of blood coagulation.

Table 17–3
Clotting Factors

International Committee Designation	Synonyms	Location
Factor I	Fibrinogen	Plasma
Factor II	Prothrombin	Plasma
Factor III	Tissue thromboplastin	Tissue cells
Factor IV	Calcium ion	Plasma
Factor V	Proaccelerin Prothrombin accelerator Accelerator globulin Labile factor	Plasma
Factor VI	Obsolete	
Factor VII	Serum prothrombin conversion accelerator (SPCA) Proconvertin Autoprothrombin I Stable factor	Plasma
Factor VIII	Antihemophilic factor (AHF) Platelet cofactor I Thromboplastinogen Antihemophilic factor A	Plasma
Factor IX	Plasma thromboplastin component (PTC) Christmas factor Platelet cofactor II Antihemophilic factor B Autoprothrombin II	Plasma
Factor X	Stuart-Prower factor Stuart factor Autoprothrombin III	Plasma
Factor XI	Plasma thromboplastin antecedent (PTA)	Plasma
Factor XII	Hageman factor Contact factor	Plasma
Factor XIII	Fibrin-stabilizing factor Plasma transglutaminase Laki-Lorand factor	Plasma
Platelet factor	Platelet factor 3	Platelets

zyme, **fibrin-stabilizing factor** (**FSF,** Factor XIII), which in the presence of calcium ions stabilizes the fibrin polymer through covalent bonding of the fibrin monomeric units.

The Retraction of a Clot

After a clot forms, it retracts (gets smaller) over a period of several hours and extrudes serum. Retraction serves to draw wound surfaces together and open the vessel if it has been occluded (blocked) by the clot, thereby increasing blood flow and promot-

ing tissue repair and clot dissolution. Retraction is caused by a platelet factor, **thrombosthenin,** a contractile protein that, to shorten, requires the presence of thrombin and adenosine triphosphate (ATP).

Fibrinolysis: The Dissolution of a Clot

Blood clots are not permanent. After hemorrhage has been checked and tissue repair has proceeded, a clot is gradually dissolved by breaking down fibrin into soluble fragments. The enzyme that degrades fibrin is called **plasmin.** Plasmin is derived from **plasminogen,** an inactive beta globulin normally present in plasma. The formation of plasmin from plasminogen requires an enzyme called **plasminogen activator.** Activator is normally absent from plasma, but its precursor, **proactivator,** is present and may be converted into activator by **cytofibrokinase** (a tissue enzyme), **staphylokinase** and **streptokinase** (bacterial enzymes), **plasma kinase** (Hageman factor), and other enzymes found in urine **(urokinase),** tears, and saliva. **Tissue plasminogen activator (TPA)** formed by endothelial cells appears to be responsible for most of the physiological fibrinolysis. TPA has been synthesized by recombinant DNA techniques and is being used to treat coronary thrombosis.

Normally, a balance between deposition of fibrin and fibrinolysis limits coagulation to the area of vascular injury. Additional inhibitors of excessive clotting include plasma enzymes, which inactivate each clotting factor shortly after it is activated, and anticoagulants.

Anticoagulants

An **anticoagulant** is a chemical that blocks or inhibits the coagulation process. Some anticoagulants are normally produced in the body and are referred to as **physiological anticoagulants;** others are chemicals, called **therapeutic anticoagulants,** that can be administered to block coagulation.

Unless there is trauma to the blood vessels or to the blood itself, coagulation does not normally occur. Although tissue breakdown and platelet destruction are normal occurrences in the absence of trauma, intravascular clotting does not usually occur because (1) the amounts of tissue and platelet procoagulants released are very small and (2) natural coagulation inhibitors such as **antithromboplastin, antithrombin,** and **heparin** are present.

Antithromboplastin inactivates tissue and plasma thromboplastin formed in small amounts, while **antithrombin** prevents thrombin from converting fibrinogen into fibrin. **Heparin,** produced by basophils and mast cells of connective tissue, prevents coagulation by acting as an antithrombin, antithromboplastin, and antiprothrombin.

Therapeutic anticoagulants, on the basis of action, may be classified into three general categories: (1) agents that depress the action of procoagulants, (2) agents that remove calcium from the blood by precipitation or chelation, and (3) agents that inhibit the synthesis of prothrombin in the liver.

In cases of excessive coagulation, heparin is widely used as a therapeutic anticoagulant. It inhib-

its the conversion of fibrinogen to fibrin and also inhibits the production of thrombin. Its effects can be countered by a protamine sulfate administration. Although heparin is produced in the body, it is normally absent from blood.

Agents that remove calcium ions from the blood are used only to prevent coagulation in blood outside the body, since removal of adequate amounts of calcium to prevent clotting interferes with the normal functioning of the heart, skeletal muscle, and nerves. **Sodium citrate** chelates (binds and removes from solution) calcium ions, whereas **sodium oxalate** precipitates calcium ions out of solution.

The presence of **vitamin K** is essential for the liver to make prothrombin. **Bishydroxycoumarin** (Dicumarol), an agent originally isolated from spoiled sweet clover, inhibits production of prothrombin by inhibiting the liver's utilization of vitamin K.

Disorders of Hemostasis

Abnormalities of hemostasis occur as a result of platelet defects, endothelial defects, coagulation defects, excessive production of anticoagulants, or a combination of these.

Thrombocytopenia is an abnormally low number of circulating platelets in blood. If the platelet count falls below 50,000/mm^3, there is an increased risk of internal hemorrhage associated with accidental trauma or surgery. Platelet counts around 20,000/mm^3 are associated with multiple small bruises **(purpura),** hemorrhagic spots **(petechiae)** in the skin, and sometimes spontaneous bleeding from mucosal surfaces. Potentially fatal hemorrhage in the intestines or brain can occur with platelet counts under 10,000/mm^3.

Thrombocytopenia can result from decreased production of platelets by the bone marrow due to toxins, radiation, or infection. It can also result from sequestration of platelets such as in **congestive splenomegaly,** or from increased destruction of platelets by autoimmune processes, as occurs in **idiopathic thrombocytopenic purpura (ITP).** Disorders characterized by increased coagulation of blood and thus increased platelet consumption, such as disseminated intravascular coagulation, also produce thrombocytopenia.

Disseminated intravascular coagulation (DIC) involves widespread coagulation that produces thrombosis in small blood vessels, increased fibrinolysis, and depletion of coagulation factors; these in turn collectively result in generalized bleeding. Causes of DIC include bacterial infections that cause widespread endothelial damage, disseminated cancers that cause increased release of procoagulants, and complications of pregnancy such as the release of amniotic fluid (procoagulant material) into the circulation.

Hemophilia is an inability of the blood to properly coagulate due to a genetic lack of a coagulation factor. **Hemophilia A** involves a deficiency in Factor VIII, and **hemophilia B** (Christmas disease) involves a deficiency in Factor IX. Hemophilias A or B are transmitted genetically and affect only males; females carry the gene but do not show symptoms. **Hemophilia C** is not sex-linked and results from a deficiency in Factor XI. About 85% of hemophilia is Type A. Hemophilia is characterized by spontaneous or traumatic subcutaneous hemorrhage, blood in the urine, and bleeding in the mouth, lips, tongue, and within the joints.

Absence or lack of adequate amounts of other factors also interferes with normal coagulation. **Afibrinogenemia,** a lack of circulating fibrinogen, may be hereditary or may result from liver disease. **Hypoprothrombinemia,** a lack of prothrombin, may be congenital, a result of liver disease, or a complication of vitamin K deficiency. Vitamin K is required for liver manufacture of prothrombin. Newborn infants, for example, lack stores of vitamin K at birth. Vitamin K is normally synthesized by bacteria in the gastrointestinal tract, but infants are born with sterile gastrointestinal tracts. Therefore, hypoprothrombinemia may develop, and life-threatening hemorrhage may occur. The disorder is called **hemorrhagic disease of the newborn.**

Hemorrhagic disease may also occur because of an increase in the amount of heparin released into the blood **(hyperheparinemia).** Allergies, collagen diseases, and disorders of the liver or bone marrow are sometimes accompanied by hyperheparinemia and a reduction in the ability of blood to coagulate effectively.

Internal Defense

Disease-producing microorganisms, called **pathogens,** are always present in our environment. We are exposed to these organisms through the food we eat and drink, the air we breathe, and the animate and inanimate objects we contact in daily life. Bodily defense against these pathogens, their toxins, and other harmful chemicals comprise both nonspecific and specific body responses.

Nonspecific defense includes mechanisms that deter a variety of pathogens by preventing their entrance into the body, as well as mechanisms that quickly destroy an invader if it gains entrance. A neutrophil encountering, phagocytizing, and destroying a bacterium is an example of general, nonspecific defense.

Specific defense includes mechanisms selectively geared toward destruction of specific pathogens or toxins. The synthesis and secretion of highly specific antibodies by a plasma cell is an example. Although other pathogens may be present in the body, the antibody reacts only with the pathogen for which it was designed, rather than reacting with any or all pathogens present.

The Recognition of "Self" and "Nonself"

All specific defenses and many nonspecific defenses require that the body be able to recognize foreign substances—that is, chemicals not usually present or a part of the body. The ability to recognize "self" and "nonself" is possible because each organism is biochemically unique; no two organisms, even if they are individuals of the same species, are chemically identical. Plasma membranes of human and other animal cells, as well as cell walls of bacteria and other forms of life, contain surface molecules of protein, glycolipid, and glycoprotein that allow the cell to be recognized as "foreign" (nonself) or "self." A molecule, or part of a molecule, recognized by the body as foreign is called an **antigen** because it stimulates specific defense mechanisms that include the production and secretion of a specific antibody. Many human cells, such as erythrocytes, have more than a dozen different antigens on their surfaces; some bacteria have more than a thousand.

Nonspecific Defense Processes

Physical and Chemical Barriers

The human skin is the first line of defense against pathogens. Most pathogens, as well as many harmful chemicals, cannot penetrate the densely packed keratinized cells of the epidermis. **Sweat** and **sebum** (a secretion of the oil glands of the skin) contain chemicals that inhibit the growth of certain pathogenic bacteria. Many nonpathogenic bacteria called **commensal bacteria** normally inhabit the skin, and their presence (owing to competition) inhibits the growth of pathogenic bacteria. Also, the pH of the skin is normally acidic because of the chemicals in sebum and sweat and the chemical products of commensal bacterial metabolism; an acid environment inhibits the growth of many pathogenic bacteria.

Mucous membranes line body cavities that open to the exterior and, like the skin, are potential sites for invasion by pathogens. The linings of the alimentary canal, respiratory passageways, urinary tract, and the reproductive tract are examples of frequently invaded sites. **Mucus** is a sticky, viscous glycoprotein secretion of the epithelial cells that form part of the mucous membrane. It traps microorganisms as well as prevents the membrane from drying out. The mucus and trapped microbes in the upper respiratory passageways are moved by **ciliated epithelial cells** toward the nasal and oral cavities from which they may be discharged or swallowed. Swallowed microbes are destroyed by the highly acidic secretions of the stomach. Microbes trapped in the mucus of the lower respiratory tract and the genitourinary tracts are usually destroyed by macrophages. In addition, the continual secretion and swallowing of saliva and the periodic flow of urine and feces help prevent pathogens from colonizing on associated mucous membranes. Millions of commensal microbes also inhabit mucous membranes and, as on the skin, they effectively reduce the chances that pathogens will multiply and invade the body.

Finally, most of the body's secretions contain antimicrobial chemicals. Sweat, saliva, tears, and urine, for example, contain **lysozyme,** an enzyme that effectively digests the cell wall of many pathogenic bacteria. Figure 17–18 summarizes primary physical and chemical defense mechanisms.

Interferons

Interferons are a group of small proteins with antiviral activities. Interferons are released by many types of cells as part of a secondary defense system that is activated when a virus invades the body. Three major groups of interferons are the **alpha-interferons,** secreted by large granular lymphocytes (NK cells) and polymorphonuclear granulocytes; **beta-interferons,** produced by fibroblasts; and **gamma-interferons** secreted by activated T cells (lymphocytes) and macrophages. Interferons are most frequently secreted by cells infected with a virus. Interferons bind to receptors on the plasma membranes of noninfected cells, triggering the cells to synthesize enzymes that break down viral mRNA, thereby making the cells virus-resistant. Interferons "interfere" with viral replication, thereby helping to prevent viral spread. Interferons also stimulate B cells (antibody-producing cells), activate macrophages, and otherwise increase the ability of cells in the immune system to recognize and destroy viruses.

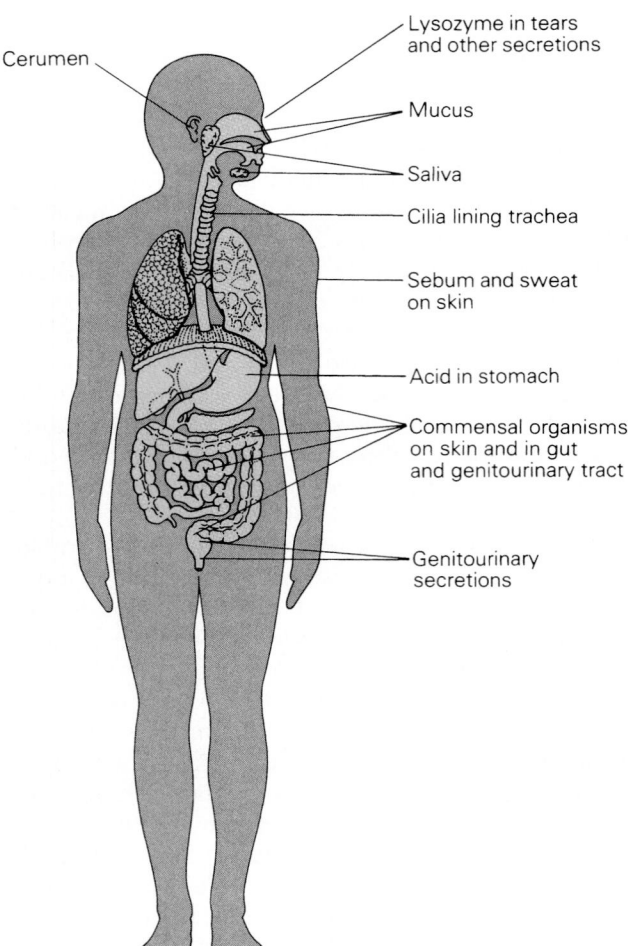

Figure 17–18
Location of various physical and chemical defense mechanisms in the body.

Complement

Complement is a set of about 20 proteins, found in normal plasma or serum, that interact in a defined sequence to promote the lysis of certain microbes and the release of chemical signals. The system is called "complement" because its functions complete—that is, are complementary to—other internal defense mechanisms. When an antibody reacts with an antigen, for example, an immune complex is formed that initiates sequential interactions between complement proteins in a manner similar to that in which chemicals initiate the reaction sequences in Stage I of blood coagulation. Activated complement proteins (Fig. 17–19) take four actions: (1) some lyse cell walls of certain microbes; (2) some increase capillary permeability by stimulating hista-

FUNCTIONS OF COMPLEMENT PROTEINS

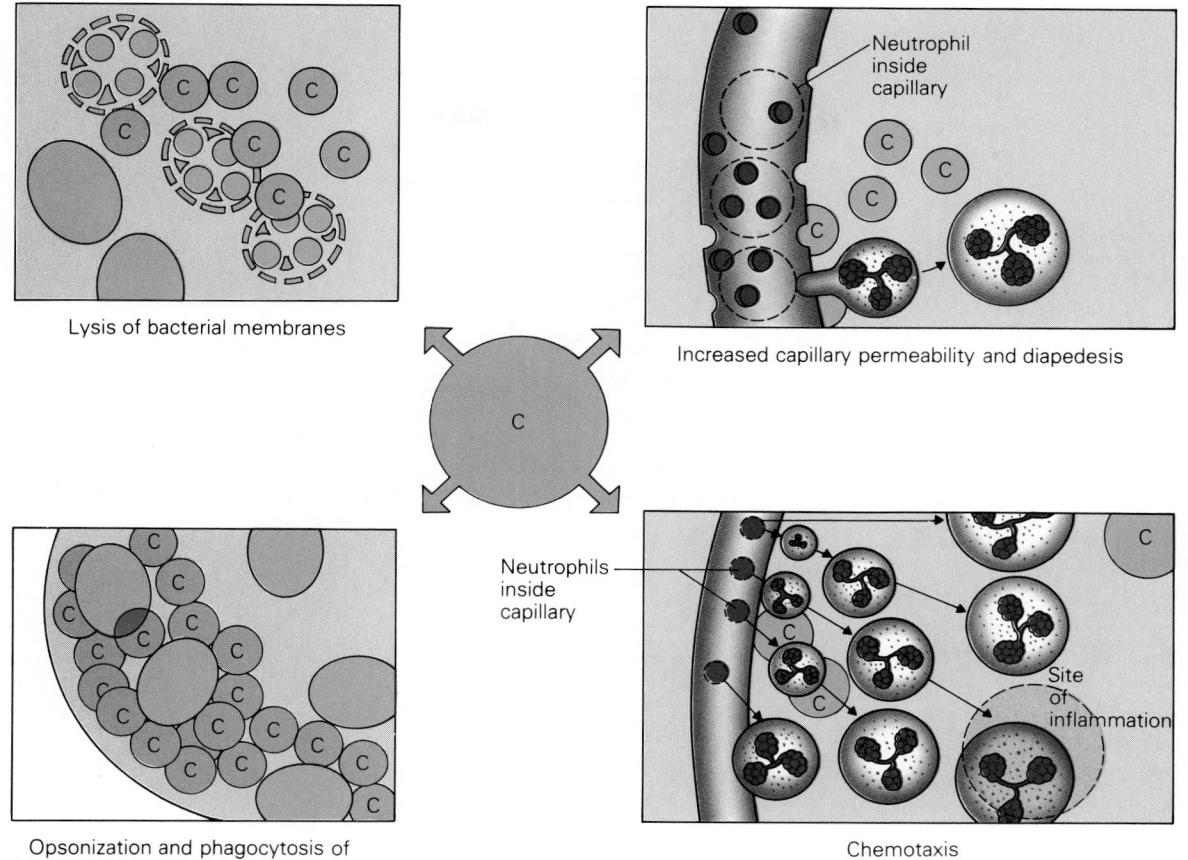

Figure 17–19
Four different ways in which complement proteins function.

mine release from tissue cells and leukocytes; (3) some attract leukocytes (by chemotaxis) to the site of the reactions; and (4) some coat the surface of the microbe, thereby making it recognizable to neutrophils and macrophages, which in turn ingest and destroy the microbe. The latter process, similar to coating distasteful medicine with candy, is called **opsonization,** and the complement proteins that function in this matter are called **opsonins.**

Inflammation

Inflammation is usually a localized body response to tissue damage or injury such as occurs when pathogens invade tissues. It signifies the body's attempt to defend itself against the pathogen and prevent it from spreading and damaging other tissues (Fig. 17–20). The clinical signs of inflammation are **redness, heat, edema,** and **pain.** They are caused by the following inflammatory responses:

1. Injured cells, basophils, mast cells, and others release **histamine** and other chemicals that dilate blood vessels and increase the blood supply to the injured area. This process makes the skin look red and feel warm.
2. Histamine and other chemicals increase the permeability of local capillaries, which allows antibody proteins and other chemicals that mediate an immune response to pass out of the capillary and enter the tissues. The increased flow of materials out of the blood and into the tissues causes local swelling or edema (fluid accumulation), which in turn, along with chemicals released from damaged cells, causes pain.
3. Neutrophils, and to a lesser extent other leukocytes, migrate out of the capillaries (diapedesis) and into the surrounding tissues (Fig. 17–21). Neutrophils and tissue macrophages are attracted to the area of tissue damage by chemicals (chemotaxis) released from damaged cells,

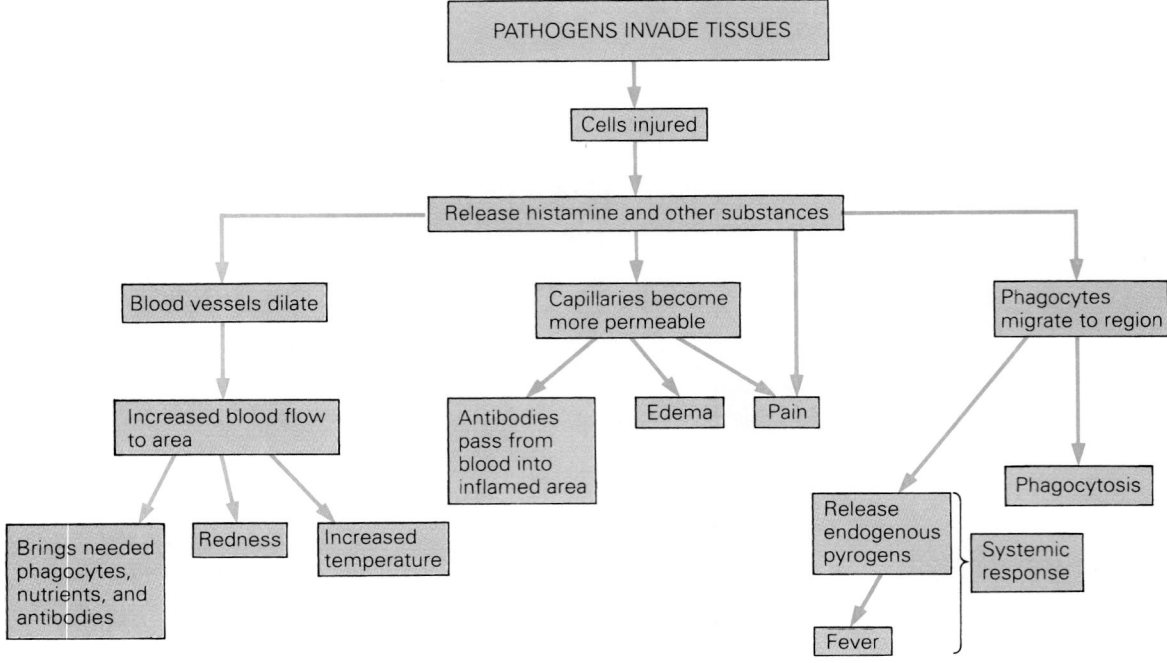

Figure 17–20
The inflammatory response.

Figure 17–21
The processes of diapedesis and chemotaxis.

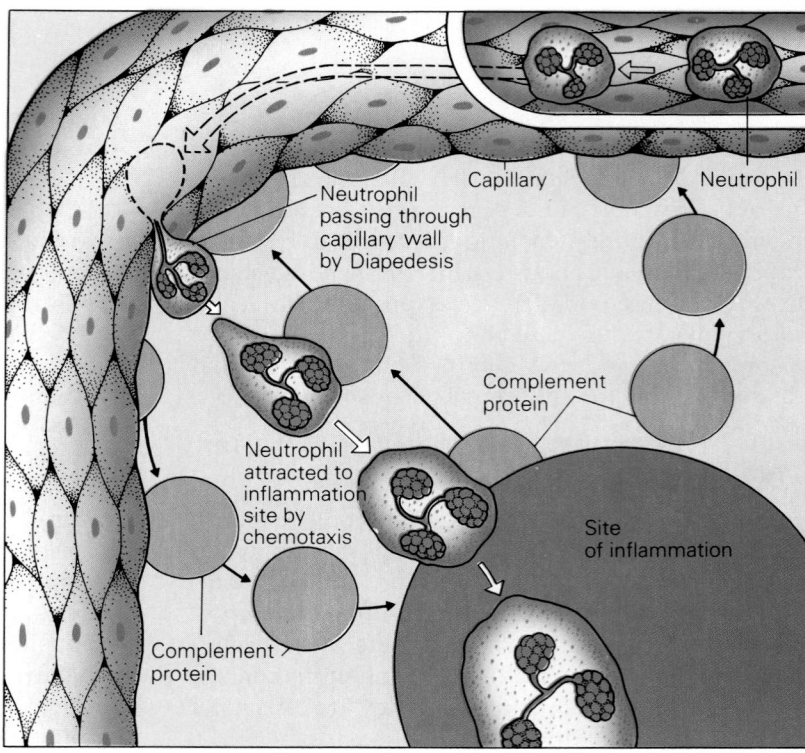

activated complement proteins, lymphokines (from lymphocytes), and other substances.

Phagocytosis

After a pathogen penetrates the body's epithelium, it encounters **phagocytes,** cells that ingest and destroy microorganisms by the processes of **phagocytosis.** Although there are several types of phagocytes, they are all derived from stem cells in the bone marrow. Phagocytes are divided into two groups: the **neutrophils,** or **polymorphonuclear leukocytes;** and the **monocyte/macrophages.**

Neutrophils, as previously discussed, are the most numerous of the leukocytes, making up about 70% of the total. They are highly phagocytic, engulfing all kinds of particles (including pathogens) and destroying them. They are short-lived, existing for one to five days, and are continually being replaced. They are the first phagocytes to arrive at an invasion site.

Monocytes migrate out of the blood and into the tissues, where they develop into macrophages. Tissue macrophages include the **Kupffer cells** of the liver, **microglia** in the brain, **alveolar macrophages** in the lung, and others. They are strategically lo-

cated where they are most likely to encounter foreign substances, including infectious agents. Some fixed macrophages, such as Kupffer cells, remain relatively stationary in the organs where they settle; others (wandering macrophages) roam about through the connective tissues.

The monocyte/macrophage differs from the neutrophil in that it is not phagocytic unless "activated" by complement proteins, antigen-antibody complexes, or other chemicals. Once activated, the macrophage becomes highly phagocytic and is known as an "angry" macrophage. Macrophages are usually slower to arrive at an invasion site but live longer (one to three weeks) than neutrophils and can ingest and destroy more material.

Phagocytosis of a bacterium by a macrophage is illustrated in Figure 17–22. The macrophage attaches to the bacterium by way of nonspecific receptors and extends pseudopodia around the attached bacterium, engulfing it and forming a membrane-bounded intracellular vesicle called a **phagosome.** Lysosomes bind to the phagosome and release digestive enzymes into it, a process known as **degranulation.** The bacterium is digested, and the end-products of the digestion are released or in some cases used by the macrophage.

Figure 17–22
Phagocytosis of a bacterium by a macrophage.

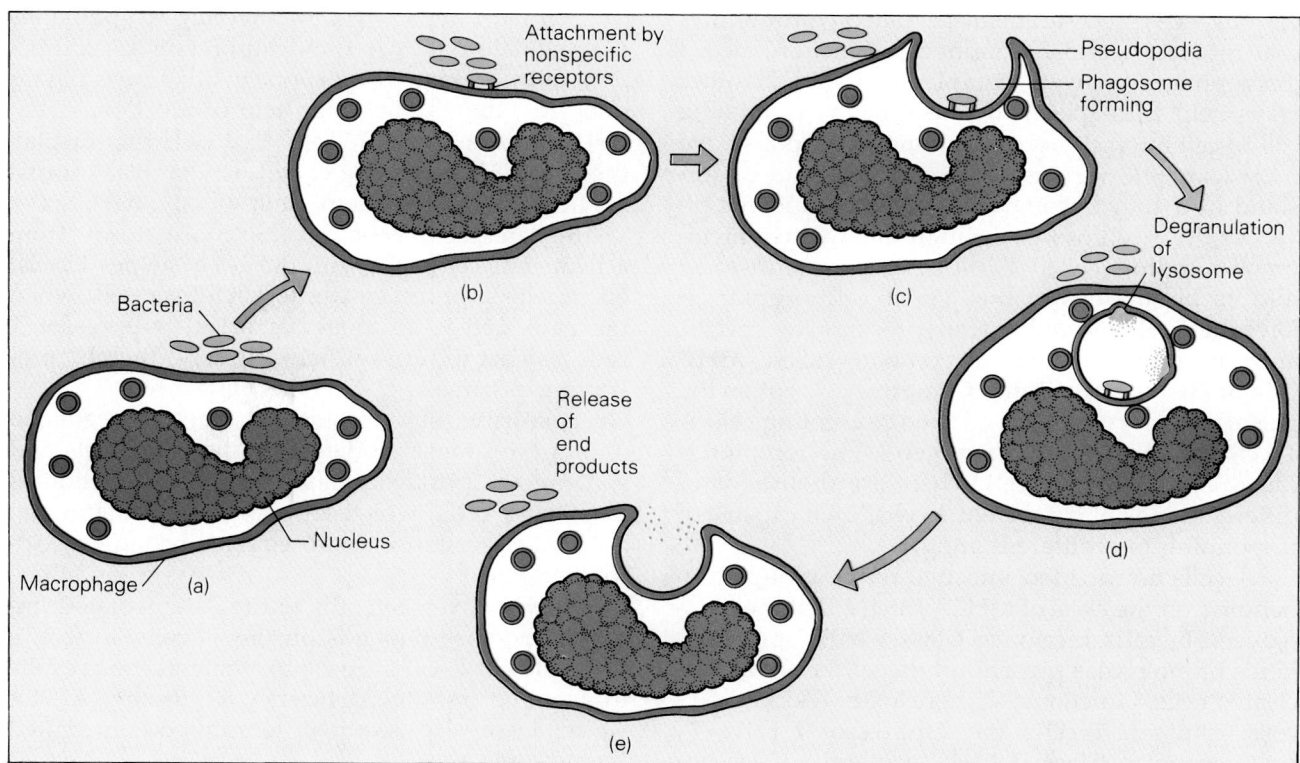

Specific Defense Processes

The term **immunity** is derived from the Latin *immunis,* meaning "to make safe." There are two kinds of immunity: **natural,** or **innate, immunity;** and **acquired,** or **adaptive, immunity.** The previously discussed nonspecific defense mechanisms, such as physical and chemical barriers, provide us with natural immunity. Specific defense mechanisms, geared toward the identification and destruction of specific pathogens, give us acquired immunity. There are two main types of specific defense responses: **cell-mediated** immune responses and **antibody-mediated** immune responses.

Cell-Mediated Immune Response

The cell-mediated immune response is the responsibility of the macrophages and the T-lymphocytes (T cells); both types of cells work together or collaborate in the direct destruction of viruses and foreign cells that enter the body. Collaboration is required because most T cells cannot by themselves recognize free antigens in tissue fluids; the antigen must be presented to the T cell by another body cell called an **antigen-presenting cell.** Usually, the antigen-presenting cell is a macrophage. Also, macrophages secrete chemicals, called **monokines,** that stimulate T cell development or help destroy a pathogen, and various T cells secrete chemicals called **lymphokines** that regulate T and B lymphocyte function, attract macrophages, activate complement, and in other ways help with specific defense. Before examining cell-to-cell interactions of the immune response, we must learn the various types of T cells and understand how they become competent to function.

As the T cell passes through the thymus, membrane receptors called T cell receptors are inserted into its plasma membrane. The T cell receptor allows the T cell to simultaneously recognize a specific antigen and a membrane protein called **MHC (Major Histocompatibility Complex)** present on the plasma membrane of the antigen-presenting cell. A T cell that possesses an antigen-MHC receptor is called a **competent T cell.** There are thousands of different kinds of competent T cells, each capable of responding to a different antigen.

T cells are divided into two major groups, depending on the class of MHC proteins that they recognize. T_8 **cells** recognize **Class I MHC proteins,** kinds of molecules present on the surface of all nucleated cells. Functionally, there are two kinds of T_8 cells: **cytotoxic T cells** and **suppressor T cells.** T_4 cells recognize **Class II MHC proteins,** which are found primarily on the surface of antigen-presenting cells. Functionally, there are two kinds of T_4 cells: **inducer T cells** and **helper T cells.** Remember that both T_8 and T_4 cells can recognize an antigen only if the antigen is presented in combination with an MHC protein on the surface of another cell.

The cell-mediated immune response depends on cooperation among the various types of T cells and macrophages. We can gain a basic understanding of the interplay between cellular elements of the cell-mediated immune response by looking at the response to bacterial infection.

When a pathogen such as a bacterium invades the body tissues, it is ingested by a macrophage, which breaks down the bacterium and displays the bacterial antigens on its own plasma membrane along with Class II MHC protein. The macrophage presents the antigen-MHC II protein to a competent T_4 cell (Fig. 17–23), which binds to the antigen-MCH II complex and becomes activated. The binding also stimulates the macrophage to secrete a hormonelike messenger called **interleukin-1 (IL-1),** a monokine that stimulates the activated T_4 cell to differentiate and reproduce, forming a clone of inducer T cells and helper T cells.

Inducer T cells secrete a **lymphokine,** called **interleukin-2 (IL-2),** which induces T_8 cells that have recognized antigen-MHC I to differentiate and reproduce, forming first a clone of cytotoxic T cells and later a clone of suppressor T cells.

Helper T cells work by secreting lymphokines that regulate cytotoxic T and suppressor T cell function, or by making direct contact with T_8 and B lymphocytes, thereby helping them to function.

Cytotoxic T cells kill infected cells that display the antigen by secreting chemicals that lyse, or disrupt, the infected cell's plasma membrane.

Suppressor T cells develop more slowly from activated T_8 cells and, with the aid of helper T cells, eventually suppress or turn off cytotoxic cells when the pathogen is no longer a threat. Suppressor T cells also act to turn off helper T cells, thereby preventing a "runaway" immune response.

Following suppression of the cell-mediated immune response, some of the helper T cells and some of the cytotoxic cells remain and function as **memory T cells,** which are capable of initiating an accelerated response to subsequent encounters with the antigen.

As we have seen, the entire cell-mediated immune response depends on the T_4 cells; in fact, a deficiency in T_4 cells caused by viral infection results in acquired immune deficiency syndrome (AIDS). T_4 cells are also required for antibody-mediated immune response.

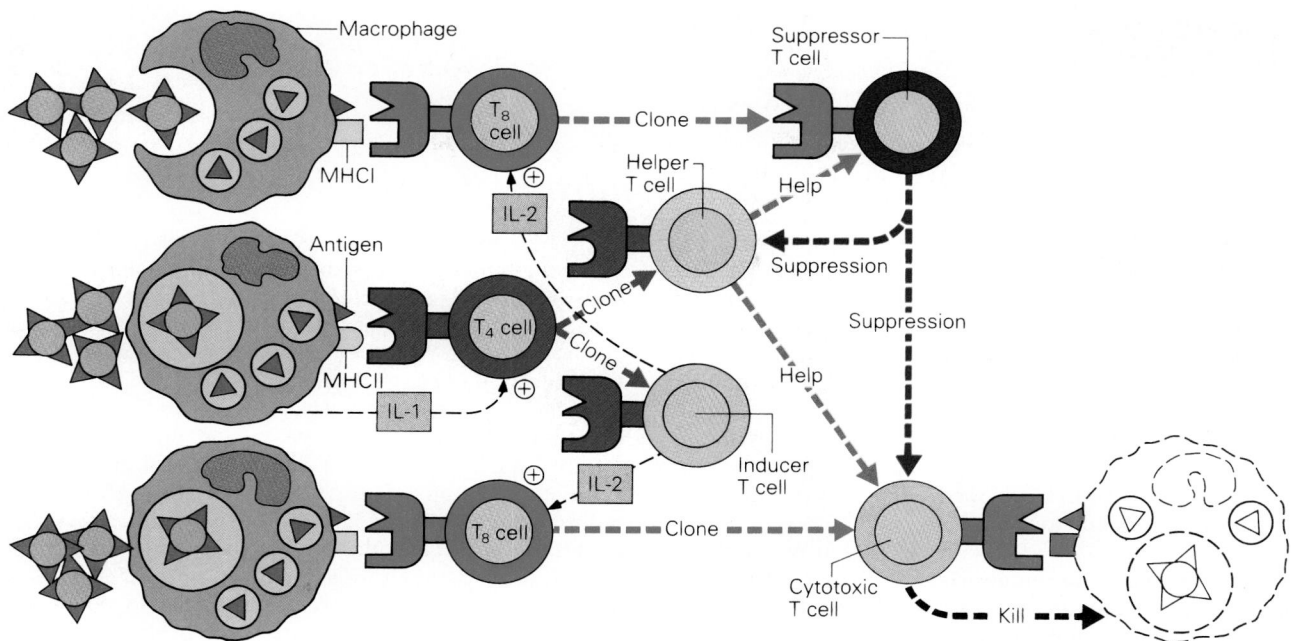

Figure 17–23
The cell-mediated immune response.

Antibodies

Antibodies are glycoproteins called **immunoglobulins** that are produced and secreted by specialized B-lymphocytes in response to specific antigens. The basic structure of all immunoglobulins is a unit consisting of four polypeptide chains and two identical light chains, stabilized and linked by disulfide (—S—S—) bonds to form a Y-shaped molecule (Fig. 17–24). The light (smaller) chains are common to all classes of immunoglobulins, whereas the heavy (larger) chains are structurally distinct for each class. Each chain has a **constant portion** and a **variable portion.**

Various **constant regions** of the heavy chains bind complement proteins and thus can initiate complement reactions. They also bind to receptors on macrophages, monocytes, neutrophils, and NK cells.

The **variable portions** of the heavy and light chains also contain the antigen-specific binding sites, one on each arm of the Y-shaped molecule. A typical antibody molecule can therefore combine with two of the same kind of antigens, forming an antigen-antibody complex (see Fig. 17–24), as well as binding to receptors on cells such as macrophages, which in turn ingest and destroy the immune complex.

Classes of Immunoglobulins. Immunoglobulins are grouped into five classes (and several subclasses)

Figure 17–24
General structure of an antibody and agglutination of red blood cells.

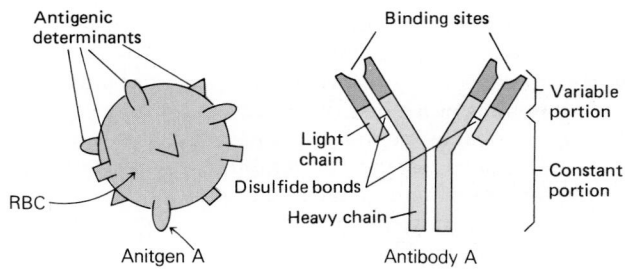

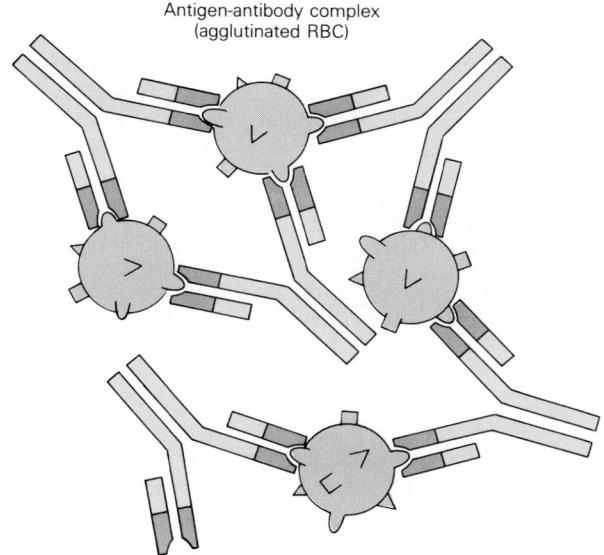

Antigen-antibody complex
(agglutinated RBC)

C L I N I C A L   F O C U S

Acquired Immune Deficiency Syndrome (AIDS)

Acquired Immune Deficiency Syndrome (AIDS) is a viral disease first recognized by the Centers for Disease Control in Atlanta in the summer of 1981. By the end of 1985, approximately 14,000 cases had been reported; by June, 1987, the number had risen to around 37,000. As of October, 1990, about 1.5 million U.S. residents and more than 6 million persons in the rest of the world are believed to be infected with the AIDS virus. No cure has been discovered, no human vaccine has been developed that will prevent infection, and the disease is fatal.

The cause of AIDS is an RNA retrovirus called the **Human Immunodeficiency Virus (HIV).** Thus far, two genetically related variants of the virus are known: HIV-1 and HIV-2. When a retrovirus infects a host cell, the viral RNA is transcribed backward ("retro") into DNA. Using the host cell's enzymes, the viral DNA directs the synthesis of new viral RNA and proteins, which are assembled into new viruses. While doing so, the virus disrupts normal cell functions, ultimately causing the cell to die and release the new viruses, which then infect more cells.

As the number of viruses and infected cells increases, the rate of cell death increases, and overt symptoms of the disease appear. In the case of AIDS, the incubation period (the time between initial infection and appearance of symptoms) may vary from several months to decades, but typically it is around seven to eight years, with death occurring nine to ten years after initial infection.

The principal (but not exclusive) target cells of HIV are the T_4 helper cells, which play central roles in orchestrating both cell-mediated and antibody-mediated immune responses (see text). T_4 helper cells bear a surface receptor, designated CD4, which HIV uses to attach to and infect the cells. A healthy adult has about 1000 CD4 positive cells per cubic millimeter of blood. HIV infection causes the number of CD4 positive cells to decline to about 800 after the first year, to about 400 after the seventh year, and to near zero after the ninth year. As the CD4 positive cell population decreases, immune defenses are gradually compromised, allowing opportunistic infections or malignancies to develop and cause death.

When the T_4 helper cell count falls to about 300/mm^3, minor skin and mucous membrane infections develop. These include fungal infections of the mouth (thrush) and feet (athlete's foot), whitish patches on the tongue (leukoplakia), and shingles, a viral infection of nerves and skin. Associated symptoms include chronic low grade fever, fatigue, diarrhea, muscle aches, night sweats, and unexplained weight loss. Collectively, these early infections and associated symptoms are known as **ARC** (AIDS-related complex). ARC signifies the interval between initial HIV infection and the development of more serious AIDS-defining infections.

When the T_4 helper cell count falls to about 200/mm^3 (about eight years after initial HIV infection), fatal opportunistic infections develop. These include *Pneumocystis carinii* pneumonia (a protozoan-caused pneumonia), toxoplasmosis (a parasitic infection of the brain), cryptococcal meningitis (a fungal inflammation of the brain), and a relatively rare form of skin cancer called Kaposi's sarcoma. Shortness of breath; dry coughing; sore throat; sharp chest pains; enlarged lymph nodes in neck, axilla, and groin; fever; and the appearance of painless purple or brownish spots on the skin or mucous membranes of the mouth and rectum may accompany the infections.

The AIDS virus does not appear to penetrate intact skin or membranes. There is no epidemiological evidence to suggest that insects, air, water, food, or casual contact spreads AIDS. The common modes of transmission are sexual contact (homosexual and/or heterosexual) and blood contact such as by infusion of infected blood, use of infected syringes, spattering of infected blood on skin cuts or abrasions, or transmission from maternal to fetal blood.

Current AIDS research is being concentrated in three major endeavors: (1) the development of a vaccine to prevent HIV infections, (2) the development of drugs to suppress and/or cure HIV infection, and (3) the development of drugs and therapies to suppress and/or cure AIDS-related infections, thereby improving survival time and the quality of the patient's life.

according to the structure of the heavy chains. Using I_g as an abbreviation for **immunoglobulin,** the five major classes are designated I_gG, I_gA, I_gM, I_gD, and I_gE.

I_gG is the major immunoglobulin in normal blood, and accounts for 70% to 75% of all immunoglobulins present. It contributes immunity against many kinds of pathogens, including bacteria, viruses, and fungi, as well as chemical defense against toxins. Usually, I$_g$G molecules are distributed evenly between the blood and extravascular fluids.

I$_g$A immunoglobulins represent about 15% to 20% of the immunoglobulin pool. They are principal antibodies found in various secretions such as mucus (respiratory and digestive tract), saliva, milk, tears, and fluids secreted into the genitourinary and digestive tracts. I$_g$A antibodies provide defense against pathogens that contact the body surface or are ingested or inhaled.

I$_g$M immunoglobulins account for approximately 10% of the body's immunoglobulins. Almost all I$_g$M molecules are in the blood. Their primary importance is in bacterial defense, working in conjunction with complement and blood phagocytes. I$_g$M molecules are also the antibodies that characterize blood type (e.g., Anti-A, Anti-B, etc.).

I$_g$D immunoglobulins make up less than 1% of the blood immunoglobulins but are present on the plasma membranes of many circulating B-lymphocytes. I$_g$D antibodies are believed to be involved in antigen-triggered lymphocyte differentiation—that is, the development of plasma cells and memory cells from B-lymphocytes.

I$_g$E immunoglobulins are rarely found as free circulating antibodies but are commonly found on the surface of the plasma membrane of basophils and mast cells of connective tissue. When engaged by an antigen, I$_g$E molecules stimulate basophils and mast cells to release histamine and other chemicals that mediate the local, often allergic, response to the presence of an antigen. Thus, I$_g$E molecules are directly involved in diseases characterized by hypersensitivity, an exaggerated immune response to certain antigens; asthma and hayfever are well-known examples.

The Antigen-Antibody Complex. All antibodies work by first combining with the specific antigen for which the antibody is designed. The formation of an antigen-antibody complex may then result in one or more of the following:

1. The antigen-antibody complex stimulates phagocytosis by neutrophils, monocytes, lymphocytes, macrophages, and other phagocytic cells.
2. In reacting with the antigen, the antibody may nullify the antigen's ability to harm. Some antibodies called **antitoxins**, for example, combine with the toxin (antigen), thereby altering its chemistry and hence its toxicity.
3. The antigen-antibody complex activates complement, which in turn attracts phagocytes, opsonizes the pathogen, lyses bacterial membranes, and stimulates other defense cells to release chemicals that mediate general and specific defensive responses to the antigen.

Antibody-Mediated Immune Response

The **B-lymphocytes,** with assistance from **T$_4$ helper cells,** are responsible for the antibody-mediated immune response. As is true of the T-lymphocytes, B-lymphocytes are able to recognize and bind to a specific antigen once they become competent and, as is also true of T-lymphocytes, there are thousands of different kinds of B-lymphocytes. In contrast to the T-lymphocyte, however, most B-lymphocytes develop competency in the bone marrow before entering the blood and lymph. As an example of the antibody-mediated immune response, let us look again at the invasion of the body by a pathogenic bacterium (Fig. 17–25).

A macrophage recognizes the bacterial antigen, binds to it, phagocytizes the bacterium, chemically removes the antigen, and displays the bacterial antigen along with Class II MHC protein on its own plasma membrane. A competent T$_4$ cell binds to the antigen-MHC II complex on the antigen-presenting cell, which secretes interleukin I (IL-1) and stimulates the T$_4$ cell to differentiate into a T$_4$ helper cell. The T$_4$ helper binds to the antigen-MHC II complex on the surface of a competent B-lymphocyte that has already bound the antigen, and begins to secrete lymphokines that stimulate the B cell to divide and differentiate. B cell division and differentiation continue as long as they are stimulated by helper T cells, resulting in the formation of two clones of cells: **plasma cells** and **memory cells.**

Plasma cells are enlarged B-lymphocytes with a highly developed and extensive endoplasmic reticulum. Plasma cells remain in the lymphatic system and secrete antibody molecules chemically identical to the antigen receptors and thus specific for the antigen. The antibodies travel via the lymph and blood to the invasion site, where they combine with the antigen, thereby marking it for destruction or reducing its capacity to cause disease.

Memory B cells function as do memory T cells; they permit a rapid mobilization of the immune response if subsequent exposure to the same antigen occurs. Memory B cells secrete small amounts of antibodies for several years after the initial infection is overcome. These antibodies circulate as part of the gamma globulin proteins in plasma and are available for immediate defense should re-infection occur. Re-infection also activates memory T cells, which in turn help memory B cells form new clones of antibody-secreting plasma cells. Memory B and T

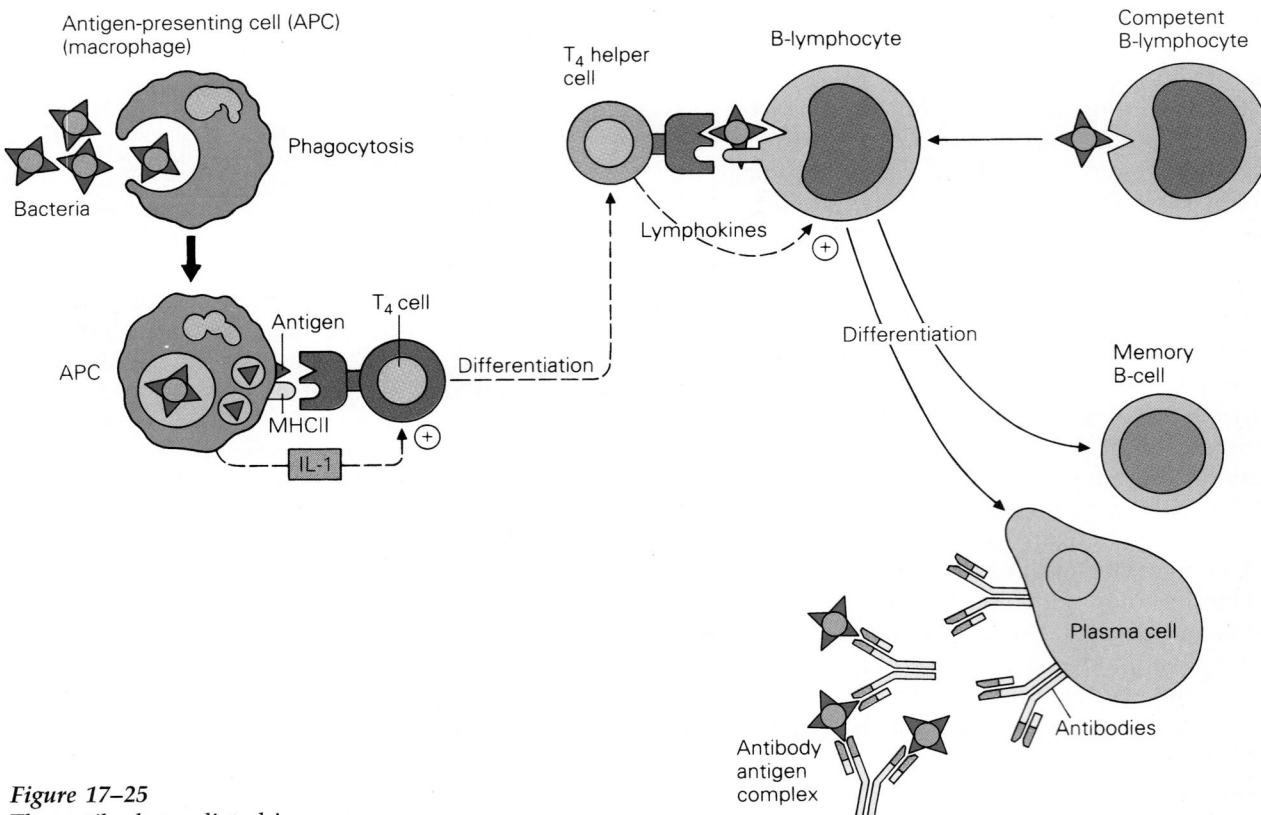

Figure 17–25
The antibody-mediated immune response.

cells form the basis of lasting immunity and often prevent us from suffering twice from the effects of the same pathogen; we may contract measles (a viral disease) as children, for example, but it is uncommon for us to suffer from it again as adults.

Blood Types

All human blood contains erythrocytes, leukocytes, platelets, and plasma; however, not all blood is the same. When some types of blood are mixed, red blood cells aggregate into clumps (a process called **agglutination**). Agglutinated cells become lodged in small vessels, blocking the flow of blood and seriously impairing tissue and organ function (e.g., kidneys, heart, brain). If agglutination is widespread in the body, death can occur. Many circumstances require transfusion of whole blood from one person to another. When transfusion of whole blood is necessary, the donor type must be carefully selected so as to avoid major agglutination in the recipient.

Incompatibilities of blood are due to the presence of antigens, called **agglutinogens,** embedded in the erythrocyte plasma membrane, and specific antibodies, called **agglutinins,** in the blood plasma. Whenever the antigen and its corresponding antibody are present in the same blood, agglutination will occur. In normal blood, if an erythrocyte antigen is present, its corresponding plasma antibody is absent, and vice versa; otherwise, agglutination of erythrocytes would occur. The presence and absence of erythrocyte antigens are determined genetically; that is, blood type is inherited.

Several types of erythrocyte antigens are genetically possible. Some of them are weakly antigenic; that is, they do not provide a strong immune response when given to a person who does not normally possess it. Others are strongly antigenic, causing the recipient to reject, by agglutination, the donated erythrocytes. Three strong antigens that may be present in erythrocytes are the **A, B,** and **D (Rh) agglutinogens.** The corresponding antibodies, normally absent if the antigen is present, are designated as **alpha (anti-A), beta (anti-B),** and **anti-D** agglutinins. The presence or absence of agglutinogens and agglutinins allows the blood to be classified into types or groups. We will discuss only the ABO and Rh blood groups.

The ABO Blood Group

The presence or absence of the A and B antigens and antibodies determines blood type in the ABO blood group as follows: **Type A** blood contains the A antigen on its red cell membranes. The plasma contains anti-B. **Type B** blood contains the B antigen on the plasma membranes of its red cells. The plasma contains anti-A. **Type AB** blood contains both the A and B antigens on its red cell membranes, but the plasma lacks both anti-A and anti-B. **Type O** blood contains neither the A nor the B antigens, but the plasma contains both anti-A and anti-B. The blood types of the ABO group and their constituents are listed in Table 17–4.

When Type B blood is transfused into a Type B recipient, no reaction occurs because the recipient's body does not recognize the blood as being foreign. When type B is transfused into a Type A recipient, two reactions occur:

1. The anti-A in the donor's plasma reacts with the A antigen on the recipient's red cells (**minor agglutination**).
2. Anti-B in the recipient's plasma reacts with the B antigen on the donor's red cells (**major agglutination**).

The first reaction is of little consequence because anti-A in the donor's plasma becomes greatly diluted on entering the recipient's blood and thus becomes ineffective. The second reaction, however, is of major consequence because of the presence of large quantities of anti-B in the recipient's blood, which react with the B antigens of the donor's cells, causing widespread agglutination. A **transfusion reaction** therefore occurs when the donor's cells are agglutinated by the recipient's antibodies.

Blood may be typed and cross-matched to determine its compatibility with other blood. **Blood typing** for the ABO group involves placing serum containing a known antibody (anti-A or anti-B) on a microscope slide and mixing it with a small quantity of the blood to be typed. The presence or absence of red cell agglutination determines the type of unknown blood (Table 17–5). For example, since its cells contain A antigens, type A blood would not agglutinate with test serum samples containing anti-B.

The blood type may be confirmed through testing of agglutinin in the serum of the blood whose type is to be confirmed. Such a procedure is called **cross-matching.** A small quantity of sera to be tested is placed on a microscope slide and mixed with red cells of known type. The presence or absence of red cell agglutination indicates the type of antibody present in the unknown sera, thus confirming the type of antigen present on unknown red cells (Table 17–6).

Table 17–5

Hemagglutination Test for Blood Type

Whole Blood	Agglutination Observed with	
Blood type	Anti-A serum	Anti-B serum
A	Yes	No
B	No	Yes
AB	Yes	Yes
O	No	No

Table 17–4

The ABO Blood Group

Blood Type	RBC Antigen (Agglutinogen)	Plasma Antibody (Agglutinin)	Population Distribution (Caucasian)
A	A	b	41%
B	B	a	10%
AB	A and B	Neither a nor b	4%
O	Neither A nor B	a and b	45%

Table 17–6

Confirmation of Blood Type by Cross-Matching

Serum Antibody	Test Cells		
	A	B	O
a	+	−	−
b	−	+	−
a + b	+	+	−
Neither a nor b	−	−	−

(+) = Agglutination.
(−) = No agglutination.

Because Type AB blood contains both the A and B antigens and neither anti-A nor anti-B, persons with Type AB blood may act as recipients for any of the other types (A, B, or O). Historically, such persons have been referred to as "universal recipients." For similar reasons, persons with Type O blood have been called "universal donors" because their erythrocytes contain neither the A nor the B antigen to oppose the recipients' anti-A and anti-B. In fact, there is no such thing as a "universal donor" or a "universal recipient." The presence of other antigens on the erythrocyte membrane or other erythrocyte antibodies in the plasma precludes this possibility. One such antigen, present in about 85% of the population, is the Rh factor. As we shall see, Type O^+ blood cannot be donated to an AB^- recipient.

The Rh Blood Group

The rhesus blood group, first described in 1940, is named after the monkey in which the antigens were discovered. As with the A and B antigen, the presence or absence of the Rh antigen is determined genetically. Most of the Rh antigens are weakly antigenic and thus of little clinical importance; however, one of the Rh antigens, called **antigen D,** is strongly antigenic. The D antigen is commonly called the **Rh factor.**

If the Rh factor is present on the red cell, the blood is said to be **positive;** if it is absent, the blood is said to be **negative.** Antibodies to the Rh factor are not normally present in the plasma of Rh-negative blood. However, on exposure to the Rh antigen (through transfusion or mixing of fetal and maternal blood during childbirth), Rh-negative persons will produce antibodies against the Rh antigen, so that the next time negative blood is exposed to positive blood, the positive cells will be destroyed. As a general rule, Rh-positive blood should never be transfused into an Rh-negative recipient; however, assuming compatibility in the ABO blood group, Rh-negative blood may be given to an Rh-positive recipient. Blood of the type A^+, for example, may not be given to an A^- recipient, but A^- could be given to an A^+ recipient.

An important consequence of the Rh blood group concerns the Rh-negative mother who bears an Rh-positive fetus for the first time. There are usually no blood complications when the mother has never been exposed to Rh-positive blood. She possesses few, if any, Rh antibodies (anti-D), and the fetal erythrocytes containing the antigen D cannot cross the placenta and enter the maternal blood to stimulate the mother's immune system. At birth, however, the placenta is destroyed, allowing fetal and maternal blood to mix. When fetal erythrocytes enter the maternal circulation, they stimulate the mother's immune system to produce Rh antibodies, which remain in her blood for several years. Upon conceiving and bearing a second Rh-positive fetus, the maternal Rh antibodies can cross the placenta, enter fetal blood, and cause the fetal erythrocytes to agglutinate, resulting in a form of hemolytic anemia called **erythroblastosis fetalis.** Unless the agglutinated fetal erythrocytes are replaced, the fetus can die from hypoxia and other complications. Fortunately, the chance of developing erythroblastosis fetalis may be significantly decreased by administration of Rh antibodies (RhoGAM) to the mother within 72 hours following birth of an Rh-positive child. The antibodies react with any positive fetal erythrocytes that have entered the mother's immune system. As a result of this procedure, called desensitization, the circulating titer (amount) of maternal Rh antibodies usually will not increase to the point of interfering with the successful carrying of a second Rh-positive fetus.

SUMMARY

General Functions of Blood

Three general functions of blood are: (1) transportation of substances from one area of the body to another, (2) defense against foreign agents, and (3) assistance in the maintenance of internal temperature and pH.

Properties of Whole Blood

Blood is a connective tissue containing erythrocytes, leukocytes, thrombocytes, and plasma. Average adult blood volume is 5 liters.

The specific gravity of normal adult blood is about 1.055, and whole blood is about 5 times as viscous as water.

A normal sedimentation rate for adult whole blood is 2 to 10 mm per hour.

A normal adult hematocrit is about 45%.

Plasma

Plasma is the intercellular fluid of the blood. Plasma is about 93% water and 7% solutes (inorganic and organic).

Plasma proteins are divided into the following fractions: albumin, globulins, and fibrinogen. Plasma proteins serve as amino acid reserves, carriers, and buffers, and also participate in blood coagulation and the regulation of fluid distribution within the body. The immune globulins play important roles in body defense. In adult blood there is about twice as much albumin as globulin. Albumin's major roles are transport and control of oncotic pressure in plasma. Globulin's major roles are transport and defense. Fibrinogen is a soluble plasma protein that is converted into insoluble fibrin when blood coagulates.

Hematopoiesis: The Generation of Blood Cells

Hematopoiesis is the generation of blood cells from primitive stem cells.

Erythrocytes: Red Blood Cells

The erythrocyte is a biconcave anucleate cell that numbers about 5 million/mm^3 of whole blood.

Erythrocytes contain hemoglobin, which transports oxygen. One gram of hemoglobin can transport 1.34 ml of O_2. The amount of hemoglobin in blood is about 15 g/dl; therefore, the oxygen-carrying capacity of blood is about 20 vol.%.

Several forms of hemoglobin exist. Type A and F are normal; other types are abnormal.

Erythropoiesis is the formation of erythrocytes. Normally, about 230 billion red cells are produced each day. Substances required for erythropoiesis include vitamin B_{12}, folic acid, iron, and protein. Erythropoiesis is controlled by a renal hormone called erythropoietin.

About 230 billion red cells are destroyed each day. When an erythrocyte is destroyed, most parts of the he-

moglobin are conserved. The remainder is excreted as bilirubin in bile.

Anemia is an abnormal reduction in the oxygen-carrying capacity of the blood. Anemias are characterized by changes in one or more of the following red cell indices: MCV, MCH, MCHC.

Polycythemia is a relative or absolute increase in the number of circulating red cells above normal. Two main types are polycythemia vera and secondary polycythemia.

Leukocytes: White Blood Cells

Leukocytes are blood cells that do not contain hemoglobin. Two principal categories are granulocytes and agranulocytes. On the average, whole blood contains about 7000 to 10,000 leukocytes/mm^3. Their principal function is body defense. Granulocytes (neutrophils, eosinophils, and basophils) contain granular cytoplasm. Their functional roles differ, but all relate to body defense. Agranulocytes (monocytes and lymphocytes) contain homogenous-appearing cytoplasm. Monocytes are phagocytic cells. Two classes of lymphocytes are the T-lymphocytes and the B-lymphocytes. T-lymphocytes confer cell-mediated immunity, and B-lymphocytes confer antibody-mediated immunity.

Leukopoiesis, the formation of leukocytes, occurs primarily in the bone marrow, although some leukocytes mature outside the marrow in lymphoid organs.

Leukocytosis is the increase above normal in the number of circulating leukocytes. Leukocytosis may be physiological or pathological.

Leukopenia is a below normal reduction in the number of circulating leukocytes and results in weakened body defense.

Thrombocytes (Platelets)

Thrombocytes, or platelets, are produced from megakaryocytes. Platelets play several roles in hemostasis.

Hemostasis: Mechanisms to Prevent Blood Loss

Local vasoconstriction helps minimize blood loss when a blood vessel is opened.

Platelets clump together, forming a plug at the site of vascular damage. Platelets also release procoagulants.

Coagulation of blood, or formation of a blood clot, at the site of vascular injury helps seal the opened vessel. Coagulation involves the formation of fibrin thread, which is the foundation of the clot. Fibrin formation occurs in three stages: (1) Stage I—formation of prothrombin-converting factor, (2) Stage II—conversion of prothrombin to thrombin, and (3) Stage III—conversion of fibrinogen to fibrin.

After formation of a clot, clot retraction helps draw wound surfaces together.

Clots are gradually dissolved by enzymes that break down fibrin.

An anticoagulant is a chemical that blocks coagulation. Anticoagulants include physiological and therapeutic types and act by blocking Stage I, II, or III.

Abnormalities of hemostasis can result from platelet defects, coagulation defects, and/or excessive production of anticoagulants. Principal abnormalities include thrombocytopenia and hemophilia.

Internal Defense

Internal defense refers to physiological mechanisms that help defend the body against toxins and pathogens.

Nonspecific defenses are general in nature and deter a wide variety of pathogens. Physical and chemical barriers against pathogen invasion include the skin, sweat, sebum, saliva, and the linings of the gastrointestinal, urinary, respiratory, and reproductive tracts. Interferons are antiviral proteins that help prevent spread of viruses and stimulate leukocytes that destroy virally infected cells. Complement is a set of proteins in the blood that interact to promote lysis of microbes and to signal leukocytes for defense. Inflammation is a localized response to an invading microbe that signals active defense; it is characterized by redness, heat, edema, and pain. Phagocytosis is a process whereby a phagocyte ingests and destroys toxins, invading microbes, foreign matter, and immune complexes. Neutrophils, monocytes, and macrophages are phagocytes.

Specific defenses are geared toward identification and destruction of specific pathogens. Two types of specific defense responses are cell-mediated and antibody-mediated immune responses. Cell-mediated immune responses are the responsibility of the macrophages and T-lymphocytes. Working together, they recognize, ingest, and destroy specific pathogens. Antibodies are proteins produced and released by B-lymphocytes called plasma cells. Antibodies are usually specific for a given antigen and help to destroy the antigen's harmful effect. Antibody-mediated immune responses are the responsibility of the B-lymphocytes, with assistance from T_4 lymphocytes.

Blood Types

The ABO blood group contains blood types based on the presence or absence of antigens A and B on the red cell membrane. Type A blood has the A antigen, Type B blood has the B antigen, Type AB has the A and B antigens, and Type O blood lacks the A and the B antigen. The plasma contains the antibodies for the absent antigen.

The Rh blood group contains blood types based on the presence or absence of antigen D. If antigen D is present on the red cell membrane, the blood type is Rh^+. If antigen D is absent, the blood type is Rh^-. The plasma of an Rh-person may or may not contain anti-D.

REVIEW QUESTIONS

Identify Each with a Term

1. The percentage of volume occupied by red blood cells in a sample of whole blood.

2. The pressure created by the presence of proteins in the plasma.

3. A solution with an osmotic pressure lower than the plasma.

4. The expansion of the red blood cell and its loss of hemoglobin.

5. The formation of blood cells from primitive stem cells.

6. A relative or absolute increase in the number of circulating red blood cells.

7. The process whereby leukocytes pass through a capillary wall.

Define Each Term

8. MCHC	**11.** anemia	**14.** phagocytosis
9. MCV	**12.** LDL	**15.** complement
10. MCH	**13.** hemostasis	**16.** I_gG

Choose the Correct Answer

17. The largest WBC, this blood cell is part of the body's macrophage defense system:

   a. neutrophil   d. basophil
   b. eosinophil   e. lymphocyte
   c. monocyte

18. The most common WBC, this highly motile and phagocytic cell forms the primary line of cell defense against bacterial invasion:

   a. eosinophil   d. monocyte
   b. neutrophil   e. basophil
   c. lymphocyte

19. The following cell recognizes an antigen presented in combination with MHC II:

   a. cytotoxic cell   d. helper cell
   b. suppressor cell   e. two of the preceding
   c. inducer cell

20. During Stage I of the blood coagulation process:

   a. thrombin is formed.
   b. fibrin is formed.
   c. prothrombin is converted into thrombin.
   d. prothrombin converting factor is formed.
   e. fibrin is dissolved.

21. Which of the following whole blood types can be safely infused into a B^+ recipient?

   a. AB^- d. AB^+
   b. A^+ e. A^-
   c. B^-

22. T$_4$ helper cells:
 a. suppress the immune response.
 b. are also known as natural killer cells.
 c. act as memory cells for antibody production and secretion.
 d. secrete interleukin I to activate neutrophils.
 e. assist T$_8$ and B cells.

23. Which of the following values is abnormal?
 a. hematocrit—43%
 b. RBC—5 million/mm^3
 c. WBC—8 thousand/mm^3
 d. O$_2$ carrying capacity—20 vol.%
 e. MCHC—29%

Calculate

24. The total blood volume in a heparinized, centrifuged test tube is 9 ml, and the packed red cell mass volume is 4 ml. Calculate the apparent hematocrit.

25. The hemoglobin content of a blood sample is 16 g/dl. Calculate the oxygen-carrying capacity of the blood.

26. Given a hematocrit of 45% and an RBC count of 5 million/mm^3, calculate the MCV.

27. Assume that 230 billion red cells are destroyed and replaced each day, that the RBC count is 5 million/mm^3, and that the total blood volume is 5 liters. How many days would the body require to replace all erythrocytes currently present?

Answer in One or Two Sentences

28. Polycythemia can be life-threatening. Why?

29. The T$_4$ helper cell plays key roles in both cell-mediated and antibody-mediated immune responses. What are the key roles?

30. A person with type AB blood used to be designated as a universal recipient. Why is this no longer true?

31. Explain the difference between major and minor agglutination with respect to blood transfusion.

SUGGESTED READINGS

Bloom, W., and Fawcett, D.W. *A Textbook of Histology*, Philadelphia: W.B. Saunders Company, 1986.

Burke, S.R. *The Composition and Function of Body Fluids*, 3rd ed. St. Louis: C.V. Mosby Company, 1980.

Kolata, G. "Immune Boosters." *Discover*, September 1987.

Laurence, J. "The Immune System in AIDS." *Scientific American*, November 1985.

Marrack, P., and Kappler, J. "The T Cell and Its Receptor." *Scientific American*, February 1986.

Mills, J. and Masur, H. "AIDS-Related Infections." *Scientific American*, August 1990.

Radetsky, P. "Closing In On An AIDS Vaccine." *Discover*, September 1987.

Roitt, I.M., Brostoff, J., and Male, D. *Immunology*, 2nd ed. St. Louis: C.V. Mosby Company, 1989.

Skikne, B.S. "The Regulation and Control of Iron Absorption: Some Recent Findings." *Laboratory Management*, 25:3, 1987.

Young, J.D., and Cohn, Z.A. "How Killer Cells Kill." *Scientific American*, January 1988.

KEY TERMS

agglutination (p. 614)
anemia (p. 592)
antibody (p. 611)
antibody-mediated immunity (p. 613)
anticoagulant (p. 604)
antigen (p. 605)
atherosclerosis (p. 585)
cell-mediated immunity (p. 610)

clot (p. 600)
complement (p. 606)
erythropoiesis (p. 585)
extrinsic clotting pathway (p. 600)
fibrinolysis (p. 603)
hematopoiesis (p. 584)
hemoglobin (p. 587)
immunity (p. 610)
immunoglobulin (p. 611)

interferon (p. 606)
interleukin (p. 610)
intrinsic factor (p. 593)
leukocyte (p. 593)
lymphocyte (p. 596)
macrophage (p. 609)
oxygen-carrying capacity (p. 588)
phagocytosis (p. 595)
plasma (p. 578)

reticulocyte (p. 589)
Rh factor (p. 616)
T cell (p. 610)
thrombocyte (p. 598)
thrombus (p. 598)

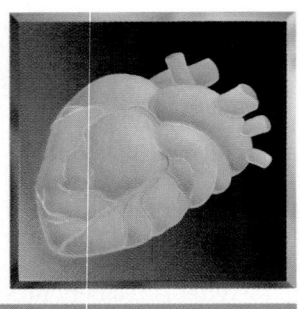

CHAPTER
18

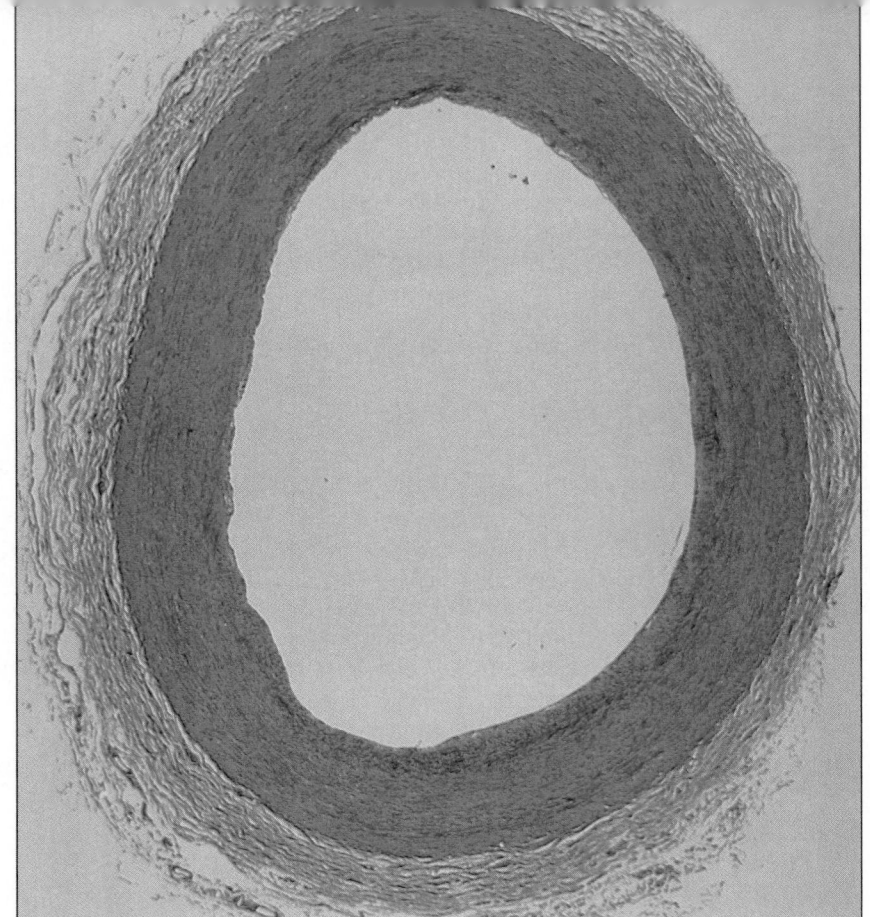

The Heart

*B*efore we begin looking in detail at the physiology of the cardiovascular system, it might be helpful to consider its overall design and function. The cardiovascular system consists of a pump (the **heart**), which propels blood throughout the body by way of a series of tubes **(blood vessels).** The **circulating blood** provides a vehicle for the rapid movement of cells and molecules from one part of the body to another. Molecules move into and out of the circulating blood by diffusion. Cells, such as white blood cells, enter or exit the blood by amoeboid movement. Once a substance has entered the blood, it is transported within seconds to any location within the body. If molecules and cells could only move through the body by diffusion and amoeboid activity, without the blood to carry them, many processes essential for the survival of a large, multicellular organism would not occur rapidly enough. For example, it would take years for nutrients to diffuse from the stomach to the brain or for gases to diffuse from our lungs to the muscles in our legs. Our immune systems would be ineffective if lymphocytes had to move by amoeboid action from the thymus to a distant site of infection, because of the length of time required for cells to travel by this method.

Overview of the Cardiovascular System

William Harvey demonstrated in the seventeenth century that the cardiovascular system forms a closed loop through which blood is pumped by the heart. As shown in Figure 18–1, this loop consists of two pumps, the **left heart** and the **right heart,** and two vascular systems, the **pulmonary circulation** and the **systemic circulation.** These four components of the cardiovascular system are arranged in **series,** which means that blood must flow through these four components in sequence, one after the other. Blood flows, via the pulmonary circulation, through the lungs, where oxygen is added and carbon dioxide is removed. This oxygenated blood then flows, via the systemic circulation, to all the cells of the body, which extract oxygen from the blood and add carbon dioxide to it. The blood then flows back to the lungs, and the cycle repeats.

A more detailed schematic of the circulatory system is shown in Figure 18–2. Notice that the sys-

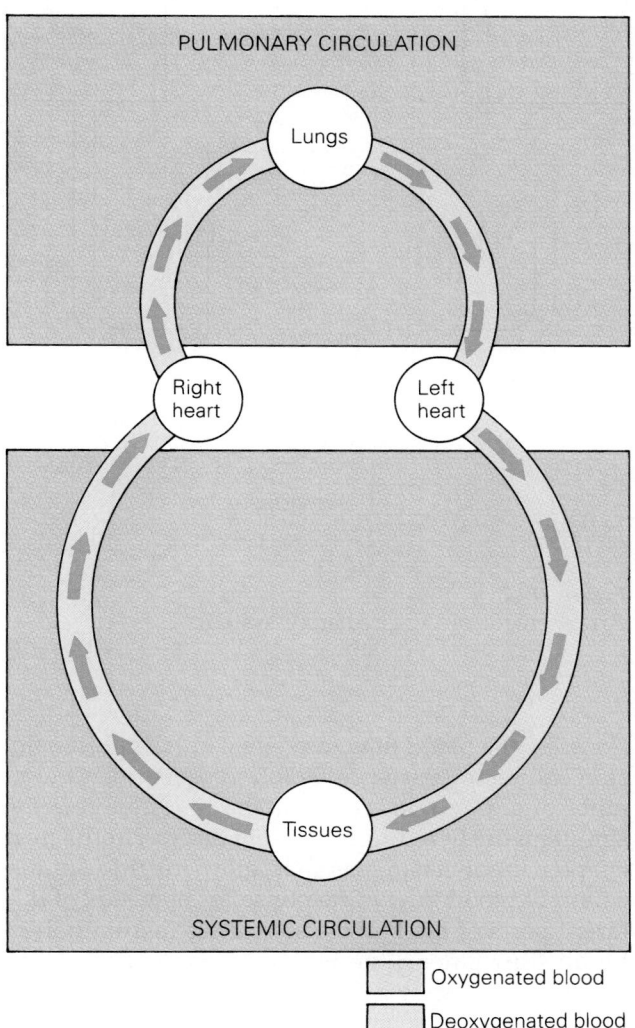

Figure 18–1
Overview of the cardiovascular system.

temic circulation consists of many separate circulatory pathways that supply blood to various organ systems within the body. Notice also that these individual **vascular beds,** which make up the systemic circulation, are arranged in **parallel** with one another. This means that the total volume of blood flowing through the systemic circulation is partitioned, or divided, among the different vascular beds. As we will see later, our bodies are able to dynamically adjust the distribution of blood flow among the various tissue beds. For example, during

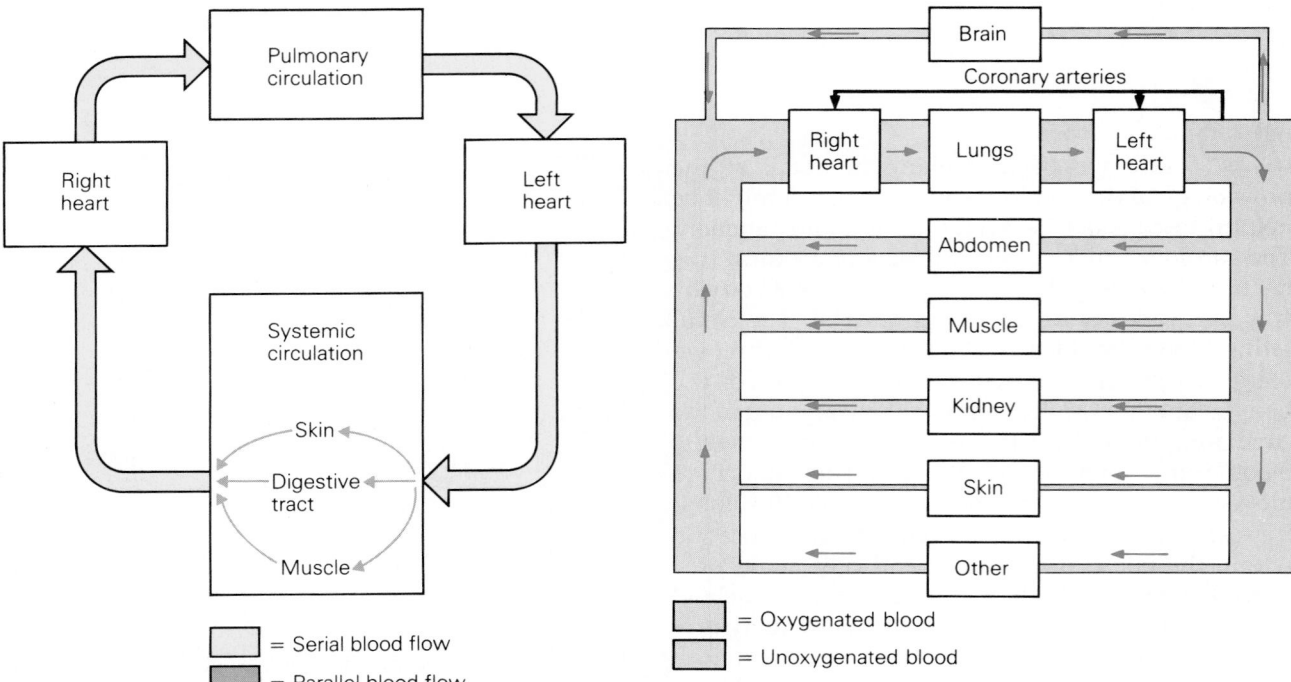

Figure 18–2
Arrangement of the circulatory system.

Figure 18–3
Relative distribution of blood flow in the cardiovascular system at rest. Vessel diameter is proportional to blood flow.

exercise, the blood flow to skeletal muscle and to the skin can be increased several fold while the flow to other organ systems, such as the digestive tract, is simultaneously reduced. We will discuss in the next chapter the mechanisms that allow for this optimization of blood flow in response to the needs of different parts of the body. The relative distribution of **blood flow** among the various organs of the body during rest is shown in Figure 18–3.

The Systemic Circulation

The vessels that carry blood *from* the heart make up the **arterial system;** the vessels that carry blood *back* to the heart make up the **venous system** (Fig. 18–4). The systemic circulation begins with a single artery, the **aorta,** which receives all the blood the heart pumps out. The aorta branches into a number of smaller **arteries,** which direct blood flow to the various organs of the body. Once it enters an organ, an artery soon branches into progressively smaller arteries, which in turn branch into smaller **arterioles.** This arteriolar tree distributes blood flow to all the cells within an organ. The arterioles then branch into an extensive network of **capillaries,** the smallest blood vessels in the body. It is here in the capil-

laries that exchange occurs between the blood and the interstitial fluid. Blood then flows from the capillaries into the venous system, which channels the blood back to the heart.

The venous system begins at the point where the capillaries re-unite to form larger vessels called **venules,** which in turn unite to form larger **veins.** All the veins from the upper portion of the body drain into a single large vein called the **superior vena cava,** and the veins from the lower portion of the body drain into the **inferior vena cava.** The two venae cavae unite to return all of the blood to the right heart.

The Pulmonary Circulation

A similar pattern of branching occurs in the **pulmonary circulation** (see Fig. 18–4). The blood leaves the right heart by way of the **pulmonary trunk,** which soon divides into the **pulmonary arteries.** The pulmonary arteries distribute the blood to the two lungs. Notice that the entire **cardiac output** passes through the lungs before being distributed throughout the systemic circulation.

As in the systemic circulation, the pulmonary arteries progressively branch into smaller and

Blood Flow and Blood Pressure: Physical Principles

We now know that the role of the cardiovascular system is to provide for a continuous flow of blood through the tissues. Let's consider for a moment what the term *flow* means and how it might be quantified.

Blood flow can be defined as the volume of blood that moves past a particular point in the cardiovascular system during a given period of time— for example, the amount that flows through the left heart in one minute. A practical example of fluid flow with which we are all familiar is the flow of water through a garden hose. If we were to collect the water coming out of a hose over a period of one minute, we could calculate the flow of water through the hose. If, after 1 minute, we had 5 liters of water in our bucket, then the calculated flow would be 5 liters in 1 minute, or 5 L/min. We have measured the volume of water that moved past a particular point in the garden hose (in this case, the nozzle) during a period of time (1 minute).

Does this number tell us anything about the flow of water elsewhere in the hose? The answer is yes, because the flow of water will be the same at all points in the hose. We can generally state that the flow in a closed system will be the same through all portions that are arranged in **series.** In terms of the cardiovascular system, this means that we could measure the flow of blood from the heart into the aorta and know that it is the same as the total blood flow through the lungs, the systemic circulation, the left heart, and the right heart, because these portions of the cardiovascular system are all connected in series and constitute a closed system.

Could we also say what the blood flow was through the kidneys on the basis of this one measurement? No, because the individual vascular beds that make up the systemic circulation are arranged in parallel, and therefore only a portion of the total systemic blood flow goes through each of these vascular beds (see Fig. 18–3). As illustrated in Figure 18–5, the combined flow through the vascular beds arranged in parallel with one another must be equal to the flow through portions of the system that are in series with these vascular beds. Furthermore, as mentioned previously, the proportion of the total systemic blood flow channeled through a particular vascular bed can change in response to the changing needs of the body, even though the total blood flow through the systemic circulation remains constant.

The force that causes blood to flow through the vessels can be measured as **blood pressure.** Blood pressure measurements are usually expressed in

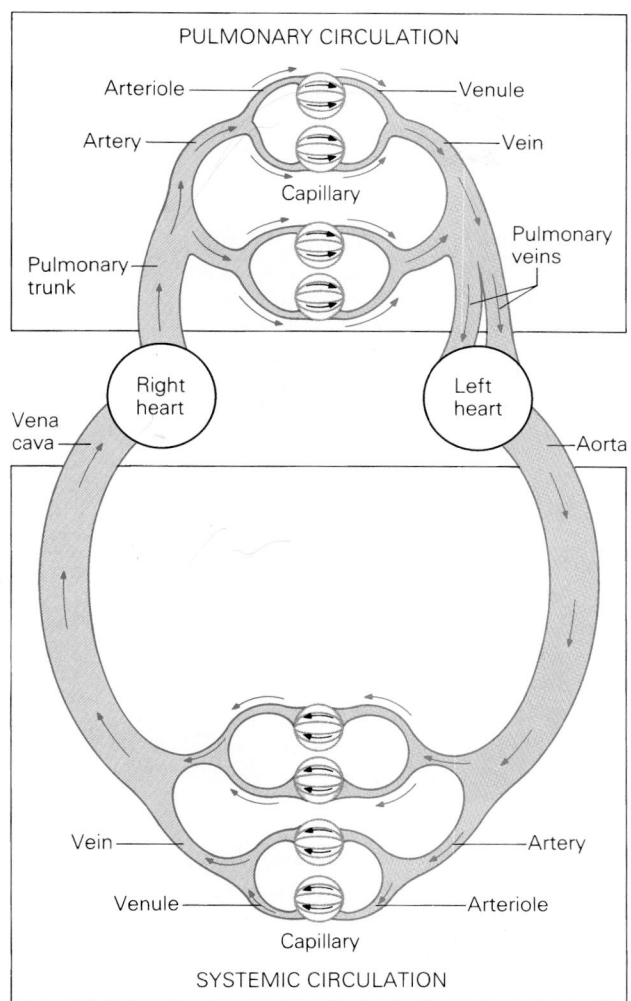

Figure 18–4
Branching pattern of blood vessels in the cardiovascular system.

smaller vessels. The **pulmonary capillaries** are the site of gas exchange between the blood and the air in the lungs. As the blood flows through the pulmonary capillaries, its oxygen content increases and its carbon dioxide content decreases as a result of diffusional exchange with the air in the lungs. (More will be said about this in Chapters 20 and 21.) This oxygenated blood then flows from the pulmonary capillaries through progressively larger venules and veins to the **pulmonary veins** and then into the left heart.

Not all arterial blood has a high oxygen content. Both the **pulmonary venous blood** and the **systemic arterial blood** are oxygenated. The distinguishing factor between arteries and veins is that arteries carry blood *from* the heart *to* capillaries, and veins carry blood *to* the heart *from* capillaries.

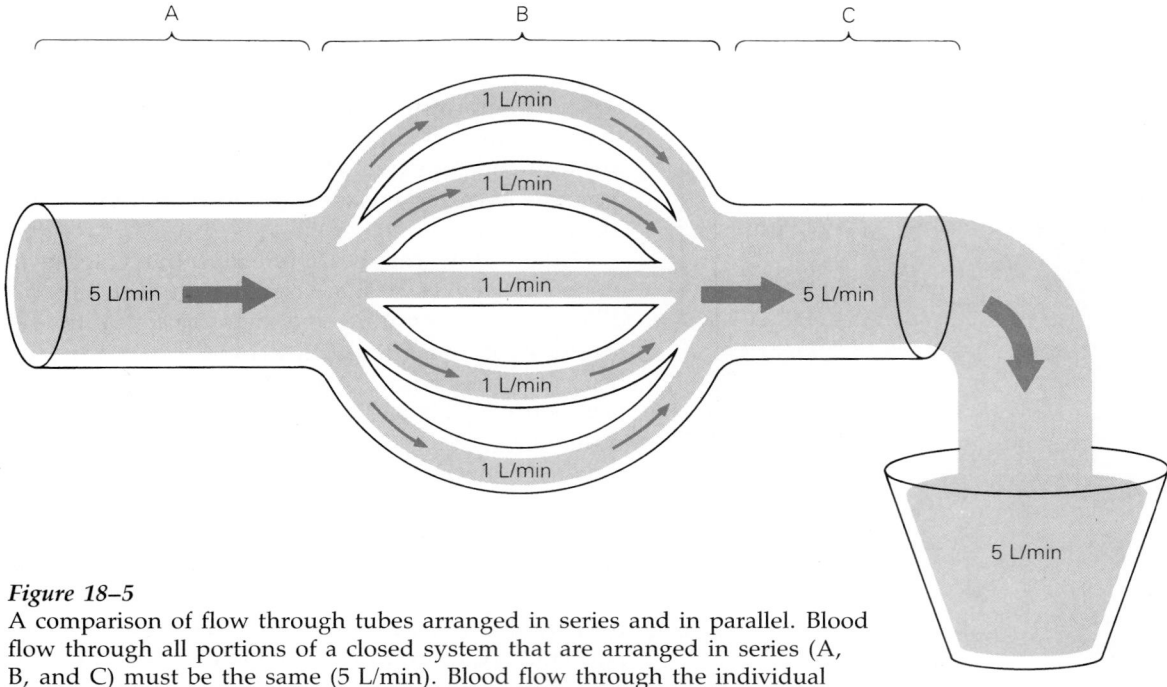

Figure 18–5
A comparison of flow through tubes arranged in series and in parallel. Blood flow through all portions of a closed system that are arranged in series (A, B, and C) must be the same (5 L/min). Blood flow through the individual vessels arranged in parallel (small vessels in section B) must add up to the total blood flow through the system (5 × 1 L/min = 5 L/min).

units of **mm Hg** (millimeters of mercury). For example, average **aortic blood pressure** is 90 mm Hg. This means that the pressure inside the aorta (relative to atmospheric pressure outside the aorta) is equal to the downward pressure exerted by a column of mercury 90 mm tall.

A more familiar unit of pressure is pounds per square inch (psi). Automobile tires, for example, have an internal air pressure of about 35 psi. A blood pressure of 100 mm Hg is equivalent to 2 psi. Blood pressure is a measure of the driving force propelling blood through the blood vessels. Just as the high air pressure within a tire will cause air to rush out of the tire through an open valve, the differences in blood pressure within the circulatory system will cause blood to flow from one point to another. Flow does not occur in response to any particular pressure (*P*), but rather to a pressure difference or **pressure gradient** ($P_1 - P_2$ or ΔP) between two points. This is analogous to net diffusion between two points; the rate of diffusion is determined by the magnitude of the **concentration gradient** and not by the absolute concentration of a diffusing substance. Thus, the blood flow between two points in the circulatory system is proportional to the *difference* in blood pressure between these same two points and not the *absolute* pressures at either point. This concept is illustrated in Figure 18–6.

The reason that a physical force or pressure is required to maintain blood flow through the circulatory system is that frictional forces tend to impede blood flow. Friction exists among the molecules that make up the blood and between the blood and the walls of the vessel. **Resistance** is a quantity that summarizes the net effect of all the frictional components that oppose blood flow in a particular situation. Thus, the greater the resistance, the lower the blood flow for any given pressure difference. The equation that describes the relationship between blood flow (*F*), driving pressure ($P_1 - P_2$ or ΔP), and resistance (*R*) is:

$$F = \frac{P_1 - P_2}{R}$$

As we will see in the next chapter, the resistance to blood flow through a vessel is determined by a number of factors, including the length and diameter of the vessel, as well as the consistency of the blood itself.

The Heart: A Muscular Pump

The major role of the heart is to maintain the flow of blood throughout the cardiovascular system by pumping blood under pressure into the vascular

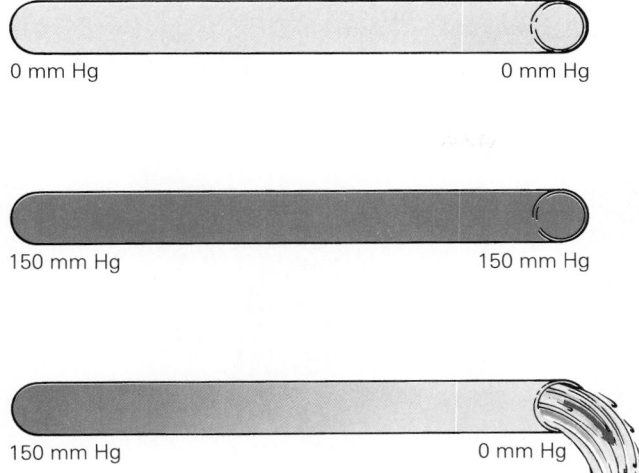

0 mm Hg   0 mm Hg

150 mm Hg   150 mm Hg

150 mm Hg   0 mm Hg

Figure 18–6
The relationship between pressure and flow. Flow
through vessels A and B is zero because there is no
pressure difference anywhere within the vessel. Fluid
flow in vessel C is driven by the pressure difference
(150 mm Hg − 0 mm Hg = 150 mm Hg) between the
ends of the vessel. The arrow indicates the direction of
blood flow.

system. The next several sections will consider the
anatomy of the heart and the manner in which cardiac muscle contraction is coordinated within the
heart to generate pressure and flow.

Chambers of the Heart

As previously mentioned, the heart consists of two
functionally distinct pumps: the left heart and the
right heart. The left and right hearts, in turn, each
contain two chambers: an atrium (plural, *atria*) and a
ventricle. The **atria** are thin-walled chambers that
receive blood returning to the heart from either the
vena cavae (in the case of the right heart) or the
pulmonary veins (in the case of the left heart). Blood
then moves from each atrium into the associated
ventricle. Ventricles have thicker, muscular walls
capable of pumping blood out of the heart at higher
pressures. Figure 18–7 shows schematically how the
chambers of the heart are organized and the direction of blood flow through the heart.

Valves in the Heart

Notice that there is a valve at the entrance and exit
of each ventricle (Fig. 18–8). These valves allow
blood to flow in only one direction. We will see later
how these one-way valves operate to ensure unidirectional flow of blood through the heart. The **atrio-**

ventricular (AV) valves are located between the
atria and the ventricles. The AV valve associated
with the right heart is called the **tricuspid valve,** and
the AV valve associated with the left heart is called
the **bicuspid** or, more commonly, the **mitral valve.**
The valves located at the junction of the ventricles
with the pulmonary and systemic arteries are called
semilunar valves. The semilunar valve between the
right ventricle and the pulmonary trunk is called the
pulmonary valve, and the valve between the left
ventricle and the aorta is called the **aortic valve.** The
opening and closing of the valves is a passive process that occurs in response to the **blood pressure
gradient** across the valve. For example, when the
pressure in the atrium is greater than that in the
adjoining ventricle, the AV valve between the two
chambers opens and allows blood to flow into the
ventricle (Fig. 18–9). If the pressure gradient is reversed—with ventricular pressure being greater
than atrial pressure—the atrioventricular valve will
be forced closed; this prevents the flow of blood
from the ventricle back into the atrium.

The Walls of the Heart

The walls of the heart are composed of three anatomically distinct tissue layers (Fig. 18–10). The
middle, muscular layer, the **myocardium,** is composed primarily of cardiac muscle cells. (The special
properties of this muscle type were discussed in
Chapter 16.) The inner surface of the myocardium,
which forms the inside wall of the atria and ventricles, is lined with a thin layer of endothelial cells,
the **endocardium.** The endothelial layer forms a
continuous lining that covers not only the inner surfaces of the heart and the heart valves but also the
entire vascular network. The outer surface of the
myocardium is covered by the **epicardium,** which
consists primarily of connective tissue and fat. The
main **coronary vessels** that supply blood to the myocardium (see Fig. 18–8) are located within this tissue
layer.

Electrical Activation of the Heart: The Cardiac Action Potential

Although the heart is composed primarily of muscle, it must function quite differently from the way
in which skeletal muscle works. Different parts of
the heart must contract at different times, and they
must do it in a proper sequence if the heart is to
pump blood efficiently. The key to understanding
much of the special function of the heart lies in a
knowledge of the electrical properties of its muscle.

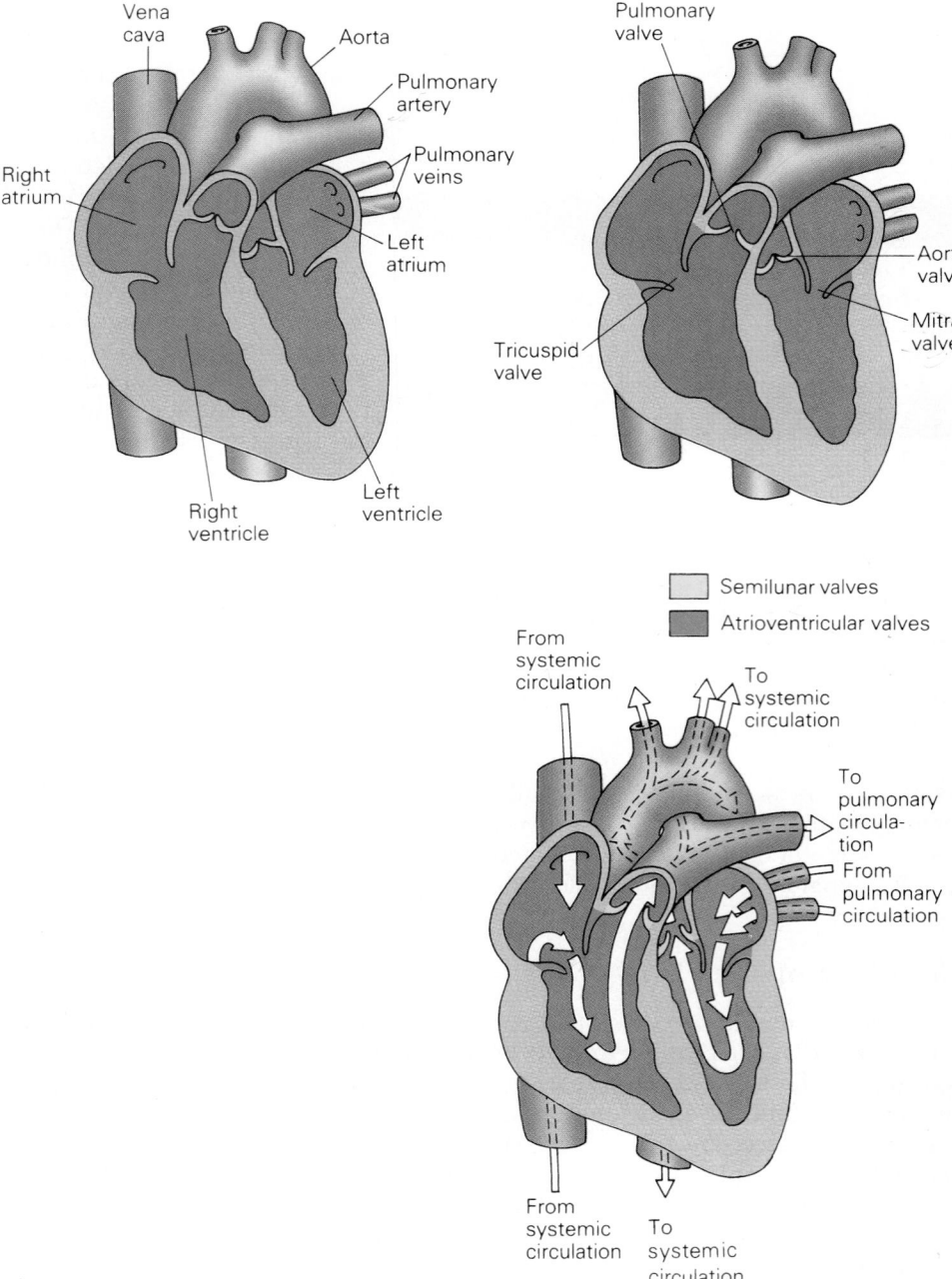

Figure 18–7
Overview of the heart. The direction of blood flow through the chambers of the heart and the major vessels leading into and out of the heart is indicated by the arrows.

Phases of the Cardiac Action Potential

As do most other muscles, the heart muscle cells have plasma membranes that can generate action potentials. It is difficult to speak of a "typical" heart muscle action potential, because each of the specialized regions of the heart has special action potentials. While they are similar in many ways to the relatively simple action potentials found in nerves and skeletal muscles, action potentials in the heart

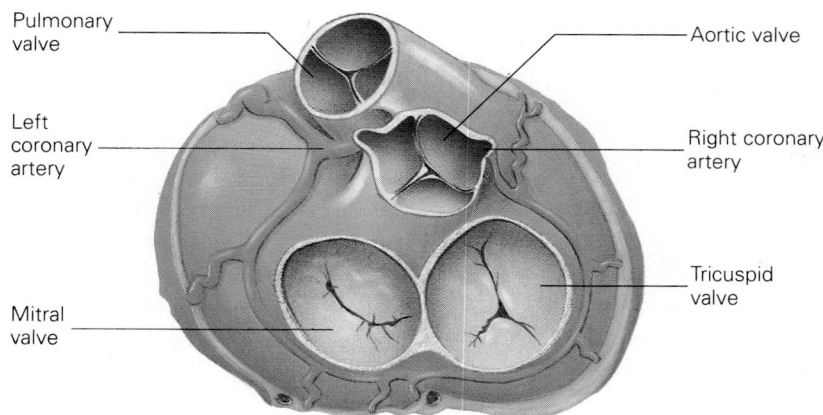

Figure 18–8
Cardiac valves. Top view of the heart with the major vessels cut away to show the location of the four cardiac valves. Both the aortic and pulmonary valves (yellow) consist of three crescent-shaped (semilunar) cusps. The left atrioventricular valve (green) consists of two valve cusps (bicuspid or mitral valve), while the right atrioventricular valve (green) consists of three valve cusps (tricuspid valve). All the heart valves are connected to a tough connective tissue framework or cardiac skeleton that lies between the atria and the ventricles.

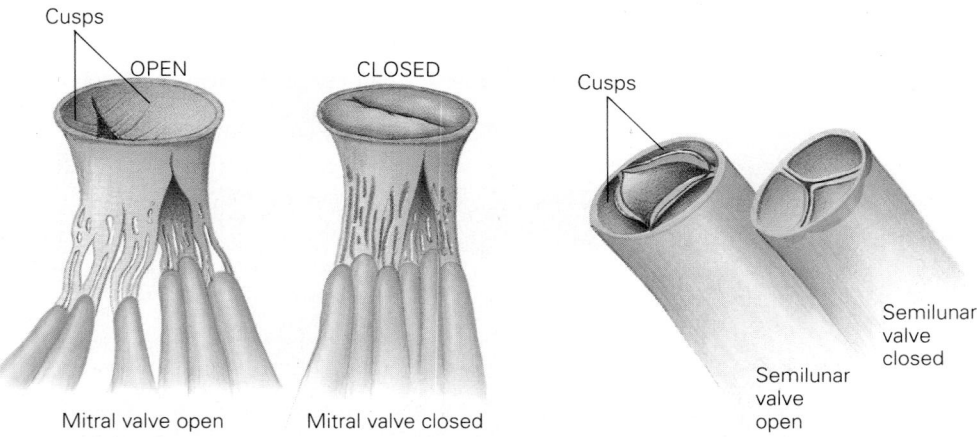

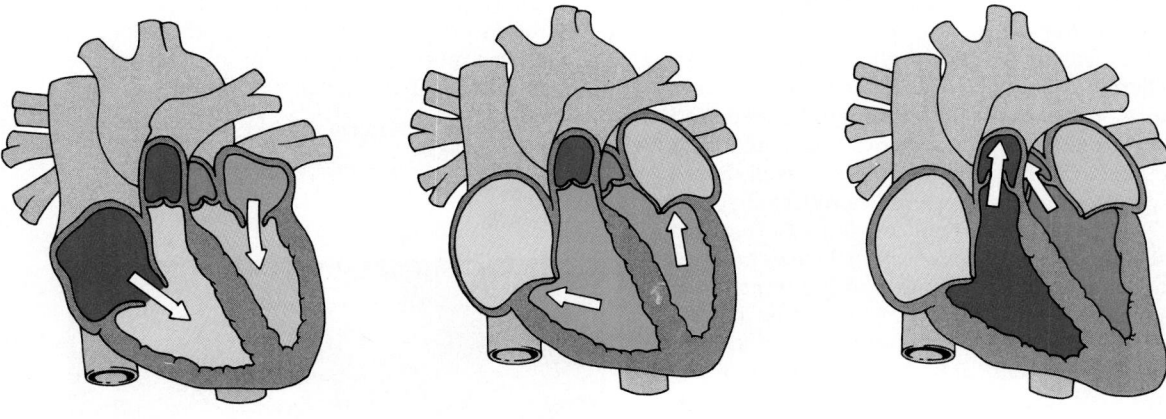

Figure 18–9
Cardiac valve operation. Valves in the heart normally open when the pressure gradient across the valve is increasing in the direction that blood normally flows through the heart (forward pressure gradient), as shown in the first drawing. In contrast, a reverse pressure gradient (see middle panel) will force the valve closed. This prevents reverse blood flow in response to reverse pressure gradients that occur as a result of the pumping action of the heart. In the third drawing, a forward pressure gradient forces the semilunar valve to open. Darker colors correspond to higher pressure.

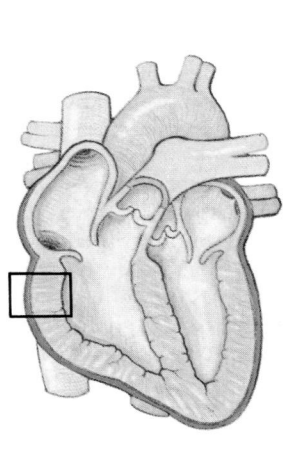
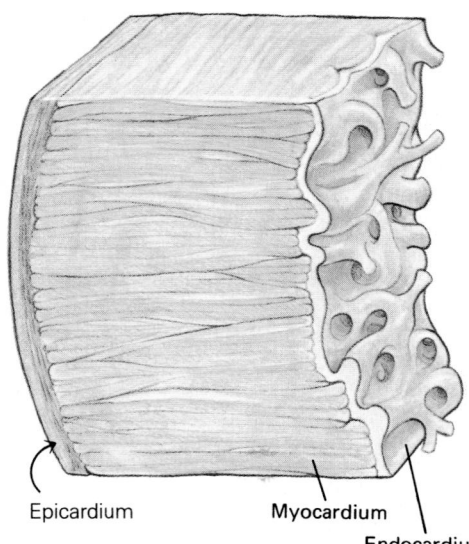

Epicardium   Myocardium

Endocardium

Figure 18–10
Anatomy of the heart wall.

C L I N I C A L   F O C U S

Cardiac Imaging

A number of techniques are available for imaging the heart chambers, heart valves, and coronary arteries. A frequently used noninvasive (not requiring surgery) technique is **echocardiography.** High frequency sound waves are used to image the chamber walls and valves in the heart. The basic principle is the same as that used for radar imaging of airplanes. An echocardiogram (see figure) provides a two-dimensional image of a cross-section through the heart and shows the same structural organization you would see if you cut a heart into two pieces. A small, hand-held device both emits and receives reflected sound waves. This device (a **transducer**) is placed against the chest wall; different cross-sectional views are generated depending on the position of the transducer. This imaging method provides information on the thickness and motion of the muscular walls, the size of the heart chambers, and the functioning of the heart valves. Absence of normal motion in a particular area of the chamber wall can be diagnostic of dead cardiac muscle **(infarct).** Excessively thickened walls frequently indicate that the heart is pumping blood against an abnormally high pressure load. Damaged (scarred, torn, or congenitally malformed) valves will frequently show an abnormal pattern of movement on the echocardiogram. Newer echocardiography techniques enable the clinician to measure the velocity of the blood as it moves through the valve orifices.

The backward leakage of blood through a damaged valve **(valvular insufficiency)** or the high velocity jet of blood moving through an obstructed valve **(valvular stenosis)** is often detected by echocardiography.

Angiography is an invasive technique that requires the insertion of a catheter into the heart **(cardiac catheterization),** followed by an injection of a radiopaque dye. The movement of the dye-bolus

Echocardiogram of a normal heart. (© Phototake.)

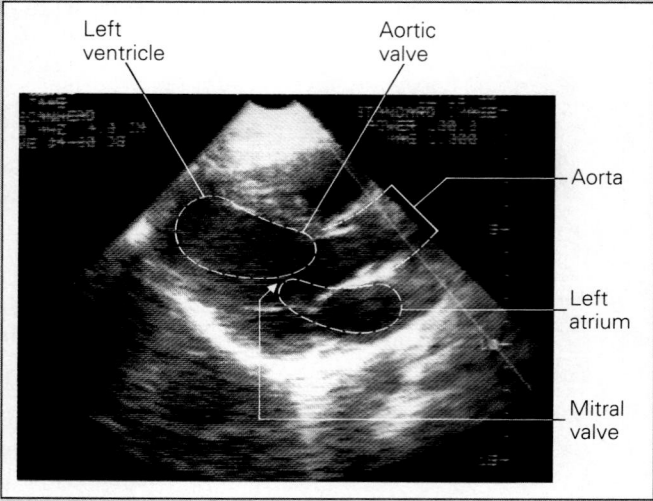

Left ventricle

Aortic valve

Aorta

Left atrium

Mitral valve

involve additional membrane processes. Cells found in the ventricular region have most of the special features that we need to discuss, so a cell of this sort can serve as our typical example.

Figure 18–11 shows such a complex action potential. This is the type of electrical activity found in the **Purkinje fibers** of the ventricles (we will discuss the function of these fibers later). Notice first of all that this action potential lasts hundreds of times longer than one from a nerve. In order to identify the extra voltage changes present, we use numbers to designate the various parts of this action potential. **Phase 0** is the rapid upstroke. This is a very fast depolarization that **overshoots** the zero millivolt level. As in nerve tissue, this rapid depolarization is due to a rapid increase in the membrane permeability to **sodium ions.** Associated with this is a reduc-

tion of the potassium permeability to a value lower than when at rest. As the high sodium permeability begins to decline, a small amount of repolarization takes place. This brief event is **Phase 1.** Shortly thereafter, the membrane potential becomes relatively steady at a value near zero millivolts. This is **Phase 2,** also called the **plateau** of the action potential. Here a slow inward membrane current due to the flow of **calcium ions** is balanced against the efflux (outward flow) of **potassium ions,** even though the potassium permeability is reduced at this time. The membrane channels through which calcium passes also allow a small amount of sodium to pass inward. Late in the plateau, the calcium permeability declines, and with the falling membrane potential, the potassium permeability begins to increase. These factors produce a **repolarization,** which is

through the heart sequentially delineates the chambers. If the catheter is inserted into one of the coronary arteries **(coronary angiography),** the size of the artery lumen can be visualized (see coronary angiography figure). Obstructions of the coronary artery

(e.g., in atherosclerosis and coronary artery spasms) can be easily visualized with this method. Recent developments in computer technology and digital image-enhancement techniques have significantly enhanced the image quality of coronary angiograms.

(*a*) Arteriogram showing a catheter inserted into the right coronary artery. (© CNRI/Phototake.)
(*b*) A schematic of the arteriogram.

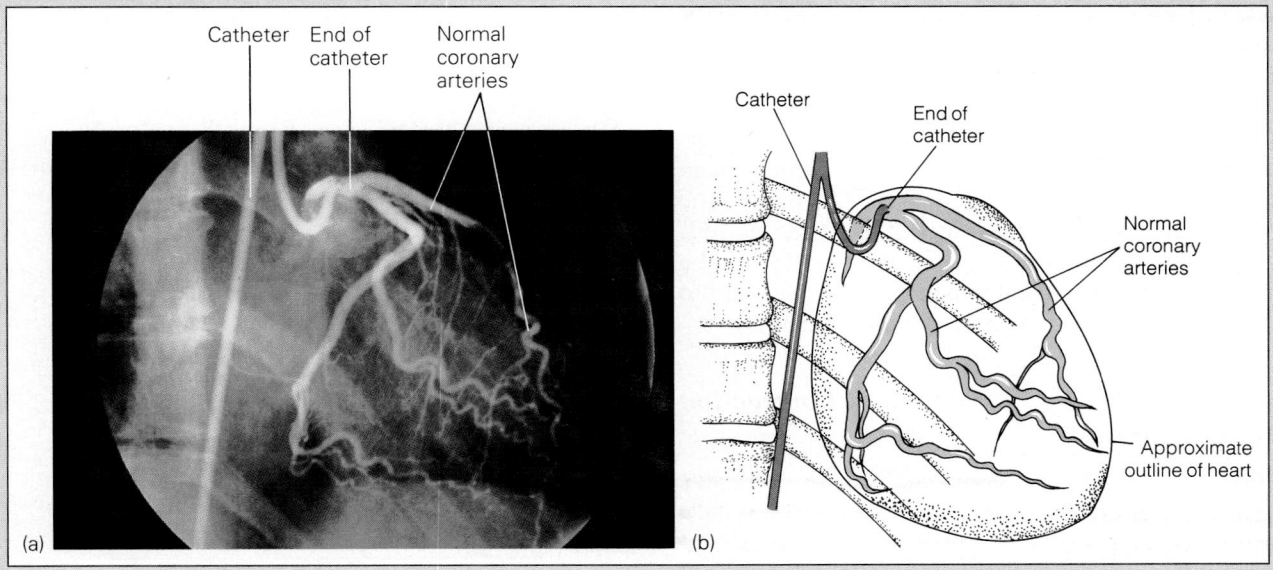

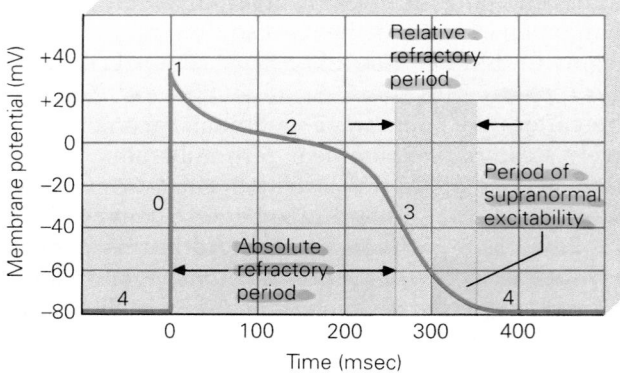

Figure 18–11
A sample action potential from heart muscle, typical of those found in the fast conduction fibers of the ventricle, such as the Purkinje fibers. The numbers above the curve designate the phases described in the text, and the approximate durations of the refractory periods are shown.

called **Phase 3.** During this time, the membrane potential falls to its original value and remains steady until the next beat. This steady portion is **Phase 4,** which is equivalent to the resting potential of other muscle cells. This sequence of membrane potential changes lasts for about 300 milliseconds, depending on the heart rate.

As in nerve and other types of muscles, these phases of the action potential are associated with refractory periods. These are diagrammed in Figure 18–12. During the **absolute refractory period,** the muscle cannot be stimulated again. The **relative refractory period** then follows, and latest of all is the period of **supernormal excitability.** Here the muscle is actually easier to stimulate because the potassium permeability is still a bit lower than normal. The action potential produced will be smaller than normal, because not all of the sodium channels from Phase 0 have had time to "re-set." Normally, the muscle is never stimulated during this time period. If such stimulation were to occur, the result could be very dangerous because the activity could spread to other tissue normally at rest at that time.

The Cardiac Action Potential and Contraction of Cardiac Muscle

Notice in Figure 18–12 that the absolute refractory period lasts almost as long as the isometric twitch. This relationship prevents the muscle from contracting again before it has relaxed, since the muscle will not respond to a second stimulus so soon after the

first. This means that a tetanus is not possible in heart muscle; such a long-lasting contraction would prevent the heart from filling with blood for the next beat, making it a poor pump indeed.

Action potentials are necessary to make heart muscle contract. In skeletal muscle, action potentials are produced as a result of stimulation by a motor nerve. Each cell in a skeletal muscle is individually supplied by its own motor nerve terminal, and coordinated activity is due to the timing of impulses from the central nervous system (CNS).

The heart, in contrast, has no motor nerve supply, and its cells are very small. A separate nerve supply to each of the millions of muscle cells in the heart would make it a huge and complex organ. Instead, the cells of heart muscle communicate with one another directly, and muscle takes over the function of nerve. This means that the heart can be stimulated to contract at a single location, and the message will be passed along to the rest of the heart muscle. Within the individual cells of heart muscle, the process of activation is similar to that in skeletal muscle. The t-tubule system of the ventricular muscle cells aids in conducting the surface action potential inward, where it causes the release of stored calcium for the activation of the troponin-tropomyosin regulation system. Significant amounts of calcium enter the cell during the plateau of the action potential, and this calcium is added to the internal stores.

Figure 18–12
The relationship between the action potential and the twitch. The isometric twitch of a sample of heart muscle is shown in relation to the action potential that caused it. To some extent the duration of the twitch is determined by the duration of the action potential. Agents that prolong the plateau of the action potential make the twitch last longer, and the force of contraction may be increased because of the greater opportunity for calcium to enter the cell during the action potential.

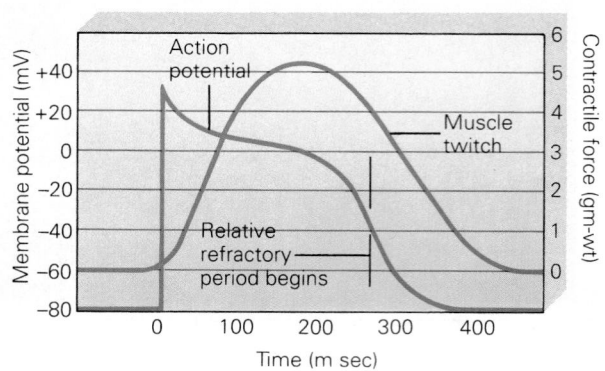

The entering calcium is also directly responsible for causing some release of internal stored calcium from the sarcoplasmic reticulum. Because of this continual entry of calcium, cardiac muscle cells must (and do) have a way of getting rid of excess internal calcium by exchanging it for incoming sodium.

Initiation of the Action Potential by Pacemaker Cells

The single location from which the stimulus for the heart action arises is called the **pacemaker.** Normally, the pacemaker region is the **sinoatrial node** (also called the **SA node**), a specialized region of muscle cells in the right atrium. A pacemaker action potential is shown in Figure 18–13. These cells do not have a steady resting potential. After each action potential, the cells repolarize to a value called the **diastolic potential.** Very soon after repolarization, the membrane begins a slow depolarization. This depolarization quickly brings the membrane potential to its **threshold,** and another action potential takes place. This action is repeated again and

again, as many as 3 billion times in a 70-year lifetime. At resting heart rates, it is repeated approximately 70 times per minute, and it can reach as high as 160 times per minute during heavy exercise. The rate of the heart is regulated by the rate at which the diastolic potential reaches the threshold.

Normally, only one pacemaker is active at one time in the heart. This is because whichever pacemaker cell "fires" first stimulates other cells that are about to produce an action potential. Therefore, the fastest pacemaker cell predominates. The heart actually contains many cells capable of becoming pacemakers, but their activity is not usually expressed because they are inherently slower. Under some conditions, if these "hidden" pacemakers do not receive a stimulus from other cells, their activity will become apparent. As we will see later, this can be an important safety device.

The rate of the SA node pacemaker is subject to control by the **autonomic nervous system.** The **parasympathetic** nervous system decreases the rate of diastolic depolarization and makes the diastolic potentials more negative (by making the cell membranes more permeable to potassium ions). This means that it takes longer for the pacemaker cells to reach threshold, and so the heart rate is decreased. The **sympathetic** nervous system speeds up the diastolic depolarization of the pacemaker cells (and the repolarization as well), and allows the cells to reach threshold sooner; this increases the heart rate. Normally, the resting rate of the heartbeat is somewhat slowed by the parasympathetic nervous system, and if parasympathetic activity is blocked, the heart will speed up. The pacemakers of heart transplant patients (whose hearts are not connected to the autonomic nervous system) can still respond to circulating hormones (e.g., epinephrine) to produce an increase in heart rate.

The rate of the heart varies in response to physiological needs, increasing during exercise and decreasing at rest. If the resting rate is very low (fewer than 60 or so beats per minute), the condition is termed *brachycardia*; an elevated resting heart rate (above 100 per minute) is called *tachycardia*. Normally, these rates are under the control of the SA node pacemaker; in other cases, blockages of conduction elsewhere in the heart may lead to a very serious brachycardia, and abnormal pacemaker activity at locations other than the SA node can lead to tachycardias with rates as high as 300 beats per minute.

Drugs and other influences that affect the rate of the heart are called *chronotropic* influences, while those agents that affect the strength of contractions are called *inotropic*.

Figure 18–13

An action potential from a pacemaker cell. The slow diastolic depolarization (Phase 4) lets the membrane potential reach the threshold voltage shortly after the cell has repolarized, resulting in another action potential. The solid line shows a pacemaker potential at a low heart rate, which is determined by the spacing between successive action potentials. In this case the low rate corresponds to a heart rate of 65 beats per minute. The dotted line shows the rate that would result from a more rapid diastolic depolarization. The action potentials occur closer together, and the new spacing corresponds to a rate of 72 beats per minute.

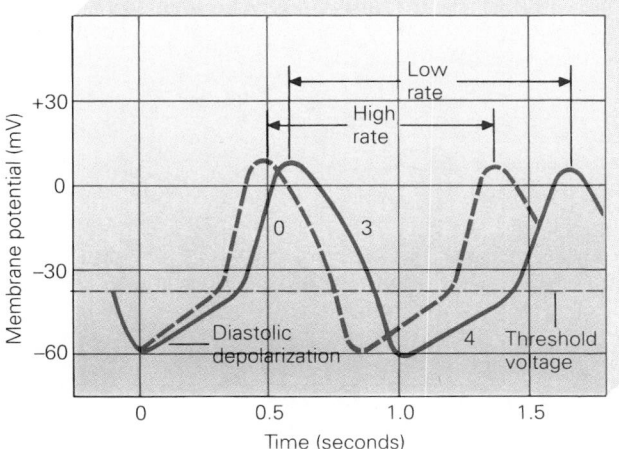

The Sequence of Cardiac Activation

Activation of the Atria.   The rhythmic activity of the SA node pacemaker is conducted first into the musculature of the right atrium and then to the left atrium (Fig. 18–14). This impulse conduction occurs when the action potential in one cell causes current to flow in an adjacent resting cell. This local current, which flows between the cells in the region of their intercalated disks (see Chapter 16), brings the membrane potential of the resting cell to its threshold. It then produces its own action potential, which in turn stimulates the next adjacent cell. This process occurs (with varying degrees of efficiency) in all regions of the heart. Conduction of action potentials between separate cells in heart muscle is similar in many ways to conduction along a nerve axon (a single cell), in which active regions stimulate the resting regions next to them. Cell-to-cell conduction in heart muscle is made possible by the low intercellular resistance associated with the intercalated disks.

Conduction through the atria is quite rapid. Atrial action potentials are conducted quickly because the muscle cells are relatively large and have a very negative resting potential. In addition, some specialized atrial conducting pathways conduct very rapidly. The rapid conduction and fast action potentials result in a very rapid Phase 0 depolarization with a large overshoot in the cells, which produces a large local current flow. These conditions favor the rapid conduction of the impulse throughout the tissue.

Figure 18–14

The succession of action potentials during the course of one heartbeat. The sequence starts with the *pacemaker potential* at the sinoatrial node. The activity spreads rapidly to the *atrial muscle*. The *atrioventricular* node action potential, similar in form to the SA node potential, is conducted slowly, and there is a longer space between this and the first ventricular potential shown, that of the *left bundle branch*. (This is similar to the slightly earlier action potential of the *Bundle of His,* which is not shown.) Action potentials of the *Purkinje fibers* and the *ventricular muscle* follow quickly in succession. The entire activation process takes approximately two-tenths of a second from the time the action potential begins in the SA node until the last ventricular muscle is activated.

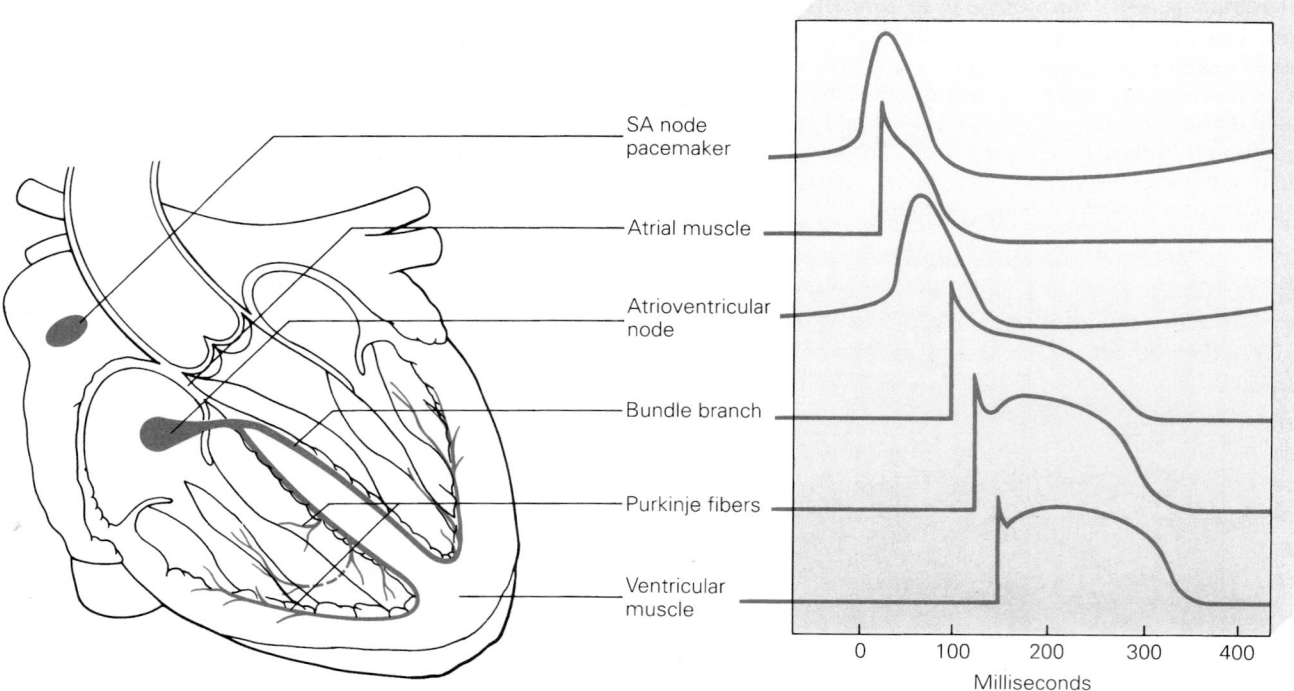

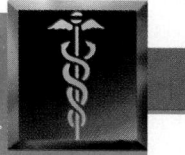

Cardiac Pacemakers

The **artificial cardiac pacemaker** is an electronic device that delivers an electrical stimulus to the heart to initiate contraction. Pacemakers are frequently used to treat patients who suffer from permanent or recurrent bradycardias (slow heart rates) with undesirable symptoms (e.g., fainting). The first totally implantable pacemakers were reported in the 1950s, and today an estimated 1 million patients worldwide have them. Implantable pacemakers are powered by lithium batteries, which may last for 6 to 10 years. Replacement of the battery requires surgical implantation of a new pacemaker.

Early pacemakers were unable to sense the electrical activity of the heart and stimulated it (either the atrium or ventricle) at a fixed rate. Modern pacemakers can sense the heart's normal electrical activity and are activated only when they fail to detect appropriate spontaneous activity.

Such **demand pacemakers** are frequently programmable to respond in a specified manner when a particular arrhythmia is detected. For example, to treat **complete heart block,** a pacemaker is used that senses the depolarization of the atria and then delivers a stimulus to the ventricles after a delay equal to the normal atrioventricular conduction time. This method of pacing retains the normal temporal relationship between atrial and ventricular contraction, and therefore preserves the normal contribution of atrial contraction to ventricular filling. In addition, by sensing the normal rate of atrial depolarization, the pacemaker can respond appropriately to increases in heart rate associated with exercise and stress. Programmable pacemakers can be reprogrammed without being removed, which allows their operating parameters to be revised to fit the patient's changing needs.

Conduction of the Action Potential from the Atria to the Ventricles. At the lower portion of the right atrium is another region of specialized cells called the **atrioventricular (AV) node.** These cells are quite small and have resting potentials that are not as negative as those of the atrial cells. As action potentials pass through this region, they become weaker and travel more slowly. This slowed impulse travel is called **decremental conduction.** If the distance is great enough, or the impulse becomes weak enough, it may die out altogether. Under normal conditions, the impulse is slowed but not stopped at the AV node. The delay this causes ensures that the atria finish their contraction before the ventricles begin theirs. The AV node is the only normal connection between the atria and the ventricles, and its delay serves as a guard against very high atrial heart rates (>200 per minute) being conducted to the ventricles.

Activation of the Ventricles. After the impulse crosses the AV node, it enters the **Bundle of His** (also called the **common bundle**). This conducting structure is composed of muscle cells specialized for very rapid conduction, and here the impulse travels at its greatest speed. The common bundle soon divides into two **bundle branches,** one for each of the ventricles. Conduction is very rapid here as well.

The bundle branches divide further in each ventricle, eventually becoming an array of **Purkinje fibers,** which form a distribution system over the endocardial surface and even penetrate slightly into the main muscle mass of the ventricles. The Purkinje fibers are also specialized muscle fibers with a high speed of conduction.

From the Purkinje fibers, the impulse passes into the muscle of the ventricle wall, stimulating contraction. The ordinary ventricular muscle fibers are somewhat slower conductors than the specialized conducting tissues, but by this time the activity is well distributed throughout the heart, and the distance remaining is relatively short. The arrangement of the specialized conducting system ensures a nearly simultaneous activation of all of the ventricular muscle.

The overall effect of the distribution of the stimulating impulse by the rapid conduction system is a wave of depolarization that sweeps over the heart from its base to its apex and from the endocardial to the epicardial surface. Because of the rapid conduction of the impulse and the fast rise of the action potential, the boundary between active and resting tissue is very sharp and moves very quickly. During repolarization, there is also an apparent movement of the active areas.

As active cells return to rest, a less-pronounced wave of repolarization sweeps over the heart in the opposite direction. However, the directionality of the repolarization is not determined by the conduction system, but by the fact that the cells activated last have the shortest action potentials and begin to repolarize first. This produces an apparent movement in the opposite direction from the wave of depolarization. Because this is not a conducted phenomenon, and because the repolarization is not nearly as rapid as the depolarization, the speed of the repolarization wave is lower and the boundary between recovering and still-active cells is not very sharp.

Abnormal Activation of the Heart. The proper function of the heart as a pump depends on the correct sequence of activation, and this in turn depends on the normal function of the conducting system. For example, failure of the AV node to conduct the impulse will cause **heart block,** in which the ventricles are not stimulated by the atrial impulse. At such times, cells other than those of the SA node can adopt a pacemaker function and provide a source of stimulation, although the rate from such auxiliary pacemakers is usually lower than normal. These "emergency" pacemakers usually lie somewhere in the conduction system, and their activity is normally suppressed by the faster pacemaker rhythm of the SA node. A typical pacemaker from the AV node might have an inherent rate of around 50 beats per minute, while a ventricular pacemaker could have a rate of 30 beats per minute.

Regular muscle tissue rarely shows any spontaneous activity. If there is such a pacemaker in the ventricular region—and if for some reason it becomes active from time to time—its impulse will be conducted both in the normal direction and "backwards" toward the atria. When it stimulates the AV node, it produces refractoriness in the nodal tissue. The next atrial impulse arriving may not be able to be conducted through the AV node, and the ventricles will "drop" a beat. This can be very apparent to the person in whom it happens, because the delay gives the ventricles a chance to become filled with extra blood, and the next heartbeat is larger than usual. Such "skipping a beat" is not rare and is usually of no concern.

However, a dangerous condition may arise if the conducting system of the ventricle is damaged so that conduction along parts of it is slowed. Such

damage is not uncommon following a heart attack (a **myocardial infarction**), because areas of the heart muscle may have been deprived of oxygen for some time. This slowed conduction can cause cells that are in the process of normal repolarization to be stimulated by a late-arriving action potential, which may arrive during the period of supranormal excitability. The resulting stimulation can give rise to even more action potentials out of their proper time sequence, which are then conducted to normally resting cells. The final result may be an uncoordinated stimulation of the ventricular muscle called **ventricular fibrillation.** Such uncoordinated contractile activity is useless for pumping blood, and death will follow shortly. These conduction and rhythm disturbances, and many others related to them, are called **arrhythmias** and are often fatal if not treated immediately.

The Electrocardiogram. The action potentials and their conduction sequence as just described can be observed only by painstaking study of exposed or isolated hearts. While the knowledge gained is important in understanding the operation of the heart, it is obviously not a realistic approach to the study or treatment of human (or veterinary) patients. Fortunately, the electrical activity of the heart may be harmlessly detected at the body surface by a technique called **electrocardiography.**

The heart may be thought of as an electrical generator lying in a conducting medium made up of the body tissue and fluids. During the phases of the cardiac cycle in which depolarization or repolarization is sweeping over the heart muscle mass, some portions are positively charged and others negatively charged. This potential difference, or **voltage gradient** (as much as 100 millivolts), causes current to flow in the external medium between these regions of the heart. These currents are strongest close to the heart, and they become weaker at greater distances from the heart. The largest voltage measured at the body surface is only around 1 millivolt; however, this is sufficient to be detected by a sensitive instrument called an **electrocardiograph.** This device converts the minute surface currents to movements of a pen on paper or a spot of light on a cathode ray tube (CRT), producing the familiar **electrocardiogram** (**ECG** or **EKG**).

Figure 18–15 is a diagram of a typical electrocardiographic recording. For convenience, the principal features of the recording have been designated by the letters P, Q, R, S, and T. The so-called **P-wave** is caused by the depolarization of the atrial muscle, and the **QRS complex** is caused by the depolariza-

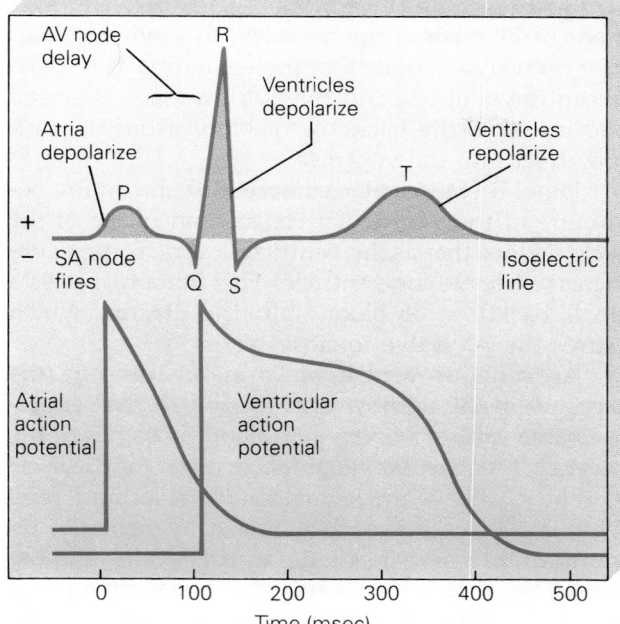

Figure 18–15
The time relationship between the action potentials and the electrocardiogram. All three traces share the same time axis, and the ECG recording may be used to infer when the actual electrical changes took place in the heart muscle.

tion of the ventricles. The **T-wave** is the result of the repolarization of the ventricles. Repolarization of the atria takes place during the ventricular depolarization, and its weak signal is lost. When the entire heart is either depolarized (active) or is resting, there is no potential difference to set up a current flow, and no deflection is seen in the ECG. The line of no deflection is called the **isoelectric line.** Figure 18–15 also shows the relationship between cellular action potentials and the surface ECG.

It is important to realize what the ECG can and cannot show. First, only electrical events are shown, and one can only *infer* that a contraction of the muscle has taken place (sometimes it has not). The ECG is only a record of **voltage** and **time.** Secondly, only the electrical events that involve large amounts of muscle show up in the ECG record. Because of the small tissue mass involved, activity of the conducting system and nodal tissues can only be inferred, but these inferences can be very revealing. For example, if a P-wave is not followed by a QRS complex, then one can assume that conduction was blocked somewhere between the atria and the ven-

tricular muscle. If the QRS complex is very broad or abnormally shaped, one can assume that some areas of the muscle of the ventricle were activated later than they should have been. This implies a defect in the ventricular conducting system. A very long delay between the P-wave and the QRS complex most probably means that the conduction in the AV node was more decremental than usual.

Further information can be gained about the direction of the spread of electrical activity by recording from electrodes located on different areas of the body. The portion of the electrical activity that the recording electrodes intercept depends on their location and orientation. Clinical recording of the ECG assumes that the heart lies at the center of an equilateral triangle and that the recording connections are made at the vertices of this triangle (Fig. 18–16). In practice, the electrical connections to the right and left arms and to the left leg are considered to connect to these vertices, and the standard **limb leads** are recordings between any two of the connections. A recording made between the right and left arms is designated as **Lead I** and is most sensitive to activity spreading laterally across the heart. **Leads II** and **III** are recordings made between the left leg and the right and left arms, respectively. Records from these leads are most sensitive to activity proceeding from the top to the bottom (base to apex) of the heart. By combining the timing and voltage information recorded from a number of leads, someone skilled in the interpretation of the ECG can build up a very complete picture of the functioning of the heart.

Figure 18–16
Standard electrocardiographic limb leads.

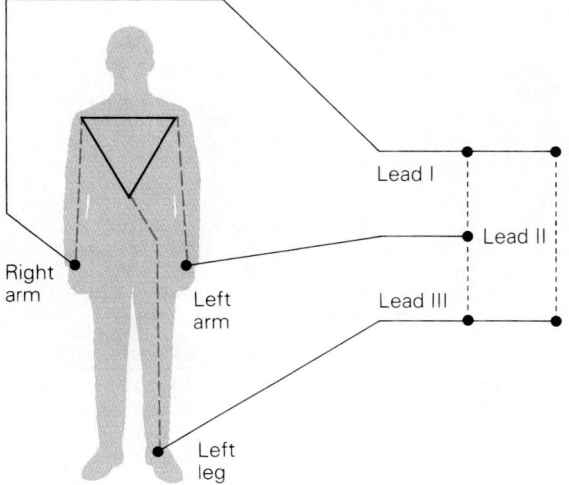

The Ventricular Pump and the Cardiac Cycle

As discussed in the previous section, the electrical activation of the left and right hearts occurs almost simultaneously. As a result, the sequence of mechanical events occurring during a single beat or cycle of the heart is very similar for both hearts. Because of this similarity, portions of the following discussion of the cardiac cycle will focus on the left heart. Keep in mind that similar events are occurring at the same time in the right heart.

The Ventricular Pump

Let us first focus on the ventricles of the heart, and examine how the ventricles function as a pump and how the heart valves ensure that blood flows in one direction. Figure 18–17 shows schematically the four basic steps in the pumping process. Beginning with Panel A **(ventricular filling),** we see that the ventricle is being filled with blood from the atrium.

Contraction of the atrial muscle facilitates the movement of blood into the ventricle through the one-way AV valve. Notice that the semilunar valve, between the ventricle and the outflow tract, is closed and prevents the backflow of blood from the outflow tract into the ventricle.

Panel B **(ventricular contraction)** shows the beginning of the ventricular contraction phase of the cycle. Notice that as the ventricle contracts, pressure increases inside the ventricle. This initially causes a small backflow of blood into the atrium, which causes the AV valve to close.

As soon as ventricular pressure exceeds the pressure in the outflow tract (Panel C), the semilunar valve is forced open, and blood flows out of the ventricle **(ventricular ejection).** In order for the cycle to repeat, the ventricle must now relax and refill with blood. As the ventricle relaxes **(ventricular relaxation),** pressure inside the ventricle falls, and the semilunar valve is forced closed as blood attempts to flow from the outflow track back into the ventricle (Panel D). When ventricular pressure falls below

Figure 18–17
The cardiac cycle. Diagrammatic depiction of a single pumping cycle of the ventricle. This illustrates the relative volume changes that occur in the atrium and ventricle during a single pumping cycle, as well as the opening and closing sequence of the inflow (atrioventricular) and outflow (semilunar) valves. Arrows indicate direction of blood flow. The intensity of blue color indicates relative pressures. Purple indicates contracting myocardium.

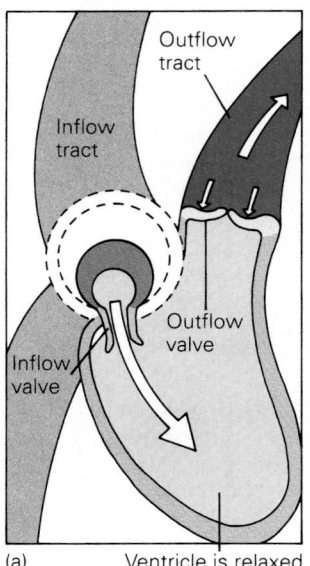

VENTRICULAR FILLING

(a)   Ventricle is relaxed and full

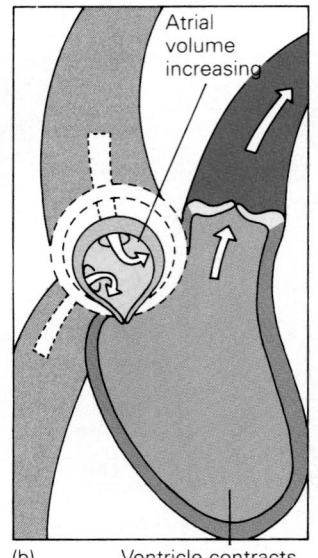

VENTRICULAR CONTRACTION

(b)   Ventricle contracts, but all valves are closed. Thus, pressure increases but volume remains constant

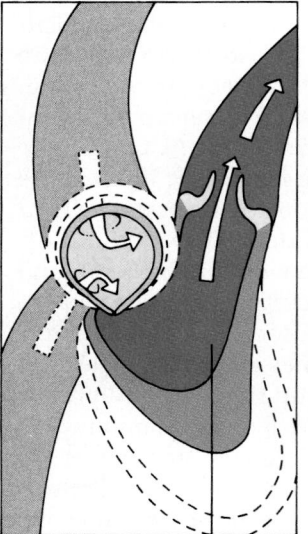

VENTRICULAR EJECTION

(c)   When ventricular pressure exceeds that in outflow tract, valves open

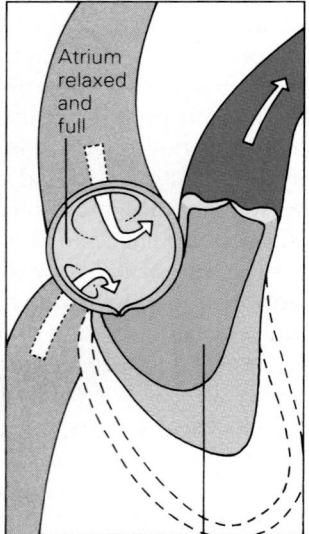

VENTRICULAR RELAXATION

(d)   Ventricle relaxes, but all valves are closed. Thus pressure decreases, but volume remains constant

that of the atrium, the AV valve opens and blood flows into the ventricle (Panel A). This completes the cycle.

The Cardiac Cycle

You will notice that the ventricle is the main pumping chamber of the heart. The atrium serves as a reservoir to receive the continuous flow of blood back to the heart. Although the atrium does contract, its contraction adds very little to the filling of the ventricle under normal resting conditions. Now let's look in greater detail at the sequence of events occurring during a single cardiac cycle as they occur in the heart. The sequence of events is depicted in Figure 18–18.

Ventricular Filling. The term **diastole** refers to the period of the cycle during which the ventricles are filling with blood, and **systole** refers to the portion of the cycle during which the ventricles are actively contracting and pumping blood out of the heart. Perhaps the easiest place to start our examination of the cardiac cycle is during the **ventricular filling** phase of diastole (see Fig. 18–18, Panel 1). At this point in the cycle, the ventricular muscle is completely relaxed, and the pressure inside the ventricles is consequently low. The blood pressure in the atria exceeds that in the ventricles because of the continuous flow of blood into the atria from the veins. This forward pressure gradient forces the AV valves open and causes blood to flow into the ventricles.

Ventricular filling occurs rapidly at first, as the accumulated blood in the atria flows into the relaxed ventricles. Atrial contraction occurs at the end of the ventricular filling phase, immediately prior to the onset of ventricular contraction. The P-wave of the ECG signals the onset of atrial depolarization. Keep in mind that the electrical depolarization precedes the mechanical contraction of the atria by a few milliseconds because of the latency period required for muscle activation. Normally, atrial contraction increases the filling of the ventricle by only about 10% to 30%. This is because adequate time is available during diastole for nearly complete filling of the ventricles prior to the onset of atrial contraction. Under conditions of high heart rate, however, the time available for ventricular filling is reduced, and the atrial contraction becomes an important contributor to complete ventricular filling (Fig. 18–19).

The end of ventricular filling corresponds to the end of diastole. The volume of blood in the ventricle at this time is referred to as the **end-diastolic volume.** As we will see in a later section, the end-diastolic volume is an important determinant of cardiac function. Notice also that there is a reverse pressure gradient across the semilunar valves during the ventricular filling phase (aortic pressure > left ventricular pressure). This reverse pressure gradient forces the semilunar valves to remain closed during this period of the cycle. This prevents blood that was pumped into the aorta during the previous cycle from flowing back into the heart.

Ventricular Contraction. The initiation of ventricular contraction marks the beginning of systole. Recall that the specialized conduction system results in rapid activation of the ventricles following atrial contraction. This electrical activation of the ventricles is reflected by the QRS complex of the ECG. Shortly after the occurrence of the QRS complex, the ventricular muscle begins to contract; this contraction results in an increase in ventricular blood pressure. Ventricular blood pressure quickly rises to a value that is greater than atrial pressure. This reverse pressure gradient causes the atrioventricular valves to close, preventing blood from being pumped out of the heart and back into the atria. During this phase of the cardiac cycle, both the AV and semilunar valves are closed. This phase is referred to as **isovolumetric ventricular contraction** (see Fig. 18–18, Panel 3) because the volume of the ventricles cannot change while both the inflow and outflow valves remain closed. If the ventricular volume cannot change, then the cardiac muscle fibers are contracting isometrically during this period. When pressure inside the left ventricle has increased to a level that exceeds that in the aorta, there is a forward pressure gradient for blood flow from the ventricle to the aorta. The aortic valve opens in response to this forward flow of blood, and the **ventricular ejection** (see Fig. 18–18, Panel 4) phase of the cardiac cycle begins. The ejection of blood from the left ventricle occurs quite rapidly at first and then slows as pressure builds up in the aorta. Just as pressure increases inside a balloon as it is inflated with air, pressure increases inside the aorta as it is inflated with blood. Keep in mind that blood is constantly flowing out of the aorta into the systemic vasculature, even as it is being pumped from the heart into the aorta. The rise in aortic pressure during ventricular ejection indicates that the flow of blood into the aorta from the heart is greater than the flow of blood out of the aorta into the systemic circulation.

Ventricular Relaxation. The onset of ventricular relaxation immediately follows ventricular repolarization. Recall that the T-wave of the ECG is a measure of the electrical repolarization of the myocar-

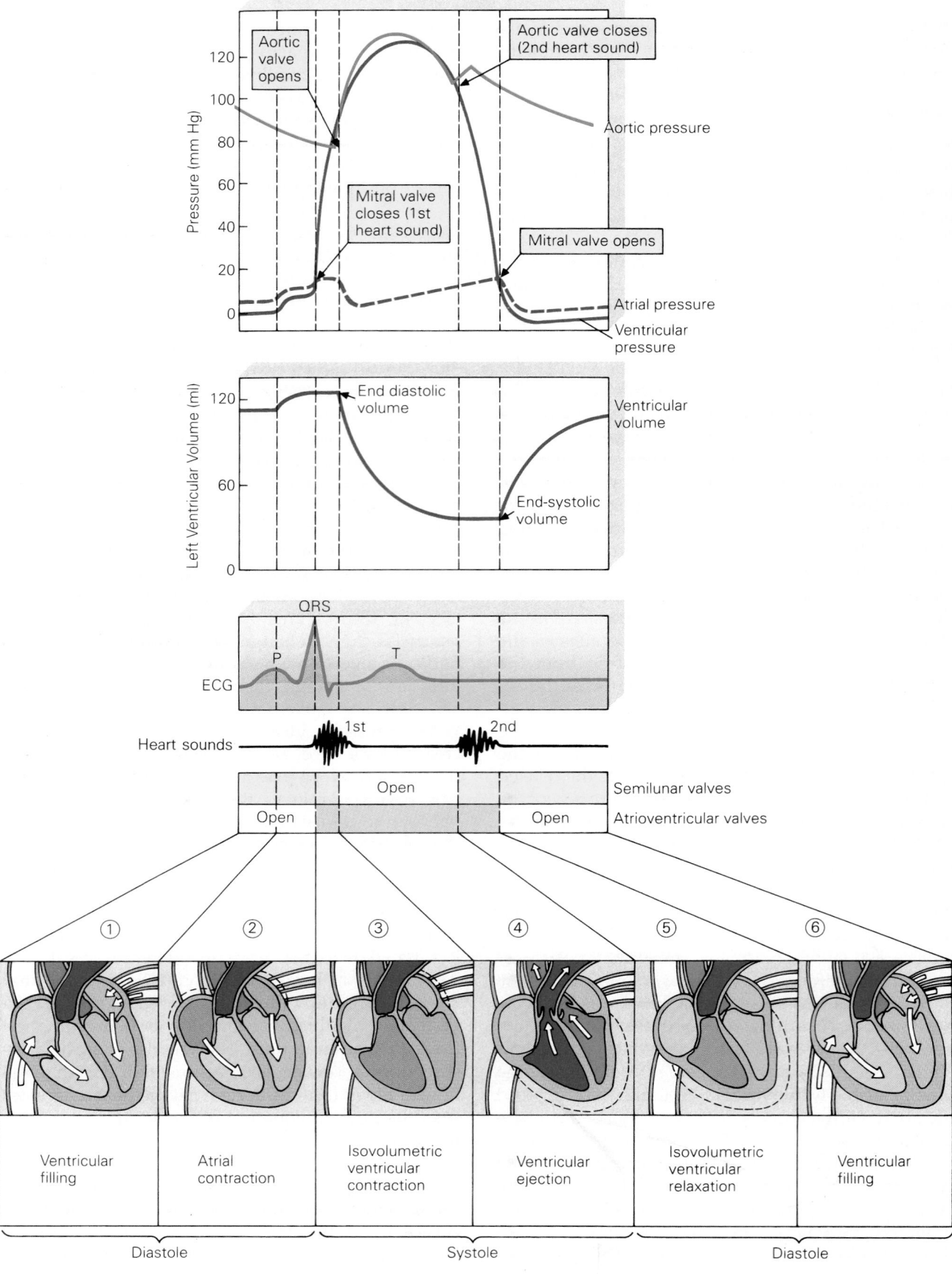

Aortic valve opens

Aortic valve closes (2nd heart sound)

Aortic pressure

Mitral valve closes (1st heart sound)

Mitral valve opens

Atrial pressure

Ventricular pressure

Pressure (mm Hg)

120 · 100 · 80 · 60 · 40 · 20 · 0

End diastolic volume

Ventricular volume

End-systolic volume

Left Ventricular Volume (ml)

120 · 60 · 0

QRS

P

T

ECG

1st   2nd

Heart sounds

Open — Semilunar valves

Open   Open — Atrioventricular valves

① Ventricular filling

② Atrial contraction

③ Isovolumetric ventricular contraction

④ Ventricular ejection

⑤ Isovolumetric ventricular relaxation

⑥ Ventricular filling

Diastole   Systole   Diastole

Figure 18–18
Events of the cardiac cycle. The graph in the upper portion of this figure shows a time-course of the pressure measured in the aorta and left ventricle during a single cardiac cycle. Also shown are the occurrence of heart sounds, left ventricular volume, and an electrocardiogram (ECG). The state of the heart valves during the cardiac cycle is shown at the bottom of the graph. The six panels at the bottom illustrate the direction of blood flow (arrows) within the heart and connecting vessels at each stage of the cardiac cycle. Higher pressures are indicated by darker colors. Purple indicates contracting myocardium.

dium. Notice in Figure 18–18 that ventricular pressure begins to fall rapidly after the occurrence of the T-wave and is soon less than the blood pressure in the aorta. This establishes a reverse pressure gradient (aortic > ventricular), which reverses the direction of blood flow in the region of the aortic valve. The aortic valve quickly closes in response to this reversal of blood flow, preventing any further flow of blood back into the heart from the aorta. Closure of the aortic valve coincides with the **diacrotic notch,** or **incisura,** of the aortic pressure tracing (Fig. 18–18). The closure of the semilunar

Figure 18–19
Effect of heart rate on diastolic filling. Atrial contraction makes only a small contribution to ventricular filling under normal conditions (upper panel). At high heart rates (lower panel), such as those occurring during heavy exercise, atrial contraction contributes significantly to ventricular filling because of the reduced time available for filling. The ventricular volume added during atrial contraction is indicated by the dark pink area.

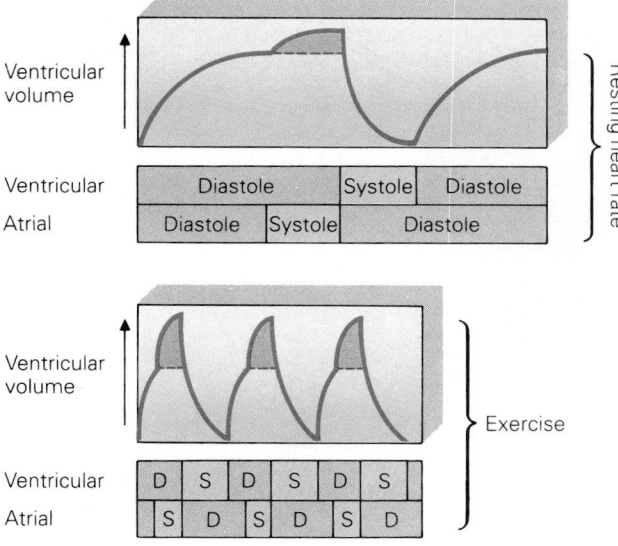

valve marks the end of systole and the beginning of diastole.

The first phase of diastole is referred to as **isovolumetric relaxation** (see Fig. 18–18, Panel 5); relaxation of the ventricles occurs at constant volume because both the AV and semilunar valves are closed. As soon as ventricular pressure falls below atrial pressure, the forward pressure gradient from the atria to the ventricles initiates blood flow into the ventricles through the one-way AV valve (see Fig. 18–18, Panel 6).

Heart Sounds

One of the physical manifestations of the intense mechanical activity occurring with each contractile cycle of the heart is a series of distinct heart sounds that correlate with specific events of the cardiac cycle. These sounds can normally be heard through a stethoscope placed against the chest, directly over the heart. Heart sounds can also be recorded electronically, as shown by the **phonocardiogram** in Figure 18–18.

Although four heart sounds are distinguishable in most healthy individuals, the first two are the most distinct. The **first heart sound** occurs at the time of closure of the AV valves, and, as previously mentioned, marks the beginning of systole. The **second heart sound** occurs at the time of closure of the semilunar valves and defines the end of systole. These sounds are the result of vibrations in the heart and chest initiated by the closure of the heart valves and the sudden cessation of blood flow. If you have ever "snapped" open a paper sack by suddenly filling it with air, you can appreciate how rapid changes in the volume of the heart might lead to audible sounds.

A number of additional heart sounds associated with abnormal conditions of the heart can occur. For example, valvular obstructions can result in **murmurs,** owing to a high velocity jet of blood flowing through the narrowed opening of an obstructed valve. Likewise, valve abnormalities that allow blood to leak backwards (e.g., from the aorta to the ventricle) are often associated with audible murmurs. The interpretation of abnormal heart sounds is based on the quality and intensity of the sound as well as on the timing of the sound relative to other normal heart sounds.

Cardiac Output

As we have already stated, the major purpose of the heart is to maintain a constant flow of blood through the pulmonary and systemic circulations. The amount of blood pumped out of the left ventricle and into the aorta during a period of one minute is defined as the **cardiac output.**

Recall that the left and right hearts are in series and must pump exactly the same volume of blood over any extended period of time. Therefore, if the average output of the left heart is 5 L/min, then the average output of the right heart must also be 5 L/min. Even though the right heart generates significantly lower pressures during systole than the left heart, the volume of blood pumped by the two hearts must be the same.

The average cardiac output for a resting male is about 5 L/min; however, this value can increase or decrease significantly in response to the changing needs of the body or in response to disease. In order to understand how the cardiac output can be altered, we must understand what determines its magnitude.

We know from our previous discussion that the heart is a pulsatile pump that ejects a volume of blood with each beat or cycle. The volume of blood ejected in a single beat is called the **stroke volume** (Fig. 18–20). The cardiac output is simply the sum of all the stroke volumes ejected from the heart over a period of one minute. If we know the stroke volume (SV) and the heart rate (HR) (number of beats in one minute), then we can calculate cardiac output (CO) as follows:

$$CO = HR \times SV$$

An average adult at rest might have a heart rate of 72 beats/min and a stroke volume of 70 ml/beat (70 ml = 0.070 L or 2.4 oz.). On this basis, the cardiac output would be:

$$CO = 72 \text{ beats/min} \times 0.070 \text{ L/beat} = 5.0 \text{ L/min}$$

The total volume of blood contained in all the blood vessels and the heart of a 70-kg male is about 5 L, or slightly more than 1 gallon. Therefore, the entire blood volume is pumped through the heart each minute. This means that, on average, a blood cell will require 1 minute to travel from the left heart, through the systemic circulation, through the right heart, through the pulmonary circulation, and back to the left heart. It is interesting to consider that over a 70-year life span the heart will pump about 200 million L of blood. This is approximately 53 million gallons—enough to fill 564 million 12-oz cans, which, if placed end-to-end, would circle the earth twice.

Figure 18–20
Stroke volume. The stroke volume is the volume of blood ejected from either ventricle during systole. The stroke volumes for both the left and right hearts must, on average, be exactly equal. Notice that a significant volume of blood remains in the heart following contraction (end-systolic volume). Purple indicates contracting muscle.

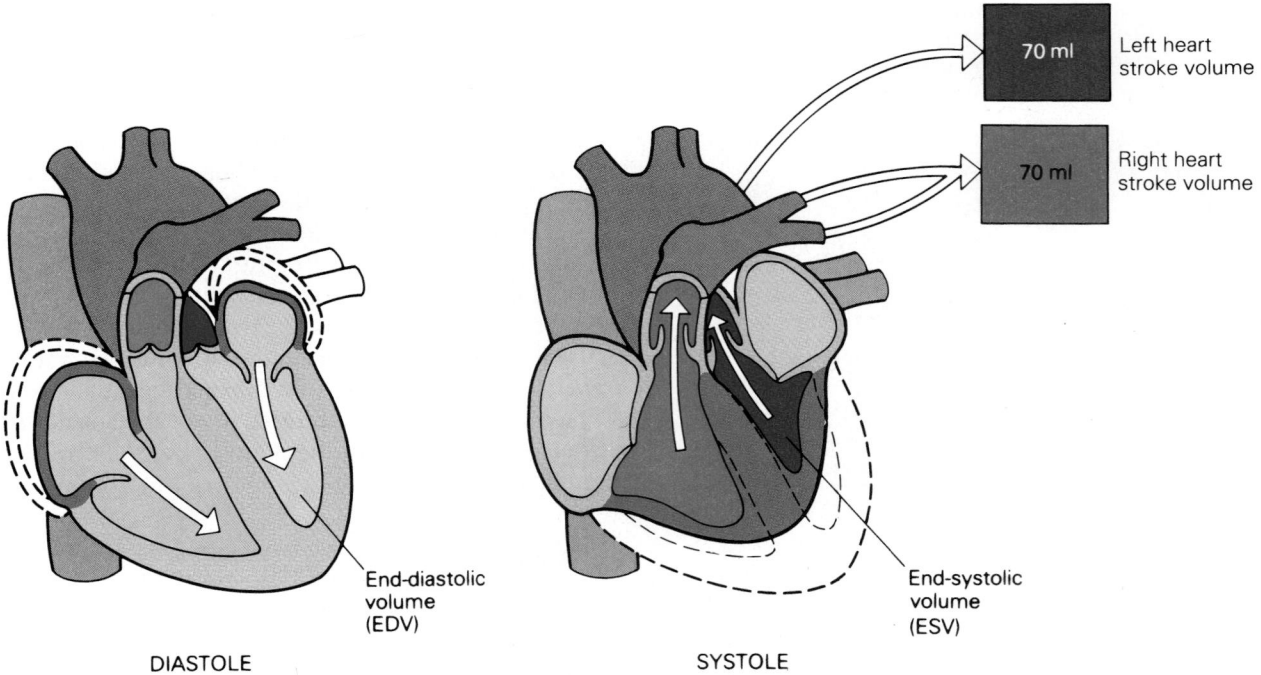

During strenuous exercise, cardiac output may increase as much as five times. From the equation for calculating cardiac output, we can see that changes in either stroke volume or heart rate can alter cardiac output. During strenuous exercise, for example, heart rate may increase to 180 beats/min and stroke volume can almost double. Let's look at each of these factors in turn and the mechanisms by which they are regulated.

Regulation of Heart Rate

You will recall from the previous discussion of the electrical activation of the heart that heart rate is determined by the rate of spontaneous depolarization of the SA node. The rate of SA nodal depolarization can be altered by a number of influences, including stimulation of autonomic nerve fibers that innervate the SA mode, circulating hormones such as epinephrine, plasma electrolyte concentrations, and body temperature.

Let's consider the effects of autonomous nerve activity on heart rate. Figure 18–21 shows how an increase in the firing rate of the sympathetic and parasympathetic cardiac nerve fibers alters the rate of spontaneous depolarization of the SA node. Norepinephrine, the sympathetic neurotransmitter, increases the rate of spontaneous depolarization of the SA nodal cells so that threshold is reached more quickly and the heart rate is increased. Acetylcholine, the parasympathetic neurotransmitter, in contrast, decreases the rate of spontaneous depolarization and, in addition, hyperpolarizes the pacemaker cells. These two effects of parasympathetic stimulation work together to increase the time required for the membrane potential to reach threshold with a resultant decrease in heart rate. Under resting conditions, both the parasympathetic and sympathetic cardiac fibers are active. If the heart is denervated, either surgically or pharmacologically, the heart rate

will increase to about 100 beats/min. The fact that the resting rate for an innervated heart is 70 beats/min indicates that under resting conditions, the parasympathetic nerves are the dominant regulatory influence on the SA node.

There are definite limits on how much cardiac output can be increased by increasing heart rate. First, the upper limit for conduction of impulses through the AV node restricts the upper heart rate to about 250 beats/min. Secondly, with rapid heart rates (tachycardia) of more than about 170 beats/min, cardiac output will begin to decrease because of inadequate time for complete filling of the ventricles during diastole. If the amount of blood that returns to the ventricle during diastole is decreased, we might suspect that the amount of blood pumped by the heart (stroke volume) would also be decreased. As we will see in the next section, stroke volume is *very* dependent on ventricular filling. Therefore, as heart rates reach very high values, the decrease in stroke volume can more than offset the effects of the increase in heart rate, resulting in a net decrease in cardiac output (see Fig. 18–25).

Changes in body temperature also produce changes in heart rate. This is of little use for normal regulation of heart rate, since body temperature remains relatively constant; however, cardiac surgeons will often decrease the body temperature to reduce the heart rate (bradycardia) and the motion of the heart during surgery.

Regulation of Stroke Volume

The volume of blood pumped by the heart with each beat is highly regulated by mechanisms that are **intrinsic** to cardiac muscle itself, as well as by **extrinsic** factors such as neural stimulation and circulating hormones. Let's begin by looking at the intrinsic mechanism for the regulation of stroke volume.

Intrinsic Regulation of Stroke Volume. The heart has the intrinsic ability to adjust its output (stroke volume) in response to changes in the input (venous return). This property of the heart is called the **Frank-Starling Law of the Heart** in honor of two physiologists, Otto Frank and Ernest Starling. As shown in Figure 18–22, there is a proportional relationship between the diastolic volume of the heart and the stroke volume over a relatively wide range of end-diastolic volumes. If the rate of blood flow to the heart increases, the end-diastolic volume of the heart will also increase, causing an increase in the stroke volume. Simply stated, the Frank-Starling law says that the heart will pump whatever volume of blood it receives.

Figure 18–21
Effect of autonomic nerve activity on spontaneous depolarization of the SA node.

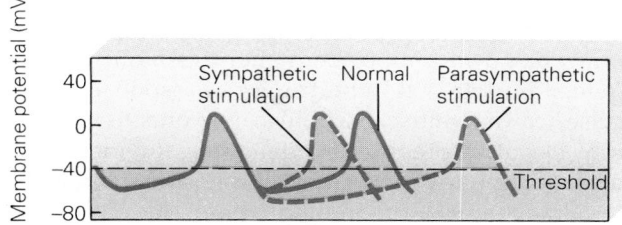

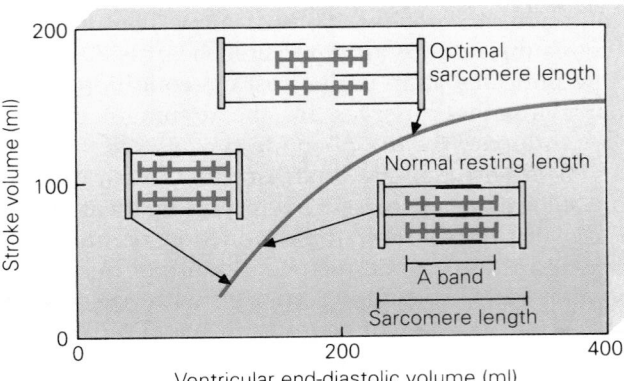

Figure 18–22
Frank-Starling law of the heart. This graph illustrates the relationship between stroke volume and changes in ventricular end-diastolic volume. The insets, showing diagrammatic sarcomeres, illustrate the relationship between end-diastolic volume and myofilament overlap. At normal resting ventricular volumes, sarcomere length is less than the optimal length for contraction.

The mechanism underlying the Frank-Starling relationship is a familiar one. When the end-diastolic volume of the heart is increased, the length of the cardiac muscle fibers in the wall of the ventricle is also increased (see Fig. 18–22). Recall from Chapter 16 that cardiac muscle, like skeletal muscle, has a well-defined length-tension relationship, which indicates that there is an optimal length for muscle contraction (see Fig. 16–13). Although skeletal muscle is generally constrained by the skeletal attachments so that it operates near the optimal length for contraction, this is not true of cardiac muscle. In fact, in a normal resting heart, the muscle fibers are *shorter* than their optimal length. Therefore, an increase in end-diastolic volume increases the muscle fiber length toward the optimal length for contraction. This increases the vigor of ventricular contraction, resulting in the expulsion of a greater stroke volume from the heart. Conversely, a decrease in end-diastolic volume results in an immediate reduction in the vigor of contraction and a decrease in the stroke volume. Therefore, the Frank-Starling mechanism ensures that the heart adjusts the vigor of contraction, on a beat-by-beat basis, in order to pump out the same volume that it receives. This mechanism regulates each ventricle independently.

Although the full significance of this mechanism for regulation of cardiac output will become apparent as we discuss the overall regulation of the cardiovascular system, one aspect can be appreciated now, based on our previous discussion of the serial arrangement of the two hearts and the systemic and pulmonary circulations. Because of this arrangement, it is vital that the two hearts each pump exactly the same volume of blood over any extended period of time. If this were not the case, excess blood would soon accumulate either in the lungs or in the systemic circulation. If the output of the right heart exceeded that of the left heart by only 1 ml/beat, within 90 minutes the entire blood volume would have accumulated in the lungs. Although such an extreme situation would never actually occur, this example demonstrates how rapidly even small imbalances in the stroke volumes of the two hearts could lead to the accumulation of blood in one portion of the cardiovascular system. An increase in the amount of blood in the pulmonary vasculature can compromise gas exchange in the lungs, with serious and often fatal results. The ability of the two hearts to independently self-regulate their output to match their input ensures that the cardiac output of the two hearts remains exactly equal. Later, we will discuss how the Frank-Starling mechanism contributes to the maintenance of normal arterial blood pressure.

The Frank-Starling mechanism also provides intrinsic regulation of cardiac output in response to changes in aortic pressure. Consider the effect on the heart of a sudden increase in mean aortic pressure from 90 mm Hg to 120 mm Hg. In order for the heart to eject blood, the ventricular pressure must now increase to 120 mm Hg before the aortic valve will open, and blood must be ejected against a greater pressure. As a result of this increased pressure load, the muscle fibers shorten more slowly, allowing for less blood to be ejected. If less blood is ejected (reduced stroke volume), more will remain in the heart at the end of systole. This extra blood in the heart, when added to the normal diastolic filling from the pulmonary vein, results in an increase in the end-diastolic volume of the heart. The Frank-Starling mechanism provides for a more forceful contraction on the next beat of the heart and results in an increase in the stroke volume back toward normal. In this manner, the Frank-Starling mechanism helps to maintain a constant cardiac output in the face of changes in aortic blood pressure.

If aortic blood pressure remains chronically elevated, the heart can self-regulate its capacity to pump blood in a more permanent fashion. Just as skeletal muscle will **hypertrophy** in response to exercise, cardiac muscle will also respond to chronically elevated pressure or demands for increased cardiac output by increasing the size and contractile protein content of the individual muscle cells of the heart.

Cardiac hypertrophy of both ventricles is readily apparent in athletes (marathon runners, for example) who perform intensive aerobic exercise. Cardiac hypertrophy can also be a response to an abnormal load placed on the heart because of disease. In many cases, this results in selective hypertrophy of a single chamber of the heart. For example, blockage of the pulmonic valve **(pulmonic stenosis)** not only results in a systolic murmur but also in hypertrophy of the right ventricle, because the right heart must generate higher-than-normal pressures to force blood through the partially occluded valve. The mechanism by which an increased load on the heart ultimately leads to increased production of contractile proteins by the myocardial cells is not currently understood.

Extrinsic Regulation of Stroke Volume.   A number of factors extrinsic to the heart can also influence the vigor of ventricular contraction without changing end-diastolic volume. Any changes in the vigor of cardiac contraction that occur independently of changes in end-diastolic volume are referred to as changes in **contractility.** A change in contractility is mechanistically different from the increased vigor of contraction seen with changes in muscle length. Changes in the contractility of the heart are the direct result of changes in the rate and extent of calcium movement into the cytoplasm, as indicated diagrammatically in Figure 18–23. Recall from Chapter 16 that the concentration of calcium ions in the cytoplasm determines the degree of muscle activation. Increased firing of cardiac sympathetic nerve

Figure 18–23
Relationship between contractility and intracellular calcium. An increase in contractility is the result of an increase in the cytoplasmic calcium concentration. This is the result of both an increased release of calcium from the sarcoplasmic reticulum and an increased influx of calcium into the cell from the extracellular space. The increased concentration of intracellular calcium results in the activation of additional crossbridges, with a resultant increase in the vigor of cardiac muscle contraction. Active crossbridges are indicated in red.

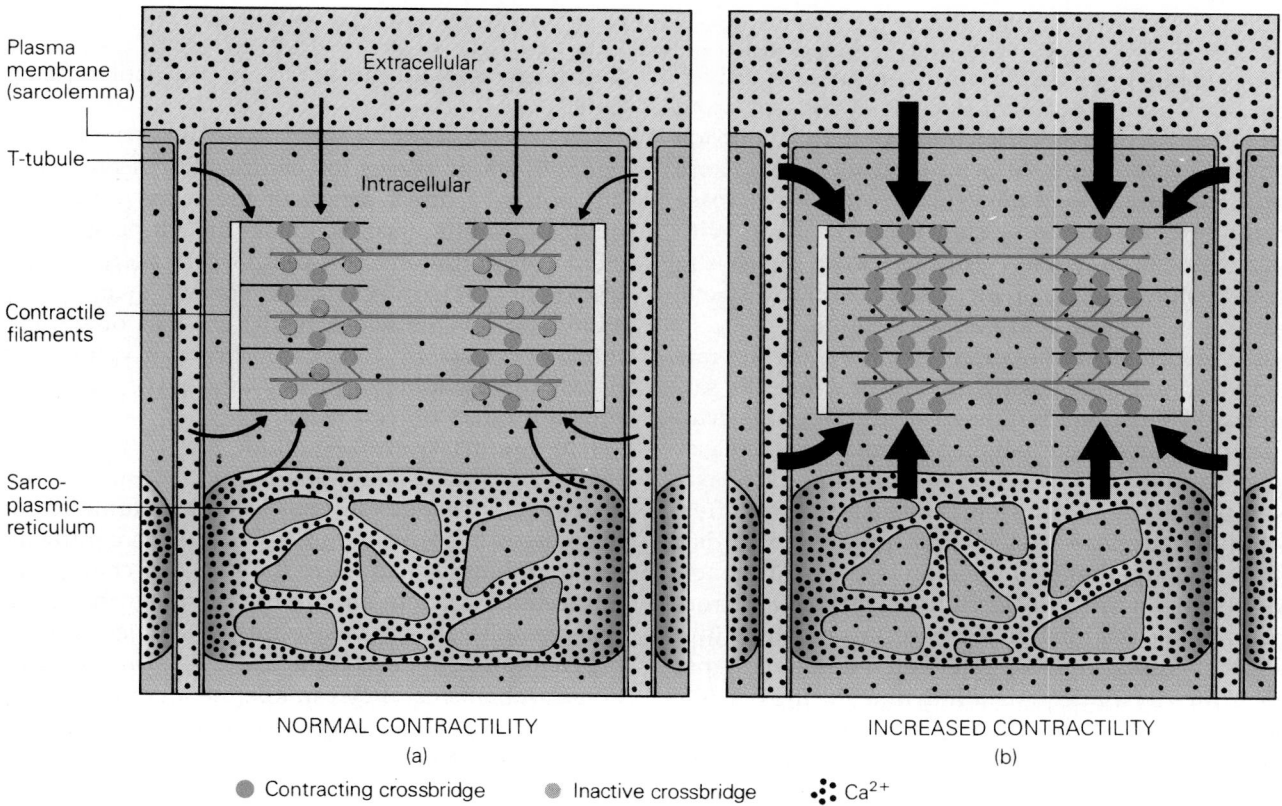

Plasma membrane (sarcolemma)

T-tubule

Contractile filaments

Sarco-plasmic reticulum

Extracellular

Intracellular

NORMAL CONTRACTILITY
(a)

INCREASED CONTRACTILITY
(b)

● Contracting crossbridge   ● Inactive crossbridge Ca^{2+}

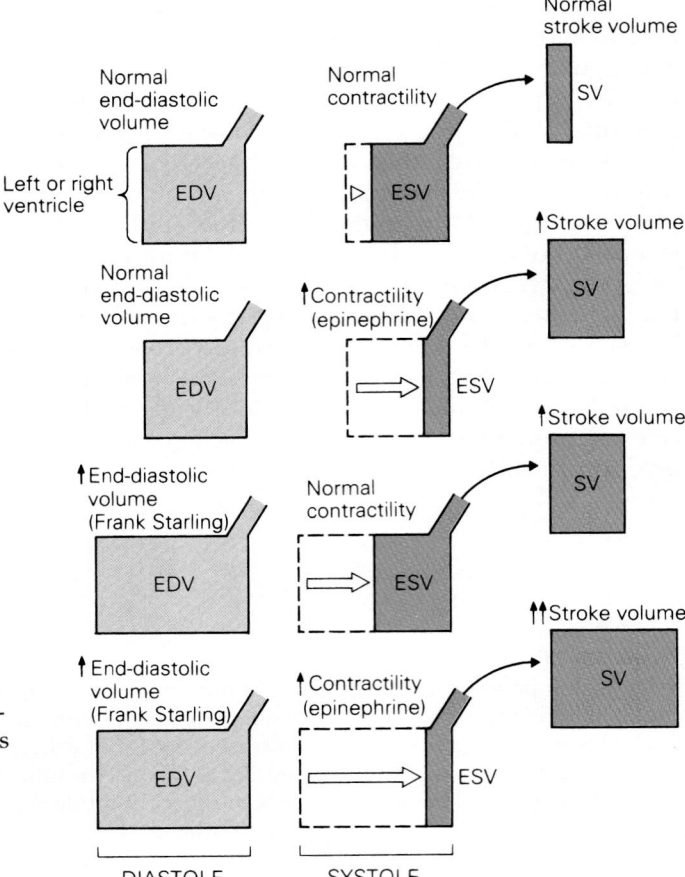

Figure 18–24
Effect of increasing stroke volume on heart volume. Changes in stroke volume due to changes in contractility are mechanistically different from those occurring as a result of increased end-diastolic volume. Therefore, the two mechanisms can operate simultaneously to increase stroke volume, as shown in the lower panel. EDV = end-diastolic volume; ESV = end-systolic volume; SV = stroke volume.

fibers results in an increase in both the rate and extent of calcium movement into the cytoplasm, from both the sarcoplasmic reticulum and from outside the cell. This results in a more rapid and a more forceful contraction of the ventricles and an increase in the stroke volume ejected from the heart with each beat. As shown in Figure 18–24, an increase in stroke volume due to an increase in contractility will result in a reduced end-systolic volume.

Sympathetic stimulation also increases the rate of relaxation by increasing the rate at which the sarcoplasmic reticulum removes calcium from the cytoplasm. Because sympathetic stimulation increases both the rate of contraction and the rate of relaxation, the duration of systole is decreased. This effect can indirectly lead to a further increase in cardiac output. Sympathetic stimulation of the heart generally results in an increase in heart rate. Recall from the previous section that increases in cardiac output due to increases in heart rate were limited at higher heart rates by inadequate filling time during diastole (see Fig. 18–25). The effect of shortening the duration of systole, at any given heart rate, is to increase

the duration of diastole and of ventricular filling time.

Increased firing of the cardiac parasympathetic nerve fibers decreases the contractility of the heart, but not by a direct action of the parasympathetic neurotransmitter (acetylcholine) on cytoplasmic calcium concentrations. It appears that acetylcholine blocks or antagonizes the effects of sympathetic stimulation of the heart. In the absence of sympathetic nerve activity, increased firing of ventricular parasympathetic nerve fibers has only a small depressant effect on contractility.

In summary, stroke volume can be altered by two basic mechanisms. The Frank-Starling law describes how changes in end-diastolic volume result in changes in stroke volume. Changes in contractility, on the other hand, are the result of changes in intracellular calcium and do not require a change in end-diastolic volume. Since these two mechanisms for changing stroke volume are based on independent mechanisms, they can operate together to produce even greater changes in stroke volume than could either mechanism alone. The interaction of

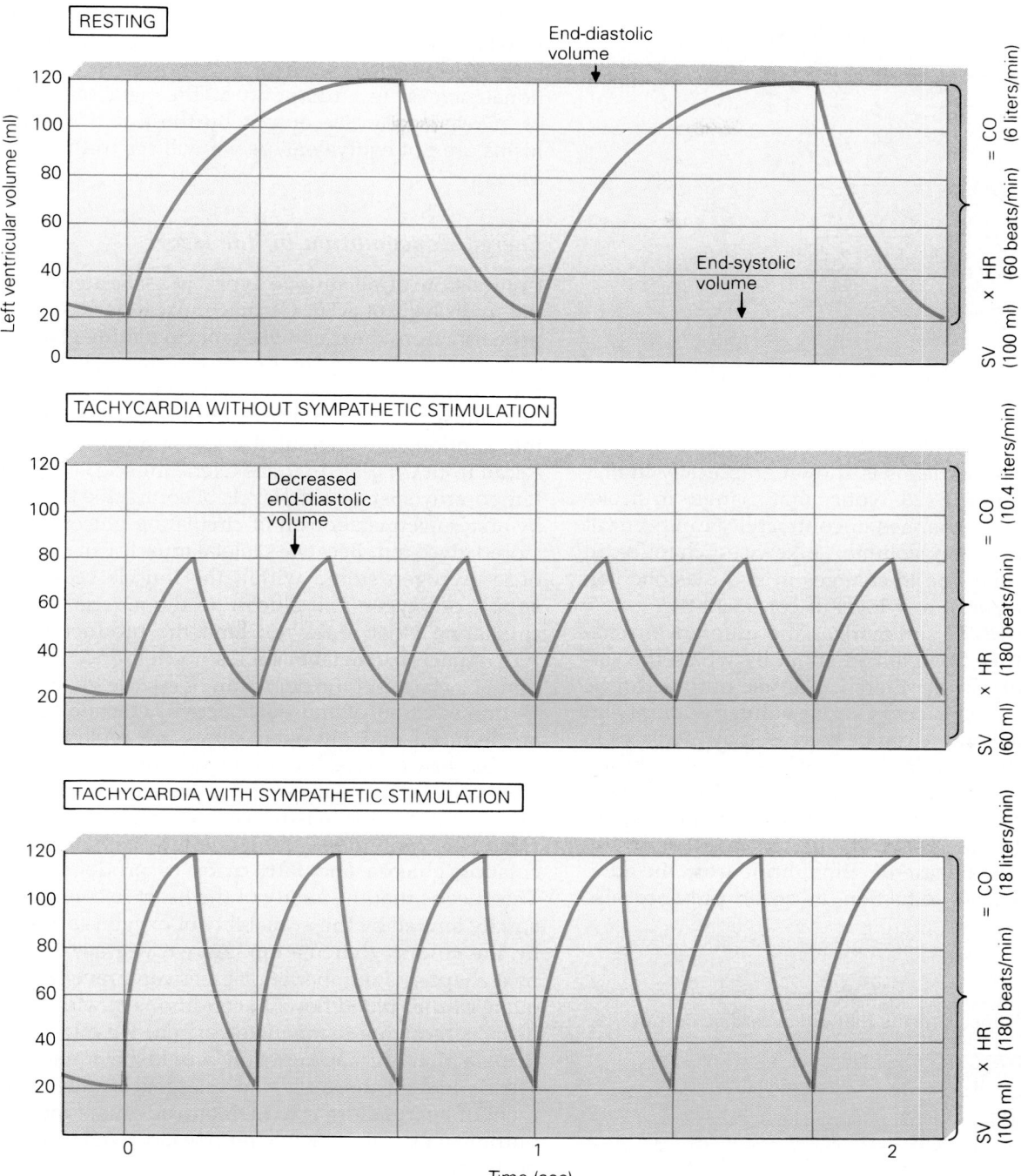

Figure 18–25
Increased cardiac output with sympathetic stimulation (effect of increasing the rate on contraction and relaxation of the heart). (*Top*) Changes in left ventricular volume in normal resting heart and calculated cardiac output (right). (*Middle*) Effect of tachycardia without sympathetic stimulation on ventricular volume and cardiac output. Tachycardia of this nature results from cardiac arrhythmias such as atrial tachycardia. (*Bottom lower*) Tachycardia induced by sympathetic stimulation and associated cardiac output. The increased rate of contraction and relaxation increases the time available for diastolic filling of the ventricle, leading to increased stroke volume and cardiac output compared to tachycardia (middle).

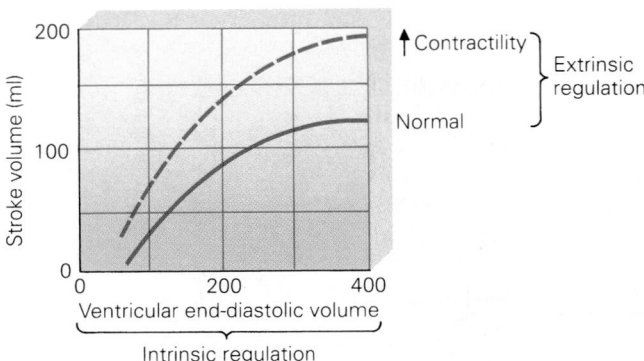

Figure 18–26
Summary of factors that affect stroke volume.

these two mechanisms is shown graphically in Figures 18–24 and 18–26. Notice that changes in stroke volume due to changes in contractility can occur at any end-diastolic volume. Likewise, changes in stroke volume due to changes in end-diastolic volume can occur at any level of contractility.

Figure 18–27 summarizes the different mechanisms that we have talked about by which the cardiac output can be altered. Cardiac output can be increased by increases in stroke volume and/or heart rate. Stroke volume can be increased by increases in end-diastolic volume (Frank-Starling law) or by increases in contractility. Both contractility and heart rate can be increased by increases in the level of sympathetic nerve activity to the heart or by increases in the release of epinephrine from the adrenal medulla. A reduction in aortic pressure also

Figure 18–27
Summary of mechanisms that affect cardiac output.

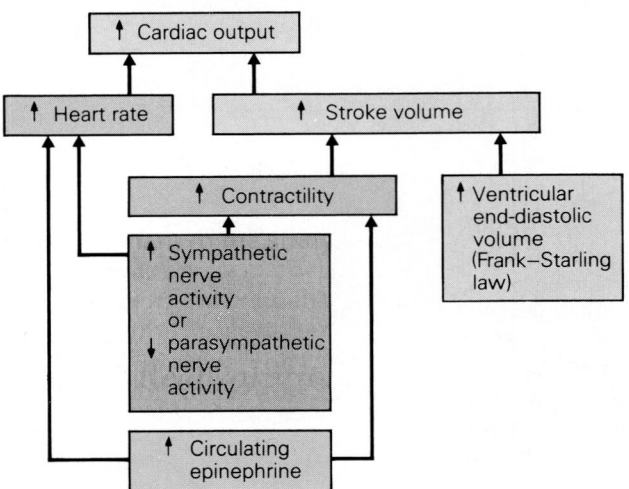

tends to increase stroke volume by reducing the load on the heart. Although an increase in either heart rate or stroke volume will result in a proportional increase in cardiac output, the metabolic costs of increasing cardiac output by these two mechanisms are not equivalent, as we will see in the next section.

Energy Consumption by the Heart

Contraction of all muscle types is associated with the hydrolysis of ATP. Cardiac muscle derives ATP primarily from the metabolism of circulating glucose and fatty acids. An important difference between cardiac muscle and other muscle types is that cardiac muscle must contract continuously, without interruption, throughout the life of an individual. Recall from Chapter 16 that skeletal muscle is able to temporarily sustain high levels of contractile activity by anaerobic metabolism of circulating glucose and stored glycogen. Because skeletal muscle can metabolize glycogen stored within the muscle cells, the supply of oxygen and glucose to the muscle by the circulating blood does not limit the production of ATP. Anaerobic metabolism results in the accumulation of lactate and the depletion of muscle glycogen. Upon cessation of muscular activity, lactate levels are reduced and glycogen stores are replenished.

Cardiac muscle, in contrast, cannot stop contracting to accumulate stored glycogen for support of anaerobic metabolism. The heart must instead rely upon continuous, oxidative metabolism of circulating glucose and fatty acids to produce ATP. This means that the ability of the heart to contract is strictly limited by the availability of oxygen supplied by the coronary circulation. As we will see in the next chapter, a number of diseases can prevent adequate cardiac blood flow. In conditions in which the heart is receiving inadequate oxygen, we might anticipate that cardiac function would be impaired. This is indeed the case, and one goal in the treatment of heart failure is to maintain adequate cardiac output while minimizing the oxygen requirements of the heart. Although at first glance this might appear to be an unresolvable situation, it turns out that the relationship between oxygen supply and cardiac output is not the same under all conditions. For example, at high heart rates and low stroke volumes, the heart uses more nutrients, and therefore more oxygen, to maintain a given cardiac output than at lower heart rates and larger stroke volumes. Blood pressure also affects the energy cost of maintaining cardiac output. The energy cost of producing a given cardiac output increases as blood pressure increases. The most economical method for increas-

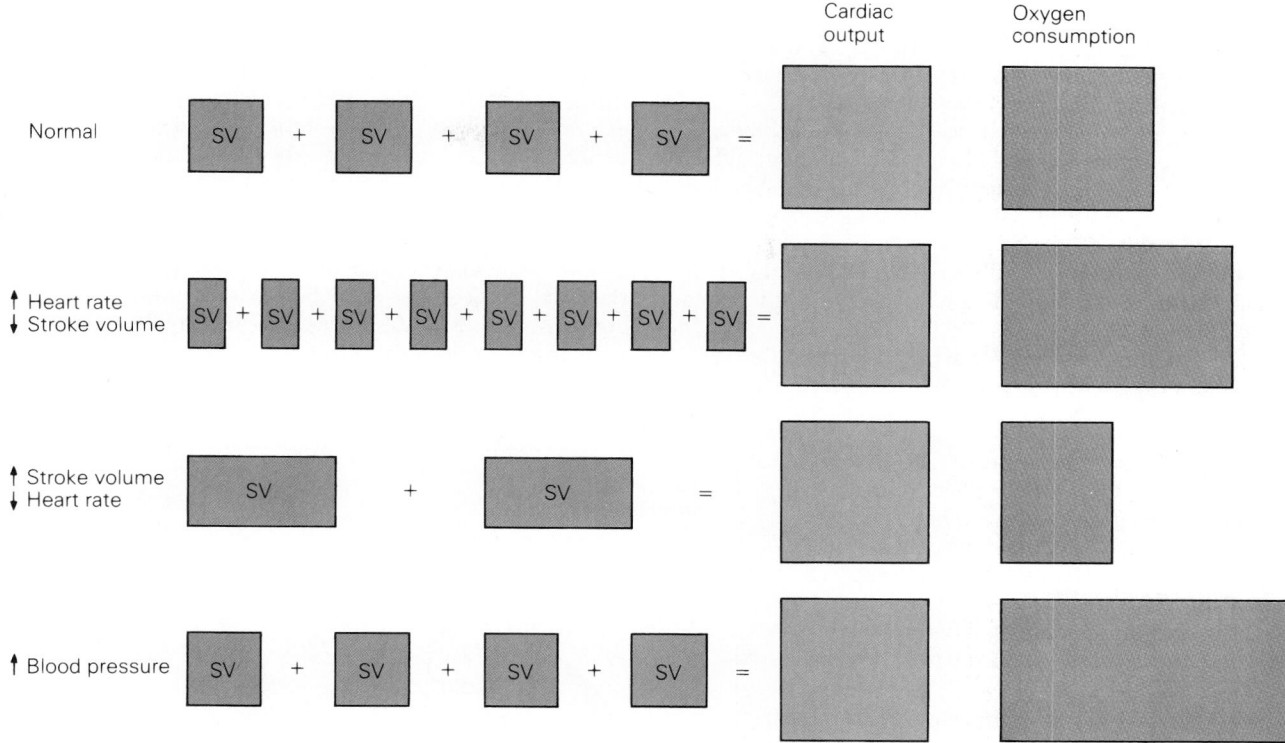

Figure 18–28
Effect of heart rate, stroke volume (SV), and arterial pressure on oxygen consumption by the heart.

ing cardiac output is to increase the end-diastolic volume (Frank-Starling law) without increasing either heart rate or blood pressure. Figure 18–28 illustrates the relationship between increases in cardiac output and increases in oxygen consumption. In the next chapter, we will see how this information can be useful in the design of effective therapies for the clinical treatment of heart failure.

R E S E A R C H   F O C U S

Cardiac Transplantation and Artificial Hearts

Cardiac transplantation was first performed in humans in 1967. As of 1987, there were over 206 surviving transplant patients. One patient has survived for more than 17 years with a transplanted heart. A major advance in the survival of transplantation patients was the introduction of the immunosuppressive agent **cyclosporine,** which reduces the possibility of rejection of the transplanted heart by the patient's immune system.

Candidates for heart transplants must have severe heart failure and a life expectancy of less than one year. After a heart has been obtained from a donor, it can be maintained for up to three hours prior to transplantation. The transplanted heart is completely denervated and lacks autonomic neural control. In spite of this, the transplanted heart shows normal contractility. Moreover, during exercise, cardiac output increases due to both an increase in stroke volume and heart rate. This is the result of an increase in the level of circulating catecholamines and the Frank-Starling properties of the heart.

In spite of the enormous cost, there has been growing interest in cardiac transplantation, which has led to a severe shortage of donor hearts. Approximately one third of patients awaiting transplant die before a suitable donor heart becomes available. One solution to this problem has been the use of the **artificial heart** as a temporary replacement prior to transplantation.

The first artificial heart to be implanted in a human donor was the **Jarvik-7.** The Jarvik-7 has two air-driven pumping chambers, which are sutured to the patient's atria and great arteries (i.e., pulmonary trunk and aorta). Pressurized air is supplied by an external pump through air tubes connected to the patient. Although the Jarvik-7 heart has successfully supported the circulation of two patients for more than a year, the formation of blood clots within the artificial heart is a serious problem. Current research is under way to develop a new heart using new materials that prevent clot formation. In addition, efforts are being made to design a fully implantable heart with compact electrical motors.

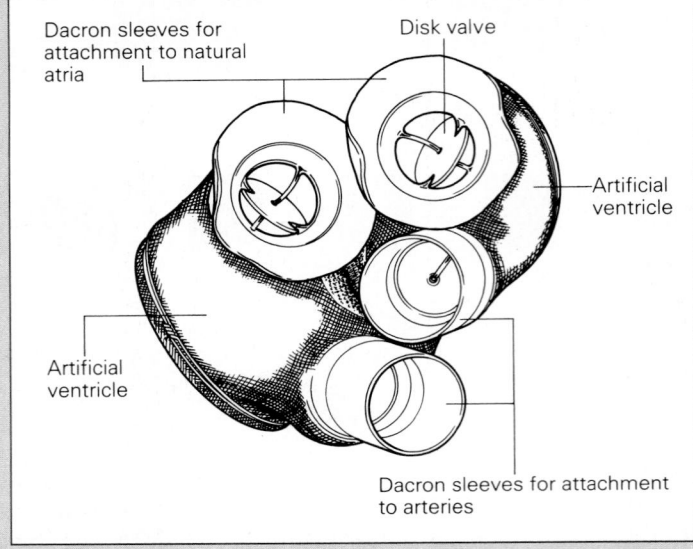

The Jarvik-7 artificial heart. The inflow tracts (*top*) and outflow tracts (*bottom*) contain artificial disk valves. The two chambers themselves contain air-driven diaphragms that simulate the pumping action of the ventricles. The patient's natural atria and arteries are attached to the inflow and outflow tracts, respectively.

SUMMARY

Overview of the Cardiovascular System

The function of the cardiovascular system is to rapidly transport blood throughout the body through an extensive network of blood vessels, which direct blood to the vicinity of all the cells in the body. Exchange of substances between the blood and the cells depends on diffusion of molecules, amoeboid activity of blood cells, and conduction of thermal energy. The cardiovascular system can be viewed as a closed loop consisting of the left and right hearts and the pulmonary and systemic circulations. The systemic circulation contains several defined vascular beds which are arranged in parallel with one another. The systemic blood flow is divided among these different vascular beds according to need. The branching pattern of the systemic blood vessels, beginning with the aorta, is as follows: aorta, arteries, arterioles, capillaries, venules, veins, and vena cava. Exchange between the blood and the cells occurs at the level of the capillaries. The branching pattern of the pulmonary vessels is as follows: pulmonary trunk, pulmonary arteries, pulmonary capillaries, and pulmonary veins.

Blood flow is the volume of blood that moves past a particular point in the cardiovascular system during a period of time. Blood flow through all portions of the cardiovascular system arranged in series is the same. The vascular beds are arranged in parallel and therefore each bed receives only a portion of the total blood flow. Blood pressure is a measure of the driving force that causes blood to flow. Blood pressure decreases as blood flows through the blood vessels because of frictional forces. Resistance is a quantity that summarizes the net effect of all the frictional forces opposing blood flow. The relationship between pressure, flow, and resistance is: Flow = ΔPressure/ Resistance.

The Heart: A Muscular Pump

The left and right hearts both consist of a thin-walled atrium and a muscular ventricle. A series of one-way cardiac valves ensures that blood flows in only one direction through the heart. Blood flows from the atria to the ventricles through the atrioventricular valves (right heart = tricuspid valve; left heart = bicuspid or mitral valve). Blood flows out of the ventricle through the semilunar valves (right heart = pulmonic valve; left heart = aortic valve). The wall of the heart consists of a middle, muscular layer (myocardium), an inner layer (endocardium), and an outer layer (epicardium). The coronary vessels supply blood flow to the heart.

The sequential activation of different parts of the heart is governed by the conduction of a cardiac action potential from the atria to the ventricles in a series of phases. The cardiac action potential is relatively long, with a duration of 300 milliseconds. Because the absolute refractory period is as long as the muscle twitch, tetanus is not possible in cardiac muscle. The action potential is initiated by pacemaker cells located in a region of the right atrium called the sinoatrial node. Spontaneous depolarization from the diastolic potential brings the nodal cells to threshold and an action potential is initiated. The rate of diastolic depolarization determines the rate at which the heart beats. The action potential is conducted from the right atrium to the ventricles via the atrioventricular node (AV node). Decremental conduction is the slowing and weakening of the action potential that occurs as it travels through the AV node. The action potential is then conducted through the Bundle of His, the bundle branches, and the Purkinje fibers to the ventricular muscle cells. This specialized conduction system ensures the nearly simultaneous activation of all of the ventricular muscle. Abnormal activation of the heart due to conduction abnormalities is referred to as an arrhythmia and may often occur following a heart attack (myocardial infarction). Electrocardiography is a method for measuring the electrical activity of the heart. The electrocardiogram (ECG or EKG) is a record made by an electrocardiograph.

The cardiac cycle is the sequence of mechanical events that occur during a single contraction of the heart. The sequence of events is the same for both the left and right hearts, which beat simultaneously. The ventricles are the primary pumping chambers of the heart. Ventricular contraction and ventricular ejection occur during systole. The systolic period begins with the occurrence of the first heart sound and ends with the occurrence of the second heart sound. Ventricular relaxation and ventricular filling occur during diastole. Diastole begins with the second heart sound and ends with the first heart sound.

The volume of blood ejected from the heart with each beat is called the stroke volume. The sum of all the stroke volumes ejected from the heart over a period of one minute is equal to the cardiac output. The body is able to alter cardiac output by changing heart rate and/or stroke volume. Heart rate is determined by both sympathetic and parasympathetic innervation of the SA node. Stroke volume is regulated by both intrinsic and extrinsic factors. The intrinsic ability of the heart to respond to an increase in end-diastolic volume with an increase in stroke volume is referred to as the Frank-Starling law of the heart, which ensures that the volume of blood pumped by the left and right hearts is equal. The heart may also increase stroke volume, on a more long-term basis, by hypertrophy of the muscle fibers. A change in the vigor of contraction that is not dependent on a change in end-diastolic volume is referred to as a change in contractility. Contractility changes are the result of changes in the concentration of intracellular calcium ion.

Cardiac muscle derives ATP from the metabolism of circulating glucose and fatty acids. The ability of the heart to contract is strictly limited by the supply of nutrients carried to the cardiac muscle by the coronary circulation. The energy cost of maintaining a given level of cardiac output is related to a number of factors, including heart rate, stroke volume, and arterial pressure.

REVIEW QUESTIONS

Identify Each with a Term

1. A change in the vigor of cardiac contraction that is independent of end-diastolic volume.

2. The volume of blood ejected from the heart during a single cardiac cycle.

3. The phrase that refers to the period of the cardiac cycle during which both the atrioventricular and semilunar valves are closed and the ventricles are contracting.

4. The volume of the ventricles immediately prior to activation of ventricular contraction.

Define Each Term

5. pacemaker
6. blood flow
7. systole
8. Frank-Starling law of the heart
9. electrocardiogram
10. arrhythmia

Choose the Correct Answer

11. Which of the following is *not* a determinant of blood flow?
    a. pressure gradient
    b. blood vessel radius
    c. colloid osmotic pressure
    d. total peripheral resistance

12. Which event occurs after the first heart sound and before the second heart sound?
    a. ventricular ejection
    b. P-wave of the ECG
    c. closure of the mitral valve
    d. atrial contraction

13. Which of the following represent parallel blood flows?
    a. total systemic blood flow and total pulmonary blood flow
    b. blood flow to muscle and blood flow to the skin
    c. right heart output and left heart output
    d. total capillary blood flow and total venous blood flow

14. An increase in the concentration of circulating epinephrine would:
    a. decrease stroke volume
    b. decrease heart rate
    c. increase cardiac output
    d. decrease contractility of the heart

15. In a normal, healthy heart, stroke volume would be increased by:
    a. increased sympathetic stimulation of the heart
    b. increased parasympathetic stimulation of the heart
    c. decreased contractility
    d. decreased end-diastolic volume

16. Contractility of the heart is *not* altered by which of the following?
    a. a change in end-diastolic volume
    b. a change in cytoplasmic calcium concentration
    c. the drug digitalis
    d. a change in the level of sympathetic stimulation to the heart

17. The exchange of nutrients and gases between the blood and the tissues occurs:
    a. in the aorta
    b. in the arteries
    c. in the arterioles
    d. in the capillaries

18. The normal location of the pacemaker in the human heart is in the:
    a. atrioventricular node
    b. sinoatrial node
    c. left ventricle
    d. Purkinje fibers

19. Cardiac muscle cannot contract in a tetanic fashion because:
    a. the long absolute refractory period prevents the muscle from becoming re-stimulated while still contracting.
    b. the action potentials travel too slowly along the conducting tissue to re-stimulate the muscle.
    c. contraction is possible only when the heart is filled with blood.
    d. the autonomic nervous system blocks rapid action potentials.

20. In the electrocardiogram, the T-wave is associated with:
    a. atrial depolarization
    b. atrial repolarization
    c. ventricular depolarization
    d. ventricular repolarization

Calculate

21. End-diastolic volume of the left ventricle is 100 ml and end-systolic volume is 30 ml. What is the stroke volume?

22. Stroke volume is 70 ml and heart rate is 80 beats/min. What is cardiac output?

23. Examination of an electrocardiogram shows that two successive QRS complexes are 0.8 seconds apart. What is the heart rate in beats per minute?

Answer Each Question in One or Two Sentences

24. How does vasoconstriction of resistance arterioles result in redistribution of blood flow to tissues?

25. Why is atrial contraction more important at elevated heart rates?

26. What is the underlying molecular mechanism associated with changes in cardiac contractility?

27. Trace the spread of activation of heart muscle, naming (in proper order) each of the structures through which the impulse passes.

28. Can an electrocardiogram be used to determine whether a change in contractility has resulted in an increased cardiac output?

SUGGESTED READINGS

Berne, Robert M., and Levy, Matthew N. *Physiology*, 2nd ed. St. Louis: C.V. Mosby, 1988.

Cantin, M., and Genest, J. "The heart as an endocrine gland." *Scientific American*, 254(2):76–81, 1986.

Eisenber, M.S., Bergner, L., Hallstrom, A.P., and Cummins, R.O. "Sudden cardiac death." *Scientific American*, 254(5):37–43, 1986.

Goerke, Jon, and Mines, Allan H. *Cardiovascular Physiology*. New York: Raven Press, 1988.

Goldstein, G.W., and Betz, A.L. "The blood-brain barrier." *Scientific American*, 255(3):74–83, 1986.

Honig, C.R. *Modern Cardiovascular Physiology*, 2nd ed. Boston: Little, Brown and Co., 1988.

Jarvik, R.K. "The total artificial heart." *Scientific American*, 244(1):74–80, 1981.

Levy, R.I., and Moskowitz, J. "Cardiovascular research: Decades of progress, a decade of promise." *Science*, 217:121–129, 1982.

Little, Robert C., and Little, William, C. *Physiology of the Heart and Circulation*, 4th ed. Chicago: Year Book Medical Publishers, 1989.

Robinson, T.F., Factor, S.M., and Sonnenblick, E.H. "The heart as a suction pump." *Scientific American*, 254(6): 84–91, 1986.

Sparks, Harvey V., and Rooke, Thom W. *Essentials of Cardiovascular Physiology*. Minneapolis: University of Minnesota Press, 1987.

KEY TERMS

aorta (p. 622)
arrhythmia (p. 634)
arterial system (p. 622)
atrioventricular node (AV node) (p. 633)
blood pressure (p. 624)
bundle branches (p. 633)
capillary (p. 622)

cardiac cycle (p. 637)
depolarization (p. 629)
diastole, diastolic pressure (p. 637)
ECG, EKG (p. 634)
endocardium (p. 625)
epicardium (p. 625)
Frank-Starling Law (p. 641)

left heart (p. 621)
myocardium (p. 625)
pulmonary circulation (p. 621)
Purkinje fiber (p. 633)
right heart (p. 621)
sino-atrial node (SA node) (p. 631)

systemic circulation (p. 621)
systole, systolic pressure (p. 637)
tachycardia (p. 631)
venous system (p. 622)
ventricle (p. 625)

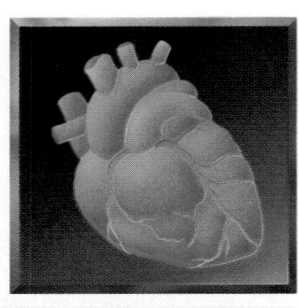

C H A P T E R
19

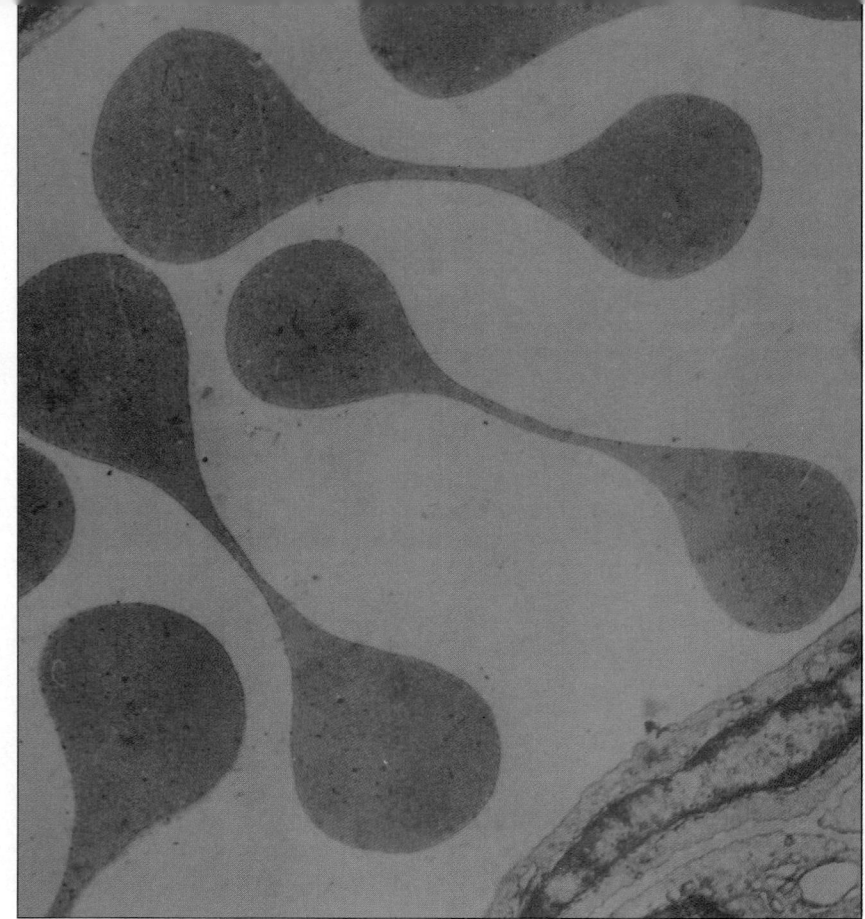

Circulation

*A*s we have previously discussed, the purpose of the cardiovascular system is to provide a flow of blood past the cells of the body. In Chapter 18, we looked at how the heart functions as a pump. The next step in our analysis will be to examine the system of blood vessels to see how they direct the flow of blood throughout the body and how exchange occurs between the blood and the interstitial fluid.

Overview of the Vascular System

Recall from our previous discussion that the major subdivisions of both the systemic and pulmonary vascular systems are: arteries, arterioles, capillaries, venules, and veins. As we will see, the exchange of substances between the blood and the interstitial fluid occurs as blood passes through the capillaries. In large part, the role of the arteries, arterioles, venules, and veins is to direct the flow of blood to and from the capillaries. Far from being passive conduits, however, these vessels also play an active role in determining what portion of the total cardiac output reaches each of the various tissue beds in the body. In addition, the venules and veins help to ensure the efficient operation of the heart as a pump, by maintaining the proper distribution of blood volume within the cardiovascular system.

The Arterial System: Carrying Blood Away from the Heart

The aorta and the arterial system are conduits that direct the flow of blood from the heart to the various tissue beds within the body. The aorta must be large enough to accommodate the combined blood flow to all the tissue beds. It is a large vessel with an average internal diameter of 2.5 cm (1 inch). The arteries that branch off the aorta to distribute blood to the various tissues have internal diameters on the order of 0.4 cm. Relative to the diameter of the arterioles and capillaries, these are large vessels that offer comparatively little resistance to the flow of blood (see Fig. 19–7).

The arterial system plays a second role in the operation of the cardiovascular system. The walls of the arteries are made of smooth muscle and elastic tissue. When the ventricle contracts and suddenly ejects a volume of blood, the aorta and arteries stretch to accommodate the ejected blood. During diastole, when the heart is no longer ejecting blood, the elastic arteries recoil and force blood through the downstream blood vessels.

Blood Pressure in the Arteries

Most people are probably aware of the fact that blood is under considerable pressure within the arterial system. Cutting a large artery results in a serious medical emergency because blood is rapidly lost from such a wound. The high pressure inside the artery forces blood out through the cut. This high arterial blood pressure is necessary for the normal operation of the cardiovascular system. As mentioned previously, blood pressure is a measure of the driving force that causes blood to flow through the blood vessels. The magnitude of the driving pressure must be adequate to overcome the **resistance** to blood flow due to friction between the flowing blood and the walls of the blood vessels, and to the friction between the individual molecules and cells that make up the blood. We all deal with frictional forces every day. It is necessary to apply pressure to one side of a heavy box to slide it across the floor because of the friction between the bottom of the box and the floor. Likewise, pressure is necessary to move blood through the vascular system.

How much pressure must the heart generate to overcome this resistance? Recall from the previous chapter the equation describing the relationship between flow (F), pressure (P), and resistance (R):

$$F = \frac{P_1 - P_2}{R}$$

Since we are looking at flow from the aorta to the vena cava, it is the pressure difference between the aorta (P_1) and vena cava (P_2) that determines flow (i.e., aortic pressure − vena cava pressure = pressure difference). We can simplify our analysis by recognizing two points. First, the pressure in the vena cava (P_2) is small compared to that in the aorta and can be considered to be zero for our purposes. Secondly, pressure within the aorta (P_1) is approximately equal to the **arterial pressure** measured in any of the major systemic arteries; this equality indicates that the resistance to flow through the arteries is relatively small. Knowing that blood pressure in the vena cava is approximately zero and that arterial pressure equals aortic pressure, we can modify the equation for blood flow as follows:

$$F = \frac{P_1 - P_2}{R}$$

$$F = \frac{(\text{aortic pressure} - \text{vena cava pressure})}{R}$$

$$F = \frac{(\text{arterial pressure} - 0)}{R}$$

$$F = \frac{\text{arterial pressure}}{R}$$

Under normal resting conditions, the arterial pressure required to provide adequate tissue flow is approximately 100 mm Hg. Figure 19–1 illustrates the values for blood pressure at various points in the circulatory system. Notice first that the systemic arterial pressure starts out about three times higher than the pulmonary arterial pressure, and yet the pressures in both systems return to very nearly the same value by the time the blood reaches the veins. The decrease in blood pressure reflects the loss of energy because of the resistance to blood flow. The greater the resistance to flow, the greater the pressure drop as blood flows through the vessels. On this basis, we can reason that the resistance to blood flow must be about three times greater in the systemic circulation than in the pulmonary circulation.

Figure 19–1
Illustration of the range of blood pressures throughout the systemic and pulmonary circulations.

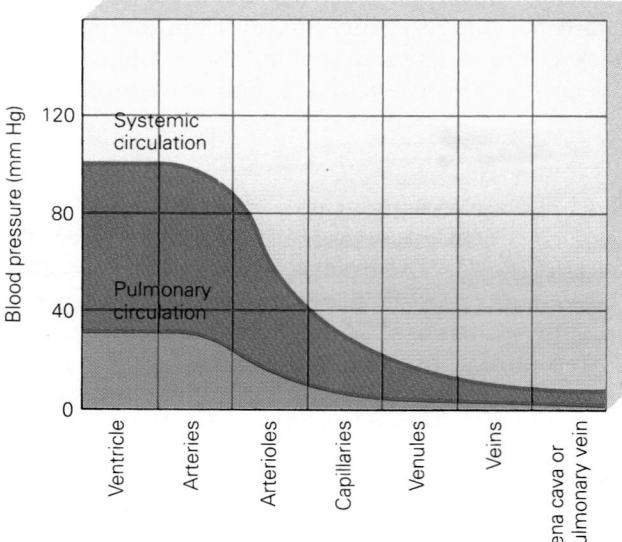

Notice that the greatest pressure drop, both in the systemic and pulmonary circulations, occurs at the level of the arterioles. This tells us that a major portion of the overall resistance to blood flow occurs at this point in the circulation. As we will see, this has important implications for the distribution of blood flow within the body.

Pulse Pressure

If we were to directly measure arterial blood pressure, the first thing we would notice is that it is pulsatile. Recall from the previous chapter that arterial pressure is lowest at the end of ventricular diastole and rapidly increases to a peak during ventricular systole. When the ventricle contracts, a volume of blood (the stroke volume) is ejected rapidly from the heart into the arterial vessels. As shown in Figure 19–2, the flow of blood into the arteries during systole is greater than the flow of blood out of the arteries into the capillaries. Blood flows out of the arteries more slowly because of the high resistance to flow through the arterioles. The elastic tissue that makes up the walls of the aorta and arteries allows them to stretch and increase in diameter to accommodate the blood ejected from the ventricle. Just as the pressure increases inside an elastic balloon as it inflates with air, the pressure inside the arterial system increases as the vessels inflate with blood. During diastole, when blood is no longer flowing into the arteries from the heart, the elastic recoil of the arteries forces blood to flow out of the arteries into the arterioles. In other words, a portion of the pressure generated by the heart during systole is stored in the stretched walls of the arteries and is then slowly dissipated during diastole as blood flows out of the arterial system. The elastic properties of the arteries help to convert the **pulsatile** flow of blood from the heart into a more *continuous* flow of blood through the rest of the circulation.

As blood flows out of the arteries, the pressure progressively falls until it reaches a minimum called **diastolic pressure** (Fig. 19–3). The peak arterial pressure, which occurs during contraction of the ventricle, is called **systolic pressure.** Your nurse or physician may have measured your blood pressure as "120 mm Hg over 70 mm Hg." These two numbers refer to the systolic and diastolic pressures respectively and are normally written as 120/70 mm Hg. The difference between systolic and diastolic pressures (120 − 70 = 50 mm Hg) is referred to as the **pulse pressure.**

As you might guess, the amplitude of the pulse pressure indicates the vigor of contraction by the ventricle; it also tells us something about the elastic-

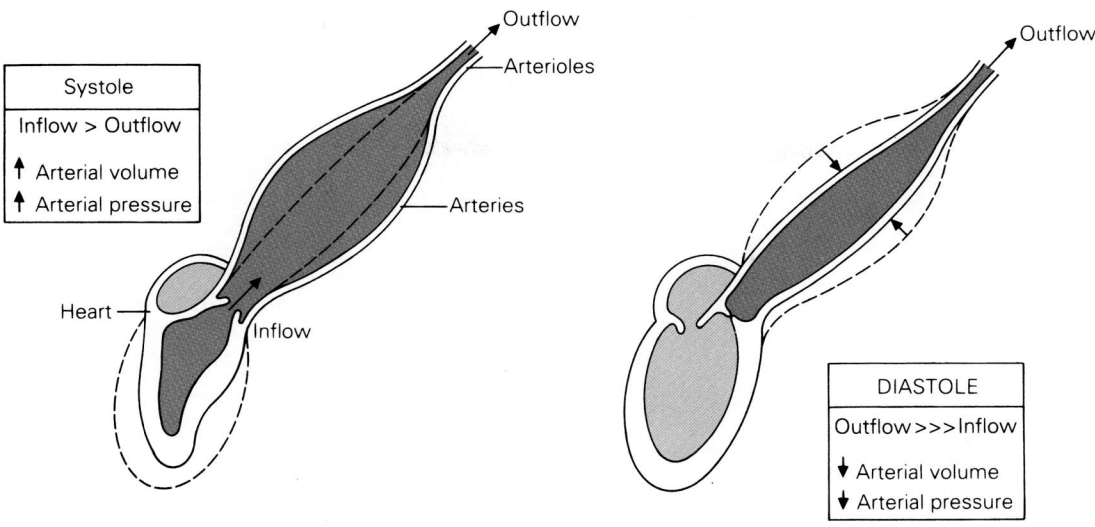

Figure 19–2
Role of the arteries as a high-pressure storage reservoir. The flow of blood into the arteries (from the heart) during systole exceeds the flow out of the arteries (through the arterioles), leading to an increase in arterial volume and pressure. During diastole, the elastic recoil of the arterial walls provides the driving force to propel blood out of the arteries.

ity of the arteries. As shown in Figure 19–4, the magnitude of the pulse pressure is determined by two factors: (1) the stroke volume and (2) the distensibility of the arteries. As we discussed in the previous chapter, the stroke volume of the heart is dynamically regulated and can change significantly from beat to beat. In order for the heart to inject an increased stroke volume into the arteries, it must generate greater pressure to further stretch the elastic wall of the artery. A sudden increase in pulse pressure therefore indicates an increase in the stroke volume ejected by the heart. Although the

distensibility of the arteries can also affect the magnitude of the pulse pressure, arteriolar distensibility is relatively constant from day to day. However, progressive arterial disease, such as atherosclerosis, results in a decrease in the distensibility of the arteries and therefore an increase in pulse pressure. Analogously, greater pressure is required to blow up a thick-walled, less distensible balloon than a thin-walled, easily distensible one. A progressive increase in pulse pressure as an individual ages can be diagnostic of arterial disease commonly referred to as "hardening" of the arteries.

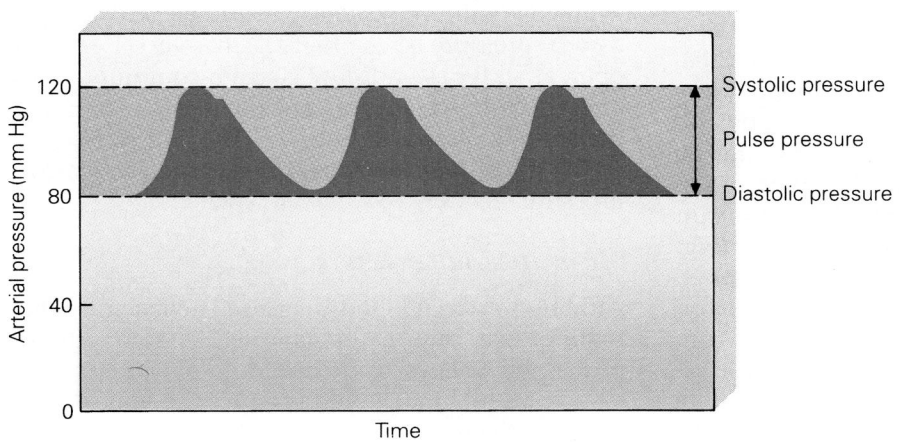

Figure 19–3
Aortic blood pressure recording, showing normal fluctuations in pressure with each cardiac cycle.

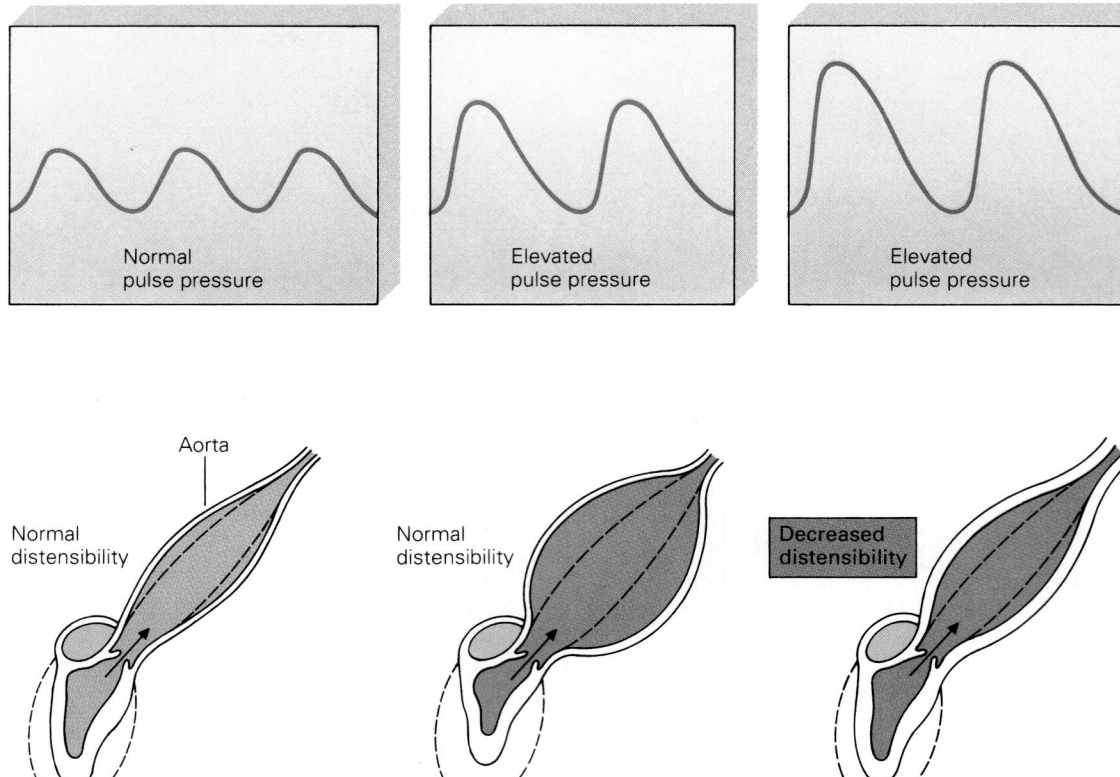

Figure 19–4
Illustration of the dependence of pulse pressure on stroke volume and aortic distensibility.

Measurement of Arterial Pressure

Everyone has had blood pressure measurements taken at one time or another. The physician or nurse places a cuff around the arm, inflates it with air, and then slowly deflates it while listening to the brachial artery with a stethoscope. Figure 19–5 illustrates how this process can measure both systolic and diastolic pressures. When the cuff is fully inflated (see Fig. 19–5a), the artery under the cuff is fully collapsed by the pressure exerted by the cuff, and no blood can flow through the artery. The pressure in the cuff is then gradually lowered. When the cuff pressure falls below the peak systolic pressure (Fig. 19–5b), blood is able to momentarily flow through the partially collapsed artery. This momentary, high-velocity spurt of blood through the artery can be heard through a stethoscope placed over the artery (downstream from the cuff) as a soft tapping sound. When this sound first occurs, the cuff pressure is recorded as systolic pressure. As the cuff pressure is progressively lowered (Fig. 19–5c), the duration of blood flow during each cardiac cycle gets longer and the nature of the sounds heard with the stethoscope changes in a characteristic manner. When the cuff pressure has decreased below the diastolic pressure (Fig. 19–5d), the vessel remains open at all times and flow is again continuous. Just after the cuff pressure falls below diastolic pressure, the sounds disappear entirely. The cuff pressure at which the sounds disappear is recorded as diastolic pressure.

Mean Arterial Pressure

Although arterial blood pressure is constantly fluctuating, we can mathematically smooth out the peaks and valleys to determine a single value for blood pressure that would produce the same flow as the pulsatile pressure that actually exists. This **mean**

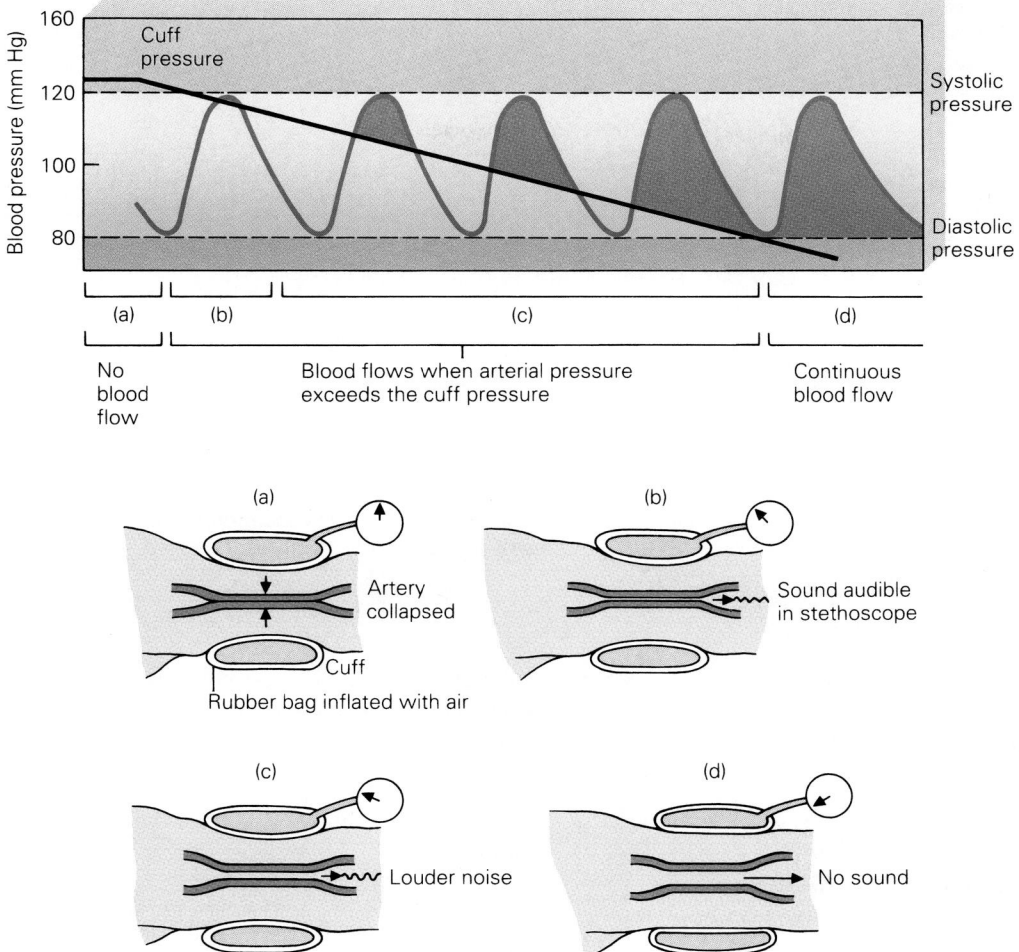

Figure 19–5
Measurement of systolic and diastolic blood pressures by sphygmomanometry. The shaded areas indicate when blood is able to flow through the brachial artery as the cuff pressure is lowered.

arterial pressure (MAP) is calculated with the following equation (Fig. 19–6):

MAP = 1/3 pulse pressure + diastolic pressure

The MAP is not the mathematical average of systolic and diastolic pressure, but rather an approximation of the geometric mean. By calculating the mean blood pressure, we can significantly simplify the calculation of blood flow because it is the MAP that is related to blood flow. Hence, we can further modify the equation for flow as follows:

$$F = \text{arterial pressure}/R = \text{MAP}/R$$

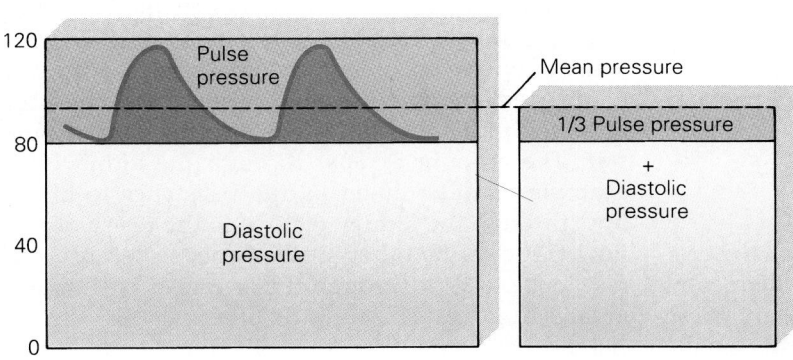

Figure 19–6
Comparison of pulsatile arterial pressure and calculated mean arterial pressure. The fluctuating pulse pressure and the constant mean arterial pressure are equivalent in that they produce equal blood flows.

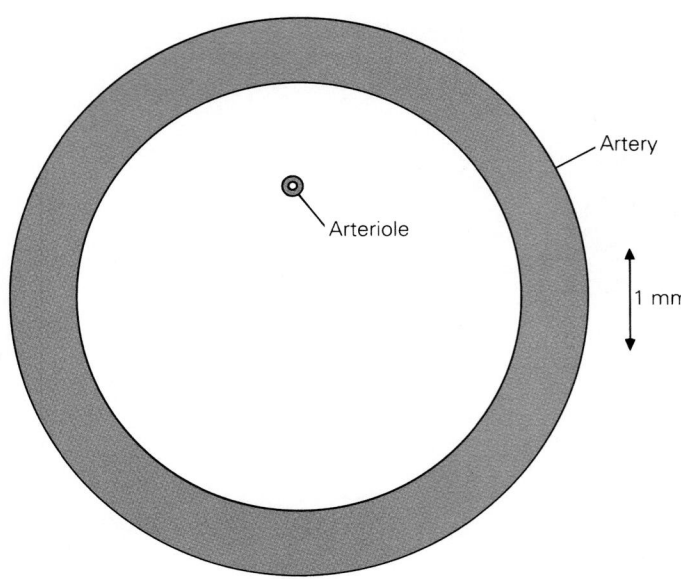

Artery

Arteriole

1 mm

Figure 19–7
Comparison of the relative diameters and wall thicknesses of arteries and arterioles.

Without calculating the MAP, it would be rather difficult, for example, to evaluate how a change in blood pressure from 120/90 to 160/70 would alter blood flow. If we calculate the MAP in each case:

MAP = 1/3(120 − 90) + 90 = 100
MAP = 1/3(160 − 70) + 70 = 100

we quickly realize that flow remains the same. As we will see in a later section, our bodies have homeostatic systems that sense MAP and then change the functional state of the heart or the vasculature, or both, to maintain MAP constant. If we recall that the purpose of the cardiovascular system is to maintain a constant **flow** of blood to the tissues, it should not be too surprising to see that our bodies regulate MAP and not pulsatile pressure.

Resistance in the Arterioles

Recall that soon after the distributing arteries reach the various tissue beds within the body, they divide into smaller arterioles. As shown in Figure 19–7, the diameter of the arteriole is considerably smaller than that of the artery. This sudden decrease in diameter results in a relatively large increase in the resistance to blood flow.

The Diameter of Arterioles

Although a number of factors contribute to the resistance of flow through blood vessels, the diameter of a blood vessel is one of the more important contributing factors, for two reasons. First, small changes in diameter cause relatively large changes in resistance. The resistance to flow through a blood vessel is inversely proportional to the radius raised to the fourth power:

$$\left(R\alpha\frac{1}{\text{radius}^4}\right)$$

Therefore, halving the diameter of an arteriole increases the resistance to flow by a factor of 16 (Fig. 19–8). Secondly, arterioles are able to dynamically adjust their diameters. The walls of arterioles contain relatively more smooth muscle cells than other blood vessels in the body. Moreover, these smooth muscle cells are arranged so that when they contract, they cause the diameter of the vessel to decrease, as illustrated in Figure 19–9 (see also Fig. 16–11). These two properties (high resistance and variable diameter) of the arterioles provide a mechanism by which the body can adjust the blood flow through different organs or tissues.

Recall from the previous chapter that the blood vessels supplying different tissues in the body are arranged into **parallel** circuits (see Fig. 18–3). Because of this arrangement, the total cardiac output is distributed among the various tissues, with the sum of all the individual flows equal to the cardiac output. The amount of flow through the kidney, for example, is determined by the resistance to blood flow through the kidney relative to the resistance to flow through the other systemic beds. Because the resistance to flow through the arterioles is so much greater than that through the other vessels, the resistance to blood flow through a tissue is deter-

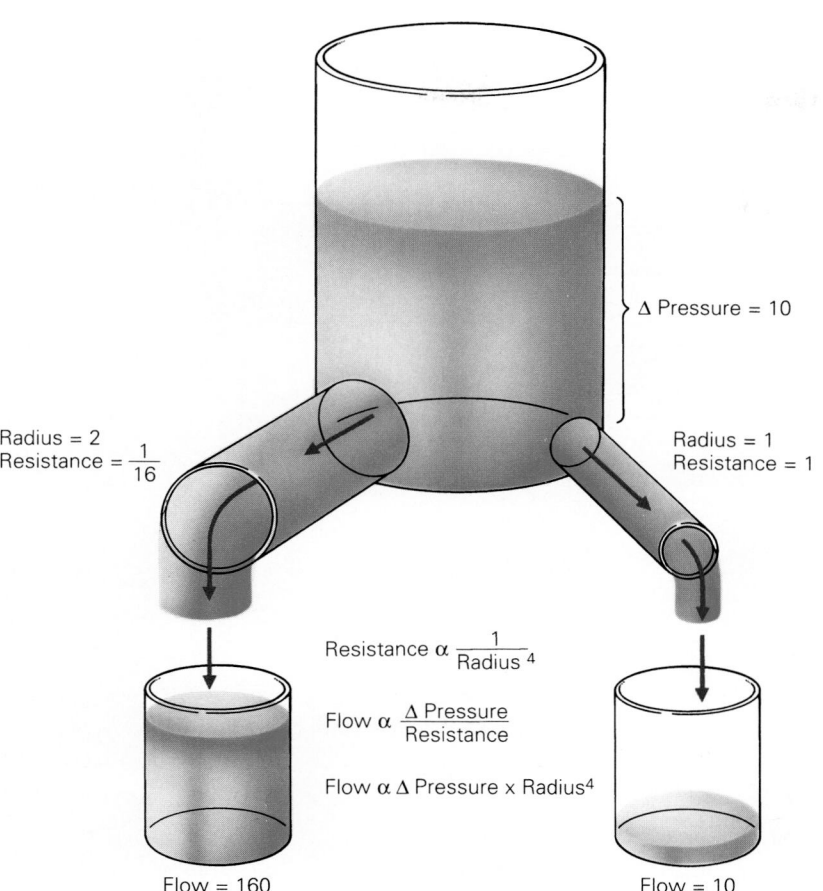

Figure 19–8
The effect of vessel radius on resistance and flow.

Δ Pressure = 10

Radius = 2
Resistance = $\frac{1}{16}$

Radius = 1
Resistance = 1

Resistance α $\frac{1}{Radius^4}$

Flow α $\frac{\Delta\ Pressure}{Resistance}$

Flow α Δ Pressure × Radius⁴

Flow = 160

Flow = 10

mined almost entirely by the resistance of the arterioles. Therefore, by adjusting arteriolar diameter, individual tissue beds can regulate the local flow of blood. Figure 19–10 shows how relative changes in arteriolar resistance can produce shifts in blood flow.

The Water System Analogy for the Regulation of Blood Flow

A good analogy for this aspect of the vascular system is a public water system. One large pipe (the aorta) carries water (the blood) from the water com-

Figure 19–9
(*a*) Smooth muscle cells wrap around the wall of an arteriole. (*b*) Changes in resistance of an arteriole due to contraction or relaxation of the smooth muscle cells in the vessel wall.

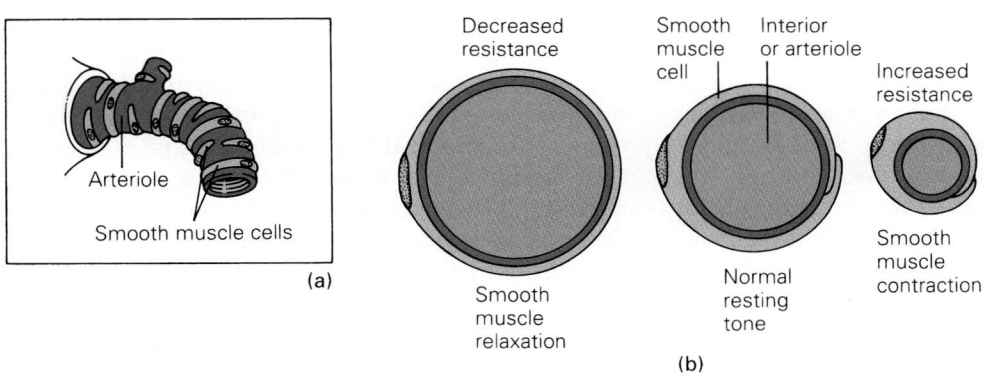

Arteriole

Smooth muscle cells

(a)

Decreased resistance

Smooth muscle cell

Interior or arteriole

Increased resistance

Smooth muscle relaxation

Normal resting tone

Smooth muscle contraction

(b)

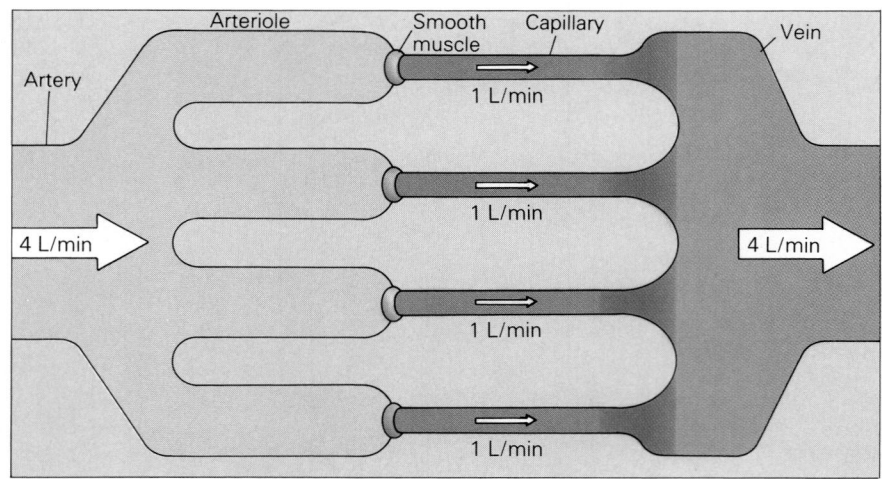

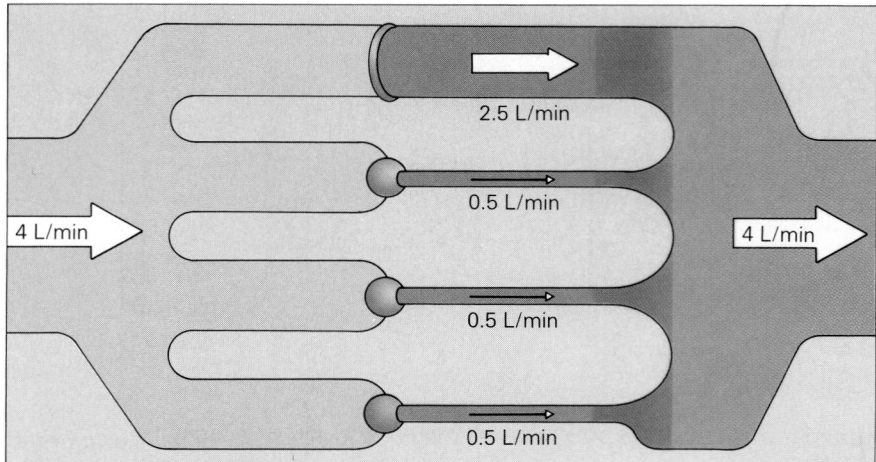

Figure 19–10
The effects of arteriolar resistance on distribution of blood flow among capillaries arranged in parallel.

pany's pump (the heart) to a series of distribution lines (arteries) that direct water to individual houses (organs). Within each house are a number of spigots (arterioles) that can be adjusted (variable resistance) to control the flow of water to different appliances (tissue beds). In this system, the distribution of water flow among the various houses and appliances is determined by local demand. The water company (the CNS) has only to ensure that the pressure in the main line (aorta) remains constant, and the flow can then be distributed among the various appliances (tissue beds) according to local demand. If you want to wash the dishes, your clothes, the dog, and the car all at the same time, you don't have to notify the water company of your intentions; you open the appropriate spigots (decrease the local resistance to flow).

Such a system makes the water company's job of providing adequate flow to all the users rather

easy because it doesn't have to monitor the usage by individuals to set the pumping rate (cardiac output). It simply monitors the water pressure (MAP) and adjusts the output of the pump (cardiac output) to keep the pressure constant. As long as the demand for water does not exceed the capacity of the water company's pump (the heart), all the consumers (the organs) will receive whatever flow they demand. If the demand on the system does exceed the capacity of the pump, then pressure will fall, and flow to the consumers will decrease. We have all experienced such a drop in water pressure during a morning shower. Since many other people are also taking showers at the same time, the demand may temporarily exceed the capacity of the pump, and both pressure and flow will decrease. If you wanted to remedy this loss of pressure and flow, you could call your neighbors (other organs) and ask them to turn off their spigots (increase local resistance to flow), while at the same time opening your spigot further (decreasing local resistance) to direct a greater portion of the limited flow to your house. This would be a reasonable thing to do if you had an urgent need (your house was on fire) compared to that of your neighbors (watering the grass). Notice that this cooperative approach to distribution of *limited* flow requires communication (the ANS) among the various users (the organs). However, as long as the capacity of the pump (the heart) exceeds the demand of the users (the organs), local regulation (by the arterioles) is adequate to adjust flow according to local needs.

Let us summarize the points made by this analogy. First, the individual organs and tissue beds can adjust local blood flow to meet demand by changing the resistance to flow through changes in the diameter of the arterioles. Second, the body can correctly adjust cardiac output to match the demand for flow simply by monitoring mean arterial pressure and adjusting the cardiac output to keep pressure constant. Third, as long as the demand for flow does not exceed the pumping capacity of the heart, individual tissue beds can function autonomously. Fourth, if demand does exceed flow, then extrinsic regulatory mechanisms must be used to redistribute flow to prevent a loss of pressure and flow to tissues with critical needs.

In order to fully understand how blood flow is distributed in the body, we must understand (1) how the resistance of the arterioles changes in response to local demand, (2) how the tissue beds in the body coordinate activity through extrinsic regulatory mechanisms, and (3) how the body monitors and adjusts cardiac output to maintain mean arterial pressure. Let's first look at the mechanisms by which local factors change arteriolar resistance.

Regulation of Blood Flow by Local Factors

Vascular Tone. The smooth muscle content of the arteriolar wall provides the means for dynamic regulation of arteriolar diameter and consequently of arteriolar resistance. Therefore, the contractile state of the smooth muscle determines the extent of local blood flow. If we were to administer a drug that relaxes vascular smooth muscle (a **vasodilator**) to a resting individual, we would find that the diameter of most of the arterioles in the body would increase. This shows that the normal resting state of the arteriolar smooth muscle is one of maintained contraction, referred to as a resting vascular **"tone."**

Resting tone is important since an increase in activity of an organ demands greater blood flow and thus relaxation of the arteriolar smooth muscle to produce dilation of the arteriole. Several factors contribute to this level of resting tone. Like the sinoatrial node of the heart, the resting membrane potential of vascular smooth muscle undergoes spontaneous depolarizations. As discussed in Chapter 16, such depolarizations are characteristic of single-unit smooth muscle and result in the initiation of a burst of action potentials, which in turn lead to the activation of muscle contraction. As we will see in the next few sections, both mechanical and chemical factors can modify this spontaneous activity.

Blood Pressure and Vascular Tone. Stretching the wall of an arteriole, and of the smooth muscle cells included in it, causes an increase in the rate of spontaneous membrane depolarizations and leads to an increased level of contraction. The increase in contractile activity in response to stretch is often referred to as **myogenic activity.** Let's examine how this might contribute to the local regulation of blood flow. An increase in arterial blood pressure causes all arteries and arterioles to increase their diameters by stretching the elastic elements (including smooth muscle cells) in the artery walls. If we recall the equation for blood flow ($F = MAP/R$), we can see that an increase in pressure increases flow not only by increasing the driving force (MAP), but also by increasing the vessel radius and decreasing the resistance to flow. Therefore, the increased myogenic activity of the smooth muscle in response to increased arterial pressure tends to minimize increases in both vessel diameter and flow. Conversely, a decrease in blood pressure, which tends to decrease flow, leads to decreased myogenic activity. This in turn leads to vasodilation, which tends to minimize the reduction in flow following a reduction in arterial pressure.

The sequence of events associated with an increase in blood pressure is shown in Figure 19–11. Experiments have shown that the blood flow to many organs remains relatively constant despite large changes in blood pressure. This ability of individual vascular beds to maintain constant blood flow is referred to as **autoregulation.** Although the mechanisms underlying autoregulation remain to by fully described, considerable evidence exists to suggest that changes in myogenic activity are a contributing factor. As we will see in the next section, a number of other factors also contribute to autoregulation of blood flow.

Local Chemical Factors and Autoregulation of Blood Flow. Blood flow to most tissues increases in pro-

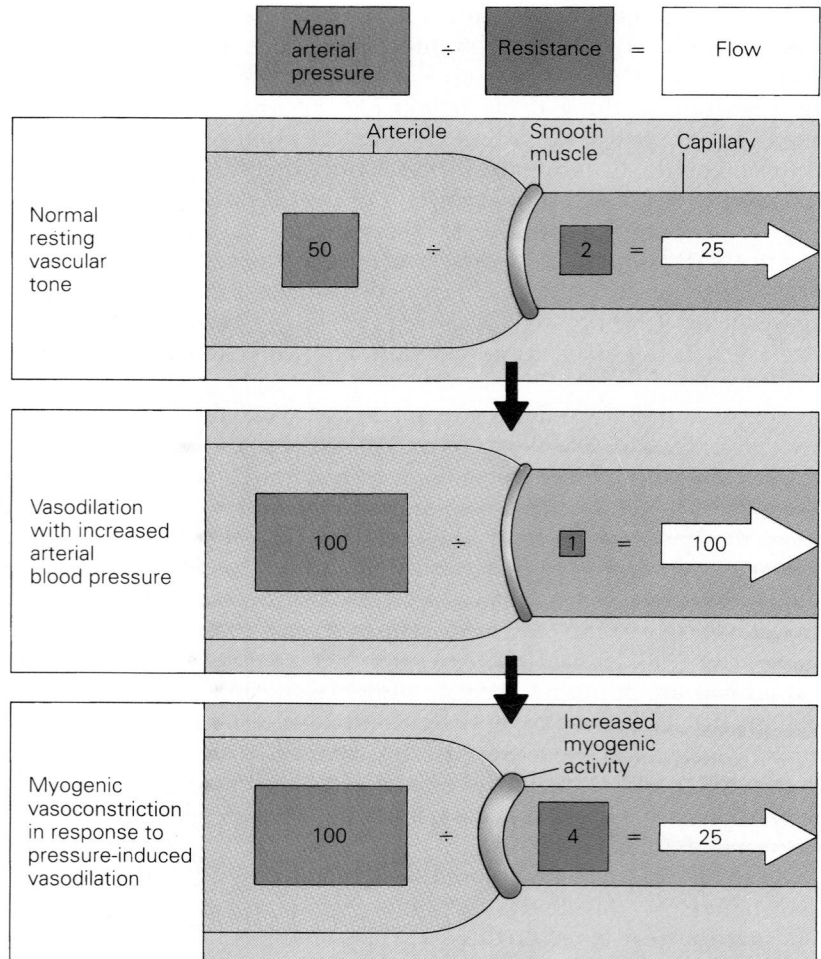

Figure 19–11
Autoregulation of capillary blood flow by myogenic activity of arteriolar and precapillary sphincter smooth muscle. Flow remains constant as perfusion pressure changes because of changes in vessel resistance due to myogenic activity.

portion to the metabolic demand of the tissue **(active hyperemia)**. This is dramatically illustrated in skeletal muscle, in which blood flow may increase as much as 20-fold in response to intense exercise. Although some of this increase is mediated by factors external to the muscle, most of it is mediated by the change in the chemical composition of the interstitial fluid.

Consider the chemical changes that occur as a result of cellular metabolism. The oxygen concentration decreases as a result of cellular use of oxygen for energy production. A wide range of metabolic end-products are generated, including carbon dioxide, adenosine, and hydrogen ions. In addition, the potassium ion concentration of the interstitial fluid increases with increased cellular metabolism because of potassium loss from inside the cells. All of these changes (increased carbon dioxide, decreased oxygen, increased adenosine, increased hydrogen ion, and increased potassium ion concentration) have been shown to promote vasodilation of the ar-

teriolar smooth muscle (Fig. 19–12). Vasodilation results in increased local blood flow, which tends to increase the oxygen concentration and reduce the metabolite concentrations in the interstitial fluid back toward resting values. Realize that local vasodilation in response to increased metabolism can be an effective method for regulating local blood flow only if the blood vessels are partially contracted at rest (resting tone).

Chemical changes such as these contribute to the autoregulatory response to changes in blood pressure that were discussed previously. For example, a decrease in arterial pressure tends to decrease tissue blood flow, promoting the accumulation of chemical substances associated with metabolism and the depletion of oxygen. These chemical changes tend to promote vasodilation and increased blood flow.

One extreme form of chemically mediated autoregulation is seen following the complete occlusion (blockage) of blood flow to a tissue. When the occlu-

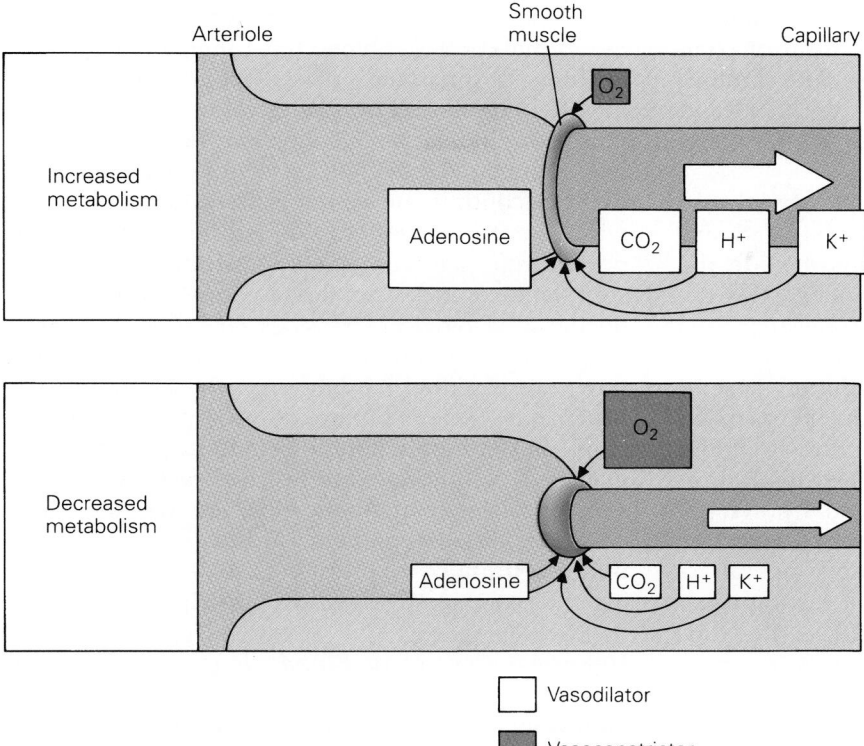

Figure 19–12
Autoregulation of capillary blood flow by local tissue factors. Relative concentrations are indicated by the size of the label.

sion is removed, an extreme reactive vasodilation occurs **(reactive hyperemia).** This is apparently due to the depletion of oxygen and the accumulation of metabolites that occur during the absence of blood flow. Next time you have your blood pressure measured, watch the changes in the blood flow to the skin in your forearm during and after inflation of the pressure cuff. Upon inflation of the blood pressure cuff, blood flow to the forearm is diminished. If the cuff is left inflated for several seconds and then deflated, the skin of the forearm will flush as a result of the reactive vasodilation of the arteriolar smooth muscle associated with the subcutaneous vasculature. As the increased blood flow restores the normal chemical balance of the interstitial fluid, vascular tone increases and blood flow to the skin returns to normal.

A number of other vasoactive substances produced locally by cells may also modify the level of vascular tone. For example, **histamine** released from mast cells during an inflammatory response (see Chapter 17) stimulates vasodilation. **Prostaglandins** may also be released from cells to cause either vasodilation or vasoconstriction, depending on the specific prostaglandin released. Vasoactive substances are also released from the endothelial cells that line the blood vessels. The specific roles of many such locally released vasoactive agents remain to be clarified.

Regulation of Blood Flow by Nerves and Hormones

Earlier in this discussion we noted that some form of communication between isolated organs and tissues is important in the distribution of blood flow, particularly under conditions in which cardiac output is limited. Massive and complete vasodilation of all the vascular beds in the body would result in a drastic fall in blood pressure because the heart would not be able to pump fast enough to maintain such a high level of flow to all the tissues (recall the water system analogy).

This is precisely what happens with heat exhaustion. The vasculature to the skin is so extensively vasodilated (to dissipate excessive heat from the body) that blood flow to other critical organs (e.g., the brain, heart, and kidneys) is seriously compromised. A less extreme example of vasodilation occurs during strenuous exercise, when blood flow to the muscles may increase to greater than 15 L/min. In both of these situations, the heart is not able to provide normal flow to the remaining tissues, and therefore redistribution of flow must occur. As pointed out in the water system analogy, redistribution of flow requires some form of communication between the various users. The communication link in our bodies involves the CNS and ANS as well as circulating hormones.

Neural Control of Blood Flow. Most arterioles in the body are richly innervated by postganglionic sympathetic nerve fibers. (Exceptions include the coronary vessels, which supply the heart, and the cerebral vessels in the brain; these two vascular beds are predominantly regulated by local metabolic factors.) The sympathetic neurotransmitter **norepinephrine** binds to **alpha receptors** on the vascular smooth muscle cells and stimulates contraction (Fig. 19–13). Under resting conditions, the systemic arterioles are constantly being stimulated by sympathetic nerve impulses originating from **vasoactive centers** in the medulla. This continuous neural activity contributes to the basal level of resting tone, which we have previously discussed.

Although the parasympathetic neurotransmitter **acetylcholine** is a vasodilator, parasympathetic nerve fibers are not generally involved in the neural regulation of blood flow. This presents us with a rather one-sided control system that can only stimulate vasoconstriction. This does not represent a serious problem for two reasons. First, in tissues such as the skin that rely heavily on sympathetic activity for normal control of blood flow, there is a high level of tonic sympathetic nerve activity and consequently a high level of resting vascular tone. Under these conditions, vasodilation can be accomplished by a reduction of the level of sympathetic nerve activity. Second, if one examines the conditions under which neural regulation of blood flow plays an important role, we see that a major function of neural stimulation is to restrict flow to nonessential organs during physiological stress. Organs whose uninterrupted function, and therefore uninterrupted blood flow, are essential for life (e.g., the brain and the heart) have less sympathetic innervation than organs that can function intermittently (e.g., skin, liver, spleen, and gastrointestinal tract). Therefore, a major role of sympathetic innervation of the resistance vessels is to prevent cardiovascular collapse, under conditions of physiological stress, by reducing the blood flow to certain noncritical organs.

Hormonal Control of Blood Flow. A number of endocrine substances have vasomotor activity. **Epinephrine,** which is released from the adrenal medulla, can bind to both alpha and **beta receptors** on smooth muscle cells. The binding of epinephrine to beta receptors of smooth muscle produces relaxation and vasodilation of the arterioles. Although

Figure 19–13
Opposing actions of the neurotransmitter norepinephrine and the hormone epinephrine on the contractile state of vascular smooth muscle. While some tissues like skeletal muscle contain both receptor types, most tissues are dominated by one receptor type. For example, the blood vessels of the skin contain predominantly alpha receptors, whereas the blood vessels of the heart (coronary arteries) contain predominantly beta receptors.

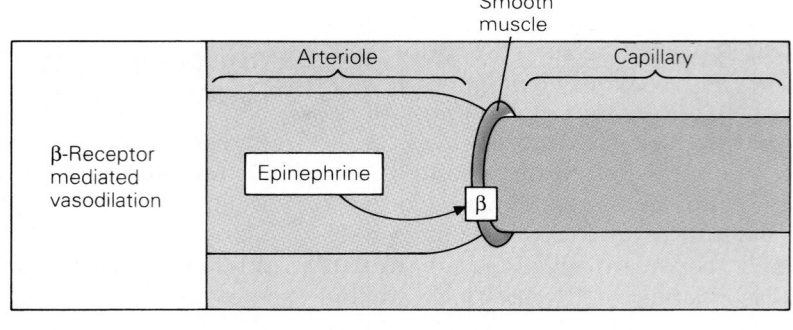

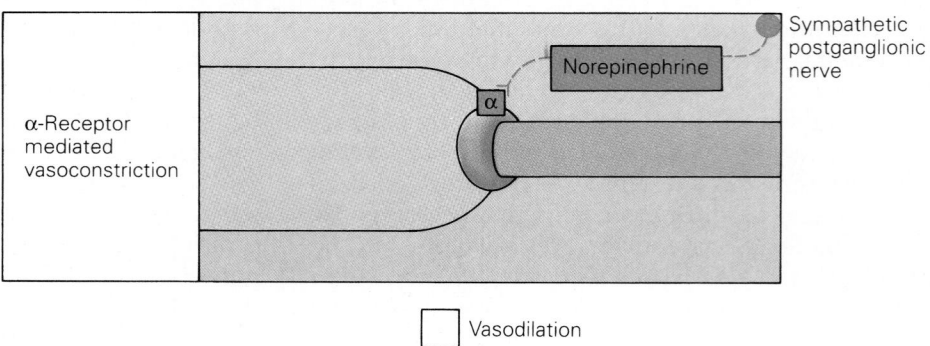

epinephrine can also bind to alpha receptors and produce vasoconstriction, this occurs only at relatively high concentrations of the hormone. The concentrations of epinephrine found in the circulating blood due to medullary stimulation are generally assumed to be too low to allow binding to alpha receptors. While all vascular smooth muscle contains alpha receptors, beta receptors are found predominantly on vascular smooth muscle cells in the coronary vessels of the heart and in the blood vessels in skeletal muscle and liver. Therefore, the effect of circulating epinephrine is to antagonize the vasoconstrictor activity of sympathetic stimulation (see Fig. 19–13). Moreover, the antagonism of nor-

epinephrine-mediated vasoconstriction is most pronounced in the heart, skeletal muscle, and liver, owing to the greater numbers of beta receptors associated with the resistance vessels of these tissues. During periods of elevated sympathetic stimulation, when both norepinephrine and epinephrine are being released, blood flow is shifted away from vascular beds containing predominantly alpha receptors and to vascular beds containing both alpha and beta receptors. Figure 19–14 summarizes the effects of these vasoactive agents on arteriolar resistance.

One of the most potent vasoconstrictor substances is **angiotensin II** (see Chapter 24). Angiotensin II may play a role in the regulation of the renal

Figure 19–14
Summary of local tissue factors that produce either vasodilation or vasoconstriction of vascular smooth muscle.

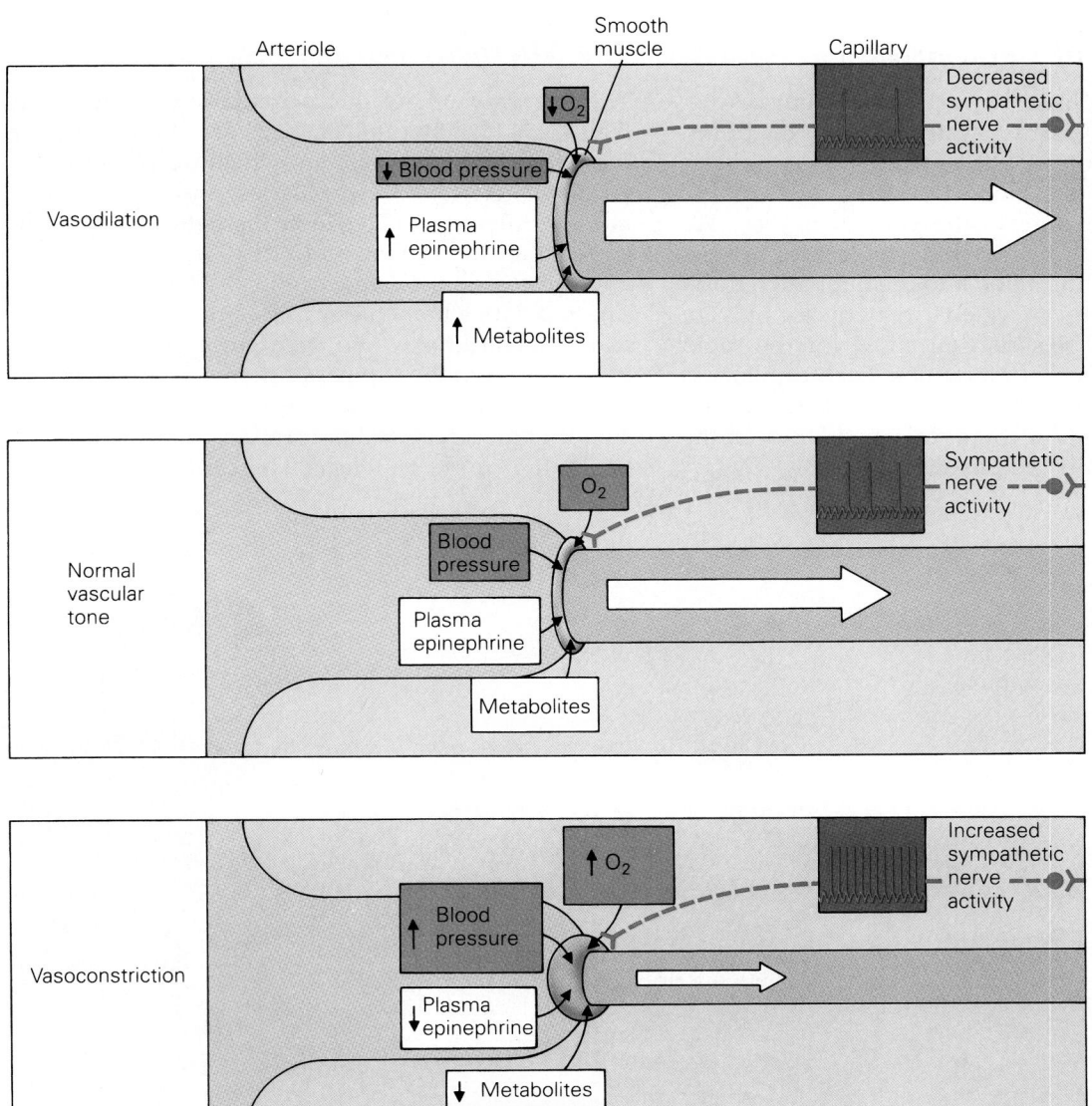

circulation and may be a contributing factor in hypertension. The major role of angiotensin II is probably the regulation of aldosterone production (see Chapter 24).

Atrial natriuretic peptide is a potent vasodilator secreted from cells in a variety of tissues, including the atria of the heart. This peptide causes both vasodilation and increased excretion of salt and water by the kidneys and may play a role in the maintenance of normal blood volume. The function of this peptide in the overall control of vascular tone, however, remains to be determined.

The Capillary System: The Exchange of Molecules

The Function of Capillaries

We have reached an important point in our discussion of the cardiovascular system. Up to this point, we have focused on how the cardiovascular system functions to generate blood flow and to distribute this flow appropriately to the various tissues. We are now ready to consider how molecules are exchanged between the blood and the interstitial fluid. The **capillaries** are an extensive network of very thin-walled blood vessels that allow for the rapid diffusional exchange of molecules. Each capillary is only about 1 mm long, and yet together they represent the major point of communication between the interstitial fluid and the blood.

The capillaries are well suited to this task. The capillary network is so extensive that most cells in the body are within 0.02 mm of a capillary. Because the distances are so small, molecules can rapidly move by diffusion from a cell to the nearest capillary. Once in the blood, molecules are then rapidly transported back to the heart.

A second aspect of the capillary that suits them to this task of molecular exchange is that the capillary walls are only one cell thick. This further facilitates rapid diffusional exchange between the blood and the interstitial fluid. The exchange processes that occur at the level of the capillary represent a dynamic balance between diffusional, osmotic, and hydrostatic forces. We will examine the dynamics of capillary exchange in some detail in the next few sections.

The Microcirculation

The anatomy of the **microcirculation** is shown in Figure 19–15. The capillary beds are at the center of the microcirculatory system, which includes the arterioles, metarterioles, capillaries, and venules. The **metarterioles** (10 to 20 μm in diameter) are slightly larger than the capillaries (5 to 10 μm in diameter) and directly connect the arterioles and the venules. The capillaries branch out from either the arterioles or the metarterioles and then empty into the venules. Although the capillaries themselves contain no smooth muscle, in some cases there is a small cuff of smooth muscle called a **precapillary sphincter** at the beginning of the capillary. The contractile state of

Figure 19–15
Diagrammatic representation of the microcirculation. The diameter of the capillaries is approximately 5 μm.

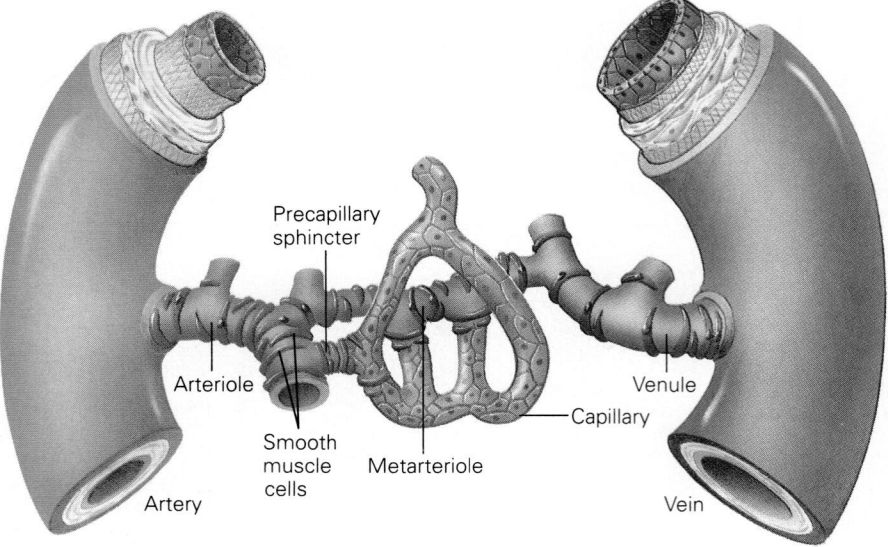

Precapillary sphincter

Arteriole

Smooth muscle cells

Metarteriole

Capillary

Venule

Artery

Vein

the arteriolar and precapillary sphincter smooth muscle determines blood flow through the capillaries. If we were to look at the flow of blood through the capillaries with a microscope, we would see that the blood flow is often sporadic. This is the result of the contraction and relaxation of the precapillary vessels **(vasomotion)**. This vasomotion is heavily influenced by the metabolic state of the tissue. For example, during exercise, the number of open capillaries in skeletal muscle will increase several-fold as precapillary sphincters vasodilate to increase blood flow to the muscle.

The Capillary Wall and Endothelial Pores

As shown in Figure 19–16, the capillary wall is made up entirely of endothelial cells and contains no smooth muscle. The flat, thin, endothelial cells are joined at their edges like patches in a patchwork quilt. Small pores **(clefts)** between the endothelial cells allow for the passage of molecules. The size and number of these **endothelial pores** vary greatly from tissue to tissue. The capillaries in the brain, for example, may not contain any pores, whereas the capillaries in the kidney contain relatively large pores to allow the passage of large molecules.

A thin basement membrane surrounds the outside of the capillaries and may serve as a molecular filter to restrict the passage of very large molecules.

Although the endothelial pores represent an obvious point of exchange, in most capillaries they represent a very small percentage of the capillary surface area. As we will see in a later section, direct diffusion of molecules across the endothelial cell membranes accounts for the major portion of molecular exchange.

Movement of Substances Across the Capillary Wall

There are three fundamentally different mechanisms by which substances can cross the capillary wall. These are **pinocytosis, diffusion,** and **bulk flow** (Fig. 19–17).

Pinocytosis

Relatively large molecules cross the capillary wall via vesicles that pinch off from the plasma membrane. Pinocytosis accounts for a very small portion of the total exchange and is probably important only for the transport of large, lipid-insoluble proteins that cannot cross the capillary wall by movement through pores.

Diffusion

By far the most important mechanism for exchange of water and dissolved substances is diffusion. We can get some idea of this exchange by comparing the

Figure 19–16
(a) Cross-sectional view of a single capillary. The capillary wall is made up of a single layer of endothelial cells. (b) Scanning electron micrograph of a single capillary surrounded by cells and interstitial space. The cells in the lumen of the capillary are red blood cells (erythrocytes). (© Don W. Fawcett/Visuals Unlimited)

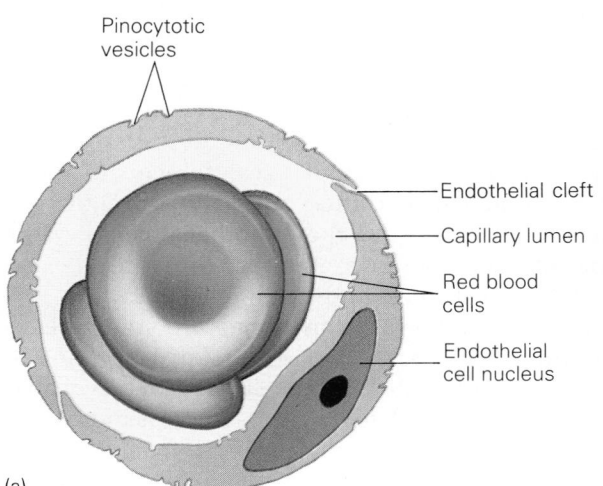
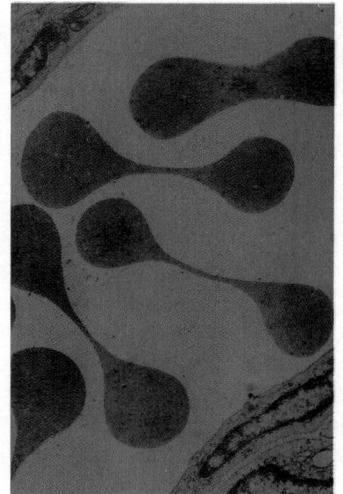

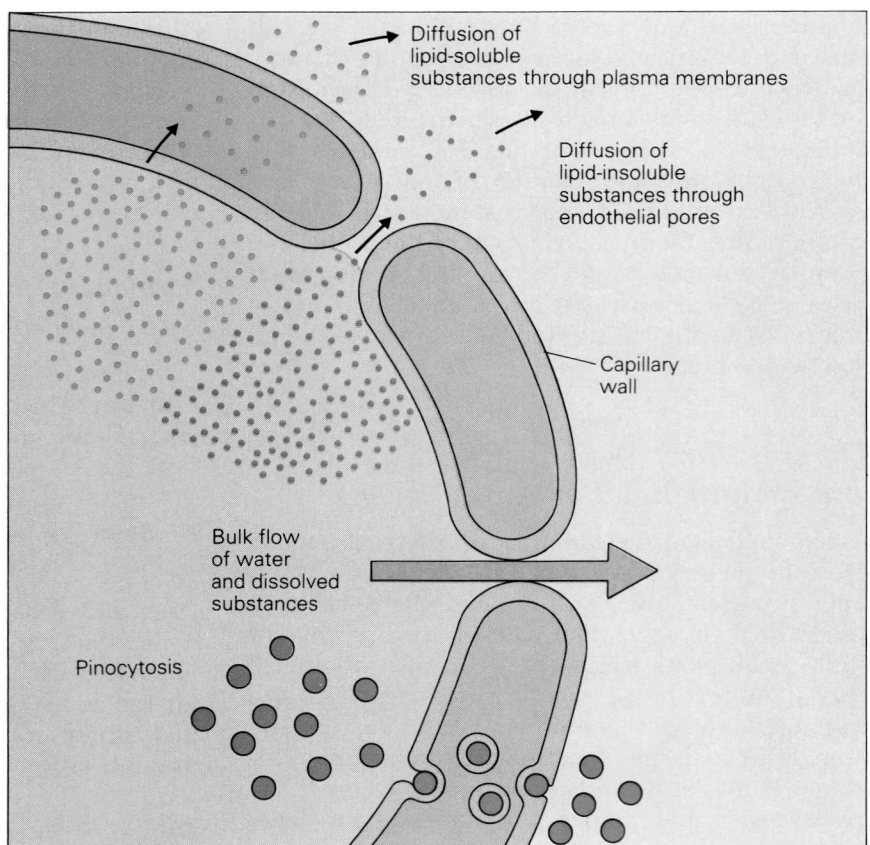

Diffusion of
lipid-soluble
substances through plasma membranes

Diffusion of
lipid-insoluble
substances through
endothelial pores

Capillary
wall

Bulk flow
of water
and dissolved
substances

Pinocytosis

Figure 19–17
Diagrammatic summary of transport pathways across the capillary wall. Movement of substances is in response to either a hydrostatic pressure gradient (bulk flow) or a diffusional gradient.

rate of blood flow through a capillary with the rate of water exchange by diffusion across the capillary wall. The rate of exchange is as much as 100 times greater than the rate of blood flow down the length of the capillary. Since water is moving across the capillary wall in both directions, there is little net gain or loss of water from the circulation by diffusion; however, the exchange rate is very high.

Of course, many solute molecules dissolved in blood and interstitial fluid are also exchanged by diffusion. Blood entering the systemic capillaries has a relatively high concentration of oxygen and glucose and a relatively low concentration of carbon dioxide. The concentrations of oxygen and glucose in the interstitial fluid are lower because these substances are being continuously taken up by the cells. Metabolism of these substances results in the production of carbon dioxide, which diffuses out of the cells into the interstitial space. As a result of these concentration gradients, there is net movement of oxygen and glucose out of the capillary and net movement of carbon dioxide into the capillary (Fig. 19–18). Of course, similar concentration gradients for other nutrients, metabolites, and hormones re-

sult in net transfer of these substances between the interstitial fluid and the blood. In all cases, the direction of the concentration gradient determines the direction of exchange. Analogously, heat generated by metabolism is transferred by convection down its thermal gradient from the cell to the blood.

Molecules can diffuse across the capillary walls either through the water-filled pores or directly through the endothelial cells. The path a molecule takes is determined by its relative solubility in water and lipids. Molecules that are only slightly soluble in lipid (e.g., Na^+, K^+, Cl^-, and glucose) will more likely diffuse through pores. Molecules such as oxygen, carbon dioxide, and urea, which are more lipid-soluble, can diffuse directly through the plasma membranes of the endothelial cells. Water molecules are able to cross either through pores or by diffusion through the endothelial cells.

These pathways for exchange are summarized in Figure 19–17. Because pores occupy less than 1% of the capillary wall area, the surface area available for diffusion of lipid-soluble substances is more than 100 times greater than that for lipid-insoluble substances. We can begin to appreciate how well suited

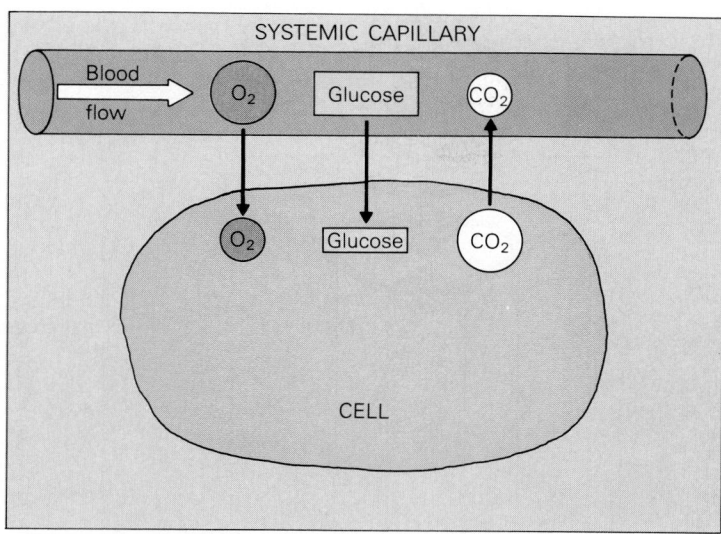

Figure 19–18
Transfer of substances between the cells and the systemic capillary blood in response to diffusional gradients. The size of the label indicates the relative concentrations of a substance.

this system is for the very rapid and continuous exchange of oxygen and carbon dioxide, which are both lipid-soluble.

A number of physical factors contribute to the rapid diffusional exchange of substances across the capillary wall through either pores or the endothelial cell. First, the distances that molecules must move are very small. Most metabolically active cells are within 20 μm of a capillary. Secondly, the surface area of the capillary walls is extensive. Both diffusion distance and surface area influence the rate of net diffusion. The total surface area of all the capillary walls may be as high as 7000 sq. ft. Imagine spreading the entire blood volume of the capillaries (approximately 250 ml or 1 cup) over an area the size of a football field; this gives you some idea of the ideal conditions for diffusion that exist in the capillaries.

A third factor is the relatively slow rate of blood flow that exists within the capillaries. The velocity of blood flow through the capillaries is greatly reduced because of the large total cross-sectional area of the capillary bed (Fig. 19–19). The velocity of blood flow is inversely proportional to the cross-sectional area of the vessel(s) through which it is flowing. We see this principle in action whenever we observe the flow of water down a river. The velocity is greatest where the stream narrows (as in rapids) and slowest where the stream widens; total flow, however, is the same at both points. When a stream empties into a lake, the velocity of water movement in the lake is imperceptible because of the dramatic increase in cross-sectional area. For precisely the same reason, the velocity of blood flow in the aorta may

be as much as 600 to 700 times greater than that in the capillaries. The average velocity of blood flow in a capillary is about 1 cm/sec (about 2 ft/min), compared to 40 cm/sec in the aorta. The dramatic reduction in flow velocity in the capillaries means that the time for exchange is substantially increased.

Bulk Flow and Oncotic Pressure

Exchange of water (and dissolved solutes) also occurs through endothelial pores by **bulk flow** in response to the pressure gradient between the inside and outside of the capillary (Fig. 19–20). As we will see, the pressure gradient is always from inside the capillary to outside, which causes water to flow out of the capillary. Although the total exchange of fluid by this means is relatively small (except in the kidney), it represents an extremely important mechanism for the maintenance of the circulating blood volume. This exchange is important because it occurs in response to a pressure gradient rather than a diffusional gradient.

The hydrostatic pressure gradient that drives water through the endothelial pores is simply the difference between the hydrostatic pressure inside the capillary (capillary blood pressure) and the hydrostatic pressure outside the capillary (interstitial hydrostatic pressure). The hydrostatic pressure outside the capillary is extremely difficult to measure, and there is some uncertainty about the actual value. Although it may in fact be slightly subatmospheric (negative), for purposes of our discussion we will assume that it is zero. Given this assumption, the hydrostatic pressure gradient (capillary hydro-

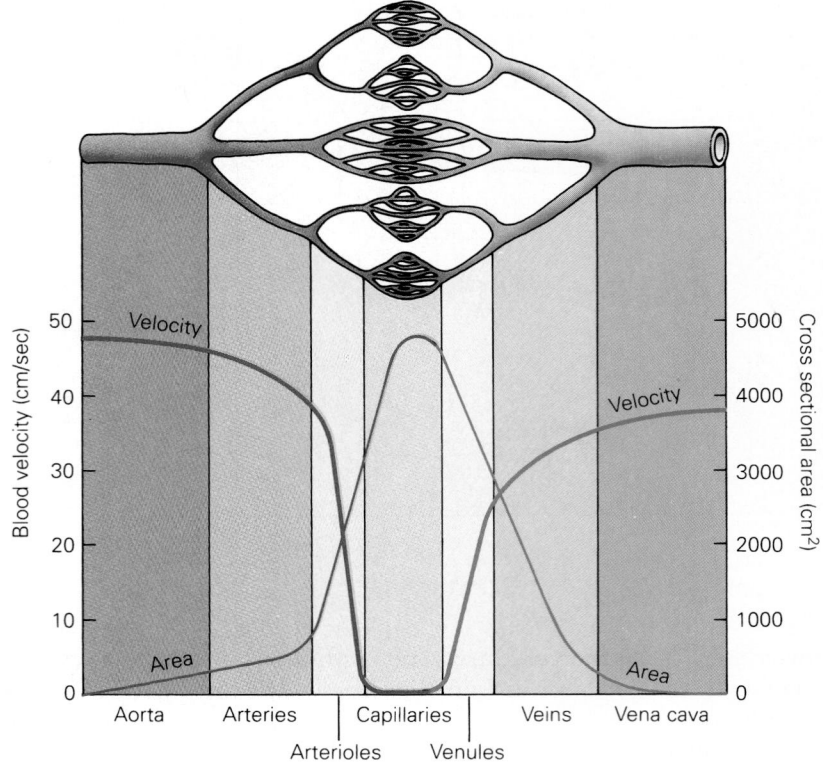

Figure 19–19
Vascular cross-sectional area and velocity of blood flow. Cross-sectional area is the total area of all the vessels added together at each level in the circulatory system. Blood velocity is that which would be measured in a single vessel at each level in the circulatory system.

static − interstitial hydrostatic pressure) is equal to the capillary hydrostatic pressure. When a blood vessel (usually a superficial vein) is cut, blood flows

Figure 19–20
Bulk flow of water and dissolved substances out of the capillary through endothelial pores in response to a hydrostatic pressure gradient. Hydrostatic pressure inside the capillary is always greater than the pressure of the interstitial fluid surrounding the capillary.

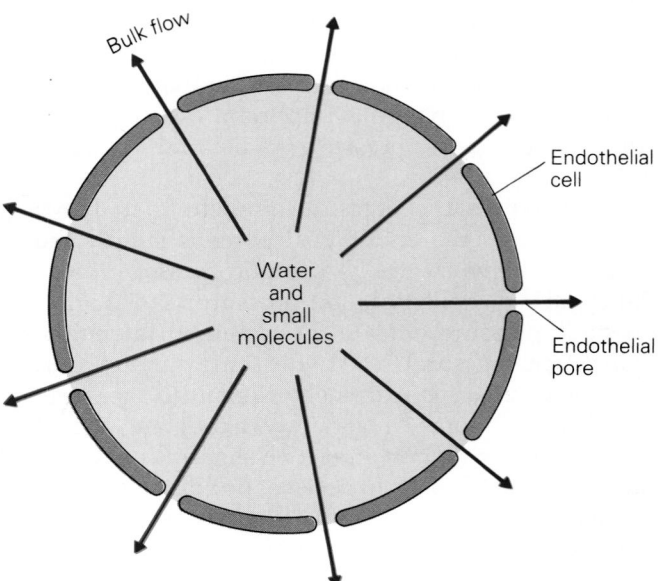

out of the vessel in response to the pressure difference between the inside and the outside of the vessel. Since water can move through the capillary pores by bulk flow, we can reason that water and dissolved solutes should be forced out of the vessels. Clearly, if this were the whole story we would have a great deal of trouble maintaining a normal circulating blood volume. There is in fact an opposing force that minimizes the loss of fluid by bulk flow.

The opposing force that prevents this loss of vascular volume is the **osmotic pressure** due to the plasma proteins (proteins in the blood plasma). Although all dissolved solutes contribute to the **total osmotic pressure** of blood, we can ignore the contribution of all solutes that are freely exchanged across the capillary wall because these solutes are at diffusional equilibrium. Only the plasma proteins that do not freely cross the capillary wall contribute to the osmotic pressure gradient between the inside and outside of the capillary. The terms **oncotic pressure** or **colloid osmotic pressure** refer to that portion of the total osmotic pressure of blood that is due to the presence of plasma proteins. Recall from Chapter 4 that net diffusion of water across a selectively permeable membrane will occur in response to a concentration gradient for water. The capillary wall functions as a selectively permeable membrane because of its limited permeability to plasma proteins. Recall that the major plasma protein is albumin.

Most plasma proteins cannot leave the capillary and therefore establish an oncotic pressure gradient that favors the net diffusion of water from the interstitial space to the intravascular space. You have probably guessed by now that the oncotic pressure exactly balances the hydrostatic pressure so that there is normally no net loss of water from the circulating blood.

Filtration and Absorption: Starling's Hypothesis

This balance of hydrostatic and oncotic pressures across the capillary endothelium was originally described by Starling in what is now called **Starling's hypothesis** (Fig. 19–21). Let's begin by looking at an idealized average capillary. (The situation in any

Figure 19–21
Summary of hydrostatic and osmotic forces causing movement of water across the capillary wall. The ideal capillary shown here represents the average of all capillaries in the body. Pressures within individual capillaries vary considerably. (Modified from *Human Physiology*, 4th Ed, 1985, ed. Vander, Sherman, Luciano. Fig. 11–53, p. 346).

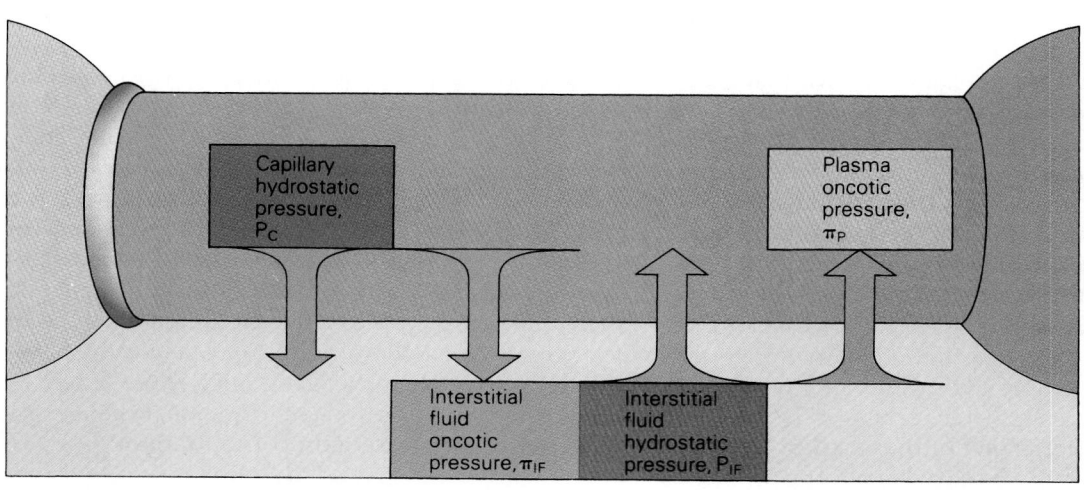

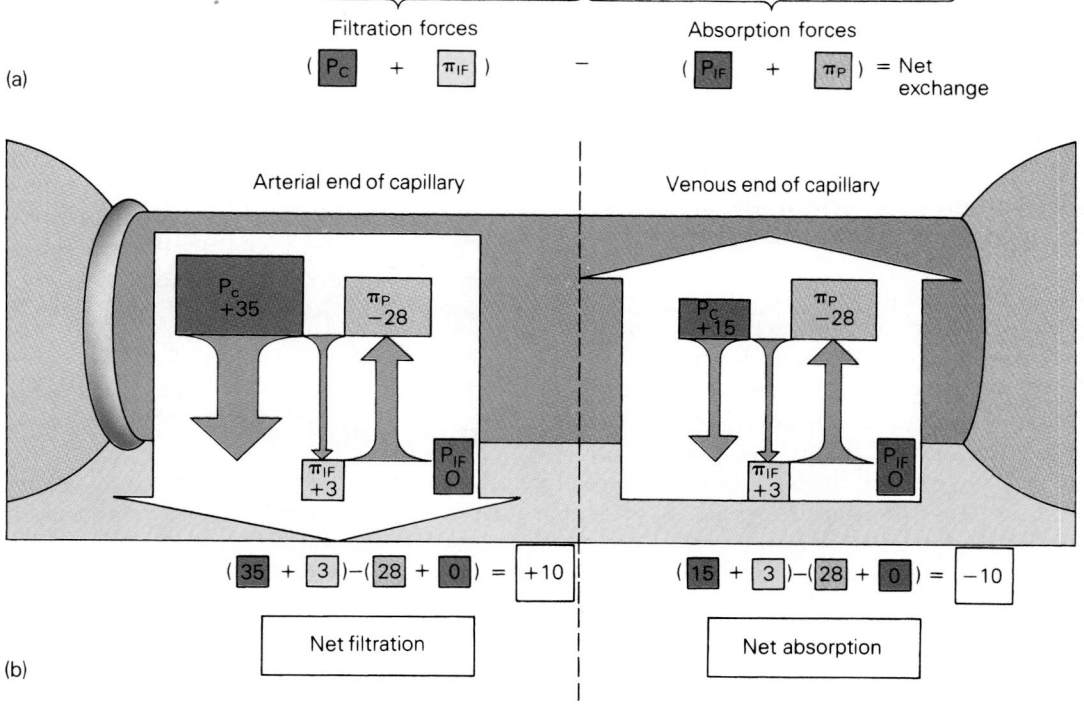

particular capillary might differ significantly from this model, but the principles remain the same.) The oncotic pressure gradient is determined by the difference between the oncotic pressure of the capillary blood and that of the interstitial fluid. The oncotic pressure of blood is normally about 28 mm Hg. Although the capillaries are relatively impermeable to plasma proteins, some proteins do leak out of the capillaries. This small amount of protein in the interstitial fluid results in a tissue oncotic pressure of about 3 mm Hg. The oncotic pressure gradient (28 − 3 = 25 mm Hg) causes a net flow of water into the capillary. Remember that the osmotic flow of water is in the direction of increasing osmotic pressure. This is easy to remember if you recall that osmosis is the diffusion of water down its concentration gradient. Water is less concentrated in blood because of the presence of plasma proteins.

If we examine a typical capillary, we will measure a hydrostatic pressure of about 35 mm Hg at the arterial end of the capillary and about 15 mm Hg at the venous end. The pressure drop along the length of the capillary reflects the resistance to flow of blood through the capillary. The hydrostatic pressure causes fluid to move out of the capillary; this is opposed by the oncotic pressure that moves water back into the capillary (Fig. 19–21). For the sake of this discussion, we can say that a positive pressure is one causing loss of capillary water and a negative pressure is one causing water reabsorption into the capillary. At the arterial end of the capillary, hydrostatic pressure (35 mm Hg) exceeds oncotic pressure (−25 mm Hg), leading to a net **filtration pressure** of 10 mm Hg. On the other hand, at the venous end of the capillary, hydrostatic pressure (15 mm Hg) is less than oncotic pressure (−25 mm Hg), leading to a net **absorption pressure** of −10 mm Hg. At the middle of the capillary, a hydrostatic pressure of 25 mm Hg is exactly balanced by an opposing oncotic pressure of −25 mm Hg. In this example, filtration exactly equals absorption, which means there will be no change in the overall capillary volume.

We stated earlier that capillary flow often changes dramatically, following the contraction of smooth muscle in the terminal arterioles and precapillary sphincters. These changes in flow are associated with similarly large changes in intracapillary pressure. As shown in Figure 19–22, when precapillary smooth muscles are fully dilated, the associated capillaries show filtration along their entire length. Fully contracted precapillary muscles lower the capillary pressure enough to cause absorption of fluid along the entire length of the capillary. Changes in arteriolar pressure can also lead to changes in capil-

lary pressure to favor either filtration or absorption. We can get some idea of what the "average" state of the capillaries is by looking at the net loss or gain of fluid from the cardiovascular system. Under normal conditions, approximately 2 to 5 L of fluid is returned from the interstitial space to the circulation by the lymphatic system during a single 24-hour period. Thus, the average situation slightly favors filtration over absorption.

Filtration and absorption provide an important mechanism for the dynamic maintenance of the circulating blood volume. Although the cardiovascular system is a closed system at the "macro" level, it is an open system at the level of the microcirculation. The constancy of the circulating blood volume in fact represents a dynamic balance between forces tending to move water into and out of the capillary. Recall that cardiac output is determined in part by filling of the heart (end-diastolic volume). Cardiac filling, in turn, is greatly influenced by the total volume of circulating blood. Any significant reduction in blood volume leads to a reduction in cardiac output. As we will see later, transfer of fluid from the extravascular space into the capillary represents an important compensatory response to blood loss (hemorrhage). The distribution of fluid can also be affected by disease. Heart failure ultimately leads to an increase in capillary pressure, causing the accumulation of fluid in the interstitial space (edema). Other diseases may also lead to edema, because of a decrease in plasma protein concentration.

The Venous System: Returning Blood to the Heart

The capillaries empty into thin-walled vessels called **venules,** which represent the beginning of the venous system. The venules contain endothelial clefts like those in the capillaries. Although fewer in number, these allow for additional exchange of substances. The capillaries and venules together are often referred to as the **exchange vessels.** The venules then converge to form progressively larger veins.

Venous Capacitance

The physical nature of the veins is considerably different from that of the arteries and arterioles. The veins are to the arteries the way a paper bag is to a balloon. The veins, like a paper bag, can accommodate large changes in volume with little change in internal pressure. In addition, the total volume of

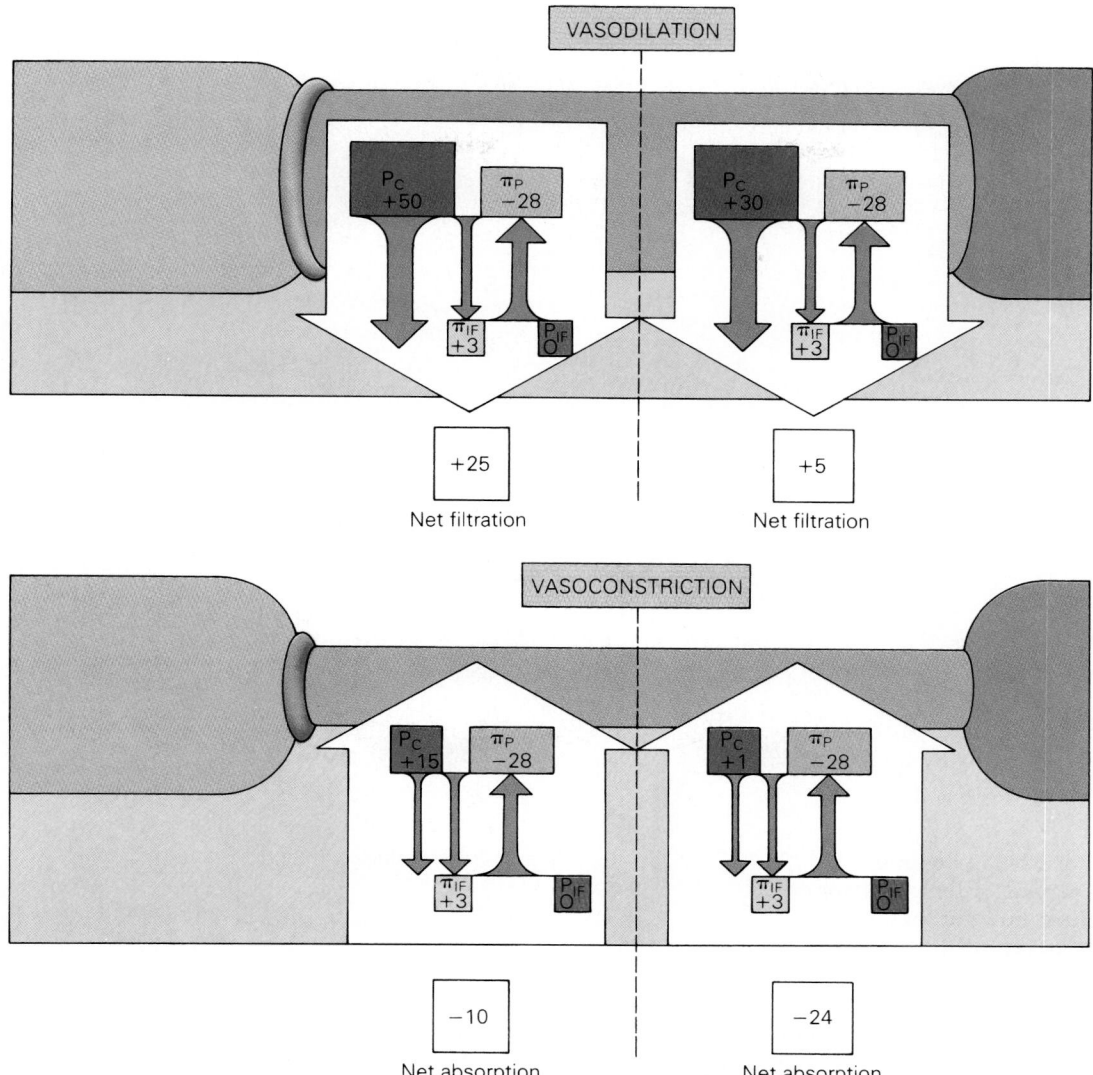

Figure 19–22
Effect of changes in arteriolar and precapillary sphincter smooth muscle contraction on the balance of hydrostatic and osmotic forces in the capillary. Vasodilation promotes filtration, and vasoconstriction promotes absorption of fluid.

the venous system is several times greater than that of the arteriolar system. These characteristics allow the veins to function as a low-pressure reservoir for blood; the veins are often referred to as the **capacitance vessels.** At any one time, approximately 75% of the total blood volume is contained within the veins, compared to 20% in the arterial system (Fig. 19–23).

This large capacity in and of itself is not significant. What is important is that the veins can reduce their capacity through contraction of the smooth muscle in their walls. This occurs in response to in-

creased sympathetic nerve activity. As the venous smooth muscle contracts, blood is transferred from the veins to the heart and the arterial system. As we will see in a later section, this ability of the veins to dynamically alter their capacity provides a mechanism by which the body can maintain end-diastolic volume of the heart. Recall that the distribution of blood volume can change dynamically as a result of changes in arteriolar resistance. For example, during exercise, the volume of blood in muscle increases as capillary beds open up in response to dilation of precapillary resistance vessels. Other

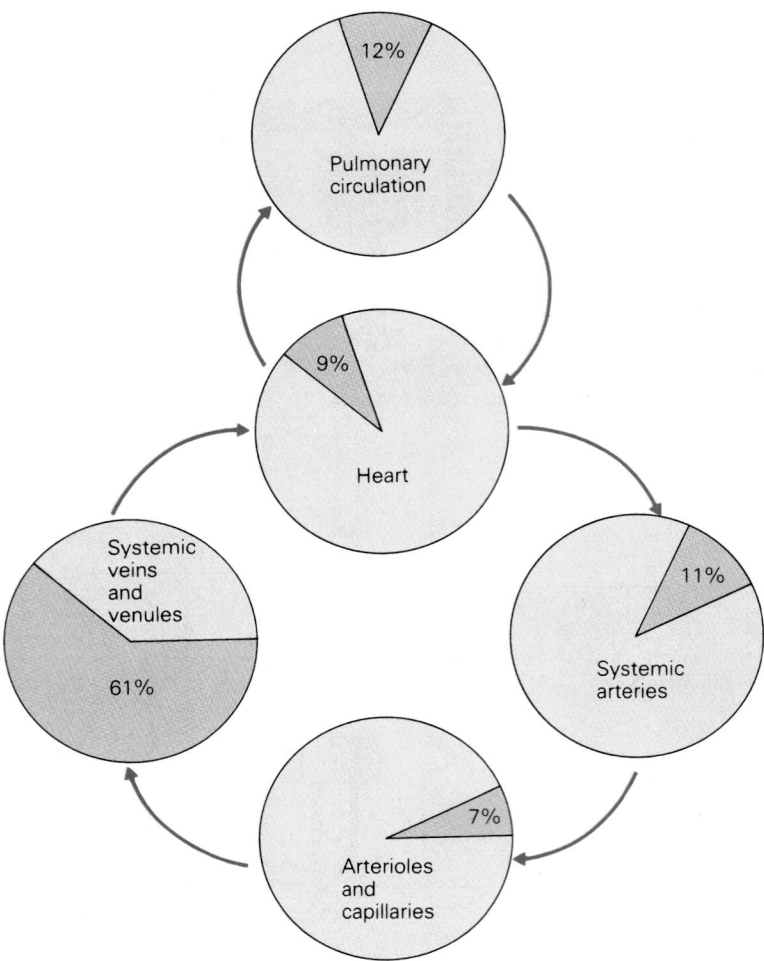

Figure 19–23
Distribution of the total blood volume among the various regions of the circulatory system. The total blood flow through each region is the same (see Fig. 18–3).

Venous Return

The pressure in the venules and small veins is only about 10 mm Hg. Since the pressure at the right atrium is essentially zero, there is a pressure drop of only 10 mm Hg between the venules and the heart, indicating that the resistance to venous blood flow is low. This low resistance is due to the large cross-sectional area of the veins.

Several other factors also contribute to the flow of blood back to the heart. Some veins contain one-way valves which function in a way similar to those in the heart (Fig. 19–24). These valves serve two major functions. First, by preventing backflow of blood away from the heart, they tend to minimize the pooling of blood in the feet and legs upon standing. Second, they allow the contractile activity of the muscles surrounding the veins to contribute to the flow of blood back to the heart. Contraction of the surrounding muscle tends to squeeze the veins. Blood is forced out of the locally constricted vein and, because of the valves, is constrained to flow toward the heart. So you can see that the combination of venous valves and skeletal muscle contraction constitutes a pump that assists the heart in maintaining the unidirectional flow of blood through the circulatory system.

An additional pumping activity further facilitates the flow of blood back to the heart. The muscular activity associated with normal respiration acts through the venous system to pump blood back to the heart. When the diaphragm descends during inspiration, the contents of the abdomen are compressed along with the abdominal veins. As before, this compression of the veins forces blood back toward the heart. During inspiration, the intrathoracic pressure decreases; this provides a negative pres-

factors such as postural changes (e.g., standing up from a recumbent position) can result in passive redistribution of blood volume in response to gravity. Contraction of the veins can compensate for these changes. We will say more about this in a later section when we discuss the overall regulation of cardiovascular function.

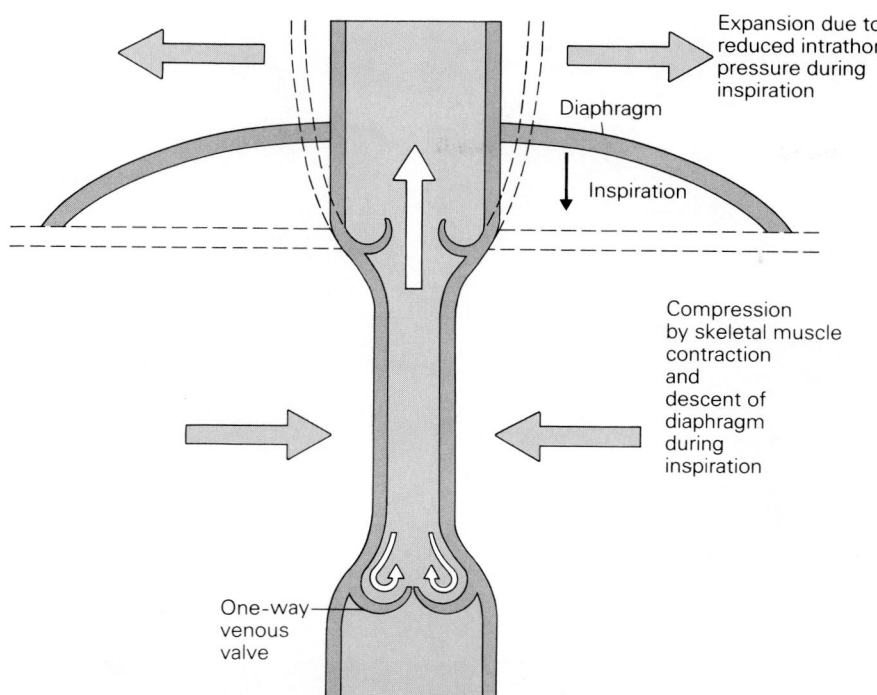

Expansion due to reduced intrathoracic pressure during inspiration

Diaphragm

Inspiration

Compression by skeletal muscle contraction and descent of diaphragm during inspiration

One-way venous valve

Figure 19–24
Factors promoting venous return to the heart. The combination of one-way venous valves and the compression of veins by skeletal muscle contraction and respiration enhance the flow of blood from the veins back to the heart.

sure that draws air into the lungs. Since the right atrium is located within the thoracic activity, this negative pressure also increases the flow of blood from the veins into the right heart.

The Lymphatic System

The lymphatic system is a second system of vessels that differ anatomically from the blood vessels (Fig. 19–25*a*). The function of the lymphatic vessels is to return whatever fluid and plasma protein has leaked out of the capillaries into the interstitial space back to the circulating blood. Despite the selectively permeable nature of the capillary wall and the balance of Starling's forces, there is a net loss of protein and fluid from the blood to the interstitial space. The extent of this loss increases significantly during exercise and with certain diseases. The normal operation of the lymphatics prevents the edema that would otherwise occur.

The Structure of Lymphatic Vessels

The lymphatic vessels, like the capillaries, are made up of a single layer of epithelial cells. As shown in Figure 19–25*b*, the lymphatic system begins as a series of sacs with low internal hydrostatic pressure. These terminal sacs have relatively large endothelial pores, allowing for the movement of both fluid and protein into the lymphatic system. The lymph capil-

laries drain into larger vessels, which themselves drain into the subclavian veins.

The Flow of Lymph

Lymph flow is driven by much the same mechanism as the one that propels venous blood back to the heart. The lymphatic vessels have an extensive series of one-way valves like those in the veins (Fig. 19–25*c*). The pumping action comes from phasic contractions of the smooth muscle associated with lymphatic vessels, as well as compression of the vessels by skeletal muscle activity. In 24 hours, the lymphatics may return a volume of fluid back to the heart equal to the total circulating blood volume (5 L). In addition, approximately one quarter to one half of the total plasma protein is returned to the blood during this period. The lymphatic system represents the only means by which plasma proteins can be returned to the blood.

The Filtration of Lymph

The lymphatic vessels perform several additional functions. The lymphatics also transport substances absorbed from the gastrointestinal tract, primarily fat, to the circulating blood. Periodic swellings of the lymphatic vessels, called **lymph nodes,** provide for the removal of microorganisms and other foreign material from the lymph.

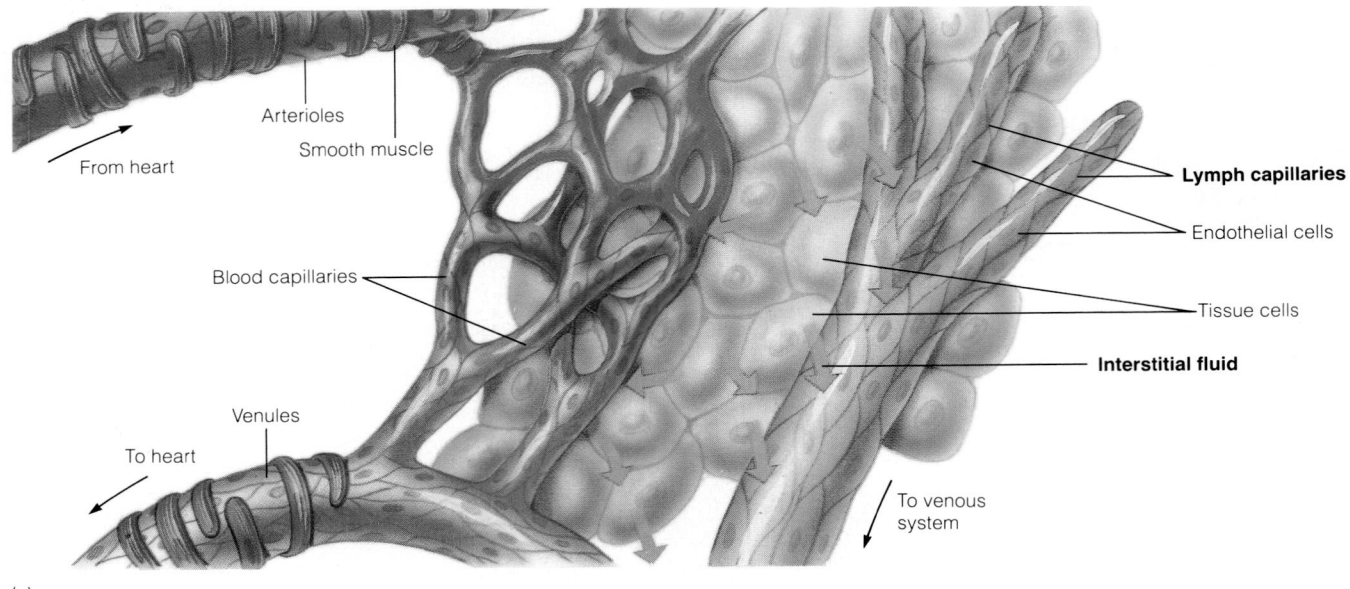

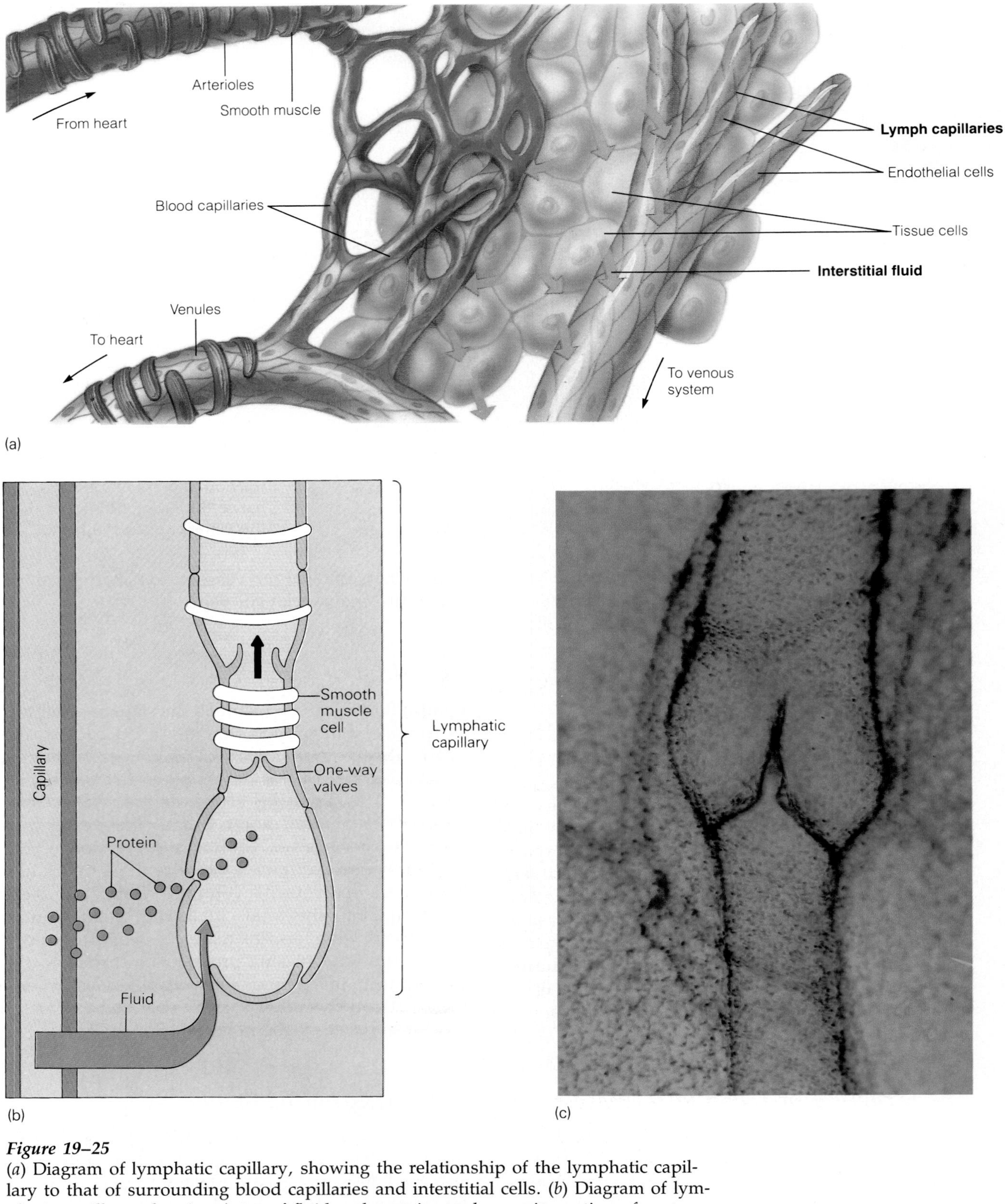

(a)

Arterioles

Smooth muscle

From heart

Lymph capillaries

Endothelial cells

Blood capillaries

Tissue cells

Interstitial fluid

Venules

To heart

To venous system

(b)

Capillary

Smooth muscle cell

One-way valves

Lymphatic capillary

Protein

Fluid

(c)

Figure 19–25
(*a*) Diagram of lymphatic capillary, showing the relationship of the lymphatic capillary to that of surrounding blood capillaries and interstitial cells. (*b*) Diagram of lymphatic capillary, showing entry of fluid and proteins and pumping action of surrounding smooth muscle and lymphatic valves. (*c*) Photomicrograph of lymphatic valve. (© John D. Cunningham/Visuals Unlimited)

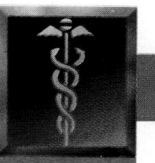

Lymphatic Pathology and Edema

Edema. Any condition which causes blockage or interruption of normal lymph flow will lead to peripheral **edema.** It results from an accumulation of protein (which has leaked out of the capillaries) in the interstitial spaces. The osmotic pressure caused by the presence of this extra protein promotes the retention of extra intravascular water, causing swelling or edema. Obstruction of lymph flow may result from injury, inflammation, parasitic infection, or surgery. Lymph nodes are surgically removed during a radical mastectomy in an attempt to arrest the spread of cancer cells from the breast to the rest of the body via the lymphatics. This may lead to temporary edema in the patient's arms; new lymph vessel growth usually alleviates the problem. Another cause of impaired lymph flow is parasitic infection, as shown in the figure on filariasis. **Filariasis** is the result of a parasitic infection of the lymphatic system. The parasitic lar-

Filariasis causes the formation of edema and swelling of the afflicted areas as illustrated here.
(© Science VU/Visuals Unlimited.)

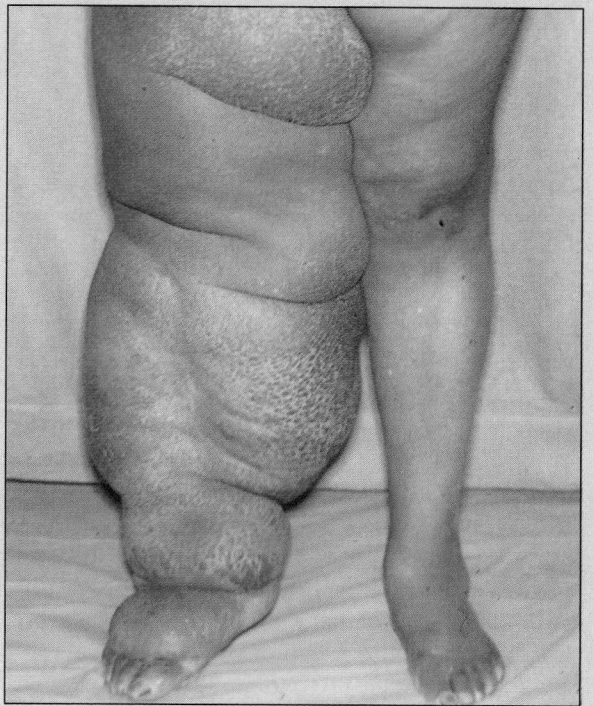

vae are transmitted by mosquitoes, and the adult worms reside in the lymph nodes. The resultant blockage of lymph flow results in progressive swelling of tissues in the affected region. As shown in the figure, affected limbs may swell to several times their normal size. The appearance of such swollen limbs has led to the more descriptive name for this disease—elephantiasis.

The other major cause of edema is accelerated leakage of fluid from the blood to the interstitial space. If the rate of leakage exceeds the rate of fluid removal by the lymphatics, edema will develop. A frequent cause of excessive lymph formation is congestive heart failure (see Fig. 19–35). When the heart fails, blood backs up in the veins and capillaries due to the reduced rate of blood transfer by the heart to the arterial system. The distension of the elastic veins causes increased venous (and therefore increased intracapillary) hydrostatic pressure, followed by increased filtration of water (see Figs. 19–21 and 19–22). Increased fluid leakage may also occur as a result of reduced concentrations of plasma proteins, secondary to either renal failure or severe protein malnutrition (see Figs. 19–21 and 19–22). With a decrease in plasma protein concentration, the edema is not localized and fluid accumulates in the most readily expandable areas of the body, such as the peritoneal cavity (see figure on kwashiorkor).

Varicose Veins. Varicose veins are usually superficial veins that have become excessively dilated, creating a distended and tortuous appearance. The primary cause for this abnormal distention is the failure of the venous valves, or **valvular incompetence.** The one-way venous valves normally facilitate the flow of blood back to the heart and prevent pooling of blood in the peripheral veins. Valvular incompetence is precipitated by factors that increase the volume of blood in the veins. Dilation of the veins by the presence of excessive blood volume impairs the proper closing of the valve leaflets; the resulting valvular incompetence compromises the flow of blood from the veins to the heart and leads to further engorgement. Prolonged distention of the veins in this fashion eventually causes permanent morphologic changes in the vessel wall, with a resultant increase in diameter and a loss in elasticity. Varicose veins can be removed surgically and this may prevent or alleviate problems such as discomfort and ulceration.

(continued)

Children suffering from kwashiorkor, San Salvador, Brazil. (© Richard Frieman/Photo Researchers, Inc.)

Numerous factors contribute to the formation of varicose veins. Prolonged standing promotes venous pooling because the weight of the column of blood between the feet and the heart pushes out on the walls of the peripheral veins, causing a progressive increase in the internal diameter; imagine filling a long narrow balloon with water and then lifting it by one end; the weight of the water and the elasticity of the balloon will allow all the water to remain near the bottom. The degree of distension will be greatest at the very bottom of the balloon because the outward pressure at any point in the balloon is directly pro-

portional to the height of the column of water above that point. Recall that skeletal muscle activity associated with walking and running normally compresses the veins and pushes blood back to the heart. This is equivalent to squeezing the bottom of the water-filled balloon with your hands to force the water back up into the elevated end. The low level of muscle activity associated with passive standing reduces the pumping effects of skeletal muscle and promotes pooling of blood in the peripheral veins. Varicose veins often form during pregnancy. The growing fetus causes compression of all the internal organs, including the inferior vena cava. The compression increases the resistance to venous return and therefore increases the pressure inside the capillaries, and inside the veins between the capillaries and the compression point; the increased internal pressure leads to venous distension. An analogous process is the increase in arterial pressure that follows constriction of the arterioles. **Hemorrhoids** are varicosities of the veins in the anal region. Formation of hemorrhoids is promoted by pregnancy for the same reasons discussed previously, and also by chronic constipation, which leads to excessive straining (attempted expiration against a closed glottis) and hence to elevated venous pressure during defecation.

Notice that no mention has been made of how gravity impedes the *flow* of blood from the lower extremities back to the heart (uphill flow of blood). This is because gravity also *assists* the flow of blood from the heart down to the lower extremities (downhill flow of blood). These two effects essentially balance one another and have *no net effect* on blood flow. In fact, if the veins and arteries were very stiff (like steel pipe), gravity would have no effect at all on the flow of blood down to the feet and back to the heart. As discussed in the accompanying section on venous return, however, the effect of gravity on the distribution of blood within the vascular system can dramatically affect cardiac output and blood flow by altering end-diastolic volume of the heart (see sections in this chapter on venous return and on disturbances of cardiovascular homeostasis: hypotension).

Regulation of Systemic Arterial Pressure

Recall from our previous discussions that the body maintains normal tissue flow by monitoring and maintaining mean arterial blood pressure (MAP). As pointed out in the water system analogy, this approach simplifies the regulatory process and at the same time allows for a great deal of local regulation of blood flow by individual tissues. You will

also recall that under conditions in which the demand for flow exceeded the capacity of the pump, it was necessary to selectively decrease the flow to certain users in order to maintain flow to others. Based on our accumulated knowledge of how the heart and vasculature function, we are now in a position to investigate the mechanisms by which the body achieves the necessary coordination of the heart and vasculature to maintain arterial pressure.

As we discuss how the body operates to maintain blood pressure, it will be easy to lose sight of the fact that the ultimate goal is to maintain blood flow, not arterial pressure. If the goal were only to maintain arterial pressure, this could be achieved with minimal effort by reducing blood flow to all the tissues to a trickle; this, of course, would not be compatible with life. Maintaining arterial pressure provides a simple method for adjusting cardiac output to meet the blood flow requirements of the tissues.

As previously noted, the relationship between cardiac output (CO), mean arterial pressure (MAP), and total peripheral resistance (TPR) is described by the equation:

$$CO = MAP/TPR$$

Algebraic rearrangement gives us the equation:

$$MAP = CO \cdot TPR$$

We can see from this equation that two factors determine the magnitude of the mean arterial pressure. One (CO) is determined by the status of the heart and the other (TPR) by the operational status of all the resistance vessels in the body. Figure 19–26 summarizes all the factors that we have discussed that can change either cardiac output or total pe-

ripheral resistance. This figure indicates only where relationships exist, without delineating how a change in one component alters related components. The ways in which these various parameters are integrated to achieve normal cardiovascular function are the subject of the next several sections.

Before proceeding, let us look briefly at the meaning of total peripheral resistance. The **total peripheral resistance** is the total resistance to the flow of blood from the aorta back to the right atrium. Keep in mind, however, that the major portion of the resistance to blood flow is posed by the arterioles and precapillary sphincters. More importantly, changes in peripheral resistance are due entirely to changes in the diameters of these vessels. Therefore, when we talk about changes in total peripheral resistance, we will be referring to changes in the overall resistance posed by all the resistance vessels in the body. Although other factors such as hematocrit can alter the viscosity of the blood—and consequently, the resistance to flow—these changes do not constitute a mechanism for the homeostatic maintenance of arterial pressure. We will therefore consider the viscosity of the blood to be constant under most conditions.

Figure 19–26
Summary of the factors that can alter cardiac output and arterial pressure.

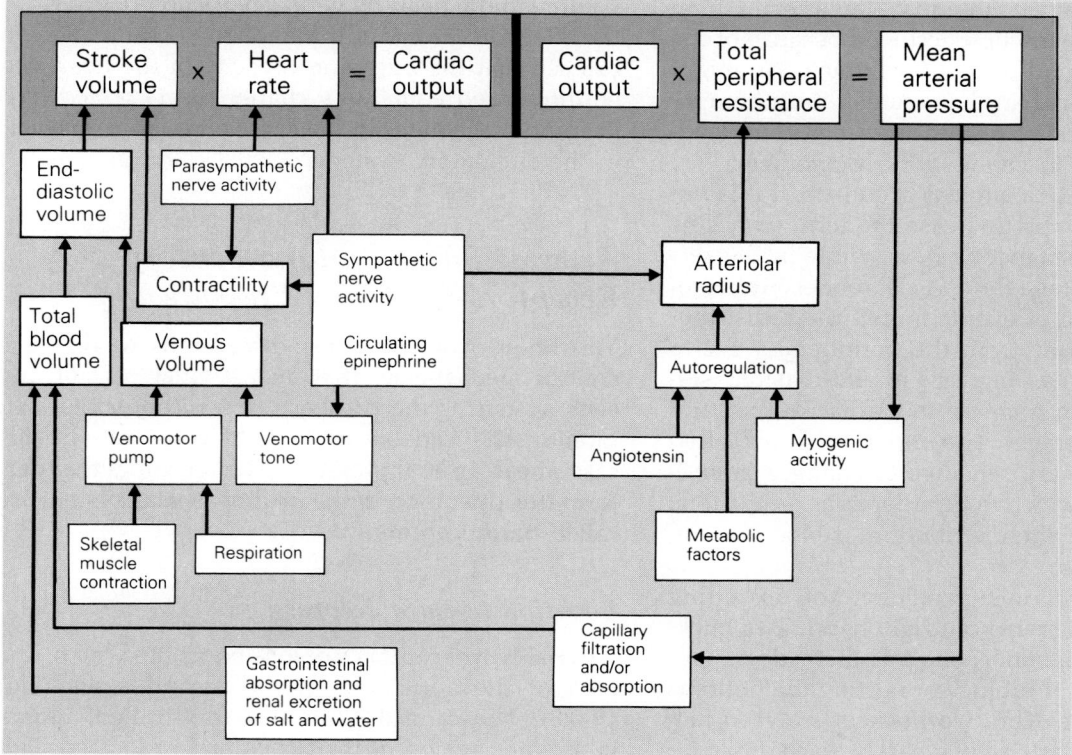

Cardiac Output, Venous Return, and End-Diastolic Volume

These terms must be defined clearly if we are to understand how the factors interact. The terms **cardiac output** and **venous return** obviously refer to blood flow from the heart to the capillaries and from the capillaries to the heart, respectively. What is not obvious is whether they refer to the average flow over a period of time or to the instantaneous flow at any given point in time.

If we use these terms to describe average flow over a period of many minutes, then cardiac output and venous return must be equal since this is a closed system. Frequently, however, we must consider transient changes in flow that occur in selected portions of the system. Consider for a moment the changes following an increase in venomotor tone. Prior to this event, cardiac output and venous return were equal. As the veins contract and shift blood volume from the peripheral veins to the heart, venous return will for a short time be slightly higher than cardiac output because of the extra volume of blood flowing back to the heart from the peripheral veins. As long as venous return to the heart exceeds cardiac output, the volume of blood in the heart will increase. The increase in end-diastolic volume will then lead to an increase in stroke volume and cardiac output (according to the Frank-Starling law). After the venoconstriction, cardiac output and venous return will again be equal (but greater than before the venoconstriction occurred because of the increased cardiac end-diastolic volume). We might then state that an increase in venous return (temporarily) produced an increase in cardiac output, leading to an increase in venous return (average).

This sounds like a circular argument if it is not made clear that in the first case the term *venous return* refers to instantaneous flow, while in the second it refers to average flow. A clearer description of the same sequence of events is that the redistribution of blood volume from the peripheral veins to the heart produces an increase in end-diastolic volume, leading to an increase in cardiac output and venous return (average). The important point is that cardiac output is very sensitive to changes in end-diastolic volume, which in turn depends on the total blood volume and the distribution of blood within the vascular system.

Consider the following analogy. You are sitting on a stool with a glass in your right hand. Two buckets are placed on the floor, one to either side of you, with a short piece of tubing connecting the bottoms of the two buckets. The two buckets are each half full of water, and you can just barely reach down far enough to fill your glass with water. Your goal is to fill the glass with water from the bucket on your right and pour it into the bucket on the left; you are pumping water from one bucket to another just as the heart pumps blood from the veins to the arteries. If you work hard enough, you will begin to accumulate more water in the bucket to your left because you are transferring water faster than it can flow back through the tubing.

But here is where you get into trouble; as you transfer water to the left bucket, the water level in the right bucket goes down. The further down the level goes, the less water you can get in your glass, and you reach a point at which the water level in the bucket sets an absolute limit on your performance. Your capabilities have not diminished, there has been no loss of water, the buckets holding the water are unchanged, but you are limited as a pump because you can no longer fill your glass.

There are two ways to remedy the problem: (1) you can put more water into the buckets to raise the water level (increased blood volume) or (2) you can raise the level of the buckets in order to reach the water in the bottom (increased venomotor tone). Whichever action you take, the results will be the same. This remedy does not require a change in the rate at which water flows back through the connecting tube (venous return). Likewise, it does not matter whether you increase end-diastolic volume by increasing total blood volume or by shifting blood volume to the heart by venoconstriction (Fig. 19–27) the effect on cardiac output will be the same. So you can see that the pumping capacity of the heart can be influenced equally by changes in *total blood volume* and by changes in the *distribution of blood volume* in the circulatory system.

Arterial Baroreceptors: Sensory Receptors for Blood Pressure

You know from previous discussions of reflexive control mechanisms that an essential part of any such system is the presence of a receptor that can monitor the variable to be held constant—in this case, mean arterial pressure. The receptors that perform this operation in the cardiovascular system are called **baroreceptors.**

Location of Baroreceptors

Arterial baroreceptors are located in the wall of the arch of the aorta and in the carotid sinus (Fig. 19–28). The **carotid sinus** is a thin-walled, highly innervated region of the carotid artery located high

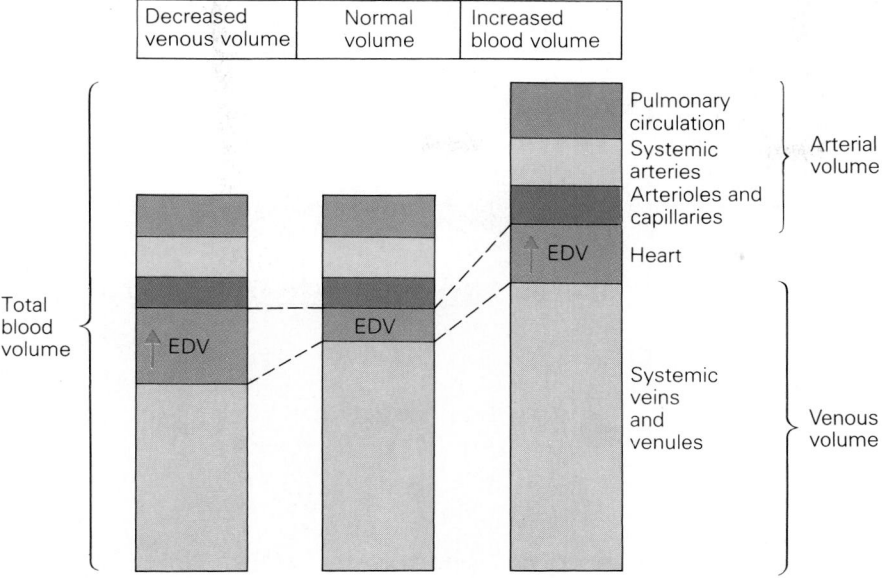

Figure 19–27
Summary of factors affecting end-diastolic volume of the heart.

in the neck at the point where the carotid artery branches into two smaller arteries.

Activation of Baroreceptors

The baroreceptor nerve endings function as stretch receptors and are distributed within the wall of the elastic arteries. You have already encountered stretch receptors in other systems in the body, and the characteristics of all stretch receptors are similar. These receptors respond to stretch with an increase in the rate of action potential firing. Because of the elastic nature of the arterial wall, an increase in arterial pressure causes the arterial wall and the associated receptors to stretch. Stretching the nerve endings initiates the firing of action potentials, which are transmitted by the associated afferent (sensory) neurons to the cardiovascular control center.

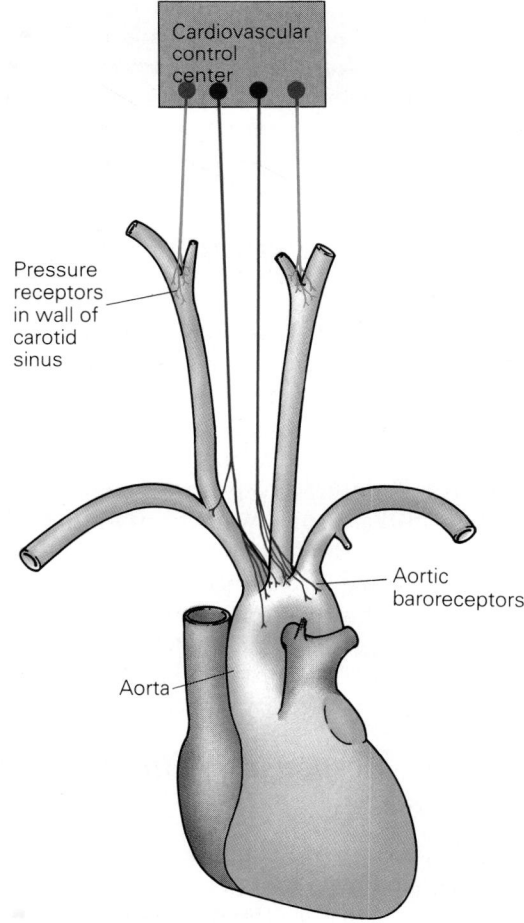

Figure 19–28
Location of the arterial baroreceptor nerve endings in the systemic arterial system. Afferent nerves from the baroreceptors travel to the medullary cardiovascular control centers.

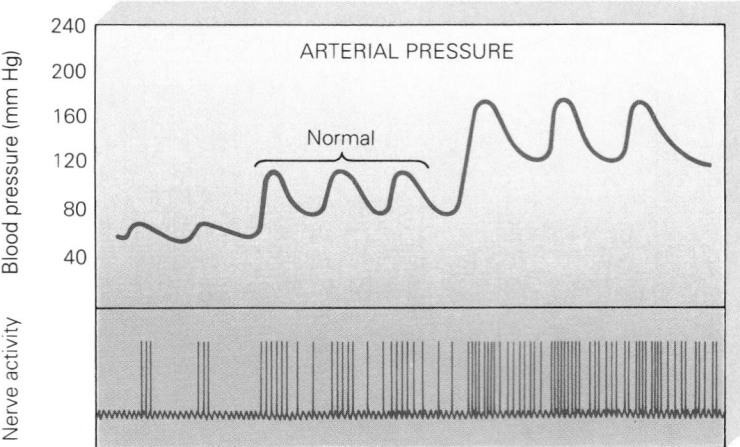

Figure 19–29
Changes in carotid sinus nerve activity as a function of changes in arterial blood pressure (modified from Berne and Levy, *Physiology*, 2nd Ed., 1988, Fig. 33–10, p. 519).

Patterns of Baroreceptor Activation by Changes in Blood Pressure

Figure 19–29 shows the pattern of action potential firing, with changes in both mean arterial pressure and pulsatile arterial pressure. Notice that action potential firing increases in response to the pulsatile pressure changes during a single heart beat. This tells us that these receptors respond very quickly to changes in pressure. Therefore, the baroreceptors are effective in monitoring acute changes in blood pressure. As mean arterial pressure increases, the overall firing rate of the receptor increases.

The Cardiovascular Control Centers of the Brain

The afferent nerves from the baroreceptors travel to the brain, where they synapse with neurons that make up the **cardiovascular control centers.** The primary control center for the baroreceptor reflex is located in the medulla. The medullary cardiovascular control center is a complex and diffuse network of neurons that integrates neural inputs from the baroreceptors with inputs from other control centers in the brain. For now we will concentrate on how changes in the level of input from the baroreceptor lead to changes in autonomic nerve activity to the heart and vasculature.

Cardiovascular Response to Blood Pressure Changes

You will recall from our previous discussions that the heart is innervated by both parasympathetic and sympathetic nerves, and that the blood vessels are innervated primarily by sympathetic nerves. Preganglionic efferent nerve fibers from the cardiovascular centers synapse with the cell bodies of both parasympathetic and sympathetic neurons that innervate the heart and blood vessels (Fig. 19–30). An increase in the rate of nerve activity from the baroreceptors leads to a decrease in the rate of stimulation of sympathetic nerves and an increase in the rate of stimulation of parasympathetic nerves. Conversely, a decrease in baroreceptor nerve activity leads to increased sympathetic and decreased parasympathetic activity. Changes in baroreceptor activity produce reciprocal changes in parasympathetic and sympathetic activity to the heart and blood vessels.

Cardiac Output and Total Peripheral Resistance

Let's now examine the overall response of the system to a change in blood pressure. If arterial pressure increases, the baroreceptor firing rate increases. This leads to a decrease in sympathetic nerve activity to the heart and blood vessels and an increase in parasympathetic nerve activity to the heart. This, in turn, causes heart rate to decrease and contractility to decrease, leading to a reduction in cardiac output. The decreased sympathetic nerve activity to the blood vessels promotes dilation of the heavily innervated vascular beds such as the skin, muscle, and splanchnic circulations, leading to a reduction of total peripheral resistance. The reduction in both cardiac output and total peripheral resistance causes a reduction of arterial pressure (MBP = CO · TPR). Clearly, the response of the system is to minimize changes in arterial pressure by changing both cardiac output and total peripheral

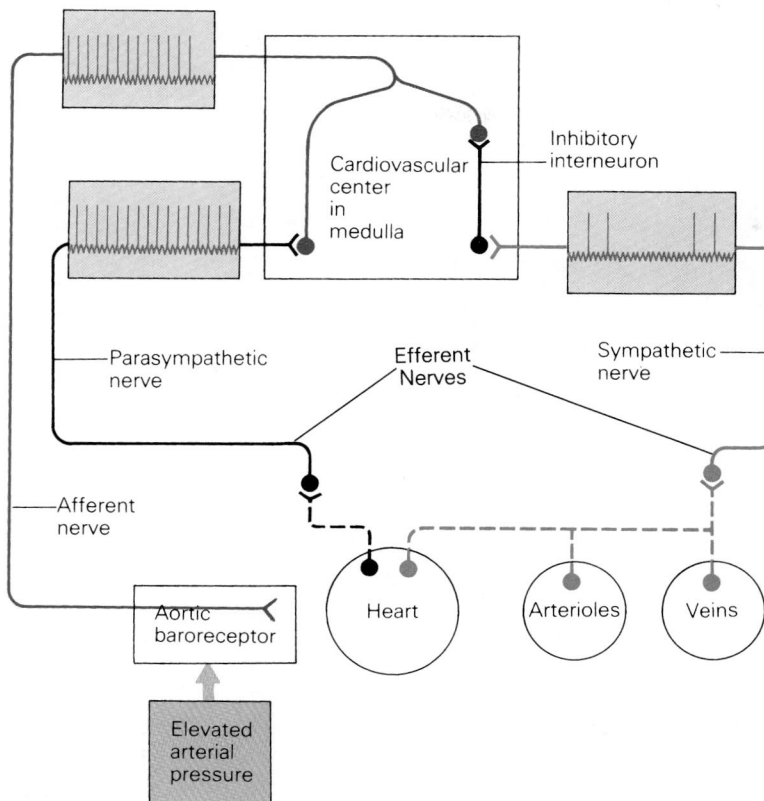

Figure 19–30
Afferent nerve activity from the baroreceptors results in reciprocal changes in efferent sympathetic and parasympathetic
nerve activity to the heart and blood vessels.

resistance in the appropriate directions. If arterial
pressure were to decrease, decreased baroreceptor
firing would lead to increased cardiac output and
increased total peripheral resistance. As before, this
compensatory response would tend to restore arterial pressure back to normal. These responses are
summarized in Figure 19–31.

Venomotor Tone

In our previous discussion of the venous system, we
pointed out that contraction of venous smooth muscle **(venomotor tone)** occurs in response to sympathetic stimulation. Changes in venomotor tone also
play a role in the baroreceptor reflex to maintain arterial blood pressure. In response to increased arterial pressure and increased baroreceptor firing,
there is a decrease in sympathetic activity to the
veins, resulting in venodilation. Venodilation results in a shift of blood volume from the heart and
arteries to the veins. This leads to a decrease in end-
diastolic volume, which decreases cardiac output
according to the Frank-Starling law. The decrease in
cardiac output then leads to a decrease in mean
blood pressure back toward normal.

Long-Term Regulation of Blood Pressure

Many physiologists contend that regulation of blood
volume is one of the most important aspects of
blood pressure regulation, particularly on a long-
term basis. Whereas reflexive mechanisms such as
those just discussed play a dominant role in short-
term pressure regulation, the maintenance of adequate circulating blood volume is critical for normal
cardiovascular function.

As you will see in later chapters, the maintenance of blood volume is primarily under the control of the kidneys (which mediate the excretion of
salt and water from the body) and the gastrointestinal system (which mediates absorption of salt and
water). You will see more clearly how important the
maintenance of blood volume is when we discuss
the ways in which hemorrhage affects cardiovascular performance. Also, we will discuss the way in
which venomotor tone and alteration of blood volume by changes in filtration and absorption in the
capillaries can temporarily compensate for inadequate blood volume. If cardiac output is compromised by inadequate end-diastolic volume, this can
be corrected by (1) increasing the total circulating

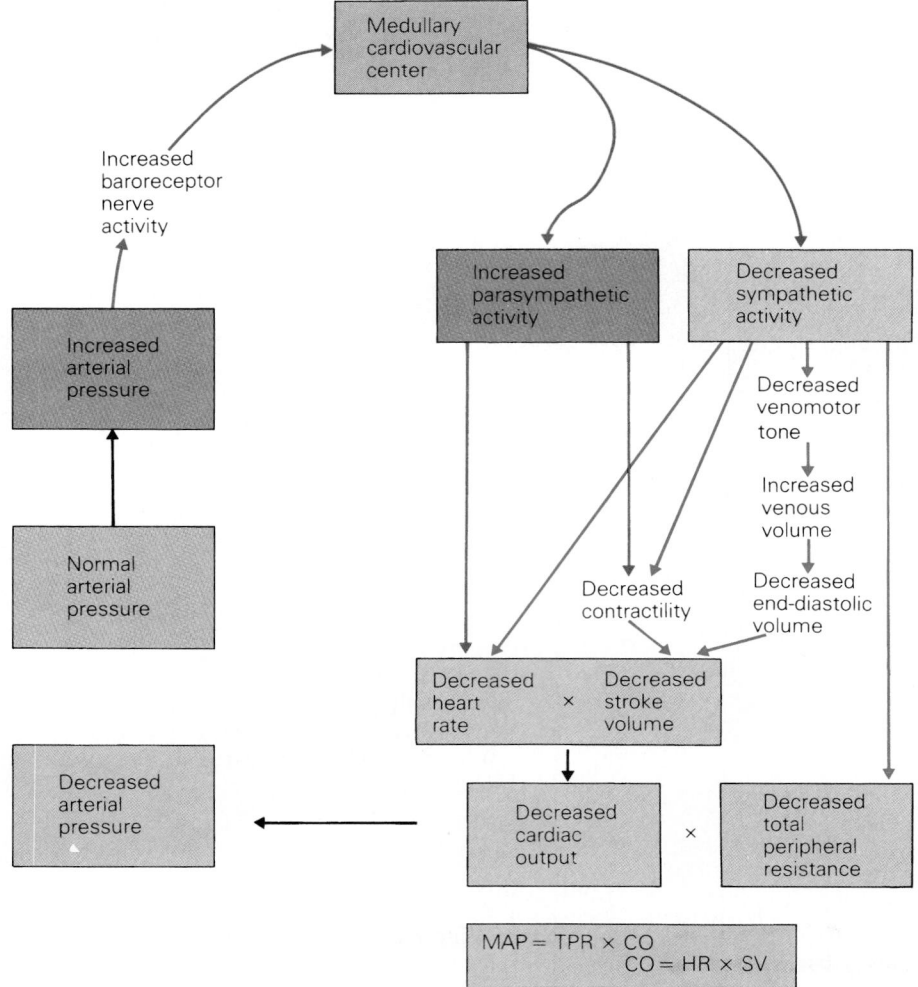

Figure 19–31
Summary of cardiovascular changes in response to increased arterial blood pressure and increased baroreceptor nerve activity. This is an example of negative feedback. An increase in blood pressure leads to compensatory changes that tend to restore blood pressure back toward normal.

blood volume, (2) shifting of blood volume from peripheral vascular beds to the heart by increasing venomotor tone, or (3) increasing the reabsorption of interstitial fluid by the capillaries (Fig. 19–22). These mechanisms are equivalent in terms of their effect on the heart. Remember, however, that they are not equivalent in terms of overall physiological homeostasis. For example, you will see that, during hemorrhage, the body compensates for loss of blood by severely decreasing flow to organs such as the kidneys. While this may temporarily serve to maintain blood pressure and blood flow to the heart and brain, the kidney can be damaged severely as a result of prolonged compensation of this nature.

Disturbances of Cardiovascular Homeostasis: Hemorrhage, Hypotension, and Exercise

We have now examined the structural and functional characteristics of the individual components of the cardiovascular system. We have discussed how the baroreceptor reflex functions to maintain arterial blood pressure by initiating changes in cardiac output and total peripheral resistance. As we have stated on several occasions, the purpose of the cardiovascular system is to maintain adequate blood flow to the tissues to support cellular metabolism. What we have begun to see is that not all tissues are

Systemic Hypertension

The term **systemic hypertension** refers to a complex disease state resulting in chronically elevated arterial blood pressure. Hypertension is the leading cause of death and disability among adults living today in the United States and most other industrialized societies. More than half of all heart attacks and two thirds of all strokes occur in hypertensive individuals.

Although the precise definition of hypertension is somewhat elusive, studies have shown that individuals with resting blood pressures between 140/90 and 160/95 have a significantly greater risk of cardiovascular disease. **Essential hypertension** is the most common form of hypertension. As the name indicates, essential hypertension is due to a primary disturbance of the cardiovascular system. Other forms of hypertension are secondary to disease processes such as renovascular disease.

Blood pressure is determined by cardiac output and total peripheral resistance. Hypertension is associated with abnormalities in both of these parameters. In many patients, cardiac output is initially high with normal peripheral resistance. With the progression of the disease, cardiac output falls and peripheral resistance increases. The initial elevation of cardiac output may have multiple causes, including abnormal retention of salt and water by the kidneys. The progressive increase in vascular resistance is most likely a compensatory response to the elevated pressure and involves structural changes in the wall of the blood vessel.

While the exact cause of hypertension is not known, a number of factors are known to contribute to the development of the disease. Hereditary predisposition to high blood pressure has been demonstrated. It is also well established that excessive dietary sodium intake is linked to hypertension, particularly in those individuals genetically predisposed to hypertension. Other factors linked to hypertension include obesity, excessive alcohol consumption, smoking, and diabetes.

Hypertension has been called the "silent killer" because the uncomplicated disease state is almost always asymptomatic. Individuals may be unaware for as long as 10 to 20 years that their elevated blood pressure is causing progressive cardiovascular damage. Frequent blood pressure measurements are necessary to detect the early onset of the disease. If left untreated, hypertension can lead to stroke, peripheral vascular disease, myocardial ischemia, and heart failure. Frequently, hypertension can be treated successfully with a combination of nondrug therapies including restriction of dietary sodium, weight reduction, exercise, and moderation of alcohol and coffee consumption. Antihypertensive drug therapy includes the use of diuretics to reduce plasma volume, adrenergic blockings drugs to reduce sympathetic stimulation of the cardiovascular system, and vasodilator agents to reduce systemic resistance.

---

equal in terms of their need for blood flow and that, under some conditions, blood flow is directed from some tissue beds to others to sustain temporarily elevated levels of metabolism (e.g., exercising muscle). Moreover, blood flow to certain tissues such as the brain and heart must be maintained, even at the expense of other tissues, because of the body's need for their uninterrupted function. Perhaps the best way to see how all of these factors operate together is to look at the response of the cardiovascular system to changes in its normal resting state. To illustrate these points we will examine three cardiovascular disturbances: exercise, postural hypotension, and hemorrhage.

Hemorrhage: Loss of Blood

The immediate effect of blood loss (hemorrhage) is a drop in arterial blood pressure. This occurs as a result of an associated decrease in cardiac output. If blood volume is removed from the vascular system, the end-diastolic volume of the heart will decrease and cardiac output will be reduced, as described by the Frank-Starling law. We know from the equation $MAP = CO \cdot TPR$ that restoration of arterial pressure requires an increase in either cardiac output or total peripheral resistance. The decrease in arterial pressure following hemorrhage immediately stimulates an increase in the rate of baroreceptor nerve activity. This leads to increases in both heart rate and contractility, due to increased sympathetic nerve activity to the heart. The associated increase in cardiac output increases arterial pressure back toward normal. The increased baroreceptor activity also results in increased sympathetic nerve activity to the resistance arterioles.

As we know from previous discussions, not all vascular beds receive the same degree of sympathetic innervation. The vascular beds in the skin, skeletal muscle, gastrointestinal tract, and kidneys are all innervated by sympathetic nerves. Arteriolar vasoconstriction of these vascular beds reduces the fall in blood pressure by increasing total peripheral resistance. The vascular resistance to the heart and brain does not increase because of the sparse sympathetic innervation and because of the high degree of local control present in these vascular beds.

We can now begin to appreciate the purpose of the selective innervation of different vasculature beds in the body. Blood flow to those tissues whose continuous function is not essential for life can be temporarily reduced so that a limited blood flow can be distributed to those organs, such as the brain and heart, that must function continuously. Loss of blood flow to the heart would weaken the heart and prevent it from participating in the restoration of blood pressure. Loss of blood flow to the brain would disrupt the function of the cardiovascular centers responsible for the mediation of the compensatory response. The kidneys can sustain a short-term reduction in blood flow, but long-term reductions lead to renal failure and death. As you will see in more detail later, the reduction of blood flow to the kidney has the added benefit of maintaining blood volume by reducing the formation of urine. Blood flow to the skin, muscles, and the gastrointestinal system can be substantially reduced with little ill effect. The net result of these changes in cardiac output and peripheral resistance is to transfer limited blood flow to the heart and brain.

Figure 19–32 provides an overview of the compensatory responses to hemorrhage. You can see that the compensatory response consists of a rapid component that includes increased heart rate, increased cardiac contractility, increased peripheral resistance, and venoconstriction. This is clearly a temporary response that sacrifices the immediate needs of some tissues to maintain those of other, more critical tissues such as the heart and brain. A second, more prolonged response consists of fluid reabsorption from the interstitial compartment to restore total blood volume. While this tends to alleviate the hemodynamic crisis, complete recovery involves several additional processes. Complete res-

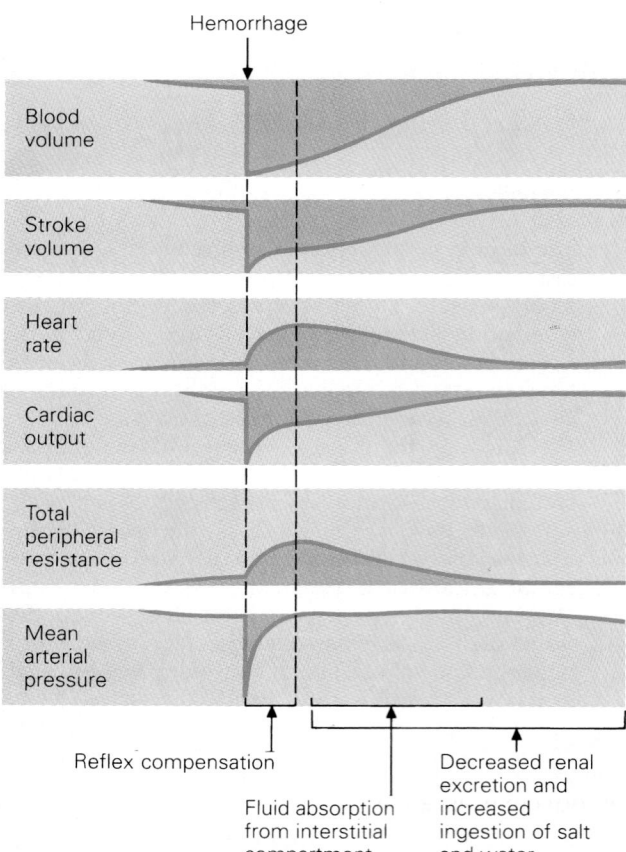

Figure 19–32
Summary of the hemodynamic responses of the cardiovascular system to mild hemorrhage. A rapid, reflexive compensation is followed by a slower response to return blood volume to normal by increasing fluid reabsorption and decreasing fluid excretion.

toration of total body water involves control of fluid ingestion and kidney function. Restoration of the blood requires stimulation of erythropoiesis to increase the hematocrit and synthesis of plasma proteins.

Reduced Stroke Volume Following Hemorrhage. One of the more pronounced changes that occurs with hemorrhage is a reduction in stroke volume. Although the increase in heart rate and contractility can partially compensate for this reduction, both of these compensatory mechanisms are limited by the diastolic filling of the heart. Recall from earlier discussions that even under normal conditions, increasing cardiac output by increasing heart rate is limited by the progressively diminished diastolic filling time. As blood is transferred to the arterial system to maintain blood pressure following hemorrhage, less blood is available for filling of the heart.

So you can see that the extent of compensation by increased heart rate, increased contractility, and increased peripheral resistance becomes self-limiting. The only way to overcome this limitation is to increase the end-diastolic volume of the heart. As we previously demonstrated with the bucket analogy, this can be accomplished by increasing the total blood volume or by redistributing blood volume.

Increased Diastolic Filling due to Increased Blood Volume, Venoconstriction, and Fluid Reabsorption to Compensate for Hemorrhage. Obviously, a blood transfusion is an effective way of returning the system to normal and is the treatment of choice for hemorrhage. There are, however, compensatory mechanisms that allow the body to recruit blood volume from other sources. One response to hemorrhage is generalized venoconstriction. Although this does not increase the total circulating blood volume, it does increase the volume of blood available to the heart. Consider what happens as the veins constrict. Because of the reduction in vessel diameter, blood shifts from the veins to the heart and arterial system. From the standpoint of the heart, this is as effective as an increase in total blood volume.

There is another mechanism by which the body is able to create a real increase in total blood volume. Consider what is happening to capillary pressure, particularly in vascular beds where arteriolar vasoconstriction has taken place (see Fig. 19–22). Mean arterial blood pressure and venous pressure have been reduced because of the hemorrhage, and capillary pressure therefore must also be reduced. The plasma protein concentration, however, has not changed. According to Starling's hypothesis, reabsorption of fluid will be favored, and interstitial fluid volume will be transferred into the capillaries. As much as 1 liter of fluid per hour can be transferred to the vascular compartment by this mechanism following severe hemorrhage.

This reabsorption of fluid does not occur immediately and may take as long as 12 to 24 hours to reach completion, depending on the extent of hemorrhage. For example, if you donate a pint of blood, within 12 hours your circulating blood volume will be restored almost to normal. The fluid transferred from the interstitial compartment is free of red blood cells, and what little plasma protein this fluid contains is filtered during reabsorption. Remember that the capillary wall is relatively impermeable to protein. As you might expect, hematocrit and plasma protein concentration are reduced as a result of fluid reabsorption. The dilutional effect of fluid reabsorption limits the extent of this compensatory mecha-

nism. As interstitial fluid is transferred into the blood, the protein concentration of the interstitial fluid increases while that of the blood decreases. This progressively reduces the osmotic gradient, which is the driving force for fluid reabsorption.

Circulatory Shock. The mechanisms just described can adequately compensate for blood losses of as much as 1 to 1.5 liters (2 to 3 pints) with few ill effects. Greater losses of blood, however, can lead to severe tissue damage, irreversible circulatory collapse, and death. This progressive deterioration of cardiovascular function is referred to as **circulatory shock.** Depending on the extent of blood loss and the elapsed time, a point is reached after which no known therapy, including massive transfusions, can prevent death. Although the details of this process are complex, it is clear that contributing factors include irreversible damage to many organ systems, including the kidneys, heart, and brain, as a result of inadequate blood flow.

Hypotension: Low Blood Pressure

From our previous discussion, we know that hemorrhage produces a form of hypotension that results from a loss of blood volume. Recall that one of the compensatory processes for hemorrhage involves the translocation of blood volume from the veins to the heart and arterial system by venoconstriction. From the point of view of the heart, this is entirely equivalent to a transfusion of an equal volume of blood. As you can see, many of the regulatory and compensatory mechanisms that we have been discussing involve blood volume redistribution.

Causes of Hypotension. Everyday examples of hypotension are fainting spells that might occur when someone stands abruptly after waking, or stands for a long time in the hot sun while wearing warm, heavy clothing. The problems here are gravity and inadequate peripheral vascular tone, which result in inadequate filling of the heart. The "buckets" (from our earlier analogy) suddenly drop out of reach.

Compensatory Responses to Hypotension. In the first example, the sleeper probably has a heart rate under 60 beats per minute, and the veins in the legs are dilated. When this person stands up, blood pools in the veins of the legs and feet, and suddenly no blood is available to fill the heart. Cardiac output drops precipitously as a result, with inadequate blood flow to the brain causing unconsciousness. In the second instance, the person is standing quietly, so that heart rate slows and blood pressure drops.

Because the person is also overheated, the subcutaneous blood vessels are dilated extensively to bring blood close to the surface of the body for heat dissipation. This extensive reduction in peripheral resistance, coupled with a lack of venomotor pumping activity, results in the transfer of blood volume away from the heart. This, in turn, leads to reduced end-diastolic volume, reduced cardiac output, reduced arterial blood pressure, and insufficient blood flow to the brain. The body compensates appropriately by adjusting its posture to promote greater blood flow to the heart and brain. Such problems are normally avoided by a combination of compensatory mechanisms. Normal levels of activity result in periodic contraction of the skeletal muscles in the legs, which prevents pooling of blood in the veins of the lower extremities. In addition, normal levels of activity, both mental and physical, are associated with slight increases in sympathetic nerve activity and therefore with increases in heart rate and venomotor tone.

Exercise and Cardiovascular Homeostasis

The cardiovascular changes occuring during exercise are in many ways the opposite of those occuring during hypotension and hemorrhage. With hypotension, the major disturbance is reduced arterial blood pressure resulting from reduced cardiac output. The compensatory response (increased cardiac output and increased vascular contraction) is mediated primarily by stimulation of the arterial baroreceptors by the reduced arterial blood pressure. During both mild and intense exercise, there is an increase in arterial pressure, which tends to decrease cardiac output and decrease constriction of systemic vascular beds via the baroreceptor reflex. Yet we know that cardiac output always increases during exercise; well-trained athletes may experience an increase in cardiac output from 5 L/min to as much as 35 L/min during maximal exercise.

It is now well accepted that the cardiovascular changes observed during exercise are not in fact mediated by the baroreceptor reflex. Although a number of local and humoral factors have been cited as contributing to vasodilation of the skeletal muscle vasculature, reflexes originating from within the contracting muscles are now thought to lead to activation of sympathetic nerves to the heart and peripheral vasculature, to further adapt the cardiovascular system to the demands of exercise.

Cardiovascular Demands During Exercise. Let us first consider what demands exercise places on the cardiovascular system. Exercise, by definition, is an

increase in the rate and extent of contraction of skeletal muscle. We can correctly assume that this is associated with a substantial increase in the metabolic rate of muscle. A good measure of the level of muscle metabolism is oxygen consumption. During maximal exertion, muscle oxygen consumption for the well-trained athlete may increase by as much as 50 to 60 times resting levels. This means that a significant increase in blood flow must occur (up to 15 times resting flow) to deliver oxygen to the muscle and to remove metabolic byproducts. If we consider that resting muscle blood flow is about 1 to 1.5 L/min, then the blood flow to skeletal muscle alone during exercise will be as high as 15 to 20 L/min. Neglecting the needs of the rest of the body, it is clear that cardiac output will have to increase substantially to support this increased demand by exercising muscle (Fig. 19–33).

Compensatory Responses to Exercise. The necessary cardiovascular changes required for exercise actually begin even before the onset of muscular ac-

Figure 19–33
Summary of the hemodynamic changes in the cardiovascular system in response to mild and intense exercise.

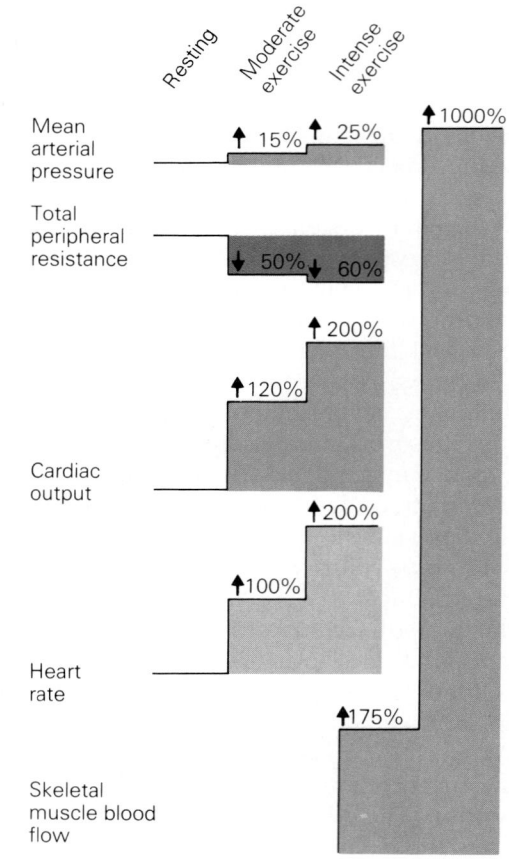

tivity. The mental anticipation of exercise leads directly to increased sympathetic activity and decreased parasympathetic activity to the heart. There is also an increase in sympathetic activity to the peripheral vascular beds such as the skin, kidney, and gastrointestinal tract, which increases vascular resistance and direct blood flow to muscle. In certain animals—and possibly humans—there may be sympathetic fibers to skeletal muscle that release acetylcholine and stimulate vasodilation of precapillary arterioles to further increase muscle blood flow before exercise.

Once muscular exercise begins, a number of additional mechanisms come into play. The release of vasodilator substances such as potassium and adenosine leads to vasodilation of the precapillary resistance vessels, allowing for a 15- to 20-fold increase in tissue blood flow. As exercise progresses and body temperature begins to increase, there is a reflexive vasodilation of the subcutaneous vessels to allow for dissipation of excess body heat. Chemoreceptors within the muscle respond to the release of metabolites and initiate a reflexive increase in sympathetic nerve activity to the heart and resistance vessels of the gastrointestinal tract and kidneys. The level of afferent nerve activity is roughly proportional to the level of muscle work being performed and therefore provides a means for matching cardiac performance with the requirements of the active muscle.

Exercise and Hemorrhage Compared. Consider for a moment how the situation during exercise differs from that during compensation for hemorrhage. In both cases, there is increased stimulation of heart rate and contractility. In both cases, blood is being shunted from inactive to active tissues. With hemorrhage, cardiac output is limited by inadequate cardiac filling; this is clearly not the case with exercise, since cardiac output may exceed 15 L/min.

The difference with exercise is that there has been no reduction of blood volume. Compensation for blood loss due to hemorrhage requires a net transfer of blood from the venous system to the heart and arterial side of the circulatory system in order to maintain adequate perfusion pressures for the heart and brain. This process is self-limiting because of decreased cardiac filling with depletion of venous blood volume. We can deduce that, during exercise, there is little change in arterial blood volume, even as cardiac output increases, because mean blood pressure increases only slightly. No transfer of blood volume from the venous to arterial systems occurs because total peripheral resistance has decreased in proportion to the increase in car-

diac output; blood flow from the arterial side to the venous side increases in direct proportion to the increase in cardiac output. If we think back to the bucket analogy, the decreased total peripheral resistance with exercise would be analogous to increasing the diameter of the connecting return tube. As long as water returns to the source bucket as fast as it is removed, the only limitation to the pumping rate is the maximum rate at which the pump can operate. This is determined by the minimum time required for contraction of the ventricle and diastolic filling. The effective upper limit for heart rate is about 180 beats/min.

Cardiovascular Disease

Congestive Heart Failure

As the name implies, heart failure is the result of the inability of the heart to maintain adequate cardiac output. Although many factors may cause failure of the heart, the resulting changes in cardiovascular performance are similar.

Cardiovascular Effects of Heart Failure

As before, let us first specify what cardiovascular parameter undergoes a change. In this case, the problem is a reduction in cardiac contractility. Recall that this means that the ability of the heart to pump blood at any end-diastolic volume is decreased (Fig. 19–34). Let us start with a normal cardiovascular system and then impose a reduction in contractility on the heart. We know that the decrease in contractility will decrease stroke volume and cardiac output. Furthermore, we can reason that a decrease in cardiac output will lead to a decrease in mean arterial pressure.

Compensation for Heart Failure

Clearly the reduced blood pressure will stimulate the baroreceptor reflex, which results in increased sympathetic and decreased parasympathetic stimulation of the heart as well as sympathetic stimulation of the vasculature. The result is increased contractility, increased heart rate, increased arteriolar constriction, and increased venomotor tone. As you know, these changes are very similar to those seen with hemorrhage. The major difference concerns the distribution of blood volume. Following hemorrhage, heart rate and vigor of contraction increase to transfer venous volume to the arterial system in

Atherosclerosis and Myocardial Infarction

Atherosclerosis is a disease process of arteries that results in the progressive occlusion of the lumen of the vessel. This process is seen most frequently in the aorta and in the cerebral and coronary arteries. Although the precise cause of atherosclerosis is still widely debated, the progression of events is well documented. The initial event probably involves damage to the endothelial lining of the vessel wall. This is followed by proliferation of the underlying smooth muscle cells that make up the muscular wall of the vessel. A progressive reduction in the size of the vessel lumen results from the thickening of the vessel wall. As the disease progresses, lipid from the blood accumulates along with fibrous tissue **(atherosclerotic plaque),** resulting in the progressive occlusion of the vessel lumen.

Occlusion of the coronary arteries in this manner **(coronary artery disease)** is the most frequent cause of cardiac dysfunction. The occlusion of the vessel lumen gradually reduces blood flow to the heart. As the disease progresses, individuals may begin to experience periodic episodes of inadequate coronary blood flow, often associated with chest pain referred to as **angina pectoris.** Episodes of angina are frequently associated with increased physical exertion or emotional stress. A heart attack is the result of an acute coronary occlusion, which may occur because of either clot formation at the site of the plaque **(thrombus)** or intense spasm of the coronary smooth muscle near the plaque. Since the heart receives its blood supply from the coronary vessels, occlusion can lead to injury and death of that portion of the

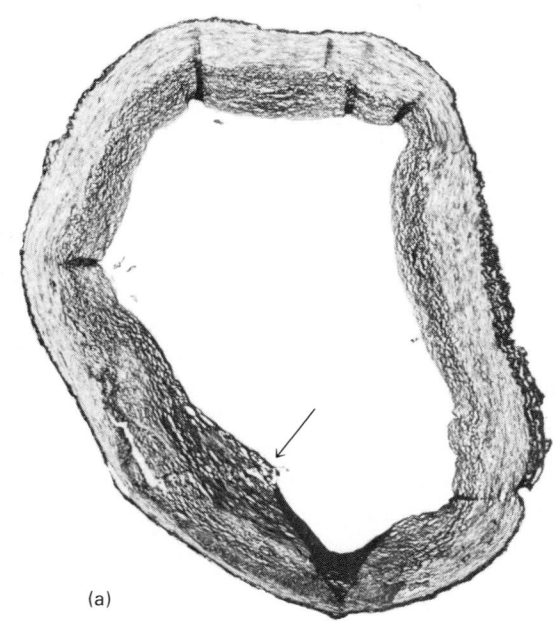

(a)

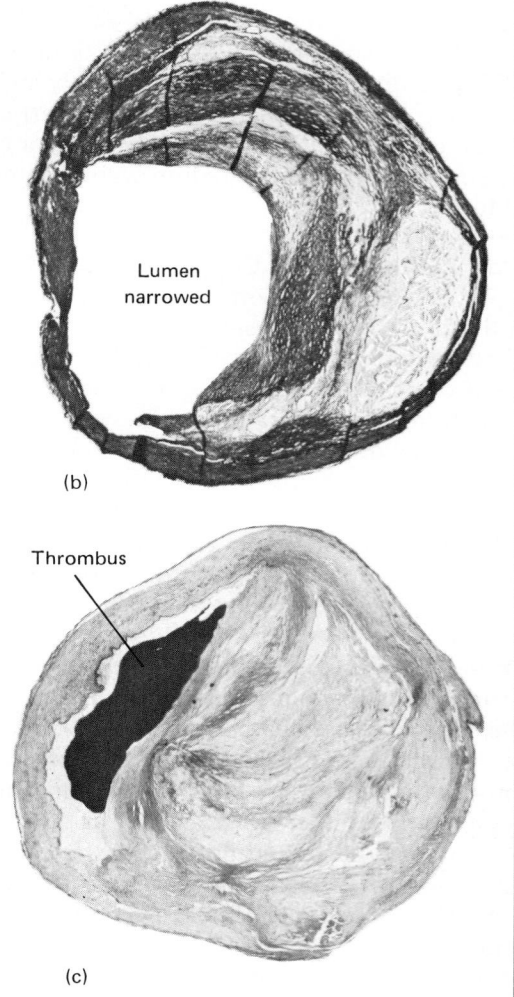

Lumen
narrowed

(b)

Thrombus

(c)

Tissue sections from an atherosclerotic vessel. (*a*) Early stage of atherosclerosis with slight thickening of the arterial wall indicated by the arrow. (*b*) Progressive accumulation of smooth muscle cells, lipid, and calcium results in plaque formation. (*c*) The progressive accumulation of plaque results in almost complete occlusion of the vessel lumen and a substantial increase in the resistance to blood flow.

myocardium supplied by the occluded vessel. The area of damaged myocardium is referred to as an **infarct.**

An infarct can reduce the heart's ability to pump blood in a number of ways. Depending on the extent and location of the infarct, the pumping capability of the heart may be compromised because of the loss of functional muscle. The infarct may also initiate cardiac arrhythmias. Either or both of these situations can lead to decreased cardiac output. This in turn may lead to reduced arterial blood pressure, depending on the body's ability to compensate by increasing peripheral resistance. Autoregulation of coronary blood flow compensates for small drops in arterial pressure; larger decreases in arterial pressure result in reduced coronary blood flow. This, in turn, further diminishes the pumping ability of the heart, leading to more reductions in cardiac output and arterial pressure. Unless halted, this downward spiral will lead to collapse of the cardiovascular system and death.

The body's compensatory responses are limited, and therefore the consequences of advanced coronary artery disease can be severe. Arterial pressure can only be increased following a heart attack by increasing cardiac contractility, heart rate, or peripheral vascular resistance. Unfortunately, these compensatory responses all require that the heart do more physical work, which in turn increases the need for oxygen delivery to the heart. Since the primary problem is inadequate blood flow and insufficient oxygen delivery to the heart, these compensatory responses may only advance the extent of myocardial damage.

As with all cardiac disorders, reversal of the primary disturbance is desirable. Several therapeutic methods are available to increase coronary blood flow. **Nitroglycerin** provides relief from angina by dilating the coronary arteries to increase blood flow. A class of drugs known as **calcium antagonists,** or **calcium channel blockers,** which block the entry of calcium into the smooth muscle cells, also provide relief by dilating the coronary arteries. Several agents that prevent formation or dissolve blood clots in the coronary arteries are being used to unblock arteries occluded by thrombus formation. If progressive occlusion of a coronary vessel is detected prior to the onset of a heart attack, measures are available to alleviate the occlusion. **Coronary bypass surgery** involves the grafting of a vein (removed from elsewhere in the patient's body) around the occluded segment of the coronary artery so that blood flow can bypass the diseased area. Another approach is to insert a catheter with a small balloon at the tip into the occluded coronary artery. When the catheter tip is

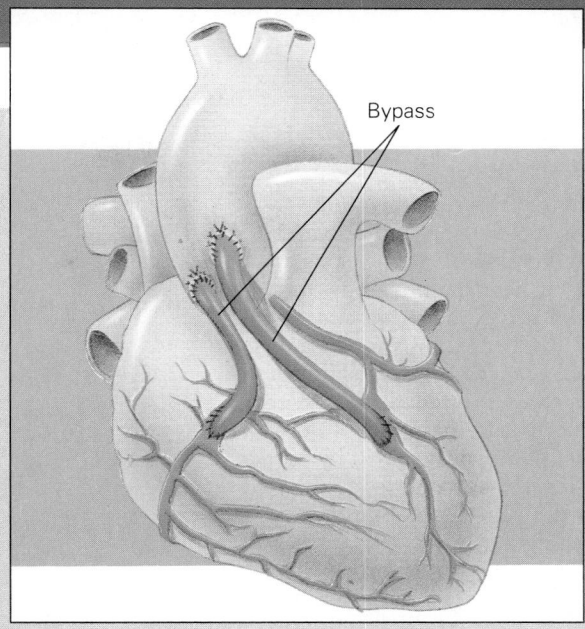

positioned within the occluded portion of the vessel, the balloon is inflated and the coronary plaque is fractured, allowing the vessel to dilate. This procedure is referred to as **coronary angioplasty.** These treatments do not stop the progression of the disease but merely relieve the symptoms temporarily.

Prevention is clearly the best treatment. Although the cause of atherosclerosis is unknown, a number of risk factors have been associated with an increased incidence of the disease. These include smoking, hypertension, elevated plasma cholesterol, and diabetes. Current evidence suggests that a reduction of cholesterol in the diet as well as regular exercise may offer some protection against the progression of the disease.

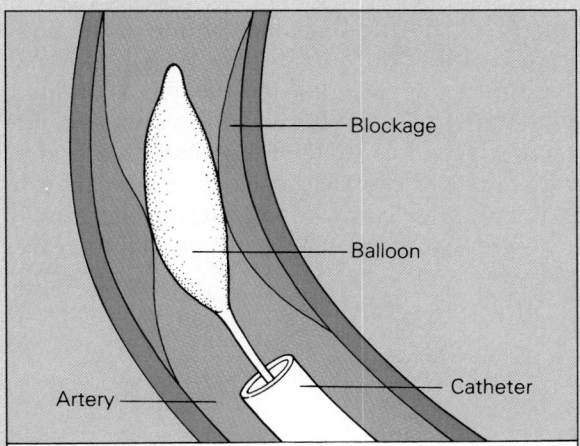

A small catheter is introduced into the blocked coronary artery. A guide wire containing a small balloon at its tip is pushed through the blockage. When the balloon is in place, it is inflated to flatten the blockage against the arterial wall.

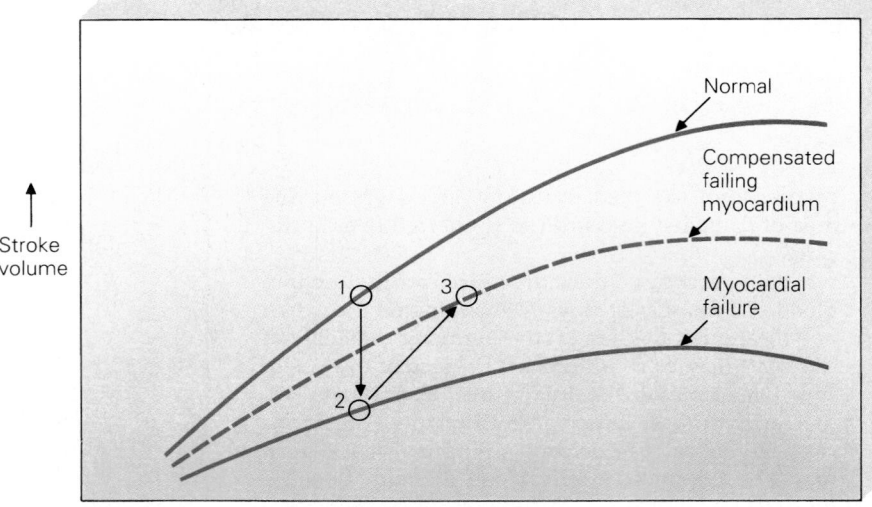

Figure 19–34
Change in the relationship between end-diastolic volume and stroke volume of the heart as a result of heart failure (1–2). Increased sympathetic nerve activity to the heart can compensate for mild failure by increasing the contractility (dotted curve). Increases in end-diastolic volume further compensate (2–3) by increasing stroke volume, as described by the Frank-Starling relationship.

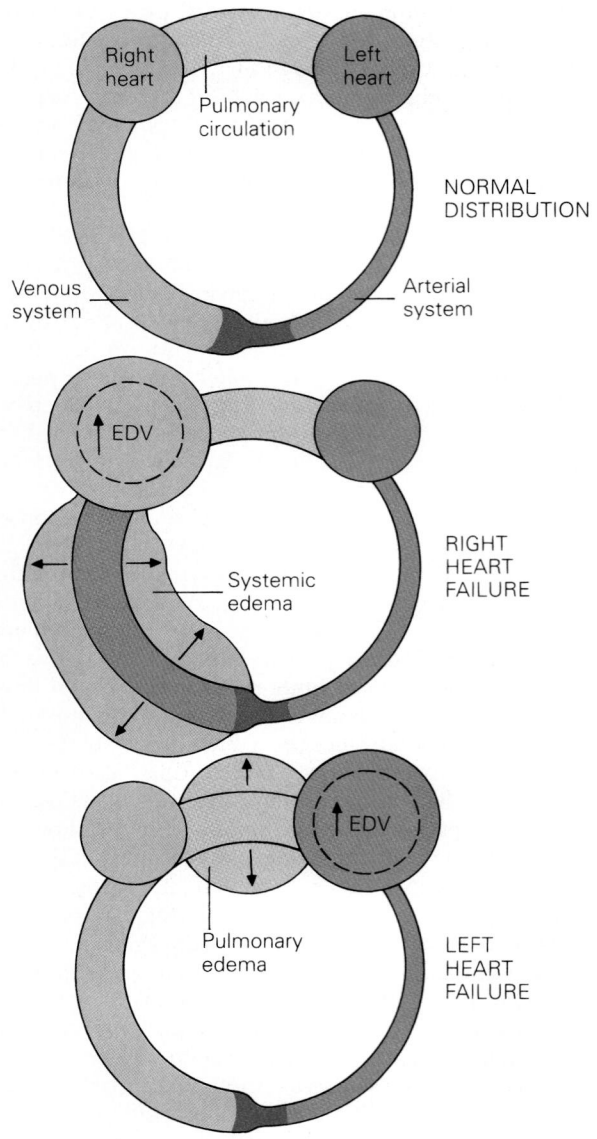

order to maintain arterial pressure. Recall that full compensation is limited by inadequate filling of the heart. With cardiac failure, just the opposite occurs. Because of the reduced pumping capacity of the heart, blood accumulates in the heart and veins. The heart fails to maintain the flow of blood to the arterial system. This, of course, results in a fall in arterial pressure and an increase in the volume of blood in the venous system and in the heart.

As the end-diastolic volume of the heart increases, the Frank-Starling mechanism tends to increase cardiac output. Recall that in hemorrhage, the kidneys respond to the reduction in blood pressure by increasing the retention of water and sodium. Exactly the same process occurs with heart failure. The increase in circulating blood volume further increases the end-diastolic volume of the heart to increase cardiac output (see Fig. 19–34). Unfortunately, this compensatory mechanism is limited. Recall that increases in end-diastolic volume lead to increased vigor of contraction because, in a normal resting heart, the cardiac muscle is shorter than its optimal length for contraction (refer to Fig. 18–14). With progressive heart failure, additional increases

Figure 19–35
Blood volume redistribution in the cardiovascular system due to failure of either the left or right heart. The increased volume in either the pulmonary or systemic veins leads to increased capillary pressure and loss of fluid to the interstitial space (edema).

in volume stretch the muscle beyond its optimal length. Once this point is reached, further increases in heart volume actually reduce the ability of the heart to pump blood. This in turn leads to further increases in venous and cardiac volume, and a vicious cycle ensues. This progressive congestion of the veins and heart with blood is described as **congestive heart failure.**

Left Heart Failure versus Right Heart Failure

Until now, we have considered the heart as a single unit. In reality, either or both ventricles can undergo failure. Failure of the right ventricle leads to an accumulation of blood in the right heart and the venous system. As blood accumulates in the veins and ultimately in the capillaries, capillary filtration increases because of the increase in capillary pressure. This leads to the accumulation of excessive fluid in the tissues in the form of edema (Fig. 19–35). The formation of edema is especially pronounced in the lower extremities because of the contributing effects of gravity.

If, on the other hand, the left heart fails, blood will accumulate in the pulmonary circulation. The formation of pulmonary edema can be particularly serious because this accumulation of fluid impairs gas exchange. This can lead to a vicious cycle; poor gas exchange will further exaggerate the loss of ventricular contractility, which in turn will lead to further edema.

Treatment of Heart Failure

How can cardiac failure be treated medically? Obviously, if contractility could be returned to normal, the remaining symptoms would be alleviated. Administration of the drug **digitalis** tends to improve contractility by increasing intracellular calcium concentrations. The excess fluid accumulation can be reversed by the administration of drugs that enhance the loss of sodium and water by the kidneys. This latter treatment must be used with caution, since excessive reduction in volume would compromise cardiac output as a result of reduced end-diastolic volume.

SUMMARY

Overview of the Vascular System

The purpose of the vascular system is to direct blood flow throughout the body and to allow the exchange of materials between the blood and the interstitial fluid.

The Arterial System: Carrying Blood Away from the Heart

The entire cardiac output flows through the aorta to the arteries, which distribute blood to the various tissues in the body. The elastic arteries serve to convert the pulsatile flow of blood from the heart into a more continuous flow of blood to the tissues. A relatively high blood pressure is required to overcome the resistance to blood flow through the blood vessels. Blood flow can be calculated as follows: Blood Flow = Arterial Pressure/Resistance. The greatest pressure drop occurs at the level of the arterioles, due to the relatively large resistance to flow at this point in the circulation. The pulsatile pressure in the arteries is calculated as the systolic pressure minus the diastolic pressure. The amplitude of the pulse pressure is affected by both the stroke volume and the distensibility of the arteries. The mean arterial pressure (MAP) is calculated as 1/3 pulse pressure + diastolic pressure. The flow of blood through the vascular system is directly related to the mean arterial pressure.

The arterioles are vessels that branch off the arteries as they enter the various tissue beds in the body. The resistance to blood flow through the arterioles is large because of their small diameter. The arterioles contain smooth muscle, which allows them to change diameter (vasodilation and vasoconstriction) and thus the resistance they provide. Altering the resistance to blood flow through the various tissue beds in the body is a way of decreasing blood flow to some tissues while increasing flow to others.

One local factor affecting blood flow is myogenic activity, a stretch-induced contraction of vascular smooth muscle that allows for autoregulation of blood flow so that it remains relatively constant in the face of changing arterial pressure. Chemical factors such as the concentrations of oxygen, carbon dioxide, adenosine, and hydrogen ion can also affect local blood flow and contribute to the overall autoregulation of flow. Reactive hyperemia represents an extreme example in which massive vasodilation follows occlusion of blood flow to a tissue. This is the result of the local depletion of oxygen and accumulation of metabolites during the absence of blood flow. Cells can release other vasoactive substances such as histamine and prostaglandins to produce a local change in blood flow.

Extrinsic regulation of blood flow provides for the coordinated distribution of the limited blood flow available from the heart to the various tissue beds. Blood vessels are innervated primarily by the sympathetic division of the autonomic nervous system (ANS). Sympathetic nerve fibers release norepinephrine, which binds to alpha receptors in the membrane of the vascular smooth muscle cells, resulting in vasoconstriction. Resting vascular tone is maintained by continuous sympathetic nerve activity originating from vasoactive centers in the medulla. Hormones also play an important role in regulation of the vasculature. Epinephrine binds predominantly to beta-receptors. The blood vessels of the heart, skeletal muscle, and liver contain the greatest numbers of beta-receptors and are therefore most sensitive to the vasodilatory activity of circulating epinephrine. Angiotensin II is a very potent vasoconstrictor agent that may play a role in the regulation of the renal circulation. Atrial natriuretic peptide, secreted from cells in a number of tissues, including the heart, is a potent vasodilator substance and also causes excretion of salt and water by the kidneys.

The Capillary System: The Exchange of Molecules

The arterioles empty into a network of short, thin-walled vessels called capillaries, through which occurs most of the exchange of substances between the blood and the interstitial fluid. At the beginning of some capillaries, there is a small cuff of smooth muscle called a precapillary sphincter, vasomotion of which regulates the flow of blood through the capillary. This activity of the sphincter is influenced heavily by the local metabolic state of the tissue.

The capillary wall is composed of a single layer of endothelial cells and contains no smooth muscle. The endothelial cells allow rapid diffusion of many substances directly across the cell. In addition, exchange of substances also occurs through endothelial pores found between the endothelial cells. Pinocytosis allows for the transport of relatively large proteins across the endothelial cell. Diffusion is by far the most important mechanism for exchange of water and dissolved substances. Lipid-soluble substances such as oxygen, carbon dioxide, and urea exchange primarily by direct diffusion through the wall of the capillary. Other, less lipid-soluble molecules diffuse through water-filled pores. Exchange of water and dissolved substances also occurs by bulk flow through endothelial pores in response to a pressure gradient between the inside and outside of the capillary. Although bulk flow through endothelial pores is not a major avenue for exchange (except in the kidneys), this method of exchange plays an important role in the maintenance of normal circulating blood volume. Bulk flow also occurs in response to osmotic gradients between the inside and outside of the capillary. The gradient of osmotic pressure due to the plasma protein difference between the intra- and extravascular compartments is referred to as the oncotic pressure gradient. Starling's hypothesis describes the balance of hydrostatic and oncotic forces across the capillary wall. The relative magnitudes of these two driving pressures determine whether net filtration or absorption of fluid will occur in a particular capillary. Under normal conditions, the hydrostatic pressure gradient tending to move fluid out of the capillary is balanced by the osmotic pressure gradient tending to move fluid into the capillary from the interstitial space, resulting in no net transfer of volume into or out of the circulation. Many processes such as hemorrhage, heart failure, and other disease processes can upset this balance, leading to a net gain or loss of circulating blood volume.

The Venous System: Returning Blood to the Heart

The venous system is made up of blood vessels that carry the blood from the capillaries back to the heart. The smallest veins, the venules, are very thin-walled like the capillaries, and allow for exchange between the blood and interstitial fluid. The capillaries and venules together are referred to as the exchange vessels. The veins are often referred to as capacitance vessels because they can accommodate a large portion of the total circulating blood volume. In addition, the smooth muscle in the walls of the veins allows them to constrict in response to sympathetic nerve activity.

This results in the transfer of blood volume to the rest of the circulatory system and contributes to the maintenance of normal venous return to the heart under a variety of circumstances. A number of other factors contribute to venous return. The veins contain one-way venous valves, which promote the flow of blood from the periphery to the heart. Contraction of skeletal muscle around the veins as well as respiratory movements of the diaphragm provide pumping activity that also increases venous return.

The Lymphatic System

The lymphatic vessels are a system of vessels that channel back into the circulation fluid and plasma proteins that have leaked into the interstitial space. The normal operation of the lymphatic system prevents edema. The lymph vessels empty into the subclavian vein. Lymph vessels have valves and rely on the pumping activity of surrounding skeletal muscle and smooth muscle in the walls of the lymph vessels to propel the flow of lymph. The lymphatics also transport fat, absorbed from the gastrointestinal tract, to the circulating blood. Lymph nodes are periodic swellings of the lymphatic vessels in which bacteria and other foreign material are removed from the lymph.

Regulation of Systemic Arterial Pressure

Normal tissue blood flow requires that the mean arterial blood pressure (MAP) be maintained. The mathematical relationship between MAP, cardiac output (CO), and total peripheral resistance (TPR) is:

$$MAP = CO \cdot TPR$$

TPR is the total resistance to the flow of blood from the aorta back to the right atrium. The major factor contributing to this overall resistance is the resistance posed by the arterioles and precapillary sphincters.

The average cardiac output must equal the average venous return. Transient changes in venous return may lead to changes in end-diastolic volume. Cardiac output is very sensitive to changes in end-diastolic volume as described by the Frank-Starling law. Changes in either the total blood volume or the distribution of blood within the circulatory system can alter end-diastolic volume and, subsequently, both cardiac output and venous return.

The arterial baroreceptors are stretch receptors located in the walls of the aorta and carotid sinuses that respond to changes in arterial pressure. An increase in pressure leads to stretching of the nerve endings, which initiates an increase in action potential firing.

Baroreceptor nerve activity is transmitted to the cardiovascular control centers in the medulla. These integrate sensory information from baroreceptors and other control centers in the brain and then initiate a change in the level of efferent parasympathetic and sympathetic nerve activity to the heart and vasculature so that arterial pressure is kept relatively constant.

An increase in blood pressure results in increased baroreceptor nerve activity, which leads to decreased sympathetic and increased parasympathetic nerve activity to the cardiovascular system. This in turn decreases heart rate and cardiac contractility, leading to a reduction in cardiac output. The changes in autonomic nerve activity also result in vasodilation and a decrease in total peripheral resistance. The combined reduction of both cardiac output and total peripheral resistance leads to a reduction of mean arterial blood pressure (MAP = CO · TPR). A reduction of arterial pressure results in the opposite changes, with the end result being an increase in blood pressure back toward normal. Contraction of venous smooth muscle (venomotor tone) also occurs in response to reduced baroreceptor stimulation as a result of an increase in sympathetic nerve activity to the veins. This causes a shift of blood volume to the heart, with a resultant increase in end-diastolic volume and subsequently an increase in cardiac output.

Regulation of circulating blood volume is one of the more important aspects of long-term blood pressure regulation. Blood volume regulation is controlled primarily by the kidneys and gastrointestinal system, which mediate the excretion and absorption of salt and water.

The operation of cardiovascular homeostatic mechanisms can be illustrated by examining the body's response to various normal and pathological cardiovascular disturbances.

Hemorrhage (loss of blood) results in an immediate decrease in cardiac output and blood pressure. The reduction in blood pressure activates the baroreceptor reflex, which stimulates an increase in both heart rate and contractility by increasing sympathetic nerve activity to the heart. Increased sympathetic nerve activity also results in vasoconstriction of resistance arterioles, particularly in the skin, skeletal muscle, gastrointestinal tract, and kidneys.

The net increase in total peripheral resistance increases arterial pressure toward normal and redirects the limited cardiac output to tissues such as the brain and heart. The constriction of resistance vessels in the kidneys reduces the formation of urine, and this helps maintain the circulating blood volume.

The effectiveness of these measures in restoring blood pressure is limited by the compromised diastolic filling of the heart. Venoconstriction results in a shift in blood volume from the veins to the heart, with a resultant increase in end-diastolic volume, cardiac output, and arterial blood pressure. The effect on cardiac output and blood pressure of venoconstriction is similar to the effect of a transfusion of blood to the hemorrhage victim. Another compensatory mechanism activated by hemorrhage is the movement of interstitial fluid into the circulation, resulting in a net increase in blood volume. Following hemorrhage, the reduction in capillary and venous pressure subsequent to constriction of the resistance arterioles shifts the balance of hydrostatic and osmotic forces to favor fluid reabsorption. The reabsorbed fluid contains very little plasma protein, thereby causing a progressive reduction in the osmotic pressure of the blood. This dilutional effect of fluid reabsorption reduces the driving force for further fluid reabsorption and limits the extent of volume expansion by this mechanism. The progressive deterioration of cardiovascular function is referred to as circulatory shock. Depending on the extent of blood loss and the elapsed time, a point is reached after which no known therapy, including massive transfusions, can prevent death.

Hypotension is a reduction in blood pressure and frequently occurs in situations not involving hemorrhage. Standing up suddenly and standing still on a warm day can both cause hypotension severe enough to produce fainting. In both cases, the problem is that blood volume has been redistributed to the veins in the lower periphery, with a resultant decrease in cardiac end-diastolic volume, cardiac output, blood pressure, and ultimately blood flow to the brain. The reduction in arterial blood pressure stimulates the carotid sinus and increases sympathetic nerve activity to the heart and vasculature. The resultant veno-constriction shifts blood volume back to the heart from the peripheral veins and increases cardiac end-diastolic volume. The cardiac stimulation increases heart rate, contractility, and, ultimately, cardiac output. The increased cardiac output, coupled with constriction of the resistance vessels, increases arterial pressure back toward normal.

During exercise, the blood flow to active skeletal muscles increases to 10 to 15 times normal. There is an increase in sympathetic nerve activity and a decrease in parasympathetic nerve activity, which begin even before the onset of physical exertion. This change in autonomic nerve activity leads to increased cardiac output and vasoconstriction of resistance vessels in the skin, kidneys, and gastrointestinal tract to increase blood flow to the skeletal muscle. After physical activity begins, metabolites (potassium and adenosine) released from the contracting muscle cells stimulate further dilation of the skeletal muscle resistance vessels. During maximal exercise, cardiac output may increase from a resting level of 5 L/min to as much as 35 L/min, and muscle blood flow may increase from 1 to 1.5 L/min up to 15 to 20 L/min.

Cardiovascular Disease

Congestive heart failure occurs when the heart is unable to maintain adequate cardiac output. Failure of the heart to maintain normal blood flow from the venous to arterial side of the system results in the accumulation of blood in the veins and heart. Compensatory responses include (1) increased cardiac output due to increased end-diastolic volume (Frank-Starling Law); (2) increased baroreceptor activity leading to increased cardiac contractility, increased heart rate, increased arteriolar resistance, and increased venomotor tone; and (3) increased retention of salt and water by the kidneys. If the right heart fails, then blood will accumulate in the systemic venous system, resulting in edema. If the left heart fails, blood accumulates in the pulmonary arteries and veins, resulting in pulmonary edema. Digitalis is often used to increase cardiac contractility. Excess fluid accumulation can be reduced by drugs that increase the rate of excretion of salt and water by the kidneys.

REVIEW QUESTIONS

Identify Each with a Term

1. The vasodilation that occurs following occlusion of blood flow to a tissue.

2. The maintained contraction of vascular smooth muscle under resting conditions.

3. The stretch-induced contraction of vascular smooth muscle.

4. The small cuff of smooth muscle at the beginning of a capillary.

5. The pressure receptor located in the carotid sinus.

Define Each Term

6. colloidal osmotic pressure
7. mean arterial pressure
8. Starling's hypothesis
9. vasomotion
10. cardiovascular control center

Choose the Correct Answer

11. Diastolic blood pressure is:
   a. equal to the cuff pressure at which flow through the brachial artery begins.

b. lower than capillary blood pressure.

c. equal to mean blood pressure minus systolic pressure.

d. equal to the cuff pressure at which flow through the brachial artery is continuous.

12. Which of the following causes vasoconstriction of the arterioles in the skin?

a. decreased arterial pressure

b. increased tissue metabolism

c. increased concentration of CO_2 in the blood

d. increased circulating epinephrine

13. In which of the following organs are the blood vessels *not* regulated by sympathetic innervation?

a. skin

b. brain

c. liver

d. gastrointestinal tract

14. A decrease in vessel radius from 10 mm to 5 mm will increase the resistance to blood flow by how much?

a. 4 times

b. 6 times

c. 8 times

d. 16 times

15. Which of the following blood vessels is *not* surrounded by smooth muscle cells?

a. artery

b. arteriole

c. capillary

d. venule

16. Which of the following vascular beds presents the greatest overall resistance to blood flow?

a. aorta

b. arterioles

c. capillaries

d. veins

17. Which of the following cause reabsorption of water into the capillary?

a. plasma proteins

b. capillary hydrostatic pressure

c. interstitial fluid oncotic pressure

d. soluble electrolytes

18. In a normal individual, the balance of filtration and absorption forces is such that over a 24-hour period:

a. filtration of water exactly equals absorption of water.

b. there is *no* net transfer of water across the capillary wall.

c. there is net absorption of water.

d. there is net filtration of water.

19. On average:

a. left heart output is slightly greater than right heart output.

b. systemic blood flow exceeds pulmonary blood flow.

c. cardiac output equals venous return.

d. resistance to blood flow is greater in the pulmonary circulation than in the systemic circulation.

20. An increase in arterial blood pressure causes which of the following?

a. a decrease in baroreceptor afferent nerve activity

b. a decrease in efferent parasympathetic nerve activity from the medullary cardiovascular control center

c. an increase in total peripheral resistance

d. a decrease in sympathetic nerve activity to the resistance arterioles

Calculate

21. An individual's blood pressure is measured as 120/90 mm Hg. What is the mean arterial pressure?

22. Mean arterial blood pressure is 100 mm Hg and total peripheral resistance is 20 mm Hg $\cdot$ min $\cdot$ L^{-1}. What is the total blood flow?

23. Capillary hydrostatic pressure is 35 mm Hg, capillary colloidal osmotic pressure is 28 mm Hg, and interstitial fluid colloidal osmotic pressure is 3 mm Hg. Calculate the net filtration or absorption pressure.

24. Pulse pressure is 50 mm Hg and systolic pressure is 120 mm Hg. What is the diastolic pressure?

Answer Each Question in One or Two Sentences

25. How do the elastic properties of the arterial blood vessels contribute to the continuous flow of blood through the capillaries?

26. What determines the distribution of blood flow among the various tissue beds in the body?

27. How does myogenic activity of vascular smooth muscle minimize the changes in tissue blood flow in response to changes in arterial pressure?

28. Exercise and hemorrhage both stimulate increased heart rate and increased contractility. Why is cardiac output much lower following hemorrhage than during exercise?

29. What cardiovascular changes occur following heart failure to compensate for the loss of contractility?

SUGGESTED READINGS

Berne, Robert M., and Levy, Matthew N. *Physiology*, 2nd ed. St. Louis: C.V. Mosby Company, 1988.

Cantin, M., and Genest, J. "The heart as an endocrine gland." *Scientific American*, 254(2):76–81, 1986.

Eisenberg, M.S., Bergner, L., Hallstrom, A.P., and Cummins, R.O. "Sudden cardiac death." *Scientific American*, 254(5):37–43, 1986.

Goerke, Jon and Mines, Allan H., *Cardiovascular Physiology*. New York: Raven Press, 1988.

Goldstein, G.W., and Betz, A.L. "The blood-brain barrier." *Scientific American*, 255(3):74–83, 1986.

Honig, C.R. *Modern Cardiovascular Physiology*, 2nd ed. Boston: Little, Brown and Co., 1988.

Jarvik, R.K. "The total artificial heart." *Scientific American*, 244(1):74–80, 1981.

Levy, R.I., and Moskowitz, J. "Cardiovascular research: Decades of progress, a decade of promise." *Science*, 217:121–129, 1982.

Little, Robert C., and Little, William, C. *Physiology of the Heart and Circulation*, 4th ed. Chicago: Year Book Medical Publishers, 1989.

Robinson, T.F., Factor, S.M., and Sonnenblick, E.H. "The heart as a suction pump." *Scientific American*, 254(6):84–91, 1986.

Sparks, Harvey V., and Rooke, Thom W., *Essentials of Cardiovascular Physiology*. Minneapolis: University of Minneapolis Press, 1987.

KEY TERMS

adenosine (p. 662)

arterial natriuretic peptide (p. 666)

arterial pressure (p. 653)

autoregulation (p. 661)

cardiovascular control center (p. 682)

carotid sinus (p. 681)

colloid osmotic pressure (p. 670)

digitalis (p. 693)

hemorrhage (p. 685)

hydrostatic pressure (p. 669)

hypotension (p. 687)

infarct (p. 691)

lymph, lymph node (p. 675)

nitroglycerin (p. 691)

oncotic pressure (p. 670)

pulse pressure (p. 654)

systemic hypertension (p. 685)

total peripheral resistance (p. 679)

venous capacitance (p. 672)

venous return (p. 674)

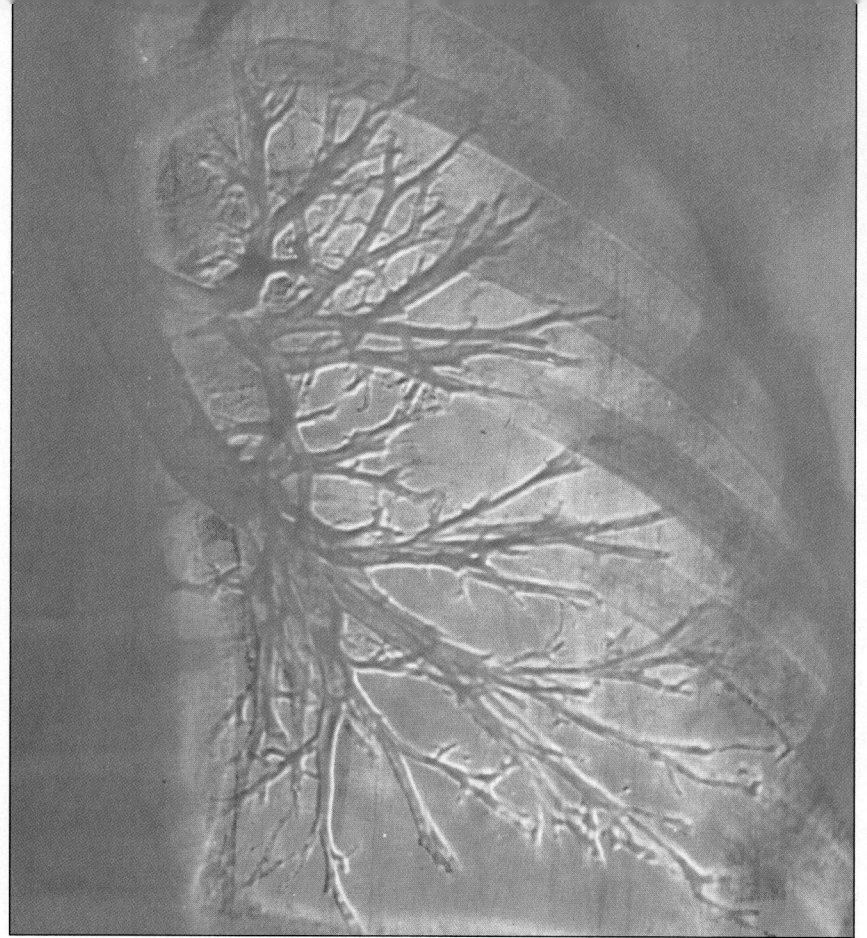

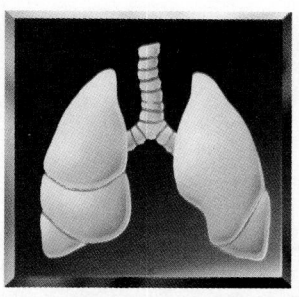

Respiration

CHAPTER OUTLINE

Most cells will die in a matter of minutes if they are deprived of oxygen or if waste products such as carbon dioxide are not removed from them. The process of exchanging oxygen and carbon dioxide between an organism and its environment is known as **respiration.** Respiration takes place in two stages. The first, known as **gas exchange,** is the process of bringing air from the external environment to the cells. More specifically, gas exchange involves the steps of bringing oxygen from the external environment and transporting it to the cell while carbon dioxide is carried from the cell to the external environment. The second stage of respiration is known as **cellular respiration,** which consists of a series of complex metabolic reactions that break down molecules of fuel, releasing carbon dioxide and energy. Recall from Chapter 6 that oxygen is required in the final step of cellular respiration to serve as an electron acceptor in the process by which cells obtain energy.

Respiration has played an important role in the success and adaptability of our species. A key physiological phenomenon in the human success story is our warm-bloodedness (homeothermy) and its impact on respiration. **Homeothermy** is a capacity we share with other mammals and birds. The term refers to the fact that the body's temperature is maintained at a high and constant level. The great advantage of warm-bloodedness is that all of the body's metabolic reactions, including the uptake and transport of respiratory gases, can take place at a fast and predictable rate. This means, for example, that the neuromuscular and cardiopulmonary systems can respond in a moment's notice to any circumstance. This is in contrast to cold-blooded animals such as the lizard, which can hardly move until the sun's rays warm its body.

Another respiratory link in the human success story is the ability of the mammalian gas exchange system to provide an abundant supply of oxygen to the millions of cells to be utilized in energy production. To drive the body's chemical reactions, especially those of muscular contraction, requires large amounts of energy. Little energy is available until the food we eat is digested and **oxidized** (burned with oxygen). Oxidation occurs when the lungs extract oxygen from the air and the blood transports that oxygen to the cells, where it can take part in cellular respiration. The mammalian respiratory system is so efficiently designed to supply oxygen and

remove carbon dioxide that gas exchange and transport rarely limit our activity.

A review of the human respiratory system can be divided into four main sections: (1) the pumping machinery, (2) the control of the pump, (3) gas uptake, and (4) gas transport. Chapter 20 discusses the "pumping machinery"—that is, the physical design of the lung, air movement in and out of the lungs, lung mechanics, and the control of breathing. Chapter 21 primarily treats gas uptake and transport—that is, circulatory design, gas diffusion between the lung and blood, and transport of respiratory gases by the blood to the cells. Chapter 22 deals with respiration in unusual environments.

The Structure of the Lung

Movement of gases in and out of cells occurs by simple diffusion. In unicellular organisms, the process by which oxygen and carbon dioxide diffuse between the cell and the surrounding environment is very simple (Fig. 20–1a). No special respiratory structures are required. In more complex organisms, however, cells deep within the body cannot exchange enough oxygen and carbon dioxide to the environment because the diffusion distance is too long. As a result, special gas exchange organs have developed. These include tracheal tubes in insects, gills in fish, and lungs in higher animals such as humans. All of the respiratory organs mentioned—from the simplest to the most complex—have three features in common. First, they all have very thin walls so that gas diffusion can readily occur. Second, they are always moist so that oxygen and carbon dioxide can dissolve in fluids. Third, they are richly supplied with blood vessels to ensure efficient gas exchange and transport of the respiratory gases to the cells.

The Internal Surface Area of the Lung

As the metabolic demands increased with larger and more complex organisms, lungs began to have a greater internal surface area for the exchange of oxygen and carbon dioxide. This was accomplished by an increase in lung size and by the formation of **septa,** or out-pockets, in the lung, which allow for more blood to flow through the lung and more air to be exposed to the blood (see Fig. 20–1b, c, d). Even

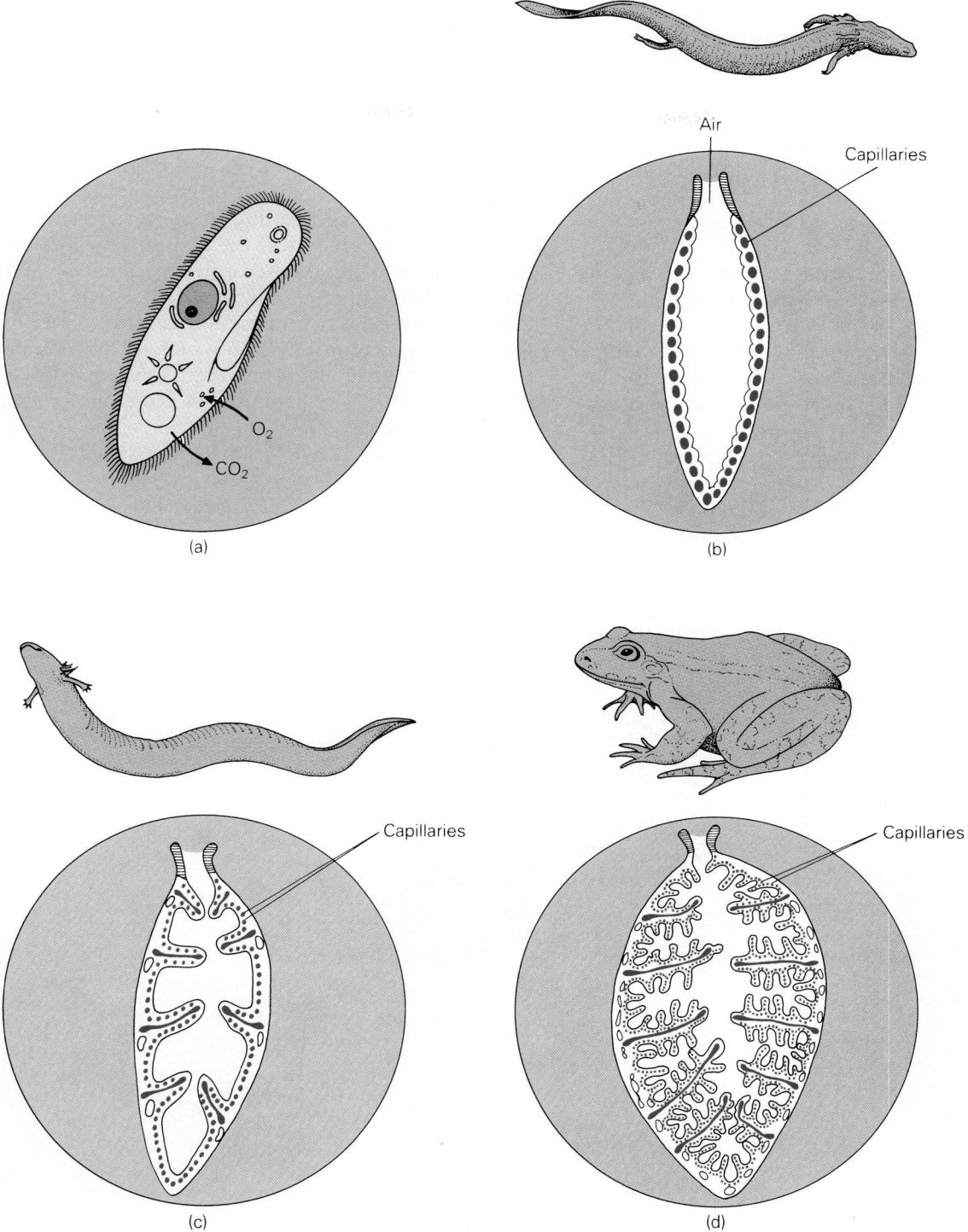

Figure 20–1
Structural design for gas exchange. (*a*) In unicellular organisms, no special organ is required. In higher animals, lungs serve as the gas exchange organ. Note the development of the multialveolar lung, which increases the internal surface area. (*b*) The uni-alveolar lung of the European olm (Proteus). (*c*) The multi-alveolar lung of the mud eel (Siren) with simple septa. (*d*) The multi-alveolar lung of the frog (Rana) with complex septa. Red dots in walls and in septa represent pulmonary capillaries.

the most primitive lung, such as that of the salamander (consisting of only a simple gas bag with numerous septa containing blood in the walls) represented a quantum jump in efficiency of gas exchange. The frog lung (Fig. 20–1*d*) has more internal surface area, which permits more blood to come into contact with air for gas exchange.

Alveoli: Microscopic Air Sacs

In the mammalian lung, these septa evolved into microscopic air sacs called **alveoli** (singular, *alveolus*) (Fig. 20–2*a, b*). Each alveolus is surrounded by a network of capillaries that bring blood into close contact with air inside the alveolus. Oxygen diffuses across the thin walls of the alveolus into the blood,

while carbon dioxide diffuses from the blood into the alveolus. The human lung contains 300 to 500 million alveoli with an internal surface area about the size of a tennis court, which is roughly 40 times greater than the external body surface area.

A simple analogy for the creation of a larger surface area is a marina. To dock more boats without enlarging the marina, smaller piers must be built. In nature, this strategy is followed as well. In species with high metabolic rates, the demand for gas exchange is also high. Accordingly, smaller and smaller alveoli are packed together in the lung, thereby increasing the internal surface area for more gas exchange. Figure 20–3, in which metabolic rate is plotted against alveolar size (diameter), shows this graphically. The manatee (sea cow), which has

Figure 20–2
(*a*) Human lung with each alveolus enmeshed in a network of capillaries.
(*b*) SEM of normal lung tissue showing an alveolar septum and the partition between the alveoli where gas exchange takes place. The capillaries are colored yellow in this image and are filled with red blood cells. (× 2200.)
(© SECCHI-LECAQUE-ROUSSEL-UCLAF/CNRI/SPL/Photo Researchers, Inc.)

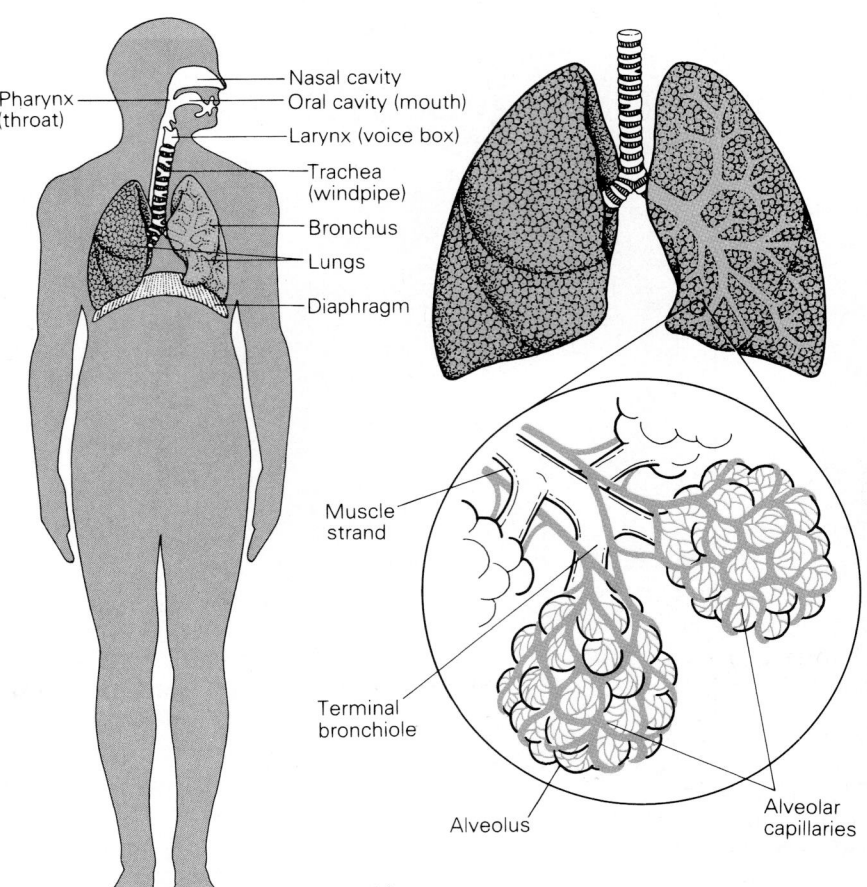

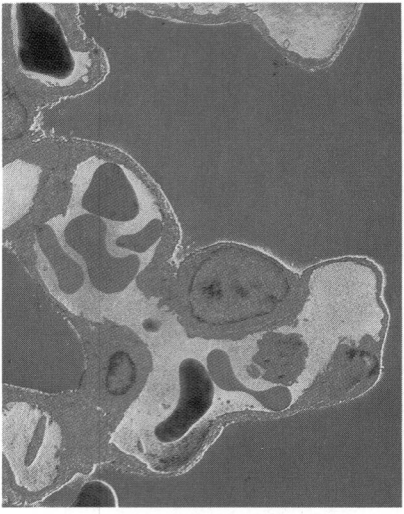

(a)

(b)

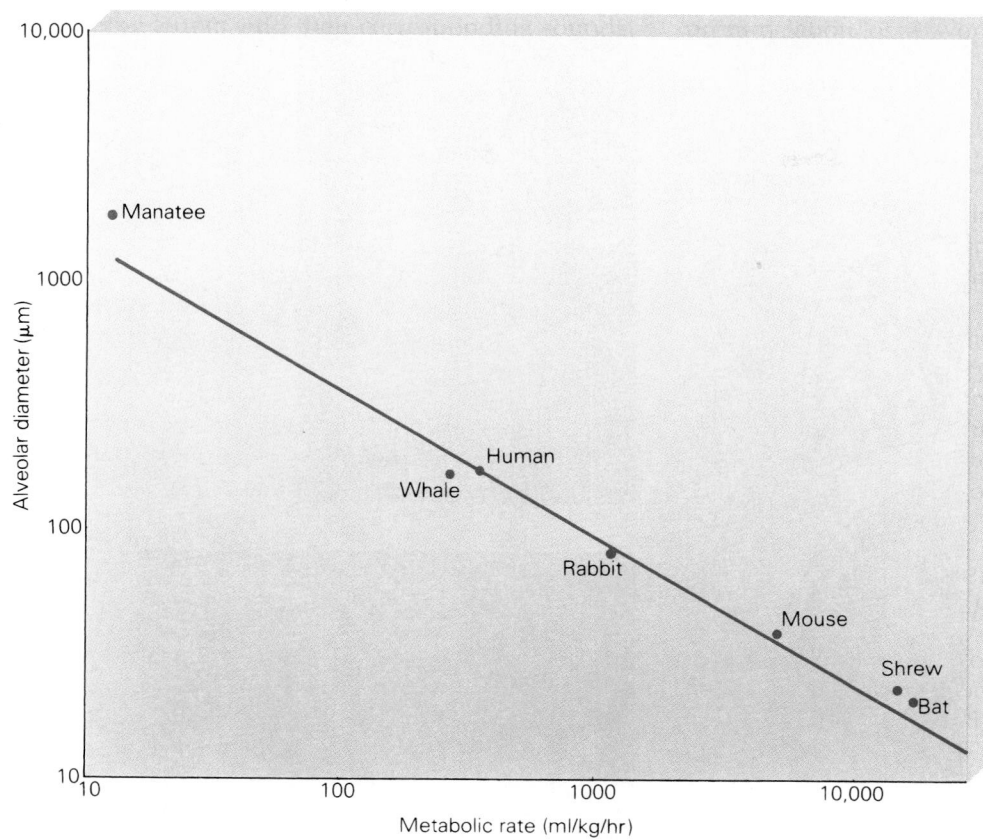

Figure 20–3

Alveolar diameter (in micrometers) plotted against resting metabolic rate (oxygen consumption in milliliters per unit body weight per unit time) in mammalian lungs of different species. Note the inverse relationship between alveolar diameter and metabolic rate.

a slow metabolic rate, has large alveoli. The shrew and bat, on the other hand, are like metabolic blast furnaces and have very small alveolar diameters.

Airways

The human gas exchange organ consists of two lungs, each divided into several lobes (Fig. 20–4). The lungs comprise two treelike structures, the **vascular tree** and the **airway tree,** which are embedded in elastic tissue. The vascular tree consists of arteries and veins tied together by capillaries. (Details of the vascular tree will be discussed in Chapter 21.) The airway tree consists of a series of highly branched hollow tubes that decrease in diameter and become more numerous at each branching. The main airway (the **trachea**) in turn branches into two **bronchi** (singular, *bronchus*), one of which enters each lung. Within each lung, these bronchi branch many times into progressively smaller bronchi, which in turn branch into terminal **bronchioles** analogous to twigs

of a tree. The trachea, bronchi, and bronchioles are part of the **conducting** zone that carries gas to and from alveoli (Fig. 20–5). The respiratory bronchioles and alveoli constitute the **respiratory zone** and are analogous to leaves on a tree having a similar function as exchange units between the organism and the environment.

The conducting units (trachea, bronchi, and bronchioles) have three important functions: (1) to warm and humidify the air, (2) to distribute air to the gas exchange surface of the lung, and (3) to serve as part of the body defense system. Since the inside of the lung is literally exposed to the external environment, the conducting unit contains a **mucociliary transport system** that keeps microorganisms, dust particles, and noxious gases from entering the alveoli (Fig. 20–6). The conducting units are lined with cilia that beat synchronously, moving foreign material and microorganisms up the trachea. The cilia are buried in a carpet of sticky mucus. At every bend in the airway, many of the heavy particles in

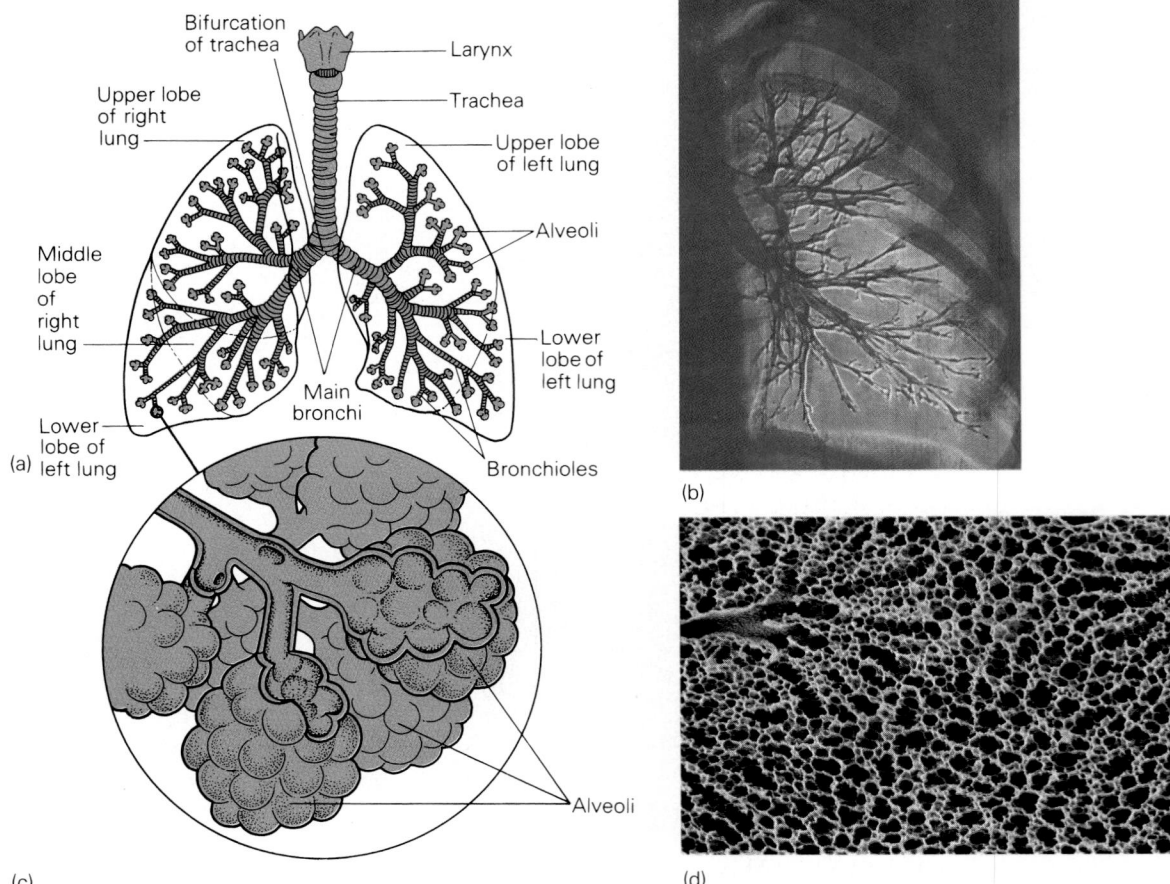

Figure 20–4
(*a*) Organization of the human lung. The right lung consists of three lobes (upper, middle, and lower) and the left lung consists of two lobes (upper and lower). (*b*) False-color arteriograph of a normal right lung showing the pattern of bronchial and bronchiolar branching. (*c*) The trachea subdivides to form bronchi, which branch repeatedly, leading to bronchioles and then to alveoli. Note that alveoli are not all the same size. (*d*) SEM of a normal human lung with numerous alveoli arising from two alvelolar ducts. (© Fawcett/ Gehr/Science Source/Photo Researchers, Inc.)

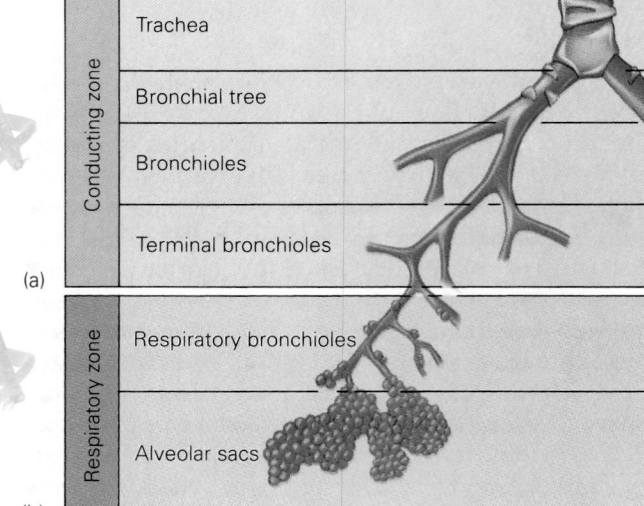

Figure 20–5
Tree-like pattern of airway, showing the conducting zone and the respiratory zone. The conducting zone (*a*) is illustrated from the trachea to the terminal bronchioles, and the respiratory zone (*b*) is illustrated from the respiratory bronchioles to the terminal bronchioles.

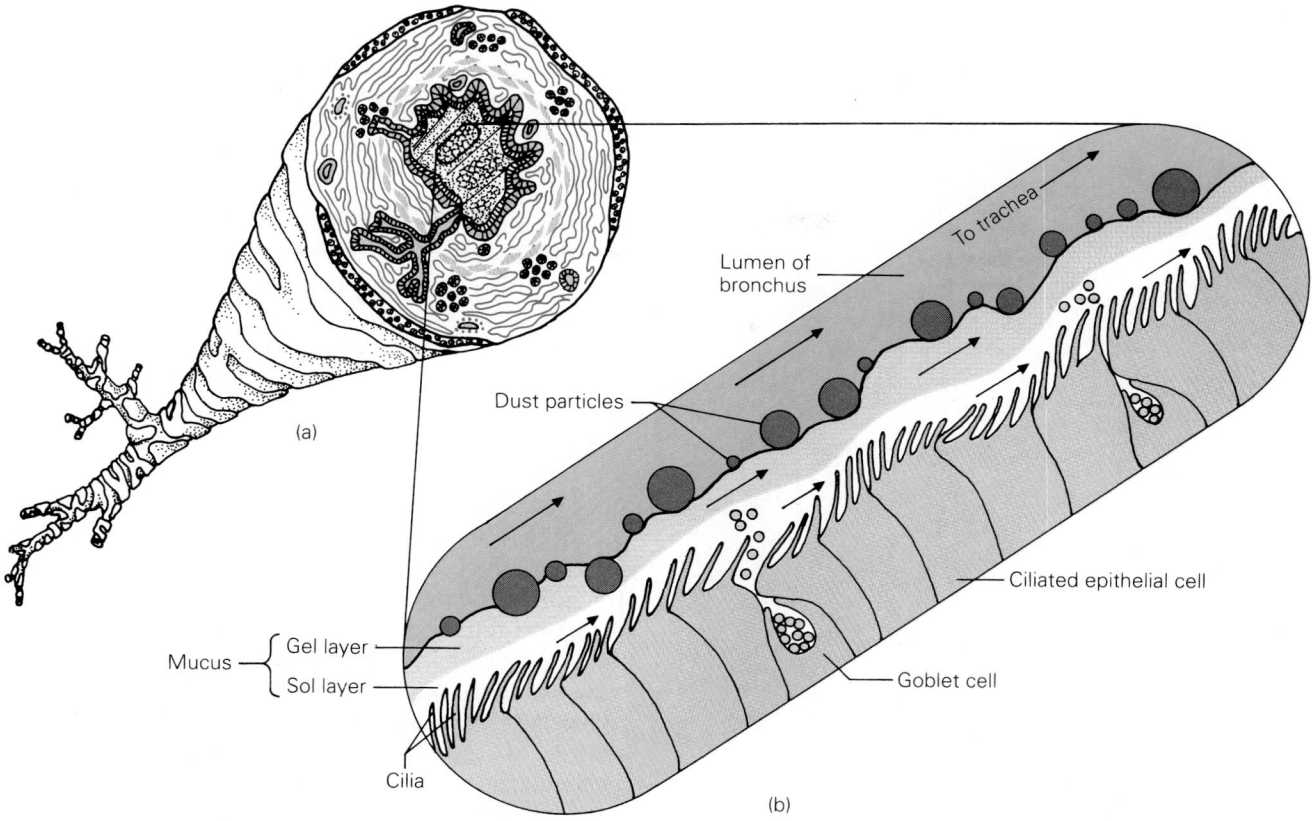

Figure 20–6
Mucociliary transport system. Mucus layer traps inhaled particles. The
mucus layer is propelled by cilia toward the trachea to remove bacteria and
particulate matter.

the inspired air fail to make the turn and stick to a
film of mucus. The mucus is coughed up and either
excreted or swallowed.

The cilia are greatly affected by tobacco smoke.
When they come in contact with it, they become
partially paralyzed, slowing the mucociliary transport system. As a result, microorganisms and excess
mucus accumulate in the lower airways (e.g., the
bronchioles). Often, the excess mucus plugs the
lower airways, and a cough is required to dislodge
and remove it. If excess mucus accumulates over a
long period of time, bacterial infection will often
develop. When this occurs, the mucus is no longer a
light yellow but instead is green as a result of the
infiltration and rupturing of white blood cells.

Lung Volumes and Ventilation

Air Flow

The process of moving air in and out of the lungs is
known as **ventilation.** When you take a breath of air
into your lungs, you can hear the sound of air rush-
ing in, especially if you breathe rapidly. The sound
comes from air flow in the airway tree. There are
two basic types of air flow.

Turbulent Air Flow

One type, occurring in the trachea and large bronchi, is **turbulent flow** (Fig. 20–7a). Turbulent flow
occurs at high flow rates and consists of completely
disorganized patterns of air flow in which the molecules literally collide with each other, resulting in a
noise that can be heard on inhalation or exhalation.
Thus, the faster you inhale or exhale air, the more
turbulent the noise.

An important consideration regarding turbulent
flow is that turbulence in the large airways is the
primary source of resistance to breathing. Turbulent
flow is also greatly affected by gas density. If you
breathe a helium-oxygen mixture, which is much
lighter than air (nitrogen and oxygen), less turbulence is produced, meaning that less resistance and
less work are involved in breathing. Gas density
also affects the vocal cords in the larynx by changing
the pitch of our voices, making us sound like Don-

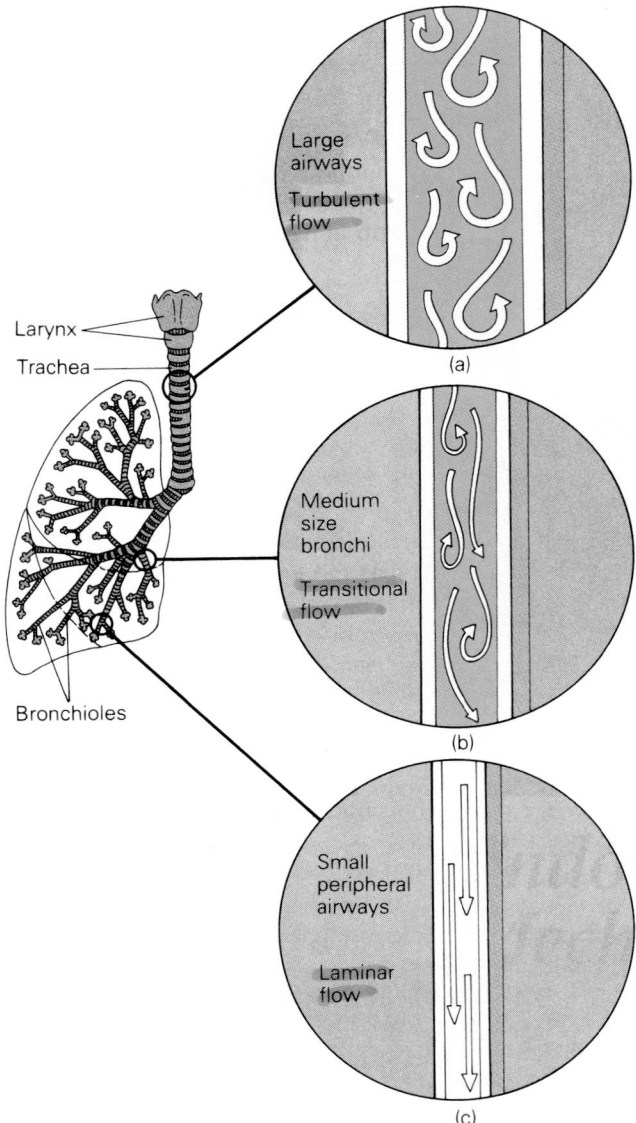

Figure 20–7
Patterns of air flow in the lung. (*a*) Turbulent flow occurs at high flow rates in large airways. (*b*) Transitional flow is a combination of laminar and turbulent flow and occurs in medium-sized bronchi. (*c*) Laminar flow occurs at low flow rates in small peripheral airways.

ald Duck when we talk after breathing a helium-oxygen mixture.

Laminar Air Flow

The second type of air flow in the lungs is **laminar air flow,** which is characterized by streamlined flow that parallels the side of the airways (Fig. 20–7*c*).

Laminar flow is silent, because air molecules slide over each other. This type of air flow occurs mainly in the small peripheral airways where air flow is extremely slow.

Laminar flow is greatly affected by airway diameter. Specifically, laminar flow is directly related to the radius raised to the fourth power (Flow $\propto$ radius^4). This means, for example, that if a peripheral airway constricts such that the radius is halved, then airflow in that tube decreases 16-fold.

$$F = 1^4 = 1$$

$$F = \left(\frac{1}{2}\right)^4 = \frac{1}{16}$$

This relationship between airflow and radius in the small airways explains why a person who suffers from asthma (a condition caused by spasmodic contraction of the bronchiole smooth muscles) has great difficulty getting air in and out of the alveoli. During asthma attacks, many asthmatics take epinephrine, a hormone that acts on the beta receptor to cause the smooth muscle to relax and in turn dilate the airways. At lower flow rates during expiration—particularly at branches of the airway tree where flow from separate tubes comes together and empties into a single airway—the air flow pattern is a mix. This is referred to as **transitional flow,** a combination of turbulent and laminar flow (Fig. 20–7*b*).

Lung Spirometry

During inspiration, the lungs are inflated, and air is drawn in. The volume of air breathed in and out of the lung is measured by a **spirometer** (Fig. 20–8). This device is a simple volume recorder consisting of a double-walled cylinder in which an inverted bell is immersed in water to form a seal. The bell is attached by a pulley to a pen that writes on a rotating drum. As air enters the spirometer from the lungs, the bell rises. Because of the pulley arrangement, the pen is lowered. Thus, a downward pen deflection represents expiration, and an upward pen deflection represents inspiration. The recording is known as a **spirogram,** which measures two factors: (1) how fast air moves in and out of the lungs (rate of air flow) and (2) the volume of air moved.

The volume of air entering or leaving the lungs during a single breath is called **tidal volume** (see Fig. 20–8). Under resting conditions, this volume is approximately 500 ml. The resting tidal volume represents only a fraction of the total air that the lungs can take in or out.

The volume of air remaining in the lungs at the end of a normal tidal volume (end of expiration) is

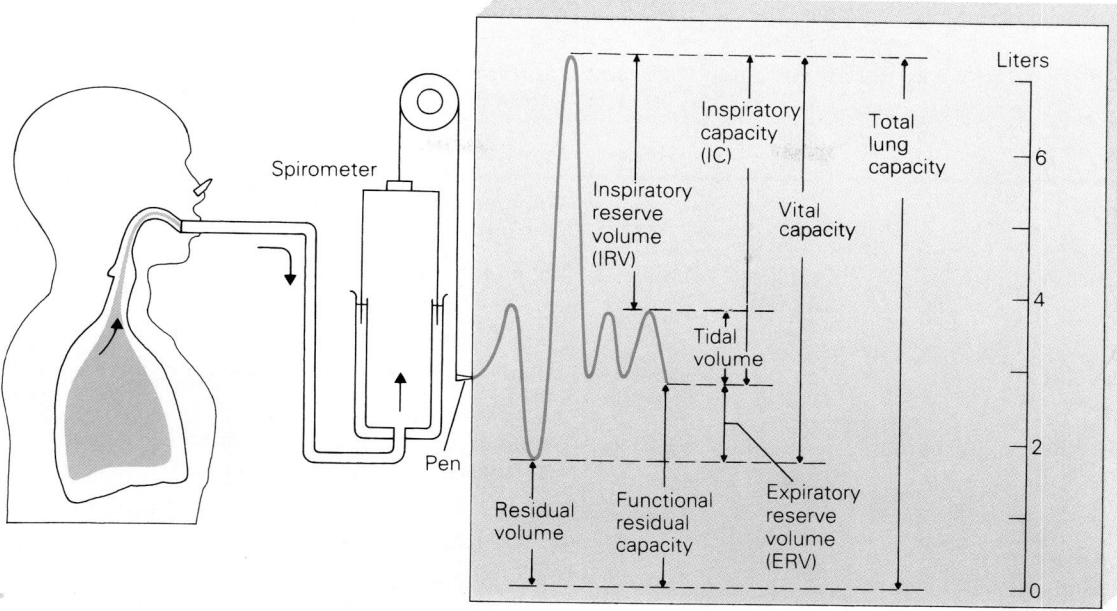

Figure 20–8

Lung volumes and capacities measured by a spirometer. With inspiration the pen shows an upward deflection on the spirograph, and with expiration a downward deflection. Note that the functional residual capacity and residual volume cannot be measured directly with a spirometer.

referred to as the **functional residual capacity (FRC).** A point of terminology is important at this juncture. When referring to the subdivisions of lung volume, the terms *volume* and *capacity* are both used. **Capacity** is used when a volume can be broken down into two or more smaller volumes; for example, **expiratory reserve volume (ERV)** and **residual volume (RV)** = FRC. These lung volumes and capacities are summarized in Table 20–1.

Forced Vital Capacity

The maximum amount of air that can be forcefully and rapidly exhaled after a deep breath (maximal inspiration) is called **forced vital capacity** (FVC) and measures approximately 5 liters for an adult. Forced vital capacity is one of the most useful tests to assess the overall ability to move air in and out of the lungs (ventilation). For example, a well-trained athlete can have a forced vital capacity of more than 7.0 L, whereas an asthmatic may have a vital capacity no greater than 3 L. The constricted bronchioles of asthmatics can collapse during forced expiration, trapping air in the lungs and leading to a decrease in forced vital capacity. A similar problem exists with individuals who smoke or have **bronchitis,** a condition in which the lining of the bronchioles becomes inflamed and swollen, thereby reducing airway diameter.

Forced vital capacity is determined primarily by three factors: (1) strength of chest and abdominal muscles, (2) airway resistance, and (3) lung volume. Any condition that decreases the strength of the respiratory muscles (e.g., poliomyelitis), decreases lung volume (e.g., tuberculosis), or increases airway resistance (e.g., bronchitis) will lead to a significant decrease in forced vital capacity.

The forced volume exhaled in one second is termed FEV_1 (Fig. 22–5). This volume has the least variability of the measurements obtained from a forced expiratory maneuver and is considered one of the most reliable measurements. Another useful way of expressing FEV_1 is as a percentage of FVC (i.e. $FEV_1/FVC \times 100$), which corrects for differences in lung size. Normally, FEV_1 is 80% of the FVC (i.e., 80% of an individual's forced vital capacity can be exhaled in the first second). Lung diseases are often classified as either *obstructive* or *restrictive disorders.* With an obstructive disorder, *expiratory flow* is obstructed, and with a restrictive disorder, *lung inflation* is restricted. In obstructive lung disorders, the ratio of FEV_1/FVC is reduced to drastically below normal (<80%). Examples include asthma and emphysema. Examples of restrictive disorders include pulmonary edema and pulmonary fibrosis.

To summarize, the measurement of FVC is most commonly and easily obtained from a forced expira-

Table 20-1

Definitions of Standard Lung Volumes, Capacities, and Pulmonary Function Tests for a Typical Young Adult

Term	Value	Definition
V_T (Tidal Volume)	0.5 liter	Volume of air inhaled or exhaled with each breath
VC (Vital Capacity)	4.8 liters	Volume of air that can be exhaled after maximal inspiration (VC = IRV + V_T + ERV)
RV (Residual Volume)	1.2 liters	Volume of air remaining in the lungs after maximal expiration
ERV (Expiratory Reserve Volume)	1.2 liters	Maximal volume of air that can be exhaled at the end of a tidal volume
IRV (Inspiratory Reserve Volume)	3.1 liters	Volume of air that can be inhaled at the end of a normal inspiration
FVC (Forced Vital Capacity)	4.8 liters	Volume of air that can be exhaled as forcibly and rapidly as possible after maximal inspiration
IC (Inspiratory Capacity)	3.6 liters	Maximal volume of air that can be inhaled after a normal expiration (IC = IRV + V_T)
FRC (Functional Residual Capacity)	2.4 liters	Volume of air remaining in the lungs at the end of a tidal volume (FRC = ERV + RV)
TLC (Total Lung Capacity)	6.0 liters	Total volume of air in the lungs after maximal inspiration (TLC = VC + RV)
RV/TLC (Residual Volume/Total Lung Capacity)	20%	Percent of total lung capacity made up of residual volumes
FEV_1 (Timed Forced Expiratory Volume in One Second)	3.8 liters	Volume of VC forcibly exhaled in 1 second
FEV_1/FVC (Timed Forced Expiratory Volume/ Forced Vital Capacity) × 100	80%	Percent of FVC that can be forcibly exhaled in one second

tory maneuver. From a FVC determination, we can readily measure FEV_1 and FEV_1/FVC ratio. These measurements, by themselves, provide a considerable amount of information about ventilatory function.

Residual Volume and Functional Residual Capacity

Even after an individual expires as forcefully and completely as possible (forced vital capacity), some air, approximately 1.0 L, remains in the lungs and is called **residual volume.** When air becomes trapped in the lungs during expiration because of airway plugging or airway closure, the residual volume becomes abnormally high.

Another often-measured lung capacity is **functional residual capacity.** This is the volume of air remaining in the lungs after a normal expiration and amounts to about 2.5 L.

Because the lungs cannot be emptied completely following forced expiration, neither residual volume nor functional residual capacity can be measured directly by simple spirometry. These two volumes are measured indirectly by a dilution technique based on the following relationship:

$$C_1 \times V_1 = C_2 \times V_2$$

in which C is concentration and V is volume. For example, if a 1-L box contains eight gas molecules (Fig. 20-9), then the box can be represented by $C_1 \times V_1$. If the box has a window that is opened to an-

Bronchitis and Emphysema

By far the most common chronic lung disease in the United States is **chronic obstructive pulmonary disease (COPD),** which includes both emphysema and chronic bronchitis. The current estimate is that over 10 million Americans suffer from one or both of these diseases. COPD is now the fifth leading cause of death in the United States (over 75,000 deaths occurred last year). The morbidity and mortality for COPD are rising at an alarming rate. Costs related to the care and treatment of COPD patients are staggering. Each year the direct costs exceed $4 billion.

The predominant clinical feature of both emphysema and chronic bronchitis is a chronic obstruction to air flow. Surprisingly, patients frequently have an abnormally high lung volume (i.e., their total lung capacity is increased) but their biggest difficulty is exhausting air from their lungs. The latter is reflected in their pulmonary function tests, which show abnormally low values for FVC, FEV_1, and FEV_1/FVC (see Table 20–1 for normal values).

Both diseases share similar symptoms: cough, wheeze, shortness of breath, and production of large amounts of mucus and sputum. Despite their similarities, the two diseases have marked structural differences and are brought about by different mechanisms. Chronic bronchitis is characterized by excessive mucus production, which plugs small airways. This comes about when the lining of the small airways becomes irritated from exposure to inhaled irritants such as tobacco smoke, which leads to excess mucus. In emphysema, alveoli become overstretched and rupture, due to excess pressure resulting from a combination of chronic coughing and plugged airways. Damage to the alveolar walls causes them to become permanently and abnormally enlarged (see inset). Late in COPD disease, patients frequently experience hypoxemia (low blood oxygen) and hypercapnia (high blood carbon dioxide). These two conditions, especially hypoxemia, lead to pulmonary hypertension (high blood pressure in the pulmonary circulation), which in turn causes right heart failure. Based on changes in blood oxygen levels, COPD patients can be categorized into two types. Those with predominantly emphysema are referred to as "pink puffers" because their oxygen levels are usually satisfactory and their skin remains pink. They develop a puffing style of breathing that is rapid and shallow. Those with predominantly chronic bronchitis are called "blue bloaters" because low oxygen levels give their skin a blue cast and because fluid retention from heart failure gives them a bloated look.

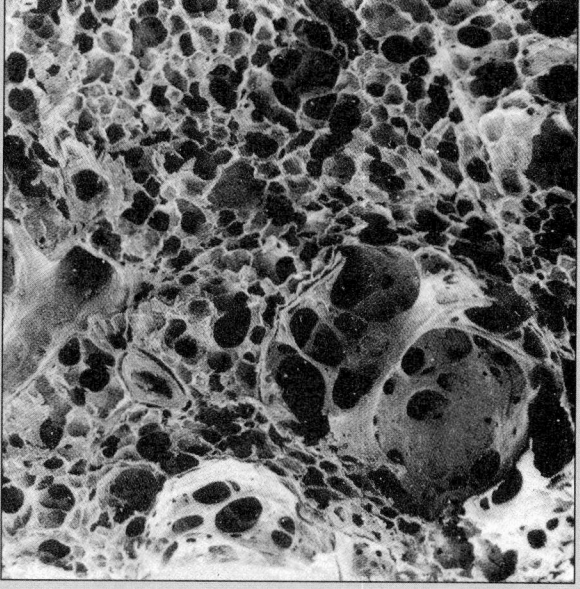

Scanning electron photomicrograph of a normal human lung showing the contrast between normal and enlarged alveolar spaces due to emphysema. The enlarged alveolar spaces are caused by over inflation. (Photo by Janice Nowell. From *Pathophysiology, 2nd ed.,* by Smith and Thier. W.B. Saunders, 1985.)

Smoking is the single most common cause of chronic bronchitis and emphysema. COPD is uncommon in nonsmokers. While pollution and occupational hazards are less important than smoking, they can act to worsen the condition. COPD is four times more common in men than in women, but as smoking habits of women change, this difference may diminish.

Recently, scientists have discovered a rare type of emphysema that occurs in families even in the absence of cigarette smoking. These individuals develop severe emphysema at a very early age (20 or 30 yrs). Scientists have discovered that this rare type of emphysema is due to the absence of a specific immunoglobin, alpha-1, in their blood. Immunoglobins are proteins that ordinarily act as antibodies, but alpha-1 also carries a substance known as alpha-1 antitrypsin, which inhibits the actions of certain enzymes (proteases) involved in the breakdown of proteins. The absence of alpha-1 antitrypsin is an inherited trait that leaves the affected individual open to the breakdown of lung proteins by his own proteases, particularly the one known as elastase. Normally there is a balance between the elastase activity and alpha-1 antitrypsin so that no detrimental breakdown of lung tissue takes place. When alpha-1 antitrypsin is present in low amounts or absent altogether, increased destruction of lung tissue leads to emphysema.

Not only can scientists explain how alpha-1 antitrypsin-deficient individuals develop early emphysema, but they have recently discovered how an alpha-1 antitrypsin-elastase imbalance is involved in emphysema in individuals who do not inherit an alpha-1 antitrypsin deficiency. Smoking can upset the balance in two ways. The first way is by increasing in

the amount of elastase. White cells are the main products of elastase. The number of white cells is significantly higher in cigarette smokers, and this leads to excess release of elastase. The second way that cigarette smoking interferes with this balance is by inactivating alpha-1 antitrypsin.

Although great strides have been taken toward conquering COPD, much research remains to be done. Future directions include the production of alpha-1 antitrypsin using recombinant DNA techniques. Another promising approach is the use of synthetic antiproteases to slow down lung damage.

A final comment: The most important therapy for COPD patients is to stop smoking.

other box with an unknown volume, then the eight gas molecules can equilibrate between the two boxes, and $C_1 \times V_1 = C_2 \times V_2$. If C_1, V_1, and C_2 are known, then V_2 can be solved to determine the unknown volume.

Since residual volume and functional residual capacity are unknown volumes, a similar dilution technique can be used. The technique used to measure these two volumes involves the dilution of he-

lium, an inert and insoluble gas. The subject is connected to a spirometer filled with 10% helium in air (see Fig. 20–9). Consequently, the initial 10% concentration of helium in the spirometer is C_1, and the volume in the spirometer is V_1. Since the lung initially contains no helium, after the subject breathes the helium-air mixture, the helium concentration in the lungs becomes the same as in the spirometer after equilibration. Because helium is insoluble and

Figure 20–9
Dilution technique to measure an unknown volume. The dilution of helium is used to measure functional residual capacity. Dots represent helium before and after equilibration.

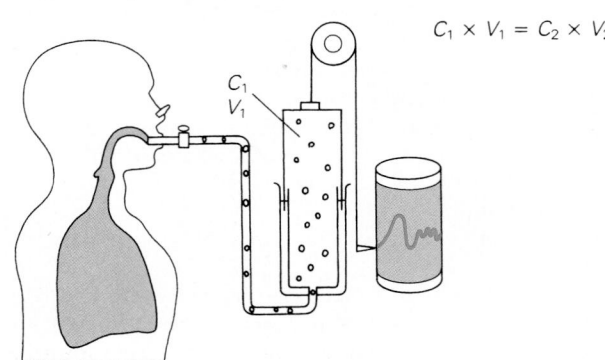

Before equilibration   After equilibration

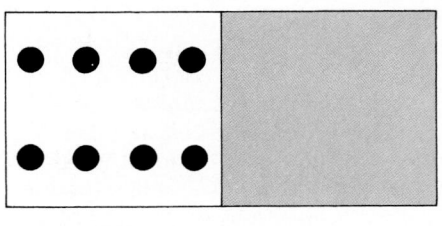

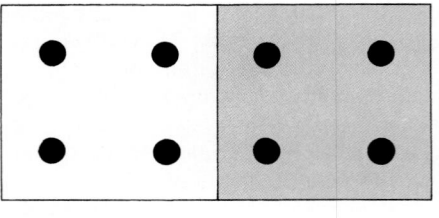

not taken up by the blood, the concentration of helium in the lung after equilibration is C_2, and the unknown volume in the lung is $V_2 - V_1$.

It is important to start the test at precisely the right time. If the test begins at the end of a normal tidal volume (end of expiration), the volume of air remaining in the lung represents functional residual capacity. If the test begins at the end of a forced vital capacity and the subject starts breathing the helium-air mixture, then the test will measure residual volume.

Ventilation Moves Air in and out of the Lungs

So far, the lung volumes discussed have been primarily static volumes. However, ventilation is a dynamic process. If 500 ml of air is brought into the lungs with each breath (tidal volume) and the breathing rate is 14 times a minute, then the total volume of inspired air that enters the lungs each minute is $500 \times 14 = 7000$ ml per minute or 7 liters per minute. This volume of air is known as **minute ventilation.**

Dead Space Volume

However, it is important to realize that not all of the minute ventilation enters the alveoli for gas exchange. The tidal volume is distributed between the conducting airways and alveoli (Fig. 20–10). Since gas exchange occurs only in the alveoli and not in the conducting airway (trachea, bronchi, bronchioles), part of the minute ventilation becomes wasted air. For each 500 ml of air inhaled, approximately 150 ml remains in the conducting airways and does not participate in gas exchange. This volume of wasted air is known as **dead space volume.** Picture, then, what occurs during a normal breathing cycle. A normal tidal volume of 500 ml is expired. During the next inspiration, another 500 ml is taken in, but the first 150 ml of air entering the alveoli is wasted air (old alveolar gas left behind). Thus, only 350 ml of fresh air reaches the alveoli, and 150 ml is left in the conducting airways.

tidal volume	=	500 ml
dead space volume	=	−150 ml
fresh air entering alveoli	=	350 ml

The normal ratio of dead space volume to tidal volume (dead space volume/tidal volume) is in the range of 0.20 to 0.35. In the example shown, the ratio (150/500) is 0.30. This means 30% of the tidal volume does not participate in the gas exchange and constitutes wasted air.

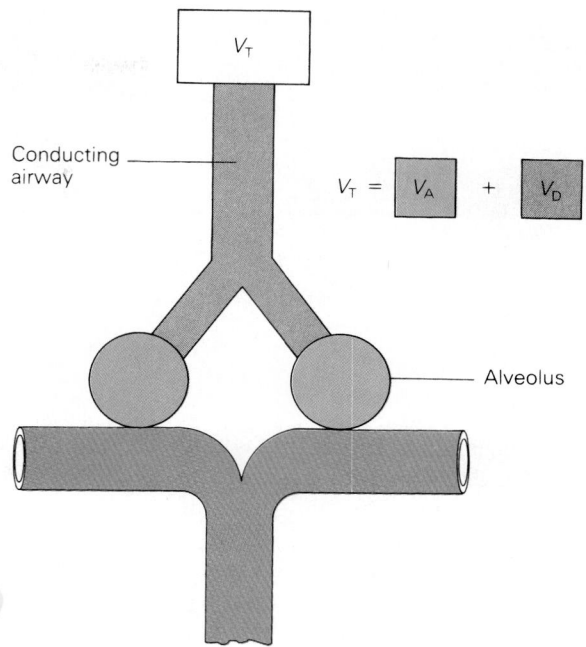

Figure 20–10
Tidal volume (V_T) is a mixture of gas from alveoli and conducting airways. The gas in the airways represents dead space volume (V_D) because this volume of gas does not participate in gas exchange. The volume of gas that participates in gas exchange in the alveoli is represented by V_A.

Alveolar Ventilation

To determine how much fresh air reaches the alveoli per minute, we subtract the dead space volume from the tidal volume and multiply the result by breathing frequency. For example, the volume of air entering the alveoli from a normal tidal volume of 500 ml with a breathing rate of 14 breaths per minute is $(500 \text{ ml} - 150 \text{ ml}) \times 14 = 4900$ ml. This represents the volume of "fresh air" available for gas exchange and is known as **alveolar ventilation.** Alveolar ventilation is the most important variable in gas exchange. This is because minute ventilation does not necessarily represent the amount of oxygen reaching the alveoli, since the amount of dead space varies. For instance, a swimmer using a snorkel breathes through a tube that increases dead space volume. Similarly, a patient connected to a mechanical ventilator also has increased dead space volume. Indeed, if minute ventilation is held constant, then alveolar ventilation is decreased with snorkel breathing or with mechanical ventilation.

The significance of dead space volume, minute ventilation, and alveolar ventilation can readily be seen in Table 20–2. In this experiment, Subject A's

Table 20–2
Influence of Breathing Patterns on Alveolar Ventilation

Subject	Tidal volume × (ml)	Frequency (breaths/min)	Minute = volume − ml/min	Dead space ventilation (ml/min)	Alveolar = ventilation (ml/min)
A	1000 ×	6	= 6000	− 900 (150 × 6)	= 5100
B	500 ×	12	= 6000	− 1800 (150 × 12)	= 4200
C	150 ×	40	= 6000	− 6000 (150 × 40)	= 0

breathing is slow and deep, while B's breathing is normal and C's breathing is rapid and shallow. Note that each subject has the same minute ventilation (i.e., each is moving the same volume of total air in and out per minute), but each has marked differences in alveolar ventilation. Subject C has no alveolar ventilation and would die in a matter of minutes, while Subject A has an alveolar ventilation greater than normal. The important message from the examples presented in Table 20–2 is that increasing the depth of breathing is far more effective in elevating alveolar ventilation than increasing the frequency or the rate of breathing. This fact is important in exercise. In most exercise situations, increased ventilation is accomplished by increases in the depth of breathing more than in the rate. For example, a well-trained athlete can often increase alveolar ventilation during moderate exercise with little or no increase in breathing frequency.

The Mechanics of Lung Action

The lungs are housed in an airtight chest cavity, the **thoracic cavity**, and are separated from the abdomen by a large, dome-shaped muscle, the **diaphragm** (Fig. 20–11). The thoracic cage is made up of 12 pairs of **ribs**, a **sternum** (breast bone), and a set of internal and external intercostal muscles that lie between the ribs. The rib cage is hinged to the vertebral column (backbone), allowing it to rise and lower during breathing. The space between the lungs and chest wall is the **pleural space,** which contains fluid and serves as a lubricant so that the lung can slide against the chest wall.

The Muscles of Breathing

Inflation of the lungs causes air to rush in (**inspiration**) and is due to an active muscular process. Consider for a moment the breathing cycle. The main muscle during the inspiratory phase is the **diaphragm.** The lungs inflate automatically when the diaphragm contracts and enlarges the thoracic cavity. This is accomplished two ways. First, when the diaphragm (which is attached to the lower ribs and sternum) contracts, the abdominal contents are pushed down, thus enlarging the pleural space. As the diaphragm pushes down on the abdominal contents, it also pushes the rib cage outward, further enlarging the pleural space. The increase in pleural space volume causes a decrease in pressure in the pleural space, which in turn causes the lung to expand and fill with air.

Muscular Movements During Inspiration

During inspiration the lungs play a purely passive role. The diaphragm is the main respiratory muscle during normal inspiration. However, during forced inspiration, in which a large volume of air is taken, the **external intercostal** muscles (see Fig. 20–11) are used as a second set of inspiratory muscles. These muscles also pull the rib cage upward and outward. To see how this works, stand sideways in front of a mirror and take a deep breath. The chest wall moves upward while the sternum moves outward, thereby enlarging the thoracic cavity.

The last group of muscles used in inspiration are the **accessory muscles.** These include the neck and chest muscles and are brought into play only during deep and heavy breathing such as during exhaustive exercise.

Muscular Movements During Expiration

Breathing out (**expiration**) is a much simpler process than inspiration. At the end of a normal inspiration, the diaphragm relaxes, the thoracic cavity decreases, and the lungs deflate. Again, the lung is passive in breathing. Although expiration is purely passive during normal breathing, with exercise or

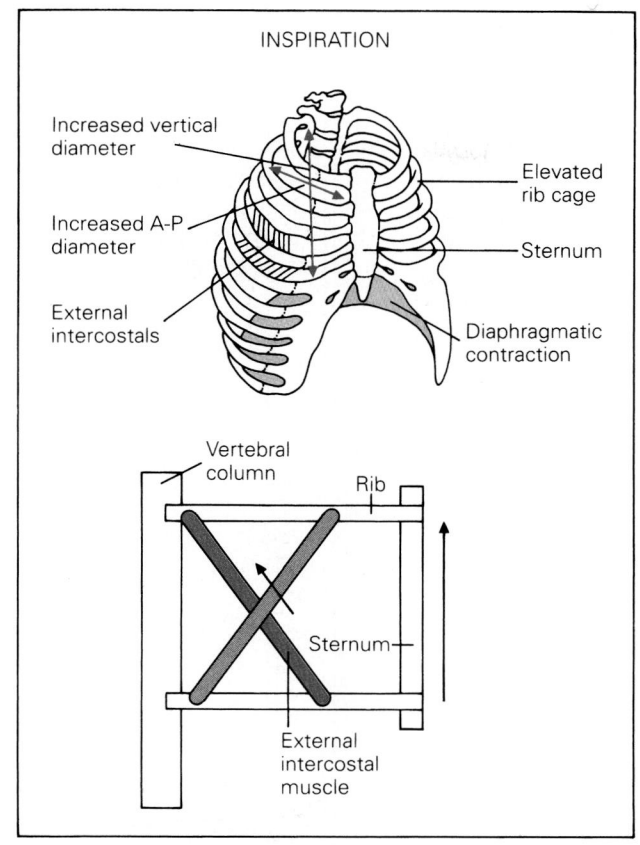

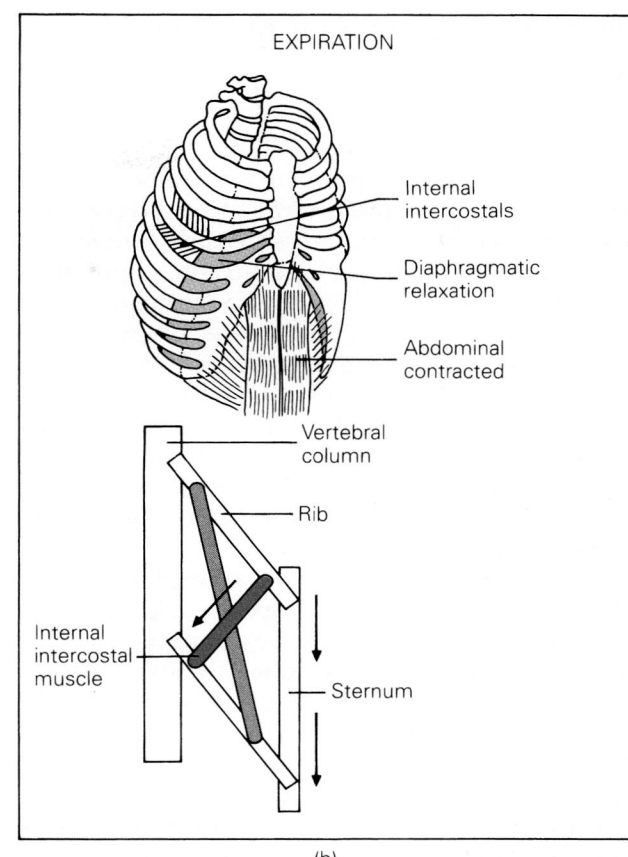

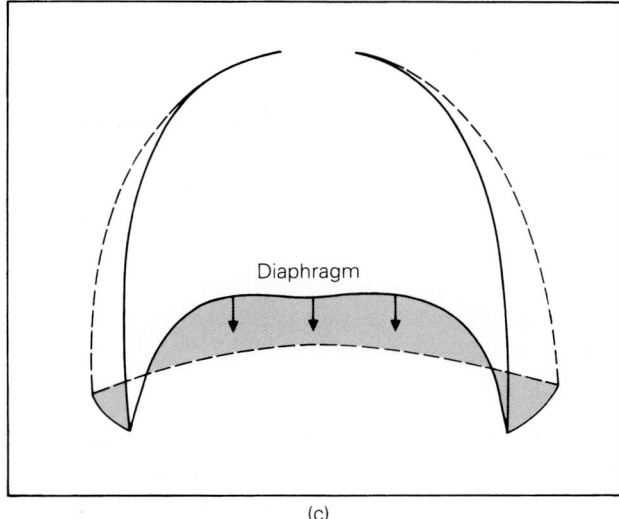

Figure 20–11
Functions of the thoracic cavity and respiratory muscles during breathing. (*a*) During inspiration, the diaphragm contracts and the rib cage is elevated, enlarging the thoracic cavity by increasing both vertical and the anterior-posterior (A-P) diameter. (*b*) During expiration, the diaphragm relaxes and the rib cage drops, thereby decreasing the space in the thoracic cavity. (*c*) The lever action of the diaphragm on abdominal contents pushes the rib cage up and outward during inspiration.

with forced expiration the expiratory muscles become active. The expiratory muscles are the muscles of the **abdominal walls** and, to a lesser extent, the **internal intercostal muscles** (see Fig. 20–11). Contraction of the abdominal muscles helps push the diaphragm up, and the internal intercostal muscles

pull the rib cage down. These expiratory muscles are important and necessary for coughing and for endurance running. This is one of the reasons that competitive long-distance runners do exercises to strengthen their abdominal muscles as part of their training.

Lung Pressures During the Breathing Cycle

In Chapter 18, which discusses the cardiovascular system, the units for pressure are millimeters of mercury (mm Hg). In respiration, however, the most convenient unit for pressure is centimeters of water (cm H_2O) because of the small pressure changes involved. In addition, respiratory physiologists commonly measure pressure relative to atmospheric pressure, or **relative pressure.** The unit for relative pressure is also cm H_2O. The atmosphere, a sea of air we breathe and live in, exerts a pressure (P) at sea level known as **atmospheric pressure** (P_{atm}) and is equal to 760 mm Hg (Fig. 20–12). Thus, relative pressure is the pressure relative to the atmospheric pressure.

For example, in the middle of expiration, the pressure inside the alveoli is 5 cm H_2O. This means the pressure is 5 cm H_2O *above* atmospheric pressure. Conversely, in the middle of inspiration the pressure inside the alveoli is −2 cm H_2O. The negative sign does not indicate that the pressure is negative but rather that it is 2 cm *below* atmospheric pressure. It is important to remember that when relative pressures are used, atmospheric pressure is set at zero. Therefore, if alveolar pressure is zero, this means alveolar pressure equals atmospheric pressure. So, unless otherwise specified, the pressures of breathing are relative pressures, and the units are cm H_2O (remember that 13.6 cm H_2O equals 1 mm Hg pressure).

Three important relative pressures involved in breathing are associated with lung inflation and air flow (Fig. 20–13). First is **pleural pressure,** which is the pressure in the pleural space between the lung and chest wall. The second, **alveolar pressure,** is the pressure inside the alveoli. The third, **transpulmonic pressure,** is the pressure gradient across the lung wall. Transpulmonic pressure (P_{TM}) is measured by subtracting pleural pressure (P_{pl}) from alveolar pressure (P_A). Thus, transpulmonic pressure ($P_{TP} = P_A − P_{pl}$) is what keeps the lungs from collapsing and is also responsible for distending (stretching) the lung during inflation. For that reason, transpulmonic pressure is often called the **distending pressure.** One way of remembering transpulmonic pressure is "In minus Out," where P_A is the pressure inside the lung and P_{pl} is the pressure outside the lung.

A key question is: why is pleural pressure negative or subatmospheric? The reason is that the lung and chest wall are both **elastic** (i.e., capable of being stretched). At the end of expiration, the lung and

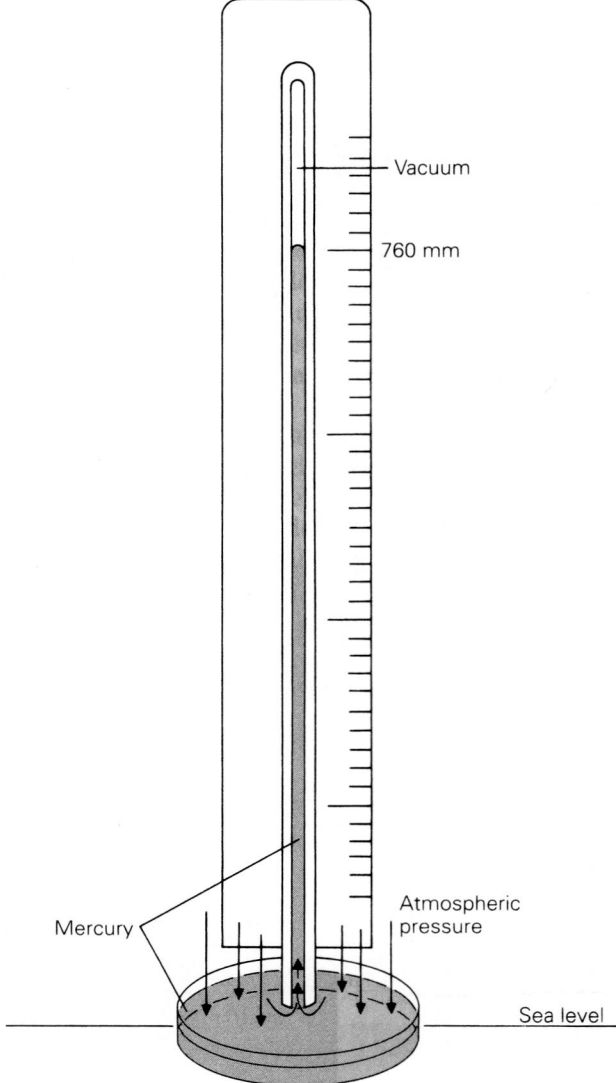

Figure 20–12
At sea level, atmospheric pressure can push a column of mercury (Hg) to a height of 760 ml. This is also known as 1 atmosphere of pressure.

chest wall are stretched in equal but opposite directions. Consequently, the elastic lung has the potential to recoil inwardly, and the elastic chest wall has the potential to recoil outwardly. These opposing forces cause the pleural pressure to be below atmospheric pressure. During quiet breathing, pleural pressure is negative and becomes more negative with deep inspiration. Only during forced expiration does pleural pressure become positive.

An example of the elastic recoil is seen when the chest is opened during surgery. The stretched lungs collapse immediately (recoil inward), and the stretched rib cage simultaneously expands outward

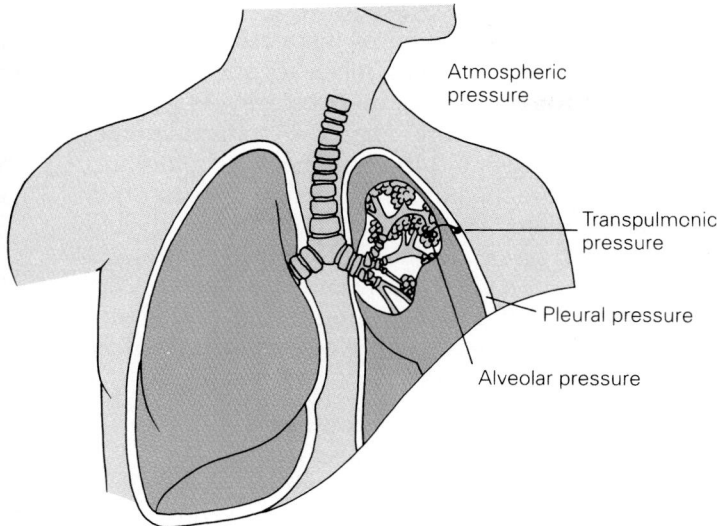

Figure 20–13
The basic pressures involved in breathing are atmospheric pressure, pleural pressure, alveolar pressure, and transpulmonic pressure.

(recoils outward). Since the pleural pressure is subatmospheric, any time the chest wall or the lung is punctured, air rushes into the pleural cavity. As a result, the pressure difference across the lung is eliminated (transpulmonic pressure equals zero), and the stretched lung collapses. When air enters the pleural space and the lungs collapse, the condition is known as a **pneumothorax** (Fig. 20–14). A pneumothorax can occur when the chest wall is punctured as with a knife or gun shot wound, or when the lung is ruptured as in an abscess and severe coughing.

Let us follow these pressures during a normal breathing cycle (Fig. 20–15). At the end of tidal volume (end of expiration), the respiratory muscles are relaxed and there is no air flow. At this point, alveolar pressure is zero (equal to atmospheric pressure) and pleural pressure is -5 cm H_2O. Transpulmonic pressure at the end of a normal tidal volume would therefore be 5 cm H_2O [$0 - (-5) = 5$]. Remember that the transpulmonic pressure is always positive in normal breathing and is the pressure difference across the lung wall ($P_A - P_{pl}$). The positive transpulmonic pressure is the distending pressure

Figure 20–14
(a) Pneumothorax. When air enters the pleural space through a hole in the chest wall or the lung, the pleural pressure equals atmospheric pressure (zero), and the lung collapses. Note that the mediasternal membrane prevents both lungs from collapsing during a pneumothorax. (b) X-ray of male lung shortly after the development of spontaneous pneumothorax. (From: *Organ Physiology: Structure and Function of the Lung,* Fraser and Paré, W. B. Saunders, 1977).

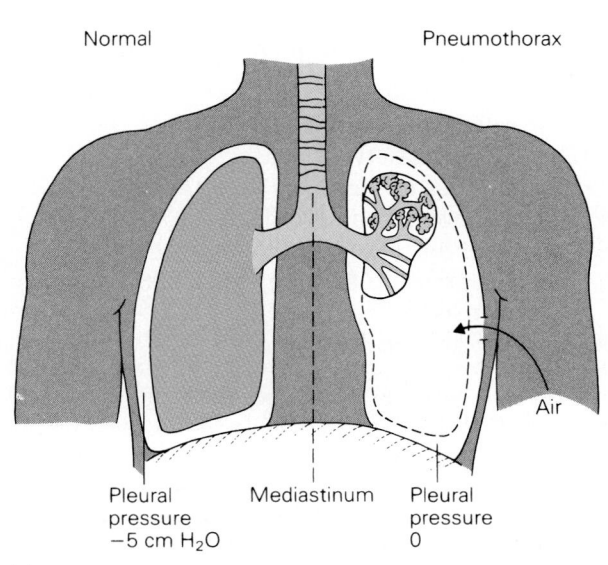

(a)

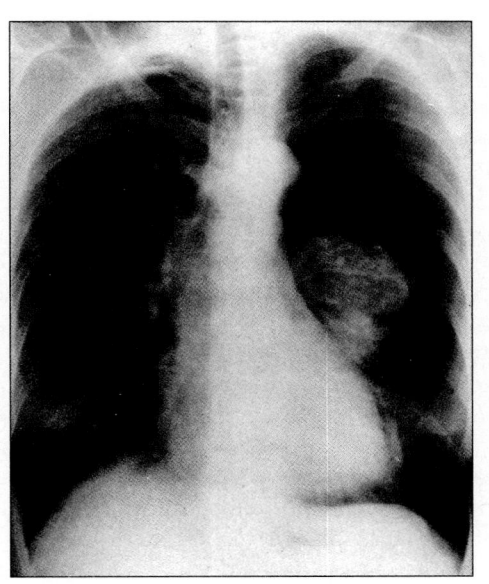

(b)

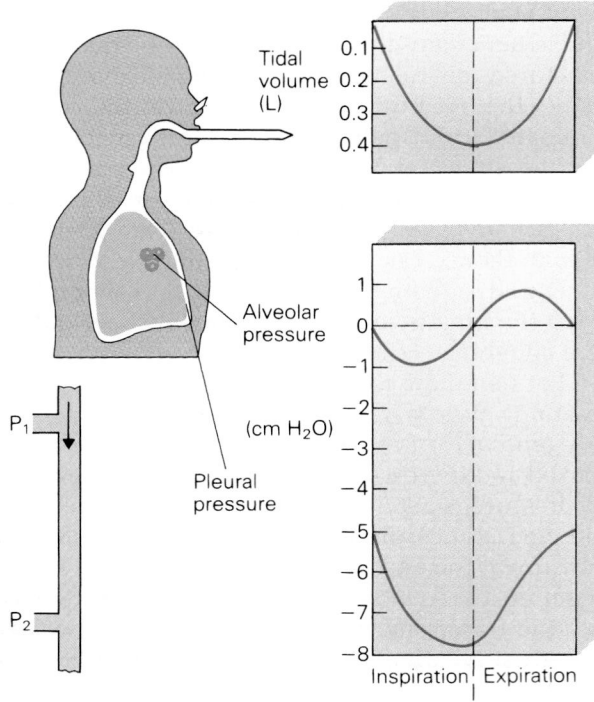

Figure 20–15
The relationship between pleural pressure, alveolar pressure, and tidal volume during a breathing cycle. The pressure gradient between mouth and alveoli is represented by P_1 and P_2.

that keeps the lungs inflated. It is important to remember that the more positive the transpulmonic pressure becomes, the more the lung inflates.

Lung Pressures During Inspiration

Inspiration is initiated by contraction of the dome-shaped diaphragm. The diaphragm moves down and at the same time pushes the rib cage out. As the thoracic cavity enlarges, the pleural pressure becomes even more negative (subatmospheric). Consequently, transpulmonic pressure increases and the lungs are inflated. As the lung inflates, the alveoli also enlarge. Alveolar pressure, therefore, drops below atmospheric pressure. This produces a pressure difference or gradient between the mouth and alveoli ($P_{atm} - P_A$), which causes air to flow down the lung to the alveoli. Air flow stops at the end of inspiration because alveolar pressure again equals atmospheric pressure (left side of Fig. 20–15).

Air flow is similar to blood flow in that air moves from a point of high pressure to one of low pressure, and flow is proportional to the pressure

difference between two points ($P_{atm} - P_A$). Thus, the greater the contraction of inspiratory muscles, the more the thoracic cavity increases and the more negative the pleural pressure becomes. As pleural pressure becomes more negative, transpulmonic pressure increases, and the lung becomes more inflated. The more the lung is inflated, the greater the alveolar diameter becomes, and the more alveoli pressure decreases, resulting in a greater pressure gradient between the mouth and the alveoli. The pressure gradient between mouth and alveoli results in air flow. The sequence of events in inspiration is as follows:

1. The inspiratory muscles contract.
2. The thoracic cavity expands.
3. Pleural pressure becomes more negative.
4. Transpulmonic pressure becomes more positive.
5. The lung inflates.
6. Alveolar pressure becomes subatmospheric.
7. Air flows into the lung.

Lung Pressures During Expiration

During expiration, the inspiratory muscles relax, pleural pressure becomes less negative, transpulmonic pressure decreases, and the lungs deflate, which is a reversal of inspiration (right side of Fig. 20–15). As the lung deflates, alveoli decrease in size. Consequently, alveolar pressure increases and exceeds atmospheric pressure, thus causing air to flow out of the alveoli through the airways. Remember, with a resting tidal volume, expiration is completely passive (muscles relax and the stretched lung recoils back to normal). It is only during forced expiration (e.g., exercise) that the expiratory intercostal and abdominal muscles contract and actively compress the thoracic cavity. The sequence of events in expiration is as follows:

1. Respiratory muscles relax.
2. The thoracic cavity decreases in size.
3. Pleural pressure becomes less negative.
4. Transpulmonic pressure decreases.
5. Lung size decreases.
6. Alveolar pressure becomes greater than atmospheric pressure.
7. Air flows out of the lung.

Typical pleural and alveolar pressures during a normal breathing cycle are shown in Table 20–3. Note from the values presented in Table 20–2 and Figure 20–15 that it takes only a small pressure gradient to move a resting tidal volume (500 ml) in and out of a normal lung. A summary of the sequential relationship between changes in transpulmonic

Table 20–3

Pressure Changes During a Normal Breathing Cycle

	Pleural Pressure (cm H$_2$O)	Alveolar Pressure (cm H$_2$O)
End expiration	−5	0
Mid inspiration	−7	−1
End inspiration	−8	0
Mid expiration	−6	+1
End expiration	−5	0

pressure, lung size, and changes in air flow during breathing is shown in Figure 20–16.

Elastic Properties of the Lung: Compliance

The lung airway and vascular trees are embedded in elastic tissue. When the lung is inflated, it must stretch to overcome these elastic components. Thus, the degree of lung expansion at any given time in the breathing cycle is proportional to transpulmonic pressure. How well a lung inflates and deflates with a change in transpulmonic pressure is a measure of its **elastic property.** Obviously, a stiff lung that does

Figure 20–16
Relationship between changes (Δ) in transpulmonic pressure, lung size, and air flow during breathing.

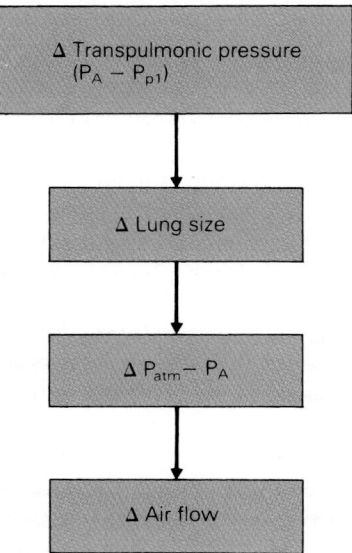

not stretch easily will not inflate to the same volume as a highly elastic lung with the same effort.

We determine the elastic properties of a lung by measuring the changes in lung volume that occur with changes in pressure, plotted on a **pressure-volume curve.** A simple analogy is inflation of a balloon with a syringe (Fig. 20–17). For each change in pressure, the balloon inflates to a new volume. We calculate the slope of the line by measuring the volume change per unit pressure change (Δ volume/Δ pressure). This value is known as **compliance** (C_L = Δ volume/Δpressure).

We can generate a similar pressure-volume curve for the human lung by measuring a change in lung volume with a spirometer and a change in pleural pressure with a pressure gauge (Fig. 20–18). In practice, a pressure-volume curve is produced by first having the subject inspire maximally to total lung capacity and then expire slowly while recording lung volume and pleural pressure measurements. Compliance (the slope of the pressure-volume curve) provides a measure of the distensibility (i.e., ability to stretch) of the lung. For example, a steep slope (high compliance) indicates that the lung is very distensible and easy to inflate. This means the lung can be inflated to a higher volume with a given amount of effort than a less distensible lung. In other words, with the same transpulmonic pressure, the lung with the higher compliance will inflate to a higher volume because it is more distensible. A lung with a low compliance (decreased slope) is stiffer or not as distensible.

What is the significance of having an abnormally high or abnormally low compliance? Low compliance, indicating a stiff lung, means more work is required to bring in a normal volume of air. Any time the lung becomes injured (by infection or by toxic environmental insults), the elastic properties are often altered, and the lung loses its distensibility (becomes stiffer). A very high compliance, however, is just as detrimental as a low compliance. Consider the disease **emphysema,** in which the lung has been overstretched from chronic coughing and congested airways. Lungs of patients with emphysema (which is strongly linked with smoking) have a high compliance (high distensibility) and are extremely easy to inflate. However, getting air out again is another matter! Lungs with abnormally high compliance (see Fig. 20–18) have poor **elastic recoil,** the property of stretched tissue that allows it to return back to normal. Thus, a lung affected by emphysema is easily distended but does not recoil back during expiration. Consequently, a lot of effort is required to get air out of the lungs. A simple analogy is a sock or stocking that has become over-

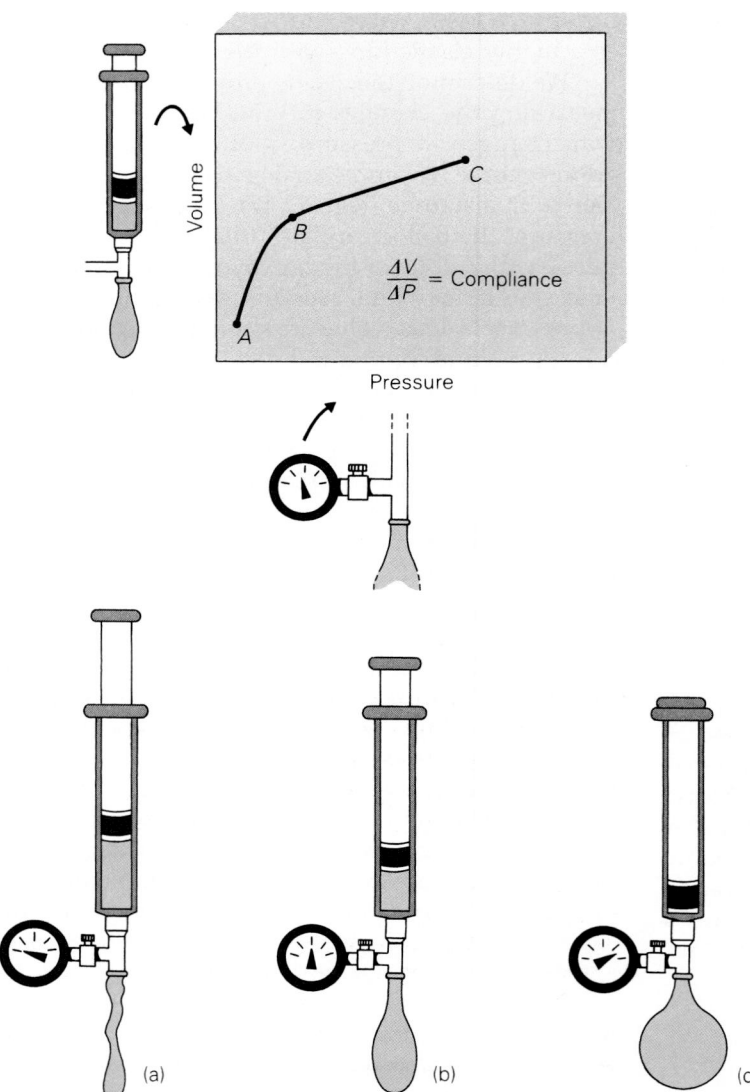

Figure 20–17
Pressure-volume curve. The slope of the line between B and C (change in volume/change in pressure) is compliance.

stretched. It can be easily stretched out but does not recoil to fit the contour of the legs and so sags and looks "baggy." Similarly, smoking-induced emphysema can lead to "baggy lungs" that retain air.

Surface Properties of the Lung: Surface Tension

Another property that markedly affects lung compliance is **surface tension** occurring within the inner surface of the alveoli at the gas-liquid interface. Surface tension is a molecular force created at the gas-liquid interface—specifically, the attractive forces between the liquid and air molecules. These forces create a surface film. Since the surface of the alveolar membrane is moist and is in contact with air, a large gas-liquid interface is produced. Surface tension in alveoli produces a force that pulls inward. This tension tends to cause alveoli to become unstable by increasing alveolar pressure proportionately more in smaller alveoli than in larger ones at low lung volumes (Fig. 20–19).

To counteract the high surface tension, specialized epithelial cells lining the alveoli secrete a detergent-like material that coats the inner surface of the alveoli and drastically reduces surface tension. This material is known as **pulmonary surfactant** and is a complex mixture of lipids (fat) and protein. The main component of pulmonary surfactant responsible for reducing surface tension is the phospholipid **phosphatidylcholine.** Thus, pulmonary surfactant lowers surface tension, prevents alveolar collapse, and makes the lung more distensible (compliant).

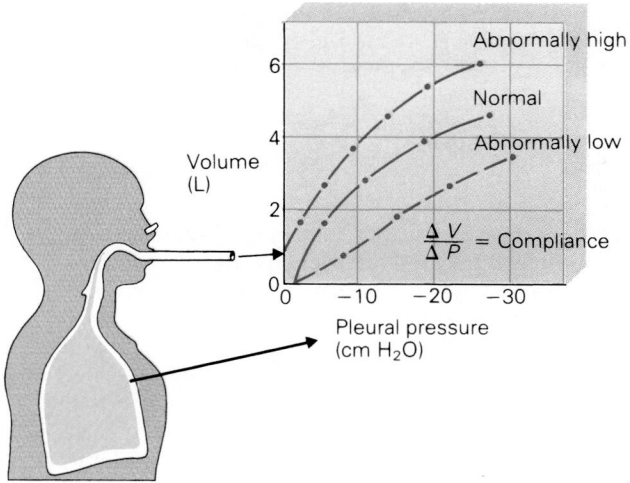

Figure 20–18
Lung pressure-volume curve and compliance. A lung with a slope greater than normal would have an abnormally high compliance. A lung with a slope smaller than normal would have an abnormally low compliance.

This is extremely important because the effects of surface tension on lung compliance are just as great as the effects of elastic forces. A striking example of what happens when inadequate amounts of surfactant are present is seen with a disorder known

Figure 20–19
The effect of surface tension on alveoli. Surface tension tends to reduce surface area and generates a pressure inside the alveoli. The smaller alveoli generate a greater pressure and cause air to flow into larger units. At low lung volumes, these smaller alveoli become unstable and collapse because of the higher surface tension. The relationship between pressure and surface tension can be expressed as $P = \dfrac{2T}{r}$ where P = pressure, T = surface tension, and r = radius.

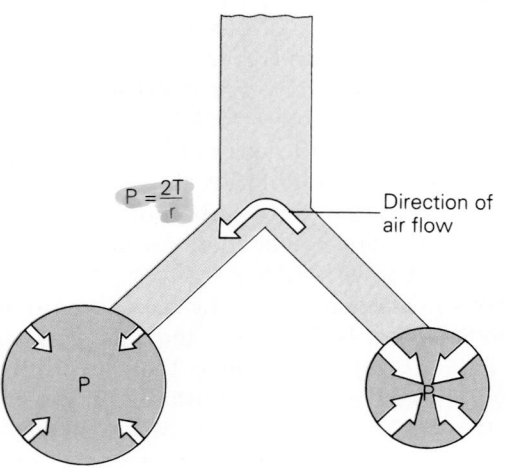

as **respiratory distress syndrome.** This disorder frequently afflicts premature neonates who are born with insufficient amounts of surfactant. These infants have extremely labored breathing because high surface tension makes it difficult to inflate the lungs. Because of the high surface tension, these infants also have alveolar collapse, a phenomenon known as **atelectasis.** Forty to fifty thousand babies are afflicted each year with infant respiratory distress syndrome. These infants are at high risk until the lung becomes mature enough to secrete surfactant. Consequently, infant respiratory distress syndrome is still the major cause of the deaths among newborns in the United States.

The Work of Breathing

During inspiration, muscular work is involved in expanding the thoracic cavity, inflating the lungs, and overcoming airway resistance. Since work can be measured as force multiplied by distance, the amount of work involved in breathing can be expressed as a change in lung volume (distance) multiplied by the change in transpulmonic pressure (force). Thus, energy is expended during muscular contraction to create a force (transpulmonic pressure) to inflate the lungs. When greater transpulmonic pressure is required to bring the same volume of air into the lungs, more muscular work and greater energy are required.

What factors affect the amount of work involved in ventilating the lungs? First is lung elasticity. As discussed earlier, lung inflation requires energy to overcome elastic forces. Thus, the more compliant the lung, the more easily it is stretched and the less energy is required in breathing. When the lung becomes stiff (less compliant), more muscular work is required to increase transpulmonic pressure in order to stretch the lungs to expansion.

A second factor affecting the work of breathing is high surface tension. As pointed out earlier, individuals who lack surfactant have high surface tension and extremely labored breathing. Some of these individuals die from exhaustion and respiratory muscle fatigue.

A third factor affecting the work of breathing is airway resistance, which can be increased in two ways: (1) by an increase in flow rate and (2) by abnormal constriction of the airways. Both of these lead to a greater energy expenditure because greater transpulmonic pressure is required to maintain adequate air flow.

How much energy is involved in breathing? In a normal person, the energy needed for breathing

C L I N I C A L F O C U S

Asthma

One of the airway diseases that many of us have encountered is asthma. Asthma is a breathing disorder brought on by narrowing of the small airways, namely the bronchi. The asthmatic person experiences labored breathing, shortness of breath, and a distressing tightness in the chest. Shortness of breath can occur suddenly and without warning. During the asthmatic attack, the person experiences a feeling of suffocation, and must sit up or stand and devote every bit of energy to breathing. The attack can last several minutes to several hours. Afterwards a cough may occur, slight and dry at first but often becoming more marked, with considerable sputum. Following an asthmatic attack, there can be considerable soreness in the chest.

The cellular mechanism of asthma, which causes the shortness of breath, wheezing, and tightness in the chest, is thought to be the release of vasoactive mediators (histamine, prostaglandins, and leukotrienes). These agents cause the bronchial smooth muscles to constrict the airways and also cause an excess secretion of mucus, which plugs the terminal airways. The mucus is usually coughed up at the end of an attack.

One form of asthma is triggered by an allergic reaction and is often called allergic asthma. This type of asthma is most prevalent in children and adolescents. In most cases, these allergic reactions occur in the bronchial tree and are triggered by a foreign protein such as pollen, animal dander, mites, mold, spores, feathers, and, less commonly, food (e.g., milk, nuts, chocolate, and seafood). Certain synthetic drugs, such as aspirin and penicillin, although not proteins, can also trigger asthma. Such agents or allergens react with antibodies and induce the release of the potent vasoactive mediators, most importantly histamine. Underlying allergic asthma is an inherited tendency to develop hypersensitivity to these allergenic agents.

A second form of asthma is idiopathic asthma, in which a specific cause cannot be identified. This form usually occurs in adults over 30 and can be triggered by exercise, emotions, inhalation of cold air, and infections.

About 1% of the entire population (1 in 100) are estimated to have asthma. Asthma is estimated to occur in 5% to 15% of children under the age of 15. It is more frequent among boys and in men 30 years and older. An important aspect of asthma treatment is the identification, when possible, of the agent that triggers the attack. Medications include oral drugs, which relax airway smooth muscle. Aerosol drugs are also used if the attack is severe.

---

represents approximately 1% of the body's total energy expenditure. Even during heavy exercise, only about 3% of the total energy expenditure is involved in breathing. Breathing is most economical when there is a balance of elastic, surface, and resistive forces. This balance requires the least amount of work. It is important to remember that the faster the breathing rate, the greater the amount of work required to overcome the increase in airway resistance. In contrast, the greater the depth of breathing, the greater the amount of work required to overcome elastic and surface forces. Thus, patients with stiff lungs will compensate by breathing small amounts of air (low tidal volume) more rapidly, while patients with severe airway obstruction tend to take deeper breaths more slowly. Both of these breathing patterns help to minimize the amount of work required for breathing under these conditions.

The Control of Breathing

Breathing is an automatic process. Ventilation occurs without any conscious effort while we are awake and asleep, and even while we are under anesthesia. We can, however, exert some conscious control over our own ventilation by voluntarily changing the rate and depth of breathing (Fig. 20–20). We can also voluntarily stop breathing for a short period of time until carbon dioxide builds up in the blood, which will then stimulate breathing regardless of how hard we try to hold our breath. Clearly, the basic automatic rhythm of breathing can be modified by both neural and chemical stimuli (see Fig. 20–20). Remember that breathing, although automatic, depends entirely on the cyclical excitation of the respiratory muscles. The diaphragm receives a respiratory signal from the lower portion of the brain stem, which also contains the cardiovascular control centers.

Inspiratory Neurons in the Medulla

Within the medulla are separate and distinct neurons, **inspiratory neurons** (Fig. 20–21), responsible for the basic rhythm. These neurons have the property of intrinsic firing (see Chapter 11).

The intrinsic rhythm pattern starts with a latent period of several seconds in which there is no activity. Action potentials begin and increase over the next few seconds. During this build-up of action potentials, inspiratory muscles are activated in crescendo-like fashion. Finally, the action potentials

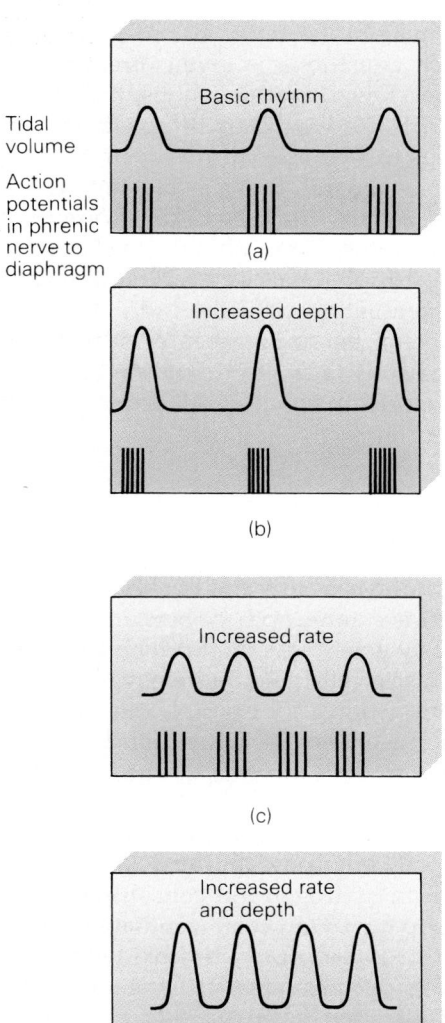

Figure 20–20
Recordings of tidal volume and phrenic nerve action potential to the diaphragm. (*a*) Normal breathing pattern with basic rhythm. (*b*) The frequency of action potentials determines depth of breathing (tidal volume), while the time interval between action potentials determines breathing rate (*c*). (*d*) Both rate and depth of breathing are increased.

phrenic nerve, which leaves the spinal cord in the upper half of the neck and passes down through the chest to innervate the diaphragm.

The Rhythm of Breathing

In a resting individual, the normal rate of breathing is approximately 14 breaths per minute. The control for the basic rhythm is housed in the **medulla,** the

Figure 20–21
The respiratory center located in the medulla. The center consists of two separate and distinct clusters of neurons: inspiratory and expiratory neurons.

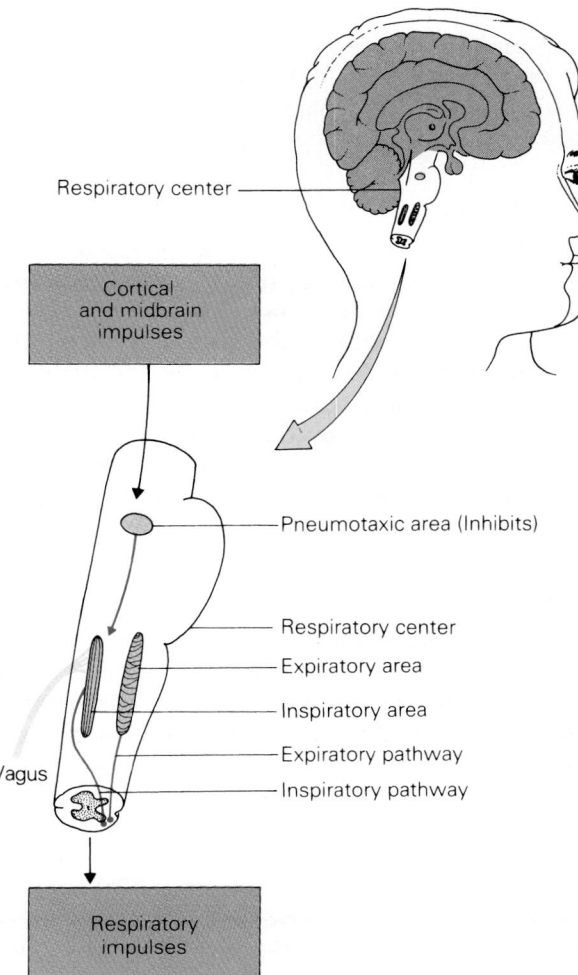

cease and the inspiratory muscles relax. The cyclic pattern then repeats itself.

The medullary center for controlling the basic rhythm also receives a rich supply of synaptic inputs from other neurons that come from the cortex and pons. Cortical input transmits the conscious effort of breathing. In automatic breathing, neuronal inputs from the pons modulate the inspiratory center and act as a "cut off" signal to inhibit inspiration.

Pulmonary Stretch Receptors

Another potential "cut off" signal comes from **pulmonary stretch receptors** embedded in the smooth muscles of the airways. When activated, these stretch receptors send impulses (action potentials) via the vagus to the medulla and inhibit the inspiratory neurons. This feedback from the airways not only helps stop inspiration but also helps prevent overinflation, which can damage the lung and alter the blood flow through the lung. In the newborn, the pulmonary stretch receptors play an important role in regulating the basic rhythm and preventing overinflation. However, in the adult, the threshold for these receptors is high and the stretch receptors play a role only under conditions in which there is a very large tidal volume (e.g., heavy exercise).

Expiratory Neurons in the Medulla

Inhibition of the inspiratory neurons is all that is required for expiration to occur, since expiration is normally a passive process. In more forceful breathing, however, expiration becomes active; **expiratory neurons** (see Fig. 20–21) stimulate the expiratory muscles. These expiratory neurons are also located in the medulla, in close proximity to the inspiratory neurons, and are quiescent (inactive) during normal breathing. They become active during forced expiration such as exercise.

Failure of the Respiratory Center

The respiratory center located in the medulla is somewhat fragile and thus can disrupt the respiratory cycle. Two of the more frequent causes of basic rhythm disruption are cerebral concussions and cerebral pressures, such as cerebral edema, created around the medulla. Both of these cause blood vessel collapse, which stops blood flow and causes the respiratory rhythm to cease functioning.

Another frequent cause of respiratory failure is barbiturate overdose. Barbiturates depress the inspiratory neurons and stop the rhythm. General anesthesia and narcotic pain relievers also act in a similar fashion, inhibiting the respiratory center. If too much anesthesia is given, breathing will stop altogether. Once breathing stops, it is extremely difficult for the body to start the cycle up again. Very few drugs can be used to reactivate the inspiratory rhythm. In general, those available to excite the respiratory centers, such as caffeine, are too weak to be of any value. The only effective treatment is to place these individuals on a mechanical ventilator to sustain ventilation until the body itself can rectify the problem. In some cases of severe respiratory depression, it may take weeks and sometimes months before the body can resume breathing on its own.

Chemical Control of Ventilation

Although the basic rhythm of ventilation is continuous, the rate and depth of breathing can vary tremendously depending on metabolic demands. First, when muscle cells need more oxygen, during muscular contractions, for example, alveolar ventilation must increase. Second, when cells become more active, they produce more CO_2. The body rids itself of the excess CO_2 by increasing alveolar ventilation. Third, in order for metabolic processes to occur, a constant pH environment must be maintained. The appropriate hydrogen ion concentrations are regulated, in part, by alveolar ventilation. These three chemicals, **oxygen, carbon dioxide,** and **hydrogen ions** (pH), can profoundly change alveolar ventilation by affecting both the rate and the depth of breathing (see Fig. 20–20).

Central and Peripheral Chemoreceptors

Two sets of **chemoreceptors** are strategically located in the body to detect changes in arterial partial pressures of oxygen and carbon dioxide and changes in hydrogen ion concentration. One set of receptors is located in the medulla. These receptors are referred to as **central chemoreceptors;** they monitor only changes in cerebral arterial CO_2 levels (Fig. 20–22).

The other set of receptors, located outside the central nervous system, are called **peripheral chemoreceptors.** These peripheral chemoreceptors are located in the arch of the aorta **(aortic bodies)** and in the neck at the bifurcation of the common carotid artery **(carotid bodies)** (Fig. 20–23). Both the carotid and aortic bodies monitor levels of all three chemicals (O_2, CO_2, and [H^+]) in the arterial blood. The peripheral chemoreceptors are made up of epithelial-like cells that contain neurons in contact with the arterial blood. The afferent (sensory) nerve fi-

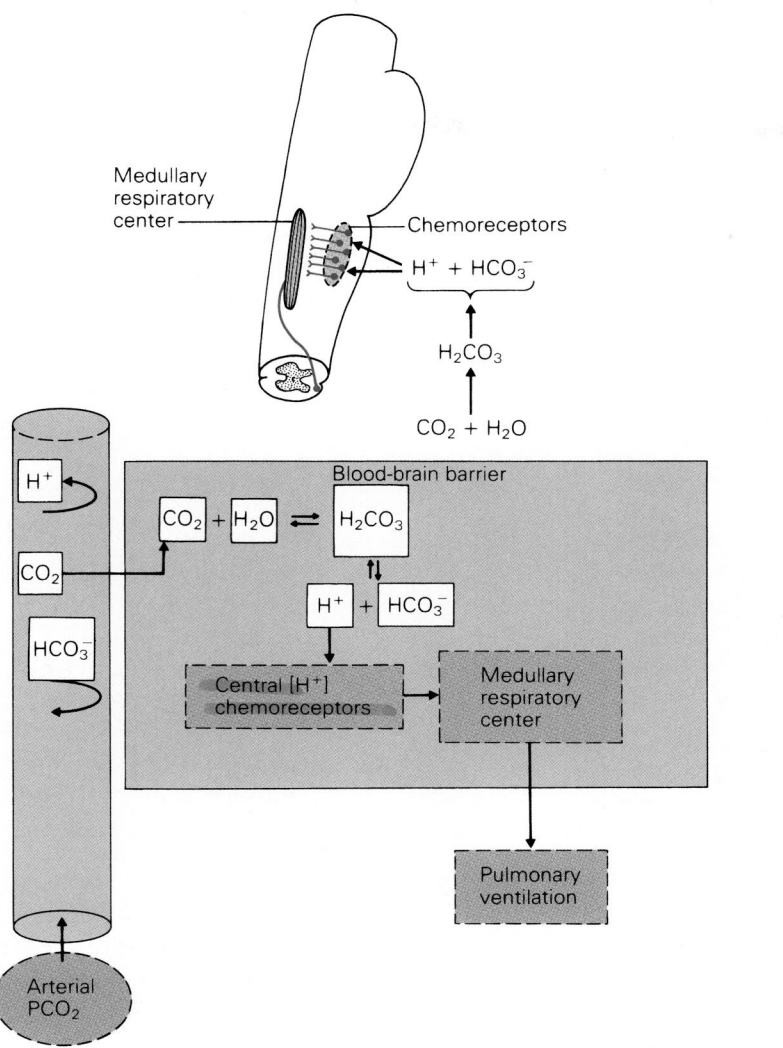

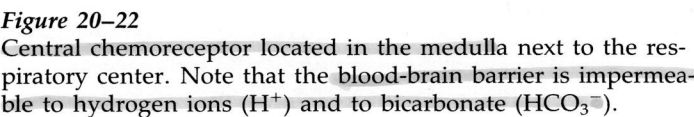

Figure 20–22
Central chemoreceptor located in the medulla next to the respiratory center. Note that the blood-brain barrier is impermeable to hydrogen ions (H^+) and to bicarbonate (HCO_3^-).

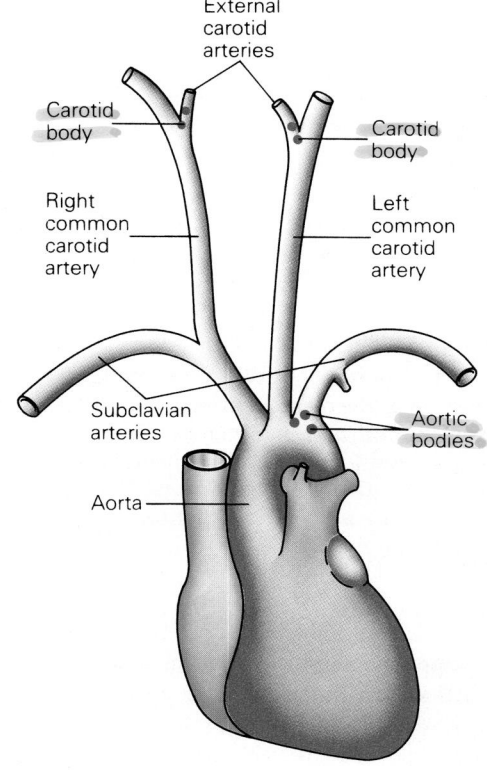

Figure 20–23
Peripheral chemoreceptors located in the aortic arch (aortic bodies) and carotid artery (carotid bodies).

bers that arise from these cells extend to the respiratory center in the medulla.

Of the three chemicals that influence alveolar ventilation, carbon dioxide is the most powerful (Fig. 20–24). When arterial carbon dioxide levels increase above normal, all portions of the respiratory center become excited. An increase of only 2 to 5 mm Hg in arterial carbon dioxide can more than double alveolar ventilation. The second most powerful chemical stimulus to ventilation is arterial pH. Arterial oxygen has the smallest effect on ventilation.

Hyperventilation and Hypoventilation: Control of Arterial CO_2

The only effective way the body can regulate short-term carbon dioxide levels in the blood is through alveolar ventilation. If arterial CO_2 levels suddenly become too high, all chemical reactions essentially stop, since most enzymatic reactions operate within a precise range of pH. To prevent this, the respiratory center is stimulated to increase alveolar ventilation, which results in "blowing off" excess CO_2 and thereby lowering arterial CO_2 back to normal. In-

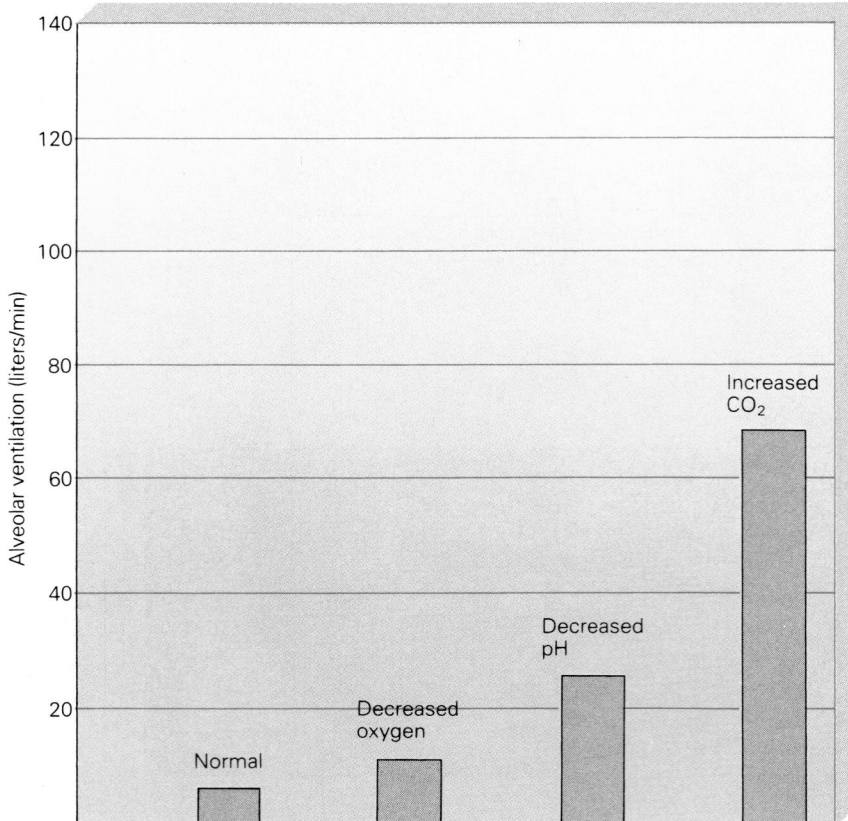

Figure 20–24
Effect of maximal decrease in arterial blood pH, decrease in arterial blood oxygen, and maximal increase in arterial blood CO_2 on alveolar ventilation. Note that elevated arterial CO_2 can increase alveolar ventilation about 15 times above normal.

creased alveolar ventilation is known as **hyperventilation.** Conversely, if arterial CO_2 becomes too low, the blood becomes too alkaline (basic), which depresses the respiratory center and causes a decreased alveolar ventilation, or **hypoventilation.**

The control of ventilation by carbon dioxide is mediated by both the central and peripheral chemoreceptors. When CO_2 diffuses across the blood-brain barrier (see Fig. 20–22), it combines with water to form carbonic acid (H_2CO_3), which rapidly dissociates into bicarbonate (HCO_3^-) and hydrogen ion (H^+) in the following manner:

$$H_2O + CO_2 \rightleftharpoons H_2CO_3 \rightleftharpoons HCO_3^- + H^+$$

Since H^+, HCO_3^-, and CO_2 are in equilibrium, any increase in arterial CO_2 will cause a profound increase in the hydrogen ion concentration in the cerebrospinal fluid (CSF). Any increase in the CSF H^+ concentration an increase in alveolar ventilation. The brain is especially sensitive to pH changes because the cerebrospinal fluid lacks a protein buffer to neutralize changes in hydrogen ion concentration. Low arterial CO_2 has an equally strong inhibitory effect on ventilation. Both the central and peripheral chemoreceptors work in concert to regulate

ventilation by CO_2, although over two thirds of the stimulation comes from the central chemoreceptor (Fig. 20–25). For further information about pH regulation and buffers, please refer to Chapter 26.

The control of alveolar ventilation by changes in arterial hydrogen ion concentration occurs exclusively through the peripheral chemoreceptors. The central chemoreceptors do not respond to arterial hydrogen ions because the blood-brain barrier is impermeable to hydrogen ions (see Fig. 20–22). In the arterial system, an increase in hydrogen ion concentration activates the peripheral chemoreceptors, which in turn stimulate the respiratory center. As a result, alveolar ventilation is increased. Similarly, the increased ventilation will blow off excess CO_2, thereby lowering arterial CO_2 and indirectly lowering the arterial hydrogen ion concentration (Fig. 20–26). The converse is true for low H^+; low arterial hydrogen ion concentration inhibits ventilation.

In many situations, a change in arterial hydrogen ion concentration is caused by factors other than altered blood carbon dioxide. For example, a decreased blood pH occurs with severe diabetes, owing to excess fatty acids in the blood, and stimulates ventilation through the peripheral chemore-

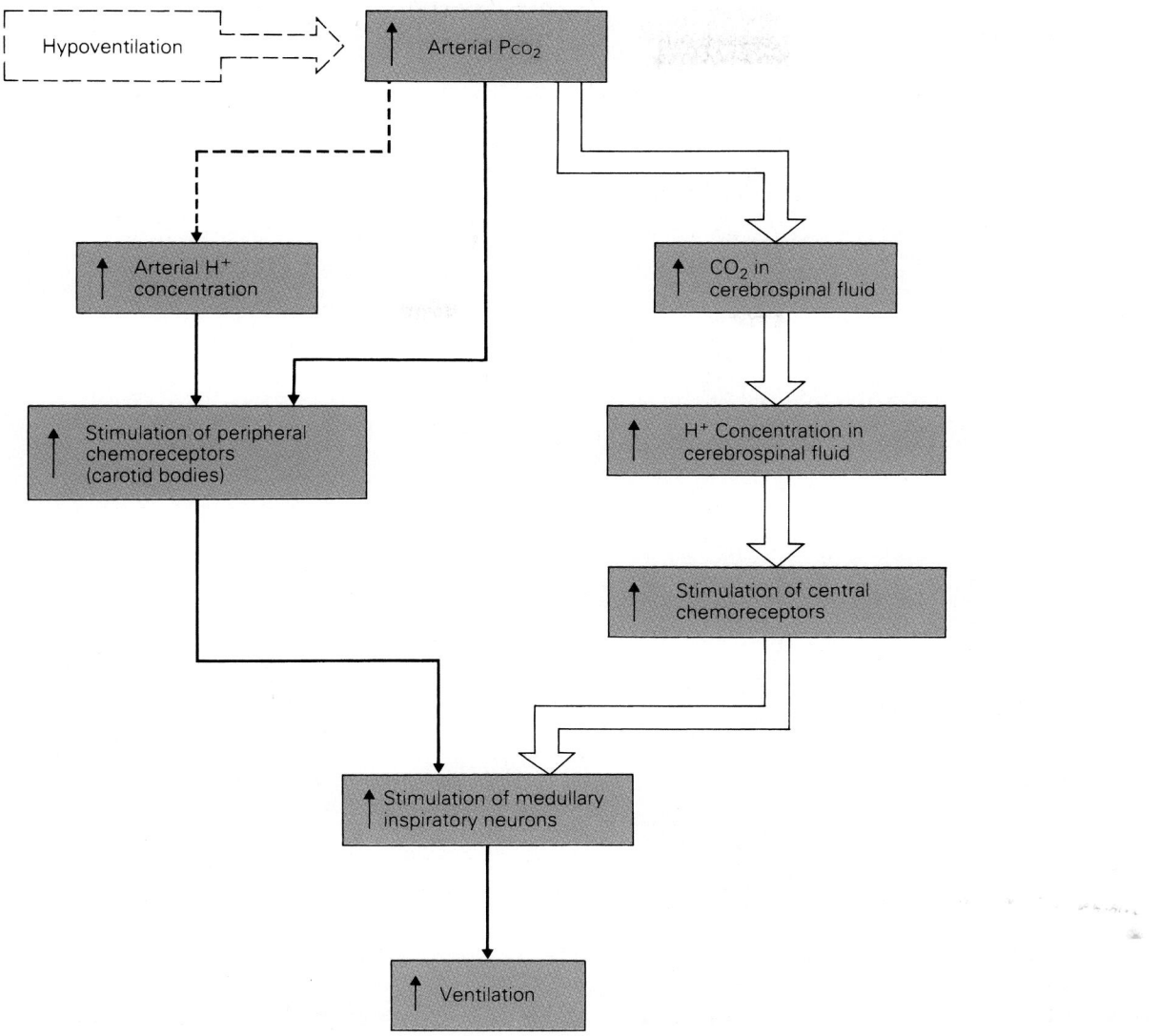

Figure 20–25
Stimulation of alveolar ventilation by carbon dioxide. The main stimulus is
via the central chemoreceptors (solid heavy line). Note that the direct effect
of arterial CO_2 on the peripheral chemoreceptors is minor (dashed line).

ceptors. Another example is during vomiting, in which hydrogen ions are lost. With the consequent decrease in arterial hydrogen ion concentration, ventilation is depressed because the peripheral chemoreceptor output is decreased.

Oxygen and Control of Ventilation

As mentioned earlier, of the three chemicals that stimulate ventilation, oxygen has the smallest effect. It may seem surprising that the control of ventilation is so insensitive to reduced oxygen, since most of us think intuitively that breathing is controlled by the body's need for oxygen. The reason that the amount of oxygen in the arterial blood has so little effect on the ventilation rate is that gas transfer in the lung is so complete that blood leaving the lung is almost fully saturated with oxygen; it is not until the oxygen in the blood decreases by 50% that delivery to the tissues is affected. Thus, sensitive control of ventilation to maintain a constant oxygen concentration in the blood is not required. Consequently, there is little increase in ventilation until the level of arterial oxygen is decreased by 50%. When the oxygen level becomes too low (i.e., below 50%) the blood becomes **hypoxemic.** The receptors that sense

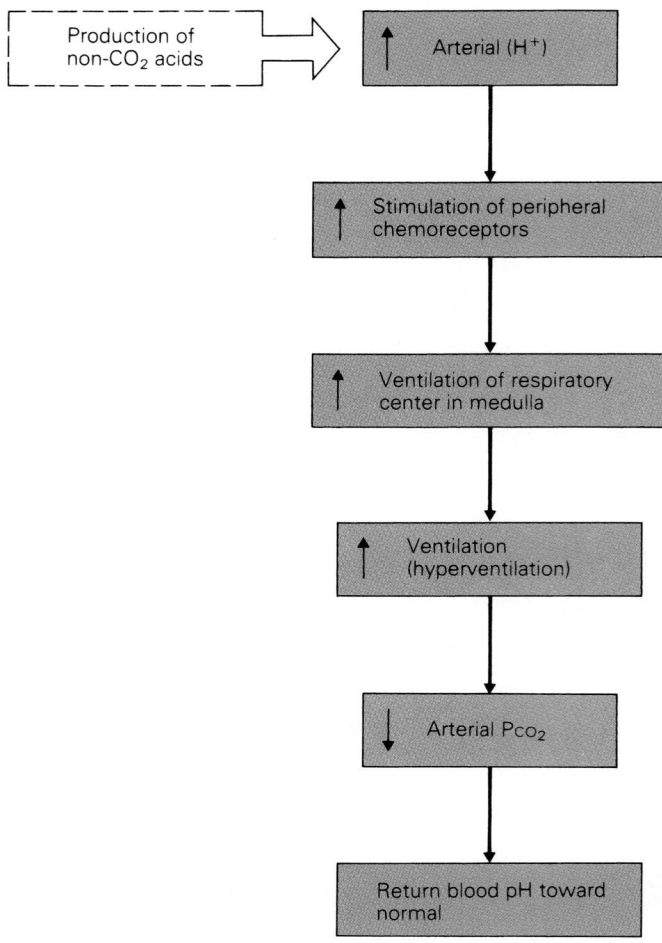

Figure 20–26
Increase in alveolar ventilation by arterial blood pH. Changes in hydrogen ion concentration can occur through the production of non-CO_2 acids such as lactic acid from muscles, or from excess fatty acid in the blood with severe diabetes.

the low O_2 are the peripheral chemoreceptors. When arterial oxygen is reduced by lung disease, hypoventilation, or by breathing at high altitude, the peripheral chemoreceptors are stimulated, which in turn results in an increase in ventilation (Fig. 20–27).

During exhaustive exercise, alveolar ventilation can increase 20-fold. The details of how exercise stimulates ventilation will be discussed in Chapter 29.

Abnormal Patterns of Breathing

Sleep Apnea

Ventilation normally decreases during sleep because the body's metabolic demand decreases. Sometimes, ventilation decreases to a far greater degree than the decrease in the body's utilization of energy. In some individuals who suffer from **sleep apnea,** a condition in which respiration stops for long periods of time (often 30 to 60 seconds), the decrease in ventilation becomes dangerously low. Sleep apnea involves obstruction of the upper airways during sleep. This obstruction often occurs when certain muscles and fatty tissue become flabby and cause airway blockage. Partial blockage produces snoring; complete blockage produces apnea (absence of breathing). When breathing stops, arterial CO_2 and O_2 become exaggerated. Eventually, it is the rising levels of arterial CO_2 that stimulate the respiratory center. The individual reacts with grunting, gasping breaths and immediately returns to a deeper sleep until the next apneic episode occurs. Sleep apnea victims may stop breathing anywhere from a dozen to a hundred times a night, although they are usually unaware of the sleep disruptions. This loss of sleep can lead to pronounced daytime drowsiness, a classic symptom of sleep apnea. Some of the victims are involved in automobile accidents because they fall asleep at the wheel. In another form of sleep

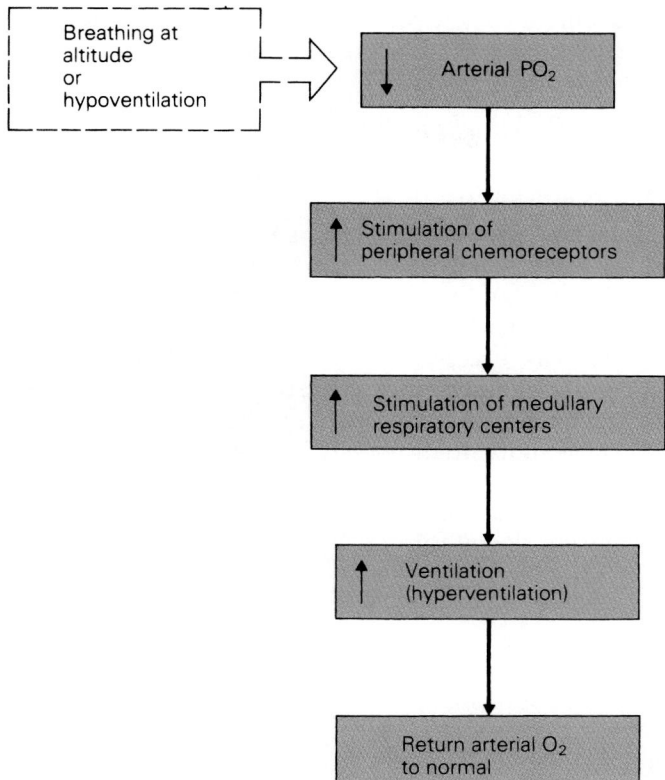

Figure 20–27
Stimulation of ventilation by hypoxemia occurs via the peripheral chemical receptors, mainly the carotid bodies.

apnea, the respiratory muscles stop working, producing a similar cycle of repeated sleep disruptions. This form often causes insomnia rather than excessive daytime sleepiness.

Sleep apnea occurs most frequently in obese people, especially men, who suffer from the obstructive form of the disease. The real danger of sleep apnea is that death from cardiac arrhythmia can occur during an episode. No one knows what causes sleep apnea. Some physiologists think the cause in the nonobstructive form may be reduced sensitivity of the chemoreceptors to carbon dioxide during certain phases of sleep. At present, sleep apnea is being studied extensively because it may be linked to **sudden infant death syndrome (SIDS)**, the disease in which a seemingly normal, healthy infant is found dead in its crib. Many of these infants, upon autopsy, have been found to have poorly developed carotid bodies.

Shallow-Water Blackout

Another type of ventilatory pattern that alters the control of breathing is **shallow-water blackout.** This occurs when a swimmer hyperventilates first before diving under water. One can hold a breath longer after first hyperventilating because hyperventilation blows off excess CO_2, which lowers arterial CO_2 and depresses breathing. When the swimmer hyperventilates, arterial CO_2 is drastically decreased, with no change in arterial oxygen levels. As a result of the low arterial CO_2, the respiratory centers become depressed. At the same time, oxygen is rapidly taken up by the muscles, lowering the arterial oxygen. However, the stimulating effect of the low oxygen is overridden by the strong inhibitory effect of the low CO_2. As a result, the brain becomes **hypoxemic.** Consequently, the swimmer can black out under water and drown.

The Diving Reflex

Another breathing pattern associated with water is the **diving reflex.** The diving reflex is most pronounced in newborns and can be elicited when the face is submerged in water. This reflex causes the individual to gasp; breathing stops (apnea), heart rate decreases, and blood flow to peripheral tissue also decreases (peripheral vasoconstriction). This shunting of blood away from the peripheral tissues allows for heart-brain perfusion. The diving reflex is a protective reflex. A number of cases have been reported in which young children have broken through ice on a pond or fallen into a lake and survived under water for 20 to 30 minutes. The gasp reflex prevents water from getting into the lungs, and oxygen is conserved, by peripheral vasoconstriction, for the brain.

The diving reflex by itself is not enough to save victims who remain under water for long periods of time. Lowered body temperature caused by the cold water is equally important. A lower body temperature means a lower metabolic rate, which means less energy is required. The reason children survive better than adults is twofold. First, they have a more pronounced diving reflex. Second, their body temperature drops at a faster rate because they have a high surface area to body weight ratio. For further

details about the relationship between body temperature and surface area, see Chapter 28.

The Pickwickian Syndrome

A fourth type of altered breathing pattern that affects ventilation is called the **Pickwickian Syndrome.** The disorder occurs in severely obese people who often suffer from hypoventilation. The Pickwickian Syndrome was named after "Joe," the boy in Charles Dickens' novel *The Pickwick Papers.* Joe was excessively sleepy during the day and snored loudly. Pickwickian patients suffer from hypoventilation because of their excess weight. These individuals usually have abnormalities in the control of breathing and often suffer from sleep apnea, which may account for the fact that they are always sleepy during the day.

SUMMARY

The Structure of the Lung

Respiration consists of gas exchange (exchange of oxygen and carbon dioxide between environment and cells) and cellular respiration (burning of fuels with oxygen to produce energy). The human respiratory system is so efficiently designed that the supply of oxygen and the removal of carbon dioxide rarely limit energy production by cells. Efficiency of gas exchange is largely dependent on the internal surface area of the lung.

In the mammalian lung, the large internal surface area is created by microscopic air sacs called alveoli.

The lung comprises two treelike structures, the vascular tree and the airway tree, which are embedded in elastic tissue. The airway consists of the trachea, bronchi, and bronchioles.

Lung Volumes and Ventilation

The process of moving air in and out of the lung is known as ventilation. Air movement in the airways results in two types of air flow. Turbulent air flow (disorganized patterns of air movement) occurs at high flow rates in large airways such as the trachea and bronchi. Laminar air flow (streamlined flow of air molecules) appears in small peripheral airways such as bronchioles and occurs under low flow rates.

Lung volumes and ventilatory function (efficiency of moving air in and out of lungs) are measured by a spirometer. Key lung volumes include tidal volume, residual volume, functional residual capacity, and forced vital capacity. Forced vital capacity is one of the most useful tests for assessing the overall ability to move air (ventilation).

The volume of air breathed per minute is known as minute ventilation. A fraction of the minute volume is not involved in gas exchange and becomes wasted air (dead space air). The volume of air that reaches alveoli for gas exchange is alveolar ventilation and represents the amount of "fresh" air entering the lung for gas exchange.

The Mechanics of Lung Action

The main respiratory muscle in ventilation is the diaphragm. The lungs are housed in an airtight cavity, the thoracic cavity, and when the diaphragm contracts, the lungs expand and air rushes in. When the diaphragm relaxes, the lung collapses and air is exhaled.

Important pressures in the breathing cycle (inspiration and expiration) are pleural, alveolar, and transpulmonic pressures. Pleural pressure is the pressure between the lung and chest wall and it changes with diaphragmatic contraction. Transpulmonic pressure is the pressure gradient across the lung wall and is what keeps the lung distended. When alveolar pressure is decreased below atmospheric pressure, a pressure gradient exists between the mouth and the alveoli that causes air to rush in. When alveolar pressure is above atmospheric pressure, the pressure gradient causes air to rush out of the lung.

Lung airways and vascular trees are embedded in elastic tissue. Thus, when the lung is inflated, it has to be stretched to overcome these elastic tissues. The elasticity (distensibility) of the lung is measured indirectly by a lung pressure-volume curve. The slope of the pressure-volume curve (compliance) provides indirect estimates of lung elasticity.

In addition to elastic properties, the lung has surface properties (surface tension) that markedly affect lung compliance. Surface tension is created in the inner surface of the alveoli at a gas-liquid interface. Alveolar epithelial cells secrete pulmonary surfactant, which reduces surface tension and makes the lung easier to inflate. Premature infants are often born with inadequate amounts of surfactant, leading to respiratory distress syndrome.

Muscular work is involved in breathing. During inspiration, energy is used to expand the thoracic cavity, inflate the lung, and overcome airway resistance. Normally, the energy required for breathing is less than 1% of the body's total energy expenditure. However, during exercise or with certain types of lung disease, energy expenditure can increase to 3%.

The Control of Breathing

Inspiratory neurons in the medulla produce a basic rhythm that provides a normal breathing rate of 14 breaths per minute.

The basic rhythm can be modified by conscious effort and chemoreceptors strategically located in the body. These chemoreceptors sense changes in arterial blood carbon dioxide, pH, and oxygen. Low oxygen (hypoxemia), low pH, or high carbon dioxide in the blood stimulates respiration and causes hyperventilation. Conversely, a high pH or low blood carbon dioxide depresses respiration and causes hypoventilation. Of the three chemical stimuli, carbon dioxide is by far the most powerful for respiration.

Sleep apnea is a condition in which respiration stops periodically during sleep. Pickwickian Syndrome is a breathing pattern disorder that leads to hypoventilation in severely obese individuals.

REVIEW QUESTIONS

Identify Each with a Term

1. The volume of air that can be forcibly exhaled after maximal inspiration.

2. The volume of air remaining in the lungs after normal expiration.

3. The amount of fresh air that reaches alveoli.

4. The pressure that keeps the lungs from collapsing.

5. A collapsed lung.

Define Each Term

6. compliance
7. pulmonary surfactant
8. atelectasis
9. hypoventilation
10. sleep apnea

Choose the Correct Answer

11. Which of the following is *not* part of the conducting zone of the airway tree?
    a. trachea
    b. bronchi
    c. bronchioles
    d. alveoli

12. Total lung capacity is the sum of:
    a. residual volume and functional residual capacity
    b. vital capacity and functional residual capacity
    c. vital capacity and tidal volume
    d. residual volume and vital capacity

13. Functional residual capacity is:
    a. the volume of air seen during normal breathing
    b. the volume of air expired after full inspiration
    c. the volume of air remaining in the lungs at the end of normal expiration
    d. the volume of air remaining in the lungs at the end of maximal expiration

14. Minute ventilation represents:
    a. total air movement in the lung
    b. total air movement minus dead space volume
    c. alveolar ventilation minus dead space volume
    d. tidal volume divided by frequency of breathing

15. Which of the following pressures is subatmospheric at rest?
    a. alveolar pressure
    b. pleural pressure
    c. transpulmonic pressure
    d. tracheal pressure

16. Pleural pressure is the most negative at:
    a. residual volume
    b. functional residual capacity
    c. end of tidal volume
    d. total lung capacity

17. A lung with an abnormally high compliance would:
    a. be easier to inflate
    b. be easier to deflate
    c. require more work to inflate
    d. have greater elastic recoil

18. Hypoxia stimulates ventilation primarily through:
    a. central chemoreceptors
    b. carotid bodies
    c. medullary chemoreceptors
    d. direct stimulation of diaphragmatic muscle

19. Control of basic rhythms of breathing is located in the:
    a. medullary center
    b. pneumotoxic center
    c. chemoreceptors located in carotid and aortic arch
    d. respiratory center of the pons

20. The most powerful blood-bone stimulus to ventilation is:
    a. bicarbonate
    b. oxygen
    c. hydrogen ion
    d. carbon dioxide

Calculate

21. A person has a forced vital capacity of 5 L and a residual volume of 1.2 L. What is the person's total lung capacity?

22. A person has a tidal volume of 500 ml, a frequency of 14 breaths/min, and a dead space volume of 150 ml. What is the person's alveolar ventilation?

23. A person has an alveolar ventilation of 5 L/min, a frequency of 10 breaths/min, and a tidal volume of 700 ml. What is the person's dead space volume?

24. A subject expires into a spirometer for 10 minutes and has an expired volume of 40 L. Her respiration rate is 12/min during the 10 minute period. a) What is the subject's minute ventilation? b) What is the subject's tidal volume?

Answer Each Question in One or Two Sentences

25. Why is forced vital capacity so useful as a test in lung spirometry?

26. How does the lung inflate/deflate itself with air during breathing?

27. Why is elastic recoil so important in breathing?

28. How does emphysema change lung compliance?

29. What is the functional importance of pulmonary surfactant?

SUGGESTED READINGS

Cherniak, N.S., Altose, M.G., and Kelsen, S.G. *Physiology*, 2nd ed., edited by R. Berne and M. Levy. St. Louis: C.V. Mosby Co., 1988.

Foster, R.E., Dubois, A.B., Briscoe, W.A., and Fisher, A. *The Lung*, 3rd ed. Chicago: Year Book Medical Publishers, 1986.

Guyton, A.C. *Textbook of Medical Physiology*, 18th ed. Philadelphia: W.B. Saunders, 1990.

Kryger, M.H. *Pathophysiology of Respiration*, edited by M.H. Kryger. New York: John Wiley & Sons, 1981.

Nunn, S.F. *Applied Respiratory Physiology*, 2nd ed. Boston: Butterworths, 1977.

Rhoades, R.A. *Physiology*, 5th ed., edited by E. Selkurt. Boston: Little, Brown, 1984.

Scarpelli, E.M., and Condorelli, S. *Pulmonary Physiology of the Fetus, Newborn and Child*, edited by E.M. Scarpelli. Philadelphia: Lea & Febiger, 1975.

West, J.B. *Respiratory Physiology*, 4th ed. Baltimore: Williams and Wilkins, 1990.

KEY TERMS

airways (p. 703)
alveolar pressure (p. 714)
alveolar surface area (p. 711)
alveolar ventilation (p. 711)
alveoli (p. 702)
aortic bodies (p. 722)

atmospheric pressure (p. 714)
breathing patterns (p. 720)
compliance (p. 717)
dead space (p. 711)
diaphragm (p. 712)
gas exchange (p. 700)

homeothermy (p. 700)
hyperventilation (p. 724)
hypoventilation (p. 724)
laminar flow (p. 706)
lung volumes (p. 705)
pleural pressure (p. 716)
pneumothorax (p. 715)

relative pressure (p. 714)
spirometry (p. 706)
surfactant (p. 718)
thoracic cavity (p. 712)
turbulent flow (p. 705)
ventilation (p. 705)

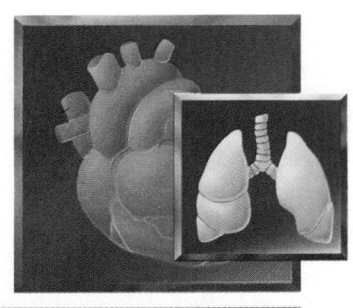

C H A P T E R
21

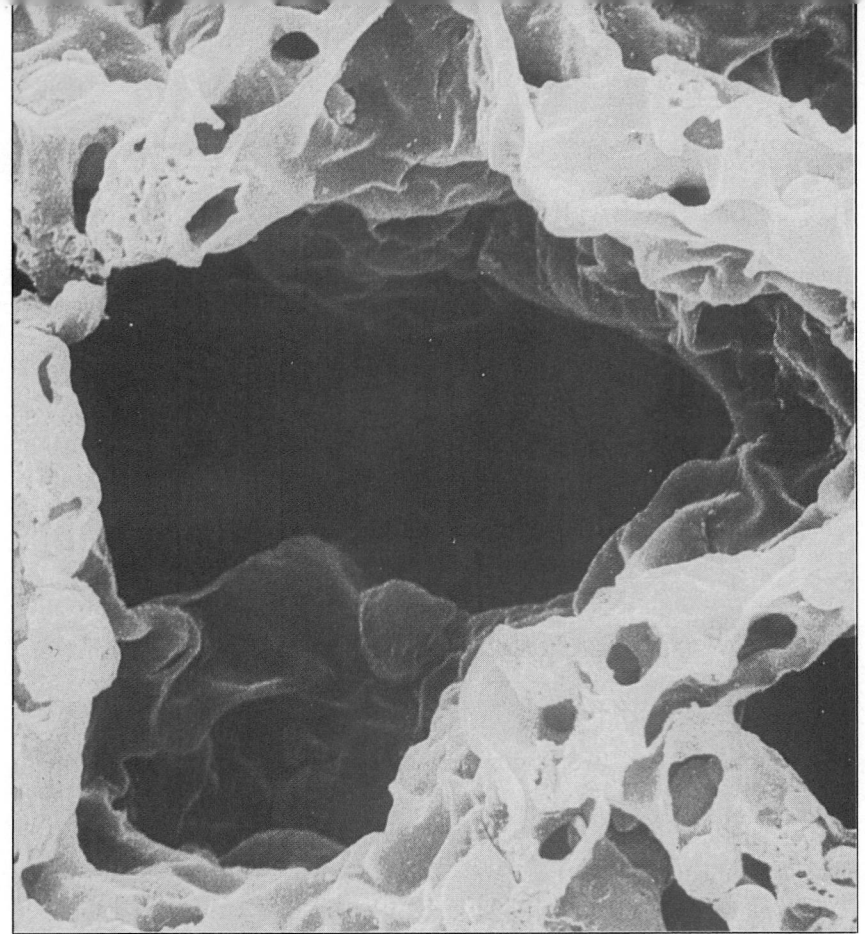

The Pulmonary Circulation and Gas Uptake

The way in which fresh air travels from the atmosphere to the alveoli was discussed in the last chapter. The other component of gas exchange, to be discussed in this chapter, is the transport of blood from the heart to the alveoli. Blood is oxygenated in the alveoli and then returned to the heart to be pumped, along with the oxygen it contains, to the rest of the body. The circulation that transports blood through the lungs, the **pulmonary circulation,** is unique because it is the only circulation in the body that must accommodate all of the cardiac output. The pulmonary vessels are designed to operate at low pressure and to have a large reserve in order to meet the increased blood flow demands of exercise.

Structure of the Pulmonary Circulation

Pulmonary Arteries

The blood leaves the right ventricle via the **pulmonary artery** and follows the same kind of treelike branching pattern found in the airways (Fig. 21–1a, b). In fact, each time the airway tree branches, the arterial tree branches also, so that the two shadow each other (Fig. 21–2).

Pulmonary Capillaries and Veins

Once the arteries reach the alveoli, the vessels break up into a fine network of tiny vessels, the pulmonary **capillaries,** which form a dense double network embedded in the alveolar wall (Fig. 21–3). This complex system results from the way that the lung develops in the embryo. As the embryonic alveoli grow, each develops its own set of capillaries. During lung maturation, the capillaries between adjacent alveoli join, but the joining is incomplete (Fig. 21–3). This leaves a complex, partially joined capillary bed with a large number of pathways through which blood can travel in each alveolar wall.

Once the blood leaves the terminal arteriole and enters the capillaries, it does not simply cross a single alveolar wall and then drain into a venule. The arterioles and venules are separated by four to eight alveoli. Thus, the blood can follow many pathways through a given alveolar wall, as well as travel over different alveolar walls; these combinations result in a huge number of possible routes by which the blood can travel across the gas exchange surface to pick up oxygen.

Each capillary is about equal to the diameter of an erythrocyte (red blood cell). This forces the erythrocytes to pass through the capillaries single file. The erythrocytes are separated from the capillary wall by a thin layer of plasma that acts as a lubricant. The distance across the capillary wall and plasma layer is short, measuring about 1 μm (in more familiar terms, a typical human hair is about 100 μm in diameter, so the distance between gas in the alveolus and the blood is 1/100th of a hair). The alveolar gas reaches the blood quickly because the distance from the gas to the blood phase is so small.

The passage of blood through the capillaries takes about one second. In that time, the blood picks up all of the oxygen it can carry. The blood then enters the smallest venules and returns through the **pulmonary venous tree** (see Fig. 21–2c) to the left side of the heart, from which the newly oxygenated blood is then pumped to the body.

Pulmonary Lymphatic Vessels

Another tree-shaped conducting system in the pulmonary system is the lymphatic system. The lymph vessels are auxiliary drainage vessels. Fluid is continually leaking from the capillaries into the interstitial space. Much of the fluid is reabsorbed by the capillaries, but part of it remains behind and must be drained away through the lymphatics. If the fluid accumulates, the distance between the alveolar gas and the capillary blood will increase and oxygen can no longer be readily absorbed. Accumulation of fluid in the lung is called **pulmonary edema.** This is a serious condition that can, in extreme cases, result in death. When edema forms, the lymphatics go into full operation in order to keep the alveoli dry. The lymphatic fluid drains into systemic veins.

Metabolic Functions in the Pulmonary Circulation

The alveoli have a large surface area, as discussed in Chapter 20. Since the capillary bed takes up about 90% of the alveolar wall, the surface area of the cap-

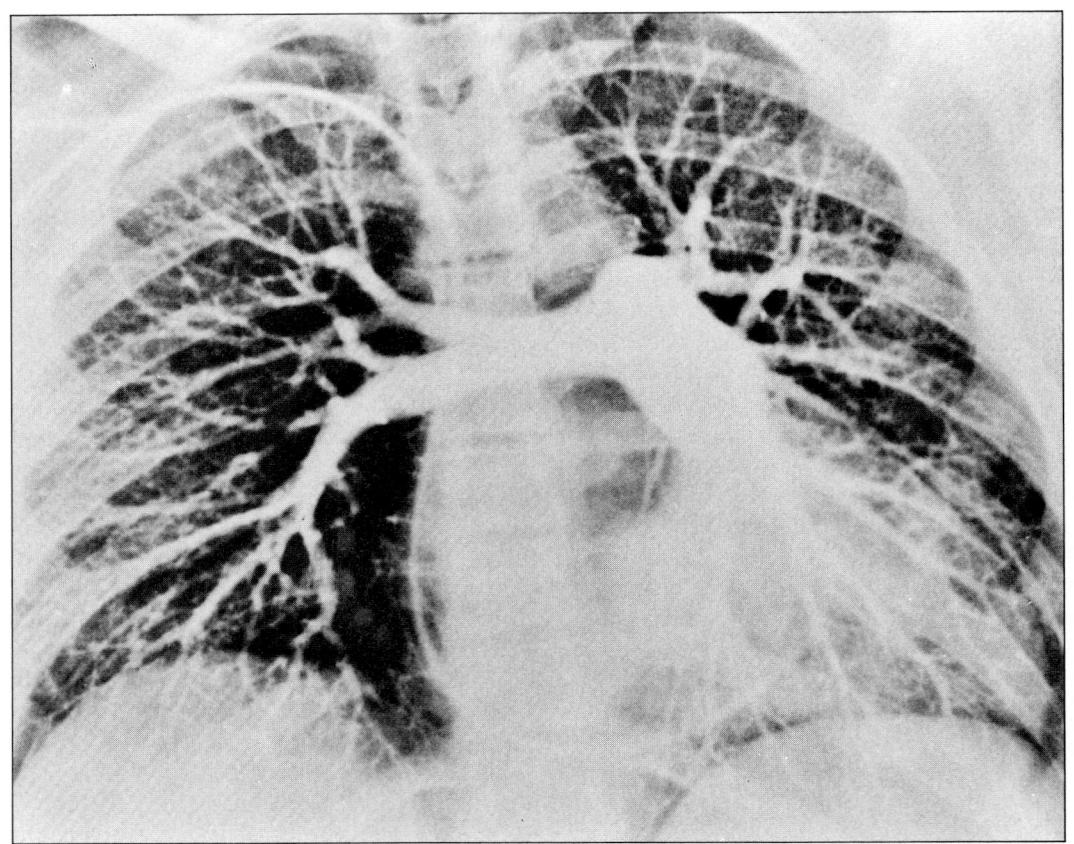

(a)

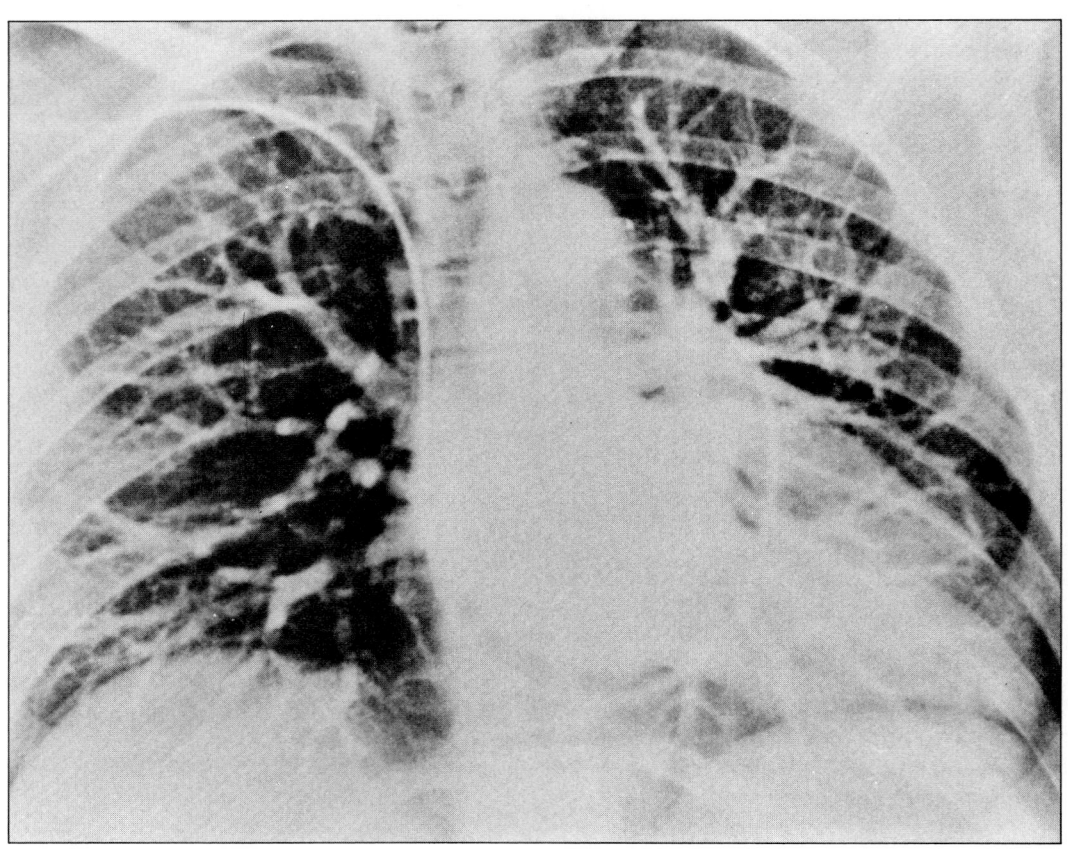

(b)

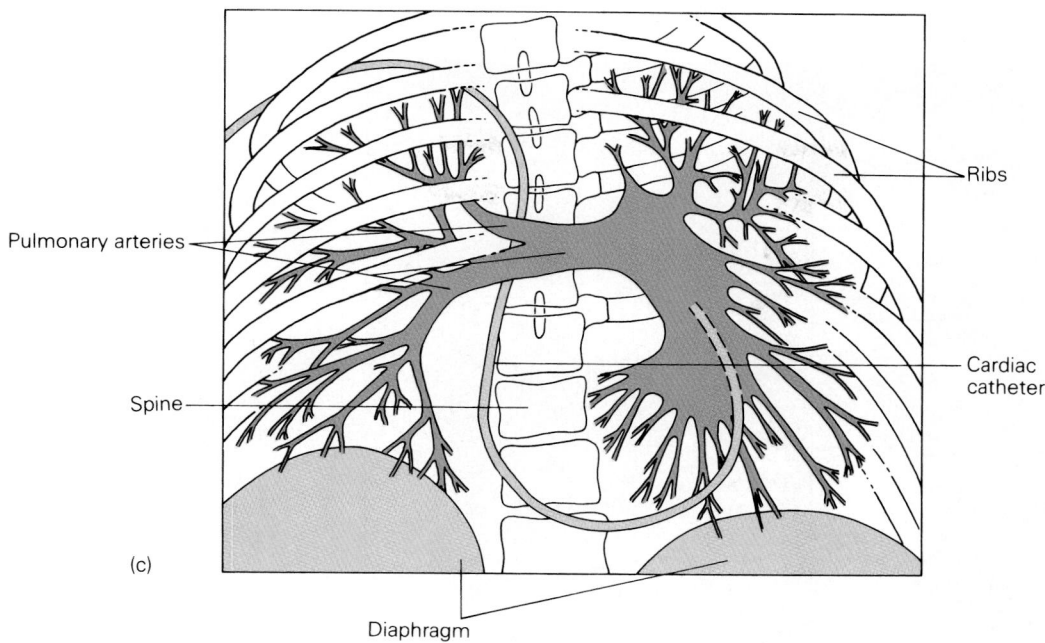

(c)

▲ *Figure 21–1*
◀ (*a*) When a dye that is opaque to x-rays is injected into the pulmonary artery and an x-ray image is made, the dye shows the pathways followed by the blood through the pulmonary arteries. (*b*) If another x-ray image is taken a few seconds later, the dye-filled veins show the pulmonary vein tree. (*c*) Artist's rendering of the arteriogram.

Figure 21–2
Photomicrograph of a thin section of lung tissue. In the center are sections of an artery (filled with blood) and an airway. Arteries and airways are almost always side by side, showing that the airway tree and the artery tree follow each other all the way to the gas-exchange region of the lung.

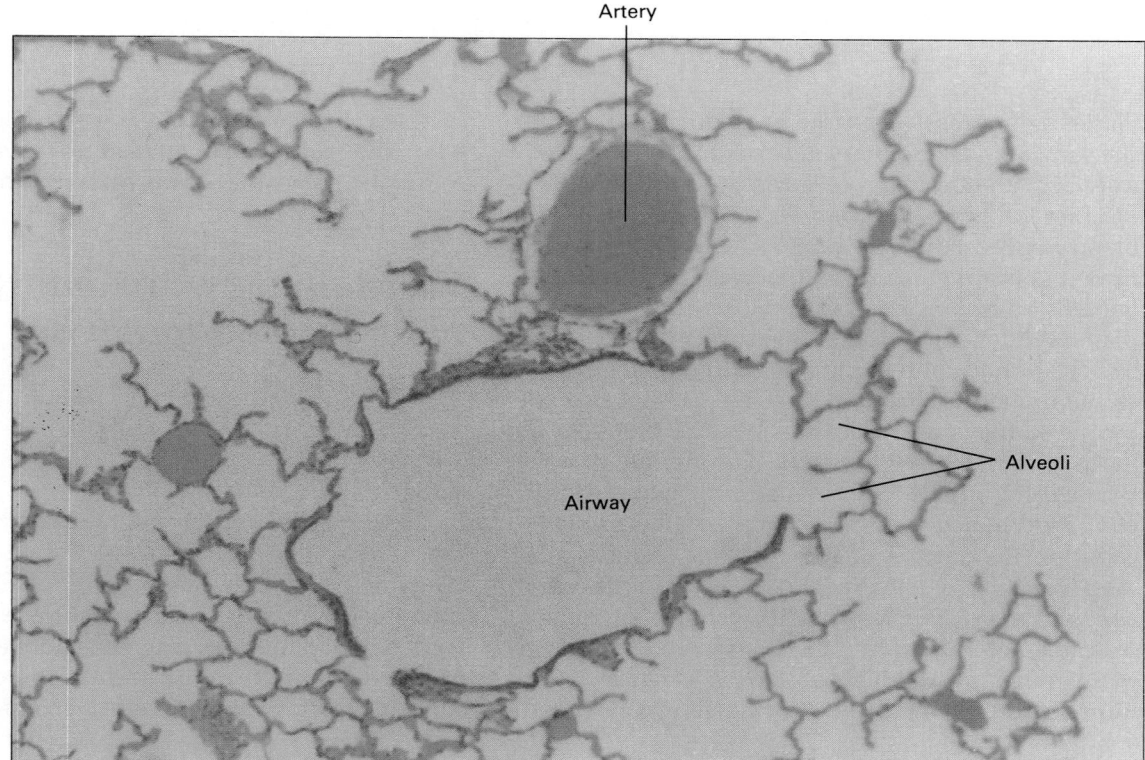

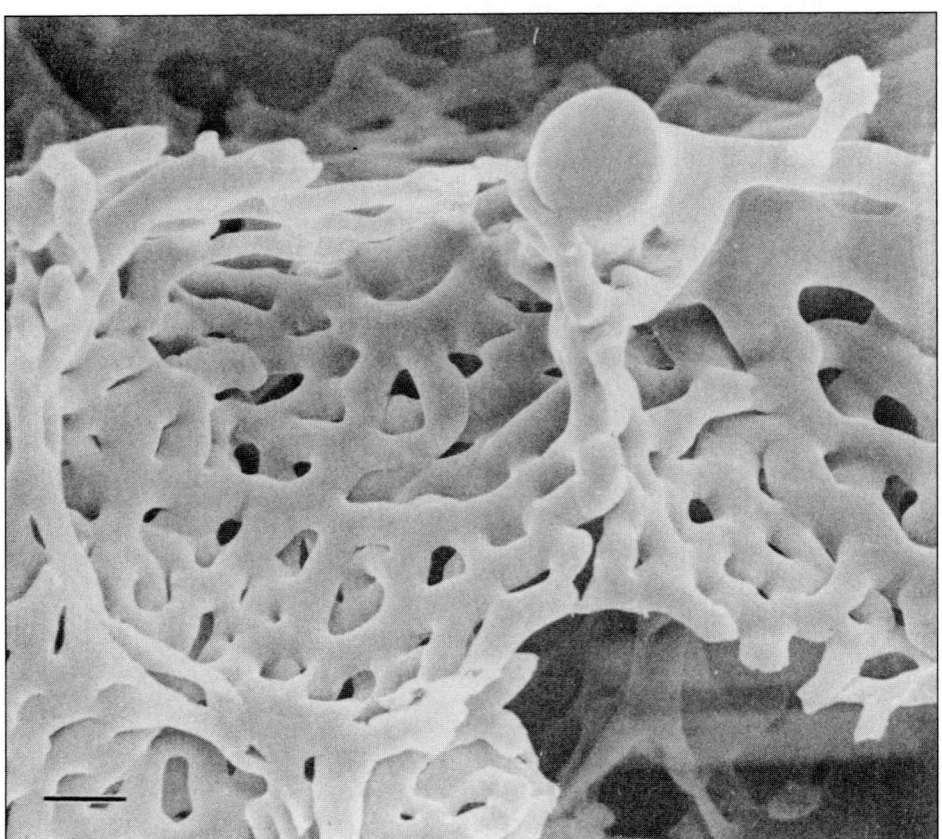

Figure 21–3
Scanning electron micrograph of the alveolar capillary bed shows an incompletely fused double network that offers the red cells a very large array of paths. Bar in lower left is 10 microns long. (With permission from Guentheroth, et al. *Journal of Applied Physiology,* 53:510–515, 1982.)

illaries is impressively large as well. The inside of the capillary wall is completely covered by metabolically active endothelial cells. Because of the central location of the pulmonary circulation, all of the blood is exposed to the capillary endothelium each time it is pumped through the body. The endothelial cells act upon a number of substances. For example, **angiotensin,** a potent vasoconstrictor, is an important hormone in the regulation of blood flow, as mentioned in Chapter 19. This material is made by the kidneys and released into the blood in an inactive form, angiotensin I. During its passage through the pulmonary circulation, angiotensin I is converted to active angiotensin II by an enzyme located on the surface of the capillary endothelium. Angiotensin II is delivered by the blood to the resistance vessels (arterioles) in the systemic circulation to cause constriction and raise blood pressure. Many other vasoactive substances, such as histamine, serotonin, and the prostaglandins, are both

released and inactivated by the endothelium. In this way, the pulmonary circulation serves important metabolic as well as gas exchange functions.

Vascular Pressure and Blood Flow in the Pulmonary Circulation

Cardiac Catheterization

Blood pressures in the pulmonary circulation are much more difficult to measure than those in the systemic circulation, where arteries and veins are conveniently located near the surface of arms and legs. The pulmonary arteries and veins lie deep within the thoracic cavity and are therefore inaccessible. **Cardiac catheterization,** a remarkable technique developed around the middle of this century, permits measurement of the blood pressures in all four chambers of the heart.

Cardiac catheterization is routine today. In this procedure, a long, narrow tube is passed into a superficial vein, often in the arm, and directed along the vein all the way to the heart. There the catheter passes through the chambers of the right side of the heart and into the pulmonary arteries. Because there are no somatic sensory nerves in the heart, the patient does not feel the catheter. Therefore, only a local anesthetic is needed where the skin is cut to isolate a superficial vein. The cardiac catheterization procedure can be performed rapidly and safely. Measurements made with this technique, along with other important technological advances such as the cardiac by-pass pump, have enabled surgeons to save countless lives.

Blood Pressures in the Pulmonary Circulation

Pressure in the pulmonary artery is much lower than in the arteries of the rest of the body. The right ventricle generates a pressure of about 25 mm Hg, which becomes the pulmonary arterial systolic pressure as the ejected blood passes through the pulmonic valve. Once the pulmonic valve closes, the pressure falls to diastolic pressure, just as it does in the systemic circulation. Pulmonary arterial diastolic pressure is about 10 mm Hg. Mean pressure for a

normal adult is 15 mm Hg. These numbers are stated the same way as systemic arterial pressure: 25/10, with a mean of 15 mm Hg (Table 21–1). Pulmonary arterial pressure is usually reported as a mean pressure. These pressures are shown in Fig. 21–4, which is a pressure tracing obtained as the tip of a catheter is passed through the right atrium and right ventricle and into the pulmonary artery. If the

Table 21–1

Pressures in the Pulmonary Artery During Various Conditions

Condition	Systolic/ Diastolic (mm Hg)	Mean (mm Hg)
Rest	25/10	15
Moderate exercise	30/15	20
Maximal exercise	50/25	33
At very high altitude	50/25	33
At very high altitude with maximal exercise	75/45	55
Severe pulmonary hypertension	175/100	125

From Groves, B.M., et al. *J. Appl. Physiol.* 63:521–530, 1987.

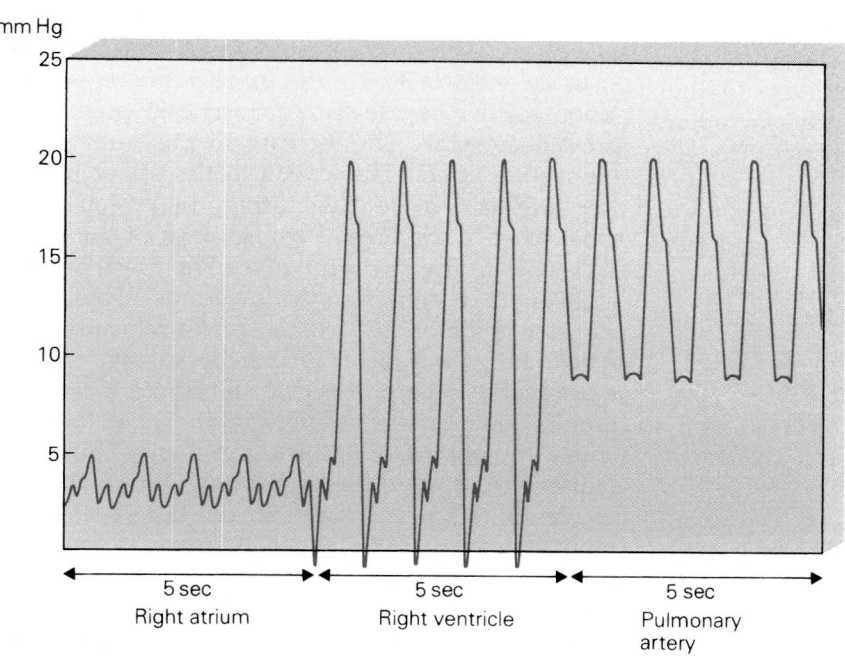

Figure 21–4
Pressure tracing from a cardiac catheter as it is passed from the right atrium into the right ventricle, past the pulmonic valve, and into the pulmonary artery.

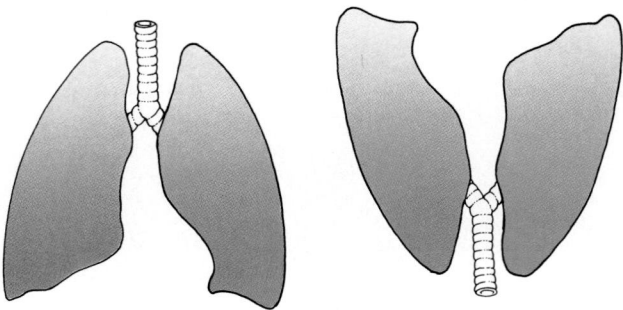

Figure 21–5
Because of the low values of both pulmonary arterial pressure and vascular resistance, most of the blood flows to the lower part of the lung during rest. If a person stands on his or her head, blood will flow to the apexes of the lung.

catheter is advanced farther and farther out along the pulmonary arteries, it will reach vessels too small to permit further passage. The catheter then becomes mechanically wedged in the vessel and pressure drops to a "wedge" pressure. In this position, the vessels downstream from the catheter act as an extension of the catheter. The wedge pressure is near left atrial pressure.

The low pulmonary arterial pressure is just enough to pump blood to the top of the lung.* Although all of the lung receives some perfusion even during resting conditions, the majority of the blood flows through the lower lung (Fig. 21–5). This creates a gradient of blood flow, with most of the flow going to the bottom of the lung and little going to the top. Almost all of the capillaries in the lower lung are open and have blood flowing rapidly through them; under these circumstances, the capillary bed is said to be highly "recruited." In the upper lung, only a few capillary pathways are open (recruited), and the blood cells flowing through them travel slowly.

Ventilation-Perfusion Balance

As long as inspired air is distributed evenly through the lung, venous blood (the blue blood in Fig. 21–6a) can flow to any alveolar region and pick up oxy-

*To calculate the height to which the right ventricle can pump the blood, consider the following. Blood has a specific gravity near 1, whereas mercury has a specific gravity of 13.6—that is, mercury is 13.6 times heavier than water with its specific gravity of 1.0. Multiplying the mean pulmonary arterial pressure of 15 mm Hg by 13.6 gives 204 mm H_2O. Thus, mean pulmonary arterial pressure is sufficient to pump the blood 200 mm high, which is approximately the height of the top of the lung above the heart.

gen from the alveolar gas; as the blood flows past the alveolar gas, it becomes **oxygenated.** Inspired air, however, is not always evenly distributed to all alveoli. This causes regional **hypoxia** (low oxygen) to exist from time to time in groups of alveoli. Of course, blood passing through such areas will not be completely saturated with oxygen (see Fig. 21–6b). This desaturated blood will mix with the rest of the oxygenated blood and reduce the total amount of oxygen that can be delivered to the body. The condition of low blood oxygen is called **hypoxemia** (-*emia* refers to blood; e.g., *anemia* refers to decreased red blood cell count).

The lung has a mechanism called **hypoxic vasoconstriction** for balancing the flow or perfusion of blood with the availability of regional alveolar oxygen. When a region of lung becomes hypoxic, the small arteries feeding that region sense the hypoxia in the alveolar gas and constrict. Their constriction causes the resistance to local blood flow to rise and forces the blood to flow away from the hypoxic region to other parts of the lung where resistance is low because oxygen is available (see Fig. 21–6c). In this way, the balance between ventilation and perfusion is maintained, and compensation for disturbances in delivery of inspired air to the alveoli can be made. When an imbalance between ventilation and perfusion occurs due to lung disease, hypoxemia, sometimes of a serious nature, can result.

Whole Lung Hypoxia

When someone experiences hypoxia throughout the entire lung—at high altitude, for example—all of the local hypoxic vasoconstrictor responses are triggered (see Fig. 21–6d). This situation causes the resistance to blood flow in the entire pulmonary circulation to increase, leading to a rise in pulmonary arterial pressure. The increase in pressure has an interesting effect. The alveoli in the upper part of the lung have more oxygen than the alveoli in the lower part of the lung, because most of the blood flow goes to the lower lung (see Fig. 21–5) and extracts more oxygen from the lowermost alveoli. The elevated pulmonary arterial pressure caused by whole lung hypoxia is useful because it enables more blood to be pumped to the upper regions of lung, where capillaries are recruited and thus increase the surface area for gas exchange. Since recruitment occurs in the upper lung where alveolar oxygen is relatively abundant, the lung becomes a more effective gas exchanger. If, however, the high pulmonary arterial pressure is sustained for long periods, there are detrimental effects, since **pulmonary hypertension** can cause vessel damage and

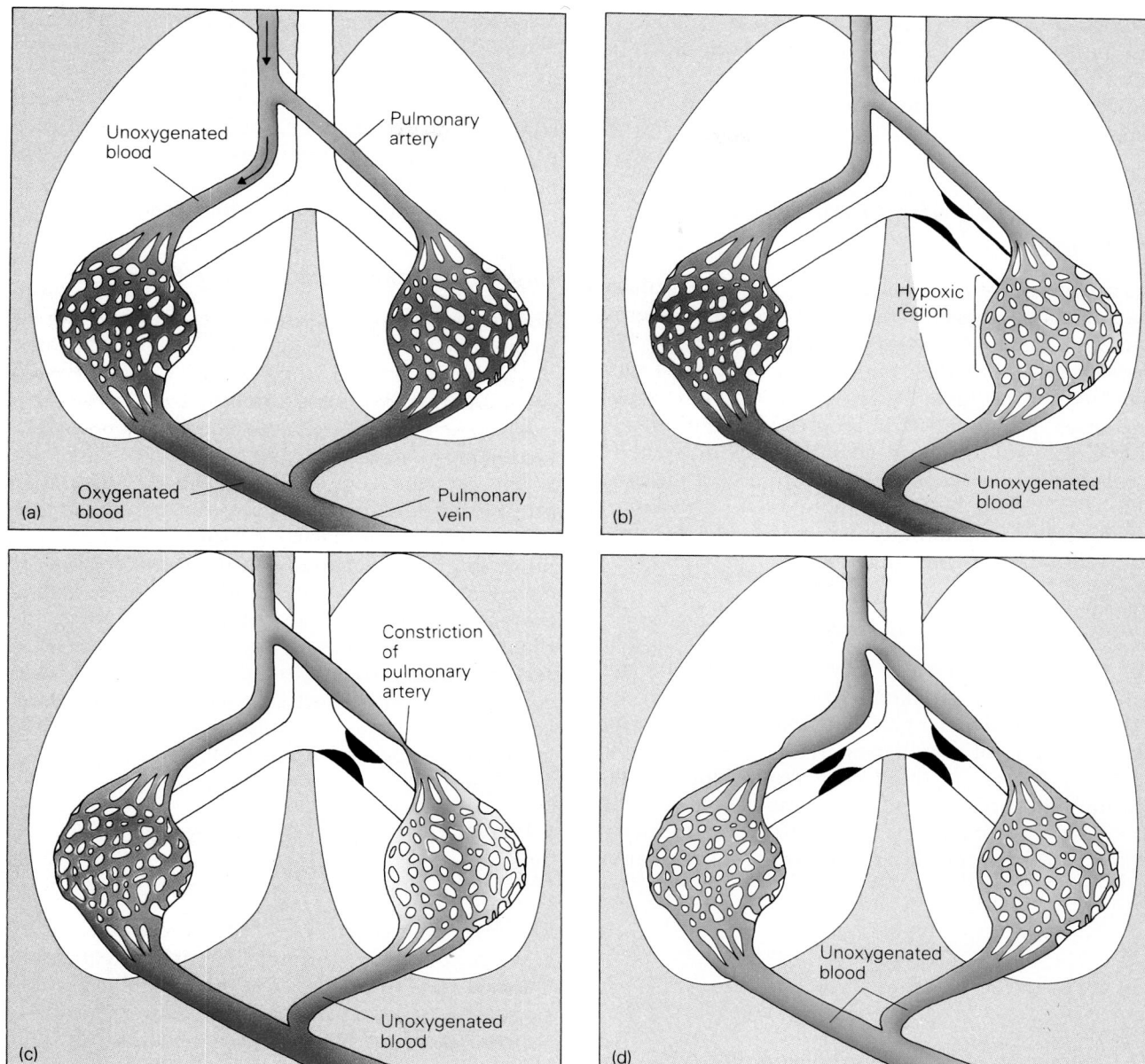

Figure 21–6

The two-alveolus model of the lung can be used to represent two alveoli or two regions of the lung with many alveoli in each region. (*a*) In this figure, all the alveoli are well ventilated so the unoxygenated (blue) blood can flow
to any region and pick up oxygen, shown here turning from blue to red. (*b*) In this case, one region of lung is hypoxic. If blood continues to flow past the hypoxic alveoli at normal rates, unoxygenated (blue) blood will enter the
pulmonary veins and reduce the proportion of oxygen in the blood. Ventilation (by air) and perfusion (by blood)
will be out of balance. (*c*) In some undiscovered way, the pulmonary artery accompanying the hypoxic airway detects the low oxygen and constricts. The narrowed vessel has increased resistance and causes local blood flow to be
reduced and directed to other, better oxygenated parts of the lung. In this way, ventilation-perfusion balance is
maintained and the amount of oxygen in the arterial blood is kept at high levels. (*d*) If all regions of the lung become hypoxic, as happens at high altitude or in certain diseases of the airways when gas movement is impeded,
the arteries detect hypoxia in the airways and constrict. Because all of the arteries constrict, the resistance to blood
flow through the lung increases and pulmonary artery pressure rises.

eventually failure of the right ventricle. Therefore, the redistribution of blood from the bottom to the top of the lung can be useful only if the resulting pulmonary hypertension is either mild or transient (see Table 21–1).

Exercise and Pulmonary Blood Pressure

As cardiac output increases during exercise, pulmonary arterial pressure rises due to the increased quantity of blood flowing through the lung. The pressure rise is modest at low levels of exercise such as walking (see Table 21–1). As the level of exercise increases, pulmonary arterial pressure continues to increase, resulting in an adequate driving force to propel the blood to the uppermost parts of the lung (Fig. 21–7). This results in more even perfusion of the lung from top to bottom and improves the ability of the lung to exchange gas.

Capillary Recruitment During Exercise

As blood flow increases, two important changes take place in the capillaries to improve oxygen uptake from the alveolar gas. The first change involves the number of capillaries with blood flowing through them. In the lower lung, blood flows through most of the capillary segments (Fig. 21–8a), even during resting conditions when cardiac output is low. In the upper lung, relatively few capillaries

are perfused during rest, which leaves capillaries in reserve that can be recruited as the demand for oxygen increases. As cardiac output rises to higher levels during exercise, pulmonary arterial pressure continues to rise as well, and more blood flow is directed to the upper lung, where reserve capillaries are recruited (see Fig. 21–8b). In this way, the surface area for gas exchange is increased.

Capillary Transit Time During Exercise

The second change that takes place during exercise to improve oxygen uptake is more rapid movement of the blood through the capillaries. Even though the normal transit time through the capillaries of one second is rapid, the blood is completely oxygenated in only a quarter of a second. Thus, a large part of the capillary transit time is wasted. As cardiac output increases with exercise, the transit time decreases until it approaches the quarter-second minimum time. This change causes more blood to be oxygenated every second during exercise. Again, there is reserve in the upper lung, where the transit times are considerably longer than one second at rest. As blood flow is shifted upward during exercise, evidence suggests that not only are there capillaries to be recruited, but also that the slowly flowing blood is speeded up as well.

Diffusion of Gases in the Alveoli

The preceding chapter described the way in which gas was conducted through the airways. Once the inspired gas approaches the alveoli, the air flow slows nearly to zero. The predominant means of oxygen transport is no longer the bulk flow that moves along a pressure gradient, but **diffusion,** which causes gas to move along a concentration gradient. In the lung, the diffusion distance across an alveolus is small, in the range of 100 to 200 μm. Gas molecules diffuse very quickly over such short distances; hence, diffusion is an effective means of moving oxygen molecules rapidly to and through the alveolar-capillary membrane.

Partial Pressures in the Alveoli

Recall from Chapter 2 that **partial pressures** of gases are used to tell us how much pressure is exerted by one particular gas in a mixture of gases. The relationship between the total pressure exerted by a mixture of gases and the pressure of individual

Figure 21–7
The distribution of blood flow during resting conditions favoring flow to the lower lung is altered by exercise when cardiac output increases and pulmonary arterial pressure rises. In these circumstances, blood flow in the upper lung increases, and perfusion from bottom to top becomes more even.

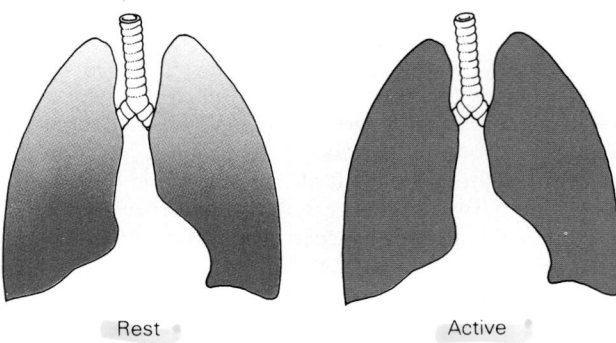

Rest   Active

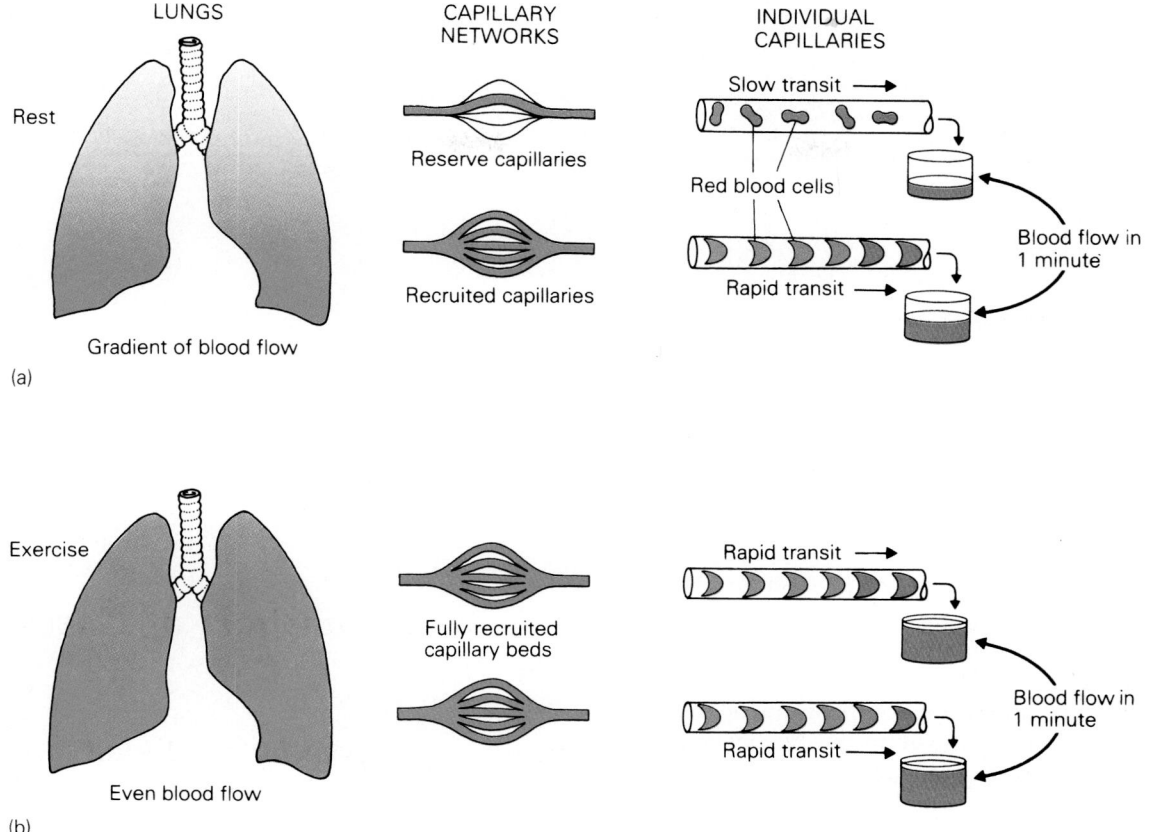

LUNGS   CAPILLARY   INDIVIDUAL
                        NETWORKS   CAPILLARIES

Rest   Reserve capillaries   Slow transit →

                                               Red blood cells

                        Recruited capillaries Rapid transit →

                                                              Blood flow in
                                                              1 minute

Gradient of blood flow
(a)

Exercise   Fully recruited   Rapid transit →
                        capillary beds
                                               Rapid transit →
                                                              Blood flow in
                                                              1 minute

Even blood flow
(b)

Figure 21–8
(*a*) During resting conditions, most of the capillaries in the lower lung are open (recruited), with blood flowing briskly through them. In the upper lung, few capillaries are recruited and what flow there is moves at a slow pace. (*b*) With the increased blood flow during exercise, more blood perfuses both the upper and lower lung. In the lower lung, the primary change is that the blood moves more rapidly through the capillaries. In the upper lung, not only does the blood move more rapidly, but there is also capillary recruitment.

gases is governed by **Dalton's law,** which states that the total pressure in a mixture of gases is equal to the sum of the partial pressures of the individual gases. The partial pressures are the pressures that the individual gases would exert if each gas were present alone in the volume occupied by the whole mixture at the same temperature. Air is made up of different gases, mostly nitrogen and oxygen, each of which exerts its own pressure, called its partial pressure. Since 21% of air is made up of oxygen, then 21% of the 760 mm Hg of air pressure at sea level must be accounted for by oxygen, and

$$0.21 \times 760 \text{ mm Hg} = 160 \text{ mm Hg}$$

is the partial pressure (P_{O_2}) exerted by oxygen. If all of the other gases in a container of air were removed, the remaining oxygen would by itself still exert a pressure of 160 mm Hg. Thus, the amount of oxygen found in air will exert a 160 mm Hg pressure whether alone or mixed with other gases. Atmospheric pressure (760 mm Hg) is the sum of all the partial pressures in air (Table 21–2). The percentages for each gas remain constant regardless of the barometric pressure.

The air in the lung is moisturized as it passes through the airways until it becomes saturated with water. The water vapor exerts a partial pressure, just as oxygen and nitrogen do. The water vapor in saturated air is directly related to the temperature. In the case of a normal body temperature of 37°C, water vapor is 47 mm Hg. Therefore, to compute the partial pressure of gases in the alveoli, 47 mm Hg must be subtracted from the barometric pressure (Table 21–1).

Table 21–2
Partial Pressures of Various Gases in Air*

	Dry Air	Moist Tracheal Air (37°C)	Alveolar Gas	Arterial Blood	Mixed Venous Blood
P_{O_2}	159.1†	149.2†	104†	100	40
P_{CO_2}	0.3	0.3	40	40	46
P_{H_2O}	0.0	47.0	47	47	47
P_{N_2}	600.6	563.5	569	573	573
P total	760.0	760.0	760	760	706

*Usual values in a resting, healthy male at sea level (barometric pressure = 760 mm Hg).

†This is an approximate value and holds only for males breathing air at sea level. Because the total atmospheric pressure at Denver, Albuquerque, or Salt Lake City is about 640 mm Hg, the partial pressure of O_2 in inspired and alveolar gas is significantly below values at sea level.

As inspired air reaches the respiratory zone of the lung, the air mixes with alveolar gas (see Chapter 20) that, being relatively rich in carbon dioxide and poor in oxygen, alters the composition of the incoming air. The final composition of alveolar gas has a partial pressure of oxygen, or **oxygen tension** as it is sometimes called, of 100 mm Hg. Even though that figure is reduced from the atmospheric partial pressure of oxygen (160 mm Hg), there is still a sufficient partial pressure gradient to drive oxygen from the alveolar gas into the capillary blood. (Alveolar oxygen tension is 100 mm Hg, while mixed venous blood oxygen tension is 40 mm Hg—see Table 21–1.) The rate of movement of molecules by diffusion is governed by Fick's Law (Chapter 2).

The Alveolar-Capillary Membrane

Once inspired air has reached the alveolus, and venous blood has reached the capillaries in the alveolar wall, gas exchange can take place. The partial pressure gradients between alveolar gas and capillary blood cause oxygen and carbon dioxide to move in opposite directions (Table 21–1). To accomplish such movement, the gases must cross the alveolar-capillary membrane. The oxygen in the alveolar gas dissolves in the membrane just as it would in water. The oxygen passes through the thin membrane and into the plasma, where it remains in solution. Diffusion continues through the plasma and then through the very thin membrane of the red blood cell. It is inside the red cell that the majority of oxygen is transported to the body.

Uptake and Transport of Gases by the Blood

Transport of Oxygen by the Blood

Oxygen Uptake During Rest and Exercise

Every 100 ml of blood that passes through the tissues delivers about 5 ml of oxygen. A normal young adult at rest requires about 250 ml of oxygen per minute. This means that the heart must pump 5 liters per minute.

A schema of the circulatory delivery is shown in Figure 21–9. During each minute, 7.5 liters of air (minute ventilation) are moved in and out of the lungs. From that, 250 ml/min of oxygen are taken up through the pulmonary capillaries and transported to the tissue capillaries, where 250 ml/min of oxygen are taken up by the tissue. The venous blood is still partially loaded with oxygen after passage through the tissue capillaries (Fig. 21–9). After returning to the right heart, the blood is circulated once again through the lungs.

With exercise, the requirements of oxygen increase considerably. During heavy exercise, a normal, physically fit adult might require 15 times the resting amount of oxygen, or 15 × 250 ml/min = 3750 ml/min. A cardiac output of 25 L/min is required to circulate enough blood to deliver 3750 ml/min of oxygen. In highly trained athletes, the oxygen demands may exceed 5000 ml/min. Cardiac outputs may be more than 30 L/min in these endurance athletes.

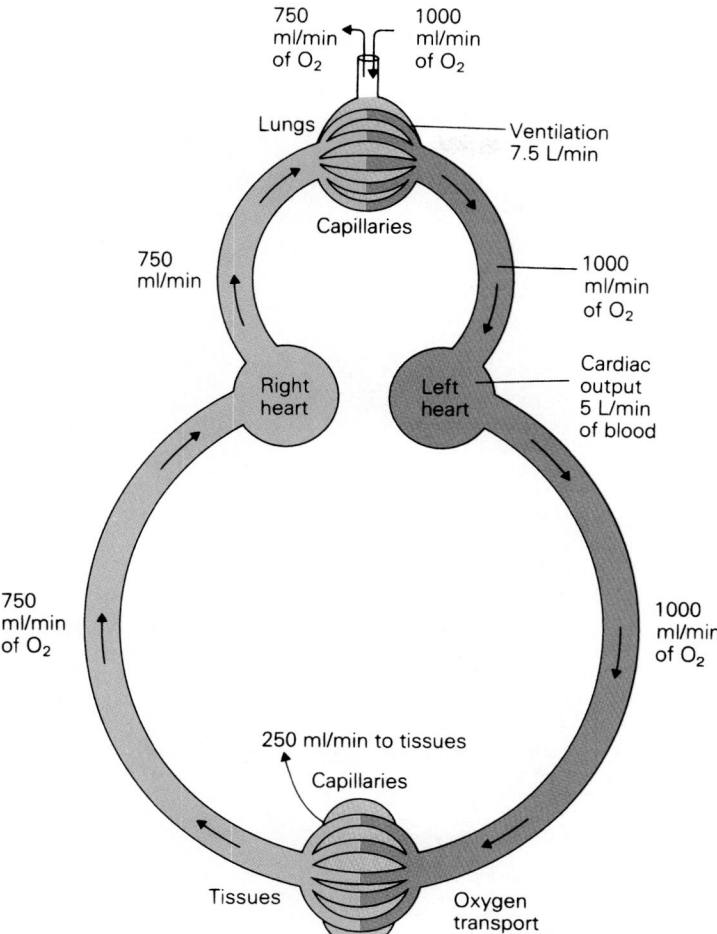

Figure 21–9
Schema of the circulation, showing oxygen deliveries.

Even though these numbers are large, they are much smaller than would be required if, in order to be transported, oxygen had to dissolve in blood. If such were the case, the cardiac output would have to be more than 1000 L/min. That flow rate would completely fill an automobile gasoline tank in 3 seconds! Such a delivery rate would require a huge heart. Each beat of the heart would cause the blood to exceed the sound barrier.

Hemoglobin in Oxygen Exchange

A remarkable molecule, hemoglobin, solves the problems of oxygen demand. Hemoglobin is a chemical that permits large amounts of oxygen to be carried to the tissues in an efficient way. The structure of hemoglobin was presented in Chapter 17. Each hemoglobin molecule contains four heme molecules, each of which contains an iron atom. Each iron atom is capable of binding with a molecule of oxygen. When oxygen combines loosely with the heme portion of hemoglobin in the lung, where the pressure of oxygen is high, it forms **oxyhemoglobin.** When the oxyhemoglobin reaches the tissue where oxygen pressure is low, the hemoglobin releases the oxygen. Because of the loose way in which O_2 binds with iron atoms, the oxygen is delivered to the tissue fluids as dissolved molecular oxygen.

The capacity of hemoglobin for oxygen is much higher than that of plasma. About 98% of the oxygen in the blood is carried by the hemoglobin. Every 100 ml of blood can combine with 20 ml of oxygen. This is written as 20 volumes percent (vol.%).

The Oxyhemoglobin Dissociation Curve

Now we can see how hemoglobin loads with oxygen in the lung. Fig. 21–10a shows an important physiological curve that is "S" shaped and called

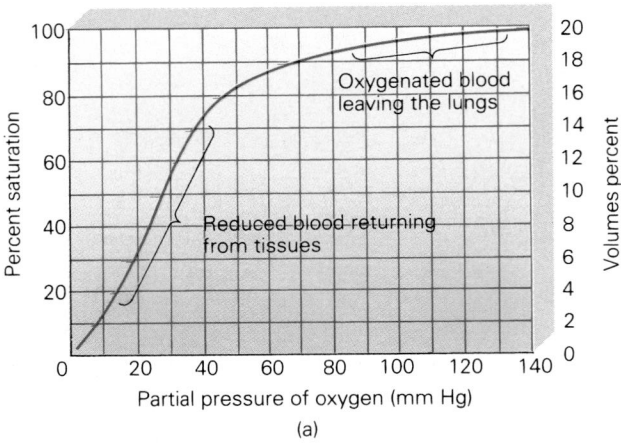

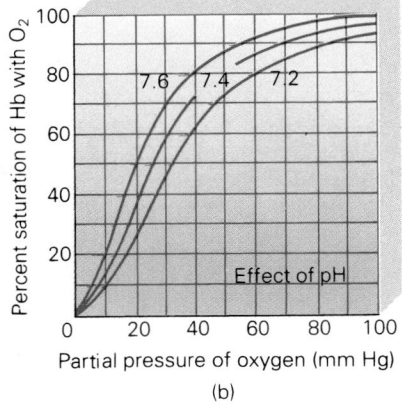

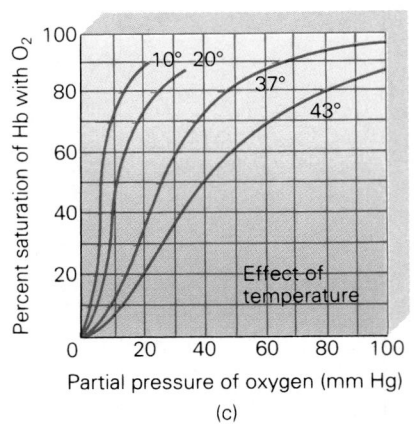

Figure 21–10

(a) The oxyhemoglobin dissociation curve, showing the way in which oxygen binds to hemoglobin in the lung and unloads oxygen in the tissues. Note that as oxygen tension increases in the mid-range, the hemoglobin loads rapidly with oxygen. As the hemoglobin molecules become saturated with oxygen, loading slows as oxygen tension increases. (b and c) Both pH and temperature cause the oxyhemoglobin curve to shift. A shift in the curve causes the hemoglobin to load or unload more easily. In the lungs, where the temperatures are relatively cool and the pH is relatively alkaline, loading is facilitated by the leftward shift of the curve. In an exercising muscle where the environment is relatively warm and acidotic, unloading is facilitated.

the **oxyhemoglobin dissociation curve.** If the blood is 100% saturated, then all of the binding sites on the hemoglobin contain an oxygen molecule. If the blood is 50% saturated, then half of the hemoglobin binding sites contain oxygen molecules. Saturation gives an indication of how completely the binding sites are filled and how many are left in reserve. Another way to measure the amount of oxygen in the blood is to measure its content of oxygen—that is, how many milliliters of oxygen are carried by each 100 ml of blood. As we have seen before, the unit for content is vol.%. That measure is found on the right-hand axis of Figure 21–10a. Under normal resting conditions, arterial blood has a partial pressure of oxygen of 100 mm Hg and a content of 20 vol.%. Venous blood has a partial pressure of oxygen of 40 mm Hg and a content of 15 vol.% (see Table 21–2).

Oxygen Delivery During Rest

Under normal conditions when alveolar tension is 100 mm Hg, the hemoglobin saturation is 97% and the oxygen content is 20 vol.%. Normal venous blood with its oxygen tension of 40 mm Hg is 75% saturated (15 vol.%). Thus 5 ml of oxygen is delivered for each 100 ml of blood pumped by the heart through the systemic capillaries, which means that the arteriovenous oxygen difference is 5 vol.%. In other words, the tissues extract 5 ml of oxygen for every 20 ml coming through them.

Oxygen Delivery During Exercise

During exercise, the oxygen tension in the muscles drops to lower values than when at rest because oxygen is being used so rapidly by the working muscle. Heavy exercise can reduce muscle oxygen tension to below 15 mm Hg. This increases the gradient that favors the movement of oxygen from the capillary blood and into the tissue. Thus the delivery of oxygen is increased because of greater extraction.

Some interesting characteristics of hemoglobin facilitate oxygen transport. Blood acidity (see Fig. 21–10b) shifts the position of the oxyhemoglobin dissociation curve to the right. For a given oxygen tension, the amount of oxygen (oxygen content) is lower. This means that for a given oxygen tension in the tissue, the hemoglobin will unload more oxygen. Exercising muscle has greater acidity owing to increased carbon dioxide and lactic acid production. The unloading of oxygen from hemoglobin is therefore facilitated. Similarly, increased temperature also causes a rightward shift of the curve (see Fig. 21–10c). Blood temperature rises when it reaches a hot exercising muscle, and oxygen unloading is enhanced.

A leftward shift in the oxyhemoglobin dissociation curve facilitates oxygen loading. At a given oxygen tension, the content and saturation are higher. As the blood enters the alveolar capillaries, carbon dioxide leaves the blood, the pH becomes more basic, and oxygen loading is enhanced. Similarly, cooler temperatures cause a leftward shift of the curve and, since the lungs are cooler than exercising muscle, a leftward shift is also promoted. Thus, the oxyhemoglobin curve shifts in a remarkable way from right to left and back to the right with each passage around the circulation in order to favor oxygen loading in the lungs and unloading in the tissues.

Transport of Carbon Dioxide by the Blood

Carbon dioxide is transported in the circulation more readily than oxygen, because carbon dioxide—being a nonpolar molecule—is highly soluble in lipid and thereby moves easily across plasma membranes. Because of this, blood carbon dioxide tension tends to remain more nearly normal than oxygen in many diseases. Since the maintenance of pH in the body is crucial to so many chemical reactions, it is important that carbon dioxide remain normal. Carbon dioxide in the blood is intimately linked to acid-base balance as described in Chapter 26.

In a normal resting adult, 4 vol.% of carbon dioxide are transported to the lungs to be exhaled each minute. Carbon dioxide is transported in the blood in three ways. All three begin with the dissolution of CO_2 in the plasma (Fig. 21–11) after it has diffused from the tissue into systemic capillaries.

Bicarbonate Ions in CO₂ Transport

The largest fraction of carbon dioxide (60% to 70%) is transported as bicarbonate ions. These ions are formed by an important series of reactions that take place inside the red blood cell.

First, carbon dioxide and water in the red blood cell combine to form carbonic acid as shown in this equation:

$$CO_2 + H_2O \xrightarrow{\text{carbonic anhydrase}} \underset{\text{carbonic acid}}{H_2CO_3}$$

Ordinarily, this is a slow reaction requiring many seconds for completion, but when catalyzed by the enzyme **carbonic anhydrase,** the reaction is accelerated by a factor of 5000 and is complete in a fraction of a second.

In the next chemical step, the carbonic acid dissociates into a bicarbonate ion (HCO_3^-) and a hydrogen ion (H^+):

$$\underset{\substack{\text{carbonic} \\ \text{acid}}}{H_2CO_3} \rightarrow \underset{\text{bicarbonate}}{HCO_3^-} + H^+$$

The bicarbonate ion moves from the red cell into the plasma, where it is readily dissolved. This leaves a hydrogen ion free in the red cell, where it rapidly combines with hemoglobin in the following way:

$$H^+ + Hb^- \rightarrow HHb$$

As the bicarbonate ions diffuse into the plasma, chloride ions diffuse into the red blood cells to replace the chloride. This exchange, known as the **chloride shift,** causes red blood cells in venous blood to have a higher chloride content than red blood cells in arterial blood (see Fig. 21–11a). These reactions are summarized in Figure 21–11.

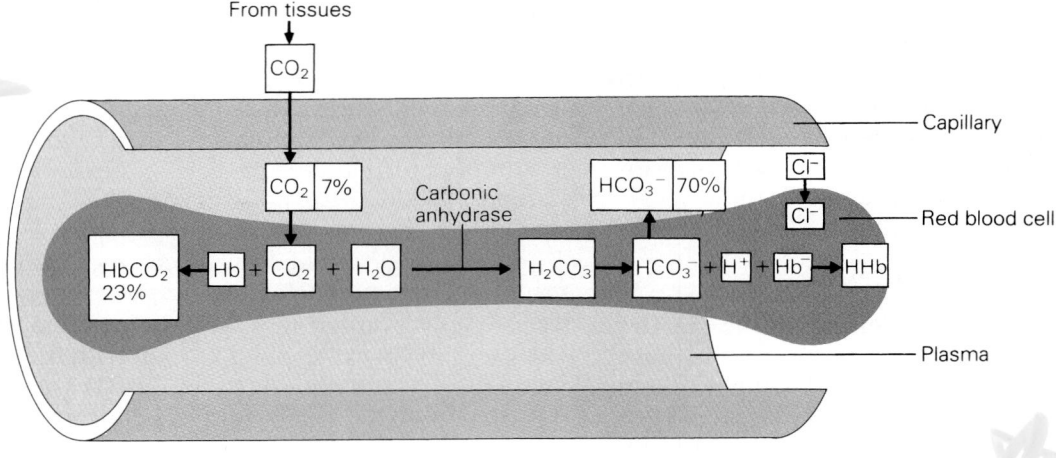

(a)

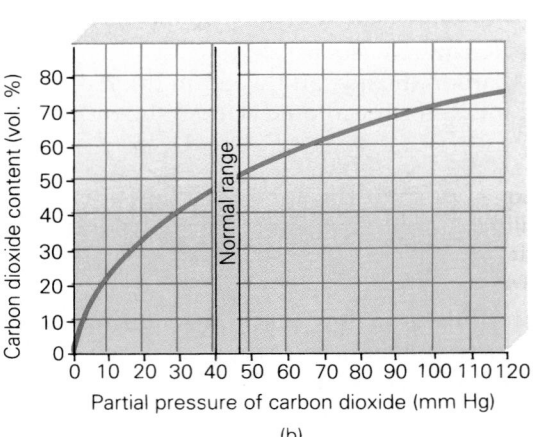

(b)

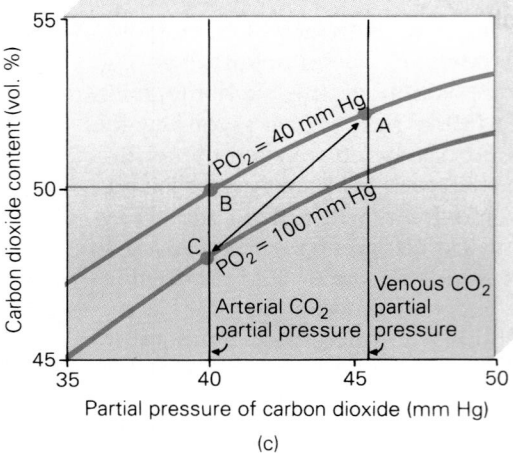

(c)

Figure 21–11

(*a*) The ways in which carbon dioxide is carried by the blood from the tissues to the lungs. (*b*) The carbon dioxide dissociation curve. (*c*) The carbon dioxide dissociation curves in the normal operating range for venous blood (blue curve) with an oxygen partial pressure of 40 mm Hg, and for oxygenated blood (red curve) with a partial pressure of 100 mm Hg. The Haldane effect is shown by the arrow.

Carbaminohemoglobin in CO_2 Transport

Carbon dioxide can combine directly with hemoglobin to form carbaminohemoglobin. The carbon dioxide does not bind at the same site on the iron atom as oxygen but instead binds loosely by a direct reaction with some of the amino groups that form the hemoglobin molecule. Between 15% and 30% of carbon dioxide transport is accomplished in this form (Fig. 21–11*a*). The hemoglobin molecule has a higher affinity for carbon dioxide when it is oxygen-desaturated—exactly the condition that exists in

venous blood, in which oxygen is low and carbon dioxide is high. The bond between hemoglobin and carbon dioxide is loose enough to be reversible in the lung. The carbaminohemoglobin reaction is much slower than the enzyme-catalyzed reaction that forms bicarbonate.

Dissolved Carbon Dioxide in CO_2 Transport

Between 7% and 10% of the carbon dioxide is transported in a dissolved state in the plasma and red blood cells. This accounts for only about 0.3 vol.% of the total carbon dioxide transport.

All three of these methods of transport are readily reversible in the pulmonary capillaries, where the concentration gradient favors the movement of carbon dioxide from the blood into the alveolar gas. Since the equations read in the reverse as well as the forward direction, the transport cycle is complete when carbon dioxide is given off in the lung.

All of the ways in which carbon dioxide can be transported in the blood are dependent on the partial pressure of carbon dioxide, just as is the case with oxygen. The **carbon dioxide dissociation curve,** with its narrow normal range of operation of between 40 and 46 mm Hg, is shown in Figure 21–11b. The capacity of blood for carbon dioxide is considerably higher than for oxygen (compare Figures 21–11b and 21–10a). As blood passes through the systemic capillaries, the carbon dioxide content rises to about 52 vol.%. As the blood passes through the pulmonary capillaries, 4 vol.% of carbon dioxide are unloaded into alveolar gas and the carbon dioxide content falls to 48 vol.%.

A physiologically important interaction, known as the **Haldane effect,** exists between the oxygen-hemoglobin reaction and carbon dioxide transport. This effect is caused by oxygenated hemoglobin becoming a stronger acid, which lessens the tendency for hemoglobin to form carbaminohemoglobin and thereby reduces that means of transport. The effect is significant, as shown in Figure 21–11c. The carbon dioxide dissociation for venous blood with an oxygen partial pressure of 40 mmHg is shown by the *blue line.* Without the Haldane effect, if the partial pressure of carbon dioxide fell from 46 to 40 mmHg, the content would change from 52 vol.% (point *A*) to 50 vol.% (point *B*). When the partial pressure of oxygen increases to 100 mmHg as the hemoglobin becomes oxygenated in the pulmonary capillaries, the Haldane effect causes the carbon dioxide dissociation curve to shift to the right (*red curve*) so that when the partial pressure of carbon dioxide is 40 mmHg, the content of carbon dioxide falls to 48 vol.% (point *C*) and a total of 4 vol.% are unloaded. This represents an additional 2 vol.% of carbon dioxide unloaded in the lungs.

It is important to remember that the blood does not empty itself of carbon dioxide in the lung or of oxygen in the tissues. Even during the most strenuous exercise—when oxygen extraction is high in the tissues and ventilation is high in the lungs—oxygen and carbon dioxide always remain in the blood. The partial pressures in mm Hg and contents in vol.% of these gases in various regions of the circulation are shown in Figure 21–12.

Disorders of the Pulmonary Circulation

Pulmonary Edema

Pulmonary edema is a condition in which an abnormal amount of fluid leaks from the pulmonary capillaries into the interstitial space and, under severe conditions, into the alveolar spaces. The condition is serious since it increases the thickness of fluid through which oxygen and carbon dioxide must diffuse for successful gas exchange to occur. In addition, edematous lungs become less compliant and impede ventilation.

The most common cause of pulmonary edema is failure of the left ventricle. When the left ventricle is unable to empty itself sufficiently during systole, diastolic pressure rises. This causes left atrial pressure to rise, which backs up the system and causes pulmonary capillary pressure to increase. If capillary pressure rises too much (past 25 to 30 mm Hg), the capillaries leak.

A second form of pulmonary edema is caused by increased permeability of the capillaries. If the capillary endothelium is damaged, then the microvessels begin to leak. Damage can occur from various toxic agents, including heroin, or it can be caused by charged oxygen molecules, called oxygen radicals, which are extremely destructive to the capillary walls. This clinical problem, called the **adult respiratory distress syndrome,** is different from infant respiratory distress syndrome described in Chapter 20.

Fetal and Neonatal Circulatory Physiology

Fetal Circulation

Before birth, the lungs have no respiratory function, since air is unavailable. The fetus must depend on the mother for gas exchange. This is accomplished through the placenta (Fig. 21–13), in which gas exchange vessels from fetus and mother pass close to each other. Such a system is less effective than lungs exposed directly to the atmosphere and results in a low fetal arterial P_{O_2}, on the order of 15 to 20 mm Hg. Blood travels to and from the placenta through the umbilical arteries and veins. The oxygenated blood flows into the right heart, but only 10% to 12% flows through the lungs. About half of the

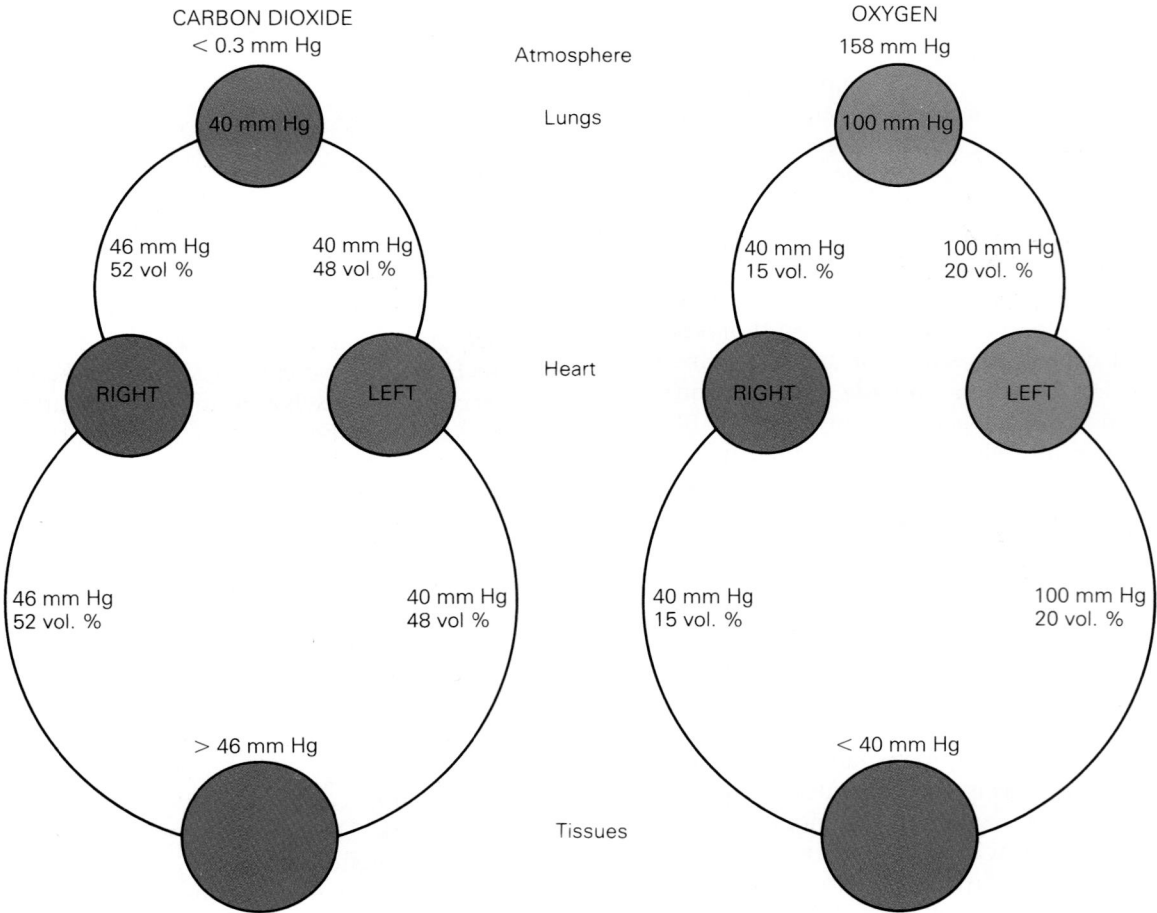

Figure 21–12
A comparison of the partial pressures of carbon dioxide and oxygen in the circulation during resting conditions. It is important to realize that arterial blood contains carbon dioxide, and that venous blood contains oxygen. That is, the blood is not scrubbed clean of carbon dioxide in the lungs, and not all of the oxygen is removed from the blood by the tissues. Partial pressures are in mm Hg and contents in Vol. %.

blood flows from the right to the left atrium through a passage called the **foramen ovale.** This oxygenated blood is pumped by the left ventricle out the aorta and perfuses mostly the head and upper body. The remainder of the blood follows the usual pathway through the right heart into the pulmonary artery. The bulk of pulmonary arterial blood, however, flows through the **ductus arteriosus** into the descending aorta to perfuse the lower body. More than half of the cardiac output passes through the umbilical artery and into the gas-exchanging vessels of the placenta (Fig. 21–13).

Circulatory Changes at Birth
The changes that occur in the circulation at birth are almost as important as the act of breathing itself. As soon as the umbilical cord is severed, the increased blood flow in the aorta immediately raises pressure. As the lungs expand, the pulmonary vessels are no longer compressed, and hypoxic vasoconstriction is relieved; these have the combined effect of lowering pulmonary vascular resistance nearly ten-fold. Since left atrial pressure is 2 to 4 mm Hg higher than right atrial pressure, the flap that lies on the left side of the atrial septum closes the foramen ovale. A few

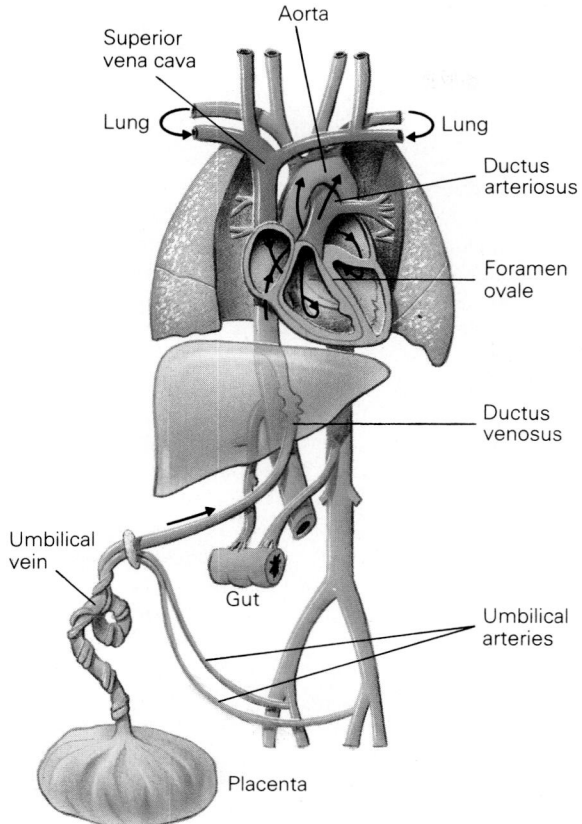

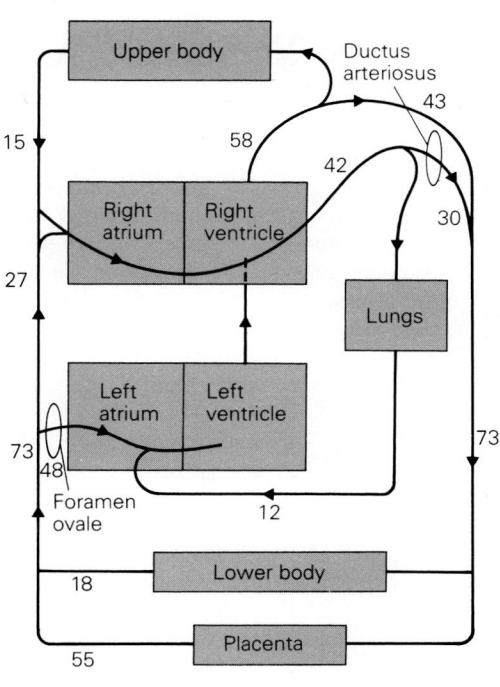

Figure 21–13

In the fetus, the circulation is different from that in the adult in a number of important ways. This figure shows the names of the vessels peculiar to the fetus and also shows how the cardiac output is distributed. Note that the partial pressures of oxygen in mm Hg (the numbers in right panel) in the fetus are very low compared to postnatal values.

hours after birth, the powerful muscular wall of the ductus arteriosus constricts and closes that vessel. Growth of fibrous tissue into the lumen of the ductus over the course of the next few months causes permanent closure.

Congenital Heart Diseases

In some newborns, the ductus does not close after birth. This condition, known as **patent ductus arteriosus** (*patent* means "open"), can lead to a left-to-right shunt of blood from the aorta into the pulmonary artery. The high aortic pressure causes **pulmonary hypertension,** which in severe cases leads to irreversible damage to the pulmonary arteries.

Sometimes, the pulmonary hypertension becomes so great that the shunt reverses and flows from right to left. Such a shunt forces venous blood into the aorta and causes hypoxemia. Patients with

patent ductus arteriosus can experience right ventricular failure and die. The patent ductus can be readily closed by surgery, and if the diagnosis is made in time, the disorder can be repaired with few untoward effects. The diagnosis is made by listening to the chest with a stethoscope. Blood coursing through the patent ductus makes a sound called a **murmur** that alerts the physician that an abnormality is present.

Numerous other defects can occur during the development of the fetus. For example, the septum between the atria can be incompletely formed, causing an **atrial septal defect.** A similar defect in the ventricles is called a **ventricular septal defect.** Although there are a number of other congenital heart defects, fortunately these disorders as a group are rare and can often be repaired by modern surgical techniques, such as those employing a heart-lung bypass machine.

Pulmonary Emboli

The design of the pulmonary circulation, with its many branching arteries all feeding into the capillary net, acts as a sieve for blood as it circulates through the body. This design has an important protective effect in that material in the circulation larger than red blood cells will plug small vessels and prevent the flow of blood. In the systemic circulation, such blockages can be life threatening if they occur in the coronary vessels (causing heart attack) or in the brain (causing stroke).

The most common abnormally large objects in the circulation are blood clots. These often form in the veins in the legs, especially when blood flow is slowed because of varicose veins or prolonged bed rest. A blood clot stuck in a vein is called a thrombus. If the clot breaks loose, it is called an embolus, or a thromboembolus. The embolus proceeds through the right side of the heart and out into the pulmonary arteries, moving farther and farther into the lung until it wedges in a branch of the pulmonary arterial tree. The resultant trapping blocks blood flow to part of the lung. Since the gas exchange surface area in the pulmonary circulation is so vast, oxygenation of blood is not adversely affected unless the blockage is massive. Sometimes the thromboembolus is so large that it can block the main pulmonary artery and cause death. Smaller clots can resolve over a period of a month or so, allowing the pulmonary circulation to return to normal.

Any large material in the venous circulation will be filtered by the pulmonary circulation. Bone marrow, from broken bones, for example, sometimes embolizes to the lung. If air enters the venous circulation, bubbles will form and cause air embolization of the lung. The air diffuses across the air-blood barrier and is exhaled with alveolar gas, so air emboli do not last long. Much air would have to be injected to cause harm.

In the accompanying figure, the pulmonary circulation has been embolized by liquid mercury. The x-ray was taken of a drug user who injected mercury intravenously for a special "kick." It is difficult to tell from an x-ray whether the mercury is in blood vessels or small airways. The differential diagnosis is made by the pool of mercury (*arrow*) in the right ventricle.

(Courtesy of Dr. William Beamish, University Hospital, Edmonton, Alberta.) Fraser and Pare: Diseases of the Chest, Saunders.

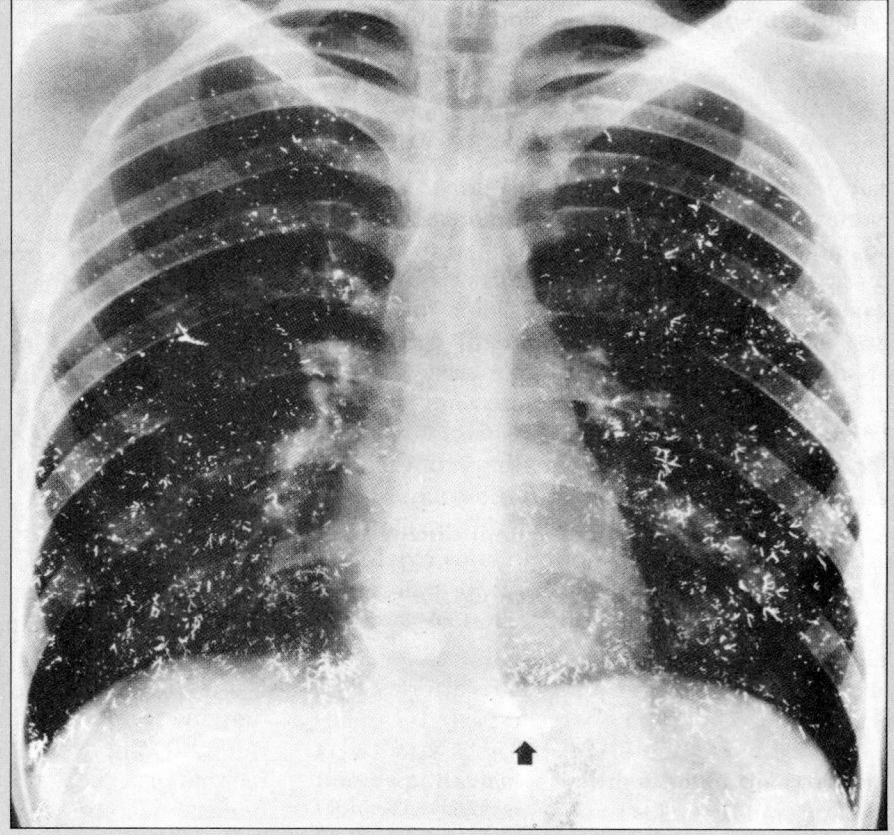

Transit of Leukocytes Through the Pulmonary Circulation

Of all the white blood cells that circulate in the blood, neutrophils make up the largest proportion and are one of the body's chief defenders against invading microorganisms. For unknown reasons, neutrophils travel through the lung far more slowly than red blood cells. This causes neutrophils to concentrate in the small pulmonary blood vessels. From that site, they can either defend the body against inhaled bacteria or move quickly through the circulation to any part of the body.

There are times, however, when neutrophils misuse their powerful bacteria-killing ability and damage the blood vessels, which causes plasma to leak into the tissue. In the lung, edema in alveoli increases the thickness of the blood-gas barrier and impedes the exchange of oxygen and carbon dioxide. In its severe form, this condition is called the Adult Respiratory Distress Syndrome.

Because neutrophils are present in large numbers in the lungs of patients with Adult Respiratory Distress Syndrome, they are implicated in the etiology of this disorder. Therefore, considerable investigative attention has been focused on what causes the neutrophils to stop under normal conditions in the pulmonary circulation. Two hypotheses are currently being investigated. The first, the mechanical impediment hypothesis, suggests that the highly viscous nature of the neutrophils impedes their travel through the pulmonary capillary bed. When the neutrophils reach capillary segments with diameters smaller than their own, they require considerable time to deform sufficiently to fit through the narrow lumens of the obstructing capillary segments. The second hypothesis, the adhesion hypothesis, is based on the fact that the neutrophil surface membrane has many adhesion glycoproteins that recognize receptors on the cells lining the blood vessel walls. There the binding of the neutrophils to the endothelial cells may slow them down. These adhesion sites are crucial in the cells' response to invading bacteria, because they anchor the neutrophils to the vessel wall, allowing them to then move through it and attack the microorganisms. Their function under normal conditions, however, is unknown.

The mechanical impediment hypothesis is being investigated by treating neutrophils with chemicals that make them either more rigid or more flexible. After treatment, the transit times of the neutrophils through the pulmonary circulation are measured to determine whether the altered viscosity affects transit. To test the adhesion hypothesis, scientists are determining the effect on neutrophil transit times of antibodies that specifically block the functioning of the adhesion molecules.

SUMMARY

Structure of the Pulmonary Circulation

The pulmonary circulation carries blood from the heart to the gas-exchange surface to be oxygenated and returns the oxygenated blood to the heart to be pumped to the rest of the body. The oxygenated blood leaves the right ventricle via the pulmonary artery and branches out into a treelike pattern of arteries that mirrors the branching of the airways.

Once the arteries reach the alveoli, they form a fine network of capillaries embedded in the alveolar wall. Gas exchange takes place when the blood is spread over the large alveolar surface area via the capillaries and the blood loads with oxygen and unloads carbon dioxide. The newly oxygenated blood returns through pulmonary veins to the left side of the heart to be pumped throughout the body.

The lymphatic system is an auxiliary network of vessels that help prevent pulmonary edema by draining excess fluid leaked from the capillaries into the interstitial space.

Endothelial cells that line the capillary wall play important roles in several metabolic functions such as the activation of angiotensin, a potent vasoconstrictor that helps regulate blood pressure.

Vascular Pressure and Blood Flow in the Pulmonary Circulation

Cardiac catheterization is a procedure by which blood pressure can be measured in the chambers of the heart and in the pulmonary arteries by threading a catheter through a superficial vein in the patient's arm or leg and into the heart.

Blood pressures in the pulmonary circulation are important measures of how the lungs are functioning. Under normal conditions, the pulmonary arterial pressure is low in comparison to arterial pressure in the rest of the body, usually 25/10 mmHg with a mean of 15 mmHg. This low-pressure, low-resistance circuit accommodates all of the cardiac output. The majority of the blood flows to the lower lung.

Sometimes, the flow of inspired air is unevenly distributed among the alveoli, causing regional hypoxia (low oxygen). The lung has a mechanism called hypoxic vasoconstriction that causes arteries in a hypoxic region of lung to constrict and direct blood flow to better-ventilated regions.

The balancing system is effective even when an individual is exposed to whole lung hypoxia, as happens at high altitudes. The ventilation-perfusion balancing system senses the low oxygen and responds by triggering vasoconstriction of the arteries, increasing pulmonary arterial pressure, and routing blood to the upper regions of the lung, where more oxygen is available.

During exercise, the increased cardiac output causes pulmonary arterial pressure to rise because of increased quantity of blood flowing through the lungs. Much of the increased blood flow perfuses the upper regions, resulting in a more even perfusion of the lungs. The body further compensates for increased oxygen demand by enhancing the oxygen uptake from the alveolar gas in two ways: (1) by increasing the number of capillaries through which the blood flows, thus increasing gas exchange surface area, and (2) by shortening the capillary transit time.

Diffusion of Gases in the Alveoli

A convenient method of measuring gas concentration is to calculate the partial pressures of a gas.

In the process of gas exchange, oxygen and carbon dioxide cross the alveolar-capillary membrane in response to their respective concentration gradients.

Uptake and Transport of Gases by the Blood

At rest, the blood delivers 5 ml of oxygen for every 100 ml of blood that circulates through the tissues. Resting cardiac output is 5 liters per minute; therefore, 250 ml/min of oxygen are delivered to the tissues. With exercise, this delivery can be increased 15 to 20 times. Hemoglobin greatly facilitates oxygen delivery to the blood by loosely binding oxygen to its four binding sites and then releasing the oxygen in the tissues. The oxyhemoglobin dissociation curve describes how hemoglobin is loaded and unloaded with oxygen at a given oxygen tension.

During rest in a normal adult, 4 vol.% of carbon dioxide are transported to the lungs to be exhaled each minute. There are three ways in which carbon dioxide is transported in the blood: bicarbonate ions, carbaminohemoglobin, and dissolved carbon dioxide.

Disorders of the Pulmonary Circulation

In pulmonary edema, fluid leaks from the pulmonary capillaries into the interstitial space and, under severe conditions, into the alveolar space, making gas exchange very difficult.

Fetal and Neonatal Circulatory Physiology

Before birth, the lungs of the fetus do not serve any respiratory function. Gas exchange takes place between the mother and the fetus through the placenta. In the fetus, blood is shunted past the lung by the foramen ovale and the ductus arteriosus.

After birth, the blood flow through the lungs increases, systemic arterial pressure rises, and the foramen ovale and ductus arteriosus close.

Examples of congenital heart disease are **patent ductus arteriosus** (ductus remains open) and **ventricular septal defect** (a hole in the ventricular septum).

REVIEW QUESTIONS

Identify Each with a Term

1. The method for measuring the pressures in the chambers of the heart and in the pulmonary artery.
2. Fluid that collects in the lung.
3. The molecule that carries oxygen in the blood.
4. Blood clots that lodge in the lung circulation.
5. The organ that the fetus uses for gas exchange.

Define Each Term

6. ductus arteriosus
7. hypoxic vasoconstriction
8. wedge pressure
9. carbaminohemoglobin
10. hypoxemia

Choose the Correct Answer

11. Which of the following causes the oxyhemoglobin dissociation shift to the right?
   a. low CO_2
   b. low pH
   c. low temperature
   d. low O_2

12. Carbonic anhydrase catalyzes the reaction between:
   a. water and carbon dioxide.
   b. the dissociation of carbonic acid to bicarbonate ions and protons.
   c. the combination of hydrogen ions and hemoglobin.
   d. carbon dioxide and hemoglobin to form carbaminohemoglobin.

13. Blood flow in the lung:
   a. is equal to the total blood flow to the rest of the body.
   b. goes mostly to the upper lung.
   c. is less per minute than in the rest of the body.
   d. is very high in the fetus.

14. Oxygen removal from the alveolar gas can be facilitated by:
   a. pulmonary edema
   b. capillary recruitment
   c. capillary transit times greater than 60 seconds
   d. anemia

15. The pulmonary and systemic circulations have the same:
   a. systolic pressure
   b. volume flow of blood per minute
   c. mean pressure
   d. diastolic pressure during rest

16. Hypoxic vasoconstriction:
   a. helps maintain ventilation-perfusion balance.
   b. directs blood away from hypoxic alveoli.
   c. causes a rise in pulmonary artery pressure at high altitude.
   d. all of the above.

17. Approximately how long does an average red blood cell spend in the pulmonary capillaries during resting conditions?
   a. 60 seconds
   b. 1 second
   c. 0.25 second
   d. 5 seconds

18. The pulmonary circulation:
   a. acts as a sieve for emboli.
   b. converts angiotensin to its active form.
   c. has a large surface area.
   d. all of the above.

19. In the fetus, which of the following is true?
   a. Arterial oxygen tension closely approximates maternal oxygen tension.
   b. All cardiac output flows through the pulmonary circulation.
   c. Blood flows from the right to the left side of the circulation through the patent ductus arteriosus.
   d. All blood flows through the placenta.

20. Carbon dioxide is transported in the blood predominantly in the form of:
   a. physically dissolved in plasma
   b. carbaminohemoglobin
   c. carbonic acid
   d. bicarbonate ions

Calculate

21. What is the partial pressure of oxygen at sea level?

22. At a resting cardiac output of 5 L/min, how much carbon dioxide is delivered to the lungs to be exhaled each minute?

23. If the top of the lung is 15 cm above the heart, what pressure in mmHg must the right ventricle generate to pump blood to the top of the lung?

24. What is the partial pressure of oxygen in Denver, Colorado?

25. If the arteriovenous oxygen difference is 7 vol.% in an individual at rest, and the cardiac output is 6 L/min, how much oxygen is being delivered to the tissue each minute?

Answer Each Question in One or Two Sentences

26. What is the function of the pulmonary lymphatics?

27. Describe the metabolic functions of the pulmonary circulation.

28. What are the blood pressures in the pulmonary circulation?

29. How does the pulmonary circulation maintain ventilation-perfusion balance?

30. What are the changes that occur in the pulmonary circulation during exercise?

SUGGESTED READINGS

Altman, L.K. *Who Goes First?* New York: Random House, 1987.

Forster, R.E., DuBois, A.B., Briscoe, W.A., and Fischer, A.B. *The Lung*. Chicago: Year Book Medical Publishers, 1986.

Guyton, A.C. *Textbook of Medical Physiology*. Philadelphia: W.B. Saunders, 1985.

Mines, A.H. *Respiratory Physiology*. New York: Raven Press, 1981.

Simon, T. *The Heart Explorers*. New York: Basic Books, 1966.

Taylor, A.E., Rehder, K., Hyatt, R.E., and Parker, J.C. *Clinical Respiratory Physiology*. Philadelphia: W.B. Saunders, 1989.

West, J.B. *Respiratory Physiology*. Baltimore: Williams & Wilkins, 1990.

KEY TERMS

alveolar-capillary membrane (p. 742)
angiogram (p. 734)
capillary recruitment (p. 740)
capillary transit time (p. 740)

carbon dioxide dissociation curve (p. 746)
carbonic anhydrase (p. 745)
chloride shift (p. 745)
gas diffusion (p. 740)
gas uptake (p. 742)
Haldane effect (p. 746)

hypoxemia (p. 738)
hypoxia (p. 738)
oxygenated blood (p. 738)
oxyhemoglobin (p. 743)
oxyhemoglobin dissociation curve (p. 743)

ventilation-perfusion balance (p. 738)
wedge pressure (p. 738)

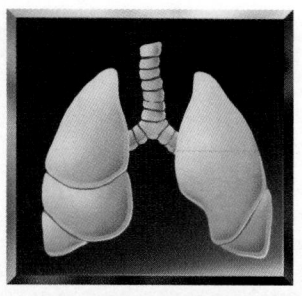

C H A P T E R
22

Respiration in Unusual Environments

For much of our history, humans were restricted to breathing atmospheric air. The largest variations in what we breathed were encountered on ascents to high altitude, where alveolar oxygen tension dropped as barometric pressure fell. There were also some cases, beginning in Grecian times, of underwater activities in primitive diving bells, in which divers must have encountered air pressures greater than one atmosphere (1 atm). In the eighteenth century, the chemistry of gases blossomed, and individual gases such as oxygen, nitrogen, and carbon dioxide were isolated. Then, the study of unusual gases began in earnest. Joseph Priestley said of pure oxygen, "Only two mice and myself have had the privilege of breathing it." Today we breathe in high-oxygen environments, low-oxygen environments, polluted environments, and artificial environments such as those in space capsules and underwater devices. Various anesthetic gases relieve pain during surgery, and we have contrived all sorts of artificial breathing devices to keep us alive in hostile environments and during times when our natural ability to breathe is inadequate. This chapter deals with this interesting area of respiratory physiology.

Respiration in Low-Oxygen Environments

The condition of low oxygen **(hypoxia)** is most commonly encountered during travel to high altitudes. As the altitude increases, the barometric pressure decreases. The percentage of oxygen in the air, however, remains constant at near 21%. The decreasing barometric pressure causes the partial pressure of oxygen to fall (Table 22–1). This fact was not appreciated until the last century, when the French physiologist Paul Bert was shocked by the deaths of balloonists who ascended to altitudes of more than 15,000 feet (Fig. 22–1). His lengthy scientific work prompted by this tragedy was published in the classic book *La Pression Barometricque*, in which he emphasized, "Oxygen tension is everything; barometric pressure in itself is nothing or almost nothing." Bert has been acclaimed as the founder of aerospace medicine; he also did important work on the effects of high pressure experienced by divers.

Figure 22–1
The first successful aerial devices were hot air balloons. Although dangerous, these balloons issued in the age of aviation.

Immediate Effects of Low Oxygen

The detrimental effects of hypoxia are exaggerated if the exposure is sudden. For example, a pilot who loses his or her oxygen supply at 30,000 feet has just over one minute to descend to lower altitude before

Table 22–1

Effects on Alveolar Gas Concentrations and on Arterial Oxygen Saturation of Acute Exposure to Low Atmospheric Pressure

Altitude (ft)	Barometric Pressure (mm Hg)	Po₂ in Air (mm Hg)	Breathing Air		Breathing Pure Oxygen	
			Po₂ in Alveoli (mm Hg)	Arterial Oxygen Saturation (%)	Po₂ in Alveoli (mm Hg)	Arterial Oxygen Saturation (%)
0	760	159	104	97	673	100
10,000	523	110	67	90	436	100
20,000	349	73	40	70	262	100
30,000	226	47	21	20	139	99
40,000	141	29	8	5	58	87
50,000	87	18	1	1	16	15

losing consciousness. In contrast, after careful acclimation, several climbers have been able to reach the summit of Mount Everest, nearly 30,000 feet, without the use of supplemental oxygen. This amazing achievement emphasizes that the body is much more able to tolerate hypoxia if the ascent is slow. The ideal of successful **acclimatization** through slow ascent is highlighted repeatedly in the literature on high altitude and mountain climbing.

Since millions of people travel to high altitudes annually for recreation, the acute effects of hypoxia are of both physiological and clinical interest. Some effects are present at even moderate elevations of 5000 feet, an altitude at which major cities are found throughout the world. There are noticeable changes in ventilation, especially during exercise, even at these relatively low elevations. Acclimatization occurs over a matter of weeks to months (Fig. 22–2). Most residents of elevations of 5000 feet soon do not notice any effect of the altitude.

Human respiration, as pointed out in Chapter 20, is very well adapted for gas exchange under normal partial pressure of oxygen. However, human respiration does not function as well as that of birds that fly at high altitudes or fish that live in low-oxygen environments. The reason is that humans have tidal flow with concurrent exchange in their lungs, while birds have continuous flow with crosscurrent exchange, and fish have continuous flow with countercurrent exchange. Humans approach continuous flow only with *hyperventilation.*

As the altitude increases, however, the body makes noticeable efforts to deliver normal amounts

of oxygen to the tissues. Chief among these responses to altitude is **hyperventilation**—that is, deeper breaths taken more rapidly. The sequence of events for hyperventilation is as follows: (1) a fall in inspired oxygen tension, (2) decreased alveolar oxygen tension, (3) decreased arterial oxygen tension, (4) increased firing of peripheral chemoreceptors (the carotid bodies), (5) hyperventilation, and (6) increased alveolar and arterial oxygen tension (see Chapters 20 and 21). The loop is regulated in part by

Figure 22–2

Approximation of the acclimatization changes that take place during a stay at an altitude of 14,000 to 16,000 feet. (From Houston, C. S. *Hypoxia: Man at Altitude.* New York, Thieme-Stratton, 1982.)

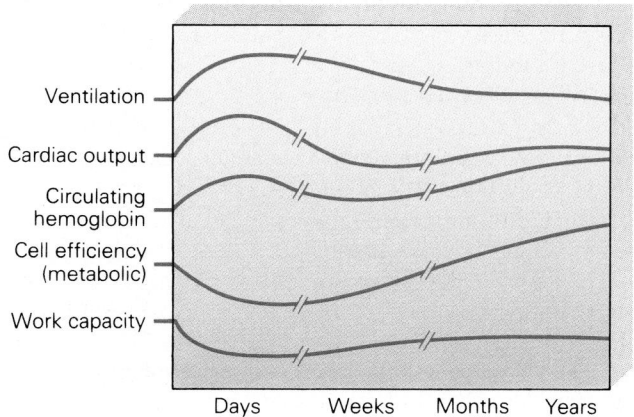

a negative feedback limb: (1) hyperventilation, (2) elimination of alveolar carbon dioxide, (3) decreased arterial carbon dioxide tension, (4) decreased arterial and cerebrospinal fluid (CSF) pH, (5) decreased firing of both central and peripheral chemoreceptors, and (6) decreased ventilation. Ventilation increases immediately upon ascent and stays elevated for many weeks (see Fig. 22–2). It is important to realize that there is an immediate increase in cardiac output to match the increased ventilation. Another long-term effect is an increase in blood volume.

Sometimes, newcomers to high altitude develop **acute mountain sickness.** This disorder is most common after rapid ascent to altitudes above 7000 or 8000 feet, when too much carbon dioxide is blown off during hyperventilation, and consists of headache, nausea, vomiting, sleep disturbances, and breathlessness (dyspnea). A drug called acetazolamide (Diamox), although still in the experimental stages, offers hope in the treatment of these symptoms. It is thought to work by interfering with the catalytic effect of carbonic anhydrase on the reaction of carbon dioxide and water in the blood (see Chapter 21), which helps overcome the alkalemia (basic conditions) of hyperventilation and ameliorates acute mountain sickness.

At higher altitudes (9000 to 10,000 feet), a rarer disorder called **high-altitude pulmonary edema** can occur. The cause of the edema is not clear, but fluid is known to leak from the pulmonary microvessels, forming edema and causing dyspnea, cough, weakness, headache, stupor, and (rarely) death. The disorder is treated with supplemental oxygen; even more important is a rapid descent to low altitude. The disease clears rapidly (in hours or days) at lower altitudes.

A more serious disorder is **high-altitude cerebral edema,** which occurs above altitudes of 10,000 to 12,000 feet. This disease is characterized by severe headache, hallucinations, problems with muscular coordination (ataxia), weakness, impaired mental ability, stupor, and death. Immediate descent is mandatory. Fortunately, high-altitude cerebral edema is rare. Altitude illnesses are summarized in Table 22–2.

Table 22–2
Altitude Illness

Acute hypoxia	Mental impairment and usually collapse after *rapid* exposure above 18,000 feet. Rare in mountains.
Acute mountain sickness	Headache, nausea, vomiting, sleep disturbance, dyspnea. Above 7000 to 8000 feet. Self-limited. Common.
High-altitude pulmonary edema	Dyspnea, cough, weakness, headache, stupor, and rarely death. Above 9000 to 10,000 feet. Requires rapid descent or treatment early. Uncommon.
High-altitude cerebral edema	Severe headache, hallucinations, ataxia, weakness, impaired mentation, stupor, and death. Above 10,000 to 12,000 feet. Descent mandatory. Uncommon.
Subacute and chronic mountain sickness	Failure to recover from acute mountain sickness may necessitate descent. Some develop dyspnea, fatigue, and heart failure after years of asymptomatic residence at altitude. Rare.
Altitude-related problems	Retinal hemorrhage, edema, thrombophlebitis and embolism, cold injury.
Chronic conditions made worse at altitude	Sickle trait, chronic cardiac or pulmonary disease.
High-altitude deterioration	Long periods spent above 18,000 to 19,000 feet cause insomnia, fatigue, weight loss, and general deterioration. Permanent residence is not possible above this altitude. Deterioration is more rapid at even higher elevations.

From Houston, C.S. In: Hypoxia in Man at Altitude. Ed. by Sutton, J.R., Jones, N.L., and Houston, C.S. New York: Thieme-Stratton, 1982.

Long-Term Effects of Low Oxygen

The body undergoes other physiological changes to adjust to low environmental oxygen. One important adaptation is that more red blood cells are manufactured, which causes elevated hematocrit. Up to a certain point, this change enables the blood to carry more oxygen to the tissues (Fig. 22–3). However, the higher the hematocrit, the more viscous (thickened) the blood becomes, a change that makes it more difficult for the heart to pump. Normal hematocrit is near 45%. In some cases, high-altitude residents develop a condition called **polycythemia** in which hematocrits reach levels of over 70%. These individuals must have blood withdrawn periodically to reduce their hematocrit to a range that permits the heart to pump effectively.

The body increases ventilation, cardiac output, and circulating hemoglobin (see Fig. 22–2) during the first few weeks to months at high altitude. Over a long period, cardiac output and ventilation tend to return to low-altitude values, while hematocrits continue to climb. The tissue cells also change to become metabolically more efficient. For example, the mitochondria increase in number, and there are alterations of metabolic pathways. Even with these changes, the maximum work capacity never reaches sea-level values. Nevertheless, the adaptation of

Figure 22–4
Sherpas are mountain-dwelling people who have highly adapted to the low oxygen environment of high altitudes (Nepal). (© Steve Covello/Mountain Stock/1990.)

Figure 22–3
The oxygen-carrying capacity of blood of mountain dwellers is greater than that of sea level natives, due in part to an increase in hematocrit. (From ''Oxygen-dissociation curves for bloods of high-altitude and sea-level residents.'' PAHO Scientific Publication No. 140, *Life at High Altitudes,* 1966.)

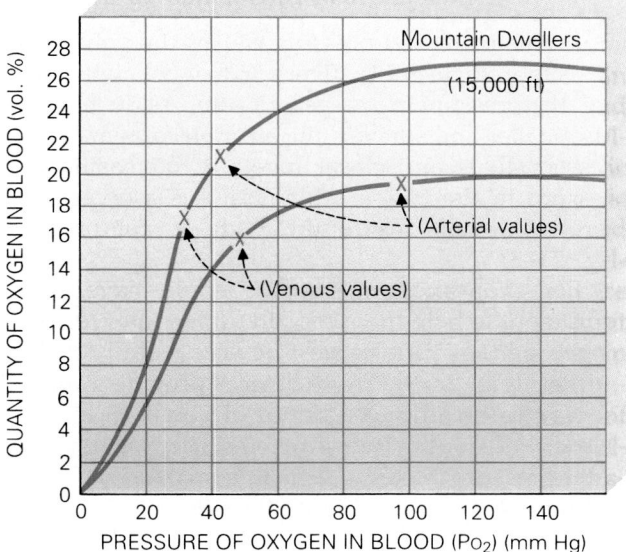

long-term residents can be impressive. In the Andes, the natives return from the high mines in the evenings to their "low"-altitude homes for a rousing soccer game at 15,000 feet; their team has an impressive home field advantage.

The best acclimatized of all high-altitude residents are the Sherpas in Nepal, whose climbing feats are legendary (Fig. 22–4). Other people—for example, the residents of high altitudes (10,000 feet) in Colorado—are often not as well adapted to low oxygen (Fig. 22–5). The percentage of Coloradans aged 60 years and older who live at high altitude is much lower than the percentage of older people who live at low altitude. The other older high-altitude residents tend to leave for health reasons, usually because of heart and lung disease (Fig. 22–6). The former residents report that their health is improved after they settle at low altitude (Fig. 22–7).

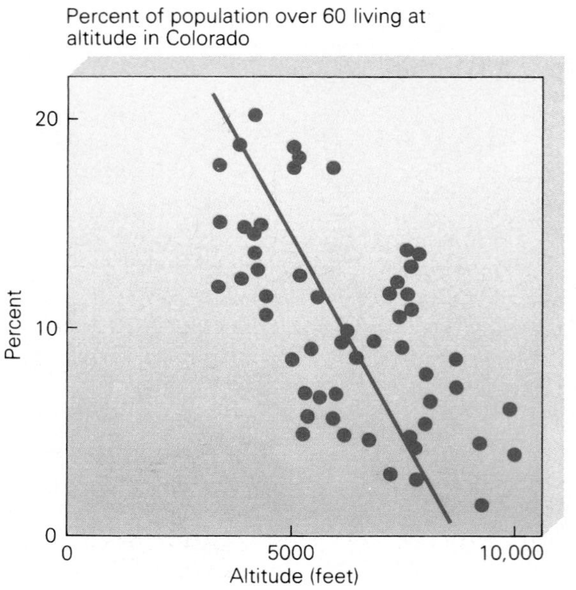

Percent of population over 60 living at altitude in Colorado

Figure 22–5
In Colorado, there is a progressive decrease in the percentage of elderly people living at high altitude. (Adapted from Regensteiner, J. G. and Moore, L. G. "Migration of the elderly from high altitudes in Colorado." *JAMA* 253:3124, 1985.)

Figure 22–6
The chief reason for migration of the elderly away from high altitudes is heart and/or lung disease. This study compared the migration patterns of a high-altitude and a low-altitude town of equal size and economic status. (Adapted from Regensteiner, J. G. and Moore, L. G. "Migration of the elderly from high altitudes in Colorado." *JAMA* 253:3124, 1985.)

People who moved because of heart and lung disease

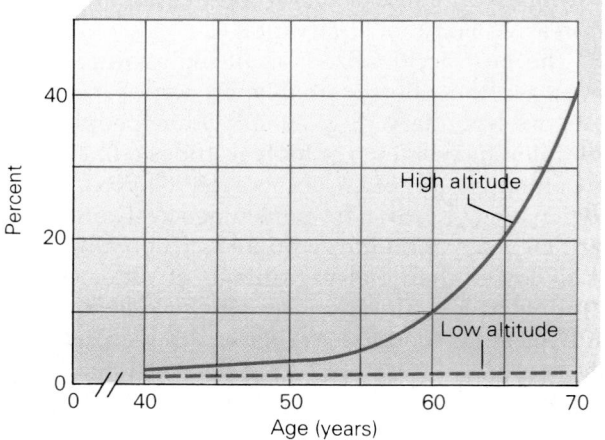

People who report health improvement after moving

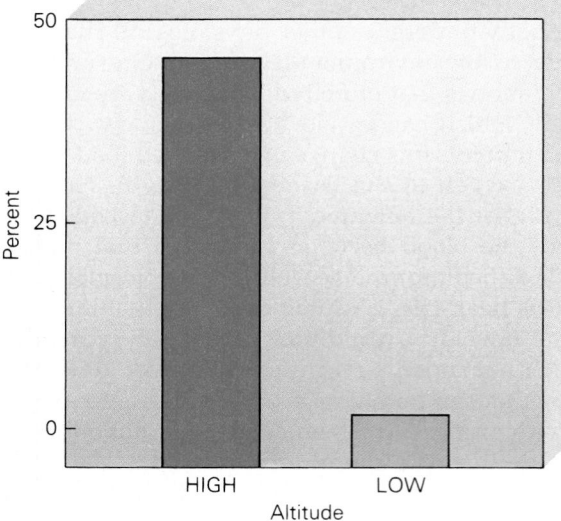

Figure 22–7
Of the high altitude migrants, about half reported improved health when they reached low altitude. (Adapted from Regensteiner, J. G. and Moore, L. G. "Migration of the elderly from high altitudes in Colorado." *JAMA* 253:3124, 1985.)

Respiration in Deep Sea Diving

Gas Pressures Under Water

As divers submerge under water, the pressure surrounding their bodies rises because of the weight of the water. To visualize how the pressure rise takes place, imagine diving with a balloon that is filled with air at a pressure of 1 atm. The deeper the balloon goes, the greater the water pressure pushing in on it. This increased pressure causes the volume of the balloon to decrease. Every 33 feet below the surface, the pressure increases by 1 atm. As the balloon gets smaller and smaller, the air molecules are compressed closer and closer together, increasing the pressure in the balloon. Similarly, as a diver submerges, the air inside the body is compressed (Fig. 22–8).

For thousands of years, people have used equipment to help them remain underwater for prolonged periods. The earliest devices were probably snorkels (Fig. 22–9). These devices provide a practical way to breathe to a depth of about 18 inches. Below that depth, two problems occur. First, the water pressure makes it difficult to expand the chest to draw in fresh air. Secondly, the snorkel adds dead space through which the diver must draw air.

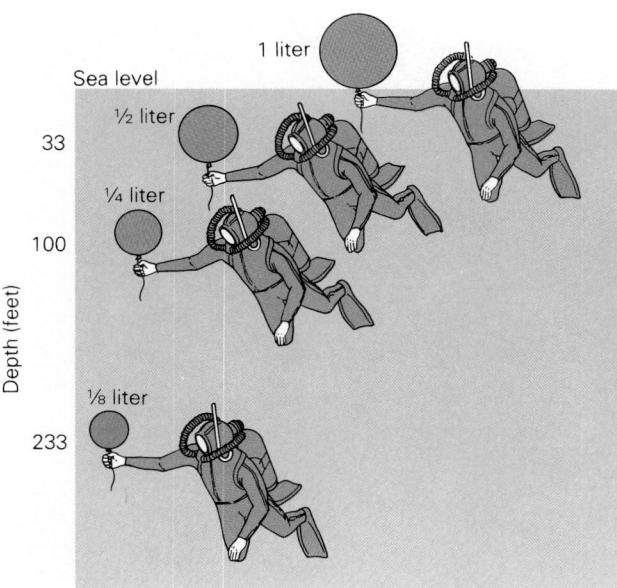

Sea level

33

100

Depth (feet)

233

1 liter

½ liter

¼ liter

⅛ liter

Figure 22–8
Air in a container is progressively compressed as the container is lowered into water, where the surrounding pressure increases.

Figure 22–9
Although a swimmer can remain submerged for long periods with a snorkel, the depth of a dive is limited because the snorkel increases respiratory dead space and because the increasing pressure on the chest makes it difficult to draw in fresh air. (© Maria Paraskevas)

If the dead space becomes too large, then the diver is simply moving air back and forth in the snorkel and does not get any fresh air.

The next apparatus to be invented was most likely the diving bell. These devices have been used for centuries. The first ones were simple, inverted chambers, open at the bottom, in which the diver stood and breathed the air in the chamber as it submerged (Fig. 22–10). These diving bells had many limitations, including air that became rapidly fouled by carbon dioxide. Also, the air volume was compressed to a smaller and smaller volume as the pressure increased during descent. Alexander the Great is said to have used a diving bell in the fourth century B.C. Modern diving bells are completely enclosed, so compressed air volume is no longer a problem. By the early nineteenth century, the enclosed diving suit had been invented (see Fig. 22–10). Fresh air was fed from the surface into the suit by a pump to sustain the diver for long periods under water.

During World War II, work began in earnest on the Self Contained Underwater Breathing Apparatus, "SCUBA" (Fig. 22–11). This kind of gear has become very popular for recreation, exploration, and work on underwater oil rigs and pipelines.

Figure 22–10
(*a*) Diving bells have been used since the time of the ancient Greeks. (*b*) During the eighteenth century, the advent of pumps allowed for the development of diving suits, which gave the diver greater mobility under water. (© The Bettmann Archive.)

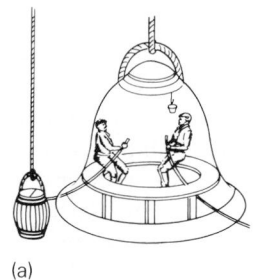

(a)

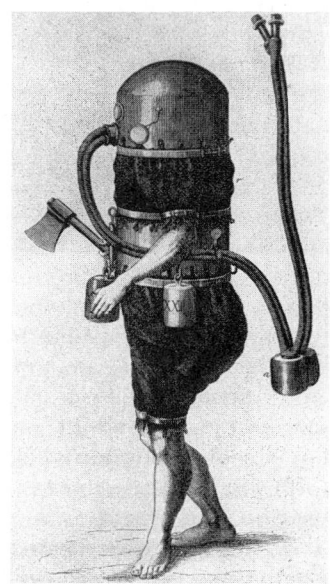

(b)

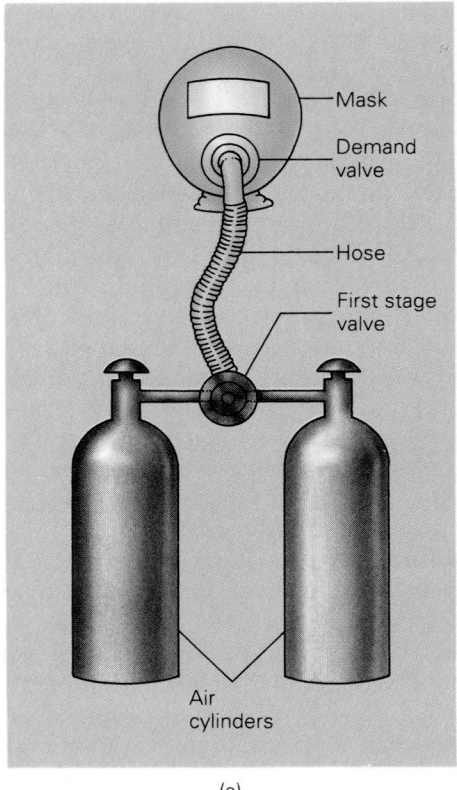

(a)

(b)

Figure 22–11
(*a*) The development of Self Contained Underwater Breathing Apparatus (SCUBA) has enabled many people to pursue underwater activities. It is critically important, however, to be cautious during deep or prolonged dives because gases such as nitrogen can be absorbed into the body and cause problems during both the dive and the ascent. (*b*) The diver in the photo makes a decompression safety stop during ascent from a dive of 120 feet. This stop facilitates the gradual release of nitrogen from the tissues, thereby preventing the bends. (Courtesy of C. Petree and J. Holtmeier)

Effects of High Pressures on Respiration

A characteristic of all these devices is that air enters the lungs at higher-than-normal pressure. Unfortunately, breathing such high pressures has some potentially serious consequences. The problems come not during the descent, but when the diver ascends. When the pressure becomes high enough, particularly at depths below 100 feet, significant amounts of nitrogen in the compressed air are absorbed into the tissue. The longer the time spent under water, the more nitrogen absorbed. As the diver rises to the surface, the nitrogen is released from the tissues. If the ascent is slow, the nitrogen can be picked up by the capillary blood and returned to the lungs to be exhaled. If the ascent is rapid, bubbles can form, both in the blood and in the tissue. When bubbles form near the joints, they cause agonizing pain. Since the diver bends over to relieve the pain, this syndrome has been called the "bends." The risk of the bends increases as both the depth of the dive and the length of time at depth increase. The treatment for the bends is to spend time in a compression chamber in which the air pressure is raised until the bubbles are compressed and the nitrogen redissolves into the tissue. Pressure in the chamber is gradually lowered, giving the nitrogen time to be exhaled through the respiratory system instead of forming bubbles in the body.

At depths below about 150 feet, the number of nitrogen molecules in the body rises to a point at which they have a narcotic effect. This causes "raptures of the deep." Under these conditions, divers can often make unintelligent, life-threatening decisions. One way they can avoid this is to switch from a nitrogen-oxygen mixture to a helium-oxygen mixture. Helium is less soluble in tissues than nitrogen, and so fewer helium molecules dissolve in the body. This reduces the narcotizing effects as well as the number of molecules that must be removed from the tissue during ascent. Also, helium has a smaller

Mask
Demand valve
Hose
First stage valve
Air cylinders

molecular size than nitrogen, so it diffuses more rapidly, which aids in elimination of the gas.

Marine Mammals: Champion Divers

Pearl divers in the Far East can work for one to two minutes at depths of 50 or 60 feet using a single breath. Yet this remarkable achievement pales beside the diving abilities of marine mammals. Seals, dolphins, and whales have developed numerous diving reflexes that assist in conserving oxygen. The blood supply to most tissues is reduced virtually to zero by powerful vasoconstriction (Fig. 22–12). The heart rate slows so that myocardial oxygen consumption decreases and permits a decrease even in coronary blood flow. Only the blood vessels that supply the brain are unaffected by the diving reflex, so that blood flow is maintained at normal levels.

Figure 22–12
When a seal submerges, the diving reflex causes intense vasoconstriction in most of the body and reduces blood flow. (Adapted from K. Schmidt-Nielsen. *Animal Physiology.* Cambridge, Cambridge University Press. 1983.)

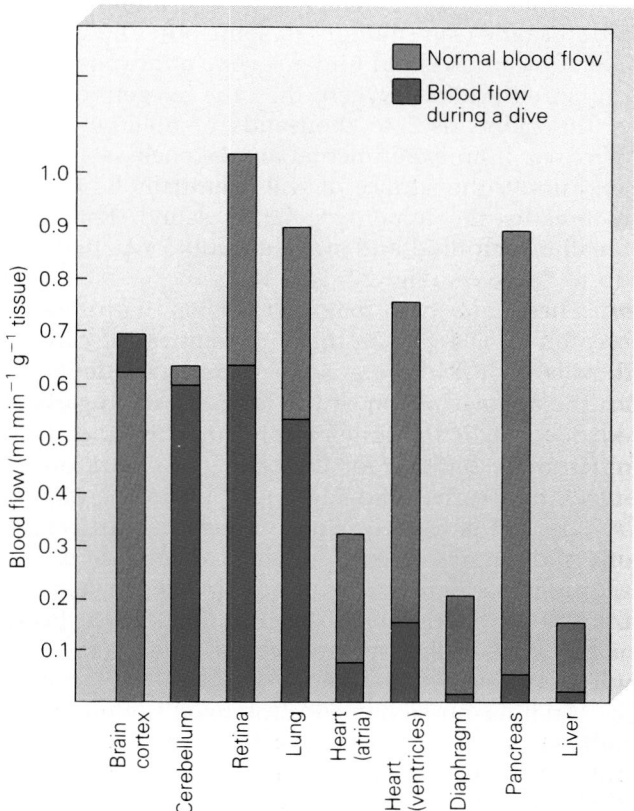

The reduced demand for blood flow permits cardiac output to be reduced; at a depth of 300 feet, the seal's heart rate is only 4 beats per minute; on the surface, its heart rate is 120 beats per minute.

Surprisingly, the seals exhale just before they dive. In Weddell seals that were monitored on free dives of up to 23 minutes and depths of 700 feet, nitrogen uptake from the alveoli was found to cease at depths of 90 feet, because the gas-containing alveoli were collapsed shut. The gas was compressed into nonexchanging airways. The seals use this technique to avoid the bends.

Whales have the best diving performance of all marine animals. These animals have been known to spend more than an hour at tremendous depths. The record is 3717 ft, recorded by an unfortunate sperm whale that became entangled in a trans-Atlantic cable broken at that exact depth. The whale dives so well because its metabolism is low and because it carries a large reserve of oxygen in its muscle. Also, the whale and other diving mammals can tolerate much higher levels of carbon dioxide than nondiving animals.

Respiration in Submarines

The first practical submarines were built in the 1800s, although crude, hand-propelled models were fabricated in the 1700s. By the start of World War I, Britain had 77 submarines, France 45, the United States 35, Germany 29, and Russia 28. The submarines in the early 1900s could only stay submerged for several hours and operated exclusively in shallow coastal waters. The effectiveness of submarines in warfare made them popular with every major power. To date, thousands of these vessels have been built.

One of the unpleasant aspects of these machines is that when they sink, they usually take all hands aboard and trap them below. That grim reality has prompted much work on methods of escape. All sorts of scuba-type gear have been developed, but usually to no avail. During World War II, only five men were saved by mechanical lungs. Ironically, a simple, excellent escape technique effective to depths of 300 feet had been known for nearly a century. It was used by Wilhelm Bauer in 1851 when his hand-propelled submarine sank in 60 feet of water (Fig. 22–13) and the hull began to collapse from the pressure. Bauer waited until the air pressure in the vessel was equalized with the water pressure and then opened the hatch. The captain and his two crew members by then were breathing air at 3 atm, but not for long enough to build up nitrogen pressure in the tissue. Thus, they could

Figure 22–13
Wilhelm Bauer's submarine sank in 60 feet of water, but the captain and crew escaped by swimming easily to the surface. (Photo courtesy of: Submarine Force Library and Museum, Groton, CT)

ascend very rapidly without experiencing the bends. As they rose to the surface, the compressed air in their lungs expanded. This caused them to exhale steadily during the ascent. They never ran out of air even though they exhaled the whole time. As Captain Bauer explained, "We came to the surface like bubbles in a glass of champagne."

This technique was ignored and forgotten. It was accidentally repeated in several instances during World War II, but was never developed. Instead, with technology gone awry, navies continued to develop impractical mechanical contrivances that did not work, while all the time the body was superbly adapted to escape. In the 1950s, the British and U.S. navies recognized that the best technology for escape was no technology. Water-filled escape training towers have been built, and unaided escapes at depths to 300 feet are a routine part of submarine training.

One step has been added to the natural escape method: sailors now inflate small life vests at the moment of departure. This extra buoyancy causes them to ascend very rapidly to the surface. The decompression is so rapid under these circumstances that the sailors exhale as fast as they can. The air they breathed while at 300 feet is so highly compressed that it expands rapidly in their lungs as they ascend and, even with continual maximal forced expiration, they never run out of air on the way up. In navy jargon, this technique is known as "blow and go."

"Homo Aquaticus"

Two methods in the research stage offer promise for underwater use. One is the actual breathing of liquid. If saline (salt solution) or some other fluid such as fluorocarbon is put under several atmospheres of pressure by 100% oxygen, then the oxygen tension of the saline rises to thousands of millimeters of mercury. If an experimental animal such as a rat is held under the surface, it will eventually be forced to breathe the hyperoxygenated saline. Rats can breathe such fluid and swim around for periods of up to 18 hours (Fig. 22–14).

There are two major problems with liquid breathing. The first is that the removal of carbon dioxide is difficult because the respiration depends on the bellows action of the lungs to wash carbon dioxide out. If the lungs are filled with fluid, it is much more difficult for the bellows to work effectively, and carbon dioxide builds up in the blood. The second is that the fluid washes the surfactant out of the lungs, leading to alveolar collapse when the animal attempts to return to air. This makes the transition from fluid back to air difficult. It has been done successfully, however, by a number of rats, mice, and even some dogs.

Another possibility for prolonged undersea activities is the development of artificial gills. This is not a small engineering feat, but it seems possible. If a gill could be developed that could trail behind a diver while being held in the mouth, its large sur-

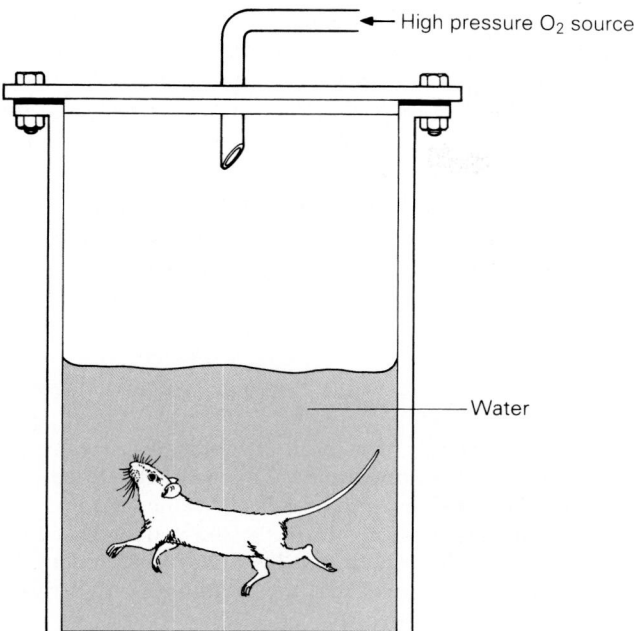

Figure 22–14
Small animals can survive for hours under water if the oxygen tension is raised to high levels.

face area could extract oxygen from the water and excrete carbon dioxide just as the gills of fish do. Animals such as the mudpuppy have external gills as well as lungs (Fig. 22–15). They use the gills to stay successfully under water for prolonged periods. Liquid breathing techniques, artificial gills, or some similar method might lead to the development

Figure 22–15
Animals like the mudpuppy have external gills as well as lungs. The external gills are used to extract oxygen from water, allowing the animal to remain under water for a longer time period (Tiger salamander larva). (© R. Calentine/Visuals Unlimited.)

of what Jacques Cousteau describes as *Homo aquaticus*, an underwater man capable of living indefinitely beneath the surface of the sea.

Respiration in High-Oxygen Environments

The usual problem in breathing is getting enough oxygen into the body. However, with modern chemistry and technology it is possible to breathe 100% oxygen and actually get too much. High-oxygen environments are often found in hospitals, where oxygen is administered to patients with oxygen deficiency, and in aviation and space flights, where the pilot or astronaut breathes air with high oxygen concentration because of extremely low oxygen tension.

Atelectasis: Collapse of Alveoli

One of the physical dangers of breathing pure oxygen is **atelectasis** (collapsed alveoli), which can occur in pilots, astronauts, and hospital patients. This physical complication occurs when small airways become blocked by mucus; the oxygen in the plugged alveoli gets absorbed, and the alveoli rapidly collapse (Fig. 22–16). The total pressure of the trapped gas is close to 760 mm Hg, but some of the individual gas pressures in the blood are much below 760 mm Hg. If gas is trapped in the course of breathing air (see Fig. 22–16), nitrogen acts to deter collapse because it is not utilized and therefore is not absorbed. However, when one is breathing pure oxygen, there is no nitrogen; hence, oxygen is continuously absorbed, and the alveoli collapse (Fig. 22–16*a*, *b*).

Postoperative atelectasis occurs commonly in patients who are ventilated with high-oxygen mixtures. Atelectasis is also prevalent in fighter pilots who are breathing high oxygen and experiencing large gravitational forces (see the section on space travel). The added gravitational effects augment alveolar collapse. Atelectasis is most likely to occur at the bottom of the lung where alveoli are poorly expanded and airways are partially closed. Reopening atelectatic areas is difficult because the surface tension effects are more pronounced on small units. Atelectasis can be reopened by deep breaths.

Toxicity Associated with Breathing High Oxygen

The second problem associated with high oxygen, much more serious than atelectasis, is toxic effects. These occur when a person breathes high concentrations of oxygen for more than one day. The primary target organ for oxygen toxicity is the lung, where damage can occur within 24 hours and can cause pulmonary edema.

Free-Radical-Induced Toxic Effects

The first site of injury is the pulmonary capillary endothelial cells, which result in leaky pulmonary capillaries. The mechanism for tissue damage is the formation of **free radicals.** These highly reactive molecules are formed as byproducts from molecular oxygen. Before molecular oxygen (O_2) can be used in the oxidation of fuels for energy, it first must be converted to "active forms." These active forms are called **oxidizing free radicals,** and the most important one in the cell is the **superoxide anion radical** ($O_2^- \cdot$). These free radicals are continuously being produced from dissolved molecular oxygen under normal oxygen tensions (tissue P_{O_2} 40 mm Hg).

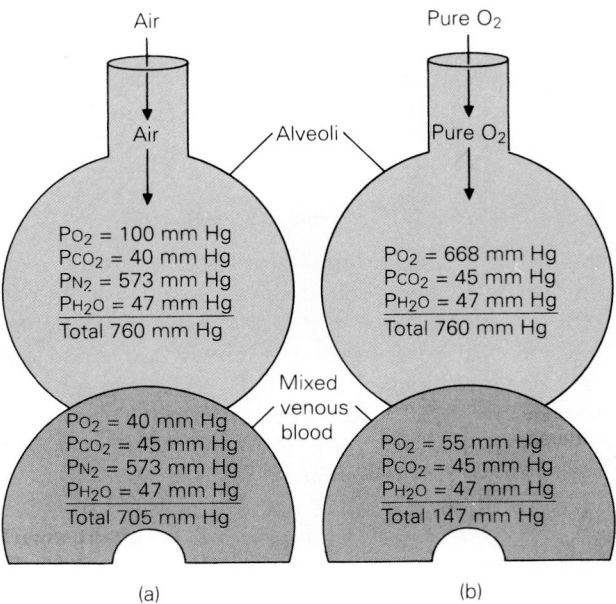

Figure 22–16
Cause of atelectasis as a result of a plugged airway. When (*a*) air is breathed and (*b*) oxygen is breathed, alveolar pressure in both cases is the same, but the sum of the gas pressures in the mixed venous blood is smaller. (*c*) Atelectasis caused by airway obstruction. Trapped air is absorbed into the pulmonary capillaries, causing alveoli to collapse.

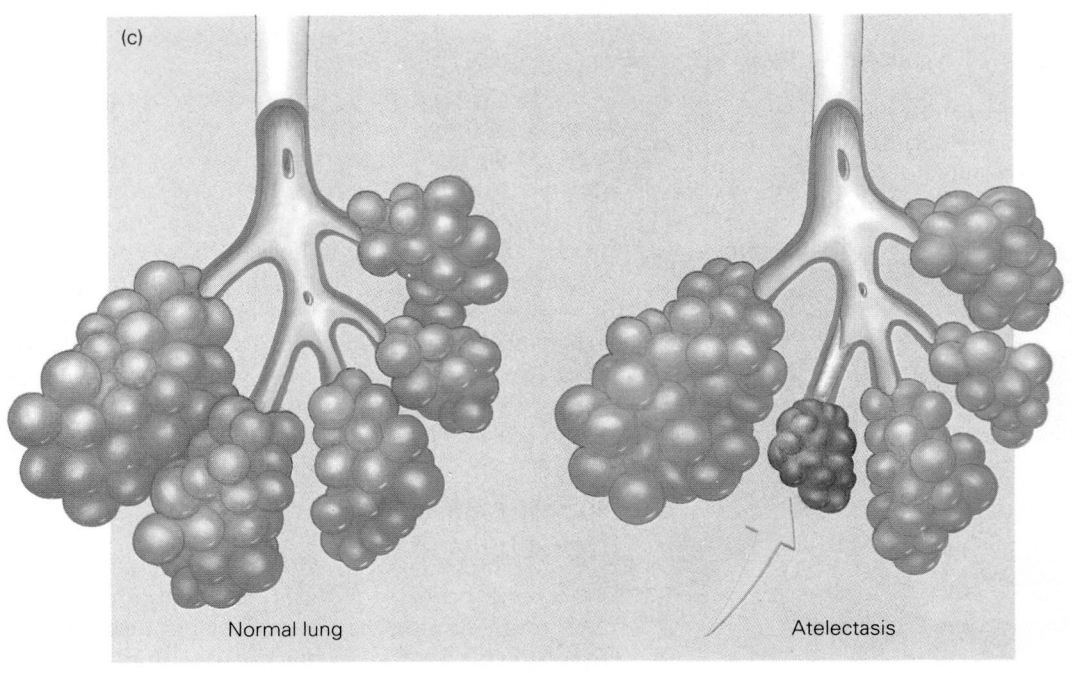

However, under these normoxic conditions, the free radicals are rapidly removed by protective cellular enzymes—namely, **superoxide dismutase, catalase,** and **peroxidase.** As long as Po_2 is normal, the oxidizing free radicals are rapidly removed and no cellular damage occurs.) When tissue Po_2 is greatly increased—such as in breathing of high oxygen—excess free radicals are formed that overwhelm the protective enzymes. As a result, these free radicals cause destructive and lethal cellular effects by (1) oxidizing membrane lipids, thus altering essential properties of membranous structures of the cells; (2) oxidizing cellular proteins and DNA; and (3) oxidizing enzymes, thus shutting down cellular respiration.

High oxygen is not the only source of oxidant damage from free radicals. Breathing ozone (O_3) or nitrogen dioxide (NO_2) from polluted air will also cause free radical injury to the lung. Some herbicides, such as paraquat, are extremely toxic because of the free-radical-induced injury. If someone smokes tobacco or marijuana that has been treated with paraquat, the paraquat in the smoke will react with lung cells to form free radicals. Crop dusters and migrant workers are also at risk because of the exposure to paraquat through the lungs or skin.

Retrolental Fibrosis

Another hazard of breathing high concentrations of oxygen is seen in intensive care units in which premature infants develop blindness from **retrolental fibrosis.** The mechanism for the blindness is deterioration of the retina and inadequate development of retinal blood vessels. The poor vasculation leads to the formation of fibrosis tissue behind the lens (retrolental fibrosis). If the Po_2 is kept below 140 mm Hg in the incubator, retrolental fibrosis can be avoided.

Toxic Effects of High-Pressure Oxygen

A third type of oxygen toxicity occurs when oxygen is breathed at pressures exceeding 760 mm Hg. Oxygen is breathed at pressures greater than 1 atm in deep sea diving or in hyperbaric (high-pressure) chambers. When oxygen is breathed at 2 to 4 atm, it stimulates the central nervous system (CNS), leading to convulsions. The exact cause for the toxic effect on brain tissue is not fully known, but the high oxygen pressure is thought to inactivate dehydrogenase enzymes. The convulsions are often preceded by twitching of the face, ringing in the ears, and nausea. Convulsions can occur when an amateur scuba diver inadvertently uses a tank filled with O_2.

Respiration in Unusual Gaseous Environments

High Concentrations of Carbon Dioxide and Respiration

Breathing high carbon dioxide levels usually does not occur. Normal Pco_2 in the air we breathe has a partial pressure of 0.3 mm Hg (0.04%), which is negligible. However, in certain types of diving gear and closed environments (e.g., submarines and spacecraft), carbon dioxide can build up and be rebreathed because of defects in equipment. Rebreathing higher-than-normal levels of carbon dioxide will stimulate ventilation. A buildup of an alveolar Pco_2 of 80 mm Hg, which is about twice normal, is the maximum that an individual can tolerate. With these high concentrations, hyperventilation is maximum, and the individual becomes severely acidotic. When Pco_2 levels are greater than 80 mm Hg, the situation becomes intolerable, and the respiratory centers are no longer stimulated but instead are inhibited. Respiration begins to fail, and acidosis becomes more severe. Eventually, the CO_2 acts like a narcotic, and the individual becomes lethargic and eventually unconscious.

Carbon Monoxide and Respiration

Carbon monoxide (CO), unlike carbon dioxide, does not stimulate respiration. The problem with carbon monoxide is that it is an odorless, colorless, and poisonous gas coming from incomplete combustion of fuels that cannot be detected by the body. Carbon monoxide becomes lethal because it competitively binds at the same site as oxygen on the hemoglobin molecule to form **carboxyhemoglobin (COHb).** The formation of COHb interferes with the transport of oxygen to the cells. The reaction (Hb + CO $\rightleftharpoons$ HbCO) is reversible and is a function of the partial pressure of CO. This means breathing a higher concentration of CO will shift the reaction to the right to form more HbCO, and breathing lower concentrations of CO will shift the reaction to the left to form less HbCO. The problem is further complicated because the binding affinity of CO for hemoglobin is about 210 times that of O_2. Because of this greater tendency for binding to Hb, even trace amounts of CO in the respired air can be extremely dangerous. For example, an alveolar Pco of 0.5 mm Hg (1/210 that of alveolar Po_2) will cause half the hemoglobin in the blood to bind with CO and half to bind with O_2. Under these conditions, the amount of oxygen carried by the blood to the cells is halved.

Thus, a Pco of 0.7 mm Hg (0.1% of air) can be lethal because the cells become **anoxic** (deprived of oxygen). The first organ to become sensitive to the lack of oxygen is the brain. Altered vision, reaction times, and coordination are the first signs of CO poisoning and often explain why there are more accidents on the freeways when CO levels rise. The real danger of CO poisoning is that it cannot be detected in the body. With hypoxemia, there are physical signs; the individual will turn blue around fingertips and lips. However, the formation of HbCO makes the blood crimson red and masks the bluish tint of hypoxemia. The second problem is that there are no chemoreceptors to detect CO. The problem is compounded because the chemoreceptors that detect changes in Po_2 are not stimulated. The reason for the latter is that Po_2 in blood is normal with CO poisoning, yet O_2 cannot bind with hemoglobin.

An individual severely poisoned with CO can best be treated by administration of 100% O_2. Administering pure oxygen elevates the blood Po_2, which displaces CO from the hemoglobin. Also beneficial is the simultaneous administration of 5% to 7% carbon dioxide. Breathing the high concentration of carbon dioxide stimulates ventilation and reduces alveolar CO, which concomitantly releases CO from the blood. When the combination of high oxygen and carbon dioxide is breathed, CO can be removed 10 to 15 times faster than when just fresh air is breathed.

Anesthetic Gases and Respiration

Without anesthesia, standard surgical procedures such as appendectomies, cardiac surgery, and neurosurgery could not be done. Until the 1930s, all anesthetics were administered via the lungs.

Prior to the introduction of anesthesia in the nineteenth century, surgical pain was relieved by alcohol or opium. For over 2000 years, opium had been known for its narcotic effects, but was first used as an anesthetic in the ninth century A.D. The patient inhaled the opium vapor, which diffused across the alveolar membrane as would oxygen to cause drowsiness and deaden the pain. However, many deaths resulted because of overdoses that depressed the respiratory centers and stopped breathing. It was not until the 1800s that effective, safe opium administration became possible, thanks to a German pharmacist who isolated the main alkaloid (morphine) from opium.

Although Joseph Priestley discovered nitrous oxide in 1772, Humphry Davy was the first to recognize its anesthetic properties in 1799; he has been given credit for the first anesthetic gas. However, it was not until 1884 that nitrous oxide was actually used as an anesthetic by Horace Wells, a dentist from Connecticut. Nitrous oxide is commonly called "laughing gas" because it literally causes individuals to laugh and feel happy. Actually, for almost one hundred years after its discovery, nitrous oxide was not taken seriously. It was very popular at parties in the nineteenth century because of its pleasurable, intoxicating effects. Another dentist in the 1800s suggested the use of a second gas, **ether,** as an anesthetic. Ether was used for the first time at Massachusetts General Hospital in 1846 and subsequently became very popular. Later, **chloroform** was introduced as an anesthetic by a British physician who used it first for women in childbirth.

The discovery of the anesthetic properties of ether, nitrous oxide, and chloroform satisfied the immediate needs of surgery despite their toxic effects and the possibility of explosion in the case of oxygen. However, in the 1950s, the fluorinated and halogenated anesthetics were discovered. These newer anesthetics were a major improvement over their predecessors because they had a less depressant effect on respiration and were neither flammable nor toxic.

There are three stages of anesthesia with inhaled anesthetics. The first stage is **analgesia,** in which the response to pain is reduced. The second stage is the **loss of consciousness.** The third stage is **surgical anesthesia,** in which all sensations are lost, motor functions are inhibited, and regular automatic breathing commences.

Respiration in Space Travel

Because of the development of spacecraft and supersonic planes, the human body can now be subjected to acceleratory forces resulting from sudden changes in velocity or direction. For example, a spacecraft must attain a speed of 25,000 miles/hr to escape the earth's gravitational field and to leave its atmosphere. High acceleration rates are required to attain these speeds. When an aircraft or spacecraft takes off, the human body inside is subjected to **positive acceleration.** (Inertial forces move in a direction that may be described as head-to-foot with respect to the pilot.) When an aircraft or spacecraft slows down at the end of a flight, **negative acceleration** occurs (inertial forces move in a foot-to-head direction).

High G Forces and the Human Body

Any time an aircraft turns, centrifugal and acceleration forces come into play as a result of gravitational attraction. The human body's sensation of weight is due to **gravitational attraction** and is referred to as **G force.** The G force is a unit used to express the magnitude of the acceleratory force. One G is equal to the force exerted by one times the normal gravitational force. For example, when an individual is sitting in a chair, the force that presses the body against the chair seat is equal to 1 G (one times the pull of gravity). If the force that presses against the seat is four times the normal weight of the body, the force is 4 G. If force is less than the body weight, then the force is a negative G. Positive and negative G forces are illustrated in Figure 22–17, in which an aircraft pulls out from a dive or goes through an outside loop so that the pilot is held down by the seat belt. In the dive, the pilot experiences positive G force, with inertial forces acting in a head-to-foot direction. In the loop, the pilot experiences a negative G force, in which the inertial forces are directed from foot to head.

The physiological effects of these acceleratory forces are due to the fact that some tissues and blood in particular are "loose" with respect to body. When a positive G force is exerted in a head-to-foot direction, blood is pulled toward the lower parts of the body. In Figure 22–17, when the pilot "pulls 4 G's," venous pressure in the leg veins is four times the normal pressure. The heart cannot pump unless venous blood is returned, and consequently there is considerable pooling of blood in the lower extremities. At 4 G, a pilot in the seated position experiences a fall in arterial systemic pressure, at the level of the heart, of approximately 40 mm Hg. Under these conditions, blood flow to the brain almost stops. Consequently, loss of vision, referred to as **blackout,** occurs. Other symptoms are also prevalent. At 3 G to 4 G, a pilot or astronaut experiences difficulty in the use of muscles. Facial tissue sags

Figure 22–17
Illustration of positive and negative G forces on an airplane.

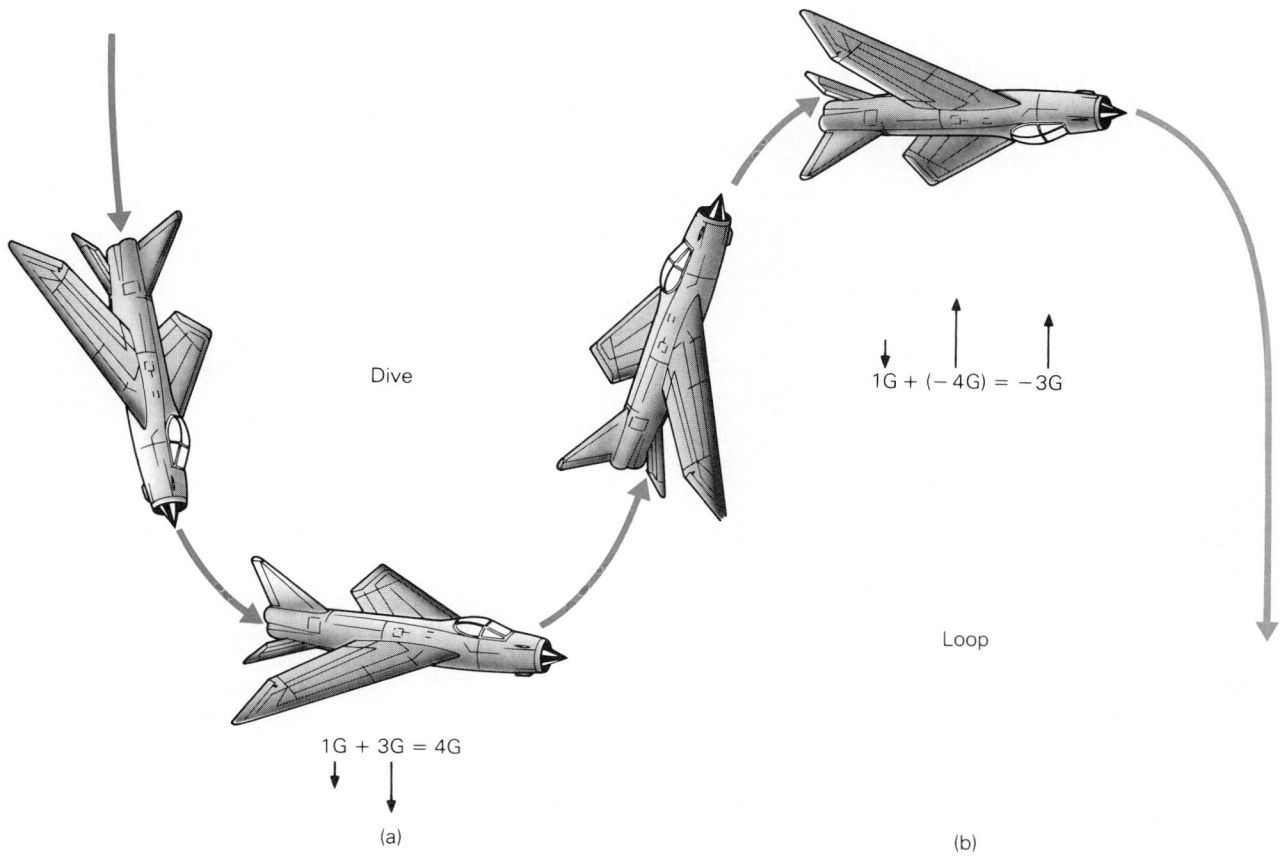

Dive

1G + 3G = 4G

(a)

1G + (− 4G) = −3G

Loop

(b)

R E S E A R C H   F O C U S

Space Shuttle Columbia: First Space Laboratory Dedicated to Life Science Research

When the United States made the decision to explore space, many Americans were very skeptical whether humans could live in outer space and many raised questions. Could cosmic particles affect the brain? Did space harbor foreign organisms that could be lethal to humans back on earth? Could humans live for long periods of time in weightlessness? Since the National Aeronautics and Space Administration (NASA) was formed over 30 years ago, we have learned that people can indeed survive, and more importantly, can live and work productively in space.

Successful exploration of space depends on the health and well-being of astronauts who travel and work there. For this reason NASA has dedicated several shuttle missions to examine how space flight affects the human body. Space-Lab Life Sciences-1 (SLS-1) is the first of these dedicated missions. It was launched as the eleventh flight of the Space Shuttle Columbia and completed its mission in June 1991. The SLS-1 was the first space laboratory dedicated to life science research. Its goal was to study how the human body adapted to microgravity.

In previous space flights, body fluid shifts, fluid losses, deconditioning of muscles, calcium loss from bones, and reduction of red blood cell production were just some of the effects observed in astronauts. What was missing from the previous studies were data on the interrelationships among different body systems in microgravity. Space Shuttle Columbia's crew of seven used themselves as human guinea pigs to enable scientists to understand how the body adapts to weightlessness. Twenty different studies were carried out on five body systems during the SLS-1 mission. These systems included: (1) cardiopulmonary, (2) renal/endocrine, (3) blood–immune, (4) musculoskeletal, and (5) neurovistibular (brain, nerves, eyes, and inner ear). Specific experiments were designed to determine how much blood was delivered to the various organs, how pulmonary function changes, and what reflex mechanisms regulate blood pressure. The problem of motion sickness in space was investigated by subjecting the crew to conflicting visual, vestibular, and tactile information to determine the effect of sense on orientation. Studies on the other systems included muscle metabolism, inhibition of bone formation, and blood and lymphocyte production.

In addition to the human studies, over 2000 jellyfish and 29 laboratory rats were carried on the mission and were used for complementary studies. In the jellyfish experiments, thousands of jellyfish polyps were encased in flasks containing artificial seawater. Early in the mission, thyroxine or iodine was injected into the flasks to induce the polyps to metamorphose into free-swimming ephyrae. The tiny jellyfish were filmed to observe their swimming motions compared to a control on the ground to determine if there is any difference between jellyfish gravity receptors that develop in space and those that develop on earth.

Figure 1.
Columbia shuttle lift-off, June 5, 1991. The Spacelab Life Sciences-1 Laboratory was the first Spacelab focusing solely on life sciences research. During the trip the crew explored physiological changes that occur during space flight, and the consequences of the body's adaptation to microgravity and the readjustment to gravity upon return to Earth. (Courtesy of NASA, John F. Kennedy Space Center.)

Figure 2.
STS-40 Mission Insignia: The STS-40 patch, like the space mission, focused on human beings living and working in space. Against a background of the universe, seven silver stars pictured around the orbital path of Columbia represent the seven crewmembers. The orbiter's flight path forms the double-helix of the DNA molecule that is common to all living creatures. In the words of a crew spokesman, "...(the helix) affirms the ceaseless expansion of human life and American involvement in space while simultaneously emphasizing the medical and biological studies to which this flight is dedicated." Leonardo's Vitruvian man, silhouetted against the blue darkness of the heavens, also serves as a powerful embodiment of the extension of human inquiry from the boundaries of Earth to the limitless laboratory of space. Drawn by artist Sean Collins, the STS-40 Space Shuttle patch was designed by the crewmembers for the flight. (Courtesy of NASA, Lyndon B. Johnson Space Center, Houston, Texas.)

The rodents were used for research on muscle, bone, and inner ear function. In one particular study, young rats in rapid growth stage were studied for alterations in bone physiology. Two scientists from Indiana University School of Dentistry, Drs. W.E. Roberts and L. Garetto, in conjunction with Dr. E.M. Holton at NASA Ames Research Center, are studying microgravity-induced osteopenia (loss of bone) in the jawbones. Under normal conditions, the conversion of the committed osteoprogenitor cell to a G_1 preosteoblast requires a certain amount of mechanical loading (stress/strain). In space, however, this conversion step is inhibited, and bone formation is decreased (see figure below). Drs. Roberts and Garetto are investigating two possible mechanisms that could inhibit this process in microgravity: First, is there a general decrease in mechanical loading, which fails to induce committed osteoprogenitor cells to differentiate into preosteoblast cells? Second, is the change in tissue geometry due to the loss of extracellular water? Differentiation of preosteoblasts is inhibited by excessive cell contact. Data from the SLS-1 mission will provide critical information on how microgravity affects bone physiology.

Figure 3.

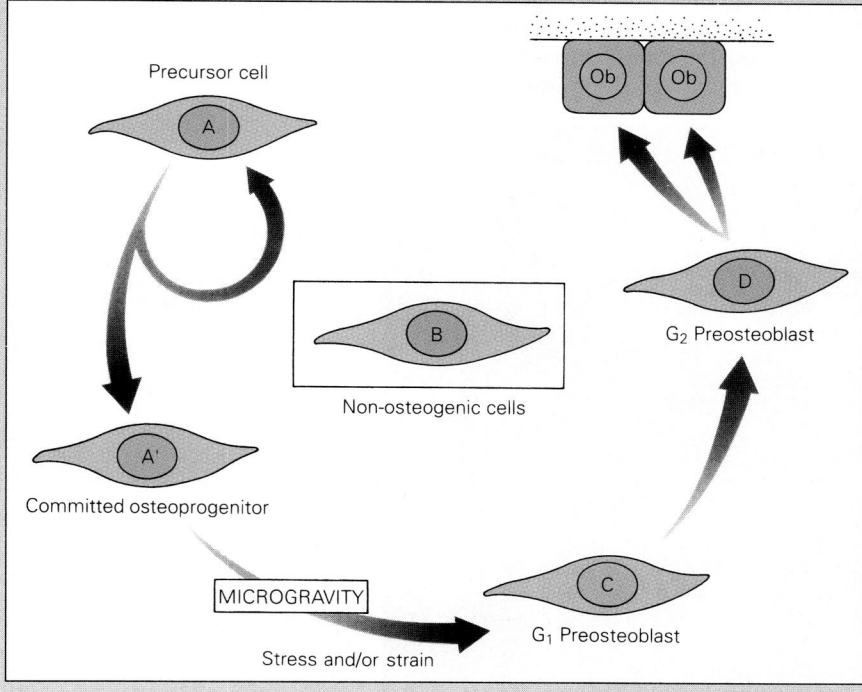

and eyelids droop. At 5 G, movement of the body is almost impossible, and respiration stops because the respiratory muscles cease to function. Between 5 G and 9 G, legs become congested and calf muscle cramps are prevalent. Pulmonary edema occurs because of venous engorgement in the base of the lung, and vision as well as hearing is lost. Finally, consciousness is lost due to cerebral hypoxia.

Most of these symptoms quickly vanish when the positive G force is removed. To minimize the G forces, astronauts lie down during take off or reentry and wear special antigravity suits. On acceleration, these suits automatically pump up, and increased pressure exerted on the body prevents pooling of the blood. With negative G force (foot-to-head), the motion of blood is toward the head. Up to −3 G, there is pressure in the eye sockets. Neck veins are overdistended and the face becomes puffy. Ringing in the ears occurs, and headaches may ensue. Above −3 G, retinal vessels and vision become engorged and vision is impaired. The light filtering through the blood gives a sensation of redness, a phenomenon known as **redout.** Often, at high negative G (greater than −3 G), retinal and cerebral hemorrhages can occur. The physiological effects of negative G are much more lasting than those of positive G, and problems with vision and headache may persist for hours following a negative G exposure.

Weightlessness and Respiration

Once a spacecraft leaves the earth's gravitational forces, astronauts are free of the problem of acceleratory forces but are faced with the problem of weightlessness. The physiological problems associated with weightlessness do not appear to be severe. One concern is the lack of hydrostatic pressures, which results in decreased blood volume, red cell mass, and cardiac output. The second physiological problem related to weightlessness is diminished physical activity, because very little muscle contraction is required for movement. The decrease in physical activity leads to a decrease in work capacity and loss of calcium from bone.

Artificial Atmospheres in Spacecraft

There is no atmosphere in space, so consequently astronauts must be provided with an artificial atmosphere that is pressurized and equipped with life-sustaining oxygen. The types of atmospheric conditions selected depend, in part, on the sophistication

Figure 22–18
Astronaut James Bagian floats through the Spacelab Life Sciences module aboard the Earth-orbiting Columbia (STS-40 mission). Bagian and six other crewmembers spent over nine days in space conducting life science research. (Courtesy of NASA, Lyndon B. Johnson Space Center, Houston, Texas.)

of the engineering. For example, the Russian space program tends to use 21% oxygen in nitrogen at a pressure equivalent to sea level, while the early Apollo missions used 100% oxygen at one third of atmospheric pressure, which gives a Po_2 equivalent of 1 atm. The reason for these two different gas mixtures was based on weight and space. In the earlier Apollo missions, NASA did not have booster rockets sufficient to launch a heavy spacecraft. Using 100% oxygen at $\frac{1}{3}$ atm gives a lighter weight load than a thick-shelled spacecraft using air at 1 atm. One of the dangers of 100% oxygen is the likelihood of an explosion or a fire. Such an accident did occur aboard one of the early Apollo crafts on the launch pad, resulting in a fatal fire and forcing NASA to change gas mixtures. Now all U.S. space flights use air as a gas mixture.

Radiation Hazards in Space

The earth's upper atmosphere is continuously being bombarded by cosmic radiation, particularly from the sun. Many of these radiation particles are trapped in the earth's magnetic field, termed the **Van Allen radiation belt.** Two major belts around the earth trap radiation particles: the inner Van Allen belt starts at an altitude of 300 miles and extends to 3000 miles; the outer belt starts at an altitude of about 6000 miles and goes to 20,000 miles. Since these two belts contain high amounts of radiation, spacecraft must not travel within them because astronauts can receive fatal doses of radiation there. At present, all spacecraft orbit below an altitude of 200 miles to avoid any radiation hazard.

SUMMARY

Respiration in Low-Oxygen Environments

At high altitudes, the barometric pressure is lower than at sea level. Since the percent of oxygen in air remains the same at near 21%, regardless of the altitude, the amount of oxygen falls as altitude increases. The body's efforts to adapt to high altitude center on compensating for the lack of oxygen (hypoxia). A person living at sea level who is suddenly exposed to a very high altitude (over 20,000 feet) will rapidly lose the ability to function. If he or she has time to *acclimatize*, the person can climb to these altitudes without the use of supplemental oxygen. The first response of the body to hypoxia is hyperventilation. Increased respiration raises arterial oxygen tension, but also blows off too much carbon dioxide, which can cause acute mountain sickness, a self-limiting disorder that is not serious, although it can be quite uncomfortable. Rapid ascent to altitudes of 10,000 feet or more can lead to high altitude pulmonary edema, a disease in which fluid collects in the lung and increases the diffusion distance for oxygen and carbon dioxide. Although serious, the edema clears rapidly on descent. Starting at about 12,000 feet, high altitude cerebral edema can occur from rapid ascent. This is an extremely serious condition that can result in death; immediate descent is mandatory.

Over the first few weeks to months at high altitude, the body makes a number of adjustments to the hypoxia. The oxygen-carrying capacity of the blood is raised by increased hematocrit. Both ventilation and cardiac output increase to improve oxygen delivery. After a longer time, ventilation and cardiac output return to normal, but hematocrit remains elevated.

The chronic effects of high altitude are pronounced in the elderly. In the United States and other countries where people have lived at high altitudes for only a few generations, there are relatively few old people adapted to high altitudes. With thousands of years to *acclimate*, the Andean natives and the Sherpas have successfully adapted to chronic hypoxia.

Respiration in Deep Sea Diving

Pressure increases rapidly as a diver submerges. For every 33 feet below the surface, the pressure increases by 1 atm. The most primitive underwater diving apparatus is the snorkel, the length of which is limited by the added *dead space* and by the increasing pressure on the chest wall.

The high pressure experienced with modern diving gear causes nitrogen to be absorbed by the blood and tissue. If the diver ascends too rapidly, the nitrogen forms bubbles and causes a painful disorder called the bends. Below 150 feet, nitrogen narcosis occurs and causes raptures of the deep, a dangerous condition that affects the diver's mental capacity.

Marine mammals are adapted for diving for long periods to depths of more than 1000 feet. Their diving reflex slows the heart and redistributes blood flow away from all organs except the brain.

In the 1950s, the Navy relearned that free ascent could be made from depths of 300 feet by simply floating up to the surface and exhaling rapidly. Since the air in the lungs expands during ascent, navy submariners never ran out of air.

Respiration in High-Oxygen Environments

Breathing high concentrations of oxygen can cause alveolar collapse (atelectasis). This occurs through irritation of the small airways, which leads to excess mucus secretion and subsequent plugging of airways. Once airways are plugged, oxygen diffuses across alveoli and, in the absence of nitrogen, rapid collapse occurs.

Breathing pure oxygen leads to pulmonary edema due to damage of pulmonary capillaries. Lung tissue is damaged by the formation of free radicals, which oxidize membrane lipids, enzymes, proteins, and DNA molecules. In addition to causing lung damage, high oxygen also causes fibrosis in the eye (retrolental fibrosis), which leads to blindness in the newborn.

Respiration in Unusual Gaseous Environments

In artificial environments, carbon dioxide can build up and be rebreathed. High concentrations of CO_2 stimulate respiration, leading to hyperventilation and acidosis. When alveolar carbon dioxide exceeds 80 mm Hg, respiration is no longer stimulated but is depressed and can lead to death.

Carbon monoxide is an odorless, colorless, and poisonous gas that cannot be detected by the body. Carbon monoxide becomes lethal because it competitively binds at the same site as oxygen on hemoglobin molecules. The problem is further complicated by the fact that CO has a much greater binding affinity for hemoglobin than does oxygen. Thus, only trace amounts of CO in the blood are necessary to cause death.

The first anesthesias were gases that had to be breathed. The first real anesthetic gas was nitrous oxide, first used in the 1800s by a dentist. Later, ether and chloroform were introduced as anesthetic gases. It was not until the 1950s that the fluorinated and halogenated anesthetics were discovered. These newer anesthetics represented a major improvement over their predecessors since they were neither toxic nor flammable.

Respiration in Space Travel

In space flights and supersonic flights, the human body can be subjected to acceleratory forces (G forces) by a sudden change in velocity or direction. When a positive G force occurs, blood is pulled toward the lower parts of the body. Consequently, venous pulling occurs, and blood flow to the brain decreases. If the G force is severe, loss of vision (blackout) and unconsciousness occur. When a negative G force is experienced, blood moves toward the head. Neck veins become overdistended, pressure builds up in the eye sockets, and retinal vision becomes blurred due to engorgement of retinal vessels (redout).

Another challenge astronauts face in space is weightlessness, the lack of gravity. At present, the physiological problems associated with weightlessness are not severe, but with future longer space flights, the problems may become more serious.

During space flights, astronauts must be provided with an artificial atmosphere suitable to sustain life. Currently, the artificial atmosphere is a mixture of air at 1 atm. However, with longer space flights, alternative gas mixtures may be required.

In addition to weightlessness and artificial atmospheres, astronauts are also potentially exposed to cosmic radiation. The earth's magnetic field (Van Allen radiation belt) traps radiation particles and prevents them from entering the earth. As long as the space flights are below the Van Allen belt (300 miles), astronauts are fairly safe from radiation exposure. However, once spacecrafts begin to venture beyond the earth's magnetic field, special precautions will need to be taken to protect astronauts.

REVIEW QUESTIONS

Identify Each with a Term

1. The change in ventilation that occurs at high altitude.

2. Fluid that collects in the lung at high altitude.

3. Loss of vision during high G force.

4. The compounds formed from breathing high concentrations of oxygen that cause oxidant damage to biological tissues.

5. Collapsed alveoli.

Define Each Term

6. acclimatization

7. high-altitude cerebral edema

8. polycythemia

9. carboxyhemoglobin

10. G force

Choose the Correct Answer

11. High-altitude pulmonary edema:
   a. is best treated by ascent to higher elevations.
   b. can be treated by supplemental oxygen.
   c. will disappear in the first year at high altitude.
   d. is a trivial condition needing no treatment.

12. Which of the following is true about underwater dives?
   a. Problems are more likely to occur on descent rather than ascent.
   b. The bends are caused by expansion of intestinal gases during rapid ascent.
   c. Free ascent from submarines at depths of 300 feet is possible with rapid exhalation during ascent.
   d. To avoid oxygen toxicity, rapid ascent is necessary after prolonged submersion at depths of 100 feet.

13. Which of the following are true about underwater physiology?
   a. The major problem with liquid breathing at hyperbaric pressures is removal of carbon dioxide.

   b. The first practical submarines were built in the 1800s.
   c. The most successful diving mammals divert blood flow to their vital organs.
   d. All of the above.

14. Which of the following are true of high altitude?
   a. With prolonged acclimatization, ventilation and cardiac output return to sea level values.
   b. There are no long-term ill effects of high altitude on North Americans, as illustrated by the large populations of retired people who move to high elevations.
   c. Acute mountain sickness is a rare disorder.
   d. With prolonged acclimatization, maximum work capacity at high altitude actually exceeds sea level work capacity.

15. For every 33 feet below sea level, barometric pressure will increase by:
   a. 1/4 atmosphere
   b. 1/3 atmosphere
   c. 1 atmosphere
   d. 2 atmospheres

16. The stimulus for hyperventilation at altitude is:
   a. low blood pH
   b. low blood oxygen tension
   c. low blood bicarbonate
   d. low blood $[H^+]$

17. Bends associated with deep sea diving are due to:
   a. excess oxygen dissolved in brain tissue.
   b. excess carbon dioxide dissolved in brain tissue.
   c. excess gas in gut tissue.
   d. excess nitrogen dissolved in fat tissue.

18. The reason carbon monoxide is so lethal is that it:
   a. has a stronger binding affinity to hemoglobin than does O_2.
   b. is taken up by brain tissue.
   c. depresses the respiratory center.
   d. is selectively toxic to heart tissue.

19. As one ascends to high altitude, which of the following decrease?

a. partial pressure of O_2
b. barometric pressure
c. partial pressure of N_2
d. all of the above

20. During acclimatization to altitude:
   a. vital capacity increases.
   b. minute ventilation increases.
   c. cardiac output increases.
   d. all of the above.

Calculate

21. What is the partial pressure of oxygen on top of Mt. Everest (barometric pressure = 231 mm Hg)?

22. Estimate how much oxygen is delivered for a cardiac output of 5 liters in a high mountain dweller (15,000 feet) if arterial oxygen tension is 40 and venous oxygen tension is 20 mm Hg.

23. A diver goes 99 ft underwater. What is the barometric pressure at this depth?

24. Calculate the total G force of a pilot who pulls a negative G force of 3.

25. What is the partial pressure of O_2 in a space cabin filled with 50% O_2 at 1/2 atm?

Answer Each Question in One or Two Sentences

26. What are the effects on respiration of acute exposure to high altitude?

27. Describe what happens to gases in the body during an underwater dive.

28. What are some of the effects of breathing high oxygen at sea level?

29. What are some of the effects of high G forces on the body?

30. Briefly describe the stages of anesthesia with inhaled anesthetics.

SUGGESTED READINGS

Altman, L.K. *Who Goes First?* New York: Random House, 1987.

Fishman, A.P., and Richards, D.W. *Circulation of the Blood.* New York: Oxford University Press, 1964.

Lambertsen, C.J. *Underwater Physiology.* Baltimore: Williams & Wilkins, 1967.

Sebel, P., Stoddart, D.M., Waldhorn, R.E., Waldmann, C.S., and Whitfield, P. *Respiration.* New York: Torstar Books, 1985.

Stoelting, R.K., and Miller, R.D. *Basics of Anesthesia.* New York: Churchill Livingstone, 1984.

KEY TERMS

acclimatization (p. 757)
analgesia (p. 768)
anesthesia (p. 768)
anoxia (p. 768)
atelectasis (p. 765)

blackout (p. 769)
carboxyhemoglobin (p. 767)
chloroform (p. 768)
diving reflex (p. 761)
free radicals (p. 766)

G force (p. 769)
hypoxia (p. 756)
oxygen radicals (p. 765)
oxygen toxicity (p. 765)

polycythemia (p. 759)
radiation (p. 773)
SCUBA (p. 761)
weightlessness (p. 772)

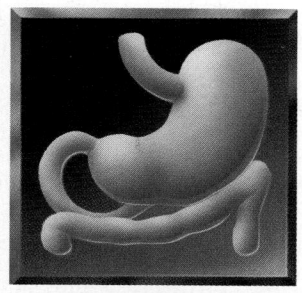

The Gastrointestinal System

CHAPTER OUTLINE

Eating is one of life's pleasures. However, there is more to eating than sensory gratification; food is needed to provide the nutrients required for the maintenance of normal structure and function of the body. Food contains water, inorganic salts, and a variety of structurally complex organic molecules such as carbohydrates, proteins, lipids, and vitamins. Carbohydrates and fats supply most of the fuel (calories) required for the activities of everyday life, whereas proteins are used mainly for the processes of growth and the constant repair of body tissue. Inorganic salts play important roles in many body processes, and vitamins are essential for normal growth and development.

Overview of the Gastrointestinal System

The **gastrointestinal system** is a collection of organs specialized to handle the food we eat (*gastro* = "relating to the stomach"). The primary function of this system is to process the materials contained in food and, in so doing, to supply the cells of the body with nutrients and energy. Four major processes are essential for the nutritive function of the gastrointestinal system: **motility, secretion, digestion,** and **absorption.** The term **motility** refers to the way in which food is moved downward through the various regions of the digestive tract at different rates depending on the functions served by the different organs of the tract. In the **secretion** process, a number of juices produced by exocrine glands enter the tract at various points along the route. The enzymes contained in some of these secretions play a primary role in **digestion,** the conversion of many large organic molecules of the diet to smaller molecules. In **absorption,** digestion products and other nutrients travel across the wall of the small intestine to the blood or the lymph, which in turn distributes them to the cells of the body. Indigestible materials and a number of waste products are moved to the end of the tract and eliminated from the body.

This chapter will consider these processes as they occur in sequence after the eating of food. However, before proceeding to a description of the handling of food by the gastrointestinal system, we will consider some of the general properties of the system as a whole.

Anatomy of the Gastrointestinal System

The gastrointestinal system (Fig. 23–1) consists of the **digestive tract and accessory structures** that lie outside the tract but are connected to it by ducts.

The digestive tract is a tube about 15 feet long that traverses the body from lips to anus; it is composed of the **mouth, pharynx, esophagus, stomach,** and **small** and **large intestines.** Both ends of the digestive tract open to the external environment, which means that the lumen of the tract is continuous with and thus a part of the external environment. The wall of the digestive tract therefore is an interface between the external and internal environments of the body. This interface permits entry of certain materials into the body and excludes others. Most nutritional materials that enter the body do so by passing through the wall of the tract.

The accessory structures that lie outside the tract are the **salivary glands, pancreas, liver,** and **gallbladder.** These structures either produce or store secretions that aid in the digestion or absorption of food; these secretions are conducted into the lumen of the tract through ducts.

Cellular Arrangements in the Gastrointestinal System

Although the microscopic structure of the digestive tract varies from one region to another, there are common features in the overall organization of that part of the tract that extends from the esophagus to the anus. These structural characteristics are illustrated diagrammatically in Figure 23–2, which shows that the wall of the digestive tract is composed of four layers. From the lumen outward, these are the **mucosa, submucosa, muscularis externa,** and the **serosa.**

From the inside out, the mucosa consists of an **epithelial lining** (invaginations of which form **exocrine glands**), the **lamina propria** (a layer of connective tissue), and the **muscularis mucosa** (a thin layer of smooth muscle).

The submucosa, a layer of connective tissue, contains blood and lymph vessels that branch off into the overlying muscularis externa and the underlying mucosa. Also located in this layer are exocrine glands and a plexus of nerve cells, the **submu-**

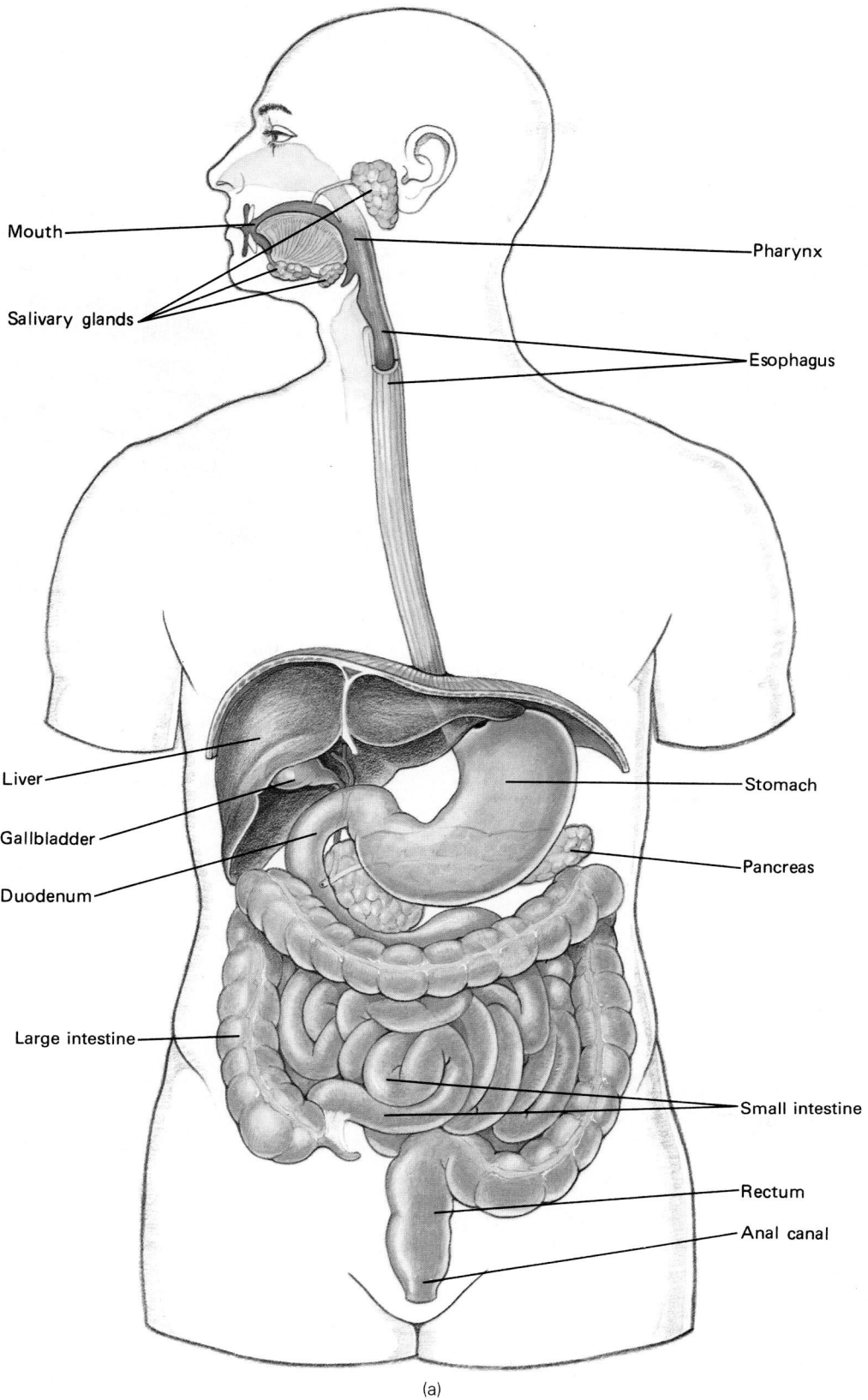

Mouth

Salivary glands

Pharynx

Esophagus

Liver

Gallbladder

Duodenum

Stomach

Pancreas

Large intestine

Small intestine

Rectum

Anal canal

(a)

Figure 23–1
(*a*) Gross anatomy of the gastrointestinal system.

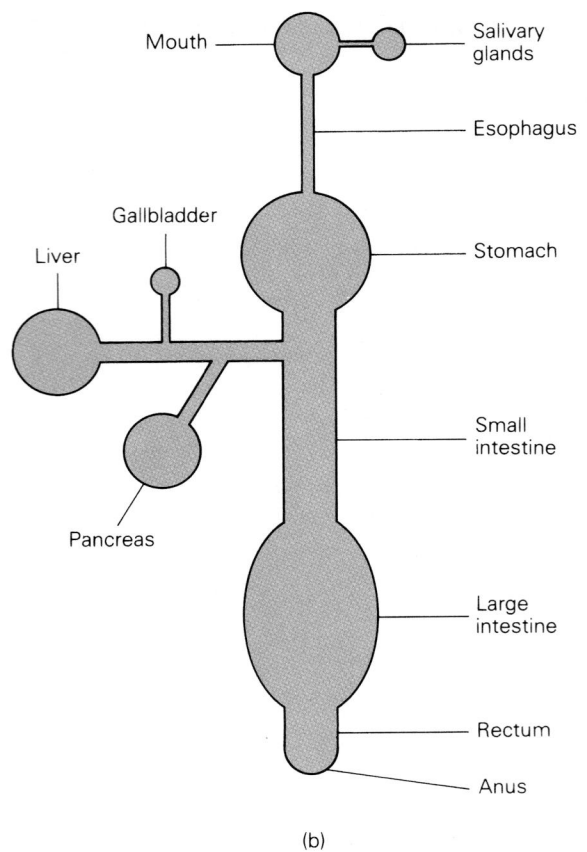

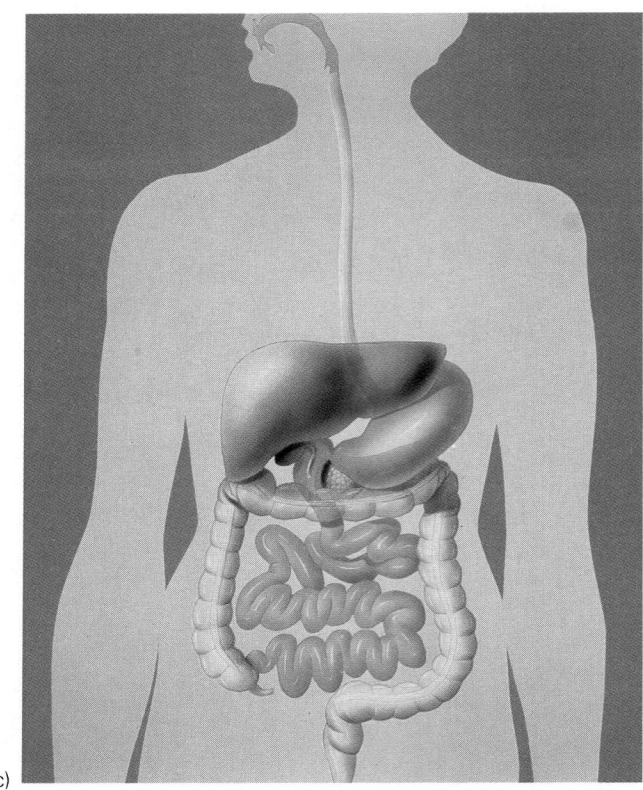

(b)

(c)

Figure 23-1 (cont'd) (*b*) Diagrammatic representation of the gastrointestinal system. (*c*) The organs of the human digestive tract arranged in their relative positions over a silhouette of the human body. (© David Gifford/Science Photo Library/Photo Researchers, Inc.)

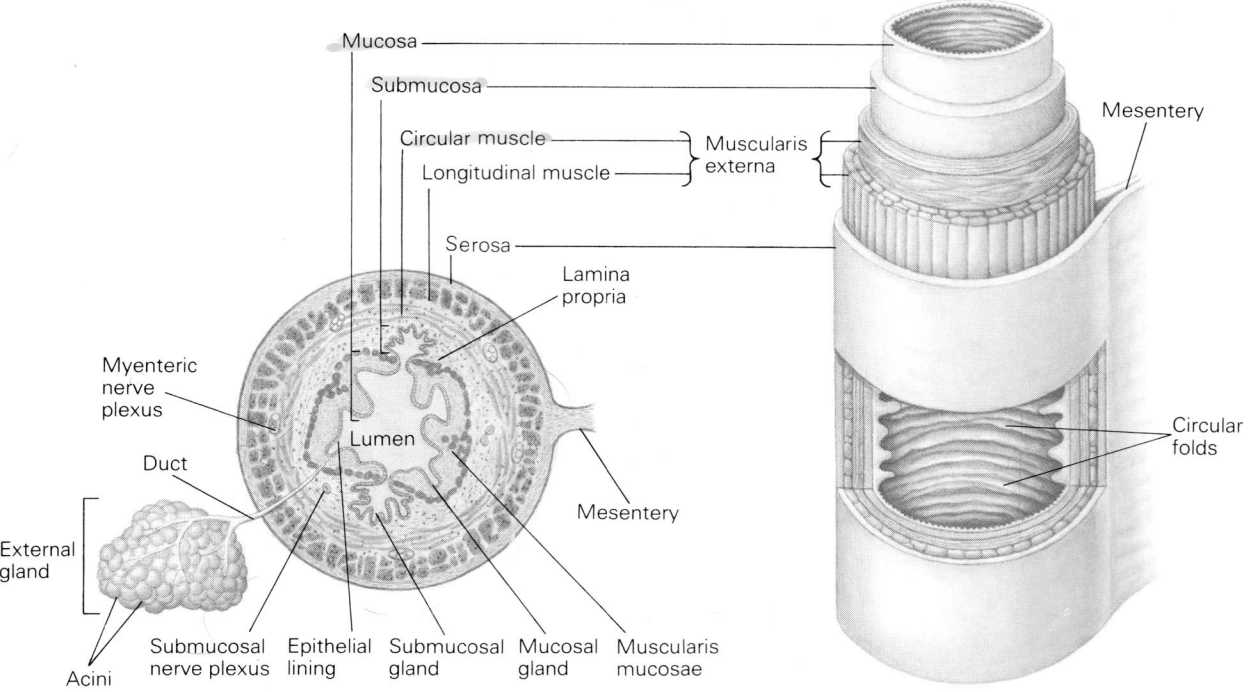

Figure 23-2
Diagrammatic representation of the structure of the gastrointestinal system, showing a cross section of the digestive tract and an exocrine gland external to the tract.

cosal plexus. The muscularis externa consists of an inner layer of **circular smooth muscle,** which encircles the tract, and an outer layer of **longitudinal smooth muscle,** which runs lengthwise along the tract. A network of nerve cells, the **myenteric plexus,** lies between these muscle layers. The serosa, a layer of connective tissue, is the outer tunic of the tract. The **mesentery** is an outward fold of the serosa that carries blood vessels, lymphatics, and nerves to and from the tract.

The salivary glands and parts of the pancreas contain clusters of exocrine secretory cells called the **acini** (see Fig. 23–2). The acini lead into a system of small ducts, which converge into larger ducts such that saliva is finally conducted into the mouth and pancreatic juice is conducted into the lumen of the upper small intestine through a main duct. A system of ducts also delivers bile (the exocrine secretion of the liver cells) into the upper small intestine.

The Gastrointestinal Circulation

The circulation of the stomach, small and large intestines, pancreas, and liver is called the **splanchnic circulation** (Fig. 23–3). The heart pumps blood to these organs through arteries that branch from the abdominal aorta. The blood that leaves the stomach, intestines, and pancreas goes next to the liver by way of the portal vein. This segment of the vasculature, which begins and ends in capillaries, is called the **portal circulation.** The significance of this vascular organization is that many absorbed digestion products are subject to metabolic processing by the liver before being distributed to the cells of the body.

The splanchnic circulation receives almost one fourth of the cardiac output. Its resting flow rate is about 1400 ml per minute, which is greater than that of any of the other major peripheral circulations.

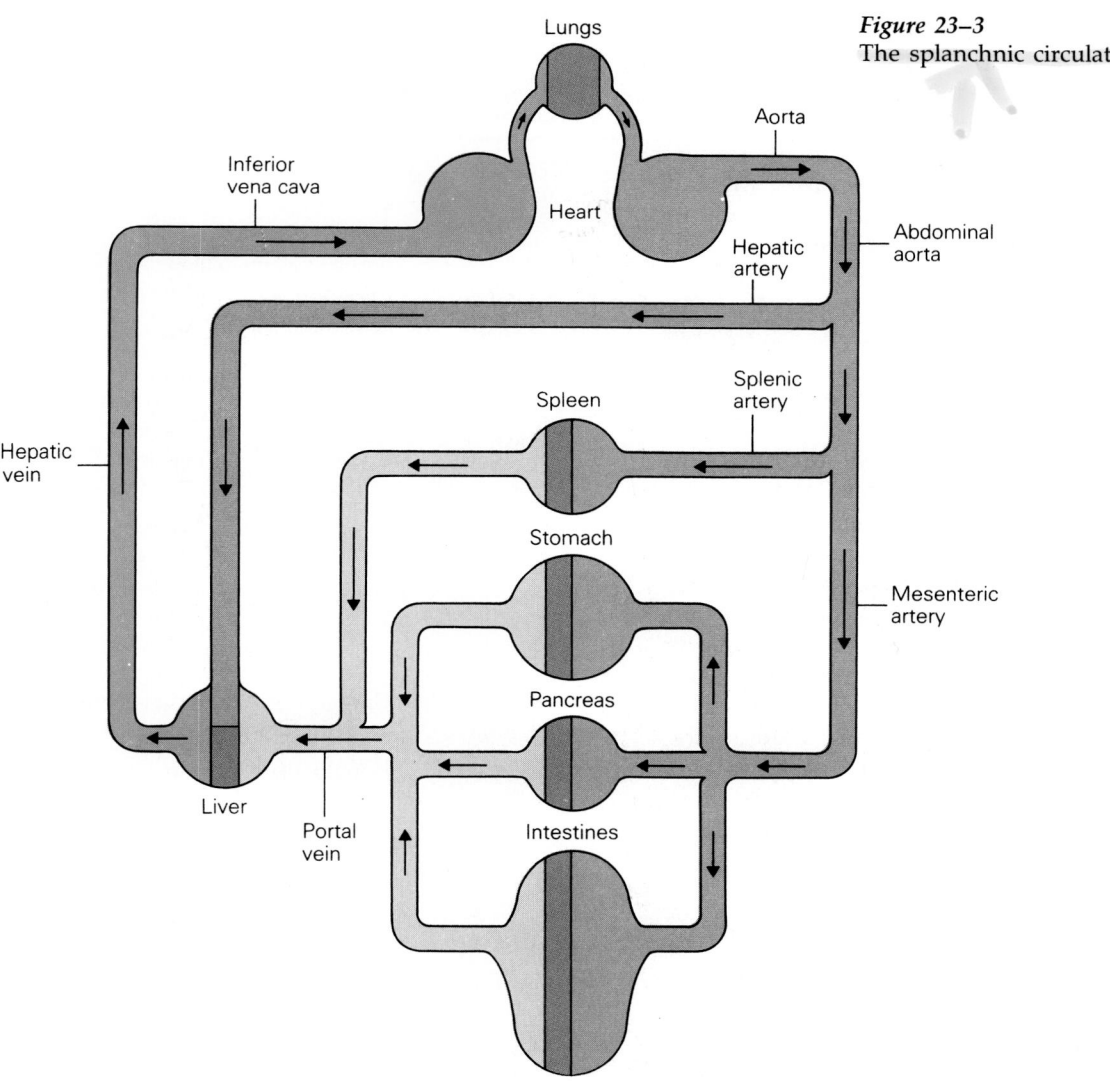

Figure 23–3
The splanchnic circulation.

Flow of blood through the splanchnic circulation increases when a meal is eaten; this increase facilitates the removal of digestion products from the digestive tract. In addition, extra oxygen and other nutrients are supplied to meet the enhanced demands for energy imposed by the increased muscular, secretory, and absorptive activities that occur after a meal.

Processes of the Gastrointestinal System

Before proceeding to a detailed discussion of motility, secretion, digestion, and absorption, we will discuss the relationships among the structures just described and the functions they serve.

Motility: Mixing Movements and Propulsive Movements

The motor functions of most of the digestive tract are performed by the smooth muscle layers that extend from the lower two thirds of the esophagus through most of the large intestine. Note, however, that skeletal muscle activity is of primary importance in the mouth and through the upper third of the esophagus at the upper end of the tract, and in the external anal sphincter at the lower end.

The two basic types of motor activity involved in the digestive and absorptive functions of the digestive tract are **mixing movements and propulsive movements.** The mixing movements of the stomach and small intestine aid digestion by mixing the digestive secretions with the food that enters from above. These movements also promote absorption from the small intestine and proximal large intestine by bringing their contents into contact with the absorbing surfaces. In the case of propulsion, the rate of movement of contents through the various regions of the digestive tract depends on the function served by each organ of the tract. For example, the transit of food from mouth through pharynx and esophagus is very rapid, because these regions simply provide a conduit for conveying food to the stomach. On the other hand, movement of contents from the stomach and through the small and large intestines is slow, to allow sufficient time for digestion and absorption.

Secretion: Inorganic Ions, Digestive Enzymes, Liver Products, and Mucin

The exocrine glands present throughout the gastrointestinal system contain a variety of secretory cells, each of which produces a characteristic secretion.

All of these secretions consist of water and inorganic ions, but many also contain organic constituents. Important inorganic ions are hydrogen ions secreted by the stomach and bicarbonate ions secreted by the pancreas. The most notable of the organic components are the **digestive enzymes** of the salivary glands, stomach, and pancreas; the **bile salts, bile pigments,** and **cholesterol** secreted by the liver; and **mucin** produced by all regions of the digestive tract. The secretory cells of the gastrointestinal system, their sites of secretion, and the major components of these secretions are listed in Table 23–1.

Blood is the ultimate source of the gastrointestinal exocrine secretions in the sense that it supplies the constituents necessary for their elaboration.

Table 23–1
Exocrine Secretions of the Gastrointestinal System

Site of Secretion	Exocrine Secretory Cell	Secretory Components
Entire gastrointestinal system	All cells	Water and inorganic salts
Salivary glands	Serous	Amylase
	Mucous	Mucin
Esophagus	Mucous	Mucin
Stomach	Parietal	HCl and intrinsic factor
	Chief	Pepsin
	Mucous	Mucin
Pancreas	Acinar	Amylase, proteases, lipase, and other digestive enzymes
	Duct epithelium	Bicarbonate
Liver	Hepatocyte	Bile salts, bile pigments, and cholesterol
	Duct epithelium	Bicarbonate
Small intestine	Mucous	Mucin
Large intestine	Mucous	Mucin

Some secretory components move directly from blood through and out of the cells at the time of secretion. Other components are synthesized by the cells from raw materials supplied by blood, and in many instances the finished products are stored in the cells until an appropriate stimulus causes them to be secreted.

Water and solutes are transferred across secretory cell membranes to the cell exterior by a variety of active and passive transport processes. Stimuli such as nerve impulses or hormones activate active transport systems in the plasma membranes of the secretory cells, and substances are actively transported from the interior to the exterior of the cell, usually into a duct. The active transport of a substance is the initial event that occurs, and water and other dissolved materials follow along passively. The overall result is the formation outside the cell of a secretion that ultimately finds its way into the lumen of the digestive tract, where it performs its function in the digestion and absorption of food.

The secretion that moves across the secretory cell membrane is known as the **primary secretion.** The composition of the primary secretion may be modified during passage through ducts as the result of movement of substances to and from blood across the duct epithelium. This is the case particularly for the salivary glands, pancreas, and liver, each of which has a relatively long duct system connecting the glandular secretory cells with the lumen of the tract.

Digestion: Breakdown of Food into Small Molecules

Carbohydrates, proteins, and fats are too large to be absorbed from the digestive tract in their ingested forms. By the process of enzymatic digestion, these large organic molecules are broken down into smaller molecules that can be absorbed. For example, complex carbohydrates such as starch are broken down to simple sugars. Proteins are degraded to small peptides and amino acids, and triacylglycerol fats are hydrolyzed to the smaller free fatty acid and monoglyceride molecules.

Digestion is carried out in the lumen of the digestive tract by enzymes contained in the secretions of the salivary glands, stomach, and pancreas, and also by enzymes in the epithelial cells lining the absorptive surface of the small intestine. The process of converting the complex molecules of food to simple molecules is a progressive one that starts in the mouth and continues in the stomach and small intestine. The small intestine is the region in which digestion is completed and most absorption occurs. Table 23–2 lists the major digestive enzymes, the substrates upon which they act, and the products of digestion.

Table 23–2

The Digestive Enzymes of the Gastrointestinal System and Their Actions

Source of Enzyme	Site of Action	Enzyme	Substrate	Products of Digestion
Salivary glands	Mouth and gastric lumen	Amylase	Polysaccharides	Oligosaccharides
Stomach	Gastric lumen	Pepsin	Proteins	Peptides
Pancreas	Small ingestinal lumen	Amylase	Polysaccharides	Oligosaccharides
		Trypsin, chymotrypsin, and carboxypeptidase	Proteins and peptides	Peptides and amino acids
		Lipase	Triacylglycerols	Monoglycerides and free fatty acids
Small intestinal mucosa	Small intestinal epithelium	Oligosaccharidases Peptidases	Oligosaccharides Peptides	Monosaccharides Amino acids

Absorption: Transfer of Nutrients to the Blood and Lymph

The absorption of water and dissolved materials is the major function of the gastrointestinal system, and the various processes of this organ system are dedicated for the most part to fulfilling this function. This transfer of water and solutes from the lumen of the digestive tract into the blood or the lymph is accomplished largely by the epithelial cells that line the inner surface of the small intestine. However, significant amounts of water and inorganic salts are absorbed from the proximal large intestine.

The luminal membranes of the intestinal epithelial lining contain a number of transport systems specialized for the absorption of specific components of the diet. Some of these transport systems are capable of moving certain digestion products and inorganic ions from lumen to blood by active transport against electrochemical gradients. Other substances move by diffusion, at varying rates depending on their molecular size and lipid solubility.

The Regulation of Gastrointestinal Function

Few mechanisms regulate the processes of digestion and absorption. Since this lack of regulation applies to those constituents of the diet that supply calories to the body, the more of these foods that are eaten, the more that will be digested and absorbed. Thus, most cases of obesity result from eating too much food too often.

In contrast to digestion and absorption, motility and secretion are highly regulated—particularly by neural and hormonal mechanisms. These processes are regulated by stimulation of receptor cells, which in turn triggers a chain of events operating through various pathways to produce specific responses of smooth muscle and exocrine gland cells.

Many of the stimuli that initiate the regulatory events are present in the lumen of the digestive tract. The most important stimuli are digestion products themselves, hydrochloric acid (HCl) secreted by the stomach, and distention by the contents of the tract. However, responses of the gastrointestinal system are not limited to stimuli applied to the tract, because stimulation of many other regions of the body can elicit responses of the gastrointestinal system. For example, both the sight and smell of food stimulate salivary, gastric, and pancreatic secretion. Another example is the profound motor activity (i.e., vomiting) that results from stim-

ulation of the vestibular apparatus during motion sickness.

The receptors of the tract include sensory nerve receptors, such as **chemoreceptors** that respond to chemical agents in the contents of the tract, and **mechanoreceptors** that respond to distention of the tract. The reflex activity that results from excitation of these nerve receptors plays an important role in the regulation of gastrointestinal secretory and motor activity. In addition, a number of different types of endocrine cells in the digestive tract serve as receptors. When excited by a variety of stimuli, these cells release hormones into the circulation. As do the nerve reflexes, these hormones also participate in the regulation of secretion and motility.

Neural Control of Gastrointestinal Processes

The neural pathways for the control of motor and secretory activity of the gastrointestinal system are summarized in Figure 23–4. Two types of neural pathways are involved in the regulation of gastrointestinal motor and secretory activity: **short reflexes,** mediated through nerve cells that lie entirely in the wall of the digestive tract, and **long reflexes,** mediated over nerves outside of the tract.

Short Reflex Regulation. Regulation by short reflexes occurs through the myenteric and submucosal nerve plexuses present in the wall of the digestive tract (see Fig. 23–2). These nerve plexuses possess all of the elements required for the local regulation of motor and secretory activity; i.e., they contain (1) sensory neurons that are excited by the intraluminal stimuli mentioned previously, (2) interneurons that conduct impulses up and down the tract, and (3) motor and secretory neurons that innervate the effector smooth muscle and secretory cells. The short reflexes that are mediated over these nerve cells give the digestive tract a degree of autonomy in the sense that muscle contraction and exocrine secretion are to an extent independent of regulatory influences originating outside the wall of the tract. These short reflexes can be either excitatory or inhibitory, depending on the mediator released at the effector cell by motor and secretory neurons. If the mediator is acetylcholine, as is usually the case, the result is excitation of exocrine secretion and contraction of visceral smooth muscle.

Long Reflex Regulation. Gastrointestinal activity is also regulated by reflexes mediated through nerves outside of the gastrointestinal system. The major function of these long reflexes is to correlate activities, not only between different regions of the gas-

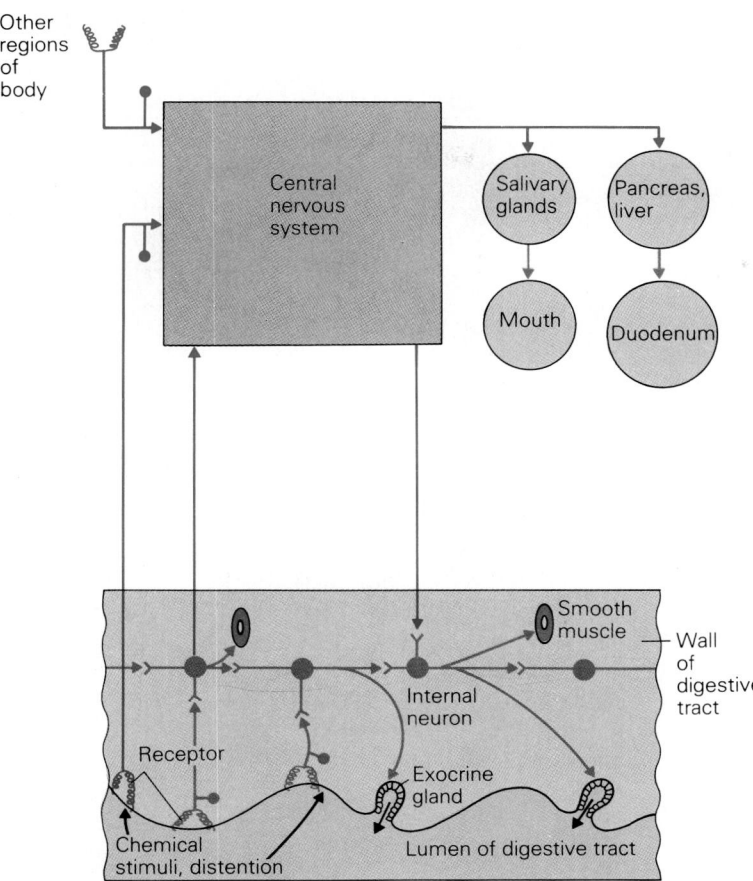

Figure 23–4
Neural pathways for the control of motor and secretory activity of the gastrointestinal system. Green lines show short reflex pathways; blue lines show long reflex pathways.

trointestinal system, but also between this system and other parts of the body. These long reflex arcs are mediated from either the gastrointestinal system or other regions of the body over afferent sensory nerves to the central nervous system (CNS). The central nervous system in turn relays impulses over efferent parasympathetic and sympathetic nerves to the gastrointestinal system. Since the efferent autonomic nerves synapse with cell bodies of the internal nerve plexuses of the digestive tract, impulses over external nerves can modify short reflex activity initiated in the wall of the tract. This modification can be either excitatory or inhibitory, depending on the type of efferent fibers involved. The vagus nerve (a parasympathetic preganglionic nerve) is the major motor and secretory external nerve of this organ system, and most vagal efferent fibers are cholinergic and excitatory. The splanchnic nerves (sympathetic postganglionic nerves) are adrenergic, and the norepinephrine they liberate depresses excitability of the nerve plexuses, thereby inhibiting short reflexes mediated through these local nerve networks.

Hormonal Control of Gastrointestinal Processes

Four hormones of gastrointestinal origin—**gastrin, cholecystokinin (CCK), secretin,** and **gastric inhibitory peptide (GIP)**—are important in the control of motility and secretion. All of these hormones are polypeptide hormones consisting of chains of amino acids of varying lengths. The four types of endocrine cells that secrete these hormones occur as single cells throughout the epithelial lining of certain regions of the tract. When stimulated, these endocrine cells release their hormones into the blood. It carries them to target organs of the gastrointestinal system, where they produce their motor and secretory effects. All four hormones are released in response to intraluminal stimuli acting directly on the endocrine cells in the luminal lining. Gastrin is also secreted in response to acetylcholine released by nerve impulses—that is, through short reflexes in the intestinal wall activated by intraluminal stimuli and through long reflexes, the efferent limbs of which are in the vagus nerve. The location of the

Table 23–3
The Gastrointestinal Hormones

	Gastrin	*CCK*	*Secretin*	*GIP*
Location of endocrine cells	Antrum of stomach	Upper small intestine	Upper small intestine	Upper small intestine
Stimuli for release	Amino acids, peptides, acetylcholine	Amino acids, peptides, fatty acids	Acid	Fatty acids, glucose
Major physiological actions				
Stomach				
Acid secretion	Stimulates		Inhibits	Inhibits
Gastric emptying	Stimulates		Inhibits	Inhibits
Mucosal growth	Stimulates			
Pancreas				
Enzyme secretion		Stimulates		
Bicarbonate secretion			Stimulates	
Growth		Stimulates		
Biliary system				
Liver bicarbonate secretion			Stimulates	
Gallbladder contraction		Stimulates		
Sphincter of Oddi		Relaxes		
Small intestine				
Ileal motility	Stimulates			
Ileocecal sphincter	Relaxes			
Mucosal growth	Stimulates			
Large intestine				
Motility	Stimulates			

gastrointestinal endocrine cells, the stimuli for the release of their hormones, and the major physiological actions of these hormones are summarized in Table 23–3.

Chewing, Salivary Secretion, and Swallowing

Chewing

The process of chewing is accomplished by the combined actions of the skeletal muscles of the jaws, lips, cheeks, and tongue. The contraction and relaxation of these muscles are coordinated by impulses over a number of cranial nerves. Although these muscles can be controlled voluntarily, the act of chewing is partly reflexive in nature. The pressure that results when the jaws close following the introduction of a bolus of food into the mouth produces a reflex inhibition of the jaw-closing muscles and a reflex contraction of the jaw-opening muscles. The mouth opens, and the resulting reduction in pressure causes a reflex contraction of closing muscles and relaxation of opening muscles. This sequence of muscular activity continues rhythmically until the bolus of food has been chewed.

A major function of chewing is to reduce a mouthful of food to particles of a size convenient for swallowing. Although this varies from one person to another, the extent to which food is chewed has little effect on digestion and absorption. These processes are completed even when large chunks of food are swallowed. Chewing, however, does aid in the digestion of food by increasing the surface area on which digestive enzymes can act. The maximal

Peptide Messengers of the Digestive Tract

In recent years, a large number of peptides present in cells of the digestive tract have been identified. Some of these peptides (**messenger peptides**) travel, by various means, from their sources to target organs in different regions of the gastrointestinal system, where they are thought to play important roles in the regulation of function. It is now possible to distinguish not only the sources, but also the targets of these physiologically important molecules, through the use of immunostaining to identify peptide-containing cells, and radioimmunoassay to measure peptides in cells and biologic fluids. It has been established that the gastrointestinal messenger peptides are of endocrine, paracrine, and neural origin.

Endocrine messenger peptides are those released from mucosal endocrine cells of the digestive tract that reach target cells via the systemic circulation. These include the well-established hormones gastrin, cholecystokinin, secretin, and gastric inhibitory peptide, all of which play important roles in the regulation of gastrointestinal function (see Table 23–3). A number of peptides produced in mucosal, endocrine-like cells may also be important in regulating gastrointestinal function, but they have not yet achieved the status of hormone. An example of such a "candidate hormone" is **motilin,** a peptide released from cells in the epithelium of the upper small intestine. It stimulates contractile activity that begins in the stomach and travels caudad down the intestine in two hour cycles after a meal. This type of muscular activity serves to cleanse the intestine of residual contents.

Paracrine peptide messengers are released from cells and either diffuse through the extracellular space or are transported by the local circulation to their target cells. **Somatostatin,** a peptide present in the wall of the stomach and small and large intestines, is of paracrine origin. This messenger peptide inhibits gastric acid secretion, gastric and intestinal motility, and the secretion of enzymes and bicarbonate by the pancreas.

Neural peptide messengers (neurocrines) are released from nerve terminals of the internal plexuses of the digestive tract and reach their target cell by diffusion across synaptic junctions. One of these peptides is **vasoactive intestinal polypeptide (VIP),** which is present in cells in the walls of the stomach and intestines. A number of functions have been ascribed to this neural peptide, including stimulation of salt and water secretion by the small intestine, inhibition of gastric acid secretion, and inhibition of gastric and intestinal motility. **Substance P,** which is distributed from the stomach through the large intestine, excites smooth muscle contraction of the digestive tract, whereas **enkephalin** is inhibitory to motor function. The neural peptide **bombesin** is a potent stimulator of gastric acid secretion.

These are just a few of the peptides that have been identified in cells of the digestive tract. As additional research is done, some will be established as playing roles in the control of gastrointestinal function, whereas others will not. Work is presently being carried out to elucidate the pathways through which these peptides manifest their actions, and as more information becomes available, it becomes increasingly apparent that regulation of gastrointestinal function is not a simple affair. Endocrine, paracrine, and neurocrine peptides are released together during the digestion of food; the evidence indicates that these molecules usually act in concert with one another as well as with the acetylcholine released by parasympathetic pre- and postganglionic nerve fibers, and with the norepinephrine released by postganglionic sympathetic nerve fibers of the digestive tract. Consequently, gastrointestinal function reflects the interplay of numerous endocrine, paracrine, and neural agents.

It is of interest that many of the peptides of the digestive tract are present in other tissues of the body, particularly the brain. Thus, these molecules are not restricted to modifying activity of the gastrointestinal system.

---

force that can be exerted during chewing is much greater than that required for the chewing of ordinary food. More important in the grinding of food is the occlusive contact area between the molars and premolars. Another function of chewing is to move food around the oral cavity and in so doing to stimulate receptors for taste and smell. This is important because much of the satisfaction of eating is derived from the perception of these sensations. In addition, stimulation of these receptors promotes secretion of saliva, which softens and lubricates the bolus of food.

Secretion of Saliva

The acinar cells of the parotid, submaxillary, and sublingual glands combine to secrete between 1 and 2 liters of **saliva** every day. The resting rate of secretion is about 0.5 ml per minute. When stimulation is maximal (by acid solutions), the secretory rate increases to about 4 ml per minute. The parotid gland contains mostly **serous cells,** which secrete a watery solution containing inorganic ions and **amylase, a** digestive enzyme that helps break down starch.

In addition to serous cells, the submaxillary and sublingual glands contain **mucous cells,** which secrete **mucin.** When mucin mixes with the watery secretion of the serous cells, a solution of high viscosity called **mucus** is produced, and this gives a thick viscous characteristic to certain types of saliva.

The composition of the saliva present in the mouth is variable because it is a mixture of serous and mucous cell secretions. The actual composition of saliva depends on the extent to which each cell type is stimulated to secrete, and this in turn depends on the nature of the stimulus and thus the function to be performed. For example, if a dog is given fresh meat, saliva that contains much mucus is secreted. This secretion lubricates the food mass and facilitates its passage into the stomach. If the same dog is given dry meat powder, a watery secretion is produced that washes the powder from the mouth.

The secretion of saliva is regulated by long reflexes. Afferent impulses originate in chemoreceptors in the mouth and nose when these receptors are stimulated by the taste and smell of food. Another sensory pathway is through stimulation of oral mechanoreceptors by food. Sight of food also stimulates salivary secretion. In all instances, impulses are sent to salivary centers in the medulla, which in turn relay impulses over the two divisions of the autonomic nervous system to the various salivary glands, causing them to secrete their specific juices. Both parasympathetic cholinergic and sympathetic adrenergic stimulation *excite* salivary secretion, in contrast to their usual opposing actions on other effector organs of the body. However, the parasympathetic nerves are the most important regulators of salivary secretion, because the mouth is dry when these nerves are nonfunctional.

Salivary secretion serves a number of functions. Saliva softens and lubricates food, which facilitates swallowing. The moistening properties of saliva facilitate chewing and speech. Saliva contains a high concentration of bicarbonate ion. This alkaline (basic) ion neutralizes acids produced by oral bacteria, thereby preventing these acids from dissolving the enamel of teeth. In addition, the decreased secretion of saliva that occurs during dehydration contributes to the sensation of thirst.

The digestive enzyme salivary amylase is synthesized by the serous cells and stored in these cells as vesicles of enzymes called **zymogen granules.** This enzyme (Fig. 23–5) degrades complex carbohydrates such as starch. Starch is a **polysaccharide** consisting of many branched and straight chains of the monosaccharide glucose. The major end-products of this enzyme are **oligosaccharides,** molecular fragments consisting of two to nine molecules of glucose. The major oligosaccharide produced is the disaccharide maltose, which is a combination of two glucose molecules. For optimal activity, salivary

Figure 23–5
Structure of part of a starch molecule and the action of amylase.

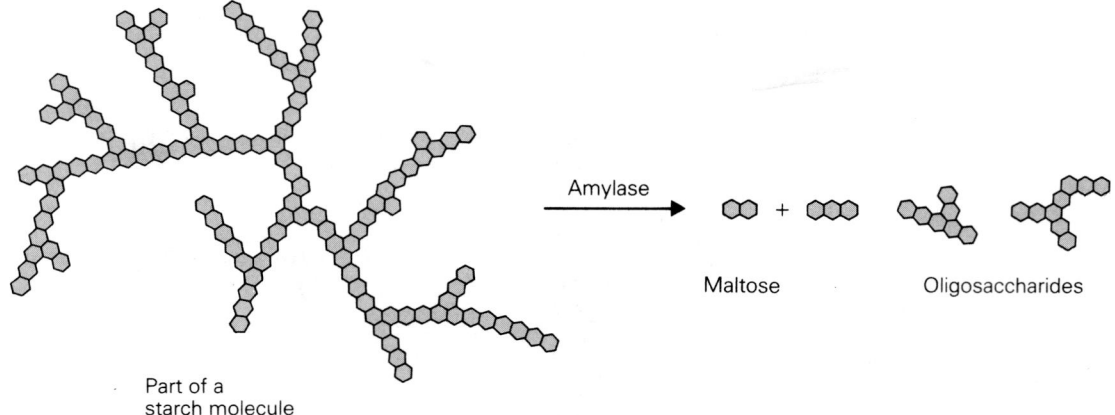

Part of a
starch molecule

Amylase

Maltose   Oligosaccharides

amylase requires a pH close to neutrality (7), as is found in the mouth. Although food remains in the mouth for only a short time, and although the contents of the stomach are highly acidic, a bolus of food remains intact for a while after entering the stomach, so that digestion can continue inside the bolus for a period of time.

Swallowing

Once a mouthful of food is chewed and mixed with saliva, it is propelled rapidly from the mouth through the pharynx and esophagus into the stomach. The food bolus moves in response to skeletal and smooth muscle contractions that occur sequentially from above to below. These muscular movements are coordinated mainly by long reflexes mediated through a swallowing center in the brain.

Swallowing Movements

Swallowing Movements in the Mouth and Pharynx. A bolus of food moves from the mouth through the pharynx into the esophagus (Fig. 23–6) in about 1 second. The movement is accomplished by the coordinated contractions of a number of skeletal muscles.

The front of the tongue is pushed up against the hard palate, and the food mass, which can be either liquid or solid, is rolled toward the back of the tongue. The bolus is then forced into the pharynx by the contraction of skeletal muscles in the back of the throat. Respiration is inhibited briefly. The soft palate is raised and closes the airways to the nose. Food is prevented from entering the trachea by closure of the glottis (the opening of the vocal cords). The glottis is closed by elevation of the larynx, the region between the base of the tongue and the trachea, and by the approximation of the vocal cords. An auxilliary mechanism that prevents food from entering the respiratory passages is that, as a bolus of food passes over the epiglottis, the bolus presses the epiglottis down over the closed glottis. As these openings close, the constrictor muscles of the pharynx contract and force the bolus into the esophagus.

The Role of the Esophagus in Swallowing. The esophagus, a muscular tube located mostly in the thorax, conducts material from the pharynx to the stomach. The major anatomical features of this organ (Fig. 23–7a) are the **upper esophageal sphincter** (a band of skeletal muscle), the **body** of the esophagus (the upper third of which is skeletal muscle and the lower two thirds of which are smooth muscle), and the smooth muscle **lower esophageal sphincter,** the last portion of which lies below the diaphragm in the abdominal cavity.

In order to understand esophageal function, it is necessary to know what pressures exist in the lumen of the different regions of the esophagus, both at rest and during a swallow. It is also important to know how these pressures relate to those in the pharynx at the upper end of the esophagus and in the stomach below, because it is the existence and direction of pressure gradients that determine not only whether contents will move through the esophagus but in which direction they will move.

Figure 23–6
Passage of a bolus of food from the mouth through the pharynx into the upper esophagus during a swallow.

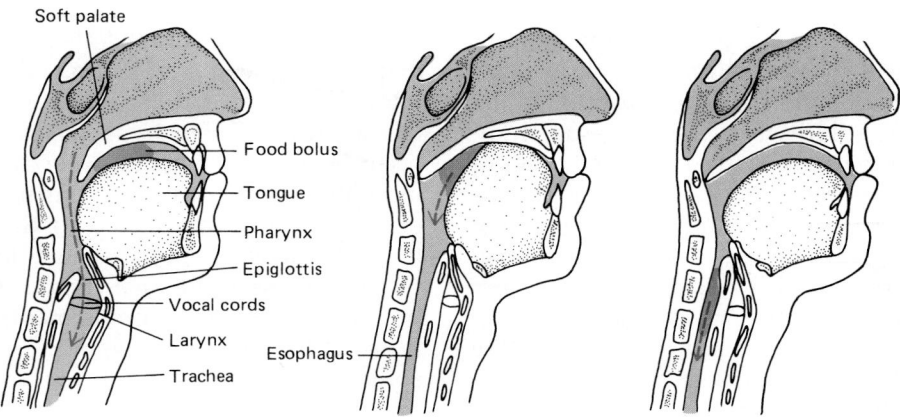

Soft palate

Food bolus

Tongue

Pharynx

Epiglottis

Vocal cords

Larynx

Esophagus

Trachea

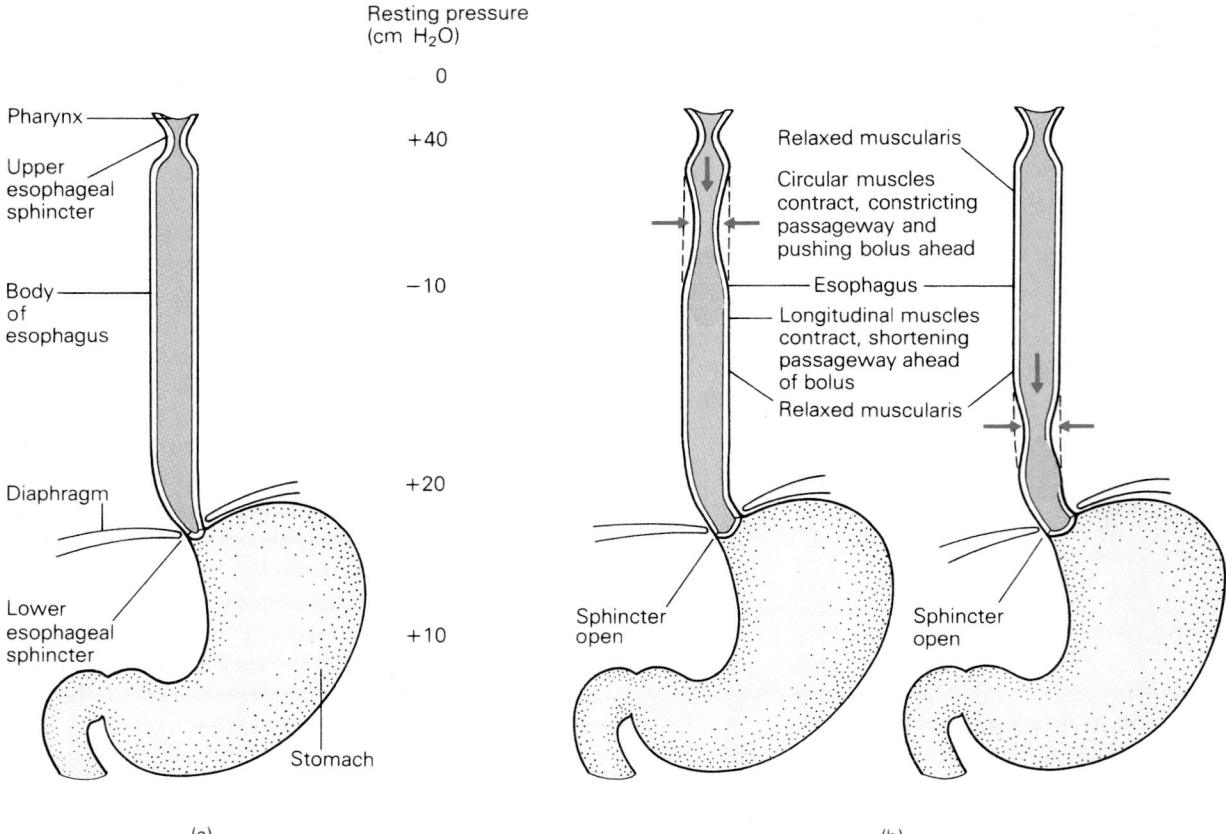

Figure 23–7

(a) Representative resting intraluminal pressures in the pharynx, upper esophageal sphincter, body of the esophagus, lower esophageal sphincter, and stomach. (b) Movement of a bolus of food through the esophagus by a peristaltic contraction.

When the esophagus is at rest (see Fig. 23–7a), pressure in the lumen of the body of the esophagus is the same as intrathoracic pressure, which is subatmospheric (see Chapter 20). On the other hand, pressures in the upper and lower esophageal sphincters are above atmospheric pressure, indicating that these sphincters are contracted when there is no swallowing activity. Since pressure in the resting pharynx is atmospheric, a pressure gradient would exist between the pharynx and the body of the esophagus if it were not for the zone of high pressure that results from the contraction of the upper esophageal sphincter. If this pressure gradient were allowed to manifest itself, large volumes of air would flow from the pharynx to the esophagus during the normal course of breathing, and much of this air would be swallowed. The above-atmospheric pressure produced by closure of the upper sphincter prevents this from happening.

At the lower end, intragastric pressure (the pressure inside the stomach) is greater than atmospheric pressure, but less than that of the contracted lower sphincter. Consequently, gastric (stomach) contents do not ordinarily reflux into the esophagus.

However, reflux is common despite the barrier imposed by the lower sphincter, and because the contents of the stomach are quite frequently acid, the esophagus is irritated, producing the unpleasant sensation known as **heartburn.** In some circumstances, such as bending or stooping, when the abdominal muscles contract forcibly, and when the diaphragm descends during deep inspirations, intra-abdominal and thus intragastric pressure increase greatly. However, reflux does not ordinarily occur in these circumstances because the increased intra-abdominal pressure is transmitted to the same extent to the stomach and to that portion of the lower esophageal sphincter that lies below the diaphragm in the abdominal cavity—that is, the same pressure barrier as before is maintained. If for any reason the subdiaphragmatic lower esophageal sphincter (and the top of the stomach) is displaced into the thoracic cavity, a condition known as **hiatus**

hernia, the terminal segment of this sphincter can no longer be assisted by changes in intra-abdominal pressure, and reflux can occur.

Soon after a swallow is initiated, pressure in the pharynx increases as a result of the skeletal muscle contraction that occurs here. Also, pressures in the upper and lower sphincters abruptly drop to atmospheric pressure, showing that these sphincters have relaxed. A bolus of food moves from the pharynx through the relaxed upper sphincter into the esophagus in response to the downward pressure gradient that now exists through these regions.

Peristalsis in the Esophagus.   After the bolus enters the esophagus, the upper sphincter closes, the glottis opens, and respiration resumes. Then pressure rises sequentially from above to below in the body of the esophagus. This sequence of pressure changes is due to the fact that a wave of contraction moves down the length of the esophagus. The contractile wave propels a semisolid food mass ahead of it (see Fig. 23–7b) and into the stomach through the previously relaxed lower sphincter. This type of propulsive activity, which is characteristic of most regions of the digestive tract, is called **peristalsis.** The esophageal peristaltic wave travels at a rate of 2 to 4 cm per second and takes about 9 seconds to traverse the esophagus. If a bolus is liquid, it is shot through the esophagus by the initial force of swallowing and travels by force of gravity to the stomach in about 1 second. Once the bolus enters the stomach, the lower sphincter contracts, preventing reflux of gastric contents into the esophagus.

The Swallowing Reflex

The all-or-none swallowing reflex is activated when a bolus of food stimulates mechanoreceptors in the pharynx. Afferent impulses are sent to a swallowing center in the medulla. This center in turn sends out, in the proper sequence, efferent impulses over somatic nerves to the skeletal muscles, and autonomic nerves to the smooth muscles involved in the swallowing response (Fig. 23–8). Motor impulses that regulate the activity of the skeletal muscles of the pharynx, larynx, upper esophageal sphincter, and upper body of the esophagus travel over various cranial somatic nerves. The esophagus is innervated primarily by the **vagus nerves.** The vagal efferent nerve fibers that supply the upper esophageal sphincter and upper body of the esophagus terminate at skeletal muscle motor end-plates and thus are somatic (and not autonomic) nerves. The vagal efferent nerve fibers that coordinate both the orderly progress of the peristaltic wave through the smooth muscle portion of the lower esophagus and the activity of the lower esophageal sphincter are autonomic nerves. These vagal preganglionic fibers synapse with cholinergic fibers of the internal nerve plexuses in the smooth muscle wall, and it is through these nerve networks that the vagus nerves manifest their motor effects. Some people lack the esophageal nerve plexuses. In such cases, the lower esophageal sphincter fails to relax following a swallow, and the resting tone of this sphincter is higher than normal. As a result, food has difficulty entering the stomach and tends to accumulate above the

Figure 23–8
Neural pathways for the regulation of esophageal motor activity.

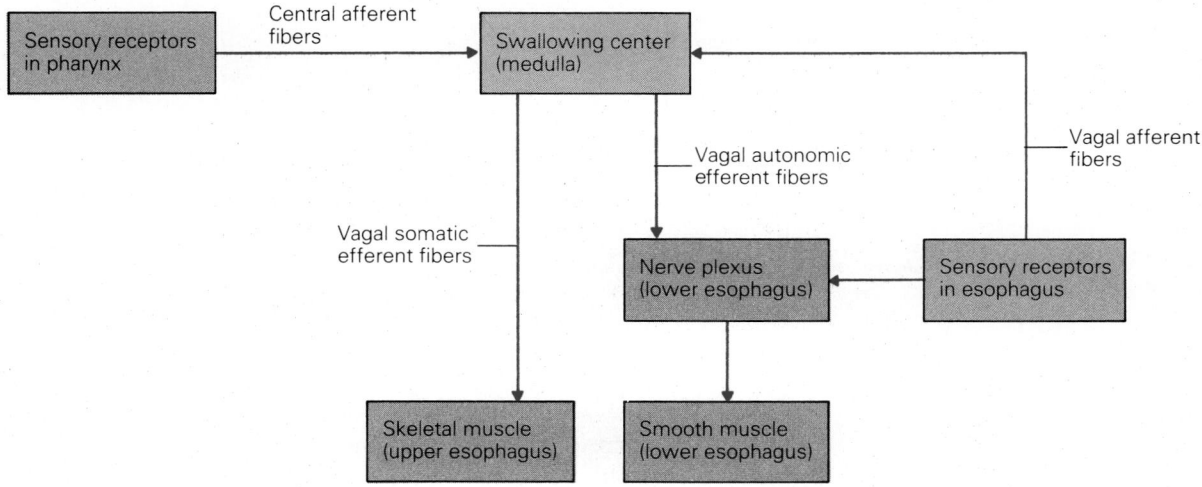

sphincter, causing the esophagus to dilate. This condition is known as **achalasia.**

The peristalsis that follows a conscious effort to swallow is **primary peristalsis. Secondary peristalsis** follows stretch of the esophageal wall without being preceded by a swallowing movement. This type of peristalsis is mediated by long reflexes over the vagus nerves from the esophagus to the swallowing center and back to the esophagus, and also by short reflexes in the esophageal wall (see Fig. 23–8). Secondary peristalsis is important because it facilitates the removal of any food that remains in the esophagus following the passage of a primary peristaltic wave.

Any one of a variety of disorders associated with the nerves and muscles involved in swallowing creates difficulty in swallowing, a condition known as **dysphagia.** One example of dysphagia is poliomyelitis, which paralyzes certain cranial nerves that are a part of the swallowing reflex. Another example of a disorder of the swallowing reflex is myasthenia gravis, a condition in which certain skeletal muscles involved in the swallowing response can contract only weakly.

The Processes of Motility and Secretion in the Stomach

The stomach (Fig. 23–9), located between the esophagus and small intestine, functions as a storage organ for food. Without the stomach, it would be necessary to consume many small meals every day, rather than a few large meals, in order to meet the energy requirements of the body. In addition to storing food, the stomach secretes HCl and the enzyme **pepsin,** both of which contribute to the digestion of food. The digestive activity, along with the muscular mixing movements of the stomach, converts large, solid food particles into a liquid suspension of finely divided particles. The small intestine can best perform its digestive and absorptive functions when presented with contents in this form. Intestinal digestion and absorption is also optimized by regulatory mechanisms that ensure that the musculature of the stomach discharges gastric contents into the small intestine slowly; that is, the small intestine is not overwhelmed by a flood of contents from above. The stomach also secretes *intrinsic fac-*

Figure 23–9
Anatomy of the stomach.

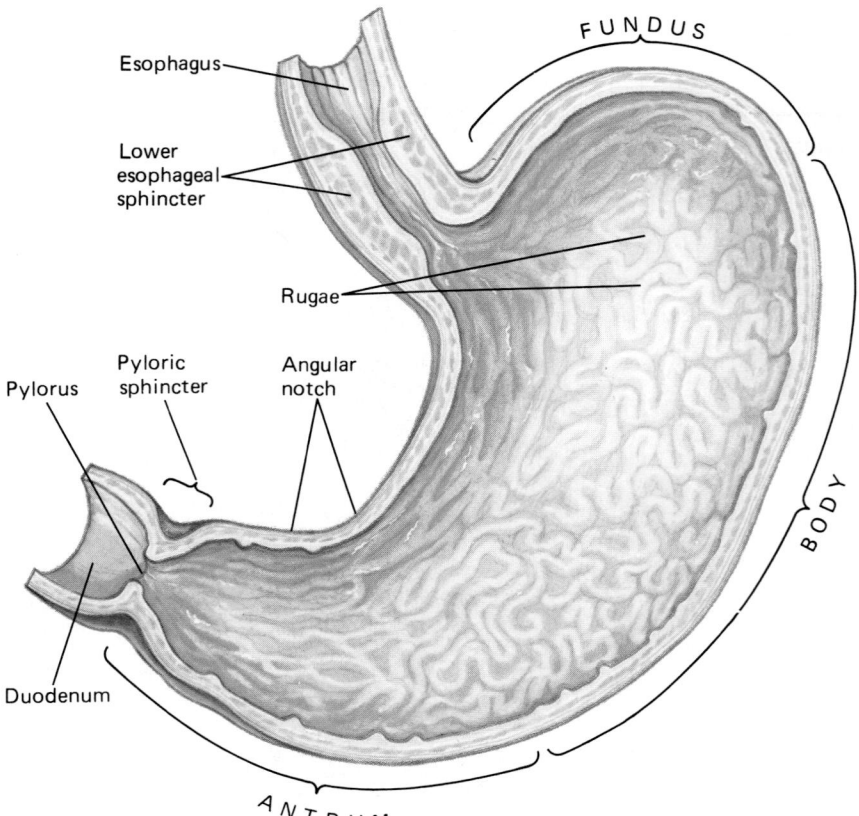

tor, a substance required for the absorption of vitamin B_{12} from the small intestine.

Motility of the Stomach

The motor activity of the stomach serves the following major functions: (1) accommodation of a meal with a relatively small increase in intragastric pressure, (2) mixing of food with gastric (stomach) secretions so that digestion can begin, (3) reduction of the size of food particles, and (4) emptying of gastric contents into the duodenum (the upper region of the small intestine).

Muscular Movements of the Stomach

Plasticity and Receptive Relaxation. The storage function of the stomach is served by the smooth muscle of the **body** and **fundus** of the stomach. These regions of the stomach can adapt to the volume of contents they contain because, as the stomach fills to capacity, intragastric pressure increases only slightly (Fig. 23–10). A property of smooth muscle that allows the stomach to accommodate large volumes with minimal changes in intragastric pressure is its ability to adjust its resting length over a wide range without large changes in resting tension (see Chapter 16). This property of **plasticity** is passive in the sense that it does not depend on neural or hormonal influences. However, a neural mechanism does contribute to the volume adaptation. As a meal is eaten, with each swallow the body and fundus of the stomach relax slightly. This **receptive relaxation** is mediated by a long reflex over the vagus nerves.

Figure 23–10
Intragastric pressure during filling of a rabbit's stomach.

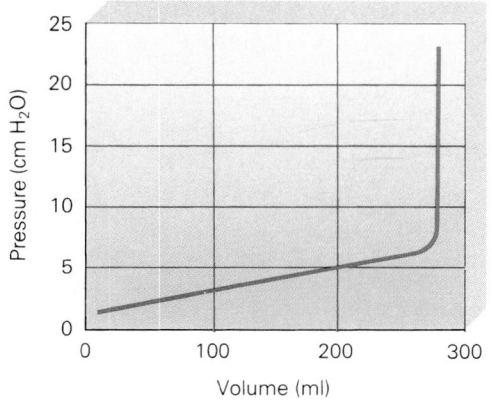

Peristalsis in the Stomach. Peristaltic contractions of the stomach mix food with gastric juice, reduce food particle size, and expel gastric contents into the duodenum. When the stomach contains food, peristaltic contractions travel over the stomach from above to below at a frequency of about three per minute. These contractions are much more vigorous in the **antrum** than in the body and fundus because antral muscle is so much thicker than elsewhere. Consequently, it is primarily the muscular activity of the antrum that generates the forces responsible for emptying the stomach. The driving force behind the emptying process is the pressure differential that develops between the antrum and the upper small intestine as a peristaltic wave travels over the antrum. At this time, pressure in the antrum rises momentarily above that in the duodenum, and a few milliliters of stomach contents move into the upper small intestine through the open **pyloric sphincter,** the thick bundle of circular smooth muscle and connective tissue separating the stomach from the small intestine. As the peristaltic wave passes over the pyloric sphincter, this structure contracts and offers enough resistance to prevent further evacuation.

Retropulsion. Another consequence of closure of the pyloric sphincter is **retropulsion,** in which stomach contents are forcibly squirted back into the body of the stomach. Retropulsion achieves a very effective mixing of food with gastric juices and also breaks lumps of food into small particles. The process of evacuation and retropulsion continues about three times a minute until the stomach is empty (Fig. 23–11).

Regulation of Stomach Motility

The Basic Electrical Rhythm of the Stomach. Peristalsis occurs at the rate of three contractions per minute because cycles of spontaneous depolarization and repolarization originate at this rate in muscle pacemaker cells at the top of the stomach. Because this activity is continuous, it has been termed the **basic electrical rhythm** of the stomach. These depolarization-repolarization cycles, which have amplitudes of 10 to 15 millivolts and durations of 1 to 4 seconds (Fig. 23–12), travel as **slow waves** through the gastric muscle from above to below as a ring around the stomach. In the absence of excitatory stimuli, the depolarizations are too small to cause the muscle membrane to reach threshold and the action potentials to be fired to elicit contraction. When excitatory influences (such as release of acetylcholine by nerve fibers or the re-

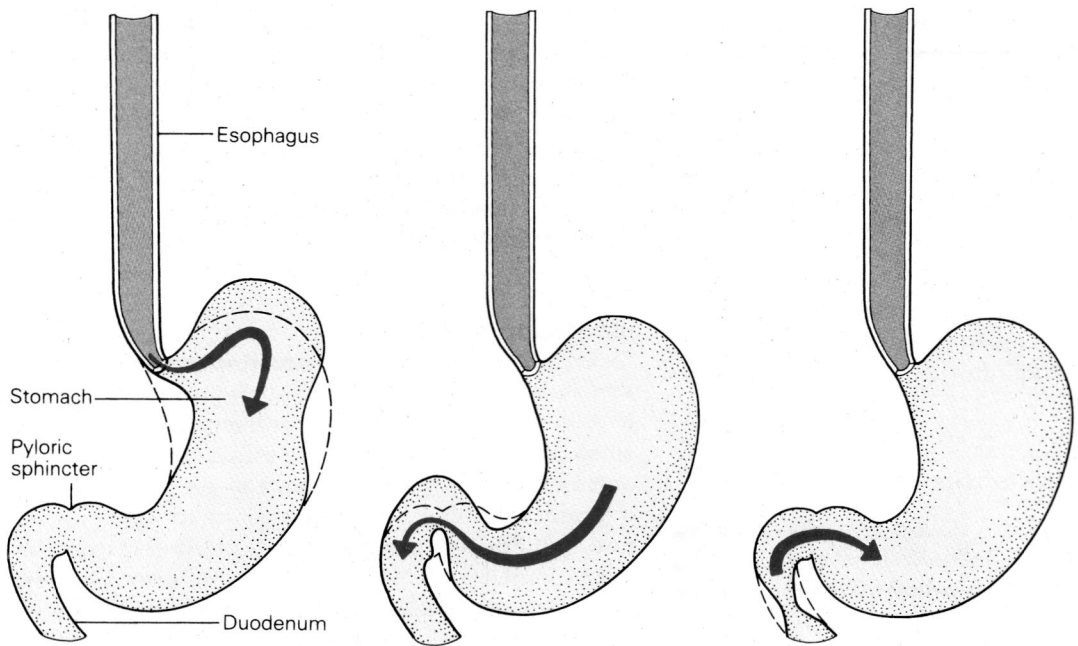

Figure 23–11
Evacuation and retropulsion of gastric contents during passage of a peristaltic wave down the stomach. (Modified from Vander, *Human Physiology*, McGraw-Hill, 1985, p. 488.)

lease of gastrin from the antral mucosa) are sufficient to produce action potentials, these occur at the peak of the slow waves because muscle excitability is closest to threshold at this time. The intensities of excitatory stimuli do not alter the frequency of the basic electrical rhythm and thus do not alter the frequency of peristalsis. On the other hand, intensity of stimulation does affect the magnitude of contraction; that is, as intensity of stimulation increases,

more action potentials are fired and contraction is more vigorous (see Fig. 23–12).

Neural and Hormonal Control. The rate at which the stomach empties is the result of the interplay between mechanisms that either excite or inhibit contraction of gastric smooth muscle. Both neural and hormonal mechanisms are involved in regulating the rate of gastric emptying.

Figure 23–12
Relationship between slow waves, action potentials, and generation of tension in gastric smooth muscle.

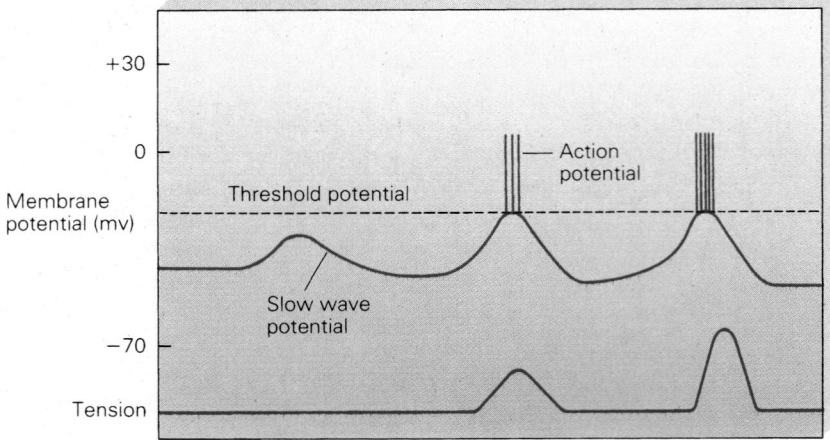

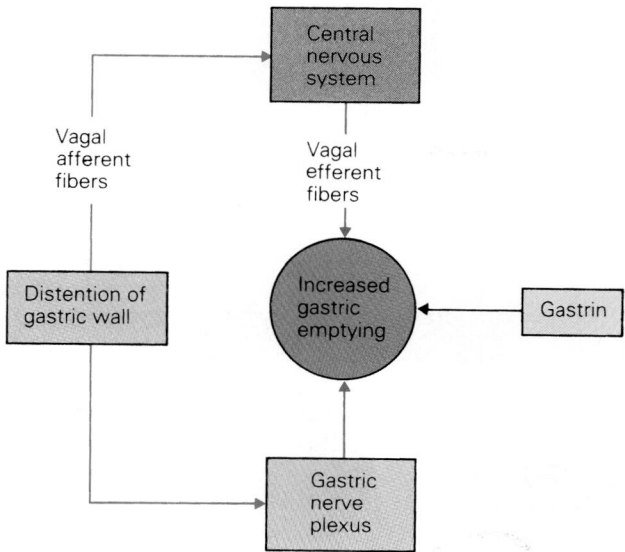

Figure 23–13
Excitation of gastric emptying.

The mechanisms that increase the vigor of gastric contractions originate in the stomach (Fig. 23–13). When the stomach is distended with food, stimulation of mechanoreceptors in the gastric wall activates short reflexes in the wall and long reflexes over the vagus nerves. Both types of reflex activity result in increased gastric motor activity. The greater the distention, the more impulses fired and the more vigorous the gastric peristalsis. Peristalsis is also augmented by the hormone **gastrin,** which is released from the antrum by intraluminal stimuli that are present during the digestion of a meal.

The excitatory influences that originate in the stomach are kept in check by inhibitory mechanisms that originate in the duodenum (Fig. 23–14a, b). These duodenal inhibitory mechanisms prevent the upper small intestine from being overwhelmed by material from the stomach. Gastric emptying is retarded through signals from the duodenum when the contents that enter this region are either high in fat, acidic (having a pH less than 3.5), hyperosmotic, or bulky enough to stretch the duodenal wall. The duodenal mucosa, when contacted by acid, releases the hormone **secretin** into the blood. This hormone is carried to the stomach, where it inhibits motor activity. **Gastric inhibitory peptide,** another hormone that inhibits gastric contractions, is released from the duodenal mucosa by the products of fat digestion. In addition, certain endocrine cells may respond to the intraluminal stimuli just listed by releasing as-yet-unidentified hormones

that inhibit gastric motility. A variety of duodenal nerve receptors, such as osmoreceptors, mechanoreceptors, and chemoreceptors, respond to these intraluminal stimuli to initiate reflex inhibition of gastric motility. These neural mechanisms, collectively called the **enterogastric inhibitory reflex,** include long reflexes over external nerves and short reflexes through the internal nerve plexuses in the duodenal and gastric walls. In general, long reflex inhibition occurs when parasympathetic (vagal) activity to the stomach is decreased or when sympathetic activity is increased. The reverse is true for excitation of gastric motility.

Although control of gastric movements largely centers on mechanisms that originate in the stomach and duodenum, gastric motility may be influenced by long reflexes initiated from any sensory region of the body. For example, stimulation of visceral and somatic pain receptors inhibits gastric motility (Fig. 23–14b). In addition, various emotional states, such as anger, fear, and depression, produce changes in gastric motor activity, although the direction of these changes is not always predictable.

Secretion of Gastric Juice

The stomach secretes about 2 to 3 liters of gastric juice every day. Most of this juice is produced by exocrine glands located in the body and fundus of the stomach. These exocrine secretory glands contain three types of secretory cells: chief cells, parietal cells, and mucous cells (Fig. 23–15).

The **chief cells,** located in the basal regions of these glands and also in the glands of the antrum, secrete **pepsinogen.** Pepsinogen is the inactive precursor of **pepsin,** an enzyme that initiates the digestion of protein in the digestive tract.

The **parietal cells** secrete the strong (highly dissociated) acid **hydrochloric acid (HCl).** Hydrochloric acid, in addition to converting pepsinogen to pepsin, also provides an optimal pH for the activity of this enzyme. This acid contributes to the breakdown of muscle fibers and connective tissue and kills bacteria that enter the digestive tract through the mouth. Hydrochloric acid is the major factor in the formation of ulcers in the upper digestive tract. The parietal cells also secrete **intrinsic factor.** This substance is necessary for the intestinal absorption of vitamin B_{12}, which is required for formation of normal red blood cells (Chapter 17).

Mucous cells are located at the necks of these glands, in the glands of the antrum, and scattered throughout the gastric epithelial lining. The alka-

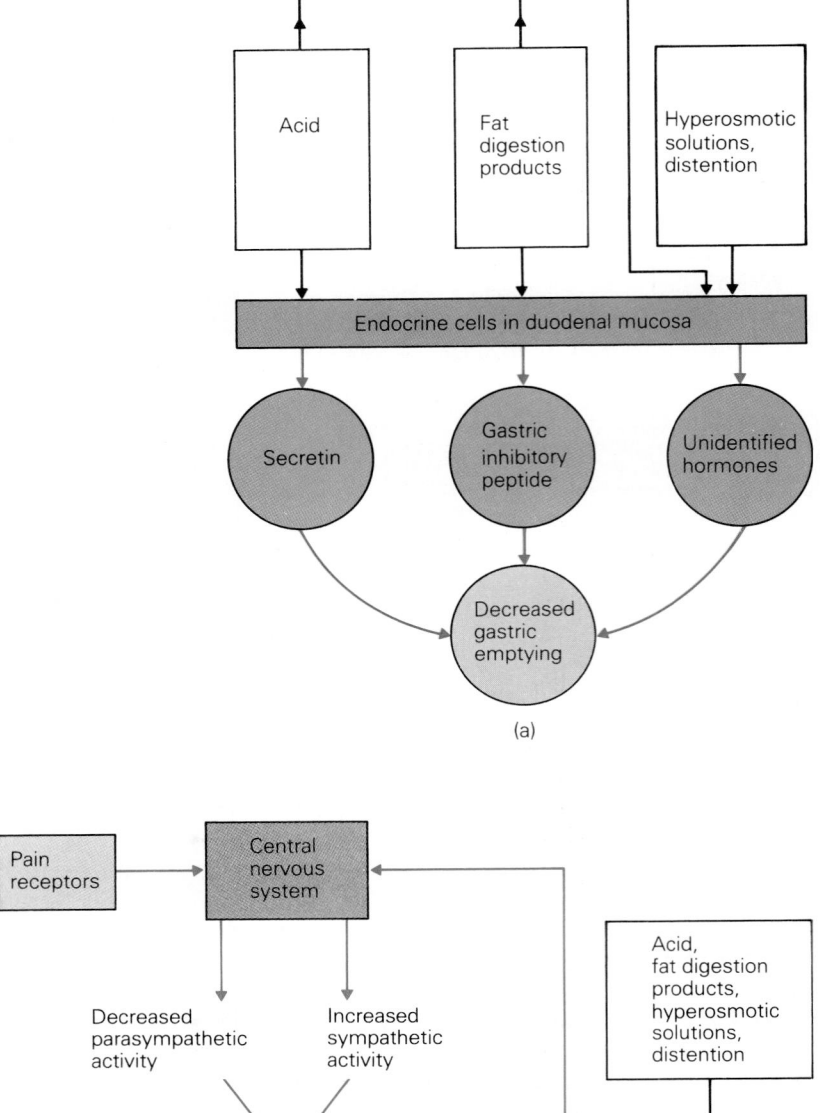

Figure 23–14
Inhibition of gastric emptying. (*a*) Hormonal pathways. (*b*) Neural pathways.

line, mucus-containing fluid that these cells secrete coats the gastric wall and protects against various forms of damage to the gastric mucosa. Its neutralizing properties provide some protection against chemical injury by HCl; the fluid also protects against mechanical injury by acting as a lubricant.

Scattered throughout the epithelial lining of the antral mucosa are endocrine cells, **the G cells,** which secrete the hormone gastrin into the portal blood. Gastrin is a potent stimulus of HCl secretion by the parietal cells.

Secretion of Acid by the Parietal Cells

The parietal cells secrete an isosmotic solution of HCl with a hydrogen ion concentration of about 150 milliequivalents per liter (pH = 0.9). Since the hydrogen ion concentration of parietal cell secretion is

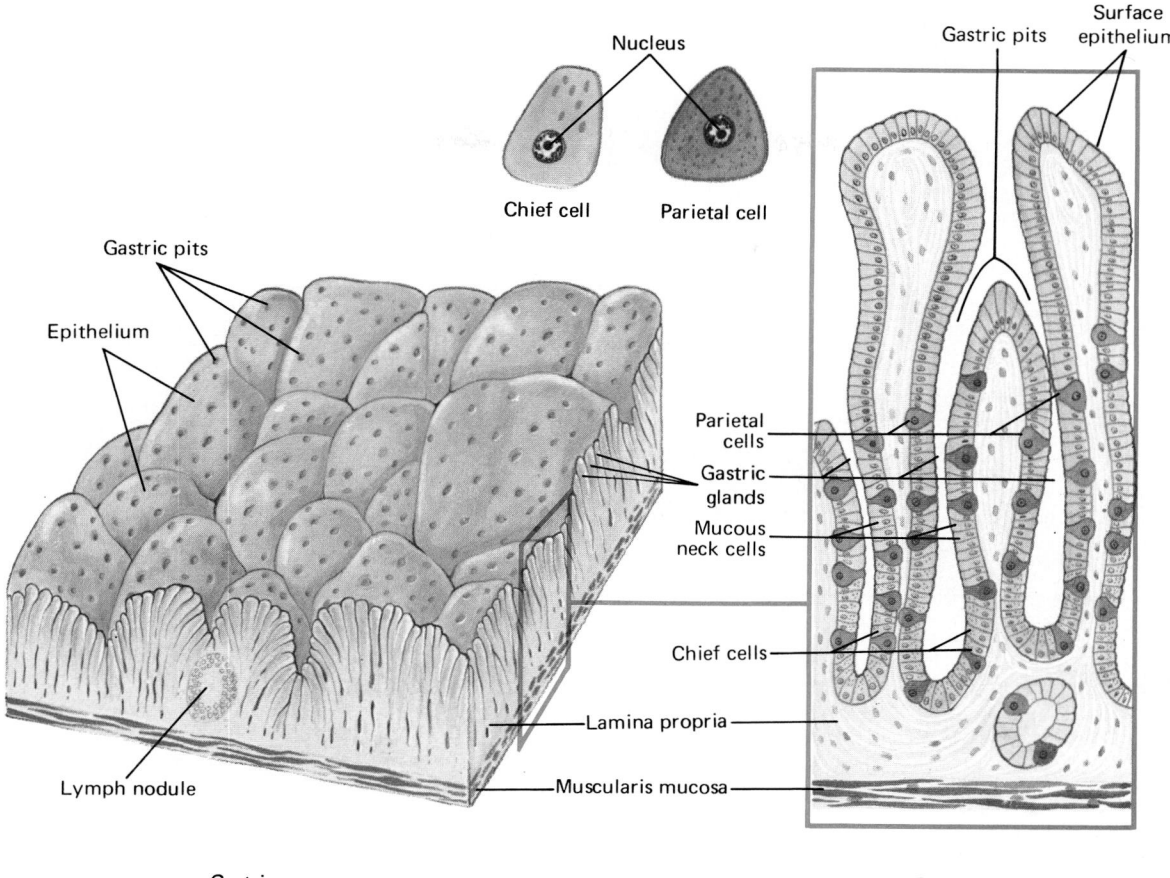

Figure 23–15
Exocrine gland from the body of the stomach.

about 3 million times that of blood, this ion moves by active transport into the gastric lumen against a large concentration gradient. One idea that has evolved concerning the mechanism of secretion of HCl is that hydrogen ions are derived from the hydrolysis of water in the parietal cells and are actively secreted into the gastric lumen by a pump located in the luminal membrane of the cell (Fig. 23–16). Energy supplied by the hydrolysis of ATP is utilized in the secretory process. For every hydrogen ion secreted, a hydroxyl ion remains in the cell, and this alkaline ion must be neutralized in order to keep intracellular pH constant. This neutralization is accomplished by the intracellular hydration of carbon dioxide to carbonic acid, a reaction catalyzed by the enzyme carbonic anhydrase, which is present in the parietal cell in high concentrations. Carbonic acid dissociates into a bicarbonate ion and a hydrogen ion, the latter combining with and neutralizing the hydroxyl ion to form water. Bicarbonate moves from

the cell interior to blood in exchange for chloride, which is actively transported from the blood to the cell interior by a pump located in the basal membrane. Chloride is then transported into the gastric lumen by a luminal membrane carrier to form HCl in the gastric lumen. Enough water is osmotically carried along with these two ions to produce an isosmotic solution of HCl.

As is the case for gastric emptying, a check-and-balance system operates in which both excitatory and inhibitory mechanisms regulate gastric HCl secretion. Excitation and inhibition of secretion occur in three phases, identified by the region in which a stimulus acts to modify the rate of secretion. These are the **cephalic, gastric,** and **intestinal** phases of gastric acid secretion.

The Cephalic Phase of Gastric Acid Secretion. During the cephalic phase (Fig. 23–17a), receptors in the head associated with the taste, smell, sight, and

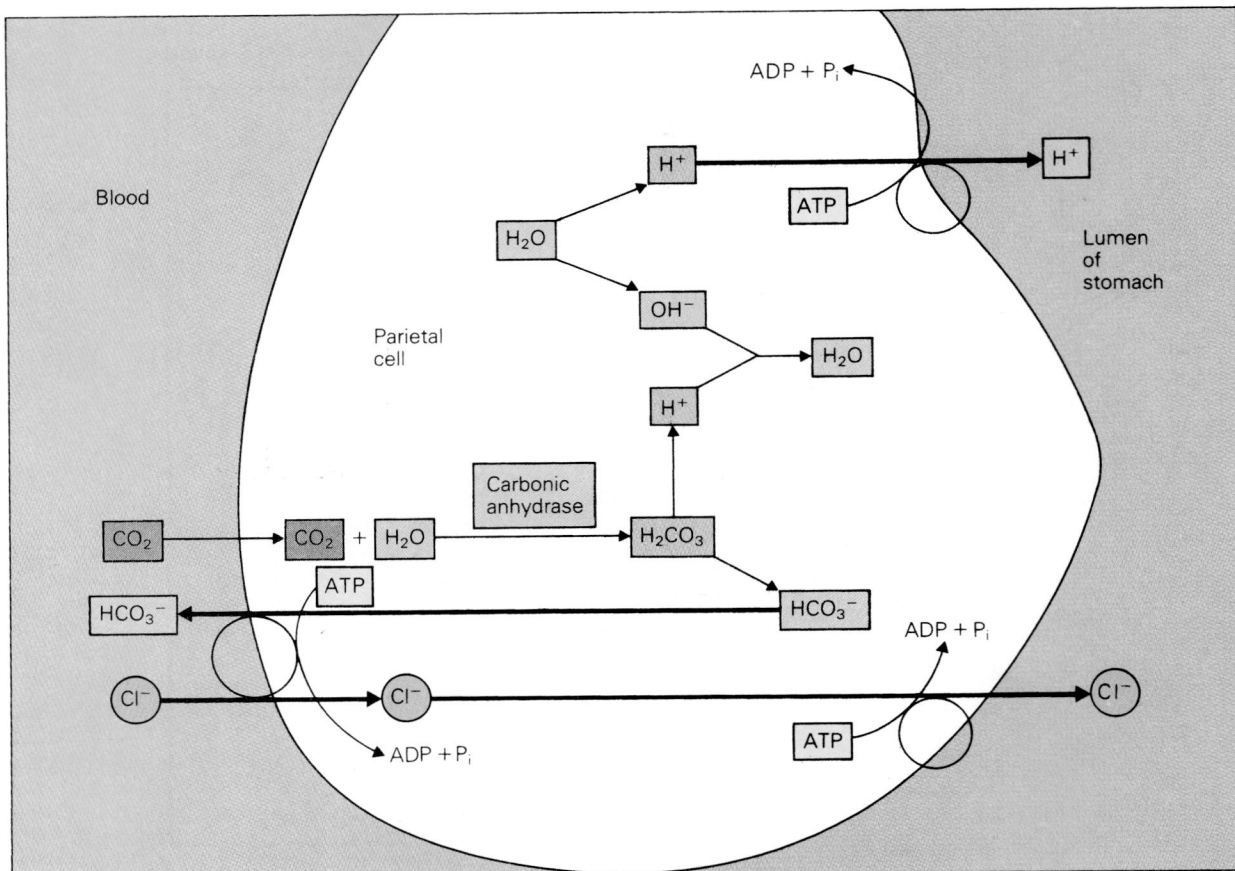

Figure 23–16
Mechanism for the gastric secretion of HCl.

chewing of food are stimulated to activate long reflexes that excite the parietal cell. The efferent fibers of these long reflex arcs are in the vagus nerves. Impulses carried over efferent vagal nerve fibers excite neurons of the internal nerve plexus. These internal neurons release acetylcholine in the vicinity of the parietal cells, stimulating these cells to secrete acid. Impulses carried over vagal nerve fibers also act through the internal nerve plexus of the antral wall to cause the release of acetylcholine at the G cell. These endocrine cells respond by secreting gastrin into the blood, which carries the hormone to the body and fundus of the stomach, where it excites secretion of acid.

The Gastric Phase of Gastric Acid Secretion. The gastric phase is triggered by certain chemical agents and by distention of different regions of the stomach when food enters (see Fig. 23–17b). For example, the products of protein digestion present in the stomach as the result of the gastric digestion of protein excite the secretion of acid by directly stimulat-

ing the parietal cells to secrete acid and also the G cells to secrete gastrin (no nerves are involved). Stimulation of mechanoreceptors in the antral wall by stretch activates short reflexes that result in the release of gastrin. Mechanoreceptors in the body and fundus are excited by stretch, resulting in stimulation of acid secretion through activation of short reflexes in the wall and long reflexes over the vagus nerves. The excitation of acid secretion that occurs as a result of the release of gastrin during the cephalic and gastric phases is checked by an inhibitory mechanism that operates during the gastric phase. Hydrochloric acid secretion is inhibited when acid contacts the antral mucosa, because HCl suppresses the release of gastrin. Thus, HCl acts as an inhibitor of its own secretion, an autoregulatory mechanism that prevents excessive secretion of acid.

The Intestinal Phase of Gastric Acid Secretion. Inhibition of secretion also occurs during the intestinal phase. Acid, various digestion products, hyperosmotic solutions, and stretch activate inhibitory

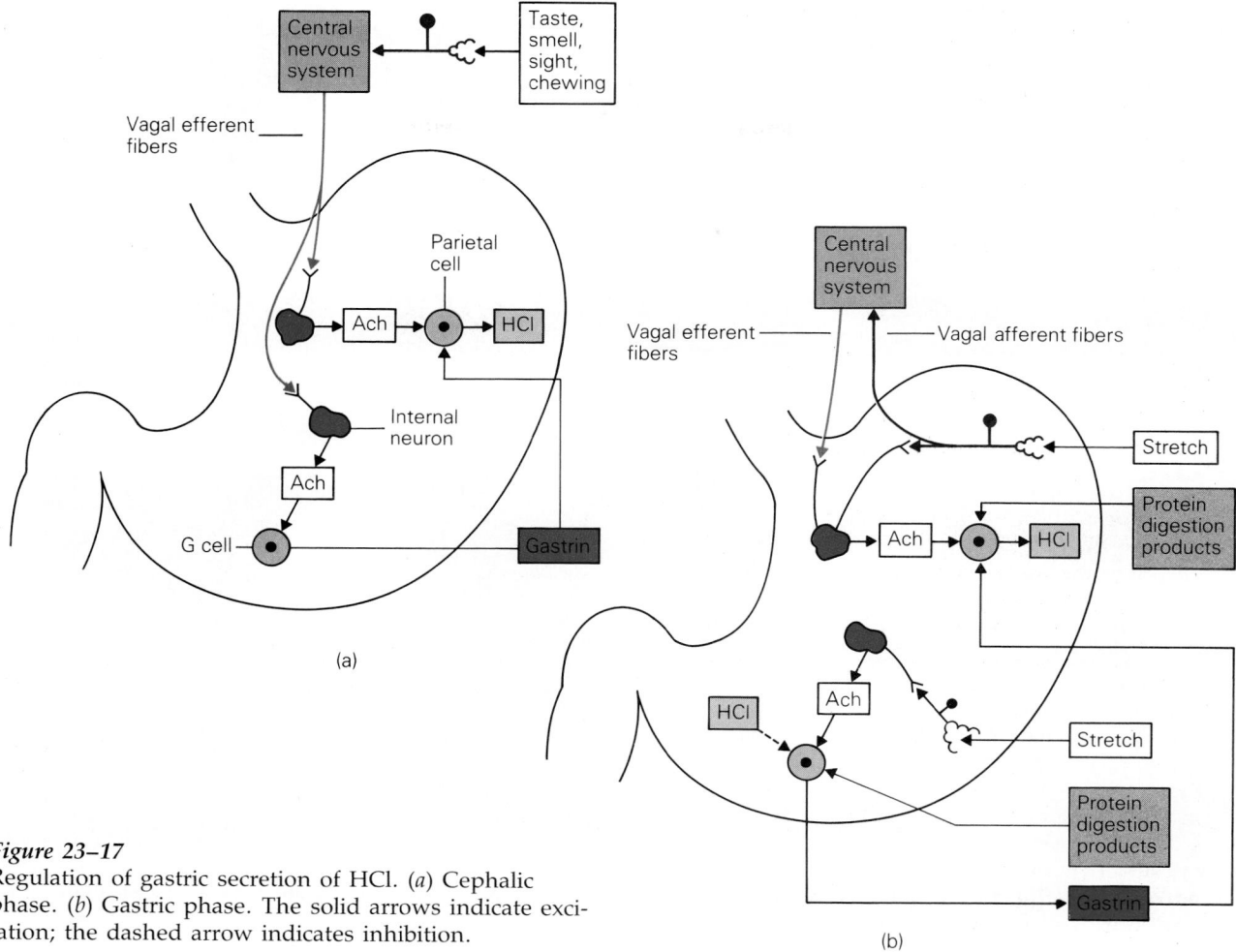

Figure 23–17
Regulation of gastric secretion of HCl. (*a*) Cephalic phase. (*b*) Gastric phase. The solid arrows indicate excitation; the dashed arrow indicates inhibition.

mechanisms that originate in the duodenum. Both neural and hormonal pathways operate in the duodenal inhibition of gastric acid secretion, and these are the same pathways involved in the duodenal inhibition of gastric motility (see Fig. 23–14*b*). These duodenal inhibitory mechanisms prevent large quantities of acid from entering the upper small intestine.

Secretion of Pepsin by the Chief Cells

Pepsinogen, the inactive precursor of the protein-digesting enzyme pepsin, is synthesized by and stored in the chief cells as packets called **zymogen granules.** Since pepsinogen cannot hydrolyze proteins, chief cells are safe from being destroyed by digestion of their own protein molecules.

Once pepsinogen enters the gastric lumen, however, both hydrochloric acid and pepsin convert it to pepsin. Hydrochloric acid initiates the conver-

sion, and the pepsin that is formed carries this process to completion very rapidly. Pepsin hydrolyzes proteins, breaking them into **peptide chains** of varying lengths. Digestion of protein by pepsin occurs most effectively at a pH of 2 to 3, an acidic environment commonly encountered in the lumen of the stomach during the digestion of a meal.

Since the most effective stimulus for pepsinogen secretion is acetylcholine, the short and long reflexes that occur during the cephalic and gastric phases of digestion and that result in the release of acetylcholine at the chief cells are important in determining pepsin levels in the gastric contents. Note that both acetylcholine and peptides, by virtue of their ability to excite HCl secretion by the parietal cell, play important roles not only in the activation of pepsin but also in the maintenance of the highly acidic environment required for optimal activity of this enzyme. The pathways in the secretion and activation of pepsinogen are summarized in Figure 23–18.

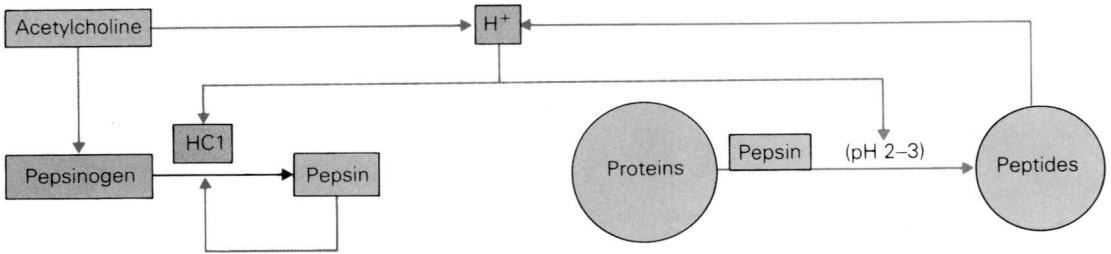

Figure 23–18
Pathways in the secretion and activation of pepsinogen.

CLINICAL FOCUS

Ulcers

An **ulcer** is an erosion of the mucosa of certain regions of the digestive tract. This erosion can extend deep into the wall of the tract, damaging blood vessels, with resultant bleeding. A **perforated ulcer** extends entirely through the wall of the tract, allowing food, gastrointestinal secretions, and bacteria to escape into the peritoneal cavity. This in turn may result in peritonitis, shock, and even death.

The dominant factors in the development of ulcers are the corrosive action of HCl and autodigestion by pepsin, as evidenced by the fact that ulcers occur mostly in regions of the digestive tract exposed to acid and pepsin. These regions are the upper small intestine (where the most prevalent type of ulcer, **duodenal ulcer,** forms), the stomach (where **gastric ulcers** occur at a frequency about one tenth that of duodenal ulcers), and occasionally the esophagus (when this organ is subjected to reflux of gastric contents for prolonged periods).

Two major reasons are generally given for the development of ulcers. One is hypersecretion of acid and pepsin, which overwhelms the natural defenses of the digestive tract. The other is lack of protection against these secretory components.

Hypersecretion of acid and pepsin may result from excessive activity of the mechanisms that excite gastric secretion. Certain surgical procedures have been used to reduce acid and pepsin secretion in ulcer patients. These procedures include vagotomy of the stomach (which eliminates the cephalic phase of gastric acid secretion), pyloric antrectomy (which removes the source of gastrin), and removal of a part of the acid- and pepsin-secreting regions of the stomach. More recently, antisecretory drugs such as **cimetidine** and **ranitidine** have been effectively used in ulcer therapy.

The most important mechanism that protects against gastric ulcer is the **gastric mucosal barrier.** The permeability characteristics of the cell membranes of the gastric mucosa restrict the entrance of

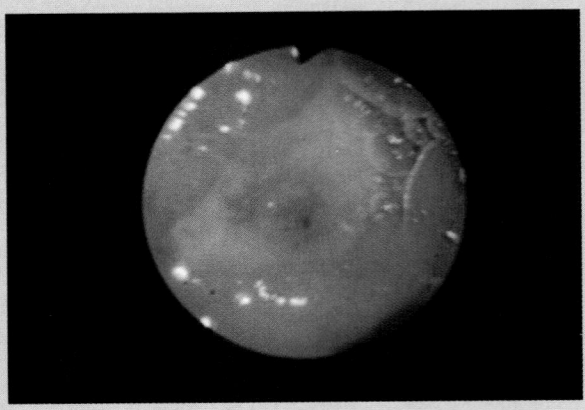

Photograph of a duodenal ulcer. (© Custom Medical Stock Photo.)

ions, even ions as small as hydrogen ions, to the interior of these cells. Thus, HCl does not usually enter these cells in quantities sufficient to destroy them. Substances such as acetylsalicylic acid (aspirin) and ethanol increase the permeability of these cell membranes, allowing acid to penetrate the cells and destroy them.

Another important protective mechanism is the neutralization of acid by food and by the alkaline secretions of the upper digestive tract. For example, the pancreas secretes an alkaline juice into the duodenum that very efficiently neutralizes acid arriving from the stomach, thereby protecting against the possibility of duodenal ulcer. The principle of neutralization is widely used in the treatment of ulcers. Various commercial alkalies (bases) and buffers are prescribed for this purpose. Another common approach is the eating of five to six well-spaced meals during the day, which, by keeping food in the stomach much of the time, keeps gastric acidity at a relatively low level.

Finally, the rapidity (every 1 to 3 days) with which cells lining this organ divide prevents lasting injury to the mucosa of the stomach.

Vomiting

Vomiting is the rapid expulsion of gastric contents through the mouth. The expulsion of gastric contents is preceded by a copious secretion of saliva, increased heart rate, sweating, pallor, and nausea, all characteristic of a general discharge of the sympathetic nervous system.

Vomiting begins with a deep inspiration and closure of the glottis and nasal passages. Contraction of skeletal muscle provides the major force for expulsion; descent of the diaphragm during a forced inspiration and contraction of the abdominal muscles raise abdominal, and thus intragastric, pressure. The increased pressure in the stomach forces gastric contents into the esophagus through the upper esophageal sphincter, which relaxes, and into the mouth. Vomiting is frequently preceded by **retching,** a process by which contents are forced into the esophagus, but not into the mouth, because the upper esophageal sphincter is not relaxed. A series of retches often leads to vomiting.

The complicated reflex act of vomiting is coordinated by a center located in the medulla. Potent stimuli arise from sensory nerve endings in the throat—for example, when a finger is stuck in the back of the throat. Other receptor areas include all regions of the digestive tract (distention of the stomach and small intestine), other abdominal viscera, and the labyrinths of the ear when rotational movements of the head result in motion sickness. Certain chemical agents, called **emetics,** cause vomiting by stimulating chemoreceptors in the stomach and small intestine or in the brain. When vomiting is prolonged and large quantities of gastric contents are lost, disturbances in water volume, electrolyte balance, and acid-base balance result (Chapter 26).

The Process of Secretion in the Exocrine Pancreas

The pancreas (Fig. 23–19a) is both an endocrine and an exocrine secretory organ. In addition to containing endocrine cells that secrete the hormones insulin and glucagon (Chapter 15), the pancreas contains exocrine secretory cells called **acinar cells** that are arranged in saclike clusters called **acini** (singular, *acinus,* "rounded lobule") (Fig. 23–19b). These cells secrete a small volume of juice containing the digestive enzymes of the pancreas. The acini are connected by small **intercalated ducts** to larger ducts that converge into one duct (sometimes two) that delivers the exocrine secretion of the pancreas

into the duodenum. The epithelial cells lining the intercalated ducts make an important contribution to the overall exocrine secretion of the pancreas. These cells secrete a relatively large volume of juice with a high concentration of the alkaline salt **sodium bicarbonate.** By neutralizing acid that arrives in the duodenum from the stomach, bicarbonate performs two functions: (1) it protects the delicate lining of the upper small intestine from damage by acid and (2) it provides an environment with the optimal pH for activity of the pancreatic digestive enzymes.

The Secretion of Pancreatic Bicarbonate

The mechanism of secretion of pancreatic bicarbonate by the intercalated duct cells of the pancreas is shown in Figure 23–20. The bicarbonate is formed when CO_2 diffuses into the cell and is converted to bicarbonate and hydrogen ions in a reaction catalyzed by the enzyme carbonic anhydrase. Bicarbonate ions are actively transported from the duct cells to the duct lumen, whereas hydrogen ions are actively transported into the interstitial fluid on the blood side of the cell. Hydrogen ions are pumped out of the cell in exchange for sodium ions that passively diffuse into the cell from interstitial fluid. Sodium passively follows the secreted bicarbonate along with enough water to form an isosmotic solution of sodium bicarbonate. Recall from the discussion of secretion of HCl by the gastric parietal cells that bicarbonate enters the blood as a part of this secretory process, and this alkalinizes the blood leaving the stomach. However, since an equivalent amount of hydrogen ion enters the blood during secretion of pancreatic bicarbonate, overall blood pH remains unchanged during the digestion of a meal.

The Secretion of Pancreatic Enzymes

Pancreatic juice is the most versatile of the digestive secretions because its enzymes are capable of nearly completing the digestion of food in the absence of all other digestive secretions. These enzymes require a pH close to neutrality for optimal activity, and this is provided by the alkaline secretion of the pancreatic intercalated duct cells. The actions of the pancreatic enzymes and the products of their digestive activity are summarized in Table 23–4.

The pancreas secretes three protein-digesting enzymes. **Trypsin** is secreted as inactive **trypsinogen,** which is converted to active trypsin by **enterokinase,** an enzyme present in the luminal membrane of the epithelial lining of the small intestine.

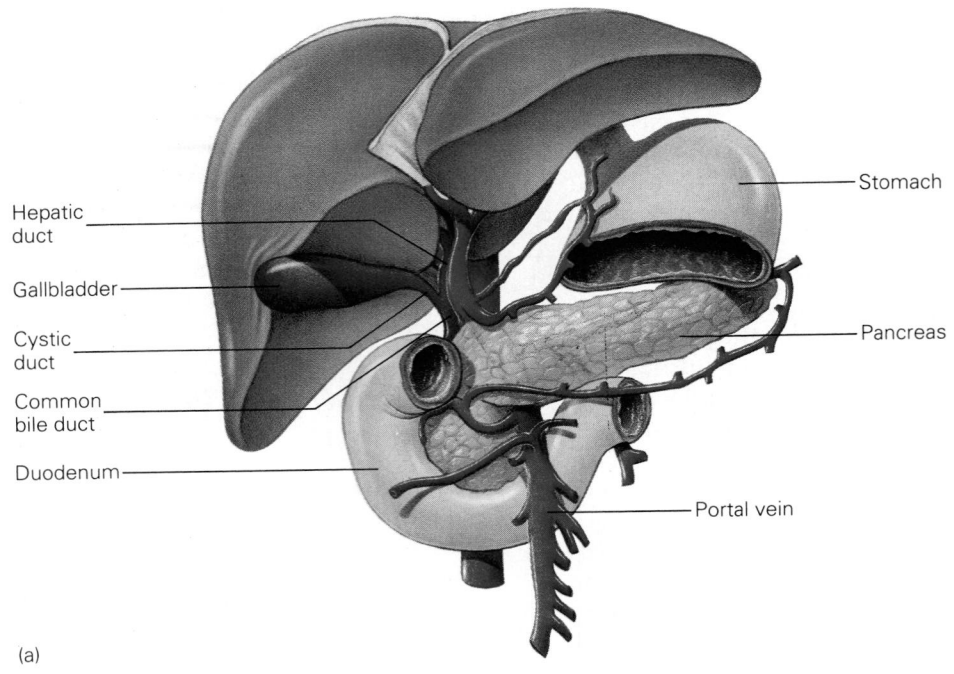

Hepatic duct

Gallbladder

Cystic duct

Common bile duct

Duodenum

Stomach

Pancreas

Portal vein

(a)

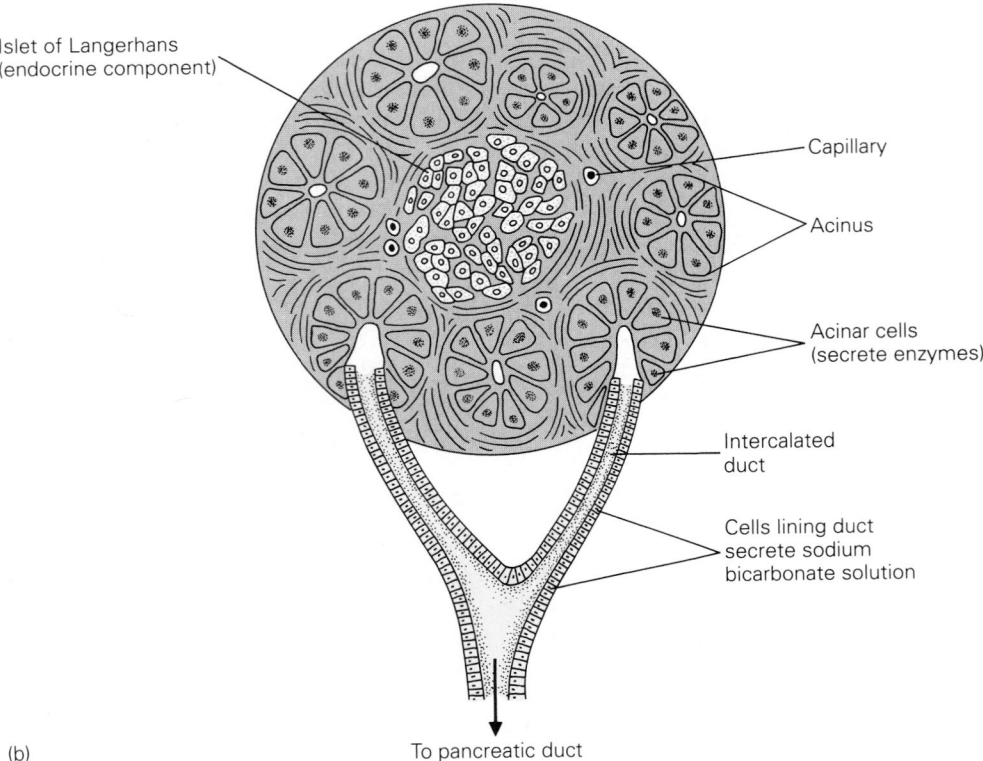

Islet of Langerhans (endocrine component)

Capillary

Acinus

Acinar cells (secrete enzymes)

Intercalated duct

Cells lining duct secrete sodium bicarbonate solution

To pancreatic duct

(b)

Figure 23–19
(*a*) Gross anatomy of the pancreas, duodenum, and biliary system. (*b*) Micro-anatomy of the exocrine pancreas.

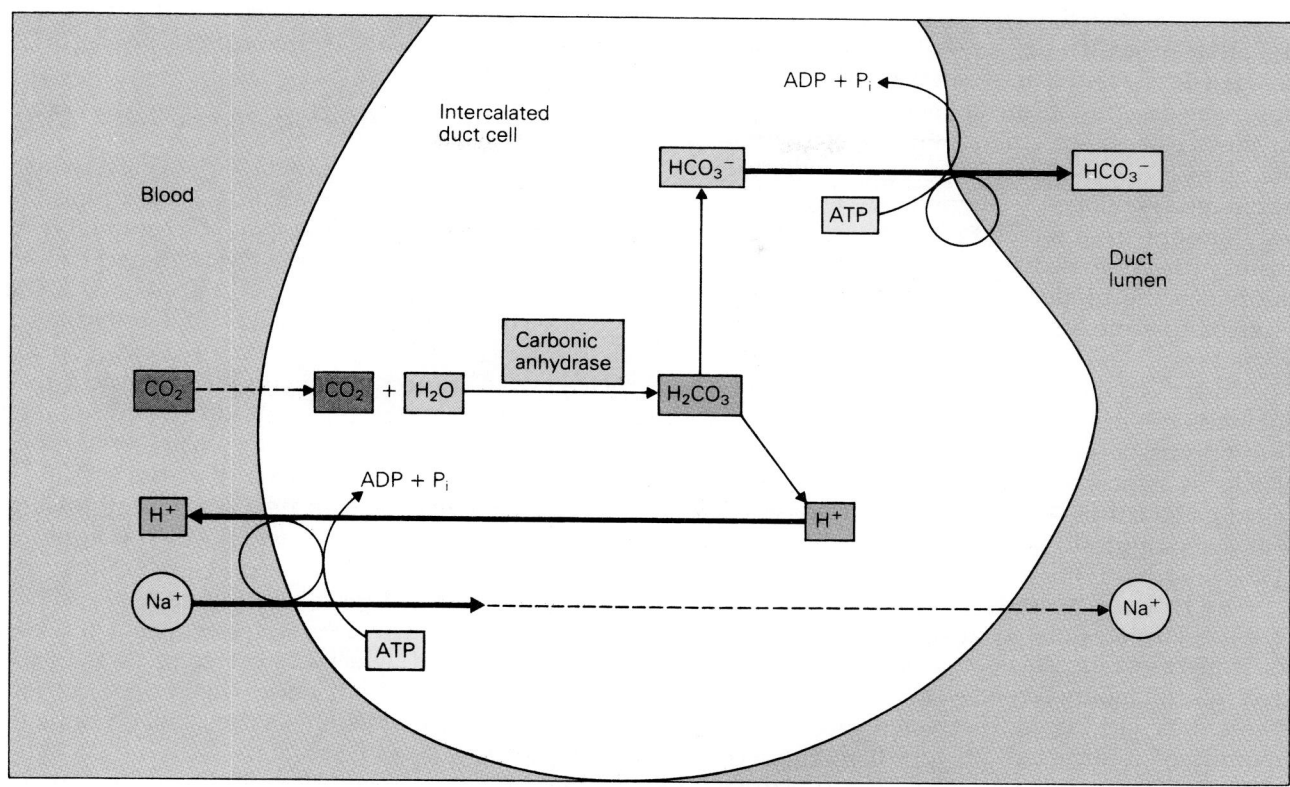

Figure 23–20
Mechanism for the pancreatic secretion of bicarbonate. Solid arrows indicate active transport and dashed arrows indicate passive transport.

Table 23–4
Digestive Enzymes of Pancreatic Juice

Enzyme	Substrate	Action	Products of Digestion
Proteolytic			
Trypsin, chymotrypsin	Proteins and peptides	Splits amino acid linkages in the interior of proteins and peptides	Small peptides
Carboxypeptidase	Proteins and peptides	Splits terminal amino acids with free carboxyl groups from the ends of proteins and peptides	Peptides and amino acids
Amylolytic			
Amylase	Polysaccharides	Splits internal glucose bonds of polysaccharides	Oligosaccharides
Lipolytic			
Lipase	Triacylglycerols	Splits two fatty acids from triacylglycerols	Free fatty acids and monoglycerides
Phospholipase	Lecithin	Splits one fatty acid from lecithin	Lysolecithin
Cholesterol esterase	Cholesterol esters	Splits the ester bond in cholesterol esters	Cholesterol and fatty acids
Nucleolytic			
Ribonuclease, deoxyribonuclease	Nucleic acids	Splits nucleic acids	Mononucleotides

Trypsin activates the remaining trypsinogen and the other proteolytic enzymes; that is, **chymotrypsinogen** is converted to **chymotrypsin** and **procarboxypeptidase** to **carboxypeptidase**. Trypsin and chymotrypsin split certain amino acid linkages in the interior of proteins and peptides, whereas carboxypeptidase liberates amino acids with free carboxyl groups from the end of proteins and peptide chains. The combined actions of gastric pepsin and the pancreatic protein-digesting enzymes degrade proteins to a mixture of small peptides and amino acids.

Under certain conditions, activated proteolytic enzymes are liberated from the acinar cells into the surrounding pancreatic tissue. These enzymes may attack the pancreas itself, producing the condition called **pancreatitis.** This self-digestion process often reduces output of both pancreatic digestive enzymes and bicarbonate.

The pancreas secretes **lipase,** the only enzyme of the gastrointestinal system capable of digesting triacylglycerol fats. This enzyme hydrolyzes triacylglycerols to monoglycerides and free fatty acids. When it is lacking, as in pancreatitis, excessive amounts of undigested fat appear in the feces, a condition known as **steatorrhea.** Other lipid-digesting enzymes are **phospholipase** and **cholesterol esterase.**

The pancreas also secretes an **amylase,** the action of which is similar to that of salivary amylase; that is, it degrades starch to oligosaccharides, mainly maltose. Other digestive enzymes secreted by the pancreas are the nucleic acid-digesting enzymes, **ribonuclease** and **deoxyribonuclease.**

The Regulation of Pancreatic Secretion

The hormones secretin and cholecystokinin, which are released from the duodenal mucosa during the intestinal phase of digestion, play the central role in the control of secretion by the pancreas. Acid is the most potent stimulus for release of secretin into the blood. This hormone stimulates the intercalated duct cells of the pancreas to secrete a bicarbonate-rich juice. This secretion neutralizes the acid that causes secretin to be released and so reduces the stimulus for additional secretin release until more acid enters the duodenum from the stomach.

The duodenal mucosa releases cholecystokinin when contacted by protein and fat digestion products. Cholecystokinin stimulates the pancreatic acinar cells to secrete digestive enzymes. As a result of the actions of these enzymes, more digestion products are produced, which in turn stimulate the pancreas to secrete more enzymes. This process continues as long as food remains in the small intestine to be digested.

Secretin and cholecystokinin have a potentiating effect on the activity of one another. For example, secretin alone is a strong stimulus for bicarbonate secretion. However, this response is increased in the presence of cholecystokinin. The pathways involved in the regulation of secretion of pancreatic juice are summarized in Figure 23–21.

Figure 23–21
Pathways in the regulation of exocrine pancreatic secretion.

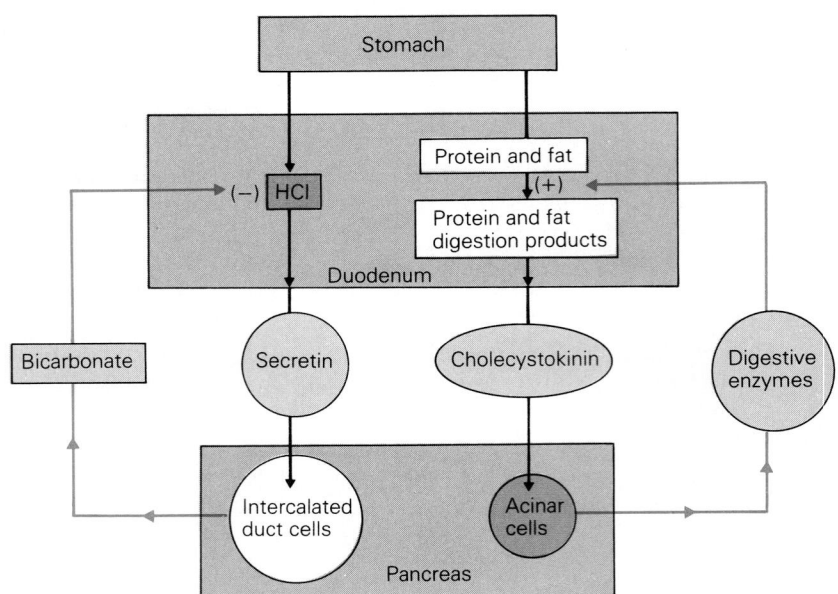

The Process of Secretion in the Biliary System

Bile is secreted by liver cells (**hepatocytes**) into minute channels called **bile canaliculi,** which drain into the **hepatic duct.** When food is not being digested, bile is diverted by way of the **cystic duct** into the **gallbladder,** a small sac that stores bile between meals (see Fig. 23–19). Following a meal, contractions of the gallbladder move bile through the **common bile duct** and the **sphincter of Oddi** into the small intestine, where it plays its role in the digestion and absorption of fat.

The molecules present in bile that are important in these processes are the **bile salts.** Other major components of bile are the **bile pigments.** These molecules are breakdown products of hemoglobin (see Chapter 17) and are for the most part excreted in the feces. Two more important organic constituents of bile are the lipid molecules **cholesterol** and **lecithin.** Inorganic ions of bile include sodium, the major cation, and chloride and bicarbonate, the predominant anions.

The Gallbladder: The Storage of Bile

Between meals, bile is prevented from entering the small intestine and is shunted into the gallbladder by the contraction of the **sphincter of Oddi,** a ring of smooth muscle surrounding the common bile duct in the region where this duct enters the duodenum. When food enters the upper small intestine, fat and protein digestion products stimulate the release of cholecystokinin from the duodenal mucosa into the blood. This hormone stimulates contraction of the gallbladder and relaxation of the sphincter of Oddi. These muscular movements result in the flow of bile into the small intestine.

While bile is stored in the gallbladder, sodium is actively transported from the gallbladder lumen to the blood by the epithelial cells that line the inner aspect of this organ. The osmotic gradient produced by the movement of sodium and accompanying anions (chloride and bicarbonate) causes water to follow across the wall of the gallbladder, with the result that the organic constituents of bile, which are not absorbed, become concentrated in the process (Table 23–5). Despite the fact that the organic con-

C L I N I C A L   F O C U S

Gallstones

The presence of gallstones in the gallbladder and bile ducts is one of the most common disorders of the gastrointestinal system. Cholesterol, a normal constituent of bile, is the major component of most gallstones. Cholesterol itself is virtually insoluble in water. However, two other constituents of bile, the bile salts and lecithin, together can dissolve cholesterol by forming water-soluble aggregates called **micelles** that contain all three of these molecules.

Whenever cholesterol precipitates out of solution, small crystals of cholesterol are formed, and these can grow gradually into large gallstones. This can happen when the capacity of the bile salts and lecithin to dissolve cholesterol is exceeded—when, for example, the liver secretes higher-than-normal quantities of cholesterol, or when the amount of bile salts secreted is lower than normal.

When the bile ducts are obstructed by a gallstone, the flow of bile into the small intestine is retarded or stopped. Since secretion of bile by the liver continues, pressure in the biliary system rises to higher-than-normal levels, and severe pain may result from stimulation of receptors in the wall of the biliary tree.

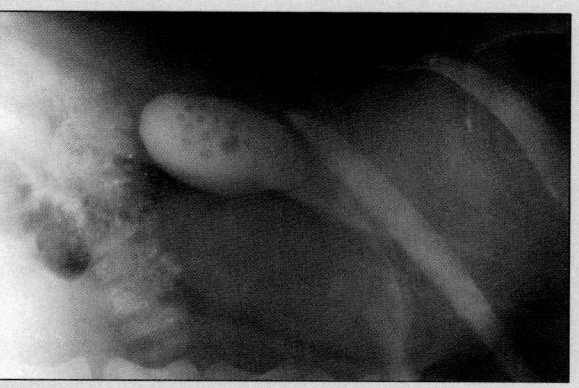

X-ray showing calcified gallstones in human gallbladder. (© Biophoto Associates/Science Source/ Photo Researchers, Inc.)

Removal of either the gallbladder (**cholecystectomy**) or of stones from the bile ducts has been the usual treatment for gallstones. More recently, substances with the capacity to dissolve cholesterol gallstones, such as the bile acid **chenodeoxycholic acid,** have been given to patients with encouraging results.

Table 23–5
Composition of Human Bile

Constituent	Liver Bile (%)	Gallbladder Bile (%)
Water	97.48	83.98
Bile pigments	0.53	4.44
Bile salts	0.93	8.70
Fatty acids	0.12	0.85
Cholesterol	0.06	0.87
Lecithin	0.02	0.14
Inorganic salts	0.83	1.02

stituents of bile are concentrated by the removal of fluid from the gallbladder, the water-insoluble lipid molecules of bile usually remain in solution.

The Bile Salts and Bile Pigments

Bile salts are synthesized by the liver hepatocytes and actively transported by these cells into the bile canaliculi. Once the bile salts enter the small intestine, they play an essential role in the digestion and absorption of fat, a topic that will be discussed later in this chapter. These molecules traverse the small intestine and are absorbed from the lower part of this organ, the **terminal ileum,** by an active process specialized for their transport. They are then returned to the liver, where the hepatocytes remove them from the blood and re-secrete them into the bile. This recycling of bile salts between liver and small intestine is called the **enterohepatic circulation of bile salts** (Fig. 23–22). The process is interrupted when the bile salts are stored in the gallbladder.

The body's pool of bile salts is circulated a number of times every day. This process is not 100% efficient, because small quantities of bile salts enter the large intestine with each circulation and are excreted in the feces. However, synthesis by hepatocytes of new bile salts from the structurally related molecule cholesterol easily makes up this loss. The amount of bile salts synthesized depends on the portal blood concentration of these molecules; that is, the rate of synthesis is high when portal blood concentration is low, and vice versa. Synthesis is regulated through inhibition by circulating bile salts of one of the steps in the series of reactions leading to the conversion of cholesterol to bile salts (see Fig. 23–22).

Bile pigments are end products of the breakdown of the heme portion of the free hemoglobin that results from the destruction of senescent red blood cells by the reticuloendothelial system (Chapter 17). The major pigment of bile is **bilirubin.** Liver hepatocytes extract this molecule from the blood

Figure 23–22
The enterohepatic circulation of bile salts as shown by the heavy solid arrows. The dashed arrow represents inhibition of bile salt synthesis.

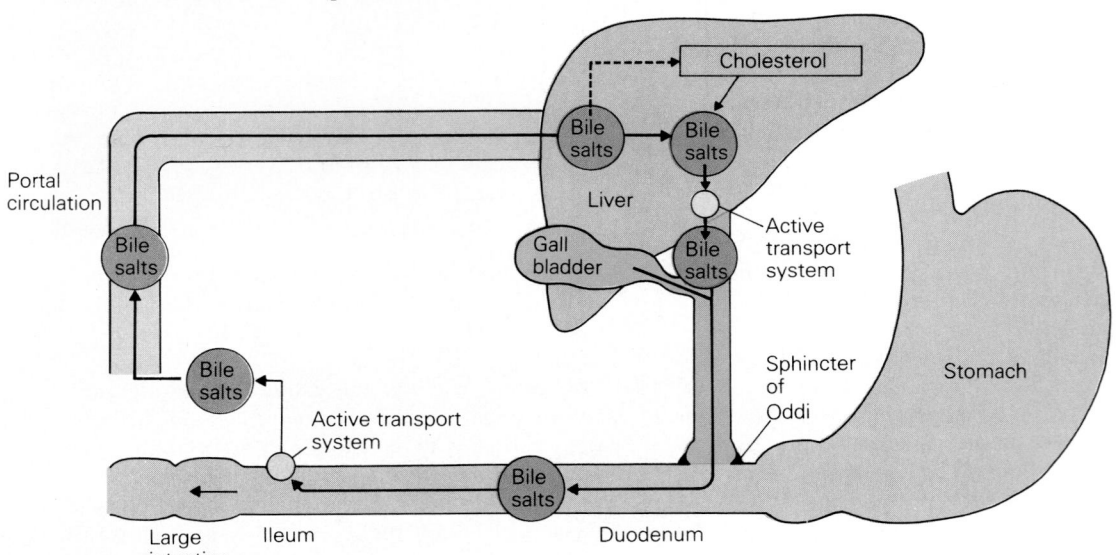

and actively transport it into the bile. Enzymes of intestinal bacteria convert bilirubin into other pigments, most of which are excreted in the feces, giving them a characteristic brown color. The remaining pigments are absorbed from the intestine and excreted by the kidneys, giving urine its yellow color. If for any reason bile pigments accumulate in the blood and tissues rather than being excreted, the skin becomes yellow, a condition called **jaundice.**

The Regulation of Bile Secretion

Bile salts are the major stimuli for the secretion of bile. As these molecules are removed from the portal circulation and actively secreted into the liver canaliculi, water and inorganic ions passively follow along in proportion to the quantity of bile salts transported. Consequently, the greater the concentration of bile salts in the portal blood, the greater the transport of bile salts and the volume of bile secreted. Portal blood concentrations of bile salts are greatest—and thus bile flow is at its highest level—when food is being digested, at which time the bile salts are being circulated between liver and small intestine. Between meals, when they are stored in the gallbladder, the concentration of bile salts in portal blood is low, and liver bile flow is minimal.

The hormone secretin also stimulates the secretion of bile, but in this instance the cells stimulated are the epithelial cells that line the bile ducts. These cells respond to secretin with an increased output of water and inorganic ions, particularly sodium bicar-

bonate. This is analogous to secretin stimulation of the pancreas in that secretin also stimulates the duct cells of this organ to secrete a bicarbonate-rich juice practically devoid of organic substances. In both instances, the resulting secretion helps neutralize acid that enters the small intestine from the stomach.

The Processes of Motility, Secretion, and Absorption in the Small Intestine

The small intestine is a tube about 3 m long that includes an initial 20- to 25-cm segment, the **duodenum.** The remainder of the tube is divided equally between **jejunum** and **ileum.** The digestion of food is completed by enzymes in the epithelial cells that line the inner surface of this organ, and the products of digestion are absorbed into the blood and the lymph. Smooth muscle contractions of the small intestine facilitate digestion and absorption and move unabsorbed food residues and other excretory products into the large intestine.

Motility of the Small Intestine

The most frequently occurring type of muscular movement of the small intestine is the **segmenting contraction,** which is illustrated diagrammatically in Figure 23–23. The segmenting contractions move the contents of different regions of the small intes-

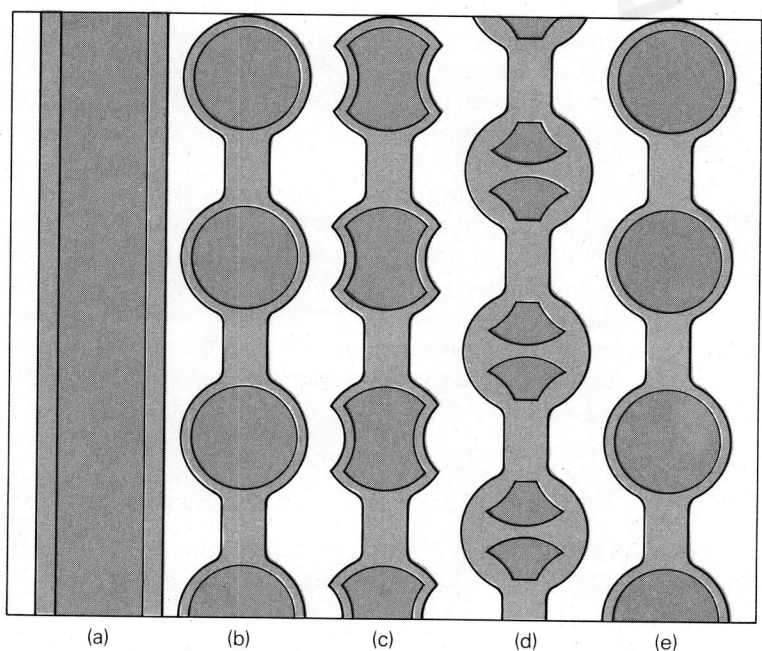

(a)   (b)   (c)   (d)   (e)

Figure 23–23
Diagram of segmenting contractions of the small intestine. (*a*) Absence of contractile activity. (*b*) The circular muscle contracts at a number of places, dividing a region of the intestine and its contents into a series of segments. (*c,d*) Each segment divides, with adjoining halves and their contents coming together to form new segments. (*e*) The original pattern is re-established.

tine back and forth over short distances. In doing so, they mix the contents with the digestive secretions and also bring the products of digestive activity into contact with the absorptive epithelium. The frequency of these movements is determined by the frequency of the slow waves initiated by pacemaker cells in the longitudinal muscle of different regions of the intestinal wall. These basic electrical rhythms progressively decrease from above to below. About 11 segmenting contractions occur every minute in the duodenum during the digestion of food, whereas only about 8 occur per minute for the ileum. This difference in the rate of contraction means that more intestinal contents are forced downward than upward, with the net effect of propelling contents toward the large intestine.

Peristalsis is the other major type of motor activity of the small intestine, and it occurs after most of a meal has been absorbed. These waves of contraction move downward slowly and travel only short distances, providing the slow transit required for the completion of the absorptive processes.

The last 2 to 3 cm of the musculature of the ileum is thicker than the rest of the ileum. This region, the **ileocecal sphincter,** is normally closed by contraction. However, during periods of increased motor activity of the ileum, the sphincter relaxes, allowing contents to enter the large intestine. It then contracts again to prevent backflow of contents into the ileum.

Chemical and mechanical stimulation of receptors in the intestinal wall by intestinal contents initiate both short and long reflexes important in the regulation of small intestinal motility. The short reflexes are mediated through the internal nerve plexuses to the smooth muscle in the intestinal wall. In the case of long reflex regulation, the generalized vagal discharge that occurs during the digestion of a meal increases vigor of intestinal contractions, whereas sympathetic stimulation, when it occurs, produces inhibition of motility. Gastrin plays a role in the propulsion of contents from the terminal ileum by increasing motor activity of the ileum and relaxing the ileocecal sphincter.

Secretion in the Small Intestine

About 1 to 2 liters of fluid move from the blood to the intestinal lumen every day. Some of this fluid movement results from the osmotic shift of water from blood to lumen in response to the presence of hyperosmotic contents in the lumen. Intestinal contents can have a high osmotic activity, either because the material that enters the small intestine from the stomach is hyperosmotic or because the rapid breakdown of large molecules to many small

molecules by enzymatic digestion produces hyperosmotic contents. Enough water moves into the lumen to produce contents that are isosmotic to blood, thereby minimizing osmotic damage to the delicate lining of the small intestine. It is also significant that gastric emptying is retarded when the contents of the duodenum are hyperosmotic. The regulatory mechanisms involved reduce the possibility of large volumes of hyperosmotic contents entering the small intestine, which in turn ensures that there will not be large shifts of water from blood to lumen. If this were to happen, circulatory collapse might occur as the result of depletion of blood volume.

The intestinal mucosa also contributes to the material contained in the intestinal lumen. Mucosal exocrine glands scattered throughout the small intestine secrete water, inorganic ions, and mucus. Local chemical and mechanical stimulation by intestinal contents are effective stimuli for secretion by these glands. In addition, epithelial cells are shed from the intestinal mucosa into the intestinal lumen. This process of extrusion of the epithelium is continuous, resulting in complete renewal of the intestinal epithelium every 1 to 2 days. The disintegration of these cells in the intestinal lumen contributes small quantities of enzymes, some organic material, and inorganic ions to the intestinal contents.

Absorption from the Small Intestine

Intestinal absorption is the movement of water and dissolved materials from the lumen of the small intestine through the intestinal epithelium into the blood and lymph. Most of the organic constituents of the contents of the small intestine are completely absorbed midway through this organ. Exceptions to this are vitamin B_{12} and the bile salts (both of which are absorbed from the terminal ileum by specific transport processes) and certain excretory products such as the bile pigments. The lower small intestine (the distal jejunum and ileum) can absorb a variety of food digestion products, and thus it acts as a reserve area of absorption in situations in which food escapes absorption upstream. Water and sodium chloride are absorbed throughout the small intestine, but significant quantities of these enter the large intestine, where their absorption is normally completed.

The Absorptive Surface of the Small Intestine

The absorptive surface of the small intestine is highly folded and, in addition, is convoluted by small fingerlike projections called **villi** (singular, *villus*) (Fig. 23–24*a*). Each villus is lined with epithelial

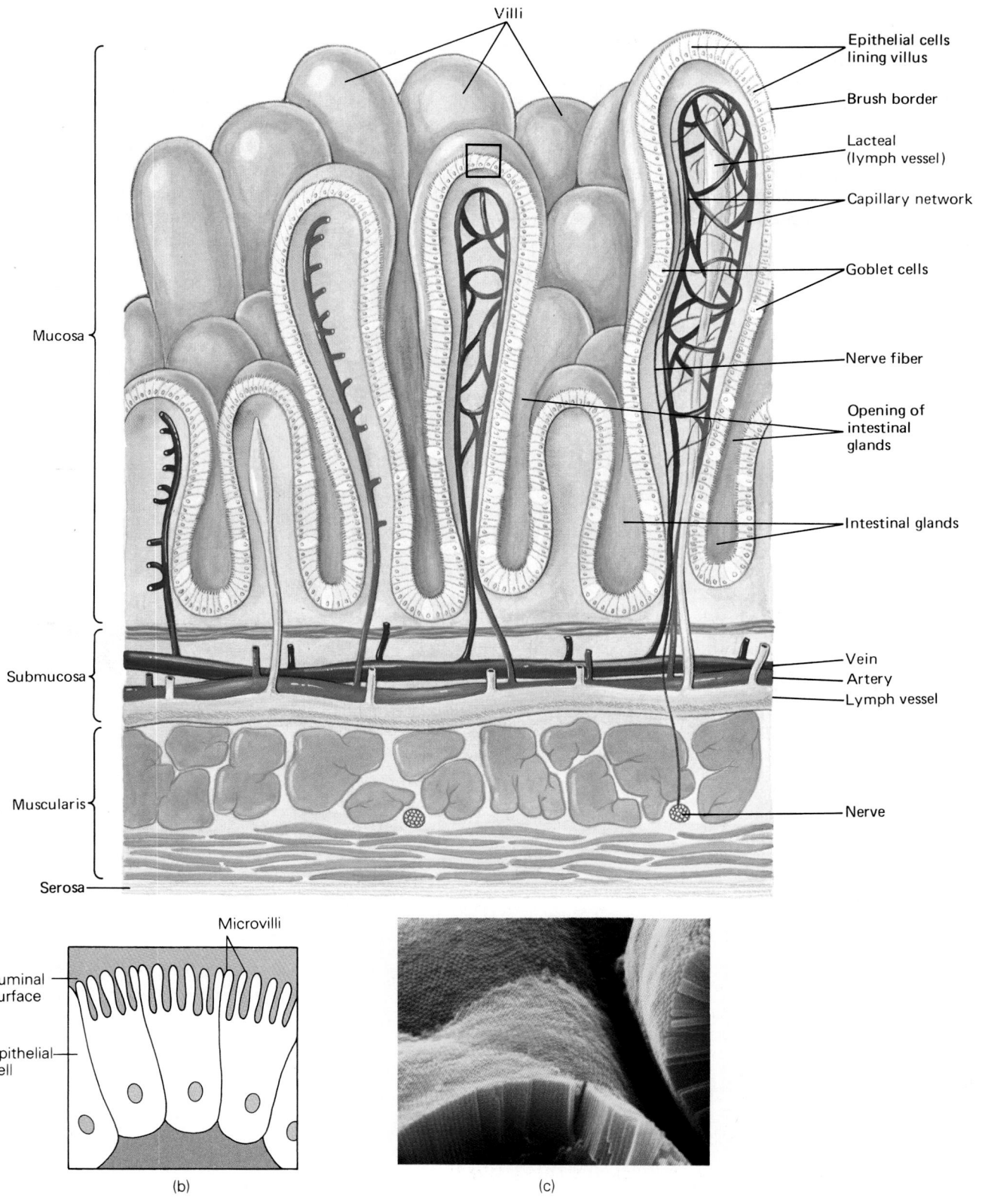

Villi

Epithelial cells
lining villus

Brush border

Lacteal
(lymph vessel)

Capillary network

Goblet cells

Nerve fiber

Opening of
intestinal
glands

Intestinal glands

Mucosa

Submucosa

Vein

Artery

Lymph vessel

Muscularis

Nerve

Serosa

Microvilli

Luminal
surface

Epithelial
cell

(b)

(c)

Figure 23–24
(*a*) Structure of the absorptive surface of the small intestine. (*b*) Microvilli on
the luminal surface of intestinal epithelial cells. (*c*) Scanning electron micrograph of the surface of an epithelial cell from the small intestine, illustrating
the microvilli. (© Custom Medical Stock Photo.)

cells and contains a rich network of blood vessels and lymph vessels. Another major feature of the absorptive surface is that the luminal aspect of each epithelial cell is covered with much smaller projections, the **microvilli** (see Fig. 23–24*b*). An epithelial cell's microvilli are the luminal membrane of that cell; collectively, the microvilli of all of the epithelial cells lining the small intestine are known as the **brush border membrane.** The small intestine is well suited for the task of absorption, because the surface area available for absorption (Fick's Law—Chapter 2) is greatly increased not only by the intestinal folds but also by the villi and the microvilli (Fig. 23–25).

The brush border membrane contains a variety of enzymes that complete the process of digestion and also a number of transport systems that move digestion products and inorganic ions from the intestinal lumen to the blood. In many instances, this is active transport against an electrochemical gradient. In other cases, molecules diffuse passively from

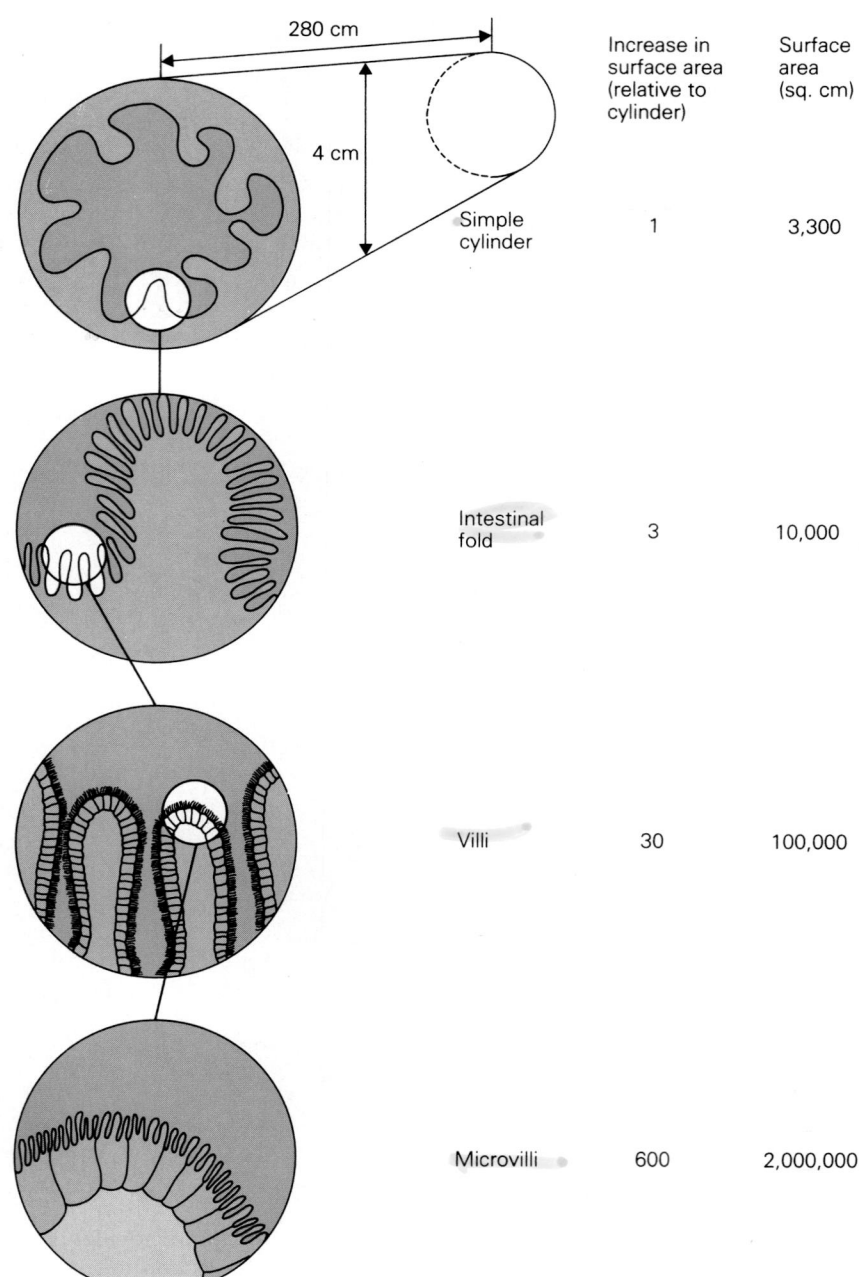

Figure 23–25
Three ways by which the surface area of the small intestine is increased.

	Increase in surface area (relative to cylinder)	Surface area (sq. cm)
Simple cylinder	1	3,300
Intestinal fold	3	10,000
Villi	30	100,000
Microvilli	600	2,000,000

280 cm

4 cm

blood to lumen or lumen to blood down concentration gradients at differing rates depending on the magnitude of the gradient and on the size and lipid solubility of the molecules.

Absorption of Water and Inorganic Ions from the Small Intestine

Approximately 9 liters of water enter the small intestine every day. About 2 liters are ingested, and the remaining 7 liters are derived from the various secretions of the gastrointestinal system. Most of this water is absorbed from the small intestine, as shown by the fact that only about 500 ml of water enters the proximal large intestine. Thus, the small intestine is about 95% efficient as a water-absorbing organ. Water is absorbed from the large intestine to such an extent that only about 100 ml is lost in the stool every day. This means that a total of 8900 ml of water is absorbed daily by the combined actions of the small and large intestines; that is, the overall efficiency of water absorption by the intestines is 99%. The handling of water by the gastrointestinal system is summarized in Figure 23–26.

The absorption of water occurs by osmosis throughout the small intestine. As solutes such as sugars, amino acids, other organic molecules and inorganic ions are transported across the intestinal epithelium into the blood, a gradient for the diffusion of water from lumen to blood is created, and water responds to this gradient by diffusing out of the intestinal lumen into the blood. Since the most abundant solute of intestinal contents is sodium, this ion is of greatest importance in the absorption of water. Sodium transport is an active, energy-requiring process and is polarized in that this ion diffuses down an electrochemical gradient from the lumen across the brush border membrane into the intestinal epithelial cell and is then extruded against an electrochemical gradient into the blood by a sodium pump located in the basolateral membrane of the cell (Fig. 23–27).

Other inorganic ions absorbed into the blood (some actively and others passively) are potassium, calcium, magnesium, iron, zinc, chloride, phosphate, bicarbonate, and iodide. The intestine possesses special mechanisms for the absorption of iron and calcium. The transport of both of these cations

Figure 23–26
Water balance of the gastrointestinal system.

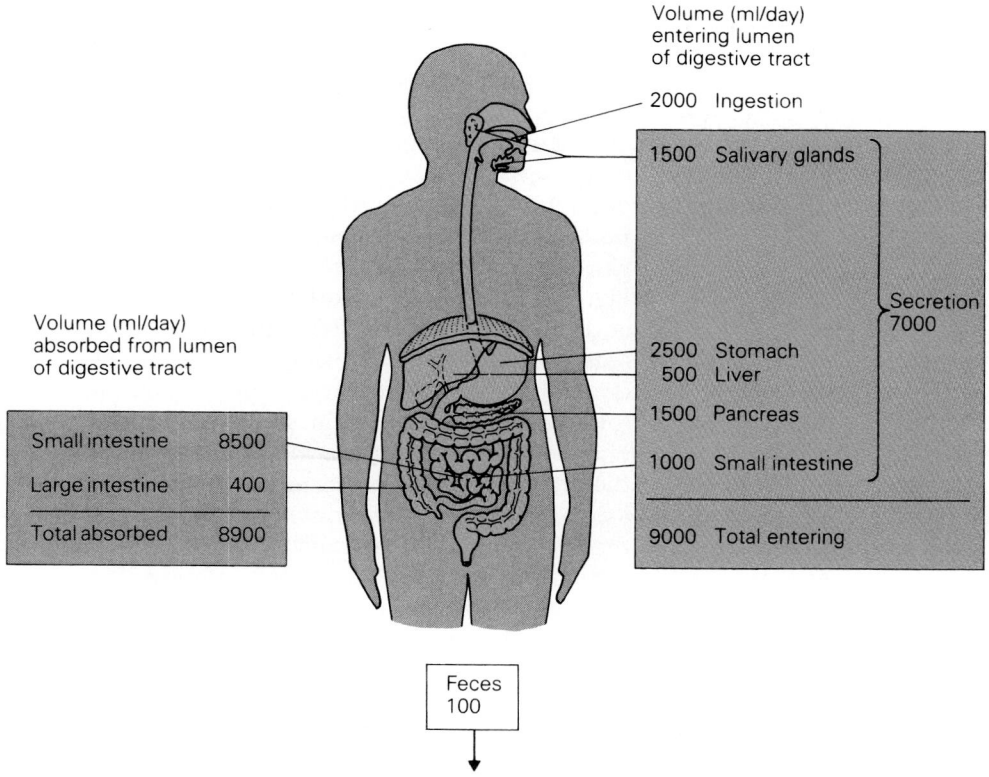

Volume (ml/day) entering lumen of digestive tract

2000   Ingestion
1500   Salivary glands
Secretion 7000
2500   Stomach
500   Liver
1500   Pancreas
1000   Small intestine

9000   Total entering

Volume (ml/day) absorbed from lumen of digestive tract

Small intestine	8500
Large intestine	400
Total absorbed	8900

Feces 100

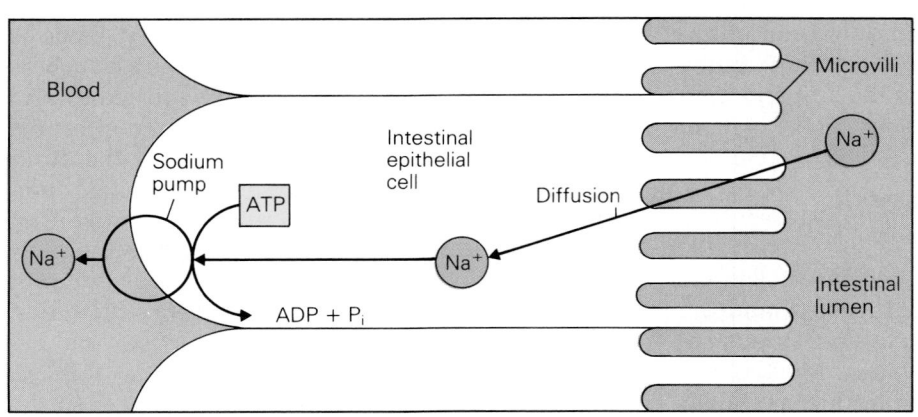

Figure 23–27
The transport of sodium across the intestinal epithelium.

is an active process and represents instances of regulated absorption. Iron is absorbed only when the body becomes deficient in this cation. Although the average daily intake of iron in the diet is about 10 to 20 mg, the amount absorbed is only about 10% of that ingested. Because of the iron lost during menstrual bleeding, the amount of iron absorbed by females is about double that of males. The iron of the mucosal epithelial cells is in equilibrium with circulating iron and with **ferritin,** a protein-iron complex in the epithelial cells. When the iron of the blood decreases below normal values, iron is liberated into the blood from the cellular stores of ferritin iron, and more iron is absorbed from the intestine to replenish these stores. Calcium requires both parathyroid hormone and vitamin D for optimal transport. A deficiency of vitamin D leads to impaired calcium absorption, and the unabsorbed calcium forms insoluble salts with phosphorus. Since these salts are not readily absorbed, the absorption of phosphorus also decreases under these conditions.

Absorption of Carbohydrates from the Small Intestine

Carbohydrates make up about one half of the caloric intake of people in the United States. Plant starch, a polysaccharide, is the major digestible carbohydrate of the diet. It is composed of many straight and branched chains of the monosaccharide glucose. Starch is hydrolyzed rapidly by salivary and pancreatic amylase to oligosaccharides, molecular fragments consisting of two to nine glucose molecules. The major oligosaccharide produced is maltose, a disaccharide that consists of two glucose molecules. Other oligosaccharides present in the diet in smaller quantities are the disaccharides sucrose, or table sugar (made up of glucose and fructose), and lactose, or milk sugar (glucose and galactose).

Another common carbohydrate in the diet is cellulose, a plant polysaccharide made up of glucose molecules but joined by linkages different from those of starch. Since humans do not secrete an enzyme that digests cellulose, this carbohydrate is excreted in the feces.

The oligosaccharides present in the lumen of the small intestine are broken down into their constituent monosaccharides by oligosaccharidases located in the microvilli of the intestinal epithelial cell. Maltose is hydrolyzed by the **disaccharidase maltase** to two glucose molecules. Lactose is degraded by lactase to one molecule of glucose and one molecule of galactose, and sucrose is broken down by sucrase to one molecule of glucose and one molecule of fructose (Fig. 23–28*a*). The higher-molecular-weight oligosaccharides produced by the action of amylase on starch are hydrolyzed in the brush border membrane to glucose molecules. Fructose is absorbed passively by facilitated diffusion. On the other hand, glucose and galactose are actively transported into the portal blood by a carrier-mediated process located in the brush border membrane (see Fig. 23–28*b*). The transport of glucose and galactose is coupled (in cotransport) with that of sodium and requires the expenditure of metabolic energy. There is a mobile carrier in the microvilli with binding sites for both sugar and sodium. It is the diffusion gradient for sodium movement into the cell that provides the driving force for moving the sugar- and sodium-loaded carrier from the outer to the inner aspect of the microvillous membrane, where both sugar and sodium are released to the cell interior. The sugar then moves from the cell interior through the basolateral membrane into the blood by facilitated diffusion. The energy in this system is expended by a sodium pump (see Fig. 23–27), which transports this ion across the basolateral membrane, thereby keeping the intracellular sodium concentration at a

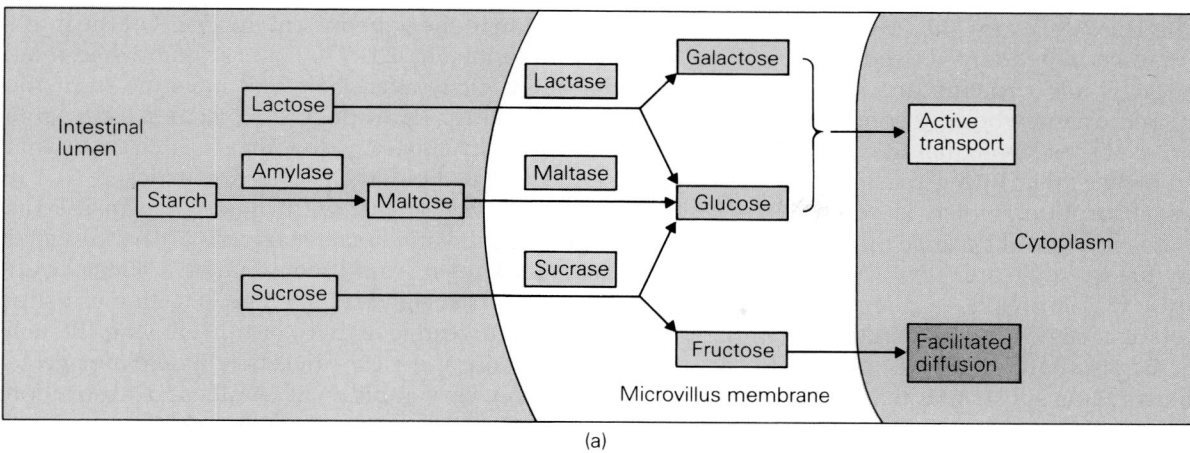

(a)

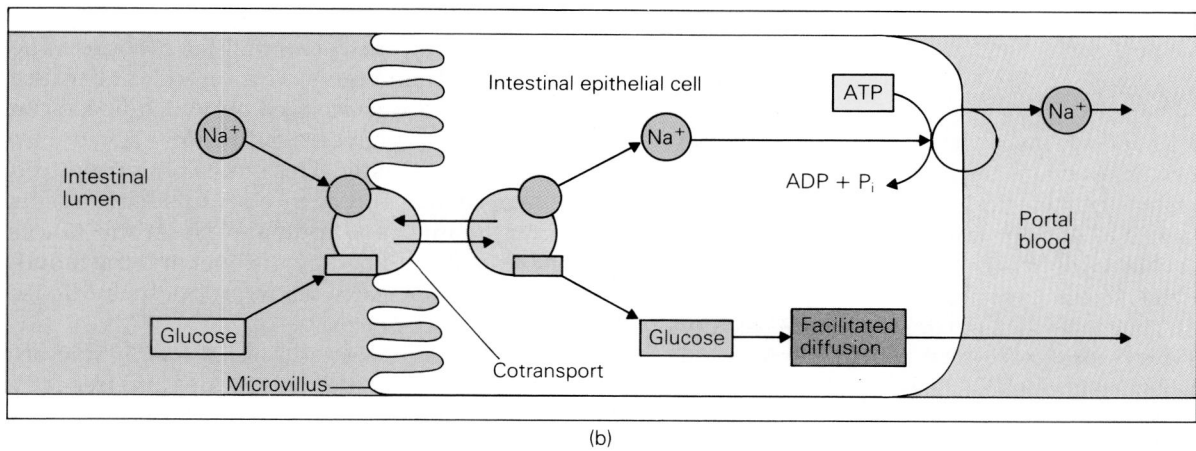

(b)

Figure 23–28
(*a*) Luminal and microvillous membrane digestion of carbohydrates. (*b*) Carrier-mediated transport of glucose and sodium across the intestinal epithelial cell.

level low enough to maintain a diffusion gradient from lumen to cell, and ensuring continued movement of the sugar-sodium carrier.

Absorption of Protein from the Small Intestine

A normal adult requires about 35 to 50 grams of protein daily in order to have essential amino acids and to achieve nitrogen balance. The average American consumes 70 to 90 grams of protein every day. However, the protein presented to the digestive tract for absorption consists not only of dietary protein but also of endogenous protein that approaches the amount of protein derived from food. This **endogenous protein** is in the form of digestive enzymes, epithelial cells, mucoproteins, and plasma proteins that leak into the tract. Most all of the dietary and endogenous protein is digested and absorbed.

The end products of the intraluminal digestion of proteins are amino acids and small peptides. A variety of enzymes that hydrolyze peptides are present in the brush border membrane, and many peptides are hydrolyzed there (by specific **peptidases**) to their constituent amino acids. Amino acids are transported into the portal blood by active processes occurring in the intestinal epithelial cell. Each class of amino acid has its own specific carrier-mediated transport system in the microvilli and, as is the case for certain sugars, amino acid transport is coupled with sodium transport. Thus, clearly there are many similarities between the absorption of carbohydrate and the absorption of protein. However, in contrast to carbohydrate absorption, in which only simple sugars are transported across the intestinal epithelium, carrier-mediated mechanisms are available in the intestinal epithelial cell for the active transport of dipeptides and tripeptides. These carrier-mediated processes for the absorption of small peptides may be dependent on sodium.

Although whole proteins are not usually absorbed as such, the newborn of some species, such as ruminants, absorb small quantities of these molecules from breast milk during the first few days of birth. This absorption, which is accomplished by pinocytosis, is important because protein antibodies present in breast milk are also absorbed by this mechanism. This provides a temporary defense against certain bacterial and viral infections until the infant can begin synthesizing its own antibodies. Breast milk contains substances that inhibit the proteolytic enzymes of the digestive tract, thereby preventing intraluminal destruction of the antibodies.

Absorption of Lipid from the Small Intestine

About 60 to 100 grams of lipid are present in the daily diet of the average adult American. Lipids are a concentrated source of calories in the diet and account for 30% to 40% of the caloric intake. About 90% of the dietary lipid is in the form of water-insoluble triacylglycerols containing mostly long-chain fatty acids (16 to 18 carbon atoms). Other dietary lipids include cholesterol esters, phospholipids, and the fat-soluble vitamins A, D, E, and K.

Both pancreatic lipase and the bile salts are required for the normal digestion and absorption of triacylglycerols. Pancreatic lipase degrades triacylglycerol to one monoglyceride and two free fatty acid molecules (Fig. 23–29). These two digestion products are the form in which fat enters the intestinal epithelial cell.

Since triacylglycerols and their digestion products are insoluble in water, special mechanisms ensure that both digestion and absorption proceed to completion in the aqueous environment of the intestinal contents (Fig. 23–30). For example, when triacylglycerols first enter the small intestine from the stomach, these hydrophobic molecules form large oil droplets by clustering together in such a way that relatively few lipid molecules are exposed to the aqueous environment, with most being inside the droplet. Pancreatic lipase acts only at the oil-water interface. Thus it would seem that only a small fraction of the triacylglycerols in the intestine are subject to the digestive activity of this enzyme. If such were actually the case, digestion would not go to completion very rapidly, if at all, and absorption would be very much limited by digestion.

In reality, co-existing in the intestinal contents with the oil droplets is a stable emulsion of triacylglycerols consisting of very small fat particles. The existence of such an emulsion is important because it provides a large surface area on which lipase can act to degrade triacylglycerol and allow digestion to go to completion rapidly. This emulsion is produced by the combined action of three emulsifying agents present in the intestinal contents. These substances are the bile salts supplied by the liver and the monoglycerides and free fatty acids produced by lipase digestion of triacylglycerols.

The bile salts play an additional role in lipid absorption. They render monoglycerides and free fatty acids water-soluble by interacting with these triacylglycerol digestion products to form very small aggregates called **micelles** (see Fig. 23–30). Each micelle contains only about 20 lipid molecules and, unlike the much larger emulsion particles, diffuses rapidly in aqueous solution.

What is the significance of micelle formation with regard to absorption of monoglycerides and free fatty acids? Both digestion products do have a limited solubility in water and are present as single

Figure 23–29
Digestion of triacylglycerols by pancreatic lipase.

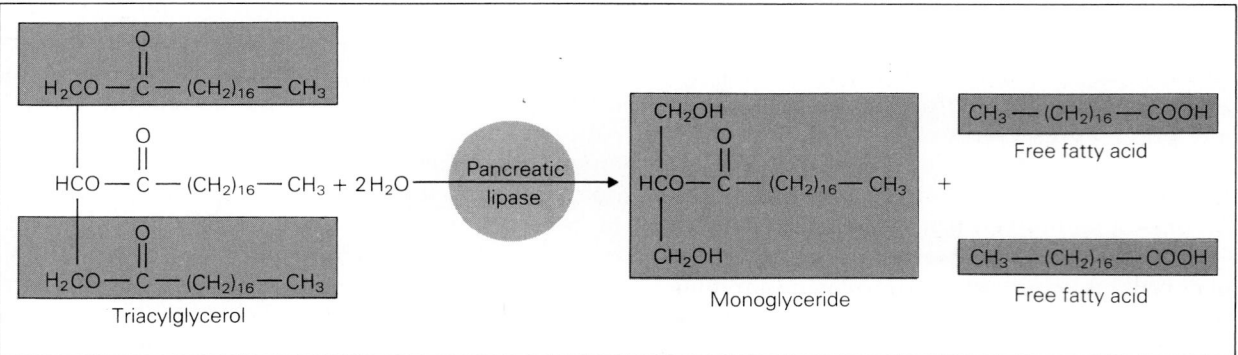

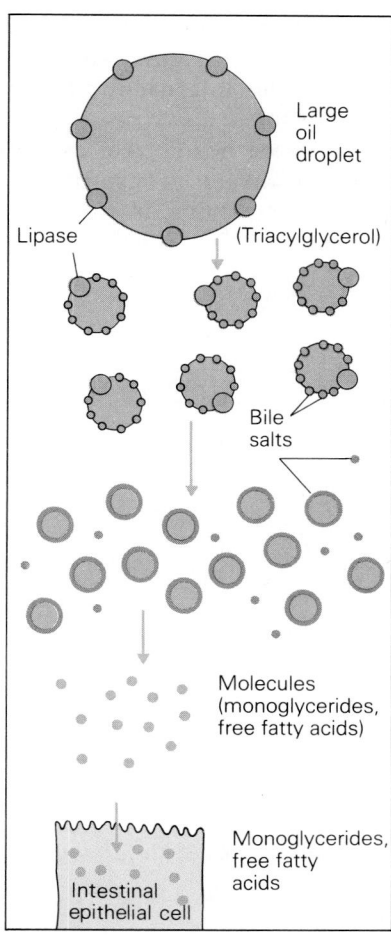

Figure 23–30
Intraluminal and cell penetration phases of triacylglycerol absorption.

glycerides and free fatty acids by movement into the free phase, the micellar phase is replenished with monoglycerides and free fatty acids from the emulsion phase, where these lipid digestion products are continuously produced.

We should point out that the micelles dissolve cholesterol and the fat-soluble vitamins within their micellar structure, thereby preparing these lipids for absorption in the same way that they prepare monoglycerides and free fatty acids. The bile salts, which are not absorbed until they reach the terminal ileum, remain in the intestinal lumen and perform their absorptive function throughout the length of the intestine until all fat is absorbed. This process is efficient because minimal quantities of bile salts are required to prepare relatively large quantities of lipid for absorption.

Once monoglycerides and free fatty acids enter the epithelial cells, they are resynthesized to triacylglycerols by intracellular enzyme systems in the smooth endoplasmic reticulum. The newly synthesized triacylglycerols are aggregated into droplets, which become progressively larger during passage through the cells. A layer of phospholipid and protein is added to each droplet, and the final product is known as a **chylomicron.** These particles enter the central lacteal of the villus and are carried by the lymphatic system to the general circulation, where they are made available to the cells of the body. The cellular phase of lipid absorption is illustrated in Figure 23–31.

Figure 23–31
Cellular phase of triacylglycerol absorption.

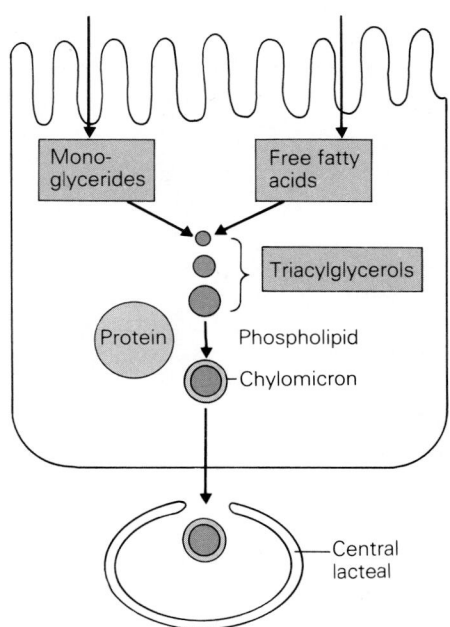

molecules in intestinal contents only in trace amounts. That is, the level of **monomolecular dispersion** for monoglycerides and free fatty acids is very low. The idea is that monoglycerides and free fatty acids are absorbed as individual molecules from the monomolecular dispersion, and that this occurs by diffusion from the intestinal lumen through the lipid portion of the intestinal epithelial cell membrane into the cell (Fig. 23–30). The function of the micelles is to maintain the monoglycerides and free fatty acids in the monomolecular dispersion phase at saturation levels by serving a reservoir function; that is, as monoglycerides and free fatty acids are removed from the monomolecular phase by absorption, they are immediately replenished by the movement of monoglycerides and free fatty acids from micelles. This allows absorption of individual lipid molecules to occur rapidly. Along this same line, as the micelles are depleted of mono-

The Processes of Absorption, Secretion, and Motility in the Large Intestine

The large intestine (the **colon**), which is about 4 feet long, extends from the ileocecal junction to the anus. This organ is subdivided into the **cecum** and its appendix, the **ascending colon**, the **transverse colon**, the **descending colon**, the **sigmoid colon**, the **rectum**, and the **anus** with its **internal** and **external anal sphincters** (Fig. 23–32).

The large intestine does not secrete digestive enzymes, and it lacks villi. Water is absorbed from the contents that enter from above; in this way, a firm fecal mass is formed. This relatively dehy-

Figure 23–32
Anatomy of the large intestine.

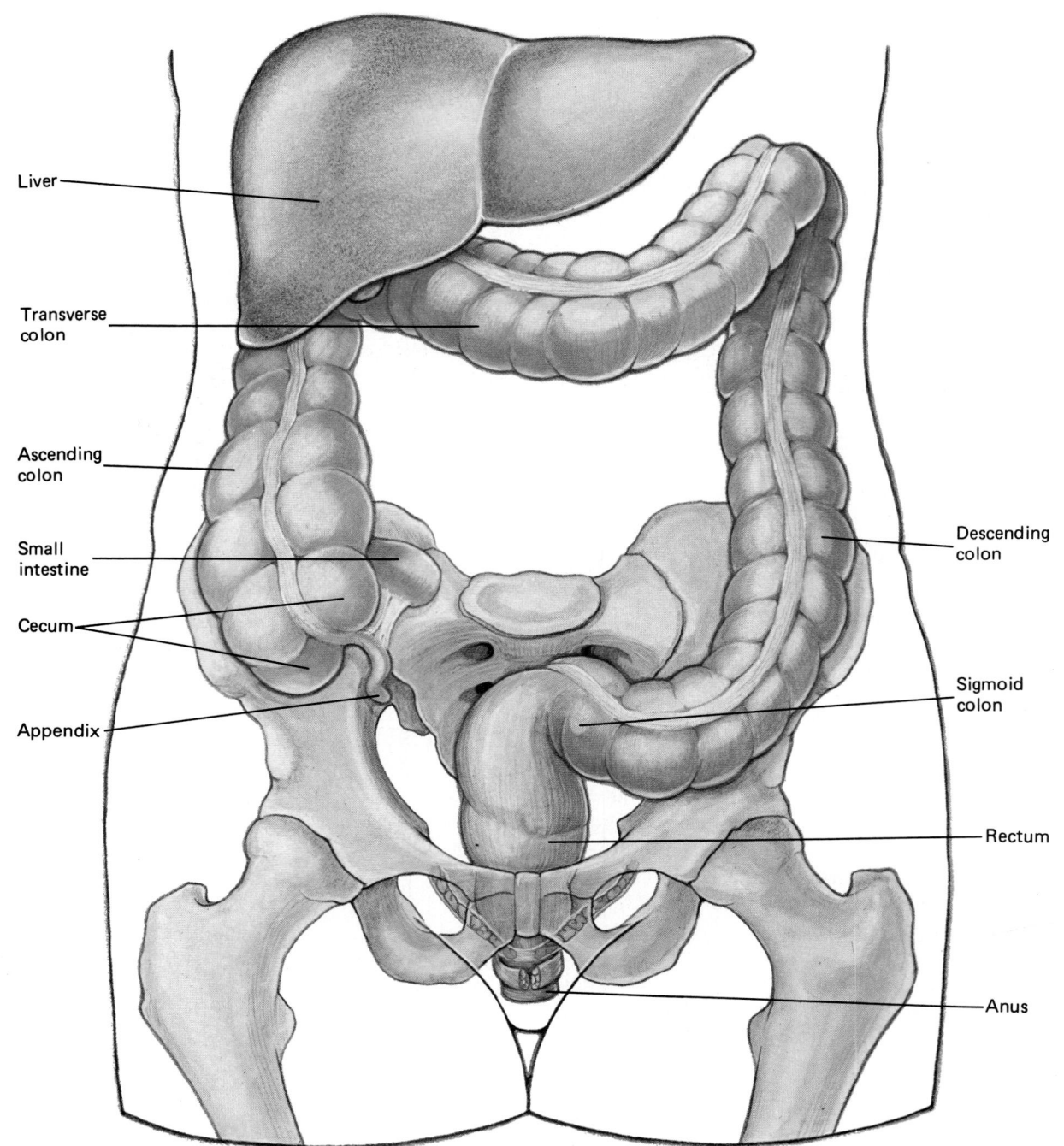

Liver

Transverse colon

Ascending colon

Small intestine

Cecum

Appendix

Descending colon

Sigmoid colon

Rectum

Anus

drated material is stored in the large intestine for variable periods of time and is ultimately evacuated to the outside of the body by defecation.

Absorption and Secretion in the Large Intestine

About 500 ml of water and a significant amount of sodium chloride enter the proximal colon every day. The primary absorptive process that occurs in the large intestine is the active transport of sodium from the colon. As in the small intestine, this is accompanied by the osmotic absorption of water. This occurs to the extent that normally only about 100 ml of water are lost to the feces every day. Carbohydrate, protein, and fat digestion products are not absorbed from the large intestine.

The bacteria present in the lumen of the large intestine synthesize certain vitamins. Although these nutrients are absorbed from the large intestine, they provide only a small part of the daily vitamin requirement. Other products of bacterial metabolism contribute to the gas (**flatus**) present in the large intestine. Flatus, which is produced at a rate of about 1 to 1.5 liters per day, consists of nitrogen and carbon dioxide, along with small amounts of hydrogen, methane, and hydrogen sulfide. Certain foods, such as beans, contain carbohydrates that cannot be digested in the small intestine. These carbohydrates are metabolized by colonic bacteria with the resultant production of large quantities of gas.

The colonic mucosa contains glands that produce an alkaline, mucus-containing secretion that protects the mucosa of the large intestine from chemical and mechanical trauma. Mucus lubricates the fecal mass, thereby facilitating the passage of feces. The bicarbonate of this secretion neutralizes irritating acids produced by bacterial metabolism. The large intestine also secretes potassium. When large volumes of fluid are abnormally excreted in the feces, body potassium can be depleted.

Motility in the Large Intestine

The musculature of the large intestine is quiescent (inactive) much of the time. Segmenting contractions occur at a much lower frequency than in the small intestine, and these produce a limited back-and-forth movement of the contents of the colon. Segmenting movements are the most frequent type of muscular activity of the large intestine, and they promote the absorption of water and inorganic ions by exposing the contents to the mucosa. A wave of

contraction passes over the proximal colon about 3 to 4 times a day and drives contents into the distal large intestine, where material is stored until defecation occurs. These propulsive **mass movements** usually occur after a meal and are often followed by the desire to defecate. Colonic movements are largely initiated locally by distention of the colonic wall, which triggers short reflexes through the nerve plexuses. However, mass movements sometimes occur when food enters the stomach. Although this has been attributed to a long reflex called the **gastrocolic reflex,** this type of muscular activity may also be stimulated by gastrin.

The contents of the distal colon are stored for varying periods of time until defecation occurs, which may be 24 hours or longer after the food is eaten. The solids of this fecal material consist of bacteria, cells shed from the intestinal mucosa, bile pigments, indigestible material such as cellulose, and inorganic salts.

The Defecation Reflex

The rectum is normally empty, but when a mass contraction moves contents into this region, the resulting distention of the rectum elicits the defecation reflex (see Chapter 10). This reflex consists of short reflexes through internal nerves and long reflexes through the sacral region of the spinal cord. The movements that result from activation of the defecation reflex suffice to propel feces through the anus and out of the body. These movements consist of increased muscular activity of the sigmoid colon, contraction of the rectum, and relaxation of the internal and external anal sphincters, which consist of smooth muscle and skeletal muscle, respectively. The anal sphincters are tonically contracted by impulses passed over external nerves during periods of nonactivity. Since the external anal sphincter consists of skeletal muscle, it is under voluntary control, and its state of contraction can be influenced by impulses from the brain. For example, a person can voluntarily keep the external anal sphincter closed. This results in postponement of defecation, because the rectum relaxes, the stimulus of distention is removed, and the desire to defecate disappears until more fecal matter enters and once again distends the rectum to elicit the defecation reflex.

In addition to the muscular movements of the distal colon that occur during defecation, contraction of the abdominal muscles and a forcible expiration with the glottis closed (**straining movements**) can supply a considerable part of the force involved in defecation. A consequence of straining at the stool is a rise in intrathoracic pressure, which causes blood pressure to rise. This is followed by a sudden drop in blood pressure as a result of decreased venous return and cardiac output. These changes in blood pressure can result in a variety of cerebrovascular accidents, such as stroke or heart attack.

Individuals with normal colonic function vary in frequency of defecation, ranging from several times a day to once every few days or so. **Constipation,** the accumulation of feces in the large intestine beyond that which is normal for each person, produces symptoms such as abdominal discomfort, headache, nausea, and lack of appetite. Many believe that these symptoms result from absorption by the large intestine of toxic products produced by bacterial metabolism. However, the primary cause is distention of the large intestine, because the sensations disappear almost immediately following relief from distention.

Constipation occurs when the muscular activity of the large intestine is decreased, as happens in certain emotional states, in older persons, or when the diet contains little **bulk** (indigestible material). Most of the bulk of the diet is derived from **dietary fiber** (cellulose and other indigestible polysaccharides) of plants. Dietary fiber helps maintain regular bowel movements by contributing to the volume of contents of the large intestine, which increases the degree of distention and thus the extent of contractile activity of this organ.

Defecation can be facilitated by the use of **cathartics (laxatives).** For example, when material is retained in the large intestine for long periods, most of the water is absorbed, and the contents become hard and dry. In this instance, ingested **mineral oil** eases defecation by lubricating the fecal mass. Unabsorbable solutes are also prescribed for people who suffer from constipation. By osmotically retaining water in the intestinal lumen, these solutes prevent the absorption of a volume of water proportional to the amount of solute administered. The excretion of this water flushes accumulated contents out of the large intestine. Typical examples of these cathartics are **magnesium sulfate (Epsom salts)** and **magnesium hydroxide (milk of magnesia).**

Diarrhea is the frequent defecation of highly fluid material or, more quantitatively, an output of fecal water greater than 500 ml per day. This condition results from poor absorption of fluid or increased fluid secretion. Either or both can occur when the intestines are subject to bacterial and viral infections. In such cases, larger-than-normal quantities of digestive contents are present in the large intestine, and the excessive distention—and thus increased motor activity—that results is the primary cause of the diarrhea. Severe or prolonged diarrhea can result in the loss of significant quantities of water and inorganic ions. One consequence is the depletion of blood volume. In addition, since the secretions of the lower digestive tract have a relatively high concentration of sodium bicarbonate, a possible result of the loss of this alkaline ion is **metabolic acidosis** (Chapter 26).

Appendicitis is inflammation of the appendix, a small, fingerlike, blind tube that extends from the cecum. This condition is preceded by obstruction of the lumen of the appendix; the infection that usually follows may result in edema, ischemia, gangrene, and perforation. Appendicitis begins with referred pain in the umbilical region, followed by nausea and vomiting. After several hours, the pain becomes severe and is localized to the right lower quadrant of the abdomen. Early removal of the appendix is recommended in all suspected cases of appendicitis.

SUMMARY

Overview of the Gastrointestinal System

The major function of the gastrointestinal system is to process food and, by doing so, provide nutrients to the cells of the body. A major component of the gastrointestinal system is the digestive tract, which consists of the mouth, pharynx, esophagus, stomach, and small and large intestines. Accessory structures connected to the digestive tract by ducts are the salivary glands, pancreas, liver, and gallbladder.

Blood flows to and from the gastrointestinal system by way of the splanchnic circulation. Blood that leaves the stomach, intestines, and pancreas goes to the liver by way of the portal vein (portal circulation) before entering the central venous system.

In the process of motility, coordinated contractions of muscle, both skeletal and smooth, move ingested material through the digestive tract and also mix this material with the juices of the exocrine glands that are secreted en route. In the process of secretion, the various exocrine glands contain a number of types of secretory cells, each of which produces a characteristic secretion. Some of these secretions consist entirely of water and inorganic ions, whereas others also contain organic compounds, including the digestive enzymes and the bile salts. In the process of digestion, the digestive enzymes catalyze the conversion of large, organic molecules of the diet to smaller molecules that are readily absorbed. In the process of absorption, substances travel from lumen to blood and lymph, mainly from the small intestine, by means of a variety of both active and passive transport processes.

Neural mechanisms for the control of gastrointestinal function include both short reflexes over nerve cells that lie entirely in the wall of the digestive tract and long reflexes over external nerves, the most important external nerve being the vagus nerve. The polypeptide hormones gastrin, secretin, cholecystokinin, and gastric inhibitory peptide also regulate gastrointestinal function.

Chewing, Salivary Secretion, and Swallowing

Chewing involves the coordinated contraction and relaxation of a variety of skeletal muscles to reduce the size of food particles and to mix them with saliva.

The long reflex secretion of saliva that accompanies chewing provides salivary amylase, an enzyme that initiates the digestion of starch, the major dietary carbohydrate.

Between swallows, sphincters at either end of the esophagus impede the entrance of air into the esophagus from above and of gastric contents from below. A swallow moves food rapidly from the mouth through the pharynx and esophagus to the stomach in response to the pressure gradients generated by the muscular contractions that occur sequentially from above to below. The characteristic propulsive muscular activity of the esophagus is peristalsis, a moving wave of contraction.

Coordination of swallowing is by long reflexes mediated through a swallowing center in the brain and short reflexes in the wall of the esophagus.

The Processes of Motility and Secretion in the Stomach

Receptive relaxation of the body and fundus, the storage regions of the stomach, allows large volumes of ingested material to be accommodated with only small increases in intragastric pressure. Peristaltic contractions of the pyloric antrum generate a pressure difference between antrum and duodenum, which is the driving force that empties the stomach.

The frequency of peristalsis is governed by the basic electrical rhythm of the stomach, a small wave of depolarization that is conducted down the gastric musculature about three times every minute. The strength of gastric peristalsis depends on a balance between the excitatory and inhibitory mechanisms that influence the gastric musculature at any given time. Excitatory mechanisms are mediated primarily through short reflexes in the gastric wall and long reflexes over the vagus nerves, both of which are triggered in response to stomach distention. Chemical and physical stimuli in the material that enters the duodenum from the stomach inhibit gastric motor activity. These stimuli produce inhibition by releasing *secretin*, *gastric inhibitory peptide*, and unidentified hormones from the duodenal mucosa, and through short and long enterogastric reflexes.

The major secretory components of gastric juice are hydrochloric acid and intrinsic factor, secreted by the parietal cells, and pepsinogen, secreted by the chief cells. HCl is secreted by active transport. When food is eaten, short and long reflexes triggered by stimulation of receptors in the head and stomach cause the release of acetylcholine by nerve fibers in the acid-secreting portion of the stomach and the release of gastrin from endocrine cells in the pyloric antrum. Both acetylcholine and gastrin excite the secretion of HCl by the parietal cell. Inhibition of HCl secretion is mediated by suppression of gastrin release by HCl and by neural and hormonal inhibitory mechanisms that originate in the duodenum. Mechanisms that protect the upper digestive tract from the corrosive actions of HCl include the impermeability of gastric cell membranes to HCl, dilution and neutralization of acid, and rapid cellular regeneration. The chief cells secrete inactive pepsinogen, which is activated by HCl to active pepsin, an enzyme that initiates the digestion of protein. The most effective stimulus for pepsinogen secretion is the acetylcholine released at the chief cell by the short and long reflexes that occur following the eating of food.

The Process of Secretion in the Exocrine Pancreas

The intercalated duct cells of the pancreas secrete a large volume of a water-electrolyte juice with a relatively high concentration of sodium bicarbonate. The sodium bicarbonate of the aqueous juice is transported actively by the duct cells and is largely responsible for the neutralization of the HCl in gastric contents.

Pancreatic acinar cells secrete a small volume of juice containing a wide variety of digestive enzymes whose combined actions almost complete the digestion of food.

Excitation of pancreatic bicarbonate secretion is largely by secretin released in response to acid stimulation of the duodenal mucosa. Excitation of enzyme secretion is largely by cholecystokinin released in response to stimulation of the duodenal mucosa by fat and protein digestion products.

The Process of Secretion in the Biliary System

Between meals, bile secreted by the liver is stored in the gallbladder, where the organic constituents of bile are concentrated by the transport of salt and water from lumen to blood. Contraction of the gallbladder and relaxation of the sphincter of Oddi result in the flow of bile into the upper small intestine in response to the release from the duodenal mucosa of cholecystokinin by fat and protein digestion products.

Bile salts, the major secretory components of bile, are synthesized and actively secreted by the cells of the liver. Following transit through the small intestine, where they perform their function in the digestion and absorption of fat, most of the bile salts are actively absorbed from the terminal ileum into the portal circulation, which returns them to the liver for resecretion (enterohepatic circulation of bile salts). The bile pigments are degradation products of hemoglobin metabolism and are excreted largely in the feces.

Circulating bile salts are the most potent stimulus for secretion of bile by the cells of the liver. The bile ducts respond to secretin stimulation by secreting a water-electrolyte fluid with a high bicarbonate content.

The Processes of Motility, Secretion, and Absorption in the Small Intestine

Segmenting contractions mix the contents of the small intestine, thereby facilitating digestion and absorption; peristalsis propels unabsorbed material into the large intestine. Short and long reflexes are largely responsible for regulating small intestinal motility.

Fluid enters the lumen of the small intestine as a result of the osmotic shift of fluid from blood to lumen, in response to the presence of hyperosmotic contents in the lumen. Mucosal exocrine glands of the small intestine are an additional source of fluid.

The major process involved in the absorption of water from the small and large intestines is the active transport of sodium, which carries water along passively from lumen to blood. Iron and calcium absorption are regulated processes mediated by special transport systems.

Digestion of carbohydrate to monosaccharides and protein to amino acids is completed by enzymes in the intestinal brush border membrane. Most monosaccharides and amino acids are absorbed into the portal circulation by sodium-dependent active transport systems.

Water-insoluble monoglycerides and free fatty acids, the end products of pancreatic lipase digestion of triacylglycerol fats, interact with bile salts to form water-soluble aggregates called micelles. Monoglycerides and free fatty acids diffuse passively from monomolecular dispersion into the intestinal epithelial cell, where they are resynthesized to triacylglycerols and passed on to the lymph as chylomicrons.

The Processes of Absorption, Secretion, and Motility in the Large Intestine

A firm fecal mass is formed by the absorption of water from the large intestine; this dehydrated material is stored for variable periods of time before defecation occurs. Products of the metabolic activities of the bacteria that reside in the lumen of the large intestine include certain vitamins and flatus. Mucus, bicarbonate, and potassium are secreted into the lumen by the colonic mucosa.

Segmenting contractions promote absorption of inorganic ions and water, and propulsive *mass movements* move contents into the distal large intestine at infrequent intervals. Distention of the rectum elicits the defecation reflex, which consists of short and long reflexes through a sacral region of the spinal cord. Although the muscular movements of the distal colon usually suffice to move feces to the outside of the body, contraction of certain skeletal muscles (straining movements) can supply a considerable amount of force in defecation. The accumulation of feces in the large intestine is known as constipation. Diarrhea is the frequent defecation of highly fluid materials.

REVIEW QUESTIONS

Identify Each with a Term

1. The transfer of water and solutes from the lumen of the digestive tract into the blood or lymph.

2. The spontaneous depolarization-repolarization cycles of pacemaker cells in the smooth muscle of the stomach and small intestine.

3. The route travelled in the body by the bile salts.

4. Very small projections of the luminal surface of the small intestinal epithelial cell.

5. The aggregates of bile salts, monoglycerides, and free fatty acids present in the lumen of the small intestine during the digestion and absorption of triacylglycerols.

6. Frequent defecation of highly fluid material.

Define Each Term

7. digestive tract
8. short reflex
9. dysphagia
10. cephalic phase
11. trypsin
12. peristalsis
13. steatorrhea
14. portal circulation
15. secretin
16. G cell
17. emetic
18. pancreatic lipase
19. ferritin
20. flatus

Choose the Correct Answer

21. The various processes of the gastrointestinal system are directly associated with certain types of gastrointestinal cells. Which of the following associations is *not* valid?
    a. motility—esophageal smooth muscle cells
    b. secretion—gallbladder epithelial cells
    c. digestion—pancreatic acinar cells
    d. absorption—small intestinal epithelial cells

22. A gastrointestinal process that is *not* regulated is:
    a. small intestinal absorption of glucose
    b. gastric secretion of HCl
    c. gastric emptying
    d. secretion of pancreatic enzymes

23. The organ of greatest importance in providing the enzymes required for the digestion of food is the:
    a. parotid salivary gland
    b. stomach
    c. pancreas
    d. liver

24. In which of the following regions is intraluminal pressure highest when there is no swallowing activity?
    a. pharynx
    b. body of the esophagus
    c. lower esophageal sphincter
    d. stomach

25. A hormone that stimulates secretion by the gastric parietal cell is:
    a. gastrin
    b. pepsin
    c. gastric inhibitory peptide
    d. secretin

26. The cephalic phase of gastric HCl secretion is *eliminated* by:
    a. surgical removal of the pyloric antrum
    b. vagal denervation of the pyloric antrum
    c. vagal denervation of the fundus and body of the stomach
    d. none of the above

27. A component of the secretion of the intercalated duct cells of the pancreas that is of major physiological significance is:
    a. trypsin
    b. $NaHCO_3$
    c. lipase
    d. bilirubin

28. A close association exists between the absorption of Na^+ and the absorption of water from the lumen of the:
    a. small intestine
    b. proximal colon
    c. gallbladder
    d. all of the above

29. The absorption of water from the digestive tract:
    a. is dependent primarily on the active transport of Ca^{2+}.
    b. occurs primarily from the small intestine.
    c. is completed during transit through the small intestine.
    d. occurs primarily from the proximal colon.

30. The intestinal *digestion* of triacylglycerols is facilitated by the incorporation of these molecules into:
    a. large oil droplets
    b. small emulsified fat particles
    c. micelles
    d. chylomicra

Answer Each Question in One or Two Sentences

31. How does the stomach adapt to the wide range of volumes it handles with only small increases in intragastric pressure?

32. What are the mechanisms by which high concentrations of HCl in the (a) stomach and (b) duodenum inhibit gastric secretion of HCl?

33. What is the functional importance of the bicarbonate ion of pancreatic juice?

34. Why are the bile salts essential for normal intestinal digestion and absorption of triacylglycerols?

35. Why is it important that our diet contains indigestible material?

SUGGESTED READINGS

Brooks, F.P., ed. *Gastrointestinal Pathophysiology*, 2nd ed. New York: Oxford University Press, 1978.

Davenport, H.W. *Physiology of the Digestive Tract*, 5th ed. Chicago: Year Book Medical Publishers, 1982.

Granger, D.N., Barrowman, J.A., and Kvietys, P.R. *Clinical Gastrointestinal Physiology*. Philadelphia: W.B. Saunders Company, 1985.

Johnson, L.R., ed. *Gastrointestinal Physiology*, 3rd ed. St.Louis: C.V. Mosby Co., 1985.

Johnson, L.R., ed. *Physiology of the Gastrointestinal System*, 3rd ed. (2 volumes). New York: Raven Press, 1987.

Makhlouf, G.M., "Neural and Hormonal Regulation of Function in the Gut." *Hospital Practice*, 25:79–98, 1990.

Moog, F. "The Lining of the Small Intestine." *Scientific American*, 245:154–176, 1981.

Shiau, Y.F. "Mechanism of intestinal fat absorption." *American Journal of Physiology*, 240:G1–9, 1981.

Thompson, J.C., ed. *Gastrointestinal Endocrinology*. New York: McGraw-Hill, 1987.

Weisbrodt, N.W. "Gastrointestinal Motility." *Annual Review of Physiology*, 43:7–51, 1985.

Williams, J.A. "Regulatory mechanisms in pancreas and salivary acini." *Annual Review of Physiology*, 46:361–365, 1984.

KEY TERMS

appendix (p. 818)
bile (p. 805)
cecum (p. 816)
colon (p. 816)
digestion (p. 778)
duodenum (p. 807)
emetic (p. 801)

epiglottis (p. 789)
esophagus (p. 789)
exocrine gland (p. 778)
gallbladder (p. 805)
gastrointestinal system (p. 778)
glottis (p. 789)

ileum (p. 807)
jejunum (p. 807)
larynx (p. 789)
microvilli (p. 810)
motility (p. 778)
mucosa (p. 778)
mucus (p. 788)

pancreas (p. 801)
pepsin (p. 795)
peristalsis (p. 791)
rectum (p. 816)
trypsin (p. 801)

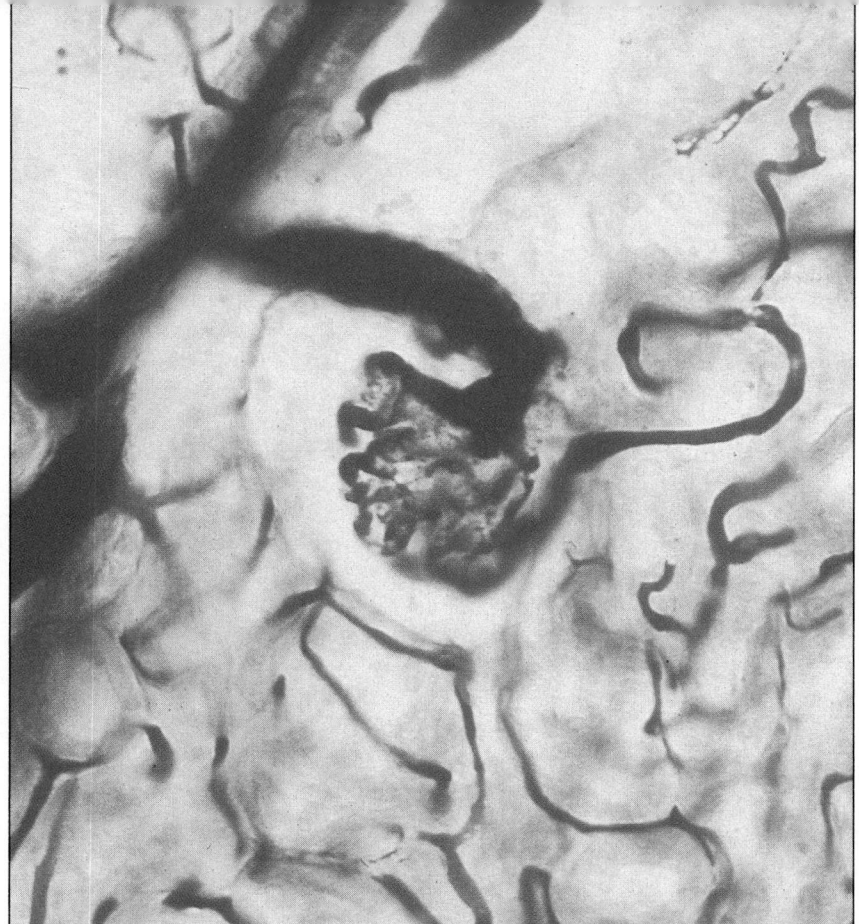

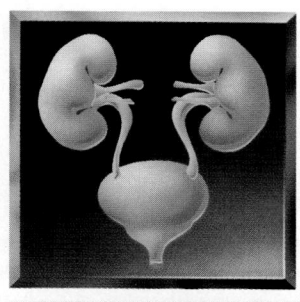

The Kidney

The kidneys play a dominant role in regulating the composition and volume of the extracellular fluids (blood plasma, interstitial fluid, and lymph). They provide a suitable "internal environment" for the body's living cells. The kidneys are commonly thought to simply dispose of waste products in the urine, but they really do much more. This chapter will consider the functional anatomy of the kidneys, renal blood flow, the processes involved in urine formation, and the ways in which the kidneys handle a variety of solutes and water. Chapter 25 will primarily consider how renal (kidney-related) and other mechanisms are controlled and integrated to accomplish the important task of regulating the internal environment. Before considering kidney function in detail, we will outline some of the functions of the kidneys.

1. The kidneys regulate the osmotic pressure of the plasma and other extracellular fluids. Most plasma membranes are highly permeable to water, and so cells have the same osmotic pressure as the extracellular fluid. By stabilizing the osmotic pressure of the extracellular fluid, the kidneys minimize shifts in water between cells and their extracellular environment. In so doing, the kidneys protect the stability of cell volume.

2. By regulating the excretion of sodium and water, the kidneys regulate the volume of the extracellular fluid.

3. The kidneys regulate the individual concentrations of numerous electrolytes in the extracellular fluid, including sodium, potassium, calcium, magnesium, chloride, sulfate, and phosphate ions.

4. The kidneys regulate the plasma bicarbonate concentration and hence the hydrogen ion concentration and so play an important role in acid-base regulation. This topic will be discussed in Chapter 26.

5. The kidneys eliminate metabolic waste products such as urea (an end product of protein metabolism), uric acid (an end product of purine metabolism), and creatinine (an end product of muscle metabolism). They also eliminate many foreign compounds from the body, including drugs such as penicillin.

6. The kidneys produce a number of special substances. These include: **erythropoietin,** a hormone that stimulates the rate of production,

maturation, and release of red blood cells from bone marrow; **renin,** a proteolytic enzyme important in the regulation of extracellular fluid volume and blood pressure; **kallikrein,** a proteolytic enzyme that leads to the formation of kinins, which are vasodilators; and various **prostaglandins** and **thromboxane,** fatty acid derivatives that act as local hormones. Prostaglandins E_2 and I_2 have several actions in the kidneys, including vasodilation, enhancement of renal excretion of salt and water, and stimulation of renin release. Thromboxane is a vasoconstrictor that may be responsible for reduced renal blood flow in a number of renal diseases.

7. The kidneys have several special metabolic functions. They are responsible for converting the inactive form of vitamin D to its active form, **1,25-dihydroxy-vitamin D_3.** The latter is a hormone that stimulates intestinal calcium absorption. The kidneys synthesize ammonia from amino acids. This is important in acid-base regulation. The kidneys can synthesize glucose from noncarbohydrate sources (e.g., amino acids), a process called **gluconeogenesis.** During a prolonged fast, the glucose added to the blood by the kidneys helps to maintain the blood sugar concentration. The kidneys are an important site of degradation (hence inactivation) of several polypeptide hormones, including insulin, glucagon, and parathyroid hormone.

Functional Anatomy of the Kidney

The functions of the kidneys can only be understood in terms of their unique anatomy. Figure 24–1 presents an overview of the urinary system and a section through the human kidney. The kidneys are bean-shaped organs that lie retroperitoneally in back of the abdominal cavity on either side of the vertebral column. They are drained by the **ureters,** which carry the urine to the **urinary bladder.** The tube draining the bladder is called the **urethra.** If a kidney is cut through (see Fig. 24–1), two parts are easily distinguished: an outer part, called the **cortex,** and an inner part, called the **medulla.** The cortex is reddish-brown and has a granular appearance. The cortex contains **glomeruli, convoluted tubules, cor-**

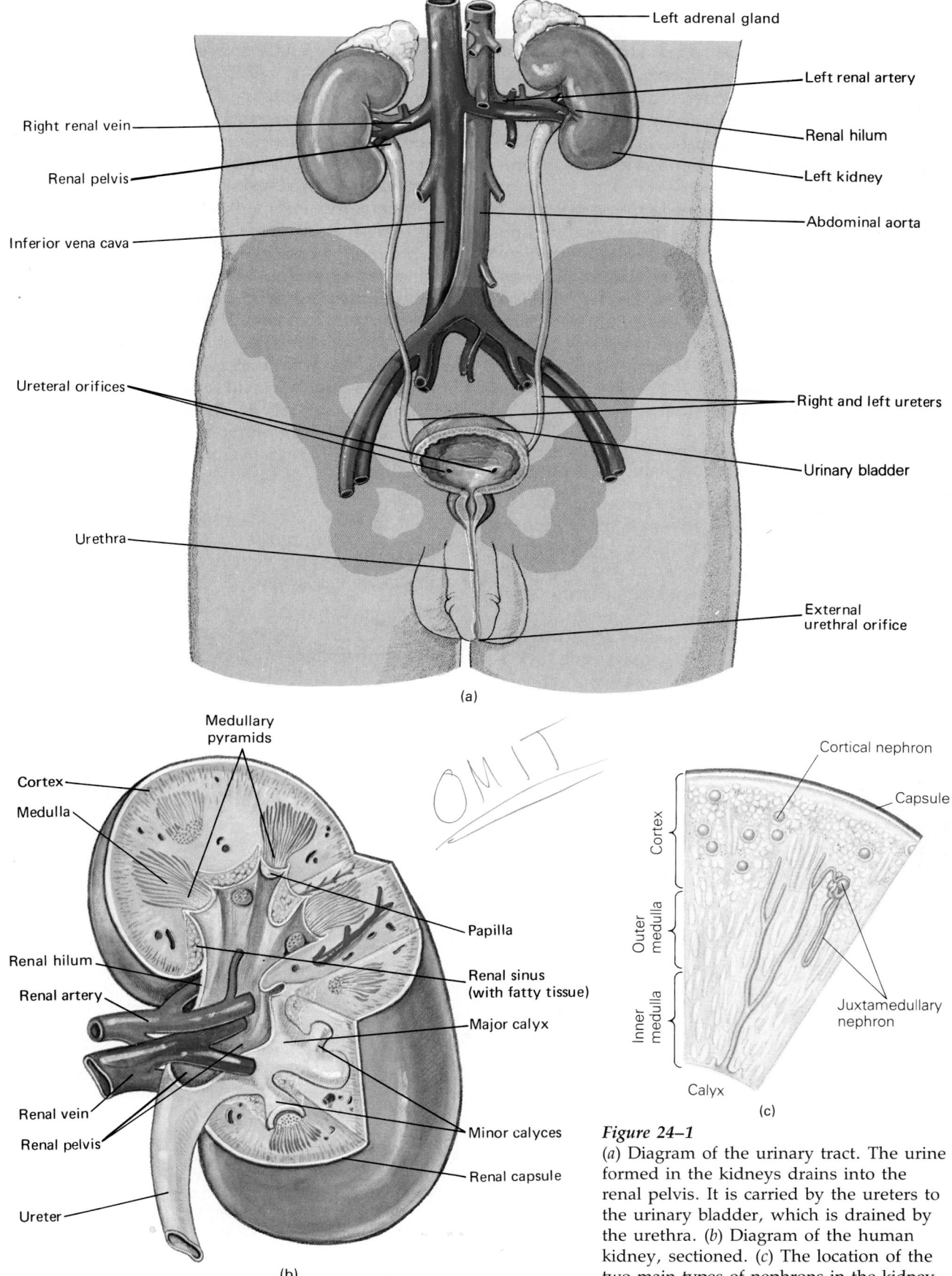

Left adrenal gland

Left renal artery

Renal hilum

Left kidney

Abdominal aorta

Right renal vein

Renal pelvis

Inferior vena cava

Ureteral orifices

Right and left ureters

Urinary bladder

Urethra

External urethral orifice

(a)

Medullary pyramids

Cortex

Medulla

Papilla

Renal sinus (with fatty tissue)

Renal hilum

Renal artery

Major calyx

Minor calyces

Renal vein

Renal pelvis

Renal capsule

Ureter

(b)

Cortical nephron

Capsule

Cortex

Outer medulla

Inner medulla

Juxtamedullary nephron

Calyx

(c)

Figure 24–1
(*a*) Diagram of the urinary tract. The urine formed in the kidneys drains into the renal pelvis. It is carried by the ureters to the urinary bladder, which is drained by the urethra. (*b*) Diagram of the human kidney, sectioned. (*c*) The location of the two main types of nephrons in the kidney.

tical collecting ducts, and associated blood vessels. The medulla is a lighter color and is striated (striped). Striations result from the parallel arrangement of **loops of Henle, medullary collecting ducts,** and blood vessels. The medulla is usually divided into an **outer medulla** (closer to the cortex) and an **inner medulla** (farther from the cortex).

The human kidney is divided into about a dozen **lobes.** Each consists of a **pyramid** of medullary tissue plus the cortical tissue overlying its base and covering its sides. The apex of a medullary pyramid forms a renal **papilla,** which drains the urine into a **minor calyx.** Minor calices unite to form a **major calyx;** these in turn lead to the expanded **renal pelvis,** which is drained by the ureter. The ureter, renal artery, renal vein, nerves, and lymphatic vessels enter or leave the kidney at the **renal hilum,** a depression of its medial surface.

The Structure of a Nephron

The basic unit of kidney structure and function is the **nephron** (Fig. 24–2). Each human kidney contains about one million nephrons. Each nephron consists of a **renal corpuscle** and its attached **renal tubule.**

The renal corpuscle consists of a tuft of capillaries, the **glomerulus,** surrounded by an expanded, double-walled cup, **Bowman's capsule.** The **urinary space** of Bowman's capsule is continuous with the lumen of the renal tubule. Traditionally, the renal tubule has been divided into several anatomically distinct segments. The first segment is called the **proximal tubule.** This is divided into two parts: the **proximal convoluted tubule,** which coils and twists (convolutes) in the neighborhood of its renal corpuscle, and the **proximal straight tubule,** which plunges toward the renal medulla. The **loop of Henle** is the portion of the nephron between the proximal convoluted and distal convoluted tubules. Its first part is the proximal straight tubule. Next is a variable **thin limb** (descending and ascending portions), and finally a **thick ascending limb** (distal straight tubule). This is followed by a short **distal convoluted tubule.** Distal convoluted tubules join **connecting tubules,** which lead to **cortical collecting ducts.** These lead to the **outer medullary collecting ducts,** which pass straight through the outer medulla. In the inner medulla, **inner medullary collecting ducts** unite to form larger and larger ducts, which convey the urine into the minor calices.

Strictly speaking, the collecting ducts are not part of the nephron. These structures have different embryological origins, and there are many more nephrons than collecting ducts. Functionally, the collecting ducts are not just conduits; they are re-

Figure 24–2
Diagram of parts of the nephron and collecting duct system. The nephron on the right is a cortical nephron, the nephron on the left is a juxtamedullary nephron. (Modified from: Kriz, W., and Bankir, L. "A standard nomenclature for structures of the kidney." *American Journal of Physiology,* 254:F1–F8, 1988.)

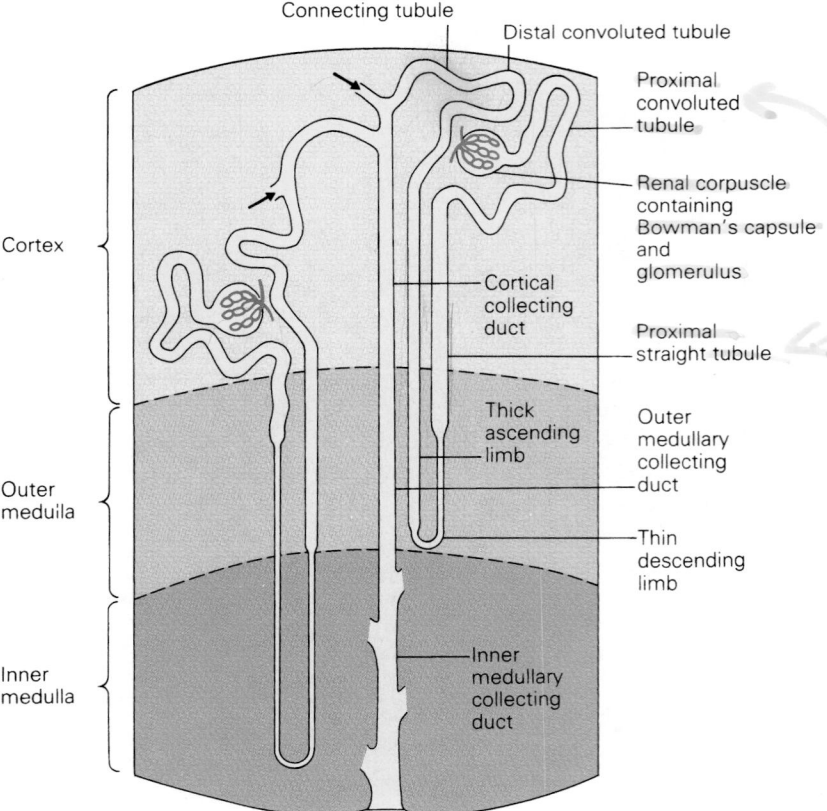

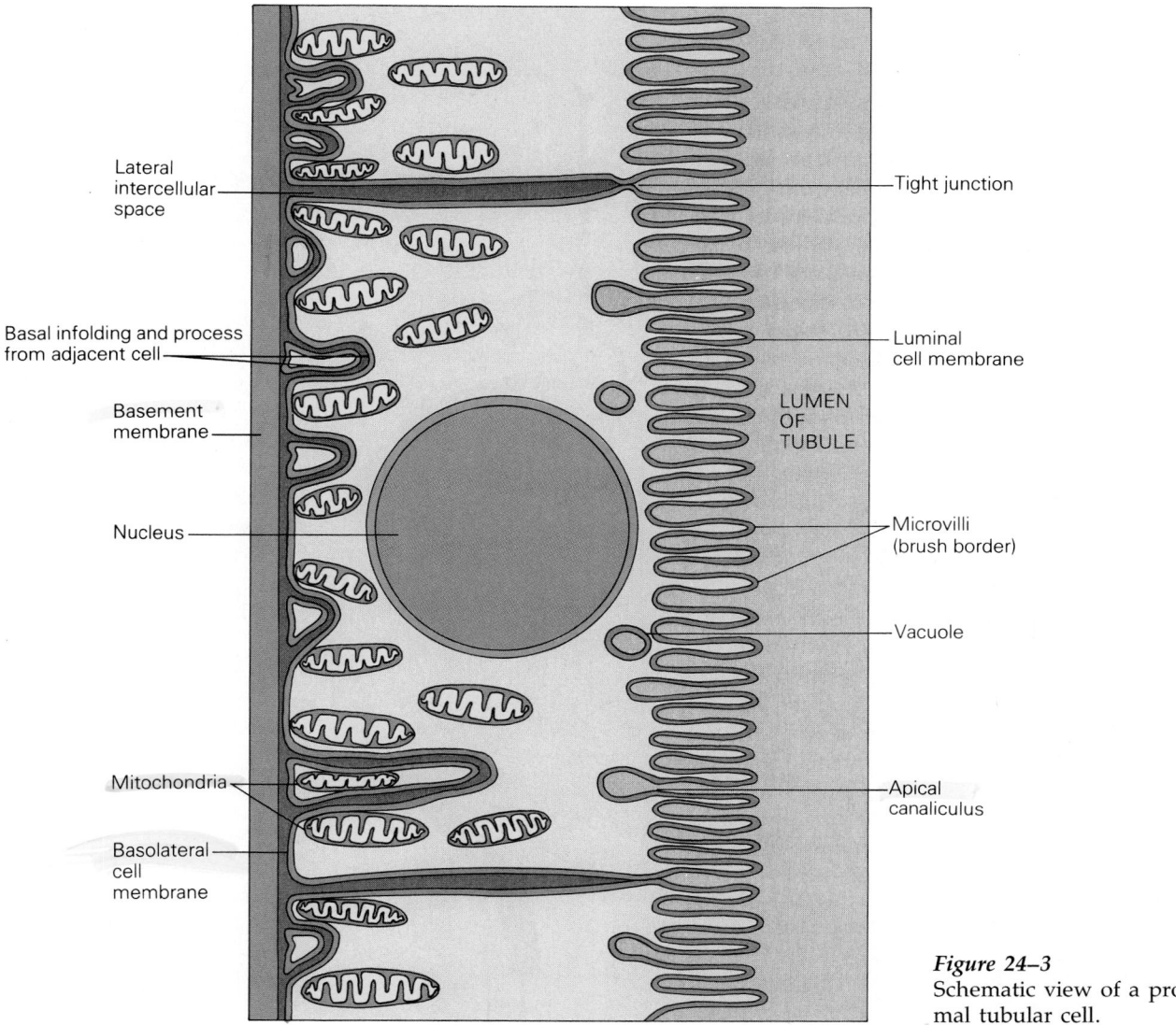

Figure 24–3
Schematic view of a proximal tubular cell.

Labels on figure:
Lateral intercellular space
Basal infolding and process from adjacent cell
Basement membrane
Nucleus
Mitochondria
Basolateral cell membrane
Tight junction
Luminal cell membrane
LUMEN OF TUBULE
Microvilli (brush border)
Vacuole
Apical canaliculus

sponsible for determining the final composition and volume of the urine. Since the collecting ducts continue the tasks performed by the nephrons, they are often considered part of the nephron unit.

Scientists are learning that there are many structural and functional differences within the more traditional nephron segments. For example, based on cell ultrastructure and functional activity rather than on form, the proximal tubule can be divided into three consecutive segments. Other nephron segments also can be subdivided; indeed, some have suggested that the nephron and collecting duct should be divided into a dozen or more distinct segments. Even in the same segment, there may be different cell types with different properties. For example, in the collecting ducts there are two different kinds of cells, the lighter-staining **principal cells** and darker-staining **intercalated cells.**

Not all nephrons are alike (see Fig. 24–2). Nephrons with renal corpuscles in the outer cortex, called **cortical nephrons,** have relatively small glomeruli, short or absent thin limbs, and loops that bend wholly within the cortex or outer medulla. Nephrons with renal corpuscles in the inner cortex (i.e., near the medulla), called **juxtamedullary nephrons,** have large glomeruli, long thin limbs, and long loops, some of which may reach the papillary tips. Functional differences between these two nephron populations include differences in filtration rate, tubular transport properties, and renin content. In the human kidney, the majority of nephrons (85%) are of the cortical type, and the rest (15%) are of the juxtamedullary type.

The renal tubule and collecting duct are composed of a single layer of epithelial cells surrounding a tubule lumen. Cell structure and function vary considerably from one segment to another. As an example of the specialized architecture of the renal tubule epithelium, we will consider a cell from the proximal tubule (Fig. 24–3). The cell surface that

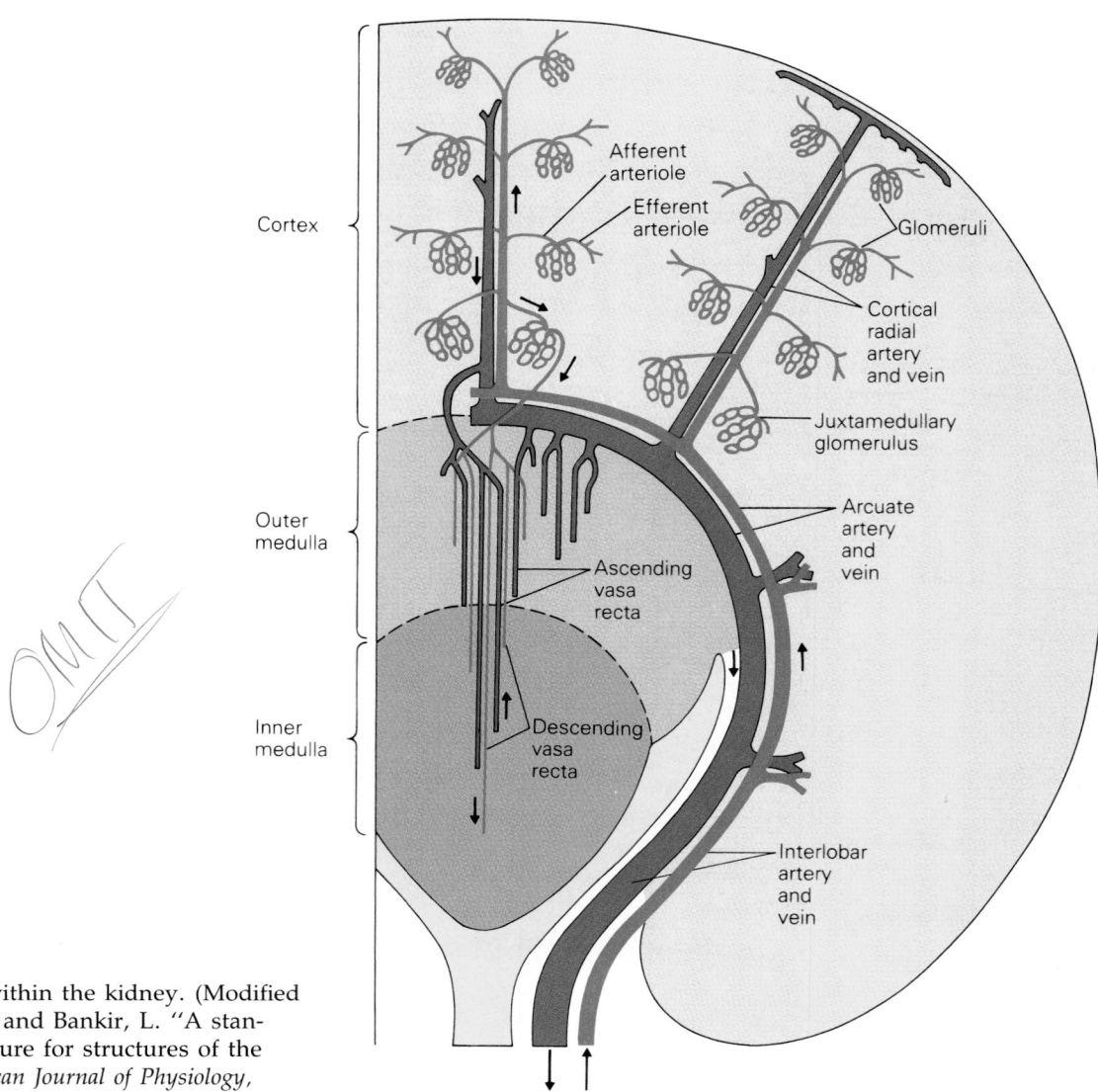

Cortex

Afferent
arteriole

Efferent
arteriole

Glomeruli

Cortical
radial
artery
and vein

Juxtamedullary
glomerulus

Outer
medulla

Ascending
vasa
recta

Arcuate
artery
and
vein

Inner
medulla

Descending
vasa
recta

Interlobar
artery
and
vein

To renal vein   From renal artery

Figure 24–4
Blood vessels within the kidney. (Modified from: Kriz, W., and Bankir, L. "A standard nomenclature for structures of the kidney." *American Journal of Physiology*, 254:F1–F8, 1988).

contacts the fluid in the tubule lumen (the **luminal** or **apical cell membrane**) has numerous **microvilli,** collectively called the **brush border.** The microvilli greatly increase the membrane surface area, thereby easing the transfer of materials between cells and tubular fluid. At the base of the microvilli are invaginations of the cell membrane **(apical canaliculi)** and associated vacuoles, thought to be involved with the reabsorption of proteins. Adjacent cells are linked to each other near the apical end of the cell by a **tight junction (zonula occludens).** In the proximal tubule, this junction permits some movement of small ions and water. The **basolateral cell membrane** surface area is considerably increased by numerous ridges and processes that are formed as adjacent cells interdigitate with each other. The space between adjacent cells, the **lateral intercellular space,** is thought to be an important route for salt and water movement across the epithelial cell layer. Within the cell, numerous mitochondria lie close to

the basolateral cell membrane and furnish the energy for an important sodium-potassium pump (Na^+/K^+-ATPase) located in these membranes. The tubule cells rest on a **basement membrane,** which acts as a supportive structure. If the cells are damaged, the basement membrane provides a framework on which regenerating cells can reline the tubule. The overall appearance of the proximal tubule cell is that of a very busy cell, and we will see later that it is heavily engaged in transport activities.

The Blood Supply of the Kidney

Each **renal artery** arises from the abdominal aorta and divides into several major branches just before entering the renal hilum. These branches further subdivide and pass into the renal substance between the medullary pyramids as **interlobar arteries** (Fig. 24–4). These give rise to the **arcuate arteries,** which arch over the base of the medullary pyramids at the junction of medulla and cortex. The arcuate arteries give rise to **cortical radial arteries,** which course toward the surface of the cortex and supply the short, muscular **afferent arterioles.** Each afferent arteriole leads to a **glomerulus.** The glomerular capillaries re-form the **efferent arteriole,** which usually has a thin muscular coat. This breaks up into the **peritubular capillaries,** which form an extensive network surrounding the tubules. The peritubular capillaries lead to veins that generally parallel the course of the arterial vessels and ultimately empty into the **renal vein.**

The renal circulation is unusual in that there are two capillary beds in series. The blood pressure is high in the glomeruli and low in the peritubular capillaries. The glomeruli are specialized for filtration, whereas the peritubular capillaries are specialized for uptake of fluid reabsorbed by the tubules. The blood supply to the kidney medulla is derived from the efferent arterioles associated with juxtamedullary nephrons. These give rise to long, straight capillaries called **vasa recta.** The vasa recta descend into and ascend from the medulla in parallel with the tubular structures of the medulla.

Nerves and Lymphatic Vessels of the Kidney

The kidneys are richly innervated. Sympathetic nerve fibers travel to the kidneys mainly in the Xth, XIth, and XIIth thoracic and Ist lumbar spinal nerves. The sympathetic efferent fibers predominantly innervate blood vessels and can cause intense vasoconstriction. Sympathetic fibers are also found adjacent to tubular cells and may increase sodium reabsorption. Afferent (sensory) nerve fibers from the renal pelvis and ureter mediate pain of renal origin. Lymphatic vessels are found in the cortex, but whether or not they are found in the medulla is not certain.

The Juxtaglomerular Apparatus

Just before it becomes the distal convoluted tubule, the thick ascending limb (distal straight tubule) contacts the afferent and efferent arterioles of its parent glomerulus (see Fig. 24–2). The tubular epithelium at this spot shows a dense crowding of cells and is called the **macula densa** (Fig. 24–5). Adjacent smooth muscle cells in the walls of the arterioles (especially the afferent arteriole) have an epithelial cell-like appearance and may contain granules. These **granular cells** synthesize, store, and release renin into the blood and interstitial fluid. In the re-

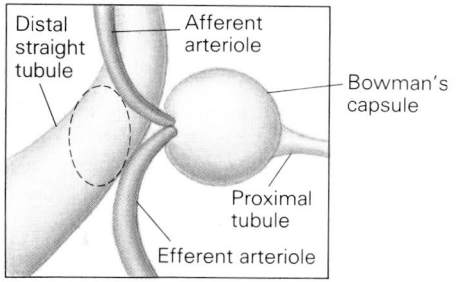

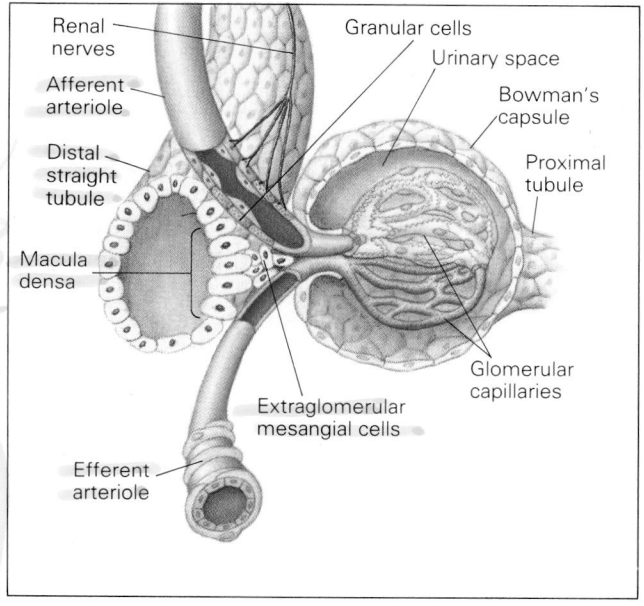

Figure 24–5
The juxtaglomerular apparatus consists of macula densa, extraglomerular mesangial cells, and granular cells. This structure may play a role in regulating glomerular blood flow and filtration rate at the level of a single nephron.

gion bounded by macula densa and arterioles are closely packed cells known as **extraglomerular mesangial cells.** These cells may act as an intermediary in transferring information from macula densa cells to cells in the blood vessel walls. The entire group of cells made up of macula densa cells, extraglomerular mesangial cells, and granular cells is known as the **juxtaglomerular apparatus.** The intimate relationship between tubule and blood vessels at this site has suggested that changes in tubule fluid composition might be signaled to the blood vessels and lead to changes in glomerular blood flow and filtration rate; this idea is known as the **tubuloglomerular feedback hypothesis.** Current evidence indicates that a sudden increase or long-term decrease in flow through the loop of Henle will cause constriction of the afferent arteriole.

The Blood Flow of the Kidney

The kidneys have an enormous blood flow. Total renal blood flow in a resting, young adult 70-kg male averages about 1200 ml per minute or 20% to 25% of the cardiac output (5 liters per minute), even though both kidneys weigh less than 0.5% of body weight (300 gm/70 kg). Blood flow per gram of kidney tissue averages 4 ml per minute (1200 ml per minute/300 gm), which is much higher than in other organs considered to be well perfused (e.g., brain, heart, liver).

Why is such a high blood flow to the kidneys necessary? It is needed to sustain a high rate of filtration of plasma in the glomeruli. Essentially all of the blood perfusing the kidneys passes through glomeruli, where it is subject to filtration. Every minute a volume of filtrate is formed that is equal to about 10% of the blood flow rate. The high renal blood flow does not reflect excessive metabolic demands. Although the kidneys use about 8% of the body's oxygen consumption in a resting adult, renal blood flow is more than adequate to deliver the needed oxygen. In fact, renal venous blood typically has a bright red color owing to its high oxygen content, a result of its low oxygen extraction.

Blood flow is not distributed uniformly within the kidney. It is highest in the renal cortex, where all of the glomeruli are located and where the blood is filtered. Less blood flows to medullary structures, and flow decreases along the depth of the medulla. The relatively low blood flow to the medulla helps to preserve higher osmotic pressures in this part of the kidney.

If arterial blood pressure is varied between 80 and 180 mm Hg, renal blood flow and glomerular filtration rate change relatively little. This is because of **renal autoregulation,** a self-regulating mechanism in the kidney that does not depend on extrinsic nerves or extrinsic hormones. Changes in the caliber of blood vessels upstream to the glomerulus (cortical radial arteries and afferent arterioles) are thought to be primarily involved. If blood pressure is increased, these vessels constrict, thereby increasing renal vascular resistance and minimizing an increase in blood flow. If blood pressure is decreased, these vessels dilate, thereby decreasing renal vascular resistance and minimizing a fall in blood flow. The autoregulatory response has a limited range, and so if blood pressure is decreased below 80 mm Hg, renal blood flow and glomerular filtration rate may fall markedly. The importance of renal autoregulation is that it prevents changes of arterial blood pressure from producing large changes in glomerular filtration rate. Such changes could lead to dramatic changes in sodium excretion, a result that might be undesirable.

Total renal blood flow is decreased in a number of stressful situations, including cold, deep anesthesia, fright, hemorrhage, pain, and strenuous exercise. This decrease is a reflex response due primarily to activation of renal sympathetic nerve fibers, which cause vasoconstriction. The increase in renal vascular resistance causes blood flow to shift to the brain and heart, which are more vital to immediate (short-term) survival.

A number of substances also affect renal blood flow. For example, angiotensin II, epinephrine, norepinephrine, and thromboxane cause renal vasoconstriction. Acetylcholine, bacterial pyrogens, kinins (e.g., bradykinin), and many prostaglandins (e.g., prostaglandins E_2 and I_2) have vasodilator effects. Some of these substances are synthesized and act locally within the kidneys. Increased amounts of vasodilator prostaglandins are produced in circumstances in which kidney blood flow is threatened—for example, by intense stimulation of renal sympathetic nerves. This production of prostaglandins opposes excessive vasoconstriction, which might otherwise damage the kidneys.

Processes Involved in Formation of Urine

Three basic processes are involved in the formation of urine: glomerular filtration, tubular reabsorption, and tubular secretion (Fig. 24–6).

Glomerular filtration is the starting point for urine formation. Because of the high hydrostatic blood pressure in the glomerular capillaries, a filtrate of plasma is pushed out of the capillaries into the urinary space of Bowman's capsule. This process of filtration is often referred to as **ultrafiltration,** and the filtrate is referred to as an **ultrafiltrate** of plasma because the glomerular capillary wall (the **glomerular filtration membrane**) behaves like an ultrafine filter. The glomerular capillary wall allows free passage of small molecules (molecules with a molecular weight below 10,000) but restricts the passage of large molecules. In particular, the large plasma proteins (serum albumin, globulins, and fibrinogen) are virtually excluded from the glomerular filtrate. Glomerular filtration is rather nonselective, and both useful substances and waste products are filtered. The final urine differs radically from the glomerular filtrate in both volume and composition. Most of the valuable constituents of the filtrate (most of the salts, water, and metabolites) are returned to the blood by the tubules.

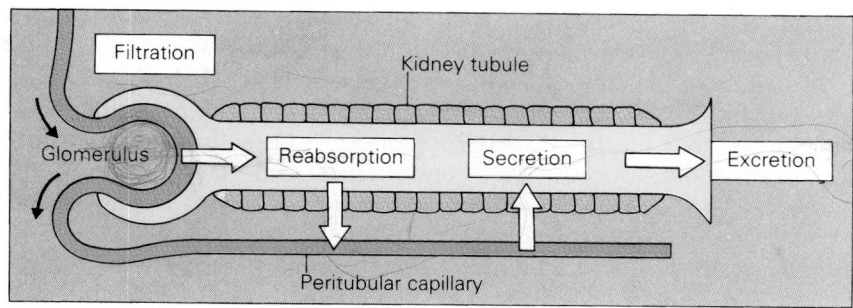

Figure 24–6
Processes involved in urine formation.

Tubular reabsorption is the process whereby substances are transferred out of the tubular fluid and returned to the peritubular capillaries surrounding the kidney tubules. In this way, substances that have been temporarily lost from the plasma in the process of glomerular filtration are returned to the circulation. Reabsorption is a selective process, and the tubules handle different substances differently.

Tubular secretion is movement of substances across the tubule epithelium in a direction opposite to that of reabsorption. Certain substances—for example, the organic anion *p*-aminohippurate (PAH)—are taken up by the tubule cells from the blood surrounding the tubules and deposited into the tubular fluid. Other substances, such as hydrogen ions and ammonia, are generated within the kidney tubule cells and then secreted into the tubular urine. The terms **reabsorption** and **secretion** simply indicate

the direction of transport across the tubular epithelium, either out of or into the tubular urine, respectively.

Excretion refers to what comes out in the urine that leaves the body. In most cases, excretion depends on both glomerular filtration and tubular transport. Many substances are filtered and reabsorbed (e.g., water); the amount excreted is equal to the difference between filtered and reabsorbed amounts. Some substances (e.g., PAH) are filtered and secreted; the amount excreted is equal to the sum of filtered and secreted amounts. A few substances are filtered, reabsorbed, and secreted (e.g., potassium); the amount excreted equals filtered minus reabsorbed plus secreted amounts. In a very few cases (e.g., inulin), a substance is only filtered; in this case, excreted and filtered amounts are equal.

Table 24–1 presents data on the daily (24-hour) filtration, reabsorption, and excretion of some nor-

Table 24–1
*Daily Filtration, Reabsorption, and Excretion of Some Normal Plasma Constituents**

Constituent	Filtered	Reabsorbed	Excreted
Water	167.5 liters	166 liters	1.5 liters
Sodium	24,000 mmoles	23,900 mmoles	100 mmoles
Potassium	720 mmoles	630 mmoles	90 mmoles
Calcium	270 mmoles	266 mmoles	4 mmoles
Magnesium	135 mmoles	120 mmoles	15 mmoles
Chloride	19,500 mmoles	19,400 mmoles	100 mmoles
Bicarbonate	4,500 mmoles	4,498 mmoles	2 mmoles
Phosphate	6 g	5 g	1 g
Glucose	150 g	150 g	0 g
Urea	50 g	25 g	25 g
Uric acid	8 g	7.2 g	0.8 g
Creatinine	1.8 g	0 g	1.8 g

*Data are for a normal, young adult, 70-kg man.

mal plasma constituents. Since filtration rate is so high (180 liters per day), enormous quantities of water, salts, and organic compounds are filtered. The bulk (more than 99%) of the filtered water, sodium, chloride, and bicarbonate is usually reabsorbed. The small fraction of these substances that is excreted is of critical importance. This is because in order for a person to stay in balance—that is, to show no net change in the amount of these substances in the body—exactly the right amount in the urine should be excreted. A small change in the percentage of reabsorption of a filtered substance may lead to a large change in the absolute excretion of the substance. For example, if the reabsorbed percentage of filtered sodium decreases from 99.6% (23,900 mmoles per day divided by 24,000 mmoles per day) to 99.2% (23,800 mmoles per day divided by 24,000 mmoles per day), the amount of sodium excreted will double from 100 to 200 mmoles per day. Clearly, tubular reabsorption of electrolytes and water must be carefully regulated to avoid excessive losses or abnormal retention.

Do not memorize the numbers in Table 24–1, but try to appreciate the magnitude of filtration, reabsorption, and excretion of different substances, and recognize that different substances are handled differently by the tubules. Notice that potassium is listed as "reabsorbed" since the excreted amount is less than the filtered amount. Actually, nearly all of the filtered potassium is reabsorbed by proximal portions of the nephron, and most of the excreted potassium is secreted by the collecting ducts. Filtered calcium is mostly reabsorbed. Dietary phosphate is largely excreted in the urine. Filtered glucose is normally completely reabsorbed; it would make little sense to lose this valuable metabolite in the urine. Filtered creatinine is excreted without reabsorption. Urea and uric acid are reabsorbed to some extent. For all three of these waste products, production rates in the body and urinary excretion are normally equal.

The rates of excretion in Table 24–1 can vary widely, depending on the diet and other circumstances. Renal handling of each of the substances listed in this table will be discussed in more detail later.

Glomerular Filtration

The Glomerular Filtration Membrane

The barrier through which the glomerular filtrate passes from capillary blood to the urinary space of Bowman's capsule is the **glomerular filtration mem-brane**. It consists of three layers: capillary endothelium, glomerular basement membrane, and podocyte cell layer (visceral epithelium of Bowman's capsule). Figure 24–7 shows part of this structure as seen with the transmission electron microscope. The endothelial cell layer lining the capillaries has large pores or windows **(fenestrae)** about 50 to 100 nanometers (nm) in diameter. The **basement membrane** consists of a meshwork of fine fibrils embedded in a gel-like matrix. The **podocytes** ("foot cells") have numerous cytoplasmic processes **(foot processes)** that rest on the basement membrane. The narrow space between adjacent foot processes is about 20 nm wide and is called the **slit pore**. Some scientists have found that a thin membrane stretches across the slit pore; this **filtration slit diaphragm** may have a fine porous structure. The glomerular filtrate must pass through all three layers of the filtration membrane. It probably passes through the endothelial cell fenestrae and the slit pores and not through the thicker cell cytoplasm.

The filtration membrane has a high permeability to water and small molecules but restricts the passage of large molecules (macromolecules). It is convenient to think that this sieving effect is due to the presence of pores in this membrane. Based on studies of the filterability of macromolecules of different molecular size, scientists have determined that the glomerular filtration membrane behaves as though it were penetrated by cylindrical pores 7.5 to 10 nm in diameter. But no one has ever seen pores of these dimensions; the fenestrae and slit pores are too big, and no pores are visible in the basement membrane. It is possible that the pores may be lost during the preparation of tissue for electron microscopy. In any case, most scientists believe that the basement membrane is probably the major size-selective barrier.

Electrical charge also affects the passage of macromolecules through the filtration membrane. The endothelial pores, basement membrane, and surface coat of the podocytes contain negatively charged glycoproteins. These negatively charged structural elements impede the filtration of negatively charged plasma proteins such as serum albumin. The layers of the filtration membrane closest to the capillary lumen, the endothelium and the innermost portion of the basement membrane, are probably the major electrostatic barriers to the filtration of negatively charged plasma proteins.

One of the hallmarks of glomerular disease is the abnormal excretion of plasma proteins. This may be due to either (1) an increase in size of glomerular pores, or (2) a loss of fixed negative charges from the glomerular filtration membrane. In either

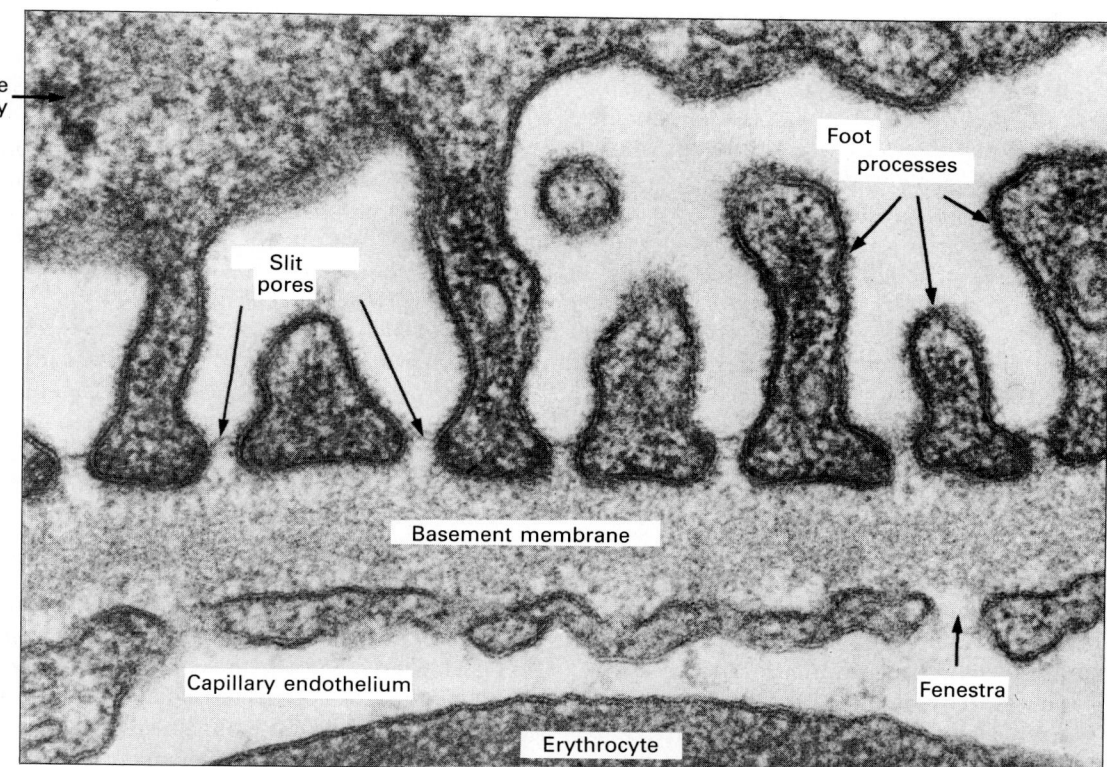

Figure 24–7
Electron micrograph of the glomerular filtration membrane. Note the pore (fenestra) in the thin capillary endothelium and the slit pores between the foot process of the podocytes. (From: Fawcett, D. W. *A Textbook of Histology*, 11th edition. Philadelphia: Saunders, 1986, p. 765; Courtesy of D. Friend).

case, abnormal quantities of protein are filtered and lost in the urine.

Factors That Affect Glomerular Filtration Rate

The factors that affect **glomerular filtration rate (GFR)** are basically the same as those that influence fluid movement across capillary walls anywhere in the body. The driving force for filtration is the glomerular capillary hydrostatic pressure (P_{GC}). This pressure ultimately depends on the energy imparted to the blood by the beating of the heart. Filtration is opposed by the hydrostatic pressure in the urinary space of Bowman's capsule (P_{BS}) and by the intracapillary colloid osmotic pressure (oncotic pressure) due to the plasma proteins (π_{GC}) (Fig. 24–8). Since the glomerular filtrate is essentially protein-free, the colloid osmotic pressure of fluid in the urinary space is zero and can be disregarded. The net filtration pressure gradient (ΔP) is equal to the dif-

ference between pressures favoring and opposing filtration. In other words,

$$\Delta P = P_{GC} - P_{BS} - \pi_{GC}$$

The glomerular filtration rate depends not only on the pressure gradient but also on the glomerular filtration coefficient K_f. We can write

$$GFR = K_f \times \Delta P$$

This equation is analogous to the equation that describes fluid flow rate along a tube, $F = \Delta P / R$ (Chapter 19), except that here we are considering fluid movement across a capillary wall. K_f can be thought of as the conductance (1/resistance) of the glomerular filtration membrane; its magnitude depends on the fluid permeability of the membrane and its surface area.

Figure 24–8 summarizes the average pressures thought to exist in the normal human kidney. Glomerular hydrostatic pressure is probably about 55 mm Hg, about twice as high as in most other capillaries in the body. The high pressure in the glo-

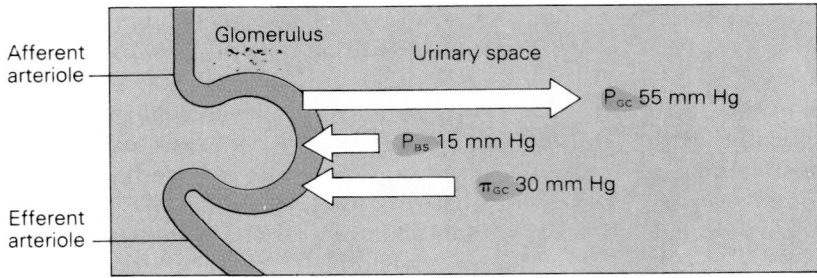

Figure 24–8

Pressures involved in glomerular filtration. The glomerular capillary hydrostatic pressure (P_{GC}) favors filtration, and the hydrostatic pressure in the urinary space of Bowman's capsule (P_{BS}) and the colloid osmotic pressure of proteins in the glomerular capillary plasma (π_{GC}) both oppose filtration. The pressures indicated are for the human kidney and are estimates based on measurements in other mammals.

merulus results from the relatively low resistance of upstream blood vessels (cortical radial artery, afferent arteriole) and high resistance of downstream blood vessels (efferent arteriole). Glomerular hydrostatic pressure probably changes very little (perhaps 2 mm Hg) along the length of the glomerular capillaries. This is because there are many (30 to 50) parallel capillary loops in a glomerulus, making the resistance to blood flow in the capillaries very low. Hydrostatic pressure in the urinary space of Bowman's capsule is about 15 mm Hg. This is the pressure left after filtration, and it opposes further filtration. It provides the driving force for fluid flow down the entire length of the nephron. The colloid osmotic pressure of glomerular capillary blood plasma averages about 30 mm Hg. This pressure rises along the length of the glomerular capillary, because a protein-free filtrate is forced out of the glomerular plasma, and the plasma proteins are left behind. Consequently, filtration rate declines from the beginning (afferent arteriole) to the end (efferent arteriole) of the glomerular capillaries. In the normal human kidney, outward filtration probably occurs along the entire length of the glomerular capillaries. The net filtration pressure gradient (ΔP), averaged over the entire glomerulus, is about 10 mm Hg (55 mm Hg − 15 mm Hg − 30 mm Hg = 10 mm Hg).

The rate of filtrate formation in the glomeruli of the kidney greatly exceeds that in all other capillary beds in the human body. The high rate of glomerular filtration in the kidneys, about 180 liters per day, is due to several factors. First, the glomerular filtration coefficient is very high, which reflects, in part, a high fluid permeability of the glomerular filtration membrane. This membrane appears to be very "porous"; it behaves as though it contains many more pores than are present in most other capillaries. In part, the high glomerular filtration coefficient also reflects a large surface area. If all of the glomerular capillaries were spread out as a flat sheet, the total surface area occupied would be about 2 square meters (roughly equal to body surface area). Second, glomerular filtration rate is high because glomerular hydrostatic pressure is unusually high for a capillary bed. Third, the high rate of renal blood flow permits a rapid rate of filtration. This effect on filtration rate is subtle. Consider what would happen, however, if glomerular blood flow were low. In this case, forcing a small quantity of protein-free filtrate out of the capillary would lead to a sharp rise in intracapillary colloid osmotic pressure early along the length of a glomerular capillary. Filtration would soon be brought to a halt; the rest of the capillary would no longer function in filtration. Therefore, a high renal blood flow is necessary for a high glomerular filtration rate.

GFR may be altered in a variety of physiological and pathophysiological conditions. The conditions that affect filtration rate do so by affecting the factors already mentioned—namely, filtration coefficient, glomerular hydrostatic pressure, hydrostatic pressure in Bowman's space, and plasma colloid osmotic pressure.

Filtration coefficient (K_f)—and with it GFR—will be decreased if there is a decrease in either filtration membrane fluid permeability (e.g., due to membrane thickening or loss of pores) or membrane surface area. K_f may be altered physiologically by a number of hormones (e.g., angiotensin II). In chronic renal failure, glomeruli are destroyed, leading to a reduction in the filtering surface and a diminished GFR.

Glomerular hydrostatic pressure (P_{GC}) depends mainly on the arterial blood pressure and the resistances of upstream blood vessels (cortical radial arteries, afferent arterioles) and downstream blood vessels (efferent arterioles). Within the range of operation of renal autoregulation (an arterial pressure of 80–180 mm Hg), glomerular hydrostatic pressure changes very little. If, however, blood pressure is reduced below this range, glomerular hydrostatic pressure and GFR may decrease substantially. At a mean arterial blood pressure of 40 to 50 mm Hg (as in shock or hypotension), GFR and, hence, urine output are practically zero, since the glomerular hydrostatic pressure is inadequate to form a filtrate.

Glomerular hydrostatic pressure may be altered by changing the diameter of afferent and efferent arterioles (Fig. 24–9); these blood vessels are affected by sympathetic nerve fibers and by the variety of vasoconstrictor and vasodilator substances discussed earlier. Constriction of afferent arterioles lowers the downstream glomerular pressure. At the same time, blood flow is decreased. For both of these reasons, GFR decreases. Dilation of afferent arterioles increases glomerular hydrostatic pressure and blood flow, which leads to an increase in GFR. Constriction of efferent arterioles leads to an increase in the upstream glomerular pressure; blood flow decreases. The effect on GFR of efferent arteriolar constriction is complex. An increase in glomerular pressure tends to increase GFR; on the other hand, a fall in blood flow tends to decrease GFR. So, depending on the degree of efferent constriction, GFR may be either increased or decreased. With modest constriction, GFR rises (the increase in glomerular pressure predominates); with severe constriction, GFR falls (the fall in blood flow predominates). Efferent arteriolar dilation leads to a decrease in glomerular pressure and an increase in glomerular blood flow; GFR falls.

If the urinary tract is obstructed (e.g., by stones, prostate enlargement), **hydrostatic pressure in Bowman's space (P_{BS})** will increase, and consequently GFR will decrease. A very high rate of urine output also may be accompanied by a rise in hydrostatic pressure in Bowman's space and a fall in GFR, because an increased pressure head is required to force a large volume flow down the tubules and collecting ducts.

A decrease in plasma protein concentration (e.g., due to intravenous infusion of a large volume of saline) decreases **plasma colloid osmotic pressure (π_{GC})** and leads to an increase in GFR.

The Clearance Concept and the Measurement of Glomerular Filtration Rate

The determination of GFR is one of the most important measurements of kidney function, both clinically and in experimental studies on renal function. This is done by applying the rather ingenious clearance concept. First we will explain what is meant by "clearance."

One can describe the operation of the kidneys as serving to clear substances from the blood plasma. Different substances have different clearance rates. The **renal plasma clearance** of a substance is defined as the rate of excretion of a substance divided by its plasma concentration. By convention, the symbol C_x indicates the clearance of a substance x; U_x the **urine concentration** of the substance x; P_x the **plasma concentration** of the substance x; and $\dot{V}$ the **urine flow rate**. The rate of excretion of a substance is equal to the product of its urine concentration and the urine flow rate, or in symbols, $U_x \times \dot{V}$. From the definition of clearance, we can write the **clearance formula:**

$$C_x = \frac{U_x \times \dot{V}}{P_x}$$

Figure 24–9
Scanning electron micrograph of a cast of a glomerulus with afferent (af) and efferent (ef) arterioles. Constriction or dilation of the arterioles will change glomerular blood flow and pressure. (From: Kimura, K., et al. Effects of atrial natriuretic peptide on renal arterioles: morphometric analysis using microvascular casts. *American Journal of Physiology* 259:F936–F944, 1990.)

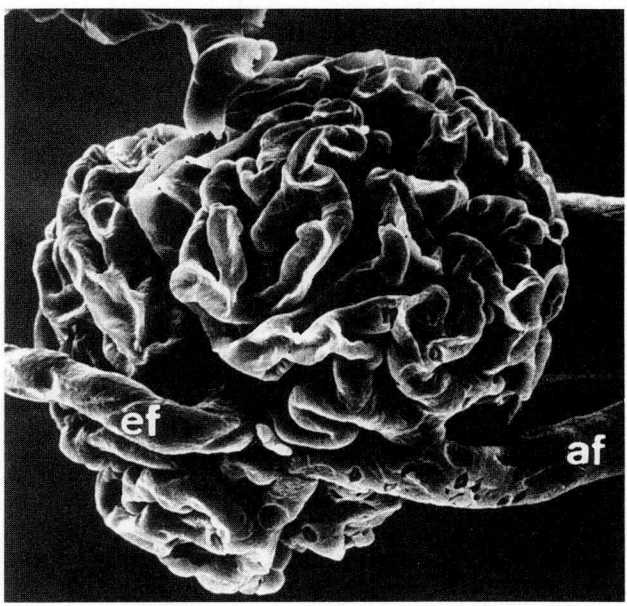

Plasma and urine concentrations of substances are often expressed in terms of milligrams per milliliter, and urine flow is given as milliliters per minute. Substituting these dimensions into the clearance formula, we see that the units of clearance are milliliters of plasma per minute.

$$C_x = \frac{\text{mg } x/\text{ml urine} \times \text{ml urine/minute}}{\text{mg } x/\text{ml plasma}}$$

$$= \text{ml plasma/minute}$$

Renal clearance may be considered as the volume of plasma per unit time that must be completely freed (cleared) of a substance to supply the quantity of a substance excreted in the urine per unit time.

To measure GFR, we need a substance that is cleared from the plasma solely by glomerular filtration. This substance, therefore, should not be reabsorbed or secreted by the kidney tubules. It should not be destroyed, synthesized, or stored by the kidneys. It should pass through the glomerular filtration membrane unhindered; in other words, it should not be a molecule that is too large or that is bound to plasma proteins. It should be nontoxic. Finally, we should be able to measure this substance in plasma and urine using simple analytical methods.

Such an ideal substance is the polysaccharide **inulin,** a fructose polymer, derived from the roots of certain plants; it has an average molecular weight of about 5000. (Do not confuse inulin with the hormone insulin!) Figure 24–10 shows that the amount of inulin (IN) filtered per unit time, $P_{IN} \times \text{GFR}$, equals the amount of inulin excreted per unit time, $U_{IN} \times \dot{V}$. The inulin clearance, C_{IN}, is defined by the expression $U_{IN}\dot{V}/P_{IN}$ and is therefore equal to the GFR.

The way in which inulin is used to measure GFR can be illustrated by an example. An inulin solution is infused intravenously into a person to achieve a constant plasma inulin concentration. A timed urine sample is collected, and average urine flow rate is calculated by dividing the urine volume by the duration of collection. In the middle of the urine collection period, a blood sample is taken. Subsequently, plasma and urine samples are analyzed for inulin. Suppose the following values are found: $P_{IN} = 0.30$ mg/ml, $U_{IN} = 30$ mg/ml, and $\dot{V} = 1.25$ ml/min. Substituting in the inulin clearance equation, $C_{IN} = U_{IN} \times \dot{V}/P_{IN} = \text{GFR}$, we get

$$\text{GFR} = \frac{30 \text{ mg/ml urine} \times 1.25 \text{ ml urine/min}}{0.30 \text{ mg/ml plasma}}$$

$$= 125 \text{ ml plasma/min}$$

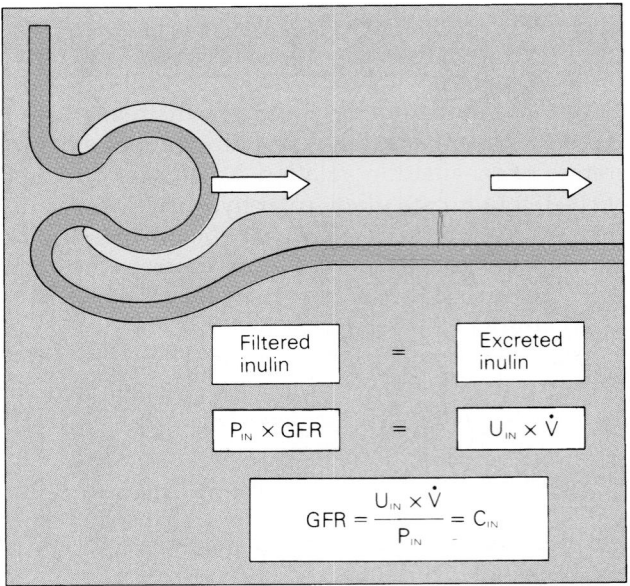

Figure 24–10
Principle behind the measurement of glomerular filtration rate (GFR). Inulin is freely filterable, so the amount filtered per unit time equals its plasma concentration (P_{IN}) multiplied by the GFR. Inulin is neither reabsorbed, secreted, synthesized, destroyed, nor stored by the kidney tubules, so filtered and excreted amounts are equal. Rearranging the equation, we find that GFR is equal to the inulin clearance (C_{IN}).

You may have noticed that in this example the inulin concentration in the urine is 100 times that in plasma. This is not due to tubular secretion of inulin; rather, it is caused by water reabsorption. As water is reabsorbed by the kidney tubules, filtered inulin is left behind and therefore becomes concentrated in a smaller volume of water. For example, if the inulin contained in 100 ml of filtrate is concentrated in 1 ml of urine, the inulin concentration will be increased 100-fold, and 99 ml of water (or 99% of the filtrate) must have been reabsorbed by the kidney tubules.

GFR in people depends on body surface area; it is usually corrected to a standard body surface area of 1.73 square meters. In normal young adult males, it averages 125 ml/min (180 liters/day), and in normal young adult females, it averages 115 ml/min (165 liters/day). GFR is very low in the newborn, 20 ml/min per 1.73 square meters of body surface area, and attains adult values (when corrected for body surface area) by about one year of age. In men, GFR is maintained at about 125 ml/min to about 45

Methods Used to Study Kidney Function

Our understanding of renal function has depended on the use of a great variety of experimental techniques. Four approaches have been of particular value: clearance methods, micropuncture, perfusion of isolated kidney tubules, and membrane vesicles.

Renal clearance methods were developed in the 1930's and still provide an excellent way to evaluate overall kidney function in an unanesthetized person or animal. This approach permits the measurement of glomerular filtration rate, renal blood flow, and the rates of tubular reabsorption or secretion of various substances.

Kidney micropuncture was developed in the 1920's in A.N. Richards' laboratory at the University of Pennsylvania. In a classic study (American Journal of Physiology, 71:209–227, 1924), Wearn and Richards punctured Bowman's space in a frog's kidney with a fine glass capillary micropipette. They analyzed the collected fluid and determined that it contained glucose and chloride but no protein. This proved that the glomerulus produced an ultrafiltrate of plasma. Since glucose and chloride were absent from the frog's bladder urine, but were present at the beginning of the nephron, it was also evident that the tubules must have reabsorbed these two substances.

The sites of reabsorption of various substances (chloride, glucose, water, urea, and inorganic phosphate) and secretion of hydrogen ions along the tubules were subsequently identified in Richards' laboratory by micropuncture in frogs and salamanders. The basic approach was to collect minute amounts of fluid from single nephrons, analyze the tubule fluid for the substance of interest, and determine the site of collection by microdissection of the nephron. This technique was later applied to the mammalian kidney in studies reported in 1941, but owing to its great difficulty and to the outbreak of World War II, the method fell into disuse. In the 1950's, Heinrich Wirz (University of Basel, Switzerland) and Carl Gottschalk (University of North Carolina) revived the method and used it to prove the existence of the countercurrent mechanism for concentrating the urine. Micropuncture continues to be an extremely valuable method to localize and study tubule functions and to measure pressures within the kidney. A limitation, however, is that not all nephron segments are accessible, because some lie deep within the kidney substance.

A method for isolating and perfusing single tubule segments in vitro was developed by Maurice

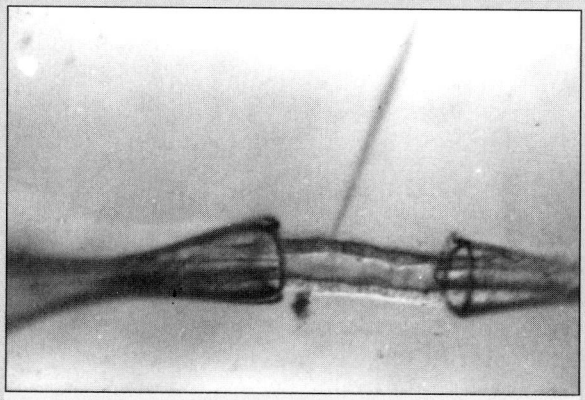

An isolated, perfused collecting duct from a salamander kidney. (Photograph by G. Tanner and J.-D. Horisberger)

Burg and his colleagues at the National Institutes of Health in 1966. With this approach any tubule segment can be dissected out of the kidney. One end of the tubule is cannulated with a glass micropipette to perfuse the lumen, and the other end is connected to another micropipette to collect the fluid which has passed through the lumen. A great advantage of this technique is that lumen and bath compositions can be changed at will, allowing the effects on tubule transport processes to be studied. Measurements of cell membrane potentials and of cell ion concentrations can be made with microelectrode or fluorescence techniques.

The membrane vesicle technique can be used to study the properties of transporters in the luminal and basolateral cell membranes of tubular epithelial cells. Basically, kidney tissue is ground up, and then biochemical techniques (precipitation, centrifugation, etc.) are used to separate luminal and basolateral cell membrane preparations. The membranes actually round up to form tiny vesicles, with a space in the middle. Uptake into the vesicle space is conveniently studied by using radioactively labeled substances. This technique has been used, for example, to demonstrate that active uptake of glucose by luminal cell membranes is dependent on sodium cotransport.

This combination of techniques permits the analysis of kidney function at whole organ, individual nephron, cellular, and subcellular levels.

to 50 years of age, after which it declines about 13 ml/min per decade.

Inulin is the "gold standard" for measuring GFR, but it is not often used clinically for this purpose. Infusing this substance is somewhat inconvenient. Instead, the **endogenous creatinine clearance** is used. Creatinine does not have to be infused since it is normally produced in the body. Its plasma levels are normally quite constant, and it can be easily measured in plasma and urine. In people, creatinine is actually secreted in addition to being filtered, but the error introduced by tubular secretion is usually small, and the endogenous creatinine clearance does provide a clinically valuable estimate of GFR.

If GFR falls, substances (including creatinine and urea) produced in the body that depend primarily on glomerular filtration for their elimination from the body will accumulate, and their plasma concentrations will rise. The extent to which the plasma concentration of creatinine is elevated may serve as an index of the degree of impairment of filtration rate. Likewise, the degree to which blood urea nitrogen (BUN) is elevated may reflect the degree of impairment of filtration rate; however, urea is less well suited for this assessment than creatinine, because BUN also depends on other factors, such as the rate of protein catabolism in the body and the urine flow rate.

We have devoted considerable attention to the measurement of GFR because it is so important in assessing kidney function. Glomerular filtration is a major process by which the kidneys regulate the volume and composition of the body fluids. A reduction in GFR to far below normal will impair a person's ability to excrete various waste products and the proper amounts of water and mineral electrolytes. In patients, the measurement of GFR provides the most valuable indicator of the severity of renal failure. Also, in studying tubule transport functions, a topic we will turn to next, the measurement of GFR is important. In order to tell how much of a substance is reabsorbed, we must first know how much was filtered. To tell whether a substance is secreted by the kidney tubules, we cannot simply measure the amount excreted; we must know how much was also filtered. Measurements of GFR are thus important to the study of transport by the kidney tubules.

Tubular Reabsorption

In the process of glomerular filtration, large quantities of mineral electrolytes, water, and organic compounds are presented to the tubules. Varying amounts of some of these substances are excreted (see Table 24–1). Most of the glomerular filtrate is not excreted, because the tubules selectively reabsorb the constituents of the glomerular filtrate.

Reabsorption may be either active or passive. **Active reabsorption** (active transport) is a process requiring the local expenditure of metabolic energy by the tubular epithelium, and it can effect the net movement of a substance against concentration or electrical gradients, or both. Examples of actively reabsorbed substances are sodium, glucose, and phosphate. **Passive reabsorption** (passive transport) does not depend directly on the expenditure of metabolic energy, and movement occurs down concentration, electrical, or osmotic gradients. Passively reabsorbed substances include urea, chloride, and water. In this section, we will discuss the mechanisms for reabsorbing various organic compounds. Later we will consider the reabsorption of various mineral electrolytes and water.

Reabsorption of Glucose

Glucose is filtered by the glomeruli and actively reabsorbed by the tubules. The glucose reabsorptive mechanism, which appears to be confined to the proximal tubule, is so effective that normally all of the filtered glucose is reabsorbed, with none excreted in the urine.

The cellular mechanism for glucose reabsorption by the proximal tubule is depicted in Figure 24–11. Glucose is cotransported with sodium across the luminal (brush border) cell membrane. This movement of glucose is energetically "uphill" and leads to accumulation of glucose in the cell. The energy required is derived from the sodium gradient across the luminal cell membrane. The cell sodium concentration is below that in the luminal fluid and is maintained at a low level by the Na^+/K^+-ATPase in the basolateral cell membrane. Also, the cell interior is electrically negative (by about 70 mv) compared to the fluid in the tubule lumen. As sodium moves into the cell down its concentration and electrical gradients, it lifts up glucose against its concentration gradient. (Think of an analogous situation: when one child on a see-saw falls down, the other is lifted up.) Glucose leaves the cell by moving down a concentration gradient (cell-to-peritubular capillary blood); this movement is via a carrier-mediated, sodium-independent mechanism in the basolateral cell membrane. Overall, glucose reabsorption is an example of "secondary" active transport coupled to "primary" active transport of sodium.

The ability of the tubules to reabsorb filtered glucose is limited (Fig. 24–12). At normal plasma

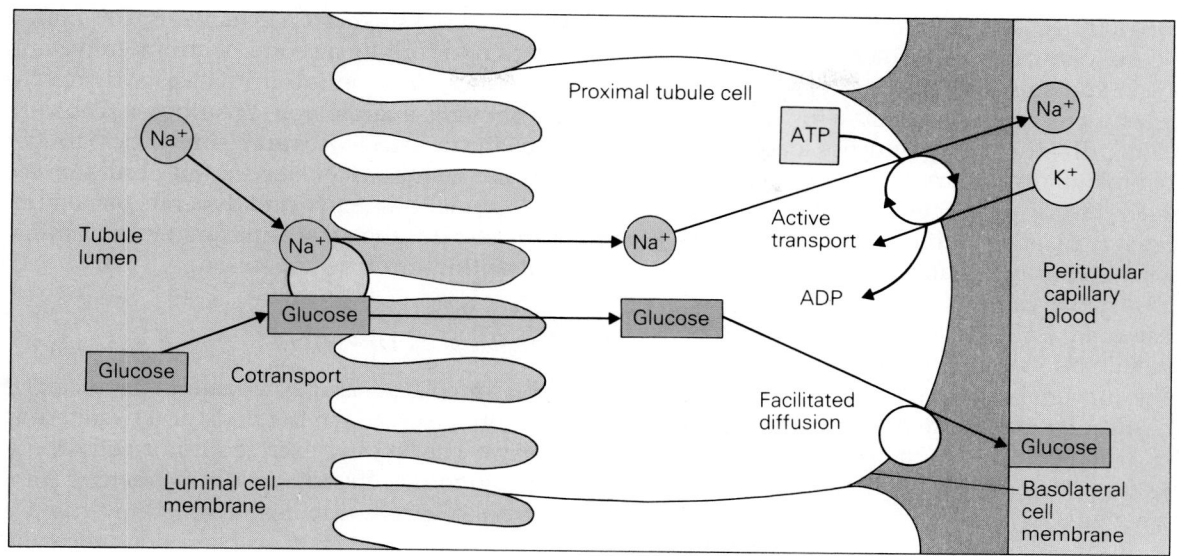

Figure 24–11

A model of glucose reabsorption by the proximal tubule cell. Glucose is co-transported with sodium across the luminal cell membrane, resulting in a high intracellular glucose concentration. Glucose passes through the basolateral cell membrane via facilitated diffusion and then diffuses into the peritubular capillary blood.

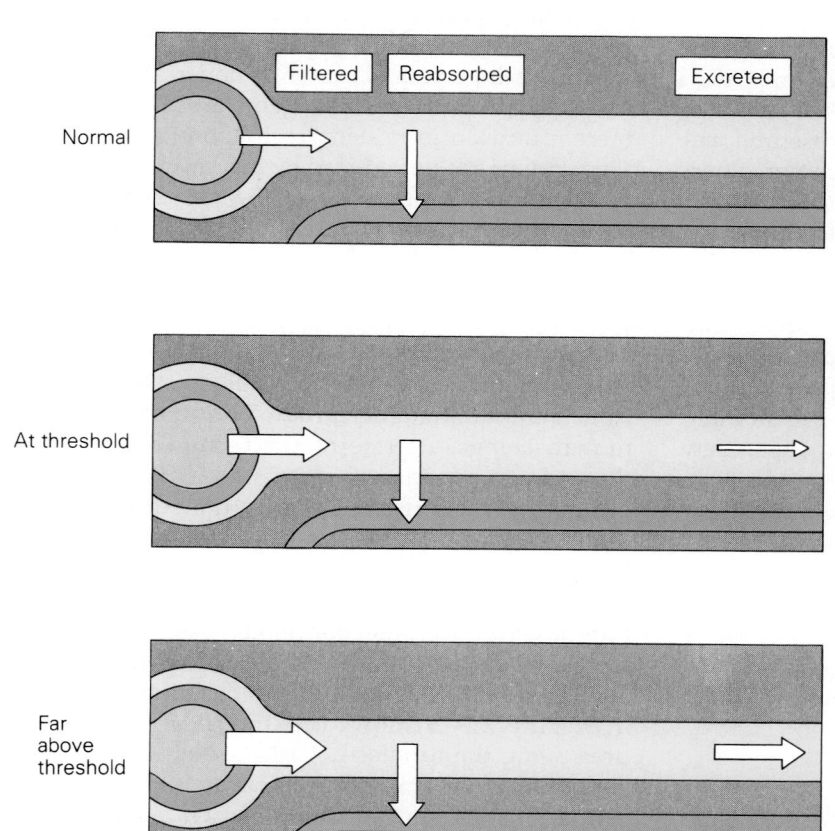

Figure 24–12

Effect of increasing the filtered amount of glucose on glucose excretion. In the normal condition (*a*), all of the filtered glucose is reabsorbed, and none is excreted. At threshold (*b*), the amount of glucose filtered is increased to a sufficiently high level (by increasing the plasma glucose concentration) so that not all of the filtered glucose is completely reabsorbed. (*c*) Far above threshold, further increases in the amount of glucose filtered result in increased glucose excretion, because the reabsorptive capacity of the tubules is operating at a maximal rate (glucose Tm). Note that the arrows for reabsorbed glucose in B and C are identical in size.

glucose levels (65 to 90 mg/dl, fasting values), the filtered glucose load can be completely reabsorbed by the tubules, and so no glucose is excreted. If the plasma level is increased sufficiently—for example, to 180 to 200 mg/dl—the filtered glucose load can no longer be completely reabsorbed. The plasma level at which glucose first appears in the urine in appreciable amounts is called the **glucose threshold.** Glucose appears in the urine because the tubules have a limited reabsorptive capacity. At high enough filtered glucose loads, all of the glucose carriers become saturated, and any extra filtered glucose is excreted. The maximal rate of glucose reabsorption is called the **tubular transport maximum (Tm) for glucose (G),** usually abbreviated Tm_G. Tm is analogous to the V_{max} (maximum velocity) of enzyme-catalyzed reactions.

Glucose in the urine **(glucosuria)** is often a sign of uncontrolled **diabetes mellitus.** In this condition, a deficiency in the insulin system results in an increased plasma glucose concentration. Filtered glucose exceeds the reabsorptive capabilities of the tubules, and so glucose is excreted in the urine. Many of the early symptoms of diabetes mellitus are due to this phenomenon. Excretion of glucose causes loss of water and salt in the urine. The result is a high urine flow rate (increased frequency of urination), dehydration, and thirst.

Glucose in the urine is not necessarily a sign of uncontrolled diabetes mellitus. Glucosuria may also occur because of a high blood glucose level after eating a heavy meal or in conjunction with emotional stress (which results in increased sympathetic nervous system activity). There are also certain kidney tubule disorders in which the ability of the tubules to reabsorb glucose is diminished.

The glucose reabsorptive mechanism reabsorbs other simple sugars such as galactose, xylose, and possibly fructose. Since the capacity of the transporting mechanism is limited, two sugars will depress each other's reabsorption if they are simultaneously presented to the tubules. One sugar may displace another from their common carrier, depending on their relative concentrations and affinities for the carrier. This phenomenon is called **competition** for transport. Saturation of transport (existence of a Tm) and competition for transport are closely linked. It appears to be a universal law of nature that when there is a limited amount of something, there is bound to be competition.

Reabsorption of Amino Acids

Plasma amino acids are filtered by the glomeruli and extensively reabsorbed by the renal tubules. Normally, only traces appear in the urine. There are several distinct transport mechanisms for amino acids, each reabsorbing a group of structurally similar compounds. The amino acid transport mechanisms, although separate from the glucose reabsorbing mechanism, show several similarities to the latter. Thus, amino acid reabsorption occurs in the proximal tubule. It is active and depends on sodium transport, usually showing a maximal rate of transport and within-group competition.

Reabsorption of Uric Acid

Uric acid, an end product of purine metabolism, is continuously produced in the body and excreted by the kidneys. The amount excreted is normally about 10% of the amount filtered. Hence, uric acid is reabsorbed by the kidney tubules. The kidney tubules also secrete uric acid, but reabsorption normally predominates. Both reabsorption and secretion occur in the proximal tubule by carrier-mediated processes.

Gout is a condition in which plasma uric acid is elevated and uric acid crystals precipitate in the joints, causing inflammation and painful swelling of the affected joints (often in the great toe). Gout may be treated by **uricosuric agents,** drugs that increase uric acid excretion. Probenecid (Benemid), sulfinpyrazone (Anturane), and high doses of aspirin inhibit tubular reabsorption of uric acid. This leads to an increase in uric acid excretion and a fall in plasma uric acid. When uricosuric agents are administered, there is a risk of uric acid precipitation in the urine. This danger can be reduced by increasing urine flow rate (drinking large amounts of water) and by making the urine alkaline (through ingestion of sodium bicarbonate).

Reabsorption of Urea

Urea, formed chiefly in the liver, is the major nitrogen-containing end product of metabolism in human beings. Its rate of production in the body depends on the rate of protein breakdown. Under ordinary conditions, this is determined by the dietary intake of protein. It is important to recognize, however, that protein catabolism, and hence urea production, can also be increased by tissue trauma, internal bleeding, infection, fever, and other conditions. Urea elimination from the body depends on an adequate rate of glomerular filtration.

Figure 24–13 shows that the renal handling of urea along the tubules is fairly complex. The numbers in this figure refer to relative amounts, not concentrations, present at different sites. The amount of urea filtered per unit of time is taken as ''100.'' Fifty percent of the filtered urea is left at the end of

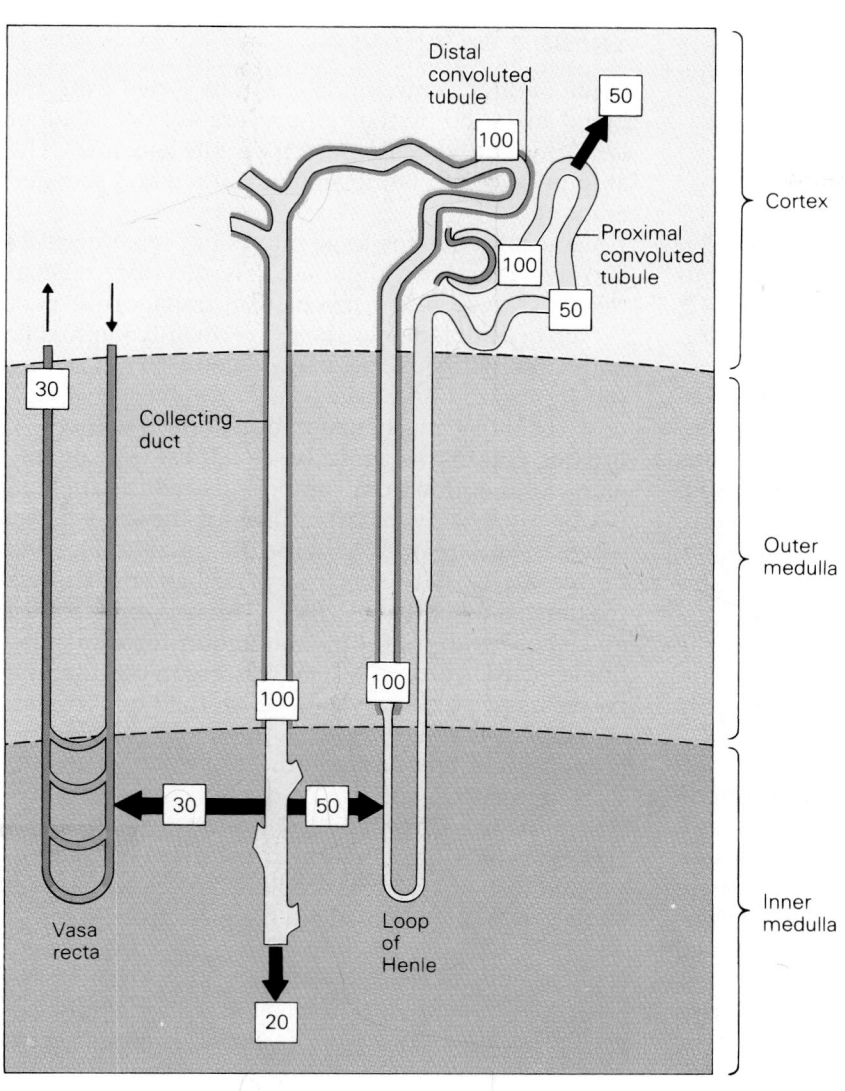

Figure 24–13
Movements of urea along the nephron. The numbers indicate relative amounts (100 = filtered urea). The heavy green outline indicates relative impermeability of these nephron segments to urea. Urea is reabsorbed by the inner medullary collecting duct; most of this urea re-enters the loop of Henle, and some is removed by the medullary blood vessels (vasa recta).

the proximal convoluted tubule. The explanation is as follows: as water is reabsorbed, filtered urea becomes more concentrated in the tubular fluid. Since the proximal tubule epithelium is somewhat permeable to urea, urea will diffuse out of the tubular fluid back into the blood surrounding the tubules. Some 50% of the filtered urea is reabsorbed in this portion of the nephron. If we examine fluid from the distal convoluted tubule, we see that more urea is present here than was present at the end of the proximal convoluted tubule (see Fig. 24–13). Hence, urea was added (or secreted) into tubular fluid in the loop of Henle; again, this is thought to be a passive movement. A very high concentration of urea exists in the medulla, and so there is a gradient for urea diffusion into the loop of Henle. Notice that the thick ascending limb, distal convoluted tubule, cortical collecting duct, and outer medullary collecting duct are quite impermeable to urea. Water is reabsorbed by the

cortical and outer medullary collecting ducts, and so a concentrated urea solution is delivered to the inner medullary collecting duct. The inner medullary collecting duct is permeable to urea, so urea diffuses out of the duct and is added to the inner medulla. As we will discuss later, this addition of urea is important in terms of the urinary concentrating mechanism. Of the quantity of urea entering the inner medullary collecting duct, 80% is reabsorbed (50% re-enters the loops of Henle to be recycled, and 30% is carried away by the blood vessels of the medulla), and 20% is excreted. In this illustration, there is a high degree of reabsorption of filtered urea (80%), reflecting a high degree of water reabsorption. If, however, water reabsorption is diminished, then urea reabsorption is also lower because a reduction in tubular water reabsorption results in less of a concentration gradient for diffusion of urea from tubule fluid back into the blood. Also, less time is available for back-diffusion when the urine flow rate is high. With urine flow rates greater than 2 ml/min, about 40% of filtered urea is reabsorbed and 60% is excreted.

Reabsorption of Proteins

The glomerular filtrate is nearly, but not completely, protein-free. Quantitatively speaking, serum albumin is the major protein in plasma, and so we will consider this substance. Serum albumin concentration in the glomerular filtrate is probably less than 1 mg/dl; compare this to a plasma concentration of 4.5 g/dl. Clearly, the glomerular filtration membrane is a very effective barrier to the passage of this macromolecule. If we consider the high rate of filtrate formation per day, the amount of serum albumin filtered per day (approximately 1 mg/dl × 180 liters/day = 1.8 gm/day) is not trivial. Normally, less than 0.03 gm of serum albumin is excreted per day. This comparison suggests that the kidney tubules reabsorb filtered serum albumin. Indeed, there is good evidence that the proximal tubules reabsorb a variety of proteins by pinocytosis. The urine normally contains less than 0.15 gm of protein per 24-hour urine sample. **Proteinuria** (excess protein in the urine) is usually due to an abnormal leakiness of the glomerular filtration membrane. Heavy proteinuria (more than 4 g/day) is almost always the result of glomerular disease.

Many other organic compounds are reabsorbed by the kidney tubules. Specific tubular reabsorptive mechanisms are known for the Krebs cycle acids (citric, alpha-ketoglutaric, and malic acids), lactic acid, ascorbic acid (vitamin C), and the ketone body acids (acetoacetic and 3-hydroxybutyric acids). Most of these mechanisms have properties similar to the glucose-reabsorbing mechanism.

Tubular Secretion

Some organic compounds are transported from the blood surrounding the kidney tubules into the tubular urine by specific secretory mechanisms. The amount excreted is the sum of filtered and secreted amounts.

In contrast to the large number of separate reabsorptive processes, there are few secretory mechanisms. One secretory mechanism transports a variety of organic anions; these are mainly carboxylic and sulfonic acids, which at blood pH (7.4) are mostly ionized and bear a negative charge. A second secretory mechanism transports a variety of organic cations (organic bases); these are mainly amine and ammonium compounds, which at blood pH bear a positive charge. Both of these secretory mechanisms are present in the proximal tubule, and both are active. The secretory mechanisms show a maximal transport rate (Tm) at high plasma levels, and show within-group competition for transport. Some of the more lipid-soluble organic acids and bases may also be passively reabsorbed or secreted.

Table 24–2 lists a few of the many secreted organic compounds. Many of these substances are foreign to the body and can be of medical importance. By supplementing filtration, tubular secretion results in rapid excretion. Urinary excretion is an important consideration in adjusting the dosage of some drugs. In patients with renal failure, drug dosage may have to be reduced to avoid excessively high plasma levels.

PAH (*p*-aminohippuric acid) is so avidly secreted by the kidney tubules that it is almost completely cleared (removed) from all of the plasma in one passage of blood through the kidneys. Therefore the clearance of PAH can be used to estimate the rate of plasma flow through the kidneys. The

Figure 24–14
The pH indicator dye chlorophenol red is avidly secreted by kidney proximal tubules. The photograph shows part of a killifish proximal tubule in vitro; the dye has been actively transported into the tubule lumen and reaches a concentration more than a 100 times higher than in the bathing medium. (Photograph by G. Tanner)

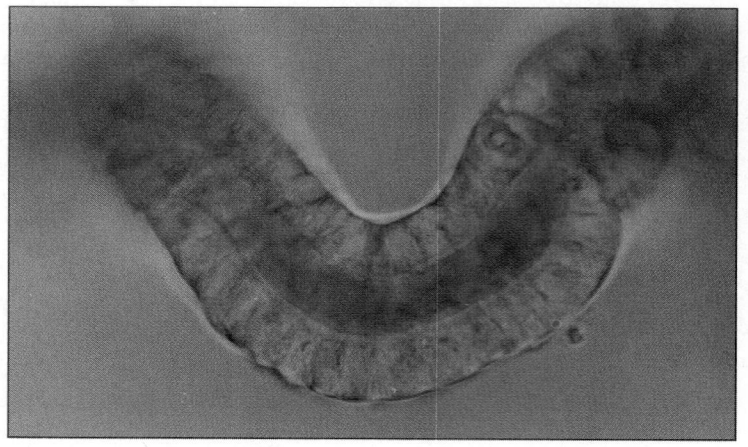

Table 24–2

Some Organic Compounds Secreted by the Kidney Tubules

Compound	Use
Organic Anions	
Phenol red (phenol-sulfon-phthalein, PSP)	pH indicator dye
p-aminohippuric acid (PAH)	Measurement of renal plasma flow and tubular secretory activity
Penicillin	Antibiotic
Probenecid (Benemid)	Inhibitor of penicillin secretion and uric acid reabsorption
Chlorothiazide (Diuril)	Diuretic drug
Acetazolamide (Diamox)	Carbonic anhydrase inhibitor
Creatinine*	Normal end product of muscle metabolism
Organic Cations	
Histamine	Vasodilator, stimulator of gastric acid secretion
Norepinephrine	Neurotransmitter
Quinine	Antimalarial drug
Quinidine	Antiarrhythmic drug
Tetraethyl-ammonium (TEA)	Ganglionic blocking agent
Creatinine*	Normal end product of muscle metabolism

*Creatinine is an unusual compound, since it is secreted by both the organic anion and cation mechanisms. This is probably because of negatively and positively charged groups in the molecule at physiological pH.

explanation for this follows: the amount of PAH per unit of time delivered to the kidneys in the renal blood flow is equal to the product of the renal plasma flow (RPF) and the plasma PAH concentration (P_{PAH}). (The amount of PAH in red blood cells is negligible.) The amount of PAH per unit of time leaving the kidneys, assuming that all is extracted from the renal plasma and excreted in the urine, is

equal to $U_{PAH} \times \dot{V}$ (the PAH excretion rate). From the principle of conservation of matter, we can write:

$$input = output$$
$$RPF \times P_{PAH} = U_{PAH} \times \dot{V}$$

Rearranging,

$$RPF = U_{PAH} \times \dot{V}/P_{PAH}$$

By definition, the expression on the right is the PAH clearance (C_{PAH}). Therefore, the renal plasma flow equals the PAH clearance. To determine the renal blood flow, we need to correct the renal plasma flow for the blood hematocrit (hct), using the following equation:

$$RBF = RPF/(1 - hct)$$

Renal blood flow can be readily measured in unanesthetized human beings or animals by applying the PAH clearance method. One simply infuses PAH so as to achieve a steady, low plasma level of this compound, collects a timed urine sample and a blood sample, measures the hematocrit, analyzes plasma and urine for PAH, and uses the equations just given. In this example, we assumed that PAH is 100% extracted from the plasma flowing through the kidneys, which, in the normal human kidney, is almost true (90% is closer to the truth). To measure renal blood flow more accurately, it would be best to determine the actual extraction by also measuring the PAH concentration in renal venous blood.

Tubular Transport of Mineral Electrolytes

The inorganic salts of the minerals sodium, potassium, calcium, magnesium, and phosphate are completely ionized in aqueous solution and can conduct electricity, so they are often referred to as mineral electrolytes. Next, we will discuss the renal handling of sodium, potassium, calcium, magnesium, and phosphate ions. Although we will consider the various ions separately, recognize that these ions do not exist in isolation. Positively or negatively charged ions are always accompanied by ions of opposite charge because the principle of electrical neutrality in solutions must be adhered to.

Tubular Transport of Sodium

The renal handling of sodium is important for several reasons. First, the kidneys are major regulators of the amount of sodium in the body. Since we take in variable amounts of sodium in our diets, the kidneys must adjust the excretion of sodium to maintain balance. We will consider this topic in detail in the next chapter. Second, sodium reabsorption by the kidney tubules is the major driving force for water reabsorption. Sodium is mostly actively reabsorbed; water follows passively. Third, sodium reabsorption is coupled to the reabsorption of both organic (e.g., glucose, amino acids) and inorganic substances (e.g. chloride, phosphate). Fourth, sodium reabsorption is intimately related to the secretion of hydrogen and potassium ions. Thus, it affects acid-base balance and potassium balance. Finally, in terms of energy demands, reabsorption of sodium is the major operation in the kidneys. Sodium and its accompanying anions (chloride and bicarbonate) constitute the solutes reabsorbed in greatest amounts. Almost all of the filtered sodium is actively reabsorbed by the kidney tubules. Normally, 80% of the oxygen consumed by the kidneys is relegated to sodium reabsorption.

Later, we will consider how sodium reabsorption is affected by various factors and how it is controlled in order to keep the body in sodium balance. For now, we will discuss how sodium is reabsorbed in various parts of the nephron. Figure 24–15 presents a summary of the percentages of filtered sodium (and water) that are reabsorbed under normal conditions by various nephron segments.

Sodium Transport in the Proximal Convoluted Tubule

The logical place to start is the proximal convoluted tubule. This segment reabsorbs 70% (in other words, the greatest fraction) of the filtered sodium and water. The cellular mechanisms involved are depicted in Figure 24–16. Sodium enters the cell across the luminal membrane (brush border) and moves down chemical and electrical gradients. The intracellular sodium concentration is low (kept low by the basolateral membrane Na^+/K^+-ATPase), and the inside negativity of the cell also favors sodium entry. The entry of sodium into the cell is coupled to the movement of other solutes. As was discussed earlier, sodium enters the cell together with glucose and amino acids. It is also accompanied by other organic solutes and phosphate. A luminal membrane Na^+/H^+ exchanger results in sodium reabsorption and acidification of the tubular urine. Sodium is actively pumped out the basolateral cell membrane by its Na^+/K^+-ATPase. This is an example of primary active transport. The energy needed to move sodium out of the basolateral cell membrane, against combined concentration and electrical gradients, is directly derived from the hydrolysis of ATP.

Figure 24–15
Percentages of filtered sodium and water reabsorbed by different nephron segments under normal conditions.

	Proximal convoluted tubule	Loop of Henle	Distal convoluted tubule and collecting duct
Percent of filtered sodium reabsorbed	70	20	9
Percent of filtered water reabsorbed	70	10	19

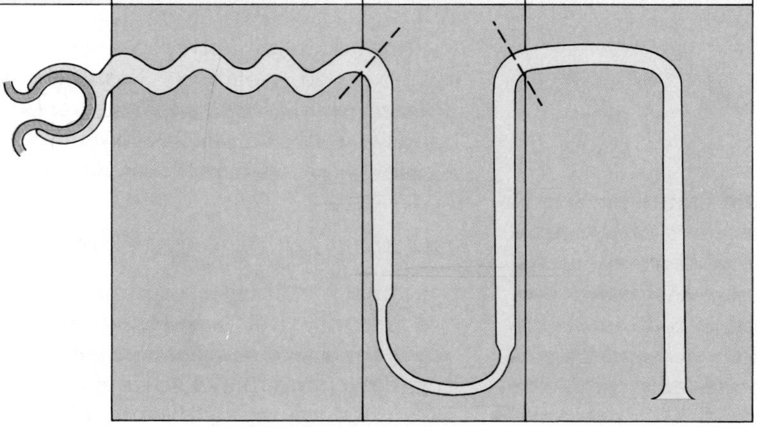

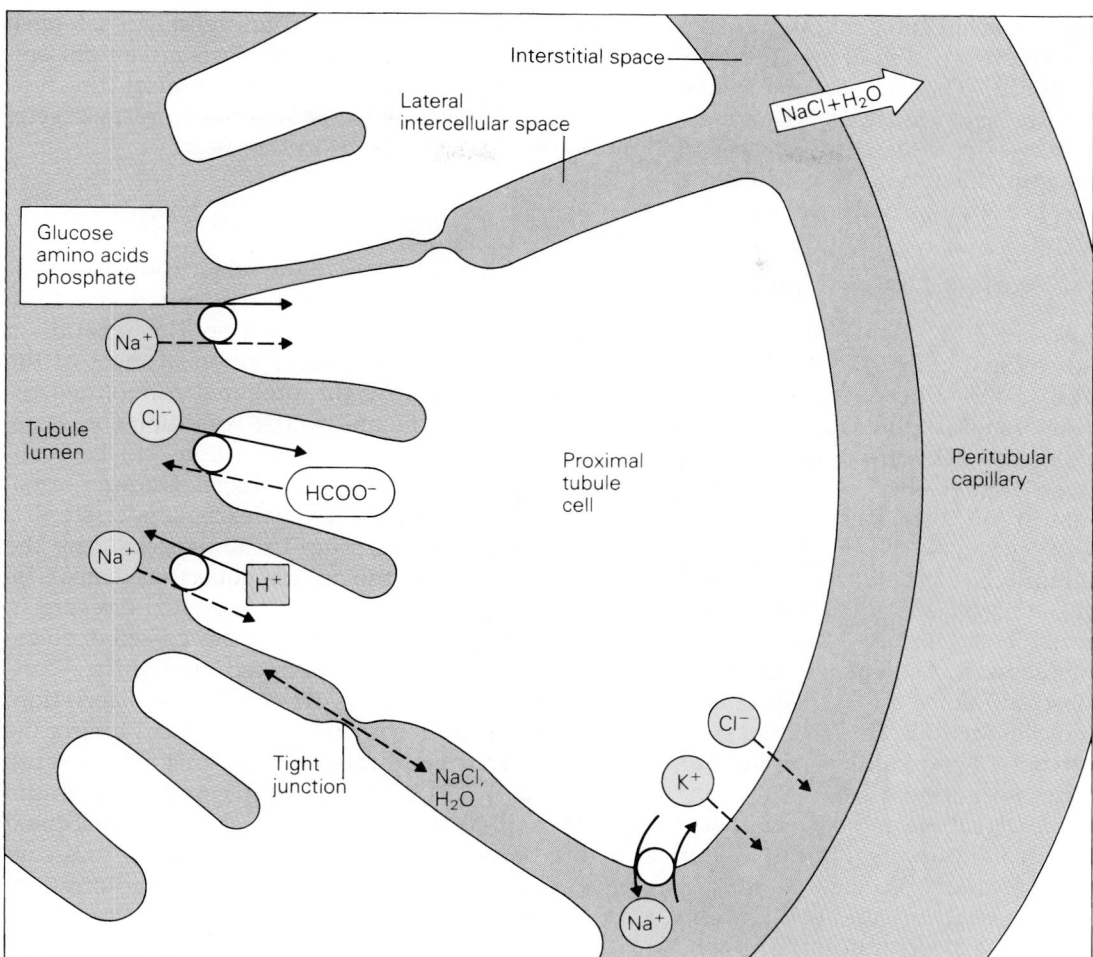

Figure 24–16
Model of ion and water transport in the proximal convoluted tubule. Solid arrows indicate active transport; dashed arrows indicate passive transport. A luminal membrane Cl^--formate ($HCOO^-$) exchanger participates in chloride reabsorption.

The reabsorption of sodium and accompanying solutes results in the reabsorption of water. The epithelium of the proximal convoluted tubule has a high water permeability, and so only a very small osmotic gradient is necessary to account for the observed rate of water reabsorption. Removal of solute is thought to lead to an effective osmotic pressure difference of about 2 to 5 mosm/kg H_2O across the tubular wall; this is considered sufficient to drive water reabsorption. Since the gradient is so small, proximal tubular fluid is essentially isosmotic to plasma. Sodium and its accompanying anions compose the major osmotically active solutes in the glomerular filtrate and proximal tubule fluid. Since 70% of the filtered sodium is reabsorbed, 70% of the filtered water is also reabsorbed in the proximal convoluted tubule.

The final step in the reabsorption of sodium is the uptake of the reabsorbed fluid by the peritubular capillaries. The forces involved here are usually referred to as **peritubular capillary Starling forces** and involve the hydrostatic and colloid osmotic pressure differences that affect fluid movement across any capillary wall (see Chapter 19). The hydrostatic pressure in these capillaries is low, since the blood has passed through several upstream resistance vessels (cortical radial arteries, afferent arterioles, efferent arterioles). Also, the colloid osmotic pressure of the peritubular capillary plasma is high because the plasma proteins were concentrated by glomerular filtration. Consequently, there is a net pressure gradient favoring uptake of reabsorbed fluid. Under some circumstances (e.g., loading a person with isotonic saline), peritubular capillary pressures are altered in a direction that diminishes the uptake force. The reabsorbed fluid then accumulates in the interstitium. The lateral intercellular spaces widen, and the tight junctions become expanded and very leaky. This results in back-leak of sodium and water into the tubule lumen and an overall reduction in net sodium reabsorption.

Sodium Transport in More Distal Nephron Segments

The loop of Henle reabsorbs about 20% of the filtered sodium and 10% of the filtered water (see Fig. 24–15). Note that more sodium than water is reabsorbed here because the ascending limb is impermeable to water but reabsorbs salt. The transcellular transport of sodium in the thick ascending limb is accomplished by a coupled $1Na^+/1K^+/2Cl^-$ co-transporter in the luminal cell membrane (which is inhibited by "loop" diuretic drugs such as furosemide and bumetanide) and a Na^+/K^+-ATPase in the basolateral cell membrane. As will be discussed later, the removal of salt along the ascending limb and its deposition in the medulla are essential steps in the mechanism whereby the kidneys produce an osmotically dilute or concentrated urine.

The distal convoluted tubule and collecting duct accomplish the reabsorption of most, but not all, of the remaining salt and water (see Fig. 24–15). Sodium reabsorption here is active, and sodium ions can be reabsorbed against steep concentration and electrical gradients. For example, if a person is depleted of salt, the collecting ducts can lower the final urine sodium concentration to 1 mEq per liter. Aldosterone, a hormone secreted by the adrenal cortex, stimulates sodium reabsorption by the collecting ducts. Most of the reabsorbed sodium is accompanied by chloride; some reabsorbed sodium may be exchanged for potassium or hydrogen ions. The collecting duct epithelium has an intrinsically low water permeability. Antidiuretic hormone (ADH), from the posterior pituitary, increases collecting duct water permeability in a graded fashion. The collecting ducts are the final site of regulation of sodium and water excretion, so it is not surprising that reabsorption there is under hormonal control.

To summarize and compare reabsorption in the early (proximal convoluted tubule) and late (collecting duct) portions of the nephron, we note that in the proximal tubule, large quantities of salt and water are reabsorbed along small gradients. The proximal tubule is a leaky epithelium and cannot sustain large gradients for small ions or water. Reabsorption here can be considered as a bulky, coarse operation. By contrast, the collecting ducts have a smaller capacity for reabsorption and can establish large gradients. The collecting ducts are a tight epithelium, and are the site of fine control; the hormones aldosterone and ADH regulate, respectively, sodium and water reabsorption here.

Tubular Transport of Potassium

Normally, the kidneys are the major route of potassium excretion from the body and serve as an important site of potassium regulation. Renal handling of potassium involves reabsorption of most of the filtered potassium by the proximal convoluted tubule and loop of Henle. Under conditions of potassium depletion, potassium continues to be reabsorbed along the collecting ducts. Under normal conditions or with potassium excess, potassium is secreted by the collecting ducts, especially by the earliest portions of the cortical collecting ducts. The collecting duct principal cells appear to be responsible for this secretion. Most of the excreted potassium usually represents secreted potassium.

A number of factors affect potassium secretion, most of which can be understood by examining the cell model for a cortical collecting duct principal cell in Figure 24–17. According to this model, potassium is taken up by the cell via a Na^+/K^+-ATPase located in the basolateral membrane. This leads to an increase in intracellular potassium ion concentration. Potassium can then diffuse out of the cell into the lumen and in this way be secreted. The intracellular potassium concentration is a key element in determining the rate of potassium secretion. An increase in extracellular potassium concentration or increased plasma levels of aldosterone enhance cell uptake of potassium and overall secretion. Aldosterone, in addition, increases the luminal membrane permeability to both potassium and sodium, which also favors potassium secretion and sodium reabsorption. There appears to be a **transtubular potential difference** across many nephron segments. If we advance a potential-measuring microelectrode through the tubule wall, we first record a negative deflection of 70 mv when the microelectrode enters the cell. (The reference potential [zero] is always on the blood side of the tubules.) A negative deflection of about 50 mv is recorded when the microelectrode penetrates the lumen of the early cortical collecting duct. Therefore, the electrical potential opposing potassium movement out of the cell is much smaller at the luminal, -70 mv $- (-50$ mv$) = -20$ mv, than at the basolateral side of the cell, -70 mv. Consequently, there is less force opposing movement into the lumen, and potassium secretion is favored. Increased amounts of poorly reabsorbed anions in

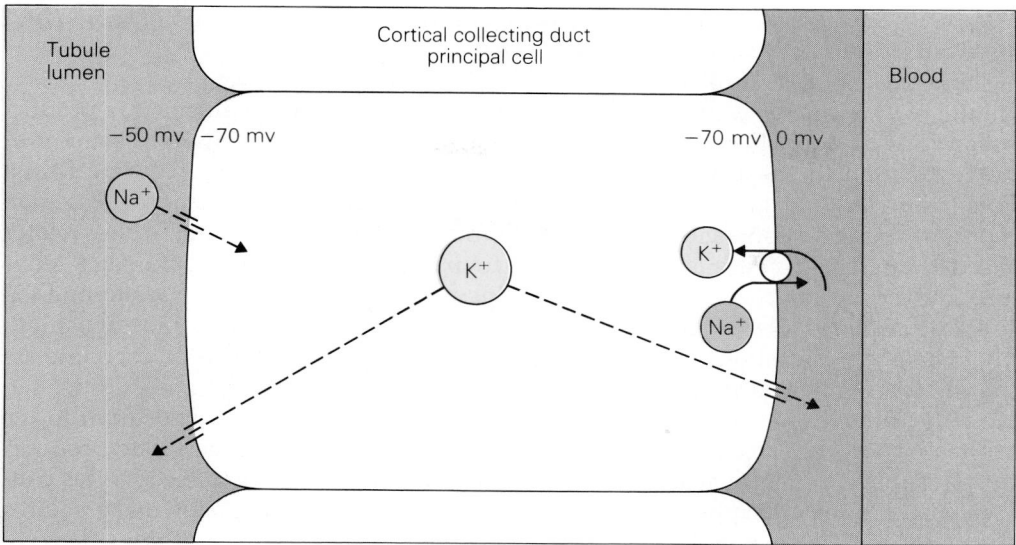

Figure 24–17

A model of potassium and sodium transport by a principal cell in the cortical collecting duct. The two steps in potassium secretion are (1) uptake into the cell via the basolateral membrane Na$^+$/K$^+$-ATPase and (2) passive diffusion across the luminal cell membrane. Sodium reabsorption involves passive diffusion into the cell through a sodium-selective ion channel in the luminal cell membrane, and active pumping of sodium across the basolateral cell membrane.

the tubule lumen increase the transtubular potential difference (make the lumen more negative) and so enhance potassium secretion, as does an increased rate of flow through the collecting duct lumen, due to a "sweeping away" effect that maintains a concentration gradient favoring potassium secretion. This effect is probably the main reason that most diuretic drugs increase potassium excretion.

Tubular Transport of Calcium

Renal excretion of calcium is important, but represents only about 10% of the total daily output of calcium from our bodies on a normal diet. About 90% of our intake is excreted in the feces. The regulation of plasma calcium involves control of intestinal calcium absorption, renal excretion, and exchanges with bone. The parathyroid hormone and vitamin D play important roles in this regulation. Calcium balance will be discussed in Chapter 27.

Calcium is filtered and reabsorbed by the kidneys. Since about 40% of plasma calcium is bound to plasma proteins, only the nonbound portion (60%) is filtered. About 50% to 60% of the filtered calcium is reabsorbed in the proximal convoluted tubule, and most of the rest in the thick ascending limb, distal convoluted tubule, and collecting ducts. The quantity excreted is about 1% to 2% of the filtered load. Regulation of urinary excretion of calcium occurs primarily in the distal portions of the nephron. The parathyroid hormone stimulates calcium reabsorption and so acts to conserve filtered calcium for the body.

Tubular Transport of Magnesium

About two thirds of the magnesium in our diet is normally excreted in the feces, and one third in the urine. Approximately 30% of plasma magnesium is bound to plasma proteins and so is not filterable at the glomeruli. The proximal convoluted tubules reabsorb about 20% to 30% of filtered magnesium, and 50% to 60% is reabsorbed in the loop of Henle. Only 3% to 5% of the filtered magnesium is excreted under normal conditions. The kidneys play a major role in regulating the plasma magnesium concentration. Excesses of magnesium are rapidly excreted in the urine; in magnesium-deficient states, magnesium is avidly reabsorbed by the kidney tubules and virtually disappears from the urine.

Tubular Transport of Phosphate

The kidneys play an important role in regulating the plasma concentration of inorganic phosphate. We usually excrete in the urine most of the phosphate ingested in our diet. Filtered phosphate is reabsorbed by sodium-dependent, secondary active transport in the proximal tubule. The reabsorptive mechanism has a limited rate, or *Tm*. The *Tm* is normally exceeded by the quantities of phosphate filtered, and so phosphate is excreted in the urine. The situation here differs from that of glucose, where the *Tm* is reached only when the plasma glucose concentration is raised dramatically. Because the filtered phosphate load and *Tm* are normally so close to one another, the kidneys can participate in regulating the plasma phosphate concentration.

If phosphate intake is increased, plasma phosphate levels will rise, and the filtered phosphate load will be increased. The *Tm* will be exceeded more than usual, and so phosphate excretion will increase, thereby ridding the body of the extra phosphate. Conversely, if phosphate intake is decreased, plasma phosphate and filtered phosphate amounts will decrease, and all of the filtered phosphate will be reabsorbed, thus conserving phosphate for the body. Phosphate excretion is also regulated by changing the *Tm*. For example, the parathyroid hormone inhibits phosphate reabsorption, thus promoting phosphate loss in the urine.

Tubular Reabsorption of Water

The rate of water excretion in the urine depends on the rates of filtration of water (GFR) and tubular water reabsorption. Since the urine volume is mostly water, we can assume that the renal water excretion rate is simply equal to the urine flow rate. Urine flow rate can vary widely, depending on conditions. If a person drinks a large quantity of water, urine flow rate may be as high as 20 ml/min. The osmolality* of this urine may be as low as 30 to 40 mosm/kg H_2O, much more dilute than plasma (approximately 300 mosm/kg H_2O). On the other hand, in a dehydrated person, urine flow rate may be as low as 0.3 ml/min. The osmolality of this urine may be as high as 1200 to 1400 mosm/kg H_2O, or approximately 4 to 5 times the normal plasma osmolality. A "normal" urine flow rate is about 1 ml/min (1.5 liters/day), and the urine is usually modestly concentrated (600 to 800 mosm/kg H_2O).

---

*The terms *osmolality* and *osmolarity* (Chapter 4) are often confused. Both are measures of solute concentration. Osmolality is expressed as osmoles per kg H_2O, and osmolarity as osmoles per liter solution. Biological fluids are mostly water, so the two terms are practically identical.

Production of Hyperosmotic Urine to Save Water

We will consider now the mechanisms for producing an osmotically concentrated (hyperosmotic) or dilute (hypo-osmotic) urine—that is, a urine with a total solute concentration greater than or less than that of plasma. If the body contains excess water (e.g., due to drinking a lot of liquid), it makes sense for the kidneys to get rid of the extra water and excrete the urinary solutes in a large volume of osmotically dilute urine. The importance of excreting an osmotically concentrated urine is more subtle. Recall that the kidneys are always called upon to excrete solutes. These solutes include waste products of metabolism and various mineral electrolytes consumed in the diet. The excretion of these solutes requires the excretion of water, since this is the vehicle in which these substances are dissolved. Suppose someone had to excrete 600 mosm of solutes per day. Suppose also that the person could form only an isosmotic urine—that is, a urine with an osmolality of 300 mosm/kg H_2O. How much water would have to be excreted? It would take 2 kg (or 2 liters) of water to excrete 600 mosm. Suppose now that the same amount of solutes could be concentrated in a urine with an osmolality of 1200 mosm/kg H_2O. How much water would have to be excreted? One-half kg (or 0.5 liter) would contain 600 mosm solute as a 1200 mosm/kg H_2O solution. We can see that by excreting the urinary solutes in an osmotically concentrated (hyperosmotic) urine, the kidneys save water for the body. In the example given, by excreting the same amount of solutes in urine with an osmolality of 1200 mosm/kg H_2O instead of 300 mosm/kg H_2O, the kidneys, in effect, saved 1.5 liters of pure water for the body.

Since water is often scarce, the ability to form an osmotically concentrated urine is an important adaptation to a terrestrial environment. By getting rid of the urinary solutes in a hyperosmotic urine, the need for water intake is diminished. In the animal kingdom, only birds and mammals can produce a urine that is osmotically more concentrated than the blood. These animals are the only classes that have loops of Henle. Furthermore, in general, those mammals with relatively long loops of Henle are able to produce the most concentrated urines. For example, the kangaroo rat, a small rodent that inhabits the southwestern deserts of the United States, has an exceptionally long loop of Henle and can produce urine with an osmolality of 5500 mosm/kg H_2O, some 18 times that of plasma. This mammal's high metabolic rate (due to its small size) results in especially vigorous sodium transport by the thick ascending limb of Henle's loop, a key step in

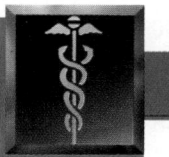
Kidney Stones

A kidney stone is a hard mass that forms in the urinary tract. At least 1% of Americans develop kidney stones at some time during their lives. Stone formation occurs more commonly in men than in women and most often affects men between 30 and 60 years of age. A stone lodged in the ureter causes severe pain. The condition causes considerable suffering and loss of time from work, and it may lead to kidney damage. Once stone formation occurs in a person, it often recurs.

Stones form when poorly soluble substances in the urine precipitate out of solution, causing crystals to form, aggregate, and grow. Most kidney stones (75% to 85%) are made up of calcium salts (calcium oxalate and calcium phosphate). Stones may form from precipitated ammonium magnesium phosphate, uric acid, or cystine. Since a low urine flow rate raises the concentration of poorly soluble salts in the urine, thereby favoring precipitation, an important key to preventing stone formation is to drink plenty of water and maintain a high urine output.

Fortunately, most stones are small enough to be passed down the urinary tract and spontaneously eliminated. Microscopic and chemical examination of the eliminated stones helps in identifying their nature, in considering possible causes of stone formation, and in selecting the appropriate treatment. Sometimes a change in diet is recommended to reduce the amount of potential stone-forming material (e.g., calcium, oxalate, or uric acid) in the urine.

If the stone is not passed, several options are available. The classical treatment has been surgery to remove the stone, but such operations are not without risks. A recently developed treatment for kidney stones that does not require surgery is called **extracorporeal shock-wave lithotripsy.** A device called a **lithotriptor** is used. The patient is placed in a tub of

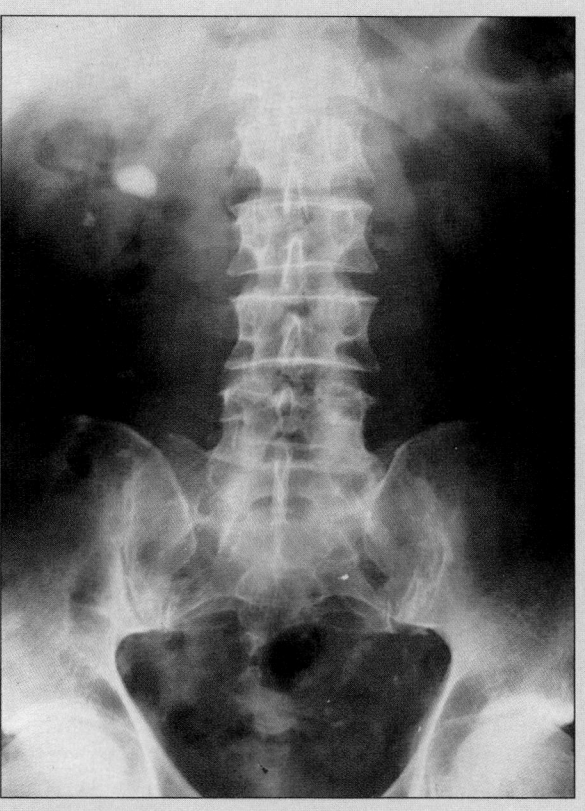

X-ray of kidney stones in the human urinary tract. (© Custom Medical Stock Photo.)

water, and the stone is localized by X-ray imaging. Shock waves generated in the water by electric discharges are focused on the stone through the body wall. The shock waves fragment the stone so that it can be passed down the urinary tract and eliminated.

concentrating the urine. These facts from comparative anatomy and physiology suggest an association between loops of Henle and the ability to produce a concentrated urine.

The Countercurrent Hypothesis for the Formation of Osmotically Concentrated Urine

The **countercurrent hypothesis** is now universally accepted as the explanation for the mechanism whereby an osmotically concentrated urine is

formed. Scientists are still not sure about certain aspects of this hypothesis—in particular, how the osmotic gradient is established in the inner medulla. The term *countercurrent* indicates that fluid flows in opposite directions in adjacent limbs of tubules and blood vessels in the kidney medulla.

There are actually two countercurrent mechanisms. The loops of Henle act as **countercurrent multipliers;** they set up a gradient of osmolality that increases from the junction of the cortex and medulla to the tip of the renal papillae. The blood vessels of the medulla (vasa recta) act as passive **countercurrent exchangers;** they help to preserve the osmotic gradient in the medulla. The collecting ducts act as **osmotic equilibrating devices;** it is here that the urine finally becomes osmotically concentrated. Depending on the plasma level of ADH, the fluid in the collecting ducts tends to equilibrate osmotically with the surrounding interstitium.

If kidney tissue from a thirsted animal is taken from various levels of the medulla, a progressively higher osmolality is found with increasing depth of the medulla. The highest osmolalities are found at the papillary tip. All of the structures in the medulla participate in this increasing gradient. Also, in a thirsted animal, with high plasma levels of ADH, the final urine has the same osmolality as tissue fluid at the tip of the kidney papilla.

This gradient is established by countercurrent multiplication in the loops of Henle. Figure 24–18 shows a simplified model for the countercurrent multiplication process. Consider a loop filled with fluid isosmotic to plasma (300 mosm/kg H_2O). Assume that the membrane separating the two loops is impermeable to water. Also assume that the loop is able to establish an osmotic gradient of 200 mosm/kg H_2O between the two limbs of the loop at any level (see Fig. 24–18b). This step is often called the **single effect.** This gradient could be produced by transport of salt out of the ascending limb and its deposition in the descending limb. Water is left behind (since the intervening membrane is water-impermeable), and so the osmolality of the ascending limb fluid decreases and that of the descending

limb increases by the same amount. Next, we add new fluid to the loop (see Fig. 24–18c), as fresh fluid flows in from the proximal convoluted tubule. Again allow the 200 mosm/kg H_2O gradient to be established (Fig. 24–18d). Repeat the process several times (Fig. 24–18e to h). What is the final result? Notice that at any level of the loop a gradient of 200 mosm/kg H_2O exists. Along the length of the loop, however, a larger osmotic gradient (about 400 mosm/kg H_2O) has been established. This is what is meant by countercurrent multiplication; namely, a gradient existing at any level of the loop is multiplied by countercurrent flow so as to produce a larger gradient along the length of the loop.

The model just discussed should not be taken too literally. For example, flow through the loop of Henle is not a discontinuous process. Also, ascending and descending limbs do not share a common membrane; rather, there is an intervening interstitial space. The model is realistic, however, in that salt is transported out of the ascending limb across a water-impermeable membrane. The descending limb is water-permeable, and its osmolality is increased primarily by water removal, not by active deposition of salt into its lumen. Notice that fluid leaving the ascending limb is osmotically dilute, even in a kidney that will be forming a concentrated urine. This is so because salt has been removed along the ascending limb.

Another important feature of countercurrent multiplication is that it is an energy-consuming pro-

Figure 24–18
The principle of countercurrent multiplication, based on the assumption that, at any level along the loop, an osmotic gradient of 200 mosm/kg H_2O can be established by salt transport between ascending and descending limbs. (Modified from: Pitts, R. F. *Physiology of the Kidney and Body Fluids,* 3rd edition. Chicago: Year Book, 1974.)

Stepwise shift of fluid

A
| 300 | 300 | 300 | 300 | 300 | 300 | 300 | 300 |
| 300 | 300 | 300 | 300 | 300 | 300 | 300 | 300 |

C
| 200 | 200 | 200 | 200 | 400 | 400 | 400 | 400 |
| 300 | 300 | 300 | 300 | 400 | 400 | 400 | 400 |

E
| 150 | 150 | 300 | 300 | 300 | 300 | 500 | 500 |
| 300 | 300 | 350 | 350 | 350 | 350 | 500 | 500 |

G
| 125 | 225 | 225 | 225 | 225 | 400 | 400 | 600 |
| 300 | 325 | 325 | 425 | 425 | 425 | 425 | 600 |

Development of "single effect"

B
| 200 | 200 | 200 | 200 | 200 | 200 | 200 | 200 |
| 400 | 400 | 400 | 400 | 400 | 400 | 400 | 400 |

D
| 150 | 150 | 150 | 150 | 300 | 300 | 300 | 300 |
| 350 | 350 | 350 | 350 | 500 | 500 | 500 | 500 |

F
| 125 | 125 | 225 | 225 | 225 | 225 | 400 | 400 |
| 325 | 325 | 425 | 425 | 425 | 425 | 600 | 600 |

H
| 112 | 175 | 175 | 225 | 225 | 313 | 313 | 500 |
| 312 | 375 | 375 | 425 | 425 | 513 | 513 | 700 |

cess. In other words, in order to establish a gradient, there must be an energy source. The energy source for operating the countercurrent multiplier is ultimately active sodium transport, which is linked to ATP hydrolysis. Note that the magnitude of the gradient established depends on the size of the single effect and the length of Henle's loop. If the single effect is reduced, or if the loops are short, then a large gradient along the length of the loop cannot be established.

Figure 24–19 shows a model for the operation of the countercurrent mechanism in the kidney. In this model, we are assuming that a maximally concentrated urine is formed. The numbers in blue give the osmotic concentration in mosm/kg H_2O. The boxed numbers in green give the relative amounts of filtered water present at various points along the nephron. The heavy blue shading along the ascending limb indicates a low water permeability. A juxtamedullary nephron, with a long loop of Henle extending to the tip of the papilla, is depicted.

Seventy percent of the filtered sodium and water is reabsorbed along the proximal convoluted tubule. This reabsorption is essentially isosmotic.

Figure 24–19
Summary of movements of water, ions, and urea in the kidney during elaboration of a maximally concentrated urine (1200 mosm/kg H_2O). Numerals boxed in blue give osmolality in mosm/kg H_2O. Numbers boxed in green give the relative amount of water present at each level of the nephron. Solid arrows indicate active transport; dashed arrows passive transport. The heavy blue outline along the thin ascending and thick ascending limbs of Henle's loop indicates that these segments are relatively water-impermeable.

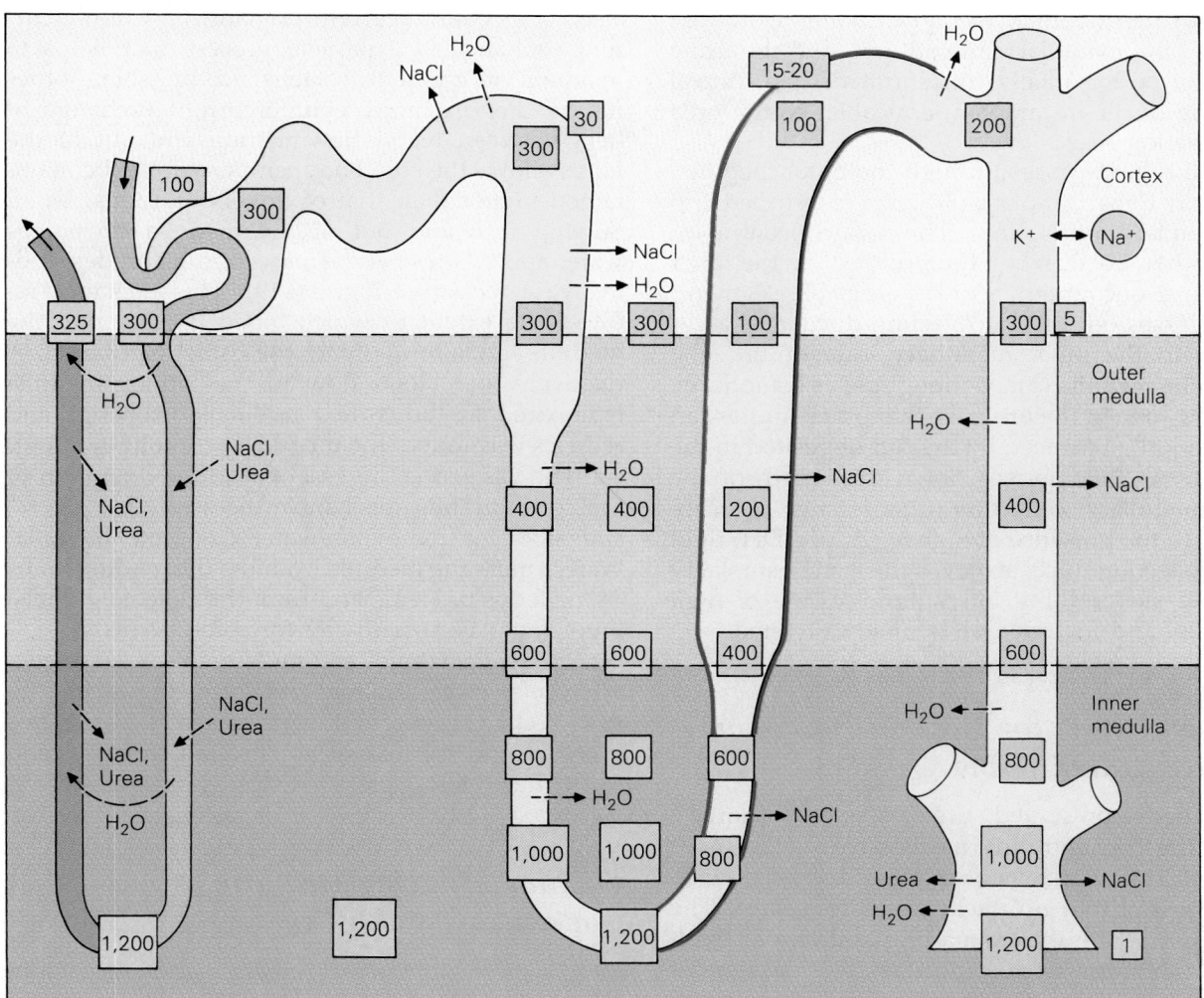

Fluid entering the descending limb of Henle's loop has an osmolality of about 300 mosm/kg H_2O. The descending limb is water-permeable, and since the interstitium surrounding the tubules is hyperosmotic, water leaves the descending limb and the osmolality increases. Since NaCl is the main solute in the descending limb fluid, its concentration rises.

In the thick ascending limb of Henle's loop, there is a vigorous Na^+/K^+-ATPase whose activity results in sodium chloride deposition in the outer medulla. An osmotically dilute fluid, which contains mainly NaCl and urea, leaves the loop and enters the distal convoluted tubule. As the fluid traverses the cortical collecting ducts in the presence of ADH, water is reabsorbed. This is because water moves along its osmotic gradient from the tubule fluid into the cortical interstitium, where it is carried away by a high blood flow. Before the tubule fluid once again re-enters the outer medulla, it becomes isosmotic, and its volume is substantially reduced. The reabsorption of dilute fluid in the cortex is important because if a large volume of water were presented to the medulla, too much water would be added to the medullary interstitium, and the urine could not be maximally concentrated. The cortical collecting ducts are urea-impermeable, so the urea concentration rises.

As the fluid passes through the outer medulla, it becomes hyperosmotic as water is reabsorbed into the medullary interstitium. The urea concentration rises further. Next, when the urine enters the inner medulla, it encounters a urea-permeable segment. Urea diffuses out of the collecting ducts and accumulates in the inner medullary interstitium. The urea in the medulla can be thought of as osmotically balancing urea in the urine, so that other solutes can be osmotically balanced by the salt deposited in the medulla. As fluid moves down the length of the inner medullary collecting duct, water is reabsorbed. In the presence of high levels of ADH, fluid in the collecting ducts achieves the same osmolality as in the surrounding interstitium. Urine of high osmolality and low volume is finally excreted.

A Special Role for Urea in Producing a Concentrated Urine

In the model presented, urea plays an important role in the concentrating mechanism in the inner medulla. This is a region of the kidney in which there are thin (but no thick) ascending limbs. The epithelium of the thin ascending limb is flat, with few mitochondria. It does not look like the type of epithelium usually associated with vigorous salt transport. Urea is added to the inner medulla from the collecting ducts and from the calyceal urine that bathes the papilla. The osmotic pressure due to urea causes the removal of water along the descending limb and thereby concentrates NaCl in this descending limb fluid. When the NaCl-enriched fluid enters the NaCl-permeable (but water-impermeable) thin ascending limb, a concentration gradient for outward, passive movement of NaCl is established. This outward movement drives the countercurrent multiplier in the inner medulla. It results in net addition of solute to this region of the kidney, which is important in establishing the gradient in the medulla that will be used to concentrate urine in the collecting ducts.

The Role of the Vasa Recta in Producing a Concentrated Urine

The blood vessels of the medulla, the vasa recta, supply the nutritional needs of the medulla. They also act as countercurrent exchangers. Countercurrent exchange is a passive process that helps to maintain a gradient established by some other means. For example, countercurrent exchange of heat between blood flowing into and out of the limbs allows the core body temperature to be maintained higher than that of fingers and toes. Blood flowing into and out of the medulla exchanges water and solutes between ascending and descending vasa recta (see Fig. 24–19). This exchange reduces the extent to which blood flowing into the medulla tends to dissipate the osmotic gradient. A relatively low blood flow to the kidney medulla (compared to the cortex) facilitates exchange and reduces washout of the medullary osmotic gradient.

The vasa recta are essential to the operation of the concentrating mechanism because they are responsible for removing water from the medulla. Water enters the medulla from the descending limbs of the loops of Henle and from the collecting ducts; there must be a pathway for removal of water—otherwise, it would accumulate in the medulla. There is a force for fluid uptake in the ascending vasa recta because of the plasma colloid osmotic pressure and because of an elevated concentration of small solutes.

Factors Affecting Urinary Concentrating Ability

Many factors influence the ability to form an osmotically concentrated urine (Table 24–3).

Table 24–3

Factors Affecting Urinary Concentrating Ability

Antidiuretic hormone (ADH)

Delivery of NaCl to ascending limb of Henle's loop

Reabsorption of NaCl by ascending limb

Delivery of fluid to medullary collecting ducts

Medullary blood flow

Urea

Length of Henle's loop

Antidiuretic hormone is necessary for the production of an osmotically concentrated urine. In the absence of ADH, the urine is not isosmotic to plasma, but is actually markedly hypo-osmotic. The range of urine osmotic concentration, from very dilute to very concentrated urine, is normally due to varying plasma levels of ADH. **Diabetes insipidus** is a condition in which there is an ADH deficiency, and as much as 20 liters of dilute urine is excreted in a day. To survive, a person with a disorder of this magnitude would have to drink 80 glasses of water a day and would spend a large part of the day and night urinating and drinking. In **central (hypothalamic, pituitary) diabetes insipidus,** synthesis or secretion of ADH is inadequate. In **nephrogenic diabetes insipidus,** ADH is present, but the collecting ducts are unresponsive.

An adequate delivery of salt to the ascending limb is necessary for maximal urine concentration. If glomerular filtration rate is abnormally low—for example, following severe hemorrhage—then concentrating ability is impaired.

Not only must delivery of salt be adequate, but the rate of reabsorption of salt along the ascending limb must be maintained. After all, this is the single effect essential for countercurrent multiplication. The "loop" diuretic drugs inhibit salt reabsorption in the thick ascending limb; they decrease concentrating ability and cause a marked increase in salt and water excretion.

The urine can be maximally concentrated only if a trickle of fluid flows through the collecting duct system. If fluid reabsorption is impaired in early portions of the nephron, as occurs during an osmotic diuresis, then a maximal urine osmolality cannot be achieved.

Medullary blood flow can affect the osmotic gradient in the medulla and thus affect concentrating ability as well. An excessive blood flow to this region tends to wash out the gradient and thereby reduces the maximum urine osmolality that can be attained.

Urea plays a critical role in the concentrating mechanism, especially in the establishment of an osmotic gradient in the inner medulla. With a low-protein diet and the resulting decrease in urea production, concentrating ability is impaired. The deficient protein intake of people in drought-stricken areas results in a greater need for water, because of impaired urine concentrating ability caused by less urea in the medulla.

Finally, the length of Henle's loops affects concentrating ability. Some desert-living mammals have very long loops for their body sizes and a correspondingly enhanced ability to concentrate the urine and save water.

Production of an Osmotically Dilute Urine

In principle, the production of an osmotically dilute urine is quite straightforward. All that is needed is the reabsorption of solute across a relatively water-impermeable membrane, with water being left behind. In the thick ascending limb, active reabsorption of sodium (accompanied by chloride) results in considerable dilution of the tubular fluid, as we have already seen. This segment is often called the **diluting segment.** In the absence of antidiuretic hormone, the collecting ducts are very water-impermeable. Even though the medullary interstitium is hyperosmotic (but not as much as in a concentrating kidney), collecting duct water mostly stays in the duct lumen. Continued reabsorption of salt results in further dilution of the urine along the collecting ducts. The end result is the excretion of a large volume of hypo-osmotic urine.

SUMMARY

Functional Anatomy of the Kidney

The kidneys play a dominant role in regulating the volume and composition of the extracellular fluid, the environment of the living body cells. The basic unit of kidney structure and function is the nephron.

Each kidney is usually supplied by a single renal artery. Important blood vessels in the kidney include interlobar arteries, cortical radial arteries, afferent arterioles, glomeruli, efferent arterioles, peritubular capillaries, and vasa recta.

The kidneys are richly innervated by sympathetic efferent fibers and by afferent fibers that mediate pain. The kidney cortex contains lymphatic vessels.

The juxtaglomerular apparatus consists of specialized cells thought to play a role in regulating glomerular blood flow and filtration rate.

The kidneys have a huge blood flow because of the need to sustain a high rate of plasma filtration.

Processes Involved in the Formation of Urine

Urine formation starts with the filtration of plasma in the kidney glomeruli. Glomerular filtration is favored by the high hydrostatic pressure of the blood in the glomerular capillaries and is opposed by the hydrostatic pressure in the urinary space of Bowman's capsule and by the glomerular capillary colloid osmotic pressure. Glomerular filtration is rather nonselective; proteins are mostly retained in the plasma by the glomerular filtration membrane, but all low-molecular-weight substances are freely filtered. Glomerular filtration rate is most accurately measured by determining the inulin clearance.

Glucose is reabsorbed by the proximal tubule. Reabsorption is active and sodium-dependent and shows a maximal rate (*Tm*). Normally, all of the filtered glucose is reabsorbed. Urea is filtered by the kidney glomeruli and is passively reabsorbed by the kidney tubules. Active reabsorption of sodium provides the driving force for tubular reabsorption of water, glucose, amino acids, chloride, and phosphate.

Some organic compounds are secreted from the blood surrounding the kidney tubules into the tubular urine.

Tubular Transport of Mineral Electrolytes

Most of the filtered sodium is reabsorbed by the kidney tubules; the quantity of sodium excreted plays an important role in body sodium balance. The proximal convoluted tubule reabsorbs the greatest fraction (70%) of the filtered sodium and water; the tubule fluid stays essentially isosmotic to plasma in this nephron segment. The loop of Henle reabsorbs about 20% of the filtered sodium and 10% of the filtered water. The distal convoluted tubule and collecting ducts reabsorb about 9% of the filtered sodium and 19% of the filtered water. The collecting ducts are the site of final regulation of sodium and water excretion; aldosterone and antidiuretic hormone (ADH) increase sodium and water reabsorption, respectively, by the collecting ducts.

Potassium is filtered, reabsorbed, and secreted by the kidney tubules. The cortical collecting duct is an important site in the determination of potassium secretion, and hence of excretion.

About 10% of dietary calcium is excreted by the kidneys.

The kidneys play a major role in regulating the concentration of magnesium in the plasma.

The kidneys are also important in the regulation of the plasma concentration of inorganic phosphate.

Tubular Reabsorption of Water

The mammalian kidney can form a urine that is osmotically more dilute or concentrated than plasma. By forming a dilute urine, the kidneys get rid of extra water from the body; by forming a concentrated urine, the kidneys save water for the body. In the absence of ADH, an osmotically dilute urine is excreted. ADH results in a more concentrated urine by increasing the water permeability of the kidney collecting ducts, thereby allowing the collecting duct urine to equilibrate with the osmotically concentrated medullary interstitium.

The formation of an osmotically concentrated urine depends on the establishment of an osmotic gradient in the medulla by the loops of Henle. These structures act as countercurrent multipliers.

The addition of urea to the inner medulla allows for efficient operation of the urinary concentrating mechanism.

The vasa recta act as countercurrent exchangers and remove water from the kidney medulla.

Factors affecting urinary concentrating ability are ADH, salt reabsorption by the ascending limb of Henle's loop, tubule fluid and blood flow, urea, and the length of Henle's loops.

Production of an osmotically dilute urine involves the reabsorption of solute across a membrane, with water left behind.

REVIEW QUESTIONS

Identify Each with a Term

1. The basic unit of kidney structure and function.

2. The liquid collected when plasma is pushed through a very fine filter.

3. The barrier separating glomerular capillary blood from the urinary space of Bowman's capsule.

4. The volume of plasma per unit of time needed to supply the quantity of a substance excreted in the urine per unit of time.

5. The plasma level at which glucose first appears in the urine in appreciable amounts.

6. The maximal rate of PAH secretion.

7. The process that leads to establishment of an osmotic gradient in the kidney medulla.

8. The disorder in which ADH synthesis or secretion is inadequate.

Define Each Term

9. afferent arteriole
10. juxtaglomerular apparatus
11. proteinuria
12. urethra
13. glomerular filtration rate (GFR)
14. tubular reabsorption
15. tubular secretion
16. filtered load
17. renal autoregulation
18. juxtamedullary nephron

Choose the Correct Answer

19. Which of the following produces renin?
   a. granular cells
   b. intercalated cells
   c. macula densa
   d. podocytes

20. Which of the following produces an increase in renal blood flow?
   a. angiotensin II
   b. norepinephrine
   c. prostaglandin I_2
   d. thromboxane

21. Approximately what percentage of filtered water is usually reabsorbed by the kidney tubules?
   a. 1%
   b. 20%
   c. 70%
   d. 99%

22. In which nephron segment is PAH secreted?
   a. collecting duct
   b. distal convoluted tubule
   c. proximal tubule
   d. thick ascending limb
   e. all of the above

23. Which of the following substances normally has the highest renal plasma clearance?

   a. glucose
   b. inulin
   c. PAH
   d. urea

24. Active reabsorption of glucose across the luminal cell membrane of proximal tubule cells is accomplished by:

   a. a glucose pump
   b. facilitated diffusion
   c. glucose-sodium co-transport
   d. simple diffusion

25. About 70% of filtered sodium is reabsorbed in the:

   a. collecting duct
   b. distal convoluted tubule
   c. loop of Henle
   d. proximal convoluted tubule

26. Which of the following is associated with an increased ability to concentrate the urine osmotically?
   a. administration of a "loop" diuretic drug
   b. a low protein diet
   c. an increased medullary blood flow
   d. long loops of Henle

Calculate

27. The following values were obtained in a renal clearance study on a young, 50-kg woman:

Plasma creatinine concentration	0.010 mg/ml
Urine creatinine concentration	0.80 mg/ml
Urine volume	720 ml
Duration of urine collection	12 hr

What is her glomerular filtration rate in ml plasma/min?

28. If glomerular filtration rate is 125 ml/min, glomerular hydrostatic pressure is 55 mm Hg, hydrostatic pressure in Bowman's space is 15 mm Hg, and mean glomerular capillary colloid osmotic pressure is 30 mm Hg, what is the glomerular filtration coefficient?

29. The following data were collected in a renal clearance study on a young, 70-kg man:

Plasma inulin concentration	0.30 mg/ml
Urine inulin concentration	7.50 mg/ml
Plasma glucose concentration	4.00 mg/ml
Urine glucose concentration	25.0 mg/ml
Plasma PAH concentration	0.02 mg/ml
Urine PAH concentration	2.64 mg/ml
Urine flow rate	5.00 ml/min
Blood hematocrit	0.45

   a. What is the glomerular filtration rate?
   b. What is the glucose clearance?
   c. What is the rate of tubular reabsorption of glucose?

d. What is the PAH clearance?

e. Assuming that the PAH clearance is equal to 90% of the renal plasma flow, what is the true renal plasma flow?

f. What is the renal blood flow?

Answer Each Question in One or Two Sentences

30. Draw a nephron and collecting duct system and the associated blood vessels; label afferent arteriole, Bowman's capsule, connecting tubule, cortical collecting duct, distal convoluted tubule, distal straight tubule (thick ascending limb), efferent arteriole, glomerulus, inner medullary collecting duct, juxtaglomerular apparatus, loop of Henle, outer medullary collecting duct, peritubular capillaries, proximal convoluted tubule, proximal straight tubule, and thin limb.

31. Why does a loss of fixed negative charges from the glomerular filtration membrane lead to proteinuria?

32. What factors account for the very high rate of plasma filtration in the glomeruli of the human kidney?

33. How will the following affect the rate of glomerular filtration: (1) a decrease in K_f, (2) a decrease in glomerular capillary hydrostatic pressure, (3) an increase in hydrostatic pressure in the urinary space of Bowman's capsule, (4) an increase in plasma colloid osmotic pressure, and (5) afferent arteriolar constriction?

34. Why does a fall in glomerular filtration rate lead to a rise in plasma creatinine concentration?

35. Why is uncontrolled diabetes mellitus often accompanied by glucosuria?

36. Why can the clearance of PAH be used to estimate renal plasma flow?

37. What are the pathways for sodium reabsorption across the proximal tubule epithelium?

38. How do sodium and water reabsorption differ in the proximal convoluted tubule and collecting duct?

39. How do the following affect potassium excretion: (1) an increase in plasma aldosterone concentration, (2) an increase in plasma potassium concentration, (3) an increase in transtubular potential difference in the cortical collecting duct (lumen more negative), and (4) an increase in tubule fluid flow rate in the cortical collecting duct?

40. What are the functions of the vasa recta?

41. What special role does urea play in the operation of the countercurrent mechanism in the inner medulla?

SUGGESTED READINGS

Brenner, B., Coe, F.L., and Rector, F.C., Jr. *Renal Physiology in Health and Disease*. Philadelphia: W.B. Saunders Company, 1987.

Gottschalk, C.W., Berliner, R.W., and Giebisch, G.H. *Renal Physiology: People and Ideas*. Bethesda: American Physiological Society, 1987.

Hladky, S.B., and Rink, T.J. *Body Fluid and Kidney Physiology*. London: E. Arnold, 1986.

Kriz, W., and Bankir, L. "A standard nomenclature for structures of the kidney." *Am J Physiol* 254:F1–F8, 1988.

Marsh, D.J. *Renal Physiology*. New York: Raven Press, 1983.

Pitts, R.F. *Physiology of the Kidney and Body Fluids*, 3rd ed. Chicago: Year Book Medical Publishers, 1974.

Seldin, D.W., and Giebisch, G.H., ed. *The Kidney: Physiology and Pathophysiology*. New York: Raven Press, 1985.

Smith, H.W. *From Fish to Philosopher*. Boston: Little, Brown and Company, 1953.

Sullivan, L.P., and Grantham, J.J. *Physiology of the Kidney*, 2nd ed. Philadelphia: Lea and Febiger, 1982.

Valtin, H. *Renal Function: Mechanisms Preserving Fluid and Solute Balance in Health*, 2nd ed. Boston: Little, Brown and Company, 1983.

Vander, A.J. *Renal Physiology*, 4th ed. New York: McGraw-Hill Book Company, 1991.

KEY TERMS

countercurrent exchange (p. 849)

countercurrent multiplication (p. 849)

diabetes insipidus (p. 853)

excretion (p. 831)

glomerular filtration (p. 830)

glomerular filtration coefficient (p. 833)

glucose threshold (p. 840)

glucosuria (p. 840)

juxtaglomerular apparatus (p. 829)

juxtamedullary nephron (p. 827)

nephron (p. 826)

proteinuria (p. 842)

renal autoregulation (p. 830)

renal plasma clearance (p. 835)

tubular reabsorption (p. 831)

tubular secretion (p. 831)

tubular transport maximum (p. 840)

tubuloglomerular feedback hypothesis (p. 829)

ultrafiltration (p. 830)

vasa recta (p. 829)

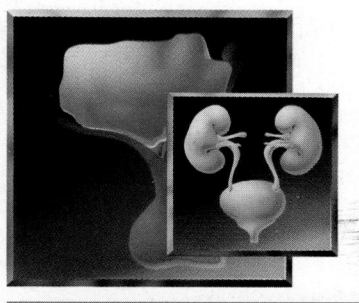

C H A P T E R
25

Regulation of Fluid and Electrolyte Balance

857

*I*n the last chapter, we discussed how various substances were handled by the kidneys. We considered the processes of glomerular filtration, tubular reabsorption, and tubular secretion. In this chapter, we will first describe in more detail the volume and composition of the various body fluids. Then we will discuss how we keep water, and sodium and potassium ions, in balance. We will indicate some of the body disturbances that occur when our kidneys fail and discuss some of the methods for treating renal failure. Finally, we will briefly consider the functions of the ureters and urinary bladder.

Body Fluids and Fluid Compartments

The fluids of the body can be divided into two categories: **intracellular fluid** (fluid within cells) and **extracellular fluid** (fluid outside of cells). These fluids differ strikingly in composition and in volume. Because most cell membranes are water permeable, however, the osmotic pressure inside and outside of our cells is the same.

Body Water and Its Divisions

The body fluids contain both solutes and water, but in terms of amount or volume occupied, water is the major constituent. Therefore, we will consider the water content of the body first, and later cover the various solutes dissolved in the body fluids.

The percentage of total body weight that is water varies from 45% to 75% in different individuals. This range mostly reflects differences in the amount of body fat. The water content of adipose (fat) tissue is about 10% by weight; most other tissues contain 70% to 75% water by weight. An obese individual has a low percentage of body weight made up of water, and a lean individual has a high percentage. Young adult males average about 60% water by weight; young adult females average about 50% water. The difference is due to females' smaller muscle mass and greater amount of subcutaneous fat. With aging, the percentage of body weight that is water decreases because water-rich muscle tissue tends to be replaced by water-poor adipose tissue.

In a healthy person, the water content of the body hardly changes from day to day. A simple way to evaluate *changes* in total body water is to weigh a person. An increase in body weight may signal abnormal fluid (water) retention; a loss of weight may indicate abnormal fluid (water) loss.

Total body water is distributed in several divisions or compartments (Fig. 25–1). Approximately two thirds of body water is contained in the **intracellular fluid compartment,** and one third in the **extracellular fluid compartment.**

The intracellular fluid compartment is not one continuous space, but is made up of trillions of cells. The intracellular space is separated from the extracellular space by cell plasma membranes. If we assume an idealized, young adult, 70-kg male, in whom total body water is 60% of body weight, we can calculate that total body water is 42 kg (or 42 liters), intracellular water is 28 liters, and extracellular water is 14 liters.

The extracellular fluid can be further divided into two major subcompartments, separated from each other by the endothelium of the blood vessels. Blood vessels contain **blood plasma,** the part of the blood exclusive of blood cells and platelets. The plasma is about 93% water (by weight or volume). Plasma water accounts for about one quarter of the extracellular water (3.5 liters in a 70-kg, young adult male). Outside of the blood vessels are the **interstitial fluid** and **lymph.** Interstitial fluid is the fluid directly bathing most body cells, and lymph is the fluid in lymphatic vessels. The interstitial fluid and lymph together contain three quarters of the extracellular water (10.5 liters in a 70-kg male). Blood plasma, interstitial fluid, and lymph are nearly identical in chemical composition, except for a higher protein concentration in plasma.

Figure 25–1 does not show an additional small extracellular fluid compartment, the **transcellular fluid.** This fluid compartment includes specialized fluids such as **cerebrospinal fluid,** the **aqueous humor** of the eye, secretions of the digestive glands, sweat, and renal tubular fluid and urine. These

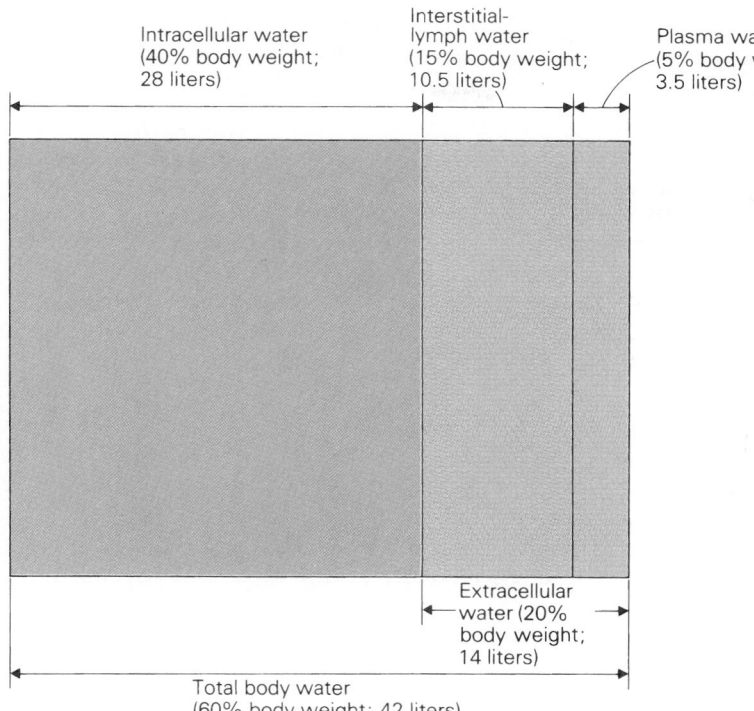

Figure 25–1

Distribution of water in the body of an average, young adult, 70-kg male.

fluids are separated from the plasma by an endothelium as well as by a continuous epithelial cell layer. This epithelium modifies the chemical composition of the transcellular fluid, so it is not a simple ultrafiltrate of plasma, as is interstitial fluid. Transcellular fluid amounts to only 1% to 3% of body weight but is extremely important, physiologically speaking. The transcellular fluids, in general, are continuously formed, and abnormal loss of these fluids or blockage of fluid drainage may have serious consequences for the organism.

Electrolyte Composition of the Body Fluids

The body fluids contain many uncharged organic substances (e.g., glucose, urea), but in terms of concentration, electrolytes are quantitatively more important. Electrolytes contribute most to the osmotic pressure of the body fluids and consequently are crucial in the distribution of body water.

The concentrations of various electrolytes in plasma, interstitial fluid, and intracellular fluid are summarized in Table 25–1. The intracellular fluid values are based on measurements in skeletal muscle cells. These cells were chosen as a representative cell type because they account for about two thirds of the cell mass in the human body. Concentrations

are expressed in terms of milliequivalents per liter solution or per kg H_2O. An **equivalent** is equal to the product of the moles and valence of an ion; one **milliequivalent (mEq)** is equal to 1/1000 Eq. For singly charged (**univalent**) ions, 1 mEq is the same as 1 millimole (mmole). Thus, we can express plasma sodium concentration as 142 mEq per liter or 142 mmole per liter. For doubly charged (**divalent**) ions, 2 mEq is the same as 1 mmole. For example, a plasma Ca^{2+} concentration of 5 mEq per liter is the same as 2.5 mmole per liter. Some electrolytes, such as proteins, are **polyvalent**, so there are several milliequivalents per millimole. The virtue of expressing concentration in terms of milliequivalents per liter is that in any fluid compartment the sum of the positive ions (cations) must equal the sum of the negative ions (anions), because of the requirement of electroneutrality in solutions. In Table 25–1, note that in every compartment the total equivalents of cation and anion are equal.

Plasma concentrations are listed in the first column of Table 25–1. Sodium is the major cation in plasma, and chloride and bicarbonate are the major anions. The plasma proteins (mostly serum albumin) bear a net negative charge at physiological pH. The electrolytes are actually dissolved in the watery phase of plasma, so concentrations in plasma water are presented in the second column of Table 25–1. These values were calculated by assuming a plasma

Table 25–1
Electrolyte Composition of the Body Fluids

Electrolytes	(1) Plasma (mEq/liter)	(2) Plasma Water (mEq/kg H$_2$O)	(3) Interstitial Fluid (mEq/kg H$_2$O)	(4) Intracellular Fluid (Skeletal Muscle) (mEq/kg H$_2$O)
Cations				
Na$^+$	142	153	145	10
K$^+$	4	4.3	4	159
Ca^{2+}	5	5.4	3	1
Mg^{2+}	2	2.2	2	40
Total	153	165	154	210
Anions				
Cl$^-$	103	111	117	3
HCO$_3$$^-$	25	27	28	7
Protein	17	18	—	45
Others	8	9	9	155
Total	153	165	154	210

water content of 93%. For example, for Na$^+$, 142 mEq per liter plasma divided by 0.93 liter water per liter plasma equals 153 mEq per liter water. Since a liter of water weighs 1 kg, the concentrations could also be expressed per kg of water.

The interstitial fluid (third column of Table 25–1) is an ultrafiltrate of plasma. It contains essentially the same concentrations as plasma of all low-molecular-weight substances, but little protein. The differences in concentrations between the small ions in plasma water and interstitial fluid (compare columns two and three in Table 25–1) arise because of differences in protein concentration between these two fluids. Two factors are involved. The first is the so-called **Donnan distribution** effect. Because the large plasma proteins are negatively charged and kept within the circulation, this leads to an excess of small cations and a deficit of small anions in the circulation. In other words, the concentrations of small cations in interstitial fluid will be lower than in plasma, and the concentrations of small anions in interstitial fluid will be greater than in plasma. The Donnan factor is 0.95 for univalent cations (Na$^+$, K$^+$) and is 1.05 for univalent anions (Cl$^-$, HCO$_3$$^-$). For example, if the plasma sodium concentration is 153 mEq per kg H$_2$O, then the interstitial fluid sodium concentration will be 0.95 × 153 mEq per kg H$_2$O or 145 mEq per kg H$_2$O. The Donnan factor for divalent cations (Ca^{2+}, Mg^{2+}) is 0.90.

The differences between plasma and interstitial fluid concentrations of Ca^{2+} and Mg^{2+} are larger than can be accounted for by the Donnan effect

alone. Approximately 40% of plasma calcium and 30% of plasma magnesium are bound to plasma proteins. It is only the non-protein-bound Ca^{2+} and Mg^{2+} that are in equilibrium across capillary walls. Note that in most instances, the differences between plasma and interstitial fluid are small. For practical purposes, we can assume (except for plasma proteins) that electrolyte concentrations in plasma and interstitial fluid are the same.

If we examine intracellular fluid composition (column four of Table 25–1), major differences between this fluid and the extracellular fluids will be apparent. The concentrations of potassium, magnesium, and protein in the cell are much higher than in the surrounding interstitial fluid. The concentrations of sodium, calcium, chloride, and bicarbonate are much lower in the cell. The anions in muscle cells labeled "Others" are mainly proteins and organic phosphate compounds (e.g., creatine phosphate, ATP), and various other anions to which the cell membrane is mostly impermeable.

The high internal potassium concentration and low internal sodium concentration are maintained by activity of the cell membrane Na$^+$/K$^+$-ATPase, which extrudes sodium ions and takes up potassium ions. The lower intracellular concentrations of chloride and bicarbonate can be mostly accounted for by the membrane potential difference, approximately −90 mv, which favors the outward movement of these negatively charged ions.

The distribution of ions within the cell is not uniform; different organelles may have different ion concentrations. For example, the cell nucleus has a higher sodium concentration than the cell as a whole. In muscle, most of the intracellular calcium is in the sarcoplasmic reticulum and mitochondria, and the cytoplasmic calcium concentration is very low.

Do not be misled by the higher number of milliequivalents per kg H_2O in the cell than in the interstitial fluid (compare totals in columns three and four of Table 25–1). A single protein molecule or phosphate compound may bear several negative charges, so the number of milliequivalents in the cell is greater than the number of milliosmoles. Also, magnesium is divalent (40 mEq per kg H_2O corresponds to 20 mmoles per kg H_2O) and is largely protein-bound within the cell, so it is not effective osmotically. It is generally accepted that the total solute concentrations (osmolalities) in the cell and surrounding extracellular fluid are identical. Water moves freely through most plasma membranes, so there is a condition of osmotic equilibrium between cellular and extracellular fluids.

Distribution of Water Between Cellular and Extracellular Fluids

Osmotic pressure is of prime importance in determining the distribution of water between cells and extracellular fluid. Osmotic pressure is directly related to total solute concentration (osmolality). An increased solute concentration lowers the activity (or concentration) of water; water moves from a region of lower osmolality to one of higher osmolality. Such water movement could be prevented by applying a hydrostatic pressure to the more concentrated solution. Animal cells cannot employ such a mechanism, since a hydrostatic pressure difference of a few centimeters of H_2O leads to rupture of the plasma membrane. In animals, osmotic pressures inside and outside of cells are equal. If solute or water is added to or lost from the extracellular fluid, osmotic equilibrium will be temporarily upset. Water will move into or out of the cells until a new osmotic equilibrium is achieved.

Most of the volume in any body fluid compartment is occupied by water. At a given osmolality, the total volume of (or amount of water in) a compartment is directly related to the amount of solute in the compartment. This important relationship follows from the equation that defines the term *concentration*—namely, concentration = amount/volume. If solute concentration (osmolality) is maintained constant, then the volume must change directly with the amount of solute.

In the extracellular fluid, the amount of sodium present is a major determinant of the volume of water present. This relationship results because sodium and its accompanying anions, chloride and bicarbonate, are the major osmotically active solutes present in extracellular fluid. If we add up the concentrations of these ions in interstitial fluid (see Table 25–1), we get 145 + 117 + 28 = 290 mmoles per kg H_2O. The corresponding osmolality is about 270 mosm per kg H_2O. (This figure is lower than 290 because of interactions among ions in solution; it is only in infinitely dilute solutions that ions behave as completely independent particles.) Since interstitial fluid (or plasma) osmolality averages 287 mosm per kg H_2O (approximately 300 mosm per kg H_2O), it is apparent that sodium salts account for more than 90% of the osmolality. The osmolality of the extracellular fluid is closely regulated by the thirst sensation and control of renal excretion of water by ADH. Consequently, if osmolality (sodium concentration) is kept constant, it follows that the amount of sodium in the body will determine the amount of water in the body. To express it another way, if a person takes in extra salt (NaCl), thirst will result, water will be ingested, and extracellular fluid volume will be increased. If a person loses salt, extracellular fluid volume will be decreased.

Cell volume, also mostly water, is similarly influenced by the amount of contained solute. In the cell compartment, potassium is the major osmotically active solute. Hence, a loss of potassium from the cell results in a decrease in cell volume, and a gain in potassium results in an increase in cell volume. The large amount of impermeable anions (e.g., proteins, organic phosphates) within cells could affect cell volume. In theory, these solutes would cause a redistribution of small, permeable ions so that an excess of solute would be present in the cell (the Donnan distribution effect), and, therefore, water would tend to enter the cell. This tendency is, however, counteracted by the activity of the Na^+/K^+-ATPase, which effectively excludes sodium and thereby lowers the total amount of solute in the cell. In this way, the Na^+/K^+-ATPase plays a key role in the regulation of cell volume. If Na^+/K^+-ATPase activity is decreased (e.g., by low temperature, oxygen lack, metabolic poisons, or cardiac glycosides), cells gain sodium and water, and they swell.

An interesting phenomenon has been recently described: when certain (e.g., proximal tubule) cells are placed in a hypo-osmotic medium, they initially swell (as expected). Soon, however, they shrink back to their original size. This response is due to loss of potassium, chloride, and water and appears to be caused by activation of membrane ion transport processes. Opposite changes are observed in hyperosmotic media (cells initially shrink and then increase back to normal size). These responses demonstrate that cells have special mechanisms for regulating and preserving their volumes.

The normal distribution of water between cellular and extracellular compartments changes in a variety of circumstances. Figure 25–2 provides some examples. Note that the height of the boxes corresponds to solute concentration. The width of the boxes corresponds to volume. The area of each box (height × width) corresponds to the total amount of solute in a compartment (concentration × volume = amount). The dotted lines represent the normal condition and are the same in all four parts of this figure. The solid lines represent conditions after a new osmotic equilibrium has been achieved. Note that the height of the boxes is always identical in intra-

cellular and extracellular fluids, since osmotic equilibrium is attained.

Figure 25–2*a* shows the normal condition. Osmolalities are identical in cellular and extracellular compartments (300 mosm per kg H_2O). The volume of the intracellular compartment (40% of body weight) is twice the volume of the extracellular compartment (20% of body weight).

In Figure 25–2*b*, pure water is added to the extracellular fluid; ingestion of water would have the same effect. Since the extracellular fluid osmolality is lowered, water moves into the cells until the osmolality in both compartments is lowered to the same level. The entry of water into the cells increases their volume. Since the cells contain two thirds of the body fluid solutes, two thirds of the added water enters the cells, and one third remains in the extracellular space. The decline in plasma osmolality suppresses secretion of ADH, which per-

Figure 25–2
The effects of adding pure water (*b*), isotonic saline (*c*), and pure NaCl (*d*) to the extracellular fluid on osmolalities and volumes of the intracellular fluid (ICF) and extracellular fluid (ECF). (*a*) is the normal condition.

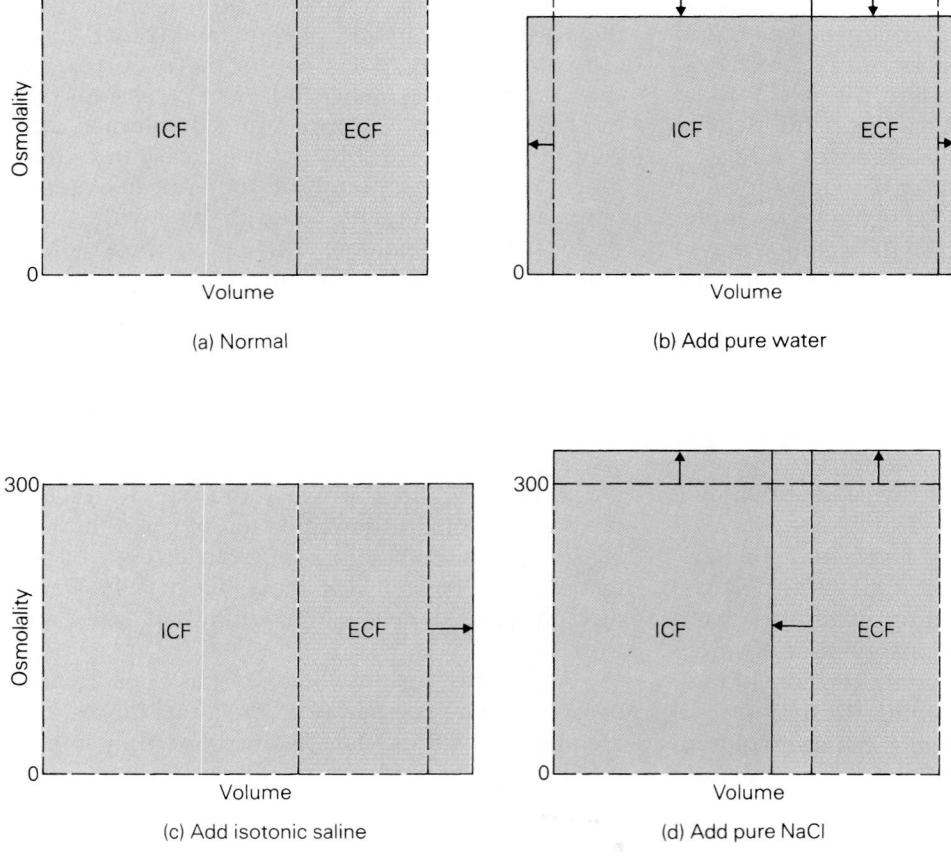

mits rapid excretion of the ingested water by the kidneys, thereby restoring body fluid osmolality and volume back to normal.

In Figure 25–2c, isotonic saline is added to the extracellular fluid. This can be done by intravenous infusion or ingestion of a 0.9% NaCl solution. Note that the final osmolality is not different from normal. This is not surprising, since isotonic saline is also isosmotic. Note also that cell volume does not change, since, by definition, cells do not shrink or swell when exposed to an isotonic solution. All of the isotonic saline is retained in the extracellular fluid and accordingly increases only this volume. In a normal person, the expanded extracellular fluid volume leads (by mechanisms we will discuss later) to an increase in renal excretion of salt and water, eventually restoring normal conditions.

Although it is not illustrated, a loss of isotonic fluid from the extracellular compartment has predictable consequences: no change in body fluid osmolality or in cell volume and a decrease in extracellular fluid volume. Hemorrhage and diarrhea are two cases in which an isotonic fluid is lost from the body. The body acts to restore depleted extracellular fluid by renal retention of salt and water and by increased water intake (due to thirst).

Figure 25–2d shows the effect of adding sodium chloride without water to the extracellular compartment. Extra salt intake (e.g., eating salted potato chips) would be a corresponding circumstance. The added salt increases the osmolality of the extracellular fluid, which results in water movement out of the cells. Extracellular fluid volume is increased at the expense of cell volume. At equilibrium, osmolality is uniformly increased. When pure NaCl (or a hypertonic NaCl solution) is added to the extracellular fluid, the NaCl is diluted by the total body water, even though all of the NaCl stays in the extracellular fluid. The body's response to NaCl excess involves (1) thirst (stimulated by cellular dehydration) and increased ingestion of water, and then (2) increased renal excretion of salt (stimulated by the expanded extracellular fluid volume).

Distribution of Water Between Blood Plasma and Interstitial Fluid

The relative volumes of plasma and interstitial fluid are primarily determined by the Starling forces that operate across capillary walls. As discussed in Chapter 19, movement of fluid across capillary walls is determined by the balance between capillary and tissue hydrostatic and colloid osmotic pressures. The hydrostatic pressure in the capillaries tends to push fluid out of the capillaries; the colloid osmotic pressure due to the plasma proteins acts to retain fluid within the vascular system. An increase in capillary hydrostatic pressure, a decrease in plasma colloid osmotic pressure, or an increase in capillary membrane permeability to proteins would favor net movement of fluid out of the vascular system and result in a decreased plasma volume and an increased interstitial fluid volume (edema). On the other hand, a decrease in capillary hydrostatic pressure or an increase in plasma colloid osmotic pressure would favor movement of interstitial fluid into the circulation. In most vascular beds, the net force favoring outward movement of the fluid at the arterial end of the capillaries is roughly the same as the net force favoring inward movement of fluid at the venous end of the capillaries. Therefore, little fluid leaves the circulation. Extra fluid that has left the circulation is returned by the lymphatics. Inadequate lymphatic drainage can lead to an appreciable increase in size of the interstitial space.

The Concept of Fluid and Electrolyte Balance

Despite varying daily intake of food and water, the volume and composition of body fluids remain constant. This suggests that we stay in balance with respect to many substances. A person in a **stable balance** maintains the same amount of a specified substance in the body over a period of time. The rates of gain or input (due to intake or production in the body) and loss or output (due to excretion or destruction in the body) of a particular substance are exactly equal, so there is no net accumulation or loss.

If fluids and electrolytes are not in balance, two alternative conditions occur. If input exceeds output, then accumulation of a substance in the body results, and a **positive balance** exists. For example, if the intake of salt exceeds the rate at which it is excreted, then salt and water will accumulate in the body and extracellular fluid volume and body weight will increase. If output exceeds input, then net loss of a substance from the body results, and a **negative balance** exists. For example, if urinary excretion exceeds the daily intake of potassium, a fall in body potassium and plasma potassium concentration results.

The kidneys play a major role in maintaining balance with respect to water, mineral electrolytes, hydrogen ions, and various organic compounds. They accomplish this by adjusting the output of these substances in the urine to match the rate of addition to the body. Renal excretion of water is under the continuous control of the antidiuretic hormone and is normally adjusted to maintain water balance. The kidneys are the primary site of control of electrolyte output. The mineral electrolytes, such as sodium, potassium, and phosphate, are ingested in variable amounts. They are not produced or destroyed in the body, and so, in a balanced system, the amounts taken in are excreted, mostly in the urine. Some substances, such as hydrogen ions or urea, are produced by metabolic reactions in the body. Since these substances should not normally accumulate, they are excreted at a rate matching their production rate.

The Regulation of Water Balance

Table 25–2 presents a balance chart for water for an average 70-kg male. We assume that the person is in water balance—that is, total input and output are equal (2500 ml per day). The input of water is normally derived from the diet. In a hospital setting, intravenous infusions may also be a source of water (and electrolytes). The various beverages we drink contain water. This intake of water is largely conditioned by habit but can also be stimulated by the thirst sensation. Water is also contained in solid food. For example, most fruits and vegetables are 80% to 90% water, and a hamburger is about 55% water. Water is also produced by oxidation of foodstuffs in the body. For example, for each mole of glucose oxidized, six moles of water are produced. (Recall from an earlier discussion the following reaction: $C_6H_{12}O_6 + 6\ O_2 \rightarrow 6\ CO_2 + 6\ H_2O$).

On the output side, we need to consider loss of water by way of the skin, lungs, gastrointestinal tract, and kidneys. Some water is always lost by way of the lungs and skin. This loss is usually not sensed and so is called **insensible water loss**. On a cold day, however, it is possible to perceive this water loss, because exhaled air is saturated with water vapor, and the water condenses to form visible droplets in cold air. Large quantities of water also can be lost from the skin through sweating. In a hot environment or during heavy exercise, as much as 4 liters of water per hour can be lost in perspiration. When we perspire, we lose not only water, but electrolytes, especially sodium chloride.

Gastrointestinal losses of water are normally small. The salivary glands, stomach, liver, pancreas, and intestine add about 7 liters of fluid to the gastrointestinal tract every day, but almost all of this is absorbed, and only 100 ml of water is normally lost in the feces (Chapter 23). With vomiting or diarrhea, large losses of water (and electrolytes) can result.

Finally, the kidneys are a site of water loss. Normally, renal water loss is regulated to maintain water balance. If there is a water deficit, renal excretion of water is decreased. If there is an excess of water in the body, renal excretion of water is increased. Renal excretion of water is controlled by the antidiuretic hormone (ADH).

The data for water balance illustrated in Table 25–2 approximate the magnitude of water inputs and outputs in a normal adult. You should recognize, however, that water intake and output vary widely, even in normal people, depending on conditions such as climate and activity. Water balance in an infant or young child is quantitatively different. Because a child has a greater surface area per unit volume and a greater metabolic rate per kilogram of body weight than an adult, the water needs of children are relatively high. Infants and young children are more vulnerable to water deficits and dehydration than are adults.

The regulation of body water balance involves two basic mechanisms: (1) the thirst sensation and (2) control of renal water excretion by the antidiuretic hormone.

Table 25–2
Daily Water Balance in an Average 70-kg Male

Input		Output	
Water in beverages	1000 ml	Skin and lungs	900 ml
Water in food	1200 ml	Gastrointestinal tract (feces)	100 ml
Water of oxidation	300 ml	Kidneys (urine)	1500 ml
Total	2500 ml	**Total**	2500 ml

Thirst and the Regulation of Fluid Intake

Thirst is a conscious sensation associated with a craving for drink and is ordinarily interpreted as a desire for water. The function of the thirst mechanism is to repair a fluid deficit.

People drink largely by habit, and such drinking normally covers water needs. Thirst is an emergency mechanism not ordinarily experienced. In other words, the thirst threshold is usually not reached. Constant thirst in temperate climates and with moderate activity may indicate an underlying disease or disorder.

The thirst sensation is interesting because—unlike many sensations—it reaches the level of consciousness. With thirst, we can sense how body mechanisms generally act to bring about normalization of body conditions. The greater the need for water, the greater the intensity of thirst and the tendency for corrective action.

In human beings, thirst is primarily associated with dryness of the mouth and throat, due to a reflex decrease in secretion by salivary and buccal (cheek) glands. A dry mouth plays a role in the conscious definition of thirst. It is also important in metering water intake and so helps us know when to stop drinking. A dry mouth is not essential in regulating water intake.

Thirst is primarily controlled by centers in the hypothalamus. The hypothalamic centers relay information to the cerebral cortex, where thirst becomes a conscious sensation. Interestingly, the hypothalamus also controls release of ADH and is therefore concerned with regulating both water intake and output. The hypothalamus is concerned with several other functions related to water balance: temperature regulation, food intake, and cardiovascular function.

There are two major stimuli for thirst: cellular dehydration and extracellular dehydration.

Cellular dehydration means a decrease in water within cells, or cell shrinkage. If the effective osmotic pressure of the plasma is increased, water will move out of all body cells. In the anterior hypothalamus are **osmoreceptor cells;** when these neurons shrink, they signal the cerebral cortex to give rise to the thirst sensation. Not all solutes are effective stimuli for the osmoreceptor cells. For example, urea and ethanol are ineffective because they readily penetrate the osmoreceptor cells and therefore do not cause them to shrink. Sodium chloride is an effective stimulus. An increase in plasma osmolality of 1% to 2% is needed to reach the thirst threshold.

Extracellular dehydration, which is synonymous with **hypovolemia,** or a decrease in blood volume, stimulates thirst. To be more exact, a decrease in **effective arterial blood volume** or **effective circulating blood volume** is the stimulus here. The effective arterial blood volume is the volume of blood pumped by the heart into the arterial system that perfuses the tissues. People with severe congestive heart failure may experience intense thirst, even though their blood volume is typically greater than normal, because, due to inadequate cardiac function, their effective arterial blood volume is decreased. Thirst is stimulated by loss of isosmotic fluid such as occurs with hemorrhage, diarrhea, or vomiting.

There appear to be several receptors for hypovolemia. The various baroreceptor areas in the cardiovascular system (Chapter 19), such as the arterial baroreceptors in the carotid sinus and aortic arch, and the stretch (volume) receptors in the cardiac atria and great veins in the thorax, are involved. These receptors transmit impulses to the thirst control centers in the hypothalamus. The kidneys may also act as volume receptors. When blood volume is decreased, the kidneys release renin into the circulation. This results in production of angiotensin II, which acts directly on the brain to stimulate thirst. This last pathway indicates that the kidneys not only participate in water balance by adjusting water output, but also by affecting water intake.

Extracellular dehydration has to be considerable before thirst is stimulated. Blood donors usually do not experience thirst after donating 500 ml of blood (10% of blood volume). If a person loses a lot of blood, however, intense thirst results.

The two stimuli, cellular dehydration and extracellular dehydration, usually work together and reinforce each other. For example, if you have been active in hot weather and have lost a lot of water by sweating, plasma osmolality will be increased and blood volume will be decreased; thirst will be stimulated by both pathways. In some situations, however, the two signals may oppose each other. For example, as stated before, people with congestive heart failure experience thirst because of a decrease in effective arterial blood volume. When they drink water, their kidneys tend to retain it in the body because ADH secretion is also stimulated. The result is that the body fluids become hypo-osmotic. Because the plasma sodium is diluted by excess water, the condition that develops is often called **hyponatremia** (low sodium concentration in the blood). In this situation, plasma osmolality is lower than normal, yet thirst persists. Apparently, the body's attempt to restore an effective arterial blood volume is a more powerful drive, and so plasma osmolality is sacrificed.

Antidiuretic Hormone and the Regulation of Fluid Output

Antidiuretic hormone (ADH) plays a key role in regulating water excretion by the kidneys by facilitating the reabsorption of water by the collecting ducts (Chapter 24). This peptide hormone is produced in the hypothalamus and released from the posterior pituitary into the blood (Chapter 13). The two major factors affecting ADH release are the same as those affecting thirst: (1) cellular dehydration and (2) extracellular dehydration.

The main stimulus under ordinary conditions is cellular dehydration, caused by an increase in effective plasma osmolality. When plasma osmolality is increased, osmoreceptor cells in the anterior hypothalamus (distinct from those that cause thirst) shrink and signal the nearby ADH-producing cells to release ADH from their nerve terminals in the posterior pituitary.

The threshold for causing ADH release is normally exceeded, and so the plasma contains ADH at a measurable level (about 2.5 pg per ml). This plasma level is associated with a modestly concentrated urine (urine osmolality about 600 mosm per kg H_2O). This means that plasma ADH can be induced to fall (by lowering the plasma osmolality) or increased (by raising the plasma osmolality). Thus,

Figure 25–3
Schema showing sites controlling thirst and ADH release. The cell bodies of the ADH-producing cells are located in the supraoptic and paraventricular nuclei of the hypothalamus, and their nerve terminals are located in the posterior pituitary.

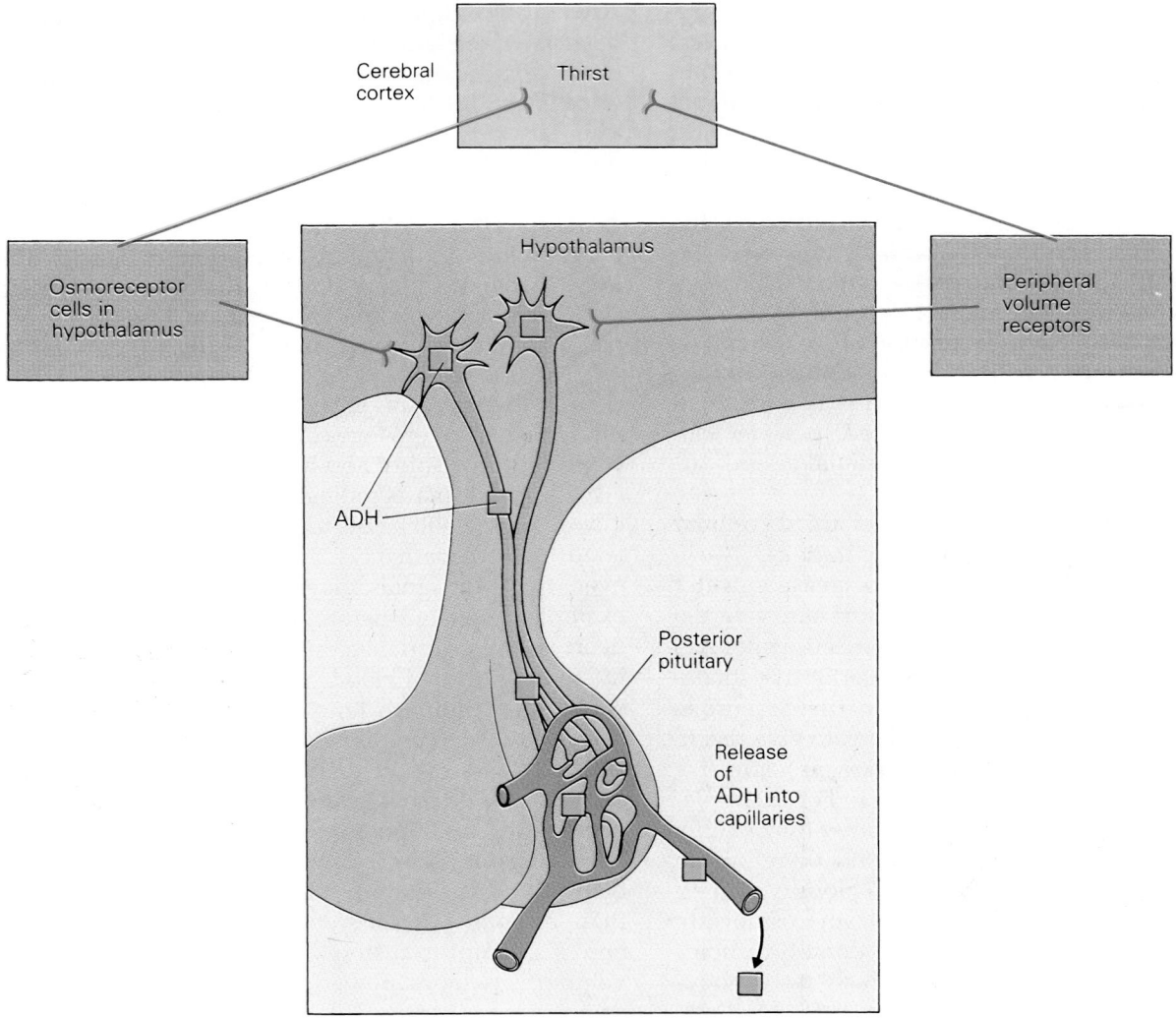

changes in plasma osmolality are continuously translated into changes in plasma ADH levels and, in turn, into changes in renal water excretion. The fact that the osmotic stimulus threshold for ADH release is considerably lower than that for thirst is consistent with the view that thirst is mainly an emergency mechanism.

Extracellular dehydration (hypovolemia or a decrease in effective arterial blood volume) stimulates ADH release. The receptors include stretch receptors in the left atrium of the heart and in the pulmonary veins within the pericardium; signals are transmitted to the brain via vagal afferents. An increase in intrathoracic blood volume inhibits ADH release and produces an increase in urine flow. The baroreceptors in the carotid sinus and aortic arch also affect ADH release. A fall in pressure at these sites stimulates ADH release. Relatively large changes in blood volume (about 10%) are required to affect ADH release. A large blood loss (e.g., 15% to 20% of blood volume), however, produces very large increases in plasma ADH. With severe hemorrhage, the high circulating levels of ADH may cause a significant rise in blood pressure due to the vasoconstrictor properties of this peptide.

The two stimuli, cellular and extracellular dehydration, often work synergistically in promoting ADH release. But (as discussed previously) in certain conditions (e.g., congestive heart failure) they may act in opposite directions. Other stimuli, primarily unrelated to water balance control, may also affect ADH release, and were discussed in Chapter 13. Figure 25–3 is a schema for the control of thirst and ADH release by osmo- and volume receptors.

The primary target for ADH action in the kidney is the collecting duct epithelium (Fig. 25–4). In the absence of ADH, the luminal membrane is water impermeable, and fluid in the lumen is hypo-osmotic to the cell or blood. ADH combines with a receptor in the basolateral cell membrane. By way of a guanine nucleotide stimulatory protein, the membrane-bound enzyme adenylate cyclase is activated. This enzyme catalyzes the formation of cyclic AMP from ATP. In turn, cyclic AMP activates a protein kinase that phosphorylates other proteins. The subsequent steps are not entirely clear, but the final result is an increase in water permeability of the luminal cell membrane, apparently due to an increase in the number of pores. Electron microscope studies have shown that in response to ADH, particle ag-

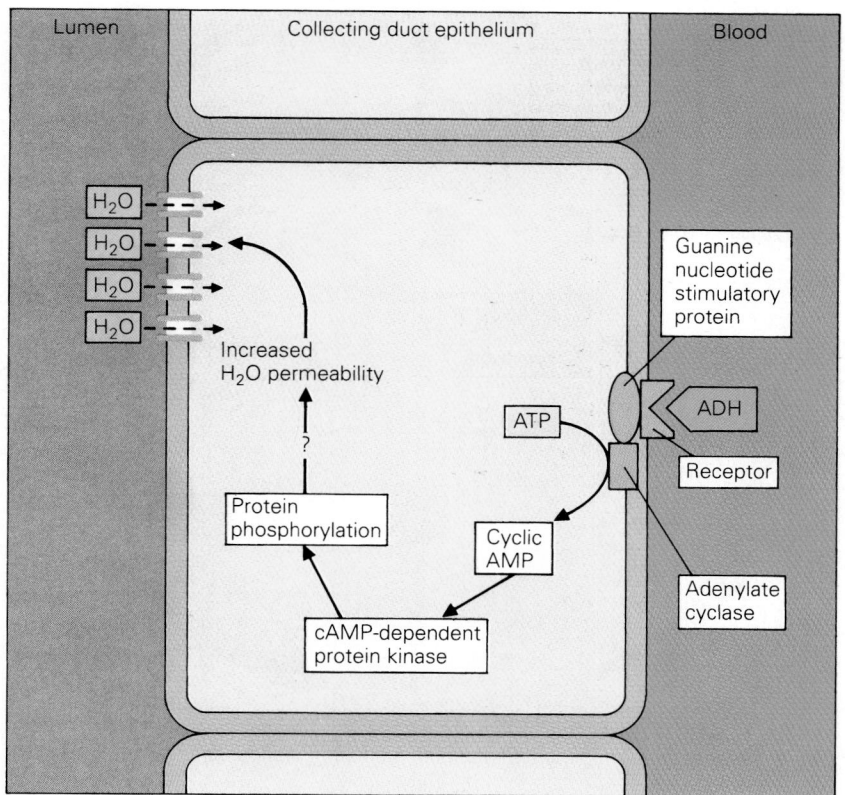

Figure 25–4
Mechanism of action of ADH on the collecting duct epithelium.

gregates (water channels?) appear in the luminal cell membrane. When the collecting duct epithelium becomes water permeable, fluid in the duct lumen tends to become concentrated by osmotic withdrawal of water.

Renal Responses to Extra Water Intake or to Dehydration

Renal excretion of water is continuously adjusted by changes in plasma ADH levels, so that plasma osmolality and extracellular fluid volume are kept at normal levels. To illustrate this adjustment, we will consider the response to two extreme states: (1) ingestion of a liter of tap water and (2) dehydration due to sweating.

Figure 25–5*a* summarizes the response to ingesting a liter of tap water. In this case, the body is threatened by dilution and excess volume, and the

appropriate renal response is excretion of a large volume of osmotically dilute urine, called a **water diuresis.** The ingested water is absorbed chiefly in the small intestine, which leads to a fall in plasma osmolality. The fall in plasma osmolality causes osmoreceptors in the hypothalamus to swell, resulting in decreased release of ADH from the posterior pituitary. The increased volume of the extracellular fluid, resulting from the water ingestion, also acts (by way of stretch receptors in the cardiovascular system) to inhibit ADH release. (This pathway appears to be less important under normal conditions.) When ADH release is decreased, plasma ADH falls, because this hormone is continuously destroyed in the body. With a low plasma ADH level, the collecting duct expresses its intrinsically low water permeability. Filtered water is then not reabsorbed at this site. The result is the excretion of a large volume (up to 20 ml per minute) of osmoti-

Figure 25–5
(*a*) Response to excess hydration. (*b*) Response to dehydration.

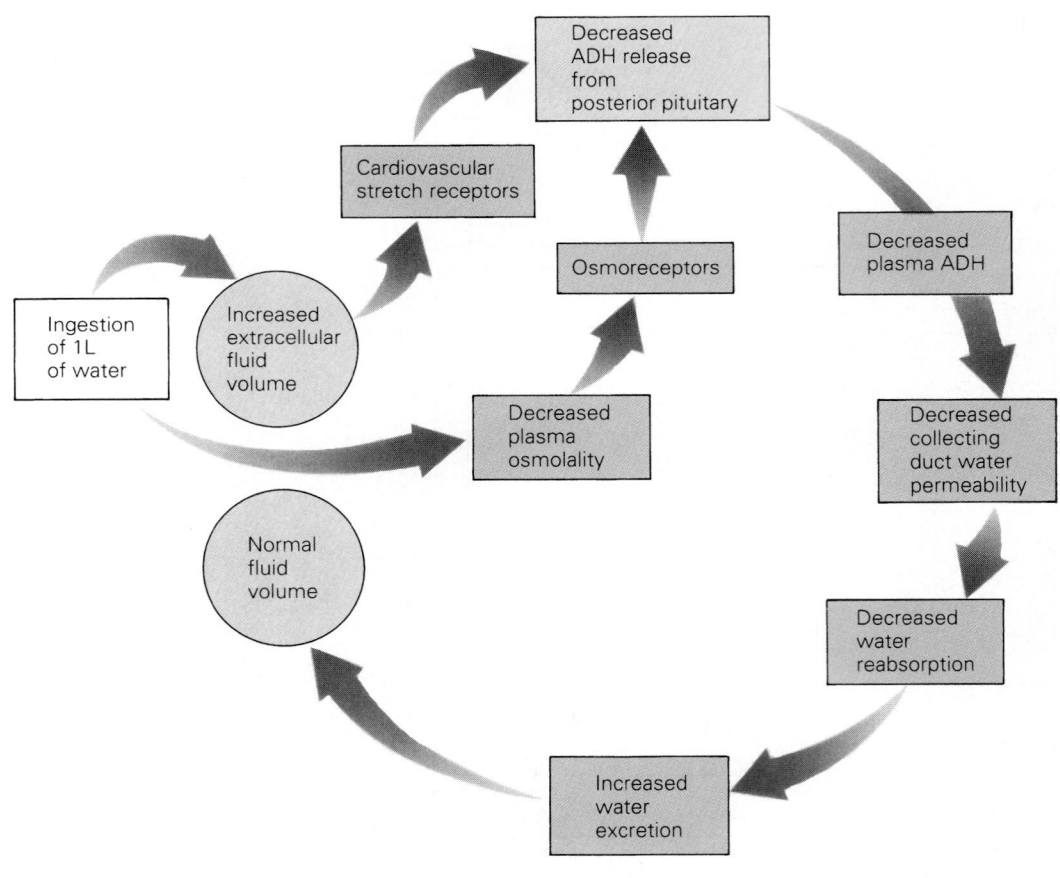

(a)

cally dilute urine. Urine osmolality may be as low as 30 to 40 mosm per kg H_2O. The increase in urine flow starts a few minutes after water ingestion, and the entire extra load of water is excreted in a few hours. By getting rid of the extra water, the kidneys are able to restore plasma osmolality and volume back to normal.

Consider now what happens in the opposite case, dehydration (see Fig. 25–5b). Suppose you exercise in hot weather, and you sweat a lot. The sweat you lose is a hypo-osmotic fluid, and so the body fluids become relatively hyperosmotic. In addition to water lost in sweat, there is also insensible water loss from the lungs and skin, and loss of water in the urine. These losses lead to a rise in plasma osmolality, stimulation of hypothalamic osmoreceptors, and increased release of ADH. Also, a fall in extracellular volume causes the cardiovascular stretch receptors to increase ADH release reflexly. The rise in plasma ADH levels causes increased water reabsorption from the collecting ducts, with resulting excretion of the urinary solutes

in a hyperosmotic urine. By excreting the urinary solutes in a minimum volume of urine, the body saves water, which can delay or minimize the effects of dehydration. To correct the dehydration, the thirst mechanism, stimulated by the same changes that affect ADH release, is engaged. Thirst stimulates water intake, and so the water deficit is repaired.

The Regulation of Sodium Balance

Sodium Input and Output

In a normal person on a normal diet, the amount of sodium ingested is closely matched by the amount of sodium lost from the body. Figure 25–6 summarizes sodium balance relationships. On the input side, sodium intake in the diet may be quite variable, but in an average American diet, sodium amounts to about 100 to 300 mEq per day (6 to 18 g NaCl per day). In a hospital setting, intravenous fluids can also be a source of sodium.

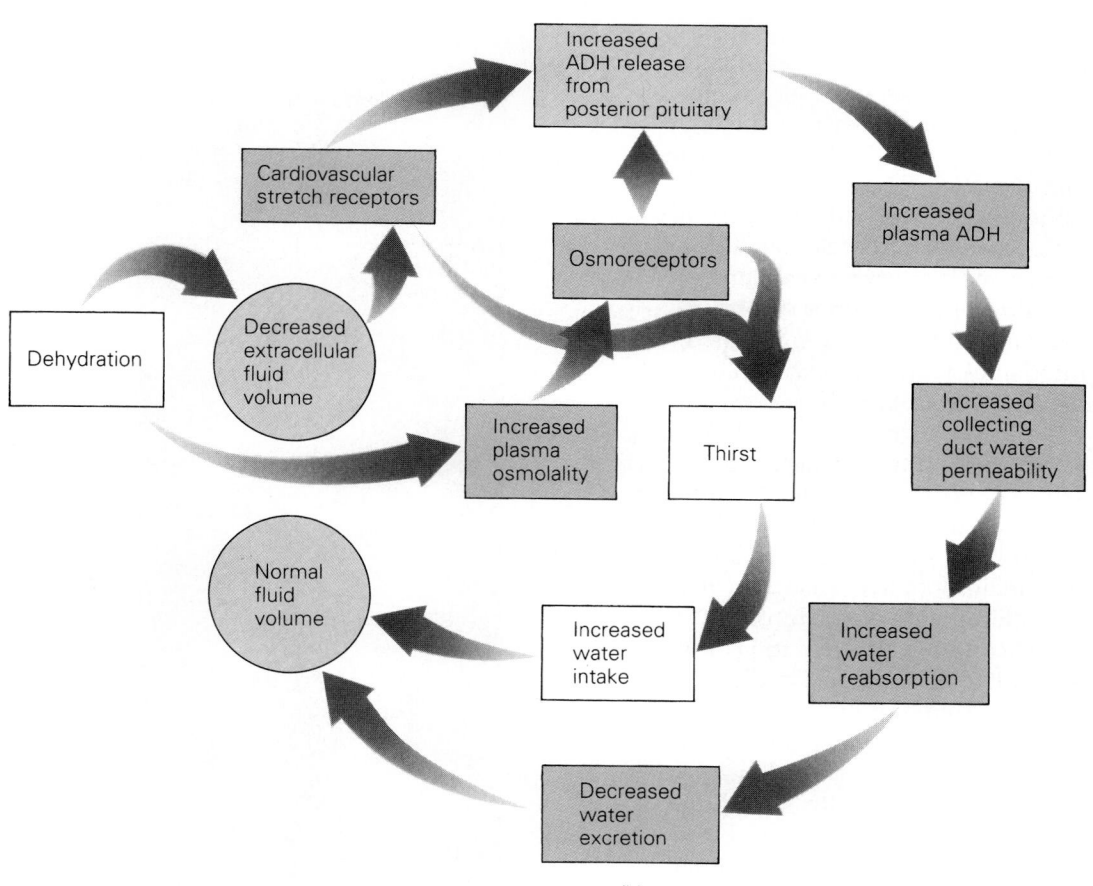

(b)

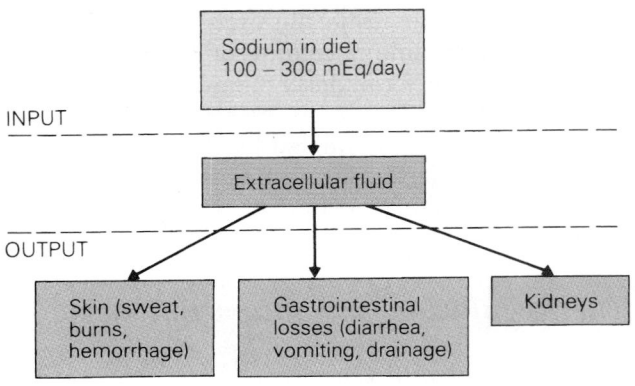

Figure 25–6
Sodium balance. The kidneys are usually responsible for 95% of the sodium output.

to a reduction of extracellular fluid volume. In turn, this produces a decrease in plasma volume and consequently may lead to an inadequate circulation, hypotension, and even death.

It should be clear that sodium balance is closely related to regulation of extracellular fluid volume. This is so because sodium salts are the major solutes of the extracellular fluid, as discussed earlier. Since the osmolality of the extracellular fluid is closely regulated by ADH, the kidneys, and thirst, it follows that retention or loss of sodium from the body will be accompanied by parallel changes in the amount of water in (and hence volume of) the extracellular fluid. The regulation of sodium balance falls largely upon the kidneys, although there is some evidence

On the output side, sodium may be lost from the skin, gastrointestinal tract, or kidneys. Loss of sodium from the skin may occur with sweating, burns, or hemorrhage. Sweat contains 5 to 80 mEq of sodium per liter, and so with heavy exercise or in warm environments, an appreciable amount of sodium may be lost in sweat. Gastrointestinal losses of sodium are normally small, but with diarrhea, vomiting, or external drainage of gastrointestinal secretions, large losses may result. The urine normally accounts for about 95% of the total output of sodium from the body. The kidneys constitute the major site of control of sodium output and normally adjust excretion to maintain balance.

A positive sodium balance results if sodium input exceeds output. Since water is retained along with sodium, and since sodium is primarily an extracellular ion, extracellular fluid volume will be increased. If sufficient salt and water are retained, expansion of the interstitial fluid space will lead to generalized edema. This condition is associated with weight gain (due to the extra water) and skin puffiness, especially in the feet and ankles (due to the effects of gravity). A positive sodium balance and generalized edema occur in a number of clinically important conditions, including congestive heart failure, hepatic cirrhosis (a liver disease), and nephrotic syndrome (a kidney disease). In order to reduce the edema, it is common practice to (1) put the patient on a low salt diet and (2) use diuretic drugs to promote sodium excretion by the kidneys. In this way, the physician tries to relieve the edema and bring the patient back into sodium balance.

A negative sodium balance results if sodium output exceeds input. This may occur if there is excessive loss of sodium via the skin, gastrointestinal tract, or kidneys. A negative sodium balance leads

Figure 25–7
Generalized edema (swelling) in the nephrotic syndrome. Due to abnormal glomerular permeability to proteins, the blood plasma protein concentration falls. This favors movement of fluid out of the capillaries into the tissue spaces. The kidneys are stimulated to retain sodium and water and thereby play an essential role in the development of the massive edema seen in this child. (From: *Renal Dysfunction: Mechanisms Involved in Fluid and Solute Imbalance, by Heinz Valtin. Little, Brown, 1979.*)

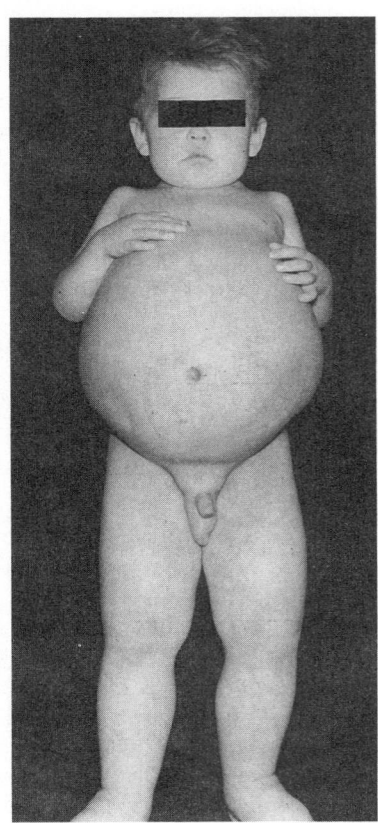

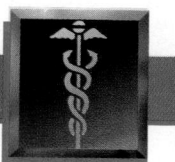

Sodium Appetite

Mammals generally maintain the amount of sodium in their bodies at a constant level. There are, however, obligatory losses of sodium in the urine and in other body secretions (e.g., sweat, saliva), for which the body must compensate. In carnivores, replacement is not a problem, since meat contains large amounts of sodium. For herbivores, however, there may be a serious lack of sodium in the diet, since most vegetation has a relatively low salt content.

Sodium intake in people is quite variable. Some people automatically salt their food without tasting it, and other people never add salt to their food. A typical American diet contains more than enough sodium.

A craving for salt in response to a real need for sodium sometimes occurs. Interestingly, many patients with Addison's disease show an unusual desire for salt. In Addison's disease there is deficient adrenal production of aldosterone, so that the patient tends to lose excessive amounts of sodium in the urine. Increased salt intake in this disease is of obvious value in maintaining a normal extracellular fluid volume. The medical literature contains many descriptions of excessive salt intake in people with Ad-

dison's disease. For example, one 34-year-old man with marked Addison's disease is reported to have put approximately a 1/8-inch layer of salt on his steak, used nearly 1/2 a glass of salt for his tomato juice, used salt on oranges and grapefruit, and made lemonade with salt!

Laboratory rats presented with two drinking bottles, one containing tap water and the other a 1% solution of sodium chloride, fail to discriminate between the two solutions and drink about the same amount from both bottles. After removal of both adrenal glands, however, they show a marked preference for the salt-containing solution.

It is well known that herbivores, such as deer and cattle, are attracted to salt-enriched mineral deposits (salt licks). The need for salt is greatly increased during pregnancy or lactation.

The physiological basis for sodium appetite remains incompletely understood. A decreased blood volume, resulting from diminished extracellular fluid sodium, may be a critical internal signal. Some studies have suggested that angiotensin II and mineralocorticoids are involved in sodium appetite, but more research needs to be done on this subject.

---

for a **sodium appetite** (which stimulates sodium intake) when a sodium deficit exists. Since the kidneys are intimately concerned with control of sodium balance and regulation of extracellular fluid volume, it should not be surprising that extracellular fluid volume is a major determinant of renal sodium excretion. We will elaborate on this later.

There is compelling evidence that hypertension (high blood pressure) may often be due to a disturbance in sodium (salt) balance. Excessive dietary intake of salt or inadequate renal excretion of salt tends to increase intravascular fluid volume; somehow this is translated into an increase in blood pressure. Avoidance of a high salt intake may help to prevent high blood pressure. A reduced salt intake and diuretic drugs are used to keep blood pressure under control in people with hypertension.

The Renal Responses to Changes in Dietary Intake of Sodium

Let us consider the renal response to changes in dietary intake of sodium. Figure 25–8 illustrates a balance study. A person was put on a fixed sodium

intake of 100 mEq per day. During the first three days of the study (the control period), the person was in a stable balance, as indicated by the fact that urinary excretion matched the sodium intake (in this study, we will ignore extrarenal losses of sodium, which are usually small). On day 4, the sodium intake was increased to 300 mEq per day and was maintained at that level until the end of day 10. What happened? Sodium excretion rose, but for the first few days (days 4 to 7) was less than the sodium input. Consequently, there was a phase of positive sodium balance. During this time, the person retained salt, and since the subject was allowed free access to water, water intake increased, and salt and water were retained in isosmotic proportions. If we measured plasma sodium concentration, we would find that it was essentially constant throughout this study. The sodium content of the body, the extracellular fluid volume, and body weight, however, increased during days 4 to 7. During days 8 to 10, the subject was back in a stable balance; that is, input and output of sodium were once again equal. When the person switched to a low sodium diet (10 mEq per day), sodium excretion fell. During days 11 to

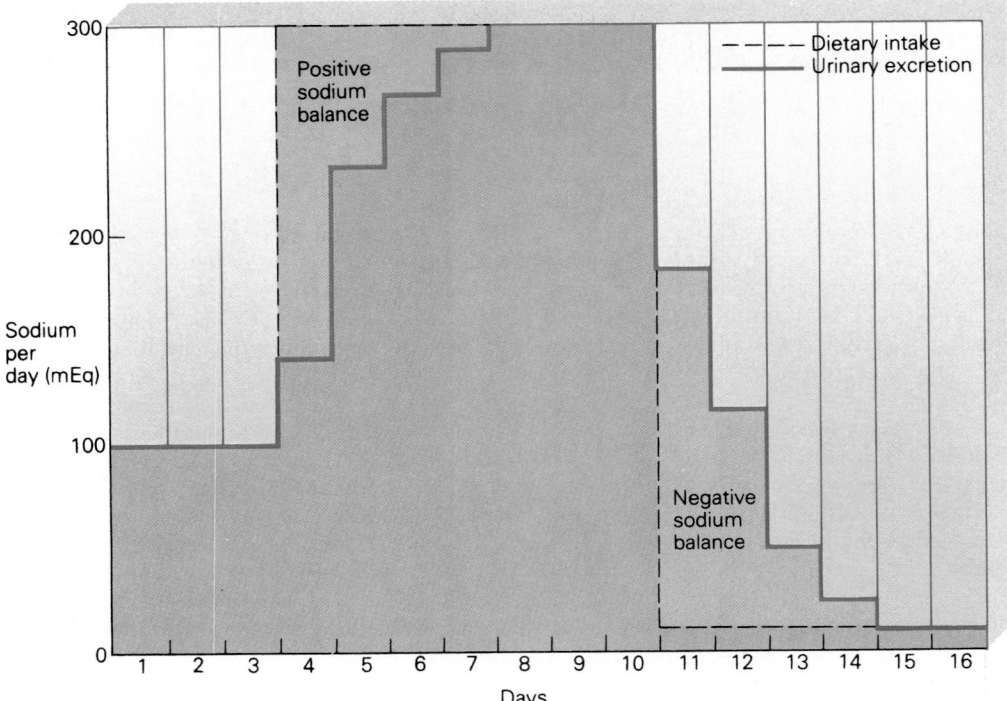

Figure 25–8
A sodium balance study. Dietary intake of sodium was initially 100 mEq per day, was then increased to 300 mEq per day (days 4 to 10), and was finally reduced to 10 mEq per day (days 11 to 16). Daily urinary excretion of sodium (solid blue lines) was measured.

14, sodium excretion was higher than sodium intake, and so a negative sodium balance resulted. During this time, extracellular fluid volume and body weight decreased. Finally, a new stable balance on the low salt diet was established (days 15 and 16).

This study demonstrates that changing the dietary sodium intake leads to appropriate changes in sodium excretion by the kidneys. An increase in sodium intake leads to increased urinary sodium excretion, and a decrease in sodium intake results in decreased sodium excretion. The kidneys can adjust sodium excretion over a wide range. Note, however, that the renal response is sluggish, and so it takes a few days to reach the appropriate rate of renal sodium excretion. During the time that renal sodium excretion is less than sodium intake (days 4 to 7), extracellular fluid volume is increasing. Extracellular fluid volume is stable, but higher than the control volume, during days 8 to 10. The increased extracellular fluid volume may be thought of as the stimulus that promotes renal sodium loss. When the sodium intake is decreased, a fall in extracellular fluid volume may be thought of as the cause of the reduced sodium excretion.

Effector Mechanisms Activated by Altered Extracellular Fluid Volume

What mechanisms allow the kidneys to adjust their output of sodium to maintain balance and relative stability of extracellular fluid volume? Figure 25–9 summarizes how sodium excretion may be controlled by a negative feedback system. According to this scheme, the regulated variable is extracellular fluid volume. Changes in extracellular fluid volume are sensed by cardiovascular volume receptors and by the kidneys. The effector mechanisms include changes in (1) glomerular filtration rate, (2) plasma aldosterone levels, (3) peritubular capillary Starling forces, (4) renal sympathetic nerve activity, (5) intrarenal blood flow distribution, and (6) plasma atrial natriuretic peptide. There may be additional effector mechanisms. Changes in these mechanisms lead to changes in sodium excretion and bring extracellular fluid volume back to normal.

We will first discuss the six effector mechanisms just listed and then the nature of the sensors and what they may actually be sensing. Sodium excretion represents the difference between filtered and reabsorbed amounts, and so factors that affect so-

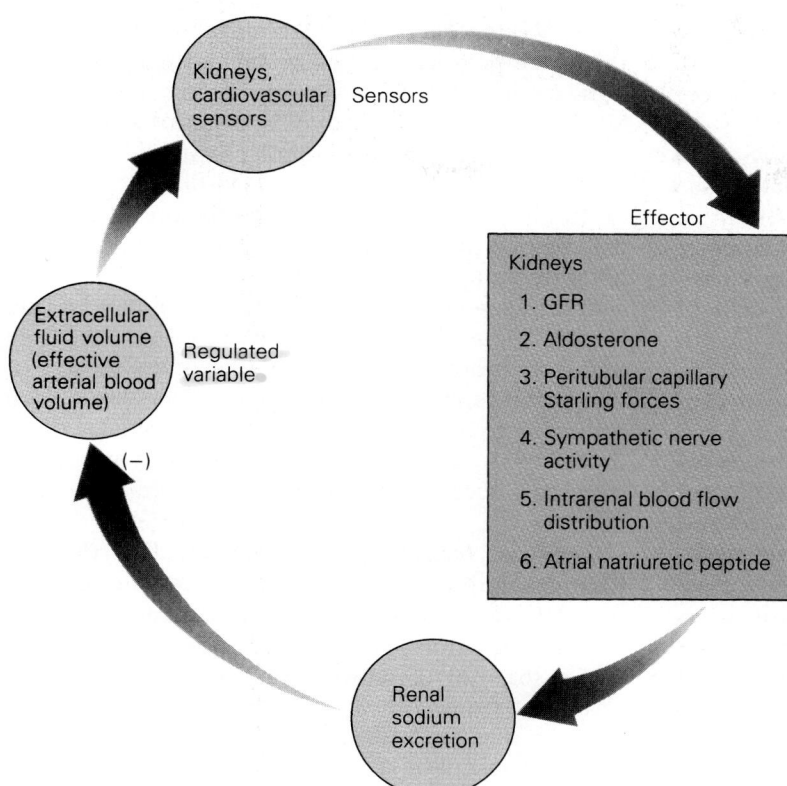

Figure 25–9
Regulation of extracellular fluid volume (or effective arterial blood volume) by a negative feedback control system. Changes in effective arterial blood volume are sensed by cardiovascular and renal sensors, and are translated into changes in renal effector pathways, leading to changes in sodium excretion that bring effective arterial blood volume back to normal.

dium excretion may act on the glomeruli, kidney tubules, or both.

Glomerular Filtration Rate.

Glomerular filtration, as discussed earlier, delivers an enormous amount of sodium to the tubules. Most of this is reabsorbed. Increases or decreases in glomerular filtration rate (GFR) lead to directionally similar changes in urinary sodium excretion. Expansion of the extracellular fluid volume with isotonic saline tends to raise GFR and so may contribute to increased sodium excretion. An increase in GFR actually leads to only a slight increase in sodium excretion, due to a phenomenon called **glomerulotubular balance.** In response to an increase in GFR, the proximal convoluted tubule and loop of Henle increase their rates of sodium reabsorption so that excessive loss of sodium does not result.

Severe hemorrhage is accompanied by a fall in blood pressure and renal vasoconstriction and hence a fall in GFR. The kidney tubules can then reabsorb the reduced filtered sodium load more completely, and sodium excretion therefore decreases. This is a valuable adjustment, because, with hemorrhage, the kidneys ideally should retain

sodium to aid restoration of the depleted extracellular volume.

Aldosterone and the Renin-Angiotensin System.

Aldosterone is the most important mineralocorticoid hormone in human beings. It is a salt-retaining steroid produced by the zona glomerulosa region of the adrenal cortex (Chapter 14). It increases the reabsorption of sodium (and secretion of potassium) by a variety of epithelial tissues, including sweat glands, salivary glands, and intestine. In the kidney, it acts chiefly on the collecting ducts. In the absence of aldosterone (in animals whose adrenal glands have been removed), the collecting ducts fail to lower the sodium concentration in the urine, and an abnormal quantity of sodium is lost. Instead of normally reabsorbing 99.6% of the filtered sodium load, 98.0% of the filtered sodium load may be reabsorbed in the absence of aldosterone. This may seem like a small difference. But remember that if 24,000 mmoles of sodium are filtered in a day, a difference of 1.6% amounts to 384 mmoles sodium, roughly the quantity of sodium in 2.5 liters of extracellular fluid. Maintained excretion of this amount of sodium over a period of days would result in serious (and even fatal) depletion of the extracellular fluid. To some extent, the effects of such salt loss in animals without their adrenal glands can be reversed by a high sodium chloride intake.

The most important pathway controlling aldosterone secretion in response to volume changes is the renin-angiotensin system (Chapter 14). Briefly, a decrease in extracellular fluid volume (or to be more exact, a decrease in effective arterial blood volume) stimulates renin release from the granular cells of the juxtaglomerular apparatus (see Fig. 24–5). Increased renin release may be due to three mechanisms: (1) a decrease in blood pressure at the level of the afferent arteriole (renal baroreceptor mechanism), (2) an increase in activity of renal sympathetic nerves that innervate the granular cells, and (3) a decrease in sodium chloride delivery to and transport by the macula densa cells. All three mechanisms are activated by volume depletion—for example, due to hemorrhage. **Renin,** a proteolytic enzyme, acts on a plasma protein, **angiotensinogen,** to produce **angiotensin I** (Fig. 25–10). Angiotensin I is converted to the active peptide, **angiotensin II,** by an enzyme called **angiotensin converting enzyme,** mainly as blood passes through the lungs. Angiotensin II has several important actions. First, it is a powerful vasoconstrictor and so raises blood pressure. Second, it increases the synthesis and release of aldosterone by a direct action on the adrenal cortex. Aldosterone causes enhanced sodium reabsorption by distal parts of the nephron. Third, angiotensin II acts on the thirst center in the hypothalamus, stimulating increased water intake. Fourth, angiotensin II stimulates ADH release from the posterior pituitary, resulting in renal water retention. In all of these actions, angiotensin II helps restore a normal blood pressure and extracellular fluid volume. In the day-to-day control of sodium balance, changes in aldosterone secretion, mediated by changes in renin release, are probably most important in bringing about changes in sodium excretion. Aldosterone secretion is also stimulated directly by a fall in plasma sodium concentration of the blood perfusing the adrenal cortex.

Peritubular Capillary Starling Forces. In Chapter 24 we said that changes in hydrostatic and colloid osmotic pressures in the peritubular capillaries affect the net reabsorption of salt and water by the proximal convoluted tubule. Specifically, an increase in hydrostatic pressure or a decrease in colloid osmotic pressure reduces fluid uptake. Such a change occurs, for example, when a large volume of isotonic saline is added to the body. The result is an increase in sodium excretion, an appropriate response.

Renal Sympathetic Nerve Activity. Stimulation of renal sympathetic nerves results in a decrease in renal sodium excretion. If stimulation is intense, the fall in sodium excretion may be explained largely by hemodynamic changes—that is, by a fall in glomerular filtration rate and renal blood flow. Low levels of renal sympathetic nerve stimulation, at intensities below those needed to produce measurable hemodynamic changes, also result in decreased sodium excretion. This latter response may be due to direct stimulation of tubular cells, so that they reabsorb more sodium. There is anatomical evidence that renal sympathetic nerve fibers lie in close proximity to tubular cells. Inhibition of renal sympathetic nerve activity—for example, by volume expansion—accordingly promotes sodium excretion.

Change in Intrarenal Blood Flow Distribution. Changes in filtration rate and blood flow in different populations of nephrons could conceivably alter sodium excretion. This idea is consistent with the notion that not all nephrons in the kidney are alike, so a shift from cortical to juxtamedullary nephron populations could result in different sodium excretion. At present, it is not certain whether this mechanism is important.

Atrial Natriuretic Peptide. Recent studies indicate that the atria of the heart produce a peptide called **atrial natriuretic peptide (ANP).** ANP is released by the atria in response to mechanical stretch—for example, by an increase in blood volume—and it increases sodium excretion.

Clearly, multiple effector mechanisms influence sodium excretion. The relative importance of these mechanisms under normal conditions and in various disease states is not fully understood.

The Nature of the Stimuli and the Location of Receptors Affecting Sodium Excretion

What exactly is the sensed variable that leads to changes in renal sodium excretion? We have said that changes in extracellular fluid volume cause appropriate changes in renal sodium excretion. The idea that extracellular fluid volume is the regulated variable seems a satisfactory explanation in a normal person. There are, however, a number of disease states in which generalized edema develops because the kidneys retain salt and water. Clearly, sodium is conserved despite an expanded extracellular fluid volume. The argument has been presented that it is not extracellular fluid volume per se, but a component thereof, the plasma volume, that is regulated. But this argument is not convincing, because blood volume is commonly increased in patients with con-

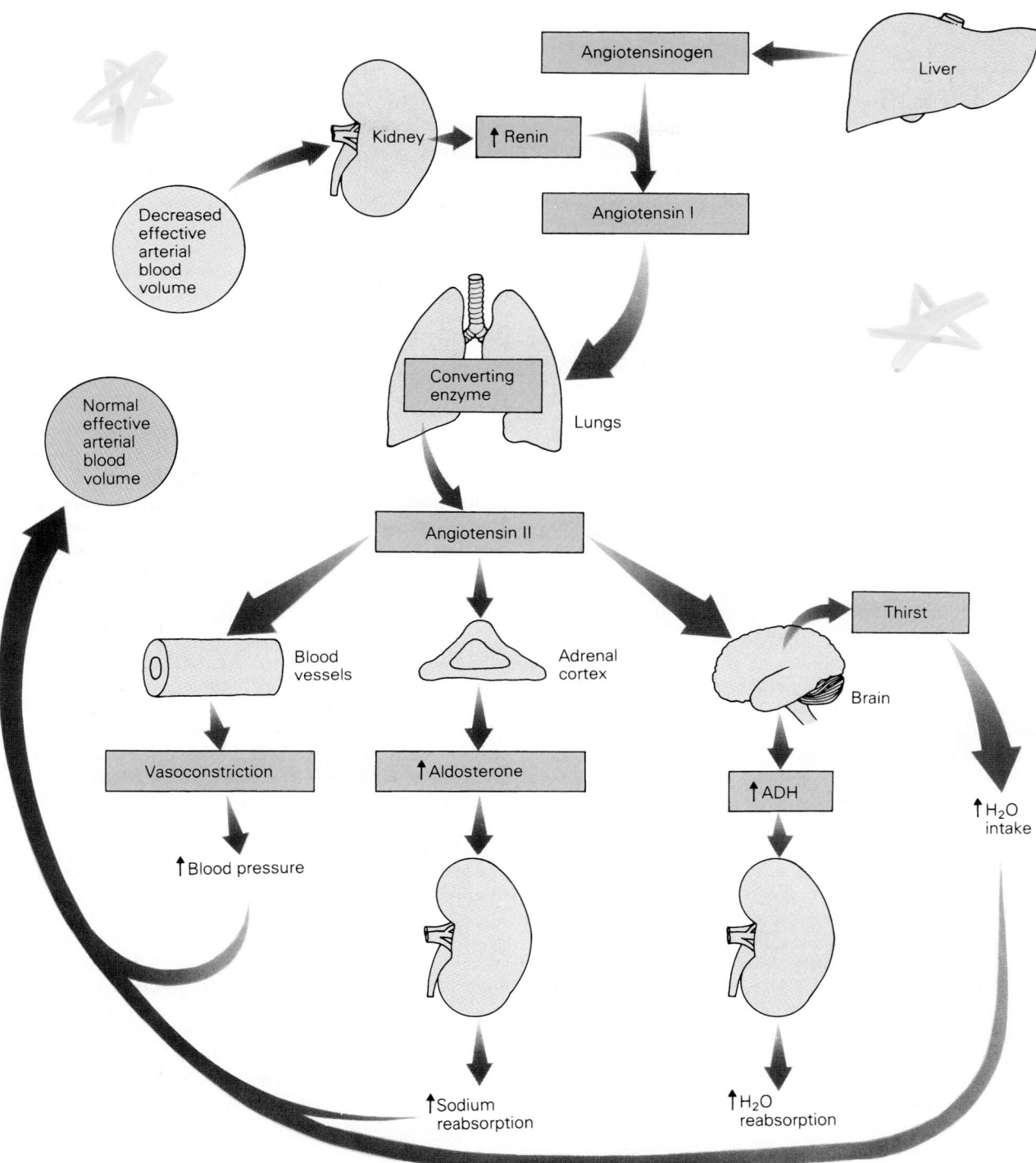

Figure 25–10
The renin-angiotensin system helps maintain normal blood pressure and extracellular fluid volume.

gestive heart failure whose kidneys avidly retain sodium.

To explain these difficulties, researchers have postulated that the **effective arterial blood volume** is the regulated variable that affects sodium excretion. The effective arterial blood volume is an index of the degree of filling of the systemic arterial tree. It determines whether blood flow will be adequate to meet the metabolic demands of the tissues. In congestive heart failure, the heart muscle is weak and does not adequately transfer blood from the venous to the arterial side of the circulation. Blood tends to

Atrial Natriuretic Peptide: A Hormone from the Heart

One of the most exciting recent developments in medicine has been the discovery of a new hormone, atrial natriuretic peptide (ANP). This hormone, which increases sodium excretion and lowers blood pressure, is produced by the cardiac atria. It now appears that the heart, in addition to pumping the blood, is an endocrine organ. The hormone it produces has important effects on the kidneys and vascular system, and thereby influences cardiac function.

The original observations were made by a group of investigators in Canada, led by Adolpho de Bold. In 1981, they reported that intravenous injection of an extract of rat atria into anesthetized rats produced a dramatic increase in renal excretion of sodium and water. Extracts of rat ventricle were without effect. Injection of crude tissue extracts is a classic method that has led to the discovery of a number of hormones.

Within two years of this discovery, the peptide sequence of the natriuretic factor was determined. The circulating hormone contains 28 amino acids. In a short time, the peptide was chemically synthesized, and rat, mouse, and human genes for ANP were cloned and their nucleotide sequences determined. The extreme rapidity of these developments made this hormone available for study all over the world. Because many different groups had simultaneously isolated it, a variety of names were applied to ANP, including atrial natriuretic factor (ANF), atrin, atriopeptin, auriculin, and cardionatrin.

ANP is stored in secretory-like granules within ordinary atrial muscle cells. These granules were discovered more than thirty years ago; however, their function was not appreciated until recently. The number of atrial granules changes in response to changes in body fluid balance; this finding supports the idea that ANP production or secretion is related to fluid balance.

The main stimulus causing ANP release into the circulation is stretching of the cardiac atria. So, for example, if there is too much blood in the thorax, and the atria are stretched, ANP is released into the circulation. It acts on the kidneys to increase excretion of salt and water, and on arterioles, which it dilates. The hormone is part of a negative feedback control system, and helps to restore blood volume and blood pressure to normal.

ANP has several actions, all consistent with its overall effect of promoting sodium excretion and lowering blood pressure. It increases renal blood flow and glomerular filtration rate. It also inhibits sodium reabsorption by inner medullary collecting ducts. It inhibits the release of aldosterone by acting directly on the adrenal cortex and by inhibiting renin release. It relaxes arteriolar smooth muscle, thereby causing vasodilation and a fall in blood pressure.

The role of ANP in the normal control of renal sodium excretion and in a variety of pathophysiological states is being intensively investigated. Simply eating a bag of lightly salted potato chips leads to a rise in plasma ANP levels, which suggests that plasma levels change appropriately in response to small forcings. Intravenous infusion of very low doses of ANP into people produces a sustained increase in sodium excretion and a fall in blood pressure. These observations support a physiological role for this hormone. In congestive heart failure, plasma ANP levels may be markedly increased; this suggests that salt retention in this disorder is not due to deficient ANP secretion. Deficiency of ANP may explain some cases of hypertension. ANP, an ANP analogue, or a drug that inhibits the breakdown of ANP might be useful in treating congestive heart failure and other edema-forming states, or hypertension.

A problem in elucidating the role of ANP is that it is only one of many factors that affect renal sodium excretion. There is growing evidence that several other hormones, still incompletely characterized, also affect sodium excretion.

"dam up" in back of a failing ventricle, which decreases effective arterial blood volume. Effective arterial blood volume is also decreased in other conditions in which generalized edema occurs, such as liver cirrhosis, malnutrition, arteriovenous fistula, and nephrotic syndrome. Effective arterial blood volume or extracellular fluid volume is decreased with hemorrhage, diarrhea, severe sweating, and low salt intake. Therefore, all of these conditions are accompanied by renal salt retention.

Where are the volume receptors? There is convincing evidence that various cardiovascular stretch receptors can affect sodium excretion. These include stretch receptors in the wall of the left atrium of the heart and in intrathoracic veins. An increase in volume of blood in the thorax is a well known stimulus that increases sodium excretion. The arterial baroreceptors may also function as stretch receptors affecting sodium excretion.

The kidneys may act as volume sensors. When effective arterial blood volume is decreased, renal blood flow is threatened. The kidneys act to retain salt and water in an attempt to improve their blood flow. How might changes in effective arterial blood volume be translated into the appropriate renal response? Three possible pathways may be involved. In the first one, increases or decreases in effective arterial blood volume lead to increases or decreases, respectively, in glomerular filtration rate, due to changes in glomerular blood flow and capillary pressure. Directionally similar changes in sodium excretion ensue. In the second, there is an intrarenal baroreceptor in the afferent arteriole. A decrease in pressure at this site causes increased renin release and ultimately increased sodium reabsorption mediated by aldosterone. Finally, in the third, changes in peritubular capillary Starling forces could result directly from changes in extracellular fluid volume and once again would lead to appropriate changes in sodium excretion.

Re-examination of Figure 25–9 will help to review some of the foregoing points. When effective arterial blood volume is decreased (e.g., after hemorrhage), renal sodium excretion is decreased. This makes sense, since renal conservation of sodium aids in restoring effective arterial blood volume to normal. The reduction in renal sodium excretion may be due to a reduced glomerular filtration rate, increased plasma aldosterone level, decreased peritubular capillary hydrostatic pressure, increased renal sympathetic nerve activity, changes in intrarenal blood flow distribution, and decreased plasma levels of the atrial natriuretic peptide. When effective arterial blood volume is increased (e.g., because of increased dietary sodium intake), sodium excretion is enhanced because of an increase in glomerular filtration rate (GFR), decreased plasma aldosterone level, increased hydrostatic pressure and decreased colloid osmotic pressure in the peritubular capillaries, reduced sympathetic nerve activity, changes in intrarenal blood flow distribution, and increased plasma levels of the atrial natriuretic peptide. The relative importance of the various effector mechanisms varies with the circumstances and is not always known. Since regulation of extracellular fluid volume or effective arterial blood volume is so important, it is not surprising that multiple mechanisms, which interact and often overlap, control renal sodium excretion.

Other Factors That Influence Sodium Excretion

A number of other factors are known to affect sodium excretion. These include (1) glucocorticoids, (2) estrogens, (3) osmotic diuretics, (4) poorly reabsorbed anions, and (5) diuretic drugs. You should learn how all of these affect sodium excretion—that is, whether they increase or decrease sodium excretion.

Glucocorticoids. The glucocorticoids (e.g., cortisol) are steroid hormones produced by the adrenal cortex (Chapter 14). They increase tubular sodium reabsorption but also cause an increase in GFR, which may mask the tubular effect. In general, glucocorticoids cause a decrease in sodium excretion.

Estrogens. Estrogens may increase sodium reabsorption by the renal tubules. High plasma levels of estrogens, together with numerous other hormonal and hemodynamic changes, contribute to sodium retention during pregnancy.

Osmotic Diuretics. Osmotic diuretics are solutes excreted in the urine that obligate the excretion of water. Urine flow rate, sodium excretion, and potassium excretion are all increased.

Clinically important osmotic diuretics include (1) mannitol, (2) glucose, and (3) urea. Mannitol is a six-carbon sugar alcohol filtered by the glomerulus but not reabsorbed. Increased excretion of mannitol leads to increased excretion of salt and water. A concentrated mannitol solution is sometimes given intravenously to promote urine output. In uncontrolled diabetes mellitus, more glucose is filtered than can be reabsorbed by the tubules. Again, increased excretion of solute (glucose) leads to enhanced excretion of salt and water. The loss of salt leads to a fall in extracellular fluid volume and so stimulates thirst.

Urea excretion also promotes renal excretion of salt. This fact may be important in sustaining sodium excretion in chronic kidney disease.

Poorly Reabsorbed Anions. Poorly reabsorbed anions result in increased sodium excretion. Because of the requirement for electroneutrality in solutions, whenever more anion is excreted, more cation must also be excreted. So if there is increased excretion of phosphate, bicarbonate, ketone body acids (as occurs in uncontrolled diabetes mellitus), or sulfate, sodium excretion will be increased. To some extent, sodium can be replaced by other cations such as potassium, ammonium, and hydrogen ions.

Diuretic Drugs.   Diuretic drugs inhibit tubular reabsorption of sodium. The increase in urine volume that ensues is basically due to impaired sodium reabsorption. Remember that tubular reabsorption of water depends on sodium reabsorption.

There are probably additional factors that can affect sodium excretion. Some investigators offer evidence of a circulating natriuretic factor distinct from the atrial natriuretic peptide. Various substances formed within the kidneys (e.g., prostaglandins, kinins) also affect sodium excretion.

The Regulation of Potassium Balance

Most of the body's potassium is found within cells. The concentration of potassium in cells is typically about 150 mEq per liter of cell water. In extracellular fluid, the potassium concentration averages about 4.5 mEq per liter.

Potassium plays a number of important roles in the body. First, it influences the electrical excitability of cells. As discussed in Chapter 7, the resting membrane potential of cells is largely determined by the unequal distribution of potassium inside and outside of cells and by a high membrane permeability to potassium ions. Disturbances in potassium balance lead to altered excitability of nerves and muscles.

Second, potassium is the major osmotically active solute in cells, and so the amount of potassium in cells affects cell volume. We are already familiar with the idea that the amount of sodium influences the volume of the extracellular compartment. Loss of potassium from cells leads to a decrease in cell volume.

Third, potassium balance is intimately related to acid-base balance. The relationships are complex, but there is abundant evidence to suggest that disturbances in acid-base status can affect potassium balance, and vice versa.

Finally, intracellular potassium affects cell metabolism. Tissue growth and repair require potassium. Conversely, tissue breakdown or increased protein catabolism leads to release of cell potassium into the extracellular fluid.

The normal range for plasma potassium concentration is 3.5 to 5.5 mEq per liter. A plasma potassium concentration below 3.5 mEq per liter is called **hypokalemia.** This condition may cause skeletal muscle weakness and even paralysis. A plasma potassium concentration above 5.5 mEq per liter is called **hyperkalemia.** Plasma potassium concentrations above 7 mEq per liter demand immediate attention because of possible cardiac irregularities. A

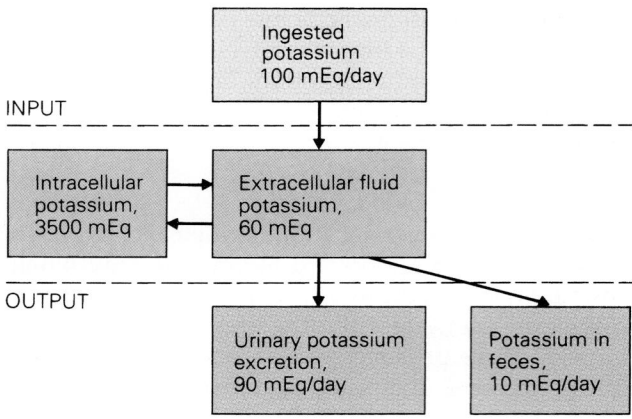

Figure 25–11
Potassium balance (input-output) and the distribution of potassium between the two major body fluid compartments.

plasma potassium concentration greater than 10 to 12 mEq per liter is usually fatal; the cause of death is cardiac arrhythmia or arrest. The changes in skeletal and cardiac muscle function that occur with abnormal plasma potassium levels result from the influence of potassium on cell membrane potentials.

Figure 25–11 summarizes potassium balance. Normally, we ingest about 100 mEq per day of potassium. The amount we take in can be quite variable, depending on what we eat and drink. About 10% of the daily potassium intake is excreted in the feces, and about 90% in the urine. Loss of potassium in sweat (which contains 5 to 10 mEq potassium per liter) is usually negligible. The kidneys are clearly the major site of potassium loss from the body, and they are the major site of control of potassium balance.

Figure 25–11 also re-emphasizes that most of the body's potassium is inside cells. The amount of potassium in extracellular fluid can be calculated from its concentration (4.5 mEq per liter) and the extracellular fluid volume (about 14 liters), and is equal to about 60 mEq. This amounts to only 2% of the body's potassium.

Several factors affect the distribution of potassium between cells and extracellular fluid. First, activity of the cell membrane Na^+/K^+-ATPase, which pumps potassium into cells, is of key importance. Impaired cell metabolism or cardiac glycosides (e.g., digitalis) inhibit the pump, and tend to produce lower cell and higher extracellular potassium levels. Second, acid-base status affects the distribution of potassium between intracellular and extracellular compartments. A fall in plasma pH causes a shift of hydrogen ions into cells in exchange for cell potassium ions. The consequence is a rise of plasma potassium concentration. Third, the availability of insulin is an important factor affecting potassium distribution. Insulin promotes the movement of potassium into skeletal muscle and liver cells; lack of insulin causes more of the body's potassium to stay outside of cells. Fourth, shifts of potassium from cells to extracellular fluid occur with cell breakdown, due to tissue trauma, infection, ischemia (inadequate blood flow), or heavy exercise. Since so much of the body's potassium is within cells, a small loss from or gain by this compartment could profoundly affect the plasma potassium concentration.

Plasma potassium must be closely regulated. Part of this regulation involves the hormones insulin, epinephrine, and aldosterone, which stimulate the uptake of potassium by cells. The kidneys also play an important part in regulating plasma potassium. Normally, when increased potassium loads are presented to the body, the kidneys can rapidly excrete excess potassium.

In Chapter 24, we discussed some of the factors that affect potassium excretion by the kidneys. Re-call that most of the filtered potassium is reabsorbed by early portions of the nephron and that the collecting ducts secrete potassium. Some of the factors that affect potassium excretion include intracellular potassium concentration, aldosterone, excretion of anions, and urine flow rate.

Figure 25–12 shows the mechanisms whereby an increase in potassium intake in the diet leads to increased renal potassium excretion. Increased potassium intake tends to raise the plasma potassium concentration. By a direct effect on the adrenal cortex, an elevated extracellular potassium concentration stimulates the release of aldosterone. Aldosterone then travels in the blood to the kidneys, where it acts on the principal cells of the collecting duct. It increases activity of the Na^+/K^+-ATPase in the basolateral cell membrane and potassium permeability of the luminal membrane of these cells, thereby favoring potassium secretion (see Figure 24–17). An increased potassium intake raises the potassium concentration in most body cells, including the principal cells, and so potassium secretion is directly increased in this way.

The opposite changes occur if potassium intake is reduced. Plasma and cell potassium levels tend to

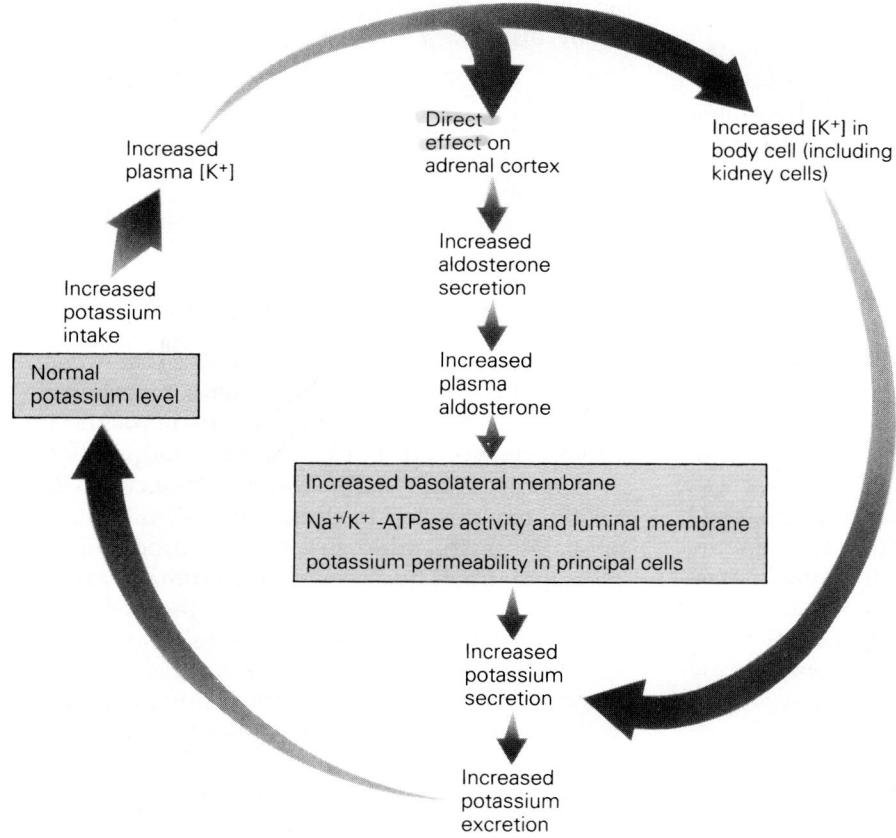

Figure 25–12
Pathways by which an increase in dietary potassium intake leads to increased renal potassium excretion.

fall, aldosterone secretion is inhibited, and secretion of potassium by the collecting duct is diminished. The result is a decreased rate of potassium excretion.

The kidneys normally do a remarkable job of maintaining potassium balance. But abnormal renal excretion of potassium (either too much or too little) is a common cause of disturbed potassium balance.

Too little renal excretion of potassium causes hyperkalemia. Most foods are rich in potassium, so continued food intake with inadequate renal excretion leads to a positive potassium balance. In acute renal failure—that is, with sudden renal shutdown—a life-threatening hyperkalemia may develop. Often compounding inadequate renal excretion (in patients with acute renal failure) are tissue trauma, infection, and acidosis, all of which tend to raise the plasma potassium level. Chronic renal disease leads to hyperkalemia when the filtration rate falls below 15 to 20 ml/min. Up to that point, hyperkalemia does not develop because each surviving nephron can excrete a greater-than-normal quota of potassium. In adrenocortical insufficiency (Addison's disease), secretion of aldosterone is inadequate. A lowered plasma aldosterone level results in decreased stimulation of renal potassium secretion and excretion, so that ingested potassium tends to be retained in the body.

Too much excretion of potassium by the kidneys leads to hypokalemia. With adrenocortical hyperfunction, aldosterone levels are abnormally high, and potassium secretion is stimulated. Treatment with glucocorticoids can also result in excessive potassium excretion. In a number of kidney diseases, the ability of the tubules to reabsorb filtered potassium is impaired, and excessive potassium excretion results. Excessive renal potassium loss occurs in uncontrolled diabetes mellitus, for several reasons: (1) an osmotic diuretic effect of glucose (increased urine output produces potassium loss), (2) increased production of ketone body acids (excretion of anions results in increased excretion of cations), and (3) increased plasma aldosterone levels (secondary to volume contraction caused by excessive loss of salt and water in the urine). Diuretic drugs (used in the treatment of hypertension and edema) are the most common cause of abnormally high renal potassium loss. This effect is probably due mostly to the increased rate of fluid flow through the collecting ducts that these agents produce. Patients taking diuretics are often advised to eat bananas, a rich source of potassium.

Excessive amounts of potassium can also be lost from the body through the gastrointestinal tract. Diarrheal fluid may have a potassium concentration of 50 to 60 mEq per liter, so a disturbance of the lower gastrointestinal tract may rapidly lead to potassium depletion. Potassium depletion also develops during vomiting, but the reasons are complex. First, gastric juice contains some potassium (about 5 to 10 mEq per liter), which accounts for some of the loss. Second, food (potassium) intake is reduced. Finally, the kidneys actually increase potassium excretion. This is due in part to the metabolic alkalosis that develops. It is also due to volume depletion, which stimulates aldosterone secretion via renin and angiotensin II. This promotes sodium conservation by the kidneys but can lead to enhanced potassium excretion.

Kidney Failure and Its Treatment

Kidney failure can occur at any age. It may develop suddenly and lead to death in a few days, or it may be due to a chronic renal disease that progresses slowly over decades. In chronic kidney disease, there is a progressive loss of nephrons. Kidney disease has a variety of causes, among which are bacterial infections, congenital defects, immunological injury, impaired renal blood flow, toxins, tumors, and urinary tract obstruction. The specific type of kidney disease can usually be diagnosed from the patient's history and clinical tests. In all forms of renal failure, however, the signs and symptoms are very similar. This symptom complex has been called **uremia** (urine in the blood). Basically, uremia indicates a disturbed internal environment, due to renal failure, that impairs the function of all body cells.

Disturbances in Renal Failure

Following are some of the disturbances that occur in uremia. This alphabetical list is not comprehensive but should serve to re-emphasize many of the important functions of normal, healthy kidneys.

Anemia, with a fall in blood hematocrit to 20% to 35%, may develop in chronic kidney failure. The anemia is mainly due to deficient production of **erythropoietin,** a hormone produced mainly by the kidneys. Erythropoietin stimulates the bone marrow to produce red blood cells (erythropoiesis). Human erythropoietin has been synthesized in the laboratory by recombinant DNA techniques. When administered to renal failure patients, it cures the anemia.

Azotemia, the accumulation of nitrogenous waste products in the blood, develops as a consequence of a markedly decreased GFR. Plasma urea, creatinine, and uric acid levels are elevated.

Calcium and phosphate metabolism are altered with renal failure. In uremia, plasma calcium is typically decreased, and plasma phosphate is increased. Secondary hyperparathyroidism, deficient synthesis of the active form of vitamin D by the kidneys, and acidosis all lead to bone disease.

Hypertension (high blood pressure) is often present in renal failure. Its origin is not always known, but there is evidence for several possible pathways. The diseased kidney may produce excessive amounts of renin. Renin causes increased production of angiotensin II (a potent vasoconstrictor) and aldosterone (which favors salt retention). The inability of the kidneys to excrete a normal amount of sodium could also lead to hypertension, owing to excess volume. Finally, the diseased kidneys may fail to produce vasodilator substances.

Metabolic acidosis typically accompanies severe renal failure. Acid end products of metabolism are not excreted at a normal rate and hence accumulate in the body. Impaired renal synthesis of ammonia contributes to the decreased ability to excrete hydrogen ions.

Potassium balance is usually well maintained until GFR falls below 15 to 20 ml/min. Retention of potassium can cause plasma levels to rise to lethal levels.

Sodium balance is typically impaired in patients with severe renal failure. On a normal sodium diet, excessive salt and water may be retained, leading to generalized edema. On a low-salt diet, patients may become salt-depleted.

Urinary concentrating ability is impaired with renal failure. Urine with an osmolality close to that of plasma is excreted.

Numerous other cardiovascular, gastrointestinal, muscular, and nervous abnormalities occur in the uremic state. In many instances, it is not understood why renal failure causes these disturbances.

Treatment by Dialysis

Most of the signs and symptoms of uremia can be relieved by dialysis. **Dialysis** is the separation of smaller from larger molecules in solution by diffusion of the small molecules through a selectively permeable membrane. In treating uremia, two methods of dialysis may be used.

In **peritoneal dialysis,** about 2 liters of a balanced salt solution are introduced into the abdominal cavity. The peritoneum acts as a dialyzing membrane. Small molecules and ions (e.g., urea,

potassium) diffuse into the injected solution, which is then drained and discarded. This procedure is used by some patients with chronic renal failure on an outpatient basis and is called **continuous ambulatory peritoneal dialysis.** The patient can perform self-dialysis and usually does this several times in a day.

Hemodialysis (Fig. 25–13) is a more efficient process involving an **artificial kidney.** The patient's blood is pumped through cellophanelike tubing immersed in a bath of balanced salt solution. Urea, potassium, and other substances in excess in the body diffuse through the tubing membrane out of the blood into the bath. Patients are usually dialyzed three times a week, and they can often undergo this procedure at home for 4 to 6 hours while watching television or reading.

Dialysis can allow patients with otherwise fatal chronic renal failure to live useful and productive lives. It has some drawbacks, however. It does not correct the anemia. It controls hypertension only with difficulty. Uremia develops between the periods of dialysis. There is a constant risk of infection and, with hemodialysis, hemorrhage. Dialysis does not maintain normal growth and development in children. Finally, it is costly. Dialysis is, however, invaluable in keeping patients alive and functioning until a suitable kidney transplant become available.

Transplantation of the Kidney

Renal transplantation is the only ''cure'' for patients with terminal chronic renal disease. It may restore complete health and function. In 1989, almost 9000 renal transplant operations were performed in the United States. At present, about 90% of kidneys grafted from a living donor related to the patient function for one year; grafts derived from an unrelated person who has just died (cadaver donors) have a one-year survival of about 80%.

The main problem in transplantation is the immunological rejection of the transplanted kidney by the host. A better understanding of immunology should lead to greater success with organ transplantation.

Functions of Ureters and Urinary Bladder

The kidneys form urine constantly. Urine is conveyed to the urinary bladder by the ureters and stored there until the bladder is emptied by way of the urethra.

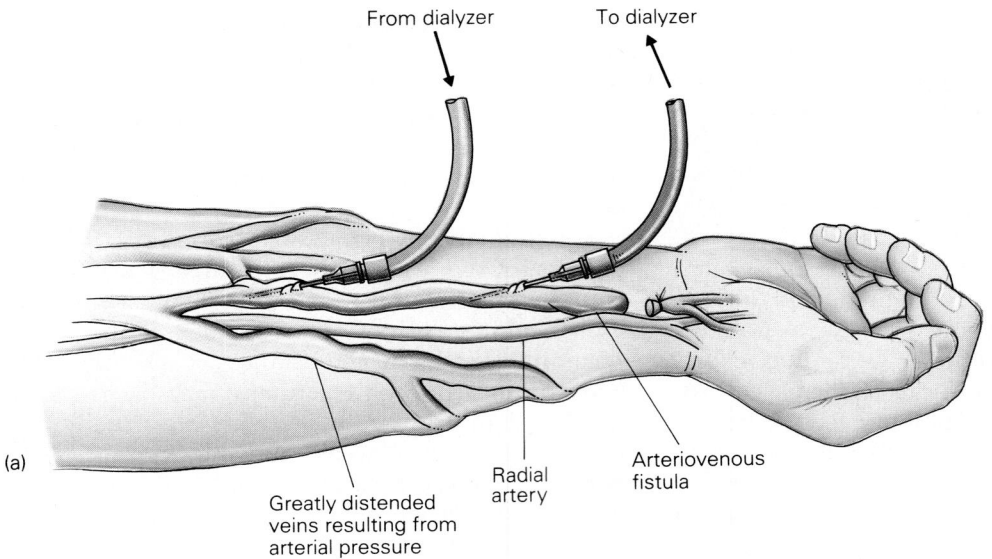

(a)

From dialyzer

To dialyzer

Greatly distended
veins resulting from
arterial pressure

Radial
artery

Arteriovenous
fistula

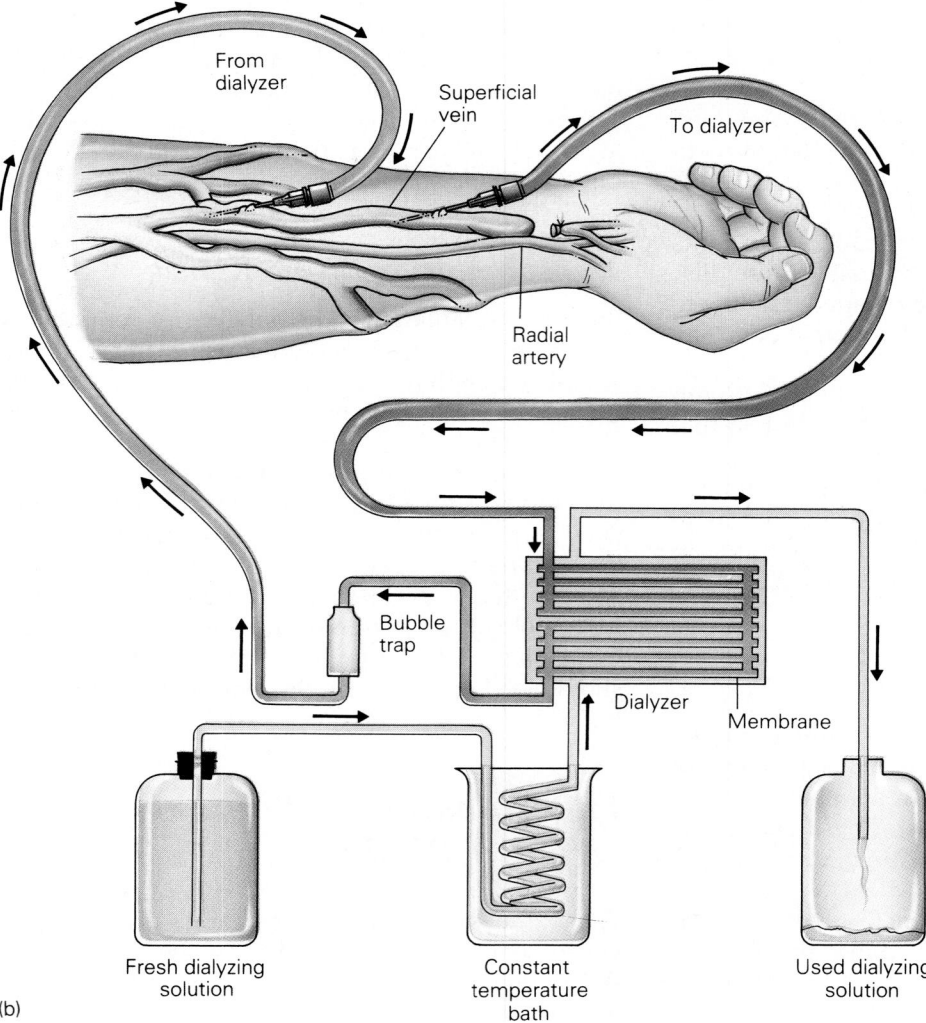

(b)

From
dialyzer

Superficial
vein

To dialyzer

Radial
artery

Bubble
trap

Dialyzer

Membrane

Fresh dialyzing
solution

Constant
temperature
bath

Used dialyzing
solution

Figure 25–13
Arrangement for hemodialysis. (*a*) An arteriovenous fistula is created between the
radial artery and a nearby
superficial wrist vein. The
vein is punctured with nee-
dles, which allows blood to
be carried to and from the
dialyzer. (*b*) In the dialyzer
("artificial kidney") the blood
is separated from dialysis
fluid by a thin membrane.
Urea, other waste products,
and potassium ions diffuse
out of the blood into the dialysis fluid. The blood may
also be ultrafiltered, to remove excess fluid, by applying a pressure gradient across
the membrane.

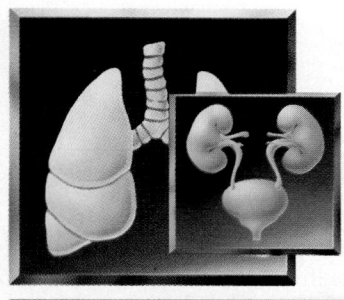

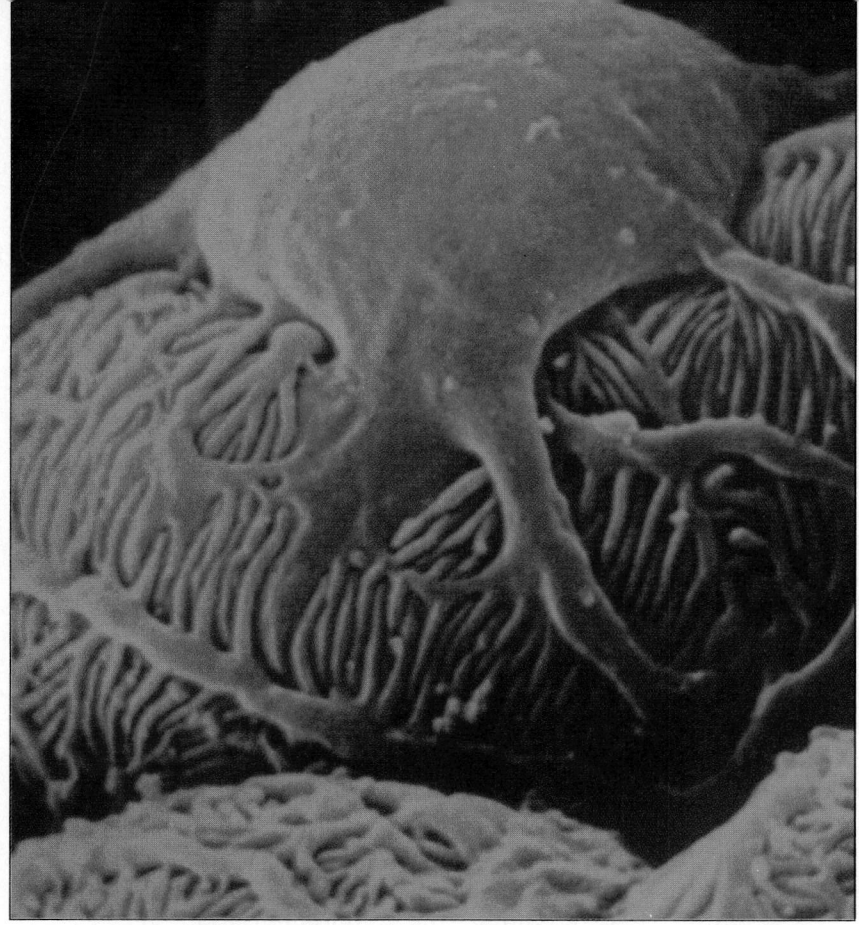

Regulation of Acid-Base Balance

CHAPTER OUTLINE

The **hydrogen ion concentration** $[H^+]$ of the extracellular fluid is one of the most important of all homeostatic states. Many abnormalities in the functioning of cells, tissues, and organs result if the extracellular fluid is too acidic or too alkaline (basic). This fact is not surprising, since hydrogen ion concentration affects the electrical charge and hence the configuration and binding properties of proteins. Hydrogen ion concentration, typically expressed in terms of pH, influences (1) the speed of chemical reactions in the body (catalyzed by protein enzymes), (2) cell permeability properties, and (3) cell structure.

This chapter will focus on the regulation of the pH of the extracellular fluid. It is obviously important for a multicellular organism to provide a suitable "internal environment," a controlled environment for its cells. As you know, the extracellular fluid is this internal environment, and pH is one of the most important conditions in that environment.

In physiology and clinical medicine, **systemic arterial blood** is used to evaluate acid-base status. The pH of whole blood, measured with a pH meter, is actually the pH of the plasma (not the cells) and so is a measurement of extracellular fluid pH.

In terms of the functioning of individual cells, the pH within cells is probably even more important than the extracellular pH. Usually, though not always, pH within cells and pH outside of cells change in the same direction. By maintaining a normal extracellular pH, the body provides for some stability of intracellular pH.

We will first review some fundamental definitions of acid-base chemistry, expanding on the introduction provided in Chapter 2. Then we will consider how changes in hydrogen ion concentration are minimized by buffers in the body. These buffers include chemical buffers in extracellular and intracellular fluids and in bone, and the physiological buffers, the lungs and kidneys. We will discuss briefly how the lungs participate in acid-base regulation by controlling the partial pressure of CO_2 (Pco_2) in arterial blood. Then we will consider how the kidneys eliminate extra acid or base from the body. We will describe the four fundamental types of acid-base disturbances, their characteristics and causes, and ways in which the body attempts to compensate for disturbances of pH. Finally, we will comment briefly on regulation of intracellular pH in a special Focus Box.

Principles of Acid-Base Physiology

Definition of Acid and Base

As first presented in Chapter 2, **an acid** is a substance that can donate hydrogen ions. A hydrogen ion (H^+) consists of a bare proton, a hydrogen atom without its orbiting electron. Examples of acids include **hydrochloric acid (HCl), sulfuric acid (H_2SO_4), nitric acid (HNO_3), phosphoric acid (H_3PO_4), ammonium ion (NH_4^+), lactic acid, acetic acid,** and **carbonic acid (H_2CO_3).** An acid will donate its hydrogen ion to a base.

Accordingly, a **base** is a substance that can accept or bind hydrogen ions. Examples of bases include **sodium hydroxide (NaOH), potassium hydroxide (KOH), ammonia (NH_3),** and **lactate, acetate,** and **bicarbonate (HCO_3^-)** ions.

Not all acids have the word *acid* in their names. For example, the ammonium ion (NH_4^+) is an acid. If we mix aqueous solutions of the salts ammonium chloride (NH_4Cl) and sodium bicarbonate ($NaHCO_3$), the following reversible reaction takes place:

$$NH_4^+ \quad Cl^- + Na^+ \quad HCO_3^- \quad \rightleftharpoons$$

$$NH_3 \quad + \quad H_2CO_3 \quad + Na^+Cl^-$$

In this reaction (left to right), the ammonium ion is an acid; it donates a hydrogen ion to the base bicarbonate.

The Neutralization Reaction

In a **neutralization reaction,** an acid reacts with a base to form a salt and water. For example, in the reaction

$$HCl \quad + \quad NaOH \quad \longrightarrow NaCl + H_2O$$

the acid HCl donates its hydrogen ion to the hydroxide ion of NaOH to form water and sodium chloride (a salt).

Quantifying Acids and Bases: Milliequivalents

The amount of an acid or base is often expressed in terms of **equivalents** or **milliequivalents (mEq)** (1 mEq is 1/1000 of an equivalent). One equivalent of acid will neutralize one equivalent of base. One mole of HCl contains one equivalent of acid. One mole of H_2SO_4 contains two equivalents of acid because sulfuric acid can donate two hydrogen ions. Therefore, two moles (two equivalents) of NaOH are required to neutralize one mole (two equivalents) of sulfuric acid, according to the following two steps:

$$H_2SO_4 + NaOH \longrightarrow NaHSO_4 + H_2O$$

$$NaHSO_4 + NaOH \longrightarrow Na_2SO_4 + H_2O$$

Amphoteric Substances

Some chemical substances can function as both an acid and a base and are referred to as **amphoteric substances** (*amphoteros* in Greek means "pertaining to both"). These include amino acids and proteins. For example, the amino acid **glycine**, $^+H_3N—CH_2—COO^-$, acts as an acid in the following reaction:

$$NaOH + {}^+H_3N—CH_2—COO^- \longrightarrow$$

$$H_2N—CH_2—COO^-Na^+ + H_2O$$

The ammonium group of glycine donates a hydrogen ion to the hydroxide ion of NaOH. Glycine acts as a base in the following reaction:

$$HCl + {}^+H_3N—CH_2—COO^- \longrightarrow$$

$$Cl^-{}^+H_3N—CH_2—COOH$$

The carboxylate group of glycine accepts a hydrogen ion from HCl. Proteins contain many different side groups that can act as acids or bases.

The Acid Dissociation Constant

When an acid (generically indicated as **HA**) is added to water, the following reversible reaction takes place:

$$HA + H_2O \rightleftharpoons H_3O^+ + A^-$$

The species H_3O^+ is a hydrated hydrogen ion and is called the **hydronium ion.** For simplicity, it is common practice to ignore water in this reaction and to write

$$HA \rightleftharpoons H^+ + A^-$$

The reaction going from left to right is called the **dissociation reaction,** and the reaction going from right to left is the **association reaction.** The rate of the dissociation reaction is equal to the product of the concentration of HA and the **dissociation rate constant k_1,** a specific value for this reaction. The rate of the association reaction is equal to the product of the concentrations of H^+ and A^- and the **association rate constant k_2,** a specific value for this reaction. At equilibrium, the rates of dissociation and association reactions are equal. We can therefore write:

$$k_1 \times [HA] = k_2 \times [H^+] \times [A^-]$$

Rearranging, we get $[H^+] \times [A^-]/[HA] = k_1/k_2$. We define a new constant K_a as equal to this ratio k_1/k_2. K_a is the **equilibrium constant** for this reaction and is called the **dissociation constant** or **ionization constant** for the acid.

The higher the acid dissociation constant, the more completely an acid is dissociated or ionized and, consequently, the more acidic the solution, because more free (unbound) hydrogen ions are present. Acids with high dissociation constants are **strong acids.** These include hydrochloric, sulfuric, nitric, and phosphoric acids. Hydrochloric acid, for example, is essentially completely dissociated into hydrogen ions and chloride ions in dilute aqueous solution. There is very little undissociated HCl. In a 0.1 M HCl solution, the free hydrogen ion concentration is nearly 0.1 M.

A low acid dissociation constant is characteristic of **weak acids.** In other words, a weak acid is incompletely dissociated in solution. For example, the acid dissociation constant of acetic acid is equal to 1.8×10^{-5}. If we have a 0.100 M solution of acetic acid in water, the concentration of undissociated acid (CH_3COOH) is 0.0987 M, and the concentrations of free hydrogen and acetate (CH_3COO^-) ions are both equal to 0.0013 M. In other words, nearly 99% of the acetic acid molecules are not dissociated in this solution. Note especially that the concentration of free hydrogen ions (or acidity) is low for a solution of a weak acid.

Acid dissociation constants are usually small and awkward to manipulate mathematically, so they are often presented in a logarithmic form. The pK_a is defined as the **negative logarithm** to the base 10 of K_a or

$$pK_a = -\log K_a$$

For example, for acetic acid, $pK_a = -\log 1.8 \times 10^{-5} = 4.7$. The negative logarithm of a number is equal to the logarithm of the inverse of the number. So the equation can be written as

$$pK_a = \log (1/K_a)$$

This equation shows more clearly that pK_a and K_a are inversely related. A low pK_a corresponds to a high dissociation constant (strong acid), and a high pK_a corresponds to a low dissociation constant (weak acid).

pH Values of Aqueous Solutions

As discussed in Chapter 2, we can also define **pH** as the negative logarithm to the base 10 of the molar activity (concentration) of hydrogen ions. In equation form:

$$pH = -\log [H^+] = \log (1/[H^+])$$

The pH of a solution is inversely related to the free hydrogen ion concentration. A low pH indicates a high hydrogen ion concentration, and a high pH indicates a low hydrogen ion concentration. The pH scale is useful because it compresses the wide range of hydrogen ion concentrations possible in aqueous solutions into a more manageable scale. The pH scale usually goes from 0 to 14. A 1 M solution of the strong acid HCl has a pH of about 0 ($\log 1 = 0$). A 1 M solution of the strong base NaOH has a pH of about 14 ($[H^+] = 10^{-14}$ M). A pH of 7 corresponds to a $[H^+]$ of 10^{-7} M. If the pH of a solution is below 7, it is considered to be **acidic;** if equal to 7, **neutral;** and if greater than 7, **alkaline** or **basic.**

The Henderson-Hasselbalch Equation

The equilibrium equation for dissociation of an acid is often expressed in logarithmic form. We can rearrange the equation

$$[H^+] \times [A^-]/[HA] = K_a$$

and solve for $[H^+]$:

$$[H^+] = K_a \times [HA]/[A^-]$$

Taking the logarithms of this equation, we get:

$$\log [H^+] = \log K_a + \log ([HA]/[A^-])$$

Multiplying both sides of the equation by -1, we get:

$$-\log [H^+] = -\log K_a + \log ([A^-]/[HA])$$

Since $pH = -\log [H^+]$ and $pK_a = -\log K_a$, we can write the equation

$$pH = pK_a + \log ([A^-]/[HA])$$

This logarithmic form of the acid dissociation equation is known as the **Henderson-Hasselbalch equation** and is very useful, as we will see later. From this equation, we can see that when $[A^-] = [HA]$ (i.e., when half of the acid is present in its dissociated form), then the pH of a buffer solution equals the pK_a (since the logarithm of one is zero).

The Ion Product of Water

Water reversibly dissociates into hydrogen and hydroxide ions according to the reaction:

$$H_2O + H_2O \rightleftharpoons H_3O^+ + OH^-$$

In this reaction, water is acting as both an acid (proton donor) and base (proton acceptor). It is customary to simplify this reaction and write:

$$H_2O \rightleftharpoons H^+ + OH^-$$

The dissociation constant (K) for this reaction is equal to

$$K = [H^+] \times [OH^-]/[H_2O]$$

The tendency of water to dissociate is very small. The concentration of water ($[H_2O]$) is large and is essentially constant. We can therefore define a new constant, K_w, as equal to $K \times [H_2O]$, and write

$$K_w = [H^+] \times [OH^-]$$

This equation indicates that in any aqueous solution, the product of the hydrogen and hydroxide ion concentrations is constant. This means that $[H^+]$ and $[OH^-]$ are inversely related. K_w, called the **ion product of water,** has a value of 1×10^{-14} M^2 at room temperature (25°C). In *pure* water, $[H^+] = [OH^-]$. Therefore, $[H^+]^2 = 1 \times 10^{-14}$ M^2 or $[H^+] = 10^{-7}$ M (pH 7). Hence, the pH of pure water is 7. If we add a strong base (NaOH) to water, OH^- ions will combine with hydrogen ions supplied by water, thereby shifting the equilibrium toward undissociated water and lowering the free hydrogen ion concentration. For example, in a 1 M solution of NaOH, $[OH^-] = 1$ M, and so $[H^+] = 10^{-14}$ $M^2/1$ M = 10^{-14} M (pH 14). In a 0.1 M solution of HCl, $[H^+]$ is essentially 0.1 M, so $[OH^-] = 1 \times 10^{-14}$ $M^2/0.1$ M or 1×10^{-13} M.

pH Values for Solutions of Acids and Bases, Body Fluids, and Beverages

Table 26–1 presents pH values for some aqueous solutions of chemicals, body fluids, and beverages. Notice that the pH of various solutions of acids depends not only on the concentration of acid, but also on the dissociation constant of the acid. Strong acids produce solutions with a lower pH than do the same concentrations of weak acids, because the stronger acid is dissociated (ionized) more and so yields a greater concentration of free hydrogen ions in solution.

Table 26–1
pH Values of Aqueous Solutions of Chemicals, Body Fluids, and Beverages

Chemicals	
0.100 M hydrochloric acid	1.1
0.010 M hydrochloric acid	2.0
0.001 M hydrochloric acid	3.0
0.100 M lactic acid (pK$_a$ = 3.9)	2.5
0.100 M acetic acid (pK$_a$ = 4.7)	2.9
Pure water	7.0
0.100 M sodium bicarbonate	8.4
0.100 M ammonia	11.1
0.100 M sodium hydroxide	13.0
Body fluids	
Arterial blood	7.40
Venous blood	7.35
Cerebrospinal fluid	7.35
Skeletal muscle cell cytoplasm	6.9
Saliva	5.8–7.1
Gastric juice	0.7–3.8
Gallbladder bile	5.6–8.0
Pancreatic juice	7.5–8.8
Intestinal juice	7.0–8.0
Feces	5.9–8.5
Urine	4.5–8.0
Beverages	
Lemon juice	2.2–2.4
Sodas	2.8–3.7
Orange juice	3.7
Beer	4.4
Coffee	5.0
Milk	6.7
Tea	6.9

The pH of arterial blood averages 7.40, which corresponds to a [H$^+$] of 4×10^{-8} M or 40 nanomolar. The pH of venous blood is slightly lower (average 7.35), owing to its higher content of carbon dioxide (hence the acid carbonic acid). Most body fluids, with the exception of gastric juice, have pH values in the range 5 to 8. Cell cytoplasm is more acidic than extracellular fluid; the pH of skeletal muscle cell cytoplasm averages 6.9. pH may differ strikingly from one organelle to another within a cell; for example, in lysosomes, which contain enzymes that operate best in an acidic environment, pH is about 4.5. Gastric juice is quite acidic, owing to its high concentration of hydrochloric acid. Pancreatic juice is slightly more alkaline than blood because of its high bicarbonate concentration. The pH of urine can vary between 4.5 and 8.0.

Many foods are acidic or alkaline. When we consume food, the load of acid or base depends not only on the free hydrogen ion concentration, but also on how much undissociated acid (or base) is present. An undissociated acid may liberate hydrogen ions (or a base may bind hydrogen ions) at the pH of body fluids. The effect of a chemical substance on acid-base balance also depends on how the substance is metabolized in the body, a subject we will discuss later.

Buffer Systems: General Principles

A buffer, by definition, is something that minimizes a shock and thereby promotes relative stability. A **pH buffer,** as described in Chapter 2, minimizes the change in pH when either acid or base is added to a solution.

A pH buffer *does not prevent* a pH change. Whenever an acid is added, pH will fall. If a base is added, pH will rise. The extent of change in pH depends on the amount and nature of added acid or base, and on the amount and nature of the pH buffer.

Previously, we wrote the equation for dissociation of an acid: HA⇌H$^+$ + A$^-$. In this reaction, HA is the acid and A$^-$ is the **conjugate base.** *Conjugate* means "joined in a pair." A **chemical buffer** consists of a weak acid and its conjugate base (or a weak base and its conjugate acid). Following are some examples of conjugate acid-base buffer pairs:

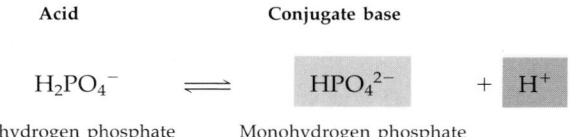

Acid		Conjugate base		
H$_2$PO$_4^-$	⇌	HPO$_4^{2-}$	+	H$^+$
Dihydrogen phosphate		Monohydrogen phosphate		

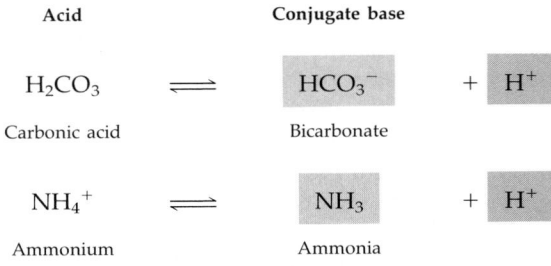

Acid   **Conjugate base**

$$H_2CO_3 \rightleftharpoons HCO_3^- + H^+$$
Carbonic acid   Bicarbonate

$$NH_4^+ \rightleftharpoons NH_3 + H^+$$
Ammonium   Ammonia

The equilibrium expression for dissociation of an acid can be written in the Henderson-Hasselbalch equation form:

$$pH = pK_a + \log ([\text{conjugate base}]/[\text{acid}])$$

For example, for a mixture of monohydrogen and dihydrogen phosphate,

$$pH = 6.8 + \log ([HPO_4^{2-}]/[H_2PO_4^-])$$

This equation demonstrates that if the ratio of conjugate base to acid is defined for a buffer of known pK, then the pH is automatically determined. This principle is useful in making up pH buffer solutions.

What, for example, is the pH of an aqueous solution containing a mixture of 0.1 M Na_2HPO_4 (sodium monohydrogen phosphate) and 0.1 M NaH_2PO_4 (sodium dihydrogen phosphate)? Since the ratio $HPO_4^{2-}/H_2PO_4^-$ is 1, and the logarithm of 1 is 0, the pH is 6.8 + 0 = 6.8. If the ratio of $HPO_4^{2-}/H_2PO_4^-$ is 4, then the pH of this solution is 6.8 + log 4 = 6.8 + 0.6 = 7.4.

Chemical buffers stabilize the pH of aqueous solutions. For example, if we add strong acid (HCl) to a phosphate buffer solution, the following reaction takes place:

$$HCl + Na_2HPO_4 \longrightarrow Na H_2PO_4 + NaCl$$

In this reaction, the base component of the buffer pair (HPO_4^{2-}) binds hydrogen ions. The strong acid (HCl) is, in effect, converted into a weak acid ($H_2PO_4^-$), thereby minimizing the increase in free hydrogen ion concentration (fall in pH). If we add strong base (NaOH) to the phosphate buffer solution, the following reaction takes place:

$$NaOH + Na H_2PO_4 \longrightarrow Na_2 HPO_4 + H_2O$$

The acid component of the buffer pair ($H_2PO_4^-$) liberates hydrogen ions, thereby diminishing the rise in pH caused by addition of strong base. The strong base (NaOH) is, in effect, converted into a weak base (HPO_4^{2-}).

Figure 26–1 shows a titration curve for a phosphate buffer solution. The y axis gives the pH. The x axis gives the amount of strong acid or base added to the solution. Going from left to right along the curve, $H_2PO_4^-$ is converted to HPO_4^{2-} by addition of strong base. Going from right to left along the curve, HPO_4^{2-} is converted to $H_2PO_4^-$ by the addition of strong acid. The slope of the titration curve is an index of the effectiveness of the buffer in resisting a change in pH. The slope of the curve is flattest when the pH is equal to the pK_a of the phosphate buffer. This means that for a given amount of added acid or base, the pH change will be least near the pK_a of the buffer. As we move away from the pK_a, the effectiveness of the buffer declines. In choosing a buffer to stabilize a certain pH, it is always best to select a buffer pair with a pK_a close to the desired pH.

The effectiveness of a buffer in minimizing pH changes depends on two factors. The first factor is

Figure 26–1

Titration curve for inorganic phosphate buffer. HPO_4^{2-} predominates at alkaline pH's. When strong acid (HCl) is added, the reversible reaction $H^+ + HPO_4^{2-} \rightarrow H_2PO_4^-$ takes place. When strong base (NaOH) is added, $H_2PO_4^-$ is converted to HPO_4^{2-}; $OH^- + H_2PO_4^- \rightarrow HPO_4^{2-} + H_2O$. The change in pH produced by a given amount of acid or base is least when the solution pH is at the pK_a (6.8) of the buffer pair $HPO_4^{2-}/H_2PO_4^-$.

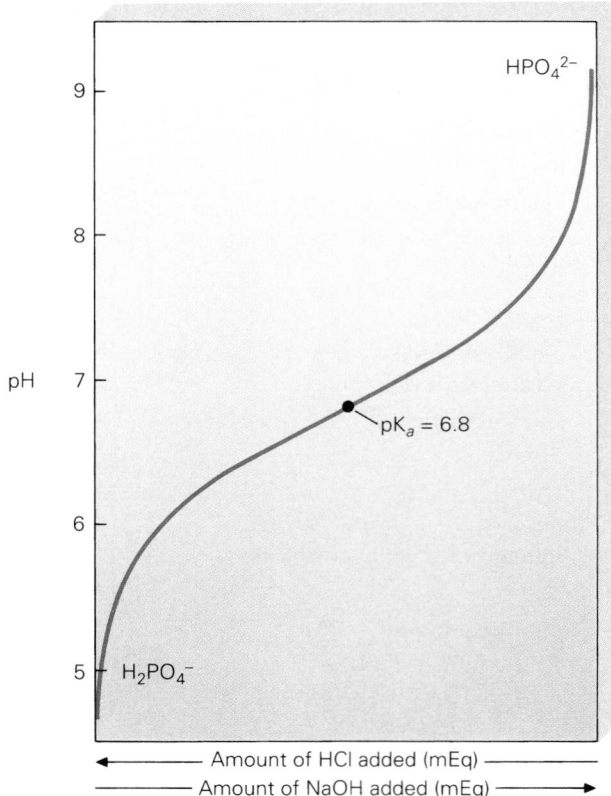

the pK_a of the buffer in relation to the desired pH. The pK_a of the imidazole group of histidine (an amino acid found in hemoglobin) is close to 7.4, and hence this is an ideal buffer in blood. The pK_a of phosphate (6.8) is also effective for buffering blood. The second factor that determines the effectiveness of a buffer is the amount (or concentration) of buffer present. The greater the concentration of buffer, the greater the ability to bind (when acid is added) or release (when base is added) hydrogen ions. If no buffers are present, as in pure water or in an aqueous solution of pure NaCl, large pH changes will be produced by addition of very small amounts of acid or base.

Sources of Hydrogen Ions for the Body

The metabolism of food consumed in our diet is usually the major source of hydrogen ions. For most people eating a mixed diet of meat and vegetables, metabolism results in net addition of acid to the body. Vegetarians are faced with the opposite situation, net addition of base. Since the usual problem is one of getting rid of acid, we will emphasize that situation for most of this chapter.

Metabolic Production of CO_2 as a Source of Hydrogen Ions

In the course of metabolism, a normal adult produces about 300 liters of carbon dioxide per day. Carbon dioxide, added to the blood in the tissues, reacts with water to produce an acid, carbonic acid:

$$CO_2 + H_2O \rightleftharpoons H_2CO_3 \rightleftharpoons H^+ + HCO_3^-$$

In the lungs, these reactions are reversed, since carbon dioxide is expired. No acid burden is imposed on the body, as long as carbon dioxide is expired by the lungs as rapidly as it is produced. Carbon dioxide production and expiration are usually matched so that arterial blood carbon dioxide tension and pH do not change.

Metabolism of Carbohydrates and Fatty Acids as a Source of Hydrogen Ions

Most carbohydrates (e.g., glucose) and fatty acids are normally completely oxidized to carbon dioxide and water. No net addition of acid to the body results, provided that carbon dioxide does not accumulate.

Incomplete oxidation of carbohydrates and fats produces nonvolatile acids. If glucose is not oxi-

dized completely, then lactic acid is produced. With severe exercise or an inadequate circulation, insufficient delivery of oxygen to the tissues results in lactic acid production and an acidic blood pH. If fatty acids are incompletely oxidized, as occurs in uncontrolled diabetes mellitus, starvation, and alcoholism, ketone body acids (acetoacetic acid, 3-hydroxybutyric acid) are produced. This situation also results in an acidic blood pH.

Metabolism of Proteins as a Source of Hydrogen Ions

The metabolism of dietary proteins results in production of strong acids. Sulfuric acid is produced by the oxidation of sulfur-containing amino acids (cysteine, cystine, methionine). Hydrochloric acid is produced by oxidation of cationic amino acids (e.g., lysine, arginine). Phosphoric acid is produced by the oxidation of phosphorus-containing proteins.

To some extent, these acid-producing reactions are neutralized by acid-consuming reactions. Thus, the oxidation of dietary organic anions (e.g., citrate, lactate, acetate) uses up hydrogen ions. For example, citrate, which is contained in fruits and vegetables, is oxidized in the body according to the following simplified reaction:

$$citrate^- + O_2 + H^+ \longrightarrow CO_2 + H_2O$$

In this reaction, a hydrogen ion is consumed. The net result of acid-producing and acid-consuming reactions on the usual American diet is such that an excess of about 50 to 100 mEq of hydrogen ions is produced in a day.

Regulation of Extracellular pH: Buffering Mechanisms in the Body

Despite the metabolic production of acids in the body, a normal pH in blood and interstitial fluid is preserved in a healthy person. The pH is stabilized by buffers in the body. The first line of defense of pH consists of chemical buffers in extracellular and intracellular fluids and in bone. The second is the lungs, which buffer blood pH by disposing of carbon dioxide as rapidly as it is formed. When there is excessive production of acids (or bases) in the body, lung ventilation changes in order to reduce the deviation from a normal blood pH. The third is the kidneys, which buffer blood pH by excreting hydrogen ions in the urine. The kidneys are in charge of eliminating excess hydrogen ions and anions from the body. Figure 26–2 sketches the pathways for maintaining a stable blood pH despite the acid threat posed by cell metabolism of foodstuffs.

Chemical pH Buffers

We will first consider chemical pH buffers in the body. These include (1) phosphates, (2) proteins, and (3) the bicarbonate-CO_2 system. Chemical buffers consist of conjugate acid-base pairs that bind hydrogen ions (when acid is added) or release hydrogen ions (when base is added).

Phosphate as a Chemical Buffer

Phosphate, by virtue of its pK of 6.8, is a good buffer for stabilizing the pH of the blood close to its normal value of 7.4. The concentration of inorganic phosphate in plasma or extracellular fluid, however, is rather low (about 1.0 mmole per liter). Because of its low concentration, inorganic phosphate has a limited capacity to buffer added acid or base. There are, however, large quantities of phosphate salts in bone, so bone can act to buffer pH. Also, the cellular compartment contains many organic phosphate compounds, such as adenosine triphosphate (ATP), adenosine diphosphate (ADP), and creatine phosphate. These compounds primarily serve in energy

Figure 26–2
Overall schema for maintenance of acid-base balance. On the usual mixed diet, pH is threatened by production of strong acids (sulfuric, hydrochloric, phosphoric), which result mainly from protein metabolism. These strong acids are buffered by chemical buffers in the body. Removal of extra hydrogen ions and the accompanying anions from the body is accomplished by renal excretion. When the kidneys excrete hydrogen ions, they add new bicarbonate to the blood, thereby restoring depleted body buffer bases. The respiratory system eliminates CO_2 produced by metabolism. CO_2 is not a threat to acid-base balance, provided it is expired at the same rate it is produced.

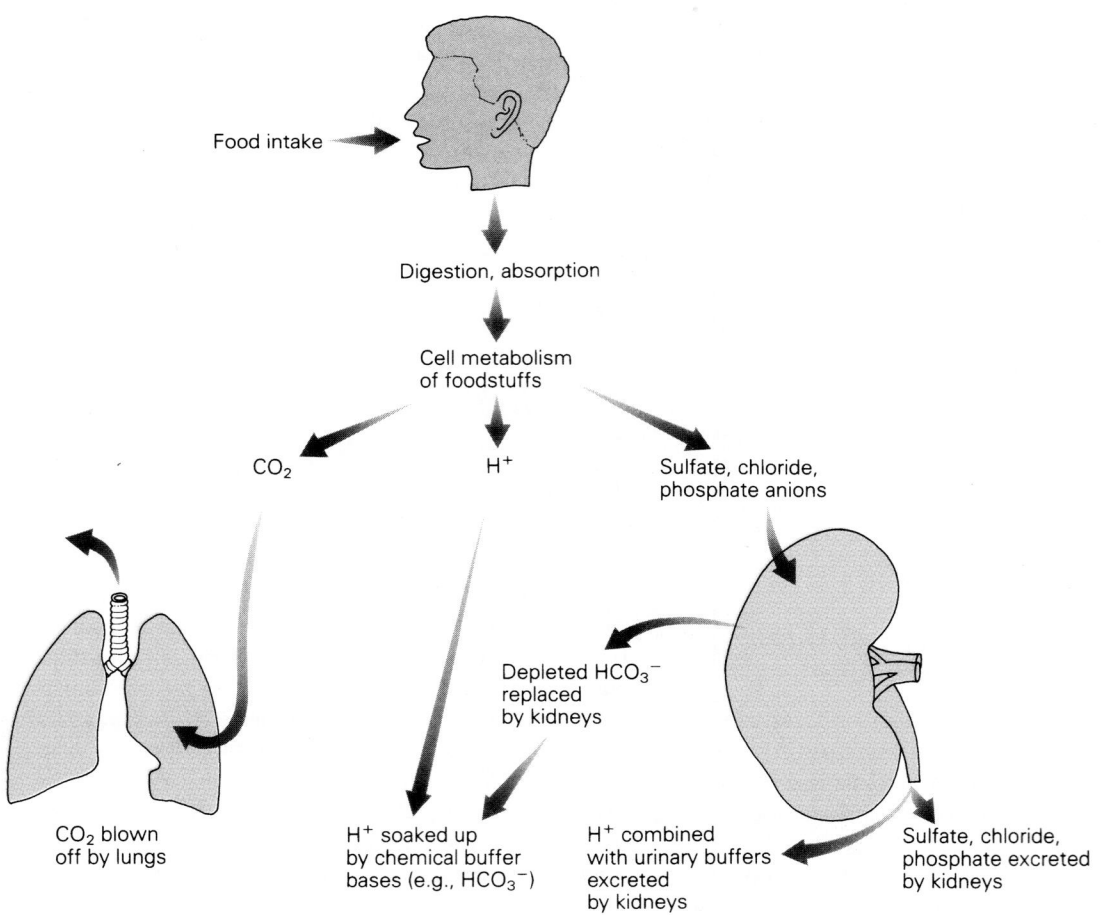

Food intake

Digestion, absorption

Cell metabolism of foodstuffs

CO_2 H^+ Sulfate, chloride, phosphate anions

Depleted HCO_3^- replaced by kidneys

CO_2 blown off by lungs

H^+ soaked up by chemical buffer bases (e.g., HCO_3^-)

H^+ combined with urinary buffers excreted by kidneys

Sulfate, chloride, phosphate excreted by kidneys

metabolism, but they also bind or release hydrogen ions in response to excess acid or base. Inorganic phosphate is a significant buffer in the urine.

Proteins as Chemical Buffers

Proteins are the largest pool of buffer in the body and are excellent pH buffers. Proteins contain many different ionizable groups, which can donate or accept hydrogen ions. Some of these—for example, the imidazole group of histidine—have pKs close to the pH of plasma and so are especially effective buffers.

Serum albumin and plasma globulins act as pH buffers in blood plasma. The large amounts of proteins in cells also participate in buffering hydrogen ions. An important example of an intracellular protein buffer is hemoglobin, found in red blood cells. When carbon dioxide is added to the blood in the tissues, hydrogen ions liberated from carbonic acid are mainly bound by hemoglobin. This combination with hydrogen ions, incidentally, also favors the release of oxygen from hemoglobin (Chapter 21). These reactions are summarized in Figure 26–3.

The Bicarbonate-CO_2 System as a Chemical Buffer

The bicarbonate-CO_2 system is of special importance in pH buffering in extracellular fluid for several reasons. First, the components of this buffer system are present in large quantities. A practically limitless supply of CO_2 is available from metabolism. There is an appreciable concentration of bicarbonate in extracellular fluid, about 24 mEq per liter. Second, despite a pK of 6.10, which is somewhat removed from the desired pH of 7.40, this buffer pair is quite effective, as will be explained later. Third, the physiological buffers, lungs and kidneys, affect blood pH by acting on components of this buffer pair.

We will now examine the bicarbonate-CO_2 relationship in the light of acid-base chemistry. We can write the following reversible reactions:

$$CO_2(g) \rightleftharpoons CO_2(d)$$
$$+$$
$$H_2O \rightleftharpoons H_2CO_3 \rightleftharpoons$$

Carbonic
anhydrase

$$H^+ + HCO_3^-$$

$$\Updownarrow$$

$$H^+ + CO_3^{2-}$$

in which $CO_2(g)$ represents CO_2 in the lung alveoli, or gaseous CO_2, and $CO_2(d)$ represents CO_2 dissolved in pulmonary capillary blood. CO_2 in the lung alveoli ($CO_2(g)$) equilibrates with CO_2 dissolved in pulmonary capillary blood. $CO_2(d)$ reacts with water in a **hydration reaction** to form carbonic acid. The reverse reaction, the formation of $CO_2(d)$ and water from H_2CO_3, is called the **dehydration reaction.**

Figure 26–3
Buffering of carbonic acid by hemoglobin. The hydrogen ion formed by dissociation of carbonic acid combines with oxyhemoglobin (abbreviated HbO_2^-), resulting in formation of reduced hemoglobin (HHb) and release of oxygen. C.A. is the enzyme carbonic anhydrase. Bicarbonate exchanges with chloride across the red cell membrane via an anion exchanger. The reverse reactions occur when Pco_2 is lowered, as happens in the capillaries of the lungs.

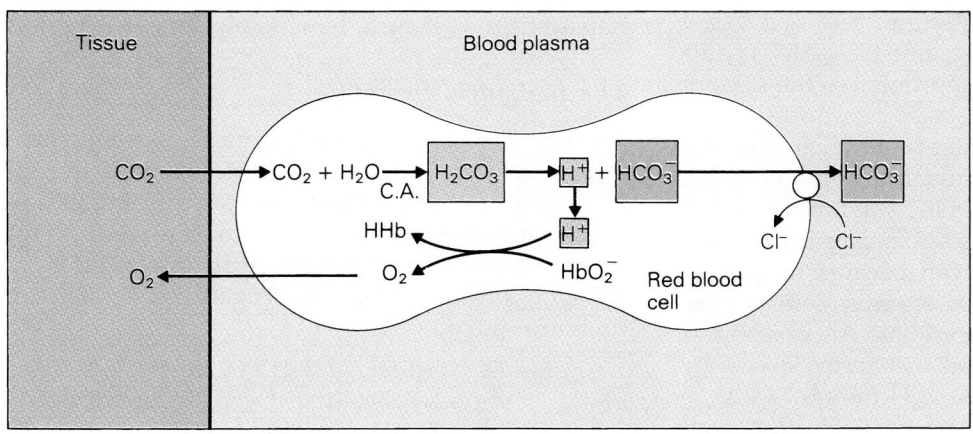

Both of these reactions are slow if uncatalyzed. As mentioned in previous discussions, they are speeded up by the zinc-containing enzyme **carbonic anhydrase.** This enzyme is present in red blood cells and in cells of the kidneys, pancreas, stomach, ciliary body of the eye, and choroid plexuses of the brain. The dissociation of carbonic acid to hydrogen ions and bicarbonate or the association of hydrogen ions and bicarbonate to form carbonic acid occur essentially instantaneously. The dissociation of bicarbonate ions into hydrogen and carbonate (CO_3^{2-}) ions occurs to an appreciable extent only at pH values much more alkaline than those encountered in the body fluids, and so we will not discuss this last reaction. It should be noted, however, that bone contains considerable amounts of calcium carbonate, which is important in buffering acids.

For the bicarbonate-CO_2 system, we can write the Henderson-Hasselbalch equation as

$$pH = 6.10 + \log ([HCO_3^-]/0.03 Pco_2)$$

Carbonic acid does not appear explicitly in this expression, but is implicit. The denominator of the log term is the concentration of dissolved CO_2. In blood plasma at a temperature of 37°C, the concentration of dissolved CO_2 is related to the Pco_2 by the equation

$$[CO_2(d)] = 0.03 \times Pco_2$$

In this equation, $CO_2(d)$ is in mmoles per liter, Pco_2 is in millimeters of Hg, and 0.03 is the solubility coefficient in mmoles per liter per millimeter of Hg. $CO_2(d)$ and H_2CO_3 are usually in equilibrium (remember carbonic anhydrase). In blood plasma at 37°C, the ratio of these two chemical species is about 400 to 1 (1.2 mmoles per liter versus 3 μmoles per liter). Because the H_2CO_3 concentration is so low and hard to measure, $CO_2(d)$ is substituted for H_2CO_3 in the Henderson-Hasselbalch equation just given. The constant (6.10) incorporates both the acid dissociation constant for H_2CO_3 and the constant for the H_2CO_3-$CO_2(d)$ equilibrium. The real source of hydrogen ions is H_2CO_3, not CO_2; the H_2CO_3 concentration is directly proportional to the Pco_2 by way of dissolved CO_2.

As stated previously, the pK of the bicarbonate-CO_2 buffer pair, 6.10, is remote from the pH of 7.40 that we want to maintain in blood. From this alone, we might conclude that this buffer pair is not effective. This is not true, however, because it operates in an "open system." In an open system, a component can be removed or added. An example of how this works is presented in Figure 26–4. Suppose we had blood containing 24 mmoles of bicarbonate per liter plasma and a Pco_2 of 40 mm Hg

($CO_2(d)$ = 1.2 mmoles per liter). The pH of the blood can be calculated as follows: pH = 6.10 + log ([HCO_3^-]/0.03Pco_2) = 6.10 + log (24/[0.03 × 40]) = 6.10 + log 20 = 6.10 + 1.30 = 7.40. Suppose we now add 10 mmoles of strong acid to each liter of blood. For simplicity, we will ignore the fact that bicarbonate is not the only buffer base that combines with the added hydrogen ions. According to the following reactions, 10 mmoles of dissolved CO_2 will be formed:

$$\boxed{H^+} + \boxed{HCO_3^-} \longrightarrow \boxed{H_2CO_3} \longrightarrow H_2O + CO_2$$

In a closed container (Fig. 26–4*b*), from which none of the CO_2 formed can escape, the resulting pH is equal to 6.10 + log ([24 − 10]/[1.2 + 10]) = 6.20. In an open system (see Fig. 26–4*c*), the CO_2 produced can be blown off as it is formed, and Pco_2 can be maintained constant at 40 mm Hg ($CO_2(d)$ = 1.2 mmoles per liter). In this case, the pH is equal to 6.10 + log ([14]/[1.2]) = 7.17, which is much better than a pH of 6.20 (a lethal pH). A fall in blood pH causes hyperventilation so that the Pco_2 is not maintained constant, but actually falls—for example, to 30 mm Hg ($CO_2(d)$ = 0.90 mmoles per liter) (see Fig. 26–4*d*). In this case, the pH is equal to 6.10 + log ([14]/[0.90]) = 7.29. We can see, therefore, that the ability to blow off CO_2 and to lower its level in the blood by hyperventilation diminishes the fall in pH produced by adding strong acid.

The bicarbonate-CO_2 buffer system is "open" in other respects, too. There is a continuous source of CO_2 from metabolism, which can replace CO_2 consumed when strong base is added to the body. The kidneys can change the amount of bicarbonate in the extracellular fluid by excreting bicarbonate in the urine (when there is excess base in the body) or by synthesizing new bicarbonate and adding it to the blood (when there is excess acid in the body). The ability of the body to change the amounts of components of this particular buffer pair makes the bicarbonate-CO_2 system a remarkably effective buffer.

The Isohydric Principle

We have discussed the various buffer systems separately, but in reality they all work together. In fact, in any fluid compartment, all are in equilibrium with the same hydrogen ion concentration. This idea is called the **isohydric principle.** The term *isohydric* means "same hydrogen ion." We can therefore write:

$$pH = 6.8 + \log([HPO_4^{2-}]/[H_2PO_4^-])$$
$$= pK_{protein} + \log ([proteinate^-]/[H-proteinate])$$
$$= 6.1 + \log ([HCO_3^-]/0.03 Pco_2)$$

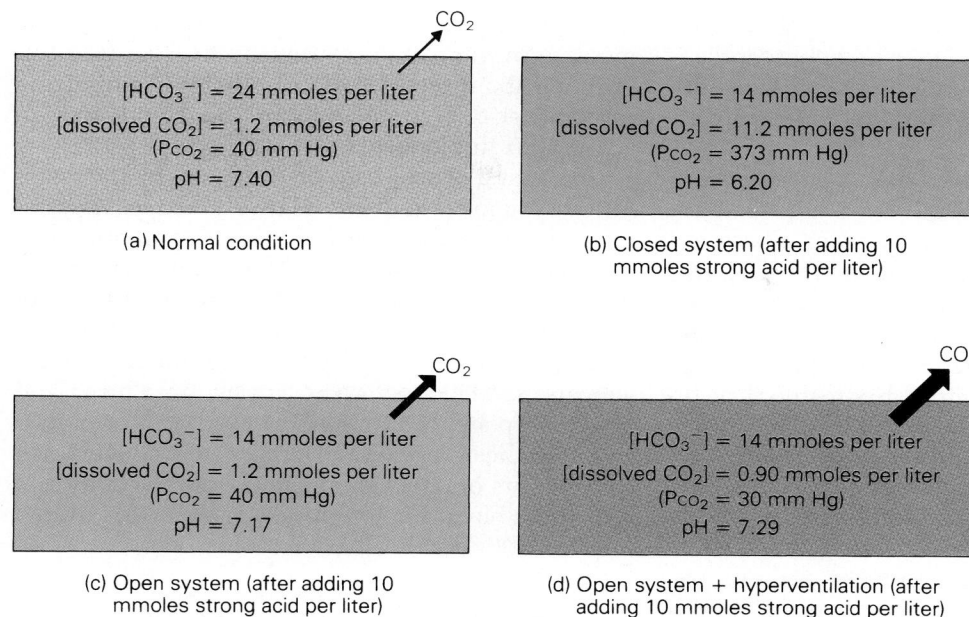

$[HCO_3^-] = 24$ mmoles per liter

$[dissolved\ CO_2] = 1.2$ mmoles per liter
($P_{CO_2} = 40$ mm Hg)

pH = 7.40

(a) Normal condition

$[HCO_3^-] = 14$ mmoles per liter

$[dissolved\ CO_2] = 11.2$ mmoles per liter
($P_{CO_2} = 373$ mm Hg)

pH = 6.20

(b) Closed system (after adding 10
mmoles strong acid per liter)

$[HCO_3^-] = 14$ mmoles per liter

$[dissolved\ CO_2] = 1.2$ mmoles per liter
($P_{CO_2} = 40$ mm Hg)

pH = 7.17

(c) Open system (after adding 10
mmoles strong acid per liter)

$[HCO_3^-] = 14$ mmoles per liter

$[dissolved\ CO_2] = 0.90$ mmoles per liter
($P_{CO_2} = 30$ mm Hg)

pH = 7.29

(d) Open system + hyperventilation (after
adding 10 mmoles strong acid per liter)

Figure 26–4
Schema to illustrate the effectiveness of the bicarbonate/CO_2 system in buffering added strong acid. Note particularly how in an open system (c and d), which allows removal of CO_2, the fall in pH is much less than in the closed system (b). (a) is the normal condition.

When an acid or base is added to the body, all buffer pairs participate in buffering. The importance of each depends on its pK, amount, and accessibility. The expression just given re-emphasizes the point that the ratio of conjugate base to acid for any buffer pair of known pK defines the pH. For the remainder of this chapter, we will stress the bicarbonate-CO_2 system and its effects on blood pH. Recognize, however, that other buffers are also important and interact with the bicarbonate-CO_2 buffer pair. Changes in the ratio HCO_3^-/CO_2 indicate that other body buffers are changing too. The reason for emphasizing the bicarbonate-CO_2 system is mainly that the physiological buffers act on its components. The respiratory system controls arterial blood pH by regulating the P_{CO_2}, and the kidneys control blood pH by regulating plasma bicarbonate concentration.

Physiological Buffers

The Respiratory System as a Buffer System: Control of the P_{CO_2}

The respiratory system controls the partial pressure of CO_2 in arterial blood. Recall from Chapter 20 that the level of alveolar ventilation is a determinant of the P_{CO_2} in the alveolar spaces of the lung. CO_2 pressures in alveoli and arterial blood are equal because of equilibration of CO_2 between alveolar gas and pulmonary capillary blood. For a constant rate of CO_2 production, alveolar ventilation and alveolar P_{CO_2} are reciprocally related. The higher the rate of alveolar ventilation, the lower the alveolar P_{CO_2}, and vice versa. Normally, CO_2 is expired at the same rate that it is produced from metabolism, and the P_{CO_2} remains near 40 mm Hg.

In a state of hyperventilation, CO_2 is flushed out of the alveolar spaces at a rate greater than the rate at which it is produced and added to the alveolar spaces. Consequently, the P_{CO_2} of alveolar gas and arterial blood falls to a new lower-than-normal level, and the reactions

$$CO_2 + H_2O \rightleftharpoons H_2CO_3 \rightleftharpoons H^+ + HCO_3^-$$

are pulled to the left. The hydrogen ion concentration of the blood falls; in other words, the blood becomes more alkaline. During hypoventilation, CO_2 added to the alveolar spaces is not removed at an adequate rate; P_{CO_2} in the alveoli and arterial blood rises. The above reactions are pushed to the right, so that a more acidic blood pH results. From these two examples, it should be clear that abnormal respiratory system function—that is, hyperventilation or hypoventilation—produces disturbances in blood pH. We will return to this topic toward the end of this chapter.

The respiratory system normally acts to minimize pH changes in the blood. Such changes might be produced by adding a "fixed" acid or a base to the blood or by adding or removing too much CO_2.

A **fixed acid** is an acid other than carbonic acid. In contrast to the concentration of carbonic acid, the concentration of a fixed acid in blood is not affected by lung function. If, for example, hydrochloric acid is added to or produced in the body, it will be neutralized by chemical buffers, and then the hydrogen ions (combined with urinary buffers) and chloride will be excreted by the kidneys. A fixed acid cannot be blown off by the lungs.

If the blood is made more acidic by addition of fixed acid, pulmonary ventilation is increased. The receptors that sense the fall in pH and reflexly increase ventilation are mainly the **peripheral chemoreceptors,** the carotid and aortic bodies. A large pH drop may also stimulate central chemoreceptors or the medullary respiratory center itself. Increased ventilation serves to lower the arterial blood P_{CO_2} and carbonic acid concentration, and so reduces the acidic shift in blood pH. Figure 26–4*d* illustrates how hyperventilation in response to added acid minimizes a fall in blood pH.

If the blood is made more alkaline by infusing base, the opposite changes occur. In this case, hypoventilation results, which leads to retention of CO_2, a higher carbonic acid concentration in the blood, and a less alkaline shift in blood pH. Hypoventilation is limited because it causes retention of CO_2 and a fall in arterial oxygen, and both of these changes stimulate ventilation.

If excess CO_2 accumulates in the body because of inadequate alveolar ventilation or breathing of CO_2-enriched air, ventilation will be stimulated. CO_2 is a very powerful stimulus to ventilation; an increase in arterial blood P_{CO_2} of 2 to 3 mm Hg causes the rate of ventilation to double. The receptors mediating this increase in ventilation are chiefly **central chemoreceptors** in the medulla of the brain. Carbon dioxide diffuses from the blood into brain interstitial and cerebrospinal fluids, where it forms carbonic acid, which liberates hydrogen ions. It is generally believed that chemoreceptors respond to hydrogen ion concentration (or pH) and not to molecular CO_2 directly. The peripheral chemoreceptors are also stimulated by an increase in arterial blood P_{CO_2}, but are less important than the central chemo-

receptors in the ventilatory response to CO_2. The increase in ventilation caused by CO_2 is useful in bringing about removal of CO_2, thereby minimizing carbonic acid accumulation in the blood and an acidic shift in blood pH.

Buffering of acid-base disturbances by the respiratory system is rapid but relatively coarse. Respiratory responses begin within minutes and are maximal in about 12 to 24 hours. The respiratory system brings blood pH back closer to normal, but cannot eliminate a fixed acid or base from the body, and so cannot restore a normal pH. By contrast, renal responses to acid-base disturbances are slower and more complete. If an excess of fixed acid or base is added to the body, normal kidneys will slowly (over many hours or days) eliminate it from the body and return the blood pH to normal. The kidneys exert a fine control over acid-base balance.

The Kidneys as a Buffer System: Acidification of the Urine

The kidneys play a major role in acid-base regulation by getting rid of acid or base in the urine when there is an excess or deficit of hydrogen ions in the body. As mentioned previously, the usual condition is one of excess acid, since metabolism of food produces about 50 to 100 mEq of strong acid in a day. This acid is first neutralized by bicarbonate (and other chemical buffer bases) in the body. Then the kidneys get rid of the acid in the urine. In the process of excreting acid, the kidneys add new bicarbonate to the blood and so bring the blood pH back to normal.

Most of the hydrogen ions excreted in the urine are combined with urinary buffers. The limits of urine pH are from a maximum of 8.0 to a minimum of 4.5. Suppose we excrete 1.5 liters of pH 4.5 urine in a day. How many *free* hydrogen ions are excreted? This can be calculated by multiplying the urine volume (1.5 liters) by the hydrogen ion concentration ($10^{-4.5}$ mEq per liter) and amounts to about 0.05 mEq. This amount is clearly inadequate to take care of the amount of acid that needs to be excreted. Most of the hydrogen ions excreted in the urine are not free but are combined with buffers. These buffers include titratable acids (e.g., inorganic phosphate) and ammonia.

Titratable acid is measured in the clinical laboratory by determining how many milliequivalents of strong base (NaOH) are needed to bring a urine sample back to the pH of arterial blood (usually 7.4). Most of the titratable acid is normally inorganic phosphate buffer. Depending on the circumstances, other substances in the urine (creatinine, various organic acids) may also be included. In the test tube, the following reaction takes place:

$$H_2PO_4^- + OH^- \longrightarrow HPO_4^{2-} + H_2O$$

This reverses the reaction that took place along the kidney tubules when hydrogen ions were added to the tubular urine:

$$H^+ + HPO_4^{2-} \longrightarrow H_2PO_4^-$$

Ammonia also buffers secreted hydrogen ions in the urine, according to the following reaction:

$$H^+ + NH_3 \rightleftharpoons NH_4^+$$

The pK_a for the ammonium ion (about 9.0) is so high that it is not included when titratable acid is measured. In blood plasma and urine, most ammonia is in the form of ammonium ion and not NH_3. Of the hydrogen ions excreted in urine, normally nearly one third are in the form of titratable acid, close to two thirds are present as ammonia (ammonium ion), and a trivial amount is free.

The urine is acidified by secretion of hydrogen ions by the tubular epithelium (Fig. 26–5). The nature of the secretory processes, the amounts of hydrogen ions secreted, and the pH changes produced differ in different segments of the nephron. For example, most of the hydrogen ions are secreted by the proximal convoluted tubule. This secretion is accomplished by a brush border membrane carrier that exchanges one sodium for one hydrogen ion (Na^+/H^+ exchanger), an example of a secondary active transport process. Because the glomerular filtrate contains large amounts of buffer (especially bicarbonate) and because the proximal convoluted tubule cannot develop steep H^+ gradients, the fall in tubular fluid pH along the proximal convoluted tubule is modest (e.g., to pH 6.8). More distal portions of the nephron (distal convoluted tubule, collecting duct) secrete smaller amounts of hydrogen ions. Distal hydrogen ion secretion involves primary active transport by a H^+-ATPase in the luminal cell membrane. The distal portions of the nephron can establish steep transepithelial pH gradients. A final urine pH of 4.4, compared to a blood pH of 7.4, means that the urine [H^+] is 1000 times higher than in plasma. For simplicity, we treat the tubular epithelium as a whole in the cell model in Figure 26–5 and do not consider differences in different nephron segments.

Three processes are involved in urinary acidification: (1) reabsorption of filtered bicarbonate, (2) excretion of titratable acid, and (3) excretion of ammonia (see Fig. 26–5). All three involve hydrogen ion secretion; furthermore, all three processes are associated with addition of bicarbonate to the peritubular capillary blood by the kidney tubules. Hydrogen ion secretion into the lumen and addition of bicarbonate ion to the blood by the tubules are best viewed as opposite sides of the same coin. Reabsorption of filtered bicarbonate is important in conserving the normal amount of bicarbonate in the extracellular fluid. In the process of glomerular filtration, large quantities (4500 mEq per day) of bicarbonate are filtered, and only traces (e.g., 2 mEq per day) are excreted in the usual acidic urine. Reabsorption of filtered bicarbonate reclaims bicarbonate for the body. Most of the hydrogen ions secreted are used to bring about reabsorption of filtered bicarbonate and are not excreted in the final urine. Those that are excreted are combined with phosphate (titratable acid) and ammonia. In these latter processes, new bicarbonate is generated in the kidney tubule cells. This new bicarbonate is added to the blood and replaces bicarbonate consumed in the neutralization of strong acids produced in the body.

Reabsorption of Filtered Bicarbonate in Urinary Acidification. Figure 26–5a depicts the reabsorption of filtered bicarbonate. This occurs mainly in the proximal convoluted tubule. Hydrogen ions are secreted into the tubular urine in exchange for sodium ions (Na^+/H^+ exchanger) and combine with bicarbonate that has been filtered by the glomeruli. This leads to formation of carbonic acid in the tubular urine. The dehydration of this acid to CO_2 and water is catalyzed in the proximal convoluted tubule lumen by carbonic anhydrase in the brush border membrane. Pco_2 values in tubular urine, cells, and blood are essentially identical since CO_2 diffuses rapidly through cell membranes. In the cell, CO_2 combines with water to form carbonic acid. This reaction is catalyzed by carbonic anhydrase inside the cell. Carbonic acid dissociates into hydrogen ions and bicarbonate. The hydrogen ions are secreted into the lumen. Bicarbonate ions exit from the basolateral side of the cell along with sodium ions. In the proximal tubule, this appears to involve a coupled Na-HCO_3 transport mechanism in the basolateral cell membrane. In this scheme, it should be apparent that filtered bicarbonate is not reabsorbed by direct transport out of the tubular urine; rather, filtered bicarbonate is reabsorbed indirectly by hydrogen ion secretion.

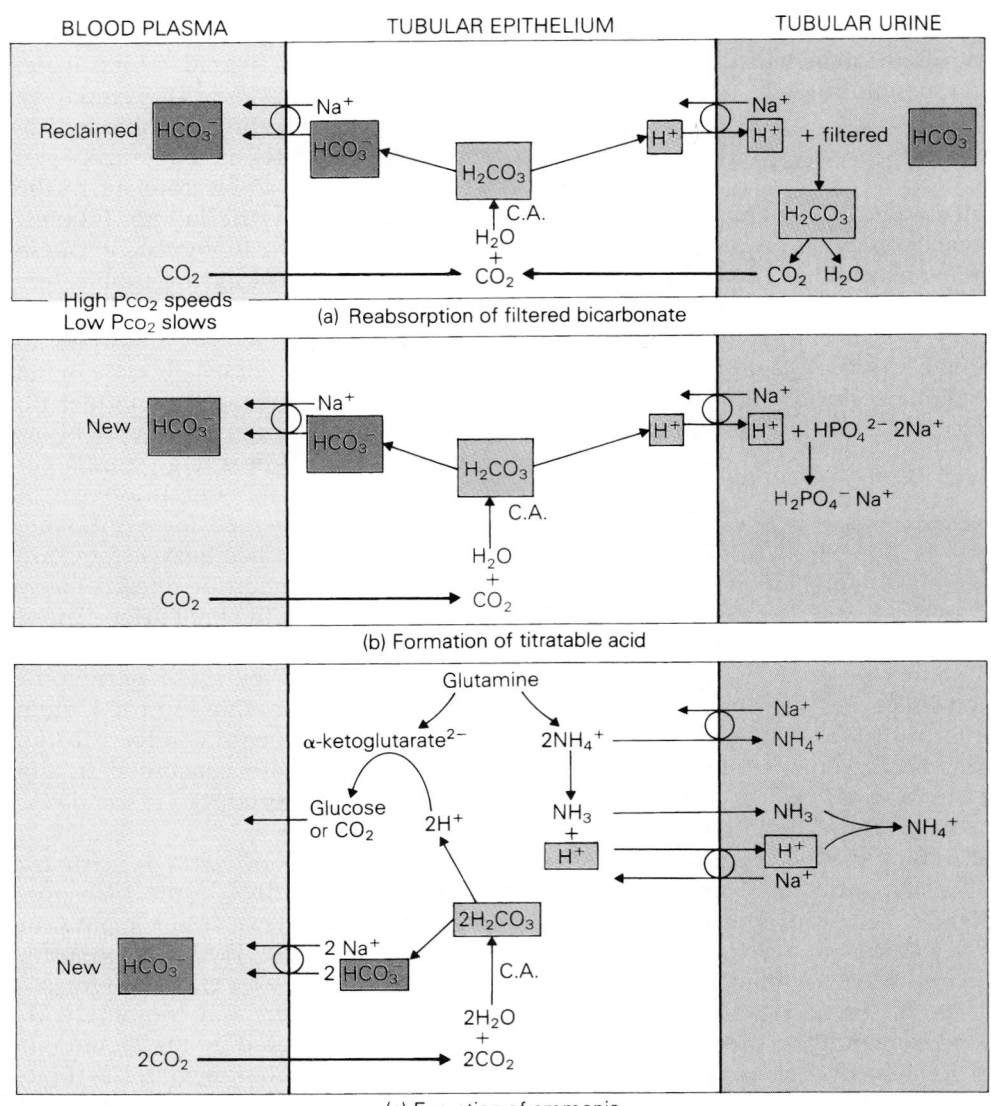

BLOOD PLASMA   TUBULAR EPITHELIUM   TUBULAR URINE

(a) Reabsorption of filtered bicarbonate

(b) Formation of titratable acid

(c) Excretion of ammonia

Figure 26–5
Cell models of the three processes involved in urinary acidification: (*a*) reabsorption of filtered bicarbonate, (*b*) formation of titratable acid, and (*c*) excretion of ammonia. C.A. = carbonic anhydrase.

Excretion of Titratable Acid in Urinary Acidification. Normally, we ingest phosphate in our diet. Filtered phosphate is incompletely reabsorbed by the kidney tubules, and the unreabsorbed phosphate acts as a pH buffer in the urine. The cell scheme for formation of titratable acid is shown in Figure 26–5*b*. Hydrogen ions are secreted into the tubular urine, and they titrate buffers in the urine in the acidic direction. For example, the basic form of phosphate (HPO_4^{2-}) is converted to the acidic form ($H_2PO_4^-$). One of the sodium ions accompanying HPO_4^{2-} is reabsorbed. The intracellular and membrane events are similar to those described in Figure 26–5*a*. For each mEq of hydrogen ion excreted as titratable acid, a *new* bicarbonate ion is added to the blood. Excretion of titratable acid, therefore, eliminates hydrogen ions from the body and, at the same time, restores depleted plasma bicarbonate reserves.

Excretion of Ammonia in Urinary Acidification. Figure 26–5*c* summarizes the role of ammonia in urinary acidification. Ammonia is synthesized by kidney tubule cells (mainly in the proximal tubule). The metabolism of glutamine results in formation of two ammonium ions and one alpha-ketoglutarate molecule. Each ammonium ion yields one molecule of free base ammonia (NH_3) and one hydrogen ion. Ammonia is secreted into the urine by two mechanisms. In the proximal tubule, ammonia is actively secreted by the Na^+/H^+ exchanger (operating in a Na^+/NH_4^+ exchange mode). NH_3 also diffuses readily through cell membranes into the urine and combines with a secreted hydrogen ion to form an ammonium ion. Recall from Chapter 4 that, mainly because of its charge, the ammonium ion does not readily diffuse through the lipid phase of plasma membranes, and so is in effect "trapped" in an acidic urine, a process called **diffusion trapping.** Diffusion trapping of ammonia is especially important in the medullary collecting ducts because they have the lowest intraluminal pH. The ammonium ion is accompanied by an anion (e.g., chloride) in the urine and allows the body to conserve a sodium ion, which it replaces. In the cell, hydration of CO_2 produces hydrogen ions and bicarbonate ions, as described previously. These hydrogen ions are consumed when alpha-ketoglutarate is metabolized to form glucose or CO_2. The bicarbonate ions are reabsorbed together with sodium ions. Note that for each hydrogen ion excreted in the urine as ammonia (ammonium ion), an equivalent quantity of *new* bicarbonate is added to the blood. The kidneys' ability to synthesize ammonia from glutamine is increased under conditions of excess acid in the body. This adaptive increase in ammonia synthesis takes several days to develop fully but allows the body to dispose of large acid loads.

Effects of CO_2 on Urinary Acidification.   The effects of CO_2 on urinary acidification are also indicated in Figure 26–5. A high blood P_{CO_2}, by mass action, favors generation of hydrogen ions in the cell. This results in more complete reabsorption of filtered bicarbonate and increased excretion of hydrogen ions combined with titratable acids and ammonia. Conversely, a low blood P_{CO_2} results in a lower rate of hydrogen ion secretion; filtered bicarbonate is less completely reabsorbed, and excretion of an alkaline (bicarbonate-rich) urine is favored. Fewer hydrogen ions will be excreted in combination with titratable acids and ammonia. These effects of CO_2 on kidney tubule hydrogen ion secretion are key to understanding the renal compensations for respiratory disturbances of acid-base balance.

If we reconsider the Henderson-Hasselbalch equation pH = 6.10 + log ($[HCO_3^-]/0.03$ P_{CO_2}), we can see that blood pH is affected by changes in plasma bicarbonate concentration. In cases of acid excess, the kidneys increase acid excretion in the urine, and raise the plasma $[HCO_3^-]$. This helps to make the blood less acidic; that is, it raises the pH toward a normal value. In cases of base excess, the kidneys increase the output of bicarbonate in the urine, thereby lowering the plasma bicarbonate concentration toward normal. This helps to make the blood less alkaline; that is, it lowers the pH toward a normal value. From these relations, we can see that it is useful to think of the kidneys as regulating plasma pH by changing the plasma bicarbonate concentration in the appropriate direction. We will discuss later the renal compensation for various acid-base disturbances.

Disturbances of Acid-Base Balance

In numerous conditions, the pH of arterial blood may not be within the normal range, 7.35 to 7.45. If arterial blood is more acidic than pH 7.35, **acidemia** exists. If the blood is more alkaline than pH 7.45, **alkalemia** exists. The range of pH values compatible with life is approximately 6.8 to 7.8.

There are four fundamental types of acid-base disturbance: (1) **respiratory acidosis,** (2) **respiratory alkalosis,** (3) **metabolic acidosis,** and (4) **metabolic alkalosis.** This classification is based on the nature of the process leading to the disturbed pH, and on the direction of pH change. If the problem is too much or too little CO_2, then a **respiratory disturbance** of acid-base balance is present. If the problem is an excess or deficit of a fixed acid or base, then a **metabolic (nonrespiratory) disturbance** of acid-base balance is present. Acidosis leads to acidemia, and alkalosis leads to alkalemia. These four disturbances of acid-base balance are each accompanied by characteristic changes in arterial blood pH, P_{CO_2}, and $[HCO_3^-]$ (Table 26–2). We will describe for each disturbance the nature of the abnormal process, characteristic changes in blood chemistry, some common causes, and how buffering mechanisms in the body minimize the deviation from normal pH.

Respiratory Acidosis: Accumulation of CO_2

Respiratory acidosis is an abnormal process characterized by accumulation of CO_2. Most often it is due to failure to expire (blow off) metabolically produced CO_2 at an adequate rate. In other words, alveolar ventilation is inadequate, and CO_2 accumulates in

Table 26–2

Summary of Directional Changes in Blood Values in the Four Fundamental Types of Acid-Base Disturbance

Disturbance	Arterial Blood			Compensatory Response
	pH	Plasma [HCO$_3^-$]	P$_{CO_2}$	
Respiratory acidosis	↓	↑	⇑	Kidneys increase H$^+$ excretion (increase plasma [HCO$_3^-$])
Respiratory alkalosis	↑	↓	⇓	Kidneys increase HCO$_3^-$ excretion (decrease plasma [HCO$_3^-$])
Metabolic acidosis	↓	⇓	↓	Alveolar hyperventilation
Metabolic alkalosis	↑	⇑	↑	Alveolar hypoventilation

Heavy arrows indicate the main change.

the blood. The problem may be caused by (1) insufficient neural drive for ventilation, (2) inadequate movements of the respiratory muscles or thoracic cage, (3) airway obstruction, or (4) pulmonary disease. Respiratory acidosis is also produced if a person breathes CO_2-enriched air. If CO_2 is allowed to build up in the arterial blood, the following reactions are pushed to the right:

$$CO_2 + H_2O \rightleftharpoons H_2CO_3 \rightleftharpoons H^+ + HCO_3^-$$

Clearly, retention of CO_2 (the primary chemical change) leads to an increase in [H$^+$] and so to a fall in pH. The plasma [HCO$_3^-$] will also be elevated, since body buffer bases (e.g., proteins and phosphates) combine with hydrogen ions generated in the previous reactions, leaving behind bicarbonate. The increase in [HCO$_3^-$] is proportionately less than the increase in P$_{CO_2}$. (Otherwise, the pH would not fall. See the Henderson-Hasselbalch equation.)

Buffering of hydrogen ions produced by excess CO_2 is first accomplished by chemical buffers in the body. These buffers include the large amounts of protein and organic phosphate buffers in cells and phosphate salts in bone. For example, hemoglobin in red blood cells binds hydrogen ions, thereby reducing the increase in free [H$^+$].

Respiratory acidosis is most often due to inadequate ventilation. The retention of CO_2 and fall in blood pH that ensue, however, both act as important stimuli to enhance ventilation. In addition, hypoxia due to inadequate ventilation may also be present, and a low arterial P$_{O_2}$ acts on the peripheral chemoreceptors to stimulate ventilation. The respiratory system is the problem, however, so it cannot return blood pH to normal.

The kidneys compensate for respiratory acidosis by increasing the output of hydrogen ions in the urine. As was mentioned, a high P$_{CO_2}$ stimulates the renal tubular epithelium to secrete hydrogen ions into the tubular urine. The result is that all of the filtered bicarbonate is reabsorbed, and increased amounts of hydrogen ions, combined with urinary buffers, are excreted in the urine. By getting rid of hydrogen ions in the urine, the kidneys raise the plasma bicarbonate concentration. In other words, they add buffer base (bicarbonate) to the blood, diminishing the severity of the acidemia. The renal response takes a few days, so with chronic respiratory acidosis, higher plasma levels of bicarbonate are achieved than those seen with respiratory acidosis of short duration.

Respiratory Alkalosis: Loss of CO_2

Respiratory alkalosis is easily understood as the opposite of respiratory acidosis. In this case, we have an abnormal process leading to excessive loss of CO_2. This causes a fall in arterial blood P$_{CO_2}$, a fall in plasma [H$^+$] (a rise in pH), and a decline in plasma [HCO$_3^-$]. Respiratory alkalosis is produced by alveolar hyperventilation, which results in excessive removal of CO_2 from the blood. Hyperventilation may be caused by voluntary effort, anxiety, or stimulation of respiratory centers owing to some abnormal influence acting directly on the brain (e.g., aspirin intoxication, fever, meningitis). Hyperventilation may also be produced reflexly. For example, hyperventilation and respiratory alkalosis are observed at high altitudes because of reflex stimulation of ventilation by low P$_{O_2}$.

The alkaline shift in pH produced during respiratory alkalosis is diminished by the action of buf-

Figure 26–6
High altitude results in respiratory alkalosis, since breathing is stimulated by the low oxygen tension. The photograph shows a scientist at the summit of Mount Everest collecting an alveolar gas sample. (© Dr. John B. West, M.D./University of California at San Diego.)

fers in the body. Chemical buffers (e.g., hemoglobin) are titrated in the alkaline direction and liberate hydrogen ions, thereby tending to diminish the fall in free $[H^+]$. As is the case for respiratory acidosis, most ($> 95\%$) of the chemical buffering occurs within cells. Although the origin of respiratory alkalosis is hyperventilation, the resulting loss of CO_2 and rise in pH limit the extent of hyperventilation. CO_2 and hydrogen ions are two important stimuli for ventilation, and a decrease in these stimuli leads to diminished ventilatory drive.

The kidneys compensate for respiratory alkalosis by excreting bicarbonate in the urine. A low P_{CO_2} leads to decreased hydrogen ion secretion by the tubular epithelium. The result is that filtered bicarbonate is incompletely reabsorbed, and so is excreted. The loss of bicarbonate in the urine results in a further decrease in plasma bicarbonate. In other words, the kidneys reduce the amount of base (bicarbonate) in the blood and so diminish the severity of the alkalemia. The kidneys work relatively slowly, so it takes them a few days to compensate fully for a persistent respiratory alkalosis.

Metabolic (Nonrespiratory) Acidosis: Gain of Fixed Acid or Loss of Bicarbonate

Metabolic acidosis is an abnormal process characterized by gain of fixed acid (i.e., an acid other than H_2CO_3) or loss of bicarbonate. Many conditions can produce metabolic acidosis. In renal failure, the acid

end-products of metabolism are not excreted at a rate that keeps up with production. In uncontrolled diabetes mellitus, large quantities of ketone body acids are produced. If the circulation is inadequate—for example, due to cardiogenic or hemorrhagic shock—glucose is converted anaerobically to lactic acid. In heavy exercise, too, lactic acid production is increased. Ingestion of certain acidifying agents (e.g., ammonium chloride) produces a metabolic acidosis. Loss of bicarbonate from the body, due to (1) diarrhea (loss of alkaline intestinal fluids) or (2) abnormal excretion of bicarbonate in the urine, also results in metabolic acidosis.

If strong acid is added to the body, the reactions

$$H^+ + HCO_3^- \rightleftharpoons H_2CO_3 \rightleftharpoons H_2O + CO_2$$

are pushed to the right. We can then correctly predict that in this type of metabolic acidosis, the hydrogen ion concentration should be increased and the plasma bicarbonate concentration decreased. Further, we might predict that P_{CO_2} should be elevated (but recall that CO_2 operates in an open system). P_{CO_2} is increased only transiently during acute infusion of large amounts of strong acid; in established metabolic acidosis, P_{CO_2} is actually below normal due to respiratory compensation (Table 26–2).

Buffering during metabolic acidosis is accomplished by chemical buffers present in extracellular fluid, cells, and bone. Roughly one half of the buffering occurs in extracellular fluid, where bicarbonate is the principal buffer. Due to the acidic blood pH, the respiratory system is stimulated and lowers the P_{CO_2}, and hence $[H_2CO_3]$, in the blood, thereby making the blood less acidic (more alkaline). Respiratory compensation for a metabolic acidosis is prompt (minutes to hours), but usually does not restore the blood pH to normal. Figure 26–4 illustrates how the bicarbonate-CO_2 system participates in buffering strong acid.

Renal compensation is slower (takes days) but more complete than respiratory compensation. The kidneys increase the output of hydrogen ions in the urine. Since the plasma bicarbonate is mainly decreased, the filtered bicarbonate load is reduced and all of the filtered bicarbonate is reabsorbed. Increased amounts of hydrogen ions, together with urinary buffers (phosphate, ammonia), are excreted in the urine, so new bicarbonate is added to the blood by the kidneys to replenish depleted bicarbonate reserves. If the underlying cause of the metabolic acidosis is corrected, then normal kidneys bring the pH of the blood back to normal.

Metabolic Acidosis in Diabetes Mellitus

Diabetes mellitus ("diabetes") is a common disorder in which there is insufficient secretion of insulin, or in which the target tissues are resistant to this hormone. In uncontrolled diabetes, a severe metabolic acidosis may develop.

The following arterial blood values were measured in a person with uncontrolled diabetes (normal values are given in parentheses).

pH	7.15 (7.40)
plasma [HCO_3^-], mEq/L	4 (24)
Pco_2, mm Hg	13 (40)
plasma glucose, mg/dl	700 (100)

Acidosis occurs because insulin deficiency leads to disturbed carbohydrate metabolism, diversion of metabolism toward utilization of fatty acids, and overproduction of ketone body acids (acetoacetic acid and 3-hydroxybutyric acid). These fairly strong acids (pK_as close to 4) are neutralized in the body by bicarbonate and other buffers. The production of these acids leads to a fall in plasma bicarbonate concentration and in blood pH.

The kidneys compensate for the metabolic acidosis by reabsorbing all of the filtered bicarbonate. They also increase excretion of titratable acids. Part of the titratable acid is composed of ketone body acids. These acids, however, can only be partially titrated to their acid form, since the urine pH cannot go below 4.5. The result is that the ketone body acids are excreted largely in their anionic form; because of the requirement of electroneutrality in solutions, increased urinary excretion of sodium and potassium results. An important compensation for the acidosis is increased renal synthesis and excretion of ammonia. This allows the person to get rid of more hydrogen ions (combined with ammonia as NH_4^+), or (in what amounts to the same effect) allows the kidneys to add new bicarbonate to the blood. The ammonium in the urine can replace sodium and potassium ions, thereby resulting in conservation of these valuable cations.

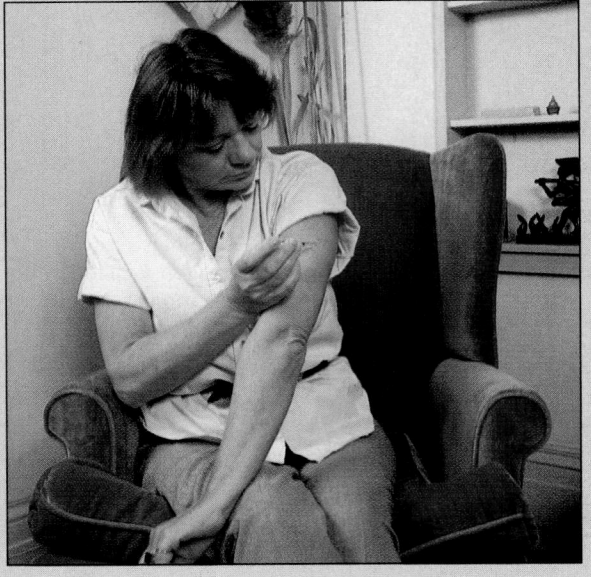

A woman injects herself with insulin to control her diabetes. (© David York/Medichrome.)

The increased acidity of the blood stimulates pulmonary ventilation, resulting in a compensatory lowering of alveolar and blood Pco_2. The lowering of blood H_2CO_3 acts to bring the blood pH back toward normal. The labored, deep breathing that accompanies severe, uncontrolled diabetes is called **Kussmaul's respiration.**

The severe acidosis, electrolyte disturbances, and volume depletion that accompany uncontrolled diabetes mellitus may lead to coma and death. Correction of the acid-base disturbance is best achieved by correcting the underlying cause, rather than just treating the symptoms. Hence, administration of a suitable dose of insulin is usually the key element of therapy. In some patients with severe acidemia, sodium bicarbonate solutions may be infused to speed recovery. Losses of sodium, potassium, and water should be replaced.

Metabolic (Nonrespiratory) Alkalosis: Gain of Strong Base or Bicarbonate or Loss of Acid

Metabolic alkalosis is an abnormal process characterized by gain of strong base or bicarbonate or loss of acid (other than H_2CO_3). Ingestion of excess sodium bicarbonate (used in treating gastric ulcers) results in a metabolic alkalosis. Oxidation of organic anions (e.g., citrate, lactate, acetate) has the same effect. A common cause of metabolic alkalosis is vomiting of acidic gastric juice. Every time a hydrogen ion is secreted into the lumen of the stomach, a bicarbonate ion is added to the blood (Chapter 23).

Intracellular pH

Normally, cytoplasmic pH is closely regulated. Many of the mechanisms involved in this regulation have been elucidated in the last decade.

Cytoplasmic pH is important because cell pH affects the activity of enzymes that control metabolic reactions. Cell pH changes affect many physiological activities, including (1) muscle contraction, (2) the opening and closing of cell membrane ion channels, (3) secretion by endocrine gland cells, and (4) cell division and proliferation.

Cell cytoplasm is generally more alkaline than we would predict if hydrogen ions were passively distributed across the cell plasma membrane. With a cell membrane potential of 60 mv, inside negative, passive distribution of H^+ across the cell membrane would lead to a cell $[H^+]$ 10 times higher than that in extracellular fluid. In other words, if the extracellular $[H^+]$ is 40 nM (pH = 7.4), intracellular $[H^+]$ should be 400 nM (pH = 6.4) at equilibrium. Intracellular pH in skeletal muscle cells, as determined with pH-sensitive microelectrodes, is about 6.9 ($[H^+]$ = 125 nM). Clearly, hydrogen ions are not distributed passively across the cell membrane, but are maintained in cells *below* the equilibrium value. Cells tend to gain H^+ because of metabolic production of acids and because hydrogen ions diffuse into the cells from the extracellular fluid along a favorable electrical gradient.

Intracellular $[H^+]$ is kept low by various membrane transport mechanisms that remove hydrogen ions (or add bicarbonate). One such mechanism that appears to be present in nearly all body cells is an **Na^+/H^+ exchanger.** This exchanger exchanges one H^+ for one Na^+ and is electrically neutral. Transport of H^+ out of the cell is uphill and is coupled to the downhill movement of sodium into the cell. Thus, the Na^+/H^+ exchanger is a secondary active transport system.

Activity of the Na^+/H^+ exchanger is increased when the cell cytoplasm is made more acidic. In part, this reflects an increase in substrate $[H^+]$ concentra-

tion. In addition, however, H^+ also stimulates the exchanger by combining with an activator site on its cytoplasmic side. This makes the exchanger more effective in pumping H^+ out into the extracellular fluid so that the cell can counteract the threat of intracellular acidosis.

The Na^+/H^+ exchanger is not only activated in defense of cell pH, but also, in some cells, when cell volume is threatened by shrinkage. If these cells are placed in a hypertonic environment, parallel Na^+/H^+ and Cl^-/HCO_3^- exchangers in the cell membrane are engaged, which leads to net uptake of NaCl. This, in turn, causes water to enter the cells, restoring normal cell volume. The Na^+/H^+ exchanger is also stimulated by many hormones and growth factors. This leads to cytoplasmic alkalinization, a change that affects cell metabolism and activity. Finally, as discussed in this chapter, the Na^+/H^+ exchanger in the kidney tubules plays a key role in acidifying the urine.

Many cells also possess various cell membrane bicarbonate transporting mechanisms—for example, sodium-independent and sodium-dependent Cl^-/HCO_3^- exchangers. These mechanisms remove bicarbonate if the cell becomes too alkaline, or add bicarbonate if the cell becomes too acidic. Different pathways are activated in response to a rise or fall in pH.

Cells have additional means to deal with excesses of acid or base. They contain large stores of protein and organic phosphate buffers that can bind or release H^+. Various biochemical reactions in cells use up or release H^+. For example, conversion of lactic acid to CO_2 and water, or to glucose, effectively disposes of acid. Various cell organelles may sequester hydrogen ions. For example, an H^+-ATPase in endosomes and lysosomes pumps H^+ out of the cytoplasm into these organelles. All of these processes assure a relatively constant cytoplasmic pH.

Loss of the stomach contents through vomiting results in net addition of bicarbonate to the body. The characteristic arterial blood changes include a primary increase in plasma $[HCO_3^-]$, a rise in pH, and a compensatory rise in P_{CO_2} (see Table 26–2).

Chemical buffers in the body act to limit the alkaline shift in pH by liberating hydrogen ions as

they are titrated in the alkaline direction. Respiratory compensation for a metabolic alkalosis is alveolar hypoventilation, brought about by the alkaline pH of the blood (inhibits ventilation). Respiratory compensation is limited, however, because hypoventilation leads to a rise in P_{CO_2} and a fall in P_{O_2}, both of which stimulate ventilation.

Renal compensation involves increased excretion of bicarbonate in the urine. Since the plasma bicarbonate is primarily elevated, more bicarbonate is filtered than can be reabsorbed by the tubules. Increased excretion of bicarbonate in the urine leads to a fall in plasma bicarbonate concentration and a return of blood pH back toward normal.

Analysis of Acid-Base Data from Blood

We have described individually each of the four fundamental disturbances of acid-base balance. Proper identification of an acid-base disturbance in a patient requires clinical evidence (medical history and examination) and supporting laboratory studies. From blood acid-base data (pH, P_{CO_2}, and plasma $[HCO_3^-]$), it is often possible to suggest what type of disturbance might be present.

Suppose, for example, that the following systemic arterial acid-base data were obtained for a patient who had been vomiting: pH 7.48, P_{CO_2} 45 mm Hg, and plasma $[HCO_3^-]$ 32 mEq per liter. What type of acid-base disturbance is present? We already know that vomiting of acidic gastric juice produces a metabolic alkalosis. To evaluate the blood chemistry values, we have to know normal values for these blood variables. These are:

pH	7.35–7.45
P_{CO_2}	35–45 mm Hg
$[HCO_3^-]$	22–26 mEq per liter

Clearly, a pH of 7.48 is above normal, and so an alkalosis is present. If it were a respiratory alkalosis, P_{CO_2} would be below normal. In this case it is not, so we conclude that a metabolic alkalosis is present. The fact that plasma bicarbonate is above normal confirms our conclusion. The P_{CO_2} on the high side of normal is probably due to respiratory compensation.

Sometimes, two independent disturbances of acid-base balance, usually one respiratory and the other metabolic, are present together in the same person. This is called a **mixed acid-base disturbance.** For example, a patient with chronic pulmonary disease may develop diarrhea. The mixed respiratory acidosis and metabolic acidosis in this patient could result in severe acidemia. Sometimes, two primary acid-base disturbances are simultaneously present and offset each other. For example, a mixed respiratory acidosis and metabolic alkalosis might be accompanied by a normal pH. Treatment of mixed acid-base disturbances is more complicated than when a single disorder is present.

SUMMARY

Principles of Acid-Base Physiology

An acid is a substance that can donate hydrogen ions. A base is a substance that can accept or bind hydrogen ions. In a neutralization reaction, an acid reacts with a base to form a salt and water.

An acid in aqueous solution reversibly dissociates into a hydrogen ion and conjugate base. A high acid dissociation constant (low pK_a) is characteristic of a strong acid; a low acid dissociation constant (high pK_a) is characteristic of a weak acid.

The pH of a solution is equal to the negative logarithm to the base 10 of the molar hydrogen ion activity (concentration). A pH value of 7 is neutral; below 7, acidic; and above 7, alkaline. Arterial blood pH averages 7.40; various body fluids have different pH values.

pH buffers minimize changes in pH produced when acid or base is added to a solution.

The body obtains H^+ ions from the metabolism of foods, and from carbonic acid formed when CO_2 reacts with water.

Regulation of Extracellular pH: Buffering Mechanisms in the Body

Chemical buffers consist of a weak acid and its conjugate base (or a weak base and its conjugate acid). Important chemical buffers in the extracellular fluid include the bicarbonate-CO_2 system, plasma proteins, and inorganic phosphate. Large amounts of protein and organic phosphate buffers are present in cells; bone (due to its phosphate and carbonate salts) also participates in pH buffering in the body.

The Henderson-Hasselbalch equation for the bicarbonate-CO_2 system is $pH = 6.10 + \log([HCO_3^-]/0.03 \times P_{CO_2})$. The bicarbonate-$CO_2$ buffer system has special importance in acid-base regulation, because the physiological buffers (lungs and kidneys) act on the components of this buffer pair. The respiratory system influences plasma pH by regulating the partial pressure of CO_2 (P_{CO_2}) in arterial blood by varying the level of alveolar ventilation. The kidneys influence plasma pH by getting rid of acid or base in the urine—in other words, by adding new bicarbonate to the blood or getting rid of bicarbonate.

On the usual mixed diet, people are challenged by excess production of nonvolatile (fixed) acids. Although chemical buffers can minimize a fall in pH, the burden of eliminating hydrogen ions from the body falls upon the kidneys. Urinary acidification involves three processes: reabsorption of filtered bicarbonate, excretion of titratable acid, and excretion of ammonia. All involve hydrogen ion secretion by the kidney tubule cells. Renal reabsorption of filtered bicarbonate returns to the blood bicarbonate that was temporarily lost in the process of filtration. Excretion of hydrogen ions combined with urinary buffers (titratable acids, ammonia) leads to addition to the blood of new bicarbonate, which replenishes depleted bicarbonate reserves. The respiratory system participates in pH buffering by expiring CO_2 as rapidly as it is produced. Ventilation is stimulated by a rise in arterial P_{CO_2} or a fall in pH.

Disturbances of Acid-Base Balance

Respiratory acidosis is characterized by accumulation of CO_2 and a fall in blood pH. The renal compensation is enhanced excretion of hydrogen ions in the urine, which results in addition of bicarbonate to the blood, thereby reducing the acidic shift in blood pH.

Respiratory alkalosis is characterized by excessive loss of CO_2 and a consequent rise in blood pH. The renal compensation is increased excretion of bicarbonate in the urine, which lowers the blood pH.

Metabolic acidosis involves an abnormal gain of fixed acid or loss of bicarbonate, and is characterized by a fall in plasma bicarbonate and pH. Respiratory compensation lowers the blood P_{CO_2}, making the blood less acidic.

Metabolic alkalosis is associated with an abnormal gain of base or loss of fixed acid, and is characterized by a rise in plasma bicarbonate and pH. Respiratory compensation raises the blood P_{CO_2}, bringing the blood pH closer to normal.

Acid-base data from the blood are used to diagnose these disturbances.

Intracellular pH is stabilized by cell buffering mechanisms and by ion transport processes (e.g., Na^+/H^+ exchanger) in plasma membranes.

REVIEW QUESTIONS

Identify Each with a Term

1. An enzyme that catalyzes the reversible reaction $CO_2 + H_2O \rightleftharpoons H_2CO_3$.

2. An arterial blood pH less than 7.35.

3. A substance that can accept or bind hydrogen ions.

4. An agent that minimizes the change in pH when either acid or base is added to a solution.

5. A chemical substance that can function as both an acid and a base.

6. An acid with a high pK_a.

Define Each Term

7. pH

8. Henderson-Hasselbalch equation

9. open system

10. respiratory acidosis

11. respiratory alkalosis

12. metabolic acidosis

13. metabolic alkalosis

Choose the Correct Answer

14. Which of the following is an acid?

  a. HCO_3^-

  b. NaOH

  c. NH_3

  d. NH_4^+

15. Most of the hydrogen ions secreted by the kidney tubules are:

  a. consumed by filtered bicarbonate

  b. excreted as ammonium ions

  c. excreted as free hydrogen ions

  d. excreted as titratable acid

16. Which process in the kidney leads to generation of *new* bicarbonate to replenish depleted bicarbonate reserves?

  a. excretion of ammonia

  b. excretion of titratable acid

  c. reabsorption of filtered bicarbonate

  d. a and b

  e. all of the above

17. The nephron segment which secretes the greatest number of hydrogen ions is the:

  a. collecting duct

  b. distal convoluted tubule

  c. proximal convoluted tubule

  d. thick ascending limb

18. The nephron segment that can establish the steepest pH gradient (urine-to-blood) is the:

  a. collecting duct

  b. distal convoluted tubule

  c. proximal convoluted tubule

  d. thick ascending limb

19. A person has been vomiting for several days. Arterial blood values are: pH = 7.53, Pco_2 = 50 mm Hg, and plasma $[HCO_3^-]$ = 40 mEq/liter. What type of acid-base disturbance is present?

  a. respiratory acidosis

  b. respiratory alkalosis

  c. metabolic acidosis

  d. metabolic alkalosis

20. A person on a mountain-climbing expedition has been living at high altitude for several days. Arterial blood values are: pH = 7.48, Pco_2 = 25 mm Hg, and plasma $[HCO_3^-]$ = 18 mEq/liter. What type of acid-base disturbance is present?

  a. respiratory acidosis

  b. respiratory alkalosis

  c. metabolic acidosis

  d. metabolic alkalosis

21. The renal compensation for chronic respiratory alkalosis is:

  a. increased excretion of bicarbonate

  b. increased excretion of titratable acid

  c. increased synthesis and excretion of ammonia

  d. both b and c

  e. all of the above

22. A person with uncontrolled diabetes mellitus also has emphysema (pulmonary disease). Arterial blood values are: pH = 7.21, Pco_2 = 60 mm Hg, and plasma $[HCO_3^-]$ = 23 mEq/liter. This subject has a:

  a. metabolic acidosis

  b. mixed metabolic acidosis and respiratory acidosis

  c. mixed metabolic acidosis and respiratory alkalosis

  d. respiratory acidosis

23. Which of the following plays a role in the regulation of intracellular (cytosolic) pH?

  a. cell membrane Na^+/H^+ exchanger

  b. cell membrane bicarbonate transport mechanisms

  c. chemical buffering by intracellular proteins and phosphates

  d. metabolic reactions that consume or liberate hydrogen ions

  e. all of the above

Calculate

24. What is the approximate pH of a 0.01 M HCl solution?

25. What is the approximate pH of a 0.01 M NaOH solution?

26. A person with severe, uncontrolled diabetes mellitus is found to have an arterial plasma pH of 7.15 and a Pco_2 of 13 mm Hg. What is the plasma bicarbonate concentration?

Answer Each Question in One or Two Sentences

27. Does a 0.1 M hydrochloric acid solution have the same pH as a 0.1 M acetic acid solution? Why or why not?

28. What two factors determine the effectiveness of a buffer?

29. Why is a person who eats a mixed diet (meat and vegetables) threatened by metabolic acidosis?

30. What are the important chemical buffers in (1) extracellular fluid, (2) intracellular fluid, (3) bone, and (4) urine?

31. Why is the bicarbonate-CO_2 buffer pair, despite a low pK of 6.10, an effective buffer in defending pH?

32. What is the isohydric principle?

33. What is the location of the receptors that stimulate ventilation when CO_2 accumulates in the body?

SUGGESTED READINGS

Davenport, H.W. *The ABC of Acid-Base Chemistry*, 6th ed. Chicago: University of Chicago Press, 1974.

Frelin, C., Vigne, P., Ladoux, A., and Lazdunski, M. "The regulation of intracellular pH in cells from vertebrates." *European Journal of Biochemistry*, 174:3–14, 1988.

Knepper, M.A., Packer, R., and Good, D.W. "Ammonium transport in the kidney." *Physiological Reviews*, 69:179–249, 1989.

Madshus, I.H. "Regulation of intracellular pH in eukaryotic cells." *Biochemical Journal*, 250:1–8, 1988.

Rose, B.D. *Clinical Physiology of Acid-Base and Electrolyte Disorders*, 3rd ed. New York: McGraw-Hill Book Company, 1989.

Valtin, H., and Gennari, F.J. *Acid-Base Disorders. Basic Concepts and Clinical Management.* Boston: Little, Brown and Company, 1987.

Winters, R.W., and Dell, R.B. *Acid-Base Physiology in Medicine. A Self-Instruction Program*, 3rd ed. Boston: Little, Brown and Company, 1982.

KEY TERMS

acid dissociation constant (p. 889)
acidemia (p. 901)
alkalemia (p. 901)
chemical pH buffer (p. 891)

Henderson-Hasselbalch equation (p. 890)
intracellular pH (p. 905)
isohydric principle (p. 896)
metabolic acidosis (p. 903)

metabolic alkalosis (p. 904)
pH (p. 890)
respiratory acidosis (p. 901)

respiratory alkalosis (p. 902)
titratable acid (p. 898)
weak acid (p. 889)

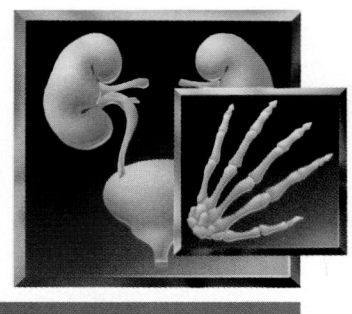

CHAPTER
27

Calcium, Phosphate, and Bone Metabolism

As described in many of the earlier chapters, a number of different homeostatic mechanisms are at work in the body, the primary focus of these being the maintenance of a constant environment in which cells can develop and function. If we were to monitor a number of different physiological parameters in the body—including, for example, the blood sugar concentration, core body temperature, and blood carbon dioxide concentration—the results would show that one of the most closely regulated of all parameters is the plasma calcium concentration. Such strict regulation usually suggests that the factor in question plays a vital role in one or more processes in the body. Although not as closely regulated, phosphate also plays several important roles in the body.

Overview: Functions of Calcium and Phosphate

Calcium, or more correctly the Ca^{2+} ion, is involved in a number of different physiological processes, several of which are listed in Table 27–1. The first of these, the effect of calcium on nerve function, is probably the most important with regard to an acute dependency on normal calcium ion concentrations.

External calcium ions dramatically influence the permeability of nerves to sodium ions. Therefore, the maintenance of extracellular calcium concentrations is crucial if the resting membrane potential of nerves is to be maintained and the nerve impulse transmission is to occur normally. When extracellular calcium ion concentrations are significantly decreased, spontaneous action potentials can be generated in the nerve. If a motor neuron is involved, the result can be **tetany** of the muscles in that motor unit. Figure 27–1 illustrates the classic clinical sign indicating the first stages of **hypocalcemic (low-calcium) tetany.** Muscles in the hand and forearm contract to produce the characteristic appearance shown. With severe, prolonged hypocalcemia, muscles of the larynx become tetanized to the extent that the individual dies from suffocation.

Other functions of calcium ion in the body are also listed in Table 27–1. At the neuromuscular junction, an influx of calcium into the nerve ending serves to trigger the release of acetylcholine. As described in Chapter 16, calcium fluxes in muscle provide the signal for excitation-contraction coupling.

Figure 27–1
Typical position of the hand in hypocalcemic tetany. The spastic contractions of muscles in the forearm and hand are the result of hyperexcitability in nerves, which occurs when the plasma calcium concentration drops below normal.

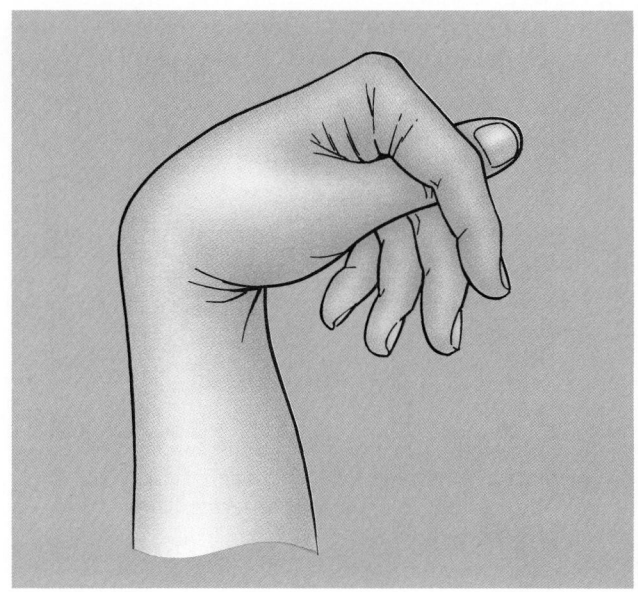

Table 27–1

Some of the Physiological Actions of Calcium

1. Required for the maintenance of normal sodium permeability in nerves
2. Involved in triggering the release of acetylcholine from nerve endings at the neuromuscular junction
3. Involved in excitation-contraction coupling in muscle cells
4. Serves as an intracellular signal for some hormones
5. Required by some enzymes for normal activity
6. Required for blood clotting to occur normally
7. Required for protein secretion
8. Constituent of bone

In other cell types, calcium influx provides a signal analogous to a second messenger; thus, calcium ion is important in the mechanism of action of some hormones. In addition, a number of cellular enzymes and proteins are known to require calcium to function properly, and the process of protein secretion is also known to depend on the presence or calcium ion.

A specialized case in which calcium ion is needed for enzymatic activity involves the enzymes responsible for blood clotting (see Table 27–1). Some of the compounds used clinically to prevent clotting in blood samples taken from patients produce their effect by binding calcium and preventing its participation in the clotting reactions.

Finally, calcium is a constituent of bone, and thus in this regard it serves an important structural role in the body. As later sections will describe, the bones provide an important reservoir of calcium and therefore play an important role in the regulation of extracellular calcium ion homeostasis.

Phosphorus is also present in considerable amounts in bone. As will be described later, the majority of phosphorus in the body exists in the form of the **phosphate ion (PO_4)**. The extracellular concentration of phosphate is not as strictly regulated as is that of calcium, however. As indicated in Table 27–2, phosphate serves the important role of providing a major portion of the intracellular buffering capacity. In addition, when coupled in organic form, phosphate is a component of a number of important cellular constituents and macromolecules. Nucleic acids, phospholipids, and various metabolic intermediates, to name a few, all contain phosphate. Chapter 5 introduced another important aspect of phosphate metabolism involving the regulation of certain metabolic processes by phosphorylation or dephosphorylation of key regulatory enzymes.

Table 27–3

Typical Body Content and Tissue Distribution of Calcium and Phosphorus

	Calcium	Phosphorus
Total body content*	~1000 g	~500 g
Tissue distribution	Relative distribution (% of total body content)	
Bone	99	85
Muscle	0.3	6
Other tissues	0.7	9

*Based on a body weight of 150 lbs and an average body composition.

The Distribution of Calcium and Phosphate in Tissue

The relative distributions of calcium and phosphate among bone and other tissues in the body are presented in Table 27–3. As this table indicates, approximately 99% of the total calcium in the body is located in bone. Even muscle, in which calcium plays such a vital role in excitation-contraction coupling, contains only about 0.3% of the total. The remaining 0.7% is distributed between the extracellular fluid and the rest of the cells in the body. Phosphate is also located predominantly in bone, although not nearly to the extent that calcium is; about 85% of the total body phosphate is present in bone.

Although the majority of body calcium and phosphate is present in bone, these are not the only constituents found in relatively high abundance there. As Table 27–4 indicates, considerable amounts of total body carbonate, magnesium, and sodium are also located in the bones. The large amount of carbonate present is of particular importance in that bone can contribute this carbonate to the extracellular fluid to combat acidosis, and thus provide a third line of defense in acid-base regulation along with the lungs and kidneys. Longstanding, uncorrected acidosis can lead to a considerable loss of bone mineral for this very reason. Perhaps surprisingly, bone also contains a considerable amount of water. About 20% of the weight of bone is actually due to water, and thus about 9% of all the water present in the body is located in bone (see Table 27–4).

Table 27–2

Some of the Physiological Actions of Phosphate

1. Functions as part of the intracellular buffer system
2. Important constituent of a variety of macromolecules, such as nucleic acids, phospholipids, metabolic intermediates, and phosphoproteins
3. Constituent of bone

Table 27–4

Major Inorganic Constituents of Bone

Constituent	Total Body Content Present in Bone (%)
Calcium	99
Phosphate	85
Carbonate	80
Magnesium	50
Sodium	35
Water	9

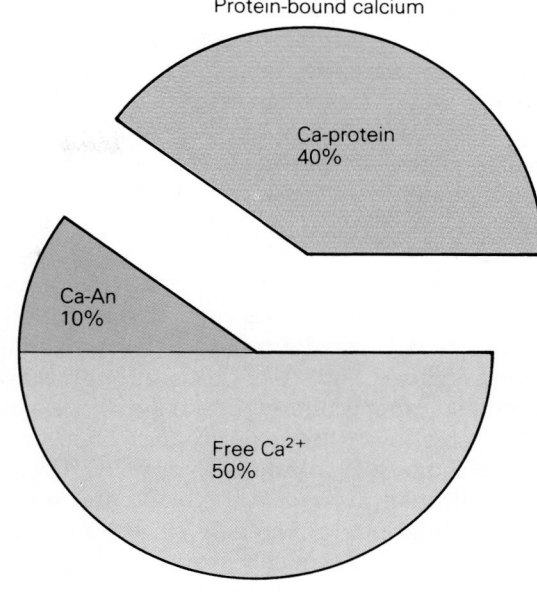

Figure 27–2
Relative distribution of different forms of calcium in the plasma. Calcium bound to plasma proteins is represented by "Ca-protein." Calcium bound to an anion, such as bicarbonate or citrate, is represented by "Ca-An," and the free ionized calcium in the plasma is represented by "Free Ca^{2+}."

The Forms of Calcium and Phosphorus in the Blood

Forms of Calcium in the Blood

As indicated previously, the calcium content of the extracellular fluid is maintained within fairly narrow limits. The total calcium content of the plasma in humans is about 10 mg/100 ml. Of this total, calcium can exist in one of several different forms. Generally, there are two major forms of calcium in the plasma. As indicated in Figure 27–2, these are the **filterable calcium** and the **protein-bound** calcium. Each of these forms behaves differently in physiological systems.

Filterable Calcium: Ionized or Complexed. The **filterable calcium** represents about 60% of the total calcium in the plasma (see Fig. 27–2 and Table 27–5) and includes those forms of calcium that are readily filtered in the glomerulus of the kidney. This calcium must therefore be reabsorbed in later segments of the renal tubules if it is not to be lost in the urine. Filterable calcium can be further subdivided into the **ionized** and the **complexed** forms.

The free Ca^{2+} ion constitutes roughly half of the total calcium present in the plasma (see Table 27–5). Ionized calcium is the **physiologically active** form of calcium. As the free ion, it is able to diffuse across plasma membranes and participate in processes such as those listed in Table 27–1. Note that this is the form of calcium that is strictly regulated in the plasma.

A smaller portion of the filterable calcium is calcium that is complexed with an anion (see Table 27–

Table 27–5

Distribution of Plasma Calcium

Form	Amount (mg/100 ml)	Percentage of Total (approximate)
Filterable		
Ionized calcium	4.8	50
Complexed, ultra-filterable, bicarbonate (0.5), citrate (0.3), phosphate (0.4), and others	1.2	10
Protein-bound	4.0	40
Albumin (3.0)		
Globulins (1.0)		
Total	10.0	100

Numbers in parentheses represent the contribution (in mg/100 ml) of each of several forms within a category.

5). The **complexed calcium** normally represents about 10% of the total calcium present in the plasma. As indicated in Table 27–5, bicarbonate is the major anion to which calcium is bound, although significant amounts are also found in association with citrate. A variety of other anions, including phosphate, make up the remainder of this category.

Protein-Bound Calcium. The second major category is that of **protein-bound calcium** (see Fig. 27–2 and Table 27–5). Calcium bound to relatively large plasma proteins cannot freely pass through plasma membranes, nor is it filtered in the kidney. As indicated in Table 27–5, the majority of protein-bound calcium is bound to **albumin.** Normally the most abundant protein present in the blood and interstitial fluid, albumin is a relatively large protein produced by the liver. A smaller proportion of the protein-bound calcium is associated with various globulins in the blood.

Forms of Phosphorus in the Blood

In contrast to the plasma calcium concentration, which is normally regulated within very narrow limits, the phosphate concentration in the blood may fluctuate on a daily basis from as low as one half of the normal concentration up to one and a half times the normal concentration. A plasma concentration ranging from 3.0 to 4.5 mg/100 ml (expressed as milligrams of **phosphorus**) is considered to be normal in humans.

In the plasma, phosphorus exists primarily as **orthophosphate,** or **PO_4.** At a normal blood pH, about 80% of the phosphate is present as HPO_4^{2-} and about 20% as $H_2PO_4^-$. When present as the phosphate ion, the majority of the phosphorus readily diffuses across membranes and is also filterable in the glomerulus of the kidney. In addition to the phosphate ion, small amounts of phosphorus are also present in the blood in organic form, such as in hexose or triose phosphates, but these represent a relatively small proportion of the total.

Pathways of Calcium and Phosphate Homeostasis

The diet is, of course, the ultimate source of both calcium and phosphate in the body. Once these ions are ingested, the **intestines, kidneys,** and **bone** are primarily involved in determining their plasma concentrations. As we will see, however, the overall patterns of movement in the body of the two ions differ considerably.

Pathways of Calcium Homeostasis

Figure 27–3 shows both the tissue distribution and the average daily flow of calcium among different tissues. In this figure, the quantity of calcium present is indicated by the blue numerals, and the daily flow of calcium among tissues is indicated by the red numerals.

The average calcium intake is about 1 gram per day (see Fig. 27–3). Most normal people can consume up to twice that amount without any deleterious effects. However, excessive calcium intake can result in problems with deposition of calcium in tissues (soft tissue **calcification**) or kidney stone formation. Of the 1 gram of calcium that is ingested, normally about one third to one half actually is absorbed from the intestines (see Fig. 27–3). The percentage can change with differing calcium intake and with differing plasma calcium status, however. For example, in growing children and pregnant or nursing women, this percentage tends to be somewhat higher, while in people of advanced age it tends to be lower.

Note also in Figure 27–3 that about 150 mg of calcium enters the intestines from the body each day. This occurs in part as a result of the sloughing of cells that line the digestive tract and also because of secretion of fluids and various digestive enzymes into the intestines. This component of calcium flux into and out of the intestines tends to change relatively little with varying calcium intake or plasma calcium status. Therefore, as Figure 27–3 indicates, the major route of excretion of *ingested* calcium is via the feces, simply because the majority of ingested calcium is not absorbed. However, as we will see later, regulation of calcium absorption from the intestines is an important aspect of overall calcium homeostasis.

Figure 27–3 also indicates that the kidneys excrete about 150 mg of calcium into the urine each day. This represents only about 1% of the total amount of calcium filtered through the glomeruli of the kidneys, as the remaining 99% is reabsorbed and returned to the blood. A small change in calcium reabsorption by the kidneys can therefore dramatically affect calcium homeostasis.

Although we may think of bone as being relatively inert, bone mineral is constantly being turned over and remodeled to meet differing mechanical needs, as well as contributing to calcium homeostasis on both a short-term and long-term basis. Roughly 500 mg of calcium both enter and leave

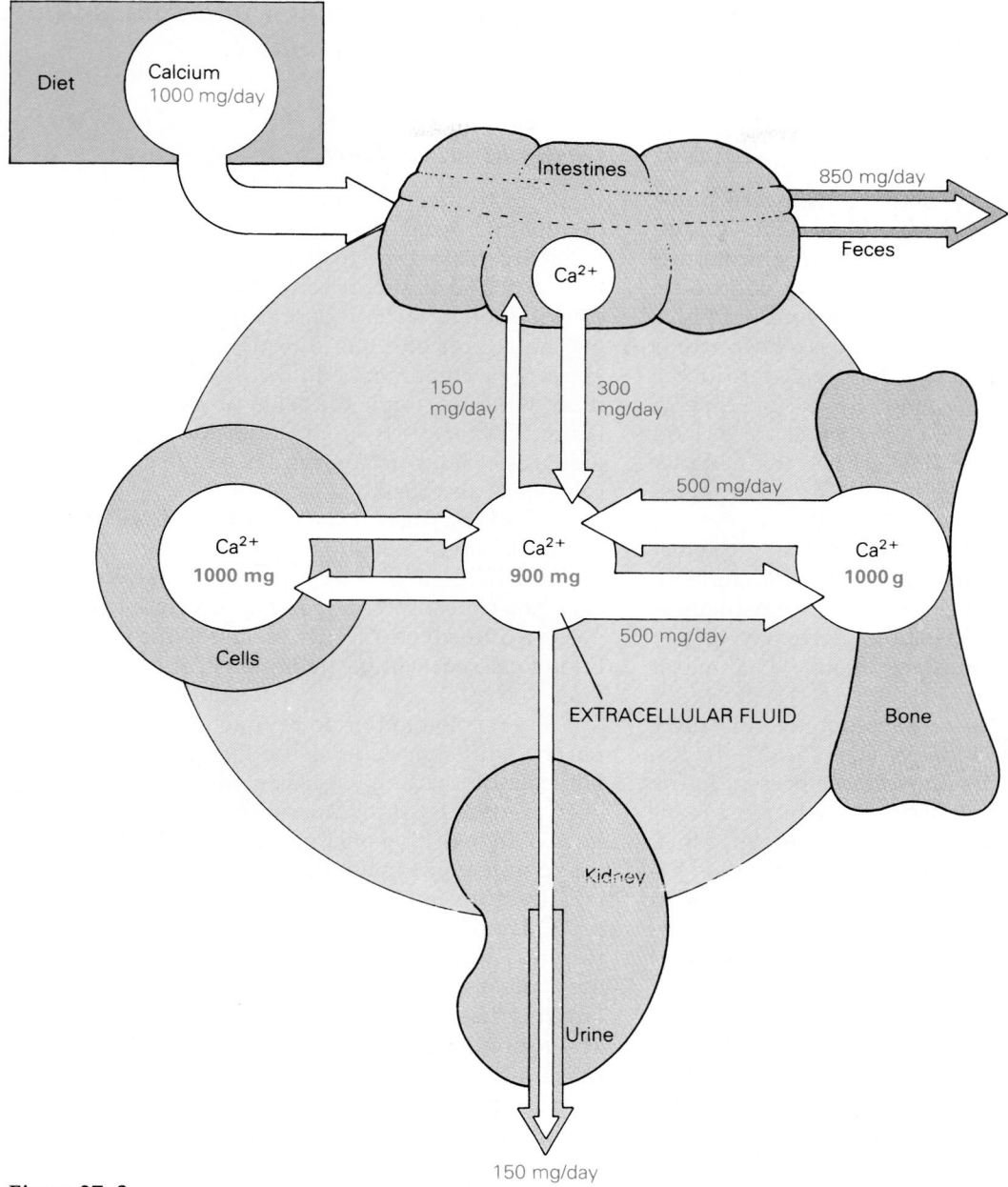

Figure 27–3
Diagram of the typical daily exchanges of calcium that occur between tissue compartments of a normal adult. Numbers shown in blue represent the amount of calcium in each compartment in *grams* or *milligrams*. Numbers in red represent the exchange of calcium between compartments in *milligrams/day*. Note that most calcium ingested each day in the diet is excreted in the feces; also note the high daily exchange of calcium between bone and the extracellular fluid. [1000 milligrams (mg) = 1 gram]

bone each day (see Fig. 27–3), indicating that bone mineral turnover is rather extensive. In a nongrowing adult, the fluxes of calcium into and out of bone will of course be equal, as indicated in Figure 27–3. In growing children, the rate of calcium deposition

is greater than the rate of calcium loss, and as bone mineral is lost with advancing age, the converse becomes true.

Each of these processes—calcium absorption from the intestines, calcium reabsorption by the kidneys, and bone mineral formation and resorption—is precisely controlled. The mechanisms involved in this control will be discussed in later sections.

Pathways of Phosphate Homeostasis

Figure 27–4 shows the various pathways of phosphate metabolism in the body. As previously indicated, the extracellular phosphate concentration is not regulated as tightly as the calcium concentration. Normally, about 1.4 grams (1400 mg) of phosphorus is ingested each day. Because phosphate is relatively abundant in most foods, a nutritional phosphorus deficiency is generally not possible as long as sufficient food is taken in to meet caloric and protein needs. As Figure 27–4 shows, the majority of ingested phosphate is absorbed from the intestines. This is in direct contrast to calcium metabolism, since as shown in the previous figure, the majority of ingested calcium is not absorbed. As with calcium, however, a small amount of phosphate is also secreted into the intestines each day. Thus, there is a *net* uptake of approximately 1100 mg of phosphorus each day.

The primary route for excretion of this daily phosphate load is the kidneys (see Fig. 27–4). Remember that most of the phosphorus present in the plasma exists as the phosphate ions, which are readily filtered in the glomeruli of the kidneys. Phosphate would thus be readily lost from the body if not for mechanisms that reabsorb and return it to the plasma. As will be described later, the reabsorption of phosphate by kidney tubule cells is a major mechanism of the hormonal regulation of phosphate homeostasis.

As with calcium, considerable amounts of phosphate flow both into and out of bone each day. Again, this reflects the very active turnover and remodeling of bone mineral that normally occurs.

Mechanisms Involved in Calcium and Phosphate Homeostasis

The previous sections referred to the processes of absorption, reabsorption, and resorption of calcium and phosphate as they occur in the gastrointestinal tract, kidney, and bone, respectively. We will now describe in more detail specific mechanisms involved in handling calcium and phosphate in each of these tissues.

Absorption of Calcium and Phosphate in the Gastrointestinal Tract

Calcium Absorption in the Gastrointestinal Tract

In the gastrointestinal tract, calcium is absorbed predominantly from the small intestine. Calcium is taken up from the small intestine by both active transport and simple passive diffusion. The relative contribution of each of these two processes depends on the region of small intestine. Active transport appears to predominate in the duodenum and jejunum, whereas simple diffusion predominates in the ileum. Because it is longer than the other two regions, the ileum is the region where most of the calcium is absorbed.

As do the other active transport processes, calcium transport exhibits characteristics of saturability. It requires energy and can move calcium against a concentration gradient. Calcium uptake by passive diffusion is directly related to the concentration of calcium within the lumen of the small intestine.

The relative importance of active transport and passive diffusion depends on a number of factors. For example, with low calcium intake, active transport predominates because uptake by diffusion is low. With ingestion of large amounts of calcium, the active transport system becomes saturated, and uptake by diffusion predominates.

Calcium uptake from the intestines is regulated through regulation of active calcium transport mechanisms. As we will see later, metabolites of vitamin D that function as hormones regulate calcium absorption by the intestinal mucosal cells. Specific calcium-binding proteins involved in calcium transport are present in intestinal mucosal cells. Under the influence of vitamin D metabolites, these mucosal cells produce more calcium-binding proteins, thereby increasing their capacity for calcium transport.

Phosphate Absorption in the Gastrointestinal Tract

Phosphate absorption also occurs by both active transport processes and passive diffusion, with active transport predominating. The small intestine is the primary site of phosphate absorption from the gastrointestinal tract. As Figure 27–4 indicates, phosphate is very efficiently absorbed from the intestines, as usually 70% to 90% of ingested phosphate is eventually absorbed. Active transport of

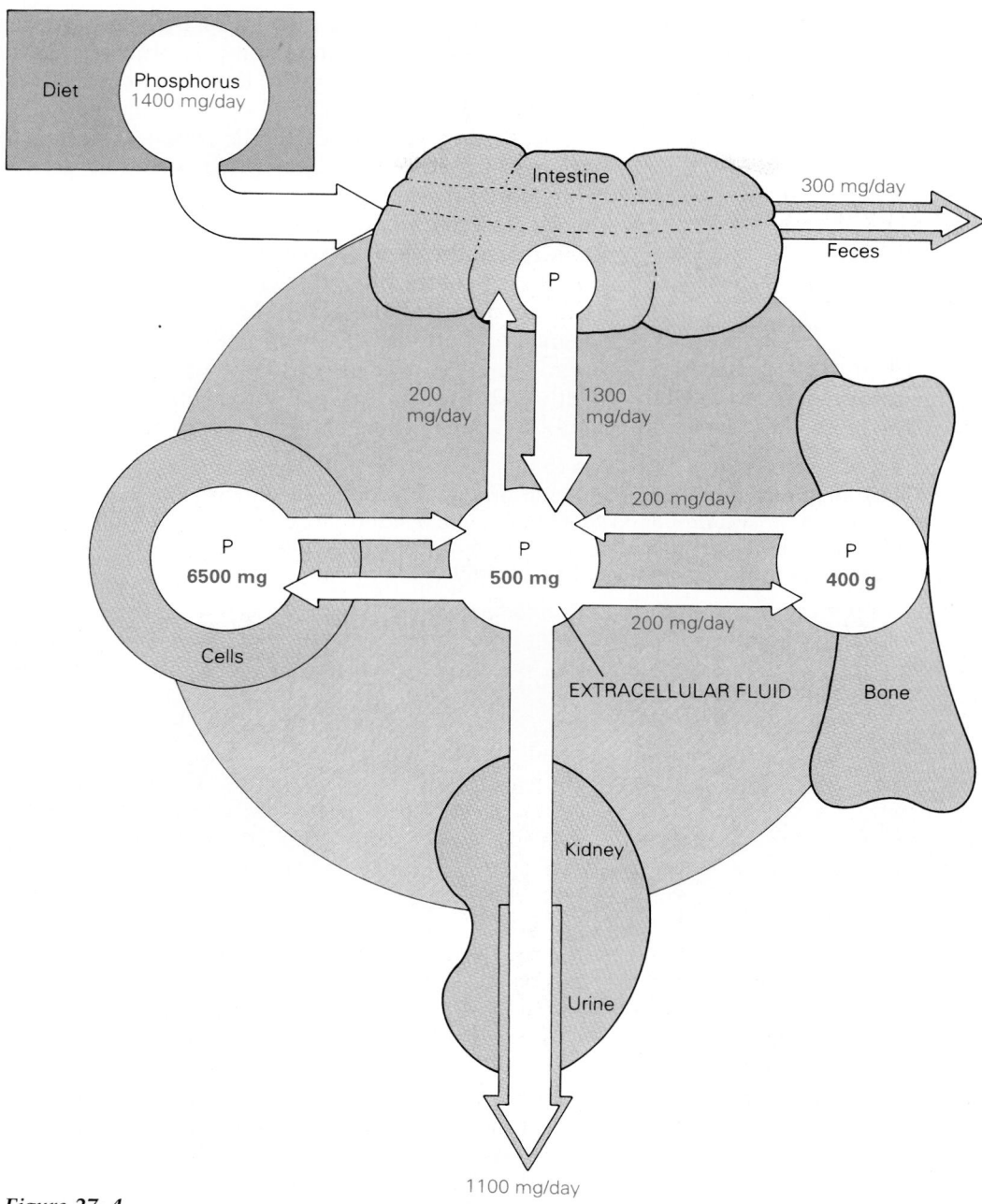

Figure 27–4

Diagram of the typical daily exchanges of phosphate that occur between tissue compartments in a normal adult. As in the previous figure, blue numbers represent amounts of phosphate in each compartment, and red numbers represent the exchange of phosphate between compartments. Note that most of the phosphate ingested each day in the diet is absorbed from the gastrointestinal tract and is excreted in the urine. Values are given in terms of grams or milligrams of phosphorus.

phosphate in the small intestine is in part coupled to calcium transport. Therefore, in situations in which active transport of calcium is low (e.g., with vitamin D deficiency) phosphate absorption is also low. The

coupling of phosphate transport to calcium transport offers certain physiological advantages that we will discuss later in this chapter. However, phosphate absorption from the small intestine differs from calcium absorption in that it is only slightly regulated, if at all. As will be described subsequently, the primary site of regulation of phosphate metabolism is the kidney.

Reabsorption of Calcium and Phosphate in the Kidney

Calcium Reabsorption in the Kidney

Recall from Table 27–5 and Figure 27–2 that filterable calcium consists of free calcium ion and calcium bound to filterable anions. Together, these make up about 60% of the total calcium in the plasma; the remaining 40% circulates bound to protein and is not filtered in the glomeruli of the kidneys. Ordinarily, about 99% of the filtered calcium is eventually reabsorbed by the kidney tubule cells and does not appear in the urine. Reabsorption occurs by active transport processes in the proximal and distal tubules. The nature of calcium reabsorption (active or passive) in the loop of Henle is not well established. About 60% of the filtered calcium is reabsorbed from the proximal tubule, about 30% from the loop of Henle, and another 9% from the distal tubule. The remaining 1% is excreted in the urine (Fig. 27–5). Although the calcium reabsorbed in the distal tubule is a small percentage of the total reabsorbed

Figure 27–5
Diagram illustrating calcium filtration and reabsorption in a kidney nephron. Note that only free ionized calcium or calcium bound to an anion is filtered in the glomerulus. Calcium is reabsorbed by active transport processes in both the proximal and distal tubules. In the distal tubule, calcium reabsorption is stimulated by parathyroid hormone.

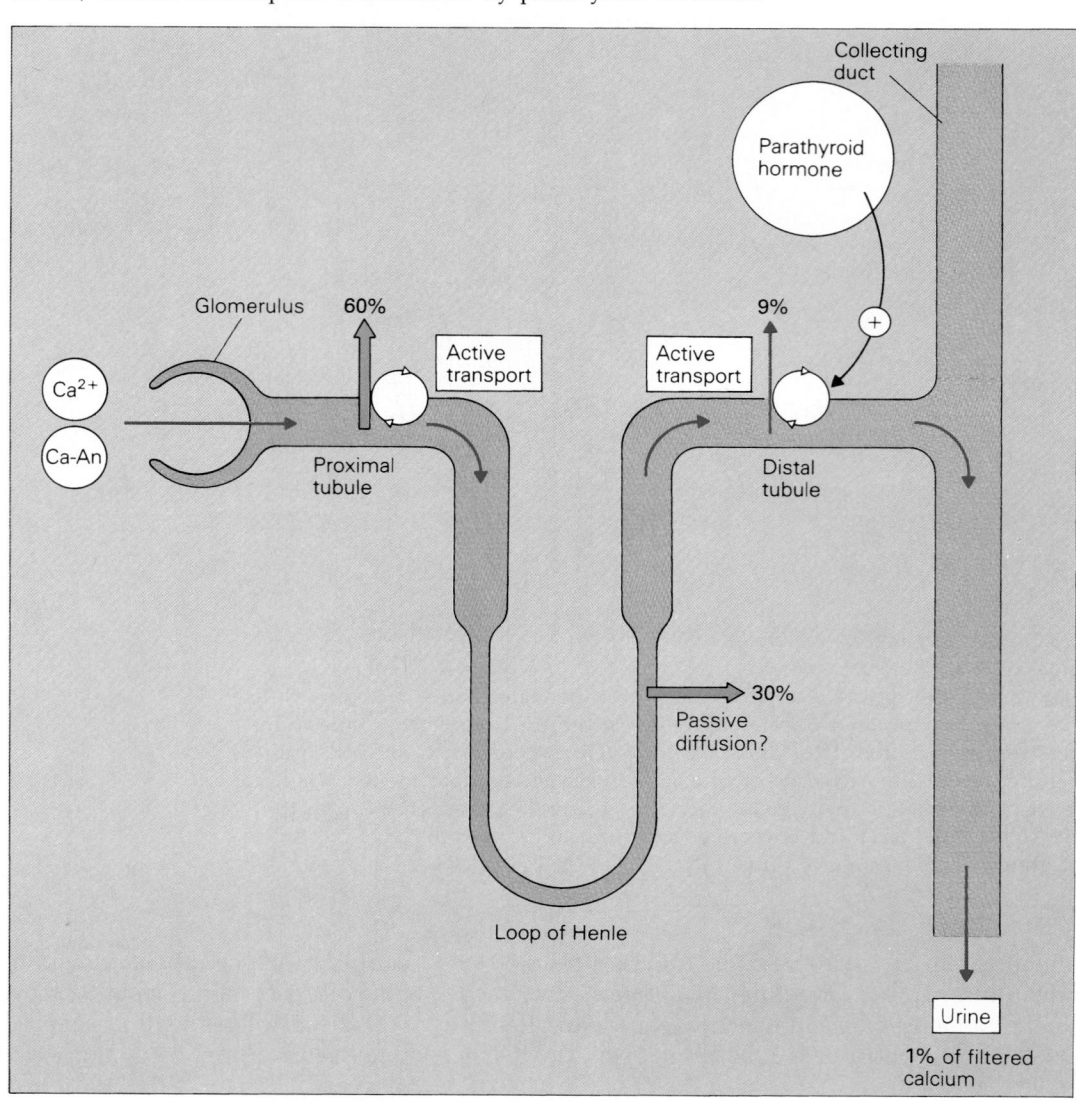

calcium, major adjustments in calcium excretion can occur at this site as a result of an action of parathyroid hormone to stimulate calcium reabsorption. Parathyroid hormone is one of three hormones involved in regulating calcium and phosphate homeostasis, as later sections of this chapter will describe. The effect of parathyroid hormone to increase the transfer of calcium from the tubular fluid to the plasma is mediated by an increase in cAMP in the distal tubule cell (refer back to Table 12–3).

Phosphate Reabsorption in the Kidney

As was indicated in Figure 27–4, the major route by which ingested phosphate leaves the body is via the urine. The kidney is therefore an important site of regulation of the extracellular phosphate concentration. Ordinarily, about 85% of the filtered phosphate is reabsorbed, and the remaining 15% is excreted in the urine. The reabsorption of phosphate is due to active transport processes. The principal site of phosphate reabsorption is in the proximal tubule (Fig. 27–6), but some reabsorption occurs in the distal tubule as well (see Fig. 27–6). Parathyroid hormone exerts a major regulatory effect on phosphate excretion by inhibiting phosphate reabsorption in the proximal tubule. As a result, parathyroid hormone causes an increase in phosphate excretion in the urine, with an accompanying decrease in plasma phosphate concentrations.

Figure 27–6
Diagram illustrating the filtration and reabsorption of phosphate in a kidney nephron. Most phosphate reabsorption occurs by active transport in the proximal tubule. As indicated, this process is inhibited by parathyroid hormone.

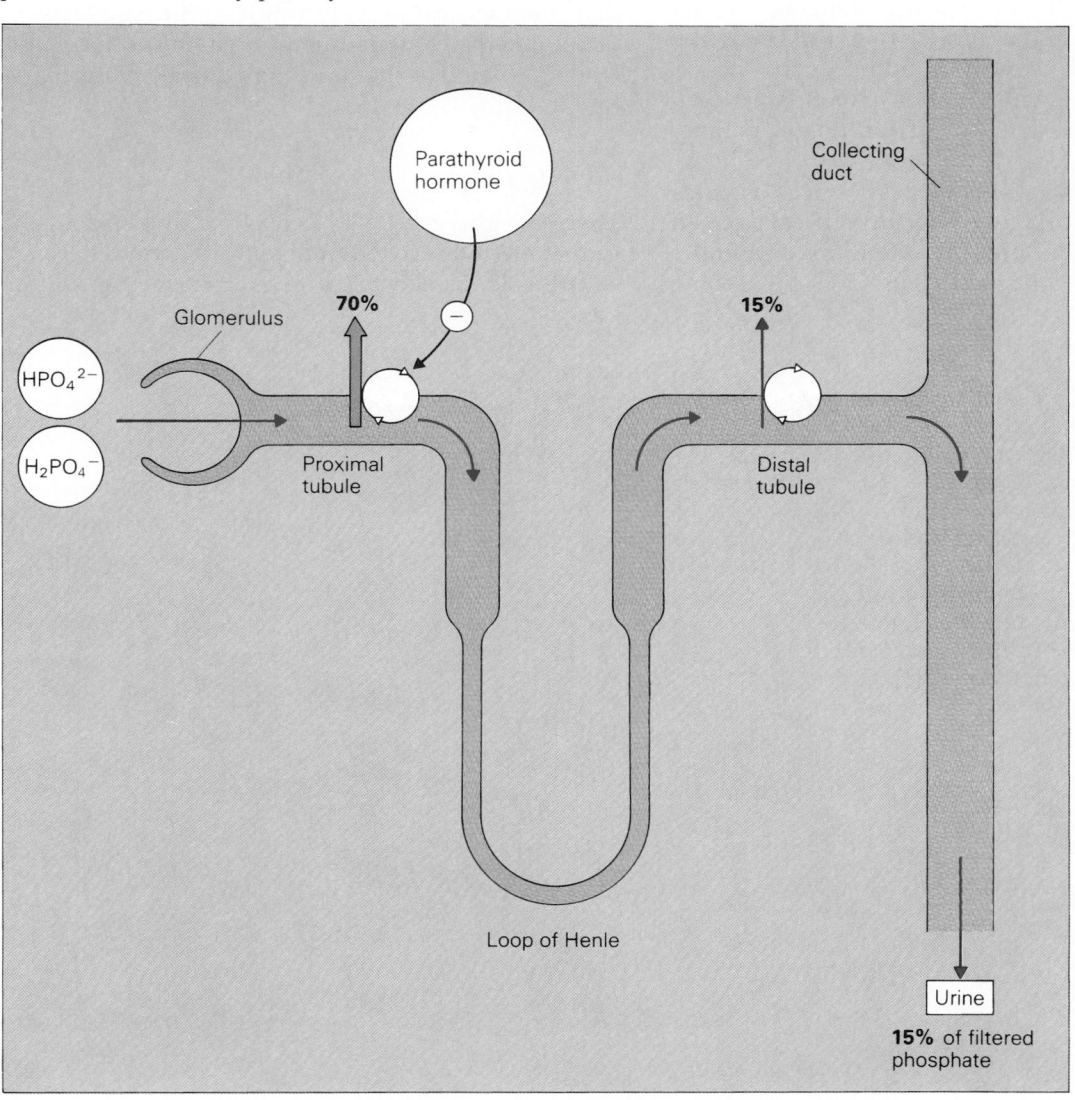

Homeostasis of Calcium and Phosphate in the Bone

The Composition of Bone

Simply described, mature bone consists of inorganic bone mineral deposited on a framework of organic support material. Imagine what would happen if you dipped a sponge into wet plaster and then allowed the plaster to harden. The sponge is analogous to the organic support material, and the hardened plaster is analogous to the bone mineral.

The mineral fraction of bone is composed largely of calcium phosphate, in the form of what are known as **hydroxyapatite crystals.** The hydroxyapatite crystals in bone have the general formula:

$$Ca_{10}(PO_4)_6(OH)_2$$

Normally, the mineral portion of bone makes up about one quarter of the volume of the bone, but because of its high density, the mineral fraction makes up about one half of the weight of bone. Recall from Table 27–4 that, in addition to calcium phosphate, the inorganic matter of bone also contains a considerable amount of carbonate, magnesium, and sodium.

The organic matrix of bone is known as **osteoid,** 95% of which is **collagen.** The type of collagen in bone is similar to that present in tendons and in the skin, although the collagen in bone does have some different biochemical properties that give it a greater degree of mechanical strength than other types of collagen. The remaining 5%, or noncollagen portion of organic matter, is known as **ground substance,** which consists of a mixture of various **proteoglycans,** high-molecular-weight compounds made up of carbohydrate and protein.

If one closely examines bone structure using the electron microscope, one sees needlelike hydroxyapatite crystals lying alongside the collagen fibers and surrounded by the somewhat amorphous ground substance. The association of hydroxyapatite with the collagen fibers is responsible for the strength and hardness characteristic of bones. If a bone is **demineralized**—that is, if all the inorganic material is removed—the shape of the bone will be preserved because of the collagen that remains, but it will be as flexible as a tendon. Conversely, removal of the organic matter will also leave the bone with its original shape, but in this case the mineral that remains will be very brittle and easily broken. Thus, it is the combination of organic and inorganic material that gives bone its high degree of mechanical strength.

Bone Cells

There are three primary cell types involved in the formation and resorption of mature bone. These are **osteoblasts, osteocytes,** and **osteoclasts** (Fig. 27–7).

Figure 27–7
Diagram illustrating the relationships among different types of bone cells. Osteoblasts line the surface of bone and secrete osteoid into the space adjacent to the bone. For the sake of simplicity, only a few of the canaliculi that interconnect osteocytes and osteoblasts are shown near the bottom of the figure. An osteoclast actively engaged in bone resorption is represented by the large cell on the right. Note the ruffled appearance of the osteoclast membrane on the side next to the bone surface.

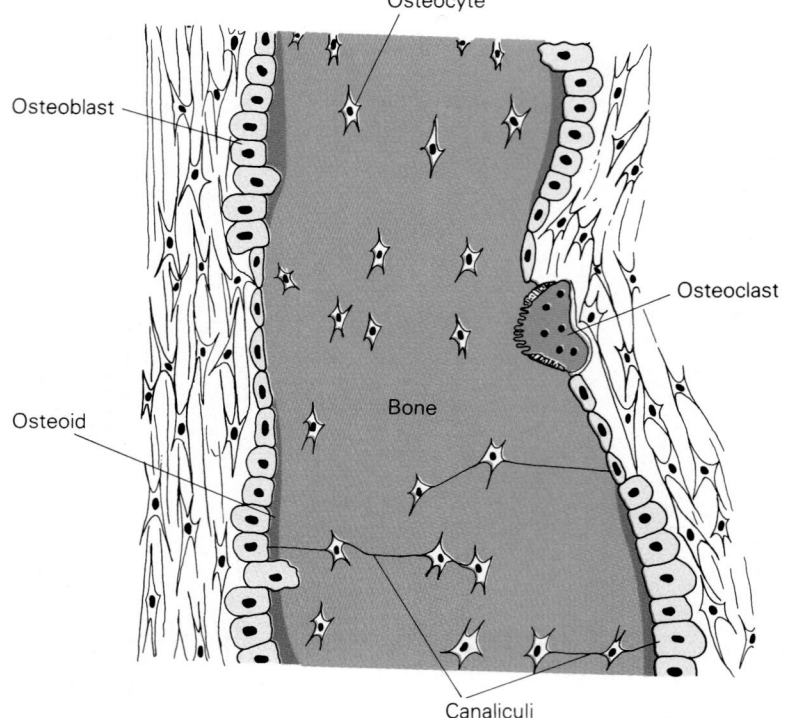

Osteoblasts.   Located side by side on the surfaces of bone, **osteoblasts** are the cells responsible for the synthesis of osteoid. Osteoblasts that are actively producing osteoid have a cuboidal shape and exhibit an abundant rough endoplasmic reticulum and Golgi complex, two characteristics typical of cells actively synthesizing proteins for export. Osteoblasts not actively engaged in bone formation have a considerably flattened appearance. Osteoblasts also have numerous cytoplasmic processes that serve to bring them into contact with neighboring osteoblasts.

As Figure 27–7 indicates, osteoid secreted from the osteoblast is deposited on the surface of the bone. Bone mineral is then deposited on the organic osteoid matrix, and as this process occurs, the osteoblasts eventually become surrounded by mineralized bone.

Osteocytes.   At this stage, osteoblasts progressively lose their bone-forming capability and are termed **osteocytes.** The cell-to-cell connections that were present among osteoblasts are maintained through the osteocyte stage of development. These interconnections are channels known as **canaliculi** in the bone and represent direct contacts between cells. This coupling among cells provides a "bucket brigade" mechanism whereby nutrients and ions can travel from blood vessels to distant osteocytes deeply embedded in bone. Hormones that control bone growth and development are also believed to pass from one osteocyte to another in this way.

Osteoclasts.   The process of bone resorption is carried out by **osteoclasts.** These are large, multinucleated cells, each containing about 4 to 6 nuclei. The surface of the osteoclast adjacent to the bone has a very ruffled appearance (see Fig. 27–7). This ruffling increases surface area and allows the cell to perform the task of bone resorption more effectively. When stimulated, osteoclasts are thought to secrete both acids and proteolytic enzymes into the extracellular space adjacent to the bone. The acidic environment increases the solubility of bone mineral, while the proteolytic enzymes attack the organic matrix of the bone. Together, these two factors promote the process of bone resorption.

The Formation of Bone

During early fetal development, the skeleton consists solely of cartilage, which is laid down in the shape and form of the bony skeleton. This cartilage "model" serves as the template for the development of the bony skeleton and is eventually replaced by bone.

The Epiphyseal Plate.   The process of cartilage replacement begins in the middle or center of the cartilage and progresses out toward each end of what will later form the bone (Fig. 27–8*a–c*). As this progression occurs, the bone increases in both length and thickness. Figure 27–8*d* illustrates the typical structure of a long bone in a growing adolescent. The histological appearance of the bone is the result of an extension of the series of events shown in parts *a* to *c* of Figure 27–8. The area of the bone designated as the **epiphyseal plate** is of particular interest, since it is here that elongation or growth of the bone occurs after birth.

Chondrocytes.   As indicated in Figure 27–9, the region of the epiphyseal plate shows considerable differences between its leading and trailing edges. On the leading edge are cells known as **chondrocytes,** which are actively engaged in the synthesis of the cartilage of the epiphyseal plate. As the cartilage is formed, it surrounds the chondrocytes and gradually becomes calcified. The trapped chondrocytes are replaced by new cells on the surface of the cartilage, so that active synthesis of cartilage continues. The calcified cartilage that is formed *does not* itself become bone, but is later replaced by bone. As the cartilage becomes increasingly calcified, the chondrocytes embedded in it begin to die off, and the calcified material begins to erode. Osteoblasts then move into the area, and the actual process of bone formation begins. As a result of the continuing processes of cartilage synthesis, calcification, erosion, and osteoblast invasion, the zone of active bone formation moves outward from the center toward the ends of the bones.

Hormones in Bone Formation.   As described earlier in Chapter 13, chondrocytes of the epiphyseal plates are under hormonal control. In particular, **IGF-I,** a somatomedin produced by the liver in response to growth hormone (GH), is the primary stimulator of chondrocyte activity. This effect of IGF-I accounts for the role of GH as an important regulator of growth in young children. Other hormones such as insulin and thyroid hormones also tend to promote chondrocyte activity and therefore promote bone growth as well.

Beginning around the time of puberty, the epiphyseal plates of the long bones, such as those in the thighs or arms, begin to "close," or become unresponsive to hormonal stimuli. In most individuals, this process is complete by about age 20, so that after this time linear growth is no longer possible. Bones of the fingers, feet, and skull remain hormonally responsive, however, which accounts for the skeletal changes that occur with GH oversecretion

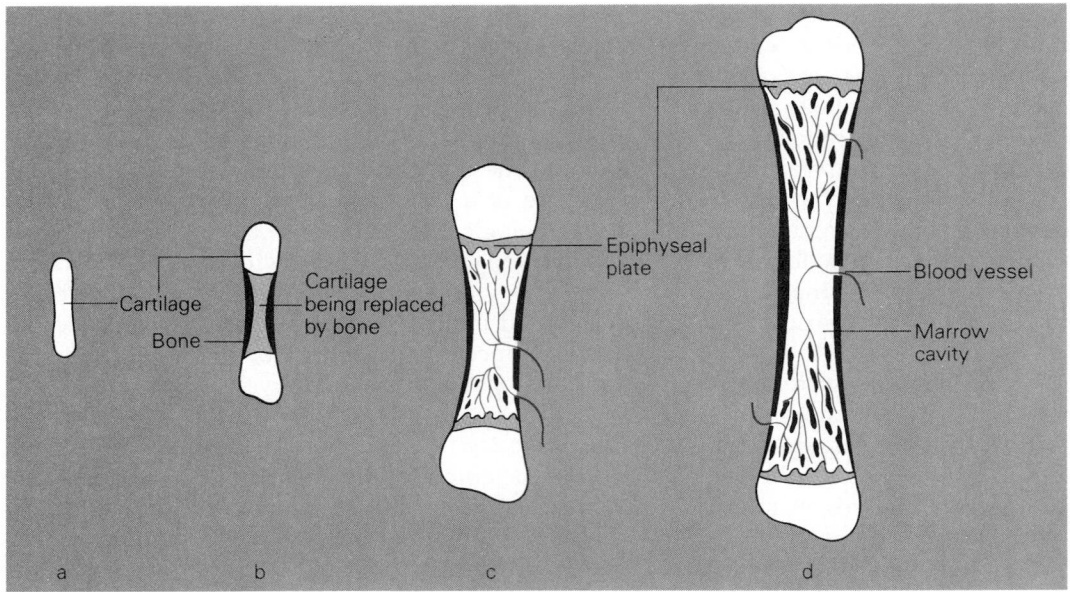

Figure 27–8
The process of growth and elongation of a typical long bone. (*a*) During fetal development, cartilage "models" of the bone are formed. (*b, c*) The cartilage is then replaced by mineralized bone, beginning in the middle of the bone and progressing towards each end. In the process, the bone becomes thicker and longer. (*d*) The center of the bone shaft becomes hollowed to form the marrow cavity. The epiphyseal plate is the zone where elongation of bone occurs.

in the adult, the condition known as acromegaly (see Chapter 13).

Bone Remodeling. As indicated earlier, bone is continually undergoing synthesis and resorption, even after linear growth has ceased. This is reflected in the high amounts of both calcium and phosphate that flow into and out of bone each day, as shown in Figures 27–3 and 27–4. This process of continued bone turnover is termed **remodeling.** Synthesis or

Figure 27–9
The appearance of an epiphyseal plate. On the right, zones of different cellular activities within the epiphyseal plate are illustrated. Active chondrocytes secrete collagen, which eventually becomes calcified. The calcified cartilage erodes and is replaced by osteoblasts, which then form new bone. The entire sequence is repeated many times as the zone of new bone formation gradually moves upward.

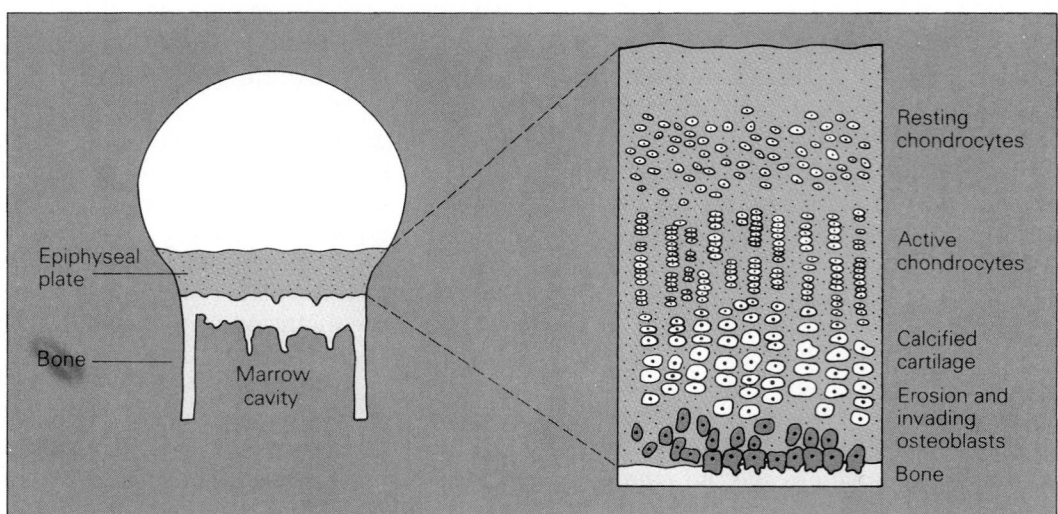

SUMMARY

Overview: Functions of Calcium and Phosphate

Because calcium ion is involved in a number of important physiological processes in the body, the concentration of calcium in the extracellular fluid must be maintained within fairly narrow limits. Phosphate also serves several important functions, but the regulation of plasma phosphate concentrations is not as critical as that of calcium. Approximately 99% of the total calcium in the body is located in bone. The majority (85%) of phosphate is also present in bone. Bone contains significant amounts of carbonate, magnesium, and sodium. About 20% of the weight of bone is due to water.

The total calcium concentration in plasma is normally about 10 mg/100 ml. The two primary forms of calcium in plasma are filterable calcium and protein-bound calcium. Filterable calcium consists of free calcium ion and calcium complexed with an anion, such as bicarbonate. Free calcium is the physiologically active form of calcium. Protein-bound calcium is not filtered in the kidney and consists of calcium bound to large plasma proteins, such as albumin. Most phosphorus in the plasma exists as phosphate (PO_4). Normal plasma concentrations range from 3.0 to 4.5 mg of phosphorus/100 ml.

Pathways of Calcium and Phosphate Homeostasis

Three tissues are primarily involved in regulating calcium and phosphate metabolism: the gastrointestinal tract, the kidneys, and bone. The majority of calcium taken in with the diet is not absorbed and leaves the body in the feces. The absorbed calcium is eventually excreted in the urine. A considerable amount of calcium both enters and leaves bone each day, reflecting the extensive turnover of bone mineral.

The majority of ingested phosphate is absorbed from the gastrointestinal tract. The primary route for excretion of this daily phosphate load is via the urine. Significant amounts of phosphate enter and leave bone each day.

Mechanisms Involved in Calcium and Phosphate Homeostasis

Calcium absorption from the gastrointestinal tract occurs primarily in the small intestine and involves both active transport and passive diffusion. Calcium absorption is regulated by changes in active transport. Phosphate absorption from the small intestine occurs by active transport and passive diffusion and does not appear to be regulated.

Approximately 99% of the calcium filtered by the kidneys is reabsorbed. Reabsorption involves both active transport and passive diffusion. Active transport in the distal tubule is stimulated by parathyroid hormone. Normally, about 85% of the filtered phosphate is reabsorbed. Reabsorption primarily occurs by active transport processes. The primary site of reabsorption in the nephron is the proximal tubule. Parathyroid hormone inhibits phosphate reabsorption by proximal tubule cells.

Bone consists of inorganic bone mineral deposited on an organic support matrix. The mineral fraction is composed primarily of hydroxyapatite; the organic matrix is termed osteoid and consists primarily of collagen. The three primary cell types in mature bone are osteoblasts, osteocytes, and osteoclasts. Osteoblasts and osteocytes are involved in bone formation and bone maintenance. Osteoclasts are responsible for bone resorption. In growing bone, chondrocytes of the epiphyseal plate are under hormonal control. Stimulation of chondrocytes leads to bone growth. After puberty, the epiphyseal plates of the long bones become unresponsive to hormonal stimulation.

Regulation of Calcium and Phosphate Concentrations in the Plasma

About 1% of the calcium in bone is in a readily exchangeable pool in rapid equilibrium with extracellular free calcium.

Three hormones are involved in the regulation of plasma calcium and phosphate concentrations, parathyroid hormone, calcitonin, and vitamin D metabolites.

Parathyroid hormone is secreted in response to decreased plasma calcium. Its net effect is to increase calcium and decrease phosphate in the plasma. Its target tissues are kidney, bone, and, indirectly, the gastrointestinal tract.

Calcitonin plays only a minor role in the regulatory process. It is secreted in response to increased plasma calcium. The net effect of calcitonin is to decrease both calcium and phosphate in the plasma. Calcitonin has its primary effect on bone.

Vitamin D_3 is converted into 1,25-OH-vitamin D_3 by sequential conversion in the liver and then the kidney. Parathyroid hormone stimulates formation of the active hormone by stimulating the renal enzyme responsible for the final step in vitamin D activation. The net effect of 1,25-OH-vitamin D_3 action is to increase the plasma concentration of both calcium and phosphate. The hormone acts primarily on the intestines to increase calcium absorption. It also promotes parathyroid hormone action in bone.

Abnormalities of Bone Mineral Metabolism

The most common abnormality of bone metabolism is a group of disorders known as metabolic bone disease. The first major category of metabolic bone disease is osteoporosis, which is characterized by an overall loss of bone mass with equal loss of bone mineral and organic matrix. Osteoporosis is most commonly seen in elderly women, as a result of hormonal changes at menopause that lead to postmenopausal osteoporosis.

Osteomalacia and rickets make up the second major category of metabolic bone diseases. These are characterized by inadequate mineralization of bone, most commonly caused by a deficiency in vitamin D activity, which can be due to a number of causes.

REVIEW QUESTIONS

Identify Each with a Term

1. The form of plasma calcium composed of ionized and complexed calcium.

2. The crystals of bone mineral, composed mainly of calcium phosphate.

3. The organic matrix of bone.

4. Cells of bone responsible for the synthesis of osteoid.

5. The cell type responsible for the synthesis and secretion of calcitonin.

6. The condition in which a reduction in overall bone mass occurs, with equal losses of both mineral and organic matrix.

7. Cells of the epiphyseal plate responsible for the synthesis of collagen.

8. The ongoing processes of bone turnover.

Define Each Term

9. hypocalcemia
10. ionized calcium
11. ground substance
12. osteoclasts
13. parathyroid hormone
14. 1,25-OH-vitamin D_3
15. ergocalciferol

Choose the Correct Answer

16. The physiologically active form of calcium in the body is:
   a. calcium bound to albumin
   b. calcium citrate
   c. calcium ion (Ca^{2+})
   d. unable to cross membranes

17. At a pH of 7.4, most of the phosphate in the blood will be present as:
   a. hydroxyapatite
   b. HPO_4^{2-}
   c. hexose phosphate
   d. $H_2PO_4^{-}$

18. The majority of ingested calcium is excreted in the:

   a. feces
   b. urine
   c. sweat
   d. plasma

19. Parathyroid hormone acts to _____ the reabsorption of calcium in the _____ convoluted tubules of the kidney.
   a. increase, proximal
   b. decrease, proximal
   c. increase, distal
   d. decrease, distal

20. The conversion of 7-dehydrocholesterol to vitamin D_3 occurs in the:
   a. epiphyseal plate
   b. kidney
   c. liver
   d. skin

21. The kidney enzyme 1α-hydroxylase catalyzes the conversion of:
   a. calcium to hydroxyapatite
   b. 25-OH-vitamin D_3 to 1,25-OH-vitamin D_3
   c. vitamin D_2 to vitamin D_3
   d. cholecalciferol to dehydrocholesterol

22. A deficiency in vitamin D action can lead to:
   a. epilepsy
   b. osteomalacia
   c. osteoporosis
   d. postmenopausal osteoporosis

Answer Each Question in One or Two Sentences

23. What is the mechanism of hypocalcemic tetany?

24. What are the major forms of calcium in the plasma and what percent of the total does each of these compose?

25. Explain how kidney disease can result in osteomalacia and rickets.

26. What are the actions of parathyroid hormone that tend to raise plasma calcium and lower plasma phosphate?

27. What conditions lead to increased secretion of parathyroid hormone? of calcitonin?

SUGGESTED READINGS

Fregley, Melvin J., and Luttge, William G. *Human Endocrinology: An Interactive Text*. New York: Elsevier Biomedical, 1982.

Ganong, William F. *Review of Medical Physiology*, 12th ed. Los Altos: Lange Medical Publications, 1985.

Goodman, H. Maurice. *Basic Medical Endocrinology*. New York: Raven Press, 1988.

Guyton, Arthur C. *Human Physiology and Mechanisms of Disease*, 4th ed. Philadelphia: W.B. Saunders Company, 1987.

Hedge, George, A., Colby, Howard D., and Goodman, Robert L. *Clinical Endocrine Physiology*. Philadelphia: W.B. Saunders Company, 1987.

Norman, Anthony W., and Litwack, Gerald. *Hormones*. Orlando: Academic Press, 1987.

Reichel, Helmut, Koeffler, H. Phillip, and Norman, Anthony W. "The Role of the Vitamin D Endocrine System in Health and Disease." *New England Journal of Medicine*, 320:980–991, 1989.

Tepperman, Jay, and Tepperman, Helen M. *Metabolic and Endocrine Physiology*, 5th ed. Chicago: Year Book Medical Publishers, Inc., 1987.

Wilson, Jean D., and Foster, Daniel W., eds. *Williams' Textbook of Endocrinology*, 7th ed. Philadelphia: W.B. Saunders Company, 1985.

KEY TERMS

C-cells (p. 924)
calcitonin (p. 923)
calcium (p. 911)
canaliculi (p. 920)
chondrocytes (p. 921)
epiphyseal plate (p. 921)

ground substance (p. 920)
hydroxyapatite (p. 920)
metabolic bone disease (p. 928)
osteoblast (p. 920)
osteoclast (p. 920)

osteocyte (p. 920)
osteoid (p. 920)
osteomalacia (p. 931)
osteoporosis (p. 929)
parathyroid gland (p. 923)
phosphate (p. 912)

phosphaturia (p. 928)
proteoglycan (p. 920)
rickets (p. 931)
tetany (p. 911)
vitamin D_3 (p. 925)

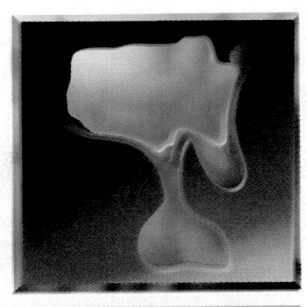

C H A P T E R
28

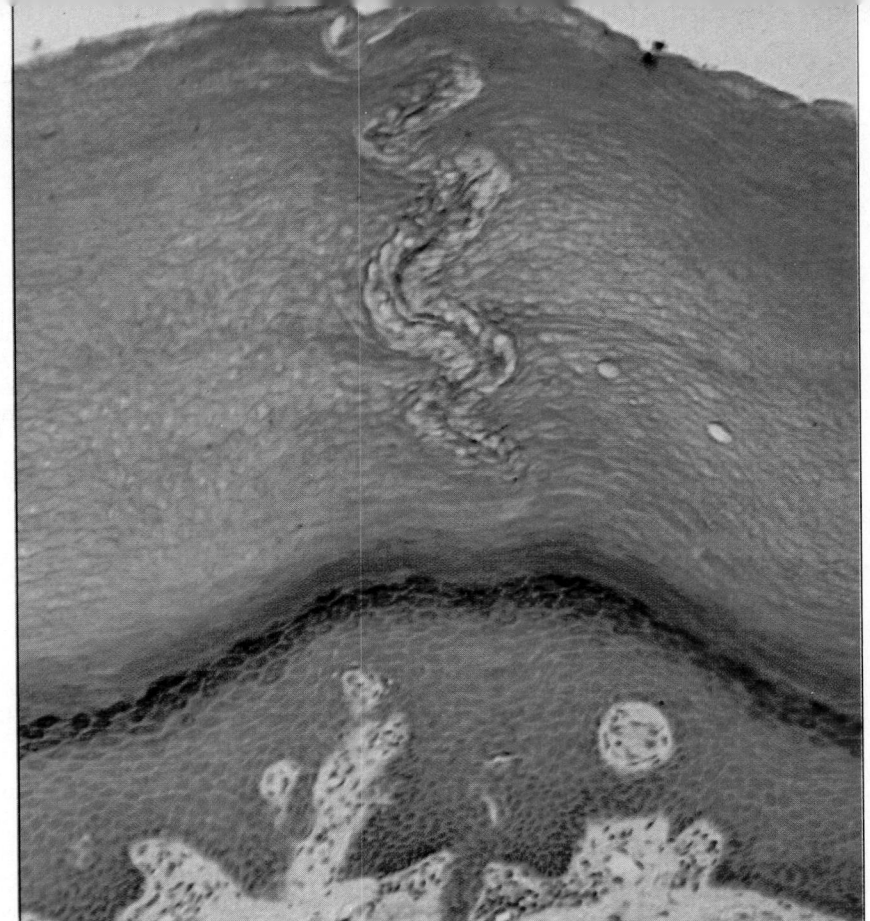

Regulation of Body Temperature

Humans are **homeotherms.** We usually regulate deep "core" body temperature within a relatively narrow range (36 to 39°C) when exposed to a wide range of climatic conditions. We share this characteristic with all mammals and birds. Other animals—for example, fish and reptiles—are **poikilotherms.** Their body temperatures may fluctuate with changes in ambient temperature.

A nude human can be exposed to temperatures as low as 12°C or as high as 60°C in dry air and still maintain a body temperature of 36 to 39°C. Higher body temperatures of 40 to 41°C, such as those associated with strenuous exercise, fever, and hyperthermia, can be tolerated for only short periods. Body temperatures higher than 42°C are associated with a rapid breakdown of vital cellular proteins and death. Humans, therefore, regulate body temperature only a few degrees below the point of thermal death.

The regulation of body temperature in humans near the higher level of thermal tolerance ensures optimal cellular function. All cellular, biochemical, and enzymatic reaction rates are temperature dependent. At normal body temperature, cellular enzymatic reactions work well. If the cellular temperature were to drop by 10°C, however, these reaction rates would decrease by a factor of 2.5. Human metabolic activity, therefore, changes approximately 12% for every degree (Centigrade) of change in body temperature. Thus, the optimal regulation of body temperature is necessary in homeotherms to ensure the capacity for rapid metabolism that makes intense physiological activity such as exercise possible. In addition, the regulation of body temperature by humans and other homeotherms ensures a near-constant internal thermal environment and provides considerable independence from any possible detrimental effects of atmospheric temperature.

Normal Body Temperature

The metabolic heat produced within the body is normally lost to the environment at the surface. The physical laws of heat flow therefore dictate that the temperature of parts of the body near the surface must be lower than the temperature of the central parts. Furthermore, the fluctuations in body temperature produced by changes in atmospheric temperature are greater at the body surface. The tem-

perature gradients of the human body can be conceived as a **"poikilothermic" shell** at the surface and a **"homeothermic" core** in the center (Fig. 28–1). The core temperature is kept constant over a wide range of atmospheric temperatures, whereas the temperature of the shell varies with atmospheric temperature.

Core Temperature

No single body temperature can be specified as *normal* since the human body has both **longitudinal** and **radial temperature gradients.** Even the temper-

Figure 28–1
Core-shell concept in the human body. The temperature of the surface and underlying areas (shell) varies with atmospheric temperature. The shell is "thicker" in a cold environment (*a*) and "thinner" in a warm one (*b*).

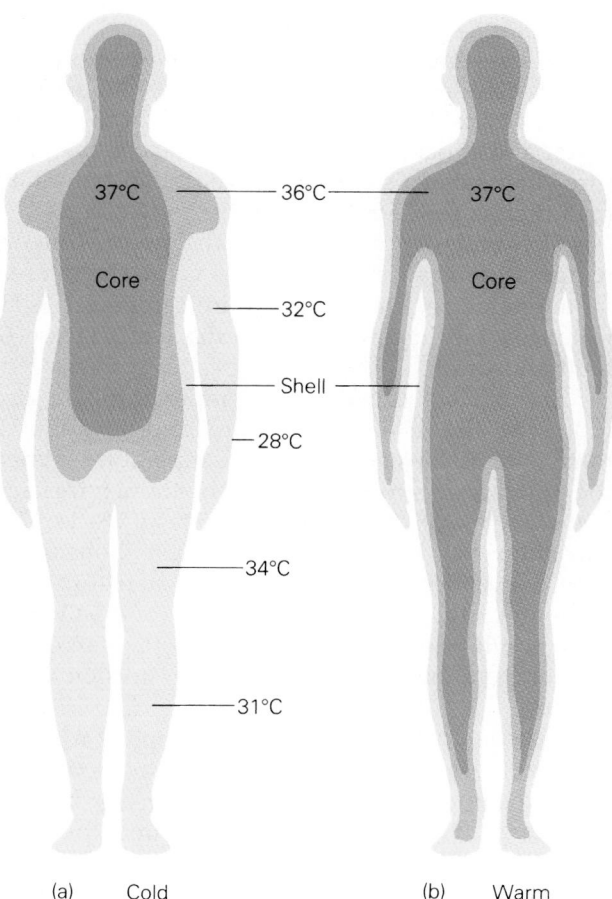

(a)   Cold

(b)   Warm

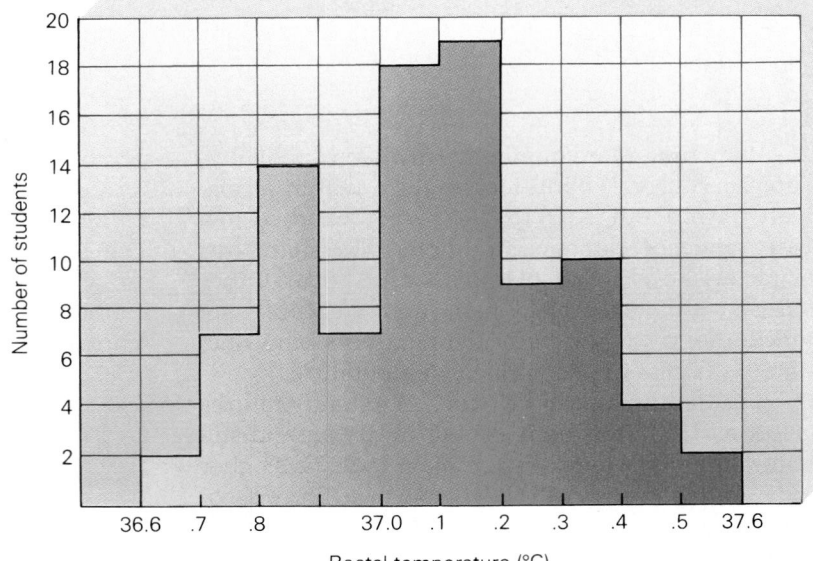

Figure 28–2
Range of rectal temperatures measured under well-controlled conditions in a group of healthy young students. The mean rectal temperature is 37.1°C.

ature of the core has temperature gradients from 0.2 to 1.2°C. For practical and comparative purposes, however, the most widely accepted measures of "normal" body temperatures are **rectal, oral,** and, in some cases, **axillary temperatures.**

In a resting human, the rectal temperature obtained at a depth of 5 to 10 cm is slightly higher than that of arterial blood, slightly lower than that of the brain, and equal to that of the liver. Rectal temperature obtained at a standard depth, therefore, can be regarded as representative of the body core.

The oral or **"sublingual"** (below-the-tongue) temperature is approximately 0.6°C lower than rectal temperature. Oral temperature can be easily affected by the inspired air and the temperature of food and drink.

The axillary (under-arm) temperature can also represent the core temperature and is approximately 0.7°C lower than rectal temperature. When the upper arm is held tightly to the thorax, the core boundary moves to the armpit (see Fig. 28–1). This shift, however, requires that the shell regions of the body gain considerable amounts of heat before a stable temperature is reached. At least one half hour must be allowed for equilibration to ensure that axillary temperature truly represents the core body temperature.

Another problem with the concept of a single value for normal body temperature is that body temperature varies considerably among resting healthy individuals (Fig. 28–2). At any given time, the rectal temperature during well-controlled conditions can vary by 1.1°C in a group of normal, resting, healthy

students. The variation is even greater in less controlled situations.

Normal body temperature should be conceived, therefore, as a *range* of core temperatures that may be expected in a group of normal subjects under various conditions (Fig. 28–3). It is perfectly normal, for example, for an athlete who has just run a race to have a rectal temperature of 40°C. The same rectal temperature in the same person resting the next day

Figure 28–3
Estimated range of body temperature in human subjects.

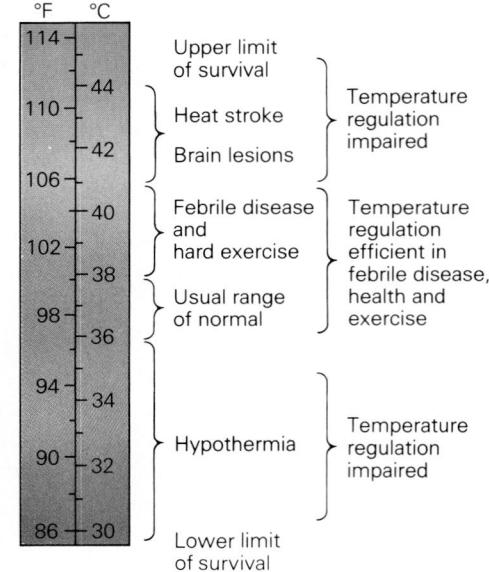

in a physician's office would be abnormal and could be considered a possible index of infection and fever. The range of body temperature observed in normal humans has a lower limit of 35°C and an upper limit of 41°C. Body temperatures outside of this range are associated with an impaired ability to effectively regulate body temperature.

Skin Temperature

Skin temperature is the accepted representative of the temperature of the body shell. The skin temperature, which is easily measured, is very different at different body sites and at different atmospheric temperatures. Surfaces covering the areas of high resting heat production have the highest skin tem-

peratures (Fig. 28–4). In a person in a comfortable room at 24 to 25°C, the skin temperature of the head, chest, and abdomen is about 35°C; the surfaces covering the large muscle masses of the arms and legs have a lower skin temperature of 31°C. Areas that cover very little muscle, such as the hands and feet, have the lowest average skin temperature of 29°C.

Mean Skin Temperature

For practical purposes, the **mean skin temperature** is regarded as representative of the temperature of the body shell and is calculated as the mean of the temperature at several representative body sites. In this calculation, the individual temperatures are

Figure 28–4
(*a*) Regional skin temperatures and heat production at rest in a comfortable ambient temperature of 25°C. (*b*) Thermograph of a woman standing in front of a tub of hot water. Color coding runs from white (hottest) through to blue and black (coldest.) (© Dr. Ray Clark & Mervyn Goff/Science Photo Library/Photo Researchers, Inc.)

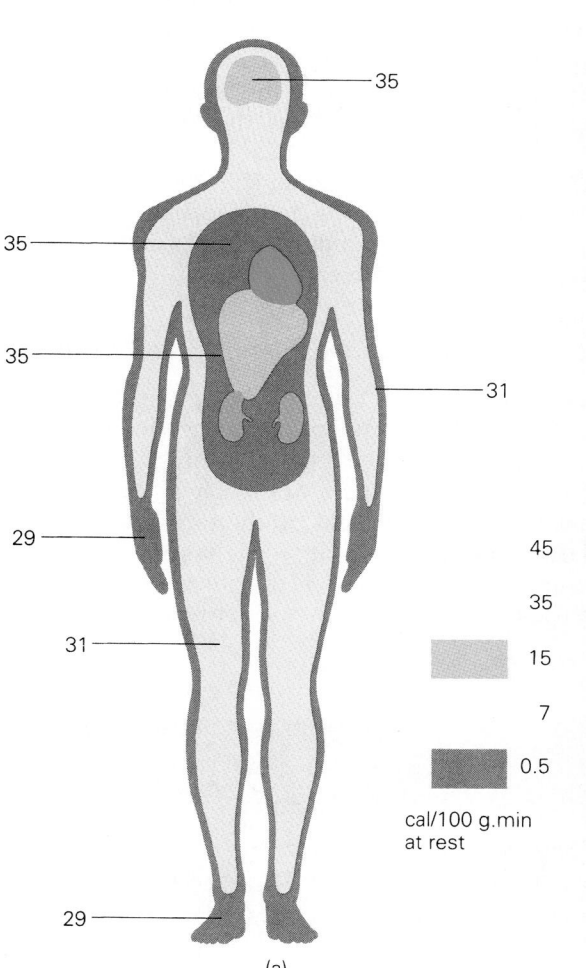

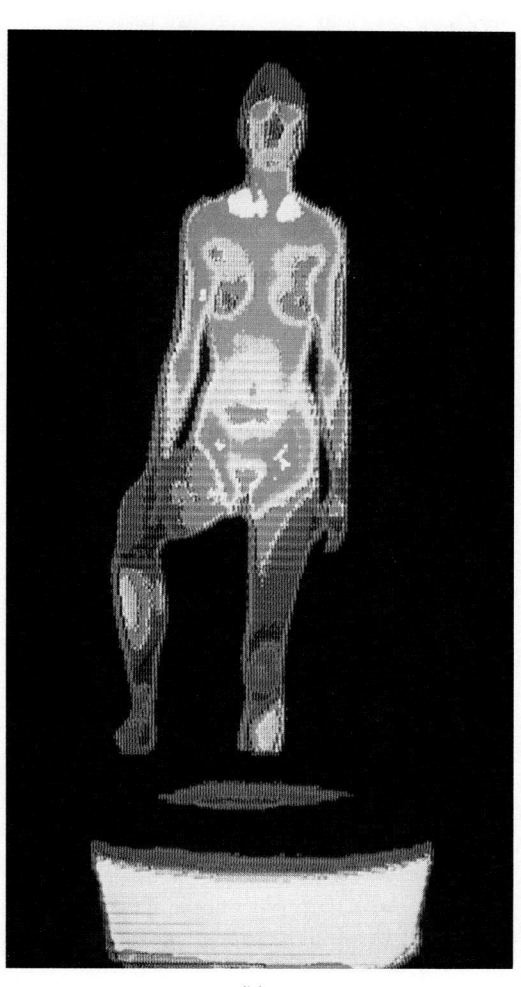

(a)

(b)

weighted in accordance with the surface area they represent. One such equation is as follows:

Mean skin temperature =
$$0.3 \text{ (chest + arm temperature)} + 0.2 \text{ (thigh + calf temperature)}$$

The mean skin temperature calculated using this equation for a person in a comfortable atmospheric temperature is about 33 to 34°C.

Thermography

Thermography is a technique for measuring mean skin temperature in humans. It consists of taking pictures of the skin with an infrared camera. The apparatus scans the skin for several minutes and records the emitted infrared rays, which are directly related to the temperature of the skin. The integrated thermogram that is generated provides an accurate measure of mean skin temperature. Thermography has also proved to be a useful clinical tool for detecting areas of abnormal growth under the skin. For example, tumors near the skin are associated with skin temperatures several degrees higher than those in corresponding normal areas.

Mean Body Temperature

The **mean body temperature** of a human resting in a comfortable temperature is a complex function of the relatively high temperature of the internal organ areas, the intermediate temperature of muscle areas, and the low temperature of the skin (see Fig. 28–4). For practical purposes, the mean body temperature is estimated by the fractional sum of the core and the shell. The core temperature is most often represented by rectal temperature, and the temperature of the shell is most often represented by mean skin temperature. For example, the mean body temperature of a person in a comfortable atmospheric temperature can be estimated from the following equation:

Mean body temperature =
$$0.3 \times \text{mean skin temperature} + 0.7 \times \text{rectal temperature}$$

The equation is based on the observation that, in a comfortable temperature, approximately two thirds of the body mass is at core temperature, and one third is at the temperature of the shell (see Fig. 28–1). Appropriate adjustments in the weighting factors for rectal and mean skin temperature are made to account for the relative changes in the ratio of the core temperature to shell temperature in per-

sons exposed to hot or cold atmospheric temperatures.

The concept of mean body temperature is important because it allows estimation of the amount of heat gained or lost from the body when exposed to hot or cold atmospheric conditions. Furthermore, it can be used to estimate the rate of heat transfer in the skin associated with the regulation of body temperature.

Factors Affecting Body Temperature

Many factors affect the body temperature of an individual, as illustrated in Table 28–1, including exercise, circadian rhythms, menstrual cycles, age, and disease.

Exercise and Body Temperature

Depending on how vigorously the muscles are working and generating metabolic heat, **exercise** can increase the core temperature by 2°C or more. During prolonged, exhausting exercise such as marathon running, rectal temperatures of 41°C have been measured. The mean skin temperature, by contrast, falls because while the muscles are generating increased metabolic heat, the sweat glands are secreting sweat, which cools the skin. Control mechanisms for the adjustments in the regulation of body temperature during exercise have not been fully elucidated and remain an area of intensive investigation and controversy.

Circadian Rhythms and Body Temperature

Recall from Chapter 11 that the body temperature of a human reaches a minimum during sleep and increases slightly in the early morning hours (see Fig. 11–2). Body temperature reaches a maximum, sometimes with two peaks, in the early afternoon or midafternoon when activity is highest. The increase in temperature, however, is not simply the result of increased activity. Even in a sedentary person with

Table 28–1
Some Factors Affecting Body Temperature

Exercise	Shivering
Diurnal rhythms	Menstrual cycle
Fever	Sweating
Age	Shift in blood distribution

all external entraining clues removed, the body temperature continues to cycle in a **circadian** or **diurnal (daily) rhythm,** with fluctuations in temperature that average 1°C.

The Menstrual Cycle and Body Temperature

The changes in temperature associated with the **menstrual cycles** of most women are superimposed on the diurnal temperature changes (Fig. 28–5).

Several days before menstruation, the body temperature usually drops about 0.6°C and is maintained at the lower level until just before ovulation, when it may drop an additional 0.2°C. After ovulation, body temperature quickly rises approximately 1°C and remains at this normal level until the onset of the next cycle.

The magnitude of the oscillations in body temperature associated with the menstrual cycle varies with the individual. Therefore, measurement of

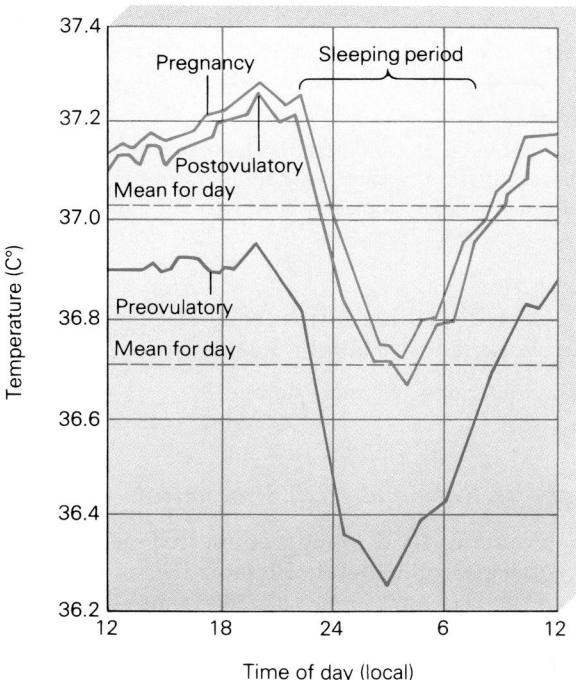

Figure 28–5
Daily variation in body temperature of normal female subjects during the menstrual cycle and pregnancy.

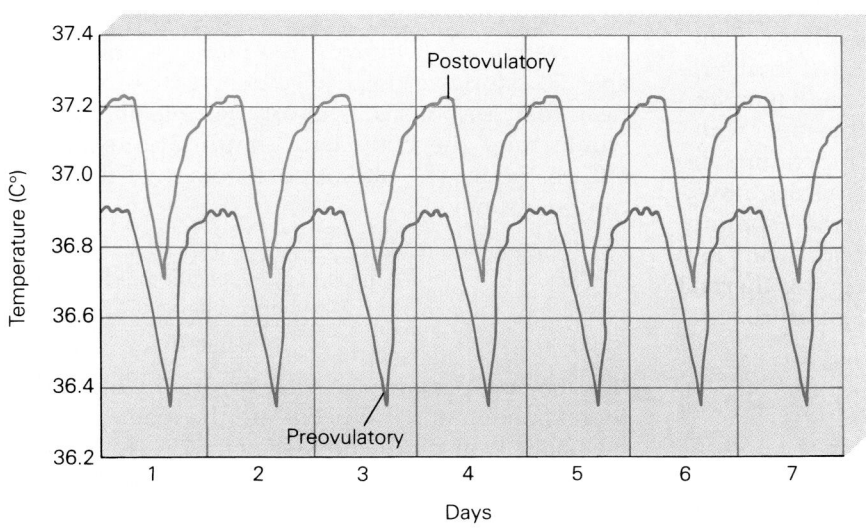

core temperature as a method of contraception has not been consistently effective. If pregnancy occurs, the premenstrual drop in body temperature is inhibited, and body temperature may actually increase to above-normal levels during pregnancy.

Age and Body Temperature

The increased body temperature of children, who tend to have higher body temperatures than adults, is the product of increases in both their physical and metabolic activity. Premature babies do not have well-developed control systems for body temperature regulation, and so their temperatures tend to fluctuate with atmospheric temperature. They are, therefore, routinely kept in incubators controlled at optimal temperature and humidity. The elderly also have body temperatures that tend to fluctuate with atmospheric temperature. Their temperatures are more variable because of age-related deficiencies in the cardiovascular responses to thermal stress.

Disease and Body Temperature

Disease states have traditionally been associated with abnormal body temperature. Fever, for example, has long been a common diagnostic index of infectious disease. High body temperatures are also produced by nonfebrile disease states, including hyperthyroidism and tumors of the adrenal gland that increase catecholamine secretion.

The Balance Between Heat Production and Heat Loss

Human body temperature is maintained at a relatively constant level because of a balance that exists between heat production and heat loss. This balance is maintained by a combination of physical and physiological mechanisms (Fig. 28–6). If heat production is greater than heat loss, the body temperature increases; conversely, when heat loss is greater than heat production, the body temperature decreases. In humans and other homeotherms, physiological regulatory mechanisms operate continuously to keep heat production and heat loss approximately equal, thereby maintaining **thermal homeostasis** and a normal body temperature.

Heat Production by the Body

Heat is being produced continuously in the body as a by-product of the **catabolism** of food. The energy liberated by catabolic processes within the body can

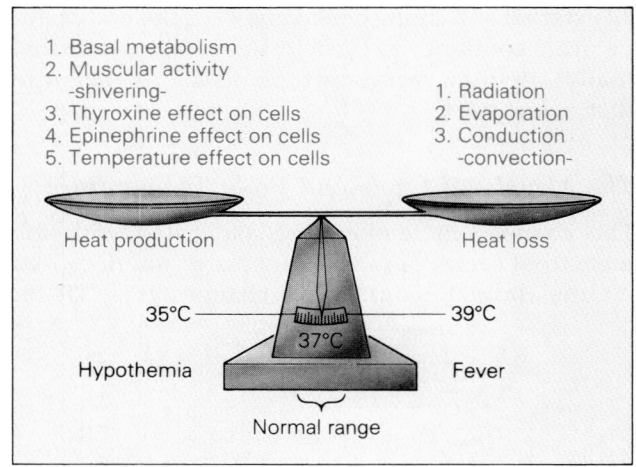

Figure 28–6
Balance between factors affecting heat production and heat loss, illustrating that body temperature remains in the normal range as long as these two are equal.

be converted to three forms: **chemical storage, external work,** and **metabolic heat.**

Food energy = chemical storage +
external work + metabolic heat

Metabolic Rate and Heat Production

The amount of food energy converted per unit of time is the person's **metabolic rate.** The energy systems of the body are about 20% efficient. That is, at maximum efficiency, 20% of the chemical energy of food is available to do external work, and 80% appears as heat (Fig. 28–7). In a resting human doing no external work, essentially all of the chemical energy of food appears as heat.

Energy = Heat

In a resting adult human, the metabolic rate averages 70 calories of heat per hour. If there were no heat loss responses, the resting metabolic rate would raise the body temperature approximately 1°C per hour. The rate of heat storage of the body can be estimated by:

$$\text{Heat storage (Cal/hr)} = \frac{\text{mass (Kg)} \times \text{specific heat (0.83)} \times \text{temp. change (°C)}}{\text{Time (hr)}}$$

In a resting person, the vital organs in the head, thorax, abdomen, and pelvis are the major areas of metabolic heat production (see Fig. 28–4). The pattern changes during work, when the muscle mass becomes a major source of heat production.

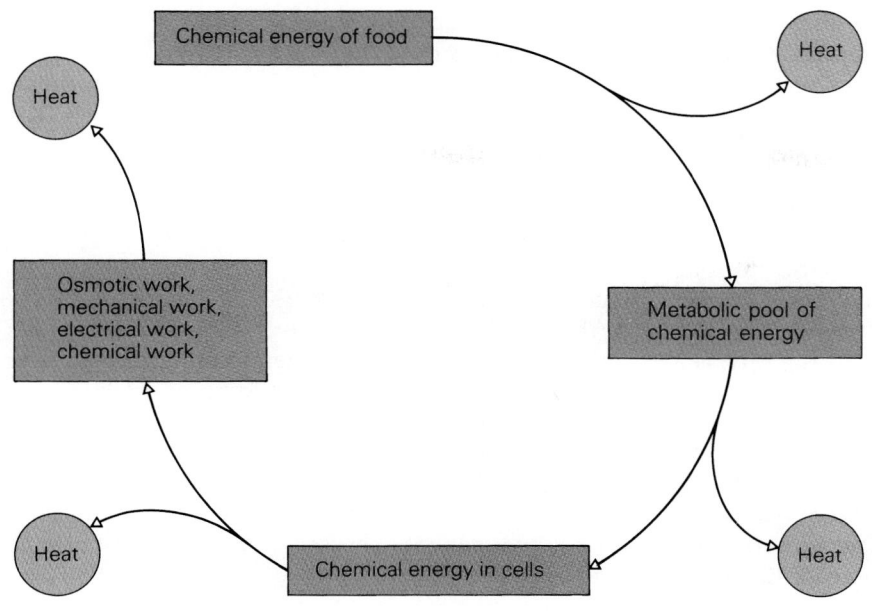

Figure 28–7
Summary of transformation of food energy within the body and its transfer as heat energy to the environment.

Factors Affecting Metabolic Rate

Any factor that increases or decreases the energy released from food also affects a person's metabolic rate. Some of the most important are listed in Table 28–2.

Exercise and Metabolic Rate. **Exercise,** the most powerful stimulus for increasing body heat production, has a dramatic effect on metabolic rate. Muscle contraction requires chemical energy, which, during strenuous exercise in a trained athlete, can increase heat production from a resting level of 1500 to 1700 calories per day to 20 times that amount for several minutes. Prolonged human activities that are classified as **light work** (e.g., writing) raise metabolic rates by approximately 500 calories per day, while those classified as **moderate work** (e.g., walking) raise metabolic rates by 1000 calories per day. **Heavy**

work (e.g., bicycling) raises metabolic heat production by 2000 calories per day, while the **heavy** occupational **physical labor** (e.g., lumbering) on the average requires 4800 calories per day and corresponds to roughly three times the resting metabolic rate. If no heat were lost from the body during the heaviest work, the heat produced would be sufficient to raise the body temperature by 4°C per hour. The limit for maximal metabolic rate for women is lower (around 3700 Cal/day) because of their lower average body weight and muscle mass.

Specific Dynamic Action and Metabolic Rate. The metabolic rate of a resting person increases following the ingestion of food and remains elevated for several hours. The increased heat production that results from processing a meal is called the **specific dynamic action** of foods. A meal containing a large quantity of carbohydrates and fats increases the metabolic rate by only 4%. A meal containing large quantities of proteins, however, can increase metabolic rate by as much as 30%. The major source of the increased heat appears to be the liver and its metabolism. This consensus is supported by the observation that metabolic rate increases in dogs when amino acids are infused intravenously. If the liver is removed prior to infusion, then heat production increases only slightly.

Hormones and Metabolic Rate. Many **hormones** influence energy metabolism. Of particular importance are the **catecholamines** and **thyroxin.** Stimula-

Table 28–2	
Some Factors Affecting Metabolic Rate	
Exercise	Shivering
Ingestion of food	Body size
Age	Gender
Temperature	Climate
Various hormones	Emotional state
Disease	Fever

tion of the sympathetic nervous system causes the release of norepinephrine and epinephrine. These catecholamines increase the breakdown of glycogen to glucose. They also increase the rates of enzymatic reactions, which increase cellular metabolism. Maximal stimulation of the sympathetic system can increase metabolic rate by 100%, but it remains elevated for only a few minutes.

Maximal secretion of thyroxin by the thyroid gland also increases most cellular enzymes. Cellular metabolism increases to about 100% above normal. Unlike the effects of catecholamines, the thyroid hormones continue to act for several weeks. Complete lack of thyroid hormones, on the other hand, can reduce metabolic rate by as much as 50%.

Many other hormones have minor effects on metabolic rate. For example, testosterone from the testes, insulin from the pancreas, and growth hormone from the pituitary gland can increase the overall metabolic rate of a person by only 5% to 15%.

Body Size and Metabolic Rate. Because of his greater tissue mass, the total energy exchange of an adult male weighing 160 kilograms is greater than that of an adult male weighing 65 kilograms. Metabolic rate, however, is not directly related to body weight, but is roughly proportional to surface area. Consequently, metabolic rate is commonly ex-

pressed as a function of body surface area (Cal/m²/hr). When heat production is expressed in this manner, adults of different body sizes have similar metabolic rates. The rationale for this observation is that metabolic heat is lost at the surface of the body and that an amount of heat equivalent to what is lost must be produced in order to maintain thermal homeostasis and a constant body temperature.

Age, Sex, and Metabolic Rate. Metabolic rate (Cal/m²/hr) reaches a peak at birth and gradually declines with age in both males and females (Fig. 28–8). The metabolic rate of growing children is higher than that of adults because large amounts of chemical energy are needed during growth for the synthesis of carbohydrates, fats, and proteins. Simultaneously, more energy is liberated as heat by the catabolic reactions that necessarily accompany synthetic reactions.

The metabolic rate of females is 5% to 10% lower than that of males of the same size and age. The cause for this difference is not certain, but it may be related to a greater proportion of relatively inactive tissue such as subcutaneous fat in females compared to males. The metabolic rate of the female increases during pregnancy as a result of the increased synthetic reactions associated with growth of the fetus.

Figure 28–8
Average basal metabolic rates in normal males and females at different ages.

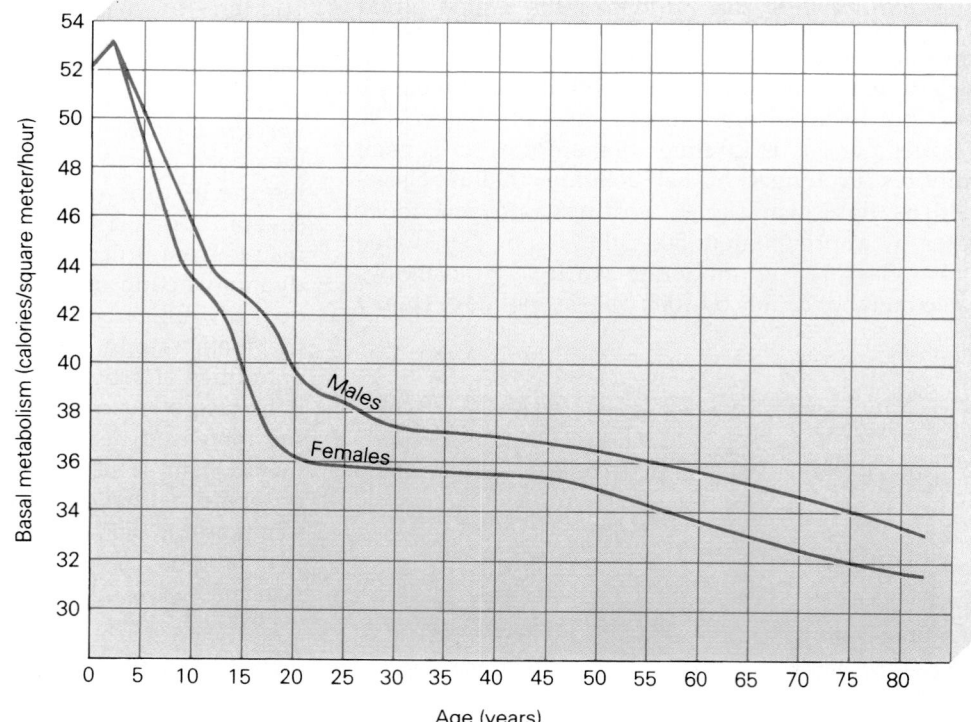

Temperature and Metabolic Rate. Fever, regardless of its cause, increases metabolic rate. This is because, as previously mentioned, all chemical reactions are temperature-dependent.

Acute exposure to cold induces shivering, which consists of a series of rapidly oscillating muscle tremors. Because no external work is performed, all of the energy liberated appears as internal heat. Shivering can increase metabolic rates to about five times above resting levels within minutes. Metabolic rates in studies of long-term residents in arctic regions have been shown to be 10% to 20% higher than the rates of residents in tropical areas. The cause for the difference is not clear, since in these studies the effects of atmospheric temperature cannot be isolated from the possible effects of other variables such as diets and light-dark cycles. Controlled studies of monkeys, however, also show that long-term exposure to cold produces a 20% to 30% increase in metabolic rate (Fig. 28–9). The increase is caused, in part, by adaptation of the thyroid gland. Thyroxin secretion increases in cold climates and decreases in hot climates.

The Basal Metabolic Rate

The concept of **basal metabolic rate (BMR)** has evolved because of our increasing knowledge of the many factors that can influence a person's metabolic rate. BMR is a standard method for determining metabolic rate that allows comparison between individuals. To record BMR, a set of standard conditions has been established to measure the so-called basal conditions. Actually, the BMR is not truly "basal," since metabolic rate during sleep is lower than BMR. BMR is more appropriately described as the "metabolic cost of living." The following four conditions are usually specified for measurements of BMR.

1. The person must have fasted for 12 hours.
2. The measurement is usually made in the morning after a night of restful sleep.
3. The person must remain at complete rest for 30 to 60 minutes before the test.
4. The atmospheric temperature must be comfortable, about 25°C.

Measurements of Metabolic Rate

The metabolic rate is a measure of the total energy used by the body per unit of time. In a person completely at rest and sitting in a comfortable temperature, all of the metabolic energy is transformed to body heat. Furthermore, a normal person under these conditions keeps body temperature constant

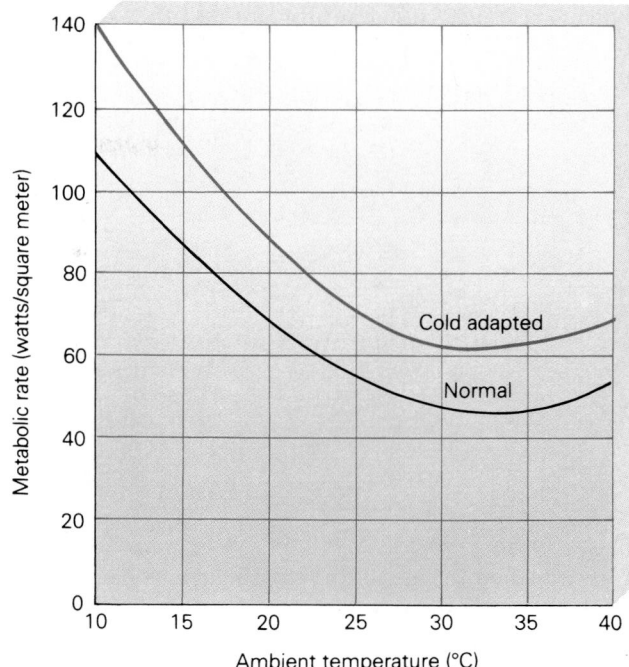

Figure 28–9
Effect of ambient temperature on resting metabolic rate before and after chronic cold exposure.

by releasing an amount of body heat into the environment equal to the heat produced inside the body. A person's metabolic rate, therefore, can be determined by measuring the heat given off from the body in a known period of time.

Direct Calorimetry. A person's metabolic rate can be measured by direct or indirect methods. In the direct method, a person is placed in a large insulated chamber called a **human calorimeter** (Fig. 28–10). The temperature of the calorimeter is kept constant by a water cooling radiator system that picks up the heat lost from the person's body. The rate of heat gain by the radiator system can be accurately measured with a thermometer and is equal to the rate of heat liberated from the person's body. Although direct calorimetry is a very accurate technique since it measures the heat output of the body directly, it is a time-consuming, expensive, and cumbersome method. It is therefore used most commonly for research purposes and to validate less costly and more convenient methods for measuring metabolic rate.

Indirect Calorimetry. A commonly used indirect method for measuring metabolic rate is based on the determination of a person's **oxygen consumption**

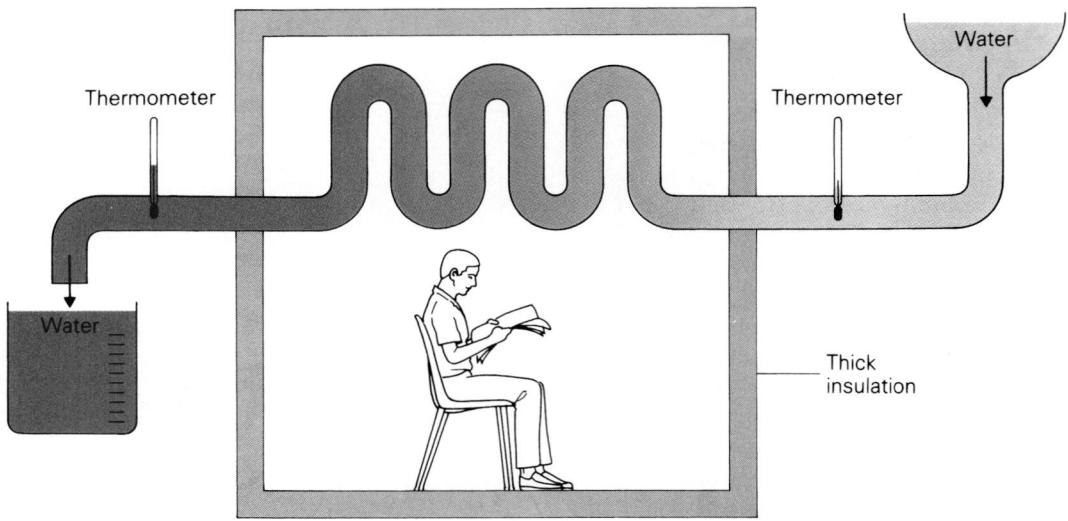

Figure 28–10
Direct method for measuring metabolic rate. The amount of heat produced
by the person's body is calculated from the total volume of water and the
difference between inflow and outflow temperature.

per unit of time. This method is based on the concept that since 95% to 100% of the energy available
to the body is derived from cellular reactions with
oxygen, metabolic rate can be accurately estimated
from oxygen use. Studies comparing metabolic rates
of humans using both indirect and direct methods
have shown good agreement, with differences of
less than 5%. The errors in the indirect method are
not large enough to outweigh its advantages.

The Energy Equivalent of Oxygen.   In practice, the
indirect method for measuring metabolic rate is dependent on the concept of the energy equivalent of
oxygen and a special instrument for the accurate
measurement of oxygen consumption called a **respirometer** (Fig. 28–11). The energy equivalent of oxygen is the quantity of energy liberated when the
body uses one liter of oxygen to produce metabolic
energy. For example, the amount of energy released

Figure 28–11
A respirometer for measuring the rate of oxygen consumption. The subject inspires oxygen from a bell gas-
ometer through a mouthpiece and valve. CO_2 is removed from the expired air by soda lime before it
returns to the bell.

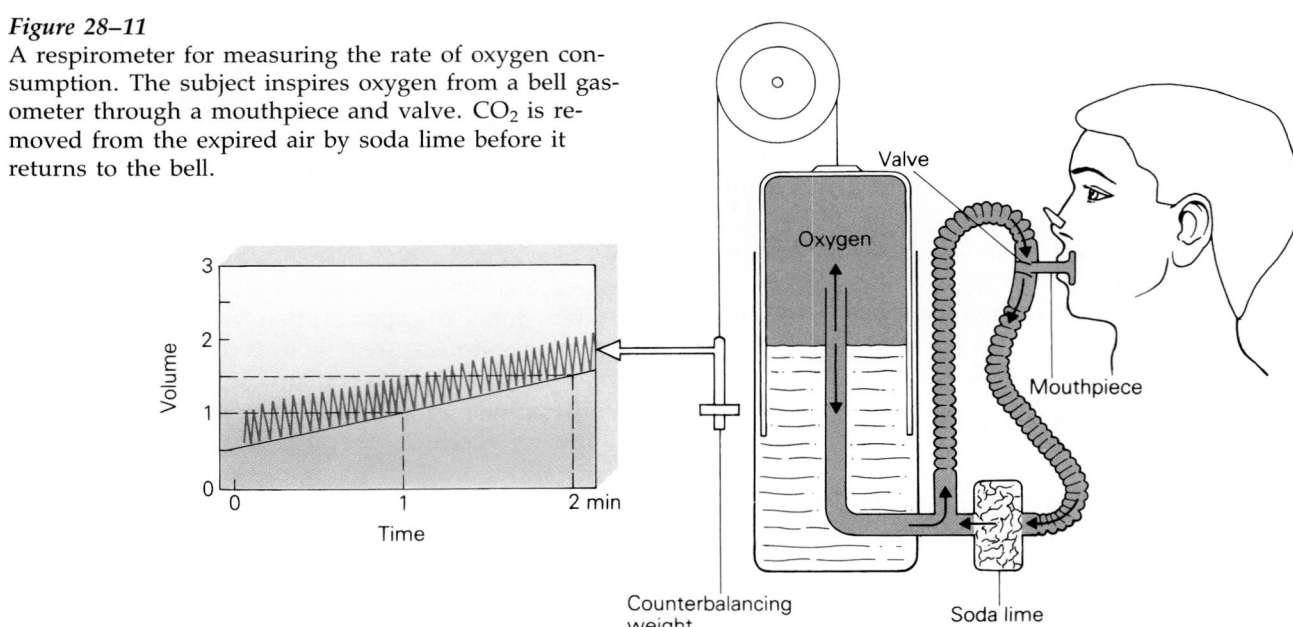

when the body uses one liter of oxygen to oxidize **carbohydrates** is 5.0 calories. When the body uses one liter of oxygen to burn **fats,** the energy released is 4.7 calories. The corresponding value for **protein** is 4.6 calories. Since the energy equivalent for all three food types is about the same, an approximate average of these values is commonly used and is equal to *4.825* calories of heat produced per liter of oxygen consumed. Metabolic rate, therefore, is commonly determined by the following equation:

$$\text{Metabolic rate (Cal/m}^2\text{/hr)} =$$
$$\text{liters of O}_2 \text{ consumed} \times$$
$$4.825 \text{ calories per liter of O}_2 \div$$
$$\text{surface area (meters}^2)$$

Heat Loss from the Body

Humans and other homeotherms must continuously exchange heat with their environment to maintain thermal homeostasis. If the heat produced by metabolism were not continually released, the body temperature would rise indefinitely. The body exchanges heat with its surroundings through four physical processes: **radiation, conduction, convection,** and **evaporation.**

Heat Loss by Radiation

Radiation is heat transfer by electromagnetic radiation between objects that are not in contact (Fig. 28–12). The greater the temperature difference, the larger the heat transfer from the warmer to the cooler object. Approximately 60% of the heat from a nude person sitting in a comfortable 21°C temperature is lost by radiation. At high atmospheric temperatures (35°C), the body may actually gain heat by radiation. The position of the body is an important factor in determining radiant heat exchange. When a person assumes the fetal position, radiant heat exchange is reduced because less surface area is exposed to the environment, whereas when a person is standing with arms and legs held away from the body, maximum heat is exchanged. Typically, most people consider air temperatures between 21 and 28°C to be comfortable because these temperatures permit the body to radiate heat at approximately the rate of metabolic heat production, thus keeping body temperature constant.

Heat Loss by Conduction

Conduction is heat exchange in the form of kinetic energy between molecules of objects in direct contact. As in radiation, the greater the temperature dif-

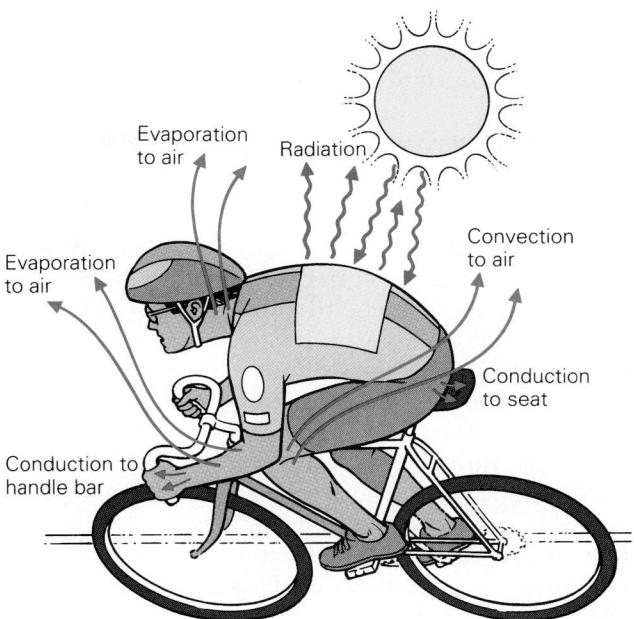

Figure 28–12
Mechanisms of heat exchange between the body and the environment.

ference, the larger the heat transfer by conduction from warmer to cooler objects. In humans, only relatively small amounts of heat are usually lost from the body by direct contact with other objects (see Fig. 28–12). Heat loss or gain by conduction, however, is very sizable in persons exposed to cold water because water is 20 times more effective than air as a conductor of heat. The rapid loss of body heat in cold water and subsequent hypothermia can lead to life-threatening conditions in boating and shipping accidents.

Heat Loss by Convection

Convection is the transfer of heat between the body and a moving gas or liquid. In fact, conduction is part of the convective process. For example, when air temperature is lower than that of the skin, heat is transferred by conduction to the cool air adjacent to the skin. The heated air expands, becomes less dense, rises, and is replaced by cooler air in a continuous process called **free convection. Forced convection** considerably increases the rate of heat exchange by the body. When the body is exposed to wind, for example, the air adjacent to the body is replaced by new air much more rapidly than normal, increasing convective heat loss. The cooling effect of wind is approximately proportional to the square root of the velocity of the wind and accounts for the "wind chill factor," which can lead to the freezing of tissue at higher ambient temperatures than occurs in calm air.

Heat Loss by Evaporation

Evaporation is the process of converting water from a liquid to a gas. Thermal energy is required to transform water from a liquid to a gas. When water evaporates from the surface of the body, 0.58 calories of heat are removed from the skin for each gram of water that evaporates. Evaporation is a continuous heat loss mechanism, since small amounts of water are continuously diffusing from the skin, the respiratory passages, and the mucous membranes of the mouth. In a person resting in a comfortable atmosphere, the total water loss from evaporation of diffused water is small and equals approximately 600 ml per day. Evaporative heat loss under these conditions is relatively small and averages about 20% of the total heat exchange (see Fig. 28–12). Since we are usually unaware of this loss, it is referred to as heat loss due to **insensible perspiration.**

Heat Loss from Sweating

Sweating, in contrast with passive diffusion of water through the skin, involves the active secretion of water by specialized sweat glands in the skin (Fig. 28–13). Sweat secretion is activated by sympathetic nerves in response to overheating, and increases linearly with abnormal increases in body temperature (Fig. 28–14). Evaporation of sweat is the only way for the body to lose heat when it is exposed to hot environments, since radiation and conduction at air temperatures greater than skin temperature actually transfer heat into the body.

Under extreme conditions the body can secrete about 1.5 liters of sweat per hour. For sweating to be an effective cooling process, sweat must rapidly evaporate from the skin. Under optimal conditions, the evaporation of 1.5 liters of sweat can remove as much as 800 calories of body heat per hour or about 12 times the rate of basal heat production.

Heat Transfer Within the Body

To keep the body temperature constant, the heat produced by metabolism must be transferred from the core cells that are producing it to the body shell, where it can be transferred to the environment. Within the body, the transfer of heat is accomplished by a combination of **internal conduction** and **circulatory convection.**

Figure 28–13
(*a*) A human sweat gland innervated by a sympathetic nerve. A protein-free filtrate is formed by the secretory coil, and most of the electrolytes are reabsorbed by the excretory duct, producing a dilute, watery sweat. (*b*) Photomicrograph of a human sweat gland. (× 35.) (© John D. Cunningham/Visuals Unlimited.)

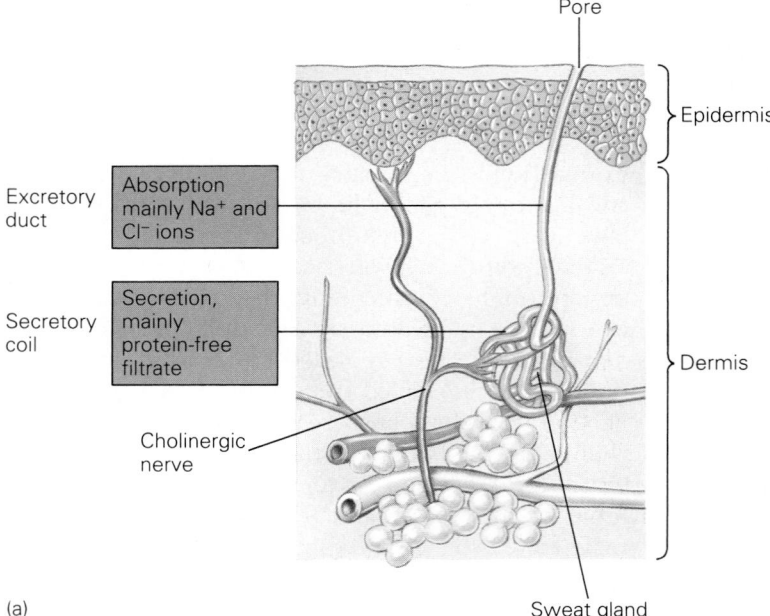
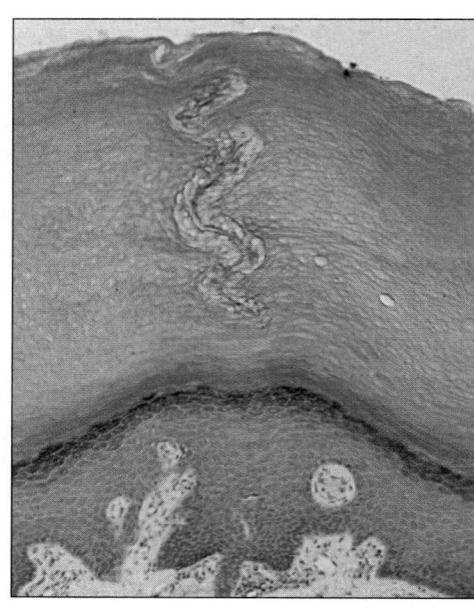

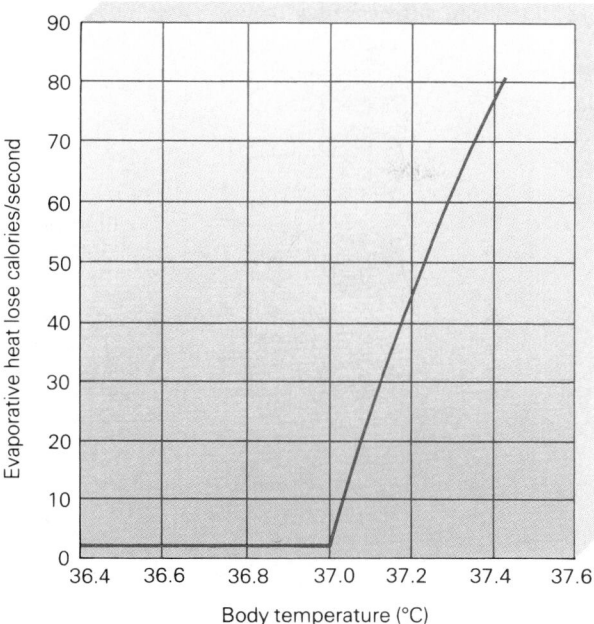

Figure 28–14
Effect of changes in body temperature on evaporative heat loss due to sweating. This figure illustrates the critical temperature at which sweating begins, and the linear increases that occur with increasing body temperature.

Heat Transfer by Internal Conduction

The rate of heat transfer due to internal conduction is slow because of the small temperature gradient that exists between adjacent cells. In addition, most of the tissues of the body are relatively poor heat conductors. The thermal conductivity of unperfused tissue is equal to that of cork, a good insulator and poor conductor. Thus, internal conduction alone is not adequate to maintain a balance between the production of heat and the loss of heat needed to maintain a constant body temperature.

Heat Transfer by Circulatory Convection

Convective heat transfer occurs within the body with the movement of body fluids. Forced convection by the circulatory system is very important for thermal balance. Cooler blood, returning from the shell, is warmed by the core tissues that are producing metabolic heat; the warmed blood returns to the periphery and is cooled by the tissues of the shell, which are in contact with atmospheric air. As previously mentioned (see Fig. 28–1), vasomotor responses that control peripheral blood flow determine the **effective insulation** of the body by

determining the body proportions that are maintained at core and shell temperatures. At comfortable temperatures, vasomotor responses regulate peripheral blood flow to achieve thermal balance without sweating or changing metabolic heat production. At the lower limit of the thermal comfort zone, peripheral vessels constrict and decrease blood flow and heat flow to a minimum (Fig. 28–15). At the upper limits of the thermal comfort zone, peripheral vessels dilate and expose large quantities of warm blood to peripheral tissues, increasing heat loss severalfold. The skin therefore can be a heat radiator for the body, and the blood flow to the skin determines the rate of heat transfer by circulatory convection from the body core to the skin.

Regulation of Body Temperature

The regulation of body temperature in humans is a classic example of a precise biological control system that has evolved to maintain thermal homeostasis. The control of body temperature within narrow limits is dependent upon the coordinated activities of several subsystems of the body, including sweating, vasomotor control, and metabolic rate. The physiological control system that coordinates the responses for effective regulation of body temperature

Figure 28–15
Effect of changes in environmental temperature on the heat conductance from the body core to the skin surface.

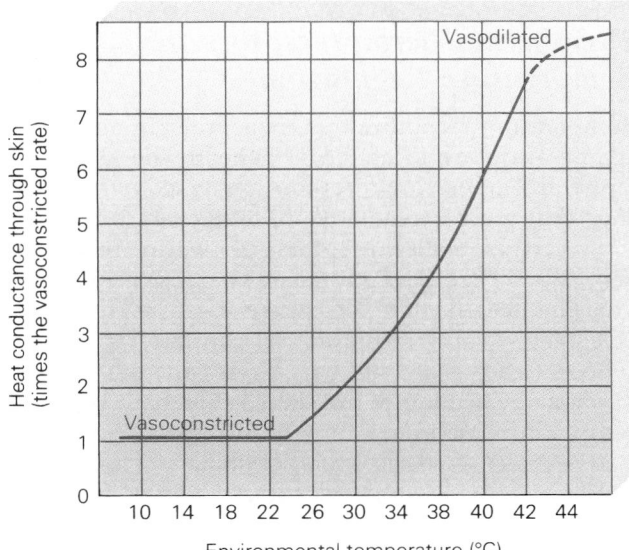

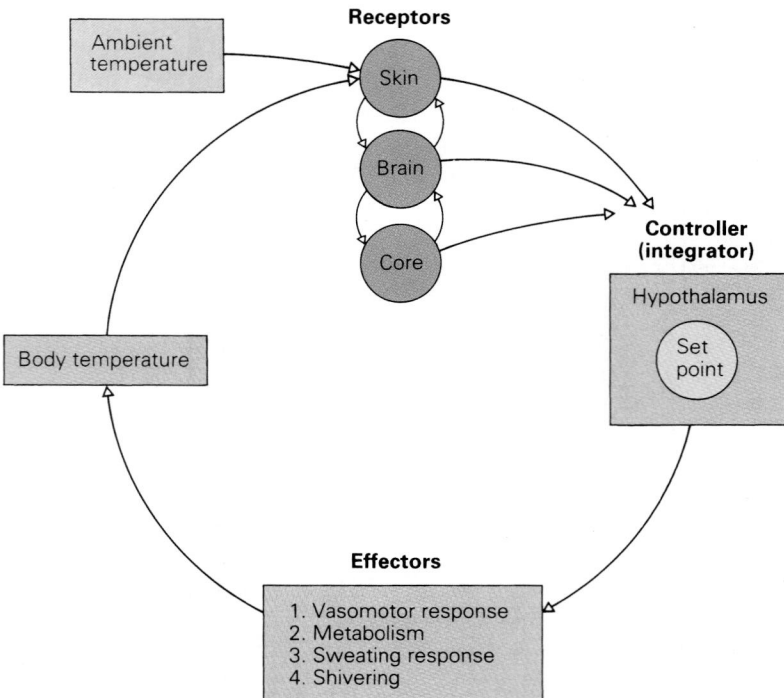

Figure 28–16
Basic components of the negative feedback control system for regulation of body temperature.

is a **negative feedback system** with at least three major components (Fig. 28–16). It has **temperature receptors** that sense the existing body temperature, **effector organ systems** that determine heat production or heat loss, and an **integrator** or **controller** that compares the sensed temperature to a "normal" or "reference" temperature to determine whether it is too high or too low. It then activates the appropriate effector systems, which return the body temperature back to normal.

Peripheral Thermoreceptors in Temperature Regulation

Temperature receptors that measure skin and shell temperature are located just beneath the skin. Peripheral temperature receptors are naked nerve endings that are very sensitive to temperature and are classified as **cold receptors** or **warm receptors** (Fig. 28–17). Cold receptors are characterized by increasing steady-state discharge rates as the skin is cooled. Warm receptors, in contrast, are those whose steady-state discharge rates increase in response to warming of the skin. While both types of temperature receptors are found throughout the body surface, cold receptors are about 10 times more numerous than warm receptors. Nerve impulses from peripheral receptors enter the spinal cord at all levels and ascend to the brain.

Central Thermoreceptors in Temperature Regulation

Central thermoreceptors are found in deep body areas including the hypothalamus, spinal cord, abdominal viscera, and great veins. Central thermoreceptors include both **cold receptors** and **warm re-**

Figure 28–17
Static discharge frequency of cold and warm nerve fibers as a function of skin temperature.

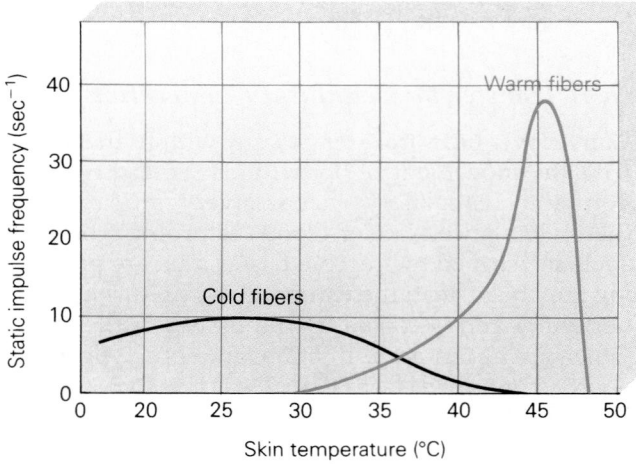

ceptors that are uniquely sensitive to decreases or increases in body core temperature. Nerve impulses from core receptors feed into the brain, where they are integrated with thermal information from peripheral nerves.

Central Integration of Thermal Information

Animal experiments have shown that the hypothalamus is the brain center for body temperature regulation. This area, through its various interactions with other brain areas as well as with central and peripheral temperature receptors, acts as a **thermostat** or controller to prevent both excessive body heating and cooling. Stimulation of the hypothalamus by overheating initiates **antirise responses** such as vasodilatation and sweating. Conversely, stimulation of the hypothalamus by excessive cooling stimulates **antidrop responses** such as vasoconstriction and shivering.

The neural pathways that deliver the hypothalamic signals to the effector organ systems include the **somatic** and **autonomic** portions of the nervous system. The sympathetic portion of the autonomic nervous system plays an important role because it controls vasomotor responses and sweating. The somatic portion, on the other hand, controls gross voluntary muscle movements such as adding or removing clothing and the small muscle movements associated with increased muscle tone and shivering.

Diseases That Affect Temperature Regulation

Many disease states cause **fever,** which is a **regulated increase** in body temperature beyond the normal range. Therefore, measurement of a person's body temperature is a common way to assess the severity of an illness. Some pathophysiological conditions produce either **hypothermia** or **hyperthermia,** which are **unregulated** decreases or increases, respectively, in body temperature.

Fever: A Regulated Increase in Body Temperature

Fever is usually caused by chemical agents called **pyrogens** released into the body during disease processes. The exact way that these agents work is not known, but they appear to **reset** the controlled level of the hypothalamic thermostat at a higher temperature level (Fig. 28–18). The hypothalamic integrating center compares existing peripheral and core temperature thermoreceptor information with the new elevated reference setting, activating heat production and conservation responses and inhibiting heat loss responses. Simultaneously, the person feels cold and may react behaviorally by turning up the heat or putting on more clothing. After several hours, the body temperature will rise to the new reference temperature, at which point the hypothalamus will equalize heat loss and heat conservation

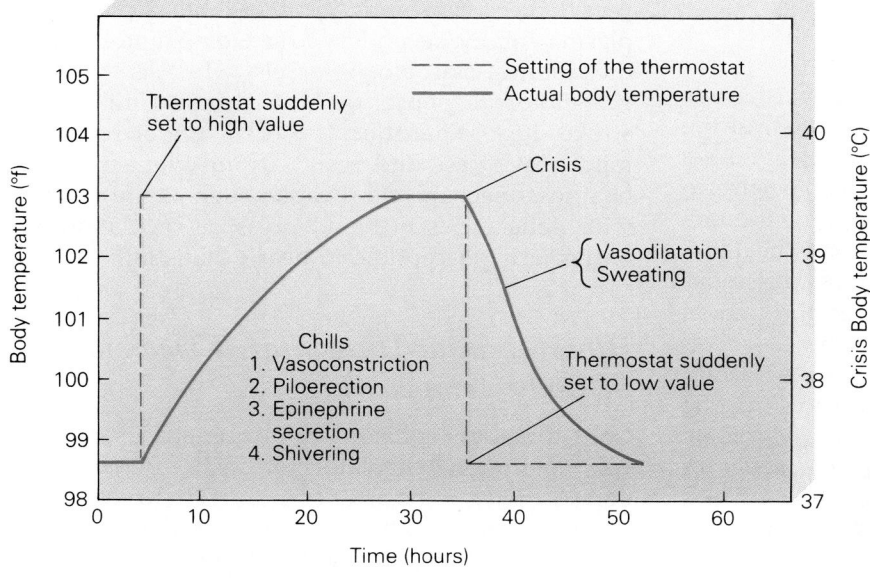

Figure 28–18
Time course of typical febrile episode. Effects of changing the setting of the "hypothalamic thermostat."

Blood-Brain Barrier (BBB) Permeability

Environmental hyperthermia and *E. coli* infection have both been implicated as conditions that increase blood-brain barrier (BBB) permeability. Lipopolysaccharide, a component of the *E. coli* cell wall, has been shown to produce a regulated fever in rats. BBB permeability to ^{14}C-sucrose (molecular weight: 340 daltons) and ^{3}H-dextran (molecular weight: 70,000 daltons) was determined in afebrile and febrile rats by calculating the permeability-surface area product (PA) for several brain regions. Rectal temperature was monitored continuously throughout the course of the experiment. PA values for afebrile rats were calculated for the cortex, cerebellum, medulla, and hypothalamic brain regions, and averaged 7.22, 7.18, 6.00, and 6.51 for ^{14}C-sucrose. The corresponding PA values for ^{3}H-dextran were 1.07, 1.24, 0.95, and 0.99, respectively. *E. coli* lipopolysaccharide injected intravenously (50 μg/kg) produced a significant fever averaging 1.10°C ($p < 0.01$). PA values for febrile rats were not significantly different from afebrile rats. These results indicate that a fever-producing dose of lipopolysaccharide does not alter BBB permeability and is consistent with the hypothesis that the BBB is maintained during a regulated febrile state.

responses. The body temperature continues to be regulated at the elevated level as long as the pyrogen is present. When the pyrogen is no longer effective or present, the so-called **crisis** occurs, and the reference temperature returns to the lower, normal range. When this happens, the hypothalamic integrator activates heat loss and inhibits heat gain processes that gradually return the body temperature to normal levels.

Hyperthermia: Unregulated Increases in Body Temperature

Heat Stroke

Heat stroke is a form of unregulated hyperthermia or heat illness associated with a breakdown of the body's temperature-regulating system. Heat stroke usually occurs during heavy work or exposure to hot, humid conditions. In these situations, the rate of heat gain can exceed the maximal rate of heat loss so that the body temperature continues to rise unless the level of work is decreased or the person moves to a cooler, less humid atmosphere.

If, however, the body temperature rises beyond a critical temperature in the range of 41 to 42°C, the person is likely to develop heat stroke. Symptoms include dizziness, abdominal distress, absence of sweating, sometimes delirium, and eventually loss of consciousness and death if the body temperature is not decreased quickly. Some of the symptoms result from circulatory shock or insufficiency brought about by fluid and electrolyte loss before the onset of the symptoms. The high temperature itself may damage body tissues, particularly in the brain, leading to loss of sweating, and a vicious cycle is started in which core temperature is raised continually toward lethal levels.

Heat Exhaustion

Heat exhaustion, a less serious heat illness caused by **circulatory insufficiency** and **hypotension,** often takes the form of fainting due to decreased blood pressure. This is brought about by depletion of plasma volume secondary to sweating, and by vasodilatation of skin blood vessels, which is a normal vasomotor response to heat stress. Unlike heat stroke, heat exhaustion can occur under relatively mild heat stress and results from over-activity of heat loss mechanisms. People, such as the elderly, with deficiencies in blood pressure regulation are particularly susceptible to heat exhaustion.

Hypothermia: Unregulated Decreases in Body Temperature

Cold illnesses or disorders in humans usually involve peripheral circulatory or cardiac defects. Hypersensitivity of the arteries in the skin to cold (Raynaud's disease) can reduce blood flow to such critical levels that pain and local asphyxia may occur

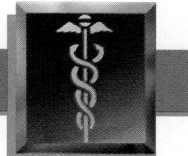

Malignant Hyperthermia

Malignant hyperthermia is the result of an inherited defect in skeletal muscle metabolism. The disorder is inherited as an autosomal dominant trait. In genetically disposed persons, several other disorders are seen, such as hypoxia, hypotension, and acidosis. Susceptibility to the condition appears to involve a defect in the uptake of calcium by the sarcoplasmic reticulum; its actual manifestation is triggered by anesthetic agents. The probability that the condition will develop in susceptible individuals increases with successive exposure to anesthetics.

In malignant hyperthermia, the calcium concentration of the sarcoplasm becomes abnormally high, and the skeletal muscles undergo prolonged contractions. The high metabolic rate of the continuously contracting muscles increases metabolic heat production and produces hyperthermia. The hyperthermia is not of CNS origin, and there are no signs of coordinated rhythmic shivering. The resulting high body temperature is a major contributor to the mortality of this disorder. Temperatures in excess of 45°C are often reported.

The preferred treatment consists of immediate removal of heat with a cooling blanket and removal of the anesthetic agents. Respiratory support is also important to provide for the high consumption of oxygen and production of carbon dioxide. Bicarbonate helps to offset the metabolic acidosis caused by the high rate of lactate production through glycolysis. The intravenous infusion of iced lactated Ringer's solution also helps to remove body heat and supports renal function. Once a person is known to be susceptible to malignant hyperthermia, general anesthesia should be avoided.

if the stress is continued. When the body mechanisms for heat production and conservation are exceeded by exposure to severe cold, **hypothermia** results, and body temperature drops below the normal range. At body temperatures of about 26 to 28°C, death can occur due to **myocardial fibrillation.** Older persons are particularly susceptible to hypothermia, since in some elderly individuals, the hypothalamic reference temperature can be reset to a lower level. Some elderly people can maintain a core temperature of 35°C or even less without the onset of shivering. In the elderly, this appears to be a regulated hypothermia, and to some extent this condition can be considered a counterpart of fever.

SUMMARY

Normal Body Temperature

Humans are homeotherms. They usually regulate their deep body temperature within a narrow range when exposed to a wide range of climatic conditions. The regulation of body temperature near the upper limit of heat tolerance ensures both rapid metabolism and a near-constant and optimal internal thermal environment.

The body has both longitudinal and radial temperature gradients. Normal body temperature should be conceived as a range of temperatures. The most widely accepted measures of body core temperature are rectal, oral, and axillary temperatures.

The mean skin temperature is the accepted measure of the temperature of the body shell.

The mean body temperature is obtained by weighing mean temperatures of the core and the skin.

Many factors, including exercise, diurnal rhythms, the reproductive cycle, age, and disease, can affect body temperature.

The Balance Between Heat Production and Heat Loss

Heat is produced continuously in the body as a by-product of catabolism. Both physiological and physical processes are continuously operating to balance heat production with heat loss, thereby maintaining thermal homeostasis and a relatively constant body temperature. The metabolic rate is the amount of food energy converted to heat per unit of time. Many factors can affect metabolic rate, including exercise, food, hormones, age, sex, size, and temperature. Basal metabolic rate (BMR) is a measure of the metabolic cost of living measured under standard conditions. It allows comparisons between individuals. A person's metabolic rate can be measured by direct or indirect calorimetry.

The metabolic heat produced must be continually removed from the body to maintain thermal balance. The body can exchange heat with its surroundings through radiation, conduction, convection, and evaporation. The evaporation of *sweat* is the only process for heat loss from a body exposed to hot environments. Sweating is the active secretion of mostly water by specialized glands in the skin. Sweat must be rapidly evaporated from the skin for cooling to occur.

To keep the body temperature relatively constant, the heat produced in the core must be transferred to the body shell, where it can be exchanged with the environment. Heat is transferred by internal conduction and circulatory convection. The rate of heat transfer due to internal conduction is too slow to maintain a balance between heat production and heat loss. Circulatory convection by the blood, therefore, is of prime importance for thermal balance. The vasomotor responses that control peripheral blood flow determine the effective insulation of the body and the rate of heat transfer by circulatory convection from the body core to the skin.

Regulation of Body Temperature

A negative feedback system is activated by temperature receptors that coordinate the responses for effective regulation of body temperature. Peripheral temperature receptors measure skin and shell temperature and are located just beneath the skin. They are nerve endings classified as cold or warm receptors. Nerve impulses from peripheral receptors enter the spinal cord at all levels and ascend to the brain.

Central thermoreceptors measure the temperature of the body core. They are located in deep body areas and include both cold and warm receptors. They send impulses into the brain, where integration with peripheral thermal information takes place.

The hypothalamus—through interactions with other brain areas, as well as with central and peripheral temperature receptors—acts as a central integrator and thermostat to prevent excessive body heating or cooling. Stimulation of the hypothalamus by overheating initiates antirise responses such as vasodilatation and sweating. Stimulation of the hypothalamus by excessive cooling stimulates antidrop responses such as vasoconstriction and shivering. The regulation of body temperature is a classic example of a precise negative feedback control system that has evolved to maintain thermal homeostasis.

Diseases That Affect Temperature Regulation

Fever is a regulated increase in body temperature, usually caused by the release of chemical agents called pyrogens into the body during disease processes. Pyrogens appear to reset the control level of the hypothalamic thermostat at a higher body temperature level. Fever, therefore, is not a loss of effective regulation of body temperature.

Heat stroke is a form of unregulated heat illness associated with a failure of the body's temperature regulating system. Heat stroke usually occurs during heavy work and/or exposure to hot, humid conditions in which the rate of heat gain can exceed the maximal rate of heat loss. Under these conditions, the body temperature continues to rise beyond the critical temperature of 41 to 42°C typically associated with heat stroke. The high temperature itself may damage brain tissue, leading to failure of the temperature-regulating system and death. Heat exhaustion is caused by circulatory insufficiency and often results in fainting due to decreased blood pressure. It occurs as a result of overactivity of heat loss mechanisms, including fluid loss due to excessive sweating and peripheral vasodilatation. People with deficiencies in blood pressure regulation such as the elderly are particularly susceptible to heat exhaustion.

Cold illnesses in humans usually involve peripheral circulatory or cardiac defects. Hypersensitivity to cold of the arteries in the skin can reduce blood flow to levels that induce local asphyxia and pain. In severe cold, heat loss may exceed the body's maximal ability for heat production, and its temperature will continue to fall, resulting in hypothermia. During severe hypothermia, in which body temperature drops to about 26 to 28°C, death can occur owing to myocardial fibrillation.

REVIEW QUESTIONS

Identify Each with a Term

1. The increased heat production associated with ingestion of food.

2. The exchange of heat using electromagnetic energy.

3. The brain center for integration of body temperature regulation.

4. The disease state characterized by unregulated hyperthermia.

5. Organisms whose body temperatures fluctuate with changes in the ambient temperature.

Define Each Term

6. homeotherm

7. fever

8. mean skin temperature

9. thermography

10. catabolism

Choose the Correct Answer

11. Which of the following is *not* a homeotherm?
   a. dog
   b. cat
   c. snake
   d. horse

12. The process by which a human loses heat when the environmental temperature exceeds the body temperature is:
   a. convection
   b. conduction
   c. evaporation
   d. radiation

13. Malignant hyperthermia is associated with a defect in the uptake of:
   a. potassium
   b. sodium
   c. calcium
   d. iron

14. Metabolic rate is *not* increased by:
   a. exercise
   b. eating
   c. age
   d. catecholamines

15. How many calories per hour does the average adult's metabolic rate burn?
   a. 300
   b. 40
   c. 70
   d. 120

Calculate

16. A person in a comfortable room has a mean skin temperature of 34°C and a rectal temperature of 38°C. What is the mean body temperature?

17. A person in a comfortable room has the following skin temperatures: arm—31°C, chest—35°C, calf—30°C, and thigh—31°C. What is the mean skin temperature?

18. A 70-kilogram adult with a specific heat of 0.83 walks from a comfortable room into a hot room. In one hour, core body temperature increases by 0.5°C. What was the rate of heat storage during that hour of heat exposure?

19. A 70-kilogram person at rest is consuming 0.25 liter of oxygen per minute. What is the metabolic rate in calories per hour?

20. A person in a hot environment evaporates 700 grams of water as sweat per hour. What is the rate of heat exchange per hour?

Essay

21. What are the conditions for the measurement of BMR?

22. What is the role of the hypothalamus in the control of body temperature?

23. Explain the major elements of a negative feedback system for control of body temperature as one goes from a comfortable room of 25°C to a hot room of 35°C.

24. Define the term *normal body temperature* as it applies to humans.

25. Describe the mechanisms that the body uses to exchange thermal energy with the environment.

SUGGESTED READINGS

Bligh, J. "Regulation of body temperature in man and other mammals." In Shitzer, A., and Eberhart, R., Eds. *Heat Transfer in Medicine and Biology*, Vol. 1. New York: Plenum Press, 1985, p. 15.

Crawshaw, L.I. "Temperature regulation in vertebrates." *Annual Review of Physiology*, 42:473, 1980.

Elizondo, R.S. "Temperature regulation in primates." In Robertshaw, D., Ed. *International Review of Physiology: Environmental Physiology II*, Vol. 15. Baltimore: University Park Press, 1977, p. 71.

Elizondo, R.S. and Johnson, G.S. "Peripheral effector mechanism of temperature regulation." In Szelenyi, Z., and Szekely, M., Eds. *Advances in Physiological Sciences*, Vol. 32. New York: Pergamon Press, 1980, p. 397.

Gordan, C.J. and Heath, J.E. "Integration and central processing in temperature regulation." *American Review of Physiology*, 48:595, 1986.

Heller, H.C., et al. "The thermostat of vertebrate animals." *Scientific American*, 239 (2):102, 1978.

Hensel, H. "Neural processes in thermoregulation." *Physiology Review*, 53:948, 1973.

Hulbert, A.J., and Else, P.L. "Evolution of Mammalian Endothermic Metabolism." *American Journal of Physiology*, 256:R63–R69, 1989.

Schonbaum, E. and Lomax, P. "Temperature regulation and drugs: an introduction." *In Thermoregulation, Physiology, and Biochemistry.* New York: Pergamon Press, 1990, P.I.

Veale, W.L. Ruwe, W.D., and Cooper, K.E. "Mechanism of fever and antipyresis." In Khogali, M., and Hales, J.R.S., Eds. *Heat Stroke and Temperature Regulation.* New York: Academic Press, 1983, p. 79.

KEY TERMS

basal metabolic rate (p. 945)
carbohydrates (p. 947)
circadian rhythms (p. 941)
conduction (p. 947)
convection (p. 947)

evaporation (p. 947)
fever (p. 951)
hyperthermia (p. 951)
hypotension (p. 952)
hypothermia (p. 951)

mean body temperature (p. 940)
mean skin temperature (p. 939)
menstrual cycle (p. 941)
poikilotherms (p. 937)

radiation (p. 947)
respirometer (p. 946)
sublingual (p. 938)
thermography (p. 940)
thermostat (p. 951)

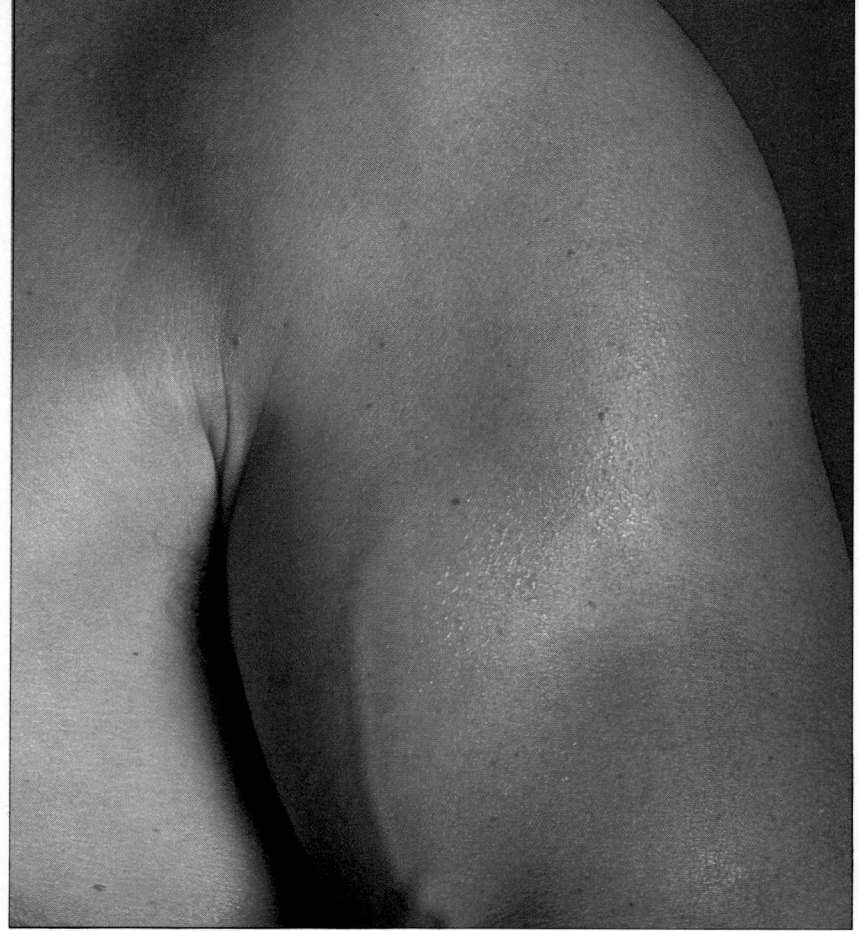

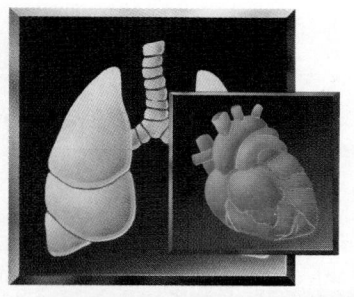

Physiology of Exercise

In all types of exercise, the final common **physiological event** is the **contraction** of skeletal muscle. But exercise also places demands on various other body systems that play a supportive role. The basic biochemical alterations at the muscle cell level that convert chemical energy into mechanical energy are accompanied by increases in the integrated activity of many organ systems, including the **central nervous system** (CNS), which initiates coordinated skeletal muscle activity. The **respiratory system** and **circulatory system** increase their activity during exercise to provide increased oxygen and nutrients to active muscle cells and to remove CO_2, other metabolites, and heat. Contracting muscle cells may increase heat production to 20 times that of resting levels; thus, exercise can also place extra demands on the **temperature regulating system.** The extent of the physiological adjustments in organ systems in response to exercise is often a direct measure of the **strain** or **stress** of exercise on the body.

The basic concepts of exercise physiology, which this chapter will discuss, have been derived mostly from studies on young adult male subjects, because relatively complete measurements of the physiology of exercise have traditionally been made only on this group of athletes. Recent studies, how-

(a)

(b)

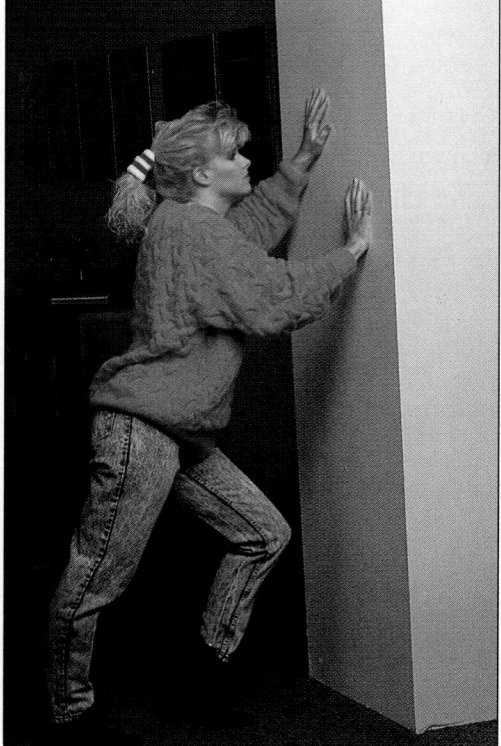

(c)

Figure 29–1
(a) Climbing a mountain—positive dynamic exercise (Mt. Everest). (© John Roskelly/ Photo Researchers, Inc.) (b) Descending a mountain—negative dynamic exercise (Alps, France). (© Didier Givois/Photo Researchers, Inc.) (c) Isometric exercise. (© David Weinstein & Associates/Custom Medical Stock Photo.)

ever, in which measurements have been made on females, indicate that the same basic principles apply. However, some **quantitative differences** are found, such as differences in body size, body composition, and sex hormones. In general, most quantitative values for women related to muscle mass, such as ventilation, cardiac output, and muscle strength, will vary from 65% to 75% of the values recorded in males. Differences in athletic performance between the sexes, on the other hand, are smaller because the female body is on average smaller, requiring smaller muscles to perform the same activities.

Exercise is often classified as either **dynamic exercise** or **static exercise.** Dynamic exercise is performed anytime muscle length changes. In a physical sense, resistances are overcome along a certain distance. In this type of exercise, which includes such activities as bicycling, running, stair climbing or mountain climbing, exercise is quantified in physical units such as watts. In **positive dynamic exercise,** such as ascending a mountain, the muscles act as a "motor." In **negative dynamic exercise,** such as descending a mountain, muscles act as "brakes." Static exercise is performed during **isometric muscle contraction.** During isometric exercise, shortening of the muscle does not occur and no distance is traversed, so it cannot be defined as work in the physical sense. Nevertheless, static exercise is associated with increased metabolic demands. The **elevated metabolic demand** is the common feature of all types of exercise, and, in most instances, this is more important for physiological adjustments than any other factor.

Metabolic Aspects of Exercise

Energy Requirements of Muscular Contraction

The primary energy source for skeletal muscle contraction, as for other types of cellular work, is high-energy phosphate compounds such as **adenosine triphosphate (ATP).** These are produced by both anaerobic and aerobic metabolic processes (Fig. 29–2). Three important metabolic systems that supply energy for muscle contraction are (1) the phosphagen system, (2) the glycogen–lactic acid system, and (3) the aerobic system **(the citric acid cycle).**

The Phosphagen Energy System

The term *phosphagen* is another name for a high-energy phosphate molecule, one that serves as a reservoir of phosphate bond energy. The **phosphagen energy system** of muscle includes ATP and **phosphocreatine.** When one phosphate group is removed from ATP (producing ADP), approximately 11,000 calories of energy become available for the contractile process. When a second phosphate is removed, leaving AMP, an additional 11,000 calories become available. The amount of energy present as ATP in the muscle is enough to maintain maximal exercise for only 5 to 6 seconds.

Phosphocreatine, however, has slightly more energy than the bond energy of ATP and can reconstitute ATP by donating phosphate groups to AMP and ADP. Muscle cells have two to three times more phosphocreatine than ATP, and together the phosphagens ATP and phosphocreatine can maintain maximal muscle contraction for only 10 to 15 seconds. Thus, the energy supply from the phosphagen system is used for short bouts of maximal exercise.

The Glycogen–Lactic Acid System

The glycogen–lactic acid system in muscle can also supply energy for contraction. The stored muscle glycogen can be split into glucose **(glycogenolysis)** and then used for energy. Recall from Chapter 6 that glycolysis occurs in the absence of oxygen and is therefore **anaerobic.** During glycolysis, each glucose molecule is converted into two molecules of pyruvic acid, and energy is released to form two molecules of ATP. If oxygen is available, pyruvic acid enters the citric acid cycle, and much more ATP

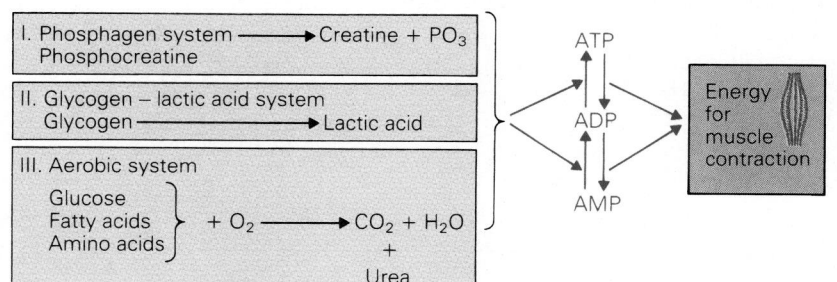

Figure 29–2
Summary of the three primary metabolic pathways that supply energy for muscle contraction.

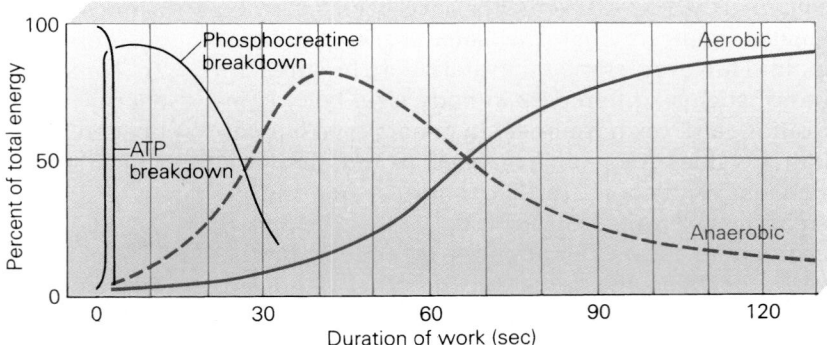

Figure 29–3
Relative contribution of various energy sources to the total energy used by the muscle at the beginning of light exercise.

is formed. If there is insufficient oxygen to support the oxidative stage of glucose metabolism, however, the pyruvic acid is converted to lactic acid. The glycogen–lactic acid system can produce ATP molecules more rapidly than the oxidative mechanisms in the mitochondria and is used as a rapid source of energy for muscle contraction. The glycogen–lactic acid system can provide the energy for maximal exercise for only 30 to 40 seconds.

The Aerobic Energy System

Recall from Chapter 6 that the aerobic energy system involves the use of oxygen within the mitochondria for the metabolism of glucose, lipids, and proteins in the citric acid cycle and electron-transport chain to produce 36 ATP molecules for every molecule of glucose. Only 20% of the energy released, however, is in the form of the mechanical energy of contraction; 80% is released as heat energy. The heat released during exercise places additional demands on the cardiovascular system and the sweat response, which play major roles in temperature regulation and the transfer of heat from the body to the environment (see Chapter 28).

Aerobic metabolism is normally the source of energy for prolonged light work. During light exercise, energy is obtained anaerobically only during the first few minutes when muscle blood flow is gradually increasing (Fig. 29–3). Thereafter, metabolism is entirely aerobic, utilizing glucose, fatty acids, and amino acids. During heavy exercise, by contrast, part of the energy is always obtained anaerobically, and if large amounts of lactic acid are produced, muscle fatigue and cramps result.

Oxygen Consumption During Exercise

During exercise, oxygen consumption increases by an amount corresponding to the intensity of the exercise. As illustrated in Figure 29–4, oxygen con-

sumption increases during the first few minutes of exercise and then reaches a "steady state" in which the oxygen uptake corresponds to the demands of the tissues. The average person in the basal state consumes about 0.25 liter (L) of oxygen per minute. An untrained person can increase O_2 consumption threefold (0.75 liter/min) during light work and eightfold to tenfold (2–3 liters/min) during heavy exercise. Well-trained athletes can increase their O_2 consumption to over 20 times (5–6 liters/min) above basal level. Oxygen consumption is often standardized in units of body weight (O_2 ml/kg/min), and averages 44 to 51 ml/kg/min for untrained males. Values between 50 and 60 ml/kg/min are often obtained in young, well-trained males. Untrained fe-

Figure 29–4
Oxygen uptake during increasing levels of steady-state exercise.

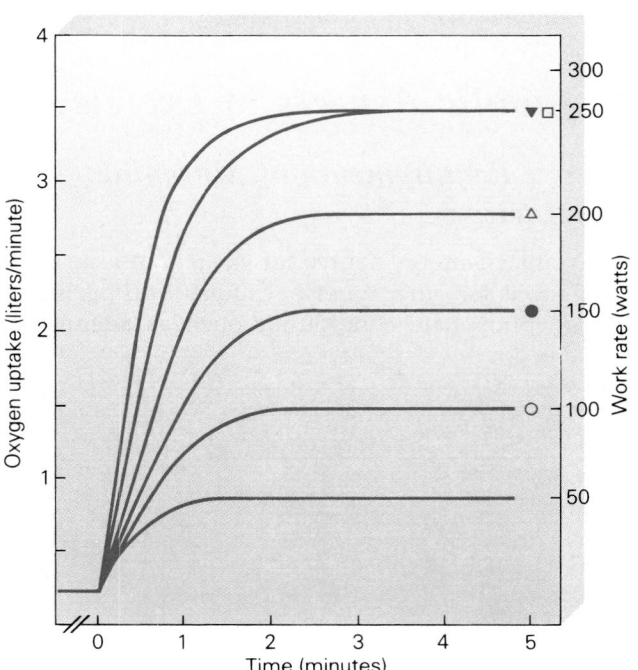

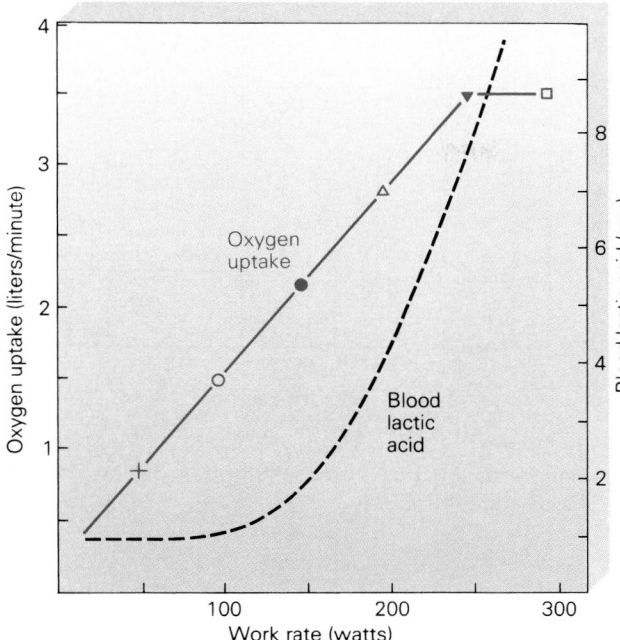

Figure 29–5
Oxygen uptake and blood lactic acid during increasing levels of exercise.

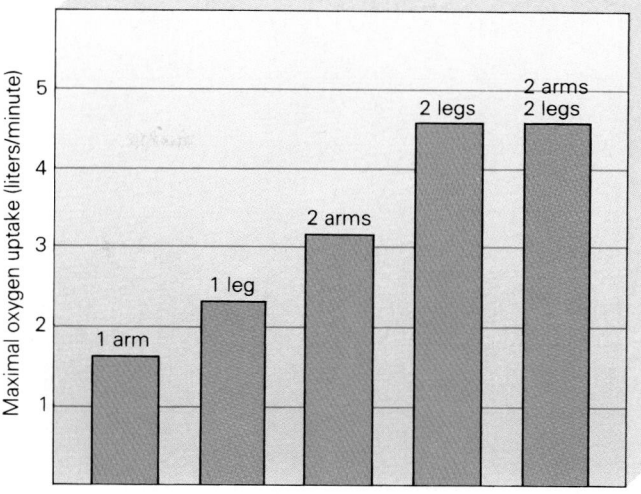

Figure 29–6
Peak oxygen consumption during maximum exercise using different sizes of muscle mass.

males normally have O_2 consumption during exercise that is about 20% less than that of untrained males.

Oxygen consumption increases linearly with increasing levels of exercise up to a point at which the maximum for oxygen transport and uptake in the tissues appears to be reached, and a plateau, or the so-called **maximal oxygen consumption (VO$_2$max),** is attained (Fig. 29–5). The VO$_2$max is generally assumed to represent a person's **maximal aerobic power** and thus is an important factor in determining one's ability to sustain high-intensity exercise. Three main criteria have been established to certify that a measurement of VO$_2$max actually represents a person's maximal aerobic power: (1) there is no further increase in oxygen consumption despite further increases in the rate of exercise, (2) the blood lactate concentration exceeds 8 mM, and (3) large muscle groups are exercised for at least 3 minutes. Figure 29–6 illustrates that the VO$_2$max peak obtained during maximal exercise is maximal only when a relatively large muscle mass (i.e., two legs) is being exercised.

Numerous factors can affect an individual's VO$_2$max. With age, VO$_2$max decreases in both males and females. As shown in Figure 29–7, maximal uptake in liters per minute increases dur-

ing the growth stages of the early years and reaches a peak between 18 and 25 years of age. This early increase in oxygen consumption is directly related to the increase in body mass that occurs at this age. Thus, the apparent increase in aerobic capacity is essentially eliminated when the VO$_2$max is expressed relative to body weight as ml/kg/min. After age 25, the VO$_2$max terms of liters per minute, as well as relative to body weight, steadily declines so that at age 55 it is about 72% below values obtained for 20-year-olds. While active adults tend to retain relatively higher VO$_2$max than inactive adults, a

Figure 29–7
Maximum oxygen consumption as a function of age in males and females.

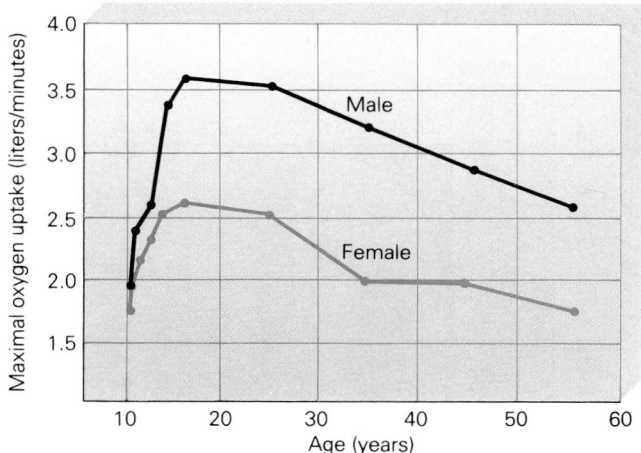

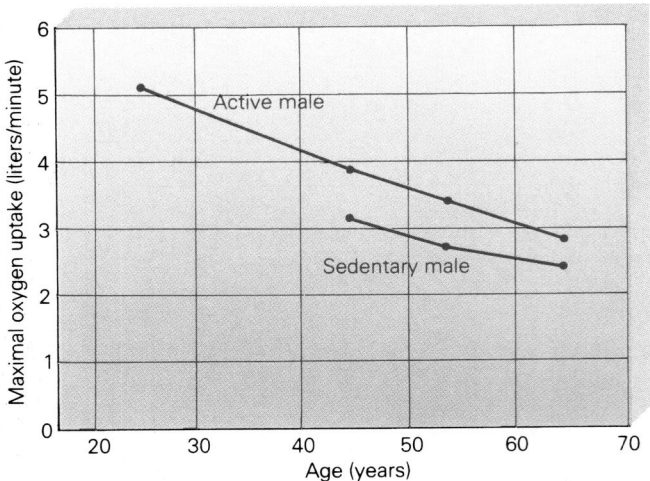

Figure 29–8
Maximum oxygen consumption as a function of age
and level of activity.

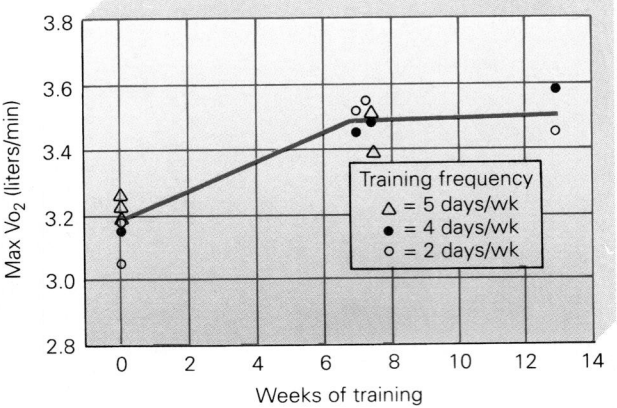

Figure 29–9
Change in maximal oxygen consumption during 7 to 13
weeks of athletic training.

progressive decline in this physiological capacity
also occurs with advancing age (Fig. 29–8).

Training increases VO_2max and can partially
counteract the effects of aging (Fig. 29–9). For example, the mean VO_2max for nine men on an initial
exposure to running 7 mph (9% grade) on a treadmill averaged 53 ml/kg/min. After 2 months of training by running on a treadmill at a rate of 8 mph
(zero incline) for 10 to 15 minutes, their VO_2max
was increased to 56 ml/kg/min. The physiological
factors that may limit human O_2 consumption include (1) the rate of cardiovascular transport to active tissue, (2) the rate of O_2 use by active tissue,
and (3) the O_2 diffusion capacity of the lungs. There
appears to be a dual basis for an increased VO_2max
with training, including (1) an increase in maximal
cardiac output and (2) an increase in blood arterial-
venous oxygen difference. Training, on the other

hand, has a relatively small effect on pulmonary
function.

Oxygen Debt During Exercise

When exercise begins, the demand for oxygen increases immediately. However, tissue blood flow
and aerobic oxidative metabolism require some time
to adjust to the new demands, and an **oxygen debt**
is incurred as oxygen is used from existing body
stores (Fig. 29–10). During light work, the oxygen
debt remains constant after a steady state is reached
in which oxygen uptake and consumption are in
balance. During heavy exercise, the oxygen debt
continues to build up until the exercise ends. At the
end of exercise, although the demand for oxygen by
contracting tissue decreases abruptly, the rate of
oxygen uptake remains above the resting level for
several minutes as the oxygen debt is being repaid.
The oxygen debt is often defined as the amount of

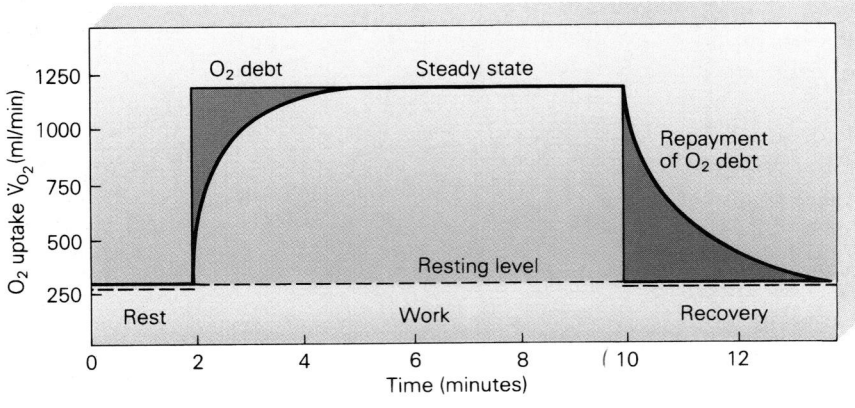

Figure 29–10
Oxygen consumption during light
steady-state exercise.

oxygen above the pre-exercise or resting level that must be taken into the body after exercise. The oxygen debt repaid after the end of exercise is often greater than that incurred during the work itself and is affected by other factors such as increases in body temperature and hydrogen ion concentration. After light work, the oxygen debt amounts to as much as 4 liters, and after heavy exercise it can be as high as 20 liters.

The repayment of the oxygen debt occurs in two stages (Fig. 29–11). The first stage is called the **alactacid oxygen debt,** the term referring to the oxygen needed to replenish the oxygen stores of the body as well as the phosphagen system. The alactacid debt is usually repaid within 2 to 3 minutes after exercise. The second stage is called the **lactic acid oxygen debt** and requires the removal of lactic acid from the body. This stage is repaid very slowly and requires at least 1 hour. During this time, some of the lactic acid is converted to pyruvic acid and then metabolized by body tissues. Also, much of the lactic acid is converted by the liver to glucose, which is used to replenish the glycogen stores in the muscle.

The Use of Nutrients During Exercise

Carbohydrates, fatty acids, and (to a lesser degree) amino acids can be utilized as fuel during exercise in proportions that vary depending on the intensity and duration of the exercise and the availability of stored fuels. At low rates of exercise of short duration, such as short walks, lipids supply a large portion of the required energy. With increasing intensity and duration of exercise, there is a gradual change toward a proportionally greater share of energy yield from carbohydrates. However, if moderately heavy exercise continues for 4 to 5 hours, the glycogen stores of the muscle become depleted; the muscle then depends on the limited supply of glucose that can be absorbed from the gut, or on energy from other sources, mainly fats and (to a much lesser extent) protein. Clearly, the choice of fuel for exercising muscle is mainly limited to carbohydrates and fat.

The relative usage of carbohydrates and fat for energy yield during prolonged, exhaustive exercise is illustrated in Figure 29–12. Note that in an individual on a mixed diet, most of the energy at this level of exercise is derived from carbohydrates during the first few minutes, but as the exercise continues there is a gradual shift toward fat use, and at exhaustion as much as 60% to 70% of the energy is being derived from fats rather than carbohydrates. It is also clear that diet can influence the relative proportion of stored fuels and thus influence the relative usage of fat and carbohydrates. A high-carbohydrate diet is associated with a relatively greater share of the energy yield from carbohydrates and longer endurance, whereas a high-fat diet results in a proportionally greater share of the energy yield from fat and relatively shorter endurance.

Diet can also affect glycogen recovery following exhaustive exercise. Figure 29–13 shows the rate of

Figure 29–11
The rate of oxygen uptake during and following four minutes of maximal exercise. Following exercise, oxygen uptake remains above resting levels as the alactacid and lactic acid debt is repaid.

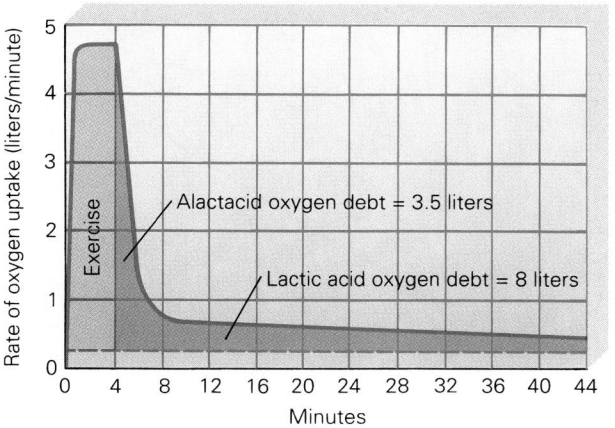

Figure 29–12
The relative use of fats and carbohydrates as a function of diet and duration of exercise.

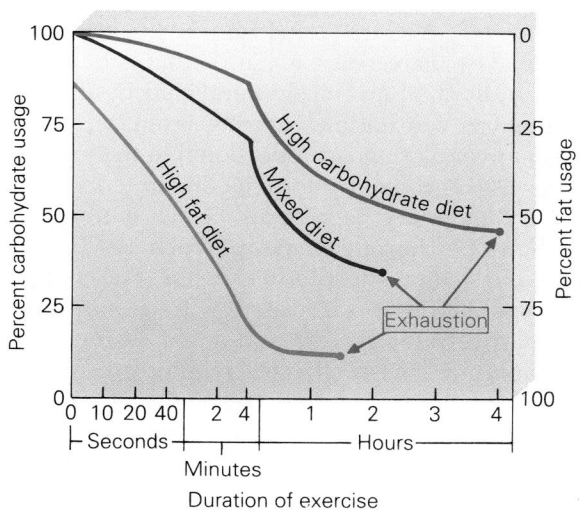

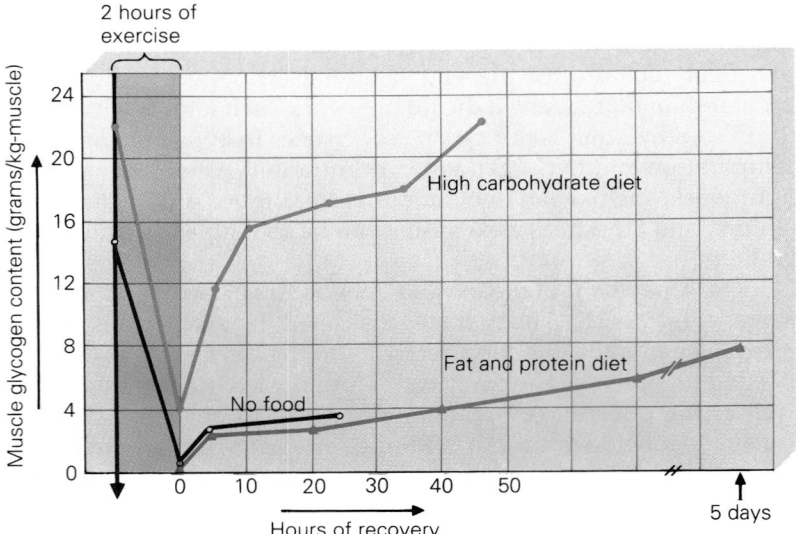

Figure 29–13
Rate of replenishment with different diets
following prolonged exercise.

muscle glycogen replenishment in three different situations: a high-carbohydrate diet, a high-fat/high-protein diet, and starvation. Note that even in individuals on high-carbohydrate diets, muscle recovery is a relatively slow process, lasting approximately two days. Individuals on high-fat and high-protein diets, or persons eating no food at all, showed very little recovery even after 5 days. It seems clear that to achieve optimal performance, approximately 48 hours of recovery after exhausting exercise are essential.

Respiration During Exercise

Pulmonary Ventilation During Exercise

The increased aerobic metabolism in muscle during exercise places extra demands on the respiratory system. The increased oxygen demands of an exercising individual are nicely paralleled by increased **pulmonary ventilation.** During exercise, the increase in ventilation is proportional to the metabolic demands (O_2 consumption), as illustrated in Figure 29–14. Pulmonary ventilation, which is normally 5 to 6 liters per minute, may approach 150 liters per minute in severe, short-term exercise. The change in total ventilation is associated with a fourfold increase in respiratory rate and a sixfold increase in tidal volume. These changes require greater work by the respiratory muscles, and the work of breathing is increased by one-hundredfold (Fig. 29–15). Note, however, that maximal breathing capacity in a healthy young individual is approximately 170 liters

per minute and is therefore 50% greater than the pulmonary ventilation achieved during maximal exercise. The point is that the respiratory system is not normally the most important factor limiting the delivery of oxygen to the muscles during maximal exercise.

The Diffusing Capacity of the Lung During Exercise

The **diffusing capacity of the lung** is defined as the milliliters of a gas diffusing across the pulmonary membrane per minute per millimeter of mercury pressure difference between the alveolar air and the pulmonary blood. The diffusing capacity of the lung

Figure 29–14
Effect of exercise on oxygen uptake and total ventilation.

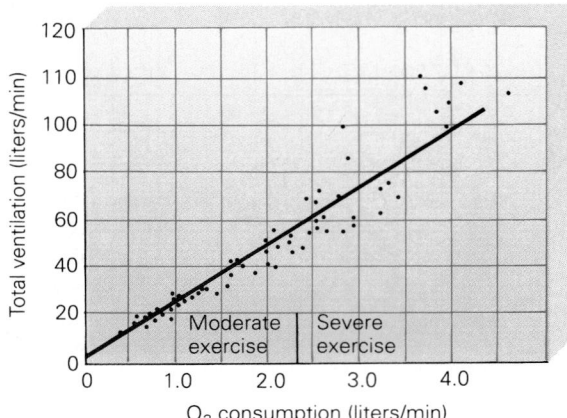

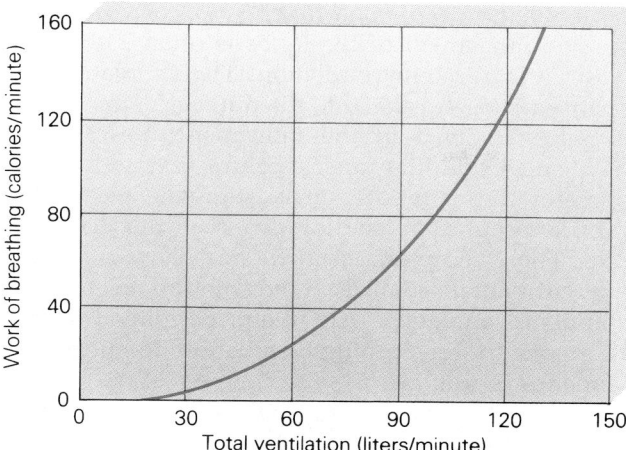

Figure 29–15
Work of breathing as a function of minute volume (pulmonary ventilation).

for the respiratory gases provides an index of the dimensions of the pulmonary capillary bed, the pulmonary membrane, and the overall efficiency of the system in the exchange of respiratory gases. The diffusion capacity of oxygen and carbon dioxide increases during exercise. Oxygen diffusion capacity in a highly trained athlete increases from an average of 23 ml/mm Hg at rest to about 80 ml/mm Hg at maximum exercise. The increased diffusing capacity in the lung is largely the result of the increased pulmonary blood flow that accompanies exercise and ensures maximum perfusion of pulmonary capillaries, thus providing a greater surface area for gas diffusion.

Control of Pulmonary Ventilation During Exercise

Pulmonary ventilation at rest exhibits inherent rhythmicity. Its control involves interactions between CNS inspiratory and expiratory centers, pulmonary stretch receptors, and peripheral and central chemoreceptors sensitive to blood gas pressure and H^+ concentrations.

The mechanisms that control ventilation at rest cannot entirely explain the increased ventilation or **hyperpnea** during exercise. For example, artificial changes in Po_2, Pco_2, and acidity in the body at rest do not lead to the high levels of ventilation observed during exercise. How this increase in ventilation is elicited during exercise is unknown. Several theories have been advanced that suggest interactions between numerous stimuli falling into two main categories: **chemical** or **humoral** (fluid-related) **stimuli,**

and **neural stimuli.** Figure 29–16 illustrates the time course of ventilation changes before and during moderate steady state exercise. The first component (Phase I) is a rapid rise that occurs just before the exercise starts. Ventilation increases so rapidly that it may occur within a single respiratory cycle. The second component (Phase II) is a slower rise, and the third component (Phase III) is associated with the development of steady state ventilation. The fast component of the response is believed to be associated with neural stimuli, since they occur sooner than the metabolic changes. The slow components are believed to be related to the slower humoral stimuli that alter CNS respiratory center activity. While the exact nature of the neural and chemical mechanisms that control ventilation during exercise is not known, it is clear that they interact to provide the proper increase in pulmonary ventilation to meet the metabolic demands of the exercise and thus keep blood respiratory gases almost normal. Some of the factors that have been studied as possible stimuli for the exercise hyperpnea will now be briefly discussed.

Chemical Stimulation of Pulmonary Ventilation

Increased Arterial Carbon Dioxide Concentration During Exercise. The close correlation between ventilation rate and metabolic rate during steady

Figure 29–16
The time course of change in ventilation before and during moderate exercise.

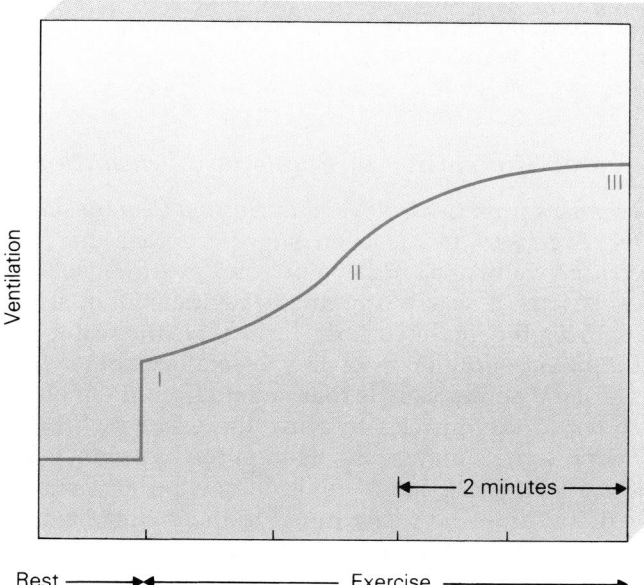

state exercise suggests that the ventilatory drive may be associated with a chemical constituent that is altered at rates proportional to metabolism. Carbon dioxide production is greatly increased in muscle during exercise and could theoretically increase arterial P_{CO_2} enough to stimulate CNS respiratory centers. Actual measurements of arterial P_{CO_2} indicate that the increases are nonexistent during moderate exercise and that a sizable decrease can occur during very heavy exercise. Much larger increases in arterial P_{CO_2} than have been measured would have to occur if P_{CO_2} alone were the primary stimulus.

Increased Arterial Hydrogen Ion Concentration During Exercise.

Lactic acid production is greatly increased in muscles during exercise and could lead to an increase in arterial hydrogen concentrations sufficient to stimulate CNS respiratory centers. Actual measurements have shown that significant changes in arterial lactic acid concentration occur only during very strenuous exercise. Furthermore, at the end of exercise, lactic acid levels may increase slightly while ventilation rate rapidly decreases. It is therefore unlikely that the blood hydrogen concentration alone provides the major stimulus for exercise hyperpnea.

Decreased Arterial Oxygen Concentration During Exercise.

Arterial **hypoxemia** can stimulate respiration by eliciting chemoreceptor activity from the carotid and aortic bodies. The threshold for these receptors, however, is an arterial blood oxygen pressure of approximately 60 mm Hg. Even during strenuous exercise, arterial oxygen pressure usually remains absolutely constant, and this mechanism alone cannot provide the primary stimulus for the increased ventilation during exercise.

Neural Stimulation of Pulmonary Ventilation

Impulses from Central Nervous System Centers During Exercise.

It has been suggested that the increased ventilation just before and at the onset of exercise is at least partly due to neural stimuli arising from the motor cortex. The effect is the result of the direct stimulation of the respiratory center by the same neural signals that radiate from the motor cortex to the muscles to cause the exercise. While neural signals may be involved in the hyperpnea of exercise, it is difficult to envision how purely neural activity can be so finely tuned to the metabolic demands of the exercise.

Impulses from Peripheral Receptors During Exercise.

Passive movement of the limbs is often associated with an increase in ventilation. The ventilatory response is closely related to the number of joints and muscles involved in the movement. Researchers have suggested that muscle contraction and movement of the extremities may stimulate peripheral proprioreceptors in the joints and in muscle spindles. These receptors, in turn, are believed to send afferent neural signals to the respiratory center and stimulate ventilation. The role of peripheral neural receptors in fine tuning ventilation to metabolic demands is unknown.

Impulses from Lung Stretch Receptors During Exercise.

Lung stretch receptors are known to be stimulated by filling of the lungs and to send afferent neural signals to the respiratory center, resulting in an inhibition of inspiration and initiation of expiration. Lung stretch receptors may play a role in the hyperpnea of exercise by increasing respiratory rate. This mechanism, however, is not considered the primary stimulus for exercise hyperpnea, since voluntary increases in respiratory rate are not self-perpetuated, as is required during exercise.

Increased Sensitivity of the Respiratory Centers During Exercise.

An increased sensitivity of the respiratory centers to arterial P_{CO_2} could account for the hyperpnea of exercise. The change in sensitivity, however, would have to be very large, since even during strenuous exercise arterial O_2 and P_{CO_2} change very little. The fact remains, however, that when breathholding time is measured during moderate exercise, rebreathing occurs at a lower alveolar CO_2 and a higher O_2 than occurs at rest. It is not possible to ascribe this effect only to increased receptor center sensitivity, because the neural factors described previously may be operating to increase the activity of the respiratory center.

Increased Body Temperature During Exercise.

An increase in body temperature increases ventilation and could play a role in exercise hyperpnea. The increase in ventilation during exercise, however, is much greater and occurs before any significant rise in body temperature. While an increase in body temperature is not the primary stimulus for exercise hyperpnea, the possibility remains that the hyperthermia of exercise may have a direct effect on the activity of joint and muscle receptors as well as on the respiratory center during exercise.

We must conclude that neither neural nor chemical stimuli alone can explain the hyperventilation of exercise. Furthermore, while neurogenic factors cannot be considered as a regulating stimulus, they may act as an activator of the response. The humoral factors, on the other hand, appear to be critical for the finer adjustments of ventilation to keep the blood respiratory gases almost normal during exercise.

Circulation During Exercise

Muscle Blood Flow During Exercise

One of the most important functions of the cardiovascular system during exercise is to deliver oxygen and other nutrients to the working muscles and remove CO_2 and other metabolic by-products. As illustrated in Figure 29–17, this function is facilitated by a marked increase in muscle blood flow during dynamic exercise. Muscle blood flow can increase approximately twenty-fivefold during maximal exercise. Approximately 50% of the increase in flow is the result of local vasodilatation caused by the direct effects of the increased muscle metabolism. The rest of the increase in muscle blood flow during exercise is the result of several factors, an important one

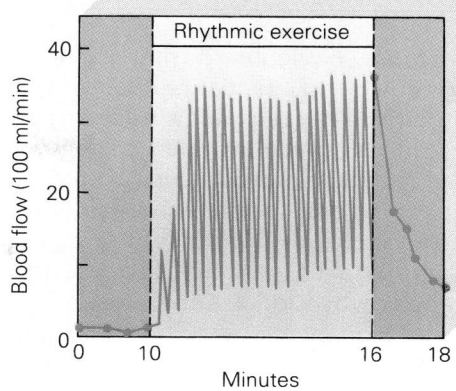

Figure 29–17
Effect of rhythmic muscle contraction on human calf muscle blood flow.

being a moderate (about 30%) increase in arterial blood pressure. Figure 29–17 also shows that during contraction, muscle blood flow temporarily decreases as the intramuscular vessels are compressed. Strong tonic contractions are known to cause rapid muscle fatigue, owing to the inadequate delivery of oxygen and nutrients during the continuous contraction.

C L I N I C A L   F O C U S

Exercise and Immunity

A relationship between exercise and immunity was first demonstrated in 1920, when it was shown that intense exercise caused leukocytosis, an increase in white blood cell (WBC) count from between 5000 and 7000 WBC per milliliter of blood to between 25,000 and 30,000 WBC per milliliter. This increase in WBC count cannot be explained by hemoconcentration alone. It is also due to the washing out of WBCs from their storage places in the bone marrow, spleen, liver, and lungs, which receive increased blood flow as a result of the increased cardiac output that occurs with exercise.

Recent studies have shown that there are many similarities in the response of the immune system to the stress of exercise and to infection. During both exercise and infection, body temperature increases, which increases the rate of all chemical reactions, in-

cluding those associated with the immune system. Other specific immune responses to exercise seen in infection include (1) leukocytosis, (2) increased natural killer cell activity, (3) increased plasma levels of acute phase reactants such as C-reactive protein, and (4) increased levels of plasma interferon. These responses to exercise are mediated by the secretion by macrophages of the polypeptide hormone interleukin-1 (IL1). IL1 is a prominent member of a group of polypeptide mediators now called cytokines, and is one of the key mediators of the body's response to microbial invasion, inflammation, immunological reactions, and tissue injury. The specific stimulus for the secretion of IL1 during intense exercise is unknown but may be related to local muscle inflammation or tissue injury.

Cardiac Output During Exercise

Cardiovascular regulation during exercise is designed to provide an adequate cardiac output to ensure an adequate oxygen and nutrient supply to the working muscles. As illustrated in Figure 29–18, there is a remarkable consistency between cardiac output, oxygen consumption, and work output. As work output increases during light or moderate exercise, there is a proportional increase in oxygen consumption, which dilates the muscle blood vessels. This dilation increases venous return, resulting in a linear increase in cardiac output. There is a deviation from linearity as oxygen consumption approaches VO$_2$max related to thermoregulation at high work rates. As body temperature rises during strenuous exercise, skin blood flow must be increased so that the high internal heat loads can be transferred to the skin and dissipated to the environment.

The increased blood flow to the skin under these conditions places an additional demand on the available cardiac output. Typical cardiac outputs for nontrained and highly trained athletes are illustrated in Table 29–1. The table shows that during maximal exercise, an untrained individual can increase cardiac output to approximately four times the resting level, while the well-trained person can increase cardiac output to about six times the resting level. The greater cardiac output in trained athletes is mainly due to **hypertrophy** (enlargement) of the heart, which results in an increase in both the size of the chambers and the mass of the heart. Hypertro-

Table 29–1

Comparison of Stroke Volume and Heart Rate in a Trained Athlete and a Nonathlete

State	Stroke Volume (ml)	Heart Rate (beats/min)
Resting		
Nonathlete	75	75
Athlete	105	50
Maximum exercise		
Nonathlete	110	195
Athlete	162	185

phy of the heart, however, occurs only in endurance-type training and has no important clinical significance.

Stroke Volume and Heart Rate During Exercise

Table 29–1 also compares stroke volume and heart rate in nonathletic and endurance-trained individuals at several levels of exercise. It shows that both at rest and during maximum exercise, the trained individual has a significantly greater stroke volume and a lower heart rate than the untrained person. Because of the greater pumping effectiveness of the heart, the trained individual can achieve the same cardiac output both at rest and during exercise with a decreased heart rate. Figure 29–19 illustrates the interrelationships between stroke volume, heart rate, and cardiac output in a trained person with a resting cardiac output of 5.5 liters per minute and a maximum exercise output of 30 liters per minute. The heart rate increases more than 250%, while stroke volume increases about 50% and reaches a maximum when the cardiac output approaches 15 liters per minute. Further increases in cardiac output are the result of increases in heart rate. Therefore, in strenuous exercise, the increase in heart rate accounts for the major increase in cardiac output.

During maximum exercise, cardiac output increases to about 90% of the maximum that the person can achieve. Therefore, the capacity of the cardiovascular system is normally a limiting factor in terms of maximal oxygen consumption and exercise performance. Maximum cardiac output decreases with age and significantly contributes to the decrease in the maximal exercise performance of the elderly.

Figure 29–18
Relationship among cardiac output, oxygen consumption, and work output during different levels of exercise.

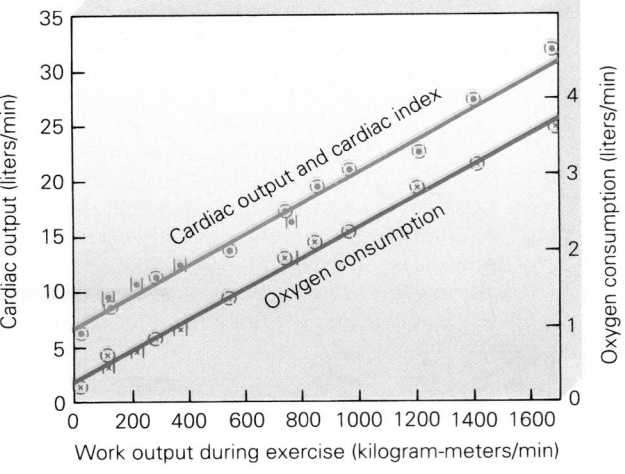

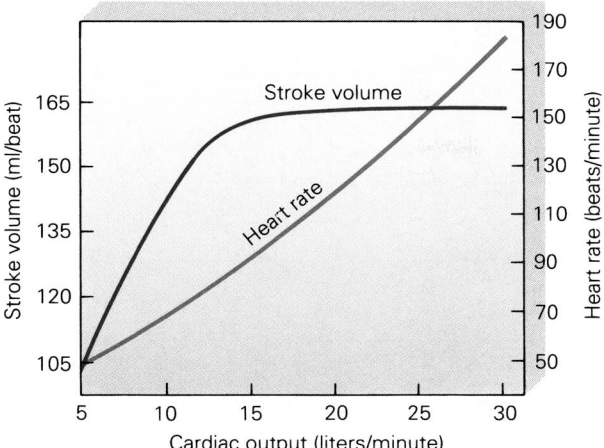

Figure 29–19
Relationship among stroke volume, heart rate, and different levels of cardiac output during exercise in an endurance-trained athlete.

Local Blood Flow to Organs During Exercise

The active muscles, including the respiratory muscles, receive the bulk of the increased cardiac output during exercise. The increase in muscle blood flow is associated with drastic alterations in the blood flow distribution to other organs. Blood flows and the percentage of the cardiac output perfusing various tissues during rest and strenuous exercise are illustrated in Table 29–2. As the table shows, brain blood flow is maintained in heavy exercise. Splanchnic and renal blood flow, on the other hand, are decreased, while skeletal muscle, heart, and skin blood flow are increased. Exercise has a dual effect on skin blood flow. At the start of exercise, there is a decrease in cutaneous blood flow that facilitates increased muscle blood flow. However, as exercise continues, skin blood flow increases above resting levels to permit effective transfer and dissipation of body heat. Clearly, during heavy exercise some of the large vascular beds in the viscera receive reduced blood flow so that the flow through more vital and active areas can be maintained or increased.

Body Temperature in Exercise

Prolonged, strenuous exercise markedly increases body temperature. Since the mechanical efficiency of the human body is less than 25%, more than 75% of the metabolic energy of exercise is converted to heat. The greater the exercise intensity, the greater the total amount of heat produced, and at maximal exercise it can be 10 to 20 times greater than resting levels. The excessive heat must be removed and dissipated from the body to prevent overheating and tissue damage from excessive hyperthermia.

Figure 29–20 illustrates the relationship between muscle and rectal temperature as a function of the percentage of a person's maximal oxygen uptake. As expected, the highest temperatures occur in the exercising muscles, where most of the increased heat is being produced. It should be noted that the temperatures do not depend on the absolute magnitude of the energy output, but rather on the level of metabolism relative to the individual's

Table 29–2
Blood Flow Distribution During Rest and Exercise in an Athlete

Area	Rest		Heavy Exercise	
	ml/min	%	ml/min	%
Splanchnic	1400	24.0	300	1.0
Renal	1100	19.0	900	4.0
Brain	750	13.0	750	3.0
Coronary	250	4.0	1000	4.0
Skeletal muscle	1200	21.0	22,000	85.5
Skin	500	9.0	600	2.0
Others	600	10.0	100	0.5
Total cardiac output	5800	100.0	25,650	100.0

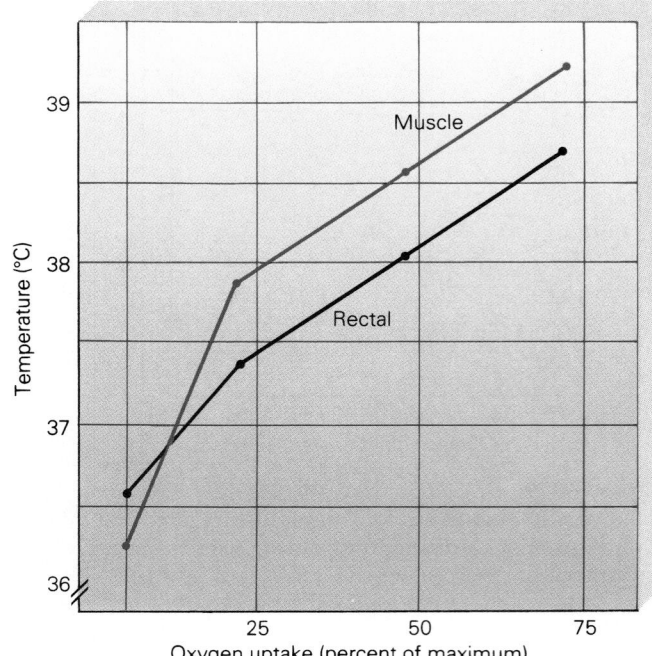

Figure 29–20
Change in exercising muscle and rectal temperature in relation to the oxygen uptake as a percentage of a person's maximum.

maximal oxygen consumption. Rectal temperatures may reach 40 to 41°C in highly trained athletes during prolonged exertion. Body temperature increases during exercise are independent of environmental temperature except at very extreme levels. Normally, internal body temperature levels off at a few degrees above normal regardless of the atmospheric temperature. The increase in body temperature during exercise is not due to a failure in the mechanisms for effective temperature regulation. On the contrary, it is a well-regulated response that occurs even during exercise in a cold environment. The primary stimulus for the increase in body temperature during exercise is not known, but it may involve a change in the set point for body temperature regulation.

A key advantage of allowing body temperature to rise during exercise is that less strain is placed on the thermoregulatory mechanisms. Maintenance of a resting body temperature during exercise would necessitate a greater increase in skin blood flow and could reduce essential blood flow to the working muscles. The rate of sweating would also have to be increased, resulting in possible dehydration earlier in prolonged exercise. These considerations do not mean that heat is always beneficial to the athlete. In very hot and humid environments, or great excesses in athletic clothing, nonregulated body temperature increases exceeding 42°C can easily occur. When this happens, multiple symptoms begin to appear, including extreme weakness, dizziness, decreased

sweating, headache, nausea, confusion, collapse, and unconsciousness. This whole complex is called **heat stroke** (see Chapter 28), and failure to treat by appropriate body cooling can lead to death.

Hormonal Responses to Exercise

Exercise is known to stimulate the secretion of numerous hormones. With few exceptions, however, their specific role in the physiology of exercise is unknown. Two hormonal systems deserve mention in this regard: (1) secretion of catecholamines by the adrenal medulla and (2) secretion of adrenocortical-stimulating hormone by the anterior pituitary.

During exercise, catecholamine secretion by the adrenal medulla system increases in response to stimulation by the sympathetic nervous system. Frequently, the secretion of catecholamines (epinephrine and norepinephrine) begins before the onset of exercise as an anticipatory response, and then catecholamine secretion coincides with the onset of work. The total amount of catecholamines secreted is related to the intensity of the exercise. Among other actions, catecholamines mobilize the fat and glycogen deposits of the body and thus play an important role in regulating the availability of nutrients during exercise. Catecholamines also enhance cardiac activity and play an important role in the cardiovascular adjustments in exercise.

Interleukin-2

The effects of prolonged exercise on interleukin-2 (IL2) induction in mitogen-stimulated mononuclear cells and the role of core temperature in that response were studied. Eight men exercised on a bicycle ergometer at 60% VO_2max for one hour, and on a different day underwent passive heating. Venous blood was sampled before, during, and 30 minutes after exercise and heating. The white blood cells (WBC) were separated and incubated with mitogen. WBC supernatant was collected 48 hours post-incubation and measured for IL2 activity. The exercise test produced IL2 levels of 1.64, 2.37, and 2.15 units/ml; heating produced levels of 2.47, 3.18, and 2.99 units/ml for rest, experiment, and recovery, respectively. These results were analyzed by a two-way ANOVA with repeated measurements on both factors. During exercise and heating, IL2 levels were significantly higher ($p < .05$) than resting levels. Exercise enhanced IL2 induction by 44% above resting levels, and heating by 30% above resting levels. These data suggest that exercise may stimulate cell-mediated immunity, and that elevated core temperature may partially account for the enhanced IL2 induction.

During exercise, there is also an increase in pituitary secretion of hormones that stimulate the adrenal cortex. Specifically, after exercise starts and with a latency of about 2 minutes, the anterior pituitary increases its secretion of ACTH. This increased ACTH then stimulates the release of corticosteroids from the adrenal cortex. The significance of the increase in corticosteroids during exercise is not fully understood, but their effect on the mobilization of glycogen stores is well known. As do catecholamines, corticosteroids probably play an important role in regulating blood nutrients during exercise.

SUMMARY

Metabolic Aspects of Exercise

Three important metabolic systems supply energy for contraction: the phosphagen system, the glycogen–lactic acid system, and the aerobic system. During prolonged light exercise, energy is obtained from the aerobic system. During heavy exercise, by contrast, part of the energy is always obtained from the anaerobic system. If large amounts of lactic acid are produced, muscle fatigue results.

Oxygen consumption increases linearly with increasing levels of exercise until a plateau or the so-called maximal oxygen consumption (VO_2max) is reached. The VO_2max represents an index of a person's maximal aerobic power and the ability to sustain high-intensity exercise. Factors including age, sex, and level of training can affect an individual's VO_2max.

An oxygen debt is incurred at the beginning of exercise as oxygen is used from existing body stores. The difference between the oxygen requirement and the actual oxygen consumed is called the oxygen debt. At the end of exercise, oxygen consumption remains elevated above resting levels as the oxygen debt is repaid. The repayment of the oxygen debt has two stages: the alactacid oxygen debt and the lactic acid oxygen debt.

Carbohydrates, fatty acids, and, to a lesser extent, amino acids can be used as fuel during exercise. The relative usage of the three nutrients during exercise varies depending on the intensity and duration of the exercise as well as on the diet, which determines the relative availability of stored fuels for exercise. To achieve optimal performance, approximately 48 hours are required to replenish fuel stores after exhausting exercise.

Respiration During Exercise

The increased oxygen demands of an exercising person are paralleled by increased pulmonary ventilation. The increased ventilation is brought about by increases in both respiratory rate and tidal volume. The maximal breathing capacity in a healthy young person is approximately 50% greater than the pulmonary ventilation achieved during maximal exercise. The respiratory system, therefore, is not normally a limiting factor in the delivery of oxygen to the muscles during maximal exercise.

The diffusing capacity of the lung provides an index of the dimensions of the pulmonary capillary bed, the pulmonary membrane, and the overall efficiency of the system in the exchange of respiratory gases. The diffusion capacity of the lung for oxygen and carbon dioxide increases during exercise, largely because of an increase in pulmonary blood flow.

The mechanisms that control ventilation at rest cannot entirely explain the increased ventilation or hyperpnea during exercise. How the increase in ventilation during exercise is elicited is not known. Several theories suggest interactions between chemical/humoral stimuli such as PCO_2, PO_2, and pH, and neural stimuli from both central and peripheral neural receptors. Neither neural nor chemical stimuli alone can entirely explain the hyperpnea of exercise. Furthermore, neurogenic factors may act as an activator of the response, while chemical factors, on the other hand, are very important for the finer adjustments to ventilation that keep blood respiratory gases almost normal during exercise.

Circulation During Exercise

One of the most important functions of the cardiovascular system during exercise is to deliver oxygen and nutrients to the working muscles and remove CO_2 and other metabolic by-products. This function is facilitated by a marked increase in muscle blood flow during exercise. Blood flow to the muscle approaches a twenty-fivefold increase during maximal exercise.

Cardiovascular function is regulated during exercise to ensure adequate blood flow to the working muscles. This results in a remarkable coordination among cardiac output, oxygen consumption, and work output. During maximal exercise, an untrained person can increase cardiac output to approximately four times the resting level, while an endurance-trained individual can increase cardiac output to about six times the resting level, mainly because of hypertrophy of the heart.

Both at rest and during maximum exercise, the trained individual has a significantly greater stroke volume and a lower heart rate than the untrained person. Furthermore, in heavy exercise, an increase in heart rate accounts for the major increase in cardiac output. During maximal exercise, cardiac output increases to about 90% of the maximum that the person can achieve and is normally a limiting factor in terms of maximal oxygen consumption and exercise performance.

The active muscles receive the bulk of the increased cardiac output during exercise. The increase in muscle blood flow is associated with drastic alterations in the blood flow distribution to other organs. In heavy exercise, blood flow to the brain is maintained; however, some of the large vascular beds in the viscera have reduced blood flow, while skeletal muscle, heart, and skin blood flow are reduced.

Body Temperature in Exercise

Since the mechanical efficiency of the body is less than 25%, more than 75% of the metabolic energy of exercise is released as heat. Core temperature normally increases during exercise, with the magnitude of the rise determined by the relative work rate compared to VO_2max. Body temperature may reach 40 to 41°C in highly trained athletes during prolonged exercise. The well-regulated increase in core temperature during exercise probably creates an optimal thermal environment for increased physiological and metabolic function. Exercise in hot, humid environments or with excessive athletic clothing can produce nonregulated increases in body temperature exceeding 42°C, and heat stroke.

Hormonal Responses to Exercise

Exercise stimulates the secretion of numerous hormones; however, their specific role in the physiology of exercise remains largely unknown. Catecholamines play an important role in the regulation of the cardiovascular adjustments to exercise, as well as the mobilization of nutrients during exercise. The role of corticosteroids during exercise is not fully understood, but, as is true of catecholamines, they probably play a role in regulating blood nutrients during exercise.

REVIEW QUESTIONS

Identify Each with a Term

1. The type of exercise in which shortening of the muscles does not occur.

2. A person's maximal aerobic power.

3. The increased ventilation that occurs during exercise.

4. The enlargement of muscle tissue that occurs with repeated exercise.

5. The increase in white blood cells stimulated by exercise.

Define Each Term

6. oxygen debt

7. cytokines

8. mechanical efficiency

9. proprioreceptors

10. phosphagen

Choose the Correct Answer

11. Proportionately, the largest drop in blood flow during maximal exercise occurs to which area?
   a. renal
   b. brain
   c. skin
   d. splanchnic

12. Approximately what percent of the energy released during muscle contraction is available to do work?
   a. 70
   b. 33
   c. 20
   d. 80

13. During maximal exercise, pulmonary ventilation may exceed how many times the resting rate?
   a. 50
   b. 30
   c. 10
   d. 100

14. Muscle blood flow during exercise can increase to how many times the resting level?
   a. 50
   b. 25
   c. 10
   d. 100

15. The major factor associated with the increase in muscle blood flow during exercise is:
   a. increased vasodilation
   b. increased heart rate
   c. increased body temperature
   d. increased blood pressure

16. Maximum oxygen uptake would be the same during maximum exercise involving:
   a. two arms versus two legs
   b. one arm versus one leg
   c. two legs versus two arms plus two legs
   d. one leg versus two legs

17. During exercise, which area shows an initial decrease in blood flow, followed by an increase?
   a. skin
   b. renal
   c. brain
   d. muscle

18. The phosphagen system does not include:
   a. ATP
   b. phosphocreatine
   c. glucogen
   d. ADP

19. Glycolysis of a single molecule of glucose produces how many molecules of ATP?
   a. 36
   b. 2
   c. 6
   d. 20

20. When one phosphate group is removed from ATP, approximately how many calories of energy are released?
   a. 25,000
   b. 5,000
   c. 11,000
   d. 100,000

Essay

21. What are the factors that control pulmonary ventilation during exercise?

22. What are the relationships between exercise and the immune system?

23. Compare and contrast the cardiovascular responses to exercise of a nonathlete and an endurance-trained athlete.

24. What are the criteria for the determination of maximum oxygen consumption?

25. What is the relationship between rectal temperature and level of exercise in humans?

SUGGESTED READINGS

Adams, G.M. *Exercise Physiology Primer*. Cosa Mesa, Calif.: Custom Publishing Co., 1983.

Astrand, P.O., and Rodahl, K. *Textbook of Work Physiology*. New York: McGraw-Hill, 1986.

Bell, A.W. and Hales, J.R.S. "Cardiac Output and its distribution during exercise and heat stress." In *Thermoregulation, Pathology, Pharmacology, and Therapy*. New York: Pergamon Press, 1991, p. 105.

Bogert, L.J., Briggs, G.M., and Calloway, D.H. *Nutrition and Physical Fitness*. Philadelphia: W.B. Saunders Co., 1984.

Brengelman, G.L. "Circulatory adjustments to exercise and heat stress." *Annual Review of Physiology*, 45:191, 1983.

Dempsey, J.A., Vidruk, E.H., and Mastenbrook, S.M. "Pulmonary control systems in exercise." *Federation Proceedings*, 39:1498, 1980.

Lamb, D.R. *Physiology of Exercise Responses and Adaptations*. New York: Macmillan Publishing Co., 1984.

McManus. *Cardiovascular Complications of Exercise*. Chicago: Year Book Medical Publishers, 1985.

Murase, U., Kobayashi, K., Kamei, S., et al. "Longitudinal study of aerobic power in superior junior athletes." *Medicine Science Sports and Exercise*, 13:180, 1981.

Noble, B.J. *Physiology of Exercise and Sport*. St. Louis: Times Mirror/Mosby College Publishing, 1986.

Saltin, B., and Rowell, L.B. "Functional adaptations to physical activity and inactivity." *Federation Proceedings*, 39:1509, 1980.

Teitz, C.C., ed. *Scientific Foundations of Sports Medicine*. Philadelphia: Decker, Inc. 1989.

KEY TERMS

alactacid oxygen debt (p. 963)

cytokins (p. 967)

dynamic exercise (p. 959)

humoral stimuli (p. 965)

hyperpnea (p. 965)

hypertrophy (p. 968)

isometric muscle contraction (p. 959)

lactic acid (p. 963)

leukocytosis (p. 967)

maximal aerobic power (p. 961)

pulmonary stretch receptors (p. 965)

pulmonary ventilation (p. 964)

pyruvic acid (p. 959)

respiratory center (p. 966)

static exercise (p. 959)

stretch receptors (p. 965)

watts (p. 959)

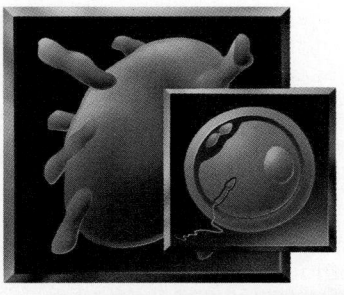

C H A P T E R
30

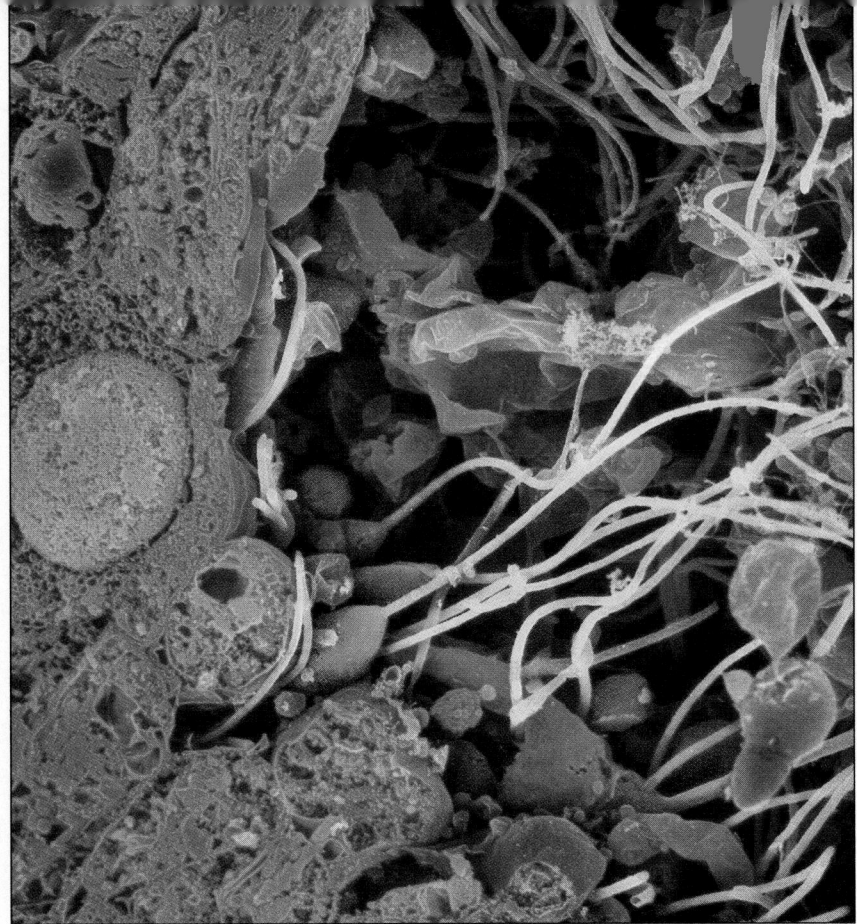

Reproductive Physiology

In preceding chapters we have examined the structural adaptations that allow cells to carry out particular physiological functions. We have seen that cell specialization produces a complex interdependence among the different cell types that control various processes within the human body. We have often observed that a common purpose of many of these processes is homeostasis, or the maintenance of a stable internal environment.

In contrast, the physiological processes we are about to discuss center on cells from two separate individuals that interact to form a third individual. The sperm cell is structurally and chemically designed to leave the male and join with the female ovum. The ovum, while remaining inside the female, is designed to react with the sperm that originates in the male. These are the only two human cell types that are so adapted.

In Chapter 30 we will first examine the male and female anatomy involved in reproduction. Then we will discuss how the two major integrative organ systems of the body, the nervous and the endocrine systems, act together to control reproductive functions. In Chapter 31, we will examine how fertilization (the union of sperm and ovum) occurs in the female. We will follow the alterations that this union produces in the female nervous and endocrine systems, which then allow for the development of the fetus. The special physiological features of the developing fetus will also be examined. In Chapter 32, we will examine sexual physiology as separate from reproductive physiology, and look at the factors that determine an individual's sex.

Reproductive Physiology in the Male

The major male reproductive functions are (1) secretion of sex hormones, (2) production of sperm, and (3) transport of sperm from the male to the female. The first and second functions are carried out by the testes, while the third function is performed by a network of ejaculatory ducts together with the penis.

Male Reproductive Anatomy

Anatomy of the Male External Genitalia

The external reproductive organs in the male are the **penis** and **scrotum** (Fig. 30–1). The penis is covered by thin, loose skin and contains the **urethra, erectile**

Figure 30–1
Side view of the male reproductive tract. (Redrawn from Masters and Johnson, *Human Sexuality, Second Edition.*)

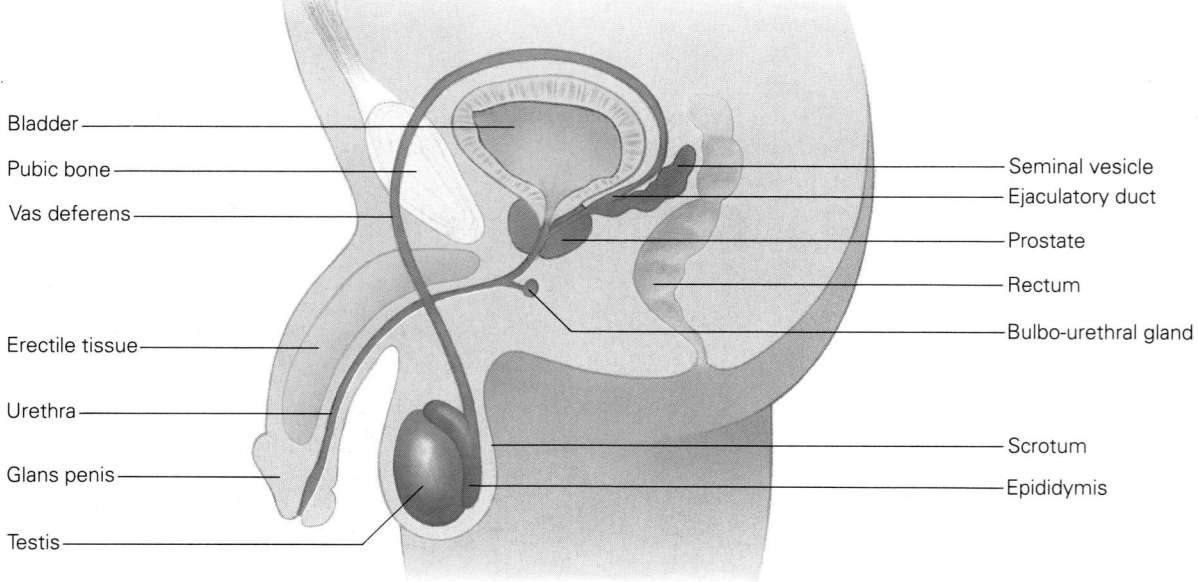

Bladder

Pubic bone

Vas deferens

Erectile tissue

Urethra

Glans penis

Testis

Seminal vesicle

Ejaculatory duct

Prostate

Rectum

Bulbo-urethral gland

Scrotum

Epididymis

tissue, nerves, and many blood vessels. The scrotum is a pouch of skin made of connective tissue and muscle. This pouch contains the **testes,** the **epididymis,** and the **spermatic cord,** which contains the vas deferens. The location of the testes in the scrotal compartment outside the body is essential for normal sperm production, which occurs only at temperatures below normal body temperature. In response to cold, the muscles of the scrotum contract involuntarily and lift the testes closer to the body to provide warmth. In warm weather, the muscles of the scrotum relax and allow the testes to move away from the body, which reduces heat. These muscular reflexes ensure that the testes are maintained at a temperature conducive to spermatogenesis, or the formation of spermatozoa.

As we will learn in Chapter 32, both the penis and the scrotum contain a dense supply of sensory nerves that make these external organs very sensitive to touch and pressure.

Anatomy of the Male Internal Genitalia

The testes have two major functions: production of sperm and production of male sex hormones. Each testis contains a large number of tightly coiled tubules called **seminiferous tubules** (Fig. 30–2). The seminiferous tubules contain sperm-producing or **spermatogenic cells,** also referred to as **germinal epithelium.** The cells that synthesize the male sex hormone **testosterone** are called **Leydig cells** or the **interstitial cells of Leydig** because they are located in the interstitial areas between the seminiferous tubules in the testes. In addition to the Leydig cells, the interstitial space also contains lymphatics, blood vessels, nerve fibers, connective tissue, and extra-

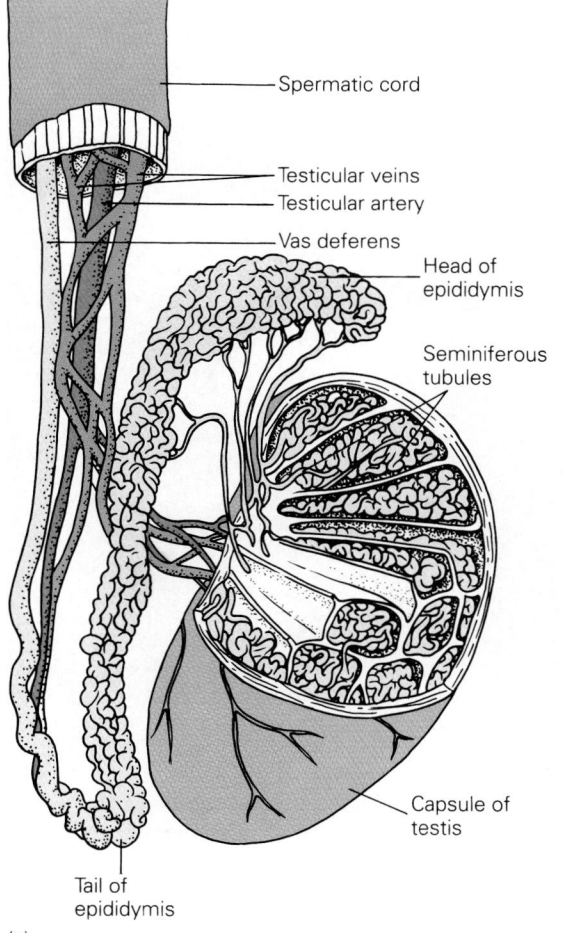

(a)

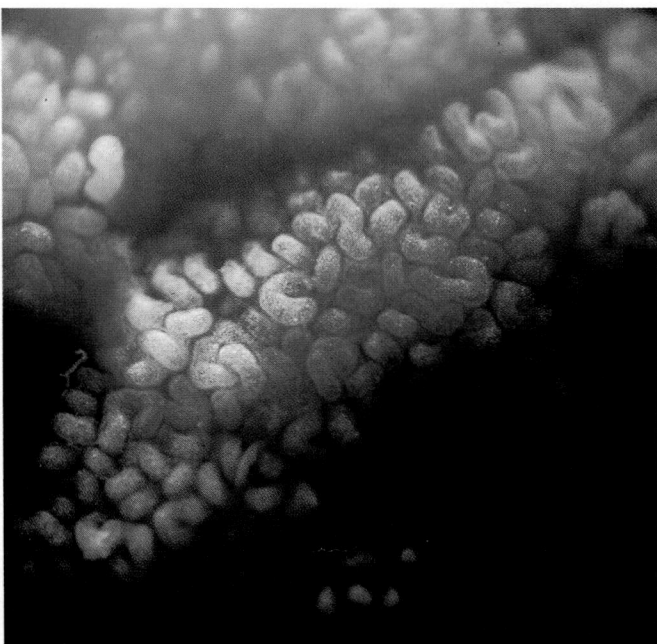

(b)

Figure 30–2
(*a*) A cross-section of the internal structure of the testis, illustrating compartmentalization of the seminiferous tubules. (*b*) The convolutions of a single seminiferous tubule. (© Lennart Nilsson, *A Child Is Born*, Dell Publishing Company, 1977.)

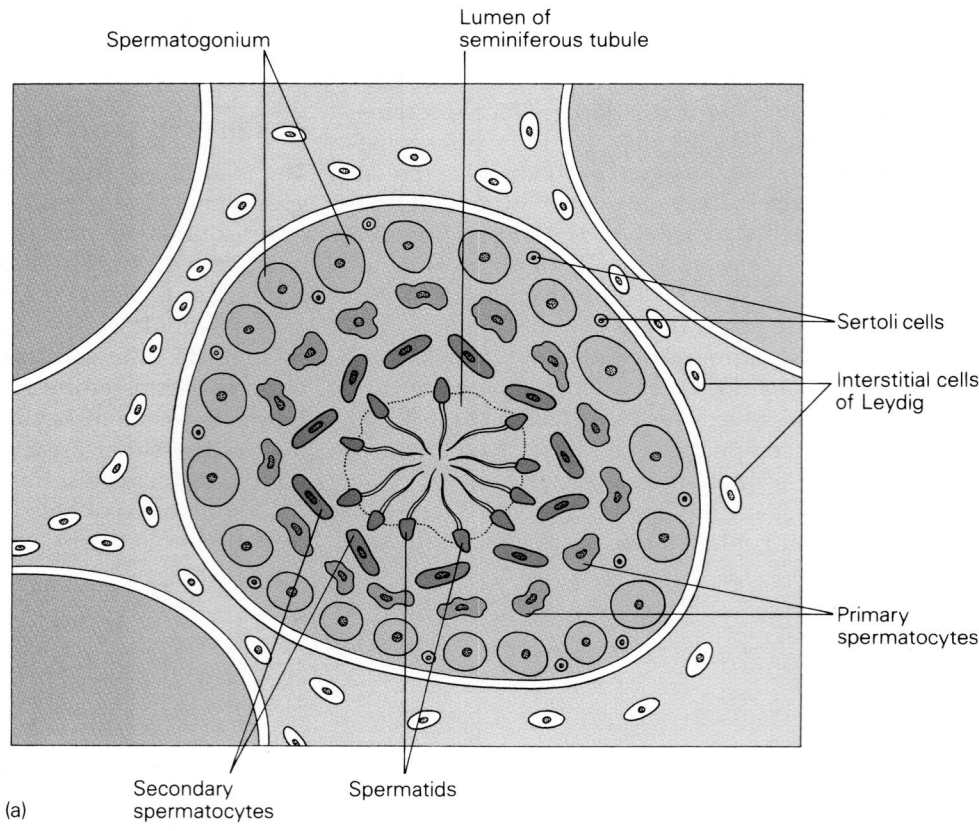

(a)

Spermatogonium

Lumen of seminiferous tubule

Sertoli cells

Interstitial cells of Leydig

Primary spermatocytes

Secondary spermatocytes

Spermatids

cellular fluid. The nerve fibers and blood vessels enter and leave the testes via the spermatic cord.

Spermatogenesis: The Formation of Sperm

The process of sperm production takes place within the coiled **seminiferous tubules** of the testes. Although primitive germ cells are present in the testes from an early age, active spermatogenesis does not begin until puberty. Once begun, the process of sperm production continues into old age.

During spermatogenesis, germ cells undergo a series of divisions and differentiations to become mature sperm. This process takes about 70 to 80 days. Spermatogenesis is constantly being repeated, and all stages of sperm cell development are present in the seminiferous tubules at any given time.

The magnitude of sperm production in the testes is impressive. As many as several hundred million sperm may be produced daily. However, of this enormous number, only a few hundred may reach the distal portion of the female fallopian tube,

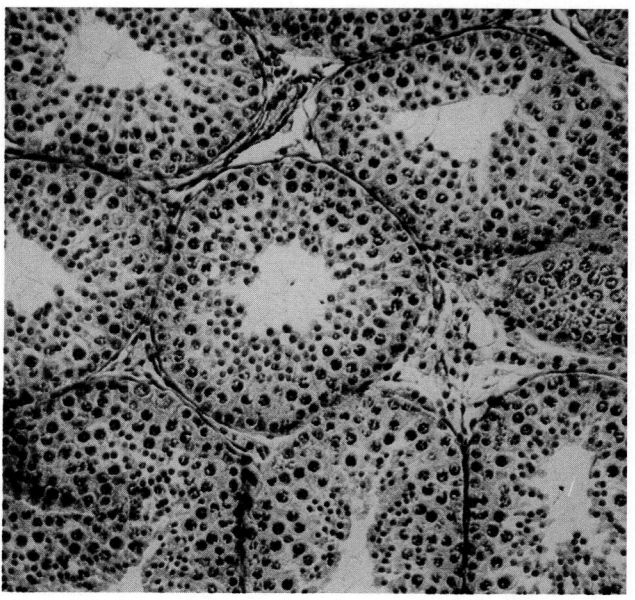

(b)

Figure 30–3
(*a*) A cross-section of a seminiferous tubule in the testes, illustrating the anatomical relationships among different cell types. (*b*) Photomicrograph of cells of the seminiferous tubules. (© John D. Cunningham/Visuals Unlimited.)

where fertilization occurs. A very high rate of sperm production appears to be necessary to overcome the odds against fertilization.

Differentiation of Germ Cells

Two types of cells are involved in spermatogenesis: the **spermatogenic cells,** from which mature sperm arise, and the **Sertoli cells,** which provide mechanical and nutritional support for the spermatogenic cells. Both spermatogenic and Sertoli cells are located inside the seminiferous tubules. Each seminiferous tubule is surrounded by a basement membrane that separates the tubule from the surrounding Leydig cells and connective tissue. The

anatomical relationship among the spermatogenic, Sertoli, and Leydig cells is illustrated in Figure 30–3.

The immature or primitive germ cells are found along the outer border of each seminiferous tubule in contact with the basement membrane. At puberty, the primitive germ cells undergo a mitotic division and differentiate into **spermatogonia** (singular, *spermatogonium*). In the adult, spermatogonia represent the first cell stage of spermatogenesis. Spermatogenesis begins when some of the spermatogonia differentiate into **primary spermatocytes.** As the spermatogonia differentiate, they increase in size, undergo a second mitotic division, and move away from the basement membrane towards the center of the seminiferous tubule (Fig. 30–4). The

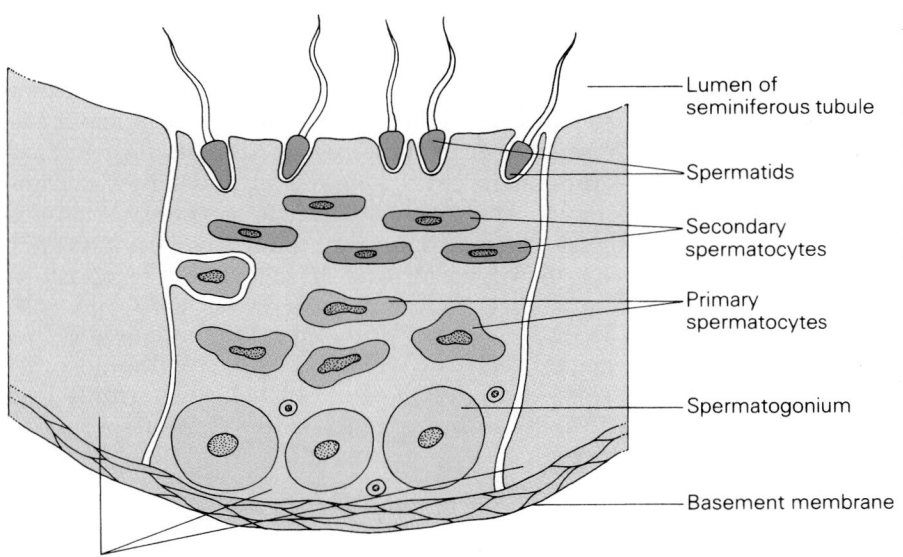

Lumen of seminiferous tubule

Spermatids

Secondary spermatocytes

Primary spermatocytes

Spermatogonium

Basement membrane

Sertoli cells

(a)

Figure 30–4
(a) Cross-section of the seminiferous tubule, illustrating the location of the spermatogenic cells.
(b) Mature sperm that have been pushed into the central duct of a seminiferous tubule. (© Lennart Nilsson, *A Child Is Born*, Dell Publishing Company, 1977.)

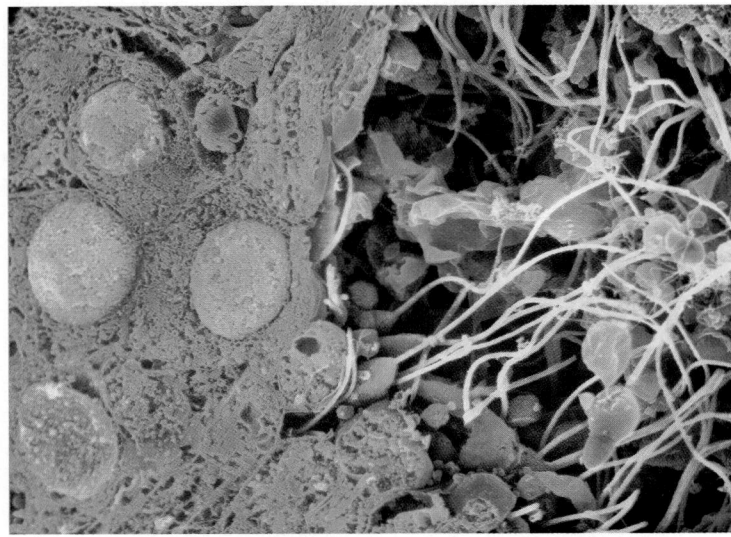

(b)

remaining spermatogonia continue to divide mitotically and provide a constant source of new germ cells that differentiate into generation after generation of primary spermatocytes.

Mitotic Divisions During Spermatogenesis. Spermatogonia, in common with all other cells of the body, contain 46 chromosomes, 23 contributed by the mother and 23 by the father. During mitotic division, the chromosomes are duplicated before cell division. When the cell divides, each of the two new cells receives one copy of each of the 46 chromosomes present in the parental cell. Therefore, primary spermatocytes, formed during mitotic division of spermatogonia, have chromosomes identical to those of the spermatogonia that produced them (23 maternal and 23 paternal chromosomes).

Meiotic Divisions During Spermatogenesis. Primary spermatocytes undergo two more divisions through a special form of cell division known as

Figure 30–5
(*a*) "Crossing over," or exchange of chromosomal segments between homologous chromosomes, during meiotic division of the spermatocyte. (*b*) Structure of the human sperm.

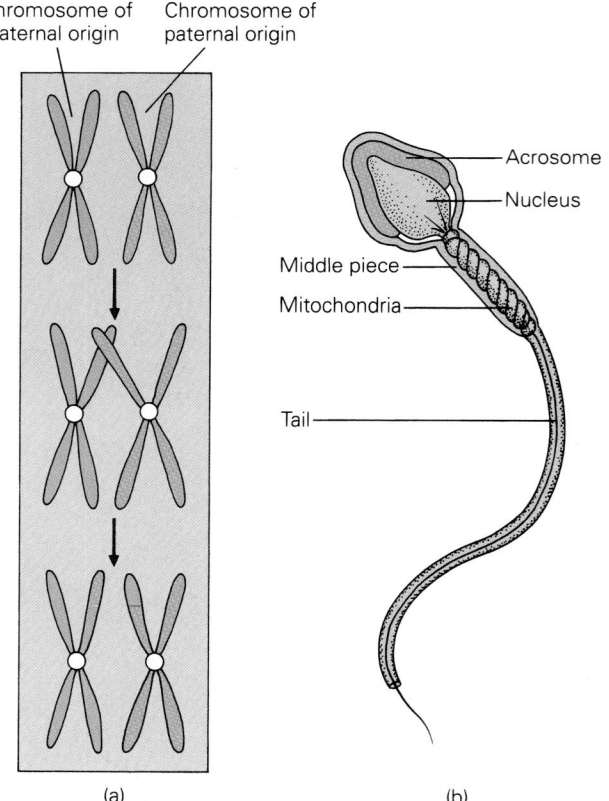

Chromosome of maternal origin   Chromosome of paternal origin

Acrosome
Nucleus
Middle piece
Mitochondria
Tail

(a)   (b)

meiosis. Each meiotic division reduces the number of chromosomes in the cell by half. During meiotic division there is a random assortment and recombination of maternal and paternal chromosomes. In addition, homologous chromosomes (similar chromosomes, one of maternal origin and the other of paternal origin) exchange chromosomal segments in a process known as **crossing over** (Fig. 30–5*a*). This process ensures that genetic material is recombined in an infinite number of ways, leading to infinite individual variation.

Before the first meiotic division, DNA synthesis increases within the primary spermatocyte, and each of the 23 maternal and 23 paternal chromosomes in the cell is duplicated. The homologous maternal and paternal chromosomes pair with each other to form 23 four-stranded chromosomes. The primary spermatocyte then undergoes two very rapid meiotic divisions that do not involve chromosomal duplication. The **secondary spermatocytes,** formed after the first meiotic division, receive a random mixture of maternal and paternal chromosomes. Therefore, individual secondary spermatocytes differ from each other and from primary spermatocytes in terms of their genetic composition.

The secondary spermatocytes then undergo a second meiotic division and differentiate into **spermatids,** each of which contains 23 chromosomes. The process of meiotic cell division ensures great diversity in the genetic composition of sperm.

Physiology of the Mature Sperm

As primary spermatocytes differentiate into spermatids, they migrate away from the basement membrane of the seminiferous tubule toward the fluid-filled lumen of the tubule (see Fig. 30–4). Near the lumen, the spermatids undergo the final phase of spermatogenesis and differentiate into mature **spermatozoa** or **sperm,** which enter the lumen in a process called **spermiation.** From the lumen, the sperm leave the seminiferous tubule and enter the epididymis.

During the final phase of spermatogenesis, the spermatids do not undergo cell division but rather go through a complex series of cytological changes. The **chromatin,** or genetic material of the spermatid, is contained within the nucleus and is surrounded by a membrane. As the spermatid differentiates into a mature sperm, it increases in size and becomes elongated (Fig. 30–5*b*). The membrane surrounding the nucleus contracts, and the nucleus condenses and becomes part of the head of the mature sperm. The anterior half of the head of the sperm is covered by a small, cap-like structure called the **acrosome,**

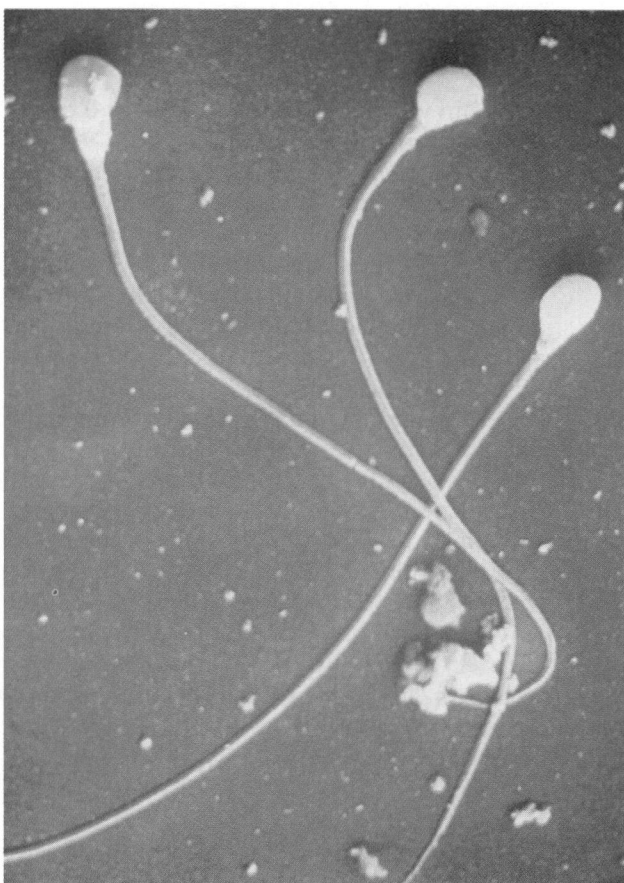

Figure 30–6
Photograph of human sperm taken with a scanning electron microscope. (× 5,000.) (© David M. Phillips, Visuals Unlimited.)

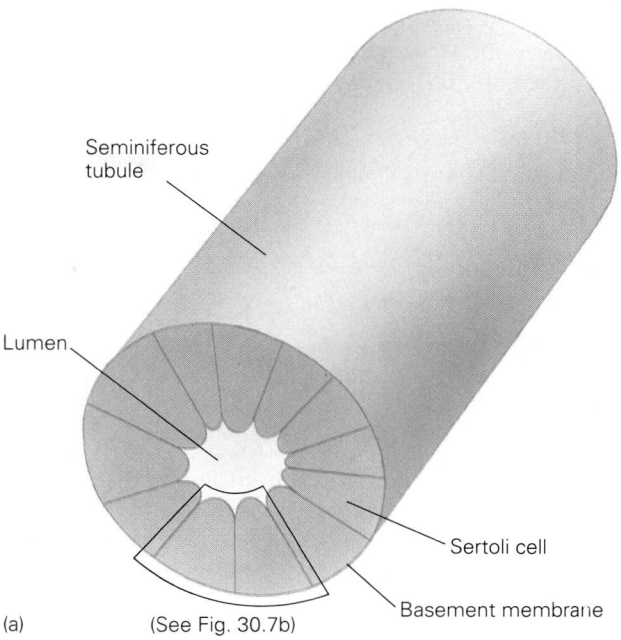

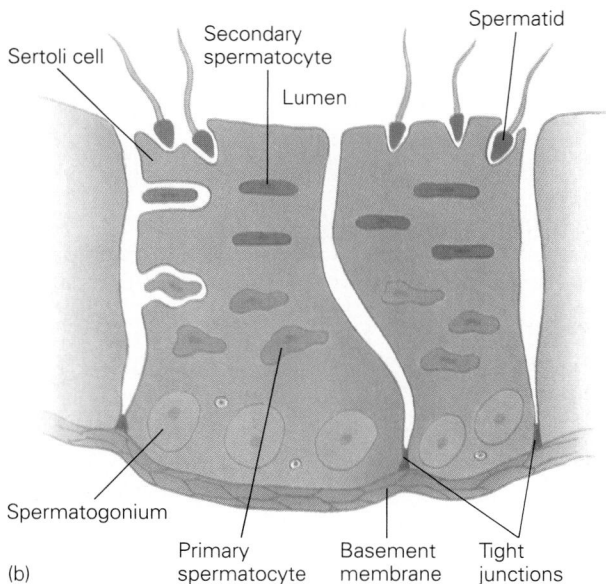

Figure 30–7
(*a*) Cross-section of a seminiferous tubule, illustrating the ring formed by Sertoli cells. (*b*) Two Sertoli cells that surround the spermatogenic cells in the seminiferous tubule.

which contains enzymes that aid the sperm in penetrating the ovum during fertilization. The midpiece of the sperm contains a spiral arrangement of mitochondria that generate the energy necessary for sperm movement. The tail of the sperm contains pairs of microtubules that run down its length and produce a back-and-forth movement in the tail, which propels the sperm forward. Mature human sperm are shown in Figure 30–6. Movement of sperm through various portions of the reproductive tract is discussed in an upcoming section.

Functions of the Sertoli Cells

The **Sertoli cells** are located inside the seminiferous tubules. Each Sertoli cell extends from the basement membrane all the way to the fluid-filled lumen of the seminiferous tubule (Fig. 30–7*a*). The Sertoli cells are joined to one another by specialized attachments called **tight junctions.** As a result, the Sertoli

cells form a ring around the inner diameter of the seminiferous tubule. All the spermatogenic cells, in different stages of differentiation, are interspersed between and surrounded by the Sertoli cells (see Figure 30–7*b*). This anatomical relationship between the spermatogenic and Sertoli cells is important for a number of reasons (Table 30–1).

Table 30–1
Functions of Sertoli Cells

Maintenance of the blood-testes barrier

Nourishment of developing germ cells

Production of seminiferous tubular fluid

Removal of damaged germ cells

Synthesis of androgen-binding protein

Synthesis of inhibin

The Sertoli cells form a **blood-testes barrier,** which prevents many substances from entering or leaving the seminiferous tubules. For instance, the Sertoli cell barrier excludes from the tubule a variety of proteins, hormones, ions, and drugs, some of which could adversely affect the developing spermatogenic cells. Conversely, this barrier keeps the sperm in the tubules from diffusing into the blood. Mature sperm are very immunogenic. Thus, if they entered the blood, they would be recognized as foreign antigens, which would result in the production of antibodies specific to them. The blood-testes barrier develops at the time of puberty, just before spermatogenesis begins, and provides a protected environment in which sperm maturation and development can occur.

Sertoli cells secrete most of the fluid in the lumen of the seminiferous tubules. This fluid is rich in proteins, enzymes, nutrients, androgens, and certain ions such as potassium and bicarbonate. These chemicals stimulate and nourish the developing sperm. The fluid produced by the Sertoli cells also provides a drive for flushing sperm out of the tubules into the epididymis. Another role of the Sertoli cells is the engulfment and removal of any damaged germ cells in the seminiferous tubules.

Sertoli cells synthesize two proteins of particular importance. The first is called **androgen-binding-protein,** or **ABP,** which is secreted into the seminiferous tubular fluid. ABP binds testosterone and increases the concentration of testosterone within the seminiferous tubules of the testes. The second protein is called **inhibin,** which is secreted into the blood and transported to the pituitary gland, where it regulates the secretion of **follicle-stimulating hormone (FSH).** As we will see, both of these proteins are important for spermatogenesis.

Figure 30–8
Reproductive duct system of the male, illustrating sperm pathway and sphincter contraction in the prostate during ejaculation.

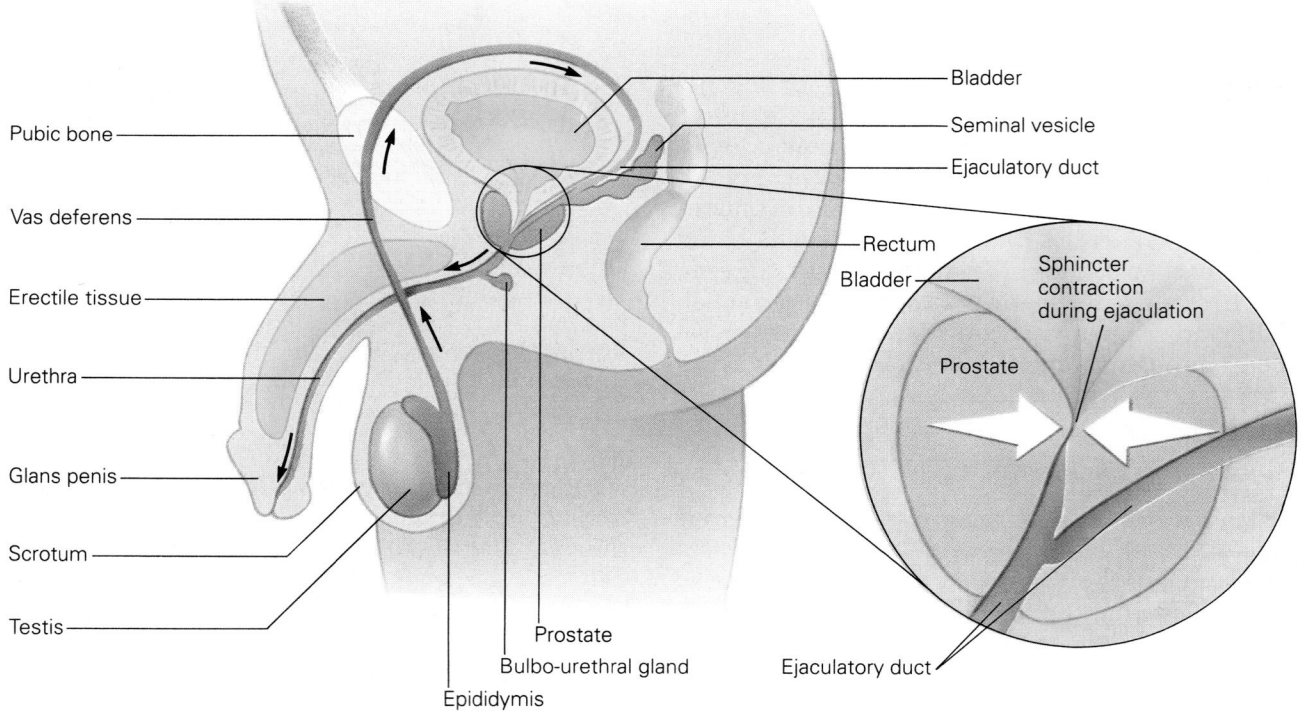

Sperm Transport Through the Internal Reproductive Structures

Maturation of Sperm in the Epididymis

The seminiferous tubules meet at the top of the testes in a network of ducts called the **rete testis** (see Fig. 30–2). Sperm move from the tubules through the rete testis into the epididymis as a result of pressure created by the continuous formation of fluid in the tubules. The **epididymis** lies along the back portion of each testis and is lined with smooth muscle. The muscle undergoes rhythmic peristalsis, which transports the sperm through the epididymis towards the vas deferens (see Fig. 30–2). The duct system involved in sperm transport is illustrated in Figure 30–8.

When sperm enter the epididymis, they are immature, nonmotile, and incapable of fertilization. The epididymis secretes a fluid containing nutrients, enzymes, and hormones that aid in sperm maturation. Sperm take about two weeks to pass through the epididymis, during which time they become mature and mobile. From the epididymis, they enter the vas deferens. As the sperm pass through the epididymis, about 90% of the fluid surrounding them is reabsorbed, and hence the sperm entering the vas deferens are highly concentrated.

Storage of Mature Sperm in the Vas Deferens

A small number of sperm are stored in the epididymis, but most of them travel to the vas deferens, where they can be stored for several months in the absence of ejaculation. The **vas deferens (ductus deferens),** components of the spermatic cords, arise from the tail of the epididymis. The vas deferens are composed of long slender tubes, or ducts, which run from the scrotum up into the abdominal cavity and behind the bladder, where they empty into the **ejaculatory duct** (see Fig. 30–8). The ejaculatory duct runs through the center of the **prostate gland,** located at the base of the bladder surrounding the urethra (see Fig. 30–1). The thick wall of the vas deferens is composed of smooth muscle and innervated by sympathetic nerves. The importance of this innervation for ejaculation will be discussed in Chapter 32. As the vas deferens enter the ejaculatory duct, they are joined by the secretory ducts of the seminal vesicles.

Addition of Nutrient Fluid by the Seminal Vesicles

The **seminal vesicles** are composed of two sacs, one on either side of the bladder. The seminal vesicles secrete a viscous material rich in **prostaglandins,** fructose, and other nutrients. The prostaglandins in seminal vesicle fluid stimulate contractions of the uterus in the female; these contractions play a role in moving the sperm through the female reproductive tract. Nutrients in seminal vesicle fluid serve as an energy source for the sperm following ejaculation. Seminal vesicle fluid accounts for approximately 60% of the total volume of ejaculated semen. During ejaculation, the contents of the vas deferens are emptied into the ejaculatory duct immediately before the contents of the seminal vesicles. Seminal vesicle fluid washes sperm out of the ejaculatory duct.

Addition of Alkaline Fluid by the Prostate Gland

The **prostate gland** secretes a liquid that contains citric acid, calcium, and various enzymes. Fluid from the prostate gland adds to the volume of the semen. Prostatic fluid is more alkaline than the secretions of the vagina, which may be very important in protecting sperm fertility. Secretions of the vagina are acidic, and sperm mobility is inhibited in acidic environments. **Prostatic fluid** may neutralize vaginal acidity and increase sperm mobility and fertility.

The prostate gland contains an elaborate system of valves. During ejaculation, a valve in the prostate opens, and the semen is ejected into the urethra. A sphincter in the prostate contracts during ejaculation and seals off the entrance to the bladder, which prevents the passage of urine from the bladder into the urethra (Fig. 30–8). Thus, although urine and semen share a common anatomical pathway to the outside of the body, the valves in the prostate prevent these two fluids from mixing. During ejaculation, semen passes from the ejaculatory duct into the urethra, where it is joined by secretions from the bulbo-urethral glands, also called Cowper's glands (see Fig. 30–8).

Addition of Lubricant Fluid by the Bulbo-Urethral Glands

The **bulbo-urethral glands** are located on each side of the urethra. They secrete a clear viscous liquid that adds to the volume of the semen. During sexual arousal, a small amount of fluid from the bulbo-urethral glands may appear at the tip of the penis before ejaculation. This fluid, termed **pre-ejaculatory fluid,** may act as a lubricant during sexual contact. Pre-ejaculatory fluid occasionally contains a small number of live sperm cells, which makes the practice of withdrawal prior to ejaculation an unreliable method of contraception.

The **urethra** transports the semen through the shaft of the penis to the outside of the body during ejaculation (Fig. 30–8). Stimulation of the sympathetic nerves of the epididymis, vas deferens, prostate, and seminal vesicles during sexual arousal results in smooth muscle contractions in these organs and expulsion of the fluids that compose the semen.

The average volume of semen ejaculated is 3 to 5 milliliters (5 milliliters = about one teaspoonful). The normal concentration of sperm in the ejaculate is highly variable (40 to 120 million per milliliter) and depends in part on frequency of ejaculation. Once sperm are ejaculated in the semen, their maximum life span is only 24 to 72 hours at body temperature.

The major organs involved in sperm production, maturation, and transport are described in Table 30–2.

Neuroendocrine Control of Reproduction in the Male

The **pituitary gland,** which lies at the base of the brain, was described in Chapter 13. Two decades ago, it was thought to be the source of the hormonal signals that control reproduction. We now know that the pituitary gland acts as a relay station, receiving neural and hormonal input from the brain and relaying this information, in the form of hormonal messages, to other organs of the body.

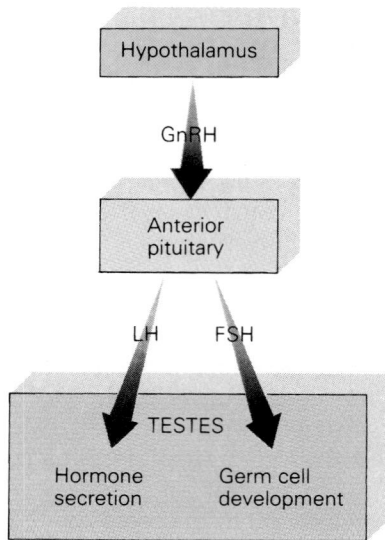

Figure 30–9
Relationship among the hypothalamus, pituitary gland, and testes.

Neuroendocrine Control of Spermatogenesis

As described in Chapter 13, the hypothalamus secretes a releasing factor called **gonadotropin releasing hormone,** or GnRH (Fig. 30–9). GnRH is transported to the anterior pituitary gland via a complex set of blood vessels known as the **hypothalamic-pituitary portal vessels** (see Fig. 13–3, Fig. 13–4). When GnRH reaches the anterior pituitary in the male, it stimulates the release of two major reproductive hormones, termed the **gonadotropins.** One of the gonadotropins, **luteinizing hormone (LH),** is secreted by the anterior pituitary into the blood and is transported to the testes. There it stimulates the Leydig cells to synthesize and secrete testosterone. The other gonadotropin, **follicle-stimulating hormone (FSH),** is secreted by the anterior pituitary into the blood and is also transported to the testes, where it binds to specific receptors on the plasma membranes of Sertoli cells. FSH stimulates the conversion of spermatogonia into spermatocytes in the seminiferous tubules of the testes. Both LH and FSH are required for spermatogenesis (see Fig. 30–9).

Feedback Regulation of Neuroendocrine Function

The secretion of GnRH, LH, and FSH is fairly constant from day to day in the male, in marked contrast to the cyclic secretion of these hormones in the female. The constancy of hormone secretion in the male is due to the presence of a continuous inhibi-

---

Table 30–2

Major Organs Involved in Sperm Production, Maturation, and Transport

Testes: Sperm production: mitosis, meiosis, and differentiation

Epididymis: Sperm transport and maturation— motility and fertility

Vas deferens: Sperm storage

Seminal vesicles: Production of seminal fluid containing nutrients, fructose, and prostaglandins

Prostate: Production of prostatic fluid that is alkaline and contains calcium and citric acid

Bulbo-urethral gland: Production of "pre-ejaculatory" fluid

Penis: Erection and ejaculation

tory **feedback loop** between the hypothalamus, pituitary gland, and testes (Fig. 30–10).

The hypothalamus is sensitive to changes in the circulating levels of sex hormones. It monitors the level of these hormones and responds by increasing or decreasing the production of its releasing factor, GnRH. For example, if the level of circulating testosterone in the blood is too high, testosterone inhibits further production of GnRH in the hypothalamus, leading to a drop in the secretion of LH and FSH from the anterior pituitary. A decrease in LH in the blood results in reduced production and secretion of testosterone by the Leydig cells of the testes, and hence blood levels of testosterone fall.

Conversely, if the circulating level of testosterone is too low, the hypothalamus increases synthesis and release of GnRH, which stimulates the anterior pituitary to secrete additional LH. The increase in circulating LH stimulates the Leydig cells of the

testes to increase the synthesis and release of testosterone into the blood. In addition to this feedback loop between testosterone and the hypothalamus, testosterone can also act directly on the anterior pituitary to inhibit the release of LH in response to any given level of GnRH secreted by the hypothalamus.

Another hormone secreted by the testes exerts an inhibitory feedback effect on the release of FSH from the pituitary. This hormone, called **inhibin,** is secreted by the Sertoli cells in the seminiferous tubules of the testes. Although inhibin is known to directly inhibit the release of FSH from the anterior pituitary, it is not known whether this hormone can inhibit the release of GnRH from the hypothalamus as well (see Fig. 30–10).

Control of Neuroendocrine Function by Sensory Stimuli

The hypothalamus receives neural input from a variety of brain areas involved in processing sensory stimuli. The amount of GnRH released by the hypothalamus is controlled by the excitatory and inhibitory neural signals that the hypothalamus receives. Visual, olfactory, or tactile stimuli are capable of altering GnRH release and hence the release of LH, FSH, and testosterone. In addition, thoughts, moods, and stress can also alter the secretion of these sex hormones. Chapter 32 discusses the role of the central nervous system in mediating the sexual response to direct or imagined stimulation.

Effects of Testosterone in the Male

Testosterone is the major **androgen,** or male sex hormone, produced in the testes. Testosterone is a steroid formed from the steroid precursor cholesterol. The general features of testosterone biosynthesis were discussed in Chapter 12. The Leydig cells do not store testosterone, but they do store large quantities of the cholesterol precursor in the form of lipid droplets. Following synthesis, testosterone is secreted into the extracellular fluid surrounding the seminiferous tubules, where it diffuses into the tubules to produce local effects on sperm production. Testosterone is also transported from the extracellular fluid into the blood vessels of the testes and from there to the general circulation, where it produces a variety of systemic effects.

When testosterone enters the blood, it is bound to albumin and to another plasma protein, **sex steroid-binding globulin (SSBG)** or **testosterone-estradiol-binding globulin (TeBG).** As discussed in

Figure 30–10
Hormonal feedback loops within the hypothalamic-pituitary-gonadal system. The minus sign indicates inhibition.

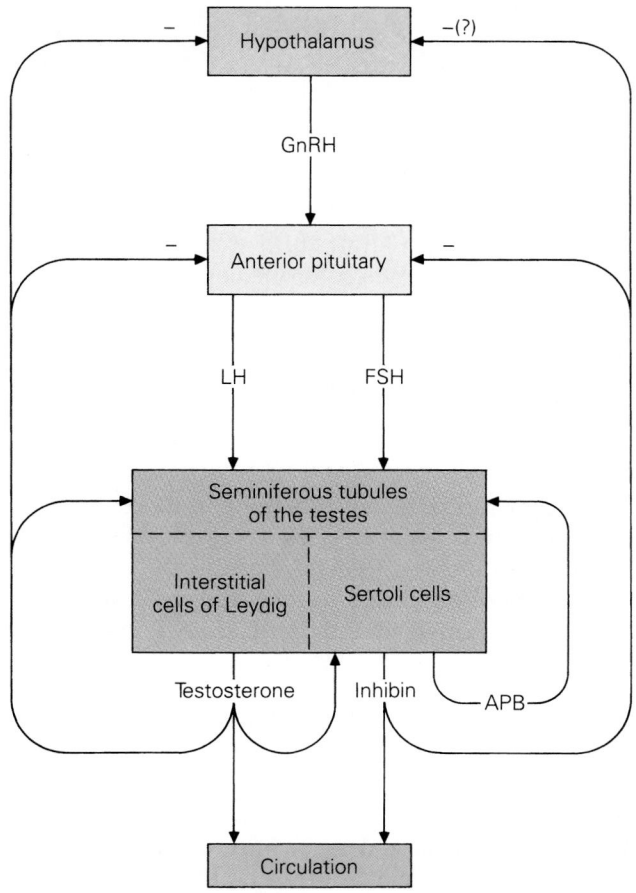

Chapter 12, hormones bound to plasma proteins are not metabolized by the body as quickly as unbound hormones. A certain proportion of all steroid hormones are "stored" in the circulation in their bound form, while the free or unbound hormone acts on target tissues.

Testosterone has some biological effect on almost every tissue of the body. Unaltered testosterone has direct effects on target tissues such as muscle, kidney, and bone. In other tissues, testosterone is transformed into another steroid by the target tissue before it exerts its effect. For instance, in the prostate and seminal vesicles, testosterone is transformed into **dihydrotestosterone** before acting on these tissues. Note that the action of testosterone on the brain is due to the conversion of testosterone to estrogen. As we will see in Chapter 32, this conversion of a "male" hormone to a "female" hormone in the central nervous system (CNS) is very important in sexual differentiation.

Effects of Testosterone at Puberty

Before puberty, the hypothalamic-gonadal system is relatively dormant. Puberty begins with the secretion of increasing amounts of GnRH from the hypothalamus, which stimulates the release of LH and FSH from the pituitary. FSH produces enlargement of the testes, and LH stimulates the secretion of testosterone from the Leydig cells. This increase in testosterone secretion at puberty results in the enlargement of the penis, scrotum, and testes, and stimulates the development of the male secondary sex characteristics. The **testosterone-dependent changes** that occur at puberty are detailed in Table 30–3. Testosterone stimulates muscular development, bone growth, thickening of the skin, and growth of facial and body hair. Testosterone also produces growth and thickening of the vocal cords and enlargement of the larynx, thereby lowering the pitch of the voice. These anatomical changes, stimulated by testosterone, mark the onset of puberty. If testosterone secretion is interrupted prior to puberty as a result of destruction or removal of the testes, a male will not undergo the anatomical changes characteristic of puberty.

Effects of Testosterone on Spermatogenesis

Spermatogenesis requires the presence of high levels of testosterone in the seminiferous tubules. A decrease in testosterone production can result in sterility. Testosterone is thought to stimulate both the formation of spermatogonia and the second meiotic division that results in the formation of sperma-

Table 30–3
Testosterone-Dependent Changes at Puberty

1. Penis and testes enlarge and become pigmented; the scrotal skin becomes wrinkled.

2. Growth of facial hair results in the development of mustache and beard, and scalp line recedes.

3. Pubic hair grows in an upward pattern on the abdomen; hair appears on the chest, armpits, and extremities.

4. Growth of the long bones increases to a rate of about 3 inches per year. Bones increase in size and strength due to an increase in the quantity of bone matrix and increased calcium retention.

5. Vocal cords and larynx enlarge, thereby lowering the pitch of the voice.

6. Muscle mass and skin thickness increase, and the skin deepens in hue due to melanin deposition. Secretions from the sebaceous glands of the face increase, resulting in the development of acne.

tids in the seminiferous tubules. An interesting interdependence exists between testosterone and the Sertoli cells (see Fig. 30–10). Testosterone stimulates protein synthesis and fluid secretion from the Sertoli cells, and the Sertoli cells in turn secrete androgen-binding-protein (ABP), which binds testosterone and keeps testosterone levels high in the seminiferous tubules (see Fig. 30–10).

Reproductive Physiology in the Female

The major reproductive functions in the female are (1) secretion of sex hormones, (2) production of ova, and (3) gestation and delivery of the fetus. The first two functions are carried out by the ovary, while the third is performed by the uterus and cervical canal.

Female Reproductive Anatomy
Anatomy of the Female External Genitalia

The external genitalia of the female, which are collectively called the **vulva**, include the **labia majora**, the **labia minora**, the **clitoris**, and the **vaginal open-**

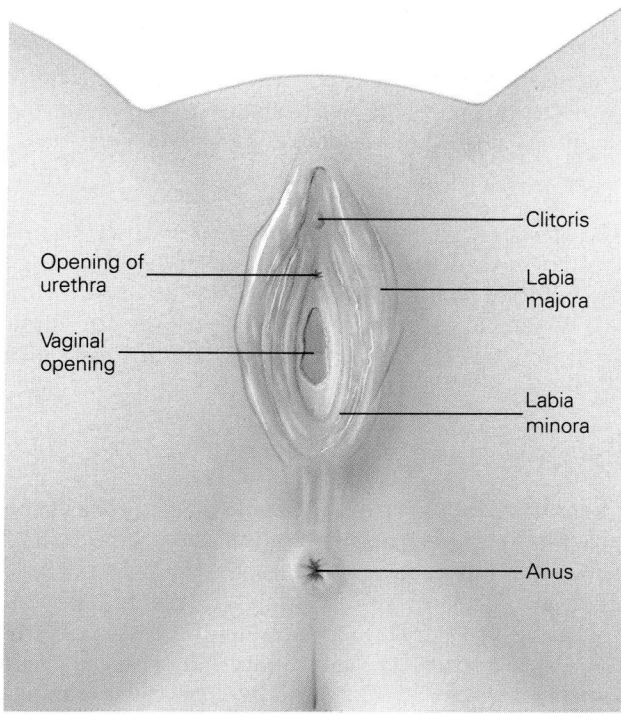

Figure 30–11
External genitalia of the female. (Redrawn from Masters and Johnson, *Human Sexuality, Second Edition.*)

ing (Fig. 30–11). The labia majora consist of folds of skin that cover fat and smooth muscle. The labia majora meet on the midline and protect the urinary opening and the entrance to the vagina. The labia minora are composed of skin folds, rich in small blood vessels, that meet above the clitoris and provide mechanical protection for the clitoris. The only visible part of the clitoris is termed the **glans clitoris.** This structure contains a dense supply of nerve endings, which makes it highly sensitive to touch, pressure, and temperature. Note that while the penis and the clitoris are derived embryologically from the same tissue, the clitoris is the only tissue in either sex that has no known reproduction function. The uniquely sexual function of the clitoris will be discussed in Chapter 32.

Anatomy of the Female Internal Genitalia

The principal internal reproductive organs of the female include the **vagina, uterus, ovaries,** and **fallopian tubes** (Fig. 30–12). In contrast to those in the male, the urinary and reproductive tracts in the female are anatomically separate.

Ova (singular, *ovum*) are produced in the ovaries. During the middle of each ovulatory cycle, a single ovum is expelled by one of the two ovaries into the abdominal cavity. The ovum passes through one of the fallopian tubes into the uterus. If the ovum has been fertilized by a sperm, it implants

Figure 30–12
Side view of the female reproductive tract. (Redrawn from Geer, J., Heiman, J., and Leitenberg, H. *Human Sexuality.*)

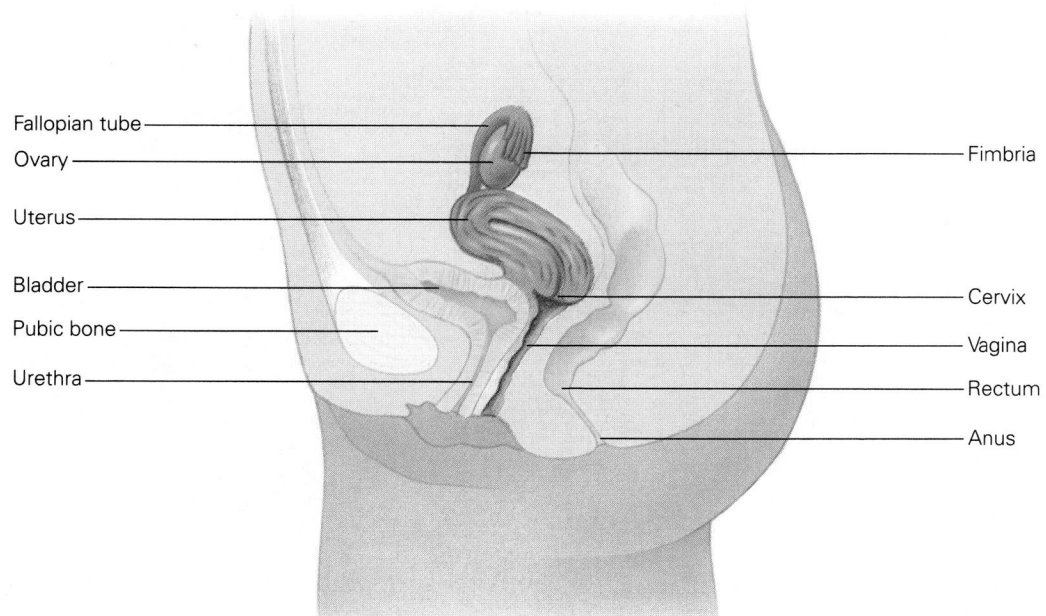

in the uterus. If it has not been fertilized, it leaves the uterus along with the menstrual flow.

The Vagina.   The vagina is a muscular organ that extends from the external opening in the vulva to the uterus. The vaginal walls have the ability to contract and expand, allowing for the passage of a baby during childbirth. The vagina is lined with a membrane called the **vaginal mucosa,** which secretes a fluid during sexual arousal that acts as a lubricant. The vagina contains few sensory nerves and is relatively insensitive to touch and pressure.

The Uterus.   The uterus is a hollow organ with walls composed of smooth muscle. The term **cervix** applies to the portion of the uterus that extends into the vagina. The lining of the uterus, termed the **endometrium,** changes in thickness during the menstrual cycle. At the beginning of pregnancy, the fertilized ovum implants in the endometrium of the uterus. The muscular walls of the uterus, called the **myometrium,** expand and contract during labor and delivery.

The Fallopian Tubes.   The **fallopian tubes,** or **oviducts,** arise from the uterus and extend into the abdominal cavity (Fig. 30–13a). These muscular,

tubelike structures terminate near the ovaries. The end of each fallopian tube is surrounded by long, fingerlike extensions called the **fimbria.** The fallopian tubes are lined with cilia, which propel the ovum through the fallopian tube to the uterus (Fig. 30–13b).

The Ovaries.   The ovaries are oval-shaped organs located in the pelvic cavity below the open ends of the fallopian tubes (see Fig. 30–13). The ovaries and the testes have similar embryonic origins and similar functions: (1) the production of germ cells and (2) the production of hormones.

Oogenesis: The Formation of Ova

The **oocytes** or **ova** are formed from germ cells in the ovary called **primordial follicles.** The production of oocytes in the female contrasts sharply with the production of sperm in the male. While millions of sperm may be produced daily in the testes during the entire reproductive lifespan of the male, all of the follicles that a female will ever produce, a maximum of about 7 million, are formed in the ovary during the first six months of embryonic life. Most of these follicles will disappear. By the time a female is born, only about 2 million follicles remain in the

Figure 30–13
(*a*) Front view of the female reproductive tract. (Redrawn from Masters and Johnson, *Human Sexuality, Second Edition.*) (*b*) An unfertilized ovum in the folds of the fallopian tube. Photograph taken with an electron microscope. (× 100.) (© Lennart Nilsson, *A Child Is Born*, Dell Publishing Company, 1977.)

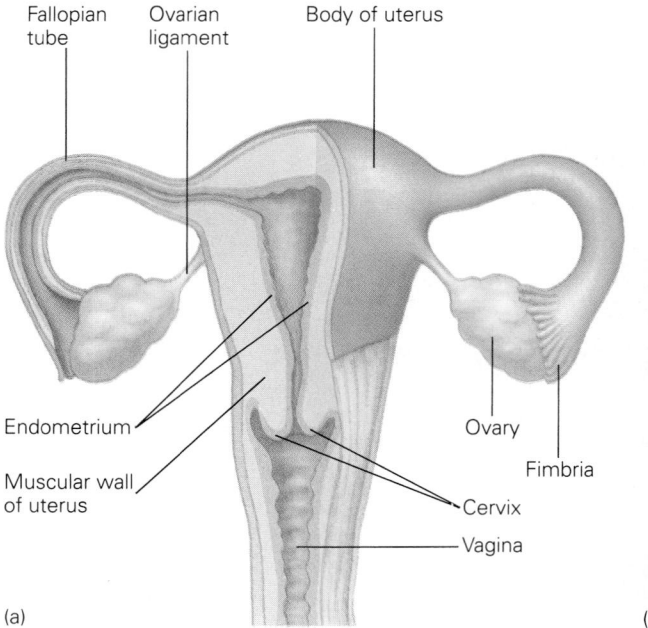

Fallopian tube
Ovarian ligament
Body of uterus
Endometrium
Muscular wall of uterus
Ovary
Fimbria
Cervix
Vagina

(a)

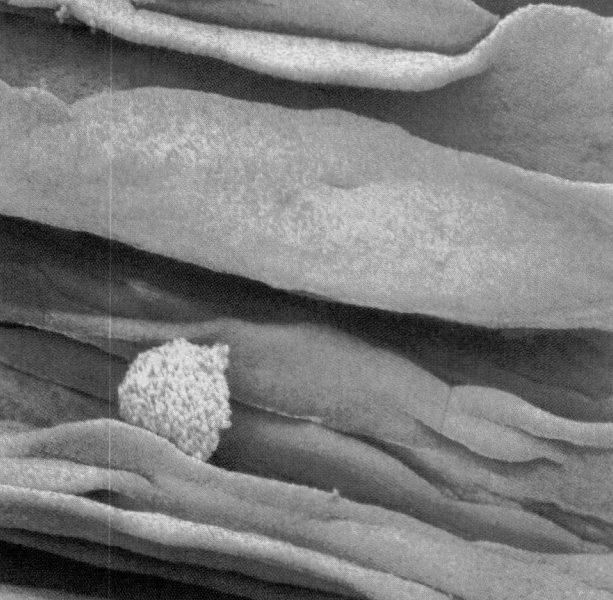

(b)

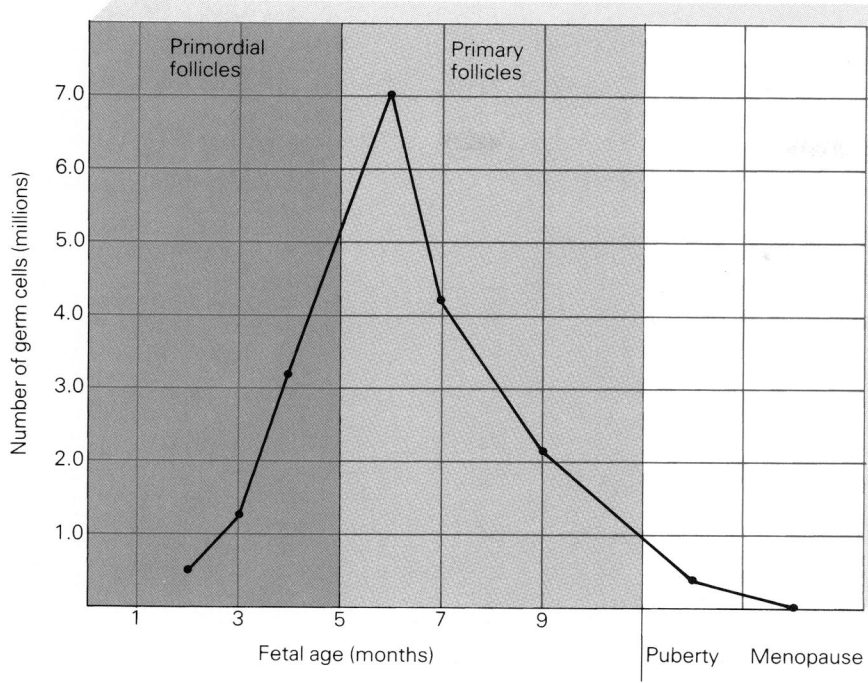

Figure 30–14
Number and type of follicles in the ovary during development.

ovary, and at the onset of puberty this number has fallen to about 400,000 (Fig. 30–14). When the female reaches menopause at approximately 50 years of age, few follicles are left in the ovary, and reproduction ceases. Only one ovum is released from one of the two ovaries each month during the reproductive lifespan of the female, which begins at the onset of puberty and ends with menopause. So of the 7 million follicles formed in the ovaries during embryonic development, only about 400 to 500 will ever develop into mature ova and be released from the ovaries.

There are major differences between the growth patterns of germ cells in the male and female. In the male, as we have seen, the spermatogenic cells are constantly undergoing a complicated structural and biochemical transformation into mature sperm in the testes. In contrast, in the female, meiotic division of the germ cells begins during embryonic development but then stops until puberty, when ovulation begins.

The primary germ cells in the female are termed **oogonia.** During the second and third months of embryonic development, meiosis begins in some oogonia, converting them first to primary oocytes and then to primordial follicles. The **primary oocytes** increase in size and are surrounded by a single layer of **spindle cells,** the precursors of **granulosa cells.** The primary oocyte is then called a **primordial follicle.** Primary oocytes that are not converted to primordial follicles degenerate.

During the fifth and sixth months of fetal development, the number of primordial follicles in the ovary increases, and some primordial follicles differentiate into **primary follicles** (see Fig. 30–14). The primordial follicles increase in size and are surrounded by spindle-shaped cells, which divide to create several layers of granulosa cells. The granulosa cells secrete mucopolysaccharides that form a coat, called the **zona pellucida,** around the developing follicle. The granulosa cells provide nutrients and chemical signals that stimulate further maturation of the follicle.

As do the Sertoli cells in the testes, the **granulosa cells** form a barrier around the developing follicle that prevents the entrance of substances such as proteins, ions, and drugs, which could destroy the developing germ cell. During the fifth and sixth months of gestation, the connective tissue around the follicle differentiates to form a highly vascular cell layer, the **theca interna,** which surrounds the follicle. The primary follicles remain in this first stage of meiotic division until puberty, when ovulation begins. The first meiotic division of the oocyte is completed just before ovulation; the second meiotic division does not occur unless and until the germ cell is fertilized by a spermatozoon.

Ovulation

Once each month, from puberty until menopause, 6 to 20 primary follicles enter the next stage of follicu-

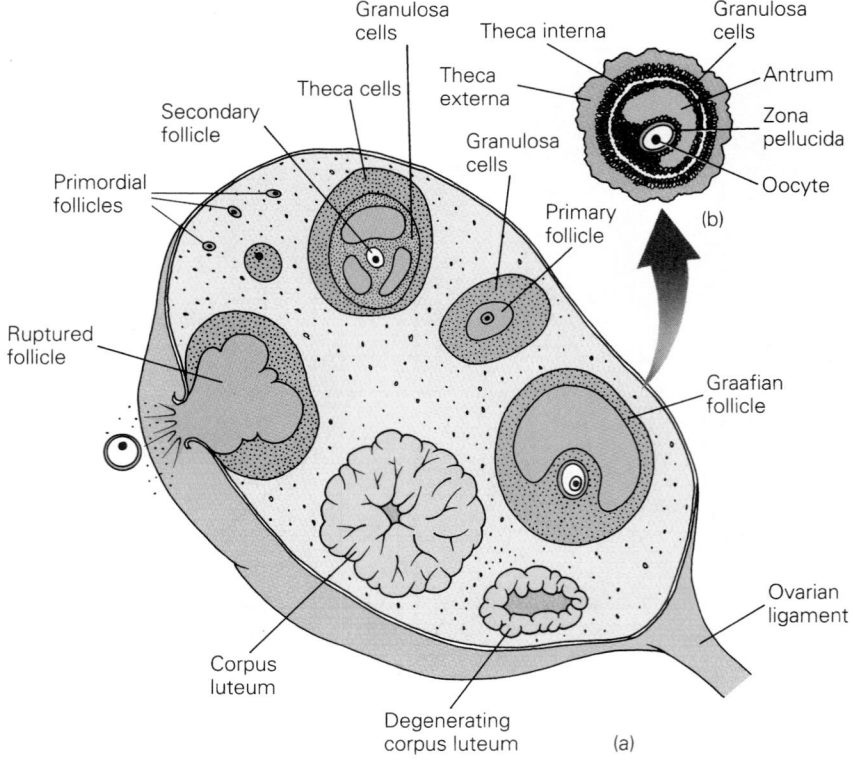

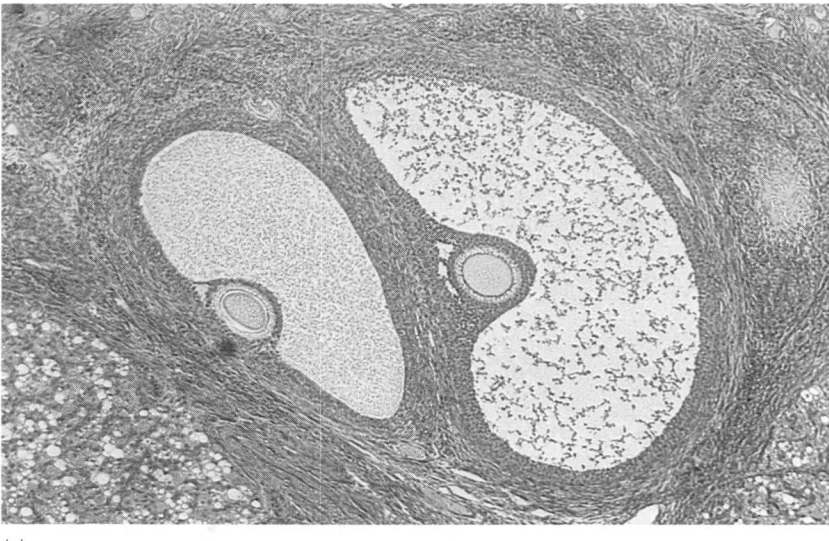

Figure 30–15
(*a*) Cross-section of the ovary, containing follicles in different stages of development. (*b*) Cross-section of a Graafian follicle. (*c*) Photomicrograph of Graafian follicles in the ovary. (× 4.5.) (© Cabisco/Visuals Unlimited)

(c)

lar development, usually resulting in the release of a single ovum from one of the two ovaries.

Growth of the Follicle

Each month, the spindle cells surrounding 6 to 20 of the primary follicles proliferate to form the **theca externa.** These mature follicles are called **Graafian follicles.** Figure 30–15*a* illustrates the structure of

the ovary, which contains follicles in different stages of development. Although 6 to 20 follicles may begin to develop in the ovary during each month, only one reaches maturation. As one follicle increases in size faster than the others, the remaining follicles in the ovary degenerate in a process known as **atresia.** Atresia of the remaining follicles can occur at any stage of follicular development, and atretic follicles can always be found in the ovary. As we will see, hormone levels in the ovary may determine whether a particular follicle will ovulate or undergo atresia.

Several days before ovulation, a single Graafian follicle increases dramatically in size to reach 10 to 20 mm in diameter within 48 hours. The granulosa cells surrounding this follicle begin to secrete increasing amounts of fluid that contains proteins, electrolytes, polysaccharides, and hormones. The follicular fluid in the Graafian follicle collects in a single central area called the **antrum** (see Fig. 30–15b). The fluid-filled antrum displaces the ovum to the side of the follicle, and the follicle itself forms a bulge on the outer surface of the ovary. The follicle secretes proteolytic enzymes that break down the surface of the ovary, and the antral fluid and the ovum are released from the follicle into the peritoneal cavity in a process known as **ovulation** (see Fig. 30–15a). Just before ovulation, the first meiotic division of the ovum is resumed and, following ovulation, the ovum enters the fallopian tube.

Growth of the Corpus Luteum

During the first few hours after ovulation, the remaining granulosa cells of the follicle undergo complex physical changes, and the follicle becomes the **corpus luteum** (see Fig. 30–15a). The cells of the corpus luteum increase in size and develop a rich vascular supply for 7 to 8 days following ovulation. If fertilization and implantation of the ovum do not occur, the corpus luteum degenerates between days 8 and 21 following ovulation. The corpus luteum secretes large amounts of **progesterone** and smaller amounts of **estrogen,** which interact with pituitary hormones to retard the growth of new follicles in the ovary. When the corpus luteum degenerates, new follicles begin to grow again in the ovary, marking the beginning of a new monthly ovarian cycle.

Neuroendocrine Control of Reproduction in the Female

Neuroendocrine control of reproduction in the female is dependent on a continuous interaction between the hypothalamus, the pituitary gland, and the ovaries. The major difference between hormonal control of reproduction in the male and female is that hormone secretion and reproductive processes in the female are cyclic rather than continuous.

In the female, as in the male, the hypothalamus secretes gonadotropin-releasing hormone, or GnRH, which is transported to the anterior pituitary, where it stimulates the release of LH and FSH.

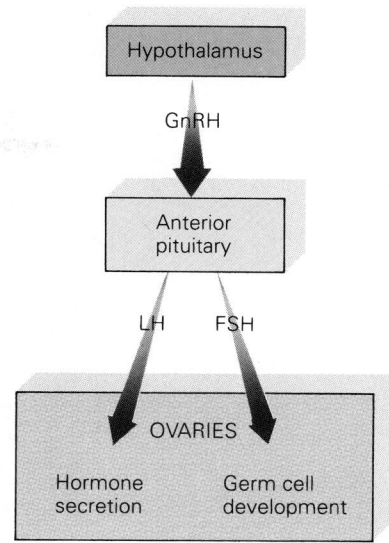

Figure 30–16
Relationship among the hypothalamus, pituitary gland, and ovaries.

LH and FSH stimulate germ cell development and hormone secretion in the ovaries (Fig. 30–16).

In the female, the effects of LH and FSH on the ovary vary during different phases of the ovarian cycle. The **ovarian cycle,** also called the **menstrual cycle,** is approximately 28 days long. This cycle can be divided into three phases: (1) the follicular phase, which lasts for approximately 12 days; (2) the luteal phase, which lasts for approximately 11 days; and (3) the menstrual phase, which lasts for approximately 5 days.

During the **follicular phase,** a single follicle in one of the two ovaries matures and is released during ovulation. During the **luteal phase,** which immediately follows ovulation, the post-ovulatory follicular cells in the ovary differentiate into the corpus luteum, which functions for a few days and then degenerates. During the **menstrual phase,** there is little activity in the ovary but much activity in the uterus, where the lining of the uterus, the endometrium, is shed in a process called **menstruation.** Figure 30–17 illustrates the cyclic hormonal, ovarian, and uterine changes that occur in the female during the ovarian cycle.

Neuroendocrine Control of the Follicular (Pre-ovulatory) Phase

Once each month, just after menstruation ceases, growth and development of follicles begin in the ovary. Follicular growth is stimulated by the pitui-

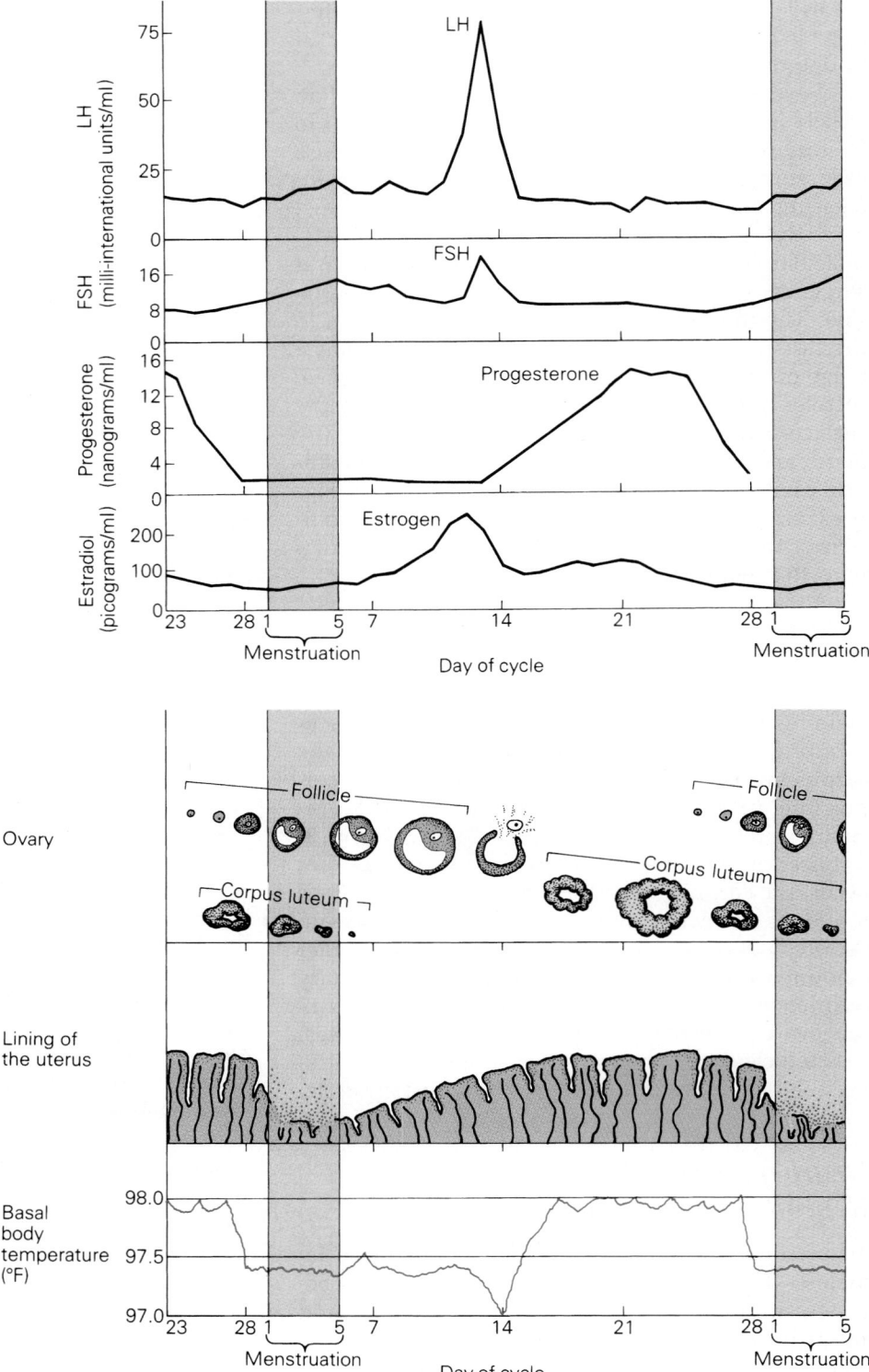

Figure 30–17
Hormonal, ovarian, and uterine changes that occur during the ovarian cycle. (Courtesy of Dr. Resco, Oregon Health Sciences University).

tary hormone FSH. FSH stimulates both hypertrophy (increase in cell size) and **hyperplasia** (increase in cell number) of the granulosa cells that surround the primary follicle. In response to FSH, the granulosa cells synthesize estrogen from androgen pre-

cursors. Estrogen secretion from the granulosa cells results in rising levels of circulating estrogen (see Fig. 30–17). Estrogen stimulates further GnRH secretion from the hypothalamus and additional LH and FSH secretion from the pituitary (Fig. 30–17).

The hormonal interactions between the pituitary and the ovary during the follicular phase of the ovarian cycle are represented in Figure 30–18. **Estrogen,** which is secreted by the granulosa cells of the follicle, acts on the granulosa cells to increase the number of receptors in the follicle for both estrogen and FSH. As the number of estrogen receptors in the follicle rises, more estrogen is retained in the follicle, and follicular estrogen content rises. The local action of estrogen, which results in an increase in the number of FSH receptors on the granulosa cells, augments the growth-promoting effect of FSH on the follicular granulosa cells. As the follicle grows, it secretes more estrogen, which continues to augment the action of FSH in the follicle.

Follicular growth and development in the ovary can be viewed as a self-perpetuating mechanism that involves a complex interaction between the pituitary gland and the ovary, resulting in a steady rise in follicular growth and estrogen secretion. As discussed in Chapter 12, when a hormone acts on the cell from which it was secreted, it is said to have an **autocrine action.** Estrogen, secreted by the granulosa cells of the follicle, acts on these same cells to augment the further secretion of estrogen and to augment the action of FSH. Therefore, estrogen exerts an autocrine action on the granulosa cells of the developing follicle. FSH secreted by the pituitary and estrogen secreted by the follicle together increase the number of LH receptors on the granulosa cells of the follicle. This increase, as we will see, plays an important role in the formation of the corpus luteum.

During the follicular phase, another interaction between the pituitary gland and the ovaries occurs (see Fig. 30–18). The increasing estrogen, secreted by the granulosa cells of the follicle, stimulates the pituitary to synthesize and secrete additional LH. This rise of LH in the blood stimulates the theca cells of the follicle to synthesize and secrete two androgens: testosterone and **androstenedione.** These androgens exert a **paracrine** action on neighboring cells in the follicle. The androgens diffuse into the follicular granulosa cells and serve as precursors for estrogen production in the granulosa cells.

The rising blood levels of LH stimulate the granulosa cells of the follicle to secrete progesterone. Progesterone enters the circulation, and blood levels of progesterone begin to rise (see Fig. 30–17). In addition, some of the progesterone from the granu-

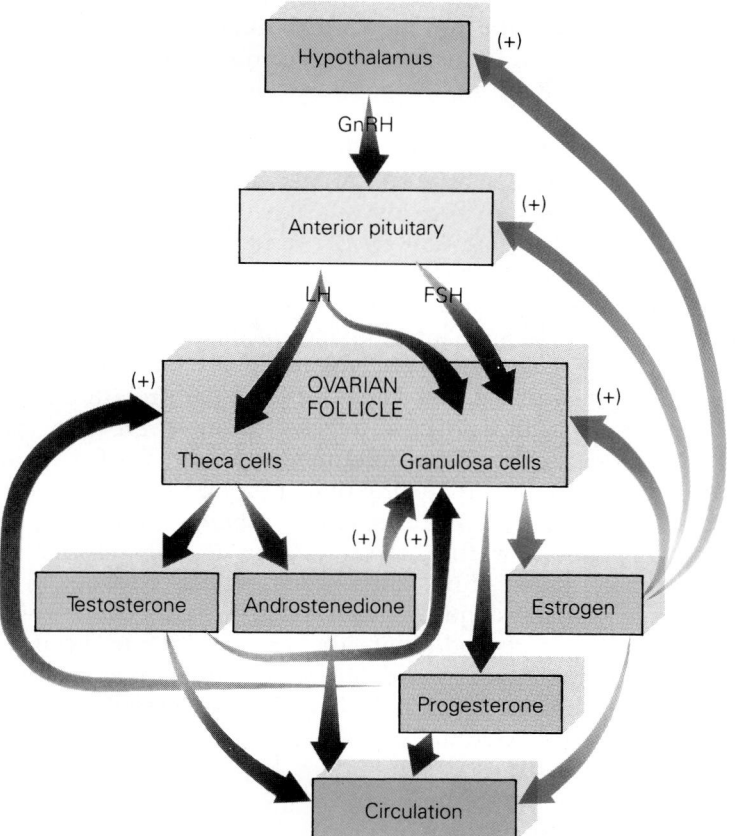

Figure 30–18
Hormonal interactions between the pituitary gland and ovary during the follicular phase of the ovarian cycle. The plus sign indicates stimulation.

losa cells diffuses back into the theca cells and acts as a substrate for further androgen synthesis by the theca cells (see Fig. 30–18).

As we have seen, as many as 20 follicles may increase in size each month during the follicular phase in the ovary, but only one follicle reaches maturity. The others undergo atresia. The mechanisms that control the selection of a single follicle for full maturation are not completely understood. It appears that complex endocrine events in the ovary control the growth and atresia of the various follicles. For example, atresia may result from an insufficient number of FSH receptors in some follicles, or from an inability of the granulosa cells of certain follicles to produce adequate amounts of estrogen from androgen precursors. Fluid from atretic follicles contains lower levels of both FSH and estrogen than fluid from mature follicles.

Neuroendocrine Control of Ovulation

During the follicular phase of the ovarian cycle, increased release of estrogen from the ovary stimulates LH and FSH release from the anterior pituitary, which results in progressive follicular growth for approximately 10 days (see Fig. 30–17). On day 14, just prior to ovulation, blood levels of LH and FSH are higher than at any other time in the ovarian cycle. These peaks in blood levels of LH and FSH are referred to as the **preovulatory surges** of LH and FSH. The surge of LH induces changes in follicular structure that result in ovulation (see Fig. 30–17). As we have discussed, these LH-induced changes include an increase in the production of antral fluid, compression of the ovum toward the wall of the follicle, protrusion of the follicle against the wall of the ovary, and, finally, enzymatic degeneration of the ovarian wall and release of the antral fluid and ovum into the abdominal cavity during ovulation. One or two days before ovulation, the follicle stops secreting estrogen, and blood levels of estrogen begin to fall (see Fig. 30–17). The mechanism that controls the cessation of estrogen secretion by the follicle is not known. Ovulation marks the end of the follicular phase of the ovarian cycle in the female.

In many women ovulation is accompanied by a slight upward shift of approximately 0.5 to 1.0 degree in basal body temperature, which persists throughout the luteal phase (see Fig. 30–17). The exact cause of this rise in temperature is not known. However, since ovulation can occur in the absence of a detectable change in body temperature, measurements of body temperature are not reliable indicators of ovulation.

Neuroendocrine Control of the Luteal (Postovulatory) Phase

Immediately following the release of the ovum from the follicle on day 14 of the ovarian cycle, the remaining granulosa cells of the follicle increase in size and number and undergo chemical changes to form the corpus luteum (see Fig. 30–17). This process is called **luteinization.** Differentiation and growth of the corpus luteum are stimulated by LH from the anterior pituitary. In fact, LH derives its name, "luteinizing hormone," from its action on the corpus luteum. The role of FSH in the luteal phase is unknown. Although the presence of LH is necessary for the formation of the corpus luteum, the ovary is thought to secrete an additional substance that inhibits the formation of the corpus luteum in the mature follicle prior to ovulation, as well as in follicles that do not ovulate.

During the luteal phase, the corpus luteum secretes a large amount of **progesterone** and a smaller amount of **estrogen,** increasing blood levels of these two hormones (see Fig. 30–17). The estrogen secreted by the corpus luteum is structurally identical to the estrogen secreted by the follicle during the follicular phase. However, while estrogen stimulates release of GnRH, LH, and FSH during the follicular phase, the estrogen and progesterone secreted during the luteal phase *inhibit* GnRH secretion from the hypothalamus and LH and FSH secretion from the pituitary (Fig. 30–19). During the luteal phase, blood levels of LH and FSH fall (see Fig. 30–17). As we will see in Chapter 32, the efficacy of the "birth control pill" is based on this inhibitory action. The opposite effect of estrogen and progesterone on GnRH, LH, and FSH release during the ovarian cycle is due to a change in hypothalamic sensitivity to estrogen and progesterone during the follicular and luteal phases of the cycle.

As LH and FSH blood levels fall during the luteal phase, the signal for follicular growth diminishes, and no new follicles begin to grow in the ovary. The corpus luteum increases in size for about 7 to 8 days following ovulation and then begins to degenerate. The mechanism responsible for degeneration of the corpus luteum is not completely known. Falling blood levels of LH may not provide an adequate signal for luteal growth (see Fig. 30–17). In addition, the corpus luteum itself is thought to secrete hormones, known as **prostaglandins,** which induce further degeneration of the corpus luteum.

As the corpus luteum degenerates, it secretes less progesterone and estrogen, so that blood levels of these hormones begin to fall (see Fig. 30–17). The

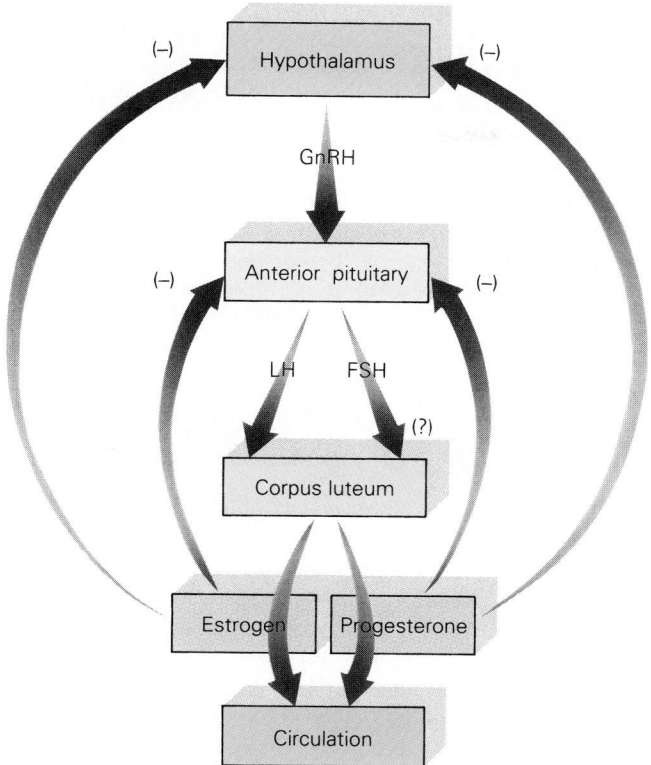

Figure 30–19
Hormonal interactions between the pituitary gland and ovary during the luteal phase of the ovarian cycle. The minus sign indicates inhibition.

inhibition of GnRH, LH, and FSH release, which was maintained by progesterone and estrogen from the corpus luteum, is now removed, so that blood levels of LH and FSH begin to rise again. Rising levels of LH and FSH in the blood stimulate growth of another set of follicles in the ovary, which marks the beginning of a new follicular cycle (see Fig. 30–17).

Neuroendocrine Control of the Menstrual Phase

Dramatic changes in the structure and secretory capacity of the endometrium of the uterus occur during the ovarian cycle in the female. These changes are induced by both estrogen and progesterone. As we have seen, during the follicular phase of the ovarian cycle, the growing follicle secretes large amounts of estrogen. This estrogen stimulates proliferation of the endometrium of the uterus (see Fig. 30–17). In addition, estrogen increases the number of progesterone receptors in the uterus. At the end of the follicular phase in the ovary, just before ovulation, the follicle stops secreting estrogen, resulting in a drop in circulating levels of estrogen (see Fig. 30–17).

Following ovulation, during the luteal phase of the ovarian cycle, estrogen and progesterone (se-

creted by the corpus luteum) stimulate continued growth and proliferation of the coiled blood vessels of the endometrium (see Fig. 30–17). The tubular glands of the endometrium fill with glycogen, and enzymes accumulate in the connective tissue of the endometrium. During the luteal phase, the endometrium swells and becomes approximately 4 to 6 mm thick, owing to the progressive growth of glands, blood vessels, and connective tissue.

The function of the proliferation of endometrial cells is to provide a continuing source of nutrients for the ovum if it is fertilized and implants in the uterus. If the ovum is not fertilized, the corpus luteum begins to degenerate on day 8 following ovulation and stops secreting progesterone and estrogen.

A drop in circulating levels of estrogen and progesterone resulting from degeneration of the corpus luteum causes **hemorrhagic changes** in the uterine endometrium that result in the onset of menstruation (see Fig. 30–17). In the absence of stimulation from estrogen and progesterone, the endometrium becomes **necrotic.** Arterioles of the endometrium constrict, which results in a slowing of the circulation. Blood enters the vascular layer of the endometrium, resulting in blood pooling. Gradually, the outer layers of the endometrium separate from the uterus at the site of hemorrhage.

Chemical Termination of Early Pregnancy

Termination of pregnancy normally requires surgical intervention, which is accompanied by the risks of infection and postsurgical trauma. Medical researchers have recently developed a new drug, mefipristone or RU 486, that represents a noninvasive alternative to surgical termination of early pregnancy. RU 486 is a synthetic steroid derivative with a high binding affinity for progesterone receptors. It blocks the action of progesterone and hence has been termed an *antiprogesterone*.

Progesterone, secreted by the corpus luteum during the ovarian cycle and by the placenta after the sixth week of pregnancy, is essential for the maintenance of pregnancy. Progesterone stimulates growth and proliferation of the endometrial lining of the uterus and strongly inhibits the uterine contractions.

In the absence of stimulation from progesterone, the endometrium undergoes hemorrhagic changes; the outer layers of the endometrium separate from the uterus and are discharged, which represents the onset of menstruation.

Oral administration of RU 486, together with prostaglandin, induces uterine contractions and the onset of menstruation, resulting in evacuation of the contents of the uterus whether or not a fertilized ovum is present.

RU 486 has been found to be both safe and effective in terminating early pregnancy, and is currently being tested as a possible contragestive agent for fertility control. RU 486 has been authorized for clinical use in France but has not yet been approved for clinical use in the United States.

Table 30–4
Summary of the Hormonal, Ovarian, and Uterine Changes That Occur During the Ovarian Cycle

1. During the early part of the follicular phase, blood levels of LH and FSH begin to rise.

2. This rise in LH and FSH stimulates growth of 6 to 12 primary follicles and the full maturation of a single Graafian follicle.

3. The growing follicles secrete large amounts of estrogen, lesser amounts of progesterone, and small amounts of testosterone and androstenedione.

4. Rising blood estrogen stimulates growth of the endometrium of the uterus.

5. Rising blood estrogen also stimulates further secretion of GnRH, which results in the pre-ovulatory surges of LH and FSH.

6. Just prior to ovulation the follicle stops secreting estrogen and blood estrogen levels fall.

7. The surges of LH and FSH at midcycle induce ovulation.

8. Following ovulation, the remaining follicular cells in the ovary become the corpus luteum.

9. The corpus luteum secretes large amounts of progesterone and lesser amounts of estrogen. Blood levels of these hormones begin to rise.

10. Progesterone and estrogen from the corpus luteum stimulate further growth of the endometrium of the uterus.

11. Progesterone and estrogen also inhibit secretion of LH and FSH, and blood levels of these pituitary hormones begin to fall.

12. As blood levels of LH and FSH fall, the signal for follicular growth diminishes and no new follicles begin to grow.

13. On day 8 following ovulation, the corpus luteum begins to degenerate and secretes less progesterone and estrogen.

14. Falling blood levels of progesterone and estrogen result in hemorrhagic changes in the uterine endometrium, which culminate with the onset of menstruation.

15. As blood levels of progesterone and estrogen fall at the end of the luteal phase, the inhibition of GnRH, LH, and FSH, which was exerted by progesterone and estrogen, is removed. Blood levels of LH and FSH begin to rise, thus marking the beginning of another follicular phase (see point 1).

Prostaglandins, produced by the uterus, stimulate the uterine smooth muscles to contract rhythmically. These uterine contractions result in the discharge of blood and endometrial cells that constitutes the **menstrual flow.** The menstrual flow lasts for approximately 5 to 7 days (see Fig. 30–17), during which the uterus expels an average of 20 to 200 milliliters of blood. An overproduction of prostaglandins in the uterus during the menstrual phase of the ovarian cycle can lead to excessive uterine contractions in many women, which result in menstrual cramps, a condition called **dysmenorrhea.** Prostaglandin action on smooth muscles in other areas of the body can also result in the nausea, vomiting, and headache that sometimes accompany the menstrual phase.

A summary of the hormonal, ovarian, and uterine changes that occur during the ovarian cycle is contained in Table 30–4.

Effects of Estrogen and Progesterone in the Female

The major sex hormones in the female are **estrogen** and **progesterone.** As described in Chapter 12, both of these hormones are steroids. In addition to their roles in regulating the ovarian cycle, estrogen and progesterone have widespread effects on many other physiological and metabolic processes. The effects of estrogen are summarized in Table 30–5.

Effects of Estrogen at Puberty

Estrogen stimulates the development of the sex organs and the appearance of the secondary sex characteristics in the female. Therefore, the effects of estrogen in the female are analogous to the effects of testosterone in the male. At puberty, estrogen stimulates growth of the vagina, uterus, and fallopian tubes as well as the external genitalia, breasts, and body hair. Estrogen stimulates linear growth but also accelerates closure of the epiphyses of the long bones, which results in a shorter height in women than in men. Stimulation by estrogen results in the particular pattern of distribution of body fat around the hips, buttocks, and abdomen that characterizes the female form. When the ovary is absent or when it secretes inadequate amounts of estrogen, puberty either fails to occur or progresses very slowly.

Metabolic Effects of Estrogen

Estrogen interacts with a variety of other hormones that control metabolic processes. For instance, estrogen decreases the sensitivity of peripheral tissues to

Table 30–5
Effects of Estrogen

Puberty

Stimulates growth of internal reproductive organs, external genitalia, and breasts

Stimulates growth of body hair

Stimulates growth of long bones and early closure of the epiphyses

Stimulates "female" pattern of fat distribution

Hormonal

Stimulates secretion of GnRH, LH, and FSH during the follicular phase and inhibits secretion of these same hormones during the luteal phase

Decreases sensitivity of peripheral tissues to insulin

Increases circulating levels of renin and angiotensin II

Metabolic

Protects against bone loss

insulin, which can result in decreased glucose tolerance. Estrogen increases the levels of plasma proteins that bind steroids such as thyroxine, cortisol, and progesterone. In addition, estrogen increases circulating levels of renin and thus angiotensin II, which can result in hypertension.

Evidence is accumulating that estrogen protects against bone loss. For instance, removal of the ovaries, or inadequate ovarian secretion of estrogen, results in the acceleration of bone loss known as **osteoporosis.** In such cases, replacement of estrogen can dramatically slow the rate of bone loss. Postmenopausal women, in whom ovarian secretion of estrogen is very low, exhibit osteoporosis most often. Note that, although men have very low circulating levels of estrogen, they are not particularly prone to osteoporosis. However, impairment of gonadal function in the male, a condition called **hypogonadism,** is often accompanied by low bone mass. It is possible that androgens may protect against bone loss in the male.

Effects of Progesterone

Progesterone is the most active naturally occurring antagonist of androgens. Progesterone antagonizes the effects of both testosterone and dihydrotestosterone by competing for the androgen-binding sites

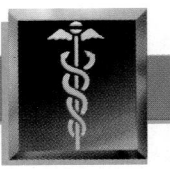

Menopausal Osteoporosis

Menopause is accompanied by decreased ovarian production of estrogen. One of the long-term metabolic effects of decreased estrogen is a reduction of total bone mass or bone density. Following menopause, women lose approximately 1% to 2% of their bone mass per year. This reduction of bone mass is caused by the fact that more bone matrix erodes than is being replaced. Consequently, bones become thin and brittle, and fractures occur easily. Bone pain is often the first symptom indicating the presence of osteoporosis. Intense pain may develop suddenly because of compression fractures of the vertebrae or fractures of the long bones. In x-ray images, the bones appear abnormally lucent (transparent), owing to thinness, and the spine is curved because of compression. Such curvature of the spine may result in decreased height over a period of years.

Treatment of menopausal osteoporosis is difficult, because no pharmacologic agent has been discovered that is both safe and effective in inducing the formation of new bone. At present, the best treatment of menopausal osteoporosis involves maintaining estrogen levels through hormone replacement therapy (HRT). When begun at the onset of menopause, estrogen replacement decreases the rate of bone loss that normally occurs following menopause.

Estrogen is usually administered in the lowest dose found to relieve symptoms, since high doses are associated with increased risk for certain types of cancer.

In treatment of menopausal osteoporosis, replacement of estrogen is accompanied by maintenance of an adequate dietary intake of calcium. However, excessive calcium intake is of limited value since absorption of calcium from the gastrointestinal tract is never complete. Dietary calcium is derived from dairy products such as milk, cheese, and eggs, as well as from white bread. Adults on normal diets consume approximately 1000 mg of calcium daily. Of this amount, less than half is absorbed from the gastrointestinal tract, and the rest is lost in the feces and urine. Most women with menopausal osteoporosis excrete normal amounts of calcium into the urine.

Although a cure for menopausal osteoporosis remains elusive, prevention involves ensuring that excessive bone loss does not occur before the onset of menopause. Prevention also involves slowing the rate of bone loss following menopause. Maintenance of physical activity and an adequate diet throughout life, together with estrogen replacement at the onset of menopause, will decrease bone loss in later years.

---

present in androgen-dependent tissues. Progesterone prevents target tissues from responding to the normally low levels of circulating androgens in the female. Progesterone and its derivatives are very useful in treating abnormally heavy hair growth in the female, a condition termed **hirsutism,** which is thought to be due to an overproduction of androgens.

Progesterone, secreted by the corpus luteum, may also promote the edema (salt and fluid retention) that occurs in many women at the end of the luteal phase of the ovarian cycle prior to the onset of menstruation. Progesterone is thought to increase circulating levels of renin, which stimulates aldosterone secretion and results in salt and fluid retention. However, the severity of premenstrual fluid retention does not appear to correlate with circulating levels of renin or aldosterone. The role of ovarian

hormones in controlling fluid and electrolyte levels in the female is not completely understood.

Menopause

Menopause, the cessation of menstruation in the female, occurs on the average at age 50. For several years before the onset of menopause, menstruation occurs less frequently and at variable intervals. As previously discussed, the number of follicles in the ovary declines with age, and hence estrogen secretion declines as well. When all of the follicles have completely disappeared, the ovary stops secreting estrogen. The low circulating levels of estrogen found in postmenopausal women are due to secretion of estrogen precursors by the adrenal glands, which are then converted to estrogen in the periphery.

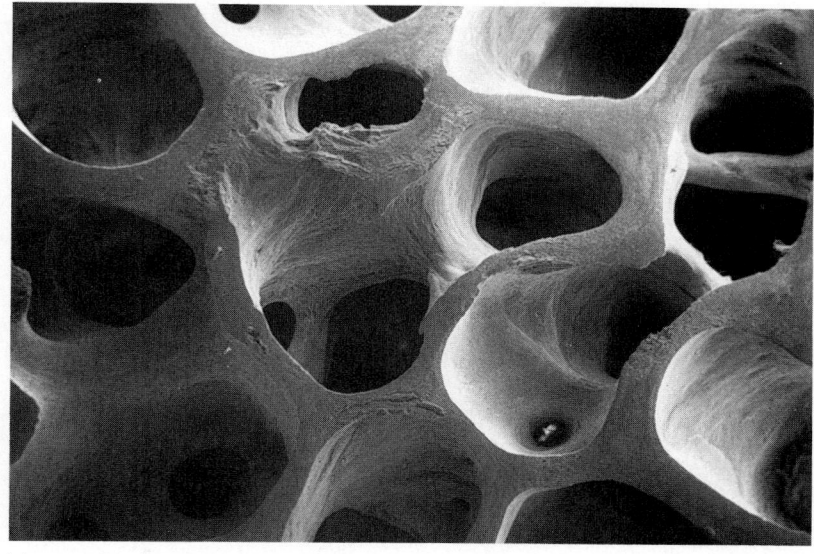

(a)

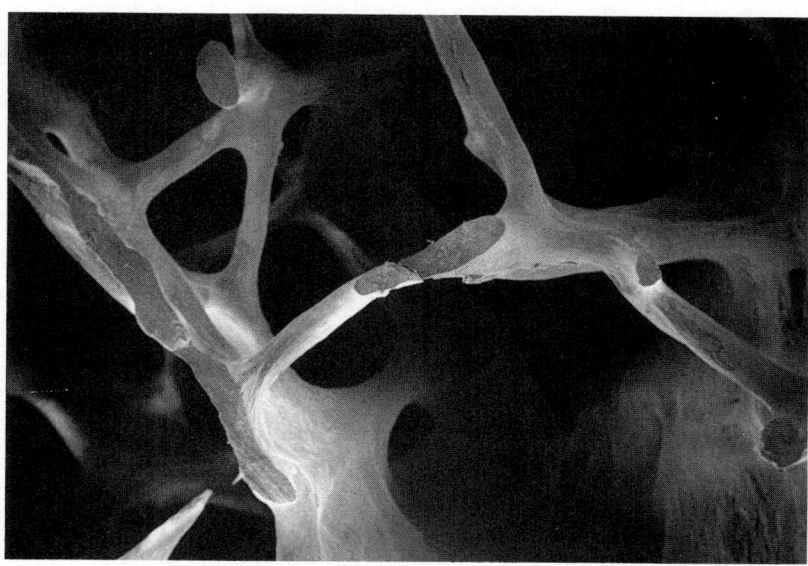

(b)

Figure 30–20
Photograph of (*a*) normal bone matrix and (*b*) erosion of bone matrix which characterizes osteoporosis. (With permission from Dempster et al., J. Bone Min. Res. *1*, 15–21, 1986.)

During menopause, the remaining follicles in the ovary become less sensitive to LH and FSH, and to compensate, blood levels of these pituitary hormones begin to rise. The decline in circulating estrogen during menopause results in a loss of vaginal epithelium, a decrease in breast mass, and an increase in bone loss. Other symptoms that accompany menopause include vascular flushing ("hot flashes"), rapid shifts in mood and emotion, and an increase in coronary vascular disease.

SUMMARY

Reproductive Physiology in the Male

The external reproductive organs in the male are the penis and scrotum. The internal reproductive organs include the testes, epididymis, vas deferens, seminal vesicles, and prostate gland.

Spermatogenesis takes place in seminiferous tubules of the testes. Spermatogenesis begins at puberty and is constantly repeated throughout the reproductive years of the male. During spermatogenesis, germ cells undergo a series of divisions and differentiate from spermatogonia into mature spermatozoa. The Sertoli cells, which form a protective blood-testes barrier around the inside of the seminiferous tubules, ensure that the sperm develop in a protected environment.

Sperm mature in the epididymis and are stored in the vas deferens. The seminal vesicles, prostate gland, and bulbo-urethral gland contribute important fluids that nourish sperm and add volume to the ejaculated semen.

Neuroendocrine control of reproduction in the male involves the secretion of gonadotropin releasing hormone from the hypothalamus, which in turn stimulates the release of luteinizing hormone and follicle-stimulating hormone from the pituitary. LH stimulates the Leydig cells of the testes to secrete testosterone. FSH stimulates spermatogenesis in the testes. Release of GnRH, LH, and FSH is controlled through feedback loops involving both testosterone, secreted by the Leydig cells, and inhibin, secreted by the Sertoli cells within the seminiferous tubules of the testes.

Testosterone stimulates the onset of puberty and promotes spermatogenesis in the testes.

Reproductive Physiology in the Female

The external reproductive organs in the female are the labia majora, the labia minora, and the clitoris. The internal reproductive organs include the vagina, uterus, ovaries, and fallopian tubes. In contrast to the male, the urinary and reproductive tracts in the female are anatomically separate.

Oogenesis takes place in the ovaries. In contrast to the male, germ cell division in the female begins during embryonic development but is then arrested until puberty. Oogenesis begins with the growth of primary follicles in the ovary. Once each month beginning at puberty, one of the primary follicles in the ovaries grows at a rapid rate and differentiates into a mature Graffian follicle containing an ovum.

At midcycle, this mature follicle releases its ovum in a process termed ovulation. Following ovulation, the ovarian follicle becomes the corpus luteum.

Neuroendocrine control of reproduction in the female involves the secretion of GnRH from the hypothalamus and LH and FSH from the pituitary. During the follicular phase of the ovarian cycle, FSH stimulates the growth of ovarian follicles, which in turn secrete estrogen. The luteal phase, which begins following ovulation, is marked by growth of the corpus luteum in response to LH stimulation. The corpus luteum secretes both estrogen and progesterone, which in turn stimulate growth and proliferation of the endometrium of the uterus. As the corpus luteum begins to degenerate at the end of the luteal phase, secretion of estrogen and progesterone diminishes; this results in necrotic changes in the endometrium that culminate in the onset of menstruation.

Estrogen, secreted first by the developing follicles and later by the corpus luteum, has a wide variety of effects on other endocrine systems and on many tissues of the body, including bone.

Menopause, the cessation of menstruation, marks the end of the reproductive years in the female. As ovarian follicles begin to degenerate during menopause, follicular secretion of estrogen diminishes. The decline in circulating levels of estrogen during menopause results in partial atrophy of tissues that depend on estrogen stimulation, including the vaginal epithelium and breasts.

REVIEW QUESTIONS

Identify Each with a Term

1. The process resulting in a reduction of total bone mass or bone density.

2. The tightly coiled tubes in the testes that contain sperm-producing cells.

3. The ovarian follicle after ovulation.

4. The lining of the uterus.

Define Each Term

5. crossing over
6. menopause
7. mitotic division
8. spermatogonia

Choose the Correct Answer

9. Which of the following is *not* an action of Sertoli cells?

    a. synthesis of inhibin
    b. removal of damaged germ cells
    c. synthesis of testosterone
    d. production of seminiferous tubular fluid

10. The acrosome contains:

a. enzymes
b. mitochondria
c. chromatin
d. secondary spermatocytes

11. Sperm become mobile in the:
a. seminiferous tubules
b. epididymis
c. urethra
d. vas deferens

12. At the onset of puberty, the ovary contains the following number of follicles, or germ cells:
a. 400
b. 4,000
c. 400,000
d. 400,000,000

13. Mature sperm are produced in:
a. one day
b. one week
c. one month
d. three months

Answer Each Question in One or Two Sentences

14. Why is the location of the testes in the scrotal compartment important for sperm production?

15. What are the two most important functions of the blood-testes barrier?

16. Why does the endometrium thicken during the luteal phase of the ovarian cycle?

SUGGESTED READINGS

Hedge, G.A. *Clinical Endocrine Physiology*. Philadelphia: W.B. Saunders Co., 1987.

Jacobs, H.S., ed. *Reproductive Endocrinology*. Philadelphia: W.B. Saunders Co., 1985.

Mishell, D.R. and Davajan, V., eds. *Infertility, Contraception and Reproductive Endocrinology*. Medical Economics Books, 1986.

Shearman, R.P. *Clinical Reproductive Endocrinology*. Churchill Livingstone, 1985.

Warshaw, J.B. *The Biological Basis of Reproductive and Developmental Medicine*. New York: Elsevier Science Publishing Co., 1983.

Williams, R.H. *Textbook of Endocrinology*. Philadelphia: W.B. Saunders Co., 1981.

Yen, S.S.C. and Jaffe, R.B., eds. *Reproductive Endocrinology*. Philadelphia: W.B. Saunders Co., 1986.

KEY TERMS

androgen-binding protein (ABP) (p. 982)
corpus luteum (p. 991)
crossing over (p. 980)
dysmenorrhea (p. 997)
endometrium (p. 988)

epididymis (p. 983)
estrogen (p. 991)
fallopian tube (p. 988)
gonadotropin releasing hormone (p. 984)
inhibin (p. 985)

luteinization (p. 994)
meiotic division (p. 980)
menopause (p. 998)
ovary (p. 987)
ovum (p. 987)
progesterone (p. 991)

puberty (p. 986)
spermatogenesis (p. 978)
testis (p. 977)
vas deferens (p. 983)

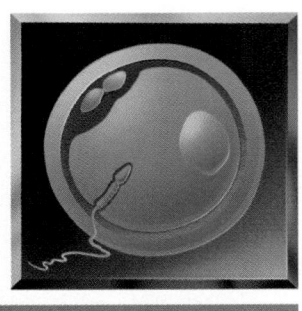

C H A P T E R
31

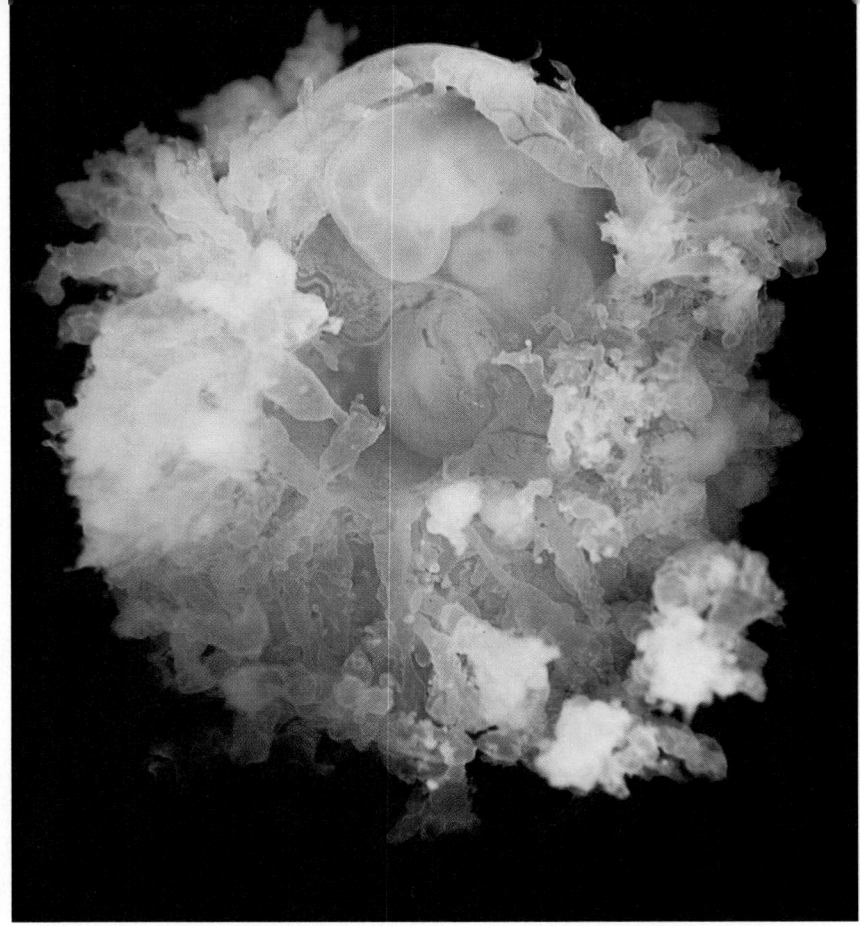

Pregnancy, Fetal Development, and Lactation

In Chapter 30, we investigated the cyclic hormonal and physiological changes that occur in the female, culminating in the release of an ovum once each month during the years between puberty and menopause. We also examined the processes that control production and expulsion of sperm in the male. We will now investigate the processes of fertilization of the ovum and maturation and development of the embryo, as well as the physiological changes that occur in the female during pregnancy, parturition (childbirth), and lactation (milk production and release).

The Process of Fertilization

Because both sperm and ova are viable only for very short periods of time, fertilization of an ovum will occur only when sperm are deposited in the vagina close to the time of ovulation. Following ejaculation, sperm normally retain their capacity to fertilize an ovum for approximately 24 to 72 hours in the female reproductive tract, although longer periods of sperm viability have been reported. Following ovulation, the ovum is "receptive" to fertilization for approximately 10 to 15 hours. Therefore, in order

for the ovum to be fertilized, sperm must be present in the female reproductive tract during the 72 hours before or the 15 hours after ovulation. This restricted time interval may make the odds against fertilization appear rather high. However, one out of every four women becomes pregnant after only one month of repeated intercourse without contraception.

Transport of the Ovum to the Fallopian Tube and Uterus

As described in Chapter 30, ovulation occurs in response to stimulation from high blood levels of LH and FSH approximately 14 days after the beginning of the previous menstrual cycle. Except in rare instances, ovulation occurs from only one of the two ovaries each month.

During ovulation, a single ovum is released from an ovary into the abdominal cavity at a point directly below the open end of the fallopian tube (Fig. 31–1). The open end of the fallopian tube is composed of long, fingerlike projections called **fimbriae.** The fimbriae and fallopian tubes are lined with **cilia,** which move or sway in a rhythmic pat-

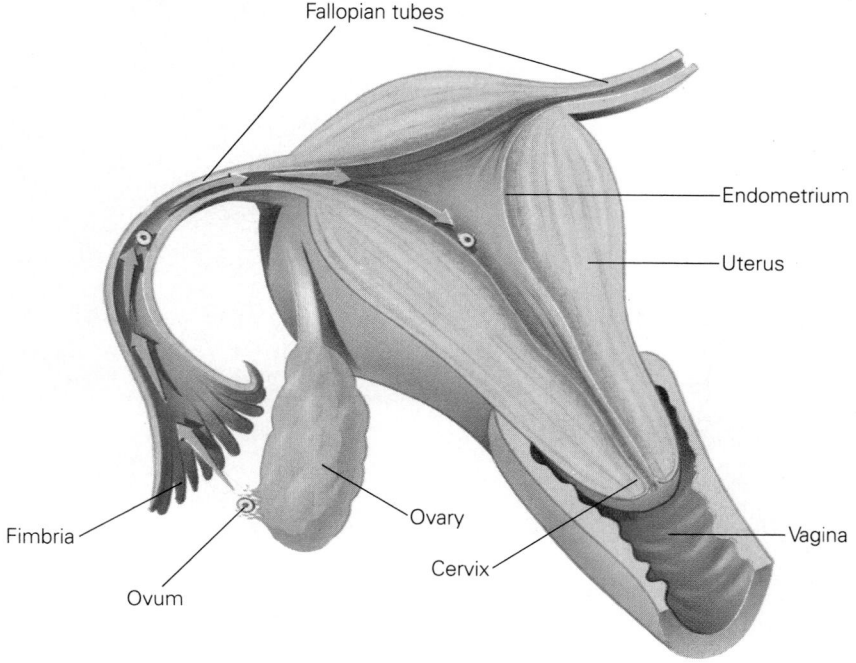

Fallopian tubes

Endometrium

Uterus

Fimbria

Ovary

Ovum

Cervix

Vagina

tern toward the uterus and induce wavelike motions on the inner surface of the fallopian tube. The uterus and cervix are also lined with cilia, but not as many as are found in the fallopian tubes. Following ovulation, the ovum is drawn from the abdominal cavity into the fallopian tube by the sweeping motion of the cilia (see Fig. 31–1).

After entering the fallopian tube, the ovum is transported to the uterus by the wavelike motion of the cilia. The ovum takes approximately 3 to 5 days to travel from the ovary through the fallopian tube to the uterus. Because the ovum is receptive to fertilization only for approximately 10 to 15 hours following ovulation, fertilization normally takes place in the fallopian tube before the ovum has reached the uterus.

Transport of Sperm to the Fallopian Tube

Following ejaculation into the vagina, sperm are transported by way of the vagina, cervix, and uterus into the fallopian tubes (Fig. 31–2a). The time re-

quired for sperm to reach the fallopian tubes is not precisely known, but it is thought to be a few hours, with some sperm reaching the fallopian tubes within minutes following ejaculation.

Sperm are propelled through the female reproductive tract toward the fimbrial ends of the fallopian tubes by wavelike movements in the tail portion of the sperm and by contractions of the muscles of the uterus and fallopian tubes (Fig. 31–2b). In the female, vaginal and cervical stimulation during intercourse results in the release of **oxytocin** from the pituitary, which in turn may stimulate uterine contractions during intercourse. In addition, seminal fluid contains a high concentration of **prostaglandins,** which also act to stimulate contractions of the uterus.

Cervical Mucus and Sperm Transport

The ease with which sperm can migrate through the cervix into the uterus is partially determined by secretions of the cervix. The cervix contains glands that secrete a mucus composed of glycoproteins, salts, and water. The volume and consistency of the cervical mucus change during the ovarian cycle in

Figure 31–2
(*a*) Pathway of sperm through the vagina, cervix, and uterus into the fallopian tubes. (*b*) A sperm cell migrates through the fallopian tube towards the ovum. (© Lennart Nilsson from *Behold Man*, published by Little, Brown and Company.)

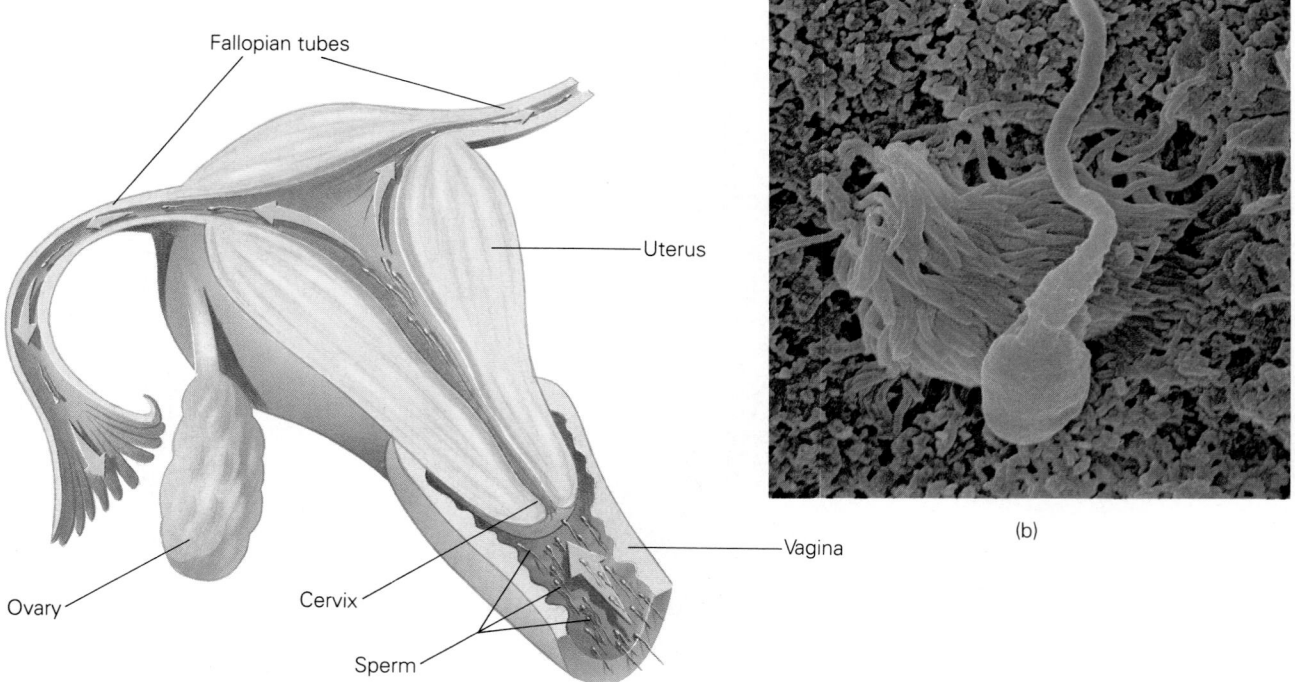

response to changes in the circulating levels of estrogen and progesterone.

Cervical Mucus Secretion at Midcycle.

As was discussed in Chapter 30, estrogen secretion by the ovary increases at midcycle just prior to ovulation. Estrogen stimulates the cervix to secrete large amounts of mucus, which is clear, watery, and nonviscous. Upon estrogen stimulation, the glycoproteins in the cervical mucus assemble to form elongated fibers arranged in channels that allow sperm to penetrate the mucus and pass upward through the cervix (Fig. 31–3). These changes in the fluidity and glycoprotein structure of cervical mucus at midcycle facilitate sperm migration through the cervix into the uterus at a time that coincides with ovulation. Note that many structurally abnormal sperm do not penetrate the cervical mucus at midcycle. Whether this cervical exclusion of abnormal sperm is due to decreased motility of the sperm or to immunological factors present in the cervical mucus is not known.

Cervical Mucus Secretion before and after Midcycle.

Prior to and following ovulation, circulating levels of progesterone are higher than at midcycle. **Progesterone** stimulates the cervix to secrete a thick, viscous, sticky mucus that lacks glycoprotein channels. This mucus plugs the cervical opening into the uterus and acts as a barrier that impedes sperm migration into the uterus at times other than at midcycle. As we will see in Chapter 32, the efficacy of many oral contraceptives containing steroids is due in part to steroid-induced alterations of the physical and chemical properties of cervical mucus.

Sperm Loss in the Female

Although the number of sperm found in ejaculated semen depends on many factors—including the quantity of the ejaculate and the time elapsed since the previous ejaculation—an average of 100 to 500 million sperm are deposited in the vagina during intercourse. Since only one sperm can fertilize the ovum, this number may seem much larger than is necessary to ensure fertilization. However, of this enormous number, only a few thousand sperm reach the fallopian tubes, and only a few hundred reach the vicinity of the ovum.

Sperm loss is high for many reasons. Up to 50% of the sperm in a normal ejaculate may be incapable of fertilization because of abnormal shape or reduced motility. When deposited into the vagina, some sperm are destroyed by the acidic secretions of the vagina. Upon reaching the cervix, sperm entry into the uterus may be blocked by the cervical mucus at certain times in the ovarian cycle, as we have previously seen. Of the sperm that manage to reach the uterus, about half will enter the fallopian tube that does not contain an ovum.

Although these factors result in a large reduction in the number of sperm that reach the ovum, only one sperm is required for fertilization of the ovum.

Figure 31–3
Photograph of the structure of human cervical mucus, taken with a transmission electron microscope. At midpoint of the ovarian cycle, coinciding with the time of ovulation, large channels are formed that permit sperm migration through the cervix. (×26,000.) (© David Phillips/Visuals Unlimited.)

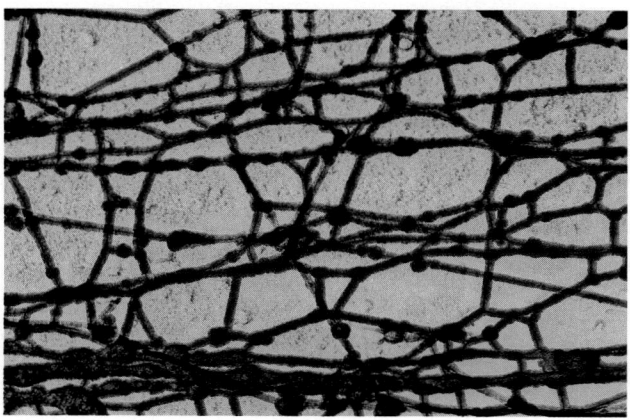

Capacitation and Activation of Sperm

As stated in Chapter 30, sperm are not mobile when they leave the seminiferous tubules of the testes and enter the epididymis. It is in the epididymis that sperm mature and acquire the ability to move. However, mature sperm from the epididymis are not capable of fertilizing an ovum. Sperm must first undergo a process termed **capacitation.** Although capacitation is not well understood, it appears to involve at least two steps. First, a physiological change in the plasma membrane of the sperm takes place that allows the sperm to penetrate the surface membrane of the ovum. Secondly, the movements of the tail of the sperm become faster and more pronounced, resulting in an increase in sperm motility. Capacitation normally takes place when the sperm are in the female reproductive tract, before the fusion of sperm and ovum.

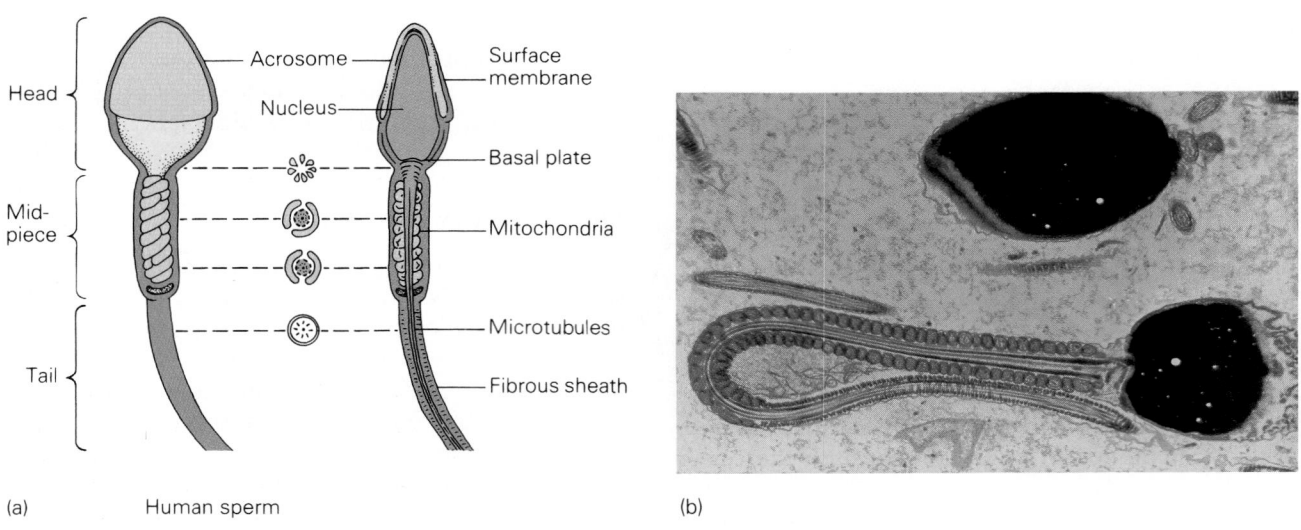

(a)   Human sperm   (b)

Figure 31–4
(*a*) Structure of a human sperm. (Redrawn from Warsaw, J.B., Ed., *The Biological Basis of Reproductive and Developmental Medicine*, 1983.) (*b*) Transmission electron micrograph of a human sperm showing the nucleus, acrosome, midpiece, mitochondria, and flagella. (© David M. Phillips/Visuals Unlimited.)

The Fusion of Sperm and Ovum

Before we examine the process of sperm-ovum fusion, let us review the anatomy of the mature sperm and ovum. Figure 31–4*a* illustrates the general structure of a mature sperm (also see Fig. 30–6). The head of the sperm is composed of a dense, compact nucleus, an acrosome that caps the nucleus, and a surface membrane. The **acrosomal portion** of the sperm head contains enzymes that play an important role in the process of fertilization. The midpiece of the sperm contains spirals of mitochondria, while the tail of the sperm is composed of microtubules surrounded by a fibrous sheath (Fig. 31–4*b*).

The organization of the mature human ovum prior to fertilization is illustrated in Figure 31–5. The mature ovum is surrounded by the **vitelline membrane.** On top of the vitelline membrane is the vitelline space, which separates the vitelline membrane from the next cell layer, the zona pellucida. The **zona pellucida** is composed of a layer of cells that form a jellylike coating around the ovum. The zona

Figure 31–5
Organization of the mature human ovum prior to fertilization.

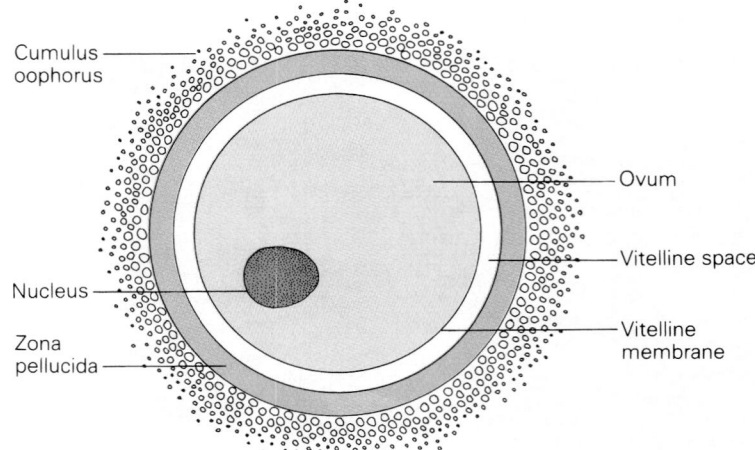

pellucida itself is surrounded by several layers of granulosa cells called the **cumulus oophorus.**

The Acrosomal Reaction

The term *fertilization* refers to the fusion of the sperm and ovum. The fertilized ovum, or zygote, is the first cell of a new organism to divide and eventually differentiate to become an embryo. However, before the sperm can fertilize the ovum, the head of the sperm must undergo a controlled disruption termed the **acrosome reaction,** (Fig. 31–6a) which allows the sperm to penetrate the cumulus oophorus and the zona pellucida of the ovum. The acrosome reaction (Fig. 31–6) involves a series of fusions between the outer membranes of the acrosome, resulting in the formation of channels in the head of the sperm through which proteolytic acrosomal enzymes are released. These enzymes create a pathway through the cumulus oophorus and the zona pellucida, which surround the ovum (Fig. 31–6b).

The Incorporation of the Sperm by the Ovum

Very soon after penetrating the zona pellucida, the head of the fertilizing sperm attaches to and fuses with the vitelline membrane of the ovum. At this point, the tail of the sperm stops moving, and the ovum gradually engulfs the sperm in a process that resembles phagocytosis. Therefore, after reaching the inner portion of the zona pellucida the sperm does not "penetrate" further into the ovum but rather is "incorporated" into it. Figure 31–7 illustrates the initial meeting of sperm and ovum.

Prevention of Polyspermy by the Ovum

Fertilization of the ovum by more than one sperm, a condition called **polyspermy,** is lethal for the fertilized ovum. The prevention of polyspermy is therefore very important for reproduction. Although many sperm may attach themselves to the surface of the ovum, normally only one sperm penetrates to the inner portion of the zona pellucida surrounding the ovum.

The ovum has efficient ways of blocking the entry of more than one sperm. First, when a sperm attaches to the ovum, there is thought to be a brief electrical block, lasting about 30 seconds, on the surface of the ovum. In addition, a large structural change in the ovum occurs shortly after sperm-ovum fusion. Fusion of the sperm and ovum results in the exocytosis of cortical granules from the periphery of the ovum into the space between the

Figure 31–6
(*a*) Successive stages of the acrosome reaction of the human sperm (Redrawn from *Atlas of Fine Structure of Human Sperm Penetration, Eggs, and Embryos Cultured in Vitro*, Sathananthan et al., 1986.) (*b*) The cap of the sperm gradually dissolves during the acrosome reaction, allowing enzymes to be released, which help the sperm to penetrate the ovum. (© Lennart Nilsson from *A Child Is Born*, Dell Publishing Company.)

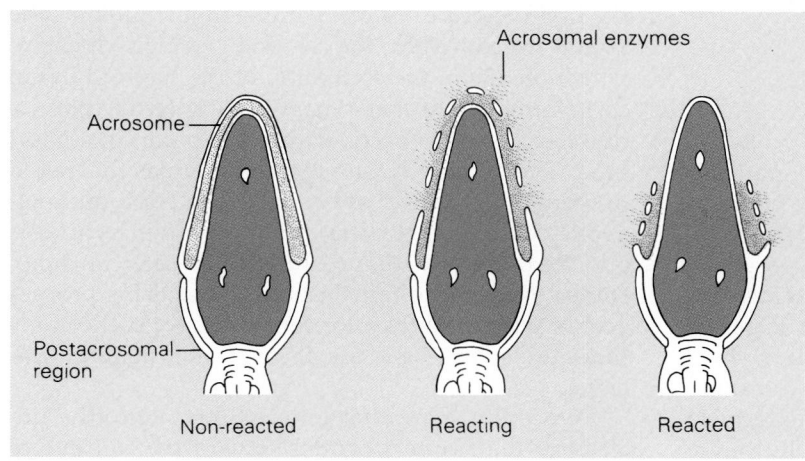

(a)

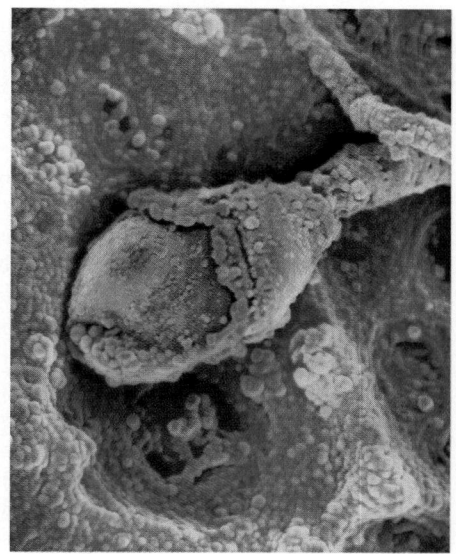

(b)

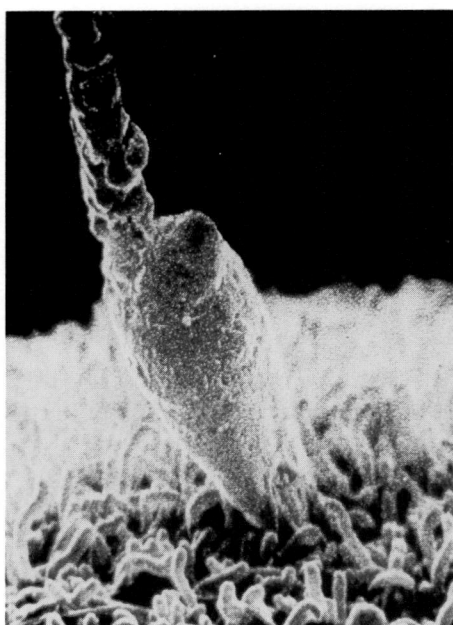

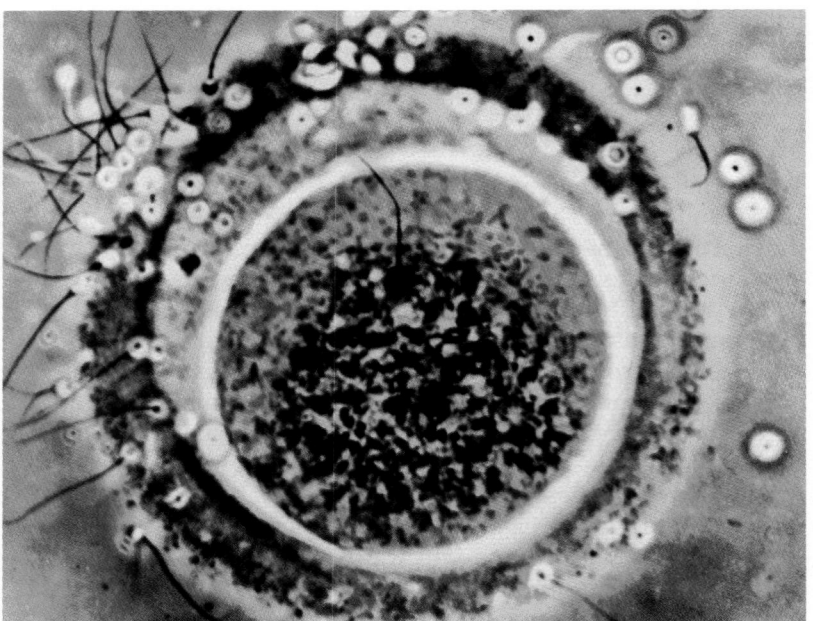

Figure 31–7
Photograph of a sperm beginning to penetrate an ovum. (© David M. Phillips/Visuals Unlimited.)

Figure 31–8
Fertilization of the human ovum. Note that as one sperm enters the ovum, the entrance of all other sperm is blocked, which prevents polyspermy. Phase contrast microscopy (original magnification: ×400.) (© Dr. Landrum B. Shettles.)

plasma membrane and the zona pellucida. These granules alter the surface membrane of the ovum, making penetration by additional sperm more difficult. In addition, the cortical granules released from the ovum during fertilization alter the binding sites for sperm on the surface of the zona pellucida and the plasma membrane of the ovum. These processes prevent additional sperm from entering the zona pellucida to fuse with the ovum. As Figure 31–8 illustrates, after one sperm enters the inner portion of the zona pellucida, all other sperm are virtually stopped before they reach the inner zona pellucida.

Multiple Births: Fertilization of More Than One Ovum

Normally only one ovum is released at midcycle each month. However, occasionally two or more ova may be released during ovulation. Fertilization of each of these ova by individual sperm results in the formation of **dizygotic** (or **fraternal**) **twins,** triplets, or quadruplets. Dizygotic siblings are as genetically different from each other as any other siblings. **Monozygotic** (or **identical**) **twins,** triplets, or quadruplets, on the other hand, result from the fertilization of a single ovum by a single sperm followed by a special type of cell division that results in a single

duplication or multiple duplications of the fertilized ovum prior to implantation. Monozygotic siblings are genetically identical. Multiple births may be monozygotic, dizygotic, or a combination of both.

Cell Division and Implantation Following Fertilization

As Chapter 30 discussed, when the ovum is released from the follicle at midcycle, it has completed the first meiotic division. Fertilization stimulates the ovum to complete the second meiotic division, which results in the formation of the **haploid ovum** containing 23 chromosomes. During fertilization, a number of events occur within the ovum in a short time. The ovum releases cortical granules that act to prevent polyspermy, as described. In addition, metabolic changes occur that initiate protein synthesis and the transformation of the haploid set of chromosomes to form the **female pronucleus.** This process occurs within hours after fertilization. At the same time, the sperm head swells to form a **male pronucleus.**

Once the sperm has entered the ovum, the nucleus of the sperm migrates toward the nucleus of the ovum, and the paternal and maternal nuclei fuse. The two sets of unpaired chromosomes, 23 from the ovum and 23 from the sperm, align themselves to form a complete set of 46 chromosomes. This alignment is such that each parent contributes half of the genetic material that will determine the characteristics of the embryo.

Stages in Early Embryonic Development

The Zygote

Initially, the fertilized ovum is a single cell containing 46 chromosomes and is called the **zygote.** The zygote begins to divide mitotically about 30 hours after fertilization. Cell division, which begins prior to implantation, marks the beginning of embryonic development. Note that embryonic cells are the only cells, other than cancer cells, that divide on their own initiative. All other cells divide only in response to a need for tissue replacement. The signal for division of embryonic cells is contained in the genetic material of the zygote.

The zygote first divides into two cells, then these two cells divide into four cells, and so on as illustrated in Figure 31–9. As cell division progresses, the size of each individual cell decreases. During mitotic division, each of the 46 chromosomes in the dividing cell duplicates itself before division, which is why all the cells of the body contain the same 46 chromosomes.

When the zygote is composed of 8 or more cells, it is called a **morula** (see Fig. 31–9). As the morula continues to divide, a hollow area begins to form in the center of the cell mass. When the hollow inner portion becomes filled with fluid, the morula is called a **blastocyst** (see Fig. 31–9). Note that the cell divisions resulting in the formation of the zygote, morula, and blastocyst occur while the fertilized ovum is being transported through the fallopian tube.

The Blastocyst

The blastocyst is the precursor of the embryo. Blastocysts are composed of two major cell types. The outer cells, which maintain the fluid-filled cavity of the blastocyst, are called **trophoblasts.** The trophoblasts are tightly connected to each other by membranes and hence form a protective shell around the blastocyst. Trophoblasts are involved in the initiation of implantation. These cells are sticky and adhere to the endometrium of the uterus. The trophoblasts will eventually differentiate into the placenta.

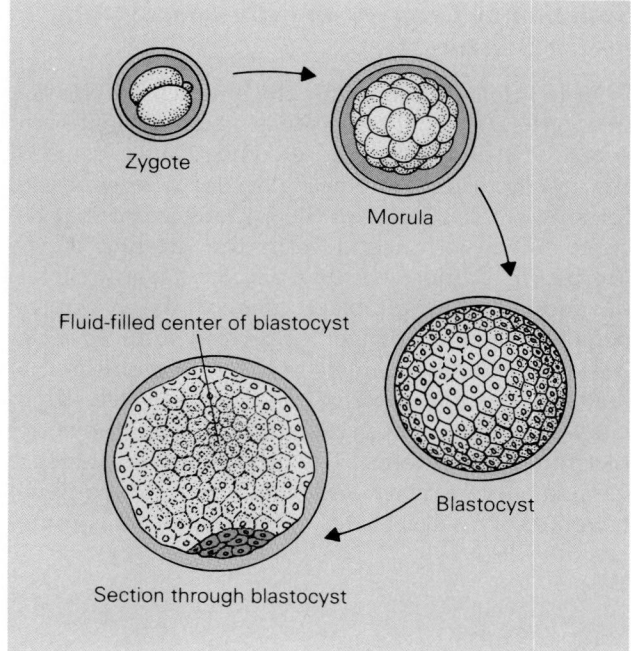

Figure 31–9
Stages in the early development of the fertilized ovum. (Redrawn from Masters, Johnson, and Kolodny, *Human Sexuality*, Little, Brown, 1985.)

The internal cells of the blastocyst are called **embryoblasts** because they give rise to the embryo itself. Hence, both the placenta and the embryo arise from cells of the fertilized ovum.

Implantation of the Blastocyst

As stated earlier, the fertilized ovum, while undergoing cell division, is transported through the fallopian tube to the uterus over the course of approximately 3 to 5 days. By the time the fertilized ovum has reached the uterus, it has undergone many cell divisions and has reached the blastocyst stage of development. The blastocyst is moved by contractions of the fallopian tube and wavelike motions of the cilia through the fallopian tube and into the uterus, where it remains and undergoes continuous cell divisions for an additional 2 to 4 days before becoming implanted in the endometrium. Once the blastocyst has fully implanted in the endometrium, at approximately 2 weeks after fertilization, it is called an **embryo.** After the second month of development, the embryo is called a **fetus.**

Secretion of Enzymes and Hormones by the Blastocyst

Prior to implantation, while the blastocyst is floating free in the uterus, it begins to secrete hormones and proteolytic enzymes. The enzymes cause some of the neighboring endometrial cells to degenerate, forming small cavities in the endometrium that become filled with blood from the arteries of the uterus. In addition, within 6 or 7 days following fertilization, the trophoblast cells of the blastocyst begin to secrete estrogen. Estrogen induces localized permeability changes in the endometrium that are necessary for implantation. Estrogen also stimulates the secretion of prostaglandins from the endometrium of the uterus. The prostaglandins increase vascular permeability and swelling of the endometrium. As we will see later, prostaglandins also stimulate the maternal pituitary gland and corpus luteum to secrete hormones necessary for the continuation of pregnancy.

Formation of Embryonic Membranes

During implantation, the trophoblast cells form microvilli on the outer surface of the blastocyst. These microvilli begin to invade the endometrium of the uterus. The trophoblasts penetrate to the level of the first muscle layer, and the blastocyst becomes buried in the endometrium. The trophoblasts differentiate into two cell layers that will form the amnion and the chorion. The amnion is a tough extraembryonic membrane that surrounds and protects the developing embryo. The chorion, which is the outermost extraembryonic membrane, gives rise to the placenta (Fig. 31–10).

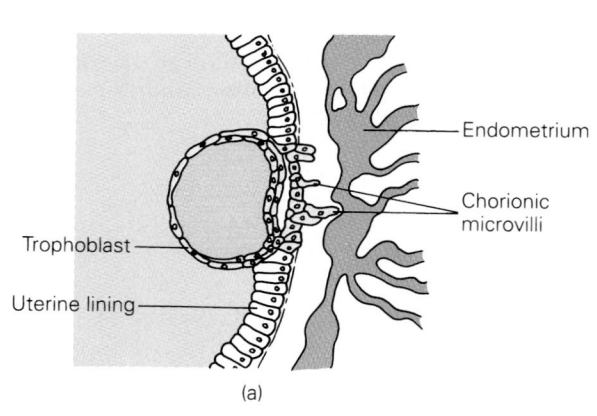

(a)

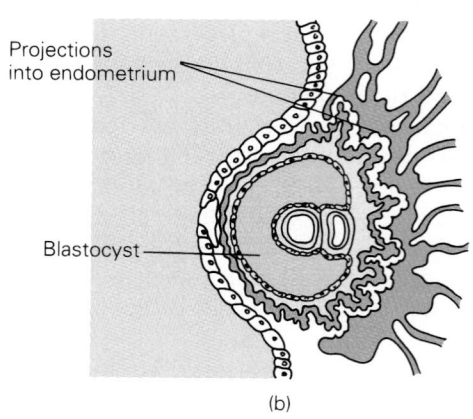

(b)

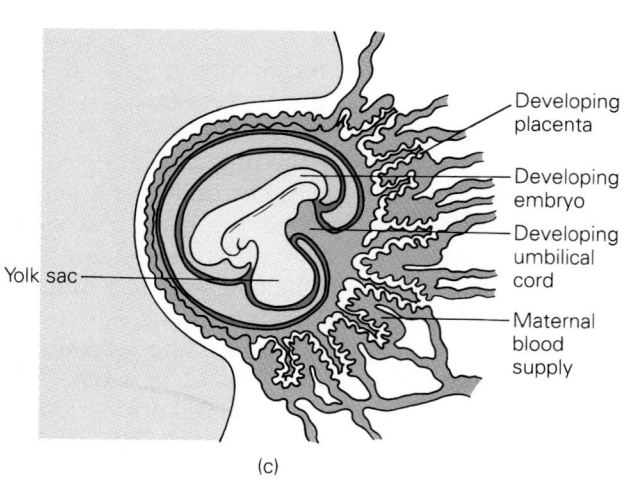

(c)

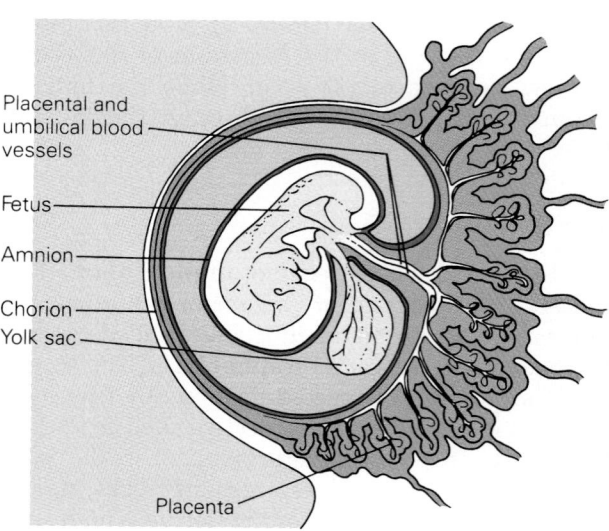

(d)

Secretion of Progesterone by the Corpus Luteum

Implantation is completed around day 7 following ovulation, which corresponds to day 21 of the ovarian cycle. As we learned in Chapter 30, on day 21 the corpus luteum is secreting large amounts of progesterone. In response to progesterone stimulation, the endometrial cells of the uterus have proliferated, and the endometrium has become thick and vascular in preparation for receiving and nourishing the blastocyst. Progesterone not only prepares the uterus for implantation but also helps stimulate cell division of the fertilized ovum. The heightened secretion of progesterone following ovulation is an example of the importance of timing in the reproductive cycle.

Abnormal Implantation

Implantation is not accompanied by any physical sensations, so it is impossible for the female to know when it has occurred. Occasionally, transport of the fertilized ovum from the fallopian tube to the uterus is blocked or slowed down, and the fertilized ovum does not reach the uterus and instead becomes implanted in the fallopian tube. Abnormal ovum transport can result from anatomical abnormalities, from the presence of scar tissue in the fallopian tube following infections or surgery, or from tumors located in the fallopian tube. In rare cases, the fertilized ovum may be expelled from the fallopian tube back into the abdominal cavity and may become implanted there. Both of these cases are called **ectopic pregnancies;** the term *ectopic* refers to implantation outside of the uterus.

◀ *Figure 31–10*

Sequential development of the human placenta. (*a*) Initial implantation of the blastocyst in the wall of the uterus. (*b*) Blastocystic invasion of the endometrium. (*c*) Development of the placenta and the umbilical cord, which connects the fetus and the endometrium. (*d*) Formation of the blood vessels of the placenta and the umbilical cord. (Adapted from "The placenta," Beaconsfield, Birdwood, and Beaconsfield, *Scientific American*, 243:94, 1980.) (*e*) On the twelfth day following fertilization, the blastocyst is securely anchored in the endometrium. (© Lennart Nilsson from *A Child Is Born*, Dell Publishing Company.) (*f*) The embryo at 4 to 5 weeks of development. The chorion, which is shaggy in appearance, and the amniotic sac have been opened to visualize the embryo. (© Lennart Nilsson from *A Child Is Born*, Dell Publishing Company.)

(e)

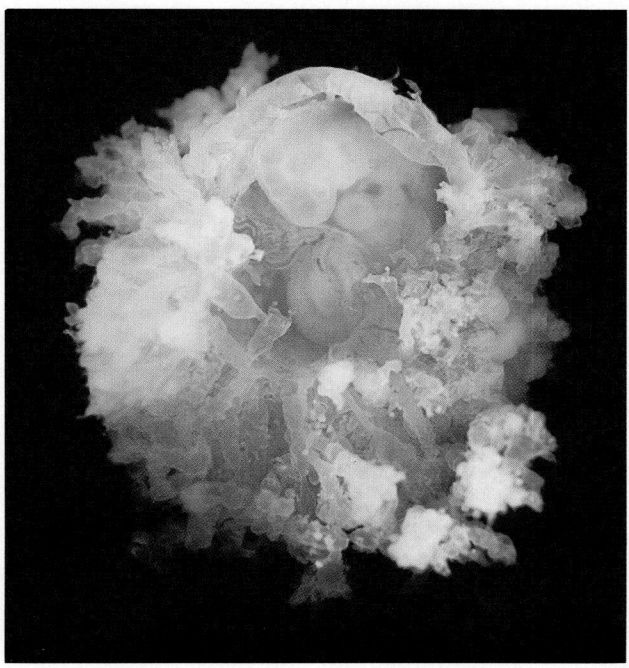

(f)

Ectopic pregnancy occurs in approximately one out of every hundred pregnancies. About 97% of ectopic pregnancies occur in the fallopian tubes and are termed **tubal pregnancies.** Tubal pregnancies do not continue to term because there is not enough space in the fallopian tube for fetal growth. Most tubal pregnancies abort at a relatively early stage. When they do not, surgical removal may be necessary in order to reduce the risk of tubal rupture and bleeding. In contrast, implantation of the fertilized ovum in the abdomen may proceed to term, in which case delivery is accomplished surgically.

The Role of the Placenta

Following implantation, the embryo is no longer independent. From this point on, it must take everything it needs—including oxygen, nutrients, salts, and fluid—from the maternal circulation. The embryo also depends on the maternal circulation for removal of its waste products.

The needs of the embryo, and later of the fetus, are met by the **placenta,** which delivers oxygen and nourishment, absorbs and removes waste products, delivers immunity to various diseases, and secretes hormones necessary for both the maintenance of pregnancy and the expulsion of the fetus at term (Table 31–1).

Note that while the fetus is totally dependent on the placenta for survival, the placenta is not dependent on the presence of the fetus for survival. The placenta can continue to function in the uterus even after a fetus dies or is removed. However, the placenta is normally expelled along with the fetus during delivery and hence is the only organ of the body with a life span of less than one year.

Table 31–1
Functions of the Placenta

Exchange of gases between fetus and mother

Delivery of nutrients from mother to fetus

Delivery of antibodies from mother to fetus

Removal of fetal waste

Secretion of hormones including human chorionic gonadotropin (hCG), progesterone, estrogen, and human chorionic somatomammotropin (hCS)

Development of the Placenta

The **placenta** is a flat organ that is connected to the uterus on one side and to the embryo, via the umbilical cord, on the other side. The **umbilical cord** is formed from trophoblast cells that surround the blastocyst. The placenta is formed from trophoblast cells and from membranes, called **decidua,** that arise from the endometrium. When the blastocyst implants in the endometrium, the tissue and blood vessels of the endometrium break down at the site of implantation to form small spaces in the endometrium that fill with maternal blood. The developing embryo sends out rootlike villi into the pools of maternal blood. In the first few weeks following implantation, the villi are composed of thick columns of cells that contain capillaries. These capillaries serve as the route for exchange of nutrients from the mother to the fetus and of waste products from the fetus to the mother. These small vessels become the arteries and veins of the umbilical cord and placenta. Within 5 weeks after implantation, the vasculature of the umbilical cord and placenta is well established. Figure 31–10 illustrates the different stages in the development of the placenta.

The placenta is heavily supplied with fetal blood vessels on one side and with maternal blood vessels on the other side, but the two blood supplies normally never mix. The placenta anatomically separates the circulatory systems of the fetus and mother and acts as an **interface** between these two circulatory systems. As shown in Figure 31–11*a*, blood from the mother enters the placenta through the uterine artery, perfuses the placenta, and exits through the uterine vein. Fetal blood enters the placenta through the umbilical artery, perfuses the placenta, and exits through the umbilical vein. As illustrated in Figure 31–11*a*, the side of the placenta that faces the fetus is smooth and highly vascularized, with membranes that surround the fetal arteries and veins. The maternal side of the placenta, on the other hand, is divided into thick, highly vascularized lobes that provide a very large surface area for exchange of materials between the fetal and maternal blood supplies. The exchange of all materials between the fetus and the mother takes place through the placenta.

Exchange of Materials Through the Placenta

Placental Exchange of Gases

Oxygen diffuses from the maternal blood into the placenta, where it is picked up by the fetal blood and carried through the umbilical cord to the fetus

(Fig. 31–11b). Fetal blood is capable of carrying more oxygen than maternal blood, for at least two reasons. First, the concentration of hemoglobin is 50% higher in fetal blood than in maternal blood. Second, fetal hemoglobin can carry 20% to 30% more oxygen than maternal blood.

Carbon dioxide, which is continually formed in fetal tissues, enters the fetal blood and is transported to the placenta, where it diffuses into maternal circulation.

Placental Delivery of Nutrients

Some substances that the fetus needs pass from maternal blood to fetal blood by simple diffusion in the placenta. These substances include glucose, sodium, potassium, and chloride. Other materials, such as proteins, calcium, and iron, are actively transported across the placental membranes.

The fetus grows at a more rapid rate than any other tissue in the adult body. To sustain this rapid growth, the fetus must be supplied with a variety of structural proteins that will be incorporated into fetal tissues. Some of these intact protein molecules are transported by the placental membranes from the maternal circulation to the fetal circulation and are made available to the developing fetus. However, not all of the proteins needed by the fetus are available in maternal blood. The placenta extracts amino acids and enzymes from the maternal blood and synthesizes additional proteins, which are then transported into the fetal circulation and delivered to the developing fetus. In fact, the placenta synthesizes proteins at a higher rate than any other organ of the body. Although the demands of the developing fetus may vary throughout gestation, on average the placenta uses approximately one third of all the oxygen and glucose that the maternal blood sup-

Figure 31–11
(a) Cross-section of placental tissue illustrating the directions of fetal and maternal blood flow. (Adapted from "The placenta," Beaconsfield, Birdwood, and Beaconsfield, *Scientific American*, 243:100, 1980) (b) Fetal blood cells in a U-shaped capillary protruding from the placenta. The fetal blood cells (blue) are separated from the maternal blood cells (red) by a thin membrane that transfers oxygen and nutrients from the maternal blood to the fetal blood. As fetal blood cells are oxygenated they turn from blue to red. (© Lennart Nilsson from *A Child Is Born*, Dell Publishing Company.)

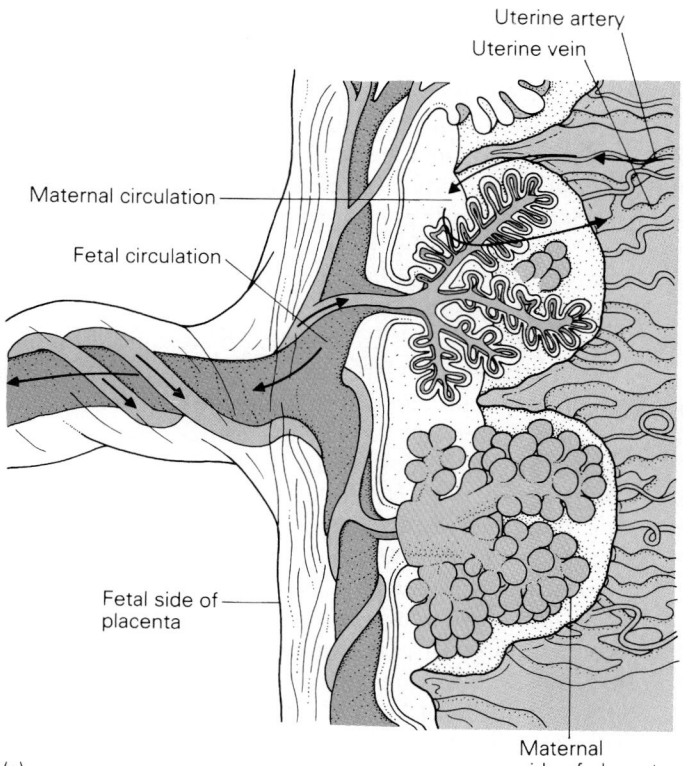
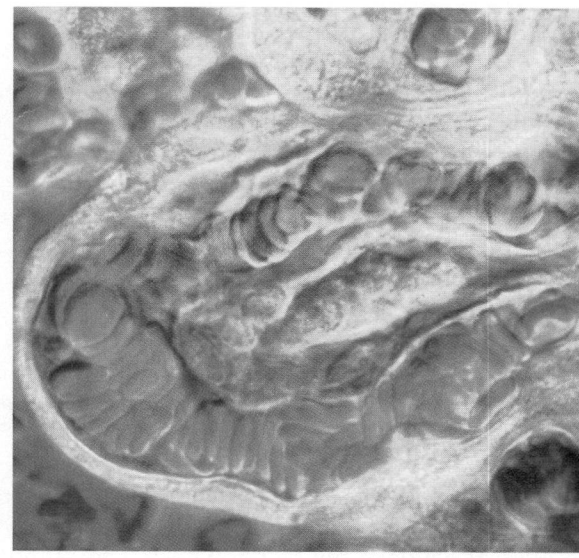

plies in order to meet the metabolic needs of the fetus and the placenta itself.

Placental Removal of Excretory Products

Fetal waste products, including urea, uric acid, and nonprotein nitrogens, pass by simple diffusion from fetal blood into maternal blood in the placenta. These fetal waste products are then excreted by the mother along with her own wastes.

Placental Transport of Drugs

While the efficiency of the transport system from mother to fetus is essential for nutrient delivery to the fetus, this same transport system exposes the fetus to many of the drugs and pollutants that the pregnant woman is exposed to. Such exposure is particularly dangerous during the first three months of pregnancy, when fetal tissue is still differentiating and is especially susceptible to damage. Alcohol, aspirin, and the chemicals contained in tobacco can reach the fetus via the placenta and influence its growth and development. Perhaps the most striking example of this transport of drugs to the fetus is seen in the offspring of women who use heroin during pregnancy. These babies experience severe heroin withdrawal symptoms after birth.

Placental Transport of Antibodies

Another function of the placenta is to provide the fetus with maternal antibodies. The mother has formed these antibodies during her contact with a variety of different antigens. Antibodies are too large to diffuse across the placenta, but the placenta has highly selective active transport systems that convey these large antibodies to the fetus. In this way, the newborn infant is protected against a variety of infections until its immune system can begin to function on its own.

Protection of the Placenta and Embryo from Immunological Rejection

Embryonic and placental tissue (which arise from the embryo) are immunologically foreign to the mother and hence might be expected to stimulate the synthesis of antibodies by the maternal system and induce tissue rejection of the embryo. This does not occur for a variety of reasons. First, during the early stages of development the blastocyst does not express antigens that would stimulate the production of antibodies by the mother. Second, the pla-

centa secretes many hormones, such as **human chorionic gonadotropin (hCG),** that suppress lymphocytes, the cells that induce tissue rejection. Third, the embryo is physically protected from the maternal system by membranes that surround it, and, as noted, the fetal and maternal blood supplies do not mix.

Hormones of the Placenta

During pregnancy, the placenta secretes large amounts of four important hormones that regulate the physiological changes associated with early pregnancy: human chorionic gonadotropin (hCG), estrogen, progesterone, and human chorionic somatomammotropin (hCS), also called human placental lactogen (hPL).

Placental Secretion of Human Chorionic Gonadotropin

Effects of hCG on Endometrial Growth. As discussed in Chapter 30, in the absence of fertilization of the ovum, the corpus luteum secretes progesterone for about 7 days following ovulation, and then progesterone levels fall as the corpus luteum degenerates. The rise in progesterone, secreted by the corpus luteum, stimulates growth of the endometrium of the uterus, and the fall in progesterone following degeneration of the corpus luteum results in necrosis of the endometrium and the onset of menstruation.

If these events were to occur following fertilization of the ovum, the fertilized ovum would be expelled from the uterus along with the menstrual flow. These events are prevented by the secretion of **human chorionic gonadotropin** (Table 31–2). hCG is a glycoprotein, which has a molecular structure very similar to that of luteinizing hormone (secreted by the pituitary). hCG is secreted first by trophoblast cells of the blastocyst before implantation in the endometrium and later by the placenta. hCG diffuses from fetal blood into maternal circulation and elevates maternal plasma hCG content as shown in Figure 31–12. hCG stimulates the corpus luteum of the pregnant woman and prevents the degeneration and involution of the corpus luteum that normally occur during the ovarian cycle, thus ensuring the continued secretion of progesterone. In response to hCG, the corpus luteum doubles in size during the first 2 months of pregnancy. As pregnancy continues, hCG stimulates the corpus luteum to secrete large amounts of both estrogen and progesterone. These two hormones stimulate continued growth of

Table 31–2

Physiological Functions of Human Chorionic Gonadotropin

Prevents degeneration of the corpus luteum

Stimulates the corpus luteum to secrete estrogen and progesterone, which in turn stimulate continued growth of endometrium

Stimulates steroid synthesis in the developing fetal adrenals

Stimulates fetal gonads, especially testosterone production by the fetal testes

Suppresses maternal lymphocytes and reduces the possibility of immunorejection of the fetus

the endometrium, which is necessary for adequate development of placental and fetal tissue. If the corpus luteum is removed during the first 7 to 10 weeks of pregnancy, spontaneous abortion will occur because progesterone and estrogen are no longer stimulating the endometrium. During the second month of pregnancy, the placenta becomes capable of secreting estrogen and progesterone on its own, and the corpus luteum is no longer needed. At this time, the placenta stops secreting hCG, and as maternal plasma levels of hCG fall (see Fig. 31–12), the corpus luteum becomes less functional. During the remaining months of pregnancy, the high levels of estrogen and progesterone in the maternal circulation are mainly the result of secretion by the *placenta* (see Fig. 31–12).

Effects of hCG on Fetal Adrenal Glands and Testes.
It is thought that hCG also stimulates steroid synthesis in the developing fetal adrenal gland (see Table 31–2). In the male fetus, hCG stimulates the synthesis of testosterone by the fetal testes. Fetal testosterone stimulates the expression of secondary sex characteristics in the male fetus.

Placental Secretion of Progesterone

The placenta begins to secrete progesterone at about the sixth week of pregnancy, and by weeks 12 to 14 the placenta produces enough progesterone to replace that previously secreted by the corpus luteum

Figure 31–12
Concentration of hormones in maternal plasma during different stages of pregnancy.

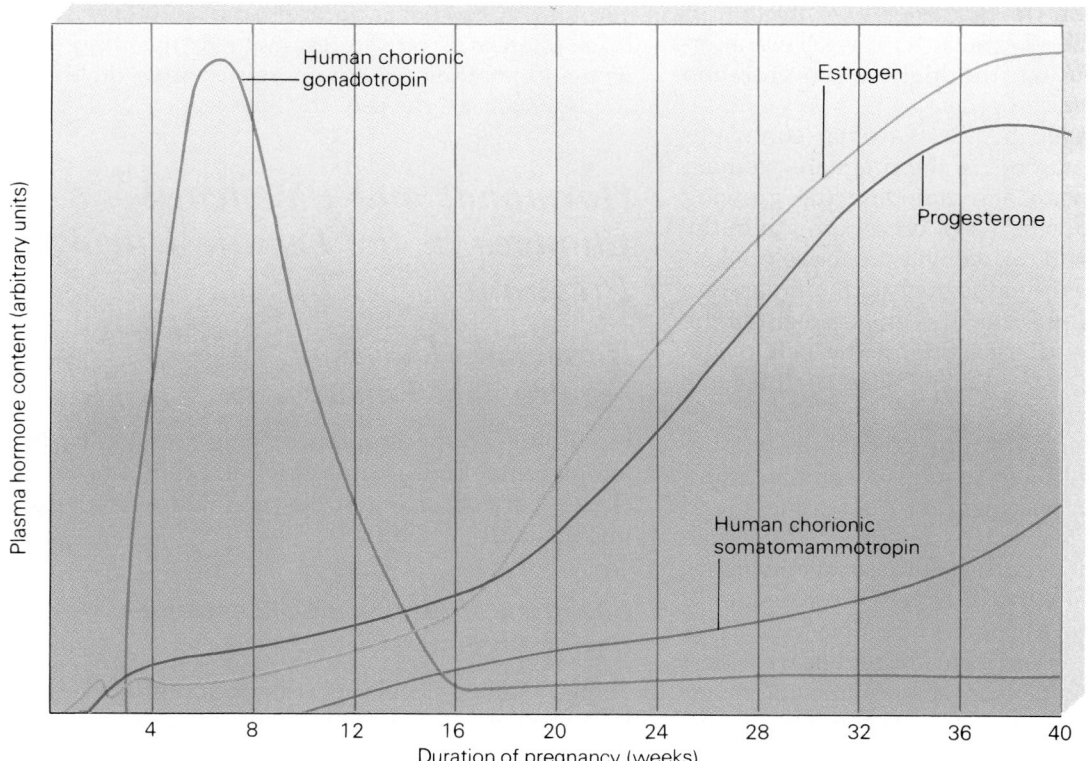

during the ovarian cycle. In the later stages of pregnancy, the placenta secretes 10 times as much progesterone as the corpus luteum is capable of producing during the ovarian cycle (see Fig. 31–12). The placenta extracts cholesterol from the maternal circulation and uses this cholesterol to form progesterone, which then diffuses back into maternal circulation.

Progesterone, secreted by the placenta, is essential for the maintenance of pregnancy for many reasons. First, progesterone stimulates growth of the endometrium and secretion of nutrients from glands in the endometrium, which nourish the developing embryo. Second, progesterone inhibits contractions of the uterus, which prevents premature expulsion of the fetus. Third, as we will see, progesterone stimulates the breast in preparation for lactation (milk production) and release following delivery. Fourth, placental progesterone serves as a precursor for the synthesis of estrogen by the fetoplacental unit.

Placental Secretion of Estrogen

During the first month of pregnancy, the corpus luteum secretes estrogen in response to hCG stimulation. After the first month, hCG begins to stimulate estrogen secretion from the placenta (see Table 31–2). As shown in Figure 31–12, circulating levels of estrogen in maternal plasma continue to rise up to the time of delivery. In fact, the concentration of estrogen in maternal plasma at term is several hundred times higher than the highest concentration seen during the ovarian cycle.

Estrogen, secreted by the placenta, stimulates enlargement of the uterus in the pregnant woman, which is necessary to accommodate the growing fetus. Estrogen also stimulates breast growth in the pregnant woman and, as we will see, development of the duct system within the breasts. In preparation for delivery, estrogen stimulates enlargement of the external genitalia and relaxation of the pelvic ligaments, which facilitates the passage of the fetus through the birth canal.

Estrogen production in the placenta is unique. The synthesis of estrogen requires a complex interaction between the placenta and the adrenal gland of the fetus. As we learned in Chapter 14, the formation of estrogen involves the synthesis of androgens from cholesterol, followed by the conversion of these androgens to estrogens. The placenta lacks the enzymes necessary to form androgens from cholesterol but contains those necessary to convert certain androgens into estrogens. On the other hand, the fetal adrenal cortex contains the enzymes needed to form androgens from cholesterol (specifi-

cally, **dehydroepiandrosterone-sulfate,** or **DHEA-sulfate**). The placenta and the fetal adrenal work together to produce estrogens in the following way: the fetal adrenal synthesizes the androgen DHEA-sulfate from cholesterol. This androgen is then transported to the placenta, where it is converted to **estriol,** a form of estrogen. Estriol is then released from the placenta into maternal circulation. Since estrogen synthesis is dependent on the participation of both the fetal adrenal and the placenta, it is sometimes said to be secreted by the **fetoplacental unit.**

Placental Secretion of Human Chorionic Somatomammotropin

Another hormone that begins to be secreted by the placenta at about four weeks of gestation is **human chorionic somatomammotropin (hCS).** Placental secretion of hCS rises steadily during pregnancy (see Fig. 31–12). hCS is a protein with a molecular structure similar to that of human growth hormone. Human chorionic somatomammotropin is believed to be involved in metabolism and nutrition of both the pregnant woman and her fetus. It has weak growth-promoting effects that may play a role in fetal development and maternal breast development. It also has effects on both glucose and fat metabolism in the pregnant woman. For instance, it decreases maternal utilization of glucose, making more glucose available to the developing fetus. In addition, it promotes the release of free fatty acids from maternal fat stores, which the pregnant woman may then use for energy, sparing other nutrients, such as glucose, for the fetus.

Hormonal and Physical Changes in the Female During Pregnancy

Hormonal Changes in the Pregnant Female

In addition to the hormones being secreted by the fetoplacental unit, a number of other hormonal changes are occurring in the pregnant woman during gestation of the fetus.

Inhibition of Pituitary Gonadotropins During Pregnancy

High circulating levels of estrogen and progesterone in maternal plasma during pregnancy inhibit the secretion of **gonadotropin-releasing hormone (GnRH)** from the maternal hypothalamus and hence inhibit release of **luteinizing hormone** and **follicle-stimulating hormone** from the maternal pituitary. Low circulating levels of LH and FSH during pregnancy are not adequate to stimulate growth of ovarian follicles. Therefore, follicular development, ovulation, and menstruation are all eliminated for the duration of pregnancy.

Stimulation of Prolactin During Pregnancy

The maternal pituitary secretes an abundance of prolactin during pregnancy. As we will see, prolactin is essential for estrogen and progesterone stimulation of maternal breast development and for the synthesis of breast milk. Circulating prolactin levels rise progressively throughout pregnancy and reach very high concentrations at birth.

Stimulation of Relaxin During Pregnancy

During the first two months of pregnancy, **relaxin** is secreted by the corpus luteum in response to hCG stimulation. Later in pregnancy, relaxin is secreted by the endometrium of the uterus. Relaxin, as its name suggests, relaxes pelvic ligaments in preparation for parturition. Relaxin, together with progesterone, inhibits uterine contractions and prevents premature expulsion of the fetus.

Stimulation of Insulin During Pregnancy

Secretion of **insulin** from the maternal pancreas increases after the third month of pregnancy in response to decreased maternal sensitivity to insulin. Insulin insensitivity during pregnancy may become so severe that it produces a temporary form of diabetes, called **gestational diabetes.** After delivery, maternal sensitivity to insulin usually returns to normal.

Stimulation of Aldosterone During Pregnancy

Increased activity within the renin-angiotensin system during pregnancy results in increased secretion of **aldosterone** from the cortex of the maternal adrenal gland. Hypersecretion of aldosterone is needed during pregnancy to maintain adequate sodium in the maternal circulation, much of which is being provided to the developing fetus. Sodium retention and accompanying water retention during pregnancy help to maintain high maternal and fetal plasma volumes.

Stimulation of Cortisol During Pregnancy

Levels of both free and protein-bound **cortisol** increase in the maternal plasma during pregnancy. These changes are due in part to increased secretion of cortisol by the maternal adrenal gland, and also to a threefold increase in **cortisol binding globulin (CBG)** in maternal plasma. The liver increases synthesis of CBG during pregnancy in response to estrogen stimulation. Increased plasma cortisol contributes to an increase in maternal adipose tissue volume and development of the mammary glands.

Stimulation of Thyroxine (T_4) and Triiodothyronine (T_3) During Pregnancy

Maternal thyroid function is increased during pregnancy, as evidenced by increased thyroid gland size, thyroid secretion of T_3 and T_4, basal metabolic rate, and resting pulse rate. T_3 and T_4 increase cardiopulmonary function in the mother, which results in an increase in the rate of delivery of oxygen and nutrients to the developing fetus.

Physical Changes in the Pregnant Female

Pregnancy begins when a fertilized ovum becomes implanted in the endometrium of the uterus. Pregnancy, which usually lasts for approximately 266 days, is divided into three-month periods called **trimesters.** The first trimester is composed of the first three months after fertilization of the ovum. The second trimester includes months four to six of pregnancy, and the third trimester refers to the time between month seven of pregnancy and parturition or delivery.

Physical Changes During the First Trimester

Tiredness, nausea, and vomiting often accompany the first trimester of pregnancy. These symptoms, referred to as **morning sickness,** usually disappear within the first few months of pregnancy. Other physical changes that occur during this time are frequent urination due to compression of the bladder by the enlarging uterus, breast swelling and tenderness, and increased vaginal secretions. A pregnant woman cannot feel the changes occurring in her uterus during the first trimester.

Physical Changes During the Second Trimester

During the second trimester of pregnancy, a woman becomes very aware of the changes in her uterus, because the fetus begins to move. Toward the end of the fourth month, the fetus begins to twist and kick, and these movements intensify during the following weeks. In response to aldosterone and estrogens, most women begin to retain fluid during the second trimester. Increased fluid retention results in swelling, or edema, of the hands and feet. Increased production of red blood cells by maternal bone marrow, together with fluid retention, results in a large increase in maternal blood volume. The increased nutritional demands of the fetus during this time increase the mother's appetite. An increased demand for oxygen by the fetus is accompanied by an increase in maternal blood flow to the placenta and an increase of 30% to 40% in maternal cardiac output.

Growth of the Uterus. A pregnant woman's uterus grows at a rapid rate during the second trimester. Expansion of the uterus results in many physical changes, including protrusion of the abdomen and widening of the waist. Rapid growth of the abdomen can cause red or pink stretch marks to form in the skin covering the abdomen. As the uterus grows, it compresses other internal organs such as the stomach and intestines, which can lead to indigestion, constipation, and hemorrhoids.

Development of the Breasts. In response to estrogen and progesterone, secreted by the placenta, and prolactin, secreted by the maternal pituitary, the breasts increase dramatically in size during the second trimester. This increase in size occurs so quickly that the skin covering the breast, like that covering the abdomen, is stretched, which can result in the formation of pink or red marks on the surface of the breasts. Breast development during pregnancy will be discussed in more detail later. During the second trimester of pregnancy, the breasts begin to secrete a thin, yellowish fluid called **colostrum,** which is the precursor of breast milk.

Physical Changes During the Third Trimester

The largest weight gain of pregnancy usually occurs during the third trimester. Although individual weight gains vary widely, the average woman gains approximately 20 to 25 pounds during pregnancy. The fetus, placenta, and amniotic fluid, all of which are expelled during delivery, account for approximately 10 to 15 pounds of this added weight.

During the third trimester, the uterus reaches its largest size. In a standing position, a woman's uterus protrudes forward, which alters her center of gravity. To compensate for this, a pregnant woman often walks with head and shoulders held back, which can result in backache.

The increased size of the uterus puts constant pressure on the bladder, which results in the need to urinate four to five times during the night, disrupting sleep patterns and increasing fatigue. Increased pressure on the blood vessels of the lower part of the body results in muscle cramps in the legs, and compression of the diaphragm can cause shortness of breath.

During the third trimester, the muscles of the uterus undergo short periods of spasmodic contractions, called **Braxton-Hicks contractions.** The fetus changes position in the uterus during the last few weeks of pregnancy, and the head of the fetus commonly becomes oriented toward the pelvis in preparation for parturition.

Growth and Development of the Fetus

Prenatal development, which occurs throughout the nine months of pregnancy, can be divided into three stages: (1) the **germinal stage,** the first two weeks of development following fertilization of the ovum, (2) the **embryonic stage,** weeks 2 to 8 following fertilization, and (3) the **fetal stage,** the time between week eight following fertilization and birth.

The Germinal Stage of Development (Weeks 1 and 2)

During the first two weeks of pregnancy, the fertilized ovum, though microscopic in size, has reached the blastocyst stage of development. Two major fetal membranes arise from the blastocyst, the inner **amnion** and the outer **chorion.** The amnion forms a thin sac of tissue that surrounds the fetus and is filled with a liquid called **amniotic fluid.** Amniotic fluid is constantly absorbed by the placenta and replaced with new fluid every 2 to 3 hours throughout pregnancy. This fluid-filled sac provides a constant temperature for fetal development and serves as a shock absorber to protect the fetus against physical injury. The chorion gives rise to the placenta.

The inner cells of the blastocyst differentiate into three cell layers. The outer layer of cells will become the skin, sense organs, and nervous system of the developing fetus. The middle layer will form the skeleton, muscles, heart, and blood vessels. The inner cell layer will develop into the liver, pancreas, and digestive system of the fetus.

The Embryonic Stage of Development (Weeks 2 Through 8)

This is a period of remarkable growth for the developing embryo even though it is less than an inch long by the end of the eighth week of development

(Table 31–3). During this time, all of the embryonic organs become partially formed, although cellular development is far from complete. By the end of the first month the primitive heart has begun to beat; the digestive system, brain, spinal cord, and nervous system are developing; and buds of tissue that will become the arms and legs begin to form (see Fig. 31–13). By the sixth week the head of the embryo is quite large, and the outlines of eyes can be seen. During the sixth and seventh weeks, the eyes and ears develop more fully, and the teeth, bones, and facial muscles begin to form.

The Fetal Stage of Development (Week 8 Through Birth)

At about 9 weeks of development, the embryo is now called a **fetus.** At the beginning of the ninth week of development, the head makes up almost half the total length of the fetus. All primitive organs and structures become more differentiated.

By the second month, distinct hands and feet have formed, and major blood vessels are forming (see Fig. 31–13). By the third month, the fetal stomach is producing digestive fluids. The heart is pumping blood through the fetal arteries and veins, the kidneys have begun to function, and the bone marrow is manufacturing red blood cells. By the third month of development, the external genitalia of the fetus appear, and the sex of the fetus becomes anatomically distinguishable.

By the fourth month, the fetus has developed vocal cords, lips, fingernails, toenails, and hair on its head (see Fig. 31–13). Sucking motions and respiratory movements begin during the fourth month, and the fetus swallows and absorbs amniotic fluid. Many peripheral reflexes are well developed by this time, though the central nervous system is still very underdeveloped and incapable of higher functions.

During the fifth month, the fetal heartbeat can be heard through the maternal abdomen. The body of the fetus becomes covered with fine, downy hair called **lanugo.** At the end of the sixth month, the fetus is only about 12 inches long and weighs about 1.5 pounds (see Table 31–3). If the fetus is born prematurely at 6 months of development, its chances for survival are very poor.

During the seventh month, the brain and nervous system become more developed, although many important higher functions of the central nervous system are still undeveloped. Fatty tissue forms under the skin of the fetus, and the downy hair covering its body begins to disappear. During the eighth month, many fetal organs become mature, and the fetal lungs produce a substance called **surfactant** (see Chapter 21), which is important for survival of the fetus after birth. Surfactant is produced by epithelial cells in fetal lung alveoli. This substance, which has a very low surface tension, coats the inside of the lungs and prevents the collapse of the lungs at low lung volumes. When fetal lungs do not contain enough surfactant, the infant is prone to **respiratory distress syndrome (RDS).** RDS is characterized by decreased lung compliance and instability of lung alveoli, which cause prolonged periods of respiratory depression that can result in death. Infants born prematurely during the eighth month of pregnancy have a good chance of survival because the fetal lungs have already begun to produce surfactant.

During the ninth month of pregnancy, the fetus migrates to a lower part of the uterus and normally positions itself head downward in preparation for parturition. The average fetus is 20 inches long and weighs 7 pounds at birth (see Table 31–3).

Parturition

As stated earlier, weak, spasmodic contractions of the uterine muscles occur infrequently throughout the third trimester of pregnancy. As parturition approaches, these contractions become stronger, more regular, and more frequent. Delivery of the fetus and expulsion of the placenta result from strong contractions of the uterus that push the fetus and the placenta through the cervix and out of the vagina. **Labor** is said to begin when uterine contractions start to occur once every 10 to 15 minutes. In some women, the amnionic membrane surrounding the fetus ruptures, and amniotic fluid flows out of the vagina before delivery begins. But in most women the amnionic membrane ruptures during delivery (Fig. 31–14).

Table 31–3
Average Fetal Growth Patterns During Gestation

Age of Fetus (in weeks)	Crown to Rump Length (in inches)	Weight (in pounds)
5–8	0.2–10	0.01–0.03
13–16	3.5–5.5	0.12–0.43
21–24	7.9–9.1	1.06–1.81
29–32	11.0–11.8	3.10–4.60
37–40	13.8–14.2	6.60–7.50

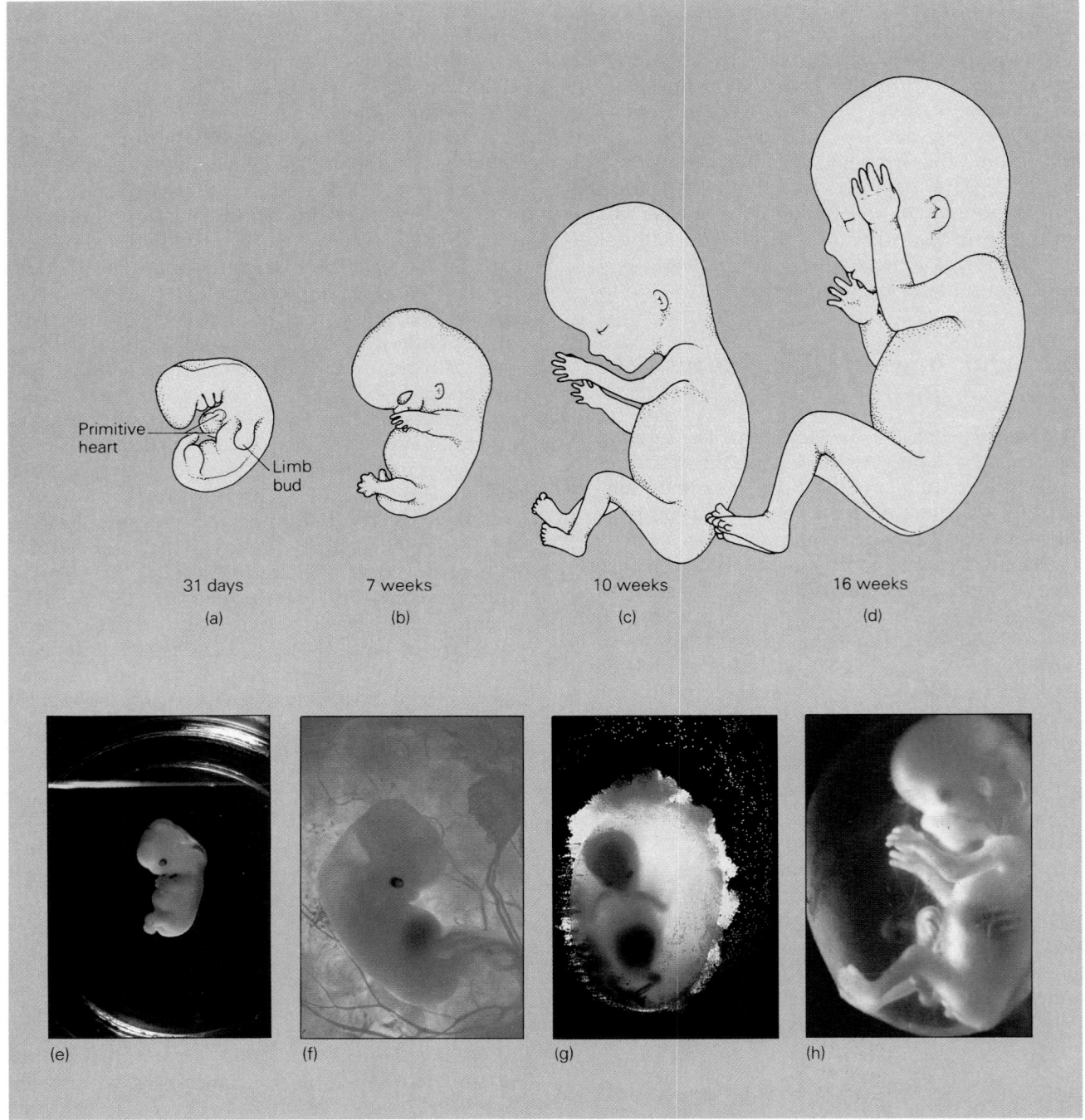

Figure 31–13
Appearance of the embryo (*a* and *b*) and the fetus (*c* and *d*) at various times during gestation. (*e*) The embryo at 31 days. (© Dr. Landrum B. Shettles.) (*f*) The embryo at 7 weeks. (© Petit Format and Guigoz.) (*g*) The fetus at 10 weeks. (© CNRI/Phototake, N.Y.) (*h*) The fetus at 16 weeks. (© Petit Format/ Science Source.)

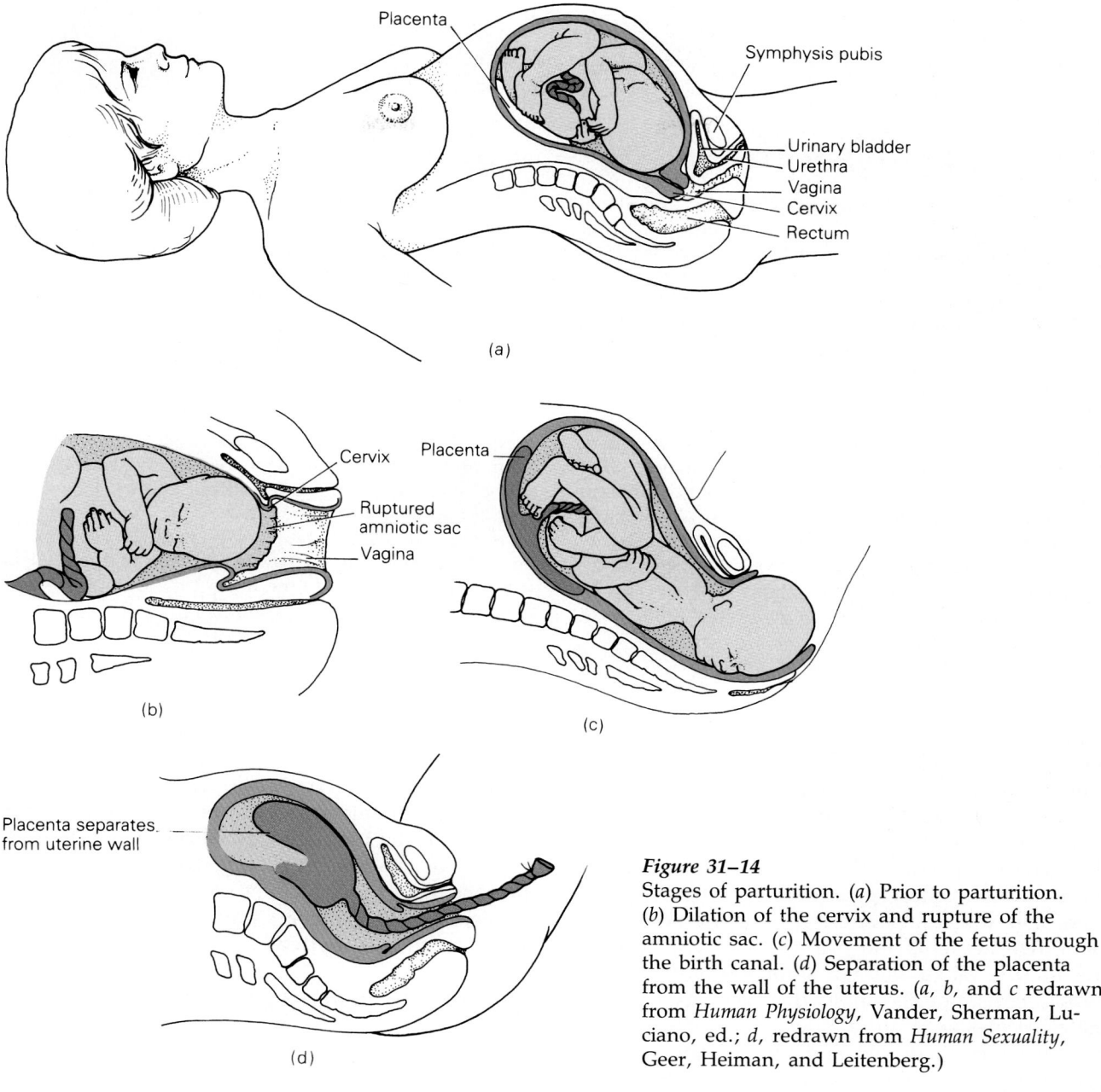

Figure 31–14
Stages of parturition. (*a*) Prior to parturition.
(*b*) Dilation of the cervix and rupture of the
amniotic sac. (*c*) Movement of the fetus through
the birth canal. (*d*) Separation of the placenta
from the wall of the uterus. (*a*, *b*, and *c* redrawn
from *Human Physiology*, Vander, Sherman, Luciano, ed.; *d*, redrawn from *Human Sexuality*,
Geer, Heiman, and Leitenberg.)

Uterine Contractions During Parturition

Although the smooth muscles of the uterus are capable of autonomous contractions, uterine contractions at delivery are facilitated by a number of hormones, which will be discussed shortly. Muscular
contractions begin in the upper portion of the uterus
and move down toward the cervix. These contractions force the fetus against the cervix. Under hormonal stimulation, and with pressure from the
fetus, the cervix widens or dilates to a maximum
diameter of about 10 cm. **Cervical dilation,** combined with pressure generated by the mother, forces
the fetus and placenta through the cervix and out of
the vaginal canal (see Fig. 31–14). At birth, the infant is still connected by the umbilical cord to the
placenta. Within minutes following delivery, the

Alcohol Drinking During Pregnancy

It has long been recognized that drinking alcohol during pregnancy is harmful to the developing fetus. The most severe hazard associated with heavy drinking by pregnant women is the development of fetal alcohol syndrome (FAS). FAS is characterized by growth retardation and abnormalities of facial development, as well as central nervous system dysfunction including hyperactivity, tremors, and impairment of intellectual development.

The exact relationship between the amount of alcohol consumed by the mother and the degree of impairment of the fetus is hard to establish because of the difficulty in separating the effects of alcohol from the contribution of other factors such as smoking, drugs, and malnutrition, which can also increase the risk of fetal impairment. However, the effects of alcohol on the fetus are commonly regarded as a continuum from severe impairment associated with heavy drinking to minor impairment associated with light drinking.

Heavy alcohol drinking as early as around the time of conception can produce craniofacial deformity and impairment of the central nervous system. Therefore, although discontinuation of heavy drinking after the first trimester of pregnancy will be beneficial to fetal growth, it cannot fully overcome the increased risk of congenital abnormalities.

Maternal alcohol drinking is also associated with increased risk of spontaneous abortion, premature labor, and neonatal death. The precise way in which alcohol damages the fetus is still not known. However, it has been determined that alcohol can impair transport of amino acids and glucose to the fetus, and can induce fetal acidosis and hypoxia through constriction of blood flow to the fetus.

maternal and infant blood vessels in the placenta and umbilical cord constrict, and the infant is separated from the maternal blood supply. The placenta becomes separated from the wall of the uterus, and continued uterine contractions expel the placenta through the birth canal (see Fig. 31–14). Because the placenta leaves the vagina after the infant is delivered, it is sometimes referred to as the **afterbirth.**

When the fetus and placenta are expelled from the uterus, all of the hormones secreted by the fetoplacental unit begin to decrease markedly in maternal plasma. Within 3 days after delivery, the concentration of steroid and protein hormones in maternal plasma returns to nonpregnant levels.

Hormonal Control of Parturition

Many hormones, including estrogen, progesterone, relaxin, oxytocin, and the prostaglandins, have been found to play important roles in the initiation of parturition (Fig. 31–15).

Effects of Oxytocin on Parturition

Oxytocin stimulates contraction of uterine muscles. Dilation of the cervix ("cervical stretch") before delivery provides afferent neural input to the hypothalamus, which results in the release of oxytocin from the posterior pituitary gland. In addition—and perhaps more importantly—in response to estrogen and relaxin stimulation, the number of receptors for oxytocin in the uterus increases progressively during the later part of pregnancy. This increases uterine sensitivity to oxytocin at the end of pregnancy. In response to oxytocin, uterine contractions become strong and rhythmic.

Effects of Prostaglandins on Parturition

Prostaglandins, specifically those of the PGF and PGE series, also stimulate contractions of uterine smooth muscle. In response to oxytocin, the myometrial or smooth muscle cells of the uterus secrete prostaglandins, which, through a paracrine action, stimulate further uterine contractions. The prostaglandins are so powerful in stimulating uterine contractions that exogenous administration of prostaglandins can induce labor and terminate pregnancy at almost any stage of gestation.

Effects of Progesterone on Parturition

As stated earlier, progesterone strongly inhibits uterine contractions. In some women, circulating levels of progesterone decline just prior to delivery,

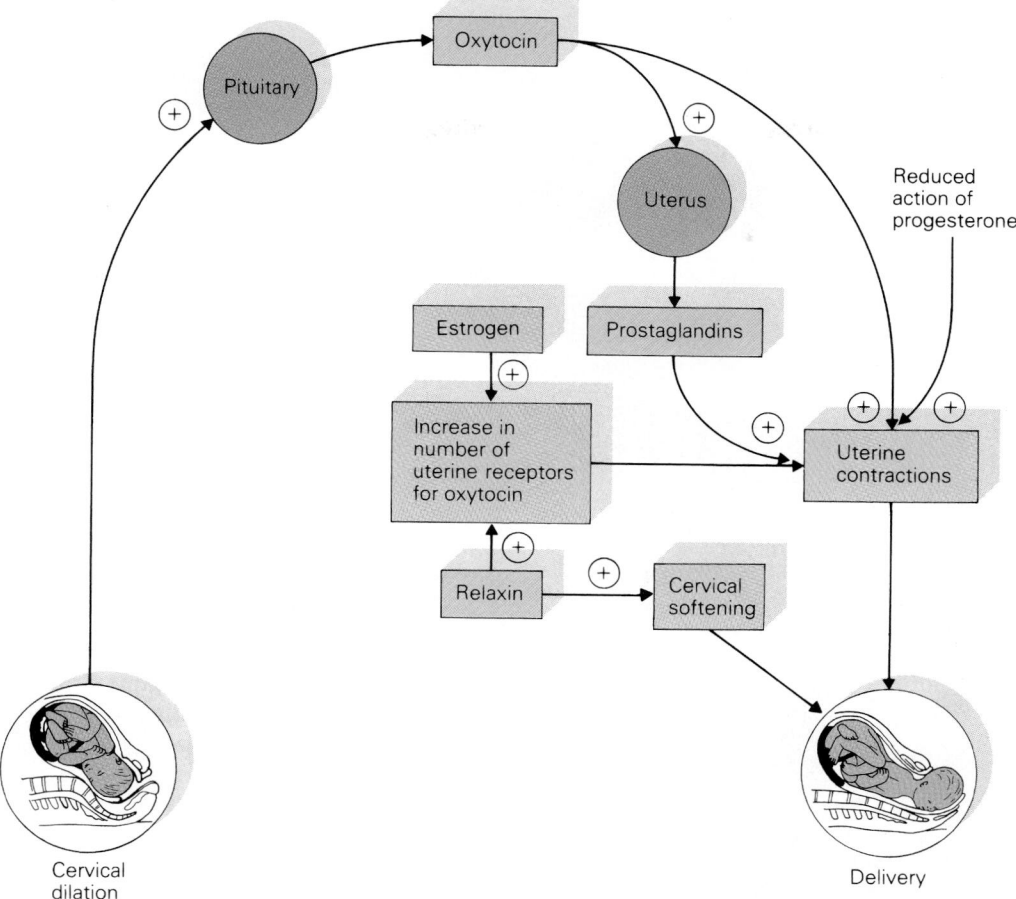

Figure 31–15
Hormonal factors that contribute to the onset of parturition. The plus sign indicates stimulation.

which removes progesterone inhibition of uterine contractions and facilitates the onset of labor. However, a decrease in total circulating progesterone prior to delivery has not been found in all women. Recent evidence indicates that the placenta may secrete a progesterone-binding protein before delivery, which reduces the amount of free progesterone in maternal circulation. This could serve functionally to reduce the action of progesterone on uterine muscle and facilitate delivery.

Effects of Relaxin on Parturition

Relaxin, secreted first by the corpus luteum and later by the endometrium, may facilitate delivery in at least three ways. First, as previously stated, relaxin increases the number of oxytocin receptors in the uterus and makes the uterus more sensitive to the effects of oxytocin at the end of pregnancy. Second, relaxin softens the cervix, making it more pliable and facilitating passage of the fetus through the cervix during delivery. Third, as previously mentioned, relaxin causes relaxation of the pelvic ligaments, easing passage of the fetus through the birth canal.

Lactation

Anatomy of the Breast

The outer surface of the female breast is illustrated in Figure 31–16a. In the center of the breast is a pigmented area, called the **areola,** which surrounds the nipple. The internal structure of the breast is illustrated in Figure 31–16b and c. The mammary or

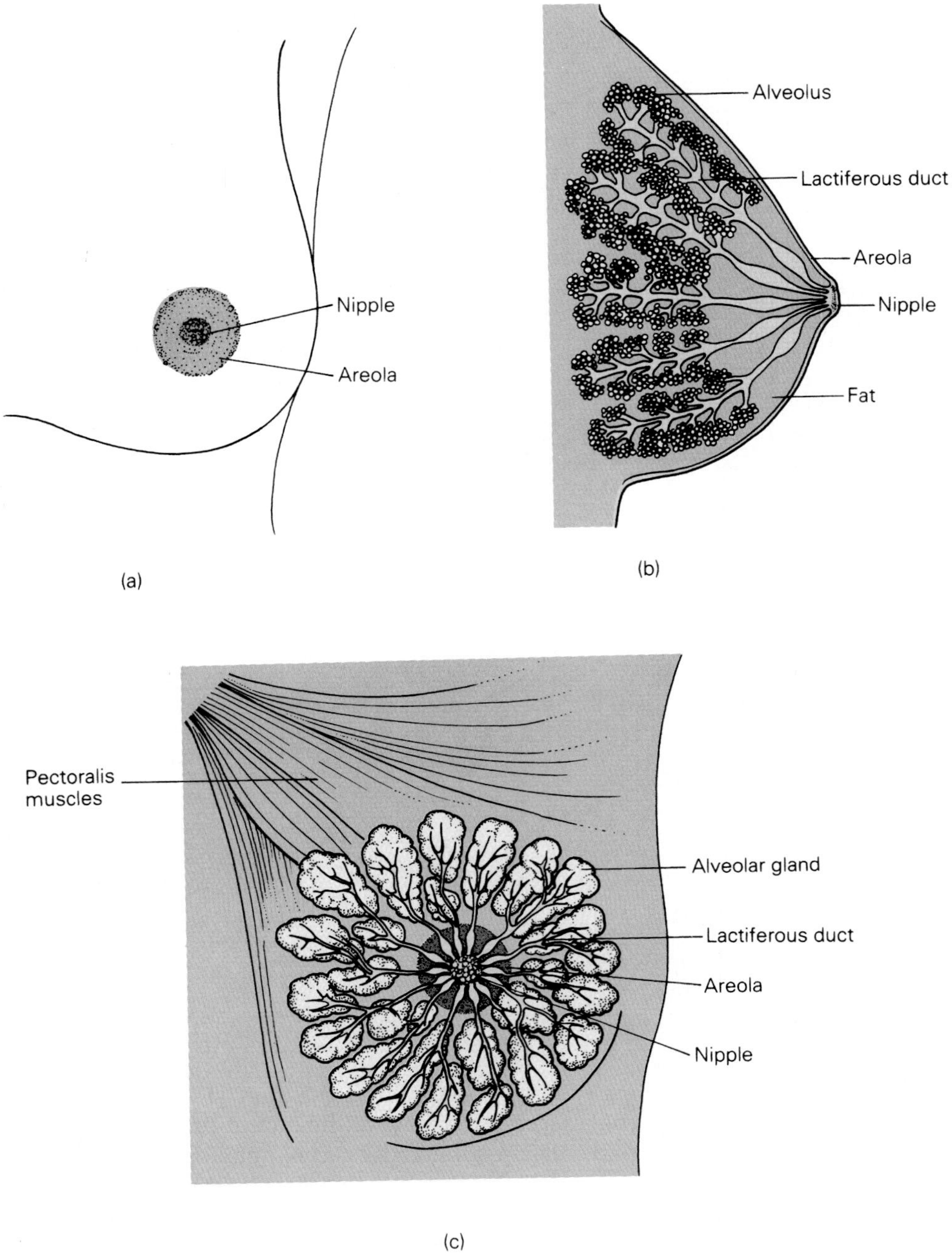

(a)

(b)

(c)

Figure 31–16
Anatomical features of the human breast. (*a*) External view. (*b*) Internal side
view. (*c*) Internal front view. (*a* is redrawn from Masters, Johnson, and
Kolodny, *Human Sexuality*. *b* and *c* are redrawn from Worthington-Roberts,
Nutrition in Pregnancy and Lactation.)

breast tissue lies directly over the pectoralis major
muscle and is separated from this muscle by a layer
of fat (see Fig. 31–16*c*). The breast is composed of
glandular epithelium and a **ductal system** embedded in interstitial tissue and fat. The ducts of the

breast, which terminate in the nipple, branch
through the breast tissue and are connected to
glands called the alveoli (see Fig. 31–16*b*). The **alveoli,** which secrete breast milk, have been described
as resembling clusters of grapes on stems. The alve-

oli are surrounded by contractile cells called **myoepithelial cells,** which, as we will see, are involved in the expulsion of milk from the breasts.

Development of the Breast

A number of hormones, including estrogen, progesterone, human chorionic gonadotropin, human chorionic somatomammotropin, prolactin, and growth hormone, stimulate breast development during pregnancy. Under the influence of these hormones, the breasts increase in size, and the ductal system and alveolar structures become more complex. A summary of various hormones and their effects on breast development prior to parturition is shown in Table 31–4.

Hormonal Control of Lactation

Under hormonal stimulation, the alveolar cells of the breast extract a variety of substances, including glucose, amino acids, fatty acids, and glycerol, from the maternal circulation and convert these materials into breast milk. During lactation, the breasts may produce as much as 1.5 liters of milk each day.

Effects of Estrogen and Progesterone on Lactation

High circulating levels of estrogen and progesterone, secreted by the placenta during pregnancy, stimulate breast development. Estrogen also stimulates prolactin release; prolactin in turn stimulates additional breast growth. However, both estrogen and progesterone inhibit milk production in the breast prior to parturition by blocking the milk-inducing effects of prolactin on the breasts. After the placenta is expelled from the uterus at parturition, the concentrations of estrogen and progesterone in maternal plasma fall (see Fig. 31–18). The removal of estrogen and progesterone leaves the breasts under the influence of prolactin and oxytocin. As we will see, prolactin, together with oxytocin, controls the production and release of milk from the breasts following parturition (Fig. 31–17).

Effects of Prolactin on Lactation

Prolactin stimulates synthesis of milk in the breast and secretion of milk into the alveoli of the breasts (see Fig. 31–17). High circulating levels of estrogen during pregnancy stimulate the release of prolactin from the maternal pituitary gland. Prolactin levels in maternal plasma rise during pregnancy and reach very high concentrations at birth (Fig. 31–18). After birth, plasma prolactin levels begin to decline, but during this time the hypothalamus is very sensitive to neural signals it receives from the breasts. Breast stimulation during **suckling** stimulates afferent neural impulses that travel from the breast, through the spinal cord, to the hypothalamus and stimulate release of a large amount of prolactin from the mater-

Table 31–4

Hormones Involved in Breast Development During Pregnancy

Hormone	Origin	Function During Pregnancy
Prolactin	Anterior pituitary	Stimulates mammary growth
Oxytocin	Posterior pituitary	Generally no effect on mammary growth, but sensitivity of myoepithelial cell to oxytocin increases during pregnancy
Estrogen	Ovary and placenta	Stimulates proliferation of glandular tissue and ducts of the breast; stimulates prolactin release but blocks action of prolactin on the breast
Progesterone	Ovary and placenta	Stimulates proliferation of glandular tissue and ducts of the breast; blocks action of prolactin on the breast
Human chorionic somatomammotropin (hCS)	Placenta	Stimulates mammary growth
Human chorionic gonadotropin (hCG)	Placenta	Stimulates mammary growth

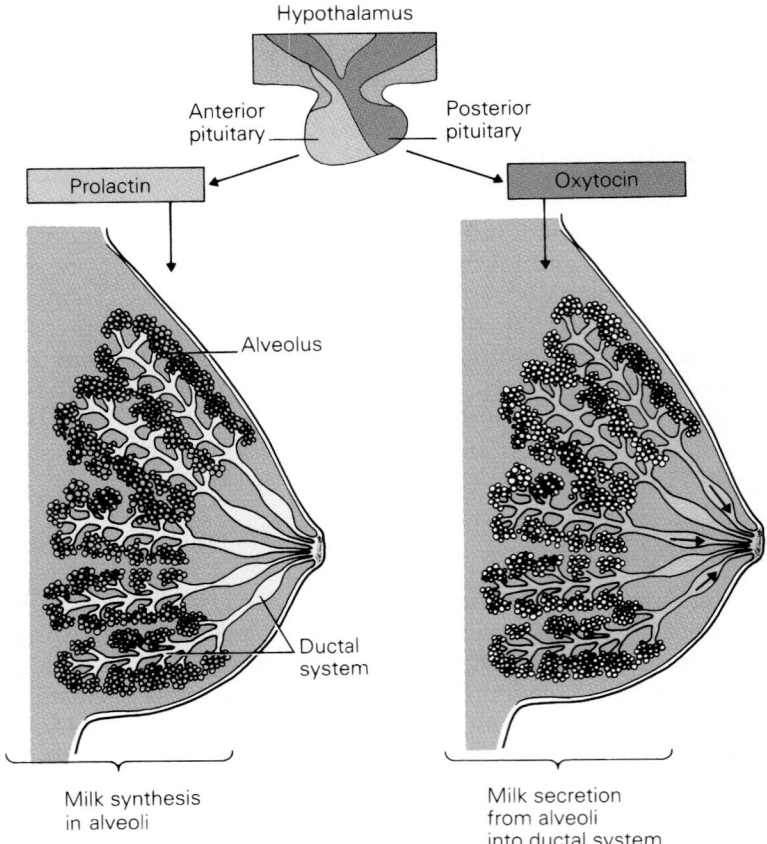

Figure 31–17
Hormonal control of milk production and
expulsion in the breast.

nal pituitary. Suckling results in a tenfold rise in
plasma prolactin that lasts for about 1 hour, after
which plasma prolactin levels return to normal.
Therefore, prolactin concentrations in maternal
plasma rise and fall with every episode of nursing
(see Fig. 31–18). Prolactin, released in response to a
single nursing episode, stimulates **milk production**
in the breast, which makes additional milk available
for the next nursing episode. If an infant is not
breastfed or when nursing is terminated, there is no
stimulus for prolactin release; hence, milk synthesis
in the breast is not stimulated, and the breast stops
secreting milk within a few days. With prolonged
nursing, milk production in the breast begins to decline within 7 to 9 months after birth, but the breast
can still produce significant quantities of milk for
several years.

As we discussed in Chapter 13, prolactin inhibits the release of GnRH from the hypothalamus and
hence inhibits LH and FSH release from the pituitary. Low levels of LH and FSH result in the absence
of follicular growth and ovulation. Therefore,
breastfeeding, which stimulates the episodic release
of prolactin, tends to inhibit ovulation and reduce

fertility. However, there is a large amount of individual variation in the time during which prolactin
inhibits GnRH release with prolonged nursing;
breastfeeding is thus not a reliable method of birth
control.

Effects of Oxytocin on Lactation

Oxytocin stimulates the release of milk from the
breast, a process termed **milk let down**. Oxytocin
induces milk let down by stimulating contraction of
the myoepithelial cells that surround the outer walls
of the alveoli. This pressure forces milk from the
alveoli into the ductal system of the breast (see Fig.
31–17). Once milk reaches the ductal system, it can
be pulled from the breast by the negative pressure
that results from suckling.

As with prolactin, oxytocin is released in response to breast stimulation during nursing. Suckling stimulates sensory nerves in the breast that
transmit afferent neural impulses to the spinal cord
and from there to the hypothalamus, resulting in
the release of oxytocin from the pituitary gland. As
with prolactin, oxytocin is released episodically in

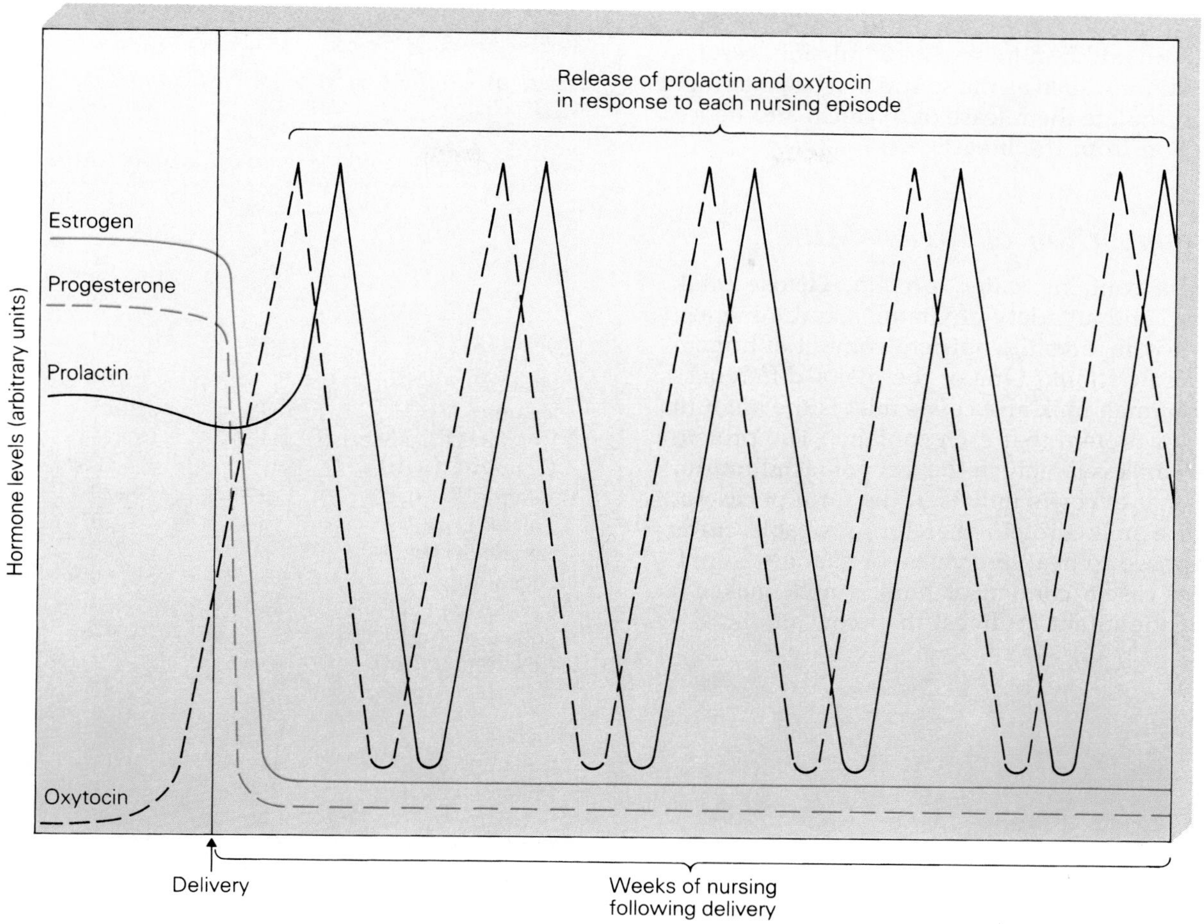

Release of prolactin and oxytocin
in response to each nursing episode

Estrogen

Progesterone

Prolactin

Hormone levels (arbitrary units)

Oxytocin

Delivery

Weeks of nursing
following delivery

Figure 31–18
Changes in maternal hormones following delivery and during episodes of
nursing.

C L I N I C A L   F O C U S

Contaminants of Breast Milk

Many substances to which a lactating woman is exposed are transferred to her breast milk. Drugs, caffeine, viruses, environmental pollutants, and alcohol can pass into breast milk by simple diffusion, by carrier-mediated diffusion, or by active transport. The entry of these substances into breast milk depends on many factors, including the molecular weight and solubility properties of the particular substance. In general, the amount of a drug that enters breast milk is not more than 2% of the maternal dose. Although moderate amounts of many substances pose no risk to the suckling infant, some drugs can be harmful. For instance, many sedatives, including lithium, reserpine, and Valium (diazepam), produce drowsiness and lethargy in breast-fed infants. Anticoagulants can induce bleeding problems in the infant. More powerful narcotics, such as heroin, or the painkiller Darvon (dextropropoxyphene), can lead to addiction in breast-fed infants.

response to nursing episodes, as illustrated in Figure 31–18. In addition to breast stimulation, certain auditory stimuli, such as the sound of a baby crying, can also stimulate the release of oxytocin and hence milk ejection from the breast.

The Composition of Breast Milk

Breast milk contains water, protein, **lactose** (milk sugar), fat, and a variety of vitamins and minerals. Table 31–5 compares the nutrient content of human milk and cow's milk. One of the major differences between human milk and cow's milk is the amount and type of protein that each contains. The protein in human milk is largely in the form of **lactalbumin,** whereas that of cow's milk is in the form of **casein.** Casein is a milk protein that forms sizable curds when exposed to heat, enzymes, or changes in pH. The lower casein content of human milk makes it easier for the infant to digest than cow's milk.

Table 31–5

Nutrient Content of Human Milk and Cow's Milk

Constituent (per liter)	Human Milk	Cow's Milk
Energy (kcal)	690	660
Protein (g)	9	35
Fat (g)	40	38
Lactose (g)	68	49
Vitamins		
Vitamin A (IU)	1898	1025
Vitamin D (activity)	40	14
Vitamin E (IU)	3.2	0.4
Vitamin K (μg)	34	170
Thiamin (μg)	150	370
Riboflavin (μg)	380	1700
Niacin (mg)	1.7	0.9
Pyridoxine (μg)	130	460
Folic acid (μg)	41–84.6	2.9–68
Cobalamine (μg)	0.5	4
Ascorbic acid (μg)	44	17
Minerals		
Calcium (mg)	241–340	1200
Phosphorus (mg)	150	920
Sodium (mg)	160	506
Potassium (mg)	530	1570
Chlorine (mg)	400	1028
Magnesium (mg)	38–41	120
Sulfur (mg)	140	300
Iron (mg)	0.56–0.3	0.5
Iodine (mg)	200	80
Manganese (μg)	5.9–4.0	20–40
Copper (μg)	60	110
Zinc (mg)	4–0.5	3–5
Selenium (μg)	20	5–50
Fluoride (mg)	0.05	0.03–0.1
Chromium (μg)	4	2

Adapted from Worthington-Roberts, B. S., Vermeersch, J., and Williams, S. R. *Nutrition in Pregnancy and Lactation*. St. Louis: Times Mirror/Mosby, 1985, p. 257.

SUMMARY

The Process of Fertilization

Ovulation occurs approximately 14 days after the beginning of the last menstrual cycle in response to stimulation from high circulating levels of LH and FSH. Following ovulation, the ovum enters one of the two fallopian tubes and is transported to the uterus over the course of 3 to 5 days.

Following ejaculation into the vagina, sperm are transported through the vagina, cervix, and uterus into the fallopian tubes. Sperm loss following ejaculation is very high, owing to many factors, including abnormal morphology, reduced motility, destruction by acidic secretions of the vagina, blockade by cervical mucus, and entry of sperm into the fallopian tube that does not contain an ovum.

Capacitation, a process that prepares the sperm to fertilize the ovum, normally occurs once the sperm enter the female reproductive tract.

When sperm reach the vicinity of the ovum, they must undergo the acrosomal reaction before they can fertilize the ovum. The acrosomal reaction results in the release of proteolytic enzymes from the head of the sperm that dissolve a pathway through the outer layer of the ovum and allow the sperm to penetrate the ovum.

Sperm-ovum fusion results in the release of cortical granules from the ovum, which alter the surface membrane of the ovum and block entry of more than one sperm into the ovum (polyspermy). Fertilization normally occurs in the fallopian tube before the ovum has reached the uterus.

Cell Division and Implantation Following Fertilization

Initially, the fertilized ovum is a single cell, called a zygote, which contains 46 chromosomes. The zygote begins to divide mitotically about 30 hours after fertilization while it is still in the fallopian tube. Cell division marks the beginning of embryonic development. By the time the fertilized ovum reaches the uterus, it has divided many times and is called a blastocyst. The blastocyst secretes estrogen and proteolytic enzymes that act on the surface of the endometrium to increase endometrial permeability.

At the same time, the blastocyst sends out tissue projections that invade the endometrium, resulting in the implantation of the blastocyst in the endometrium. When the blastocyst is fully implanted in the endometrium (around day 7 following ovulation), it is called an embryo. After the second month of development, the embryo is called a fetus.

The Role of the Placenta

The placenta, which is formed from cells of the early embryo, acts as an interface between the fetal and maternal blood supplies.

The embryo is dependent on maternal circulation for its supply of oxygen, fluid, nutrients, salts, and antibodies, as well as for the removal of carbon dioxide, urea, and nitrogen. All substances that pass between fetal and maternal blood pass through the placenta. The efficiency of the placental transport system ensures removal of fetal waste products and delivery of needed substances, but it can also result in the delivery of drugs to the fetus if they are present in maternal circulation.

A variety of mechanisms protect embryonic and placental tissue from immunological rejection by the mother.

hCG is secreted by the blastocyst prior to implantation and then by the placenta during the first two months of pregnancy. hCG stimulates growth of the corpus luteum during early pregnancy and ensures continued secretion of estrogen and progesterone from the corpus luteum. Estrogen and progesterone stimulate continued growth of the endometrium of the uterus. During the second month of pregnancy, the placenta secretes estrogen and progesterone on its own, and the corpus luteum is no longer needed. As the placenta secretes more estrogen and progesterone, it stops secreting hCG, leading to the degeneration of the corpus luteum. Throughout the remaining months of pregnancy, the placenta continues to secrete estrogen and progesterone.

Progesterone secreted by the placenta is essential for the maintenance of pregnancy. It stimulates endometrial growth and secretion of nutrients from the endometrium, which nourish the developing embryo. Progesterone inhibits uterine contractions, helping to prevent premature birth. Progesterone also stimulates breast development in preparation for nursing.

Estrogen secreted by the placenta stimulates enlargement of the uterus to accommodate the growing fetus, and development of the ducts in the breasts in preparation for nursing. At birth, estrogen induces relaxation of pelvic ligaments, facilitating delivery. hCS secreted by the placenta decreases maternal utilization of glucose and promotes release of free fatty acids from maternal fat stores, increasing the availability of glucose and free fatty acids to the developing fetus.

Hormonal and Physical Changes in the Female During Pregnancy

High circulating levels of estrogen and progesterone in maternal plasma during pregnancy inhibit the secretion of hypothalamic gonadotropin-releasing hormone (GnRH), and hence pituitary secretion of luteinizing hormone (LH) and follicle stimulating hormone (FSH). In the absence of stimulation from LH and FSH, no new follicles develop in the ovaries, resulting in the inhibition of ovulation and menstruation during pregnancy.

Maternal plasma prolactin levels rise progressively throughout pregnancy, resulting in growth of the breasts.

Maternal secretion of insulin, aldosterone, cortisol, and thyroxine also increases during pregnancy. These hormones control maternal metabolic rate as well as the levels of fluid, salt, and glucose in maternal plasma.

Pregnancy, which usually lasts nine months, is divided into three-month periods called trimesters. Major physical changes occur in the pregnant woman during the second and third trimesters of pregnancy. The second trimester is marked by increased fluid retention, urination, appetite, and cardiac output. Both the uterus and the breasts enlarge dramatically during this time. The largest weight gain during pregnancy occurs in the third trimester and is accounted for, in part, by the increased weight of the fetus, placenta, and amniotic fluid.

Growth and Development of the Fetus

The germinal stage of fetal development is marked by the formation of the blastocyst and implantation of the blastocyst in the endometrium. Membranes of the amnion, which surround the fetus, also form during this period and are filled with amniotic fluid, which protects the fetus from physical injury.

During the embryonic stage of development, internal organs including the heart, digestive system, and nervous system, as well as external organs including the eyes and ears, begin to form in the developing embryo.

During the fetal stage of development, the fetus develops from a partially formed organism that is completely dependent on the maternal blood supply into an organism capable of sustaining life outside the uterus. During the fetal stage, all internal and external organs undergo rapid growth and differentiation of function.

Parturition

During delivery, uterine contractions force the fetus through the dilated cervix and through the birth canal.

As delivery approaches, oxytocin released by the pituitary and prostaglandins secreted by the uterus stimulate uterine contractions and cervical dilation. In addition, relaxin, secreted by the endometrium of the uterus, softens the cervix and relaxes pelvic ligaments.

Lactation

The breast is composed of glands called alveoli, which synthesize milk, and a ductal system, which secretes milk.

A variety of hormones including prolactin, estrogen, progesterone, hCG, and hCS stimulate breast development during pregnancy in preparation for nursing.

After delivery, production and release of milk from the breasts are largely controlled by prolactin and oxytocin, which are secreted by the anterior and posterior lobes of the pituitary, respectively. Prolactin stimulates both the synthesis of milk in the breasts and the secretion of milk into the breast alveoli. Oxytocin stimulates the release of milk from the breasts by inducing contractions of the myoepithelial cells in the breasts. Both prolactin and oxytocin are released in response to neural signals generated by breast stimulation during nursing. Therefore, repeated nursing episodes after delivery result in repeated cycles of milk synthesis and release from the breasts.

Breast milk contains a highly digestible protein called lactalbumin as well as fat and vitamins.

While breast milk is very nutritious, it can also act as a vehicle for transport of drugs, viruses, and environmental pollutants to the nursing infant from the maternal circulation.

REVIEW QUESTIONS

Identify Each with a Term

1. Fertilization of the ovum by more than one sperm.

2. Implantation of a fertilized ovum outside of the uterus.

3. The contractile cells in the breast that are responsible for milk expulsion.

Define Each Term

4. dizygotic twins
5. zygote
6. chorion

Choose the Correct Answer

7. Fertilization of the ovum normally occurs in the:
   a. vagina
   b. abdominal cavity
   c. uterus
   d. fallopian tube

8. Polyspermy is prevented by:
   a. mitotic division of the ovum prior to fertilization
   b. secretion of estrogen by the ovum
   c. a brief electrical block on the surface of the ovum
   d. formation of the zygote

9. The time between fertilization of the ovum and implantation of the fertilized ovum in the uterus is approximately:
   a. 5 to 10 days
   b. fewer than 24 hours
   c. 3 weeks
   d. 2 days

10. Which of the following is *not* an action of the placenta?

a. delivery of waste from fetus to mother
b. exchange of gases between fetus and mother
c. delivery of antibodies from mother to fetus
d. delivery of nutrients from fetus to mother

11. An increase in which of the following hormones would serve to inhibit the onset of parturition?

a. relaxin
b. progesterone
c. oxytocin
d. prostaglandins

Answer Each Question in One or Two Sentences

12. Why can sperm enter the uterus more easily at midcycle?

13. Why is the acrosomal reaction important for fertilization?

14. How does sucking increase milk availability in the breast?

SUGGESTED READINGS

Beaconsfield, P., Birdwood, G., and Beaconsfield, R. "The placenta." *Scientific American*, 243:94–102, 1980.

Briggs, G.G., Freeman, R.K., Yaffe, S.J., eds. *Drugs in Pregnancy and Lactation*. Baltimore: Williams and Wilkins, 1986.

Diamond, M.P., and Decherney, A.H., eds. *Ectopic Pregnancy*. Philadelphia: W.B. Saunders Co., 1991.

Lagercrantz, H., and Slotkin, T.A. "The 'stress' of being born." *Scientific American*, 254:100–107, 1986.

Neville, M., and Daniel, C.W., eds. *The Mammary Gland*. Plenum Press, 1987.

Wilson, J.D., and Foster, D.W. *Williams' Textbook of Endocrinology*. Philadelphia: W.B. Saunders Co., 1985.

Worthington-Roberts, B.S., Vermeersch, J., and Williams, S.R. *Nutrition in Pregnancy and Lactation*. St. Louis: Times Mirror/Mosby, 1985.

KEY TERMS

acrosome reaction (p. 1007)
amniotic fluid (p. 1018)
areola (p. 1023)
blastocyst (p. 1009)
Braxton-Hicks contractions (p. 1018)
colostrum (p. 1018)
dizygotic twins (p. 1008)
embryo (p. 1009)
fertilization (p. 1007)
fetus (p. 1009)
lactation (p. 1023)
monozygotic twins (p. 1008)
parturition (p. 1019)
placenta (p. 1012)
polyspermy (p. 1007)
pregnancy (p. 1016)
prolactin (p. 1025)
relaxin (p. 1023)
umbilical cord (p. 1012)
zygote (p. 1009)

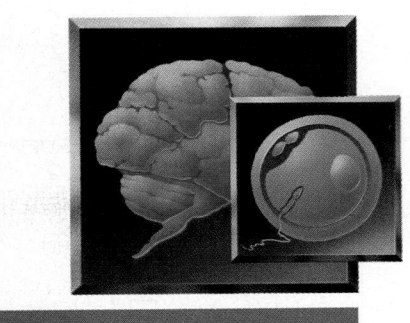

CHAPTER
32

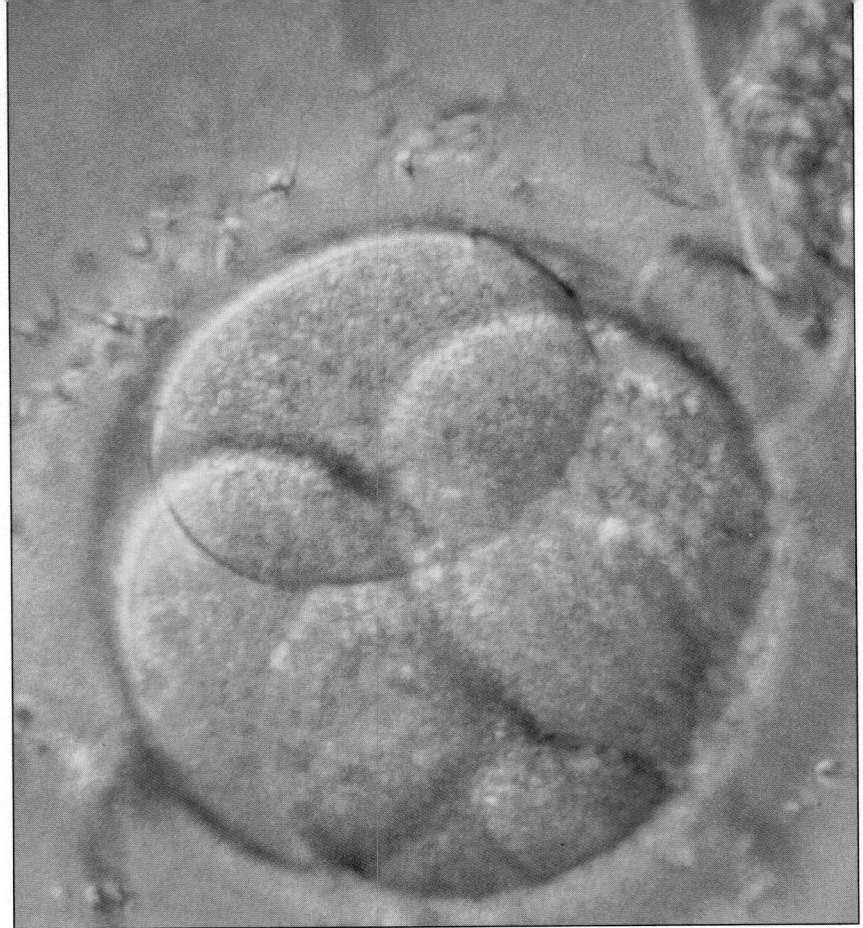

Sexual Physiology

In Chapter 30, we discussed the physiological processes that control reproduction in the male and female. In Chapter 31, we examined the events that occur during pregnancy, fetal development, parturition, and lactation (Fig. 32–1). We will now discuss the physiological processes that control human sexuality, beginning with an examination of how the biological sex of an individual is established. We will also investigate the physiological basis of the human sexual response, sexual dysfunction, and how contraceptives work.

Sexual Determination and Differentiation

When we consider the number of differences in the anatomy, physiology, and endocrinology of the male and female, it is surprising that male and female embryos are identical during the first six weeks of life. The characteristics that define an individual's sex are established through the processes of **sexual determination** and **sexual differentiation.** Determination of sex begins with the fusion of sperm and ovum, which establishes the **chromosomal (genetic,** or **genotypic) sex** of the zygote. The genetic information contributed to the zygote determines whether the **primitive gonads** will differentiate into **testes** or **ovaries,** which constitute the **gonadal sex** of the individual. The differentiated gonads, in turn, control development of the genital ducts, external genitalia, secondary sex characteristics, and endocrine profiles that constitute the **phenotypic sex** of the individual. Therefore, "maleness" or "femaleness" begins with chromosomal determination of sex, followed by gonadal differentiation and the development of phenotypic sex characteristics.

Sexual Determination: The Sex Chromosomes

Every cell in the body, except the sperm and ovum, contains 23 pairs of chromosomes. Twenty-two of these chromosomal pairs are termed **autosomes;** the twenty-third pair are the **sex chromosomes.** Autosomes provide the genetic information for the development and differentiation of most of the cells in the body. The sex chromosomes provide the genetic information that directs the differentiation of gonadal cells into ovaries or testes; hence, the sex chromosomes determine the gonadal sex of the individual.

Meiosis and the Sex Chromosomes

The two sex chromosomes in the female are designated **XX;** the two sex chromosomes in the male are designated **XY.** As discussed in Chapter 30, prior to fertilization the male germ cells (sperm) and the female germ cells (ova) undergo a special form of cell division known as **meiosis.** During meiotic division, each daughter cell receives only 23 chromosomes, one from each of the 23 chromosomal pairs. Given that the twenty-third pair (the sex chromosomes) in the male are XY, half of the sperm formed during meiosis will receive 22 autosomes and an X sex chromosome, while the other half will receive 22 autosomes and a Y sex chromosome. In contrast, the twenty-third pair of chromosomes in the female are XX, so all the ova formed during meiosis will each receive 22 autosomes and an X sex chromosome. The Y chromosome carries genes that direct the gonads to differentiate as testes and hence determines "maleness."

Figure 32–1
Human zygote at the 4-cell stage of cell division. (© CNRI/Phototake, NYC.)

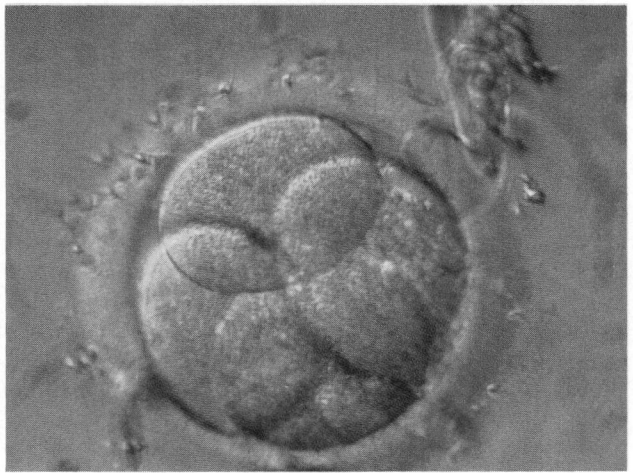

The Sex Chromosomes in Fertilization

During fertilization, when the sperm and ovum fuse, the 23 chromosomes from the sperm (22 autosomes and an X or Y sex chromosome) and the 23 chromosomes from the ovum (22 autosomes and an X sex chromosome) combine to produce a single-celled zygote that contains a total of 46 chromosomes. When a sperm containing an X chromosome fertilizes the X-bearing ovum, the chromosomal sex of the zygote will be XX, or **female.** When a sperm containing a Y chromosome fertilizes the X-bearing ovum, the chromosomal sex of the zygote will be XY, or **male.** If an ovum is not fertilized by a Y-bearing sperm, the testes, male genital ducts, and external genitalia will not develop. Therefore, the sex of the zygote is determined by the genetic material contributed by the father.

The Role of the Y Chromosome and the H-Y Antigen

A question of current interest to molecular geneticists is how the Y chromosome directs the embryonic gonad to differentiate as a testis. Although this process is not completely understood, it appears that the Y chromosome contains a genetic locus (or loci) which either codes for, or regulates the expression of, a specific protein found on the membrane of male cells. This protein is antigenic and induces immunological rejection in females. Hence, it is termed the **histocompatibility-Y antigen** or **H-Y antigen.** The H-Y antigen is thought to be responsible for inducing testicular differentiation in the primitive gonad. The chromosomal location and the mechanism of expression of the H-Y structural gene (or genes) that codes for the H-Y antigen are not known. Current evidence suggests that the H-Y antigen, disseminated by cells in the developing gonads, binds to H-Y receptors in the gonads and induces gonadal cells to differentiate as testes.

In the female, both of the X chromosomes are essential for normal differentiation of the ovaries. Whether a counterpart to the H-Y antigen exists on the X chromosomes that might direct differentiation of the ovaries is not known. Currently, it is thought that the absence of the H-Y antigen, and receptors for the antigen, predisposes the embryonic gonads to differentiate into ovaries. Therefore, the absence of the Y chromosome and consequent absence of the H-Y antigen in the female embryo results in the development of ovaries rather than testes.

Differentiation of the Internal Genitalia

During the first six weeks of embryonic development, the gonads of a genetically male or female embryo are anatomically and morphologically identical. During the third to the seventh week of embryonic development, two sexually undifferentiated gonads (one on each side) develop, which will differentiate into ovaries in the female or testes in the male (Fig. 32–2).

The Embryonic Ductal Systems

In addition, two distinct ductal systems develop in the embryo. One ductal system, termed the **Mullerian duct system,** is the precursor of the internal genitalia of the female. The other duct system, called the **Wolffian duct system,** is the precursor of the internal genitalia of the male (see Fig. 32–2). In the genetically female embryo, the Mullerian duct system will differentiate into fallopian tubes, uterus, cervix, and vagina, and the Wolffian duct system will regress (see Fig. 32–2). In the genetically male embryo, the Wolffian duct system will differentiate into the epididymis, vas deferens, seminal vesicles, prostate, and ejaculatory duct, and the Mullerian duct system will regress (see Fig. 32–2). As we will see, differentiation and development of one or the other of the duct systems is due to the presence or absence of hormones secreted by the developing gonads.

Differentiation of Internal Genitalia in the Male

The timing of gonadal differentiation in the male embryo is different from that in the female. Presence of the H-Y antigen in the male stimulates differentiation of the testes at 6 to 7 weeks of gestation, while in the female, the ovaries do not begin to differentiate until week 12 of gestation.

Mullerian Inhibiting Substance and Gonadal Differentiation. As the testes develop in the male embryo, under the direction of the H-Y antigen, the Sertoli cells of the testes begin to secrete a substance called **Mullerian Inhibiting Substance (MIS).** MIS is a glycoprotein that has local effects, through paracrine action, on cells surrounding the gonads. MIS induces regression, or involution, of the Mullerian duct system. The paracrine action of MIS is evi-

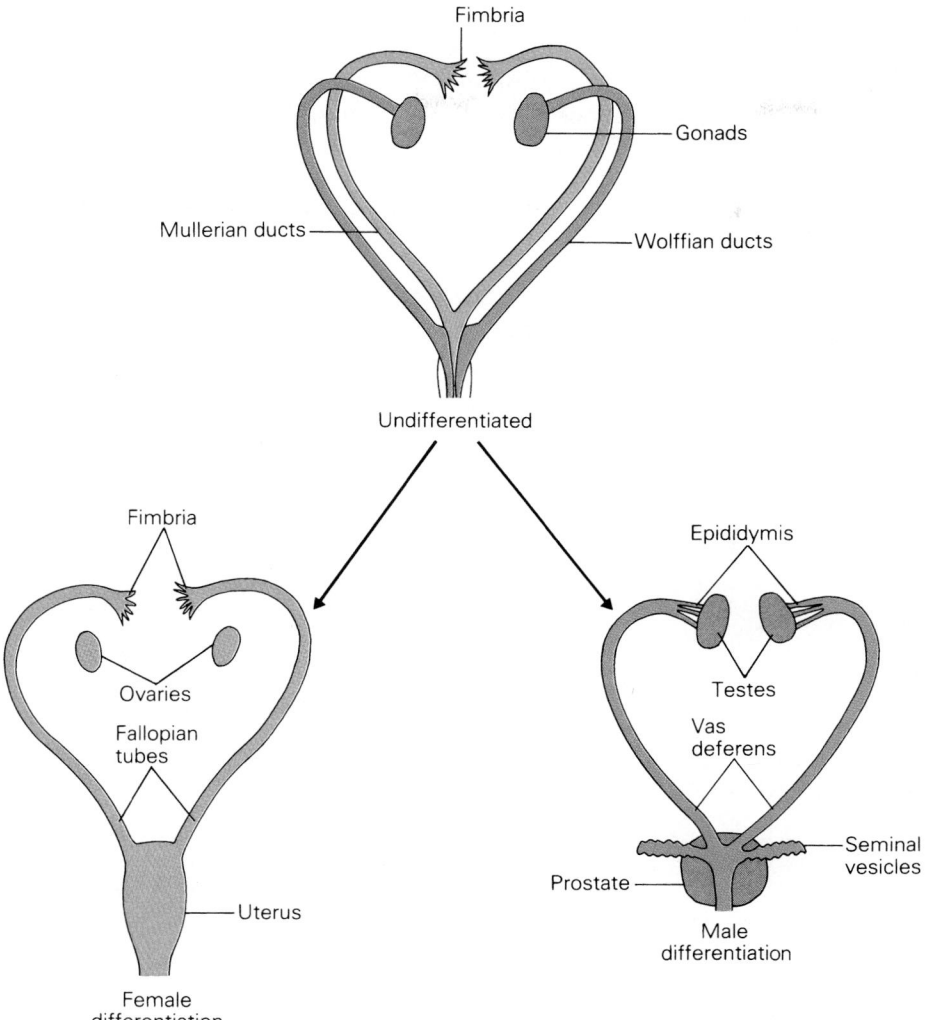

Fimbria

Gonads

Mullerian ducts

Wolffian ducts

Undifferentiated

Fimbria

Ovaries

Fallopian tubes

Uterus

Female differentiation

Epididymis

Testes

Vas deferens

Prostate

Seminal vesicles

Male differentiation

Figure 32–2
Embryonic differentiation of the male and female internal genitalia.

denced by the fact that if one of the two testes is removed at an early stage of development, Mullerian regression will occur only on the side of the body that has an intact testis. Although the Sertoli cells secrete MIS for about the first two years of life, secretory rates are much higher during the first year, when regression of the Mullerian ducts is required in the male.

Testosterone and Gonadal Differentiation. Although MIS has been identified as the factor responsible for regression of the Mullerian ducts, it does not stimulate the Wolffian ducts to differentiate into the internal genitalia of the male. Differentiation of the male internal genitalia is under the control of

testosterone. Leydig cells of the developing testes begin to secrete testosterone at about 7 to 9 weeks of gestation. Testicular content and secretion of testosterone rise from week 6 to 14 of gestation and then begin to decline during weeks 15 to 20 (Fig. 32–3). However, even at birth, when testosterone levels are lower than during early development, the male fetus is exposed to considerably higher concentrations of testosterone than the female. Testosterone, which diffuses from the testes, acts locally to stimulate differentiation of the male or Wolffian duct system during weeks 9 to 16 of embryonic development. Therefore, along with MIS, testosterone has paracrine actions on the developing gonads. While MIS induces Mullerian duct regression, testosterone

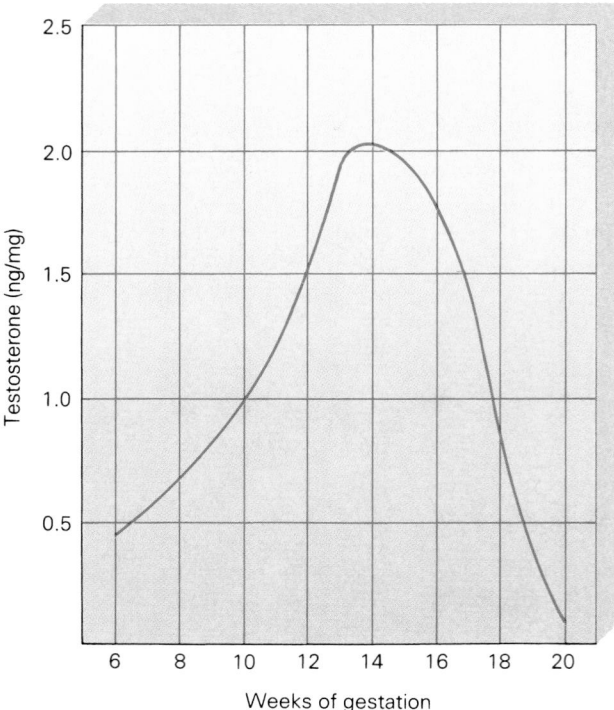

Figure 32–3
Testicular content of testosterone in the male fetus during gestation. (Adapted from Shearman, *Clinical Reproductive Endocrinology*, 1985, p. 349, originally reproduced from Tapanainen, et al., 1981.)

stimulates Wolffian duct development (see Fig. 32–2). As we will see, when testosterone biosynthesis is deficient in the developing testes, or when the Wolffian ducts do not respond to testosterone, the internal genitalia of the male embryo do not develop, but remain rudimentary, resulting in a condition called **testicular feminization.**

Differentiation of Internal Genitalia in the Female

Ovarian differentiation in the female embryo begins long after the testes have differentiated in the male embryo. At about 12 weeks of gestation, the germ cells (oogonia) in the ovary begin to divide meiotically to form oocytes, marking the beginning of ovarian differentiation. By week 20 to 25 of gestation, the female gonads can be structurally characterized as ovaries. During the third month of gestation, the Mullerian duct system in the female embryo undergoes further development, while the Wolffian duct system degenerates, owing to the absence of testosterone stimulation (see Fig. 32–2).

The presence of the ovaries is not necessary for Mullerian duct development in the female embryo, since differentiation of the uterus and fallopian tubes occurs even when no ovaries are present. Therefore, it appears that the ovaries (unlike the testes) do not contribute hormonal factors that govern differentiation of the female genital tract.

Differentiation of the External Genitalia

As we have seen with the internal genitalia, the external genitalia of both the male and female embryo are identical during the first 6 weeks of gestation.

Figure 32–4
Embryonic differentiation of the male and female external genitalia.

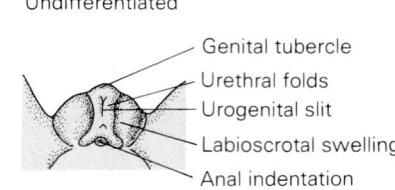

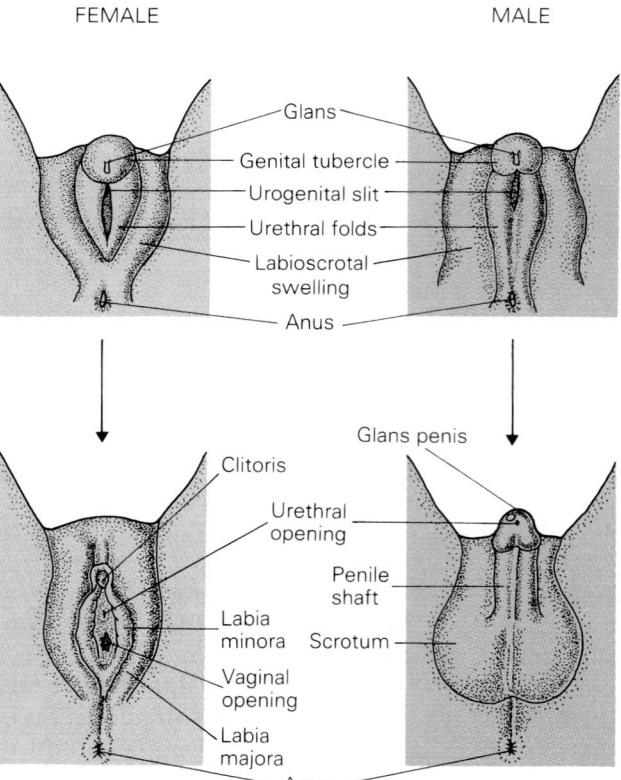

Table 32–1
Structural Origin of Male and Female External Genitalia

Undifferentiated Embryonic Structure	Differentiation in the Female	Differentiation in the Male
Genital tubercle	Clitoris	Glans of penis
Urethral folds	Labia minora	Shaft of penis
Labioscrotal swellings	Labia majora	Scrotum
Urogenital slit	Urethra and vaginal opening	Urethra
Anal indentation	Anus	Anus

The external genitalia consist of a urogenital slit surrounded by urethral folds, which are in turn surrounded by the labioscrotal swellings (Fig. 32–4). A genital tubercle is located above the urogenital slit, and the anal indentation is located below. Sexual differentiation of these embryonic structures into the external genitalia of the male and female is summarized in Table 32–1.

Differentiation of External Genitalia in the Female

Differentiation of the external genitalia in the female occurs because of the absence of testosterone. In the female embryo, the urethral folds and the labioscrotal swellings remain separate and differentiate into the labia minora and labia majora, respectively (see Fig. 32–4). The genital tubercle differentiates first into the glans and then into the clitoris (see Fig. 31–4). The urogenital slit differentiates into the vaginal opening and the urethra (see Fig. 32–4).

Differentiation of External Genitalia in the Male

Secretion of testosterone by the developing testes stimulates differentiation of the external genitalia in the male embryo. Testosterone reaches the external genitalia via the systemic circulation and hence exerts an endocrine action on differentiation of the external genitalia. Note that the external genitalia are androgen-sensitive tissues that contain the enzyme **5-alpha-reductase,** which converts testosterone to **dihydrotestosterone,** or **DHT.** Therefore, while testosterone directly stimulates differentiation of the Wolffian ducts, DHT, rather than testosterone itself, stimulates sexual differentiation of the external genitalia in the male. Deficiencies of testosterone secretion, 5-alpha-reductase activity, or receptors for

DHT in external genital tissues can, as we will see, result in impaired masculinization of the male external genitalia.

In response to stimulation from dihydrotestosterone, the urethral folds fuse and surround the urethra. Fusion of the urethral folds forms the shaft of the penis (see Fig. 32–4). Dihydrotestosterone also stimulates fusion of the labioscrotal swellings on the midline, forming the scrotum and the skin that covers the ventral portion of the penis (see Fig. 32–4). Although the clitoris and penis originate from similar embryonic tissue, the penis grows 3 to 4 times faster than the clitoris. Therefore, the sex of the embryo can be determined by the size of the glans as well as by the presence or absence of urethral fold fusion. In the male embryo, dihydrotestosterone also induces differentiation of the prostate and inhibits the development of a vagina.

The factors controlling sexual determination and differentiation are schematically illustrated in Figure 32–5.

Abnormalities of Sexual Differentiation

Normal patterns of sexual differentiation can be altered because of abnormalities at the level of chromosomal sex (genotype), gonadal sex, internal genital sex, or external genital sex (phenotype). Although many variations and degrees of variation in patterns of sexual differentiation are possible, very few are common. We will discuss only those that are best understood.

Chromosomal Abnormalities in the Female

As we have previously discussed, the normal female embryo has 23 pairs of chromosomes (22 autosomal pairs and one pair of sex chromosomes, XX)

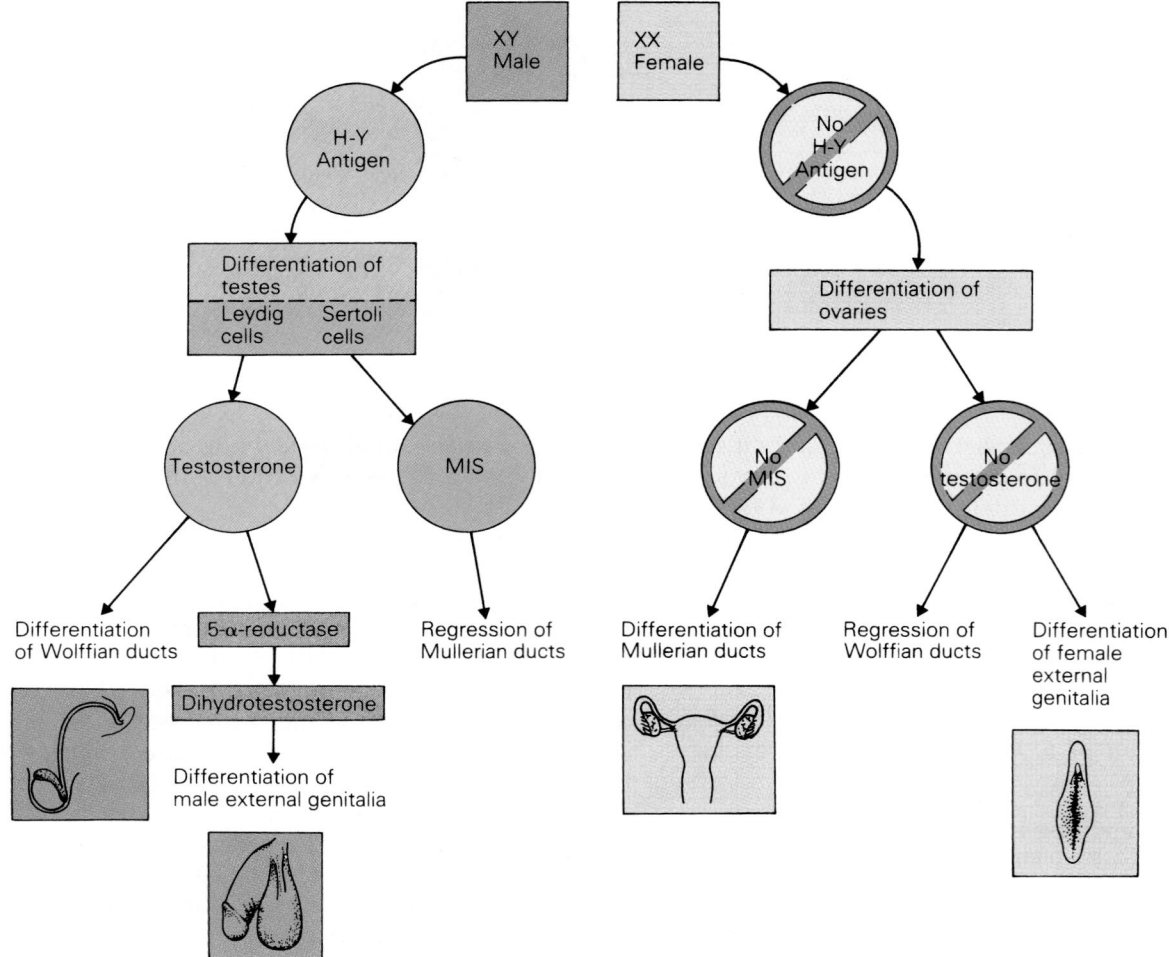

Figure 32–5
Schematic illustration of sexual differentiation.

(Fig. 32–6). Both of the X chromosomes are essential for normal differentiation of the ovaries, but only one X is necessary for differentiation of the Mullerian ducts and the female external genitalia. One chromosomal abnormality that has been identified in the female is the presence of only one X sex chromosome, designated as XO, which is termed **Turner's syndrome** (Table 32–2). This chromosomal deficiency is characterized by impaired development of the female gonads, which results in the appearance of an incompletely formed gonad, termed a **vestigal gonadal streak.** The uterus, fallopian tubes, and external female genitalia develop normally.

Chromosomal Abnormalities in the Male

The normal male embryo has 23 pairs of chromosomes (22 autosomal pairs and one pair of sex chromosomes, XY) (see Fig. 32–6). Some genetically

male individuals receive an extra X chromosome (XXY), which results in **Klinefelter's syndrome** (see Table 32–2). In these cases the presence of the Y chromosome directs differentiation of the testes, but they are reduced in size. As in normal males, testosterone secretion from the testes in XXY males stimulates development of the Wolffian duct system and development of the male external genitalia, while MIS secretion from the testes induces Mullerian regression. However, the presence of the extra X chromosome leads to testicular dysfunction, owing to impaired development of the seminiferous tubules and a deficiency in spermatogenesis.

Enzyme Deficiencies in the Female

As we have previously learned, the presence of androgens (testosterone and dihydrotestosterone) during early embryonic development results in the

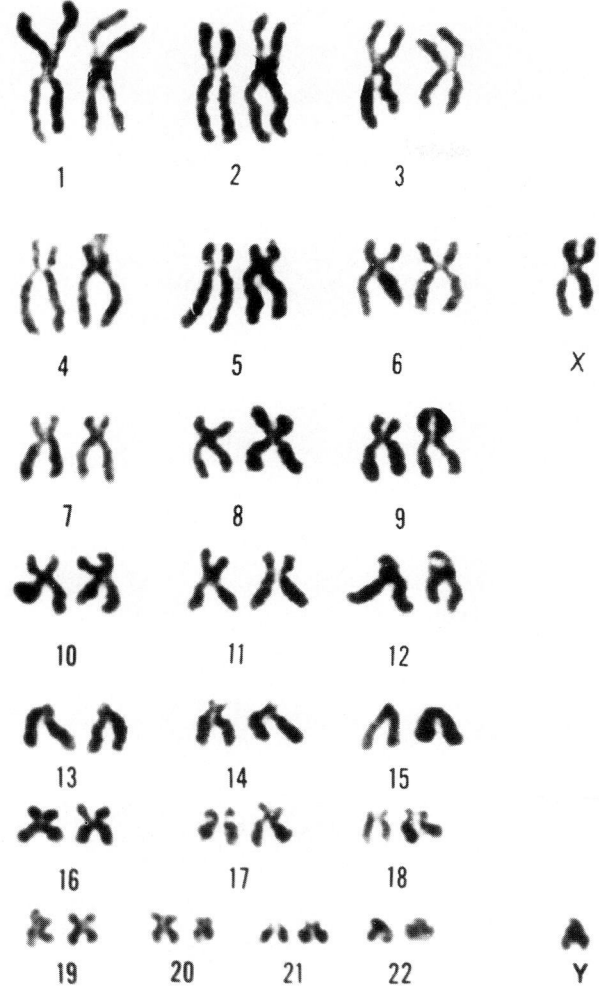

Figure 32–6
Photograph of normal human karyotype. (© Leonard Lessin/Photo Researchers, Inc.)

differentiation of male external genitalia. Therefore, it is not surprising that exposure of a genetically female embryo (XX) to androgens during embryonic development can result in **masculinization,** or **virilization,** of the female embryo's external genitalia.

Some genetic females (XX) lack the enzymes **11-hydroxylase** or **21-hydroxylase** in their adrenal glands. These enzymes are necessary for the synthesis of gluco- and mineralocorticoids. In the absence of these enzymes, the adrenal glands use the steroid precursor cholesterol to synthesize androgens. Hypersecretion of adrenal androgens results in the development of the **adrenogenital syndrome,** also called **congenital adrenal hyperplasia** (Table 32–2). These individuals have normal ovaries, owing to a full chromosomal XX complement. The absence of a Y chromosome in these individuals precludes the development of testes and hence testicular secretion of MIS. The absence of MIS ensures normal development of the Mullerian duct system (including the uterus and fallopian tubes). However, hypersecretion of adrenal androgens can result in masculinization of the external genitalia. The degree of masculinization depends on the severity of the adrenal enzymatic deficiency; it can range from mild hypertrophy of the clitoris to fusion of the urethral folds and the labioscrotal swellings, which then resemble a penis and scrotum. Severe masculinization of the female external genitalia, a condition termed "female pseudohermaphroditism," can lead to mistakes in identifying the gender of the infant.

Enzyme and Receptor Deficiencies in the Male

Genetic males (XY) who lack androgen receptors cannot respond to testosterone or dihydrotestosterone, which results in a condition known as **testicular feminization syndrome** (see Table 32–2). In these individuals, the presence of the Y chromosome results in development of the testes, and testicular secretion of MIS induces regression of the Mullerian ducts. However, androgen insensitivity results in a lack of development of the Wolffian duct system and the absence of masculinization of the external genitalia. Therefore, these individuals have no internal genitalia at all, because MIS induces regression of the Mullerian ducts while the lack of androgen receptors precludes development of the Wolffian ducts.

As with androgen receptor deficiencies, a deficiency of the enzymes necessary for testosterone biosynthesis also results in varying degrees of Wolffian duct underdevelopment and reduced masculinization of external genitalia (see Table 32–2). In such cases, the presence of the Y chromosome and of MIS ensures the development of the testes and regression of the Mullerian ducts. A deficiency of the enzyme 5-alpha-reductase, necessary for converting testosterone to dihydrotestosterone, results in reduced masculinization of the external genitalia (which respond to dihydrotestosterone). However, this enzyme deficiency does not interfere with the formation of normal testes due to the presence of the Y chromosome. Mullerian regression occurs because of the presence of MIS, and the Wolffian duct system develops in response to testosterone stimulation (see Table 32–2).

A summary of the important events that control sexual determination and differentiation is given in Table 32–3.

Table 32–2
Abnormalities of Sexual Differentiation

Chromosomal Complement	Enzyme or Receptor Defects	Gonads	Internal Genitalia	External Genitalia
XX Normal female	—	Ovaries	Female	Female
XO Turner's syndrome	—	Ovaries with impaired development—"streak gonads"	Female	Female
XXY Klinefelter's syndrome	—	Testes with impaired development of seminiferous tubules and deficient spermatogenesis	Male	Male
XX Adrenogenital syndrome	Adrenal 11- or 21-hydroxylase deficiency	Ovaries	Female	Male/Female
XY Testicular feminization	Lacks androgen receptors	Testes	None	Female
XY	Lacks ability to synthesize testosterone	Testes	Male with impaired development	Female/Male
XY	5-Alpha-reductase deficiency	Testes	Male	Female/Male

Table 32–3
Summary of Events Controlling Sexual Determination and Differentiation

Female	*Male*
1. Two X chromosomes are needed for normal differentiation of the ovaries.	1. The Y chromosome, which codes for or regulates the expression of the H-Y antigen, is required for differentiation of the testes.
2. Only one X chromosome is required for differentiation of the Mullerian ducts and the female external genitalia.	2. Mullerian Inhibiting Substance, secreted by the Sertoli cells of the testes, induces regression of the Mullerian ducts.
3. The absence of ovaries will not preclude development of the Mullerian ducts or the female external genitalia.	3. Testosterone, secreted by Leydig cells of the testes, stimulates Wolffian duct development but has no effect on the Mullerian ducts.
4. Exposure of the genetically female embryo to androgens during early development will result in varying degrees of masculinization of the external genitalia, depending on when during development the exposure occurs. Exposure to androgens will not alter development of the ovaries or the Mullerian ducts.	4. Testosterone is converted to dihydrotestosterone (DHT) in tissues of the external genitalia, and DHT stimulates differentiation of the male external genitalia. An absence of DHT formation results in reduced masculinization of the external genitalia but does not affect normal development of the Wolffian ducts, which respond directly to testosterone.

Human Sexual Response

With the increased availability and popularity of contraceptives, it is now useful to draw a distinction between reproductive and nonreproductive, or sexual, physiology. Note, however, that sexual and reproductive physiology are inextricably intertwined. Most of the reproductive organs in both sexes serve both sexual and reproductive functions. In addition, sexual arousal in the male is necessary for reproduction since ejaculation, which results in sperm transmission, does not occur in the absence of sexual arousal.

Physiological Changes During the Sexual Response Cycle

Sexual arousal, like other physiological responses, can result from either **physical stimulation** or **mental stimulation.** For instance, salivation can occur in response to the thought or sight of food as well as to eating. In much the same way, sexual arousal can result from mental stimulation provided by movies or books as well as from physical stimulation such as touching or kissing. Therefore, the human sexual response can be significantly altered by thoughts, feelings, and learned values. However, the physiological responses to sexual arousal are the same whether arousal is due to physical or mental stimulation. Sexual arousal, which occurs throughout life from infancy into old age, involves activation of the nervous system, the endocrine system, and the sex organs.

The physiology of the human sexual response in the male and female can be divided into four stages of sexual arousal, as defined by Masters and Johnson: **excitement, plateau, orgasm,** and **resolution.** While distinctions among the four stages are not clearcut and individual variations are many, all four stages of sexual arousal occur in both males and females and in both heterosexuals and homosexuals.

Excitement Phase in the Male

Sexual arousal during the excitement phase increases parasympathetic nerve activity, which causes the arteries in the penis to dilate, resulting in increased blood flow to the penis. As blood becomes concentrated in the tissues of the penis, a process called **vasocongestion,** the penis increases in diameter, length, and firmness, which defines **penile erection** (Fig. 32–7). Although the penis of some lower animals contains a bone, human males do not have a penile bone. The human penis does contain a thin skeletal muscle, but this muscle does not contribute significantly to penile erection, as evidenced by the fact that normal erection occurs in the absence of penile muscle activity following certain spinal cord injuries. Instead, penile erection depends on increased blood flow to the penis and increased intrapenile pressure. Penile vasoconstriction during sexual arousal in the male is not constant. Decreased blood flow to the penis can result in a reduction in penile size and firmness even though sexual stimulation and neuromuscular tension have not diminished. Such fluctuations in blood flow to the penis are usually short-lived.

Sexual arousal also results in increased blood flow to the testes, causing an increase in testicular size (see Fig. 32–7). During the excitement phase, the muscles surrounding the vas deferens contract, lifting the testes closer to the body and smoothing the scrotal skin (see Fig. 32–7). An increase in general muscle tone or tension, termed **myotonia,** also accompanies sexual arousal. Erection of the nipples of the breast occurs in some (but not all) men.

Excitement Phase in the Female

One of the physiological responses to sexual arousal during the excitement phase in the female is **vaginal lubrication** (Fig. 32–8). Both penile erection in the male, and vaginal lubrication in the female, occur in response to vasocongestion. Sexual arousal results in increased blood flow to the vagina, and vasocongestion of the vagina leads to release of moisture from the vaginal lining, a process termed **transudation.** As vaginal secretion of moisture increases, some moisture may reach the vaginal opening and the surrounding labia. Vaginal secretion of moisture lubricates the vagina and facilitates insertion of the penis without discomfort. Note that there are great individual variations in the amount of moisture secreted by the vagina during sexual arousal, and the amount of moisture present cannot be used as an index of sexual arousal. Heightened sexual arousal may be accompanied by sparse vaginal lubrication, and vaginal secretion of moisture can occur in the absence of sexual arousal.

During the excitement phase, the vagina expands, and the cervix and uterus are lifted upward away from the vagina (see Fig. 32–8). Increased blood flow to the external genitalia results in increased size of the clitoris, labia minora, and labia majora (Fig. 32–9a). Muscle contractions in the breast tissue often result in nipple erection, and increased blood flow to the breast causes breast enlargement (see Fig. 32–9b). As in the male, there is

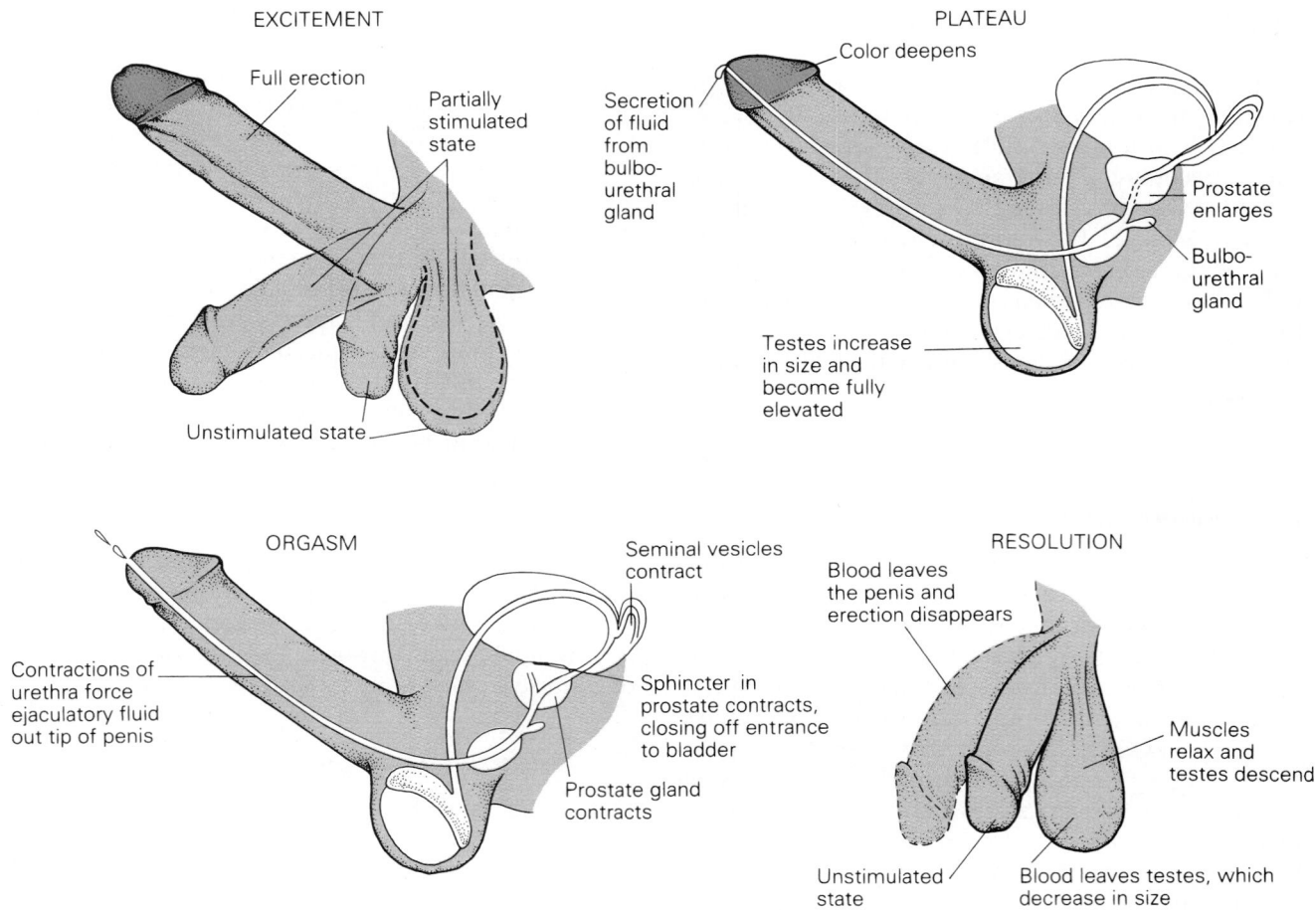

Figure 32–7
Changes in the external and internal genitalia of the male during the sexual response cycle. (Redrawn from Masters, Johnson, and Kolodny, *Human Sexuality*, Little, Brown, 1985.)

an increase in general muscle tension, or myotonia, in the female during the excitement phase.

Plateau Phase in the Male

The plateau phase of sexual arousal precedes orgasm. Time spent in the plateau phase varies widely within and among individuals. Men who ejaculate quickly spend little time in this phase. During the plateau phase, blood pooling in the head of the penis causes this area to enlarge and become darker in color (see Fig. 32–7). The testes continue to increase in size because of increased blood flow, and muscle contractions in the scrotum lift the testes up and back toward the anus (see Fig. 32–7). As was discussed in Chapter 30, pre-ejaculatory fluid from the bulbo-urethral glands, which contains viable sperm, is often secreted at the tip of the penis during the plateau phase of sexual arousal (see Fig. 32–

7). Heart rate, blood pressure, respiration rate, and muscle tension all increase significantly during this phase.

Plateau Phase in the Female

In the female, blood flow to the vagina increases, resulting in swelling of the walls of the vagina, which reduces the size of the vaginal opening (see Fig. 32–8). This increases the vaginal surface area that contacts the penis during intercourse. The uterus is elevated away from the vagina in a process termed *tenting* (see Fig. 32–8). The clitoris, which is highly sensitive to touch during this phase, is retracted against the pubic bone and becomes partially covered by the skin of the clitoral hood, which reduces direct physical contact of the clitoris (see Fig. 32–9*a*). Increased blood flow to the labia produces a deepening in the color of the labia minora and labia

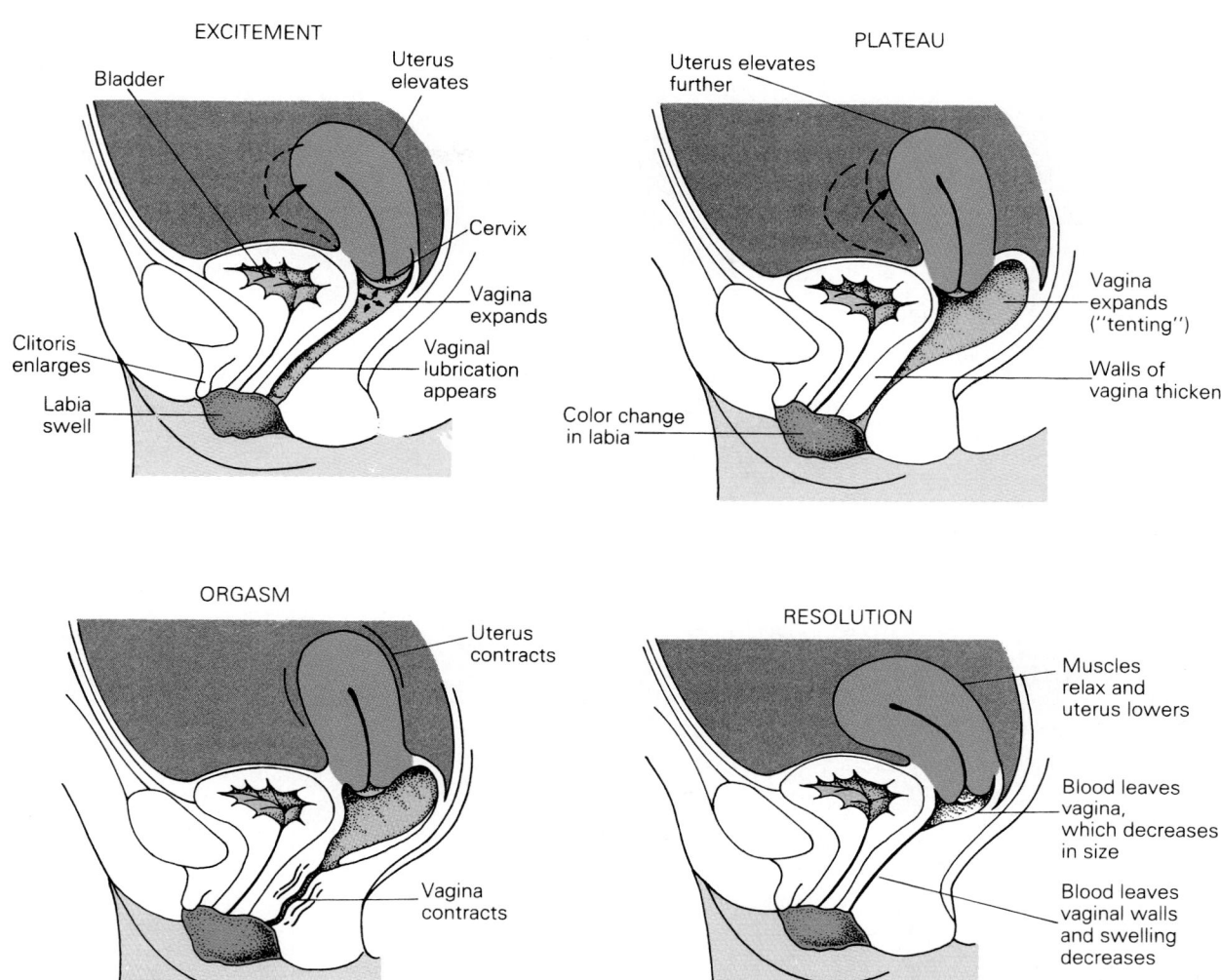

Figure 32–8
Changes in the external and internal genitalia of the female during the sexual response cycle. (Redrawn from Masters, Johnson, and Kolodny, *Human Sexuality*, Little, Brown, 1985.)

majora. Blood flow to the breasts results in increased size of the areoli and reddening of the skin color in areas on the breasts, referred to as the "sexual flush" (see Fig. 32–9b). Reddening of the skin on other areas of the body may occur as well. As is seen in the male, heart rate, blood pressure, respiration rate, and muscle tension all increase during the plateau phase in the female.

Orgasmic Phase in the Male

Ejaculation in the male involves two processes: the **emission** and the **expulsion** of ejaculatory fluid (see Fig. 32–7). During emission, increased sympathetic nerve activity stimulates rhythmic muscular contractions of the vas deferens, seminal vesicles, and prostate, forcing ejaculatory fluid into the urethra.

Once these contractions have begun, they cannot be stopped until ejaculation is complete. During the expulsion phase, rhythmic muscular contractions of the urethra force ejaculatory fluid out of the tip of the penis. These two phases of ejaculation usually last only a few seconds. As was discussed in Chapter 30, a sphincter in the prostate contracts during ejaculation, sealing off the entrance to the bladder and preventing urine from mixing with ejaculatory fluid (see Fig. 32–7).

Orgasmic Phase in the Female

Orgasm in the female is characterized by rhythmic muscular contractions of the uterus and vagina that may last for as long as 15 seconds (see Fig. 32–8). As in the male, muscles in various parts of the body

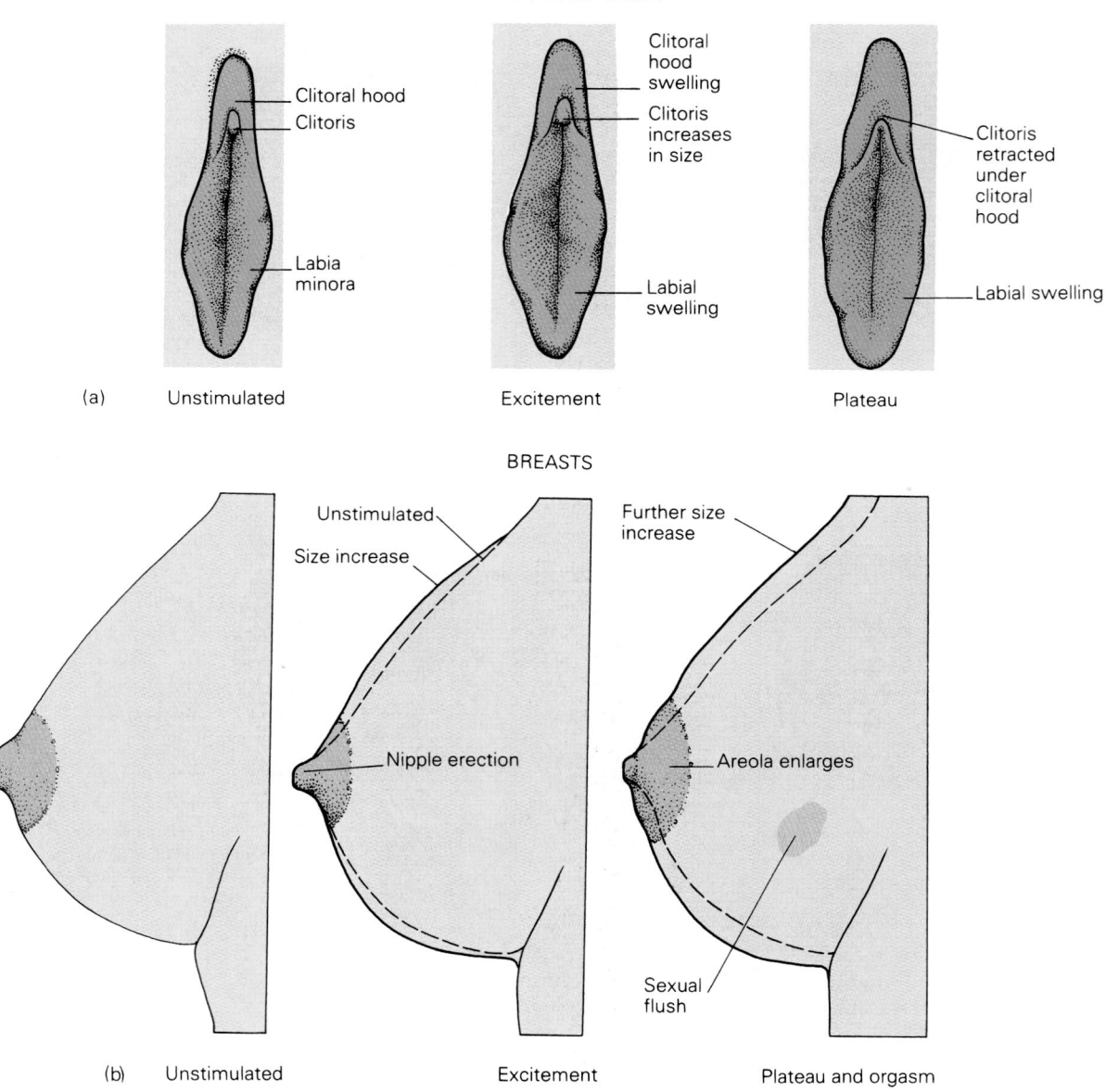

Figure 32–9
Changes in the (*a*) clitoris, labia, and (*b*) breasts during the female sexual response cycle. (Redrawn from Masters, Johnson, and Kolodny, *Human Sexuality*, Little, Brown, 1985.)

involuntarily contract during female orgasm, resulting in muscular rigidity and spasms.

Some controversy surrounds the importance of clitoral stimulation in inducing orgasm in the female, and whether females ejaculate. It appears that most women, while capable of achieving orgasm in response to vaginal stimulation, experience more intense orgasms when clitoral and vaginal stimulation are combined. Although most females do not ejaculate fluid from the urethra during orgasm, ejaculation-like responses have been reported by some females. At present, it is not known whether the composition of the fluid expelled from the urethra during orgasm in these women resembles that of prostatic fluid or urine.

Note that female orgasm is not essential for reproduction, since insemination of the female can occur in the absence of female sexual arousal.

Resolution Phase in the Male

Following ejaculation, most males enter a refractory period during which no further orgasm or ejaculation is possible. The length of the refractory period is highly variable (from a few minutes to many hours) and is prolonged by fatigue, time elapsed since the last orgasm, and age. During the refractory period, increased sympathetic nerve activity constricts the arteries in the penis, and blood leaves the penis through the veins, resulting in a loss of penile erection (see Fig. 32–7). Decreased blood flow to the testes also results in a decrease in testicular size. As muscles relax, the testes are lowered away from the body (see Fig. 32–7). Heart rate, blood pressure, respiration, and muscle tension all return to normal during the resolution phase. When intense sexual arousal is not followed by orgasm, the length of the resolution phase is prolonged and can be accompanied by aching sensations in the testes due to continued vasoconstriction.

Resolution Phase in the Female

Many females, unlike most males, are **multiorgasmic** (capable of experiencing more than one orgasm before entering the resolution phase) if intense sexual arousal is maintained. During the resolution phase, muscular relaxation results in a lowering of the uterus and cervix toward the vagina, which increases the probability of cervical contact with any ejaculatory fluid present in the vagina (see Fig. 32–8). Blood flow away from the breasts and genitals results in a reduction in the size of the breasts, clitoris, labia, and vagina. As in the male, the female's heart rate, blood pressure, respiration, muscle tension, and skin flushing all return to normal during the resolution phase. Breast and genital tissues are extremely sensitive to touch following orgasm in both sexes, and continued physical stimulation of these tissues during the resolution phase can be irritating or painful.

Sexual Response During Infancy

Sexual arousal in response to physical stimulation occurs even before birth. Penile erections have been detected by ultrasound in the male fetus during the last few months of gestation. At birth, many male infants exhibit penile erections, while female infants manifest vaginal lubrication and clitoral enlargement. Following birth, certain forms of physical contact, including breastfeeding, can result in penile erection in male infants and vaginal lubrication and enlargement of the clitoris in female infants. Al-though both males and females are capable of achieving orgasm from birth, the ability of the male to ejaculate does not occur until the internal and external genitalia have matured at puberty.

Sexual Response During Aging

Although physiological changes in the sexual response occur in both males and females during aging, the ability to become sexually aroused does not normally disappear.

Aging and the Female Sexual Response

In the female, the breasts, clitoris, and vagina remain sensitive to stimulation in old age. However, after menopause, the ovaries stop secreting estrogen, which leads to changes in the vagina. A low level of circulating estrogen results in diminished blood flow to the vagina during sexual arousal, which in turn leads to decreased vaginal lubrication. Use of an artificial lubricant can circumvent the pain that may result from intercourse in the absence of sufficient vaginal lubrication, a condition termed **dyspareunia**. Some evidence suggests that regular sexual activity following menopause slows the decline in vaginal lubrication that occurs during aging. A decrease in the elasticity of the vaginal walls in aging women leads to reduced vaginal expansion during sexual arousal. In addition, enlargement of the breast, development of the sexual flush, and increased muscle tension during arousal are significantly reduced in older women. Changes in the female response during aging are summarized in Table 32–4.

Aging and the Male Sexual Response

Although circulating testosterone levels begin to decline gradually after age 55 to 60 (see Chapter 30), there is no dramatic drop in testosterone levels during aging in the male that parallels the rapid decline in circulating estrogen levels in the female following menopause. Healthy males can maintain the capacity for penile erection throughout their entire life spans, and sperm production continues till age 80 or 90, although the production rate slows down after age 40.

However, physiological changes in the male sexual response do occur with aging. The sensitivity of penile sensory nerves to tactile stimulation decreases with age, and hence more prolonged and direct penile stimulation may be needed to induce erection. Blood flow to the penis during sexual arousal may also be reduced with age, and so the

Table 32–4
Changes in the Sexual Response During Aging

Female	Male
1. A decrease in circulating estrogen following menopause results in decreased blood flow to the vagina, which in turn results in diminished production of vaginal lubrication. Decreased vaginal lubrication can result in painful intercourse, or dyspareunia.	1. Penile erection occurs more slowly and may require more prolonged stimulation.
2. The vaginal walls lose elasticity, and sexual arousal produces less vaginal expansion.	2. The penis is less firm when fully erect.
3. Many responses to sexual arousal—including enlargement of the breast, reddening of skin color (sexual flush), and muscle tension—are significantly reduced.	3. Time spent in the refractory period following orgasm, during which no erection or ejaculation is possible, is prolonged.
	4. The volume of ejaculatory fluid expelled at orgasm is decreased.
	5. Testicular size and elevation during sexual arousal are reduced.
	6. Many responses to sexual arousal—including reddening of skin color (sexual flush) and muscle tension—are reduced.

penis may be less firm when fully erect. Testicular elevation during sexual arousal is slower and less complete in the older male. The amount of fluid ejaculated at orgasm is often reduced, and the refractory period following orgasm is prolonged. As in the female, muscle tension during sexual arousal in the older male is also reduced. A summary of the changes in the male sexual response during aging is contained in Table 32–4.

Sexual Dysfunction

The term *sexual dysfunction* refers to any condition in which the normal physical responses to sexual stimulation are impaired. Sexual dysfunction can be a source of anxiety, frustration, and distress, disrupting personal relationships and family life.

Endocrine Factors in Sexual Dysfunction

Estrogen Deficiencies in the Female

Estrogen plays an important role in maintaining vaginal lubrication. Estrogen stimulates proliferation of the vascular bed beneath the vaginal epithelium and increases blood flow to the vagina, which results in secretion of moisture from the vaginal walls. Inadequate estrogen stimulation can result in **atrophic vaginitis,** which is characterized by dryness and thinning of the vaginal muscosal mem-

brane and decreased elasticity and muscle tone of the vagina. Following menopause, when circulating estrogen levels are decreased, vaginal elasticity and vascularization are reduced, and blood flow to the vagina can be decreased by as much as 80%. Estrogen replacement therapy following menopause can increase blood flow to the vagina by fourfold.

Premenopausal women with very low levels of circulating estrogen (for instance, following ovarian disease or surgical removal of the ovaries) can also experience atrophic vaginitis with accompanying loss of vaginal lubrication and elasticity. Inadequate vaginal lubrication can produce discomfort or pain during sexual intercourse (dyspareunia). Women with very low levels of circulating estrogen can also experience symptoms of peripheral neural dysfunctions, which can include slowed response to tactile stimulation or loss of clitoral sensation. Estrogen replacement therapy can reduce the severity of these symptoms.

Testosterone Deficiencies in the Male

Some evidence suggests that testosterone is necessary for normal sexual activity in the male. Although males with low circulating levels of testosterone can attain penile erection, the capacity to ejaculate during orgasm may be delayed, diminished, or absent. In addition, low levels of testosterone are often associated with reduced sexual desire or sexual libido. Androgen replacement therapy often increases both **ejaculatory capacity** and **sexual**

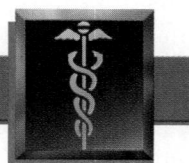
Amenorrhea

Amenorrhea, or the absence of menstruation, is classified as either primary or secondary. Primary amenorrhea is the absence of menstruation in females who have never menstruated. Secondary amenorrhea is the cessation of menstruation in females who have previously menstruated. Primary amenorrhea is rare and usually results from developmental abnormalities of the ovaries or the reproductive tract, or from the formation of scar tissue in these structures in response to physical injury or infections occurring before the first menstrual cycle would normally begin.

Secondary amenorrhea is more common. Diagnosis can be problematic, because the length of the menstrual cycle varies greatly among women. A general rule used to diagnose this disorder is the absence of menstruation for an interval greater than three times the individual's normal cycle length.

Secondary amenorrhea can result from malfunctions at the level of the ovary, uterus, pituitary, or hypothalamus. During adulthood, autoimmune diseases, radiotherapy, chemotherapy, and surgery or infections that cause scar tissue can result in ovarian failure and a reduction in the number of viable follicles. At the level of the pituitary, the most common cause of secondary amenorrhea is the presence of a prolactin-secreting tumor, which results in hyperprolactinemia and consequent suppression of GnRH release (see text). At the level of the hypothalamus, many factors can contribute to a suppression of GnRH release and the onset of secondary amenorrhea. Stress, which often accompanies intense pro-

grams of physical exercise, can suppress GnRH release and result in secondary amenorrhea. Extreme weight loss, which can occur during dieting; in response to drugs such as alcohol, narcotics, or amphetamines; and in the condition termed "anorexia nervosa," can also suppress GnRH release and result in secondary amenorrhea.

Extreme weight gain can result in secondary amenorrhea through a different mechanism. An increase in adipose tissue results in increased conversion of adrenal androgens to estrogen, and this increase in circulating estrogen disrupts the normal feedback signals to the hypothalamus (see text). Psychological stresses of many types also result in amenorrhea.

Secondary amenorrhea can be seen following discontinuation of birth control pills in approximately 2% of the women who have taken the pills for two or more months, a condition termed "post-pill amenorrhea." It is not clear whether this type of amenorrhea is caused by a hypothalamic disturbance in response to the steroid content of the pills, or whether the women displaying post-pill amenorrhea have a prior history of menstrual irregularities.

Secondary amenorrhea is often reversible and tends to disappear with prolonged discontinuation of oral contraceptives, when normal dietary patterns are resumed, or when the source of stress is removed. It should be noted that the most common nonpathological condition that results in secondary amenorrhea is pregnancy.

desire. The exact mechanism by which testosterone facilitates ejaculation or sexual desire is not known. It has been proposed that androgens are involved in sensory perception and the capacity for development of muscle tension within the pelvis.

Neural Disorders and Sexual Dysfunction

Afferent sensory impulses from the internal and external genitalia enter the spinal cord and travel to the brain. Efferent nerve impulses from the brain leave the spinal cord and activate muscles and blood vessels of the internal and external genitalia. The excitement phase of sexual arousal in both males

and females involves increased afferent neural input from sensory nerves in the genitals and efferent neural induction of vascular changes in the internal and external genitalia. Orgasm involves increased efferent motor activity, which induces muscular contractions in the internal and external genitalia. Therefore, both mechanical and disease-related injuries to the spinal cord or brain can result in impairment of sexual arousal and orgasm in both males and females. Interruption of afferent sensory pathways can interfere with sexual arousal. Interruption of efferent pathways may disrupt increased blood flow to the genitals during sexual arousal and hence may reduce penile erection in the male and vaginal swelling and lubrication in the female, as well as muscle contractions during orgasm in both sexes.

Multiple sclerosis, a disease characterized by patches of demyelinization in the spinal cord and brain, can be accompanied by a reduction or elimination of sexual arousal and orgasmic capacity in both males and females because of lesions in the spinal cord that disrupt afferent and efferent nerve transmission.

Severe **diabetes mellitus** can be accompanied by dysfunctions of the sensory fibers that innervate the penis and clitoris, resulting in a reduction or absence of the genital sensations necessary for sexual arousal and orgasm. Diabetic neural dysfunction can also result in decreased blood flow to the genitals, which can result in insufficient penile rigidity to achieve vaginal penetration.

Drugs and Sexual Dysfunction

Although small amounts of **alcohol** are often used to reduce sexual inhibitions, large amounts have a depressant effect on the central nervous system, which can interfere with sexual arousal. In addition, chronic and excessive consumption of alcohol, as seen in alcoholics, can lead to damage of the CNS, which results in alcoholic neuropathy in males and females. Alcohol-induced injury to the somatic and autonomic nerves in the periphery, spinal cord, or brain can result in a reduction or absence of sexual arousal and orgasmic capacity.

Sedatives (e.g., barbiturates, chloral hydrate), **narcotics** (e.g., codeine, morphine), and **tranquilizers** (e.g., diazepam, meprobamate), when taken in high doses, can retard or inhibit sexual arousal and orgasm in both males and females because of a depression of nerve conductance in the central nervous system.

Some drugs used to treat hypertension, such as reserpine and alpha-methyldopa, can alter neural control of blood flow to the internal and external genitalia in both males and females, resulting in decreased vaginal swelling and lubrication in the female and difficulty in achieving erection in the male.

Some antihistamines that contain ephedrine decrease blood flow to the vagina and vaginal lubrication, resulting in dyspareunia.

Other Sources of Sexual Dysfunction

Anorgasmia, Infection, and Vaginismus

Difficulty in reaching orgasm, which is referred to as **anorgasmia,** represents the largest category of sexual dysfunction in the female. However, organic disorders that can interfere with orgasm, such as hormone deficiencies, alcoholism, diabetes, or neurological disorders, account for fewer than 5% of the cases of anorgasmia. Anorgasmia appears to be largely due to psychological factors, which may include sexual inhibitions acquired during early training, unrealistic performance expectations for oneself or one's partner, difficulty with sexual communication, and negative self-image. These issues are beyond the scope of this chapter. For further information, consult the volumes on human sexuality listed at the end of this chapter.

Vaginal infections, allergic reactions, or chemical irritations can produce a condition termed **nonatrophic vaginitis.** This condition, which is quite common and usually of short duration, can result in vaginal dryness or tenderness.

Endometriosis (see "Infertility") and **pelvic inflammatory disease** can both cause pain during sexual arousal and intercourse. This is due to the fact that vasocongestion and pelvic muscle contractions during sexual arousal place increased pressure on diseased tissues, in the case of endometriosis, or on chronically inflamed tissues, in the case of pelvic inflammatory disease. Further pressure due to intercourse can result in deep pelvic pain. Pain associated with sexual arousal and intercourse can result in decreased vaginal lubrication, loss of orgasmic capacity, and a loss of sexual desire, or sexual aversion.

Vaginismus is defined as involuntary contractions and spasms of the vaginal (or pubococcygeal) muscles, which make vaginal penetration very difficult or impossible (Fig. 32–10). Vaginismus may stem from emotional problems or traumatic or painful sexual experiences; or it may arise as a protective reflex to avoid pain in the case of pelvic disease. Prolonged dyspareunia, or painful intercourse, can sometimes also lead to vaginismus. Note that dyspareunia and/or vaginismus in the female can cause sexual problems in male partners when they feel responsible for unsuccessful attempts at intercourse. Feelings of inadequacy and guilt can result in erectile difficulty and loss of sexual desire.

Erectile Dysfunction, Premature Ejaculation, and Dyspareunia

Erectile dysfunction, or **impotence,** is defined as the inability to achieve penile erection sufficient for vaginal penetration. Erectile dysfunction is classified as either primary or secondary. **Primary erectile dysfunction,** which is very rare, defines the male who has never been able to achieve erection sufficient for vaginal penetration. **Secondary erectile**

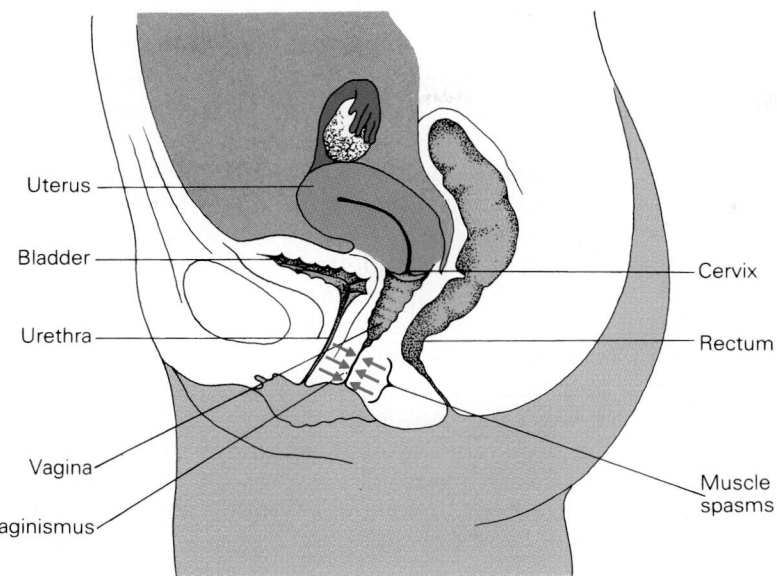

Figure 32–10
The involuntary muscle spasms of vaginismus. (Redrawn from Masters, Johnson, and Kolodny, *Human Sexuality*, Little, Brown, 1985.)

dysfunction is defined as a period of inability to achieve or maintain erection. Secondary erectile dysfunction can occur in response to anxiety, stress, illness, fatigue, lack of privacy, alcohol consumption, or change of sexual partner. Episodes of secondary erectile dysfunction are usually short-lived, and isolated episodes occur in most men.

Premature ejaculation can be defined as consistent, unintentional ejaculation prior to vaginal penetration. When defined in this way, very few men experience premature ejaculation. More typically, loss of ejaculatory control results in ejaculation soon after vaginal penetration. The probability of premature ejaculation increases with time elapsed since last orgasm, novelty of the sexual partner, decreased age, and prolonged stimulation prior to vaginal penetration. Manual application of pressure to the penis for a few seconds, from front to back just below the coronal ridge or at the base of the penis, can reduce the urge to ejaculate and help to control premature ejaculation (Fig. 32–11). Application of pressure to the penis may also cause a partial loss of erection that is only temporary.

Dyspareunia in the male is classified as painful erection, intromission, or ejaculation. Painful erection and/or pain on intromission can result from many factors, including inflammation of the foreskin of the penis (termed **balanitis**), reduced elastic-

Figure 32–11
Application of manual pressure to the penis, used to prevent premature ejaculation. (Redrawn from Masters, Johnson, and Kolodny, *Human Sexuality*, Little, Brown, 1985.)

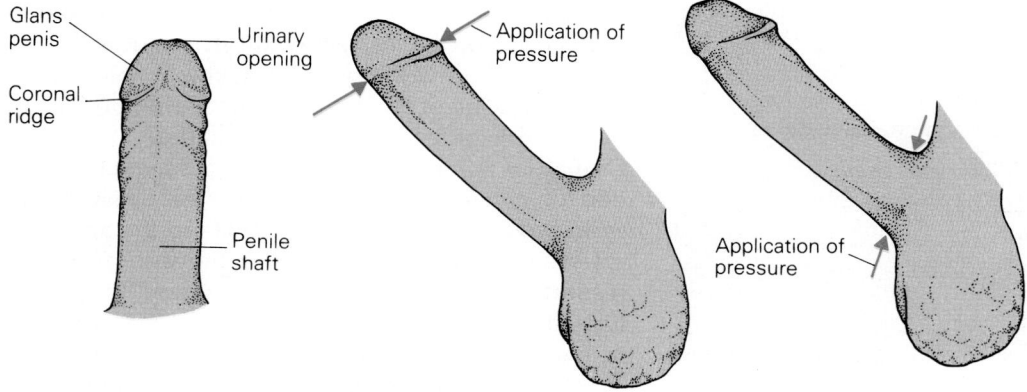

ity of the skin covering the penis, and penile scarring. These conditions can result from physical trauma to, or infection of, penile tissue. Painful ejaculation can result from physical, infectious, or chemical irritation of the urethra, termed **urethritis.**

Sources of sexual dysfunction in the male and female are summarized in Table 32–5.

Table 32–5
Origins of Sexual Dysfunction

Male

1. Neurological disorders resulting from physical or disease-related injury (multiple sclerosis, diabetes mellitus) to the brain, spinal cord, or peripheral nerves.

2. Testosterone deficiency characterized by decreased penile erection, capacity to ejaculate, or sexual desire.

3. Excessive use of drugs including alcohol, sedatives, narcotics, tranquilizers, antihistamines, and antihypertensive agents.

4. Primary or secondary erectile dysfunction resulting from physiological or psychological causes.

5. Premature ejaculation.

6. Dyspareunia due to physical, chemical, or infectious damage to the penis or urethra.

Female

1. Neurological disorders resulting from physical or disease-related injury (multiple sclerosis, diabetes mellitus) to the brain, spinal cord, or peripheral nerves.

2. Chronic estrogen deficiency that results in atrophic vaginitis characterized by vaginal dryness, thinning, and decreased elasticity.

3. Acute estrogen deficiency that results in nonatrophic vaginitis characterized by vaginal dryness.

4. Excessive use of drugs including alcohol, sedatives, narcotics, tranquilizers, antihistamines, and antihypertensive agents.

5. Dyspareunia due to physical, chemical, or infectious damage to the vagina, clitoris, or labia.

6. Endometriosis and pelvic inflammatory disease.

7. Involuntary contractions of the vaginal muscles (vaginismus).

Infertility

Infertility, or the inability to reproduce, can be due to anatomical, hormonal, chromosomal, immunological, or psychological factors.

Fertility in the male depends on several factors: adequate secretion of the gonadotropins that act to stimulate spermatogenesis; testes capable of responding to gonadotropins by producing sperm; glandular production of seminal fluid; an intact ductal system for sperm delivery; and an intact nervous system for the control of penile erection and ejaculation. A disruption of normal function at any of these levels can result in infertility.

In addition, three aspects of the sperm themselves are important for sperm penetration and fertilization of the ovum: sperm number, sperm morphology (shape and structure), and sperm motility. Although it is always difficult to establish "normal" ranges for physiological processes, it is generally accepted that 15 to 250 million sperm per milliliter of semen represents a normal sperm count. A further definition of sperm normality requires that at least 60% of the sperm have a normal appearance and that at least 60% are active or motile.

A low sperm count and/or decreased sperm motility can result from gonadotropin deficiencies or from damage to the testes. Testicular damage can result from infections, inflammation, impaired circulation, and certain diseases such as mumps. In addition, a variety of environmental factors such as fever, prolonged heat exposure, chemical agents, and irradiation can result in testicular damage and impaired spermatogenesis. The extent of testicular damage produced by these agents is determined by the intensity and duration of testicular exposure. Abnormal sperm shape or structure, which includes a non-oval head section, a broken midpiece, or a bent or coiled tail, can result in infertility by preventing sperm migration through the cervical mucus or penetration of the ovum. Infection and inflammation of the epididymis, vas deferens, or ejaculatory ducts can reduce or prevent sperm transport, which can also result in infertility.

As in the male, infertility in the female can be due to a number of factors. One of the most common causes of female infertility is failure to ovulate. Ovulatory failure can result from inadequate secretion of gonadotropins, failure of the ovary to respond to gonadotropins, variations in the anatomical structure of the ovaries, and ovarian damage resulting from infection or inflammation. Other factors that can cause female infertility include overproduction of thick cervical mucus, which can block sperm entry into the uterus, and infection or disease of the uterus or fallopian tubes, with accompanying scarring that can block ovum transport to, or implantation in, the uterus. For instance, **endometriosis** is a disease characterized by the presence of endometrial-like tissue in locations other than the uterus, such as the fallopian tubes or ovaries. Tissue formation, which may appear as cysts or nodules, can reduce the diameter of the fallopian tubes and block ovum transport to the uterus (Fig. 32–12).

Immunological factors may also contribute to infertility. Antisperm antibodies have been found in both males and females. Antibodies, produced by the male, coat the surface of the sperm and reduce its ability to penetrate the cervical mucus. Antibodies produced by the female can induce **agglutination,** or aggregation of sperm into clumps, which also reduces their ability to penetrate the cervical mucus. Whether immunological factors play an important role in infertility remains to be determined.

Figure 32–12
Radiographic image of the uterus and (*a*) normal bilateral fallopian tubes and (*b*) bilateral fallopian tube blockage. (© *Hysterosalpingography: A Text and Atlas* by Ott and Fayez. Published by Urban & Schwarzenberg, 1991.)

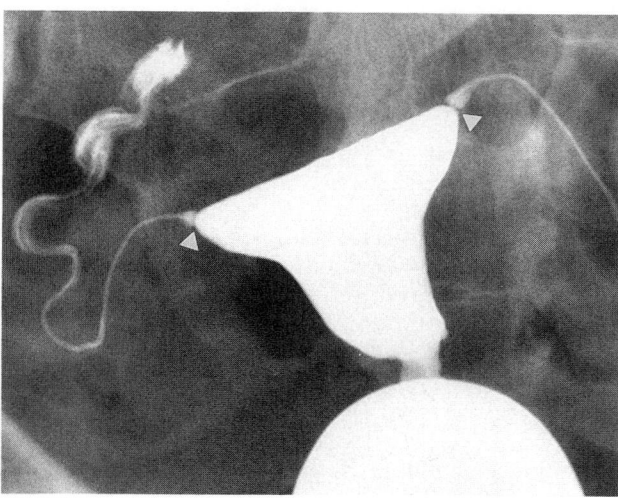

(a)

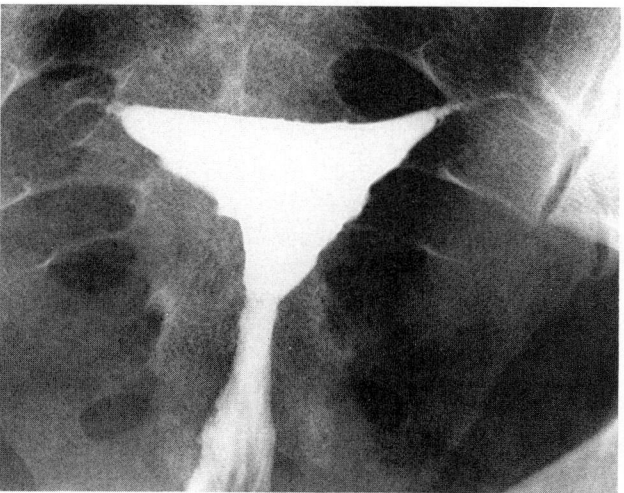

(b)

Contraception

Being able to block conception ("contra-ception") has resulted in increased independence of human sexuality from reproduction. Various methods are used to interfere with conception, but all of them fall within four basic categories: **barrier methods, chemical methods, surgical methods,** and **abstinence.** In addition, all methods of contraception work by disrupting one of three major steps in the reproductive process: ovulation, sperm and ovum transport, or implantation of the fertilized ovum in the uterus. The sites of actions of various methods of contraception are illustrated in Figure 32–13.

Barrier Methods

All barrier methods of contraception prevent sperm transport. Barrier methods include condoms, diaphragms, cervical caps, and contraceptive sponges. Condoms are thin sheaths of rubber that fit over the penis and prevent sperm from entering the vagina. Diaphragms, cervical caps, and sponges are rubber or polyurethane devices that cover the entrance to the cervix and prevent sperm from entering the uterus. Condoms, diaphragms, and sponges are used once and then must be replaced (or, in the case of the diaphragm, re-inserted) prior to each sexual encounter. Cervical caps, which work similarly to the diaphragm, can be worn for longer periods prior to removal; they are held in place by suction.

Barrier methods of contraception vary in effectiveness and are often combined with chemical methods to increase their effectiveness. For instance, the contraceptive sponge contains **spermicides,** chemicals that kill sperm. Spermicides, contained in cream or jelly, are also applied to the inside of the diaphragm to kill sperm that may pass the rim of the diaphragm.

Barrier methods of contraception have minimal side effects, which include the possibility of an allergic reaction to the spermicide or to the material used in construction, and a chance of introducing infection into the vagina.

Infertility

A variety of methods are currently available for the diagnosis and treatment of infertility. When male infertility is suspected, a microscopic examination of a sperm sample can be used to determine if infertility is due to a low sperm count, abnormal sperm morphology, or decreased sperm motility. When sperm counts are low, sperm can be collected and introduced directly into the uterus or fallopian tubes via a catheter to avoid the sperm loss which normally occurs during passage of the sperm through the cervix and uterus. This process, termed artificial insemination, can also be used to circumvent immunological infertility, which can occur when cervical mucus induces an allergic reaction in sperm resulting in sperm death.

Many cases of female infertility are due to damage or obstruction of the fallopian tubes resulting from previous infections. Reconstructive surgery has been successful in repairing damage to the fallopian tubes and restoring the capacity to conceive in approximately a third of these cases. Dilation of the cervix, a relatively simple operation, can also be used to treat infertility resulting from cervical constriction. Another cause of female infertility is inadequate secretion of gonadotropins from the pituitary gland, which results in a failure to ovulate. In such cases, oral or intramuscular administration of drugs can be used to induce gonadotropin secretion by the pituitary, which in turn stimulates estrogen secretion from the ovary and facilitates follicular maturation and ovulation. Hormonal induction of ovulation is often so successful that multiple births result.

In vitro or "test-tube" fertilization is an approach used to treat both male and female infertility. This method involves removing one or more ova from a woman's ovary by means of aspiration through a needle inserted into the ovarian follicle. Collection of ova is carefully timed to coincide with ovulation. The conditions used for storage and treatment of ova and sperm are designed to approximate the conditions normally found in the woman's body following ovulation. For example, following collection, the ova are stored for a few hours in a nutrient solution at 37° C (body temperature) prior to the introduction of sperm. The ova and sperm are then incubated together at body temperature for two days prior to implantation of the fertilized ovum in the uterus. *In vitro* fertilization is useful in treating male infertility resulting from a low sperm count because it requires far fewer sperm than does fertilization resulting from sexual intercourse. In vitro fertilization can also be used to circumvent fallopian tube obstruction in the female. Research efforts are currently being directed towards developing a method of injecting a single sperm directly into an ovum. This process, if feasible, would eliminate many cases of infertility that arise from low sperm number, poor sperm motility, fallopian tube damage, or immunological blockade.

Figure 32–13
Sites of action of various methods of contraception. Minus sign indicates inhibition.

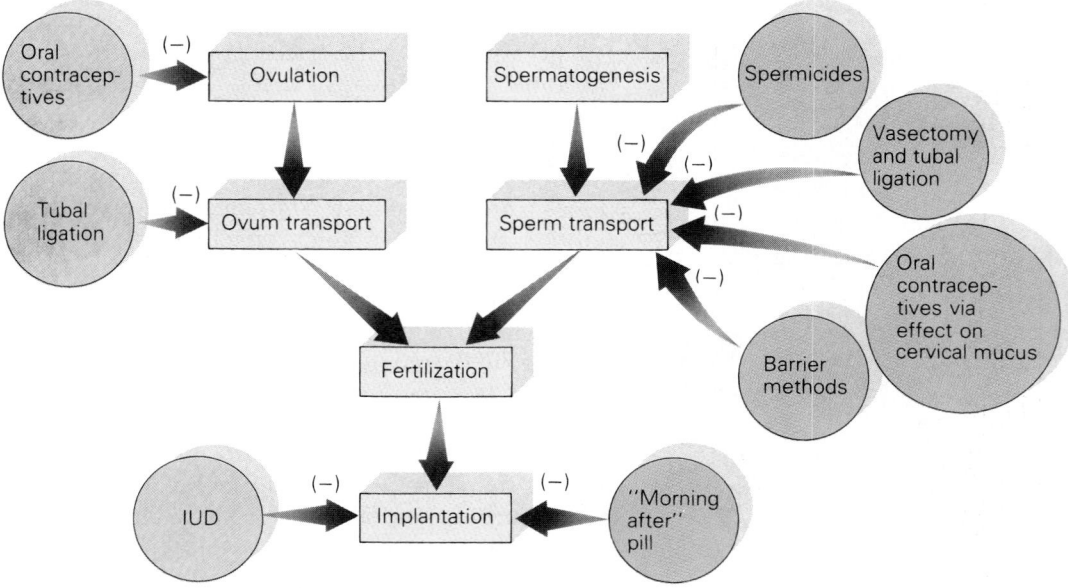

Chemical Methods

The most popular method of contraception, the **birth control pill,** is taken orally and acts to prevent pregnancy by blocking ovulation (see Fig. 32–13). There are many types of birth control pills, but they all contain synthetic variations of the gonadal hormones estrogen and/or progesterone, and they act to inhibit the release of follicle-stimulating hormone (FSH) and luteinizing hormone (LH) from the pituitary. As we learned in Chapter 30 (see Fig. 30–17), a rise in circulating levels of LH and FSH at the midpoint of the ovarian cycle stimulates maturation of the ovarian follicle, resulting in ovulation. When LH and FSH release are suppressed by oral contraceptives, as illustrated in Figure 32–14, follicular maturation does not occur, and ovulation is inhibited. Continuous use of oral contraceptives will inhibit follicular maturation, ovulation, and the onset of menstruation. Discontinuation of oral contraceptives for one week each month results in a drop in circulating levels of estrogen and progesterone, which stimulates the onset of menstruation. The

contraceptive efficacy of birth control pills does not require monthly discontinuation of use.

Progesterone, contained in oral contraceptives, also interferes with conception in two additional ways. First, progesterone increases the viscosity of cervical mucus, which, as we have seen in Chapter 31 (see Fig. 31–3), blocks the entry of sperm into the uterus. Second, progesterone interferes with implantation of the ovum in the uterus by inhibiting estrogen-induced proliferation of the endometrial lining of the uterus, which is required for implantation.

Birth control pills can produce a variety of side effects, including nausea, constipation, elevated blood pressure, skin rashes, and salt retention with accompanying weight gain. However, the most serious side effect is increased risk of cardiovascular disease, which becomes significant in women over 35 who use both oral contraceptives and tobacco.

Another type of oral contraceptive, termed the **morning-after pill,** contains a very high level of **estrogen,** which alters the endometrial lining of the uterus and prevents implantation of the fertilized ovum. However, the high levels of estrogen contained in the morning-after pill often induce severe nausea and vomiting, which have limited the use of this type of oral contraceptive.

Another type of contraceptive method is the **intrauterine device (IUD),** which can be worn for extended periods. The IUD is made of plastic or metal and is inserted by a physician into the uterus through the vagina and cervix. Although the mechanism of action of the IUD is not completely understood, it is thought to induce a local inflammatory reaction in the endometrium of the uterus, which prevents implantation of the ovum. Side effects of the IUD include uterine bleeding, spasms and cramping of uterine muscles, pelvic infection, and perforation of the uterine wall.

Surgical Methods

The most effective form of birth control, with the exception of abstinence, is surgical sterilization. Sterilization procedures, which include vasectomy in the male and tubal ligation in the female, have enjoyed a recent increase in popularity because they are safe, effective, and permanent.

Vasectomy

Vasectomy is a simple surgical procedure performed within 15 to 20 minutes under local anesthesia. As illustrated in Figure 32–15, vasectomy consists of a small surgical incision on each side of the scrotum to expose a small section of the vas deferens. The vas

Figure 32–14
Plasma levels of gonadotropins during the ovarian cycle in women (*a*) not using oral contraceptives and (*b*) using oral contraceptives.

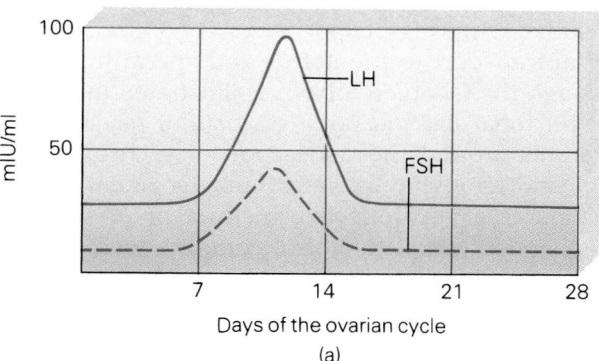

(a)

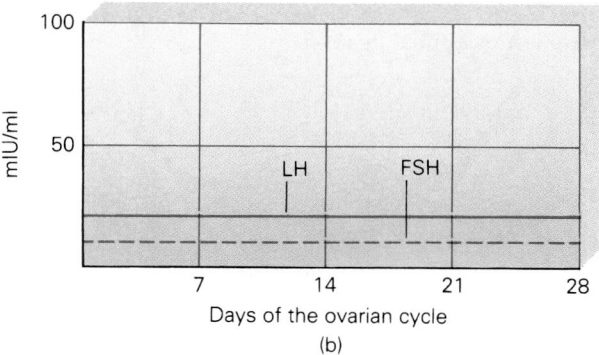

(b)

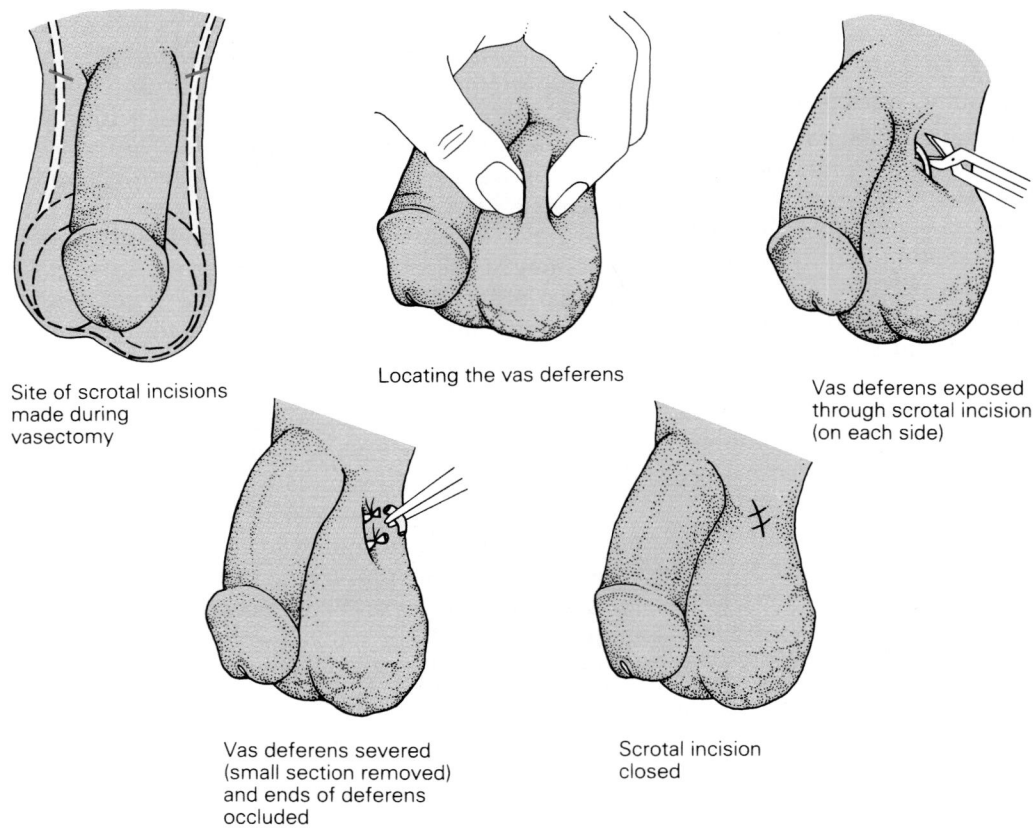

Site of scrotal incisions
made during
vasectomy

Locating the vas deferens

Vas deferens exposed
through scrotal incision
(on each side)

Vas deferens severed
(small section removed)
and ends of deferens
occluded

Scrotal incision
closed

Figure 32–15
Steps involved in a vasectomy.

deferens are severed, and the ends are tied, clamped, or cauterized to prevent sperm transport.

Vasectomy does not interfere with sperm production. Following vasectomy, sperm accumulate in the epididymis of the testes, where they are engulfed and destroyed by phagocytosis. Vasectomy does not interfere with the synthesis or release of testosterone from the testes and does not affect erection, ejaculation, or orgasm in any way. Vasectomy is a very effective method of contraception, with failure rates below 0.15%. Postsurgical complications, which can include bleeding, infection, and swelling of scrotal tissue due to sperm leakage, are rare. No long-range health risks appear to be associated with vasectomy.

Tubal Ligation

Tubal ligation consists of an abdominal incision to permit insertion of an instrument called a **laparoscope,** which is used to locate and sever the fallopian tubes on each side of the body. As illustrated in Figure 32–16, the cut ends of the fallopian tubes are occluded with rings, clips, or cautery, blocking both ovum transport to the uterus and sperm transport through the fallopian tubes. As illustrated in Figure 32–16, tubal ligation often consists of placing two ligatures around the midsection of each fallopian tube and removing the intervening segment. Tubal ligation is a very effective method of preventing pregnancy, and postsurgical complications, which can include bleeding and infection, are rare. As with vasectomy, no long-range health risks appear to be associated with tubal ligation.

Figure 32–16
Illustration of a tubal ligation.

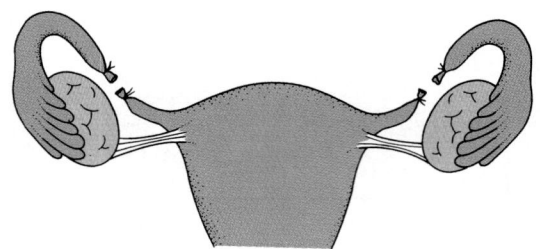

Abstinence

Complete abstinence, which is the only completely effective form of birth control, is unpopular for obvious reasons. However, many people practice **periodic abstinence** (termed the **rhythm method**) during the time of ovulation when fertility is the greatest. The efficacy of this method depends on the accuracy of predicting the time of ovulation.

Three methods are used to predict ovulation. The first involves monitoring the ovarian cycle and length of menstruation for at least 6 months and avoiding intercourse on the days preceding and following ovulation (at midcycle). This method has a high failure rate because many women exhibit a large variation in length of monthly cycles. The second method involves daily monitoring of basal body temperature and avoiding intercourse on the days preceding and following ovulation. As discussed in Chapter 30 (Fig. 30–17), body temperature decreases prior to ovulation and rises again following ovulation. This method also has a high failure rate because the temperature shifts that occur during the ovarian cycle are small (usually less than 1 degree) and are difficult to identify accurately. The third method of identifying the time of ovulation requires monitoring the consistency of cervical mucus. As described in Chapter 31, cervical mucus becomes clear, thin, and stringy around the time of ovulation. Predicting ovulation on the basis of consistency of cervical mucus has a high failure rate because of the difficulty of accurately identifying these changes.

Future Trends

Several new approaches to contraception are currently being developed and clinically tested. The important variables that should be identified for all new methods of contraception are efficacy, duration of action, reversibility, ease of use, and incidence of both short- and long-term side effects. In the female, research efforts are being directed toward blocking ovulation and implantation. In the male, the focus is on blocking production, maturation, motility, and/or transport of sperm.

Female Contraception

Long-acting forms of estrogen and progesterone are being administered by subcutaneous implantation and by injection, which can prolong the effectiveness of these substances for months. Other hormonal analogs that suppress LH and FSH release are also being investigated, as are substances that antagonize the actions of progesterone and human chorionic gonadotropin (hCG) on the endometrium and hence interfere with implantation of the ovum. Synthetic analogs of the prostaglandins, which induce uterine contractions, are being investigated for their efficacy in inducing the onset of menstruation, which would evacuate the contents of the uterus whether or not a fertilized ovum is present.

Male Contraception

Combinations of estrogen, progesterone, and testosterone are currently being examined for their ability to transiently block sperm production. Other naturally occurring substances, such as an extract from cottonseed oil (termed *gossypol*), that are capable of immobilizing or killing sperm are also being investigated. Efforts are focused as well on identifying and testing substances, such as inhibin, that inhibit FSH release from the pituitary and hence remove the hormonal signal for sperm production in the testes.

SUMMARY

Sexual Determination and Differentiation

The genetic information contributed by the parents determines the chromosomal or genotypic sex of the zygote. The chromosomal sex of the zygote determines the gonadal sex, and the gonads in turn determine the phenotypic sex of the individual. The presence or absence of a Y chromosome determines the sex of the zygote. The Y chromosome codes for, or regulates the expression of, the H-Y antigen, a protein that binds to receptors in the gonads. The H-Y antigen induces gonadal cells to develop into testes. In the absence of a Y chromosome (XX), the gonads develop into ovaries.

During the first six weeks of embryonic development, the gonads in all embryos, whether genetically male or female, are anatomically and morphologically identical. Hormones secreted by the developing gonads direct gonadal differentiation and the development of the internal genitalia. In the male embryo, the developing testes secrete Mullerian Inhibiting Substance (MIS), which induces regression of the female internal genitalia. Testosterone secreted by the developing testes stimulates development of the male internal genitalia (the Wolffian duct system). In the absence of the H-Y antigen, the gonads of the female embryo develop into ovaries rather than into testes. The developing ovaries do not secrete testosterone or MIS. In the absence of MIS, the Mullerian duct system develops in the female embryo.

The external genitalia in the male and female differentiate in response to the presence and absence of testosterone, respectively.

Exposure of the female embryo to androgens during development does not alter differentiation of the female internal genitalia, which occurs normally in the absence of Mullerian inhibiting factor. However, exposure to testosterone during development may produce masculinization of the female external genitalia.

Abnormal patterns of sexual differentiation can occur at four levels: chromosomal sex (genotype), gonadal sex, internal genital sex, and external genital sex. Chromosomal abnormalities can alter gonadal differentiation as well as enzyme activity and hormone receptor activity, resulting in altered hormone secretion from the gonads or altered action in hormone-sensitive tissues.

Human Sexual Response

The ability to become sexually aroused is present before birth and remains throughout the life span of both males and females. Sexual arousal in the adult, which can result from physical or mental stimulation, is divided into four phases: excitement, plateau, orgasm, and resolution. The excitement phase of sexual arousal is marked by increased blood flow to the genitals, which results in penile erection in the male and transudation in the female, as well as increased muscle tension or myotonia in both sexes. During the plateau phase of sexual arousal, heart rate, blood pressure, respiration rate, muscle tension, and genital vasocongestion increase in both sexes. The orgasmic phase is marked by muscular contractions of the internal and external genitalia of both sexes. In the male, these contractions result in emission and expulsion of ejaculatory fluid from the penis. During the resolution phase, increased sympathetic nerve activity in both sexes reduces blood flow to the internal and external genitalia, and blood leaves the genitals through the genital veins. The dimensions and position of the penis, vagina, and cervix return to those of the unaroused state. Heart rate, blood pressure, respiration, and muscle tension all return to normal as well. The external genitalia of both sexes are extremely sensitive to touch during the resolution phase.

Sexual Dysfunction

Endocrine deficiencies can result in reduced sexual arousal and decreased capacity for orgasm. Insufficient estrogen secretion in the female results in reduced vaginal lubrication and painful intercourse, or dyspareunia. Chronic estrogen deficiencies can lead to atrophic vaginitis. Testosterone deficiencies in the male can reduce ejaculatory capacity and sexual desire.

Mechanical or disease-related damage to the peripheral nerves, spinal cord, or brain can result in impaired sexual arousal and/or impaired capacity for orgasm in both sexes. Both multiple sclerosis and diabetes mellitus can produce neuropathies that can compromise sexual arousal and orgasm.

Alcohol, antihistamines, sedatives, narcotics, and tranquilizers can retard or inhibit sexual arousal and orgasm by altering blood flow to the genitals or the neural transmission of sexually arousing stimuli.

Other conditions that interfere with sexual arousal and/or orgasm include anorgasmia, endometriosis, and vaginismus in the female, and erectile dysfunction, urethritis, and premature ejaculation in the male.

Infertility can result from hormonal deficiencies, destruction of gonadal tissue resulting from inflammation or disease, and the production of antisperm antibodies.

Contraception

Barrier methods of contraception, including diaphragms, cervical caps, contraceptive sponges, and condoms, interfere with sperm transport. The efficacy of barrier methods in preventing pregnancy is increased by spermicides.

Chemical methods of contraception include hormones, such as estrogen and progesterone, which are taken orally in the form of birth control pills. Estrogen and progesterone inhibit the secretion of pituitary hormones that control ovulation. Progesterone also increases the viscosity of cervical mucus, impeding sperm entry into the uterus. In addition, progesterone inhibits proliferation of the endometrial lining of the uterus, interfering with implantation of the ovum in the uterus. The morning-after pill contains a high level of estrogen, which prevents implantation of the ovum in the uterus.

Surgical methods of contraception include vasectomy in the male, which interferes with sperm transport, and tubal ligation in the female, which interferes with both sperm and ovum transport.

The rhythm method of contraception refers to periodic abstinence from intercourse at midcycle, around the time of ovulation. The efficacy of this method depends on the accuracy of predicting the time of ovulation in the female. Changes in body temperature, consistency of cervical mucus, and length of the ovarian cycle are used to predict time of ovulation, with varying accuracy.

Future trends in contraception include the development of long-acting natural and synthetic chemicals that inhibit ovulation, spermatogenesis, sperm transport, ovum transport, and/or implantation of the ovum in the uterus.

REVIEW QUESTIONS

Identify Each with a Term

1. The release of moisture from the lining of the vagina during sexual arousal.

2. The elevation of the uterus away from the vagina during sexual arousal.

3. Any condition in which the normal physical responses to sexual stimulation are impaired.

4. The protein thought to be responsible for inducing testicular differentiation in the primitive gonad.

Define Each Term

5. vaginismus
6. impotence
7. Mullerian Inhibiting Substance (MIS)
8. myotonia
9. Turner's syndrome
10. amenorrhea

Choose the Correct Answer

11. Which of the following does *not* result from testosterone stimulation?
    a. masculinization of the female embryo's external genitalia
    b. stimulation of Mullerian duct regression
    c. masculinization of the male embryo's external genitalia
    d. stimulation of Wolffian duct development

12. The chromosomal make up of an individual with Klinefelter's syndrome is:
    a. XO
    b. XYY
    c. XXY
    d. XX

13. Which of the following can help control premature ejaculation?
    a. testosterone therapy
    b. increasing the time since last orgasm
    c. vasectomy
    d. application of pressure at the base of the penis

14. Which of the following is *not* a barrier method of contraception?
    a. birth control pill
    b. condom
    c. diaphragm
    d. contraceptive sponge

Answer Each Question in One or Two Sentences

15. What is the importance of the sex chromosomes?

16. How can intense programs of physical exercise result in amenorrhea?

SUGGESTED READINGS

Acosta, A.A., ed. *Human Spermatozoa in Assisted Reproduction.* Baltimore: Williams and Wilkins, 1990.

Allgeier, E.R., and Allgeier, A.R., eds. *Sexual Interactions.* Lexington, Mass.: D.C. Heath and Co., 1984.

Geer, J., Heiman, J., and Leitenberg, H., eds. *Human Sexuality.* Englewood Cliffs, N.J.: Prentice Hall, Inc., 1984.

Glover, T.D., ed. *Human Male Fertility and Semen Analysis.* Academic Press, 1990.

Masters, W.H., Johnson, V.E., and Kolodny, R.C., eds. *Human Sexuality.* Boston: Little, Brown and Co., 1985.

Reinisch, J.M. *The Kinsey Institute New Report on Sex.* St. Martin's Press, 1990.

Shearman, R.P., ed. *Clinical Reproductive Endocrinology.* Churchill Livingstone, 1985.

KEY TERMS

amenorrhea (p. 1047)
autosomes (p. 1033)
chromosomal sex (p. 1033)
clitoris (p. 1044)
contraception (p. 1051)
dyspareunia (p. 1045)

ejaculation (p. 1043)
endometriosis (p. 1051)
gonadal sex (p. 1033)
histocompatibility-Y
  antigen (H-Y antigen)
  (p. 1034)

impotence (p. 1048)
infertility (p. 1050)
Mullerian duct system
  (p. 1034)
orgasm (p. 1041)
spermicides (p. 1051)

transudation (p. 1041)
vaginismus (p. 1048)
vasectomy (p. 1053)
Wolffian duct system
  (p. 1034)

Glossary

A-band A transverse band of thick filaments of cardiac and skeletal muscle located in the center portion of the sarcomere. (Ch. 16)

absolute refractory period The period of time in which an axon is unable to generate an action potential regardless of the applied stimulus strength. (Ch. 7)

absorption (ab-**sorp**-shun) The movement of digestion products and other nutrients across the wall of the small intestine to the blood or lymph. (Ch. 23)

acclimatization The process of becoming accustomed to a new environment. (Ch. 22)

A-cell A glucagon-secreting cell of islets of Langerhans; also called alpha-cell. (Ch. 14)

acetone (**ass**-ih-tone) A volatile, sweet-smelling ketone body produced from acetyl CoA during prolonged fasting or untreated severe diabetes mellitus. (Ch. 26)

acetyl coenzyme A (as-ee-til koh-**ehn**-zym) An important intermediate in the Krebs cycle and the main precursor to lipids; it is formed in the mitochondrial matrix when all of the acetyl groups produced from food molecules are coupled to the terminal —SH group of coenzymes. (Ch. 6)

acetylcholine (as-ee-til-**koe**-leen) (Often abbreviated as Ach.) A low-molecular weight neurotransmitter; found in the central and peripheral nervous systems. (Ch. 10)

acetylcholinesterase (as-ee-til-**koe**-leen-s-tur-aze) Enzyme essential for breaking down Ach to an inactive form once it has engaged a receptor. (Ch. 16)

acid (**as**-id) Any substance whose dissociation in water releases hydrogen ions. (Ch. 2)

acidemia (as-i-**dehm**-ee-uh) An arterial blood pH less than 7.35. (Ch. 26)

acidosis (as-i-**doe**-sis) An abnormal process that tends to produce acidemia. (Ch. 26)

acinus (**as**-ih-nuhs) Saclike clusters of exocrine secretory cells in the pancreas that secrete digestive enzymes. (Ch. 23)

acromegaly (ak-roe-**meg**-ah-lee) A disfiguring condition resulting from the overproduction of GH in adults whose epiphyses have closed; stimulates excessive growth of soft tissues and transverse diameter of bones leading to bony thickening and deformity. (Ch. 13)

acrosome reaction (ak-roe-sowm) A controlled disruption of the head of a sperm, which allows it to penetrate the cumulus oophorus and the zona pellucida of the ovum. (Ch. 31)

actin (ak-tin) A globular protein to which myosin cross bridges bind; located in muscle thin filaments and microfilaments of cytoskeleton. (Ch. 16)

actin-linked regulation Muscle regulation that is accomplished through some action on the thin filaments. (Ch. 16)

action potential The mechanism by which impulses are transmitted, thereby carrying information along the nerve fiber membranes. The action potential is a rapid change in the voltage of the membrane potential that occurs when adequate stimulus is applied. The all-or-none response causes the voltage to change from its negative resting state to a positive value. (Ch. 7)

activation energy (ac-ti-**va**-tion **en**-er-gy) The initial energy required to break the existing bonds in a molecule and form new ones. (Ch. 2)

activation state One of three states of the sodium channel; if the nerve cell is stimulated and the membrane potential is depolarized to the threshold value, the sodium channel switches from its resting state to its activating state; sodium ions will then flow until the inactivating gate closes the channel. (Ch. 7)

active site Areas on the surfaces of enzymes that permit reactant molecules to bind. (Ch. 2)

active transport A situation in which a solute must move from a less concentrated region outside the cell to a more concentrated region inside, against the prevailing gradient and thus requiring energy. When the movement of solute is coupled directly to an energy-yielding reaction, the process is termed **primary active transport;** when the active transport of a solute is not coupled directly to the energy-yielding reaction, it is described as **secondary active transport.** (Ch. 4)

active zone Restricted areas of the presynaptic terminal. (Ch. 7)

acute mountain sickness An illness typically seen in high-altitude mountain climbers, in which too much carbon dioxide is blown off during hyperventilation. (Ch. 22)

Addison's disease (**add**-i-sonz) A condition caused by an overall deficiency in steroid production by the adrenal cortex. It is a progressive disease with symptoms such as diarrhea, hypotension, increased skin pigmentation, fatigue, and digestive difficulties. (Ch. 14)

adenohypophysis (ad-e-no-high-**pof**-i-sis) The anterior lobe of the pituitary gland. (Ch. 13)

adenosine (a-**den**-oh-seen) A nucleoside that has been shown to promote vasodilation of arteriolar smooth muscle. (Ch. 19)

adenosine diphosphate (ADP) (a-**den**-oh-seen die-**fos**-fate) An important molecule that is formed from ATP when the terminal phosphate is removed. This reaction produces the energy that drives other energy-requiring reactions. (Ch. 6)

adenosine monophosphate (AMP) (a-**den**-oh-seen mah-noe-**fos**-fate) A molecule that is formed from ATP when pyrophosphate is removed (also known as adenosine phosphate). (Ch. 6)

adenosine triphosphate (ATP) (a-**den**-oh-seen try-**fos**-fate) The universal energy compound that is produced in all cells and is the source of energy in most of the energy-requiring reactions of metabolism. ATP is produced by adenosine—plus three phosphate groups attached. (Ch. 6)

adenylate cyclase (a-**den**-ih-late **sye**-klase) An enzyme located in the inner surfaces of cell membranes that catalyzes the conversion of ATP to cAMP following the binding of a signal molecule to a membrane-bound receptor. (Ch. 12)

adequate stimulus The type of stimulus to which a sensory receptor is most sensitive. (Ch. 8)

adipose tissue (**ad**-i-poz) Tissue composed mainly of fat-storing cells. (Ch. 11)

adrenal cortex (ad-**ree**-nal **kore**-teks) The outer two-thirds shell of the adrenal gland that produces the steroid hormones—cortisol, aldosterone, and androgens. (Ch. 12)

adrenal glands A pair of endocrine glands, one located above each kidney: Each consists of outer adrenal cortex and inner adrenal medulla. (Ch. 12)

adrenal medulla (ad-**ree**-nal meh-**dul**-ah) The inner core of the adrenal gland that secretes the catecholamines, epinephrine and norepinephrine. (Ch. 12)

adrenergic (ad-ren-**er**-jik) Pertaining to epinephrine or norepinephrine. (Ch. 10)

adrenocorticotropic hormone (ACTH) (ad-*ree*-no-kore-ti-kow-**trope**-ik) A polypeptide hormone that is synthesized, stored, and released from the anterior pituitary gland. It stimulates the production and secretion of cortisol by the zona fasciculata of the adrenal cortex. (Ch. 11)

adrenogenital syndrome (ad-ree-no-*gehn*-ih-tahl) A genetic disorder in females in which hypersecretion of adrenal androgens results in the development of female internal genitalia but male/female external genitalia. (Ch. 32)

adult respiratory distress syndrome A form of pulmonary edema caused by increased permeability of the capillaries; damage is caused by leakage of oxygen radicals. (Ch. 21)

aerobic (air-**oh**-bik) Any oxygen-requiring process. (Ch. 6)

afferent (**af**-er-ent) Carrying toward. (Ch. 9)

afferent arteriole (**af**-er-ent ar-**tee**-ree-ole) The arteriole that carries blood toward the glomerulus. (Ch. 24)

afferent nerve fiber Nerve fiber that transmits stimuli to the CNS. (Ch. 9)

affinity chromatography (ah-**fih**-ih-tee kro-muh-**tog**-raph-ee) A technique that uses the specificity of an antigen-antibody reaction to study antibodies. (Ch. 3)

afterload The force a weight on a lever system will provide during an isotonic contraction. (Ch. 16)

agglutination (a-*glue*-ti-**nay**-shun) The process that is described by the aggregation or clumping of red blood cells which is caused by the attachment of substances to cell surfaces that, in turn, bind the cells together. (Ch. 17)

agglutinin (a-**glue**-ti-nin) Specific antibodies in the blood that cause clumping or agglutination. (Ch. 17)

agglutinogen (a-**glue**-ti-nin-oh-gin) Specific antigens on red blood cell membrane surfaces that causes the production of an agglutinin-type of immunoglobulin. (Ch. 17)

agonist (ah-**goh**-nist) Any ligand capable of both binding to a plasma membrane receptor and evoking a physiologic response. (Ch. 5)

agranulocyte (a-**grahn**-you-low-site) Leukocyte that does not have a granular cytoplasm: monocyte and lymphocyte. (Ch. 17)

airway A series of highly branched hollow tubes within the lung that carry gas to and from the alveoli; trachea, bronchii, bronchioles. (Ch. 20)

alactacid oxygen debt The first stage of repayment of the oxygen debt during exercise; refers to the oxygen needed to replenish the oxygen stores of the body as well as the phosphagen system. (Ch. 29)

albumin (al-**byou**-min) The most common plasma protein; produced in the liver. It regulates plasma osmotic pressure. (Ch. 12)

aldosterone (al-**dos**-ter-own) A mineral-ocorticoid (steroid) hormone produced by the zona glomerulosa of the adrenal cortex that helps to regulate body fluid volume by controlling sodium chloride and potassium excretion from the kidney tubular cells. (Ch. 12)

alkalemia (*al*-kuh-**lee**-me-uh) An arterial blood pH greater than 7.45. (Ch. 26)

alkaline (**al**-kuh-line) Basic; a pH greater than 7. (Ch. 26)

alkalosis (*al*-kuh-**low**-sis) An abnormal process that tends to produce alkalemia. (Ch. 26)

allosteric enzyme (*al*-o-**steer**-ik **en**-zime) A catalytic protein that has two reactive sites. One site activates the enzyme. The other one, the allosteric site, can enable a substance other than the normal substrate to bind to the enzyme molecule, thereby changing the shape of the molecule and activity of the enzyme. (Ch. 6)

alpha adrenergic receptor (**al**-fah ad-ree-**ner**-jik) A type of receptor site for norepinephrine and epinephrine that, when activated, opens a calcium channel in the plasma membrane where it is located. (Ch. 7)

alpha cells of the pancreas One cell type found in the islets of Langerhans; produces glucagon and is responsible for regulation of blood glucose concentrations and fuel homeostasis. (Ch. 15)

alpha motor neuron An efferent neuron originating in the CNS that synapses with extrafusal muscle fibers. (Ch. 9)

alpha-helix A type of conformation that results when a single polypeptide chain coils about itself to form a uniform cylindrical structure. (Ch. 2)

alveolar capillary membrane (al-**vee**-oh-lar) The membrane through which gas exchange takes place between the alveolus and the blood in the pulmonary capillaries. (Ch. 21)

alveolar pressure The pressure of air inside the alveoli. (Ch. 20)

alveolar surface area The site of gas exchange in the lung; roughly 80

times greater than the external body surface. (Ch. 20)

alveolar ventilation The volume of atmospheric air entering alveoli each minute. (Ch. 20)

alveolus (al-**vee**-oh-lus) A microscopic air sac in the lung; effects gas exchange (plural: alveoli). (Ch. 20)

ameboid movement (a-**mee**-boyd) Movement similar to that of an amoeba, in which the cell projects protoplasmic extensions ahead and then follows them. (Ch. 17)

amenorrhea (a-men-oh-**ree**-ah) The absence of menstruation. (Ch. 32)

amino acid The basic unit of protein molecules whose chemical substrate can be written as R—$CHNH_2COOH$. There are 20 naturally occurring amino acids that appear in biological proteins. (Ch. 2)

amino group —NH_2. (Ch. 2)

aminoacyl attachment site of tRNA A feature of transfer RNA in which an amino acid destined to join a growing polypeptide chain attaches for transport to a ribosome. (Ch. 5)

ammonia NH_3. (Ch. 26)

amniotic fluid (am-**nee**-ot-ik) The liquid that fills the amnion. (Ch. 31)

amphetamine (am-**fet**-uh-meen) A sympathomimetic drug that stimulates the CNS, increases blood pressure, and reduces the appetite. (Ch. 11)

amphipathic molecule (*am*-fee-**path**-ik) Molecule that has a polar region at one location and a nonpolar region at another location, making it both hydrophobic and hydrophilic. Examples of amphipathic molecules are the phospholipids in cell membranes. (Ch. 4)

amphoteric (am-**fo**-tear-ik) Capable of acting as either acid or base. (Ch. 26)

amylase (**am**-uh-laze) A digestive enzyme that helps break down starch. (Ch. 23)

amylin (**am**-uh-lyn) Amylopectin; the insoluble constituent of starch. (Ch. 15)

anabolic steroid (an-a-**ball**-ik) A testosteronelike agent that increases protein synthesis. (Ch. 14)

anabolism (biosynthesis) (a-**nab**-o-lizm) A series of steps in which the energy extracted in catabolism is used to assemble building block molecules into proteins, nucleic acids, lipids, and other macromolecules for the cell. (Ch. 6)

anaerobic (an-air-**oh**-bik) Refers to processes that can occur in the absence of molecular oxygen. (Ch. 6)

analgesia (an-al-**jee**-zee-ah) The first stage of anesthesia; a reduced response to pain without loss of consciousness. (Ch. 22)

anatomical planes Imaginary lines used to divide the body into parts. (Ch. 1)

androgen (**an**-drow-jen) Any chemical with testosteronelike action. (Ch. 11)

androgen-binding-protein (ABP) A protein synthesized by Sertoli cells that binds testosterone and increases the concentration of testosterone in seminiferous tubules of the testes. (Ch. 30)

anemia (ah-**nee**-mee-uh) A condition characterized by an abnormally low oxygen-carrying capacity of the blood, resulting from marked deficiencies in the number of circulating red blood cells, the amount of hemoglobin, or both. (Ch. 17)

anesthesia (*an*-es-**thee**-zee-ah) Loss of feeling or sensation. (Ch. 22)

angina pectoris (an-**jie**-nah **pek**-toe-ris) Chest pain due to hypoxia of the myocardium that may occur because of overexertion, stress, or excitement. (Ch. 19)

angiogram (**an**-jee-oh-gram) A roentgenogram of blood vessels filled with a contrast medium. (Ch. 21)

angiotensin I (*an*-jee-o-**ten**-sin) A compound in the blood formed by the action of renin, an enzyme secreted by the kidney. Angiotensin I becomes angiotensin II by means of a converting enzyme of the pulmonary capillaries. (Ch. 10)

angiotensin II A hormone that causes the hypothalamus to release ADH into the circulatory system; stimulates aldosterone secretion from adrenal cortex, vascular smooth muscle contraction, and thirst. (Ch. 10)

angiotensin-converting enzyme An enzyme that converts angiotensin I into angiotensin II through the removal of two amino acids. (Ch. 14)

angiotensinogen A plasma protein synthesized by the liver and secreted into the blood; precursor for angiotensin I. (Ch. 14)

angular gyrus The association cortex thought to be involved with the association of visual, auditory, and tactile sensations. (Ch. 11)

anion (**an**-eye-on) An atom that gains one or more electrons, thus acquiring a negative charge. (Ch. 2)

anorgasmia (*an*-ore-**gaz**-me-uh) Difficulty in reaching orgasm. (Ch. 32)

anoxia (ah-**nok**-see-uh) Reduced oxygen supply. (Ch. 22)

antagonist (an-**tag**-o-nist) A ligand that binds with high affinity to a receptor but evokes no response. (Ch. 5)

antagonist muscle A muscle that opposes an agonist muscle at a joint where two bones meet, contraction of the agonist muscle results in the extension of the antagonist and vice versa. (Ch. 9)

anterior hypothalamus The anterior portion of the hypothalamus; appears to be the center for sexual drive in both males and females. (Ch. 11)

anterior pituitary gland Anterior portion of the pituitary gland. (Also called *adenohypophysis*.) (Ch. 13)

anterograde fast transport The movement of organelles and materials from the soma to the axon terminal of a nerve cell. (Ch. 7)

anterolateral pathway One of two major pathways for the transmission of somatic sensory information from the surface of the body to the neocortex. (Ch. 8)

antibiotic (*an*-ti-bye-**ot**-ik) Molecules produced by microorganisms that kill or inhibit the growth of other microorganisms. (Ch. 5)

antibody (**an**-ti-*bod*-ee) A highly specific protein produced by plasma cells in response to the introduction of foreign matter (antigens). The antibody reacts against the antigen to remove or destroy it. The new term for antibody is *immunoglobulin*. (Ch. 3)

antibody-mediated immunity Humoral immunity; major protection against bacteria and viruses in extracellular fluid. (Ch. 17)

anticoagulant (*an*-ti-koh-**ag**-you-lant) A chemical that blocks or inhibits the coagulation process. (Ch. 17)

anticodon A set of three nucleotides in tRNA complementary to a codon on an mRNA molecule. (Ch. 5)

antidiuretic hormone (ADH) (an-ti-die-**your**-et-ik) A hormone that directs kidneys to increase the absorption of water; see also vasopressin. (Ch. 10)

antidrop response Vasoconstriction and shivering due to stimulation of the hypothalamus by excessive cooling. (Ch. 28)

antigen (**an**-tih-jen) Any molecule that stimulates the production of antibodies. (Ch. 3)

antiparallel DNA strands When the 3′ ends of the two complementary nucleic acid strands of the DNA molecule are at opposite ends of the double helix. (Ch. 5)

antiport The movement of solutes in the opposite direction of Na^+. (Ch. 4)

antirise response Vasodilation and sweating due to stimulation of the hypothalamus by overheating. (Ch. 28)

antithrombin A natural coagulation inhibitor. Prevents thrombin from converting fibrinogen to fibrin. (Ch. 17)

antithromboplastin A natural coagulation inhibitor. Inactivates tissue thromboplastin. (Ch. 17)

antrum (ahn-**trum**) The lower portion of the stomach closest to the pyloric sphincter. (Ch. 23)

aorta (ay-**or**-tah) The artery that receives all the blood the left heart pumps out; branches into a number of smaller arteries that direct blood flow to organs. (Ch. 18)

aortic arch The part of the aorta in which baroreceptors that sense blood pressure are located. (Ch. 10)

aortic bodies Chemoreceptors of the peripheral nervous system within the wall of the aorta; sensitive to levels of O_2 and CO_2 in blood. (Ch. 10)

aortic valve The valve between the left ventricle and the aorta. (Ch. 18)

aphasia (ah-**fay**-zee-ah) Disorders of language resulting from damage to the brain. (Ch. 11)

appendicitis An inflammation of the appendix; usually preceded by obstruction of the lumen of the appendix. (Ch. 23)

appendix A small, fingerlike blind tube that extends from the cecum. (Ch. 23)

aprosodia Damage to areas of the nondominant hemisphere of the brain; results in an inability to understand or express intonation. (Ch. 11)

aqueous (**ak**-wee-us) Prepared with water. (Ch. 2)

areola (ah-**ree**-o-lah) The pigmented area of skin that surrounds the nipple of the breast. (Ch. 31)

arm projection neuron A neuron of the posterior parietal cortex that generates action potentials when the arm reaches for an object. (Ch. 9)

arrhythmia (ah-**ryth**-mee-ah) Conduction and rhythm disturbances of the heart. (Ch. 18)

arterial baroreceptors (ar-**teer**-ee-al) Nerve endings sensitive to stretch or distortion produced by arterial blood pressure changes; located in carotid sinus or aortic arch arteries. (Ch. 19)

arterial pressure The pressure within arteries. (Ch. 19)

arteriole (ar-**tee**-ree-ole) The smaller blood vessels that branch off an artery once the artery has entered an organ. (Ch. 18)

arteriosclerosis (ar-**teer**-ee-o-skleh-roe-sis) A group of diseases characterized by thickening and loss of elasticity of arterial walls. (Ch. 17)

artery (ar-**ter**-ee) A blood vessel that transports blood away from the heart to the various organs of the body. (Ch. 1)

associative learning A type of learning in which a subject acquires an understanding of the relationships between stimuli or between a behavior and subsequent reward or punishment. (Ch. 11)

asthma (**az**-muh) A breathing disorder brought on by narrowing of the bronchi. (Ch. 20)

astrocytes Star-shaped glial cells of the CNS that form the blood-brain barrier and provide nutrients, support, and insulation for CNS neurons (also called *astroglia*). (Ch. 7)

atelectasis (*at*-ee-**lek**-tah-sis) Alveolar collapse due to high surface tension in the lungs. (Ch. 20)

atherosclerosis (*ath*-er-or-skleh-**roe**-sis) A disease process of arteries that results in the progressive occlusion of the lumen of the vessel. (Ch. 17)

atherosclerotic plaque Accumulation of lipid and fibrous tissue inside blood vessels resulting in narrowing of the vessel lumen. (Ch. 19)

atmospheric pressure Air pressure surrounding the body; equal to 760 mm Hg at sea level. (Ch. 20)

atom (**a**-tom) The smallest unit of an element that retains the properties of the element. (Ch. 2)

atomic number (a-**tom**-ic **num**-ber) A specific number of protons in the nucleus of an atom, by which it can be identified. (Ch. 2)

ATP synthetase An enzyme that transforms the energy of the proton electrochemical gradient into chemical energy in the form of ATP. (Ch. 6)

atrial natriuretic peptide (ay-**tree**-al) A peptide secreted from cells in the walls of the atria; causes both vasodilation and increased excretion of salt and water by the kidneys. (Ch. 19)

atrial septal defect A congenital birth defect in which the septum between the atria is not completely formed. (Ch. 21)

atrioventricular (AV) node (ay-tree-o-ven-**trik**-you-lar) A region of specialized cells located in the right atrial wall near the interventricular septum. It relays the contractile stimulus to the bundle of His, the Purkinje fibers, and the ventricular myocardium. (Ch. 18)

atrioventricular valve Valve located between an atrium and a ventricle. The left AV valve is called the bicuspid or mitral valve. The right AV valve is called the tricuspid valve. (Ch. 18)

atrium (**ay**-tree-um) A thin-walled chamber of the heart that receives blood returning to the heart from either the vena cava or the pulmonary vein. (Ch. 18)

atrophic vaginitis (a-**tro**-fik) Drying and thinning of vaginal mucosal membrane and decreased elasticity and muscle tone of the vagina due to inadequate estrogen stimulation. (Ch. 32)

atrophy (**at**-roe-fee) Wasting away. (Ch. 16)

auditory cortex (aw-di-**toe**-ree **kor**-teks) That part of the neocortex that detects sounds via the ear. (Ch. 7)

auditory sensory aphasia A condition produced by lesions in the auditory association cortex or Wernicke's area, in which voices can be heard but the words appear to have no meaning. (Ch. 8)

auditory system The system of the body that pertains to the sense of hearing. (Ch. 8)

autocrine (*aw*-**toe**-krin) Mechanism by which a hormonelike substance is released into the interstitial fluid surrounding the cells in which it is produced, and engages receptors on the same cells. (Ch. 12)

autonomic nervous system (aw-toe-**nom**-ik) The portion of the nervous system that controls the visceral functions of the body such as regulation of the cardiac and smooth muscles and glands. The ANS comprises the sympathetic and parasympathetic divisions. (Ch. 10)

autophagy (*aw*-toe-**fay**-jee) A process in which lysosomes break down the cell's own organelles and other matter; can be a response to cell injury or can occur during remodelling of the cell or during adverse nutritional conditions. (Ch. 3)

autoradiography (*aw*-toe-ray-**dee**-og-ra-fee) A technique that locates the radioactive molecules within cells or tissues; has been useful in discovering the secretory pathways in cells and in determining the role of the Golgi complex in this process. (Ch. 3)

autoreceptor Alpha-2 and beta-2 receptors located on the membrane of the presynaptic terminal that appear to regulate the amount of NE released. (Ch. 7)

autoregulation The ability of individual vascular beds to maintain constant blood flow despite changes in arterial blood pressure; the intrinsic ability of the heart to regulate the strength of ventricular contraction. (Ch. 19)

autosome (**aw**-toe-sowm) Chromosomes that provide genetic information for the development and differentiation of most of the cells of the body. (Ch. 32)

auxotonic contraction (ox-o-tohn-ik) Muscle movement in which the force continually increases as the muscle shortens. (Ch. 16)

axial muscles (**ak**-si-al) Muscles that produce movement of the skull, backbone, and ribs. (Ch. 9)

axoaxonic synapse (**ak**-so-ak-son-ik) Synaptic contact made between two axons. (Ch. 7)

axodendritic synapse Synaptic contact made between axons and dendrites. (Ch. 7)

axon (**ak**-son) A long fiberlike structure that extends from the soma to make contact with other nerve cells. (Ch. 7)

axon terminal The ends of axons; the point of contact with other nerve cells at synapses to transmit information. (Ch. 7)

axonal transport The movement of cellular components from the cell body to the terminals. (Ch. 7)

axosomatic synapse Synaptic contacts made between axons and somata. (Ch. 7)

Babinski sign (**bah**-bin-skee) Extension of the big toe and fanning of the other toes in response to stroking the bottom of the foot; due to upper–motor-neuron lesions. (Ch. 9)

bacteria (bak-**teer**-ee-ah) Unicellular organisms that have an outer cell wall and a plasma membrane. (Ch. 17)

baroreceptor (*bar*-oh-re-**sept**-tor) A stretch receptor sensitive to pressure and to rate of change in pressure. (Ch. 10)

basal (**bay**-sal) Refers to resting level. (Ch. 3)

basal ganglia (**bay**-sal **gang**-lee-ah) A group of nuclei in the brain involved in the programming of motor patterns. (Ch. 9)

basal metabolic rate (BMR) Also called metabolic cost of living, it is the minimal energy cost to any individual under controlled resting conditions. (Ch. 28)

base Any substance that can accept or bind hydrogen ions. (Ch. 2)

basement membrane The thin proteinaceous layer of extracellular material upon which epithelial and endothelial cells fit. (Ch. 3)

basilar membrane (**bas**-ih-lar) Thin membrane that divides the scala tympani from the cochlear duct in the inner ear. It supports the organ of Corti and helps to record the frequency of sound waves. (Ch. 8)

basolateral membrane The side of the epithelial cell facing away from the lumen. (Ch. 23)

basophil (**bay**-so-fil) The rarest type of leukocyte; found in greater numbers outside the blood in loose connective tissue. (Ch. 17)

belly The wide or thick portion of a muscle. (Ch. 16)

beta cells of the pancreas One cell type found in the islets of Langerhans; produces insulin and is responsible for regulation of blood glucose concentrations and fuel homeostasis. (Ch. 15)

beta receptor (**bay**-tah) One of two types of plasma membrane receptor sites for norepinephrine and epinephrine. When stimulated, the effects include accelerated heart rate, dilation of bronchioles, and glycogenolysis. (Ch. 10)

beta sheet A secondary structural conformation of a protein molecule that results when the polypeptide chain runs back and forth upon itself. (Ch. 2)

beta-lipotropin A large polyprotein that breaks down into a number of opiates that can inhibit the sensation of pain. (Ch. 7)

bicuspid valve (bye-**kus**-pid) The atrioventricular valve located between the atrium and the ventricle of the left heart; also called the mitral valve. (Ch. 18)

bile (byl) The fluid secreted by the liver; contains bicarbonate, bile salts, cholesterol, lecithin, and bile pigments. (Ch. 23)

bile canaliculi Small ducts next to liver cells into which bile is secreted. (Ch. 23)

bile salt An organic component secreted by the liver; promotes solubilization and digestion of fat in the small intestine. (Ch. 23)

bilirubin (*bil*-ee-**roo**-bin) A red-colored product of the breakdown of hemoglobin by the liver cells. It is then excreted as waste in the bile. (Ch. 17)

biliverdin (*bil*-ee-**vur**-din) A green-colored product of the breakdown of hemoglobin by the liver cells. It is the first product of catabolism and is then converted to bilirubin. It is excreted as waste. (Ch. 17)

bipolar cell A type of nerve cell found in the retina. (Ch. 8)

bipolar neuron A type of neuron that has one axon and one dendrite, each extending from opposite sides of the soma. (Ch. 7)

birth control pill An oral contraceptive containing synthetic estrogen and/or progesterone that prevents pregnancy by blocking ovulation. (Ch. 32)

blackout Loss of vision; often due to reduced blood flow to the brain. (Ch. 22)

blastocyst (**blas**-toe-sist) The precursor of the embryo; composed of a ball of developing cells surrounding a central cavity. (Ch. 31)

blood group A system of related blood types, such as the ABO blood group or the Rh blood group. (Ch. 17)

blood pressure The force that causes the blood to flow through the vessels; usually expressed in units of mm Hg. (Ch. 18)

blood type Method of classifying blood by determining the presence or absence of the specific antigens on the red blood cells. (Ch. 17)

blood vessels A series of tubes that convey blood from the heart throughout the body and back to the heart. (Ch. 18)

blood-brain barrier A group of anatomical barriers and transport systems that controls kinds of substances entering the brain extracellular space from blood and their rates of entry. (Ch. 7)

B-lymphocyte A blood cell that develops in the bone marrow and that is the basis of humoral or antibody-mediated immunity. (Ch. 17)

bolus A rounded mass of food. (Ch. 23)

Bowman's capsule A double-walled cup that surrounds the glomerulus and collects the glomerular filtrate. (Ch. 24)

bradykinin (**brad**-i-kin-in) A hormone produced by the kidneys, whose primary biological effect is vasodilation. (Ch. 10)

brain The organ of the nervous system encased in the skull. (Ch. 7)

brain stem The section of the brain consisting of the medulla oblongata, pons, and midbrain; located between the spinal cord and the forebrain. (Ch. 7)

Braxton-Hicks contractions Spasmodic muscle contractions of the uterus that occur during the third trimester of pregnancy. (Ch. 31)

breast alveolus A gland located in the breast that stores and secretes breast milk. (Ch. 31)

Broca's area (**broe**-kaz) Region of the neocortex responsible for the motor control of speech. Usually located in the left frontal lobe. (Ch. 7)

Brodmann area 3 The location of cells in the somatosensory cortex that exhibit the annular receptive field. (Ch. 8)

bronchiole (**brong**-kee-ol) A small division of a bronchus. (Ch. 20)

bronchitis (**brong**-kye-tis) A condition in which the lining of the bronchioles becomes inflamed and swollen, thereby reducing airway diameter. (Ch. 20)

bronchus A large-diameter airway branching off the trachea and entering the lung. (Ch. 20)

brush border membrane The microvilli of all the epithelial cells lining the small intestine. (Ch. 23)

buffer An agent that minimizes the change in pH when either acid or base is added to a solution. (Ch. 2)

bulk Material that cannot be digested. (Ch. 23)

bulk flow The movement of fluids or gases from a region of higher pressure to one of lower pressure. (Ch. 19)

bundle branches Right and left branches of the AV bundle going to each ventricle. (Ch. 18)

bundle of His Group of modified muscle cells between the right atrium and the interventricular septum specialized for very rapid conduction of impulses between the AV node and the Purkinje network; also called the *atrioventricular bundle (AV bundle)*. (Ch. 18)

calcitonin (*kal*-sih-**toe**-nin) A peptide hormone produced by c-cells of the thyroid gland; secreted in response to increases in plasma calcium concentration causing increased deposition of calcium in bones. (Ch. 27)

calcium antagonist (**kal**-see-um) A class of drugs that block the entry of calcium into the smooth muscle cells and dilate coronary arteries. (Ch. 19)

calcium ion A chemical that acts as a second messenger to initiate the intracellular reactions that modify the behavior of the target cell. (Ch. 5)

calmodulin (kal-**mod**-you-lin) An intracellular calcium-binding protein that mediates many of calcium's second-messenger functions. (Ch. 11)

caloric density (kal-**or**-ik) The amount of calories per gram of basic foodstuff. (Ch. 15)

calyx (**kayl**-ikz) An extension of the renal pelvis; collects urine from the renal papillae. (Ch. 24)

cAMP *See* cyclic AMP.

canaliculus (*kan*-ah-**lik**-you-lus) A narrow tubular passage or channel between cells. (Ch. 27)

capacitation A process involving physiological change in the plasma membrane of the sperm that increases motility and permits penetration of the surface of an ovum. (Ch. 31)

capillary (**kap**-ih-lar-ee) The smallest type of blood vessel between arteriole and venule. (Ch. 1)

capillary recruitment An increase in the number of capillaries that have blood flowing through them. (Ch. 21)

capillary transit time The amount of time necessary for blood to move through the capillaries; normal transit time is 1 second. (Ch. 21)

carbaminoglobin (*kar*-bam-**in**-oh-globe-in) The binding of carbon dioxide and hemoglobin to facilitate the transport of carbon dioxide in venous blood. (Ch. 21)

carbohydrate (*kar*-boe-**hye**-drayt) A class of organic compounds containing carbon, hydrogen, and oxygen in a ratio of 1:2:1. Sugars, starches, and cellulose are carbohydrates. This class is divided into monosaccharides, disaccharides, and polysaccharides. (Ch. 2)

carbon monoxide CO, a colorless, odorless gas that readily binds to hemoglobin, thereby decreasing blood-oxygen capacity. (Ch. 22)

carbonic anhydrase (kar-**bon**-ik an-hi-draz) An enzyme that catalyzes the reversible reaction $CO_2 + H_2O \leftrightarrow H_2CO_3$. (Ch. 7)

carboxyhemoglobin (COhb) (kar-**bok**-see-hee-moh-**glo**-bin) The binding of carbon monoxide, rather than oxygen, to hemoglobin, interfering with transport of oxygen to the cells. (Ch. 22)

carboxyl group —COOH. (Ch. 2)

carboxypeptidase (car-*bok*-see-**pep**-tih-daze) An enzyme secreted into the small intestine as a precursor of procarboxypeptidase by the exocrine pancreas; breaks the peptide bond at the carboxyl end of the protein. (Ch. 23)

carcinogen (car-**sin**-oh-jin) Any agent that induces cancerous transformation of cells. (Ch. 3)

cardiac cycle   One heartbeat sequence of the heart. (Ch. 18)

cardiac muscles   Striated, involuntary muscles that form the myocardium of the heart. (Ch. 9)

cardiac output   The amount of blood pumped out of one ventricle in 1 minute. (Ch. 18)

carotid body (kah-**rot**-id)   A chemoreceptor located at the bifurcation of the carotid artery. It is sensitive to changes in levels of oxygen, carbon dioxide, and pH in the blood. (Ch. 10)

carotid sinus   The dilated portion of the internal carotid artery located just above the branching of the common carotid artery. Pressure receptors that help regulate blood pressure are located here. (Ch. 10)

carrier proteins   Membrane proteins that allow ions and large water-soluble molecules to pass through the lipid bilayer via facilitated diffusion. (Ch. 4)

catabolism (kah-**tab**-o-lizm)   The breakdown of organic molecules into simpler compounds accompanied by the release of energy. (Ch. 6)

catalase (**kah**-tuh-laz)   A cellular enzyme that catalyzes the decomposition of hydrogen peroxide. (Ch. 22)

catecholamine (*kat*-e-**kole**-ah-mean)   A hormone that acts on the heart to increase heart rate and the volume of blood pumped, thereby increasing blood pressure: dopamine, epinephrine, and norepinephrine. (Ch. 10)

cathartic   A purgative; facilitates defecation by increasing bulk. (Ch. 23)

catheter (**kath**-eh-ter)   A tube that is inserted into a body cavity or along a blood vessel for the introduction of body fluids. (Ch. 24)

cation (**cat**-eye-on)   An atom that loses one or more electrons, thus acquiring a positive charge. (Ch. 2)

caudal   Anatomical term of position that means nearer to the tail, or further from the head; also called *inferior*. (Ch. 1)

caudate (**kaw**-dayt)   One type of basal ganglia; receives inputs from the neocortex, thalamus, and substantia nigra of the brain stem. The **caudate nucleus** is one pair of basal ganglia associated with the reward system in the brain. (Ch. 9, Ch. 11)

c-cells   Parafollicular cells in the thyroid gland; cells that produce the peptide hormone calcitonin. (Ch. 27)

cecum (**see**-come)   The blind sac that marks the first part of the large intestine. (Ch. 23)

cell adhesion molecule (CAM)   A glycoprotein important in nervous system development. (Ch. 7)

cell cycle   The period of time from the beginning of one cell division to the beginning of the next division. (Ch. 3)

cell differentiation   *See* differentiation.

cell fractionation   The separation of different organelles of the cell in order to study each on its own in a strictly controlled system. (Ch. 3)

cell theory   The broad basis for all modern biology and physiology; key concepts are (1) all organisms are made up of cells; (2) new cells arise only from preexisting cells; (3) all cells have the same fundamental chemical makeup and metabolic processes; and (4) activities and processes of an organism result from interdependent and cooperative workings of groups of cells. (Ch. 1)

cell-mediated immunity   A type of specific defense response wherein several kinds of T-cells and macrophages work together to destroy an antigen. (Ch. 17)

cellular respiration   The process by which organic molecules are catabolized by reactions that require atmospheric O_2, producing CO_2 and H_2O. The reverse of photosynthesis. (Ch. 6)

cellulose   A polysaccharide composed of glucose subunits; found in plant cells. (Ch. 23)

central nervous system (CNS)   The brain and spinal cord. (Ch. 7)

centrifuge   A machine that allows materials to be spun at high speeds, thus separating them by density. (Ch. 3)

cerebellar anterior lobe (*ser*-eh-**bel**-ar)   The portion of the cerebellum involved with planning, initiation, and control of limb movement. (Ch. 9)

cerebellar flocculonodular lobe   The portion of the cerebellum involved with the maintenance of equilibrium and posture. (Ch. 9)

cerebellar granule cell   Fibers in the cerebellum that synapse with the dendrites of the Purkinje cells, which in turn generate simple patterns of action potentials. (Ch. 9)

cerebellar posterior lobe   The portion of the cerebellum involved with planning, initiation, and control of limb movement. (Ch. 9)

cerebellum (*ser*-eh-**bel**-um)   The area of the brain responsible for locomotor coordination. (Ch. 7)

cerebrospinal fluid (CSF) (se-*ree*-broe-**spy**-nal)   The fluid that fills the cerebral ventricles and the subarachnoid space surrounding the brain and the spinal cord; provides an extracellular environment for brain cells. (Ch. 10)

cervical mucosa (**ser**-vih-kul mew-**ko**-sah)   Secretions of the cervix composed of glycoproteins, salt, and water. (Ch. 31)

cervix (**ser**-viks)   The lower portion of the uterus; connects uterus and vagina. (Ch. 31)

channel gating   Mechanism by which the opening and closing of ion channels is regulated. Channel gating is regulated by channel proteins. (Ch. 4)

channel protein   Membrane proteins that allow ions and large water-soluble molecules to pass through the lipid bilayer via facilitated diffusion. (Ch. 4)

characteristic frequency, f_o The frequency that most effectively activates a particular auditory nerve fiber. (Ch. 8)

chemical energy   The energy of the cell used to do work; released by reactions that take place within the cell. (Ch. 2)

chemical equation   A representation using chemical symbols to show a chemical reaction. (Ch. 2)

chemical pH buffer   A weak acid and its conjugate base. (Ch. 26)

chemical reaction   Process in which chemical bonds are broken down and recombine into new compounds. (Ch. 2)

chemical stimuli   Humoral or fluid-related stimuli. (Ch. 29)

chemical symbol   An abbreviation for an element. (Ch. 2)

chemiosmotic synthesis of DNA (kee-me-oz-*moh*-tik)   The overall process by which oxidative phosphorylation generates ATP. (Ch. 6)

chemoreceptor   An afferent nerve ending (or cells) that is sensitive to concentrations of certain chemicals. (Ch. 10)

chemotaxis (**kee**-moe-*tak*-sis)   The attraction by and subsequent movement toward a chemical by a leukocyte or macrophage. (Ch. 17)

chloride shift   The exchange of chloride ions for bicarbonate ions by the erythrocyte as carbon dioxide enters and leaves the erythrocyte. (Ch. 21)

chloroform   An inhalation anesthetic, first used for women in childbirth. (Ch. 22)

cholecalciferol (koe-lee-**kal**-sif-er-al)   Natural vitamin D. (Ch. 27)

cholecystectomy (koe-lee-sis-**tek**-toe-mee)   The surgical removal of the gall bladder. (Ch. 23)

cholecystokinin (CCK)   A polypeptide hormone produced by the gastrointestinal tract; stimulates enzyme secretion and growth in the pancreas, stimulates gallbladder contraction, and relaxes the sphincter of Oddi. (Ch. 22)

cholesterol (koe-**les**-ter-ol)   The most abundantly found steroid located in cell membranes; it is actually classified as a lipid. Cholesterol is a component of gallstones and is a precursor in the production of many steroid hormones. (Ch. 4)

cholinergic receptor (koe-lin-**er**-jik)   A receptor of acetylcholine. (Ch. 7)

chondrocyte (**kon**-droe-site)   A cell involved in cartilage formation. (Ch. 13)

chromaffin cell (kroe-**maf**-in) The functional unit of the adrenal medulla; functions as a neuroendocrine cell. A cell that is readily stained with solutions containing chromium salts. (Ch. 14)

chromaffin granules Dense, membrane-bound vesicles in chromaffin cells that contain epinephrine. (Ch. 14)

chromatid (**kroe**-mah-tyd) One of two identical strands of chromatin that results from DNA replication during meiosis and mitosis. (Ch. 3)

chromatin (**kroe**-mah-tin) A combination of DNA and nuclear proteins that is the principal component of chromosomes. (Ch. 3)

chromosomal sex The genotypic sex of an individual. (Ch. 32)

chromosome (**kroe**-moe-sowm) A highly coiled, condensed form of chromatin formed in the cell nucleus during mitosis and meiosis. (Ch. 3)

chylomicron (kie-loe-**my**-kron) An aggregate of triglycerides, cholesterol, and phospholipids surrounded by protein that is absorbed from the intestinal cells into the lacteal of a villus. (Ch. 17)

chymotrypsin (kie-moe-**trip**-sin) An enzyme secreted by the exocrine pancreas; breaks down certain peptide bonds in proteins and polypeptides. (Ch. 23)

circadian rhythms (ser-**kay**-dee-an) Rhythms of the body that vary on a daily basis. (Ch. 11)

circulatory convection A method of transfer of heat within the body via the movement of body fluids. (Ch. 28)

circulatory shock The progressive deterioration of cardiovascular function; generally caused by great losses of blood, severe tissue damage, and irreversible circulatory collapse. (Ch. 19)

citric acid cycle A series of reactions occurring in the mitochondrion that represent the final common pathway for the breakdown of all the fuel molecules in the cell. Also called the Krebs cycle or the tricarboxylic acid cycle. (Ch. 6)

classical conditioning One type of associative learning in which an animal acquires an understanding about the relationship between one stimulus and another. (Ch. 11)

clearance The volume of plasma per unit time needed to account for the rate of removal of a particular substance. (Ch. 24)

clitoris (**kli**-tor-iss) A small body of erectile tissue in the female external genitalia; homologous to the penis in males. (Ch. 32)

cloning vector Plasmids used in DNA cloning studies. (Ch. 3)

clot A semisolid mass similar in consistency to gelatin. (Ch. 17)

clot retraction The decrease in the size of a clot; caused by thrombosethenin. (Ch. 17)

CNS Central nervous system. (Ch. 7)

CO₂ fixation The series of steps in which CO_2 from the atmosphere is used in the creation of carbohydrates and other high-energy organic compounds. (Ch. 6)

coagulation (koh-*ag*-you-lay-shun) The process of clot formation. (Ch. 17)

cocaine (koh-**cane**) A drug that enhances release of dopamine and augments the rate of self-stimulation. (Ch. 11)

cochlea (**koke**-lee-ah) A spirally coiled, fluid-filled inner tube of the inner ear; contains the organ of Corti, where nerve impulses are generated in response to sound waves. (Ch. 8)

cochlear hair cells Cells in the cochlea that transduce the vibrations of the basilar membranes into action potentials. (Ch. 8)

codon (**koh**-don) Three-base sequence in a polynucleotide of DNA or mRNA that specifies the location of a single amino acid in a polypeptide chain. (Ch. 5)

coenzyme (koe-**en**-zime) Complex organic cofactors that serve to transfer atoms and small groups from one enzymatic substrate to another acceptor. Vitamins function as coenzymes. (Ch. 2)

coenzyme A A coenzyme also known as pantothenic acid; involved in many reactions, such as the oxidation of pyruvate or fatty acids to acetyl CoA in the citric acid cycle. (Ch. 6)

cofactor (**koe**-fak-tor) The nonprotein component that some enzymes need to function as catalysts. (Ch. 2)

cold receptors Peripheral temperature receptors that are naked nerve endings sensitive to temperature; characterized by increasing steady-state discharge rates as the skin is cooled. (Ch. 28)

collagen (**kol**-a-jen) A fibrous protein found in collagen fibers that is the principal support of many connective tissues. (Ch. 27)

colloid (**kah**-loyd) A mixture in which there is suspended material of large molecular size. The protein-containing material found in the interior of the follicles of the thyroid gland. (Ch. 13)

colloidal osmotic pressure (also called **oncotic pressure**) Refers to the portion of the *total osmotic pressure* that is due to the presence of plasma proteins. (Ch. 19)

colon (**koe**-lon) Part of the large intestine between the cecum and the rectum; subdivided into the ascending colon, transverse colon, descending colon, and sigmoid colon. (Ch. 23)

color-blind Refers to a visual defect in which an individual is unable to distinguish between certain colors. (Ch. 8)

colostrum (koe-**las**-strum) Thin yellowish fluid excreted by the breast; precursor to breast milk. (Ch. 31)

combined white blood cell count The number of leukocytes per cubic millimeter of whole blood. (Ch. 17)

common bundle *See* bundle of His.

complement (**kom**-plih-ment) A system of proteins that are normally present in plasma and that are necessary for the lysis of foreign cells in immunological defenses. (Ch. 17)

compliance The change in volume of a hollow organ that results from a change in pressure within the organ. (CL = delta volume/delta pressure.) (Ch. 20)

compound A combination of two or more elements. (Ch. 2)

concentration gradient The gradation in concentration that occurs between two regions having different concentrations of the same chemical. Expressed as a difference divided by the distance that separates the two regions. (Ch. 7)

conditioned response The behavior that results from repeated use of a conditioned stimulus. (Ch. 11)

conditioned stimulus The stimulus used to create an association with another stimulus to acquire the desired response. (Ch. 11)

conductance The ease with which ions can flow through an ion channel; measured in Siemens. (Ch. 7)

conduction Heat exchange in the form of kinetic energy between molecules of objects that are in direct contact. (Ch. 28)

conductive hearing loss Hearing loss resulting from damage to the middle ear. (Ch. 8)

cone photoreceptor A light-sensitive cell located in the retina that is cone-shaped; responsible for color vision. (Ch. 8)

congestive heart failure Heart failure that results from the inability of the heart to maintain adequate cardiac output due to progressive congestion of the veins and heart. (Ch. 19)

conjugation (*kon*-ja-**gay**-shun) To join together. For example, a process by which a steroid is coupled to a polar sulfate or glucuronide group to make it more water-soluble to facilitate elimination. (Ch. 12)

connective tissue A loosely arranged group of cells, separated by an intercellular matrix, that support, bind, protect, and partition all bodily components. (Ch. 1)

connexin A protein channel that spans the gap between presynaptic and postsynaptic cells. (Ch. 7)

connexon A protein subunit of connexin; there are six, arranged in hexagonal assembly. (Ch. 7)

constipation (*kon*-stih-**pay**-shun) The accumulation of feces beyond what is normal in the large intestine. (Ch. 23)

continence (**kon**-tih-nence) The ability to keep urine in the bladder and to retain fecal material within the colon and rectum. (Ch. 10)

contraception (*kon*-trah-**sep**-shun) The ability to block or prevent conception. (Ch. 32)

contractility (*kon*-trak-**til**-ih-tee) The ability of muscle to shorten or contract. (Ch. 16)

contraction (kon-**trak**-shun) Any form of muscle activity in response to stimulation. (Ch. 16)

control systems Mechanisms by which the body maintains homeostasis. (Ch. 1)

convection (kon-**vek**-shun) The transfer of heat between the body and a gas or liquid moving over the body surface. (Ch. 28)

coronary angioplasty A procedure in which a catheter is inserted into the occluded portion of a coronary artery and inflated, compressing the plaque and expanding the vessel's diameter. (Ch. 19)

coronary bypass surgery Surgery in which a vein from elsewhere in the body is grafted around the occluded segment of the coronary artery so that blood flow can bypass the diseased area. (Ch. 19)

coronary circulation The arrangement of blood vessels and their distribution through the myocardium. (Ch. 18)

corpus luteum (**kor**-pus **loo**-tee-um) An ovarian structure formed from the follicle after ovulation; secretes estrogen and progesterone. (Ch. 30)

cortex (**kor**-teks) The outer part of an organ. (Ch. 24)

cortical efferent zone (CEZ) Functional columns of cells in the motor cortex; all the cells in a given zone are involved in the contraction of a given muscle. (Ch. 9)

corticobulbar pathway (kor-ti-koe-**bul**-bar) The avenue by which information from the motor cortex is transmitted to the brain stem. (Ch. 9)

corticospinal pathway (kor-ti-koe-**spy**-nal) The major motor pathway within the brain and spinal cord, used for voluntarily activating skeletal muscle. (Ch. 9)

corticospinal tract lesion A damaged area of the spinal cord that disrupts inputs to motor neurons; characterized by loss of strength and movement in muscle groups, loss of strength in voluntary muscles, and the Babinski sign. (Ch. 9)

corticosteroid-binding globulin (CBG) (kor-tih-koe-**stear**-oid) A protein that binds with cortisol to facilitate transport in blood. (Ch. 14)

corticotropin-releasing hormone (CRH) A hormone produced by the hypo-thalamus that regulates the secretion of ACTH. (Ch. 10)

cortisol (**kor**-ti-sol) A glucocorticoid secreted by the adrenal glands. (Ch. 12)

cortisol-binding globulin (CBG) A carrier protein that binds cortisol for transport in the blood. (Ch. 12)

cortisone (**kor**-tih-zone) A hormone regulated by the hypothalamus. (Ch. 11)

countercurrent exchange The passive exchange of solutes, water, or heat between two adjacent streams flowing in opposite directions, as in the vasa recta of the kidney medulla. (Ch. 24)

countercurrent multiplication An energy-demanding process that sets up a solute (osmotic) gradient along the length of two streams flowing in opposite directions, as in Henle's loop in the kidney medulla. (Ch. 24)

coupled reactions Endergonic reactions that are paired with exergonic reactions to obtain the free energy necessary to proceed. (Ch. 6)

covalent bond (koe-**vay**-lint) The chemical bond formed when two or more atoms share electrons. (Ch. 2)

c-peptide A peptide involved in insulin synthesis. (Ch. 15)

cranial An anatomical term of position that means nearer to the head; also called superior or caudal. (Ch. 1)

creatine phosphate (CP) A high-energy compound capable of transferring its phosphate and energy to ADP to create ATP. (Ch. 6)

creatinine (kree-**at**-ih-nine) An end product of muscle metabolism excreted in the urine; its renal clearance can be used to estimate glomerular filtration rate. (Ch. 24)

crenation The shriveling of an erythrocyte as it loses an excessive amount of water. (Ch. 17)

cretinism (**kree**-tin-izm) A mental handicap that results from an untreated thyroid hormone deficiency; linear growth is also impaired, resulting in dwarfism. (Ch. 13)

cross bridge Myosin molecule head that projects from the ends of the thick myofilaments within a muscle cell; can bind to an active site on a thin myofilament in the presence of calcium, causing the filaments to slide past one another. (Ch. 16)

crossing over A process in mitosis in which homologous chromosomes exchange chromosomal segments, ensuring the infinite recombination of genetic material. (Ch. 30)

crossmatching A process by which agglutinin in the serum of blood is tested to confirm blood type; the presence or absence of red cell agglutination indicates the type of antibody present, thus confirming the type of antigen present on unknown red cells. (Ch. 17)

current flow The movement of electric charge; in biologic systems the current flow is the movement of ions. (Ch. 7)

Cushing's syndrome A condition resulting from excessive secretion of the adrenal cortex. General appearance of the patient is thin arms and legs, increased fat in the face, trunk, shoulder blades, and base of the neck, as well as masculinization of the female. (Ch. 14)

cutaneous (cue-**tay**-nee-us) Pertaining to the skin. (Ch. 8)

cyclic AMP (cAMP) Cyclic 3',5'-adenosine monophosphate; the compound derived from nucleotide that mediates the action of many hormones in cells. (Ch. 2)

cyclosporine (*sye*-cloe-**spore**-in) An immunosuppressive agent that reduces the possibilities of rejection of foreign substances by the immune system. (Ch. 18)

cystic fibrosis (*sis*-tik fi-**broe**-sis) An inherited genetic disorder in which secretions of body glands are abnormally thick and sticky, clogging glands and preventing them from working properly. (Ch. 4)

cytochemistry (*sye*-to-**khem**-is-tree) A research technique used with microscopy to produce color or enhanced contrast at the sites of specific molecules. (Ch. 3)

cytochrome (*sye*-toe-krome) A protein that serves as an electron carrier in oxidative phosphorylation. (Ch. 6)

cytokine (*sye*-toe-keen) A general term for a protein messenger secreted by macrophages and monocytes or by lymphocytes. (Ch. 29)

cytokinesis (*sye*-toe-kih-**nee**-sis) The process by which the cytoplasm divides, each half taking with it one of the two new nuclei to form a new cell. (Ch. 3)

cytoplasm (*sye*-toe-plazm) One of two major compartments of the cell; composed of the cytosol and the organelles. (Ch. 3)

cytoskeleton (*sye*-toe-skel-eh-ton) The internal structure of the cell responsible for maintaining cell shape. (Ch. 3)

cytosol (*sye*-toe-sol) The fluid surrounding the nucleus. (Ch. 3)

data interpretation The comparison of amounts, rates, times, and other measurable evidence of the body processes that reveals relationships between function and structure. (Ch. 1)

dead space air The amount of air in conducting airways, approximately 150 ml for each 500 ml inhaled, that does not participate in gas exchange. (Ch. 20)

deamination of amino acids (*dee*-am-in-**ay**-shun) Removal of an amino acid group (NH_2) from a molecule. A process by which amino acids are broken down to acetyl compounds so they can be used to generate ATP. (Ch. 6)

decremental conduction The slowed travel of a pacemaker signal as it passes through the atrioventricular node. (Ch. 18)

defecation (deaf-ih-**kay**-shun) The removal of fecal material from the colon and rectum. (Ch. 10)

dehydration (dee-hi-**dray**-shun) A deficiency of water in the body. (Ch. 15)

7-dehydrocholesterol A precursor to cholecalciferol, which itself is derived from cholesterol; changes to cholecalciferol through the action of ultraviolet light. (Ch. 27)

dehydroepiandrosterone (DHEA) (dee-hy-dro-ep-i-an-dro-stir-**own**) A steroid hormone secreted by the adrenal cortex. (Ch. 12)

dehydrogenation reaction The removal of an electron during oxidation through the loss of an entire atom, almost always hydrogen. (Ch. 6)

deiodinase A compound involved in the breakdown of thyroglobulin through the removal of iodide from MIT and DIT to produce iodine and tyrosine. (Ch. 13)

delta cells of the pancreas Cells found in the islet of Langerhans responsible for the production of somatostatin. (Ch. 15)

dendrite (**den**-drite) A fiberlike structure that branches out from the cell body. (Ch. 7)

dentate nucleus A deep cerebellar nucleus that forms synaptic connections with the axons of Purkinje cells. (Ch. 9)

deoxyribonucleic acid (DNA) (dee-ok-see-*ri*-boe-nyoo-**klee**-ik) A double-helical molecule consisting of two strands of four different nucleotides. Contains the genetic information that is duplicated in cell division. (Ch. 3)

deoxyribonucleic (DNA) ligase An enzyme that joins together broken strands of different DNA molecules, allowing any DNA fragment to be grafted into the DNA of a bacterium. (Ch. 3)

dependent variable The variable that changes as a result of changes in the independent variable; usually represented on the vertical axis. (Ch. 1)

depolarization (dee-*poe*-lar-i-**zay**-shun) The gradual movement of the membrane potential toward a more positive value. (Ch. 8)

dermatome A particular area of the skin whose receptors send sensory information to specific spinal cord segments. (Ch. 8)

dermis (**der**-mis) The thick layer of skin located beneath the epidermis;

composed of irregular, dense connective tissue. (Ch. 28)

desmosome (**dez**-moe-zome) Type of cell-to-cell junction where membrane differentiation takes place. (Ch. 16)

diabetes insipidus (*dye*-ah-**bee**-teez in-**sip**-ih-des) A disease due to a lack of ADH; marked by great thirst and excretion of a large volume of dilute urine. (Ch. 13)

diabetes mellitus (**mel**-ih-tus) A disease in which plasma glucose control is defective because of insulin deficiency or decreased target cell response to insulin. (Ch. 15)

diabetic neuropathy A secondary complication of diabetes involving the impairment of nerve function, resulting in abnormalities in the bladder or gastrointestinal tract function, impotence, or loss of feeling in the lower limbs. (Ch. 15)

diabetic retinopathy (*reh*-tin-**op**-uh-thee) A secondary complication of diabetes that results in the deterioration of blood vessels that nourish the retina, eventually resulting in blindness. (Ch. 15)

diacylglycerol (DAG) (dye-*ace*-ill-**glis**-er-all) The second messenger that activates the protein kinase C, which then phosphorylates a large number of other proteins. (Ch. 5)

dialysis (dye-**al**-ih-sis) The separation of smaller from larger molecules in solution by diffusion of the small molecules through a selectively permeable membrane. (Ch. 25)

diapedesis (*dye*-ah-peh-**dee**-sis) The process by which leukocytes pass through the capillary wall to enter spaces around blood vessels. (Ch. 17)

diaphragm (**dye**-ah-fram) A large, dome-shaped skeletal muscle that separates the chest cavity from the abdomen. (Ch. 20)

diarrhea (*dye*-ah-**ree**-ah) Frequent defecation of highly fluid material. (Ch. 23)

diastole (dye-**as**-toe-lee) The period of the cardiac cycle during which the ventricles are filling with blood. (Ch. 18)

dichromat (dye-**kroh**-mat) Males who are color-blind due to possessing only two types of cones; a sex-linked trait. (Ch. 8)

dichromatic color vision Color vision that relies on two types of cones. (Ch. 8)

diencephalon (die-en-**sef**-ah-lon) The core of the anterior part of the brain; lies beneath the cerebral hemispheres and contains the thalamus and the hypothalamus. (Ch. 9)

dietary fiber Cellulose and other indigestible polysaccharides found in plant food sources. (Ch. 23)

differential white blood cell count The number of each kind of white

blood cell, usually expressed as a percentage of the total or combined white blood cell count. (Ch. 17)

differentiation (dif-er-en-she-**aye**-shun) The process by which a single cell type develops into many different cell types. (Ch. 1)

diffuse synapse A chemical synapse in which the neurotransmitter release is not limited to specific active zones. (Ch. 7)

diffusing capacity of the lung The milliliters of a gas diffusing across the pulmonary membrane per minute per millimeter of mercury pressure difference between the alveolar air and the pulmonary blood. (Ch. 29)

diffusion (dif-**you**-zhun) A process that involves the random movement of particles, such as ions and molecules in solution, such that the substances become evenly mixed. (Ch. 2)

diffusion coefficient $D = 1.0 \times 10^{-5}$ cm^2/sec. (Ch. 4)

diffusion trapping A phenomenon in which the neutral form of a weak base will diffuse across a cell membrane, whereas the charged form will not. (Ch. 4)

digestion (dye-**jes**-chun) The conversion of large organic molecules in the diet to smaller molecules. (Ch. 23)

digestive enzymes Organic compounds that break down large organic molecules into simpler, smaller components. (Ch. 23)

digestive tract A tube about thirty feet long that traverses the body from lips to anus and is composed of the mouth, pharynx, esophagus, stomach, and small and large intestines; also called *alimentary canal*. (Ch. 23)

digitalis A drug that tends to increase the contractility of cardiac muscle by increasing intracellular calcium concentrations. (Ch. 19)

dihydrotestosterone (DHT) (dye-*hi*-droe-tes-**tos**-ter-own) A steroid that stimulates sexual differentiation of the external genitalia in the male. (Ch. 32)

1,25 dihydroxyvitamin Vitamin D_3. (Ch. 12)

diiodotyrosine (DIT) An iodinated compound found primarily within the thyroid follicle cell; used to make T_3 and T_4. (Ch. 13)

direct calorimetry A method of measuring metabolic rate that uses a large insulated chamber to determine the heat loss from a person's body. (Ch. 28)

disaccharidase (dye-**sak**-er-ide-aze) An enzyme that can hydrolyze disaccharides. (Ch. 23)

disaccharide (dye-**sak**-er-ide) A carbohydrate molecule composed of two monosaccharides. (For example, maltose = glucose-glucose.) (Ch. 2)

discrete synapse   A chemical synapse in which the chemical neurotransmitter is released from restricted areas of the presynaptic terminal into a small synaptic cleft. (Ch. 7)

distal (**dis**-tal)   Refers to the insertion, or less stationary point of attachment, of a muscle. (Ch. 16)

distal convoluted tubule   A short nephron segment between the thick ascending limb of Henle's loop and the connecting segment. (Ch. 24)

distal muscles   Muscles involved with the movement of the skeleton, primarily the limbs. (Ch. 9)

distal tubule   The distal convoluted tubule, connecting segment, and first part of the cortical collecting duct. (Ch. 24)

diuresis (dye-you-**ree**-sis)   An increased urine flow rate. (Ch. 24)

diuretic (dye-you-**ret**-ik)   An agent that increases urine output. (Ch. 25)

diurnal (**die**-urn-al)   Occurring on a 24-hour cycle. (Ch. 11)

diving bell   A chamber, either open at the bottom or enclosed, which holds air for a diver to breathe. (Ch. 22)

diving reflex   A protective reflex, elicited when the face is submerged in water, that shunts blood away from peripheral tissues and allows for heart–brain perfusion. (Ch. 20)

dizygotic twins   Siblings resulting from fertilization of two ova by individual sperm. (Ch. 31)

DNA polymerase   An enzyme that during DNA replication forms a new DNA strand by joining together nucleotides already base-paired with an existing DNA strand. (Ch. 5)

DNA replication   The copying of the chromatin in the nucleus. (Ch. 5)

DNA template (**tem**-plit)   A pattern. (Ch. 5)

DNA transcription   The formation of mRNA containing, in linear sequence of its nucleotides, genetic information in a specific gene; the first stage of protein synthesis. (Ch. 5)

dominant hemisphere   The side of the brain in which the language centers are located. (Ch. 11)

dopamine (**dope**-ah-mean)   A catecholamine synthesized from tyrosine; a precursor of epinephrine and norepinephrine. (Ch. 7)

dopamine receptor   A receptor that causes the hyperpolarization of the postsynaptic potential through the activation of chemically sensitive potassium ion channels. (Ch. 7)

dorsal column nuclei   A collection of neurons located on the dorsal column. (Ch. 8)

dorsal column pathway   A major pathway for the transmission of somatic sensory information from the surface of the body to the neocortex. (Ch. 8)

dorsal hypothalamus   A central component of the autonomic nervous system. (Ch. 10)

dorsal root ganglion   A cluster of sensory cell bodies located along the posterior root of a spinal nerve. (Ch. 8)

double helix   A configuration in which two nucleotide strands of DNA are coiled around each other. (Ch. 2)

down regulation of receptor number   A sustained elevation of the concentration of an agonist circulating in the blood, leading to a decrease in the number of receptors present in target cells. (Ch. 5)

ductus arteriosus (**duk**-tus ar-*tee*-ree-**o**-sus)   A fetal bypass vessel that connects the pulmonary artery to the aorta. (Ch. 21)

duodenum (**doo**-o-dee-num)   The initial 30 cm segment of the small intestine. (Ch. 23)

dynamic exercise   Activities such as bicycling or running in which the length of muscles changes and which can be quantified in physical units such as watts. (Ch. 29)

dynamic neurons   Neurons that change their activity when there is an increase in the level of force applied. (Ch. 9)

dynein   A protein from the microtubules of cilia and flagella; functions as an ATP-splitting enzyme. (Ch. 7)

dyslexia (dis-**lek**-see-ah)   A congenital disorder that affects an individual's ability to read, write, or spell. (Ch. 11)

dysmenorrhea (*dis*-men-o-**ree**-ah)   Excessive uterine contractions or menstrual cramps. (Ch. 30)

dyspareunia (*dis*-**par**-ee-oo-nee-ah)   The absence of sufficient vaginal lubrication. (Ch. 32)

dysphagia   Difficulty in swallowing. (Ch. 23)

E_m The difference in electrical potential across a membrane. (Ch. 7)

ear drum   The membrane located between outer and middle ear that vibrates in response to sound waves; also called *tympanic membrane*. (Ch. 8)

ectopic pregnancy (ek-**top**-ik)   The implantation and development of a fetus at a site other than the uterus. (Ch. 31)

edema (e-**dee**-mah)   An abnormal accumulation of fluid in the interstitial fluid spaces. (Ch. 24)

EEG   *See* electroencephalogram.

efferent (**ef**-er-ent)   Carrying away from. (Ch. 9)

efferent arteriole   The arteriole that carries blood away from the glomerulus. (Ch. 24)

efferent neuron   Nerve cell that carries signals *away* from the CNS. (Ch. 7)

efficiency   The amount of work done for a certain amount of energy. (Ch. 16)

ejaculation (e-jak-yoo-*lay*-shun)   The emission and expulsion of ejaculatory fluid. (Ch. 32)

elastic recoil   The property of stretched tissue that allows it to return to a prestretched state. (Ch. 20)

elasticity   The ability of stretched material to return to its unstretched position. (Ch. 20)

electrical potential (E)   The difference in net charge between the inside and the outside of a cell. (Ch. 7)

electrocardiogram (ECG or EKG) (e-*lek*-troe-**kar**-dee-o-gram)   A record of the electrical activity of the heart. (Ch. 18)

electrochemical gradient   The combination of chemical and electrical gradients of a solution that determines the flux of positive and negative ions. (Ch. 4)

electroencephalogram (EEG) (e-**lek**-troe-en-**sef**-ah-loe-gram)   A record of the frequency and amplitude of the brain's electrical activity. (Ch. 11)

electrolyte (e-**lek**-troe-lite)   An ionized substance in solution; capable of conducting electricity. (Ch. 24)

electromagnetic radiation   A continuum of electromagnetic energy of varying wavelengths, ranging from radio waves to X-rays. (Ch. 8)

electron   Negatively charged particle of an atom. (Ch. 2)

electron carrier   A molecule that donates electrons to proteins. (Ch. 6)

electron transport chain (respiratory chain)   The transfer of electrons from reduced coenzymes to O_2, a process that is coupled with phosphorylation of ADP to ATP. (Ch. 6)

element   A substance that cannot be broken down into simpler substances by chemical reactions. (Ch. 2)

embolus (**em**-boe-lus)   A foreign object floating in the blood; for example, a clot (thromboembolus) or an air bubble (airembolus). (Ch. 17)

embryo (**em**-bree-o)   A blastocyst after it has fully implanted in the endometrium of the uterus. (Ch. 31)

embryonic stage   Weeks 2 to 8 following fertilization. (Ch. 31)

emetic (**em**-et-ik)   Chemical agent that causes vomiting by stimulating chemoreceptors in the stomach and small intestine or brain. (Ch. 23)

emphysema (**em**-fih-**see**-mah)   A lung disease that is characterized by destruction of the alveolar wall and consequent impairment of gas exchange. (Ch. 20)

emulsion (ee-**mul**-shun)   A preparation of one liquid distributed in globules throughout a second liquid. (Ch. 23)

en passant   An overlapping system of synapses in passage. (Ch. 7)

end-diastolic volume (EDV) dye-ah-**stol**-ik)   The volume of blood in the ventricle at the end of diastole. (Ch. 18)

end-plate current A small local current that flows across the postsynaptic membrane at the neuromuscular junction. (Ch. 16)

end-plate potential The depolarization of motor end-plate of skeletal muscle fiber in response to acetylcholine; initiates action potential in muscle plasma membrane. (Ch. 16)

end-systolic volume The volume of blood in the ventricle at the end of systole. (Ch. 18)

endergonic reaction A reaction that cannot proceed without the input of free energy. (Ch. 6)

endocardium (en-doe-**kar**-dee-um) A thin layer of endothelial cells that lines the inner surface of the myocardium, which forms the inside walls of the atria and ventricles. (Ch. 18)

endocrine gland (**en**-doe-krin) Any of several ductless-type glands that secretes hormones into the blood stream, thereby influencing the activity of cells located away from the gland. (Ch. 12)

endocrine pancreas Small portion of the pancreas that secretes insulin and glucagon and other hormones (also called *islets of Langerhans*). (Ch. 15)

endocrine system All of the body's hormone-secreting glands. (Ch. 12)

endocytosis (*en*-doe-sye-**toe**-sis) The process that allows the entrance into a cell of substances that are otherwise unable to penetrate the cell membrane; includes pinocytosis and phagocytosis. (Ch. 3) **Receptor-mediated endocytosis** involves larger peptide hormones that are internalized by their receptor tissue cells. (Ch. 12)

endogenous (en-**doj**-e-nus) Produced within the body or due to internal causes. (Ch. 24)

endogenous protein Digestive enzymes, epithelial cells, mucoproteins, and plasma proteins that leak into the digestive tract and are absorbed. (Ch. 23)

endometriosis (*en*-doe-**mee**-tree-oh-sis) A disease characterized by the presence of endometrial-like tissue in locations other than the uterus, such as the fallopian tubes or ovaries. (Ch. 32)

endometrium (*en*-doe-**mee**-tree-um) The lining of the uterus. (Ch. 30)

endomysium (*en*-doe-**meez**-ee-um) The connective tissue sheath surrounding individual muscle fibers. (Ch. 16)

endoplasmic reticulum (ER) (*en*-doe-**plaz**-mik re-**tik**-you-lum) A network of membranous tubules, flattened sacs, and vesicles located in the cytoplasm that function in intracellular transport, synthesis, storage, packaging, and secretion. (Ch. 3)

endorphin (en-**dor**-fin) Any of the "endogenous-morphine" family of peptides; they function as a neurotransmitter at synapses activated by opiate drugs and as a paracrine and hormone. (Ch. 12)

endothelial clefts (or pores) (*en*-doe-**thee**-lee-al) Small openings between endothelial cells that allow for the passage of molecules. (Ch. 19)

enkephalin (en-**kef**-ah-lin) A peptide of the nervous system that affects the perception of pain. (Ch. 8)

enterohepatic circulation of bile salts (*en*-teh-roe-he-**pat**-ik) Recycling of bile salts between liver and small intestine. (Ch. 23)

enterokinase (*en*-teh-roe-**kin**-aze) An enzyme present in the luminal membrane of the epithelial lining of the small intestine. (Ch. 23)

enzyme (**en**-zime) A natural catalyst that increases the rate of chemical reactions in the body. (Ch. 2)

eosinophil (ee-o-**sin**-o-fil) A leukocyte whose primary function is detoxification of foreign proteins and other substances. (Ch. 17)

epicardium (ep-i-**kard**-ee-um) The outer surface of the myocardium, which consists primarily of connective tissue and fat. (Ch. 18)

epididymis (ep-i-**did**-i-mis) The portion of the male reproductive duct system in which sperm mature between the seminiferous tubules and the vas deferens. (Ch. 30)

epiglottis (ep-ih-**glot**-is) Cartilage guarding the superior opening into the larynx. (Ch. 23)

epimysium (ep-ih-**meez**-ee-um) A connective tissue sheath that covers the entire surface of a muscle. (Ch. 16)

epinephrine (ep-ih-**nef**-rin) One of the two catecholamines that is produced by the adrenal medulla; stimulates the sympathetic nervous systems, increasing blood pressure, stimulating the cardiac muscle, and causing glycogenolysis. Also known as *adrenaline*, it acts together with norepinephrine to prepare the body for "fight or flight" response. (Ch. 14)

epiphyseal plate (*ep*-ih-**fizz**-e-al) The region near the end of a bone where linear growth occurs. (Ch. 13)

epithelial tissue Closely packed cells arranged in flat sheets ranging from one to several layers in thickness that cover exposed body surfaces and line body cavities, tubes, and organs. (Ch. 1)

equilibrium A state at which the rate of formation of products of a chemical reaction is exactly balanced by the rate of reformation of the original reactants. (Ch. 2)

equilibrium potential The voltage of electrical force required to draw ions across a cell membrane at the same rate as they pass because of the concentration gradient. (Ch. 7)

ergocalciferol Vitamin D_2, the form of vitamin D found in plants and yeasts. (Ch. 27)

erythrocytosis (eh-*rith*-roe-sigh-**toe**-sis) An increase in the number of red blood cells in response to low oxygen environments and other states in which oxygenation of blood is reduced. (Ch. 17)

erythropoiesis (eh-*rith*-roe-**poy**-ee-sis) Erythrocyte formation. (Ch. 17)

erythropoietin (eh-*rith*-roe-**poy**-ih-tin) A glycoprotein hormone, produced chiefly by the kidneys, that stimulates the rate of production, maturation, and release of red blood cells from bone marrow. (Ch. 24)

esophagus (eh-**sof**-ah-gus) The muscular tube located mostly in the thorax that conducts material from the pharynx to the stomach. (Ch. 23)

essential amino acids Amino acids that cannot be synthesized and must be ingested as food. (Ch. 6)

essential hypertension Chronically elevated arterial blood pressure due to a primary disturbance of the cardiovascular system. (Ch. 19)

estradiol (**es**-trah-die-ol) The principal sex steroid in females. (Ch. 13)

estrogen (**es**-trow-jen) A hormone secreted by the ovaries and the placenta that stimulates the growth and development of tissues in the reproductive tract. (Ch. 24)

ether An anesthetic gas. (Ch. 22)

eukaryotic cell A cell that has a well-defined nucleus contained within a nuclear membrane. (Ch. 3)

evaporation The process of converting a liquid to a gas. (Ch. 28)

excision repair of DNA A process under enzyme control by which a section of damaged DNA is selectively removed and replaced. (Ch. 5)

excitatory postsynaptic potentials (EPSP) Potentials that depolarize the postsynaptic membrane and tend to excite nerve cells to discharge action potentials. (Ch. 7)

excretion (eks-**kree**-shun) The elimination of a substance from the body in the urine or feces. (Ch. 24)

exergonic reaction A chemical reaction that releases free energy and is likely to proceed spontaneously. (Ch. 6)

exocrine gland (**ek**-so-krin) A type of gland that secretes its products by way of ducts that open onto lining or covering epithelium. (Ch. 23)

exocrine pancreas A portion of the pancreas responsible for the secretion of fluid and various enzymes involved in the process of digestion. (Ch. 15)

exocytosis (**eks**-oh-sih-**toe**-sis) A process in which a secretory vesicle fuses with the plasma membrane and materials are expelled into the extracellular fluid. (Ch. 3)

exogenous (eks-**oj**-en-us) Foreign to the body. (Ch. 24)

experimental design Construction of experiments so that specific conditions are created deliberately to produce the observations in question, ideally so that only one explanation exists for the observation. (Ch. 1)

expiration (eks-pih-ray-shun) Movement of air out of the lungs. (Ch. 10)

expiratory neurons Neurons in the medulla that stimulate expiratory muscles during exercise. (Ch. 20)

extension The movement of one body part away from another; for example, movement of the forearm away from the arm when extension occurs at the elbow. (Ch. 16)

extensor A muscle that, by contracting, causes extension; for example, the triceps brachius is an extensor of the forearm. (Ch. 9)

external Of two sites, the one farther from the center of the body. (Ch. 1)

external anal sphincter Skeletal muscles at the end of the large intestine that are relaxed during defecation. (Ch. 23)

external intercostal muscles A set of inspiratory muscles that pull the rib cage upward and outward. (Ch. 20)

external sphincter A group of striated muscles that close off the exit from the bladder. (Ch. 10)

extracellular fluid The fluid outside of cells. (Ch. 24)

extrafusal fiber Skeletal muscle fibers that form the bulk of skeletal muscles. Contraction is controlled by alpha motor neurons. (Ch. 9)

extrapolation The extension of a relationship beyond the range of available experimental data. (Ch. 1)

extrinsic (eks-**trin**-sik) Having an external origin; arising from outside the body. (Ch. 1)

extrinsic (peripheral) proteins Membrane proteins that do not penetrate the lipid bilayer. (Ch. 4)

extrinsic clotting pathway The process by which clotting begins after blood comes in contact with injured tissue. (Ch. 17)

F cells of the pancreas Cells that produce pancreatic polypeptide. (Ch. 15)

facilitated diffusion (**fah**-sil-ih-**tay**-ted) Downhill movement of substances, in or out of the cell, that is enabled by the presence of a carrier protein. (Ch. 4)

fallopian tubes (fal-**low**-pee-an) Muscular, tubelike structures that arise from the uterus and extend into the abdominal cavity to terminate near the ovaries. (Ch. 30)

fasciculi Bundles of individual muscle fibers. (Ch. 16)

fast axonal transport The movement of newly produced membranous organelles down the axon. (Ch. 7)

fast skeletal muscles Muscles that contract rapidly but fatigue quickly because they are poorly vascularized. (Ch. 9)

fastigial nucleus One of three deep cerebellar nuclei involved in the maintenance of balance and posture. (Ch. 9)

fat cells Adipose tissue that stores triglycerides, a major source of energy for the body. (Ch. 11)

fats Lipids that are poorly soluble in water and highly soluble in organic liquids; *see also* triacylglycerol. (Ch. 2)

fatty acid A long chain of covalently linked carbon atoms to which only hydrogen atoms are attached, except for a carboxylic acid (—COOH) group at one end. (Ch. 2)

fatty acid oxidation The breakdown of fatty acids in the mitochondrial matrix. (Ch. 5)

feedback excitation A process by which the activation of beta-2 receptors increases the release of norepinephrine from the presynaptic terminal. (Ch. 7)

feedback inhibition A process by which the activation of alpha-2 autoreceptors inhibits the release of norepinephrine from the presynaptic terminal. (Ch. 7)

female pseudohermaphroditism Severe mascularization of the female external genitalia. (Ch. 32)

ferritin (**fer**-ih-tin) An iron-protein complex that stores iron in the liver. (Ch. 17)

fertilization The union of sperm and ovum. (Ch. 31)

fetal stage The period between week 8 after fertilization and birth. (Ch. 31)

fetoplacental unit (fee-toe-plah-*sen*-tahl) The fetal adrenal gland and placenta, which together stimulate estrogen synthesis. (Ch. 30)

fetus (**fee**-tus) An embryo after the second month of development. (Ch. 31)

fever (**fee**-ver) A regulated increase in body temperature beyond the normal range. (Ch. 28)

fibrin (**fye**-brin) An insoluble protein derived from fibrinogen, which forms the foundation of a blood clot. (Ch. 17)

fibrinogen (fye-**brin**-o-jin) A soluble plasma protein manufactured in the liver. (Ch. 17)

fibrinolysis The enzymatic dissolution of a blood clot. (Ch. 17)

"fight or flight" response The physiological changes that prepare the body for the stress of threatening situations. (Ch. 10)

filterable calcium One of two major forms of calcium in extracellular fluid. (Ch. 27)

filtered load The quantity of a substance, per unit of time, that is filtered by the glomeruli and is therefore presented to the tubules; for a freely filterable substance, this is equal to the product of the plasma concentration and the glomerular filtration rate. (Ch. 24)

filtration (fil-**tray**-shun) The passage of a liquid through a membrane because of a hydrostatic pressure gradient. (Ch. 24)

first heart sound The closure of the atrioventricular valves that marks the beginning of systole. (Ch. 18)

flatus (**flay**-tus) Gas present in the large intestine. (Ch. 23)

flavin adenine dinucleotide (FAD) A coenzyme derived from riboflavin. (Ch. 5)

flexion (**flek**-shun) A muscle action that causes a joint to close and a limb to assume a more withdrawn position. (Ch. 16)

flexor (**flek**-sor) A muscle that by contracting causes a joint to close up. (Ch. 9)

fluid mosaic model A generalized model of the structure of cell plasma membranes. (Ch. 4)

follicle stimulating hormone (FSH) A glycoprotein hormone, created by the anterior pituitary, that stimulates growth and development of the gonads. (Ch. 12)

foramen ovale (foe-**ray**-men o-**va**-lee) A passage that allows blood to flow between the right and left atria in the fetal heart. (Ch. 21)

forced vital capacity The maximum amount of air that can be forcefully and rapidly exhaled after a deep breath. (Ch. 20)

fovea (**foe**-vee-ah) The most sensitive part of the retina. (Ch. 8)

Frank-Starling law of the heart The ability of the heart to adjust its output (stroke volume) in response to the input (venous return). (Ch. 18)

free energy The energy available to be put to work. (Ch. 2)

free radicals Highly reactive molecules formed as byproducts from molecular oxygen. (Ch. 22)

frequency code of contractile force The control of muscle tension as determined by the frequency of action potentials generated by upper motor neurons. (Ch. 9)

frequency code of stimulus intensity The number of action potentials generated per unit of time as a function of the intensity of a stimulus. (Ch. 8)

frontal plane The plane that divides the body into front and back portions (also called the coronal plane). (Ch. 1)

functional residual capacity The volume of air remaining in the lungs after a normal expiration; the sum of expiratory reserve volume and residual volume. (Ch. 20)

gallbladder A small sac that stores bile between meals; sends bile to the small intestine, where it emulsifies fat. (Ch. 23)

gallstones Small crystals of cholesterol that form in the gallbladder; this formation occurs when the capacity of bile salts and lecithin to dissolve cholesterol is exceeded. (Ch. 23)

gamma amino butyric acid (GABA) A neurotransmitter found throughout the CNS; a potent inhibitory neurotransmitter synthesized from glutamate by glutamic acid decarboxylase. (Ch. 7)

gamma motor neuron Motor neurons that synapse with intrafusal muscle fibers. (Ch. 9)

ganglion cells (gang-lee-on) Nerve cells in the retina that generate action potentials even in the absence of input from the bipolar cells. (Ch. 8)

gap junction The intercellular junction that allows ions and small molecules to flow between the cytoplasm of adjacent cells; the two adjacent plasma membranes are joined by small tubes. (Ch. 5)

gas diffusion The net movement of gas molecules from an area of higher to lower concentration. (Ch. 21)

gastric chief cells (gas-trik) Cells, located in the basal region of exocrine secretory glands, that secrete pepsinogen, the precursor to pepsin, the enzyme that initiates the digestion of protein in the digestive tract. (Ch. 23)

gastric inhibitory peptide (GIP) A hormone of gastrointestinal origin that is important in the control of motility and secretion; released in response to intraluminal stimuli acting directly on the endocrine cells in the luminal lining. (Ch. 23)

gastric mucosal barrier The ability of the cell membranes of the gastric mucosa to restrict entrance of hydrogen ions to the interior of these cells because of the low permeability of the cell membrane to hydrogen ions. (Ch. 23)

gastric parietal cells Cells that secrete an isosmotic solution of HCl. (Ch. 23)

gastrin (gas-trin) A hormone released by the stomach mucosa in response to stretching or the presence of substances such as caffeine and partially digested proteins. Stimulates the release of gastric juice from the gastric glands. (Ch. 7)

gastrointestinal hormone Gastrin, cholecystokinin, secretin, and gastric inhibitory peptide. (Ch. 23)

gastrointestinal system (gas-tro-in-testin-al) A collection of organs specialized to handle food through the processes of motility, secretion, digestion, and absorption. (Ch. 23)

G-cells Cells in the epithelial lining of the antral mucosa, which secrete the hormone gastrin into the portal blood. (Ch. 23)

gene Each section of chromosomal DNA that carries the information necessary for synthesis of a single polypeptide. (Ch. 3)

generator potential A change in the membrane potential of a sensory receptor cell that generates action potentials. (Ch. 8)

genetic code A reference to the way in which information about protein synthesis is stored in DNA, translated into RNA, and then used to make proteins. (Ch. 5)

genetic expression The phenomenon of the fact that only certain genes are tapped for information about how to make specific proteins they describe. (Ch. 5)

genome The sum of all the chromosomal genes of a cell. (Ch. 3)

germinal stage In prenatal development, the first two weeks of development following fertilization of the ovum. (Ch. 31)

gestation The period of development of the young from the time of fertilization of the ovum until birth. (Ch. 11)

gestational diabetes A transient form of diabetes that occurs in about 3% of pregnant women. (Ch. 15)

G-force The human body's sensation of weight due to gravitational attraction. (Ch. 22)

gigantism *See* acromegaly.

glia (glee-al) Nonneuronal cellular elements of nervous tissue; provide physical and functional support to adjacent neurons. (Ch. 7)

glial scar Formed by reactive astrocytes in conjunction with fibroblasts after a brain injury; the scar blocks the regeneration of new axons. (Ch. 7)

globulin (glob-you-lin) A type of plasma; classified into alpha, beta, and gamma groups. (Ch. 17)

glomerular filtration (glo-mer-yoo-lar) The process in the kidney glomeruli in which an ultrafiltration of plasma is pushed out of the capillaries into the urinary space of Bowman's capsule. (Ch. 24)

glomerular filtration rate (GFR) The rate at which the plasma is filtered by the kidney glomeruli; usually @ 125 ml/min. (Ch. 24)

glomerulotubular balance The proportionate change in sodium reabsorption by the proximal convoluted tubule and Henle's loop when glomerular filtration rate is changed. (Ch. 24)

glomerulus (glo-**mer**-you-lus) In the kidney, the tuft of capillaries, surrounded by Bowman's capsule, where the blood is filtered. (Ch. 24)

glottis (glot-iss) The true vocal cords and the opening between them. (Ch. 23)

glucagon (**gloo**-kuh-gon) Hormone released by the islets of Langerhans of the pancreas; elevates the level of blood sugar. (Ch. 15)

glucagonlike peptide Peptides that are synthesized in the small intestine and hypothalamus. (Ch. 15)

glucocorticoid (*gloo*-koe-**kor**-tih-koyd) A steroid hormone produced by the adrenal cortex and having major effects on glucose metabolism. (Ch. 10)

gluconeogenesis (*gloo*-koe-*nee*-oh-**jen**-ih-sis) The formation of glucose by the liver and kidney from noncarbohydrate precursors. (Ch. 6)

glucose threshold The plasma level at which glucose first appears in the urine. (Ch. 24)

glucosidase An enzyme that splits a glucoside. α-Glucosidase (maltase) occurs in intestinal juice and β-glucosidase (cellobiase) in the kidney, liver, and intestinal mucosa. (Ch. 2)

glucosuria (*gloo*-koe-**soo**-ree-uh) A condition in which glucose is excreted in the urine. (Ch. 15)

glutamate (**gloo**-tah-mate) One of the most potent excitatory neurotransmitters in the nervous system. (Ch. 7)

glycerol-phosphate A three-carbon molecule that is combined with free fatty acids in triacylglycerol synthesis. (Ch. 15)

glycogen (gly-ko-jen) A polysaccharide carbohydrate resembling starch that occurs in many tissues, particularly the liver and skeletal muscles; a polymer of glucose. (Ch. 6)

glycogenolysis (*gly*-koe-jeh-**nol**-ih-sis) The process by which glycogen is broken down into units of glucose-1-phosphate, which enter gluconeogenesis after conversion to glucose-6-phosphate. (Ch. 6)

glycolipid (*gly*-koe-**lip**-id) Lipids, occurring in cell membranes, that are derived from sphinogosine and contain one or more monosaccharide units. (Ch. 2)

glycolysis (gly-**kol**-ih-sis) A process by which glucose obtained from food is converted to pyruvic acid. (Ch. 6)

glycophorin An intrinsic glycoprotein in the red blood cell membrane. (Ch. 4)

glycoprotein (*gly*-koe-**pro**-teen) A member of a class of conjugated proteins composed of protein with an attached carbohydrate group. (Ch. 2)

goiter Any abnormal enlargement of the thyroid gland. (Ch. 13)

goitrogen An agent that inhibits iodide uptake or organification reactions, resulting in a blockage of thyroid hormone formation and the formation of a thyroid goiter. (Ch. 13)

Goldman equation The relationship between the intra- and extracellular concentrations of Na^+, K^+, and Cl^- ions, as well as their membrane permeabilities, P_{Na}, P_K, and P_{Cl}, and membrane potential. (Ch. 7)

Golgi complex (**goal**-jee) Part of the cell responsible for modifying macromolecules synthesized in the ER; processes and packages the macromolecules inside vesicles to be transported to the cell surface or cytosol. (Ch. 3)

Golgi tendon organ A type of muscle receptor located in series with the extrafusal muscle fiber and that transmits information about the force or tension produced by the contraction of a muscle. (Ch. 9)

gonadal sex The presence of testes or ovaries that determines the gender of the embryo. (Ch. 32)

gonadotropin (*gon*-ad-oh-**trow**-pin) Luteinizing hormone and follicle-stimulating hormone; hormones whose target tissue is the gonads. (Ch. 13)

gonadotropin-releasing hormone (GnRH) A hormone secreted by the hypothalamus that stimulates the release of luteinizing hormone and follicle-stimulating hormone. (Ch. 30)

gout A condition in which plasma uric acid is elevated and uric acid crystals precipitate in the joints, causing inflammation and painful swelling of the affected joints (often in the big toe). (Ch. 24)

G-proteins A group of integral membrane proteins that bind and require guanosine triphosphate to function. (Ch. 12)

gradient (**gray**-dee-ent) The continuous decrease or increase of a variable over a distance. (Ch. 2)

granulocyte (**gran**-yoo-low-site) A leukocyte that contains large granules and has a complex lobulated nucleus; also called polymorphonuclear granulocyte. (Ch. 17)

graph A diagram that expresses a relationship between two or more quantities. (Ch. 1)

Graves' disease Hyperthyroidism that occurs due to an abnormality in the immune system that leads to the production of antibodies that recognize and bind to sites on the surface of thyroid follicle cells. (Ch. 13)

ground substance The noncollagenic portion of organic matter in bone; consists of a mixture of various proteoglycans. (Ch. 27)

growth cone A specialized enlargement at the top of the growing neuron. (Ch. 7)

growth hormone A large peptide hormone that stimulates body growth; also called somatotropin. (Ch. 12)

gustatory receptors Taste receptors; a type of epithelial cell clustered with other cells to compose a taste bud. (Ch. 8)

habituation (hah-bit-you-**a**-shun) A type of behavior in which the individual learns to decrease a response to a repeated stimulus. (Ch. 11)

hair receptor A mechanoreceptor that is a hair follicle on the surface of the skin and that has a nerve fiber wrapped around its base. (Ch. 8)

Haldane effect Describes the amount of carbon dioxide the blood can load, which is greater the more the hemoglobin is deoxygenated. (Ch. 21)

hand manipulation neurons Neurons in the posterior parietal cortex that increase their firing rate when the hand explores an object. (Ch. 9)

hand–eye coordination neurons Neurons in the posterior parietal cortex that are most active when the eye fixates on an object that is being touched. (Ch. 9)

haptoglobin A protein that binds with hemoglobin; prevents hemoglobin from being excreted in the urine. (Ch. 17)

heart block A situation in which the ventricles are not stimulated by the atrial impulse; due to failure of the AV node or AV bundle to conduct the impulse. (Ch. 18)

heart failure The inability of the heart to maintain adequate cardiac output. (Ch. 19)

heart sounds A series of distinct sounds that correlate with specific events of the cardiac cycle; four heart sounds are distinguishable in most healthy individuals. (Ch. 18)

heartburn An unpleasant sensation due to irritation of the esophagus caused by reflux of gastric acid. (Ch. 23)

heat exhaustion An illness caused by circulatory insufficiency and hypotension; takes the form of fainting due to decreased blood pressure. (Ch. 28)

heat stroke A form of unregulated hyperthermia or heat illness associated with a breakdown of the body's temperature-regulating system. (Ch. 28)

hematocrit (hee-**mat**-o-krit) The volume occupied by packed red blood cells in a centrifuged 100-ml sample of whole blood. (Ch. 17)

hematopoeisis (*hem*-ah-toe-poy-**ee**-sis) The formation of blood cells in the red bone marrow. (Ch. 17)

hemocytoblast (*hee*-moh-**sigh**-tow-blast) A colony-forming unit or stem cell of the bone marrow that gives rise to several types of committed stem cells. (Ch. 17)

hemoglobin (*hee*-moh-**glo**-bin) A pigment molecule that gives the red blood cell its color; can take up oxygen or release it. (Ch. 17)

hemolysis (he-**mol**-ih-sis) The destruction of erythrocytes and the resultant escape of hemoglobin. (Ch. 17)

hemolytic disease of the newborn Hypoprothrombinemia causing possible life-threatening hemorrhage in newborns. (Ch. 17)

hemophilia (*hee*-moh-**fil**-ee-ah) A disease in which one of the clotting factors is absent due to a genetic mutation. (Ch. 17)

hemorrhage (**hem**-or-ij) Loss of blood. (Ch. 19)

hemostasis (*hee*-moh-**stay**-sis) Mechanisms that minimize or prevent the loss of blood when a blood vessel is opened. (Ch. 17)

Henderson-Hasselbalch equation The logarithmic form of the acid dissociation equation: $pH = pK_a + \log([A^-]/[HA])$. (Ch. 26)

heparin (**hep**-ah-rin) A mucopolysaccharide acid, abundant in the lungs and liver; inhibits the clotting process. (Ch. 17)

hiatus hernia When the subiaphragmatic lower esophageal sphincter is displaced into the thoracic cavity; the terminal segment of this sphincter can no longer be assisted by changes in intraluminal pressure and reflux can occur. (Ch. 23)

hierarchical organization The range of complexity from lowest level to highest; also called serial organization. (Ch. 9)

high altitude cerebral edema Cerebral edema that occurs at altitudes of 10,000 to 12,000 feet; characterized by severe headache, hallucinations, ataxia, weakness, impaired mental ability, stupor, and death. (Ch. 21)

high altitude pulmonary edema Edema that occurs at altitudes of 9000 to 10,000 feet; characterized by dyspnea, cough, weakness, headache, stupor, and rarely death. (Ch. 21)

high energy phosphate compound A compound that has a strong tendency to transfer its phosphate group to an acceptor molecule; the transfer is accompanied by the release of a large amount of free energy. (Ch. 6)

histamine A vasoactive substance released from mast cells during an inflammatory response that stimulates vasodilation. (Ch. 19)

histocompatibility-Y antigen (H–Y anti-

gen) An antigenic protein that induces an immunologic rejection in females; thought to be responsible for inducing testicular differentiation in the primitive gonad. (Ch. 32)

histone A specialized protein that binds to DNA and organizes its structure in the nucleus. (Ch. 3)

homeostasis (*ho-me-o-sta-*sis) The stable state of the internal environment, which is maintained by regulatory physiologic processes, despite changes that may occur in the external environment. (Ch. 1)

homeothermy (*ho-mee-oh-thur-*me) The maintenance of a constant body temperature despite changes in environmental temperature. (Ch. 20)

"homo aquaticus" Jacques Cousteau's description of an underwater human capable of living indefinitely beneath the surface of the sea due to liquid breathing techniques or artificial gills. (Ch. 21)

hormone (**hoar-**moan) A chemical messenger that helps regulate the activity of other tissues and organs; secreted by endocrine or ductless glands. (Ch. 3)

hormone receptor A hormone that allows the signaling molecule to recognize and bind to the target cell. (Ch. 12)

hormone sensitive lipase An enzyme that catalyzes the breakdown of triglycerides into their component fatty acids and glycerol. (Ch. 15)

human calorimeter A large insulated chamber used for measuring metabolic rate. (Ch. 28)

human chorionic gonadotropin (hCG) (*ko-*ree-**on-**ik *gon-*ah-do-**trow-**pin) A glycoprotein that is secreted by cells of the trophoblast, that signals the corpus luteum that pregnancy has begun. (Ch. 12)

human chorionic somatomammotropin (hCS) A peptide hormone secreted by the placenta during pregnancy. (Ch. 13)

humoral stimuli Chemical or fluid-related stimuli. (Ch. 29)

hydration The addition of water molecules to an ion. (Ch. 7)

hydrochloric acid (HCl) A highly dissociated acid (dissociates to H and chloride ions) produced by parietal cells in the gastric gland. HCl provides an optimal pH for the activity of the pepsin enzyme and is, therefore, an aid in digestion. HCl kills bacteria that enter the digestive system through the mouth. It contributes to the breakdown of muscle fiber and connective tissue and is a major factor in the formation of ulcers in the upper digestive tract. (Ch. 23)

hydrogen bond Similar to van der Waals bond, but slightly stronger; forms when the negatively charged electron cloud formed by an asymmetrical orbit of electrons in one atom is attracted to the positively charged nucleus of a neighboring atom. (Ch. 2)

hydrogen ion A hydrogen atom without its orbiting electron; a proton. (Ch. 2)

hydrogenation reaction The reduction process in which a hydrogen atom is gained. (Ch. 6)

hydrophilic compound (*high-*droe-**fil-**ik) (literally, "water loving") A compound that readily absorbs water or readily dissolves in water. (Ch. 2)

hydrophobic compound (*high-*droe-**foe-**bik) (literally, "water fearing") A compound that is insoluble in water; repels water molecules. (Ch. 2)

hydrostatic pressure (*high-*droe-**stat-**ik) Pressure created by a fluid that is either moving or stationary. (Ch. 19)

hydroxyapatite Calcium phosphate deposits usually found in crystalline form in the bone. (Ch. 27)

α-hydroxylase A specific enzyme in the proximal tubule cells of the kidneys that catalyzes the conversion of 25-hydroxy-vitamin D_3 to 1,25-dihydroxy-vitamin D_3. (Ch. 27)

5-hydroxytryptamine (5-HT) *See* serotonin.

hyperalgesic zones (*high-*pur-al-**jee-**zik) Painful areas on the surface of the body that are associated with inflammations of the internal organs. (Ch. 10)

hyperbilirubinemia An elevation of blood bilirubin above normally low levels. (Ch. 17)

hypercalcemia (*high-*pur-kal-**see-**mee-ah) An excessive amount of calcium in the blood. (Ch. 27)

hyperchromic A morphologic description of red cells; indicates an erythrocyte that is deeper in color than normally. (Ch. 17)

hyperglycemia (*high-*pur-gly-**see-**me-ah) An above-normal increase in the blood glucose concentration due to the absence of either insulin secretion or insulin action. (Ch. 15)

hyperkalemia (*high-*pur-kah-**lee-**me-ah) A plasma potassium concentration greater than 5.5 mEq/liter. (Ch. 25)

hypernatremia (*high-*pur-nah-**tree-**me-ah) A plasma sodium ion concentration greater than 145 mEq/liter. (Ch. 25)

hyperpnea Increased ventilation. (Ch. 29)

hyperpolarization (high-per-*pohl-*er-ih-**zay-**shun) When a membrane becomes more negative on its inner surface than it was in the resting state. (Ch. 7)

hypertension (*high-*pur-**ten-**shun) A chronically elevated blood pressure above 145 systolic or 95 diastolic. (Ch. 25)

hyperthermia (*high-*pur-**ther-**mee-ah) An elevation above normal body temperature. (Ch. 28)

hyperthyroidism The excess secretion of thyroid hormones that results in markedly increased heart rate, considerable weight loss, and high responsiveness to external stimuli. (Ch. 13)

hypertonic (*high-*per-**ton-**ik) Having an osmotic pressure or solute concentration greater than that of some other solution, which is taken as standard. (Ch. 17)

hypertrophy (high-**per-**trow-fee) An increase in the mass of a muscle without an increase in the actual number of cells. (Ch. 16)

hyperventilation (*high-*pur-ven-tih-**lay-**shun) Increased alveolar ventilation. (Ch. 20)

hypervolemia (high-pur-vo-**lee-**mee-ah) An abnormal increase in the volume of circulating blood in the body. (Ch. 19)

hypocalcemia (*high-*poe-kal-**see-**me-ah) A reduction in the blood-calcium level to below normal. (Ch. 27)

hypochromic A morphologic description of red cells; indicates an erythrocyte that is paler in color than normally. (Ch. 17)

hypoglycemia (*high-*poe-gly-**see-**me-ah) A decrease in the blood glucose concentration to below normal. (Ch. 15)

hypokalemia (*high-*poe-kahl-**ee-**me-ah) A plasma potassium concentration level less than 3.5 mEq/liter. (Ch. 25)

hyponatremia (*high-*poe-nay-**tree-**me-ah) A plasma sodium concentration less than 137 mEq/liter. (Ch. 25)

hypoosmotic Having an osmolality less than a reference solution (usually plasma). (Ch. 25)

hypophysis (high-**pof-**ih-sis) The pituitary gland; located at the base of the brain. (Ch. 13)

hypotension Low blood pressure; less than 100 systolic, 60 diastolic. (Ch. 19)

hypothalamic-pituitary portal system A complex of blood vessels that consists of the primary capillary network, the secondary capillary network, and the vessels that connect them; carries factors that regulate the secretion of anterior pituitary hormones from the hypothalamus to the anterior pituitary. (Ch. 13)

hypothalamic-pituitary-adrenal axis The relationship between CRH, ACTH, and the adrenal cortex. (Ch. 13)

hypothalamus (high-poe-**thal-**ah-mus) A portion of the brain that regulates the pituitary gland, the autonomic nervous system, emotional response, body temperature, water balance, and appetite. (Ch. 10)

hypothermia (*high-*poe-**ther-**mee-ah)

The lowering of body temperature below normal. (Ch. 28)

hypothesis (high-**poth**-e-sis) A statement that tentatively explains an observation. (Ch. 2)

hypothyroidism A condition that exists when there is a deficiency in thyroid hormone production; affected individuals cannot tolerate cold temperatures and exhibit lethargy; sometimes accompanied by myxedema. (Ch. 13)

hypotonic (high-poe-**ton**-ik) Refers to a solution whose osmotic pressure or solute content is less than that of some standard of comparison. (Ch. 17)

hypoventilation (*high*-poe-ven-tih-**lay**-shun) Decreased alveolar ventilation below normal. (Ch. 20)

hypovolemia (**high**-poe-vo-**lee**-mee-ah) An abnormally decreased volume of circulating blood in the body. (Ch. 19)

hypoxemia (high-pock-**see**-me-ah) The condition of low blood oxygen. (Ch. 21)

hypoxia (high-**pock**-see-ah) Oxygen deficiency. (Ch. 21)

hypoxic pulmonary vasoconstriction A mechanism in the lung that balances the flow or perfusion of blood with the availability of regional alveolar oxygen. (Ch. 21)

H-zone A somewhat lighter region in the sarcomere; found in the center region of the A-band where the thin filaments of the I-band have not penetrated. (Ch. 16)

I-band A region of a muscle where only thin filaments are present. (Ch. 16)

ileocecal sphincter (ill-ee-oh-**see**-kal) The last 2–3 cm of the musculature of the ileum; controls passage of material to the large intestine. (Ch. 23)

ileum (**il**-lee-um) The distal portion of the small intestine; extends from the jejunum to the cecum. (Ch. 23)

immune response (ih-**mewn**) Any reaction that is designed to defend the body against pathogens or any other foreign substances. (Ch. 17)

immunity (ih-**mew**-nih-tee) Either nonspecific (natural) or specific (adaptive) physical and chemical barriers or defense mechanisms geared toward the identification and destruction of specific pathogens. (Ch. 17)

immunoglobulin (Ig) (*im*-yoo-no-**glob**-yoo-lin) Any glycoprotein secreted by B-lymphocytes in response to specific antigens. (Ch. 17)

immunosuppression The inhibition of the body's ability to mount an effective immune response; may be due to certain drugs or exposure to ionizing radiation. (Ch. 14)

implantation (*im*-plan-**tay**-shun) The penetration of the epithelial lining of the uterus by a blastocyst to embed in the endometrium and form an embryo. (Ch. 31)

impotence (**im**-poe-tence) The inability to achieve penile erection sufficient for vaginal penetration. (Ch. 32)

in utero (**yoo**-ter-oh) Within the uterus. (Ch. 31)

in vitro (**vee**-tro) In an artificial environment, such as of cells grown in culture dishes. (Ch. 2)

in vivo (**vee**-vo) Within the body. (Ch. 2)

inactivating state One of three states of operation of the sodium channel of membranes; the time-dependent gate is closed. (Ch. 7)

incontinence (in-**kon**-tih-nence) The inability to retain urine or feces due to loss of sphincter control. (Ch. 23)

incus The middle of the three ossicles of the ear; serves to conduct vibrations from the tympanic membrane to the inner ear; also known as the anvil. (Ch. 8)

independent variable A datum plotted on the horizontal axis of a line graph. (Ch. 1)

indirect calorimetry An estimate of metabolic rate based on oxygen consumption. (Ch. 28)

infarct (**in**-farkt) An area of tissue death resulting from circulatory obstruction, usually a thrombus or embolus. (Ch. 19)

inferior (in-**fe**-ree-or) Nearer to the lower end of the body. (Ch. 1)

inferior vena cava The large vein that returns blood from the lower part of the body to the heart. (Ch. 18)

infertility A diminished or absent ability to reproduce. (Ch. 32)

inflammation (*in*-flah-**may**-shun) A localized body response to tissue damage or injury; the clinical signs are redness, heat, edema, and pain. (Ch. 17)

inhibin (in-**hib**-in) A protein secreted by Sertoli cells of the seminiferous tubules that regulates the secretion of follicle stimulating hormone (FSH). (Ch. 30)

inhibitory postsynaptic potential (IPSP) A potential that hyperpolarizes the membrane potential and inhibits the nerve cell from generating an action potential. (Ch. 7)

initial segment The first portion of an axon plus part of the cell body where the axon joins. (Ch. 7)

inner ear The part of the ear containing receptors for hearing and equilibrium. (Ch. 8)

inner mitochondrial membrane A substrate for proteins in the electron transport chain of oxidative phosphorylation. (Ch. 6)

inositol triphosphate (IP$_3$) One of a pair (with diacylglycerol) of intracellular second messenger molecules. (Ch. 5)

inotropic agent A compound that affects the contractility of heart muscle. (Ch. 16)

insensible perspiration Water loss by evaporation from the lungs and skin. (Ch. 28)

insertion (in-**sir**-shun) The distal, or less stationary, point of attachment of a skeletal muscle. (Ch. 16)

insomnia (in-**som**-nee-uh) The inability to sleep. (Ch. 11)

inspiration (*in*-spih-**ray**-shun) The act of drawing air into the lungs. (Ch. 10)

inspiratory neurons Neurons within the medulla oblongata responsible for rhythmic breathing when the body is at rest. (Ch. 20)

insulin (**in**-suh-lin) A double-chain protein hormone produced by the beta cells of the islets of Langerhans in the pancreas that allows cells to take in glucose in response to an increase in blood sugar levels. (Ch. 15)

insulin resistance A lack of response by tissues to insulin in the blood. (Ch. 15)

insulin-dependent diabetes mellitus (IDDM) A disease caused by inadequate insulin secretion; type I diabetes. (Ch. 15)

insulin-like growth factor Somatomedins produced by the liver; **IGF-I** stimulates skeletal growth; **IGF-II** stimulates tissue growth and repair. (Ch. 13)

integument (in-**teg**-yoo-ment) Any covering, especially the skin. (Ch. 28)

intercalated disc (in-**ter**-kah-lay-ted) A specialized structure that provides mechanical and electrical connection between cells of cardiac muscle. (Ch. 16)

intercalated duct A system that delivers exocrine secretion of the pancreas to the duodenum. (Ch. 23)

intercellular Between cells. (Ch. 24)

interferon (*in*-ter-**feer**-on) A group of proteins secreted by cells that have been infected by a virus, inducing noninfected cells to form antiviral proteins that inhibit viral multiplication. (Ch. 17)

interleukins (*in*-ter-**loo**-kins) Hormone-like messengers that stimulate T lymphocytes; **IL-1** stimulates activated T$_4$ lymphocytes to differentiate and reproduce, forming a clone of inducer T cells and helper T cells; **IL-2** induces T$_8$ lymphocytes to differentiate and reproduce, forming cytotoxic T cells and suppressor T cells. (Ch. 17)

internal Nearer the center of an organ, cavity, or part of the body. (Ch. 1)

internal anal sphincter The smooth muscle that is relaxed during defecation. (Ch. 23)

internal environment The extracellular

fluid that surrounds the cell, outside the cell but inside the organism. (Ch. 1)

internal intercostal muscle The expiratory muscle that contracts to pull the rib cage down. (Ch. 20)

internal sphincter A group of smooth muscles that close off the exit from the bladder. (Ch. 10)

interneuron Any neuron in a chain of neurons situated between the sensory and the final motor neuron. (Ch. 7)

interoceptor (*in-ter-o-sep-ter*) A sensory receptor that transmits information from viscera to CNS. (Ch. 8)

intermediolateral lateral nuclei (ILN) The location in the spinal cord of the preganglionic nerve cells of the sympathetic nervous system. (Ch. 10)

interphase (*in-ter-faze*) The period between the end of one cell division and the beginning of the next. (Ch. 3)

interpolation Prediction of data values that fall between existing data points. (Ch. 1)

interstitial fluid (*in-ter-stish-al*) The fluid outside blood vessels that directly bathes most body cells. (Ch. 25)

interstitium (*in-ter-stish-ee-um*) The fluid-filled space between tissue cells. (Ch. 24)

intracellular (*in-tra-sell-yoo-lar*) Within a cell. (Ch. 24)

intracellular pH The pH level within the cell; generally lower than extracellular pH. (Ch. 26)

intrafusal muscle fibers Fibers in the inside of skeletal muscles; arranged parallel to extrafusal fibers. The fibers are part of the neuromuscular spindle and supplied by gamma motor neurons. (Ch. 9)

intrauterine device (IUD) (*in-trah-yoo-ter-in*) A contraceptive device inserted into the uterus to prevent implantation of the ovum. (Ch. 32)

intrinsic clotting pathway The initiation of clotting in the absence of tissue damage; can occur inside or outside the body. (Ch. 17)

intrinsic factor A glycoprotein, secreted by the lining of the stomach, that is essential for absorption of vitamin B_{12} in the small intestine. (Ch. 17)

intrinsic (integral) protein (*in-trin-sik*) A protein that partially or completely penetrates the lipid bilayer of a plasma membrane. (Ch. 4)

inulin A fructose polymer whose renal clearance is used to measure glomerular filtration rate. (Ch. 24)

inverse myotatic reflex A reflex that inhibits skeletal muscle contraction when the Golgi tendon organ detects excessive force that might damage the muscle. (Ch. 9)

inverse relationship The relationship between data such that y-values get smaller as x-values get larger. (Ch. 1)

involuntary muscle The smooth or cardiac muscle whose actions are controlled by the autonomic nervous system in response to some internal reflex. (Ch. 16)

iodide trap The transport process that pumps iodide into cells of the thyroid. (Ch. 13)

ion channel (*eye-on*) The pores in a nerve cell membrane that allow passage of certain ions. A **passive ion channel** allows a specific ion to pass through; a **chemically activated ion channel** can be opened or closed by a chemical transmitter; a **voltage-activated ion channel** is opened when it detects a certain voltage. (Ch. 7)

ionic bond The equal and opposite electrical charge that holds two atoms together. (Ch. 2)

ionic current The flow of ions. (Ch. 7)

ionization (*eye-on-eye-zay-shun*) The dissociation of a substance in solution into ions. (Ch. 26)

ionized calcium A free Ca^{2+} ion, the physiologically active form of calcium. (Ch. 27)

ischemia (*is-kee-me-ah*) Reduced blood flow to an organ due to obstruction of a functional blood vessel. (Ch. 17)

islets of Langerhans (*eye-lits of Lahng-er-hanz*) Small groups of cells interspersed within the cells of the exocrine pancreas; secrete insulin, glucagon, and somatostatin. (Ch. 15)

isoelectric line A line of an electrocardiograph with no deflection. (Ch. 18)

isohydric principle The idea that all buffer systems are in equilibrium with the same hydrogen ion concentration. (Ch. 26)

isometric contraction A contraction in which a muscle does not shorten but develops force or tension while pulling against immovable attachments. (Ch. 26)

isosmotic (*eye-soz-mat-ik*) Having the same osmotic pressure. (Ch. 24)

isotonic (*eye-se-tahn-ik*) Having the same osmolality, equal to that of a reference solution. (Ch. 16)

isotonic contraction (*eye-so-ton-ik*) A contraction in which the force in a muscle remains constant while the muscle shortens. (Ch. 16)

isotope (*eye-so-tope*) An atom of an element that has a differing number of neutrons. (Ch. 2)

isovolumetric ventricular contraction Early phase of the cardiac cycle in which both the atrioventricular valves and the semilunar valves are closed and ventricular volume does change. (Ch. 18)

isovolumetric ventricular relaxation Early phase ventricular relaxation when atrioventricular and aortic valves are closed.

jaundice (*jawn-dis*) A condition in which the skin is yellowish due to hyperbilirubinemia and deposition of bile pigment in the skin. (Ch. 17)

jejunum (*jeh-joo-num*) The middle portion of the small intestine; extends from the duodenum to the ileum. (Ch. 23)

joint The site of junction or union between two or more bones of the skeleton. (Ch. 16)

junctional complex The cellular structure that prevents the lateral diffusion of proteins in the plasma membrane. (Ch. 4)

junctional folds A series of deep grooves of the muscle plasma membrane.

juxtaglomerular apparatus (*juks-tah-glo-mer-you-lar*) A structure in the kidney consisting of macula densa, extraglomerular mesangial cells, and granular cells, located in the region where the distal straight tubule touches afferent and efferent arterioles of its parent glomerulus. (Ch. 4)

juxtamedullary nephron Nephrons with renal corpuscles in the inner cortex; have large glomeruli, long thin limbs, and long loops; 15% of all nephrons are of this type. (Ch. 24)

kainate One of three subtypes of glutamate receptors. Its activation results in excitatory postsynaptic potentials by opening ion channels that increase in Na+ and K+ conductance. (Ch. 7)

kallikrein A proteolytic enzyme that leads to the formation of kinins. (Ch. 24)

keloid (*key-loyd*) An elevated, enlarging scar due to excessive amounts of collagen in the dermis during connective tissue repair. (Ch. 8)

keto acid (*key-toe*) Acid formed by the removal of an amino group from amino acid. (Ch. 16)

ketoacidosis An acid/base imbalance caused by an increase in the concentration of ketones in the blood. (Ch. 15)

ketogenesis The formation of ketones in the liver; stimulated by glucagon. (Ch. 15)

ketone bodies (*key-tone*) Four-carbon compounds that are released by and accumulate in the blood during starvation and untreated diabetes mellitus; acetoacetate, acetone, or β-hydroxybutyrate acid. (Ch. 15)

ketosis A condition characterized by an abnormally elevated concentration of ketone bodies in body tissue and fluids; a complication of diabetes mellitus and starvation. (Ch. 15)

kinesthesia (*ki-nee-steez-ee-ah*) The sense of movement. (Ch. 8)

kinetic energy Energy in use that is producing motion. (Ch. 2)

kinin (*Kye*-nin) An endogenous peptide derived from kallikrein that causes vasodilation. (Ch. 24)

knee jerk reflex A stretch reflex activated by tapping the patellar tendon below the knee. (Ch. 9)

Krebs cycle Aerobic cycle of respiratory chemical reactions that occur in the mitochondria and produce cellular ATP (*also called the citric acid cycle*). (Ch. 3)

labor Strong contractions of the uterus that occur with increasing frequency and intensity during the latter part of pregnancy; results in expulsion of the fetus. (Ch. 31)

lactase (*lak*-tase) An β-Galactosidase. An enzyme that digests lactose into glucose and galactose. (Ch. 23)

lactation (lak-*tay*-shun) The production and secretion of milk by the breast. (Ch. 23)

lacteal (*lak*-tee-al) Pertaining to milk; any of the intestinal lymphatics that transport chyle. (Ch. 31)

lactic acid (*lak*-tik) A three-carbon molecule formed as a byproduct of anaerobic metabolism of the energy pathways of cells. (Ch. 16)

lactic acid oxygen debt The second stage of repayment of the oxygen debt that involves removal of lactic acid from the body. (Ch. 29)

lactose (*lak*-tose) A disaccharide composed of one molecule of glucose and one molecule of galactose. (Ch. 23)

laminar air flow The streamlined flow of air in the lungs that parallels the sides of the airways. (Ch. 20)

larynx (*lar*-inks) The upper part of the airway between the pharynx and the trachea that contains the vocal cords. (Ch. 23)

latent period Period of time (several milliseconds) between stimulus delivery and muscle contraction. (Ch. 16)

lateral (*lat*-er-al) Position farther from the midline or midsagittal plane. (Ch. 1)

lateral reticulospinal tract A descending pathway from the brain stem to the spinal cord; used to control flexor muscles involved in the maintenance of posture. (Ch. 9)

laxative A cathartic such as mineral oil that hydrates or lubricates material in the large intestine, promoting evacuation of the bowel. (Ch. 23)

L-dopa Dopamine precursor that is often administered to patients with Parkinson's disease. (Ch. 9)

lead I An electrocardiograph recording made with electrical connections between the right and left arms; **lead II** with electrical connections between the right arm and the left leg; **lead III** with electrical connections between the left leg and the left arm. (Ch. 5)

length-tension curve The relationship between the force capability of a muscle and its length. (Ch. 18)

leukemia (loo-*key*-mee-ah) A malignant disease characterized by proliferation of hematopoietic cells that are immature and therefore functionally impaired. (Ch. 16)

leukocyte (*loo*-ko-site) A white blood cell (WBC). (Ch. 17)

leukocyte inducing factor A specific factor in plasma that can be demonstrated when leukocytes are depleted; stimulates leukopoiesis. (Ch. 17)

leukocytosis (*loo*-ko-sigh-*toe*-sis) An abnormal increase in the total number of circulating leukocytes. (Ch. 17)

leukopenia (*loo*-ko-*pee*-nee-ah) An abnormal reduction in the total number of circulating leukocytes. (Ch. 17)

leukopoiesis (*loo*-ko-*poe*-ee-sis) The development of white blood cells. (Ch. 17)

ligand (*lig*-and) Any compound that binds with a high degree of specificity to a receptor. (Ch. 5)

limb leads An electrocardiograph recording made between any two of the standard connections: the right and left arms and the left leg. (Ch. 5)

limbic system Parts of the prefrontal cortex, hippocampus, amygdala, and hypothalamus that collectively function in the expression of emotional behavior. (Ch. 18)

linear relationship A relationship that can be described by a straight line. (Ch. 8)

lipase (*lye*-pase) An enzyme produced by the pancreas that digests triglycerides and phospholipids. (Ch. 23)

lipids A group of fat and fatlike substances that are water-insoluble and highly soluble in organic liquids. (Ch. 2)

lipogenesis (*lipe*-oh-*jen*-eh-sis) The production of lipids from precursors such as glucose and amino acids. (Ch. 15)

lipolysis The breakdown of triglycerides to fatty acids and glycerol. (Ch. 14)

lipoprotein lipase An enzyme that promotes the uptake of lipoproteins from the blood. (Ch. 14)

lipoproteins A complex of lipids and proteins, synthesized in the endoplasmic reticulum lumen, that has a critical role in the transport and distribution of lipids. (Ch. 3)

local chemical mediator A chemical signal released by a cell that affects only nearby cells and that is rapidly destroyed if not used. (Ch. 3)

long-term memory Memory that is stable, less volatile, and long-lasting. (Ch. 5)

loop of Henle The hairpin portion of the nephron situated between proximal convoluted and distal convoluted tubules. (Ch. 24)

lordosis A receptive posture adopted by female animals during mating. (Ch. 24)

low density lipoprotein A lipoprotein aggregate that transports most of the cholesterol to the cell. (Ch. 11)

lower esophageal sphincter A ring of smooth muscle that acts as a valve controlling the entrance to the stomach. Also called the cardiac sphincter. (Ch. 23)

lower-motor-neuron lesion Occurs with spinal cord or brain stem injuries that affect motor neurons directly; characterized by ipsilateral hypoactive reflexes, paralysis, and limp, atrophic muscles. (Ch. 9)

lumen (*loo*-men) The cavity within a tubelike structure such as a blood vessel, lymphatic vessel, or intestine. (Ch. 24)

luminal membrane The cell surface that contacts fluid in the tubule lumen of the kidney. (Ch. 24)

luteinization The formation of the corpus luteum from the remnants of the ovarian follicle after ovulation. (Ch. 30)

luteinizing hormone (LH) (*loot*-eh-*nigh*-zing) Gonadotropin hormone released by the anterior pituitary that initiates ovulation. (Ch. 11)

lymph (limph) A transparent, slightly yellow fluid of alkaline reaction that is found in lymphatic vessels and derived from tissue fluids. (Ch. 11)

lymph node Small organelle located along lymph vessels; filters lymphocytes, microorganisms and other foreign material. (Ch. 11)

lymphatic system (lim-*fat*-ik) A system of vessels that directs plasma protein from interstitial spaces to the blood. (Ch. 19)

lymphocyte (*lim*-foe-site) An agranular leukocyte, the second most common type of white blood cell. (Ch. 17)

lymphokines (*lim*-foe-kines) Chemicals secreted by lymphocytes that regulate T- and B-leukocyte function, attract macrophages, activate complement proteins, and help with specific defense. (Ch. 17)

lysis (*lye*-sis) The destruction of cells by a specific antibody. (Ch. 15)

lysosome (*lye*-so-sowm) A cytoplasmic, membrane-bound organelle that contains digestive enzymes used for intracellular digestion. (Ch. 4)

macrocytosis (mak-roe-sye-*toe*-sis) The production of red blood cells that are larger than normal, also called macrocythemia. (Ch. 17)

macromolecule A very large molecule (Ch. 17)

having a polymeric chain structure; all contain carbon and assume huge proportions and a great variety of shapes. (Ch. 2)

macrophage (mak-roe-faje) A large phagocytic cell that forms part of the body's defense against microorganisms and harmful chemicals. (Ch. 17)

macula densa (mak-you-la den-sa) Portion of the straight distal tubular epithelium that helps make up the juxtaglomerular apparatus of the kidney. (Ch. 24)

major histocompatibility complex II (MHCII) A protein present on the plasma membrane of antigen-presenting cells; involved in cell-mediated immune response. (Ch. 17)

malignant hyperthermia An inherited defect in skeletal muscle metabolism. (Ch. 28)

malleus (mal-lee-us) The largest of the auditory ossicles, the one attached to the tympanic membrane; also known as the hammer. (Ch. 8)

maltase (mawl-tase) α-Glucosidase. A digestive enzyme that breaks down maltose into glucose. See also glucosidase.

mass movement A propulsive contraction of the colon, moving material into the distal large intestine. (Ch. 23)

mass number The total number of protons and neutrons in the nucleus of an element. (Ch. 2)

mast cell A specialized type of cell in connective tissue; contains histamine and is important in allergic reactions. (Ch. 5)

mastocytosis (mas-toe-sigh-toe-sis) A condition in which greatly elevated numbers of mast cells accumulate in body tissues. (Ch. 5)

matrix (ma-triks) The intercellular substance of a tissue or the tissue from which a structure develops. (Ch. 1)

maximal aerobic power A means of determining an individual's ability to sustain high-intensity exercise; the maximal oxygen consumption is an indicator of this state. (Ch. 29)

maximal oxygen consumption ($\dot{V}O_{2max}$) The state at which the maximum oxygen transport and uptake in the tissues are reached and a plateau is attained. (Ch. 29)

mean arterial pressure A calculation of a single value for blood pressure that factors both the pulse and diastolic pressures. (Ch. 19)

mean body temperature An estimate of the fractional sum of core and mean skin body temperatures. (Ch. 28)

mean corpuscular hemoglobin (MCH) An index used to determine the concentration of hemoglobin in red blood cells. (Ch. 17)

mean corpuscular volume (MCV) A quantitative assessment of the volume of red blood cells, reported in mm³. (Ch. 17)

mean skin temperature The temperature of the body shell, calculated as the mean of the temperatures at several representative parts of the body. (Ch. 28)

mechanoreceptor A sensory receptor that is excited by mechanical pressures or distortions, such as bending, twisting, or compressing. (Ch. 8)

medial (mee-dee-al) A term of position; literally means nearer to the mid-sagittal plane. (Ch. 1)

medial lemniscal pathway A pathway in the brain stem through which sensory fibers of the dorsal column ascend toward the thalamus. (Ch. 8)

medial reticulospinal tract A component of the ventromedial pathway; originates in the reticular formation and is involved in maintaining posture by activation of extensor muscles. (Ch. 9)

median eminence A specialized region in the hypothalamus where arterial blood vessels branch into a network of capillaries. (Ch. 13)

medulla (mah-dull-ah) The innermost portion of an organ or structure. (Ch. 10)

megakaryocyte (meg-ah-kare-ee-oh-site) A type of blood cell from which thrombocytes, or platelets, are derived. (Ch. 17)

meiotic division (my-ah-tik) A division of germ cells that reduces the number of chromosomes in the cell by half; random assortment and recombination of maternal and paternal chromosomes occur as well as crossing over. (Ch. 30)

meiotic muscle contraction (my-oh-tahk-ik) A contraction in which the force lessens as a muscle shortens. (Ch. 16)

melatonin (mel-ah-toe-nin) A hormone derived from tryptophan; suspected to have a role in puberty onset and control of body rhythms. (Ch. 12)

membrane conductance The ease with which ions can flow through channels in the plasma membrane. (Ch. 7)

membrane potential Voltage difference between inside and outside of cell due to an ionic concentration gradient. (Ch. 7)

memory cell A B-lymphocyte cell that permits rapid mobilization of immune response upon subsequent exposure to an antigen. (Ch. 17)

menopause (men-oh-pawz) The cessation of menstruation in the human female; occurs on the average at age 50. (Ch. 30)

menstrual cycle (men-stroo-al) The approximately 28-day ovarian cycle; consists of the follicular phase, the luteal phase, and the menstrual phase. (Ch. 30)

menstruation (men-streh-way-shun or men-stray-shun) Normal flow of menstrual fluid from the uterus that occurs approximately once a month in most nonpregnant females. (Ch. 30)

mesentery (mez-en-ter-ee) A membrane attaching various organs to the body wall, especially the peritoneum connecting the intestine to the posterior abdominal wall. (Ch. 30)

messenger RNA The transcribed segment of DNA containing the information specifying the amino acid sequence of a polypeptide chain. (Ch. 5)

metabolic acidosis An abnormal process characterized by a gain of acid which leads to a fall in blood pH; due to any cause other than elevated carbon dioxide. (Ch. 22, 25, 26)

metabolic alkalosis An abnormal process characterized by a gain of strong base, which leads to a rise in blood pH; due to any cause other than a decrease in blood carbon dioxide levels. (Ch. 22, 25, 26)

metabolic bone disease A group of disorders in which there is generalized abnormality of the ongoing process of bone formation and bone resorption. (Ch. 27)

metabolic pathway The sequence of enzyme-mediated chemical reactions by which molecules are synthesized and broken down in cells. (Ch. 6)

metabolic rate Total energy production of an individual per unit time. (Ch. 3)

metabolism (meh-tab-oh-lizm) The sum of all physical and chemical processes by which living organized substance is produced and maintained; also the transformation by which energy is made available for the organism's uses. (Ch. 3)

metarteriole Blood vessels that are slightly larger than capillaries.

micelle (my-sel) Small aggregates of monoglycerides and free fatty acids that can diffuse in an aqueous solution. (Ch. 23)

microcirculation The circulation of blood through the arterioles, metarterioles, capillaries, and venules. (Ch. 19)

microcytosis (my-kroe-seh-toe-sis) The production of red blood cells that are smaller than normal; also called microerythremia. (Ch. 17)

microfilament A component of cytoskeletal structure; a rodlike cytoplasmic protein believed to have a supportive and cytoskeletal function and/or mediate movement of the cell and the organelles within it. (Ch. 3)

microglia (my-krog-lee-ah) Small, non-neural, interstitial cells of mesodermal origin that form part of the

microtubule (my-krow-too-bewl) A tubular cytoplasmic filament that provides internal support for cells; allows change in cell shape and organelle movement in cell. (Ch. 7)

microtubule-associated protein (MAPS) Accessory proteins that are thought to be responsible for the distribution of material to dendrites and axons. Dendrites have high-molecular-weight MAPS, while axons have low-molecular-weight MAPS. (Ch. 7)

microvilli (my-krow-vill-ee) Small projections that cover epithelial cells of the digestive tract. (Ch. 23)

micturition (mik-tu-rish-un) Emptying of the bladder. (Ch. 10)

middle ear The portion of the ear between the tympanic membrane and the cochlea; contains the ossicles and internal auditory meatus (eustachian tube). (Ch. 8)

midsagittal plane An imaginary plane that passes directly along the midline of the body, dividing it into right and left halves. Also called median sagittal plane. (Ch. 1)

mineral A nonorganic homogeneous solid substance, usually a constituent of the earth's crust. (Ch. 2)

mineralocorticoid (min-er-al-oh-kor-ti-koyd) A steroid hormone produced by the adrenal cortex; involved in regulation of electrolyte and water balance. (Ch. 12)

minute ventilation The total volume of inspired air that enters the lungs per minute; equals tidal volume times frequency. (Ch. 20)

mitochondrion (my-tow-kon-dree-un) An organelle of the cell enclosed in a double membrane; the site of ATP production through the process of aerobic respiration. (Ch. 3)

mitosis (my-tow-sis) A process of cell division whereby each of the two daughter cells receives a complete set of chromosomes. (Ch. 3)

mitral cell (my-tral) Cells located in the olfactory bulbs that form the olfactory tract. (Ch. 8)

mitral valve The atrioventricular valve on the left side of the heart; also called the bicuspid. (Ch. 18)

M-line A structure in muscle cells that holds the thick filaments in side-by-side alignment. (Ch. 16)

molality A unit of concentration; the number of moles of solute per kilogram of solvent in a solution. (Ch. 2)

molarity A unit of concentration; the number of moles of solute per liter of solution. (Ch. 2)

molecular formula A symbol that shows the number of each type of atom in a molecule or group of atoms. (Ch. 2)

molecular mass The sum of the atomic masses of the elements in a compound. (Ch. 2)

monochromatic color vision Color vision that relies on one type of cone. (Ch. 8)

monoclonal antibody A specific antibody produced in vitro by hybridoma cell lines; the hybridoma cell lines are produced through fusion of activated B cells with cancer cells. (Ch. 3)

monocyte (mon-oh-site) One of two types of agranulocyte white blood cells; migrate out of blood and into tissues to form macrophages. (Ch. 3)

monoiodotyrosine (MIT) A compound found within the thyroid follicle cell; used in the synthesis of T_3 and T_4. (Ch. 17)

monokine A chemical secreted by macrophages that stimulate T-lymphocyte development. (Ch. 13)

monosaccharide (mon-oh-sak-ah-ride) A simple sugar; cannot be degraded by hydrolysis to a simpler sugar. (Ch. 17)

monozygotic twins Twins that result from the fertilization of a single ovum by a single sperm, followed by cell division that results in a single duplication or multiple duplicaté ovum prior to implantation; also called identical twins. (Ch. 31)

mossy fiber Neurons that originate in the neocortex and spinal cord; sensory stimuli and voluntary movements about muscle activity. (Ch. 9)

motility The ability to move spontaneously. (Ch. 23)

motor cortex The area of the brain that controls movement. (Ch. 7)

motor end-plate The connection between a muscle fiber and the myelinated axon of a motor neuron; also called myoneural junction. (Ch. 7)

motor neuron A neuron that carries instructions about muscle movements from the CNS in response to sensory stimuli; also called an efferent neuron. (Ch. 7)

motor neuron pool A functional grouping in the spinal cord of motor neurons that innervate a single muscle. (Ch. 9)

motor system Collectively, the areas of the brain and spinal cord that control and coordinate skeletal muscle movement. (Ch. 9)

motor unit A motor neuron plus the muscle fibers it innervates. (Ch. 9)

mucin (mew-sin) A nonenzymatic protein found in mucus; lubricates and protects the lining of the digestive tract. (Ch. 23)

mucosa (mew-ko-sah) A mucous membrane. (Ch. 23)

mucus (mew-kus) A viscous mixture of mucin and watery secretions from mucus and serous cells of salivary glands. (Ch. 23)

müllerian duct system An embryonic ductal system that is the precursor of the internal genitalia of the female. (Ch. 32)

müllerian inhibiting substance (MIS) A substance secreted by the Sertoli cells of the testes that induces regression or involution of the müllerian duct system. (Ch. 32)

multipolar neuron A type of neuron with several short dendrites and one long axon. (Ch. 7)

multiunit smooth muscle A class of muscles that are closely controlled by the nervous system. (Ch. 16)

muscarinic acetylcholine receptor A category of cholinergic receptor; a receptor on target cells such as muscle. (Ch. 10)

muscle (mus-ul) A number of muscle fibers bound together by connective tissue. (Ch. 9)

muscle contraction The shortening of muscle tissue. (Ch. 9)

muscle fiber A type of contractile cell that is elongated in shape; also called muscle cell. (Ch. 9)

muscle relaxation The lengthening of muscle tissue. (Ch. 9)

muscle spindle A receptor complex attached to somatic sensory neurons that detect the length of the muscle and its velocity of contraction. (Ch. 9)

muscle spindle receptor A capsule-enclosed arrangement of afferent nerve fiber endings in skeletal muscle; sensitive to stretch. (Ch. 8)

muscle tissues Tissues whose cells are elongated and contain many myofibrils; specialized for contraction; functions to accomplish the organism's movement. (Ch. 1)

muscle tone The amount of resistance of muscle to passive stretch. (Ch. 9)

myasthenia gravis A disorder of neuromuscular function characterized by muscle weakness; may affect any muscle of the body, but especially those of the eye, face, lips, tongue, throat, and neck. (Ch. 16)

myelin (my-eh-lin) An insulating lipid elaborated by oligodendrocytes (centrally) or Schwann cells (peripherally) that surrounds the fiber extension of neurons in order to speed the transmission of information. (Ch. 7)

myenteric plexus A network of nerve cells that lie between the two layers of the muscularis externa of the digestive tract. (Ch. 7)

myocardial infarction (my-oh-kar-dee-al in-fark-shun) A heart attack; necrosis of the myocardium when a coronary blood vessel is occluded. (Ch. 23)

myocardium (my-oh-kar-dee-um) The middle and thickest layer of the heart; composed of cardiac muscle. (Ch. 18)

myofibril (my-oh-fye-bril) A functional unit of muscle cells formed by hun-

dreds of sarcomeres connected in series. (Ch. 16)

myofilament (*my*-oh-**fil**-ah-ment)   A protein filament that provides for contraction of skeletal, cardiac, or smooth muscle. (Ch. 16)

myogenic activity   The increase in the contractile activity of smooth muscle cells in response to stretch. (Ch. 19)

myoglobin (*my*-oh-**glow**-bin)   A respiratory pigment that transports oxygen and imparts a red color to muscle tissue. (Ch. 16)

myoneural junction (*my*-oh-**new**-ral)   The connection between a muscle fiber and the myelinated axon of a motor neuron; also called motor end-plate. (Ch. 9)

myosin (**my**-oh-sin)   A very large protein that composes the thick filaments of muscle cells; 68% of protein in muscle is myosin. (Ch. 16)

myosin light-chain kinase (MLCK)   An enzyme involved in phosphorylation and the regulation of muscle contraction. (Ch. 16)

myosin-linked regulation   The regulation of muscle contraction that takes place via thick filaments. (Ch. 16)

myotonia (*my*-oh-**toe**-nee-ah)   A condition that is characterized by tonic spasms or temporary rigidity of muscle. (Ch. 32)

myxedema (*mik*-seh-**dee**-mah)   The accumulation of water in the skin, leading to a thickened and puffy appearance. (Ch. 13)

narcolepsy   A severe disorder of sleep in which an individual suddenly lapses into sleep during the middle of the day; thought to result from the activation of the REM-sleep generator. (Ch. 11)

natriuretic   An agent that promotes the excretion of sodium in urine. (Ch. 24)

negative balance   The condition in which there is net loss of a substance from the body. (Ch. 24)

negative feedback   A class of control systems active in the body; works to restore normal values of a variable and thus to exert a stabilizing influence. (Ch. 1)

neocortex   The outer gray matter of the cerebrum. (Ch. 7)

nephron (**nef**-ron)   The basic unit of kidney structure and function; consists of a renal corpuscle and its attached tubule. (Ch. 24)

Nernst equation   A chemical equation that allows the equilibrium potential for any ion to be calculated. (Ch. 7)

nerve   A group of many nerve fibers traveling together in the peripheral nervous system. (Ch. 7)

nerve cell   A cell specialized to initiate,

integrate, and conduct electric signals; also called neuron. (Ch. 7)

nervous tissue   One of four primary types of tissue; made up of neurons and glial cells. (Ch. 1)

neural fat   *See* triacylglycerol.

neuroendocrinology   An area of study that covers the overlap between the nervous and endocrine systems. (Ch. 12)

neurofilament   A rodlike cytoplasmic protein that forms a major component of the cytoskeleton of neurons. (Ch. 7)

neurohormone   Chemical messenger released by a neuron and carried by the blood to its target cell. (Ch. 12)

neurohypophyseal peptide (*new*-row-high-**pawf**-is-*ee*-al)   A neuropeptide released from the posterior pituitary that functions to regulate plasma osmolarity and lactation. (Ch. 7)

neuromuscular junction   The location of synaptic connections between alpha motor neurons and skeletal muscles where acetylcholine is released from the nerve fiber terminal. (Ch. 9)

neuron (**new**-ron)   *See* nerve cell.

neuropeptide transmitter   A family of at least 50 neurotransmitters composed of two or more amino acids; often function as chemical messengers in nonneural tissues. (Ch. 7)

neurotransmitter   A chemical messenger used by neurons to communicate with each other or with effectors. (Ch. 5)

neutron (**new**-tron)   The electrically neutral particle found in the nucleus of the atom. (Ch. 2)

neutrophil (**new**-tro-fil)   A polymorphonuclear granulocytic leukocyte whose granules show preference for neither eosin nor basic dyes; functions as a phagocyte and releases chemicals involved in inflammation. (Ch. 17)

nexus (**nek**-sus)   A structure that connects numerous portions of smooth muscle cells; involves a fusion of the outer layers of the membrane of each cell so that there is no extracellular space at the region between cells; also called gap junction. (Ch. 16)

NH$_3$ Ammonia; a base. (Ch. 24)

NH$_4$$^+$ Ammonium ion; an acid. (Ch. 24)

nicotinamide adenine dinucleotide (NAD)   A coenzyme that is most frequently employed as a hydrogen acceptor in metabolic reactions. (Ch. 6)

nicotinic   An cholinergic receptor that also responds to nicotine; located on postganglionic autonomic neurons. (Ch. 10)

nitroglycerin (*nigh*-troh-**glis**-eh-rin)   A therapeutic drug that provides relief from angina by dilating the coronary arteries to increase blood flow. (Ch. 19)

nitrous oxide   A colorless, sweetish gas used as an anesthetic; commonly called laughing gas. (Ch. 22)

NMDA (*N*-methyl-*D*-aspartate)   One of three subtypes of glutamate receptors; its activation results in an increase in Ca^{2+} conductance. It is both a ligand and a voltage-gated channel. (Ch. 7)

nociceptor (*no*-sih-sep-tor)   A receptor that detects painful stimuli; a **mechanical nociceptor** is activated by intense mechanical stimulation such as a slap to the head; a **heat nociceptor** responds to temperatures above 45°C. (Ch. 8)

nodes of Ranvier   A space between adjacent myelin-forming cells along a myelinated axon where axonal plasma membrane is exposed to extracellular fluid. (Ch. 7)

nonassociative learning   A type of learning in which we adapt to a single type of stimulus according to its relevance to our survival. (Ch. 11)

nonatrophic vaginitis   Vaginal dryness or tenderness due to allergic reactions or chemical irritations. (Ch. 32)

non-insulin-dependent diabetes mellitus (NIDDM)   A form of diabetes in which the target tissues of insulin lose their responsiveness to the hormone; also known as type II diabetes. (Ch. 15)

norepinephrine (*nor*-ep-ih-**nef**-rin)   A catecholamine neurotransmitter released from the adrenal medulla and at most sympathetic postganglionic endings and in many CNS regions. (Ch. 7)

normovolemia   The normal blood volume. (Ch. 17)

nuclear bag fiber   Intrafusal fibers that help make up muscle spindle; innervated by type Ia nerve fibers that transmit information about muscle length and velocity of contraction to the CNS. (Ch. 9)

nuclear chain fibers   Intrafusal fibers that help make up muscle spindle; innervated by type II nerve fibers that transmit information about muscle length to the CNS. (Ch. 9)

nuclear envelope   The double membrane that surrounds the cell nucleus. (Ch. 3)

nucleic acid (new-**klee**-ik)   A nucleotide polymer in which phosphate of one nucleotide is linked to the sugar of the adjacent one; stores and transmits genetic information; includes DNA and RNA. (Ch. 2)

nucleolus (new-**klee**-oh-lus)   Densely stained structure within the nucleus that contains the DNA that codes for rRNA; site of ribosome assembly. (Ch. 3)

nucleotide   A compound consisting of a pentose sugar, one of a set of special groups called bases, and one to three phosphate groups. (Ch. 2)

nucleus (new-**klee**-us)   A large membrane-bound organelle that contains the cell's DNA; (neural) cluster of neuron cell bodies in the CNS. (Ch. 3)

nucleus accumbens The lower forebrain structure that is part of the reward system. (Ch. 11)

nucleus of the vagus A region of the brain stem that, when activated, directly reduces the heart rate. (Ch. 10)

ocular dominance column (ah-kew-lar) The organization within the visual cortex in which columns of cells respond to light impinging on either the right or left eye. (Ch. 8)

"off" center and "on" surround cells Ganglion cells for which the annulus is inhibitory and periphery is excitatory. (Ch. 8)

olfactory bulb (ohl-**fak**-toe-ree) The site of the mitral cells of the olfactory system. (Ch. 8)

olfactory receptor cells Cells within the nasal cavity that are activated by odors; enervate mitral cells in the olfactory bulb. (Ch. 8)

olfactory system The physiological system that provides the ability to smell odors. (Ch. 8)

olfactory tract A bundle of fibers that lead from mitral cells in the olfactory bulb and innervate areas of the cortex. (Ch. 8)

oligodendroglia (ohl-ih-go-den-**dro**-glee-ah) One of two types of glial cells that wrap around the fiber extensions of neurons to form myelin. (Ch. 7)

oligosaccharide A carbohydrate that on hydrolysis yields a small number of monosaccharides. (Ch. 23)

oliguria (ohl-ih-**goo**-ree-ah) The excretion of a small amount of urine in relation to fluid intake. (Ch. 25)

"on" center and "off" surround cells Ganglion cells for which the center of the annulus is excitatory and the periphery is inhibitory. (Ch. 8)

oncotic pressure (on-**kot**-ik) The osmotic pressure due to the presence of colloids in a solution; in the case of plasma interstitial fluid interaction, it is the force that tends to counterbalance the capilary blood pressure. (Ch. 17)

operant conditioning A type of associative learning that involves the relationship between a behavior and a subsequent reward or punishment. (Ch. 11)

opioid (oh-pea-oid) One of a family of neuropeptides that is structurally similar to opiates and that modulate the perception of pain. (Ch. 7)

opsonin (op-**so**-nin) A complement protein that is part of the internal defense mechanism; coats bacteria and other cells so that they are susceptible to phagocytosis. (Ch. 17)

ordinate The vertical axis of a graph; used to plot a dependent variable. (Ch. 1)

organ (or-gan) A group of tissues that have been combined for the performance of a specific function or a series of related functions. (Ch. 1)

organelle An organized, reproducible, permanent subcellular structure, membranous or nonmembranous, that provides for specific functions of the cell. (Ch. 3)

orgasm (or-gazm) The apex and culmination of sexual excitement; characterized in the male by ejaculation and in the female by rhythmic muscular contractions in the uterus and vagina. (Ch. 30)

orientation column Nerve cells of the primary visual cortex that are sensitive to bars of light oriented at different angles. (Ch. 8)

origin The more stationary and proximal point of the attachment of a muscle. (Ch. 16)

orthophosphate PO_4; the primary form of phosphorus in plasma. (Ch. 27)

osmolality Total solute concentration, expressed as osmoles per kg of H_2O. (Ch. 24)

osmolarity The concentration of osmotically active particles in solution, expressed as osmoles of a solute per liter of solution. (Ch. 2)

osmoreceptor Specialized neurons in the anterior hypothalamus that are stimulated by increased osmolality of the extracellular fluid; their excitation promotes the output of antidiuretic hormone and stimulates the thirst sensation. (Ch. 9)

osmosis (oz-**mow**-sis) The net movement of solvent (usually water) from a solution of lesser to one of greater solute concentration when the two solutions are separated by a membrane that selectively prevents passage of solute molecules, but is permeable to the solvent. (Ch. 2)

osmotic fragility (oz-**mot**-ik) The ability of a cell to withstand an osmotic stress. (Ch. 4)

osmotic pressure A property of a solution that is proportional to the solute concentration; a cause of diffusion across membranes. (Ch. 2)

ossicle (**os**-ih-kul) A small bone within the inner ear. (Ch. 8)

osteoblast (**os**-tee-oh-*blast*) A cell that arises from a fibroblast and which at maturity is associated with the production of bone. (Ch. 27)

osteoclast (**os**-tee-oh-*klast*) A cell that carries out the process of bone resorption. (Ch. 27)

osteocyte (**os**-tee-oh-*site*) An osteoblast that has progressively lost its bone-forming capability. (Ch. 27)

osteoid (**os**-tee-oyd) The organic matrix of bone. (Ch. 27)

osteomalacia (**os**-tee-oh-mal-*ay*-see-ah) A metabolic bone disease in which

there is softening of the bone due to inadequate mineralization with excess accumulation of osteoid. (Ch. 27)

osteoporosis (*os*-tee-oh-poe-**row**-sis) A metabolic bone disease in which there is reduction of overall bone mass. (Ch. 27)

otolithic membrane (oh-toe-*lith*-ik) A membrane in which the stereocilia of hair cells in the utricle of the inner ear are embedded; contains calcium carbonate crystals called statoconia. (Ch. 9)

otolithic organ The utricle and saccule of the inner ear; detects linear acceleration. (Ch. 8, 9)

otosclerosis (oh-toe-skler-oh-sis) Abnormal growth of bones in the middle ear, causing conductive hearing loss. (Ch. 8)

outer ear The portion of the ear that channels sound waves to the ear drum. (Ch. 8)

oval window A membranous structure of the inner ear that vibrates with movement of the ossicles. (Ch. 8)

ovarian cycle (oh-var-ee-an) The approximately 28-day cycle of the female reproductive system; also called the menstrual cycle. (Ch. 30)

ovary (**oh**-var-ee) The female gonad; an oval-shaped organ located in the pelvic cavity below the open ends of the fallopian tubes; produces germ cells and hormones. (Ch. 30)

Overton's rules Relationships that characterize the permeability of cell membranes to polar and nonpolar solutes. (Ch. 4)

ovulation (*oh*-vu-**lay**-shun) The release of a single ovum from one of the two ovaries. (Ch. 30)

ovum (**oh**-vum) The female reproductive cell, which after fertilization develops into an embryo. (Ch. 30)

oxidation reaction (oks-ih-**day**-shun) A chemical reaction in which electrons are lost by a reactant. (Ch. 6)

oxidation-reduction reaction A class of chemical reactions in which one reactant accepts electrons that are removed from another. (Ch. 2)

oxidative phosphorylation The chemical pathway that provides ATP to the cell by phosphorylation of ADP in the mitochondrion. (Ch. 6)

oxygen An element with a high affinity for electrons; the primary receptor of electrons in the mitochondrial respiratory chain. (Ch. 6)

oxygen debt The amount of oxygen required to metabolize excess lactic acid formed in skeletal muscle during exercise. (Ch. 16)

oxygen radicals Highly reactive byproducts of molecular oxygen that are toxic to cells. (Ch. 22)

oxygen toxicity Damage to the lung and other body parts that occurs after breathing high concentrations of oxygen for more than 1 day. (Ch. 22)

oxygenated blood Blood that has passed through the pulmonary capillaries and carries about 20 ml of O_2 per deciliter of whole blood. (Ch. 21)

oxygen-carrying capacity The amount of oxygen that the hemoglobin in 100 ml of whole blood is capable of transporting. (Ch. 17)

oxyhemoglobin (*ok*-see-**he**-mow-glow-bin) Hemoglobin that is saturated with oxygen after passing through the lungs. (Ch. 17)

oxyhemoglobin dissociation curve An S-shaped curve that illustrates the way that hemoglobin binds to oxygen in the lung and unloads oxygen in the tissues. (Ch. 17)

oxytocin (*ok*-see-**toe**-sin) A hormone secreted by the posterior pituitary; stimulates contraction of uterine muscles during labor and milk ejection actions. (Ch. 13)

pacemaker Neurons that set the rhythm of biological clocks independent of external cues; any nerve or muscle cell that has an inherent autorhythmicity and determines activity patterns of other cells. (Ch. 18)

pacemaker cell A cardiac cell that regulates heart rate. (Ch. 10)

Pacinian corpuscle A somatic sensory mechanoreceptor, located beneath the skin and in walls of viscera, that responds to pressure. (Ch. 9)

pain receptor A mechanical or heat nociceptor. (Ch. 9)

palmitic acid A fatty acid; one of the most common types found in phospholipids. (Ch. 4)

pancreas (**pan**-kree-as) An organ of the digestive system; located in the abdomen near the stomach; connected by ducts to the small intestine; contains endocrine gland cells, which secrete insulin, glucagon, and somatostatin into the bloodstream, and exocrine gland cells, which secrete digestive enzymes and bicarbonate into the intestine. (Ch. 15, 23)

pancreatic polypeptide A polypeptide produced by the F-cells of the pancreas; its physiological function is not fully characterized. (Ch. 15)

pancreatitis (*pan*-kree-ah-**tie**-tis) An acute or chronic inflammation of the pancreas; due to autodigestion of pancreatic tissue by its own enzymes. (Ch. 23)

paracellular transport pathway Junctions that are relatively permeable and allow water and ions to move between the epithelial cells. (Ch. 4)

paracrine (**pear**-ah-krin) The mechanism whereby a hormone secreted by one cell affects an adjacent or nearby target cell. (Ch. 12)

parafollicular cells Thyroid cells that produce calcitonin; also called C-cells. (Ch. 27)

parallel organization The arrangement of the motor system whereby information from multiple sources can simultaneously and independently affect the same target. (Ch. 9)

parasympathetic nervous system The portion of the autonomic nervous system that coordinates the body's vegetative activities. (Ch. 10)

parathormone *See* parathyroid hormone.

parathyroid gland (*par*-ah-**thy**-roid) Two pairs of glands located on the dorsal surface of each of the two lobes of the thyroid glands. (Ch. 27)

parathyroid hormone (PTH) A polypeptide protein produced by the parathyroid glands. (Ch. 27)

paraventricular nucleus The location of osmoreceptors in the hypothalamus. (Ch. 11)

partial pressure That part of the total gas pressure that is due to molecules of one gas species; a measure of concentration of a gas in a gas mixture. (Ch. 2)

parturition (*par*-too-**rish**-in) Birth; the delivery of an infant and placenta. (Ch. 31)

patellar tendon (pah-**tell**-ar) A tendon of the anterior thigh muscles that passes through the knee, contains the patella (knee cap), and is inserted into the anterior tibia of the leg. (Ch. 9)

pathogen (**path**-oh-jen) Any disease-producing organism or agent (Ch. 17)

P_{CO_2} Carbon dioxide partial pressure or tension. (Ch. 29)

penile erection (**pee**-nile) The increase in diameter, length, and firmness in the penis; due to vasocongestion. (Ch. 32)

pentose phosphate pathway The pathway for oxidation of glucose in the cytosol; also called the hexose monophosphate shunt. (Ch. 6)

pepsin (**pep**-sin) An enzyme, secreted by the stomach, that initiates digestion of protein. (Ch. 23)

peptidase Any enzyme that hydrolyzes specific peptides to their constituent amino acids. (Ch. 23)

peptide (**pep**-tide) A compound that is composed of two or more amino acids; forms the constituent parts of proteins. (Ch. 7)

peptide bond A polar covalent chemical bond joining two amino acids; forms the protein backbone. (Ch. 2)

perfusion Capillary blood flow. (Ch. 19)

perimysium (*per*-ih-**mis**-ee-um) A type of connective tissue that groups bundles of muscle fibers. (Ch. 16)

peripheral nervous system (PNS) Nerve fibers that emerge from the brain stem and spinal cord; composed of somatic and autonomic portions. (Ch. 7)

peripheral proprioceptors Neuroreceptors in joints and muscle spindles that detect movement of limbs. (Ch. 29)

peripheral thermoreceptors Temperature receptors that detect changes in skin and shell temperature; located just beneath the skin. (Ch. 28)

peristalsis (*per*-ih-**stal**-sis) The propulsive movement of a semisolid food mass through the digestive system by a wave of muscle contractions. (Ch. 23)

peritoneum (*per*-ih-tow-**nee**-um) A membrane that lines the abdominal cavity and covers some of the abdominal viscera. (Ch. 23)

peritubular capillary The capillary network surrounding the tubules of the kidney. (Ch. 24)

permeability coefficient A number expressing the ease with which a solute can penetrate a plasma membrane; governed by the thickness of the membrane and the lipid solubility of the solute; can be expressed as *P*. (Ch. 4)

permissive action The amplifying effect of one hormone on another. (Ch. 14)

peroxidase (per-**ox**-ih-dase) A cellular enzyme that readily reacts with oxidizing free radicals. (Ch. 22)

peroxide An oxide of any element that contains more oxygen than any other. (Ch. 22)

peroxisome (per-**ox**-ih-zome) A small, membrane-bounded bag, found in a cell, that is involved in digestion and food waste; also important in detoxifying a number of molecules. (Ch. 3)

pH A measure of the acidity or alkalinity of a solution; it is equal to the negative logarithm (to base 10) of the hydrogen ion concentration (the latter measured in moles/liter). (Ch. 2)

phagocytosis (*fag*-oh-sigh-**toe**-sis) A special type of enocytosis involving internalization of particulate matter. (Ch. 3)

pharynx (**fare**-inks) The portion of the digestive tract between the mouth and the esophagus. (Ch. 23)

phase contrast microscope A microscope that enhances the contrast between the internal organelles so that structures can be seen clearly in living, unstained, and nonfixed cells. (Ch. 3)

phasic contraction (**faze**-ik) A contraction that occurs intermittently. (Ch. 16)

phasic discharge A very brief burst of action potentials by a proprioceptor. (Ch. 9)

phenotype (**fee**-no-type) The visible expression of the genotype; physical or other characteristics, which are produced by genes and influenced by the interaction between genes and environmental factors. (Ch. 11)

pheromone A substance secreted to the outside of the body by an individual and perceived (as by smell) by a second individual. (Ch. 10)

phonocardiogram The electronic recording of heart sounds. (Ch. 18)

phosphagen system The energy system of muscles based on ATP and phosphacreatine; used for short bouts of maximal exercise. (Ch. 29)

phosphate PO_4. (Ch. 27)

phosphatidylinositol A minor phospholipid constituent of the plasma membrane. (Ch. 12)

phosphatidylinositol 4,5-bisphosphate (PIP2) The immediate precursor for the generation of two intracellular second messengers, derived from phosphatidylinositol. (Ch. 12)

phosphaturia (fos-fa-**tur**-ee-ah) Increased phosphate excretion in the urine. (Ch. 27)

phosphocreatine A high-energy phosphate molecule used as an energy resource in the metabolism of muscle cells. (Ch. 29) *See also* phosphate.

phosphodiesterase An enzyme present in the disc membrane of photoreceptors; catalyzes the conversion of cyclic AMP to AMP. (Ch. 9)

phospholipase C (fos-fo-**lie**-pace) A receptor-controlled plasma-membrane enzyme that catalyzes phosphatidyl bisphosphate breakdown to inositol trisphosphate and diacylglycerol. (Ch. 5)

phospholipid (fos-fo-**lie**-pid) A lipid subclass similar to tracylglycerol except that a phosphate group and a small nitrogen-containing molecule are attached to the third hydroxyl group of glycerol; a major component of the cell membrane. (Ch. 2)

phosphoprotein phosphatase An enzyme that removes phosphate from protein, restoring the protein's original shape. (Ch. 5)

phosphorylation The addition of a phosphate group to an organic molecule. (Ch. 6)

photon A particle of light; photons have wavelengths and energies associated with different colors. (Ch. 9)

photoreceptor cones (fo-toe-he-**sep**-tor) Conically shaped photoreceptors that are the basis of color perception and day vision. (Ch. 9)

photoreceptor rods Cylindrically shaped photoreceptors that are very sensitive to low levels of illumination. (Ch. 9)

photoreceptors Receptor cells of the retina that absorb photons. (Ch. 9)

photosynthesis The use of radiant energy to manufacture ATP, which drives biosynthesis in plant cells. (Ch. 6)

phrenic nerve (**free**-nik) The nerve that sends respiratory messages to the diaphragm. (Ch. 20)

physiology (*fiz*-ee-**ol**-oh-jee) A branch of biology dealing with the mechanisms by which the body functions. (Ch. 1)

Pickwickian syndrome A type of altered breathing pattern that affects ventilation; occurs in severely obese individuals who often suffer hypoventilation. (Ch. 20)

pituitary dwarfism (pit-**too**-ih-*terr*-ee) Dwarfism that is due to a deficiency in GH secretion, which occurs as the result of a defect in pituitary function. (Ch. 13)

pituitary gland The endocrine gland that lies in a bony pocket below the hypothalamus; includes the anterior, middle, and posterior pituitary. (Ch. 13)

pituitary stalk A thin segment of tissue that connects the pituitary gland to the hypothalamus. (Ch. 13)

placenta (plah-**sen**-tah) An interlocking fetal and maternal organ of molecular exchange between fetal and maternal circulation. (Ch. 31)

plasma (**plas**-muh) The liquid portion of blood; a component of extracellular fluid. (Ch. 17)

plasma cell A cell that differentiates from activated B-lymphocytes and secretes antibodies. (Ch. 17)

plasma membrane The membrane that forms the outer surface of the cell and separates the cell's contents from extracellular fluid. (Ch. 1)

plasmid (**plas**-mid) A small accessory DNA molecule used for DNA cloning. (Ch. 3)

plateau The phase of sexual excitement that precedes orgasm. (Ch. 32)

platelet (**plate**-let) Thrombocyte; a cell fragment, suspended in plasma, that functions in blood clotting. (Ch. 17)

platelet plug An aggregation of platelets at the site of vascular damage; effectively plugs the defect end. (Ch. 17)

pleural pressure (**ploor**-ahl) The pressure in the pleural fluid between the lung and the chest wall. (Ch. 20)

pneumatology An early theory of physiology put forward by Galen; involved the belief that the blood took on natural spirits and carried them through the veins to the bodily organs. (Ch. 1)

pneumothorax The condition when air enters the pleural space and the lungs collapse. (Ch. 20)

P_{O_2} Partial pressure of oxygen. (Ch. 29)

poikilocytosis A variation in the shape of an erythrocyte. (Ch. 17)

poikilotherm (poy-**kee**-lo-therm) An animal whose body temperature may fluctuate with ambient temperature. (Ch. 28)

polycythemia (*pol*-ee-sie-**thee**-me-ah) An increase in the number of red blood cells in the blood. (Ch. 17)

polypeptide A polymer consisting of amino acid subunits joined by peptide bonds; also called peptide and protein. (Ch. 2)

polysome A structure composed of ribosomes attached to a single strand of mRNA, playing a role in peptide synthesis; also called a polyribosome. (Ch. 5)

polyspermy Fertilization of the ovum by more than one sperm; lethal for the fertilized ovum. (Ch. 31)

polyuria (*poly*-ee-**yoo**-ree-ah) The excessive excretion of urine; characteristic of diabetes. (Ch. 15)

population code A method of informing the brain about the intensity of a stimulus; the strength of a stimulus is related to the number of receptor cells that are activated. (Ch. 9)

population code of contractile force A means of control of the amount of force that a muscle generates during a contraction; an increase in the number of motor neurons and motor units can increase a muscle's force of contraction. (Ch. 9)

portal circulation Blood flow from the stomach, intestines, and pancreas through the portal vein and to the liver. (Ch. 23)

positive balance The condition in which there is net gain of a substance in the body. (Ch. 25)

positive feedback A class of control system that destabilizes normal values of a variable and thus has limited value in maintaining homeostasis. (Ch. 1)

posterior (pos-**teer**-ee-or) A descriptive term of position that means nearer to the back of the body. (Ch. 1)

posterior parietal cortex An area of the brain that processes sensory stimuli, leading to purposeful movement. (Ch. 9)

postganglionic cell (post-gang-glee-**on**-ik) The autonomic neuron that directly innervates the effector cell. (Ch. 10)

postmenopausal osteoporosis The rapid loss of bone due to decreasing estrogen levels in postmenopausal women. (Ch. 27)

post-translational modification of proteins A series of reactions in which newly synthesized RNA molecules are covalently modified; also known as RNA processing. (Ch. 5)

potential energy Stored energy. (Ch. 2)

power output The rate of doing work. (Ch. 16)

presynaptic inhibition (*pre*-sin-**ap**-tik) The action of an interneuron to inhibit the release of substance P by the release of the peptide enkephalin. (Ch. 9)

precapillary sphincter (**sfink**-ter) A smooth muscle ring around the arteriole that regulates the flow of blood into the capillaries supplied by that arteriole. (Ch. 19)

preganglionic cells Efferent sympathetic and parasympathetic neurons whose cell bodies are located within the CNS but whose fibers leave the CNS and synapse within peripheral autonomic ganglia. (Ch. 10)

pregnancy The condition of having a developing embryo or fetus within the body. (Ch. 31)

pregnenolone A steroid hormone that is an early constituent of the synthesis of all steroids; its precursor is cholesterol. (Ch. 12)

premature ejaculation Consistent, unintentional ejaculation prior to vaginal penetration; more commonly defined as loss of ejaculatory control resulting in ejaculation soon after vaginal penetration. (Ch. 32)

premise Given information stated in the form of an absolute rule using a word such as ''all'' or ''always.'' (Ch. 1)

premotor cortex A component of the frontal, cerebral cortex where movements are planned and programmed. (Ch. 9)

preoptic area An area of the hypothalamus in which electrical stimulation induces non-REM sleep; destruction of this area of the brain causes insomnia in laboratory rats. (Ch. 11)

pressure gradient $P_1 - P_2$, or ΔP; the difference in pressure between two points. (Ch. 18)

presynaptic facilitation An underlying mechanism that enhances an animal's response to a particular stimulus; causes sensitization to the stimuli. (Ch. 11)

primary somatosensory area A narrow strip of cortical tissue, the location of cells in the somatosensory cortex that exhibit the annular receptive field; also called Brodmann area 3. (Ch. 9)

primitive gonads Gonads found in the zygote before they have differentiated into testes or ovaries. (Ch. 32)

procarboxypeptidase A inactive precursor of carboxypeptidase, which is a proteolytic enzyme secreted by the pancreas. (Ch. 23)

progesterone (pro-**jes**-ter-own) A steroid hormone secreted by the corpus luteum and placenta; stimulates uterine-gland secretions, inhibits uterine smooth-muscle contractions, and stimulates breast growth. (Ch. 12)

progestin (pro-**jes**-tin) One of the major categories of steroid hormones; involved in the maintenance of pregnancy as well as being produced by the ovaries and the placenta. (Ch. 12)

proglucagon (pro-**glue**-ka-gon) The immediate precursor of glucagon; synthesized in the pancreas, small intestine, and hypothalamus, but only converted to glucagon in the pancreas. (Ch. 15)

proinsulin A large precursor molecule from which insulin is derived; synthesized in the rough endoplasmic reticulum of the beta cells in the pancreas. (Ch. 15)

prokaryotic cells Cells in which the DNA is not in a separate compartment from the rest of the cell. (Ch. 3)

prolactin (pro-**lak**-tin) A peptide hormone secreted by the anterior pituitary; stimulates milk secretion in the mammary glands. (Ch. 13)

promoter site The place where transcription of a given gene begins when the RNA polymerase binds on a DNA template. (Ch. 5)

pro-opiomelanocortin (POMC) A large protein precursor for ACTH, endorphins, and several other hormones. (Ch. 13)

proprioception (*pro*-pre-oh-**sep**-shun) The sensation of a change in the position of the limbs. (Ch. 9)

proprioceptor (*pro*-pre-oh-**sep**-tor) A somatic sensory receptor activated by movement of the limbs. (Ch. 9)

prosomatostatin The immediate precursor to somatostatin. (Ch. 15)

prostaglandins (*pros*-tah-**glan**-dins) A group of naturally occurring hydroxy fatty acids derived from arachidonic acid; affect a wide variety of physiological processes. (Ch. 24)

protease (**pro**-tee-ase) A general term for a proteolytic enyzme; *see* peptidase. (Ch. 23)

protein (**pro**-teen) A large polymer consisting of one or more sequences of amino acid subunits joined by peptide bonds. (Ch. 2)

protein asymmetry in membranes The condition when extrinsic and intrinsic membrane proteins are distributed unequally between the two halves of the bilayer. (Ch. 4)

protein kinase One of a family of enzymes that phosphorylate certain other proteins by transferring them to a phosphate group from ATP. (Ch. 5)

protein-bound calcium One of two major forms of calcium in plasma; calcium bound to relatively large plasma proteins that cannot pass through plasma membranes. (Ch. 27)

proteinuria (pro-tee-**nyoo**-re-ah) The presence of an excess of proteins in the urine. (Ch. 24)

proteoglycan High-molecular-weight compounds that form the noncollagen portion of bone. (Ch. 27)

prothrombin (pro-**throm**-bin) An inactive precursor of thrombin; produced by the liver and normally present in plasma. (Ch. 17)

prothrombin converting factor The activated Stuart factor, in the presence of calcium ions; forms complexes with accelerin on phospholipid micelles provided by tissue thromboplastin; converts prothrombin to thrombin. (Ch. 17)

proton A positively charged subatomic particle. (Ch. 2)

protoplasm A complex mixture of chemicals that displays the attributes of life; that is, is organized into specific kinds of structural units, has the ability to enter into chemical activities that include the transformation of energy and the maintenance or synthesis of protoplasm, the ability to respond to changes in the environment, and the ability to grow and reproduce. (Ch. 1)

proximal Nearer, closer to a reference point; the opposite of distal. (Ch. 16)

proximal convoluted tubule The first part of the renal tubule. (Ch. 24)

pseudopodia Footlike extensions of the cytoplasm. (Ch. 17)

puberty The attainment of sexual maturity, when conception becomes possible; as commonly used, refers to 3 to 5 years of sexual development that culminates in sexual maturity. (Ch. 30)

pulmonary artery (**pul**-mow-*ner*-ee) The large vessel leading from the right ventricle that carries systemic venous blood to the lungs. (Ch. 18)

pulmonary circulation The circulation through the lungs; the portion of the cardiovascular system between the pulmonary trunk, as it leaves the right ventricle, and the pulmonary veins, as they enter the left atrium. (Ch. 18)

pulmonary edema An abnormal accumulation of fluid in the lung interstitium and alveoli. (Ch. 21)

pulmonary embolism Blockage in the pulmonary circulation due to emboli which are often blood clots (thrombi) that have formed in the systemic venous system. (Ch. 21)

pulmonary hypotension High pulmonary artery pressure; can cause vessel damage and eventual failure of the right ventricle. (Ch. 21)

pulmonary stretch receptors Afferent nerve endings lying between airway smooth muscle cells and activated by lung inflation. (Ch. 20)

pulmonary trunk A large artery that carries blood from the right ventricle of the heart. (Ch. 18)

pulmonary valve The valve between the right ventricle of the heart and the pulmonary trunk. (Ch. 18)

pulmonary veins Veins that take oxygenated blood into the left heart. (Ch. 18)

pulmonary ventilation Tidal volume × respiratory rate; normally 5 to 6 liters per minute at rest. (Ch. 29)

pulmonic stenosis Narrowing of the right atrioventricular orifice or pulmonary trunk resulting in hypertrophy of the right ventricle. (Ch. 18)

pulse pressure The difference between systolic and diastolic arterial blood pressures. (Ch. 19)

Purkinje cell (per-**kin**-jee) The most prominent nerve cell in the cerebellum. (Ch. 9)

Purkinje fiber A specialized myocardial cell that constitutes part of the conducting system of the heart; conveys excitation from bundle branches to ventricular muscle. (Ch. 18)

putamen One of the basal ganglia involved with involuntary control of skeletal muscle. (Ch. 9)

P-wave A component of an electrocardiogram that reflects atrial depolarization. (Ch. 18)

pyloric sphincter (pie-**lor**-ik) A thick bundle of circular smooth muscle and connective tissue that controls the passage of chyme from the stomach to the small intestine. (Ch. 23)

pylorus (pie-**lor**-us) The distal funnel-shaped end of the stomach; stomach contents are emptied through it to the duodenum. (Ch. 23)

pyramidal tract Nerve fibers that descend from the motor cortex to the brain stem and down the lateral corticospinal tract. (Ch. 9)

pyrogens Chemical agents that cause fever. (Ch. 28)

pyruvic acid A three-carbon intermediate in the glycolytic pathway that, in the absence of oxygen, forms lactic acid or, in the presence of oxygen, enters the Krebs cycle. (Ch. 29)

QRS complex The component of an electrocardiogram corresponding to ventricular depolarization. (Ch. 18)

qualitative Pertaining to quality. (Ch. 1)

quantal release Refers to the release of a fixed number of neurotransmitter molecules from synaptic vesicles. (Ch. 7)

quantitative Denoting or expressible as a quantity; relating to the proportionate quantities or the amount of the constituents of a compound. (Ch. 1)

quisqualate One of three subtypes of glutamate receptors that, when activated, results in excitatory postsynaptic potentials by opening ion channels that increase in Na^+ and K^+ conductance. (Ch. 7)

radiation Heat transfer by electromagnetic radiation between objects that are not in contact. (Ch. 28)

radioimmunoassay A procedure used to measure very small amounts of molecules with great accuracy; uses radioisotopes to label cyclic AMP in a sample. (Ch. 3)

radioisotope An unstable isotope of an element that undergoes radioactive decay. (Ch. 3)

raphe nucleus A portion of the brain stem that is responsible for the generation of REM sleep. (Ch. 11)

rapid eye movement (REM) sleep A phase of sleep during which eye movement becomes rapid, the heart rate and blood pressure increase, and gastrointestinal motility decreases. (Ch. 11)

rate-limiting enzyme An enzyme in the metabolic pathway that is most easily saturated with substrate; determines the rate of the entire metabolic pathway. (Ch. 6)

rate-limiting reaction The slowest reaction in a metabolic pathway. (Ch. 6)

Rathke's pouch A pocket of cells from the primitive oral cavity that are involved in the embryological development of the pituitary. (Ch. 13)

ratio An expression that compares two numbers or quantities by division. (Ch. 1)

reactive hyperemia (hi-per-ee-me-ah) Extreme reactive vasodilation. (Ch. 19)

receptive field The small area of the skin that can activate the receptor of a nerve cell in the dorsal root ganglion (Ch. 8)

recombinant DNA (re-**kom**-bih-nent) A DNA molecule created by combining pieces of DNA from markedly different organisms. (Ch. 3)

rectum (**rek**-tum) The portion of the large intestine between the sigmoid colon and the anus. (Ch. 23)

redout A loss of vision due to high negative G-forces. (Ch. 22)

redox potential The oxidation–reduction potential; a measure of the affinity of molecules to give up or accept electrons. (Ch. 6)

reduction reaction A chemical reaction in which a molecule is reduced by gaining electrons. (Ch. 6)

referred pain Pain that is attributed to the surface of the body rather than its internal source. (Ch. 10)

reflex (**ree**-flex) An automatic, involuntary action. (Ch. 9)

refractory period (re-**frak**-tor-ee) The period between the peak of an action potential and the return of a membrane to the resting state. (Ch. 16)

regenerative process The basis for the generation of an action potential; the depolarization of a membrane

after a threshold is reached. (Ch. 7)

relative pressure Pressure relative to the atmospheric pressure. (Ch. 20)

relative refractory period The period near the end of an action potential during which the threshold voltage needed to initiate a second action potential is large. (Ch. 7)

relaxin (re-**lax**-in) A hormone that relaxes pelvic ligaments in preparation for parturition. (Ch. 31)

REM generator *See* raphe nucleus.

REM sleep *See* rapid eye movement sleep.

renal (**ree**-nul) Pertaining to the kidneys. (Ch. 24)

renal hilum The indentation of the medial aspect of the kidneys. (Ch. 24)

renal pelvis The expanded upper portion of the ureter that collects the urine from the renal calyces. (Ch. 24)

renin (**reh**-nin) A proteolytic enzyme, produced by granular cells in the kidney, that is involved in the regulation of thirst, osmolarity, and water volume. (Ch. 10)

replication fork A structure that is produced by the separation of the two original DNA strands at the site of active synthesis. Many replication forks can exist on a DNA molecule at once. (Ch. 5)

repolarization The period during an action potential in which the membrane potential is restored to its original value. (Ch. 18)

residual volume The amount of air remaining in the lungs after a person expires as forcefully as possible. (Ch. 20)

respiratory acidosis An abnormal process characterized by CO_2 accumulation, which leads to a fall in blood pH. (Ch. 26)

respiratory alkalosis An abnormal process characterized by the loss of too much CO_2, which leads to a rise in blood pH. (Ch. 26)

respiratory center The portion of the medulla that is responsible for the control of breathing. (Ch. 20)

respiratory distress syndrome Labored breathing that afflicts premature neonates born with insufficient amounts of surfactants in the lungs. (Ch. 20)

respirometer An instrument for the accurate measurement of oxygen consumption. (Ch. 28)

restriction endonuclease An enzyme that can break a strand of DNA at a specific site. (Ch. 3)

retching A process by which the stomach contents are forced into the esophagus but not expelled. (Ch. 23)

reticular formation (reh-**tik**-yoo-lur) A diffuse network of neurons in the brain stem; involved in maintaining posture by the activation of extensor muscles. (Ch. 9)

reticulocytes Stem cells that have lost their nuclei but are not yet mature erythrocytes. (Ch. 17)

retina (**ret**-ih-na) The interior surface of the eye; contains photoreceptors and other visual system neurons. (Ch. 8)

retinotopic map The organization of the visual cortex in the form of a map of the external visual field. (Ch. 8)

retrograde transport (**ret**-row-grade) Transport against the main direction of flow. (Ch. 7)

retrolental fibrosis Blindness due to deterioration of the retina and inadequate development of the retinal blood vessels; the formation of fibrosis tissue behind the lens. (Ch. 22)

retroperitoneal (*reh*-trow-*per*-ih-tow-nee-al) External to the peritoneal lining of the abdominal cavity. (Ch. 23)

retropulsion The closure of the pyloric sphincter and forcible squirting of stomach contents back into the body of the stomach. (Ch. 23)

re-uptake A mechanism by which transmitter molecules are returned into the presynaptic terminal. (Ch. 7)

reward system Brain structures that provide pleasurable inner emotions in response to certain behaviors and drugs. (Ch. 11)

Rh factor Antigen D, which is genetically determined to be present or absent on red blood cells. (Ch. 17)

rhodopsin (row-**dop**-sin) A visual pigment; a photosensitive purple-red chromoprotein in the retinal rods that is bleached to visual yellow by light. (Ch. 8)

ribonucleic acid (RNA) (rye-bo-new-**clay**-ik) A single strand of nucleotides that represent a copy of a limited region of the nucleotide sequence of a strand of DNA. (Ch. 3)

ribosomal RNA (rRNA) (rye-bo-**zome**-al) One of three kinds of RNA synthesized; the nucleotide sequence on rRNA molecules provides information for protein synthesis. (Ch. 5)

ribosome (**rye**-bo-sowm) Cellular organelles that make proteins according to the instructions carried by RNA. (Ch. 3)

rickets (**rik**-ets) A metabolic bone disease of children, characterized by inadequate mineralization of new bone matrix; caused by deficiency of vitamin D. (Ch. 27)

right heart The chambers of the heart that receive systemic venous blood and pump it to the lungs. (Ch. 18)

rigor mortis After death, a stiffness of skeletal muscles resulting from loss of ATP. (Ch. 16)

RNA polymerase Enzyme that synthesize RNA during the transcription of DNA. (Ch. 5)

RNA processing (post-transcriptional modification) A series of reactions that modify newly synthesized RNA before it leaves the nucleus. (Ch. 5)

RNA splicing A type of RNA processing in which axon segments are joined together to form functional mRNA. (Ch. 5)

RNA translation The translation of codons on mRNA in the synthesis of proteins. (Ch. 5)

rod photoreceptors Cylindrically shaped photoreceptors that contain one type of pigment capable of absorbing photons representing a broad range of wavelengths. (Ch. 8)

Ruffini capsule A slowly adapting mechanoreceptor sensitive to the stretching of skin. (Ch. 8)

saccule (**sak**-yool) One of two sacs in the membranous labyrinth of the inner ear. (Ch. 9)

saltatory conduction (**sal**-tah-*tow*-ree) Conduction of a nerve impulse (action potential) by a myelinated fiber in which the impulse jumps from one node of Ranvier to the next. (Ch. 7)

sanguine Abounding in blood. (Ch. 1)

sarcolemma (*sar*-koe-**lem**-ma) The plasma membrane of a muscle cell, especially a skeletal muscle cell. (Ch. 16)

sarcomere (**sar**-koe-mere) A repeating structural unit of myofibril; composed of thick and thin filaments and extending between two adjacent Z lines. (Ch. 16)

satiety (sah-**tie**-ih-tee) A feeling of fullness or of not being hungry. (Ch. 11)

saturation The degree to which protein-binding sites are occupied by ligands. (Ch. 2)

scanning electron microscope A microscope that images the deflection of electrons by a specimen. (Ch. 3)

Schwann cell (shvon) A type of cell that forms the myelin sheath around a motor axon. (Ch. 7)

scientific method A form of inquiry that involves posing questions and then searching for answers to those questions through experimentation and testing to validate answers. (Ch. 1)

scientific notation Quantities expressed as products of numbers plus a power of ten. (Ch. 1)

second heart sound A heart sound that occurs at the time of closure of the semilunar valves and defines the end of systole; the sound is caused by the vibrations in the heart and chest that are initiated by the closure of the heart valves and the sudden cessation of blood flow. (Ch. 18)

second messenger An extracellular signal that travels from a cell to a target cell, where it reacts with receptor molecules. (Ch. 5)

secondary somatosensory area The area of the somatosensory cortex that does not exhibit an annular receptive field; also called Brodmann areas 1 and 2. (Ch. 8)

secretin (**seek**-reh-tin) A strongly basic polypeptide hormone secreted by the upper small intestine; stimulates the pancreas to secrete bicarbonate into the small intestine. (Ch. 12)

secretion (se-**kree**-shun) The elaboration and release of organic molecules, ions, and water by cells in response to specific stimuli. (Ch. 23)

sedimentation rate The rate at which red cells settle; the normal rate is between 2 and 10 mm/hr. (Ch. 17)

segmenting contraction The most frequently occurring type of muscular movement in the small intestine; moves the contents of different regions of the small intestine back and forth over short distances. (Ch. 23)

semicircular canal The three sensory organs of the vestibular apparatus of the ear that detect position and motion of the head in space; the **superior, inferior,** and **horizontal semicircular canals** are each situated perpendicular to the other two. (Ch. 9)

semiconservative replication A process in DNA replication in which the structure of the parent DNA molecule is disrupted but the structure of the nucleotide sequence of each strand is conserved. (Ch. 5)

semilunar valve (*sem*-ee-**loo**-nar) A valve located at the junction of the ventricle with the pulmonary and systemic arteries; consists of flaps shaped like half-moons. (Ch. 18)

sense organ Any organ of the sensory system, such as the eyes or ears. (Ch. 8)

sensitization An increase in the response to a stimulus. (Ch. 11)

sensor A mechanism that detects changes to bodily or environmental conditions. (Ch. 1)

sensory adaptation The gradual loss of perception of a stimulus that is continuously applied. (Ch. 8)

sensory neural hearing loss Hearing loss due to the damage to hair cells located between the basilar and tectorial membranes of the ear. (Ch. 8)

sensory neuron A neuron that carries information about external or internal stimuli from sense receptors to the central nervous system; also called an afferent neuron. (Ch. 7)

sensory transduction The conversion of energy (light, heat, sound, etc.)

into nerve impulses by a sensory receptor. For example, the process by which the sensory receptors of the eye convert various wavelengths of light into electrical impulses that nerve cells uses to transmit information to the thalamus. (Ch. 8)

serotonin (sair-oh-**toe**-nin) A vasoconstrictor synthesized in the brain stem; inhibits gastric secretion, stimulates smooth muscle, and serves as a central neurotransmitter. (Ch. 7)

sex chromosomes The X and Y chromosomes; females have two X chromosomes and males have one X and one Y; provide the genetic information for the development and differentiation of most of the cells in the body. (Ch. 32)

sex steroid binding globulin (SSBG) A plasma protein that binds with testosterone in the blood; aids in storing hormones in the blood. (Ch. 30)

sexual dysfunction Any condition in which the normal physical responses to sexual stimulation are impaired. (Ch. 32)

sexual flush A darkening of the body's skin color during sexual intercourse. (Ch. 32)

shivering Involuntary trembling or quivering of the body caused by contraction or twitching of the muscles; a physiologic method of heat production. (Ch. 28)

short portal system A vascular system that connects the posterior and anterior lobes of the pituitary; may carry substances secreted by the posterior lobe to the anterior lobe where they regulate hormone secretion from the anterior pituitary. (Ch. 13)

short-reflex A regulatory reflex that is mediated over sensory neurons, interneurons, and motor and secretory neurons that lie entirely within the wall of the digestive tract. (Ch. 23)

short-term memory A section of the brain's memory bank that is easily disrupted and short-lived. (Ch. 11)

Siemens Dimensional units of conductance. (Ch. 7)

single channel current The number of ions that flow through a channel per unit of time. (Ch. 7)

sino-atrial node (SA node) (sigh-no-**ay**-tree-al) The region, in the right atrium of the heart, containing specialized cardiac cells that depolarize spontaneously faster than other heart cells; determines heart rate; also called pacemaker of the heart. (Ch. 18)

skeletal muscle The muscle that produces voluntary movement of body parts. (Ch. 9)

sleep apnea A condition in which respiration can stop for long periods of time (often 30–60 seconds) during sleep. (Ch. 20)

slow axonal transport The movement of proteins along the axon from the soma to the terminal region. (Ch. 7)

slow skeletal muscle The muscle that contracts slowly and is more resistant to fatigue. (Ch. 9)

slowly adapting tactile mechanoreceptor A type of receptor that continues to generate action potentials as long as a stimulus is applied. (Ch. 8)

small intestine A portion of the gastrointestinal system consisting of the duodenum, jejunum, and ileum; digestion is completed here by enzymes in the epithelial cells that line the inner surface, and the products of digestion are absorbed into the blood and lymph. (Ch. 23)

smooth muscle A muscle type associated with blood vessels and other organ systems. (Ch. 9)

sodium pump A primary active transport system by which a cell is able to maintain a high K^+ concentration and a low Na^+ concentration in the cytosol; also called the Na^+/K^+-ATPase pump. (Ch. 4)

soft tissue calcification The deposition of calcium in tissues. (Ch. 27)

solute (**sol**-yoot) A substance that is uniformly distributed in a solvent to form a solution. (Ch. 2)

solution A type of mixture in which one substance is uniformly distributed throughout another substance, which is usually a liquid. (Ch. 2)

solvent (**sol**-vent) A substance in which a solute is uniformly distributed to form a solution. (Ch. 2)

soma The body and metabolic center of a neuron. (Ch. 7)

somatic Pertaining to the skin and skeletal muscles of the body. (Ch. 7)

somatic nervous system (so-**mat**-ik) Sensory nerves that transmit information from sense receptors to the central nervous system and motor nerves that transmit instructions about appropriate responses of skeletal muscles. (Ch. 7)

somatic sensory cortex The portion of the neocortex that responds to sensations felt on the surface of the skin and the position of limbs relative to the body. (Ch. 7)

somatomedin (so-mah-tow-**me**-din) A growth factor found in many tissues, including the liver, that mediate the effect of growth hormone on cartilage. (Ch. 13)

somatostatin (so-mat-oh-**stat**-in) A hypothalamic hormone that inhibits growth hormone and TSH secretion by the anterior pituitary; a possible neurotransmitter; also in the stomach and pancreatic islets. (Ch. 13)

somatotropin *See* growth hormone.

spatial summation The additive effects of two simultaneously active synaptic inputs. (Ch. 7)

specific dynamic action The increased heat production that results from processing food. (Ch. 28)

specific gravity The ratio between the weight of a given volume of a liquid and the weight of an equal volume of pure water. (Ch. 17)

specificity The ability of a binding site to react with only one (or a limited number of) type of molecule. (Ch. 5)

spectrin An extrinsic protein found only on the cytoplasmic side of the red blood cell membrane; thought to be important in determination of red blood cell shape. (Ch. 4)

sperm loss Nonviable sperm. (Ch. 31)

sperm viability The ability of a sperm to fertilize an ovum. (Ch. 32)

spermatogenesis (spur-mah-tow-**jen**-eh-sis) The formation of sperm. (Ch. 30)

spermicide (**spur**-muh-side) A chemical that kills sperm; one form of contraception. (Ch. 32)

spherical micelle A small sphere formed by phospholipids in solution. (Ch. 4)

sphincter of Oddi (sfink-ter) A smooth muscle ring surrounding the bile duct at its entrance into the duodenum. (Ch. 23)

spinal autonomic reflex arc (**spy**-nal) A loop of nerve cell activity; begins with the detection of visceral information by autonomic sensory receptors in an organ that in turn activates interneurons within the spinal cord. These neurons synapse with the cells of the IML; the IML cells integrate incoming information and discharge to activate postganglionic neurons. These neurons convey information to the target organ, causing it to make the appropriate response. (Ch. 10)

spinal cord The part of the central nervous system that transmits information to and receives information from most of the body's internal organs, muscles, and skin; located caudal to the brain stem. Consists of 31 segments, each with a pair of spinal nerves. (Ch. 7)

spirometry The measurement of pulmonary volumes and capacities. (Ch. 20)

splanchnic circulation The circulation in the stomach, small and large intestines, pancreas, and liver. (Ch. 23)

stable balance The condition in which there is no net gain or loss of a substance. (Ch. 3)

stance phase The phase of the walk cycle in which the foot is on the ground and supporting the weight of the body. (Ch. 9)

standard free energy change $\Delta G°$, or the difference in total free energy between reactants and products of a chemical reaction under a set of standard conditions: temperature of 25°C, pressure of 1 atmosphere, and a concentration of 1.0 M for all reactants and products. (Ch. 6)

stapes (**stay**-peas) The innermost of the auditory ossicles, shaped somewhat like a stirrup; also called stirrup. (Ch. 8)

start codon A codon that initiates translation of nucleotide triplets. (Ch. 5)

stasis (**stay**-sis) A slowing down or stoppage of flow, such as of blood or other body fluid. (Ch. 17)

static exercise Exercise that involves isometric muscle contractions. (Ch. 29)

statistics The science that deals with the collection, tabulation, and analysis of numerical facts. (Ch. 1)

statoconia Calcium carbonate crystals contained within the otolithic membrane. (Ch. 9)

steady state A condition unvarying with respect to time. (Ch. 5)

steatorrhea The appearance of excessive amounts of undigested fat in the feces; usually due to the absence of lipase in the gastrointestinal system. (Ch. 23)

stereocilia Fiberlike elements that protrude from hair cells located between the basilar and tectorial membranes of the ear. (Ch. 8)

steroid (**steer**-oid) A lipid subclass; the molecule consists of four interconnected carbon rings to which polar groups may be attached. (Ch. 11)

stomach (**stum**-ak) The portion of the gastrointestinal system located between the esophagus and the small intestine; functions as a storage organ for food. (Ch. 23)

stop codon A codon that terminates translation of nucleotide triplets. (Ch. 5)

stretch receptors Afferent nerve endings that are depolarized by stretching; *see also* muscle spindle receptor. (Ch. 29)

stretch reflex Reflex contraction of skeletal muscle in response to its stretch. (Ch. 9)

stria A narrow bandlike structure; a general term for such longitudinal collections of nerve fibers in the brain. (Ch. 14)

striate muscle (**stri**-ate) *See* skeletal muscle.

striatum An area beneath the neocortex that is involved in the initiation and coordination of movement. (Ch. 7)

stroke volume The blood volume ejected by a ventricle during one heartbeat. (Ch. 18)

strong acid An acid with a high dissociation constant, which consequently is highly ionized in solution. (Ch. 26)

structural formula A formulaic illustration of the arrangement in space of the atoms of a molecule. (Ch. 2)

subcutaneous (*sub*-kyoo-**tay**-nee-us) Beneath the skin. (Ch. 28)

sublingual (sub-**ling**-gwal) Below the tongue. (Ch. 28)

submucosa (sub-myoo-**koe**-sah) A connective tissue layer under the mucosa in the gastrointestinal tract. (Ch. 23)

submucosal plexus A nerve-cell network in the submucosa of the esophageal, stomach, and intestinal walls. (Ch. 23)

substantia gelatinosa A group of nerve cells located in the dorsal horn of the spinal cord. (Ch. 8)

substantia nigra An area of the brain associated with the reward system. (Ch. 11)

substrate-level phosphorylation (sub-straight) The direct formation of high-energy phosphate compounds such as ATP by a chemical reaction that incorporates a phosphate group. (Ch. 6)

suckle To derive or to provide nourishment by feeding. (Ch. 32)

sucrase An enzyme that digests sucrose to glucose and fructose. (Ch. 23)

sudden infant death syndrome (SIDS) A disease in which a seemingly normal, healthy infant is found dead in its crib; sleep apnea is being studied extensively as a causative link. (Ch. 20)

sulfonylureas A class of drugs that promote insulin secretion from the beta cells; widely used in the treatment of diabetes mellitus. (Ch. 15)

summation The cumulative effects of a number of stimuli applied to a muscle nerve or reflex arc. (Ch. 16)

superior Literally, nearer to the head. (Ch. 1)

superior vena cava (**vee**-nah **kay**-vah) A large vein that carries blood from the upper half of the body to the right atrium of the heart. (Ch. 18)

superoxide anion radical An active form of free radical. (Ch. 22)

superoxide dismutase A protective cellular enzyme that removes free radicals. (Ch. 22)

supplemental motor cortex An area of the brain that designs the necessary movements required to accomplish a task; also called Brodmann area 6. (Ch. 9)

suprachiasmatic nucleus The nucleus of the hypothalamus, thought to be responsible for coordinating various biological rhythms. (Ch. 11)

supraoptic nucleus Cells in the hypothalamus that secrete vasopressin into the blood in response to increases in osmolarity. (Ch. 10)

surface tension A molecular force created at the gas–liquid interface; specifically, the attractive forces between the liquid and air molecules. (Ch. 20)

surfactant (ser-**fak**-tant) A lipoprotein produced by pulmonary alveolar cells; a detergent-like material that reduces surface tension of fluid film lining alveoli. (Ch. 20)

swallowing The propulsion of a food bolus through the pharynx and esophagus into the stomach; the bolus moves in response to skeletal and smooth muscle contractions that occur sequentially from above to below. (Ch. 23)

sweating The active secretion of water and electrolytes by specialized sweat glands in the skin for the purpose of increasing evaporative heat loss; activated by sympathetic nerves in response to overheating; increases linearly with abnormal increases in body temperature. (Ch. 28)

swing phase The phase of the walk cycle in which the foot is off the ground and swinging forward. (Ch. 9)

sympathetic nervous system (*sim*-pah-**thet**-ik) One portion of the autonomic nervous system; preganglionic fibers leave the CNS at the thoracic and lumbar portions of the spinal cord. (Ch. 10)

sympathoadrenal system (*sim*-pah-thow-ah-**dree**-nal) The sympathetic nervous system and the adrenal medulla together. (Ch. 14)

sympathomimetic (*sim*-pah-thow-my-**met**-ik) Adrenergic; producing effects resembling those produced by the sympathetic nervous system. (Ch. 10)

symport The movement of solutes in the same direction as Na^+. (Ch. 4)

synapse (**sin**-aps) The junction between nerve cells; **electrical synapses** consist of gap junctions that permit direct passage of ions, and **chemical synapses** prevent the direct passage of ions. (Ch. 7)

synaptic cleft (sin-**ap**-tik) In a chemical synapse, the space, between one cell and another, that prevents the direct passage of ions. (Ch. 7)

synaptic transmission The process by which there is communication between nerve cells. (Ch. 7)

synaptic vesicles Small structures filled with neurotransmitting chemicals that fuse with the presynaptic membrane. (Ch. 7)

syncytium (**sin**-she-um) A multinucleate mass of protoplasm produced by the merging of cells. (Ch. 16)

synergistic muscle (sin-er-**jis**-tik) A muscle that works together with another to produce a given movement. (Ch. 9)

systemic circulation (sis-**tem**-ik) The circulation from the left ventricle through all the organs except the lungs and back to the heart. (Ch. 18)

systemic hypertension A complex disease state resulting in chronically elevated arterial blood pressure. (Ch. 19)

systole (**sis**-toe-lee) A period of ventricular contraction. (Ch. 18)

T cell A lymphocyte derived from a precursor that differentiated in thymus; *see also* cytotoxic T cell, helper T cell, suppressor T cell. (Ch. 17)

tachycardia (*tak*-ih-kar-**dee-ah**) An abnormally rapid heart rate, usually defined as more than 100 beats per minute. (Ch. 18)

tactile receptor (**tak**-tile) A mechanoreceptor that detects touch, pressure, and vibrations applied to the skin. (Ch. 8)

target cell A cell whose activity is affected by a specific hormone. (Ch. 5)

taste bud A cluster of gustatory receptors, basal cells, and supporting cells. (Ch. 8)

tectorial membrane (tek-**toe**-ree-al) A membranous structure in which the hair cells of the cochlea are embedded. (Ch. 8)

tectospinal tract A component of the ventromedial pathway that originates in the tectum. (Ch. 9)

tectum (**tek**-tum) A structure involved in the coordinated control of head and eye movements. (Ch. 9)

temperature The metabolic heat produced within the body. (Ch. 28)

temperature regulating system A system that works to maintain thermal homeostasis in the body. (Ch. 29)

temporal summation The addition of postsynaptic potentials that arise from activation at one synaptic input. (Ch. 7)

tendon A collagen fiber bundle that connects muscle to bone and transmits muscle contractile force to the bone. (Ch. 16)

tenting The elevation of the uterus away from the vagina as the blood flow to the vagina increases during the plateau phase of sexual arousal. (Ch. 32)

terminal cisterna Enlarged end portions of the sarcoplasmic reticulum; also called lateral sacs. (Ch. 16)

terminator site on DNA A sequence of nucleotides immediately following the 5′ end of the gene; acts as a recognition site for the ending of the RNA strand. (Ch. 5)

testicular feminization syndrome A condition in which the internal genitalia of the male do not develop; usually due to a deficiency of testosterone biosynthesis or when Wolffian ducts do not respond to testosterone. (Ch. 32)

testis The male gonad. (Ch. 11)

testosterone (tes-**tos**-teh-rone) A steroid hormone produced in interstitial cells of the testes; essential for spermatogenesis and maintaining growth and development of reproductive organs and secondary sexual characteristics of males. (Ch. 12)

tetanic fusion frequency The lowest frequency that will produce a fused

tetanus; around 20 to 60 times per second for most skeletal muscles. (Ch. 16)

tetanus (tet-ah-nus) A maintained mechanical response of muscle to high-frequency stimulation; the disesase lockjaw. (Ch. 16)

tetany (tet-ah-nee) A syndrome manifested by sharp flexion of the wrist and ankle joints; occurs in response to low concentrations of extracellular calcium ion. (Ch. 27)

thalamus (**thal**-uh-mus) An area of the brain that receives all types of sensory information. (Ch. 8)

thalassemia (thal-ah-**see**-me-ah) A group of inherited hemolytic anemias characterized by decreased synthesis of one or more hemoglobin polypeptide chains. (Ch. 17)

theory A formulated hypothesis. (Ch. 1)

thermal homeostasis Physiological regulatory mechanisms continuously operating to keep heat production and heat loss approximately equal. (Ch. 28)

thermal receptor A sensory receptor that responds to temperature. (Ch. 8)

thermography A technique for measuring mean skin temperature in humans. (Ch. 28)

thermoreceptor (*ther*-mow-ree-**sep**-tor) A receptor that detects changes in temperature. (Ch. 1)

thermostat A controller to prevent excessive body heating or cooling. (Ch. 28)

thick myofilament A 12- to 18-nm myosin filament in a muscle cell. (Ch. 16)

thin myofilament A 5- to 8-nm myosin filament in a muscle cell; consists of actin, troponin, and tropomyosin. (Ch. 16)

thirst A conscious sensation associated with a craving for drink; ordinarily interpreted as a desire for water. (Ch. 25)

thoracic cavity (thow-**ras**-ik) The chest cavity. (Ch. 20)

threshold The point when a membrane potential is depolarized sufficiently to produce an action potential. (Ch. 18)

thrombin An enzyme that catalyzes conversion of fibrinogen to fibrin. (Ch. 17)

thrombocyte (**throm**-bow-site) A fragment of cytoplasm, enclosed in plasma membrane, that plays a role in blood clotting; found in the circulation. Also called platelet. (Ch. 17)

thrombocytopenia A deficiency in circulating platelets; results in excessive bleeding. (Ch. 17)

thromboplastin (*throm*-bow-**plas**-tin) A complex of several phospholipids and a proteolytic enzyme released from damaged tissue; important in the clotting process. (Ch. 17)

thrombosis (throm-**bow**-sis) The formation or presence of a clot in an un-

broken blood vessel or heart chamber. (Ch. 17)

thrombus A blood clot. (Ch. 17)

thyrocalcitonin A peptide hormone, secreted by the parafollicular cells of the thyroid, that causes increased deposition of calcium in bones; also called calcitonin. (Ch. 27)

thyroglobulin (*thy*-row-**glob**-yoo-lin) A large protein to which thyroid hormones bind in the thyroid gland; a storage form of thyroid hormones. (Ch. 13)

thyroid follicle (**thy**-roid) The internal structure of the thyroid gland as a collection of closely packed follicles; normal thyroid contains about 3 million follicles. (Ch. 13)

thyroid hormone (TH) A collective term for amine hormones released from the thyroid gland; that is, thyroxine (T4) and triiodothyronine (T3). (Ch. 13)

thyroid-stimulating hormone (TSH) A glycoprotein hormone secreted by the anterior pituitary; induces secretion of thyroid hormone. Also called thyrotropin. (Ch. 13)

thyrotropin-releasing hormone (TRH) A hypothalamic hormone that stimulates thyrotropin and prolactin secretion by the anterior pituitary. (Ch. 13)

thyroxin (thy-**rok**-sin) A hormone, secreted by the thyroid gland, that accelerates the rate at which cells consume oxygen, metabolize glucose, and produce heat. (Ch. 11)

thyroxine-binding globulin (CBG) A carrier protein that binds and carries mainly thyroxine in the blood. (Ch. 12)

tidal volume The volume of air inhaled or exhaled during a single breath. (Ch. 20)

"tight" epithelia Epithelia that do not permit leakage between the cells. (Ch. 4)

tight junction A specialized attachment that connects Sertoli cells to one another inside the seminiferous tubules; the area where plasma membranes are joined. (Ch. 30)

time period The time between the peaks of a sound wave of a tone; the inverse of the frequency of a sound wave. (Ch. 8)

tinnitus (tie-**nie**-tis) A constant ringing in the ear due to the inappropriate discharge of action potentials by particular nerve fibers associated with specific frequencies. (Ch. 8)

tissue An aggregate of differentiated cells of similar type united in the performance of a particular function; also denotes the general cellular fabric of a given organ. (Ch. 1)

titratable acid A solution whose acid concentration can be determined by measuring the milliequivalents of strong base needed to bring the pH to a certain level; for example, the amount of base needed to bring a urine sample back to the pH of arterial blood. (Ch. 26)

tone *See* muscle tone.

tonic contraction A long, sustained contraction in response to single or continued stimulation by nerves, drugs, or hormones. (Ch. 16)

tonic discharge An action potential generated at one specific frequency in response to the position of a joint. (Ch. 8)

tonotopic map The organization of auditory nerve fibers such that fibers with a high characteristic frequency innervate hair cells at the base of the cochlea, while those with a low characteristic frequency innervate hair cells near the apex of the cochlea. (Ch. 8)

tonus The constant steady force of contraction of most smooth muscles; also present in skeletal muscles. (Ch. 16)

total peripheral resistance (TPR) The total resistance to flow in systemic blood vessels from the beginning of the aorta to the end of the venae cavae. (Ch. 19)

trachea (tray-kee-ah) The main airway of the respiratory system, which branches into two bronchi. (Ch. 20)

transcellular transport pathway The movement of water and solutes through epithelial cells; entry and exit pathways are on opposite sides of the cell. (Ch. 4)

transcription (tran-**skrip**-shun) The synthesis of mRNA using a DNA template; the first step in the transfer of genetic information from DNA to proteins. (Ch. 2)

transducin A protein present in the disc membrane of photoreceptors. (Ch. 8)

transfer RNA (tRNA) A type of RNA; different tRNAs combine with different amino acids and with codon on mRNA specific for that amino acid, thus arranging amino acids in sequence to form specified proteins. (Ch. 5)

transferrin An iron-binding protein carrier for iron in plasma. (Ch. 17)

translation The synthesis of a polypeptide using mRNA as a template. (Ch. 2)

transmission electron microscope A microscope that uses magnetic lenses to focus a beam of electrons in much the same way that glass lenses in ordinary microscopes focus a beam of light; specimens must be fixed and stained to be seen under a transmission electron microscope. (Ch. 3)

transmitter *See* neurotransmitter.

transpulmonic pressure The pressure across the lung; measured by subtracting the pleural pressure from the alveolar pressure. (Ch. 20)

transudation The vasoconstriction of the vagina, leading to the release of moisture from the vaginal lining; lubricates the vagina to facilitate insertion of the penis without discomfort. (Ch. 32)

transverse plane An anatomical position that describes a plane dividing the body into upper and lower portions. (Ch. 1)

transverse tubule (t-tubule) A tubule extending from striated muscle plasma membrane into the fiber, passing between opposed sarcoplasmic-reticulum segments; conducts muscle action potential into muscle fiber. (Ch. 16)

triacylglycerol (tri-ace-il-**gliss**-er-al) A subclass of lipids composed of glycerol and three fatty acids; also called fat, neutral fat, or triglyceride. (Ch. 6)

triad The region of contact in skeletal muscles of one t-tubule and the terminal cisternae from two adjacent sarcomeres. (Ch. 16)

trichromatic color vision The ability to discriminate between different colors on the basis of three types of photoreceptors: blue, red, and green cones.

tricuspid valve The valve between the right atrium and the right ventricle of the heart. (Ch. 18)

triglyceride *See* triacylglycerol.

triiodothyronine (T3) (*try*-eye-oh-dow-**thy**-row-neen) An iodine-containing amine hormone secreted by the thyroid gland. (Ch. 13)

trimesters The three 3-month periods of pregnancy. (Ch. 31)

trophic Of or pertaining to nutrition. (Ch. 7)

tropomyosin A regulatory protein of the I-band that inhibits contraction unless its position is modified by troponin so that myosin molecules can make contact with actin. (Ch. 16)

troponin A regulatory protein bound to actin and tropomyosin of striate-muscle thin filaments; a site of calcium binding that initiates contractile activity. (Ch. 16)

trypsin (**trip**-sin) A protein-digesting enzyme secreted by the pancreas. (Ch. 2)

tryptophan (**trip**-toe-fan) An essential amino acid; a precursor to serotonin. (Ch. 7)

tubal ligation (lie-**gay**-shun) The cutting and tying-off of both fallopian tubes, resulting in sterilization. (Ch. 32)

tubular reabsorption The transport of materials by the kidney tubule epithelium out of the tubular urine. (Ch. 24)

tubular secretion The transport of materials by the kidney tubule epithelium into the tubular urine. (Ch. 24)

tubular transport maximum (T_m) The maximum rate at which a particular substance is reabsorbed (e.g., glucose) or secreted (e.g., PAH) by the kidney tubules. (Ch. 24)

turbulent flow Completely disorganized patterns of air flow, which occur at high flow rates. (Ch. 20)

Turner's syndrome A chromosomal abnormality in females in which there is only one X sex chromosome; characterized by impaired development of female gonads, resulting in the appearance of an incompletely formed gonad. (Ch. 32)

T-wave A component of an electrocardiogram, corresponding to ventricular repolarization. (Ch. 18)

twitch The mechanical response of muscle to a single maximal stimulus. (Ch. 16)

tympanic membrane (tim-**pan**-ik) A thin, semitransparent membrane that stretches across the ear canal; separates outer and middle ear. (Ch. 8)

type I diabetes Diabetes in which insulin is completely or almost completely absent; also called insulin-dependent diabetes. (Ch. 15)

type II diabetes Diabetes in which insulin is present at near-normal or even above-normal levels, but the tissues do not respond optimally to it; also called insulin-independent diabetes. (Ch. 15)

tyrosine An amino acid; the precursor of catecholamines and thyroid hormones. (Ch. 12)

tyrosine hydroxylase An enzyme that converts tyrosine to dihydroxy-phenylalanine. (Ch. 14)

ulcer (**ul**-ser) An erosion of the mucosa of certain regions of the digestive tract. (Ch. 23)

ultrafiltrate (ul-tra-**fill**-trate) An essentially protein-free fluid formed from plasma as it is forced through capillary walls by a pressure gradient. (Ch. 24)

ultrafiltration Filtration through a membrane with exceedingly fine pores. (Ch. 24)

umbilical cord (um-**bil**-ih-kal) An artery or vein that transports blood between fetus and placenta. (Ch. 31)

unconditioned stimulus A stimulus that is paired with a conditioned stimulus to evoke a conditioned response. (Ch. 11)

unitary postsynaptic potential A change in the membrane potential due to the flow of synaptic current resulting from the release of transmitters. (Ch. 7)

unitary smooth muscle The smooth muscle that makes up the bulk of the visceral organs; less strictly controlled by the nervous system; a large number of its cells function together as a single unit. (Ch. 16)

upper esophageal sphincter A band of skeletal muscle between the pharynx and the body of the esophagus. (Ch. 23)

urea (yoo-**ree**-ah) The major nitrogen-containing end product of metabolism; formed chiefly by the liver. (Ch. 24)

urea cycle A cyclic series of reactions that produce urine; a major route for removal of ammonia. (Ch. 14)

uremia (yoo-**ree**-mee-ah) The symptom that results from renal failure. (Ch. 24)

ureter (yoo-**ree**-ter) The tube that carries urine from the kidney to the urinary bladder. (Ch. 24)

urethra (yoo-**ree**-thra) The tube that carries urine from the urinary bladder to the outside. (Ch. 24)

uric acid An end product of purine metabolism, secreted by the kidneys. (Ch. 24)

uricosuric An agent that promotes excretion of uric acid in the urine. (Ch. 24)

urinary bladder A hollow, muscular organ that collects the urine and is periodically emptied. (Ch. 24)

utricle (**yoo**-trih-kul) The largest of the two divisions of the membranous labyrinth of the inner ear. (Ch. 9)

vacuole (**vac**-you-ohl) A cavity enclosed by a membrane and located in cytoplasm. (Ch. 3)

vaginismus Involuntary contractions and spasms of the vaginal or pubococcygeal muscles; makes vaginal penetration difficult or impossible. (Ch. 32)

Van Allen belt Particles of cosmic radiation trapped in the earth's magnetic field. (Ch. 22)

van der Waals bond Weak attractive forces between nonpolar regions of molecules. (Ch. 2)

variables The related quantities displayed on a graph. (Ch. 1)

vas deferens (**def**-ur-enz) One of the paired male reproductive ducts that connect the epididymis of testes to the urethra; also called ductus deferens. (Ch. 30)

vasa recta The blood vessels of the medulla that supply the nutritional needs of the kidney medulla as well as act as countercurrent exchangers. (Ch. 24)

vascular bed (**vas**-kyoo-lar) Many separate circulatory pathways that supply blood to various organ systems within the body. (Ch. 18)

vasectomy (vah-**sec**-tow-mee) The cutting and tying-off of both ductus deferens, which results in sterilization of the male without loss of testosterone. (Ch. 32)

vasocongestion The concentration of blood in the tissues of the penis,

which causes an increase in diameter, length, and firmness. (Ch. 32)

vasodilator (vas-oh-die-uh-**lay**-tor) An agent that increases blood-vessel diameter due to vascular smooth muscle relaxation. (Ch. 19)

vasopressin (vas-oh-**press**-in) A hormone, secreted by the paraventricular nucleus of the hypothalamus, that causes the kidneys to reabsorb more water. (Ch. 10)

vein (vane) Any vessel that returns blood to the heart. (Ch. 1)

venomotor tone The contraction of venous smooth muscle. (Ch. 19)

venous capacitance The total blood volume in the veins. (Ch. 19)

venous return The rate (ml/min) of blood return to the right atrium; usually equal to cardiac output. (Ch. 19)

venous system The vessels that carry the blood back to the heart; begins at the point where the capillaries reunite to form venules. (Ch. 18)

ventilation The process of moving air in and out of the lungs. (Ch. 20)

ventilation–perfusion balance A mechanism for balancing regional blood flow with regional alveolar ventilation. (Ch. 21)

ventral An anatomical position that describes the area of the body; specifically, toward, or at the front of, the body. (Ch. 1)

ventral corticospinal tract Fibers from the motor cortex that descend to the spinal cord without crossing the midline of the body; primarily innervate motor neurons in the medial region of the ventral horn associated with axial muscles of the body. (Ch. 9)

ventral posterior lateral (VPL) thalamus The region of the thalamus that receives somatic sensory information. (Ch. 8)

ventral posterior medial (VPM) thalamus The region of the thalamus that receives information from taste receptors. (Ch. 8)

ventricle (**ven**-trih-kul) Any cavity. (Ch. 18)

ventricular ejection Part of the pumping process of the heart when the semilunar valve is forced open and blood flows out of the ventricle; this occurs when ventricular pressure exceeds the pressure in the outflow tract. (Ch. 18)

ventricular fibrillation The uncoordinated stimulation of the ventricular muscle; useless for pumping blood. (Ch. 18)

ventricular filling Part of the pumping process of the heart when the ventricle is being filled with blood from the atrium. (Ch. 18)

ventricular relaxation Part of the pumping process of the heart when the ventricle relaxes and pressure inside the ventricle falls; the semilunar valve is forced closed as blood

attempts to flow from the outflow track back into the ventricle. (Ch. 18)

ventromedial pathway A pathway that descends along the ventral and medial region of the spinal cord. (Ch. 9)

venule (**ven**-yule) A small vessel that carries blood from the capillary network to a vein. (Ch. 18)

vestibular nerve fiber (ves-**tib**-you-lar) A nerve fiber that innervates the hair cells of the vestibular apparatus of the ear. (Ch. 9)

vestibulospinal tract A component of the ventromedial tract that originates in the vestibular nucleus and carries information for the reflex control of equilibrium. (Ch. 9)

villus A small, fingerlike projection that creates the convoluted surface of the small intestine. (Ch. 23)

viscera (**vis**-ur-uh) Organs of the thoracic and abdominal cavities. (Ch. 10)

visceral afferent fiber (**vis**-er-al) Nerve fibers that carry information from an organ to the spinal cord. (Ch. 10)

visceral efferent fiber Nerve fibers that carry information from the spinal cord to an organ. (Ch. 10)

visceral muscles Muscles that line the walls of the internal organs. (Ch. 16)

viscosity (vis-**kos**-ih-tee) A property of fluid that makes it resist flow. (Ch. 17)

visual agnosia Difficulty recognizing familiar objects or colors due to lesions in the secondary visual cortices. (Ch. 8)

visual cortex The area of the neocortex that responds to visual images seen by the eye. (Ch. 7)

visual field The area that can be seen through either the right or the left eye. (Ch. 8)

visual orientation columns Columns of nerve cells in the visual cortex that are sensitive to bars of light oriented at different angles. (Ch. 8)

visual pigment The photoreceptor pigment that absorbs photons. (Ch. 8)

visual system The physiological system that perceives light to determine the shape, color, and grouping of objects in the environment. (Ch. 8)

visuotopic map A precise mapping of the visual field onto the visual cortex. (Ch. 8)

vitamin A general term for a number of unrelated organic molecules that occur in many foods in trace amounts and that are necessary for normal growth and development. They may be water-soluble or fat-soluble. (Ch. 27)

volt A unit of electrical potential. (Ch. 7)

voluntary muscles Muscles that move in response to conscious effort. (Ch. 16)

vomiting The rapid expulsion of gastric contents through the mouth. (Ch. 23)

warm receptors Peripheral temperature receptors sensitive to temperature; their steady-state discharge rates increase in response to warming of the skin. (Ch. 28)

warm-blooded *See* homeothermy.

wavelength A characteristic of electromagnetic radiation that is unique for each part of the continuum of radio waves to X-rays. (Ch. 8)

weak acid An acid that does not dissociate completely when dissolved in water. (Ch. 4)

weak base A base that does not dissociate completely when dissolved in water. (Ch. 4)

"wedge" pressure A pressure measurement made during cardiac catheterization; near-left atrial pressure. (Ch. 21)

weightlessness Lacking gravitational pull. (Ch. 22)

Wernicke's area An area of the neocortex responsible for the interpretation of language. (Ch. 7)

Wolffian duct system A duct system in the embryo that is the precursor to the internal genitalia of the male. (Ch. 32)

***x*-axis** The abscissa or horizontal axis of a graph. (Ch. 1)

***y*-axis** The ordinate or vertical axis of a graph. (Ch. 1)

Z-disc Another term for Z-line to emphasize its two-dimensional nature. (Ch. 16)

Z-line A structure running across a myofibril at each end of a striated muscle sarcomere; anchors one end of the thin filaments. (Ch. 16)

zona fasciculata (**zow**-nah fah-*sik*-you-**lay**-tah) The thick middle layer of the adrenal cortex that secrets glucocorticoid hormones. (Ch. 14)

zona glomerulosa (glow-*mer*-you-**low**-sah) The outer layer of the adrenal cortex that secretes mineralocorticoid hormones. (Ch. 14)

zona pellucida (peh-**loo**-sih-da) The thick, clear layer separating the ovum from surrounding granulosa cells. (Ch. 31)

zona reticularis (reh-**tik**-you-lar-is) The inner layer of the adrenal cortex that secretes sex hormones, mainly androgens. (Ch. 14)

zygote (**zye**-goat) A fertilized ovum; the cell resulting from the fusion of male and female gametes. (Ch. 31)

zymogen granules (**zye**-mow-jen) Vesicles of the enzyme salivary amylase that are stored in the serous cells. (Ch. 23)

Answers to Review Questions

Chapter 1

1. mm Hg
2. exponents
3. metric system
4. extracellular fluid
5. chemoreceptors
6. tissue
7. abscissa or x-axis
8. homeostasis: a term used to describe a constant internal environment despite changes in the external environment
9. hypothesis: an idea proposed to explain a scientific observation
10. scientific method: a sequence of steps used to discover how processes work
11. proximal: a term used to describe a structure that is nearer to the origin or point of attachment
12. distal: a term used to describe a structure that is farther from the origin or point of attachment
13. *y*-axis: the vertical axis on a graph; the ordinate
14. speculation
15. sagittal
16. Aristotle
17. Priestley
18. Harvey
19. Malpighi
20. Cannon
21. a. 1×10^6 cells   b. 5×10^{-4} mm
22. a. 0.005 moles/liter   b. 0.0001 μm
23. 12×10^{-6} kg
24. one to ten or 1/10
25. The concept that cells originate from other cells and not from nonliving matter laid the foundation for the idea that all cells have the same fundamental chemical makeup and that all organs are made up of cells.
26. The scientific method is a quantitative approach that allows scientists to devise experiments to validate their answers. The scientific method also allows other scientists to reproduce their results.
27. All enzymatic reactions and protein conformation are dependent on a constant environment (i.e., temperature, pH, ionic concentration).
28. Graphs allow physiologists to learn about relationships between variables.
29. Scientific notation is a convenient mechanism which allows scientists to express a quantity as the product of a number and a power of ten.
30. There are basically two types of control systems: negative feedback and positive feedback.

Chapter 2

1. exergonic
2. DNA (deoxyribonucleic acid)
3. Vmax (maximum velocity)
4. glycoprotein
5. flux
6. molecular formula: shows the number of each type of atom present in a molecule
7. atomic number: the number of protons in the nucleus of an atom
8. covalent bond: a very strong bond formed when two or more atoms share electrons
9. solution: usually a mixture of a liquid (solvent) containing another substance (solute), which is distributed uniformly
10. hexose: a monosaccharide containing six carbon atoms
11. a
12. c
13. d
14. d
15. c
16. 18 daltons
17. 1×10^{-5} moles/liter
18. 75 mOsm

19. A peptide bond is formed between the alpha carboxyl group (—COOH) of one amino acid and the alpha amino group (—NH$_2$) of the other amino acid. The peptide bond has the structure —CONH—.

20. (a) Hydrogen bond formation between two adjacent peptide bonds or between polar side chains of amino acids. (b) Hydrophobic interactions between nonpolar side chains of amino acids. (c) Disulfide bonds between —SH groups of cysteine amino acids.

21. The two DNA strands in the double helix are not identical. The base adenine in one strand always lies opposite thymine in the other strand. Likewise, cytosine in one strand is always opposite guanine in the other strand.

22. The crystal structure of KCl, a hydrophilic compound, is stabilized by ionic bonds. Small polar water molecules enter the crystal structure and form shells around the K$^+$ and Cl$^-$ ions. This breaks the ionic bonds, and the K$^+$ and Cl$^-$ ions separate or dissociate from one another.

23. An enzyme is a protein which lowers the activation energy level for a chemical reaction. It provides binding sites (active sites) on its surface so that reactant molecules can be brought close together and in the proper orientation for the reaction to occur. These encounters between reactant molecules do not occur at a high rate in the absence of the enzyme.

Chapter 3

1. cytosol
2. chromatin
3. exocytosis
4. cytoskeleton
5. cytokinesis
6. eukaryotic cell: type of cell characterized by the presence of a well-defined nucleus confined within a nuclear membrane
7. genome: the sum of all the different chromosomal genes
8. autoradiography: a technique for determining the location of radioactive molecules in cells and tissues
9. density gradient centrifugation: a procedure that allows separation of cell organelles based on differences in their densities
10. recombinant DNA: a DNA molecule formed experimentally by combining pieces of DNA from different organisms
11. d

12. b
13. a
14. b
15. c
16. c
17. (a) Degradative enzymes can be stored safely within lysosomes. (b) Physical separation of synthetic and degradative pathways allows independent control. (c) Some newly synthesized molecules can be processed further and sorted according to final destinations.
18. DNA binds to histone proteins, and parts of the double helix wrap around the histones to form a nucleosome, the basic packing unit. Close packing of nucleosomes forms a chromatin fiber. This, together with futher levels of folding, allows the double helix to be packed in a highly organized and compact way.
19. (a) Ribosomal RNA joins with other proteins to form ribosomes. (b) Messenger RNA contains information about the sequence of amino acids which must be joined together at the ribosomes. (c) Transfer RNA brings the required amino acids to the ribosomes.
20. Aerobic respiration is the main metabolic pathway for extracting the energy stored in fuel molecules. When the energy is released, it is trapped in the form of ATP. Aerobic respiration is completed in the mitochondrion, which is the major site of ATP production within the cell.
21. Primary cultures are formed by transferring cells directly from a tissue sample to culture medium. Most of the cells will divide and grow only a limited number of times in culture before dying. In contrast, cells from a cell line will divide and grow indefinitely in culture.

Chapter 4

1. integral (or intrinsic)
2. osmosis
3. facilitated diffusion
4. antiport
5. regulatory volume decrease
6. amphipathic molecule: a molecule, such as a phospholipid or intrinsic membrane protein, that has both hydrophobic and hydrophilic regions in its structure
7. Overton's rules: (a) Membrane permeability to small nonpolar solutes is directly proportional to their lipid solubility. (b) Membrane permeability to polar solutes is inversely proportional to their molecular size.

8. diffusion trapping: the neutral form of a weak base (or acid) diffuses across a cell membrane, but the charged form will not. The neutral form can be "trapped" on one side of a membrane if it reverts to its charged form after crossing the membrane.

9. secondary active transport: carrier-dependent transport of solute against the electrochemical gradient. Solute transport is not directly coupled to the energy-yielding reaction.

10. tight epithelium: an epithelium that does not permit leakage of water and solutes between the cells. This is because the cell junctions are highly impermeable.

11. c

12. a

13. d

14. d

15. b

16. a

17. (a) The core of the plasma membrane is a double layer of phospholipid molecules. (b) Proteins are closely associated with, and often embedded in, the lipid bilayer. (c) The interior of the lipid bilayer is like a viscous fluid. (d) Both phospholipid and protein molecules move laterally within the bilayer. (e) Both phospholipids and proteins are distributed asymmetrically between the two halves of the bilayer.

18. Amino acids form only linear chains, but monosaccharides can be assembled in branched chain structures. This provides almost unlimited possibilities for making a membrane receptor that will bind only one specific type of extracellular molecule.

19. The normal erythrocyte has a biconcave shape. When water enters from a hypo-osmotic solution, the cell can swell to become more like a sphere. In this way the cell can accommodate a large increase in its normal volume.

20. This describes the response of mammalian cells to an increase in intracellular volume. The increase in volume activates mechanisms which, in most cells, increase the efflux of K^+ and Cl^- from the cell. This decreases intracellular osmolarity and stimulates efflux of water, returning cell volume to normal.

21. The tight junctions between the cells of a leaky epithelium allow passage of small ions and water. This "leak" pathway allows paracellular transport across the epithelium in addition to transcellular transport. The tight junctions of a tight epithelium do not permit leakage between the cells. Transcellular transport is the only pathway across a tight epithelium.

Chapter 5

1. degenerate

2. semiconservative

3. intron

4. gap junction

5. ligand

6. start codon: the mRNA codon that is involved in initiating translation of the nucleotide triplets in the mRNA molecule

7. RNA polymerases: the group of enyzmes that use a DNA template to synthesize molecules of RNA

8. post-translational modifications: alteration of newly synthesized polypeptides, e.g., to produce a biologically active polypeptide from a larger precursor

9. antagonist: a ligand that binds with high affinity to a receptor but produces no response

10. cyclic AMP phosphodiesterase: the enzyme that rapidly breaks down intracellular cyclic AMP

11. c

12. d

13. b

14. d

15. a

16. a

17. c

18. This is possible because the genetic code is a triplet code. The four different nucleotides, when combined in groups of three, allow 64 different combinations (4^3). A doublet code would provide only 16 different combinations (4^2), which is not sufficient for 20 amino acids.

19. (a) RNA synthesis requires ribose instead of deoxyribose, (b) uracil instead of thymine, (c) and RNA polymerase instead of DNA polymerase. (d) RNA synthesis produces a single nucleotide strand instead of two, and (e) the DNA template is not incorporated into the RNA product.

20. The reason is that expression of the genetic information in a cell is controlled so that different genes are expressed in different cell types. Most of the genetic information in any cell remains suppressed and is not used. The genes which are being expressed, in the form of specific protein products, may change during the life cycle or in response to a specific stimulus such as a hormone.

21. An agonist is any ligand which is capable of both binding to a receptor and producing a physiological response. An antagonist will bind to the same receptor but cannot evoke a response. By occupying the receptors, the antagonist will block the action of the agonist.

22. (a) Local mediators (e.g., prostaglandins), (b) neurotransmitters, and (c) hormones.
23. A common mechanism is to activate a protein kinase, an enzyme which will catalyze addition of phosphate groups to specific protein targets. Phosphorylation often changes the function of the protein. For example, if the protein target is another enzyme, it may be activated (or inactivated) by the phosphorylation step.

Chapter 6

1. anabolism or biosynthesis
2. endergonic reaction
3. oxidized nicotinamide adenine dinucleotide (NAD^+)
4. gluconeogenesis
5. urea
6. chemiosmotic mechanism: synthesis of ATP by a process that is driven by the flow of protons across the inner mitochondrial membrane
7. pentose phosphate pathway: a cytosolic pathway for glucose oxidation, which is important in cells heavily involved in biosynthesis. The pathway produces no ATP but generates NADPH to drive biosynthetic reactions.
8. oxidative phosphorylation: synthesis of ATP by coupling oxidation to phosphorylation of ADP. The oxidation steps occur in the mitochondrial electron transport chain as electrons are transferred from reduced coenzymes to O_2.
9. allosteric inhibitor: a compound that inhibits an enzyme activity by binding to a site on the enzyme which is separate from the active site. Binding of the inhibitor produces a conformational change in the active site which impairs its function.
10. high–energy phosphate compound: a compound with a strong tendency to transfer a phosphate group to an acceptor. This transfer is accompanied by release of a large amount of free energy.
11. a
12. d
13. c
14. c
15. b
16. (a) 0   (b) 12
17. (a) 31   (b) 15
18. NADH is oxidized by the mitochondrial electron transport pathway to drive synthesis of ATP. Oxidation of NADPH is used to drive reactions in various biosynthetic pathways in the cytosol, e.g., synthesis of fatty acids and steroids.
19. Catabolism of proteins, carbohydrates, and fats. Biosynthesis of fatty acids, cholesterol, and steroids.
20. When the supply of ATP is high, ATP acts as an allosteric inhibitor of citrate synthetase to reduce the rate of formation of citric acid. This reaction is the entry point in the cycle for acetyl groups from acetyl coenzyme A.
21. (a) Plants use light from the sun to make ATP by photosynthesis. (b) ATP is used by plant cells during CO_2 fixation. (c) This process produces high-energy organic compounds and O_2. (d) These organic "fuel" molecules are a source of chemical energy and building-block molecules for animals that eat the plants. (e) In animals the molecules are catabolized by cellular respiration, a process which requires O_2, and the chemical energy is trapped as ATP. (f) Water and CO_2 are produced as by-products.
22. A mobile animal, unlike a plant, must be able to carry its energy store with it. The amount of ATP produced by a given weight of fat far exceeds that produced by the same weight of carbohydrate. In addition, more fat can be stored than carbohydrate. Thus, it is much more efficient for an animal to use fat as an energy store.

Chapter 7

1. cation
2. anion
3. depolarization
4. hyperpolarization
5. membrane potential: the voltage across the membrane that results from an imbalance in electrical charge between the inside and outside of the cell
6. equilibrium potential: the potential at which the influx of the ion across the membrane is equal to its efflux so that there is no net current flow
7. action potential: an electrical impulse generated by nerve cells and some muscle cells. It is characterized by a depolarization that quickly reaches a peak positive value, which is rapidly followed by a repolarization of the membrane.
8. saltatory conduction: the apparent jumping of the action potential from one node of Ranvier to the next as it propagates along the axon
9. d
10. d
11. c
12. b
13. −85.7 mV
14. 75 meters per second

15. During the absolute refractory period the potassium channels are in the open state, causing the membrane to be hyperpolarized. In addition, the sodium channel inactivation gates are in the closed state. Together these mechanisms prevent the depolarization of the membrane to threshold levels.

16. Increases in membrane conductance to Na^+ ions and decreases to K^+ can cause EPSPs.

17. Increases in membrane conductance to K^+ and Cl^- ions as well as decreases to Na^+ can cause IPSPs.

18. Temporal summation is the process by which postsynaptic potentials are added together over a period of time. If the additive postsynaptic changes result in a depolarization of the membrane above threshold levels, an action potential will be initiated.

19. Ions will move from a region of high concentration to a region of lower concentration. Anions will move to regions that are positively charged, while cations will move to negatively charged regions.

Chapter 8

1. ocular dominance columns
2. proprioceptor
3. slowly adapting receptor
4. somatic sensory map
5. generator potential: a change in the membrane potential of a sensory cell in response to a sensory stimulus
6. conductive hearing loss: results from damage to the bones of the middle ear
7. fovea: the most sensitive part of the retina where all of the photoreceptors are cones
8. labeled-line code of sensory stimuli: different types of sensory information are transmitted along different pathways.
9. a
10. d
11. c
12. b
13. The transduction of sensory information requires an adequate or effective stimulus, which causes an alteration in membrane conductance. The change in membrane conductance, in turn, produces a generator potential, leading to the generation or inhibition of action potentials.
14. Stimulus intensity is encoded in the nervous system by a frequency code and by a population code.
15. Receptive fields are regions on a transducing surface that activate and/or inhibit a neuron with the presence of the sensory stimulus upon the transducing surface.
16. The trichromatic theory of color vision states that the total spectrum of colors that we perceive can be explained by the differential absorption of photons by three types of photoreceptors, which optimally absorb photons with wavelengths of 443 nm, 535 nm, and 570 nm, respectively.

Chapter 9

1. flexor
2. cortical efferent zone
3. muscle spindle
4. semicircular canal
5. extensor: a muscle that increases joint angle
6. alpha motor neuron: a nerve cell in the spinal cord that directly innervates and activates skeletal muscles
7. Babinski sign: a fanning of the toes when the sole of the foot is stroked; is an indication of a lesion of the corticospinal tract
8. utricle: otolithic organ that detects linear acceleration and is also capable of detecting the position of the head with respect to gravity
9. a
10. b
11. c
12. c
13. Alpha motor neurons can increase the force of muscle contraction by increasing their frequency of discharge and/or by increasing the number of alpha motor neurons that are activated.
14. Alpha motor neurons that innervate fast skeletal muscles have rapid conduction velocities and high frequencies of action potential discharge. Those that innervate slow skeletal muscles have slower conduction velocities and lower discharge rates.
15. Gamma motor neurons maintain muscle spindle sensitivity as the extrafusal muscle fiber is shortening.
16. Motor units are composed of a single alpha motor neuron and the number of individual muscle fibers that it innervates. A motor neuron pool is a cluster of alpha motor neurons that innervate the same skeletal muscle.

Chapter 10

1. "fight-or-flight" response
2. epinephrine and norepinephrine
3. acetylcholine

4. norepinephrine
5. sympathetic nervous system: a subcomponent of the autonomic nervous system that coordinates the body's response to stress
6. parasympathetic nervous system: coordinates the body's response for vegetative and digestive functions
7. referred pain: the sensation of pain that appears to come from the surface of the body but the source of the pain originates from an internal organ
8. viscera: internal organs of the body
9. b
10. d
11. c
12. a
13. The optimal conditions for cellular function occur at a temperature of 37°C, pH 7.4, and osmolarity of 300 mOsm. These conditions are maintained by central autonomic regulatory systems in the hypothalamus and brain stem. Thermo- and osmoreceptors in the hypothalamus detect changes in temperature and osmolarity, while chemoreceptors in the brain stem monitor pH.
14. The urge to urinate is detected by stretch receptors in the bladder. Information conveyed by these receptors enters the spinal cord and activates parasympathetic pelvic nerve fibers, resulting in the contraction of the smooth muscles of the bladder wall. In addition there is a coordinated inhibition of sympathetic and somatic outflow, resulting in the relaxation of the internal and external sphincters, respectively.
15. The urge to consume water is initiated by osmoreceptors in the hypothalamus, which monitors plasma osmolarity. Increases in osmolarity above 300 mOsm activate somatic motor responses, leading to drinking behavior and autonomic responses resulting in the retention of water.
16. The hypothalamus coordinates the physiological changes seen in the "fight-or-flight" response. These bodily alterations act in concert to provide more oxygen to the skeletal muscles and brain by increasing heart rate and contractility and the depth of respiration. Changes also occur to provide more fuel and energy for critical organ systems.

Chapter 11

1. dominant hemisphere
2. planum temporale
3. narcolepsy

4. raphe nucleus
5. synaptic facilitation: the process that increases neurotransmitter release
6. synaptic habituation: the process that decreases neurotransmitter release
7. associative learning: the process by which an understanding is acquired between two stimuli, as in classical conditioning, or a behavioral response and a reward, as in operant conditioning
8. REM sleep: a period of sleep associated with the state of dreaming during which the eyes are rapidly moving
9. b
10. d
11. c
12. a
13. There are five stages of sleep: Stages I–IV and REM. REM sleep is associated with dreaming.
14. Several areas of the brain are involved in the sleep process. The preoptic area of the hypothalamus controls non-REM sleep, while the raphe nucleus controls REM sleep. The suprachiasmatic nucleus, the body's biological clock, has inputs to the preoptic area to control the wake–sleep cycle.
15. Prenatal exposure to androgens during critical periods of development can alter adult sexual behavior. Fetal male brains that have not been exposed to androgens result in "feminized" sexual responses. Fetal female brains that have been exposed to androgens result in "masculinized" sexual responses.
16. Short-term memory can be easily disrupted and limited in time, while long-term memory is more stable and long-lasting. In addition, long-term memory and its consolidation from short-term memory require protein synthesis.

Chapter 12

1. autocrine, or paracrine
2. neuroendocrine cell
3. cholesterol
4. radioimmunoassay, or RIA
5. adenylate cyclase
6. inositol trisphosphate (IP_3) and diacylglycerol (DAG)
7. hormone receptor: a chemical entity, usually a protein, that is present either on or in a cell that specifically recognizes a hormone. Binding of the hormone to the receptor triggers a series of events that result in the specific biological effects characteristic of that hormone.
8. neuroendocrinology: that branch of science that

deals with the study of the interaction of the nervous and endocrine systems

9. mineralocorticoids: a class of steroid hormones produced by cells of the adrenal cortex that act to regulate sodium balance by the kidneys

10. conjugation of steroids: the process by which steroid metabolites are coupled to a sulfate or glucuronide group, resulting in improved solubility of the conjugated steroid and facilitation of its excretion.

11. G-proteins: a group of GTP-binding proteins that act to couple cell surface receptors with their respective intracellular signal transduction systems. In particular, stimulation of G_s proteins by the appropriate receptor results in activation of adenylate cyclase.

12. receptor-mediated endocytosis: the process by which peptide hormones are taken up by their target cells after binding to specific cell surface receptors. Typically, receptor-mediated endocytosis results in hormone degradation.

13. phosphodiesterase: a cellular enzyme that is responsible for the conversion of cAMP to AMP and thus the termination of intracellular signals mediated by an increase in cAMP

14. b
15. c
16. d
17. d
18. b
19. e
20. a
21. b
22. d
23. e
24. b

25. Receptors act as signal transducers in the endocrine system by converting extracellular hormone signals into intracellular signals that ultimately result in altered cell function. Specificity results from the fact that receptors typically only recognize and bind one hormone, thus preserving the faithfulness of the signalling process.

26. Steroid and thyroid hormones readily enter their target cells and bind to specific receptors in or on the nucleus. The hormone-receptor complex interacts with DNA and alters the transcription of a few specific genes coding for proteins responsible for the actions of that particular hormone.

27. Signal amplification is important in hormone action because hormones circulate in the blood in exceedingly low concentrations. Amplification occurs in many hormone action pathways as a result of a cascade of enzymatic steps in which the number of signal molecules increases at each step due to enzyme activation.

28. Hormones that modulate intracellular cAMP interact with cell surface receptors that are functionally coupled to adenylate cyclase through either stimulatory (G_s) or inhibitory (G_i) G-proteins. Adenylate cyclase forms cAMP from ATP, and cAMP alters cell function by activating specific cAMP-dependent protein kinases. Several forms of cAMP-dependent protein kinase are known, each with its own particular protein substrates. Changes in intracellular cAMP in different cells may lead to different effects, depending on the particular form of cAMP-dependent protein kinase that is present and on the particular protein substrates that are present.

29. Activation of certain cell surface receptors results in the activation of a form of phospholipase C which cleaves phosphatidylinositol 4, 5-bisphosphate (PIP_2) into diacylglycerol (DAG) and inositol trisphosphate (IP_3). IP_3 acts as a second messenger inside the cell by promoting the release of calcium from intracellular stores, and DAG activates a membrane-bound protein kinase, C-kinase, which catalyzes the phosphorylation of specific cellular proteins, leading to alterations in their activity and specific changes in cell function.

Chapter 13

1. Rathke's pouch
2. median eminence
3. somatostatin
4. diabetes insipidus
5. acromegaly
6. thyroglobulin
7. the hypothalamic-pituitary portal system: a vascular system consisting of a primary capillary network in the median eminence of the hypothalamus and a secondary capillary network in the anterior pituitary, which are connected by long portal veins. The pituitary portal system carries factors from the hypothalamus to the anterior pituitary that serve to regulate the secretion of anterior pituitary hormones.
8. corticotrophs: ACTH-secreting cells of the anterior pituitary; normally they comprise about 20% of the cells of the gland
9. Gonadotropins: a collective term that refers to both LH and FSH, two hormones secreted by the anterior pituitary, which are of primary importance in regulating gonadal function

10. Somatomedins: a collective term referring to growth factors that are responsible for many of the growth-promoting actions of growth hormone. The principal somatomedins in the body are IGF-I and IGF-II.
11. "iodide trap": the active transport mechanism present on the basal surface of thyroid follicle cells, which concentrates iodide within the thyroid gland
12. a
13. b
14. a
15. d
16. c
17. d
18. c
19. d
20. b
21. The posterior pituitary secretes two hormones (oxytocin and ADH), which are synthesized in neuroendocrine cells whose cell bodies are located in the hypothalamus. These hormones are carried within the axons of the neuroendocrine cells from the hypothalamus to the anterior pituitary where they are released from the nerve terminals. From there, they enter capillaries and are carried to the general systemic circulation. Anterior pituitary hormone secretion is under the control of hypothalamic release/inhibiting factors carried from the median eminence of the hypothalamus to the anterior pituitary via a vascular system referred to as the hypothalamic-pituitary portal system.
22. The two posterior pituitary hormones, oxytocin and ADH, are synthesized in cell bodies of neuroendocrine cells in the hypothalamus and are transported via their axonal projections to the posterior pituitary. Upon appropriate stimulation of these neuroendocrine cells, the hormones are released from nerve termini in the posterior pituitary, where they enter capillaries and are carried to the general systemic circulation.
23. The two primary stimuli for ADH secretion are (1) an increase in osmolality of the blood or extracellular fluid and (2) a large decrease in blood volume. The primary effect of ADH is to increase water retention by the kidneys.
24. The hypothalamic-pituitary-thyroid axis consists of three hormones: TRH, TSH, and T_4. Hypothalamic TRH stimulates TSH secretion by the anterior pituitary; TSH in turn stimulates T_4 secretion by the thyroid gland. In addition to its biological effects, T_4 has a negative feedback effect on the anterior pituitary to inhibit TSH secretion.
25. The shortened stature associated with dwarfism results from a deficiency in GH secretion or action prior to puberty. Gigantism results when increased GH secretion occurs during childhood; acromegaly results when overproduction of GH begins after puberty. Persons exhibiting gigantism have disproportionately long arms and legs, owing to the overstimulation of the bones in the limbs. Persons with acromegaly typically exhibit an enlarged jawbone, hands, and feet, owing to the continued stimulation of these bones, even after the closure of the epiphyseal plates of the long bones.
26. Thyroid hormones generally promote glycogen formation in liver and glucose uptake into adipose tissue and muscle. Thyroid hormones promote all aspects of lipid metabolism but in excess tend to promote the loss of most fat stores in the body. The effects of thyroid hormones on protein metabolism are biphasic: At normal physiological concentrations, thyroid hormones increase protein synthesis and promote an overall accumulation of protein in the body; when present in excess, thyroid hormones promote an overall loss of protein.

Chapter 14

1. zona fasciculata
2. aldosterone
3. renin
4. Cushing's syndrome
5. tyrosine hydroxylase
6. chromaffin cells
7. zona glomerulosa: the outermost zone of the adrenal cortex. It is the predominant site of synthesis of aldosterone in the body.
8. corticotropin-releasing hormone: a hypothalamic peptide responsible for the regulation of ACTH secretion by the anterior pituitary
9. pharmacological effects of glucocorticoids: those produced by high concentrations of the hormone; these effects can be classified primarily as being immunosuppressive or anti-inflammatory in nature.
10. angiotension II: an eight amino acid peptide hormone formed from angiotensin I, which acts to regulate aldosterone secretion from the zona glomerulosa of the adrenal cortex
11. Addison's disease: a disorder resulting from insufficient steroid hormone production by the adrenal cortex

12. lipolysis: the process whereby stored triglycerides are broken down into their component glycerol and free fatty acids
13. renin-angiotensin system: a multi-step process that serves to couple changes in blood pressure or volume to appropriate changes in aldosterone secretion. Several tissues, enzymes, and peptide or protein factors are involved.
14. c
15. b
16. d
17. b
18. b
19. d
20. c
21. "Tropic" effects refer to the fact that ACTH promotes the growth of all three zones of the adrenal cortex. The steroidogenic effect of ACTH involves the regulation of cortisol secretion by cells of the zona fasciculata.
22. Corticotropin-releasing hormone is produced by the hypothalamus and regulates ACTH secretion by the anterior pituitary; ACTH in turn regulates cortisol secretion by the adrenal cortex. Various inputs from higher brain centers serve to regulate CRH production by the hypothalamus. Negative feedback regulation by cortisol occurs at the level of the hypothalamus and anterior pituitary to inhibit CRH and ACTH secretion, respectively, and ACTH itself has negative feedback effects in the hypothalamus to inhibit CRH release.
23. Addison's disease results from a deficiency in hormone secretion by the adrenal cortex and can arise as a result of a lesion in the adrenal cortex itself or from a defect in the hypothalamus or anterior pituitary. In those individuals in whom the problem resides in the adrenal cortex and pituitary function is normal, ACTH secretion will increase owing to the lack of negative feedback effects of cortisol on ACTH secretion. Because ACTH has some homology to MSH, high circulating levels of ACTH will stimulate melanocytes of the skin and excess pigmentation can occur.
24. The adrenal medulla and the sympathetic nervous system, sometimes referred to collectively as the sympathoadrenal system, have influence over a number of organ systems in the body and are responsible for activation of the "fight or flight" responses. Functionally, cells of the adrenal medulla are analogous to postganglionic fibers of the sympathetic nervous system, but rather than secreting catecholamines into a synapse, these cells secrete epinephrine directly into the bloodstream.
25. The protein-catabolic and lipid-mobilizing effects of cortisol lead to a loss of muscle mass and fat in the extremities with a redistribution of fat to the trunk, shoulder blades, and face, which result in the physical appearance of thinned arms and legs, a large abdomen, and a round face. The catabolic effects of excess cortisol also leads to a loss of connective tissue in skin and vascular beds, resulting in the formation of striae, reddening of the cheeks, and easy bruising.

Chapter 15

1. islets of Langerhans
2. beta cells of the pancreas
3. proinsulin
4. lipoprotein lipase
5. gluconeogenesis
6. Type II diabetes, or NIDDM
7. exocrine pancreas
8. C-peptide: the peptide segment of proinsulin that connects what will eventually form the A-chain and B-chain of insulin. It is removed from proinsulin by proteolytic cleavage and is secreted from beta cells in equal amounts as insulin.
9. hormone-sensitive lipase: the enzyme in adipose tissue and liver that breaks triglycerides into fatty acids and glycerol. It is inhibited in adipose tissue by insulin and stimulated in liver by glucagon.
10. glycogenolysis: the process of breaking down stored glycogen into glucose-6-phosphate, which can be used for energy or converted into glucose and released into the blood.
11. ketogenesis: the name for the process of ketone body formation from fatty acids. It occurs primarily in liver and is stimulated by glucagon.
12. caloric density: the number of calories in a substance per unit weight of the substance
13. Type I diabetes mellitus (IDDM): the form of diabetes generally thought to be caused by an autoimmune disorder that leads to a lack of circulating insulin. Patients with Type I diabetes are typically treated with insulin injections.
14. sulfonylureas: a group of oral drugs often used in the treatment of Type II diabetes mellitus (NIDDM). Sulfonylureas promote insulin action in target tissues and insulin secretion from the pancreas.
15. hyperglycemia: the term used to describe a higher than normal blood glucose concentration

16. b
17. b
18. a
19. d
20. d
21. a
22. d
23. d
24. The insulin-secreting beta cells are most numerous and are located toward the center of the islet; glucagon-secreting alpha cells are located around the outside of the islet; and somatostatin-secreting delta cells tend to be located between alpha and beta cells. This arrangement of cells may serve to allow paracrine interactions among pancreatic hormones, as in the inhibition of insulin and glucagon secretion by somatostatin.
25. In muscle, insulin stimulates amino acid uptake, stimulates protein synthesis, and inhibits protein degradation. Its overall effect is therefore to promote accumulation of muscle protein.
26. Pure triglycerides have about twice the caloric density as does glycogen. In addition, glycogen is hydrophilic, and thus considerable water is found associated with stored glycogen, greatly reducing the number of calories for a given weight of stored fuel as compared to triglycerides.
27. Type I diabetes is characterized by little or no circulating insulin as a result of autoimmune destruction of beta cells. Type II diabetics have apparently adequate circulating insulin levels, but the target cells do not respond to the insulin that is present. Both forms of diabetes are characterized by hyperglycemia.
28. Acute complications of diabetes include hyperglycemia with glucosuria, increased urine volume and frequency of urination, and primarily in Type I diabetics also ketoacidosis and electrolyte imbalance.
29. In insulin target tissues such as adipose and muscle, glucose transporters exist both in an active form in the plasma membrane and in an inactive pool inside of the cell. Insulin stimulates the movement of glucose transporters from their intracellular site to the membrane, thus making a greater number of active transporters available and increasing glucose uptake as a result.

Chapter 16

1. sarcomere
2. myofilaments

3. smooth muscle (visceral muscle, involuntary muscle)
4. isometric contraction
5. myoneural junction (motor end-plate)
6. ATP (adenosine triphosphate)
7. force-velocity curve
8. twitch: a single brief contraction of a skeletal muscle in response to a single stimulus
9. tetanus: a series of twitch contractions that occur so close together in time that relaxation between them does not take place, and the force produced is a summation of the forces of the individual twitches
10. sarcoplasmic reticulum: the membranous cellular organelle in muscle that is responsible for the release and uptake of the calcium ions that regulate the activity of the contractile proteins
11. troponin: a complex of three proteins that can bind calcium ions and regulate the interaction of the contractile proteins in skeletal and cardiac muscle
12. crossbridges: the projections of the myosin molecules from the thick myofilaments that make contact with the thin (actin) myofilaments. The relative motion of the crossbridges and the actin myofilaments produces the muscle contraction, and the association between the actin and myosin molecules is the site where the energy of ATP is released.
13. end-plate potential: the electrical response in the postsynaptic (end-plate) membrane of skeletal muscle. It occurs when the membrane channels become open to sodium and potassium ions through the action of the acetylcholine that has been released from the presynaptic (nerve) membrane.
14. acetylcholine esterase: an enzyme located in the postsynaptic region of the myoneural junction that hydrolyzes acetylcholine and terminates its action on the receptor sites of the ion channels
15. rigor mortis: a rigidity of muscle that occurs when ATP is not available to bind with myosin crossbridges to allow their release from the actin filaments. It occurs progressively after death as energy stores become depleted.
16. myoglobin: a protein found predominantly in red skeletal muscle fibers that is capable of binding oxygen in a manner similar to that found in hemoglobin
17. disuse atrophy: a progressive loss of muscle tissue that occurs when a muscle is prevented from contraction by immobilization, denervation, and so on.
18. b
19. b

20. a
21. a
22. a
23. b
24. c
25. c
26. b
27. b
28. a
29. c
30. The three types of muscle are (a) skeletal (striated, voluntary), (b) cardiac (heart), and (c) smooth (nonstriated, involuntary, visceral).
31. Skeletal muscle undergoes hypertrophy in response to a long-term increase in its work load or contractile activity, and atrophy occurs when the muscle is used very little or is immobilized.
32. The speed of contraction of cardiac muscle lies between that of the other two types; it is slower than most skeletal muscles and faster than smooth muscle.
33. Curare mimics acetylcholine in that it can bind to the postsynaptic receptors and prevent the binding of any other molecule; however, it cannot cause the channels to open, and it is not broken down by acetylcholinesterase.
34. During a tetanus a skeletal muscle produces more force because relaxation from the individual twitches is prevented, and the force due to subsequent stimulations adds to the force already present.
35. In all types of muscle, relaxation is associated with a reduction in the number of calcium ions that are free to interact with the regulatory proteins. In some cases the calcium is taken back up into the sarcoplasmic reticulum, while in other cases it is actively transported out of the cells.

Chapter 17

1. hematocrit
2. colloidal osmotic pressure
3. hypotonic solution
4. hemolysis
5. hematopoiesis
6. polycythemia
7. diapedesis
8. Mean Corpuscular Hemoglobin Concentration: an index of the proportion of hemoglobin (weight/volume) measured per average red cell in a sample of blood. Normal adult value is 34 ± 2%.
9. mean corpuscular volume: an index of average red cell volume. Normal adult value is 87 ± 5 μm^3.

10. Mean Corpuscular Hemoglobin: an index of the average concentration of hemoglobin per red cell. Normal adult value is 29 ± 2 μg.
11. anemia: a condition in which the number of circulating red cells and/or the amount of hemoglobin in the blood are below normal, resulting in reduced oxygen-carrying capacity of the blood
12. Low Density Lipoproteins: transport proteins in the plasma that favor the deposition of cholesterol in arterial walls and thereby contribute to the development of vascular disease
13. hemostasis: refers to mechanisms that minimize or prevent the loss of blood. Four interrelated events constitute hemostasis: (1) local vasoconstruction, (2) platelet clumping, (3) formation of a clot, and (4) clot retraction and dissolution.
14. phagocytosis: a process whereby a cell, called a phagocyte, recognizes, ingests, and enzymatically breaks down (digests) foreign substances
15. complement: a set of plasma proteins that interact in a defined sequence to promote lysis of microbes and release of chemical signals that trigger other defensive reactions in response to foreign substances
16. IgG: the major immunoglobulin in the plasma; helps provide immunity against bacteria, viruses, and fungi, as well as chemical defense against toxins
17. c
18. b
19. e
20. d
21. c
22. e
23. e
24. $4/9 \times 100 = 44\%$
25. 16 g/dl × 1.34 ml O_2/gm = 21.4 ml O_2/dl
26. $(45 \times 10) \div 5 = 90 \ \mu$m^3
27. (5 million/mm^3)(1000 mm^3/cm^3)(1000 cm^3/liter) (5 liters) = 15×10^{12} cells $(25 \times 10^{12}$ cells) ÷ $(23 \times 10^{10}$ cells/day) = 109 days
28. Polycythemia increases blood viscosity, which increases cardiac work and blood pressure, thereby increasing the possibilities of hemorrhage and cardiac failure.
29. In the cell-mediated immune response, the helper cell assists in the development of cytotoxic and suppressor cells. In the antibody-mediated immune response, the helper cell assists in the development of plasma cells and memory cells.
30. The presence of antigens other than A and B on the erythrocyte membrane precludes the "uni-

versal recipient'' designation. For example, an AB$^-$ recipient cannot receive A$^+$, B$^+$, O$^+$, or AB$^+$ whole bloods, because the recipient lacks the Rh antigen (D).

31. Minor agglutination occurs when a recipient receives antibodies from a donor's blood that react with the recipient's red cells antigens; for example, an A$^+$ recipient receiving O$^+$ blood containing anti-A. Major agglutination occurs when a recipient receives antigens on donor red cells for which the recipient has antibodies in the plasma; for example, an A$^+$ recipient receiving B$^+$ blood.

Chapter 18

1. contractility or change in contractility
2. stroke volume
3. isovolumetric contraction
4. end-diastolic volume
5. pacemaker
6. blood flow: the volume of blood that moves past a particular point in the cardiovascular system during a given interval of time

$$flow = volume/time$$

7. systole: that portion of the cardiac cycle that occurs after the first heart sound and before the second heart sound
8. Frank-Starling law of the heart: An increase in end-diastolic volume will result in an increase in stroke volume, whereas a decrease in end-diastolic volume will decrease stroke volume.
9. electrocardiogram (also ECG or EKG): an electrical recording, made at the body surface, that provides a record of the electrical activity of the heart. It is produced by very small electrical currents caused by the action potentials of the atrial and ventricular muscles.
10. arrhythmia: a disturbance of the normal rate or pathway of the heartbeat
11. c
12. a
13. b
14. c
15. a
16. a
17. d
18. b
19. a
20. d
21. stroke volume = end-diastolic volume − end-systolic volume

    70 ml = 100 ml − 30 ml

22. cardiac output = stroke volume × heart rate

    5600 ml/min = 70 ml/beat × 80 beats/min

23. If the interval between beats is 0.8 second, the rate is 1 beat/0.8 second, or 1.25 beats/second. Since there are 60 seconds in a minute, the rate is:

    1.25 beats/second × 60 seconds/minute,

    or 75 beats/minute.

24. Vasoconstriction increases the resistance for blood flow to a particular tissue bed and therefore decreases blood flow to that tissue. If blood pressure is below normal, as with hemorrhage, vasoconstriction would increase total peripheral resistance and consequently increase blood pressure; this would then increase flow to tissues such as the heart and brain.
25. At normal heart rates, there is sufficient time for complete filling of the ventricle before the atria are activated. At higher heart rates, ventricular filling time is reduced and atrial contraction occurs before the ventricles have filled completely. The atrial contraction accelerates the flow of blood into the incompletely filled ventricle.
26. Contractility changes are primarily the result of changes in the intracellular calcium concentration.
27. SA node to atrial muscle to AV node to common bundle (or Bundle of His) to right and left bundle branches to Purkinje fibers to ventricular muscle.
28. No, because the EKG measures electrical activity only and cannot directly determine anything about the mechanical function of the heart; however, it can provide information by which a physician can make inferences about cardiac function.

Chapter 19

1. reactive hyperemia
2. tone or vascular tone
3. myogenic activity
4. precapillary sphincter
5. carotid baroreceptor
6. colloidal osmotic pressure: that portion of the total osmotic pressure of blood due to plasma proteins that cannot readily diffuse across the capillary wall. This is also referred to as the "oncotic pressure."
7. mean arterial pressure: a calculated pressure that provides the same effective driving force as a particular pulsatile pressure. For example, the blood flow through a particular artery in response to a pulsatile pressure of 120/90 mm Hg

would be about the same as the calculated mean pressure to 100 mm Hg.

mean pressure = 1/3 pulse pressure + diastolic pressure

8. Starling's hypothesis: a mathematical description of the balance of hydrostatic and osmotic pressures across the capillary endothelium

9. vasomotion: the contraction and relaxation of arteriolar and precapillary sphincter smooth muscle. Vasomotion is strongly influenced by local tissue factors.

10. cardiovascular control center: region in the medulla that coordinates the actions of the heart and blood vessels to maintain a constant blood pressure. Afferent inputs from the baroreceptors initiate appropriate changes in sympathetic and parasympathetic efferent nerve activity to the heart and blood vessels to maintain normal arterial blood pressure.

12. d
13. b
14. d
15. c
16. b
17. a
18. d
19. c
20. d

21. mean pressure = 1/3 pulse pressure + diastolic pressure

100 mm Hg = 1/3 30 mm Hg + 90 mm Hg

22. total blood flow = MAP/TPR = 100 mm Hg/20 mm Hg · min · liter^{-1} = 5 liters/min

23. net pressure = capillary hydrostatic + interstitial fluid colloidal osmotic pressure − capillary colloidal osmotic pressure
= 35 mm Hg + 3 mm Hg − 28 mm Hg
= +10 (positive = net filtration)

24. pulse pressure = systolic pressure − diastolic pressure

diastolic pressure = systolic pressure − pulse pressure

diastolic pressure = 120 mm Hg − 50 mm Hg = 70 mm Hg

25. During systole the left ventricle rapidly ejects blood into the elastic aorta, which expands to accommodate the increased volume. During diastole, when the ventricle is no longer pumping blood, the walls of the arteries recoil and pump blood, momentarily stored in the arterial system, to the capillaries.

26. The distribution of blood is determined by the relative resistances of the blood vessels leading to any particular bed. Tissue beds with higher resistances will receive less blood than those with less resistance.

27. An increase in arterial pressure will stretch the arterioles and associated smooth muscle. The stretch will activate smooth muscle contraction (myogenic activity), which will cause the vessel diameter to decrease until flow is returned to its normal value.

28. During hemorrhage, the circulating blood volume is reduced. This leads to reduced end-diastolic volume and low cardiac output in spite of increased heart rate and contractility. During exercise, end-diastolic volume is maintained, and therefore the increased heart rate and increased contractility lead to a proportional increase in cardiac output.

29. The weakened pumping action of the heart leads to accumulation of blood in the venous circulation and in the heart. The increased end-diastolic volume of the heart results in increased vigor of contraction as described by the Frank-Starling law of the heart.

Chapter 20

1. forced vital capacity
2. residual volume
3. alveolar ventilation
4. transpulmonary pressure
5. pneumothorax
6. compliance: the slope of a lung pressure-volume curve
7. pulmonary surfactant: a lipoprotein that coats the inner surface of the alveoli and acts as a surface reducing agent
8. atelectasis: a condition in which alveoli become unstable, causing smaller alveoli to collapse
9. hypoventilation: a condition in which alveolar ventilation is below normal, causing high CO_2 in the blood (hypercapnia)
10. sleep apnea: a condition in which breathing stops for a long period of time (30–60 seconds) during sleep
11. d
12. a
13. c
14. a
15. b
16. d
17. a
18. b
19. a
20. d

21. 6.2 liters
22. 6.2 liters/min
23. 200 ml
24. a. 4.0 liters/min b. 0.333 liter
25. FVC is one of the most useful measurements in lung spirometry because it is a dynamic measurement that provides important information about airflow and elastic properties of the lung.
26. The lung inflates when the chest cavity expands, causing pressure inside the alveoli to drop below atmospheric pressure and thereby creating a pressure gradient between mouth and alveoli that causes air to flow into the lung. Deflation occurs when the chest cavity decreases in size, causing alveolar pressure to rise above atmospheric pressure and creating a pressure gradient that causes air to flow out of the lung.
27. Lung elastic recoil is essential for the exhalation of air from the lung during expiration. During inflation the lung is stretched and during deflation the elastic recoil from the stretched lung causes alveolar pressure to be greater than atmospheric pressure, thus causing airflow out of the lungs.
28. Emphysema causes destruction of alveoli and connective tissue of the lung. As a result, the lung is not as stiff. Thus lung compliance is abnormally high in an emphysema patient.
29. Surfactant lowers surface tension in the alveoli, making the lung more compliant and thereby reducing the work required to inflate it. Also by reducing surface tension, surfactant prevents edema by reducing the tendency for the alveoli to absorb fluid from the pulmonary capillaries. At low lung volumes, surfactant promotes alveolar stability.

Chapter 21

1. cardiac catheterization
2. pulmonary edema
3. hemoglobin
4. embolus
5. placenta
6. ductus arteriosus: a fetal blood vessel connecting the pulmonary artery directly to the descending aorta
7. hypoxic vasoconstriction: vasoconstriction of the pulmonary artery caused by alveolar hypoxia
8. wedge pressure: pressure measured at the tip of a catheter that is wedged in a small pulmonary vessel
9. carbaminohemoglobin: the chemical term for carbon dioxide bound to hemoglobin

10. hypoxemia: abnormally low P_{O_2} in the blood
11. b
12. a
13. a
14. b
15. b
16. d
17. b
18. d
19. c
20. d
21. 159.6 mm Hg (760 mm Hg × 0.21 = 159.6 mm Hg)
22. 200 ml/min. In a normal resting individual, 4 vol% (4 ml/100 ml blood) are transported to the lungs each minute (40 ml/liter × 5 liters/min = 200 ml/min).
23. 11 mm Hg $\left(15 \text{ cm} = 150 \text{ mm} \times \dfrac{1 \text{ mm Hg}}{13.6 \text{ mm}}\right)$
24. 134.4 mm Hg (640 mm Hg × 0.21 = 134.4 mm Hg)
25. 420 ml/min (70 ml/liter × 6 liters/min = 420 ml/min)
26. Pulmonary lymphatics are auxiliary drainage vessels to remove excess interstitial fluid.
27. The metabolic function of the lung includes the metabolism (synthesis, release, and conversion) of vasoactive mediators such as angiotension, histamine, prostaglandin, and serotonin.
28. Mean pulmonary arterial pressure at rest is 15 mm Hg, systolic is 25 mm Hg, and diastolic is 10 mm Hg.
29. Ventilation-perfusion balance is maintained by directing blood away from poorly ventilated regions in the lung by hypoxic vasoconstriction.
30. Major changes in pulmonary circulation include an increase in blood flow, increased pulmonary arterial pressure, a decrease in capillary transmit time, and capillary recruitment.

Chapter 22

1. hyperventilation
2. edema
3. blackout
4. free radicals
5. atelectasis
6. acclimatization: the ability to tolerate hypoxia if the ascent is slow so that the body can become accustomed to the low oxygen
7. high altitude cerebral edema: an abnormal collection of fluid in the cerebellum due to hypoxia exposure above altitudes of 12,000 feet
8. polycythemia: an increase in red cell production

9. carboxyhemoglobin: a compound formed when blood is exposed to carbon monoxide
10. G-force: the unit used to express the amount of gravitational force on the body
11. b
12. c
13. d
14. a
15. c
16. b
17. d
18. b
19. d
20. d
21. 48.6 mm Hg
22. 1000 ml
23. 4 atm
24. −2 G
25. 190 mm Hg
26. Acute effects of altitude on respiration include increased ventilation, respiratory alkalosis, and increased pulmonary arterial vasoconstriction.
27. Nitrogen gases during diving dissolve into the tissue. If the ascent is too rapid, bubbles form in both blood and tissue, causing "chokes" and "bends."
28. One problem associated with breathing high concentrations of oxygen is atelectasis, and the other is the toxic effects of free radical formation.
29. High G-forces alter blood hemodynamics in the eye, causing either blackout or redout.
30. Three stages of anesthesia occur with inhaled anesthetics. These include analgesia (reduced pain), loss of consciousness, and surgical anesthesia (loss of motor function).

Chapter 23

1. absorption
2. basic electrical rhythm
3. enterohepatic circulation
4. microvilli
5. micelles
6. diarrhea
7. digestive tract: the hollow tube extending from the lips to the anus that is composed of the mouth, pharynx, esophagus, stomach, and small and large intestines
8. portal circulation: the segment of gastrointestinal circulation (portal vein) that carries blood from the stomach, intestines, and pancreas to the liver
9. short reflex: a reflex that is mediated over sensory neurons, interneurons, and motor and secretory neurons that lie entirely within the wall of the digestive tract
10. secretin: a peptide hormone, secreted by endocrine cells in the upper small intestine, that inhibits gastric HCl secretion and motility and stimulates pancreatic and biliary secretion of bicarbonate
11. dysphagia: any condition in which there is difficulty swallowing
12. G-cell: endocrine cells in the mucosa of the gastric pyloric antrum that secrete the hormone gastrin
13. cephalic phase: the regulation of gastrointestinal function through neural and hormonal mechanisms that are activated by stimulation of receptors in the head
14. emetic: a chemical agent that causes vomiting by stimulation of chemoreceptors in the stomach and small intestine or in the brain
15. trypsin: a digestive enzyme, secreted by the pancreas, that breaks certain amino acid linkages in proteins and peptides
16. pancreatic lipase: a digestive enzyme, secreted by the pancreas, that splits triacylglycerols into monoglycerides and free fatty acids
17. peristalsis: a wave of smooth muscle contraction that moves along the wall of the digestive tract, propelling the contents in the lumen of the tract ahead of it
18. ferritin: an iron-protein complex that stores iron in intestinal epithelial cells
19. steatorrhea: the condition in which there are excessive amounts of fat in the feces
20. flatus: gas that is present in, or expelled from, the large intestine
21. b
22. a
23. c
24. c
25. a
26. d
27. b
28. d
29. b
30. b
31. When stretched as a meal enters the stomach, the smooth muscle of the body and fundus of the stomach adjusts its resting length without large changes in resting tension (plasticity). In addition, with each swallow as a meal is eaten, a long reflex over the vagus nerves causes a slight relaxation of the body and fundus.
32. HCl in the gastric pyloric antrum inhibits release of gastrin, and HCl in the duodenum

causes release of secretin, a hormone that inhibits secretion by the gastric parietal cells.

33. Bicarbonate of pancreatic juice neutralizes HCl entering the duodenum from the stomach and in so doing (1) protects against damage by HCl to the lining of the upper small intestine and (2) increases the pH of the duodenal contents to an optimal level for the activity of pancreatic digestive enzymes.

34. Bile salts (1) promote digestion of triacylglycerols by participating in the formation of an emulsion, consisting of very small fat particles, that provides a large surface area on which lipase can act to split triacylglycerols and (2) facilitate absorption by participating in the micellar solubilization of monoglycerides and free fatty acids, the digestion products of triacylglycerols.

35. Dietary fiber (bulk) helps maintain regular bowel movements by contributing to the volume of contents of the large intestine, which increases the degree of distention and thus the contractile activity of this organ.

Chapter 24

1. nephron
2. ultrafiltrate
3. glomerular filtration membrane
4. renal plasma clearance
5. glucose threshold
6. tubular transport maximum for PAH (Tm_{PAH})
7. countercurrent multiplication
8. central diabetes insipidus
9. afferent arteriole: the arteriole that carries blood toward the glomerulus
10. juxtaglomerular apparatus: the region in which the distal straight tubule touches afferent and efferent arterioles of its parent glomerulus; consists of macula densa, extraglomerular mesangial cells, and granular cells
11. proteinuria: the presence of an excess of proteins in the urine
12. urethra: the tube that carries urine from the urinary bladder to the outside of the body
13. glomerular filtration rate (GFR): the rate at which plasma is filtered in the kidney glomeruli
14. tubular reabsorption: the transport of substances out of the tubular urine
15. tubular secretion: the transport of substances into tubular urine
16. filtered load: the quantity of a substance, per unit of time, which is filtered by the glomeruli
17. renal autoregulation: the maintenance of a relatively constant renal blood flow and glomerular filtration rate despite changes in arterial blood pressure
18. juxtamedullary nephron: a nephron whose glomerulus lies deep in the cortex, adjacent to the medulla; such nephrons typically have long loops of Henle.
19. a
20. c
21. d
22. c
23. c
24. c
25. d
26. d
27. 80 ml plasma/min
28. 12.5 ml/min per mm Hg
29. a. 125 ml plasma/min   b. 31 ml plasma/min c. 375 mg/min d. 660 ml plasma/min e. 733 ml plasma/min   f. 1.33 liters/min
30. See Figures 24–2 and 24–4.
31. Serum albumin and many other plasma proteins bear a net negative charge at physiological pH. Passage of these macromolecules through the glomerular filtration membrane is normally impeded by fixed negative charges in the glomerular capillary membrane. Loss of this electrostatic barrier leads to filtration and hence abnormal excretion of plasma proteins (proteinuria).
32. The high rate of filtration of plasma in the human kidney is explained by (a) a high filtration coefficient (high fluid permeability, large surface area), (b) a high hydrostatic pressure in the glomerular capillaries, and (c) a high renal blood flow.
33. All of the changes will result in a decrease in GFR.
34. Creatinine is continuously produced in the body as an end-product of muscle metabolism. In a steady state, its production and urinary excretion rates are equal. The excretion of creatinine depends on filtration of plasma in the kidneys. If GFR falls, creatinine production will exceed excretion, and so the plasma creatinine concentration will rise.
35. In uncontrolled diabetes mellitus, due to a deficiency in the insulin system, plasma glucose levels are abnormally increased. This results in an increase in the amount of glucose filtered in the kidney glomeruli. When the filtered glucose load exceeds the reabsorptive capacity of the tubules for glucose, some of the filtered glucose will be excreted in the urine (glucosuria).
36. PAH, in addition to being filtered by the kidney glomeruli, is so vigorously secreted by the kid-

ney (proximal) tubules that it is nearly completely cleared from all of the plasma in one passage of blood through the kidneys. Its renal plasma clearance, therefore, approximates the renal plasma flow.

37. Sodium enters proximal tubule cells by symport with glucose, amino acids, other organic solutes, phosphate, and chloride, and by exchange with cellular hydrogen ions. Sodium is pumped out of the basolateral cell membrane by its Na^+/K^+-ATPase. Sodium ions also traverse the epithelium by passive movement through leaky "tight junctions."

38. The proximal tubule reabsorbs large quantities of sodium and water as an isosmotic solution. The collecting ducts reabsorb smaller amounts of sodium and water. Unlike the proximal tubule, they can establish steep concentration gradients for sodium. Collecting duct fluid can be rendered hypo-osmotic (due to the low intrinsic water permeability of its epithelium, expressed in the absence of ADH) or hyperosmotic (due to equilibration of the collecting duct urine with the medullary interstitium, in the presence of ADH). ADH and aldosterone increase, respectively, water and sodium reabsorption by the collecting ducts but do not affect reabsorption by the proximal tubule.

39. All of these factors increase renal potassium excretion.

40. The vasa recta (a) supply the nutritional needs of the medulla, (b) act as countercurrent exchangers, and (c) remove water from the medulla.

41. Urea, added to the inner medulla by its collecting ducts and from the calyceal urine, osmotically promotes withdrawal of water along the descending thin limb of Henle's loop. This results in a rise in NaCl concentration in this fluid. When the tubule fluid moves up the ascending thin limb, sodium and chloride move passively into the medullary interstitium. This comprises a "single effect" and permits countercurrent multiplication in the inner medulla. The concentrated urea solution which enters the medulla furnishes the potential energy for establishing an osmotic gradient in the inner medulla.

Chapter 25

1. hypovolemia
2. osmoreceptors
3. water diuresis
4. sodium appetite

5. dialysis
6. renin: a proteolytic enzyme, produced by granular cells in the kidney, that splits off angiotensin I from angiotensinogen
7. converting enzyme: an enzyme that catalyzes the conversion of angiotensin I to angiotensin II
8. extracellular fluid: the fluid outside of cells
9. negative balance: the condition in which there is net loss of a substance from the body
10. insensible water loss: water loss by evaporation from the lungs and skin
11. hypokalemia: a plasma potassium concentration below the normal range, i.e., below 3.5 mEq/liter
12. micturition: urination
13. b
14. b
15. d
16. a
17. b
18. d
19. d
20. d
21. e
22. c
23. e
24. d
25. (a) Intracellular water = 1110 ml. (b) Interstitial fluid-lymph water = 1335 ml.
26. 4.2 kg (or 9 lbs)
27. Interstitial fluid-lymph is an ultrafiltrate of plasma and has practically the same composition as plasma except for the higher protein concentration in plasma and the effects this has on the distribution of small ions across capillary walls. The concentrations of small cations are slightly higher in plasma than in interstitial fluid-lymph, and the opposite is true for small anions.
28. Transcellular fluids (e.g., cerebrospinal fluid, digestive secretions, tubular urine) can differ strikingly in composition from plasma, because these fluids are formed by or acted upon by epithelial cells, which transport various ions.
29. Potassium, magnesium, and proteins are present at higher concentrations in cells than in extracellular fluid.
30. We drink mostly by habit, and this intake covers our water needs under ordinary circumstances. We usually operate below the threshold for thirst. In the same way, when it is cold we regulate our body temperature by behavioral adjustments and by cutaneous vasoconstriction; rarely do we get cold enough to shiver.
31. Drinking a large volume of water results in a fall

in plasma osmolality, which is detected by osmoreceptors in the anterior hypothalamus. They cause a decrease in release of ADH from the posterior pituitary, which allows the collecting ducts to express their intrinsically low water permeability. Tubular urine becomes osmotically dilute in the thick ascending limb; continued reabsorption of solute (mostly NaCl) out of the collecting ducts, without water following, leads to excretion of a large volume of dilute urine.

32. ADH combines with a receptor in the basolateral cell membrane of collecting duct cells. Via a guanine nucleotide stimulatory protein, adenylate cyclase is activated. This enzyme catalyzes the formation of cyclic AMP, which in turn activates a protein kinase. The subsequent steps are not entirely clear, but the final result is an increase in water permeability of the luminal cell membrane.

33. (a) an increase in GFR, (b) a decrease in plasma aldosterone levels, (c) an increase in peritubular capillary hydrostatic pressure or decrease in peritubular capillary colloid osmotic pressure, (d) a decrease in renal sympathetic nerve activity, (e) a change in intrarenal blood flow distribution, and (f) an increase in plasma atrial natriuretic peptide level.

34. Angiotensin II causes (a) vasoconstriction, (b) increased aldosterone secretion, (c) thirst, and (d) increased ADH secretion.

35. Effective arterial blood volume is an index of the degree of filling of the systemic arterial system, which determines whether blood flow needs of the tissues will be met. It is sometimes called the "effective circulating blood volume."

36. (a) Reduced plasma potassium concentration and (b) reduced plasma aldosterone level.

37. The urinary bladder (a) stores the urine and (b) is emptied when this is appropriate.

Chapter 26

1. carbonic anhydrase
2. acidemia
3. base
4. pH buffer
5. amphoteric substance
6. weak acid
7. pH: $-\log [H^+]$
8. Henderson-Hasselbalch equation: $pH = pK_a + \log ([A^-]/[HA])$
9. open system: a system in which a component can be added or removed

10. respiratory acidosis: an abnormal process characterized by CO_2 accumulation, which leads to a fall in blood pH

11. respiratory alkalosis: an abnormal process characterized by the loss of too much CO_2, which leads to a rise in blood pH

12. metabolic acidosis: an abnormal process characterized by a gain of acid (other than H_2CO_3) or loss of bicarbonate, which leads to a fall in blood pH

13. metabolic alkalosis: an abnormal process characterized by a gain of strong base or bicarbonate or a loss of acid (other than H_2CO_3), which leads to a rise in blood pH

14. d
15. a
16. d
17. c
18. a
19. d
20. b
21. a
22. b
23. e
24. 2
25. 12
26. 4.4 mEq/liter
27. No. Hydrochloric acid, a strong acid, will be essentially completely dissociated (ionized); a 0.1 M solution has a pH of 1. Acetic acid is a weak acid and so will be incompletely dissociated; a 0.1 M solution has a pH of 2.9.
28. The effectiveness of a buffer depends on (a) how close the pK_a of the buffer is to the desired pH and (b) on the amount of buffer present.
29. Metabolism of proteins produces strong acids (sulfuric, hydrochloric, and phosphoric) that are not matched by sources of base in the diet.
30. (a) bicarbonate-CO_2, plasma proteins, inorganic phosphate; (b) proteins, organic phosphates, bicarbonate-CO_2; (c) carbonate and phosphate salts; (d) bicarbonate-CO_2, ammonia, inorganic phosphate.
31. The bicarbonate-CO_2 buffer pair operates in an open system. The respiratory system can change arterial blood P_{CO_2} and the kidneys can change the plasma bicarbonate concentration.
32. The isohydric principle states that in any fluid compartment all buffer pairs are in equilibrium with the same hydrogen ion concentration.
33. The main receptors which cause an increase in ventilation in response to elevated P_{CO_2} are central chemoreceptors located in the medulla of the brain. Peripheral chemoreceptors (carotid and aortic bodies) play a lesser role.

Chapter 27

1. filterable calcium
2. hydroxyapatite
3. osteoid
4. osteoblasts
5. parafollicular, or C-cells
6. osteoporosis
7. chondrocytes
8. remodelling
9. hypocalcemia: the condition in which there is a lower than normal plasma calcium concentration
10. ionized calcium: this form of calcium represents about 50% of the total calcium in the blood and is the physiologically active form of calcium
11. ground substance: this comprises about 5% of the organic material of bone and is composed primarily of proteoglycans
12. osteoclasts: large, multinucleated cells that secrete acids and proteolytic enzymes and are responsible for bone resorption. The activity and maturation of osteoclasts are stimulated by parathyroid hormone.
13. parathyroid hormone: a peptide hormone secreted by the parathyroid glands in response to decreased plasma calcium. It acts on bone, kidney, and indirectly on the gastrointestinal tract; its overall effects are to increase plasma calcium and decrease phosphate.
14. 1,25-OH-vitamin D_3: the hormonally active form of the vitamin. It is formed in the kidney and acts primarily on the gastrointestinal tract, where it promotes calcium absorption.
15. ergocalciferol (or vitamin D_2): found in many plants and yeasts. It is metabolized, has actions similar to those of vitamin D_3, and is often used for the supplementation of foods with vitamin D.
16. c
17. b
18. a
19. c
20. d
21. b
22. b
23. Hypocalcemia results in altered sodium permeability in nerves and the generation of spontaneous action potentials. This can lead to continuous stimulation and tetanization of the muscles of the motor unit.
24. In the plasma, calcium consists primarily as free ionized calcium ion (Ca^{2+}), calcium complexed to an anion, and calcium bound to plasma proteins. The first two forms comprise the filterable calcium and represent 50% and 10% of the total, respectively; calcium bound to plasma proteins is not filterable and constitutes approximately 40% of the total.
25. The kidney contains the enzyme 1 α-hydroxylase, which is responsible for the formation of the hormonally active form of vitamin D_3. In certain kidney diseases, the action of this enzyme can be impaired, leading to a deficiency of vitamin D action and osteomalacia or rickets.
26. In bone, parathyroid hormone stimulates osteoclasts, leading to bone resorption. In the kidney, parathyroid hormone stimulates calcium reabsorption in the distal tubule and inhibits phosphate reabsorption in the proximal tubule. The hormone also stimulates the enzyme responsible for the final step of vitamin D activation in the kidney, leading to an increase in calcium and phosphate absorption by the gastrointestinal tract.
27. Parathyroid hormone and calcitonin secretion are reciprocally regulated in response to changes in plasma Ca^{2+} concentrations; a decrease in calcium stimulates parathyroid hormone secretion, and an increase stimulates calcitonin secretion. Several gastrointestinal hormones, particularly gastrin, also promote calcitonin secretion.

Chapter 28

1. specific dynamic action
2. radiation
3. hypothalamus
4. heat stroke
5. poikilotherm
6. homeotherm: an organism that regulates its body temperature within a relatively narrow range
7. fever: a regulated increase in body temperature associated with disease
8. mean skin temperature: a representative temperature of the body shell
9. thermography: a technique using infrared ray detection to measure skin temperature
10. catabolism: biochemical reactions that break down complex substances into simpler compounds
11. c
12. c
13. c
14. c
15. c

16. 36.8°C
17. 32.0°C
18. 29.1 cal/hr
19. 72.4 cal/hr
20. 406 cal/hr
21. The conditions for measuring BMR are:
   a. after a person has fasted for 12 hours
   b. measurement usually made in morning
   c. measurement made after 30 minutes of complete rest
   d. measurement made in a comfortable temperature
22. The hypothalamus is the central controller and integrator of all the body responses for effective regulation of body temperature.
23. As an individual goes from an ambient temperature of 25°C to 35°C, the skin temperature increases. This increase is detected by thermoreceptors in the skin, which send neural information to the hypothalamus. Additionally, core temperature may increase slightly. This increase is detected by the central thermoreceptors, which also send neural information to the hypothalamus. The hypothalamus, acting as a central controller, integrates the neural information from thermoreceptors and activates antirise responses, which correct the initial rises in skin and core temperatures.
24. Normal body temperature in humans is a range of body temperatures between 35°C and 41°C.
25. The mechanisms that the body uses to exchange thermal energy include:
   a. radiation—heat transfer by electromagnetic radiation
   b. conduction—transfer of heat by direct contact
   c. convection—transfer of heat to a moving liquid or gas
   d. evaporation—transfer of heat by conversion of water to a gas

Chapter 29

1. isometric exercise
2. maximal oxygen consumption
3. hyperpnea
4. hypertrophy
5. leucocytosis
6. oxygen debt: the use of oxygen from body stores, associated with the beginning of exercise
7. cytokines: cellular polypeptides which mediate tissue responses to infection and injury
8. mechanical efficiency: the fraction of the available chemical energy converted to mechanical energy during exercise

9. proprioreceptors: receptors in joints and muscle spindles, which detect motion of the extremities
10. phosphagen: high-energy phosphate molecule
11. d
12. c
13. b
14. b
15. a
16. c
17. a
18. c
19. b
20. c
21. The factors that control pulmonary ventilation during exercise include chemical or humoral stimuli and neural stimuli from body receptors.
22. Exercise causes an increase in white blood cells and the release of cytokines, which mediate immune responses to infection and tissue injury.
23. Both at rest and during maximal exercise, the trained individual has a greater stroke volume and lower heart rate, which are the result of a greater pumping effectiveness of the heart of the trained individual.
24. The criteria for determination of VO_2 may include (a) no further increase in oxygen consumption with increases in the rate of exercise; (b) blood lactate in excess of 8 mM; (c) large muscle mass is exercised for at least 3 minutes.
25. The body temperature during steady-state exercise is not related to the absolute level of exercise but instead is related to the level of exercise relative to the person's maximum exercise capacity.

Chapter 30

1. osteoporosis
2. seminiferous tubules
3. corpus luteum
4. endometrium
5. crossing over: the exchange of maternal and paternal homologous chromosomal segments during meiotic division.
6. menopause: the phase in a woman's life when menstruation ceases.
7. mitotic division: duplication of chromosomes prior to cell division, which results in the formation of two cells, each of which have one copy of the chromosomes present in the original cell.
8. spermatogonia: the first cell stage of spermatogenesis in the adult male. At puberty, the primitive germ cells undergo a mitotic division and differentiate into spermatogonia.

9. c
10. a
11. b
12. c
13. d
14. Sperm production occurs only at temperatures below normal body temperature. The musculature of the scrotum moves the testes away from the body in warm weather and lifts the testes closer to the body in cold weather to ensure that the temperature in the testes is conducive to spermatogenesis.
15. (a) The blood-testes barrier prevents substances which could destroy the developing spermatogenic cells from entering the seminiferous tubules. (b) The blood-testes barrier keeps sperm from diffusing from the seminiferous tubules into the blood and hence prevents the production of antibodies specific to sperm.
16. During the luteal phase of the ovarian cycle the corpus luteum secretes estrogen and progesterone, which stimulate growth and proliferation of blood vessels and accumulation of glycogen and enzymes in the endometrium, which results in an increase in endometrial thickness.

Chapter 31

1. polyspermy
2. ectopic pregnancy
3. myoepithelial cells
4. dizygotic twins: dizygotic or fraternal twins result from the fertilization of two ova by individual sperm.
5. zygote: the single-cell fertilized ovum, which contains 46 chromosomes.
6. chorion: trophoblast cells which form a membrane that completely surrounds the developing embryo.
7. d
8. c
9. a
10. d
11. b
12. Secretion of estrogen by the ovary increases at midcycle, and estrogen stimulates the formation of glycoprotein channels in cervical mucus, which facilitates sperm migration through the cervix into the uterus.

13. The acrosomal reaction results in the release of enzymes from the head of the sperm, which allows sperm to penetrate the ovum during fertilization.
14. Suckling stimulates afferent neural impulses that travel from the breast, through the spinal cord, to the hypothalamus and stimulate prolactin release from the pituitary. Prolactin stimulates milk production in the breast.

Chapter 32

1. transudation
2. tenting
3. sexual dysfunction
4. histocompatibility-Y (H-Y) antigen
5. vaginismus: involuntary contractions of the vaginal muscles, which make vaginal penetration very difficult or impossible
6. impotence: also termed erectile dysfunction; the inability to achieve penile erection sufficient for vaginal penetration
7. Mullerian Inhibiting Substance (MIS): a glycoprotein, secreted by the Sertoli cells of the testes, which induces regression of the Mullerian duct system of the embryo
8. myotonia: an increase in general muscle tone or tension
9. Turner's syndrome: a chromosomal abnormality in the female characterized by only one X sex chromosome (XO), which results in the development of an incompletely formed gonad, termed a vestigal gonadal streak
10. amenorrhea: the absence of menstruation
11. b
12. c
13. d
14. a
15. The sex chromosomes (XX in the female and XY in the male) provide the genetic information that directs the differentiation of gonadal cells into ovaries or testes and hence determine the gonadal sex of the individual.
16. Stress, which often accompanies intense programs of physical exercise, can suppress GnRH release and hence LH and FSH release, which can result in secondary amenorrhea.

Normal Physiological Values for the Adult

Total Body Water (% body weight)

Female	50
Male	60

Core Temperature

	°C	°F
	36–38	96.8–100.4

Blood

Hematocrit (volume % RBC in blood)	37–52
Hemoglobin content (g/dl)	12–18
Erythrocytes (million/mm^3)	4.6–6.2
Leukocytes (thousand/mm^3)	7–10
Plasma glucose (mg/dl)	80–120
Plasma calcium (ionized—mEq/liter)	2.0–2.5
Plasma iron (μg/dl)	50–150
Plasma sodium (mEq/liter)	136–145
Plasma osmolarity (mOsm/kg H_2O)	290–300

Cardiovascular

Resting heart rate (beats/min)	65–75
Systolic pressure (mm Hg)	100–140
Diastolic pressure (mm Hg)	60–95
Mean systemic arterial pressure (mm Hg)	80–100
Mean pulmonary arterial pressure (mm Hg)	10–15
Cardiac output (liter/min)	5–8

Renal

Glomerular filtration rate (ml/min)	115–130
Urine output (ml/day)	600–6000
Urine pH	4.5–8.0
Urine osmolarity (mOsm/kg H_2O)	500–800

Respiratory

Alveolar ventilation (liter/min)	4.2
Respiratory rate (breaths/min)	12–15
Tidal volume (ml)	500
Total lung capacity (liter)	6
Vital capacity (liter)	4.8
Oxygen uptake (ml/mm)	240
Carbon dioxide production (ml/min)	192
Respiratory exchange ratio (CO_2 output/O_2 uptake)	0.8
Alveolar gas tension	
O_2 (mm Hg)	104
CO_2 (mm Hg)	40
Arterial blood gases	
pH	7.35–7.45
O_2 saturation (% saturation of HB with O_2)	95–98
O_2 tension (mm Hg)	85–100
CO_2 tension (mm Hg)	35–45
Oxygen content (ml/dl)	20
Carbon dioxide content (ml/dl)	50
Bicarbonate (mEq/liter)	23–28

Medical Terminology

Prefixes

Prefix	Meaning (and Example)
a-, ab-	away from (*abnormal*)
a-, an-	negative, not, without (*anaerobic*)
abdomin-	pertaining to abdomen (*abdominal*)
ad-	toward, adhere to (*adduction*)
aer-	pertaining to air (*aerobic*)
af-	movement toward a point (*afferent nerve*)
algi-	pain (*algesic*)
all-	other, different (*allele*)
amb-	both, on both sides (*ambidextrous*)
amph-	on both sides (*amphiarthrosis*)
ana-	up (*anaphylaxis*)
angio-	pertaining to blood or lymph vessels (*angiogram*)
ante-	before (*antebrachium*)
anti-	against (*antibiotic*)
apo-	from, opposed to (*apocrine*)
aqua-	water (*aqueous*)
arche(i)-	first (*archetype*)
arthr-	joint (*arthritis*)
aud-	pertaining to hearing (*audiology*)
auto-	pertaining to self (*autoimmunity*)
bi-	double (*binocular*)
bio-	life, living (*biofeedback*)
blast-	germinal, bud (*blastocyst*)
brachi-	arm (*brachium*)
brachy-	short (*brachycephalic*)
brady-	slow (*bradycardia*)
bucc-	cheek (*buccinator muscle*)
cac-	bad, evil (*cachexia*)
calc-	pertaining to calcium, lime, or stone (*calcification*)
carcin-	cancer (*carcinogen*)
cardi-	heart (*cardiac*)
cata-	down, lower (*catatonia*)
caud-	tail (*caudal*)
centri(o)-	center, middle (*centrilobular*)
cephal-	pertaining to head (*cephalomeningitis*)
cerebro-	brain (*cerebrovascular accident*)
chol-	bile (*cholangiography*)
chondr-	cartilage (*chondrocranium*)
chromo-	color (*chromosome*)

Prefix	Meaning (and Example)
circum-	around (*circumcision*)
co-	together (*coarctation*)
coel-	hollow cavity (*coelom*)
com-	together (*communicable*)
con-	together, with (*congested*)
contra-	against (*contraceptive*)
corn-	hardened, horny (*cornea*)
corp-	body (*corpus cavernosum*)
cryo-	pertaining to cold (*cryopreservation*)
crypt-	hidden (*cryptocephalus*)
cyan-	blue in color (*cyanotic*)
cyst-	bladder, baglike (*cystocarcinoma*)
cyto-	cell (*cytoplasm*)
dacry-	tears, lacrimal (*dacryostenosis*)
dactyl-	pertaining to fingers (*dactylospasm*)
de-	from, not, down (*decongestant*)
deci-	tenth (*decibel*)
dent-	pertaining to teeth (*dentition*)
derm-	skin (*dermatitus*)
di-	two, double (*digastric*)
dia-	through, between, separate (*diastema*)
dipl-	double (*diplococcus*)
dis-	apart, away from, negative (*dislocation*)
duct-	to lead or conduct (*ductus arteriosus*)
dys-	difficult, bad, painful (*dysmenorrhea*)
ecto-	external, outside (*ectoderm*)
ef-	out (*effusion*)
endo-	internal, inside (*endoderm*)
entero-	intestine (*enteroscope*)
ento-	within (*entocyte*)
epi-	upon, over, in addition to (*epicardium*)
erythro-	red (*erythrocyte*)
eu-	well, healthy, normal (*euphoria*)
ex-	from, out of (*excrement*)
exo-	outside (*exoskeleton*)
extra-	beyond, outside of (*extraembryonic*)
fasci-	band (*fasciculus*)
febr-	pertaining to fever (*febricide*)
feto-	pertaining to a fetus (*fetoscope*)
fibr-	fibrous (*fibroma*)

Prefix	Meaning (and Example)
for-	opening (*foramen rotundum*)
fore-	in front of (*forebrain*)
galact-	pertaining to milk (*galactosemia*)
gastr-	pertaining to the stomach (*gastroenteritis*)
gloss-	tongue (*glossectomy*)
gluco-	pertaining to sugar (*glucogenesis*)
glyco-	pertaining to sugar (*glycosemia*)
gonado-	pertaining to the gonads (*gonadotropin*)
gran-	particle, grain (*granulocyte*)
gravi-	heavy (*gravimetric*)
gyn-	female sex (*gynecologist*)
haplo-	simple, single (*haploid*)
hema(o)-	blood (*hemapoiesis*)
hemato-	blood (*hematologist*)
hemi-	half (*hemisphere*)
hepat-	liver (*hepatitis*)
hetero-	different, other (*heterosexual*)
histo-	tissue (*histocompatibility*)
holo-	whole (*holocrine*)
homo-	same (*homosexual*)
hydro-	water (*hydrocephalic*)
hyper-	excessive, above (*hypertension*)
hypo-	under, below (*hypoglycemia*)
idio-	individual, distinct (*idiopathic*)
ilei(o)-	pertaining to the ileum (*ileitis*)
ilio-	pertaining to the ilium (*iliosacral*)
in-	in, within, negative (*incontinence*)
infra-	beneath (*infraorbital*)
inter-	between, among (*intercoastal*)
intra-	within, inside (*intramuscular*)
irid-	pertaining to the iris (*iridectomy*)
iso-	like, equal (*isometric*)
karyo-	nucleus (*karyotype*)
kerato-	pertaining to tough substances, cornea (*keratosis*)
keto-	pertaining to ketone bodies (*ketosis*)
kine-	movement (*kinesiology*)
kypho-	humped (*kyphosis*)
labi-	lip (*labia majora*)
lacri-	tears (*lacrimal duct*)
lacto-	pertaining to milk (*lactogenic*)
laparo-	pertaining to loin or abdomen (*laparoscopy*)
laryngi(o)-	pertaining to larynx (*laryngoscope*)
later-	side (*lateral*)
leuk-	white (*leukocyte*)
lipo-	pertaining to fat (*lipocyte*)
lith-	stone (*lithodialysis*)
lymph-	pertaining to lymphatic system (*lymphoma*)
macro-	large (*macrophage*)
mal-	abnormal, bad, ill (*malnutrition*)
mamm-	pertaining to breast (*mammography*)
mast-	pertaining to breast (*mastectomy*)

Prefix	Meaning (and Example)
medi-	middle (*mediastinum*)
mega-	large (*megakaryocyte*)
mela-	black (*melanoma*)
meningo-	meninges (*meningopathy*)
meso-	middle (*mesoderm*)
meta-	beyond, after (*metacarpal*)
metra(o)-	pertaining to uterus (*metrocystosis*)
micro-	small (*microcephaly*)
mio-	less, smaller (*mioplasmia*)
mito-	thread (*mitosome*)
mono-	one, alone (*mononucleosis*)
morph-	shape (*morphogenesis*)
muco-	pertaining to mucous (*mucoenteritis*)
multi-	many (*multicellular*)
myelo-	pertaining to bone marrow (*myeloma*)
myo-	muscle (*myocardium*)
narc-	numbness, stupor (*narcolepsy*)
naso-	nose (*nasoscope*)
necro-	dead, corpse (*necrosis*)
neo-	young, new (*neonatal*)
nephro-	kidney (*nephropathy*)
neuro-	nerve (*neurosurgery*)
nucleo-	nucleus (*nucleotide*)
ob-	against, in front of (*obstruction*)
oc-	against (*occlusion*)
oculo-	eye (*oculomotor*)
odont-	tooth (*odontoblast*)
oligo-	small, few (*oligospermia*)
omo-	shoulder (*omoclavicular*)
oo-	egg (*oocyte*)
oophoro-	ovary (*oophorotomy*)
opisth-	backward, behind (*opisthotic*)
orchi-	testis (*orchidectomy*)
ortho-	straight, normal (*orthopedics*)
osteo-	bone (*osteocyte*)
oto-	ear (*otopathy*)
ovi(o)-	egg (*oviduct*)
pachy-	thick, heavy (*pachyderma*)
pan-	all (*panarthritis*)
para-	beside, near (*paranasal*)
path-	disease (*pathology*)
pedia-	children (*pediatrics*)
per-	through (*perfusion*)
peri-	surrounding, near (*periosteum*)
phag-	pertaining to eating (*phagocyte*)
pharmaco-	drugs, medicine (*pharmacotherapy*)
phlebi(o)-	vein (*phlebitis*)
photo-	pertaining to light (*photoreceptor*)
platy-	flat, broad (*platycephaly*)
pleur-	pertaining to pleura (*pleurisy*)
pneumo-	air, lung (*pneumonia*)
pod-	foot (*podiatry*)
poly-	much, many (*polydactyly*)
post-	after, behind (*postmortem*)
pre-	before (*prenatal*)

Prefix	Meaning (and Example)
pro-	before (*prochondral*)
proct-	anus (*proctology*)
proto-	first (*protoplasm*)
pseudo-	false (*pseudogout*)
psycho-	mental (*psychogenesis*)
pyo-	pus (*pyodermatitis*)
quad-	pertaining to four (*quadriplegia*)
re-	again, back (*reabsorb*)
reno-	pertaining to the kidney (*renography*)
rete(i)-	network (*reticulocyte*)
retro-	behind, backward (*retroperitoneal*)
rhino-	nose (*rhinopathy*)
saccharo-	sugar (*saccharuria*)
sarco-	flesh (*sarcolemma*)
sclero-	hard (*scleroderma*)
semi-	half (*semilunar*)
steno-	narrow (*stenosis*)
sub-	under, below, beneath (*subcutaneous*)
super-	beyond, above, upper (*supervirulent*)
supra-	over, above (*supramandibular*)
sym-	joined together (*symphysis*)
syn-	joined together (*synarthrosis*)
tachy-	fast (*tachycardia*)
tele-	far (*telencephalon*)
tetra-	four (*tetraploid*)
therm-	pertaining to heat (*thermogenesis*)
thorac-	chest (*thoracic cavity*)
thrombo-	clot, lump (*thrombosis*)
tox-	poison (*toxicology*)
trans-	over, across (*transfuse*)
tri-	three (*triceps brachii*)
tropho-	pertaining to nourishment (*trophoblast*)
ultra-	excess, beyond (*ultrasonogram*)
uni-	one (*unicellular*)
uro-	urinary organs, urine (*urogenital*)
vas-	vessel (*vasodilation*)
viscer-	internal organs (*visceroreceptor*)
vit-	life (*vitality*)
xanth-	pertaining to yellow (*xanthoderma*)
zoo-	animal (*zoology*)
zygo-	union, join (*zygote*)

Suffixes

Suffix	Meaning (and Example)
-able	capable of (*digestable*)
-a(e)sthesia	sensation (*anesthesia*)
-algia	pain (*gastralgia*)
-ase	enzyme (*lactase*)
-asis	condition (*homeostasis*)
-cele	tumor, cyst (*hydrocele*)
-cide	destroy (*fungicide*)
-coele	enlarged cavity, swelling (*blastocoele*)
-cyte	cell (*erythrocyte*)
-dynia	pain (*ophthalmodynia*)
-ectomy	surgical removal (*appendectomy*)
-emesis	vomiting (*hyperemesis*)
-emia	pertaining to the blood (*glycemia*)
-ferent	moving, carrying (*afferent*)
-form	shape (*penniform*)
-fuge	drive away (*centrifuge*)
-gen	causative agent (*pathogen*)
-genic	causing (*carcinogenic*)
-gram	recording, record (*electrocardiogram*)
-ia	condition (*uremia*)
-iatrics	medical specialties (*geriatrics*)
-ism	condition (*rheumatism*)
-itis	inflammation (*arthritis*)
-kinesis	movement (*orthokinesis*)
-lith	stone (*otolith*)
-logy	study of (*osteology*)
-lysis	dissolve, break apart (*hemolysis*)
-megaly	enlarged (*acromegaly*)
-oid	resembling (*nephroid*)
-oma	tumor (*melanoma*)
-ory	pertaining to (*sensory*)
-ose	full of (*adipose*)
-osis	condition (*scoliosis*)
-pathy	disease, abnormality (*colonopathy*)
-penia	deficiency (*leukopenia*)
-phil	attraction to (*neutrophil*)
-phobia	abnormal fear (*hematophobia*)
-phylaxis	protection (*prophylaxis*)
-plasty	to reconstruct (*rhinoplasty*)
-plegia	stroke, paralysis (*paraplegia*)
-pnea	pertaining to breathing (*apnea*)
-poiesis	formation of (*hematopoiesis*)
-rrhage	overflow (*hemorrhage*)
-rrhea	discharge, flow (*menorrhea*)
-sclerosis	hardness, dryness (*arteriosclerosis*)
-sis	action, process (*dialysis*)

Suffix	Meaning (and Example)
-tomy	cut, incision (*tracheostomy*)
-trophy	pertaining to nutrition (*hypertrophy*)
-tropic	changing, influencing (*gonadotropic*)
-uria	pertaining to urine (*glycosuria*)

Latin and Greek Terms for Body Parts

ankle: (L): *tars-, tarsi-, -tarsus*

anus: (L): *ano-, -anus*; (G): *procto-, -proctus*

arm: (G): *brachi-, -brachium*

back: (L): *dors-, dorsi-, -dorsum*; (G): *noto-, -notus*; (L): *terg-, -tergum*

bag: See *bladder*.

beak: (G): *rhyncho-, -rhynchus*; (L): *rostr-, -rostrum*

belly: (L): *-venter, ventr-, ventro-*

bladder: (G): *asco-, -ascus*; (G): *cysto-, -cystis*; (G): *-physa, physo-*; (L): *-vesica, vesico-*

blood: (G): *-haema, haemato-, haemo-*; (L): *sanguini-, -sanguis*

body: (L): *corporo-, -corpus*; (G): *-soma, somato-*

bone: (L): *-os, ossi-*; (G): *osteo-, -osteum*

border: (G): *chilo-, -chilus*; (G): *craspedo-, -craspedum*

brain: (L): *cereb-, cerebr-, -cerebrum*; (G): *encephalo-, -encephalus*

breast: (L): *pector-, -pectus*; (L): *stern-, -sternum*; see also *chest*.

bristle: (G): *-chaeta, chaeto-*; (L): *-seta, seti-*

cartilage: (G): *chondro-, -chondrus*

cell: (L): *-cella, celli-*; (G): *cyto-, -cytus*

cheek: (L): *bucc-, -bucca*; (L): *gen-, -gena, geno-*

chest: (G): *stetho-, -stethus*; see also *breast* and *thorax*.

claw: (G): *chel-, -chela, chelo-*; (G): *onycho-, -onyx*; (L): *ungui-, -unguis*

crest: (L): *crist-, -crista*; (G): *lopho-, -lophus*

crown: (L): *coron-, -corona*; (G): *stephano-, -stephanus*

digit: (G): *dactylo-, -dactylus*; (L): *digiti-, -digitus*

ear: (L): *auri-, -auris*; (G): *otido-, oto-, -ous*

egg: (G): *oo-, -oum*; (L): *ovi-, -ovum*

eye: (L): *oculi-, -oculus*; (G): *-omma, ommato-*; (G): *ophthalmo-, -ophthalmus*; (G): *opo-, -ops, opto-*

eyelash: (G): *-blepharis, blepharo-*

eyelid: (L): *cili-, -cilium*

face: (L): *faci-, -facies*; (G): *-ops*

feather: (L): *-pinna, pinni-*; (L): *plum-, -pluma, plumi-*; (G): *-pteryla, pterylo-*; (G): *ptilo-, -ptilum*

finger: See *digit*.

flesh: (L): *carni-, -caro*; (G): *sarco-, -sarx*

foot: (L): *pedi-, -pes*; (G): *podo-, -pus*

forehead: *-frons, front-*; (G): *metopo-, -metopus, -metopius*

gill: (G): *branch-, -branchium, brancho-*

gland: (G): *-aden, adeno-*; (L): *glandi-, -glans*

groin: *-inguen, inguini-*

hair: (L): *capill-, -capillus*; (L): *crini-, -crinis*; (L): *pil-, pili-, -pilus*; (G): *-thrix, tricho-*

hand: (G): *-chir, chiro-*; (L): *mani-, manu-, -manus*

head: (L): *capit-, capiti-, -caput*; (G): *-cephala, cephalo-*

heart: (G): *cardi-, -cardia*; (L): *-cor, cordi-*

heel: (L): *calcan-, calcane-, -calcaneum*; (L): *talari-, tali-, -talus*

horn: (G): *-cera, cerato-*; (L): *corn-, -cornus*

jaw: (G): *genyo-, -genys*; (G): *gnatho-, -gnathus*; (L): *maxill-, -maxilla*

joint: (G): *arthro-, -arthrum*; (L): *articuli-, articulus, -artus*

kidney: (G): *nephro-, -nephrus*; (L): *ren-, -ren, reni-*

knee: (L): *genu-, -genu*; (G): *gony-, gonyo-, -gonys*; (G): *-gonatium, gonato-*

knuckle: (G): *condylo-, -condylus*

leg: (G): *cnemi-, -cnemis*; (L): *crur-, -crus*; (G): *-scelis, scelo-, scelido-*

lip: (G): *chilo-, -chilus*; (L): *labi-, labio-, -labium, labr-, -labrum*

liver: (G): *-hepar, hepato-*; (L): *jecori-, -jecur*

lung: (G): *-pneuma, pneumo-*; (L): *pulmo-, -pulmo, pulmono-*

membrane: (G): *chorio, -chorium*; (G): *-hymen, hymeno-*; (L): *membran-, -membrana*; (G): *meningo-, -meninx*

mouth: (L): *ora-, ori-, -os*; (G): *-stoma, stomato-*

mucus: (G): *blenno-, -blennus*

muscle: (G): *myo-, -mys*; see also *flesh*.

neck: (G): *-auchen, aucheno-*; (L): *cervic-, -cervix*; (L): *coll-, -collum*; (G): *-dera, dero-*; (G): *trachelo-, -trachelus*

nose: (L): *nasi-, -nasus*; (G): *rhino-, -rhis*

rib: (L): *cost-, -costa, costi-*; (G): *scelido-, -scelis*

rump: (G): *gluteo-, -gluteus*; (G): *-pyga, pygo-*

scale: (G): *lepido-, -lepis*; (L): *squam-, -squama, squami-*

shell: (G): *-concha, concho-*; (G): *ostraco-, ostracum*

shoulder: (G): *omo-, -omus*

skin: *-byrsa, byrso-*; (G): *chorio-, -chorium*; (L): *cutan-, cuti-, -cutis*; (G): *derm-, -derma, dermo-, dermato-*; (G): *scyto-, -scytus*

skull: (G): *cranio-, -cranium*

snout: See *beak*.

sperm: (L): *-semen, semin-*; (G): *-sperma, spermato-*

spine: (G): *-acantha, acantho-*; (G): *rhachi-, -rhachis*; (L): *-spina, spini-*

stomach: (G): *-gaster, gastro-*; (L): *-venter, ventr-*

suture: (G): *-rhapha, rhapho-*

tail: (L): *caud-, -cauda*; (G): *cerco-, -cercus*; (G): *-ura, uro-*

thigh: (L): *femor-, -femur*; (G): *mero-, -merus*

thorax: (G): *thoraco-, -thorax*

throat: (L): *gula-, -gula*; (L): *guttur-, -guttur*; (G): *laemo-, -laemus*; (G): *pharyngo-, -pharynx*; (G): *trachelo-, -trachelus*

tissue: (G): *histo-, -histus*; (L): *tel-, -tela, teli-*

toe: See digit.

tongue: (G): *-glossa, glosso-, -glotta, glotto-*; (L): *-lingu, -lingua*

tooth: (L): *-dens, dent-, denti*; (G): *odonto-, -odous, -odus*

vein: (G): *phlebo-, -phleps*; (L): *ven-, -vena, veni-*

windpipe: (G): *broncho-, -bronchus*; (G): *trache-, -trachea*

wing: (L): *ala-, -ala, ali-*; (G): *ptero-, -pterum*; (G): *pterygo-, -pteryx*

wrist: (L): *carpo-, -carpus*

Index

Page numbers in bold type indicate definitions.
Page numbers followed by "F" indicate illustrations.
Page numbers followed by "T" indicate tables.

Chapter Opening Photo Captions

Chapter 1 MRI of a herniated brain. (© Howard Sochurck/Medichrome/The Stock Shop)

Chapter 2 A computerized model of an enzyme (protein) that is unattached to its substance (left) and one that is attached (*right*). (© Visuals Unlimited)

Chapter 3 False-color transmission electron micrograph of a section through a single dictyosome forming part of the Golgi apparatus in a mammalian cell. The dictysome is the stack of paired membranes (green with red outlines) curving across the center of the image (original ×31,800). (© Dr. Gopal Murti/SPL/Photo Researchers, Inc.)

Chapter 4 An electron micrograph of a kidney proximal tubule cell showing microvilli of the apical plasma membrane, extensive infoldings of basal and lateral membranes, and mitochondria between the basolateral infolding. (© C. Craig Tischer, M.D./University of Florida)

Chapter 5 Micrograph of an unwound human DNA molecule. (© Dr. John A. Olschowka, Ph.D./Dr. Richard Insel, M.D./Dr. Louise Chow/Phototake)

Chapter 6 Mitochondria. (© Don W. Fawcett/Visuals Unlimited)

Chapter 7 Scanning electron micrograph of a human neuron in tissue culture. (© Petit Format/Science Source)

Chapter 8 A scanning electron micrograph of the inner hair cells of the human ear. (© Dr. G. Bredberg/SPL/Photo Researchers)

Chapter 9 Neurons of the spinal cord. (© Cabisco/Visuals Unlimited)

Chapter 10 The medulla of the human brain (silver preparation). (© Cabisco/Visuals Unlimited)

Chapter 11 Human brainwaves during REM sleep (dreaming state). (© SIU/Visuals Unlimited)

Chapter 12 Scanning electron micrograph of the progisterone crystal (original ×3000). (© David M. Phillips/Visuals Unlimited)

Chapter 13 Photomicrograph of a section through the thyroid gland (original ×425). (© Bruce Iverson/Visuals Unlimited)

Chapter 14 The zona fasiculations of the adrenal cortex (original ×450). (© R. Calentine/Visuals Unlimited)

Chapter 15 Photomicrograph of an islet of Langerhans. (© David M. Phillips/Visuals Unlimited)

Chapter 16 Transmission electron micrograph of cardiac muscle showing the intercattel discs. (© Don W. Fawcett/Visuals Unlimited)

Chapter 17 A scanning electron micrograph of an angry macrophage engulfing a yeast cell. (© Biology Media/Science Source/Photo Researchers, Inc.)

Chapter 18 A normal coronary artery. (© Cabisco/Visuals Unlimited)

Chapter 19 A single capillary surrounded by cells and interstitial space. The cells in the lumen of the capillary are red blood cells. (© Don W. Fawcett/Visuals Unlimited)

Chapter 20 A false-color arteriograph of a normal right lung showing the pattern of bronchial and bronchiolar branching. (© CNR/SPL/Photo Researchers, Inc.)

Chapter 21 Scanning electron micrograph of a lung alveolus. (© David M. Phillips/Visuals Unlimited)

Chapter 22 Take-off of the space shuttle Columbia, the first space mission dedicated to life science research. (Courtesy of NASA)

Chapter 23 Scanning electron micrograph of the surface of an epithelial cell from the small intestive illustrating the microvilli. (© Custom Medical Stock Photo)

Chapter 24 An isolated, perfused proximal convoluted tubule. (© Biophoto Associates/Photo Researchers, Inc.)

Chapter 25 Light microscopy of sodium chloride crystals. (© Cabisco/Visuals Unlimited)

Chapter 26 Scanning electron micrograph of two adjacent renal glomerular capillaries. (© F. Spinelli, Ciba-Geigy/Visuals Unlimited)

CLINICAL ABBREVIATIONS

a.c.	before eating	CF	complement fixation; cystic fibrosis	ESRD	end-stage renal disease
abs. feb.	without fever			ESV	end-systolic volume
ACTH	adrenocorticotrophic hormone	CHD	congenital heart disease	F	Fahrenheit
		CHF	congestive heart failure	F1, F2	first, second filial generation
ad effect.	until effective	cm	centimeter		
ad lib.	as desired	CNS	central nervous system	FDA	Food and Drug Administration
ad us.	as customary	CO	cardiac output		
ad us. ext.	for external use	CoA	coenzyme A	FEV	forced expiratory volume
ADH	antidiuretic hormone	COAD	chronic obstructive airways disease	fl oz	fluid ounce
ADP	adenosine diphosphate			FSH	follicle-stimulating hormore
adst. feb.	when fever is present	comp	compound		
ag. feb.	when fever increases	contra	against	FUO	fever, unknown origin
AIDS	acquired immuno-deficiency syndrome	COPD	chronic obstructive airways disease	G6PD	glucose 6-phosphate dehydrogenase
alt. dieb.	every other day	CPAP	continuous positive airway pressure	GABA	gamma-aminobutyric acid
alt. hor.	every other hour			GFR	glomerular filtration rate
ANS	autonomic nervous system	CPR	cardiopulmonary resuscitation	GI	gastrointestinal
				GnRH	gonadotrophin-releasing hormone
AQ	aqueous	Cr	creatinine		
ARC	AIDS-related complex	CRF	chronic renal failure	gr	grain
ARDS	adult respiratory distress syndrome	CRH	corticotrophin-releasing hormone	GTP	guanosine triphosphate
				gtt	drops
ARF	acute respiratory failure; acute renal failure	CSF	cerebrospinal fluid	GU	genitourinary
		CT	computed tomography	Hb	hemoglobin
AS	aortic stenosis	CV	cardiovascular	Hcy	hematocrit
atm	atmosphere	CVA	cerebrovascular accident	HD	Huntington's disease
ATP	adenosine triphosphate	CVP	central venous pressure	HDL	high density lipoprotein
AV	atrioventricular	CVS	chorionic villus sampling	hgb	hemoglobin
BBB	bundle branch block; blood-brain barrier	d	dexto (right); day	HGH	human growth hormone
		D & C	dilation and curettage	HIV	human immunodeficiency virus
bid	twice daily	da	day		
bis in 7d	twice a week	DBP	diastolic blood pressure	HLA	human leukocyte group A
BMR	basal metabolic rate	def	defecation	HTLV-III	human T-lymphotrophic virus Type III
BP	blood pressure	DH	delayed hypersensitivity		
BSA	body surface area	DHAP	dihydroxyacetone	hypo	hypodermic
BUN	blood urea nitrogen	dl	deciliter	HZ	hertz (cycles per second)
C	celsius; complement; gallon	DNA	deoxyribonucleic acid	IBS	irritable bowel syndrome
		dr	dram	ICF	intracellular fluid
c̄	with	DTP	diphtheria-tetanus-pertussis	IDL	intermediate density lipoprotein
C_cr	renal creatinine clearance				
C1, etc.	first cervical vertebra, etc.	ECF	extracellular fluid	lg	immunoglobulin
ca	about	ECG	electrocardiogram	IM	intramuscular
CAD	coronary artery disease	ECT	electroconvulsive therapy	inf	infusion
cal	calorie	EDV	end-diastolic volume	inhal	inhalation
Cal	kilocalorie	EEG	electroencephalogram	inj	injection
cAMP	cyclic AMP	EM	electron micrograph	IPPB	inspiratory positive pressure breathing
CAPD	continuous ambulatory peritoneal dialysis	EMG	electromyogram		
		ENS	enteric nervous system	IPSP	inhibitory postsynaptic potential
CAT	computerized axial tomography	ENT	ear, nose, and throat		
		EPSP	excitatory postsynaptic potential	IRV	inspiratory reserve volume
CBC	complete blood count				
cc	cubic centimeter (cm^3)	ER	endoplasmic reticulum	IU	international unit
CCK,	cholecystokinin-	ERV	expiratory reserve volume	IUD	intrauterine device
CCK-PZ	pancreozymin			IV	intravenous
CDC	Centers for Disease Control	ESR	erythrocyte sedimentation rate	IVP	intravenous pyelogram
				s̄	without